Carbon Management Technology Conference 2015

Sustainable and Economical CCUS Options

Sugar Land, Texas, USA
17-19 November 2015

Volume 1 of 2

ISBN: 978-1-5108-1886-6

Printed from e-media with permission by:

Curran Associates, Inc.
57 Morehouse Lane
Red Hook, NY 12571

Some format issues inherent in the e-media version may also appear in this print version.

Printed by Curran Associates, Inc. (2016)

AIChE
120 Wall Street, FL 23
New York, NY 10005-4020

Phone: (800) 242-4363
Fax: (203) 775-5177

www.aiche.org

Additional copies of this publication are available from:

Curran Associates, Inc.
57 Morehouse Lane
Red Hook, NY 12571 USA
Phone: 845-758-0400
Fax: 845-758-2633
Email: curran@proceedings.com
Web: www.proceedings.com

TABLE OF CONTENTS

Driving Innovation ♦ Delivering Results

Coal with CCS to Electricity: Baseline Performance, Cost and Sensitivities

(DOE/NETL-2016/1741)

Tim Fout, Kristin Gerdes, Travis Shultz
NETL
Marc Turner, Vincent Chou, Norma Keuhn, Arun Iyengar, Mark Woods - BAH

U.S. DEPARTMENT OF ENERGY | National Energy Technology Laboratory

Presentation Outline

- Objective of Bituminous Baseline Study
- Pulverized Coal Supercritical Steam with CO_2 Capture (Case B12B) Walkthrough by Plant Section
- Summary of B12B Efficiency and Costs
- Sensitivities
- Conclusions

Objective

- **Determine cost and performance estimates of near-term commercial offerings for power plants, both with and without current technology for CO_2 capture**
  - Consistent design requirements
  - Up-to-date performance and capital cost estimates
  - Technologies built and deployed in the near-term
- **Provides baseline cost and performance**
  - Compare existing technologies
  - Guide R&D for advancing technologies within the DOE Office of Fossil Energy (FE), NETL Programs

Reports referenced:

U.S. DOE, NETL, July 2015, *Cost and Performance Baseline for Fossil Energy Plants, Volume 1a: Bituminous Coal (PC) and Natural Gas to Electricity, Revision 3*, Pittsburgh, PA : Department of Energy, 2015. Report DOE/NETL-2015/1723

U.S. DOE, NETL, June 2015, *Cost and Performance Baseline for Fossil Energy Plants Supplement: Sensitivity to CO_2 capture rate in coal-fired power plants*, Pittsburgh, PA : Department of Energy, 2015. Report DOE/NETL-2015/1720

Pulverized Coal Power Plant
Case B12B – with CO_2 Capture

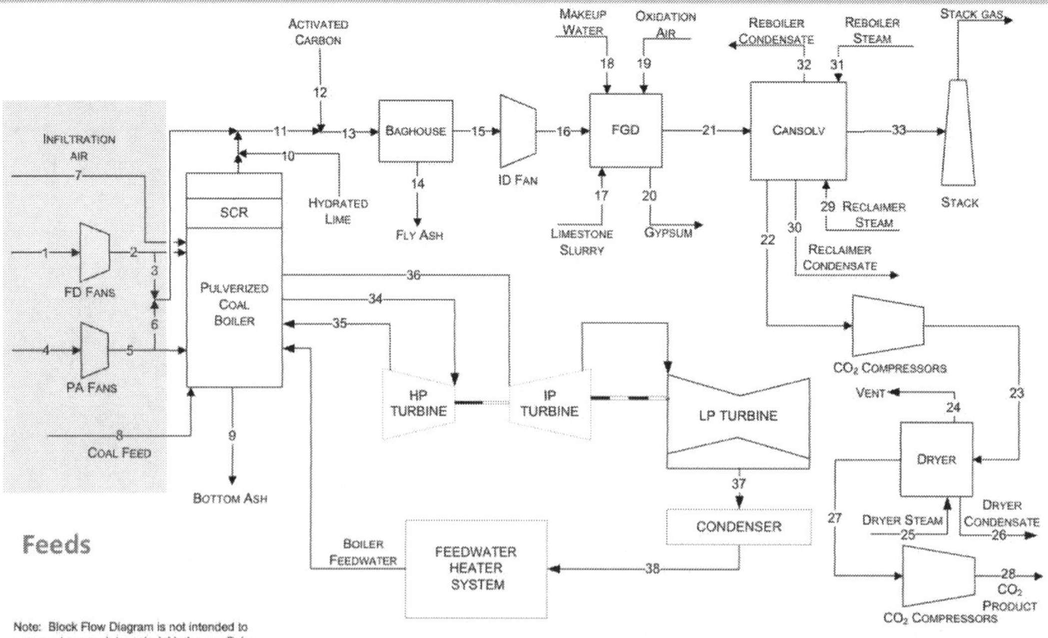

Note: Block Flow Diagram is not intended to represent a complete material balance. Only major process streams and equipment are shown.

Case B12B Breakdown – Feeds

NETL

- **Illinois #6 Coal**
  - 11,666 Btu/lb as received, HHV
  - $2.94/MM Btu
- **Train delivery**
- **Conveyors**
- **Mills (Pulverizers)**
  - 72% passing 200 mesh, 0.5% remaining on 50 mesh
- **Air Fans**
- **Air Preheaters (Ljungstrom)**

Illinois #6 Coal Ultimate Analysis (weight %)		
	As Rec'd	Dry
Moisture	11.12	0
Carbon	63.75	71.72
Hydrogen	4.50	5.06
Nitrogen	1.25	1.41
Chlorine	0.29	0.33
Sulfur	2.51	2.82
Ash	9.70	10.91
Oxygen (by difference)	6.88	7.75
	100.0	100.0
HHV (Btu/lb)	11,666	13,126

Pulverized Coal Power Plant
Case B12B – with CO_2 Capture

NETL

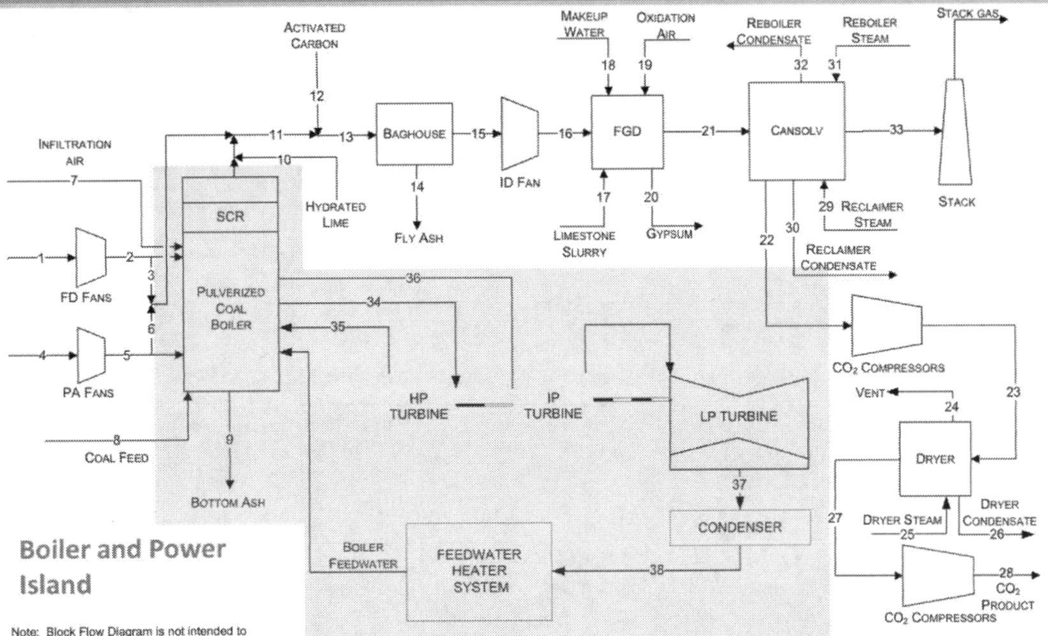

Note: Block Flow Diagram is not intended to represent a complete material balance. Only major process streams and equipment are shown.

Case B12B Breakdown – Boiler and Power Island

- **Boiler**
  - 24 to 36 coal nozzles at multiple elevations
  - Low NOx burners with over-fire air
  - Oil-fired ignition, warm-up and flame stabilization at start-up and low loads
  - Soot blowers (50 to 150 retractable nozzles and lances) clean furnace walls and convection surfaces with HP steam
- **Steam Turbine**
  - Tandem compound (HP-IP-two LP sections)
  - Hydrogen-cooled generator
  - 6 steam extractions with capture
  - 3,500 psig / 1,100F / 1,100 F
- **Balance of Plant**

Pulverized Coal Power Plant
Case B12B – with CO_2 Capture

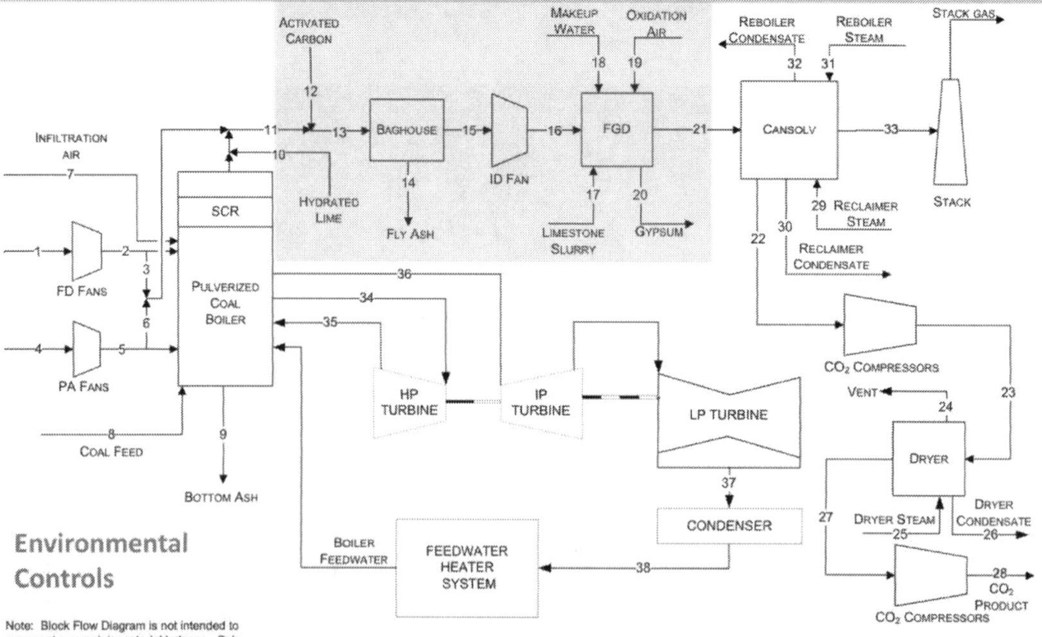

Note: Block Flow Diagram is not intended to represent a complete material balance. Only major process streams and equipment are shown.

Case B12B Breakdown – Environmental Controls

NETL

- **Hg and HCl targets based on March 2013 update to Utility Mercury and Air Toxics Standards for PC and IGCC Technologies**
- **Particulate Matter (PM), SO_2 and NOx targets based upon February 2013 New Source Performance Standards for PC, IGCC and NGCC Technologies**
- **Emissions lower in CO_2 capture cases due to requirements/performance of capture system**
- **NO_X**
  - Low NO_x burners with over fire air
  - Selective Catalytic Removal (SCR) (83 – 85 % reduction)
    - Selective Non-catalytic Removal (SNCR) not used due to relatively low NO_x out of boiler
- **SO_2**
  - Wet limestone Flue Gas Desulfurization (FGD)
  - 98% removal
- **Hg**
  - Activated carbon injection (ACI) removed in fabric filter
  - Direct Sorbent Injection (DSI) for SO_3 removal to improve effectiveness of activated carbon injection
  - Total removal 90% co-benefit of SCR, fabric filter and wet FGD

Case B12B Breakdown – Environmental Controls

NETL

- **PM**
  - Fabric filter (or baghouse)
  - High air-to-cloth ratio design with pulse jet online cleaning
  - Hg control (DSI and ACI) increase particulate matter by 26%
- **Summary of Environmental Control**

Pollutant	PC (lb/MWh-gross)	Control Technology
SO_2	1.00	Wet limestone scrubber
NOx	0.70	Low NOx burners, overfire air and SCR
PM (Filterable)	0.09	Fabric filter
Hg	$3x10^{-6}$	Co-benefit capture, dry sorbent injection[A], activated carbon injection
HCl	0.010	SO_2 surrogate[B]

[A] Limits SO_3 levels and their detrimental effects on activated carbon injection (See Report Section 3.1.6)
[B] SO_2 may be utilized as a surrogate for HCl measurement if the EGU utilizes wet FGD. (Report reference 17)

Pulverized Coal Power Plant
Case B12B – with CO_2 Capture

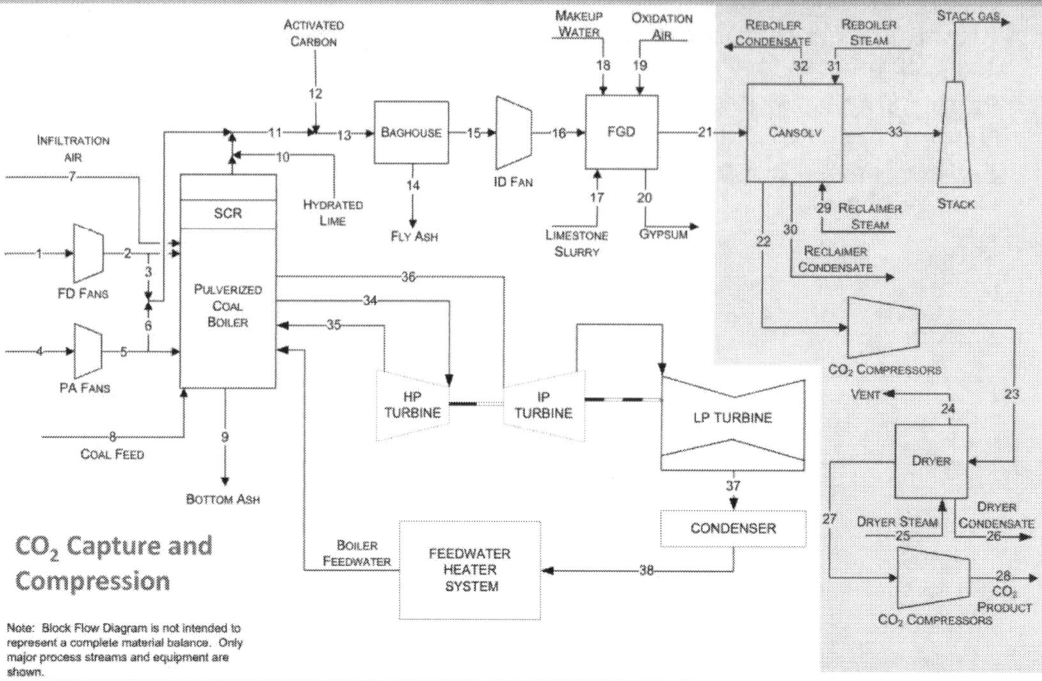

Note: Block Flow Diagram is not intended to represent a complete material balance. Only major process streams and equipment are shown.

Case B12B Breakdown – CO_2 Capture and Compression

- **Pre-treatment**
  - Lowers SO_x to ~ 1 ppmv from ~40 ppmv out of FGD
- **Cansolv CO_2 Capture Process**

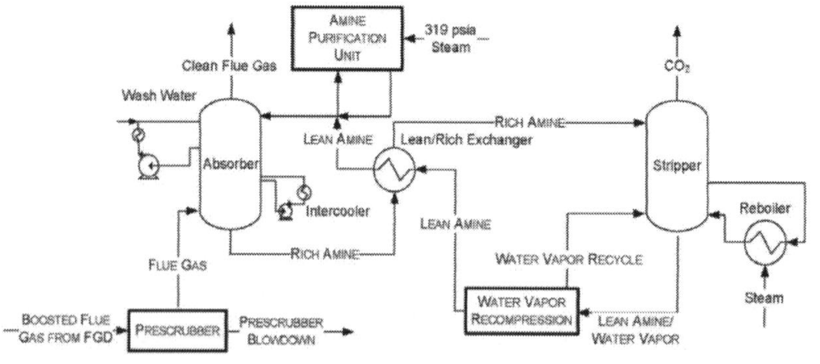

Case B12B Breakdown – CO_2 Capture and Compression

- **Cansolv CO_2 Capture Process (con't)**
  - 90 % CO_2 capture
  - Steam extraction from crossover pipe between IP and LP sections of steam turbine
  - Product CO_2 ~ 30 psia
- **CO_2 Compression System**
  - CO_2 compressed to 2,200 psig
  - 8 stages (2.23 to 1.48 stage pressure ratios)
  - Intercooling in each stage
    - Water knockout in first 3 stages
  - TEG dehydration unit between stages 4 and 5
    - 300 ppmw H_2O in CO_2 product

CO_2 Transport and Storage – Case B12B and Sensitivities

- **CO_2 Transportation and Storage (T&S) Costs[1]**
  - CO_2 transported 100 km, stored in the Illinois Basin due to Midwest plant location and monitored for 80 years
  - CO_2 T&S costs added on to the cost of electricity (COE)

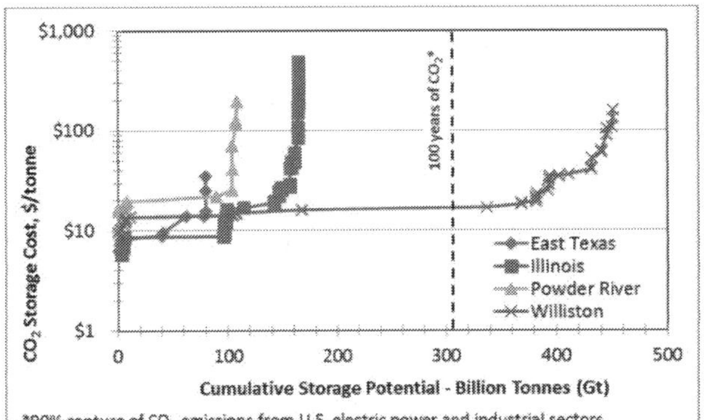

*90% capture of CO_2 emissions from U.S. electric power and industrial sectors

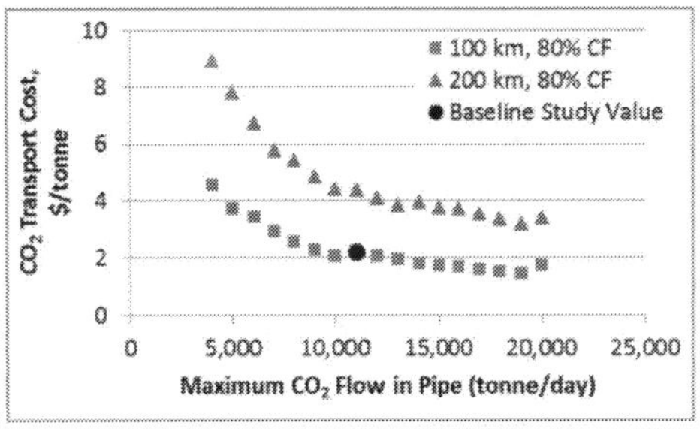

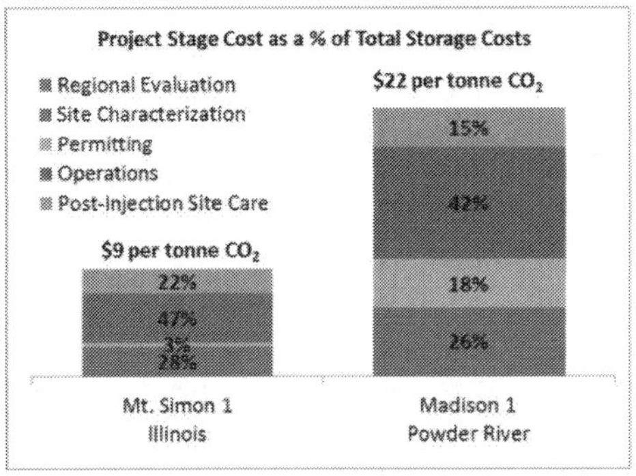

[1]**National Energy Technology Laboratory.** *Quality Guidelines for Energy System Studies: Carbon Dioxide Transport and Storage Costs in NETL Studies.* Pittsburgh, PA : Department of Energy, 2014. DOE/NETL-2014/1653

Case B12B Breakdown – Financial Parameters

- **2011 First Year of Capital Expenditure & Year Dollars**
- **30 year economic plant life**
- **85% Capacity Factor for PC & NGCC**

	5 Year Construction Period		3 Year Construction Period	
	High Risk	Conventional	High Risk	Conventional
Applicability	PC-CCS	PC	NGCC-CCS	NGCC
Capital Charge Factor	12.4%	11.6%	11.1%	10.5%

Considerations on Reported Costs

Capital Costs:

- **Estimated to be -15% to + 30% (AACE Class 4)[1]**
- **Intended to be next commercial offering**
- **Rely on vendor cost estimates for component technologies**
- **Attempts to apply contingencies (challenge for emerging technologies)**

Costs of mature technologies and designs:

- **PC and NGCC Plants without CO2 Capture**
  - Estimates reflect nth-of-a-kind
  - Costs have dropped over time due to "learning by doing" and risk reduction that result from serial deployments as well as continuing R&D

Costs of emerging technologies and designs:

- **PC and NGCC Plants with CO2 Capture**
  - Use same cost estimating methodology as mature plant designs
  - Does not fully account for the unique cost premiums associated with the initial, complex integrations of emerging technologies in a commercial application
  - Initial deployments may incur costs higher than those reflected in these reports

[1]**National Energy Technology Laboratory.** *QGESS: Cost Estimation Methodology for NETL Assessments of Power Plant Performance.* Pittsburgh, Pa : U.S. Department of Energy, April 2011. DOE/NETL-2011/145

Considerations on Reported Costs (con't)

NETL

Other Factors:

- **Actual reported project costs for all of the plant types are also expected to deviate from the cost estimates in this report due to project- and site-specific considerations**
  - i.e. contracting strategy, local labor costs, seismic conditions, water quality, etc

Future Cost Trends:

- **Continuing research, development, and demonstration (RD&D) is expected to result in designs that are more advanced than those assessed by this report, leading to costs that are lower than those estimated**

U.S. DEPARTMENT OF ENERGY | National Energy Technology Laboratory 17

Summary Performance and Cost Results

NETL

Case	B12A	B12B
CO_2 Capture	0%	90%
Gross Power (MW)	580	642
Auxiliary Power Summary		
Balance of Plant	27	35
Flue Gas Cleanup	3	4
CO_2 Capture	-	16
CO_2 Compression	-	36
Total Aux. Power (MW)	**30**	**91**
Net Power (MW)	550	550
Heat Rate (Btu/kWh)	8,400	10,500
Efficiency (HHV)	41	33
Energy Penalty[1]	-	8

Case	B12A	B12B
CO_2 Capture	0%	90%
Total Plant Cost, in Millions 2011$[2]		
Base Plant	947	1,111
Gas Cleanup (SOx/NOx/Hg/HCl/PM)	167	197
CO_2 Capture	-	484
CO_2 Compression	-	98
Total	**1,114**	**1,890**
Cost of Electricity, $/MWh (2011$)		
Capital	39	72
Fixed	10	15
Variable	9	15
Fuel	25	31
CO_2 T&S	-	10
Total[3]	**82**	**143**
CO_2 Captured (w/o T&S), $/tonne (2011$)		
Compared to SCPC or NGCC	-	58

[1] CO_2 Capture Energy Penalty = Percent points decrease in net power plant efficiency due to CO_2 Capture

[2] Total Plant Capital Cost (Includes contingencies and engineering fees but not owner's costs

[3] 85% Capacity Factor

U.S. DEPARTMENT OF ENERGY | National Energy Technology Laboratory 18

Cost of Electricity Breakdown – Overall Study

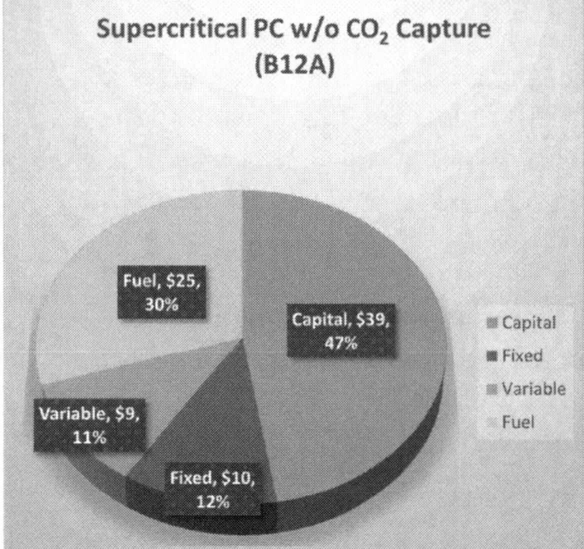

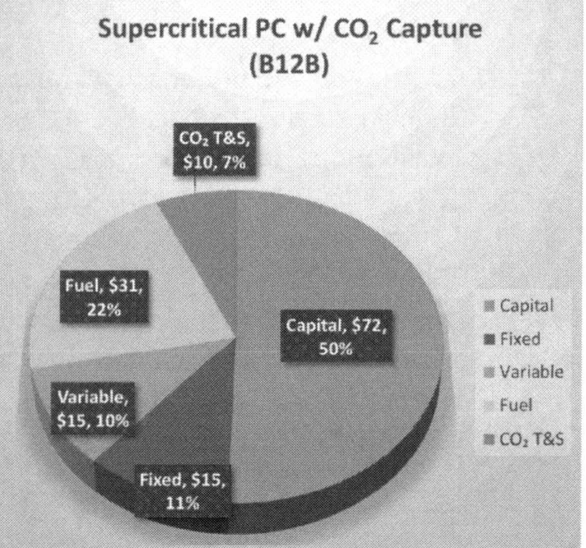

All costs in $ /MWh ($2011)

U.S. DEPARTMENT OF ENERGY | National Energy Technology Laboratory

19

Cost of Capture Breakdown – B12B

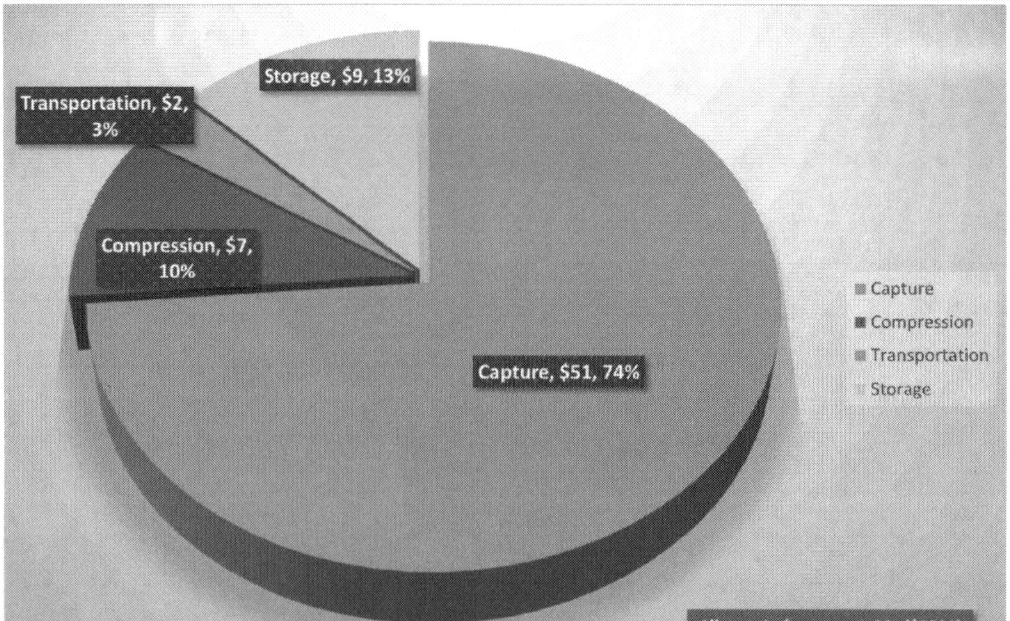

All costs in $ per tonne CO_2 ($2011)

U.S. DEPARTMENT OF ENERGY | National Energy Technology Laboratory

20

Cost of Capture Breakdown – B12B

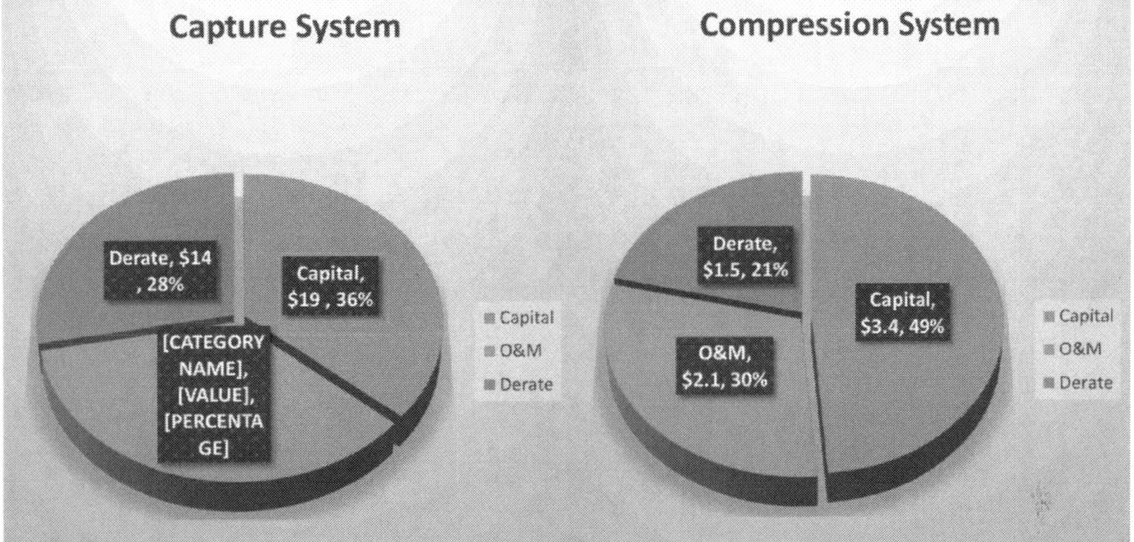

All costs in $ per tonne CO_2 ($2011)

Sensitivity of COE to CCS Parameters

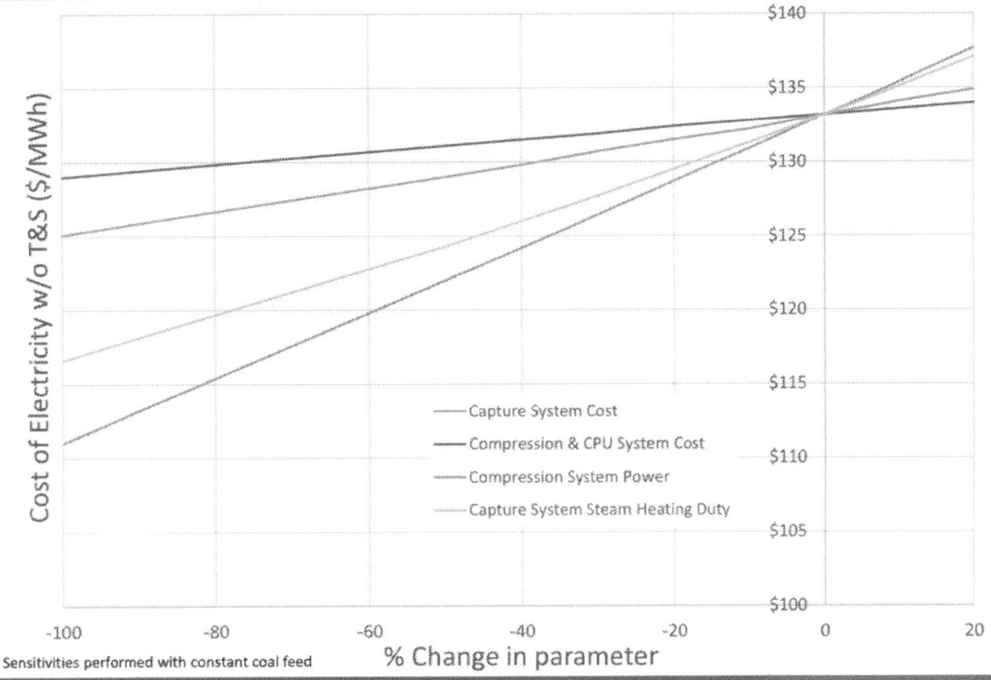

Sensitivities performed with constant coal feed

R&D Trajectories to Meet Program Goals

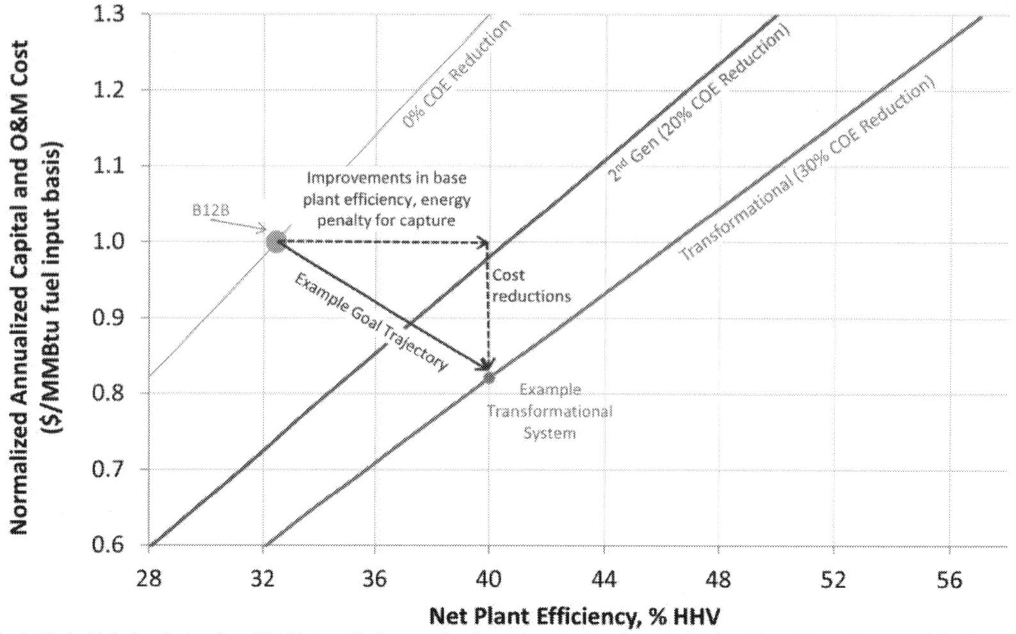

State-of-the-Art Coal with Carbon Capture from NETL "Cost and Performance Baseline, Volume 1a: Bituminous Coal (PC) and Natural Gas to Electricity" (July 2015)
http://www.netl.doe.gov/research/energy-analysis/energy-baseline-studies

COE Sensitivity to Financial Assumptions

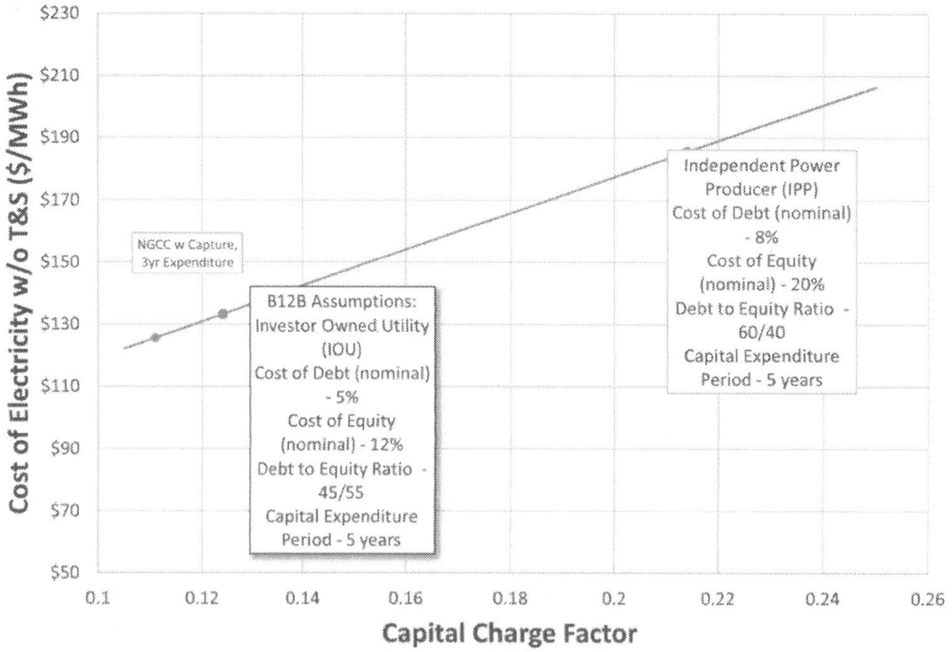

CO_2 sales price

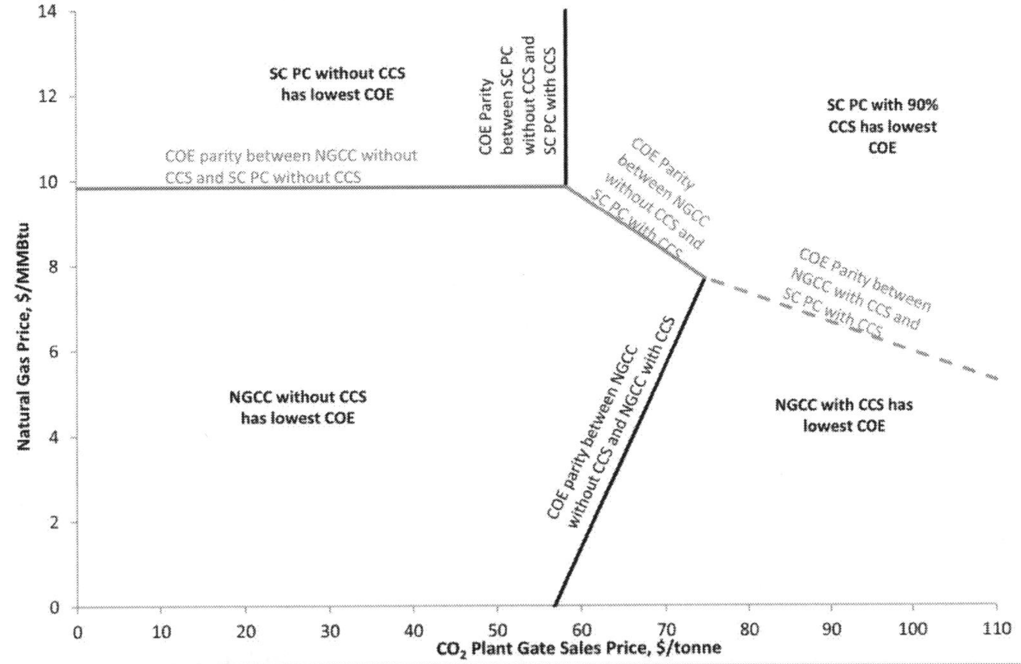

BB SCPC Case (B12B) Modified for Partial Capture

- **Evaluate the cost and performance of a <u>BB SCPC plant (Case B12B)</u> that is modified for partial CO_2 capture with all other design assumptions remaining constant**
  - An appropriate portion of the flue gas stream exiting the desulfurization step is directed towards the Cansolv CO_2 capture system to evaluate systems with CO_2 emissions ranging from ~ 1700 – 210 lb/MWh-net
  - The remaining flue gas stream is allowed to bypass the CO_2 capture system and is sent to the stack without further treatment
  - Cost and performance scaled from 90% capture Cansolv quote and accordingly sized compression quote

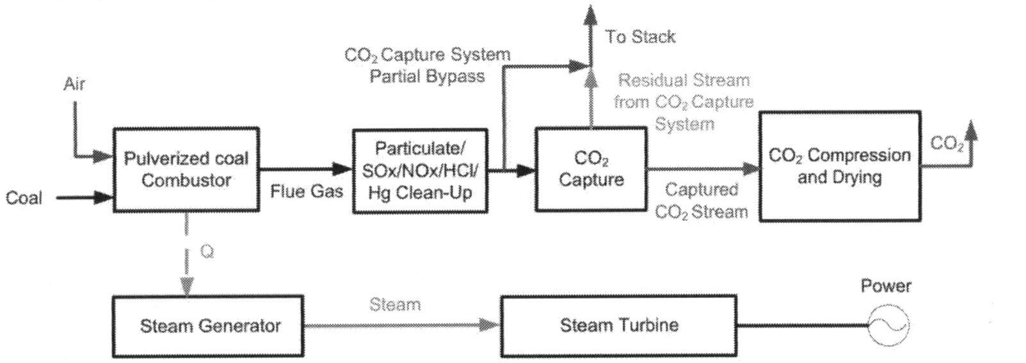

Source: NETL

Partial CO_2 Capture
Efficiency vs CO_2 Emissions

NETL

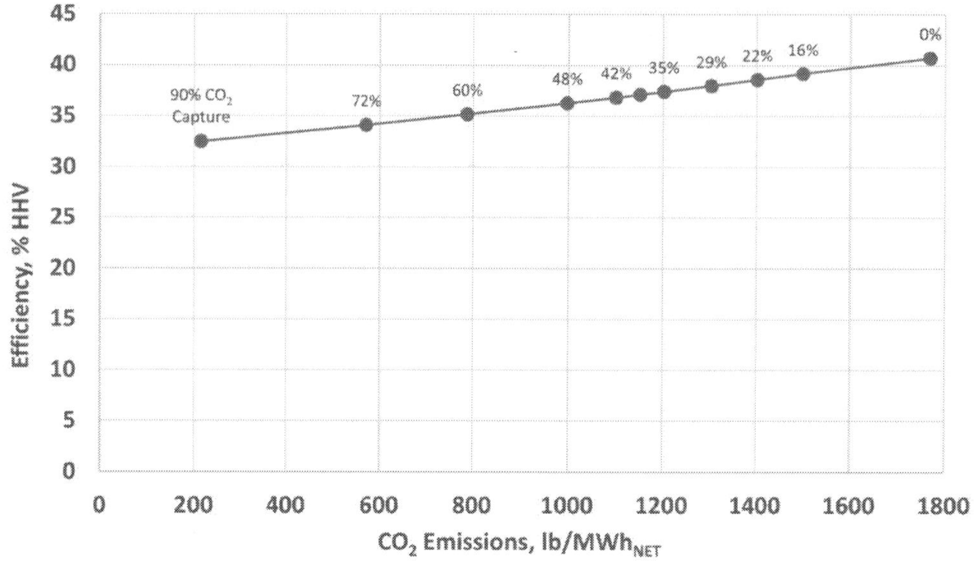

U.S. DOE, NETL, June 2015, Cost and Performance Baseline for Fossil Energy Plants Supplement: Sensitivity to CO_2 capture rate in coal-fired power plants, Report DOE/NETL-2015/1720

Partial CO_2 Capture
Capital Cost vs CO_2 Emissions

NETL

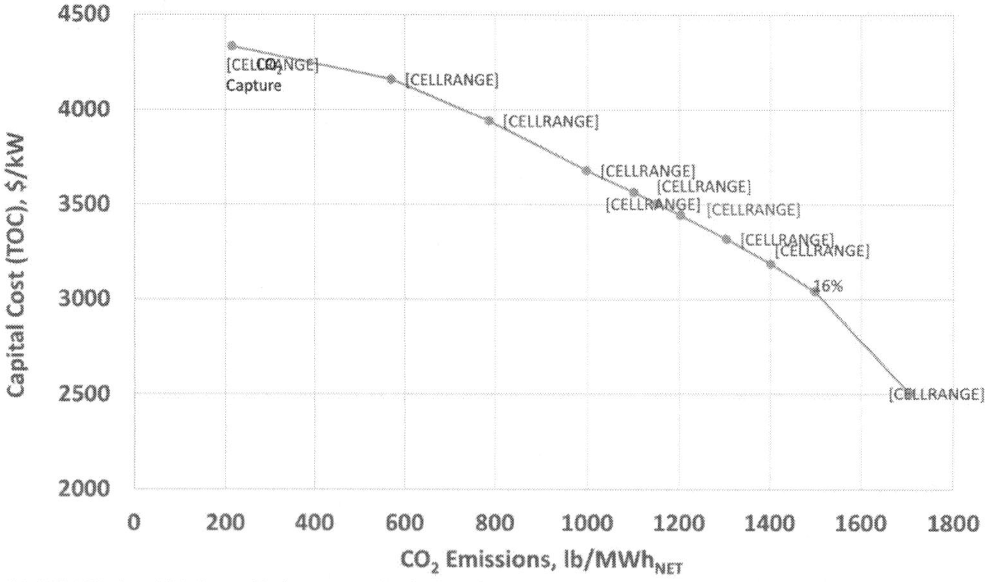

U.S. DOE, NETL, June 2015, Cost and Performance Baseline for Fossil Energy Plants Supplement: Sensitivity to CO_2 capture rate in coal-fired power plants, Report DOE/NETL-2015/1720

* Fully-loaded design rates; does not account for start-up, shutdown, performance degradation between maintenance, part-load operation, etc.

BB SCPC Case for Various Levels of Capture
COE Including T&S

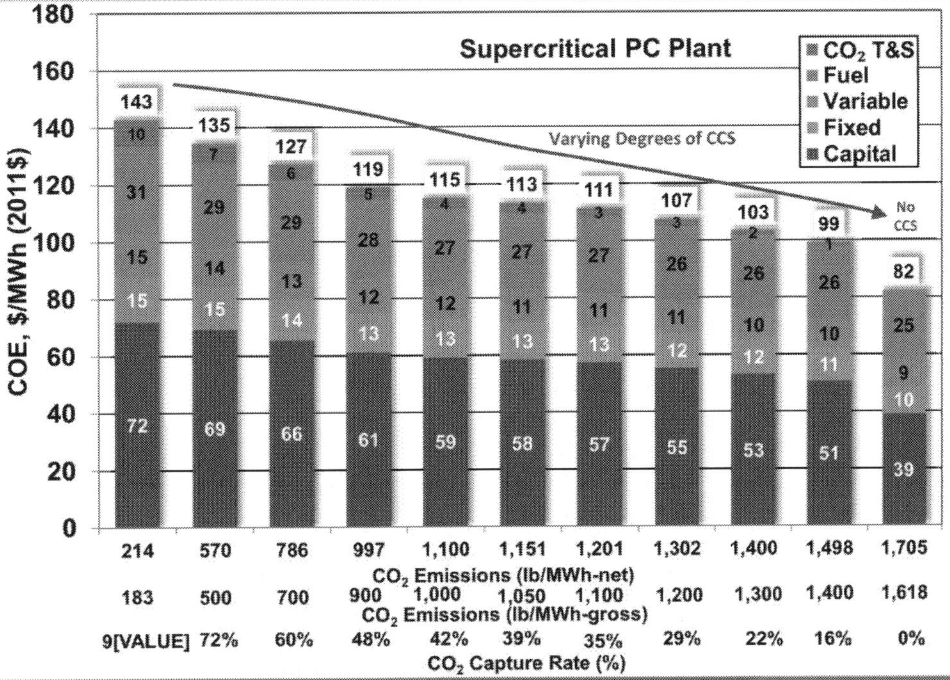

*Note that the non-capture coal plant and capture coal plant cases use different finance structures

BB SCPC Case for Various Levels of Capture
Cost of Capture, excluding T&S

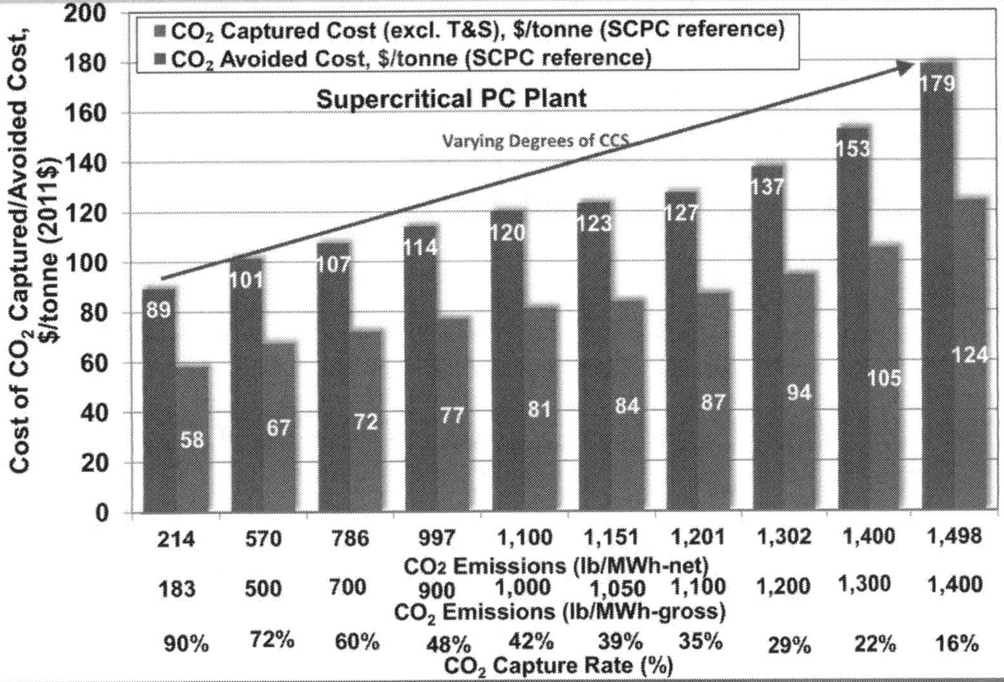

Summary

- **Most up-to-date performance and costs available in public literature to date**
- **Establishes baseline performance and cost estimates for current state of technology**
  - Coal-based plants using today's technology are efficient and clean
    - PC: 40%, HHV (without capture on bituminous coal)
    - Meet or exceed current environmental regulations
    - Today's capture technology can remove 90% of CO_2, but at significant increase in COE
- **Coal with CCS COE preferred with > \$70/tonne CO_2 sales price and > \$8/MMBtu NG**
  - As NG prices decline, higher CO_2 sales prices required
  - At reference price of \$6.13/MMBtu, \$90/tonne CO_2 sales price required

Summary

- **Lower capture rates result in lower COE**
  - Lower capital and operating costs
  - Lower solvent regeneration energy steam use
- **CO_2 capture cost (\$/tonne CO_2) higher at lower capture rates**
  - Economies of scale
  - Less CO_2 available for sale
- **Bulk of the cost to capture and store CO_2 is in the CO_2 capture system**
  - Compression, storage, and transportation for Midwest location total < 25%
- **Capture system capital cost and solvent regeneration energy significantly impact CO_2 capture cost**
  - R&D needs to address both
- **Compression capital cost ~ 50% of compression impact to CO_2 capture cost**

Online Reports Available

- **Cost and Performance Baseline for Fossil Energy Plants, Volume 1a: Bituminous Coal (PC) and Natural Gas to Electricity, Revision 3 (DOE/NETL-2015/1723)**
- **Cost and Performance Baseline for Fossil Energy Plants Supplement: Sensitivity to CO_2 Capture Rate in Coal-Fired Power Plants (DOE/NETL-2015/1720)**
  - Includes PC and IGCC cases
- **Cost and Performance Baseline for Fossil Energy Plants, Volume 1b: Bituminous Coal (IGCC) to Electricity, Revision 2 – Year Dollar Update (DOE/NETL-2015/1727)**

http://www.netl.doe.gov/research/energy-analysis/energy-baseline-studies

It's All About a Clean, Affordable Energy Future

For More Information, Contact NETL

the ENERGY lab

Delivering Yesterday and Preparing for Tomorrow

CARBON MANAGEMENT TECHNOLOGY CONFERENCE

CMTC-440206-MS

Process Engineering Perspectives on CO_2 Capture Technology Development

Andrew Sexton, Austyn Douglas, Kevin Fisher, and Ray McKaskle, Trimeric Corporation

This paper was prepared for presentation at the Carbon Management Technology Conference held in Sugarland, Texas, USA, 17–19 November 2015.

Abstract

This paper provides a broad overview of CO_2 capture technologies from the perspective of the process engineer who is challenged to develop, evaluate, and scale these technologies up from laboratory, bench, and pilot-scale to full-scale, commercial applications. The CO_2 capture problem is stated in process engineering terms, and implications of the source characteristics including composition, temperature, pressure, flow, and the variability of these characteristics - and corresponding CO_2 purity/delivery specifications for various end users – are analyzed. The process engineer's tasks are described in terms of the wide range of process engineering activities that are involved, ranging from experimental data collection and process simulation, process and equipment design, and economic analyses, along with the challenges that each task presents. The nature and importance of the universal thermodynamic constraints and resulting minimum energy requirements that govern all CO_2 capture process are described, and the need to integrate the CO_2 capture process with the CO_2 source process and CO_2 use or storage locations is explored. CO_2 capture technologies are categorized and compared in terms of the key process design parameters that affect each group of processes in a different way, revealing the various process advantages and disadvantages of the different approaches. Capture approaches that are discussed include chemical and physical absorption, chemical and physical adsorption along with temperature, pressure, and vacuum regeneration approaches, organic and inorganic membranes, cryogenic separation, as well as emerging approaches including chemical looping, electrochemical, and microbial capture technologies. Equipment challenges such as the physical size of mass and heat transfer devices, mechanical limitations on compression technology, and materials of construction issues are described. Conventional methods of estimating and comparing capital and operating costs among CO_2 capture technologies are reviewed.

Introduction

In recent years, many researchers around the world have been engaged in efforts to develop improved technologies for capturing carbon dioxide (CO_2) from streams that would otherwise be vented to the atmosphere. The scope of the task is enormous and involves the combined efforts of many scientific and engineering disciplines. This paper discusses this challenge from the specific perspective of the process engineer tasked with developing, evaluating, and scaling up CO_2 capture processes for commercial applications. CO_2 source characteristics including composition, temperature, pressure and flow, as well as the end user specifications for the captured CO_2, are reviewed. CO_2 capture technologies further along in the development stages are categorized and compared. Process approaches discussed include chemical and physical absorption, chemical and physical adsorption, membranes, and cryogenic separation.

The CO_2 Capture Process Engineering Challenge

Looking at CO_2 capture from the view of the process engineer presents the challenge of removing CO_2 from streams that would otherwise be vented to the atmosphere. These CO_2 carrying streams come from various sources including stationary sources such as power generation flue gas from coal or natural gas fired boilers, process heaters, cement kilns, chemical manufacturing processes, ethanol production, and mobile sources like cars, trucks, trains, and planes. The scope of this paper only addresses possible CO_2 capture solutions for the stationary sources.

The sources of CO_2 emissions vary in terms of the amount of CO_2 emitted per year, pressure of the emitted CO_2, and composition of the emitted stream (or purity of CO_2 in the stream). Different removal technologies will suit different CO_2 sources better than others based on the source characteristics and the removal process. The figures below compare CO_2 sources and the amount of CO_2 each source generates. Electricity from coal is the largest producer of CO_2 at roughly 1,800 million metric tonnes of CO_2 per year.

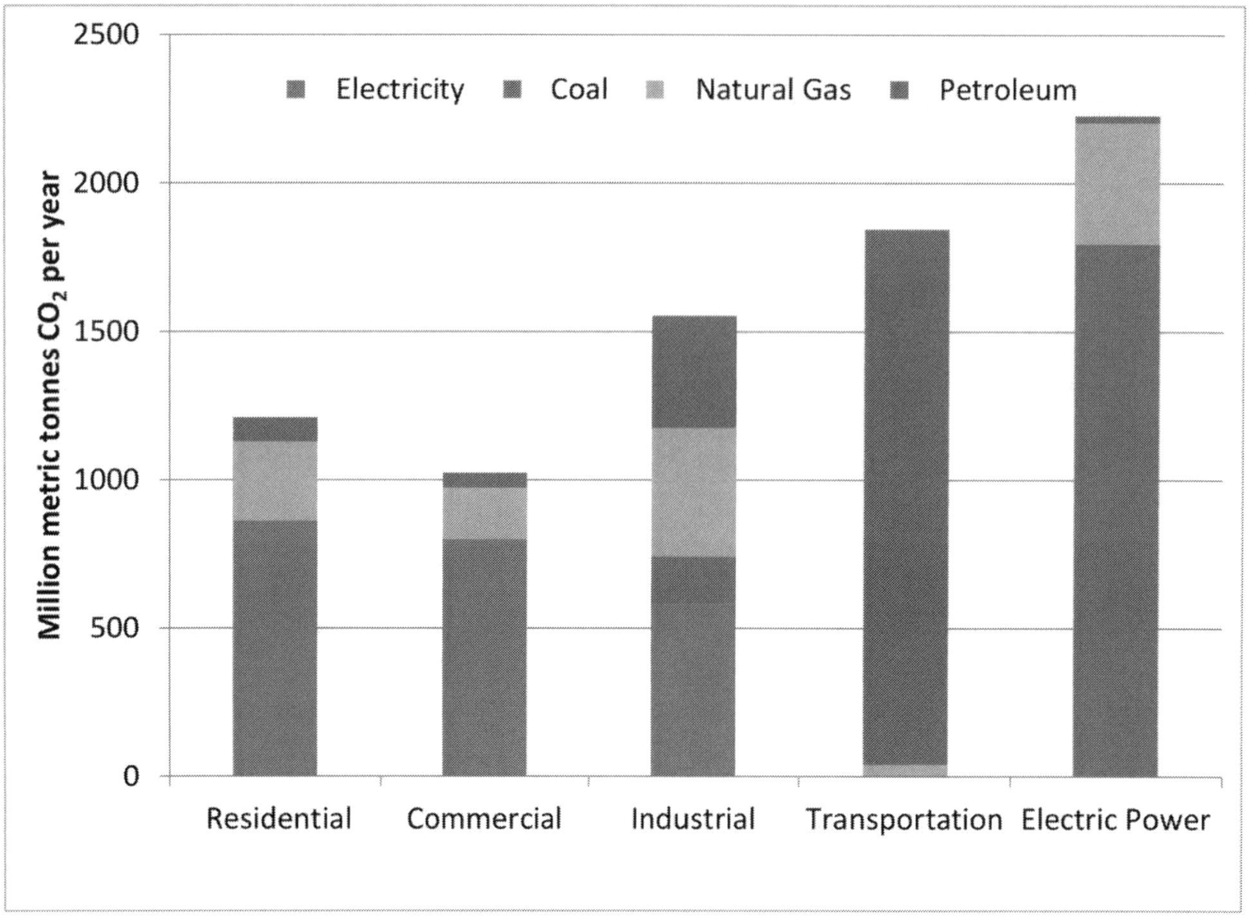

Figure 1. US CO_2 Emission Sources 2011 (EIA, 2013)

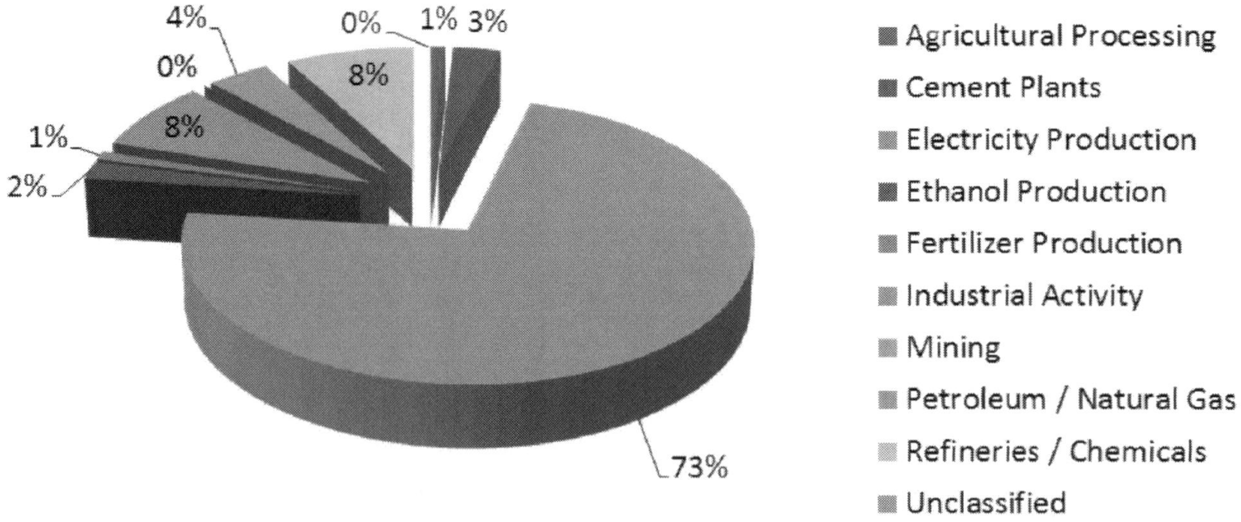

Figure 2. US Stationary CO_2 Emission Sources – 2012 (DOE, 2012)

As seen in Figures 1 and 2 above, electricity from coal produces more CO_2 than other sources on a mass basis. However, it is not the source with the highest concentration of CO_2. Figure 3 below shows the concentrations of CO_2 in flue gas from various sources and the pressure of the flue gas. Higher concentrations of CO_2 generally result in easier separation and a more efficient capture process. Ethanol production yields a high concentration CO_2 stream; however, CO_2 production from ethanol only makes up 2% of the CO_2 production from stationary sources, so while the process to capture this CO_2 is extremely efficient, it only addresses a small fraction of CO_2 emitted from stationary sources. Removing CO_2 from power plants has a greater overall impact on reducing the total mass of CO_2 emitted from stationary sources; however, it is not very cost effective to capture the CO_2 as the concentration of CO_2 is low and CO_2 removal requires more energy compared to processes that produce a gas stream with a higher concentration of CO_2.

The pressure of the CO_2 source also affects the efficiency of removal. The efficiency of different CO_2 capture technologies can be affected by the feed pressure. In addition, most CO_2 end-use applications require high pressure CO_2 which takes additional energy and equipment to compress from lower pressure sources of CO_2. The following figure shows the CO_2 purity and pressure for common sources.

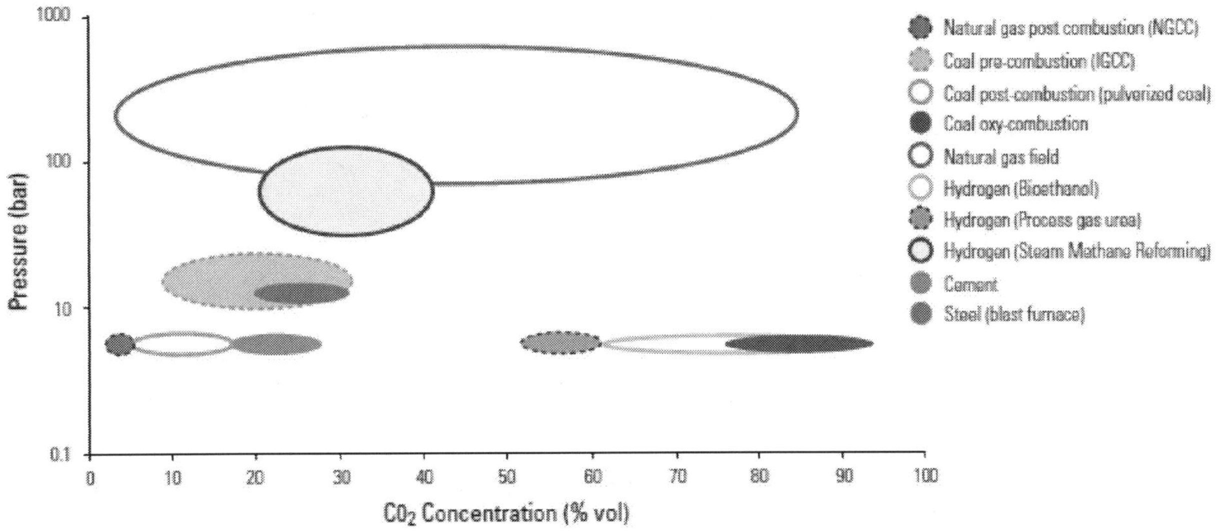

Figure 3. CO_2 Purity and Pressure for Common Sources (KAPSARC, 2012)

Flue gas generated from the combustion of coal, which is a major contributor to CO_2 emissions from the power generation sector, is near atmospheric pressure and typically has a CO_2 concentration of around 14% (by volume). Low pressure and low concentration sources of CO_2 are not optimal conditions for CO_2 removal from an energy and cost perspective. Figure 4 below shows CO_2 emissions sources in terms of their CO_2 purity and volume of CO_2 emitted. As demonstrated in Figure 4, the higher the concentration of CO_2 and the higher the amount of CO_2 mass emissions in metric tonnes (MT)/day, the higher the value or cost-effectiveness of CO_2 removal.

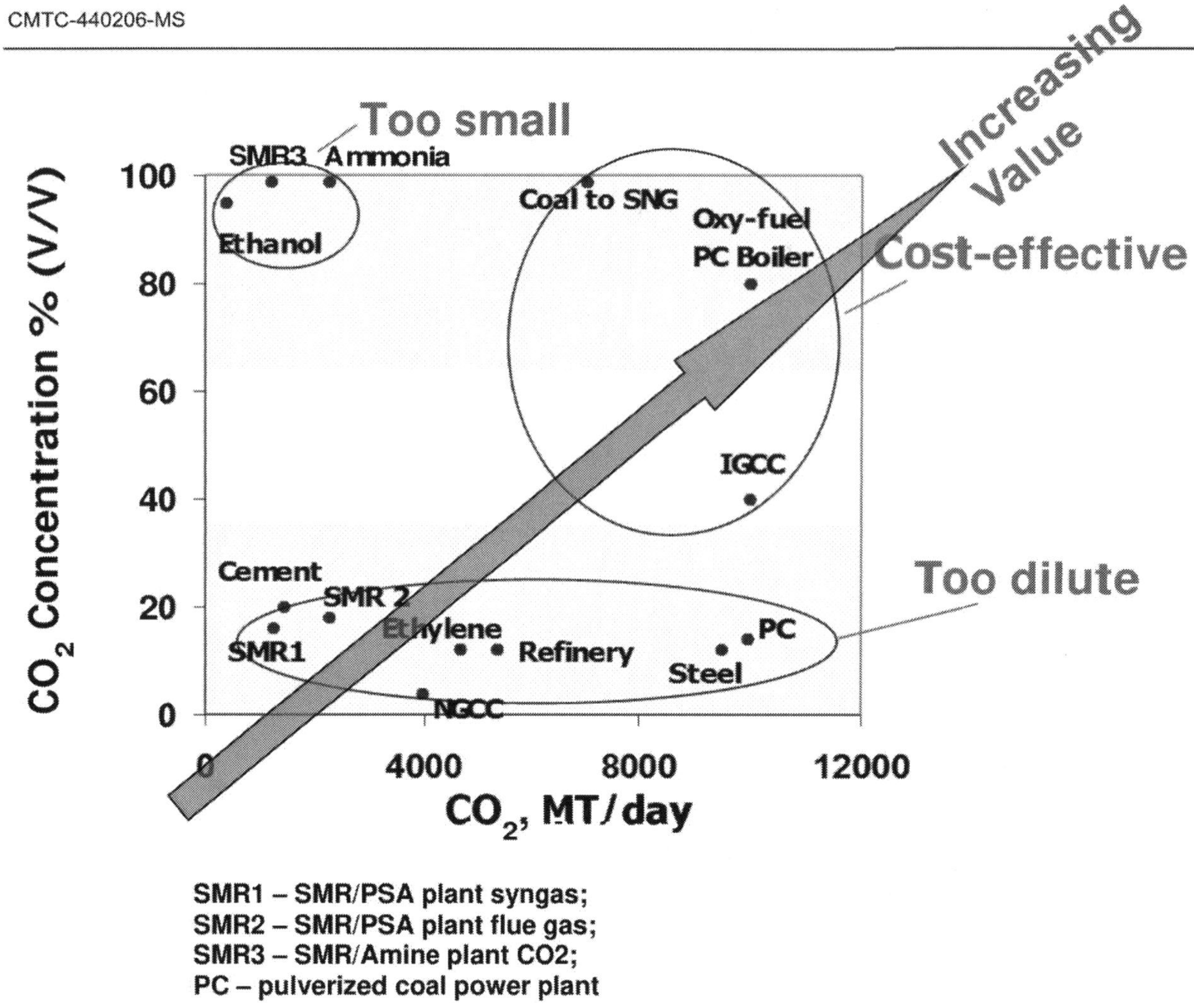

Figure 4. Source Purity vs. Volume (Praxair, 2006)

The specifications for the captured CO_2 stream depend on the end user. Some examples of CO_2 capture end users include sequestration, enhanced oil recovery, food/beverage industry, and other industrial uses. Typically the CO_2 will need to be transported as CO_2 sources are rarely located at the same site as the end use. The large amounts of CO_2 for sequestration or enhanced oil recovery are typically transported via pipeline.

There is a smaller demand for CO_2 used in beverages or specialty chemicals which may be liquefied and transported via truck or rail. High pressure CO_2 is required for transportation via pipelines and conversion to the refrigerated liquid state is needed for truck or rail transport. Other uses may convert CO_2 into another form, e.g. making urea for fertilizer or mineralization of the CO_2 into solid products. Tables 1 and 2 provide example CO_2 product specifications for an industrial pipeline, a pipeline as specified in a DOE report, and for injection into a formation for EOR. In all cases the stream has a minimum purity of 95% CO_2.

Table 1. Example Pipeline / EOR Specifications (Havens, 2008 and NETL, 2013)

Component	Example Industrial Pipeline (Ref. 1)	DOE Pipeline Design Guideline (Ref. 2)	DOE EOR Design Guideline (Ref. 2)
CO_2	≥ 95 mol. %	≥ 95 vol. %	≥ 95 vol. %
H_2O	≤ 30 lb / MMcf (633 ppmv)	≤ 500 ppmv	≤ 500 ppmv
Oxygen	≤ 10 ppmw	≤ 10 ppmv	≤ 10 ppmv
H_2S	≤ 20 ppmw	≤ 100 ppmv	≤ 100 ppmv
Total Sulfur	≤ 35 ppmw		
Nitrogen	≤ 4 mol. %	≤ 4 vol. %	≤ 1 vol. %
Hydrocarbons	≤ 5 mol. %	(Table 2)	(Table 2)
Glycol	≤ 0.3 gal / MMcf	≤ 46 ppbv	≤ 46 ppbv

Table 2. Example Pipeline / EOR Specifications (NETL, 2013)

Component	DOE Pipeline Design Guideline (Ref. 2)	DOE EOR Design Guideline (Ref. 2)
Ar	≤ 4 vol. %	≤ 1 vol. %
H_2	≤ 4 vol. %	≤ 1 vol. %
CH_4	≤ 4 vol. %	≤ 1 vol. %
C_2H_6	≤ 1 vol. %	≤ 1 vol. %
C_3+	< 1 vol. %	< 1 vol. %
CO	≤ 35 ppmv	≤ 35 ppmv
NO_x	≤ 100 ppmv	≤ 100 ppmv
SO_2	≤ 100 ppmv	≤ 100 ppmv
COS	Trace	≤ 5 ppmv

Overview of Processes Used or Under Development for CO_2 Capture

This section describes various categories of CO_2 capture processes such as absorption (including amines, chilled ammonia, physical solvents and ionic liquids), adsorption, membranes and cryogenic separation, and provides brief comparisons of the technologies including key process criteria and state of commercialization or development.

Absorption. Absorption processes have had a long history of use for gas separations, and this technology has been proven on a large scale. There are existing commercial facilities for CO_2 capture using absorption processes, and other advanced CO_2 removal technologies are often measured against absorption processes for comparison to a benchmark technology. In an absorber, lean solvent contacts the gas stream to selectively remove the component of interest (in this case CO_2). The component is released from the rich stream in a regenerator. The flow diagram below shows a generic absorption process for removing CO_2 from flue gas.

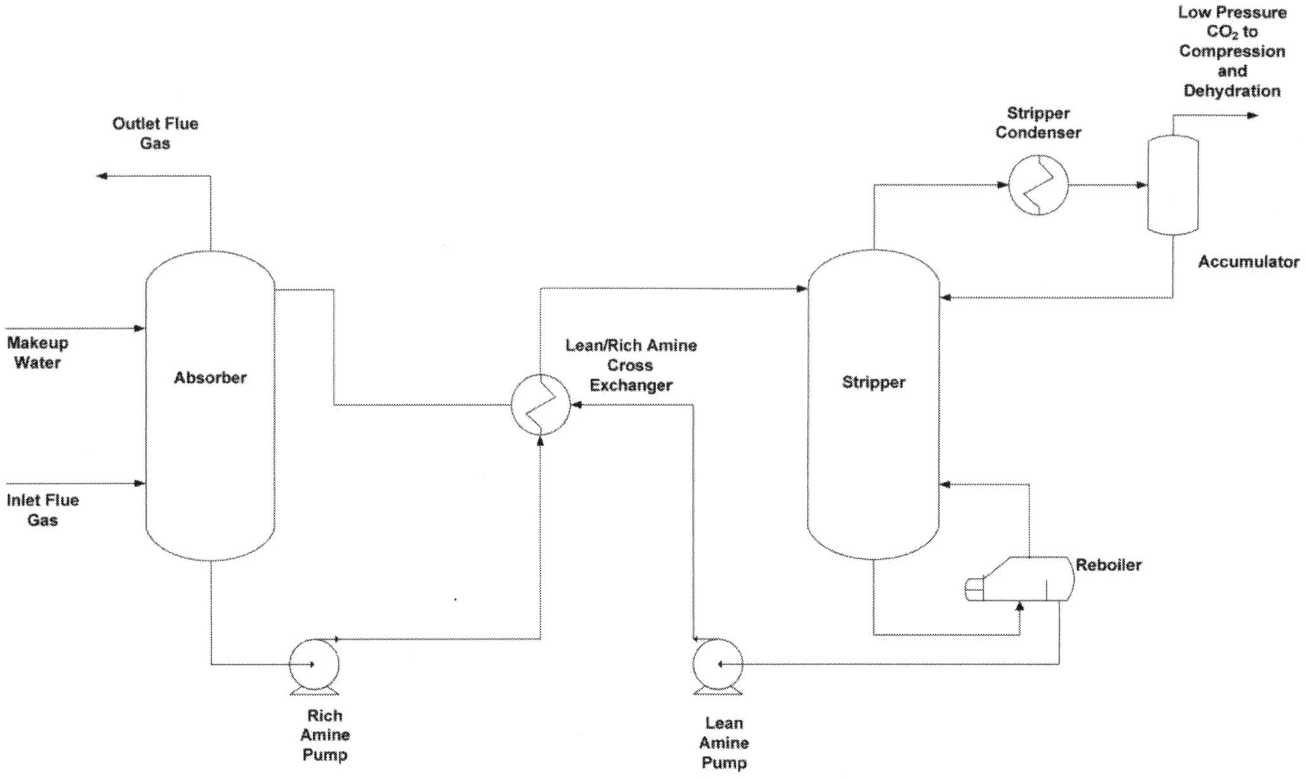

Figure 5. General Absorption Process Flow Diagram

Some important parameters for deciding whether an absorption process is a suitable technology for CO_2 removal include the source pressure and temperature and solvent characteristics – such as loading at the process operating conditions of interest and the solvent heat of absorption. Loading is defined as the maximum amount of solute (CO_2) that the solvent can absorb under the process operating conditions of interest. The solvent either chemically reacts with the CO_2 stream to remove CO_2 or physically interacts with the stream for CO_2 removal. Chemical solvents generally have a higher CO_2 solubility at lower pressures and temperatures when compared to physical solvents.

In coal gasification applications, the flue gas feeding the CO_2 removal process is at an elevated pressure which makes it a better application for physical solvents, while flue gas at ambient pressure and ambient temperature would usually be considered to be a better application for chemical solvents. A low heat of absorption may in some cases reduce the amount of heat input required in the stripper reboiler, although a lower heat of absorption does not always result in lower energy requirements for the overall capture and compression process. The ability to selectively remove CO_2, or selectivity, for chemical solvents is high while selectivity for physical solvents is moderate (Ciferno, 2011). Some examples of absorption technologies that are discussed below in greater detail include amine scrubbing, chilled ammonia, and ionic liquids.

Amine Scrubbing. Amine scrubbing is the most proven CO_2 capture approach for power plant flue gas applications. There have been dozens of demonstration units in sizes up to around 140 MWe (Boundary Dam); Boundary Dam uses a Shell Cansolv proprietary solvent. A 25 MWe demonstration at

Southern Company Plant Barry used the MHI KS-1 solvent. A 240 MWe NRG/Petranova WA Parish Station capture plant, scheduled for operation in 2016, will also use the KS-1 solvent. There are numerous suppliers of commercially available solvents including Fluor, Mitsubishi, and Shell Cansolv. With amine solvents, there is also some potential for simultaneous SO_2 scrubbing.

Amine scrubbing also has its challenges. There is a high energy penalty driven primarily by the energy required to regenerate the solvent. Further amine solvent development and technology improvements are focused on producing reductions in energy requirements, and many strategies aimed at reducing the energy requirement involve a fairly complex integration with the host power plant utilities including steam, water, and other utilities.

Additionally, there are concerns with amine emissions and oxidative and thermal degradation byproducts. Amines have the potential to produce submicron aerosol particles that are entrained with the treated flue gas stream leaving the absorber resulting in losses of solvent which are emitted to the atmosphere. Potential for nitrosamine formation from the reaction between NO_X and some amine solvents is also an ongoing concern – both for the recirculating solvent and for emissions from the absorber and regeneration columns – as some nitrosamines are classified as carcinogens. Other concerns include fly ash and metals accumulating in the solvent reclaimer waste and the potential for these wastes to be classified as hazardous waste.

Chilled Ammonia. Chilled ammonia is another example of a chemical solvent. Ammonia is readily available in bulk, is inexpensive, and there are less degradation concerns with using ammonia. The chilled ammonia process generally follows the most established absorption and stripping CO_2 capture process flow scheme. Similar to amines, there are concerns about ammonia slip emissions into the treated gas stream. There is also a high energy requirement to cool the flue gas for this low temperature absorption process.

Ionic Liquids. Ionic liquids can function as either a chemical or a physical solvent depending on their specific chemical composition. One benefit of using ionic liquids is that they generally have no vapor pressure, which means that atmospheric absorber emissions due to evaporation are minimized. However there is still a possibility of emissions and associated solvent losses due to mechanical entrainment. Another benefit with some ionic liquids is that they have a very high thermal stability.

However, ionic liquids are in a more developmental stage than conventional amine solvents and the availability of the ionic liquid solvents may be more limited and come at a higher cost. Degradation mechanisms and the potential impacts on solvent replacement costs are not yet well understood. In addition, the high viscosity of some ionic liquids presents a challenge. High viscosity increases pumping power requirements, reduces heat transfer (resulting in higher required surface areas for heat exchangers), and reduces the mass transfer of bulk CO_2 into and out of the ionic liquid solvent.

Adsorption. In chemical and physical adsorption processes, the gas stream is usually contacted with a solid sorbent which selectively adsorbs the gas component of interest (see Figure 6). Desorption of the CO_2 out of the sorbent can be a batch, cyclical process in the same vessel or the adsorbent can be fluidized and moved between adsorption and desorption vessels in a continuous process. Adsorption processes can operate at high temperatures. Similar to absorption processes, chemical adsorption has high selectivity, while physical adsorption has moderate selectivity. Adsorption processes tend to scale more linearly with size. The surface area required for contacting the flue gas generally increases linearly with the flue gas flowrate as does the equipment required for the adsorption process.

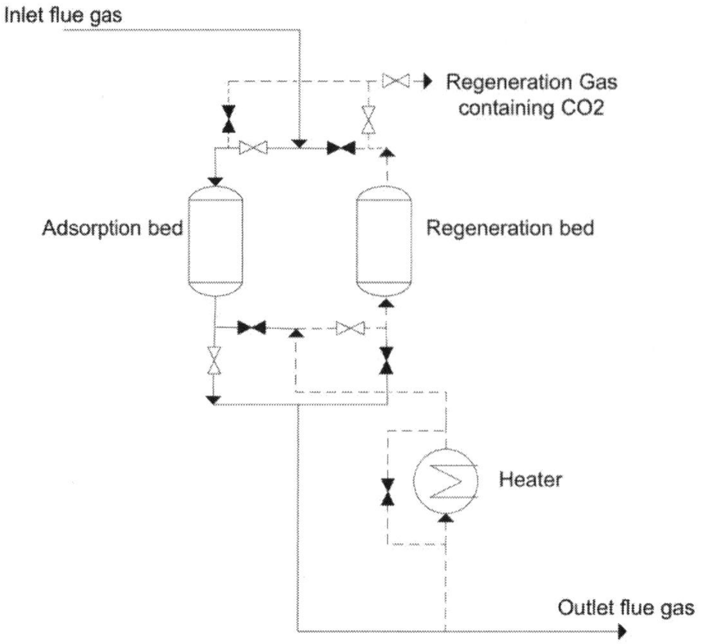

Figure 6. Example Adsorption Process Flow Diagram

Adsorption eliminates the need to pump around a solvent and may eliminate the concerns associated with atmospheric absorber emissions and disposal of waste solvent. However, adsorption media can be prone to fouling, contamination, or mechanical damage, which may result in reduced media performance. Depending on the process configuration, the purity of CO_2 achieved by the adsorption process may not meet pipeline specifications for oxygen and other impurities. Rapid-cycle adsorption may require a fairly complex process configuration with switching valves and controls.

Membranes. Figure 7 below shows some example membrane process flow diagrams. Separation with organic and inorganic membranes occurs by differing partial pressures of the retentate and permeate on opposite sides of the membrane; as a result higher pressures are beneficial (Ciferno, 2011). A high concentration of the gas component to be removed is preferred. Some important parameters for deciding to use membranes include source temperature and pressure, membrane permeance, and selectivity of all chemical species present. However, there is a tradeoff between permeance and selectivity. The more permeable the membrane, the less selective the removal will be. Selectivity can be high for dense inorganic membranes and microporous membranes, but is variable for polymeric membranes. Like the adsorption process, membranes have a more linear scale-up as the surface area of the membranes and therefore equipment costs scale linearly with flowrate.

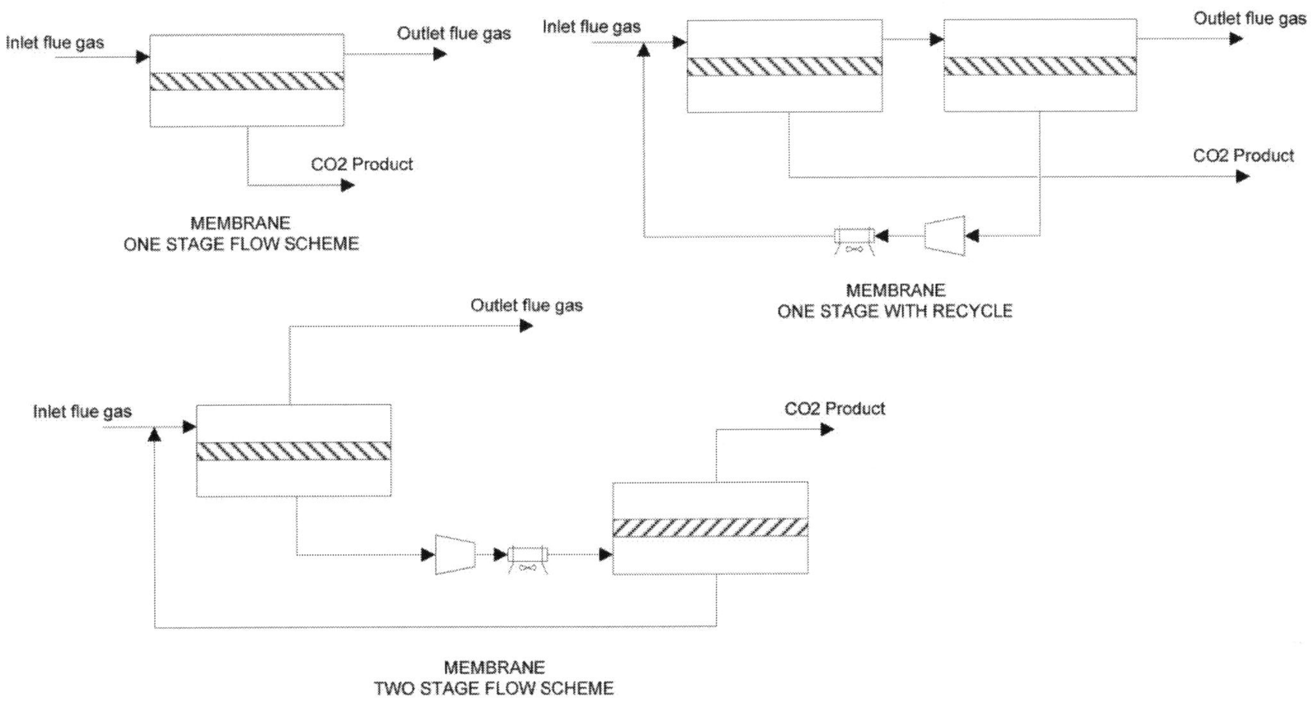

Figure 7. Example Membrane Process Flow Diagrams

Membranes require very little integration with the host site power plant. This is more of a "bolt-on" capture approach than the absorption and stripping capture concepts. However, because membranes require a high pressure differential, the process may require compression of the corrosive flue gas stream to force CO_2 to permeate through the membrane. This has two potential negative impacts – special materials required for compression of wet, corrosive flue gas and the energy impact of compression of the entire flue gas stream (as opposed to compression of only the CO_2 product). Alternatively, some processes produce a captured CO_2 permeate stream at sub-atmospheric pressure. The equipment and operating costs required to compress this low pressure, corrosive and high actual volumetric flow stream from sub-atmospheric pressure to pipeline/sequestration pressure are higher than for other capture approaches. Another cost challenge for membrane processes is that in order to produce CO_2 that meets the purity specifications, multiple stages of membranes may be required.

Cryogenic Separation. Cryogenic separation is separation by low temperature distillation or a similar process. Distillation relies on difference in relative volatilities (boiling points) of the compounds to be separated. Packing, trays, or other distillation column internals provide gas-liquid contact area and ensure proper gas and liquid distribution within column. Heat is applied at the bottom of the column to generate stripping gas that travels up the column while cooling is applied to the overhead stream to condense some of the less volatile components (CO_2 in this instance) and return a reflux liquid stream to travel back down the column. Low temperatures are required for distillation of CO_2 from flue gas, so refrigeration and heat integration are critical components that significantly impact the performance and capital and operating costs of this process.

Cryogenic processes typically produce a liquid CO_2 product that is easier to pump to pipeline pressures than gaseous CO_2 is to compress from other capture processes. In other low temperature separation processes, CO_2 is removed by solidifying it instead of liquefying it in a distillation process. Another advantage with the cryogenic separation processes is that no regeneration of solvents/sorbents is required.

Some cryogenic separation processes generate solids, which are inherently more difficult to handle than gases and liquids. Cryogenic separation processes are not typically favored for low partial pressure or high temperature applications, but selective removal can be achieved and cryogenic separation processes have the potential to be cost effective as a large-scale flue gas CO_2 capture application.

Other Processes. There are other processes not covered in this paper that are being developed for CO_2 capture. Some of these include, but are not limited to, chemical looping, mineralization, electrochemical and microbial processes, and processes that produce fuels and chemicals from CO_2.

Process Engineering Tasks

Simply stated, the ultimate goal of the process engineer is to provide a detailed and validated process design that will define how the final commercial process will work, and how much it will cost to build and operate.

The process engineering activities will naturally vary depending on the technical nature of the process being considered for development, and the tasks will be iterative to some extent as new information becomes available. These tasks include:

- Develop a conceptual design and flowsheet for the process, identifying all chemical and physical processes to be used.
- Develop material and energy balances. At this point, chemical and physical property data gaps are identified.
- Obtain fundamental experimental data needed to support material and energy balances from a chemical engineering thermodynamics standpoint. This is often lacking for emerging technologies.
- Develop and/or obtain information needed to size the major equipment and to estimate operating costs.
- Perform preliminary economic analyses with available data at hand. This generally has to be refined as new data become available, and may lead to recommendations to change one or more aspects of the process.
- Evaluate key aspects of the process that could lead to incorrect estimates of energy requirements or costs.
- Execute a pilot-scale test campaign and revise material and energy balances and capital and operating cost estimates.
- Prepare detailed material and energy balances and capital and operating cost estimates.

Conceptual Design. The first task for the process engineer is to identify the chemical and/or physical processes to be used, develop a conceptual design for the process and create a flowsheet based on the design. This task may seem trivial, but it is extremely important that the technology developer is able to effectively explain their process concept and illustrate it in a flow diagram.

A chemical process involves a chemical reaction that changes the composition of the inlet stream compared to the outlet stream (in this case removing CO_2). A physical process involves a change in temperature, change in pressure, physical separation of phases, system holdup, fluid phase changes, or a change in composition due to physical diffusion, for example. For each identified step in the process, all inlet and outlet process streams and all inlet and outlet utility streams are identified and defined. All of the potential unit operations that could achieve the desired chemical process, physical process, or combination thereof are also identified. Then the appropriate unit operation is selected to represent each process on the flowsheet, and a written process description is developed to accompany the flowsheet that describes the objective of each unit operation.

Material and Energy Balances. During this step, chemical and physical property data gaps may be identified. This is a key task, but it is sometimes overlooked by the technology developer. It often requires multiple iterations between the technology developer and process engineer to obtain a satisfactory and valid material and energy balance. The material and energy balances can be done manually, but in many cases it is preferable to use a commercial process simulator.

Process simulation models require a thorough cross-check of the simulator physical property, thermodynamic property, and chemistry packages via off-line hand calculations using literature references. Specific examples include heat capacities, thermal conductivities, densities, activity coefficients, Henry's constants, vapor pressures, solubilities, reaction equilibrium constants, and enthalpies of reaction. Some process simulators have regression tools that allow for literature data to be input and then the process simulator will regress the data and develop a model to fit the data. This is a useful feature; however, it must be used with caution and the regressed models must be closely checked. In some cases multiple property packages may be required for a single process simulation. Simulation chemistry packages need to have the correct order of reaction and all species (including reactants and products) in the database. A couple of examples of things to watch out for include using the Peng-Robinson equation of state around the CO_2 critical point as it has been an inaccurate representation in the past, and limited ability of process simulations to accurately represent solutions with high ionic strength.

A complete material and energy balance should be performed around each unit operation, in addition to the material and energy balances around the entire process. Key physical and chemical property data that are missing and key assumptions used in the absence of these data should be noted.

Obtain Missing Data. Experimental data are often lacking for emerging technologies. Based upon material and energy balance items evaluated in the previous step, a list of experimental data required to refine the process simulation is created, along with any key assumptions used in place of the data.

Example data needed to represent solvent-based processes includes:

- Pure component molecular weight
- Pure component vapor pressure data
- Binary solvent/CO_2 vapor-liquid equilibrium at absorber and regenerator conditions (temperatures, partial pressures) – this is extremely important data from the perspective of cross-checking circulation rate and the enthalpy of reaction as a function of absorber and regenerator temperatures
- Binary solvent/water vapor-liquid equilibrium
- Ternary solvent/water/CO_2 vapor-liquid equilibrium
- Henry's constant for inert gases in solvent at equilibrium
- Liquid/liquid or liquid/solid phase equilibria data (if applicable)
- Reaction stoichiometry
- Density (CO_2 loaded and unloaded solutions)
- Viscosity (CO_2 loaded and unloaded solutions)
- Heat capacity (CO_2 loaded and unloaded solutions)
- Solvent degradation rate, including effects of impurities (O_2, NO_X and SO_X, for example)

Example for data needed to represent adsorption processes includes:

- Bulk density of adsorbent material
- Heat capacity of adsorbent material
- Geometry of adsorbent material
- Pressure drop characteristics through bed of adsorber material
- Adsorption isotherms with relationship between sorbent mass loading of CO_2 versus temperature and CO_2 partial pressure
- Equilibrium capacity for water and other impurities
- Effects of impurities

Equipment Sizing. Part of this task involves the development of a matrix of technical challenges for each unit operation, which can be helpful in drafting test plans for bench-scale or pilot-scale testing that will be conducted to obtain key information.

These types of technical challenges include:

- Minimum and Maximum design pressures and temperatures (compared to operating pressures and temperatures)
- Materials of construction
- Mass or heat transfer limitations (on CO_2 removal rate)
- Pressure drop
- Avoiding undesirable phase changes or precipitation
- Minimizing atmospheric emissions
- Identifying opportunities for heat recovery or integration
- Identifying opportunities to recycle process and utility streams (and avoid makeup requirements)
- Selecting a continuous versus a batch process, and defining cycle times for batch processes
- Selecting a heat transfer medium

Key information is needed for specific equipment sizing. The following types of example information are typically needed for equipment sizing:

- Blower – efficiency, pressure rise, flow rate
- Compressors – efficiency, number of stages, flow rate, maximum discharge temperature
- Turbine – efficiency, flow rate
- Pump – efficiency, pressure rise, flow rate, fluid density
- Heat Exchanger – estimated heat duty process and utility temperatures (supply and return), overall heat transfer coefficient, heat exchanger type
- Separator – residence time
- Column Diameter – liquid rate, gas rate, packing type, approach to flood
- Column Height – rate of chemical reaction between solvent and CO_2
- Tank – residence time
- Membrane Module – specific surface area, pressure drop, permeance
- Adsorption Module – heat of sorption, sorbent loading, sorbent capacity
- CO_2 Dehydration Unit – dewpoint specification for water, dehydration medium
- Reclaimer – rate of heat stable salt removal

Preliminary Economic Analysis. The preliminary economic analysis is generally refined as new data become available and may lead to recommendations to change one or more aspects of the process. Below is a list of relevant literature from the Department of Energy that can assist in performing an economic analysis. Note: this is not an all-inclusive list.

- DOE/NETL-2010/1397 "Cost and Performance Baseline for Fossil Energy Plants Volume 1a: Bituminous Coal (PC) and Natural Gas to Electricity (Revision 3, July 2015)"
- DOE/NETL-2015/1726 "Quality Guidelines for Energy Systems Studies: Performing a Techno-Economic Analysis for Power Generation Plants"
- DOE/NETL-341/013113 "Quality Guidelines for Energy Systems Studies: Capital Cost Scaling Methodology (January 2013)"
- DOE/NETL-2011/1455 "Quality Guidelines for Energy Systems Studies: Cost Estimation Methodology for NETL Assessments of Power Plant Performance (April 2011)"
- DOE/NETL-2012/1557 "Current and Future Technologies for Power Generation with Post-Combustion Capture (March 2012)"
- DOE/NETL-2009/1366 "Research and Development Goals for CO_2 Capture Technology"

For the economic analysis, the energy requirements for CO_2 capture and compression are estimated. The energy requirements include electricity from blowers, pumps, and compressors and thermal input for regeneration. Energy requirements for CO_2 capture from power plants are described in terms of reducing or derating the energy produced by the plant. This energy requirement is sometimes referred to as the parasitic energy requirement for the capture process.

The heat input required for the process, or thermal derating, is often expressed in terms of an equivalent electrical derating. For example, the steam required for solvent/sorbent media regeneration is converted to an equivalent electrical derating from the power generated by the power plant by estimating the equivalent amount of electric power that could have been produced using the amount of steam required for the capture process in a steam turbine.

The temperature and pressure of the steam are important factors to determine the equivalent electrical derating – the higher the temperature and pressure, the larger the equivalent electrical energy penalty for steam usage. If there is a considerable gap between the delivery pressure and temperature of the steam and the required operating pressure of the steam for process regeneration, it may be justified to add a power recovery turbine to recover some additional electricity from the superheated steam prior to use in the CO_2 capture process.

In the case of a power plant CO_2 capture process, the energy requirements must be accounted for in conjunction with the overall power plant. When the CO_2 capture process is a retrofit to an existing power plant, the equivalent electrical derating for capture and compression is subtracted from the gross electrical capacity of the power plant. When part of a greenfield, or new power plant carbon capture facility, an escalation factor is applied to the base power plant based on the energy requirements for the capture facility such that after the parasitic energy requirements are applied to the "escalated" plant, the net electrical output meets the required value.

Selected equipment from the process flowsheet are sized and costed. Resources such as vendor quotes or internally developed cost curves can be used to estimate the purchased equipment costs based upon equipment sizing. For critical equipment or equipment with high purchased equipment costs, it is preferable to work with vendors to estimate purchased equipment costs. Once the equipment costs have been developed, preliminary total plant costs can be estimated using a standard factored approach to account for labor, engineering, and process/project contingencies.

Evaluation of Critical Process Criteria. A critical task for the process engineer to perform during the early stages of development is to evaluate the process for any potential flaw that can be identified and corrected prior to getting too far into the development process. While the identification of such deficiencies is not welcome news for the technology developer, the cost of correcting them becomes much more expensive in the later stages if they are not addressed as early as possible in the capture technology design processes. Several examples that might be overlooked early are given below:

- High sorbent degradation or loss rates that, when coupled with a high unit cost for the sorbent, result in prohibitively high costs per ton of CO_2 captured
- Underestimating the overall energy requirements for the process. The system boundaries must be properly defined so that all energy requirements of the capture process are properly accounted for in the system energy balance.
- Underestimating the impact of various factors which can result in the equipment being prohibitively large (e.g. slow absorption or desorption kinetics, high viscosity solvents, low capacity solvents)
- Failure to compare the estimated costs with the costs of the current state-of-the-art processes
- Underestimating the impact of exotic materials requirements on the equipment capital cost
- Overlooking features of the process that will prevent the process from producing product streams that meet purity specifications, which could result in additional costs from additional unit operations necessary to further process the CO_2 stream (or other end product stream if some other CO_2 derivative product is produced like fertilizer, methane, etc.)
- Assuming that the economics will be "better" if the sorbent has a low heat of absorption for CO_2
- Failure to identify thermodynamic inconsistencies

The last two items are addressed in more detail below.

It has been commonly assumed by many that having a lower heat of absorption/desorption for CO_2 always leads to lower overall energy requirements for the entire capture process. However, this is deceptive for a couple of reasons. Perhaps the best example is pure water. The heat of absorption of CO_2 in water is approximately 19 kJ/mole CO_2 at ambient temperature (Gill and Wadso, 1982, Caroll 1991), whereas for MEA it is approximately 80 to 85 kJ/mole over the expected range of operational loadings (Kim, 2007). However, using water as the solvent at typical operating conditions for CO_2 capture would obviously be uneconomic due to the high circulation rates that would be required, which would result in prohibitively large equipment and associated capital costs.

A second reason is that a low heat of absorption results necessarily (for thermodynamic reasons discussed later) in a low temperature-dependence of the CO_2 partial pressure. This means that in general a higher temperature (or lower partial pressure of CO_2) is required to regenerate the sorbent when compared to the same sorbent/process if the heat of absorption is higher. Furthermore, at a fixed temperature, an increase in the heat of absorption leads to a higher required heat input, but also a higher pressure in the regenerator; as a result, the power required to compress the CO_2 from regenerator pressure to pipeline pressure decreases. In some cases, the benefit of decreased compression work outweighs the increase regenerator thermal input and secondary effects on circulation rate. The trade-off and interaction between heat of absorption and the overall energy requirements is complex and has been extensively researched by multiple groups (Rochelle, 2007).

A common thermodynamic inconsistency arises when the specified value for the heat of absorption for the sorbent is inconsistent with the process conditions reported for regenerating the sorbent. For example, there are rigorous thermodynamic constraints (e.g. the van't Hoff equation) that relate the temperature dependence of the partial pressure of CO_2 over a loaded sorbent to the heat of absorption. A consequence of this constraint is that it is not possible to have a sorbent that has a low heat of absorption and yet is easily regenerated without raising the temperature or reducing the CO_2 partial pressure. Usually, after properly accounting for these constraints, the temperatures required for regeneration are higher than originally estimated (resulting in higher energy costs) and/or the CO_2 product pressure is lower – resulting in higher CO_2 compression costs or process steam requirements.

Another example is what is known as a "temperature cross" or "pinch". When two process streams are being processed in a heat exchanger to recover heat, it is not uncommon to have preliminary calculations showing that a large amount of energy recovery is feasible. CO_2 capture processes usually require very close temperature differences between the streams being exchanged, and a more rigorous analysis of the heat exchanger is required, leading in some cases to the identification of a situation where the maximum amount of heat that can be recovered is limited due to the presence of the temperature cross or pinch. The revised calculations usually reveal that the energy requirements are higher than originally estimated due to the thermodynamic constraints on the heat exchanger that limit the amount of useful heat that can be recovered.

Despite having access to state-of-the-art commercial process simulation software, these types of errors can still occur. It is not uncommon for emerging CO_2 capture processes to use chemical and physical processes that are not adequately represented by the "default" properties packages available with the simulation software. Frequently, these properties are estimated using a variety of methods (e.g. using UNIFAC to estimate vapor-liquid equilibrium). Systems involving concentrated electrolyte solutions are also notoriously difficult to model properly due to both the lack of specific property data and the inability of current models to accurately represent concentrated electrolytes. The only known way to address these problems is to have all of the required experimental data in hand and then perform separate calculations to verify all process simulation results.

Another important thermodynamic constraint that process engineers must be aware of is the concept of minimum work. There are many textbooks with excellent discussions of this topic, so these are not going to be repeated in detail in this paper (Smit 2014, Wilcox 2012, McGlashan 2007). The primary constraint that applies in the context of most CO_2 capture processes is the minimum work required to "de-mix" or separate the CO_2 from the flue gas, and then minimum work required to raise the pressure of the pure CO_2 to meet pipeline specifications. These minimum thermodynamic requirements are path-independent, and apply to any CO_2 capture process.

The minimum work required to separate CO_2 from other gases can readily be calculated from basic thermodynamic relationships. As a simple example illustration of this, the minimum work required to separate a mixture of 12 mole % CO_2 and 88 mole % nitrogen into pure streams of CO_2 and nitrogen, at ambient pressure and 25 °C, is approximately 7 kJ/mole. Similarly, the minimum energy requirement needed to raise the pressure of the product CO_2 from ambient pressure to 2,200 psig is approximately 11 kJ/mole, which results in a total minimum work requirement of approximately 18 kJ/mole.

This simple example shows that any estimates of energy requirements lower than this cannot be correct. In practice, the calculations are more complex and have to consider the process specific temperatures, any waste heat integration features, and the specific compositions of the feed and product streams, but this type of analysis has proven useful in establishing a bound beyond which further energy reductions cannot be achieved. The concept of the minimum theoretical work has also been addressed by DOE in some of their work. (DOE, 2011).

Another thermodynamic constraint specific to absorption and adsorption processes is the tradeoff between the heat of absorption and the circulation rate (which is directly correlated to the solvent capacity, defined as the difference between the lean and rich loadings). There is a trade-off between these two properties, and they cannot be examined individually.

The heat of absorption has a direct effect on the energy requirements for regeneration, which affects the parasitic load of the capture plant; this in turn has a direct effect on the size of the power plant and the capture facility.

The solvent capacity has a secondary effect on energy requirements, as the circulation rate affects pumping power requirements and the latent heat required to heat and cool the solvent; it also has a direct effect on the sizing of any process equipment involved in the liquid circulation loop.

Figure 8 shows an example series of isotherms (constant temperature curves) that plots solvent loading as a function of CO_2 partial pressure (hypothetical pressure if CO_2 gas was the only component in the mixture) over a range of temperatures, and how this information can be used to estimate the operating capacity of the solvent as a function of operating conditions. The working capacity is fixed once operating conditions are chosen for a solvent.

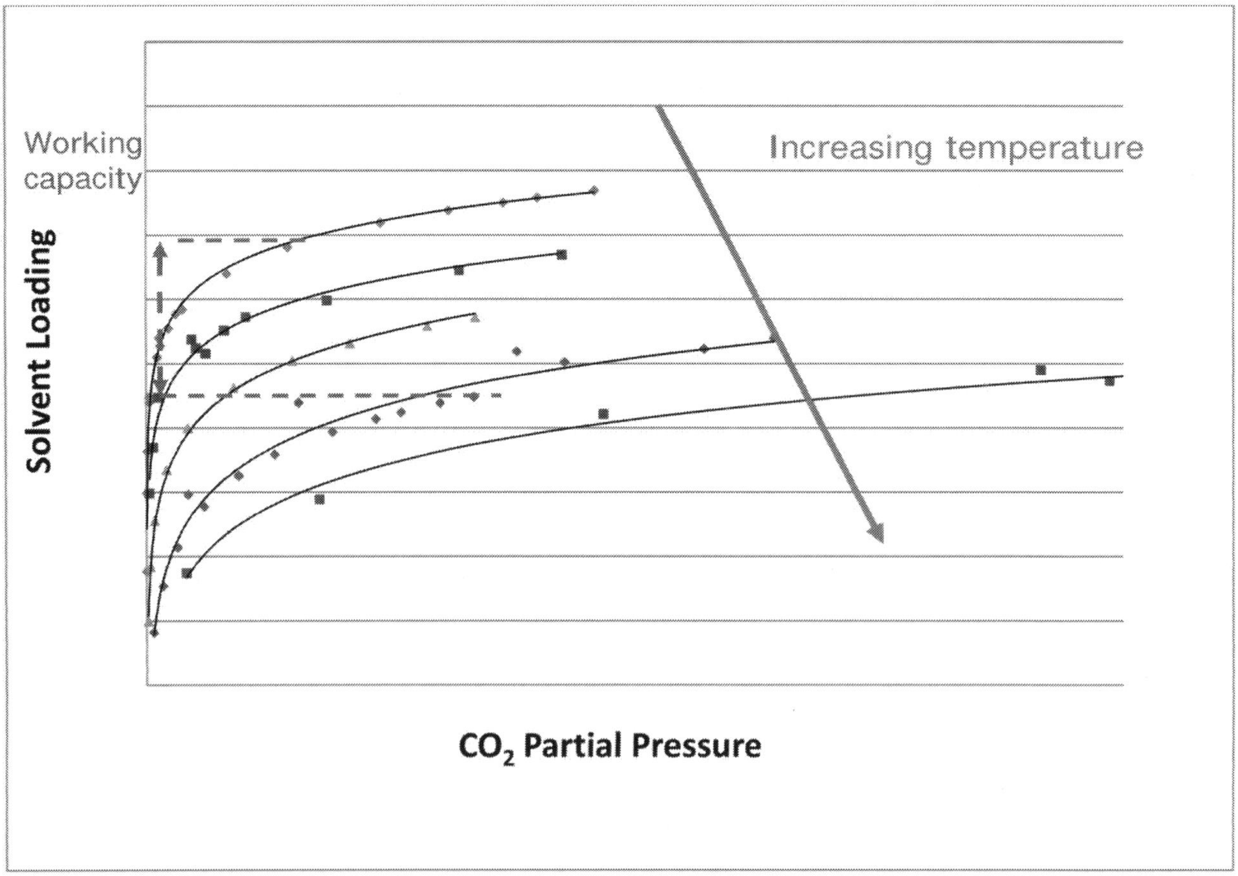

Figure 8. Operating Capacity of Solvent as a function of Operating Conditions

Pilot-scale testing. The objectives of pilot testing should be defined and used to develop a scope for the test plan. The test plan ensures that sufficient data are collected to satisfy objectives and address process design challenges. This is the time to address data gaps which were previously identified in the chemical and physical property data that were not resolved with experimental bench-scale tests.

It is desirable to use equipment for pilot-scale testing that is as similar as possible to equipment that would be utilized in full-scale plant.

In addition to developing the items previously discussed such as process description, process flow diagram, material and energy balances, equipment list, etc., the following may need to be developed to support pilot-scale testing. (Note: exact tasks will depend upon whether an existing pilot facility is used or a new pilot facility is constructed).

- Control strategy with key control loops
- Cause and effect diagrams
- Table of alarms and interlocks
- Startup, shutdown, and operating procedures
- Piping and materials specifications
- Control valve/safety valve sizing
- Define utility requirements
- Develop process and instrumentation diagrams (P&IDs)
- Instrument and line lists
- Develop an equipment layout including estimated weight and dimensions

Detailed material/energy balances and cost estimates. As previously stated, this task is contingent upon the collection of sufficient operating data during bench-scale and pilot-scale testing. This requires operating for a long enough period of time to collect sufficient data on operating conditions, energy performance, and impact on selected materials of construction and solvent stability. A process simulator can be used to revise the material and energy balances and note any major changes from the initial simulation and assumptions. The processes previously outlined in this paper can be used to revise the cost estimate.

Conclusions

The process engineer's task for CO_2 capture is broad and complex, requiring careful and rigorous process engineering work. This is good news for the process engineer looking for interesting work, but it may pose challenges for the technology developer or investor that needs results quickly. CO_2 capture process technology development presents many challenges and must include the acquisition of fundamental physical/chemical data through laboratory and pilot-scale testing, the preparation of detailed material/energy balances and economic analyses, and careful evaluation of the impact of thermodynamic constraints, flue gas impurities, and CO_2 source and product requirements.

References

Carroll, John J., John D. Slupsky, and Alan E. Mather. "The solubility of carbon dioxide in water at low pressure." *Journal of Physical and Chemical Reference Data* 20.6 (1991): 1201-1209.

Ciferno, Jared P., John J. Marano, Ronald K. Munson, "Technology Integration Challenges", *Chemical Engineering Progress*, August 2011.

DOE NETL 2012 Carbon Utilization and Storage Atlas.

EIA/Annual Energy Outlook 2013 Early Release.

Gill, S. J., and I. Wadsö. "Flow-microcalorimetric techniques for solution of slightly soluble gases. Enthalpy of solution of oxygen in water at 298.15 K." *The Journal of Chemical Thermodynamics* 14.10 (1982): 905-919.

Havens, Ken, "CO_2 Transportation", Kinder-Morgan CO_2 Company, June 5, 2008.

KAPSARC, "Carbon Capture and Storage: Technologies, Policies, Economics, and Implementation Strategies", 2012.

Kim, Inna, and Hallvard F. Svendsen. "Heat of absorption of carbon dioxide (CO_2) in monoethanolamine (MEA) and 2-(aminoethyl) ethanolamine (AEEA) solutions." *Industrial & Engineering Chemistry Research* 46.17 (2007): 5803-5809.

Matuszewski, Michael, J. Ciferno, and S. Chen. "Research and Development Goals for CO_2 Capture Technology." *US Dept of Energy, National Energy Technology Laboratory* (2012).

McGlashan, N. R., and A. J. Marquis. "Availability analysis of post-combustion carbon capture systems: minimum work input." *Proceedings of the Institution of Mechanical Engineers, Part C: Journal of Mechanical Engineering Science*221.9 (2007): 1057-1065.

NETL (National Energy Technology Laboratory), Quality Guidelines for Energy System Studies, CO_2 Impurity Design Parameters, August 2013.

Praxair, Overview of Industrial CO_2 Sources, EOR Carbon Management Workshop, December 2006.

Rochelle, Gary T. "Amine scrubbing for CO_2 capture." *Science* 325.5948 (2009): 1652-1654.

Smit, Berend, Jeffrey R. Reimer, and Curtis M. Oldenburg. *Introduction to Carbon Capture and Sequestration*. Vol. 1. London: Imperial College Press, 2014.

Wilcox, Jennifer. *Carbon Capture*. Springer Science & Business Media, 2012.

Oil Price: Effect on Carbon Emission

Enobong Bassey
School of Energy, Construction and Environment
Faculty of Engineering, Environment and Computing
Coventry University
Coventry
United Kingdom

Coventry University

Presentation Outline

- Crude Oil and Gas Profile
- Carbon Emission Scenarios
- Primary Drivers of Carbon Emission
- Carbon Emission Models
- Oil Price and CO_2 Emission Correlations

Crude Oil Price

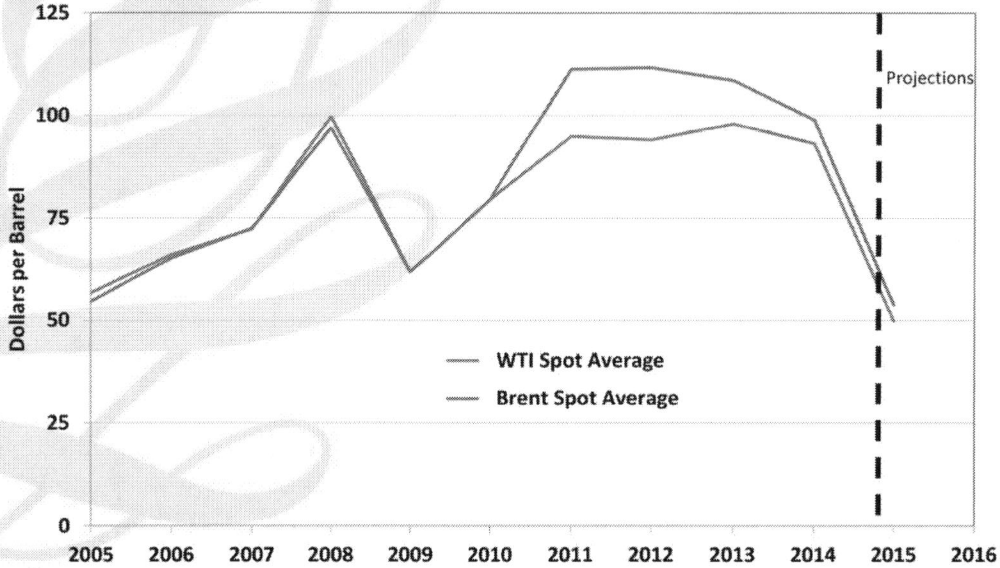

eia Source: Short-Term Energy Outlook

Natural Gas Price

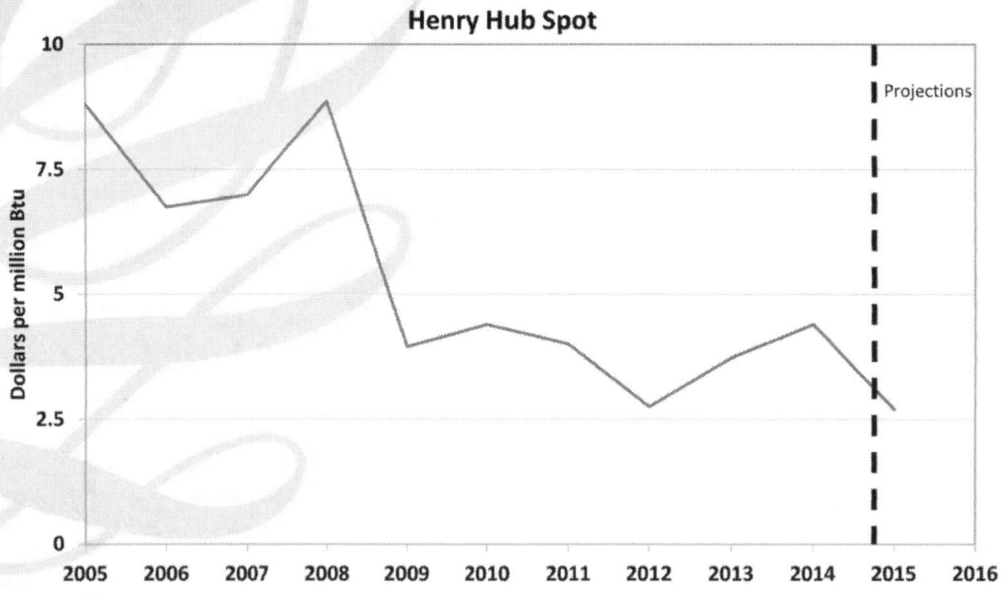

eia Source: Short-Term Energy Outlook

Total Petroleum and Other Liquids Production 2014

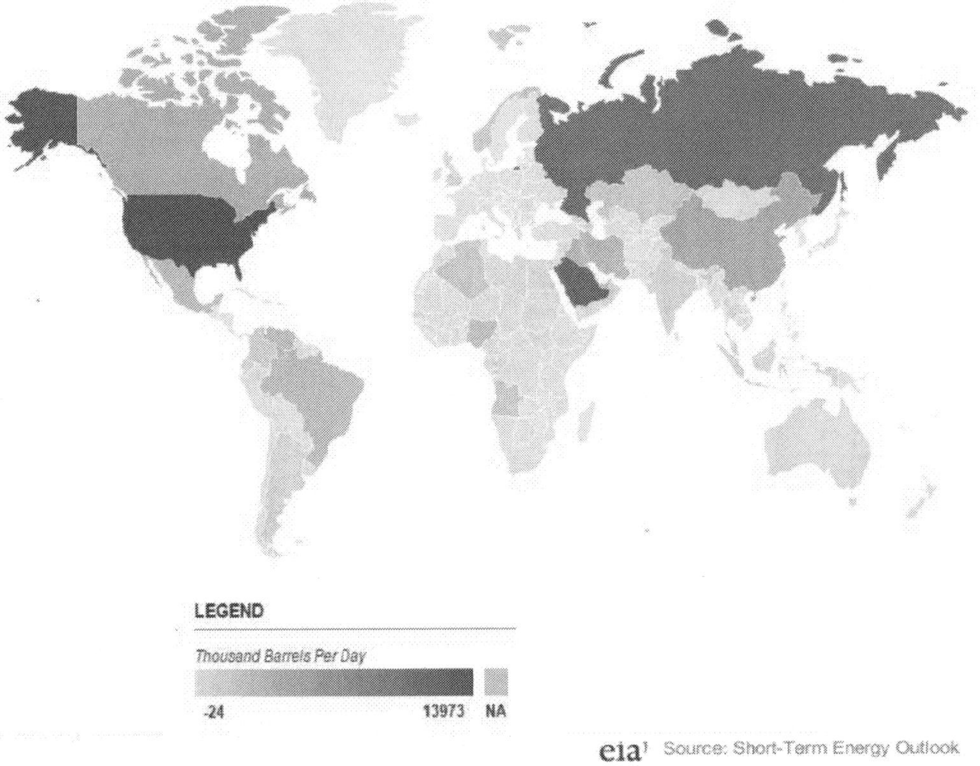

GHG Emissions Scenarios

The principal emission scenarios depend on:

- GDP
- Population
- Resource base
- Lifestyle

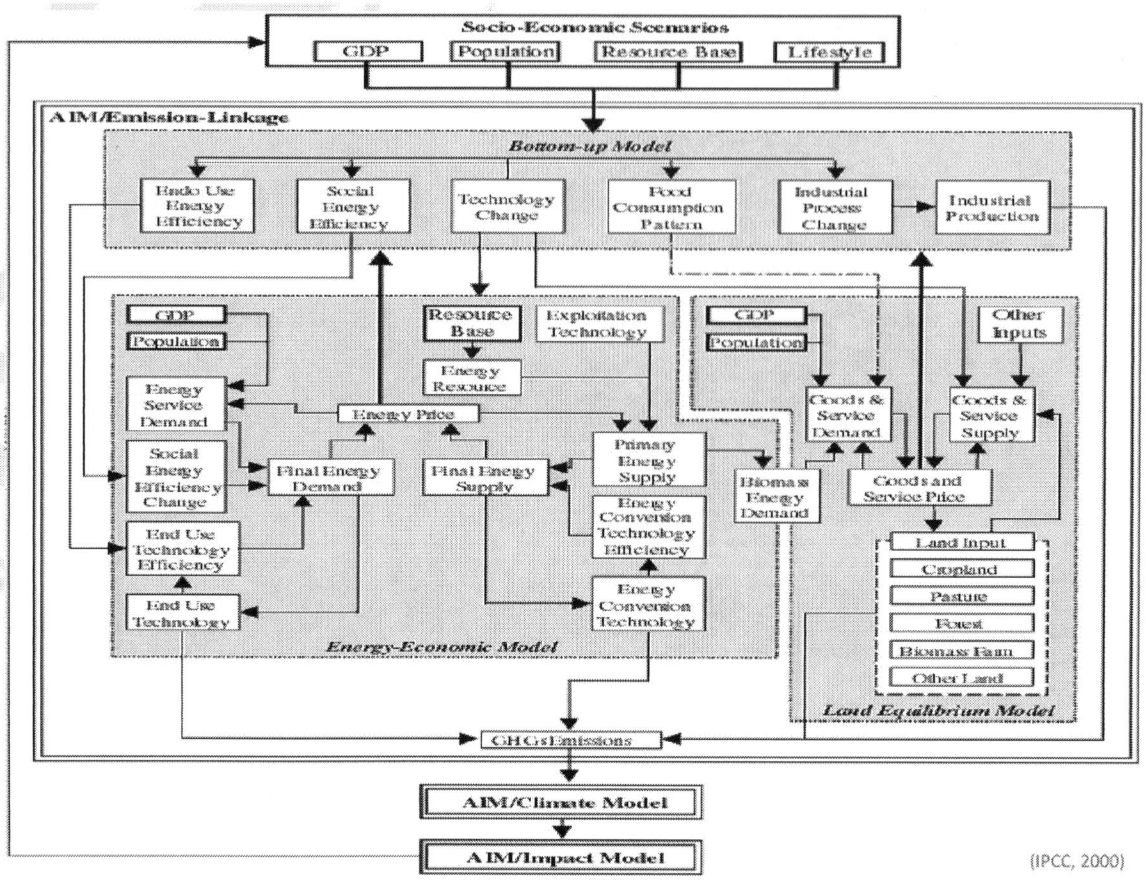

(IPCC, 2000)

Carbon Emission

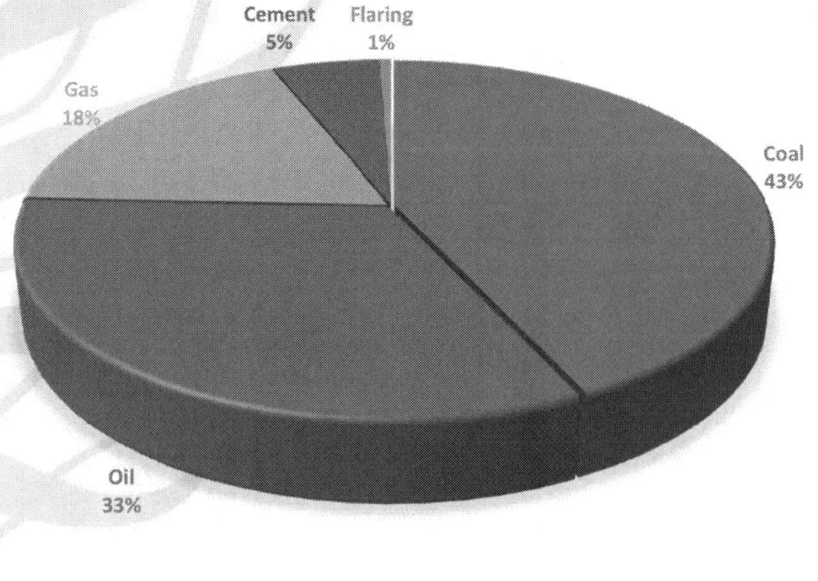

Source: CDIAC (2014)

GHGs Emission

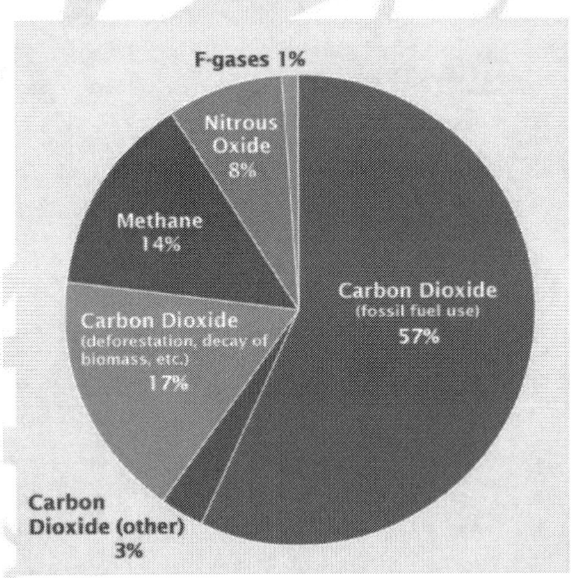

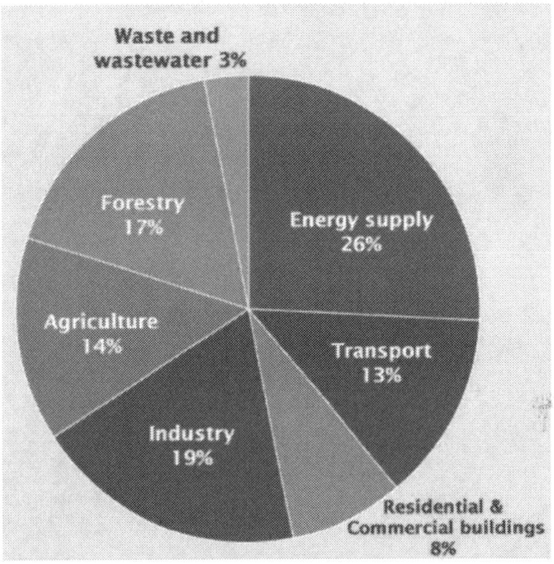

Source: IPCC (2007)

Global Carbon Emission

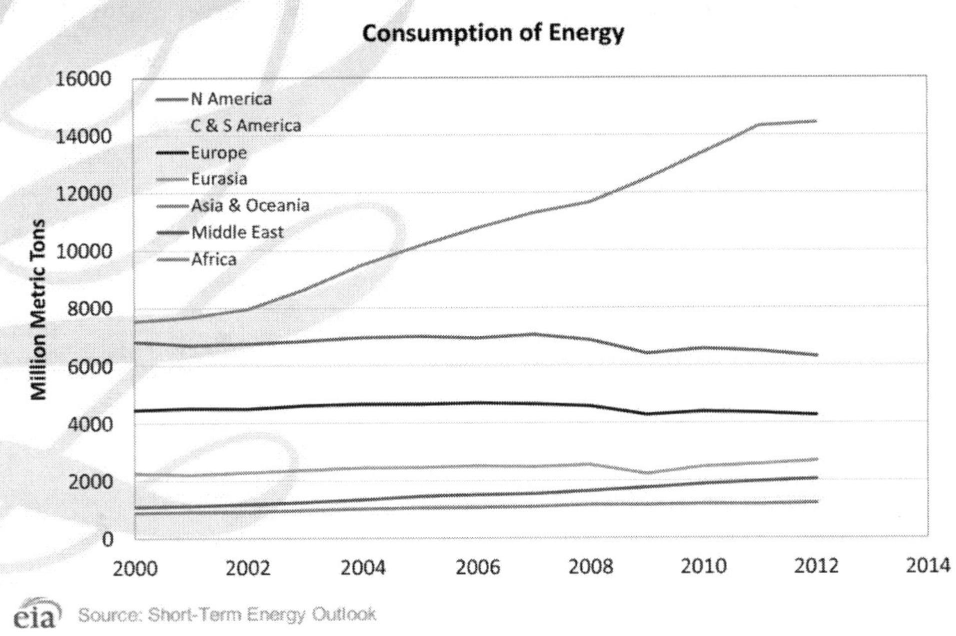

Carbon Emission: Major Contributors

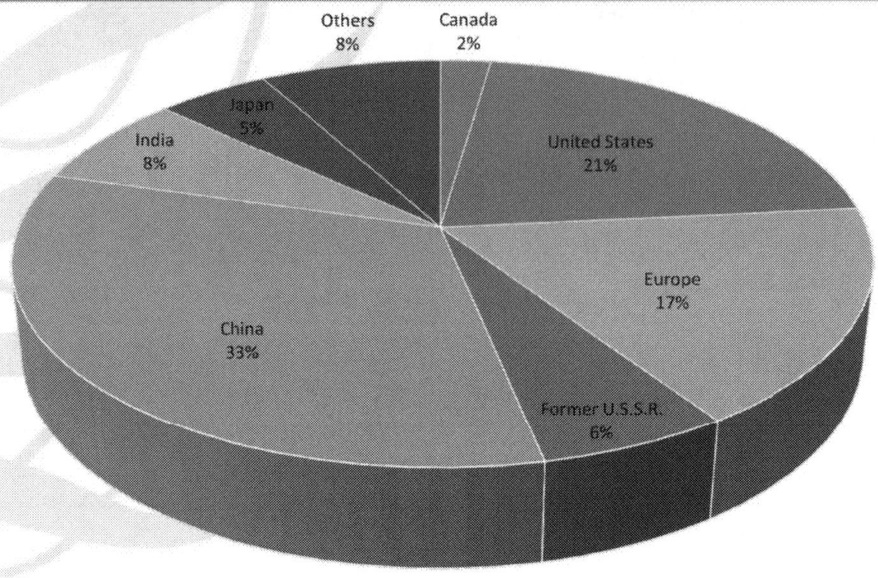

eia Source: Short-Term Energy Outlook

Carbon Emission

CO_2 emissions from fossil-fuel use and cement production in the top 5 emitting countries and the EU

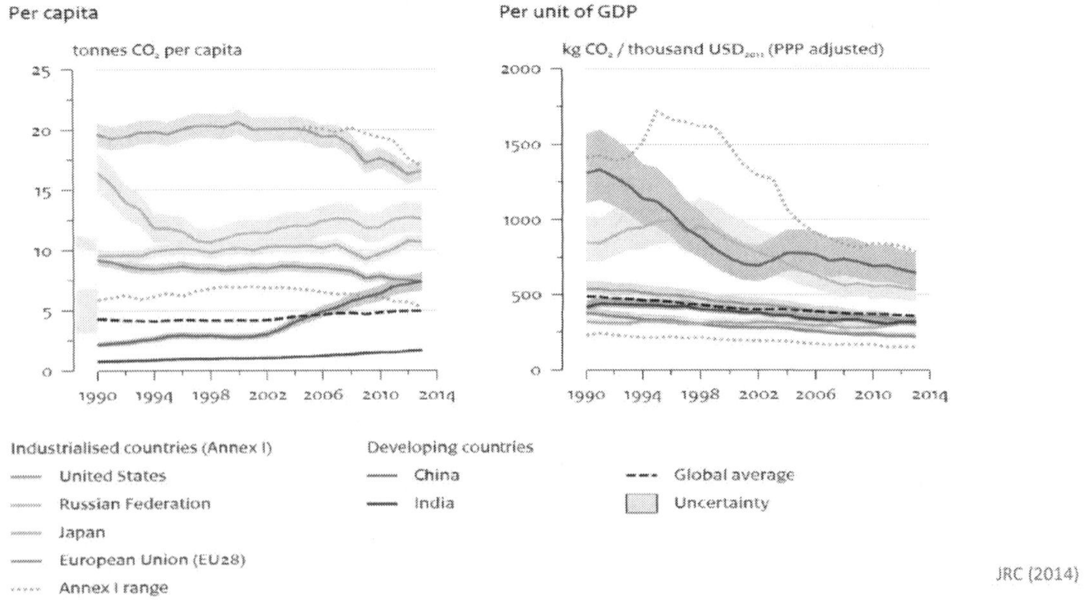

JRC (2014)

Carbon Emission per unit of Total Primary Energy Supply (TPES) in the top 5 emitting countries and the EU

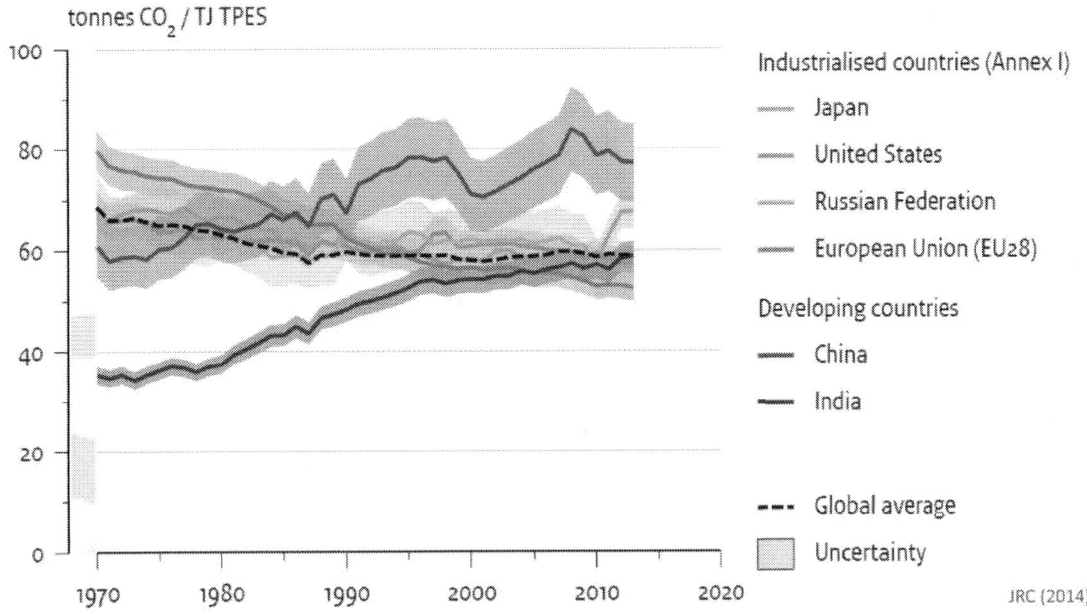

JRC (2014)

Carbon Emission: Total Primary Energy Supply by Type

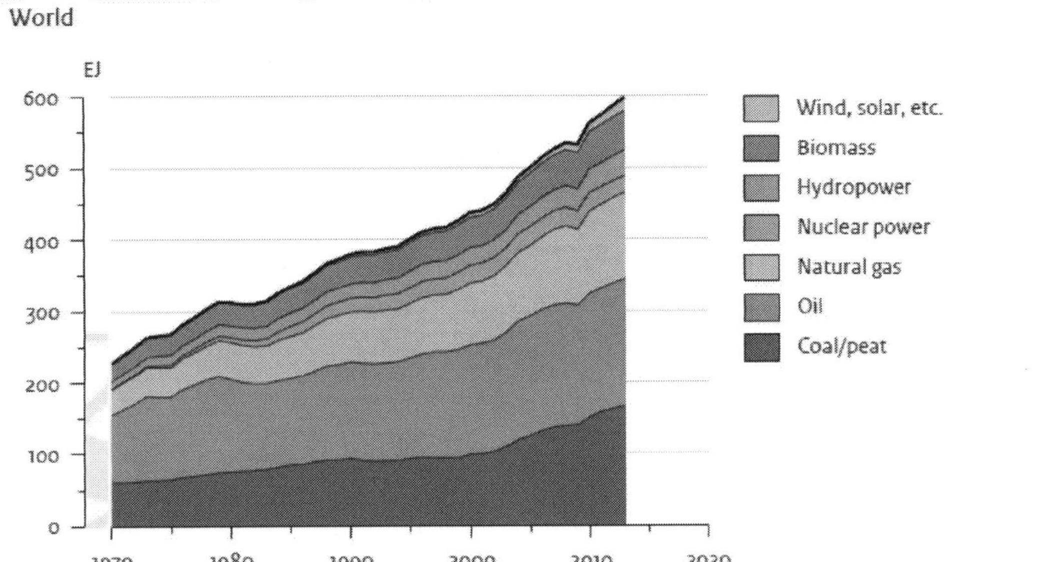

JRC (2014)

Carbon Emission: Total Primary Energy Supply by Type

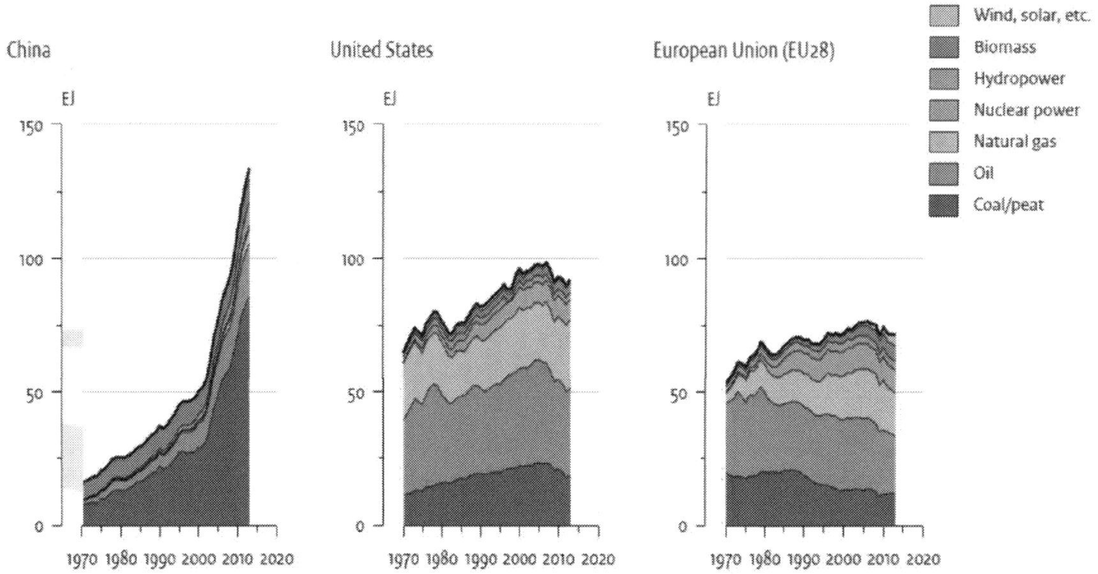

JRC (2014)

CO_2 Emissions (USA)

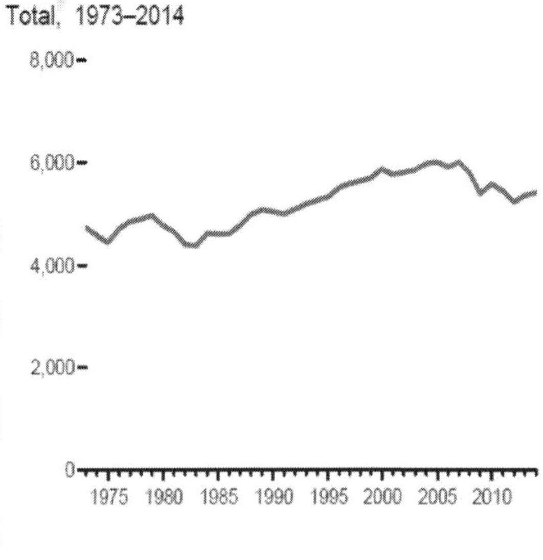

eia Monthly Energy Review, Oct 2015

CO_2 Emissions (USA)

Million Metric Tons of CO_2

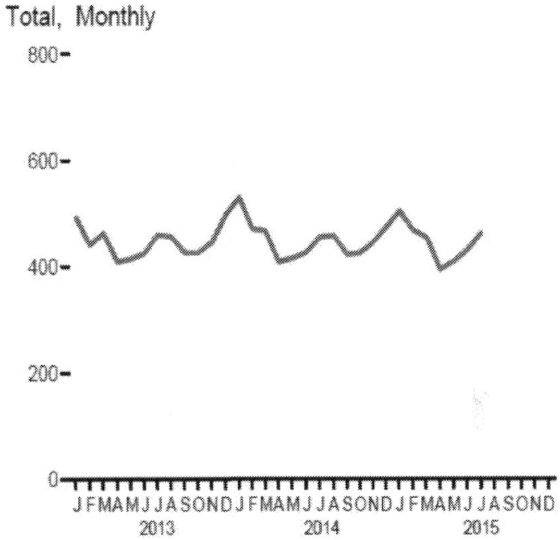

eia Monthly Energy Review, Oct 2015

CO_2 Emissions From Energy Consumption by Source (USA)

Million Metric Tons of CO_2

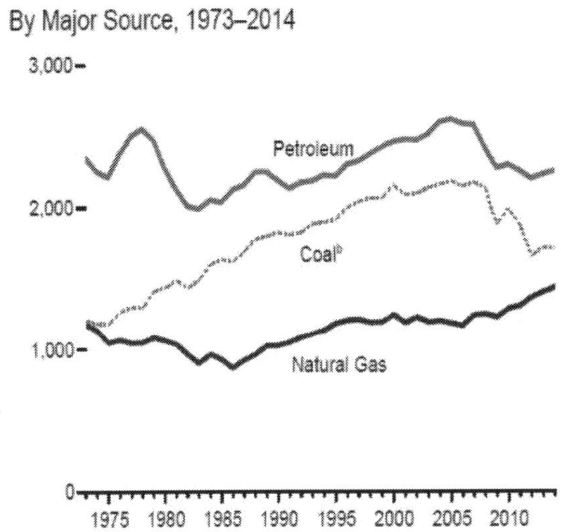

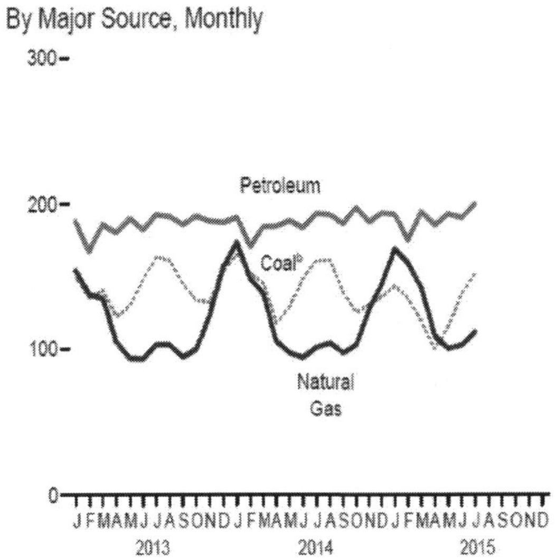

eia Monthly Energy Review, Oct 2015

CO_2 Emissions From Energy Consumption by Source (USA)

Million Metric Tons of CO_2

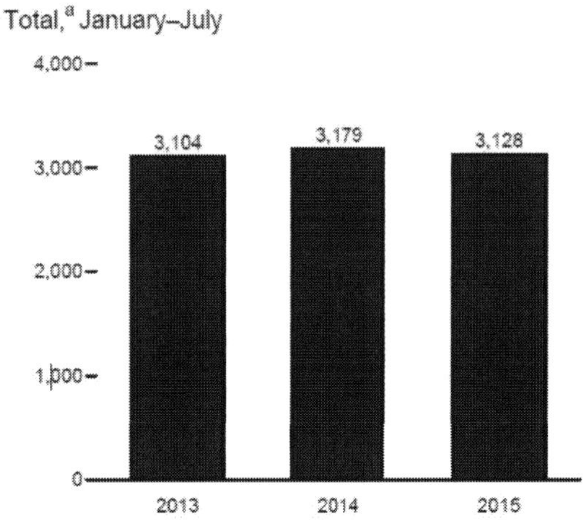

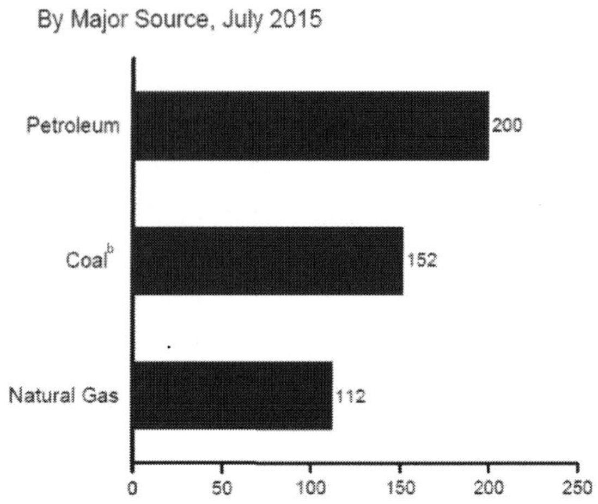

[a] Excludes emissions from biomass energy consumption.
[b] Includes coal coke net imports.

Monthly Energy Review, Oct 2015

CO_2 Emissions From Energy Consumption by Sector (USA)

Million Metric Tons of CO_2

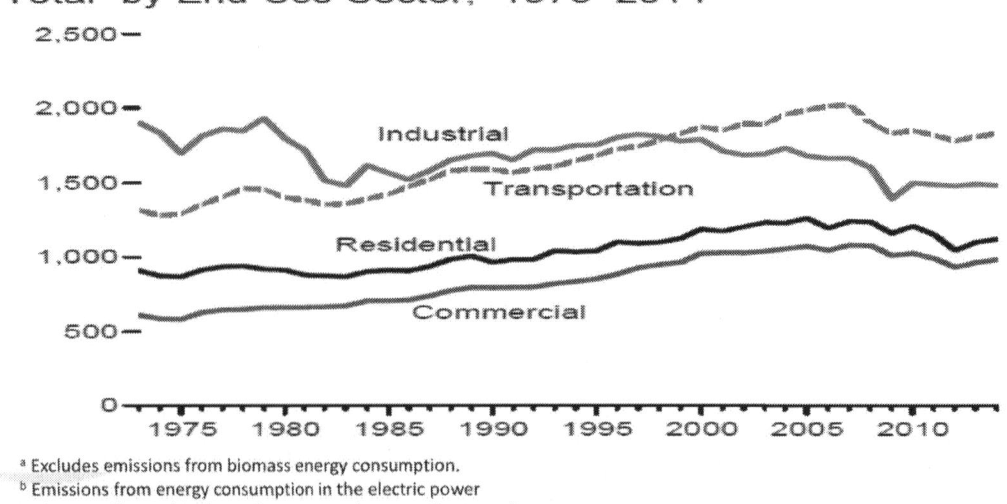

[a] Excludes emissions from biomass energy consumption.
[b] Emissions from energy consumption in the electric power sector are allocated to the end-use sectors in proportion to each sector's share of total electricity retail sales.

Monthly Energy Review, Oct 2015

CO_2 Emissions From Energy Consumption by Sector (USA)

Million Metric Tons of CO_2

Residential Sector by Major Source, 1973–2014

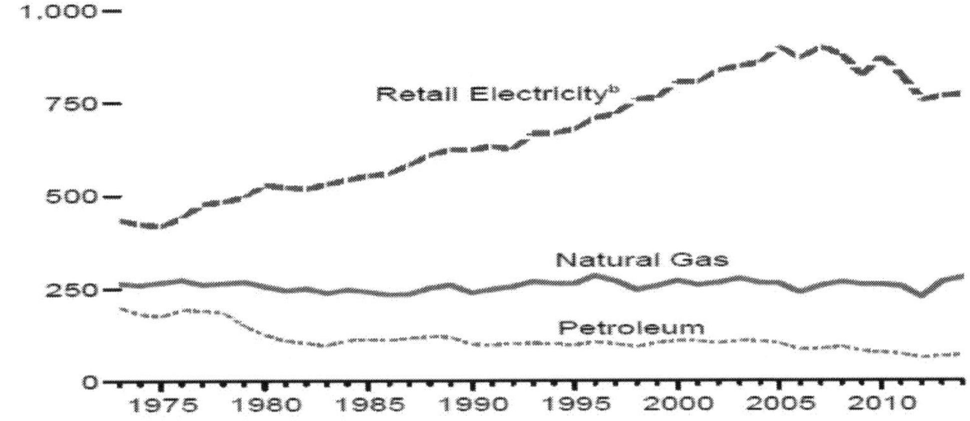

[b] Emissions from energy consumption in the electric power sector are allocated to the end-use sectors in proportion to each sector's share of total electricity retail sales.

eia Monthly Energy Review, Oct 2015

CO_2 Emissions From Energy Consumption by Sector (USA)

Million Metric Tons of CO_2

Commercial Sector by Major Source, 1973–2014

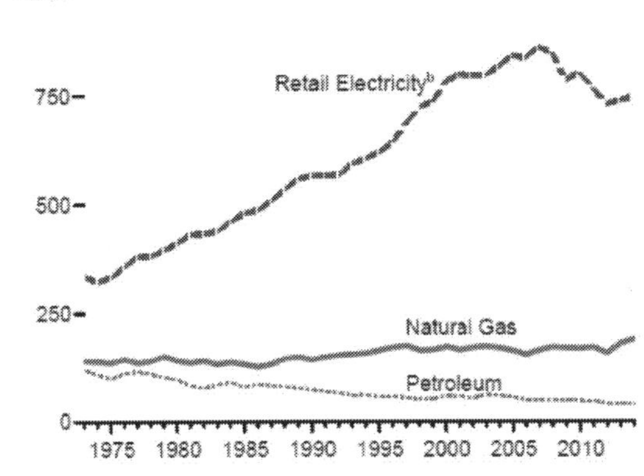

Industrial Sector by Major Source, 1973–2014

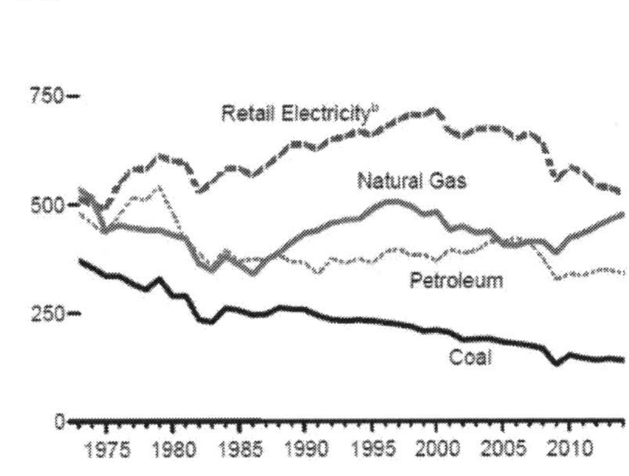

[b] Emissions from energy consumption in the electric power sector are allocated to the end-use sectors in proportion to each sector's share of total electricity retail sales.

eia Monthly Energy Review, Oct 2015

CO_2 Emissions From Energy Consumption by Sector (USA)

Million Metric Tons of CO_2

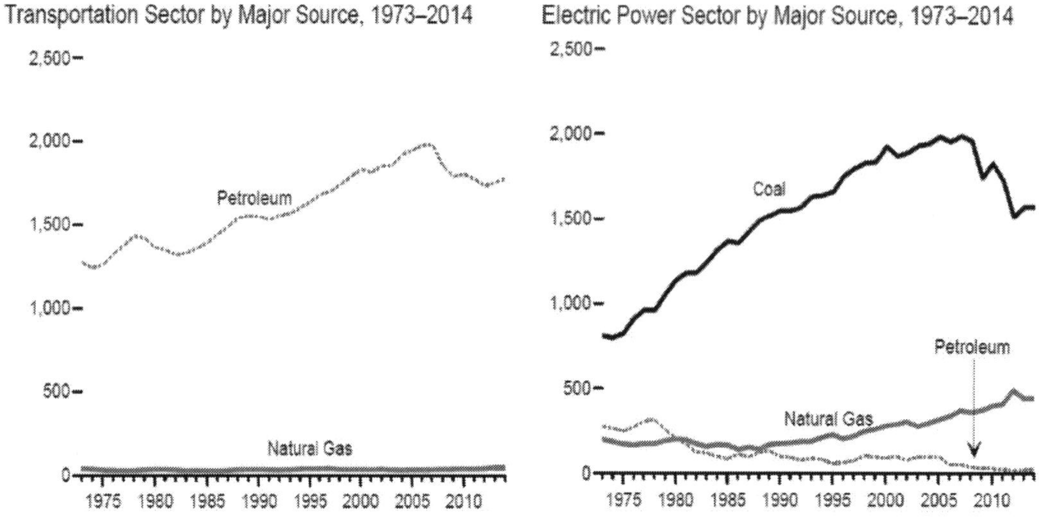

eia Monthly Energy Review, Oct 2015

Carbon Emission Models

- IPAT model (Ehrlich and Holdren 1971):

  $I = PAT$

  where,

  I = impact, P = population,

  A = affluence (per capita GDP), T = technology

- STIRPAT model known as Stochastic Impacts by Regression on Population, Affluence, and Technology (Dietz and Rosa 1997):

  $I_i = aP_i^b A_i^c T_i^d e_i$

  where, a = b = c = d = e =1

The Market Trend

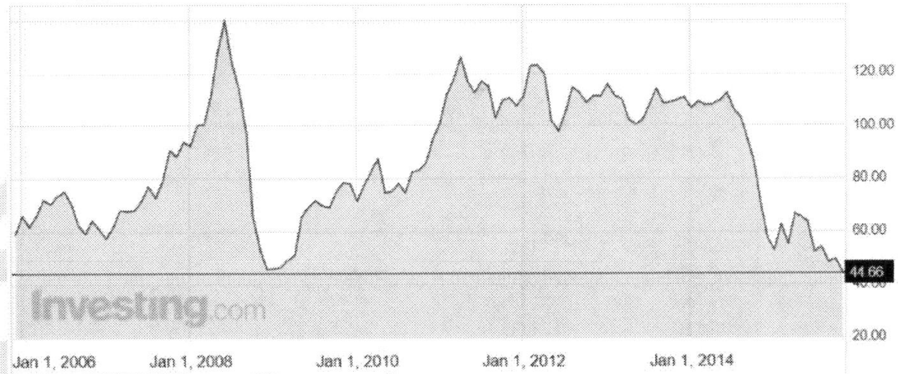

Brent Crude Oil

CO_2 Emission

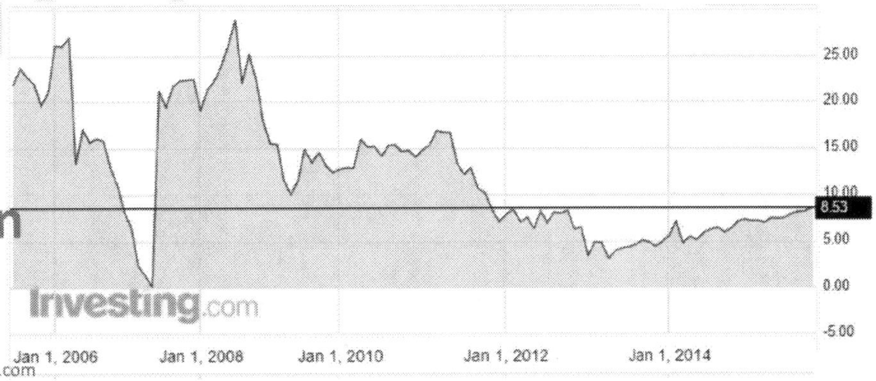

Source: www.investing.com

Brent Crude Oil Future

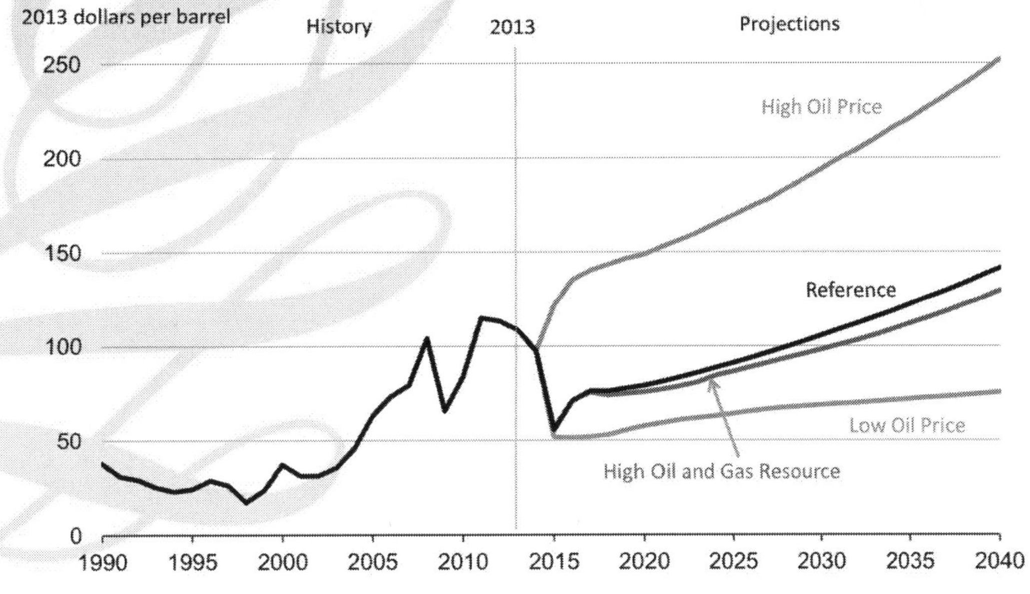

Source: EIA, Annual Energy Outlook 2015

Carbon Emission Projection

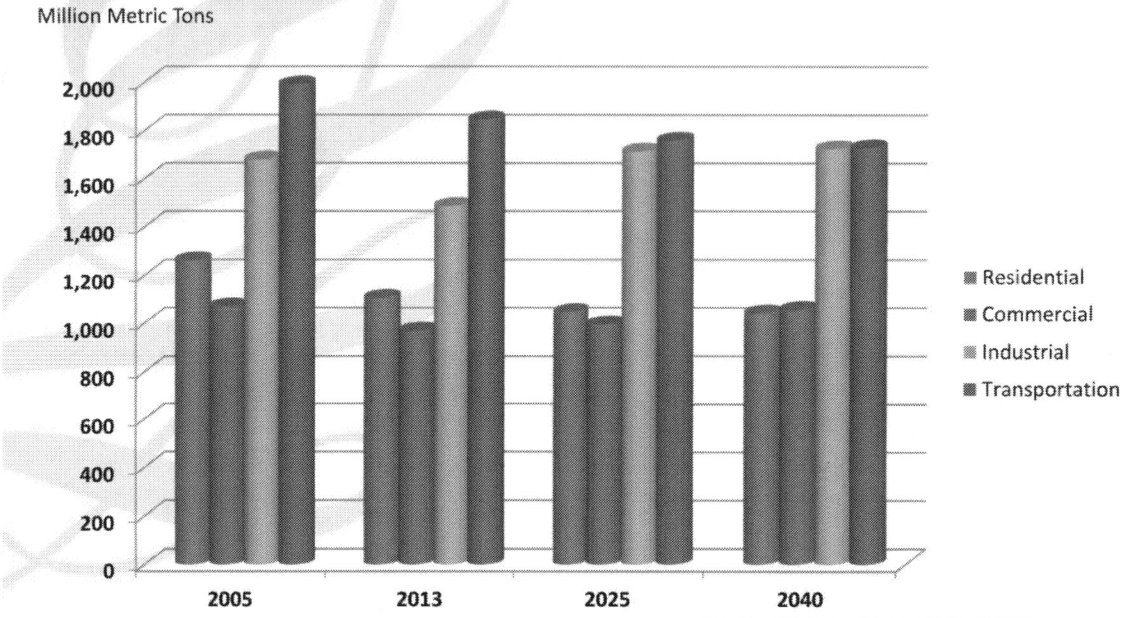

Source: EIA, Annual Energy Outlook 2015

Carbon Emission Projection

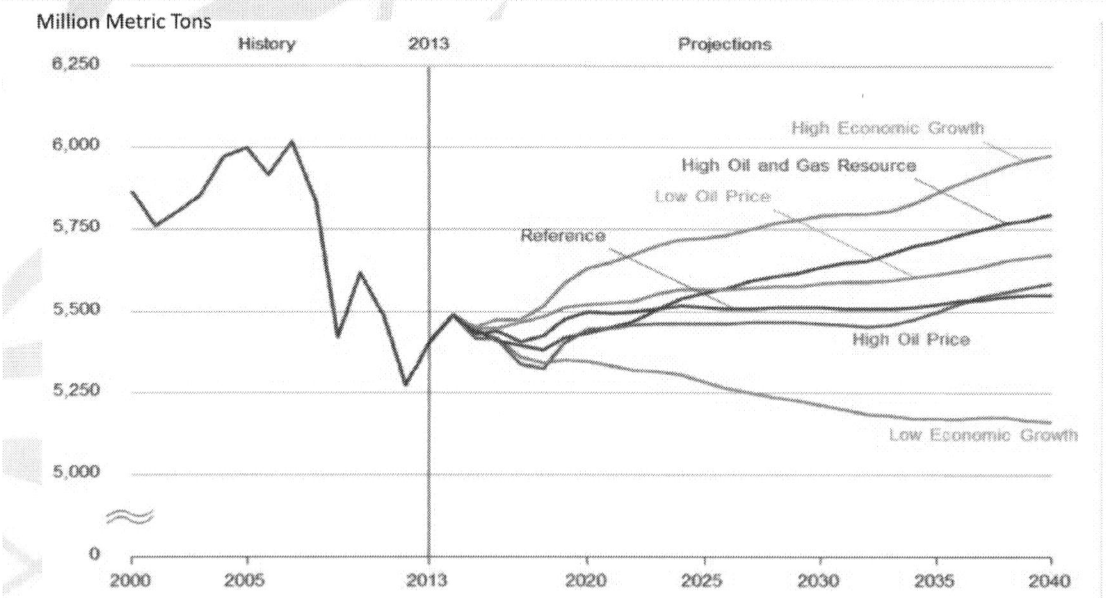

Source: EIA, Annual Energy Outlook 2015

Thank you

Coventry University

CCSI
Carbon Capture Simulation Initiative

Predictive Dynamic Model of a Carbon Capture System: Pilot Scale Validation at National Carbon Capture Center

Anderson Soares Chinen[a], Joshua C. Morgan[a], **Benjamin Omell[a]**, Debangsu Bhattacharyya[a], David C. Miller[b], John Wheeldon[c]

[a] *Department of Chemical Engineering, West Virginia University, Morgantown, WV 26506, USA*
[b] *National Energy Technology Laboratory, 626 Cochrans Mill Road, Pittsburgh, PA 15236, USA*
[c] *National Carbon Capture Center, 31800 Highway 25, North Wilsonville, AL 35186, USA*

CMTC 2015

NETL West Virginia University U.S. DEPARTMENT OF ENERGY

CCSI For Accelerating Technology Development

Rapidly synthesize optimized processes to identify promising concepts → Better understand internal behavior to reduce time for troubleshooting → Quantify sources and effects of uncertainty to guide testing & reach larger scales faster → Stabilize the cost during commercial deployment

National Labs	Academia	Industry
Berkeley Lab, Pacific Northwest National Laboratory, NETL, Los Alamos National Laboratory, Lawrence Livermore National Laboratory	Carnegie Mellon, Princeton University, West Virginia University, Boston University, The University of Texas at Austin	ADA, Alstom, B&W, GE, EPRI Electric Power Research Institute, Fluor, Phillips 66, AEP American Electric Power, Southern California Edison, Southern Company, ExxonMobil, Chevron, URS, Air Products, Eastman, aspentech, ANSYS, Schneider Electric, PSE

CCSI Carbon Capture Simulation Initiative NETL Berkeley Lab Lawrence Livermore National Laboratory Los Alamos Pacific Northwest U.S. DEPARTMENT OF ENERGY 2

Motivation

- Development of a Gold Standard baseline MEA model
  - Open source
  - Validation framework
  - Well documented
  - Uncertainties quantified
- Demonstrate as a Framework for proprietary systems
  - Methodology for robust, predictive models
- Steady state validation
- **Dynamic validation**

Deficiencies in Existing Steady State Models

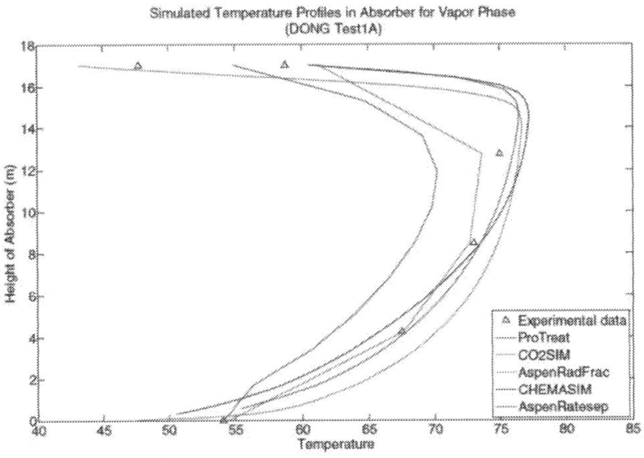

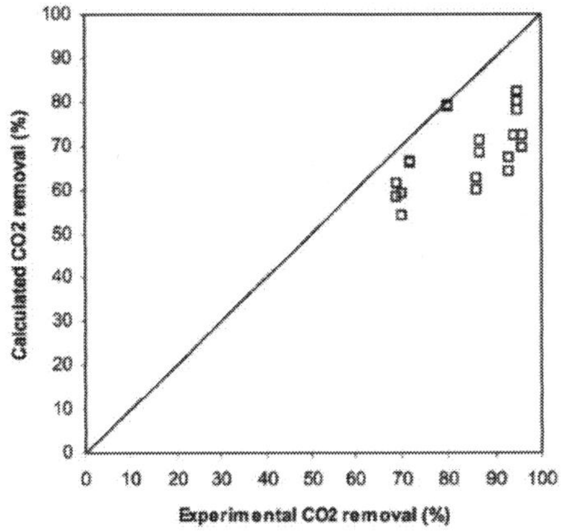

ProTreat-Optimized Gas Treating, Inc.; CO2SIM-NTNU/SINTEF CHEMASIM-BASF SE; AspenRatesep-modified by IFP

Luo et al., "Comparison and validation of simulation codes against sixteen sets of data from four different pilot plants", Energy Procedia, 1249-1256, 2009

Zhang, et al., Rate-Based Process Modeling Study of CO_2 Capture with Aqueous Monoethanolamine Solution, Ind. Eng. Chem Res., 48, 9233-9246, 2009

Deficiencies in Existing Dynamic Models

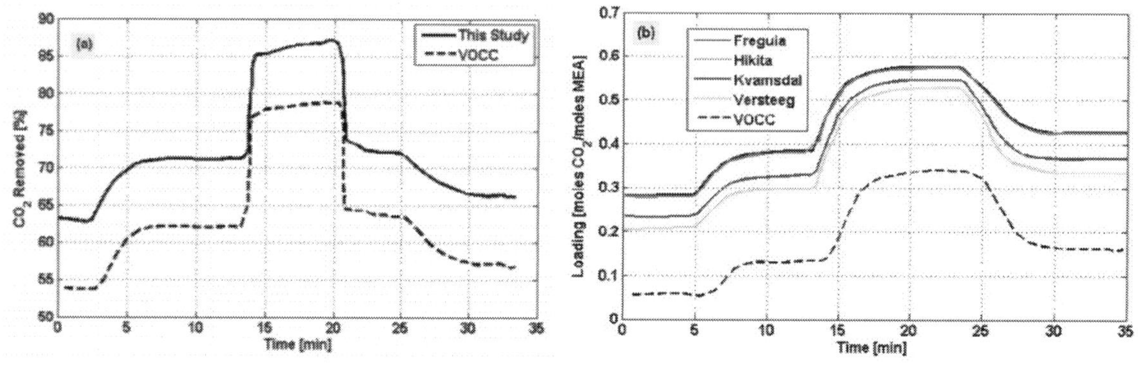

***Data from NTNU/SINTEF**

Hanne M. Kvamsdal, Actor Chikukwa, Magne Hillestad, Ali Zakeri, Aslak Einbu, A comparison of different parameter correlation models and the validation of an MEA-based absorber model, Energy Procedia, 4, 1526-1533, 2011

Outline

- Steady state model
- Dynamic model using Aspen Dynamics
- Test conditions
- Dynamic data reconciliation
- Results
- Conclusion

How to Develop a Gold Standard Model

- Property models
  - Valid for absorber and stripper operating conditions
- Hydraulic and mass transfer models
  - Developed simultaneously with relevant properties models using both WWC and packing data
- Steady State Validation
- Dynamic Validation

Physical Property Model Development

- Initial framework based upon the "Phoenix" model*
  - Developed by Prof. Rochelle's Group at UT, Austin
- Independent property models
  - Viscosity
  - Density/Molar Volume
  - Surface Tension
- Thermodynamic framework
  - Vapor-Liquid Equilibrium
    - Binary MEA-H_2O system
    - Ternary MEA-H_2O-CO_2 system
  - Heat Capacity
  - Heat of Absorption
  - Reaction Kinetics
    - Model developed for consistency with reaction equilibrium constants

***Jorge Mario Plaza, Ph.D. Dissertation, UT Austin, May 2012**

Integrated Mass Transfer Model Development

- Properties (such as diffusivity, viscosity, surface tension) as well as interfacial area, and mass transfer coefficients all affect mass transfer
- Data from both wetted wall column and packed column considered
- In Aspen Plus, simultaneous regression of these models not possible; thus solution can be sub-optimal
- FOQUS has the capability of simultaneous regression

Usual approach: Sequential regression

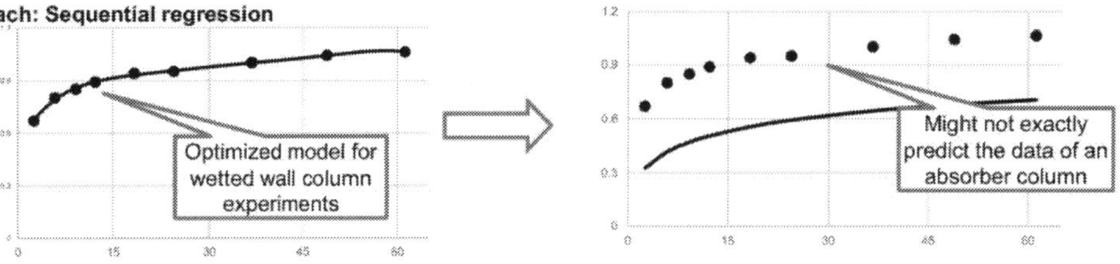

FOQUS capability: Simultaneous regression

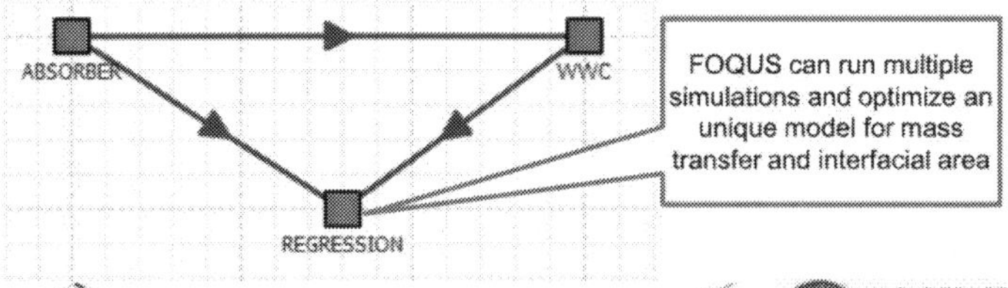

CCSI Carbon Capture Simulation Initiative | NETL | Berkeley Lab | Lawrence Livermore National Laboratory | Los Alamos National Laboratory | Pacific Northwest National Laboratory | U.S. DEPARTMENT OF ENERGY

9

Integrated Mass Transfer Model Results

- Final model form for hydraulics and mass transfer models:
  - Pressure drop: Billet and Schultes (1999)
  - Holdup: Tsai (2011)
  - Mass transfer coefficients: Billet and Schultes (1993)
  - Interfacial area: Tsai et al. (2012)
- Model parameters regressed for Mellapak plus^{TM} 252Y

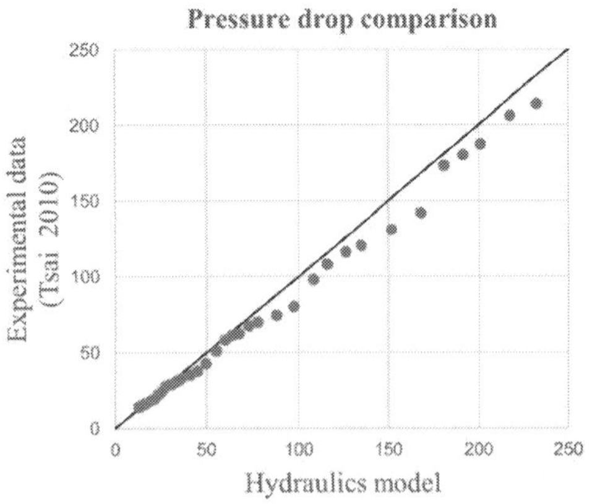

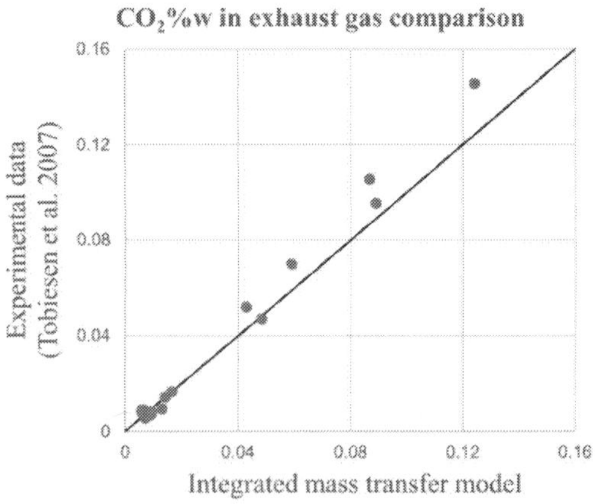

CCSI Carbon Capture Simulation Initiative | NETL | Berkeley Lab | Lawrence Livermore National Laboratory | Los Alamos National Laboratory | Pacific Northwest National Laboratory | U.S. DEPARTMENT OF ENERGY

10

Steady State Validation

Absorber Validation

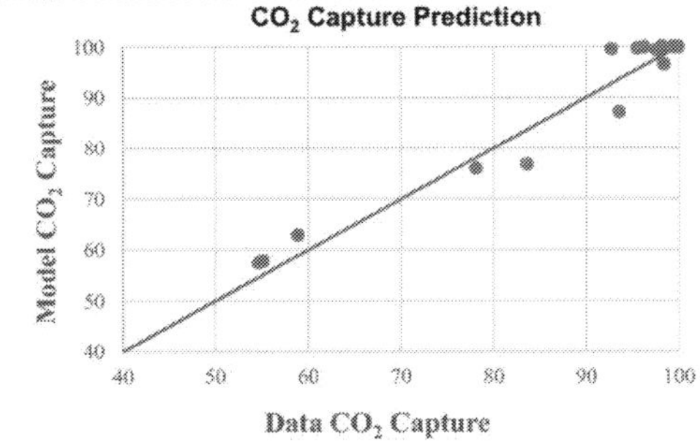

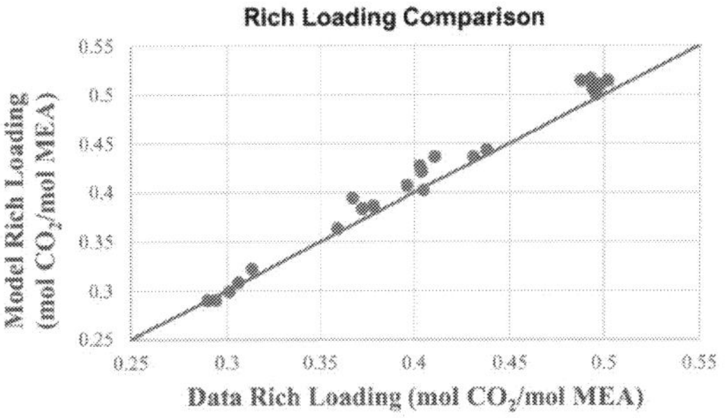

Regenerator Validation

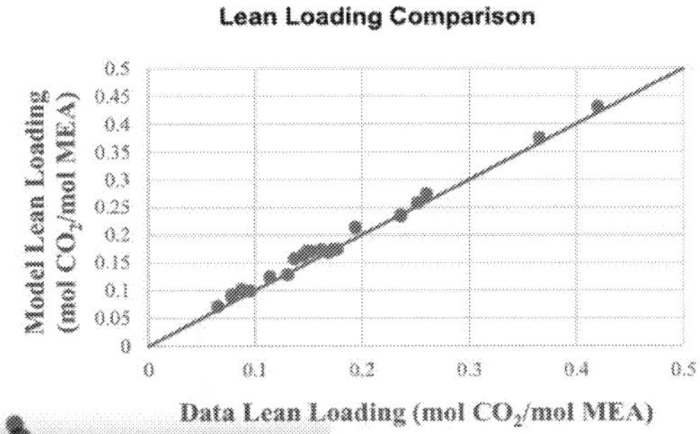

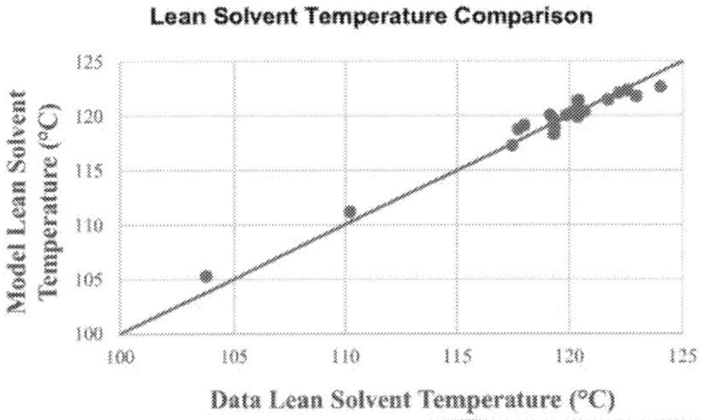

Aspen Dynamics

- **Capability of using a steady-state model to generate a dynamic model**
- Properties model are shared, including user models
- Absorber and regenerator can only be solved using an equilibrium assumption
- Rate-based results can be approximated by Murphree efficiencies*

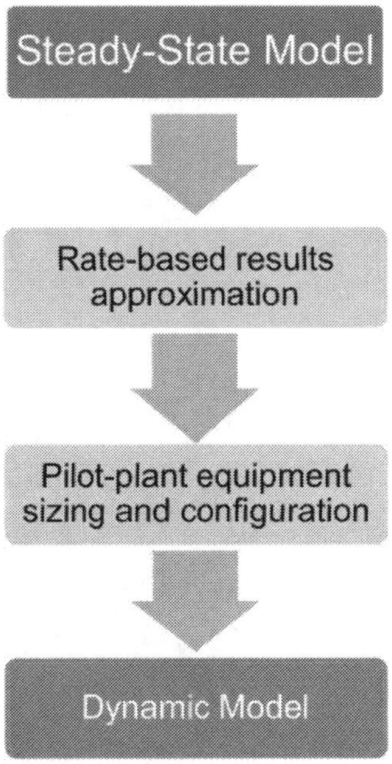

*Zhang et al. "Modeling and model predictive control of a MEA-based post-combustion CO_2 capture process". Industrial Engineering Chemistry Research 2015.

Dynamic Model Development

Efficiency Model

$$\varepsilon = A\left(\frac{F_L}{F_{Lo}}\right)^B \left(\frac{F_V}{F_{Vo}}\right)^C \left(\frac{CO_{2\,load}}{CO_{2\,load,o}}\right)^D \left(\frac{MEA}{MEA_o}\right)^E$$

Conditions	Absorber		Regenerator	
	Max	Min	Max	Min
Liquid flowrate (kg/h)	12961	5390	6503	4981
Gas flowrate (kg/h)	2325	2133	623	441
MEA (%w)	25.41	11.92	0.27	0.24
CO_2 loading (mol/mol)	0.25	0.12	0.47	0.15

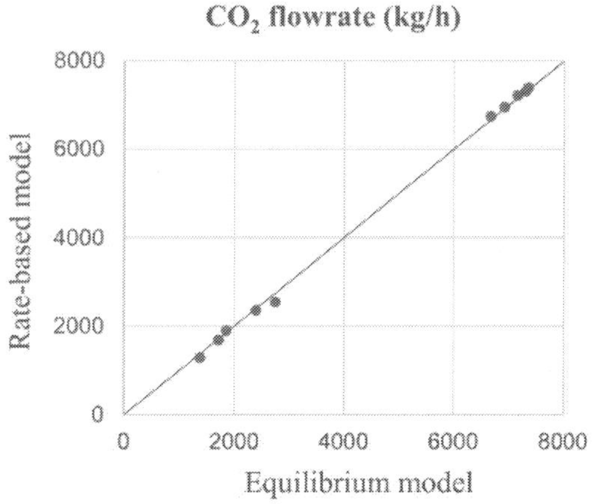

Correlated component efficiency implemented in Aspen Dynamics

CCSI team conducted tests at NCCC

Dynamic Test Conditions

- Dynamic tests capture nonlinearity
- Persistence of excitation
- Step test conducted
  - Solvent flow (lb/hr); x_1=6, datum= 12,500
  - Inlet flue gas(lb/hr); x_2=10, datum= 5,000
  - Reboiler Steam Flow(lb/hr); x_3=6, datum = 5,000

Test#	Test Condition
1	datum
2	+x% of datum
3	-x% of datum
4	+2x% of datum
5	-2x% of datum
6	+x% of datum
7	-x% of datum
8	datum

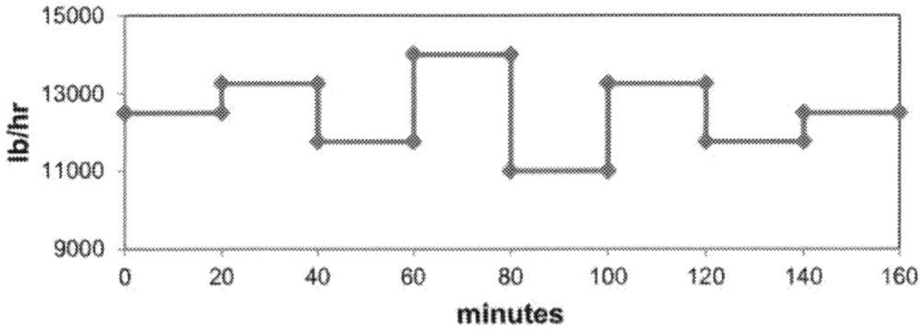

Time periods as well as x_1, x_2, and x_3 determined by conducting initial step tests and recording sensitivities in outputs

Challenges of Dynamic Validation

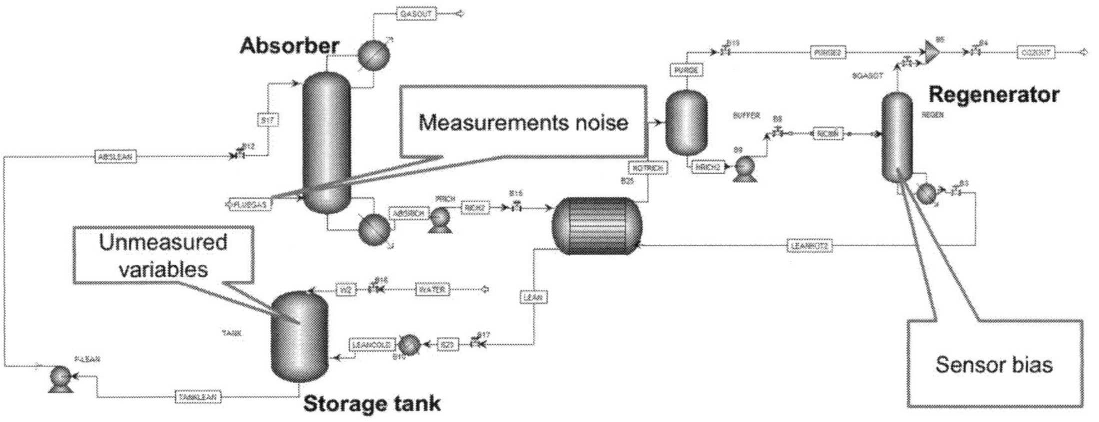

Dynamic data can contain noisy, inaccurate and missing measurements

Dynamic Data Reconciliation

- Noisy, inaccurate, and missing measurements
- Data reconciliation guarantees mass and energy conservation in the dynamic data

$$\min \ (y-\eta)'\Sigma^{-1}(y-\eta)$$

s.t.

$$\dot{\eta} = f(\eta, u, \theta)$$

$$g(\eta, u, \theta) \le 0$$

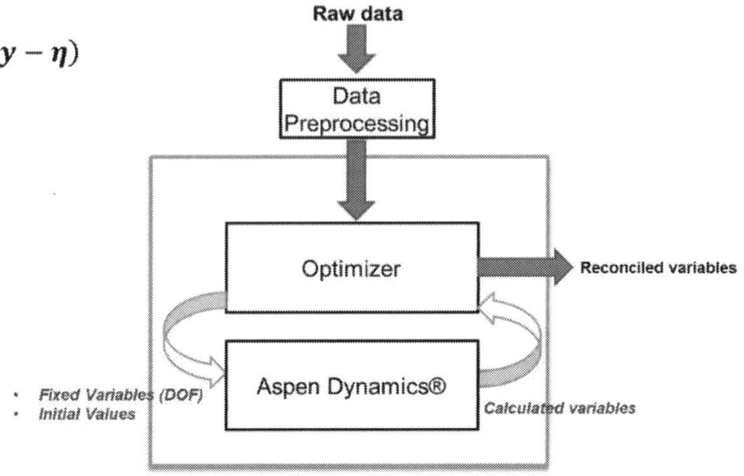

Absorber Validation with DDR

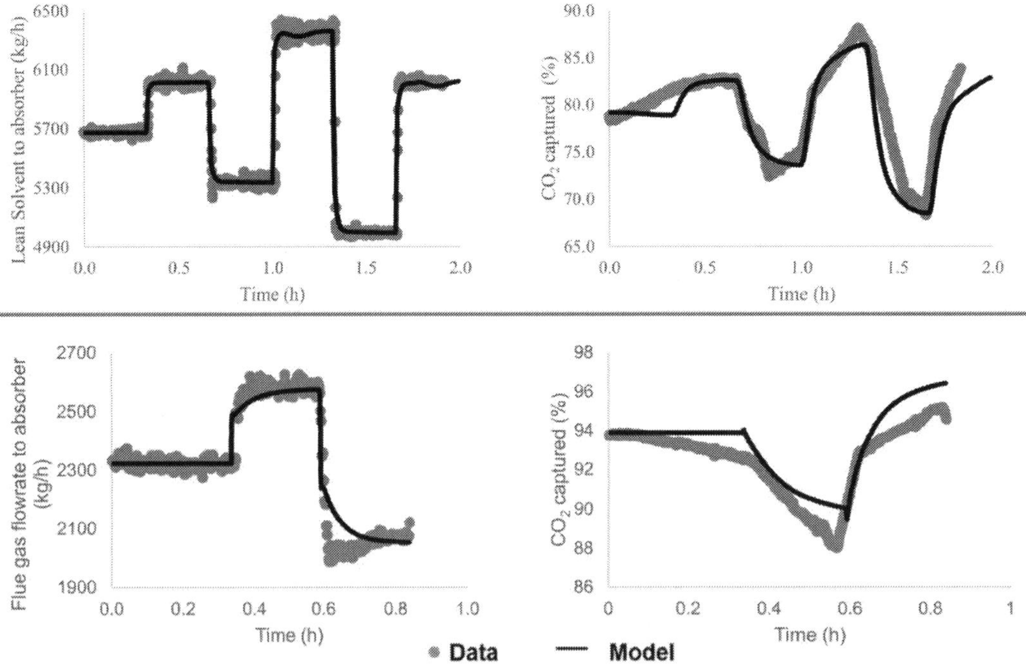

Regenerator Validation with DDR

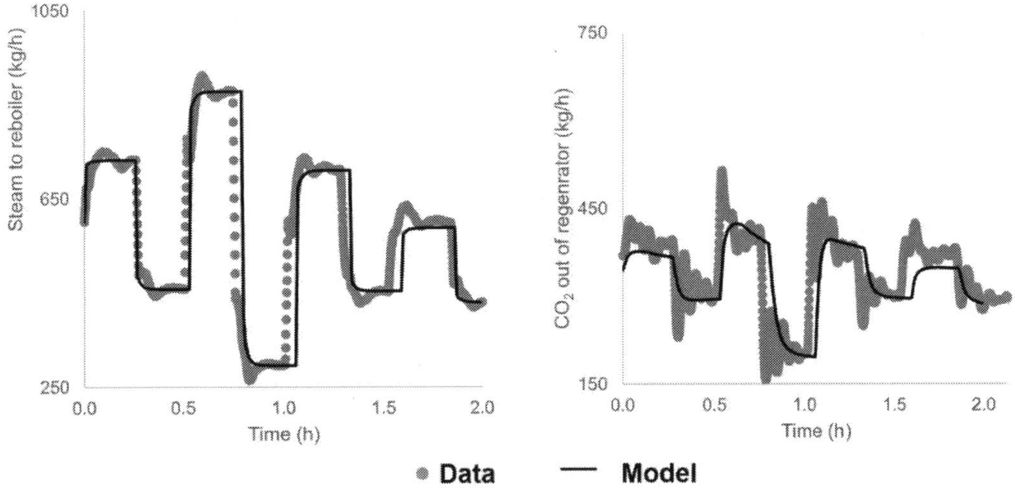

CCSI Carbon Capture Simulation Initiative | NETL | Berkeley Lab | Lawrence Livermore National Laboratory | Los Alamos National Laboratory | Pacific Northwest National Laboratory | U.S. Department of Energy

19

Conclusions

- Efficiency-based dynamic model captures most of behavior in steady state rate-based model
- Dynamic data reconciliation enables best use of noisy inaccurate, and missing data
- Dynamic model predicts gain & time constant of process
- Demonstrates how dynamic data can be used for model validation
- Accuracy of dynamic model might allow its use for control applications

CCSI Carbon Capture Simulation Initiative | NETL | Berkeley Lab | Lawrence Livermore National Laboratory | Los Alamos National Laboratory | Pacific Northwest National Laboratory | U.S. Department of Energy

20

Thank you!

Acknowledgements

This research was conducted through the Carbon Capture Simulation Initiative (CCSI), funded through the U.S. DOE Office of Fossil Energy.

A portion of this work was conducted as part of the National Energy Technology Laboratory's Regional University Alliance (NETL-RUA), a collaborative initiative of the NETL; this technical effort was performed under the RES contract DE-FE0004000.

Disclaimer

CCSI Carbon Capture Simulation Initiative | NETL | Lawrence Livermore National Laboratory | Los Alamos | Pacific Northwest | U.S. DEPARTMENT OF ENERGY

Challenges for a Gold Standard Model

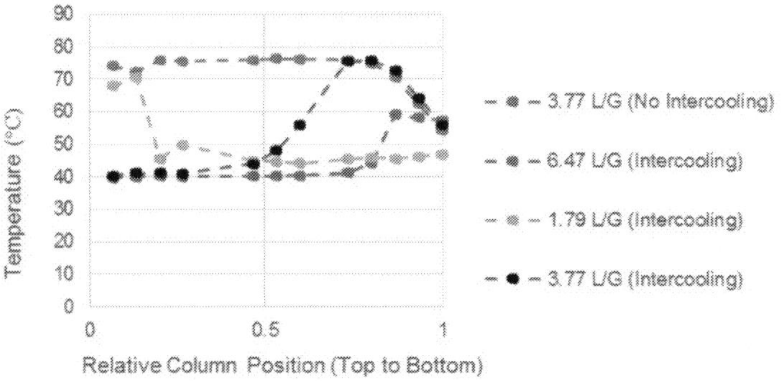

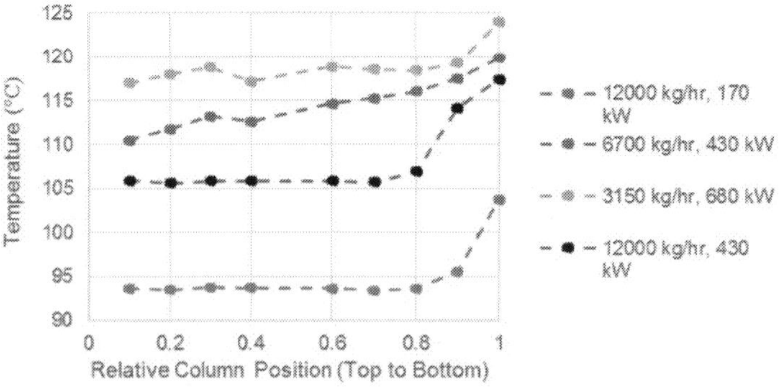

CCSI Carbon Capture Simulation Initiative | NETL | Lawrence Livermore National Laboratory | Los Alamos | Pacific Northwest | U.S. DEPARTMENT OF ENERGY

Heat of Absorption Comparison

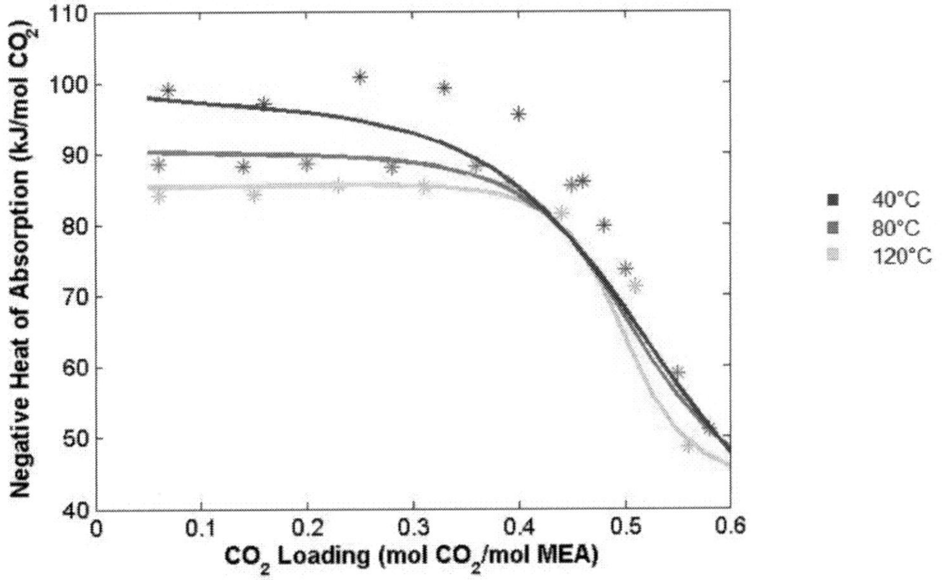

Data from: Kim et al., Energy Procedia,2014;63:1446-1455

VLE Ternary Data Model Fit (30 wt%)

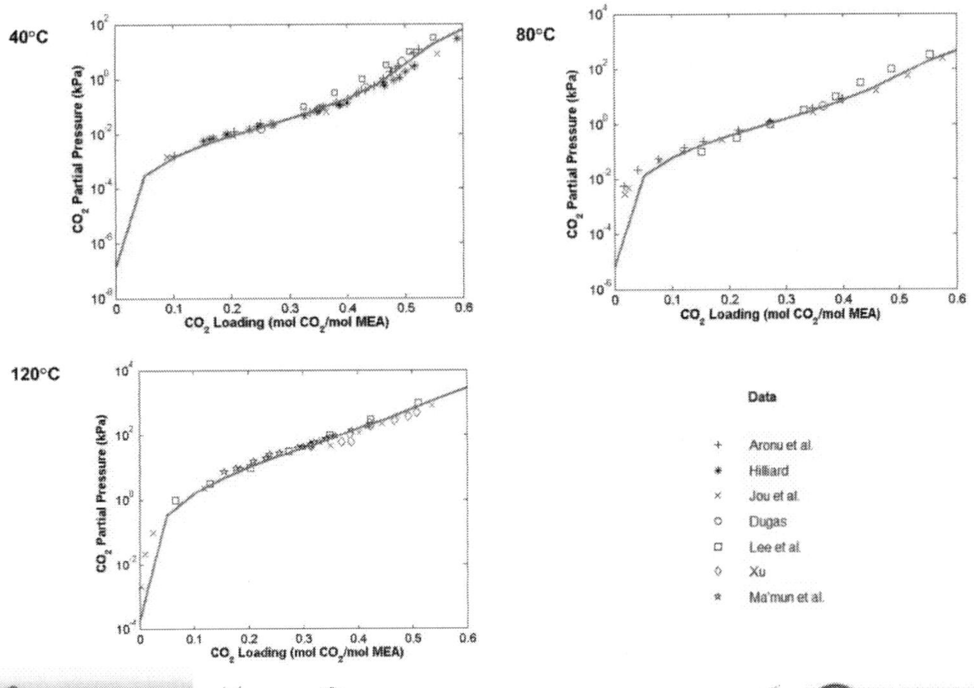

VLE Data Binary Data Model Fit

Txy Diagrams (data from Cai et al.)

P = 101.33 kPa

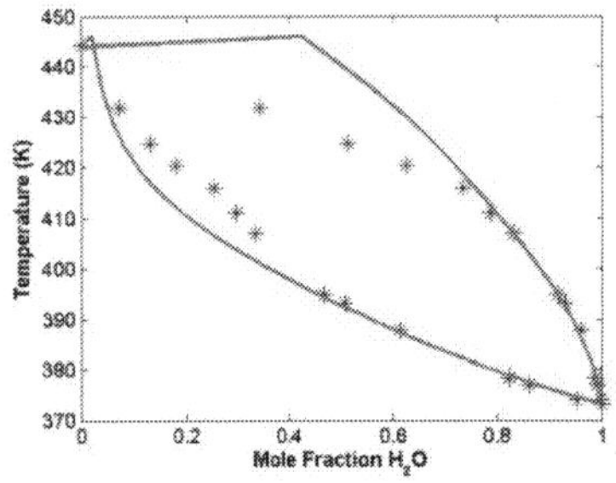

P = 66.66 kPa

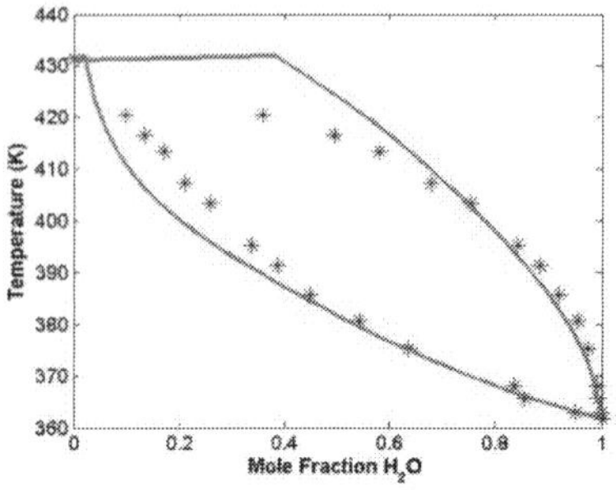

Pxy Diagrams (data from Tochigi et al.)

T = 363.15 kPa

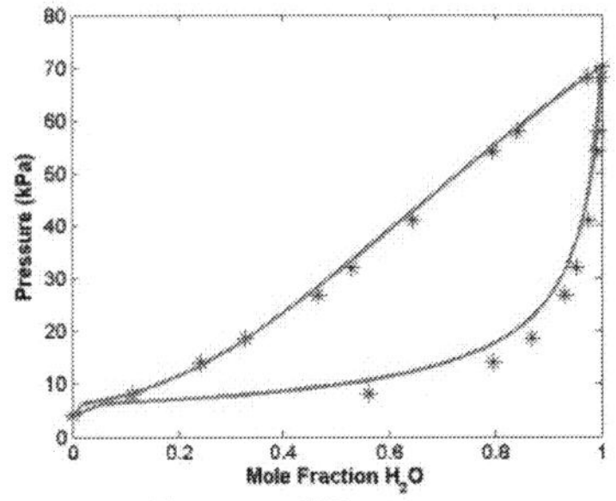

Cai et al., J Chem Eng Data,1996;41:1101-1103
Tochigi et al., J Chem Eng Data, 1999;44:588-590

CCSI Carbon Capture Simulation Initiative | NETL | Berkeley Lab | Lawrence Livermore National Laboratory | Los Alamos National Laboratory | Pacific Northwest National Laboratory | U.S. DEPARTMENT OF ENERGY 25

NCCC vs Other Pilot Plants

	CO_2 Capacity (tpd)	Source of Flue Gas	Absorber		Regenerator	
			Diameter (cm)	Height (m)	Diameter (cm)	Height (m)
UT, Austin	3.0	Non-coal	42.7	6.1	42.7	6.1
NTNU/ SINTEF	0.3	Non-coal	15.0	4.4	10.0	3.9
ITC, Regina	1.0	Non-coal	33.0	7.1	33.0	10.0
ITT, Stuttgart	0.3	Non-coal	12.5	4.2	12.5	2.5
Esbjerg CASTOR	24.0	Coal	110.0	17.0	110.0	10.0
NCCC (PSTU)	10.0	Coal	64.1	18.5	59.1	12.1

Intercooler and flexibility of number of beds also differ

CCSI Carbon Capture Simulation Initiative | NETL | Berkeley Lab | Lawrence Livermore National Laboratory | Los Alamos National Laboratory | Pacific Northwest National Laboratory | U.S. DEPARTMENT OF ENERGY 26

Steady-State Test Runs

Operating Conditions	Range
Solvent Flow (lb/hr)	7,000-26,000
Inlet Flue Gas (lb/hr)	5,000-6,500
Reboiler Steam Flow (lb/hr)	600-2,500
Inlet FG CO_2 vol%	9-11%
# of beds	1-3
Intercooler	no - yes

- **All possible combinations of different operating conditions tested**

Steady-State Test Matrix

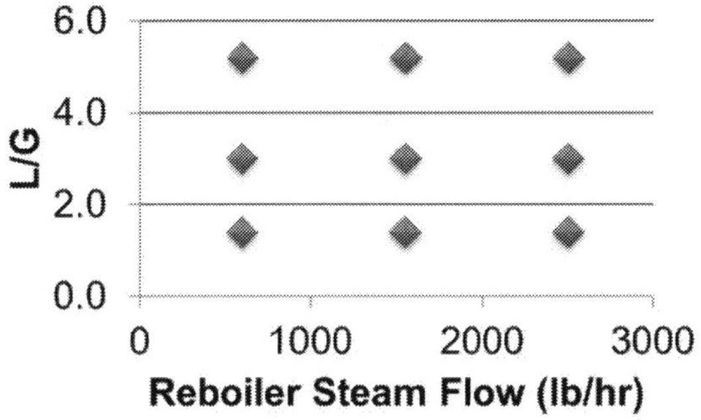

Steady State Absorber Validation

CO_2 Capture Prediction

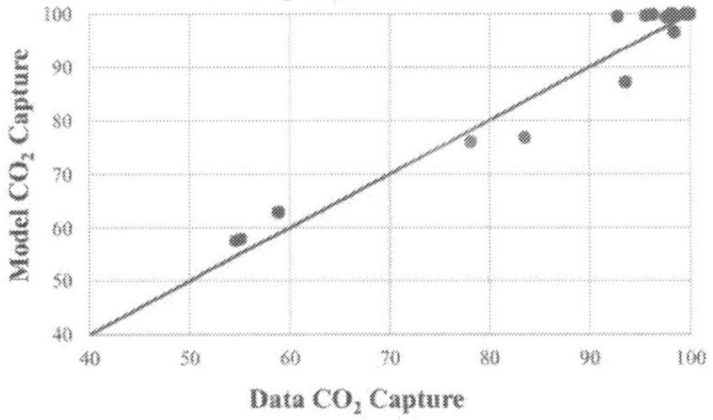

Rich Loading Comparison

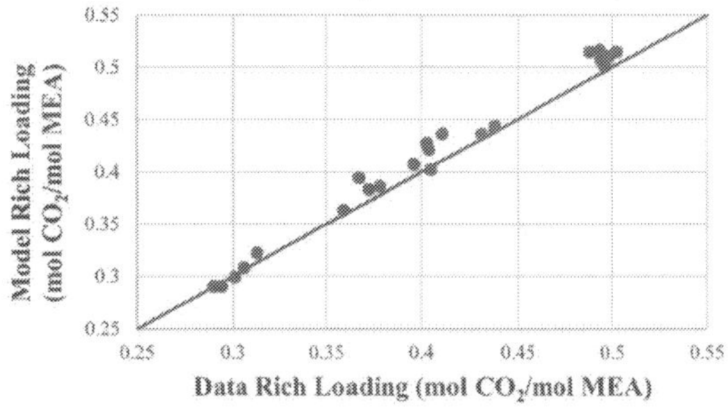

Percent Deviation Between Data and Model Values (Summary)

	Data CO_2 Capture-Liquid vs. Gas Discrepancy	CO_2 Capture-Gas Side	CO_2 Capture-Liquid Side	Rich Loading
Maximum	9.19	8.09	10.84	7.36
Average	3.62	2.69	3.97	2.69

CCSI Carbon Capture Simulation Initiative | NETL | Lawrence Livermore National Laboratory | Los Alamos | Pacific Northwest | U.S. DEPARTMENT OF ENERGY

Steady State Absorber Validation

Sample Temperature Profiles No parameter tuned

Case K3

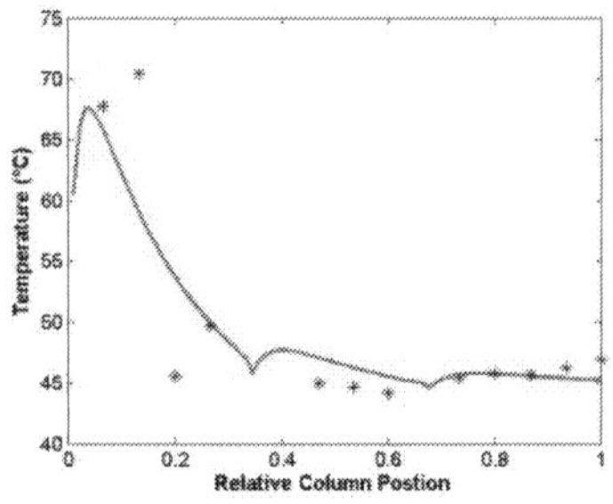

Case K6

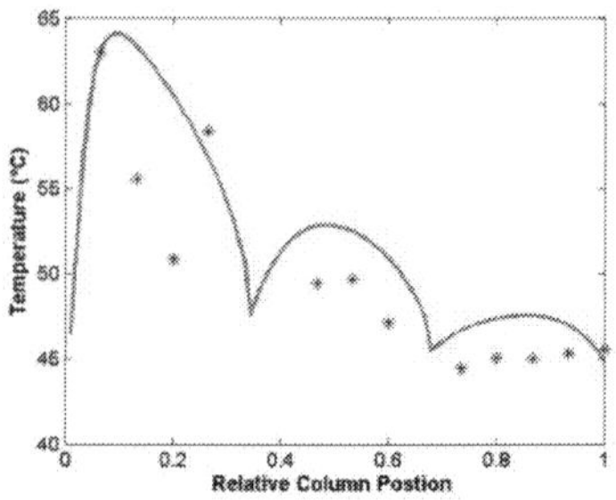

Case K20

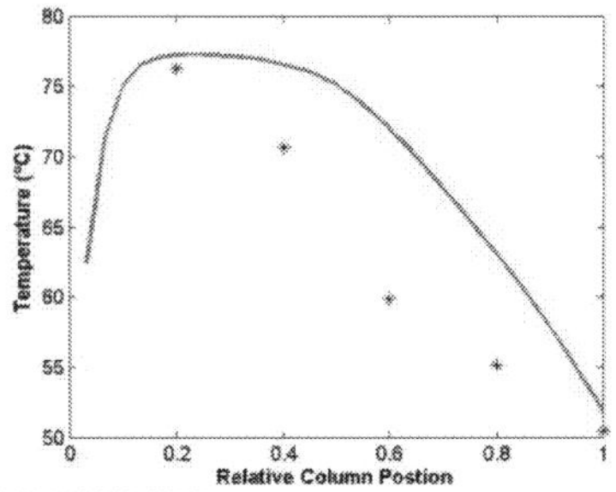

Relative column positions of 0 and 1 correspond to top and bottom of column, respectively

Case	L/G (mass)	Beds/Intercooling	Lean Loading (mol CO_2/mol MEA)
K3	1.41	3/Yes	0.091
K6	3.02	3/Yes	0.347
K20	2.38	1/No	0.075

CCSI Carbon Capture Simulation Initiative · NETL · Berkeley Lab · Lawrence Livermore National Laboratory · Los Alamos · Pacific Northwest National Laboratory · U.S. DEPARTMENT OF ENERGY 29

Steady State Regenerator Validation

Lean Loading Comparison

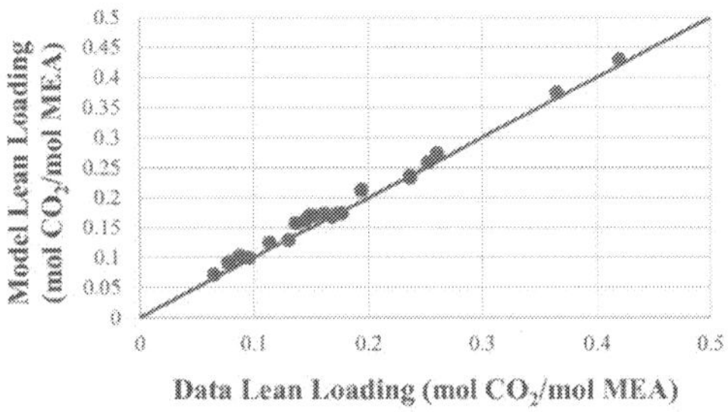

Lean Solvent Temperature Comparison

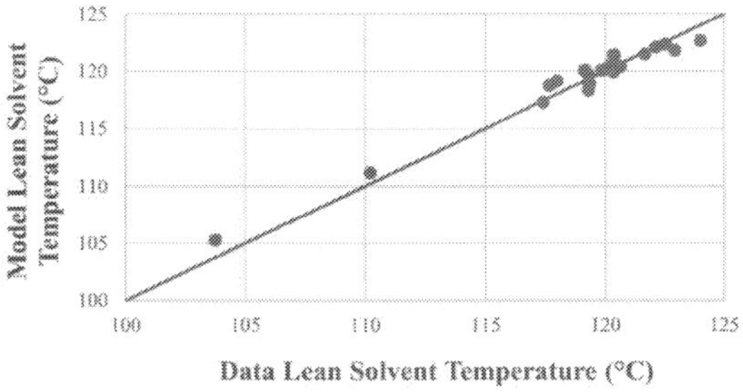

Percent Deviation Between Data and Model Values (Summary)

	Lean Loading	Lean Solvent Temperature
Maximum	16.53	1.14
Average	6.39	0.48

CCSI Carbon Capture Simulation Initiative · NETL · Berkeley Lab · Lawrence Livermore National Laboratory · Los Alamos · Pacific Northwest National Laboratory · U.S. DEPARTMENT OF ENERGY 30

Regenerator Validation

No parameters tuned

Sample Temperature Profiles

Case K1

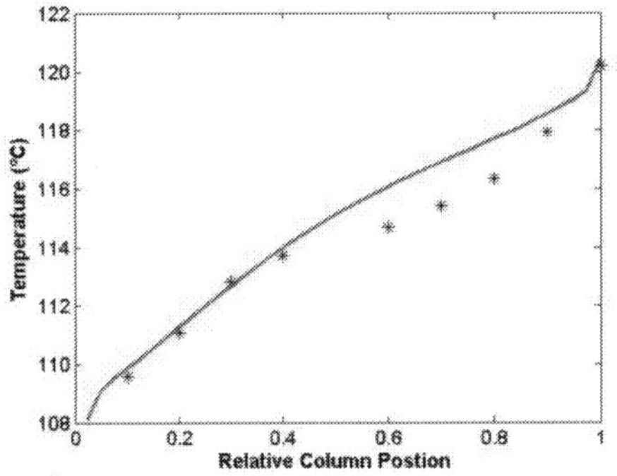

Case K9

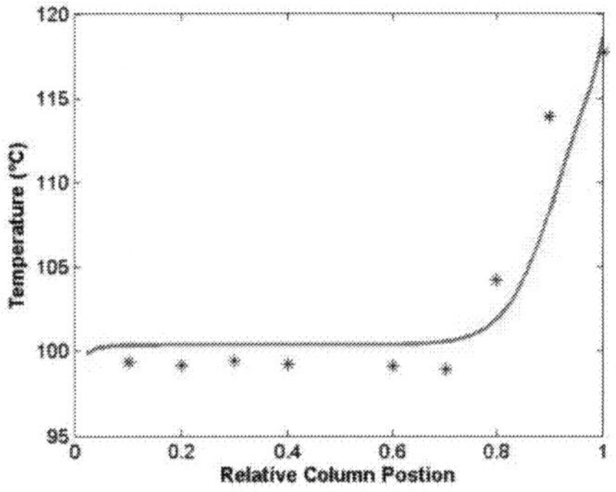

Case K10

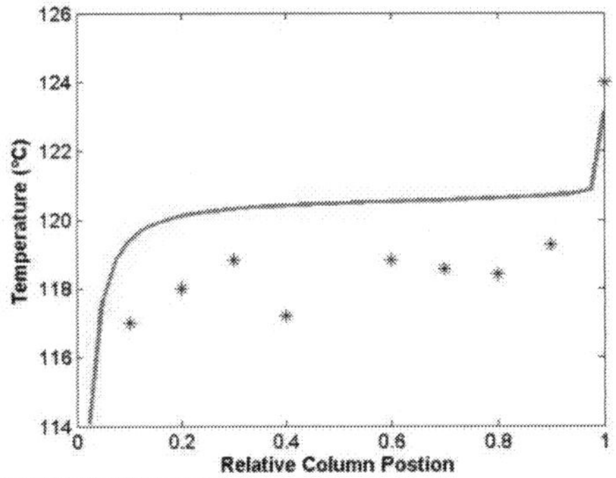

Case	Rich Solvent Flow (kg/hr)	Reboiler Duty (kW)	Rich Loading (mol CO_2/mol MEA)
K1	7242	430.61	0.384
K9	3337	165.74	0.474
K10	3358	670.62	0.477

CCSI Carbon Capture Simulation Initiative NETL Lawrence Livermore National Laboratory Los Alamos Pacific Northwest National Laboratory U.S. DEPARTMENT OF ENERGY 31

National Risk Assessment Partnership

NRAP leverages DOE's capabilities to help quantify uncertainties and risks necessary to remove barriers to full-scale CO_2 storage deployment.

Building toolsets and improving the science base to address...

- Potential impacts related to release of CO_2 or brine from the storage reservoir
- Potential ground-motion impacts due to injection of CO_2

Technical Team

Stakeholder Group

Wade, LLC

NRAP's approach to quantifying performance relies on reduced-order models to probe uncertainty in the system.

A. Divide system into discrete components

B. Develop detailed component models that are validated against lab/field data

C. Develop reduced-order models (ROMs) that rapidly reproduce component model predictions

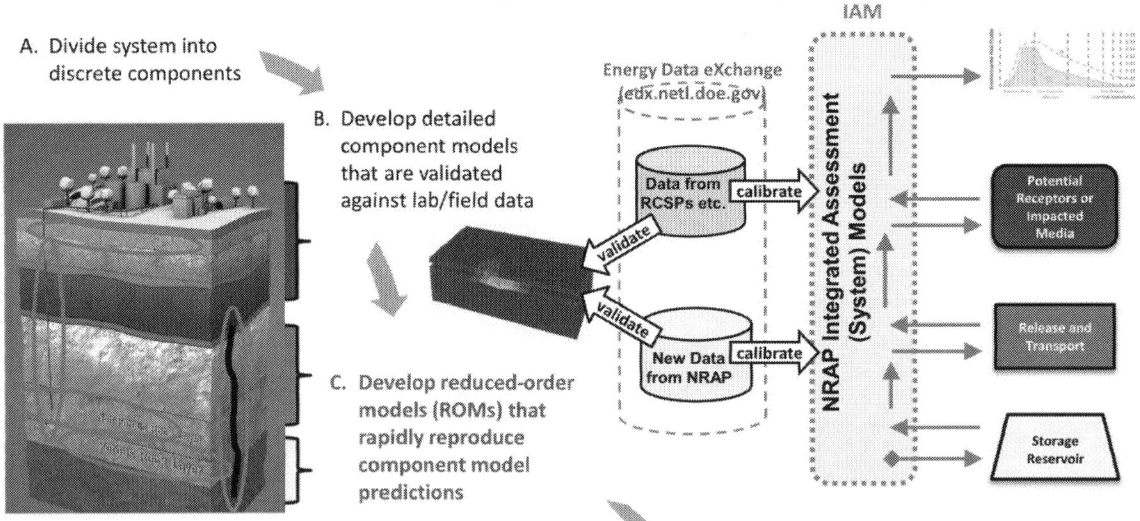

D. Link ROMs via integrated assessment models (IAMs) to predict system performance & risk; calibrate using lab/field data from NRAP and other sources

E. Develop strategic monitoring protocols that allow verification of predicted system performance

Several NRAP products have been released for beta testing.

- Integrated Assessment Model – Carbon Storage (IAM-CS)
- Reservoir Evaluation and Visualization (REV) Tool
- Wellbore Leakage Analysis Tool (WLAT)
- Natural Seal ROM (NSealR)
- Aquifer Impact Model (AIM)
- Design for Risk Evaluation and Monitoring (DREAM)
- Short Term Seismic Forecasting (STSF)

Available now at
www.edx.netl.doe.gov/nrap

U.S. DEPARTMENT OF ENERGY NETL Los Alamos Pacific Northwest NRAP

NRAP's Integrated Assessment Model for Carbon Storage NRAP-IAM-CS

- Simulates long-term full system behavior (reservoir to aquifer/atmosphere)
- Generates risk profiles (time-lapse probability of leakage and GW impact)
- Estimates storage permanence quantitatively amidst system uncertainty
- Identifies key drivers of risk amidst system uncertainty

- Integrates ROMs of system components including: storage reservoir, cemented and open wellbores, faults, and groundwater aquifer
- Quantifies flux of CO_2 and brine to overlying receptors (groundwater and atmosphere), and impacts to groundwater aquifers
- Monte-Carlo simulation allows robust, time-dependent uncertainty quantification
- Uses built-in and user-defined models

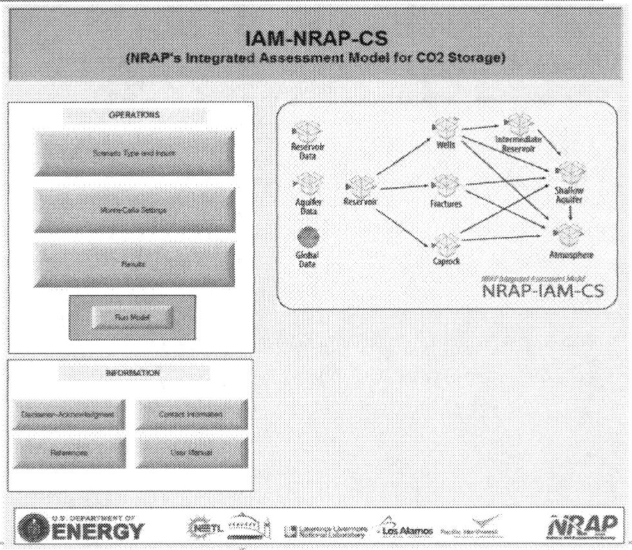

U.S. DEPARTMENT OF ENERGY NETL Los Alamos Pacific Northwest NRAP

Reservoir Evaluation & Visualization (REV) Tool

- Generates pressure and CO_2 plumes sizes over time
- Suitable for Area of Review (AoR) determination
- Visualizes reservoir behavior probabilistically

- Uses pressure and saturation values from simulation software (modular design accommodates different file types).
- Outputs plume sizes through time and pressure values in specified grid blocks at each time step.
- Functions for a single reservoir or accepts multiple simulations and outputs probabilistic values for defined thresholds.

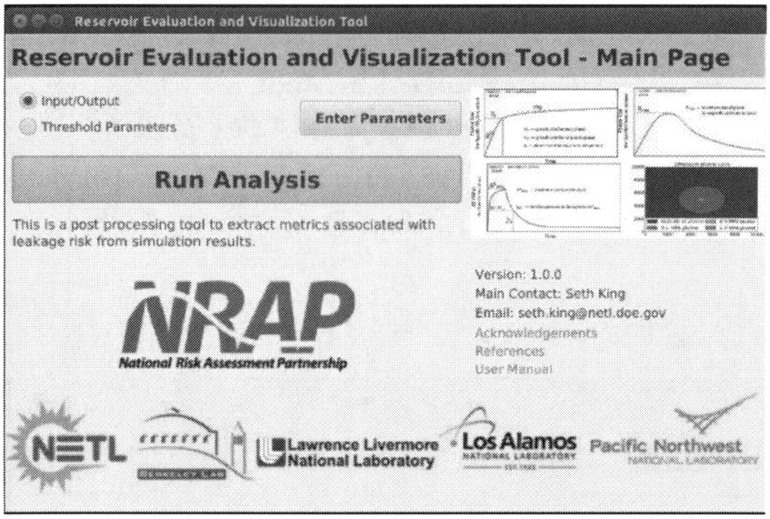

Wellbore Leakage Analysis Tool address vertical migration for a variety of conditions.

- Evaluate existing wells for leakage potential
- Explore leakage response as a function of well disposition
- Evaluate the implications of permeable overburden zones

- **Models migration of brine and/or CO_2 outside of storage reservoir**
- **Inputs of reservoir pressures and saturations**
- **Predicts flowrate into thief zone and groundwater aquifer**
- **Incorporates chemistry to identify flowrate changes as a function of time**

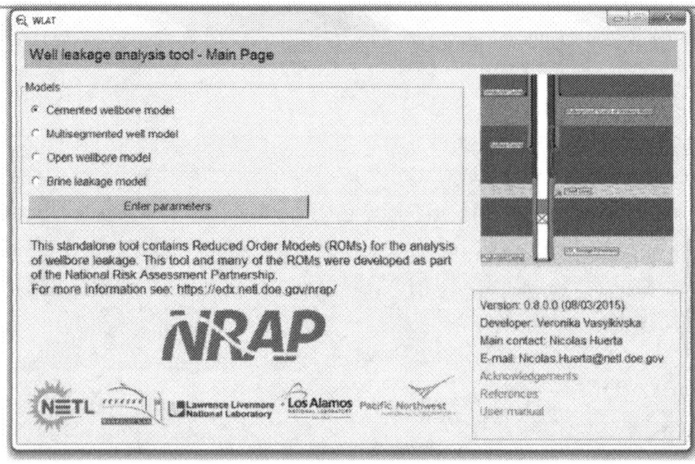

Tool for estimating leakage through fractured seal (*NSealR*)

- Estimate flux through a fractured or perforated seal
- Account for storage outside of primary target zone

- Uses inputs of pressure and saturation at the reservoir/seal interface
- Computes two-phase (brine and supercritical CO_2) flux and Includes fluid thermal/pressure dependence
- Module to compute leakage through a Barrier (Seal) Layer
- Various levels of complexity to model barrier response
- Accounts for effective stress dependence of aperture

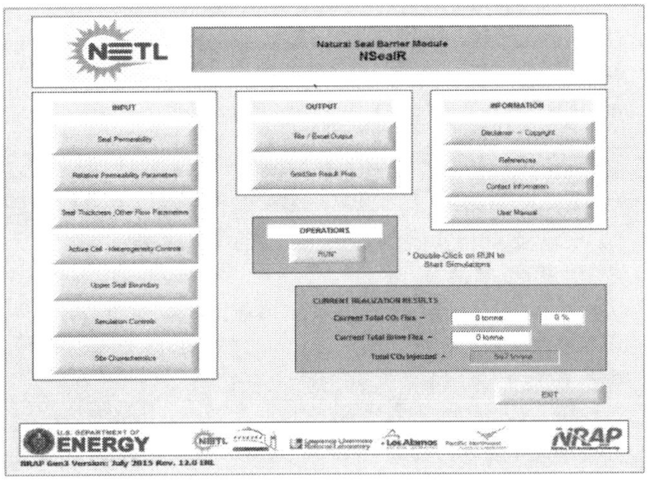

Aquifer Impact Model (AIM)

- **Rapid estimation of aquifer volume impacted by a leak**
- **Distinguishes between CO_2 and brine leaks**
- **Used to determine impact of threshold criteria.**

- Inputs migration rate and concentrations from wellbore or similar models
- Includes two different end member aquifer types
- Incorporate flow and chemistry
- Metrics include: pH, TDS, metals concentrations, organics concentrations

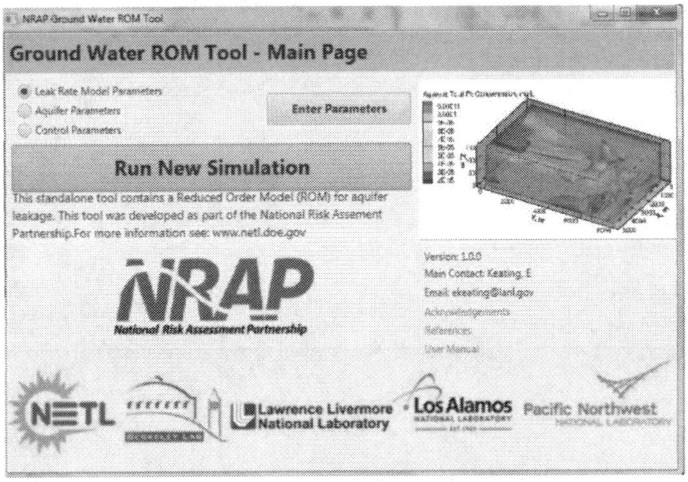

Prototype DREAM monitoring optimization tool has been developed.

- Estimate time to detection for a monitoring system
- Evaluate and select optimal monitoring designs

- **Optimizes subsurface monitoring design for a specified CCS site**
- **Finds monitoring design (well location and depth, sensor type) that yields minimum expected time to first detection of CO_2 leakage (E[TFD])**
- **Can incorporate budget and operational constraints**
- **Uses a collection of realizations of a subsurface simulation**
- **User defined alarm and inference criteria**
  - Sensor detection threshold
  - How many sensors imply a leak

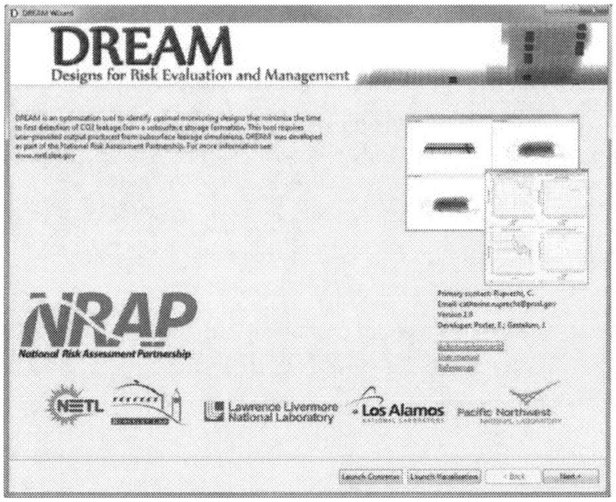

Short Term Seismic Forecasting (STSF) Tool

- Forecasts seismic event frequency over the short term
- Potential to complement stoplight approach for induced seismicity planning and permitting

- **Based on Gutenburg and Omori laws**
- **Originally an aftershock models**
- **Reads a seismic event catalog**
- **Forecasts seismic frequency for a window of a few days**

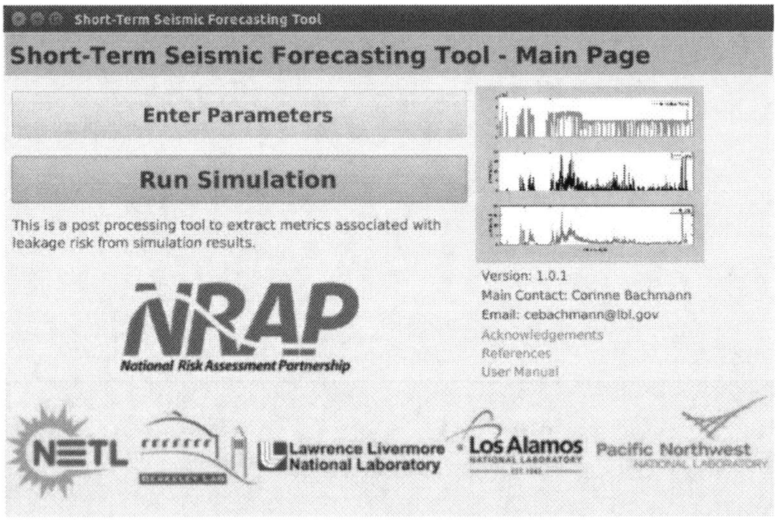

NRAP's IAM can used to quantify risks and study impact of subsurface parameters.

- Quantitative risk profiles with realistic storage conditions
- Importance diagrams identify parameters that do and don't have an impact on leakage
- Can evaluate impacts in any part of the storage-containment system
- Greenfield vs. brownfield conditions influence likelihood of failure

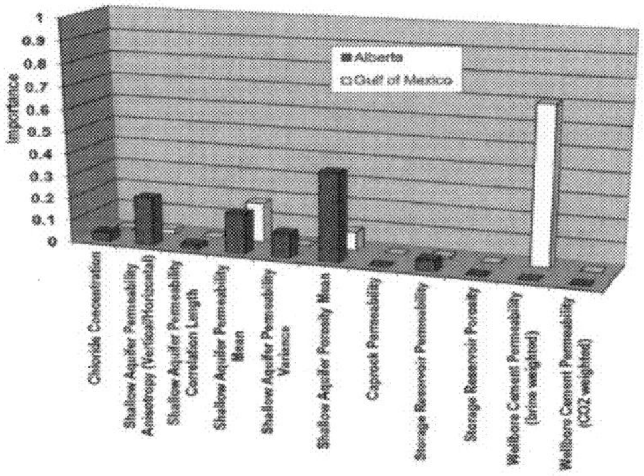

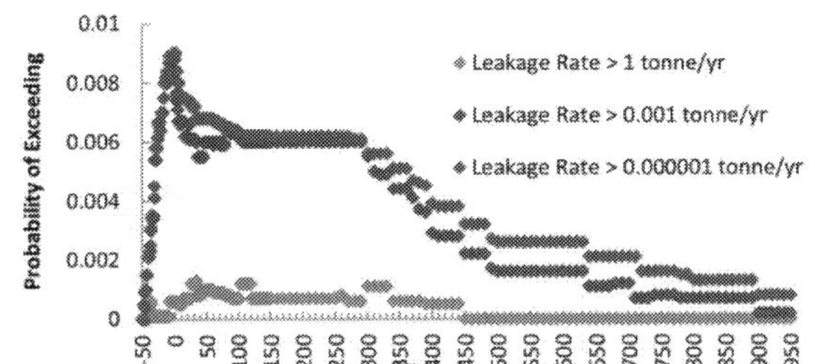

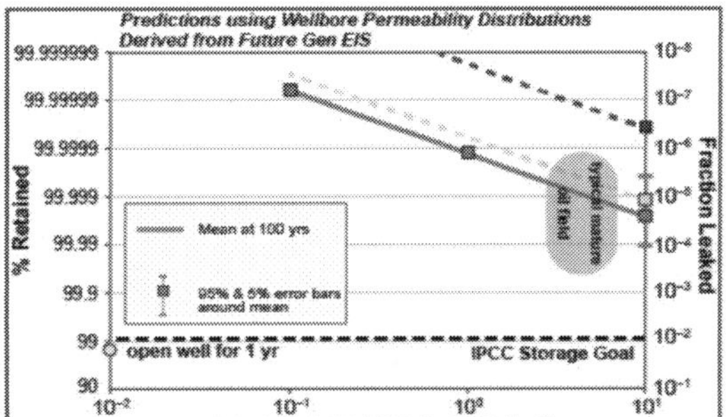

Ref: Pawar et al, GHGT-12, 2014

NRAP analyzed key risk-based metrics for the reservoir component of the storage system using tools.

- Size of CO_2 plume injection
  - Rate of growth for early phase
  - Rate of growth for long-term phase
  - Plume radius at end of injection
- Size of pressure plume
  - Maximum size of plume
  - Various pressure thresholds, relevant
    - Brine rise
    - Fault-slip criteria
- Pressure at a location
  - Maximum pressure increase

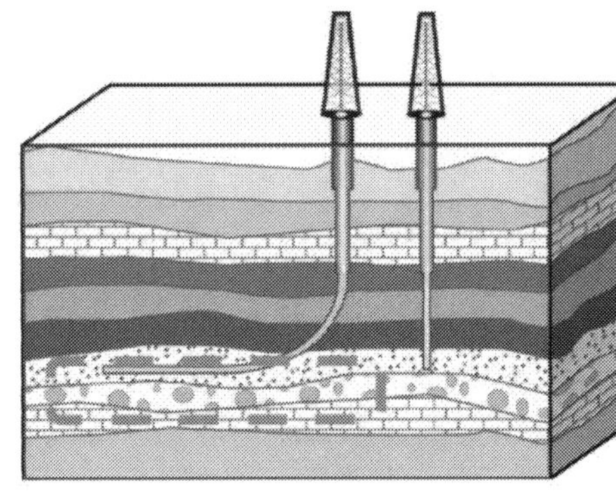

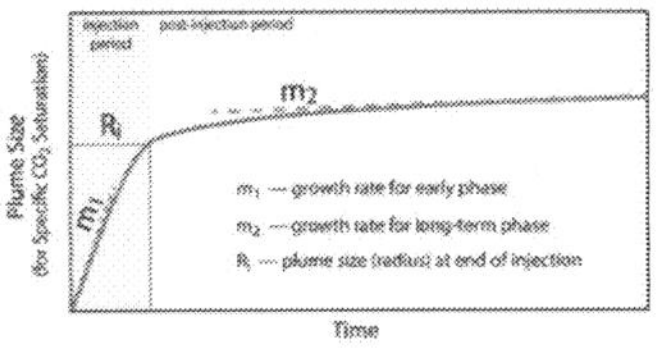

Size of CO_2 Plume

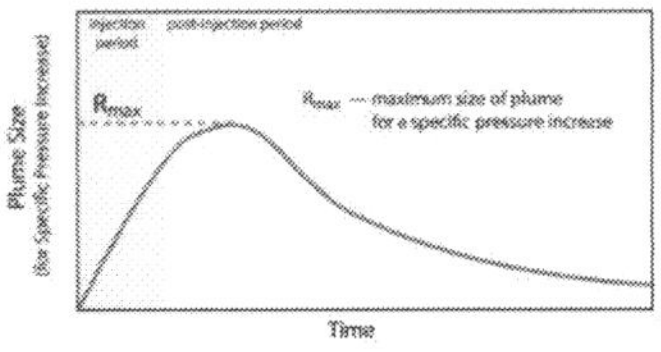

Size of Pressure Plume

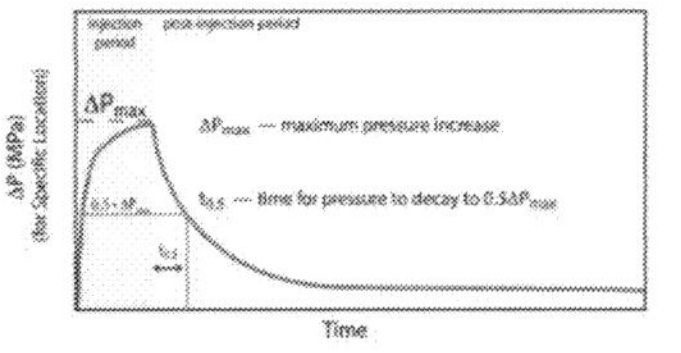

Pressure at at a Location

Growth of pressure plume has a characteristic behavior, albeit details vary between sites.

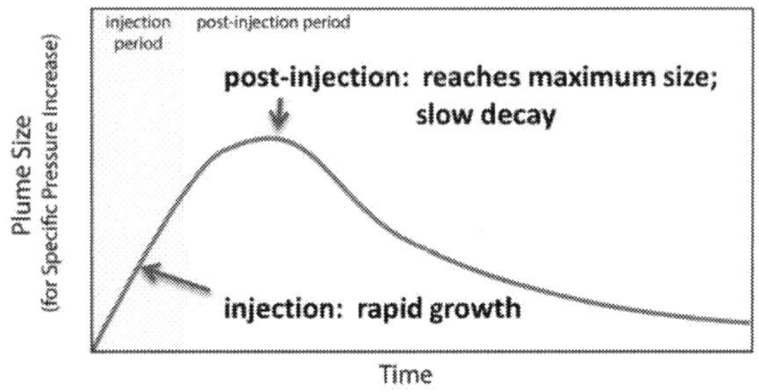

Unbound Sandstone, Regional Dip (0° dip)

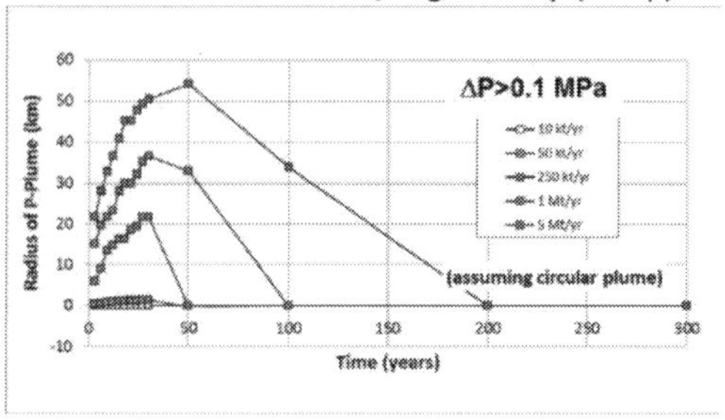

Domal Sandstone-Limestone

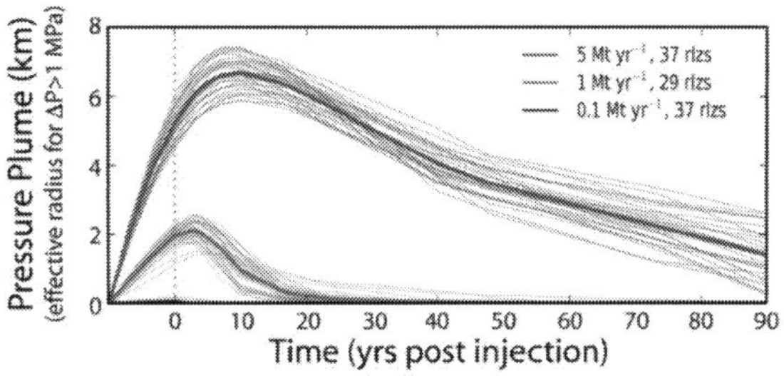

U.S. DEPARTMENT OF ENERGY NETL Los Alamos Pacific Northwest NRAP

Amount of CO_2 injected and geologic details each impact the area/distance for pressure effects.

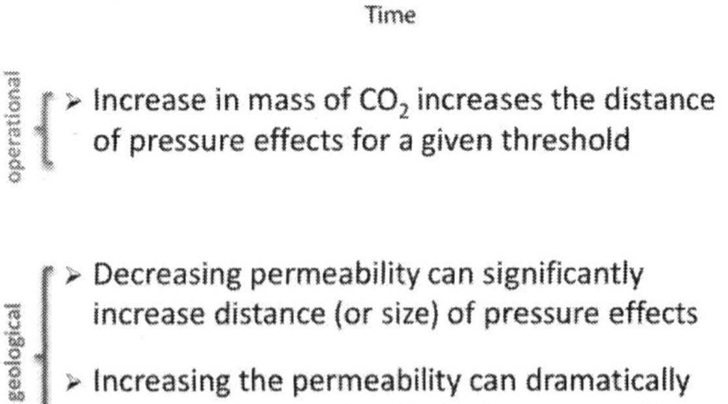

operational
- Increase in mass of CO_2 increases the distance of pressure effects for a given threshold

geological
- Decreasing permeability can significantly increase distance (or size) of pressure effects
- Increasing the permeability can dramatically decrease distance (or size) of pressure effects

Unbound Sandstone, Regional Dip (0° dip)

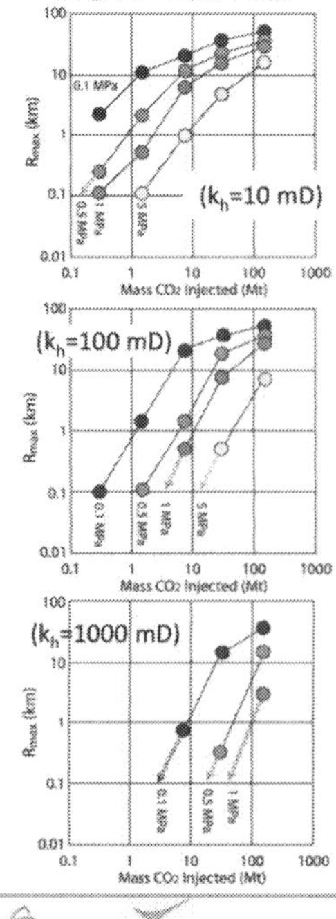

Domal Sandstone-Limestone

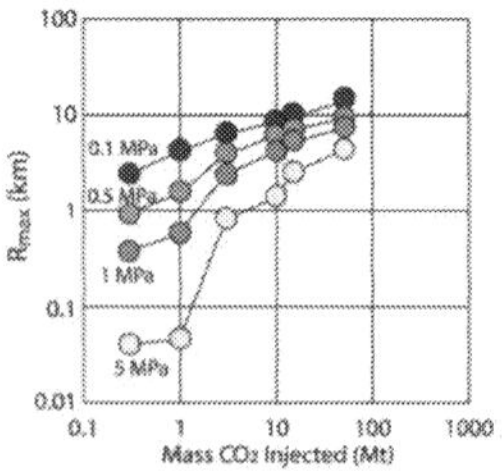

Multi-layer Sandstone (stochastic realizations)

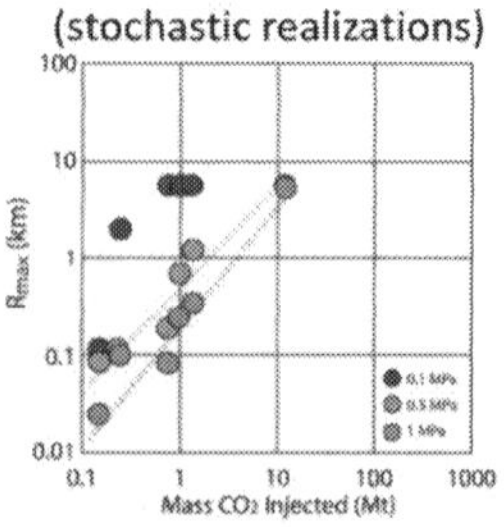

U.S. DEPARTMENT OF ENERGY NETL Los Alamos Pacific Northwest NRAP

WLAT tool can help determine what existing wells should be addressed and when.

- Relationship between well age, completion type, and cement permeability
- Distance has the greatest impact on breakthrough time
- Well age trumps when similar distances

➢ Importance of permeability in Well 5

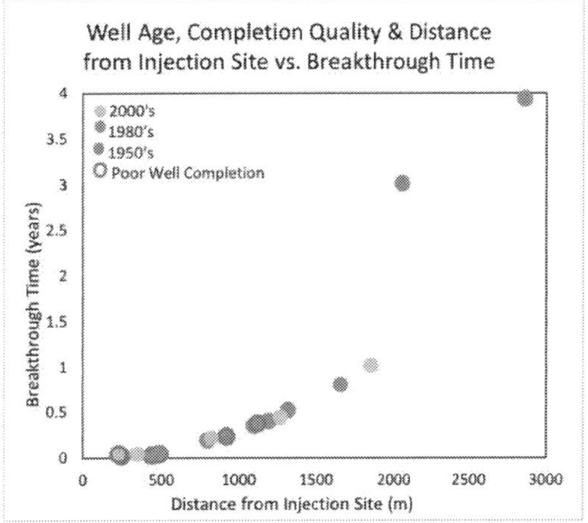

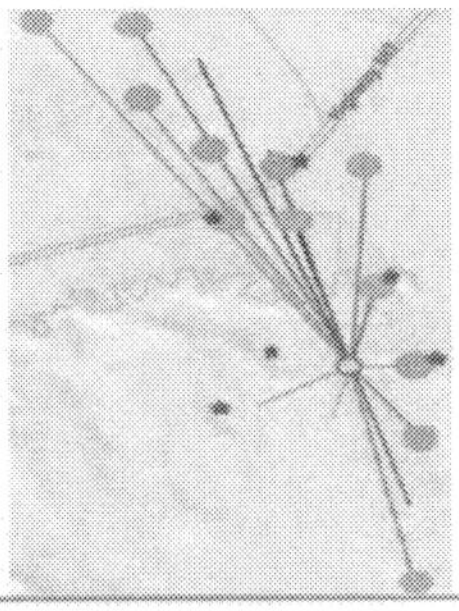

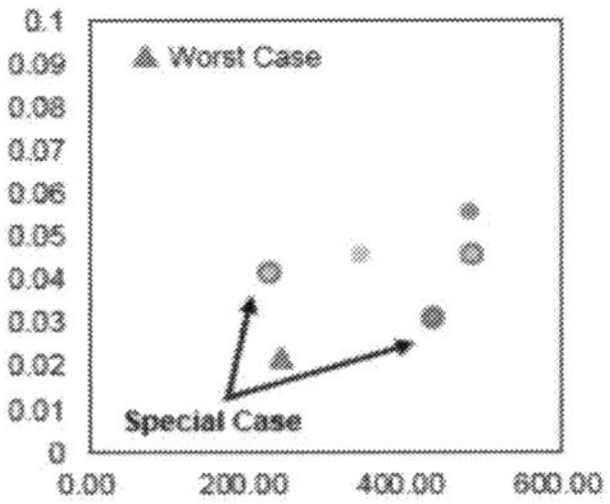

Permeability Values (m^2):

	2000	1980	1950
Good Well Completion	9.869×10^{-17}	9.869×10^{-14}	9.689×10^{-13}
Poor Well Completion	9.689×10^{-15}	4.935×10^{-13}	4.935×10^{-12}

U.S. DEPARTMENT OF ENERGY NETL Los Alamos Pacific Northwest NRAP

NRAP models are underpinned with supporting laboratory and field science.

- **Established "no-impact" threshold values for two major classes of groundwater aquifers**
- **Identified key reservoir metrics for risk**
  - Reservoir response varies with geologic type
  - Determined comparable impact between reservoir properties and geologic variables
- **Quantifying behavior of fractured well as impacted by chemistry**
- **Measure changes in permeability of fractures after slip events**

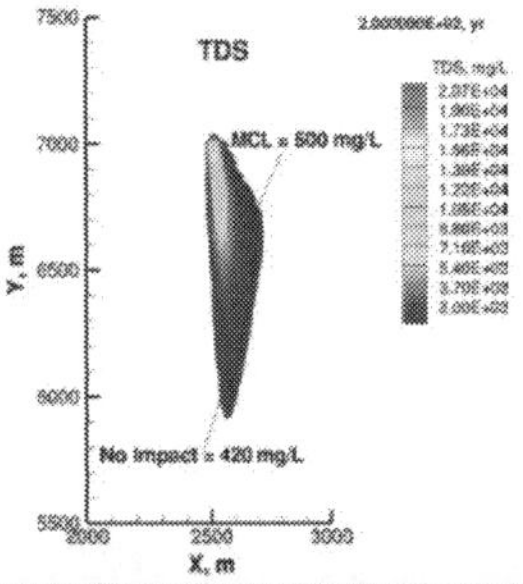

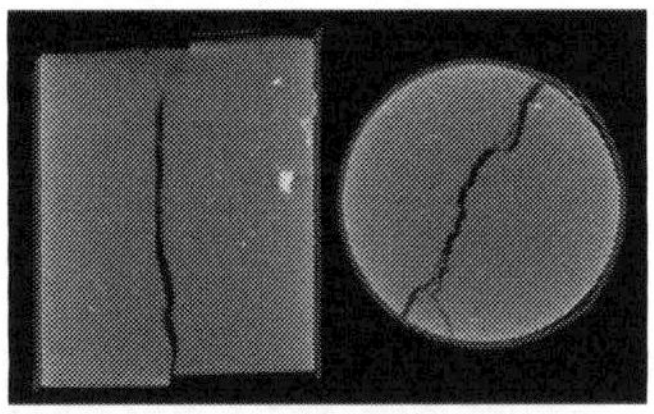

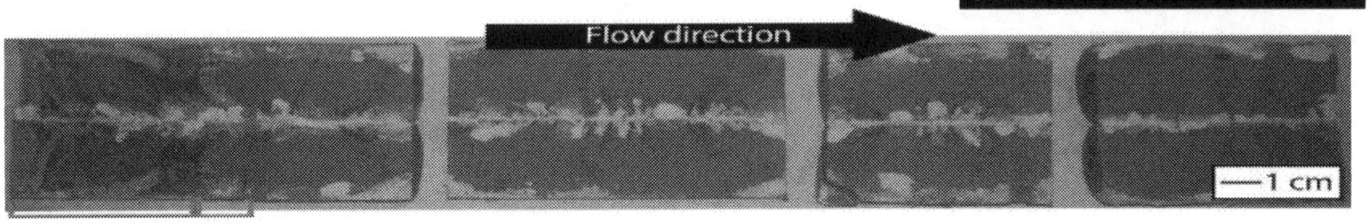

U.S. DEPARTMENT OF ENERGY NETL Los Alamos Pacific Northwest NRAP

How to access the NRAP tools

- **Where are they located?** http://edx.netl.doe.gov/nrap
- **Is a license needed?** No, but registration is required.
- **What platform is needed?** Varies. All work on Windows. Some also work on Linux or Mac.
- **What info is needed to install?** Installation is generally self-explanatory. Some codes (NRAP-IAM-CS and NSealR) require a copy of the Goldsim player (free at http://www.goldsim.com/Web/Downloads/)
- **Are there manuals or tutorials?** Manuals can be found on the web site. A series of webinars are being given as brief tutorials for each tool. Recordings of the webinars can be found on the NRAP web site on the "Beta Testing Tools" page.
- **What if there are questions or bugs?** Feedback forms are available with the download. Address questions to NRAP@netl.doe.gov

U.S. DEPARTMENT OF ENERGY — NETL — Los Alamos — Pacific Northwest — NRAP

Tools and manuals can be downloaded.

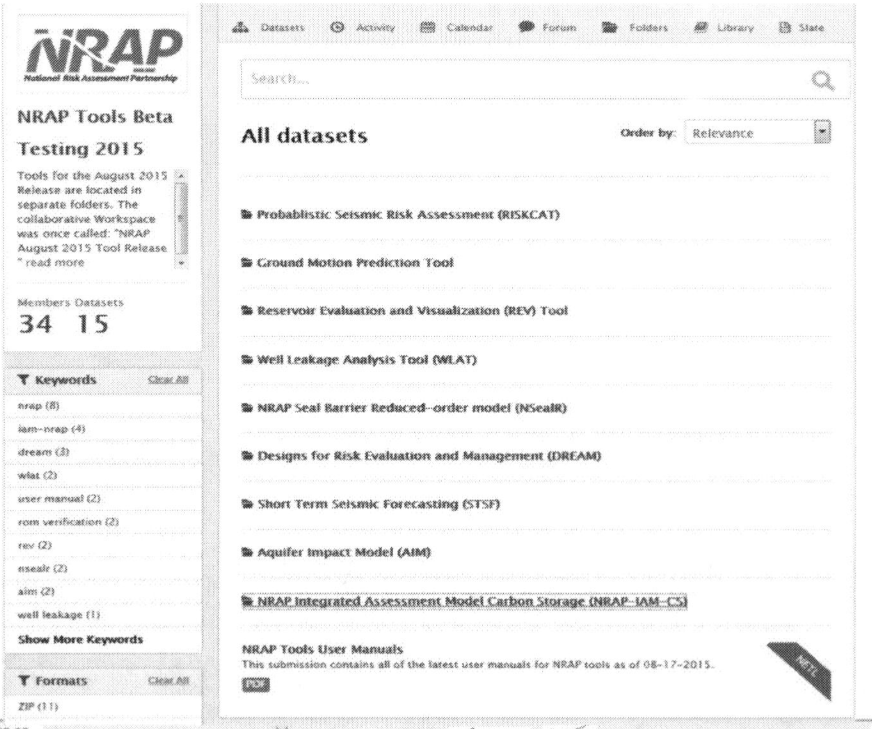

U.S. DEPARTMENT OF ENERGY — NETL — Los Alamos — Pacific Northwest — NRAP

Schedule for NRAP Tool Webinar Series

NRAP Beta Tool Webinars

Scheduled Date/Time	NRAP Tool	Presenter(s)	Recorded Webinar Link	
			SLIDES	VIDEO
October 13, 2015 Time: 1:00 - 3:30 PM EDT	Integrated Assessment ModelCarbon Storage (NRAP-IAM-CS)	Rajesh Pawar		
October 19, 2015 Time: 1:00 - 2:00 PM EDT	Natural Seal ROM (NSealR)	Ernest Lindner	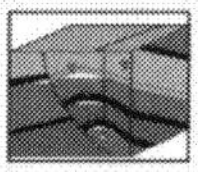	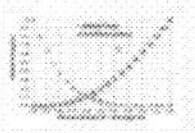
October 26, 2015 Time: 1:00 - 2:00 PM EDT	Reservoir Evaluation and Visualization (REV) Tool	Seth King	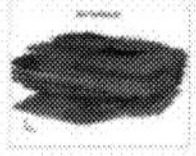	
November 2, 2015 Time: 1:00 - 2:30 PM EST	Wellbore Leakage Analysis Tool (WLAT)	Nicolas Huerta		
November 9, 2015 Time: 1:00 - 2:00 PM EST	Aquifer Impact Model (AIM)	Diana Bacon	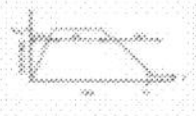	
November 16, 2015 Time: 1:00 - 2:00 PM EST	Design for Risk Evaluation and Monitoring (DREAM)	Catherine Ruprecht		
November 30, 2015 Time: 1:00 - 2:00 PM EST	Short Term Seismic Forecasting (STSF)	Josh White, Corinne Bachman		
December 7, 2015 Time: TBD	NRAP-IAM-CS and REV			

Back to top

Check for updates at www.edx.netl.doe.gov/nrap

U.S. DEPARTMENT OF ENERGY — NETL — Los Alamos — Pacific Northwest — NRAP

Thank you!

www.edx.netl.doe.gov/nrap

NRAP@netl.doe.gov

NRAP National Risk Assessment Partnership

National Risk Assessment Partnership

NETL

Home | Initiative | Approach | Research Products | Tools | Team | TOOL BETA TESTING

Become an NRAP Tool Beta Tester!

NRAP leverages DOE's capabilities to help quantify uncertainties and risks necessary to remove barriers to full-scale CO_2 storage deployment.

NETL's Office of Research and Development (ORD) is leading a multi-laboratory effort that leverages broad technical capabilities across the DOE complex into a mission-focused platform that will develop the integrated science base that can be applied to risk assessment for long-term storage of CO_2: the National Risk Assessment Partnership (NRAP). NRAP involves five DOE national laboratories: NETL, Los Alamos National Laboratory, Lawrence Berkeley National Laboratory, Lawrence Livermore National Laboratory, and Pacific Northwest National Laboratory.

The motivating goal of NRAP is to develop science-based methodologies and tools for calculating risks at any CO_2 storage site while providing necessary scientific and technological advances to support that methodology. Fiscal Year 2016 will span a period of transition for the NRAP research program, with Phase I of this multi-year research effort expected to come to completion by June 2016, and Phase II research beginning concurrently. Phase I is focused on assessment of risk associated with large-scale CO_2 storage, and with quantifying uncertainties associated with those assessments; Phase II will focus on management of risk associated with large-scale CO_2 storage, and with reducing associated uncertainties.

CONTACTS

Cynthia Powell
Director
Office of Research and Development
National Energy Technology Laboratory

Grant Bromhal
Technical Director, NRAP

Tom Richard
Deputy Technical Director, NRAP

Robert Dilmore
NETL Technical Portfolio Lead

George Guthrie
Chair, Executive Committee

Traci Rodosta
Technology Manager, Carbon Storage

Mark Ackiewicz
Director
Division of Carbon Capture and Storage

U.S. DEPARTMENT OF ENERGY — NETL — Los Alamos — Pacific Northwest — NRAP

Supporting Slides

Current work includes developing a risk-based approach to evaluate the AoR

- Build on a 3-tier methodology to define the AoR proposed by *Birkholzer et al. (2013)*:
  - ✓ **Tier 1:** CO_2 leakage risk
  - ✓ **Tier 2:** Open borehole brine leakage risk
  - ✓ **Tier 3:** Plugged and abandoned wellbore brine leakage risk
- Spatial extents of Tiers 1 and 2 are easily defined. *Tier 3 is needed for* quantitative understanding of risks
- Probabilistic methodology to generate "risk maps" related to tier 3 AoR

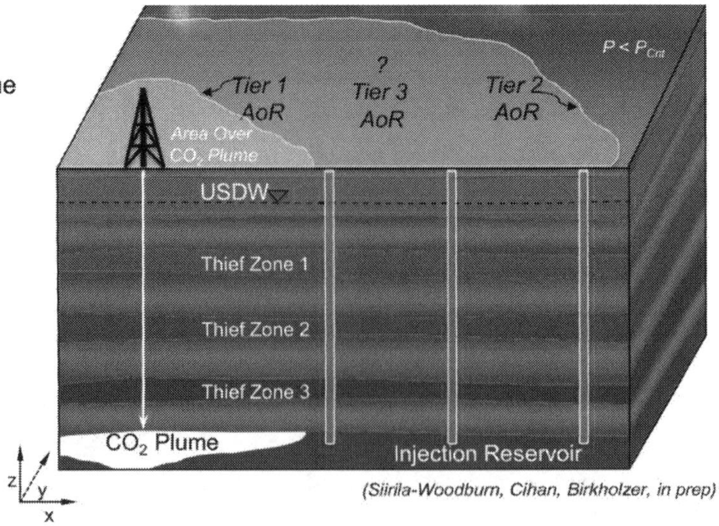

(Siirila-Woodburn, Cihan, Birkholzer, in prep)

U.S. DEPARTMENT OF ENERGY NETL Los Alamos Pacific Northwest NRAP

Methodology to Generate Risk Maps

Two greatest sources of uncertainty in P&A well leakage: ***1) well location and 2) well permeability***

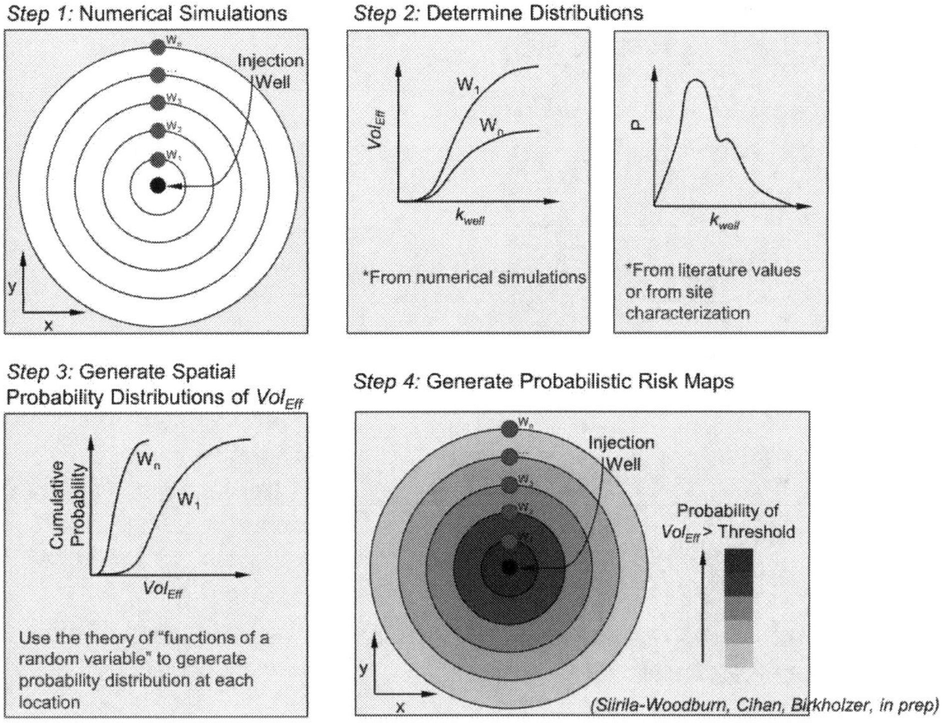

Community Data Sets

- Build from "real" and "synthetic" data
- Detailed model of the large system (reservoir to receptors)
- To use to test models and monitoring technologies
- Still in the thinking stage

- Concept: We are generating a model with data sets to support our research. Can we do a little bit more work and make the data sets valuable to the broader CCS community?

U.S. DEPARTMENT OF ENERGY

NRAP

An example application of an adaptive algorithm

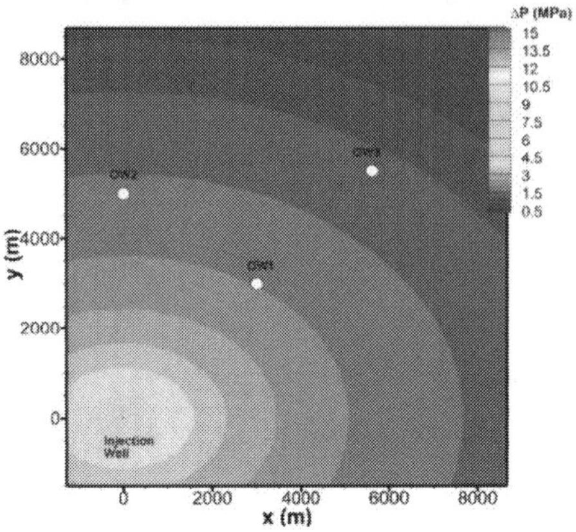

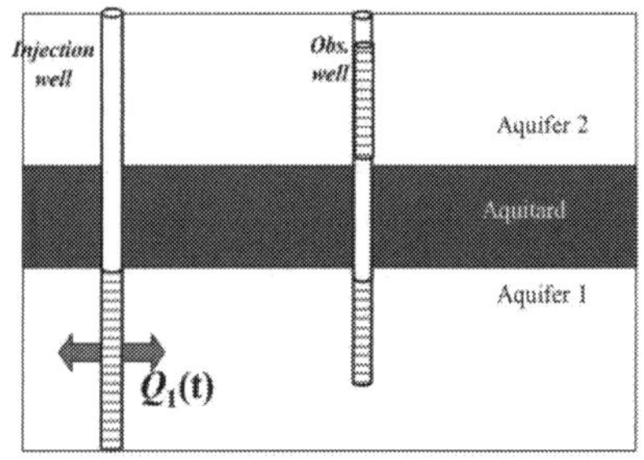

Reservoir parameters assumed to be unknown	Actual values
K1 (m/d)	1.015E-01
S1 (Specific Storage Coef, 1/m)	1.693E-06
Horizontal Anisotropy Ratio	2.000E+00
K2 (m/d)	1.037E-01
S2 (1/m)	1.704E-06
K' (aquitard)	1.028E-06
S' (aquitard)	1.988E-06

- Fluid injection rate is 1Mt/yr over 30 yrs
- An adaptive algorithm involving monitoring + model update is applied to test after how many years (or period of observations) the model results match the actual values, generated hypothetically.
- No risk assessment is implemented

Conformance of Reservoir Model Predictions

- Initial model predictions will carry significant uncertainty due to incomplete knowledge, and model results can deviate significantly from actual observations.
- Conformity of long-term model results with the actual system behavior can only be approached reasonably close by an adaptive management strategy.

An Adaptive Strategy for Conforming Reservoir Predictions with Actual System Behavior

Questions planned to be addressed:

- What measurements and at what resolution (spatial and temporal) are needed to test/update reservoir models for a specific site for making accurate risk assessments?
- How closely do the model results match the field observations for reliable predictions of future risks?
- How long may it take for the reservoir models under realistic reservoir conditions to reach a conformance with actual data?

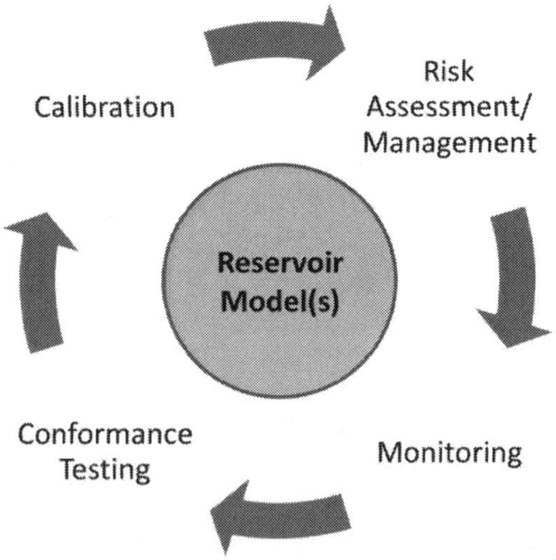

Methodology to Generate Risk Maps

*Two greatest sources of uncertainty in P&A well leakage: **1) well location and 2) well permeability***

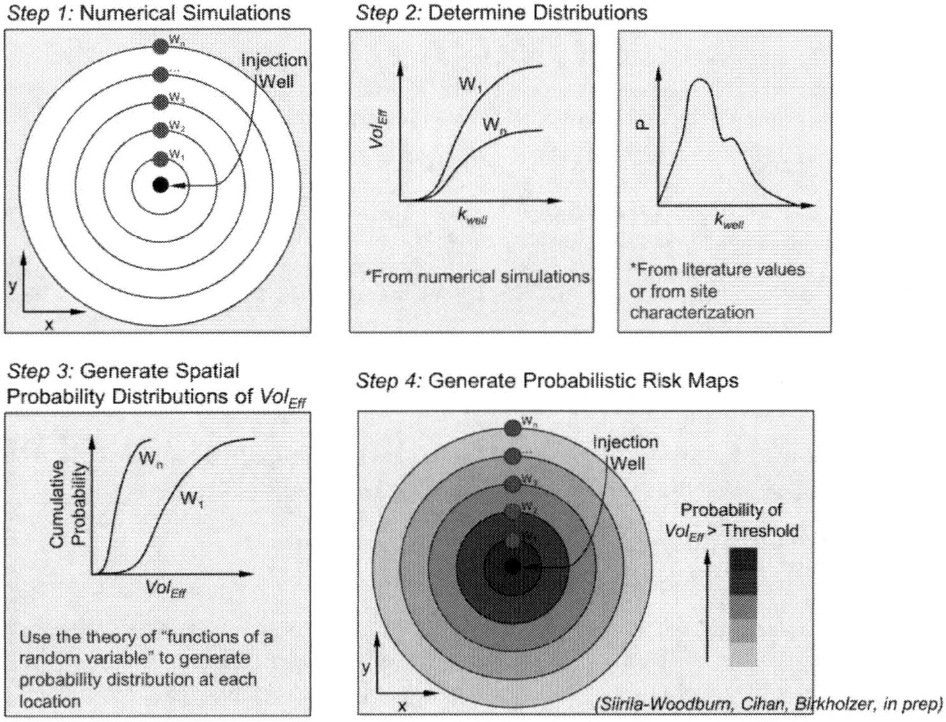

(Siirila-Woodburn, Cihan, Birkholzer, in prep)

Using Science-Based Prediction to Probe Reservoir Behavior: General Observations

Two questions

- How does reservoir performance change as a function of injection volumes & rates?
- How does a reservoir respond as a function of time when injection stops?

Impact of operational variables...

- Reservoir response varies with amount of injected CO_2
  1. Size of CO_2 Plume—rapid growth during injection; slower growth post injection
  2. Size of Pressure Plume—rapid growth during injection; reaches maximum and decays post
  3. Change in Pressure over Time—rapid growth during injection; rapid decline post injection
- Potential to exceed some pressure thresholds depends on size of injection

Impact of geologic variables...

- Reservoir response varies with geologic type
- Rate of CO_2 plume growth depends on reservoir porosity
- Potential to exceed some pressure thresholds depends on reservoir permeability

Impact of geologic variables is on the same order as operational variables.

U.S. DEPARTMENT OF ENERGY

NRAP

Using Science-Based Prediction to Probe Reservoir Behavior: General Focus/Approach

What we're doing

- Developing a synthetic database of reservoir behavior to inform two questions:
  - How does reservoir performance change as a function of injection volumes & rates?
  - How does a reservoir respond as a function of time when injection stops?

What we've done...

- Explored behavior of a few prototype reservoirs as a function of operational and geological variables using physics-based simulations
- Used three reservoir performance metrics that relate to potential risks of concern (i.e., not a direct measure of risk)
  1. Size of CO_2 Plume
  2. Size of Pressure Plume
  3. Change in Pressure over Time (at a Point)
- Monitored metrics using simple parameters; threshold values for pressure reflect a range from small increase through thresholds with potential to cause impact

What we have NOT done...

- Calculate risk—current effort probed only one component of risk (reservoir behavior)
- Calculate behavior of specific reservoirs—current effort explored only general behavior of a few common prototypes; injection only (no co-production of fluids)
- Calculate all physical phenomena—current effort focused only on fluid flow (conservative)

ENERGY NETL Los Alamos Pacific Northwest NRAP

Growth of CO_2 plume has a characteristic behavior, albeit details vary between sites.

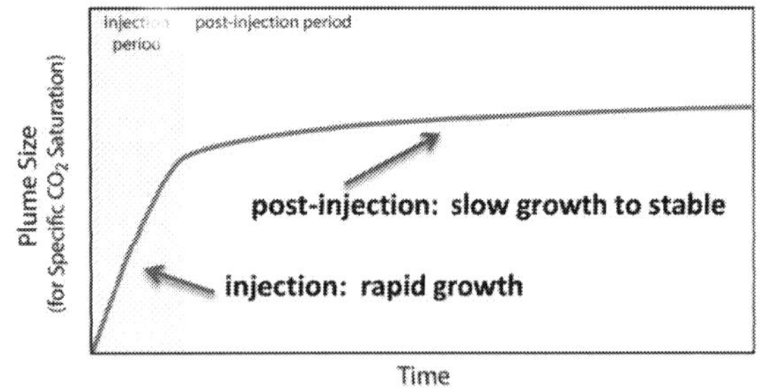

Unbound Sandstone, Regional Dip
(0° dip; k_h=100 mD; ϕ=0.2)

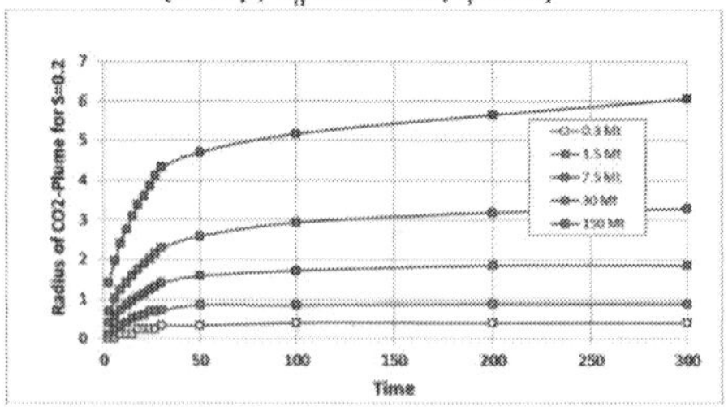

Unbound Sandstone, Regional Dip
(0° dip; k_h=1000 mD; ϕ=0.2)

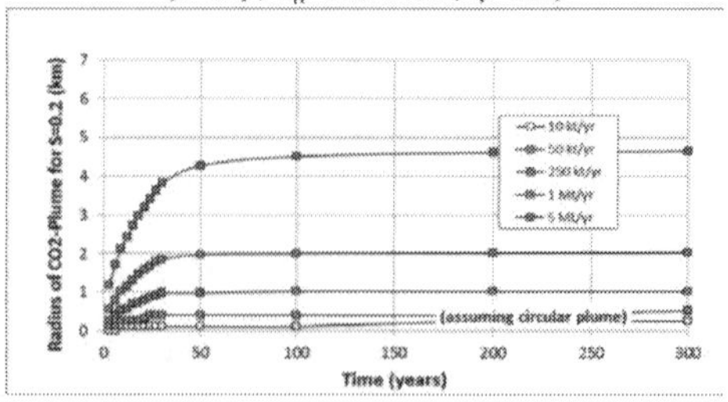

Multi-layer, Domal Sandstone
(3-yr injection; 0.1-10 Mt)

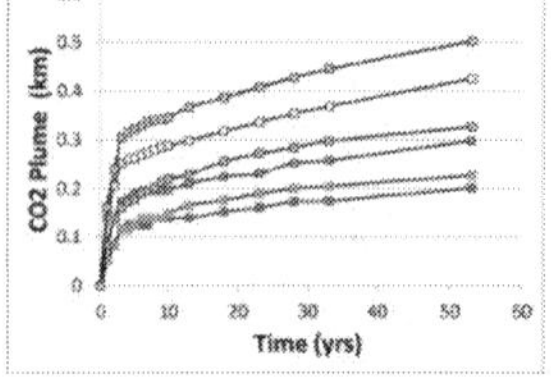

Domal Sandstone-Limestone
(10-yr injection; 1-50 Mt)

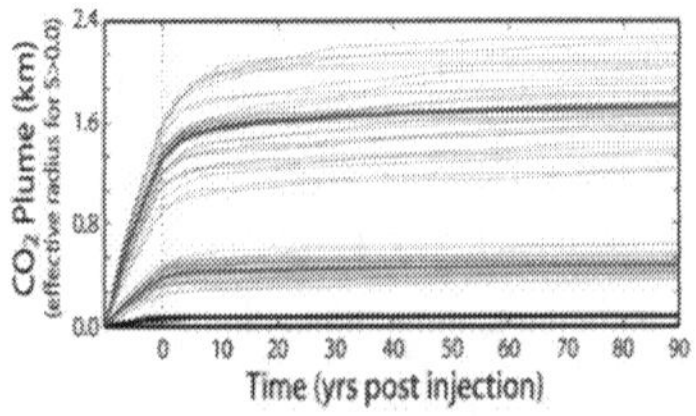

U.S. DEPARTMENT OF ENERGY NETL Los Alamos Pacific Northwest NRAP

Amount of CO_2 injected is primary factor in the growth rate and size of CO_2 plume.

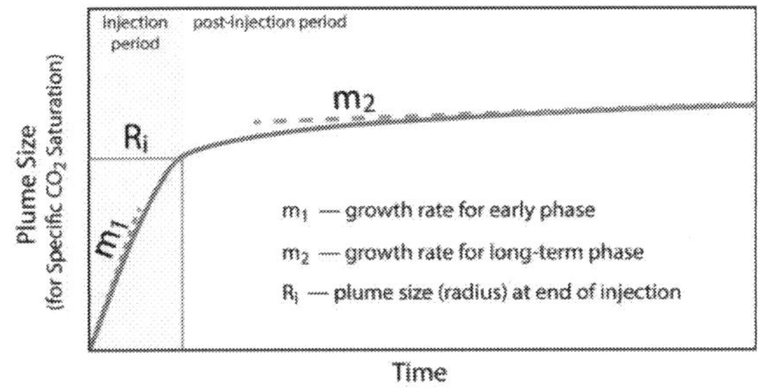

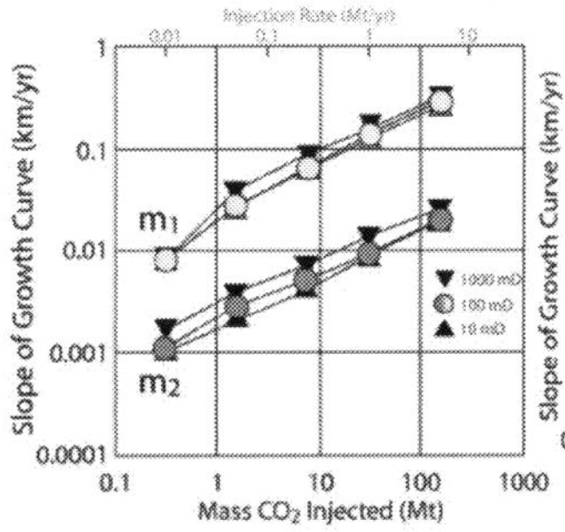

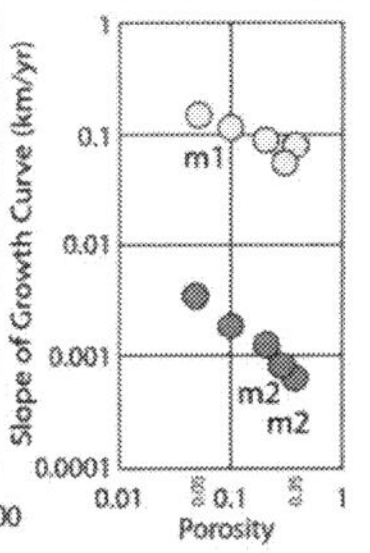

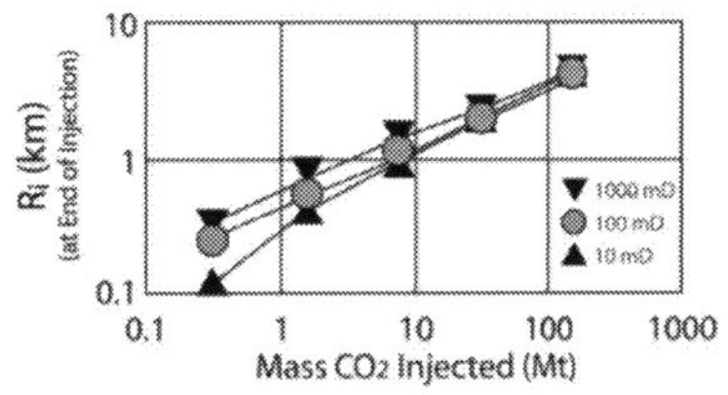

operational
- Increase in mass of CO_2 increases the growth rate for the CO_2 plume (both early and long-term rates)

geological
- Increase in porosity decreases growth rate
- Permeability has minimal impact on growth rate and small impact on size of CO_2 plume

U.S. DEPARTMENT OF ENERGY NETL Los Alamos Pacific Northwest NRAP

Pressure increase in reservoir has a characteristic behavior, albeit details vary between sites.

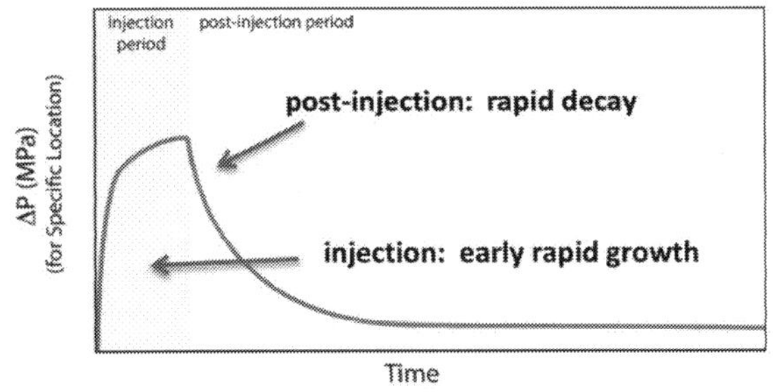

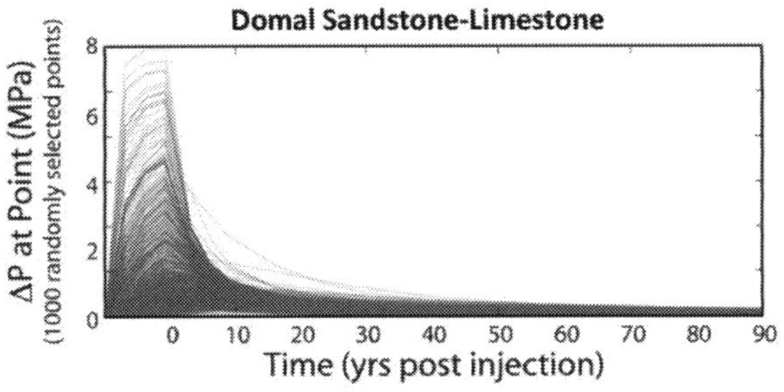

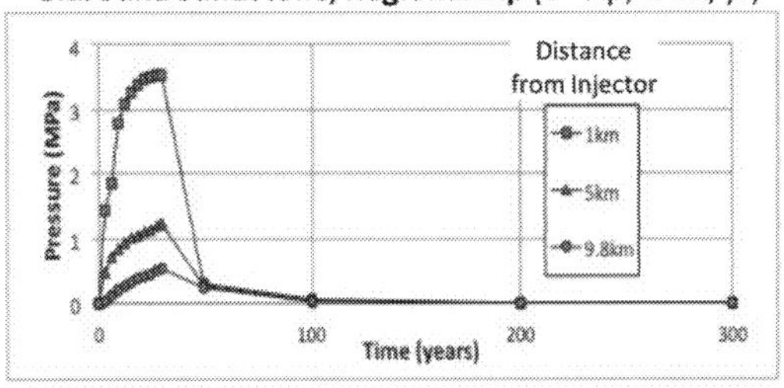

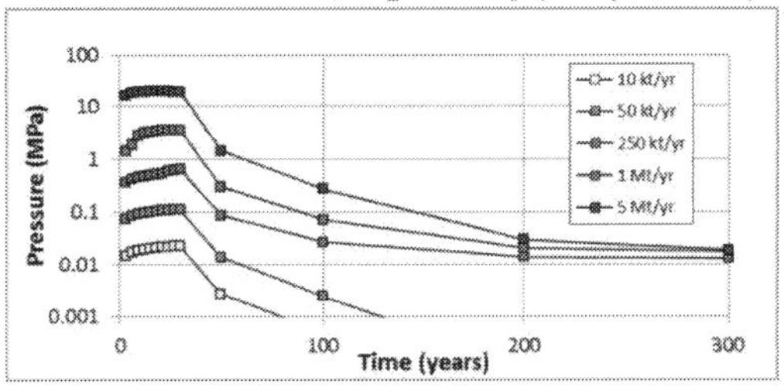

U.S. DEPARTMENT OF ENERGY NETL Los Alamos Pacific Northwest NRAP

Amount of CO_2 injected and geologic details each impact the magnitude of pressure increase.

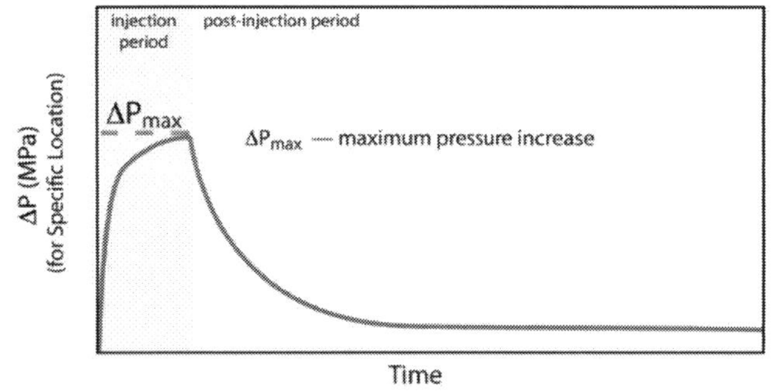

Unbound Sandstone, Regional Dip (0° dip)

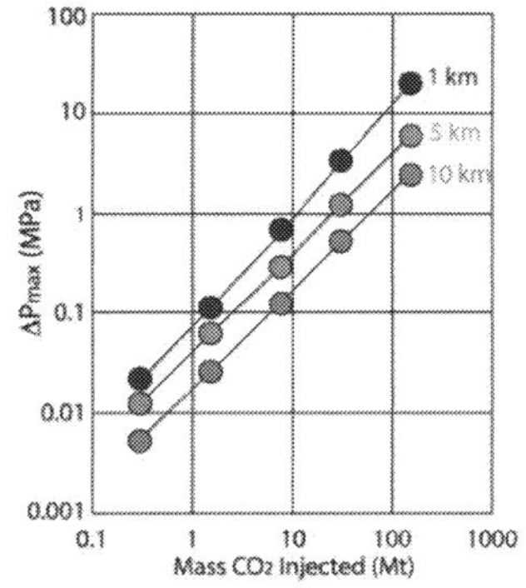

operational
- Increase in mass injected increases maximum ΔP at a given distance from the injection well

geological
- Increase in permeability decreases maximum ΔP at a given distance from the injection well

Unbound Sandstone, Regional Dip
(1° dip; 1 Mt/yr)

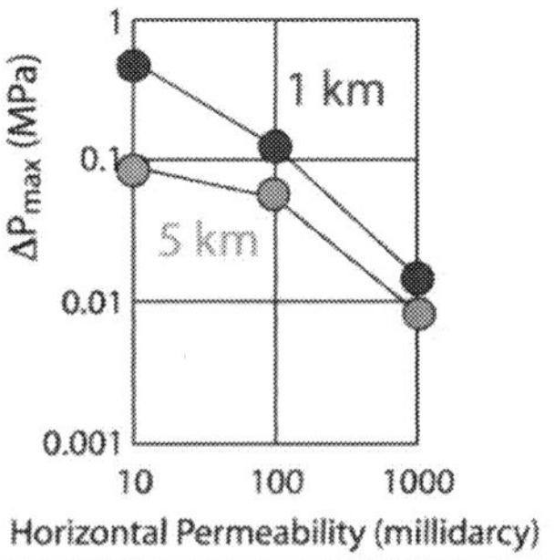

U.S. DEPARTMENT OF ENERGY
NRAP

Techno-Economic Analysis of Coal Direct Chemical Looping Combustion Using Aspen Plus

Originally presented on: 11/19/2015 13:10:00 - 13:35:00

Chemical Looping Combustion (CLC) is considered as one of the most efficient technologies for carbon capture from coal directed power plants. The effectiveness of CLC in terms of efficient power generation and almost pure CO_2 capture has been well documented; it has the least energy penalty among several other combustion technologies currently being considered for CO_2 capture, e.g. the oxy-combustion. For CLC, the concept of using a multi-staged model has also been studied by the authors recently; the purpose of multi-staging is to increase the rate of fuel conversion in the char gasification step in CLC. This increased rate of conversion supposedly would not only make the process faster, but would most likely also increase the energy output of the CLC system. Although the benefits of CLC and multi-staged CLC in terms of efficiency and energy output have been well studied, the techno-economic analysis of a CLC based power plant has been scarce in the literature. In this paper, techno-economic analysis of a coal direct multi-stage CLC system is conducted using the software ASPEN Plus. The relative cost and benefit analysis of single-stage, two-stage and four-stage CLC system is performed. The results of techno-economic simulations show that the coal direct chemical looping combustion (CD-CLC) can be an economically viable technology for power generation when compared to the traditional power plants. The results also indicate that the two-stage system CLC is the most economically viable system among the single-, two- and four-stage CLC systems considering both the capital and the operating costs.

CARBON MANAGEMENT TECHNOLOGY CONFERENCE

CMTC-439569 -MS

A Proposed Alternative Measure for Climate Change Potential

F. DeGroff CCSF

This paper was prepared for presentation at the Carbon Management Technology Conference held in Sugarland, Texas, USA, 17–19 November 2015.

Abstract

There currently exists no comprehensive metric to measure and value anthropogenic changes in carbon flux between all geospheric carbon sinks. We propose that changes in carbon residence time within geospheres be used as a metric to assess anthropogenic changes in carbon flux, and the term 'carbon quality' (cq) be used to describe such changes. Carbon residence time represents the inverse of carbon flux; as carbon flux increases, the corresponding cq will decrease, and vice versa.

Focusing on atmospheric carbon emissions as a measure of anthropogenic activity on the environment ignores the fungible characteristics of carbon that are crucial in both the biosphere and the worldwide economy. The ubiquitous carbon molecule enables the enormous diversity in the biosphere, as well as the widespread, strategic economic presence of carbon in the world economy. Focusing on a single form of inorganic carbon as a proxy metric for the plethora of anthropogenic activity and carbon compounds will prove inadequate, convoluted, and unmanageable. A broader, more basic metric is needed to capture the breath and scope of carbon activity.

We propose a logarithmic vector scale for cq to measure anthropogenic carbon flux. The distance between vector points, e.g. the starting and ending residence times, would represent the change in cq. A base-10 logarithmic scale would allow the addition and subtraction of exponents to calculate changes in cq.

As carbon moves between carbon reservoirs, the change in cq is measured as: cq = b ($\log_{10}$ [mean carbon residence time]) where b represents the carbon price coefficient for a particular country. For any country, cq measures the climate change potential for any organic carbon when converted to inorganic CO_2, or to any lower residence time carbon state. The greater the carbon fees for a country, the larger the b coefficient would be, and the greater the import fees would be to achieve carbon parity on imports from countries with lower carbon fees.

By assessing embodied carbon within imports for carbon parity with domestic production, cq would eliminate the incentives to use spatial shifts in carbon emissions to avoid carbon fees. Similarity, cq would modulate the incentives to use temporal displacement of carbon emissions, such as with biomass or CCS, to reduce carbon fees.

Introduction

The carbon molecule is the single most flexible and economically important molecule in the biosphere and the economy. Excess carbon pollution is the most pressing aspect of anthropogenically driven climate change. Yet, the current metric to measure and assess carbon based climate change focuses on the atmospheric emissions of a single form of inorganic carbon, CO_2. This singular focus will, over time, limit the effectiveness and efficiency of any climate change mitigation strategies and agreements that rely solely on CO_2 emissions.

The current methodology for measuring the contribution of anthropogenic activity on climate change uses an assessment of the equivalent quantity of CO_2 emitted within a specific boundary or envelope during the life of a good or service. The Life Cycle Assessment (LCA) is a summation of all equivalent CO_2 (CO_2e) that is emitted during certain anthropogenic activities. If there are no CO_2e or equivalents emitted, no climate change has occurred.

The proposal is to measure the effect of anthropogenic activity on carbon-related climate change by the difference between the temporal proximity of carbon sequestered from the atmosphere that is provided by the ecosystem, and that resulting from any human activity.

Proposal

The proposed metric is defined as:

> Temporal Proximity to Atmosphere (TPtA) as the non-anthropogenic length of time that carbon, in a particular state, is sequestered from the atmosphere.

The proposal for measuring the carbon related climate change potential would require the development of an empirically based scale of the temporal distance, of any and all forms of carbon, from the earth's atmosphere where the largest immediate risk of climate change exists.

In simple terms, the proposed metric measures how nature balances the movement, known as flux, of all forms of carbon both to and from the earth's atmosphere. This natural carbon flux to (and from) the atmosphere is then compared to the anthropogenic carbon flux; the difference is referred to the climate change potential.

Since a major part of all anthropogenic activity that drives carbon based atmospheric climate change is due to the acceleration of carbon flux "toward" the atmosphere, the empirical measurement of the natural, non-anthropogenic carbon flux toward the atmosphere would be the focus of the empirical effort.

Such empirical effort would initially measure the natural flux rates for the type of carbon that have the greatest affect on the atmosphere climate change, such as fossil fuels and other forms of carbon that typically form long-term carbon storage away from the atmosphere. The units of measurement for the temporal distance to the atmosphere would simply be time, e.g. years.

The new metric recognizes the natural carbon cycling in the environment as an environmental resource or an ecosystem[1] service[2]. What nature provides is a balance of carbon flux both to and from the atmosphere, so that each and every carbon pathway together forms a balance that has existed over much of human development. That balance has now changed. The increased extraction of fossil fuels, along with land use changes and other human activity has increased the level of CO_2 in the atmosphere, beyond what evidence has shown has existed over the last 400K years.

[1] "Ecosystem means a dynamic complex of plant, animal and micro-organism communities and their non-living environment interacting as a functional unit." Fao.org,. (2015). Natural Resources and Environment: Ecosystem Services. Retrieved 18 November 2015, from http://www.fao.org/nr/land/sustainable-land-management/integrated-approach-to-slm/ecosystem-services/en/

[2] Ecosystem services: All benefits that humans receive from ecosystems, including the sequestration of carbon from the atmosphere to balance carbon flux into the atmosphere. Ibid.

The new metric is comprehensive. It applies to changes in the flux of all forms of carbon including fossil fuels, feedstocks, biomass, etc. at any time, at any location on the planet.

Focusing on a single form of inorganic carbon, atmospheric CO_2, Life Cycle Assessment, accounts for all emissions of CO_2 within a life cycle envelope. To compare the affect of producing products on climate change, LCA requires the complete knowledge of life cycle pathway. The smaller the envelope, the less complete the assessment of climate change. A full assessment of the CO_2e for a particular good or service produced is only available at the end of product life cycle.

The emphasis on CO_2 emissions as a primary measure of climate change can present accounting challenges. The felling of a tree, for example, may focus on the ancillary human activities, such as transportation, embodied carbon in the ax used to fell the tree, and food for the logger. If the logger used a recycled ax rather than a chain saw and walked rather than drove to fell the tree, however, the site specific CO_2 emissions would be minimal, perhaps missing the shift in carbon flux from 100+ years sequestration as woody biomass to significantly less time as paper or other biomass products, with little or no emissions from any burning as fuel. LCA and CO_2e are extremely helpful as tools to understanding how human activities produce goods and services, but they do not inform as to how all human activity affects the natural carbon cycle.

The felling of a tree demonstrates the limitations of LCA/ CO_2e, resulting in the need for adjunct credits for emissions avoidance activities, such as land use changes that result in climate change but little or no CO_2 emissions, or that occur outside the LCA envelope. Emphasis on emissions, rather than on changes to the natural carbon flux, often results in a limited, oblique view of the full effect of anthropogenic activities that result in climate change.

Table 1: Distinctions Between LCA and TPtA

	LCA	TPtA
Metric	CO_2e	Carbon Quality
Measure	Scalar	Vector
Units	Mass	Mass & Time
Analysis	Path	State
Envelope	Yes	No
Scope	Full Life Cycle envelope	Empirically Determined State
Emphasis	Downstream	Upstream
Credits	Avoidance	None
Accounting	Verification ?	GAAP/ IFRS
Frequency	Every Transaction	Every State
Implementation	Regulatory	Economic

1. Metric: LCA uses CO_2e, TPtA uses the new property carbon quality (cq)
2. Measure: LCA is a scalar measure of climate change, TPtA is a vector with both magnitude + direction, from one carbon form to another
3. Units: LCA – Mass, TPtA – Mass + Time
4. Analysis: LCA is a path dependent function, TPtA is a state function
5. Envelope: TPtA requires no envelope or boundary
6. Scope: LCA requires full life cycle accounting (or within a defined envelope envelope), TPtA is an empirically determined *state* and requires no envelope

7. Emphasis: LCA emphasis is on the downstream, or total CO_2 emissions, TPtA accounts for climate change in all human activities, upstream and downstream
8. Credits: LCA requires credits for activities that avoid climate change, No such credits are required for TPtA since all human activities are included in the climate accounting
9. Accounting: Verification of avoidance activities can be problematic, TPtA does not need nor use credits
10. Frequency: LCA must be calculated for each distinct or unique economic transaction, TPtA requires that the climate potential for each carbon state be empirically determined
11. Implementation: LCA primarily through regulation, TPtA would be economically driven

Definitions

Mean Temporal Proximity to Atmosphere (mean TPtA):

The mean (non-anthropogenic) time necessary for carbon in a particular state to cycle through *to the atmosphere* as CO_2

Carbon quality (cq):

$cq_{(c\ state \rightarrow atmos)} = f\ (\log_{10}\ [\text{mean TPtA}]_{(c\ state \rightarrow atmos)})$[3]

$\Delta cq_{(c\ state \rightarrow atmos)} = f\ (\Delta \log_{10}\ [\text{mean TPtA}]_{(c\ state \rightarrow atmos)})$

Carbon toll (ct) for country X with country coefficient β:

$ct_X = \beta_X\ \Delta cq_{(c\ state \rightarrow atmos)} = \beta_X\ (\Delta \log_{10}\ [\text{mean TPtA}]_{(c\ state \rightarrow atmos)}$

Carbon avidity:

The affinity of a company, country or other entity for carbon, especially high energy or value fossil fuels, in lieu of other alternatives due to economic or infrastructural inertia, natural resource availability, societal or political structure, etc.

Carbon parity:

When disparate carbon sources have received the same economic assessments for equivalent changes in carbon quality

Carbon cognates:

Economic substitutes for CO_2 emissions, aka alternative forms of carbon exhaust as a way of avoiding fees on atmospheric CO_2 emissions

[3] For mean TPtA > 1 year; for mean TPtA $\leq$ 1 year cq=0

CMTC-438982-MS

GHG Mitigation Strategies (Kyoto Protocol)

Nnamdi A. Nwaka, University of Ibadan

Abstract

Global warming and climate change have become major concerns to the Global Community over the past two decades. The aftermath of these two adverse phenomena pose a threat to human existence and the entire Biological Ecosystem. Over 150 years of industrialization have produced significant levels of waste gases (greenhouse gases) from fossil fuels use. The greenhouse effect, resulting from these emissions (chiefly CO_2), have led to drastic upsurge in temperatures and other adverse climate changes. This paper, as part of the global efforts to curb greenhouse gas (GHG) emissions, reviews the already set out GHG mitigation strategies by the Kyoto Protocol and its consequences. It seeks to apply the ideas of the protocol to this part of the world for more energy and environmental security and economic development. Global successes and challenges of the protocol and highlighted also and the concept of carbon credits is concisely discussed. This paper shows that the protocol has generally been successful and could be more successful. It also shows that the application of the strategies in a country would provide better living conditions.

Introduction

Global warming is the continuous increase in the average temperature of the Earth's atmosphere and its oceans, a change that is seen to be permanently changing the Earth's climate. The rise in the volumes of carbon dioxide and other greenhouse gases (GHGs) released by the use of fossil fuels, land clearing, agriculture, and other human activities, are believed to be the principal sources of the global warming that has occurred over the past 50 years [Intergovernmental Panel on Climate Change, 2007]. Evident changes resulting from global warming include: sea levels rise due to melting of the polar ice caps, change in the amount and pattern of precipitation as well a probable desert expansion, frequent and severe storms, ocean acidification and species extinction due to shifting temperature region, decreasing crop yield - a threat to food security, inundation leading to habitat loss, drought and other severe events.

Substantial scientific evidence indicates that an increase in the global average temperature of more than 3.6 degrees Fahrenheit (°F) (or 2 degrees Celsius [°C]) above pre-industrial levels poses severe risks to natural systems and to human health and well-being (Union of Concerned Scientists, 2007).
The United Nations, through its arm: United Nations Framework Convention on Climate Change (UNFCCC), in the bid to reverse the situation has set out mitigation strategies embodied in the international agreement known as the Kyoto Protocol. These strategies aim to reduce emissions primarily through the use of renewable sources of energy and carbon sequestration. The protocol recognizes that the developed countries are principally responsible for the current high levels of GHG emissions in the atmosphere as a result of more than 150 years of industrial activity. Developing countries do not have binding commitments under the Kyoto Protocol as regards to their emissions but are encouraged to embark on emissions reduction projects for a more habitable environment.
In Nigeria, our primary source of energy is the fossil fuel. Most power plants, industries and other institutions make use of generators. These are potential global warmers and with our extensive solar and wind endowment and human resources, these should not be the case. In as much as we do not have any binding targets under the protocol, there is need to engage ourselves in the use of renewables for a more sustainable and environmentally friendly energy source. It is important also to consider the effects of certain emission practices such as gas flaring and enact laws to combat it.

The Mechanisms

The protocol has employed basically three strategies to see to emissions reduction. These strategies involve interactions between two or more countries.

Clean Development Mechanism (CDM)

The Clean Development Mechanism (CDM), allows a country with an emission-reduction or emission-limitation commitment under the Kyoto Protocol (Annex I Party) to implement an emission-reduction project in developing countries (United Nations, 1998). Such projects can earn carbon credits in form of saleable certified emission reduction (CER) credits, each equivalent to one ton of $CO2$, which can be counted towards meeting Kyoto targets. A CDM project activity might involve, for example, a rural electrification project using solar panels or the installation of more energy-efficient boilers.

This mechanism stimulates sustainable development and emission reductions, while giving industrialized countries some flexibility in how they meet their emission reduction or limitation targets (United Nations, 1998).

International Emissions Trading

Every country with a commitment under the Kyoto Protocol has a quota of allowed emissions referred to as "assigned amount units" (AAUs). Emissions trading allows countries that have emission units to spare to sell their excess capacity to countries that are over their targets.

Thus, a new commodity created in the form of emission reductions or removals. Since carbon dioxide is the principal greenhouse gas, carbon like any other commodity is now followed and traded. Hence the term - "carbon market".

Joint Implementation

This is similar to the CDM, but differs only in the host country. In Joint Implementation (JI), any industrialized country can invest in an emission reduction project (referred to as a "Joint Implementation Project") in any other industrialized country as an alternative to reducing emissions domestically (United Nations, 1998). In this way countries can lower the costs of complying with their Kyoto targets by investing in projects that reduce greenhouse gas emissions in an Annex I country where reducing emissions may be cheaper, and then using the resulting Emission Reduction Units (ERUs) towards their commitment goal. A JI project might involve, for example, replacing a coal-fired power plant with a more efficient combined heat and power plant.

The Reward - Carbon Credits

A carbon credit implies any tradable certificate or permit representing the right to emit one ton of or the mass of another GHG equivalent i.e. equivalent to one ton of CO_2. Consequently, one carbon credit is equal to one metric ton of CO_2. Carbon credits create a viable market for curbing greenhouse emissions by assigning monetary value to the cost of air pollution. Emissions, therefore and reasonably, become a variable cost of doing business and are recorded on the balance sheet alongside other liabilities or assets. For example, consider an establishment that owns a factory putting out 10,000 tons of greenhouse gas emissions in a year. Its government is of a country that enacts a law to limit the emissions that the business can produce. So the factory is given a quota of say 8,000 tons per year. The factory either reduces its emissions to 8,000 tons or buys carbon credits to offset the excess. After opportunity cost evaluations, the business may decide that it is not economical to invest in new machinery for that year. Instead it may decide to buy carbon credits on the open market from organizations that have been approved as being able to sell legitimate carbon credit, physically sequester its produced waste gases or use renewables.

Discussions

The idea of the adopted mechanisms can therefore be applied in solving our national and global issues. In Nigeria, the International Oil Companies (IOCs) could get involved in electricity generation by investing in power plants. These power plants can be of course run on fuels from refined crude, which would be cheaper for the IOCs if refined locally. The IOCs can therefore enact better control of emissions through bulk carbon sequestration. Through these, the IOCs would have: provided cheaper electricity for people, created a wider range of employment from the mineral exploitation through power generation and bulk sequestration, insured the profitability of oil business, provided cheaper raw materials for industries, participated actively in the carbon market and most importantly, reduced emissions.

In the past 10 years, CDM projects alone have delivered 110,000 MW of renewable energy capacity globally, which is roughly equivalent to the total power generation capacity of Africa. There are registered CDM projects in 83 developing countries, ranging from projects that reduce emissions by replacing inefficient wood stoves, to solar power projects that displace fossil fuels, to large industrial projects that destroy extremely potent greenhouse gases (GHGs) (United Nations Framework Convention on Climate Change, 2015a). Examples include (according UNFCCC's report on CDM projects):

1. Phuong Mai 3 Wind Power Project; CDM project 7279, in Vietnam, by Switzerland, with expected reduction of 31,978 metric tons CO2 equivalent per annum.

2. Cebu CTRADE Biogas to Energy Project; CDM Project 9223, in Philippines, with expected reduction of 46,914 metric tons CO2 equivalent per annum.

3. Mitigation of Methane Emissions in the Charcoal Production of Plantar; CDM Project 1051, in Brazil, by Japan et al., with expected reduction of 16,098 metric tons CO_2 equivalent per annum.

A perfect example of an emissions trading system is the EU emissions trading system (EU ETS) which covers more than 11,000 power stations and industrial plants in 31 countries, as well as airlines.

The JI projects are not left out either with the following examples (according to UNFCCC's report on JI projects):
1. BE1000314: YARA Tertre Uhde 2 abatement project in Belgium, by France, with an expected reduction of 240337.0 tons CO_2 equivalent.

2. Project 0221: YARA Köping S2 N_2O abatement project in Sweden, by Netherlands, with an expected reduction of 159,070 tons CO_2 equivalent.

3. FR1000029: Thermal oxidation of the gas effluents of the Trifluoroacetic Acid generating station at the Salindres factory (Gard) in France, by Japan, with an expected reduction of 507806.0 tons CO_2 equivalent (United Nations Framework Convention on Climate Change, 2015b).

Conclusion

On the basis of the extensive reviews, the following conclusions have been reached:

1. Clearly, many nations have followed through on their agreement and cut greenhouse gas emissions, an example is the European Union who dropped their emissions by 5%. (British Petroleum, 2014)

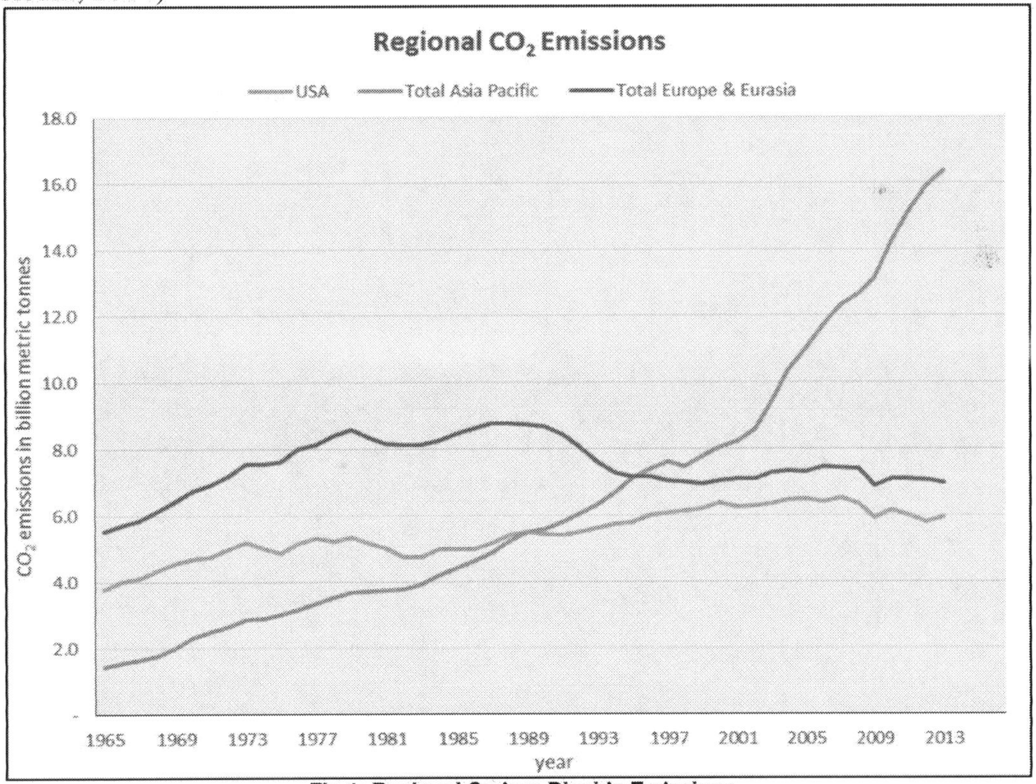

Fig 1: Regional Carbon Dioxide Emissions
Data Source: BP's Statistical Review of World Energy, 2014.

2. A major challenge of the protocol is non-ratification by highly emitting countries.
3. The mechanisms adopted by the protocol are major tools for global development through more energy and environmental security.

4. There is an urgent need for further research in the application of renewable sources of energy within our own locality.

References

British Petroleum (2014). *BP's Statistical Review of World Energy 2014.* London: Author.

Intergovernmental Panel on Climate Change. (2007). *Climate Change 2007: IPCC Fourth Assessment Report.* Retrieved from https://www.ipcc.ch/ipccreports/far/wg.../ipcc_far_wg_II_full_report.pdf

Union of Concerned Scientists. (2007). *How to Avoid Dangerous Climate Change: A Target for U.S. Emissions Reductions.* Cambridge, England: Author.

United Nations. (1998). *Kyoto Protocol to the United Nations Framework Convention on Climate Change.* Bonn, Germany: Author.

United Nations Framework Convention on Climate Change. (2015a). *CDM Projects.* Retrieved from http://cdm.unfccc.int/Projects/DB/DNV-CUK1150124653.47/view

United Nations Framework Convention on Climate Change. (2015b). *JI Projects.* Retrieved from *http://ji.unfccc.int/JI_Parties/DB/05WMEGGC4NX26HD3IPXIOHNQSIDXDL/viewDFP*.

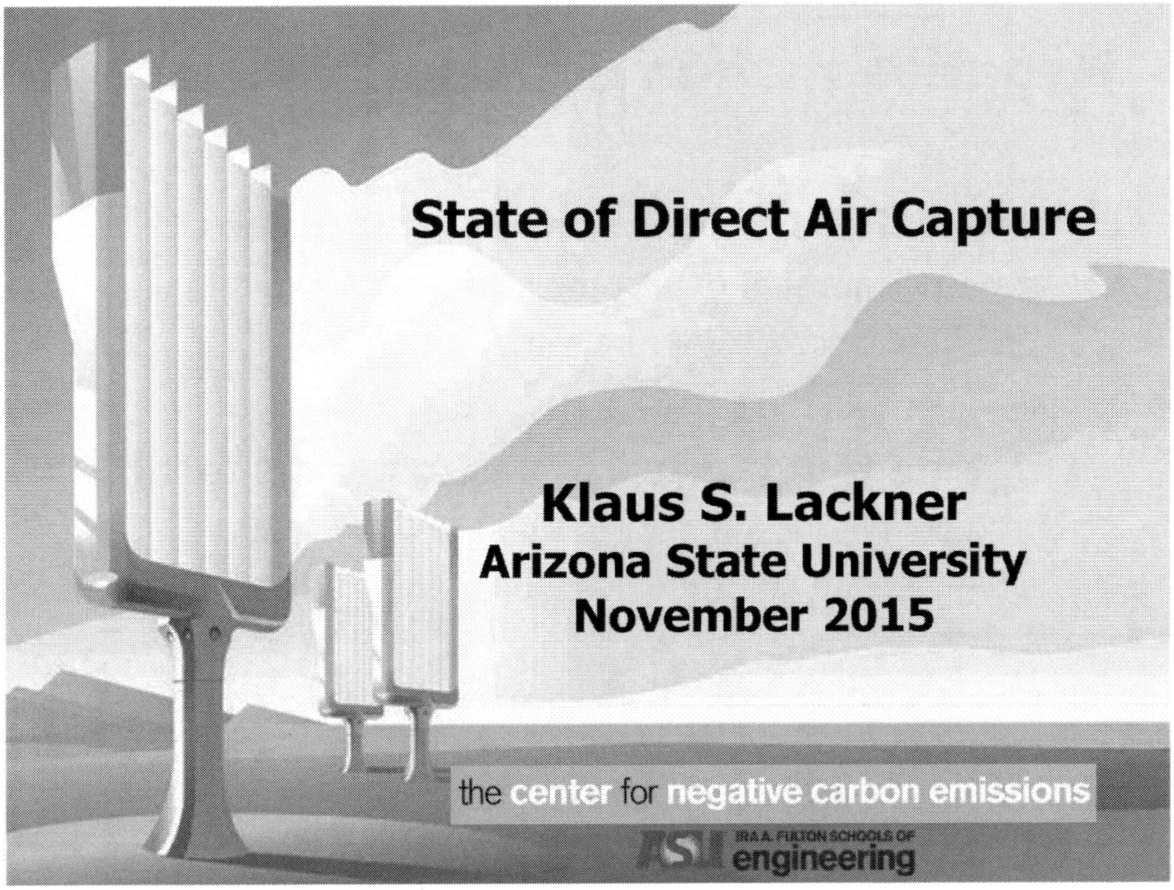

IPCC calls for negative emissions

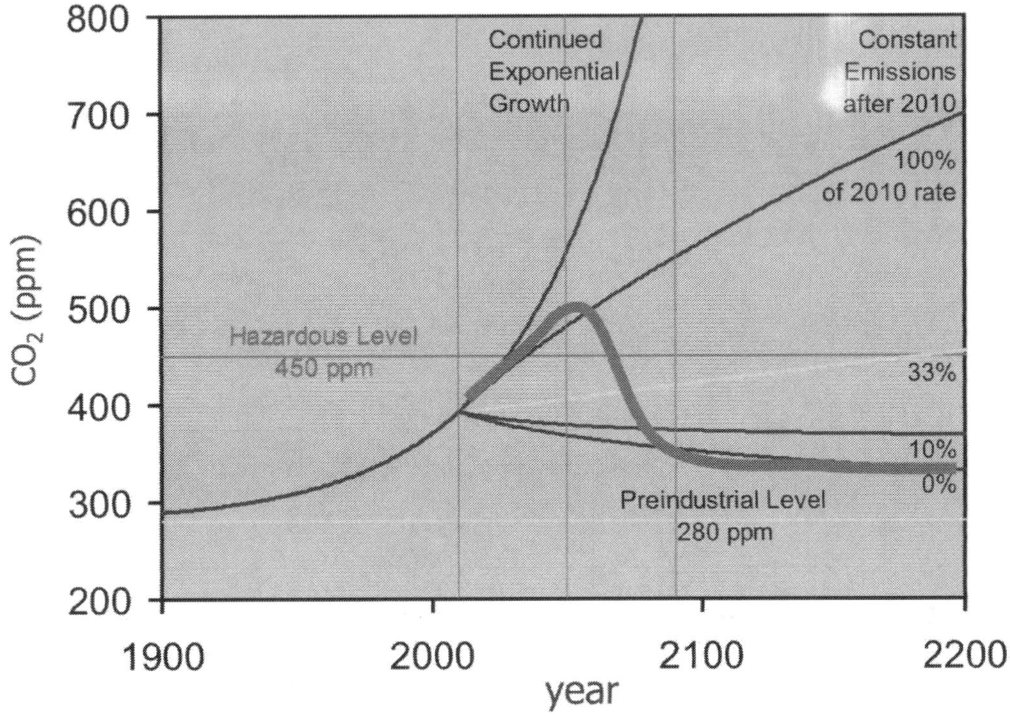

Negative carbon emissions

- **Retrieval of carbon from environment**
  - Direct removal of CO_2 from air or ocean
  - BECCS, Direct Air Capture, etc.

- **Large demand for carbon dioxide disposal**
  - 100 ppm reduction implies ~400 Gt C or ~1500 Gt CO_2
    - Ocean will return its carbon

Even zero emission must eliminate all emissions, not just 90% of coal emissions

3

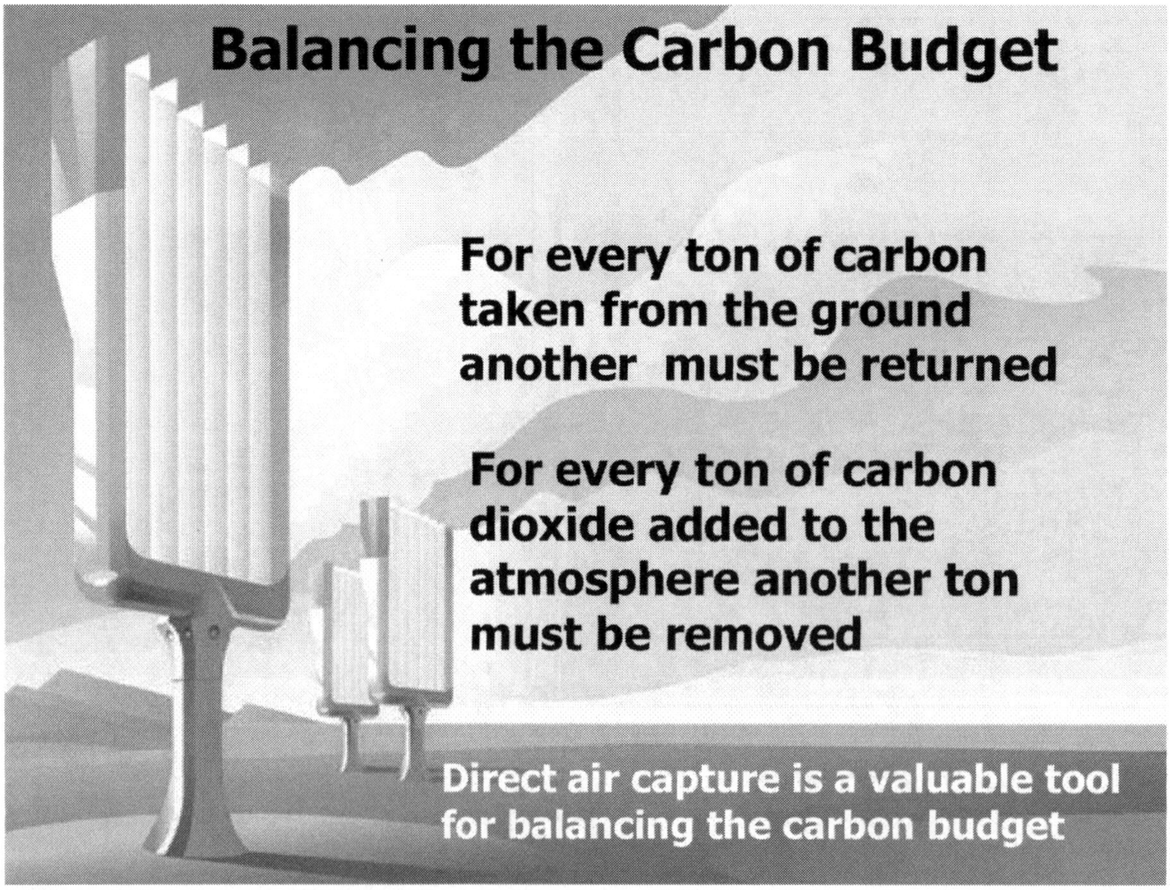

Air capture is the capture of last resort

- can cancel out all unaddressed emissions
- can recover past emissions
- sets upper limit on the cost of carbon
- allows for zero or negative carbon scenarios
- encourages point source capture

5

Wikipedia picture

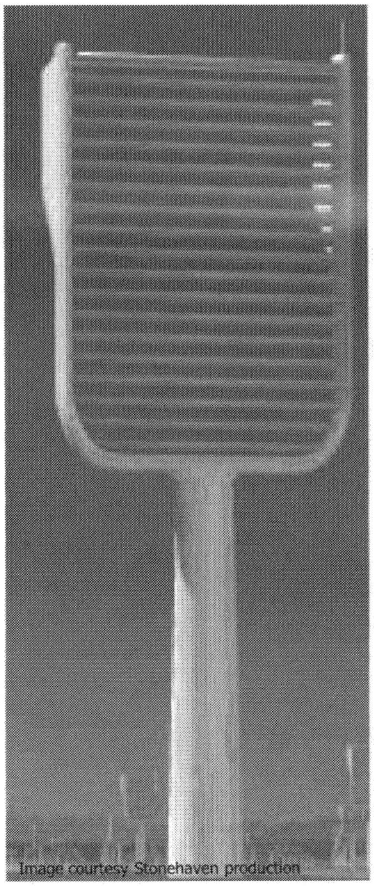

Why air capture?

- **Air capture eliminates exceptions**
  - No emission source can be exempt
  - Separates sources from sinks
  - Carbon democratization
- **Air capture for drawing down CO_2**
  - Negative emissions are already unavoidable
  - Requires vast CO_2 storage capacity
  - Negative carbon emissions
- **Air capture with non-fossil liquid fuels**
  - Synthetic fuel production from CO_2 and H_2O
  - Provides energy storage & liquid fuels
  - Requires cheap non-fossil energy
  - Carbon recycling
- **Air capture with fossil liquid fuels**
  - Carbon balanced by sequestration
  - Requires cheap CO_2 storage
  - Carbon balancing

Three Rules for Technological Fixes

D. Sarewitz and Richard Nelson:

Three rules for technological fixes, *Nature,* 2008*, 456*, 871-872

I. **The technology must largely embody the cause-effect relationship connecting problem to solution.**

II. **The effects of the technological fix must be assessable using relatively unambiguous or uncontroversial criteria.**

III. **Research and development is most likely to contribute decisively to solving a social problem when it focuses on improving a standardized technical core that already exists.**

In contrast, direct removal of CO_2 from the atmosphere — air capture — satisfies the rules for technological fixes. Most importantly, air capture embodies the essential cause–effect relations — the basic go — of the climate change problem, by acting directly to reduce CO_2 concentrations, independent of the complexities of the global energy system (Rule I). There is a criterion of effectiveness that can be directly and unambiguously assessed: the amount of CO_2 removed (Rule II). And although air-capture technologies have been remarkably neglected in both R&D and policy discussions, they nevertheless seem technically feasible (Rule III).

Thermodynamics checks out

Theoretical minimum free energy requirement for the regeneration is the free energy of mixing

Gas pressure P_0
CO_2 partial pressure P_x
Denoted as (P_0, P_x)

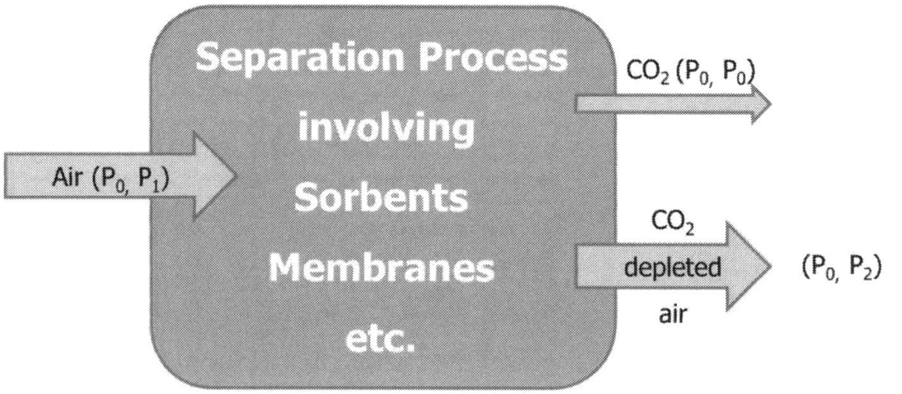

$$\Delta G = RT\left(\left(\frac{P_0 - P_2}{P_1 - P_2}\right)\frac{P_1}{P_0}\ln\frac{P_1}{P_0} - \left(\frac{P_0 - P_1}{P_1 - P_2}\right)\frac{P_2}{P_0}\ln\frac{P_2}{P_0} + \left(\frac{P_0 - P_1}{P_0}\right)\left(\frac{P_0 - P_2}{P_0}\right)\frac{P_0}{P_1 - P_2}\ln\frac{P_0 - P_1}{P_0 - P_2}\right)$$

Specific irreversible processes have higher free energy demands

22 kJ/mol

8

Feasibility & Affordability?

CO_2 in air is dilute and air is full of water

Sherwood's Rule:

Separation costs often scale linearly in dilution

Air Capture is an exception

Sherwood's Rule is not a Law

A challenge for dilute values

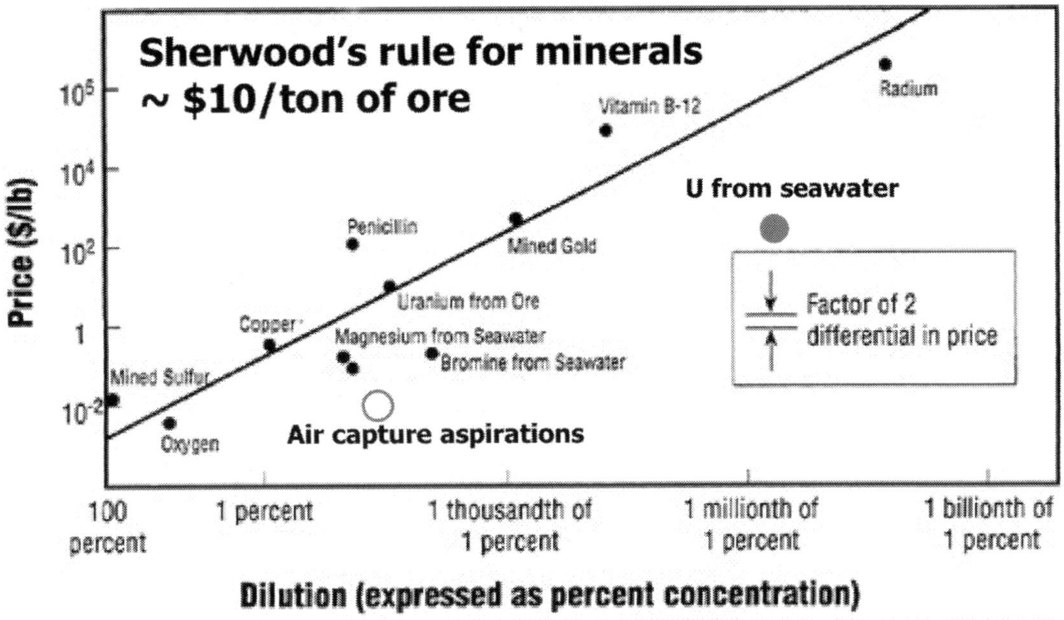

SOURCE: National Research Council (1987)

Artificial kelp to absorb uranium from seawater

- **Passive, long term exposure to water**
  - Braids of sorbent covered buoyant plastic
  - Anchored to the floor
  - Replace [illegible] active systems
- **Low energy sorbent**
  - Laminar flow over sorbent
  - Uptake is limited by boundary layer transport
- **Regeneration**
  - After harvesting the strings
- **Gross viola[illegible]**
  - Cost estimates range from $200 to $1200/kg
  - Sherwood $3 million/kg

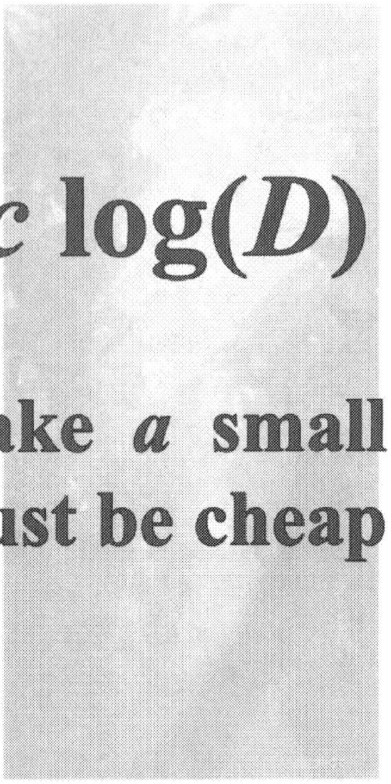

$$Cost = a\,D + b + c\log(D)$$

must make *a* small

Air contacting must be cheap

Air capture is sorbent based

- **Sorbent based separation of CO_2 from air**
  - Concentration ratio is 1 : 2500
    - Eliminates all options that perform significant work on the air
    - Sorbents postpone work to the regeneration step
- **All sorbents are chemical sorbents**
  - At 400 ppm physisorption is too weak
  - Minimum free energy of binding: $\Delta G > 22$ kJ/mol
- **Sorbents exploit carbonate chemistry**
  - Alkali hydroxides
  - Weak and strong based amines
  - Thermal, vacuum and reaction based recovery
    - Humidity swing takes advantage of $H_2O - CO_2$ – sorbent reactions

12

ASU Strategy for cost reductions

- **Use passive collectors to avoid Sherwood's Rule**
  - Cost need not scale linearly with dilution
  - no extrapolation from flue gas streams
  - Passive collectors
- **Match sorbent material to the problem**
  - Match swing to ambient conditions
    - 1:2500 CO_2
    - 1:100 H_2O
  - Minimum binding energy, low cost energy, good kinetics, ...
- **Harness cost reductions from mass production**
  - Learning curves can give orders of magnitude cost reductions
  - This makes a priori cost estimates virtually impossible
- **Take advantage of the resources in the air**
  - Wind energy, moisture content, temperature, etc.

Anionic Exchange Resins

Solid carbonate "solution"
Quaternary ammonium ions form strong-base resin

Type I Strong Base Resins

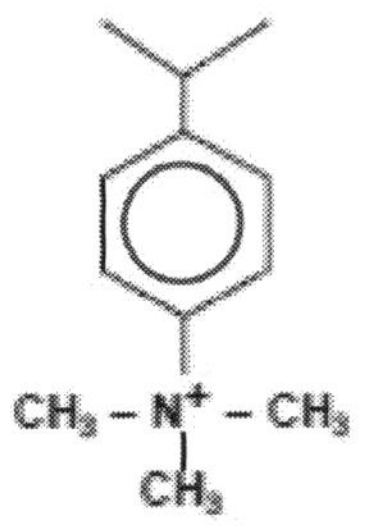

- Positive ions fixed to polymer matrix
  - Negative ions are free to move
  - Negative ions are hydroxides, OH^-
- Dry resin loads up to bicarbonate
  - $OH^- + CO_2 \rightarrow HCO_3^-$ (hydroxide → bicarbonate)
- Wet resin releases CO_2 to carbonate
  - $2HCO_3^- \rightarrow CO_3^{--} + CO_2 + H_2O$

Novel Moisture driven CO_2 swing

The Moisture Swing

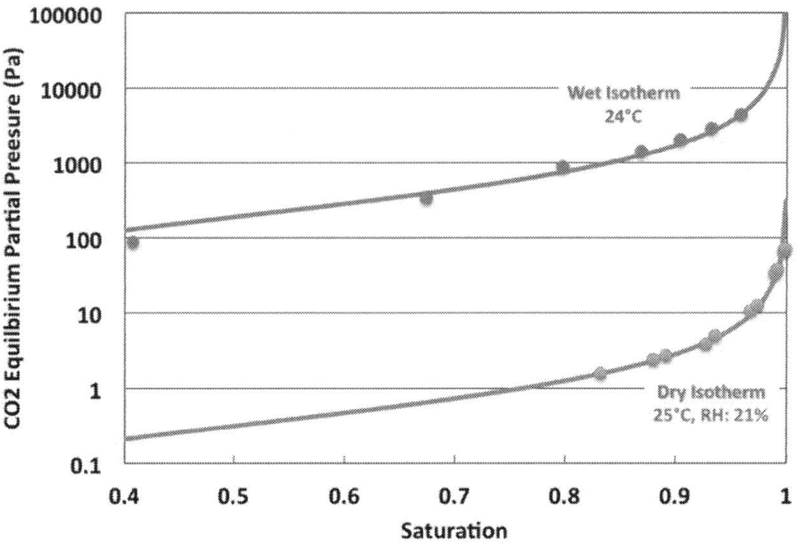

Tao Wang et al

15

Free energy from water evaporation

Water evaporation can drive CO_2 capture

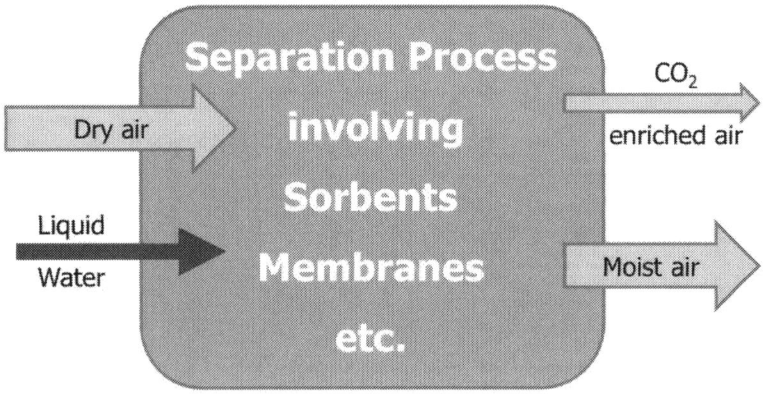

Free energy of water evaporation at a relative humidity *RH:*
$\Delta G = RT \ln(P/P_{sat}) = RT\ ln(RH)$

Ball park estimate: 2.5 kJ/mol
140 MJ/m^3
@ 20¢/m^3 0.5¢/kWh

16

Many other alternatives to DAC

- **Field without public support moved to start-ups**
  - Looking for CO_2 markets
  - ASU and its new center aim for public demonstration
    - Working with private companies and other universities
- **Many different approaches are being pursued**
  - Liquid and solid sorbents
  - Inorganic and organic
  - Passive and active air flow (low pressure drop)
  - Thermal swings
  - Moisture driven swings
  - Vacuum swings
  - Combinations of different swings
  - Shaped after large industrial processes
  - Emulating the mass production paradigm

17

CO_2 Markets

- **Merchant CO_2**
  - Markets are small and distributed
- **Chemical commodities**
  - E.g. urea, often attached to point sources
- **Biomass production**
  - Greenhouse, algae reactors
    - Limit water consumption, produce food/fuels
    - Operate with CO_2 enriched air
- **Enhanced oil/gas recovery**
  - Air capture aims at small fields, exploratory work in the absence of pipelines
- **Synthetic renewable fuels**
  - Input is excess, intermittent renewable power, often distributed
  - Energy rather than CO_2 drives cost
- **Sequestration**
  - Remote locations engender less resistance

Air capture has a competitive advantage in satisfying small, distributed or remote demands

18

Developments and Research

- **Private companies capturing CO_2**
  - Carbon Engineering
  - Climeworks
  - Global Thermostat
  - Infinitree
  - Skytree
- **Sorbent Research**
  - ASU, Georgia Tech, Columbia, Zhejiang and more
- **Systems Research**
  - Technical challenges are system's challenges
  - Systems constraint will drive component research

19

Center for Negative Carbon Emissions

Demonstrating a new approach to carbon management

Demonstration	Basic Science	Policy-Outreach
Field deployed prototypes	Sorbent chemistry, physics and thermodynamics	Study policy implications of air capture technology
Establish rapid prototyping capability	Separation Science	Move policy makers to a zero-emission world
Continuous improvement and continuous operation	System engineering, scaling and automation, techno-economics	Connect to industry and venture capital
Learning by doing in multiple iterations	Sustainability science, human interfaces	Seek support from foundations & sovereign wealth funds
		Creating IP Pool

20

Low cost comes with experience

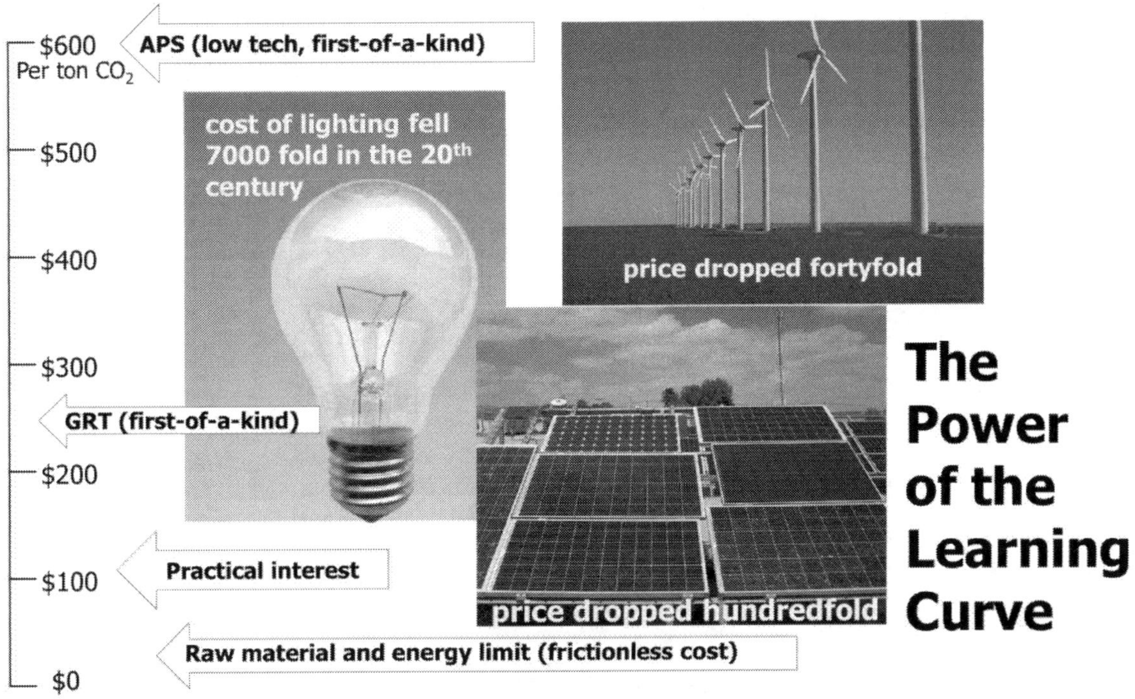

Ingredient costs are already small – small units: low startup cost

Wikipedia pictures

Finding Market Niches

Electrochemical Conversion of Air CO_2 to Dissolved Mineral Bicarbonate with the Production of C-Negative H_2

Greg H. Rau[1,2], Susan A. Carroll[2]

[1]IMS, University of California, Santa Cruz (off-campus)

[2]Energy and Environment Directorate
Lawrence Livermore National Laboratory

rau4@llnl.gov; carroll6@llnl.gov

CMTC meeting, Nov. 19, 2015

Overview

- Why air CO_2 removal?
- Description of novel, hybrid, electrogeochemical process that when powered by non-fossil energy:
  - Consumes atmospheric CO_2
  - Generates C-negative H_2 relevant as energy storage, a transportation fuel, or a chemical feed stock
  - Has high capacity - uses globally abundant raw materials and stranded energy
  - Costs <$100/ tonne CO_2 consumed, comparable to CCS
- Applications, capacity, economics and other issues requiring further R&D.

2

Why Air Carbon Dioxide Removal - CDR?

- In current models (IPCC et al.), CO_2 emissions reduction alone is insufficient to stay below 2°C global warming: 100's to 1,000's of Gt CO_2 will need removing
- Emissions reduction itself does not address remediation/mitigation of long-live legacy CO_2 in the atmosphere.
- <u>Good news</u> – natural CDR (marine and land; bio and geochemical) currently consume more than half of our emissions ≈ 20 Gt CO_2/yr.
- Hence, rather than inventing completely new CDR processes, we should consider enhancing/modfying proven, high capacity, natural CDR.

3

Natural Carbon Dioxide Removal in Action

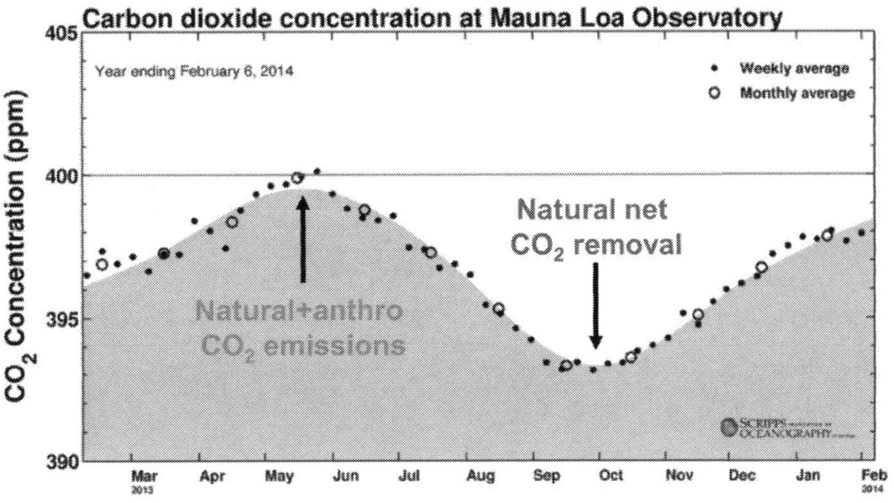

For Example: Enhance CO_2 Weathering of Base Minerals

Atmospheric or Waste CO_2

↓

Mineral Weathering Reactions

***Silicate*: CO_2 + Ca/MgSiO_3 ---> Ca/MgCO_3 + SiO_2**

***Carbonate*: $CO_2 + H_2O + CaCO_3$ ---> Ca^{2+} + 2HCO_3^-**

- The Good News - Such reactions are the primary way that natural or unnatural increases in air CO_2 are consumed and stored at global scales.
- The Bad News - Very slow reaction rates under natural conditions; Takes 100's kyrs to naturally consume excess global CO_2.

5

Acids Accelerate Mineral Dissolution

acid dissolution examples:

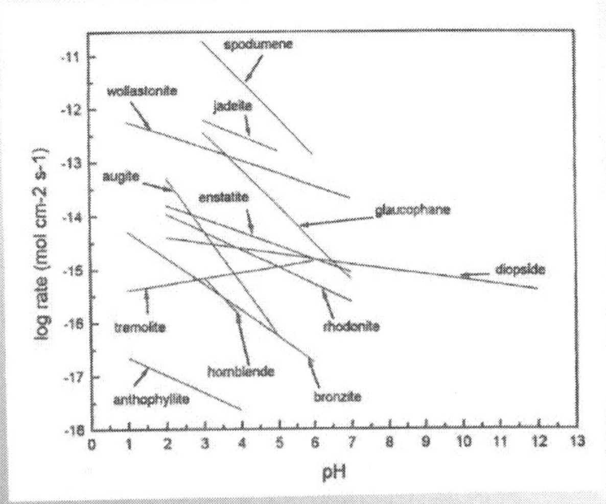

from Brantley and Chen, 1995

acid and base dissolution examples:

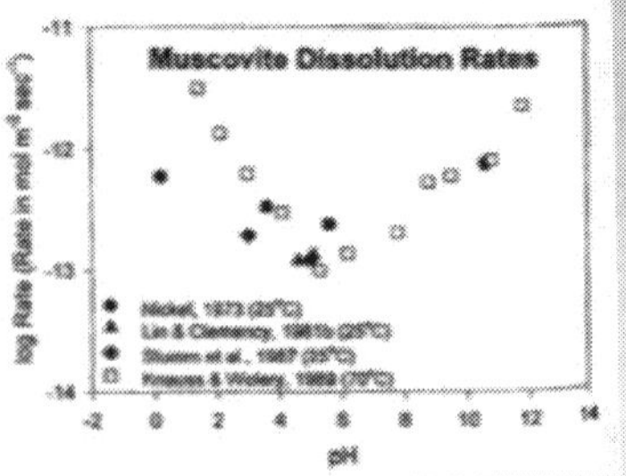

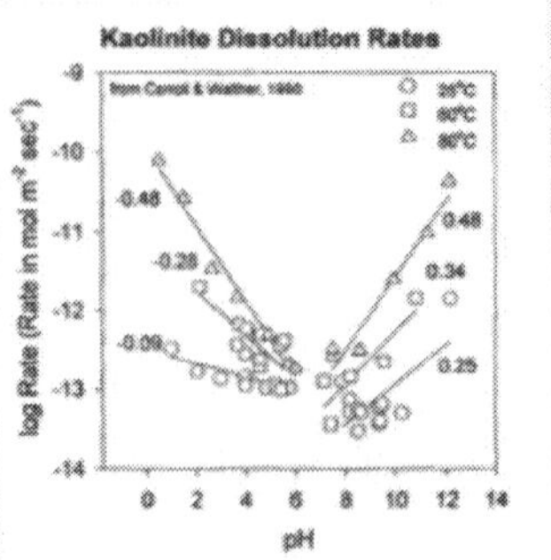

from Nagy, 1995

6

A Solution:
Integrate Mineral Weathering into Saline Water Electrolysis:

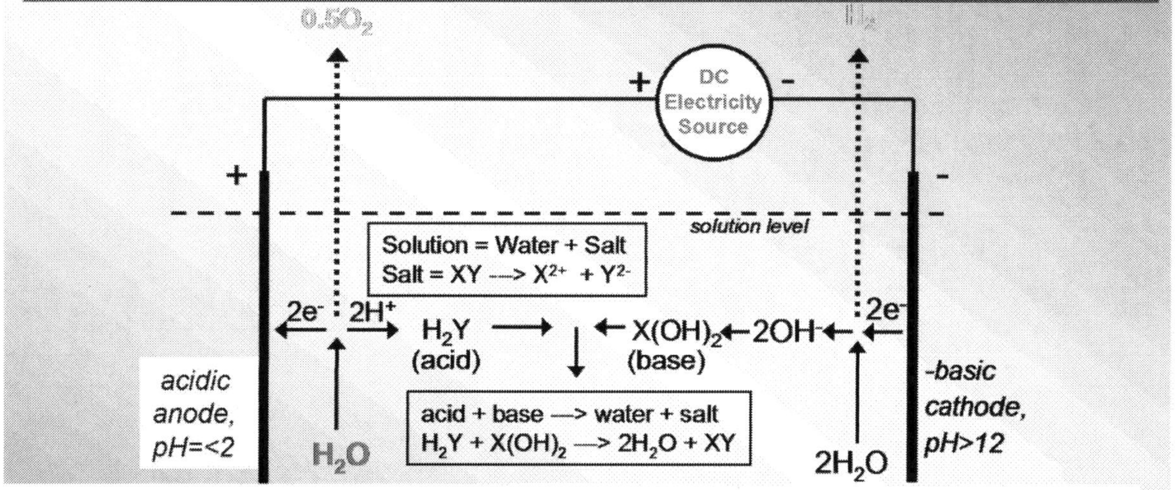

Net reaction: $H_2O + V_{DC}$ ----> $0.5O_2 + H_2$ Δ° G = $\underline{+}$237.1 kJ/mol H_2

No net gain or loss of salt, acidity, or hydroxide

But significant anode-cathode pH gradient

Innovation: ***Insert base minerals (carbonates or silicates) in the acidic anolyte to effect acid neutralization and mineral dissolution causing in hydroxide accumulation in the bulk electrolyte.***

7

So React Anodic Acid With Base Minerals to Produce Excess Soluble Base/Hydroxide

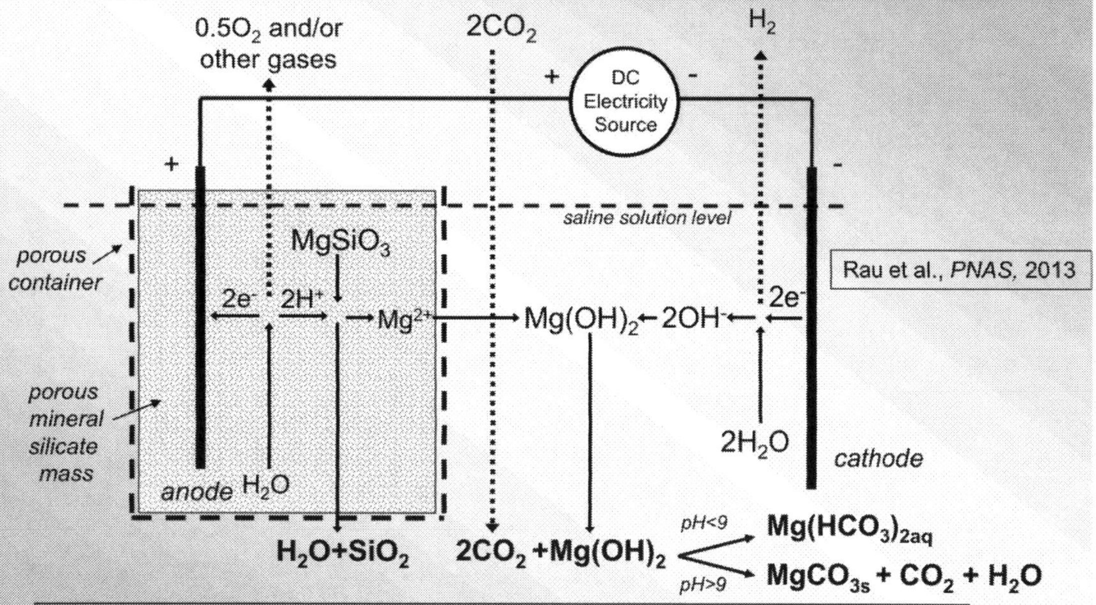

Δ° G = +246 kJ/mol H_2 or +123 kJ/mol CO_2

standard electro H_2, Δ° G = +$\underline{237}$ kJ/mol H_2

marginal energy cost ΔΔ° G = +9 kJ/mol H_2 or +4.5 kJ/mol CO_2

8

Experimental Setup at LLNL:

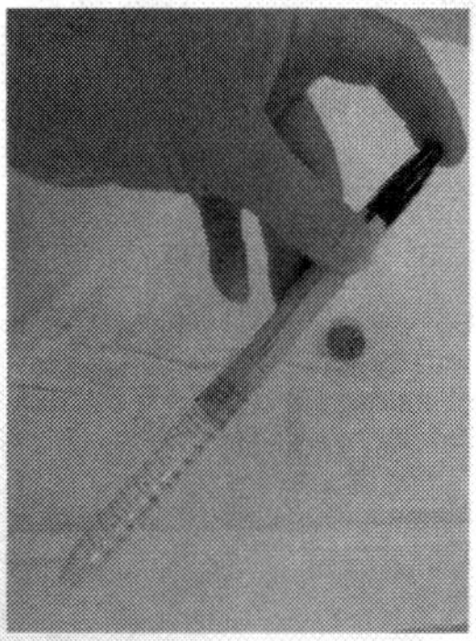

→

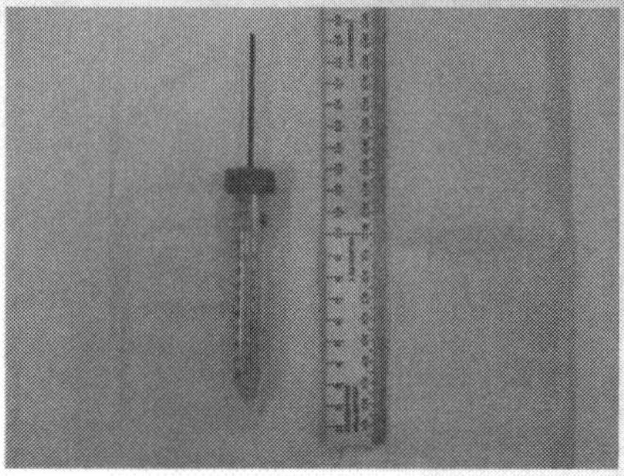

↓

←

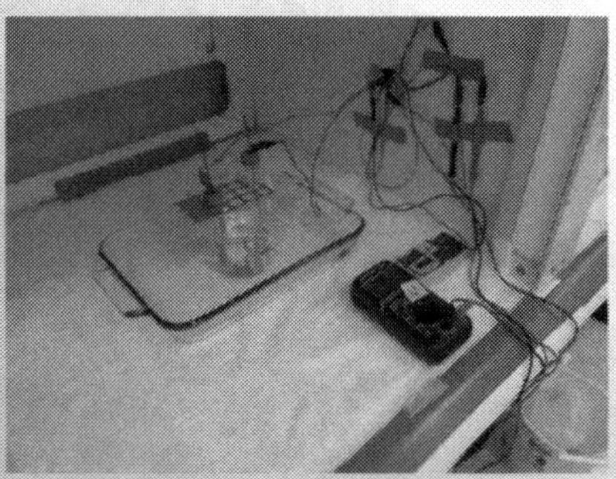

9

Results: pH response to electrolysis then aeration

Proof: saline water + mineral + V_{dc} → concentrated OH^-

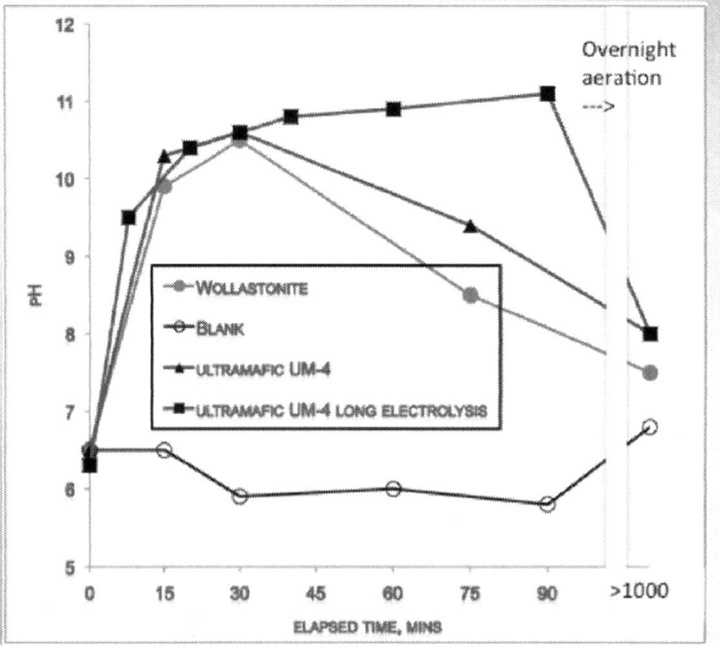

10

Results: Air CO_2 absorbed and converted to HCO_3^- in solution

Proof: concentrated $OH^- +$ air $CO_2 \rightarrow HCO_3^- (+ CO_3^{2-})$

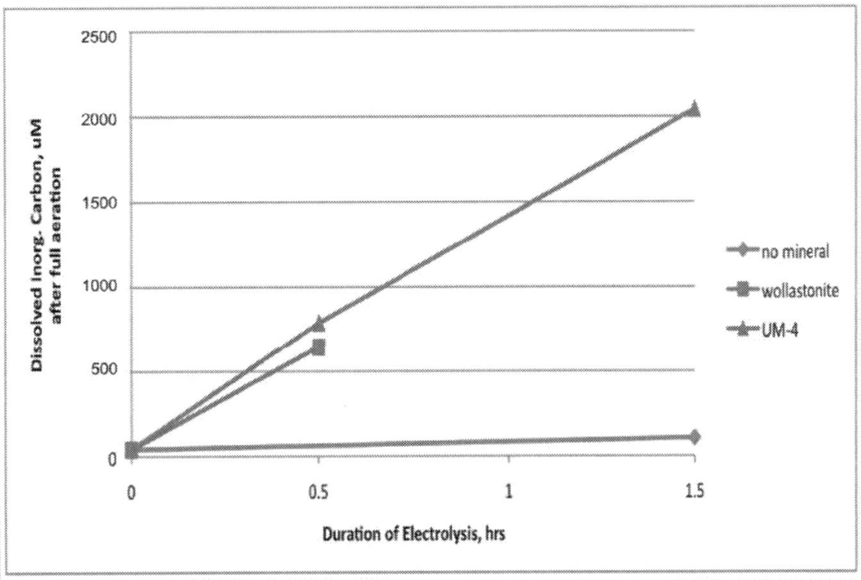

11

Expermiental conclusions:

- Saline water electrolysis in the presence of silicate (or carbonate) minerals dramatically increased hydroxide production relative to controls
- Subsequent aeration dramatically increased capture and storage of air CO_2 by as much as 45X relative to controls.
- Dissolved mineral bicarbonate was the primary carbon storage product.

Further details:

Rau et al. *PNAS* 2013 - http://www.pnas.org/contentearly2013/05/30/1222358110.full.pdf

Rau *ES&T* 2008 - http://pubs.acs.org/doi/pdf/10.1021/es800366q

12

Implications:

Use globally abundant saline/brine water, carbonate/silicate minerals and renewable energy to produce carbon-negative H_2 at large scales:

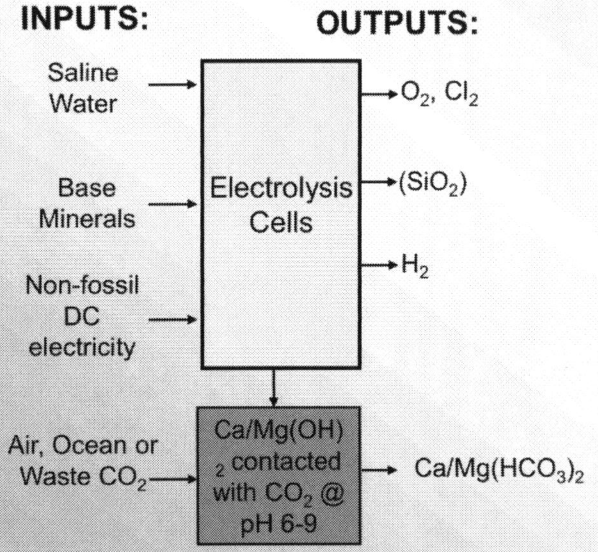

13

Integration into Renewable Energy Systems:

Use off-peak, off-grid, and/or stranded renewable electricity to produce C-negative H_2 for: energy storage, transportation fuels or chemical feed stock

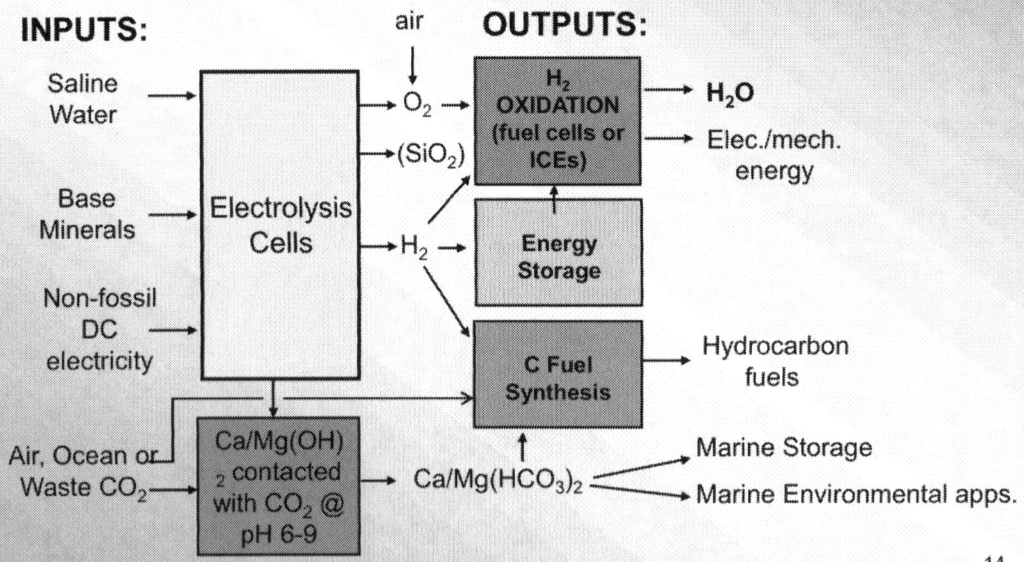

14

Energy, Reactant and Product Quantities

(tonnes/tonne CO_2 consumed)

Reactants:

(1.20) $MgSiO_3$-->

(0.41) Water -->

Non Fossil DC Electricity:

1,364 kWh_e/tonne CO_2 -->

Saline Electrolysis cell

Products:

Atmos. (1) CO_2

--> (0.66) $Mg(OH)_2$ --> (1.66) $Mg(HCO_3)_2$

--> (0.02) H_2 (0.4 tonnes CO_2 avoided)

--> (0.59) SiO_2

--> (0.36) O_2

Net CO_2 mitigated:
1 tonnes CO_2 consumed + 0.4 tonnes CO_2 emissions avoided
= 53 tonnes CO_2 mitigated per tonne H_2 generated

15

Estimated CO_2 Economics

($/tonne CO_2 mitigated)

Costs:

($24.09) $MgSiO_3$ -->

($0.33) Water -->

Non fossil Electricity:
($95.45)@$0.07/kWh_e-->

Capital, Operation, And Maintenance:
($14.05)

Saline Electrolysis cell

Revenues:

Atmos. CO_2

--> $Mg(OH)_2$ ----> $Mg(HCO_3)_2$ ($?)

--> H_2 ($34.09 @ $1.50/kg H_2)

--> O_2 ($?)

--> SiO_2 ($?)

$133.92 (gross cost) - $34.09 (revenue) = $99.83, net cost/t CO_2 consumed
i.e. within the range of CCS: $50 - $150/tonne CO_2 avoided

16

Roadmap to Low Carbon Fuels and Low Emissions Vehicles

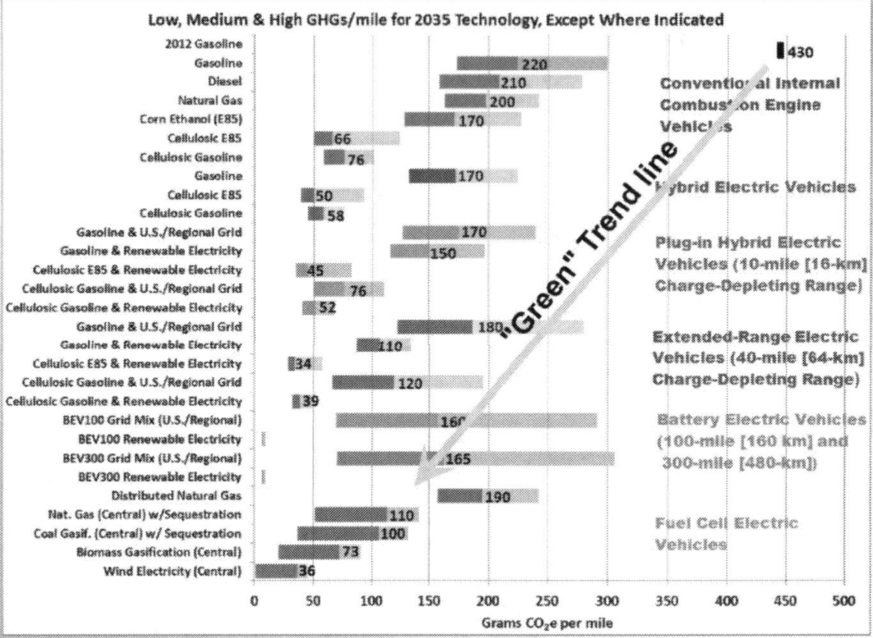

http://www.iea.org/media/workshops/2014/hydrogenroadmap/7DOEEricMiller.pdf

17

Roadmap to Low Carbon Fuels and Low Emissions Vehicles

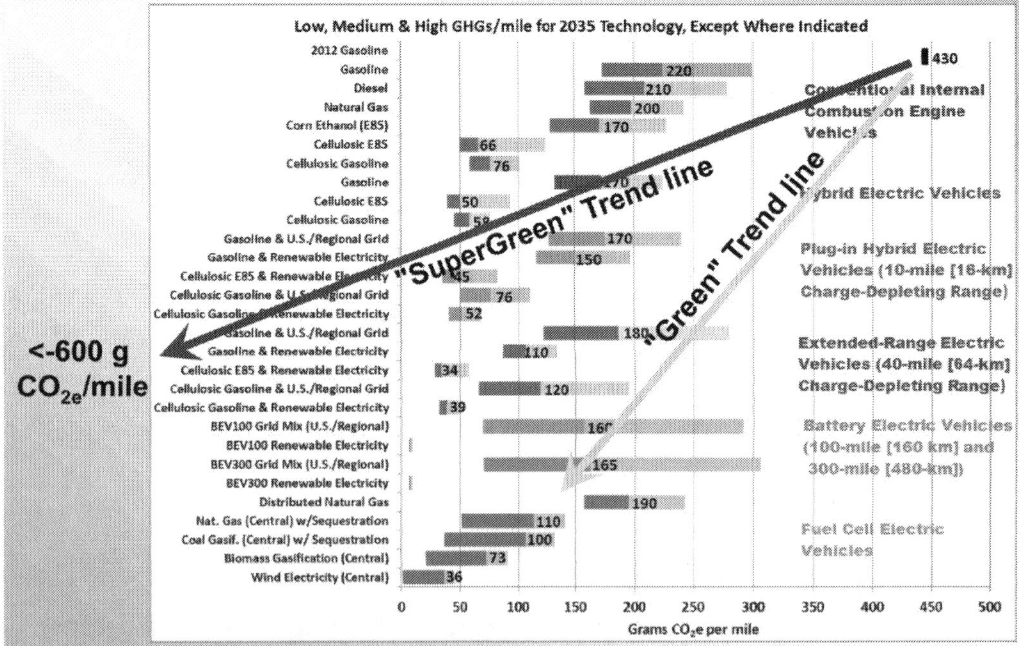

http://www.iea.org/media/workshops/2014/hydrogenroadmap/7DOEEricMiller.pdf

18

Estimated H_2 Economics:

($/kg H_2 produced)

Atmos. CO_2

Costs:

($1.06) $MgSiO_3$ -->

($0.09) Water -->

DC Electricity:
($4.20)@$0.07/kWh_e-->

Capital, Operation, And Maintenance:
= ($0.62)

Saline Electrolysis cell

Revenues:

($2.20 CO_2 **removal** @$0.05/kg CO_2)

--> $Mg(OH)_2$ ----> $Mg(HCO_3)_2$ ($?)

--> H_2 ($0.88; CO_2 avoidance @ $0.05/kg CO_2)

--> ($?) O_2

--> ($?) SiO_2

$5.89 (gross cost) - $3.03 (revenue) = <u>$2.81, net cost/kg H_2 produced</u>

EERE: 2015 target = $3/kg; 2020 target = $2/kg (assuming $0.03/$kWh_e$)

19

Benefit of Adding Carbonate Alkalinity to Ocean: Preservation of corals and shellfish

Ω = $CaCO_3$ saturation state $\propto$ [Ca^{2+}] x [CO_3^{2-}]

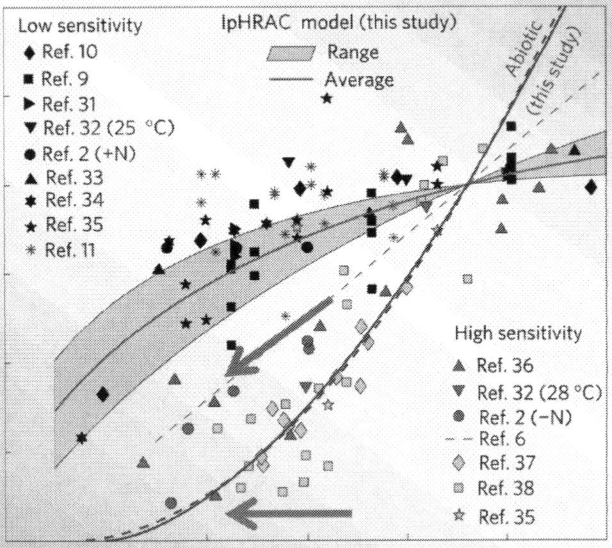

SEE MY POSTER

Economic impact:
$20B/yr (?)
$1T/yr by 2100 (?)

(Mod. from McCulloch et al. 2012)

<---increasing CO_2, decreasing pH, Ω

20

Benefit of Adding Carbonate Alkalinity to Ocean: Preservation of corals and shellfish

$CO_2 + H_2O + CaCO_{3s} \rightarrow Ca(HCO_3)_{2aq}$ $(+ CaCO_{3aq})$, $\Omega \propto [Ca^{2+}] \times [CO_3^{2-}]$

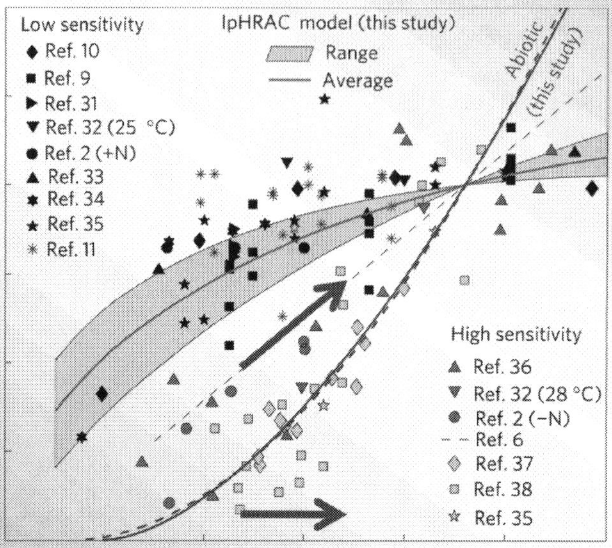

SEE MY POSTER

Helps avoid economic impact:
$20B/yr (?)
$1T/yr by 2100 (?)

(Mod. from McCulloch et al. 2012)

alkalinity addition, increasing pH, $\Omega \rightarrow$

21

Large-Scale Applications

Site large-scale electrolysis operations at the convergence of abundant water/brine, non-fossil energy, and base mineral resources, e.g.:

The Ocean -

- Massive potential electrolyte
- Massive (stranded) renewable energy ≈ 2×10^6 TWh/yr
- Massive mineral resources
- Massive CO_2 absorber; 70% of Earth's surface; already takes up 10 Gt CO_2/yr

22

For Example: Ocean-based, carbon-negative wind hydrogen

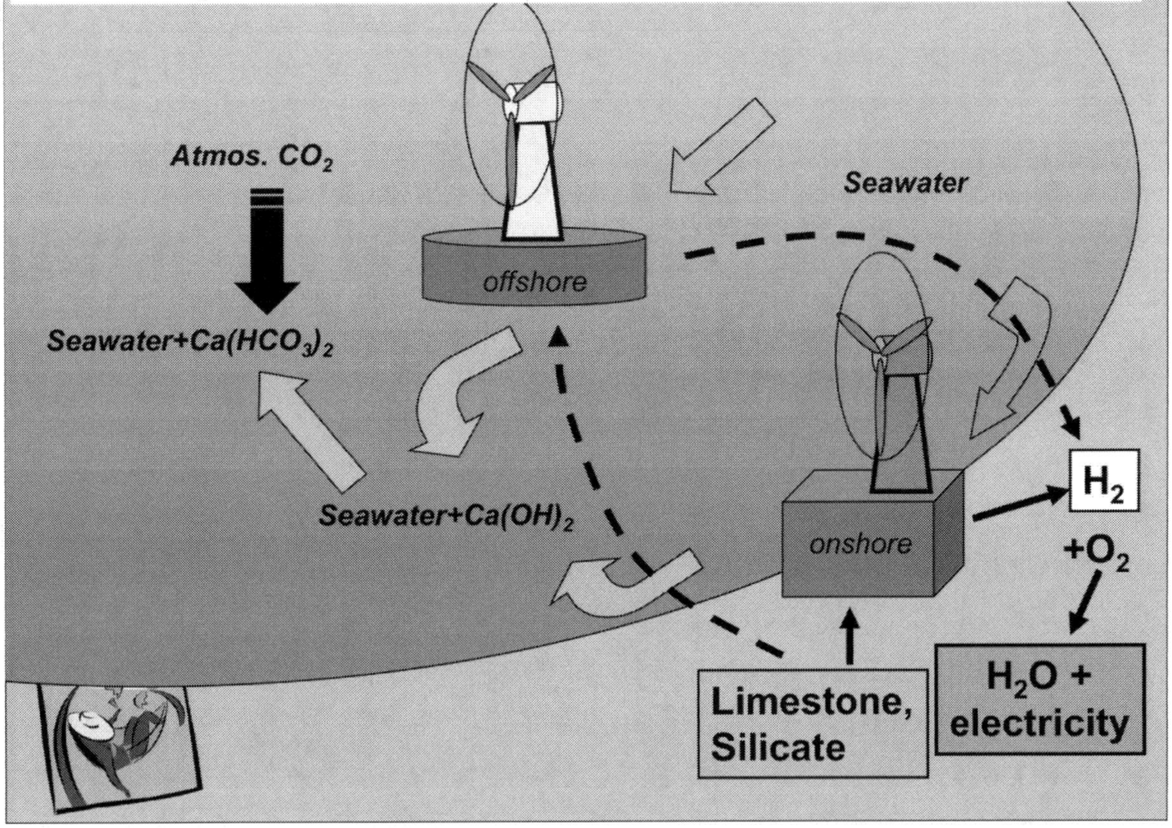

Comparison: BECCS vs C-Negative H_2

	CDR Hybrid System Type	Non-Fossil Energy Input	Energy Output	C Product	C Storage	Capacity	Cost, \$/tonne CO_2
BECCS	biology + com-bustion + CCS	land biomass (marine biomass)	electricty, (H_2, C fuels)	supercrit. CO_2	underground	limited by land use	<\$100
C-Negative Electrolytic H_2	electrochemisty + geochemistry	cheapest source	H_2, (electricity, C fuels)	bicarbonate, carbonate	surface ocean, (saline aquifers)	massive	<\$100

- C-Neg H_2 provides additional and complementary CDR to BECCS
  Increases CDR capacity, co-benefits, and flexibility

Bottom line: ***At this early stage of R&D and policymaking, let's keep an open mind about what the lowest cost, highest capacity <u>portfolio of CDR options</u> might look like***

24

Summary

- A novel electrogeochemical method of air CO_2 removal and storage has been demonstrated that:
  - when powered by non-fossil energy produces carbon-emissions-negative H_2 and other products whose value significantly reduces net air CO_2 removal cost, est. =<$100/tonne CO_2, competitive with CCS.
  - has high capacity due to global abundance of raw materials and off-peak or stranded renewable energy
  - can provide additional environmental and economic benefits e.g., using the produced alkalinity to counter ocean acidity
- **Approaches like C-negative electrolytic H_2 deserves to be part of the CDR R&D portfolio going forward in order to maximize our chances of effectively managing atmospheric CO_2 on human rather than geologic time scales.**

25

CO_2 and Air Capture Using Physical Adsorbents at Moderate Temperatures. Are Metal-Organic Frameworks Becoming the New Benchmark Materials?

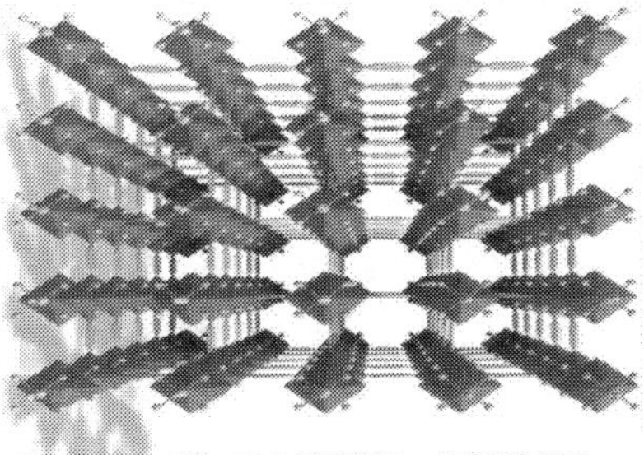

Nugent, P, Belmabkhout, Y et al (2013) Nature 494 (7439) ,80-84
Shekhah, O, Belmabkhout, Y et al (2014) Nature Commun. DOI: 10.1038/ncomms5228

Youssef Belmabkhout

Functional Materials Design, Discovery and Development (FMD^3)
King Abdullah University of Science and Technology KAUST (KSA)

Carbon Management Technology Conference 2015, Sugar Land, TX Nov, 19th 2015

FMD^3 Functional Materials Design, Discovery and Development
King Abdullah University of Science and Technology KAUST

Research and development in material sciences for gas separation and storage

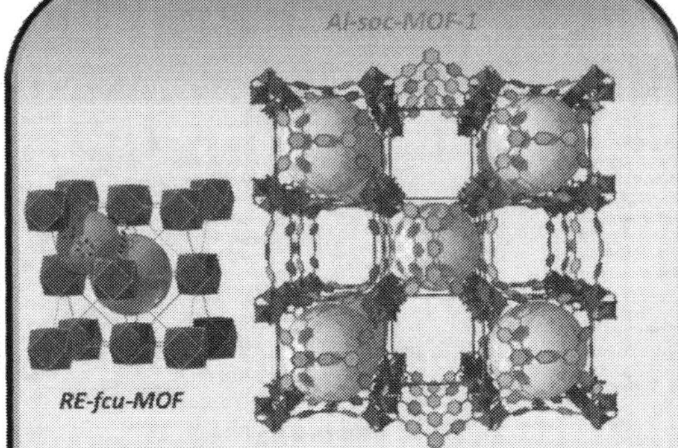

Materials Features

- Crystalline solids or amorphous
- Rational Assembly (building block approach)
- Modular, can be functionalized prior or after the assembly
- Unprecedented apparent surface areas (up to 7000 m^2/g)
- Controlled pore size (from ultra-micropores to mesopores up to 5 nm)

CO_2 Separation
- Natural gas upgrading
- Shifted syngas and Flue gas
- DAC
- Breathing systems

Other applications
- Gas /solvent dehydration
- Krypton/Xenon separation
- H_2S removal and sensing
- SOx/Nox removal
- BTEX removal from sour gas
- VOC sensing

Hydrocarbon Separation
- C_4/C_1, C_3/C_1, C_2/C_1
- Olefin / paraffin
- Paraffin/iso-paraffin
- VOCs
- Fuel upgrading

CH_4 & H_2 Storage
Mobile and stationary applications
- CH_4 : 180 v(STP)/v
- H_2: 5.5 wt% (2017)

Eddaoudi M. *et al. Science* 2002, 295, 469-471; Li, H.; Eddaoudi, M.; O'Keeffe, M.; Yaghi, O. M. *Nature 1999*, 402, 276-279 2

Quest of innovative advanced materials

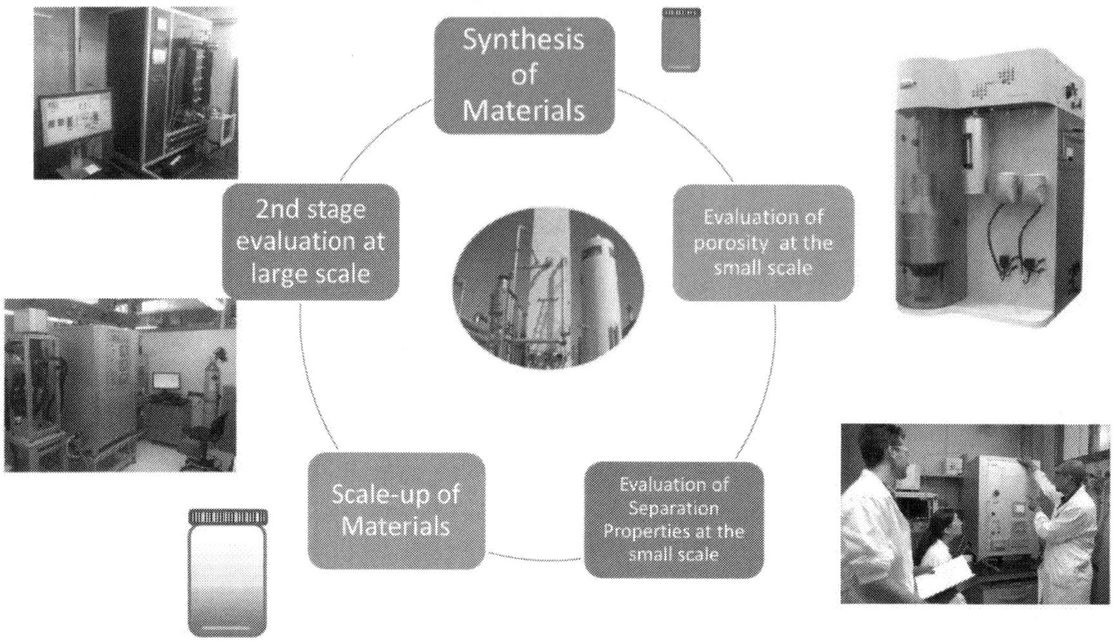

FMD³ Functional Materials Design, Discovery and Development
King Abdullah University of Science and Technology KAUST

Motivation

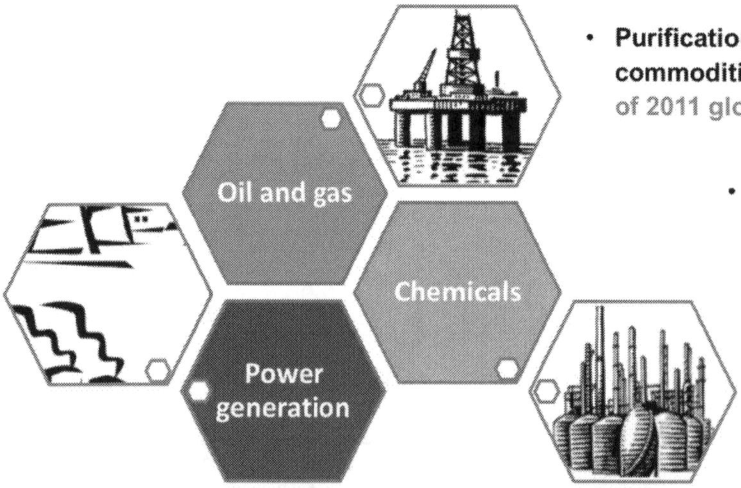

- **Purification & separation to produce commodities are energy intensive** (~15% of 2011 global energy use)
- **Global demands for commodities (water, fuel, chemicals) will** grow ~3X by 2050, **so "*business as usual*" would require an equivalent to** 45% of total 2011 energy usage for separations alone in 2050

FMD³ Functional Materials Design, Discovery and Development — King Abdullah University of Science and Technology KAUST

Motivation

Saudi Arabian Oil Production vs Consumption (thousand bbl/day)

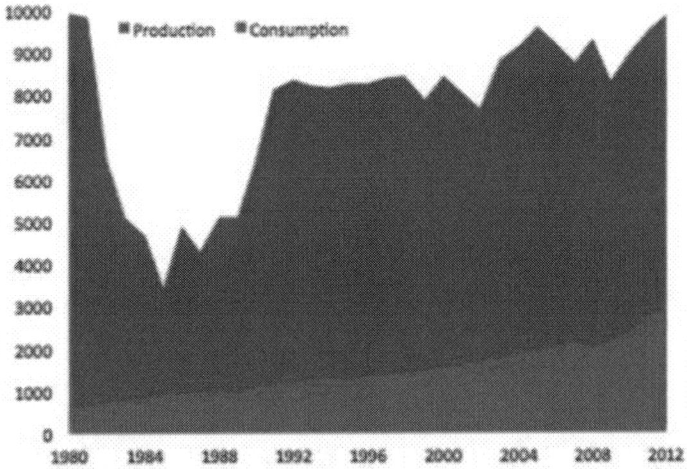

5

FMD³ Functional Materials Design, Discovery and Development — King Abdullah University of Science and Technology KAUST

Outline

- ✓ **Background: CCS and traces and low CO_2 concentration removal**
  → Chemical vs physical adsorbents
- ✓ **Understanding of structural- CO_2 adsorption properties relationships →** push the limit of gas separation with the lowest energy input using physical adsorbents

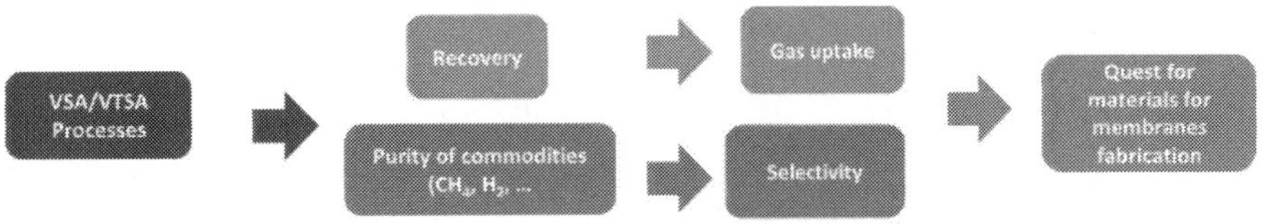

- ✓ **Conclusions from the past works (2012-2013)**
- ✓ **New finding → Shift to Membrane technology.**
- ✓ **Conclusions**

6

Evolution of CO_2 Emissions in the Middle East-Asia

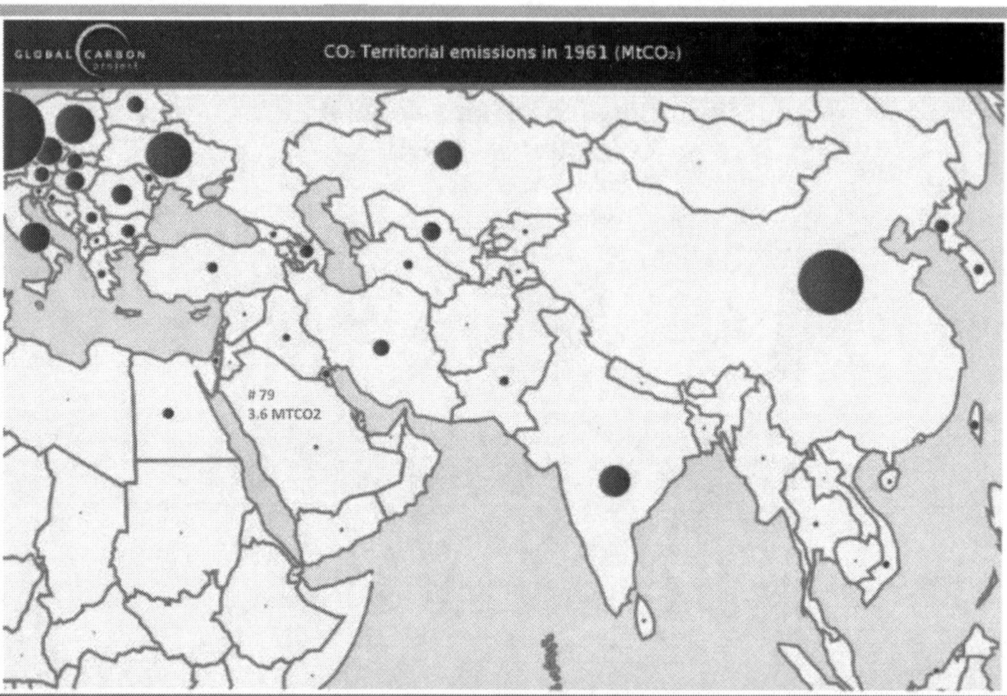

Source: http://www.globalcarbonatlas.org/?q=en/emissions

FMD[3] Functional Materials Design, Discovery and Development
King Abdullah University of Science and Technology KAUST

Evolution of CO_2 Emissions in the Middle East-Asia

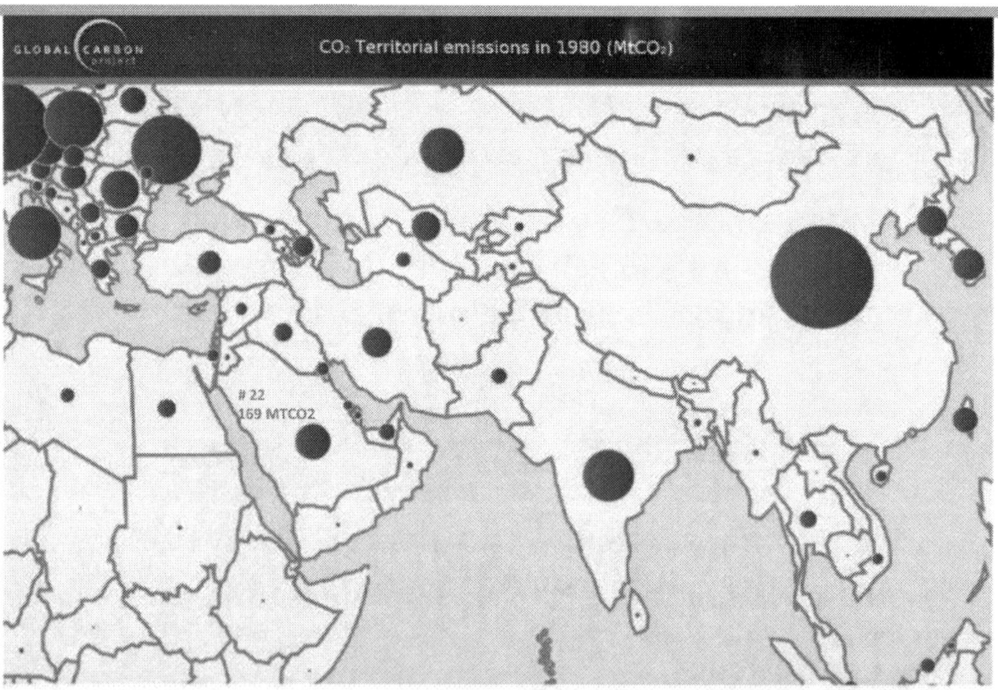

Source: http://www.globalcarbonatlas.org/?q=en/emissions

FMD³ Functional Materials Design, Discovery and Development — King Abdullah University of Science and Technology KAUST

Evolution of CO_2 Emissions in the Middle East-Asia

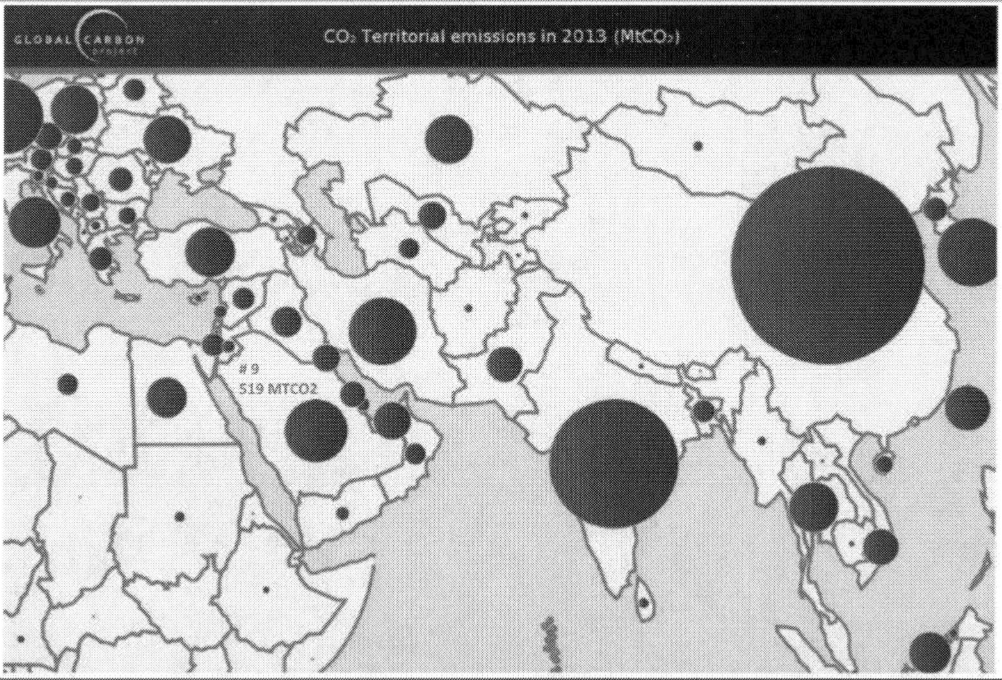

Source: http://www.globalcarbonatlas.org/?q=en/emissions

9

FMD³ Functional Materials Design, Discovery and Development — King Abdullah University of Science and Technology KAUST

Main sources of CO_2 emissions

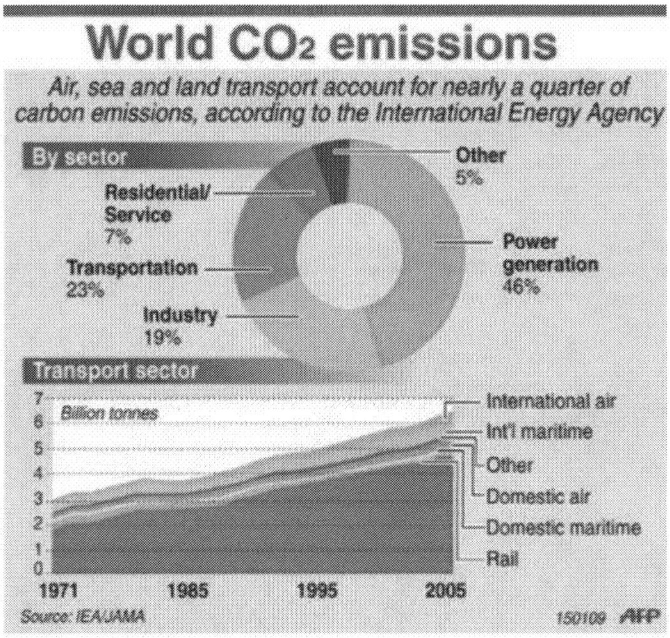

Power generation

Transportation

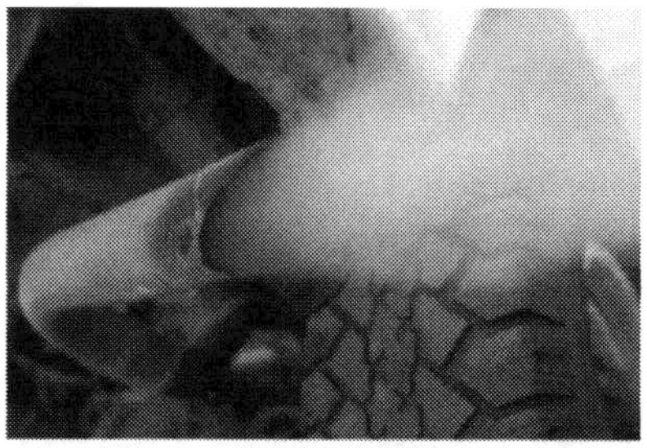

10

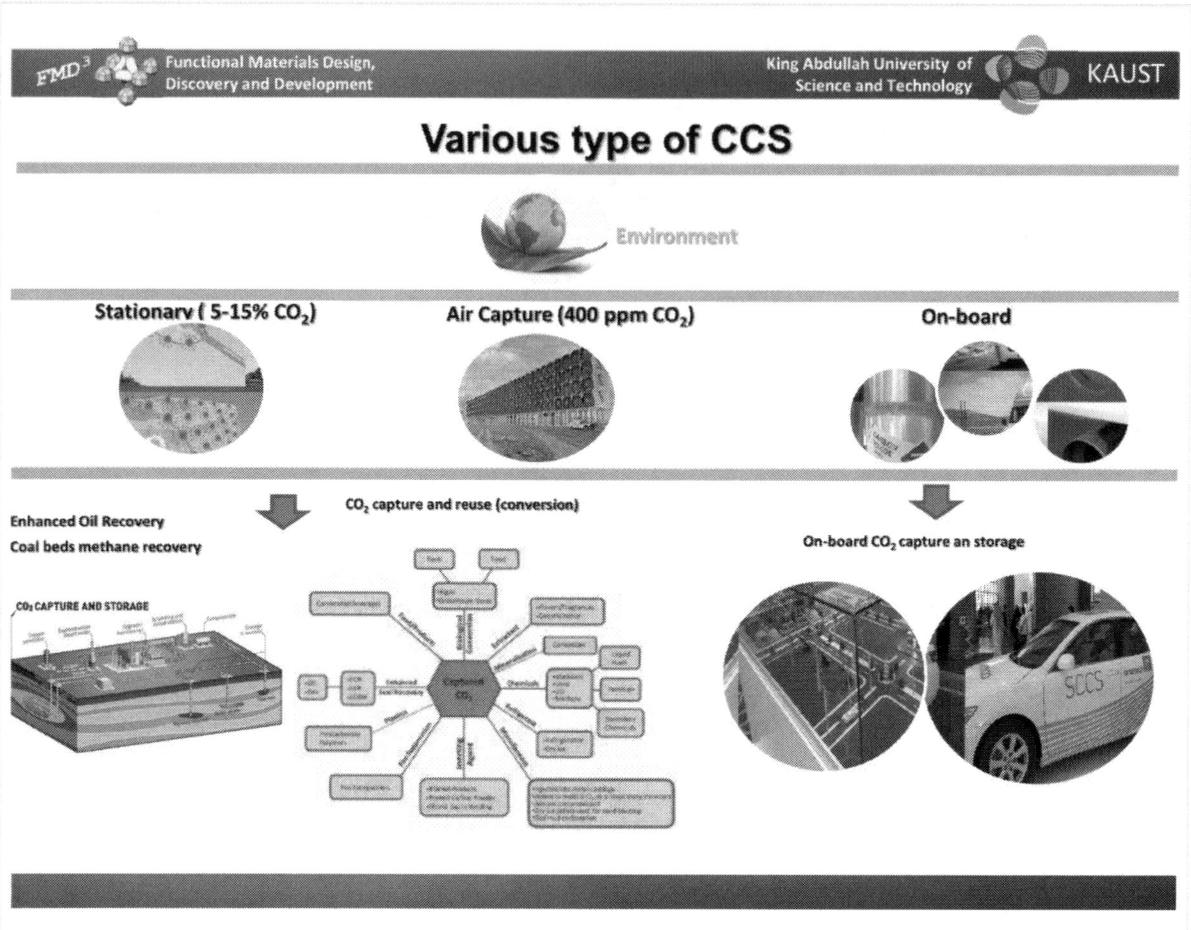
FMD3 Functional Materials Design, Discovery and Development
King Abdullah University of Science and Technology
KAUST
Various type of CCS
Environment
Stationary (5-15% CO_2)
Air Capture (400 ppm CO_2)
On-board
Enhanced Oil Recovery
Coal beds methane recovery
CO_2 capture and reuse (conversion)
On-board CO_2 capture an storage
CO_2 CAPTURE AND STORAGE
SCCS

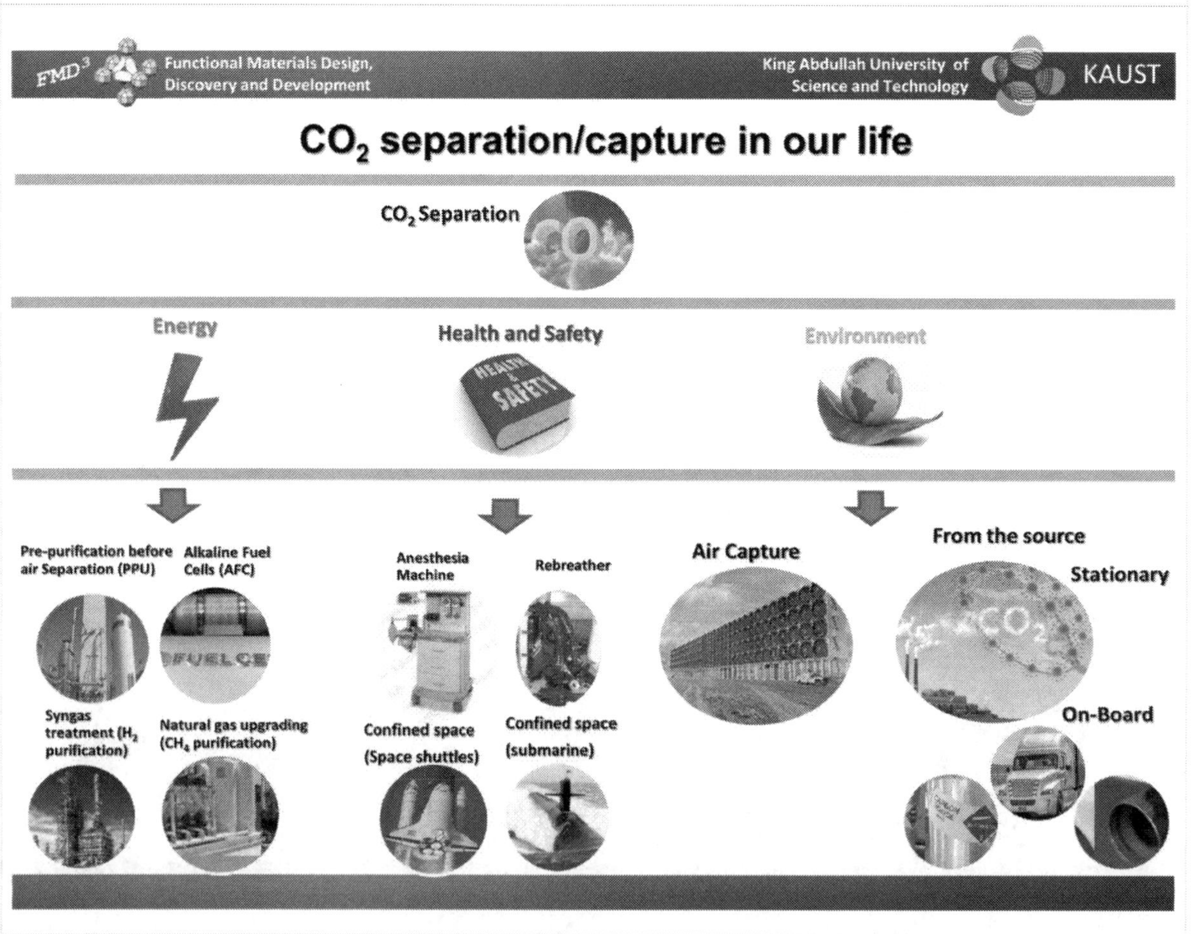
FMD3 Functional Materials Design, Discovery and Development
King Abdullah University of Science and Technology
KAUST
CO_2 separation/capture in our life
CO_2 Separation
Energy
Health and Safety
Environment
Pre-purification before air Separation (PPU)
Alkaline Fuel Cells (AFC)
Syngas treatment (H_2 purification)
Natural gas upgrading (CH_4 purification)
Anesthesia Machine
Rebreather
Confined space (Space shuttles)
Confined space (submarine)
Air Capture
From the source
Stationary
On-Board

FMD[3] Functional Materials Design, Discovery and Development — King Abdullah University of Science and Technology KAUST

Traces CO_2 and Air Capture Using Chemical Adsorbents: LiOH

Lithium Hydroxide (LiOH)

Respirator system in International Space Station (ISS)

$2LiOH + CO_2 \rightarrow Li_2CO_3 + H_2O$

LiOH canister for diving

13

FMD[3] Functional Materials Design, Discovery and Development — King Abdullah University of Science and Technology KAUST

Traces CO_2 and Air Capture Using Physical Adsorbents: 5A

Zeolite Ca^{2+} LTA (**5A**)

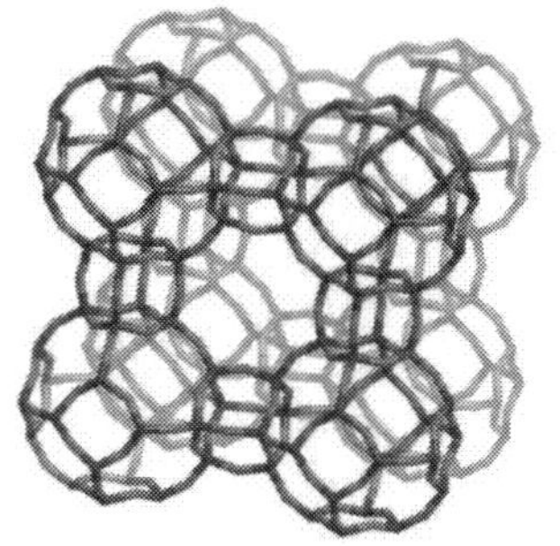

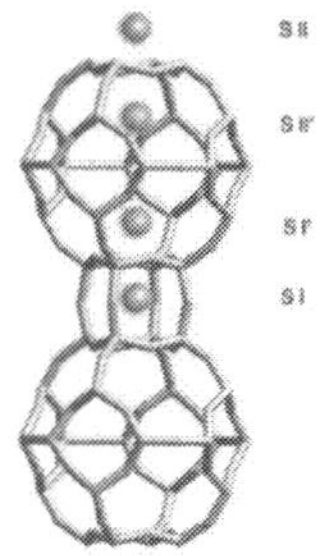

Advantage:
- ✓ 45 kJ/mol (Physical adsorption)
- ✓ recyclable

Drawbacks:
- ✓ Extremely hydrophilic
- ✓ Prior drying is needed

14

Traces CO_2 and Air Capture Using Physical Adsorbents: Strontium SAPO

Sr^{2+} SAPO-34

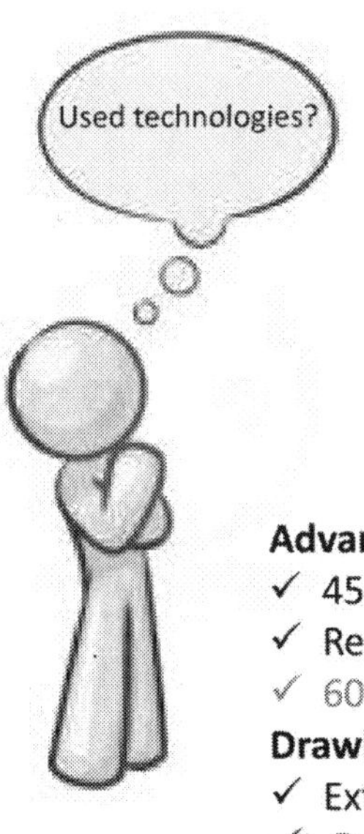

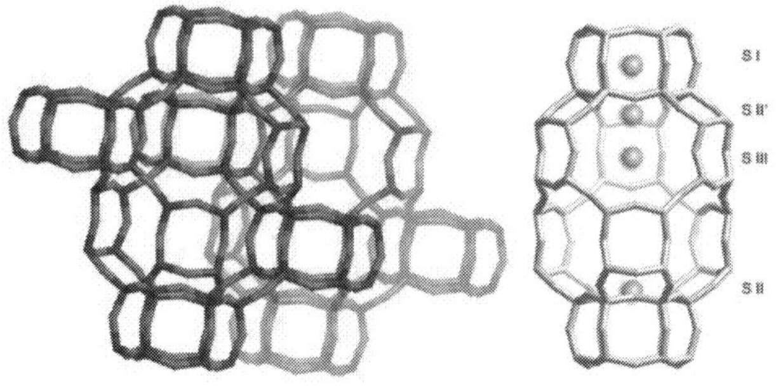

Advantage:
- ✓ 45 kJ/mol (Physical adsorption)
- ✓ Recyclable
- ✓ 60 % higher uptake (15cm^{3}STP/g) than 5A at 1300 ppm CO_2

Drawbacks:
- ✓ Extremely hydrophilic
- ✓ Prior drying is needed

Structural/chemical properties and physical adsorption properties relationships

- What is the key properties of an adsorbent to be suitable for CO_2 removal at different concentrations ?
- Which mechanism ?

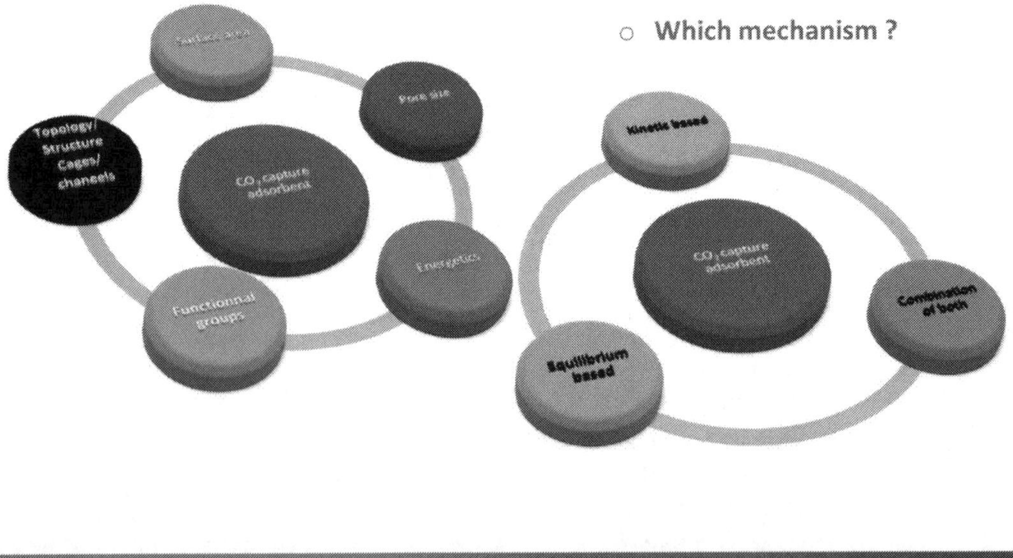

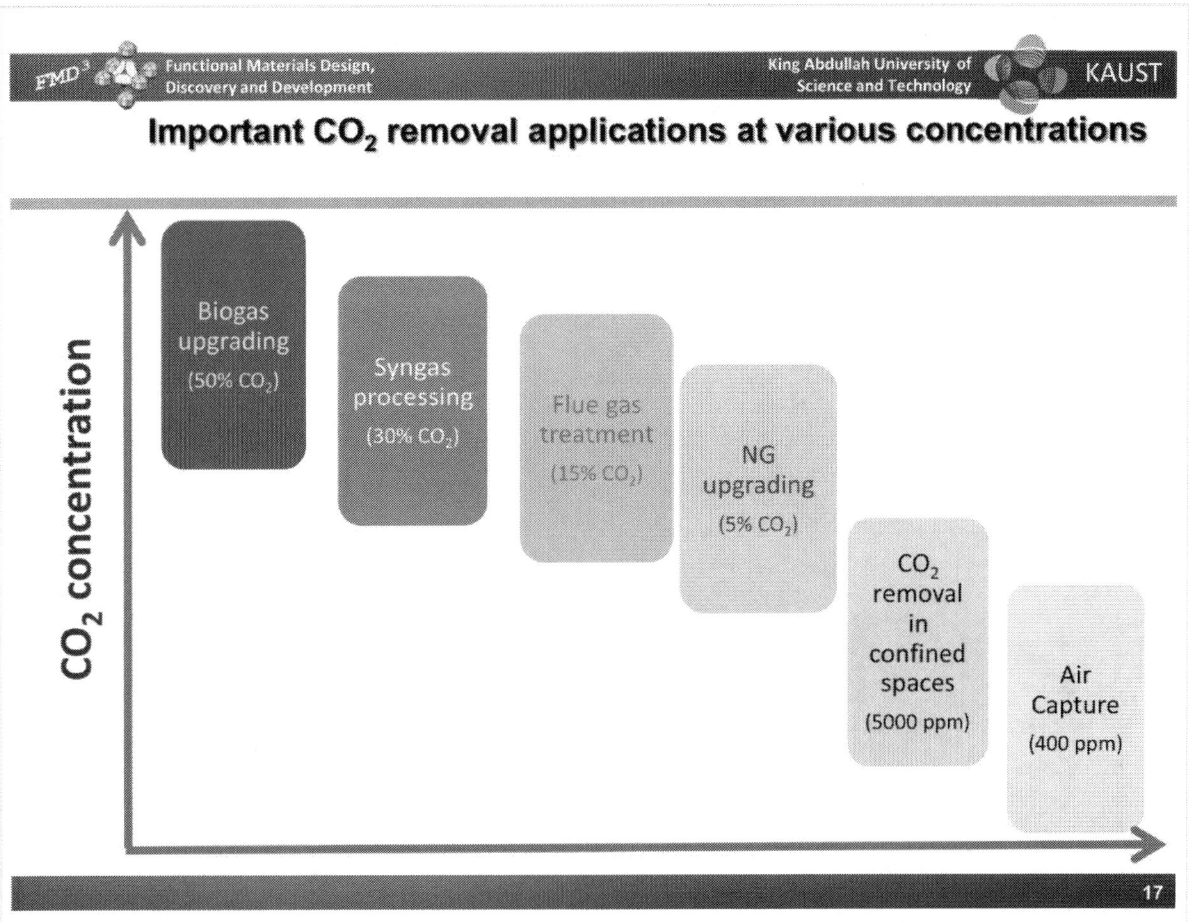
FMD³
Functional Materials Design, Discovery and Development
King Abdullah University of Science and Technology
KAUST
Important CO_2 removal applications at various concentrations
CO_2 concentration
Biogas upgrading
(50% CO_2)
Syngas processing
(30% CO_2)
Flue gas treatment
(15% CO_2)
NG upgrading
(5% CO_2)
CO_2 removal in confined spaces
(5000 ppm)
Air Capture
(400 ppm)
17

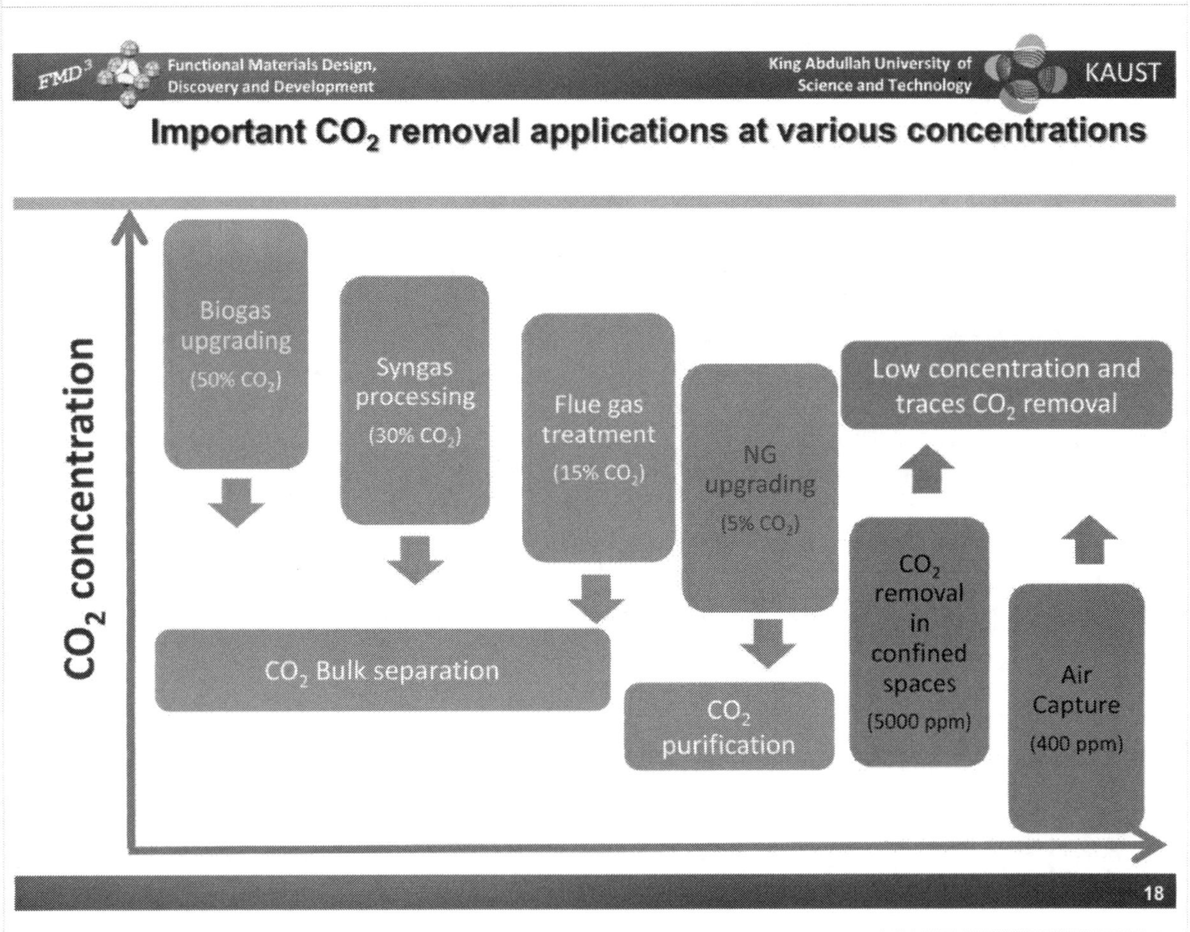
FMD³
Functional Materials Design, Discovery and Development
King Abdullah University of Science and Technology
KAUST
Important CO_2 removal applications at various concentrations
CO_2 concentration
Biogas upgrading
(50% CO_2)
Syngas processing
(30% CO_2)
Flue gas treatment
(15% CO_2)
NG upgrading
(5% CO_2)
Low concentration and traces CO_2 removal
CO_2 removal in confined spaces
(5000 ppm)
Air Capture
(400 ppm)
CO_2 Bulk separation
CO_2 purification
18

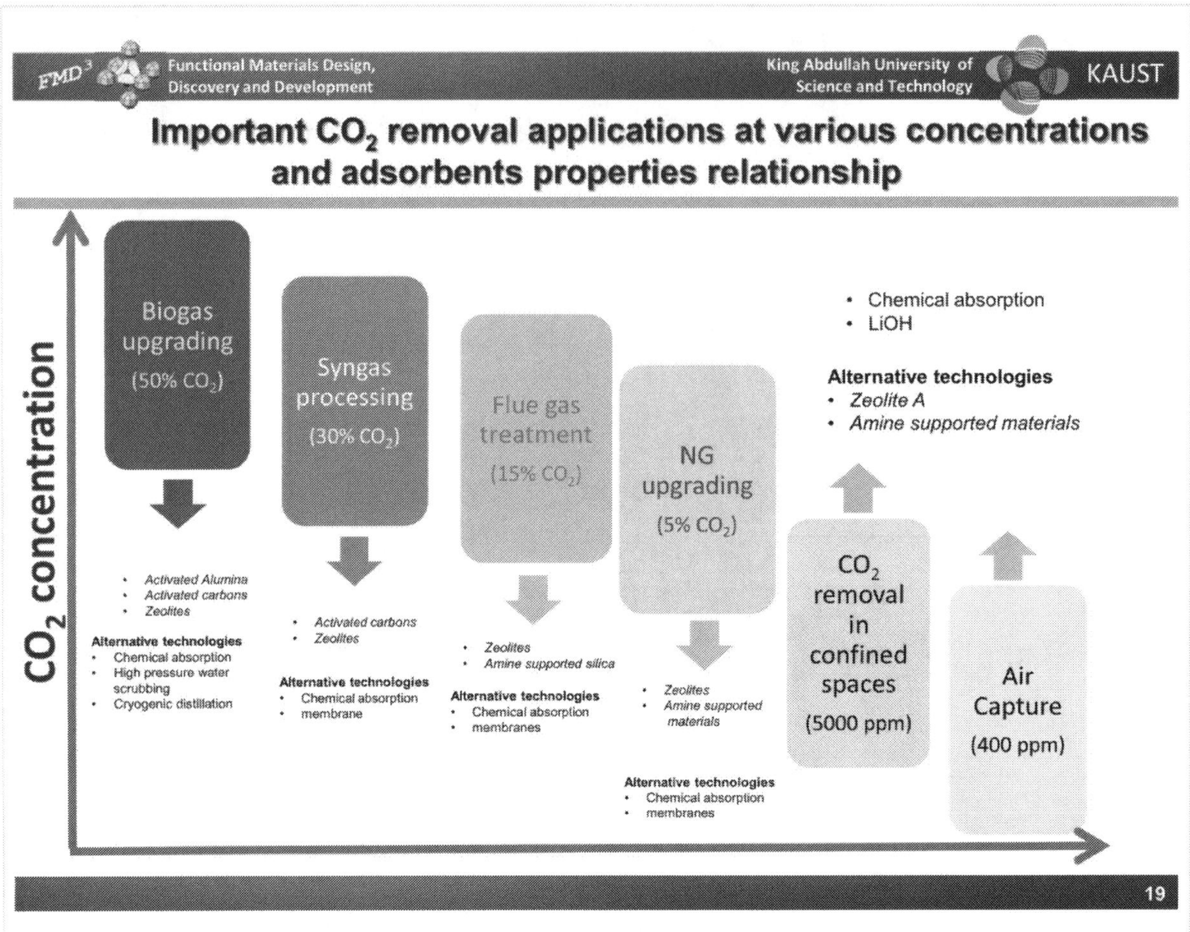
FMD³ Functional Materials Design, Discovery and Development
King Abdullah University of Science and Technology
KAUST
Important CO₂ removal applications at various concentrations and adsorbents properties relationship
CO₂ concentration
Biogas upgrading (50% CO₂)
Activated Alumina
Activated carbons
Zeolites
Alternative technologies
Chemical absorption
High pressure water scrubbing
Cryogenic distillation
Syngas processing (30% CO₂)
Activated carbons
Zeolites
Alternative technologies
Chemical absorption
membrane
Flue gas treatment (15% CO₂)
Zeolites
Amine supported silica
Alternative technologies
Chemical absorption
membranes
NG upgrading (5% CO₂)
Zeolites
Amine supported materials
Alternative technologies
Chemical absorption
membranes
CO₂ removal in confined spaces (5000 ppm)
Chemical absorption
LiOH
Alternative technologies
Zeolite A
Amine supported materials
Air Capture (400 ppm)
19

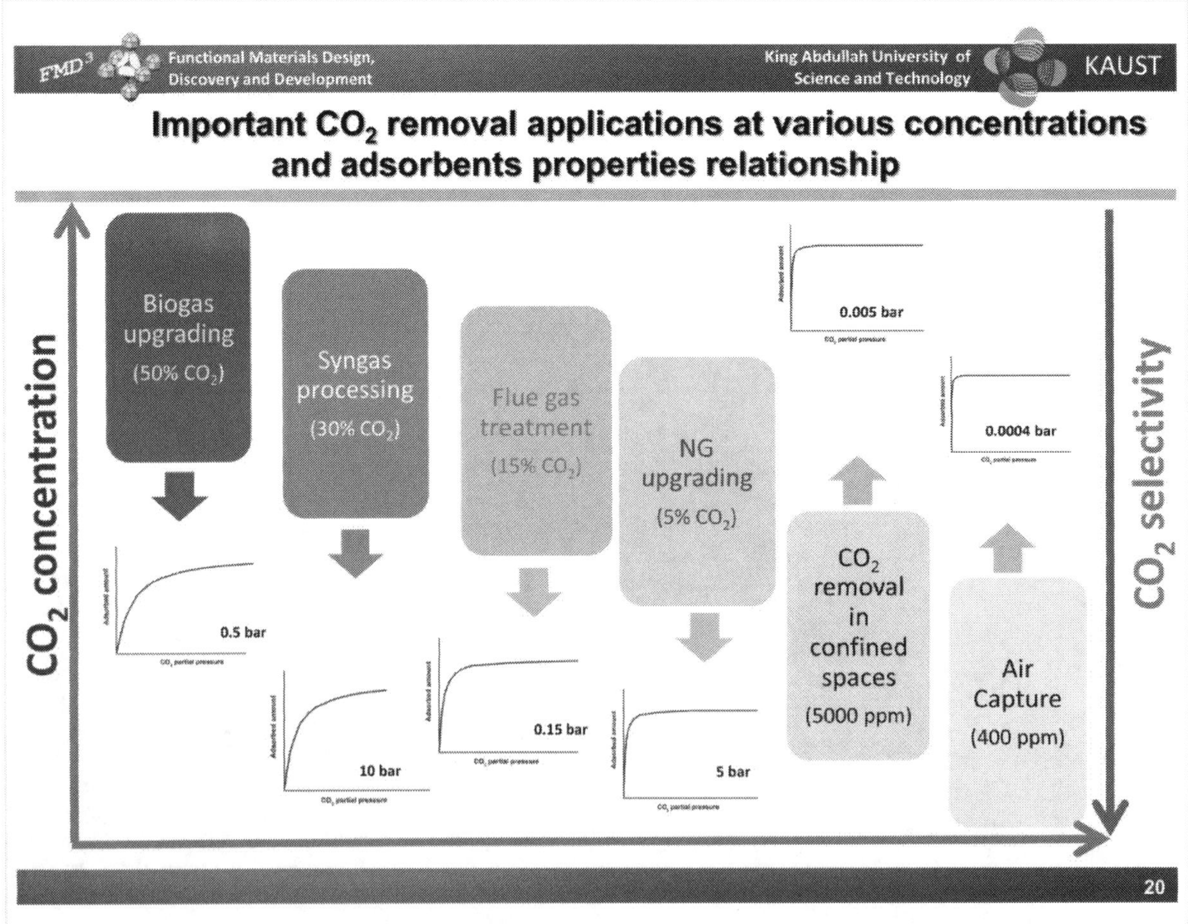
FMD³ Functional Materials Design, Discovery and Development
King Abdullah University of Science and Technology
KAUST
Important CO₂ removal applications at various concentrations and adsorbents properties relationship
CO₂ concentration
CO₂ selectivity
Biogas upgrading (50% CO₂)
0.5 bar
Syngas processing (30% CO₂)
10 bar
Flue gas treatment (15% CO₂)
0.15 bar
NG upgrading (5% CO₂)
5 bar
CO₂ removal in confined spaces (5000 ppm)
0.005 bar
Air Capture (400 ppm)
0.0004 bar
20

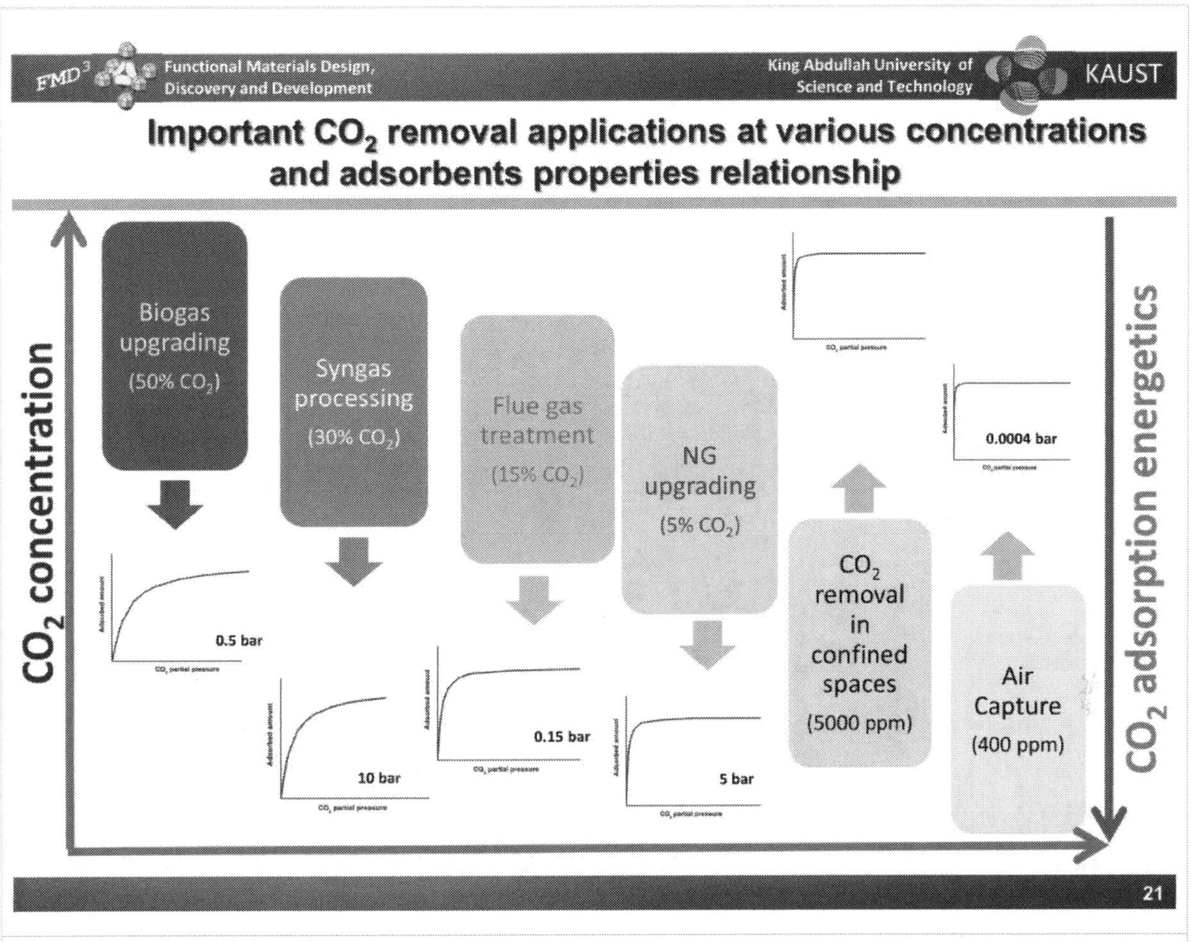
FMD3 Functional Materials Design, Discovery and Development
King Abdullah University of Science and Technology
KAUST
Important CO2 removal applications at various concentrations and adsorbents properties relationship
CO2 concentration
CO2 adsorption energetics
Biogas upgrading (50% CO2)
Syngas processing (30% CO2)
Flue gas treatment (15% CO2)
NG upgrading (5% CO2)
CO2 removal in confined spaces (5000 ppm)
Air Capture (400 ppm)
0.5 bar
10 bar
0.15 bar
5 bar
0.0004 bar
21

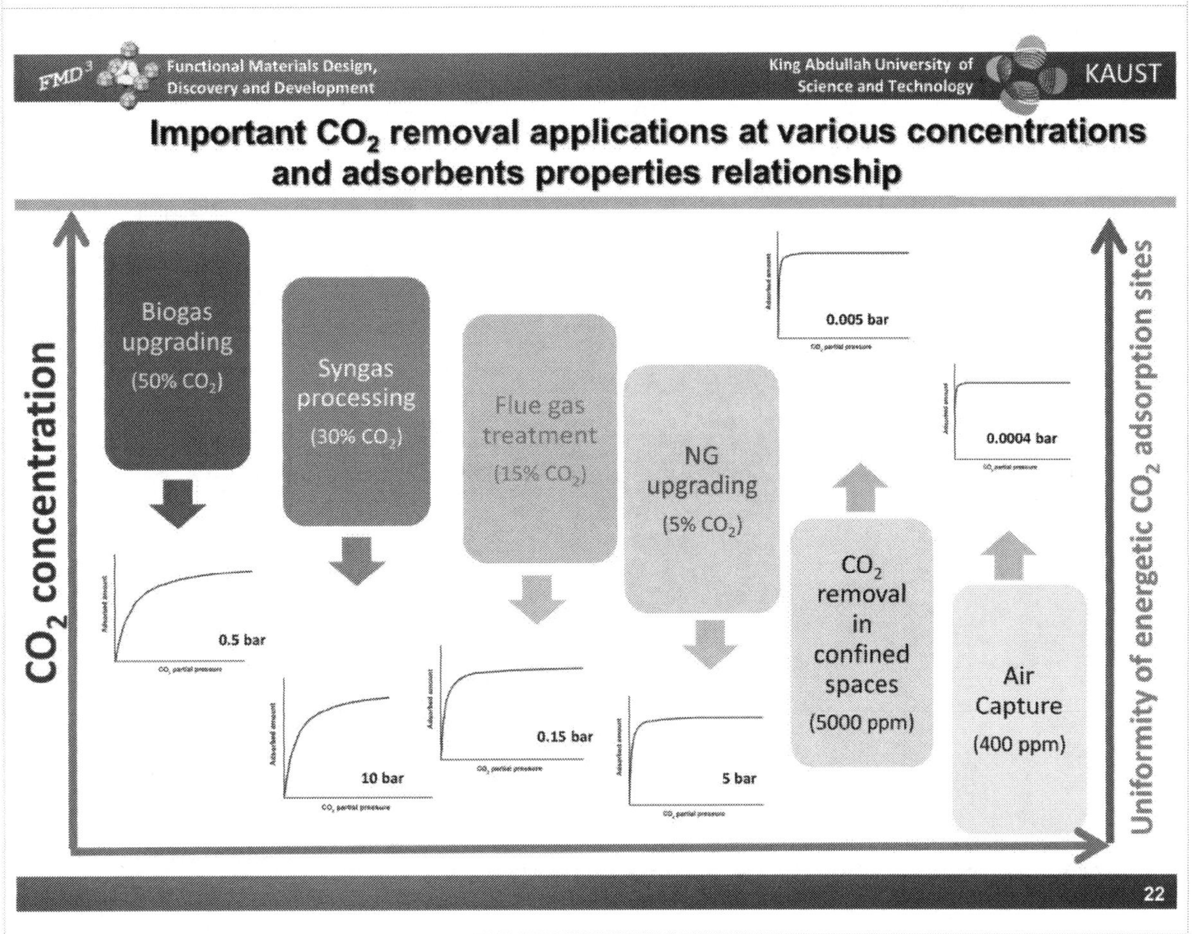
FMD3 Functional Materials Design, Discovery and Development
King Abdullah University of Science and Technology
KAUST
Important CO2 removal applications at various concentrations and adsorbents properties relationship
CO2 concentration
Uniformity of energetic CO2 adsorption sites
Biogas upgrading (50% CO2)
Syngas processing (30% CO2)
Flue gas treatment (15% CO2)
NG upgrading (5% CO2)
CO2 removal in confined spaces (5000 ppm)
Air Capture (400 ppm)
0.5 bar
10 bar
0.15 bar
5 bar
0.005 bar
0.0004 bar
22

MOFs and coordination polymers

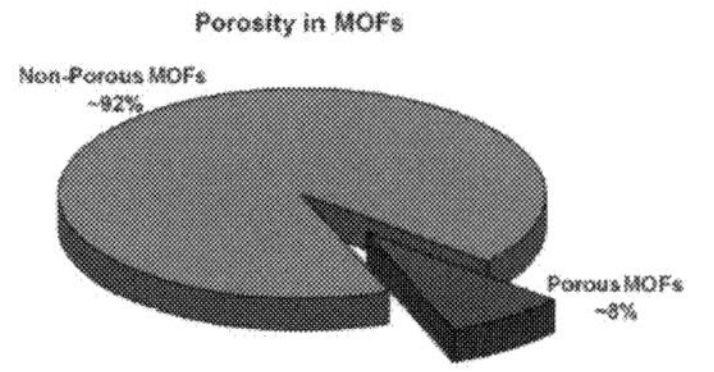

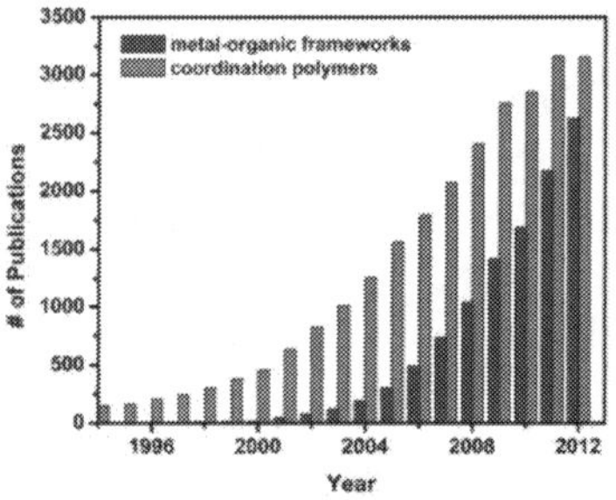

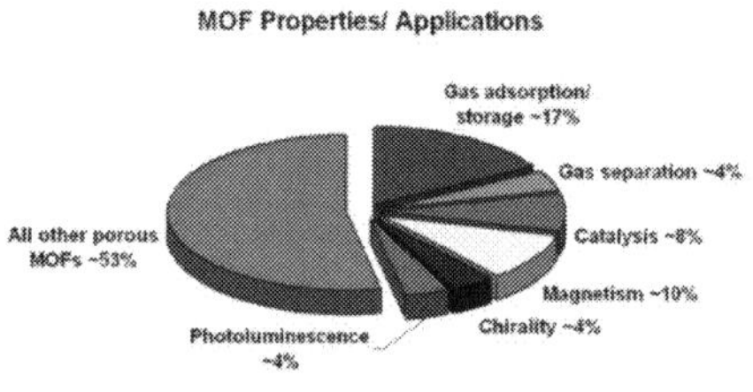

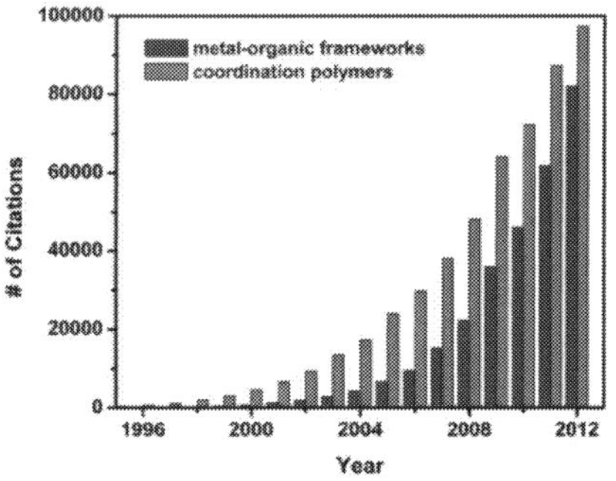

Past works (2012-2013)

SiF_6 MOF Platform

LETTER

doi:10.1038/nature11893

Porous materials with optimal adsorption thermodynamics and kinetics for CO_2 separation

Patrick Nugent[1]*, Youssef Belmabkhout[2]*, Stephen D. Burd[1], Amy J. Cairns[2], Ryan Luebke[2], Katherine Forrest[1], Tony Pham[1], Shengqian Ma[1], Brian Space[1], Lukasz Wojtas[1], Mohamed Eddaoudi[1,2] & Michael J. Zaworotko[1]

The energy costs associated with the separation and purification of industrial commodities, such as gases, fine chemicals and fresh water, currently represent around 15 per cent of global energy production, and the demand for such commodities is projected to triple by 2050 (ref. 1). The challenge of developing effective separation and purification technologies that have much smaller energy footprints is greater for carbon dioxide (CO_2) than for other gases; in addition to its involvement in climate change, CO_2 is an impurity in natural gas, biogas (natural gas produced from biomass), syngas (CO/H_2, the main source of hydrogen in refineries) and many other gas streams. In the context of porous crystalline

that are pillared via SiF_6^{2-} anions ('SIFSIX') in the third dimension to form three-dimensional nets with primitive cubic topology[13].

$[Cu(4,4'\text{-bipyridine})_2(SiF_6)]_n$, a prototypal primitive-cubic net that remains one of the best sorbents for CH_4 as measured by volumetric uptake[14], exhibits highly selective CO_2 uptake versus both CH_4 and N_2 at 1 bar and 298 K (ref. 15). In the absence of UMCs or amine groups, we attributed this behaviour to favourable interactions between CO_2 and SIFSIX. This compound, which we call here SIFSIX-1-Cu, exhibits one-dimensional square channels (pore size 9.54 Å; here and throughout this Letter, pore sizes are given as diagonal dimensions) aligned by a periodic array of SIFSIX pillars, and is prototypical for a class of

A series of channel based MOFs based on Bypiridyne/Cu 2-D periodic 4^4 square grids pillared by SiF_6 anions

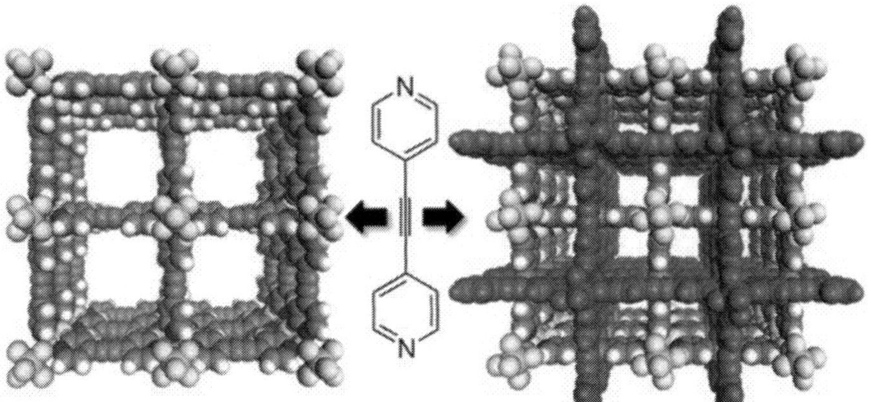

SIFSIX-2-Cu
10.5 Å pore
3140 m^2/g (N_2)

SIFSIX-2-Cu-i
4.9 Å pore
735 m^2/g (N_2)

1,2-bis(4-pyridyl)acetylene organic building block

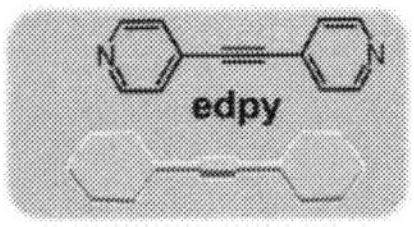

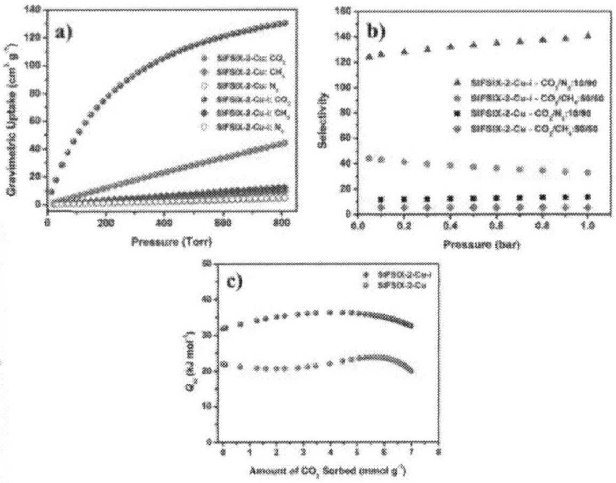

SIFSIX-2 (Cu)

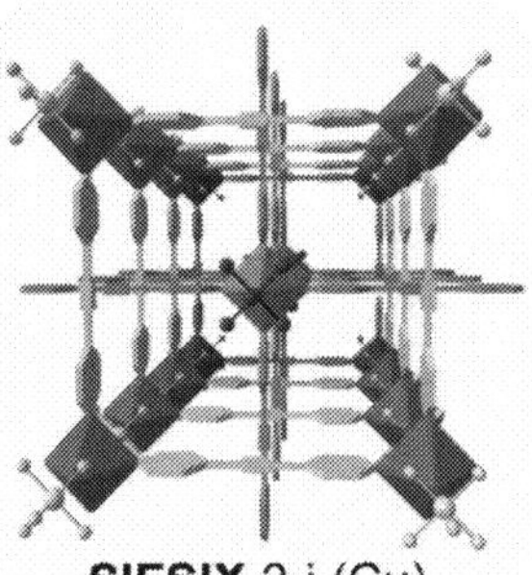

SIFSIX-2-i (Cu)

pyrazine/Zn 2-D periodic 4^4 square grids pillared by SiF_6 anions

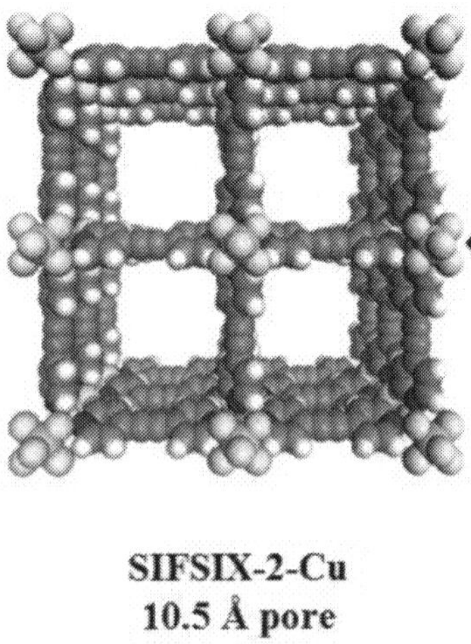

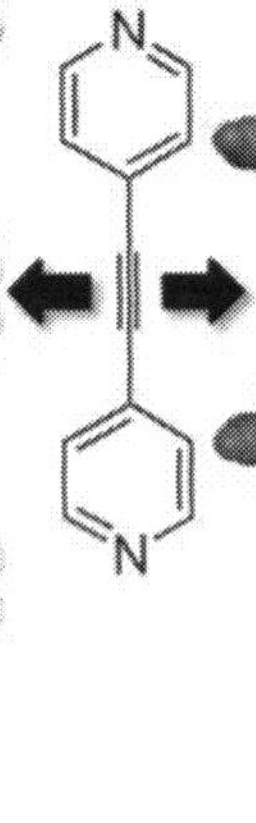

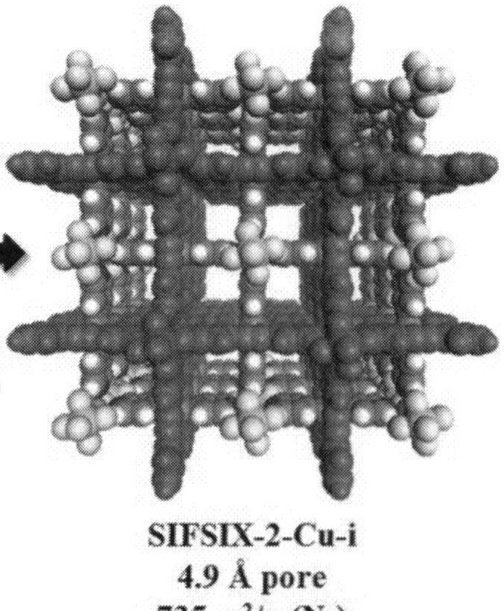

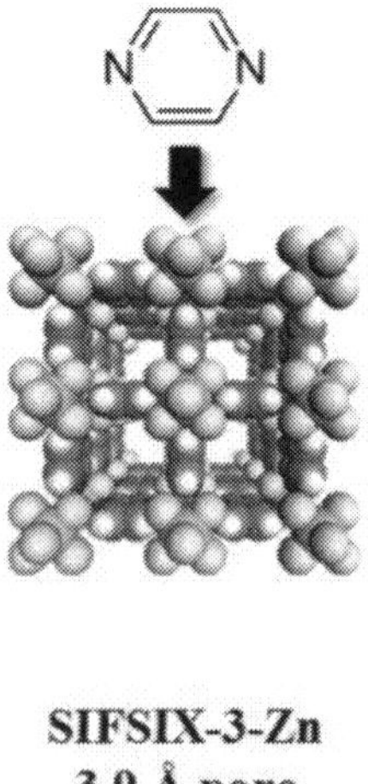

SIFSIX-2-Cu
10.5 Å pore
3140 m^2/g (N_2)

SIFSIX-2-Cu-i
4.9 Å pore
735 m^2/g (N_2)

SIFSIX-3-Zn
3.9 Å pore
250 m^2/g (CO_2)

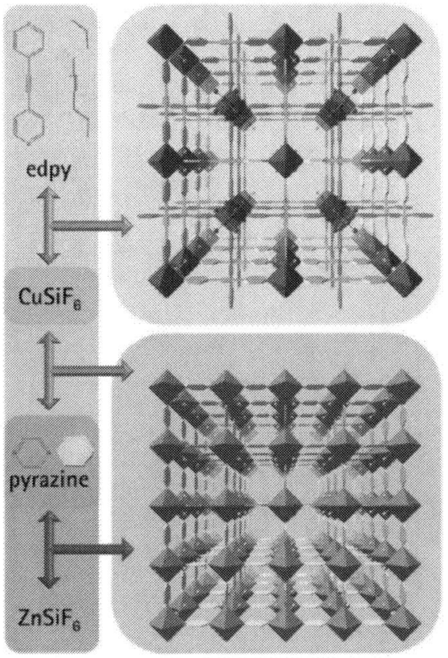

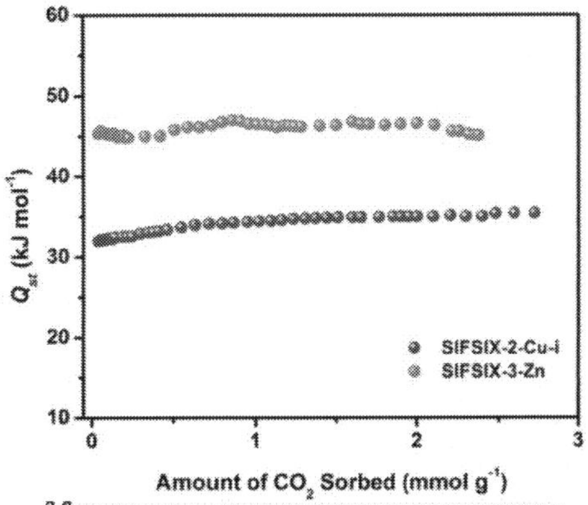

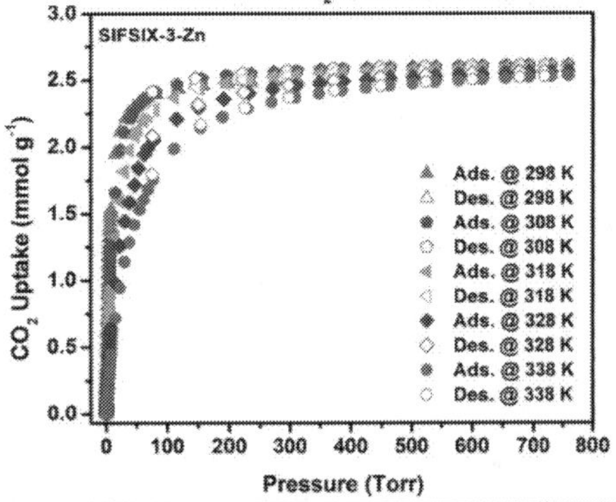

CO_2/N_2 and CO_2/H_2 separation using SIFSIX-3-Zn MOFs- Effect of humidity

CO_2/N_2 (flue gas)and CO_2/H_2 (syngas) mixtures adsorption - column breakthrough
adsorption tests in humid conditions

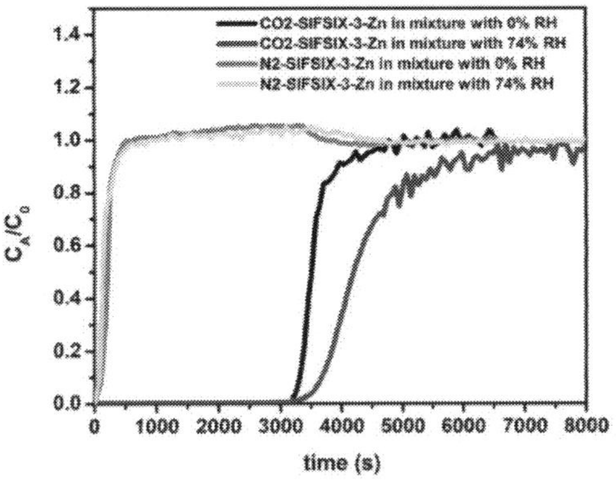

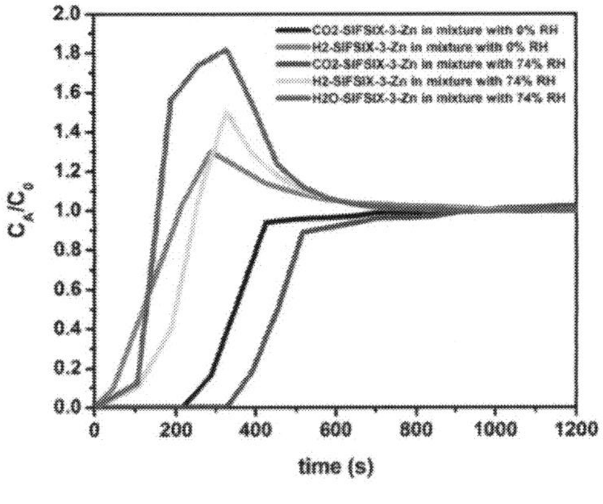

FMD³ Functional Materials Design, Discovery and Development — King Abdullah University of Science and Technology KAUST

Column breakthrough test: CO_2/N_2: 10/90 mixture

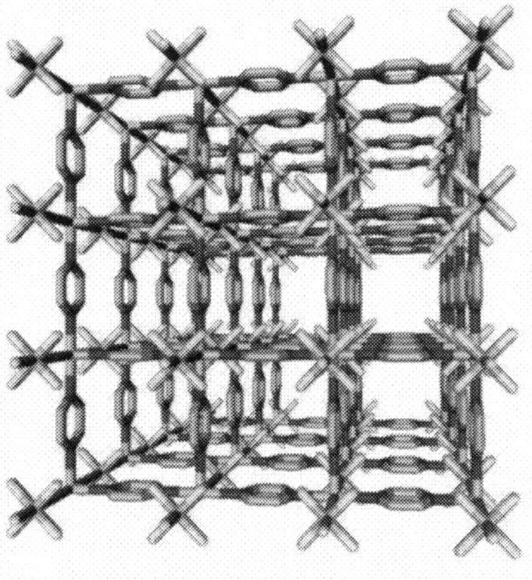

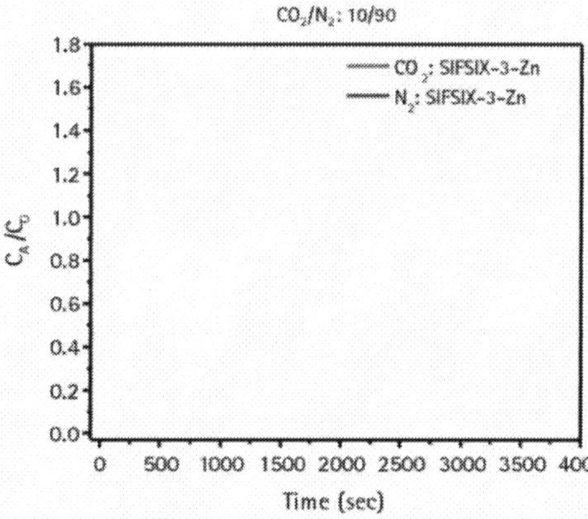

Nugent, P, Belmabkhout, Y et al (2013) Nature 494 (7439) ,80-84, Patent Pending 31

FMD³ Functional Materials Design, Discovery and Development — King Abdullah University of Science and Technology KAUST

Recent finding (2014)

nature COMMUNICATIONS

ARTICLE

Received 9 Nov 2013 | Accepted 28 May 2014 | Published 25 Jun 2014 DOI: 10.1038/ncomms5228 OPEN

Made-to-order metal-organic frameworks for trace carbon dioxide removal and air capture

Osama Shekhah[1,*], Youssef Belmabkhout[1,*], Zhijie Chen[1], Vincent Guillerm[1], Amy Cairns[1], Karim Adil[1] & Mohamed Eddaoudi[1]

32

FMD³ Functional Materials Design, Discovery and Development | King Abdullah University of Science and Technology KAUST

SIFSIX-3-Cu MOFs. Interatomic distances in SIFSIX based MOFs.

The rational is based on conventional coordination chemistry suggesting that replacement of Zn(II) by Cu(II) to form an iso-structural **SIFSIX**-3-Cu will potentially induce an additional pore contraction due to Jahn-Teller distortions of the octahedral coordination geometry of Cu(II), CuN_4F_2.

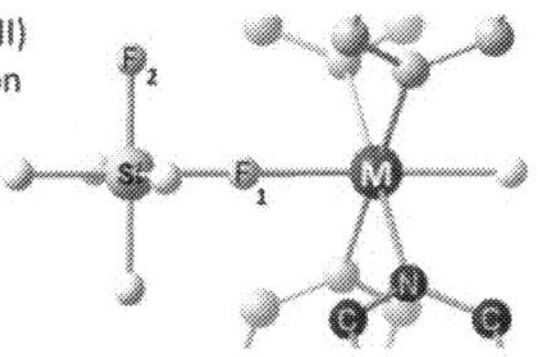

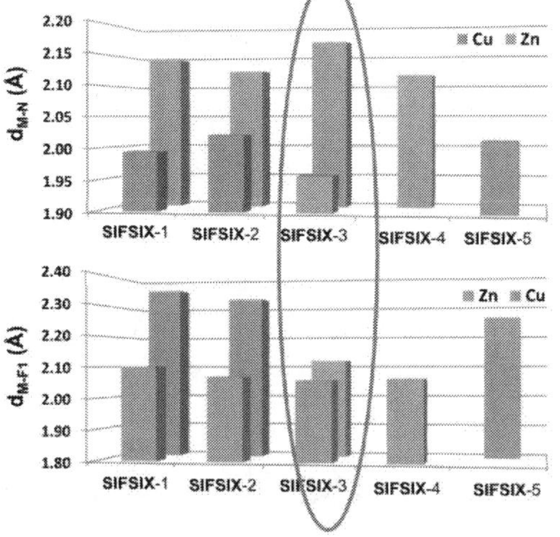

MOF	d_{M-N} (Å)	d_{M-F1} (Å)	D_{Si-F1} (Å)	D_{Si-F2} (Å)†	reference	Ligand
SIFSIX-1-Zn	2.157	2.082	1.757	1.600	[38]	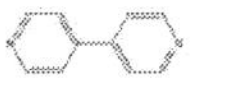
	2.131	2.122	1.720	1.650	[39]	
SIFSIX-1-Cu	2.007	2.379	1.703	1.672	[40]	
	2.009	2.357	1.698	1.609	[41]	
	1.966	2.336	1.692	1.685	[42]	
SIFSIX-2-Zn	2.125	2.069	1.698	1.668	[39]	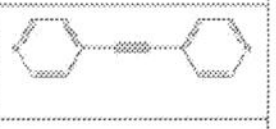
SIFSIX-2-Cu	2.027	2.300	1.684	1.684	[43]	
SIFSIX-2i-Cu	2.015	2.353	1.693	1.679	[43]	
SIFSIX-3-Zn	2.172	2.057	1.747	1.657	[44]	
SIFSIX-3-Cu	1.958	2.119	1.834	1.651	this work	
SIFSIX-2D-Cu	2.061*	2.402	1.727	1.688	[44]	
SIFSIX-2D-H_2O-Cu	2.031	2.412	1.695	1.675	[42]	
SIFSIX-4-Zn	2.117	2.062	1.712	1.653	[39]	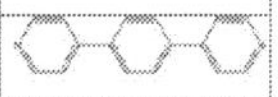
SIFSIX-5-Cu	2.012	2.258	1.727	1.681	[40]	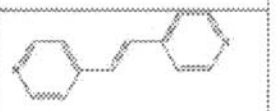

Shekhah, O, Belmabkhout, Y et al (2013). Nature Commun 2014, 5, 4228

FMD³ Functional Materials Design, Discovery and Development | King Abdullah University of Science and Technology KAUST

Enhanced charge density by slight pore contraction

Uptake at 1300 ppm

SIFSIX-3-Cu	43 cm^3 STP/g
Sr^{2+} SAPO-34	15 cm^3 STP/g
Ca^{2+} LTA (**5A**)	9 cm^3 STP/g

SIFSIX-3-M

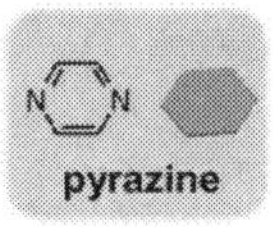

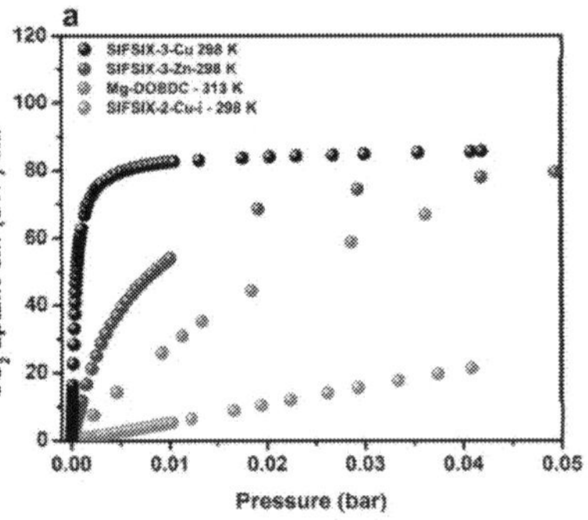

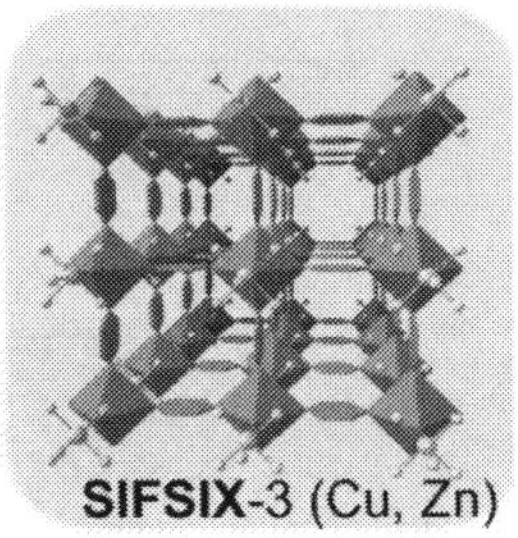

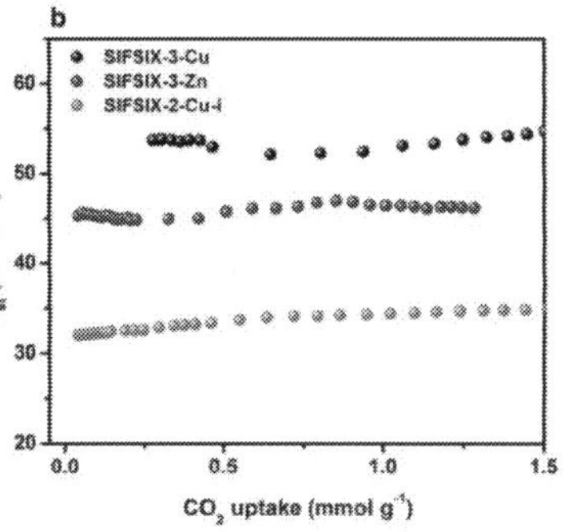

➢**High charge density + suitable pore size**

➢**Enhanced charge density by pore contraction**

Shekhah, O, Belmabkhout, Y et al (2013). Nature Commun 2014, 5, 4228, Patent Pending

FMD[3] Functional Materials Design, Discovery and Development

King Abdullah University of Science and Technology KAUST

Shift to membrane Technology

35

FMD[3] Functional Materials Design, Discovery and Development

King Abdullah University of Science and Technology KAUST

JACS
JOURNAL OF THE AMERICAN CHEMICAL SOCIETY

Communication

pubs.acs.org/JACS

Quest for Anionic MOF Membranes: Continuous sod-ZMOF Membrane with CO_2 Adsorption-Driven Selectivity

Bassem A. Al-Maythalony,[†] Osama Shekhah,[†] Raja Swaidan,[‡] Youssef Belmabkhout,[†] Ingo Pinnau,[‡] and Mohamed Eddaoudi[*,†]

[†]Functional Materials Design, Discovery & Development Research Group (FMD[3]), Advanced Membranes & Porous Materials Center (AMPMC), Division of Physical Sciences and Engineering (PSE), 4700 King Abdullah University of Science and Technology (KAUST), Thuwal 23955-6900, Kingdom of Saudi Arabia

[‡]Advanced Membranes & Porous Materials Center (AMPMC), Division of Physical Sciences and Engineering, 4700 King Abdullah University of Science and Technology (KAUST), Thuwal 23955-6900, Kingdom of Saudi Arabia

Supporting Information

ABSTRACT: We report the fabrication of the first continuous zeolite-like metal–organic framework (ZMOF) thin-film membrane. A pure phase sod-ZMOF, sodalite topology, membrane was grown and supported on a porous alumina substrate using a solvothermal crystallization method. The absence of pinhole defects in the film was confirmed and supported by the occurrence of quantifiable time-lags, for all studied gases, during constant volume/variable pressure permeation tests. For both pure and mixed gas feeds, the sod-ZMOF-1 membrane exhibits favorable permeation selectivity toward carbon dioxide over relevant industrial gases such as H_2, N_2, and CH_4, and it is mainly governed by favorable CO_2 adsorption.

based separation, preferably with a favorable permeation of the minor component (i.e., CO_2) in a given gas mixture such as CO_2/CH_4 and CO_2/H_2 relevant to natural gas upgrading and shifted syngas, respectively.

The main advantages of this emerging technology for gas separations versus conventional sorbent-based separation processes are the relatively lower energy cost, smaller footprint, and modular design.[6] The implementation of MOF membranes for gas purification and separation is in its infancy and is mainly hindered by the persisting challenge to successfully fabricate robust, continuous, and pinhole-free MOF thin-films with very good attachment to the support.[3] This challenge is reflected on the limited number of reports in the open literature pertaining to pure MOFs grown as thin films for gas separation.[7]

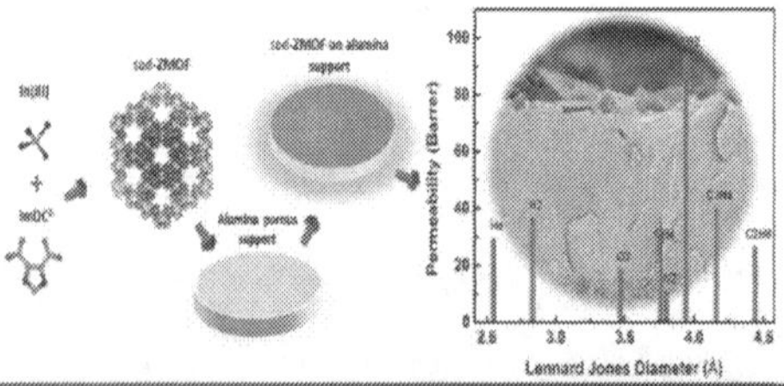

Zeolite like MOF platform. Thin film membrane

Single gas permeation

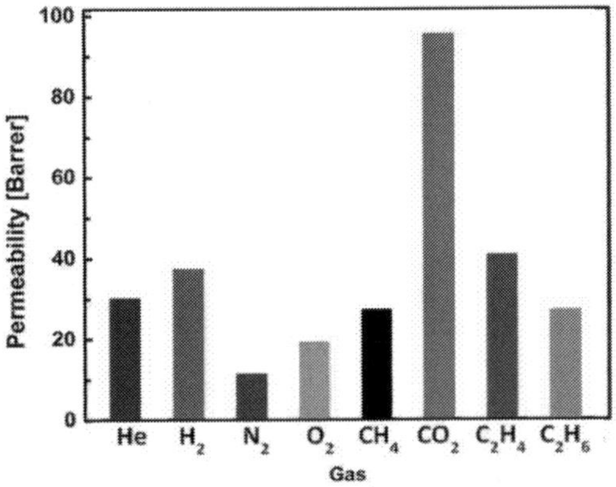

Gas mixture	Ideal selectivity
CO_2/He	3.2
CO_2/H_2	2.6
CO_2/N_2	8.7
CO_2/O_2	5.1
CO_2/CH_4	3.6
H_2/N_2	3.4
H_2/O_2	2.0
H_2/CH_4	1.4
O_2/N_2	1.7

Some historical aspects

Adsorption on Porous Zeolites (Richard M. Barrer)

1930 — 1945-1950 — 1960

Large scale by Linde and Union Carbide corporation

commercialization

Practice of mass transport principles through polymer

1900 — 1980

First Plant with polysulfone hollow fiber membranes for H_2/N_2 Permea (Monsanto)

Adsorption on Porous MOFs

1998-2000 — 2013 — 2015 — 2020-2030

Large scale-up for CH_4 storage by BASF

Start up at KAUST

commercialization

Acknowledgment

ارامكو السعودية
Saudi Aramco

مدينة الملك عبدالعزيز للعلوم والتقنية KACST

Thank you

CARBON MANAGEMENT
TECHNOLOGY CONFERENCE

Validating CO_2-EOR as a CCUS Technology

Vanessa Nuñez-Lopez, Ramon Gil-Egui
Gulf Coast Carbon Center
Bureau of Economic Geology
The University of Texas at Austin

Sugar Land, TX November 16–19, 2015

WWW.CARBONMGMT.ORG

Problem Statement

- Is CO_2-EOR a valid option for greenhouse gas emission reduction? Are geologically stored carbon volumes larger that direct/indirect emissions resulting from the CO_2-EOR operation?

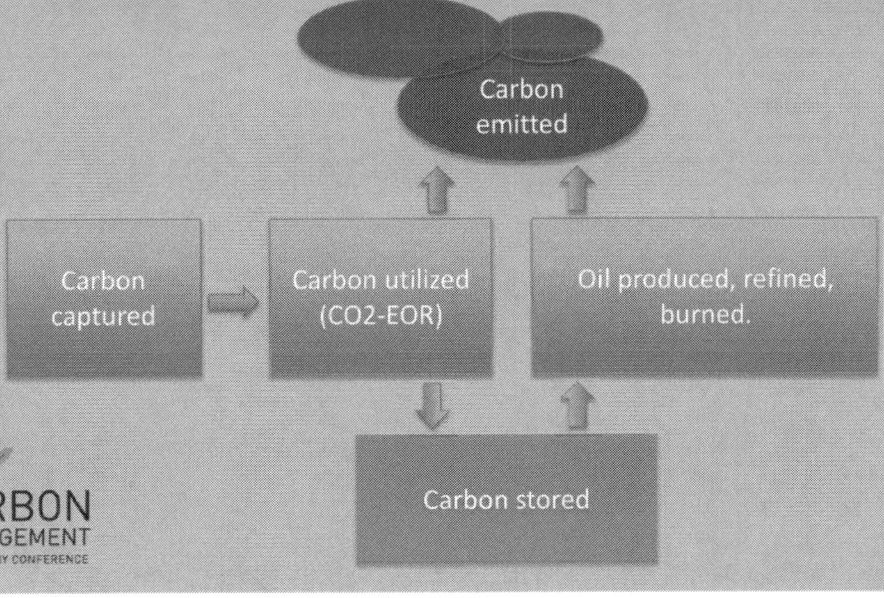

CARBON MANAGEMENT
TECHNOLOGY CONFERENCE

DOE-NETL Sponsored Project

Goal: To develop a clear, universal, repeatable methodology for making the determination of whether a CO_2-EOR operation can be classified as Net carbon Negative Oil (NCNO)

Objectives:

- Identify and frame critical carbon balance components for the accurate mass accounting of a CO_2-EOR operation.
- Develop strategies that are conducive to achieving a NCNO classification.
- Develop a comprehensive, yet commercially applicable, monitoring, verification, and accounting (MVA) methodology.

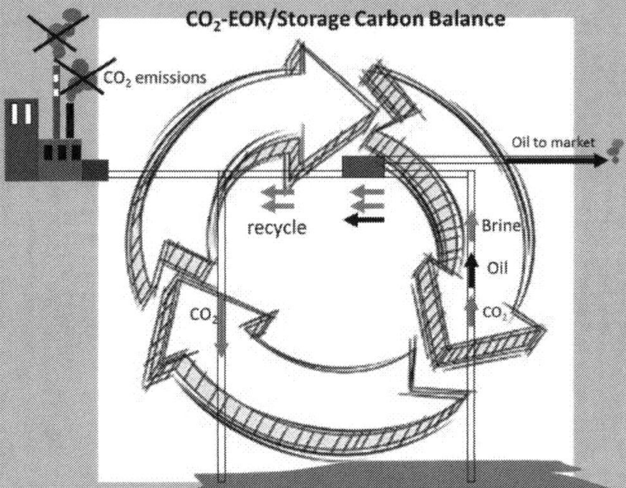

CARBON MANAGEMENT TECHNOLOGY CONFERENCE

System Boundaries of Previous Studies

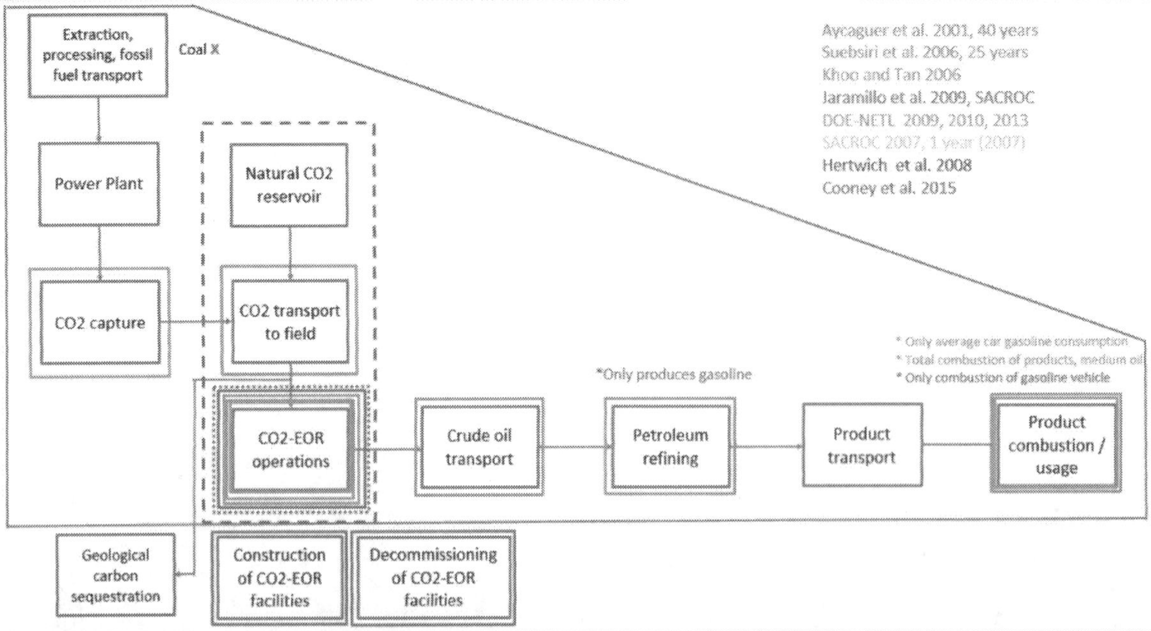

Energy intensive components within the EOR site

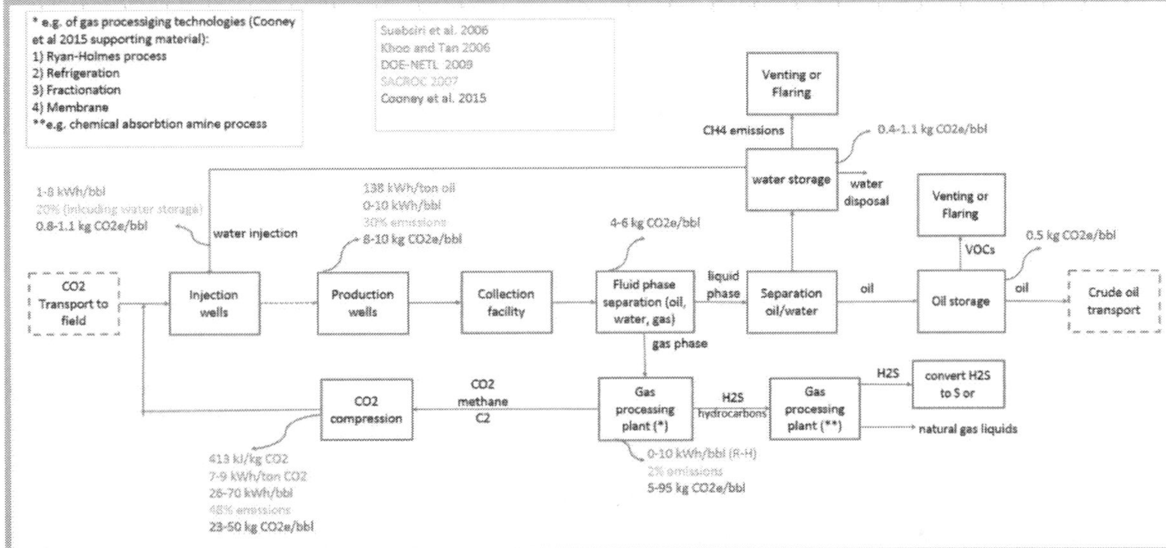

CARBON MANAGEMENT TECHNOLOGY CONFERENCE

5

Environmental Impact per Component

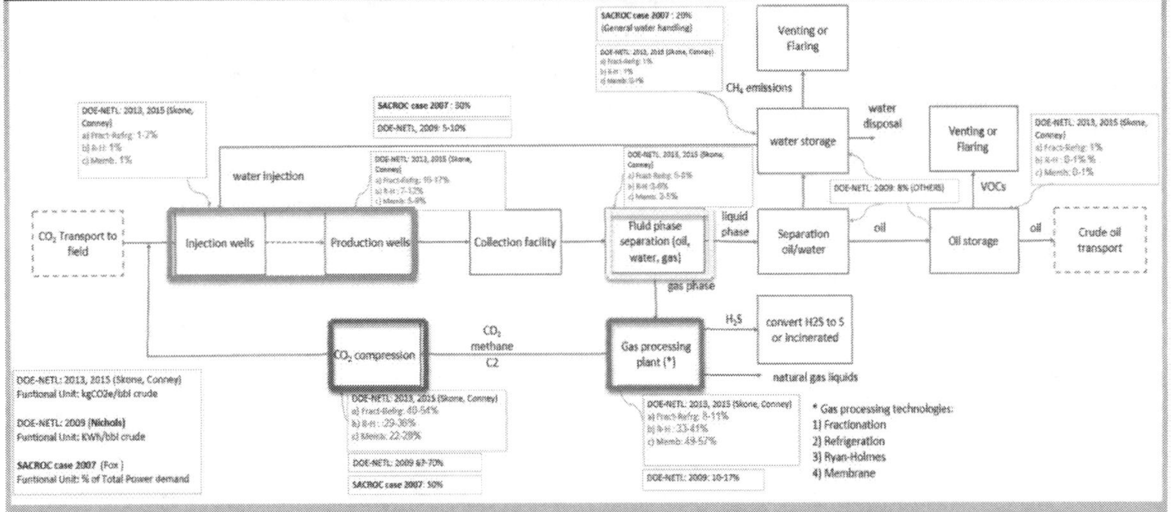

Injection/production wells | Production separation | CO_2 separation | CO_2 compression

CARBON MANAGEMENT TECHNOLOGY CONFERENCE

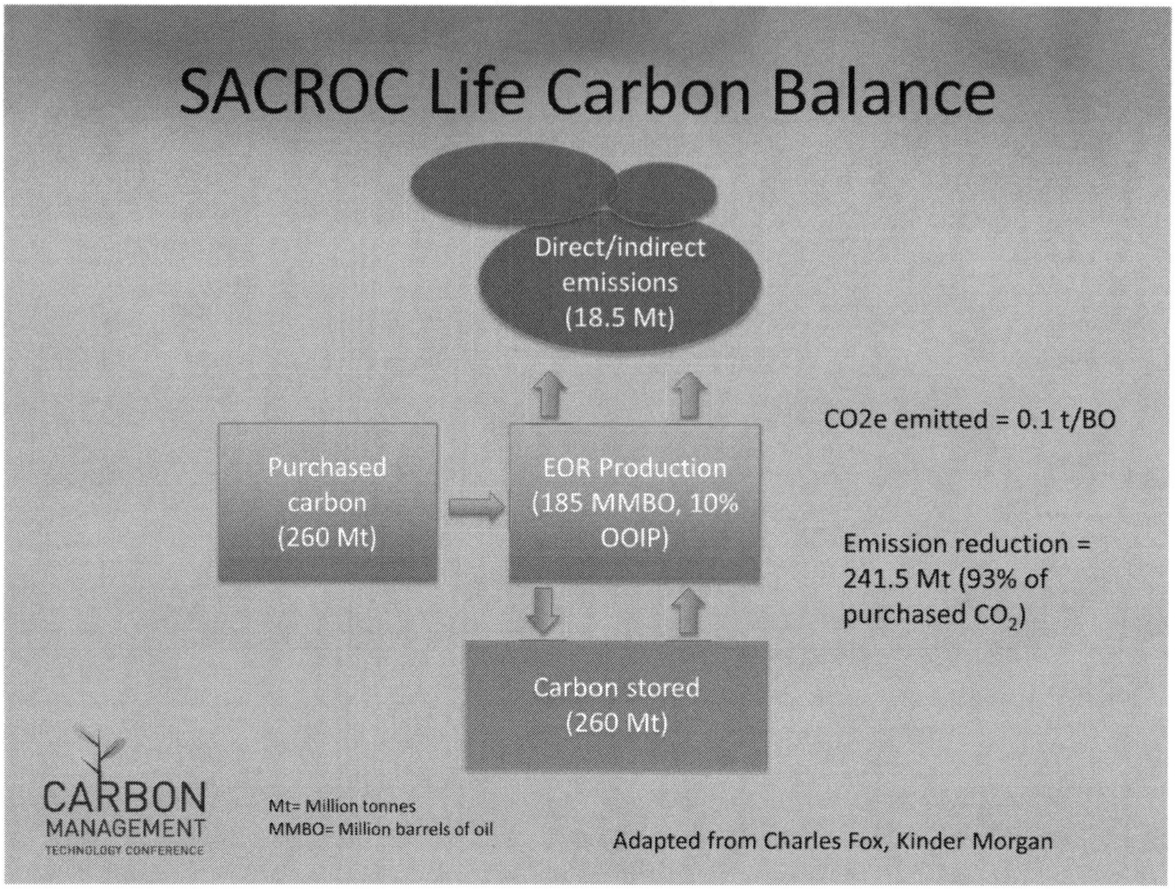
SACROC Life Carbon Balance
Direct/indirect emissions (18.5 Mt)
Purchased carbon (260 Mt)
EOR Production (185 MMBO, 10% OOIP)
Carbon stored (260 Mt)
CO2e emitted = 0.1 t/BO
Emission reduction = 241.5 Mt (93% of purchased CO_2)
CARBON MANAGEMENT TECHNOLOGY CONFERENCE
Mt= Million tonnes
MMBO= Million barrels of oil
Adapted from Charles Fox, Kinder Morgan

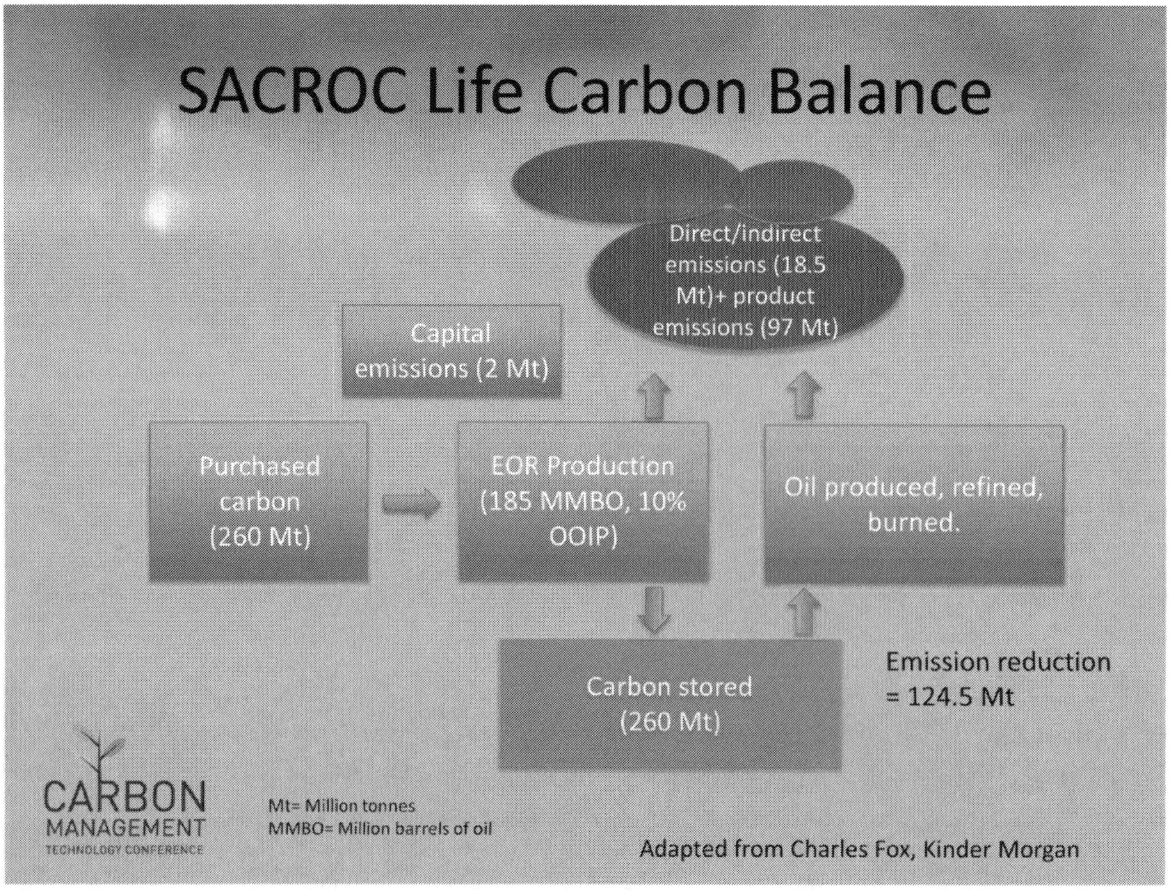
SACROC Life Carbon Balance
Direct/indirect emissions (18.5 Mt)+ product emissions (97 Mt)
Capital emissions (2 Mt)
Purchased carbon (260 Mt)
EOR Production (185 MMBO, 10% OOIP)
Oil produced, refined, burned.
Carbon stored (260 Mt)
Emission reduction = 124.5 Mt
CARBON MANAGEMENT TECHNOLOGY CONFERENCE
Mt= Million tonnes
MMBO= Million barrels of oil
Adapted from Charles Fox, Kinder Morgan

Study focus: CO_2 utilization ratios

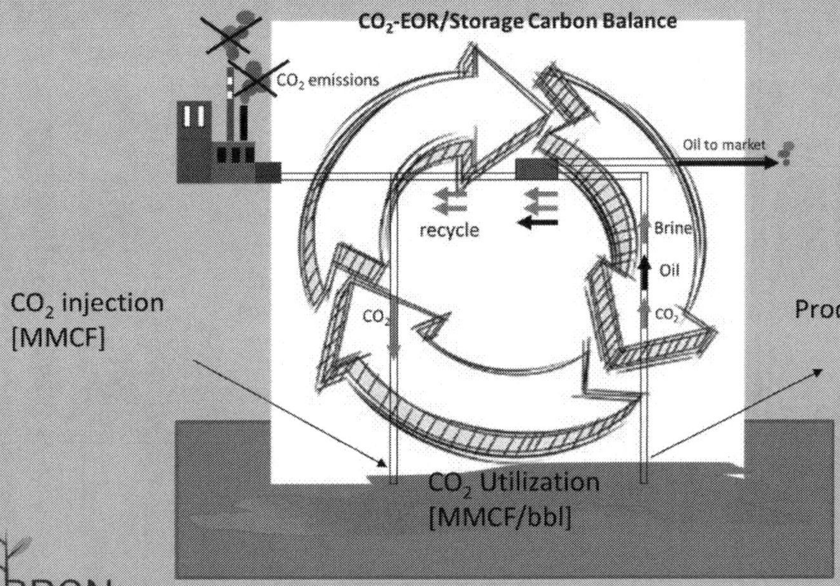

CARBON MANAGEMENT TECHNOLOGY CONFERENCE

Field Study

- (Cranfield, Mississippi)
  - It provides the optimal mass accounting data set as it was required by its comprehensive SECARB MVA program
  - It is a desirable direct injection (no WAG), which is favorable for achieving NCNO
  - Pattern geometry and operations repeated systematically around field development
  - Provides a simpler environment than many CO_2-EOR floods

CARBON MANAGEMENT TECHNOLOGY CONFERENCE

10

Methodology: Numerical Simulation

- Utilize Cranfield pattern calibrated models to:
  - Run numerical simulations for different novel and standard CO_2 injection scenarios (WAG, direct CO2 injection)
  - Evaluate how the variability of CO_2 utilization ratios for the different injection scenarios affects the identified system components and how EOR and saline storage can be co-managed as CCS matures.
  - Understand the carbon balance evolution from start of injection to completion.

CARBON MANAGEMENT TECHNOLOGY CONFERENCE

11

Development of MVA Plan

- Use predictive flow and pressure elevation results to develop a generic but comprehensive MVA plan that is based on:
  - existing regulatory monitoring requirements
  - existing best practices
  - a number of proposed and suggested processes that are currently being considered for possible future regulatory or credit trading conditions
- Integrate results to develop a universal methodology for estimating the carbon balance of a CO_2-EOR operation for making the determination of whether the operation can be classified as Net Carbon Negative Oil (NCNO)

CARBON MANAGEMENT TECHNOLOGY CONFERENCE

12

Expected Outcomes

- A comprehensive carbon balance analysis of a CO_2-EOR operation with an accurate mass accounting methodology for determining whether the operation can be classified as NCNO.
- A recommendation of CO_2 surface operation and injection strategies that are conducive to achieving a NCNO classification.
- A universal MVA methodology encompassing the entire CO_2-EOR operation and inclusive of pre CO_2 injection, injection, and stabilization periods.

CARBON MANAGEMENT TECHNOLOGY CONFERENCE

13

Comparative Analysis of EOR Technologies

- Assuming the barrel of oil will be produced regardless of the technology and of the geographic location, in order to meet a demand.

Differential approach

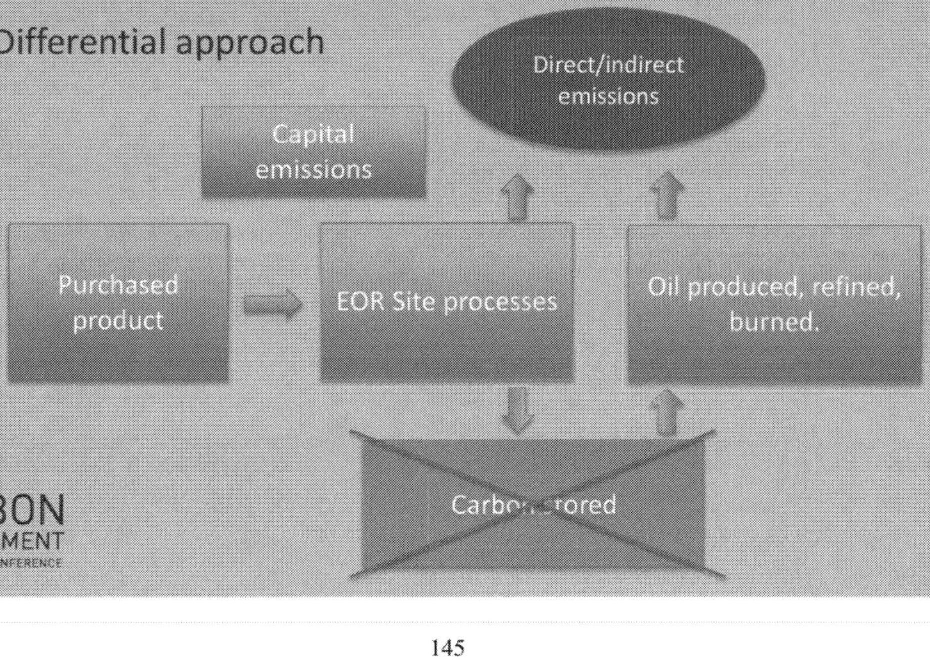

CARBON MANAGEMENT TECHNOLOGY CONFERENCE

Conclusions

- In a gate-to-gate life cycle analysis, the energy consumption (purchased and generated) is responsible for almost all the emissions associated with the EOR operation, particularly at the CO_2 separation and compression processes.
- CO_2 EOR potential for GHG emission reduction lies in the differential carbon balances between itself and other tertiary recovery technologies.

CARBON MANAGEMENT TECHNOLOGY CONFERENCE

15

CARBON MANAGEMENT TECHNOLOGY CONFERENCE

Acknowledgements:

- This material is based upon work supported by the U.S. Department of Energy, National Energy Technology Laboratory
- Cost-share and research support is provided by The University of Texas, Bureau of Economic Geology, Gulf Coast Carbon Center

NETL
BUREAU OF ECONOMIC GEOLOGY
Gulf Coast Carbon Center

WWW.CARBONMGMT.ORG

CARBON MANAGEMENT TECHNOLOGY CONFERENCE

CMTC-439549-MS

Further Advances of CO_2-EOR by Promoting a "Triple-E" Approach: Energy Security, Environmental Quality, and Economic Viability

Domenico "Mimmo" Parisi , Farshid Vahedifard, Michael S. Mazzola
National Strategic Planning & Analysis Research Center (nSPARC), Mississippi State University, Mississippi State, MS, USA.

This paper was prepared for presentation at the Carbon Management Technology Conference held in Sugarland, Texas, USA, 17–19 November 2015. This paper is drawn from a full report, "A New Framework for Energy Policy: Providing Energy Security, Environmental Quality, & Economic Viability," that can be found at nsparc.msstate.edu.

Abstract

We propose a new framework for energy policy on a national and global scale for exploring further advances of CO_2-EOR that promotes a "triple-e" approach: (1) energy security, (2) environmental quality, and (3) economic viability. Increasing the use of CO_2-EOR can reduce U.S. dependence on foreign oil, decrease greenhouse gas (GHG) emissions, create jobs, and help generate electricity to meet domestic demand. Understanding CO_2-EOR and the opportunities and challenges it brings is therefore essential for both policymakers and consumers.

Introduction

There are always two sides to a story, and carbon dioxide (CO_2) is no exception. On one side, CO_2 is presented as a greenhouse gas (GHG), and preventing its release drives the policy of some government agencies. On the other side, CO_2 is an essential nutrient necessary for plant growth and, more recently, has been recognized in the energy sector as a valuable commodity to enhance recovery in utilization operations, such as CO_2 enhanced oil recovery (CO_2-EOR), enhanced gas recovery (CO_2-EGR), and enhanced coal bed methane recovery (CO_2-ECBMR).

CO_2-EOR is a mature technology and currently generates around 300,000 barrels of oil each day in the United States. It is a contributing factor to the recent boom in U.S. oil, which has resulted in lower gasoline prices. However, CO_2-EOR's potential for energy security is largely untapped. Exploring new opportunities and technological improvements for increased CO_2 storage in EOR operations can significantly increase the volume of its generated oil and improve applicability of technology as a revenue generator for CO_2 capture and a large-scale CO_2 storage option.

Energy security calls for comprehensive use of all of our nation's energy sources, including fossil fuels. But fossil fuels, especially coal, have become controversial because of their high level of CO_2 emissions. Many power plants have replaced coal with natural gas to mitigate CO_2 emissions. The United States and the world have also pushed for renewable energy as an effort to control CO_2 emissions under the overall banner of slowing down global warming.

CO_2-EOR provides a way to create a safe, secure, and economically viable supply of fossil fuel-based energy that also reduces CO_2 emissions. Not only does the CO_2-EOR process incidentally trap CO_2 safely in the earth, but the use of anthropogenic, or man-made, CO_2 is essential to promoting and supporting CO_2-EOR, which will encourage a market where CO_2 will be stored as a commodity rather than released into the atmosphere as a waste product.

CO_2-EOR can also give the United States a major global advantage in oil production, supply, and pricing because of how it taps into oil from fields that are traditionally seen as depleted after primary and secondary recovery processes. In addition to depleted old fields, advances in CO_2-EOR also show promise for extraction of significant volumes of stranded oil in residual oil zones (ROZs) and offshore oil fields. Understanding CO_2-EOR and the opportunities and challenges it brings is therefore essential for both policymakers and consumers.

Taking advantage of the full potential of CO_2-EOR technology will only be feasible by implementing a comprehensive policy framework that concurrently considers socioeconomic interests of all parties involved. To address this need, we propose a new energy policy framework on a national and global scale for exploring further advances in CO_2-EOR that promotes a "triple-e" approach: (1) energy security, (2) environmental quality, and (3) economic viability (see Figure 1). Increasing the use of CO_2-EOR can reduce U.S. dependence on foreign oil, decrease GHG emissions, create jobs, and help generate electricity to meet domestic demand. CO_2-EOR is supported by the energy industry and many environmental groups because they understand its potential benefits.

Figure 1. Promoting CO_2-EOR advances through a "triple-e" energy policy framework: (1) energy security, (2) environmental quality, and (3) economic viability

Energy Security

Fossil fuels account for more than 80 percent of U.S. energy consumption, with 37 percent from petroleum and 27 and 20 percent from natural gas and coal, respectively. Better gasoline mileage, improved building construction, improved efficiencies of appliances, and the push for more renewable energy have significantly contributed to reduction in the energy demand rate over the past few decades. As a result, more GDP per unit of energy consumed is generated. In spite of this, fossil fuels will continue to be the most viable source of energy for many years at an expected rate of approximately 85 percent of total energy consumption (U.S. Energy Information Administration 2014).

Fossil fuels will continue to be a primary source of energy for the United States and other industrialized nations. The establishment of a reliable, inexpensive, and environmentally sustainable energy supply will be critical to ensuring social and economic success in the United States. CO_2-EOR can help maintain and promote energy security by (1) increasing domestic oil production and (2) increasing the use of clean-coal technology for electric power generation.

The U.S. Department of Energy (2012) indicates that CO_2-EOR in the United States contributes about 300,000 barrels of oil per day. The oil that could be potentially recovered with CO_2-EOR in the United States is significant -- Advanced Resources International (2010) estimates that CO_2-EOR could unlock 395 billion barrels of oil. Expansion of CO_2-EOR would push the country further ahead in oil exporting. In 2014, the United States had its greatest amount of crude oil exports in 57 years (Breul 2014).

Large-scale energy generation through renewable energy sources (e.g., wind, solar, geothermal) is uncertain at this time. Renewable energy production and use have increased in recent years but are still marginal to overall U.S. energy generation. In 2012, for example, all renewable sources accounted for only 9 percent of energy consumption, less than one-eighth of the amount accounted for by fossil fuels (U.S. Energy Information Administration 2014). The limited development of renewable energy has also been aided by substantial government subsidies. The CO_2-EOR process is a more promising approach to promoting energy security while reducing CO_2 emissions.

With current CO_2-EOR methods, an estimated additional 19 billion barrels of crude oil could be economically recovered. More advanced "next generation" methods are predicted to yield more than three times as much extra economically recoverable oil at 59.6 billion barrels. These figures are based on the assumption that oil sells at $90 per barrel, that CO_2 costs $40 per metric ton, and that there is a 20 percent before-tax rate of return (Kuuskraa 2013).

CO_2-EOR can also be used to extract additional oil from offshore sites. The offshore CO_2-EOR experience conducted in the North Sea indicates that implementation expenses tend to be greater with offshore extraction. Nonetheless, offshore CO_2-EOR is still potentially profitable. Cost projections from Advanced Resources International and Melzer Consulting (2010) suggest that employing CO_2-EOR in the Gulf of Mexico could provide a 20 percent return on investment, assuming that oil sells at $90 per barrel and that CO_2 is priced at $40 per metric ton. In addition, CO_2 pipelines to offshore fields could use the corridors that already exist for oil and gas, thereby reducing capital costs (Koottungal 2014; Mock, Williams, and Harder 2014).

Another advantage of a CO_2-EOR strategy is that it promises to reduce emissions without excluding coal from the nation's energy mix. Due to its abundance in the United States (U.S. Energy Information Administration 2015), coal remains both a central part of U.S. environmental policy debate and a very large contributor to local economies, directly through mining and transportation and indirectly through affordable electricity generation. However, recent trends have favored using natural gas rather than coal in power generation because the former results in lower CO_2 emissions.

CO_2-EOR methods can make coal power plants a more viable long-term energy solution by reducing their negative environmental impact (i.e., release of CO_2 into the atmosphere). Using CO_2-EOR on a large scale to conserve anthropogenic CO_2 emissions generated from coal-fired power plants will create a commodity market for CO_2. The incidental storage of CO_2 underground, a natural result of CO_2-EOR, would ensure that conserved CO_2 from coal power plants would not contribute to the U.S. emission rate.

Two clean-coal technology power generation plants in development – one in Mississippi and one in Texas – are used to capture chemicals that could be sold as commodities and to contribute CO_2 for use in oil production. Both the Kemper County and Petra Nova plants illustrate how CO_2 capture, storage, and utilization can help lead to energy security in the United States.

Mississippi Power's Kemper County 582-megawatt power plant is a combined power/chemical-generating gasification plant that will operate with a new approach to generating electric power. Scheduled to open in mid-2016, the plant will be fueled by low-cost lignite that will be mined on site. The Kemper County power plant will generate CO_2, sulfuric acid, and ammonia to be captured and sold. Sixty-five percent of the CO_2 to be generated by the plant will be used in CO_2-EOR operations (Mississippi Power 2015a). The CO_2 production from the plant could lead to the recovery of 2 million barrels of oil each year.

The Petra Nova project in Texas involves an international partnership between NRG Energy and JX Nippon Oil & Gas Exploration, with assistance from the U.S. Department of Energy. The project's goal is to capture CO_2 from NRG Energy's coal-fired power plant near Houston, Texas, for use in EOR. The project, scheduled to be operational by the end of 2016, is expected to capture 1.4 million tons of CO annually. It will demonstrate how the CO_2 capture technology could be commercialized, allowing existing coal power plants to both produce clean electricity and reduce the impact of CO_2 emissions (National Energy Technology Laboratory 2014).

Environmental Quality

The concentration of CO_2 in the atmosphere, the primary factor in climate change, recently reached a monthly global average of 400 parts per million (ppm), the highest level ever recorded by the National Oceanic and Atmospheric Administration (2015). Carbon capture and storage (CCS) technology has been identified as a promising solution to reduce excessive CO_2 emissions. Among 10 recent technologies listed by the International Energy Agency (2012a) for saving energy and decreasing CO_2 emissions, CCS has been identified with the highest potential and could account for up to 20 percent of cumulative CO_2 reductions in the two-degree scenario (2DS) by 2050 (Rystad 2013).

CCS is the only potential option that would enable major industries, such as steel, cement, and natural gas/oil, to reduce their emissions significantly. Despite its viability, the potential of CCS has not been fully utilized due to some economic, technical, and social challenges. These challenges and concerns have slowed the large-scale implementation of CCS as well as public acceptance of CCS technology (Mortezaei and Vahedifard 2015; International Energy Agency 2012b; Van Noorden 2010). More recently, it has been recognized that adding "utilization" to CCS and pursuing developments and opportunities with carbon capture, utilization, and storage (CCUS) technology can have a positive impact on the large-scale implementation of CO_2 capture and storage projects (McConnell 2012).

Among different emerging applications within CCUS, CO_2-EOR offers a market-driven solution with major near-term and long-term technical and economic opportunities. The CO_2-EOR process can significantly impact climate change by (1) using huge quantities of anthropogenic CO_2 emissions that would otherwise be released into the atmosphere and (2) storing residual CO_2 deep underground (Kuuskraa, Leeuwen, and Wallace 2011). One study found that expanding "next generation" uses of CO_2-EOR in the United States could bring about the associated storage of 27.8 gigatons of CO_2, or the equivalent of the CO_2 that would be produced by roughly 200 coal-fired power plants over 40 years (Kuuskraa 2013).

While there may seem to be a disconnect between using fossil fuels for energy production and reducing CO_2 emissions, both can happen simultaneously with a CO_2 capture and permanent storage strategy. The expansion of CO_2-EOR can advance such a strategy in two key ways. First, CO_2-EOR incidentally traps a significant amount of power plant-generated CO_2, which would otherwise be released into the atmosphere, within oil fields. Second, the employment of CO_2-EOR using anthropogenic supplies of CO_2 should help finance the construction of an extensive CO_2 capture and transport infrastructure that public/private partnerships can eventually use to implement a comprehensive carbon sequestration plan.

The CO_2 injected into oil fields during CO_2-EOR becomes trapped below ground. After the CO_2-EOR process, layers of strata that previously confined oil hold CO_2 in the porous rock layer. Thus, CO_2-EOR enables higher levels of oil recovery from existing fields while simultaneously enclosing the injected CO_2 in a permanent geological structure located thousands of feet below the surface (National Energy Technology Laboratory 2010). Storage occurs at a pressure that does not harm the strata housing the CO_2 or the strata above or below. As such, CO_2-EOR represents a promising avenue for avoiding CO_2 emissions.

Anthropogenic CO_2 generated by power plants can be used in CO_2-EOR operations and stored safely below ground instead of being released into the atmosphere. Technology projects such as the power plant in Kemper County, Mississippi, and the Petra Nova project in Texas provide examples of this process and demonstrate the need for future similar projects to contribute to CO_2-EOR. To expand CO_2-EOR practices, oil producers need to identify sources of anthropogenic CO_2 in a form that can be used and transported to a CO_2-EOR site. As natural supplies of CO_2 are depleted, a steady supply of anthropogenic CO_2 will increasingly become a critical need for CO_2-EOR operators (Godec, Kuuskraa, and DiPietro 2013).

Sequestration provides another opportunity to reduce CO_2 emissions. The sequestration process takes anthropogenic CO_2 as a waste stream and pumps it into a suitable underground geological formation. Unlike CO_2-EOR, sequestration does not seek to use CO_2 to recover oil or another product. But like CO_2-EOR, sequestration results in the permanent underground containment of CO_2 that would otherwise be released into the atmosphere.

CO_2-EOR could promote a full sequestration program by financing some of the requisite equipment and leaving it in place. The infrastructure necessary to acquire anthropogenic CO_2 from power plants would allow for capturing, storing, and transporting the gas (e.g., well casings, piping, drilling patterns). Oil producers' purchase of CO_2 would help finance this infrastructure and construction that could also be used in sequestration. The infrastructure would be in place in the event future regulations required the sequestering of CO_2 emissions (Leach et al. 2011).

There is additional potential for synergy between CO_2-EOR and sequestration operations. Suitable sequestration strata (e.g., deep saline aquifers) may exist below or near oil fields where CO_2-EOR will be employed (Leach et al. 2011). In such cases, infrastructure used to deliver CO_2 for CO_2-EOR could later be used to transport CO_2 for final sequestration into strata below the oil field.

Metz et al. (2005) suggest that storage of CO_2 by means of CO_2-EOR is safe and long-lasting. Geological reservoirs that are appropriately chosen and administered will likely retain 99 percent of their stored CO_2 over 1,000 years. The overwhelming majority of CO_2 will be restrained by trapping mechanisms and may be held for millions of years (Metz et al. 2005).

Economic Viability

Domestic energy generation has been tied to a panoply of economic benefits. For instance, the recent rise in U.S. energy production created an estimated 133,000 energy-sector jobs from 2010 to 2013, reduced the country's trade deficit to its lowest levels since the 1990s by decreasing petroleum imports, and increased the national gross domestic product (GDP) by 0.22 percent during 2012-2013 (Executive Office of the President of the United States 2014). It follows that in addition to helping achieve energy security and promoting emission reduction, continued development of CO_2-EOR can be expected to improve the general health of the U.S. economy.

The CO_2-EOR process and its related technological processes can create jobs, produce new revenue streams, and supply more domestic energy. Future growth in CO_2-EOR and related processes could help end the nation's dependence on oil imports and favorably impact GDP and the U.S. trade deficit. In addition, CO_2-EOR production generates revenue through several streams, including severance and ad valorem taxes, federal and state royalties, lease income, jobs, investments, and others (National Enhanced Oil Recovery Initiative 2012).

A potential major economic benefit is the future growth in oil production that could result from a sufficient supply of affordable CO_2. With expansion of CO_2-EOR, anticipated growth in demand for CO_2 supply can be estimated. Purchased CO_2 demand for EOR in the United States is forecasted to grow from 3.2 billion cubic feet a day (62 million metric tons a year) in 2012 to nearly 5.4 billion cubic feet a day (104 million metric tons a year) by 2018 (DiPietro, Balash, and Wallace 2012). Depending on CO_2 utilization rates, incremental daily U.S. oil production from CO_2-EOR will grow from 284,000 barrels in 2012 to between 480,000 and 770,000 barrels by 2018 (Godec 2014).

Projections from (Godec 2013) indicate that the country can recover an estimated 100 billion barrels of oil from conventional and ROZ fields through "next generation" CO_2-EOR. The oil and gas industry is projected to earn approximately $1.8 trillion, or 20.94 percent of the total. The power industry is expected to earn $1.3 trillion (15.53 percent of total revenues) from the sale of captured CO_2 emissions to oil producers. Private companies outside of these two industries are projected to collectively earn close to

$2.7 trillion from the provision of services and materials to CO_2-EOR-related operations; this amount represents the largest share of the revenue distribution (31.18 percent). In addition, it is estimated that federal and state government would experience a growth in revenue of nearly $2 trillion (23.29 percent of revenues). Finally, private mineral owners would gain $770 billion, or 9.06 percent of the total revenue stream. All of these estimates assume that oil sells at $85 per barrel and CO_2 costs $40 per metric ton.

Combining carbon capture and storage (CCS) with CO_2-EOR would provide more economic benefits to the U.S. economy. Revenue from CO_2 sales for the purposes of CO_2-EOR can turn CCS into an economically viable strategy and would result in more oil production that prevents CO_2 from significantly contributing to the U.S. emissions rate. CO_2-EOR could provide large new revenues to federal and state treasuries and other participants in the value chain (Godec 2014).

For example, in the case of Mississippi, a recent study found that CO_2-EOR accounts for $546 million of total disposable income and $754 million of state GDP and contributes $60 million in state and local tax revenues. The energy sector as a whole in Mississippi accounts for $3.6 billion of total disposable income and $6.7 billion of state GDP and contributes $425 million in state and local tax revenues. In addition, the energy sector as a whole supports a total of 59,734 jobs in Mississippi, while CO_2-EOR itself supports a total of 9,701 jobs in the state (NSPARC 2015).

The use of captured anthropogenic CO_2 for CO_2-EOR is transforming the view of CO_2 as a waste product from power plants to a new commodity that can be used to produce energy. New technologies are doing the same for other "waste" products, such as sulfur. The potential value of the CO_2 market in the United States and Canada has created an incentive for developing cost-effective carbon capture technologies at large stationary sources.

As the Kemper County plant project shows, clean-coal chemistry/power-generating technology can use lignite (low-cost coal) to produce other chemicals. The Kemper technology, which is being initiated in stages, will produce other chemicals in addition to CO_2, including sulfuric acid and ammonia (Mississippi Power 2015b). Plants producing both electricity and commodity chemicals can become increasingly more economically viable at larger scales (Kuuskraa 2007). The technology developed and used for low-cost lignite could be applied at extensive lignite deposits in the Southeastern United States and worldwide and also in a pure chemical manufacturing setting where all power generated from lignite is dedicated to chemicals production. Thus, this model could be sized to optimum large-scale commodity chemical production without the complications of fitting into a regulated utility business model. The initial cost of development of clean-coal plants and associated technology could be substantial, but the benefits could be wide-reaching and accomplish several objectives, including the reduction of GHG emissions and the establishment of lucrative new markets.

For example, the United States imports large amounts of sulfur and ammonia. The country is one of the world's leading producers and consumers of ammonia, which is also used in these derivatives: urea, ammonium nitrate, ammonium phosphates, nitric acid, and ammonium sulfate. Approximately 84 percent of domestic ammonia consumption is for fertilizer use. Ammonia is also used to produce plastics, synthetic fibers and resins, explosives, and numerous other chemical compounds (U.S. Geological Survey 2014). Establishing a market to satisfy the demand for these chemicals would help end the need to import them.

Growing a domestic market for chemicals – especially chemicals that are now discarded as waste products – would be beneficial to help satisfy the environmental concerns of GHG emissions and satisfy the current demand for chemicals. Other large-scale emerging markets for CO_2 still need to be identified.

Concluding Remarks

No energy technology, including CO_2-EOR, will be able to sustainably address the nation's needs in the long term without a robust policy framework. We propose a new energy policy framework on a national and global scale for exploring further advances in CO_2-EOR that promotes a "triple-e" approach: (1) energy security, (2) environmental quality, and (3) economic viability. Expanding CO_2-EOR operations can make the country's energy supply more reliable and environmentally sound by, respectively, raising domestic oil output and reducing carbon emissions from power plants.

Significant environmental and economic benefits could be achieved by using anthropogenic CO_2 to grow the CCS and CO_2-EOR markets, so those who oppose using fossil fuels to generate energy are stifling market growth.

The extent to which CO_2-EOR methods are encouraged depends on the legislative and regulatory environment. It is up to policymakers to design and promote tax and regulatory policies that allow CO_2-EOR to compete fairly with the alternatives and increase its use.

Incentives to unlock the benefits of CO_2-EOR and encourage broader implementation of its technology should include:

- Funding more research on CO_2-processing technologies to accelerate CO_2-EOR production by examining economic ways to advance CO_2 capture, separation, transport, storage, and monitoring.
- Awarding of a new federal tax credit that recognizes permanent storage of anthropogenic CO_2 during hydrocarbon recovery operations on the order of $15 per metric ton of the gas. This credit would strongly promote CO_2-EOR that relies on power plant-generated CO_2 as an input, helping offset the high cost of capturing and transporting CO_2 produced by electric facilities.
- Maintaining the well classification regulation for Class II and Class VI wells and current supervision under the Safe Drinking Water Act.

Significant costs are likely to be involved with these policy recommendations, but such changes would be justified if CO_2-EOR is posed to play a major role in shoring up the nation's future in energy security, environmental quality, and economic growth – as seems likely.

References

Advanced Resources International. 2010. *U.S. Oil Production Potential from Accelerated Deployment of Carbon Capture and Storage*. Arlington, VA: ARI. Retrieved April 29, 2015 (http://www.adv-res.com/pdf/v4ARI%20CCS-CO2-EOR%20whitepaper%20FINAL%204-2-10.pdf).

Advanced Resources International, and Melzer Consulting. 2010. *Optimization of CO_2 Storage in CO_2 Enhanced Oil Recovery Projects*. Department of Energy & Climate Change. Office of Carbon Capture and Storage. Arlington, VA: Advanced Resources International, Inc.

Breul, Hannah. 2014. "Crude Exports and Re-Exports Continue to Rise; Some Volumes Sent to Europe and Asia." U.S. Energy Information Administration. Retrieved June 15, 2015 (http://www.eia.gov/todayinenergy/detail.cfm?id=18631).

DiPietro, Phil, Peter Balash, and Matt Wallace. 2012. "A Note on Sources of CO_2 Supply for Enhanced-Oil-Recovery Operations." *SPE Economics & Management* April 2012:69-74. Retrieved May 22, 2015 (http://www.netl.doe.gov/File%20Library/Research/Energy%20Analysis/Publications/SourcesofCO2SupplyforEOR OperationsTechNote_0412.pdf).

Executive Office of the President of the United States. 2014. *The All-of-the-Above Energy Strategy as a Path to Sustainable Economic Growth*. Council of Economic Advisers. Retrieved June 10, 2015 (https://www.whitehouse.gov/sites/default/files/docs/aota_energy_strategy_as_a_path_to_sustainable_economic_growth.pdf).

Godec, Michael L. 2013. "Using the Economic Value of CO_2-EOR to Accelerate the Deployment of CO_2 Capture, Utilization and Storage (CCUS)." Presented at the 2013 Carbon Management Technology Conference, October 21, Alexandria, Virginia.

Godec, Michael L. 2014. "Industrial CO_2 Supply Crucial for EOR." *The American Oil & Gas Reporter*. February. Retrieved May 22, 2015 (http://www.aogr.com/magazine/editors-choice/industrial-co2-supply-crucial-for-eor).

Godec, Michael L., Vello A. Kuuskra, and Phil DiPietro. 2013. "Opportunities for Using Anthropogenic CO_2 for Enhanced Oil Recovery and CO_2 Storage." *Energy & Fuels* 27:4183-89.

International Energy Agency. 2012a. *World Energy Outlook 2012: Executive Summary*. Paris, France: Organisation for Economic Co-operation and Development/International Energy Agency. Retrieved June 15, 2015 (http://www.iea.org/publications/freepublications/publication/english.pdf).

International Energy Agency. 2012b. *Energy Technology Perspectives 2012: Pathways to a Clean Energy System*. Paris, France: Organisation for Economic Co-operation and Development/International Energy Agency. Retrieved May 29, 2015 (https://www.iea.org/publications/freepublications/publication/ETP2012_free.pdf).

Koottungal, Leena. 2014. "U.S. and Other Countries EOR Production." *Oil & Gas Journal*, April 7, p. 79.

Kuuskraa, Vello A. 2007. *A Program to Accelerate the Deployment of CO_2 Capture and Storage (CCS): Rationale, Objectives, and Costs*. Arlington, VA: Pew Center on Global Climate Change.

Kuuskraa, Vello A. 2013. "The Growing CO_2-EOR Oil Recovery and CO_2 Utilization 'Prize.'" Presented at the 2013 EOR Carbon Management Workshop, December 10, Midland, TX. Retrieved April 29, 2015 (http://www.co2conference.net/wp-content/uploads/2014/01/1-Kuuskraa_Midland_EOR_12_13.pdf).

Kuuskraa, Vello A., Tyler Van Leeuwen, and Matt Wallace. 2011. *Improving Domestic Energy Security and Lowering CO_2 Emissions with 'Next Generation' CO_2-Enhanced Oil Recovery*. DOE/NETL-2011/1504. Activity 04001.420.02.03. National Energy Technology Laboratory. U.S. Department of Energy.

Leach, Andrew, Charles F. Mason, and Klaas van 't Veld. 2011. "Co-optimization of Enhanced Oil Recovery and Carbon Sequestration." *Resource and Energy Economics* 33:893-912.

McConnell, Charles. 2012. "Adding 'Utilization' to Carbon Capture and Storage." U.S. Department of Energy. Retrieved June 17, 2015 (http://energy.gov/articles/adding-utilization-carbon-capture-and-storage).

Metz, Bert, Ogunlade Davidson, Heleen de Coninck, Manuela Loos, and Leo Meyer. 2005. *IPCC Special Report on Carbon Dioxide Capture and Storage*. Intergovernmental Panel on Climate Change. New York: Cambridge University Press.

Mississippi Power. 2015a. "Environmental Benefits." Retrieved June 17, 2015 (http://www.mississippipower.com/about-energy/plants/kemper-county-energy-facility/environmental-benefits).

Mississippi Power. 2015b. "Facts." Retrieved June 17, 2015 (http://www.mississippipower.com/about-energy/plants/kemper-county-energy-facility/facts).

Mock, Vanessa, Selina Williams, and Amy Harder. 2014. "Australian Repeal Deals Blow to Global Carbon-Emission Plans." *Wall Street Journal*, July 27.

Mortezaei, Kimia, and Farshid Vahedifard. 2015. "Numerical Simulation of Induced Seismicity in Carbon Capture and Storage Products." *Geotechnical and Geological Engineering* 33:411-24.

National Energy Technology Laboratory. 2010. *Carbon Dioxide Enhanced Oil Recovery: Untapped Domestic Energy Supply and Long Term Carbon Storage Solution*. U.S. Department of Energy. Retrieved April 28, 2015 (http://www.netl.doe.gov/File%20Library/Research/Oil-Gas/enhanced%20oil%20recovery/co2%20eor/NETL_CO2-EOR-Primer.pdf).

National Energy Technology Laboratory. 2014. *Petra Nova Parish Holdings: W.A. Parish Post-Combustion CO_2 Capture and Sequestration Project*. U.S. Department of Energy. Retrieved June 2, 2015(http://www.netl.doe.gov/File%20Library/Research/Coal/major%20demonstrations/ccpi/FE0003311.pdf).

National Enhanced Oil Recovery Initiative. 2012. *Carbon Dioxide Enhanced Oil Recovery: A Critical Domestic Energy, Economic, and Environmental Opportunity*. Great Plains Institute. Center for Climate and Energy Solutions. Retrieved May 22, 2015 (http://www.neori.org/NEORI_Report.pdf).

National Oceanic and Atmospheric Administration. 2015. " Greenhouse Gas Benchmark Reached." Retrieved June 17, 2015 (http://research.noaa.gov/News/NewsArchive/LatestNews/TabId/684/ArtMID/1768/ArticleID/11153/Greenhouse-gas-benchmark-reached-.aspx).

NSPARC. 2015. *The Impact of the Energy Sector and CO_2-EOR on Mississippi's Economy*. NSPARC. Starkville, MS: Mississippi State University.

Rystad, Jarand. 2013. "Petroleum Production under the Two Degree Scenario (2DS)." Rystad Energy. Retrieved June 3, 2015 (https://www.regjeringen.no/contentassets/17f83dcdadd24dad8c5220eb491a42b5/04_rystad_energy_production_under_2ds.pdf).

The Associated Press. 2014. "Southern Company Agrees to Share Kemper County Power Research with Chinese Firm." Retrieved June 15, 2015 (http://blog.gulflive.com/mississippi-press-business/2014/04/southern_company_agrees_to_sha.html).

U.S. Department of Energy. 2012. *Fossil Energy Research Benefits: Enhanced Oil Recovery*. Retrieved June 15, 2015 (http://energy.gov/sites/prod/files/eor_factcard.pdf).

U.S. Energy Information Administration. 2014. *Annual Energy Outlook 2014 with Projections to 2040*. Washington, D.C.: U.S. Department of Energy. Retrieved May 21, 2015 (http://www.eia.gov/forecasts/archive/aeo14/pdf/0383%282014%29.pdf).

U.S. Energy Information Administration. 2015. "Coal Explained: How Much Coal Is Left." Retrieved June 17, 2015 (http://www.eia.gov/energyexplained/index.cfm?page=coal_reserves).

U.S. Geological Survey. 2014. *Mineral Commodity Summaries 2014*. Washington, D.C.: U.S. Government Printing Office. Retrieved June 2, 2015 (http://minerals.usgs.gov/minerals/pubs/mcs/2014/mcs2014.pdf).

Van Noorden, Richard. 2010. "Carbon Sequestration: Buried Trouble." *Nature* 463:871-73.

Technical Economic and Regulatory Challenges Facing Large Scale Adaption of Carbon Geologic Sequestration

Tiraz Birdie, Lynn Watney, and Jennifer Hollenbach

Carbon Management Technology Conference

November 18th 2015

Sugarland, TX

KU KANSAS GEOLOGICAL SURVEY The University of Kansas

TBirdie Consulting inc

NETL

Global CO_2 Cycle

CO_2 Cycle (billion tons per year, GT)

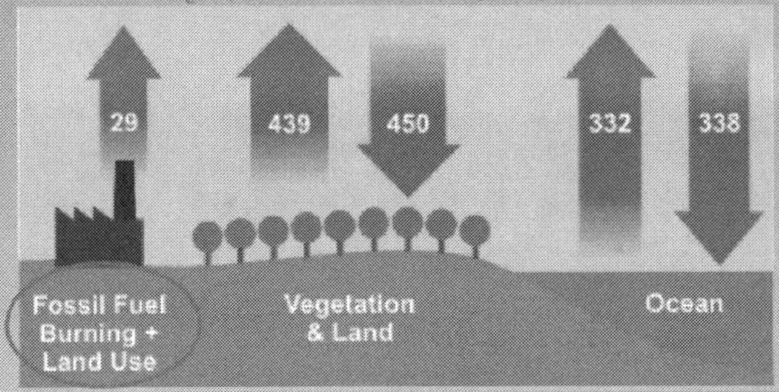

Total Emissions	**800 GT annually**
Natural Sequestration	**788 GT**
Balance	12 GT

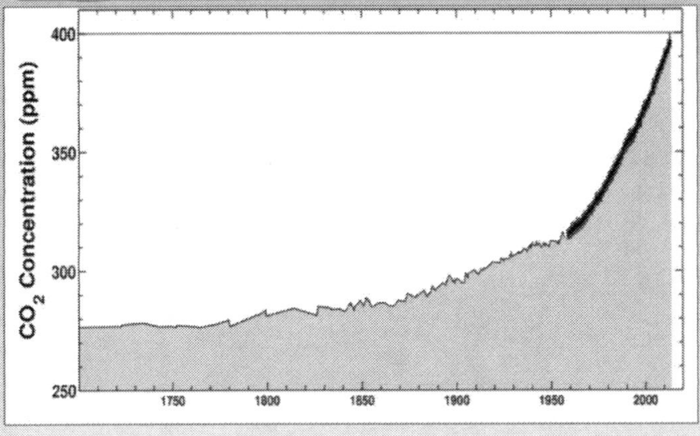

- Atmospheric CO_2 levels rising since start of industrial era.
- Present concentrations of 400 ppm CO_2 close to 2050 target of 450 ppm.

CO2 Abatement Targets

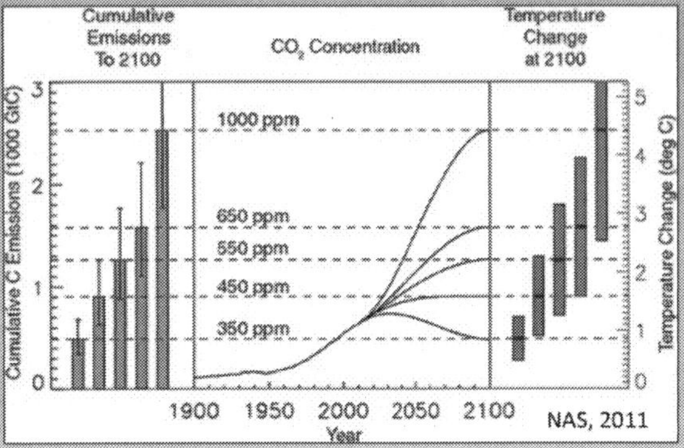

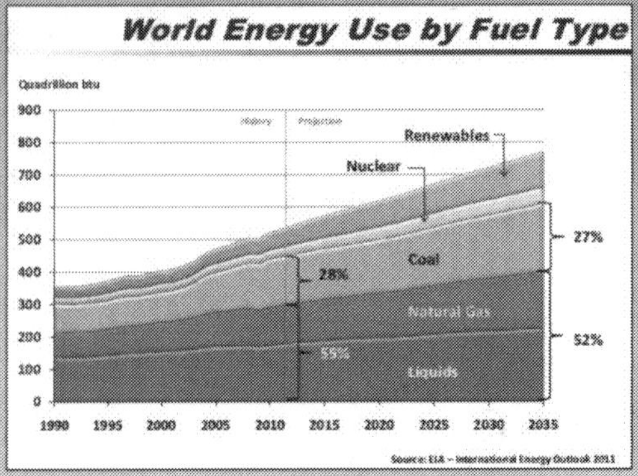

Figure 141. World energy-related carbon dioxide emissions by fuel type, 1990-2040
billion metric tons

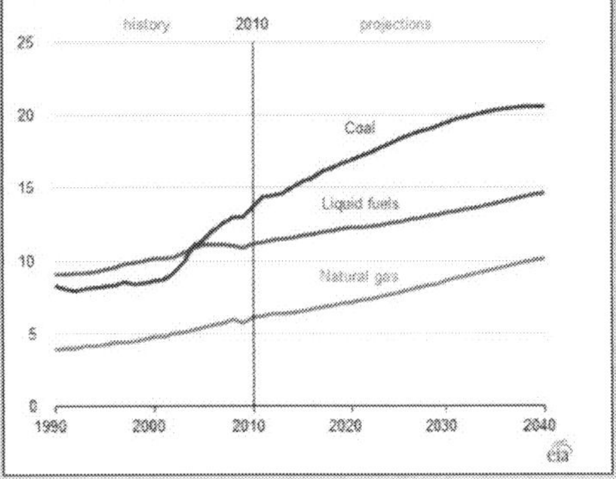

- **With business as usual, CO_2 emissions by 2035 will have reached the threshold for limiting temperature rise to 2°C this century**
- **Need to reduce emissions in the 100s-1000 GT range in this century to meet atmospheric CO_2 and temperature targets**

Geologic Sequestration Necessary to Meet Emission Targets

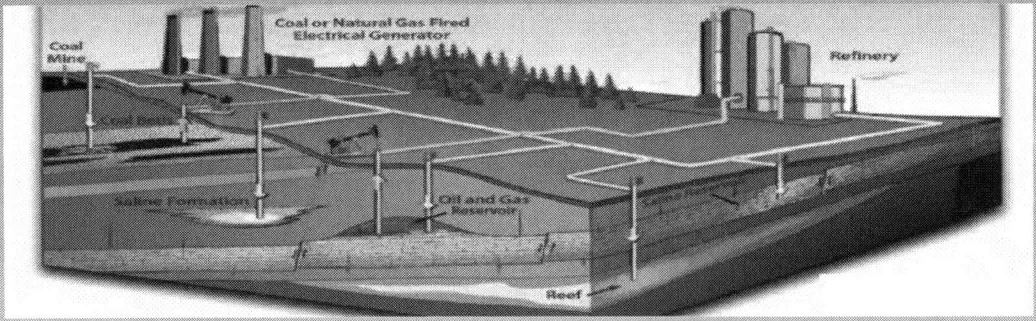

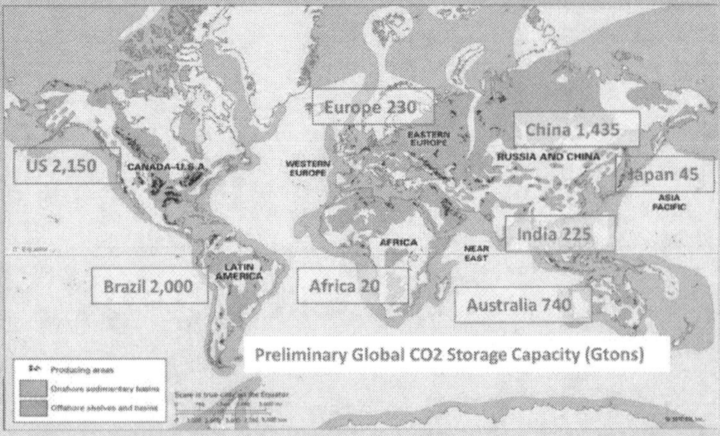

Challenges for Scaling Up CCS

- **Carbon Capture and Storage Costs**
- **Inadequate Transportation Network**
- **Storage Uncertainties including Seismic Risk and Injectivity**
- **Regulatory and Legal Issues**

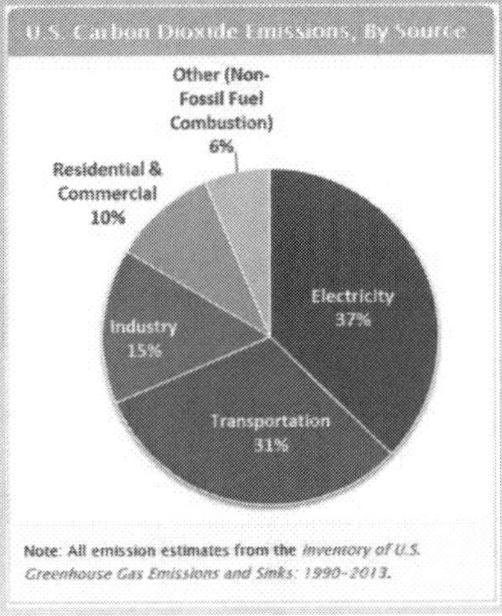

Carbon Capture Technologies

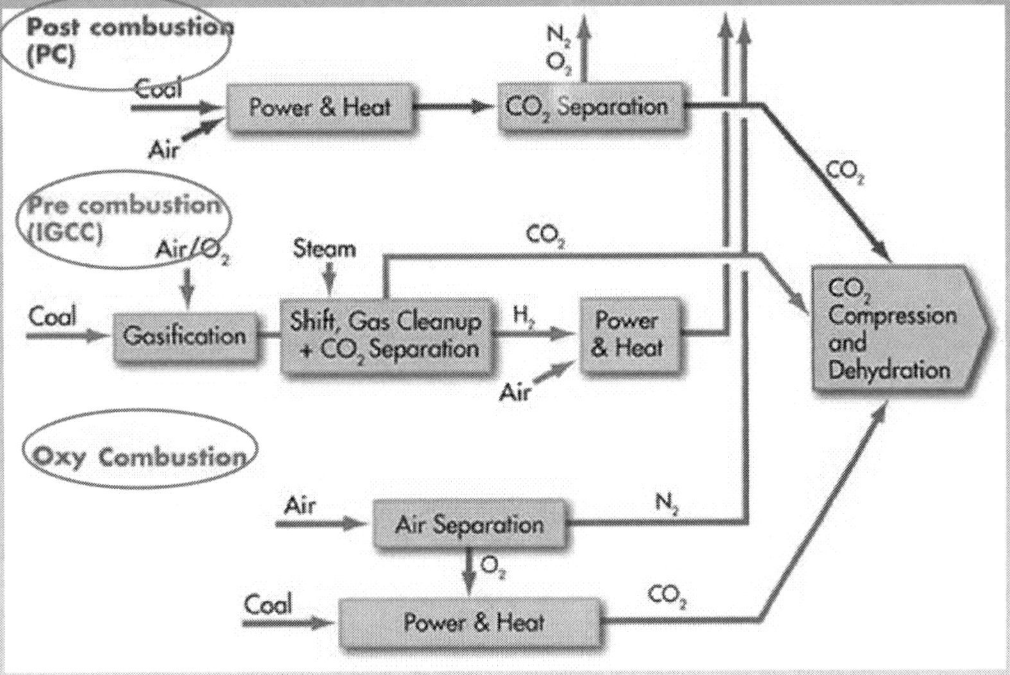

Emissions Rate by Plant Type and Capacity

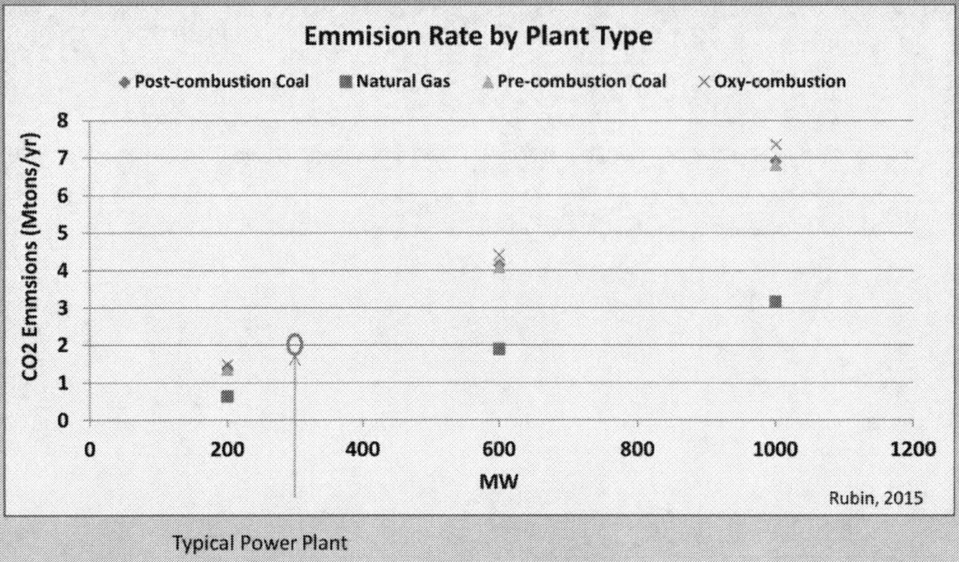

- Most coal-based plants have similar emission rates
- A typical 300 MW coal-based power plant emits ~ 2 $Mtco_2$/year
- Natural gas-based plants emit approximately half the CO_2 of coal-based plant

Cost of Carbon Capture by Plant Type

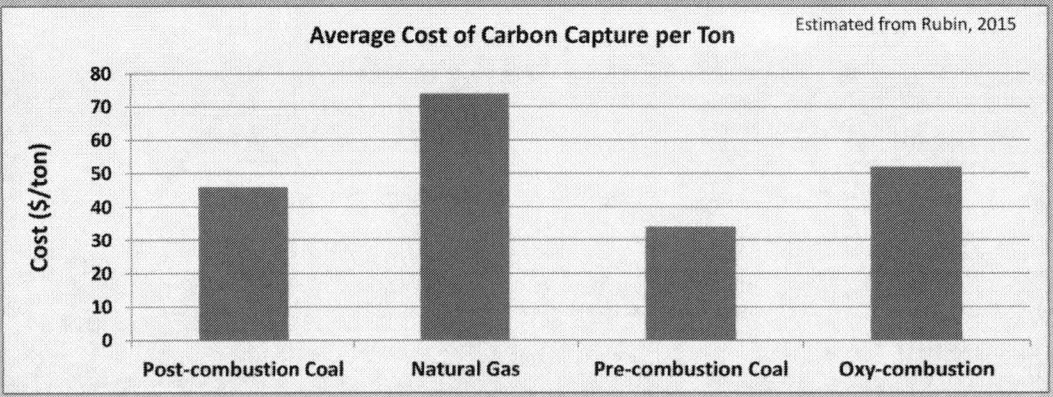

- For a typical 300 MW plant that emits emits 2 $Mtco_2$/year – annual abatement cost ~ $80M/yr

Carbon Capture and Storage Cost Components

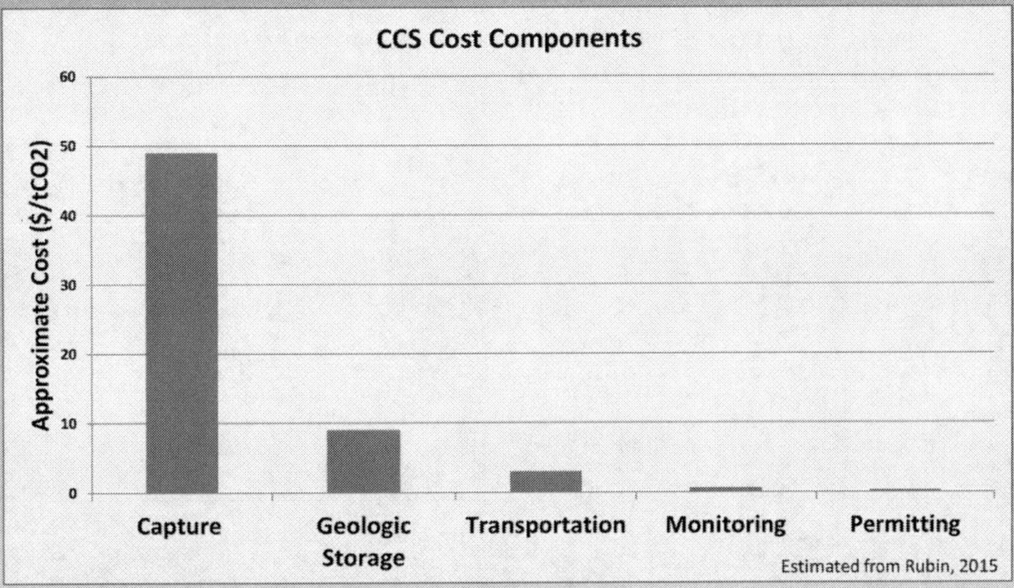

- **EOR credits ~ $15 - $40 per ton depending on oil price (Rubin, 2015)**

Cost Mitigation by Enhanced Oil Recovery (EOR)

Affordable cost for EOR ~ 2% of oil price in $/bbl. At $100,= $36/tCO2, at $50=$18.

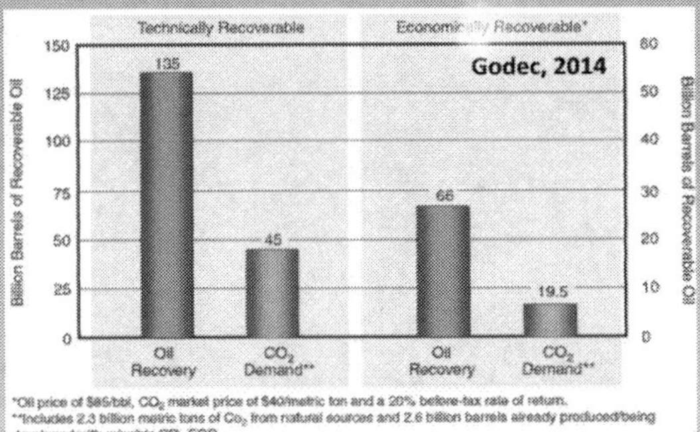

*Oil price of $85/bbl, CO_2 market price of $40/metric ton and a 20% before-tax rate of return.
**Includes 2.3 billion metric tons of CO_2 from natural sources and 2.6 billion barrels already produced/being developed with miscible CO_2 EOR.

- Large quantities of CO_2 can be used in US for EOR operations

Region	CO_2 EOR Oil Recovery (Billions of barrels)	CO_2 Storage Capacity (Billions of metric tons)
1. Asia Pacific	47	13
2. Central & South America	93	27
3. Europe	41	12
4. FSU	232	66
5. Middle East/North Africa	595	170
6. NA/Other	38	11
7. North America/U.S.	177	51
8. South Africa/Antarctica	74	21
Total	1,297	370

- ~ 370 GT can be stored globally in association with EOR

Godec, 2014

Plan Construction Cost

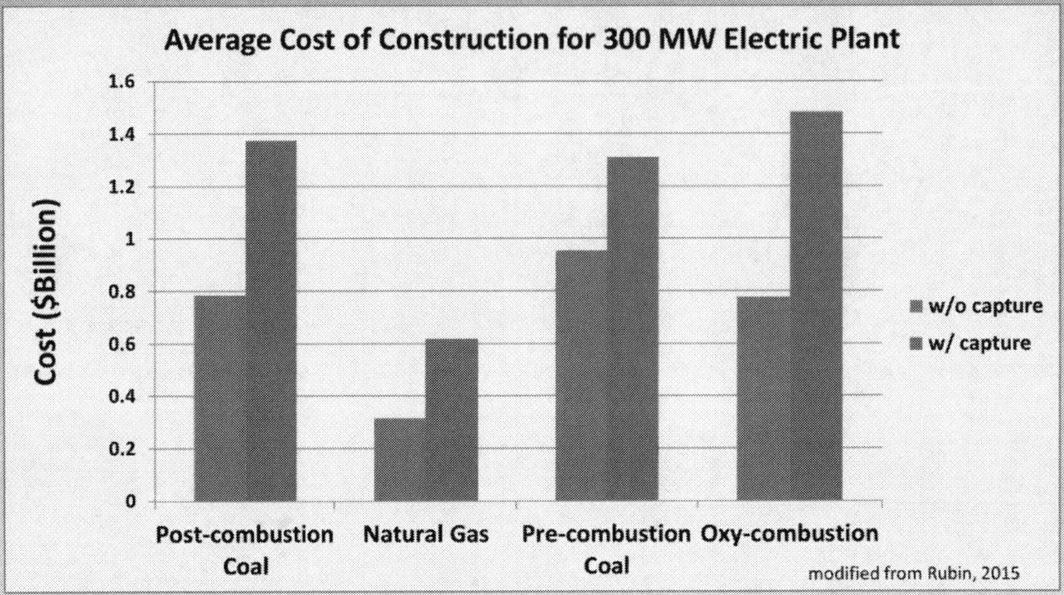

Levelized Cost of Electricity

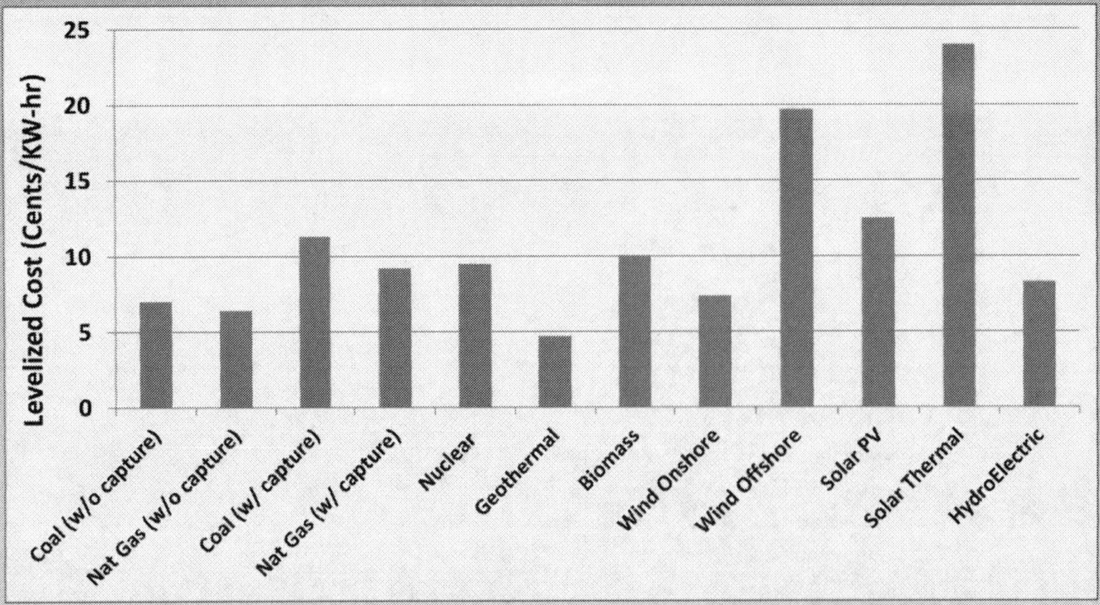

- **Coal plants competitive even with capture and storage costs**

CCS More Competitive Overseas

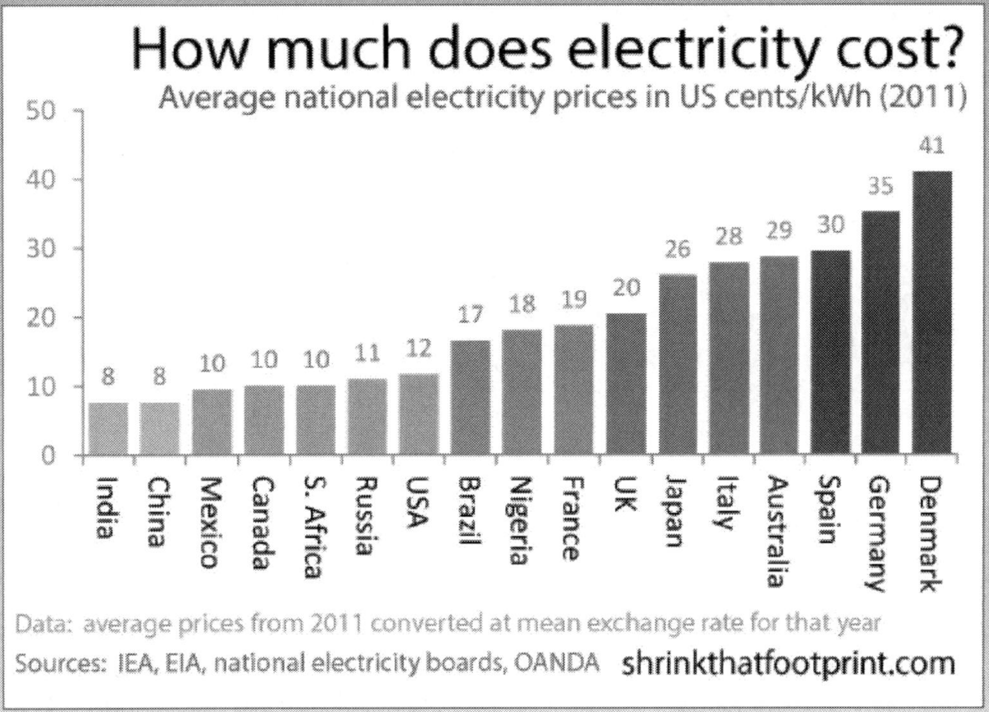

Further Incentive for CSS Adoption and Lowering of Cost with Carbon Credits

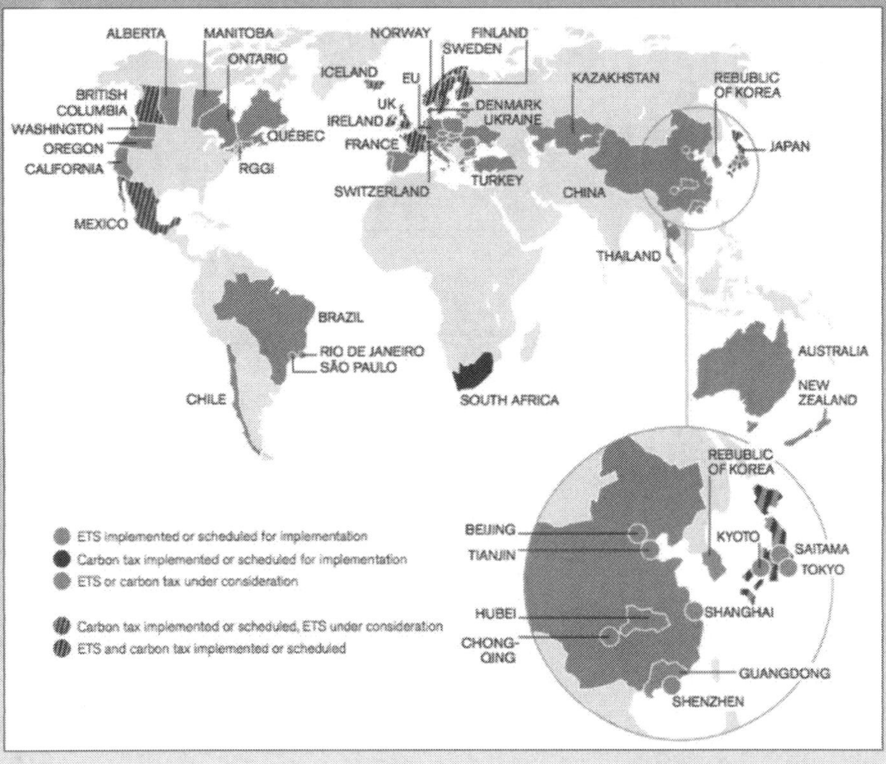

Inadequate CO_2 Pipeline Network

Uncertainty in National Carbon Policy Affecting Coal Production

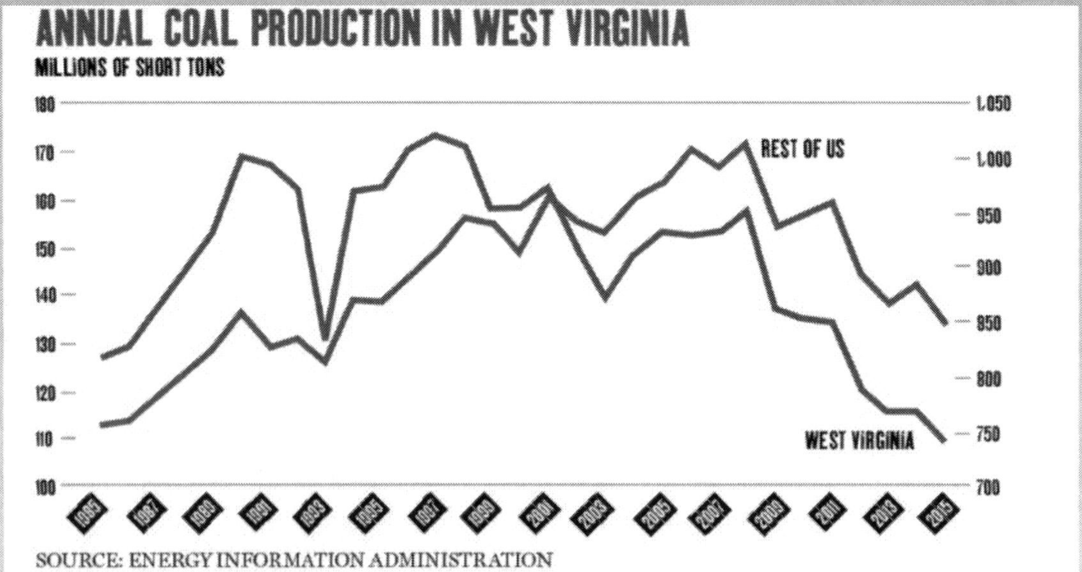

SOURCE: ENERGY INFORMATION ADMINISTRATION

Uncertainty in Carbon Policy Causing Fiscal Distress for Coal Miners

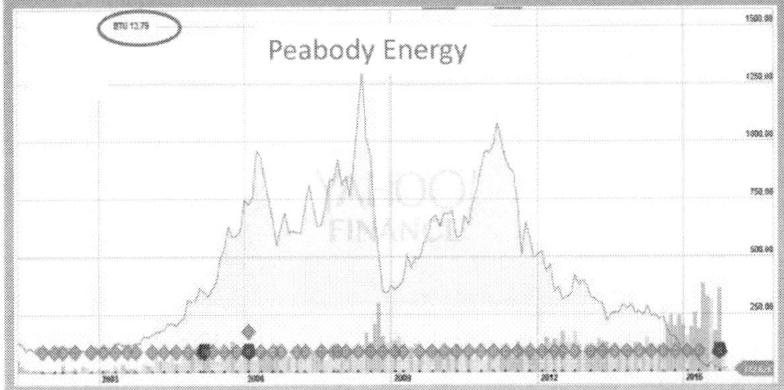

- **26 coal mining companies have filed for bankruptcy protection in the past 5 years**

Arch Coal expects to file for bankruptcy protection within months

REUTERS November 9, 2015 6:14 PM

* Out-of-court restructuring deal unlikely - sources

* Filing could take place before Dec. 15 bond payment

Storage Risk – The Case of Kansas

- **Kansas was assumed to lie in a seismically benign area**

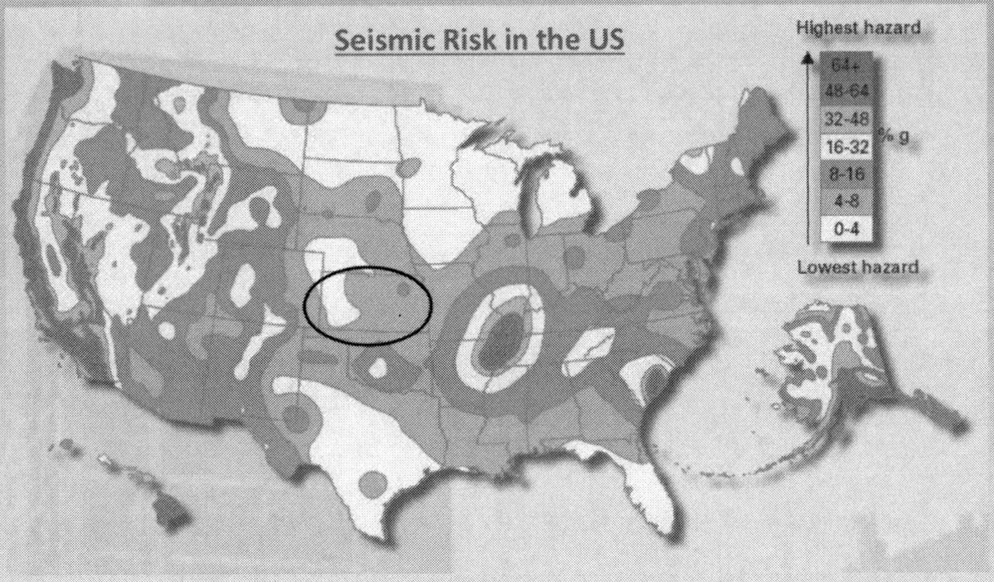

Earthquake Trends in Southern Kansas

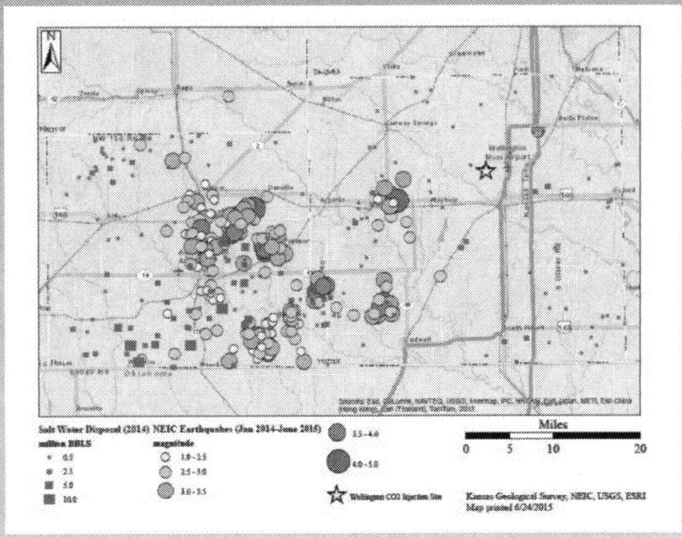

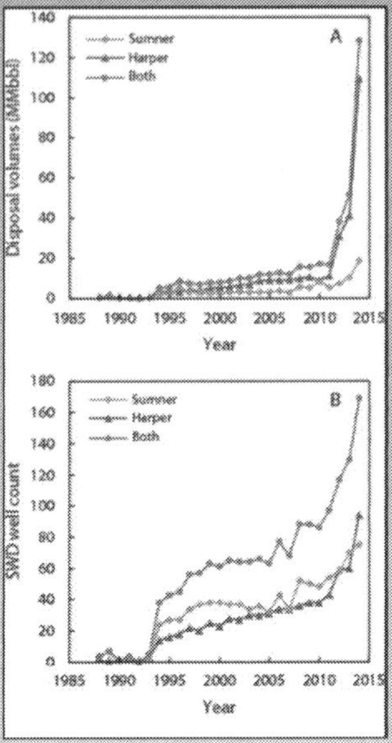

- **Large earthquakes (> M3.8) in past year associated with waste water injection in saline aquifer being considered for CO_2 injection**
- **EPA now requires seismic risk assessment at Class VI injection well sites**

Seismic Trends and Implications for CSS

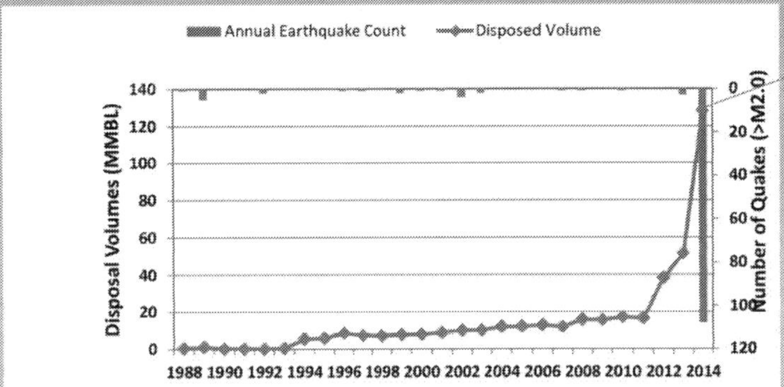

Equivalent to 16 Mtons of CO_2 (~ 8 years of CO2 emissions from a 300 MW electric power plant

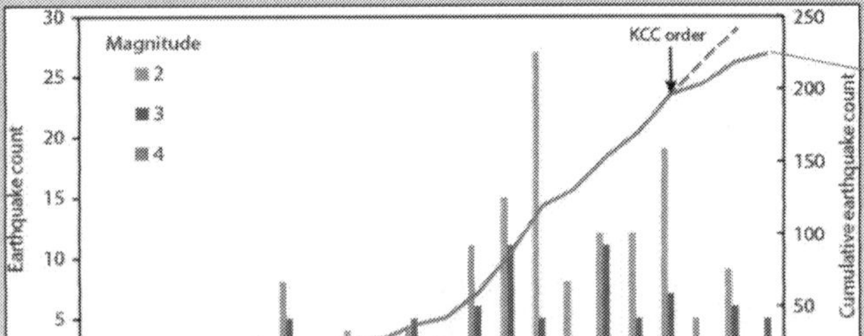

- **Some abatement in seismicity following restrictions on injection rates and volumes in Kansas**

Earthquake Trends in Central and Eastern US

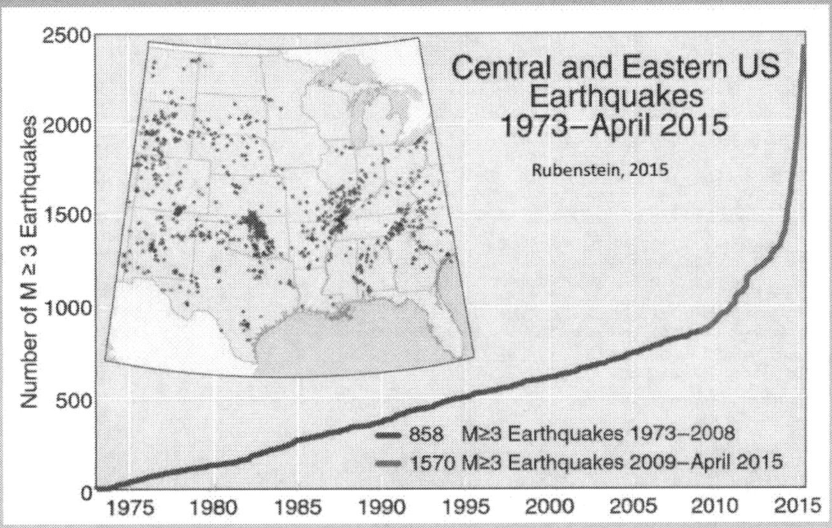

- **Induced seismicity linked to waste disposal in saline aquifers being considered for CO_2 storage**

Induced Seismicity - Physical Mechanisms

Subsurface Stress Field

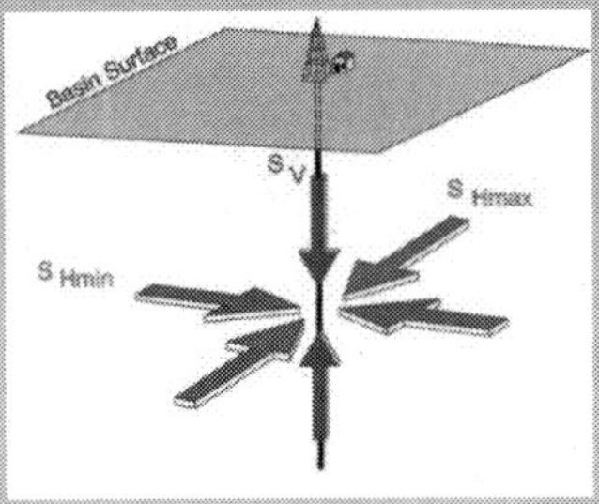

Stresses on Fault Plane

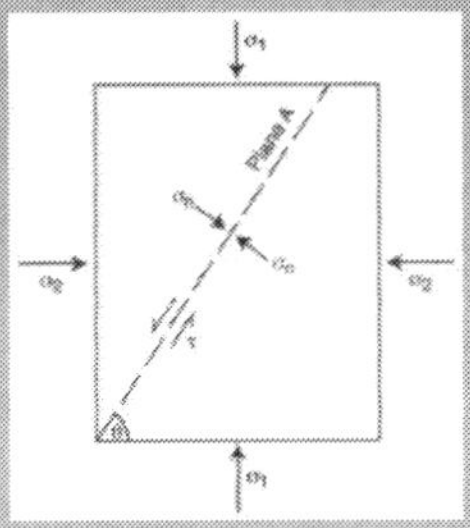

Effect of Injection on Fault Stability

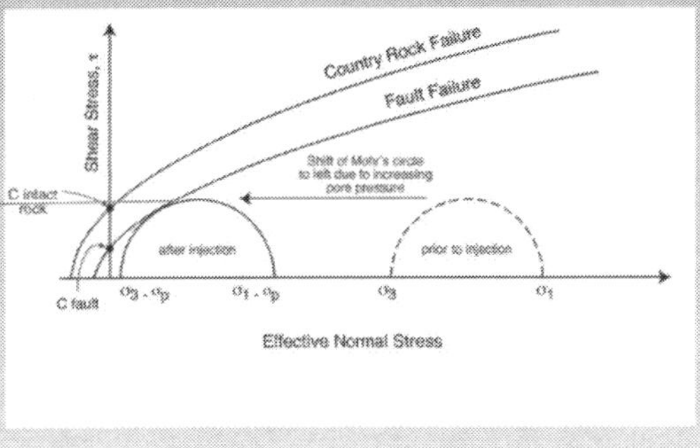

Relationship Between Earthquake Magnitude and Infrastructure Damage

Richter Magnitudes	Description	Earthquake Effects	Frequency of Occurrence
Less than 2.0	Micro	Micro-earthquakes, not felt.	About 8,000 per day
2.0-2.9	Minor	Generally not felt, but recorded.	About 1,000 per day
3.0-3.9	Minor	Often felt, but rarely causes damage.	49,000 per year (est.)
4.0-4.9	Light	Noticeable shaking of indoor items, rattling noises. Significant damage unlikely.	6,200 per year (est.)
5.0-5.9	Moderate	Can cause major damage to poorly constructed buildings over small regions. At most slight damage to well-designed buildings.	800 per year
6.0-6.9	Strong	Can be destructive in areas up to about 160 kilometres (100 mi) across in populated areas.	120 per year
7.0-7.9	Major	Can cause serious damage over larger areas.	18 per year
8.0-8.9	Great	Can cause serious damage in areas several hundred miles across.	1 per year
9.0-9.9	Great	Devastating in areas several thousand miles across.	1 per 20 years
10.0+	Epic	Never recorded; see below for equivalent seismic energy yield.	Extremely rare (Unknown)

Relationship Between Fault Plane Area and Earthquake Magnitude

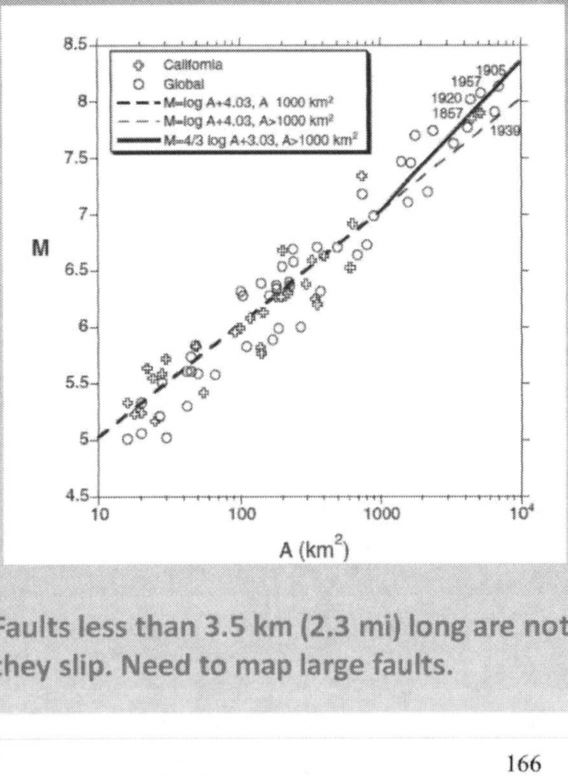

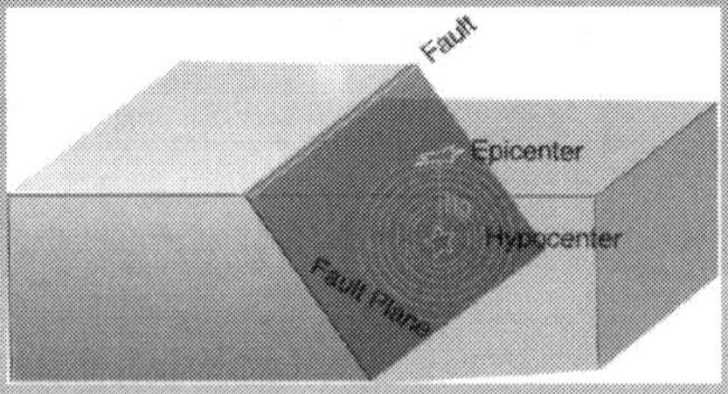

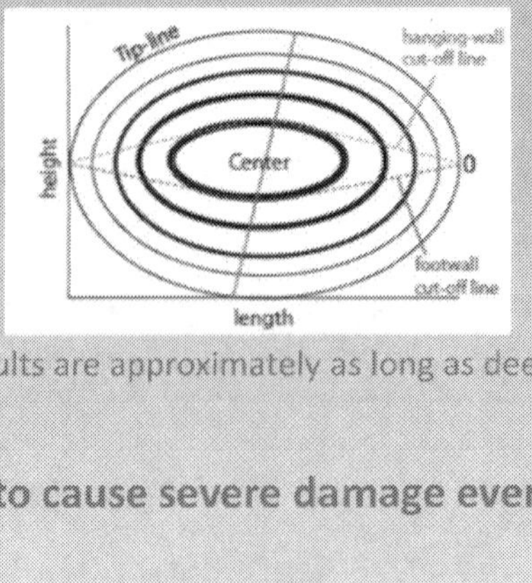

Faults are approximately as long as deep

- **Faults less than 3.5 km (2.3 mi) long are not likely to cause severe damage even if they slip. Need to map large faults.**

Inadequate Fault Mapping

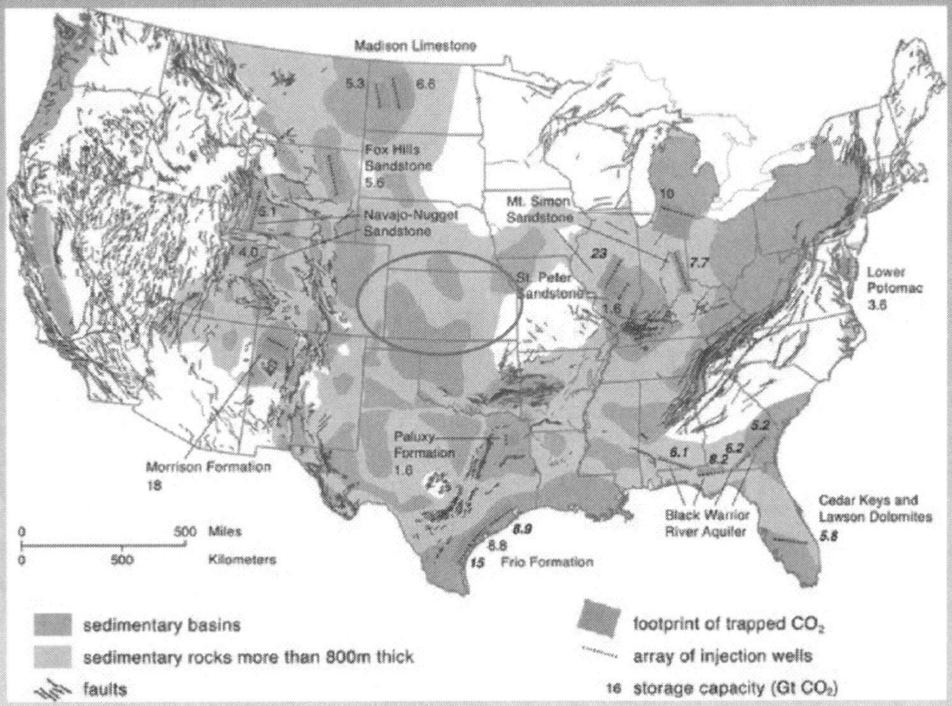

- **Faults in naturally dormant areas not adequately mapped.**

Wellington CO_2 Sequestration and EOR Site

- **Approximately 26,000 tons to be injected in the Arbuckle aquifer for CO_2 sequestration and 26,000 tons in the overlying Mississippian reservoir for EOR.**

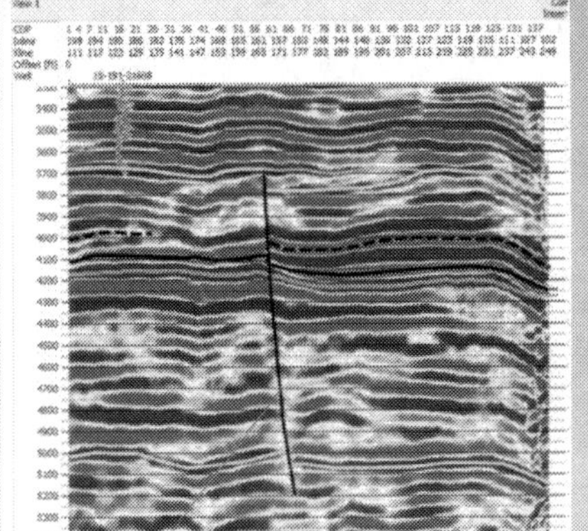

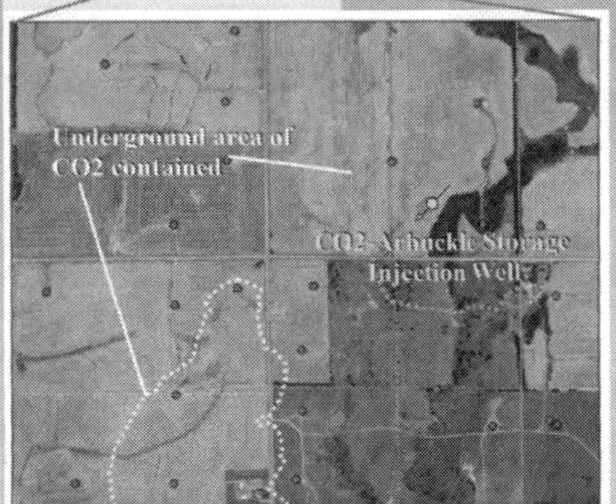

- **Extensive data acquisition required to identify faults and assess seismic risk**

Regulatory Issues

- **US EPA Class VI permit required to inject:**
  - Goals
    - a) prevent CO_2 migration into Underground Sources of Drinking Water (USDW) – TDS > 10,000 ppm (substantially higher than drinking water limit of 500 ppm)
    - b) protect caprock Integrity
    - c) prevent earthquakes
  - Expensive multi-year process requiring deep drilling, testing, and analysis to characterize the formation and perform corrective action on wells within the EPA Area of Review (AoR)
  - 50 years post-injection monitoring period
  - Between \$8 -\$70 million financial assurance (bond/insurance, @ 5% cost could be as high as \$3.5 million annually)
  - Guidelines required for defining aquifer – 2 GPD/day yield threshold for well quite restrictive
  - Fracture gradient restriction fairly stringent

Maximum Allowable (Fracture-Based) Increase in Pore Pressure (psi)

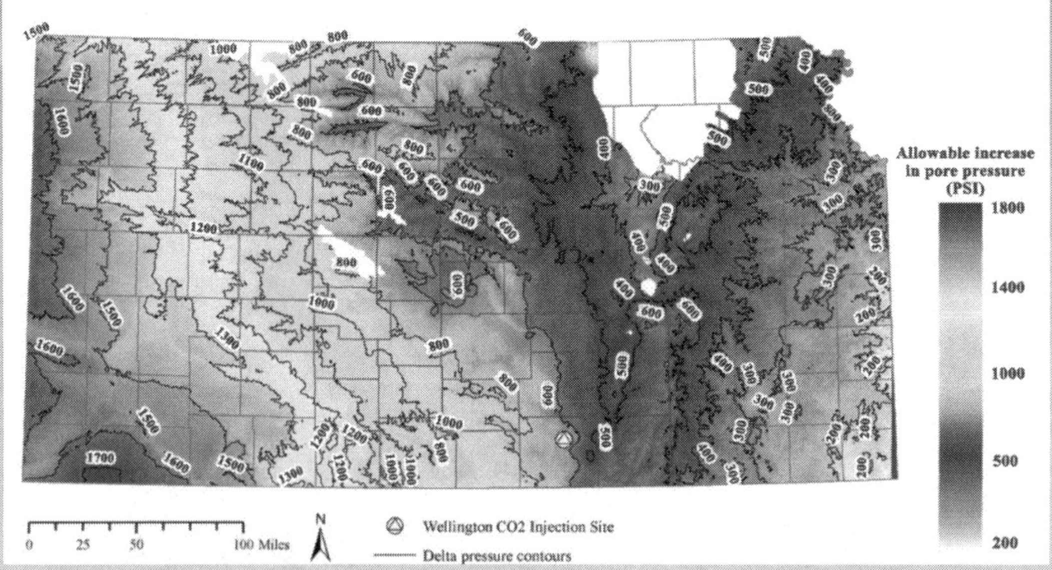

Simulated increase in pore pressures due to injection of 12 million tons/year of CO_2 over a 50 year period at 10 targeted sites in Kansas

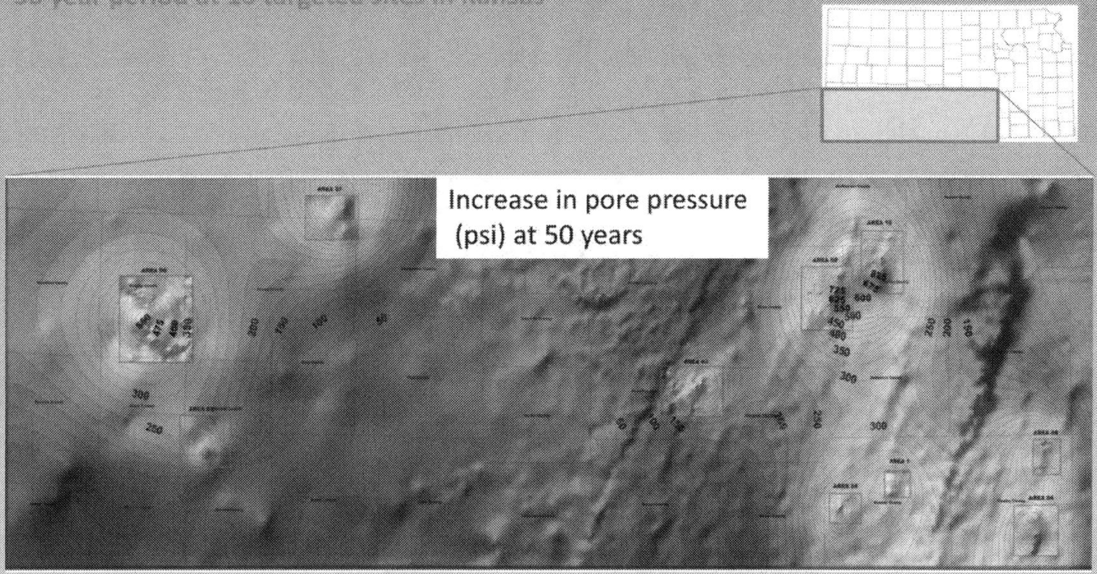

Williams, Gerlach, Fazelalavi, Doveton, et al., 2014

- Pressures greater than 700 psi can occur in the injection zone

Maximum Allowable Fracture-Based Increase in Pore Pressure

- May not be feasible to inject in large areas due to fracture gradient requirement even in the presence of tight caprock and confining layers within the injection zone

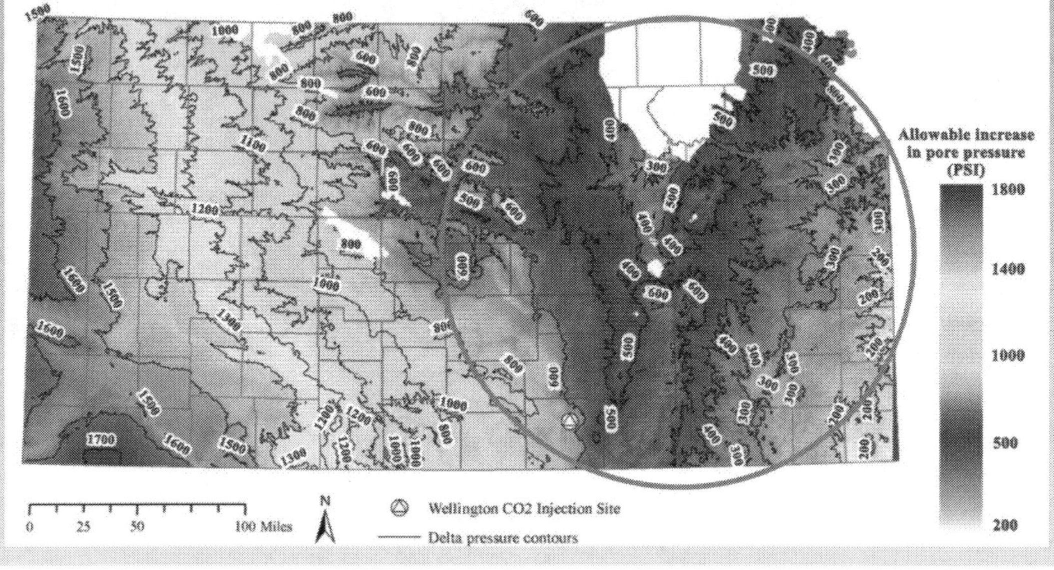

Effect of Stratification on Plume Migration and Pressure Distribution At Wellington CO2 Storage Site

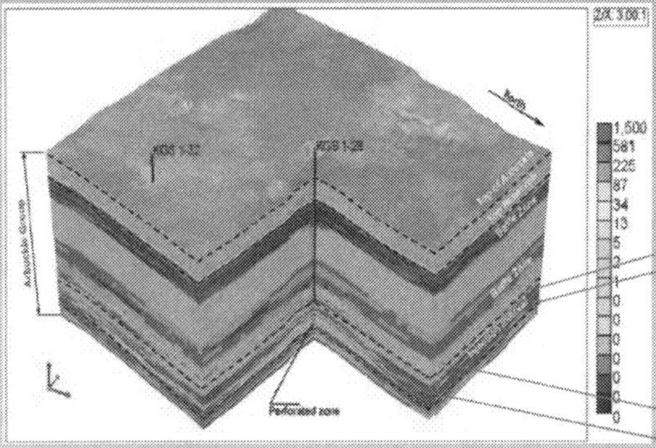

Holubnyack, Rush, Fazelalavi (KGS)

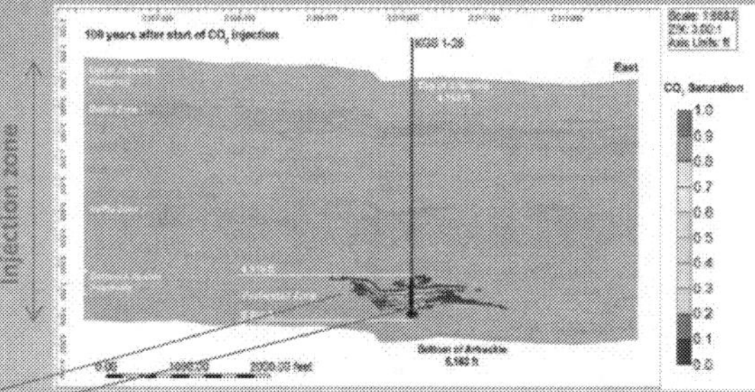

Entry pressure in caprock is 955 psi

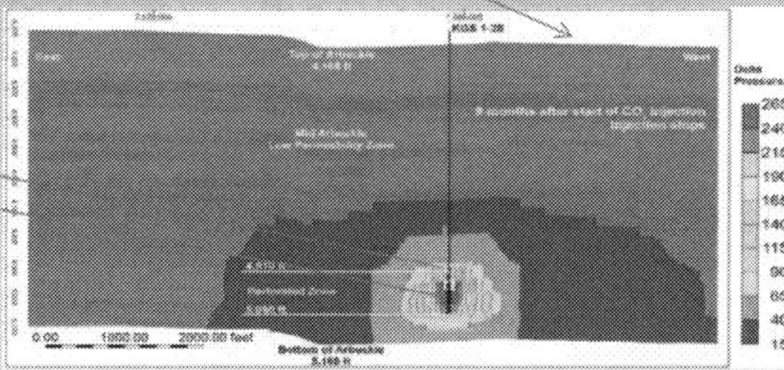

- CO_2 to remain deep in the injection zone. Negligible pressure increase at base of caprock

Simulated Sequestered Volumes of CO_2 in Kansas

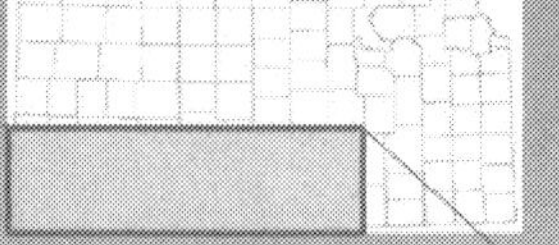

- Total sequestered volume over 50 year period ~ 0.65 GT (almost a decade of CO2 emissions in Kansas).

Other Outstanding Issues Hindering CSS Adoption

- Lack of formal national/global carbon emissions policy
- Creation of an industry-financed trust fund
- Adoption of substantive or procedural limitations on claims
- Laws and regulations regarding ownership of pore space and long-term stewardship
- Standards to account for cross-border movement of stored CO_2
- Negative public perception

NETL

Driving Innovation ♦ Delivering Results

U.S. DOE Carbon Capture R&D Program: Advancing Multiple Generations of Carbon Capture Solutions

José D. Figueroa
Carbon Capture Coordinator - Capture Division
National Energy Technology Laboratory

U.S. DEPARTMENT OF ENERGY | National Energy Technology Laboratory

Presentation Overview

NETL

- **Challenges to GHG Abatement Exist**
- **Successful implementation of GHG abatement strategies requires:**
  - Public/Private Partnerships
  - Efficient utilization of resources for an existing and growing CCS market
- **Prospective 2nd Generation technologies are moving to pilot scale, but require market-based support**
  - The search continues for a suite of economically sustainable solutions

U.S. DEPARTMENT OF ENERGY | National Energy Technology Laboratory 2

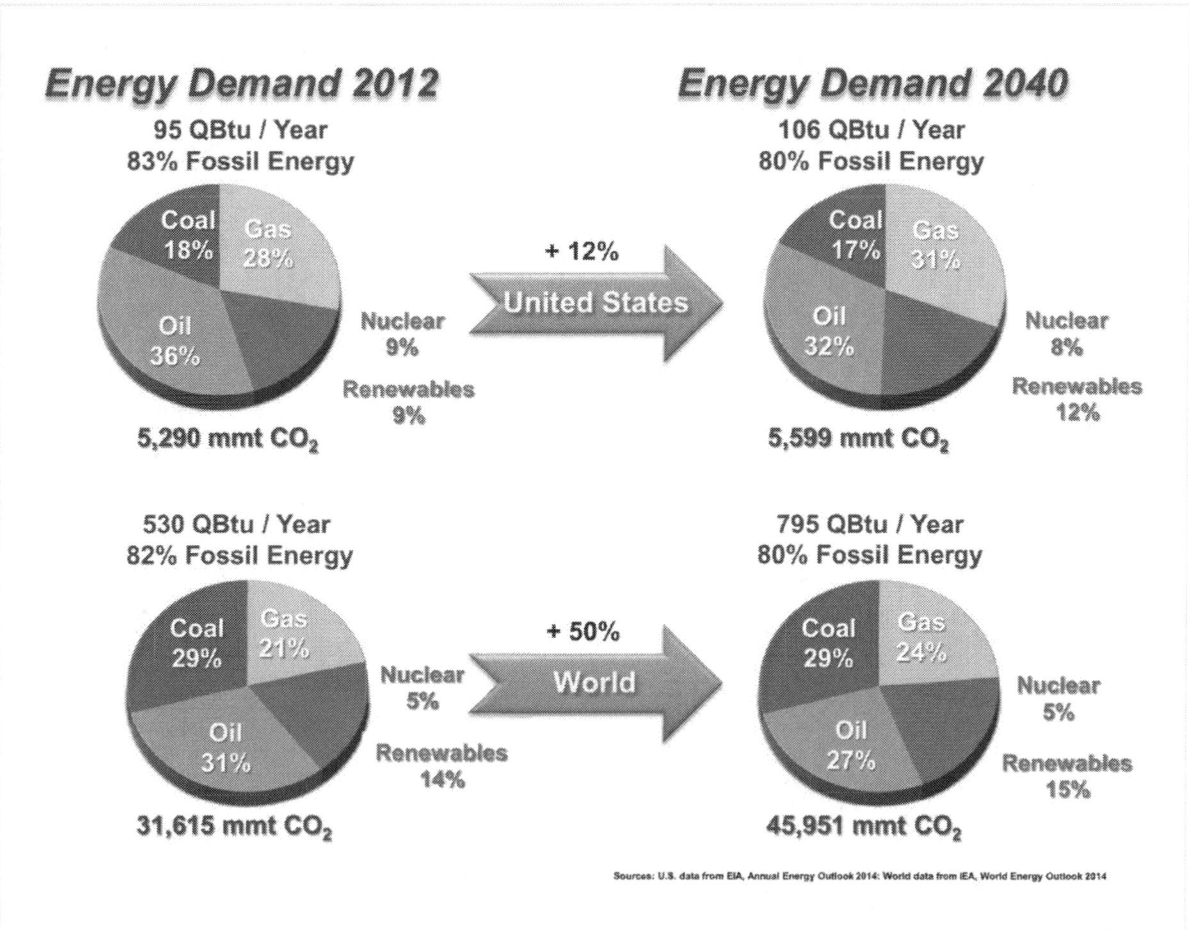

Energy Demand 2012
95 QBtu / Year
83% Fossil Energy
Coal 18%
Gas 28%
Oil 36%
Nuclear 9%
Renewables 9%
5,290 mmt CO_2
+ 12%
United States
Energy Demand 2040
106 QBtu / Year
80% Fossil Energy
Coal 17%
Gas 31%
Oil 32%
Nuclear 8%
Renewables 12%
5,599 mmt CO_2
530 QBtu / Year
82% Fossil Energy
Coal 29%
Gas 21%
Oil 31%
Nuclear 5%
Renewables 14%
31,615 mmt CO_2
+ 50%
World
795 QBtu / Year
80% Fossil Energy
Coal 29%
Gas 24%
Oil 27%
Nuclear 5%
Renewables 15%
45,951 mmt CO_2
Sources: U.S. data from EIA, Annual Energy Outlook 2014; World data from IEA, World Energy Outlook 2014

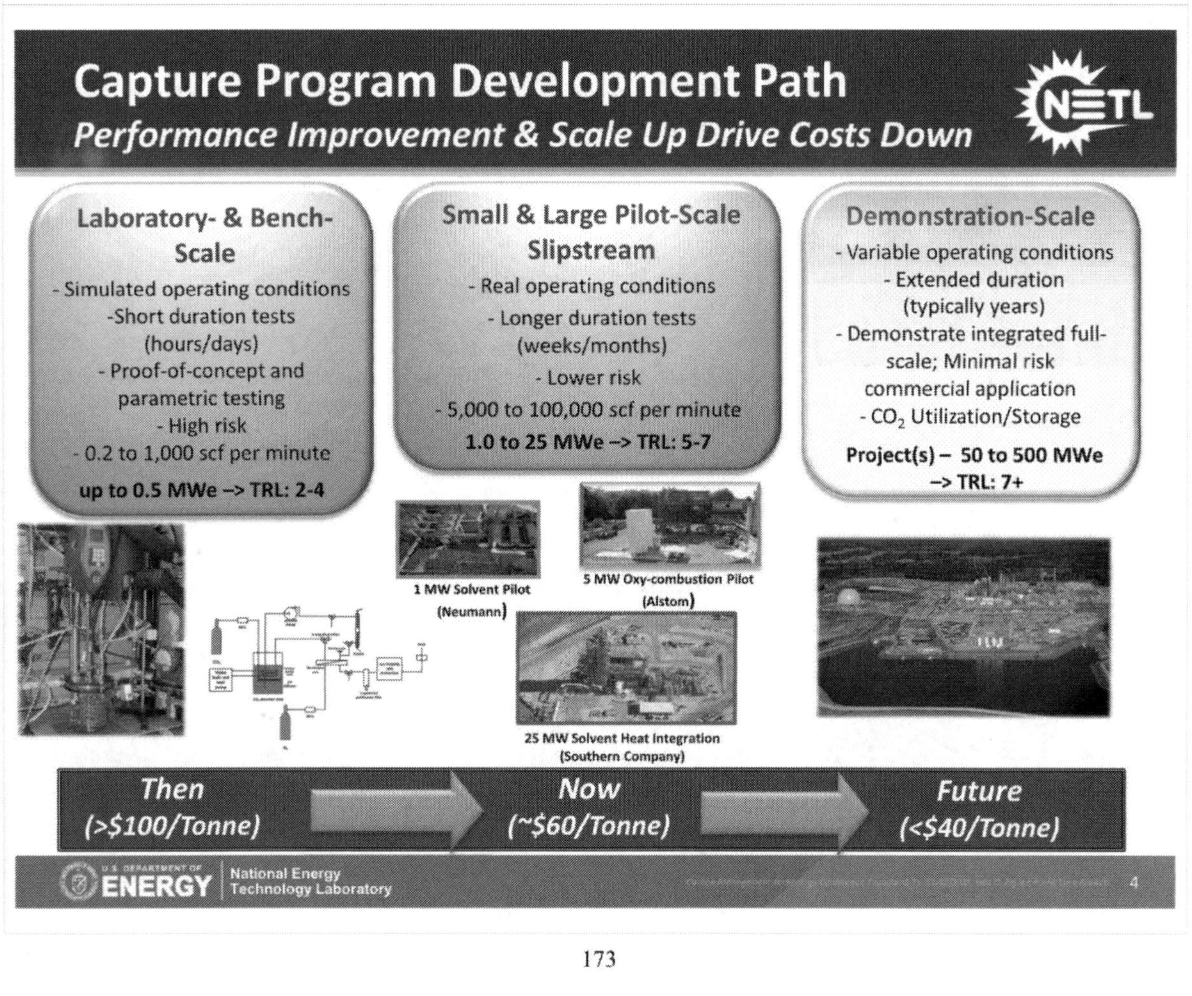

Capture Program Development Path
Performance Improvement & Scale Up Drive Costs Down
NETL
Laboratory- & Bench-Scale
- Simulated operating conditions
-Short duration tests (hours/days)
- Proof-of-concept and parametric testing
- High risk
- 0.2 to 1,000 scf per minute
up to 0.5 MWe –> TRL: 2-4
Small & Large Pilot-Scale Slipstream
- Real operating conditions
- Longer duration tests (weeks/months)
- Lower risk
- 5,000 to 100,000 scf per minute
1.0 to 25 MWe –> TRL: 5-7
Demonstration-Scale
- Variable operating conditions
- Extended duration (typically years)
- Demonstrate integrated full-scale; Minimal risk commercial application
- CO_2 Utilization/Storage
Project(s) – 50 to 500 MWe –> TRL: 7+
1 MW Solvent Pilot (Neumann)
5 MW Oxy-combustion Pilot (Alstom)
25 MW Solvent Heat Integration (Southern Company)
Then (>$100/Tonne)
Now (~$60/Tonne)
Future (<$40/Tonne)
U.S. DEPARTMENT OF ENERGY
National Energy Technology Laboratory
4

Approaches to CO_2 Capture

NETL

Solvents

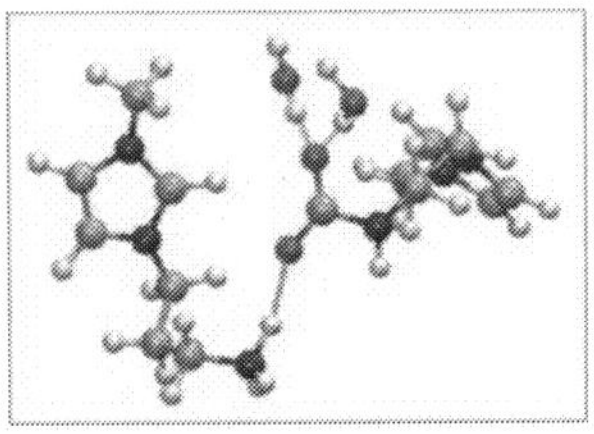

Sorbents

Membranes

Novel Concepts

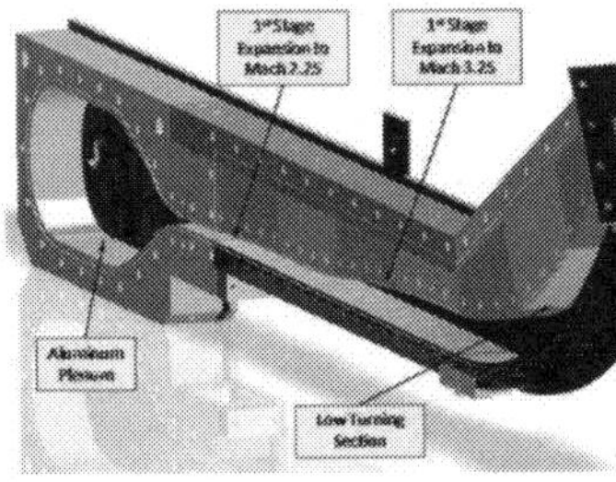

R&D Areas: CO_2 Capture

NETL

Pre-Combustion

- ☐ *Solvents*
- ☐ *Sorbents*
- ☐ *Membranes*
- ☐ *Novel Concepts*

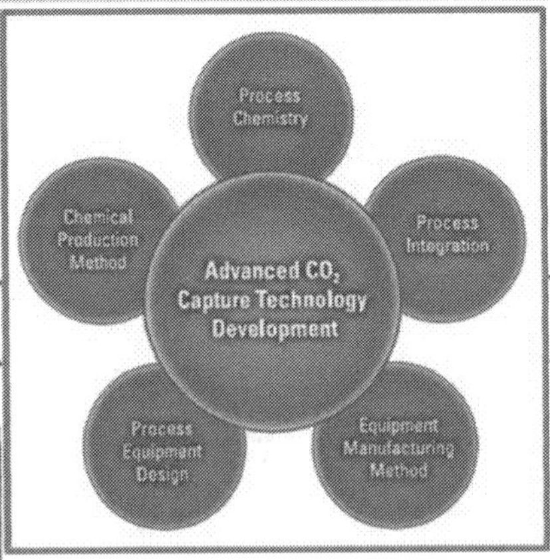

1st-Generation Technologies—*include technology components that are being demonstrated or that are commercially available.*

2nd-Generation Technologies—technologies currently in R&D scheduled to become available for large-scale testing around 2020 and available for deployment in the 2025 timeframe.

Transformational Technologies—*emerging technologies in early stages of development that offer the potential for "game-changing" improvements in cost and performance forecast to be available for large-scale testing around 2025 and available for deployment in the 2030 timeframe.*

Post-Combustion

- ☐ *Solvents*
- ☐ *Sorbents*
- ☐ *Membranes*
- ☐ *Novel Concepts*

Post-Combustion Research Focus

NETL

Key Technologies	Research Focus
Solvents	• Low-Cost, Non-Corrosive Solvents with High CO_2 Loading Kinetics, Low Capacity, Improved Reaction Regeneration Energy, and Degradation Resistance • Process Intensification/Heat integration • High Performance Functionalized Solvents • Catalyzed Absorption • Phase-Change Solvents • Hybrid Systems • Cryogenic Capture
Sorbents	• Low-Cost Base Materials, Thermal and Chemical Stability, Low Attrition Rates, Low Heat Capacity, High CO_2 Adsorption Capacity and High CO_2 Selectivity • Process Intensification/Heat integration • Novel Processes Equipment and Configurations • Structured Solid Adsorbents (eg., MOFs) • Hybrid Systems • Enhanced PSA/TSA
Membranes	• Low-Cost, Durable Membranes with Improved Permeance, Selectivity, Thermal and Physical Stability, and Tolerance to Flue Gas Contaminants • Hybrid systems • Novel Process Conditions • Nano-materials
Novel Concepts	• Supersonic Shockwave CO_2 Compression • Hybrid Systems • Cryogenic Capture

• **2nd Generation Technology** • Transformational Technology

U.S. DEPARTMENT OF ENERGY | National Energy Technology Laboratory

7

Pre-Combustion Research Focus

NETL

Key Technologies	Research Focus
Solvents	• Advanced Regeneration Process to Produce a High-Pressure CO_2 Stream • Increased Selectivity for Maximal H_2 Recovery • High temperature operation to maintain warm syngas • Dual Swing Absorption/Regeneration Cycles • Hybrid Systems
Sorbents	• Cyclic PSA Producing High-Pressure H_2 and CO_2 • WGS/CO_2 Separation Process intensification for High Efficiency Impact • Hybrid Systems
Membranes	• Membrane Materials: High-Temperature Polymer, Dual-Phase Carbonate-Ceramic, Pd, and others • Silica Molecular Sieve • Gas/Liquid Contactor • WGS/CO_2 Separation Process intensification for High Efficiency Impact • High Density and Pressure Nano-Scale Membranes • High-temperature/high-pressure seals • Process Intensification • Hybrid Systems
Novel Concepts	• WGS/CO_2 Separation Process intensification for High Efficiency Impact • Hybrid Systems

• **2nd Generation Technology** • Transformational Technology

U.S. DEPARTMENT OF ENERGY | National Energy Technology Laboratory

8

Program Goals

Driving Down the Cost of Electricity of Coal Power with CCS

NETL

Cost of Electricity Reduction Targets

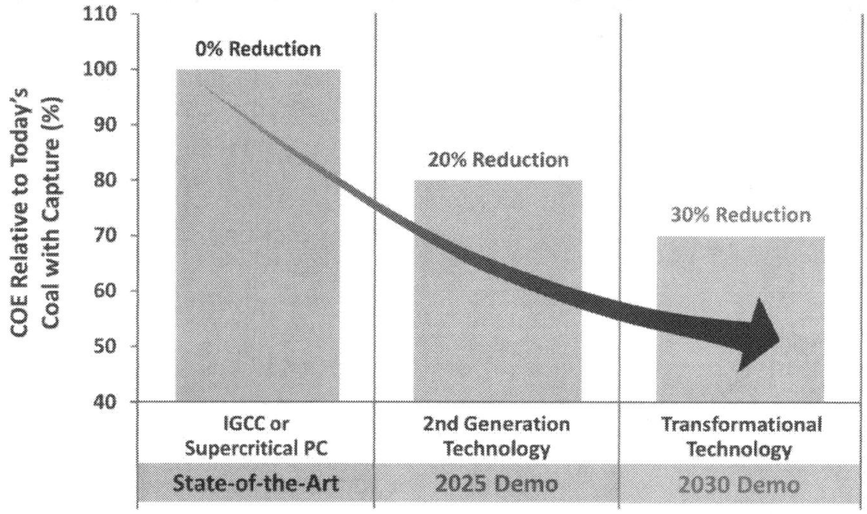

Goals are for greenfield plants. Costs include compression to 2215 psia, but exclude CO_2 transport and storage costs.

Capture Program: Active Portfolio Distribution

NETL

Program Area	Key Technology	Number of R&D Projects			Total	
		Lab/Bench	Small Pilot	Large Pilot		
Post-Combustion Capture	*Solvents*	12	4	6	**22**	
	Sorbents	8	2	1	**11**	
	Membranes	6	3	-	**9**	44
	Novel Concepts	2			**2**	
Pre-Combustion Capture	*Solvents*	-	1	-	**1**	
	Sorbents	-	1	-	**1**	9
	Membranes	5	-	-	**5**	
	Novel Concepts	2			**2**	
Compression	*Compression*	-	-	1	**1**	
Totals		**35**	**11**	**8**	**54**	

Carbon Capture Lab Scale Projects

Performer	Project Focus	Transformational
	Post-Combustion Solvents (4)	
PNNL - Pacific Northwest National Laboratory	CO2-Binding Organic Liquid (CO2BOL) Solvents	●
LLNL - Lawrence Livermore National Laboratory	Enhanced Manufacturing of Solvents	●
University of Notre Dame	Encapsulated Ionic Liquids	●
University of Illinois	Biphasic Absorption with Phase Separation	●
	Post-Combustion Sorbents (4)	
Texas A&M University	Amine-Incorporated Porous Polymer Networks (APPNS)	●
Georgia Tech Research Corporation	Novel Rapidly Cycled Pressure Swing Adsorption Process	●
Research Triangle Institute	Novel Hybrid Fluidizable Sorbents	●
Massachusetts Institute of Technology	Electrochemically Mediated Regeneration	●
	Post-Combustion Membranes (1)	
Gas Technology Institute	Go-Peek Hybrid Membranes	●
	Post-Combustion Novel Concepts (1)	
Liquid Ion Solutions LLC	Hybrid Advanced Membrane, Solvent System, and Process Integration	●
	Pre-Combustion (4)	
LANL - Los Alamos National Laboratory	Polymer-Based Membranes	
University of Southern California	A High Efficiency, Ultra-Compact Process for Pre-Combustion CO2 Capture	●
Southern Research Institute	Combined Sorbent/WGS with Integrated Heat Management	●
The Research Foundation of State University of New York	Sorption Enhanced Mixed Matrix Membranes for Hydrogen Purification and Carbon Dioxide Capture	●

Carbon Capture Bench Scale Projects (1)

Performer	Project Focus	Transformational
	Post-Combustion Solvents (8)	
URS Group Inc	Piperazine Solvent with Flash Regeneration	
William Marsh Rice University	Hybrid Absorption/Stripper Process	●
GE Global Research	Phase-Changing Absorbent	●
Research Triangle Institute	Nonaqueous Solvent	●
Akermin Inc.	Carbonic Anhydrase Catalyzed Advanced Carbonate Solvent	●
SRI International	Ammonia- and Potassium Carbonate-Based Mixed-Salt Solvent	●
University of Kentucky	Hybrid Membrane/Advanced Catalyst Solvent	●
Research Triangle Institute	Nonaqueous Solvent	●
	Post-Combustion Sorbents (4)	
Research Triangle Institute	Advanced Solid Sorbents and Processes for CO2 Capture	●
W. R. Grace & Co.	Rapid Pressure Swing Adsorption	●
ADA-ES Inc.	Cross-Heat Exchanger for Sorbent-Based CO2 Capture	
Aspen Aerogels Inc.	Advanced aerogel sorbents	●

Carbon Capture Bench Scale Projects (2)

NETL

Performer	Project Focus	Transformational
	Post-Combustion Membranes (5)	
Ohio State University Research Foundation	Inorganic/Polymer Composite Membrane	●
FuelCell Energy Inc. (FCE)	Electrochemical Membrane	●
Alliant Techsystems Operations LLC	Supersonic Inertial CO_2 Extraction System	●
American Air Liquide Inc.	Sub-Ambient Temperature Membrane	
American Air Liquide Inc.	Hollow Fiber Membrane Module	●
	Post-Combustion Novel Concepts (1)	
Membrane Technology & Research Inc.	Hybrid Membrane and Amine Absorption	
	Pre-Combustion (3)	
Media and Process Technology Inc.	Two-Stage Membrane Separation: Carbon Molecular Sieve Membrane Reactor followed by Pd-Based Membrane	●
SRI International	PBI Polymer Membrane	●
Arizona State University	Zeolite Membrane Reactor	●

Carbon Capture Small Pilot Projects

NETL

Performer	Project Focus	Scale	Cost	Construction
	Post-Combustion Solvents (5)			
Linde, LLC	Slipstream Novel Amine-Based Post-Combustion Process	1 MWe	$22.7M	Complete
Neumann Systems Group, Inc	Carbon Absorber Retrofit Equipment	0.5 MWe	$9.2M	Complete
University of Kentucky	Heat Integrated Post-combustion CO_2 Capture System Using the MHPSA Advanced Solvent	0.7 MWe	$21.4M	Complete
General Electric	Novel Aminosilicone Solvent	0.5 MWe	$6.3M	Late 2015
ION Engineering	Amine Solvent in Ionic Liquid	0.7 MWe	$10.9M	Complete
	Post-Combustion Sorbents (3)			
ADA-Environmental Solutions	Solid Sorbents as Retrofit Technology	1 MWe	$24.3M	Complete
TDA Research, Inc.	Alkalized Alumina Solid Sorbent	0.5 MWe	$5.9M	Late 2016
SRI International	Novel Solid Sorbent	1 MWe	$12.9M	Late 2016
	Post-Combustion Membranes (3)			
Membrane Technology & Research	Polymeric Membranes	1 MWe	$18.8M	Complete
Gas Technology Institute	Hollow-Fiber-Membrane Contactor with aMDEA Solvent	0.5 MWe	$12.8M	Late 2016
FuelCell Energy Inc.	Combined Electric Power and CO_2 Separation (CEPACS) System	3 MWe	$23.7M	TBD
	Pre-Combustion (2)			
SRI International	CO2 Capture Using AC-ABC Process	0.1 MWe	$6.1M	Complete
TDA, Inc.	High Capacity Regenerable Sorbent	0.1 MWe	$9.9M	Late 2015

Carbon Capture Large Pilot Projects

Performer	Project Focus	Proposed Scale
	Post-Combustion Solvents (6)	
Southern Company Services	Waste Heat Integration	25 MWe
University of Illinois	Linde/BASF CO_2 Capture Technology	25 MWe*
Alstom Power	Improvements to Chilled Ammonia Process	15 MWe*
Southern Company Services	Process Improvements/Advanced Solvent	25 MWe*
University of Kentucky	Heat Integrated Capture System	10 MWe*
General Electric	Novel Aminosilicone Solvent	10 MWe*
	Post-Combustion Sorbents (1)	
NRG CO_2NCEPT	VeloxoTherm™ Solid Sorbent Capture System	10 MWe*
	CO_2 Compression (1)	
Dresser-Rand	Supersonic CO_2 Compressor	TBD

** These projects are Phase 1 only. In FY2016, up to 2 projects will be selected for the full project (Phase 2).*

NETL's Office of Research Approach: Integrated Technology Development

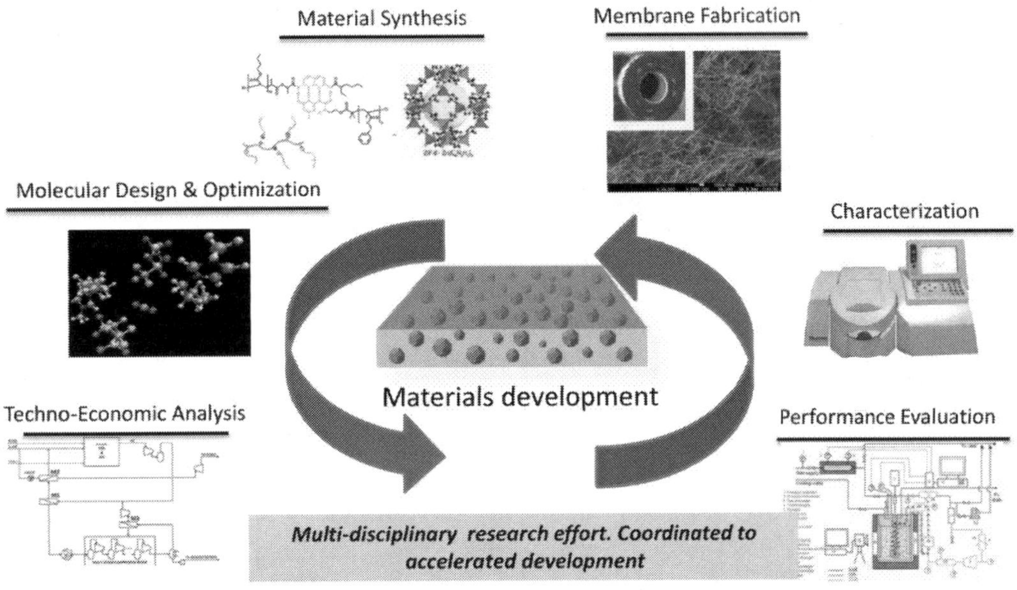

Looking Forward

- **2nd Generation:** Increased Focus on Pilot-Scale
- **Transformational:** Development at Laboratory/Bench Scale

Laboratory / Bench-Scale

PNNL - Pacific Northwest National Laboratory
CO2-Binding Organic Liquid (CO2BOL) Solvents

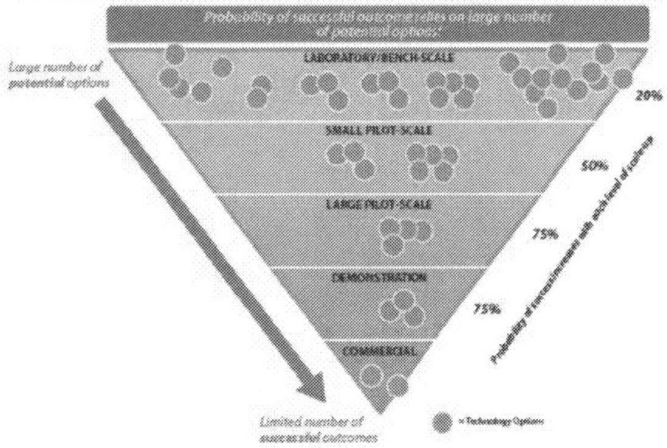

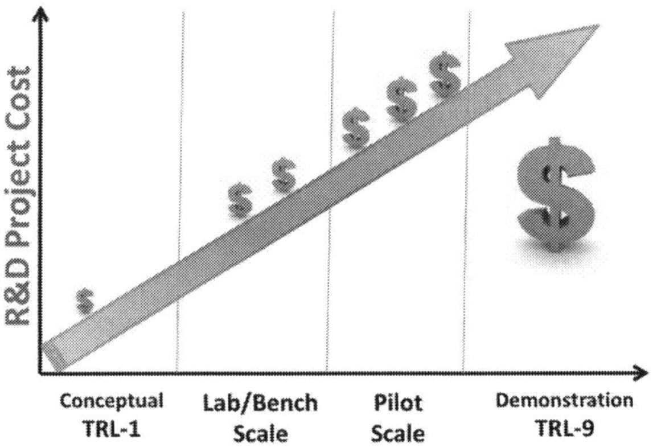

Looking Forward

- **2nd Generation:** Increased Focus on Pilot-Scale
- **Transformational:** Development at Laboratory/Bench Scale

Laboratory / Bench-Scale

LLNL - Lawrence Livermore National Laboratory
Enhanced Manufacturing of Solvents

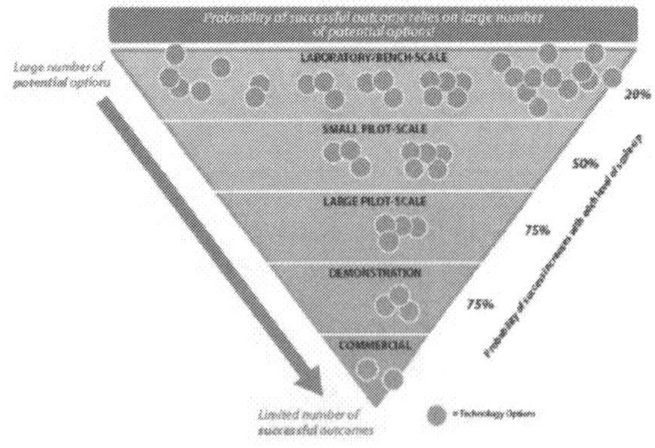

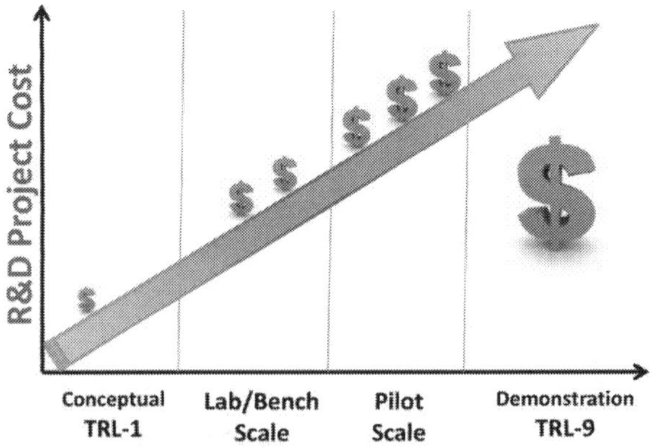

Looking Forward

- **2nd Generation:** Increased Focus on Pilot-Scale
- **Transformational:** Development at Laboratory/Bench Scale

Laboratory / Bench-Scale

University of Notre Dame
Encapsulated Ionic Liquids

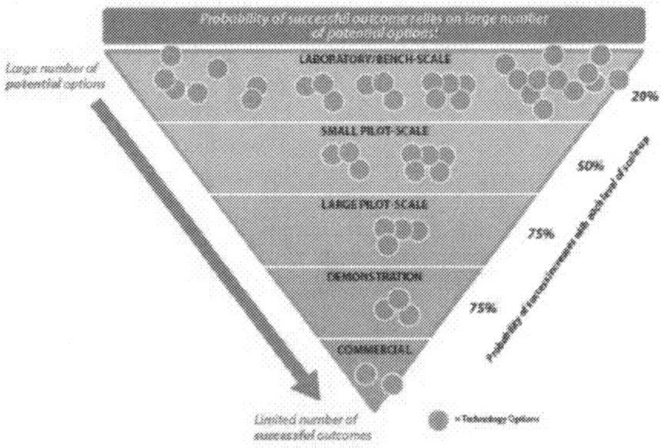

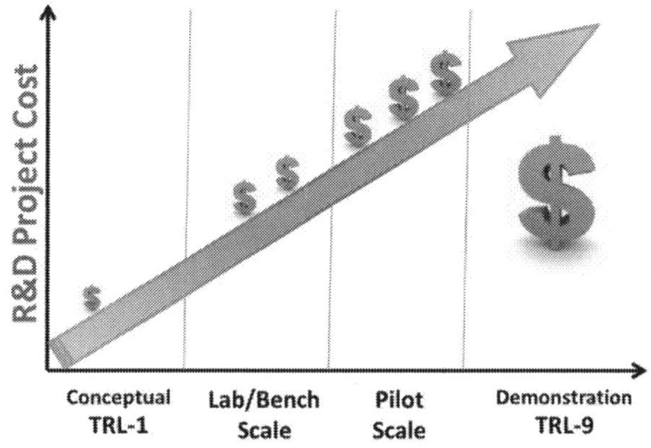

Looking Forward

- **2nd Generation:** Increased Focus on Pilot-Scale
- **Transformational:** Development at Laboratory/Bench Scale

Laboratory / Bench-Scale

University of Illinois
Biphasic Absorption with Phase Separation

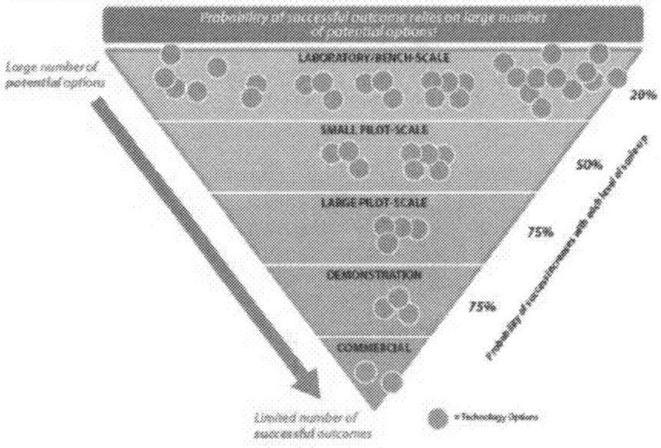

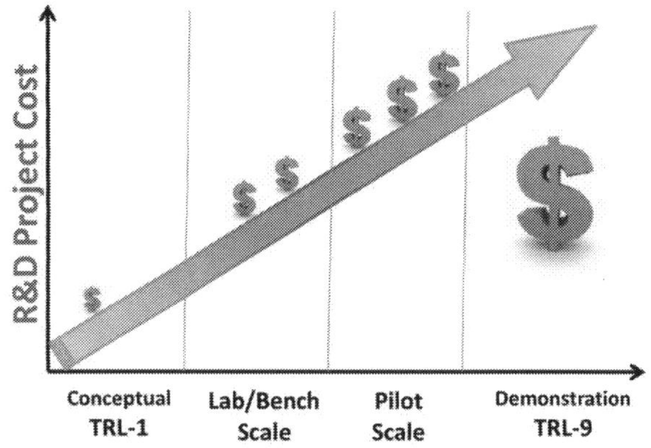

Looking Forward

NETL

- **2nd Generation:** Increased Focus on Pilot-Scale
- **Transformational:** Development at Laboratory/Bench Scale

Laboratory / Bench-Scale

Texas A&M University
Amine-Incorporated Porous Polymer Networks (APPNS)

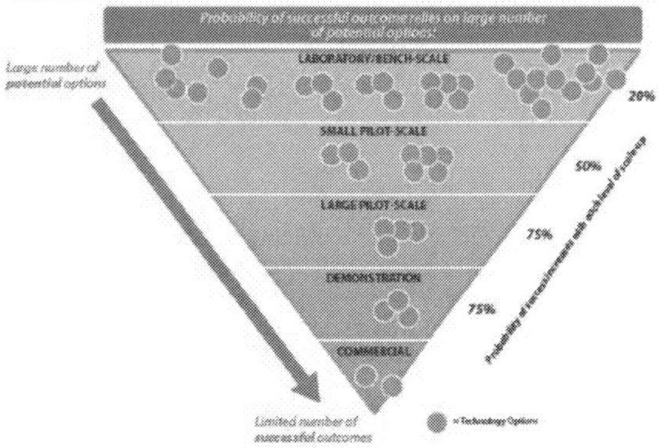

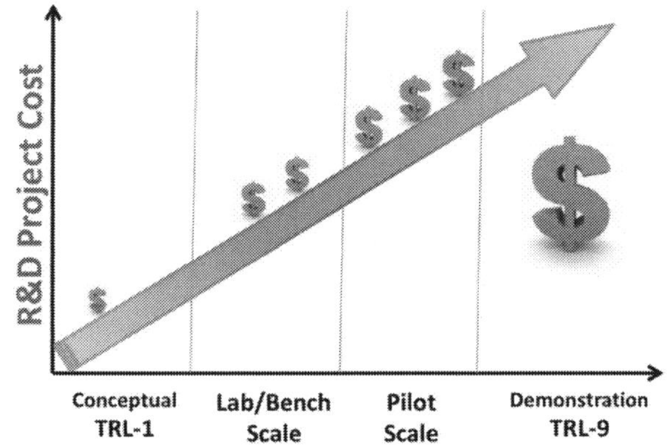

Looking Forward

NETL

- **2nd Generation:** Increased Focus on Pilot-Scale
- **Transformational:** Development at Laboratory/Bench Scale

Laboratory / Bench-Scale

Georgia Tech Research Corporation
Novel Rapidly Cycled Pressure Swing Adsorption Process

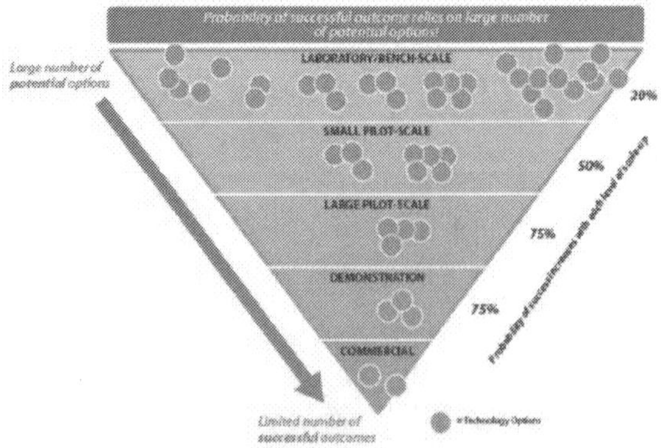

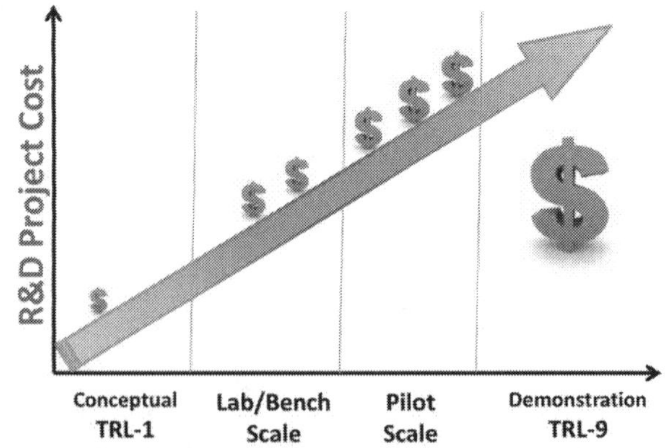

Looking Forward

- **2nd Generation:** Increased Focus on Pilot-Scale
- **Transformational:** Development at Laboratory/Bench Scale

Laboratory / Bench-Scale

Research Triangle Institute
Novel Hybrid Fluidizable Sorbents

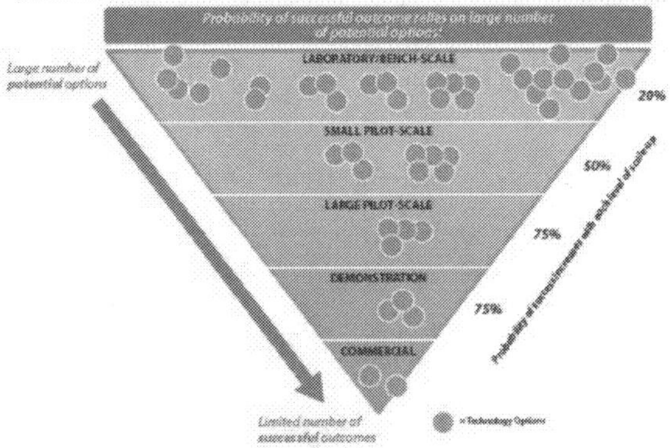

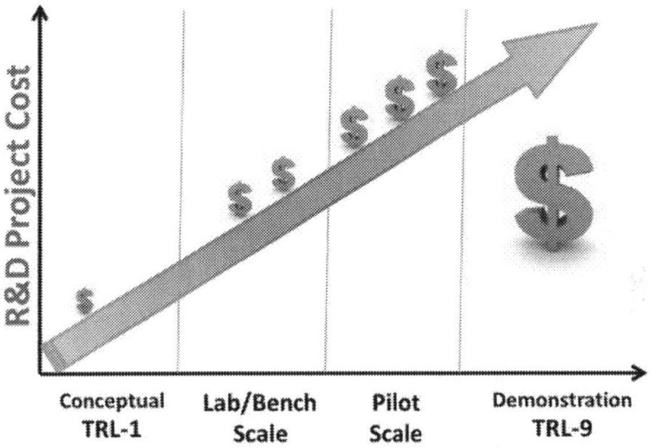

Looking Forward

- **2nd Generation:** Increased Focus on Pilot-Scale
- **Transformational:** Development at Laboratory/Bench Scale

Laboratory / Bench-Scale

Massachusetts Institute of Technology
Electrochemically Mediated Regeneration

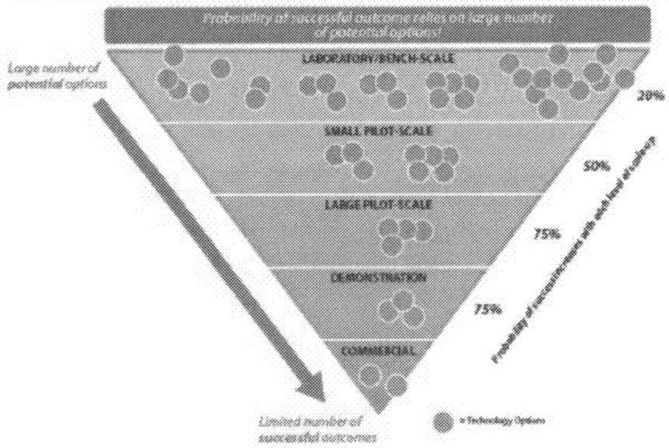

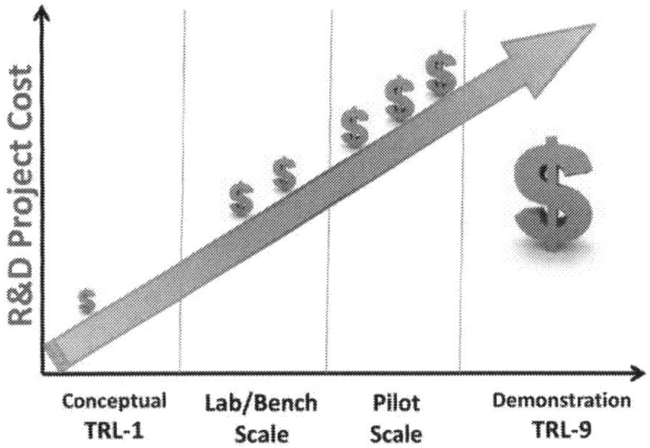

Looking Forward

- **2nd Generation:** Increased Focus on Pilot-Scale
- **Transformational:** Development at Laboratory/Bench Scale

Laboratory / Bench-Scale

Gas Technology Institute
Go-Peek Hybrid Membranes

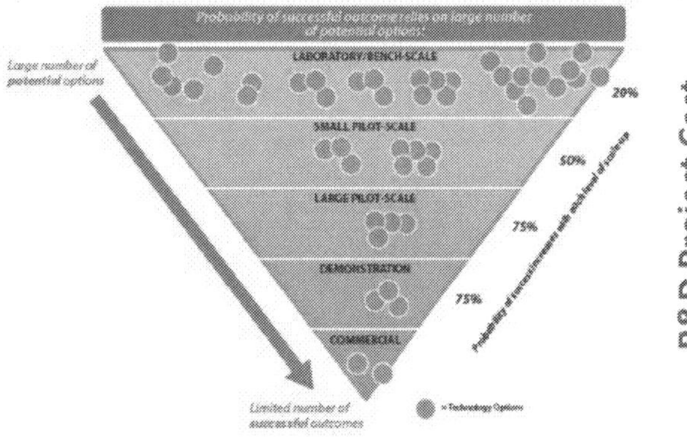

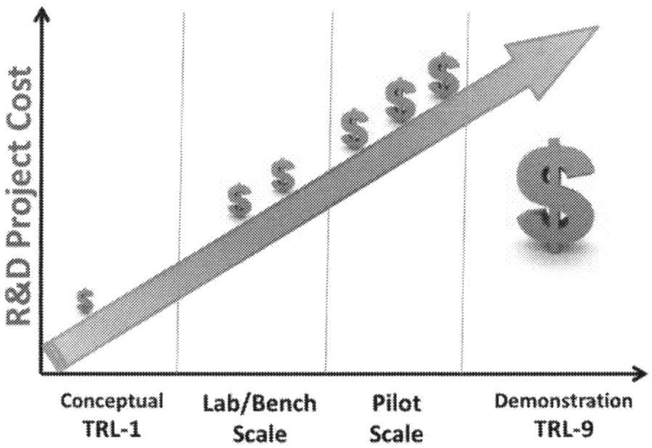

Looking Forward

- **2nd Generation:** Increased Focus on Pilot-Scale
- **Transformational:** Development at Laboratory/Bench Scale

Laboratory / Bench-Scale

Liquid Ion Solutions LLC - Hybrid Advanced Membrane, Solvent System, and Process Integration

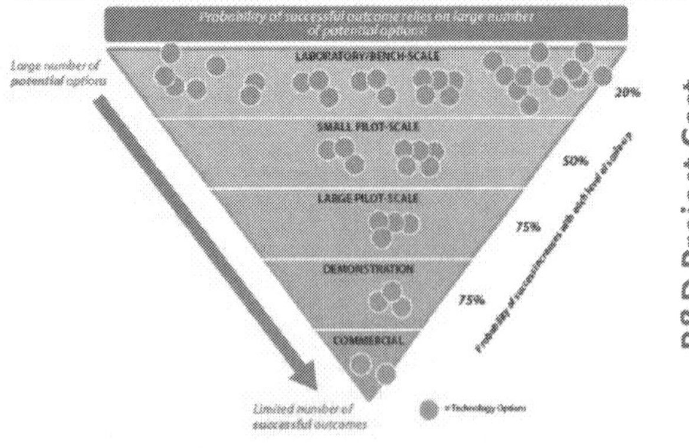

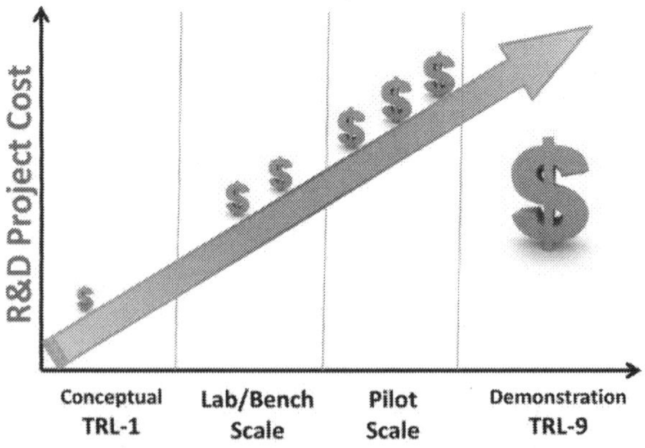

Looking Forward

NETL

- **2nd Generation:** Increased Focus on Pilot-Scale
- **Transformational:** Development at Laboratory/Bench Scale

Laboratory / Bench-Scale

LANL - Los Alamos National Laboratory
Polymer-Based Membranes

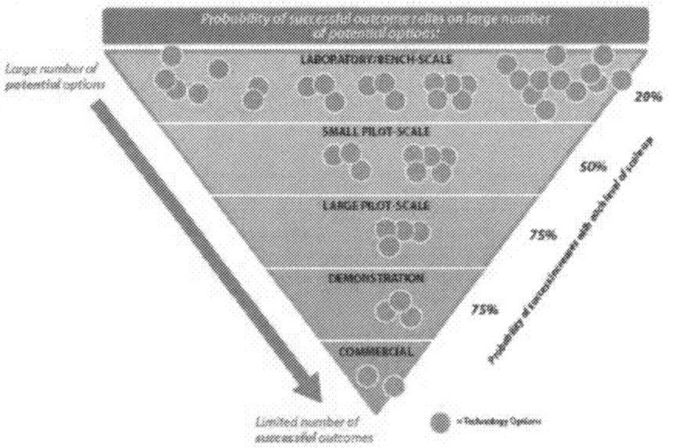

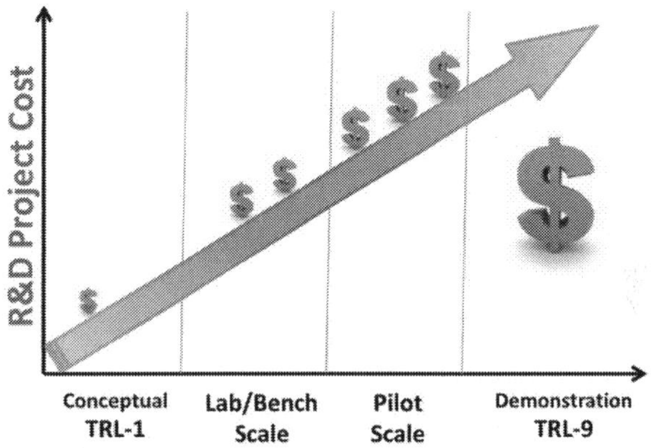

Looking Forward

NETL

- **2nd Generation:** Increased Focus on Pilot-Scale
- **Transformational:** Development at Laboratory/Bench Scale

Laboratory / Bench-Scale

University of Southern California - A High Efficiency, Ultra-Compact Process for Pre-Combustion CO2 Capture

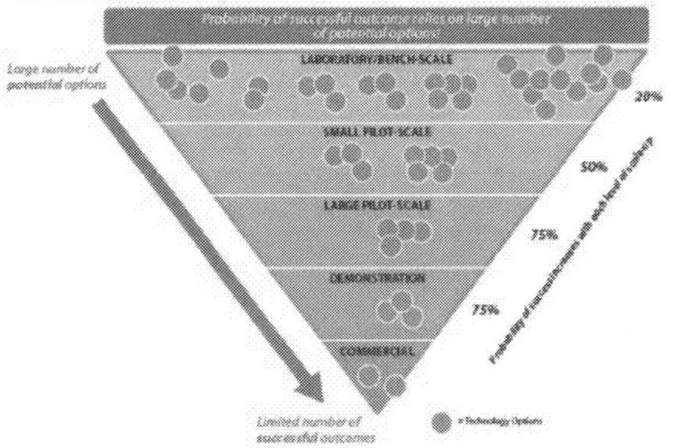

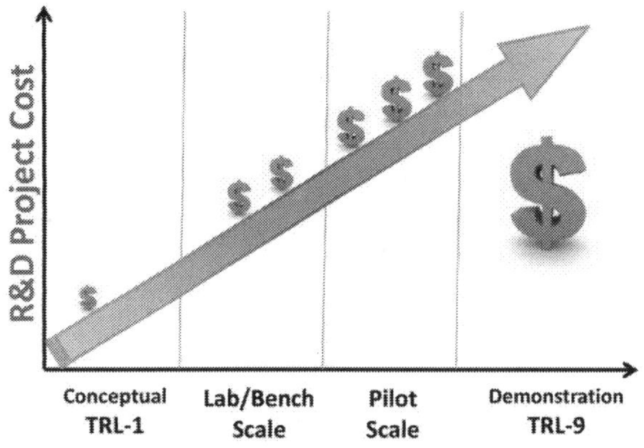

Looking Forward

- **2nd Generation:** Increased Focus on Pilot-Scale
- **Transformational:** Development at Laboratory/Bench Scale

Laboratory / Bench-Scale

Southern Research Institute
Combined Sorbent/WGS with Integrated Heat Management

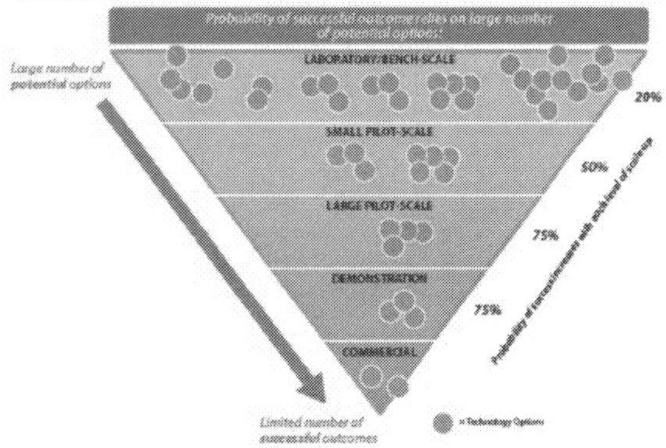

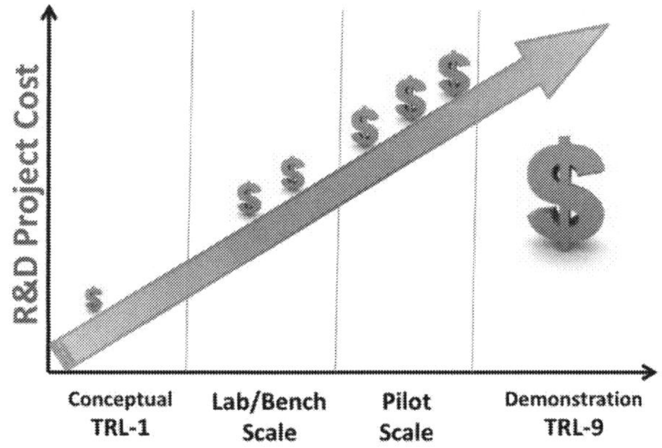

Looking Forward

- **2nd Generation:** Increased Focus on Pilot-Scale
- **Transformational:** Development at Laboratory/Bench Scale

Laboratory / Bench-Scale

The Research Foundation of State University of New York
Sorption Enhanced Mixed Matrix Membranes

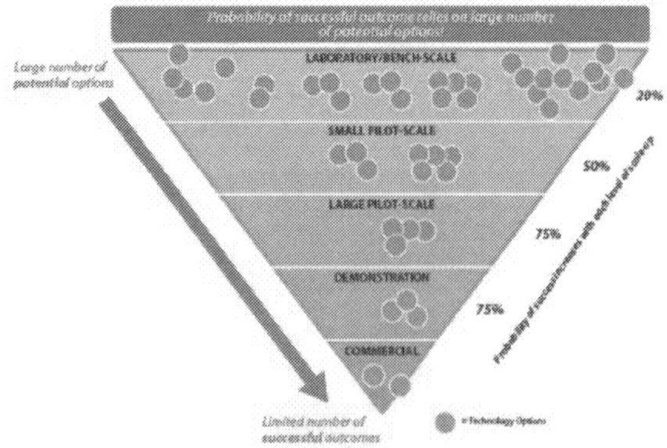

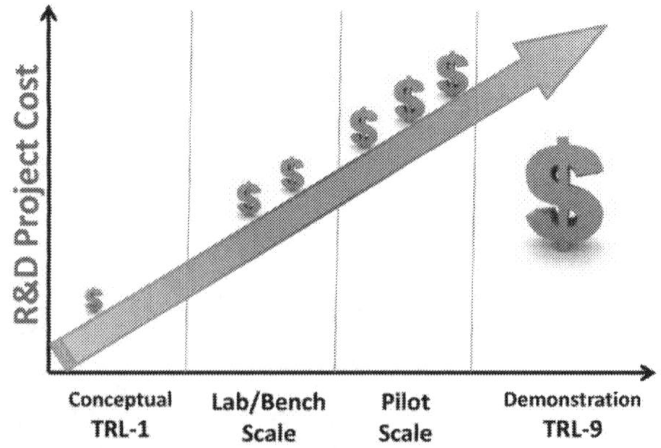

Looking Forward

- **2nd Generation:** Increased Focus on Pilot-Scale
- **Transformational:** Development at Laboratory/Bench Scale

Small Pilot-Scale

Linde, LLC
Slipstream Novel Amine-Based Post-Combustion Process

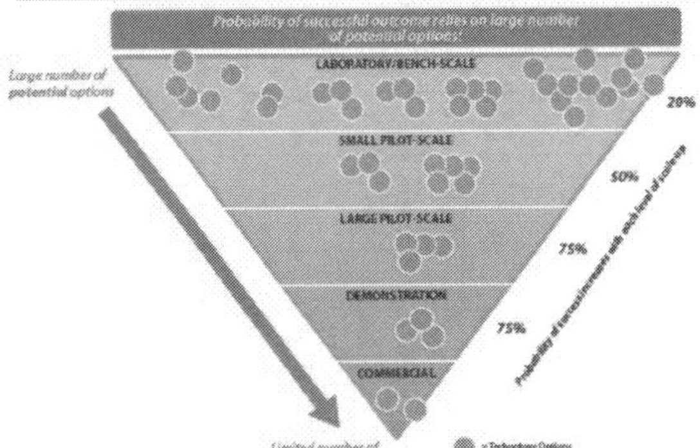

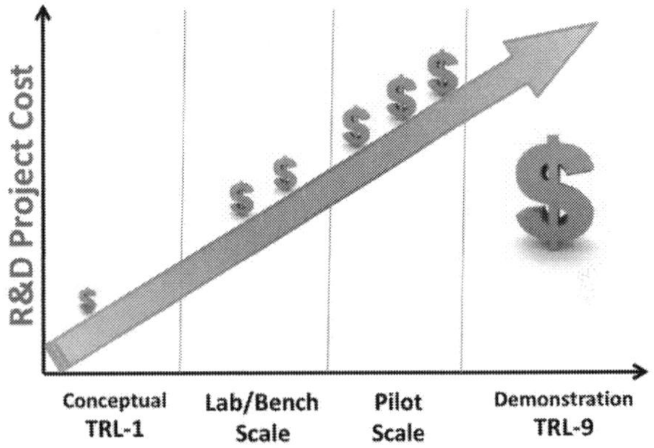

Looking Forward

- **2nd Generation:** Increased Focus on Pilot-Scale
- **Transformational:** Development at Laboratory/Bench Scale

Small Pilot-Scale

Neumann Systems Group, Inc
Carbon Absorber Retrofit Equipment

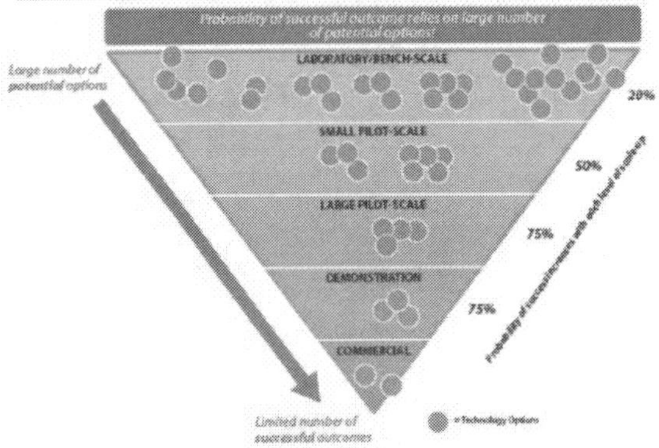

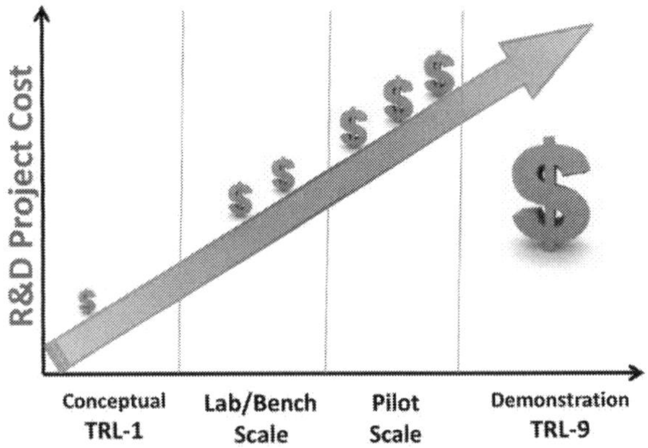

Looking Forward

- **2nd Generation:** Increased Focus on Pilot-Scale
- **Transformational:** Development at Laboratory/Bench Scale

Small Pilot-Scale

University of Kentucky - Heat Integrated Post-combustion CO_2 Capture System Using the MHPSA Advanced Solvent

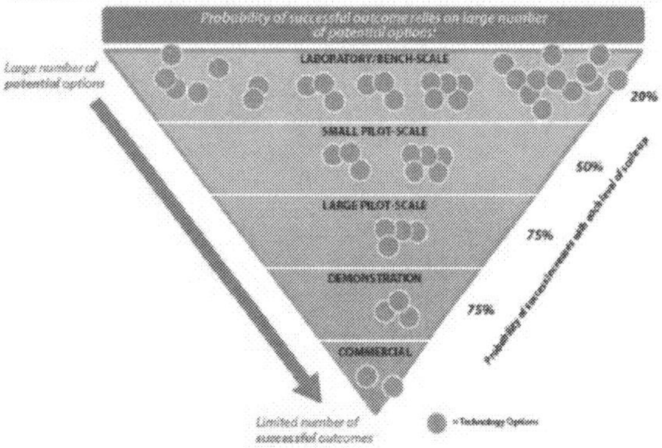

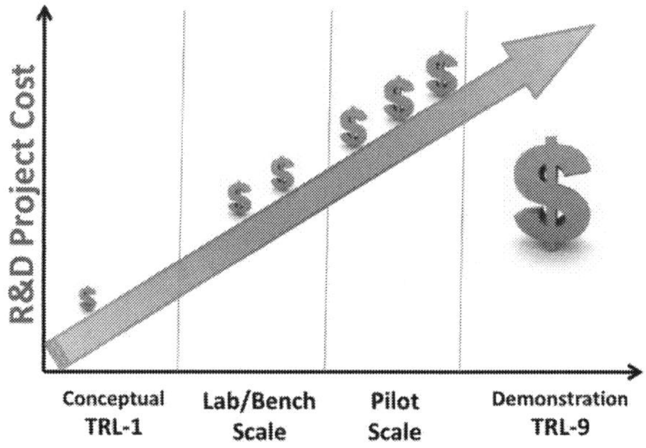

Looking Forward

- **2nd Generation:** Increased Focus on Pilot-Scale
- **Transformational:** Development at Laboratory/Bench Scale

Small Pilot-Scale

General Electric
Novel Aminosilicone Solvent

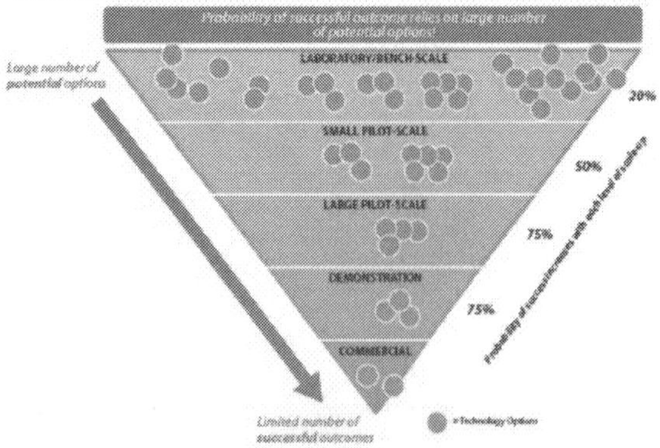

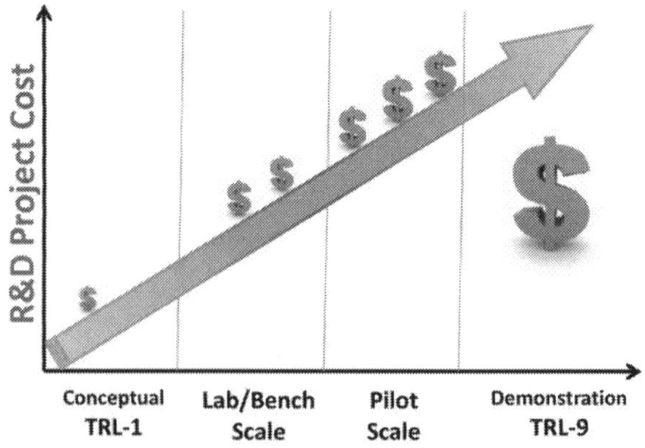

Looking Forward

- **2nd Generation:** Increased Focus on Pilot-Scale
- **Transformational:** Development at Laboratory/Bench Scale

Small Pilot-Scale

ION Engineering
Amine Solvent in Ionic Liquid

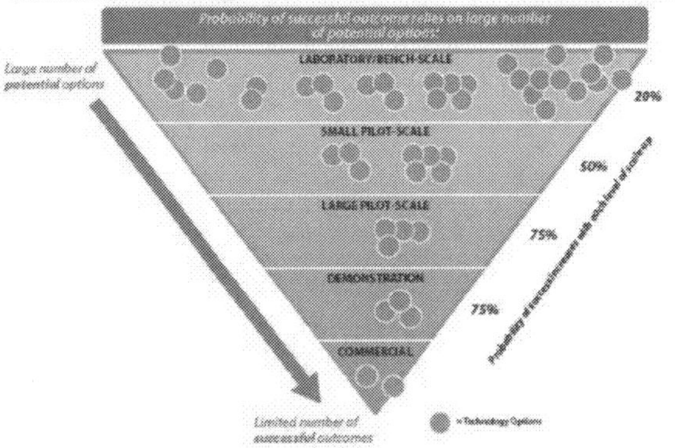

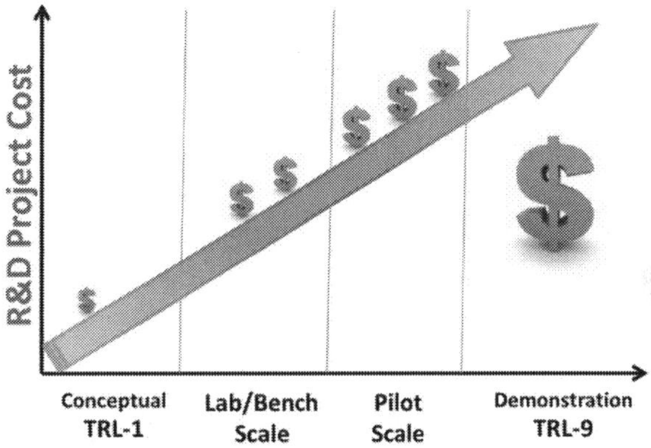

Looking Forward

- **2nd Generation:** Increased Focus on Pilot-Scale
- **Transformational:** Development at Laboratory/Bench Scale

Small Pilot-Scale

ADA-Environmental Solutions
Solid Sorbents as Retrofit Technology

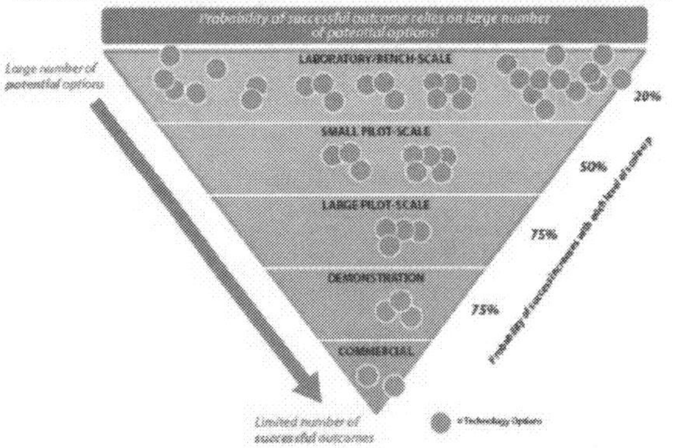

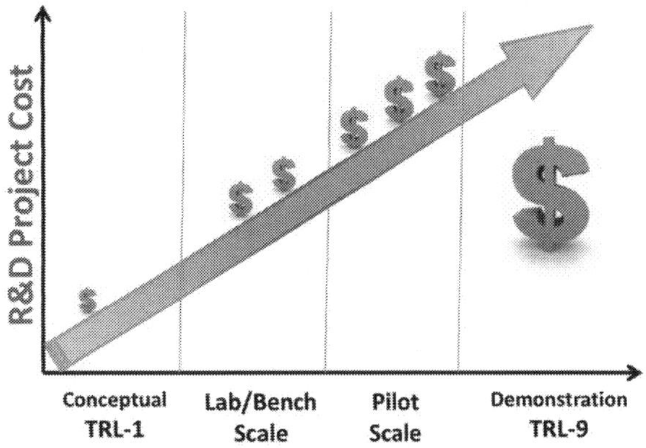

Looking Forward

NETL

- **2nd Generation:** Increased Focus on Pilot-Scale
- **Transformational:** Development at Laboratory/Bench Scale

Small Pilot-Scale

TDA Research, Inc.
Alkalized Alumina Solid Sorbent

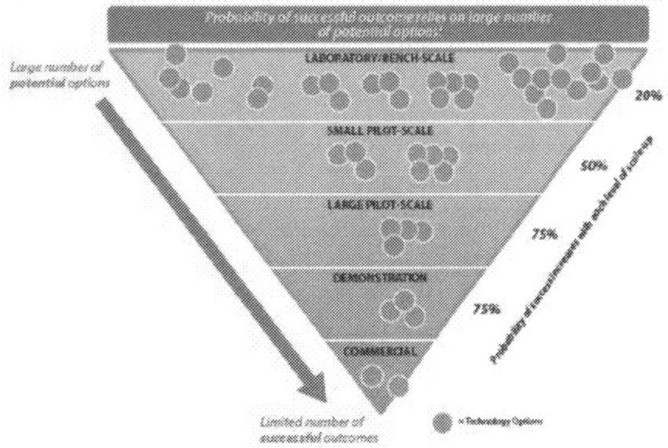

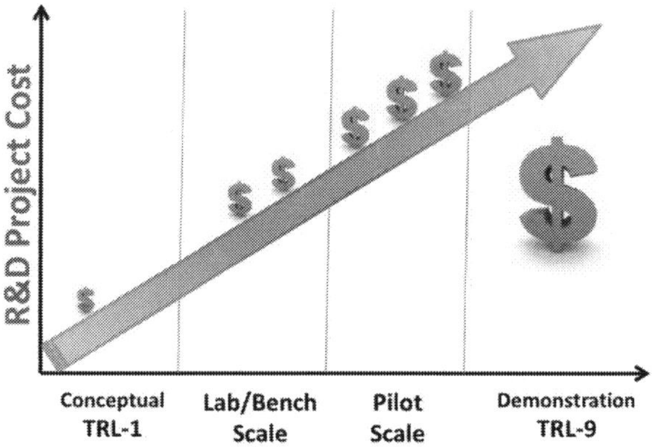

Looking Forward

NETL

- **2nd Generation:** Increased Focus on Pilot-Scale
- **Transformational:** Development at Laboratory/Bench Scale

Small Pilot-Scale

SRI International
Novel Solid Sorbent

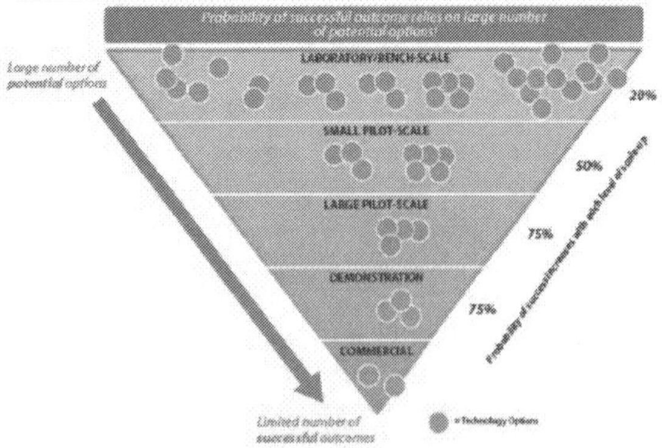

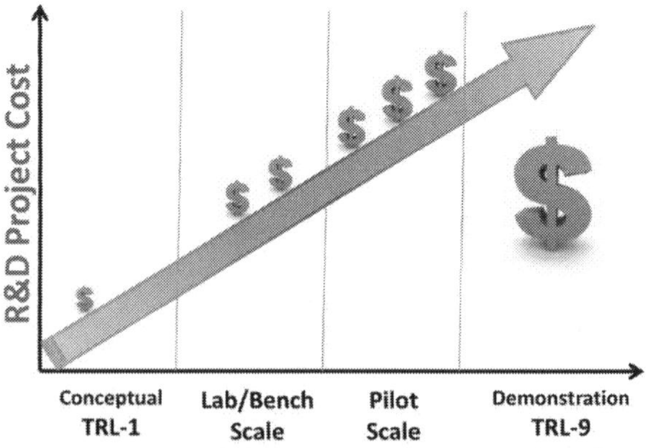

Looking Forward

- **2nd Generation:** Increased Focus on Pilot-Scale
- **Transformational:** Development at Laboratory/Bench Scale

Small Pilot-Scale

Membrane Technology & Research
Polymeric Membranes

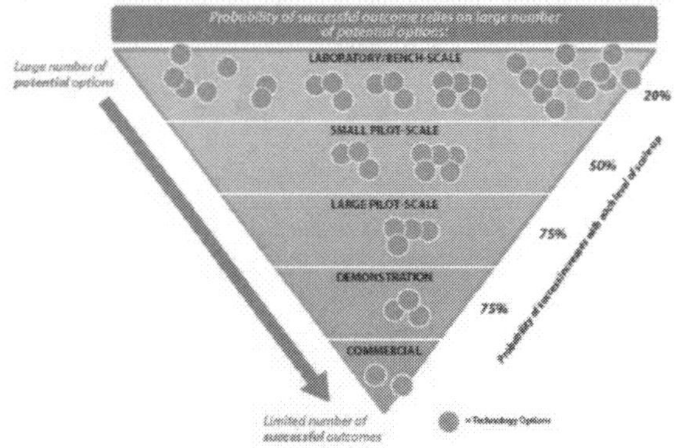

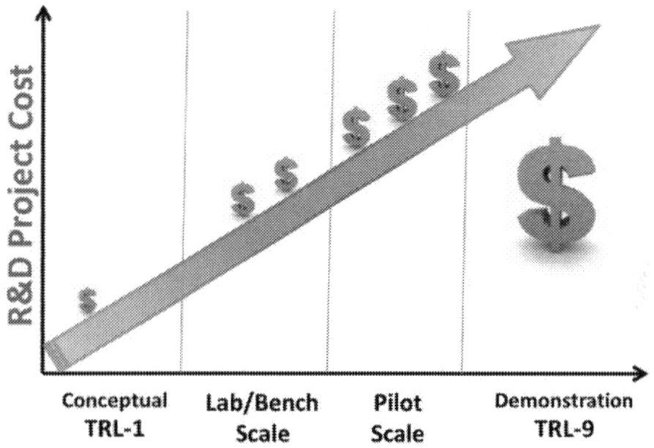

Looking Forward

- **2nd Generation:** Increased Focus on Pilot-Scale
- **Transformational:** Development at Laboratory/Bench Scale

Small Pilot-Scale

Gas Technology Institute
Hollow-Fiber-Membrane Contactor with aMDEA Solvent

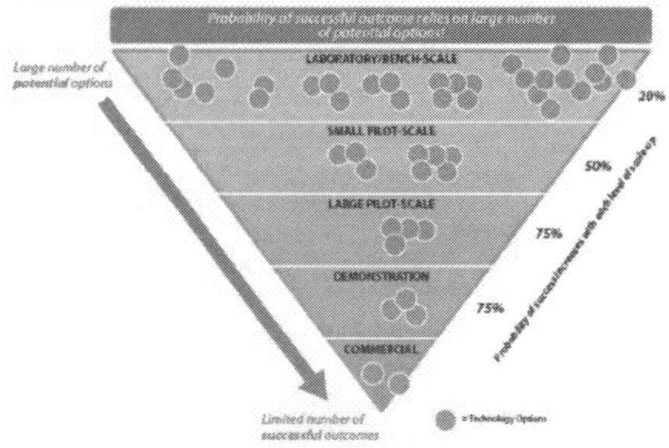

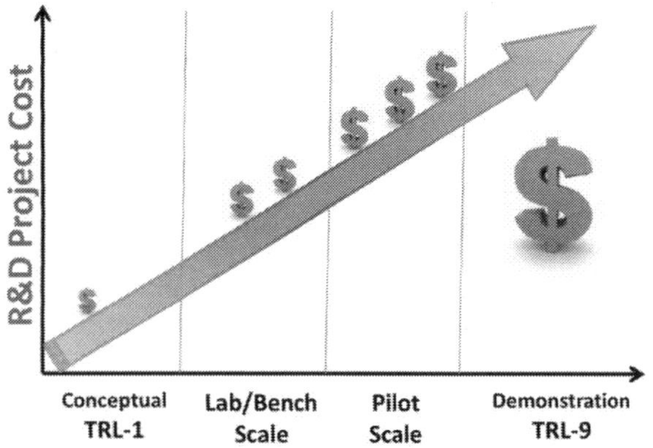

Looking Forward

NETL

- **2nd Generation:** Increased Focus on Pilot-Scale
- **Transformational:** Development at Laboratory/Bench Scale

Small Pilot-Scale

FuelCell Energy Inc.
Combined Electric Power and CO_2 Separation (CEPACS) System

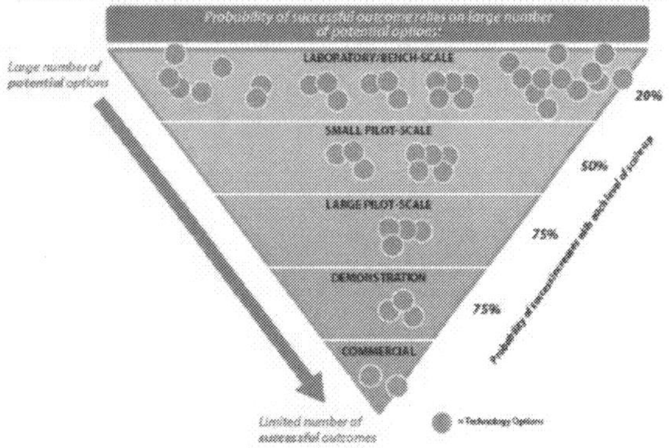

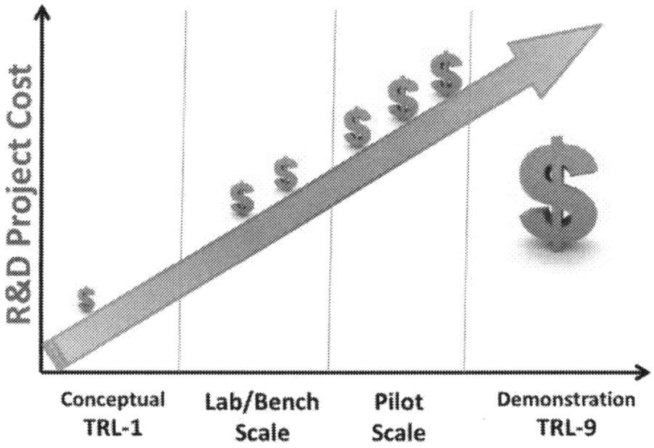

U.S. DEPARTMENT OF ENERGY | National Energy Technology Laboratory 41

Looking Forward

NETL

- **2nd Generation:** Increased Focus on Pilot-Scale
- **Transformational:** Development at Laboratory/Bench Scale

Small Pilot-Scale

SRI International
CO2 Capture Using AC-ABC Process

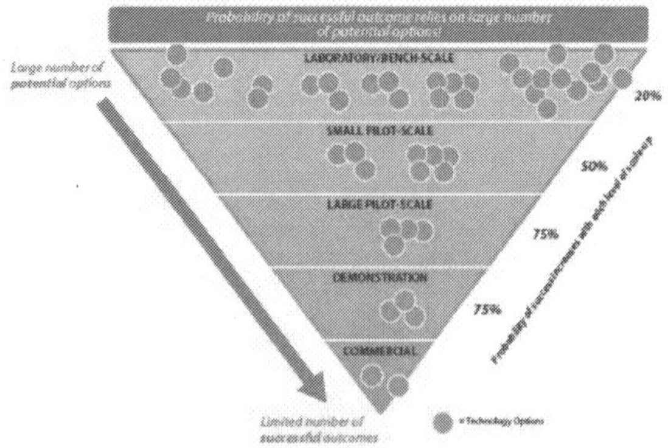

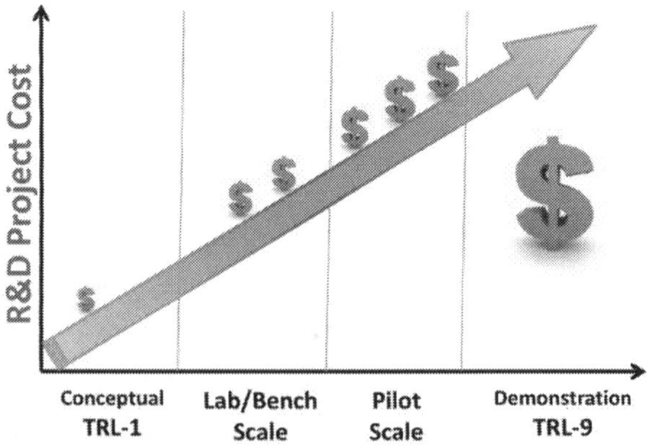

U.S. DEPARTMENT OF ENERGY | National Energy Technology Laboratory 42

Looking Forward

- **2nd Generation:** Increased Focus on Pilot-Scale
- **Transformational:** Development at Laboratory/Bench Scale

Small Pilot-Scale

TDA, Inc.
High Capacity Regenerable Sorbent

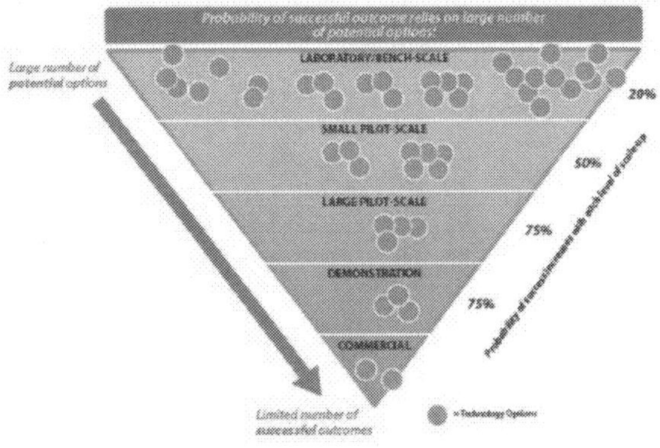

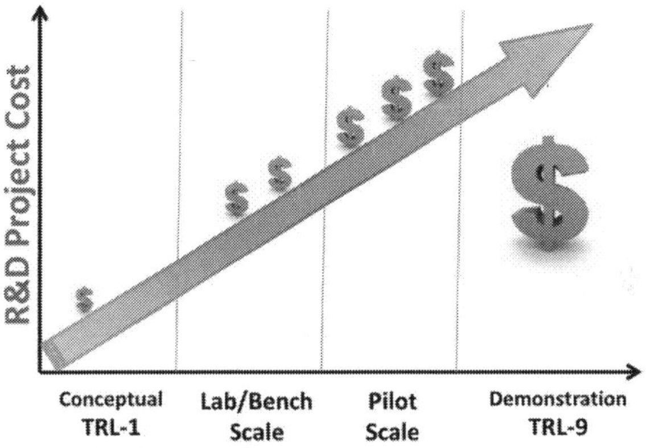

Looking Forward

- **2nd Generation:** Increased Focus on Pilot-Scale
- **Transformational:** Development at Laboratory/Bench Scale

Large Pilot-Scale

Southern Company Services
Waste Heat Integration

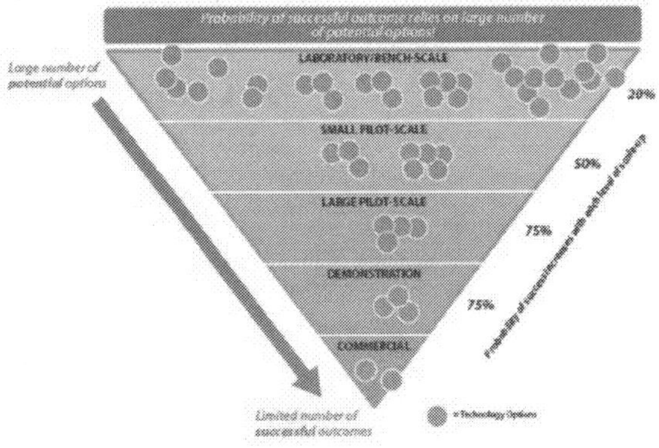

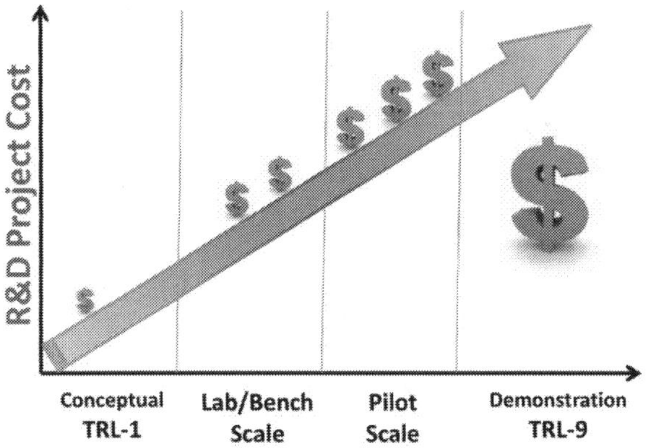

Looking Forward

- **2nd Generation:** Increased Focus on Pilot-Scale
- **Transformational:** Development at Laboratory/Bench Scale

Large Pilot-Scale

University of Illinois
Linde/BASF CO_2 Capture Technology

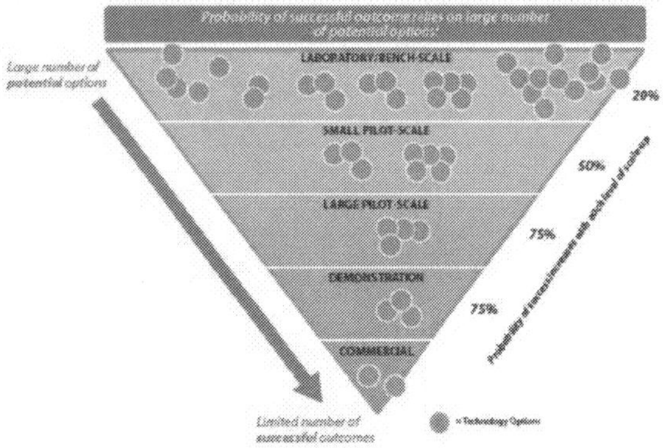

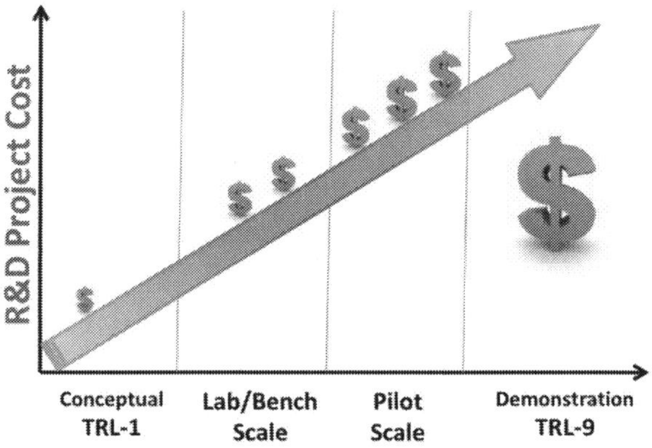

Looking Forward

- **2nd Generation:** Increased Focus on Pilot-Scale
- **Transformational:** Development at Laboratory/Bench Scale

Large Pilot-Scale

Alstom Power
Improvements to Chilled Ammonia Process

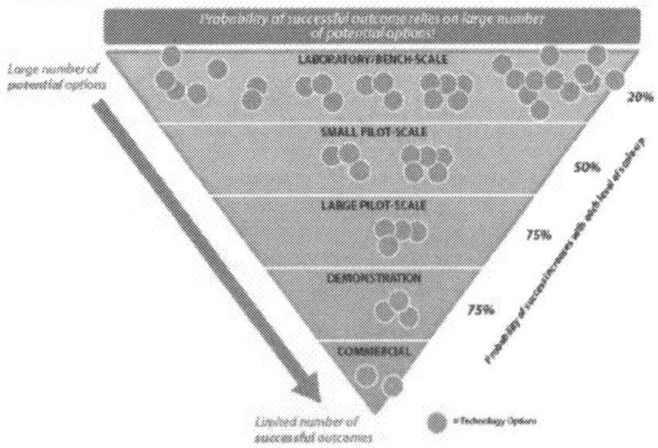

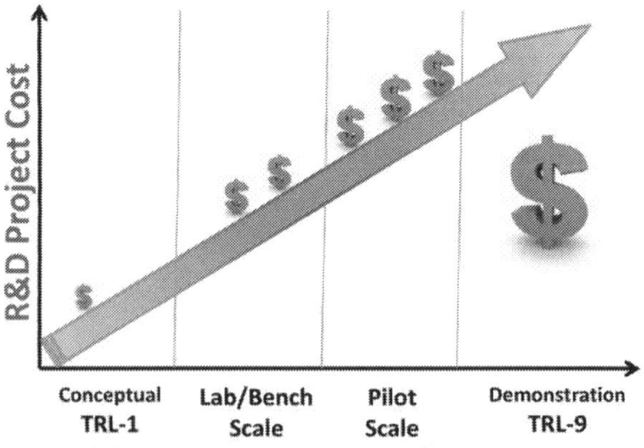

Looking Forward

- **2nd Generation:** Increased Focus on Pilot-Scale
- **Transformational:** Development at Laboratory/Bench Scale

Large Pilot-Scale

Southern Company Services
Process Improvements/Advanced Solvent

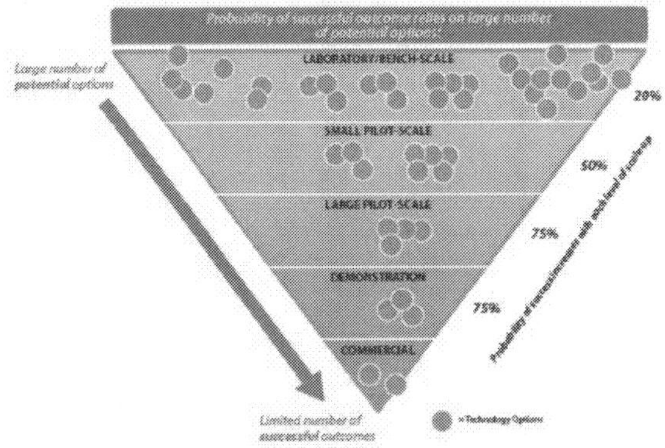

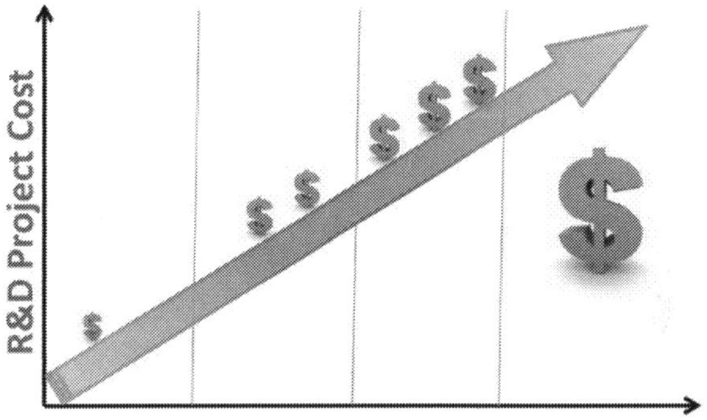

U.S. DEPARTMENT OF ENERGY | National Energy Technology Laboratory 47

Looking Forward

- **2nd Generation:** Increased Focus on Pilot-Scale
- **Transformational:** Development at Laboratory/Bench Scale

Large Pilot-Scale

University of Kentucky
Heat Integrated Capture System

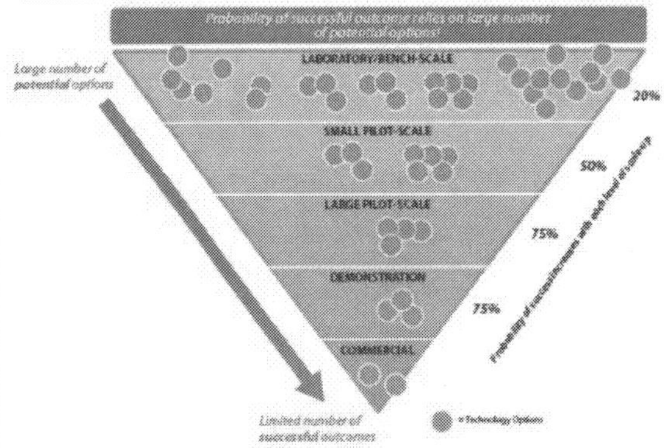

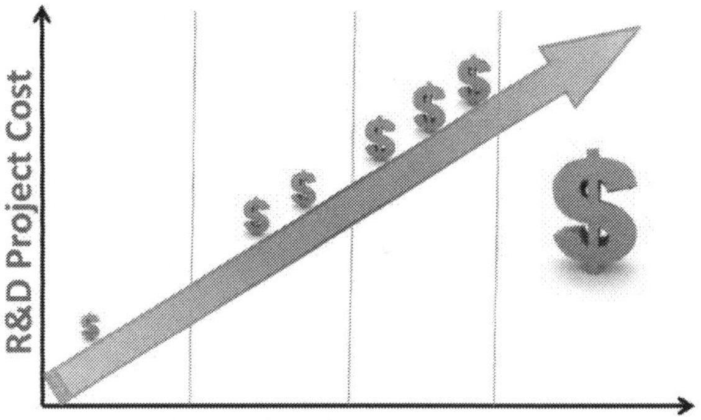

U.S. DEPARTMENT OF ENERGY | National Energy Technology Laboratory 48

Looking Forward

- **2nd Generation:** Increased Focus on Pilot-Scale
- **Transformational:** Development at Laboratory/Bench Scale

Large Pilot-Scale

General Electric
Novel Aminosilicone Solvent

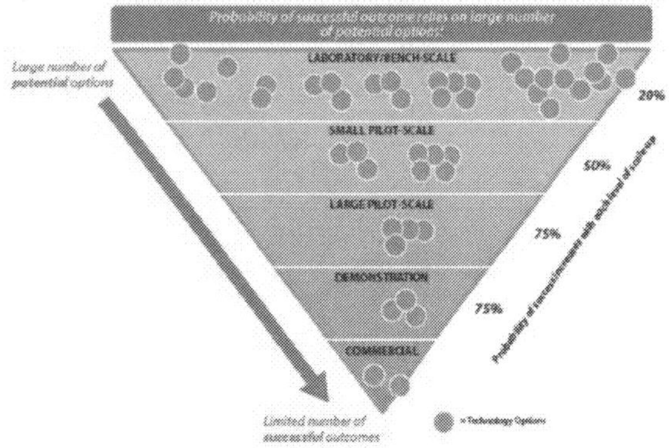

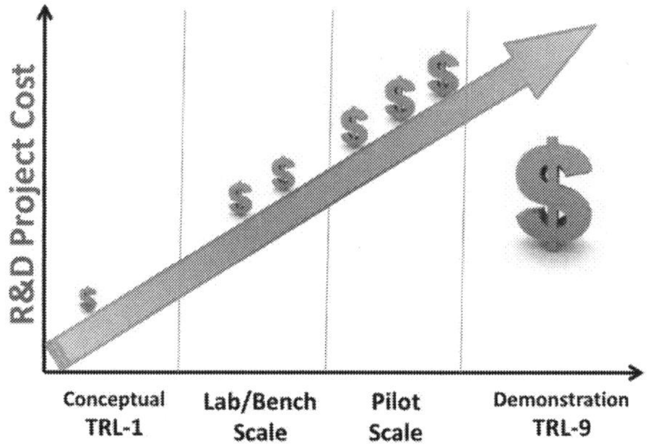

U.S. DEPARTMENT OF ENERGY | National Energy Technology Laboratory 49

Looking Forward

- **2nd Generation:** Increased Focus on Pilot-Scale
- **Transformational:** Development at Laboratory/Bench Scale

Large Pilot-Scale

NRG CO_2NCEPT
VeloxoTherm™ Solid Sorbent Capture System

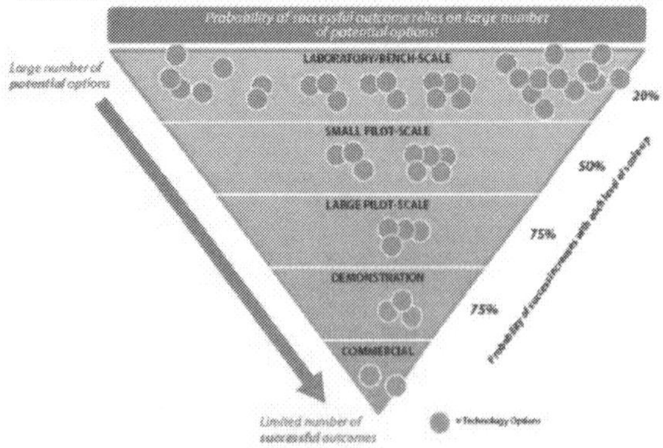

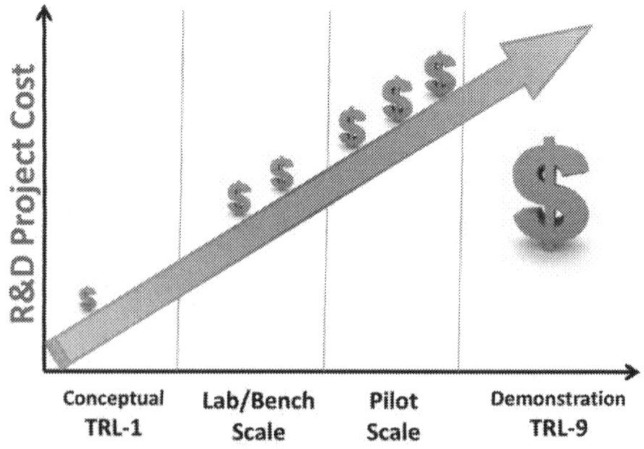

U.S. DEPARTMENT OF ENERGY | National Energy Technology Laboratory 50

Looking Forward

- **2nd Generation:** Increased Focus on Pilot-Scale
- **Transformational:** Development at Laboratory/Bench Scale

Large Pilot-Scale

Dresser-Rand
Supersonic CO_2 Compressor

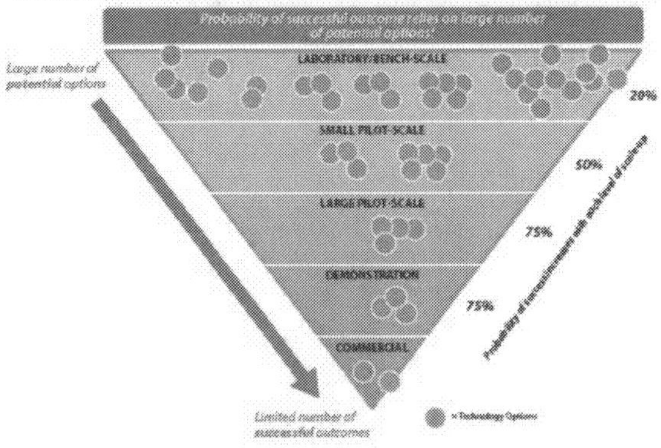

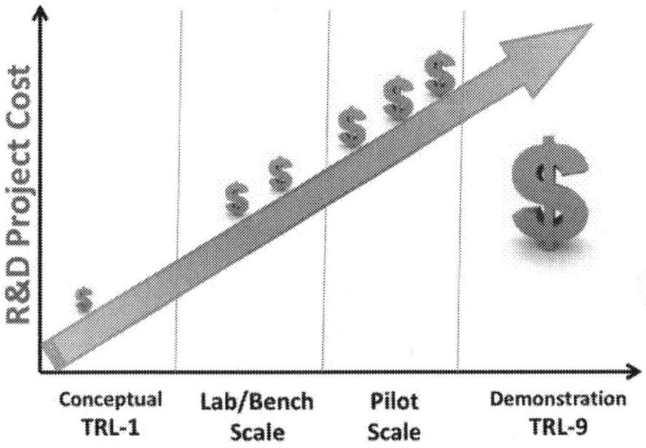

U.S. DEPARTMENT OF ENERGY | National Energy Technology Laboratory

51

Looking Forward

- **2nd Generation:** Increased Focus on Pilot-Scale
- **Transformational:** Development at Laboratory/Bench Scale

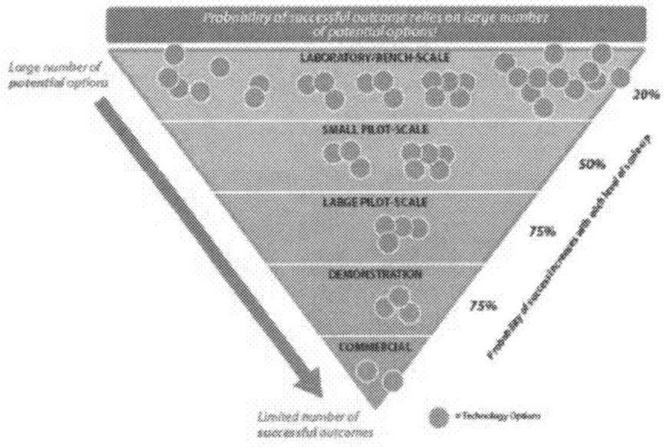

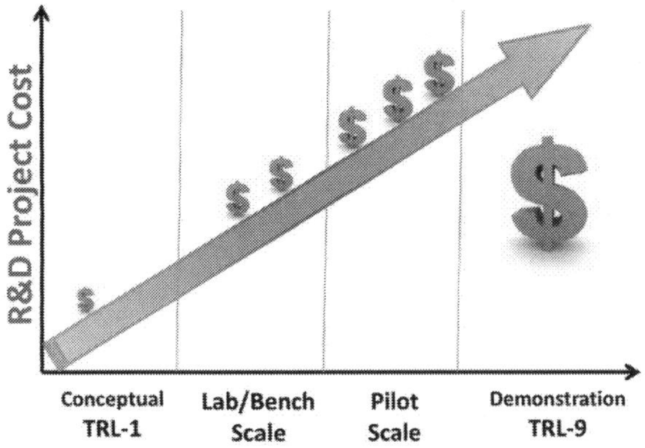

U.S. DEPARTMENT OF ENERGY | National Energy Technology Laboratory

52

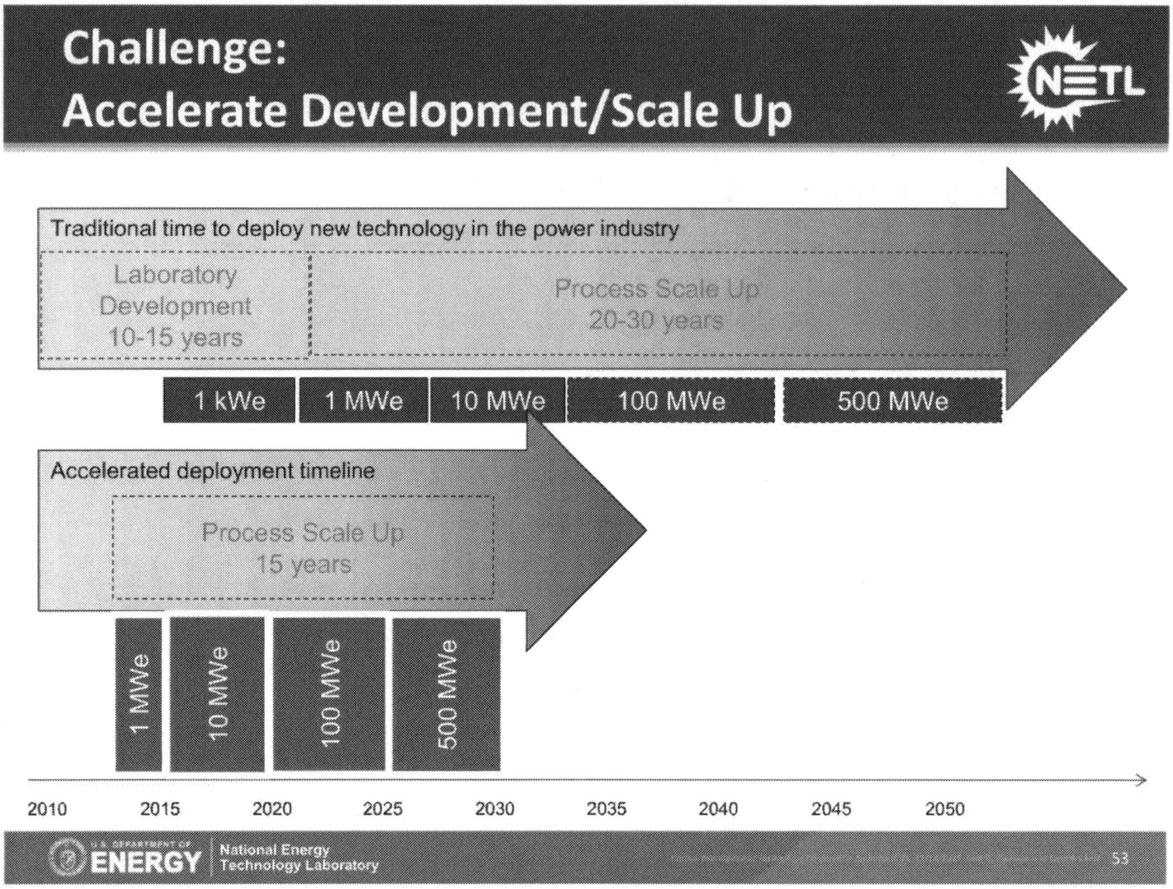
Challenge:
Accelerate Development/Scale Up
NETL
Traditional time to deploy new technology in the power industry
Laboratory Development 10-15 years
Process Scale Up 20-30 years
1 kWe
1 MWe
10 MWe
100 MWe
500 MWe
Accelerated deployment timeline
Process Scale Up 15 years
1 MWe
10 MWe
100 MWe
500 MWe
2010
2015
2020
2025
2030
2035
2040
2045
2050
U.S. DEPARTMENT OF ENERGY
National Energy Technology Laboratory
53

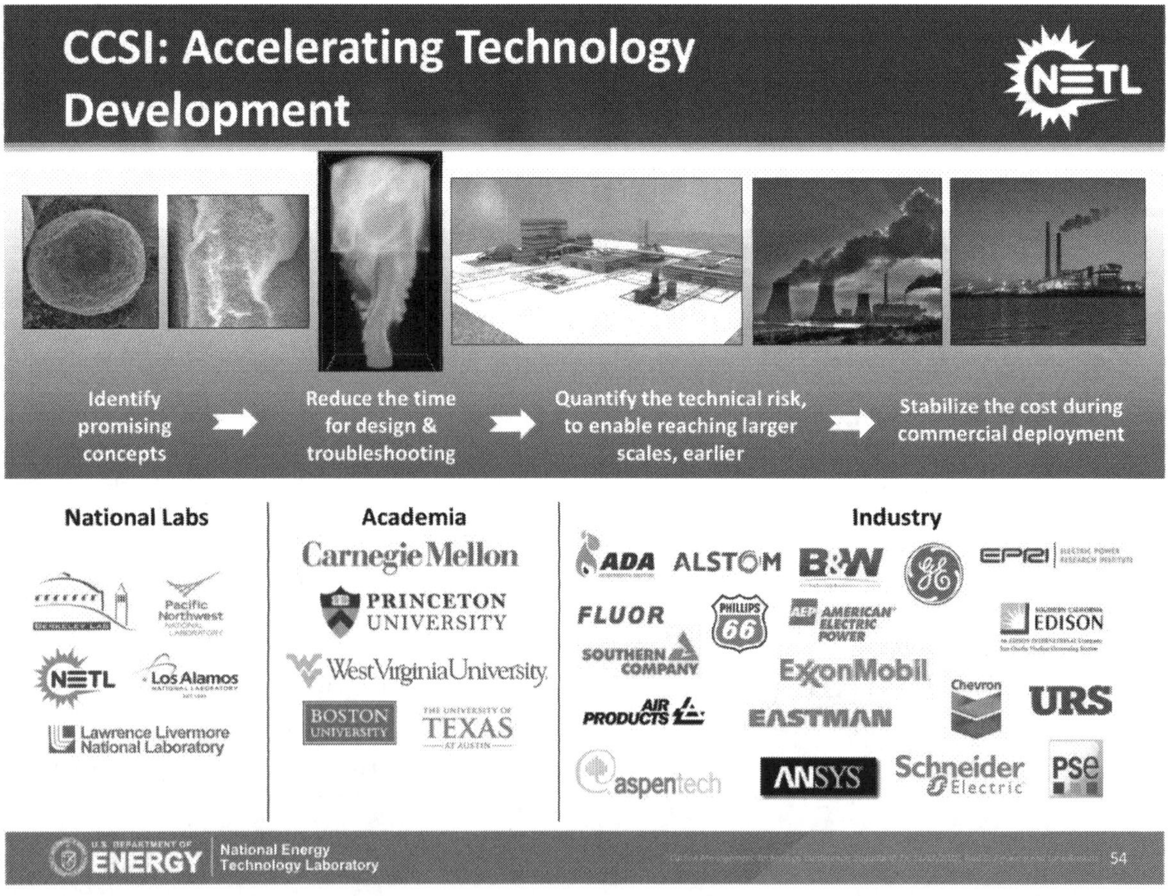
CCSI: Accelerating Technology Development
NETL
Identify promising concepts
Reduce the time for design & troubleshooting
Quantify the technical risk, to enable reaching larger scales, earlier
Stabilize the cost during commercial deployment
National Labs
Pacific Northwest National Laboratory
NETL
Los Alamos National Laboratory
Lawrence Livermore National Laboratory
Academia
Carnegie Mellon
Princeton University
West Virginia University
Boston University
The University of Texas at Austin
Industry
ADA
ALSTOM
B&W
GE
EPRI Electric Power Research Institute
FLUOR
Phillips 66
AEP American Electric Power
Edison
Southern Company
ExxonMobil
Chevron
URS
Air Products
Eastman
aspentech
ANSYS
Schneider Electric
PSE
U.S. DEPARTMENT OF ENERGY
National Energy Technology Laboratory
54

National Carbon Capture Center

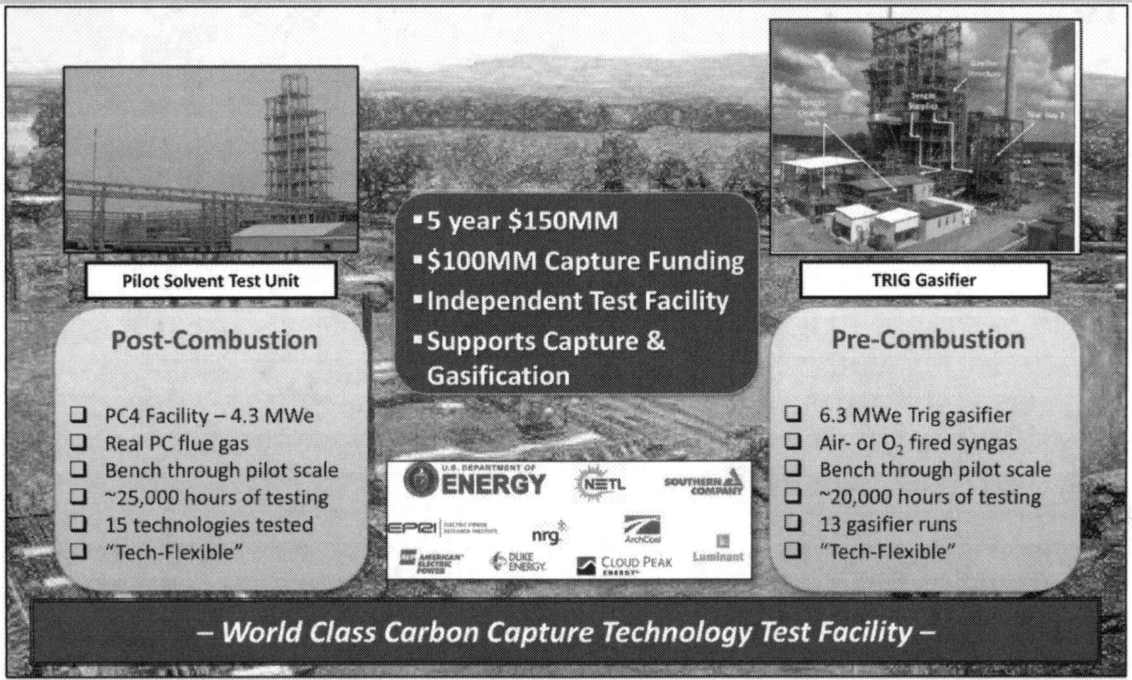

National Carbon Capture Center

Offering a **world-class neutral** test facility and a highly specialized staff, to **accelerate the commercialization** of advanced technologies and enable coal based power plants to achieve **near-zero emissions (low cost CO_2)**.

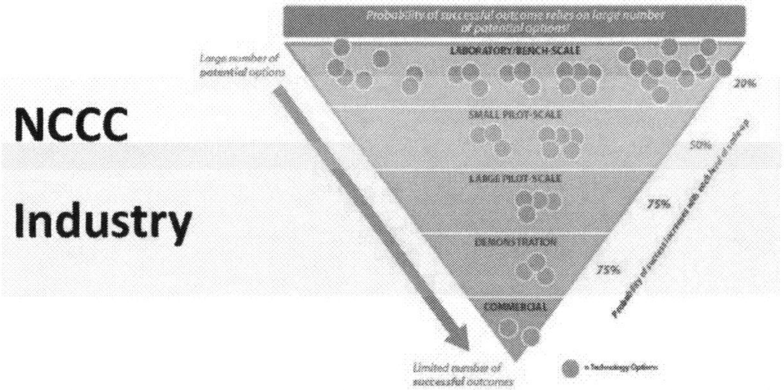

2nd Generation Small Pilot Scale Capture Plants

- **Technology scale-up is ongoing with pilot projects installed, undergoing shake down operations or testing on actual flue gas/syngas**

R&D Benefits Materializing

Figure courtesy of Southern Company, Linde, and MTR

- **Reduced Footprint**
- **Reduced Manufacturing Costs**
- **Reduced Energy Penalty**

R&D Benefits Materializing at NCCC

NETL

1 MWe : Slipstream: Novel Amine-Based Process

Linde, LLC

NETL

Project Summary

Approach

Test BASF's advanced amine-based solvent process technology and Linde's novel equipment and process innovations on a 1 MWe slipstream at the National Carbon Capture Center

Status – Operational since January 2015

Advantages

- Solvent more stable, higher capacity, faster kinetics and ~30% lower energy demand than MEA
- Projected up to 60% lower electrical energy load
- Novel intercooler
- Integrated absorber/wash unit
- High pressure regeneration
- Low cost MOC (projected ~30% capital savings)

Challenges

- Sustaining performance projections through scale up

1 MWe : Low Pressure Membrane Contactors
Membrane Technology & Research

Project Summary

Approach

Develop and test a spiral-wound membrane process on a 1 MWe slipstream at the National Carbon Capture Center

Status

Operational NC3 – January 2015 - August 2015
>11,000 hours of Polaris material tested on flue gas
Moving Unit to integrated test facility

Advantages

- Smaller footprint
- Efficient scale-up 20-25x larger than current modules
- Capital cost reduction
- Reduced process complexity

Challenges

- Overcoming sweep side pressure drop
- Effective use of all membrane area
- Maximizing packing density minimizing delta P

0.7 MWe : Process Integration to Reduce Capture Costs
University of Kentucky Center for Applied Energy Research (CAER)

Project Summary

Approach

Design, build and install an integrated CO_2 capture system into an existing PC power plant at a 0.7 MWe slipstream scale (~15 TPD CO_2) E.W. Brown Generating Station

Status – Construction Complete
Currently conducting MEA baseline testing

Advantages

- Two-stage air stripping increases solvent working capacity; CO_2-laden air sent to boiler to increase flue gas P_{CO_2}.
- Integrated cooling tower uses regenerated CO_2 waste heat to dry liquid desiccant which dries cooling tower air for improved steam turbine efficiency

Challenges

- Proving air stripping at slipstream scale
- High O_2 environment may increase solvent oxidation

3 MW_e Novel Electrochemical Membrane System for Carbon Dioxide Capture and Power Generation FuelCell Energy, Inc.

Goal: Design, fabricate, and test a 3 MW_e Combined Electric Power and Carbon-Dioxide Separation (CEPACS) system.

- Previous bench-scale (10 kW) results

Benefits: The system is anticipated to capture 90% of CO_2 with 95% purity for \$40/tonne with ≤35% increase in the COE.

- Does not impact base plant
- The ECM produces power while capturing CO_2, reduces NOx emissions

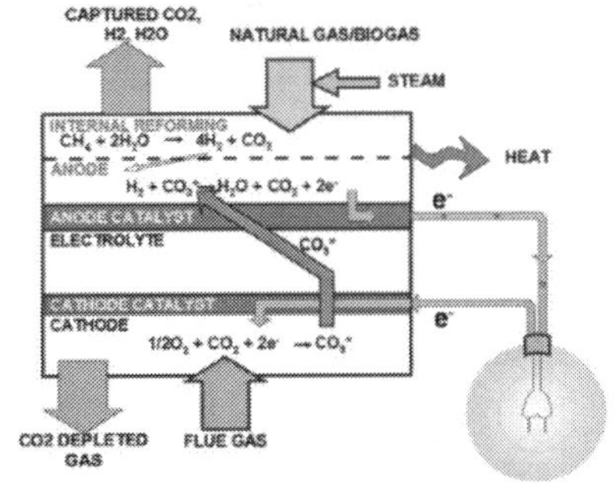

Electrochemical membrane (ECM) technology

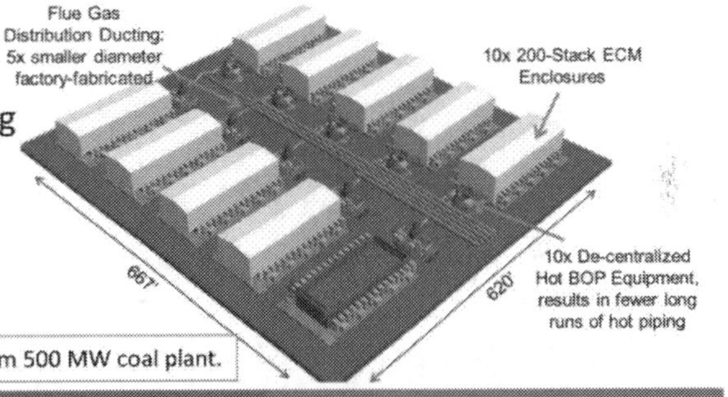

420 MW DFC-based system conceptual design for 90% CO_2 capture from 500 MW coal plant.

Lessons Learned

- **Communication is essential and required throughout project**
- **Involve ALL stakeholders early and often**
- **Scope definition with vendors, fabricators, construction contractors and project partners upfront is critical**
- **Include performance acceptance testing for equipment and pilot unit.**
  - At OEM/Fabricator site as well as host test site on actual gas
- **Confirm that host site, state, local codes, standards and procedures are well understood by all parties, since interpretations may differ.**
- **Review designs in great detail with:**
  - host site, safety personnel, fabricator, engineers, and technology developer for potential issues early and often
- **Expect unknowns will occur. Have a procedure for addressing them**
- **Delays will occur so plan for them to avoid critical path efforts that jeopardize the schedule and/or budget**

MAINTAIN A GOOD WORKING RELATIONSHIP

What's Next...

Funding Opportunity Announcement: *"Small and Large Scale Pilots for Reducing the Cost of CO_2 Capture and Compression"*

- Opened – 2/11/15; Closed – 3/16/15
- Areas of Interest:
  - AOI 1: Supersonic Compression Systems
  - AOI 2: Small Pilot Scale (0.5 – 5.0 MW) Post-combustion CO_2 Capture Development and Testing
  - AOI 3: Large Pilot Scale (10.0 – +25.0 MW) Post-combustion CO_2 Capture Development and Testing

Selected and Awarded Phase 1

Funding Opportunity Announcement: *"Lab - and Bench-Scale Applications for Research and Development of Transformational Carbon Dioxide Capture Technologies for Coal-Fired Power Plants"*

- Opened – 1/26/15; Closed – 3/30/15
- Areas of Interest:
  - AOI 1: Lab-Scale (TRL 3-4) CO_2 Capture Development and Testing
  - AOI 2: Bench-Scale (TRL 5) CO_2 Capture Development and Testing
  - AOI 3: Low Concentration, Coal-Relevant CO_2 Sources
  - AOI 4: Biological CO_2 Use/Conversion

Selections made and most awarded

FY2015 Funding Opportunity Announcements
Large Scale CO_2 Capture Projects (10+ MWe)

- **Final scale before full scale demonstration (100+ MWe)**
- **Need several pilots to support 2020 demonstrations**
- **Necessary for**
  – Validation of capture technology
  – Integration of advanced capture system components
  – Optimization of capture system for full scale demo
- **Phase I – Design (6 awards) (2015-2016)**
- **Phase II – Construction and operations (2016-2020)**

Large Scale Post-Combustion Pilots –Phase I

- **General Electric** – Aminosilicone capture process, 10 MWe+
- **University of Illinois** - Linde/BASF CO_2 capture technology at the Abbott coal-fired power plant, 25 Mwe
- **University of Kentucky** – Heat integrated capture system, 10 MWe+
- **Southern Company Services** – Process improvements/advanced solvent at Plant Barry Pilot Facility, 25 Mwe
- **Alstom Power** – Improvements to chilled ammonia process, 15 Mwe
- **NRG CO_2NCEPT**– VeloxoTherm™ solid sorbent capture system, 10 MWe

FY2015 Funding Opportunity Announcements
Transformational Capture Technologies

Post- and Pre-Combustion Projects	Performer
Development of a Novel Biphasic CO_2 Absorption Process with Multiple Stages of Liquid–Liquid Phase Separation for Post-Combustion Carbon Capture	University of Illinois
Evaluation of Amine–Incorporated Porous Polymer Networks as Sorbents for Post-combustion CO_2 Capture	Texas A&M University
Hybrid Encapsulated Ionic Liquids for Post-combustion CO_2 Capture	University of Notre Dame
Development of a Non-aqueous Solvent CO_2 Capture Process for Coal-Fired Power Plants	Research Triangle Institute
Enabling 10 mol/kg Swing Capacity via Heat Integrated Sub-ambient Pressure Swing Adsorption	Georgia Tech Research Corporation
Development of a Solid Sorbent for CO_2 Capture Process for Coal-Fired Power Plants	Research Triangle Institute
Electrochemically Mediated Sorbent Regeneration in CO_2 Scrubbing Processes	Massachusetts Institute of Technology
Testing of Next-Generation Hollow Fiber Membrane Modules	American Air Liquide, Inc.
Energy Efficient Membrane–Based "Hybrid" Hybrid Process for Post-combustion CO_2 Capture	Gas Technology Institute
Development of a Hybrid Capture System with Advanced Membrane, Solvent System, and Process Integration	Liquid Ion Solutions LLC
A High-Efficiency, Ultra-compact Process for Pre-combustion CO_2 Capture	University of Southern California
Sorption Enhanced Mixed Matrix Membranes for H_2 Purification and CO_2 Capture	University of Buffalo
Combined Sorbent/Water-Gas Shift–Based CO_2 Capture Process with Integrated Heat Management	Southern Research Institute
Zeolite Membrane Reactor for Pre-combustion Carbon Dioxide Capture	Arizona State University

FY2015 Funding Opportunity Announcements

Transformational Capture Technologies

Biological CO_2 Use/Conversion Projects	Performer
A Microalgae–Based Platform for the Beneficial Reuse of CO_2 Emissions from Power Plants	University of Kentucky Research Foundation
Microalgae Commodities from Coal Plant Flue Gas CO_2	MicroBio Engineering

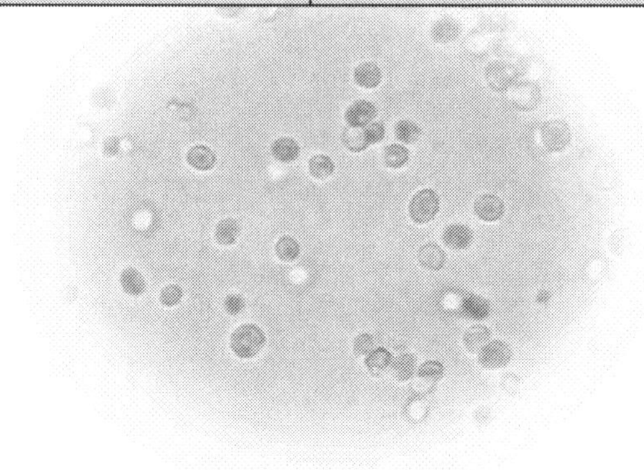

For More Information About the NETL Carbon Capture Program

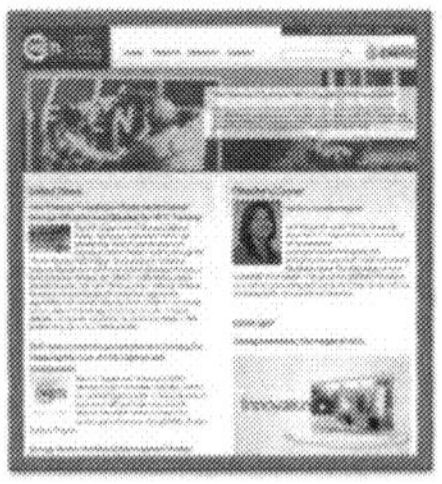

NETL
www.netl.doe.gov

facebook.com/FossilEnergy

Office of Fossil Energy
www.energy.gov/fe

twitter.com/fossilenergygov

José D. Figueroa
Carbon Capture Coordinator,
Carbon Capture Division
U. S. Department of Energy
National Energy Technology Laboratory
(Tel) 412 386-4966
Jose.Figueroa@netl.doe.gov

Lynn Brickett
Technology Manager
Carbon Capture Program
U. S. Department of Energy
National Energy Technology Laboratory
(Tel) 412-386-6574
Lynn.Brickett@netl.doe.gov

Project Update of the World's Largest Post Combustion CO_2 Capture Plant

November 18th, 2015

Mitsubishi Heavy Industries America, Inc.
Environmental & Chemical Plant Division
Dale Wilterdink

MITSUBISHI HEAVY INDUSTRIES, LTD.
Our Technologies, Your Tomorrow

Contents

MITSUBISHI HEAVY INDUSTRIES, LTD.

 2

1.1 History of Development of MHI's CO_2 Capture Technology

MITSUBISHI HEAVY INDUSTRIES, LTD.

1990	Started R&D activities with Kansai Electric Power Company (KEPCO)
1991	Started a 2 ton per day pilot plant at KEPCO's Nanko Power station
1994	Development of proprietary hindered amine solvent "KS-1™" and "KM CDR Process®" with KEPCO
1999	First commercial plant in Malaysia (200 ton per day, to enhance urea synthesis from the CO_2 recovered from a reformer flue gas)
2002	Started a pilot test for coal-fired power plant at MHI's Hiroshima R&D center
2003	High energy efficiency - Development of proprietary energy efficient process "Improved KM CDR Process"
2008	First commercial plant in Middle east (400 ton per day) which "Improved KM CDR Process" applied
2011	World's First - Started 500 ton per day fully integrated CCS demonstration plant with Southern Company for a coal-fired power plant at Alabama Power's James M. Barry Electric Generating Plant
2014	World's Largest - Received an order for a PCC plant of 4,776 ton per day for EOR sponsored by NRG Energy Inc. and JX Nippon Oil & Gas Exploration Corporation

1.2 MHI - R&D Strength & Experience

- **Deployment of R&D Facilities for specific CO_2 capture testing. More than 25 years R&D Experience**

(A) 2.0 TPD Nanko Osaka pilot plant from 1991

(B) 1.0 TPD Hiroshima pilot plant from 2002

(C) 10 TPD Matsushima pilot for testing coal fired flue gas from 2006 to 2008

(D) 500 TPD Barry CCS demonstration plant in Alabama, (25MW equivalent) from 2011

2.0TPD Nanko Pilot Plant (Kansai Electric Power Co.)

1.0TPD Hiroshima Pilot Plant (MHI' R&D Center)

10TPD Matsushima Pilot Plant (J-Power)

500TPD Barry Demonstration Plant (Southern Company)

1.3 KM CDR Process®

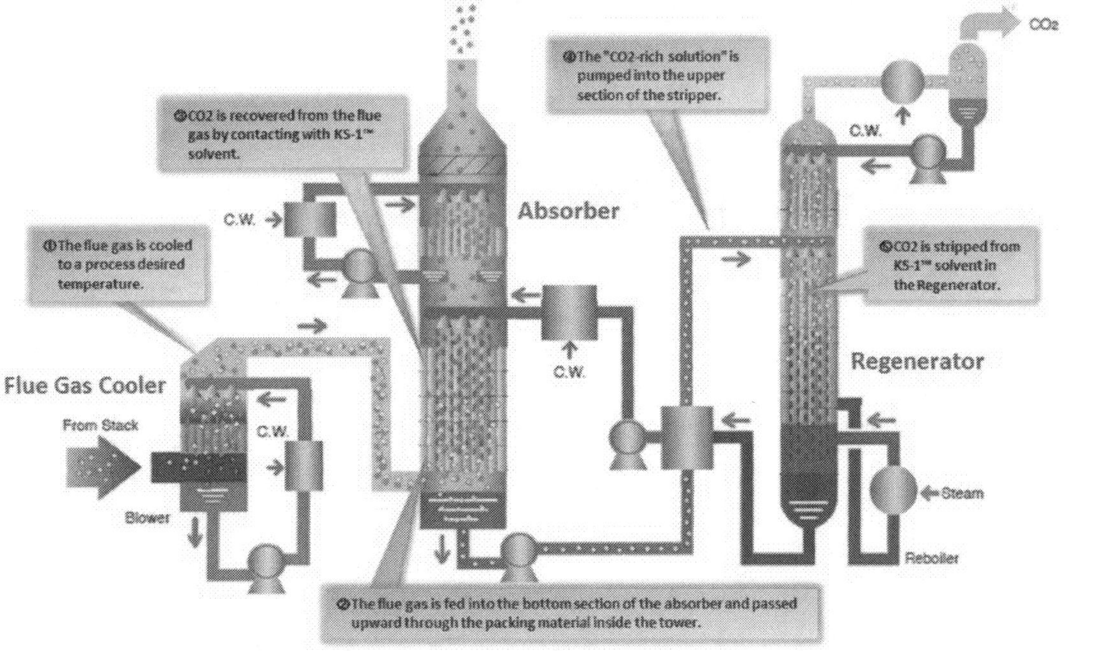

1.4 MHI CO_2 Capture Plants Commercial Experience

MITSUBISHI HEAVY INDUSTRIES, LTD.

- World leading large scale post combustion CO_2 capture technology licensor, with 11 commercial plants in operation from a variety of natural gas or heavy oil flue gas sources
- World's largest CCS plant is under construction in Texas.

1999
210 t/d Malaysia
NG Reformer
Urea expansion

2005
330 t/d Japan
Heavy oil boiler
General usage

2006
450 t/d India
NG Reformer
Urea expansion

2006
450 t/d India
NG Reformer
Urea expansion

2009
450 t/d India
NG Reformer
Urea expansion

2009
450 t/d Bahrain
NG Reformer
Urea expansion

2010
400 t/d UAE
NG Reformer
Urea expansion

2010
240 t/d Vietnam
NG Reformer
Urea expansion

2011
340 t/d Pakistan
NG Reformer
Urea expansion

2012
450 t/d India
NG Reformer
Urea expansion

2014
500 t/d Qatar
NG Reformer
Methanol expansion

- Operation Start
- CO_2 Capacity, location
- Flue gas source
- CO_2 usage

SOUTHERN COMPANY

2. 500 Metric TPD CO_2 Capture Demonstration Plant in USA

MITSUBISHI HEAVY INDUSTRIES, LTD.

MITSUBISHI HEAVY INDUSTRIES, LTD.
Our Technologies, Your Tomorrow

2.1 Specification of Demonstration Plant

Southern Company Plant Barry CO_2 Demo Plant

Item	Description
Plant location	Mobile County (Alabama, U.S.A.)
Plant owner	Southern Company subsidiary Alabama Power
Process	KM CDR Process®
Absorption liquid	KS-1™ solvent
Plant scale	Corresponding to 25 megawatts (MW)
Flue gas amount	116,800 Nm^3/h
CO_2 concentration	10.1 mol%-wet
CO_2 capture capacity	500 tonnes/day (150,000 tonnes/year)
CO_2 capture ratio	90 percent

World's First CCS Project fully integrated with Amine-based Post Combustion Process and CO_2 Injection.
Over 230,000 metric tons of CO_2 was captured and over 115,000 tons of CO_2 was injected as of August 2014.

3. Project Status of World's Largest CO_2 Capture Plant from Coal-fired Flue Gas

MITSUBISHI HEAVY INDUSTRIES, LTD.

Our Technologies, Your Tomorrow

3.1 Petra Nova CO_2 Capture Plant for CO_2-EOR (1/2)

MITSUBISHI HEAVY INDUSTRIES, LTD.

The world's largest CO_2 capture and compression plant from coal-fired power plant

- Project owner: Petra Nova, a partnership between NRG Energy, Inc. and JX Nippon Oil & Gas Exploration Corporation
- Location: NRG WA Parish Power Plant in Thompsons, TX.
- Flue gas source: Slipstream off of 650MW coal-fired boiler
- CO_2 concentration: 11.5%
- CO_2 capture capacity: 4,776 TPD (240MW equivalent)
- CO_2 capture ratio: 90%
- CO_2 Use : CO_2 EOR
- Pipeline : Approximately 81miles
- Injection Site: West Ranch oil field in Jackson County, TX
- Operation Start: 4th Quarter, 2016

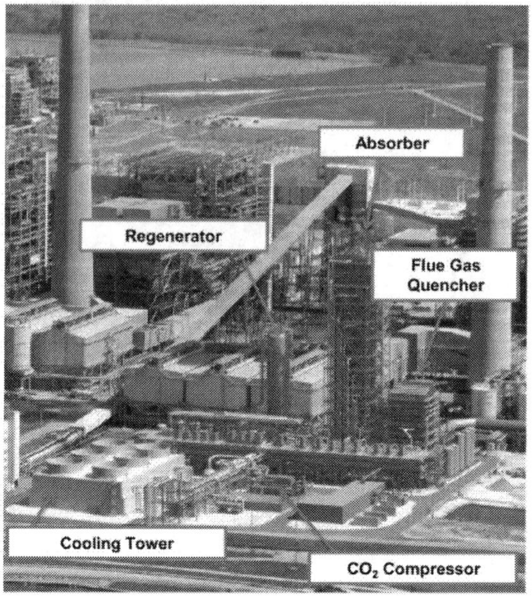

[Reference] U.S. Department of Energy "W.A. Parish Post-Combustion CO2 Capture and Sequestration Project Final Environmental Impact Statement Volume I" (Feb, 2013), DOE/EIS-0473

3.1 Petra Nova CO_2 Capture Plant for CO_2 EOR (2/2)

MITSUBISHI HEAVY INDUSTRIES, LTD.

- Captured and compressed CO_2 from W.A. Parish Power Plant will be delivered by 80 mile CO_2 pipeline to the West Ranch oil field.
- 1.4 million metric tons of greenhouse gas will be annually injected into the West Ranch formation.
- It is expected that oil production will be enhanced from 500 barrels/day to approx. 15,000 barrels/day.

The West Ranch CO_2-EOR Project — NRG FACT SHEET

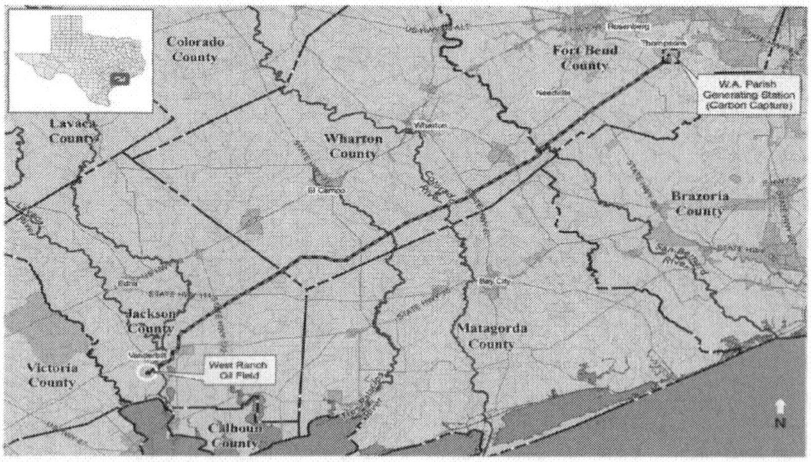

3.2 MHI's Scale-up Strategy

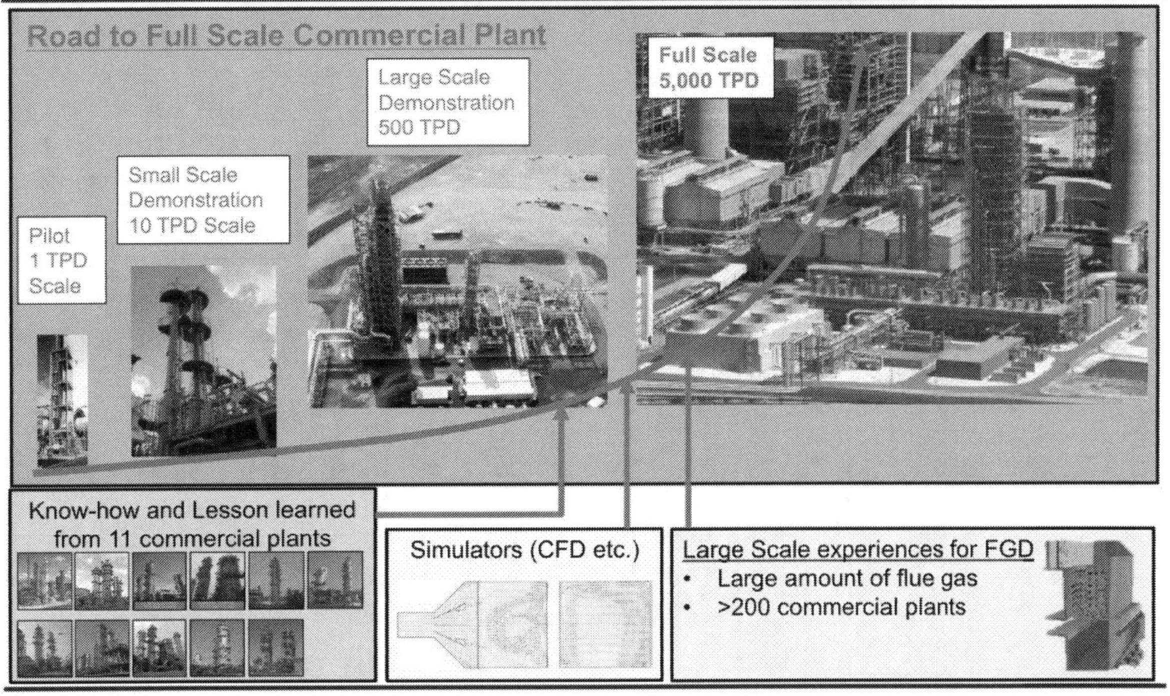

 13

3.3 Block Flow Diagram of CO_2 Capture Plant

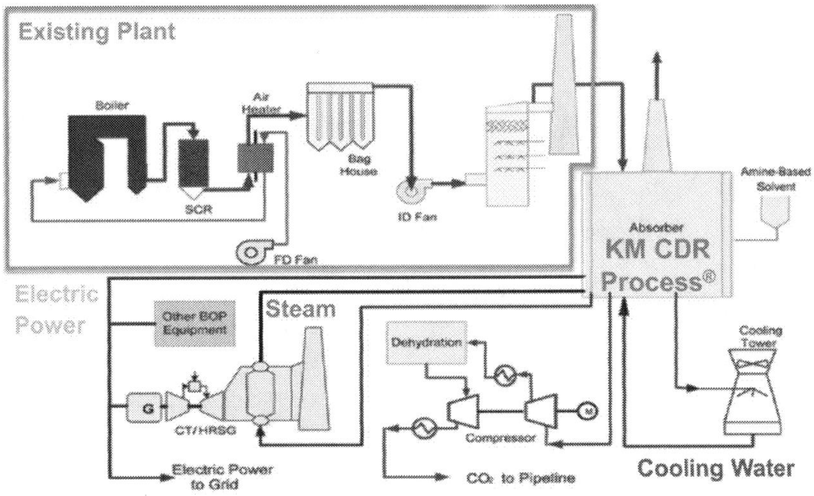

- **Power plant refurbishment is NOT REQUIRED**
- **Minimum tie-in work**
- **No parasitic load on coal unit**
- **Does not impact the price of electricity**

 [Reference] U.S. Department of Energy "W.A. Parish Post-Combustion CO2 Capture and Sequestration Project Final Environmental Impact Statement Volume I" (Feb, 2013), DOE/EIS-0473 14

3.4 Key Process Technologies for Petra Nova CO_2 Capture Plant

MITSUBISHI HEAVY INDUSTRIES, LTD.

All technologies were proven in 500 Metric TPD CO_2 Capture Demonstration Plant in USA

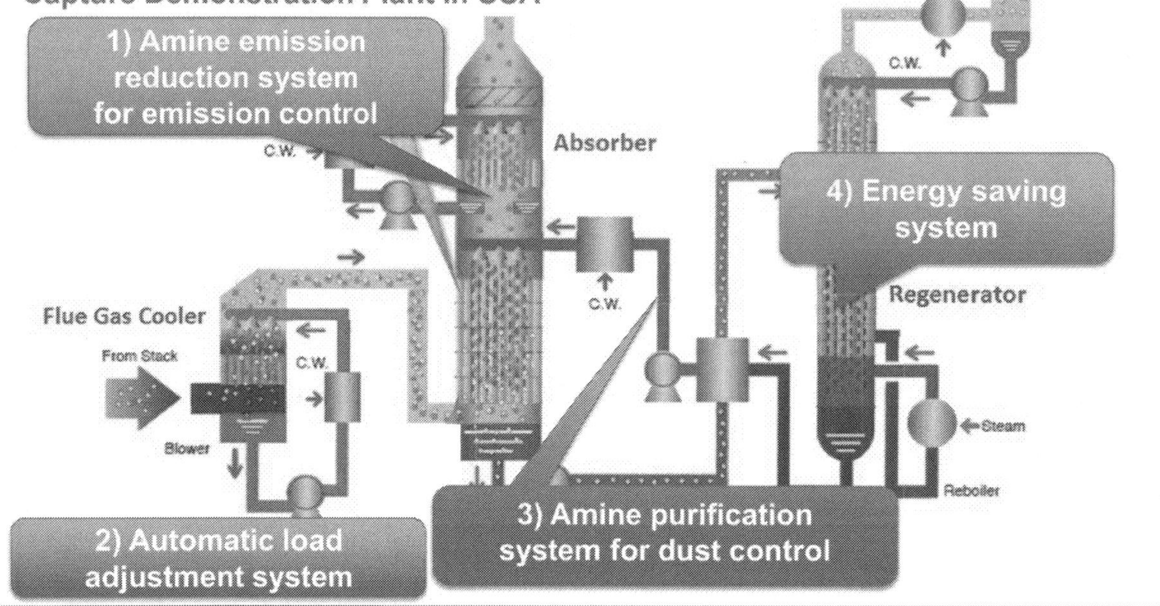

3.5 Lessons Learned from Petra Nova CO_2 Capture Plant (1) Load Change Requirement

MITSUBISHI HEAVY INDUSTRIES, LTD.

- Power Plant Load
  - ✓ CCS designed with load following flexibility under turndown conditions
  - ✓ CCS operational profile will be optimized for various load conditions
  - ✓ Wide load change is required for CO_2 Capture plant

Automated Load Adjustment System

→ Easy operation, maximum plant performance

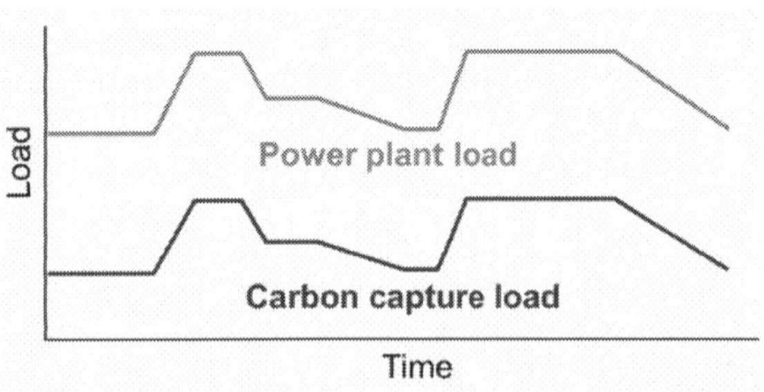

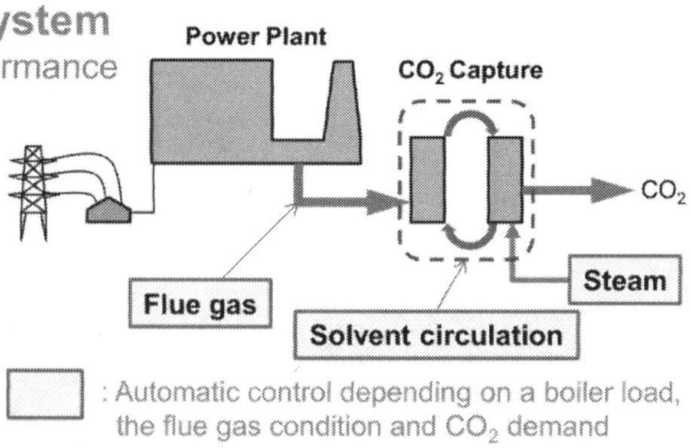

3.5 Lessons Learned from Petra Nova CO_2 Capture Plant (2) Constructability

- Construction for Quencher and Absorber

  Rectangular steel towers and Modular construction
  → Speedy Construction / High Flexibility for transportation

 17

3.6 Project Progress

- Construction: Ahead of Schedule (As of Nov. 2015)
- Planned Operation Start: 4th Quarter, 2016

 18

4. Summary

MITSUBISHI HEAVY INDUSTRIES, LTD.

MITSUBISHI HEAVY INDUSTRIES, LTD.
Our Technologies, Your Tomorrow

4.1 Summary

- The MHI's KM CDR process® is applied to eleven (11) commercial CO_2 capture plants.
- Southern Company and MHI carried out the 500 MTPD CO_2 capture and storage demonstration project at Alabama Power's Plant Barry. This project became the world's largest start to finish CCS demonstration project on coal fired flue gas.
- Based on the know-how gained from CCS demonstration project together with extensive R&D activities and commercial plant experiences, the world largest CO_2 capture plant under construction in SW Houston, Texas for NRG Energy and JX Nippon Oil & Gas Exploration Corporation.
- MHI continues extensive effort to enhance competitiveness in terms of CO_2 capture cost and to accelerates commercial viability of CCS/EOR project.

Thank you!

For more information please contact:

Dale Wilterdink – Mitsubishi Heavy Industries America, Inc.
Project Manager, PNC-EOR
Email: dale_wilterdink@mhiahq.com
Tel: +1-713-351-6405

Progress on the developments of an advanced aqueous amine-based post combustion CO_2 capture utilizing BASF's OASE® blue technology

2015 Carbon Management Technology Conference
Sugarland, Texas; November 18, 2015

Krish R. Krishnamurthy, Linde LLC, Murray Hill, NJ
Torsten Stoffregen, Linde Engineering, Dresden
Sean Rigby, BASF Corp., Houston, TX
Kevin O'Brien, ISTC, Univ. of Illinois, Champaign, IL

BASF We create chemistry
THE LINDE GROUP Linde

BASF / Linde partnership
Delivers total solutions with confidence

BASF We create chemistry
THE LINDE GROUP Linde

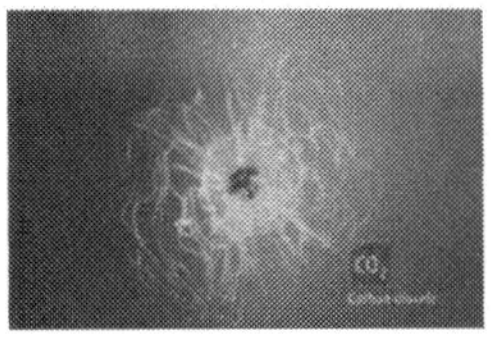

BASF Solvent/Process Expertise
Basic Design Package
Process performance
Emissions performance

Founded	1865
Sales (2014)	€74.3 billion
Employees	113,292

PCC capture

Linde Engineering Expertise
Process optimization
Basic/Detailed Engineering
Package/EPC wrap

Founded	1879
Sales (2014)	€17 billion
Employees	65,591

BASF OASE® blue Technology Development
Adopted and optimized for PCC applications

BASF We create chemistry | THE LINDE GROUP Linde

Equilibria → Kinetics → Stability

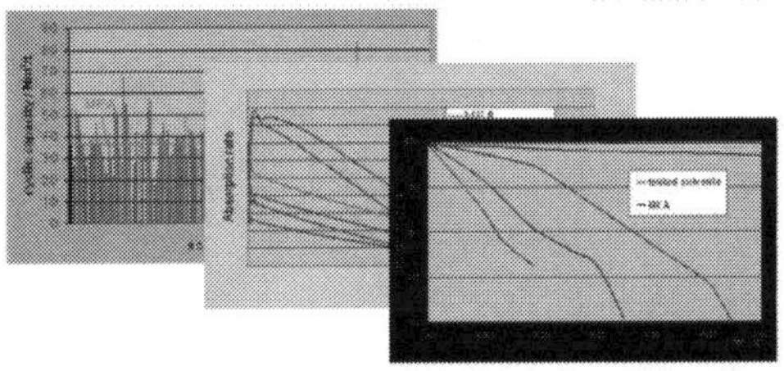

Lab scale
- Ludwigshafen, Germany
- Advanced solvent screening, development, optimization

Mini plant
- Ludwigshafen, Germany
- Solvent performance verification

Pilot: 0.45 MWe
- 2009, Niederaussem
- Process opt., materials & emissions testing

Pilot: 1.5 MWe
- 2014, Wilsonville, AL
- Design improvements, emissions confirmation

Niederaussem PCC Pilot Plant: Fact Sheet

RWE The energy to lead | BASF We create chemistry | THE LINDE GROUP Linde

- Flue gas: 1,550 Nm^3/h; CO_2 product: 7.2 t CO_2/day; capture rate 90%
- Commissioning and start-up 2009, availability of 97%
- BASF's OASE® blue was tested over 26,000 hours (> 3 years)
- nearly 7,000 t CO_2 were captured with OASE® blue

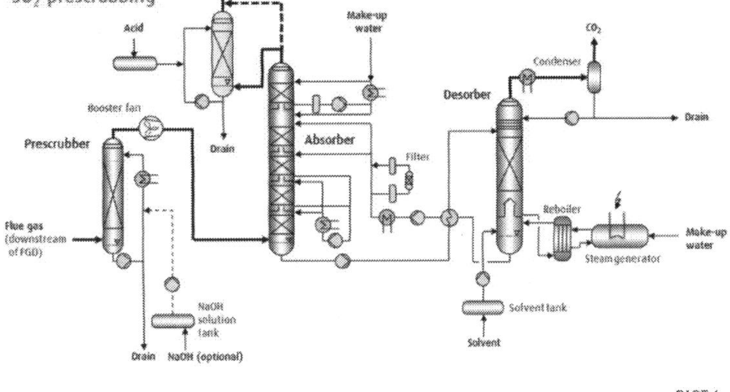

Niederaussem PCC Pilot Plant
Test Campaigns

Solvent testing

- MEA & Process
- GUSTAV200
- LUDWIG540
- ⇨ Selection optimal solvent: OASE® blue

Long-term testing, optimisation

- Modification of plant components
- Intermediate testing
- Long-term testing (FGD)
- Long-term testing (FGDplus)

Optimisation, Long-term testing

- Overall optimum emission mitigation
- Increase of O_2-content flue gas
- Variation OASE® blue
- ⇨ Optimum OASE® blue
- Long-term testing (FGD/FGDplus)

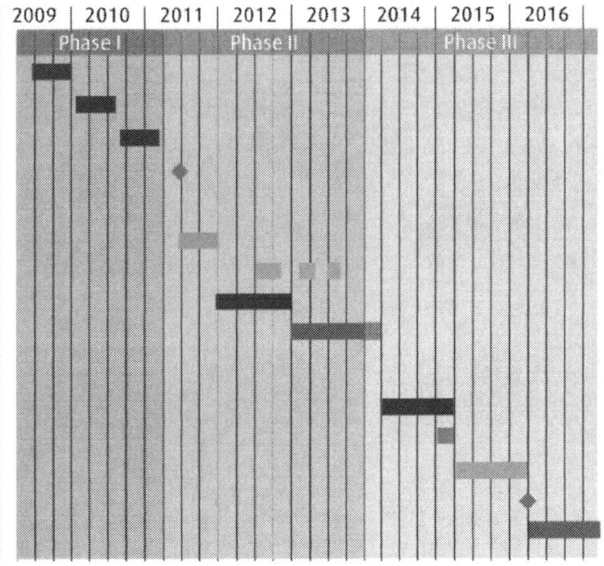

Niederaussem PCC Pilot Plant
Operational experiences and main results

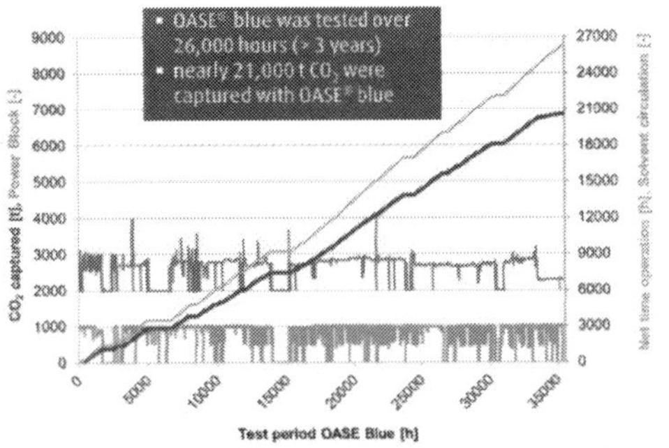

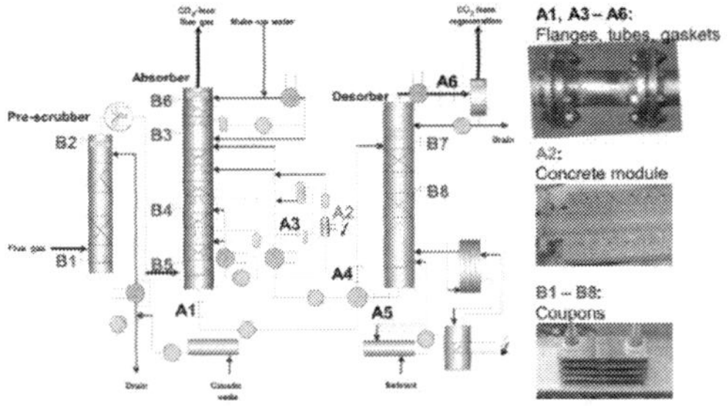

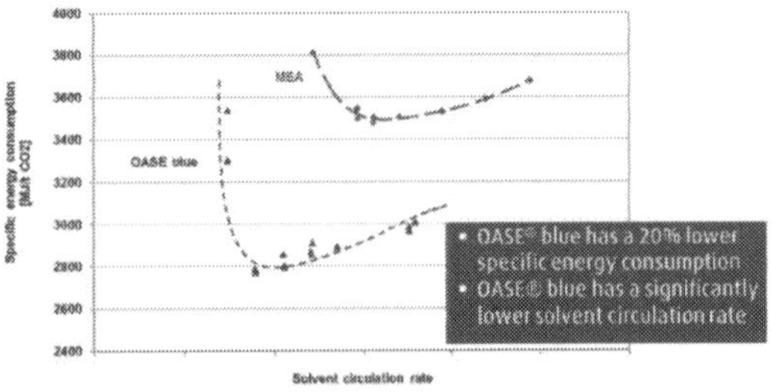

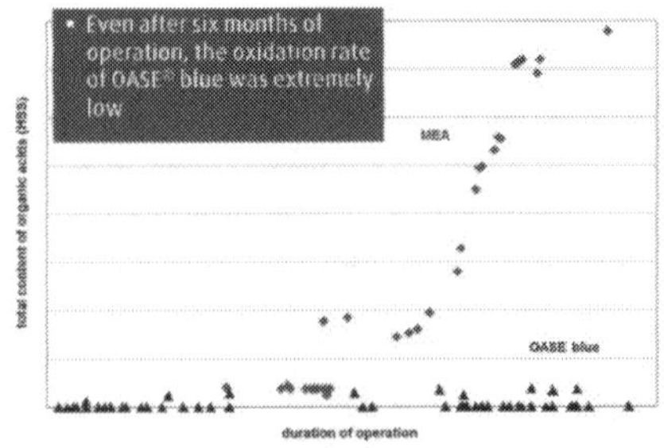

Niederaussem PCC Pilot Plant
Emission reduction measures

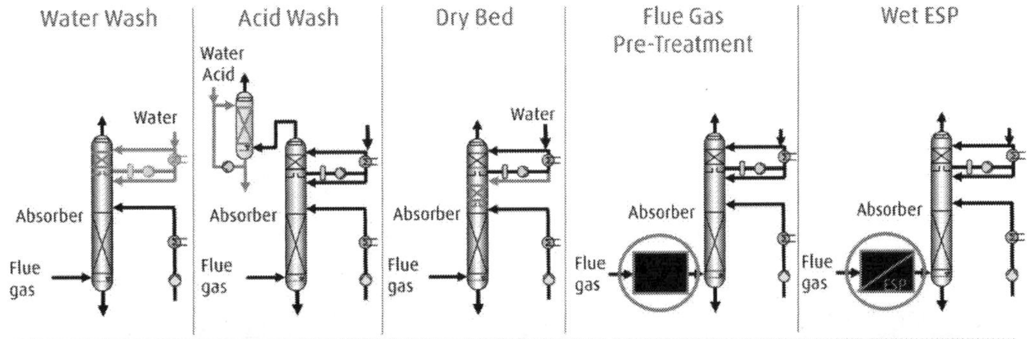

Variation of Process Configurations:
- FGDplus/pre-scrubbing (w/wo addition of NaOH)
- Number of water wash steps (1 or 2)
- Water wash with double height
- Combination water wash and dry bed
- Combination acid wash and dry bed
- Combination with wet electric precipitator

Variation of Parameters:
- Water wash temperature (40° - 60°C)
- Intercooler temperature
- pH-value acid wash
- Voltage of wet electric precipitator

Niederaussem PCC Pilot Plant
Emission reduction measures – "Dry bed"

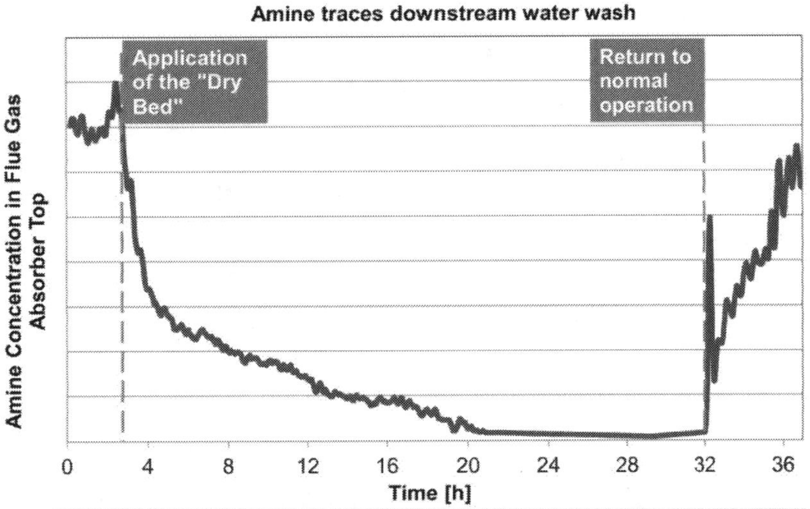

Reduction of amine emissions by an order of magnitude:
→ Proprietary process configuration „Dry Bed"

Wilsonville PCC Pilot Plant
Fact Sheet

BASF We create chemistry | THE LINDE GROUP Linde

Project essentials

- DOE-NETL funded project (\$16.2 million funding)
- Total project cost \$22.7 million
- Location: 880 MWe Gaston Power plant (operated by Southern Co.) in Wilsonville, AL
- Site of the National Carbon Capture Center
- Capacity: Up to 6,250 Nm^3/h flue gas from coal fired power plant (30 t/d CO_2); Up to 1.5 MWe
- CO_2 purity 99+ vol % (Dry basis)
- Project start: November 2011
- Start-up: January 2015
- Project Duration: 4.5 years
- Partners: Linde LLC, Linde Engineering North America, Linde Engineering Dresden, BASF, DOE-NETL, EPRI, Southern Company (Host site)

Wilsonville PCC Pilot Plant
Flow sheet - Novel features evaluated

BASF We create chemistry | THE LINDE GROUP Linde

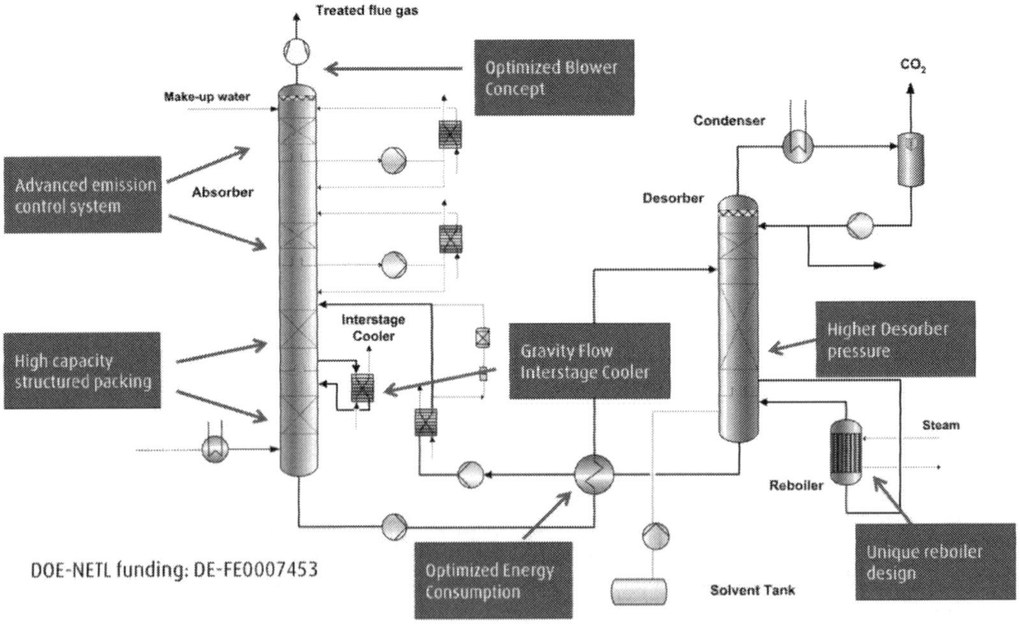

DOE-NETL funding: DE-FE0007453

Wilsonville PCC Pilot Plant
Construction

BASF We create chemistry

THE LINDE GROUP Linde

Wilsonville PCC Pilot Plant
Test operation

BASF We create chemistry

THE LINDE GROUP Linde

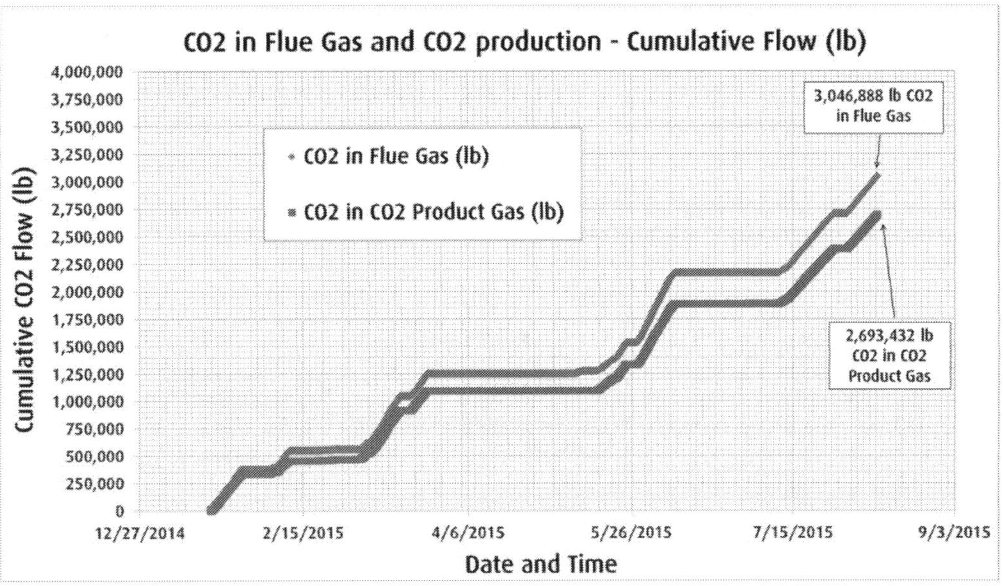

Wilsonville PCC Pilot Plant
Typical result for specific energy consumption

BASF We create chemistry | THE LINDE GROUP Linde

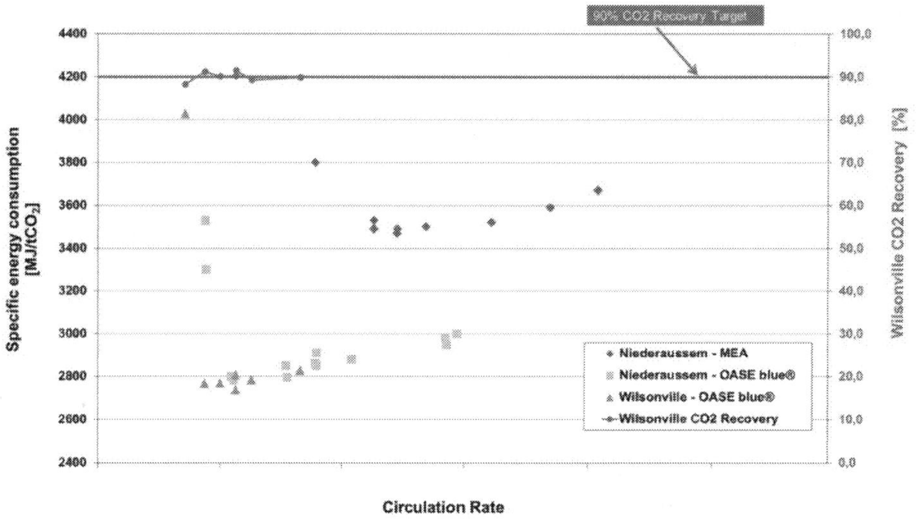

Wilsonville PCC Pilot Plant
Parametric Testing Performed

BASF We create chemistry | THE LINDE GROUP Linde

S.No.	Key variable	Status
1	Flue gas flow rate	7,500 to 15,750 lbs/hr
2	Flue gas temperature to absorber	86°F to 104°F
3	Treated gas temperature exit absorber	86°F to 115°F
4	Lean solution temperature to absorber	104°F to 140°F
5	Inter-stage cooler	On (104°F) /Off
6	Regeneration pressure	1.6 to 3.4 bars
7	Solvent circulation rate	Varied from 80 to 120%
8	CO_2 capture rate	90% typical Varied from 85% to >95%

Large pilot development at the Abbott Power Plant, University of Illinois, Champaign, IL

BASF We create chemistry | THE LINDE GROUP Linde

PILOT

- 1.5 MWe coal-fired flue gas (30 TPD CO2)
- Located at the National Carbon Capture Center, Wilsonville, AL; Host site: Southern Co.'s Alabama Power Gaston plant 880 MWe
- Project period: 2011-2016

LARGE PILOT

- 25 MWe coal-fired flue gas (500 TPD CO2)
- Located at Abbott Power Plant, Campaign, IL ; Host site: University of Illinois 35 MWe cogen plant
- US DOE funding: $1 million; Phase 1 Project definition in progress (Q4 2015 – Q3 2016)
- Phase 2 proposal & selection in Q3/Q4 2016

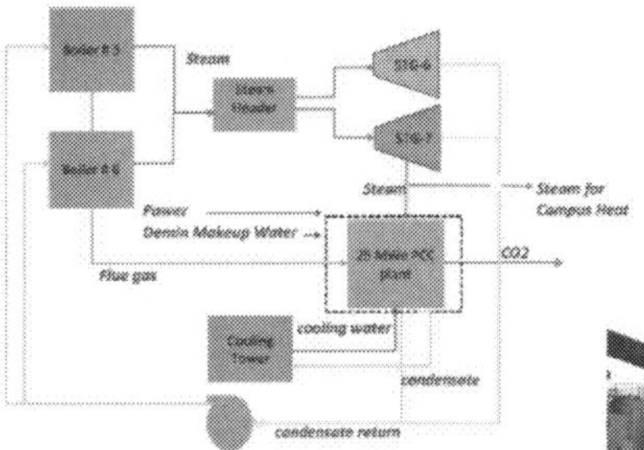

ILLINOIS
UNIVERSITY OF ILLINOIS AT URBANA-CHAMPAIGN

15 15

Large pilot development at the Abbott Power Plant, University of Illinois: Deliverables

BASF We create chemistry | THE LINDE GROUP Linde

Phase 1 (Oct. 1, 2015 – June. 30, 2016)

- Phase 1 Technology engineering and economic analysis report – March 30, 2016
- Phase 1 Technology gap analysis – March 31, 2016
- Phase 1 EH&S study – March 31, 2016
- Phase 1 Topical Report addressing technology, design basis & capabilities, schedule, host site, project participants, updated Phase 2 (detailed design, procurement, construction, operation & decommissioning) costs – March 31, 2016
- Updated project management plan – March 31, 2016
- Phase 2 Environmental Questionnaire – March 31, 2016
- Executed financial and host site agreements – June 30, 2016
- Updated representations and certifications – June 30, 2016

16

Linde-BASF PCC technology for large scale

Two prong development and commercialization approach

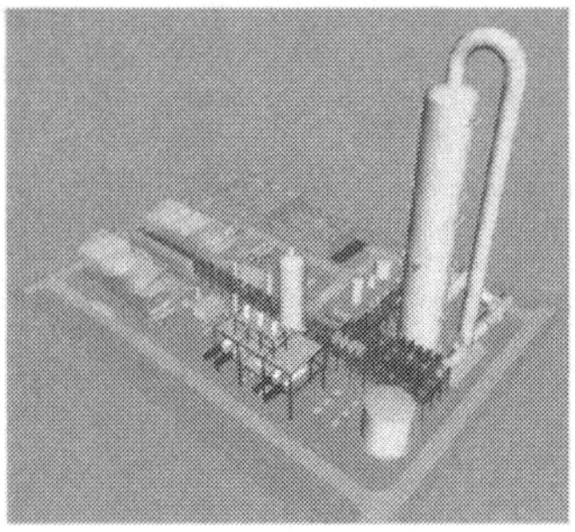
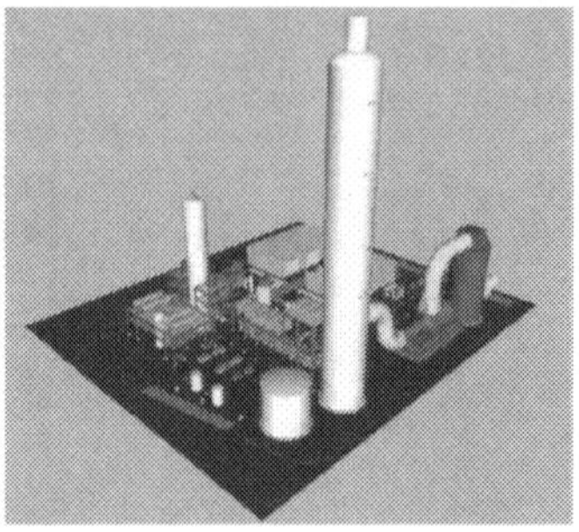
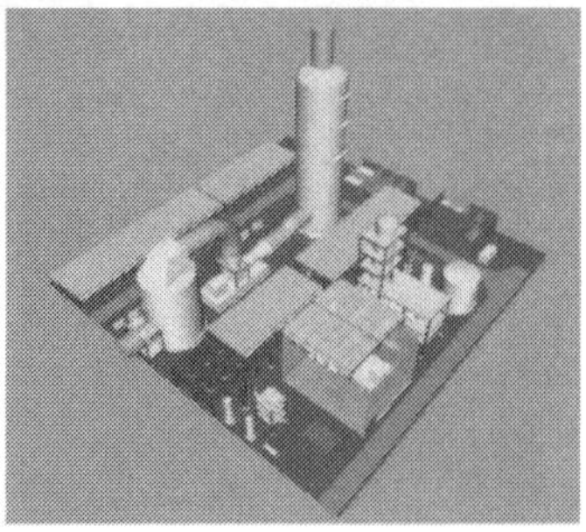

- Solvent performance
- Equipment design
- Process Design
- Emissions control
- Scale-up
- EPC at large scale

- OASE® blue Technology is ready for commercialization
- continued development for cost reduction required

Summary and conclusions

- Linde and BASF are partnering in the development of an advanced PCC technology incorporating BASF's novel amine-based process along with Linde's process and engineering innovations
- Performance demonstrated and long term stability validated on a 0.45 MWe lignite fired power plant flue gases (Niederaussem, Germany)
- The current testing campaign includes parametric testing of two new solvents from BASF, followed by a long term test for the most promising solvent
- Nominal 1 MWe pilot plant at the NCCC in Wilsonville, AL commissioned; initial operations & testing have demonstrated stable operation, validation of functional features and initial achievement of several key targets
- The current testing campaign in Wilsonville is focused on parametric tests aimed at energy optimization, emissions minimization and validation of higher pressure regenerator operation. This will be followed long duration testing (4-6 months) to demonstrate solvent stability.
- Phase 1 design and engineering evaluation for a 25 MWe large pilot implementation at the Abbott Power Plant in University of Illinois, Champaign, IL is in progress.
- Technology is ready for commercialization with continuing efforts on further development to reduce cost of capture.

Disclaimer

Disclaimer: "This presentation was prepared as an account of work sponsored by an agency of the United States Government. Neither the United States Government nor any agency thereof, nor any of their employees, makes any warranty, express or implied, or assumes any legal liability or responsibility for the accuracy, completeness, or usefulness of any information, apparatus, product, or process disclosed, or represents that its use would not infringe privately owned rights. Reference herein to any specific commercial product, process, or service by trade name, trademark, manufacturer, or otherwise does not necessarily constitute or imply its endorsement, recommendation, or favoring by the United States Government or any agency thereof. The views and opinions of authors expressed herein do not necessarily state or reflect those of the United States Government or any agency thereof."

19 19

Acknowledgements

1. This presentation is based on work supported by the Department of Energy under Award Number DE-FE0007453 (to Linde as prime for PCC Wilsonville, AL) & DE-FE0026588 (to UIUC as prime for large pilot) and BMWi under sponsorship codes: 0327793 A to I for RWE Power, Linde and BASF (PCC Niederaussem)
2. Linde: Stevan Jovanovic, Devin Bostick, Annett Kutzschbach
3. BASF: Gerald Vorberg and Gustavo Lozano
4. RWE: Peter Moser, Sandra Schmidt and Georg Wiechers
5. UIUC: Yongqi Lu
6. DOE-NETL Project Manager(s): Andrew Jones, Bruce Lani
7. SCS/NCCC: Justin Anthony and Frank Morton

ILLINOIS
UNIVERSITY OF ILLINOIS AT URBANA-CHAMPAIGN

RWE The energy to lead

SOUTHERN COMPANY

THE LINDE GROUP

We create chemistry

Thanks for your attention.

21

21

CAER
CENTER FOR APPLIED ENERGY RESEARCH

UK

The Unique UKy-CAER CO_2 Capture Process: Effectiveness of the Liquid Desiccant Air Drying System

Bethany Clontz[a], Kunlei Liu[a,b],
Michael Manahan[c], David Link[c]

[a] *Center for Applied Energy Research, University of Kentucky, Lexington, KY, USA*

[b] *Department of Mechanical Engineering, University of Kentucky, Lexington, KY USA*

[c] LG&E and KU Energy LLC, Louisville, KY USA

CAER
CENTER FOR APPLIED ENERGY RESEARCH

University of Kentucky
Center for Applied Energy Research

UK

www.caer.uky.edu

Overview

- Background Information
- Operation at Design Conditions
- Conclusions
- Future Experimentation

CMTC Conference

November 17 - 19, 2015

The Unique Uky-CAER Carbon Capture Facility

CMTC Conference

November 17 - 19, 2015

Technical and Economic Analysis

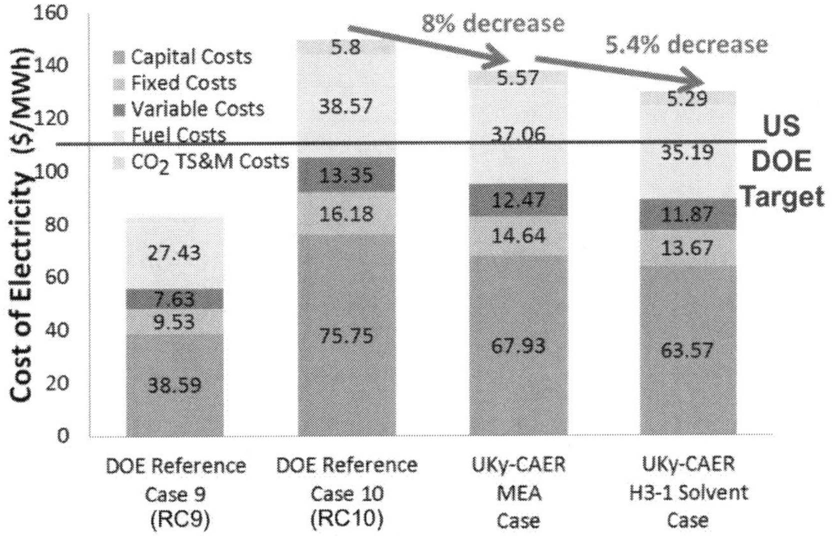

All at 90% capture rate and CO_2 compression to 2200 psia.

55.8% increase in Cost of Electricity from RC9

\$46.93/tonne CO_2 Captured, lower than RC10 by \$14.38/tonne CO_2

The Unique CAER-Uky Carbon Capture Process

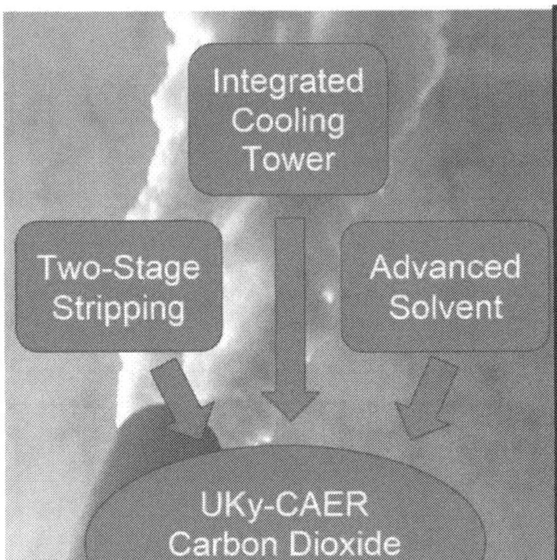

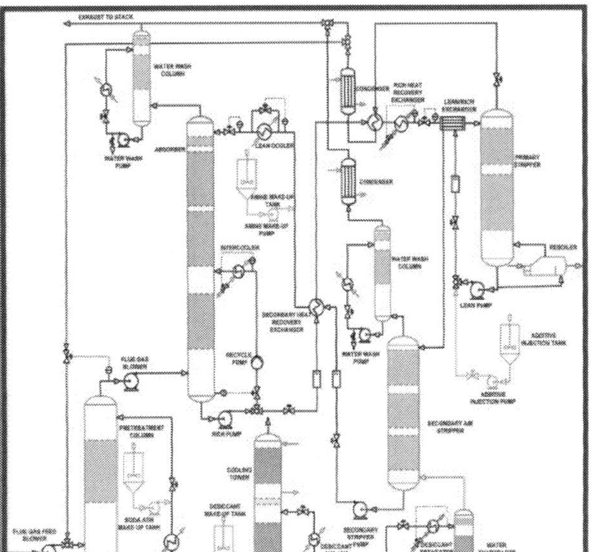

CMTC Conference November 17 - 19, 2015

Purpose of Heat Integrated Cooling Tower

- To recover waste heat otherwise rejected by the process.
- To reduce the wet bulb temperature to the top stage, which provides more efficient and effective cooling in the water stage.
- To provide warm, water saturated air to the secondary air stripper and to provide a mechanism for water make-up

Design of Cooling Tower

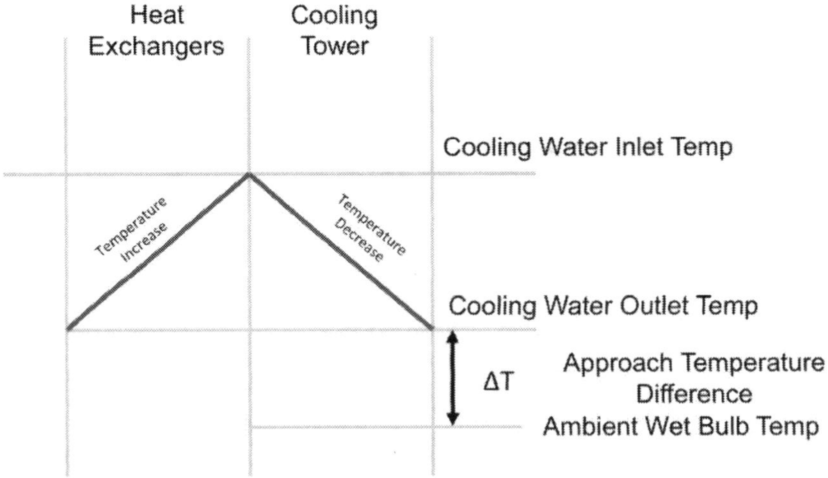

- The difference between the cooling water outlet and the wet bulb temperature is typically close to 15 ˚F.

Design of Cooling Tower

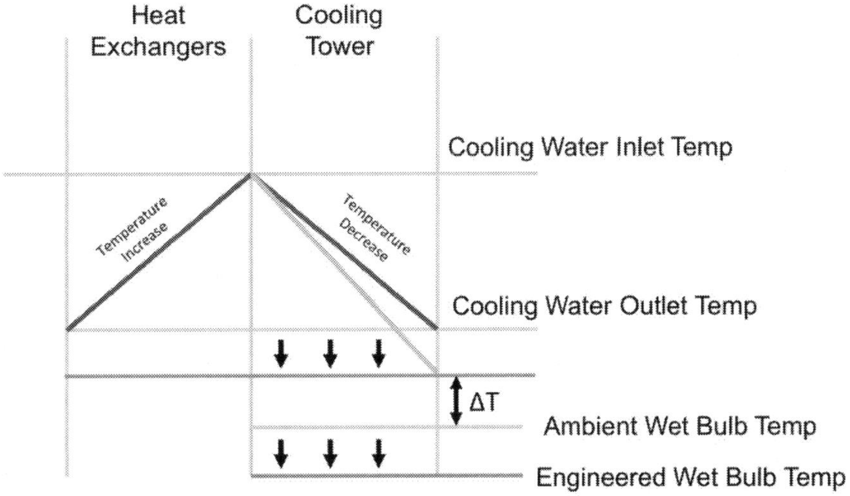

- This difference will be maintained with a shift in the wet bulb temperature, resulting in cooler water.
- The approach temperature difference is between the ambient wet bulb temperature and the cooling water outlet temperature.

CMTC Conference November 17 - 19, 2015

Region and Duration at which Benefit is Realized

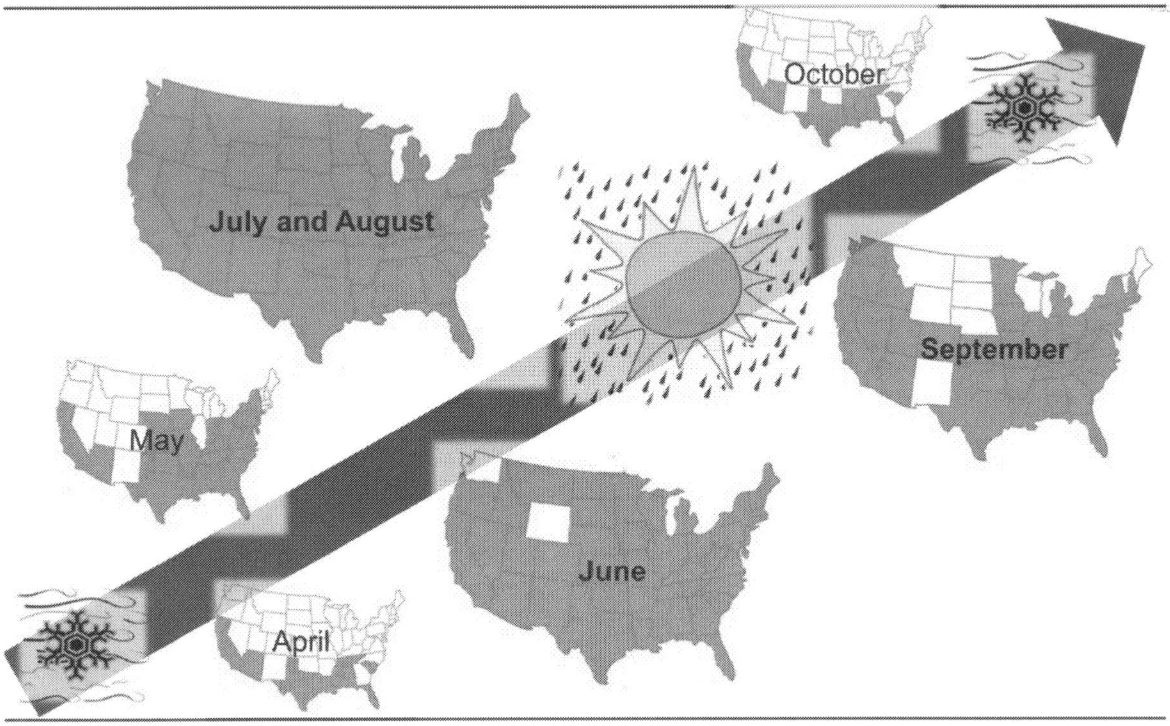

Design Conditions Operations Data

Design Operating Conditions

- Air Flowrate to the Cooling Tower: 25000 SCFM
- Liquid Desiccant Circulation Flowrate: 30 to 50 GPM
- Liquid Desiccant Preheater Temperature: 155 °F to 180 °F
- Chiller Temperature: 65 °F

Normal Operations Data

Cooling Tower
Efficiency

- Efficiency for cooling towers:

$$\mu = (T_i - T_o) / (T_i - T_{wb})*100$$

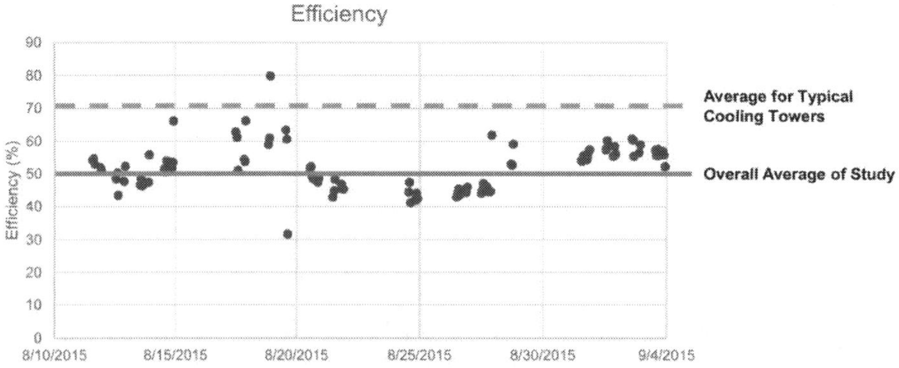

Normal Operations Data

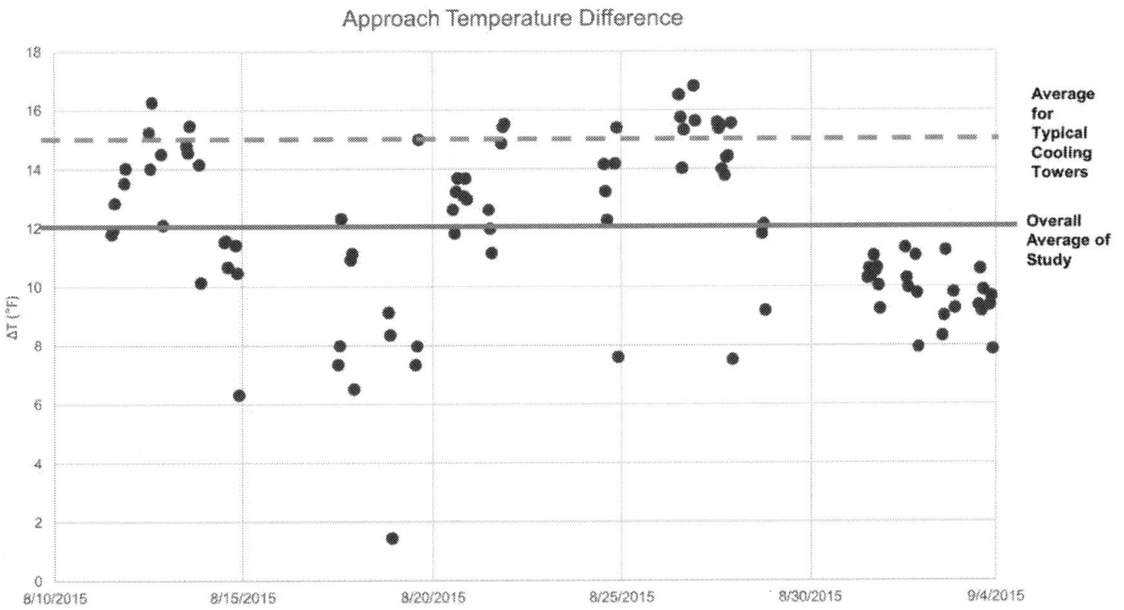

CMTC Conference

November 17 - 19, 2015

Normal Operations Data

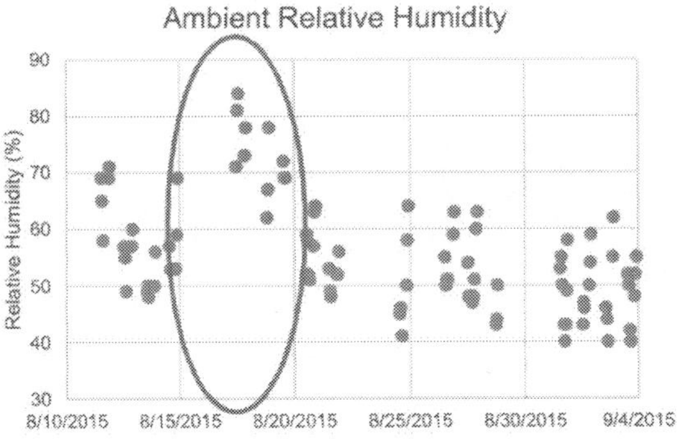

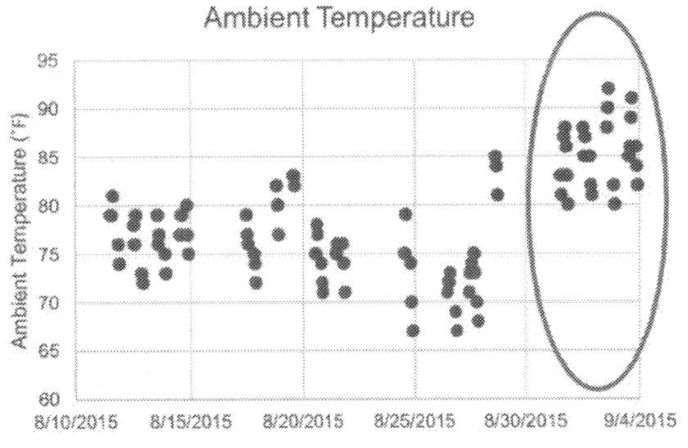

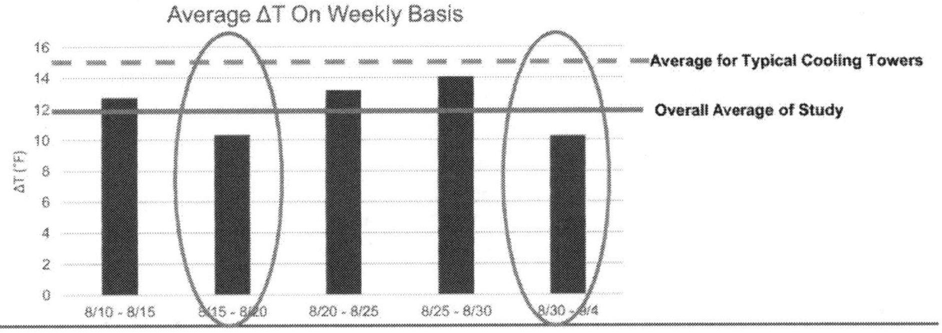

CMTC Conference

November 17 - 19, 2015

Conclusions

- At the design conditions, the approach temperature difference is less than that of a typical cooling tower.
- With the addition of heat exchangers in the amine loop, and the flow of the liquid desiccant through these heat exchangers does add collection of low grade heat.
- This collection of waste heat can be used for other purposes, such as water make-up and humid air provided to the secondary air stripper.

Technology Development Pathway

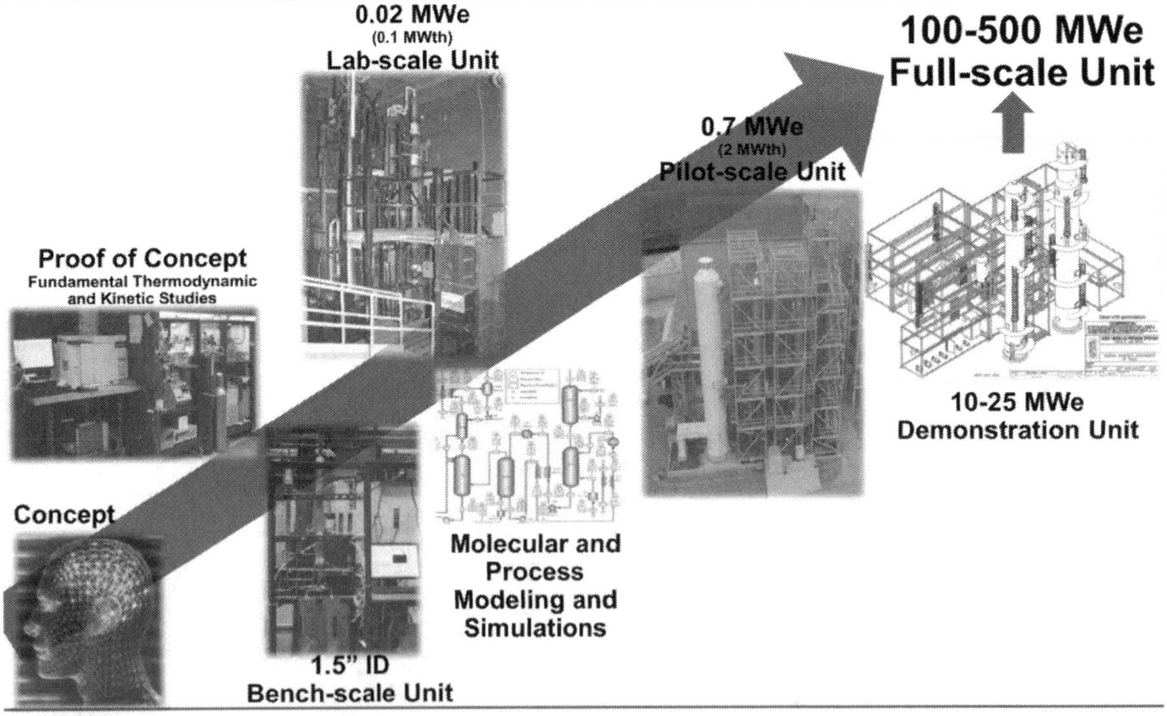

Acknowledgements

José Figueroa, DOE NETL

CMRG Members

Donnie Duncan, LG&E and KU
David Link, LG&E and KU
Michael Manahan, LG&E and KU
Jeff Fraley, LG&E and KU

UKy-CAER Slipstream Team

- **Zhen Fan**
- **Brandon Harris**
- **Kunlei Liu**
- **Marshall Marcum**
- **Heather Nikolic**
- **Jonathan Pelgen**
- **Charlie Whitenack**

GLOBAL
CCS
INSTITUTE

CCS Pilot Projects in the Asia Pacific Region

Tony Zhang, Senior Adviser – Carbon Capture

Carbon Management Technology Conference 2015, Sugar Land, Texas, USA 17-19 November 2015

Global CCS Institute

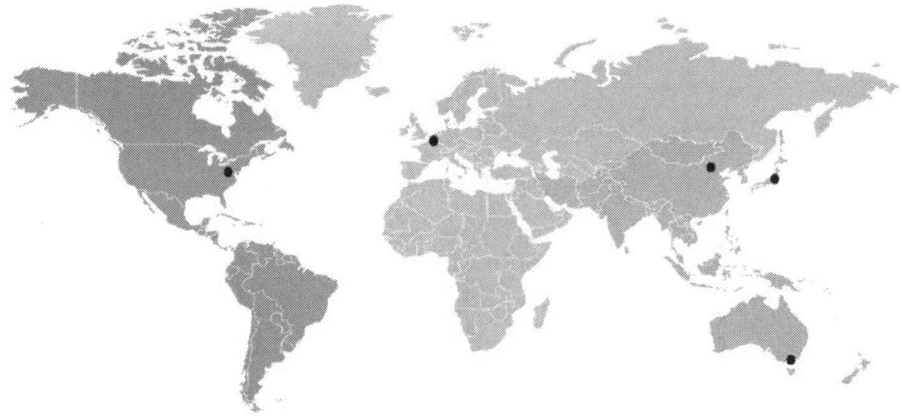

- Knowledge sharing
- R&D coordination
- International collaboration
- Global networks and regional networks
- Fact-based advice and advocacy

http://www.globalccsinstitute.com/publications

http://co2degrees.com

The cost objectives for future carbon capture technology

US DoE target

- By 2025 – 2nd Generation:
  - Commercial deployment of 2nd Gen beginning
  - A capture cost of around $40/tonne of captured CO_2.
- By 2035 – 3rd Generation:
  - 3rd Gen / Transformational technology commercial readiness
  - Less than $40/tonne of captured CO_2

K Gerdes, US DoE NETL, 2013

Pilot Projects in Asia Pacific Region

Strategic benefits of conducting pilot projects

- Capture cost reduction (material, process and equipment)
- Technical issues investigation
- Public confidence in the technology
- Personnel training
- Learning by doing
- Adaptation according to local conditions

Pilot Projects in Asia Pacific Region

List of projects (including pilot and small pilot projects)

Australia

- Alcoa Kwinana Carbonation Plant Australia
- Callide Oxyfuel Project Australia
- CO2CRC Otway Project Australia
- Hazelwood Carbon Capture and Mineral Sequestration Pilot Plant Australia
- The Australia-China Post Combustion Capture (PCC) Feasibility Study Project

Greater China Area

- Chongqing Hechuan Shuanghuai Power Plant CO_2 Capture Industrial Demonstration Project
- Datang International Power Generation NGCC Carbon Capture Project
- Huaneng Gaobeidian Power Plant Carbon Capture Pilot Project
- Huaneng GreenGen IGCC CCUS Pilot Scale System
- Huazhong University of Science and Technology Oxyfuel Project
- Jilin Oil Field EOR Demonstration Project
- Shanghai Shidongkou 2nd Power Plant Carbon Capture Demonstration Project
- Shenhua Group Ordos Carbon Capture and Storage (CCS) Demonstration Project
- Sinopec Shengli Oilfield Carbon Capture Utilization and Storage Pilot Project
- Sinopec Zhongyuan Carbon Capture Utilization and Storage Pilot Project
- ITRI Calcium Looping Pilot.pdf ITRI Calcium Looping Pilot

South Korea

- Dry-sorbent CO_2 Capture System - Hadong
- KoSol Process for CO_2 Capture (KPCC) – Boryeong

Japan

- COURSE 50 – "CO2 Ultimate Reduction in Steelmaking Process by Innovative Technology for Cool Earth 50"
- Mikawa Post Combustion Capture Pilot Plant
- Saga City Waste Incineration Plant
- The EAGLE Project
- The Osaki CoolGen Project
- Tomakomai CCS Demonstration Project

India

- Carbon Capture Solutions – Solvay Vishnu Capture Project India

Pilot Projects in Asia Pacific Region

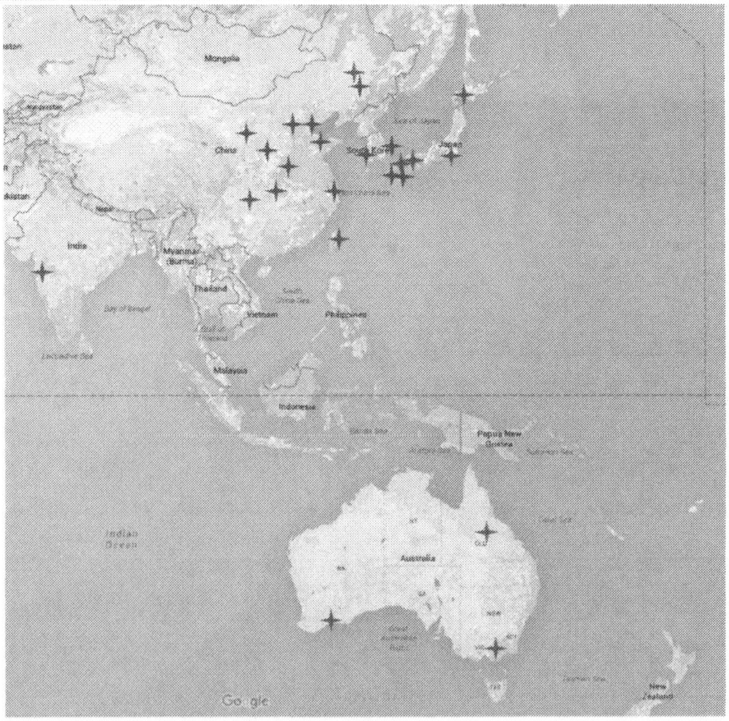

PILOT CAPTURE PROJECTS IN APAC

CNPC Jilin Oilfield CCUS Project

Key Information

- Qian'An, Jilin Province, China
- Capture Capacity: 630,000 tonnes per year
- Time: first stage commenced operation in 2008.
- CO_2 Source:
    - natural gas field; oil field associated gas
    - 21.95% - 31.92% v/v
    - 5.57 MPa
- Capture technology:
    - Amine-base solvent (activated MDEA) (1,800 tpd)
    - Adsorbent (pilot, 80,000 m^3 per day)
    - Membrane (pilot, 50,000 m^3 per day)

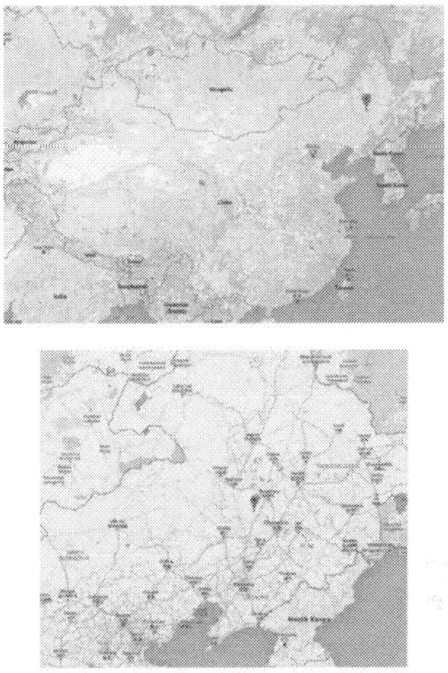

Jinlin Oilfield CCS-EOR location

Reference: 1. Ma. XH 2015. *CNPC CO2 Capture and Recycled Injection.*

CNPC Jilin Oilfield CCUS Project

Technology Comparison – CO_2 capture from natural gas

- CAPEX (Feed gas: 1.20 million Nm^3/day (42.47 MMSCFD))

    - Membrane + Amine: RMB 205 million
    - Activated MDEA: RMB 214 million (selected)
    - Pressure Swing Adsorption: RMB 190 million

Reference: 1. Ma. XH 2015. *CNPC CO2 Capture and Recycled Injection.*

CNPC Jilin Oilfield CCUS Project

Technology Comparison – CO_2 separation from EOR gas oil mixture

- Technology Choices:
  i. PSA;
  ii. Membrane;
  iii. Solvent;
  iv. Membrane + Amine;
  v. Membrane + PSA
- OPEX comparison

PSA is much cheaper to run compared with other options.

Reference: 1. Ma. XH 2015. *CNPC CO2 Capture and Recycled Injection.*

CNPC Jilin Oilfield CCUS Project

Key Performance to date

- Period of operation more than 7 years
- Accumulated injection amount of CO_2: More than 700,000 tonnes
- Oil production: 200, 000 tonnes per year
- Pipeline:
  - ✓ Block Hei#59 17.2 km (6 injection, 25 production, started April 2008) *Pilot scale*
  - ✓ Block Hei#79, 23.7 km (18 injection, 60 production, started March 2010) *Demonstration scale*
  - ✓ Block Hei#46, ~26 km (gas phase pipeline transport, supercritical injection, 31 injection, 157 production, 1,200 tpd injection) *Industry scale, August 2013 started construction, commissioned 25 September 2015*

Jinlin Oilfield Changling Natural Gas Purification Plant

Jinlin Oilfield Block Hei#46 injection station

Reference: 1. The Central People's Government of PRC 2009. <http://www.gov.cn/jrzg/2009-12/23/content_1495097.htm>.
2. CPCIA 2014. (3 September 2014) <http://www.cpcia.org.cn/html/1/20149/141438.html>.
3. ChinaIRN 2015. (22 April 2015) < http://www.chinairn.com/news/20150422/170255444.shtml >.
4. CNPC 2014. <http://news.cnpc.com.cn/system/2014/10/08/001509858.shtml>.
5. CNPC 2015.< http://news.cnpc.com.cn/system/2015/04/22/001538344.shtml >.

Datang International Power Generation NGCC Carbon Capture Project

Key Information

- Gaojing, Beijing, China
- Total investment: RMB8.1 million(~US$1.32 million)
- Capacity: 5 tonnes per day
- Construction completed July 2014
- Commissioned 27 November 2014
- Capture technology: absorption (three solvents)

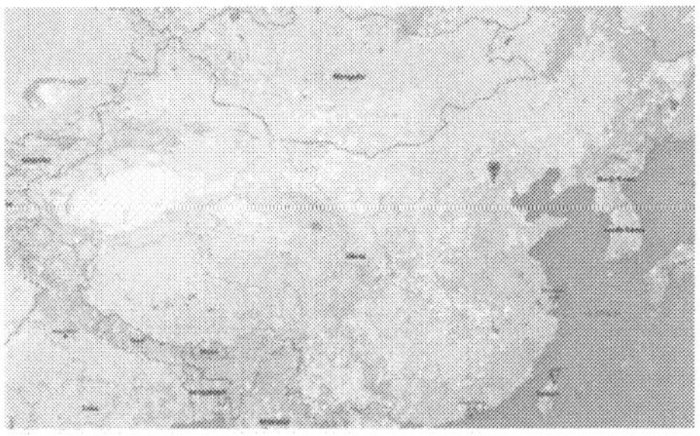

Location of Datang NGCC capture plant

Reference: 1. China Datang Corporation 2014. <http://www.china-cdt.com/dtwz/indexAction.ndo?action=showDoc&d=98B5F050-5BE4-F8D7-1DBD-B76241E1F942&t=index_news>.

Datang International Power Generation NGCC Carbon Capture Project

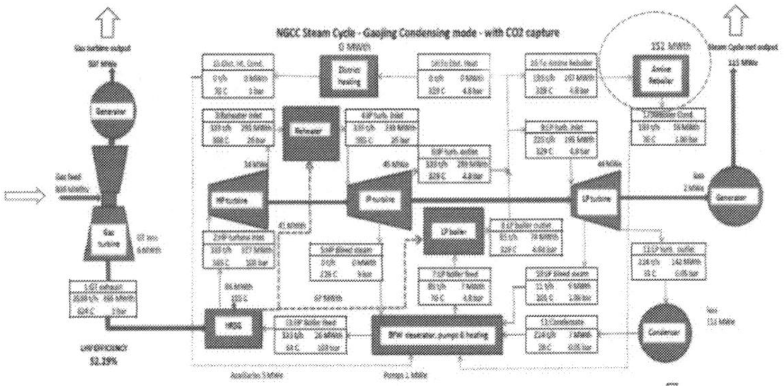

Datang Gaojing NGCC Plant

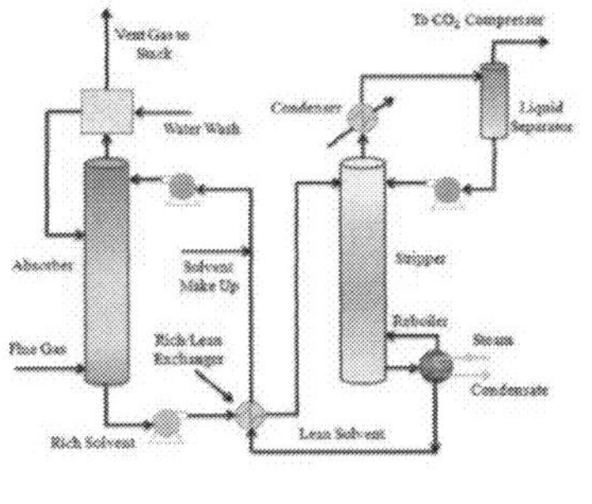

Datang Gaojing NGCC Capture Process Flow Diagram

Datang Gaojing NGCC Capture Process Pumps

Datang International Power Generation NGCC Carbon Capture Project

Key Parameters

- CO_2 concentration: 4.6% v/v
- Energy penalty: 4.6GJ/t CO_2 captured
- 1 x Absorber
- 1 x Stripper
- 1 x CO_2 Compressor

Strategic benefits for Datang

- A demonstration of Datang's commitment to low emission energy technology
- A training platform for Datang personnel
- A testing platform for CCS technology developers
- A learning platform for Datang management
- A culture transforming platform for Datang

Datang International Power Generation NGCC Carbon Capture Project

Preliminary Costs

Input	Unit	Quantity	Unit Cost		RMB/tCO_2
Steam	T/ tCO_2	2.75	192	RMB/t	528
MEA	Kg/tCO_2	2.30	20	RMB/Kg	46
Cooling water	T / tCO_2	234	0.15	RMB/t	35
Electricity	KWh/ tCO_2	200	0.564	RMB/kwh	112
Desalted water	Kg/ tCO_2	48	0.015	RMB/Kg	1
Total					722

Costs for ionic liquid option

Input	Unit	Quantity	Unit Cost		RMB/tCO_2
Steam	T/ tCO_2	1.75	192	RMB/t	336
Ionic liquid	Kg/tCO_2	0.25	80	RMB/Kg	20
Cooling water	T / tCO_2	237.6	0.15	RMB/t	36
Electricity	KWh/ tCO_2	200	0.564	RMB/kwh	112
Desalted water	Kg/ tCO_2	48	0.015	RMB/Kg	1
Total					505

Sinopec Zhongyuan Oilfield Refinery CCUS Project

Key Information

- Liutun, Puyang, He'Nan Province
- Capacity: 100,000 tonnes per year
- Total investment RMB150 million
- Sinopec Group approval: ~January 2014
- Environmental Impact Assessment approval August 2014
- Construction started September 2015
- Commissioned 30 June 2015
- Capture technology: amine-base solvent

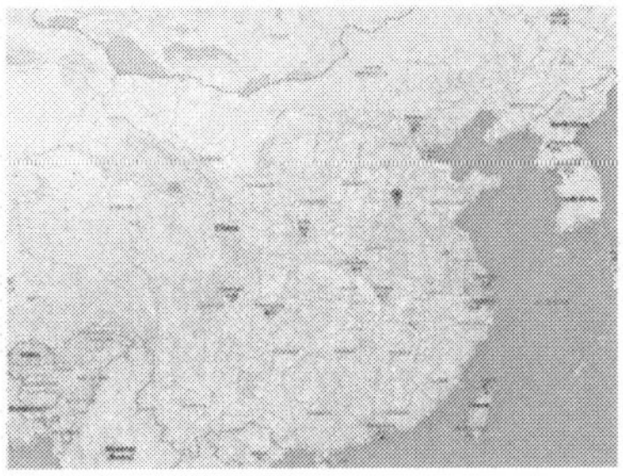

Sinopec Zhongyuan Oilfield CCS-EOR Project Location

Sinopec Zhongyuan Oilfield CCS-EOR Project Photos

Reference: 1. Sinopec 2015. <http://www.sinopecnews.com.cn/news/content/2015-07/01/content_1538865.shtml>
2. Li, H 2015. China Petrochemical Newspaper, 18 August 2015. <http://202.149.227.159/zgshb/html/2015-08/18/content_629767.htm?div=0>
3. China Energy Newspaper, 13 April 2015. <http://paper.people.com.cn/zgnyb/html/2015-04/13/content_1553914.htm>.

Sinopec Zhongyuan Oilfield Refinery CCUS Project

Catalytic cracker → De-NOx → Dust Removal → De-SOx → Carbon capture → Liquefaction → Liquid Storage → EOR and storage

65,000 Nm³/hr
~12% CO_2

12.50 tonne per hour CO_2

- 1 x Absorber (51 m in height)
- 1 x Stripper
- 2 x CO_2 Compressor
- 2 x Ammonia Refrigerator
- 2 x Spherical storage 650m³
- Loading dock

Sinopec Zhongyuan Oilfield CCS-EOR Project - Absorber

Reference: 1. Li, H 2015. China Petrochemical Newspaper, 18 August 2015. <http://202.149.227.159/zgshb/html/2015-08/18/content_629767.htm?div=0>
2. GMW 2015. 2 June 2015. <http://pic.gmw.cn/cameramanplay/599905/3094179/3826596.html>..

Sinopec Zhongyuan Oilfield Refinery CCUS Project

Performance highlights

- Environmental benefits
  - ✓ SO_2 reduction 90%
  - ✓ NOx reduction 70%
  - ✓ Dust reductions
  - ✓ Permanent storage of CO_2

  Meet high standards by China *Environment Protection Act 2015* revision

- Economic benefits
  - RMB390 per tonne CO_2 market price
  - Saving of RMB39 million per year (100,000 tpa injection)
  - Revenue from increased oil output
    - 33,000 tonne oil per year
    - US$10,885,050 per year
    - RMB69,664,320

  Assuming: 1 tonne oil requires 3 tonne CO_2, oil price of US$45/barrel, US$1= RMB6.40.

- 101,800 tonnes CO_2 injected in first half of 2015.
- 113,800 tonnes CO_2 injected in second half of 2015

Reference:

1. Li, H 2015. China Petrochemical Newspaper, 18 August 2015. <http://202.149.227.159/zgshb/html/2015-08/18/content_629767.htm?div=0>

Time is of the essence for technology development

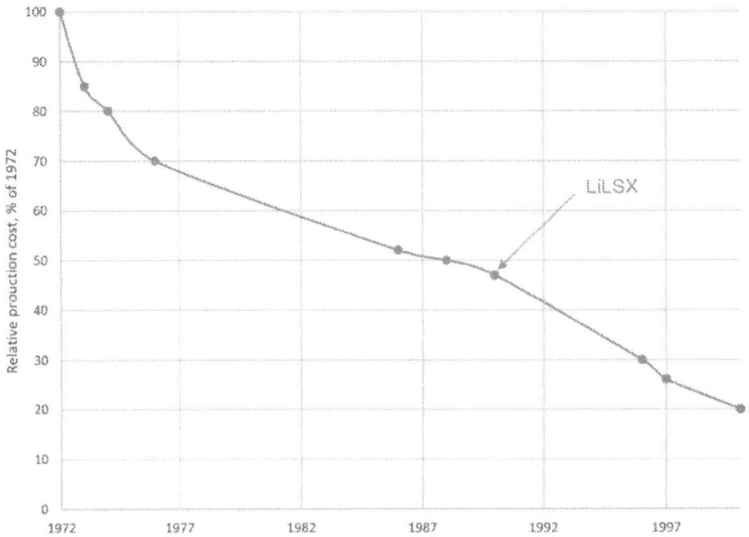

Oxygen production cost using adsorption technology

Reproduced based on data from R T Yang, *Adsorption*, 2003

Knowledge sharing and international collaboration are critical

A holistic approach to cost reduction is needed.

For APAC region

- Continue demonstration and pilot testing are needed.
- New technologies will continue to be developed and tested.
- More capacity development are needed to promote the technology and to conduct more personnel training.
- More knowledge sharing is needed to avoid unnecessary repeats.
- International collaborations are needed both for R&D and large scale projects.

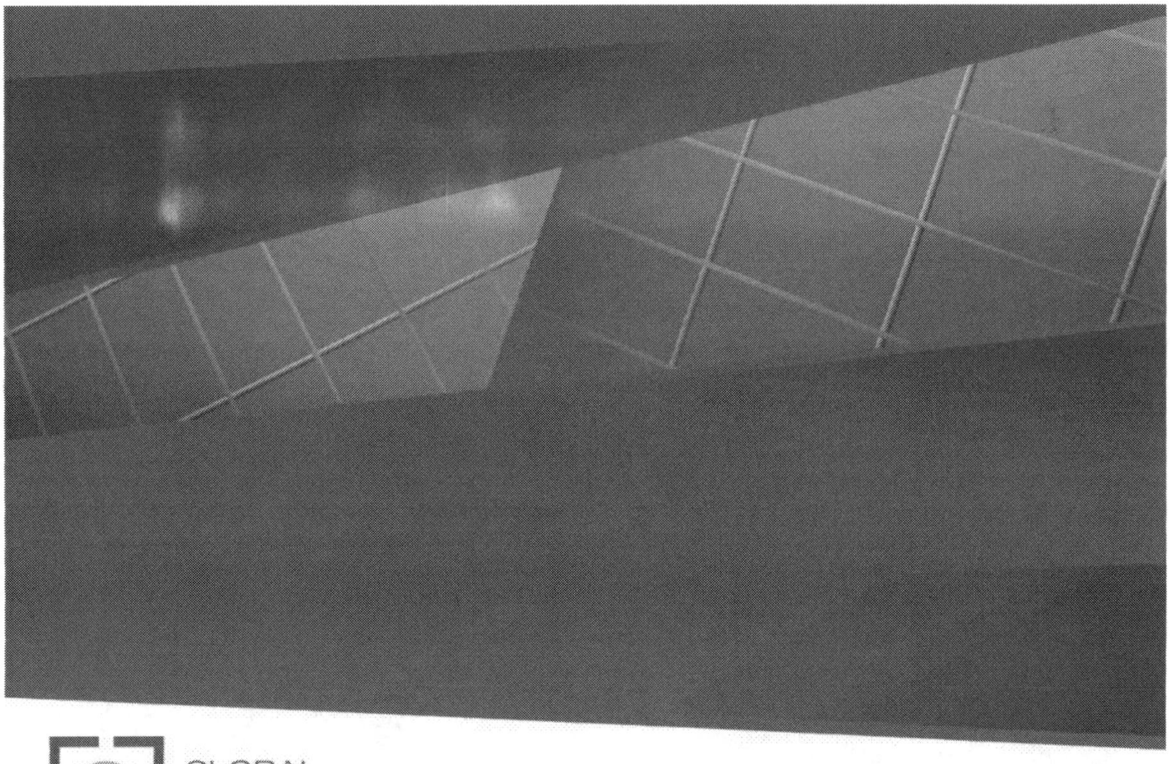

GLOBAL
CCS
INSTITUTE

GLOBALCCSINSTITUTE.COM

FuelCell Energy
Ultra-Clean, Efficient, Reliable Power

Progress in Development of Electrochemical Membrane Technology for Post-Combustion Carbon Capture

Hossein Ghezel-Ayagh and Stephen Jolly

Carbon Management Technology Conference 2015 (CMTC 2015)
November 17-19, 2015
Sugar Land, TX

Ultra-Clean | Efficient | Reliable Power

FuelCell Energy
Ultra-Clean, Efficient, Reliable Power

Integrated Fuel Cell Company

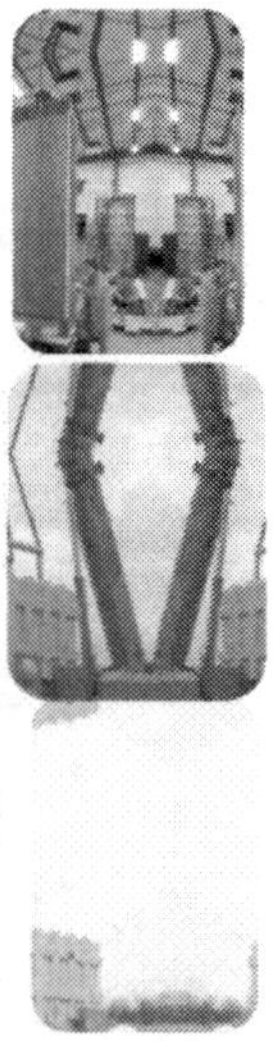

Research & Development

- *Global fuel cell technology platform*
- *Robust intellectual property portfolio*
- *Leveraging core technology for new market opportunities*

Sales, Manufacture & Project Execution

- *Project development – Direct Sales*
- *Global manufacturing (200+ MW capacity)*
- *Engineering, Procurement and Construction*

Services

- *Operate & Maintain power plants*
- *100+ DFC® plants operating at 50+ sites globally*
- *>4 billion kWh ultra-clean power produced*
- *> 300 MW installed/backlog*

1.4 MW power plant

11.2 MW fuel cell system

2.8 MW power plant

5 unit fuel cell park

2

FuelCell Energy
Ultra-Clean, Efficient, Reliable Power

Global Manufacturing Footprint

North America

Manufacturing
Torrington, CT
- Module Assembly & Stacking
- 65,000 ft^2 facility *(pre-expansion)*
- Opened 2001

Corporate
Danbury, CT
- Research labs
- Engineering design
- Global Service center
- Conditioning

SOFC Research
Littleton, Colorado
Calgary, Canada
- Research labs

Asia & Europe

Manufacturing
Pohang, South Korea
Capacity for Asian market via partner, POSCO Energy

Manufacturing
Ottobrunn, Germany
Capacity for European market via FCES, GmbH

3

FuelCell Energy
Ultra-Clean, Efficient, Reliable Power

World's Largest Fuel Cell Park

- **Scalable consisting of 21 DFC3000® power plants**
  - Power Rating: 59 MW
- **Supplying electric grid and district heating system**
- **Constructed in only 14 months**
- **Adequate to power ~ 140,000 S. Korean homes**

4

Electrochemical Membrane (ECM) Technology

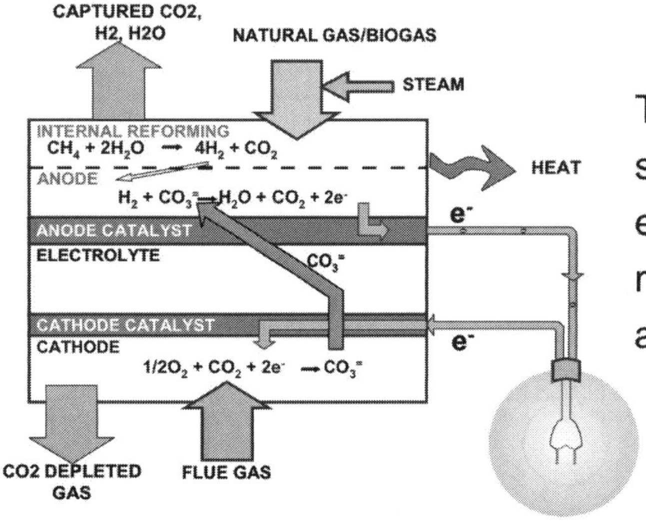

The driving force for CO_2 separation is electrochemical potential, not pressure differential across the membrane

Net Results

- Simultaneous Power Production and CO_2 Separation from Flue Gas of an Existing Facility
- Excess Process Water Byproduct
- Complete Selectivity towards CO_2 as Compared to N_2

5

FuelCell Energy
Ultra-Clean, Efficient, Reliable Power

Modular Technology

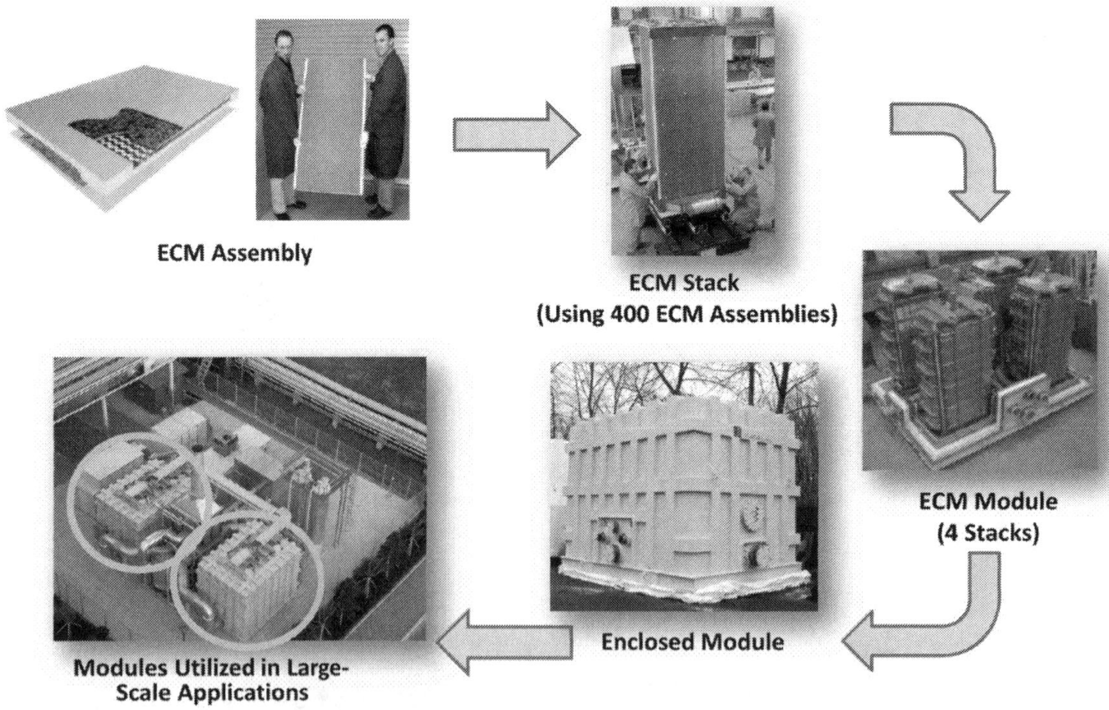

6

FuelCell Energy
Ultra-Clean, Efficient, Reliable Power

System Block Flow Diagram

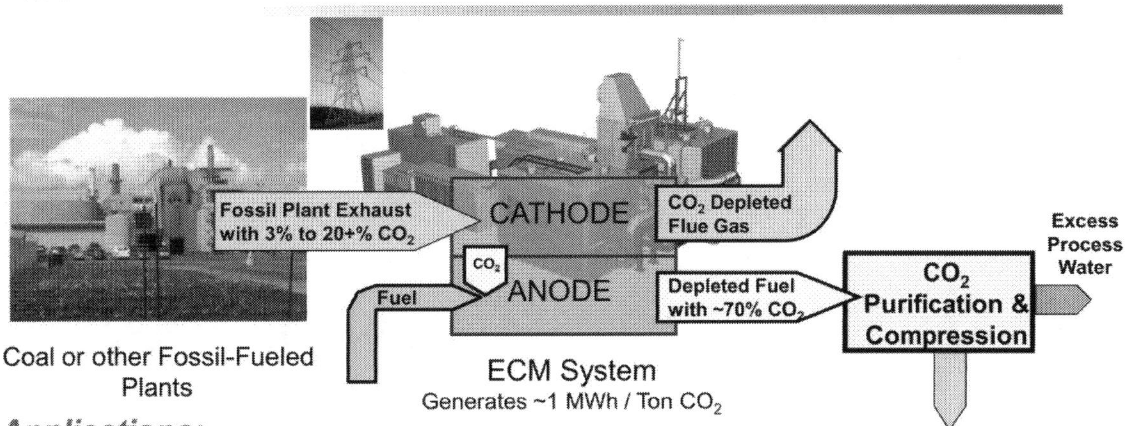

Applications:

- **Large-scale carbon capture from coal-fired plants**
  - Ultimate objective of DOE-supported development
- **Carbon capture natural gas systems**
  - Near term application can provide low-carbon baseload or peaking systems
- **Carbon capture from industrial processes**
  - Reduced carbon footprint from processes such as cement production, which also have high on-site power needs
- **Enhanced Oil Recovery (EOR)**
  - On-site generation from associated gas with CO_2 capture for EOR

7

FuelCell Energy
Ultra-Clean, Efficient, Reliable Power

Coal Power Plant Application

- **420MW ECM-based system would capture 90% of CO_2 from 550MW Pulverized Coal (PC) plant**
- 4.9 Million tons of CO_2 capture per year
- 3.2GWh ultra-clean power generated per year
- System designed for achieving high availability (>90%) for capture from large scale coal plants
- Cost of CO_2 captured less than $40/ton, or less than $0.02/kWh

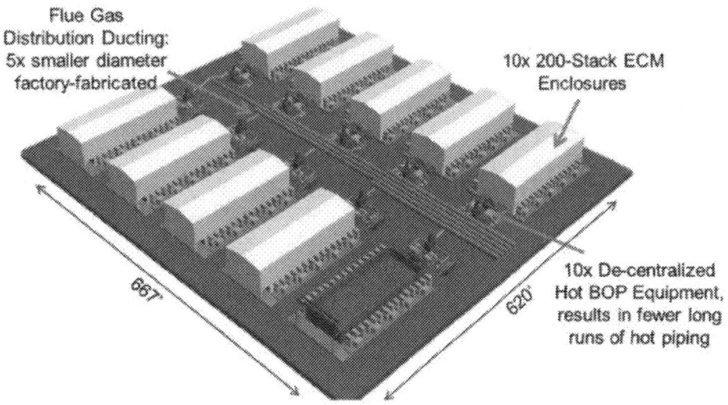

8

FuelCell Energy
Ultra-Clean, Efficient, Reliable Power

ECM Flue Gas Contaminant Tolerance: SO_2

To simulate flue gas cleanup system upsets, ECM response to spikes of SO_2 concentrations was studied:

Pacific Northwest
NATIONAL LABORATORY

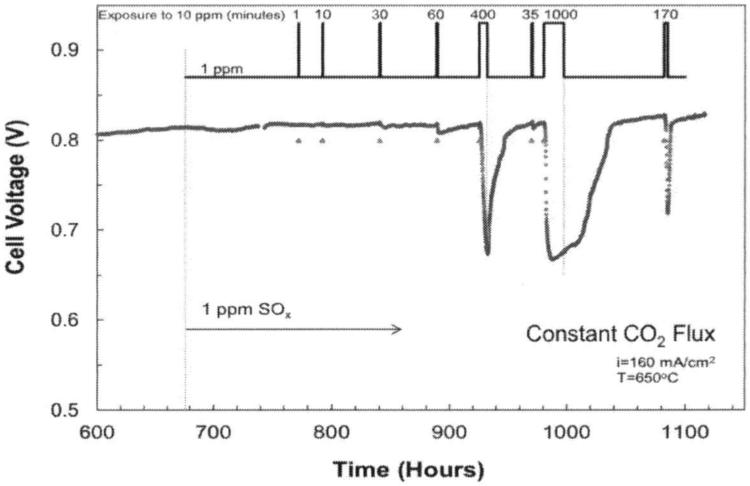

- ECM performance is stable using a polishing equipment which reduces upstream SO_2 concentration in the flue gas (cathode gas) to <1 ppm
- Performance loss was fully recoverable after exposing ECM to 10 ppm transients SO_2 of varying lengths with recovery time proportional to length of transient

9

FuelCell Energy
Ultra-Clean, Efficient, Reliable Power

ECM Flue Gas Contaminants Tolerance: Summary

Flue Gas Contaminant	Concentration in Cathode Inlet Gas After Polishing FGD, Estimated by AECOM		Highest Concentration Tested by PNNL, with low/no power degradation		Notes
SO_2	0.18	ppmv	1	ppmv	Performance losses due to short-term SO_2 exposure up to 40ppm were fully reversible
Se	0.30	ppbv	10	ppbv	No apparent degradation over 860 hours.
Hg	0.08	ppbv	250	ppbv	Expected form is predominantly elemental Hg. No apparent degradation over 1100 hours.
HCl	12.7	ppbv	200	ppbv	No apparent degradation over 900 hours.

- Based on PNNL testing and AECOM performance estimates, a polishing wet-FGD scrubber is designed to sufficiently clean flue gas for ECM operation

10

FuelCell Energy
Ultra-Clean, Efficient, Reliable Power

ECM NO_x Removal Mechanism

- Based on FCE's prior experience:
  - ECM materials are not expected to be degraded by NOx in flue gas
  - CEPACS system offers co-benefit of NOx reduction

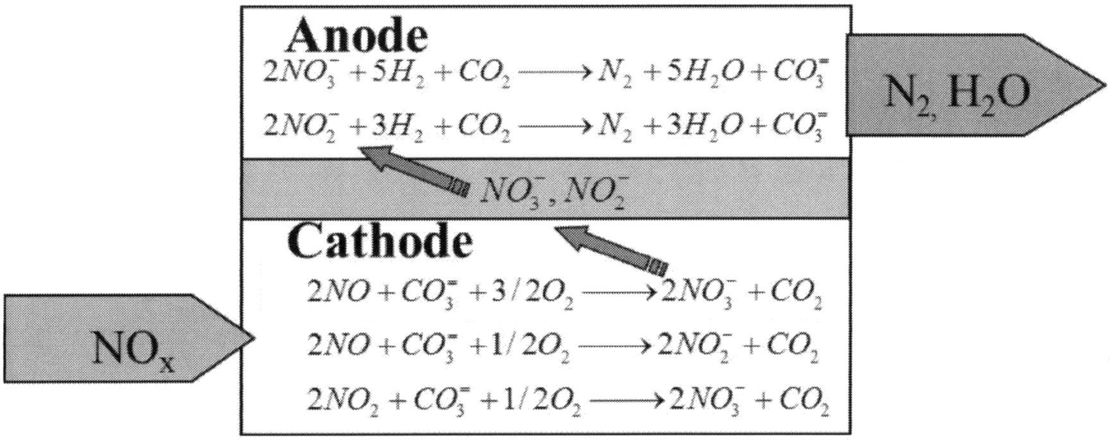

Reaction Mechanism by which NOx is removed from the Flue Gas (cathode-side), transferred to the anode-side along with CO_2, and subsequently destroyed

11

FuelCell Energy
Ultra-Clean, Efficient, Reliable Power

ECM NO_x Removal Capabilities

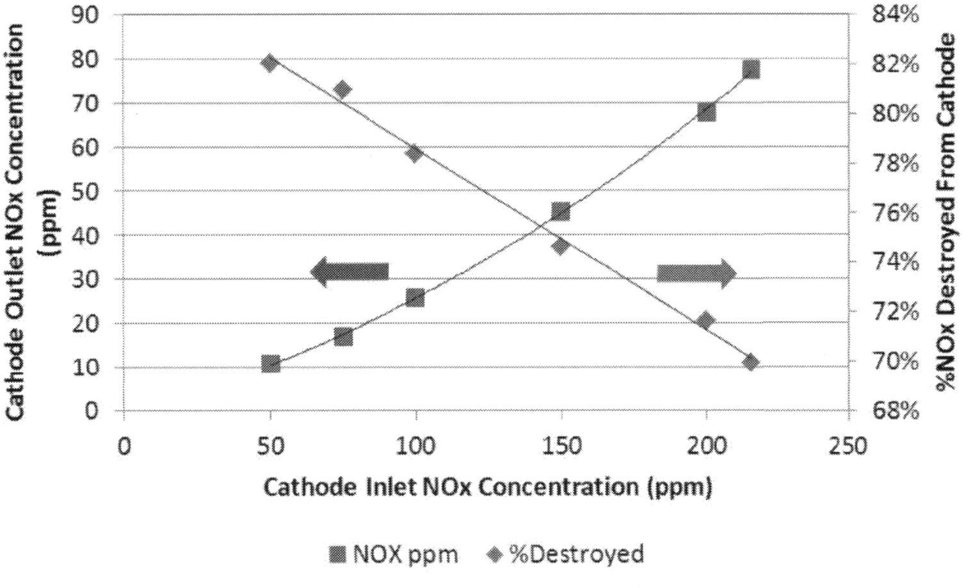

- ECM Capability for NO_x Destruction Remains > 70% at High Inlet NO_x Concentration (200 ppm) During Carbon Capture under System Conditions

12

FuelCell Energy
Ultra-Clean, Efficient, Reliable Power

Bench-Scale Demonstration

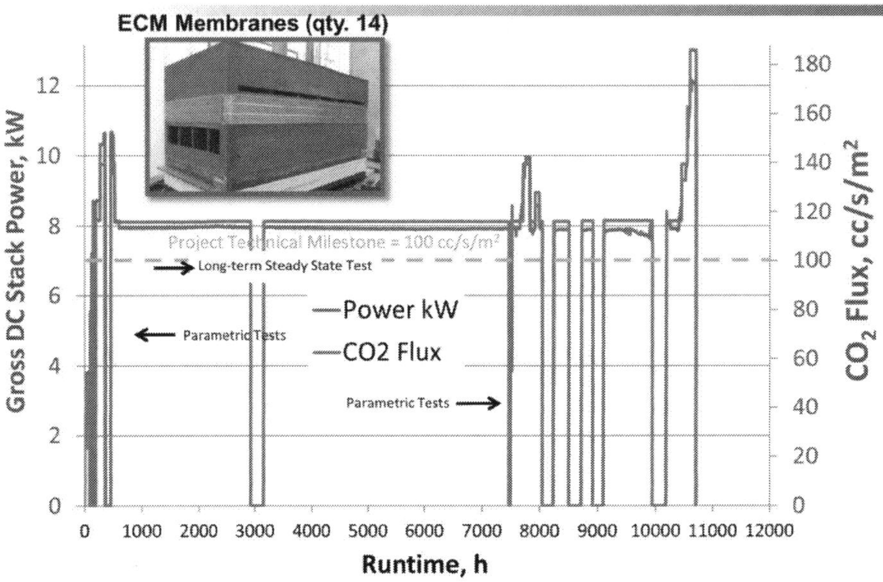

CEPACS demonstration system for PC flue gas currently undergoing testing at FCE:

- 100 tons/year liquid CO_2 product
- >10 kW peak power production
- Bench-scale CEPACS test results verified high CO_2 flux and stable operation for >10,000 hours

13

FuelCell Energy
Ultra-Clean, Efficient, Reliable Power

ECM Performance: Effect of Flue Gas Composition

ECM cell performance data for NGCC and PC plant flue gases at 93% carbon capture:

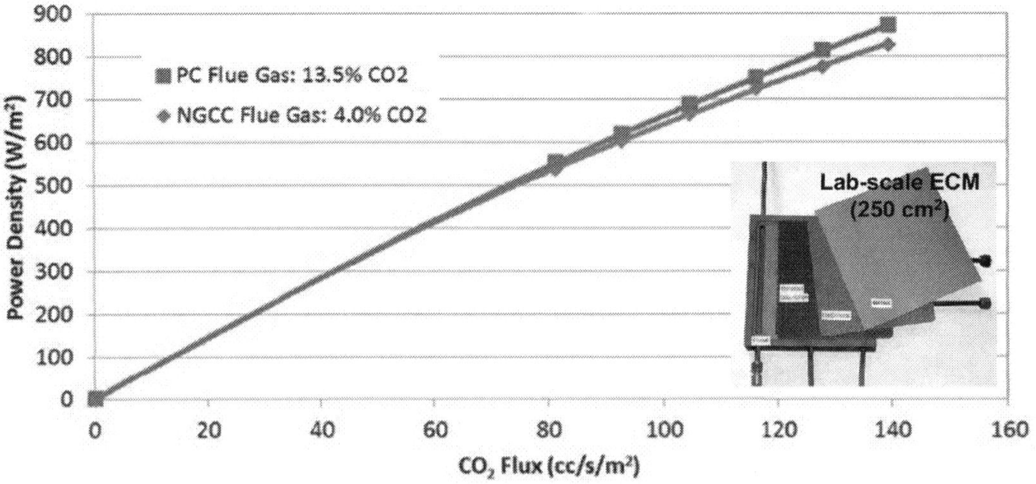

- High cell power densities at high CO_2 flux is observed in ECM tests
- ECM is capable of operating on flue gases with a wide range of CO_2 partial pressure
  - Pulverized coal-fueled boilers
  - Natural gas-fueled boilers
  - Natural gas turbine and combined cycle plants

14

FuelCell Energy
Ultra-Clean, Efficient, Reliable Power

Case Study: ECM for CO_2 Capture from SAGD Bitumen Extraction

Jacobs Consultancy developed an independent analysis* of an ECM system for 90% CO_2 capture applied to a 33,000 BOPD Steam Assisted Gravity Drainage (SAGD) facility under a study by Alberta Innovates (Alberta, Canada)

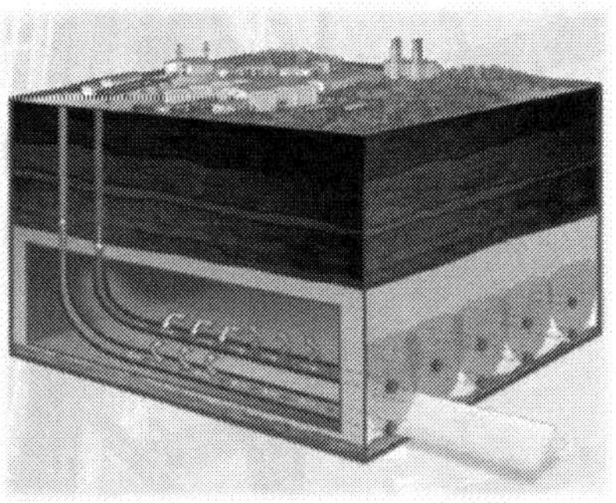

ECM system:

- Captures 90% of CO_2 from SAGD NG-fired Once Through Steam Generator (OTSG)
- Produces 62 MWe net, enough to cover all SAGD power requirements and export 48 MWe
- Reduces SAGD facility makeup water requirements by 44% (compared to without CCS)

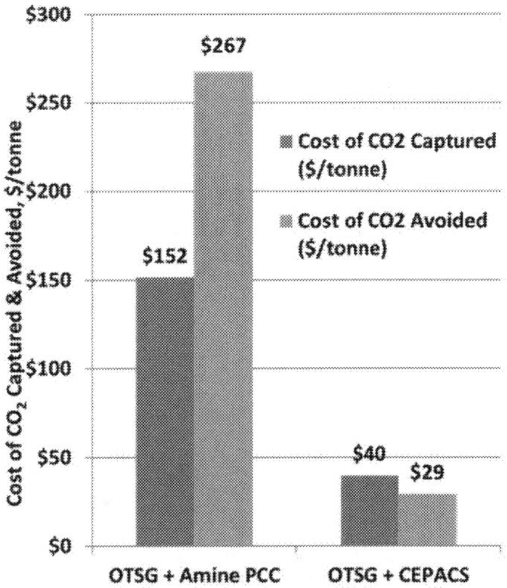

Alberta Innovates Energy and Environment Solutions

*available: http://www.ai-ees.ca/media/14423/ecm_evaluation_study_report-final_all.pdf

15

FuelCell Energy
Ultra-Clean, Efficient, Reliable Power

Pilot –Scale Project

- Recently awarded Co-operative Agreement with DOE (DE-FE0026580)
- Pilot ECM system to capture >90% of CO_2 from a 3 MWe equivalent slipstream of a PC plant flue gas
- Objective to confirm high purity (>95%) CO_2 capture with a cost of electricity <30% of state-of-the-art capture technologies in large-scale applications
- 42 months project starting 10/1/2015
- Host site is to be confirmed

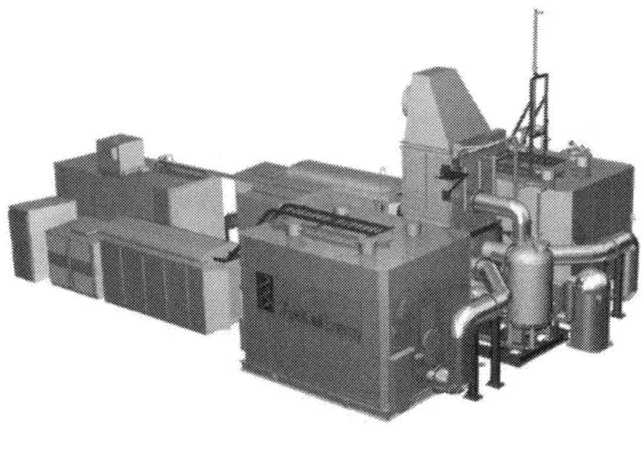

Pilot-Scale Project:
Designed for capturing **>58 tons CO_2 per day** while simultaneously producing **~40,000 kW-h per day** of ultra-clean electricity

16

DFC Carbon Capture Overview

Captures and Concentrates exhaust from
- Coal powerplant
- Natural gas powerplant
- Industrial process

Proven Technology
- Leverages commercial fuel cell technology
- Verified in pilot plant with commercial hardware

Economical
- Produces additional power vs power reduction
- Generates return on capital vs operating expense

Additional Benefits
- 70% reduction in NOx
- Clean water production

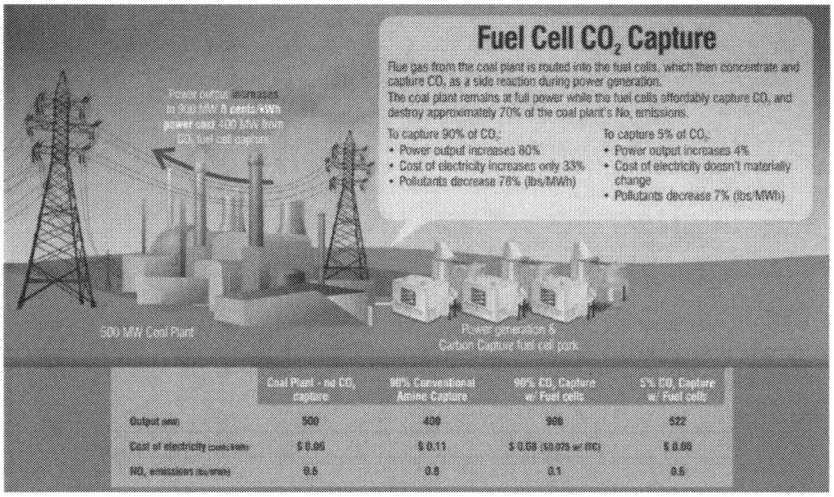

17

Acknowledgements

ECM Carbon Capture from Coal Plants supported by DOE/NETL (Co-operative Agreements: DE-FE0007634 & DE-FE0026580)

Guidance from NETL team: José Figueroa, Elaine Everitt, Lynn Brickett, John Litynski, and others at NETL/DOE

18

MTR Membrane Technology & Research

CO_2 Capture from Power Plant Flue Gas by Polaris™ Membranes: Update on Field Demonstration Tests

Carlos Casillas, Ken Chan, Don Fulton, Jurgen Kaschemekat,
Jay Kniep, Jennifer Ly, Tim Merkel, Vincent Nguyen, Zhen Sun,
Xuezhen Wang, Xiaotong Wei, and Steve White

Carbon Management Technology Conference
Sugar Land, TX
Thursday, Nov. 19, 2015

MTR Introduction

- Privately-held, 90 employees mostly located in Newark, California
- Started in 1982 funded by U.S. SBIR grants
- Commercial products in petrochemical, natural gas and refinery industries
- Provides complete turn-key solutions with about 200 membrane systems installed worldwide
- 20 person R&D group with expertise in membrane materials/formation and process design

MTR

MTR CO_2 Capture Process

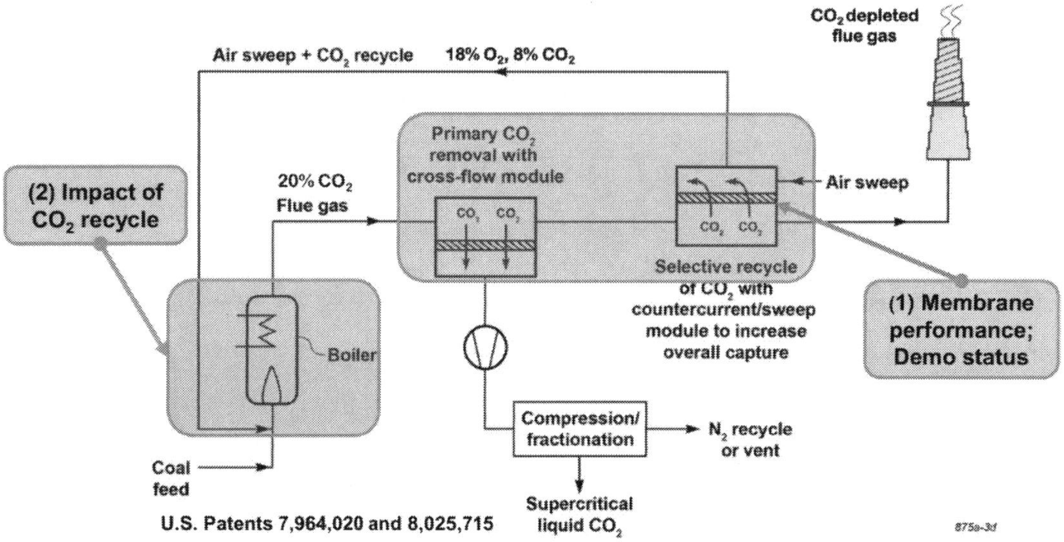

- Combustion air sweep provides driving force that lowers the capture energy
- Pre-concentrated CO_2 decreases membrane area and power required

3

MTR

MTR CO_2 Capture Development Timeline

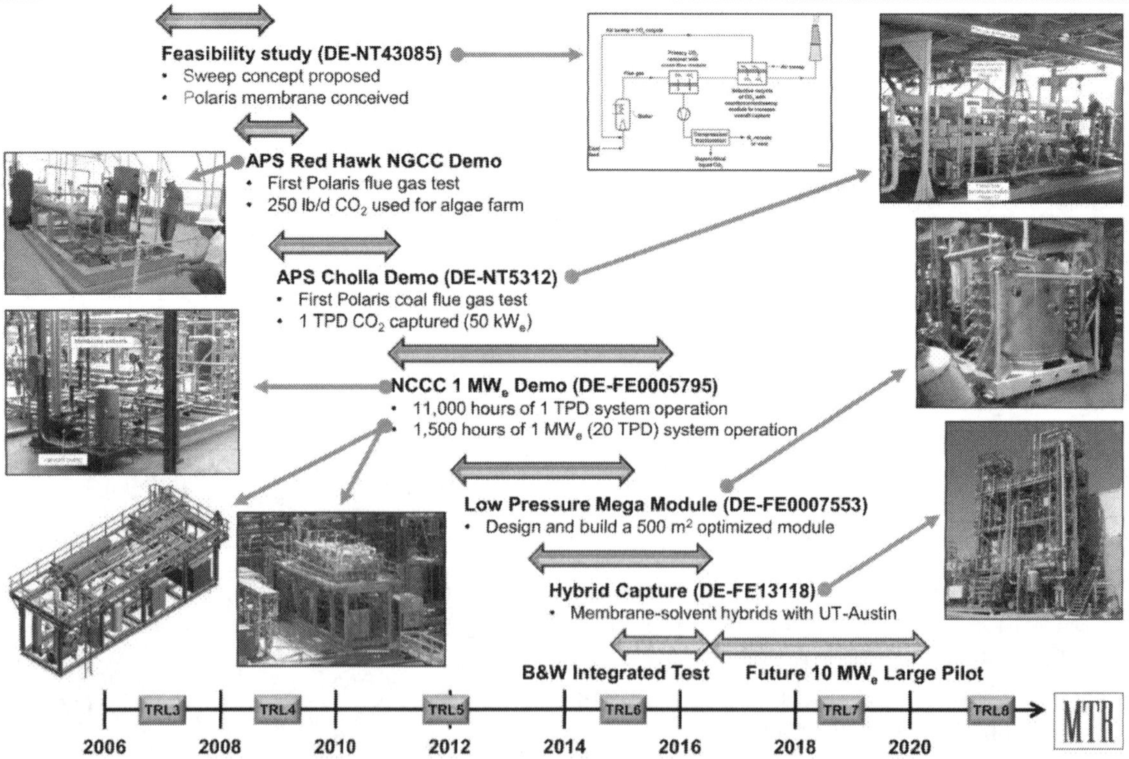

MTR

Pros and Cons of a Membrane Post-Combustion Capture Process

Benefits:

- No hazardous chemical handling, emissions, or disposal issues
- Not affected by oxygen, SO_x or NO_x
- Water use lower than other technologies (recovers H_2O from flue gas)
- No steam use → no modifications to existing boiler/turbines
- Near instantaneous response; high turndown possible
- Modular and compact; relatively simple installation/operation
- Particularly well-suited for partial capture (40-60%)

Challenges:

- Very low partial pressure driving force favors high permeance membranes
- Unknown impact of real flue gas on membrane-module lifetime
- Module pressure drop and flow distribution issues

5

MTR

Polaris™ Membranes Continue to Improve

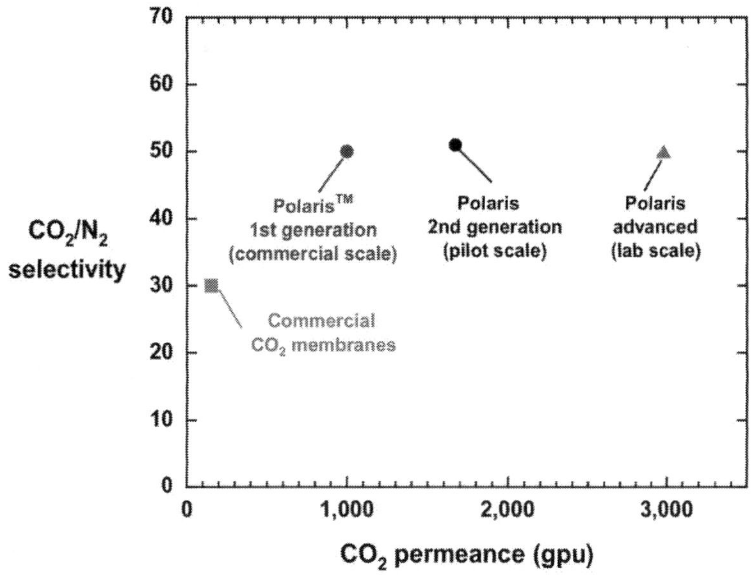

1 gpu = 10^{-6} cm^3(STP)/(cm^2 s cmHg)= 3.35 x 10^{-10} mol/(m^2 s Pa)

- In addition to lowering costs, these improvements are important to shrink the size of the capture system

6

MTR

1 TPD Testing at NCCC

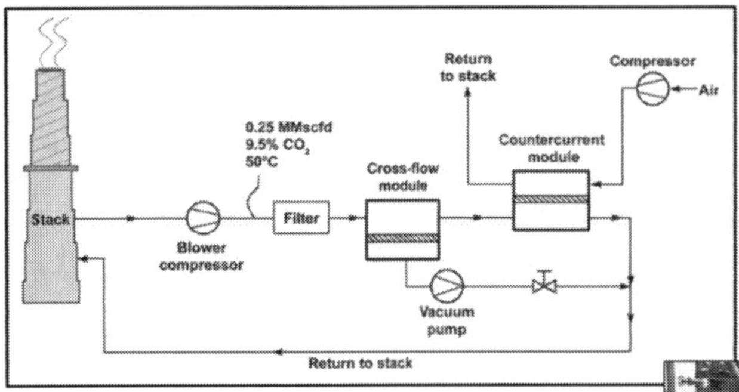

- Field laboratory that allows lifetime evaluation (>11,000 hours cumulative), and validation testing of new membranes
- Many lessons learned (i.e., ammonium sulfate deposition) applied to scale up

- System is testing vacuum and air sweep membrane steps
- Sized to capture 1 ton CO_2/day using commercial sized module
- System was in operation from Spring 2012 through Summer 2015

7

Sample 1 TPD Results from NCCC

Capture rate vs time

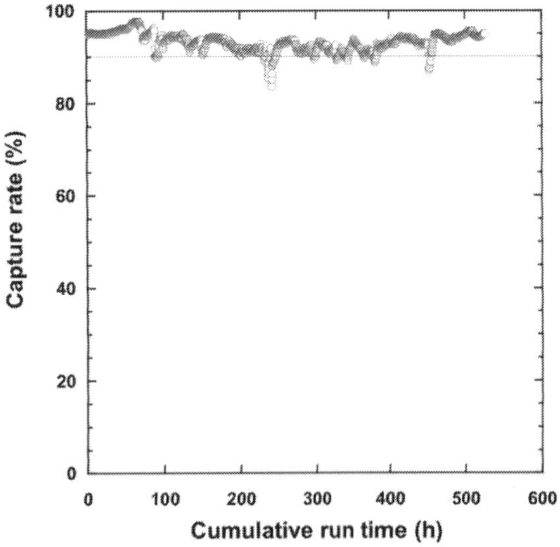

New membrane validation

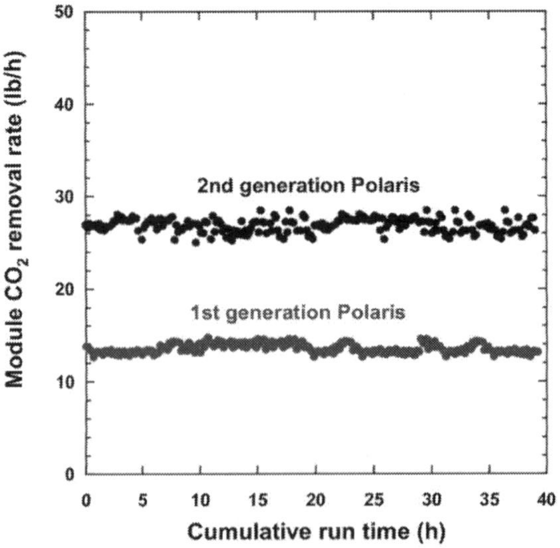

System allows more rapid materials optimization through real world testing

MTR

8

Scale up to 20 TPD Small Pilot

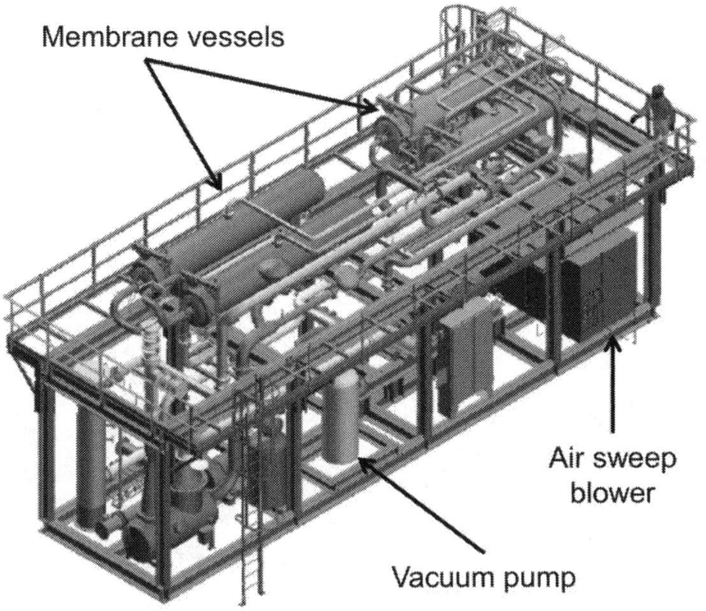

- 20 TPD skid (1 MW_e) was installed by NCCC and MTR personnel during summer 2014
- In operation from January to June 2015
- Test objective was validation of advanced modules (multi-tube and plate-and-frame) designed for low pressure drop and small footprint

MTR

Installation of 20 TPD Small Pilot at NCCC

1st floor of system arriving by truck

Crane lowering 2nd floor of system into place

10

Photos courtesy of Tony Wu

MTR

20 TPD System at PC4

- System installation completed Aug 2014
- Operation during January – June 2015
- Decommissioned in July 2015

- System showed stable performance over 1,500 hours of operation, including demonstration of prototype low-pressure drop plate-and-frame sweep modules

11

System Tests Scaled-Up Membrane Modules

Bundled spiral sweep modules | **Bundled Polaris spirals** | **Polaris plate-and-frame (designed in DE-FE0007553)**

Advanced modules demonstrate lower cost and pressure drop

12

New Modules Demonstrate Improved Pressure Drop Performance

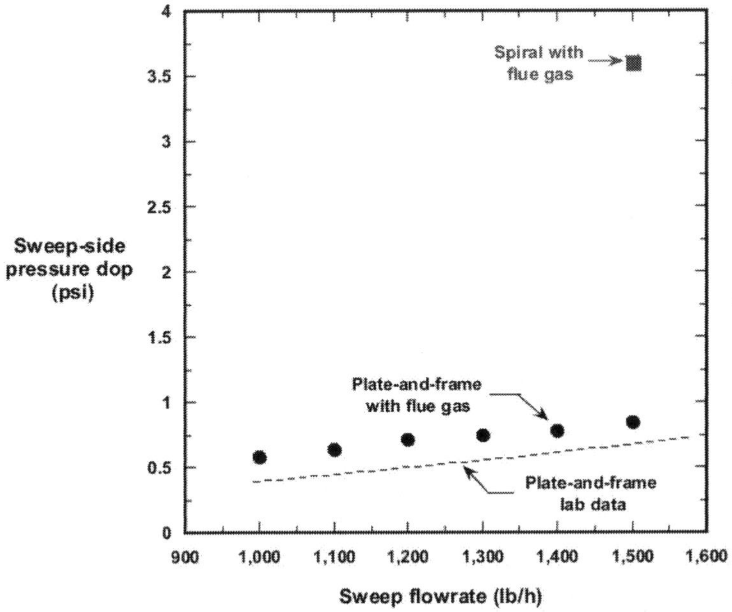

- Field data is consistent with lab results, and confirms much lower air sweep pressure drop in new modules
- At full scale, the difference in pressure drop amounts to savings of about 10 MW_e

13

MTR

20 TPD System Shows Stable Performance

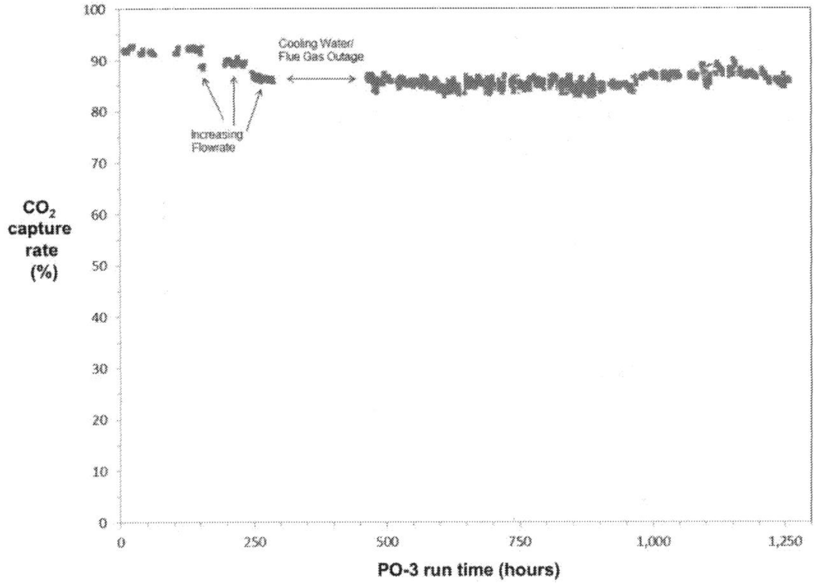

- Stable system performance through various startup / shutdowns and large variations in ambient conditions (sub-freezing in January to >95°F in June)
- System goes from cold start to steady state in ~15 minutes

Figure data from last NCCC campaign (PO-3: May to June 2015)

14

MTR

B&W Studies of CO_2 Recycle Impact on Boiler Performance

Phase I – CFD modeling

- B&W modeled 2 boiler configurations (radiant boiler firing bituminous coal and SWUP firing PRB coal) and 2 sweep recycle cases (constant secondary air flow and constant stoichiometry)
- <u>Main conclusion of modeling study</u>: secondary air laden with CO_2 appears feasible as a retrofit in either of the boiler configurations examined if oxygen mass flow to boiler is fixed

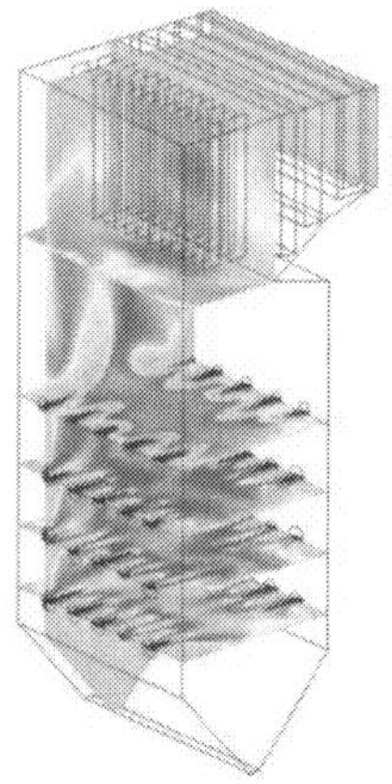

Phase II – Pilot testing

- B&W's SBS-II 1.8 MW_{th} pilot boiler operated with CO_2-laden combustion air
- Two coals evaluated: a western sub-bituminous coal and a highly volatile bituminous coal
- O_2 content of windbox air varied from 21% to 16% through CO_2 dilution
- Monitored flame stability, length, and shape; unburned combustibles in fly ash, and furnace exit gas temperature
- Radiant furnace and convective pass heat absorptions were measured
- Boiler efficiencies for air and sweep firing were determined

15

B&W Boiler Study Highlights

- Stable and attached flames with air (21% O_2) and CO_2-enriched air (16-18% O_2)
- CO_2-enriched flame was less luminous than air-fired case
- Lower furnace heat absorption but higher convection pass/air heater heat transfer for CO_2-enriched operation relative to air
- For bituminous coal, 30% lower NO_x emissions with CO_2-enriched air
- No burner modifications necessary
- Net reduction in plant efficiency of ~0.75% at 18% O_2

Flame image from combustion of PRB coal with air (21% O_2)

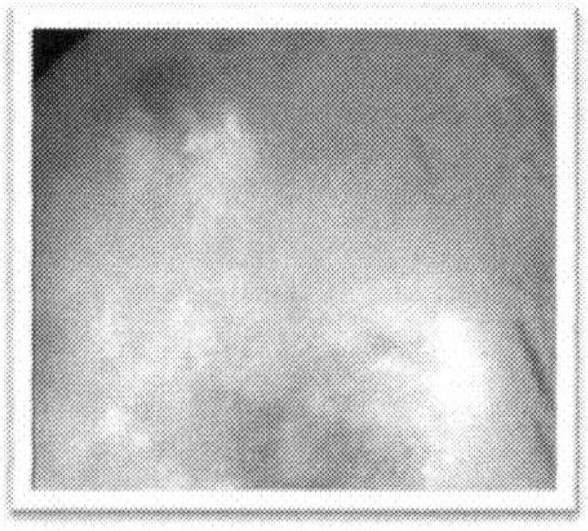

Flame image from combustion of PRB coal with CO_2–enriched (18% O_2)

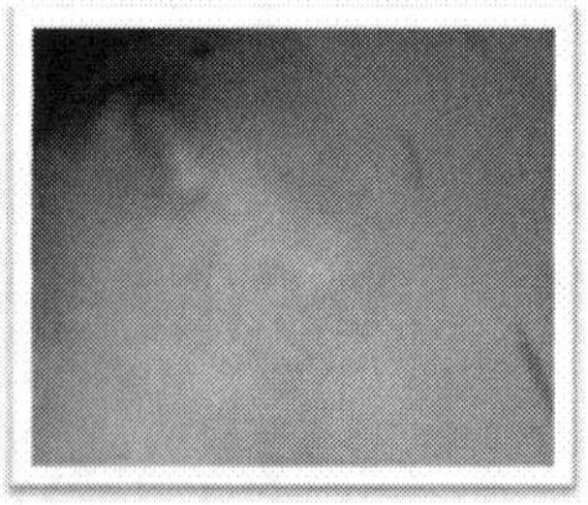

16

Systems Analysis: Membranes are Particularly Effective at Partial Capture

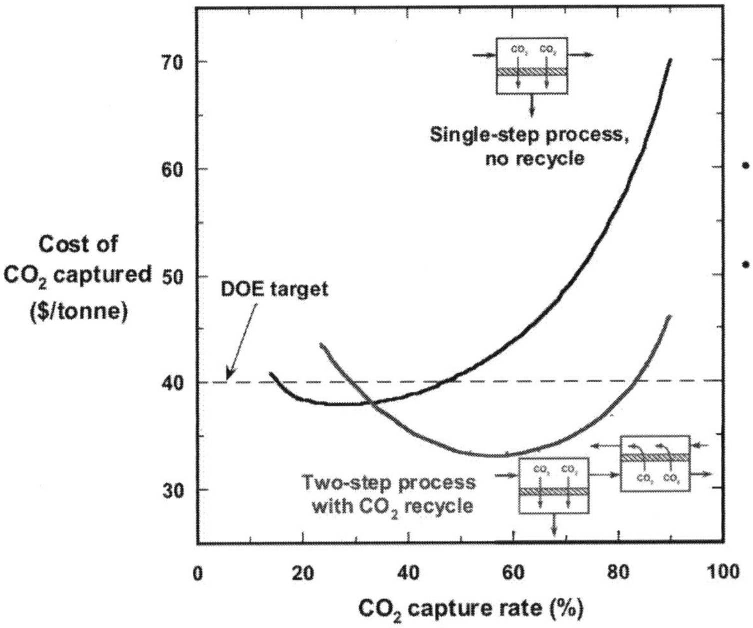

- Membranes show a minimum in capture cost
- To meet proposed EPA emission limits for coal (~40% capture), a simple system without recycle may be preferable

17

MTR

Pilot Boiler / Membrane Schematic

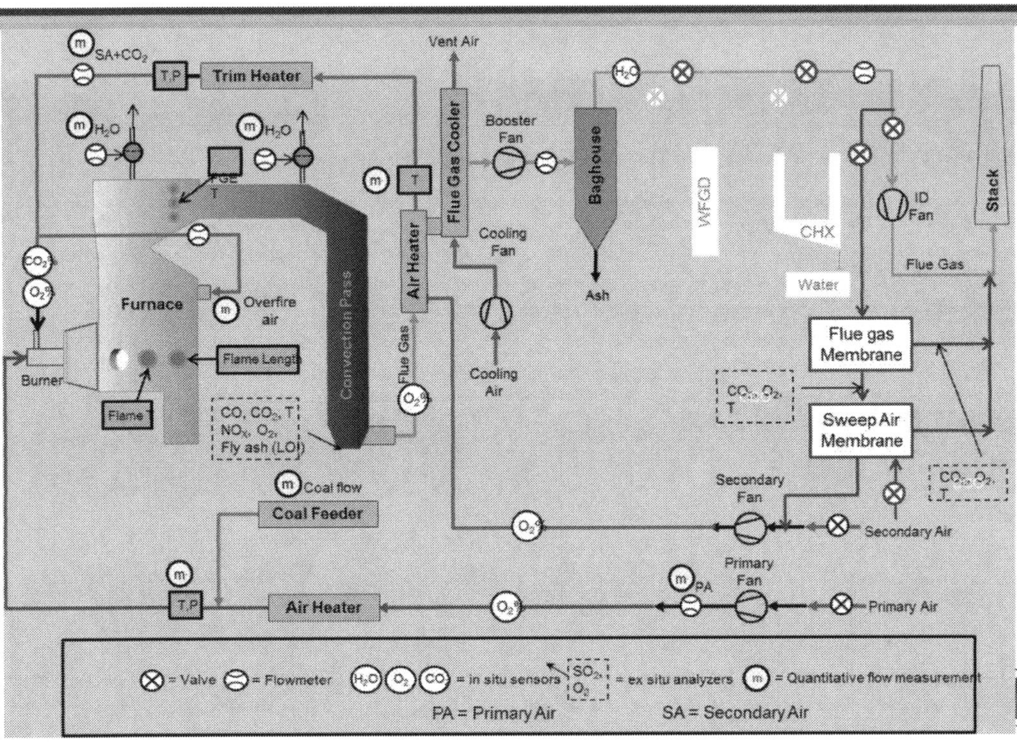

18

MTR

Summary

- CO_2 capture membrane performance continues to improve and has been validated on the 1 TPD slipstream system with over 11,000 hours of runtime at NCCC
- 20 TPD small pilot operation at NCCC is complete. Testing successfully demonstrated optimized modules (low Δp, low cost) over 1,500 hours of runtime.
- B&W boiler test suggests CO_2 recycle with sweep membrane is feasible; efficiency impact is similar to predictions
- Project for fully integrated testing of 20 TPD small pilot system with B&W coal-fired boiler is underway; tests to be conducted in 2016.

19

MTR

Acknowledgements

- **U.S. Department of Energy,**

National Energy Technology Laboratory
  - José Figueroa
  - Mike Mosser
- **Southern Company Services (NCCC)**
  - Tony Wu
- **Babcock & Wilcox**
  - Hamid Farzan

U.S. DEPARTMENT OF ENERGY

SOUTHERN COMPANY

U.S. Department of Energy National Carbon Capture Center

B&W

20

MTR

CO2 Capture for Fossil Energy Power Plant
Large Scale ASU Developments for Oxy-Combustion and Pre-Combustion CO2 Capture Applications

Alekseev, A., Linde AG
Krishnamurthy, K. R., Linde LLC
Originally presented on: 11/19/2015 14:00:00 - 14:25:00

The development of large scale ASU (Air Separation Unit) for oxy-combustion application during the period 2007-2011 was driven by extremely high expectations on the emerging carbon capture, utilization and storage market. All world-class industrial gas companies invested a lot of efforts into development of new ASU concepts for oxyfuel applications. Main target was a considerable reduction in power consumption and capital cost. As a result, a new class of cryogenic processes for production of large amount of gaseous low pressure oxygen with purity of 95-97% was developed. Some features developed include multi-column and multi-condenser processes for reducing the energy consumption, integration of the compressor heat into the steam cycle, greater operational flexibility of the ASU's, multi-train compression and air purification systems, and, new concepts for rotating equipment incorporating an axial compression stage to enable better heat integration.

As a result of the above efforts, the ASU-efficiency was considerably increased: the specific power for production of low pressure oxygen was reduced from 0.35-0.4 kWh/nm3 to 0.25-0.3 kWh/nm3. Somewhat unexpected was the minor reduction in ASU cost. It is because these new process concepts were still based on more or less conventional equipment for almost all process subunits like purification, heat exchange, rectification etc. Generally, the ASU cost and the impact of the ASU cost on power plant economics was underestimated during this time.

This rapid development has changed abruptly after in 2011 mainly because of dramatically reduced expectation on oxyfuel CCS market. Cancellation of large scale oxyfuel power plant projects in Jänschwalde in Germany and more recently Futuregen in the US were key contributing factors. Other contributing reasons were the lessons learned from other large oxyfuel projects: namely that the real installed cost of an oxyfuel power plant (not ASU only) are considerably higher than expected. This fact has direct impact on the economics of this application and makes the oxyfuel based CCS opportunity not as attractive as expected. Consequently, all development efforts were reduced to a minimum and no mentionable achievements in terms of power consumption, or any disruptive technical developments were presented during the time between 2012 and 2015. However, during this time a considerable improvement has occurred on the combustion side as well as on flue gas side of the oxyfuel power plant process: the elevated pressure boiler instead of conventional ambient pressure boiler becomes an attractive option because of higher efficiency (+2...3%) and reduced cost. These boiler options are still under development, but may become a competitive solution in the coming years. These new boiler design developments set the new requirements on ASU process design, which has to be developed in the next years. This development will be probably driven by combination of work done in academic and government institutions, as well as start-ups in addition to the industrial gases industry.

The contribution of industry in next years will be of two kinds: 1) improvements in load range and ramping rate of ASU, because of a lot of background activities in this field, 2) further increase in capacity of single ASU, which helps to reduce the number of required ASU units and improve the economics by economies of scale. For example, the Linde Group was able to realize 5 worldwide largest air separation units in Jamnagar, India, each with oxygen capacity of 5250 tpd, which is now the new benchmark.

This presentation will review the details on the past low pressure oxygen production developments and the expectations for the future high pressure oxygen plant developments.

Non-Aqueous Aminosilicone Solvent for CO_2/H_2 Separation

Dan Hancu, Surinder Singh, Ben Wood, Barbara Miebach, Irina Spiry
Laura E. Weaver, John Blieszner

November 18, 2015

imagination at work

JACOBS Consultancy

Agenda

- CO_2 Capture and Separation Technologies at GE GRC
- Aminosilicone CO_2 Capture Solvent .. Development
- CO_2 / H_2 separation by Aminosilicone solvent.. Preliminary Techno-economical analysis

CO_2 Separation Technology... Evolution

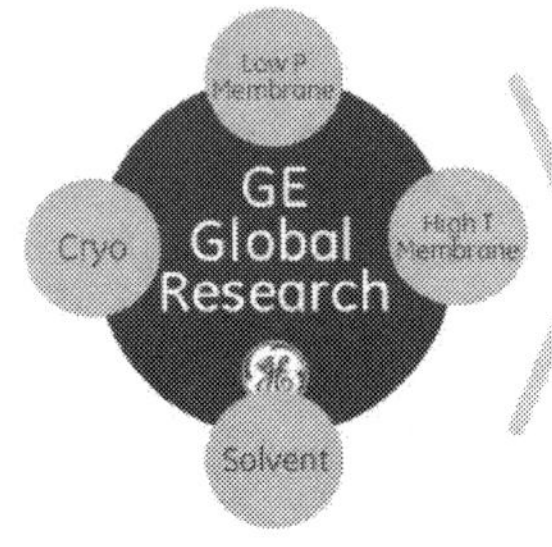

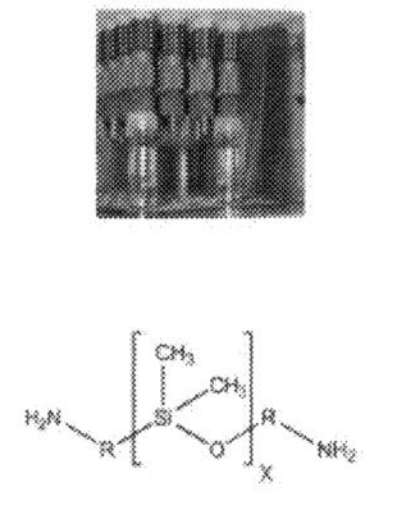

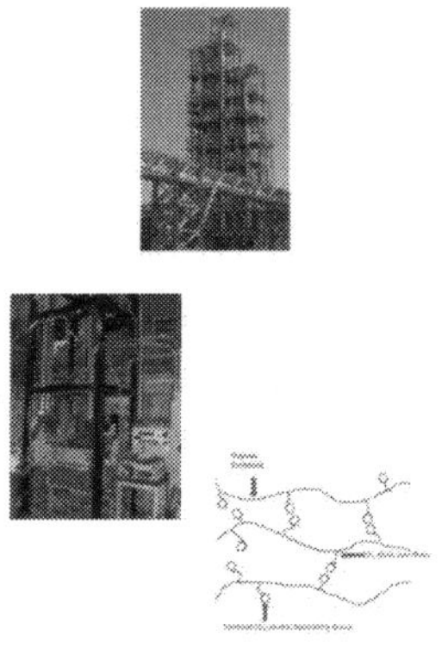

2008

Program launch for development of energy efficient pre & post combustion CO_2 Capture

2008 - 2010

Silicone POC (TRL 2)

2011-2013

Silicone: Bench (TRL 4)
Phase Change: POC (TRL 3)
Holo-fiber membrane: POC (TRL 3)

2015+

Silicone Solvent: Pilot (TRL 5)
Phase Change: Bench (TRL 4)
CO_2 *Thickener*: POC (TRL 2)
CO_2 Reuse: Bench (TRL 4)
O&G R&D: Oklahoma City

7+ years development of CO_2 Capture / Utilization Technologies

imagination at work

3
November 18, 2015

Aqueous Amine Capture Process

Postcombustion capture (absorption process)

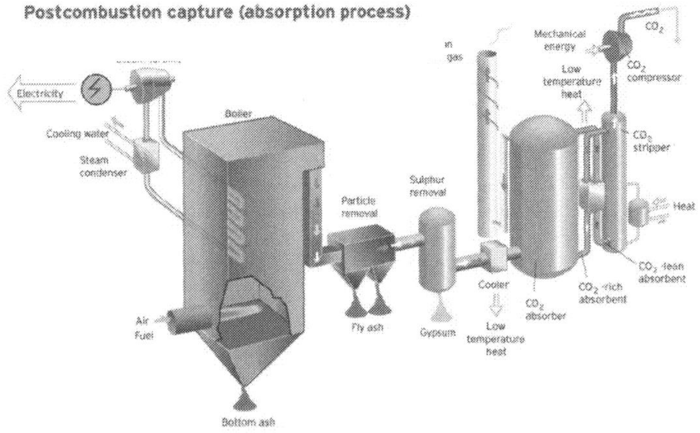

Source: http://www.industrytap.com/how-carbon-capture-technology-is-turning-co2-into-cash/[illegible]

- 80 % increase in COE.. Heating & condensing water
- MOC's... Solvent corrositvity
- Solvent Management.. Oxidation, volatility, and nitroso-amines
- Amine emissions .. Aerosols
- Large footprint.. Adsorber/ Desorber

Providers:
MHI, Fluor, Alstom, BASF, Shell, Linde

Need Solvent Redesign for Cost Effective CO_2 Capture

imagination at work

4
November 18, 2015

New Solvent System.. Amino-Silicone

Desired Solvent Properties
for cost effective CO_2 capture:

- Low/no water
- Liquid carbamate salt
- Thermal stability
- High CO_2 loading
- High desorption pressure
- Low volatility
- High reaction rates

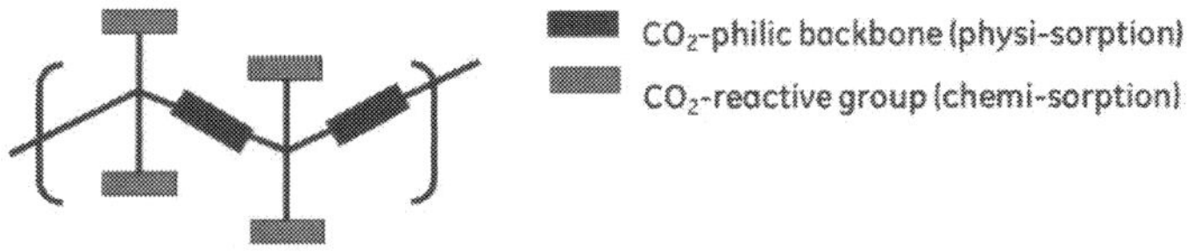

Aminosiloxanes

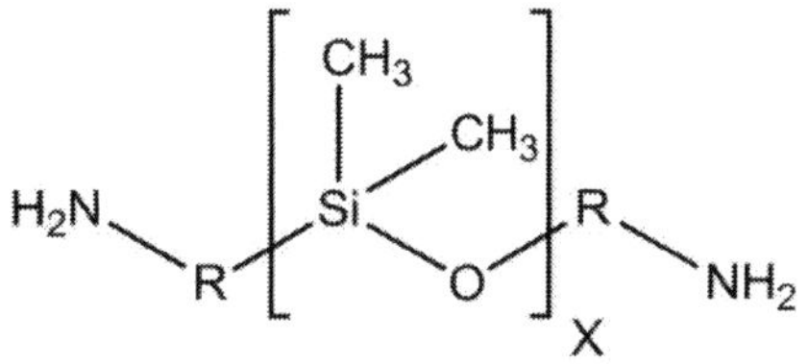

DOE Contract: DE-FE0007502

Amino-Silicone.. Working Capacity

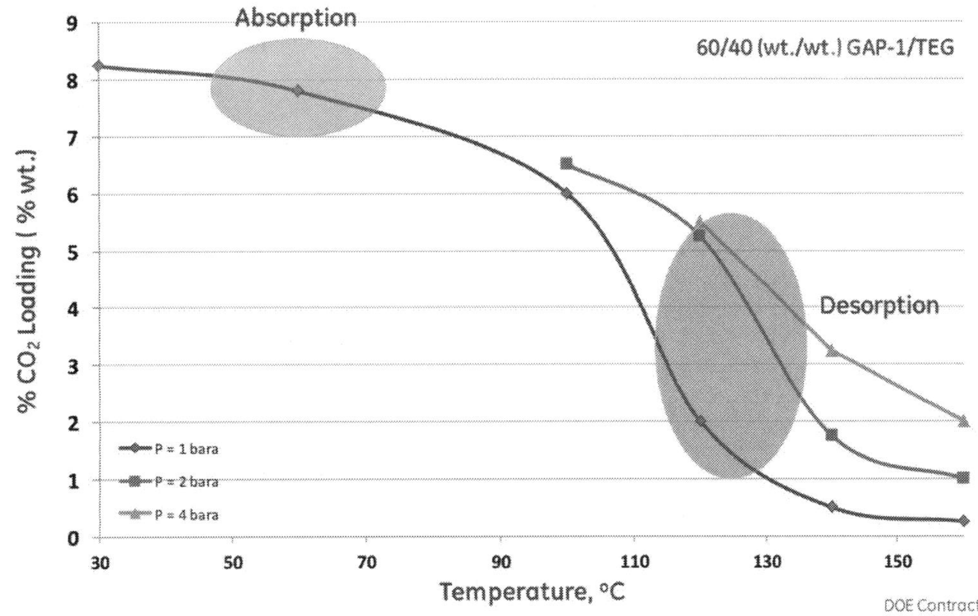

DOE Contract: DE-FE0007502

High Working Capacity... Low Recirculation Flow

Silicone technology.. Simplified PFD

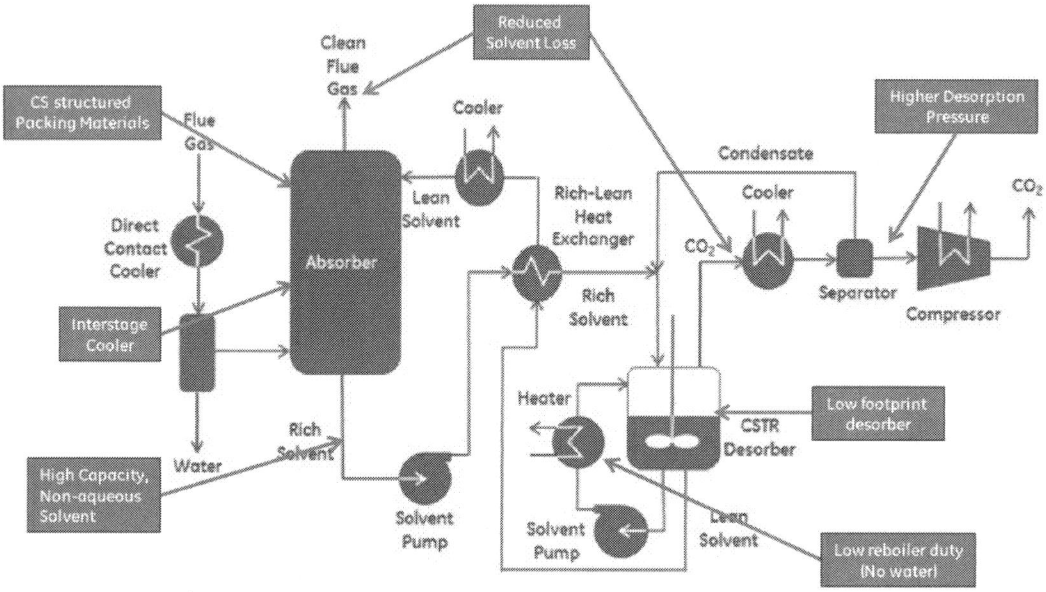

imagination at work

GE Confidential

H_2 Plant.. High Pressure Application

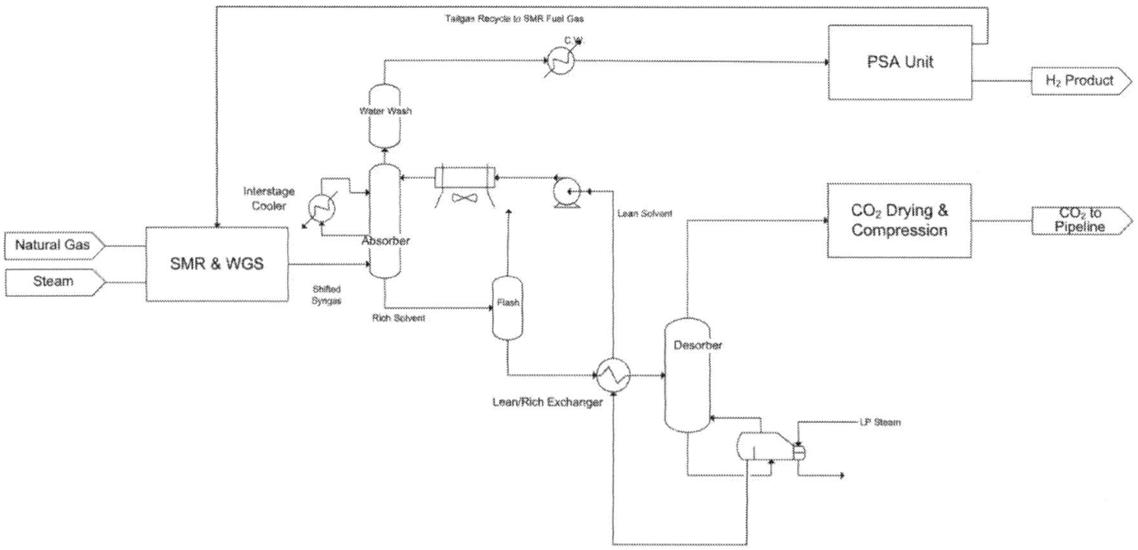

H_2 Plant vs. Coal Application:
High Pressure absorption in the presence of hydrogen and methane

imagination at work

GE Confidential

JACOBS Consultancy

Technical Approach Process & Plant Modelling

ASPEN PLUS: CO_2 Capture Island
GE Global Research

H_2 Plant
JACOBS Consultancy

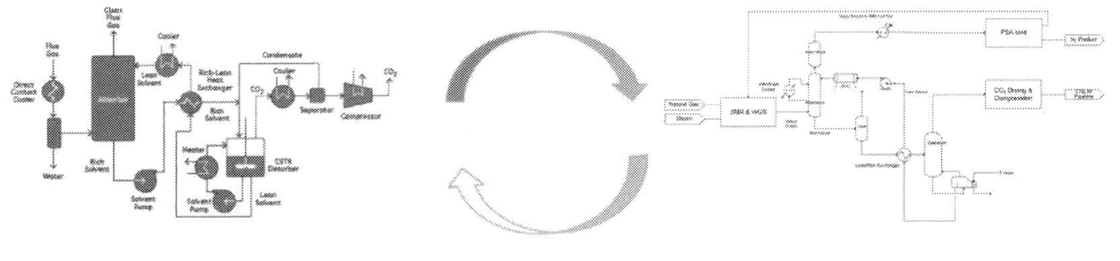

- ✓ **Physical Properties..** CH_4 / H_2 solubility
- ✓ **Absorber...** Packing, intercooling, MOC
- ✓ **Desorber...** CSTR, temperature, Pressure, MOC

- ✓ **MDEA Baseline Modeling..**
- ✓ **Techno-economic analysis..** CO_2 Cost
- ▪ **Heat Integration..** Cooling water, steam

Dynamic interplay between the CO_2 Capture Island and H_2 Plant

CO_2 Separation Process Modeling.. Gen 1 Model

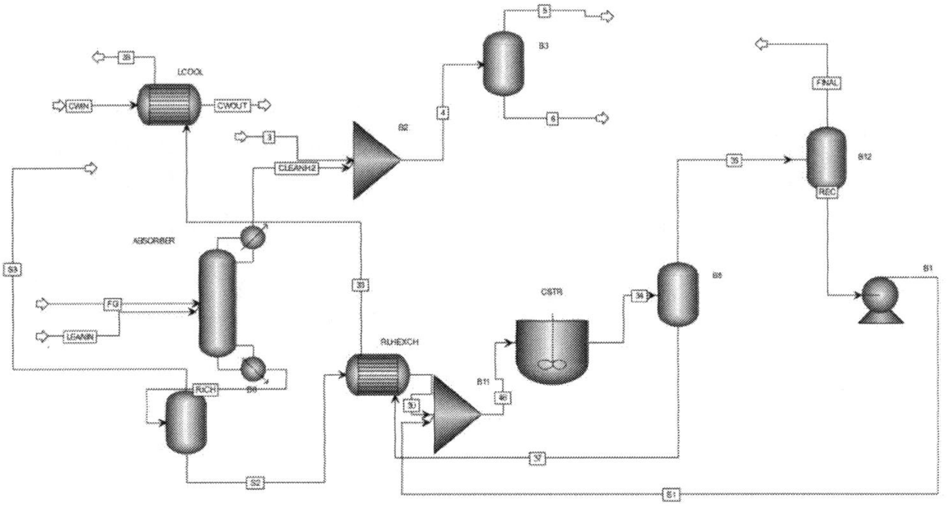

- Assumptions: H_2 and CH_4, no chemical absorption in the solvent
- Aspen Plus Models made for GE (GAP-1/TEG) solvent systems

Techno-Economic Analysis..

Gen 1 Model
(% Delta from baseline)

H_2 Production	-2 %
OPEX	-26 %
CAPEX	5 %
CO_2 Cost	-14 %

imagination at work

JACOBS Consultancy

Physical Properties Measurements.. Reactor

Overall Picture

450 mL windowed Parr reactor (2000 Psi)
ISCO syringe pump for gas delivery

Reactor Head

Reactor head fitted with 2 dip tubes, rupture disk, vacuum & vent

Gas Delivery System

Gas manifold for delivering gaseous components H_2, C_1-C_6

imagination at work

Physical Properties Measurements... H_2 Solubility

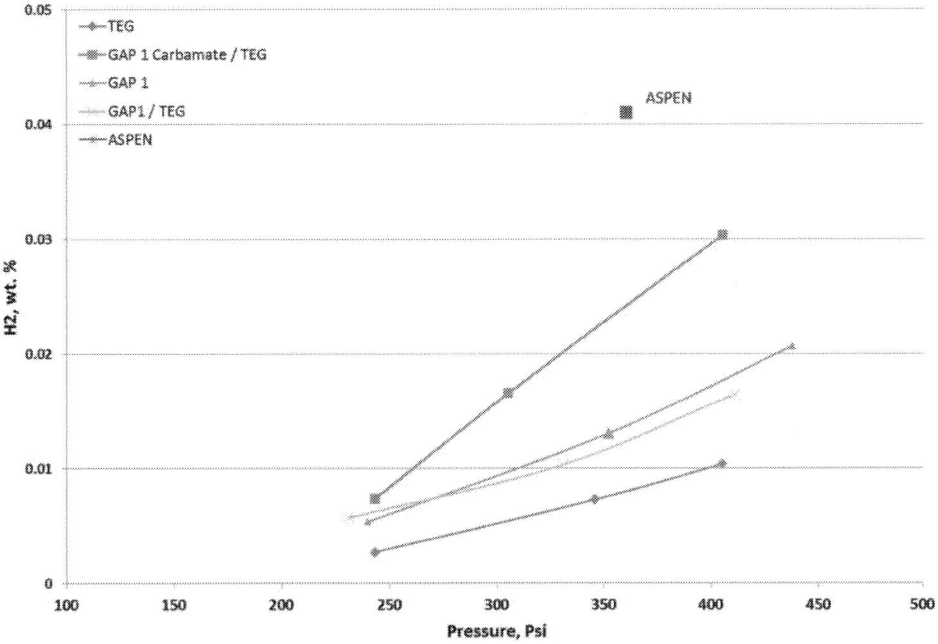

Gen 1 Model Over-predicts H_2 Solubility

imagination at work

GE Confidential

13
November 18, 2015

Techno-Economic Analysis..

	Gen 1 Model *(% Delta from baseline)*		*Gen 2 Model* *(% Delta from baseline)*
H_2 Production	-2 %		- 1 %
OPEX	-26 %	→ H_2 / CH_4 experimental solubility	- 32 %
CAPEX	5 %		- 7 %
CO_2 Cost	-14 %		- 23 %

Aminosilicone Solvent: Over 20% reduction in CO_2 Cost

imagination at work

GE Confidential

JACOBS Consultancy

14
November 18, 2015

Summary

Aminosilicone CO_2 capture technology:

- ✓ Non-aqueous solvent with improved CO_2 working capturing capacity
- ✓ Low volatility & low corrosivity solvent
- ✓ Lower footprint operations
- ✓ 20+ % reduction in CO_2 removal cost vs. baseline DMEA
- Reduce H_2 loss & solvent make-up hrough process optimization

Further technology advancements

Advanced absorber... (advanced intercooling, alternate packing)	Reduce CAPEX & OPEX
High T Desorber... (use of stabilizers)	Lower solvent recirculation
Heat Integration...	Reduce steam requirements
Gen 2 Silicone solvents... (lower heat of reaction & improve working capacity)	Reduce steam requirements and lower solvent recirculation

Acknowledgment

Authors would like to thank Suncor Energy and Canadian Natural for the financial support of the techno-economic analysis.

MTR Membrane Technology & Research

Industrial Use of Polaris™ CO_2 Capture Membranes

Karl Amo, Richard Baker, Trevor Carlisle, Zhenjie He, Jay Kniep, Jennifer Ly, Tim Merkel, Saurabh Pande, Steve White

Membrane Technology and Research, Inc.

www.mtrinc.com

CARBON MANAGEMENT TECHNOLOGY CONFERENCE

Sugarland, TX
November 18, 2015

MTR

Outline

- Introduction
- Technology background
- Current and emerging industrial use
  - Natural gas
  - Refinery/syngas
- Summary

2

MTR

MTR Introduction

- Privately-held, 90 employees mostly located in Newark, California
- Started in 1982 funded by U.S. SBIR grants
- Commercial products in petrochemical, natural gas and refinery industries
- Provides complete turn-key solutions with over 200 membrane systems installed worldwide
- 20 person R&D group with expertise in membrane materials/formation and process design

Membrane Technology Basics

- Membranes have composite structures to maximize flux

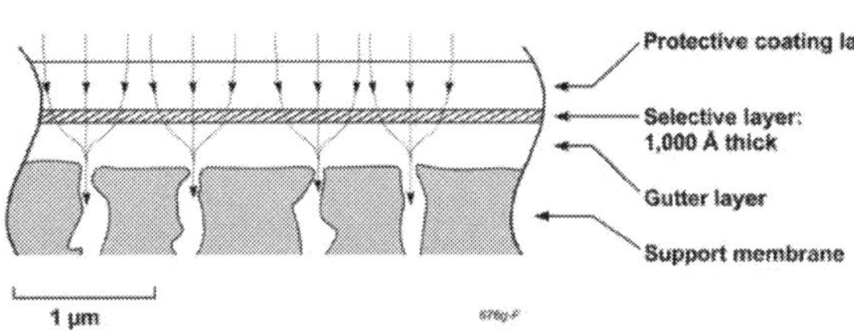

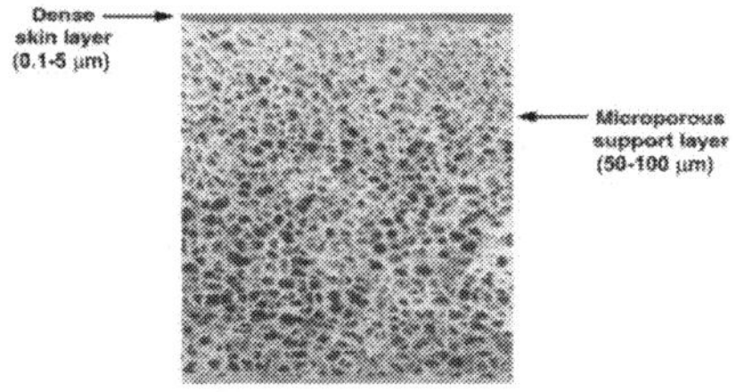

- Membranes are packaged in modules for industrial use

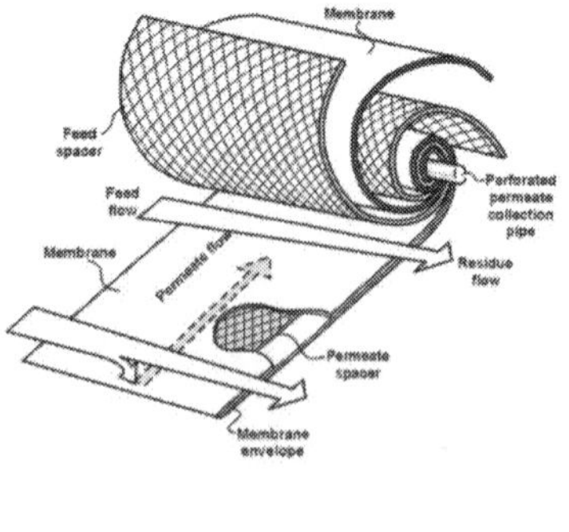

MTR CO_2 Capture Development Timeline

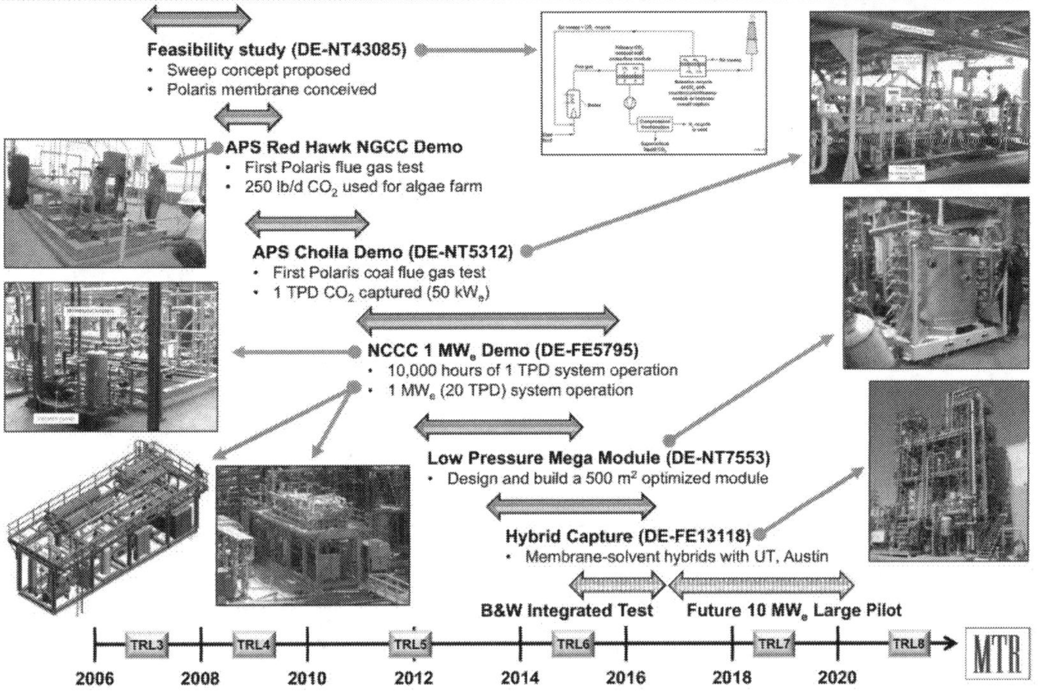

Polaris Membrane Development

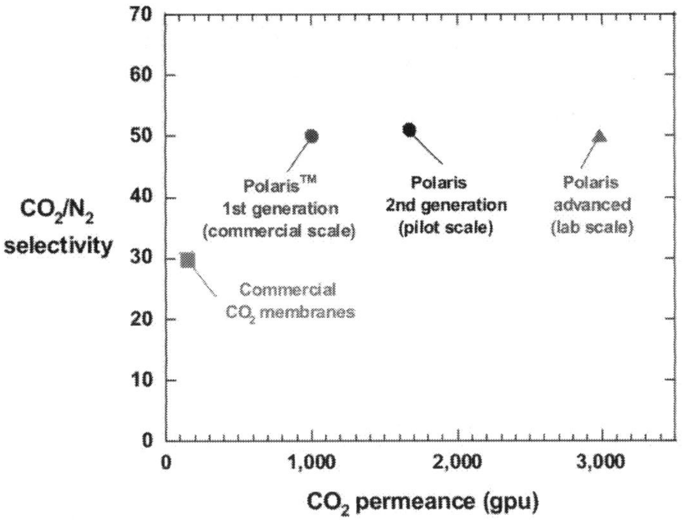

- Polaris membrane has exceptionally high CO_2 permeance and good polar/light gas selectivities
- Over the course of the DOE programs, this membrane has been scaled up from a concept to commercial production in 1000 ft. rolls

1 gpu = 10^{-6} $cm^3(STP)/(cm^2\ s\ cmHg)$ = 3.35×10^{-10} $mol/(m^2\ s\ Pa)$

6

MTR

Natural Gas Application: Membrane Fuel Gas Conditioning

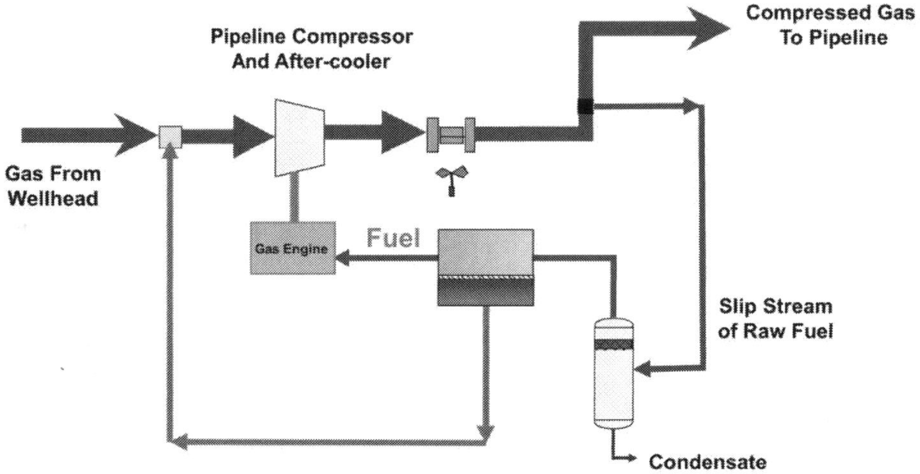

- Remove CO_2, H_2S & water from fuel
    - Recover valuable heavy HCs

7

MTR

Example of a Fuel Gas Conditioning Membrane System

Compressor station, New Mexico

Stream	Raw Feed Gas	Conditioned Fuel Gas
Methane Number	53	74
HHV (Btu/scf)	1139	1028
Component (mol%)		
Nitrogen (N_2)	1.69	2.98
Carbon Dioxide (CO_2)	4.76	1.76
Methane (C1)	76.78	87.77
Ethane (C2)	9.77	5.38
Propane (C3)	4.68	1.54
Butanes (C4)	1.43	0.38
Pentanes (C5)	0.56	0.12
Hexanes plus (C6+)	0.34	0.06

8

Over 100 installations in various North American shale plays

MTR

Polaris System Can be Used to Treat Sour Gas and Reduce Emissions

H_2S contaminated gas, Canada

Gas Stream Component	Feed Gas (mol %)	Conditioned Gas (mol %)
Propane	2.72	0.624
i-Butane	0.37	0.049
n-Butane	0.67	0.088
i-Pentane	0.18	0.018
n-Pentane	0.19	0.019
Hexane	0.16	0.010
C_{6+}	0.14	0.008
Total C_{3+} Hydrocarbons	**4.43**	**0.82**
H_2S Content	**3400 ppm**	**40 ppm**

9

Fuel Gas Conditioning: Summary of Membrane Benefits

- Acid gas and VOC (NMHC) emissions reductions – Meet strict emissions requirements
- Engine performance significantly improved, unscheduled downtime and engine maintenance costs reduced (due to reduced heavy hydrocarbons & BTU content of fuel gas)
- Added revenues through increased condensate recovery
- A membrane success story with 100+ installations

10

Future Application: Polaris System for CO_2 Reuse

- Membranes are particularly cost-effective for bulk CO_2 recovery
- Potential applications include CO_2 capture from flue gas (power, steel, cement) for reuse: algae, greenhouses etc.

- Demo system at APS Redhawk (a 1,060 MW_e NGCC power plant located near Phoenix, Arizona
- Polaris system captured 250 lb CO_2/day at ~30% CO_2 for delivery to an algae biofuel farm

MTR

Polaris Development for Refinery/Syngas

PSDF at NCCC

Pilot-scale Polaris CO_2 capture skid

Bench-scale Proteus skid

- MTR has been testing Polaris (CO_2-selective) and Proteus (H_2-selective) membranes for 6 years at the National Carbon Capture Center (NCCC)
- Membranes and semi-commercial modules are ready for scale-up and demonstration on industrial streams

MTR

Overview of Potential Industrial Syngas Applications for Polaris

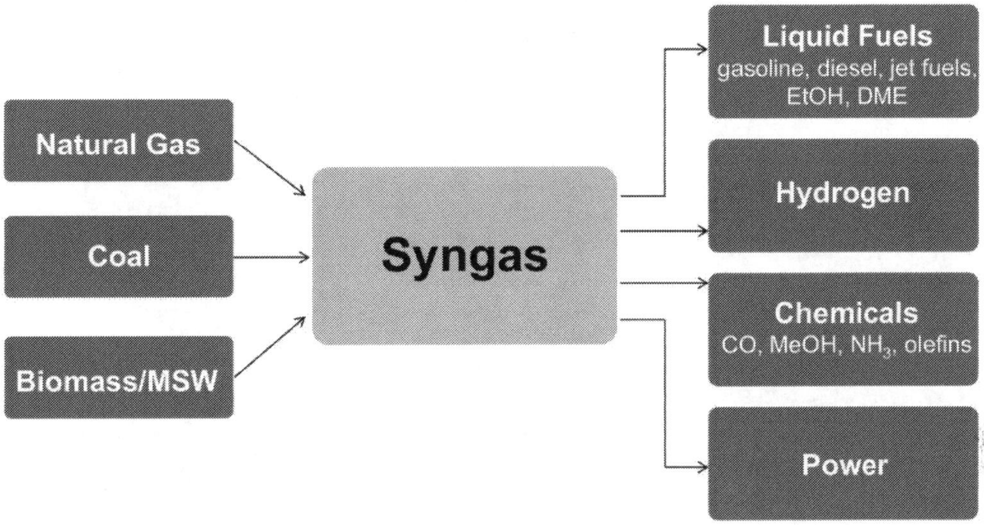

13

MTR

500 lb/h CO_2 Liquefaction Pilot System

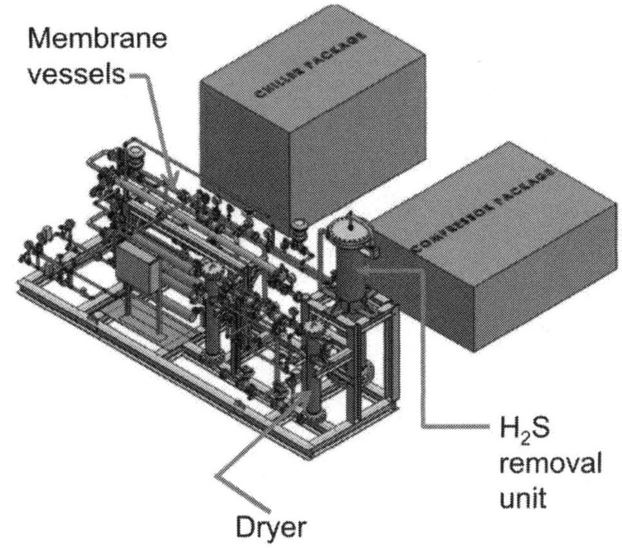

- Treats 500 lb/h of syngas; makes high-purity liquid CO_2
- Uses full-scale 8-inch diameter Polaris modules operating cold
- Commissioning in Dec 2012; performance data collected during R10 and R11 runs in 2013

14

MTR

500 lb/h Pilot Skid Installed at NCCC

- Installed in October/November 2012; commissioned in December 2012
- System uses commercial-scale CO_2-selective Polaris™ membrane modules

MTR

15

Polaris CO_2 Liquefaction Pilot System: Test Highlights

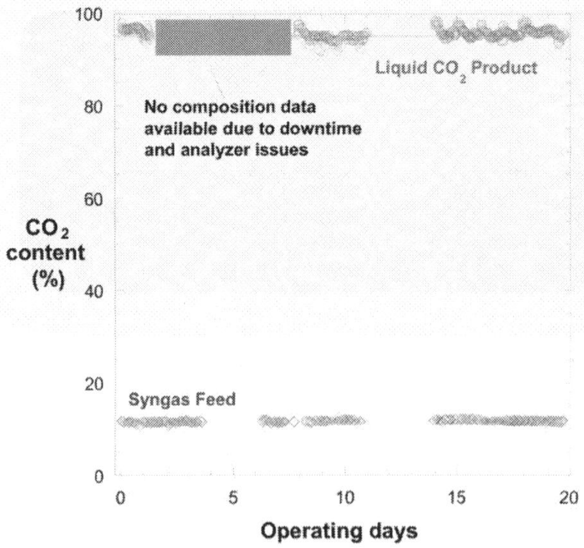

- Over 1,000 hours of skid operation during NCCC campaigns R10 and R11
- No degradation in performance of Polaris modules during testing
- Liquid CO_2 produced from ~12% syngas feed
- CO_2 purity consistently >97% at -33°C and 390 psig
  - Excellent agreement with simulated value
- Captured >900 lb/day of liquid CO_2 (400 kg/day)

MTR

16

Polaris Field Test at the Advanced Energy Research Facility (AERF)

- AERF test site is located at Edmonton Waste to Biofuels facility
- Objective is to field test Polaris (CO_2-selective) and Proteus (H_2-selective) membranes within municipal waste to biofuel process
- Funding from Alberta Innovates and DOE; project partners are City of Edmonton and Enerkem

Alberta Innovates Energy and Environment Solutions

Enerkem

The City of Edmonton Waste Management Services

Picture courtesy of City of Edmonton

MTR

Field Test Simplified Process Diagram

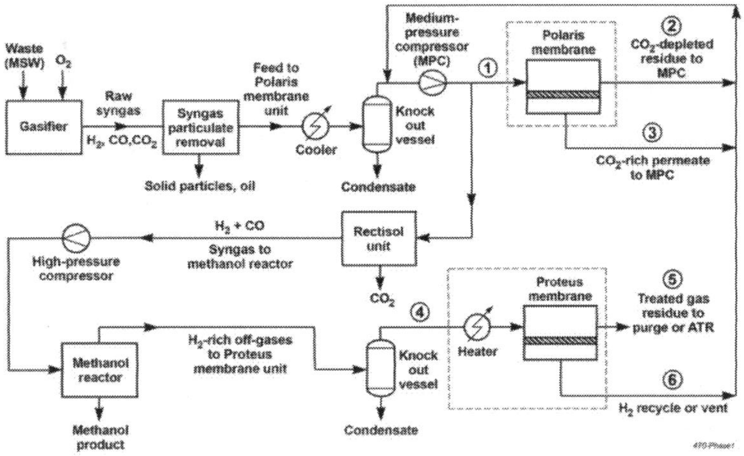

Polaris to debottleneck a Rectisol unit and Proteus to enhance methanol reactor efficiency

MTR

MTR Industrial CO_2 Capture Test Skid Installed at the AERF

Module installation

Skid installed in July, 2015

First test campaign scheduled for 1Q2016

MTR

Summary

- Polaris membranes, being developed for post-combustion CO_2 capture with DOE funding, have started to find commercial industrial use
- In some cases, like fuel gas conditioning, there are economic drivers other than CO_2 regulations/credits
- Others, like CO_2 reuse or IGCC capture, would benefit from regulations
- Membranes are particularly cost effective when partial capture and/or CO_2-enriched streams are acceptable

20

MTR

Acknowledgements

- U.S. Department of Energy,
  National Energy Technology Laboratory
  - Jose Figueroa
- Southern Company Services (NCCC)
- Alberta Innovates – Energy and Environmental Solutions
- Enerkem
- City of Edmonton

21

TSI and CO_2 Market Report

Tech Services Inc.

Carbon Management Technology Conference (CMTC Nov. 2015)

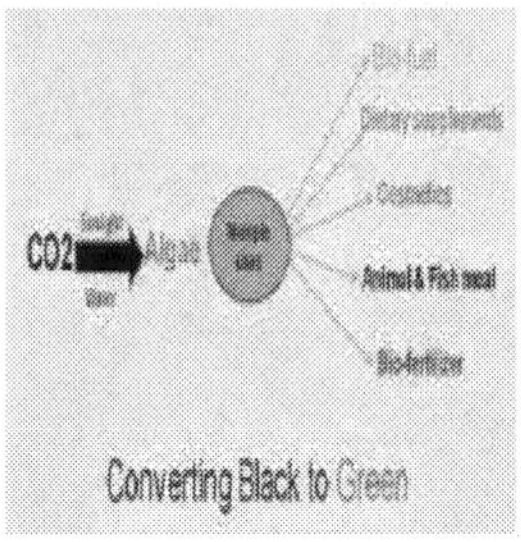

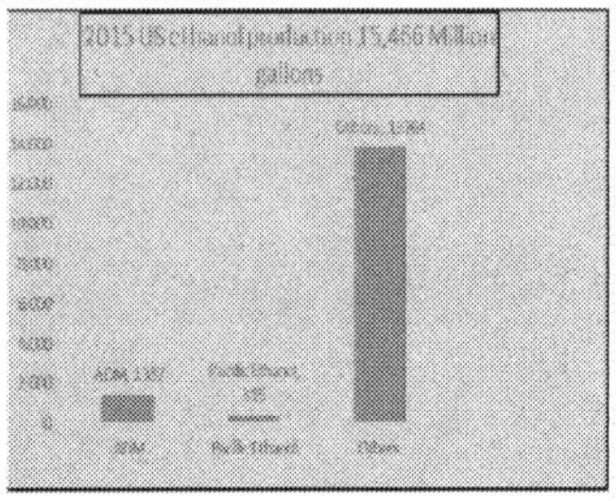

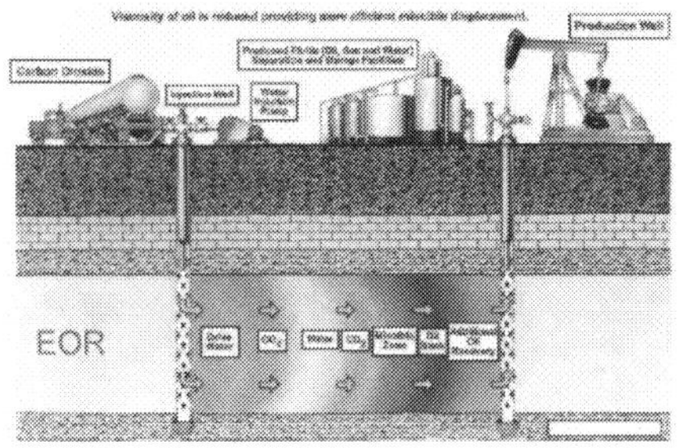

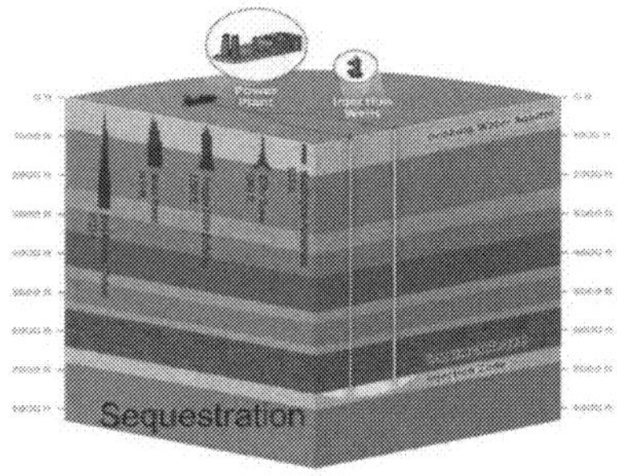

11/17/2015 **Confidential Information** Sudhir Brahmbhatt, Ph.D., MBA 1

 Providing Innovative and cost effective solutions to the industry

TSI and CO_2 Market Report

Tech Services Inc.

Topics Covered

- CO_2 Gas - A high profile **Greenhouse gas** and **valuable market commodity**
- US CO_2 Market - A changing Landscape
  - Suppliers in US, CO_2 Applic., CO_2 market Opportunities
  - CO_2 applications in the market
- US Regions and CO_2 Market
  - Market, Key Drivers and End Users
  - CO_2 supply capabilities, Regional Distribution Capabilities
- Commercial Aspect:
  - Market Pricing for US Region as a function of Volume used by the end users
- Business Options - Options and ROI calculation
  - Notes on Economic Analysis
- References

11/17/2015 2

 Providing Innovative and cost effective solutions to the industry

TSI and CO2 Market Report

Tech Services Inc.

Technology services Incorporated: The Company

Highlights:

Expert in **Industrial gases applications** in

>Chemical Industry, >Environmental Industry >Pharma and Biotech Industry
>Metal Industry >Pulp & Paper Industry >Cryogenic Applications

- Industrial gases related experience more than 35 years with top 100 corporations.
- **More than 25 patents in a variety of industries**, Commercialized new concepts successfully with very attractive ROI
- Have developed variety of markets from small portfolio to large no of clients.
- Work with end users very closely to understand their needs and prepare solutions to their needs.

Our Services:

Technology Support for productivity improvements, Yield increases in

- ★ Support in **Environmental, Chemicals**, high temperature and metal industry in productivity gain, meet EPA regulation
- ★ **Pharma/Biotech**, In processes such as fermentation, Lyophilization, Deoxygenation, Support in Global pharmacopeias requirements.
- ★ **Creating Innovative ways to optimize operations. Premier Consulting**
- ★ **Add value to your products**.

Contact Information:

Technology Services Incorporation, 1700 Country top Court, Glencoe, MO 63038 636-458-1775 (o), 314-409-1120 (cell), 636-273-6093 (Fax), Sudhir.Brahmbhatt@TSInc-US.com, www.TSInc-US.com

11/17/2015

 Providing Innovative and cost effective solutions to the industry

TSI and CO_2 EMISSIONS

Tech Services Inc.

>Average coal plant today "generates **3.5 Million tons of CO_2** per year

>In the U.S., the Obama administration has set a goal under the Clean Power Plan to slash CO_2 emissions from existing coal-fired power plants **30 percent below 2005 levels by 2030**.

>In the U.S. alone, burning coal emitted **1.87 billion tons of** CO_2 in 2011, according to the U.S. Energy Info. Admin. **Worldwide, coal-burning released 14.4 billion tons** of CO_2 in 2011.

U.S. GREENHOUSE GAS POLLUTION INCLUDES:

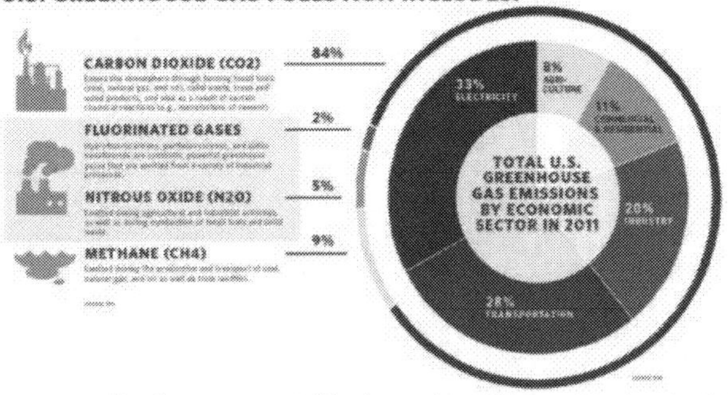

> CO_2 emissions from U.S. electricity generation by source, 2014

Source	Million metric tons	Share of total
Coal	1,562	76%
Natural gas	444	22%
Petroleum	23	1%
Other[3]	11	<1%
Total	2,043	

- Coal generates 44% of our electricity, and is the single biggest air polluter in the U.S.
- Coal pollutes when it is mined, transported to the power plant, stored, and burned

11/17/2015

The contents are strictly confidential and are not for any unauthorized person **Providing Innovative and cost effective solutions to the industry**

US Coal Fired Power plants:

Tech Services Inc.

Table A-2. The Nation's 100 Most-Polluting Power Plants, Carbon Dioxide Emissions Equivalent in Passenger Vehicles and Primary Fuel Category, 2011

Rank	State	Operator Name	Plant Name	Primary Fuel Category	Emissions (Million Metric Tons)[iv]	Emissions Equivalent in Passenger Vehicles (Millions)[iv]
1	GA	Georgia Power Co.	Scherer	Coal	21.3	4.44
2	AL	Alabama Power Co.	James H. Miller Jr.	Coal	20.7	4.3
3	TX	Luminant Generation Company, LLC	Martin Lake	Coal	18.8	3.91
4	MO	Union Electric Co. (MO)	Labadie	Coal	18.5	3.85
5	TX	NRG Texas Power, LLC	W. A. Parish	Coal	17.8	3.71
6	IN	Duke Energy Indiana Inc.	Gibson	Coal	16.9	3.53
7	OH	Ohio Power Co.	General James M. Gavin	Coal	16.6	3.46
8	PA	FirstEnergy Generation Corp.	FirstEnergy Bruce Mansfield	Coal	16.4	3.41
9	MI	Detroit Edison Co.	Monroe	Coal	16.4	3.41
10	AZ	Salt River Project	Navajo	Coal	15.9	3.32
11	IN	Indiana Michigan Power Co.	Rockport	Coal	15.4	3.22
12	KS	Westar Energy Inc.	Jeffrey Energy Center	Coal	14.7	3.05
13	GA	Georgia Power Co.	Bowen	Coal	14.2	2.97
14	WV	Appalachian Power Co.	John E. Amos	Coal	13.9	2.89
15	NM	Arizona Public Service Co.	Four Corners*	Coal	13.8	2.88
16	NC	Duke Energy Carolinas, LLC	Belews Creek	Coal	13.8	2.87
17	TX	Luminant Generation Company, LLC	Monticello	Coal	13.7	2.85
18	MT	PPL Montana, LLC	Colstrip	Coal	13.6	2.82
19	TX	NRG Texas Power, LLC	Limestone	Coal	13.3	2.77
20	LA	Louisiana Generating LLC	Big Cajun 2	Coal	13.2	2.75
21	MN	Northern States Power Co. – Minnesota	Sherburne County	Coal	13.1	2.73
22	SC	South Carolina Public Service Authority	Cross	Coal	12.9	2.69
23	WY	PacifiCorp	Jim Bridger	Coal	12.9	2.68
24	IL	Dynegy Midwest Generation Inc.	Baldwin Energy Complex	Coal	12.8	2.67
25	OH	Dayton Power & Light Co.	J. M. Stuart	Coal	12.7	2.66
26	KY	Kentucky Utilities Co.	Ghent	Coal	12.7	2.65
27	TN	Tennessee Valley Authority	Cumberland	Coal	12.4	2.57
28	WY	Basin Electric Power Coop	Laramie River Station	Coal	12.2	2.54
29	UT	Los Angeles Department of Water & Power	Intermountain Power Project*	Coal	12.0	2.51
30	KY	Tennessee Valley Authority	Paradise	Coal	12.0	2.49

*Indicates that this power plant is scheduled for retirement.[lvi]

11/17/2015

 Providing innovative and cost effective solutions to the industry

CO_2, A changing Landscape – Global Perspective

Tech Services Inc.

- CO_2 is both a high profile greenhouse gas and a valuable market commodity
- This product is found naturally from underground wells
- It is also emitted from Fossil Fuel burning industrial plants

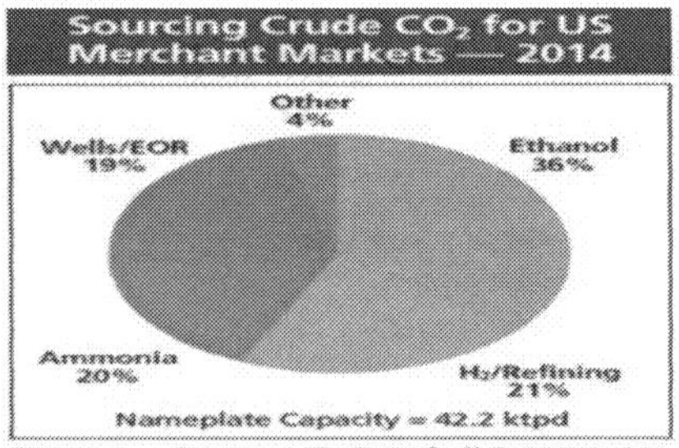

Figure 1 Source: J. R. Campbell & Assoc., Inc.

-Sourcing CO2 in US:

>For 2010 total US Merchant CO_2 production capacity was **40,000 tpd**

>Most of the Merchant CO_2 is now from Ethanol plants.

>Ammonia plants are expected to continue to lag as No. of plants shut down to high NG prices in 2002 onward. This could change due to low NG price now.

>**Ethanol plants continue to be dominant** source of CO_2 supply.

The current global demand

For CO_2 is estimated at **80Mtpa** based on 2010 data

- Of this 80Mtpa, at least **50Mtpa** is utilized for EOR, almost exclusively in North America.
- The remaining **30Mtpa** represents the global demand of all other uses, predominantly the mature industries of beverage carbonation and food industry uses.

The future demand

- Estimate (for the year 2020) for the short-listed reuse technologies is **140Mtpa**, including EOR.
- This estimate is based on a predicted growth of current technologies such as EOR and urea fertilizer and the implementation and commercialization of demonstration projects for the remaining technologies in line with their prospective development timeframes.

11/17/2015 Ref: SRI Consulting, March 2010, Chemical Economics Handbook 2010 6

 Providing innovative and cost effective solutions to the industry

CO_2 Supplies in US

Tech Services Inc.

Suppliers in US:

Major CO_2 suppliers and distributors are unique as they commonly do not produce the gas.

- **Linde is the largest supplier of CO_2 in US** with 25% of total 27,000 tpd produced.
- **Fullerton, NY Sunoco ethanol plant** started in 2010 producing 600 tpd of CO_2 and supplying to Food, beverage, chem. manufacturers including dry ice to the industry.
- **Praxair, Airgas, and Air Liquide** follow Linde.
- **Airgas is the largest producer** and whole sale distributor of dry ice in US.
- **EPCO, recently (2014)** purchased by Air Products, is large specialized producer with 10% market share.
- Pain Enterprise, Flo CO_2, and Continental Carbonic together contribute 22% in market share. Matheson Trigas bought Continental Carbonic in Feb 2014.
- 75% of CO_2 liquefier plants built in past 10 years are owned by small specialized producers.

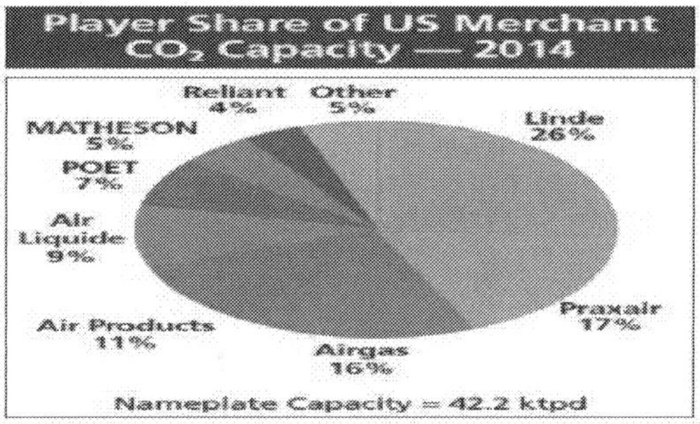

Figure 2 Source: J. R. Campbell & Assoc., Inc.

For 2010 total of 40,000 tons per day of CO_2 was the capacity. Two third, 27,000 tpd was produced and distributed in the market.

11/17/2015 7

The contents are strictly confidential and are not for any unauthorized person Providing innovative and cost effective solutions to the industry

TSI and CO_2 Applications

Tech Services Inc.

Carbon Dioxide Supplies And Users

- Gaseous CO_2: Supplied by Pipeline
- Liquid CO_2 is supplied by Merchant Channels such as tankers and Cylinders to a wide variety of industries.

>Demands are steady for :

- Food and Beverage Applications
- Laser production and welding
- Dry ice
- Pharmaceutical Industry
- Green House Gas

>Demand has surged and will remain strong in:

- Enhanced Oil recovery using pipeline CO_2 consuming 177,000 tpd of CO_2 to extract additional Oil from reservoirs

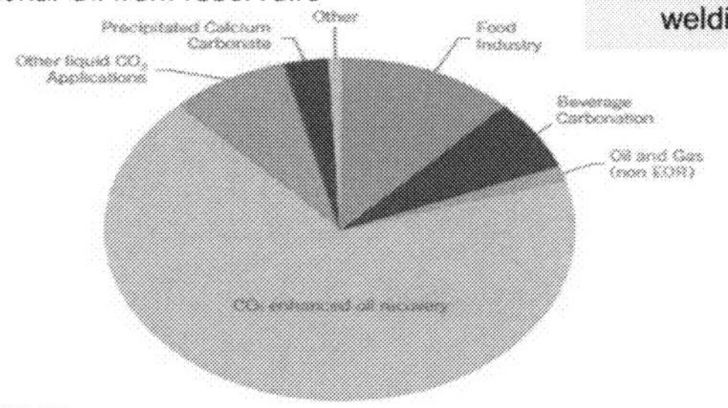

❖ Key Applications: For Dry Ice

-Food Industry: Meat processing, Short term food storage, distribution In flight catering, R & D

-Dry ice blasting: surface cleaning technique, a gentler and green process.

-Pharma and Biotech Industry

As a note:

-In Midwest region, Illinois based Continental Carbonic produces 1400 tpd of Liquid CO_2 where 80% of this product is converted to and sold as dry ice. It has 8 production facilities and 31 distribution sites. Matheson Trigas, a major industrial gases company in US, recently purchased Continental Carbonic gaining market share in dry ice business. This will also offer organic sales growth to Continental Carbonic as Matheson market share is diverse that includes welding, specialty gases and electronics.

11/17/2015 8

The contents are strictly confidential and are not for any unauthorized person Providing innovative and cost effective solutions to the industry

Some Applications of CO_2 by Industry

Tech Services Inc.

Manufacturing and Construction Uses:

>Carbon dioxide is used on a large scale as a shield gas in MIG/MAG welding, where the gas protects the weld puddle against oxidation by the surrounding air. A mixture of argon and carbon dioxide is commonly used today to achieve a higher welding rate and reduce the need for post weld treatment.

>Dry ice pellets are used to replace sandblasting when removing paint from surfaces. It aids in reducing the cost of disposal and cleanup.

Chemicals, Pharmaceuticals and Petroleum Industry Uses:

>Large quantities are used as a raw material in the chemical process industry, especially for methanol and urea production.

Rubber and Plastics Industry Uses:

>Flash is removed from rubber objects by tumbling them with crushed dry ice in a rotating drum.

Food and Beverages Uses for Carbon Dioxide:

>Liquid or solid carbon dioxide is used for quick freezing, surface freezing, chilling and refrigeration in the transport of foods.

>In cryogenic tunnel and spiral freezers, high pressure liquid CO_2 is injected through nozzles that convert it to a mixture of CO_2 gas and dry ice "snow" that covers the surface of the food product. As it sublimates (goes directly from solid to gas states) refrigeration is transferred to the product.

>Carbon dioxide gas is used to carbonate soft drinks, beers and wine and to prevent fungal and bacterial growth.

>Liquid carbon dioxide is a good solvent for many organic compounds. It is used to de-caffeinate coffee.

>It is used as an inert "blanket", as a product-dispensing propellant and an extraction agent. It can also be used to displace air during canning.

>Supercritical CO_2 extraction coupled with a fractional separation technique is used by producers of flavors and fragrances to separate and purify volatile flavor and fragrances concentrates.

>Cold sterilization can be carried out with a mixture of 90% carbon dioxide and 10% ethylene oxide, the carbon dioxide has a **stabilizing effect** on the ethylene oxide and reduces the risk of explosion.

11/17/2015 9

 Providing Innovative and cost effective solutions to the industry

TSI and CO_2 & Applications

Tech Services Inc.

•**Medicine:** In medicine, up to 5% CO_2 is added to pure oxygen. This helps in provoking breathing and to stabilize the $O2/CO_2$ balance in blood. $SCCO_2$ applic. in sterilization of drugs and Extraction. Global CO_2 market in Pharma & Biotech is $85 mm, US is $32 mm and growing.

•**CO_2 Laser:** The CO_2 laser, a common type of industrial gas laser uses CO_2 as a medium. Welding: It also find its use as **an atmosphere for welding**.

•**Oil Wells:** Carbon dioxide is commonly injected into or next to producing oil wells to draw lost traces of crude oil .

•**Metals Industry:** It is used in the manufacture of casting influences so as to enhance their hardness.

•**Fumigation:** Used as a fumigant to increase shelf life and remove infestations.

•**Enhanced Oil Recovery:**

➢Heightened interest in US in Enhanced Oil Recovery (EOR) using tertiary techniques such as EOR using $CO2$. However, Laid in cost for using CO_2 in EOR projects is driven by distance from its source to application site.

➢Danbury resources, a key CO_2 supplier in Texas via pipeline sells 400 mmscf per day $CO2$ for EOR in Southeast region.

➢Major limitation in other areas of US is CO_2 supply in scattered EOR project locations.

<u>**Carbon Sequestration and storage:**</u>

➢ Syngas and Power projects for many regions of US and Canada will provide CO_2 for EOR services in years to come.

➢Oklahoma state has numerous EOR projects where CO_2 is pipelined from Bravo Dome and others are supplied by ammonia where CO_2 is byproduct.

➢This is new area and it will take years before its activities are at peak.

11/17/2015 10

 Providing Innovative and cost effective solutions to the industry

Waste CO_2 to Valuable Products from Algae

Tech Services Inc.

- Separated CO2 from Flue emission can be further:
  - -Purified, compressed, and sold to commercial gas clients
  - -Used in agriculture e.g. algae farming.
- Algae has **found its usage in a variety of industries** as follows:
  - - Fuel: EPA 2011 renewable fuel std. volume requirements at **1.35 billion gallons of advanced biofuels.** And the U.S. Navy's goal is to operate at least 50% of its fleet on clean renewable fuel sources by 2020. Achieving this objective will require a significant use of biofuels.
  - -Renewable Oil Markets: This area is focused on **addressing the needs of chemicals, nutritionals, and skin and personal care markets using algae.**
  - -Chemicals: In 2009, about 9 million metric tons of palm kernel and coconut oils were consumed globally for food and industrial applications such as laundry detergent, soap, and shampoo, functional fluids such as airplane de-icing fluids, **Algae oils are a natural fit for these markets.**
  - -Nutritionals: The $2 trillion packaged food industry is actively seeking more natural, sustainable sources of fiber and healthful fats without compromising value, quality, taste, or nutrition. **Commercializing a suite of microalgae-derived food ingredients is in progress.**
  - -Health Sciences: Target market is global skin and personal care market, which is estimated at $323 billion. Within this market, focus area of business are
    - (1) Toiletries, makeup, and hair care, estimated at $239 billion
    - (2) skin care, estimated at $84 billion.

    Consumers are staying active and living longer, and their desire to appear youthful will continue to drive growth in the beauty and personal care markets.

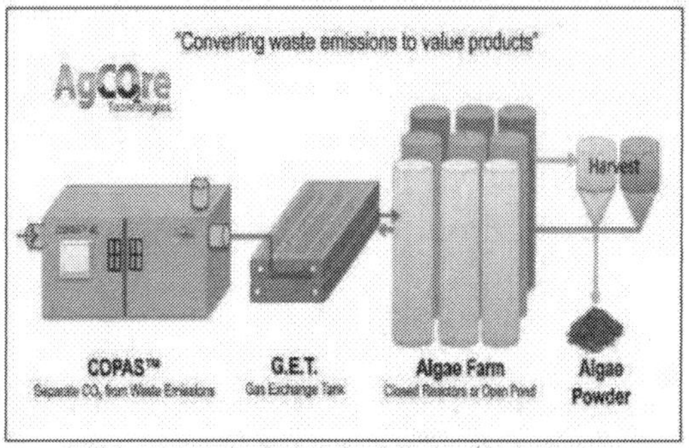

This diagram, provided by Agcore Technologies, demonstrates how its equipment is used to capture, separate, and infuse CO_2 for use in its algae farming operations.

11/17/2015 11

CO_2 Market

Tech Services Inc.

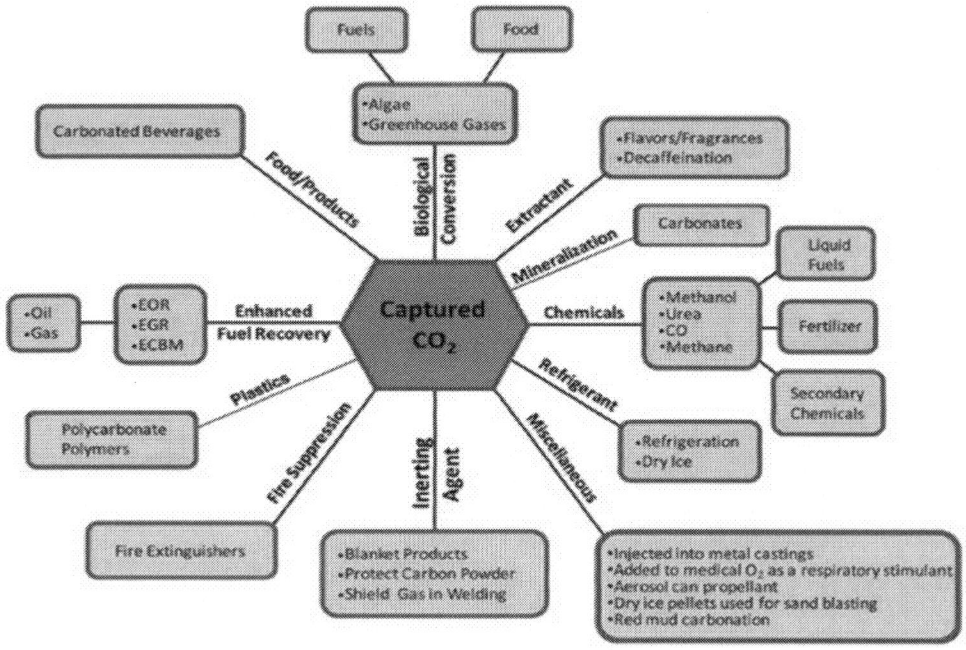

11/17/2015 12

TSI and CO_2 Market for the CO_2 Manufacturer

Tech Services Inc.

CO_2 Market Opportunities – Should it be Wholesale Raw Gas or Direct Marketing to Consumers?

- Some decades back, in the United States, many of the sources (i.e. ethanol, ammonia plant) operators and owners were and are the same parties who owned operated the CO_2 plant near the raw gas source operation and marketed.
- Today, also **numerous independents** thrive in the United States, as direct sales to consumers. Since the emergence of the major gas companies, most of the raw CO_2 is actually sold to the gas refiner – which is also the marketing operation for the merchant CO_2.
- In addition to this exception today in North America,

This direct CO_2 marketing scheme is quite common in Latin America and many other regions.

Ref: http://www.biofuelsdigest.com/bdigest/2011/11/23/carbon-dioxide-applications-%E2%80%93-a-key-to-ethanol-project-developments/

11/17/2015 13

 Providing innovative and cost effective solutions to the industry

TSI and CO_2 Market for the CO_2 Manufacturer

Tech Services Inc.

- **One fact** supporting direct sales of CO_2 by the ethanol firm to a limited merchant or niche market is a **large margin difference** between the raw gas prices to a refiner/gas company. **Raw gas prices** range **from $5 to $25/ton** v. consumer **market prices** usually averaging around **$90 to $100/ton**; and in some high priced markets with little regional competition or no local supply can be **$200 to $300/ton.**
- It will be necessary to evaluate the costs of production, distribution, and overhead, initially to consider wholesale raw gas sales or direct markets.
- Further, once markets are understood, and the costs & requirements for producing CO_2 for the **merchant trade** is known, along with distribution; then potential risks for direct marketing can be properly evaluated.
- Numerous cases exist, where a niche market or a specific region would make the most sense, in terms of directly marketing the commodity – and a true opportunity to produce much stronger revenues.

We now quickly review different US regions, their key drivers and CO_2 capabilities

11/17/2015 14

 Providing innovative and cost effective solutions to the industry

The New England Region -Market

Tech Services Inc.

Key Drivers in the Region:

- For New England Region the most important driver is the **Govt. R &D** as total revenue has been steady from \$67 billions to \$69 billions from 2006 to 2011. **Health care and related services** have grown to \$66 Billion in 2011 an increase of \$7 billion since 2006. **Computer, electronics and component mfg**. has seen largest growth since 2006 from \$20 b to \$34 billion in 2011. Chemical, Petrochemical and plastic segment has dropped more than 40% while construction more than 20% in same five years period. **Food and beverage sector** is steady \$5 Billion during the five year period.

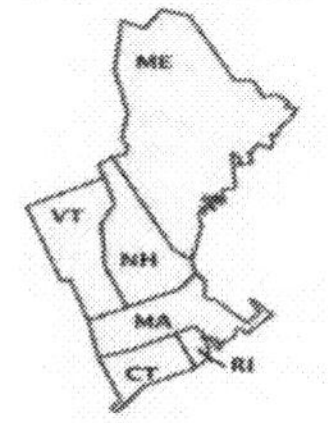

CO2 Supply capabilities:

- **No CO_2 production** facility in this region.
- Nearest CO_2 sources for New England region are from Ethanol plants operating in New York state.
- **Both CO_2 plants, one EPCO Now Air Products in Madina near Canada, and other by Linde near Albany were commissioned in 2007-2008.**
- Independent CO_2 distributors are -American Carbonation in Palmer, MA, - Esquire Gas Production now Tech Air Both companies have rail receiving Terminals.
- In addition, **Linde has rail depot with 500 t CO_2 storage capacity for beverage grade CO_2 supplied from their Volney, NY facility.**

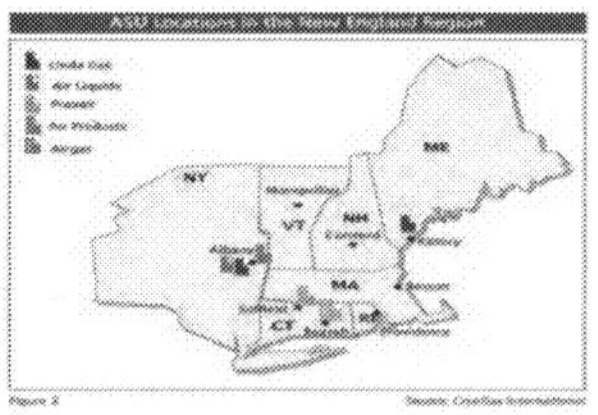

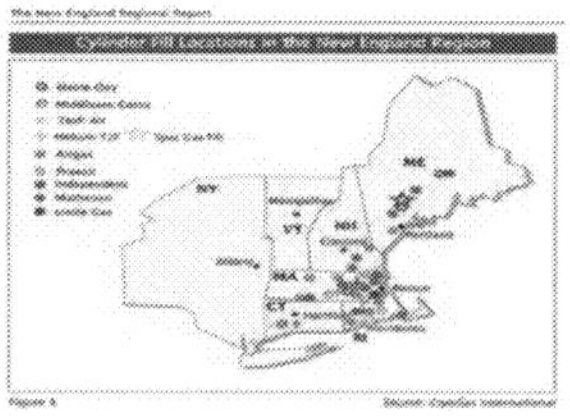

 Providing Innovative and cost effective solutions to the industry

The Mid-Atlantic Region -Market

Tech Services Inc.

Key Drivers in the Region:

- **Govt. R & D market** in this region has grown steadily from \$269 billion in 2006 to \$277 billion in 2011 roughly 3% increase. Largest potential market for gas industry.
- **Healthcare and related services** is the second growing market that grew 10% form 2006 till 2011. It is now \$188 billion market.
- **Computer, Electronics and component market** grew 65% from 2006 till 2011. It was \$38 billion in 2011.
- In contract chemical segment shrunk to \$31 billion in 2011 from \$46 Billion in 2006.
- **The Natural gas extraction** is picking up at rapid pace.

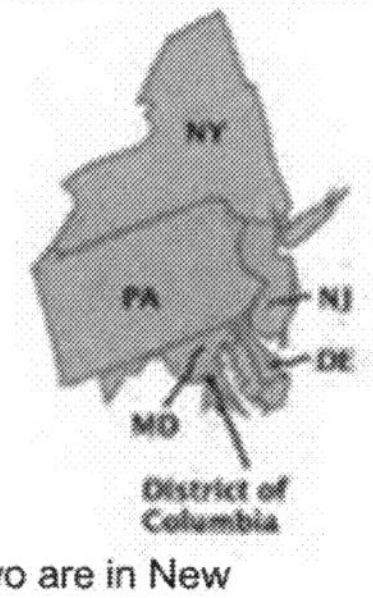

CO_2 supply capabilities:

- **Four known CO_2 plants in this region with Total Capacity of 1500 tpd of CO_2.** Two are in New York both from Ethanol plants. Medina, NY plant was built **by EPCO now APCI while other plant in Albany, NY owned by Linde.**
- **Medina plant built by EPCO also covers Canada.**
- **Linde ships product to the region by rail and has terminals in NJ and PA.** It has second LCO_2 plant in Cumberland, Maryland from cogen plant.
- Air Liquide has CO_2 plant next to the refinery in Delaware. The plant was shut down in 2010-2011.

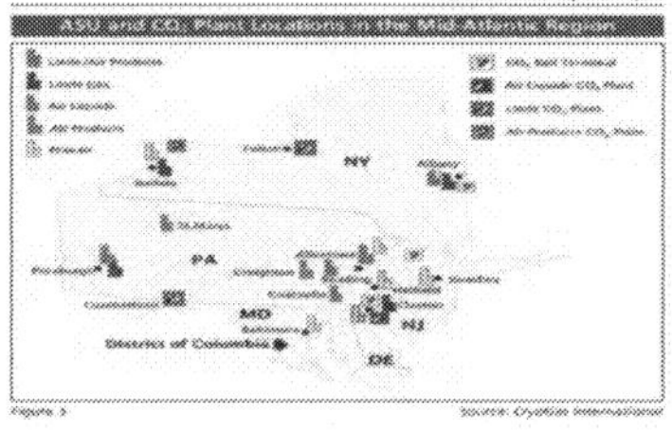

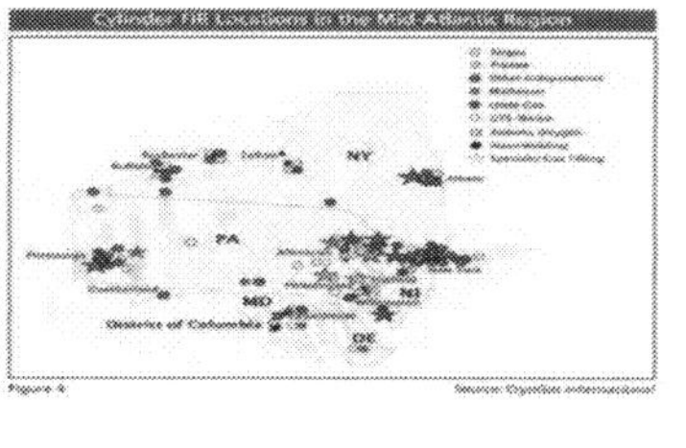

 Providing Innovative and cost effective solutions to the industry

The SouthEast Region -Market

Tech Services Inc.

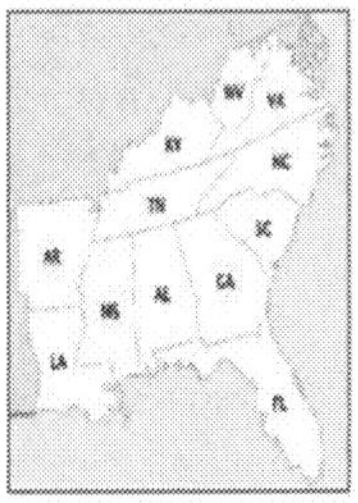

Key Drivers in the Region:

- **Large Number of Govt. R & D** in region value $403 Billion in 2011. Florida is home for NASA Operation at Kennedy space Center, while Virginia for govt. labs and TN for largest National lab for DOE.
- **Healthcare is another driver** worth $215 Billion in 2011.
- **Total manufacturing** is another key driver in the region, where food and beverage is $67 billion in 2011, automotive $22 Billion, $50 billion for computer, electronics and components.
- Construction sector has dropped 31% in 2001 from 2006 to $107 billion from $154 Billion.

CO_2 Supply Capabilities:

- **SouthEast has large infrastructure of ASUs**, 105 ASUs owned and operated by major gas companies with total capacity of 21,000 tpd of LIN/LOX which is 27% of total installed liquid capacity in US. This is large region with12 states and more than 80 million population.
- There are **22 known and identified LCO_2 production plants with total capacity of 9000 tpd.** Over 75% of CO2 is associated with Natural gas wells. Owned and operated by **Danbury resources**. 15 % is from Ammonia plants and the rest from ethanol plants and refinery.
- **Air Gas, Linde, and Praxair are major CO_2 players**, and some independent sources such as Matheson (Former Continental Carbonic) are also in this region.

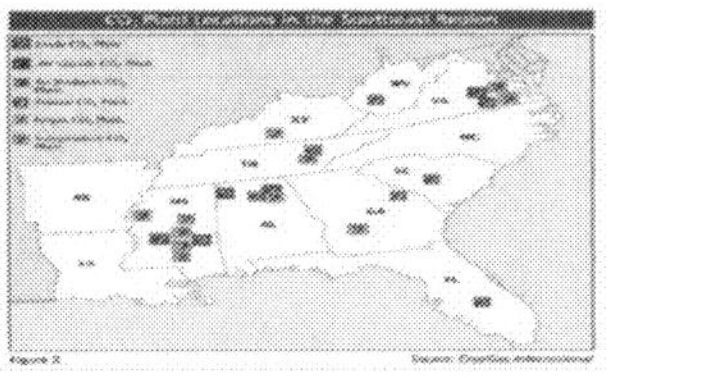

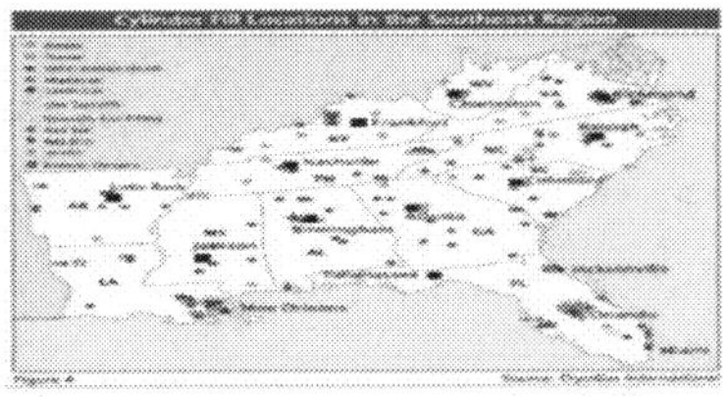

11/17/2015 17

 Providing Innovative and cost effective solutions to the industry

The Great Lakes Region -Market

Tech Services Inc.

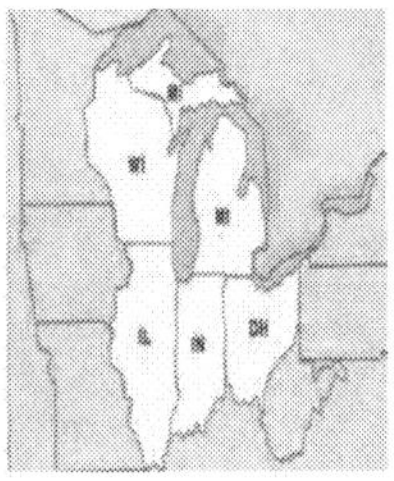

Key Drivers in the Region:

- Great lakes Region **manufacturing** represents very large share of the Industrial gases market (US gas business is $16 billion in 2011)
- **Largest Automobile related market** ($44b) in region and home for power houses like Kraft, Kellogg, Smucker and food and beverage sector ($33b)
- **Healthcare and related services** is also a bright spot ($148b)
- **Computer, electronics growing** 7% and was $21b in Great lake area in 2011.
- **Chemicals, Petroleum, plastic sector** growing 7% and was $66b in 2011.
- **Govt. R & D** represents second largest overall market in Great lake region after total mfg.

CO_2 Supply Capabilities:

- **19 CO_2 production facilities** with total installed capacity of over 6000 tpd. Mainly from IL, WI, and IN.
- Over last decade **dramatic increase in CO_2 capacity**. And it is linked to number of new ethanol plants. 75% of capacity is associated with from Ethanol plants.
- **APCI acquired EPCO in 2013 with four new CO_2 plants in the region.**
- **Air gas has important position in the region for CO_2. Linde, Praxair** have owned and operated plants for some time now.

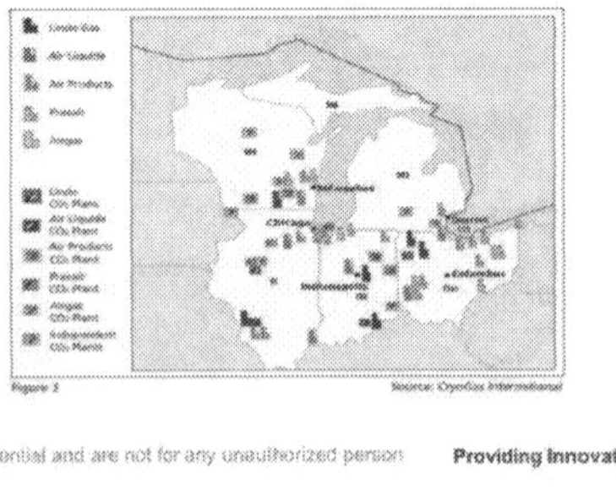

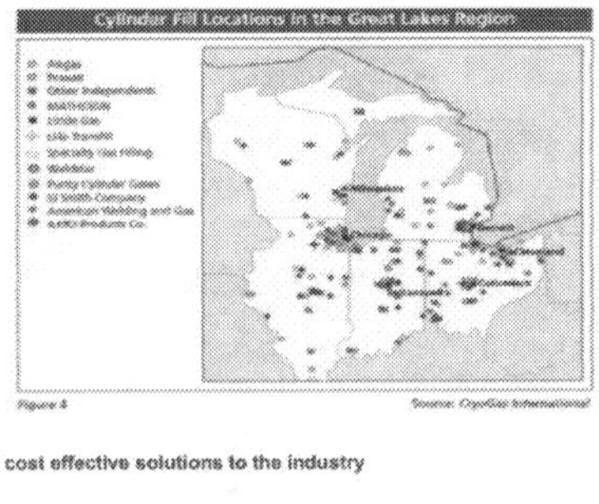

11/17/2015 18

 Providing Innovative and cost effective solutions to the industry

The Plains Region Market

Tech Services Inc.

Key Economic Drivers in the Region:

- **Agricultural segment** grew 37% between 2006 and 2011 ($30b).
- **Food and Beverage market** size is $22b in 2011 making it largest individual manufacturing segment by market share.
- **Healthcare and Govt. R & D sector** in the plains is healthy. Healthcare grew 13% to $69b from 2006 thought 2011, while govt. services sector grew 2.1% to $99b.
- **Mining, oil and Exploration** grew 62% from $2.9b in 2006 to $4.7b in 2011.
- **Computer and Electronic grew** 38% to $20b from 2006 to 2011.

CO_2 Supply Capabilities:

- Important source of CO_2 through Ammonia plants and Ethanol plants. 5000 tpd installed capacity.
- **20 Known and identified LCO_2 production facilities** Operating in the region.
- **75% of CO_2 capacity is from Ethanol plants** and 15% is from the ammonia(NH_3) plants., remaining 10% is sourced from the refineries.
- **APCI with EPCO purchase now has 10 production plants competing with Praxair who was the leader in the region before 2013. Matheson purchase** of Continental Carbonic and **Air Liquide through acquisition** has also increased their share in the market.

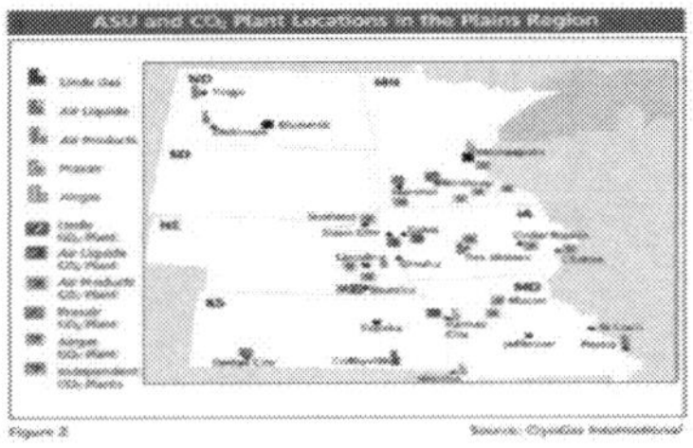

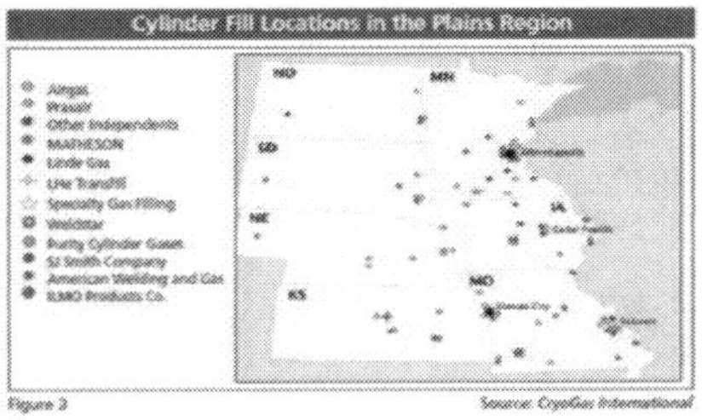

 Providing Innovative and cost effective solutions to the industry

The SouthWest Region -Market

Tech Services Inc.

Key Drivers in the Region:

- **Manufacturing segment** is key driver with $204 billion in 2011 (pertaining to the gas business). Largest is chemical, petroleum, and plastics sector with $65.5 billion in 2011.
- **Computer, Electronics, & components mfg. is second** largest with $61 Billion in 2011.
- The key driver for the gas industry is the **govt. R & D**, $183 billion with key govt. institutions such as Los Alamos lab, Sandia National Lab.

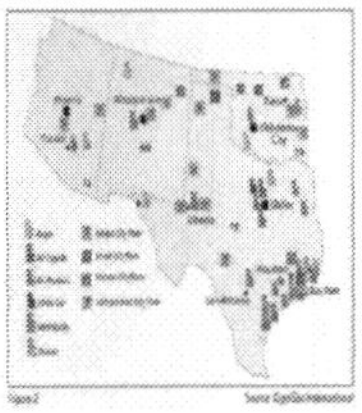

CO_2 Supply Capabilities:

- **22 known LCO_2 production facilities** with total Installed capacity of 6600 tpd of which 38% are sourced from Natural gas wells compared to 75% in SouthEast Region.
- **Ammonia plant** is the next largest source of CO_2 accounting or 28% of total supply of CO_2.
- **Petrochemicals and refining account for 22% of CO_2 sourcing** while remaining 12% is from Ethanol plant, Power plant and cogen. (CO_2 for EOR is excluded in analysis)
- **Reliant Holdings is the largest operator of CO_2 merchant capacity**. **Linde has five CO_2 plants** while **Airgas owns and operates 3 facilities**. Air Liquide and Praxair have access to some merchant CO_2.
- Continental now Matheson also has presence in this region.

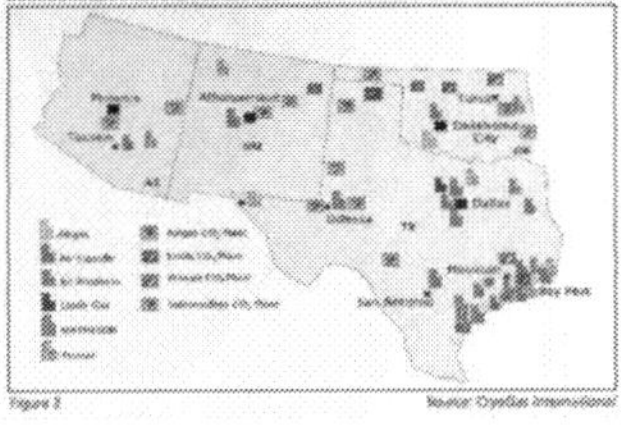

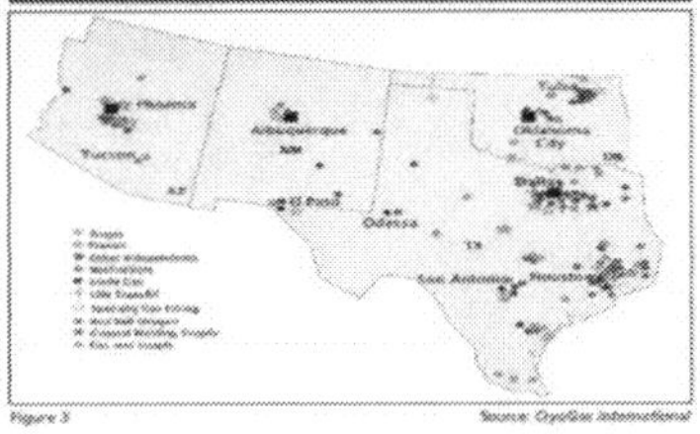

 Providing Innovative and cost effective solutions to the industry

The Rocky Mountain Region -Market

Tech Services Inc.

Key Drivers in the Region:

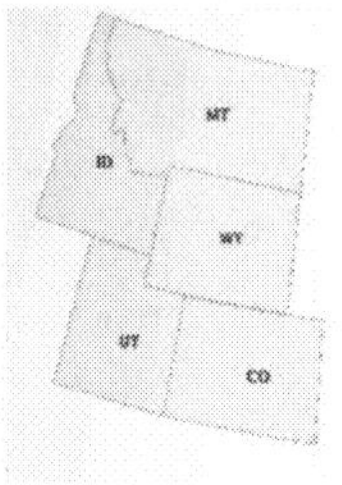

- **Govt. R & D** \$57.4 B (in 2011) is leading the key drivers for the gas Industry. Rocky Mountain Lab (RML) in Hamilton, Montana is the largest lab NIH has in this region.
- **The Mfg. segment** is \$44 B (in 2011) followed by computer and component equipment sector \$15.6 B (in 2011). The Chemicals, petroleum and plastics make up \$7.9 B in 2011.
- **The Healthcare industry** is \$28.5 b (in 2011) while Oil and mining is \$12 B (in 2011). Colorado accounts for \$ 9.0 B in Mining while Wyoming mining industry is \$8.8 B
- The construction industry is \$14 B (in 2011) that was down by 26.6% from 2006.

CO_2 Supply Capabilities:

- **11 ASUs owned by major gas companies** with total capacity of 2300 TPD LOX/LIN
- NORCO, large distributor owns a ASU in Boise, ID.
- **8 LCO2 production facilities in the region** with total installed capacity of 2000 tpd of CO_2. Over 50 % of CO_2 is from Natural gas processing. 25% of CO_2 is from the natural wells. Remaining 25 is from Ammonia in Wyoming and Ethanol plant in Colorado.
- **Significant CO_2 pipelines** in the region for EOR (Enhanced Oil Recovery).
- **Linde and Praxair are two largest CO_2 players** in the region. However, Ferus bought independent operator of pure CO2 and Reliant Holding built CO_2 recovery plant at only ethanol plant. Kinder Morgan Energy partner invested \$671 M in 2011 in CO_2 infrastructure and expand CO_2 production and 500 miles Cortez pipeline to reach from Colorado to Eastern New Mexico.

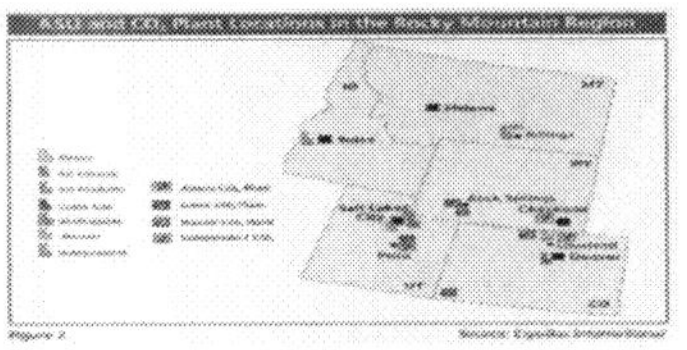

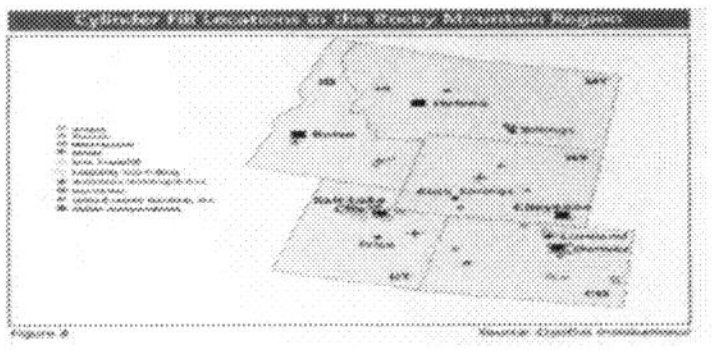

11/17/2015 21

The contents are strictly confidential and are not for any unauthorized person Providing innovative and cost effective solutions to the industry

The Far West Region -Market

Tech Services Inc.

Key Drivers in the Region:

- The large significant **manufacturing sector is computer, electronics, and component** market totaled \$152 billion in 2011 out of which \$86 B is in California and \$60.5 B in Oregon. Electronics and Semiconductor processing are large users of gases and specialized materials.
- **Healthcare and related services** is also large market share of \$154 billion in 2011.
- **Govt. R & D sector** totaled \$274 billion in 2011, while Petroleum, plastics sector totaled \$41 billions in 2011.
- **Oil and Gas mining sector** totaled \$31 billion in 2011, Alaska mining was \$10.5 b while California was \$16.4 b in 2011.
- Construction shrunk down by 36% to \$74.5 b in 2011 from 2006.

CO_2 Supply Capabilities:

- **California has largest no. of ASUs in the region**. The region has 7 merchant plants operated by Air Liquide, with two Joint ventures one with Praxair and other with Air Products. Total liquid supply capability by all ASUs in the region is 10,000 tpd
- **11 LCO_2 production facilities in the region** with capacity of 4500 tpd of CO_2 almost all associated with refinery H2 production except single facility operated by Airgas in Hawaii.
- **Praxair and Linde have similar size capacity followed by Air Liquide. Recent start of Pixley CO_2 facility by Air Liquide has further increased AL presence in the region.**

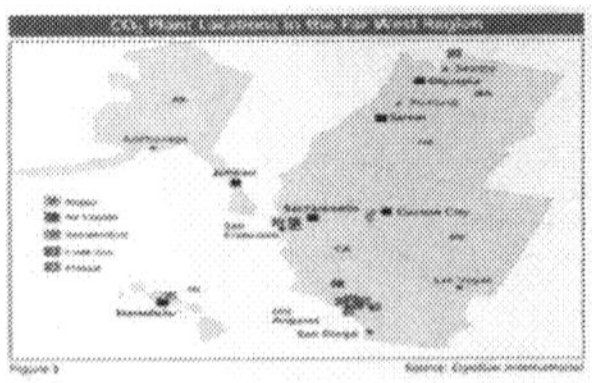

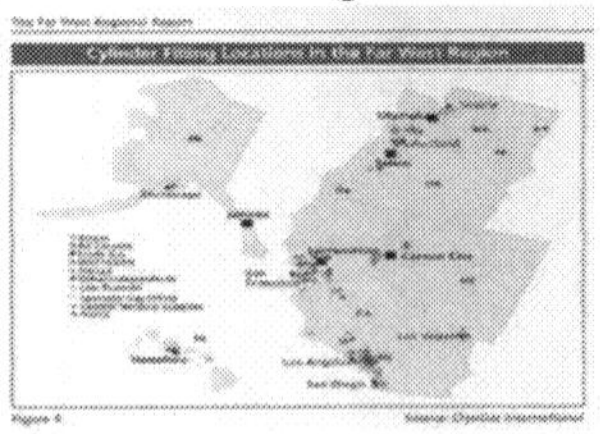

11/17/2015 22

The contents are strictly confidential and are not for any unauthorized person Providing innovative and cost effective solutions to the industry

Various Business Options for CO_2 Suppliers

Tech Services Inc.

Business Options

Option I: **Contractual relationship** with Industrial Gases Company
- Pros: Minimum investment by the supplier, long term contract
- Cons: Typical current price can be $ 8- $10 per ton.
- Comment: Contact Distributors and industrial gases companies to pursue this option.

Option II: **Install pipeline to send CO_2** to closest pipeline for Enhanced Oil Recovery project in the area.
- Pros: Minimum investment and involvement once the CO_2 is sent to the pipe for EOR and make CO_2 as a revenue stream
- Cons: The pipeline cost may be prohibitive depending on the location of the main pipeline for EOR and CO_2 source.
- Comment: Will require to ensure that a long term agreement is put in place with EOR. This may generate $10 to $12 per ton of CO_2.

Option III: **Independent distributorship**. This will require investment of capturing and testing CO_2 onsite and have delivery station to distribute the product to local distributors and demand price >$11 per ton.
- Pros: Can demand higher price that market can bear, long term contracts
- Cons: Invest in capturing CO_2 onsite and delivery station. May want to use third party transportation service to deliver product.
- Comment: This will require separate business line with sales and mgmt. team.
- Competition: This will be competing with major CO_2 suppliers such as $NuCO_2$, Linde, Praxair, Air Liquide, Airgas, Air Products,Matheson, and other independent distributors.

11/17/2015 23

The contents are strictly confidential and are not for any unauthorized person Providing Innovative and cost effective solutions to the industry

Various Business Options for CO_2 Suppliers

Tech Services Inc.

Option IV: **Make beverage and USP grade CO_2 onsite** by investing and having delivery station.
- Pros: Can provide excellent ROI based on New customer base and growing business line. Long term contracts with commitments from 5 to 7 years.
- Con: Will require investment, Sales team, distribution network
- Comment: Separate business line and portfolio.
- Competition: Airgas, Air Products, Linde, Praxair, independent distributors.

Option V: **Make selectively USP grade CO_2** for Pharma industry onsite by investing and having delivery station.
- Pros: Based on pharma industry intelligence, US supplies $32 mm USP CO_2 to the Pharma industry. Excellent opportunity
- Cons: Will require investment to make product, Sales team, distribution network.
- Competition: Major Industrial gases companies such as Airgas, Linde, Praxair, Air Liquide etc.

11/17/2015 24

The contents are strictly confidential and are not for any unauthorized person Providing Innovative and cost effective solutions to the industry

Notes on Economic Analysis

Tech Services Inc.

- **A 300 tpd plant,** the **estimated cost to clean up captured** LCO_2 to food grade and USP grade is $5 million, however, $6 million is used on calculation.
- The **operating cost used in calculation is 30% of revenue**. This number is very conservative as average production cost is $15 per ton for new plant, plus the cost of piping from the source and the cost of producing CO_2.
- The **price for CO_2 is estimated by the field sales** as stated in the analysis. If price is significantly reduced to $30, $35 and $40 for different use rate, the ROI is 3.67. This analysis is in the spreadsheet titled Analysis in Extreme case.
- It is assumed that **CO_2 available from the power plant is 72% w/w CO_2** and purified to >99% when available as USP and Food grade for use

11/17/2015 25

Economic Analysis for Options

Tech Services Inc.

	CO2 plant Business Analysis for Options III, IV, and V		
	Option II	Option III, IV, V	Extreme Case
	Pipeline		Options III, IV, V
Estimated Investment LCO_2 purification, piping	$2,000,000.00	$6,000,000.00	$6,000,000.00
Tanks, compressors investment		$5,000,000.00	$5,000,000.00
Production capability, TPD	300.00	300.00	300.00
Product spec	Industry grade	Beverage and USP grade	Beverage and USP grade
Expected ROI, Years	2.64	1.83	3.67
No of contracts	1.00	20 for 300 TPM, 16 for 100TPM, 10 for 50 TPM	20 for 300 TPM, 16 for 100TPM, 10 for 50 TPM
Contract length, Years	10.00	5.00	5.00
Distribution, $ per ton CO2 for 150 miles	$40.00	$40.00	$40.00
Delivery Charge, $ per delivery	$90.00	$90.00	$90.00
HAZMAT per delivery	$50.00	$50.00	$50.00
Tons per month, available (90% uptime)	8,183.70	8,183.70	8183.7
Tons used	8,183.70	8,100.00	8100
	Pricing, $ per ton	Pricing, $ per ton	Pricing, $ per ton
Users , 300 tpm	$11.00	$85.00	$30.00
100 tpm		$95.00	$35.00
50 tpm		$140.00	$40.00
Net Revenue (Estimate) per month	$63,015.00	$500,000.00	$250,000.00

Note: >Option II piping, Option III,IV,V have compressors, tank investments as separate item
>Except Options I, and II, estimated investment for all options is for purification equipment to meet food grade and USP grade product
>Distribution capital not included.
>This is strictly an estimate based on best available data.

26

11/17/2015

Recommendations

Tech Services Inc.

- TSI team **can help evaluating the potential of becoming supplier** of CO_2 to the market

- Depending on the option, TSI can support the **client in developing information** that can help make sound decision. This may include:

  - **Regional Market study** and identifying:
    1. Industrial gases company that may take additional product, its contact and support in **contract negotiation** if needed.
    2. TSI can help finding distributor and/or delivery net work for the product
    3. TSI can support in exploring **opportunity** in pipeline for EOR projects in the area.
    4. TSI can help come up with the **list of end users** and their volume needs with the possible availability date if option III, IV, or V is chosen and do the detailed market study.
    5. May **help secure businesses** using experienced account Managers if required.

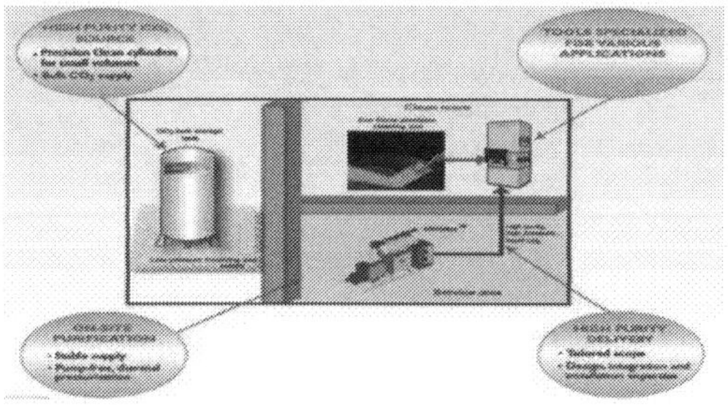

11/17/2015 27

The contents are strictly confidential and are not for any unauthorized person

Providing Innovative and cost effective solutions to the industry

CO2 Capture for the Industrial Sector

CCUS State of Readiness: Global Examples of Three Technology Developments and Demonstration Projects

Originally presented on: 11/18/2015 17:35:00 - 18:00:00

The state of readiness for implementation of three carbon capture utilization and storage technologies is reviewed. Carbon dioxide capture during hydrogen production from an industrial steam methane reformer has been implemented in PortJerome, France. At this site, the CO_2 in the PSA offgas is captured using a hybrid cryogenic distillation and hollow fiber membrane solution. Liquid CO_2 is then supplied as a product into the regional market. In Callide, Australia, CO_2 capture has been demonstrated using oxycoal combustion and cryogenic distillation to capture at a rate of 75 tons per day. Performance and reliability data covering more than 5600 hours of operation shows that the technology can provide an effective, reliable approach for the electrical power industry. At the National Carbon Capture Center in the US, a next generation hybrid cold membrane system is undergoing field tests. To date, the system has met all performance targets, demonstrating it to be a viable approach to reducing CO_2 emissions in post combustion, air fired coal power plants.

CO2 Utilization and Geological Storage

Methodology for Identification of Oil Reservoirs Suitable for CO2-EOR in Alberta, Canada, and Estimation of Their Storage Capacity Using Reserves Databases

Bachu, S.,

Originally presented on: 11/18/2015 12:45:00 - 13:10:00

Carbon dioxide capture and storage in geological media is a short-to-medium term technology that can reduce significantly atmospheric emissions of anthropogenic CO_2. Geological storage of CO_2, particularly through CO_2 enhanced oil recovery (CO_2-EOR) and in deep saline aquifers, has been identified by Alberta's government as an important component of its strategy for reducing greenhouse gas emissions. Use and storage of CO_2 in EOR operations provides the advantage that some of the costs associated with CO_2 capture and transportation are offset by the monetary value of the incremental oil produced as well as utilizing existing facilities. While oil companies may perform feasibility studies for CO_2-EOR on individual oil pools, these studies are confidential and limited in scope. Considering that there are close to 13,000 oil reservoirs in Alberta, a methodology for the quick assessment of the potential for CO_2-EOR in the province using reserves databases maintained by the provincial regulator of the oil and gas industry has been developed and applied.

A set of screening criteria for CO_2-EOR suitability was developed based on a review of literature to date and of the characteristics of more than 130 miscible CO_2-EOR operations worldwide, and on the availability of data in the reserves and production databases of the Alberta Energy Regulator. These screening criteria include oil gravity and viscosity, reservoir depth, temperature and porosity, minimum miscibility pressure, original and remaining oil in place, and some operational characteristics. Based on operational and reservoir characteristics, approximately 3,300 oil reservoirs in Alberta would be suitable for CO_2-EOR. However, when the original and remaining oil in place are taken into account, only 136 oil reservoirs in 85 oil fields are deemed as being suitable for CO_2-EOR, i.e., these oil reservoirs are of sufficient size to justify their development for tertiary oil recovery.

Based on the statistical production characteristics of 31 CO_2-EOR operations in USA which have injected to date CO_2 between 100% and more than 300% HCPV, the incremental oil recovery from these 136 oil reservoirs has been estimated. The CO_2 storage capacity in these 136 oil reservoirs has been evaluated as well based on net CO_2 utilization factors. The location and CO_2 storage capacity through CO_2-EOR of these oil reservoirs has been compared with the location of 38 large stationary CO_2 emitters in Alberta which emit individually more than 500 kt CO_2/yr and cumulatively slightly more than 100 Mt CO_2/yr. Finally, a methodology for source-sink matching is proposed, to identify the oil reservoirs best suited for any particular large CO_2 source. The results of this work will be presented at the conference.

1

Ohio Development Services Agency NETL MRCSP Battelle The Business of Innovation

Project Goals

- Develop a rigorous framework for fully evaluating the potential for CO_2-assisted EOR and geologic storage in Ohio's depleted oil fields,
- Develop process understanding and evaluate technical and economic feasibility of CCUS in the region,
- Focus on Clinton Sandstone and Knox Group Dolomite formations (under-pressured, low permeability reservoirs with poor primary recovery), and
- Provide systematic assessment of CO_2-EOR and geologic storage potential in these reservoirs.

2

Ohio Development Services Agency NETL MRCSP Battelle The Business of Innovation

Project Approach

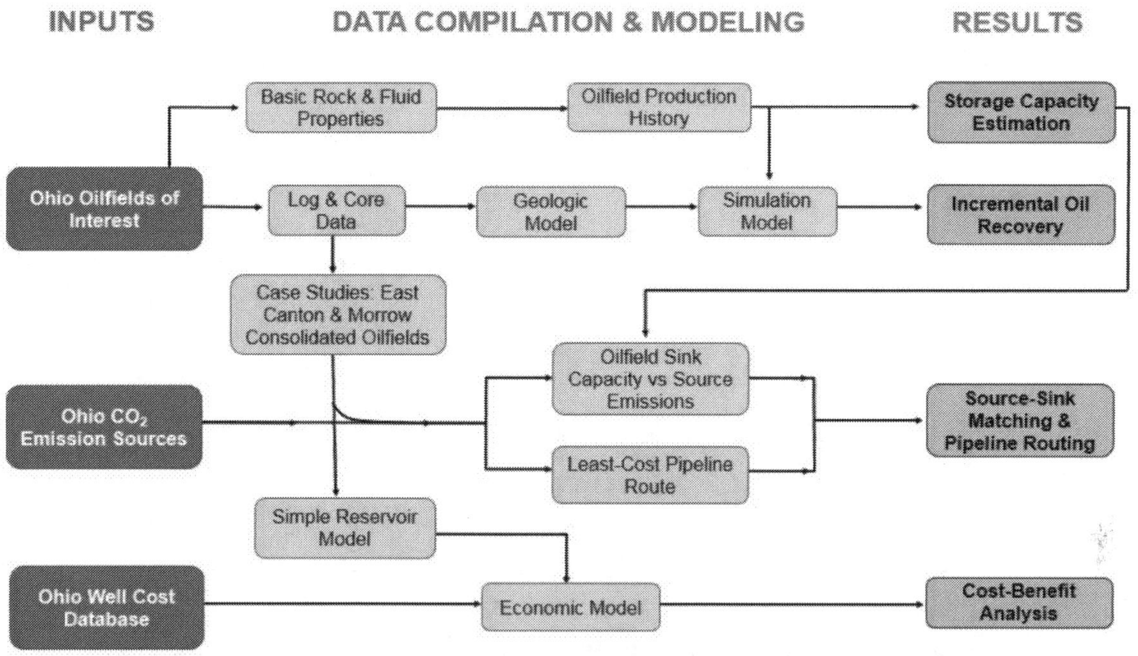

Ohio Development Services Agency | NETL | MRCSP | Battelle The Business of Innovation

Ohio's Oilfields of Interest

30 Depleted Oil Fields

8,851 Original Oil in Place [MMbbls]

1,274 Cumulative Production [MMbbls]

878 CO_2 Storage Capacity [MMt]

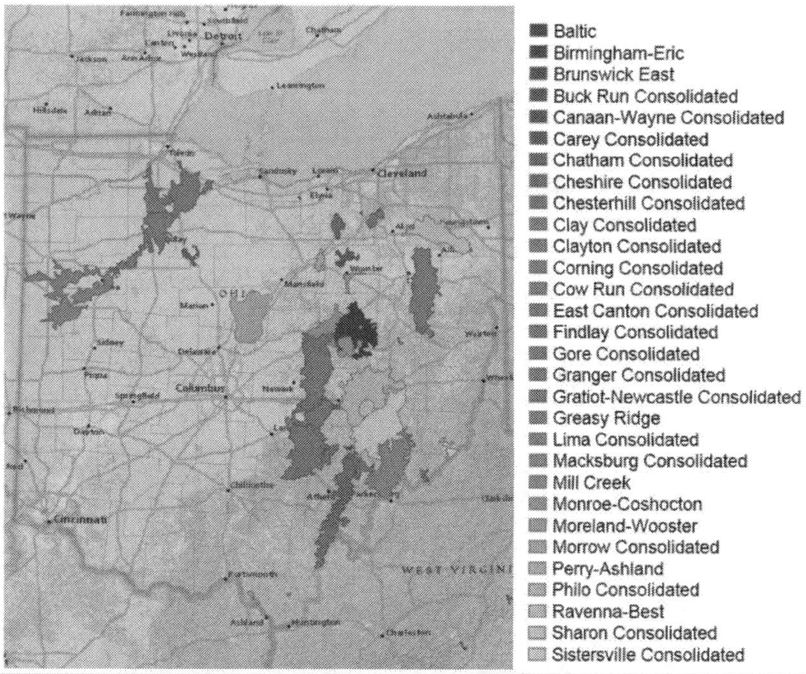

Ohio Development Services Agency | NETL | MRCSP | Battelle The Business of Innovation

Storage Capacity Estimation Method

CO_2 storage capacity was calculated using a production-based equation, which uses actual oil and gas production to evaluate storage.

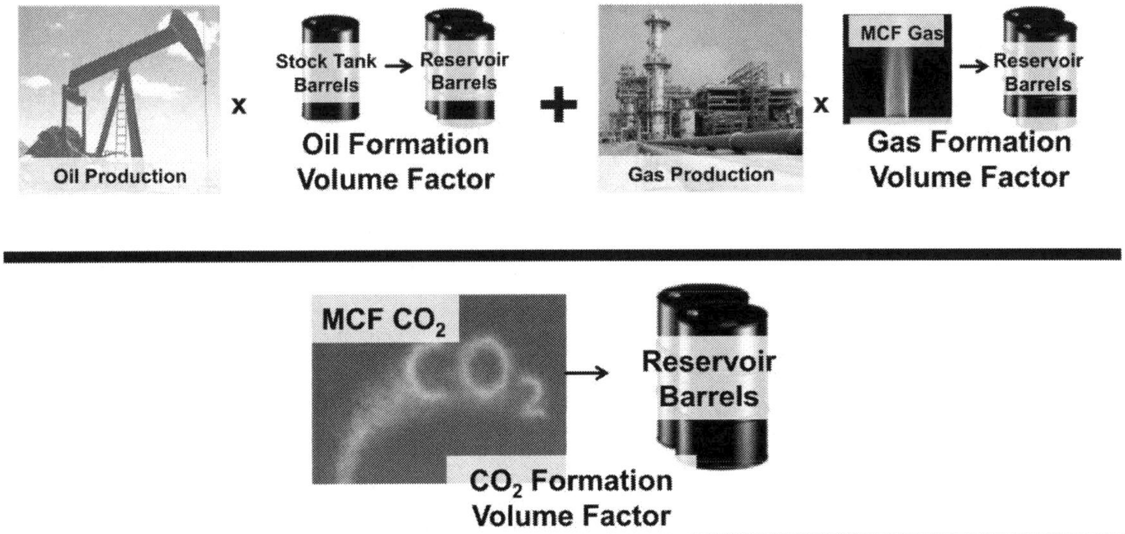

Storage Capacity Estimation Results

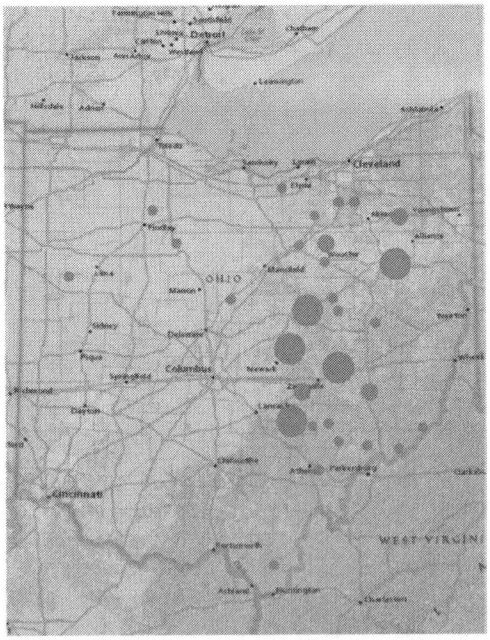

The 30 major oilfields had a combined total of 878 million metric tons of CO_2 storage capacity.

CO_2 Storage Capacity (Production Method) Million Metric Tons (MMT)

- 72 - 154
- 25 - 72
- 0.004 - 25

Representative Field Selection - ECOF

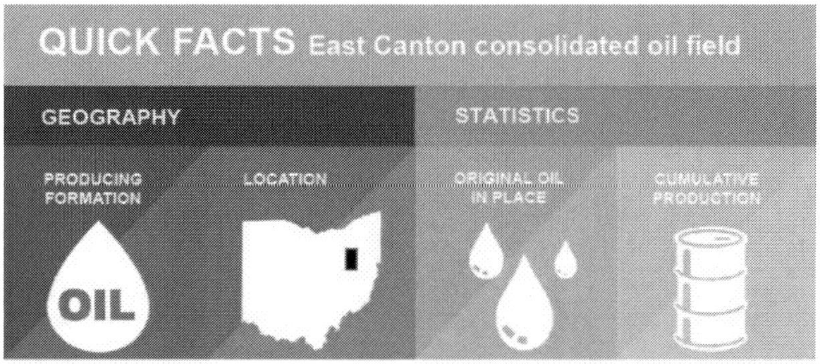

Clinton Sandstone - "Clinton" Sands – The Clinton formation, not to be confused with the overlying Clinton Group, is a sequence of interbedded sandstones, siltstones, and shales deposited from a westward prograding delta system. The "Clinton" sands is an informal driller term erroneously applied to sandstones of the Medina, as it is known in Pennsylvania. The Clinton is the most productive oil and gas unit in Ohio, and thus the most drilled.

Eastern Ohio - The coverage spans 175,000 acres in Eastern Ohio and covers Stark, Carroll, Tuscarawas, and Harrison counties. The east canton consolidated oil field was discovered in 1947.

1,537 MMBbls - Original oil in place (OOIP) refers to the total volume of hydrocarbon stored in a reservoir prior to production. Reserves or recoverable reserves are the volume of hydrocarbons that can be profitably extracted from a reservoir using existing technology.

96.0 MMBbls - The cumulative production is the sum of all oil that has ever been produced to date.

99.9 MMT - CO_2 storage capacity calculated using the Production Method

Representative Field Selection - MCOF

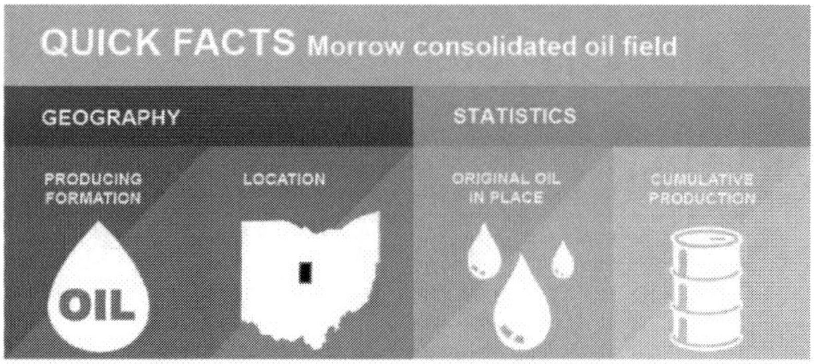

Copper Ridge Dolomite - The Copper Ridge Dolomite, also known as the Trempealeau formation, is the basal unit of the Knox Group, a heterogeneous assemblage of dolostones and dolomitic sandstones deposited in tidal flat to shallow marine environments. In the vicinity of the Morrow Consolidated, the dolostones have secondary porosity that has created discrete, highly productive oil deposits.

North-Central Ohio - The coverage spans 16,000 acres in North-Central Ohio and covers Morrow, Richland, Marion, Knox, and Delaware counties. The morrow consolidated oil field was discovered in 1959.

177 MMBbls - Original oil in place (OOIP) refers to the total volume of hydrocarbon stored in a reservoir prior to production. Reserves or recoverable reserves are the volume of hydrocarbons that can be profitably extracted from a reservoir using existing technology.

46.3 MMBbls - The cumulative production is the sum of all oil that has ever been produced to date.

25.0 MMT - CO_2 storage capacity calculated using the Production Method

Project Approach

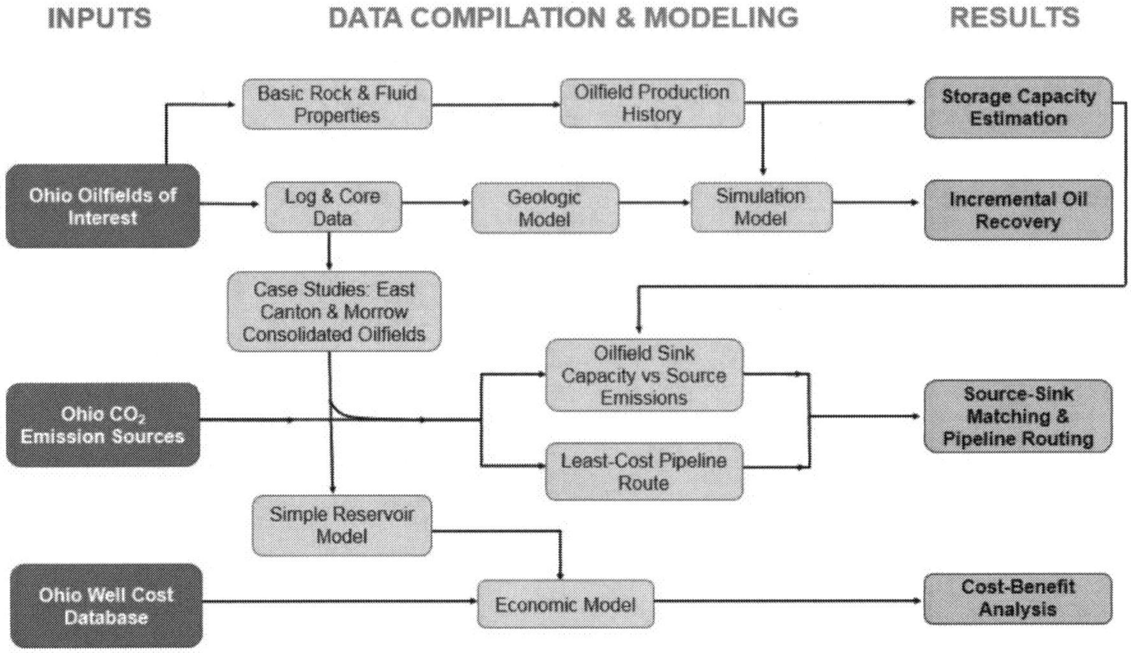

Geologic Data Synthesis

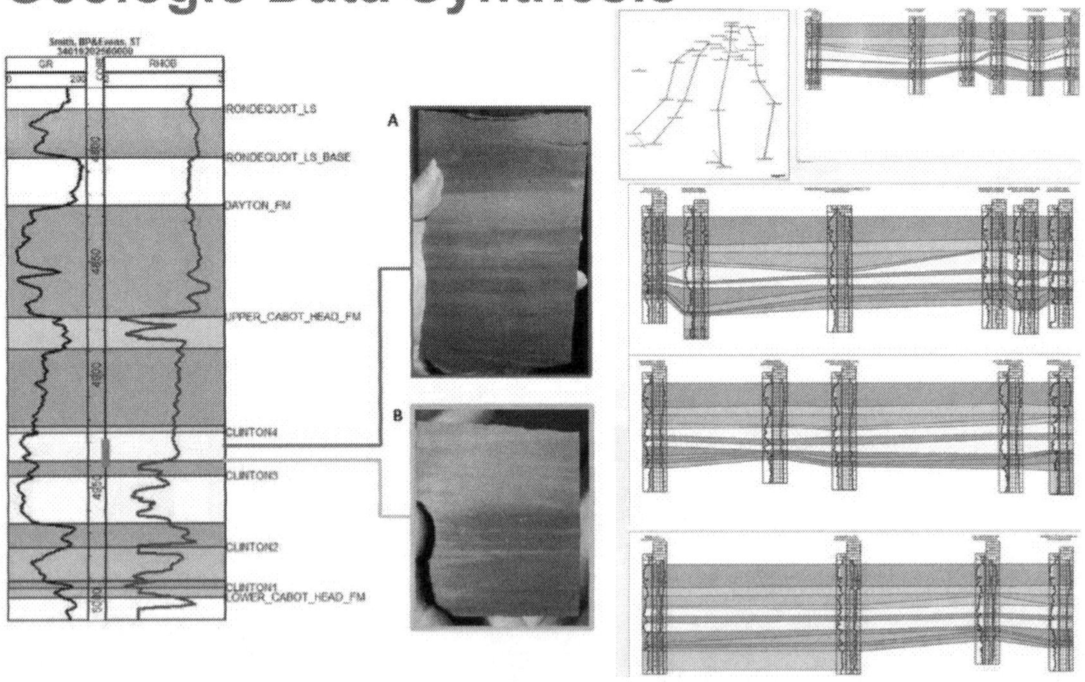

Geologic Model Building and Results

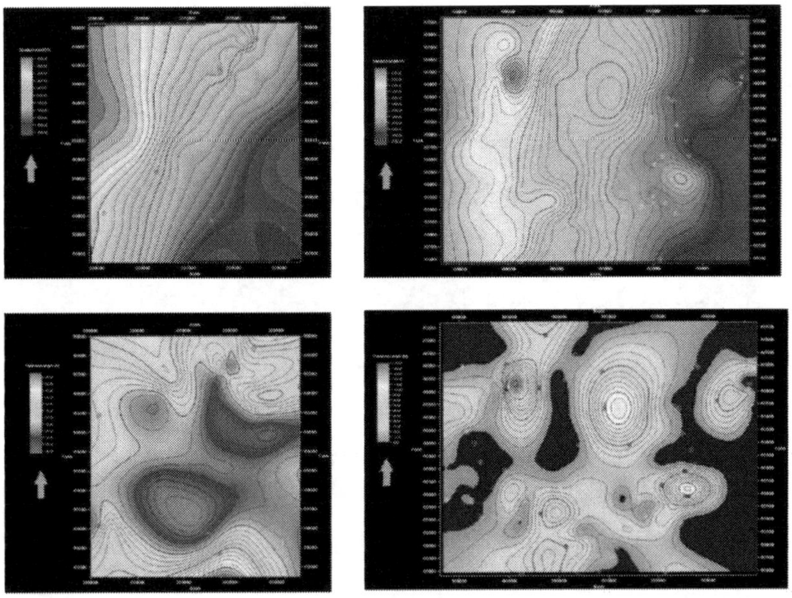

Grid Parameter	ECOF	MCOF
Overall grid size in x direction, ft.	47,060	51,279
Overall grid size in y direction, ft.	63,717	49,345
Cell grid size in x direction, ft.	300	400
Cell grid size in y direction, ft.	300	400
Number of grid cells	26,195	19,276
Minimum Well Spacing, ft.	760	565
Minimum Node Spacing	2.5	1.4

- Challenges
  - lack of information on distribution of natural fractures,
  - sparse advanced log coverage in study areas, and
  - minimal availability of core samples for developing porosity-perm relationship.

Fluid Property Prediction Toolbox

Inputs:		
Shadding Indicates Needed Input		
Input Bubble Point Property		
Select Known Bubble Point Property :	Pressure	
Pressure, pB	2017.0	psia
Additional Reservoir Properties		
Oil gravity, γ_{API}	43.6	API
Gas gravity, γ_g	0.76	-
Reservoir Temperature, T	108	°F
Single Point Pressure of Interest Evaluation at right		
Pressure, p	1250	psia
Pressure Range of Interest See next sheets for tables and plots		
Maximum Pressure	2694	psia
Min Pressure	15	psia
Brine-Gas Calculations		
Calculate Brine-Gas Properties?	Yes	
If Yes, Salinity =	1.2	Weight % Solids
CO_2 Calculations		
Calculate Pure CO_2, CO_2-Oil and CO_2-Brine Properties?	Yes	
*Pure CO2 properties only available for 100psia ≤ p ≤ 3600psia and 0°F ≤ T ≤ 350°F		
*CO2-Oil and CO_2-Brine Calculations assume dead oil		
If Yes, Avg. Oil Molecular Weight (MW) =	220	lb/lbmol
If Yes, Salinity =	1.5	Weight % Solids

Outputs at Single Point of Pressure:			
At:	Pressure, p	1250	psia
	Bubble Point Pressure, p_B	2017	psia
	Solution Gas Oil Ratio at Bubble Pt, R_{sob}	774	SCF/BBL
Oil-Gas	Formation Volume Factor, B_o	1.21	RB/STB
	Solution Gas Oil Ratio, R_{so}	439.2	SCF/BBL
	Viscosity, μ_o	0.466	cP
	Compressibility, c_o	3.41E-04	1/psi
	Density, ρ_o	46.2	lbm/ft3
Gas	Formation Volume Factor, B_g	0.0103	ft3/SCF
	deviation factor, Z	0.802	-
	Viscosity, μ_g	0.014	cP
	Compressibility, c_g	9.57E-04	1/psi
	Density, ρ_g	5.64	lbm/ft3
Brine-Gas	Formation Volume Factor, B_w	1.0097	RB/STB
	Solution Gas Water Ratio, R_{sw}	15.31	SCF/BBL
	Viscosity, μ_w	0.64	cP
	Compressibility, c_w	5.04E-05	1/psi
	Density, ρ_w	62.29	lbm/ft3
Pure CO_2	Formation Volume Factor, B_{CO2}	0.01	ft3/SCF
	Viscosity, μ_{CO2}	0.03	cp
	Density, ρ_{CO2}	21.88	lbm/ft3
	Compressibility Factor, Z_{CO2}	0.43	-
CO_2 - Oil	CO_2-Oil Solubility, Sol	0.60	mole fraction
	Solution CO2 Oil Ratio, $R_{so\text{-}CO2}$	740.8	SCF/STB
	CO_2-Oil Swelling, SF	1.29	-
	CO_2-Oil Viscosity Ratio, μ_{oCO2}/μ_{oil}	0.23	-
CO_2 -Brine	Solution CO_2-Water Ratio, $R_{sw\text{-}CO2}$	142.3	SCF/STB

Reservoir Simulation

- Goal ⇨ Evaluate the feasibility of CO_2-EOR for reference Ohio reservoirs in terms of reservoir performance following CO_2 injection
- Develop reservoir model(s) to integrate known geologic, fluid and production data
- Calibrate models to observed production history
- Quantify incremental oil recovery following CO_2 injection and amount of CO_2 stored
- Examine sensitivity of injection scenarios to geologic and engineering factors

Reservoir Simulation Results - ECOF

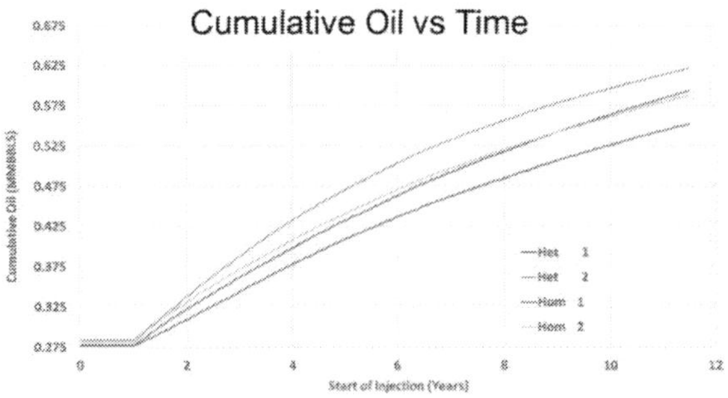

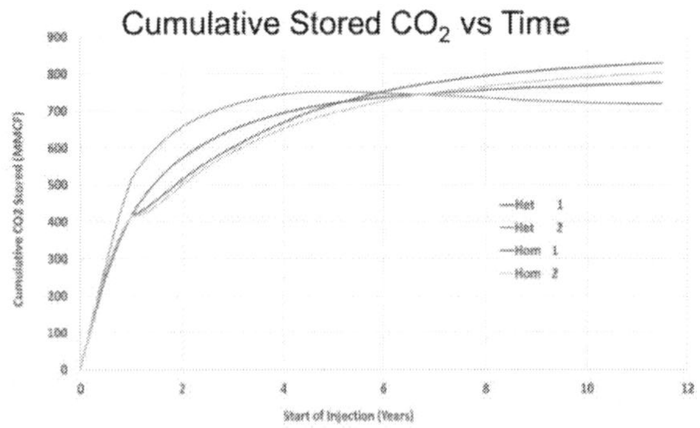

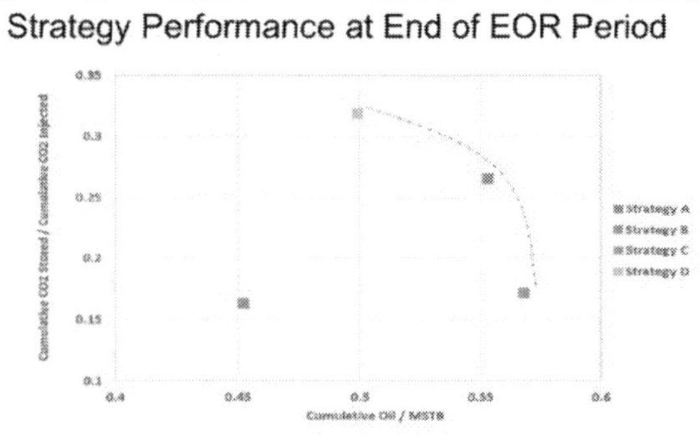

Ohio's CO_2 Emission Sources

229
Stationary Sources

129
Million Metric Tons CO_2 Emitted Per Year

22
Sources are Coal-Fired Plants

63%
CO_2 Emissions Come From Coal-Fired Power Plants

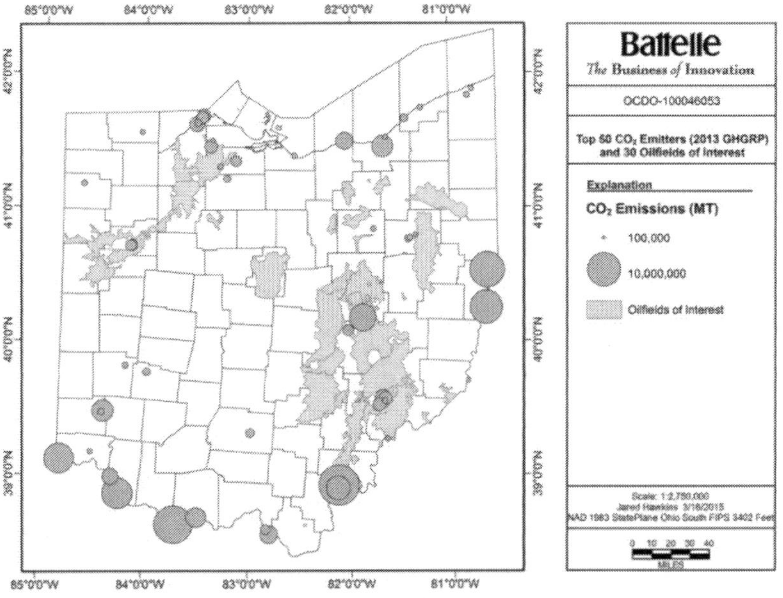

Oilfield Sink Capacity vs Emission Source Volume – Source-Sink Matching

- Buffer maps were generated to determine the emission sources in close proximity to ECOF and MCOF.
- A pipeline routing analysis was carried out to determine the optimal routes from the top 45 emission sources to ECOF and MCOF.

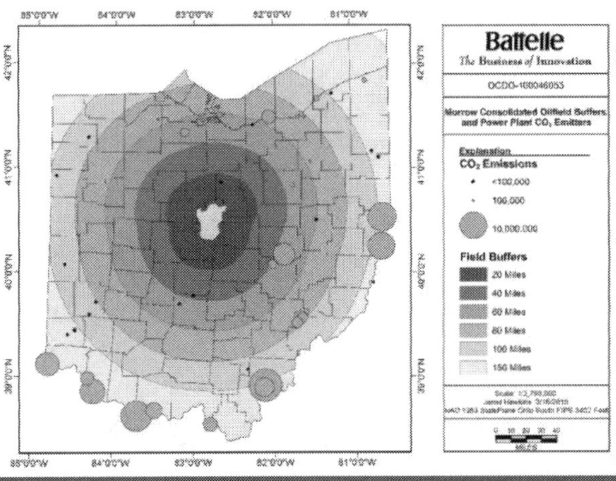

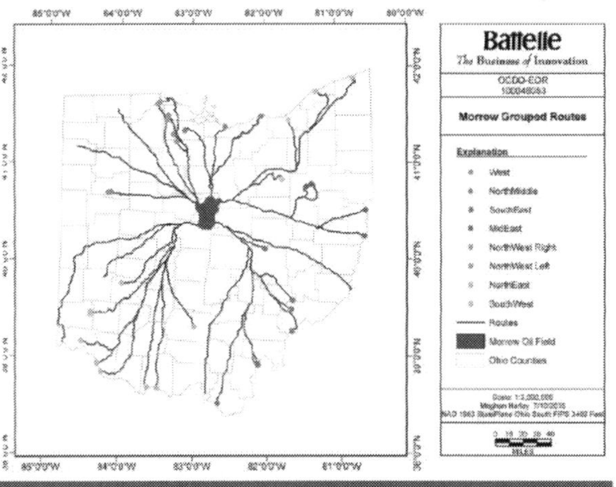

Cost-Benefit Analysis Methodology

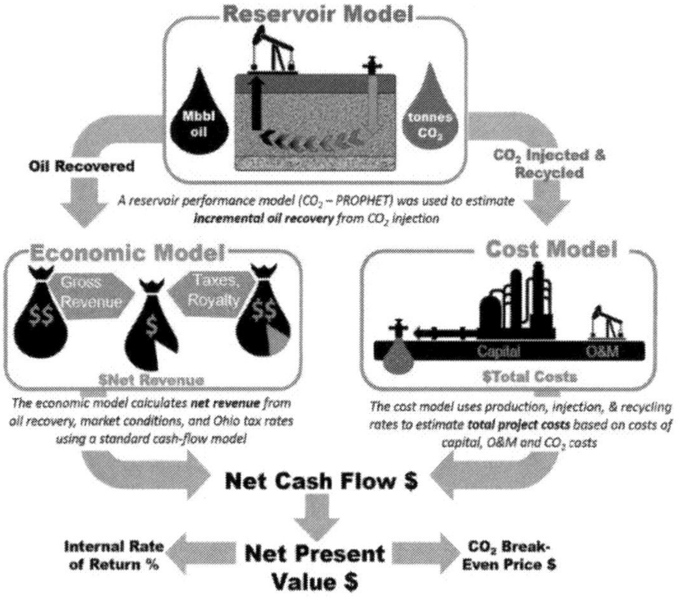

- CO2-PROPHET reservoir performance model
- Ohio-specific well cost data + cost models established in previous studies.
- Results include: NPV (net present value), IRR (internal rate of return), and a break-even cost of CO_2.
- Applied to ECOF and MCOF

Ohio Development Services Agency NETL MRCSP Battelle The Business of Innovation

Cost-Benefit Analysis Results

- Example calculations
  - For both ECOF and MCOF
  - With representative pattern elements
  - Using Oil price = [40, 70, 100] $/bbl
  - Using CO_2 cost = [40, 80, 100] $/ton
- NPV analysis shows good project economics at $100/bbl oil for most CO_2 price scenarios
- Breakeven analysis suggests that for every dollar increase in oil price, corresponding increase in break-even cost of CO_2 is ~$4 per ton

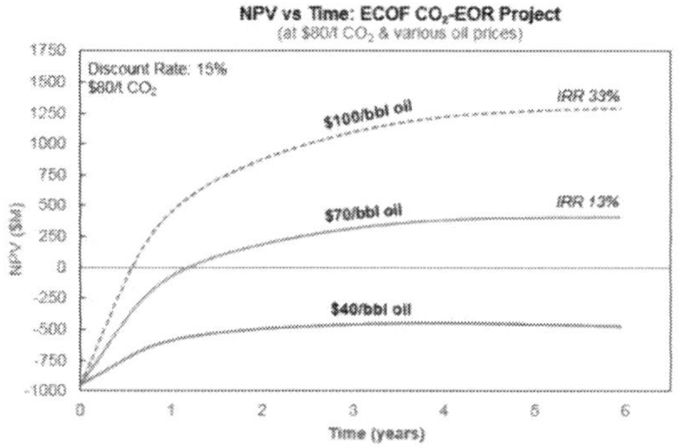

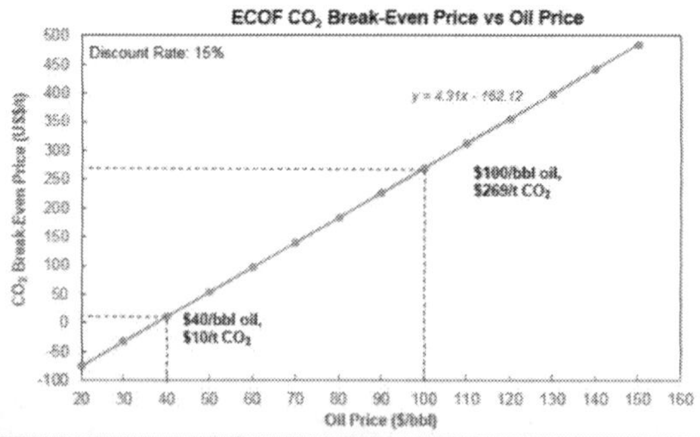

Ohio Development Services Agency NETL MRCSP Battelle The Business of Innovation

Project Key Findings

30
Depleted Oil Fields

8,851
Original Oil in Place [MMbbls]

1,274
Cumulative Production [MMbbls]

878
CO_2 Storage Capacity [MMt]

229
Stationary Sources

129
Million Metric Tons CO_2 Emitted Per Year

22
Sources are Coal-Fired Plants

63%
CO_2 Emissions Come From Coal-Fired Power Plants

70 - 120 %
Incremental oil recovery as a % of primary production with CO_2-EOR for ECOF & MCOF

5 - 12
Tons of CO_2 stored for every 100 barrels of oil produced

~4x
Change in breakeven CO_2 cost per $ change in the oil price

19

Ohio Development Services Agency | NETL | MRCSP | Battelle The Business of Innovation

Acknowledgments

Collaborators:

Neeraj Gupta	Mark Moody	Natalie Zeleznik
Mark Kelley	Rick Peterson	Jared Schuetter
Amber Conner	Samin Raziperchikolaee	Ashley Kubatko
Ola Babarinde	Priya Ravi Ganesh	Jacob Markiewicz
Jackie Gerst	Jacob Markiewicz	

Funding for this study was provided by Ohio Development Services Agency's Ohio Coal Development Office (OCDO) grant agreement OOE-CDO-D-13-24 in conjunction with the Midwest Regional Carbon Sequestration Partnership (MRCSP) grant agreement number DE-FC26-05NT42589.

Ohio Development Services Agency | Battelle The Business of Innovation | MRCSP Midwest Regional Carbon Sequestration Partnership

800.201.2011 | solutions@battelle.org | www.battelle.org

CO_2 storage capacity was calculated using the volumetric equation with three different efficiency factors.

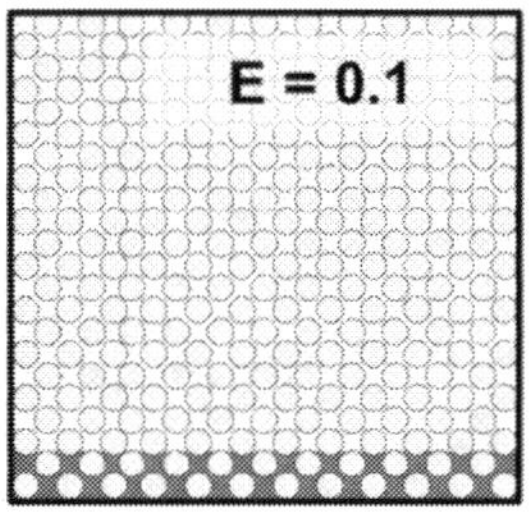

205 Million Metric Tons

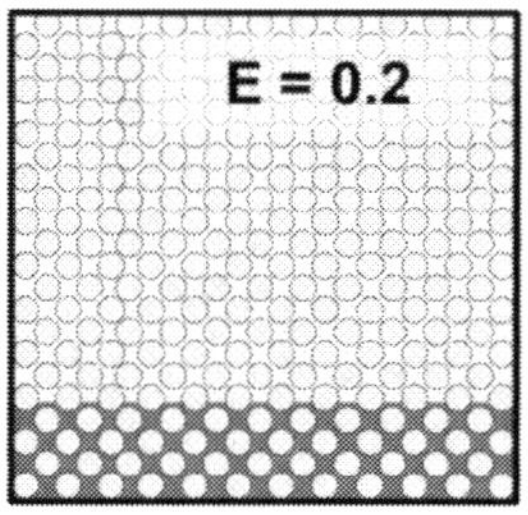

411 Million Metric Tons

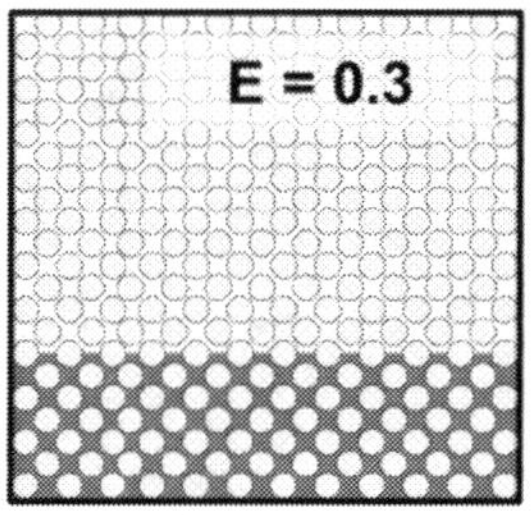

616 Million Metric Tons

The upper bound CO_2 storage capacity, calculated by setting E = 1.0, is the total storage capacity of a formation if all pore space occupied by hydrocarbons replaced by CO_2: 2,054 Million Metric Tons.

Summary of Storage Capacity Estimates

Method	Storage Capacity (Million Metric Tons)
Production-based	**878**
Volumetric-based	**205 – 616 (E = 0.1 – 0.3)**
Upper Bound (Volumetric)	**2,054 (E = 1.0)**
Wickstrom et al (2005)	**419**
DOE (2012)	**261 – 796**

1. Wickstrom, L. H. and et al., 2005, Characterization of geologic sequestration opportunities in the MRCSP region: Final report DOE cooperative agreement DE-PS26-05NT42255, 152 p.
2. Department of Energy, 2012. Carbon sequestration atlas of the United States and Canada (4th ed.): [Washington, D.C.], National Energy Technology Laboratory, 130 p., Available at: http://www.netl.doe.gov/File%20Library/Research/Coal/carbon-storage/atlasiv/Atlas-IV-2012.pdf

Ohio Development Services Agency NETL MRCSP Battelle The Business of Innovation

Geologic Model Development

- Integration of data from geologic analysis and reservoir characterization
- Building "static earth model" for each oilfield of interest
  - Create a 3D grid incorporating porosity and water saturation
- Export developed model as input to dynamic reservoir modeling effort

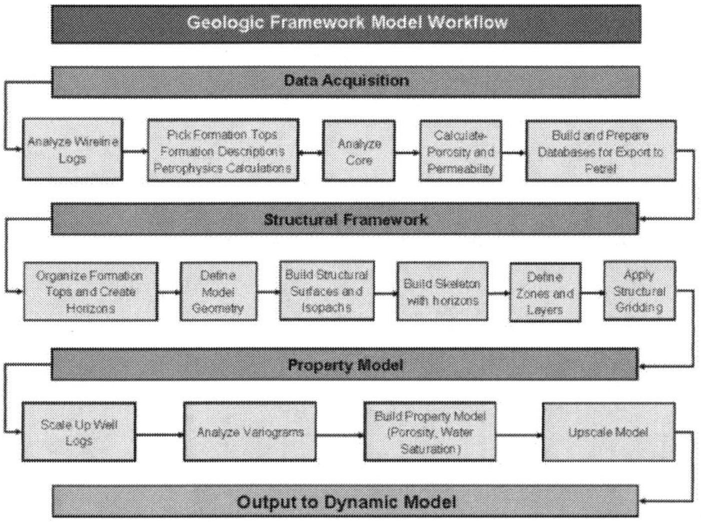

Reservoir Fluid Properties Study

- An Excel-based toolbox for predicting fluid properties of oil, gas, brine and CO_2 has been developed using literature-based and newly developed correlations
  - Formation volume factor
  - Solution-gas liquid ratio (gas in oil, CO_2 in water)
  - Density
  - Isothermal compressibility
  - Viscosity
  - Gas deviation factor
- These correlations are primarily a function of pressure, temperature, oil gravity, gas gravity and salinity (as appropriate)

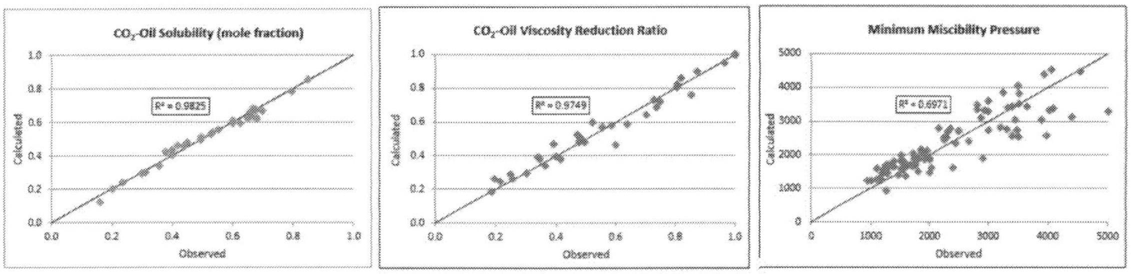

Reservoir Simulation Results - MCOF

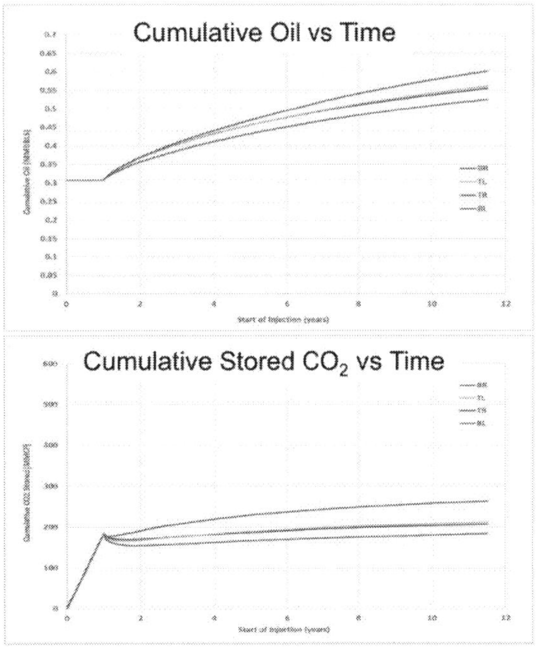

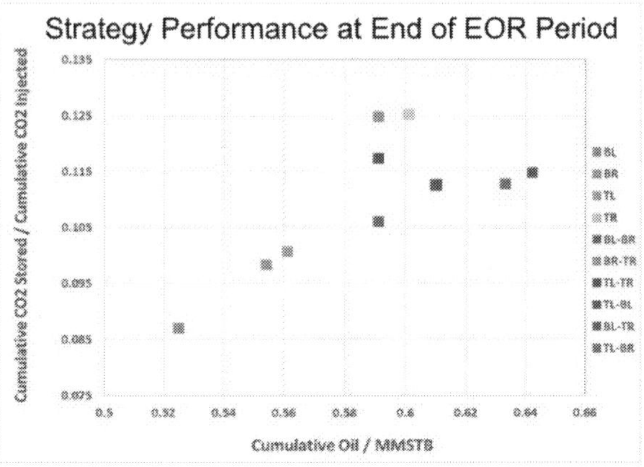

Pipeline Routing Trade-off Analysis

- Assessment of the total CO2 available and total pipeline distance of the grouped routes was performed to determine route efficiencies.
- This was done for the individual routes as well.

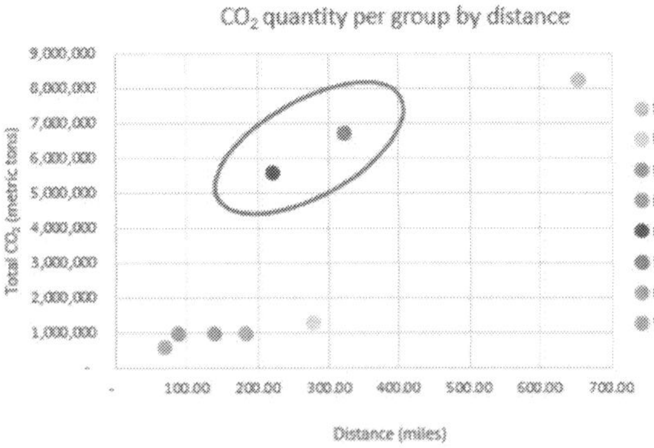

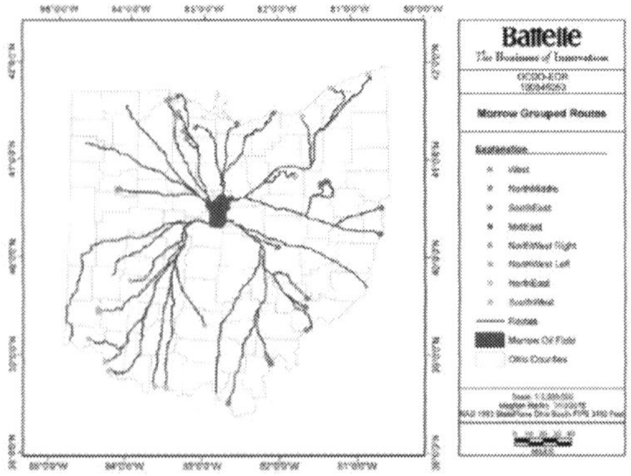

CMTC-439476-MS

Compositional Simulation of CO_2 Storage Capacity in Depleted Oil Reservoirs

William Ampomah[1], Robert S. Balch[1], Reid B. Grigg[1], Zhenxue Dai[2] and Feng Pan[3]
[1]Petroleum Recovery Research Center, [2]Earth and Environmental Sciences Division, Los Alamos National Laboratory, [3] Energy & Geoscience Institute, University of Utah

This paper was prepared for presentation at the Carbon Management Technology Conference held in Sugarland, Texas, USA, 17–19 November 2015.

Abstract

The Farnsworth Unit (FWU) of Ochiltree County, Texas operated by Chaparral Energy L.L.C. is the site of a CO_2-EOR project using anthropogenic CO_2. The Southwest Regional Partnership on Carbon Sequestration (SWP), sponsored by the Department of Energy's National Energy Technology Laboratory, is using this project to monitor CO_2 injection and movement in the field to determine CO_2 storage potential in depleted oil reservoirs. The field was discovered in 1955 with estimated initial oil in place of about 120 MMBO. The target reservoir is the Pennsylvanian-aged Upper Morrow sandstone, locally termed the Morrow "B" sandstone. CO_2 flooding was initiated by Chaparral Energy in December 2010.

This paper describes a compositional simulation of CO_2 storage in a depleted oil reservoir. Mechanisms considered for CO_2 storage include structural/ stratigraphic trapping, dissolution in formation water and oil, and residual trapping. A high resolution geological model constructed from geological, geophysical and engineering data from FWU was used for the study. FWU has no recorded oil-water contact. The model was first calibrated to the reservoir's primary and secondary recovery history performance as a benchmark for the study. Several models were constructed and the storage capacity was analyzed as a function of injection volume, time and pressure.

Numerical simulation results show that with over 25 years of WAG injection, 75% of the CO_2 was sequestered. Afterwards wells were shut-in to monitor the storage. Significant amount of stored CO_2 was dissolved in remaining oil, contributing to enhanced oil recovery from the tertiary stage of the field operations. Supercritical phase CO_2 mass within the reservoir compared to CO_2 dissolved in formation water was dependent on CO_2 injection rate. Residual trapping contribution was significant when hysteresis was modeled. Pressure, volume of reservoir fluid present, and cap rock integrity and optimized WAG injection strategies were significant parameters to determine long-term CO_2 storage capacity within FWU.

This work shows an improved strategy of maximizing CO_2 storage within a depleted oil reservoir. It was evident in this study that pressure changes within the reservoir have to be continuously monitored to enhance CO_2 storage. This study could serve as a benchmark for future CO_2-EOR projects in the Anadarko basin or geologically similar basins throughout the world.

Introduction

The purpose of carbon capture, utilization, and storage (CCUS) is to reduce the amount of CO_2 released into the atmosphere to help mitigate the effects of anthropogenic climate change (IPCC, 2005; Dai et al., 2014a; Bachu, 2015). CO_2 can be injected into structural reservoirs in deep geological formations such as depleted oil and gas reservoirs, saline aquifers and unmineable coal seems (Holloway, 1997; IPCC, 2005). Using CO_2 for enhance oil recovery (EOR) is common in the oil industry, but most of the CO_2 is usually produced from natural, geological sources. CCUS uses anthropogenic sources of CO_2 for EOR. This, together with other factors, motivated the US Department of Energy (DOE) to set up regional carbon sequestration partnerships with research institutions and operating companies to study injection of anthropogenic CO_2 for storage and EOR in the United States (Goodman et al. 2011, Dai et al., 2014b,c). Southwest Regional Partnership (SWP) is one of such groups who is working closely with an industry partner, Chaparral Energy, to ascertain the storage capacity of CO_2 and its utilization within the Farnsworth field unit (FWU) (SWP, 2013).

Numerous studies have been focused on mitigating CO_2 emissions by capturing and injecting into oil reservoirs for enhanced oil recovery (Bachu and Shaw, 2003; Qi et al., 2008; Gorecki et al., 2012; Saini and Jimenez, 2014). CO_2 storage in depleted oil reservoirs has tremendous advantages due to economic incentives over other geological sequestration mechanisms currently available. In the oil industry, CO_2 flooding has been utilized for over 35 years (GCEP Technical Report 2003) so the processes have been more thoroughly investigated than sequestration in saline aquifers or unmineable coal seams. Most often, the water alternating gas (WAG) process has been successfully used to improve sweep efficiency by delaying early CO_2 breakthrough (Lake, 1989). The typical goal of a CO_2-EOR project is to maximize tertiary oil recovery with a minimum amount of CO_2; sequestration projects obviously include making efforts to store as high a percentage as possible at conclusion of the project.

Previous research identified four main primary mechanisms by which CO_2 is stored within geological structures. Mechanisms include: structural / stratigraphic trapping, solubility trapping, residual gas trapping and mineral trapping (Nghiem et al., 2009; Pruess et al., 2003; Kumar et al., 2005). Structural trapping involves the storage of CO_2 in a supercritical state in through the effects, often combined, of reservoir statigraphy and structure in a manner analogous to trapping in a hydrocarbon reservoir. CO_2 is usually trapped below a low permeable formation such as shale or mudstones. This prevents the supercritical fluid from migrating to the surface. Trapping can also be due to impermeable zones such as sealed faults; typically a trap involves some components of both structure and stratigraphy. Prior to sequestration, cap rock and seal integrity must be assessed for long-term potential. Solubility trapping occurs when injected CO_2 dissolve in formation fluids. The dissolution can occur in both the formation water and/or oil phase. Solubility of CO_2 in formation water depends on pressure, temperature and salinity within the subject formation (Chang et al. 1998). Residual trapping involves the storage of CO_2 as an immobile phase within the porous media. This is an important mechanism since it keeps CO_2 far from the seal. In mineral trapping, formation mineralogy may react with CO_2 and induce precipitation of carbonate minerals such as dolomite, calcite and siderite (Nghiem et al., 2009). The storage security of these mechanisms is summarized in Figure 1 below (IPCC, 2005).

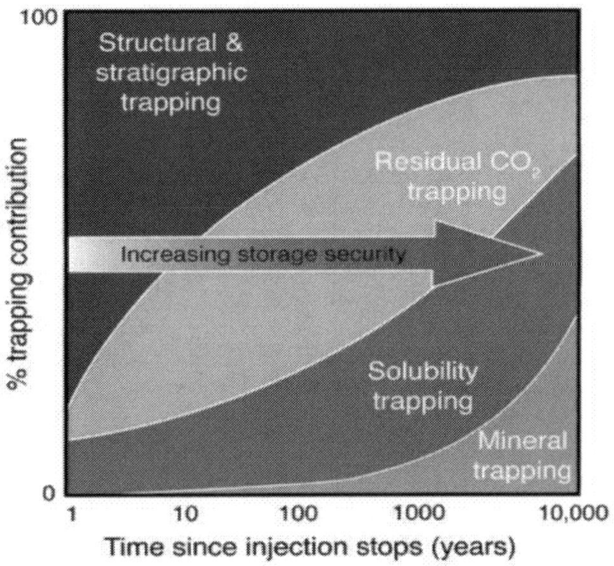

Figure 1- Increasing of storage security with time (IPCC, 2005)

This paper presents a field scale compositional numerical modeling of CO_2 storage capacity as a function of different trapping mechanisms within the Farnsworth field in the Texas panhandle. The targeted Morrow sandstone reservoir presents opportunities both to enhance oil production and sequester a large portion of injected CO_2. The trapping mechanisms discussed include structural/ stratigraphic, solubility (water and oil) and residual trapping. The effects of WAG cycle, salinity and hysteresis on CO_2 storage are studied. A cap rock integrity analysis has been conducted on FWU to access the long-term containment of supercritical CO_2. A primary, secondary and tertiary history matched model constructed from the FWU was used as the base for the study. WAG injection with CO_2 recycle was continued for a total of 25 years and wells shut-in to monitor the storage for additional 200 years as a function of different mechanisms.

Material and Methods

Study Site (Reservoir Description and History)

The Farnsworth Unit (FWU) produces from sandstones of the upper Morrow formation (Munson 1988; Gallagher, 2014). The Morrow "B" formation is an operational name for strata deposited during the Morrowan- age of the Pennsylvanian. Morrowan reservoirs are prevalent within the Anadarko basin throughout the Oklahoma and Texas panhandles, western Kansas, and southeastern Colorado (Puckette et al., 2008). The upper Morrow formation within FWU contains up to five sandstone packages separated by mudstone intervals (Munson 1988; Gallagher, 2014). Production comes from the upper most sand interval referred to as the Morrow "B" sandstone.

The Morrow B sandstone deposited in an incised valley, fluvial-estuarine setting as shown in Figure 2 (Puckette et al., 1996; Gallagher, 2014). Stratigraphy of the reservoir was controlled by a complex interplay of climactic and tectonic forces during the Pennsylvanian. Large scale glacioeustatic sea level fluctuations are thought to have caused the rapid transitions from shale to sandstone intervals (Sonnenburg et al., 1990; Krystinik and Blakeney, 1990; Al-Shaieb et al., 1995; Gallagher, 2014). Deposition was also influenced by positive elements at the time that resulted from the collision of the South American and North American plates. These positive elements include the Apishapa-Sierra Grande Uplift to the west, the Ancestral Front Range and Transcontinental Arch to the north, the Central Kansas Uplift and the Amarillo-Wichita uplifts to the south (Sonnenberg and others 1990, Puckette et al., 2008; Gallegher, 2014).

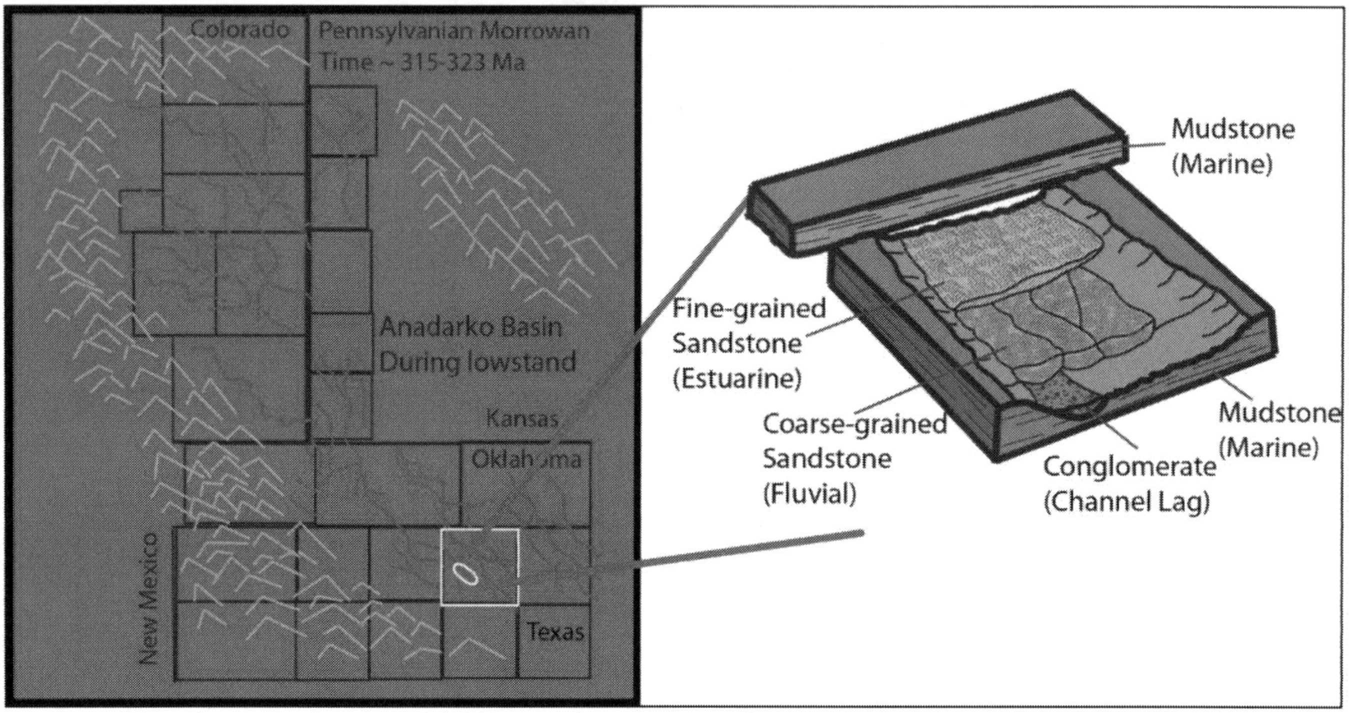

Figure 2 - Geological setting and Farnsworth field location (Rose-Coss,. 2014)

The reservoir tends to be comprised of subarkosic, moderately to poorly sorted, sub-angular to sub-rounded, upper medium to very coarse sands and fine gravels. The overall immaturity and presence of both micrographic granitic rock fragments as well as volcanic quartz suggests that the Precambrian crystalline rocks of the Amarillo-Wichita Uplift were a probable source area (Munson, 1988 Gallagher, 2014). After deposition, diagenetic events such as feldspar alteration, authigenic clay formation, dissolution/precipitation of silica, calcite cementation, and pyrite formation either enhanced or diminished primary porosity and permeability (Munson, 1988; Cather, 2014; Cather and Cather, 2015).

The first discovery well was drilled by Unocal in October 1955 in the east side of the field. It has an average connate water saturation of 31.4%. The initial reservoir pressure and temperature was 2203 psig at a datum of 4900 ft subsea and 168 °F respectively. The original bubble point pressure was 2059 psi. The original oil had a formation volume factor of 1.192 RB/STB. Also, oil initially in place was estimated at about 120 MMSTB. The primary production was drive by a solution gas mechanism since there was no reported aquifer associated with the reservoir. About 100 wells were completed by 1960. The field was unitized in 1963 with Unocal as operator. Water injection for secondary recovery started in 1964. Chaparral Energy LLC (CELLC) acquired the FWU in November 2009 and Tertiary CO_2-EOR began in December 2010. The anthropogenic CO_2 for the project is supplied from the Arkalon ethanol plant in Liberal, Kansas and the Agrium fertilizer plant in Borger, Texas as shown in Figure 3.

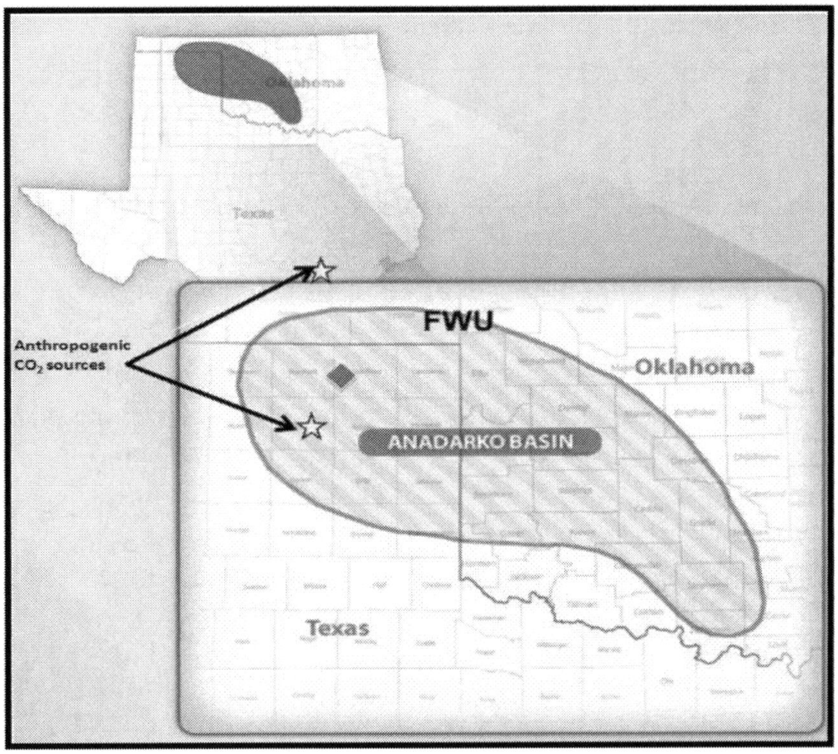

Figure 3-Location of the Farnsworth Unit (red diamond) in the northeast part of the Texas panhandle within the Anadarko Basin (shaded region). Stars indicate locations of large point sources from which anthropogenic CO_2 is scrubbed from emissions and compressed for injection. The northern star is the Arkalon Ethanol Plant in Liberal, Kansas, and the southern star is the Agrium Fertilizer Plant in Borger, Texas. Modified from Grigg and McPherson, 2012.

Trapping mechanisms

This work focuses on three storage mechanisms including structural trapping, solubility and residual trapping. Mineral trapping is examined in a separate study (Ahmmed, 2015)

Solubility Trapping

In this study, equation of state (EOS) is used to model oil and gas phase densities and fugacities. CO_2 separation between oil and gas phases is computed by a fugacity equilibration. A compressive PVT analysis on a Farnsworth field fluid sample has been presented by Gunda et al. (2015). The original minimum miscible pressure (MMP) measured from slim tube experiment was 4200 psia. Gunda et al. (2015) predicted an MMP of 4009 psia which represents less than 5% error compared to original value. Water is not allowed to dissolve in the gas or oil phased. However, CO_2 can be present in all three phases. CO_2 solubility in formation water and other aqueous phases are modeled by using Chang et al., (1998) correlation. The equation developed for distilled water solubility extended to include effects of salinity as shown in Eq (1).

$$R_{sb} = R_{sw} * e^{(-0.028 \times C \times T^{-0.12})} \qquad (1)$$

Where R_{sb} is solubility in Scf of CO_2 per Stb of formation water, R_{sw} is solubility in Scf of CO_2 per Stb of distilled water, T is temperature in °F, and C is salinity of formation water in weight percent of solid.

The salinity is available from formation water chemistry analysis. The current measured salinity for FWU is around 3500 ppm.

Residual Trapping

Residual trapping is an important CO_2 trapping mechanism (Kumar et al. 2005). Hysteresis modeling allows capillary pressures and relative pemeabilities to range between imbibition and drainage curves through scanning curves (Killough, 1976). Capillary pressure follows drainage curves for decreasing wetting-phase saturations and imbibition curves for increasing wetting-phase saturations. In case of a reversal in saturation directions, capillary pressure follows along the scanning curves (Killough, 1976). Entrapment of the non-wetting phase occurs when it is bypassed by the wetting phase thereby making it immobile. Previous research has presented several correlations in modeling hysteresis (Spiteri et al., 2005; Killough, 1976; Land 1968). In this paper the relative permeability hysteresis is based on the equation developed by Killough (1976) as shown in Eq. 2.. Sample drainage and imbibition curves used in modeling hysteresis are shown in Figure 4.

$$S_{ncrt} = S_{ncrd} + \frac{S_{hy} - S_{ncrd}}{1 + C\,(S_{hy} - S_{ncrd})} \quad (2)$$

Where

$$C = \frac{1}{S_{ncri} - S_{ncrd}} - \frac{1}{S_{n\,max} - S_{ncrd}} \quad (3)$$

S_{ncrt} is the critical trapped saturation of non-wetting phase; S_{ncrd} is the critical saturation of drainage curve; S_{hy} is the maximum non-wetting phase saturation; C is land's coefficient; S_{nmax} is the maximum non-wetting saturation; S_{ncri} is the critical saturation of the imbibition curves.

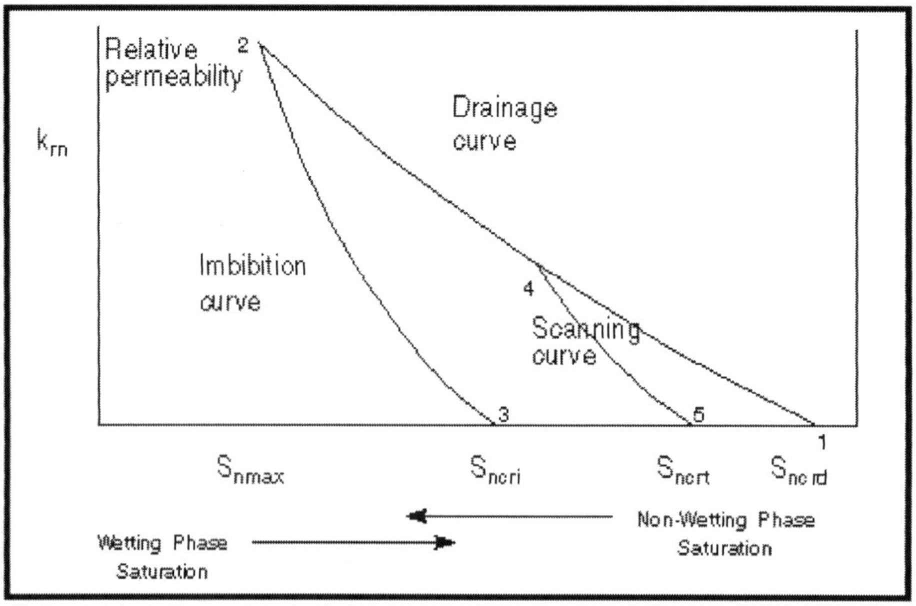

Figure 4- Drainage and imbibition curves used in modeling hysteresis effect

Structural / stratigraphic Trapping

Structural and stratigraphic trapping mechanisms are both important in the storage of CO_2 in a depleted hydrocarbon reservoir. The integrity of the caprock is vital to prevention of upward movement of supercritical CO_2 towards the surface. Seal bypass refers to the presence of fast paths for fluid transmission that have the potential to compromise the integrity of a caprock (Cartwright et al., 2007). Heath and Mozley (2015) analyzed the integrity of the caprock for the FWU using mercury porosimetry. The mercury porosimeter computes the height of supercritical CO_2 that the caprock can contain (Downey, 1984). The results from the analysis for FWU indicate that sealing formations including the overlying Morrow shale and Thirteen Finger Limestone units are excellent seals capable of providing storage of supercritical CO_2 up to a column height of 10,000 ft.

Reservoir Model Description

The static (geologic) model for the FWU was constructed from geophysical, geological and engineering data as described by Ampomah et al. (2015). This model has been used by the SWP Simulation Working Group, and serves as the basis for research activities in hydrogeology, reactive transport, and risk assessment. Stratigraphic definition of the reservoir and adjacent units was supported by a 3-D seismic survey acquired in January 2013 as part of the SWP Phase III project. Ampomah et al. (2015) presented a detailed upscaling study to decrease the number of cells in the geological model to reduce computation cost for dynamic modeling studies.

The dynamic model used for this study was extracted from the SWP geological model (Ampomah et al., 2015) which has been history matched. The model, which describes the west half of FWU, has a size of 81×77×5 grid blocks with gridblock dimensions of 200 ft × 200 ft. This model includes the Thirteen Finger limestone, Morrow shale, and Morrow "B" sandstone (Figure 5). The first two horizons serve as a caprock for the Morrow "B" sandstone reservoir. Figure 6 shows a 3D view of the model. All wells in FWU are vertical wells. The 3-D distribution of porosity and permeability for the Morrow "B" sandstone reservoir is shown in Figure 7 and 8 respectively. There was no correlation existing between porosity and permeability at the time static model was constructed (Ampomah et al., 2015).

The dynamic model had 60 production wells during primary depletion in the model. As there were no recorded gas-oil or water-oil contacts, all grid blocks were assigned an initial oil saturation of 69% and 31% connate water saturation. An initial reservoir pressure of 2217.7 psia was assigned to a datum depth of 4900 ft below subsea. The bubble point pressure of 2073.7 psia indicates that the reservoir was slightly undersaturated. Volume of oil originally in place (OOIP) was about 60 MMstb with 20.82 MMscf of dissolved gas. During waterflood, there were 47 producer wells with 13 wells converted from producers to injector wells. Six additional water injector wells were drilled during this same period. FWU has 55 years of primary and secondary depleted recovery. In December 2010, Chaparral Energy LLC commenced CO_2 flooding of the field. The detailed history matching processes are presented in another publication and are beyond the scope of this paper. The additional four years of CO_2 flood history was calibrated to the model. Oil and water saturation distribution at the end of CO_2 flood history match simulation for layer 4 is illustrated in Figures 9 and 10 respectively. Figure 11 shows a 3D distribution of amount of CO_2 sequestered in terms of lb-mol/rb at the end of the simulation run (December 2014) for layer 4 within the model. The areas in the figure with large volumes of sequestered CO_2 are the locations of CO_2 injection patterns. This figure also indicates the direction of CO_2 flow. There is less accumulation as the flow moves away from the injectors.

The FWU history matched model is extended to study the storage mechanisms associated with injected CO_2 volume in this paper.

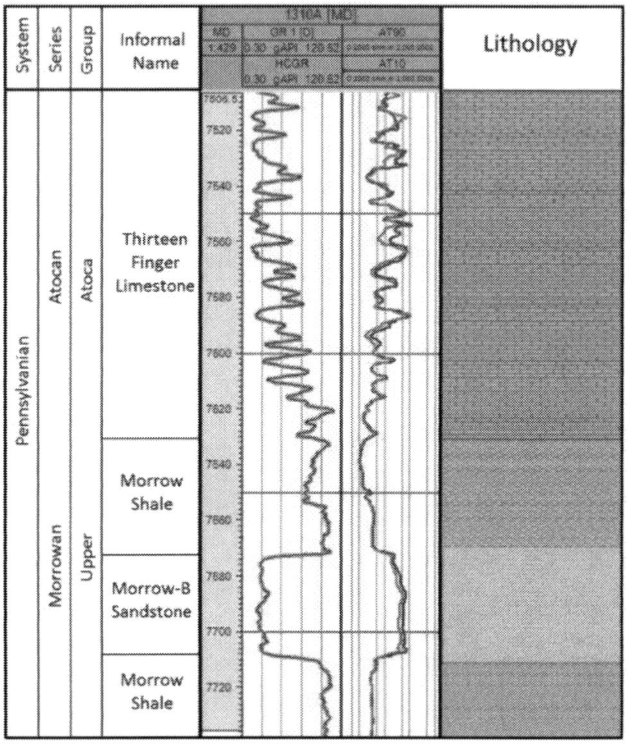

Figure 5- Stratigraphy and type log of FWU (modified from Ampomah et al. 2015; Ross-Coss, 2014)

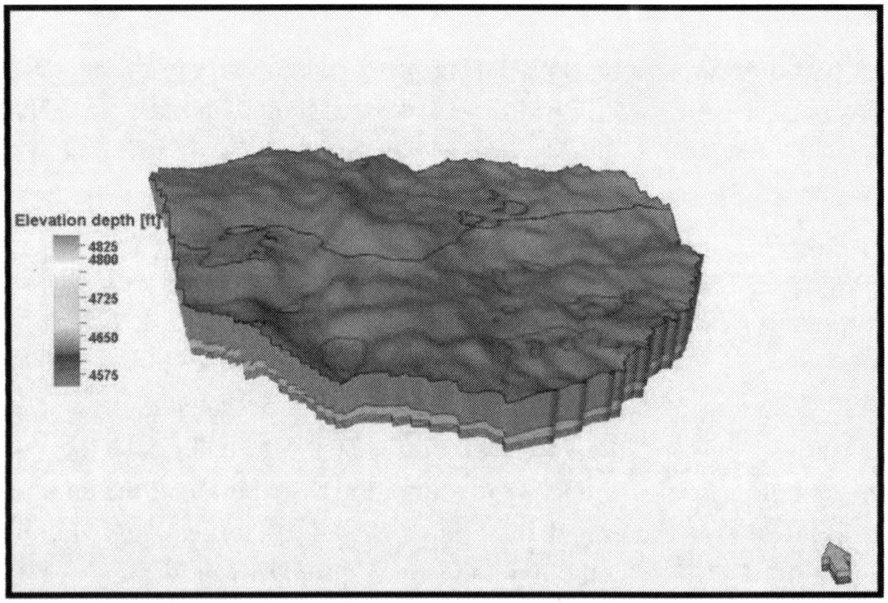

Figure 6- 3D view of west half of FWU showing different horizons in the constructed model

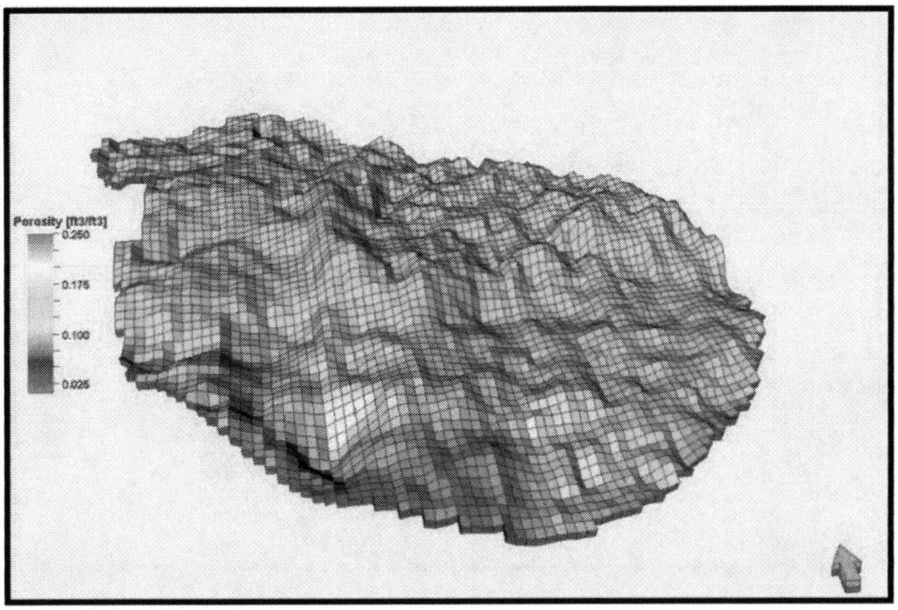

Figure 7- 3D porosity distribution for west half Morrow B formation of FWU

Figure 8- 3D permeability distribution for west half Morrow B formation of FWU

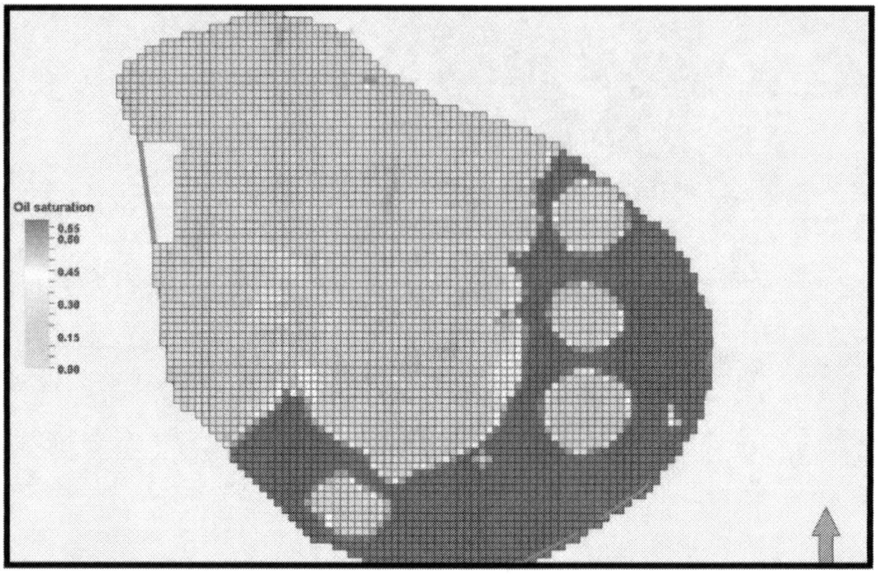

Figure 9- Oil saturation distribution of layer 4 at the end of CO_2 history match (December 2014). Lower saturations spots indicate the location of WAG injectors where CO_2 saturation is higher

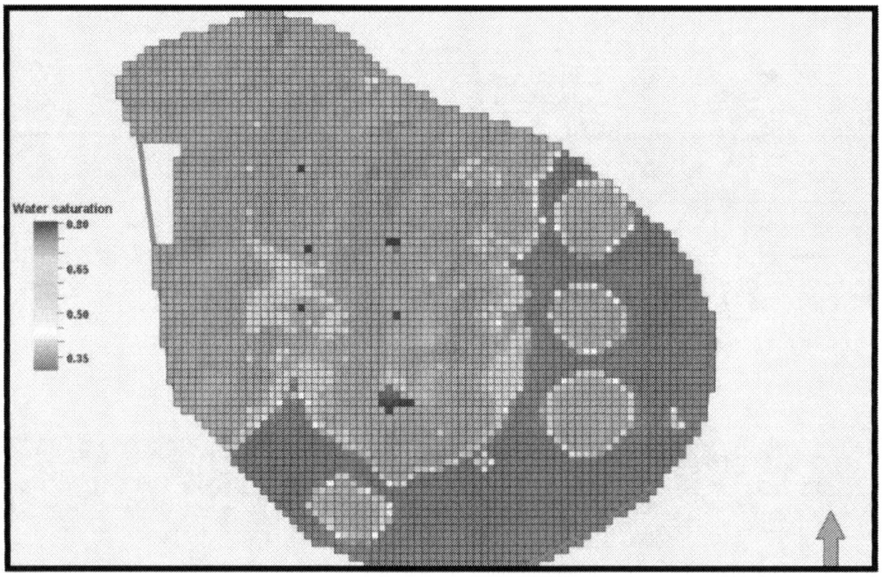

Figure 10 - Water saturation distribution of layer 4 at the end of CO_2 history match (December 2014). The interior part where WAG wells are located shows slightly lower water saturation

Figure 11- Layer 4 CO_2 sequestered within the Morrow formation expressed in lb-mol/rb

Results and Discussion

Scenario based models were run and analyzed to ascertain how CO_2 is stored within the reservoir under various operational scenarios. A total of 25 WAG wells and 35 producers were introduced into the model by the end of 2018. There is an anticipated constant CO_2 purchase of 26,350 lb-mol/d and a compressor facility to recycle produced gas for re-injection purposes. A user-defined algorithm was developed to use purchased CO_2 in addition to produced gas (recycled) as total group injection available rate for various simulation cases. A constant WAG ratio of 2:1 was used to initialize the simulation runs.

All studied cases involve a continuous WAG injection for additional 21 years beyond the history matched period. After 25 years of injection, wells were completely shutin and stored CO_2 was monitored for additional 200 years as a function of various storage mechanisms. The entire duration of study was from December 2014 to January 2236.

Injection profiles effect on CO_2 *storage capacity*

Two main development strategies are used in this work to illustrate the effect of CO_2 injection profiles on storage. Both cases have 25 WAG wells, 35 producers, CO_2 purchase and recycle. The WAG cycle for these cases is 3:1 with a water salinity of 3500 ppm.

A. With constant CO_2 purchase (26350 lb-mol/d).
B. With decreasing CO_2 purchase from 2022 to 2030 and inject only recycled gas after 2030.

Case A involves injecting continuous 26350 lb-mol/d of purchased CO2 in addition to recycle gas for the 21 year period. Case B also involves continuous injection of 26,350 lb-mol/d of purchase CO2 in addition to recycled gas until end of 2021. At year 2022, purchaseed CO_2 was cut to 21,080.03 lb-mol/d representing a 20% reduction. CO_2 purchased was subsequently reduced by 2,635 lb-mol/d each year until 2030. Afterwards, CO_2 purchase ceased and only recycled gas was injected.

Table 1 shows summary of results comparing above cases. Even though Case A had about 15% more CO_2 volume stored than Case B, a large volume of CO_2 was produced and amount of recycled percentage was low in Case A due to other constraints such as bottomhole pressures. Case B had almost the same amount of oil recovery with 40% less CO_2 purchased. Case B has a probable chance of rendering a higher rate of return if the FWU were operated under such a scenario.

In terms of long term storage of CO_2 both cases demonstrated sufficiently high pressure above the MMP even at the end of the 200 year monitoring simulation run, as shown in Figure 12. Case A has a slightly higher average pressure due to the volume of total injected fluid. The minute gradual reduction in pressure could be due to changes in total compressibility and equilibrating of the system after wells are shut in. There is a possibility of reservoir pressure lowering below the MMP which could eventually result in an increase in supercritical CO_2. However, FWU has multiple good stratigraphic seals which will serve to trap the CO_2 in the subsurface .

Case B illustrates that recycling a high percentage of produced gas, addition of wells/patterns, and reduction in purchased CO_2 after some years of operation improves oil recovery and maximizies CO_2 storage. This development strategy is used in the following sections to study effect of WAG cycles, salinity and hysteresis on different storage mechanisms.

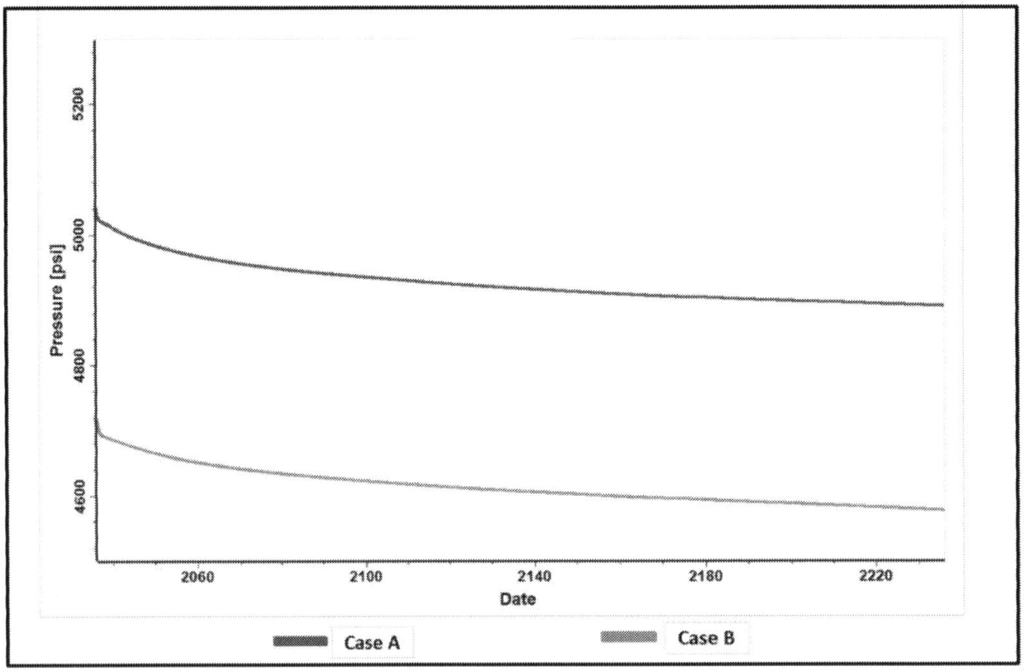

Figure 12- Pressure profile during monitoring period for Case A and Case B

Effect of WAG cycles on Storage Mechanisms

Changes in CO_2 and water cycling in a CO_2 flood can affect both the amount of CO_2 stored and the storage mechanism within the reservoir. In addition to Case B described above, a development strategy with a WAG cycle of 1:1(Case B1) was simulated to analyze its effect on CO_2 storage. CO_2 and water injection profiles for these cases are shown in Figure 13. Case B has a higher CO_2 injected volume and lower injected water volume than Case B1. The trend translated to higher oil recovery and volume of CO_2 stored for Case B as summarized in Table 2. It is therefore necessary to use optimum injection cycles to ensure higher oil recovery and maximum utilization of available CO_2. Figures 14 and 15 show CO_2 storage profiles for Case B and Case B1 respectively. During the injection period in Case B, CO_2 in supercritical phase was competing with dissolution in oil as opposed to Case B1. This was due to the larger volume of CO_2 injected and lower bottomhole pressures at the producer wells. Case B1 had slight higher average pressure due to the larger volume of water enabling extra dissolution of supercritical CO_2 as shown in Table 2. In both cases at the shut in of wells, reservoir pressure started to equilibrate and was still above MMP, ensuring mobile CO_2 stayed dissolved in the remaining oil. The solubility of CO_2 in water was nearly constant at shut-in. At the end of simulation only 4% of stored CO_2 is in the supercritical state which could be due to a gradual reduction in pressure. Though it is theoretically possible for the pressure to decrease below MMP after several years, it was not realized in this study. FWU has multiple overlying formations that act as formidable seals to contain mobile CO_2 within the reservoir. The residual trapping was not dominant in cases B and B1. The 1% residual trapping could be due to capillary forces existing within the reservoir. Trends exhibited in these figures could change based on several factors including volume of distinct fluids within the reservoir, pressure, salinity and temperature. But it is obvious dissolution of CO_2 in oil would dominate in most partially depleted oil reservoirs.

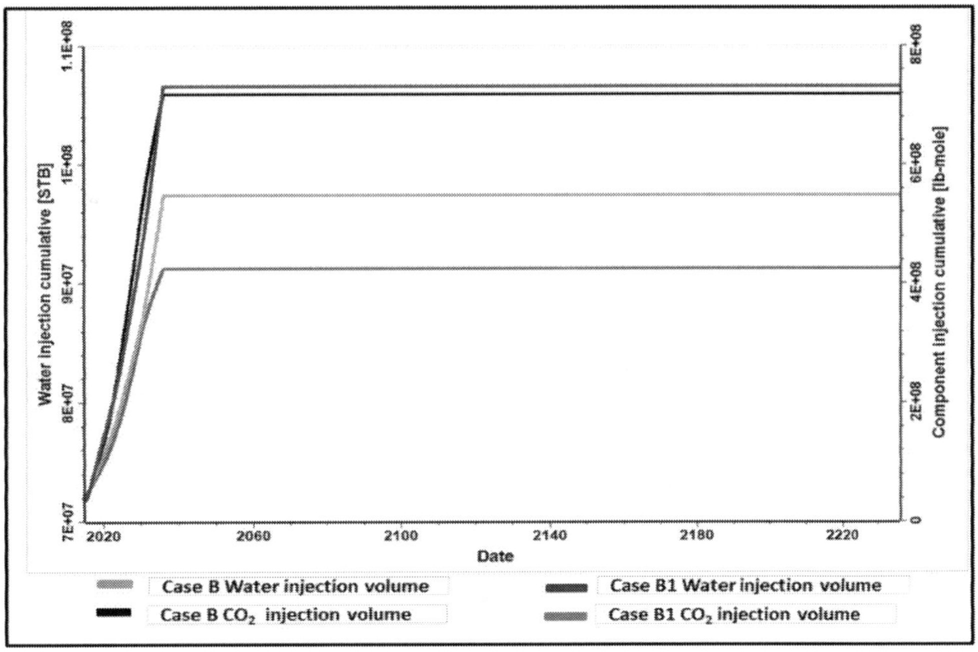

Figure 13- Injection profiles for Case B and Case B1 to study effect of WAG cycles on CO_2 storage

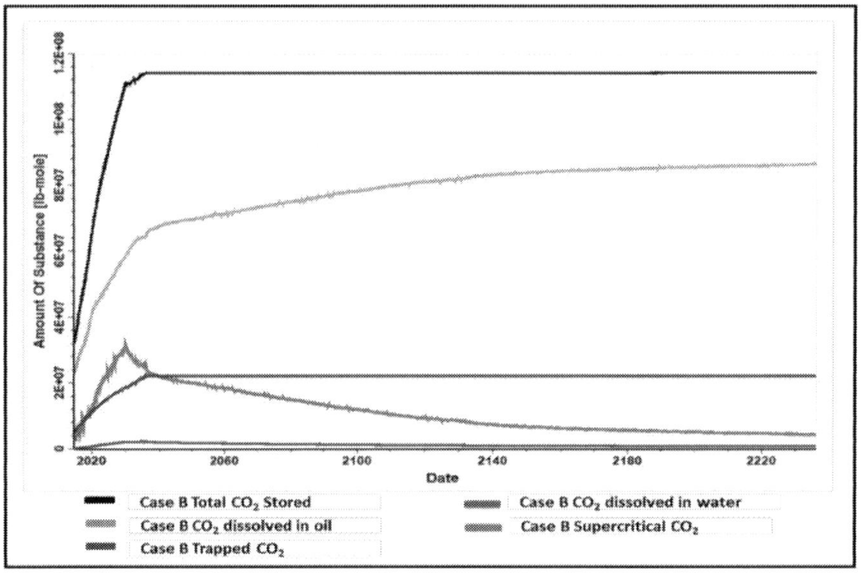

Figure 14- CO_2 storage profile based on different mechanisms within the reservoir for Case B

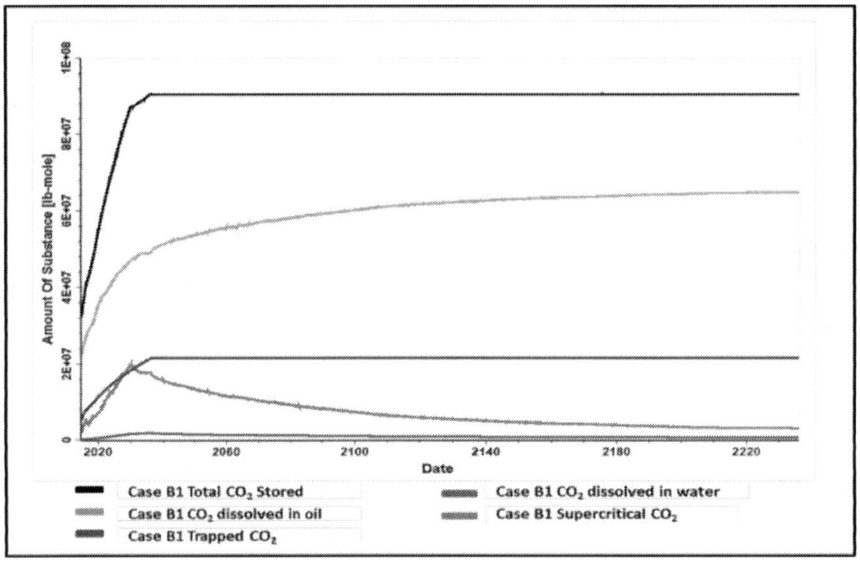

Figure 15 - CO_2 storage profile based on different mechanisms within the reservoir for Case B1

Table 1 - Summary results for Cases A and B Injection Scenarios

Results	Units	Case A	Case B
CO_2 Purchased	MMlb-mol	237.92	153.25
CO_2 Production Cumulative	MMlb-mol	720.04	604.97
CO_2 Injection Cumulative	MMlb-mol	854.43	719.22
Recycle	MMlb-mol	616.52	565.97
Total Storage*	MMlb-mol	134.39	114.25
% Storage	%	57	75
% Oil Incremental Recovery	%	34.1	33.9

**Total Storage = CO_2 Purchase – CO_2 produced + CO_2 Recycled*

Effect of Salinity on Storage Mechanisms

Prior to waterflood reservoir salinity at FWU was thought to be ~35,000 ppm. Reservoir water salinity following waterflood was 3,500 ppm due to the injection of fresh water of TDS ~500 ppm. Numerical simulations were run with these two salinities to study the effect on CO_2 storage within FWU. From simulation results, salinity changes strongly affects the solubility of CO_2. The salinity in Case B was changed to 35,000 ppm while keeping other parameters constant. Figure 16 compares CO_2 dissolved in water for salinities of 3,500 ppm and 35,000 ppm. For these two scenarios, the total CO_2 in place was equal prior to and after wells shut in. The initial case with a salinity of 3,500 ppm had nearly 2 MMlb-mol more of CO_2 dissolved in formation water as compared with the 35,000 ppm senario. This extra 2 MMlb-mol dissolved in remaining oil in the latter case. There was not significant change in oil recovery for either case.

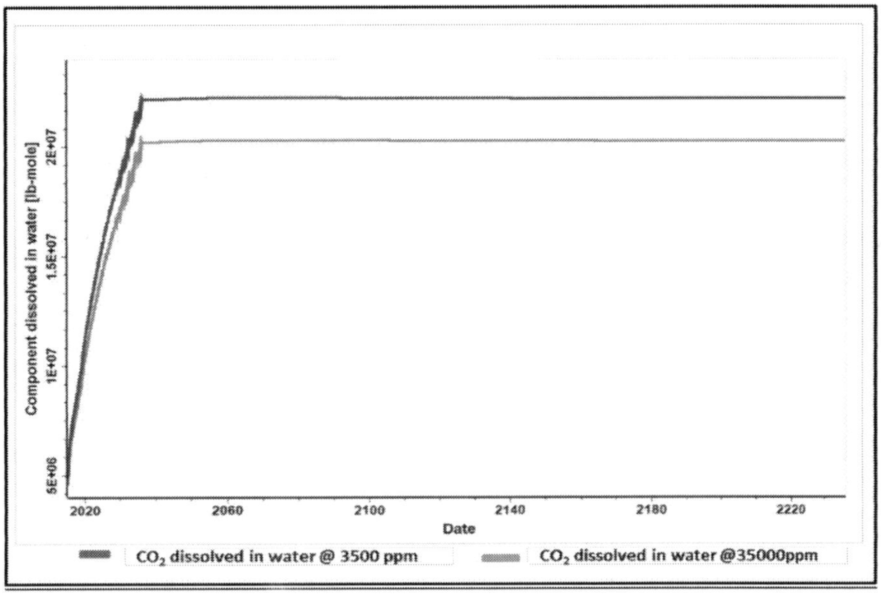

Figure 16- CO_2 dissolved in water at different salinities.

Effect of Hysteresis on Storage Mechanisms

Whenever a wetting phase fluid bypasses a non-wetting phase it often renders the non-wetting phase immobile. During the imbibition process, the relative permeability of the wetting phase increases drastically compared to the non-wetting phase thereby increasing entrapment. FWU is considered a water wet reservoir as can be seen from Figure 17. Drainage curves were measured as part of the special core analysis. Corey correlation was used to construct imbibition curves to study the hysterestic effect on CO_2 storage. The cases B and B1 described above did not include the hysteresis model that contributes to about 1% storage due to residual trapping. Case B2 is an extension to Case B and includes hysteresis. The volume of CO_2 stored in both cases is almost the same. The storage profiles for Case B and Case B2 are shown in Figures 14 and 18 respectively. Due to an increase in wetting phase relative permeability, a larger volume of water was produced in Case B2, subsequently reducing oil recovery and average reservoir pressure. The supercritical CO_2 phase also increased slightly in Case B2 due to pressure reduction. Table 3 shows a summary comparison of both cases. In Case B2, CO_2 stored due to residual trapping was pronounced. 14% of stored CO_2 was trapped due to hysteresis. This apparently affected the amount of CO_2 dissolved in oil since the percentage of CO_2 dissolved in water was almost the same in both cases.

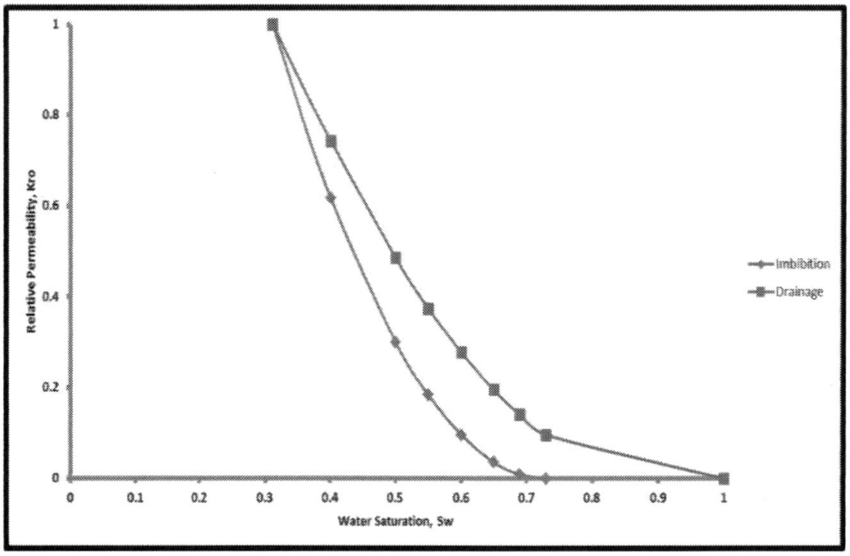

Figure 17- Drainage and imbibition curves for FWU

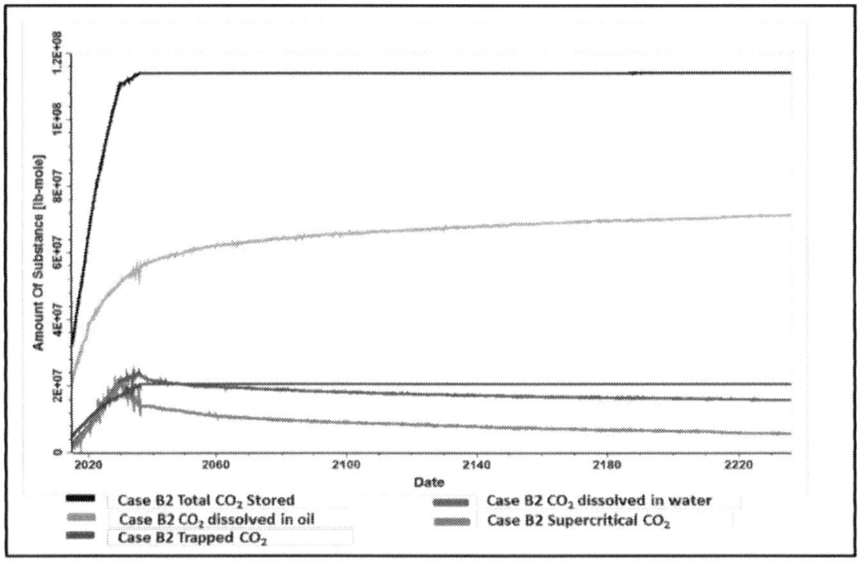

Figure 18- CO_2 **storage profile based on different mechanisms within the reservoir for Case B2 to study effect of hysteresis**

Table 2- Summary results for effects of WAG cycles on CO_2 **storage**

		Oil	Solubility		Supercritical CO_2	Residual
	WAG	Recovery	Oil	Water		
Case	Cycle	%	%	%	%	%
B	3:1	73	76	19	4	1
B1	1:1	70	72	24	4	1

Table 3- Summary results for effects of hysteresis on CO_2 storage

		Oil	Solubility		Supercritical CO_2	Residual
	Hysteresis	Recovery	Oil	Water		
Case	Model	%	%	%	%	%
B	No	73	76	19	4	1
B2	Yes	71	63	18	5	14

Conclusions

This work presents an extension of performance assessment of FWU, studying different mechanisms contributing to CO_2 storage within a depleted oil reservoir.

At FWU, the average reservoir pressure has been maintained to reduce the supercritical CO_2 phase which makes solubility (oil and water) and residual trapping the most important storage mechanisms. The gradual reduction in pressure seen in simulation results after all wells are shut-in signifies a possible increase in supercritical CO_2 after passage of significant time. However, the caprock intergrity for FWU has been thoroughly studied and demonstrates the ability of the seal to hold up to a 10,000 ft column of CO_2.

Dissolution of CO_2 in oil was found to be the predominant mechanism for storing CO_2 within FWU. The optimum WAG injection strategies are necessary to ensure high storage capacity of CO_2.Changes in salinities affects amount of potential CO_2 disolution in formation water. Residual trapping when hysteresis is modeled accurately can be a competing mechanism in storing CO_2.

This study could serve as a benchmark for future CO_2-EOR projects in the Anadarko basin or geologically similar basins throughout the world.

Acknowledgements

Funding for this project is provided by the U.S. Department of Energy's (DOE) National Energy Technology Laboratory (NETL through the Southwest Regional Partnership on Carbon Sequestration (SWP) under Award No. DE-FC26-05NT42591. Additional support has been provided by site operator Chaparral Energy, L.L.C., and Schlumberger Carbon Services and Informataion Support.

References

Ahmmed, B., 2015, Numerical modeling of CO2-water-rock interactions in the Farnsworth, Texas hydrocarbon unit, USA: Unpublished master's thesis, University of Missouri—Columbia, 63 p.

Al-Shaieb, Zuhar; Puckette, James; Abdalla, A. 1995. Influence of Sea-Level Fluctuation on Reservoir Quality of the Upper Morrowan Sandstones, Northwestern Shelf of the Anadarko Basin 249–268. Stillwater, OK.

Ampomah, W., Balch, R. S., and Grigg, R. B. (2015) Analysis of Upscaling Algorithms in Heterogeneous Reservoirs with Different Recovery Processes. SPE Production and Operations Symposium March 1-5 2015, Oklahoma USA. doi:10.2118/173588-MS

Bachu, S. Review of CO_2 storage efficiency in deep saline aquifers, *International Journal of Greenhouse Gas Control*, 2015, DOI: 10.1016/j.ijggc.2015.01.007

Bachu, S., & Shaw, J. (2003). Evaluation of the CO2 sequestration capacity in Alberta's oil and gas reservoirs at depletion and the effect of underlying aquifers. *J. Can. Petrol. Technol.*, *42*(9), 51–61.

Cather, S.M. (2014). Prelminary Petrographic Analysis of Pennsylvanian (Upper Morrow) Strata from the Farnsworth Unit 13-10A Well, Ochiltree County, Texas. Unpub. Manuscript.

Cather, S.M. and Cather, M. C. (2015) Comparative Petrography and Paragenesis of Pennsylvanian (Upper Morrow) Sandstones from the Farnsworth Unit 13-10A, 13-14, and 32-08 Wells, Ochiltree County, Texas. Unpub. Manuscript.

Chang, Y.-B., Coats, B. K., & Nolen, J. S. (1998). A Compositional Model for CO2 Floods Including CO2 Solubility in Water. *SPE Journal*. doi:10.2118/35164-PA

Dai, Z., R. Middleton, H. Viswanathan, J. Fessenden-Rahn, J. Bauman, R. Pawar, S. Lee and B. McPherson, 2014a, An integrated framework for optimizing CO_2 sequestration and enhanced oil recovery, *Environ. Sci. Technol. Lett.*, 1, 49-54.

Dai, Z., P. H. Stauffer, J. W. Carey, R. S. Middleton, Z. Lu, J. F. Jacobs, K. Hnottavange-Telleen, L. Spangle, 2014b, Pre-site characterization risk analysis for commercial-scale carbon sequestration, *Environ. Sci. Technol.* 48, 3908−3915.

Dai, Z., Viswanathan, H., Fessenden-Rahn, J., Middleton, R., Pan, F., Jia, W., Ampomah, W., Grigg, R. (2014c). Uncertainty Quantification for CO2 Sequestration and Enhanced Oil Recovery. *Energy Procedia*, *63*, 7685–7693. doi:10.1016/j.egypro.2014.11.802

Downey MW (1984) Evaluating seals for hydrocarbon accumulations. AAPG Bulletin, 68, 1752– 1763.

Heath, J. E., Dewers, T. A., Mozley, P. S. (2015). Characteristics of the Farnsworth Unit , Ochiltree County , Texas : Southwest Partnership CO2 Storage - EOR Project.

Holloway, S., 1997. An overview of the underground disposal of carbon dioxide. Energy Convers. Manag. 38, 193 – 198.

Gallagher, S. R. (2014). Depositional and Diagenetic Controls on Reservoir Heterogeneity: Upper Morrow Sandstone, Farnsworth Unit, Ochiltree County, Texas. M.S. Thesis, New Mexico Tech, Socorro NM, USA.

Goodman, A., et al. (2011). U.S. DOE methodology for the development of geologic storage potential for carbon dioxide at the national and regional scale. *International Journal of Greenhouse Gas Control*, *5*(4), 952–965. http://doi.org/10.1016/j.ijggc.2011.03.010

Gorecki, C. D., Hamling, J. A., Ensrud, J., Steadman, E. N., & Harju, J. A. (2012). Integrating CO2 EOR and CO2 Storage in the Bell Creek Oil Field. Carbon Management Technology Conference. . February 7-9 2012, Orlando, Florida USA. doi:10.7122/151476-MS

Grigg, R.B., McPherson, B.J., (2012). Phase III- Deployment Phase Farnsworth Unit CCUS, Ochiltree, Texas, National Energy Technology Laboratory Carbon Storage RD Meeting: Developing the Technologies and Building the Infrastructure for CO2 Storage. August 21-23, Pittsburgh, USA.

Gunda D., Ampomah, W., Grigg, R. B. and Balch, R. S. (2015) Reservoir Fluid Characterization for Miscible Enhanced Oil Recovery. Carbon Management Technology Conference November 16-19 2015, Sugarland, Houston-Texas USA. CMTC-440176-MS

IPCC (Intergovernmental Panel on Climate Change). IPCC special Report on Carbon Dioxide Capture and Storage. Prepared by Working Group III of the Intergovernmental Panel on Climate Change; Cambridge University Press: Cambridge, 2005, p 442.

Killough, J. E. (1976). Reservoir Simulation With History-Dependent Saturation Functions. *SPE Journal*. doi:10.2118/5106-PA

Krystinik, Lee; Blakeney, B. 1990. Sedimentology of the upper Morrow Formation in the Eastern Colorado and Western Kansas. In *Morrow sandstones of southeast Colorado and adjacent areas* 37–50. Denver, Co: Rocky Mountain Association of Geologists.

Lake .L.W. (1989). **ENHANCED OIL RECOVERY**, Prentice Hall, Inc., New Jersey (1989)

Land, C. E. "Calculation of Imbibition Relative Permeability for Two- and Three-Phase Flow from Rock Properties," paper SPE 1942, *Society of Petroleum Engineers Journal* (1968) 8, No. 2, 149-156.

Munson, T. 1988. *Depositional, Diogenetic, and Production History of the Upper Morrowan Buckhaults Sandstone, Farnsworth Field, Ochiltree County Texas*. West Texas State University.

Nghiem, L. X., Shrivastava, V. K., Tran, D., Kohse, B. F., Hassam, M. S., & Yang, C. (2009). Simulation of CO2 Storage in Saline Aquifers. SPE/EAGE Reservoir Characterization and Simulation Conference , October 19-21 2009, Abu Dhabi, UAE. doi:10.2118/125848-MS

Pruess, K., Xu, T., Apps, J., Garcı´a, J., 2003. Numerical modeling of aquifer disposal of CO2. Soc. Pet. Eng. J., 49 – 59 (March).

Puckette, Jim; Azhari, Abdalla; Aaron Rice. Al-Shaieb, Z.1996. The Upper Morrow Reserviors: Complex Fluvio-Deltaic Depositional Systems. Stillwater, OK: Oklahoma Geological Survey.

Qi, R., LaForce, T., & Blunt, M. J. (2008). Design of carbon dioxide storage in oil fields. *SPE Annual Technical Conference and Exhibition*, (September), 21–24. Retrieved from https://www.onepetro.org/conference-paper/SPE-115663-MS

Rose-Coss, D. The Morrow-B Reservoir Sandstone, Farnsworth Unit, Ochiltree County Texas. New Mexico Tech 125th Anniversary Poster Competition, October 2014 & Rocky Mountain Rendezvous Poster presentation, September 2014.

Saini, D., & Jimenez, I. (2014). Evaluation of CO2 -EOR and Storage Potential in Mature Oil Reservoirs. *SPE International, SPE-169555*(April), 16–18.

Sonnenberg, Stephen. Shannon, Lee. Rader, Kathleen. Drehle, W. 1990. Regional Structure and Stratigraphy of the Morrowan Series, Southeast Colorado and Adjacent Areas. In *Morrow sandstones of southeast Colorado and adjacent areas* .1–8. Denver, Co: Rocky Mountain Association of Geologists.

Spiteri, E., Juanes, R., Blunt, M. J., & Orr, F. M. (2005, January 1). Relative-Permeability Hysteresis: Trapping Models and Application to Geological CO2 Sequestration. SPE Annual Technical Conference and Exhibition, October 9-12, Dallas, Texas, USA. doi:10.2118/96448-MS

SWP, The Southwest Partnership on Carbon Sequestration (SWP) 2013, http://www.southwestcarbonpartnership.org/.

White, M. D., McPherson, B. J., Grigg, R. B., Ampomah, W., & Appold, M. S. (2014). Numerical Simulation of Carbon Dioxide Injection in the Western Section of the Farnsworth Unit. *Energy Procedia*, *63*, 7891–7912. http://doi.org/10.1016/j.egypro.2014.11.825

Unit Conversions

1 Mscf	= 2.635 lb-mol
1 lb-mol	= 1.977E-05 Metric tonnes

Potential Issues and Costs Associated with CO_2 Storage During and After CO_2-EOR

Prepared for:
Carbon Management Technology Conference – 2015

Prepared By:
Michael L. Godec, David Riestenberg, and Shawna Cyphers
Advanced Resources International, Inc.

November 16-19, 2015
Sugarland, Texas

Background

- U.S. EPA has promulgated Underground Injection Control (UIC) well (Class VI) requirements for geologic storage of CO_2.
- EPA also established reporting requirements under its Greenhouse Gas Reporting Program for facilities that inject CO_2 underground
  - Subpart UU for CO_2-EOR
  - Subpart RR for geologic storage
- CO_2-EOR operator injecting anthropogenic CO_2
  - Could convert to geologic storage (Class II to Class VI), or
  - Qualify for CO_2 storage during CO_2-EOR operations.
- Recent EPA guidance confirms that CO_2-EOR can result in stored CO_2, and a conversion to Class VI is not required for assuring storage.
- Many are concerned substantial legal and regulatory obstacles exist to allowing CO_2-EOR to be a viable source for CO_2 emissions reduction.

Objective of Presentation

- **Characterize potential issues and costs associated with:**
  - Ensuring CO_2-EOR project gets "credit" for stored CO_2
  - Converting a CO_2-EOR project to a CO_2 storage project
- **Based upon the EPA Class VI rule, GHGRP Subparts RR and UU, and current Class II regulations.**
- **Estimated costs based on:**
  - Cost analyses developed as a part of EPA rulemakings
  - Actual costs at DOE-funded research projects.
- **Focus on "quantifiable" costs – some potential costs are currently very difficult to quantify, and thus not included.**

Potential Issues and Costs Associated with CO_2 Storage During and After CO_2-EOR

Transitioning CO_2-EOR to CO_2 Storage

- Transition may proceed through one or more phases:
  1. Conventional CO_2-EOR
  2. CO_2-EOR with credit for stored CO_2
  3. Transition from CO_2-EOR to CO_2 storage
  4. CO_2 storage post-CO_2-EOR operations (Class VI).
- Phases 1 - 3 still represent, for the most part, traditional CO_2-EOR/Class II operations.
- In Phases 2 - 3, additional MRV activities may be necessary and/or required to verify the volumes of CO_2 stored for which credit may be desired.
- To date, most of the focus has been on the conversion of CO_2-EOR to CO_2 storage

Potential Technologies/Approaches to Ensure CO_2 Storage with CO_2-EOR

- Develop MRV plans and update models
- Convert some existing wells to monitoring wells
- Install new shallow monitoring wells
- Perform soil flux/vadose zone monitoring
- Perform micro-seismic monitoring
- Special-case plugging and abandonment (P&A)
- Drill new deep monitoring wells
- Perform 3D seismic
- Perform vertical seismic profiling (VSP)/cross-well seismic
- Perform eddy covariance/ Color Infrared Orthoimaging (CIR) monitoring
- Conduct cased hole logging
- Replace tubulars/wellhead/packers in existing wells
- Conduct mechanical integrity testing (MIT) program
- Monitor surface pressure, temperatures, rates, gas composition, and corrosion
- Monitor subsurface pressure and fluid sampling

Key Characteristics of the CO_2-EOR/CO_2 Storage Project Assumed for Estimating Costs

	Deep Saline Formation	CO_2-EOR Project
Maximum Monitoring Area	28,000 acres (44 sq m)	7,000 acres (10.9 sq m)
Well Depth	5,250 feet	5,250 feet
Capacity	74 million metric tons	18.5 million metric tons
# of wells	348 injection wells, 0 production wells	87 injection wells, 87 production wells
# of above-zone monitoring wells	20	4 converted producers
# of in-zone monitoring wells	20 converted producers, 8 new	6 converted producers, 3 new
Project Life, for annualizing costs	20 years of injection, 50 years of post-injection monitoring	20 years of injection, 10 years of post-injection monitoring

CO_2 Storage with EOR

Issues to be Considered Regarding CO_2 Storage with CO_2-EOR

- Fundamental question -- can a CO_2-EOR operator get credit for stored CO_2?
- If so, what exactly is required to ensure storage?
  - Can purely implementing GHG accounting/ reporting be sufficient (e.g., Subpart RR program)? – (Base Case)
  - Or, will CO_2-EOR operations need to accept most (but maybe not all) of the responsibilities associated with the UIC Class VI program? – (High Case)
  - ***Or, will legal and regulatory obstacles, mostly procedural, preclude CO_2-EOR from being an option for ensuring storage?***

CO_2-EOR/Class II Activities That Also Apply to Class VI/CO_2 Storage

- Most site characterization
  - Maps and cross sections
  - Geochemistry
  - Assess storage reservoir(s)/confining zone(s)
  - Risk assessment/ mitigation
- Most or all well drilled (?)
- Abandonment of unused existing wells (?)
- Most surface infrastructure
- Existing computational model(s) (reservoir simulation); as appropriately modified
- Some monitoring activities
  - Pressure, temperature, composition, fluid sampling, etc.

Monitoring Scenarios for Ensuring CO_2 Storage During CO_2-EOR

MRV Category	Base Case	High Case
Develop MRV plans and update models	YES	YES
Convert some existing producers to monitoring wells	YES	YES
Install shallow monitoring wells[1]	NO	YES
Perform soil flux/vadose monitoring	YES	YES
Perform micro-seismic install/monitoring	NO	YES
Special-case P&As	YES	YES
Drillin new deep monitoring wells	NO	YES
Perform 3D seismic	NO	YES
Perform VSP/cross-well seismic	NO	YES
Perform eddy covariance/CIR	YES	YES
Conduct cased hole logging	NO	YES
Replace tubing/wellhead/packers for injectors	YES	YES
Drill new Class VI injector(s)	NO	NO
Conduct MIT program	IN PLACE	YES[2]
Monitor surface pressure, temperatures, rates, gas composition, and corrosion	YES	YES
Monitor subsurface pressure and fluid sampling	YES	YES

[1] Additional shallow ground water monitoring wells required

[2] Assume annual MITs required, compared to every 5 years in Class II

YES = will be done to comply

NO = will not be done

IN PLACE = already in place

Costs For Ensuring CO_2 Storage During CO_2-EOR – Summary

	Total Costs ($)		Storage Costs ($/metric ton)	
Storage with CO_2-EOR	Base	High	Base	High
Site Characterization and Preparation	$8,755,818	$56,010,370	$0.48	$3.82
Injection Phase (20 years)	$12,268,567	$125,131,031	$0.67	$5.11
PISC (10 years)	$2,188,132	$610,189	$0.12	$0.98
TOTAL	**$23,212,516**	**$181,751,590**	**$1.27**	**$9.91**

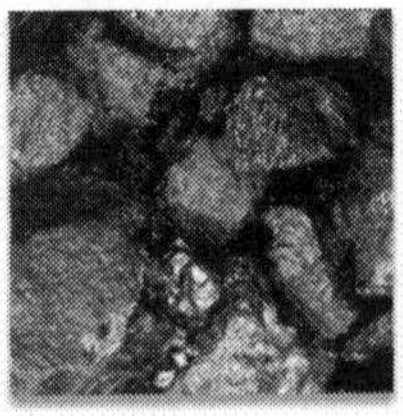

CO_2-EOR Conversion to CO_2 Storage

Transitioning CO_2-EOR to CO_2 Storage

- **May need to "prepare" CO_2-EOR reservoir for storage, prior to end of CO_2-EOR operations, to "re-optimize" for storage.**
- **May be particularly important for reservoirs that have undergone WAG processes**
  - Reduce reservoir pressure and improve CO_2 injectivity.
- **Could be facilitated by converting from a WAG to continuous CO_2 injection**
  - Producing the water (and incremental oil), but not reinjecting the water.
- **Would allow for pressure in the reservoir to decline, the injection rate for CO_2 to increase, and free up pore space in the reservoir for additional CO_2 storage.**

Transitioning CO_2-EOR to CO_2 Storage

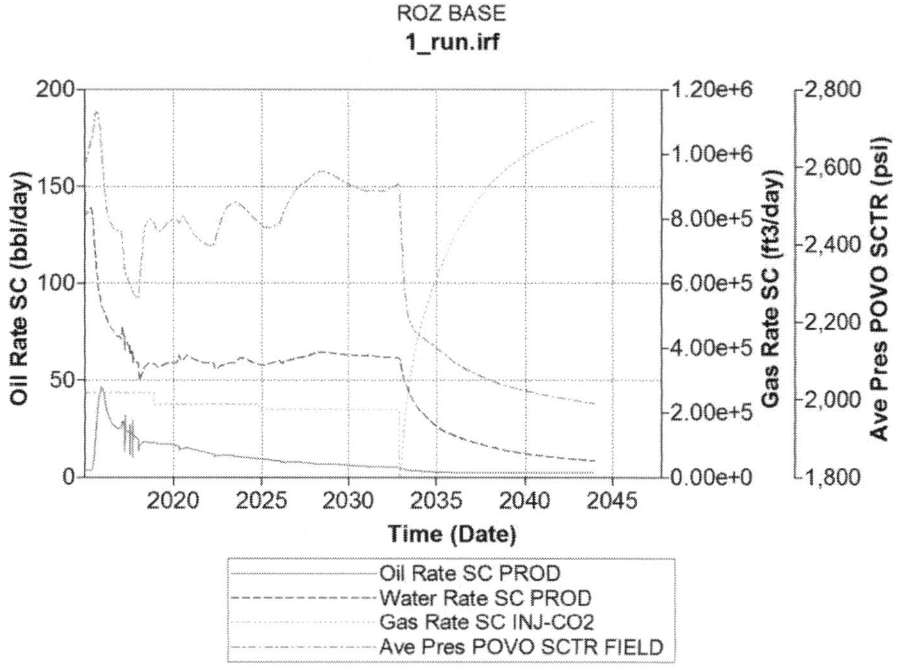

General Assumptions for Converting Class II/ CO_2-EOR to Class VI/CO_2 Storage

- Meets Class VI regs; assumes comprehensive site and geological characterization of the target reservoir available from the EOR project
- A 20 square mile AoR (EPA baseline) beyond field acreage
- One CO_2-EOR producer per injector modified for deep monitoring
- CO_2 storage spacing is 320 acres; from 40 acres for CO_2-EOR

Base Case

- Existing CO_2-EOR injector is used for injector (wells "grandfathered")
- Add shallow monitoring wells
- Microseismic only, no 3D or VSP/cross-well seismic
- Class VI monitoring plan is sufficient for Subpart RR
- 10 year PISC

High Case

- New injectors constructed to Class VI standards
- Seismic (VSP, cross-well, 3-D)
  - Additional RR requirements
  - Surface leakage monitoring
- Additional and more frequent monitoring for Subpart RR
- 50 year PISC

Costs For CO_2-EOR Conversion to CO_2 Storage – Summary

	Total Costs ($)		Storage Costs ($/metric ton)	
CO_2-EOR Conversion to Class VI	**Base**	**High**	**Base**	**High**
Site Characterization and Preparation	$21,533,656	$78,484,890	$1.17	$4.28
Injection Phase (20 years)	$89,052,618	$171,481,263	$4.85	$9.35
PISC	$22,671,365	$198,213,295	$1.24	$10.80
TOTAL	**$133,257,639**	**$448,179,448**	**$7.26**	**$24.43**
PISC Time Period (years)			10	50

Costs Not Considered for Converting Class II/ CO_2-EOR to Class VI/CO_2 Storage

- ***Legal and regulatory costs for multiple plan submittals or potential legal costs.***
- Reservoir pressure control (i.e., water withdrawal for plume management, injectivity/capacity enhancement)
- Bonding
- Pore space rights
- Land use permits
- Water discharge permits
- Air emission permits
- Taxes
- Mitigation

Potential Legal/Regulatory Obstacles under Subpart RR

- **Under EPA "Clean Power Plan," CO_2 storage w/ CO_2-EOR recognized as a potential compliance mechanism.**
- **If CO_2 storage is verified under Subpart RR, many are concerned that RR fundamentally conflicts with state-level mineral property and resource conservation law.**
- **Most significant concerns under Subpart RR include:**
  - Process for and timeliness of EPA approval of MRV plans for CO_2-EOR projects
  - What constitutes "new" activity in a CO_2-EOR project necessitating submittal of a new MRV plan
  - How does federal requirement relate to current state CO_2-EOR/Class II permitting processes.
  - The extent to which MRV plans are subject to Part 78 litigation procedures.

Critical Elements Impacting Costs

- Whether new MRV plans are required whenever "new" activity is undertaken (e.g., like converting a producer to an injector), and whether such plans are legally challenged.
- Class VI Program Director determination of "acceptability" of permit specifications
  - In particular, the acceptability of existing infrastructure and wells for CO_2 storage
  - Can existing wells be converted (and at what cost) or will new wells to Class VI standards be necessary?
  - Understanding what monitoring activities provide useful information, and which do not
- Ability to prepare CO_2-EOR operation for storage
  - May take some time to allow for cost-effectiveness
- Length of time for post-injection site care

Concluding Thoughts and Observations

- The transition from conventional CO_2-EOR to CO_2 storage post- CO_2-EOR may proceed through one or more phases.
- Many of the activities that would serve to ensure/verify CO_2 storage are already part of most CO_2-EOR operations
- Fundamental question -- can a CO_2-EOR operator get credit for stored CO_2, and if so, what exactly is required to ensure storage?
- ***If important legal and regulatory considerations can be addressed, it is conceivable that costs for ensuring storage can be relatively modest.***
- However, certain monitoring requirements, if imposed, could change this:
  - In particular, Class VI well construction required for all CO_2-EOR wells, long period of PISC, and extensive 3D seismic requirements.

Office Locations
Washington, DC
4501 Fairfax Drive, Suite 910
Arlington, VA 22203
Phone: (703) 528-8420

Houston, Texas
11931 Wickchester Ln., Suite 200
Houston, TX 77043-4574
Phone: (281) 558-9200

Cincinnati, Ohio
1282 Secretariat Court
Batavia, OH 45103
Phone: (513) 460-0360

AIChE ASME AIME IEEE ASCE

Carbon Management

Impact of thermal pressurization on fluid pressure build up in carbon capture and storage projects

Kimia Mortezaei, Farshid Vahedifard

Department of Civil and Environmental Engineering and Center for Advanced Vehicular Systems (CAVS), Mississippi State University, Mississippi State, MS, USA

Contents

MISSISSIPPI STATE UNIVERSITY

2

Geomechanical Issues with CO_2 sequestration

MISSISSIPPI STATE UNIVERSITY

CO_2 Injection

↓

Increase pore pressure
Decrease effective stress

↓

Expansion in reservoir rocks /deformation of overburden

Upward migration of CO_2	Occurrence of seismic events

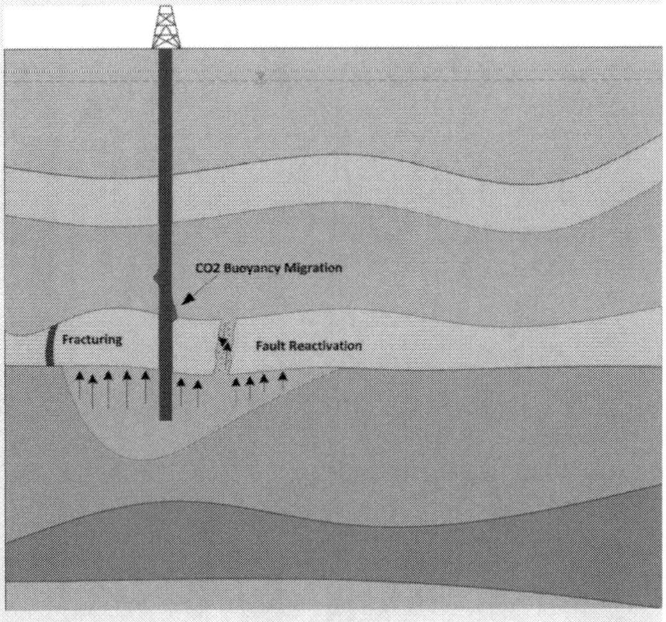

Potential stress and strain changes due to CO_2 injection

Introduction:
Thermal weakening mechanisms (at seismic slip rates)

MISSISSIPPI STATE UNIVERSITY

◈ **Flash heating of microscopic asperities**

Effective before bulk melting takes place

◈ **Thermal pressurization of pore fluid**

Effective if a fault is saturated with water

4

Introduction: Thermal pressurization

MISSISSIPPI STATE UNIVERSITY

1- Frictional sliding generates heat.

2- Pore fluid expands more than rock.

3- Pore pressure increases if rate of heat production exceeds rate of fluid and heat transport.

4- Effective normal stress decreases.

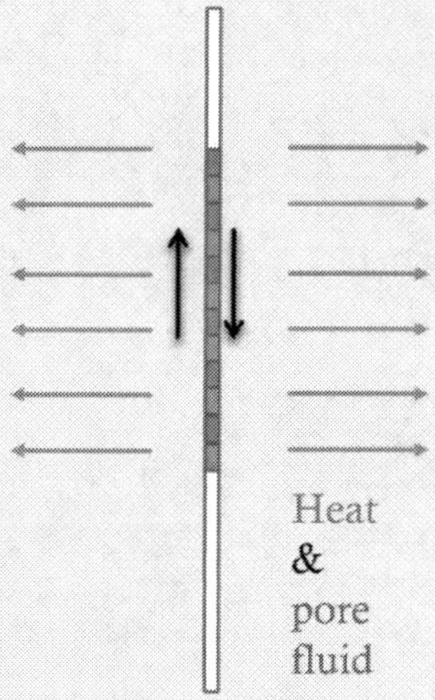

5

Objectives

MISSISSIPPI STATE UNIVERSITY

- Compare the response of the fault shear zone during CO_2 injection with and without considering the TP effect
- The effects of hydraulic diffusivity, slip velocity, shear zone thickness as well as injection rate on pore pressure, temperature and fracture energy prediction were evaluated.

6

Equations

MISSISSIPPI STATE UNIVERSITY

The following factors affect pore pressure during slip:

1) shear heating
2) porosity change
3) heat flow
4) fluid flow
5) dehydration reaction

7

Equations: Heat transport

MISSISSIPPI STATE UNIVERSITY

Based on energy conservation law and the Fourier law, the temperature change is given by:

$$\frac{\partial T}{\partial t} = \frac{\tau_f v}{\rho_{bulk} C_{bulk} w} + c_{th} \frac{\partial^2 T}{\partial x^2}$$

$$c_{th} = \frac{k}{\rho_{bulk} C_{bulk}}$$

ρ_{bulk}, the density of bulk composite	C_{bulk}, the specific heat of the bulk composite at a constant pressure	T, the temperature
t, time	τ_f, the frictional resistance, ($\tau_f = f(\sigma_n - P)$)	v, the dislocation rate
c_{th}, thermal diffusivity	k, the thermal conductivity,	x, the spatial coordinate normal to the fault

8

Equations: Fluid Transport

MISSISSIPPI STATE UNIVERSITY

Based on Darcy's law and mass conservation law, pore pressure diffusion can be expressed as:

$$\frac{\partial P}{\partial t} = \frac{\alpha_f}{\beta_f}\frac{\partial T}{\partial t} - \frac{1}{\beta_f \emptyset}\frac{\partial \emptyset}{\partial t} + \omega \frac{\partial^2 P}{\partial x^2}$$

Porosity is assumed to be constant

$$\omega = \frac{K}{\mu \beta_f \emptyset}$$

β_f, fluid compressibility
α_f, fluid expansivity,
v, the dislocation rate,
ω, is the hydraulic diffusivity,
K, the permeability,
μ, is the dynamic viscosity of the fluid,
$\emptyset$, the porosity.

9

Equations: Thermal pressurization

MISSISSIPPI STATE UNIVERSITY

The coupled temperature and pore pressure equations are:

$$\frac{\partial P}{\partial t} = \frac{\alpha_f}{\beta_f}\frac{\partial T}{\partial t} + \omega \frac{\partial^2 P}{\partial x^2}$$

$$\frac{\partial T}{\partial t} = \frac{\tau_f v}{\rho_{bulk} C_{bulk} w} + c_{th}\frac{\partial^2 T}{\partial x^2}$$

10

Equations: Fracture Energy

MISSISSIPPI STATE UNIVERSITY®

The evaluation of the fracture energy at a specific point relies on knowledge of the relationship between stress τ versus coseismic slip δ and is defined as below:

shear stress as a function of slip

$$G(\delta) = \int_0^{\delta} [\tau(\delta') - \tau(\delta)] d\delta'$$ → slip

Assuming all the energy released during earthquake slip is transformed into sensible heat

specific heat of the bulk composite — temperature

$$G = \rho_{bulk} C_{bulk} w (T - T_0) - \tau\delta$$

11

Numerical Model

MISSISSIPPI STATE UNIVERSITY®

Values of model parameters

Material	Symbol	Value	Unit
Permeability (base)	K	10^{-19}	m^2
Porosity	∅	0.025	1
Thermal expansivity of the solid	α_f	2.5×10^{-3}	K^{-1}
Fluid compressibility	β_f	4×10^{-9}	Pa^{-1}
Dynamic viscosity of the fluid	μ	10^{-4}	Pa s
Fluid density	ρ_f	340	$Kg\ m^{-3}$
Density of bulk composite	ρ_{bulk}	2700	$Kg\ m^{-3}$
Thermal conductivity	k	1.35	$J\ m^{-1}\ s^{-1}\ K^{-1}$
Fictional coefficient	f	0.5	1
Young modulus	E	10	GPa
Poisson's ratio	ν	0.25	-
Initial temperature	T_0	50	°C

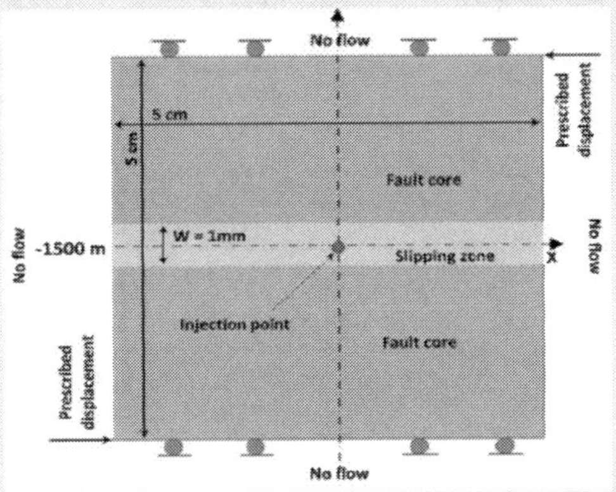

Model geometry used in numerical simulations

12

Results

MISSISSIPPI STATE UNIVERSITY®

Several factors can influence the thermal pressurization mechanism:

- Slip rate
  - Rate of frictional heating
  - Rate of fluid pressure build-up
- Depth and width of the deformation zone can
  - The hydraulic transport properties
  - Normal stress

13

Results (Effect of shear zone thickness)

MISSISSIPPI STATE UNIVERSITY®

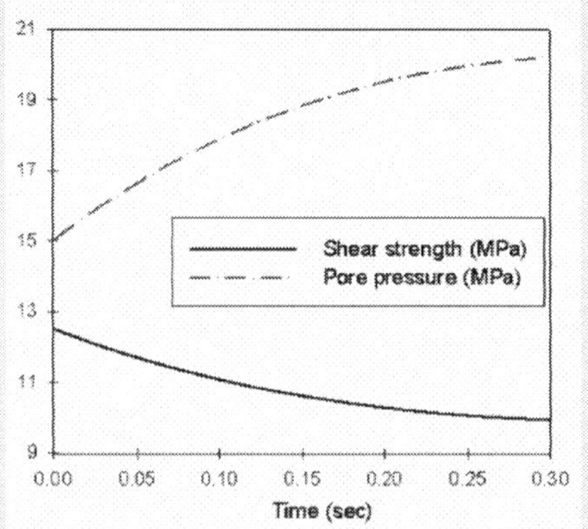

Shear resistance and pore pressure versus time in the base model (w = 1 mm)

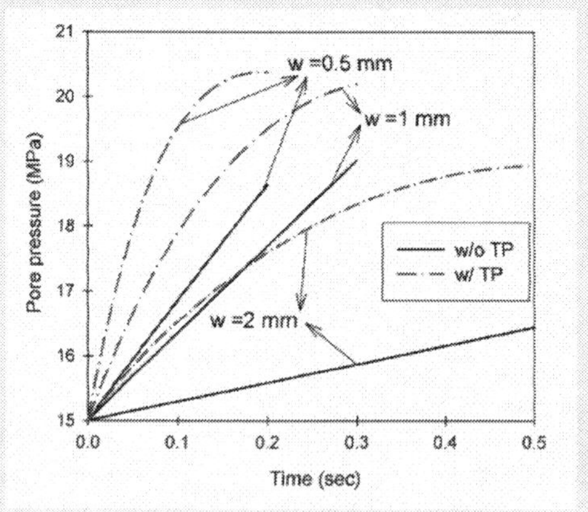

Pore pressure versus time for different slipping zone thicknesses

14

Results (Effect of shear zone thickness)

MISSISSIPPI STATE UNIVERSITY

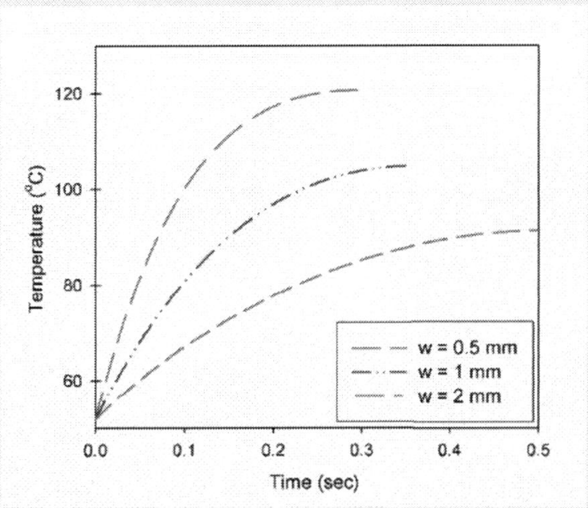

Temperature versus time for different slipping zone thicknesses.

15

Results (Effect of shear zone thickness)

MISSISSIPPI STATE UNIVERSITY

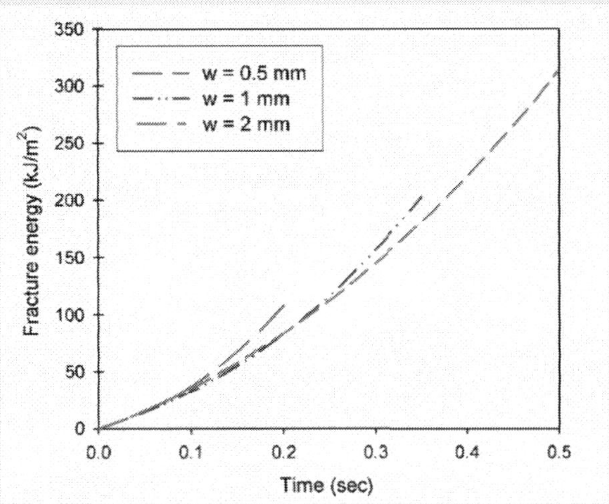

Fracture energy versus time for various shear zone thicknesses

16

Results (Effect of hydraulic diffusivity)

MISSISSIPPI STATE UNIVERSITY®

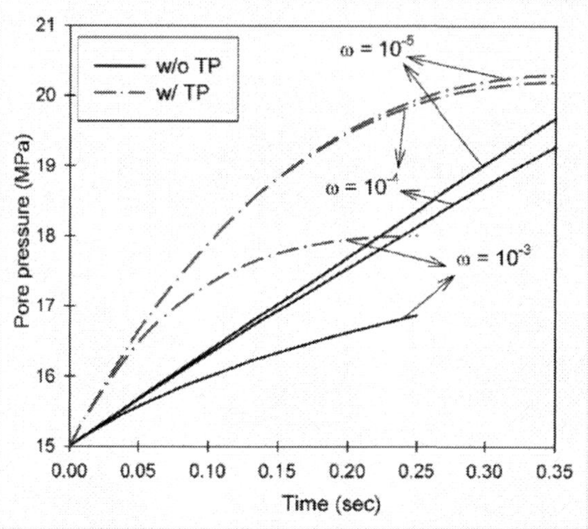

Pore pressure versus time for different hydraulic diffusivity values.

17

Results (Effect of hydraulic diffusivity)

MISSISSIPPI STATE UNIVERSITY®

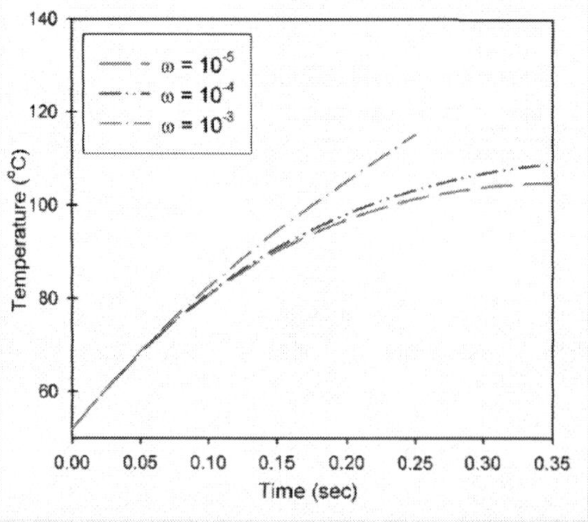

Temperature versus time for different hydraulic diffusivity values.

18

Results (Effect of hydraulic diffusivity)

MISSISSIPPI STATE UNIVERSITY®

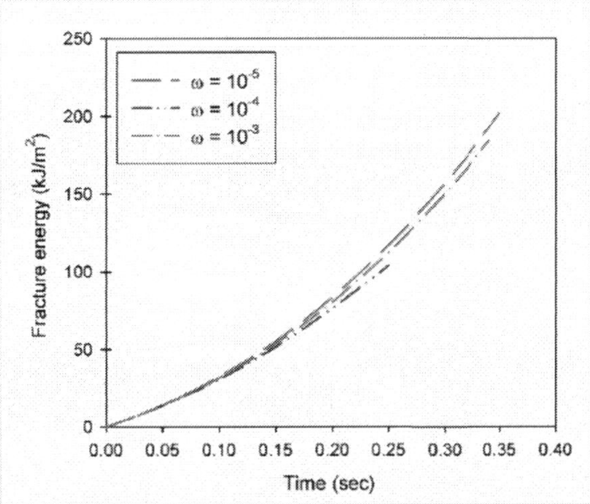

Predicted fracture energy (G) versus time for various hydraulic diffusivity values

19

Results (Effect of slip velocity)

MISSISSIPPI STATE UNIVERSITY®

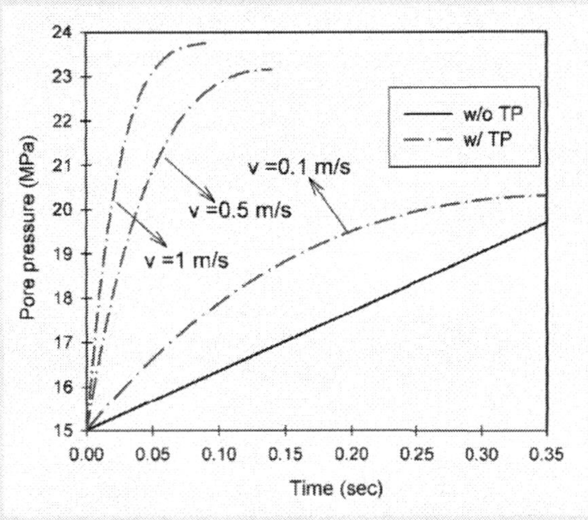

Pore pressure versus time for various slip velocity

20

Results (Effect of slip velocity)

MISSISSIPPI STATE UNIVERSITY®

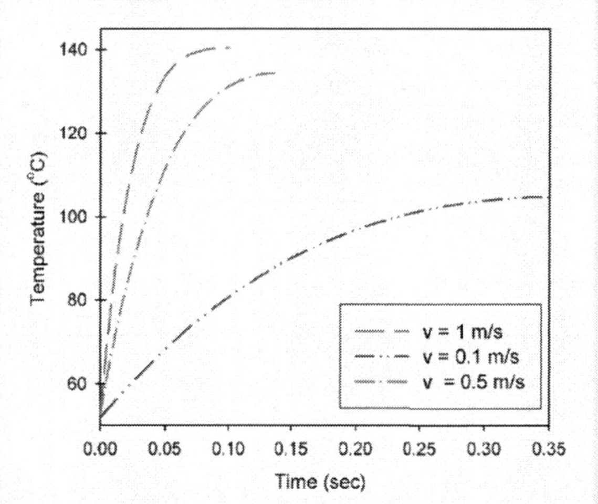

Temperature change versus time for various slip velocity

21

Results (Effect of slip velocity)

MISSISSIPPI STATE UNIVERSITY®

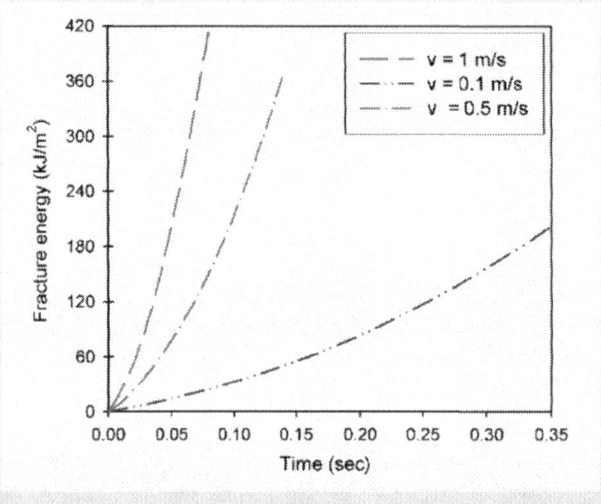

Predicted fracture energy (G) versus time for various slip velocity

22

Results (Effect of injection rate)

MISSISSIPPI STATE UNIVERSITY®

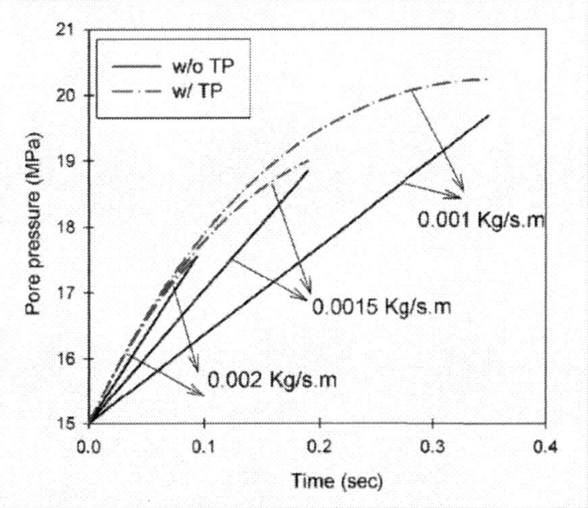

Pore pressure versus time for various injection rate

23

Results (Effect of injection rate)

MISSISSIPPI STATE UNIVERSITY®

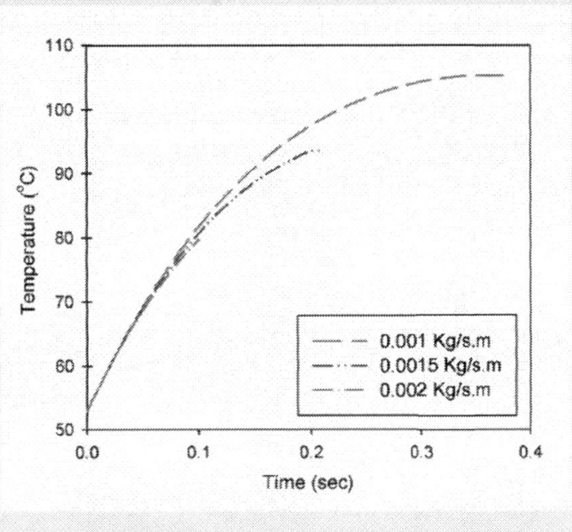

Temperature change versus time for various injection rate

24

Results (Effect of injection rate)

MISSISSIPPI STATE UNIVERSITY

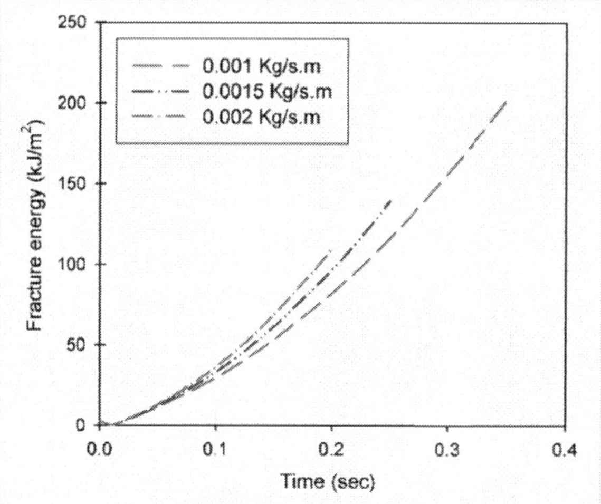

Predicted fracture energy (G) versus time for various injection rate

25

Conclusion

MISSISSIPPI STATE UNIVERSITY

- Once fault slip initiates, frictional heat due to slip is continuously generated which increases pore pressure and decreases effective normal stress.
- The comparative study of two models, with and without TP effect, indicated that in all cases ignoring the TP role can lead to a significant underestimation of pore pressure built up during CO_2 injection.
- Among the different parameters that were examined, slip velocity had the highest impact and the highest pore pressure as well as temperature were developed in the fault with largest slip velocity.
- Weakening rate and pore pressure increases as shear zone thickness (w) increases.
- As hydraulic diffusivity decreases, reduction in the effective normal stress and effect of TP becomes larger..

26

Conclusion (cont'd)

MISSISSIPPI STATE UNIVERSITY

- Regarding slip velocity, a reduction of 4 MPa pore pressure was observed by altering the velocity rate from 1 to 0.1 m/s as a result of higher frictional heating. For the velocity of 1 m/s, temperature grows three time higher than its initial value.
- Rate of pore pressure change becomes higher for higher injection rates in the model without TP effect but pore pressure evolution decreases for higher injection rate in the model considering TP.

27

Thank you for your Attention!

28

CMTC-439489-MS

Quantifying Local Capillary Trapping Storage Capacity Using Geologic Criteria

Bo Ren, The University of Texas at Austin, Steven L. Bryant, University of Calgary and Larry W. Lake, The University of Texas at Austin

This paper was prepared for presentation at the Carbon Management Technology Conference held in Sugarland, Texas, USA, 17–19 November 2015.

Abstract

The capillary pressure heterogeneity or *local capillary trapping* (LCT) determines the final distribution of CO_2 in a saline aquifer during geological carbon sequestration. This locally trapped CO_2 would not escape from the storage formation even if caprock integrity is compromised. It is, therefore, essential to predict the extent and storage capacity of LCT during the design of GCS projects.

This work employs a fast method based on the geologic criteria to estimate the structures of local capillary traps. The method assumes a critical capillary entry pressure (CCEP) and a geostatistical realization of the reservoir entry pressure field as inputs. It then finds the capillary barriers inside the domain and identifies the grid blocks beneath clusters of barriers. These grid blocks are the local capillary traps. The criterion for choosing the CCEP is important, and we suggest a criterion in this work.

We verify this algorithm by full-physics simulations in small 2D and 3D domains. We employ a large CO_2 injection rate (N_{gr}~0.1) to fully sweep the storage domain, followed by a long period of buoyant flow to allow for complete charging of those local capillary traps. We test several CCEPs to determine the most physically representative value by comparing the LCT predicted from both methods. We find that a single value of CCEP enables the geologic algorithm to give a very good approximation of LCT distribution as well as LCT volume in uncorrelated and weakly correlated porous media. This means that the concept of CCEP is a reasonable approximation to the physical process by which traps are filled.

LCT can be described in terms of percolation theory. The percolation threshold arises from the competition between connected clusters of barriers and connected clusters of local traps. We show that both the percolated CCEP (corresponding to the percolation of LCT) and optimal CCEP (corresponding to the best match between geologic criteria and full-physics simulation) change with each other in a predictable linear way for the uncorrelated capillary entry pressure field.

Introduction

Geological CO_2 sequestration (GCS) in the saline aquifer has been widely accepted as the most promising and easily accessible way to reduce the carbon emissions and global warming (Bachu, 2008). One of the main concerns is whether the CO_2 can be permanently stored in a safe way. In order to achieve this, it is essential to understand the CO_2 migration behavior, storage state and leakage characteristics.

After CO_2 injection, four widely accepted storage mechanism are responsible for immobilizing CO_2 in the storage formation. They are structural, solubility, residual, and mineral trapping. Quantitatively, the relative importance of these different trapping mechanisms over time depends on several factors, such as geometry of formations, permeability and its distribution, injection rate, and wettability (Doughty, 2007; Juanes *et al.*, 2006; Kumar *et al.*, 2005; Mo and Akervoll, 2005, Saadatpoor, 2009, Taku Ide *et al.*, 2007).

Recently, a new trapping mechanism called local capillary trapping was proposed in reservoirs exhibiting capillary pressure heterogeneity (Saadatpoor *et al.*, 2009). Specifically, it occurs during buoyancy-driven migration of bulk phase CO_2 within a saline aquifer with spatially varying petrophysical properties (permeability and capillary entry pressure). The most important benefit of LCT is that saturation of the stored CO_2 in the LCT is larger than that predicted for other trapping mechanisms, and CO_2 in these regions of locally large saturation will not escape from the formation even if caprock integrity or other geomechanical disruptions occur in the long-term sequestration period. Ren (Ren *et al.*, 2014) demonstrated that the LCT can be as significant as residual trapping in immobilizing CO_2 into the storage domain. Taking no account of the LCT would consequently overestimate the CO_2 leakage flux, amount, and risks. Additionally, these local accumulations of CO_2 patches would significantly impact the seismic response and imaging during the monitoring period (Behzadi *et al.*, 2011, Meckel *et al.*, 2015). Therefore, it is essential to predict and quantify the amount, extent and distribution of LCT during GCS in the saline aquifer.

A number of simulation studies have been conducted to investigate the impact of capillary heterogeneity (Behzadi and Alvarado, 2012; Kong *et al.*, 2014; Li and Benson, 2015; Mouche *et al.*, 2010; Saadatpoor, 2012) on the CO_2 upward migration, saturation distribution, and trapping state. The main issue of these simulations is the long run time because of the incorporation of this small-scale capillary heterogeneity.

Considering this, Saadatpoor (Saadatpoor, 2012) developed a fast algorithm to predict the LCT distribution and capacity based on the geologic model, specifically on the entry capillary pressure field which is correlated with permeability field. A critical capillary entry pressure (CCEP) is assumed that will determine the grid blocks to be barriers or flow path and then find the flow path grids surrounded by barriers. These flow path grid blocks are the LCT. The algorithm is tested by simulations of buoyant flow and it is found the algorithm always predicts some "false" LCTs not observed in the full-physics simulation because CO_2 does not sweep these grids during buoyancy-driven flow. The other issue with the algorithm is the identification of the CCEP *a priori*.

The objective of this work is to test the applicability of the geologic algorithm when the CO_2 is emplaced in the reservoir by injection and how to determine the value of CCEP. During the injection period of CO_2, viscous forces can enable the CO_2 to invade much of the pore space near the wellbore regardless of capillary heterogeneity. Thus the initial state of the buoyancy-dominated flow period differs significantly from some previous studies, and the occupancy and saturation of LCTs attained during post-injection buoyancy driven flow could therefore differ as well.

Description of Geologic Criteria Algorithm

The following subroutines can be applied to both 2D and 3D capillary entry pressure fields.

1. Given a value of critical capillary entry pressure, find all cells in the domain that have entry pressure exceeding the critical value of CCEP (=2 psi). Fig. 1a shows a sample 2D capillary entry

pressure field and Fig. 1b shows barrier grid blocks in the orange that have capillary entry pressure larger than CCEP.

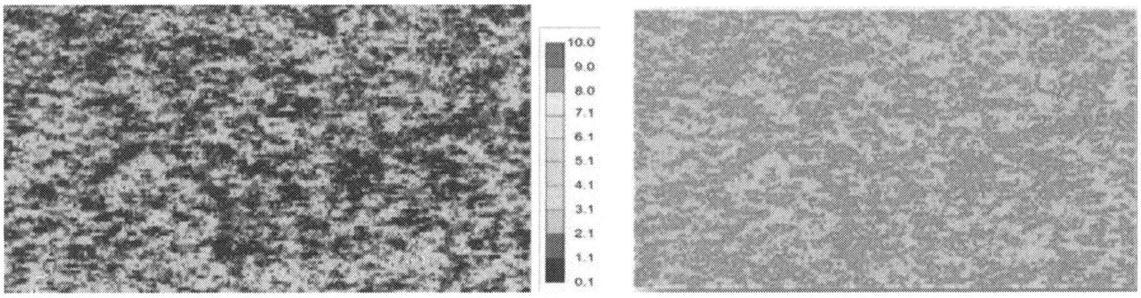

(a) (b)

Fig. 1. (a) Heterogeneous capillary entry pressure (2D sample); (b) Orange cells have capillary entry pressure larger than CCEP and are considered as barriers.

2. All the connected clusters of barriers in the set of cells from step 1.

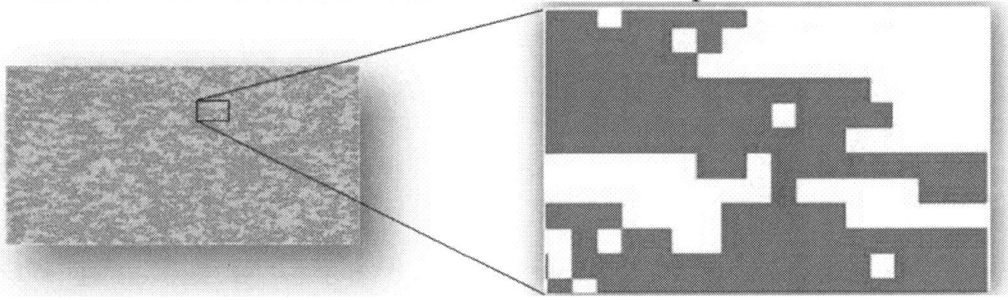

Fig. 2. The connected cluster of barriers shown in the red blocks.

3. Find the non-barrier clusters that are surrounded by the set of clusters of barriers from step 2.

Fig. 3. The local capillary traps are shown in the green blocks.

2D/3D Geologic Model Construction

We built 2D and 3D statistical geologic models. The model dimensions and properties are shown in Table 1. The porosity field is correlated with permeability by the Eq. 1 (Holtz, 2002). The capillary entry pressure field is generated by the Leverett J-function (Leverett, 1941) which correlates the permeability, porosity, and entry capillary pressure together. The Leverett J-function is shown in Eq. 2.

$$\phi = \left(\frac{k}{7E+7} \right)^{1/9.61} \qquad \text{(1)}$$

$$J(S_w) = \frac{p_c(S_w)}{\sigma_{g/w} \cos\theta} \sqrt{\frac{k}{\phi}} \qquad \text{(2)}$$

In the above equation, p_c is the capillary pressure, σ is the interfacial tension between CO_2 and brine, θ is the contact angle, k is the permeability, ϕ is the porosity. To scale the capillary pressure curve for each grid block in the aquifer, we assume that the interfacial tension and contact angle do not vary spatially. We also assume that the shape of capillary pressure curve does not vary spatially. Thus for arbitrary grid block i and j, we can transform Eq. 2 into Eq. 3.

$$p_{ci}(S_w) = p_{cj}(S_w)\sqrt{\frac{k_j}{k_i} * \frac{\phi_i}{\phi_j}} \quad \text{..(3)}$$

Subscript j denotes the properties of the rock for which we know the capillary pressure curve, and we call it reference rock. ϕ_j and k_j are the reference permeability and porosity, respectively. We assign the reference capillary pressure curve to the grids block with the average permeability, and capillary pressure curves for other grids can be calculated by Eq. 3. The generated 2D capillary entry pressure field and its histogram are shown in Fig. 4a and 4b, respectively. The corresponding 3D fields are shown in Fig. 5a and 5b.

Table 1. 2D and 3D Geologic Model Parameters

Model parameters	2D	3D
Model dimension (Nx, Ny, Nz), ft	400×100	$64 \times 32 \times 32$
Grid block size, ft	1×1	$1 \times 1 \times 1$
Permeability field	lognormal	lognormal
Auto correlation length (Lx, Ly, Lz), ft	(5,~, 0)	(5, 5, 0)
Mean of permeability, mD	194	403
Anisotropy of permeability field,	isotropic	isotropic
Standard deviation of permeability, mD	339	774
Porosity, constant	0.269	0.269
Mean of capillary entry pressure (μ), psi	2.07	2.13
Standard deviation of capillary entry pressure (σ), psi	1.36	1.43

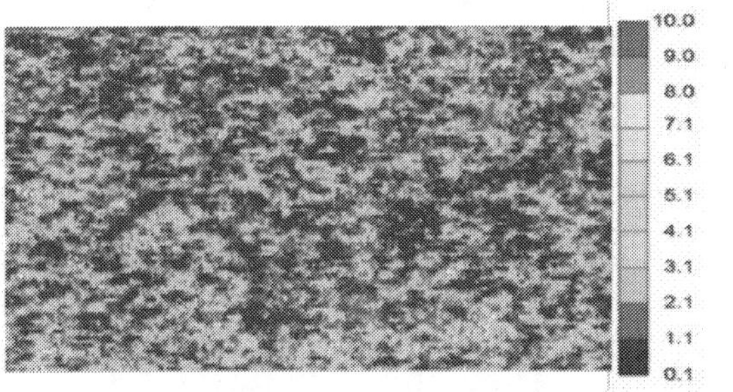

(a) 2D capillary entry pressure field (psi)

(a) 3D capillary entry pressure field (psi)

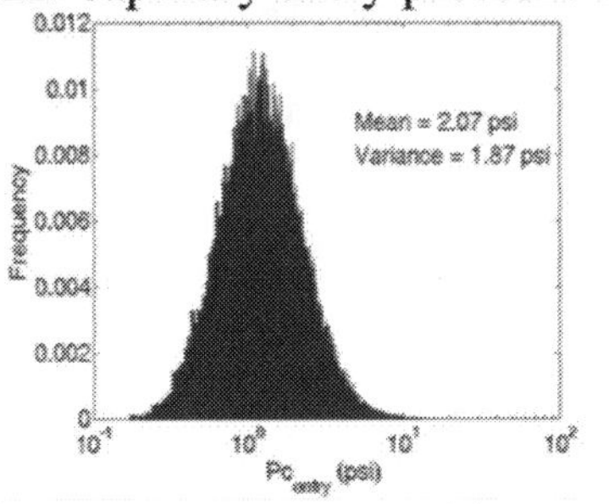

(b) Capillary entry pressure histogram in 2D

Fig. 4. 2D capillary entry pressure and histogram

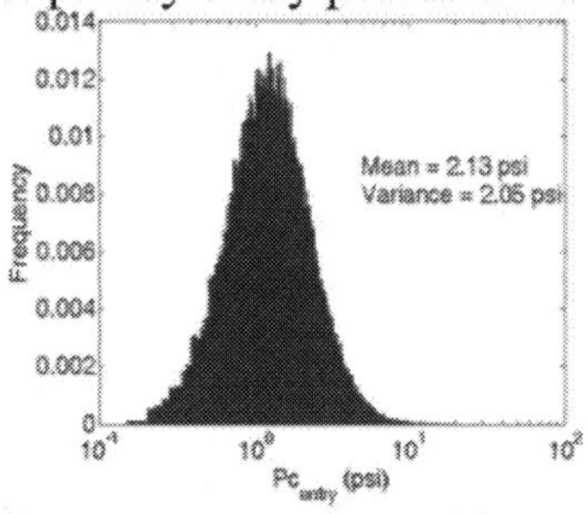

(b) Capillary entry pressure histogram in 3D

Fig. 5. 3D capillary entry pressure and histogram

Simulation Settings in the CMG-GEM

We employ the simulator of CMG-GEM (CMG, 2012) to conduct the CO_2 flow simulation in the above 2D and 3D geologic models. The fluid properties of CO_2 and brine are the same as before (Kumar, 2005). The relative permeability curve and reference capillary pressure curve are shown Fig. 6 and Fig.7, respectively. Hysteresis in the relative permeability is considered but not in capillary pressure.

The capillary pressure curve for each grid was scaled by the Leverett J-function and the reference capillary pressure curve as described above.

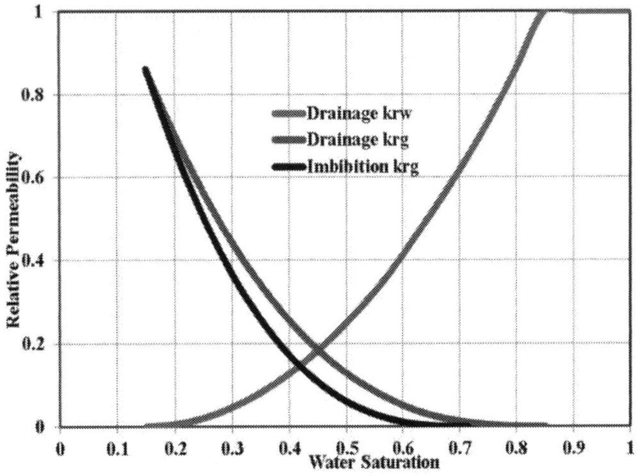

Fig. 6. Relative permeability curves

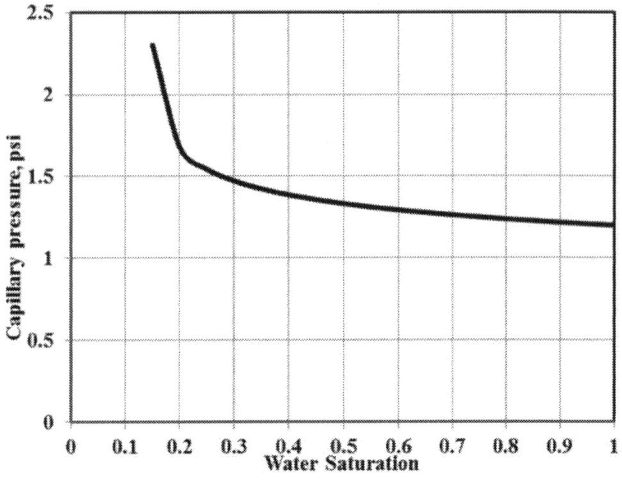

Fig. 7. Reference capillary pressure curve

The simulation injection scheme is shown in Table 2. In the CMG-GEM, we employ a very large VOLMOD in the right boundary grid blocks to mimic the open flow boundary. We select the injected volume to be enough to fill the storage domain, but it accounts for little (4.2E-5) of the total pore volume. The injection period is short considering the small storage domain, and after injection, we continue to run the simulation to allow for the buoyant flow, during which LCT would be filled, to reach a steady state. The period during injection and post-injection lasts for 50 years. The Darcy velocity along the wellbore for 2D is set to be same as that of 3D. The buoyancy number along the wellbore perforation for 2D and 3D is calculated based on the following equation and shown in Table 2.

$$N_{gr} = \frac{\Delta \rho \, g k_v L}{u \mu H} \quad \text{.......(4)}$$

In In the above equation, N_{gr} is the buoyancy number, $\Delta\rho$ is density difference between brine and CO_2 (405.9 kg/m^3), g is buoyancy constant, k_v is the average vertical permeability (see Table 1), u is the Darcy velocity of CO_2 along the wellbore, $u = q/(2\pi r_w H)$, q is the injection rate, r_w is the wellbore radius. μ is the viscosity of CO_2 (0.049 mPa·s), L is the reservoir length, H is the perforated length.

Table 2. Injection simulation scheme in the CMG-GEM

Parameters	2D	3D
Well type	Vertical	Vertical
Perforation interval	Left lower quarter	Left middle lower half
Perforation length, ft	25	16
Injection rate, Scf/d	4.9E+5	3.1E+5
Injection period, day	73	70
Simulation period, yr	50	50
Buoyancy number along wellbore	0.1	0.2

Results from Geologic Criteria Algorithm

Since the local capillary trap is surrounded by the grid blocks with the locally larger capillary entry pressure, we expect that the best CCEP will be very close to the mean of the capillary entry pressure field. Based on the histogram of capillary entry pressure as shown in the Fig. 4b and Fig. 5b, we select several values of CCEP around the mean of capillary entry pressure. They are from 0.80 through 2.0 psi with the interval of 0.05 psi.

Fig. 8a shows the variation of LCT volume fraction under different values of CCEP in the 3D. Here, the LCT volume fraction is defined as the number of LCT grid blocks divided by the total number of grid blocks of the storage domain. The CCEP is normalized by the mean and standard deviation of the capillary entry pressure. We can see from Fig. 8a that LCT volume fraction first increases with the CCEP followed by a decrease. The increasing part of the trend is simply because as the CCEP increases, more grid blocks become flow paths instead of barriers, and since most of the domain is a barrier at small CCEP, the additional flow paths are also additional LCT. However, increasing CCEP also decreases the number of barrier grid blocks, which is necessary to surround the LCT grids, and eventually the number of possible LCT must decrease. Therefore, there exists a threshold value of CCEP which gives rise to the maximum LCT volume fraction. In this example, the threshold CCEP is 1.05 psi and the corresponding LCT volume fraction is 16%.

Fig. 8b shows the LCT distribution and different sizes of LCT clusters under the threshold CCEP of 1.05 psi. Total 753 clusters of LCT is detected and the largest LCT cluster (ID=2) has 3731 grid blocks (5.69% of the whole domain). In the future publication, we will show the LCT cluster size distribution and fractal behavior analysis in detail. We label the LCT cluster ID from the bottom middle grid block, and it can be observed that even for the largest LCT cluster (ID=2), it cannot span the whole domain from the bottom to the top. This phenomenon is different from the traditional invasion percolation theory (Wilkinson and Willemsen, 1983). However, to be simple, we still call the CCEP at which gives the largest LCT fraction as the ***'percolated CCEP'***. In the following section of comparison of LCT between geologic criteria and CMG-GEM, we will show LCT for a range of closely spaced CCEP values and try to identify the optimal CCEP that could give the best match with the CMG-GEM.

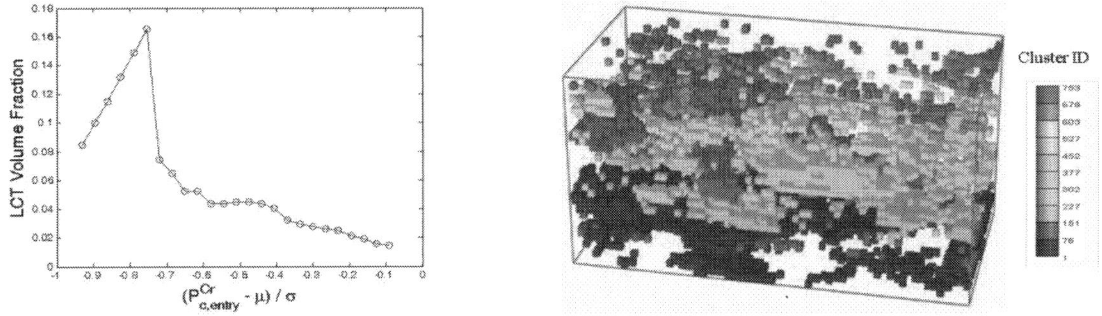

(a) (b)

Fig. 8. (a) LCT volume fraction versus CCEP in the 3D, μ=2.13 psi, σ=1.43 psi (refer to Table 1); (b) LCT clusters in the domain under the percolated CCEP 1.05 psi; colors are arbitrary.

Results from CMG-GEM

The CO_2 saturation at the end of the injection in 2D storage domain is shown in Fig. 9a. It can be seen that CO_2 is in the compact displacement under the high injection rate and CO_2 displaces brine almost everywhere, though not to the same saturation. Clearly all the local capillary traps that are intrinsic to the geologic model are occupied by CO2 at the end of injection. We then continued the simulation to allow buoyant flow. The CO_2 saturation at the end of buoyant flow is shown in the Fig. 9b, the yellow and red color grid blocks with high CO_2 saturation are the LCT. Fig. 9c and 9d show the histogram of the CO_2 saturation at the end of both injection and buoyant flow, respectively. The CO_2 saturation at the end of injection mostly follows the normal distribution with the mean value of 0.63. This is consistent with others' observation from coreflooding experiments (Kong *et al.*, 2014; Krause, 2012). However, after buoyant flow, two peaks of saturation frequency are observed, one is around the residual gas saturation (0.287), and the other peak is around 0.8 (1-S_{wirr}), which is the local capillary trapping. The corresponding results for the 3D are shown in Fig. 10. It should be noted that LCT frequency (CO_2 saturation larger than 0.75 is assumed as LCT) in 3D is half of that for 2D.

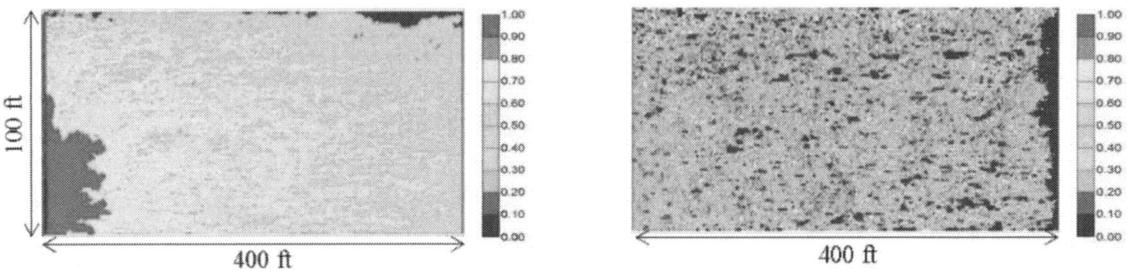

(a). CO_2 saturation at the end of injection (b). CO_2 saturation at the end of buoyant flow

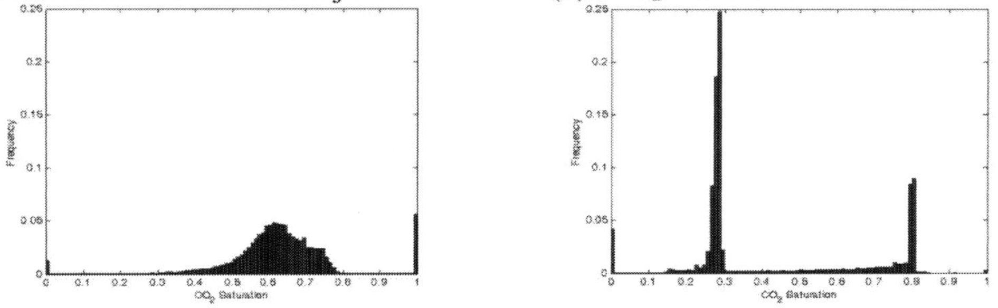

(c). CO_2 saturation histogram of (a) (d). CO_2 saturation histogram of (b)

Fig. 9. CO_2 saturation distribution and histogram in 2D model.

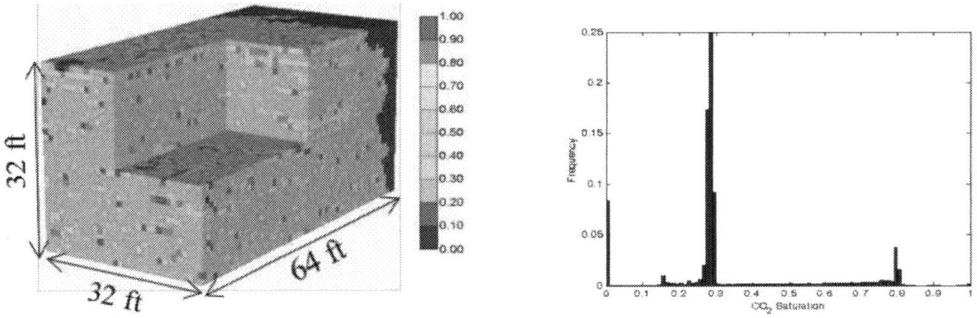

(a). CO_2 saturation distribution at the end of buoyant flow (d). CO_2 saturation histogram of (a)

Fig. 10. CO_2 saturation distribution and histogram in the 3D model.

Comparison of LCT between Geologic Criteria and CMG-GEM

After we obtain the LCT saturation distribution from CMG-GEM, we extract the LCT from the saturations map at the end of buoyant flow (see Fig. 9b and Fig. 10a) and compare them with the LCT from the geologic algorithm. Grid block with CO_2 saturation higher than 0.75 is assumed as LCT.

We define a match index (Eq. 5) to search for the ***optimal CCEP***. It is defined as the ratio of the grids predicted from both methods over the sum of the number of grids predicted only from CMG-GEM and that predicted only from geologic criteria. Therefore, the match index ranges from 0 to 1. The worst match has the match index that approaches to 0 while the best match approaches to 1. The CCEP corresponding to the peaked match index would be the optimal CCEP.

The changes of LCT volume fraction and match index with CCEP in the 2D simulation are shown in left plot of Fig. 11. It can be observed the CCEP corresponding to the peak value of LCT volume fraction is equal to 1.8 psi, while the optimal CCEP that gives the best match is 1.15 psi. We also show the comparison under the optimal CCEP in the left plot of Fig. 12. In the figure, if a grid block is predicted as LCT from both methods, we set it to be green. If the grid block is predicted as LCT only from CME-GEM and geologic criteria, we set them to be red and blue, respectively. The LCT volume fraction predicted from the geologic criteria under the optimal CCEP is 25% while it is 20% as predicted from CMG-GEM.

Similarly, we conduct a 3D analysis on the optimal CCEP and the result is shown in the right plot of

Fig. 11. The optimal CCEP is found to be the same as the percolated CCEP. The corresponding comparison of LCT between the two methods in several selected layers of the reservoir is shown in Fig. 12. It can be observed that LCT from the geologic criteria could fully and exactly represent the LCT from the CMG-GEM under the optimal CCEP. However, geologic criteria overestimates LCT as we have some "false" blue grid blocks only predicted from the geologic criteria. The LCT volume fraction from the geologic criteria under the optimal CCEP is 16% while it is 7% as predicted from CMG-GEM.

$$M.I = \frac{\text{LCT from both methods (\# green grid)}}{\text{LCT from both methods (\# green grid)} + \text{LCT from CMG−GEM only (\# red grid)} + \text{LCT only from GC only (\# blue grid)}} \quad \text{.......(5)}$$

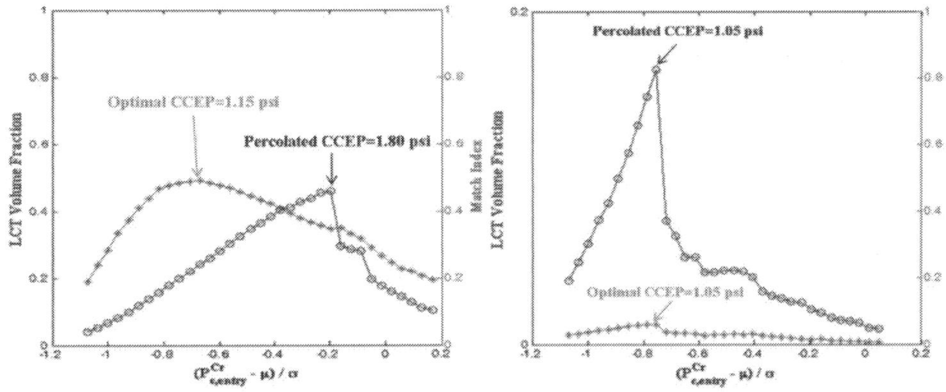

Fig. 11. LCT volume fraction and match index (Eq. 5) vary with CCEP in 2D (left) and 3D (right).

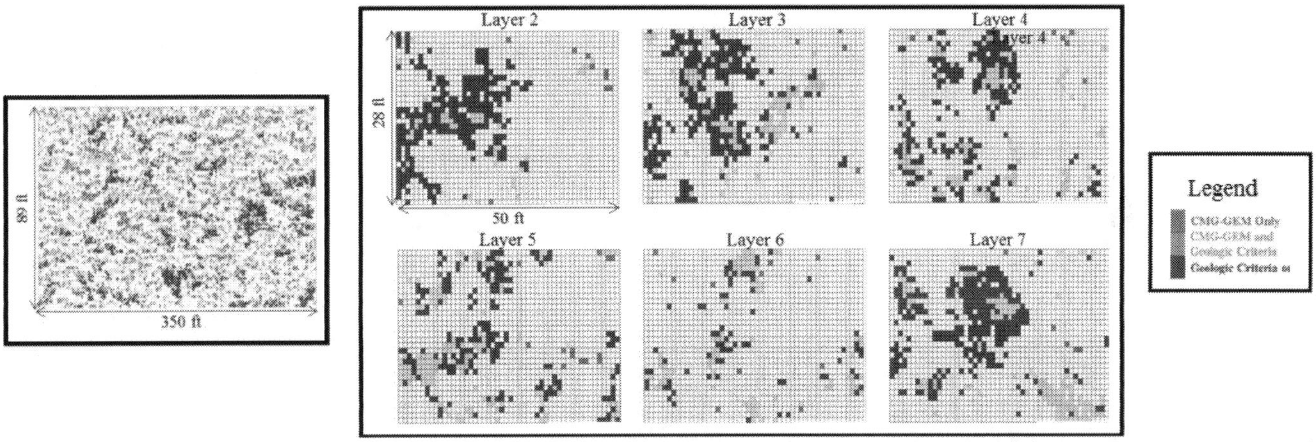

Fig. 12. Comparison of LCT between CMG-GEM and geologic criteria algorithm under the optimal CCEP. Left: 2D with the optimal CCEP=1.15 psi. For the comparison, we exclude the top 11 layer with high saturation CO_2 accumulation beneath the seal and the right 50 columns with CO_2 dissolved in brine (refer to Fig. 9b). Right: 3D in layer 2 through 7 from top to the bottom of the domain under the optimal CCEP =1.05 psi. We exclude the top four layers with high saturation CO_2 accumulation beneath the seal and the right fourteen columns with CO_2 dissolved in brine (refer to Fig. 10a).

The complicated relationship between optimal and percolated CCEP is suspected to be the interplay between LCT and the variation of the distribution and correlation length of capillary entry pressure fields. We make a further study into the effect of these parameters on the CCEP selection. Considering the final distribution of CO_2 is determined by the capillary pressure heterogeneity, we set the porosity and permeability to be constant in the following analysis. The porosity and permeability are 0.269 and 194 mD, respectively. Additionally, the mean of the capillary entry pressure (μ) keeps as constant of 2.07 psi as the change of the mean will not alter LCT distribution, but the standard deviation (σ) will be changed. The detailed parameter settings are shown in Table 3.

Table 3. Auto correlation lengths and standard deviations of capillary entry pressure fields

Model	2D
Dimensions (N_x, N_z)	(512, 128)
$(L_x,\ L_z)$, ft	(0, 0),
σ, psi	0.86, 1.36*, 1.86, 2.36
σ, psi	1.36*
$(L_x,\ L_z)$, ft	(0, 0), (5, 0), (20, 0), (50, 0)
σ, psi	1.36*
$(L_x,\ L_z)$, ft	(0, 0), (0, 5), (0, 20), (0, 50)

Note: * base case with the standard deviation same as that of 2D example study above.

Both the optimal and percolated CCEP are normalized by mean (μ) and standard deviation (σ) of the capillary entry pressure field. The normalized function is defined as $(\text{CCEP} - \mu)/\sigma$. For the capillary entry pressure field, the dimensionless horizontal and vertical correlation length is $\lambda_{Dx} = L_x/L$, and $\lambda_{Dz} = L_z/H$, respectively. The coefficient of variation is $C_v = \sigma/\mu$. The left plot of Fig. 13 shows the change of peaked match index (M.I.) with λ_{Dx}, λ_{Dz}, and C_v. For the uncorrelated capillary entry pressure fields, the peaked match index keeps well around 55% irrespective of the coefficient of variation. However, it decreases as the vertical and horizontal correlation length increase. This indicates that a single value of CCEP is not enough to characterize the LCT structures with strong correlation. In other words, the advantage gained by the geologic criteria described here is more significant for the random and weakly correlated porous media than for the strongly correlated.

Considering about this, we focus on the uncorrelated capillary pressure field with the change of coefficient of variation. We try to seek the relationship between the normalized percolated and optimal CCEP, and they are shown in the right plot of Fig. 13. Different solid square points represent different coefficients of variation with the constant mean of the capillary entry pressure field. It can be observed that the relationship between the two complexes is almost linear, which provides a potential way of determining the optimal CCEP *a priori*. It can be expected that varying the mean of the capillary entry pressure field will move this curve up and down proportionally, but the slope will be always kept the same. This is because the variation of the mean of capillary entry pressure field will not alter the LCT distribution. Therefore, for the random capillary entry pressure field, this linear curve will directly offer the right choice of the physical representative CCEP as long as the mean and standard deviation of the field are known.

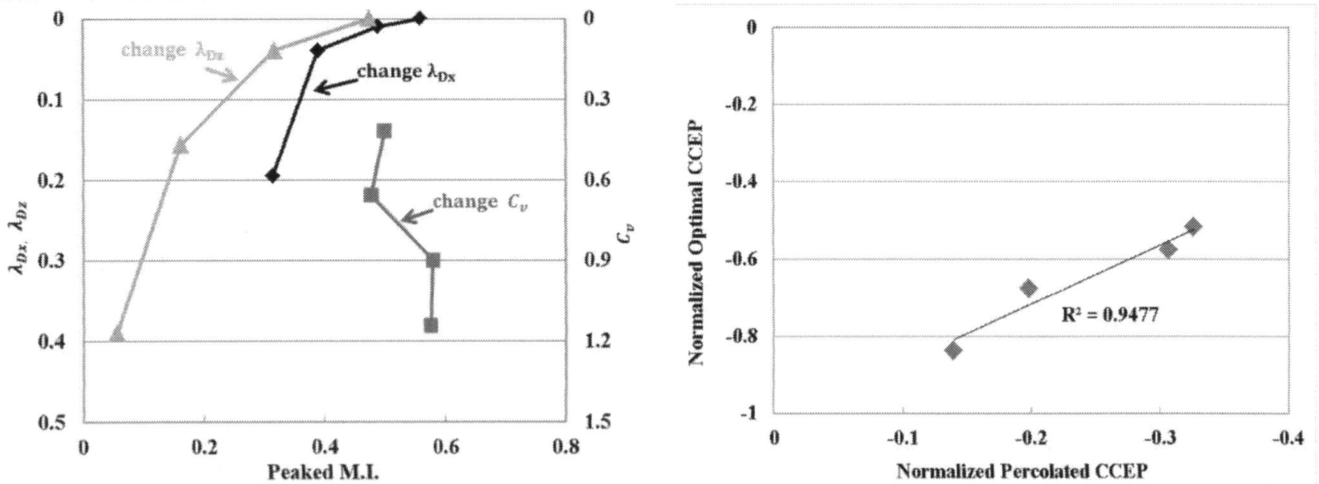

Fig. 13. Left: change of the peaked match index with the heterogeneity of the capillary entry pressure field (coefficient of variation, auto vertical, and horizontal correlation length); Right: the cross plot between the normalized percolated and optimal CCEP for the different coefficients of variation in the uncorrelated capillary entry pressure field.

Summary and Conclusion

The final distribution of CO_2 during the long-term buoyant flow of GCS is influenced by the capillary heterogeneity, and the security of storage is influenced by the local capillary trapping (LCT). In this work, we show the use of *geologic criteria* to identify the distribution of LCT based only on the capillary entry pressure field. The algorithm assumes a critical capillary entry pressure and searches for the clusters of LCT in the storage domain. By comparing with the full-physics simulation, we demonstrate that the geologic criteria algorithm is a feasible way of predicting LCT in the random and weakly correlated porous media. The uncertainty that lies in the method is the selection of CCEP. We define an optimal function that can be used to find the optimal CCEP that gives the best match between the geologic criteria and full-physics simulation by CMG-GEM. The optimal CCEP shows linear behavior with the percolated CCEP in the uncorrelated capillary entry pressure field, which provides a potential way of identifying LCT *a priori*.

Nomenclature

Roman Symbols

C_v	Coefficient of variation
g	Gravity acceleration, 9.8m/s^2
H	Reservoir height, ft
$J(S_w)$	Leverett J-function, []
k	Permeability, mD
k_v	Average vertical permeability, Darcy
k_i	Permeability in grid i, Darcy
k_j	Permeability in grid j, Darcy
k_{rg}	Relative permeability of CO_2, []
k_{rw}	Relative permeability of water, []
L	Reservoir length, ft
L_x	Correlation length in x-direction, ft
L_y	Correlation length in y-direction, ft
L_z	Correlation length in z-direction, ft
$M.I.$	Match index, []
N_{gr}	Buoyancy number, []
N_x	Number of grid blocks in x-direction,
N_y	Number of grid blocks in y-direction,
N_z	Number of grid blocks in z-direction,
$p_c(S_w)$	Capillary pressure, psi
$p_{ci}(S_w)$	Capillary pressure for grid i, psi
$p_{cj}(S_w)$	Capillary pressure for grid j, psi
$p^{Cr}_{c,entry}$	Critical capillary entry pressure (CCEP), psi
S_{grmax}	Maximum residual gas saturation, []
S_w	Water saturation, []
u_g	Darcy velocity of CO_2, ft/d

Greek Symbols

μ_g Viscosity of gas phase (CO_2), cp

μ Mean of capillary entry pressure distribution, psi

$\sigma_{g/w}$ Interfacial tension between CO_2 and water, N/m

σ Standard deviation of capillary entry pressure field, psi

θ Contact angle, degree

ϕ Porosity, []

ϕ_i, ϕ_j Porosity in grid i and j, respectively []

λ_{Dx}, λ_{Dz} Dimensionless horizontal and vertical correlation length, respectively, []

$\Delta\rho$ Density difference between brine water and CO_2, kg/m^3

Acronym

CCEP Critical capillary entry pressure

CMG Computer modeling group

GC Geologic Criteria

GCS Geological carbon sequestration

LCT Local capillary trapping

M.I. Match index

Scf Standard cubic feet

Acknowledgement

We would like to thank the sponsors of the Geological CO_2 Storage Industrial Affiliates Project at The University of Texas at Austin. Part of this work was supported by the Office of Fossil Energy, National Energy Technology Laboratory of the United States Department of Energy under DOE Award Number DE-FE0004956. Steven Bryant holds the Canada Excellence Research Chair in Materials Engineering for Unconventional Oil Resource at the University of Calgary. Larry W. Lake holds the W.A. (Monty) Moncrief Centennial Chair in the Department of Petroleum and Geosystems Engineering at The University of Texas at Austin.

Reference

Bachu, S. 2008. CO_2 Storage in Geological Media: Role, Means, Status and Barriers to Deployment. *Progress in Energy and Combustion Science* **34** (2): 254-273.

Behzadi, H., Alvarado, V., Mallick, S. 2011. CO_2 Saturation, Distribution and Seismic Response in Two-Dimensional Permeability Model. *Environmental Science & Technology* **45** (21): 9435-9441.

Behzadi, S. and Alvarado, V. 2012. New Upscaling Procedure for Capillary Dominant Buoyancy-Driven Displacement. Presented at the SPE Annual Technical Conference and Exhibition. San Antonio, Texas, USA, 8-10 October. SPE-158603-MS.

CMG. 2012. GEM technical manual: Calgary, Alberta, Canada.

Doughty, C. 2007. Modeling Geologic Storage of Carbon Dioxide: Comparison of Non-hysteretic and Hysteretic Characteristic Curves. *Energy Conversion and Management* **48** (6): 1768-1781.

Holtz, M.H. 2002. Residual Gas Saturation to Aquifer Influx: A Calculation Method for 3-D Computer Reservoir Model Construction. Presented at the SPE Gas Technology Symposium. Calgary, Alberta, Canada, 30 April-2 May. SPE-75502-MS.

Kong, X., Delshad, M. and Wheeler, M.F. 2014. History Matching Heterogeneous Coreflood of CO_2/Brine by Use of Compositional Reservoir Simulator and Geostatistical Approach. *SPEJ* **20**(2): 267-276.

Krause, M.H. 2012. Modeling and Investigation of the Influence of Capillary Heterogeneity on Relative Permeability. Presented at SPE Annual Technical Conference and Exhibition, San Antonio, Texas, USA, 8-10 October. SPE-160909-STU.

Kumar, A., Noh, M.H., Ozah, R.C., Pope, G.A., Bryant, S.L., Sepehrnoori, K. and Lake, L.W. 2005. Reservoir Simulation of CO_2 Storage in Deep Saline Aquifers. *SPE Journal* **10** (3): 336-348.

Leverett, M. C. 1941. Capillary Behavior in Porous Solids. *AIME Petroleum Transactions* **142**: 152-169.

Li, B.X. and Benson, S.M. 2015. Influence of Small-scale Heterogeneity on Upward CO_2 Plume Migration in Storage Aquifers. *Advances in Water Resources* **83**: 389-404.

Meckel, T.A., Bryant, S.L. and Ravi, G.P. 2015. Characterization and Prediction of CO_2 Saturation Resulting from Modeling Buoyant Fluid Migration in 2D Heterogeneous Geologic Fabrics. *International Journal of Greenhouse Gas Control* **34**: 85-96.

Mo, S. and Akervoll, I. 2005. Modeling Long-Term CO_2 Storage in Aquifer With a Black-Oil Reservoir Simulator. Presented at the SPE/EPA/DOE Exploration and Production Environmental Conference. Galveston, Texas, USA, 7-9 March. SPE-93951-MS.

Mouche, E., Hayek, M. and Mügler, C. 2010. Upscaling of CO_2 Vertical Migration through a Periodic Layered Porous Medium: The Capillary-free and Capillary-dominant cases. *Advances in Water Resources* **33**: 1164-1175.

Ren, B., Sun, Y.H. and Bryant, S.L. 2014. Maximizing Local Capillary Trapping During CO_2 Injection. *Energy Procedia* **63**: 5562-5576.

Saadatpoor, E. 2009. Effect of Capillary Heterogeneity on Buoyant Plumes: New Trapping Mechanism in Carbon Sequestration. MS thesis, The University of Texas at Austin, Austin, Texas.

Saadatpoor, E. 2012. Local Capillary Trapping in Geological Carbon Storage. Ph.D. dissertation, The University of Texas at Austin, Austin, Texas.

Spiteri, E.J. and Juanes, R. 2006. Impact of Relative Permeability Hysteresis on the Numerical Simulation of WAG Injection. *Journal of Petroleum Science and Engineering* **50** (2): 115-139.

Taku Ide, S., Jessen, K. and Orr, F.M. 2007. Storage of CO_2 in Saline Aquifers: Effects of Gravity, Viscous, and Capillary Forces on Amount and Timing of Trapping. *International Journal of Greenhouse Gas Control* **1** (4): 481-491.

Wilkinson, D. and Willemsen, J.F. (1983). Invasion Percolation: a New Form of Percolation Theory. *Journal of Physics A: Mathematical and General* **16** (14): 3365-3376.

CMTC-440049-MS

Investigating the Geochemical Alterations in an Aquifer Due to Long-Term Sequestration of CO_2 Using Time-Lapse Seismic Information

Sang H. Han, The University of Texas at Austin, Sanjay Srinivasan, The Pennsylvania State University

This paper was prepared for presentation at the Carbon Management Technology Conference held in Sugarland, Texas, USA, 17–19 November 2015.

Abstract

The effects of chemical interaction between injected CO_2, brine, and formation rocks are often ignored in sequestration studies because chemical reactions are assumed to be localized to carbonate rocks that make up only a small proportion of the potential reservoirs. It is conjectured in this work that long-term exposure of certain types of clays and cement material to CO_2-brine mixtures can induce chemical reactions and subsequent alteration of rock properties that can be subsequently detected in time-lapse seismic surveys. This is demonstrated using a case-study structured after the Cranfield field injection site. Geochemical alterations of the reservoir rock are quantified by performing reactive transport simulations and subsequently rock physics models are used to translate the altered petrophysical properties into seismic responses. The study quantifies the long-term geochemical effects of CO_2 injection on the seismic response and conversely, presents an approach to infer the reservoir regions contacted by the CO_2-saturated brine based on the observed seismic response and a unique model selection algorithm.

Time-lapse or passive seismic monitoring is an effective method for mapping the progress of the CO_2 plume through the subsurface. But, because of the lack of resolution of the seismic information, it is necessary to use the seismic information together with prior geologic knowledge about the surface in order to identify if there is any migration of CO_2 into regions that might be deemed sensitive e.g. overlying aquifers or faults. Because of uncertainties in the prior geologic description of the reservoir, the feasibility of implementing a model selection process is explored in this work. The model selection procedure utilizes the observed well data and reference seismic map to select a subset of models that represent the posterior geologic uncertainty. The flow simulation of CO_2 injection and forward seismic modeling were repeated on the posterior models, and the seismic responses were compared for the reaction and non-reaction cases.

The study showed that the effects of geochemical reactions and the resultant impact on petrophysical properties and fluid saturation distribution were visible in the seismic response. Major differences in seismic responses were detected in regions of the reservoir where significant amount of minerals were dissolved and precipitated. These regions were at the top of the reservoir due to the reactions caused by the buoyant CO_2 plume. The presence of carbonate facies, even in small proportion, plays an important role in geochemical reactions and their effect is manifested at the seismic scale. The results of this time-lapse study can provide new interpretation of events observed in time-lapse seismic data that might lead to a better assessment of leakage pathways and other risks.

Introduction

CO_2 storage in a deep formation requires structural and geochemical trapping mechanisms in order to ensure that the CO_2 remains sequestered in the subsurface for a long duration. Structural trapping mechanism involves physical trapping of the CO_2 plume due to buoyancy forces and the capillary barrier exerted by the reservoir seal (Lu et al., 2013; Mandalaparty et al., 2011). When CO_2 is injected, its low density relative to in situ formation fluids, typically brine, causes it to migrate upward due to buoyancy. Capillary forces due to interfacial effects between the CO_2 and the rock mineral prevent the CO_2 from entering the pore space of the formation seal until a high enough entry pressure is attained. Geochemical trapping involves chemical interactions between rocks, formation fluids, and injected CO_2 (Lu et al., 2012). When CO_2 is injected, it dissolves in brine and raises the acidity (or lowers the pH) of formation fluids, initiating geochemical alteration of reservoir rocks by mineral dissolution and precipitation. CO_2 is thereby stored in the formation in the form of new material formed as a result of precipitation.

Currently, there are various analytical models and numerical simulators like CMG-GEM and PHREEQC (Computer Modeling Group Ltd., 2013; Parkhurst and Appelo, 2013) that can account for the geologic heterogeneity in reservoirs and compute the physical behavior of the CO_2 plume. There are numerous papers examining the physics and chemistry of carbon sequestration. However, these works do not take the results pertaining to the alterations in properties of the reservoir all the way to its impact on seismic wave propagation and the consequent interpretation and modeling implications. The research presented in this paper aims to address this shortcoming.

Study Site

As a part of the Regional Carbon Sequestration Partnership Program supported by the U.S. Department of Energy and the National Energy Technology Laboratory, the Southeast Regional Carbon Sequestration Partnership (SECARB) is a $34 million project aimed at designing sequestration and monitoring techniques for long-term carbon storage using Cranfield Field in Mississippi as the test field site. The Gulf of Coast Carbon Center (GCCC) of Bureau of Economic Geology in collaboration with Denbury Resources, Sandia Technologies, Schlumberger Carbon Services, Lawrence Berkeley National Lab, and University of Mississippi are the partners on this project. The project began in 2003 and more than 3 million metric tons of CO_2 has been injected at Cranfield Field since July 2008 (Hovorka, 2013).

The reservoir model for this work is created based on the reservoir conditions of the Designed Area of Study (DAS) specified by the GCCC. Located 12 miles east of Natchez, MS, the 25m-thick amalgamated fluvial sandstone reservoir is located in the Upper Cretaceous Lower Tuscaloosa Formation at a depth that varies from 3178m to 3207m, with a dip angle of about 2° (Delshad et al., 2013; Hosseini et al., 2013; Kordi, 2013; Lu et al., 2011). One injection well (CFU 31F-1) and two observatory wells (CFU 31F-2 and CFU 31F-3) are drilled within a spacing of 100m (Ajo-Franklin et al., 2013). Hundreds of core samples have been extracted from two observatory wells and petrophysical properties such as porosity, permeability, and lithology have been determined (Lu et al., 2013). The observation wells have been used to monitor the migration of CO_2 and the changing water saturation.

Lu et al. (2012) experimentally examined the geochemical effects of carbonates (calcite and dolomite), which make up 1.5% of the reservoir in the DAS. From dissolution of carbonates, they estimated that 147 mg/L of Ca and 30 mg/L of Mg were released from the mineral samples. These effects were concluded to be "insignificant," in terms of mineral trapping. The goal in this work is to quantify the effects in petrophysical terms and then to assess whether these geochemical alterations could be assessed using seismic information.

Modeling Approach

Mineralogy interpreted from well logs were used within a geostatistical modeling scheme in order to characterize the reservoir. Because of uncertainties associated with this characterization, more models were generated and refined through a model selection process. CMG-GEM, a commercial compositional simulator, was used to simulate CO_2 injection into the reservoir, and the geochemical effects were observed for the subsequent 100 years. Subsequently, rock physics models were used to compute the seismic signatures at different time steps corresponding to the porosity and saturation changes. These computed responses were compared with the reference seismic map at the corresponding time steps in order to select the best models. This model selection work flow was first implemented to the synthetic reference models for validation. It was then applied to the real data set for a complete assessment of the carbon sequestration in Cranfield and the accompanying geochemical alterations in a long term. Figure 1 outlines this modeling scheme.

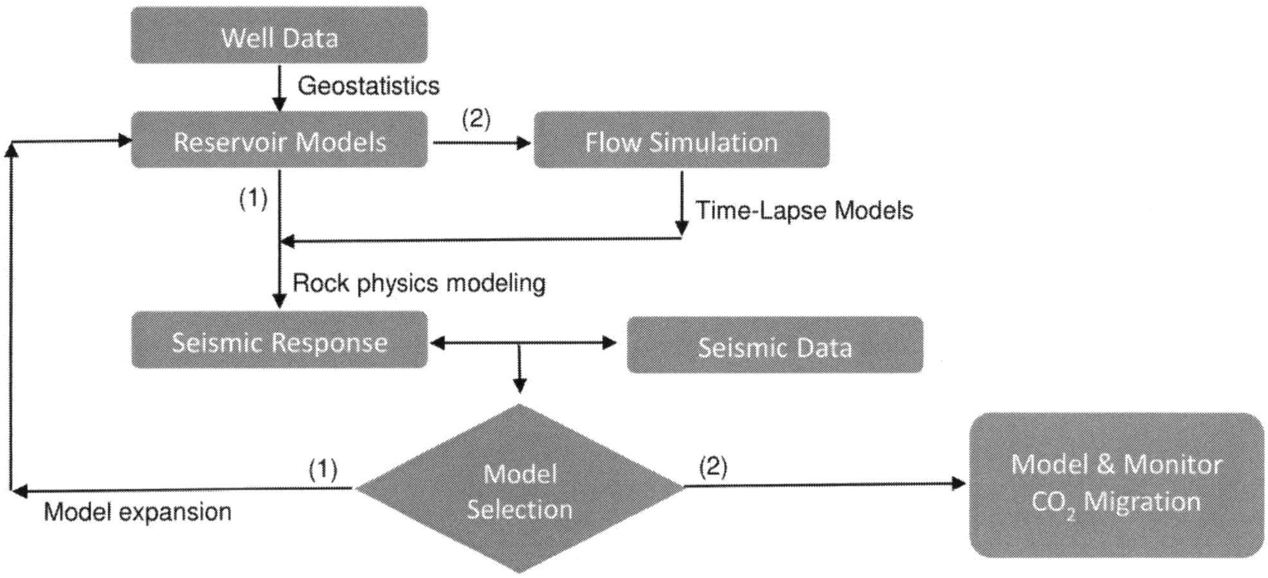

Figure 1: Proposed modeling scheme to characterize a reservoir through model selection and expansion

Modeling of Carbonate Facies (Synthetic Reference Model)

Reservoir Characterization

In the Designed Area of Study (DAS) there are three wells: 31F-1, 31F-2, and 31F-3. Mineralogy data based on quantitative log analysis is only available for the well 31F-2. The major reactive minerals in the reservoir are calcite and dolomite, that make up only 1.5% of the reservoir (Lu et al., 2012). It is essential to determine where in the reservoir these carbonates (calcite and dolomite) facies are present. To do so, the indicator coding of data is proposed. If carbonate facies is present at a location then an indicator code of "1" was specified and "0" if not.

In order to perform sequential indicator simulation using the indicator data, a two-dimensional reservoir grid covering 120-meters in width and 30-meter in height was created using SGeMS. The grid covers the DAS horizontally from 10m west of 31F-1 to 10m east of 31F-3 and vertically from a depth 3175m to 3205m. The horizontal extent of the area was divided into 600 blocks, and the vertical extent into 720 blocks in order to create a very fine grid. Figure 2 shows one of the reservoir models created using the sequential indicator simulation with the global proportion of carbonates specified to be 1.5%, which is based on the well log data observed along well 31F-2. Orange indicates regions where carbonate facies are present. The injection well (F31-1) is located at the position marked as 0m, and the two observation wells (F31-2 and F31-3) are located at 70m and 100m. This model was defined as "reference model."

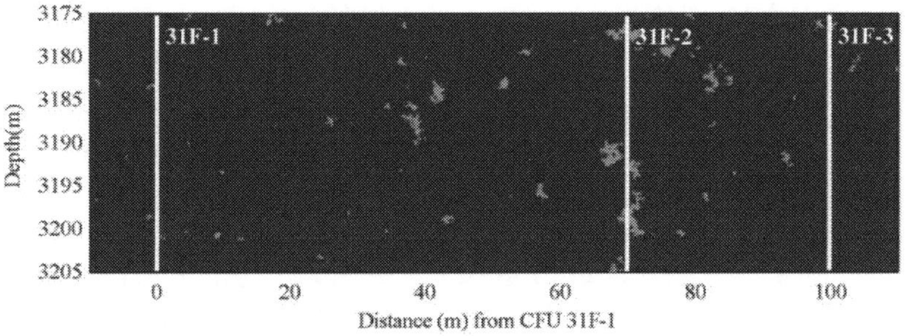

Figure 2: (a) Indicator map of carbonate facies in the DAS generated using the sequential indicator simulation using the carbonates data from well 31F-2. Orange indicates where carbonates are present. The carbonates make up 1.5% of the reservoir lithology. This model was defined as "reference model"

The fine scale model was coarsened by computing the average of the indicator data in order to estimate the percentage of carbonate facies at that location of the reservoir. Furthermore, a 2° dip of the reservoir was modeled. This process resulted in a 25-by-120 reservoir grid with 1m-by-1m blocks. Within each larger block, the mean of indicator data of carbonates was calculated in order to yield the approximate percentage of carbonates in the region. Figure 3 shows the reservoir with the carbonate facies proportion. This format of the model allow users to enter the data into CMG-GEM to perform flow simulations.

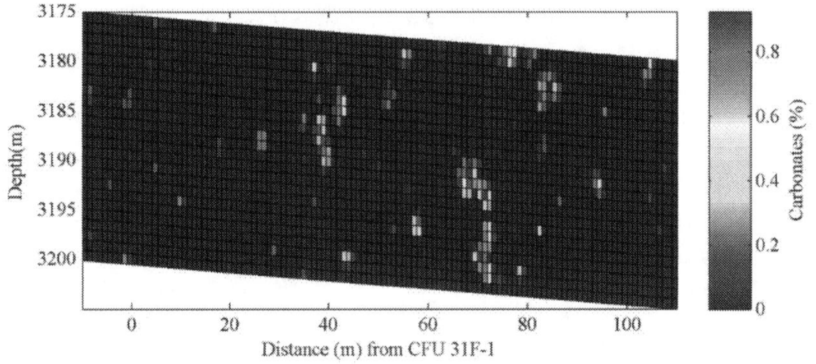

Figure 3: A reservoir model with the carbonates facies proportions. It was created based on the indicator map of carbonates created on a very fine grid using the sequential indicator simulation.

Flow Simulation

The reference rock facies data along with the reservoir conditions of the DAS, were entered into CMG-GEM in order to perform reactive-transport simulations to investigate the geochemical effects of CO_2 injection on reservoir properties. The conditions corresponding to an infinite reservoir were simulated because the reservoir extends far beyond the DAS. The initial pressure and temperature of the reservoir were 32MPa and 125°C, respectively (Delshad et al., 2013; Hosseini et al., 2012; Kordi, 2013). There are two carbonate facies: calcite and dolomite. According to the mineralogy along well 31F-2, the proportion of calcite is 1.1% while that of dolomite is 0.4% of dolomite (Lu et al., 2012).

Equations 1, 2, and 3 express the reactions involved in the flow and transport simulation. When CO_2 is injected, it dissolves in brine to release hydrogen and bicarbonate ions (1) that lowers the pH of the aquifer and initiates calcite (2) and dolomite (3) dissolution/precipitation.

$$CO_2 + H_2O \rightarrow H^+ + HCO_3^{2-} \quad \text{.......................................} (1)$$

$$CaCO_3 + H^+ \leftrightarrow Ca^{2+} + HCO_3^- \quad \text{.......................................} (2)$$

$$CaMg\left(CO_3\right)_2 + 2H^+ \leftrightarrow Ca^{2+} + Mg^{2+} + 2(HCO_3^-) \quad (3)$$

One pore volume of CO_2 was injected at the bottom interval of well 31F-1 over a two-year period. The behavior of the reservoir was observed for a period of 100 years after injection. There was a change in porosity due to calcite and dolomite dissolution and precipitation. Even after injection stopped, CO_2 continued dissolving in brine to cause reactions. Figures 4 shows the changes in porosity at the beginning of year 50, and 100, respectively. The changes in water saturation of the reservoir (Figure 5) reflects the buoyant nature of the CO_2 plume. As CO_2 continues to migrate upward, water saturation became dominant in the lower half of the reservoir. Because of the high water saturation in the surrounding saline aquifer, water tends to encroach in order to re-saturate the reservoir.

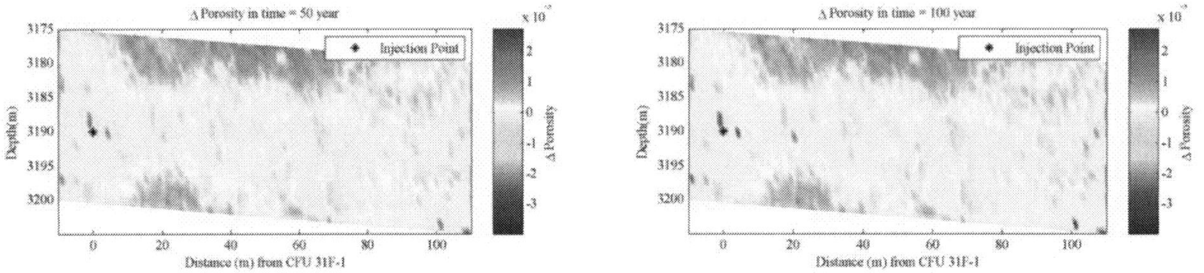

Figure 4: Porosity change in the reference model in year 50 (left) and in year 100 (right)

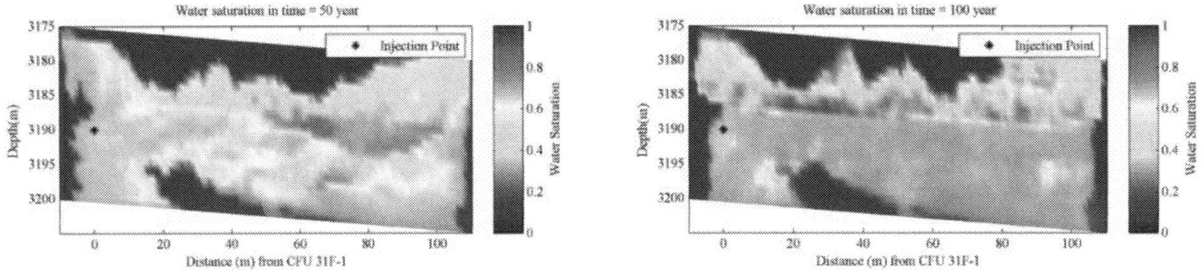

Figure 5: Water saturation in the reference model in year 50 (left) and in year 100 (right)

Rock Physics and Forward Seismic Modeling

From the flow simulation, the variation in porosity and fluid saturation with time was obtained. Subsequently, the contact cement model (Dvorkin and Nur, 1996) and Hertz-Mindlin theory (Mindlin 1949; Dvorkin and Nur, 1996) followed by Gassmann's theory (Gassmann, 1951) were used to derive the elastic moduli of the saturated reservoir rock at different time. Two different types of sorting trends are observed in the DAS (Carter and Spikes, 2013). Relative to well 31F-2, the region below 3195 m is a sandstone with a variation of poorly- and well-sorted formations, and above this region are sediments with concentric deposition of clay cement around the grains. According to this sorting trends, Carter and Spikes (2013) implemented the Hertz-Mindlin theory in the lower region and the contact cement model in the upper region. We used the same methods to characterize the elastic properties in these two regions.

In order to perform forward seismic modeling, the reflection coefficient was derived in travel time coordinates from the rock properties for each layer of the reservoir. It was then convolved with a zero-phase Ricker wavelet shown in Figure 6(a). Figure 6(b) shows a synthetic seismogram of the DAS before injection. The two distinct regions are observed. The lower region is where there is a variation of sorting trends. The upper regions is where clay-cemented quartz grains were observed. Figure 7 shows the synthetic seismogram of the DAS in years 50 and 100, respectively.

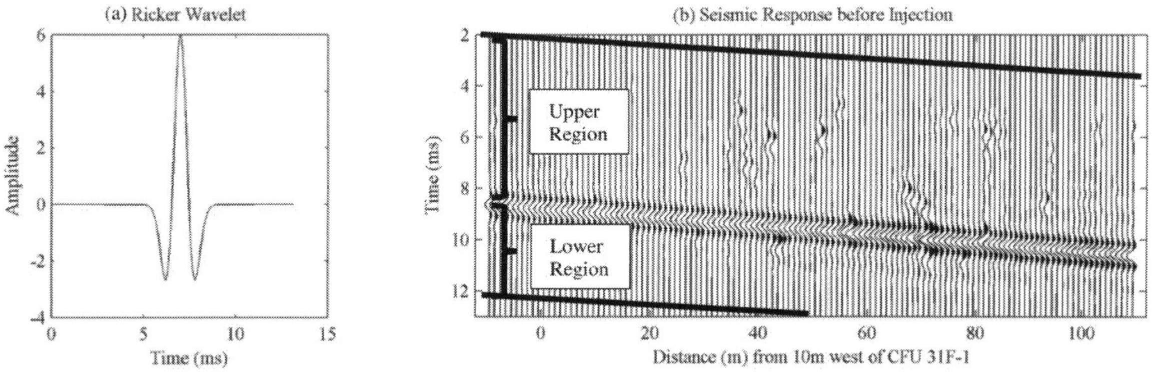

Figure 6: (a) Ricker wavelet used in forward seismic modeling. The reflection coefficient derived using rock physics models was convolved with the wavelet to compute (b) seismic response of the reference model in the DAS before injection.

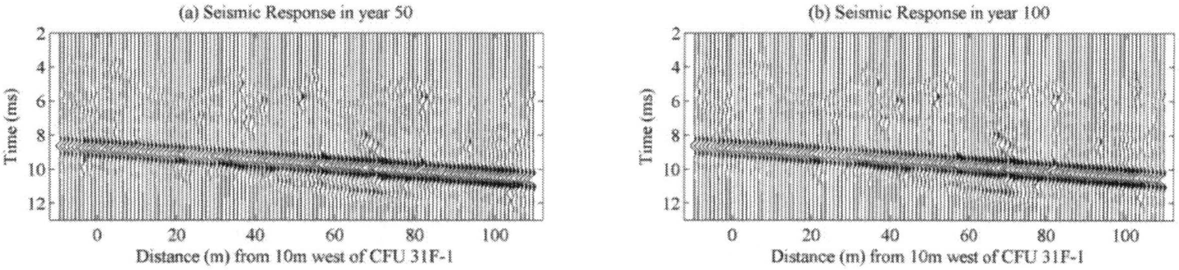

Figure 7: (a) Seismic response of the reference model in the DAS in year 50 and (b) in year 100.

Model Selection and Expansion

The reference model was created using sequential indicator simulation conditioned to very scarce mineralogy data. A limited set of conditioning data were sampled from the reference model and used to generate a suite of models reflecting the prior geologic uncertainty in the form of varying carbonate facies proportion. These models need to be refined through the model selection and expansion process in order to reduce the uncertainties, especially in the carbonate facies proportions. The mineralogy data extracted along 20 psuedo-wells within the reference model is shown in Figure 8. The rock physics models and forward models were applied to all the models before injection, and the subsequent seismic responses were compared to that of the reference model.

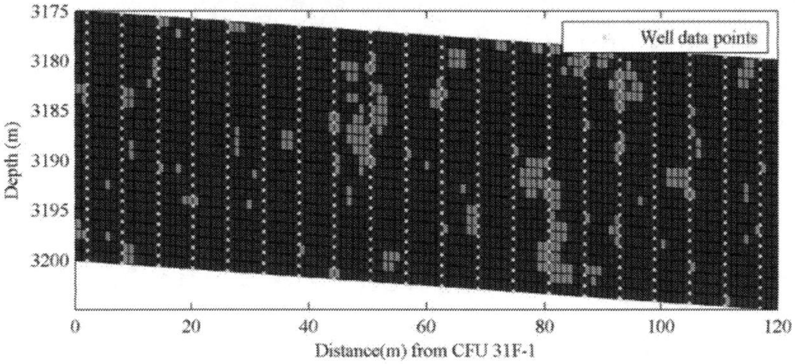

Figure 8: The mineralogy data of the reference model. The cyan points indicate the data points extracted to run the sequential indicator simulation to generate an initial suite of models.

The discrete Fréchet distance (Eiter and Mannila, 1994), which is a measure of dissimilarity between two profiles, was computed between the reference seismic response and that for the prior reservoir models prior to the start of injection. Subsequently, the models were clustered based on the distribution of the discrete Fréchet distances. The smaller the discrete Fréchet distance, the more the models match. The models in the first cluster were then resampled in order to obtain conditioning data that were used to generate an expanded posterior set of models. A common spatial correlation model reflecting the spatial characteristics of the selected set of models was used for this resampling process. Flow simulation was run on all the models in the new set of models, to compute the corresponding reservoir response after CO_2 injection. The seismic responses of the models in year 50 (48 years after CO_2 injection stopped) were compared with the seismic response of the reference model in year 50. The models in the new set were clustered based on the distribution of the discrete Fréchet distances in year 50. The models in the best group were again used to generate another set of 100 models. Repeating the process in the previous section, flow simulation was performed on all the models, and rock physics models and forward seismic models were used to compute the corresponding seismic responses in year 100. The seismic responses were compared with that of the reference model by computing the discrete Fréchet distances.

Comparing the carbonate facies proportions of the reference model and those of the selected set of models, it can be observed that the refined models were able to represent the areas of high carbonate facies proportions (Figure 9). If the model selection and expansion is repeated until the discrete Fréchet distance between the model and the reference model is minimum, the carbonate facies proportions of the model should approximately be equal to those of the reference model. It is not possible to create a reservoir model that yields seismic response that perfectly matches that of the reference model because lithology is not the only factor contributing to the changes in seismic response. The carbonates proportions, however, have a major impact on geochemical environment of the reservoir, and it is essential to have information about their locations. This method allows us to predict where carbonates are present in the reservoir with the limited data available and to better assess the risks associated with storage of CO_2-saturated brine.

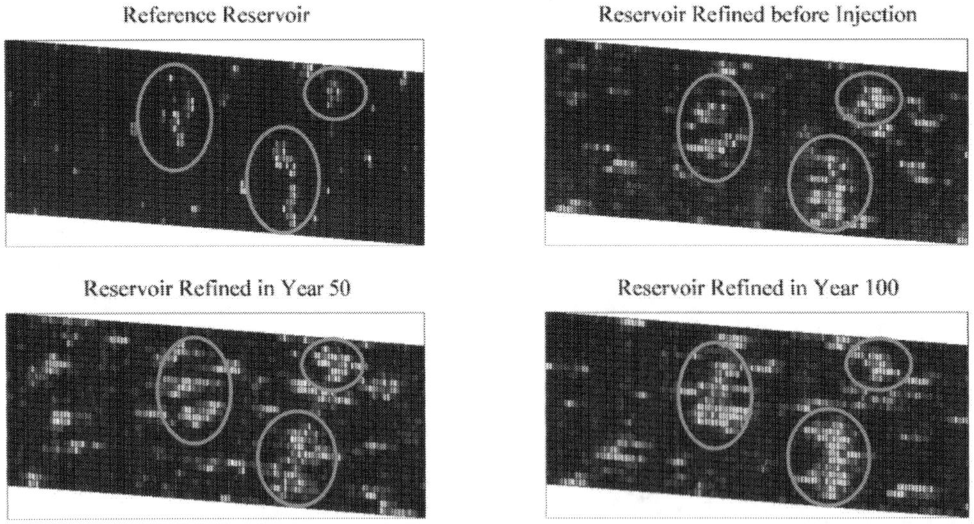

Figure 9: The concentration of carbonates of the reference model and the refined models. The refined models capture the areas of high proportions of carbonates.

Modeling of Carbonate Facies using Actual Time-Lapse Seismic Data

Inverted Time-Lapse Seismic Data

Carter (2014) performed inversion of the time-lapse seismic data of Cranfield to obtain the acoustic impedance and porosity. Figure 10 shows the acoustic impedances before injection and ten months after injection in 2010, in the region around the Designed Area of Study (DAS). The black lines indicate the approximate upper and lower boundaries of the reservoir where CO_2 was injected. The vertical brown lines from right to left are three wells in the DAS: 31F-1, 31F-2, and 31F-3, respectively. The spacing between inlines is about 25 meters.

In order to perform reactive transport simulations, it was required to transform the coordinates of the seismic data (two-way travel time and inline) into spatial coordinates. The seismic inversion results in spatial coordinates obtained after the transformation process is as shown in Figure 11(a). The vertical brown lines from right to left are the three wells 31F-1, 31F-2, and 31F-3, respectively. In addition to acoustic impedance, spatial variation in porosity is also available by Carter (2014) and is shown in Figure 11(b). However, inverted porosity is only available based on the seismic data prior to injection.

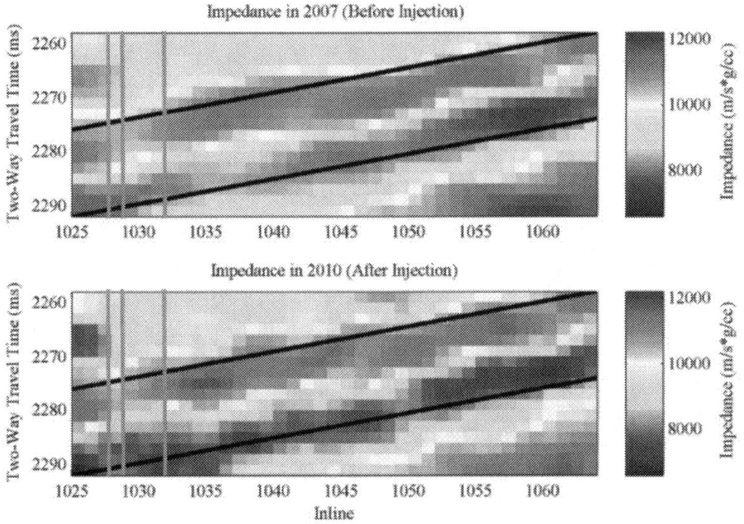

Figure 10: The inverted acoustic impedance before injection and ten months after injection (Carter, 2014). The black lines indicate the approximate upper and lower boundaries of the reservoir. The brown vertical lines are wells 31F-1, 31F-2, and 31F-3 from right to left, respectively.

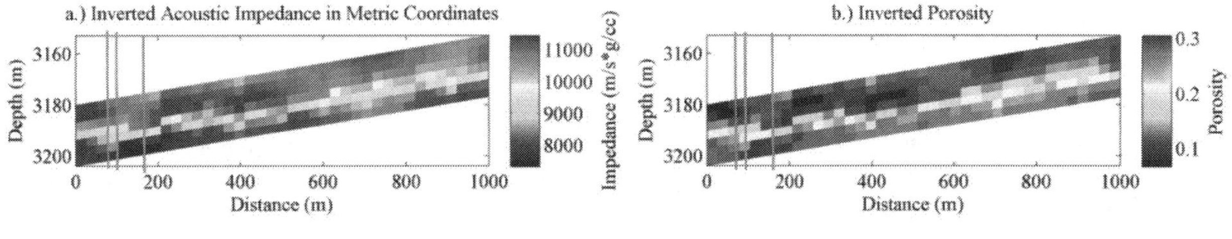

Figure 11: (a) The inverted acoustic impedance before injection and (b) inverted porosity in the DAS before injection in the metric coordinates.

Model Selection and Expansion

Based on the inverted acoustic impedance data before injection, ten wells were sampled as shown in Figure 12. The data sampled along those wells were used to create a variogram that was used to model the spatial continuity of the carbonate facies for which data was only available along well 31F-2. The variogram model for the carbonate facies was applied in the sequential indicator simulation in order to generate an initial suite of reservoir models.

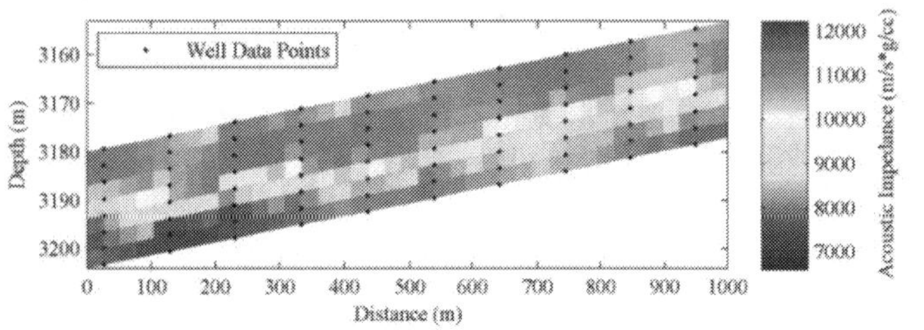

Figure 12: The sampling locations along which acoustic impedance values were sampled in order to create a variogram for the initial suite of models.

Rock physics model was applied to all prior models in order to calculate acoustic impedances and the discrete Fréchet distance was used to compare the acoustic impedances of the models with the inverted acoustic impedance before injection. In order to cluster the models based on the spatial distribution of impedances, the histogram of the discrete Fréchet distance was created with 10 bins as shown in Figure 13. Figure 13 also shows the inverted acoustic impedance and the calculated acoustic impedance for the models falling in the first bin. The carbonate facies data from the models in the first bin were resampled as conditional data chosen to create a common variogram, which was then applied in sequential indicator simulation in order to generate an expanded posterior set of reservoir models.

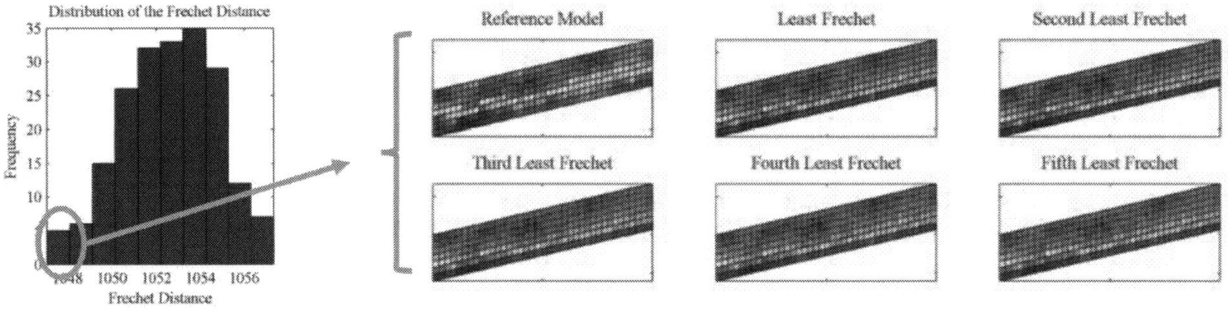

Figure 13: The distribution of the discrete Fréchet distances between the reference model and the newly generated models before injection (left) and the reference seismic response and the responses of the models that fall into the first bin of the distribution (right).

Using CMG-GEM, reactive transport simulations were run on all the models in the expanded set of posterior reservoir models. CO_2 was injected through well 31F-1 at one million tons-per-year rate for ten months from December 2009 to the end of September 2010, after which time a seismic survey was again completed. The changes in porosity and saturation over this period were simulated and then translated into seismic responses through rock physics modeling. Figure 14 (a) shows the inverted acoustic impedance in 2010 and the processed acoustic impedances of the best-refined model in the new set of models is shown in (b). Figure 15 shows the carbonate facies proportions of the refined model, which will be used for the assessment of CO_2-rock interactions and its long-term effects on sequestration.

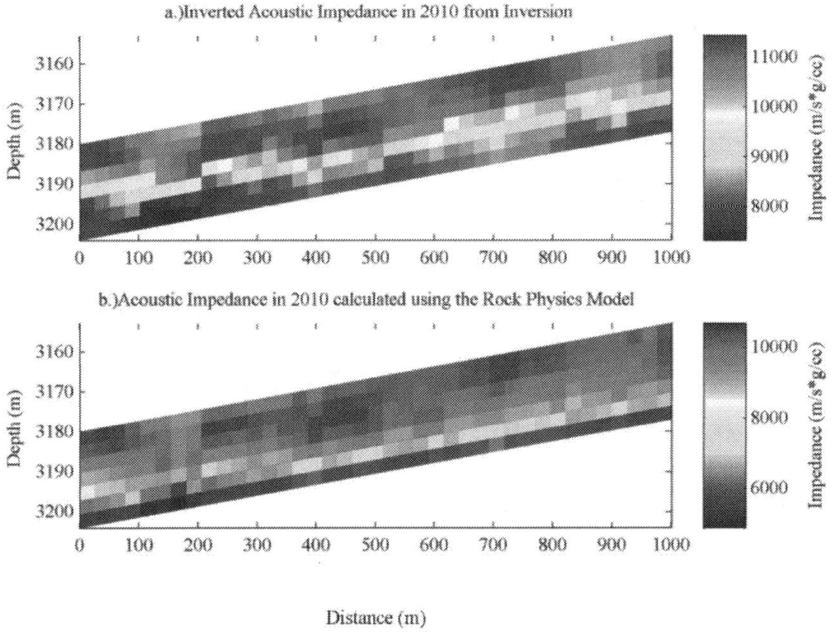

Figure 14: (a) The inverted acoustic impedance in 2010 and (b) the acoustic impedances computed using the rock physics model for the best model in the new set of models

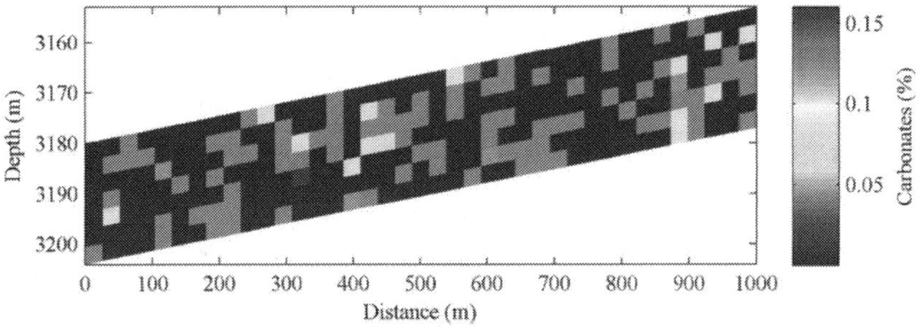

Figure 15: The carbonate facies proportions of the refined model that is used to investigate the long-term effects of CO_2 sequestration.

Modeling and Monitoring CO_2-Rock Interaction in the Selected Model

The carbonate facies model obtained in the previous section along with the inverted porosity were entered in CMG-GEM to perform reactive transport simulations in order to investigate the geochemical effects of CO_2 injection on reservoir properties. The CO_2 was injected into the bottom half of the reservoir through 31F-1 at the rate of one million tons-per-year for three years. The subsequent effects were observed for 100 years. Another simulation was run on the same model with the same injection conditions but with no reactions, and the results were compared. Acoustic impedances were calculated using the rock physics model. Figure 16 (a) shows the spatial distribution of acoustic impedance in 2010 with reactions while the difference between acoustic impedance derived with reactions and the one derived with no reactions is shown in Figure 16 (b). The difference in acoustic impedances between the two cases is partly due to the alterations in rock fabric and partly due to changes in fluid saturations. Figure 16 (c) shows the difference in pore structure/porosity while the difference in water saturation is shown in (d). This was ten months after injection. Most changes occurred near the injection zone as expected. The reactions induced rock frame alterations led to porosity changes that in turn led to changes in water saturation by affecting the fluid flow.

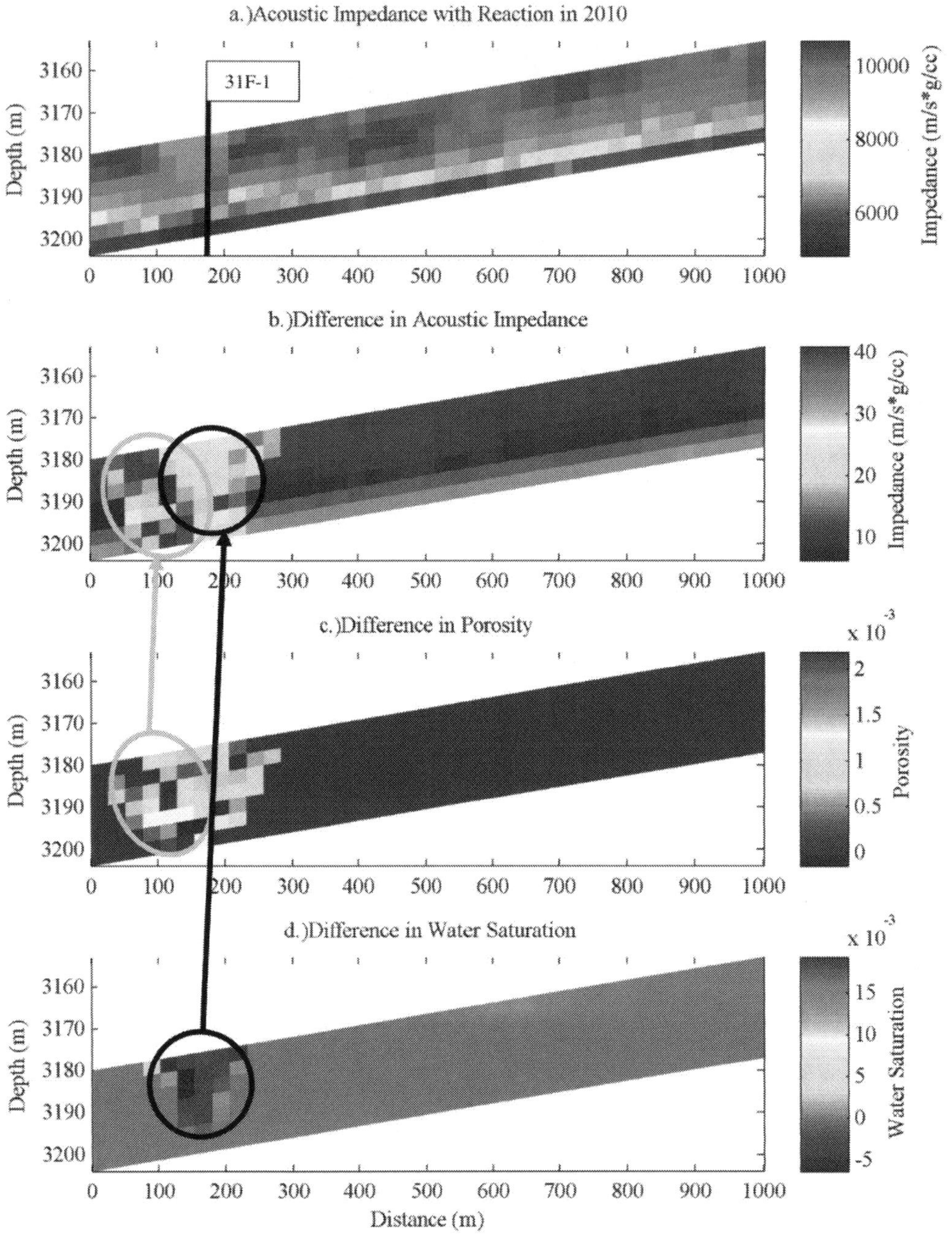

Figure 16: (a) Spatial distribution of acoustic impedances in 2010 corresponding to the case with reactions, (b) the difference between acoustic impedance derived with reactions and the one with no reactions, (c) the difference in porosity, and (d) the difference in water saturation.

The progressive alterations in models post-injection were also assessed. Figure 17 (a) shows the spatial distribution of acoustic impedances in 2059 with reactions and (b) shows the difference between acoustic impedance derived with reactions and the one derived with no reactions. The difference in porosity 50 years after injection is shown in Figure 17 (c) and the difference in water saturation is shown in (d). Because of relatively low density of CO_2 compared to water, the CO_2 plume migrated toward the up-dip east (right) side of the reservoir. Although rock frame alterations were observed throughout the reservoir, the major changes in porosity were observed toward the up-dip end at shallower depth. Major changes in water saturation were observed in the top portion of the reservoir. This is again due to the rock frame alterations that affect the flow of the buoyant CO_2. It induces a slight change in the rate at which the CO_2 migrates upward and this in turn introduces a change in acoustic impedance at the seismic scale.

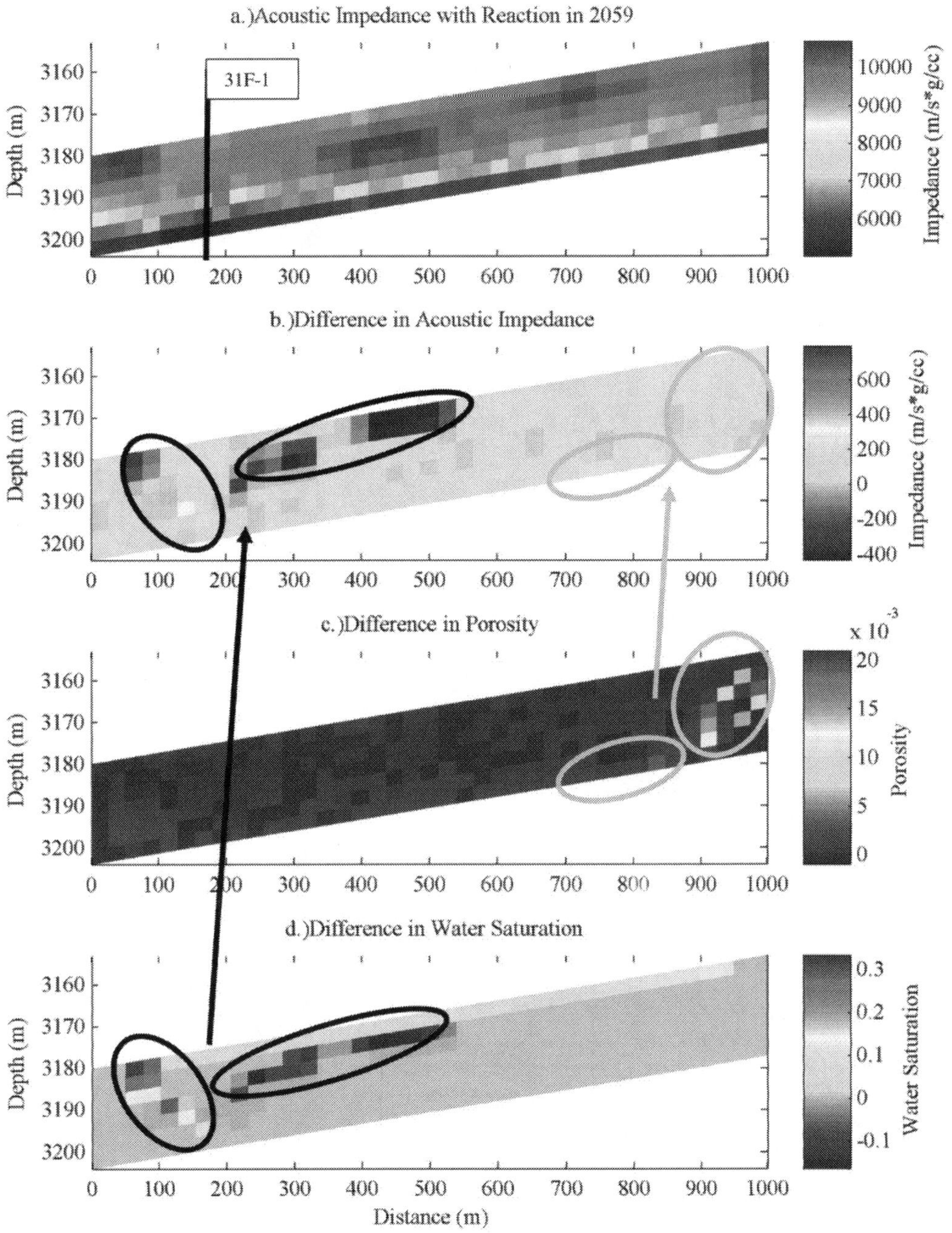

Figure 17: (a) Acoustic impedance in 2059 (50 years after injection) with reactions, (b) the difference between acoustic impedance derived with reactions and the one derived with no reactions, (c) the difference in porosity, and (d) the difference in water saturation.

Figure 18 (a) shows the spatial variation in acoustic impedance hundred years after injection (in 2109) with reactions. The difference between acoustic impedance derived in models with reactions and in models with no reactions is depicted in Figure 18 (b). The difference in porosity is shown in Figure 18 (c) while the difference in water saturation is shown in (d). Slight differences in the reservoir are observed between the models in 2059 and the one in 2109 indicating that there are ongoing reactions in the reservoir long after the injection. The results thus show the importance of modeling reactive transport when predicting reservoir fluid migration. In the short term and long term, the effects were clearly visible at the seismic scale.

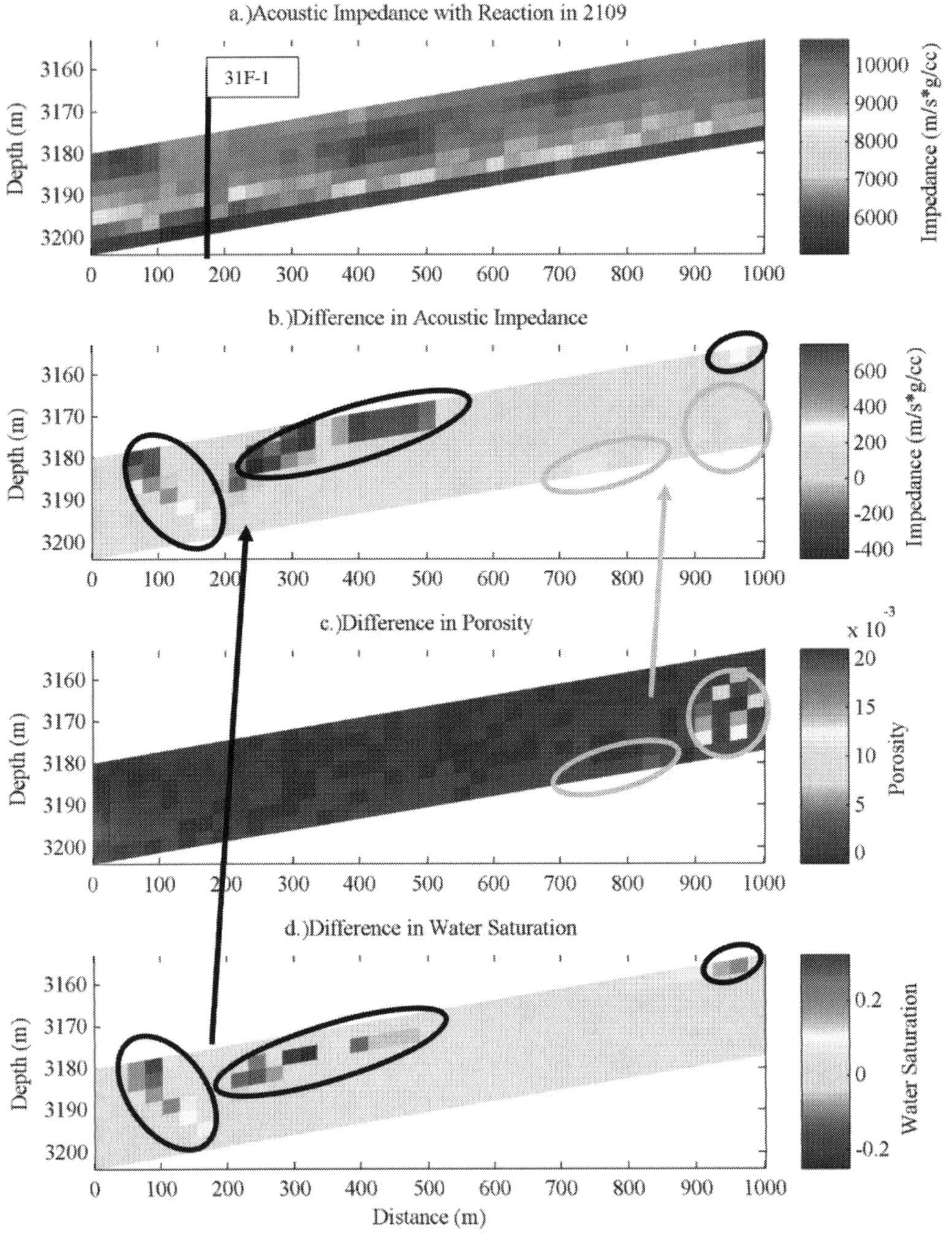

Figure 18: (a) Acoustic impedance in 2109 (100 years after injection) with reactions, (b) the difference between acoustic impedance derived with reactions and the one derived with no reactions, (c) the difference in porosity, and, (d) the difference in water saturation.

Conclusions

In this research, we investigated the geochemical effects of CO_2/brine mixtures in subsurface formations and based our study on data obtained for the siliciclastic reservoir in Cranfield, Mississippi. The reservoir has a very small proportion of carbonate facies. We investigated the behavior of the reservoir using a commercial compositional simulator CMG-GEM. We studied both reactive and non-reactive cases and the effect of rock frame alteration on the rock physics properties of the formation. With the limited data available for the Cranfield reservoir, we implemented a unique model selection and expansion scheme to progressively refine our estimates of likely carbonate facies locations within the reservoir. With the well log data, an initial suite of models was generated. The models were then processed through a flow model and rock physics model in order to transform the mechanical and flow properties of the rock into the corresponding seismic response, i.e. acoustic impedance. The models were clustered based on their similarity with the actual seismic impedance data available for the field, and the models in the closest cluster were used to generate an expanded posterior set of models. Through the model selection and expansion scheme, we were able to obtain a carbonate facies model that was similar to the reference model.

Subsequently, a model was selected from the expanded set of posterior models and was used to assess the geochemical alterations of the siliciclastic reservoir due to long-term sequestration of CO_2. The dissolution and precipitation of the carbonate facies altered the rock frames of the reservoir and affected the flow of CO2- saturated brine. It was clearly seen through this work that the effects of the presence of carbonate facies, even in small proportions, could be manifested in the reservoir response at the seismic scale and should be carefully modeled when investigating CO_2 sequestration.

References

Ajo-Franklin, J. B., J. Peterson, J. Doetsch, and T. M. Daley. 2013. "High-Resolution Characterization of a CO2 Plume Using Crosswell Seismic Tomography: Cranfield, MS, USA." *International Journal of Greenhouse Gas Control* 18. Elsevier Ltd: 497–509. doi:10.1016/j.ijggc.2012.12.018.

Carter, Russell W. 2014. "Fluid Characterization at the Cranfield CO 2 Injection Site : Quantitative Seismic Interpretation from Rock-Physics Modeling and Seismic Inversion." The University of Texas at Austin.

Computer, Modeling Group Ltd. 2013. *Compositional & Unconventional Reservoir Simulation*. Calgary.

Delshad, Mojdeh, Xianhui Kong, Reza Tavakoli, Seyyed a. Hosseini, and Mary F. Wheeler. 2013. "Modeling and Simulation of Carbon Sequestration at Cranfield Incorporating New Physical Models." *International Journal of Greenhouse Gas Control* 18. Elsevier Ltd: 463–73. doi:10.1016/j.ijggc.2013.03.019.

Dvorkin, Jack, and Amos Nur. 1996. "Elasticity of High-Porosity Sandstones: Theory for Two North Sea Data Sets." *Geophysics* 61 (5): 1363. doi:10.1190/1.1444059.

Eiter, Thomas, and Heikki Mannila. 1994. "Computing Discrete Fréchet Distance." Technische Universitat Wien.

Gassmann, Fritz. 1951. "Über Die Elastizität Poröser Medien." *Vierteljahrsschrift Der Naturforschenden Gesellschaft* 96 (1): 1–23. http://www.ngzh.ch/archiv/1951_96/96_1/96_2.pdf.

Hosseini, Seyyed Abolfazl, Hamidreza Lashgari, Jong W. Choi, Jean Philippe Nicot, Jiemin Lu, and Susan D. Hovorka. 2013. "Static and Dynamic Reservoir Modeling for Geological CO2 Sequestration at Cranfield, Mississippi, U.S.A." *International Journal of Greenhouse Gas Control* 18. Elsevier Ltd: 449–62. doi:10.1016/j.ijggc.2012.11.009.

Hosseini, Seyyed Abolfazl, Simon a. Mathias, and Farzam Javadpour. 2012. "Analytical Model for CO2 Injection into Brine Aquifers-Containing Residual CH4." *Transport in Porous Media* 94 (3): 795–815. doi:10.1007/s11242-012-0025-x.

Hovorka, Susan D. 2013. "Three-Million-Metric-Ton-Monitored Injection at the Secarb Cranfield Project—Project Update." *Energy Procedia* 37 (0). Elsevier B.V.: 6412–23. doi:http://dx.doi.org/10.1016/j.egypro.2013.06.571.

Kordi, M. 2013. "Characterization and Prediction of Reservoir Quality in Chlorite-Coated Sandstones : Evidence from the Late Cretaceous Lower Tuscaloosa Formation at Cranfield Field , Mississippi , U . S . A ." The University of Texas at Austin.

Lu, J., M. Kordi, S.D. Hovorka, T.A. Meckel, and C.A. Christopher. 2013. "Reservoir Characterization and Complications for Trapping Mechanisms at Cranfield CO2 Injection Site." *International Journal of Greenhouse Gas Control* 18 (October): 361–74. doi:10.1016/j.ijggc.2012.10.007.

Lu, Jiemin, Yousif K. Kharaka, James J. Thordsen, Juske Horita, Athanasios Karamalidis, Craig Griffith, J. Alexandra Hakala, et al. 2012. "CO 2-Rock-Brine Interactions in Lower Tuscaloosa Formation at Cranfield CO 2 Sequestration Site, Mississippi, U.S.A." *Chemical Geology* 291: 269–77. doi:10.1016/j.chemgeo.2011.10.020.

Lu, Jiemin, Kitty Milliken, Robert M. Reed, and Susan Hovorka. 2011. "Diagenesis and Sealing Capacity of the Middle Tuscaloosa Mudstone at the Cranfield Carbon Dioxide Injection Site, Mississippi, U.S.A." *Environmental Geosciences* 18 (1): 35–53. doi:10.1306/eg.09091010015.

Mandalaparty, Prashanth, Milind Deo, and Joseph Moore. 2011. "Gas-Compositional Effects on Mineralogical Reactions in Carbon Dioxide Sequestration." *SPE Journal* 16 (4). doi:10.2118/124909-PA.

Mindlin, R. D. 1949. "Compliance of Elastic Bodies in Contact." *Trans. ASME*, no. 71. New York, NY: Springer New York: A – 259. doi:10.1007/978-1-4613-8865-4_24.

Parkhurst, D.L., and C.A.J. Appelo. 2013. "Description of Input and Examples for PHREEQC Version 3—A Computer Program for Speciation, Batch-Reaction, One-Dimensional Transport, and Inverse Geochemical Calculations." In *U.S. Geological Survey Techniques and Methods*. http://pubs.usgs.gov/tm/06/a43/.

CO2 Utilization and Geological Storage

Unconventional Hydrocarbon Extraction from Geologic Formations with Carbon Storage Potential

Park, A. H. A., Columbia University
Gadikota, G., Columbia University

Originally presented on: 11/18/2015 17:35:00 - 18:00:00

Hydrocarbons from unconventional reservoirs such as shale account for an increasing fraction of the global energy portfolio. The current water-intensive hydrofracking technology for hydrocarbon extraction from shale-bearing reservoirs has limited efficiency and is environmentally unsustainable in the long run. Therefore, the use of CO_2 as an alternative fracking fluid has been proposed because it can penetrate into smaller pore spaces, and in the presence of water, CO_2 can react with shales to produce various chemical and morphological changes in these rocks. The fundamental mechanisms of shale-scCO_2-water interactions are investigated in this study. In this study, the chemical and morphological changes in shales with varying compositions of carbonates, clays, and quartz are investigated at elevated temperatures (T_{max} = 90 °C) and pressures ($P_{CO2, max}$ = 150 atm). Since the pore spaces in shales are to the order of a few nanometers, synchrotron techniques at Argonne National Laboratory—including Ultra Small Angle X-Ray Scattering and Small Angle Scattering (USAXS and SAXS)—were performed to determine the changes in the pore spaces. The morphological changes were identified using X-Ray Microtomography. Changes in the phases were also determined using Thermogravimetric Analysis (TGA) and X-Ray Diffraction (XRD). Our studies showed that CO_2 reacts with shales in the presence of varying levels of water, produces significant increases in the pore spaces and effectively disorders the structure of shales. While, the clay phases in shales were disordered in the presence of scCO_2 and water, the carbonate and quartz phases remained relatively unchanged.

Post-Injection Tracer Test to Constrain CO2 Residual Trapping and Plume Evolution

Zeidouni, M., Louisiana State University
Hovorka, S. D., University of Texas
Shi, K.,
Originally presented on: 11/18/2015 16:20:00 - 16:45:00

CO_2 residual trapping, post-injection plume extent, and time for plume stabilization of CO_2 geological storage highly depend on the hysteresis process which is the discrepancy between drainage and imbibition processes. CO_2 flow in the injection zone during the injection period is mainly controlled by the drainage process where the non-wetting gaseous phase replaces the wetting aqueous phase. However, during the post-injection period the viscous forces diminish allowing for imbibition process to control the CO_2-rich phase distribution. The post-injection plume extent and the amount of trapped CO_2 via residual trapping will then depend on both the imbibition and drainage processes. Therefore, conditioning the reservoir model using data collected over the injection period is insufficient in constraining the imbibition curve. Long-term data collection over the post-injection period may be required to determine the residually trapped CO_2 and to predict the CO_2 plume evolution and stabilization. Alternatively, a tracer test can be used to obtain such information in a timely manner. In this work, we introduce a tracer test at the end of the injection period to obtain information on the residual trapping parameters including imbibition/drainage discrepancy (hysteresis) and critical CO2-rich phase saturation. The sensitivity of the proposed tracer test to residual trapping parameters is evaluated based on the tracers' peak times and concentrations at the injection well (which serves as observation well during post-injection period). The effect of residual trapping on the plume evolution is also studied considering reservoir properties representative of Citronelle CO2 injection project.

CMTC-439481-MS

Carbon Dioxide Sequestration and Hydrocarbons Recovery in the Gas Rich Shales: An Insight from the Molecular Dynamics Simulations

Manas Pathak, University of Utah, Gorakh Pawar, University of Utah, Hai Huang, Idaho National laboratory and Milind D. Deo, University of Utah

This paper was prepared for presentation at the Carbon Management Technology Conference held in Sugarland, Texas, USA, 17–19 November 2015.

Abstract

Recent technological advances like horizontal drilling and hydraulic fracturing have made recovery of gas possible from the ultra low permeable shale plays in the United States. However, recoveries from these gas shales still tend to be in the range of single digits. It is believed that more than 80% of the generated hydrocarbons still remain in these tight formations. In order to increase the recovery from these tight shale plays, an enhanced recovery procedure using supercritical carbon dioxide as the injection fluid is recommended in this paper. In the current research, molecular dynamics simulations (MDS) have been performed on the kerogen-methane-carbon dioxide system to understand the absorption-adsorption-desorption phenomena of the super critical carbon dioxide fluid.

Previous studies have confirmed that the kerogen has a tendency to adsorb and absorb the hydrocarbons. In the current work, the type II kerogen model was chosen and annealed with the methane molecules. The Nose-Hoover style non-Hamiltonian equations of motion were used in a molecular simulator to generate positions and velocities sampled from the canonical (nvt) and isothermal-isobaric (npt) ensembles. This updates the position and velocity for atoms in the group each time step during the simulations. The pairwise distribution function and density of the mixture were calculated at the end of the simulations in order to validate the model with the experimental observations. The subsequent simulations with the carbon dioxide molecules in the periodic boundary conditions reveal that the carbon dioxide can sweep the absorbed methane from the kerogen matrix in the shales. The interactions at the interface between the carbon dioxide and the methane rich kerogen matrix are studied. These replicate the interactions at the fracture surface in the process of using super critical carbon dioxide as the fracturing fluid for enhanced gas recovery. Compared to methane, the carbon dioxide molecule has higher affinity to be adsorbed and eventually absorbed in the kerogen matrix because of its polar nature and linear shape. This allows the carbon dioxide to replace methane in the kerogen, thereby, sequestrating itself in the kerogen rich shale formations. Adsorption and absorption of the carbon dioxide in turn cause desorption of the methane from the kerogen, which aids in the gas recovery from such gas rich shales as the Marcellus Shale in the United States.

Introduction

The shale play resources have played a key role in increased oil production in last decade in United States (US). US is expected to become single largest oil producer due to unprecedented exploitation from shale reservoirs according to International Energy Agency (2001). Shale reservoir refers to the organic rich formations, which have porosity in range of nano-meters and permeabilities in the range of nano-darcies. Major US shale plays are shown in figure 1. US oil and gas production from shale plays have increased many folds in past decade with current oil production reaching over 5 million barrel oil per day and gas production 25 billion cubic feet gas per day (figure 2 and 3). However, majority of oil and gas is still left in shale reservoirs in the sub-surface and is currently not being exploited. A good understanding of dynamics at the scale of pores is required to exploit the full potential of these shale reservoirs and increase the recoveries. The current research aims to contribute in this direction so that a good understanding of the physics at the molecular scale can be developed and could be used towards better exploitation of shale play resources.

Method

The Molecular Dynamics (MD) simulation techniques were used to explore the possibility of enhanced recovery in shales through CO_2 sequestration. The MD simulations give insight into the molecular scale interaction of CO_2 with other hydrocarbons and the organic matter in shales. Kerogen is believed to be the precursor to oil and gas. During catagenesis phase of kerogen (Pathak et. al, 2015), it is converted into oil and gas. The objective of the current research was to look at the interaction of CO_2 with kerogen rich in hydrocarbons.

Kerogen model described in Ungerer, 2015 was used for the MD simulations. A type II kerogen model was used as it is usually both oil and gas prone (McCarthy et. al, 2011). The energy minimization routine was performed in LAMMPS molecular dynamics simulations package (Plimpton, 1995) using General Amber Force Field (GAFF) after Wang. J, et. al, 2004. Series of canonical (nvt) and isothermal-isobaric (npt) ensembles were run in LAMMPS, solving Nose-Hoover style non-Hamiltonian equations of motion to generate positions and velocities after each time step (in range of picoseconds). The nvt and npt ensembles were used for total time of couple of nano seconds, in order to perform simulated annealing of the kerogen structure. The nvt and npt ensembles were run with maximum T= 2000K and maximum P= 1000 atm. Final temperature and pressure while calculating properties were 300K and 300 atm, respectively, which is close to in-situ reservoir conditions. At the end of the simulated annealing, the kerogen structure attains its least energy state as in its natural occurring state. An annealed kerogen structure behaves as a bulk kerogen and could be then saturated with fluids like CH_4, CO_2 and/or Octane. The type II bulk kerogen model (figure 2) had a density over 1 gm/cc, which is a important check for the authenticity of the modeling.

In nature, kerogen behaves as a cross-linked elastomer (Ertas, 2006). This molecular structure of bulk kerogen developed in the current work would behave slightly different from the kerogen in native state with intrinsic cross linkages in the structure, However, developing a model for cross linked kerogen is beyond the scope of current work. A kerogen model with density over 1 gm/cc is good enough to measure properties of fluids in this kerogen structure. The key properties associated with the fluid kerogen interaction are diffusion coefficient of fluids like CH_4 and CO_2 in the kerogen and volumetric swelling of kerogen.

The enhanced recovery in shales through CO_2 sequestration involves adsorption of CO_2 followed by its absorption in the kerogen matrix. This in turn may lead to desorption of CH_4 from the kerogen matrix, thereby aiding towards enhanced recovery. An important property of interest is the associated volumetric strain in this process when CO_2 replaces CH_4 from the kerogen matrix. The volumetric strain may reveal the possibility of micro fracturing induced due to possible shrinkage of the matrix.

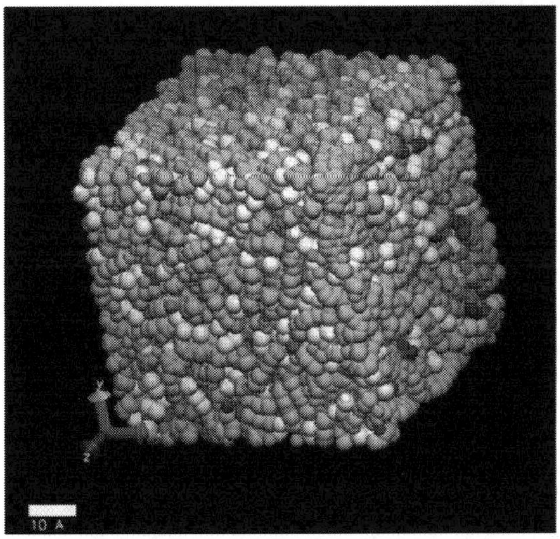

Figure 1. Molecular structure of type II kerogen after simulated annealing. Final density is 1.15 gm/cc

Discussion

In order to understand the combined mechanism of enhanced recovery and CO_2 sequestration, a hypothetical scenario was proposed in the current work. At first, the kerogen matrix was saturated with the methane gas, CH_4 (5% by mass) as shown in scenario 1 in the figure 2. Assuming a successful enhanced recovery operation, 50% of the methane in the kerogen is replaced by equal number of CO_2 molecules (scenario 3 in figure 2). This scenario represents kerogen matrix rich in both, CO_2 and CH_4. Scenario 2 represents the case where only CO_2 was allowed to interact with kerogen. We used MD simulations to measure diffusion coefficient of fluids in kerogen and the volumetric strain associated in each scenario.

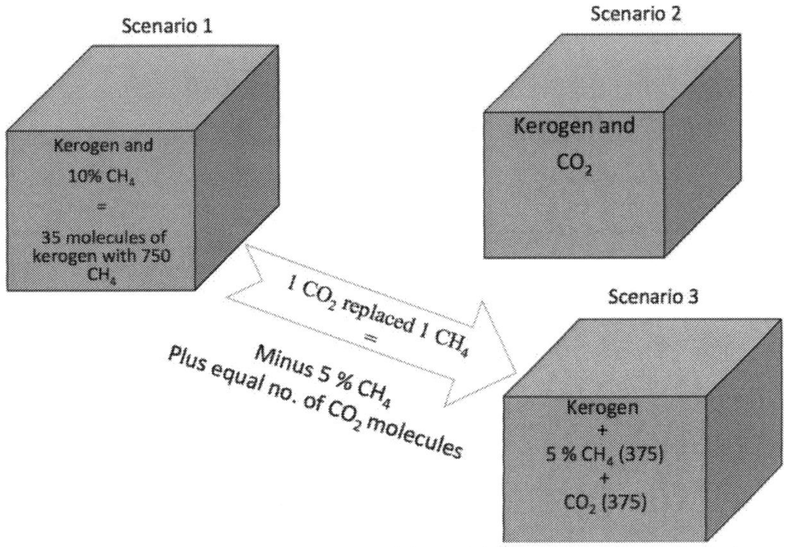

Figure 2. Three different scenarios to measure diffusion coefficient of CO_2 and CH_4 and associate volumetric strain in each scenario

The self-diffusion and inter diffusion coefficients were measured by the slope of mean square displacement and time plots. A better estimate of percentage increase in recovery of methane by CO_2 could be achieved by Grand Canonical Monte Carlo (GCMC) simulations, which is mentioned as a future work in this paper. The volumetric strain was calculated by the change in volume of scenario 2 and 3 compared to scenario 1. The volumetric strain of scenario 1 was calculated by comparing it the volume of pure bulk kerogen in its least energy state.

Results

The results are given for each of the three scenarios. The molecular diffusion coefficient and the volumetric change are calculated for each of the scenario as shown below:

1. Scenario-1. Type II kerogen with 10 % by mass of CH_4 [35 molecules of kerogen with 750 molecules of CH_4]:

- Volumetric strain due to swelling: 25%
- Molecular diffusivity of CH_4 in Type II kerogen: $2*10^{-9}$ m^2/s

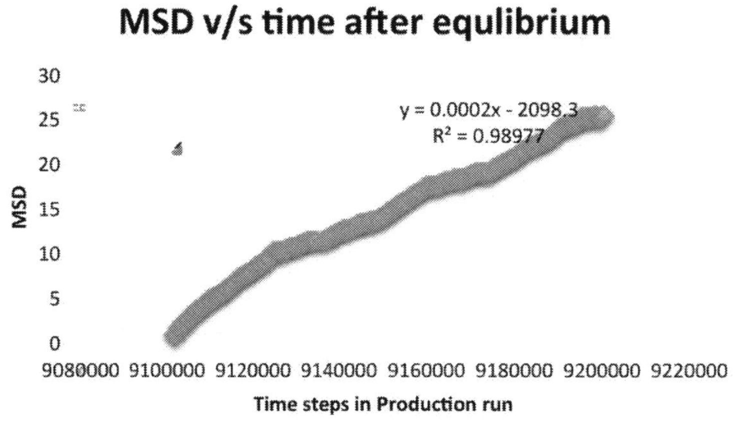

Figure 3. Mean Square Displacement (MSD) of CH_4 molecules with time plot for scenario 1

2. Scenario-2. Type II kerogen with CO_2 [35 molecules of kerogen with 750 molecules of CO_2]:

- Volumetric strain due to swelling: 18%
- Diffusion coefficient of CO_2 in Type II kerogen: $5*10^{-10}$ m^2/s

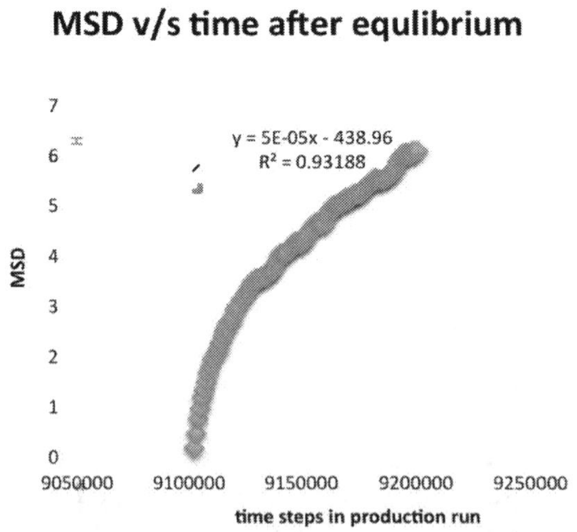

Figure 4. Mean Square Displacement (MSD) of CO_2 molecules with time plot for scenario 2

3. Scenario-3. Type II kerogen with 5% CH_4 and rest replaced by CO_2 [35 molecules of kerogen with 350 molecules of CH_4 and 350 molecules of CO_2]:

- Volumetric strain when 50% CH_4 replaced by equal no. of CO_2: -1.5%
- Diffusion coefficient of CH4 in Type II kerogen,in presence of CO_2: $2*10^{-9}$ m^2/s
- Diffusion coefficient of CO_2 in Type II kerogen, in presence of CH_4: $7*10^{-10}$ m^2/s

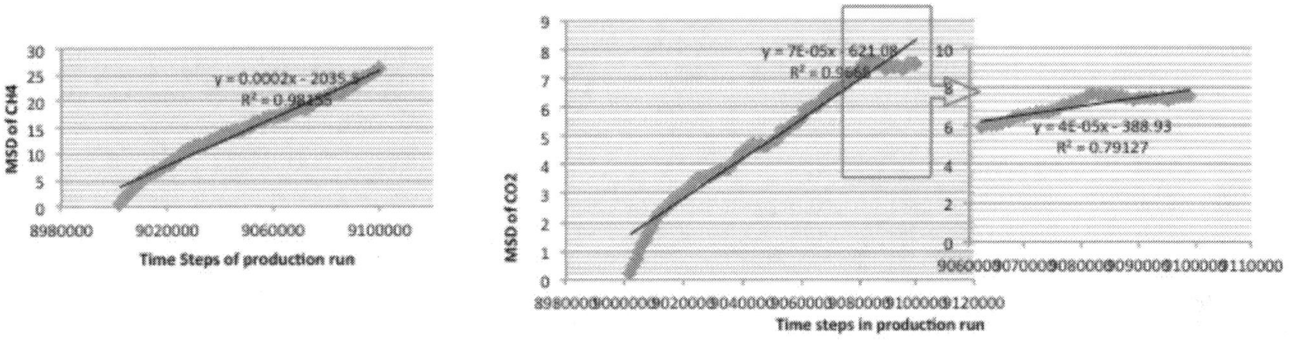

Figure 5. Mean Square Displacement (MSD) of CH_4 and CO_2 molecules with time plot for scenario 3

Conclusions

When no external force is applied, CO_2 has a lesser tendency to diffuse in kerogen compared to CH_4. This helps in retaining CO_2 in kerogen rich shales with time and in turn enhancing the CH_4 recovery. The volumetric strain associated with 50% enhanced recovery of CH_4 is -1.5%. Shrinkage of kerogen during CO_2 sequestration would help open flow pathways at kerogen-mineral grain boundary, aiding to enhanced recovery. MD simulation techniques are useful in assessing the potential of unconventional reservoirs for enhanced recovery through CO_2 sequestration

Future work

The author intends to study the absorption/ adsorption-desorption of CO_2 in the gas and liquid rich shale plays using Grand Canonical Monte Carlo Simulations (GCMC)

References

Ertas, D., Kelemen, S.R., and Halsey, T.C. 2006. Petroleum Expulsion Part 1. Theory of Kerogen Swelling in Multicomponent Solvents, Energy & Fuels 20 1), 295-300. DOI: 10.1021/ef058024k

International Energy Agency, "World Energy Outlook: Assessing Today's Supplies to Fuel Tomorrow's Growth," 2001

McCarthy Kevin, Rojas Katherine, Niemann Martin, Palmowski Daniel, Peters Kenneth, Stankiewicz Artur, 2011. Basic Petroleum Geochemistry for Source Rock Evaluation, Oilfield Review Summer 2011:23, no. 2

Pathak, M., Deo, M. D., Panja, P., & Levey, R. A., 2015. The Effect of Kerogen-Hydrocarbons Interaction on the PVT Properties in Liquid Rich Shale Plays. Society of Petroleum Engineers. doi:10.2118/175905-MS

Plimpton, S., 1995. Fast Parallel Algorithms for Short-Range Molecular Dynamics, J Comp Phys, 117, 1-19

Ungerer, P., Collell, J., and Yiannourakou,M. 2015. Molecular Modeling of the Volumetric and Thermodynamic Properties of Kerogen: Influence of Organic Type and Maturity. *Energy & Fuels 29* (1), 91-105. DOI: 10.1021/ef502154k

Wang, J., Wolf, R. M., Caldwell, J. W., Kollman, P. A. and Case, D. A. (2004), Development and testing of a general amber force field. J. Comput. Chem., 25: 1157–1174. doi:10.1002/jcc.20035

Simulation and Optimization of Enhanced Geothermal System Using CO2 As a Working Fluid

Agarwal, R., Washington University in St. Louis
Biagi, J.,
Zhang, Z.,
Originally presented on: 11/18/2015 17:10:00 - 17:35:00

Because of rising concerns about CO2 emissions from fossil fueled power plants, in recent years there has been strong emphasis on the development of safe and economical CCUS (Carbon Capture Utilization and Storage) technology. One such technology that shows some promise is EGS (Enhanced Geothermal System), where CO2 is used as a working fluid to extract heat from a geothermal reservoir. Permanent carbon sequestration is also achieved as a byproduct due to subsurface fluid losses throughout the life of the system. In this paper, numerical simulations of subsurface flow in EGS are conducted using the multiphase flow solver TOUGH2 (Transport of Unsaturated Groundwater and Heat). An optimization code based on a multi-objective genetic algorithm is combined with TOUGH2 (designated as GA-TOUGH2) and modified for EGS application. Using GA-TOUGH2, the CO2 injection rate is optimized for both constant mass and constant pressure injection scenarios to manage the production temperature profile and to ensure that the heat extraction occurs for the entire life of the system thus allowing more efficient use of CO2. The results of this study show promise of EGS technology for consideration of deployment on a commercial scale.

CMTC-439769-MS

CO_2 EOR Mechanisms in Bakken Shale Oil Reservoirs

Hui Pu, Yinghui Li*, InPetro Technologies Inc.

This paper was prepared for presentation at the Carbon Management Technology Conference held in Sugarland, Texas, USA, 17–19 November 2015.

Abstract

Preliminary studies have been done to characterize rock-fluid properties, and flow mechanisms in the shale reservoirs. Most of these studies modified parameters to be used for conventional reservoirs and failed to capture dynamic features of shale rock and fluids in confined nano-pore space. In unconventional reservoirs, interactions between the wall of organic and inorganic matter and contained fluid significantly affect phase and flow behaviors. The inability to model capillarity with the consideration of pore size distribution (PSD) characteristics using commercial software may lead to an inaccurate oil production performance in Bakken. This paper presented a novel formulation that consistently evaluates capillary force and adsorption using PSD directly from core measurements. In the meantime, an adsorption model using a local density optimization algorithm was developed to address the incapability and inaccuracy of Langmuir model for liquid-rich formations. This new adsorption model, based on exactly the same physical chemistry principles at different pressures, revolved from monolayer at low pressure to more complex multilayer model spontaneously and consistently and was imperative to CO_2 cycling. With our advances in understanding key production mechanisms of capillarity and adsorption, we are able to differentiate production driving mechanisms in unconventional reservoirs vs. conventional ones. A new compositional simulator was developed that captures those differentiations and results showed that should capillarity be consistently formulated with PSD, significant difference in production profile was observed. As shown for CO_2 huff-and-puff process in unconventional reservoirs, a smaller amount of soaking time and an 18%OOIP incremental in ultimate recovery was achieved, as compared with the case not considering capillarity and adsorption properly. The new findings could better address differences in flow mechanisms in unconventional reservoirs, and thus lead to an optimized IOR practice.

Introduction

The Bakken Formation is one of the oil-bearing strata of the Williston Basin underlying parts of North Dakota, Montana, Saskatchewan, and Manitoba (**Figure 1**). The Devonian-Mississippian age Bakken formation consists of three members: the Upper Bakken is about 20 ft thick black marine shale; the Middle Bakken is 30-80 ft thick interbedded layer of limestone, siltstone, dolomite, and sandstone;

and the Lower Bakken is 10-50 ft thick black marine shale (LeFever, 2005). The Mississippian age Lodgepole Limestone overlies and the Devonian age Three Forks formation underlies Bakken formation as depicted in **Figure 2**. Lithology of the Middle Bakken formation varies from calcareous siltstone to predominant quartz sandstone, dolostone, and silty limestone. The Upper and Lower members, as hydrocarbon source rocks, are organic-rich, with total organic carbon (TOC) content ranging from 12 to 36 wt%, average 25 to 28 wt% over large portions of the basin (Tran *et al.*, 2011). The Middle Bakken member, which is the primary oil target, is organic-poor with TOC content ranging from 0.1 to 0.3 wt% (Price, 1999). The average depth of the Middle Bakken is about 10,000 ft. The porosity of the Middle Bakken is about 6% and the permeability averages 0.001-0.01 *mD* or less. Water saturation ranges between 25% and 50% in the Middle Bakken (Cherian *et al.*, 2012). The average oil gravity is about 42 °API (Yu *et al.*, 2014), gas oil ratio (GOR) ranges from 507 to 1,712 SCF/bbl, and the bubble point pressure varies from 1,617 to 3,403 psi (Nojabaei *et al.*, 2013). The Bakken Formation is variably overpressured with pressure gradients up to 0.73 psi/ft in the central part of the basin (Meissner, 1978).

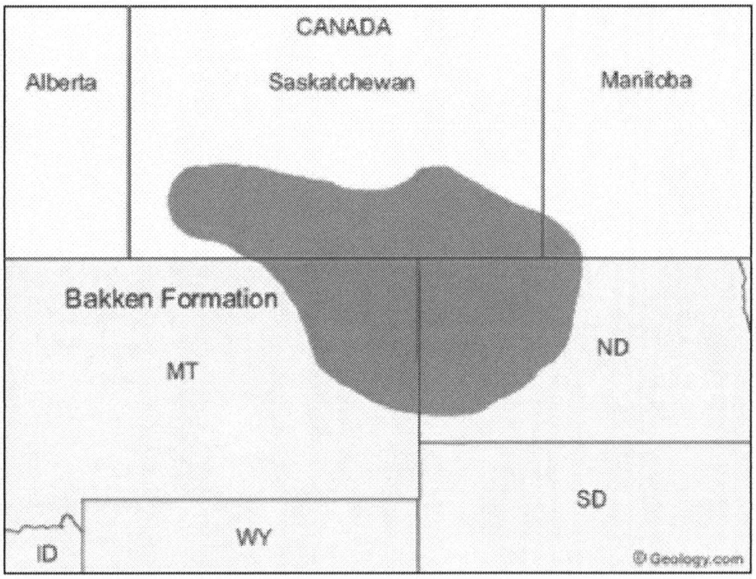

Figure 1. Map of Bakken Formation (Source: www.geology.com)

The Bakken formation is one of the largest unconventional resources in the world with approximately 7.4 billion barrels of recoverable oil as estimated by the U.S. Geological Survey in 2013, and the estimate of original oil in place is considered to be 100 to 900 billion barrel. With the advancements in horizontal drilling and hydraulic fracturing technologies, the development of Bakken unconventional resources has been very rapid, and the Bakken Formation contributes significantly to tight oil production. Although currently Bakken produced over 1 million bpd of oil due to the advances in horizontal drilling and multi-stage hydraulic fracturing, primary oil recovery factor remains very low and has been estimated to be less than 10-15% of original oil in place (OOIP) (Iwere *et al.*, 2012; Cherian *et al.*, 2012; Hoffman, 2012). Given the huge Bakken resources and low primary oil recovery, there is a need and enormous potential for enhanced oil recovery (EOR) in Bakken. Although waterflooding is the most widely used recovery method after primary production for conventional oil reservoirs, it may not be a viable option in tight and shale oil formations due to extremely low permeability/injectivity and reaction with clay content. In contrast, gas injection (i.e., CO_2) is recognized to be more suitable for enhanced oil recovery in the Bakken Formation due to beneficial properties such as low viscosity and high injectivity than water. While the use of CO_2 in conventional reservoirs is a widely applied practice, its use for enhanced oil recovery in unconventional oil reservoirs is a very opaque, albeit a few laboratory demonstrations for its success. Furthermore, in spite of great success in the development of unconventional reservoirs, the fluid flow mechanisms in shale, fluid properties in confined pore space and interactions between the wall of nano-pores and contained fluid are still not properly understood. Therefore, many unique features of Bakken formation further present a challenge for understanding, and its application, of CO_2 EOR in Bakken.

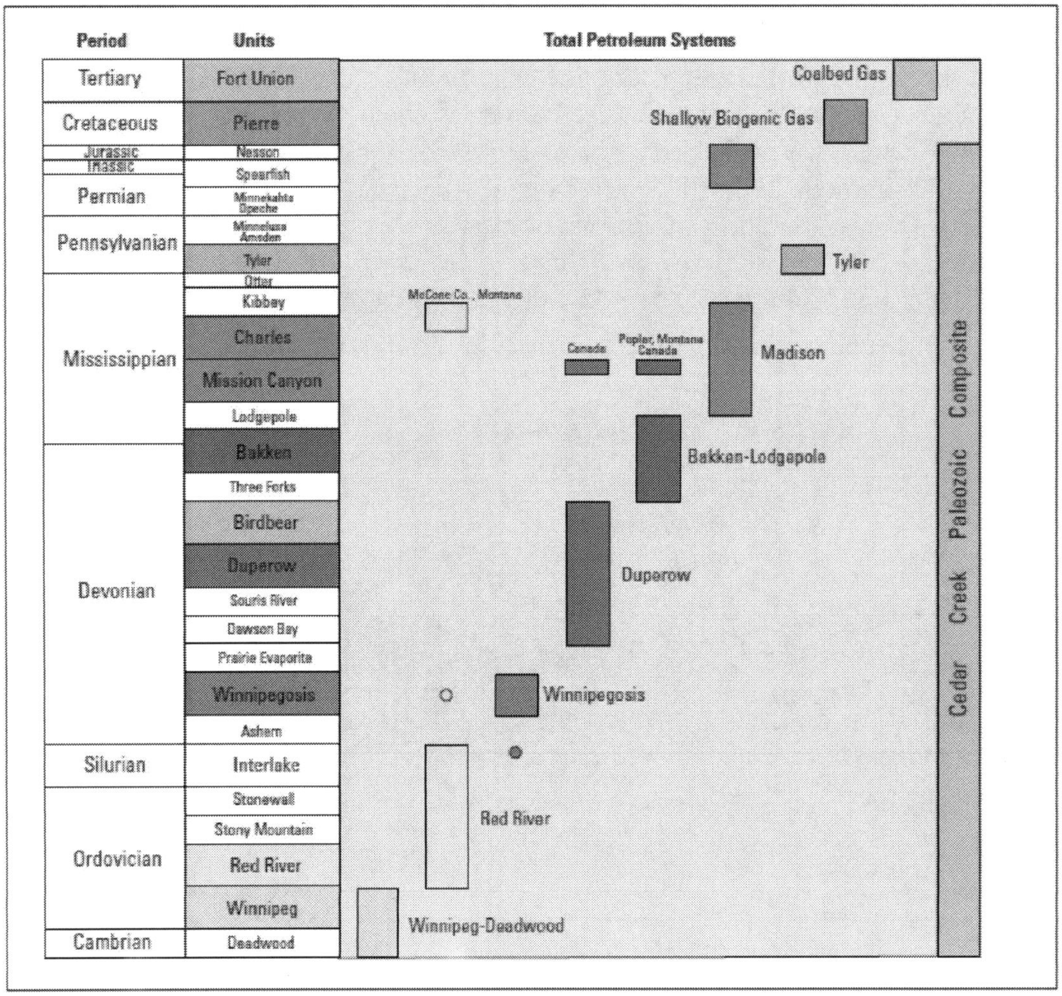

Figure 2. Generalized Stratigraphic Chart and Associated Total Petroleum System for the Williston Basin (Source: Anna et al., 2013)

The matrix in shale reservoirs consists of micropores smaller than 2nm in diameter, mesopores with diameters in the range of 2 to 50 nm, and macropores larger than 50nm in diameter (Kuila *et al.*, 2011). Those small pores on the order of nanometers lead to a significant increase in capillary pressure. The confined hydrocarbon phase behavior also largely deviates from the bulk measurements as noted by several authors and laboratory measurements (Wang *et al.*, 2013; Jin *et al.*, 2015). The effect of capillary pressure on vapor-liquid equilibrium is generally negligible for conventional reservoirs, as assumed in current commercial reservoir simulators and there is no pressure difference between vapor and liquid phases during flash calculations. Failure of accounting for large capillary pressure in nano-pores can lead to significant underestimates for ultimate recovery and saturation pressures (Nojabaei *et al.*, 2013; Jin *et al.*, 2015; Kazi *et al.*, 2015; Li *et al.*, 2015). Therefore, it is imperative to fully and correctly characterize capillary pressure for simulation in unconventional reservoirs to more guide the development in unconventional reservoirs. One of the objectives of this paper was to present a breakthrough formulation to model capillarity with the consideration of PSD directly from core measurements, and then the capillary pressure model is incorporated into an in-house compositional reservoir simulator.

Another important area that differentiated the simulation of liquid-rich formation and conventional reservoirs was an adsorption driving mechanism. Shale and tight formations commonly exhibit a high adsorption potential due to its significant internal surface area. To have a better understanding of adsorption, many isotherm models have been proposed for gas adsorption on solids surface (Kohlmann *et al.*, 2001), and Langmuir model is the most popular one, representing the adsorption of a monolayer of molecules on the inner surfaces of the material (Langmuir, 1916). Adsorption is commonly modeled using Langmuir isotherm in the commercial reservoir simulation software. However, there are several assumptions in the Langmuir model which were not held valid for wet or liquid-rich shale formations of interests, and CO_2 EOR and storage in particular (Do, 1998; Afonja *et al.*, 2012): 1). Adsorption surface is homogeneous; 2). Adsorption on surface is localized, that adsorbed atoms or molecules are adsorbed at definite/localized sites; 3). Each site can accommodate only one molecule or atom. This paper developed a new adsorption model that incorporated adsorption behavior using a local density optimization algorithm using the same PSD.

Shale Pore Size Distribution and Its Capillary Pressure

Shale is a fine-grained sedimentary rock that forms from the compaction of silt and clay-size mineral particles that we commonly call "mud". In contrast to conventional reservoir rocks, the shales typically have very fine-grained rock texture (dominant grain size ≤62.5 µm), low porosity (≤10%) and very low permeabilities (in nanodarcy range) (Kuila *et al.*, 2011). One of the unique features of shale rocks are their extremely small pore-sizes being attributed to their fine-grained nature. Quantification of pore structure attributes in shales still remains a challenge even after the advancement in the imaging techniques and other techniques. Saidian *et al.* (2015) used nuclear magnetic resonance (NMR) and other methods to measure porosity and PSD of 28 samples from Middle Bakken and Three Forks formations. **Figure 3** shows the PSD of different Bakken samples using NMR method (Saidian *et al.*, 2015). As we can see from these PSD results, most Bakken pore diameters fell in the range from 1 nm to 7 nm. Due to confinement of these extremely small pores in unconventional reservoirs, shale reservoirs exhibited flow and interaction characteristics that were uniquely related with nano-scale pore throats. Confinement in tiny pores changed the fluid behavior due to the geometric constraint of the fluid molecules and the interaction of these molecules with the pore walls (the molecule–wall interaction). Experimental results also suggested that a confined fluid had a depressed bubble point pressure and an increased density as compared with its bulk properties (Wong *et al.*, 1993, Thommes *et al.*, 1995, Hiejima *et al.*, 2004, Arakcheev *et al.*, 2008). Fluid properties and flow conditions are modified in confined pores due to pore proximity effects (Sapmanee, 2011). Given the fact of pore proximity effect and increased capillary pressure in unconventional reservoirs, it is inappropriate to ignore the large effect of capillary pressure on vapor-liquid equilibrium. Historically, mercury intrusion techniques are used for pore-size analysis of conventional reservoirs. However, for unconventional shale reservoirs, very high pressures (> 60,000 psi) would be required for mercury to access the full pore-structure (Kuila *et al.*, 2011). The mercury injection capillary pressure (MICP) result for a shale rock sample is shown in **Figure 4** (Elgmati *et al.*, 2011), which showed that the nano-scale pores lead to very high mercury injection pressure, and the major intrusion occurred between 10,600 and 60,000 psi.

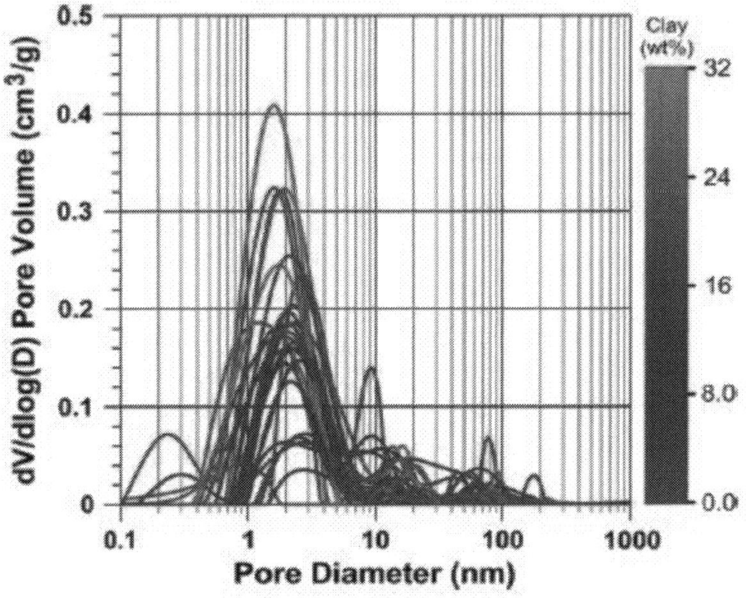

Figure 3. Pore Size Distribution of Bakken Samples using NMR (Saidian and Prasad, 2015)

Capillary pressure may be calculated by two methods. One method is Young-Laplace Equation (Eq. 1), where σ is the interfacial tension between phases j and i in dynes per cm, and r is the pore radius and contact angle is assumed to be 0 here. σ is calculated using Macleod-Sugden correlation, where ψ_i is the parachor of component i , ρ_i is the molar density of phase j in lb-mole per cubic feet.

$$P_c = \frac{2\sigma}{r}, \ \sigma_{jl}^{0.25} = 0.016018 \sum_{i=1}^{n_c} \Psi_i(\rho_j x_{ij} - \rho_l x_{il}) \quad (1)$$

Another option is to use J-Function (Eq. 2), where the parameters C_{pc} and E_{pc} need to be determined by matching gas/oil experimental capillary pressure curves. ϕ is the porosity and K is the permeability. S_{on} and S_{gn} is the normalized oil and gas saturation.

$$P_c = C_{pc}\sigma\sqrt{\frac{\emptyset}{K}}\frac{S_{gn}}{S_{on}+S_{gn}}^{E_{pc}} \quad (2)$$

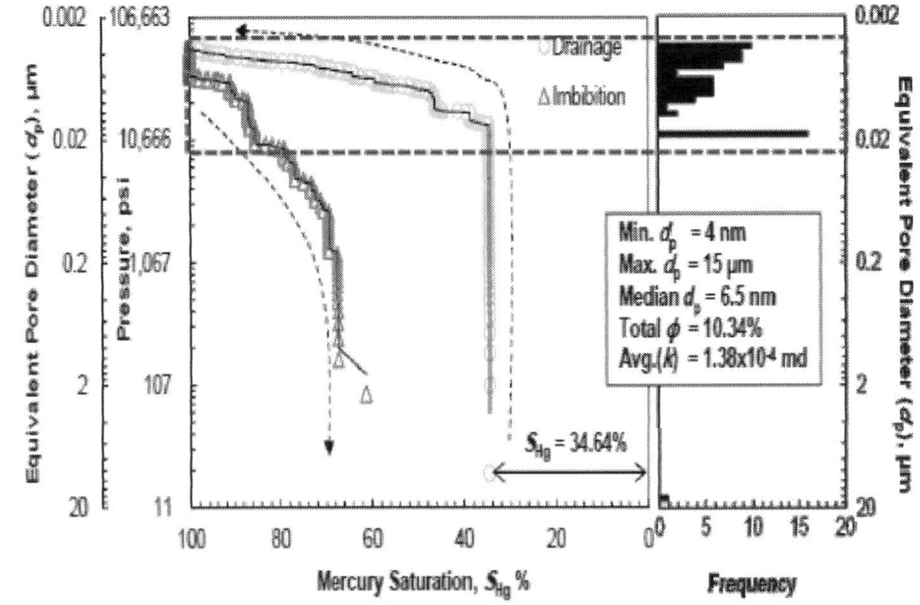

Figure 4. Mercury Injection Capillary Pressure Result for a Shale Sample (Elgmati et al., 2011)

Vapor-liquid equilibrium is an important part of phase behavior calculation using an equation of state. As mentioned above, the effect of capillary pressure on vapor-liquid equilibrium in unconventional shale reservoir shall be considered to establish a correct representation of flow mechanisms in shales (Nojabaei *et al.*, 2013). Once vapor-liquid equilibrium is established considering the effect of capillary pressure, an adsorption for the vapor phase is superimposed to account for molecular interactions with the surface and changes in storage properties, which was described in a later session. The condition of equilibrium of the two phases can be written in terms of the chemical potential, μ, at a given pressure, P:

$$\mu_i^L(T, P^L, x) = \mu_i^V(T, P^V, y) \qquad i = 1,2, \ldots, m \qquad (3)$$

Under the impact of capillary forces, the pressure exerted at the curved liquid-vapor interface, P_c, is defined as

$$P^V - P^L = P_c \qquad (4)$$

where P^L and P^V are the phase pressures of liquid and vapor phases, respectively. μ_i is the chemical potential of component i at the respective temperature, phase pressure, and mole fraction. m is the number of components in the system.

Optimized Simplified Local Density (OSLD) Model for Adsorption

Organic shale has the ability to store significant amount of gas permanently because of trapping of the gas in an adsorbed state within its finely dispersed matters. Organic-rich shales can serve as targets for sequestration of significant volumes of anthropogenic CO_2. Methane adsorption has long been seen as important for gas production in unconventional reservoir, and has been studied extensively. CO_2 adsorption, which had less attention as CH_4, were often studied in the context of enhanced gas recovery in coal seams in coal bed methane. Researchers have found significant preferential adsorption of CO_2 over CH_4 and thus great potential for enhanced recovery and sequestration (Kovscek *et al.*, 2005) in CBM and shale gas. Nuttall *et al.* (2005) studied CO_2 adsorption capacity in gas shales through a series of experiments, and found that Langmuir-like adsorption of CO_2 in gas shales was more than five times greater than the amount of CH_4 adsorption. Kang *et al.* (2010) studied CO_2 adsorption on two Barnett shale samples, finding 5-10 times greater adsorption of CO_2 than CH_4. However, when liquid-rich formation was depleted under bubble point pressure, i.e., those Bakken CO_2 EOR cases, the assumptions of using monolayer Langmuir model no longer held true, and a new multi-layer multi-component model was required to account for the large quantity of $C3^+$ components in vapor-liquid equilibrium. In this work, a Simplified Local Density (SLD) model was further improved and developed to include various pore sizes of shale samples, using Peng-Robinson Equation of State (PR EOS). Finally, the new OSLD-PR EOS is integrated to an in-house compositional simulator for study of mechanisms of CO_2 sequestration in organic-rich shales.

The SLD-PR model was originally developed by Rangarajan *et al.* (1995) to correlate adsorption isotherms of a pure component on a flat surface over a large pressure range. Chen *et al.* (1997) investigated adsorption and desorption isotherm of carbon dioxide adsorbed on activated carbon at subcritical conditions through experiments and also found that the adsorption isotherms can be represented very well by the simplified local density model. Fitzgerald *et al.* (2003) modified SLD-PR model to study the adsorption of pure gases (such as methane, nitrogen, and CO_2) on coals at near-critical and supercritical conditions, and reported that the modified SLD-PR model can correlate accurately data for supercritical adsorption systems. Ma and Jamili (2014) used SLD-PR to study the effects of confinement in shale formations on phase behavior. Density profiles of pure methane, ethane, propane, or n-butane in slit-shaped pore with with of 2.4 nm- or 5 nm under condition of 185 °F and 2600 psi were given in the paper (Ma and Jamili, 2014).

The SLD-PR model has some basic assumptions: (1) the chemical potential at any point near the absorbent surface is equal to the bulk phase chemical potential, and (2) the chemical potential at any point above the surface is sum of the fluid-fluid and fluid-solid interactions. Therefore, the equilibrium chemical potential is calculated by contributions from these fluid-fluid and fluid-solid interactions. The SLD model is based on the work of Chen *et al.* (1997). The pore geometry used in local density model is slit with a width of *L*. Let distance *z* be the normal distance from the point of interest to the surface of the solid. The chemical potential of fluid inside a porous medium at a distance *z* is the result of fluid-fluid and fluid-solid interactions:

$$\mu_{bulk} = \mu(z) = \mu_{ff}(z) + \mu_{fs1}(z) + \mu_{fs2}(L - z) \qquad (5)$$

where the subscript "*bulk*" refers to the bulk fluid, "*ff*" refers to fluid-fluid interactions, and "*fs*" refers to the fluid-solid interactions.

The chemical potential of bulk can be written in terms of fugacity:

$$\mu_{bulk} = \mu_0(T) + RTln(f_{bulk}/f_o) \qquad (6)$$

where f_{bulk} is the bulk fugacity and f_o is the fugacity at standard state.

At d distance *z* normal to the surface of pore wall, the chemical potential of fluid-fluid interactions can be defined in terms of fugacity due to fluid-fluid interaction $f_{ff}(z)$ as:

$$\mu_{ff}(z) = \mu_0(T) + RTln\big(f_{ff}(z)/f_o\big) \qquad (7)$$

The chemical potential due to fluid-solid interaction can be written as

$$\mu_{fs}(z) = N_A[\Psi(z) + \Psi(L - z)] \qquad (8)$$

where $\Psi(z)$ is the fluid-solid potential energy function. Lee's partially integrated 10-4 Lennard-Jones potential was used (Lee, 1988):

$$\Psi(z) = 4\pi\rho_{atoms}\varepsilon_{fs}\sigma_{fs}^2\left(\frac{\sigma_{fs}^{10}}{5(z')^{10}} - \frac{1}{2}\sum_{i=1}^{4}\frac{\sigma_{fs}^4}{(z'+(i-1)\cdot\sigma_{ss})^4}\right) \qquad (9)$$

where ρ_{atoms} is the number of carbon plane atoms per unit area (38.2 atoms/nm^2), ε_{fs} is the fluid-solid interaction energy parameter, σ_{fs} is the molecular diameter of the adsorbate, $\sigma_{fs} = (\sigma_{ff} + \sigma_{ss})/2$, σ_{ss} is the carbon interplanar distances which is taken to be that of graphite (0.335nm) (Subramanian *et al.*, 1995), The distance z' is the perpendicular distance between the center of a fluid molecule to the center of the first plane of solid atoms, i.e., $z' = z + \sigma_{ss}/2$.

Through combining equations (5)-(8), the relations of the bulk fugacity and fluid-fluid fugacity can be expressed as:

$$f_{ff}(z) = f_{bulk}exp\left(-\frac{\mu_{fs1}(z)+\mu_{fs2}(L-z)}{RT}\right) \qquad (10)$$

The Peng-Robinson equation of state was used to calculate the fugacity of a bulk fluid:

$$lnf_b = \frac{b\rho_b}{1-b\rho_b} - \frac{a_b\rho_b}{RT(1+2b\rho_b-b^2\rho_b{}^2)} - ln\left[\frac{(1-b\rho_b)}{RT\rho_b}\right] - \frac{a_b}{2\sqrt{2}RT}ln\left[\frac{1+(1+\sqrt{2})\rho_b}{1+(1-\sqrt{2})\rho_b}\right] \qquad (11)$$

Similarly, the expression for fugacity of fluid-fluid interaction can be written:

$$lnf_{ff}(z) = \frac{b\rho(z)}{1-b\rho(z)} - \frac{a(z)\rho(z)}{RT\{1+2b\rho(z)-b^2[\rho(z)]^2\}} - \ln\{\frac{[1-b\rho(z)]}{RT\rho(z)}\} - \frac{a(z)}{2\sqrt{2}RT}ln\left[\frac{1+(1+\sqrt{2})\rho(z)}{1+(1-\sqrt{2})\rho(z)_b}\right] \qquad (12)$$

The equations for $a(z)$ depend on the ratio of the slit width L to the molecular diameter σ_{ff}. The details for calculations of $a(z)$ are described in paper of Chen *et al.* (1997), and are shown in Appendix A. From equations (6)-(12), the local density $\rho(z)$ at any point can be calculated.

In our model, the adsorption per unit weight area of adsorbent is calculated as follows:

$$\Gamma = A \int_{\sigma_{ff}/2}^{L-\sigma_{ff}/2} [\rho(z) - \rho_b(z)]dZ$$

where A is the surface area per unit weight of the adsorbent.

The SLD-PR model is developed through considering adsorption behavior using a simplied local density optimization function, so that the characterization and understanding of multi-component interactions at adsorption sites for a full spectrum of reservoir pressures is possible.

Numerical Simulation

Our advances in modeling of aforementioned capillarity with the consideration of PSD characteristics and adsorption using a local density function were integrated to a compositional simulator. The simulator was then used to model the process of injection of CO_2 in Bakken Formation in this study, to understand a whole core lab experiment. An 1-D model was built in this work. The model size was 6.25 ft × 0.5 ft × 0.5 ft, which corresponds to length, width, and thickness, respectively. The computation domain was divided into 25 blocks in the *x*-direction. Injection and production was in the leftmost cell for huff-and-puff simulation. The basic reservoir properties data of the model was shown in **Table 1**. The reservoir was homogeneous with a permeability of 1 μD.

Table 1. Parameters for Bakken Simulation Model

Parameter	Value
Initial reservoir pressure, psi	3200
Reservoir temperature, ºF	240
Reservoir permeability, µD	1
Initial water saturation	0.1

The pressure, volume, and temperature (PVT) experimental data for Bakken crude oil samples were available and were used to define PVT relationships under reservoir condition. Constant composition expansion (CCE), differential liberation (DL) analysis, separator, swelling test and fluid compositional analysis data were also available for oil samples. Our OSLD-PR EOS model was tuned based on the available experimental PVT data to perform phase equilibrium and property calculations. The Bakken reservoir oil was carefully grouped into nine pseudocomponents, i.e., CO_2, C_1, C_2, C_3, C_4, C_5-C_6, C_7-C_{12}, C_{13}-C_{21}, C_{22}-C_{80}, and their molar fractions were 0.0001, 0.36726, 0.14885, 0.09334, 0.05751, 0.06406, 0.15854, 0.0733, and 0.03704.

Base Case (Primary Recovery)

The reservoir (or the model in the study) was first produced by primary depletion. Therefore, a baseline was established prior to assessing the effectiveness of the CO_2 huff-and-puff process. The "base cases" consisted of primary production for a period of 150 days, without injection of any fluid, at constant bottomhole pressures of 800, 1,100, 2,000, 3,000 psi, respectively. The bottomhole pressure determines the pressure drawdown at the wellbore and the pressure profile in the reservoir, which ultimately affect the drive mechanism significantly. Through comparing the oil recoveries, average reservoir pressure after primary depletion, and the field practices, BHP was selected as the well constraint for simulations.

The average reservoir pressure profile of base case is shown in **Figure 5**. It shows the reservoir was depleted after 70 days of production. As **Figure 6** shows, the oil recovery reached 11% OOIP after primary depletion, which is in agreement with numerous studies on estimating the primary recovery in the Bakken formation by various methods (Clark, 2009; Dechongkit and Prasad, 2011; Cherian *et al.*, 2012). Our model exhibits the characteristic primary depletion behavior of a shale reservoir. This primary oil recovery of base case will be directly compared with CO_2 huff-and-puff recovery in the following CO_2 huff-and-puff section to determine if CO_2 huff-and-puff is able to increase the ultimate recovery factor.

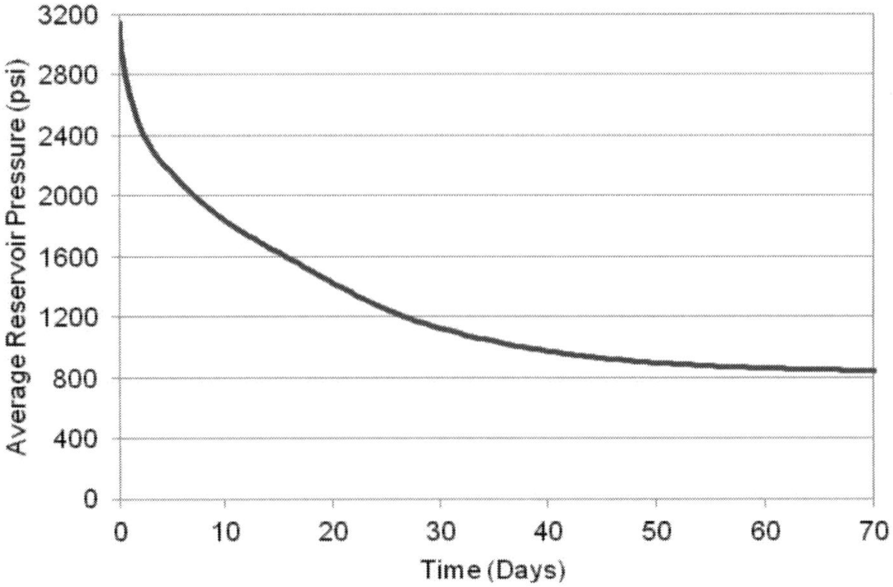

Figure 5. Average Reservoir Pressure Profile of Base Case

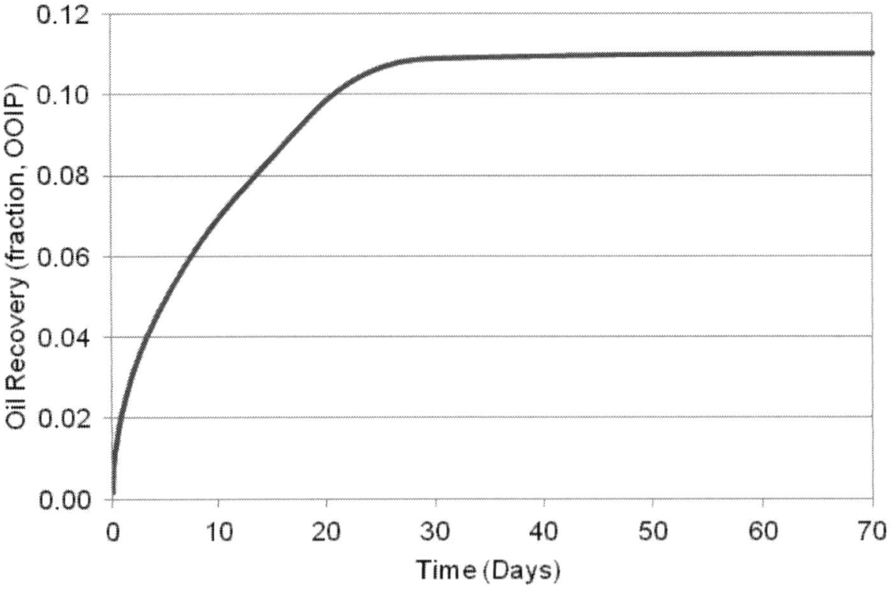

Figure 6. Oil Recovery of Base Case

The effect of capillary pressure (using either J-function or Young-Laplace equation), fixed pore radius, and pore size distribution on primary production is studied using our in-house simulator. The following simulation cases were designed and run: 1)"No P_c" case in which the effect of capillary pressure and adsorption is not included; 2)"Fixed Pore Radius" case in which the radius of pore is set as a constant value of 10 nm and the capillary pressure is calculated using Young-Laplace equation and adsorption used OSLD-PR method; 3)"J-function" case in which the capillary pressure is computed using J-function with adsorption used OSLD-PR method; 4) "PSD" case in which the pore size distribution information is considered. As can be seen from average reservoir pressure profile (**Figure 7**), the average reservoir pressure is almost identical except with small differences after the reservoir is depleted. The ultimate oil recoveries of these primary production cases vary from 11% to 12.4% OOIP after 30 days of production, therefore, the difference of primary oil recoveries is about 1% incremental whether the capillarity is considered or not, or using different method to represent the capillary pressures or adsorption (**Figure 8**). As we have a more in-depth discussion on how pressure and adsorption would affect CO_2 application in another paper, this presentation focuses on the general optimization for CO_2 huff-and-puff practices in Bakken reservoirs.

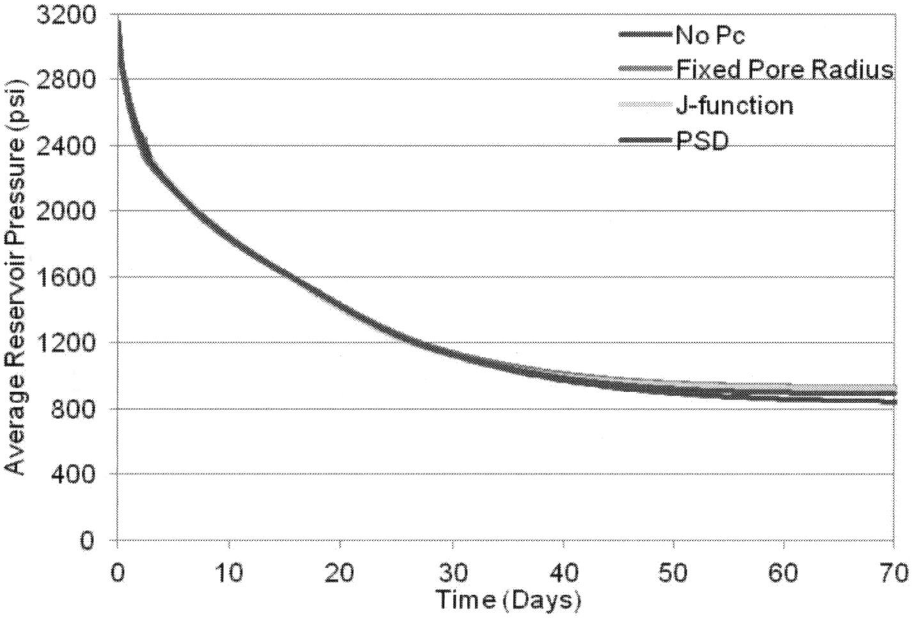

Figure 7. Average Reservoir Pressure during Primary Production

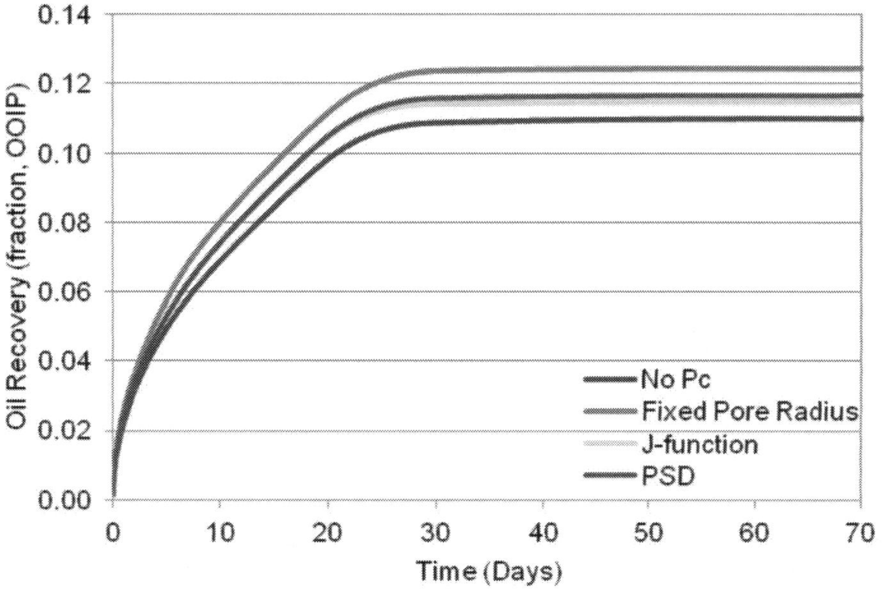

Figure 8. Oil Recovery during Primary Production

CO_2 Huff-and-Puff

Huff-and Puff is a cyclical process in which an injection-soaking-production sequence is repeated over a certain period of time for a well or a group of wells. In this work, the well initially produced for 30 days, which gave an ultimate recovery of primary production, and then it was converted to a CO_2 injector to inject CO_2 for a certain days, followed by shut-in for soaking, and lastly production. This was one cycle of CO_2 Huff-and-Puff. The length of the injection and soaking periods, and the number of cycles affect the performance of CO_2 Huff-and-Puff process. In this simulation work, different cases were designed and run to compare the CO_2 Huff-and-Puff process.

Length of the Injection Period

The length of the injection period is a key component of the Huff-and-Puff process. The CO_2 injection time and injection rate has a large impact on incremental oil recovery. The length of injection period is directly related to the volume of CO_2 injected and the degree to which Huff-and-Puff induces changes in reservoir pressure and fluid properties. Four cases were run to simulate a 1-cycle Huff-and-Puff treatment. The length of the CO_2 injection period varies from case to case (5, 10, 15, 20 days), while the soaking and overall cycle length was kept constant at 10 days and 245 days, respectively. The CO_2 injection began after 30 days of primary production, and a constant injection pressure of 4,500 psi was used. The oil recovery and average pressure of these simulations are shown in **Figures 9** and **10**. As shown in **Figure 9**, CO_2 Huff-and-Puff process yielded excellent incremental oil recovery compared to the recovery of base case of primary production. The oil recovery and incremental recovery of these cases are summarized in **Table 2**. As expected, the case with 5 days of CO_2 injection gave the smallest incremental oil recovery of 3.66% OOIP, and a higher pore-volume of CO_2 injection (i.e., 20 days core CO_2 injection) may increase oil recovery by over 18%, suggesting the great benefits for CO_2 EOR in unconventional reservoirs. However, when comparing the length of CO_2 injection between 15 days and 20 days, the extent of increase in oil recovery was only 2% OOIP. This suggests that there may have an optimal slug size for unconventional CO_2 EOR. **Figure 10** shows the average reservoir pressures for these cases with different CO_2 injection periods. The larger the volume of the injected CO_2, the higher the average reservoir pressure. The average reservoir pressure for case with 5 days of injection is lower than the bubble point pressure, and it gave the smallest incremental oil recovery. That optimal slug size may be required to increase the average reservoir pressure to above original depressed bubble point pressure.

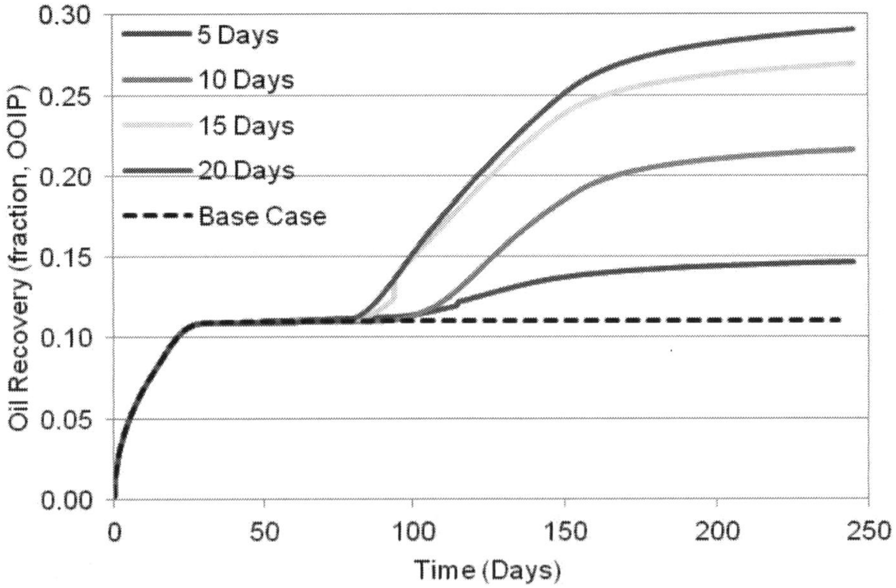

Figure 9. Oil Recoveries of Different Lengths of CO2 Injection Period

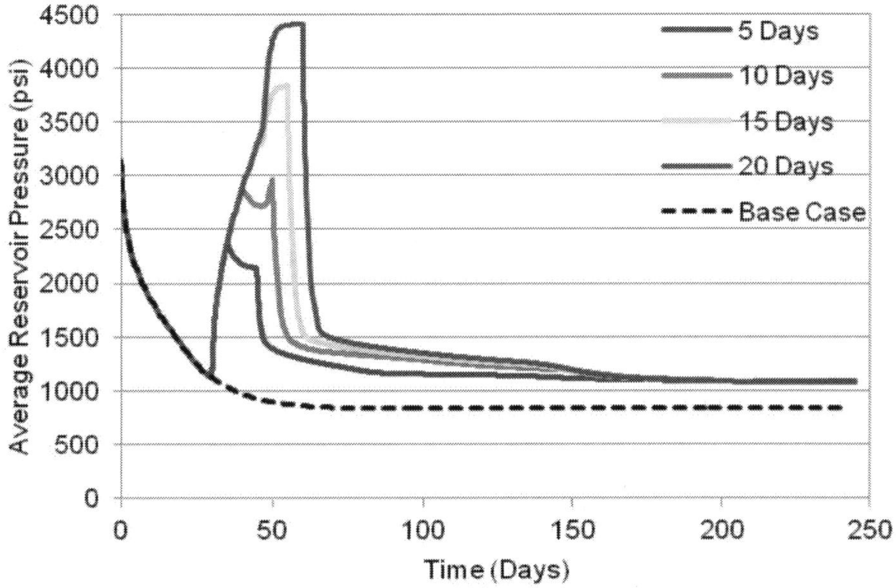

Figure 10. Average Reservoir Pressures of Different Lengths of CO2 Injection Period

Table 2. Recovery and Incremental Oil Recovery for Different Lengths of CO_2 Injection Period

Injection Time, Days	Soaking, Days	Ultimate Recovery, OOIP%	Incremental Recovery, OOIP%
5	10	14.68	3.66
10	10	21.64	10.62
15	10	26.88	15.86
20	10	29.06	18.04
Base Case(Primary)	N/A	11.02	N/A

Length of the Soaking Period

The length of the soaking period is also studied. Four simulation cases, each with one cycle CO_2 injection, were designed by varying soaking period (1, 10, 50, and 100 days), while keeping the CO_2 injection period and overall cycle length as 15 days and long enough to study the one-cycle ultimate recovery. The BHP for injection was 4,500 psi. It shows although flow in soak time is quite small, a relative short period of time may allow good incremental oil recovery (**Figure 11**). The results also show that the longest soaking period (100 days) had almost same incremental oil recoveries as cases with 10 and 50 days of soaking, in other words, longer soaking time proved to be less effective given longer waiting time delays the production. Soak time is insensitive, relatively; however, it shall not be too small and impair production performance. The simulation results show that the soaking period of 5-10 days yields the good incremental recovery compared to longer-soak-time cases. **Figure 12** shows the average reservoir pressures of these simulation cases.

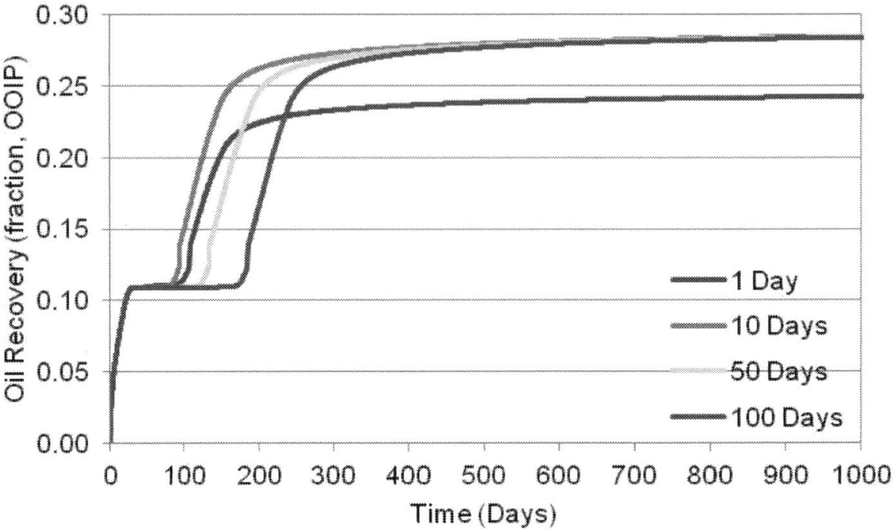

Figure 11. Oil Recoveries of Cases with Different Soaking Periods

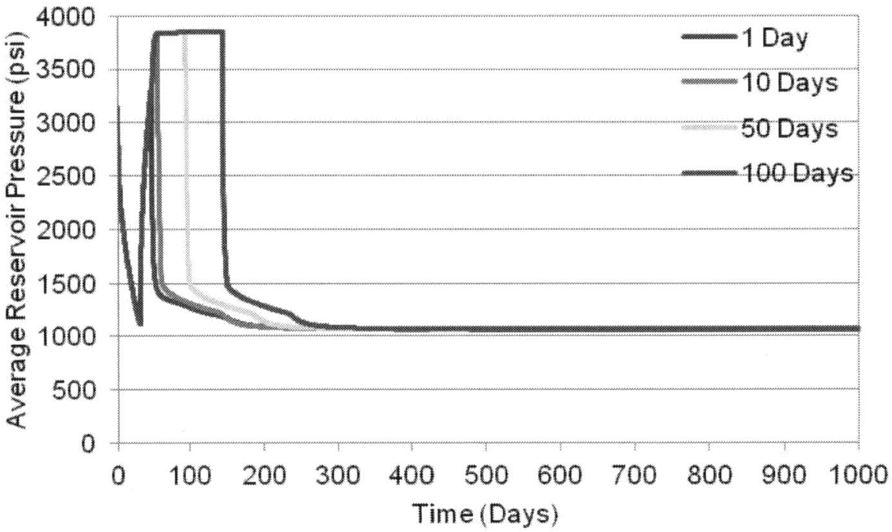

Figure 12. Average Reservoir Pressures of Cases with Different Soaking Periods

Number of Cycles

Huff-and-Puff is a multi-cyclical process conducted in the field with a goal to economically maximize economic matrices, i.e., total amount of produced oil. Given the complex geologic conditions of the reservoir, and the limitations of each EOR method, there is a limit of crude oil that could be extracted by EOR process, including CO_2 Huff-and-Puff. The number of cycles of CO_2 Huff-and-Puff process has to be studied and optimized, and we used an eight-cycle CO_2 Huff-and-Puff case for that objective. Same as previous CO_2 Huff-and-Puff process that the CO_2 injection began after a 30 days of primary recovery, each cycle of CO_2 Huff-and-Puff consisted of 15 days of injection, 10 days of soaking, and 200 days of production. The BHP constraint for injection period was 4,500 psi. **Figure 13** shows the oil production rate of 8-cycle CO_2 Huff-and-Puff process, indicating that the oil rate was significantly boosted during the first a few cycles. The last three cycles (Cycles 6, 7, and 8) basically did not show much benefits for CO_2 injection. While more cycles yield additional incremental recovery, their impact on production is progressively diminished. The "spikes" in the oil rate curve is related with the switching of well status during the simulation. The oil recovery of 8-cycle CO_2 Huff-and-Puff process is shown in **Figure 14**. The oil recovery increased by 32.71% OOIP compared to primary production (Base Case), more than quadruple primary recovery currently suffered in Bakken reservoirs. It can also be observed on oil recovery profile that the oil recovery almost does not increase after 5 cycles, and the first cycle yields the highest incremental oil recovery and is the most effective. The average reservoir pressure of 8-cycle CO_2 Huff-and-Puff process is demonstrated in **Figure 15**.

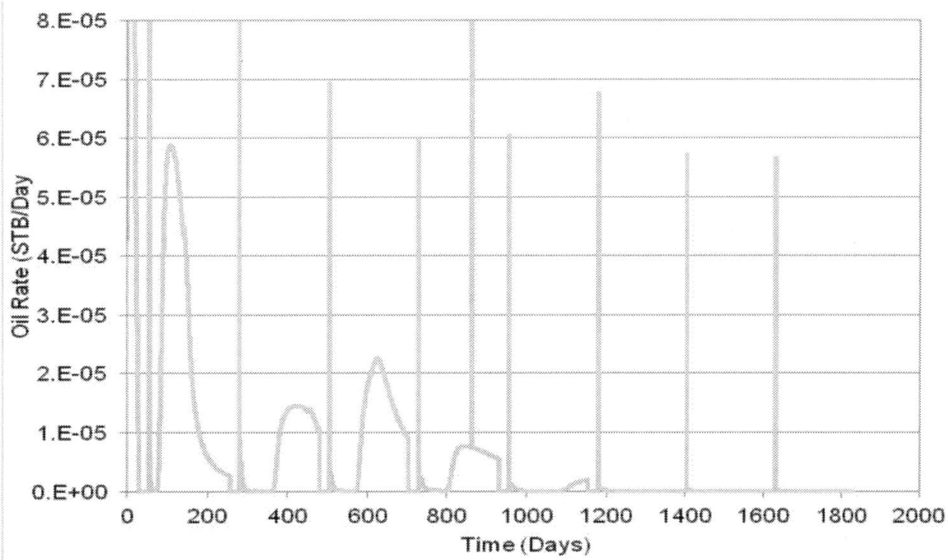

Figure 13. Oil Rate of 8-cycle CO2 Huff-and-Puff Process

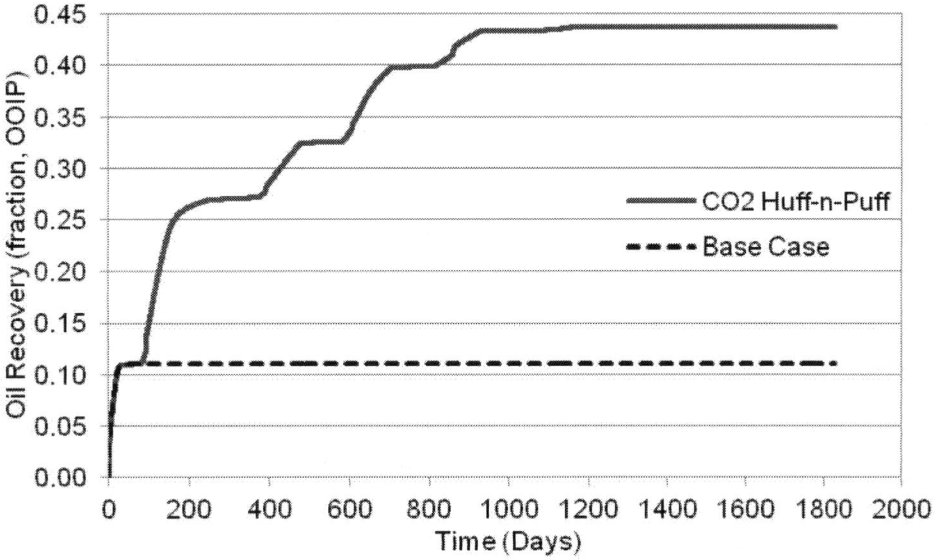

Figure 14. Oil Recovery of 8-cycle CO2 Huff-and-Puff Process

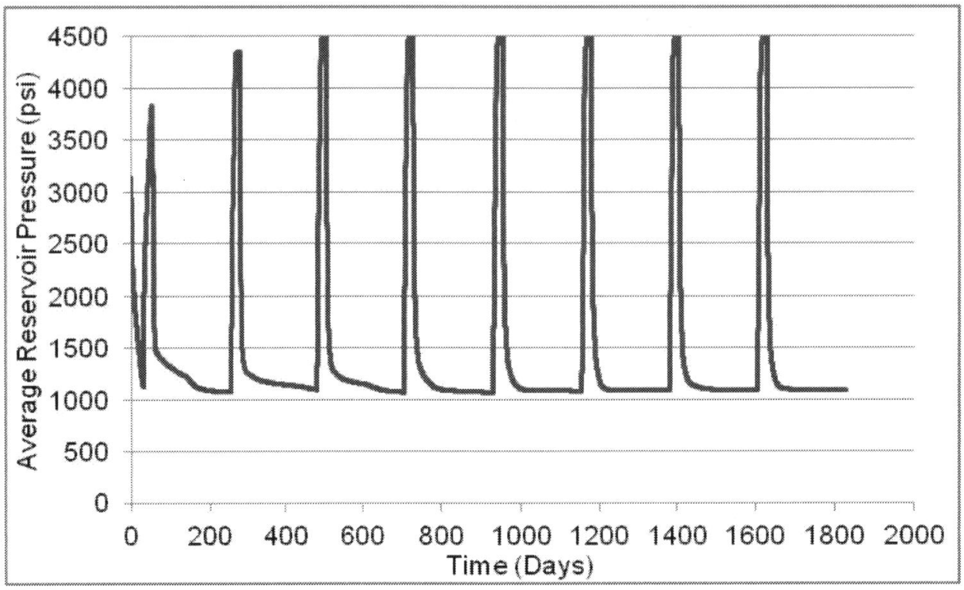

Figure 15. Average Reservoir Pressure of 8-cycle CO2 Huff-and-Puff Process

According to Sorensen (2015), five North Dakota Bakken field injection tests were conducted among which two were CO_2 tests. Field results showed that injectivity and production responses were demonstrated, but there is a gap between the simulation and field results. To illustrate the importance of optimal injection rate/length of injection period and understand the possible causes of the discrepancy, we compared 8-cycle CO_2 huff-and-puff cases with 5-day and 15-day injection period for each cycle. The oil recovery at different pore volume of CO_2 injected is shown in **Figure 16**. As observed, the ultimate recovery of 15-day injection period reached almost 44%OOIP at 0.8PV of CO_2 injected, 11% higher than the recovery of the 5-day injection case which was only 33%. This was especially true for early recovery build-up. The figure also showed a larger response for CO_2 injection by 15-day injection period case at, say, 0.1 PV CO_2 injected. 15-day injection case showed 27% oil recovery, while two 5-day-injection cycles produced 19% recovery combined. The results indicate that there might have a threshold for the minimum volume of CO_2 injected to achieve a significant incremental recovery. It was especially true when it comes to the field application which required a large volume of initial CO_2 injection.

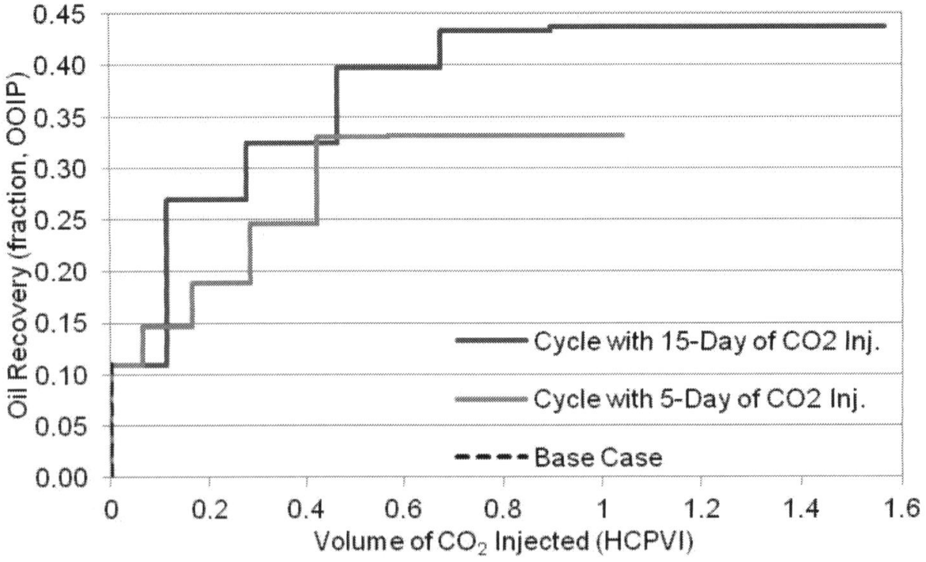

Figure 16. Effect of Length of Injection Period (5-day or 15-day) on Oil Recovery of 8-cycle CO2 Huff-and-Puff Processes

Conclusions

With our advances in understanding key production mechanisms of capillarity and adsorption, we are able to differentiate production driving mechanisms in unconventional reservoirs vs. conventional ones. Through using in-house simulator, we performed a series of simulations for CO_2 Huff-and-Puff recovery in the Bakken Formation. The following conclusions can be drawn from this work:

(1). This paper implements a novel formulation that captures capillarity pressure under pore confinement using a full spectrum of PSD characterization for shale oil core analysis.

(2). A substantial increase in oil recovery is observed for CO_2 Huff-and-Puff process in the Bakken Formation.

(3). CO_2 injection rate/length of CO_2 injection period, soaking time, and number of cycles for important parameters for CO_2 Huff-and-Puff recovery in the Bakken.

References

Afonja, G.,Hughes, R.G.,Nagineni,V.G.R., and Jin, L., 2012, Simulation Study for Optimizing Injected Surfactant Volume in a Miscible Carbon Dioxide Flood, Paper SPE 158220 presented at SPETT 2012 Energy Conference and Exhibition,Port-of-Spain, Trinidad, 11-13 June.

Anna, L.O., Pollastro, Richard, and Gaswirth, S.B., 2013, Williston Basin Province—Stratigraphic and structural framework to a geologic assessment of undiscovered oil and gas resources, chap. 2 of U.S. Geological Survey Williston Basin Province Assessment Team, Assessment of undiscovered oil and gas resources of the Williston Basin Province of North Dakota, Montana, and South Dakota, 2010 (ver. 1.1, November 2013): U.S. Geological SurveyDigital Data Series 69–W, 17 p.

Arakcheev,V.G., Valeev, A.A. Morozov, V.B. Olenin, A.N., CARS Diagnostics of Molecular Media under Nanoporous Confinement, Laser Physics 18 (2008) 1451–1458.

Chen, J. H., Wong, D. S. H., Tan, C. S., Subramanian, R., Lira, C. T., & Orth, M. 1997, Adsorption and Desorption of Carbon Dioxide onto and from Activated Carbon at High Pressures. Industrial & Engineering Chemistry Research, 36(7), 2808-2815.

Cherian, B.V., Stacey, E.S., Lewis, R., Iwere, F.O., Heim, R.N., and Higgins, S.M., 2012, Evaluating Horizontal Well Completion Effectiveness in a Field Development Program, Paper SPE 152177 presented at Hydraulic Fracturing Technology Conference, 6-8 February, The Woodlands, Texas.

Clark, A.J. 2009, Determination of Recovery Factor in the Bakken Formation, Mountrail County, ND, Paper SPE 133719 presented at SPE Annual Technical Conference and Exhibition, New Orleans, Louisiana, 4-7 October.

Dechongkit, P. and Prasad M., 2011, Recovery Factor and Reserves Estimation in the Bakken Petroleum System (Analysis of the Antelope, Sanish and Parshall fields), presented at Canadian Unconventional Resources Conference,Calgary, Alberta, Canada15-17 November.

Do, D. D., 1998, “Adsorption Analysis:Equilibria and Kinetics”.

Elgmati, M., Zhang, H., Bai. B., Flori, R., Qu, Q., 2011, Submicron-Pore Characterization of Shale Gas Plays, Paper SPE 144050 presented at North American Unconventional Gas Conference and Exhibition, 14-16 June.

Fitzgerald, J.E., Sudibandriyo, M., Pan, Z., Robinson R.L. Jr., Gasem, K.A.M. Modeling the Adsorption of Pure Gases on Coals with The SLD Model, Carbon 41 2203-2216 (2003).

Hiejima,Y., Yao, M., Phase Behaviour of Water Confined in Vycor Glass at High Temperatures and Pressures, Journal of Physics: Condensed Matter 16 (2004) 7903–7908.

Hoffman, B.T., 2012, Comparison of Various Gases for Enhanced Recovery from Shale Oil Reservoirs, paper SPE 154329 presented at SPE Improved Oil Recovery Symposium, Tulsa, Oklahoma,14-18 April.

Iwere, F.O., Heim, R.N., Cherian, B.V., 2012, Numerical Simulation of Enhanced Oil Recovery in the Middle Bakken and Upper Three Forks Tight Oil Reservoirs of the Williston Basin, paper SPE 154937 presented at SPE Americas Unconventional Resources Conference, Pittsburgh, Pennsylvania,5-7 June.

Jin, L., Pu, H., Wang, Y., Li, Y., 2015, The Consideration of Pore Size Distribution in Organic-rich Unconventional Formations May Increase Oil Production and Reserve by 25%, Eagle Ford Case Study. Paper SPE 178507/URTeC-2148314, Proceedings of the Unconventional Resources Technology Conference, San Antonio, TX, Jul. 20-22.

Kang, S.M., Fathi, E., Ambrose, R.J., Akkutlu, I.Y., and Sigal, R.F., 2010, CO_2 Storage Capacity of Organic-Rich Shales, paper SPE 134583 presented at SPE Annual Technical Conference and Exhibition.

Kazi, T., Li, Y., Jin, L., Wu, X., 2015, How Typical Langmuir Isotherm Curves Underestimate OGIP and Production by 20%, a Barnett Case Study. AAPG Datapages/Search and Discovery Article #90214, Proceedings of Southwest Section AAPG Annual Convention, Wichita Falls, TX, April 11-14.

Kohlmann, J.; Zevenhoven, R., 2000, The Removal of CO_2 from Flue Gases using Magnesium Silicates in Finland. In Proceedings of the 11th International Conference on Coal Science, San Francisco, CA, USA, 29 September–5 October.

Kovscek, A.R., Tang, G.-Q., and Jessen, K., 2005, Laboratory and Simulation Investigation of Enhanced Coalbed Methane Recovery by Gas Injection: Paper SPE 95947 presented at SPE Annual Technical Conference and Exhibition.

Kuila, U. and Prasad, M., 2011, Surface Area and Pore-Size Distribution in Clays and Shales. Paper SPE 146869 presented at the SPE Annual Technical Conference and Exhibition, Denver, Colorado, 30 October–2 November

Langmuir, I., 1916, The Constitution and Fundamental Properties of Solids and Liquids. Part I. Solids: Journal of the American Chemical Society, v. 38, p. 2221-2295.

LeFever, J. A. 2005. Montana North Dakota Middle Member Bakken Play. North Dakota Geological Survey.

Li, Y. Pu, H. 2015, Modeling Study on CO_2 Capture and Storage in Organic-Rich Shale, Paper CMTC-439561 presented at Carbon Management Technology Conference (CMTC 2015), Sugarland, Texas, November 17-19.

Ma, Y. and Jamili, A., 2014, Using Simplified Local Density/ Peng-Robinson Equation of State to Study the Effects of Confinement in Shale Formations on Phase Behavior, Paper SPE 168986 presented at SPE Unconventional Resources Conference,The Woodlands, Texas, 1-3 April.

Meissner, F.F., 1978, Petroleum Geology of the Bakken Formation, Williston basin, North Dakota and Montana, in D. Rehig., (ed.), The economic geology of the Williston basin: Proceedings of the Montana Geological Society, 24th Annual Conference, p. 207-227.

Nojabaei, B., Johns, R.T., Chu, L., 2013 Effect of Capillary Pressure on Phase Behavior in Tight Rocks and Shales, SPE Reservoir Evaluation & Engineering Journal, July 2013.

Nuttall, B.C., Eble, C.F., Drahovzal, J.A., and Bustin, R.M., 2005, Analysis of Devonian Black Shales in Kentucky for Potential Carbon Dioxide Sequestration and Enhanced Natural Gas Production.

Price, L.C., 1999, Origins and Characteristics of the Basin-Centered Continuous-Reservoir Unconventional Oil-Resource Base of the Bakken Source System, Williston Basin: The Energy & Environmental Research Center (EERC), Website accessed October 22, 2015. http://www.undeerc.org/News-Publications/Leigh-Price-Paper/Default.aspx

Rangarajan, B., Lira, C. T. and Subramanian R., 1995, Simplified Local Density Model for Adsorption over Large Pressure Ranges. AIChE Journal. 41.4: 838-845.

Saidian, M. and Prasad, M., 2015 Effect of Mineralogy on Nuclear Magnetic Resonance Surface Relaxivity: A Case Study of Middle Bakken and Three Forks Formations, Fuel, Volume 161, Pages 197-206

Sapmanee, K., 2011. Effects of Pore Proximity on Behavior and Production Prediction of Gas/ Condensate. Master Thesis, University of Oklahoma, 2011.

Sorensen, J., 2015, Bakken CO_2 Storage and Enhanced Recovery Program, Presented at U.S. Department of Energy, National Energy Technology Laboratory Carbon Storage R&D Project Review Meeting, 18-20 August.

Subramanian R., Pyada H., Lira C.T. An Engineering Model for Adsorption of Gases onto Flat Surfaces and Clustering in Supercritical Fluids, Ind. Eng. Chem. Res., 34 (11) 3830-37 (1995).

Thommes, M., Findenegg, G.H., Schoen, M., Critical Depletion of a Pure Fluid in Controlled-Pore Glass: Experimental Results and Grand Canonical Ensemble Monte Carlo Simulation, Langmuir 11 (1995) 2137–2142.

Tran, T., Sinurat, P., and Wattenbarger, R.A., 2011, Production Characteristics of the Bakken Shale Oil, Paper SPE 145684 presented at the SPE Annual Technical Conference and Exhibition, Denver, Colorado, 30 October-2 November.

U.S. Geological Survey Fact Sheet 2013-2013, U.S. Geological Survey, Denver, Colorado.

Wang, Y., Yan, B., Killough, J., 2013, Compositional Modeling of Tight Oil Using Dynamic Nanopore Properties, SPE 166267 presented at Annual Technical Conference and Exhibition, New Orleans, Louisiana, 30 September–2 October.

Yu, W., Lashgari H., Sepehrnoori, K., 2014, Simulation Study of CO_2 Huff-n-Puff Process in Bakken Tight Oil Reservoirs, Paper SPE 169575 presented at SPE Western North American and Rocky Mountain Joint Meeting, 17-18 April, Denver, Colorado.

Wong, A.P.Y, Kim, S.B., Goldburg, W.I. Chan, M.H.W. Phase-Separation, Density Fluctuation, and Critical-Dynamics of N_2 in Aerogel, Physical Review Letters 70 (1993) 954–957.

Appendix A

The calculation of $a(z)$ is described in the paper of Chen *et al.* (1997). For $L/\sigma_{ff} \geq 3$, $a(z)$ is given by

$$\frac{a(z)}{a_b} = \frac{3}{8}\left[\frac{z}{\sigma_{ff}} + \frac{5}{6} - \frac{1}{3\left(\frac{L-z}{\sigma_{ff}} - 0.5\right)^3}\right] \quad \text{for} \quad 0.5 \leq \frac{z}{\sigma_{ff}} \leq 1.5$$

$$\frac{a(z)}{a_b} = \frac{3}{8}\left[\frac{8}{3} - \frac{1}{3\left(\frac{z}{\sigma_{ff}} - 0.5\right)^3} - \frac{1}{3\left(\frac{L-z}{\sigma_{ff}} - 0.5\right)^3}\right] \quad \text{for} \quad 1.5 \leq \frac{z}{\sigma_{ff}} \leq \frac{L}{\sigma_{ff}} - 1.5$$

$$\frac{a(z)}{a_b} = \frac{3}{8}\left[\frac{L-z}{\sigma_{ff}} + \frac{5}{6} - \frac{1}{3\left(\frac{z}{\sigma_{ff}} - 0.5\right)^3}\right] \quad \text{for} \quad \frac{L}{\sigma_{ff}} - 1.5 \leq \frac{z}{\sigma_{ff}} \leq \frac{L}{\sigma_{ff}} - 0.5$$

When $2 < L/\sigma_{ff} < 3$, $a(z)$ is given by

$$\frac{a(z)}{a_b} = \frac{3}{8}\left[\frac{z}{\sigma_{ff}} + \frac{5}{6} - \frac{1}{3\left(\frac{L-z}{\sigma_{ff}} - 0.5\right)^3}\right] \quad \text{for} \quad 0.5 \leq \frac{z}{\sigma_{ff}} \leq \frac{L}{\sigma_{ff}} - 1.5$$

$$\frac{a(z)}{a_b} = \frac{3}{8}\left(\frac{L}{\sigma_{ff}} - 1\right) \quad \text{for} \quad \frac{L}{\sigma_{ff}} - 1.5 \leq \frac{z}{\sigma_{ff}} \leq 1.5$$

$$\frac{a(z)}{a_b} = \frac{3}{8}\left[\frac{L-z}{\sigma_{ff}} + \frac{5}{6} - \frac{1}{3\left(\frac{z}{\sigma_{ff}} - 0.5\right)^3}\right] \quad \text{for} \quad 1.5 \leq \frac{z}{\sigma_{ff}} \leq \frac{L}{\sigma_{ff}} - 0.5$$

When $1.5 \leq L/\sigma_{ff} \leq 2$, $a(z)$ is given by

$$\frac{a(z)}{a_b} = \frac{3}{8}\left(\frac{L}{\sigma_{ff}} - 1\right)$$

Integrated Reservoir Modeling and General Workflow for Determining Minimum Data Set Requirements for CO2 Leakage Detection and Migration Monitoring during Post CO2-EOR Storage

GAO, S.,
Sun, A., The University of Texas at Austin
Nicot, J., Bureau of Economic Geology
Originally presented on: 11/19/2015 12:45:00 - 13:10:00
Title

Integrated reservoir modeling and general workflow for determining minimum data set requirements for CO2 leakage detection and migration monitoring during post CO2-EOR storage Objective

A generally applicable workflow is developed for credible migration monitoring and leakage detection of CO2 geological storage during post CO2-EOR storage with most economical data acquisition investment. A systematic approach to determine the minimum data set is presented with an integrated reservoir modeling workflow for post CO2-EOR stage, from the 3D seismic data, geologic data, well log and core data, and field flowing history data (Figure 1) collected during CO2-EOR period. The minimum data set considers the minimum well log data types, as well as maximum vertical and lateral separations of the monitoring wells and temporal sampling intervals. Reservoir pressure evolution patterns are simulated with the reservoir modeling with the data so that CO2 migration paths are predicted.

This project advances the rich data set from Weyburn and Midale oilfields in south-eastern Saskatchewan, which apply man-made CO2 for enhanced oil recovery (EOR) for more than 10 years, and become the largest reservoir on post-EOR anthropogenic CO2 storage with most scrutinized world-leading research. The structural reservoir model is integrated in Petrel E&P platform. 3D geostatistical modeling integrating seismic data, well log data, and core data are analyzed in R. Reservoir simulation during post CO2-EOR storage is implemented in Compositional & Unconventional Reservoir Simulator CMG-GEM.

The minimum parameter set to model the reservoir hydraulic system, predict storage capacity and monitor CO2 migration that meets a certain level of risk assessment include:

(1) The types of formation petrophysical properties that are most sensitive to the CO2 geological storage prediction and migration monitoring,

(2) A largest spatial range of data acquisition that captures the possible CO2 migration range, which is normally larger than the storage complex,

(3) The smallest spatial resolution of sampling measurements, determined from

local geologic and petrophysical heterogeneity and anisotropy,

(4) The smallest temporal span of data acquisition for the purpose of reservoir monitoring.

In the exercise of reducing sampling data, the petrophysical properties of permeability and porosity are simulated in a stochastic way, and the reduction is accepted only when the pressure evolution follows the similar results as with the full dataset. In order to reduce the dataset, a hierarchical automatic clustering and geostatistic techniques are employed for hydraulic flowing facies classification with available core and well log measurements.

(1) From an extensive history matching with CMG-GEM reservoir simulation, the field's effective porosity, anisotropic permeability, and initial fluid composition at the beginning of (post-EOR) CO2 injection, are calibrated as the baseline for the minimum dataset derivation process. Permeability field is interpolated by kriging from core anisotropic permeability and porosity data, and modified to include local fractures and boundaries from the production history matching. Pressure distribution evolution is simulated with CO2 injected at various locations and different injection rates in the field within the rock's mechanic tolerance.

(2) The field of interest undergoes the primary facies classification, with effective porosity, anisotropic permeability, fluid composition, and stratigraphic formation indicator (from seismic amplitude data and interpreted by geologists) as the inputs. The spatial location of each data point is considered in the process to retain the spatial continuity of each facies. As a result, the sedimentary hydraulic flowing facies are automatically categorized honoring geologic information. For each facies, a secondary typing is implemented from effective porosity, anisotropic permeability so as to identify subfacies at a lower scale (Figure 2).

(3) To model the hydraulic flowing unit spatial heterogeneity of the field, continuous lag Markov chains based stochastic models are employed to the facies from step (2) to obtain transition probability matrix.

(4) Indicator geostatistical techniques are applied to the facies and subfacies, and transition rate matrix in 3-dimensional space are estimated.

The workflow is applied for 100 realizations for the Weyburn field, and the monitoring wells are recommended from each of the realization following by reservoir simulations in CMG-GEM. The workflow shows that seismic data takes a positive part in the facies recognization in the lateral directions, but not in the vertical part due to the limited seismic vertical resolution. The indicative kridging of permeability and porosity field provides more than 70% reduction of required well data by virtue of the facies indication.

Pressure Transient Response Analysis at the Southeast Regional Carbon Sequestration (SECARB) Anthropogenic Test Site near Citronelle, Alabama

Prepared for:

2015 Carbon Management Technology Conference
Sugar Land, TX

JohnRyan MacGregor
Advanced Resources International, Inc.

November 19, 2015

Advanced Resources International, Inc.

Project Assessment and Evaluation of the Area of Review (AoR) at the Citronelle SECARB Phase III Site Alabama, USA

Acknowledgement

This presentation is based upon work supported by the Department of Energy National Energy Technology Laboratory under **DE-FC26-05NT42590** and was prepared as an account of work sponsored by an agency of the United States Government. Neither the United States Government nor any agency thereof, nor any of their employees, makes any warranty, express or implied, or assumes any legal liability or responsibility for the accuracy, completeness, or usefulness of any information, apparatus, product, or process disclosed, or represents that its use would not infringe privately owned rights. Reference herein to any specific commercial product, process, or service by trade name, trademark, manufacturer, or otherwise does not necessarily constitute or imply its endorsement, recommendation, or favoring by the United States Government or any agency thereof. The views and opinions of authors expressed herein do not necessarily state or reflect those of the United States Government or any agency thereof.

NETL · SECARB Southeast Regional Carbon Sequestration Partnership · SOUTHERN COMPANY Energy to Serve Your World · SOUTHERN STATES ENERGY BOARD · EPRI ELECTRIC POWER RESEARCH INSTITUTE · Denbury · Advanced Resources International, Inc.

Presentation Outline

I. Introduction
II. Study Area Overview and Background
  i. Geologic Characterization
  ii. Injection History
III. Objectives and Rationale
IV. Empirical Transient Data
V. Hypothetical Pressure Transient Calculations
VI. Preliminary Findings and Next Steps

Advanced Resources International, Inc.

Pressure Data and Interference Testing

This talk details the efforts to qualitatively assess the empirical pressure transient data collected from the SECARB Anthropogenic Test site and place constraints on reservoir saturations of CO_2

Demonstrate non-endangerment

Exhibit new techniques for constraining plume location

Advanced Resources International, Inc.

Anthropogenic Test Project Background

The SECARB Anthropogenic Test is an integrated Carbon Capture and Storage project

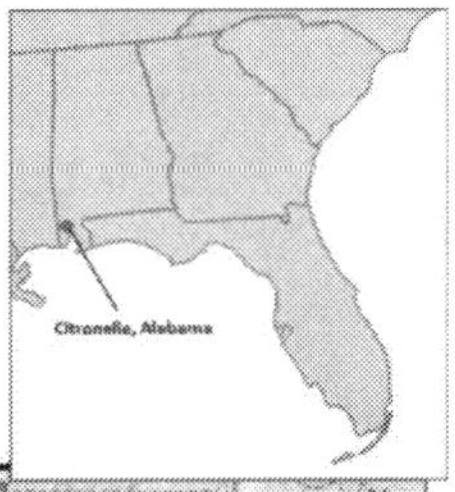

- CO_2 is captured from Southern Company's Plant Barry (25 MW slipstream from a 2,657 MW coal-fired power plant)
- A ~12 mile pipeline delivers the CO_2 to Denbury's Citronelle Oilfield
- The captured CO_2 is injected into the saline bearing Paluxy formation (~9,500 feet deep)

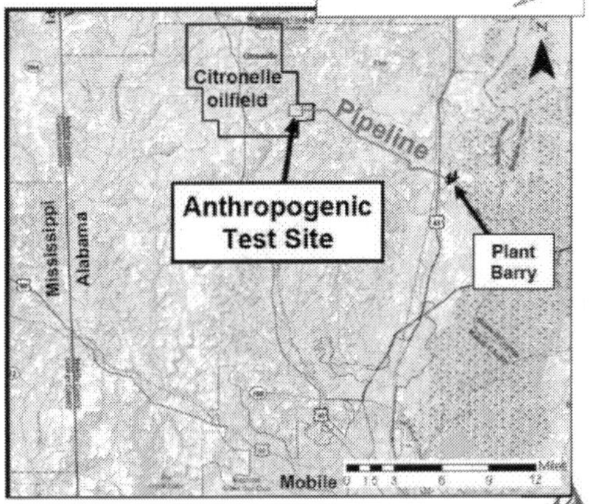

Approximately 114 thousand metric tons of CO_2 were injected between 8/12 and 9/14

Study Area Background

The Citronelle Oilfield lies atop the Citronelle Dome

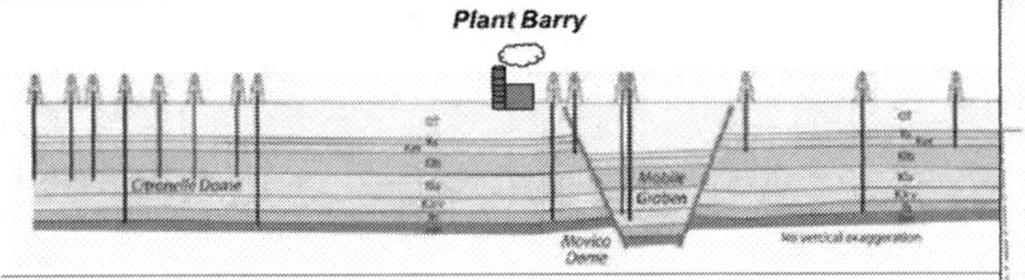

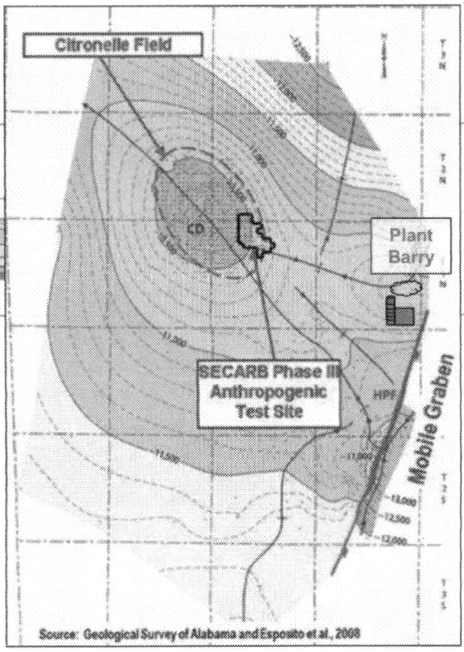

What makes the Citronelle Dome amenable to CO_2 sequestration?

- Well defined reginal geology from oil exploration and production
- Gulf coast sediments comprise excellent quality reservoirs
- A proven structural trap
- Structural closure is exhibited in all directions
- The limbs of the dome gently dip ~1°

Storage Site: The Citronelle Oilfield

CO_2 is injected at the Southeast Unit in Citronelle Oilfield

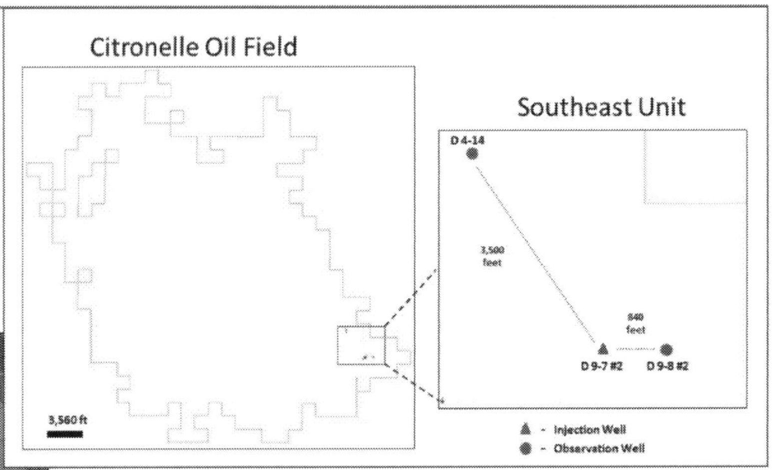

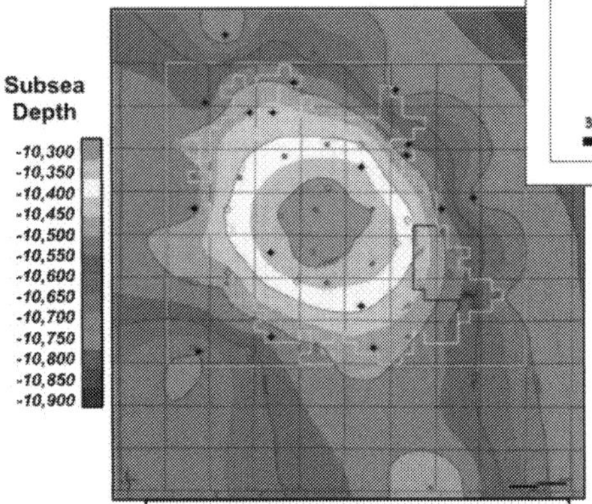

Structure Map of Ferry Lake Anhydrite

- One injection well perforated in multiple zones in the Paluxy Formation
- Two in-zone observation wells perforated in the injection zone

Project Assessment and Evaluation of the Area of Review (AoR) at the Citronelle SECARB Phase III Site Alabama, USA

Study Area Stratigraphy

Gulf coast sedimentary strata consist of extensive deposits of high porosity and permeability formations

Numerous confining units of impermeable marine shale and the Selma Group 2,000+ ft chalk unit provide confining zones to restrict CO_2 migration

CO_2 is injected into the Paluxy Formation

The field's target oil reservoir lies below the CO_2 injection zone

System	Series	Stratigraphic Unit	Major Sub Units	Potential Reservoirs and Confining Zones
Tertiary	Plio-Pleistocene		Citronelle Formation	Freshwater Aquifer
	Miocene	Undifferentiated		Freshwater Aquifer
	Oligocene		Chicasawhay Fm.	Base of USDW
		Vicksburg Group	Bucatunna Clay	Local Confining Unit
	Eocene	Jackson Group		Minor Saline Reservoir
		Claiborne Group	Talahatta Fm.	Saline Reservoir
		Wilcox Group	Hatchetigbee Sand Bashi Marl Salt Mountain LS	Saline Reservoir
	Paleocene	Midway Group	Porters Creek Clay	Confining Unit
Cretaceous	Upper	Selma Group		Confining Unit
		Eutaw Formation		Minor Saline Reservoir
		Tuscaloosa Group	Upper Tusc.	Minor Saline Reservoir
			Mid. Tusc. — Marine Shale	Confining Unit
			Lower Tusc. — Pilot Sand Massive sand	Saline Reservoir
Cretaceous	Lower	Washita-Fredericksburg	Dantzler sand	Saline Reservoir
			Basal Shale	Primary Confining Unit
		Paluxy Formation	'Upper' 'Middle' 'Lower'	CO_2 Injection Zone
		Mooringsport Formation		Confining Unit
		Ferry Lake Anhydrite		Confining Unit
		Donovan Sand	Rodessa Fm. 'Upper'	Oil Reservoir
			'Middle'	Minor Saline Reservoir
			'Lower'	Oil Reservoir

Reservoir Characteristics

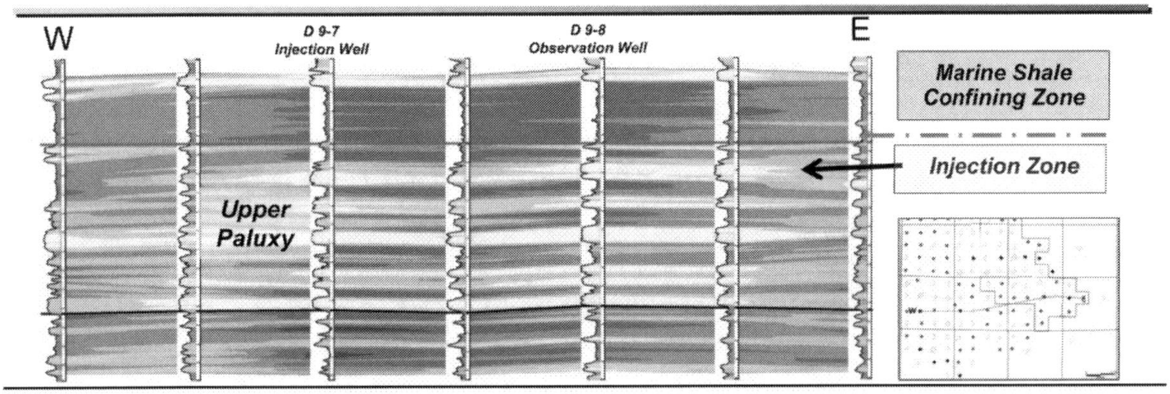

Site characterization demonstrated satisfactory geologic conditions for CO_2 storage

- The Paluxy SS reservoir is a relatively continuous saline formation
- Measured permeability up to 4,000 mD
- Porosity range: 4.4% - 26.1%
- Well established overlying cap rock

Parameters	Unit	Value
Mean Permeability, *k*	mD	373
Porosity, ϕ	%	19
D 9-8 Distance from Injector	feet	840
D 4-14 Distance from Injector	feet	3,500
Pressure	psi	4,430
Temperature	°F	224
Total Dissolved Solids (TDS)	ppm	200,000

Project Assessment and Evaluation of the Area of Review (AoR) at the Citronelle SECARB Phase III Site Alabama, USA

CO_2 Injection History

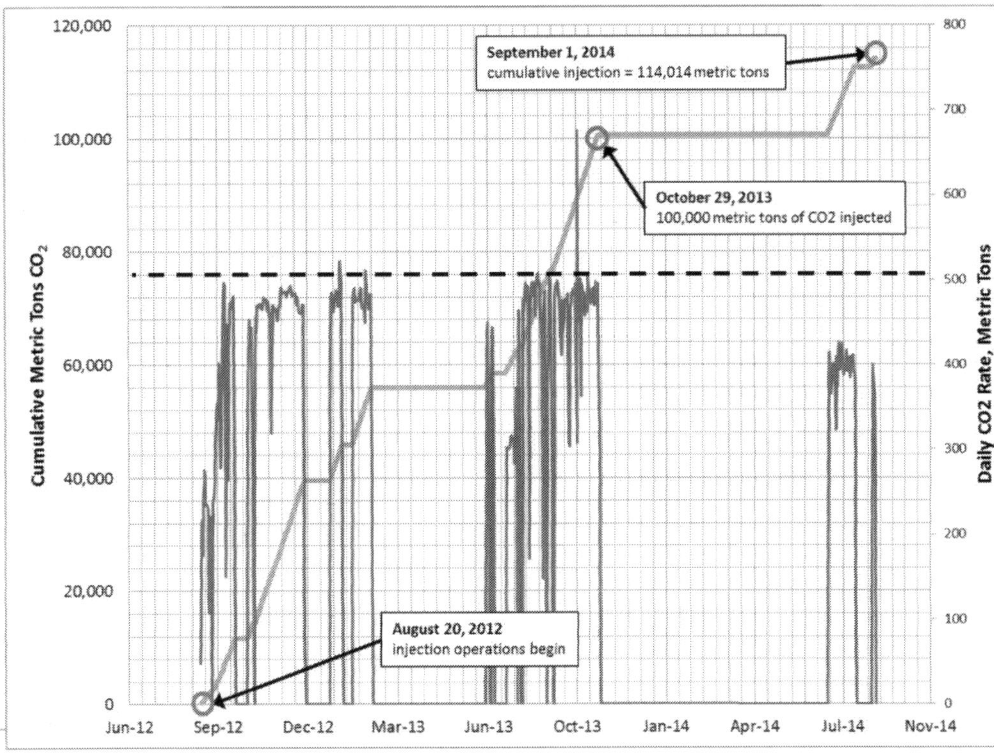

MVA Elements and Frequency

Multiple lines of evidence to confirm CO_2 containment

MVA Method	Frequency				
	Continuous	Monthly	Quarterly	Annual	Milestone *(Baseline, Injection, Post)*
Shallow					
Soil flux		■			
Groundwater sampling (USDW)			■		
PFT survey				■	
Deep					
CO2 volume, pressure & composition	■				
Reservoir fluid sampling				■	
Injection, temperature & spinner logs				■	
Pulse neutron logs				■	
Crosswell seismic					■
Vertical seismic profile (VSP)					■
Experimental					
Distributed Temperature Sensing (DTS)		■			
Comparative fluid sampling methods				■	
MBM VSP				■	
Distributed Acoustic Sensing (DAS)				■	
MBM VSP & OVSP Seismic				■	

Assure non-endangerment of USDWs

Project Assessment and Evaluation of the Area of Review (AoR) at the
Citronelle SECARB Phase III Site Alabama, USA

Objective

To detail changes in saline reservoir dynamics of the Paluxy Sandstone with the injection of CO_2 at the Citronelle Oil Field and to demonstrate non endangerment of USDW's

Rationale:

- ***Pressure data collected from two, in-zone observation wells provides a unique opportunity to establish pressure transient relationships for the reservoir***
- *Building an empirically validated model will enable the prediction of reservoir behavior with CO_2 injection, facilitating plume monitoring*

Objective

To detail changes in saline reservoir dynamics of the Paluxy Sandstone with the injection of CO_2 at the Citronelle Oil Field and to demonstrate non endangerment of USDW's

Rationale:

- *Pressure data collected from two, in-zone observation wells provides a unique opportunity to establish pressure transient relationships for the reservoir*
- ***Building an empirically validated model will enable the prediction of reservoir behavior with CO_2 injection, facilitating plume monitoring***

Methodology

- Pressure data from two in-zone observation wells were collected from downhole pressure and temperature gauges deployed in the injection zone.
- Pressure data was compared with injection volume and timing to determine pressure-transient response time.

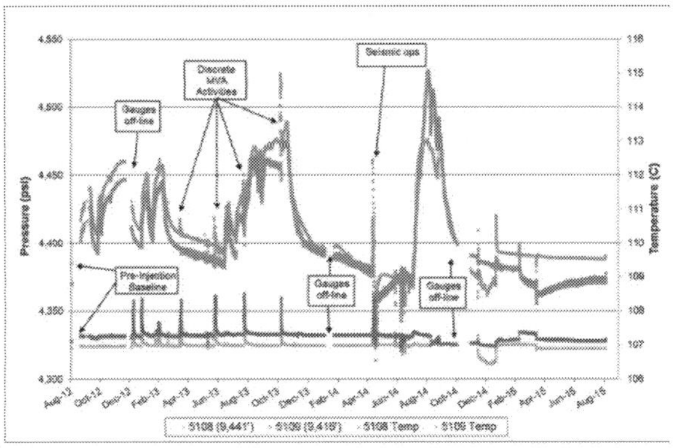

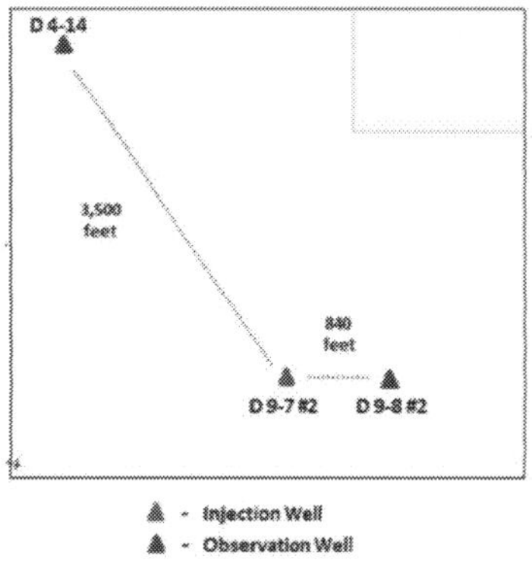

Theoretical response times were calculated using reservoir properties for several CO_2 saturations to place constraints on changes in CO_2 saturation in the system

Simple Transient Pressure Model

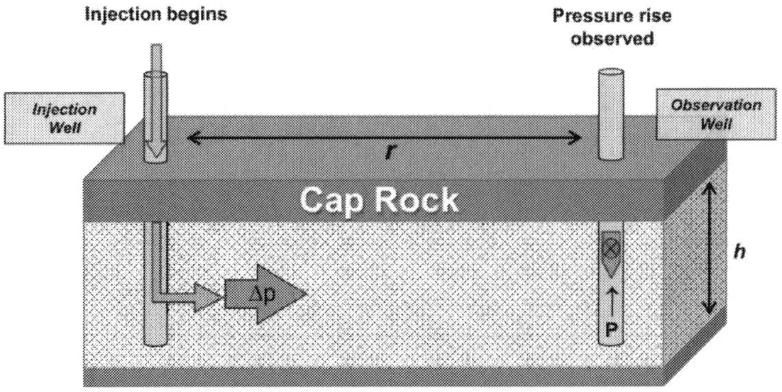

Time the pulse takes to reach the observation well is a function of reservoir characteristics

Empirical Observations

D 9-8 Pulse Testing

840 feet from injector

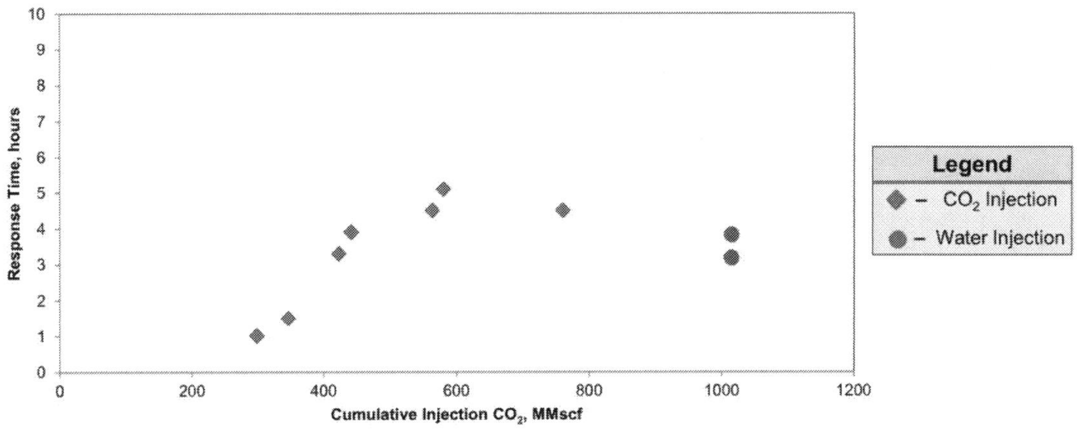

Response times recorded from downhole pressure data

- Red diamonds represent CO_2 injection starts
- Blue circles represents post-injection water pulse tests

Project Assessment and Evaluation of the Area of Review (AoR) at the Citronelle SECARB Phase III Site Alabama, USA

D 9-8 Pulse Testing

840 feet from injector

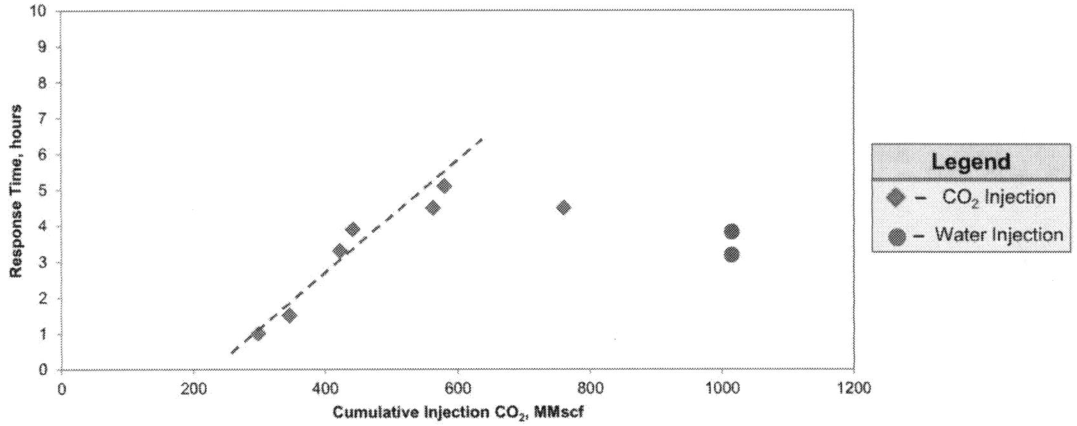

A general increase in response time is observed with an increase in cumulative injected volume for early data

D 9-8 Pulse Testing

840 feet from injector

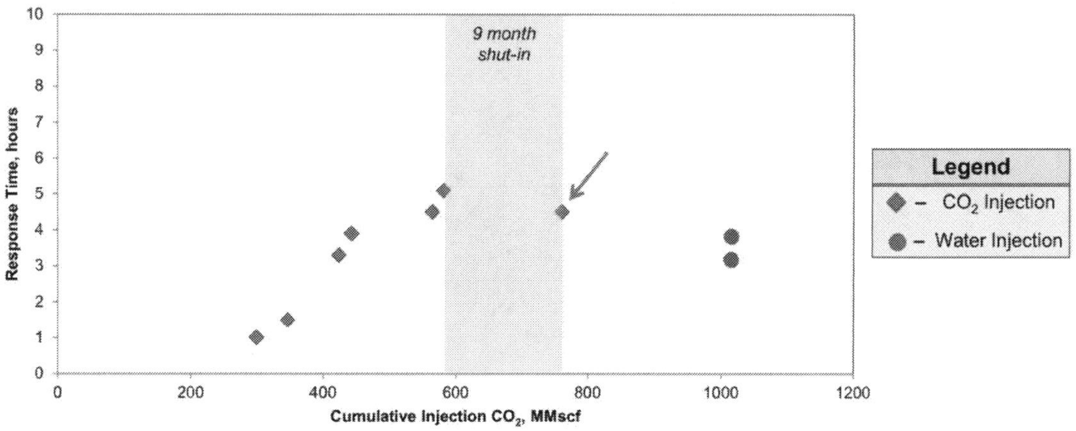

The observed drop in response time follows a 9 month shut-in period

Project Assessment and Evaluation of the Area of Review (AoR) at the Citronelle SECARB Phase III Site Alabama, USA

D 9-8 Pulse Testing

840 feet from injector

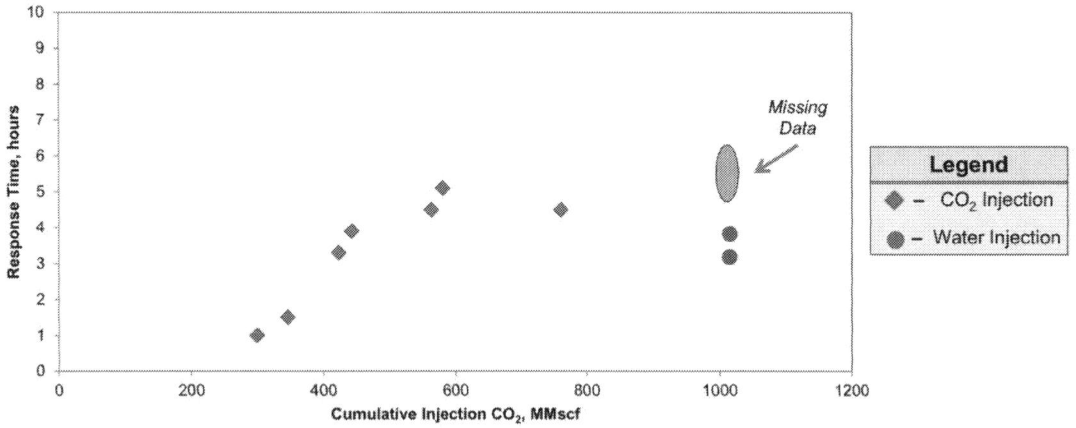

The pulse for the final phase of injection was not observed due top gauge failure

D 9-8 Pulse Testing

840 feet from injector

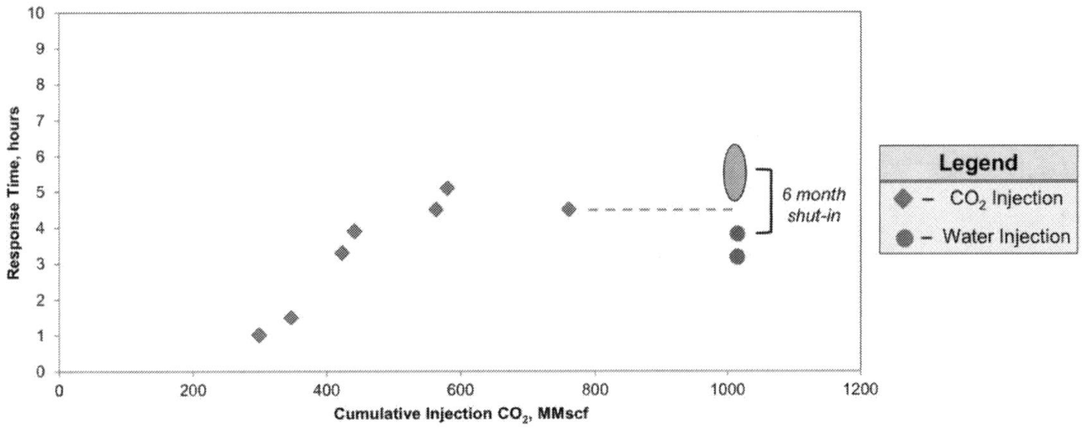

A second shut in period of 6 months yields another pressure drop

Project Assessment and Evaluation of the Area of Review (AoR) at the Citronelle SECARB Phase III Site Alabama, USA

D 9-8 Pulse Testing

840 feet from injector

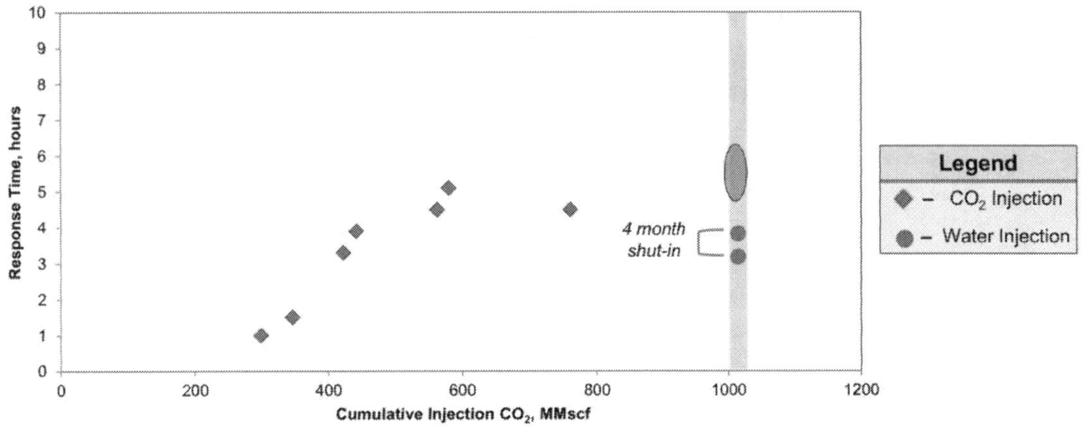

A third pressure drop is observed following 4 more months shut-in

Constraining Saturation of CO_2 in the Reservoir

Theoretical Transient Time Calculations

Utilizing the measured reservoir properties, the travel time of a pressure pulse transient from the injection well to the observation well can be solved by its relationship to dimensionless time (Ahmed, 2010)

$$t_D = \frac{\eta t}{r^2} \quad \text{which rearranges to} \quad \frac{t_D r^2}{\eta} = t$$

Where:

- η – *diffusivity constant* - **(function of: k, ϕ, fluid μ, c_t)**
- t_D – *dimensionless time*
- ***r*** – *radius of investigation*
- ***t*** – *time*

Calculating this for a range of CO_2 saturations will place constraints on the CO_2 saturation of the system

Theoretical Transient Time Calculations

The Hurst-van Everdingen exponential integral (Ei) function for the radial diffusivity equation is utilized to solve for t_D

The transient pressure rise observed at a known radius for a specific time is a function of the initial pressure, reservoir properties, and the Ei function —

$$p_{r,t} = p_i - \frac{0.141 q \mu}{kh}\left[{}^{1}/_{2}\left(-\mathrm{Ei}\frac{-1}{4t_D}\right)\right]$$

Equation 3.20a, Slider (1983)

Rearrange to solve for the Ei pressure function using known reservoir properties and Δp —

$$\left[\!\left[{}^{1}/_{2}\left(-Ei\frac{-1}{4\,t_D}\right)\right]\!\right] = (p_i - p_{r,t}) \div \left(\frac{0.141\,q\,\mu}{k\,h}\right)$$

Ei function | Δp | *Reservoir properties*

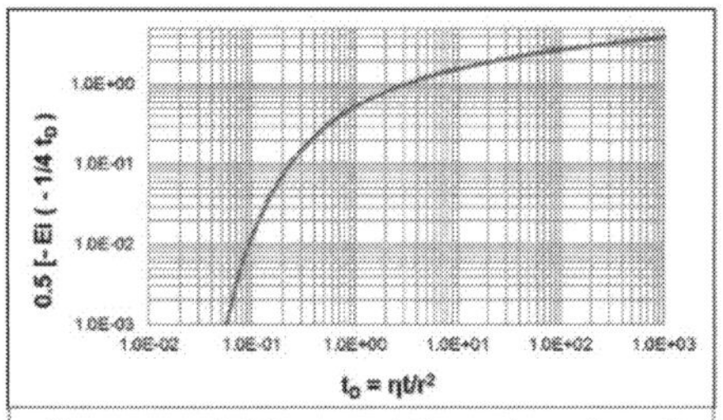

Exponential Integral Function Graph

← The Ei pressure function solution is related to t_D

t_D is used to solve for the theoretical pressure transient time at a given reservoir saturation of CO_2

$$\frac{t_D\, r^2}{\eta} = t$$

Advanced Resources International, Inc.

Transient Pressure Model

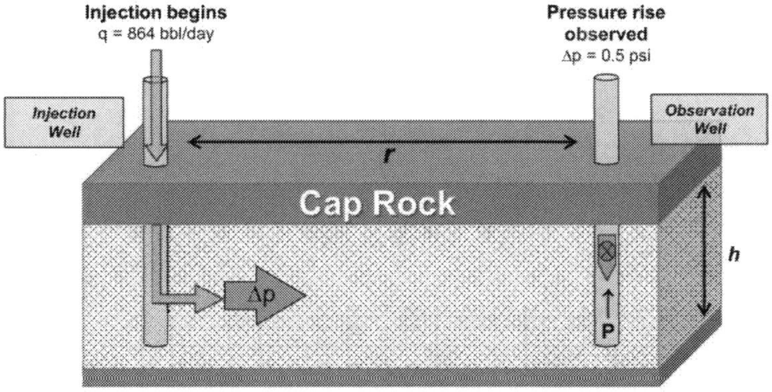

D 9-8 #2 Example (10% CO_2 Saturation)	
Fluid viscosity	0.49 cP
Net thickness, h	35 ft
Permeability, k	373 mD
Porosity, φ	19%
Radius, r	840 ft
Diffusivity C, η	1,450,474
System Compressibility	1.75E-5
Pressure gauge	

$$\left[\!\left[{}^{1}/_{2}\left(-Ei\frac{-1}{4\,t_D}\right)\right]\!\right] = 0.1093$$

$$\frac{t_D\, r^2}{\eta} = \quad t = 2.8 \text{ hours}$$

Theoretical response times for a pressure transient to travel from the injector to the observation well were calculated as a function of CO_2 saturation in the reservoir (a component of η)

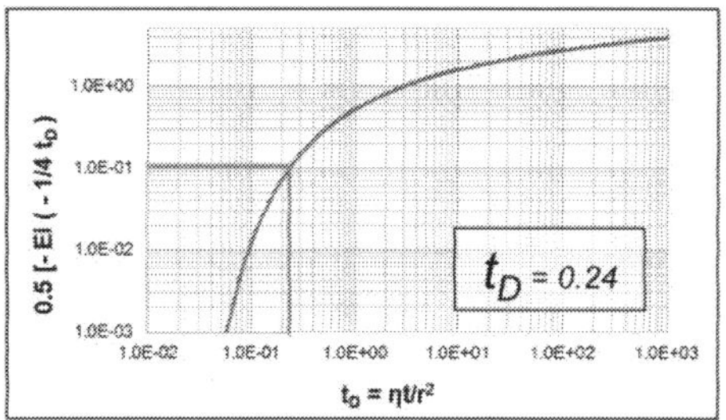

Exponential Integral Function Graph

Assumptions

- Homogenous distribution of CO_2 in reservoir
- Fixed reservoir properties

Advanced Resources International, Inc.

D 9-8 Pulse Testing

840 feet from injector

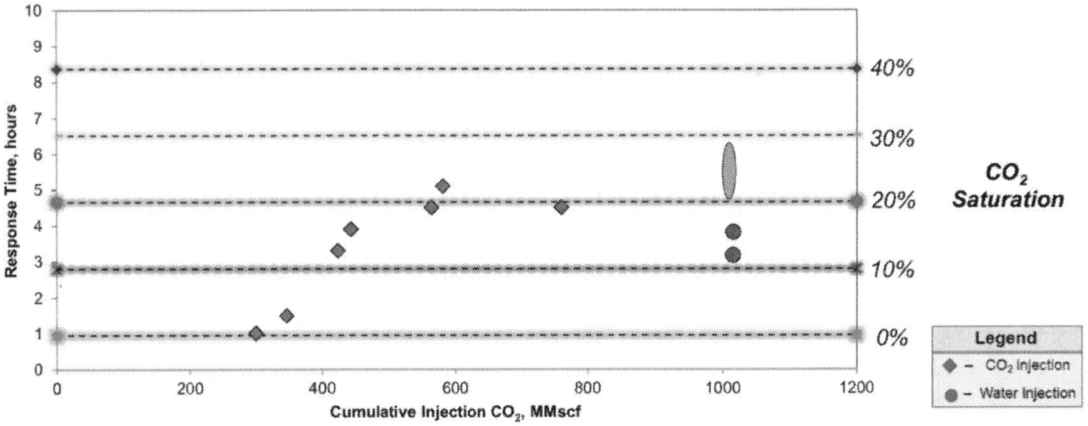

Theoretical response times are calculated from the E_i solution of the radial diffusivity equation for several saturation levels

Project Assessment and Evaluation of the Area of Review (AoR) at the Citronelle SECARB Phase III Site Alabama, USA

D 9-8 Pulse Testing

840 feet from injector

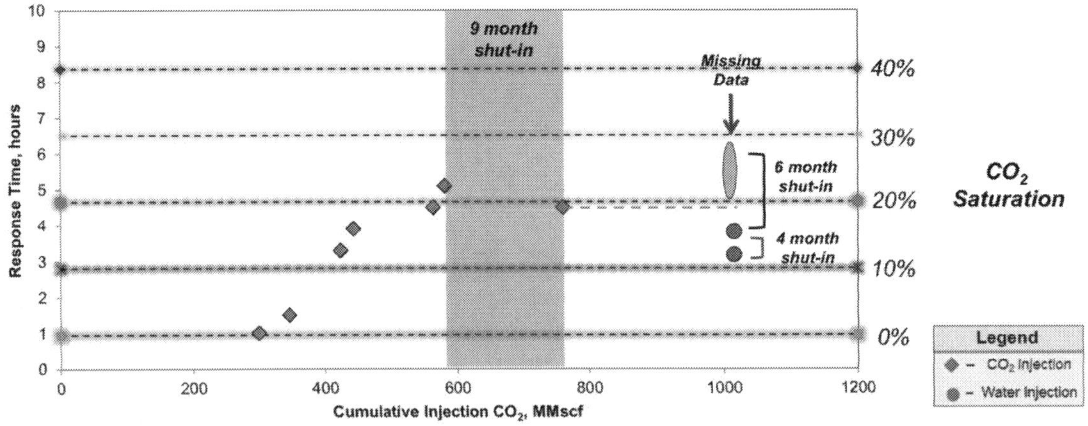

Theoretical response times are calculated from the E_i solution of the radial diffusivity equation for several saturation levels

D 4-14 Pulse Testing

3,500 feet from injector

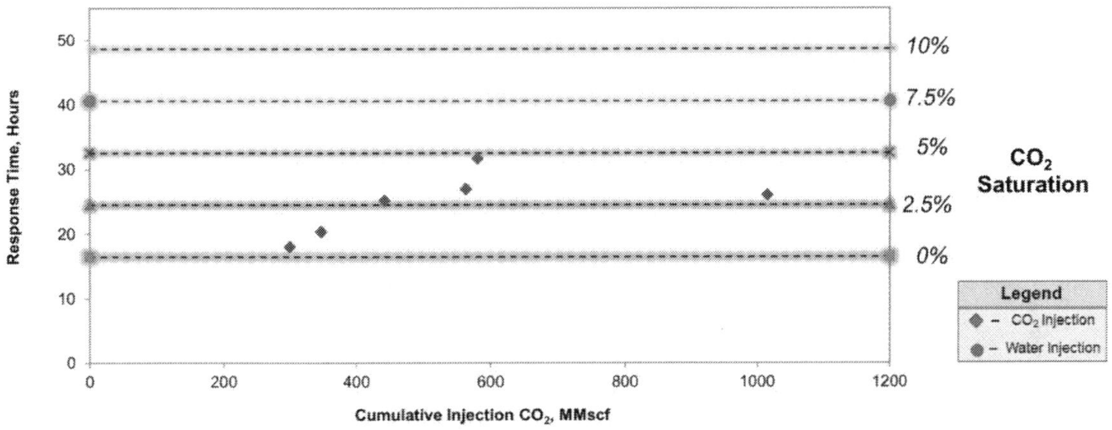

Similar observations are made for the D 4-14 well, albeit with lower saturation values due to larger distance from the injection well

Failed gauges account for the fewer data points

Project Assessment and Evaluation of the Area of Review (AoR) at the Citronelle SECARB Phase III Site Alabama, USA

D 4-14 Pulse Testing

3,500 feet from injector

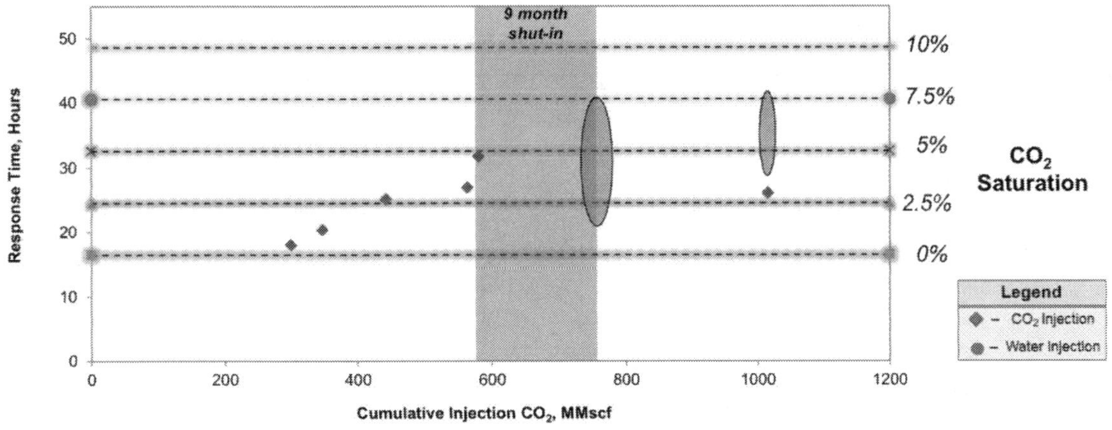

Similar observations are made for the D 4-14 well
(albeit with lower saturation values due to larger distance from the injection well)

Failed gauges account for the fewer data points

Preliminary Findings and Next Steps

Project Assessment and Evaluation of the Area of Review (AoR) at the
Citronelle SECARB Phase III Site Alabama, USA

Preliminary Findings

- The observed pressure data shows that response time of the pressure transient increases with increasing volumes of CO_2.
- The response-time increase may be attributed to an increase in the system's overall compressibility
- Drops in pressure response time following long, non-injection periods suggests that CO_2 may be going into solution with the formation water

CARBON MANAGEMENT TECHNOLOGY CONFERENCE

CMTC-439559-MS

Effectiveness of 4D Seismic Data to Monitor CO_2 Plume in Cranfield CO_2-EOR Project

Masoud Alfi, Texas A&M University, Seyyed A. Hosseini, Bureau of Economic Geology, The University of Texas at Austin, Mehrdad Alfi, Texas A&M University, and Mahmood Shakiba, Bureau of Economic Geology, The University of Texas at Austin

This paper was prepared for presentation at the Carbon Management Technology Conference held in Sugarland, Texas, USA, 17–19 November 2015.

Abstract

In this paper, we investigate the reliability of 4D seismic interpretation as a potential CO_2 injection monitoring tool by comparing its results to that of a numerical simulation in a CO_2 injection project at Cranfield, Mississippi, USA. Underground CO_2 injection, as a way to reduce greenhouse gas emission and/or improve ultimate recovery from mature oil fields, has been demonstrated to be effective and safe. This brings up the necessity of using monitoring tools, e.g. seismic methods, for leak detection, to provide information for optimizing injection plans, and to support the monitoring, verification, and accounting plans that will be required for site licensing. Using seismic surveys as monitoring methods has the advantage of covering a large area where it can provide information about the CO_2 saturation/distribution at locations in which no monitoring well exists. Comparison between CO_2 distribution in 4D seismic interpretations and numerical simulation results in Cranfield shows that although a sensible match is observed in areas of higher CO_2 saturation, central areas in the field, located along a sealing fault, as well as the areas on the edge of seismic survey limits show a less clear match. In fact, in these areas our compassion of CO_2 saturation distribution with interpreted 4D seismic data shows visible discrepancies. These discrepancies can be attributed to various challenges in Cranfield, impacting the reliability of the seismic interpretations. These challenges include, but are not limited to, masking effects due to the presence of residual gas (mainly methane) along the sealing fault in the reservoir, high depth of the formation (higher pressure makes the CO_2 density to be higher, reducing its contrast with oil and water), low thickness of formation (net sands are less than 65 ft thick), and low injection gas saturations (relative permeability curves suggest that gas saturation would barely go over 30-40% in most parts of the reservoir).

Introduction

Using carbon dioxide for enhanced oil recovery (EOR) has attracted a great deal of attention as the world grapples with the twin challenges of improving oil recovery from mature oil fields and reducing emissions of greenhouse gases. CO_2 is considered to be a favorable EOR agent as it can result in a miscible flood process while being less expensive compared to the other similar choices (solvents). Although very fascinating when expressed in terms of numbers and benefits, the life-saver CO_2 flood/sequestration projects may turn into a potential risk to the environment when not operated and monitored appropriately. The fact that many oil and gas formations have held hydrocarbons for millions of years prior to hydrocarbon production provides the fundamental support for possibility of storing CO_2 in similar formations; however, it may induce the risky conception of over-trusting these formations to securely store CO_2. Hence, aside from taking mandatory precautionary steps related with different safety aspects of CO_2 flood/sequestration (Duncan et al. 2008; He et al. 2011; Asamoto et al. 2013; Shadravan et al. 2015), monitoring the CO_2 injection process appears to be a necessity in these projects, not only because it helps to improve the efficiency of the process, but also to reduce the risk of CO_2 leaking out of the storage zone. The key factor during monitoring process is to keep track of the properties that can change as a result of CO_2 injection. Examples of such properties include, but are not limited to, pressure, saturation, temperature, density, etc. In this study, we try to evaluate the performance of the 4D seismic method, as a tool to monitor CO_2 injection process in a mature oil field. Although using seismic methods has the advantage of covering a large area, caution should be always practiced as understanding of the seismic data and dealing with unpredicted noise and related misinterpretations can pose many challenges. Reliability and repeatability of seismic interpretations not only depend on the way seismic signals have been generated and recorded, but also on reservoir related properties like geology and mineralogy, fluid and rock properties, pressure, depth and so forth (Lumley et al. 1997). As a result, while seismic methods can be considered a trustworthy source of information in some occasions, their reliability can be seriously questioned in some other circumstances.

The focus of this study is the CO_2 injection operation in the Cranfield reservoir, located in west-central Mississippi. Cranfield in known as a siliciclastic reservoir of the lower Tuscaloosa formation dated to the upper Cretaceous age. The average reservoir depth is about 10,500 ft below the sea level with an average reservoir pressure and temperature of 4,640 psi and 257 °F. Cranfield is believed to have a porosity ranging from 0 to 0.37 and permeability in the interval of 0.1 to 1,000 mD (Lu et al. 2012a; Lu et al. 2012b). The Cranfield field is a shallowly dipping (~2°) regional anticline and contains a volumetrically significant gas cap in the dome crest (Mississippi Oil and Gas Board 1966). The circular dome overlying the deepseated salt structure is slightly faulted and acts as a trap (Weaver and Anderson 1966). Figure 1 provides a structure contour map of the Cranfield reservoir, showing the location injection and production wells, along with the seismic survey region. Also shown in Figure 1, a northwest-southeast trending fault crosses the apex of the dome, offsetting the Tuscaloosa reservoir with a throw equal to the interval thickness (Mississippi Oil and Gas Board 1966). Although evidences show that the fault is non-transmissive in the horizontal direction, results of the current study suggest that residual gas updip migration and accumulation along the fault is considered to cause a masking effect, impacting the reliability of the seismic interpretations to monitor CO_2 plum during the injection process.

Cranfield reservoir was originally developed at 1940s by producing oil, gas condensate, and methane (Mississippi Oil and Gas Board 1966). A recycling plan was also implemented, by reinjection of the produced and make-up gas from Cranfield and adjacent reservoirs, to maintain pressure in the reservoir. Primary oil production was completed before mid-1960s and the gas cap was produced after that, resulting in a pressure drop followed by water encroachment from the edges. The operation at Cranfield was then stopped due to a high water cut where a cumulative amount of 37 MMbbl oil was produced up till 1965. With modern developments in enhanced oil recovery from CO_2 injection, and due to regionally available CO_2 and existing local infrastructure (Lu et al. 2012a), the site was revived for CO_2 injection on 2008. The area has been under CO_2 flood since then and by January 2015, more than 5 million metric tonnes of CO_2 were stored in the formation.

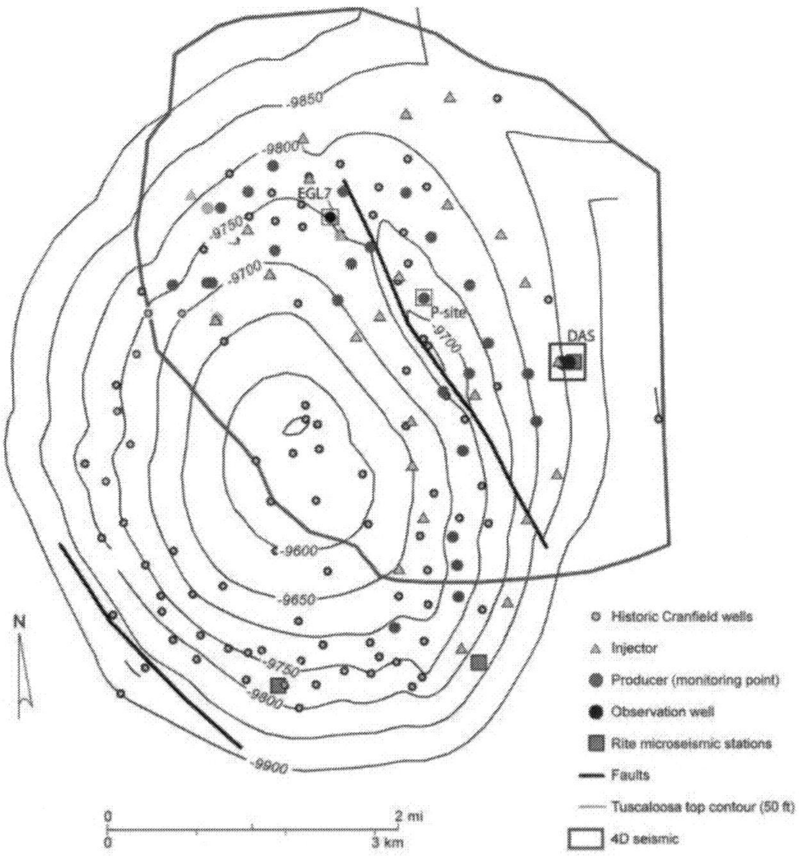

Figure 1—Map view of the top of Tuscaloosa sand at Cranfield including the well locations and data collected for characterization and monitoring through early 2011 (Hovorka et al. 2013).

Reservoir model

CMG-IMEX black oil simulator (Computer Modeling Group, Calgary, Canada) was used in this study to simulate CO_2 injection in a mature oil filed. CO_2 behavior and first- and multiple-contact miscibilities are modeled using a designated solvent phase and the concept of pseudo-miscibility (Todd and Longstaff 1972). The reservoir is simulated using a Cartesian grid system with the size of 124 by 49 by 20 blocks (i-j-k) where each grid block is 200×200×4 ft. Our simulation model covers an areal extent of 17,700 acres. The data presented by Hosseini et al. (2013) are used to generate permeability and porosity distribution maps, along with the relative permeability curves in this model. Outer boundaries of the reservoir are surrounded by an active aquifer which provides additional pressure support and maintains reservoir pressure within a certain limit (Hosseini and Nicot 2012). The numerical simulation model is tuned using pressure trends and production data and its validity is confirmed before using it for predictive simulations. History matching process was performed using 2.25 years' worth of data. Our goal was to history match cumulative oil, water, and gas production from the time CO_2 injection kicked off on July 2008 till the time of the second seismic survey (August 2010). Figures 2a, b, and c compare cumulative oil, water, and gas production from the field data and simulation results. Oil production data is used as hard constraint in the simulations. As seen from the graphs, our history matched model reasonably matches the field data. This increases our confidence to use this model to compare the output saturations with saturation maps interpreted from 4D seismic. To further verify the validity of the geological model, we are comparing the breakthrough times as well. By looking at Figure 2d, we can see that the breakthrough times predicted by our simulation model resemble well that of the field measurements.

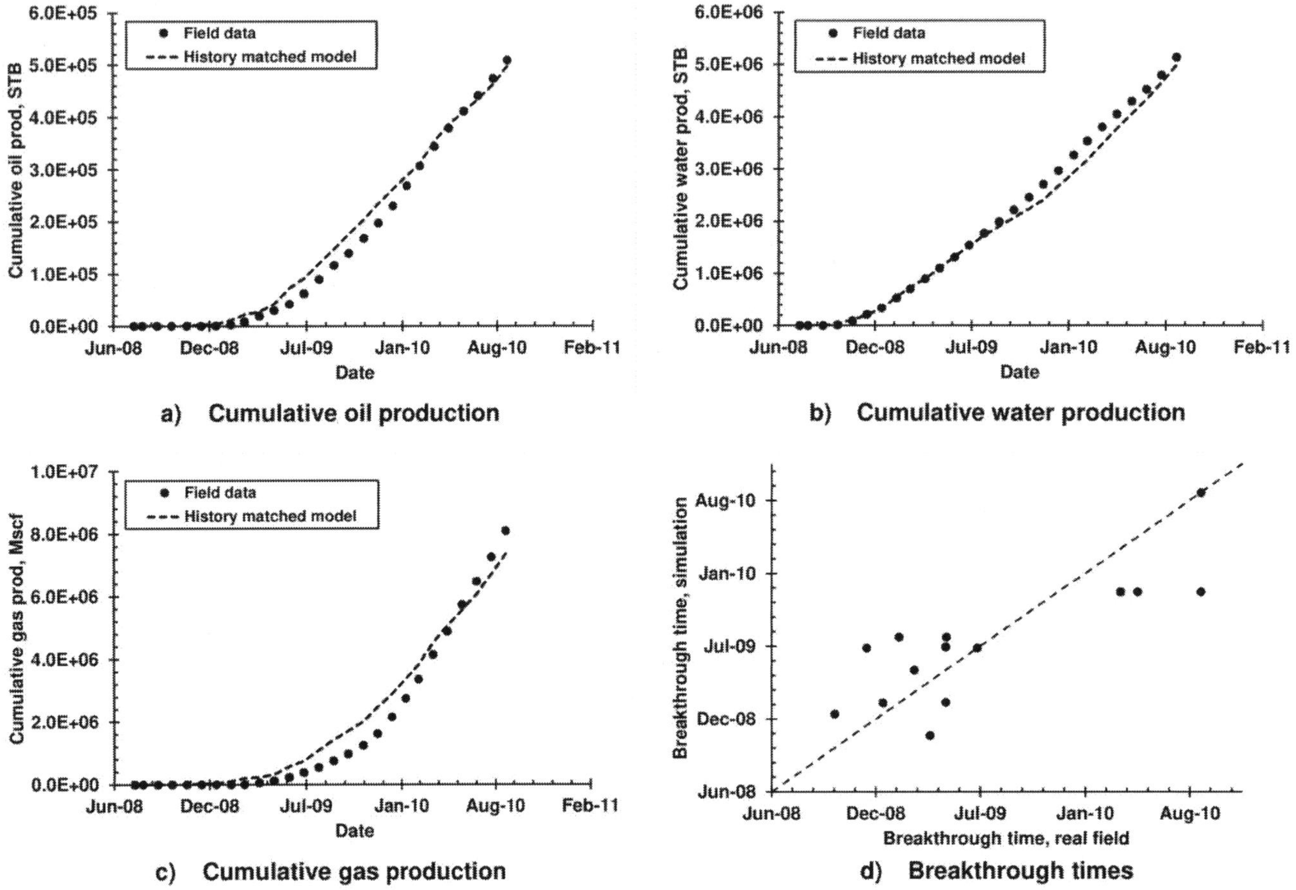

Figure 2—Simulation results of cumulative (a) oil, (b) water, and (c) gas production match that of the field data. (d) Modeled and observed breakthrough times are in an acceptable agreement as well.

Comparison of 4D seismic and simulation

Recent improvements in seismic data acquisition and interpretations have increased the demand to use such techniques for geophysical monitoring and verification purposes. Commonly known as 4D seismic, time-lapse seismic reservoir monitoring is the comparison of a 3D seismic survey with one or more repeat surveys taken in the same geographic location at different times. In fact, 4D seismic compares the changes in 3D seismic surveys over time. In this work, we have used the seismic interpretation of Cranfield reported by Carter (2014). Carter (2014) linked elastic properties of the media to the variations of CO_2 saturation, lithology, and cement content using rock-physics models to estimate a CO_2 saturation distribution map. Discussed seismic interpretation includes two vintages of 3D surface seismic data (from Cranfield, Mississippi), one for 2007, before CO_2 injection started, and a repeat from 2010, after a significant amount of CO_2 is injected into the reservoir zone. The repeat survey was collected to assess changes in saturation due to CO_2 injection around the oil ring in Cranfield. Seismic surveys were performed in the northeastern part of the reservoir and the overlapping region covers an area of approximately 3.7 miles × 3.7 miles. In both surveys the in-line and cross-line spacings were approximately 80 ft. With the rock-physics model calibrated to the reservoir, Carter (2014) found the reservoir properties (e.g. porosity) based on the P-impedance (Ip) values from the baseline impedance information. Ip values were calculated by inverting the surface seismic data using the basis pursuit inversion (BPI) algorithm (Zhang and Castagna 2011). The resulting inverted and measured porosity values were compared against the well-log data to assure the integrity of the process. Considering the porosity estimates to be fixed during the injection process, CO_2 saturation was estimated from the time-lapse Ip data. A growing amplitude anomaly with time in the targeted reservoir interval can be interpreted as CO_2 replacing oil and water in the reservoir. Figure 3a shows the interpreted CO_2 saturation map generated from the seismic surveys. The black triangles in the map show the location of injection wells. Regions denoted in red show the places where considerable amplitude changes is observed between the initial survey and the repeat survey. These areas are most likely to be influenced with injected CO_2 and alteration in reservoir fluid properties. The first glance at the saturation map confirms that areas with higher calculated CO_2 saturation are *all* located around an injection well. However, not all the injection wells are visible to the seismic survey. In other words, seismic survey does not appear to provide a comprehensive view of the CO_2 plume. It should be noted that the reliability of seismic interpretations and inversion results at the edges of the surveyed area is a place of question because of a lack of coverage. Hence, one should avoid over interpretation of such regions, as it might be just an art of fact. In an effort to compare the simulation data and seismic interpretation, Figure 3b shows the simulated average CO_2 saturation distribution in the reservoir at the same time as the repeat seismic survey is acquired.

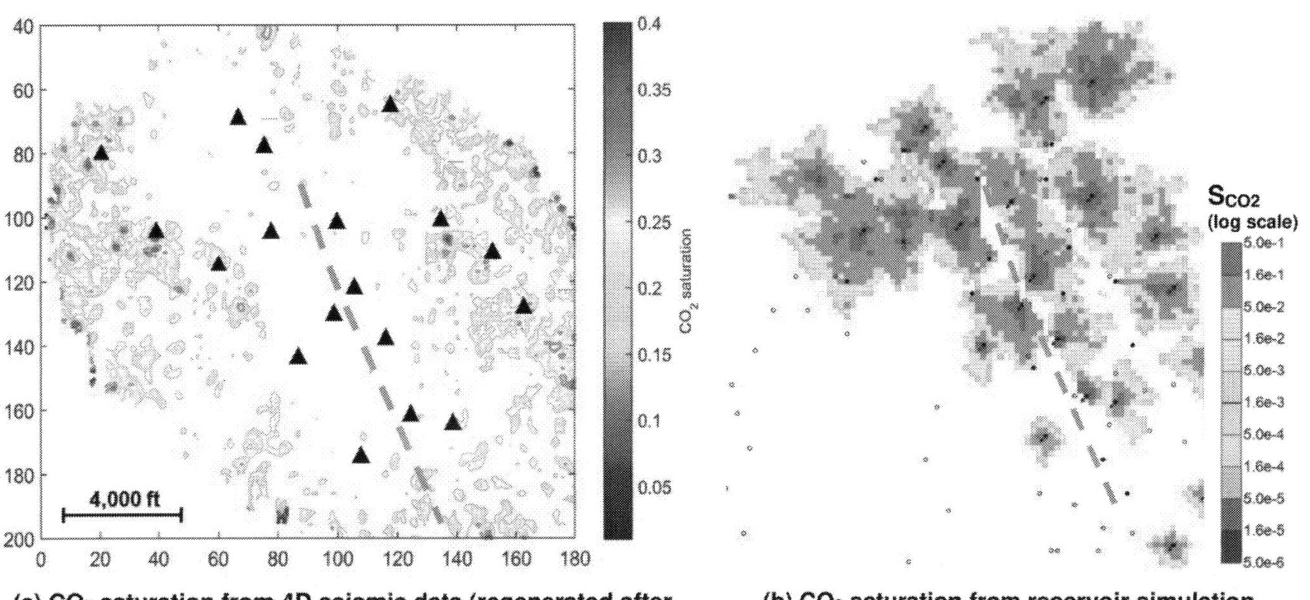

(a) CO_2 saturation from 4D seismic data (regenerated after Carter (2014))

(b) CO_2 saturation from reservoir simulation

Figure 3—Comparison of the 4D seismic interpretations and simulation results show that although seismic results are able to detect CO_2 plume closer to the edges, the central area around the sealing fault is not visible to the seismic surveys and is completely ignored by the interpretations.

According to Figure 3, although seismic data and numerical analysis share a common pattern in certain areas of the reservoir (closer to the edges), the differences between these two distribution maps is not something one can simply overlook. The uncertainty inherited in the geophysical-based data acquisition methods (e.g. signal strength, noise ratio, and repeatability) can result in the inconsistent interpretation of the seismic data in the areas distant for where any CO_2 was injected and seismic data is collected. On the other hand, field information and simulated data show that considerable amount of CO_2 is injected in the central part of the reservoir, along the sealing fault where the seismic signals look to be blind. This behavior can be probably explained by the fact that the presence of a fault would cause the residual gas in the reservoir (mainly made of methane) to migrate updip against the fault and accumulate over time. Existence of even a small percentage of high compressibility residual gas in layers above the injection zone in central areas is able to mask the acoustic change caused by CO_2 injection, reducing the sensitivity of seismic signals to CO_2 saturation alteration in this region, especially when net sand thickness is less than 65 ft. Some levels of mismatch between the observed and modeled change can also be attributed to the specific properties of the CO_2 injection site. Low net-to-gross ratio and narrow reservoir zone in some parts of Cranfield can reinforce the uncertainty related to seismic interpretations. The average reservoir depth at the injection interval of Cranfield is about 10,500 ft. When injecting in reservoirs of relatively high pressure, CO_2 will have a density close to that of oil and brine. Distinguishing the reservoir and injection fluid in such conditions poses additional interpretation challenges.

Conclusions

In this paper, we have evaluated the performance of 4D seismic methods in monitoring a CO_2 EOR project upon a nearly depleted oil reservoir to understand its potentials and drawbacks. This is achieved by comparing the saturation distribution maps generated based on interpretation of time-lapse seismic surveys with the numerical simulation results. Consistency of the numerical model used in this study is confirmed by history matching oil, water, and gas production data. CO_2 breakthrough times are also used to check the validity of our geological model. Our observations suggest that although 4D seismic can be considered as a potential monitoring tool, caution should be practiced as the accuracy and reliability of this method depends on a variety of reservoir, historical, geological, and geographical factors. To be more specific, for the case of Cranfield, presence of a sealing fault in the middle of the surveyed region can result in accumulation of the reservoir gas (gas exsolved during the historical production of the field in pressures below bubble point), which can ultimately mask the CO_2 invaded zones and intrigue the seismic signals. Low-thickness of the net pay and high depth of the reservoir are among the other factors affecting the quality of seismic incidents.

Acknowledgment

The authors thank Computer Modeling Group for providing the academic license for CMG-IMEX. Publication authorized by the Director of the Bureau of Economic Geology, the University of Texas at Austin.

References

Asamoto, S., Guen, Y.L., Poupard, O. et al. 2013. Well integrity assessment for CO_2 injection: A numerical case study on thermo-mechanical behavior in downhole CO_2 environments. *Engineering Computations* **30** (6): 842-853.

Carter, R.W. 2014. Fluid Characterization at the Cranfield CO_2 injection site: Quantitative Seismic Interpretation from Rock Physics modeling and Seismic Inversion. PhD dissertation, The University of Texas at Austin.

Duncan, I.J., Nicot, J.-P., and Choi, J.-W. 2008. Risk Assessment for Future CO_2 Sequestration Projects Based CO_2 Enhanced Oil Recovery in the U.S. Paper presented at the the 9th International Conference on Greenhouse Gas Control Technologies (GHGT-9), 16-20 November, Washington, D.C.

He, M., Luis, S., Rita, S. et al. 2011. Risk assessment of CO_2 injection processes and storage in carboniferous formations: a review. *Journal of Rock Mechanics and Geotechnical Engineering* **3** (1): 39-56.

Hosseini, S.A., Lashgari, H., Choi, J.W. et al. 2013. Static and dynamic reservoir modeling for geological

CO_2 sequestration at Cranfield, Mississippi, U.S.A. *International Journal of Greenhouse Gas Control* **18** (0): 449-462.

Hosseini, S.A. and Nicot, J.-P. 2012. Numerical modeling of a multiphase water–oil–CO_2 system using a water–CO_2 system: Application to the far field of a U.S. Gulf Coast reservoir. *International Journal of Greenhouse Gas Control* **10**: 88-99.

Hovorka, S.D., Meckel, T.A., and Treviño, R.H. 2013. Monitoring a large-volume injection at Cranfield, Mississippi—Project design and recommendations. *International Journal of Greenhouse Gas Control* **18**: 345-360.

Lu, J., Cook, P.J., Hosseini, S.A. et al. 2012a. Complex fluid flow revealed by monitoring CO_2 injection in a fluvial formation. *Journal of Geophysical Research: Solid Earth* **117** (B3).

Lu, J., Kharaka, Y.K., Thordsen, J.J. et al. 2012b. CO_2–rock–brine interactions in Lower Tuscaloosa Formation at Cranfield CO_2 sequestration site, Mississippi, U.S.A. *Chemical Geology* **291**: 269-277.

Lumley, D.E., Behrens, R.A., and Wang, Z. 1997. Assessing the Technical Risk of a 4D Seismic Project. Paper presented at the SEG Annual Meeting, 2-7 November, Dallas, Texas.

Mississippi Oil and Gas Board. 1966. Cranfield field: Cranfield unit, basal Tuscaloosa reservoir, Adams and Franklin Counties. 42-58.

Shadravan, A., Ghasemi, M., and Alfi, M. 2015. Zonal Isolation in Geothermal Wells. . Paper presented at the Fortieth Workshop on Geothermal Reservoir Engineering, 26-28 January, Stanford University, Stanford, California.

Todd, M.R. and Longstaff, W.J. 1972. The Development, Testing, and Application Of a Numerical Simulator for Predicting Miscible Flood Performance. *Journal of Petroleum Technology* **24** (7).

Weaver, L.K. and Anderson, K.F. 1966. Cranfield Field, Cranfield Unit, Basal Tuscaloosa Reservoir, Adams and Franklin Counties, Mississippi, in Oil Recovery from Gas- Cap Reservoirs: An Engineering Evaluation of Conservation Practices in Six Reservoirs. *Interstate Oil Compact Commission, Oklahoma City*: 42-58.

Zhang, R. and Castagna, J. 2011. Seismic sparse-layer reflectivity inversion using basis pursuit decomposition. *Geophysics* **76** (6): R147-R158.

CARBON MANAGEMENT TECHNOLOGY CONFERENCE

CMTC-439492-MS

Microseismic Monitoring on CO_2 Migration in a Tight Oil Reservoir during CO_2-EOR Process

Bo Ren*, Shaoran Ren, Liang Zhang, Haidong Huang**, China University of Petroleum, Guoli Chen, Hua Zhang, PetroChina Jilin Oilfield Company
* Now with the University of Texas at Austin
** Now with the Schlumberger Limited (China)

Abstract

Jilin Oilfield is conducting a large-scale demonstration project on CO_2 EOR and storage in China. CO_2 separated from a nearby natural gas reservoir (15-30 mole% CO_2) is injected into the northern part of H-59 oil block with permeability and porosity of 3.5 mD and 12.7%, respectively. After about six years of operation, nearly 0.26 million tons of CO_2 (0.32 HCPV) has been injected into the thin oil layers with well-developed natural fractures. In order to track the movement of CO_2 in the oil reservoir, a microseismic monitoring program has been implemented to map the CO_2 flow anisotropy and sweeping efficiency. Gas tracer testing has also been conducted to examine the inter-well connectivity. The production data has been analyzed in a real-time mode to monitor the dynamic response of different production wells, and meanwhile to verify the credibility of the results from microseismic and gas tracer tests. It is demonstrated that the migration or movement of CO_2 in the thin oil layers can be successfully detected by the microseismic technique, and the sweeping profiles of CO_2 obtained from the inversed microseismic are in good agreement with the temporal change of the produced CO_2 rate from producers as well as the characterization of the reservoir's petrophysical properties.

Introduction

The development of sustainable and safe technologies to mitigate emissions of greenhouse gases to the atmosphere is one of the major challenges in this century. Carbon capture and storage (CCS) has been widely accepted as the most promising and easily accessible way to reduce the carbon emission and global warming (IPCC, 2005). The captured CO_2 is injected into the subsurface reservoirs for the long-term storage/sequestration. The potential geological sites suitable for CO_2 storage include depleted oil and gas reservoirs, deep saline aquifers, unminable coal bed methane, and gas hydrate and geothermal reservoirs (Bachu, 2008). At present and for the near future, oil and gas reservoirs are the most favorable choices because of their proven trapping safety and economic potential combined with

CO_2 enhanced oil or gas recovery (Zhang *et al.*, 2015).

Although CO_2-EOR has been widely employed in the oil and gas industry for more than 60 years, many uncertainty and risks still exist in the face of the combined geological storage. One of the big concerns is the CO_2 migration in the oil reservoirs and the possible leakage from the geological bodies. From the perspective of enhanced oil recovery, the CO_2/oil miscibility, gas front migration, and injector/producer response also need to be monitored to accurately evaluate the field performance. Therefore, it is essential to deploy the monitoring, verification and accounting (MVA) techniques to understand the behavior of CO_2 in the geologic formations, track its migration/fate, and assess/prevent CO_2 leakage from the storage media (Brown *et al.*, 2009; Ren *et al.*, 2014). Meanwhile, through the MVA practice, the advantages and limitations of monitoring tools can be assessed for the future improvements (Litynski *et al.*, 2008). Additionally, this will help the operator to update the reservoir geological model and optimize the oilfield development program.

The CO_2 EOR and storage project in Jilin oilfield is the first-large scale CCS demonstration project in China. It aims at enhancing the understanding of CO_2 EOR mechanisms and ensuring the storage safety of CO_2 in the reservoir. In order to achieve these purposes, a comprehensive monitoring program has been designed and conducted during CO_2 injection into the tight oil reservoirs. The primary purpose of this study is to examine the feasibility of micro-seismic monitoring program employed in the field to track the CO_2 migration in the tight and thin oil reservoir. Detailed well fluid sampling analysis has been conducted and compared with the micro-seismic results. Gas tracer results have also been used in complimentary to the above analysis and additional insights are obtained with respect to the gas channeling/breakthrough features in the tight and heterogeneous oil reservoir.

The Jilin Oilfield CCS-EOR

Jilin Oilfield is located in the Jilin province of the Songliao basin in North-east China (Fig. 1a). The target oil reservoir of H59 block belongs to the DaQingZiJing region of the oilfield. The reservoir was selected for CO_2 EOR and storage based on its geological characterization and petrophysical properties under the reservoir conditions. The overall CCS-EOR project design and surface processing (Fig. 1b) has been described in several publications (e.g., Zhang *et al.*, 2015). The CO_2 is separated from the Changshen natural gas reservoir and then pumped via an 8 km pipeline to the H59 block. The CO_2 content in the natural gas stream is around 15-30 mole%. After the initial pilot test at the H59 oil block during the first phase, the following phases will include more than 200 patterns in other blocks nearby.

The target oil reservoir is composed of several producing zones from the Qingshankou formation formed in the age of Cretaceous. It is deposited in a delta front environment and controlled by two antithetic down faults (Fig. 2a). The oil-bearing formations are manifested with well-developed sand bodies. The gross sand layers range from 11.2 m to 18.2 m in thickness, including main layers 7, 12, 14, 15 in the Qing-I formation of the Qingshankou Formation, which take 78% of OOIP (Fig. 2b). The high-angle or vertical natural fractures are well developed in the direction of west-east in the oil reservoir, and hydraulic fracturing treatments have been conducted in several producers to boost oil production. The average porosity and permeability are 12.7% and 3.5 mD, respectively.

The CO_2 pilot area consists of 31 wells with a spacing of 440 x 140 m; 25 production wells and 6 CO_2 injectors. They are divided into 6 well groups as shown in Fig. 2a. The reservoir has been produced via primary depletion, and then water flooding was started in well H-4-2 and 6-6 from 2004 to 2008. In April 2008, CO_2 was firstly injected into wells 6-6 and 12-6, and then CO_2 injection was started with other wells over the subsequent six months. The wellhead gas injection pressure is of 11-14 MPa, and at -22 °C (liquid CO_2). Since June 2009, water alternating gas (WAG) injection was initiated in well 12-6 and 6-6, and spread to other injectors in four months. The WAG injection cannot be maintained in a mode of the strict cycle due to the poor water injectivity. Since 2011, the well pattern has been changed; H59-14-6 with high produced CO_2 content was converted to gas injector, followed by H59-10-4, H-8-8, and H-4-1. Up to June 2014, approximately 0.32 Hydrocarbon Pore Volume (HCPV) of CO_2 has been totally injected into the oil reservoir with the incremental oil recovery factor of 4.5%.

3. Field Implementation of Monitoring Techniques for Tracking CO2

3.1 Microseismic test

As shown in Fig. 3, an array of 24 three-component geophone receivers was placed around the monitoring well at the near-surface with a spacing of 150 m. The geophones were clamped at the depth of 10 to 30 meters below surface to avoid the surface noise and adsorption of the microseismic wave. The total coverage area of the sensor network is about 600 x 600 m, with each side of about twice of the injector-producer distance.

Before the microseismic test, the CO_2 injectors were shut down for more than 15 hours in order to make the existing fractures in the reservoir closed. When the wellhead pressure decreases quite low (< 2 MPa), CO_2 injection is re-started with the normal injection rate, and the injection pressure might re-open the previously-closed natural fractures and meanwhile with some micro-seismic events induced. The dense degree of micro-seismic events induced from different layers/positions is used to monitor the movement direction and sweeping profile of CO_2 in the oil reservoir.

In the H59 oil block, micro-seismic mapping has been implemented in the four CO_2 vertical injectors; H59-4-2, H59-6-6, H59-1 and H59-14-6. The detailed monitoring time, interval and totally CO_2 injected during monitoring test has been shown in Table 1. The microseismic events have been interpreted and located by the contractor (Sinopetro Longshuo Ltd). Here we discuss them in relation to the injection/production response and compare them with the gas tracer results. As seen from Table 1, most of the microseismic events were interpreted to be induced from the layer 7, which sucks most of the CO_2 as confirmed by the injection profiling test. In the following result part, we will focus on layer 7 as an example of analyzing and examining the effectiveness of microseismic monitoring.

3.2 Gas Tracer Test

The first gas tracer test in the north five well groups (Fig. 2a) has failed at the initial period of CO_2 injection due to the incomplete sampling of an unexpected early breakthrough of CO_2 tracer. Therefore, a second gas tracer test was implemented into the two middle well groups (H59-8-4 and 6-6) in Jan 2012. After a careful screening test on different kinds of gas tracers, QT-1 and QT-5 (a kind of CO_2-soluble fluorocarbon) were chosen, and a slug of the corresponding tracer with the mass of 90 kg was pumped into the Qing-I member of well H59-8-4 and 6-6, respectively. The injection rate is 10-30 L/min with the injection pressure of 15-16 MPa. After the injection, samples of casing gas or dissolution gas from produced oil were collected by the vacuum plastic bags once a day, and the collected gas volume is about 500 ml each time. The concentration of the produced gas tracer was analyzed by the gas chromatograph-electron capture detector (GC-ECD). The sampling operation lasts for 3-6 month in all the production wells of the pilot area.

4. Results and Discussion

4.1 Well Groups of H59-6-6 and H59-1

The stacking CO_2 microseismic events distributions for well H59-6-6 and H59-1 were shown in Fig. 4a and 4b. The induced microseismic events around the well by the CO_2 injection increase gradually from the inside area in pink to the outside area in red color. It can be seen that, the CO_2 preferential flow is mainly along the E-W direction for both wells, this observation is particular pronounced in well H59-6-6. As stated before, the natural fractures are developed in the east-west direction. Therefore, this would indicate the mapped microseismic events is mainly due to the natural fractures opening triggered by the injection pressure. It should be noted that the pink or red areas are not CO_2 sweeping area during the monitoring period; they are just the indicator of the injection induced pressure fronts. The CO_2 migration rate or displacement front is much less that the pressure propagation rate, and it is the former that cause the microseismic events which are used to map the potential fluid movement trend or pattern.

Since the real-time CO_2 injection pressure during microseismic test is not recorded, the long-term injection rate and pressure data was used to help to analyze subsurface CO_2 migration behavior. This analysis is reasonable and valid considering the fact that the microseismic monitoring are selectively conducted on the wells with relatively-stable injection performance and the CO_2 injection rate is kept the same as before during microseismic test. CO_2 injection pressure fluctuates along the CO_2 injection rate or WAG cycle. During the microseismic monitoring, the wellhead injection pressures of CO_2 for both wells are around 15 MPa with the measured bottomhole pressures of 38 MPa, respectively. It is suspected that this pressure is enough to open the pre-existing natural fractures, detailed study on the geomechanical properties of the reservoir and natural fractures is necessary to validate this point.

A comparison of predicted CO_2 sweeping profile between the microseismic results and gas tracer test has been conducted for well group H59-6-6. The gas tracer test result is shown in Fig. 5; the tracer was injected via H59-6-6, and detected later in the surrounding production wells. As shown from the figure, the injected CO_2 moves along the W-E direction toward to the producers, this is very consistent with the CO_2 sweeping profile as indicated from the outer boundaries of microseismic events (Fig. 4a). The gas tracer test shows fast CO_2 average movement rate between H59-6-6 and 4-6, also produced gas tracer concentration of the well is the highest among the detected six producers. This reveals very good connectivity between the two wells and, therefore, the most of CO_2 would move toward to H59-4-6. However, microseismic monitoring demonstrates that the dense flow of CO_2 is toward to the middle area of H59-6-8 and 4-6 (see the areas with blue and green colors in Fig. 4a), which is almost parallel to the natural fracture direction. This might indicate that the re-opened natural fractures by CO_2 injection dominate the flow of CO_2.

Further, it can be seen from Fig. 4b that the CO_2 sweeping profile for well group H59-1 is almost uniform, therefore it can be expected that the injection response, i.e., CO_2 breakthrough time and produced CO_2 rate, would be close to each other in the surrounded production wells. As revealed from the well group configuration of Fig. 2a, the production response of wells in the lower parts of the H59-1 group (i.e., H59-1-3, 3-3 and 3-1 of Fig. 4b) might be only affected by the CO_2 injection in well H59-1. Therefore, the CO_2 production rates along with time for these three wells are plotted and they are shown in Fig. 6. It can be seen that the temporal change of the produced CO_2 amounts from the above three wells have a very similar trend. The CO_2 breakthrough happens around almost the same time of period during September to December of 2009. This supports the credibility of microseismic interpretation on the CO_2 flooding profile.

Another very interesting observation from the two figures is that the stacked microseismic events occurrence area as well as event rates (number of microseismic events during monitoring) in well H59-6-6 is much larger than that in well H59-1. However, the microseismic monitoring or CO_2 injection lasting periods is very close for the two wells, 210 vs 190 min. And the CO_2 injection rates are 37 and 42 t/d, respectively. So the total injected CO_2 would be comparable as well as the well head injection pressures. Also, both microseismic results are interpreted from the 7 layer of Qing-I (Table 2), which shows limited difference of permeability around the two wells. All of these would make us expect that, at least, the detected microseismic coverage area would be almost similar. Further work might be required to characterize the fracture density and length variations around the two wells to help understand the reservoir deformation and induced micro-seismic events during CO_2 injection. However, it should be indicated that, especially from Fig. 4b, the CO_2 sweeping profile interpreted from microcosmic distribution is less anisotropic; this might imply that the developed E-W natural fractures have a favorable effect in improving the CO_2 sweeping efficiency in our specific case of study.

4.2 Well Groups of H-59-14-6 and H59-4-2

The distribution of the dense degree of micro-seismic events occurred around well H59-14-6 and H59-4-2 was shown in Fig. 7a and 7b, respectively. It can be seen from Fig. 7a that the microseismic events are distributed in a NW-SE strip, which is parallel to the connected line of H59-14-6 and H59-12-4. For the monitoring well H59-6-6, the micro-seismic events cluster along the NE-SW direction and they become almost uniformly distributed as moving far from the injector.

As seen from Fig. 7a, the injected CO_2 in well H59-14-6 mainly spreads towards well H-12-4, but moves to other directions weakly, such as the directions to wells H-16-8, H-14-8 and H-12-8. For the injector H59-4-2 (Fig. 7b), CO_2 moves preferentially along the strip line connecting H-59-4-1, 4-2 and 4-4 together. It should be indicated that, before the microseismic monitoring, it is expected that CO_2 plume behavior would be mainly controlled by the E-W natural fractures that are well developed in the reservoir. However, from the microseismic distribution for both injectors, it can be seen that CO_2 migration pattern interpreted from microseismic is mainly driven by the pressure gradient between injectors and producers. This might indicate that the injection pressure is not enough to open the closed natural fractures.

H-59-14-6 was converted to the CO_2 injector in October of 2012 because of the serious CO_2 breakthrough. During the microseismic monitoring, the wellhead injection pressure of CO_2 for well H59-14-6 is about 12.5 MPa, while it is 10 MPa for well H59-6-6. It is suspected that this pressure is not enough to open the pre-existing natural fractures. In order to verify our speculation, a careful comparison between CO_2 flow pattern and permeability field is conducted for each layer of H59-14-6. The basic idea for the comparison is the permeability distribution would be the dominant factor of CO_2 migration if it is driven by the pressure gradient according to the Darcy flow. Injection profile tests show that the injected CO_2 through well H59-14-6 mostly moves into the layer 7 and 12, so we superimposed the outer boundaries lines of microseismic event intensity for each layer onto the pressure field and they are shown in Fig. 8a and 8b, respectively. It can be seen that, for layer 7 and 12, CO_2 migrates toward the areas with either high permeability or low pressure (producer). Therefore, the microseismic events detected from the reservoir are very likely not caused via the stress gradients changes induced by localized fractures. Rather, they are mainly controlled by the pressure gradients from the injection to production sides.

Also, from the strip-like CO_2 flow pattern around H59-14-6 as shown in Fig. 7a, it can be predicated that, once CO_2 breakthrough occurs, the produced CO_2 from H59-14-8 (the upper blue dashed circle well in Fig. 8a) would be much smaller than H59-12-4 (the lower blue dashed circle well in Fig. 7a). However, for injector H-59-4-2, the CO_2 sweeping profile is almost uniform with slightly preferential flow toward well H59-4-4 and H-4-1, therefore, it can be expected that the produced CO_2 amount from these two wells would be marginally higher than that of wells in the perpendicular direction, e.g., H59-6-2 and 2-2. Considering about this, we plot daily-averaged CO_2 production rate for well H59-14-8 and 12-4 (in H59-14-6 group of Fig. 7a) as well as H-59-4-4 and H59-2-2 (blue dashed circle wells in H59-4-2 group of Fig. 7b), and they are shown in Fig. 9a and 9b. As seen from the figures, the difference of CO_2 production rate between well H59-14-8 and 12-4 is 4-7 times higher than that between H59-4-4 and 2-2 (note the scale of Y axis in the two figures). This observation has very good consistency with the microseismic results which shows that the CO_2 sweeping profile of well H-59-4-2 is better than that of well H59-14-6. It should be indicated that, for Fig. 9a and 9b, the produced CO_2 from these wells are not exactly from the corresponding central CO_2 injector, and they might come from other CO_2 injectors. In this case, the results of interwell connectivity test - gas tracer – could be used to quantify how much CO_2 is produced from the corresponding central CO_2 injector. Gas tracer result is available for well H59-4-4 and 2-2 and it shows that the producer H59-4-4 is only impacted by gas injector 4-2, whereas well 59-2-2 has almost equivalent connectivity with four CO_2 injectors (i.e., H59-4-2, 8-4, 10-8, and 6-6). If considering about this, it can be further derived that the CO_2 production rate of H59-2-2 caused only from injector 4-2 would be much less that of H59-4-4. Therefore, the approximation of produced CO_2 measurement from the near-by CO_2 injector analyzed in the above is reasonable in terms of comparison with microseismic results. All the above analysis and comparison with different kinds of data confirms the credibility of the microseismic test results and it can be used to support future optimization of the reservoir performance, which would be of particular interest for the CO_2-EOR project.

5. Summary and Conclusions

Different methods have been employed to monitor the CO_2 migration in thin and tight heterogeneous oil reservoirs, combining the microseismic monitoring with the gas tracer testing and produced gas analysis. Gas tracer testing is effective for detecting the orientation of gas channeling from CO_2 injectors to producers due to hydraulic or natural fractures while the microseismic technique can be effectively used for monitoring CO_2 migration and its sweeping profiles in the reservoirs.

In applying the microseismic technique, the open and closure of the pre-existing natural fractures were well corresponding with the variation of CO_2 injection pressure. Higher injection pressure can induce the re-opening of the pre-existing natural fractures, and the CO_2 flow pattern induced by the fracture network become less anisotropic. And the microseismic events detected were consistent with the orientation of the natural fractures and the stress distribution in the reservoir. However, for the lower injection pressure, the inversed microseismic profiles clearly show the preferential CO_2 flows and their anisotropy in different well groups, which are mainly dominated by the pressure gradient and the heterogeneity of reservoir permeability. The detected micro-seismicity is caused by stress gradients induced by production and injection gradient, rather than the local natural fractures.

The results of microseismic tests are in good consistency with the produced CO_2 measurements. The mapped preferential flow direction and sweeping profile of CO_2 in each injector well group can provide a good prediction for the breakthrough of CO_2 in different production wells, which is useful for conformance control and injection optimization.

6. Acknowledgements

This research is supported by the National Major S&T Project (No. 2011ZX05016-005). Permission to publish this paper by Jilin Oilfield Company, PetroChina, is gratefully acknowledged.

7. References

Bachu, S. 2008. CO_2 storage in geological media: Role, means, status and barriers to deployment. *Progress in Energy and Combustion Science* **34**:254-273.

Brown, B., Carr, T.R., Vikara, D. 2009. Monitoring, verification, and accounting of CO_2 stored in deep geologic formations. U.S. Department of Energy: National Energy Technology Laboratory.

IPCC (Intergovernmental Panel on Climate Change). 2005. IPCC special report on carbon dioxide capture and storage. Cambridge University Press, Cambridge, UK, pp. 195-276.

Litynski, J.T., Plaskynski, S., Mcilvried, H.G., Mahoney, C., Srivastava, R.M. 2008. The United States Department of Energy's regional carbon sequestration partnerships program validation phase. *Environment International* **34** (1):127-138.

Ren, S.R., Li, D.X., Zhang, L. 2014. Leakage pathways and risk analysis of carbon dioxide in geological storage. *Acta Petrolei Sinica* **35**(3):591-601.

Zhang, L., Ren, B., Huang, H.D., Li, Y.Z., Ren, S.R., Chen, G.L., Zhang, H. 2015. CO_2 EOR and storage in JiLin Oilfield China: monitoring program and preliminary results. *Journal of Petroleum Science and Engineering* **125**:1-12.

Table 1 Microseismic monitoring well information

Monitoring well	H-14-6		H-4-2		H-59-1		H-6-6
Monitoring date	5/24/2013	5/25/2013	8/6/2012	8/7/2012	8/6/2012	8/7/2012	7/2/2013
Monitoring interval depth, m	2454-2482	2422-2425	2482-2493	2430-2466	2467-2504	2436-2444	2435-2440
Corresponding layer	Qing-I: 12,14,15	Qing-I: 7	Qing-I: 14,15,16	Qing-I: 7,12	Qing-I: 12,15,16	Qing-I: 7	Qing-I: 7
Monitoring period, min	10	9.8	6.9	7.6	8.3	7.8	4.4

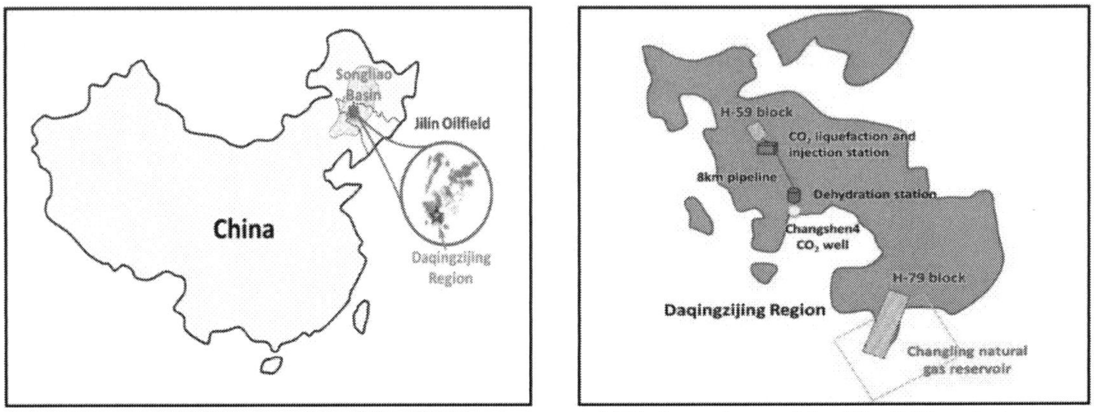

(a) Location of the Jilin oilfield (b) Project scheme of the Jilin CCS project

Fig. 1. CO_2 EOR and storage project in Jilin oilfield northeast China (adapted from Zhang *et al.*, 2015)

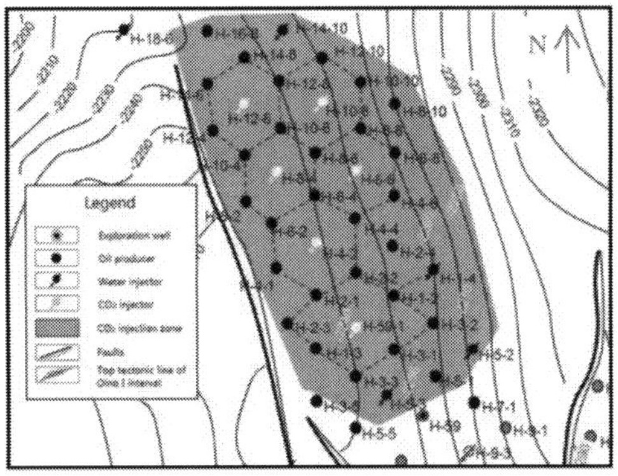

YaoJia Formation				
Cretaceous Period	QingShanKou Formation	Qing-III Member	VI Unit	-
			VII Unit	38—42
		Qing-II Member	I Unit	1—5
			II Unit	6—11
			III Unit	12—14
			IV Unit	16—21
			V Unit	22—24
		Qing-I Member	I Unit	1—3
			II Unit	4—8
			III Unit	9—12
			IV Unit	14—16
QuanTou Formation				

(a) (b)

Fig. 2. (a) is the surface layout of well location and pattern in H59 oil Block; Well H-59-1 was converted to a CO_2 injector in September, 2009. (b) is a schematic stratigraphy map with the detailed reservoir layering system. The reservoir includes Qing-I, II and III member, and the main oil production layers are # 7, 12, 14 and 15 of Qing-I member as shown in the table with the gray background color.

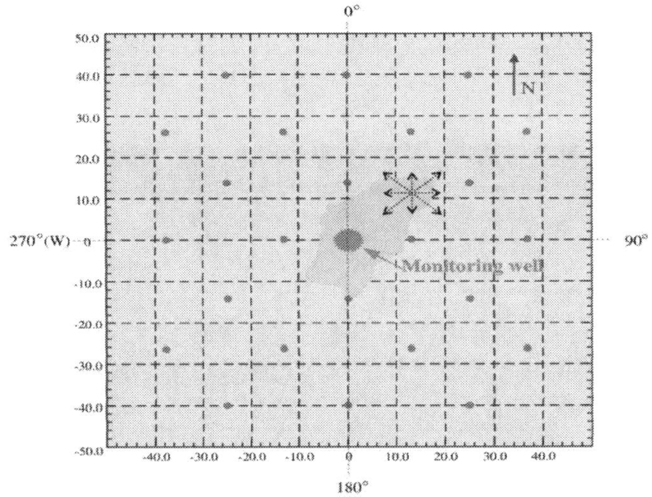

Fig. 3. Microseismic sensor network: twenty-four sensors have been employed with the 3 component geophones clamped near-surface with the depth of 10-30 m. The distance between sensors is around 150 meter with the network coverage area of 600 m x 600 m centered on the target monitoring CO_2 injector. The distance along the axes are in meter with the monitoring well as the coordinate origin, a factor of 6 should be multiplied to get the field distance.

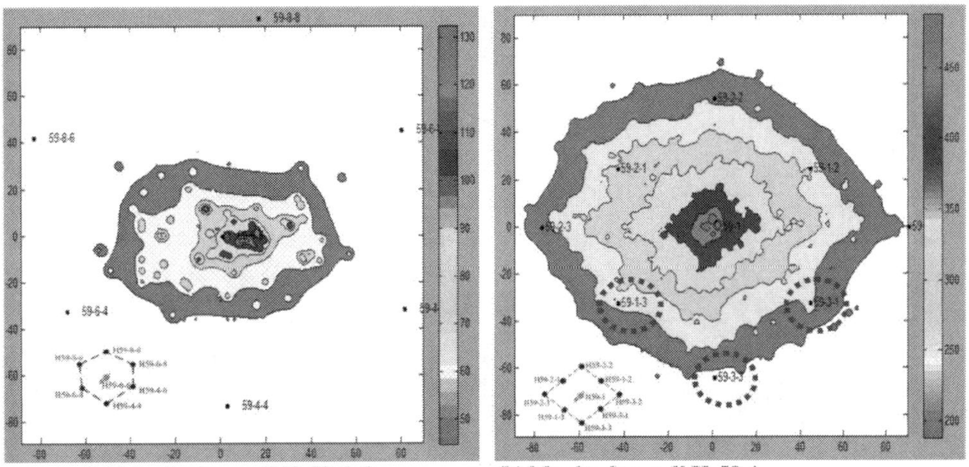

(a) Monitoring well H-59-6-6 (b) Monitoring well H-59-1

Fig. 4. Micro-seismic events stacking at the reservoir intervals of 2435-2440 m (corresponding to layer 7, 4a) and 2436m-2444 m (corresponding to layer 7, 4b) in the well H59-6-6 and H59-1, respectively. The observation time of CO_2 injection lasts for 210 and 190 min, respectively. The CO_2 injector is shown in the middle black circle with other producer surrounded shown in the black dots. The color bar shows the number of microseismic events during the observation time period. The ink and blue colors represent areas of more micro-seismic energy emissions while the red areas are with less microseismic events. The distance along the axes are in meter with the monitoring well as the coordinate origin, a factor of 6 should be multiplied to get the field distance. The elongate anomalies in the microseismic events distribution for both wells are in the W-E direction, which is parallel to the direction of maximum principal stress. The corresponding well groups are shown in the left bottom corner.

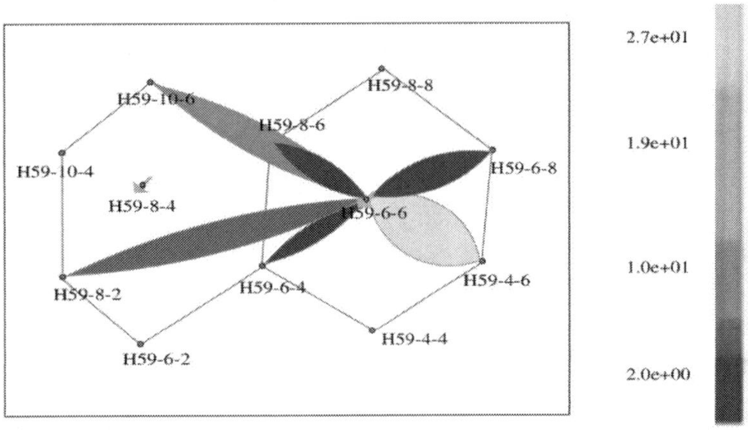

Fig. 5. Gas tracer test in well H59-6-6, the gas tracer is injected via H59-6-6. The color bar shows the CO_2 average movement rate between wells with the unit of m/d. the relative width of leave-like shapes with two ends connecting injector and producers are the relative peak concentration monitored during the sampling period. The wider of the leave, the higher the peak concentration of gas tracer produced. Note most of injected tracer moves along the E-W direction toward to the oil producers.

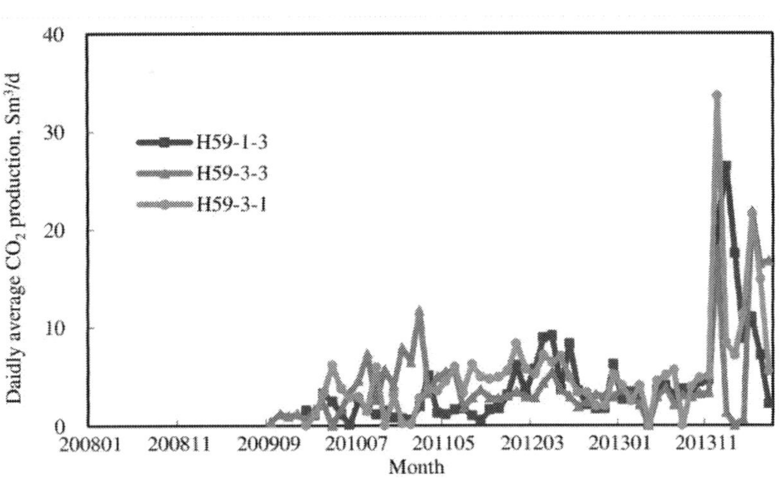

Fig. 6. Daily CO_2 production rate for well H59-1-3, 3-3 and 3-1 in the lower part of H59-1group (refer to the blue dashed circle of Fig. 4b).

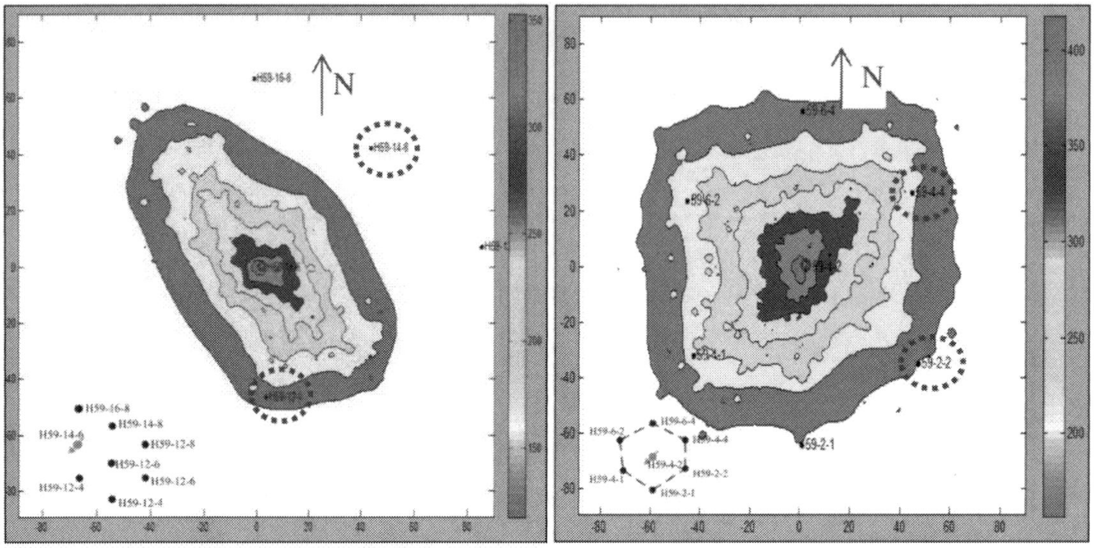

(a) Monitoring well H-59-14-6 (b) Monitoring well H-59-4-2

Fig. 7. Micro-seismic events stacking at the reservoir depth intervals of 2454-2482 m (corresponding to layer 12, 14 and 15, 7a) and 2430-2466 m (corresponding to layer 7 and 12, 7b) in the well H59-14-6 and H59-4-2, respectively. The microseismic observation time of CO_2 injection lasts for 150 and 170 min, respectively. The CO_2 injector is shown in the middle black circle with other producer surrounded shown in the black dots. The color bar shows the number of the microseismic events during the observation time period. The ink and blue colors represent areas of more micro-seismic emissions while the red areas are with less microseismic events. The distance along the axes are in meter with the monitoring well as the coordinate origin, a factor of 6 should be multiplied to get field distance. The elongate anomaly in the microseismic events distribution is in the SE-NW direction for the well H59-14-6, whereas NE-SW direction for the H59-4-2. The corresponding well group or surrounded production wells are shown in the left bottom.

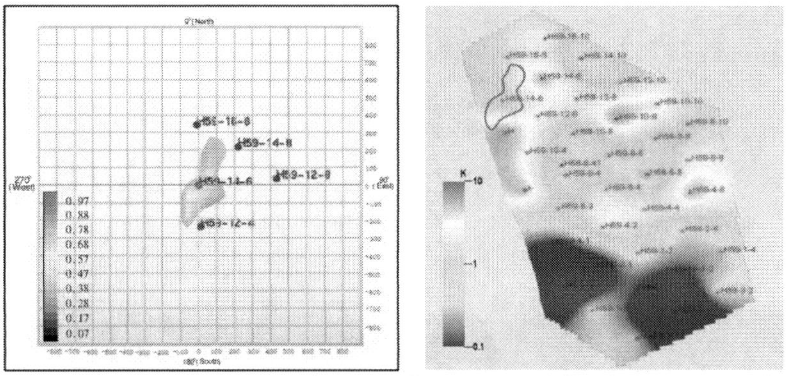

(a) Microseismic event intensity in layer 7 (left) and permeability field of layer 7 (right)

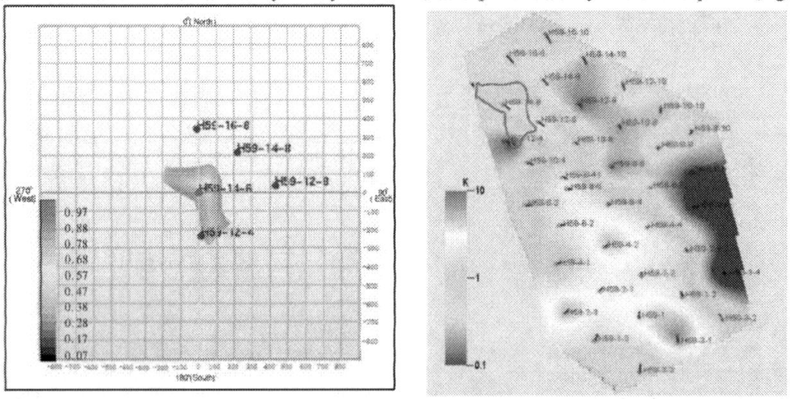

(b) Microseismic event intensity in layer 12 (left) and permeability field of layer 12 (right)

Fig. 8. Comparison between micro-seismic event intensity with the permeability field for layer 7 (a) and 12 (b) of Qing-I well H-59-14-6. In each layer comparison, microseismic event intensity is shown in the left, while the permeability field on the right. The microseismic profiles are the depth slices through a 190 min passive seismic events stack at the onset of CO_2 injection and the corresponding depth for layer 7 and 12 are 2424.7 m and 2455.9m, respectively. The blue dots are production wells and red spot in the center is the monitoring well. The color bar is the micro-seismic frequency, and hotter colors indicate areas of great emission of microseismic. The outer boundary lines of the micro-seismic profile are superimposed onto the right permeability field which is shown in the red color.

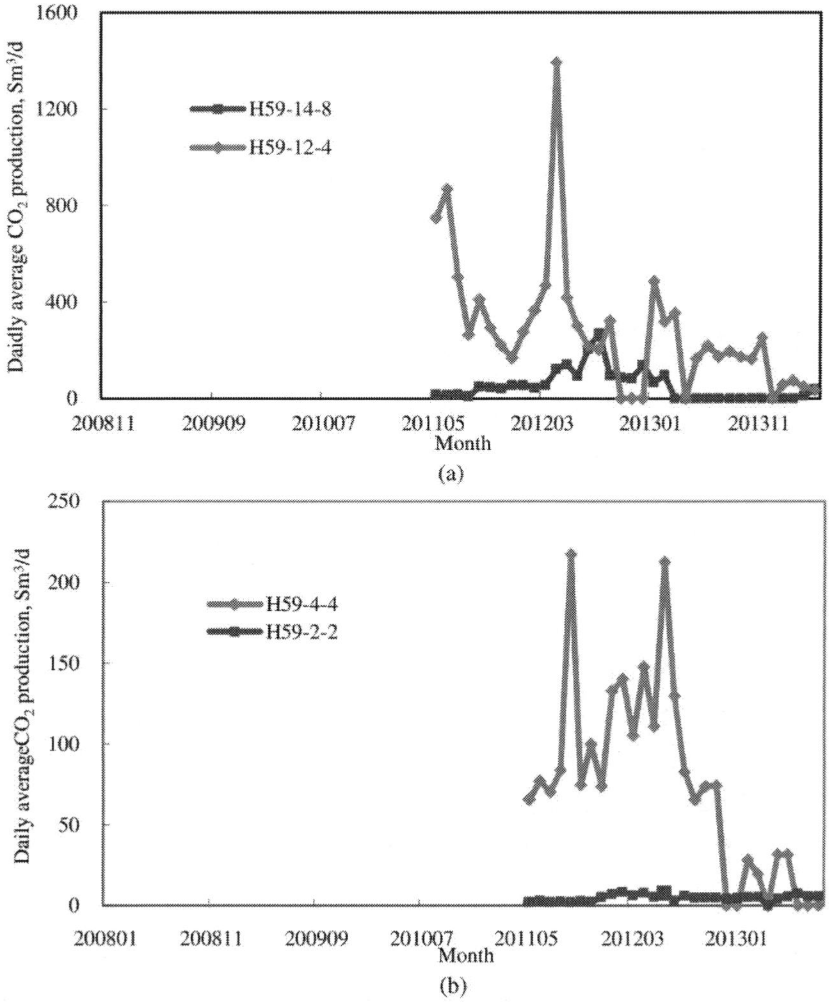

Fig. 9. Daily CO_2 production rate for: (a) well H59-14-8 and H59-12-4 (in H-15-14-6 group), H59-14-8 is along the direction of CO_2 preferential flow determined from microseismic while H59-12-4 is in the perpendicular direction (refer to the blue dashed circle of Fig. 7a). (b) for well H59-4-4 and H59-2-2 (in H-15-4-2 group), H59-4-4 is along the direction of CO_2 preferential flow determined from microseismic while H59-2-2 is in the perpendicular direction (refer to the blue dashed circle of Fig. 7b).

Demonstration of CO_2 Containment and Non-Endangerment at the SECARB Anthropogenic Test Site

Carbon Management Technology Conference 2015

David Riestenberg
George Koperna
Advanced Resources International, Inc.

Robert Trautz
Richard Rhudy
Electric Power Research Institute

November 17-19, 2015

Advanced Resources International, Inc.

Acknowledgement

This presentation is based upon work supported by the Department of Energy National Energy Technology Laboratory under **DE-FC26-05NT42590** and was prepared as an account of work sponsored by an agency of the United States Government. Neither the United States Government nor any agency thereof, nor any of their employees, makes any warranty, express or implied, or assumes any legal liability or responsibility for the accuracy, completeness, or usefulness of any information, apparatus, product, or process disclosed, or represents that its use would not infringe privately owned rights. Reference herein to any specific commercial product, process, or service by trade name, trademark, manufacturer, or otherwise does not necessarily constitute or imply its endorsement, recommendation, or favoring by the United States Government or any agency thereof. The views and opinions of authors expressed herein do not necessarily state or reflect those of the United States Government or any agency thereof.

NETL | SECARB Southeast Regional Carbon Sequestration Partnership | SOUTHERN COMPANY Energy to Serve Your World

SOUTHERN STATES ENERGY BOARD | EPRI ELECTRIC POWER RESEARCH INSTITUTE | Denbury | Advanced Resources International, Inc.

Topics of Discussion

1. Project Introduction
2. Monitoring Lines of Evidence for Non-Endangerment
3. Modeling Lines of Evidence for Non-Endangerment
4. Conclusions and Next Steps

DOE Regional Carbon Sequestration Partnerships

Large-Scale Projects

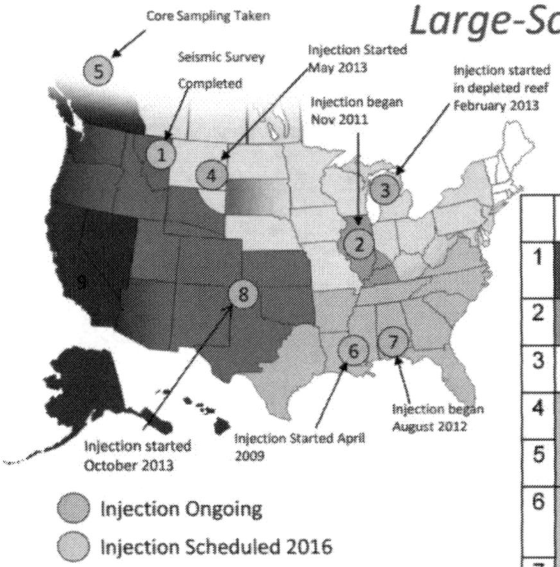

Note: Some locations presented on map may differ from final injection location

- ✓ *Large-volume tests*
- ✓ *Four Partnerships currently injecting CO_2*
- ✓ *Remaining injections scheduled 2016*

	Partnership	Field Project – Geologic Formation	Metric Tons Injected to Date
1	Big Sky	Kevin Dome- Duperow Formation	Injection starting in 2016
2	MGSC	Illinois Basin Decatur-Mt. Simon Sandstone	999,215
3	MRCSP	Michigan Basin - Niagaran Reef	324,163
4	PCOR	Bell Creek - Muddy Sandstone	1,660,570
5	PCOR	Fort Nelson - Sulfur Point Formation	Injection TBD
6	SECARB	Early Test (Cranfield Field) - Tuscaloosa Formation	4,743,898
7	SECARB	Anthropogenic Test (Citronelle Field) – Paluxy Formation	114,104
8	SWP	Farnsworth Unit -Morrow Formation	229,060
	WESTCARB	Regional Characterization	

NETL

Project Objectives

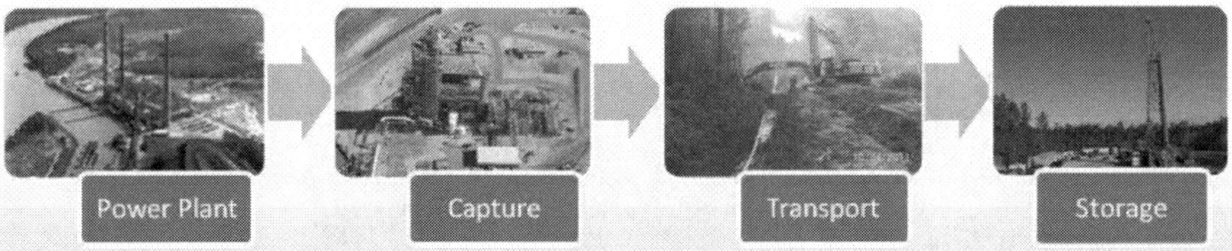

1. Support the United States' largest prototype ***coal-fired CO_2 capture and transportation*** demonstration with ***injection, monitoring and storage*** activities;
2. Test the CO_2 flow, trapping and storage mechanisms of the Paluxy;
3. Demonstrate how a saline reservoir's architecture can be used to maximize CO_2 storage and minimize the areal extent of the CO_2 plume;
4. Test the adaptation of commercially available oil field tools and techniques for monitoring CO_2 storage
5. Test experimental CO_2 monitoring activities, where such technologies hold promise for future commercialization;
6. Begin to understand the coordination required to successfully integrate all four components (capture, transport, injection and monitoring) of the project; and
7. Document the permitting process for all aspects of a CCS project.

Storage Site: The Citronelle Oilfield

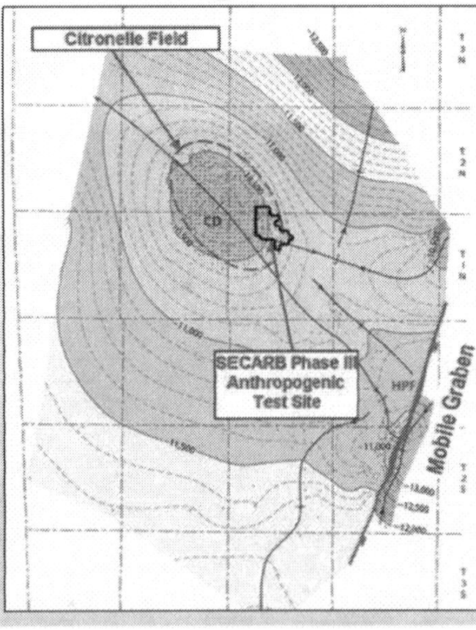

System	Series	Stratigraphic Unit	Major Sub Units	Potential Reservoirs and Confining Zones
Tertiary			Citronelle Formation	Freshwater Aquifer
		Undifferentiated		Freshwater Aquifer
			Chickasawhay Fm.	Base of USDW
		Vicksburg Group	Bucatunna Clay	Local Confining Unit
		Jackson Group		Minor Saline Reservoir
		Claiborne Group	Talahatta Fm.	Saline Reservoir
		Wilcox Group	Hatchetigbee Sand Bashi Marl Salt Mountain LS	Saline Reservoir
		Midway Group	Porters Creek Clay	Confining Unit
Cretaceous	Upper	Selma Group		Confining Unit
		Eutaw Formation		Minor Saline Reservoir
		Tuscaloosa Group		Minor Saline Reservoir
			Marine Shale	Confining Unit
			Pilot Sand Massive sand	Saline Reservoir
	Lower	Washita-Fredericksburg	Dantzler sand	Saline Reservoir
			Basal Shale	Primary Confining Unit
		Paluxy Formation	'Upper' 'Middle' 'Lower'	Injection Zone
		Mooringsport Formation		Confining Unit
		Ferry Lake Anhydrite		Confining Unit
		Donovan Sand	Rodessa Fm. Upper' 'Middle' 'Lower'	Oil Reservoir Minor Saline Reservoir Oil Reservoir

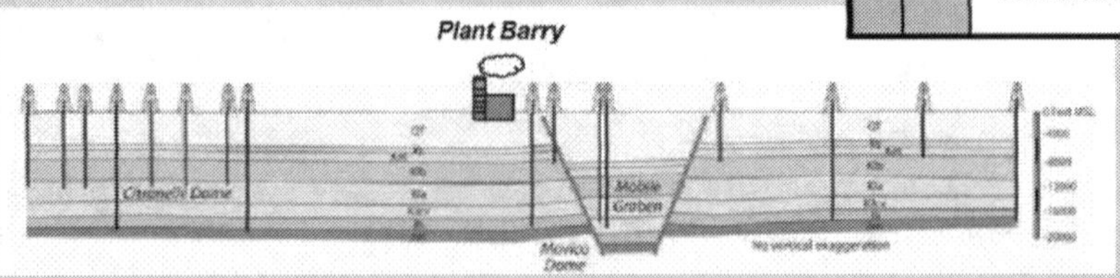

Underground Injection Control Class V Permit

- **UIV Class V Experimental Injection Well permit**
  - Short duration of injection (3 years) and modest volumes of CO_2
  - Characterization and modeling of "stacked" CO_2 storage
  - CO_2 injection under "real world" operating conditions
  - *Demonstration of experimental monitoring tools and methods*
- **Several UIC Class VI (CO_2 sequestration well) standards were applied**
  - Injection Area of Review (AOR) determined by modeling and monitoring results (Annually)
  - Deep, shallow and surface monitoring
  - Injection stream compositional monitoring
  - **Site closure based on USDW non-endangerment and CO_2 containment demonstration (5-yr renewal)**
    - *Based on monitoring and modeling results*

Storage Project Status

- ADEM granted permission to inject on August 8, 2012
  - ➢ Injection commenced on August 20, 2012
- Through September 1, 2014, approximately 114,104 metric tons of CO_2 were injected
  - Less than the proposed 500,000+ metric tons due to low dispatch of Plant Barry Unit 5
- A crosswell seismic survey acquired in June, 2014 captured a time-lapse image of the CO_2 plume
- Other testing and monitoring activities have indicated containment
- The project entered the Post-Injection Site Care Period in September, 2014
- Site closure will be based on **CO_2 containment and drinking water non-endangerment demonstration**

CO_2 Injection History

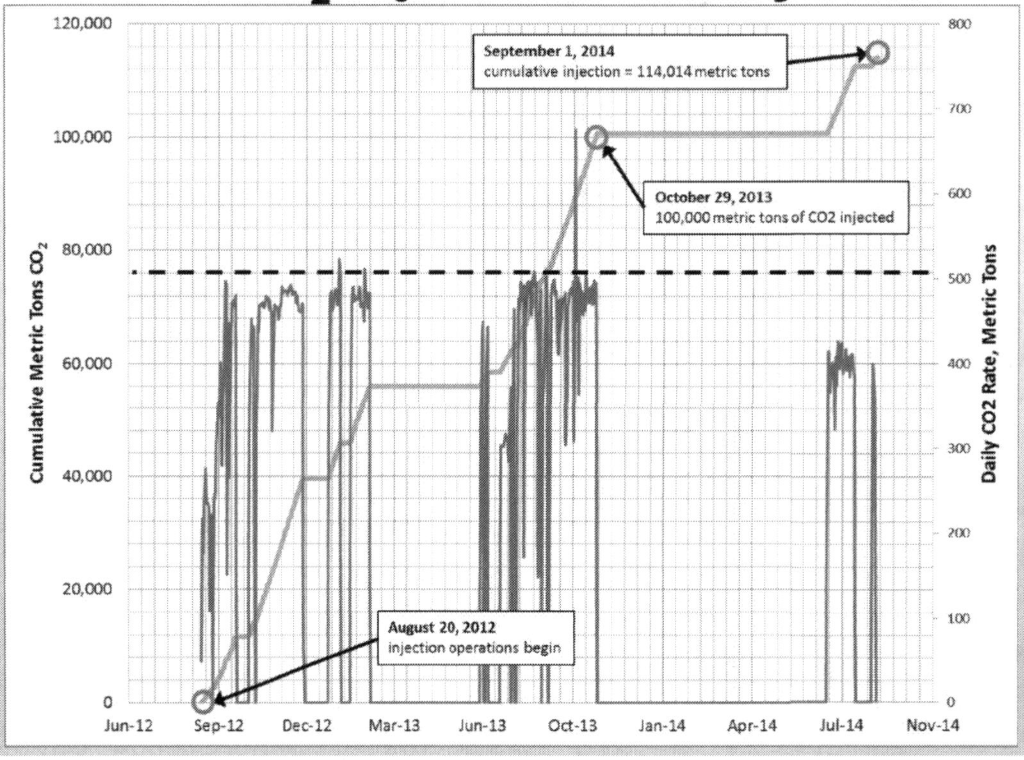

Monitoring Lines of Evidence

MVA Elements and Frequency

MVA Method	Frequency: Continuous	Monthly	Quarterly	Annual	Milestone *(Baseline, Injection, Post)*
Shallow					
Soil flux		■			
Groundwater sampling (USDW)			■		
PFT survey				■	
Deep					
CO2 volume, pressure & composition	■				
Reservoir fluid sampling				■	
Injection, temperature & spinner logs				■	
Pulse neutron logs				■	
Crosswell seismic					■
Vertical seismic profile (VSP)					■
Experimental					
Distributed Temperature Sensing (DTS)		■			
Comparative fluid sampling methods				■	
MBM VSP				■	
Distributed Acoustic Sensing (DAS)				■	
MBM VSP & OVSP Seismic				■	

Test Site Development

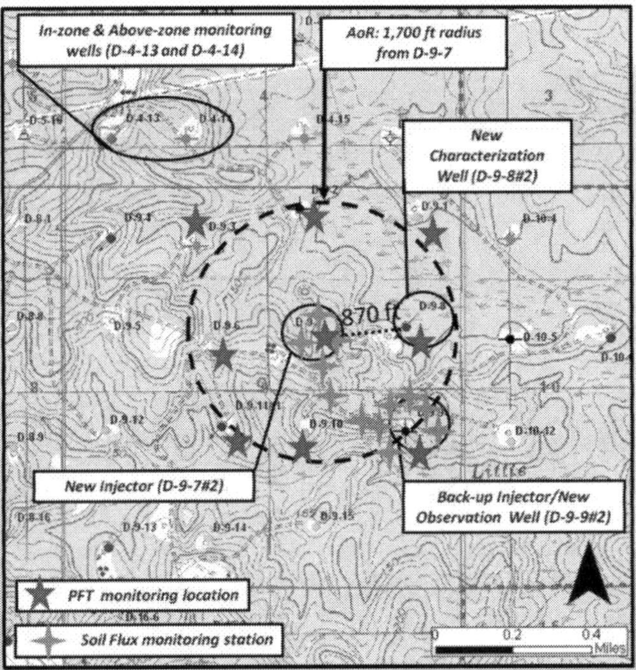

- One new injector (D-9-7 #2)
- Two new deep observation wells (D-9-8 #2 & D-9-9 #2)
- Two in-zone & above zone monitoring wells (Citronelle wells D-4-13 & D-4-14)
- 12 soil flux monitoring locations
- Tracer monitoring on nine well pads

Shallow Monitoring
Soil CO_2 Flux and Tracers

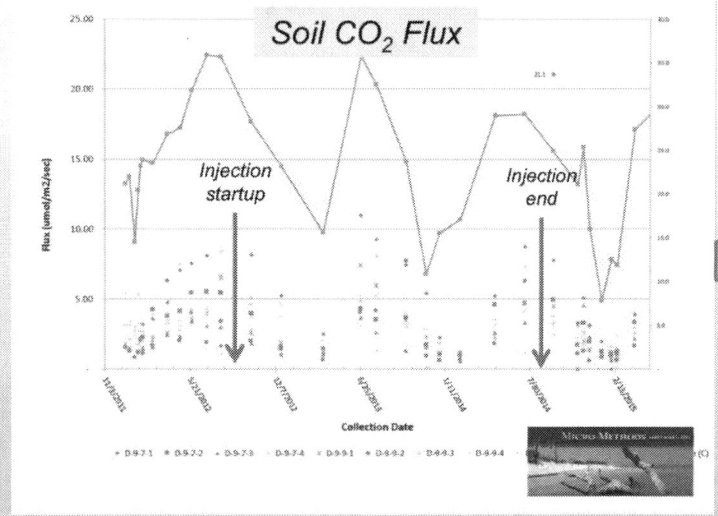

Tracer Results

	Inoculation	Testing	
Well/Sample	AUG 2012	JUN 2013	NOV 2013
D-9-1	ND	ND	ND
D-9-2	ND	ND	ND
D-9-3	ND	ND	ND
D-9-6	ND	ND	ND
D-9-7-1	ND	ND	ND
D-9-8	Invalid Data	ND	ND
D-9-9	ND	ND	ND
D-9-10	Invalid Data	ND	ND
D-9-11	ND	ND	ND
Air Blank 1	ND		
System Blank		ND	ND

PRAXAIR

Soil CO_2 results appear to vary as a function of mean temperature and tracer results have been non-detects

Deep Monitoring – Time-Lapse Crosswell Seismic

- Crosswell seismic surveys allow for high-resolution mapping of the acoustic travel time (velocity) and seismic reflectors between a pair of wells
- When CO_2 displaces water in the formation, it changes the acoustic impedance of the rock
  - Acoustic wave decreases and its direct travel time increases
- Results from "repeat" surveys performed during or after CO_2 injection can be compared to a pre-injection "baseline" survey to image the extent of the CO_2 plume (referred to as "time-lapse imaging")
- Baseline and repeat 2-D crosswell seismic surveys were performed between the injection well and the observation well

Time-Lapse Differencing Using the Baseline and Repeat Velocity Tomograms

- *First arrivals from repeat survey were of sufficient quality to produce a velocity difference image (right) showing regions where seismic velocity has changed over time*
- *Time-lapse difference image indicates a decrease in seismic velocity in the upper injection zone of up to 3%, suggesting an increase in CO_2 saturation*

More importantly, no negative velocity anomalies are observed in or above the confining unit…implying no detectable leakage out of inj. zone

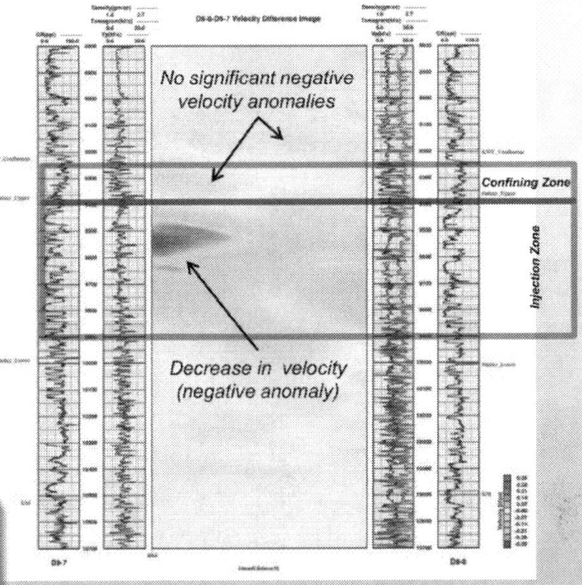

Pixelized difference tomography results without seismic reflection overlay showing positive velocity differences in warm colors and negative differences in cool colors

Deep MVA – Pressure Response

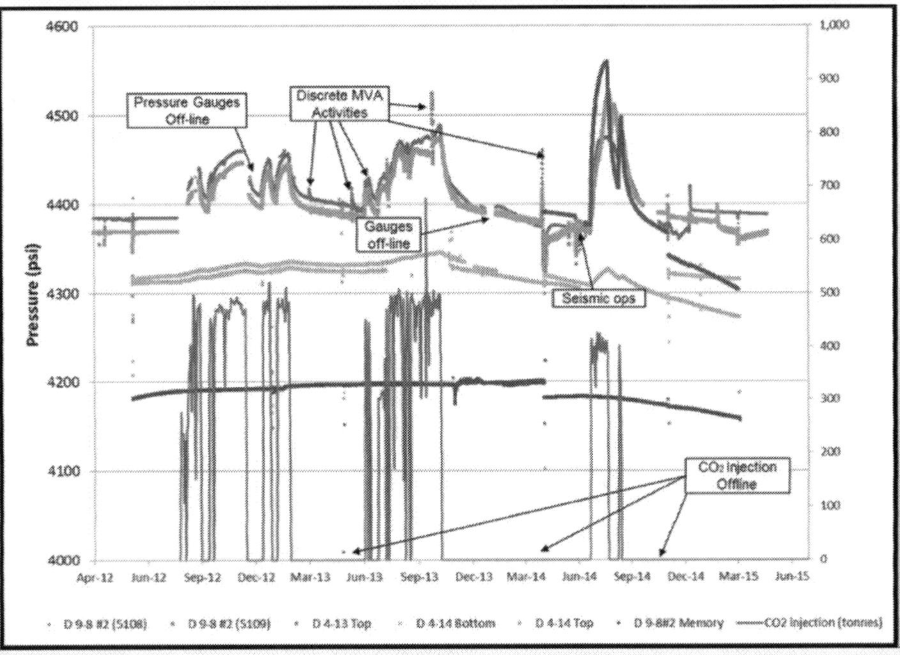

Downhole pressure data is a primary input to the history match and plume model

Transient Pressure Model

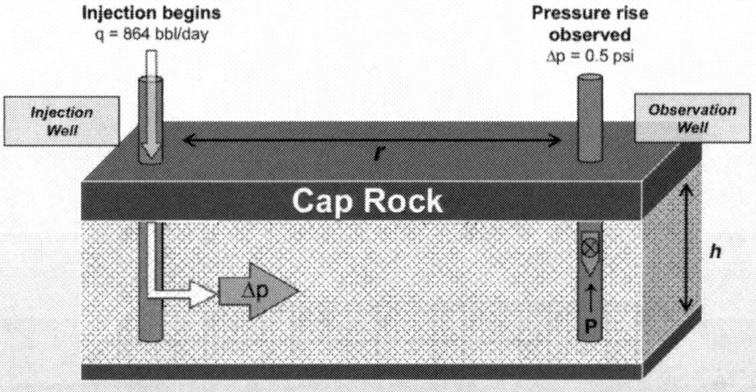

D 9-8 #2 Example
(10% CO_2 Saturation)

Fluid viscosity	0.49 cP
Net thickness, h	35 ft
Permeability, k	373 mD
Porosity, φ	19%
Radius, r	840 ft
Diffusivity C, η	1,450,474
System Compressibility	1.75E-5

Pressure gauge

$$\frac{t_D\, r^2}{\eta} = \quad t = 2.8 \text{ hours}$$

$$\left[{}^1\!/_2 \left(-Ei \frac{-1}{4\, t_D} \right) \right] = 0.1093$$

Theoretical response times for a pressure transient to travel from the injector to the observation well were calculated as a function of CO_2 saturation in the reservoir

Assumptions

- Homogenous distribution of CO_2 in reservoir
- Fixed reservoir properties

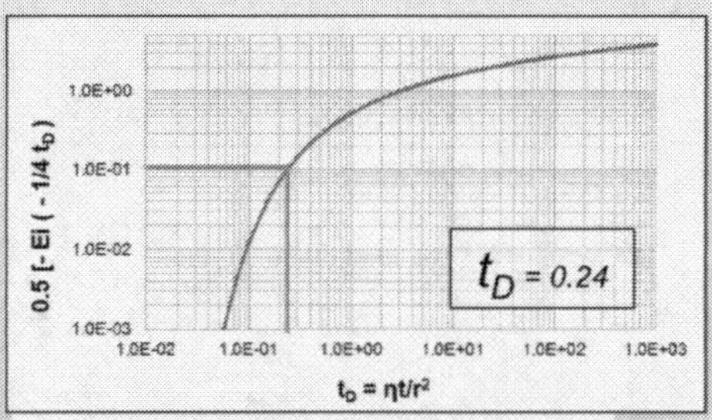

Exponential Integral Function Graph

MacGregor, 2015 CMTC

D 9-8 Pulse Testing

840 feet from injector

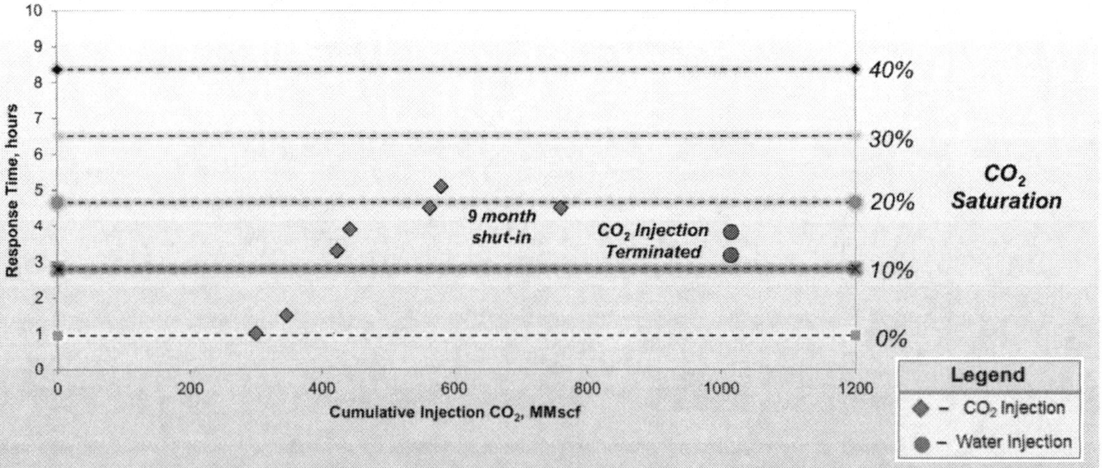

Response times increased during the CO_2 injection phase, appear to be decreasing post injection

MacGregor, 2015 CMTC

Modeling Lines of Evidence

Reservoir Modeling and Simulation

Computer Modeling Group's GEM software was used to model the injection into the upper Paluxy and forecast the subsurface movement and pressure profile of the CO_2.

Modeling was done in a two-step process. These steps were:

- History matching the entire injection period using injection rates, pressure (injection and monitoring wells), spinner surveys.
- Then, the plume was allowed to relax to understand pressure equilibration and plume stabilization.

CO_2 Plume Extent on September 1, 2014 Radius of 600 ft

- As expected, the two most porous and permeable sands accepted the bulk of the CO_2.

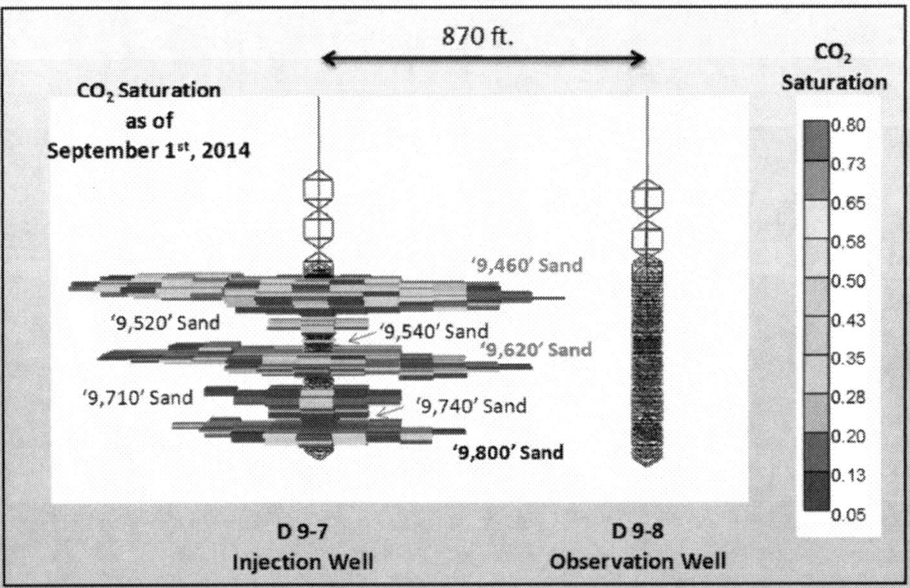

Predicted Radial CO_2 Plume Extent on October 31, 2017 remains ~600 ft

- Despite the high permeability of the formation and a gentle dip, there is not a substantial change in the plume extent in the three years following the cessation of injection.
- In fact, the plume extent has equilibrated.
- This is due to injection into multiple zones, the limited injection volume and injection shut ins which allowed for equilibration during the injection phase

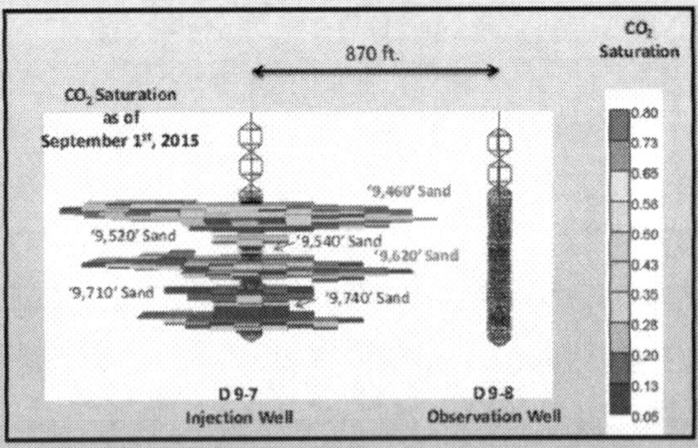

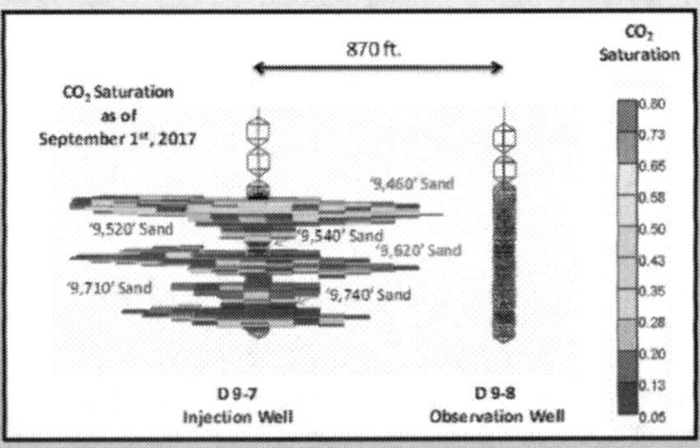

Conclusions and Next Steps

Conclusions

- Successful demonstration of integrated CO_2 capture, transport and storage
- Multiple injection interruptions limited volumes but provided opportunities for monitoring
- Multiple lines of evidence from monitoring and modeling suggest CO_2 containment in injection zone and non-endangerment of USDWs

Next Steps

- Continue non-endangerment demonstration through monitoring activities
- Close out test site including transfer of project wells to oilfield operator
- Pursue permit closure

Questions?

Advanced Formation Characterization Techniques for CO2 Geologic Sequestration

Birdie, T., TBirdie Consulting, Inc
Watney, L., Kansas Geological Survey
Scheffer-Villareal, A.,
Rush, J.,
Holubnyak, E., Kansas Geological Survey
Fazalalavi, M., Kansas Geological Survey
Doveton, J., Kansas Geological Survey
Raney, J., Kansas Geological Survey
Datta, S., Kansas State University
Hedke, D.,
Roberts, J.,

Originally presented on: 11/19/2015 15:10:00 - 15:35:00

The Arbuckle aquifer system of Cambro-Ordovician period (~ 500-1,000 million years old) in the central plains states of Kansas, Missouri, and Oklahoma, is one of the largest saline aquifer systems in North America. Based on U.S. Department of Energy estimates, approximately 89.5 billion metric tons of can be stored in this aquifer, equaling many years of total annual US CO_2 emissions of approximately 6 billion metric ton. To facilitate large scale carbon sequestration, the DOE funded a multi-year 11 million dollar study to characterize the aquifer specifically for CO_2 sequestration purposes. This effort was in support of an ongoing pilot scale carbon capture and sequestration (CCS) project at Wellington, Kansas, to demonstrate the ability of the Arbuckle aquifer in particular, and saline aquifers in general, to accept and store supercritical CO_2permanently.

The Arbuckle aquifer at the site exists between 4,000 -5,000 ft below ground surface. Shales overlying the Arbuckle Group have caprock characteristics and function as the top confining zone. Precambrian-age basement granites underlie the Arbuckle Group and provide basal confinement. For typical hydrocarbon extraction, derivation of bulk petrophysical properties are adequate to predict hydrocarbon recovery rates. However, for sequestration purposes, it is necessary to conduct advanced geologic analysis in order to account for sequestration in each of the four primary CO_2trapping mechanisms: structural, residual, solubility, and mineralogical, and to predict induced formation pressures for assessing seismic risk. Two 5,000+ feet wells were drilled into basement to derive an extensive suite of geophysical logs, cores, and swab samples, in order to be understand the geology/hydrogeology, derive petrophysical properties, and conduct hydraulic tests.

The geochemistry data including ion composition, molar ratios, biogeochemistry, isotopic characterization, was used to estimate the competence of the caprock and the hydraulic stratification within the injection zone. The biomass concentrations and microbial diversity/counts confirmed the presence of a highly stratified Arbuckle reservoir. X-Ray Diffraction and Spectral Gamma Ray Analyses (specifically the Rhomma-Umma analysis) were utilized for mineralogical characterization of the injection and confining zones, which was necessary to develop the reaction kinetics for conducting geochemical simulations and to predict sequestration quantities in the mineralogical phase and to predict changes in formation petrophysical properties such as permeability and porosity due to precipitation of minerals. Helical computerized tomography scans were used to inspect the texture of the rocks and to detect the presence of minute fractures. The Nuclear Magnetic Resonance (NMR) and sonic logs were used collectively to estimate the matrix and vuggy porosities. The acoustic measurement of porosity records the first arrival of ultrasonic compressional waves and is primarily sensitive to interparticle porosity that occurs between grains within carbonates and is often referred to as “primary” or “matrix” porosity. In contrast, the MRI, neutron, and density measurements respond to pore spaces at all scales and so provide a measure of total porosity. The difference between the acoustic porosity and the total porosity is termed the “secondary porosity” which can be interpreted to be vuggy porosity, where vugs can range in size anywhere from a dissolved grain to large cavities. The Flow Zone Interval and residual saturation information was used to develop a new technique for estimating the hydraulic conductivity, which compared favorably with core based estimates of this parameter. The T2 distribution data from the NMR logs were used to estimate the pore throat radius (as a function of capillary pressure) in order to estimate the entry pressure of the caprock. This information was necessary to demonstrate the presence of a competent seal to the U.S. Environmental Protection Agency (EPA) which regulates geologic sequestration.

The potential for induced seismicity was assessed using rock strength data from cores to construct the Mohr-Coulomb failure envelope. The regional stress field was derived from the strep-rate test (a variant of the leak-off test) and dipole-sonic data. These data sets were used to derive the fault slip and dilation tendencies.

The characterized data was the combined with a 3D geocellular geomodel (obtained from a 3D multi-component seismic volume) to develop a geostatistical multiphase flow and transport simulation models to estimate the impacts of CO_2 injection on subsurface fluid pressures and extent of plume migration. The results of the characterization resulted in a complex movement of the plume, which will be discussed.

CMTC-440095-MS

China's CCUS Progresses and a New Evaluation Method of CO_2 Storage Capacity in Coalbed Reservoirs

Kai Jiang, China University of Geosciences (Beijing); Hong'en Dou, Pingping Shen, Research Institution Petroleum Exploration & Development (RIPED); Tao Sun, RIPED Graduate School.

This paper was prepared for presentation at the Carbon Management Technology Conference held in Sugarland, Texas, USA, 17–19 November 2015.

Abstract

Carbon Capture Utilization and Storage (CCUS) is playing a significant role in dealing with the global climate change. This paper introduces policy supports of CCUS and CCUS demonstration projects in oilfields and deep saline aquifers in China. In order to study CO_2 storage capacity in coalbed reservoirs, adsorption/desorption experiments of coal rocks from different regions and coal ranks are conducted. Experimental results show that laws of desorption and absorption for two different coal ranks are obviously different and CO_2/CH_4 adsorption ratio is different along with reservoir depth. This paper points out weaknesses of evaluation methods in existing models, shows clearly what problems existing in the previous methodologies and regards the proved ratio is a key factor when evaluation. Various factors are taken into account including proved ratio, replacement ratio of CO_2 to CH_4, recovery ratio, ash content and moisture, completion factor, efficient pores volume, gas saturation and so on. Therefore, it is the first time to clearly put forward storage depth limitation. In addition, new methods for evaluating theoretical, effective, practical and matched capacity are proposed. Also, reasonable depth of CO_2 storage is determined in China according to present production technology. Finally, this paper expresses evaluation results of CO_2 storage capacity in China's coalbed reservoirs which prove the applicability of new methodologies.

Key words: CCUS progresses, CO_2 Storage projects, ECBM, Storage mechanism, Evaluation methods, Storage capacity

1 Introduction

CO_2 emission has reached 35.0 billion tons in the world in 2013, China has come to 9.6 billion tons which already occupies approximately a fourth of the world. The concentration of CO_2 is now 399 parts per million, the concentration of methane is greater than 1,774 parts per billion. For the global average, warming in the last century has occurred in two phases. Phase one, 1910s to the 1940s, temperature went up 0.35°C. Phase two, 1970s to the present, temperature went up 0.55°C. The temperature is becoming higher and higher as the concentration of CO_2 is going up sharply in recent ten years. By the end of the century, we could expect the planet's average temperature to rise anywhere between about 1.4 and 6°C. Therefore, Global warming has been become a hot topics in the world in recent year. Human being faces great challenges in reducing emissions due to its coal-dominated energy mix and large yet growing carbon dioxide emissions. CO_2 Capture and Utilization Sequestration will likely become a strategic technology option to reduce CO_2 emission in China in the future.

Coalbed reservoirs are regarded as ideal sites to sequestrate carbon dioxide as the large storage capacity. Many scholars and organizations have come up with evaluation methods of CO_2 storage capacity in coalbeds. Buchu (2002)[1] defined geological sequestration as "the capture of CO_2 directly from anthropogenic sources and disposing of it deep into the ground for geologically significant periods of time." Yu[2]; Liu and Li(2005)[3] evaluated CO_2 storage capacity in coalbed reservoirs in China, however, calculation parameters is not easy to obtain for their methods. Kaldi and Gibson-Poole (2008)[4] used the storage factor, including fraction of coal seam thickness and pore-scale displacement efficiency. Frailey and Finely (2009)[5] thought that capacity must be based on active wellbores, which have reasonable representations of the feasible injection. W.G.Allinson et al. (2014)[6] presented the logic methodology of a capacity classification system that is more closely aligned to the defined commercial, contingent and prospective storage capacity. Fang and Li (2014)[7] thought it can improve 4260 billion cubic meters methane production using CO_2-ECBM and store 9881 million tons carbon dioxide. Carbon Sequestration Leadership Forum (CSLF, 2007)[8] put forward a pyramid model that developed the understanding of different capacity considering multiple limitation of the degree of certainty, such as technology, policy, knowledge, regulatory framework, exploration progress and etc. U.S Department of Energy (USDOE, 2007)[9] exhibited storage potential system for Carbon Sequestration Atlas of the United States and Canada, the system of evaluation are lack of practical or technical limitations. The Energy and Environmental Research Center (EERC, 2009)[6] proposed a classification system dividing theoretical storage into various types based on geological, engineering, economic limitations and current conditions.

The aims of this paper are to: firstly, introduce recent CCUS progresses in China including policy support and projects; secondly, analyze weaknesses of evaluation methods in existing evaluation models and show clearly what problems in the models; thirdly, present a new methodology for predicting CO_2 storage capacity and finally, estimate the theoretical capacity, effective capacity, practical capacity and matched capacity in coalbed reservoirs of whole China.

2 China CCS Progresses

In order to achieve CO_2 emission reduction goal of 2.0 billion tons by 2050, China government has launched many policies, scientific and technological researches.

Table 1 Policies/scientific and technological researches in China to deal with CO_2 emission

Year	Orgnization/Insititution	Type	Scientific and Technological Researches/Policies
2006	China Ministry of Science and Technology, China National Petroleum Corporation	National Fundamental Research Program	Research on Utilizing Greenhouse Gas as a Resource for EOR and Geological Storage
2007	China National Petroleum Corporation	science & technology research project	Utilizing Greenhouse Gas as a Resource and Geological Storage
		pilot project	Pilot Test of CO_2-EOR and CO_2 Storage in Jilin Oil Field
2008	Chinese government	national program	CO_2 Capture and Storage for EOR
2011	China Ministry of Science and Technology	Chinese National Fundamental Research Program	A Basic Research on CO_2 Emission Reduction and Storage and Utilization
		Chinese National Major Science and Technology Project	
2011	Chinese government	policy support	The China's response to climate change policies and action
2012			12th five-year plan
2014	Chinese government with America		the Chinese government in the sino-us joint statement on climate change

Policies

China Ministry of Science and Technology and China National Petroleum Corporation (CNPC) have launched a *National Fundamental Research Program* called *Research on Utilizing Greenhouse Gas as a Resource for EOR and Geological Storage* since 2006. *CNPC* established a pilot test of CO_2-EOR and CO_2 Storage in Jilin Oil Field in 2007. After *the 2014 Beijing Asia-Pacific Economic Cooperation (APEC)* meeting, Chinese government has made a pledge of carbon emission peak by 2030 in the *US-China joint announcement on climate change* —China intends to achieve the peaking of CO_2 emissions around 2030 and to make best efforts to peak early and intends to increase the share of non-fossil fuels in primary energy consumption to around 20% by 2030. **Table 1** is launched CCUS projects and policies by China government to deal with CO_2 emission.

After nearly 20 years of practice, in support of national policies, many demonstration projects covered CCS whole process of capture and storage are established (parts of them are shown in **Figure 1**), and all of them have achieved fairly good performance.

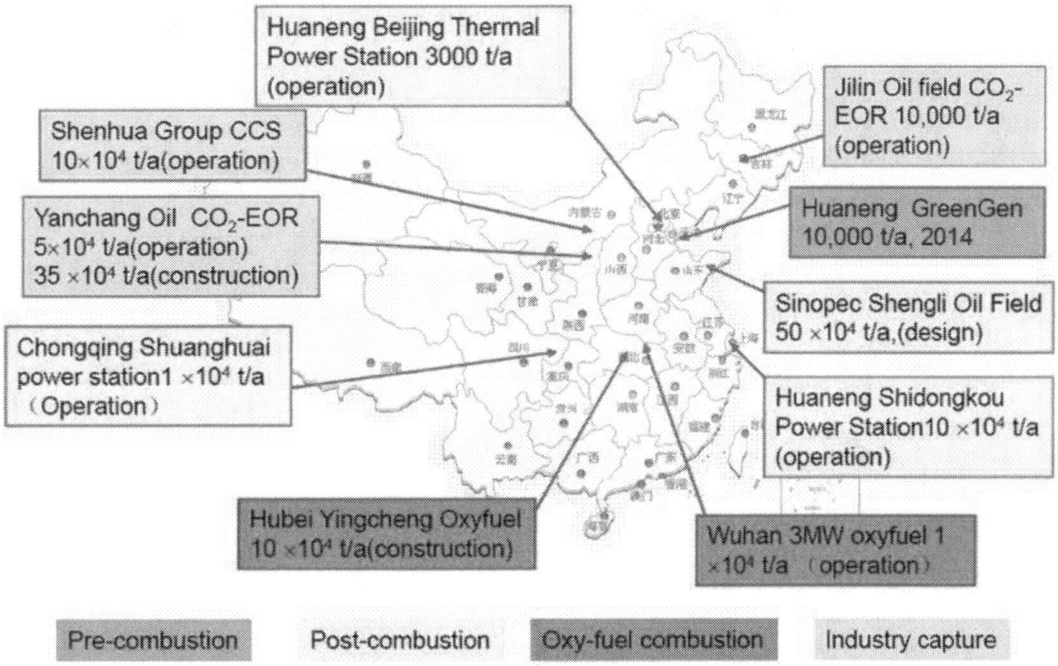

Figure 1 Parts of China CCUS projects

CO_2 Capture

China HuaNeng Group built the first CO_2 capture pilot plant in 2009, which the amount is 120 thousand per year from simulated gas of natural gas combined cycle in China, it had been continuously run for more than 3000 hours to demonstrate absorbent and process. The test shows that the designed capture process runs stably and reliably. The developed CO_2 absorbent has the advantages of high absorption efficiency, low corrosion, low degradation and low solvent loss. The novel steam-flash-mechanical vapor recompression system can reduce 10% to 25% of steam to demand and decrease 10% to 15% of total energy cost for CO_2 capture (Guo Dongfang, 2014)[10].

CO_2 Utilization and Storage

CNPC has implemented a CO_2-EOR pilot in *Hei-59 in Jilin Field.* Geological reserves is 1.02 million tons, oil reservoir is distributed at the depth of 2450m, oil-bearing area is 2.3 square kilometers, formation pressure/saturation pressure is 24.2MPa/7.0MPa, permeability is 3.0md, minimum miscible pressure is 22.3MPa. As a pilot test area, it has built 7 injection wells and 26 oil production wells. After supercritical CO_2 injected into wells at 40 tons per day, single well productivity was 4.3 tons per day which improves 1 ton per day more than gas injected, and increase 2.5 tons per day by 53% more than predicted value by water flooding (2.8 tons per day). Up to now, *Hei-59* improved production per day is 33 tons per day and improved production per year is 9.3 thousand tons. Predicted recovery of CO_2 flooding is 35.7% which increased by 12.7% more than water flooding (Sun Ruiyan, 2012)[11].

CO_2-ECBM has conducted a pilot test in *TL003 well of the southern Qinshui Basin*. In the pilot test, 19.2 tons of liquefied CO_2 were intermittently injected into No.3 coal seam. The injection rate is 0.8 MMscf/d. A set of high-quality data were obtained, including the injection rates, surface and bottom-hole pressures and temperatures during injecting CO_2 period, the bottom-hole pressures and temperatures during shut-in period, and the bottom-hole pressures and temperatures, gas and water production rates, and gas composition during production period. The history match of the dataset from the pilot was successful with the CMG simulator. The multi-well simulation and numerical prediction indicated that CO_2 was almost adsorbed into No.3 coal seam, and the coalbed methane recovery increased at least 40% more than drainage (**Figure 2**) while CO_2 could be stored in No.3 coal seam for a long term in south part of Qinshui Basin (Ye Jianping etc., 2007)[12].

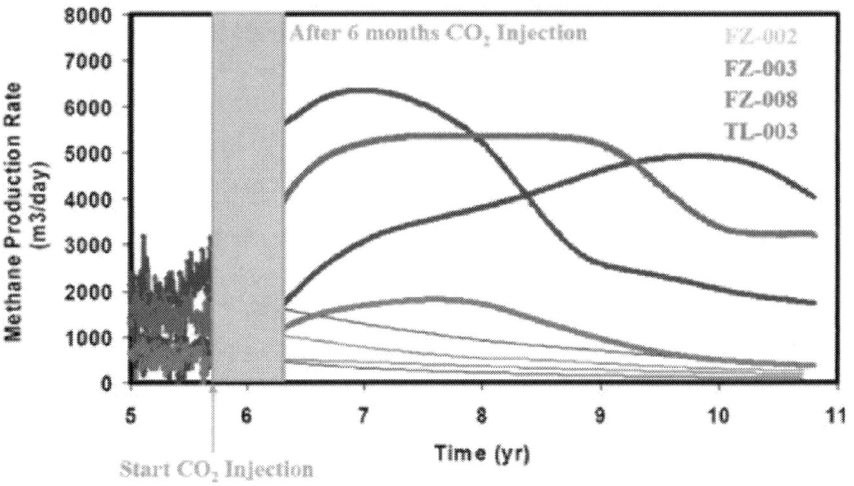

Figure 2 Enhanced coalbed methane recovery rate after CO_2 injection in Qinshui Basin (Ye Jianping)

Whole Process

In Febrary 2009, *China Shenhua Group* began to built the first whole process demonstration project in deep saline aquifer in China. The object region is saline aquifer which has low porosity and low permeability in the southeastern Erdos City. It finished 3-D seismic exploration in May 2010 and trap apparatus foundation in June 2010. The injection experiment was completed in Janary 2011. In May 2011, it officially began to inject CO_2. It has been completed by April 2015. The injection capacity per day is above 400 tons per day, injection pressure is 4.98MPa, injection rate is 17.23 cubic meters per hour and the total injection amount is 302 thousand tons. It contains capture high concentration carbon dioxide from coal chemical industry, transportation by tank trunks and CO_2 injection into saline aquifer by separate layer injection technology, multilayer injection technology and separate layer monitoring injection technology (Wu Xiuzhang, 2010[13]; Lan Zhengkai[14], 2015).

Although China has made some achievements on CCS, there are large gaps in variety and quality as compared with advanced standards (**Table 2**). Especially at the aspect of carbon dioxide transportation, it mostly by tank truck, while the USA has built up to 808 km of pipeline transportation which annual transport capacity of 20 million tons. In 2020 and 2030, the capture scale, using scale and storage capacity will increase while the energy consumption and cost will decrease (**Table 3**).

Table 2 Comparison of CCUS technology project between China and foreign countries

Technology Stage		Engineering Quantity		Largest Project Scale		Longest-running	
		Foreign	Inland	Foreign	Inland	Foreign	Inland
Capture	post-combustion	>5	4	10×10^4t/a	12×10^4t/a	2 years	2 years
	Pre-combustion	>2	-	$5\times10_4$t/a	$6\sim10\times10^4$t/a under construction	< 6 months	-
	Oxygen-enriched Combustion	>3	-	6×10^4t/a	5×10^4t/a under construction	3 years	-
Transportation	CO_2 Pipeline Transportation	>15	2	808km, quantity delivered is 2×10^7t/a	short-distance pipework	40 years	10 years
	CO_2 Large-scale Storage	/	4	$3000m^3$/pot	$1000m^3$/pot	40 years	2 years
Utilization	CO_2-EOR	>100	10	120×10^4t/a	10×10^4t/a	nearly 40 years	6 years
	CO_2-ECBM	>5	1	$>20\times10^4$t/a	$1\sim2\times10^4$t/a	7 years	4 years
	CO_2 Chemical Utilization	>10	>5	5×10^4t/a		>5 years	>3 years
	CO_2 Bioconversion	>5	3	biodiesel 10^4t/a	biodiesel 10^3t/a	/	under contruction
Storage	Saline Aquifer onshore	>2	1	100×10^4t/a	10×10^4t/a	7 years	6 months
	Saline Aquifer under the Sea	2	-	100×10^4t/a	-	15 years	-
	Acid Gas Reinjection	>60	-	48×10^4t/a	-	21 years	-
	Depleted Gas and Oil Fields	1	-	1×10^4t/a	-	7 years	-

Table 3 Visions and goals of China's CCUS technologies in 2015, 2020 and 2030

Stage	Details	2015	2020	2030
Whole Process	System Scale ($\times 10^4$t/a)	>30	100	>100
	Increase in Energy Consumption (%)	<25	<20	<17
	Cost (RMB/t)	350	300	240
Capture	Capture Scale ($\times 10^4$t/a)	30~100	100	>100
	Increase in Energy Consumption (%)	<20	<15	<12
	Cost (RMB/t)	210	180	140
Transportation	Pipe Transportation ($\times 10^4$t/a)	>80	200	>=1000
	Cost (RMB/t)	90	80	70
	Gas Deliverability (km)	30	>100	—
Utilization	Utilization Scale in Total ($\times 10^4$t/a)	100	200	>200
	Oil Production ($\times 10^4$t/a)	30	60	100
	Dynamic Storage Rate (%)	40~50	50~60	60
Storage	Storage Capacity ($\times 10^4$t/a)	30	100	>100
	Cost (RMB/t)	50	40	30

3 New CO_2 Storage Capacity Evaluation Methodology in Coalbed Reservoirs

Many researchers[2],[8],[9] proposed evaluation methodologies, however there are drawbacks listed as follows:

(1) The formations worldwide are different. Take China and the USA for example, the coal permeability in the U.S. is one or two orders of magnitudes more than them in China. Else, the enrichment area of CBM resources in America is always continuous reservoir underground, but in China the potential regions and target areas are inconsecutive and fragmented geographically.

(2) All methods do not take security and available resources into account. Because of physical characteristics of CO_2, residents live near storage sites would suffer from mortal danger if CO_2 leaks. Besides, coals in the depth from 0m to 1000m can be mined. CO_2 Injected into these layers would lead to energy waste.

(3) Microscopic characteristics are indispensable elements. Microscopic properties of coals such as areal and vertical as well as pore-scale displacement efficiency can influence the capacity of storage capacity.

(4) CO_2/CH_4 displacement ratio. One of the critical factors in estimating storage capacity is how to ensure CO_2/CH_4 replacement rate change with coal rank. Reeves (2003)[15] gave a relationship presented in **Figure 3**. However, according to classification standards of coal rank (**Table 4**), the vitrinite range Reeves adopted belongs to bituminous coal (medium rank coal). It aims at coal beds in America. Therefore, all results based on Reeves's conclusion are worth discussing.

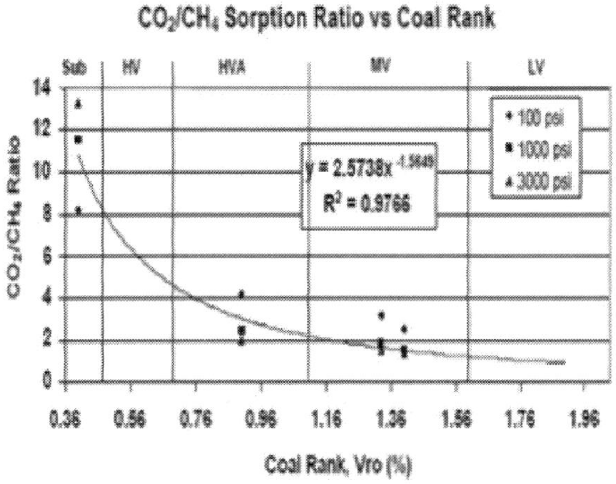

Figure 3 CO_2/CH_4 replacement ratios vs. coal rank by Reeves (2003)

Table 4 Classification standards of coal rank

Rank		Classification Standards
Low-rank	Low-rank C (Lignite C)	35%<Bed Moisture <75% and R<0.4%
	Low-rank B (Lignite B)	Bed Moisture<=35% and R<0.4
	Low-rank A (Subbituminous coal)	0.4%<=R<0.5%
Medium-rank	Medium-rank D (Bitumite D)	0.5%<=R<0.6%
	Medium-rank C (Bitumite C)	0.6%<=R<1.0%
	Medium-rank B (Bitumite B)	1.0%<=R<1.4%
	Medium-rank A (Bitumite A)	1.4%<=R<2.0%
High-rank	High-rank C (Anthracite C)	2.0%<=R<3.0%
	High-rank B (Anthracite B)	3.0%<=R<4.0%
	High-rank A (Anthracite A)	4.0%<=R<6.0%

CO_2 is injected at pressures over 8MPa, where it exists as a supercritical fluid with a density of 100kg/m^3, which is less dense than most formation brines and oils (James P. Verdon, 2012)[16]. In coals, the storage mechanism is chemical trapping. It occurs when CO_2 dissolves in subsurface fluids (solubility trapping) and may then be involved in chemical reactions with the rock matrix (mineral trapping), or becomes adsorbed onto mineral surfaces (adsorption trapping). Therefore, this paper considers Adsorption Storage Capacity, Dissolution Storage Capacity and Mineralization Storage Capacity. Therefore, it should consider these matters when begin to estimate:

(1) In coal seams, the free gas accounts 10% ~ 20%, dissolved gas can be neglected, the rest are adsorbed gas. So storage potential can be divided into three parts: adsorption storage quantity more than dissolution storage quantity more than mineralization storage amount.

(2) The treatment of the mined-out area mainly has three ways: full caving method, the filling-in method and roof control by wall-shape pillar. The caving method is simple, reliable, less cost and the most commonly used. CO_2 injection easily caused leakage risk, so this paper believes mined-out area has no CO_2 storage potential.

(3) Coal bed methane mining method is drainage decompression or depletion development. After drainage of coal seam in above 1000m, dissolution storage capacity needs to refer to a given actual production data. This evaluation is about theoretical potential.

3.1 Adsorption Storage Capacity

Carbon dioxide storage in coal beds occurs when CO_2 is preferentially adsorbed onto coal. Coal contains a natural system of fractures called cleats, which imparts some permeability to the system. Between the cleats, the solid does not contain macropores through which fluids can flow, but does contain a very large number of micropores into which gas molecules can diffuse from the cleat. The combined surface area of the micropores, which form adsorption sites for gas molecules, is not saturated, and the non polar gas molecules (such as CO_2) can produce Van Der Waal's Force (Sun Jiaopeng et al. 2009)[17], the adsorbed molecules can be very tightly packed. Coal can physically adsorb many gases, for which it has different affinities (**Figure 4**). Coal has higher affinity for gaseous CO_2 than for methane. When CO_2 is injected into coal seams, there exists the competing adsorption phenomenon between CO_2 and CH_4. In the meantime, the pore pressure increases due to CO_2 injection. Compared with CH_4, CO_2 is prone to super-critical state when the CO_2 has the characteristics of high density, low viscosity, low diffusion and high solubility. Thus, CO_2 storage in coal beds is based on that the injected CO_2 will replace the methane in coal and stay adsorbed onto the coal surface as long as the coal is undisturbed.

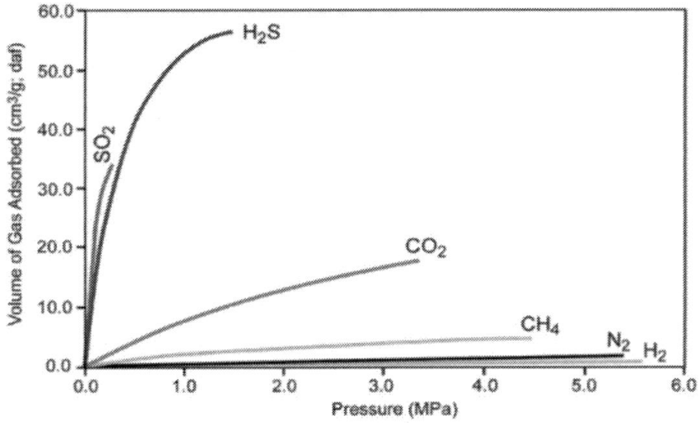

Figure 4 Adsorption of various gases on coal (Chikatamarla & Bustin, 2003)[18]

Apart from factors which previous methods mentione, effective stress should be considered while estimating. During coalbed methane production process, porosity and permeability decrease with actual stress increasing. This phenomenon is confirmed by experiments and theoretical derivation (Equation 1 to 5). In addition, permeability of different coal ranks reduces in the form of negative exponential function with the increase of pressure. Therefore, porosity change is regarded as an essential element in this evaluation.

Actual stress equals overburden pressure minus reservoir pressure. Effective stress has a relation with permeability (K) as equation:

$$K = 1.03 \times 10^{-0.31\sigma} \tag{1}$$

Mckee gave an equation:

$$K = K_0 \exp(-3C_p \Delta\sigma) \tag{2}$$

$$C_p = \frac{1}{\phi} \times \frac{d\phi}{dp} \tag{3}$$

By evaluating the integral of equation 3, it has:

$$\phi = \phi_0 \exp(-C_p \Delta\sigma) \tag{4}$$

$$\Delta\sigma = \sigma - \sigma_0 \tag{5}$$

where K_0 is the initial permeability, mD; K is the permeability, mD; σ is the effective stress, MPa; C_p is the pore volume compressibility, MPa^{-1}; p is the fluid pressure, MPa; σ_0 is the initial pressure, MPa; ϕ is the porosity impacted by effective stress, %; ϕ_0 is the initial porosity, %.

Besides, the storage capacity of CO_2 in coal seams includes two parts: one is the ability of remaining space after drainage gas recovery technique, the other part is the ability which through CO_2 flooding. Learned from America experience, there has 75 percent resources left as the basis of CO_2-ECBM.

This new method takes fully into account. Equation 7 is to compute production using drainage technique. Equation 8 and Equation 9 is to calculate production which through carbon dioxide flooding. It should be noted that coalbed methane in deep depth cannot depend on drainage gas recovery because of low permeability without fracture.

$$N_{proved} = N_{geological} \cdot \xi \tag{6}$$

$$N_{CBM} = N_{proved} \cdot \eta \tag{7}$$

$$N_{ECBM} = N_{proved} \cdot (\eta_{CO_2-ECBM} - \eta) \tag{8}$$

$$N'_{ECBM} = N'_{proved} \cdot \eta'_{CO_2-ECBM} \tag{9}$$

where N_{proved} is the proved resources, m^3; $N_{geological}$ is geological resources, m^3; ξ is the proved ratio, %; N_{CBM} is the methane production using drainage method, m^3; η is the recovery rate, %; η_{CO_2-ECBM} is the recovery factor of carbon dioxide flooding, %, N_{ECBM} is the improved recoverable reserve in shallow depth, m^3; N'_{ECBM} is the improved recoverable reservoir which in deep depth, m^3. is not only up to gas content, law of adsorption and desorption, reservoir pressure as well as temperature, but also strongly depends on drilling, completion and production technologies. is the product of areal and vertical displacement ratio.

The theoretical capacity is:

$$M_{CO_2 t} = \rho_{CO_2} \cdot N_{ECBM} \cdot (1 - f_a - f_m) \cdot D_R \tag{10}$$

The effective capacity is:

$$M_{CO_2 e} = M_{CO_2 t} \cdot R_f \cdot E \cdot \lambda \tag{11}$$

$$\lambda = \frac{\phi}{\phi_0} \tag{12}$$

The practical capacity is:

$$M_{CO_2 a} = M_{CO_2 e} \cdot C \tag{13}$$

The matched capacity is:

$$M_{CO_2 m} = M_{CO_2 a} \cdot E_q \tag{14}$$

where $M_{CO_2 t}$ is the theoretical storage capacity, t; $M_{CO_2 e}$ is the effective storage capacity, t; $M_{CO_2 e}$ is the practical storage capacity, t; $M_{CO_2 a}$ is the matched storage capacity, t; ρ_{CO_2} = 1.813kg/m^3 is CO_2 density at standard conditions; f_a and f_m are fraction of the ash and moisture, %; D_R is the CO_2/CH_4 displacement ratio, decimal; R_f is the recovery factor, %; E is the storage factor is storage factor between 28% and 40% according to Monte Carlo simulations (**CSLF, 2008**)[19], %; λ is the rate of porosity change, decimal; C is the completion factor, %, E_q is the equipment factor, %.

3.2 Dissolution Storage Capacity

After CO_2 injected into coal beds, the thick shale and clay rock above coal beds prevent CO_2 flow upward. CO_2 shows a higher solubility in the formation water. If CO_2 dissolved in the formation water, the density of CO_2 is increasing so that CO_2 will sink in the coal layer structure other than float to the surface. However, CH_4 dissolves in the formation water is more easily to escape in the same pressure conditions.

CO_2 can dissolve into water after injection. Carbon dioxide dissolubility has positive correlation with pressure, and is inverse ratio with temperature and salinity. **Table 5** is the solubility of CO_2 in water under different pressure and temperature. It is going to distinguish different values in different depth using geothermal gradient and formation pressure gradient.

Table 5 Solubility of CO_2 in water under different pressure (unit: dm^3/kgH_2O)

Pressure (MPa)	Temperature(℃)										
	0	10	20	30	40	50	60	70	80	90	100
0.2		2.3	1.72	1.23							
0.4		4.7	3.4	2.5							
0.6		6.75	4.83	3.7							
0.8		8.85	6.4	4.8							
1		10.7	7.85	5.83							
2.5	30.46	20.8	18.82	14.54	11.62	9.71	8.51	7.44	6.36	5.62	5.36
5	36.89	36.68	30.52	24.65	20.35	17.25	14.48	12.78	12.04	11.11	10.18
10	49.13	38.39	33.26	30.43	27.81	25.63	23.95	21.94	19.64	18.09	17.67
20	47.33	41.03	36.32	33.04	30.71	29.14	29.14	27.25	26.29	25.74	25.69
30	45.11	41.84	38.69	35.72	33.22	31.34	31.34	29.6	29.51	29.62	29.53
40			40.09	37.08	34.87	33.29	33.29	31.24	32.03	32.42	32.39
50			41.22	38.82	36.73	31.75	31.75	13.16	6.27	9.98	34.4

Adsorption and desorption experiments are also done to study rules at different temperature, pressure and moisture. Rock cores experiments of Huo-test1 well, Jin7 wells and Chang-test1 well show that high temperature, low pressure and high viscosity lead to drastic thermal motion of molecular gas; and methane adsorption of coal is an exothermic process, temperature elevating lead to a decrease in adsorb amount(**Figure 5**). In aquifers, water molecules occupy the adsorption sites which belong to methane molecules before so that methane adsorption capacity reduces. Dry coal cores and coal cores with balanced water show great difference in adsorption and desorption curves for one same rock core (**Figure 6**).

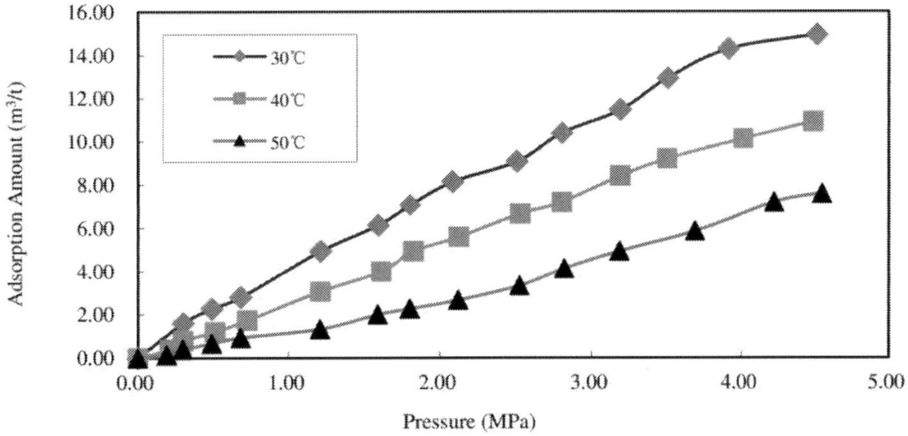

Figure 5 Adsorption capacity influenced by temperature and pressure

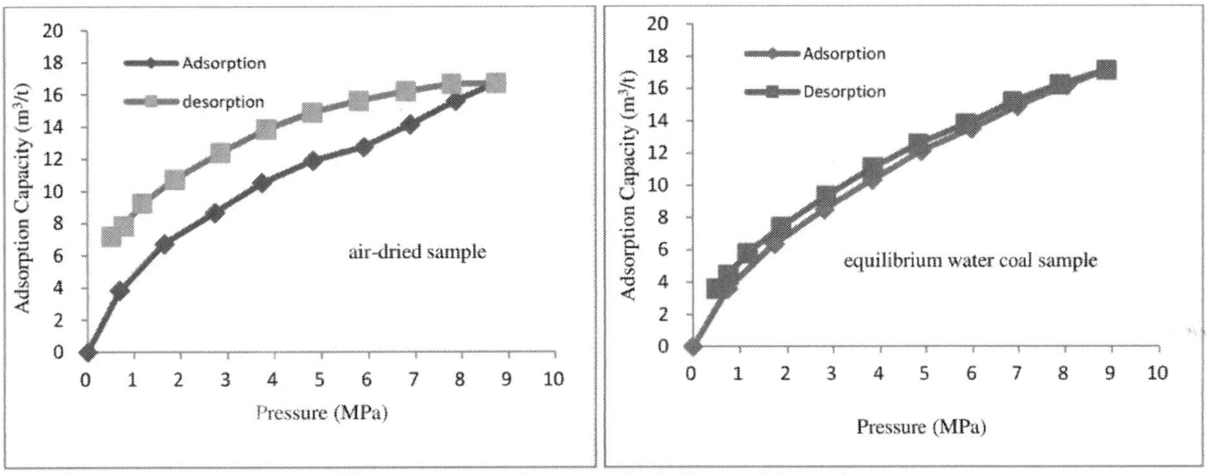

Figure 6 Adsorption capacity influenced by moisture

According to the industrial standard of *China's oil and gas industry standards (CN-PC)*, using volumetric method to calculate coalbed methane geological reserves as follows:

$$G_i = 0.01AhDC_{ad} \tag{15}$$

$$\text{Or } G_i = 0.01AhD_{daf}C_{daf} \tag{16}$$

where $C_{ad} = 100C_{daf}(100 - f_a - f_m)$, G_i is the CBM geological reserves, $10^8 m^3$; A is the gas bearing area, km^2; h is the seam thickness, m; D is the density of air dried basis, t/m^3; C_{ad} is the gas content of air dried basis, m^3/t; D_{daf} is the mass density of dry ash-free basis, t/m^3; C_{daf} is the gas content of dry ash-free basis, m^3/t.

So the dissolution storage capacity is:

$$M_s = \rho_{CO2} \cdot M_{coal} \cdot S \cdot f_m \tag{17}$$

where M_s is the dissolution storage capacity, t; M_{coal} is the quality of aqua coal , t; S is dissolubility, %.

3.3 Mineralization Storage Capacity

With the chemical reaction between CO_2 and bedrocks, the mechanism of the mineral trapping is beginning to play a role. In addition, the dissolved CO_2 set off a chemical reaction with minerals in rocks to form the formation of ionic substances and carbonate minerals (Equation 1 to 5).

$$CaCO_3 + H^+ \Leftrightarrow Ca^{2+} + HCO_3^- \tag{18}$$

$$CaAl_2SiO_8 + 8H^+ \Leftrightarrow 4H_2O + Ca^{2+} + 2Al^{3+} + 2SiO_2(soluble) \tag{19}$$

$$Al_2O_3 + 2SiO_2 + 2H_2O + 6H^+ \Leftrightarrow 5H_2O + 2SiO_2(soluble) + 2A^{3+} \tag{20}$$

$$FeCO_3 \Leftrightarrow Fe^{2+} + CO_3^{2-} \tag{21}$$

$$K\{(Al, Fe, Mg)_2[AlSiO_{10}](OH)_2\} + 10H^+ \Leftrightarrow K^+Fe^{3+} + Fe^{2+} + 2Mg^{2+} + 3Al^{3+} + 3SiO_2(soluble) + 6H_2O \tag{22}$$

There are numerous factors impacting mineralization storage capacity such as reservoir water and mineral rock compositions, system temperature and pressure, interfacial tension of solid-liquid, fluid velocity and so on. All play roles in the mineralization process. But until now, no study for this kind has been systematically developed. Thus this paper cannot give quantitative descriptions of mineralization storage.

4 Application

There are 36810 billion cubic meters coalbed methane geological resources within 2000m and 10870 billion cubic meters coalbed methane geological resources within 1500m in 45 major gas-bearing basins in China. The average proved ratio is 9‰. The average recovery rate is 35% by 2013.

Buried Depth

China puts coal mining depth into four categories: less than 400m is shallow coal mine, 400m to 800m is mid-deep coal mine, 800m to 1200m is deep mine, more than 1200m is special deep mine. Because of high stress, high geothermal, high risk and high cost, several regions stop to approve deeply mining. Deeply mining wells are distributed in the east of TaiHang Mountain, in the south of northeast and eastern of Inner Mongolia region (**Figure 7**). So this paper believes that the buried depth of coalbeds in China should be more than 1000m. Considering the permeability decreases with depth increases and present exploration situations, the maximum depth should be less than 2000m.

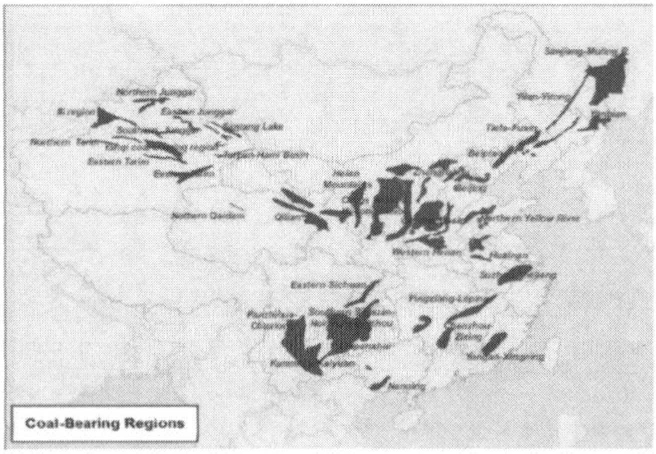

Figure 7 Mine wells which depth more than 1000m (red points) distributed in China. The blue areas are gas-bearing basins.

CO_2/CH_4 displacement

Bustin (2002)[20] insisted that the replacement ratio of CO_2/CH_4 was highest for low rank coals, and it decreased with coal ranks increasing (**Figure 8**). Quantitatively, the replacement ratio decreased from the range of 10:1 for sub-bituminous coals to 1:1 for LV coal. It is noted that the ratios are influenced by pressure. Lower pressure can lead to a less apparent ratio gap while higher pressure seems to be the operating range of most sequestration projects.

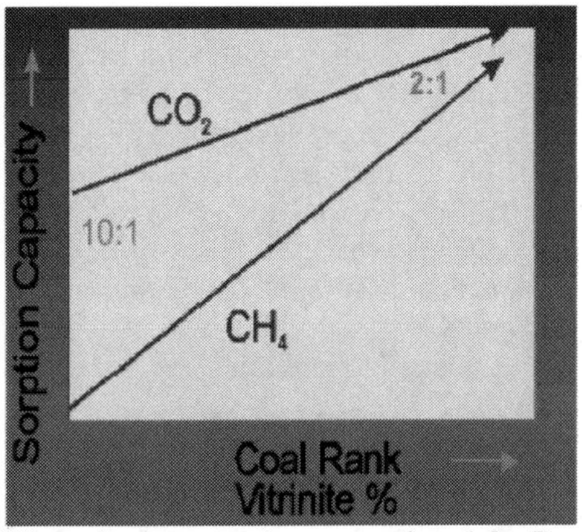

Figure 8 CO_2/CH_4 replacement ratios vs. coal rank by Bustin (2002)

To ensure the precision of CO_2/CH_4 displacement value, metathesis experiments of cores at the bury depth of 1000m (from Qinshui Basin, Erdos Basin and Xinjiang Basin etc.) have been accomplished. The value range is from 2 to 6. Besides, according to Hilt Law, the deeper the buried depth is, the higher the metamorphic grade is, and the smaller the CO_2/CH_4 displacement is. As a result, the displacement value in 1000m to 2000m is 0.2 bigger than it in 1500~2000m.

Table 6 Recoverable reserve and CO_2 adsorption theoretical storage capacity of coalbed reservoirs in China

Basin	Recoverable Reserve(10^8m^3)		Theoretical Storage Capacity(10^8t)		
	Converntional Technology	CO2-ECBM	1000~1500m	1500~2000m	Total
A	796.00	1903.03	4.15	3.34	7.49
B	34.70	36.88	0.32	0.00	0.32
C	147.33	156.59	1.20	0.00	1.20
D	62.27	191.42	0.54	0.51	1.06
E	53.40	164.17	0.34	0.32	0.66
F	33.64	103.40	0.35	0.33	0.68
G	1507.90	6080.56	16.98	23.63	40.61
H	87.61	218.93	0.42	0.27	0.69
I	192.88	653.84	0.98	1.03	2.01
G	229.38	777.56	2.31	2.52	4.82
K	70.48	238.92	0.58	0.62	1.20
L	280.06	909.11	2.35	2.40	4.75
M	276.29	1139.26	1.92	2.75	4.66
N	79.76	192.97	0.34	0.21	0.54
O	254.12	628.85	0.93	0.59	1.52
p	207.62	702.15	1.16	1.23	2.39
In Total	4313.45	14097.64	34.86	39.74	74.61

Table 7 CO_2 adsorption effective, practical, matched and missolution storage capacity of coalbed reservoirs in China

Basin	Effective Storage Capacity(10^8t)			Practical Storage Capacity(10^8t)			Matched Storage Capacity(10^8t)			Dissolution Storage Capacity (10^8t)		
	1000~1500m	1500~2000m	Total	1000~1500m	1500~2000m	Total	1000~1500m	1500~2000m	Total	1000~1500m	1500~2000m	Total
A	1.73	1.38	3.10	1.38	1.10	2.48	1.04	0.83	1.86	0.17	0.16	0.33
B	0.13	0.00	0.13	0.11	0.00	0.11	0.08	0.00	0.08	0	0	0
C	0.50	0.00	0.50	0.40	0.00	0.40	0.30	0.00	0.30	0.02	0	0.02
D	0.23	0.21	0.44	0.18	0.17	0.35	0.14	0.13	0.26	0.01	0.01	0.01
E	0.14	0.13	0.27	0.11	0.10	0.22	0.09	0.08	0.16	0.01	0.01	0.02
F	0.15	0.14	0.28	0.12	0.11	0.23	0.09	0.08	0.17	0.01	0.01	0.03
G	7.06	9.73	16.79	5.65	7.79	13.44	4.24	5.84	10.08	0.29	0.44	0.73
H	0.17	0.11	0.28	0.14	0.09	0.23	0.10	0.07	0.17	0.01	0	0.01
I	0.41	0.43	0.83	0.33	0.34	0.67	0.24	0.26	0.50	0.08	0.1	0.18
J	0.96	1.04	2.00	0.77	0.83	1.60	0.58	0.62	1.20	0.04	0.05	0.09
K	0.24	0.26	0.50	0.19	0.21	0.40	0.14	0.15	0.30	0.03	0.03	0.06
L	0.98	0.99	1.97	0.78	0.79	1.57	0.59	0.59	1.18	0.21	0.24	0.45
M	0.80	1.13	1.93	0.64	0.90	1.54	0.48	0.68	1.16	0.07	0.11	0.19
N	0.14	0.08	0.23	0.11	0.07	0.18	0.08	0.05	0.14	0.01	0.01	0.02
O	0.39	0.24	0.63	0.31	0.19	0.50	0.23	0.15	0.38	0.03	0.02	0.06
P	0.48	0.51	0.99	0.39	0.41	0.79	0.29	0.30	0.59	0.03	0.03	0.06
In Total	14.50	16.37	30.87	11.60	13.10	24.69	8.70	9.82	18.52	1.03	1.22	2.25

Evaluation Results

Taking equation 6 and equation 7 to calculate CO_2 storage capacity in each coal area. Middle value results and dissolution storage capacity are shown in **Table 6** and **Table 7**. The average recovery rate is enhanced from 35% to 72% after utilizing CO_2-ECBM technology. Recoverable reserve can be improved 1409.8 billion cubic meters. CO_2 theoretical storage capacity of coalbeds in China is 7461million tons, effective storage capacity is 2480 to 3748 million tons, practical storage capacity is 1736 to 33.73 million tons, matched storage capacity is 1042 to 3036 million tons. The potential between 1000m to 1500m is larger than it between 1500m to 2000m. And the dissolution storage capacity in total is 2250 million tons. Hence deep buried depth has a enormous potential to store carbon dioxide with the development of exploration and production technologies.

Moreover, **Figure 9** is map of CO_2 point sources. Combining with the evaluation areas, storage fields in China are conformity with the main emission regions.

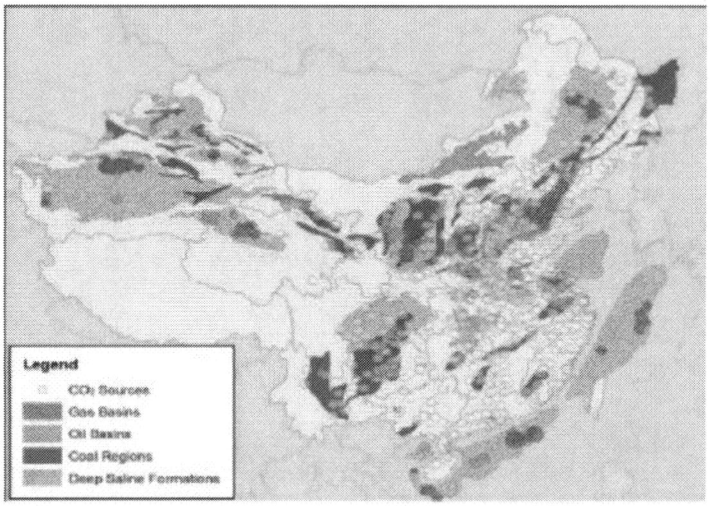

Figure 9 Map of large CO_2 point sources with candidate geologic storage formations (Source: USDOE, 2009).

The basins can be classified by capacity range, namely more than 1000 million tons, ranging from 100 to 1000 million tons and from 0 to 100 million tons, as shown in **Table 8**. There are five basins which possess capacities lager than 200 million tons, and the total capacity of these five basins is 2578 million tons, accounting for 83.5% of the whole China. These five basins are the ideal basins for large-scale CO_2 storage projects implemented in the future.

Table 8 Basin classified by storage capacity

Storage Capacity Range（10^8t）	Number of Basins	Storage Capacity（10^8t）	Percentage of Total Effective Capacity(%)
0-1	11	5.08	16.5
1-10	4	8.99	29.1
>10	1	16.79	54.4

It can be easily concluded from **Table 9** that the productivity calculated by the prediction equations is higher than the actual value by a huge difference. Considering overall evaluation methods deeply and comprehensively, the results should be revalued. But the economic and operating limitations of storage capacity are basic coincidence.

Table 9 Results comparison in different ways

Index	Yu(2005)	Liu and Li(2005)	Fang and Li(2014)	This Method
Buried Depth(m)	0～2000m	300～1500m	1000～2000m	1000～2000m
Geological CBM Resources(10^8m^3)	143369.43	276100	225410.3	225410.33
Conventional Gas Production(10^8m^3)	70790.83	9666.69	—	5225.4
CO_2-ECBM Improved Production(10^8m^3)	37443.46	16321.94	42600	12885.69
CO_2 Minimum Value(10^8t)	1417.13	121	98.81	24.8
CO_2 Maximum Value(10^8t)		484		37.48
Considering the Economic and Operating Conditions, CO_2 Storage Capacity(10^8t)	—	10	—	10.42

5 Conclusion

(1) China's environmental management is to build a comprehensive management system that takes into account all kinds of pollutants and pollution sources, and has launched a series CCS polies and regulations to deal with global climate change. In China, ECBM policies were performed as practical and economical ways for reducing CO_2 emissions, which function is one stone two birds.

(2) New evaluation methodologies of CO_2 storage capacity of the theoretical storage potential, effective storage potential are built, also, CO_2 geological storage potential and coal recoverable reserves by CO_2-ECBM both are estimated in China. We think that the reasonable buried depth in China should be more than 1000m and less than 2000m. China has a huge theoretical storage capacity in Erdos Basin, Zungar Basin, Tarim Basin, Tuha Basin and Qinshui Basin, where have not only great geological conditions but also multiple types of emission sources. Therefore, these basins are ideal regions to promote CCUS project.

(3) We need to develop practical technology for CCS/CCUS because the cost of carbon capture is very high for now. For CCUS project is a system engineering, including capture, transportation and storage, finance is very complicated, it is suggested that emission enterprises will afford the cost of CO_2 capture, and both emission enterprises and customers will afford the cost of transportation, and storage will be afforded by customers.

(4) For CCU project, emission enterprises capture CO_2, will consider selling it to customers, such as food plant and industrial plant of requiring CO_2.

(5) In order to acquire more accurate CO_2 storage capacity, further studies could be considered in future: study of micro flow characteristics in different coal ranks research of CO_2 flow in coalbed reservoirs and make quantitative analysis on coal rock's physical-chemical reaction with CO_2 and economic evaluation for CCS and CCUS-CO_2-ECBM project.

Acknowledgements

This work is supported by A Basic Research on CO_2 Emission Reduction and Storage and Utilization (No. 2011CB707302) which is a part of National Science Foundation of China.

References

[1] Buchu, S. Energy Convers. Manage. 2002, 43, 87-102.
[2] Yu Hongguan. Study of Characteristics and Prediction of CH_4, CO_2, N_2 and Binary Gas Adsorption on Coals and CO_2/CH_4 Replacement [D]. Doctor of Philosophy. Shandong University of Science and Technology. 2005.
[3] Liu Yanfeng, Li Xiaochun and Bai Bing. Preliminary estimation of CO_2 Storage Capacity of Coalbeds in China [J]. Chinese Journal of Rock Mechanics and Engineering. 2005.
[4] Kaldi, J. and Gibson-Poole, C.M. (Eds). Storage Capacity Estimation, Site Selection and Characterization for CO_2 Storage Projects," Cooperative Research Center for Greenhouse Gas Technologies, Canberra, Australia. CO2CRC Report RPT08-1001. 2008.
[5] Frailey, S.M., and Finley, R.J. Classification of CO_2 Geologic Storage: Resource and Capacity [J]. Energy Procedia1 (1):2623-2630. 2009.
[6] W.G. Allinson; Y. Cinar; P.R. Neal and J. Kaldi. CO_2- Storage Capacity- Combining Geology, Engineering and Economics [J]. SPE-133804-PA. 2014.
[7] Fang Zhiming, Li Xiaochun. A Preliminary Evaluation of Carbon Dioxide Storage Capacity in Unmineable Coalbeds in China [J]. Acta Geotechnica (2014) 9: 109-114.
[8] CSLF (Carbon Sequestration Leadership Forum). Estimation of CO_2 Storage Capacity in Geological Media, June 2007, 43p.
[9] USDOE (U.S. Department of Energy, Office of Fossil Energy). Carbon Sequestration Atlas of United States and Canada, 86 p. 2007.
[10] Guo Dongfang, Liu Lianbo, Wang Jinyi, etc. Pilot-scale Experimental Study of 1000t/a CO_2 Capture From Flue Gas of Natural Gas Power Station [J]. Proceedings of CSEE. 2014,34 (23):3849-3855.
[11] Sun Ruiyan, Wang Xianzhong, Ma Xiaohong etc. CO_2 flooding Ground Engineering Technology in Hei-59 [J]. Planning and Design. 2012, 31(2):37-38.

[12] Ye Jianping, Feng Sanli, Fan Zhiqiang, etc. Micro-pilot Test for Enhanced Coalbed Methane Recovery by Injecting Carbon Dioxide in South Part of Qinshui Basin [J]. ACTA PETROLEI SINICA. 2007, 28(4):77-80.
[13] Wu Xiuzhang, Cui Yongjun. Study on 0.1Mt/a CO_2 Sequestration in Saline Formation [J]. ACTA PETROLEI SINICA (PETROLEUM PROCESSNG SECTION). 2010,10:236-239.
[14] Lan Zhengkai, Li Xue'e. Three-dimensional Geological Modeling in Shenhua's CCS Demonstration Area [J]. Inner Mongolia Petrochemical Industry. 2015,8:40-42.
[15] Reeves S. Assessment of CO_2 Sequestration and ECBM Potential of U.S. Coalbeds. U.S. Department of Energy, DE-FC26-00NT40924, February, 2003.
[16] James P. Verdon. Microseismic Monitoring and Geomechanical Modeling of CO_2 Storage in Subsurface Reservoirs [D]. Doctoral Thesis accepted by University of Bristol, United Kingdom.
[17] Sun Jiaopeng, Guo Wei, Cui Lu etc. The Meaning and Theory Analysis of CO_2 injection into Coal Seams [J]. Science and Technology Innovation Herald. 2009(22):116.
[18] Chikatamarla, L. and M.R. Bustin, 2003: Sequestration potential of acid gases in WesternCanadian Coals. In: Proceedings of the 2003 International Coalbed MethaneSymposium, University of Alabama, Tuscaloosa, Alabama, 5-8 May, 2003, 16 p.
[19] CSLF (Carbon Sequestration Leadership Forum). Comparison between Methodologies Recommended for Estimation of CO_2 Storage Capacity in Geological Media. 2008.
[20] Bustin, M. Research Activities on CO_2, H_2S and SO_2 Sequestration at UBC. 2002.

THE UNIVERSITY OF TEXAS AT AUSTIN

WHAT STARTS HERE CHANGES THE WORLD

Improving the Ease and Accuracy of Environmental Monitoring at CCS Sites

Katherine Romanak,
Gulf Coast Carbon Center
Bureau of Economic Geology
The University of Texas at Austin

Carbon Management Technology Conference
Sugar Land TX, USA
November, 2015

THE UNIVERSITY OF TEXAS AT AUSTIN

WHAT STARTS HERE CHANGES THE WORLD

Why Do We Need Improvement?

- Research is showing us…
  - We need to change our thinking
  - We have the methods that allow new thinking
- Its important to get it right
  - Weyburn/Kerr Allegation
- Simple, fast, accurate

THE UNIVERSITY OF TEXAS AT AUSTIN

WHAT STARTS HERE CHANGES THE WORLD

Transition to Industrial Monitoring

- Minimalistic approach relative to research-oriented
- All questions will not be answered
- All tools will not be used
- Balance between regulatory and technical goals
- Balance between cost effective and accurate data collection

THE UNIVERSITY OF TEXAS AT AUSTIN

WHAT STARTS HERE CHANGES THE WORLD

Questions To Be Answered

- How can data acquisition, reduction and interpretation be simplified?
  - E.g. "so simple my mother could do it" *[Julio Friedman, Deputy Assistant Secretary for Clean Coal, U.S. Department of Energy, Future of Carbon Storage Research, NETL Geological Storage Program Review, August 12th, 2014, Pittsburgh, PA]*
- How to balance cost with areal coverage?
- What amount of data is necessary to address situations where interpretation becomes problematic?
- How can we respond to public claims quickly, accurately and economically?
- How can we quantify a leak if it happens?

Environmental Monitoring

1) Find an anomaly that could signal a leak to the surface
   - Detected during routine monitoring
   - Environmental Impact
   - Public claims of leakage

2) Attribute its source
   - Natural variation or leakage?

THE UNIVERSITY OF TEXAS AT AUSTIN

WHAT STARTS HERE CHANGES THE WORLD

Baselines For Attribution

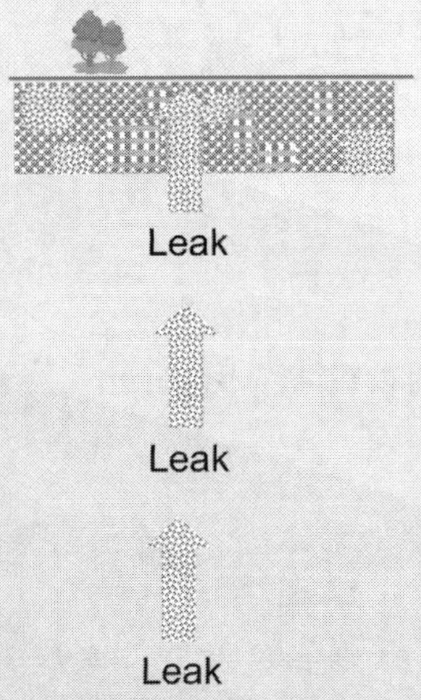

- Measure "baseline" CO_2 for 1-3 years before project start to define seasonal variability.
- Monitor CO_2 during project and compare to baseline.
- Significant increase from baseline during a project could signal a leak

THE UNIVERSITY OF TEXAS AT AUSTIN — WHAT STARTS HERE CHANGES THE WORLD

Challenges

- Ease
  - Long lead time (1-3 years)
  - Complex statistical analysis and comparison with weather parameters required
- Usefulness
  - Background cannot be measured across all potential leak points
  - Baseline measurements can only signify a change in CO_2 concentrations; they cannot attribute the source of the change.
- Accuracy
  - Baseline will be dynamic over the project lifetime
    - Climate
    - Land use

THE UNIVERSITY OF TEXAS AT AUSTIN — WHAT STARTS HERE CHANGES THE WORLD

Lessons from Weyburn and Kerr Leakage Claim

Weyburn Project

- European research team
- Rigorous environmental research over many seasons,
- Many geochemical parameters,
- Dense sampling grids (offsite background location & onsite location)
- Rainfall and temperature data
- Detailed statistical analysis
- High spatial and temporal density (Beaubien et al., 2014).

Kerr Claim

- Three independent research teams (European team, operator, independent team)
- Landowner claim of environmental damage from leakage at the Weyburn site.

THE UNIVERSITY OF TEXAS AT AUSTIN — WHAT STARTS HERE CHANGES THE WORLD

Learnings from IEAGHG Weyburn-Midale Monitoring and Storage Project

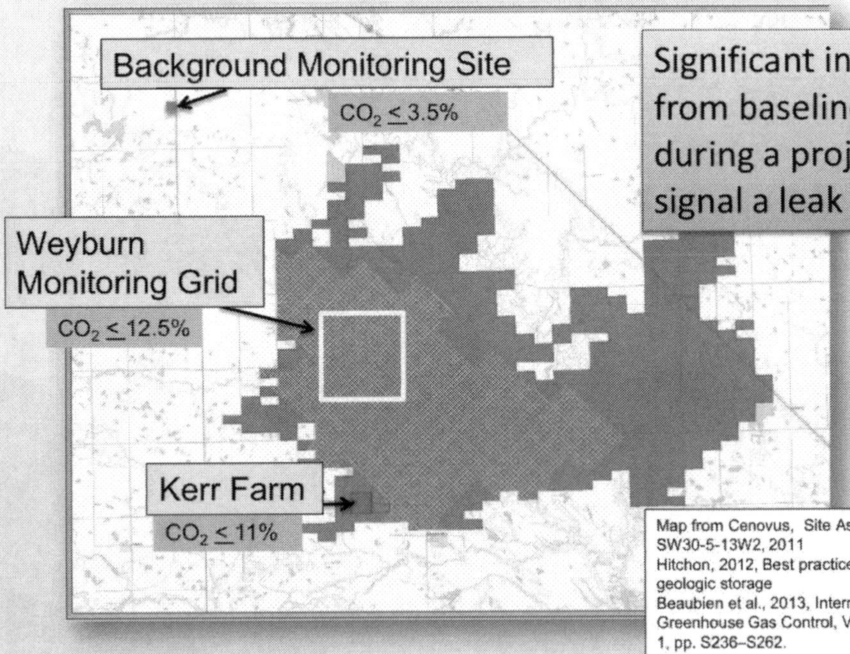

Significant increase from baseline range during a project could signal a leak

Map from Cenovus, Site Assessment Weyburn Unit SW30-5-13W2, 2011
Hitchon, 2012, Best practices for validating CO_2 geologic storage
Beaubien et al., 2013, International Journal of Greenhouse Gas Control, Volume 16, Supplement 1, pp. S236–S262.

THE UNIVERSITY OF TEXAS AT AUSTIN — WHAT STARTS HERE CHANGES THE WORLD

Soil Gas CO_2 (October 2011)

Statistical distribution of at sites within the Weyburn-Midale site (Beaubien et al., 2014).

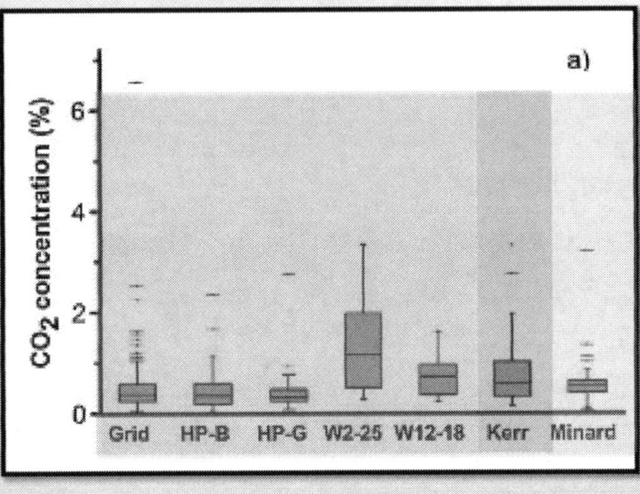

Leakage refuted based on complex reasoning and statistical analysis of multiple geochemical parameters (Hitchon, 2012; Beaubien et al., 2014).

THE UNIVERSITY OF TEXAS AT AUSTIN — WHAT STARTS HERE CHANGES THE WORLD

Ratios for Attribution: Kerr Farm

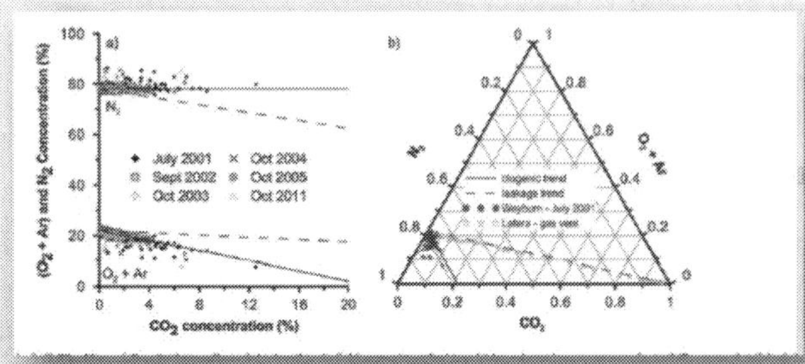

Beaubien et al., 2014

Kerr Farm, Saskatchewan, Canada

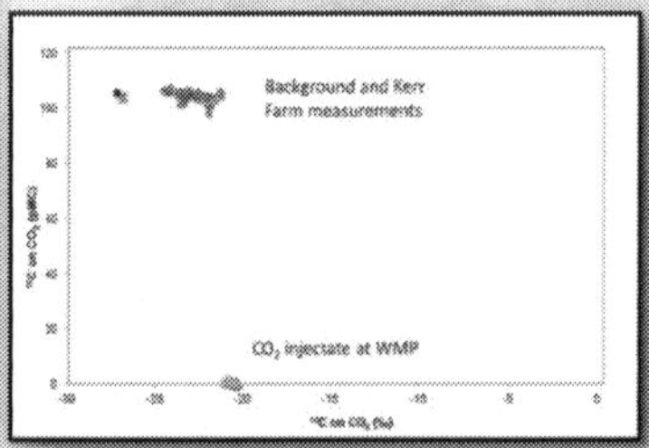

Trium, 2013

Ratios that define processes

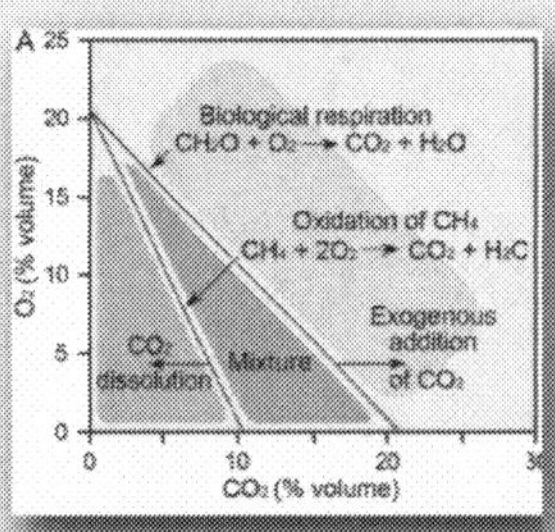

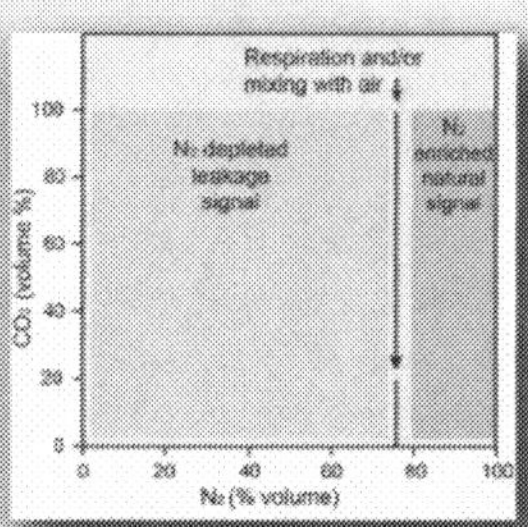

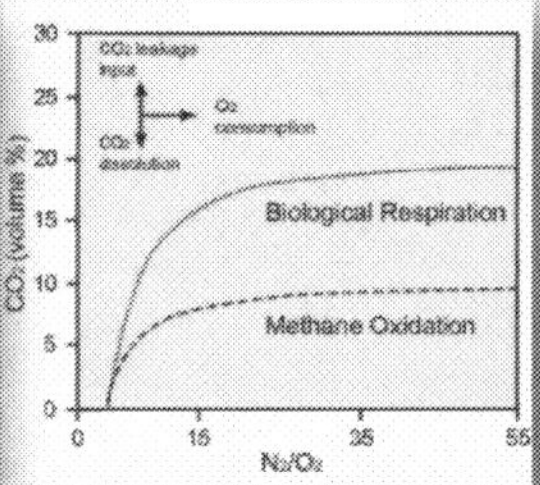

Romanak et al., 2012, 2014

THE UNIVERSITY OF TEXAS AT AUSTIN — WHAT STARTS HERE CHANGES THE WORLD

Process-Based Ratios

- Uses simple gas ratios to identify **processes**
  - Biologic respiration
  - Methane oxidation
  - Dissolution
  - Leakage
- No need for years of background.
- Method can be applied in any environment regardless of variability.

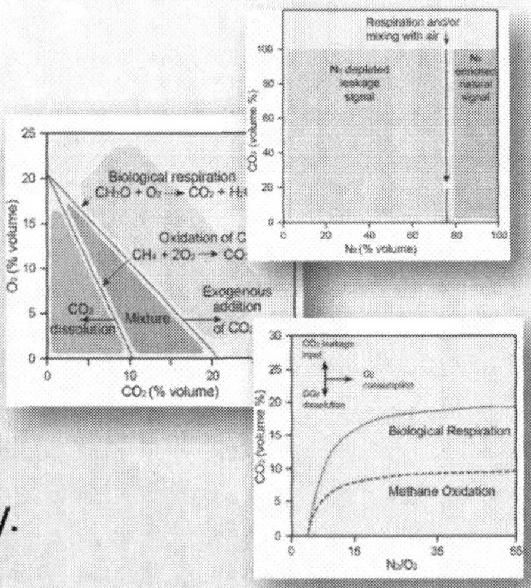

Dixon and Romanak, in press, Improving Monitoring Protocols for CO_2 Geological Storage with Technical Advances in CO_2 Attribution Monitoring, International Journal Greenhouse Gas Control.
Romanak et al., 2014, International Journal of Greenhouse Gas Control, 30, 42-57
Romanak et al., 2012, Geophysical Research Letters, *39* (15).

THE UNIVERSITY OF TEXAS AT AUSTIN

WHAT STARTS HERE CHANGES THE WORLD

Process-Based Ratios: Kerr

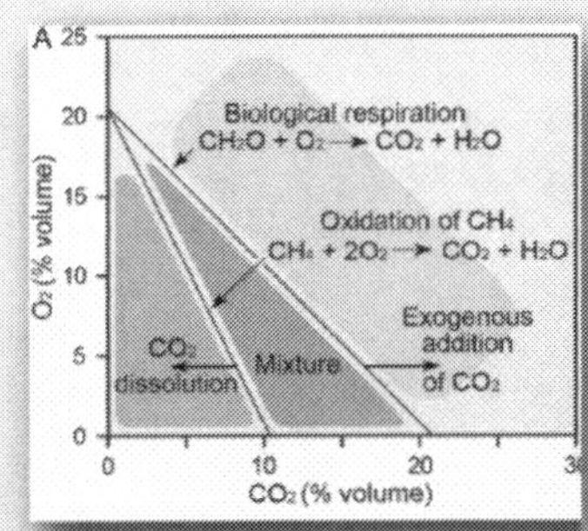

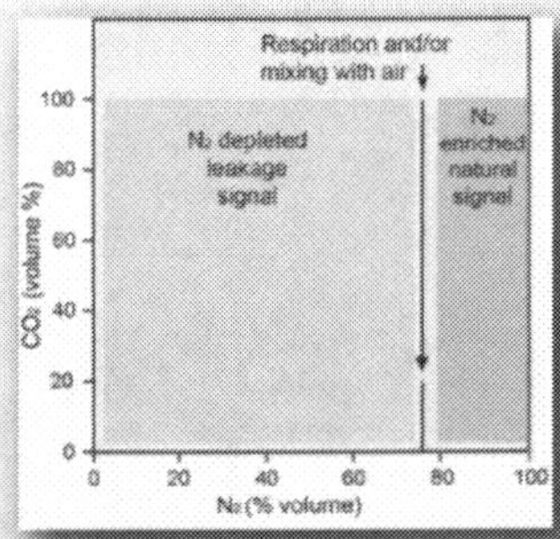

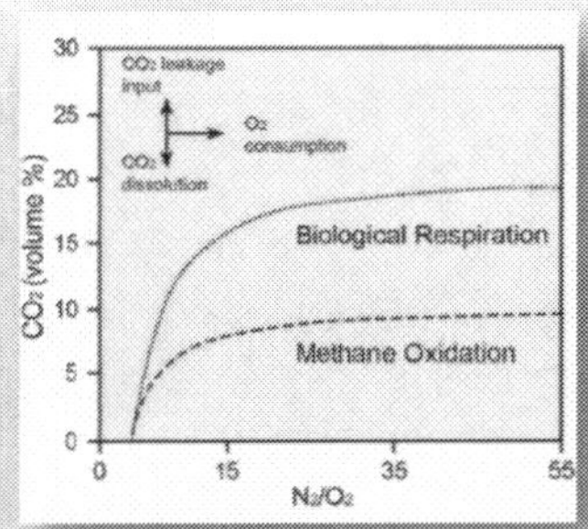

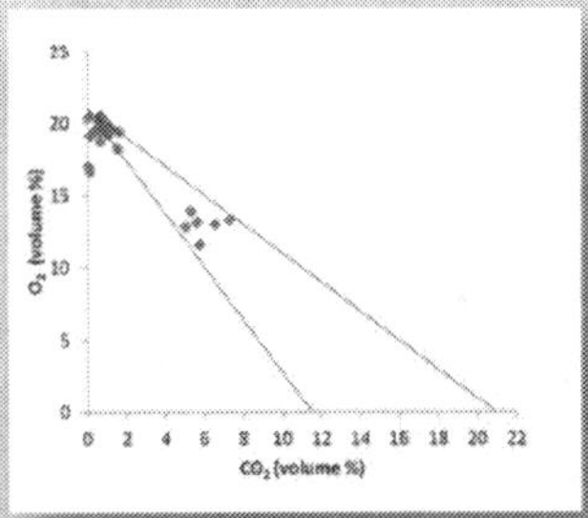

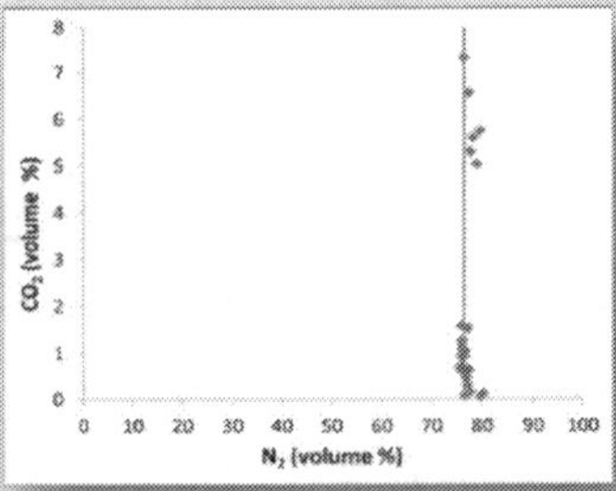

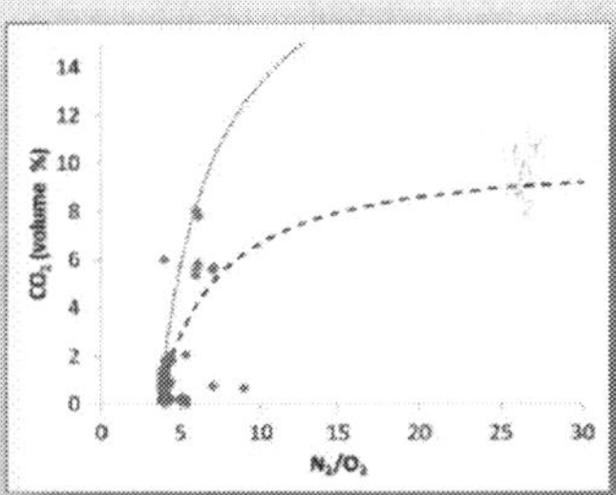

THE UNIVERSITY OF TEXAS AT AUSTIN

WHAT STARTS HERE CHANGES THE WORLD

Process- Based: ZERT

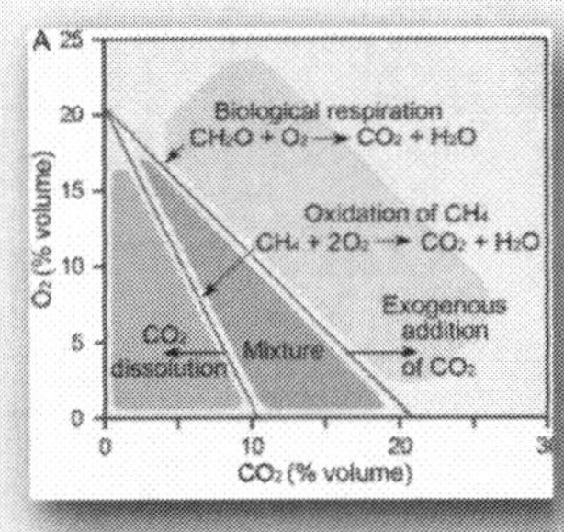

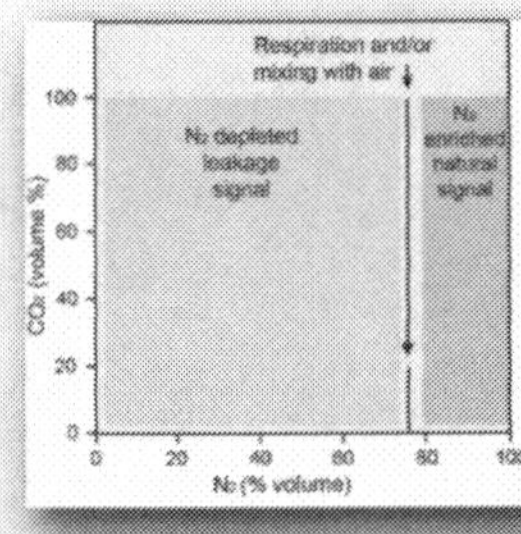

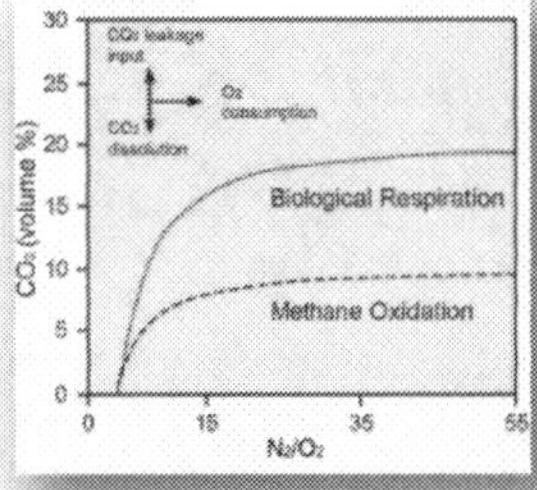

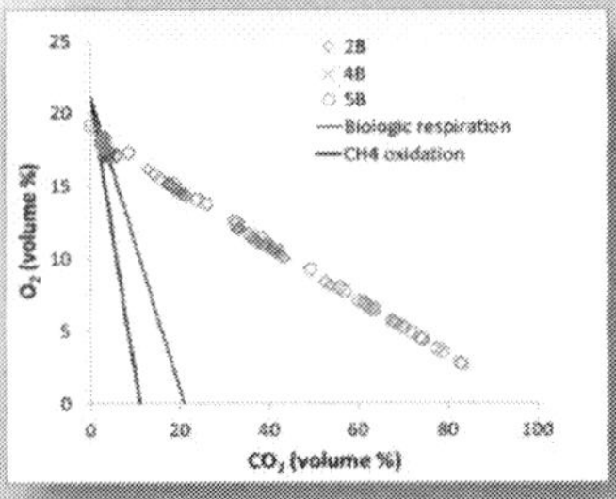

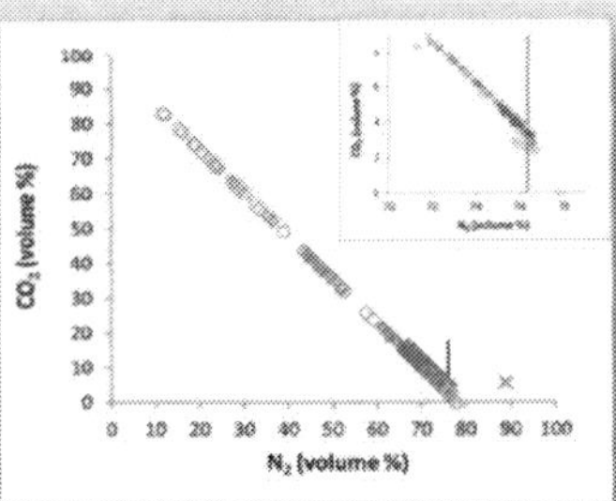

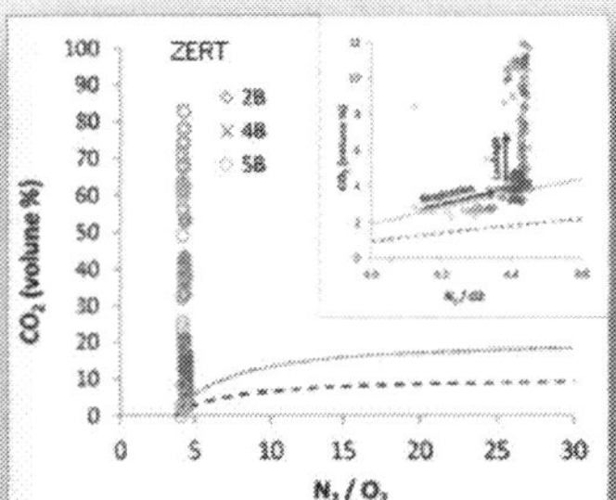

THE UNIVERSITY OF TEXAS AT AUSTIN

WHAT STARTS HERE CHANGES THE WORLD

Controlled Laboratory Pilot Experiments

THE UNIVERSITY OF TEXAS AT AUSTIN

WHAT STARTS HERE CHANGES THE WORLD

"Characterization" Instead of Baselines

Methods that don't require definition of temporal variability provide a subtle but powerful shift in how we approach environmental attribution.

- Spatial variability is important instead of temporal variability.
- Characterization is a one-time baseline measurement.
- Allows leakage signal to be predicted
- Simplified and more accurate
- Respond to public claims quickly, accurately and economically.
- Quantify a leak if it happens

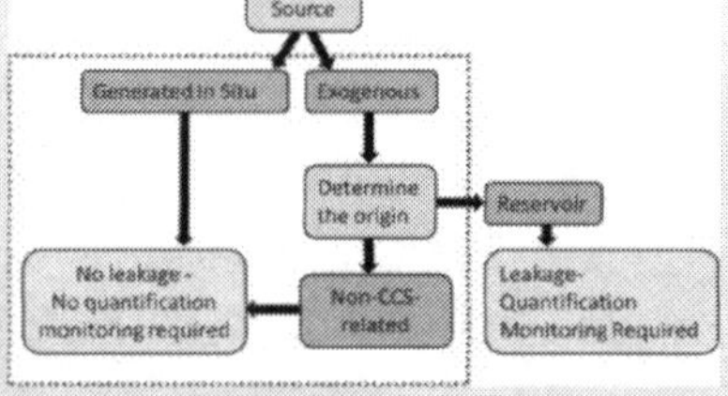

What if?

- We simplified our approach to locating and attributing anomalies.
- A simplified approach was more accurate than a complex one?
- We had monitoring approaches that didn't rely on years of background data?
- We could replace current labor intensive monitoring methods with dependable real-time continuous smart data collection?
- Monitoring was so simple Julio's mother could do it?

THE UNIVERSITY OF TEXAS AT AUSTIN
WHAT STARTS HERE CHANGES THE WORLD

Thank you!

Katherine Romanak,
katherine.romanak@beg.utexas.edu
+1 512-471-6146
www.beg.utexas.edu/gccc/

Dixon and Romanak, 2015,

International Journal of Greenhouse Gas Control

journal homepage: www.elsevier.com/locate/ijggc

Improving monitoring protocols for CO_2 geological storage with technical advances in CO_2 attribution monitoring

Tim Dixon[a], Katherine D. Romanak[b,*]

DEVELOPING A DATA MANAGEMENT SYSTEM FOR THE AQUISTORE CO_2 STORAGE MONITORING PROGRAM

Peter Anderson, SGS

November 19th 2015

Carbon Management Technology Conference 2015

Sugar Land, Texas

WHEN YOU NEED TO BE SURE SGS

SGS

OUTLINE

- Introduction to SGS
- Why Data Management Systems for CCS?
- SGS E-Data Management System
- Aquistore Data Management Project
- Next Steps in Project

SGS GLOBAL SERVICES TAILORED TO INDIVIDUAL INDUSTRIES

INSPECTION
TESTING
VERIFICATION
CERTIFICATION
TRAINING
CONSULTANCY
OUTSOURCING

SGS GLOBAL REACH AND LOCAL SUPPORT

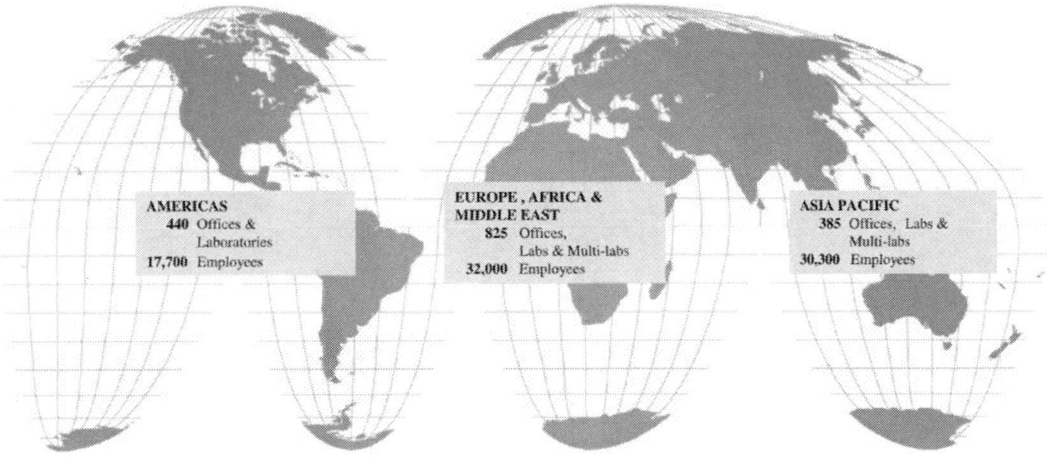

- 80,000 employees
- Over 1,650 offices and laboratories operating all over the world

SGS

CCS CHAIN AND SERVICES

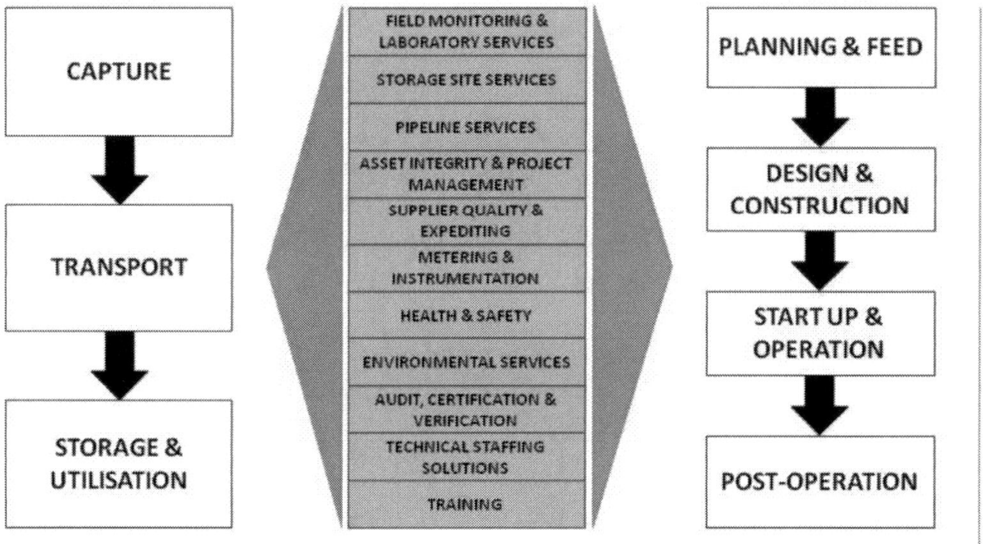

SGS

THE DATA CHALLENGE IN CO_2 STORAGE MONITORING

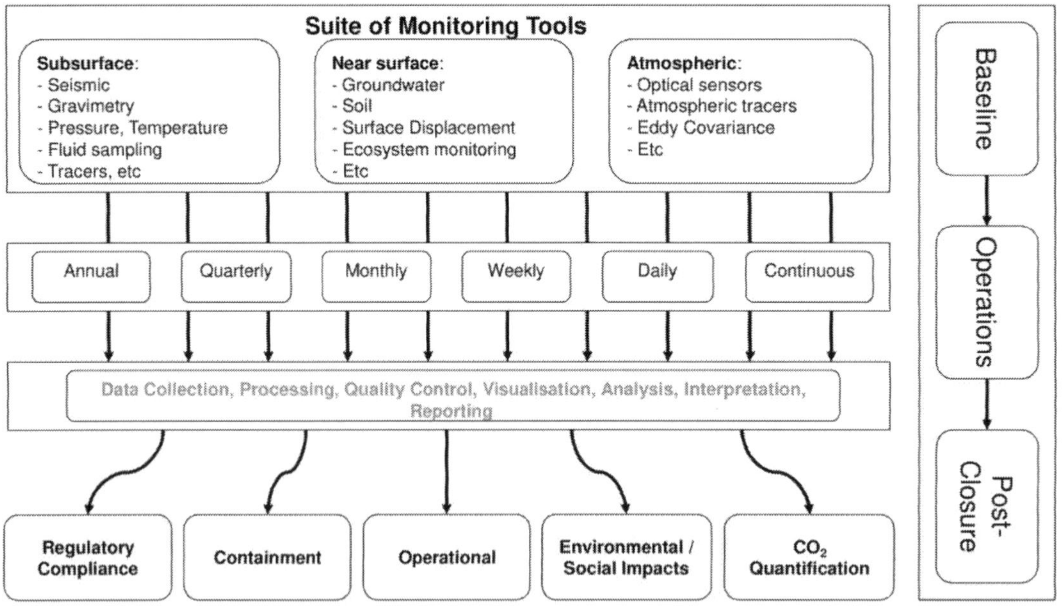

SGS E-DATA FOR CCS

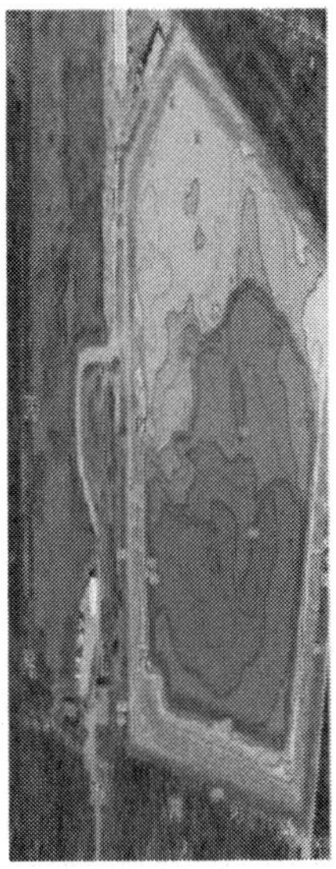

- Ideal solution for handling large volumes of data from MMV programs
- A secure client-accessible web-portal for monitoring and sampling projects
- Capability to manage diverse data streams from historical lab results to real-time data from sensors
- Fully flexible and customisable system enabling project data to be presented in a variety of analysis tools (maps, graphs and charts, tables, statistics, etc)
- Supported by a back-end database with integrated QA/QC protocols
- Compare data directly against regulatory compliance limits and instantly identify and be alerted to exceedences and issues
- Analyse and interpret data to enable better adaptive management strategies

SGS E-DATA SYSTEM WORKFLOW

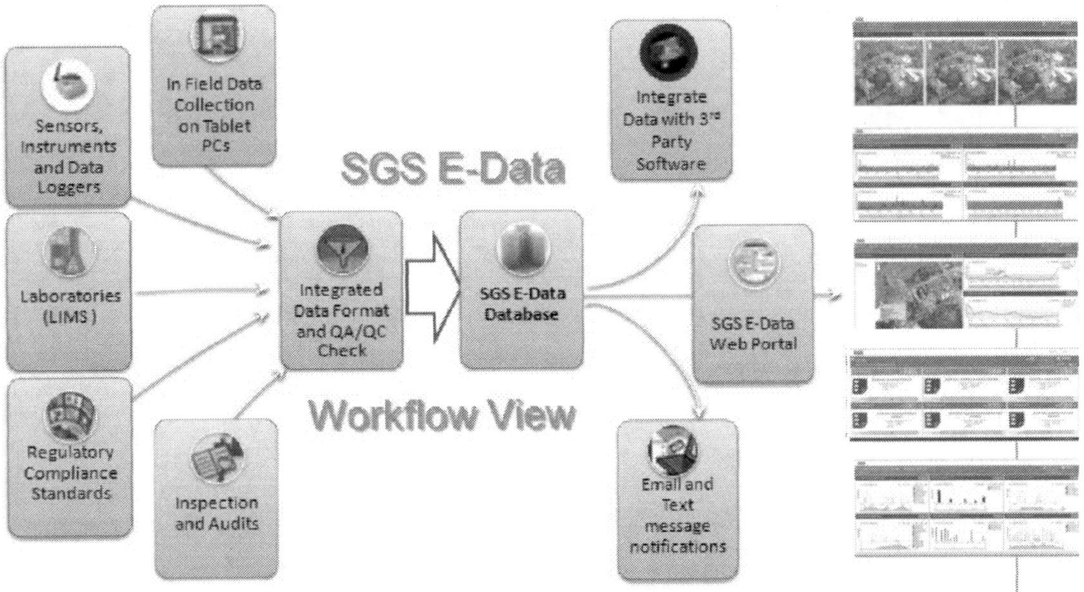

AQUISTORE DATA MANAGEMENT SYSTEM PROJECT

- Aquistore
  - Located in Saskatchewan, Canada, managed by Petroleum Technology Research Centre (PTRC)
  - Canada's first deep saline CO_2 project injecting CO2 captured from Boundary Dam CCS project
  - Research project enabling testing of wide range of monitoring tools
- SGS and PTRC working to develop a data management system for Aquistore using the SGS E-Data platform that can aid project partners to manage and interpret the wide array of data produced
- Developed, tested and approved prototype using historical groundwater data

PROTOTYPE DASHBOARD

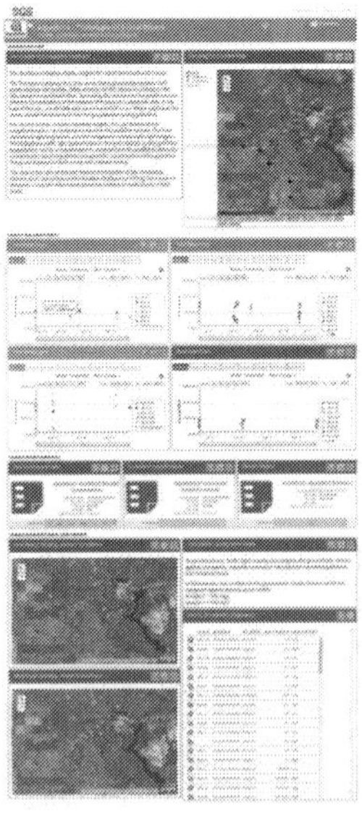

PROTOTYPE DASHBOARD

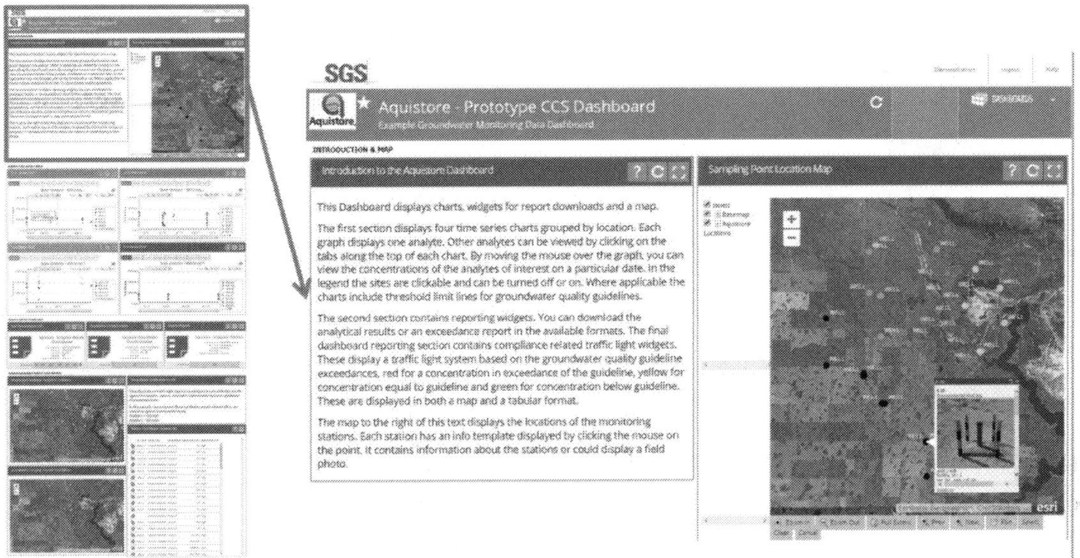

PROTOTYPE DASHBOARD

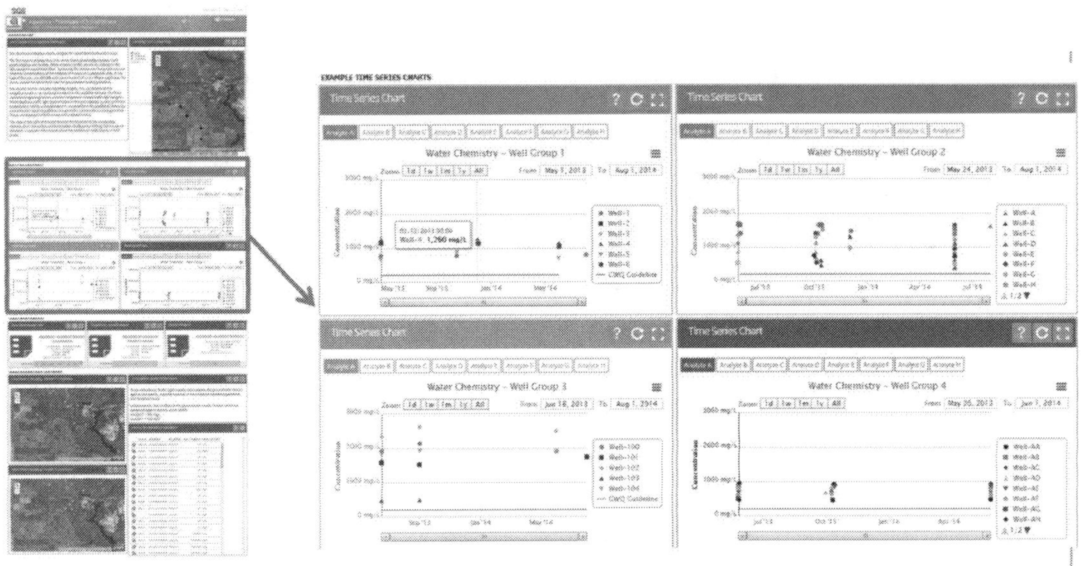

SGS PROTOTYPE DASHBOARD

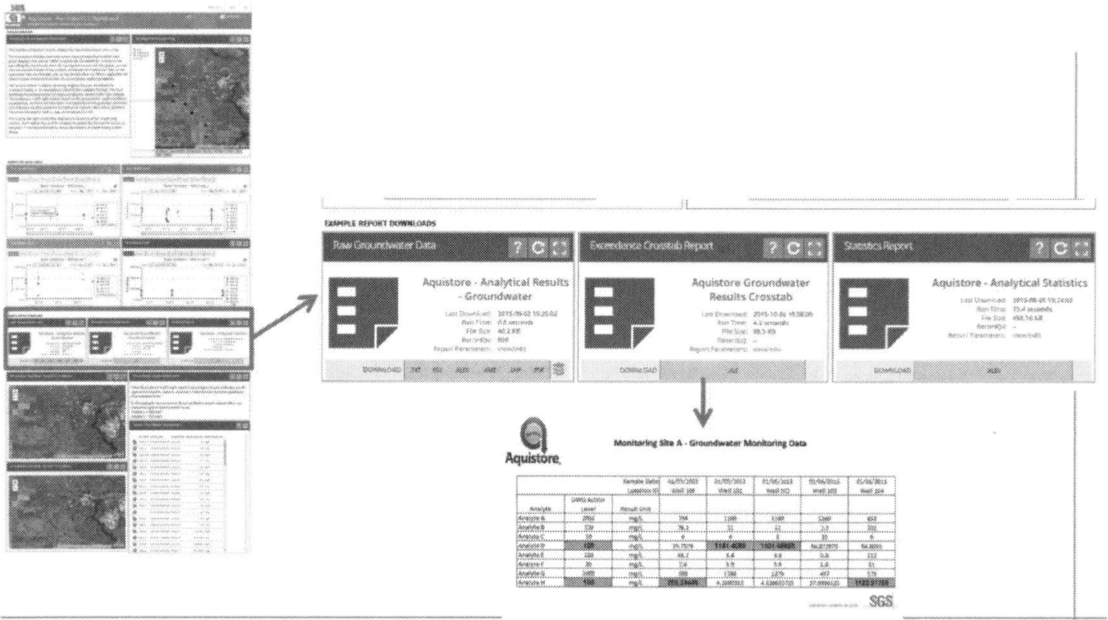

SGS PROTOTYPE DASHBOARD

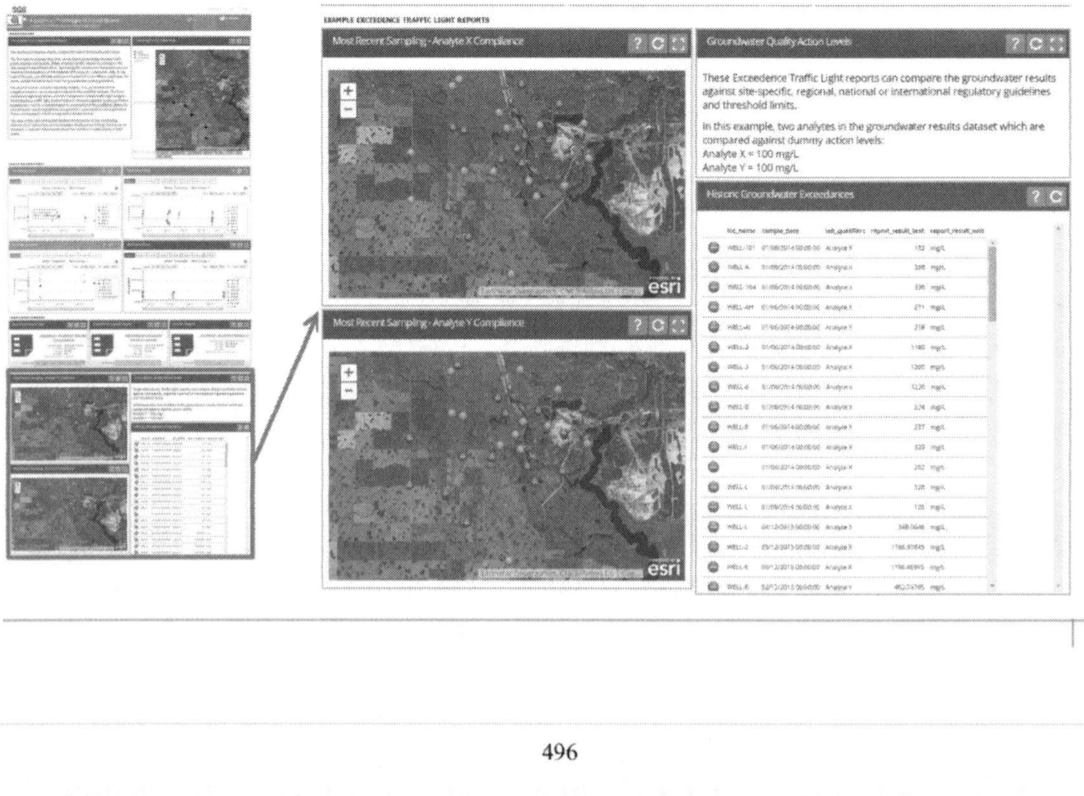

SGS NEXT STEPS IN PROJECT

- Initial dashboard developed to test system and tailored to meet needs of Aquistore team
- Next step to extend dashboard to incorporate range of monitoring data:
  - Surface monitoring (tiltmeters, inSAR, GPS)
  - Shallow subsurface (groundwater, soil-gas, atmospheric)
  - Injection, pressure, temperature, fluid samples
  - Seismic program
- Configure the e-mail alert system
- Refine system to best suit needs of project – how best can information be utilised, presentation format, reports for internal / external stakeholders, etc....developing a bespoke system

SGS SGS E-DATA – ENVIRONMENTAL DATA MANAGEMENT SYSTEM

REPORTING & ANALYSIS PLATFORM

EXAMPLE DASHBOARD IN SGS E-DATA WEB PORTAL

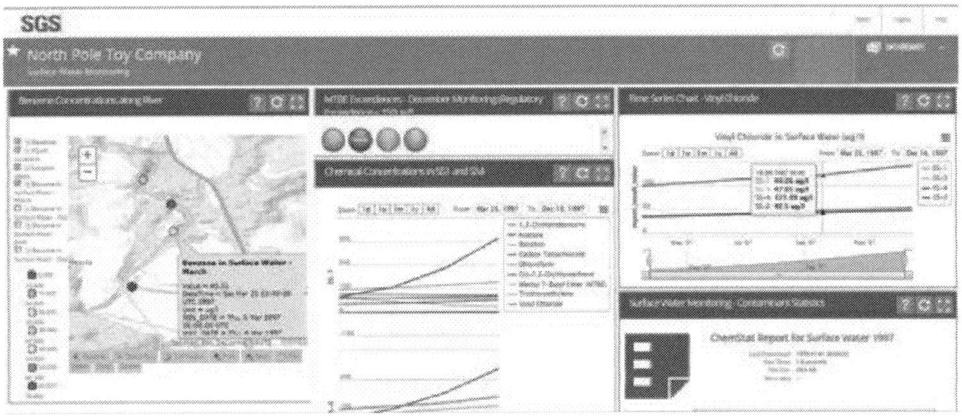

HTTPS://EDATA.SGS.COM/DEMO

Username = Demonstration Password = Geneva1

SGS E-DATA – ENVIRONMENTAL DATA MANAGEMENT SYSTEM

SGS DATA MANAGEMENT CLIENTS

For further information and a full demonstration of SGS E-Data capabilities, please contact:

Peter Anderson (Global CCS Manager)
peter.anderson@sgs.com

Karl Daines (Global Data Management & Interpretation Manager)
karl.daines@sgs.com

Physical Sciences Inc.

VG-2015-137

Laser-Based Sensors and Algorithms for GCS MVA

Mickey Frish and Richard Wainner
Physical Sciences Inc.

2015 Carbon Management and Technology Conference
Sugar Land, TX

19 November 2015

Physical Sciences Inc. (PSI)

PSI is:

- **A 42 year-old company of ~150 scientists, engineers, and administrative personnel**
- **Headquartered in Andover, MA with eight satellite locations in the US**
- **Employee-owned via the PSI Employee Stock Trust**
- **Three wholly-owned subsidiaries: Q-Peak (Bedford, MA), Research Support Instruments (Lanham, MD; Princeton, NJ), Faraday Technology (Dayton, OH)**

PSI does:

- **Applied contract research and development for all major agencies of the U.S. government**
- **Technology transition and product development for government and industrial customers**
- **Prototype product development for industry and commercial applications**
- **Components, systems, and instrumentation for industry and government sales**
- **Technology and product licensing**

PSI Industrial Sensors:

- **Interdisciplinary combination of science and engineering skills with specific strengths in development and commercialization of photonic sensors and instrumentation**
- **Product development from concept to manufacturing prototype**
- **Go to market via direct sales, strategic partnerships, pilot scale manufacturing, and licensing**
- **Serving oil & gas and other energy industry needs since 1994**

PSI Leak Detector Background

Physical Sciences Inc.

VG-2015-137-2

- **Spectrum Diagnostix, Inc. (1992 – 1996)**
  - **Developed the first commercial *Tunable Diode Laser Absorption Spectroscopy (TDLAS)* products: SpectraScan®**
  - **Collaboration with Petroleum Environmental Research Forum (PERF), including Exxon, Mobil, BP, Shell, Amoco for refinery leak detection**

- **Invented, developed and licensed RMLD™**
  - **Heath Consultants product released in 2005, R&D 100 Award**

- **Conducts advanced TDLAS R&D**
  - **> 150 custom TDLAS systems delivered to government and industrial customers**

Absorption Spectroscopy

Physical Sciences Inc.

VG-2015-137-3

- **Gas molecules absorb light at specific colors ("absorption lines")**

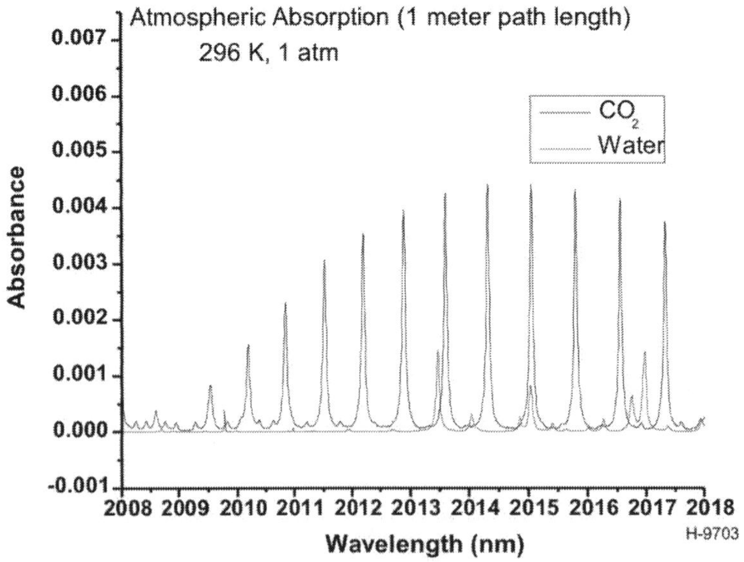

TDLAS

PSI Physical Sciences Inc.

VG-2015-137-4

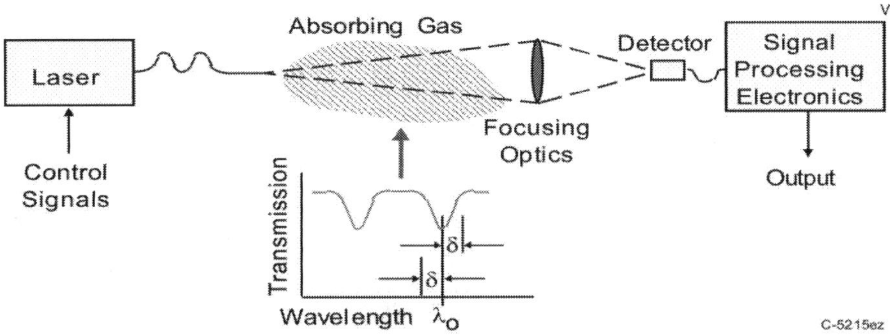

- **A frequency agile (i.e. tunable) laser beam transits an analyte gas sample**
- **Laser wavelength scans repeatedly across absorption line unique to analyte gas**
- **Received signal processed to deduce analyte concentration**
- **Laser sources**
  - SWIR – Distributed Feedback (DFB) laser or
  - SWIR – Vertical Cavity Surface Emitting Laser (VCSEL)
  - MWIR – Interband Cascade Lasers
  - LWIR – Quantum Cascade Lasers

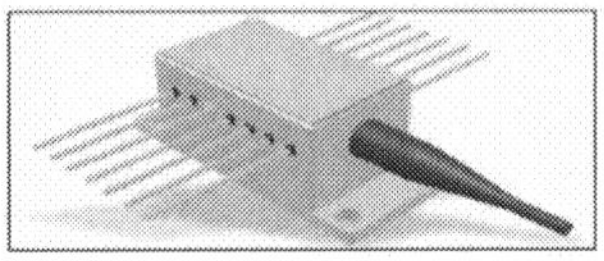

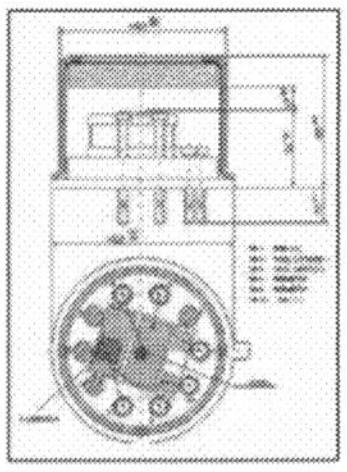

Laser-based Sensors for CO_2 MVA and Safety

PSI Physical Sciences Inc.

VG-2015-137-5

- **PSI envisions suites of affordable continuous monitoring sensors comprising:**
  - Open-path CO_2 gas sensors (red)
  - Handheld /mobile leak survey tools for pipeline inspection
  - Shallow in-ground CO_2 gas point sensors (blue)
  - Well-depth liquid CO_2 sensors (green)

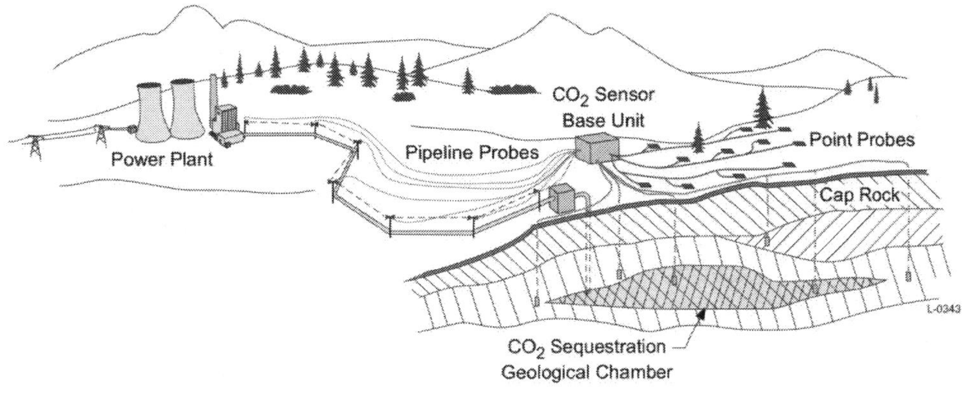

Portable Standoff near-IR TDLAS for Leak Survey

PSI Physical Sciences Inc.

VG-2015-137-6

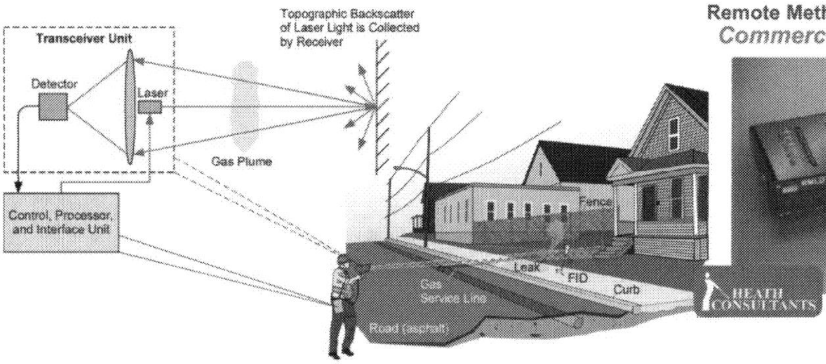

Remote Methane Leak Detector (RMLD™)

Commercial product (since 2005)

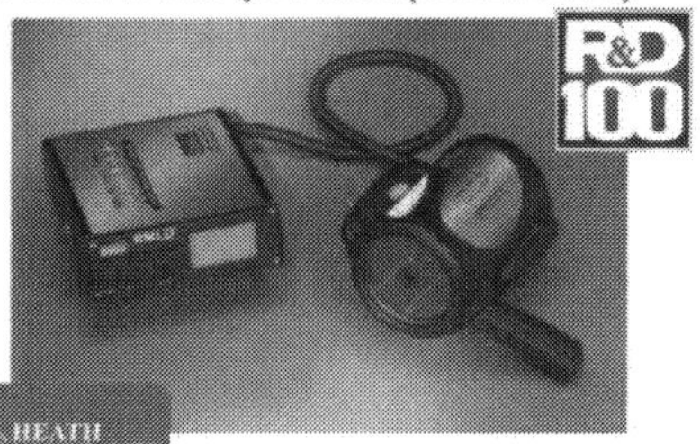

- **Laser beam illuminates a distant surface**
- **Senses analyte gas between transceiver and illuminated surface**
  - ***Standoff range ~100 ft with handheld transceiver***
- **Scanning laser beam across plume results in rapidly changing analyte gas measurement**
- ***~3000 RMLDs™ in use for natural gas leak surveying***

Remote Carbon-dioxide Leak Detector (RCLD)

PSI Physical Sciences Inc.

VG-2015-137-7

- Compact, portable, personnel-wearable laser module with hand-held transceiver
- Battery powered with optional data logging via RS-232 port
- The RCLD (photo below) was laboratory tested to show its response to CO_2 plumes and field tested to illustrate its effectiveness around a CCS site

ILLINOIS STATE GEOLOGICAL SURVEY
PRAIRIE RESEARCH INSTITUTE

CO_2 version demonstrated at CCS test site (IBDP) wellhead during maintenance

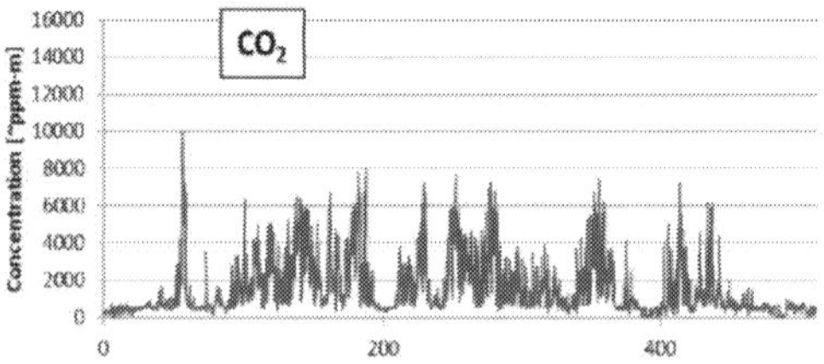

J.W. Zimmerman, R.A. Locke II, C.S. Blakley, M.B. Frish, M.C. Laderer, R.T. Wainner, *"Initial testing of prototype tunable diode laser absorption spectrometers for CO_2 monitoring applications at the Illinois Basin - Decatur Project"*, 13th Annual Carbon Capture, Utilization and Storage Conference, Pittsburgh PA (April 30, 2014)

Open-Path Pipeline Monitors

PSI Physical Sciences Inc.

VG-2015-137-8

- **Permanent laser-based open-path alarms for use along pipelines and wellhead infrastructure**
  - Wireless
  - Solar-powered
  - Easy installation & alignment
  - Real-time alarm notification
    - *Operator alert within one minute of urgent leak detection*
    - *Hourly notification of non-urgent leaks, enabling proper operator assessment and response*
  - ***Enabled by proven, industry-accepted RMLD™ technology***

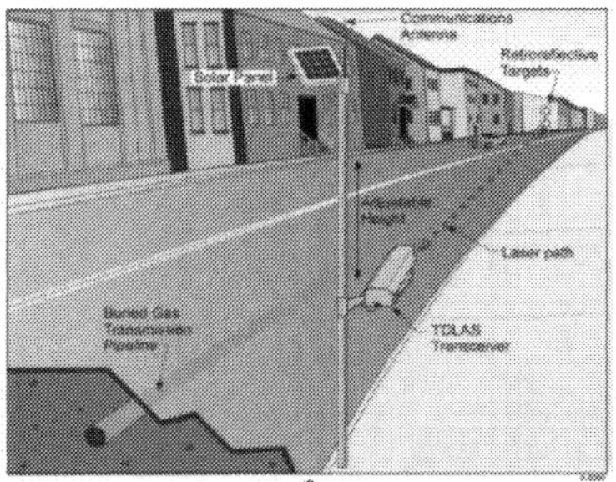

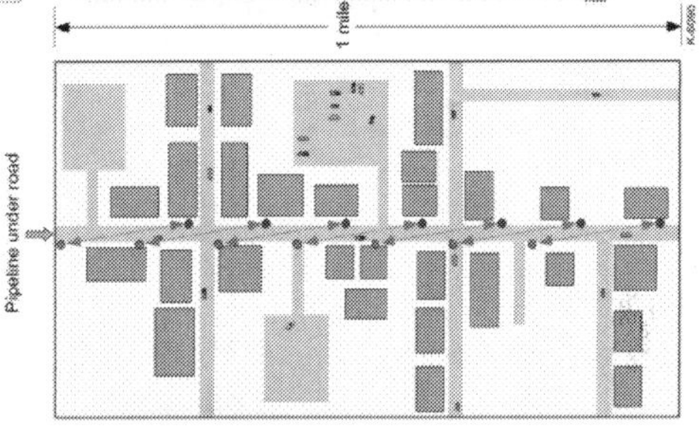

- **CH_4 version demonstrated at PGE Livermore CA Training Center**
  - 580' ft path
  - Months of maintenance-free operation
- **CO_2 version tested for more than two years at PSI and Illinois-Basin Decatur Project (IBDP) GCS Site**

PG&E — HEATH CONSULTANTS — NETL — ILLINOIS STATE GEOLOGICAL SURVEY, PRAIRIE RESEARCH INSTITUTE

Open-Path CO_2 Measurements and Alarms

PSI Physical Sciences Inc.

VG-2015-137-9

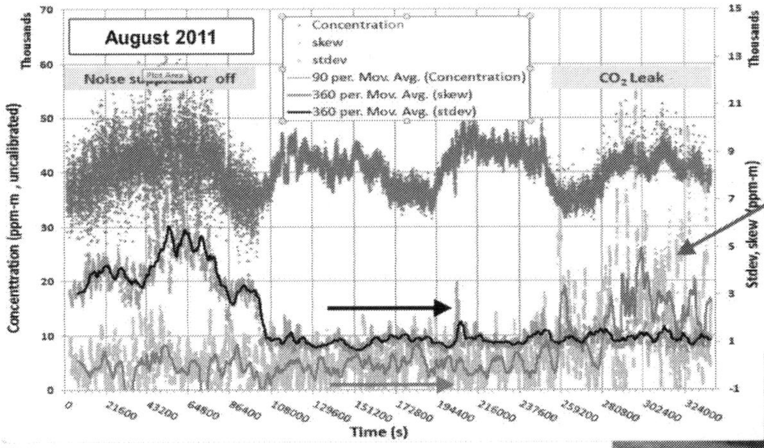

U.S. DEPARTMENT OF ENERGY

Statistical extraction of signal from background noise

Skew: 2nd central moment of temporal sample

- **Continuous operation (PSI site (Andover, MA))**
  - ***No adjustment, calibration, or zeroing***
- **Diurnal variations correlate with point sensor measurements**
- ***Leaking CO_2 yields statistically distinct signature***

Open Path CO_2 – Illinois Basin Decatur Project

Carbon Sequestration Injection Wellhead

PSI Physical Sciences Inc.

ILLINOIS STATE GEOLOGICAL SURVEY
PRAIRIE RESEARCH INSTITUTE

VG-2015-137-10

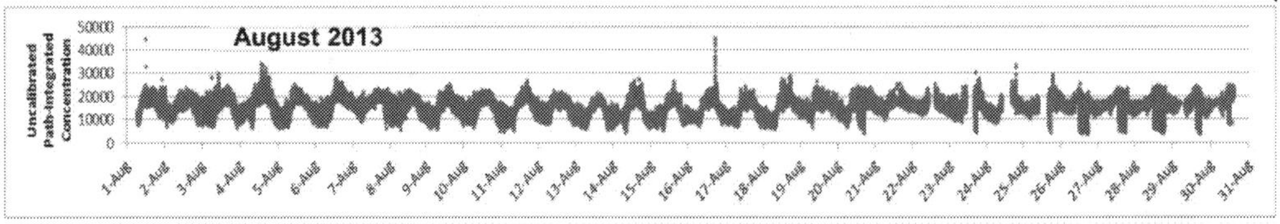

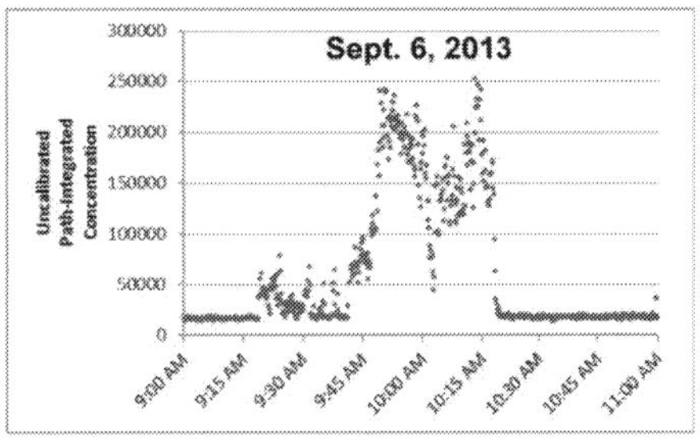

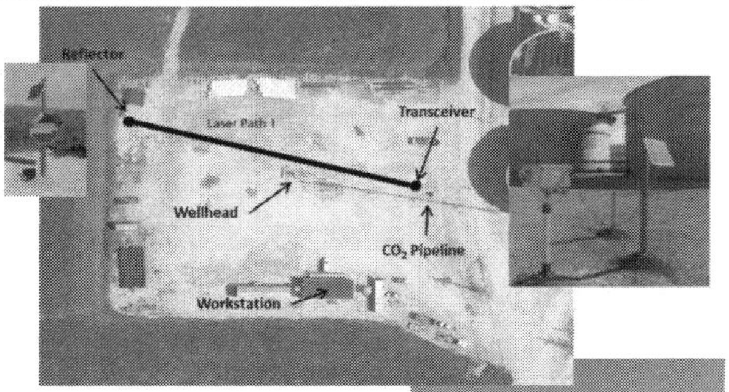

- **Continuous CO_2 Monitoring**
- **Detected CO_2 trap exhaust event during routine wellhead pipeline blowdown maintenance with injection stopped**

Algorithms for Enhanced Leak Detection

PSI Physical Sciences Inc.

VG-2015-137-11

- **Need: Signal processing technology to discriminate a GCS leak to atmosphere from natural (diurnal, seasonal, annual) variation and local environmental effects**
  - Desire large area coverage, improved accuracy, continuous (real-time) monitoring capabilities, and long-term durability
- **Approach: Statistics-based algorithms using unsupervised machine learning to automatically recognize temporal and frequency signal features of importance for classifying "background" and "abnormal" contributions.**
  - Leak-free training period and then continual expansion/improvement of background signature library
- Currently applying algorithms to previously-acquired open-path sensor data
- Planning to build and field test an array of subsurface (and OPS) sensors at a validation test site to enable quantifiable leakage detection performance

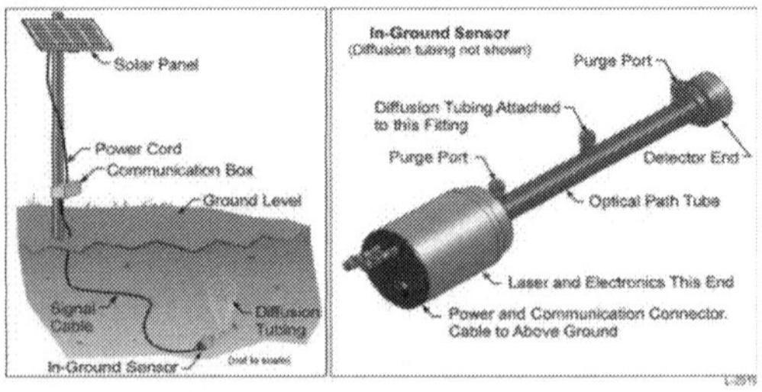

Deep Downhole Sensors

Physical Sciences Inc.

VG-2015-137-12

- **Need: Reliable and cost effective monitors to confirm permanent storage of CO_2 in geologic formations.**
  - Ensure that CO_2 remains within the injection formation as modeled
  - Protect potable groundwater sources and sensitive ecosystems as required by the Safe Drinking Water Act and Clean Air Act

- **Desired solution: Technologies involving field-based MVA hardware that quantify CO_2 leakage from geologic storage fields in the unlikely event that CO_2 migrates out of the injection zone.**
  - New or enhanced MVA tools that can cover a large area for monitoring supercritical CO_2 in the deep subsurface environment
  - Improved accuracy, continuous (real-time) monitoring capabilities, and/or reduced cost over current methodologies

Deep Downhole CO_2 Sensors

Physical Sciences Inc.

VG-2015-137-13

- **Motivation:**
  - Assure reservoir efficacy and aquifer safety
  - Account for stored gas
  - Continually monitor migration of stored CO_2. Discover, in real-time, unexpected migration paths
  - Improve understanding of CO_2 storage and transport processes (model validation)
- **Current methods for tracking sequestered CO_2 plumes include:**
  - Direct: Pressure. Geochemical sampling in the reservoir
  - Indirect: Downhole and surface seismic, electrical, and gravitational data
  - Near well bore: Reservoir Saturation Tool (RST) (pulsed neutrons)
- **Current methods for detecting CO_2 leaks include:**
  - Direct: Pressure fall-off. Geochemical sampling at "monitor" wells
  - Indirect: Surface / USDW monitoring. Acoustic.
- **Shortcomings:**
  - Complex
  - Slow
  - Poor Sensitivity
  - Poor chemical specificity
  - Expensive

Sleipner Field Seismic Map **(source: CO2ReMoVe)**

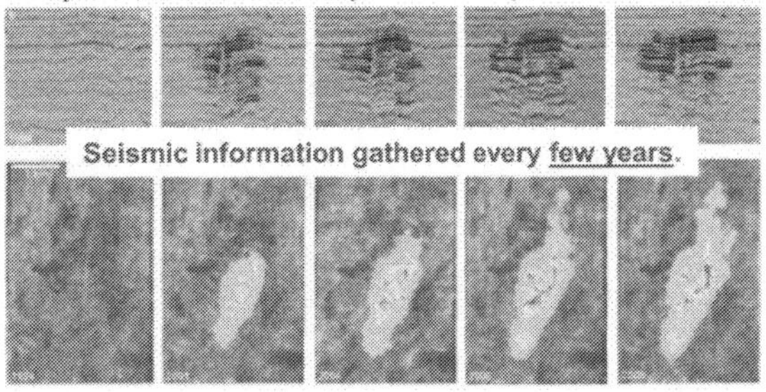

Technical Approach

Physical Sciences Inc.

VG-2015-137-14

- A network of permanently-installed, continuously operating CO_2 sensors deployed several thousand feet below ground within 1-3" diameter boreholes.
- Comprised of:
  1. Near-infrared (NIR) fiber-coupled laser source
  2. Sensing by spectroscopic measurement of supercritical CO_2
  3. Passive optical heads suitable for downhole insertion
- Functions:
  - Monitoring plume progression in the reservoir; A tool to improve reservoir fluid dynamics models.
  - Monitoring for leakage outside the reservoir; An alarm to protect neighboring property and aquifers.
  - Semi-quantitative *in situ* CO_2 measurements; A tool to support seismic or other sensors:

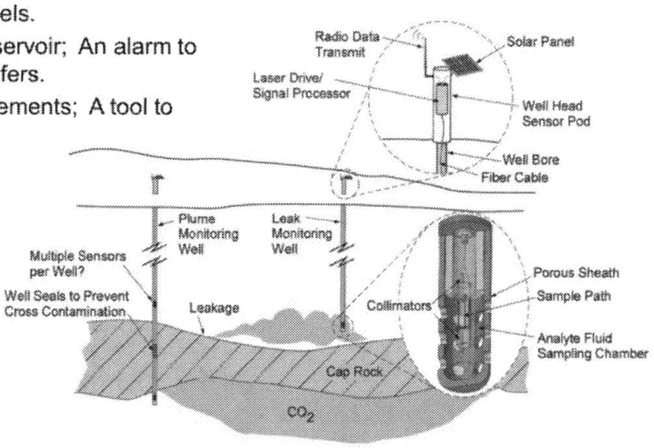

Sensor Goals

Physical Sciences Inc.

VG-2015-137-15

- **Applicable to 3000m depth**
  - Typically 1000-3000m depth ⇒ 10-30MPa, 45-100°C
- **Measure 0-100% CO_2 in brine with <1% CO_2 resolution**
- **~$15k unit price**

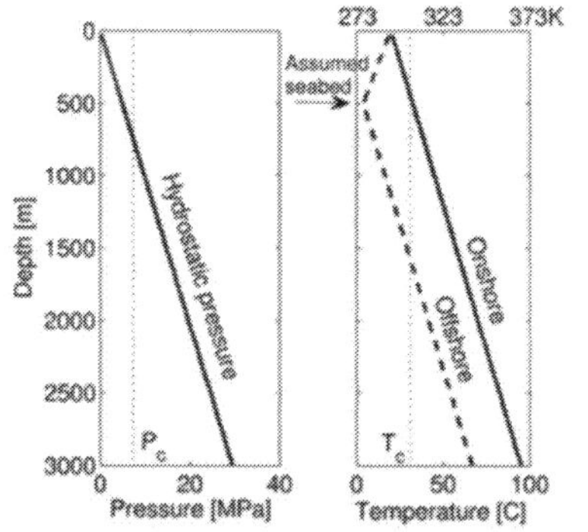

- **Passive probe head characteristics**
  - **Simple construction**
  - **Inserts easily down a small (~ ≥1 inch) borehole**
  - **Survives a harsh and corrosive environment**
  - **Attaches simply to a TDLAS Control Unit**
  - **Maintenance free**
  - **Provides appropriate measurement sensitivity**

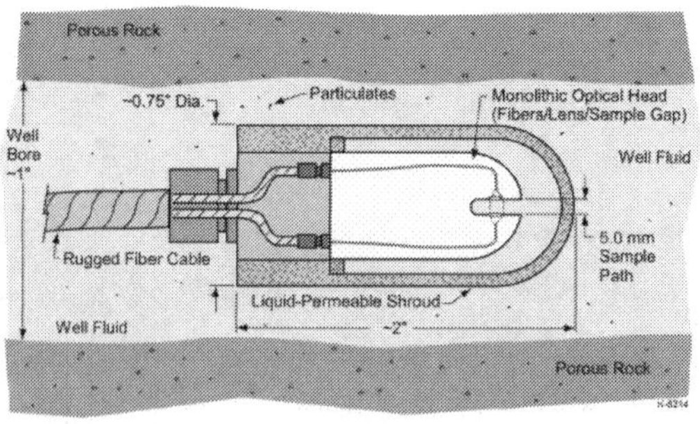

Extension to Supercritical CO_2

Physical Sciences Inc.

VG-2015-137-16

Spectral features of atmospheric CO_2 (1 atm; 293 K; 12 mm path).

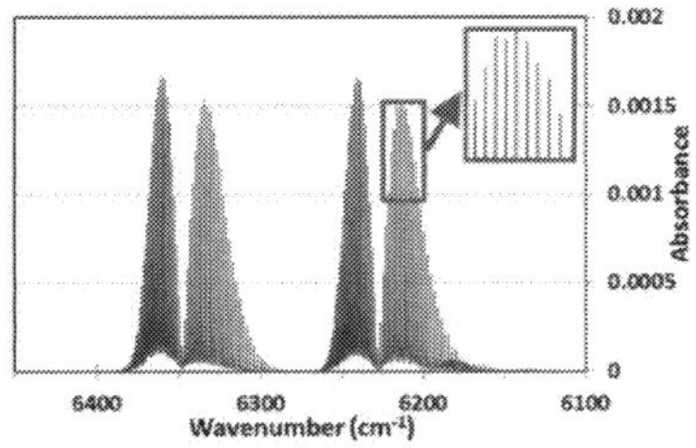

Spectral features of supercritical CO_2 (160 atm; 333 K; 12 mm path).

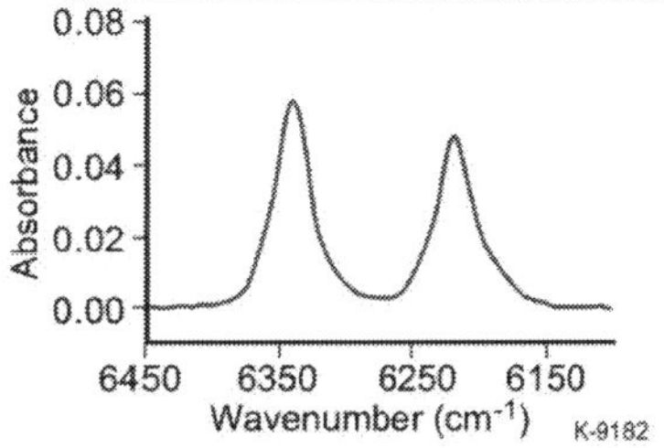

K-9182

- **At depth, spectral features span wavelengths 200x broader than 1 atm gas phase**
  - Linewidths > 30cm^{-1}
- **Modified laser modulation technique enables ~30cm^{-1} low-speed (Hz) spectral sweep**
  - Acceptable for monitoring slowly varying phenomena
  - Suitable for the supercritical CO_2 feature
  - Expected detection limit ~1% CO_2

Results vs. P and Mixture Fraction

Physical Sciences Inc.

VG-2015-137-17

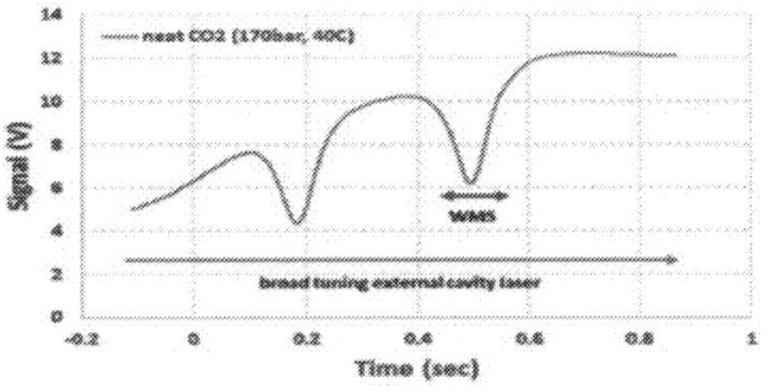

Recorded transmission spectrum through 5cm supercritical CO_2.

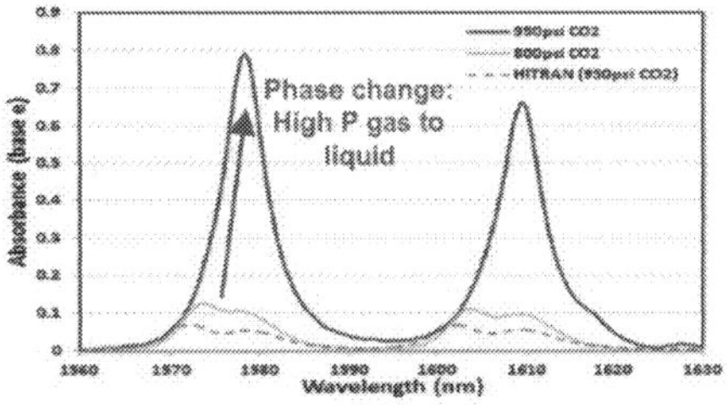

Absorbance signals for pure CO_2 vs. pressure near phase change boundary. For reference, a calculation for high pressure gas is also included.

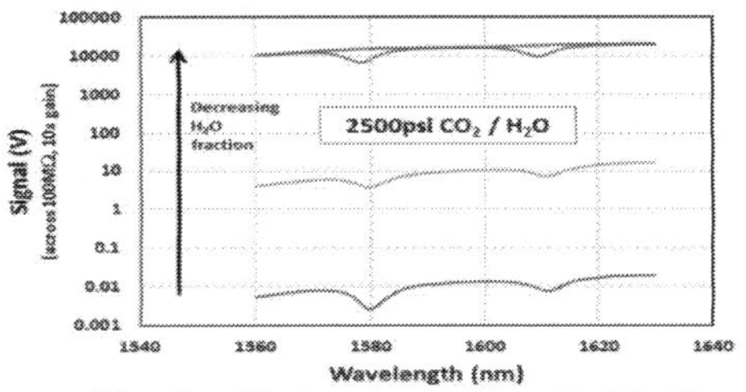

Temporal signal traces versus wavelength for varying CO_2 / mixtures.

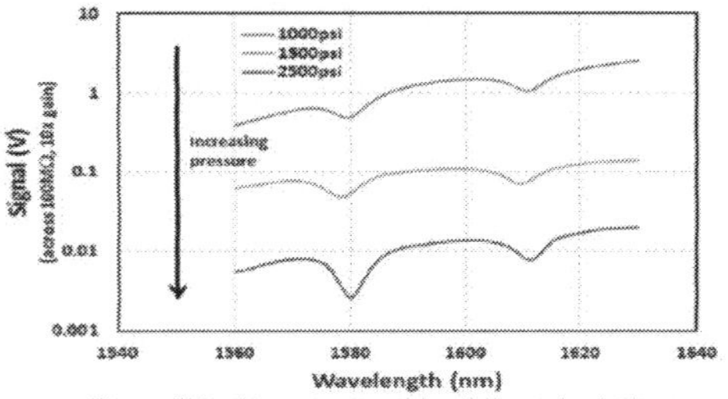

Temporal signal traces versus wavelength for varying fluid pressure.

- Assembled laboratory apparatus to create and interrogate supercritical CO_2
- Scanned wavelength with a broadly-tunable commercial laser source
- Varied pressure and CO_2/H_2O
- ***10^6 dynamic range demonstrated***

Summary

PSI Physical Sciences Inc.

VG-2015-137-18

- **Cost-effective laser-based sensors are being developed for sequestration MVA and safety**
- **Handheld and long open-path sensors are proven reliable and sensitive monitors ready for deployment as wellhead and pipeline leak detectors and alarms**
  - Research is underway to discern reservoir leakage by applying advanced statistics-based algorithms to data from open-path and in-ground sensors
- **Deep-downhole supercritical CO_2 sensors are being developed to monitor reservoir plumes and verify containment**

Acknowledgment

PSI Physical Sciences Inc.

VG-2015-137-19

- **R&D funding provided by US Department of Energy**
- **Instrumentation provided by Heath Consultants Inc.**
- **Collaboration with Illinois State Geological Survey, Illinois Basin Decatur Project, Midwest Geological Sequestration Consortium, and Berkeley National Laboratory is gratefully acknowledged**
- **Technical advice from Schlumberger Carbon Services is appreciated**

CMTC-439342-MS

A Chemical CO_2 Sensor Monitoring CO_2 Movement Under Reservoir Conditions

Sai Wang, Jingshan San, Ning Liu
Petroleum Recovery Research Center, New Mexico Tech

This paper was prepared for presentation at the Carbon Management Technology Conference held in Sugarland, Texas, USA, 17–19 November 2015.

Abstract

A dowhole CO_2 sensor can continuously collect real-time data about CO_2 movement and concentration changes at subsurface conditions. These data are very valuable for better understanding of subsurface uncertainties and quality-controlling theoretical studies such as reaction, transport, and mechanics in oil and gas formations. This paper describes the development of a downhole CO_2 sensor tested under high pressure and reservoir conditions to monitor aqueous CO_2 concentration change. The CO_2 sensor developed is a Severinghaus-type sensor, which includes a metal-oxide electrode, a gas-permeable membrane, a porous steel cup, and a bicarbonate-based internal electrolyte solution. The CO_2 sensor thus prepared 0.7 in. in diameter and 1.5 in long. A linear correlation was observed between a change in sensor output potential and dissolved CO_2 in water under 1,000 psi pressure. CO_2/brine coreflooding tests were performed to simulate the CO_2 storage process and the sensor was deployed to monitor CO_2 movement. The results indicated that the CO_2 sensor could monitor CO_2 movement in-situ in CO_2 storage processes.

Key words Chemical sensor CO_2 storage monitoring

Introduction

Geologic sequestration of CO_2 involves putting CO_2 into long-term storage in geologic zones at subsurface conditions [1]. Such sites as deep saline aquifers and unmined coal seams onshore, and depleted oil or gas formations both onshore and offshore have been recommended for further serious consideration [2]. Thus far, in various regions of the world (Pacific Ocean, Gulf of Mexico, North Sea, Chinese East sea, and the Atlantic Ocean), a large part of research studies and pilot projects have looked at the feasibility of geological sequestration of CO_2 [3] . The first commercial project occurred in Norway in1996, in which CO_2 was captured from natural gas streams and around 1 million tons of CO_2 per year were into the Utsira formation [4]. At Weyburn, a CO_2 monitoring and storage project supported by the International Energy Agency was developed to investigate the fate of CO_2 storage particularly in depleted oil and gas reservoirs [5]. In conjunction with enhanced oil recovery, the project was expected to establish a reliable method for predicting the fate of CO_2 injected into a formation, and to develop practical application of CO_2 monitoring methods to track CO_2 movement under reservoir conditions. To date, around 1.9 billion m^3 of CO_2 have already been injected into the Weyburn field. In recent decades, more and more research has focused on the storage of supercritical CO_2 in deep-water environments, especially since the International Maritime Organization gave permission for CO_2 storage under the deep seabed in 2007 [6]. N. Pilisi *et al.* had worked on a feasibility study of CO_2 sequestration under deep-water conditions in the Gulf of Mexico [7]. Based on available technologies, he demonstrated the probability of CO_2 sequestration under sub-seabed formations where the water depth is greater than 10,000 feet, which theoretically supports reliable carbon sequestration in deep-water formations. Furthermore, numerous pilot scale projects have been led by the National Energy Technology Laboratory of the U. S. Department of Energy, to demonstrate the capability of storing CO_2 under different types of geological formations, and to provide insight into CO_2 migration [8]. All these pilot studies, located in Kansas, Virginia, West Virginia and Canada, are either under consideration or have been initiated.

Due to the rapid development of geological CO_2 storage, more and more researchers are focused on the behaviors of CO_2 during geological sequestration. In order to assure the safety of storage and provide real-time information about CO_2 plume migration, downhole CO_2 sensor which can monitor subsurface CO_2 movement in situ is highly demanded. At present, a broad range of technologies based on geophysical, microbiological, geoelectrical and geochemical principles have been applied [9]. At Sleipner, a time-lapse gravity measurement method was used at a CO_2 injection site for understanding the physical properties and phase behavior of injected CO_2 between 2002 and 2005 [10]. The gravity change proved to be related to density variations in the CO_2. Allan Mathieson *et al.* proved the reliability of time lapse satellite imaging, which detects deformation of the ground and evaluates CO_2 movement in the subsurface environment [11]. Another novel technology was reported in German that combined 3D reflection seismic imaging with surface–downhole measurement and crosshole tomographic survey [12]. A geoelectrical method which works through the relationship between CO_2 saturation and electrical resistivity was also applied for monitoring CO_2 movement at the same site [13]. At Weyburn, in a particularly geochemical aspect, a carbon isotope (^{13}C) was identified as a valuable precursor to track the spread and movement of the CO_2 plume [14].

Compared with those technologies mentioned, which may have time-consuming or costly requirements, a preferred method for in-situ underground CO_2 migration monitoring is to install down-hole CO_2 sensors directly underground. However, such a sensor must be able to resist high temperatures, high salinity and high pressures. This paper recounts the development and testing of a robust, novel downhole CO_2 sensor which can monitor CO_2 concentration changes continuously at subsurface conditions. Differing from the traditional sensor, this type of the CO_2 sensor uses a metal-oxide electrode instead of a glass electrode to improve resistance to the harsh downhole environments encountered in the oilfield. To date, this Severinghaus-type sensor will be the first downhole chemical CO_2 sensor for monitoring CO_2 movement. This method will prove invaluable for large-scale storage of CO_2. Compared to all other monitoring methods, this small chemical sensor is easy to handle and can monitor data continuously, accurately, and in a timely manner.

Experimental

CO_2 sensor design and construction

Figure 1 shows the diagram of the CO_2 sensor fabricated in our research. The principle of the sensor is based on the measurement of the pH change between the electrodes in the electrolyte solution, which is caused by hydrolysis of CO_2. The CO_2 molecules diffuse into the internal solution, where they react with water to form carbonic acid, then subsequently dissociate into bicarbonate and proton ions (as seen in Equation 1). Consequently, the produced proton ions decrease the pH of the electrolyte, which is detected by the internal electrodes [15].

$$CO_2\ (aq) + H_2O \leftrightarrow H_2CO_3 \leftrightarrow H^+ + HCO_3^- \qquad [1]$$

First, the prerequisite of the sensor was assembled through two electrodes, which are the working and reference electrodes. The working electrode was generated by thermal decomposition of an iridium salt (TIROF) [16]; the ultra-sonically cleaned iridium fiber was positioned with Li_2CO_3 at a high temperature of 870°C temperature for oxidation, then a uniform black layer was attached on the surface of the bare iridium wire. The iridium oxide thus generated was entwined with the gold wire and the combined nut and gold wire were insulated through a plastic heat-shrink tubing. Next, an electrochemical method was applied for the reference electrode preparation [17]. The bare silver wire was electroplated in 0.1-M HCl electrolyte solution to form a thin coating of silver chloride on the silver wire surface. Along this route the silver wire was used as the anode while the bare copper wire was treated as the cathode. Also, this silver/silver chloride electrode was insulated by the plastic heat-shrink tubing. Both electrodes were fixed through a stainless steel cap machined at PRRC's machine shop and stabilized by silicone sealant. The stabilized electrodes were dried for two days under air conditions at room temperature (25 ± 2°C).

A porous stainless steel unit was selected to support the whole frame, which enables the sensor to work under harsh downhole environments, especially under high-pressure conditions. This support unit was coated by silicone sealant onto the inner and outer surface to prevent contacting with electrodes. Then the stabilized electrodes were inserted into the support unit that was filled with a bicarbonate-based internal electrolyte solution ($NaHCO_3$-NaCl electrolyte solution). All the joints were sealed with silicone sealant in order to preventing leakage of the internal solution.

Another critical material used in the sensor construction was a gas-permeable membrane, purchased from Cole-Parmer Inc. The component of the membrane was silicon formed polymer, which can permit gases to pass through but not fluids. The prepared CO2 sensor, as shown in Fig. 1b, is small and compact, 0.7 in. in diameter and 1.5 in. long.

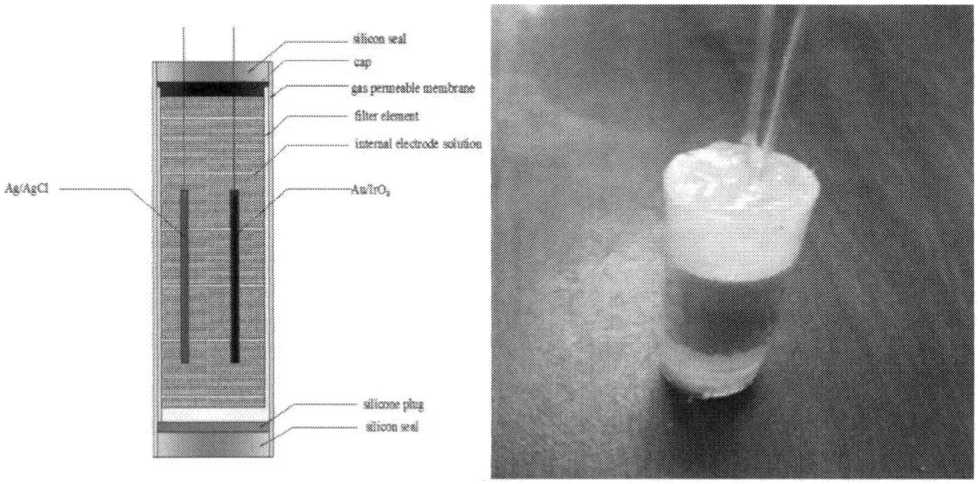

(a) (b)

Fig. 1 Schematic structure (a); and picture (b) of the CO_2 sensor

CO_2 sensor performance test under high pressure in synthesized produced water

Table 1 indicates the composition of salts used to make the synthetic produced water. As per reported composition of Permian Basin produced water, the top seven ions, Na^+, Mg^{2+}, Ca^{2+}, Cl^-, K^+, HCO_3^- and SO_4^{2-}, were all considered to investigate the potential response of CO_2 sensor. Synthesized produced water was prepared by dissolution of $CaCl_2{\cdot}2H_2O$, $MgCl_2$, NaCl, $NaHCO_3$, KCl and Na_2SO_4 in DI water.

Table 1 The Composition of Synthetic Permian Basin Produced Water

g/100 ml (±0.001)	$CaCl_2{\cdot}2H_2O$	$MgCl_2$	NaCl	$NaHCO_3$	KCl	Na_2SO_4
Permian Basin	1.280	1.380	15.700	0.018	0.034	0.240

The high-pressure plant, which was purchased from the High Pressure Equipment Company, includes the followed five units: cover, cap, "O" ring, tank body and high-pressure screw. This plant can withstand high pressures up to 20000 psi and retains its resistance ability under high salinity conditions.

Figure 2 shows the schematic structure of the high pressure apparatus. First all the units were combined and connected with the CO_2 chemical sensor. Then the pump was brought into play to verify that the whole system was leakproof by increasing the pressure by steps to 500 psi, 1000 psi, 1500 psi, 2000 psi, 2500 psi and 3000 psi, maintaining each pressure level for 24 hours to observe the pressure change. Then the performance of the CO_2 sensor was tested. All potentiometric measurements were carried out at room temperature of 20°C ±2°C. Before carrying out the tests, the base potential of the CO_2 sensor was measured by maintaining the CO_2 sensor in a mixed citric acid-produced water system solution. After the potential reached a stable value, a certain volume of the 0.1M $NaHCO_3$ solution was injected into the tank and then the tank was immediately sealed. Then the ESCO pump was started to increase the tank pressure. The sensor output signal was measured by a digital multi-meter. The CO_2 concentration of the solution was increased by additions of $NaHCO_3$ solution. Under acidic conditions, all bicarbonate and carbonate are converted into CO_2 by the reverse process of reaction, so the CO_2 concentration could be calculated from the added amount of $NaHCO_3$. The whole process was performed under pressure of 1000 psi.

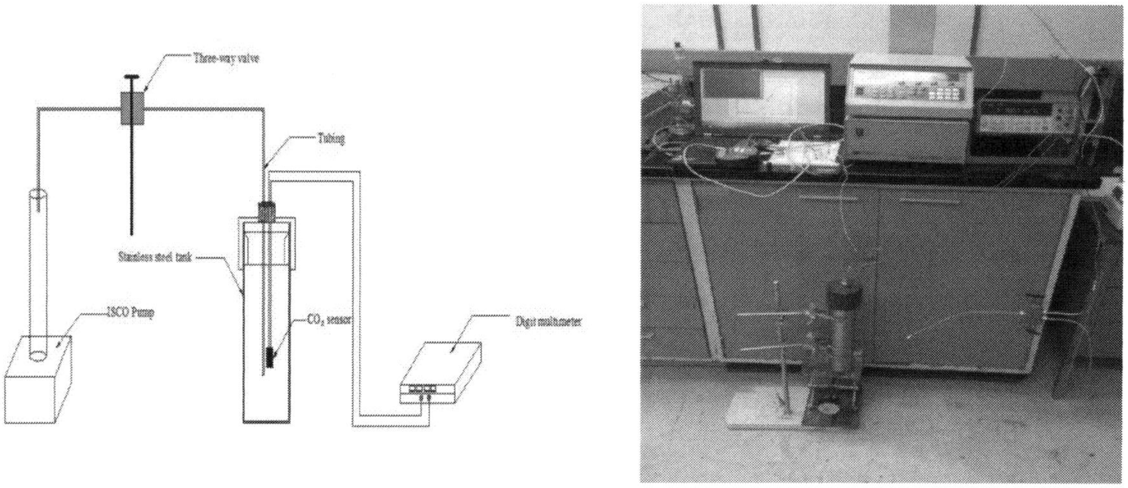

(a) (b)

Fig. 2 Schematic structure (a); and picture (b) of the high pressure plant.

CO_2 sensor performance test under simulated reservoir conditions

This series of tests was performed to investigate the performance of the sensor in a simulated geological environment. The stable pressure was 1250 psi. Table 2 shows the properties of the core used.

Table 2 Core Properties

Diameter, D/inch	1.47
Length, L/in	9.5
Pore Volume, PV/ml	57.71
Porosity, Φ	21.85%
Permeability, k/mD	143

The schematic of the CO_2 sensor setup is shown in Figure 3. The core sample and the spacer were first set into the core sleeve and then placed into the core holder. Then the sensor was inserted into the spacer and sealed completely by the plugs. The other end of the sensor was connected to the digital multi-meter for the purpose of potential data monitoring. An external overburden pressure of 3,750 psi was then set around the core.

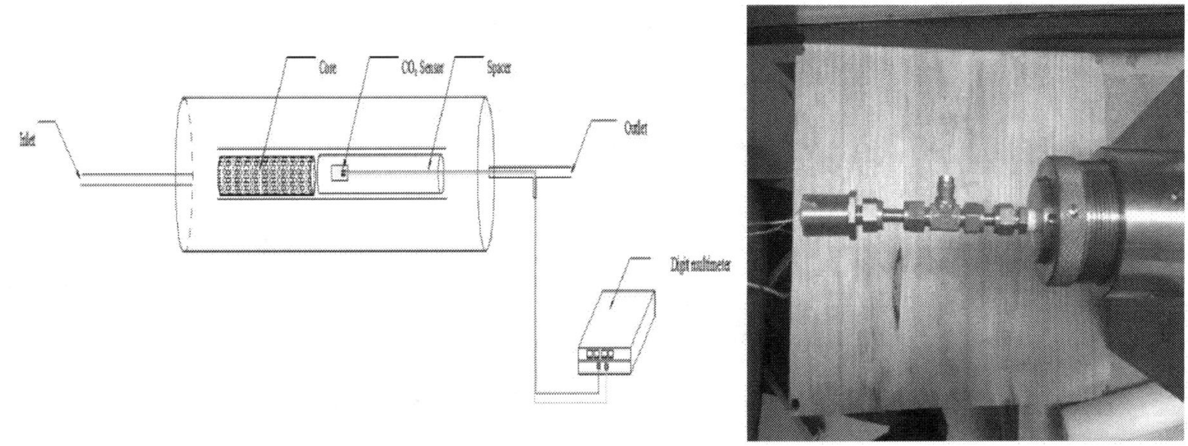

Fig. 3 Schematic structure and picture of CO_2 sensor setup.

The schematic diagram of the coreflooding device is shown in Figure 4. 2% brine and CO_2 were loaded into two accumulators, respectively. Three backpressure regulators (BPR) were set in the system to control the experimental pressures. The pressure drop along the core during flooding was measured by a differential pressure transducer, and the digital multi-meter was used to synchronously monitor the potential change of the CO_2 sensor. All potentiometric measurements were carried out at room temperature of 20°C±2°C in the test.

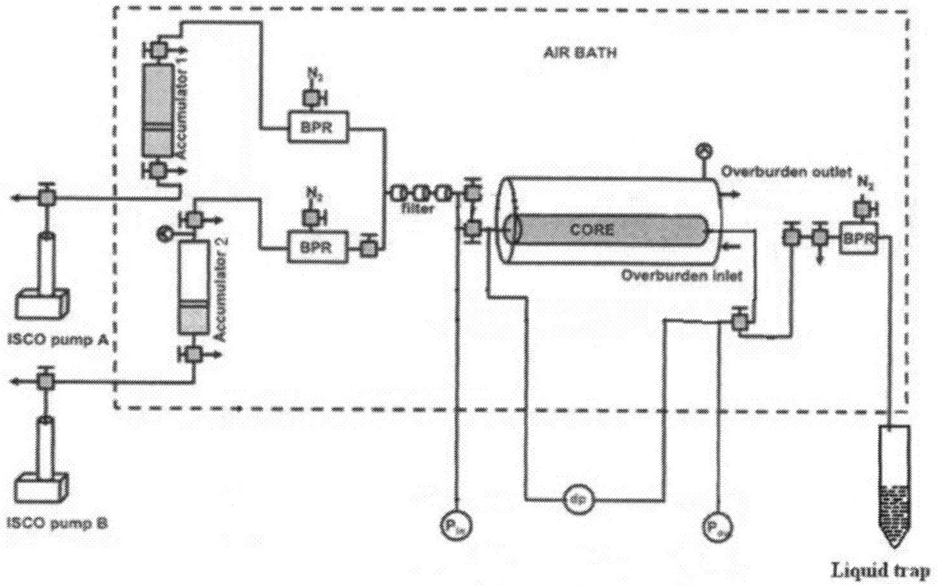

(a)

(b)

Fig. 4 Schematic diagram (a); and picture (b) of core flooding devices.

After the device was set up, the experiment proceeded to investigate the performance of the CO_2 sensor under simulated high pressure and high salinity conditions. First, baseline experiments were performed. Brine was injected into the core at flow rate of 100mL/hour. After 1 pore volume (PV) of brine was injected, the CO_2 concentration of the solution was increased by co-injection of CO_2 and brine into the core at the same injection rate and a fixed CO_2–brine volumetric injection ratio at 1:1. Under this condition, the geological formation was simulated and the performance of the CO_2 sensor under the harsh downhole environment was tested.

Results and Discussion

CO_2 sensor performance test under high pressure in produced water

Response time t and potential differential ΔE are two important parameters to evaluate the CO_2 sensor sensitivity. The potential differential was defined as: $\Delta E = E\ (CO_2) - E\ (\text{base})$, the response potential caused by CO_2. Therefore, the CO_2 responses for each sensor were evaluated by using ΔE and response time t in this experiment.

Figure 5 illustrates a typical potential response of the CO_2 sensor. The measurement was divided into three stages: 1) the potential stabilized at 80 mV in the bicarbonate-based internal electrolyte solution; 2) the CO_2 diffused through the gas permeable silicone membrane to the internal electrolyte solution. As a result, the potential dramatically increased from 80 mV to 126 mV; 3) the reaction reached the equilibrium and the potential stabilized at 126 mV.

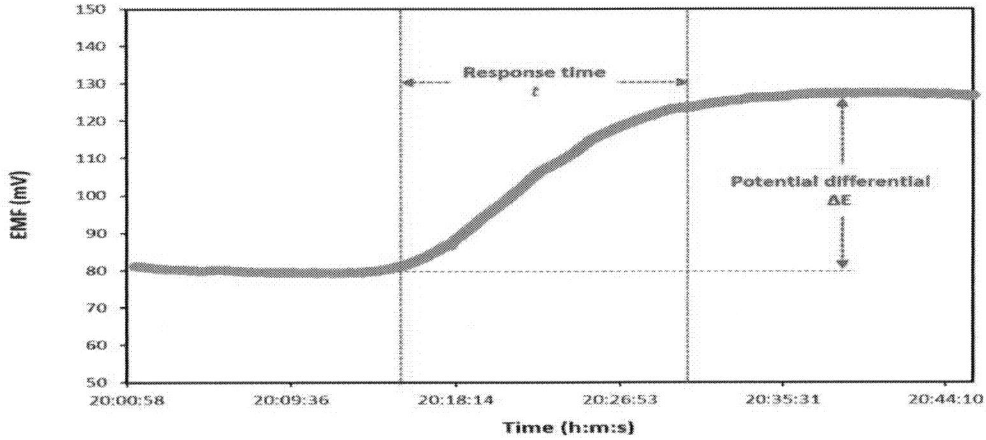

Fig. 5 A typical potential response of the CO_2 sensor.

Generally, an excess of citric acid was present in the system solution, All the $NaHCO_3$ reacted with the acid and transferred to the CO_2 following equation [2]. Thus a certain quantity of CO_2 can be generated and employed automatically.

$$2HCO_3^- + 2H^+ = 2CO_2 + 2H_2O \qquad [2]$$

By adding certain amounts of $NaHCO_3$, the fixed quantities of CO_2 were generated with the reaction. Figure 6 shows the potential response of the CO_2 sensor under variable CO_2 concentrations ranging from 1mM to 6 mM. As shown in (a), the potential increased from 62 mV to 84mV with the generated 1mM CO_2. While generated CO_2 concentration increased to 2mM, the potential increased from 64mV to 109mV. All four reaction graphs followed the same increase tendency dramatically.

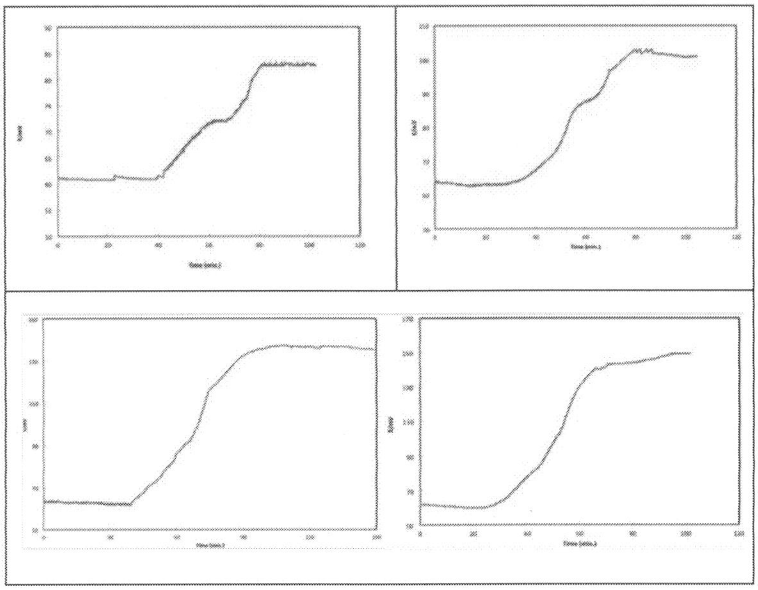

Fig. 6 Potential response of the CO_2 sensor with time in synthetic produced water, pressure=1000psi, generated CO_2 concentration: (a) 1mM; (b) 2mM; (c) 4mM; (d) 6mM.

Figure 7 shows the potential differential ΔE of the sensor for different CO_2 concentrations in the produced water with pressure of 1000 psi. The sensor showed good linearity with the sensitivity (slope of the line) as 89.209 mV/ Log [CO_2], and correlation factor of 0.981. Although a variety of ions exist in the outside environment, they cannot easily transfer into the internal electrolyte solutions through the gas-permeable silicone membrane. Thus, there is no salt ions influence on sensor sensitivity.

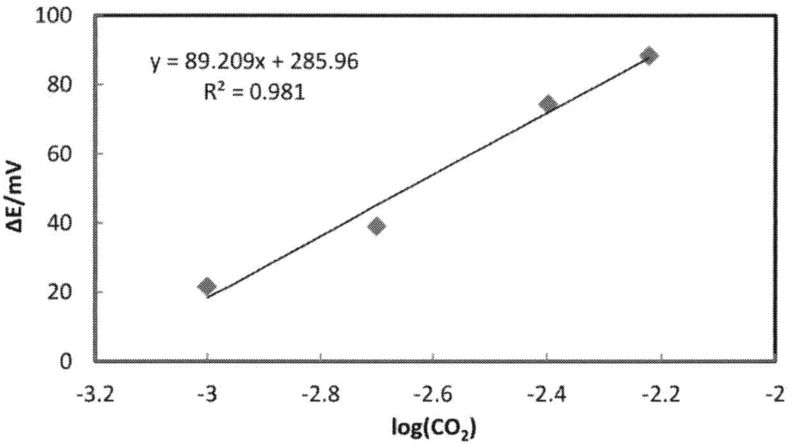

Fig. 7 The relationship between ΔE and log (CO_2), 1000 psi.

Table 3 presents the response time of the CO_2 sensor in produced water under high pressure. With the CO_2 concentration increase, the response time increased slightly. It can also be seen that the response time is about 1 hour in the produced water with the CO_2 concentration ranging between 1 mM and 6 mM.

Table 3 Response Time of CO_2 Sensor in Produced Water.

CO_2 concentration (mM)	Log [CO_2]	Response Time (s)
1	-3	3094
2	-2.70	3491
4	-2.40	3852
6	-2.22	4479

CO_2 sensor performance test under simulated reservoir conditions

Figure 8 illustrates the real potential response of CO_2 sensor. When the co-injection volume reached to 1.25 PV, the gaseous CO_2 can be observed from the outlet, showing that the CO_2 sensor was already submerged with CO_2 and started to work. As described above, this potential vs. time graph was also divided to three stages in the measurement. Stage 1 showed the base potentials of the CO_2 sensor. At stage 2, the CO_2 diffused through the gas permeable silicone membrane to the internal electrolyte solution and decreased the pH of the solution, which finally caused the potential to increase dramatically.

Stage 3 showed the potentials were stabilized again. Based on the potential change of the sensor, the CO_2 movement along the core can be estimated. The response time of the sensor, which is around 1400 seconds, showed that the CO_2 sensor can present the real data quickly.

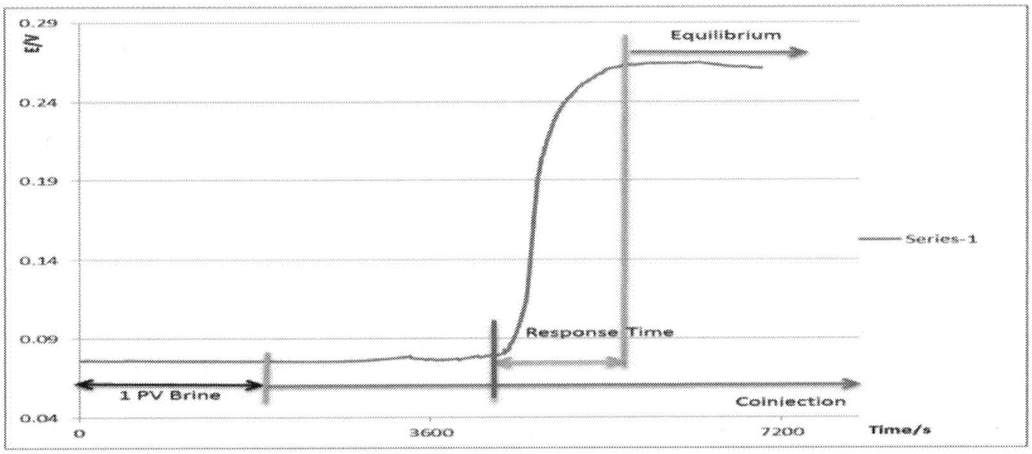

Fig. 8 Potential response of CO_2 sensor vs. time
(Brine injection rate: 100mL/hour; Co-injection rate: 100 mL/hour).

Conclusion

An iridium oxide electrode-based CO_2 sensor for downhole monitoring CO_2 *in-situ* was designed and constructed. The sensor was tested under high pressure and reservoir conditions to monitor CO_2 concentration change. The CO_2 sensor displayed a very good sensitivity and reasonable response time in synthesized produced water under high pressure (1000 psi). The sensitivity was 89.2 mV/ Log [CO_2] and the response time was around 1 hour. The performance of the sensor was also investigated under simulated geological CO_2 storage conditions. The results indicated that the prepared CO_2 sensor could monitor CO_2 concentration change in a CO_2/brine coreflooding test. The short response time, which is around 20 minutes, demonstrated that the sensor could detect CO_2 movement real time.

Acknowledgements

This work was supported by the Department of Energy through the National Energy Technology Laboratory under contract number DE-FE0009878. The author also gratefully acknowledges the support of Petroleum Recovery Research Center at New Mexico Tech.

Reference

[1] Pacala, S., and Socolow, R., 2004, Stabilization wedges-solving the climate problem for the next 50 years with current technologies: Science, v. 305, p. 968-972.

[2] Nicolas Pilisi, Davood Ghorbani, CO_2 Sequestration Studies in Ocean and Deep Sea Sub-seabed Formations. AAPG Annual Convention and Exhibition, Long Beach, California, Apr 22-25, 2012

[3] D. Ghorbani, N. Pilisi, S. Vasantharajan, CO_2 Sequestration in Deepwater Sediments: Optimization and Parametric Studies: Carbon Management Technology Conference, Feb 7-9, 2012, Orlando, Florida, USA

[4] Korbol, R. and A. Kaddour: Sleipner West CO_2 disposal: injection of removed CO_2 into the Utsira formation. Energy Conversion and Management, 36(6-9), 509-512

[5] D.J. White, G. Burrowes, T. Davis, Z. Hajnal, K. Hirsche, I. Hutcheon, E. Majer, B. Rostron, S. Whittaker, Greenhouse gas sequestration in abandoned oil reservoirs: The International Energy Agency Weyburn pilot project: GSA today, 14(7), 4-11

[6] M.A. de Figueiredo,H.J. Herzog, P.L. Joskow, K.A. Oye, D.M. Reiner, Regulating Carbon Dioxide Capture and Storage, CEEPR WP-2007-003, Apr 2007.

[7] Nicolas Pilisi, Ismail Ceyhan, Sriram Vasantharajan, CO_2 Sequestration in Deepwater Subseabed Formations. SPE International Conference on CO_2 Capture, Storage, and Utilization, New Orleans, Louisiana, USA, Nov 10-12, 2010

[8] NETL, Fit For Purpose Projects, [DB/OL] http: // www.netl.doe.Gov/research/coal/ carbon-storage /carbon-storage-infrastructure/fit-for-purpose,

[9] Giese, R., Henniges, J., Lüth, S., Morozova, D., Schmidt-Hattenberger, C., Würdemann, H., Zimmer, M., Cosma, C., Juhlin, C., Monitoring at the CO_2 SINK site: a concept integrating geophysics, geochemistry and microbiology. Energy Procedia 1 (1), 2251–2259.

[10] Scott Nooner, Mark Zumberge, Ola Eiken, Torkjell Stenvold, Sylvain Thibeau, Constraining the density of CO_2 within the Utsira formation using time-lapse gravity measurements. Proceedings of the 8th International Conference on Greenhouse Gas Control Technologies, Trondheim, Norway, Jun 19-22, 2006.

[11] Allan Mathiesona, Iain Wrighta, David Robertsa, Philip Ringrose, Satellite imaging to monitor CO_2 movement at Krechba, Algeria. The 9th International Conference on Greenhouse Gas Control Technologies (GHGT-9), 1-1, 2201–2209, Washington DC, USA, Nov 16–20, 2008

[12] Stefan Lüth, Peter Bergmann, Calin Cosma, Nicoleta Enescu, Rüdiger Giese, Julia Götz, Alexandra Ivanova, Christopher Juhlin, Artem Kashubin, Can Yang, Fengjiao Zhang, Time-lapse seismic surface and down-hole measurements for monitoring CO_2 storage in the CO_2 SINK project (Ketzin, Germany). 10th International Conference on Greenhouse Gas Control Technologies, Volume 4, 2011, Pages 3435–3442

[13] Dana Kiesslinga, Cornelia Schmidt-Hattenbergera, Hartmut Schuetta, Frank Schillinga, Kay Kruegera, Birgit Schoebela, Erik Danckwardtd, Juliane Kummerow, Geoelectrical methods for monitoring geological CO_2 storage: First results from cross-hole and surface-downhole measurements from the CO_2SINK test site at Ketzin (Germany). International Journal of Greenhouse Gas Control, 4: 816 – 826, Sep, 2010.

[14] D.J. White, G. Burrowes, T. Davis, Z. Hajnal, K. Hirsche, I. Hutcheon, E. Majer, B. Rostron, S. Whittaker, Greenhouse gas sequestration in abandoned oil reservoirs: The International Energy Agency Weyburn pilot project: GSA today, 14(7), 4-11

[15] N. Liu, Y. Li, L.X. Li, R. Grigg, B. McPherson, and R. Lee, Development of a CO_2 chemical sensor and sensor strategy for CO_2 monitoring in carbon sequestration, Seventh Annual Carbon Capture and Sequestration Conference, Pittsburgh, PA, 2008.

[16] S. Yao, M. Wang, M. Madou, A pH Electrode Based on Melt-Oxidized Iridium Oxide, Journal of The Electrochemical Society, 148 (2001) H29-H36.

[17] Jianjia Yu, Munawar Khalil, Ning Liu, Robert Lee, Iridium oxide-based chemical sensor for in situ pH measurement of oilfield-produced water under subsurface conditions, Ionics, Mar 2015, Volume 21- 3, 855-861

GLOBAL
CCS
INSTITUTE

The Impact of CO_2 Capture on Water Requirements of Thermal Power Plants

Carbon Management Technology Conference, November 2015, Texas

Guido Magneschi, Ron Munson, Tony Zhang

The Global CCS Institute

Our Vision for CCS:
CCS is an integral part of a low-carbon future

- We are an international membership organisation.
- Offices in Washington DC, Brussels, Beijing and Tokyo. Headquarters in Melbourne.
- Our diverse international membership consists of:
  - governments,
  - global corporations,
  - small companies,
  - research bodies, and
  - non-government organisations.
- Specialist expertise covers the CCS/CCUS chain.

The Global Status of CCS: 2015

The Institute's key publication

Summary Report, Key Findings and other advocacy materials can be found at:

www.globalccsinstitute.com

Full report is available online at the Institute's Members Portal.

Content

1. Introduction: water requirements of post combustion systems
2. Where is water needed?
3. Literature numbers and observations
4. How to reduce water consumption?
5. Conclusions

Water requirements in CO_2 capture systems

Generally, a CO_2 capture system requires water, raising the water demand of the power plant or industrial facility where it is applied.

Two types of water requirements for capture systems:

- Water for cooling
- Water for process make up.

Most of the water is required for cooling - make-up is a smaller portion.

Literature data

Author	Year	Post				Pre	Oxy	
		SUB	SC	USC	NGCC	IGCC	SC	USC
Zhai & Rubin	2010	X	X	X				
IEAGHG	2011			X		X		X
DOE [1]	2012						X	
DOE	2013		X		X	X		
ROAD	2014			X				
DOE	2015		X		X	X		

SUB = subcritical PC, SC = supercritical PC, USC = Ultra-supercritical PC, NGCC = natural gas combined cycle

Post-combustion: cooling requirements

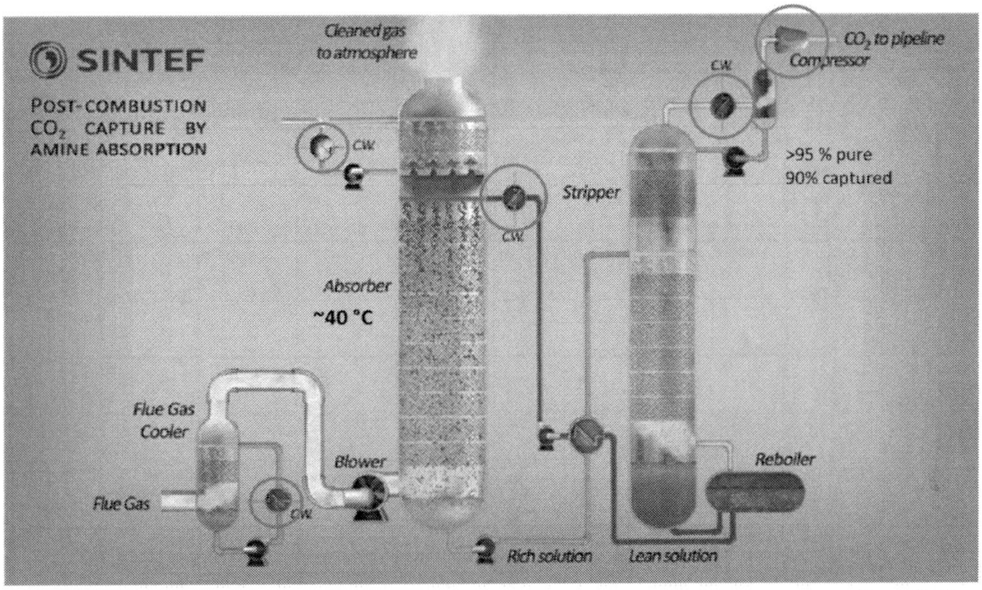

Source: Sintef

Post-combustion: water balance (make-up)

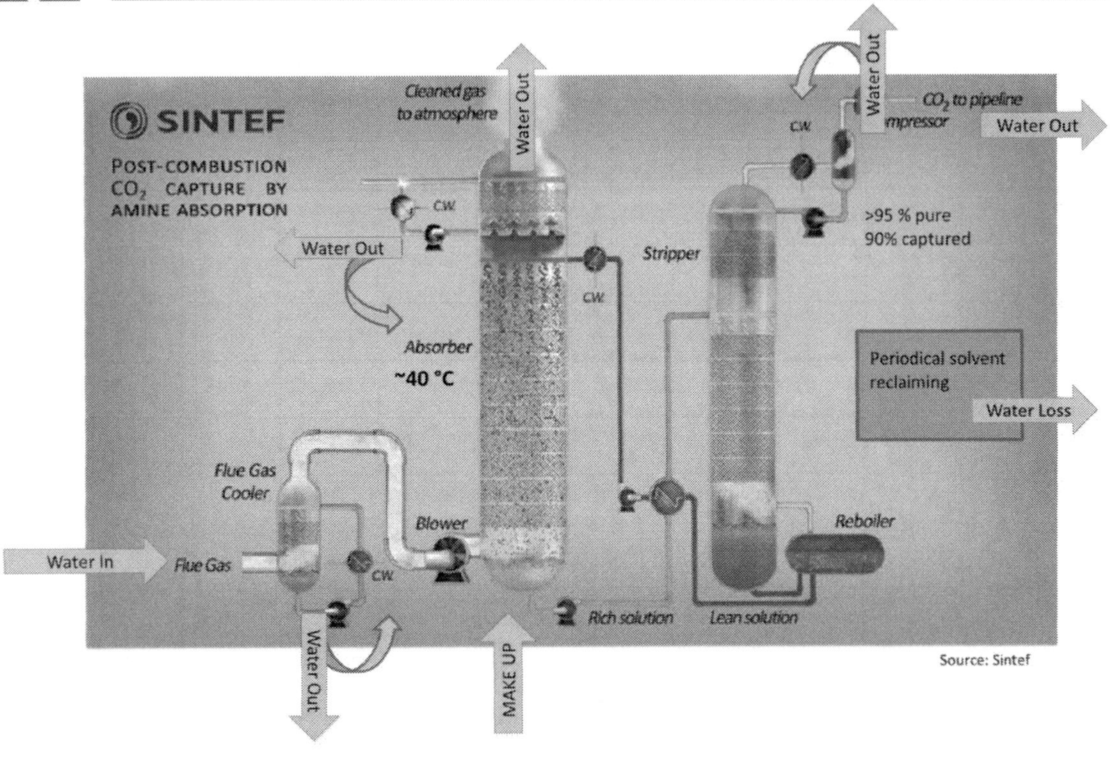

Source: Sintef

Estimations for post-combustion systems

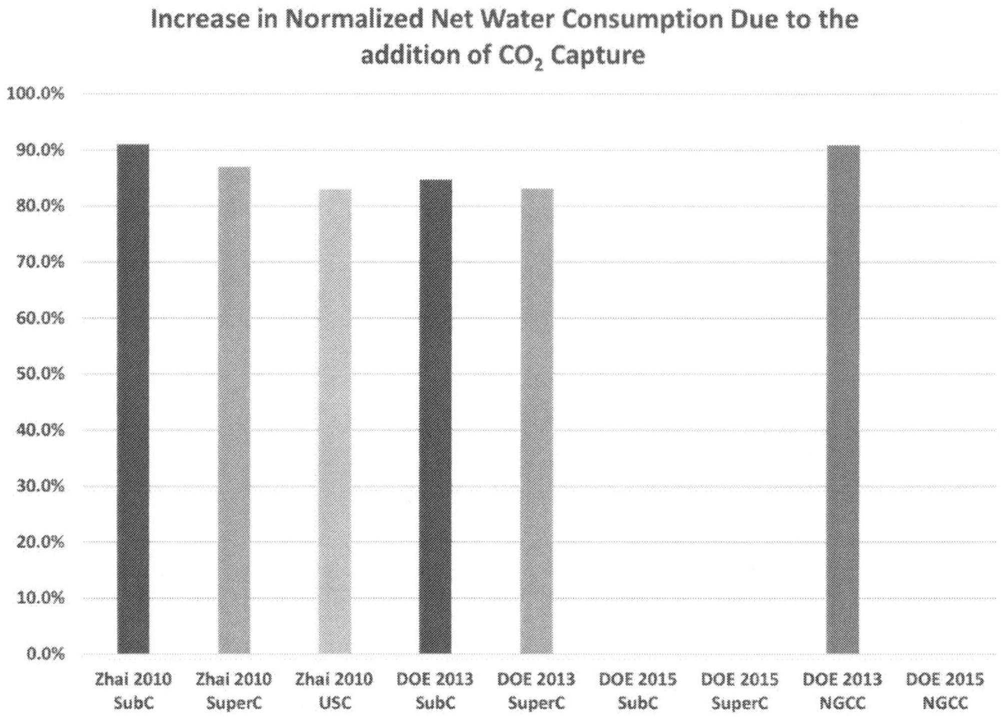

Estimations for post-combustion systems

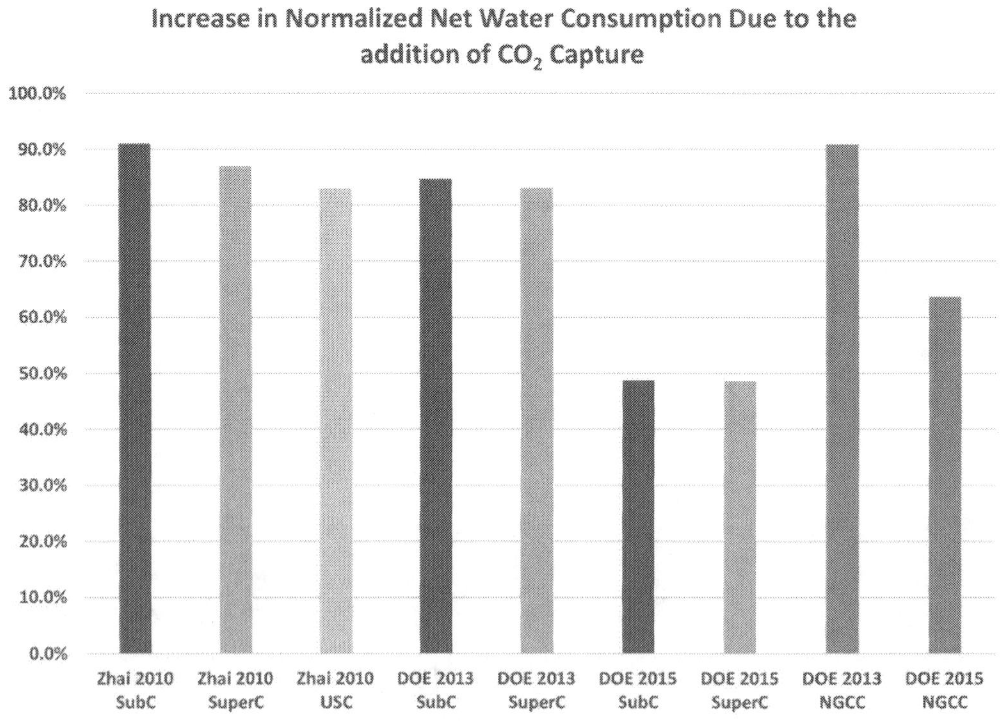

Estimations for post-combustion systems

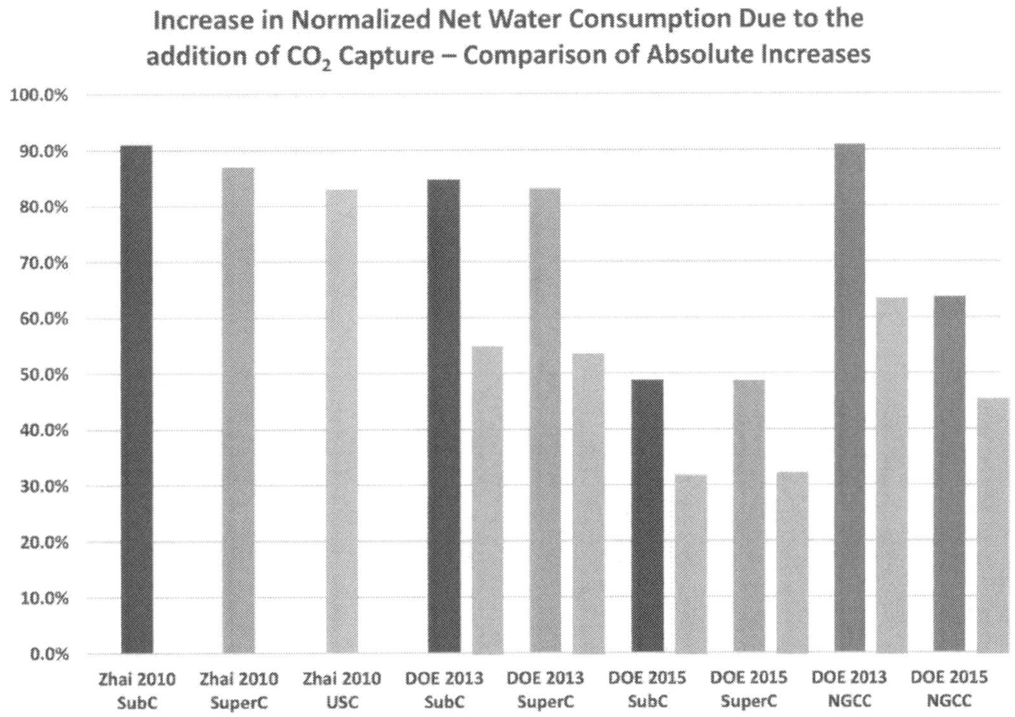

Estimations for post-combustion systems

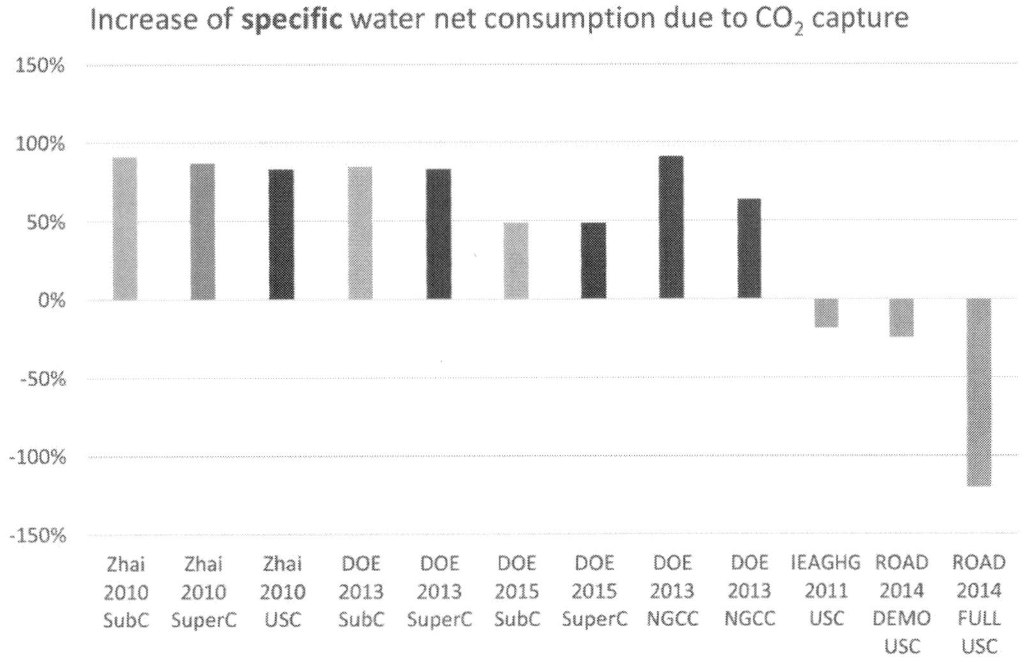

Options to reduce water use in CO_2 capture systems

For cooling water requirements:

- Waste heat integration
- Increase performance of capture systems
- Use of low temperature solvents
- Use dry cooling systems

For process (make-up) water requirements:

- Optimize water balance management
- Reduce reclaiming frequency

The benefit of waste heat integration

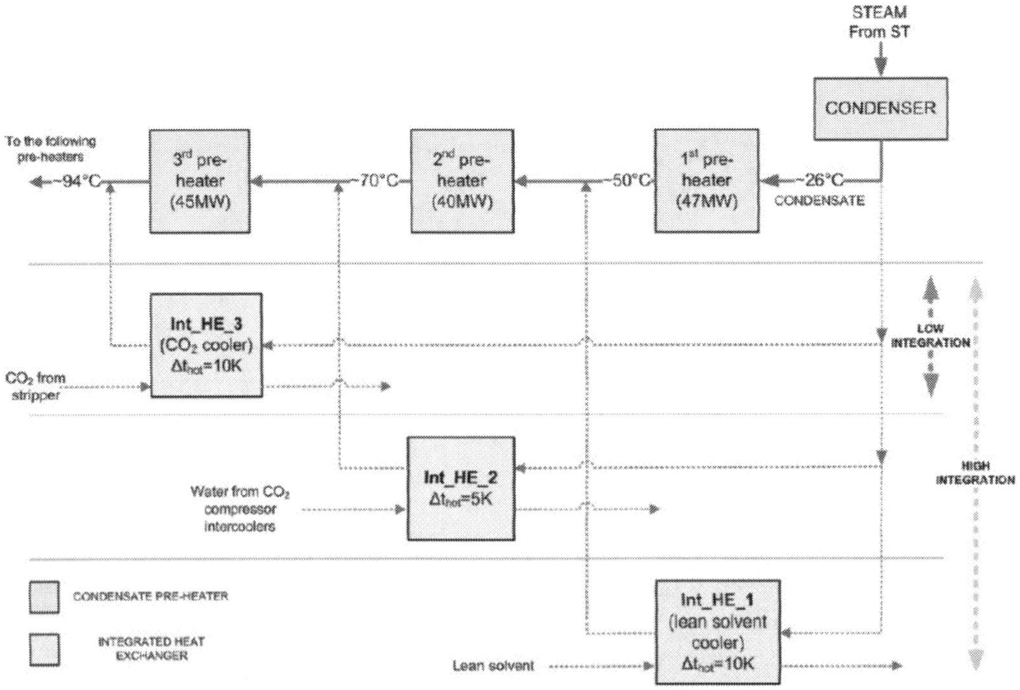

The benefit of waste heat integration

Integrating waste heat from the capture plant into the power plan steam cycle has also the effect to reduce cooling water requirements.

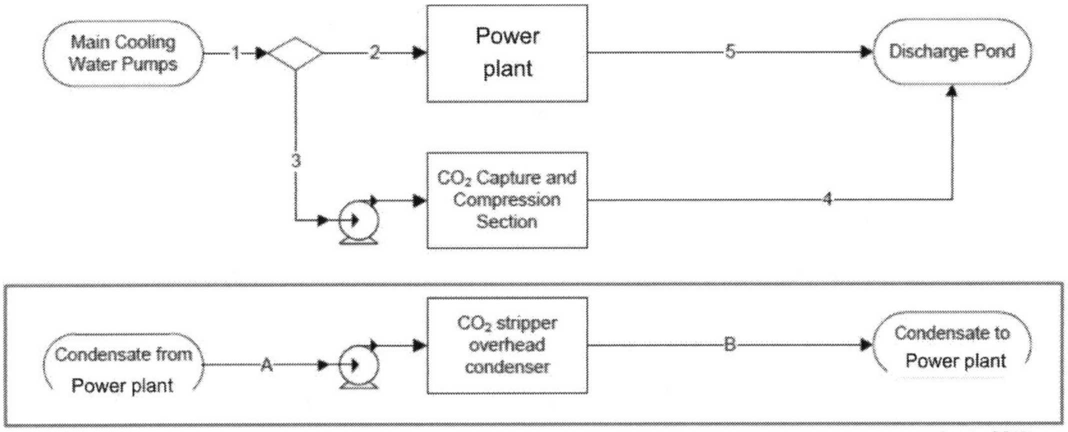

Waste heat integration

Source: ROAD

Other Options

- Dry cooling can be employed, although it has higher costs and power demand
- Newer technologies like *membranes* or *adsorption systems,* that requires less water, might be more suitable for region with scarce water availability - but not mature yet
- Enhanced Water Recovery EWR (after proper water treatment) could help to cooling water in arid regions (e.g. China)

Conclusions

- State of art post-combustion systems increase water consumption of PC plants around 50% (on a t/MWh basis)
- This correspond to about 32% increase in absolute value (t/h)
- NGCC systems requires around half of the water required in PC plants
- Water consumption increase is mainly a concern for wet recirculating cooling systems as they dissipate water
- For once-through PC the addition of a CO_2 capture systems can actually reduce fresh water demand
- Waste heat integration helps to reduce water consumption

Report under preparation

The institute is currently preparing a report on this topic that includes also oxy-combustion and pre-combustion systems. The report is expected to be ready Q1 2016.

For enquiries, suggestions and clarifications please contact:
guido.magneschi@globalccsinstitute.com

References

- **[Zhai, 2010]** Zhai, H. and Rubin E. S., 2010, Carbon capture effect on water use at coal fired power plant. GHGT-10 proceedings.
- **[DOE, 2013]** DOE/NETL, 2013, Cost and Performance Baseline for Fossil Energy Plants Volume 1: Bituminous Coal and Natural Gas to Electricity. Revision 2a, September 2013. DOE/NETL-2010/1397
- **[DOE, 2015]** DOE/NETL, 2015, Cost and Performance Baseline for Fossil Energy Plants Volume 1a: Bituminous Coal (PC) and Natural Gas to Electricity. Revision 3, July 2015. DOE/NETL-2015/1723
- **[IEAGHG, 2011]** IEAGHG, 2011, water usage and loss analysis for bituminous coal power plant with CO2 capture, report 2010/05.
- **[ROAD, 2014]** Hylkema, H. and Read, A., 2014, Reduction of freshwater usage of a coal fired power plant with CCS by applying a high level of integration of all water streams. GHGT-12 proceedings.

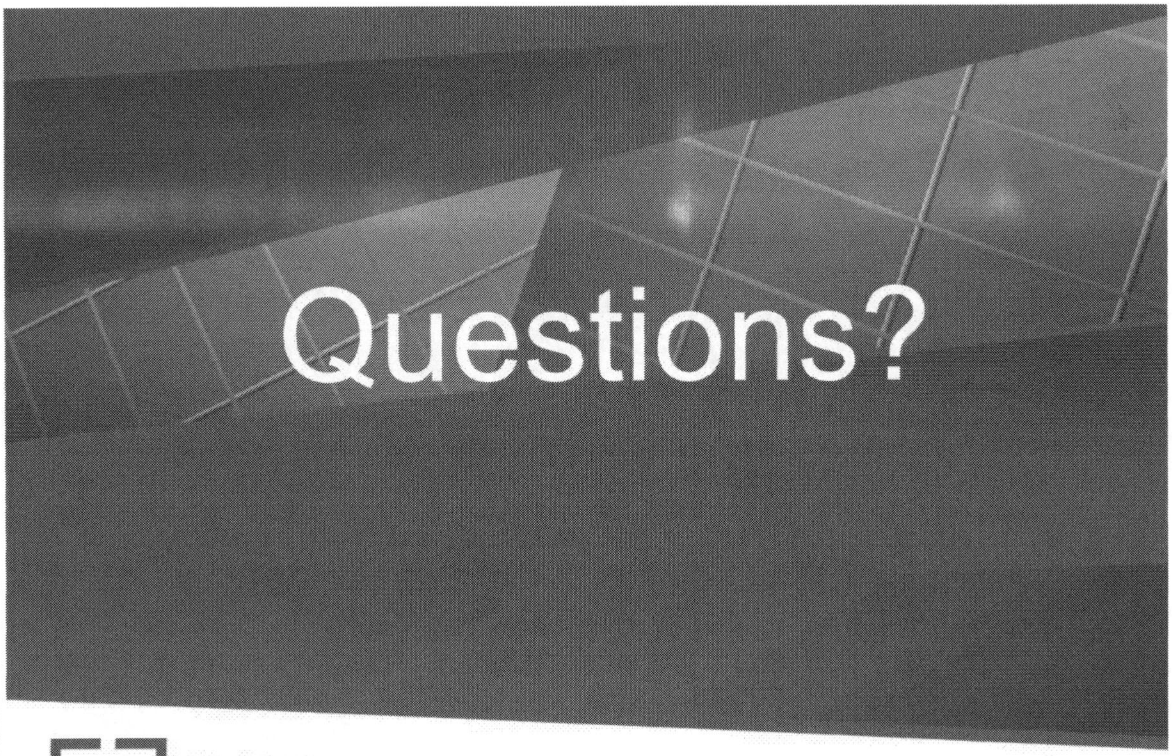

GLOBAL CCS INSTITUTE

GLOBALCCSINSTITUTE.COM

Session 19 Energy/Water Nexus
Carbon Management Technology Conference 2015

Improving the Physical and Economic Performance of CO_2 Capture, Utilization and Storage in Saline Reservoirs by Producing Brine

Thomas A. Buscheck[1], Jeffrey M. Bielicki[2], Joshua A. White[1], Yunwei Sun[1], Yue Hao[1], William L. Bourcier[1], Susan A. Carroll[1], and Roger D. Aines[1]

[1]Lawrence Livermore National Laboratory, P.O. Box 808, L-223, Livermore, CA USA
[2]Dept of Civil, Environmental, & Geodetic Engineering and John Glenn College of Public Affairs, The Ohio State University, Columbus, OH, USA
buscheck1@llnl.gov

November 19, 2015

This work was sponsored by the US DOE National Energy Technology Laboratory, managed by Traci Rodosta and Andrea McNemar. We acknowledge Statoil and the Statoil Production License for use of data from the Snøhvit CO_2 Storage Project and Philip Ringrose for useful discussions.

LLNL-PRES-679315
This work was performed under the auspices of the U.S. Department of Energy by Lawrence Livermore National Laboratory under contract DE-AC52-07NA27344. Lawrence Livermore National Security, LLC

Outline

- Introduction/Motivation
- Objectives
- Approach
- Model results
  - testing the efficacy of brine removal for a real geologic setting: Tubåen Fm. at Snøhvit
  - brine reinjection options
- Conclusions
- Final thoughts

Introduction/Motivation

- Major challenges for saline-reservoir CCS
  - pressure buildup drives storage risks (induced seismicity, caprock fracture, leakage) that can limit storage capacity and permanence
  - until large amounts of fluid move into and/or out of a reservoir, estimates of storage capacity and permanence are subject to uncertainty
  - reducing this uncertainty could be necessary prior to securing financing for CO_2 capture and transportation infrastructure
- We present an approach called ***pre-injection brine production*** that
  - first removes large amounts of brine out of a reservoir
  - uses individual wells to produce brine prior to injecting CO_2
  - may generate water for beneficial use

Introduction/Motivation

- This approach, sometimes called enhanced water recovery (EWR), shares key advantages with CO_2 EOR
  - generate a product
  - ample reservoir characterization prior to CO_2 injection
  - depleted reservoir pressure prior to CO_2 injection

Objectives

- Site selection and characterization: assess storage capacity and permanence prior to CO_2 injection
- Increase storage capacity and efficiency per well and per unit of brine removed
- Maximize resource utilization (increase source-to-sink ratio) to better leverage fixed storage costs
- Reduce (1) area of review (AoR), (2) interference with neighboring subsurface operations, and (3) duration of site care and monitoring
- Generate water prior to, during, and after CO_2 injection

The physical and economic performance of CCS can be improved with brine production

Approach

Pre-injection brine production

- Monitor pressure drawdown to assess
  - CO_2 storage capacity and compartmentalization
  - CO_2 leakage potential
- Reservoir diagnostics where needed most: at the center of CO_2 storage

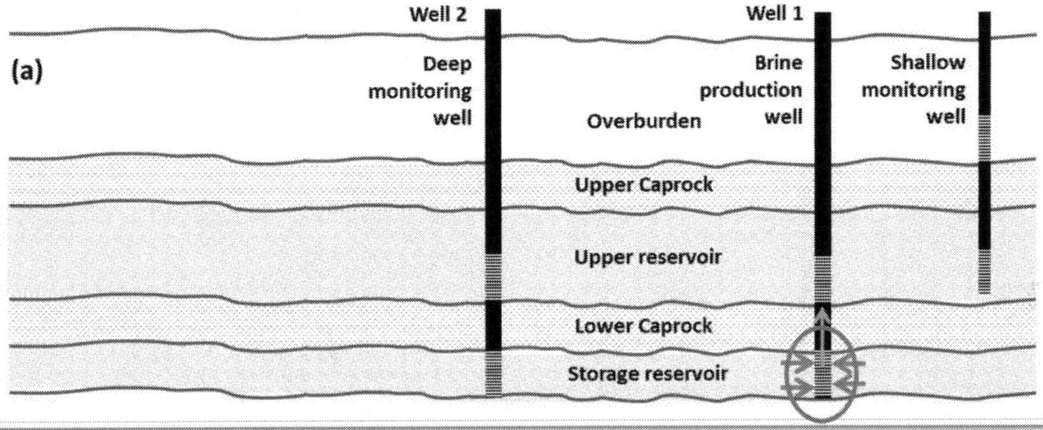

Lawrence Livermore National Laboratory LLNL-PRES-679315 7

Pre-injection brine production and reinjection

- Some of the produced brine may be reinjected in a nearby reservoir
- Measuring pressure buildup from brine reinjection can determine
  - brine storage capacity
  - brine leakage potential

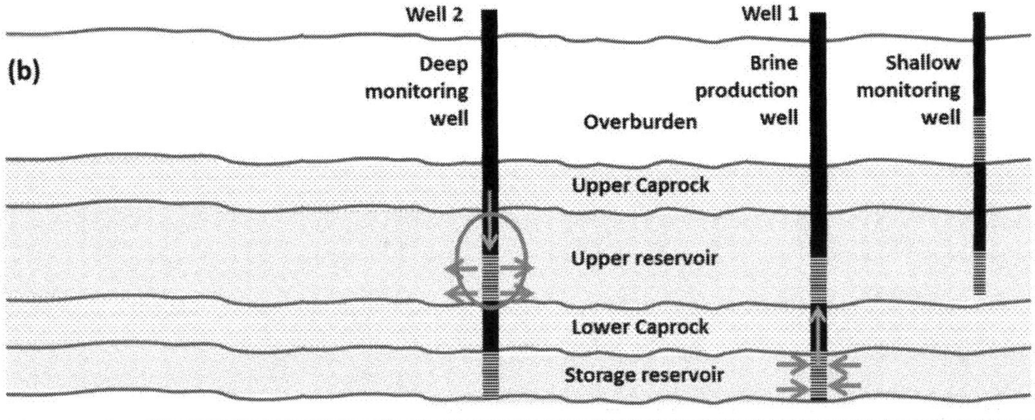

Lawrence Livermore National Laboratory LLNL-PRES-679315 8

Co-injection brine production

- Pre-injection pressure drawdown allows
  - greater spacing between injectors and producers (fewer wells overall)
  - ongoing pressure-management planning
- Pressure relief is greatest where needed most: at the center of CO_2 storage

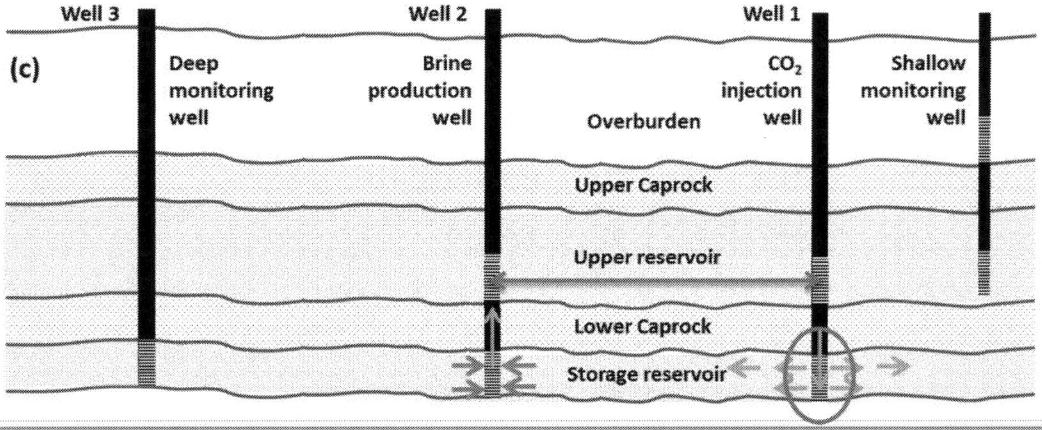

Co-injection brine production and reinjection

- Brine that is not consumed for use can be reinjected in a nearby reservoir
- Reinjecting in an overlying reservoir can reverse the overpressure gradient and reduce CO_2 leakage potential

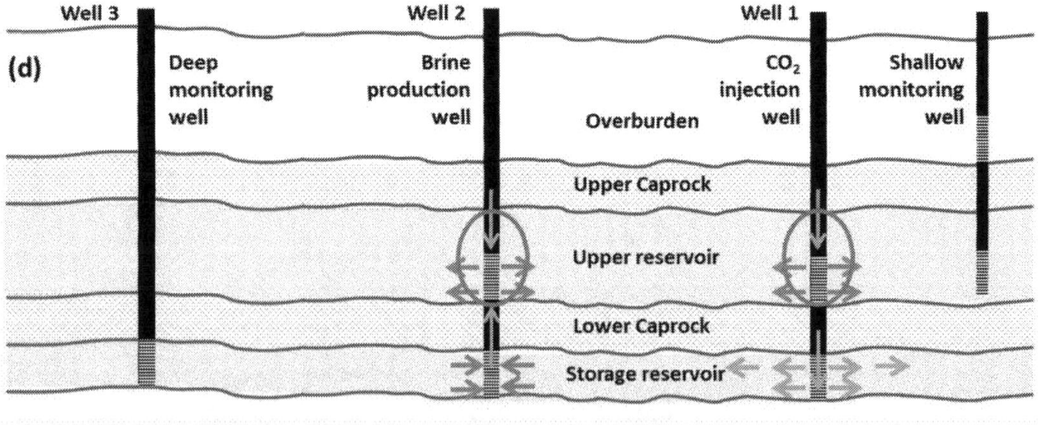

Co-injection brine production

- After CO_2 breakthrough, a brine producer can become a CO_2 injector
- Each successive well goes through three stages:
  - monitoring
  - brine production
  - CO_2 injection

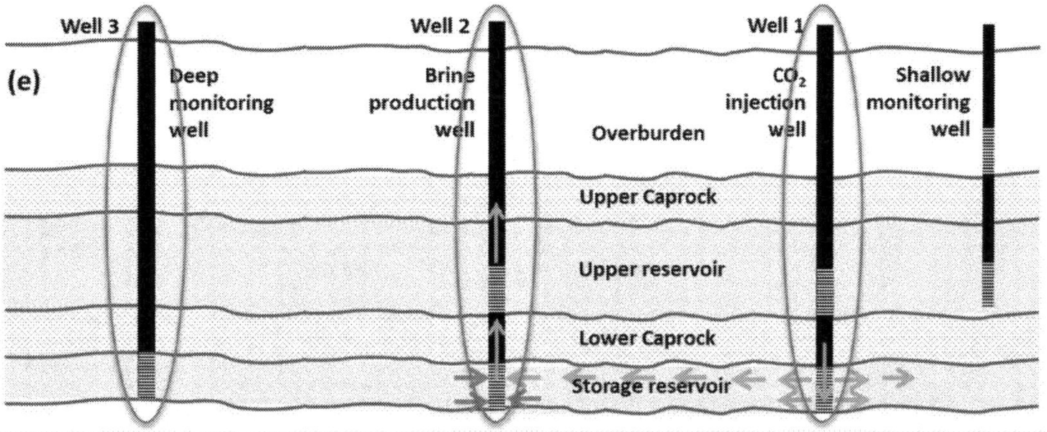

Co-injection brine production

- Each additional well can be located and operated with the most information
- CO_2 storage operations can be managed proactively and more efficiently

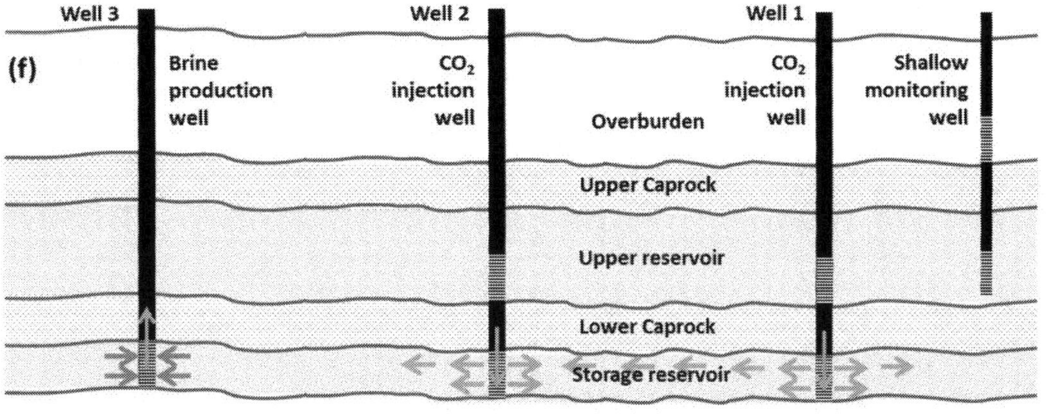

Co-injection brine production and reinjection

- Isolating well intervals with packers can enable simultaneous
  - CO_2 injection
  - brine production or reinjection
- Greater value can be leveraged out of each well

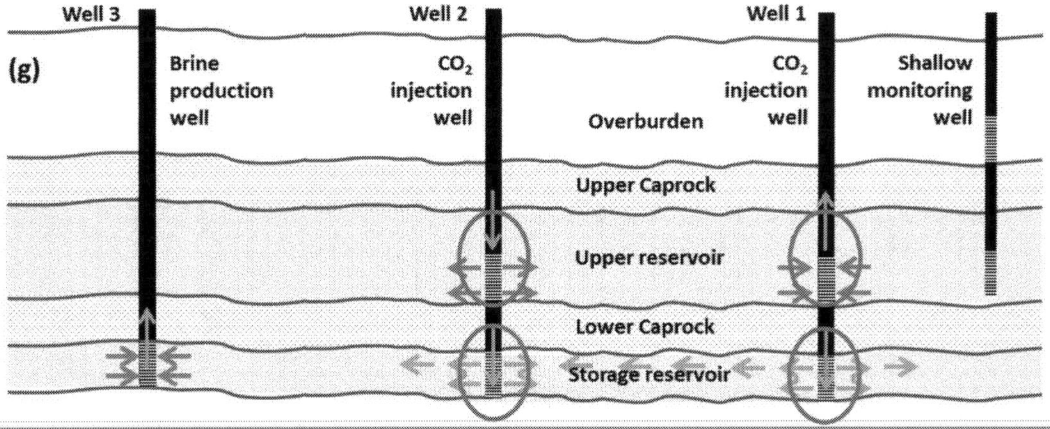

Post-injection brine production

- Brine production may continue <u>post-injection</u> to nullify overpressure
  - reducing the areal and temporal extent of site care and monitoring

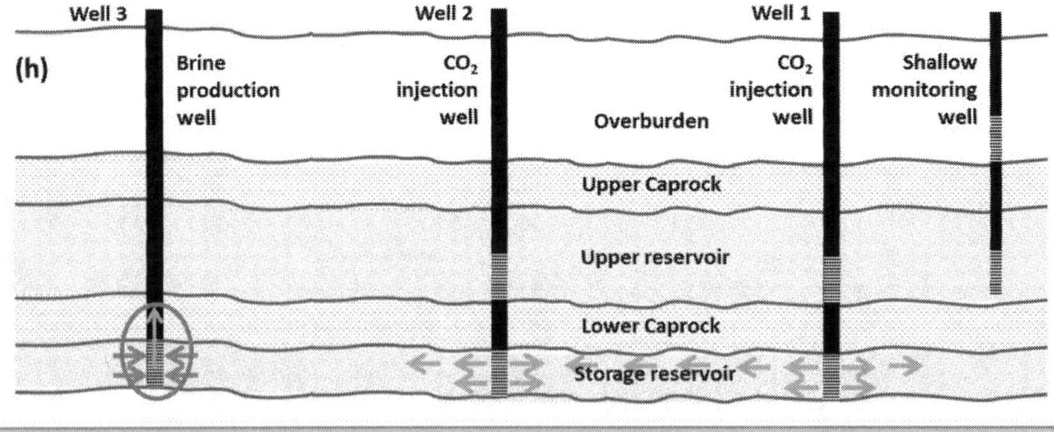

Testing the efficacy of brine removal for a real geologic setting: Tubåen Fm. at Snøhvit

Calibrated reservoir model of Snøhvit CO_2 project

- Reservoir model, using NUFT code, was calibrated for 3 yr of injection of 1.09 MT of CO_2 into Tubåen Fm with data provided by Statoil
  - thickness, porosity, and permeability
  - CO_2 injection-rate and bottom-hole pressure for injection well
  - structural geology and 4-D seismic difference amplitude maps, which show overpressure and CO_2 migration controlled by long fluvial channels in 3 permeable subunits of the Tubåen
  - production logging tool (PLT) data show 80% of injection going into lower perforated zone

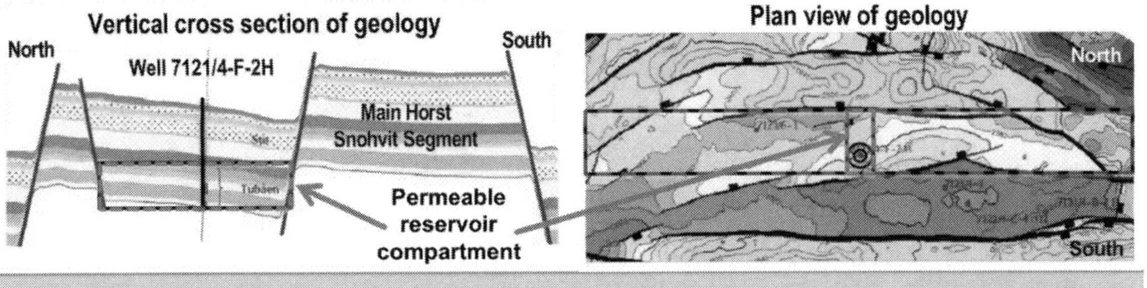

Calibrated reservoir model of Snøhvit CO_2 project

- Initial conceptual model was a 230 m x 2000 m compartment, with leaky lateral boundaries, bounded by low-permeability seal units
  - after calibration, final conceptual model has a 230 m x 2300 m "leaky" compartment
  - after calibration, a permeability ratio of 0.01 between the far field and permeable compartment was determined

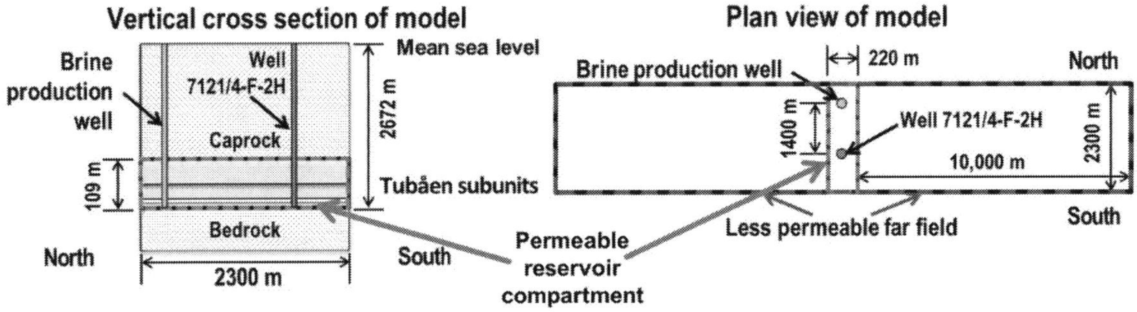

Calibrated reservoir model of Snøhvit

Injection-well overpressure history for the Phase I Snøhvit Project

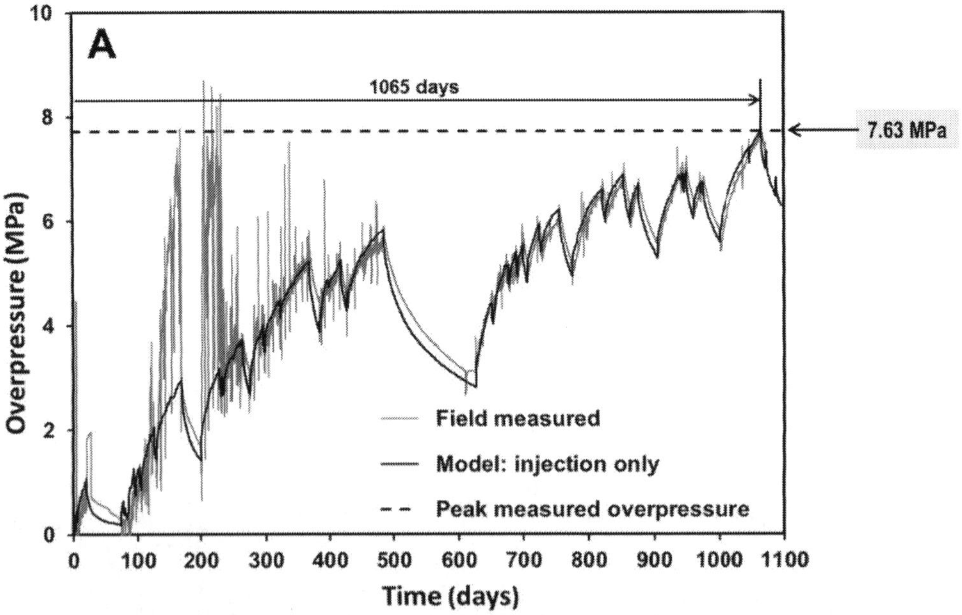

Snøhvit study: Pre-injection brine production

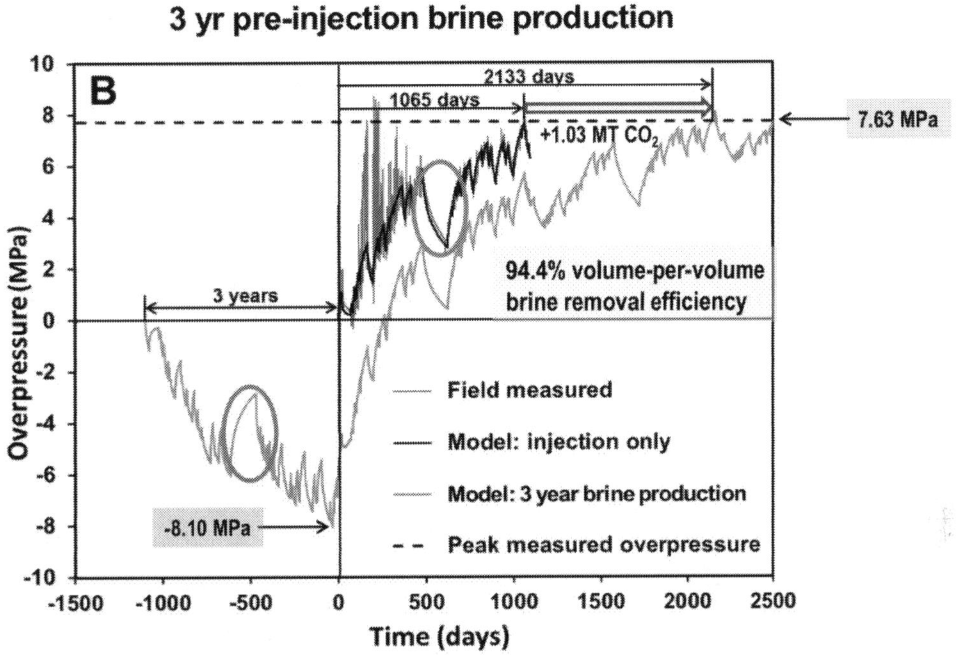

Snøhvit study: Pre-injection brine production

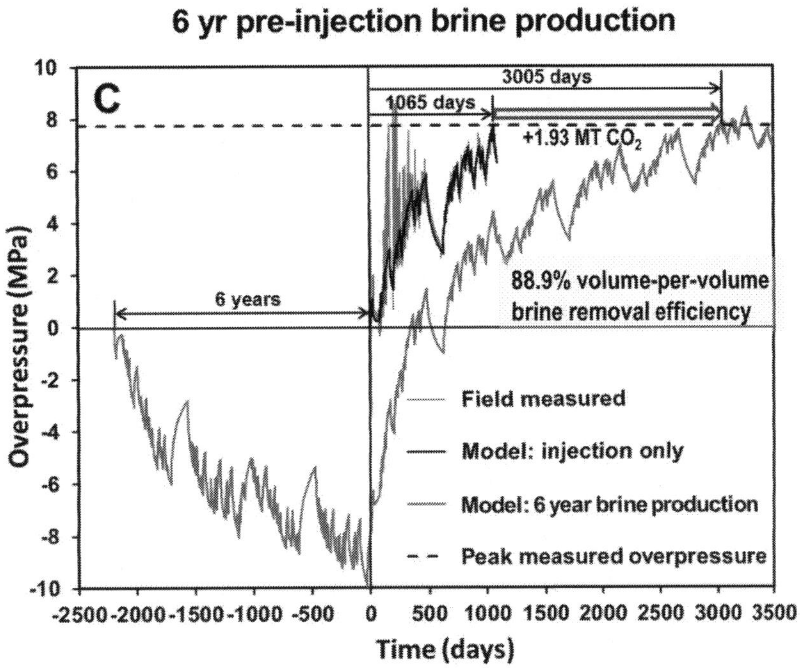

Snøhvit study: Pre-/co-injection brine production

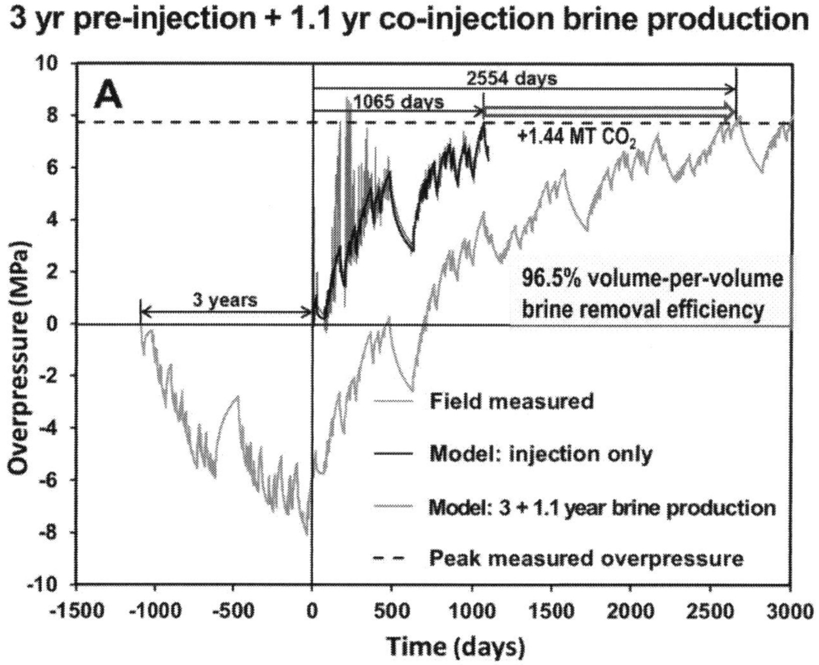

Snøhvit study: Pre-/co-injection brine production

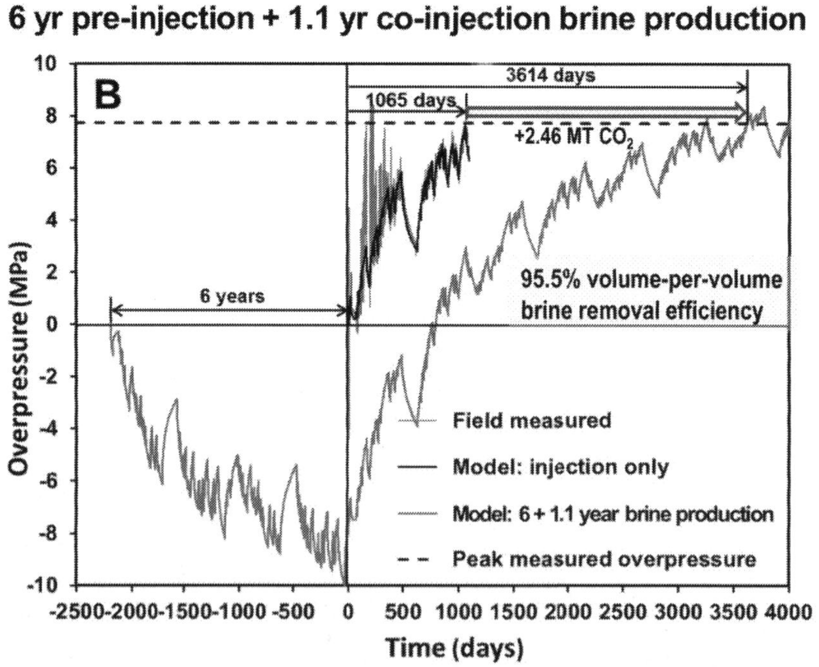

Snøhvit study: Effectiveness of brine removal

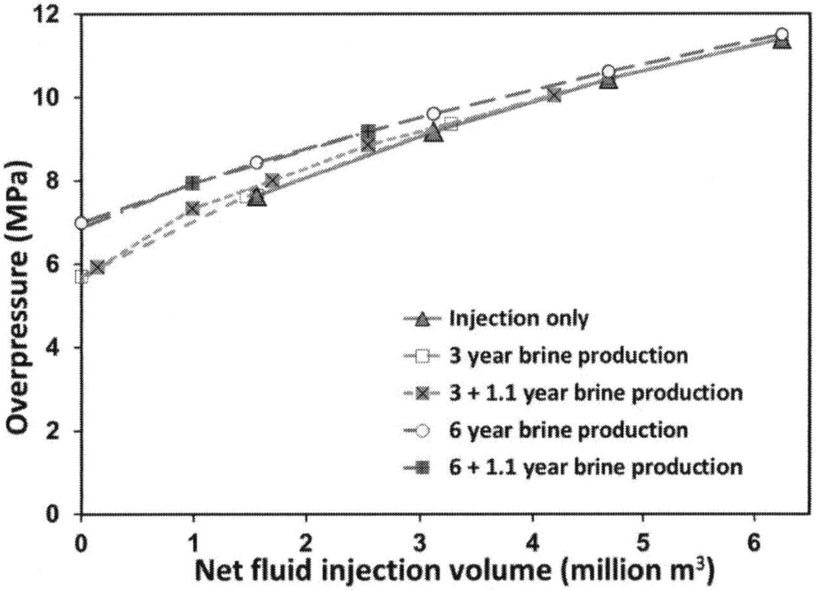

Snøhvit study: Post-injection pressure decline

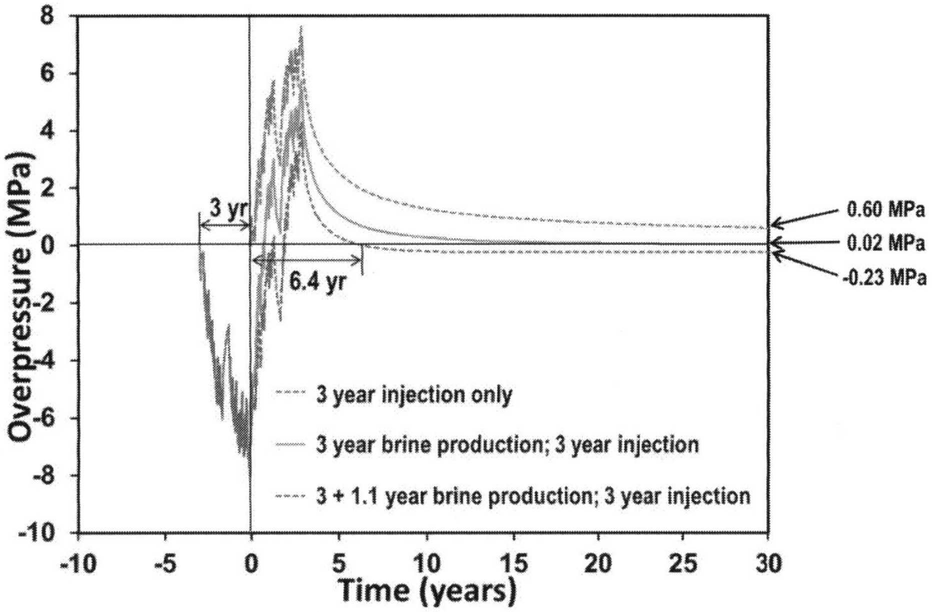

Snøhvit study: Post-injection pressure decline

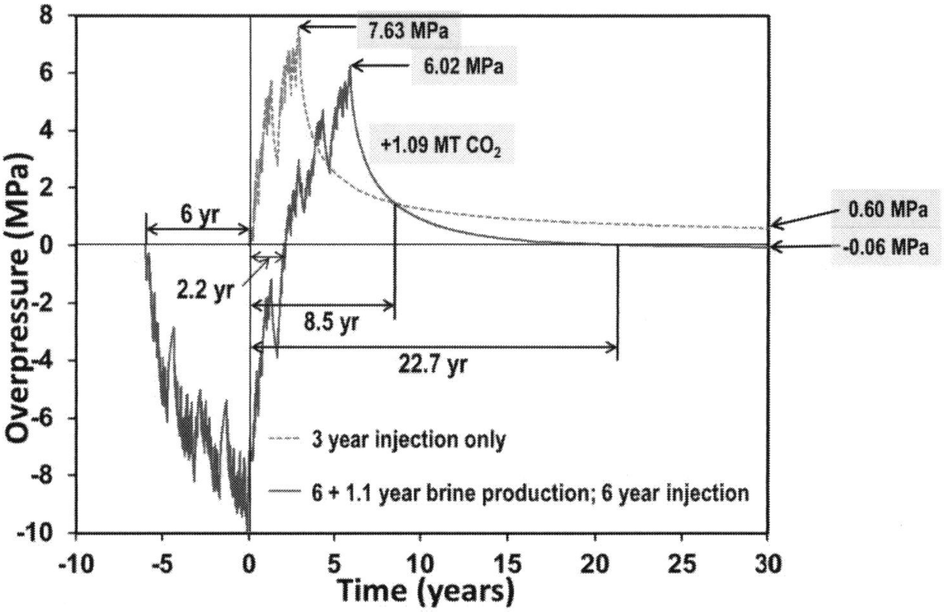

Snøhvit study: Effectiveness of brine removal

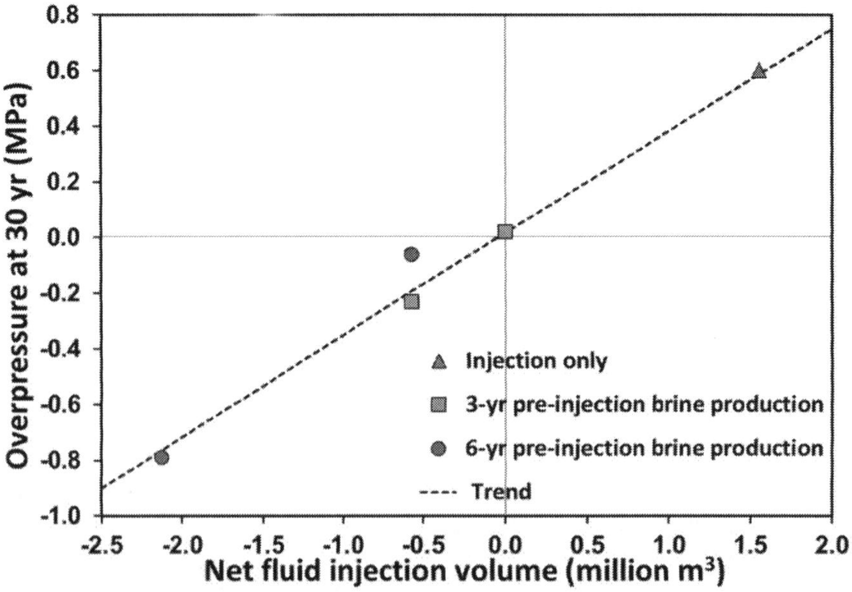

Brine Reinjection Options

Injection only

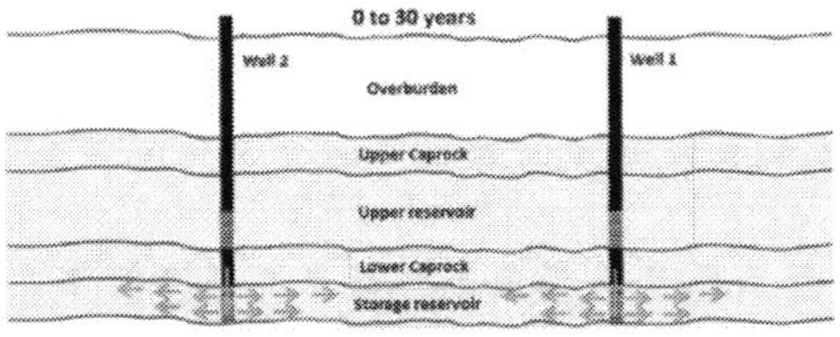

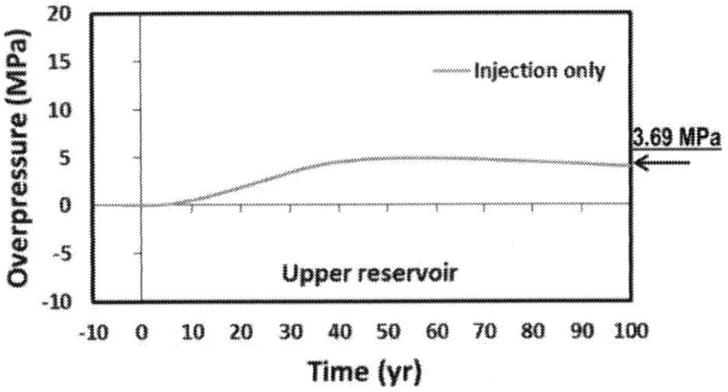

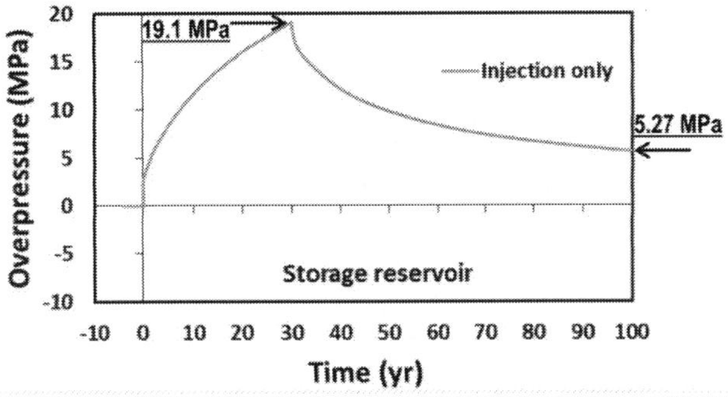

Brine disposition: 0% consumption, 100% reinjection

0.0 brine-removal to CO_2 injection ratio

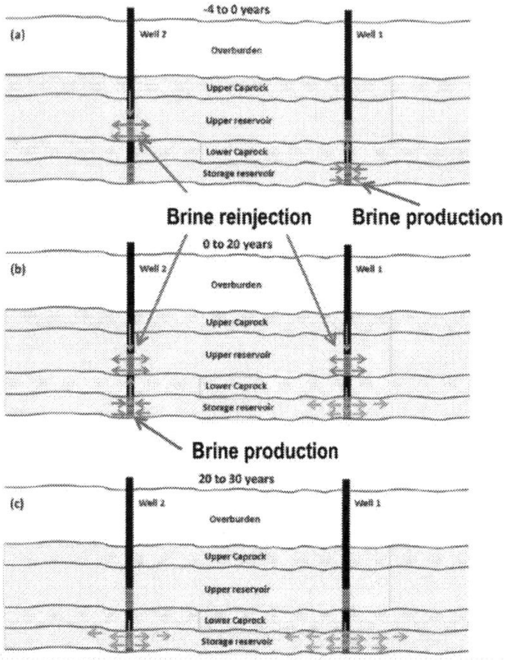

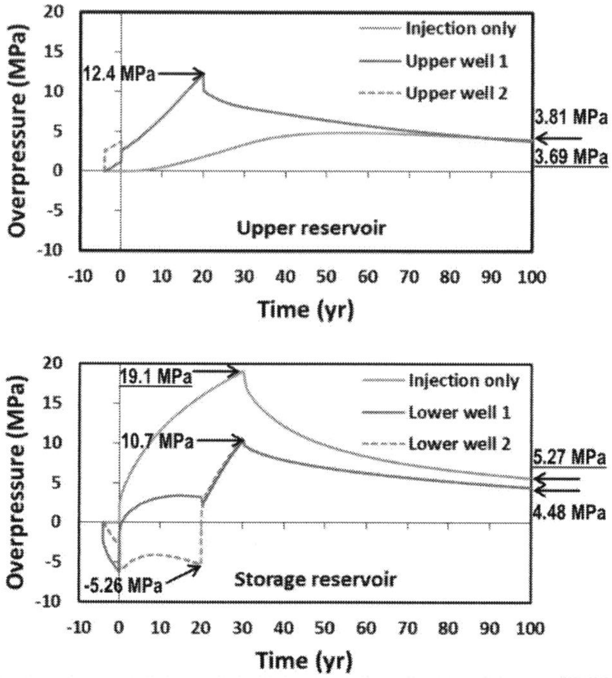

Brine disposition: 50% consumption, 50% reinjection

0.35 brine-removal to CO_2 injection ratio

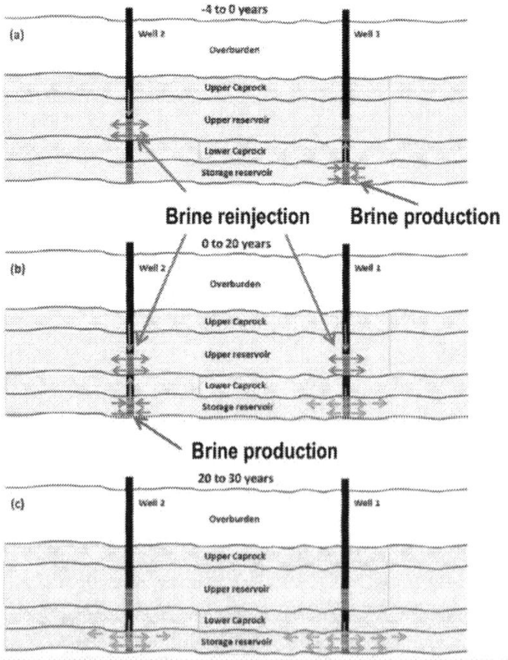

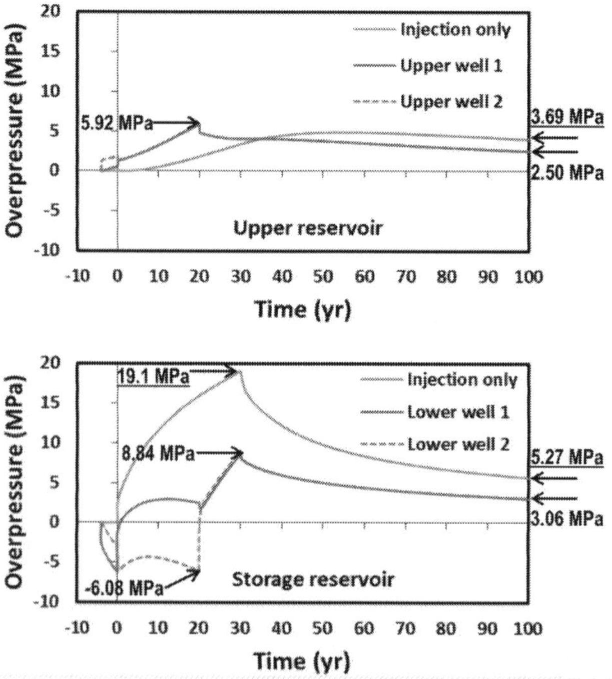

Brine disposition: 100% consumption, zero net injection

0.70 brine-removal to CO_2 injection ratio

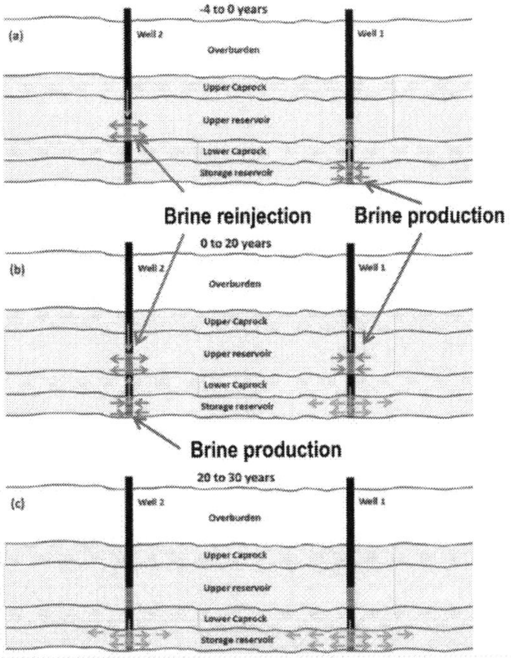

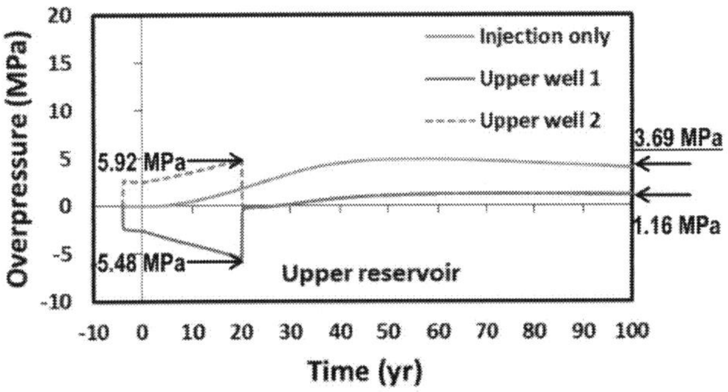

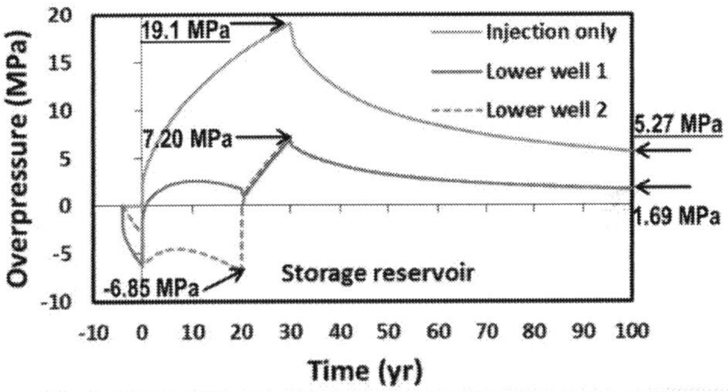

Conclusions

- Pre-injection brine-production can provide
  - determination of storage capacity and permanence prior to CO_2 injection
  - better assurance that storage integrity can be managed
  - pressure relief and reservoir diagnostics where needed most
- This approach can reduce project cost and environmental risk by
  - staying ahead of pressure buildup and minimizing related storage risks
  - minimizing the number or wells required for CO_2 storage
  - maximizing storage capacity and resource utilization
  - reducing cost of financial security to cover environmental risk
  - generating water
- Brine disposition options span from 100% reinjection to zero net injection

Final thoughts

- This approach effectively prepares a saline aquifer storage formation to perform like a depleted oil and gas reservoir
  - regulators, stakeholders, and local populations are more likely to be comfortable with such operations
- This approach can provide additional assurance to regulators, investors, and stakeholders that a well-informed reservoir-management plan is in place prior to CO_2 injection operations
- Because only one well is required, it is ideal for a pilot project

Thank you

TECHNOLOGY FRONTIERS

Reducing Energy's Footprint by Producing Water and Storing CO_2

By Thomas A. Buscheck
Group Leader,
Geochemical, Hydrological, and Environmental Sciences,
Physical and Life Sciences Directorate,
Lawrence Livermore National Laboratory

Jeffrey M. Bielicki
Assistant Professor,
Department of Civil, Environmental, and Geodetic Engineering
and John Glenn College of Public Affairs,
The Ohio State University

The global energy sector faces many challenges, perhaps the two most important of which are reducing greenhouse gas emissions from fossil fuels, which made up over 86% of primary energy consumption in 2014,[1] and addressing the growing challenge of water scarcity. One key aspect of the landmark U.S.-China Joint Announcement on Climate Change is the consideration of a collaborative effort to engage on both challenges simultaneously through research, development, and demonstration of a CO_2 capture and storage (CCS) project that would produce freshwater.[2]

The U.S.-China Clean Energy Research Center's (CERC) Advanced Coal Technology Consortium (ACTC) is an opportunity to leverage years of experience and research to investigate an emerging CO_2 capture, utilization, and storage (CCUS) technology called enhanced water recovery (EWR). To date, CCUS has been deployed by injecting CO_2 into petroleum reservoirs for enhanced oil recovery (CO_2-EOR). While CCUS/CO_2-EOR can make early contributions to reducing CO_2 emissions and raising revenue for first mover CCS demonstrations, the total scope is limited. For example, in the U.S., in the year 2013 alone, emissions were 6.7 billion tonnes of CO_2. For comparison, total storage capacity using "next generation" CO_2-EOR in the U.S. was estimated to be 45 billion tonnes of CO_2, with less than half (20 billion) considered to be economic at an oil price of $85/bbl, which would address the equivalent of three years of CO_2 emissions.[3,4]

"Producing brine has a number of operational benefits that enhance the efficacy of CO_2 storage, while simultaneously producing water that may help alleviate the stress in the water–energy nexus."

Unlike CO_2-EOR, which has limited deployment potential in many regions of the world, EWR can be deployed in saline aquifers that are well distributed and close to CO_2 sources (e.g.,

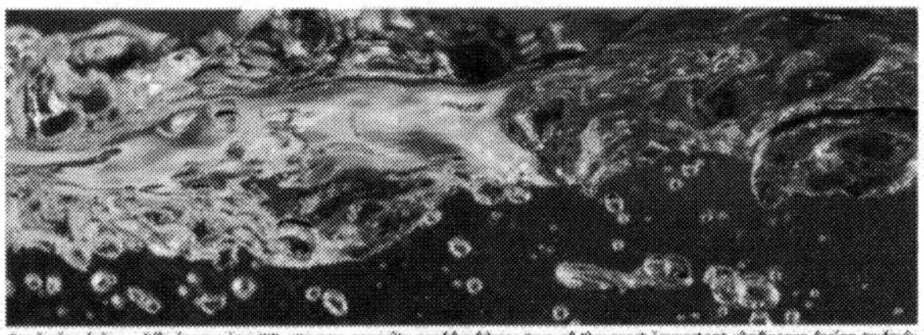

Producing brine while increasing CO_2 storage capacity could address two of the most important challenges facing today's energy industry.

Water for Energy
Reducing the Carbon Water Footprint

Carbon Management Technology Conference 2015
(CMTC2015)

Russell Lefevre
Director, Metropolitan Water District of Southern California
Adj. Prof. Physics-University of North Dakota
IEEE-USA Energy Policy Committee

r.Lefevre@ieee.org

Agenda

- Water for green energy
- Conventional Hydropower
- Small Hydro
- Geothermal
- Marine and Hydrokinetic Power
- Osmotic Power
- Tidal Lagoon Power

Water for Green Energy

- Water driven power plants have virtually no carbon footprint after deployment.
- Water driven power plants generate baseload renewable power with no requirement for nature driven processes e.g. wind and sun. (However drought has a significant effect.)
- Many water driven power plants are "semi-green". (To be explained)

Hydropower

- Hydropower (HP) is the United States largest source of renewable energy
  - 79GW of generating assets
  - 22GW of pumped storage assets
- HP provided 50% of renewable energy in the US in 2014
- HP has contributed 10.5% of US net power generation over 6.5 decades
- Potential for new hydropower-80GW (US DoE)
  - San Diego County Water Authority/San Diego currently planning 500MW plant at San Vicente Reservoir

Metropolitan Water District

- Regional water wholesaler
- 26 Member Agencies
- 6 counties
- Serving approximately 19 million residents
- 5,200 square mile service area
- $1 trillion economy

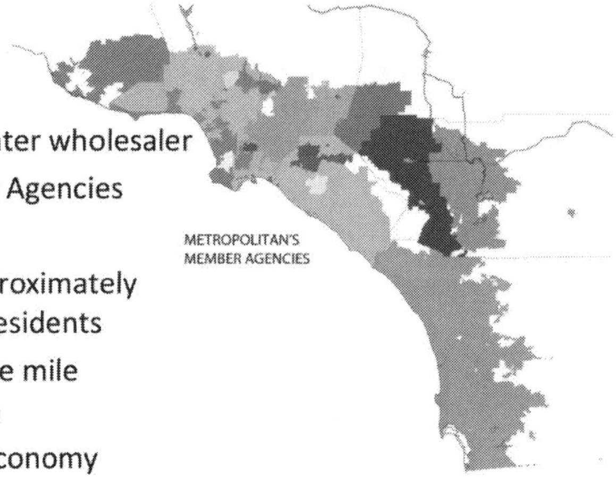

Semi-Green Hydropower MWD examples

- Hydropower can be generated from sources that were installed for other reasons.
  - California State Water Project (SWP) has the Edmonston Pumping Plant to bring SWP water from northern California over the Tehachapi mountains. It lifts water 1926 feet. When it arrives in the MWD Service Area it is used to deliver drinking water and also to generate hydropower. (See next slides)
  - MWD constructed Diamond Valley Lake reservoir to hold water for drought situations. When the water is delivered it is used to generate hydropower. The Wadsworth plant has a Nameplate Capacity of 30MW.
  - MWD pumps water from Lake Havasu to southern California and generates power when it reaches SC.

WP&S Committee Item 7a April 13, 2015

Metropolitan Hydroelectric Power Plants

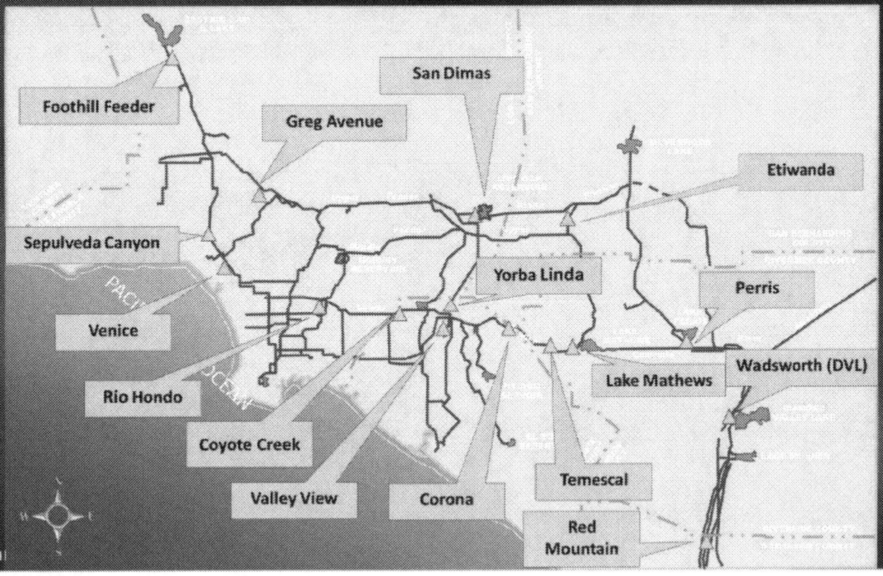

E&O Com

MWD Hydroelectric Power Capacity MW

- Greg Avenue 1
- Lake Mathews 5
- Foothill Feeder 9
- San Dimas 10
- Yorba Linda 5
- Sepulveda Canyon 9
- Venice 10
- Temescal 3
- Corona 3
- Perris 8
- Rio Hondo 2
- Coyote Creek 3
- Red Mountain 6
- Valley View 4
- Etiwanda 24
- Wadsworth 30

- Total 131MW

Small Hydro Systems

- Small Hydropower-100kW to 30MW
  - Microhydro <100kW
- Small Hydropower capacity worldwide-85GW in 2008
  - China 65GW
  - Japan 3.5GW
  - US 3GW
  - India 2GW
- Uses
  - Connect to conventional electrical distribution systems
  - Isolated areas
  - No national electrical distribution system
  - Capacity limit
  - Do It Yourself (<~5kW)

There is a substantial commercial enterprise addressing this specific market

Small Hydro/Pumped Hydro/Microgrid

- El Hierro, Canary Islands, Spain-100% of energy from renewable
- Installed capacity 35,000 kW
  - During abundant wind immediate needs are met
  - Excess energy used to pump water up to a large reservoir-700 Meters
  - When wind dies down hydropower turbines supply needed power
- Isle of Eigg, Scotland, United Kingdom-87% renewable
  - Wind/Solar/Hydro

Renewable Microgrids Rocky Mountain Institute and Carbon War Room

Hydropower Drought Effect-California

Hydropower is often considered baseload power, however:

- Hydropower average of total California electricity from 1983 to 2013 = 18%
- From 10/2011 to 9/2014 the average was less than 12%
- Approximately 34,000GW lower than long term average
- Increase in electricity costs of $1.4B

- MWD hydroelectric power generation 2011/12=393.5MW
- MWD hydroelectric power generation 2013/14=216.2MW

Geothermal Power

- Geothermal contributes <1% of total U.S. electricity generation
- 3000MW on line
- Demand is tempered by high up-front costs and early-stage risk
- However, California
  - Geothermal = 50% of renewable power production
  - 5% of total power production
  - Salton Sea leaders view Geothermal as a funding vehicle for restoration
- And, Nevada
  - >10% of in-state electricity

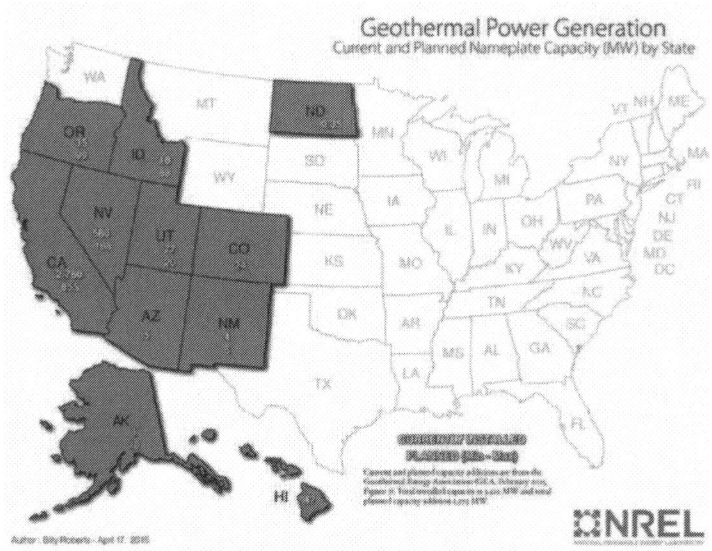

Marine and Hydrokinetic Power (MHK)

- Generate energy from:
  - Waves 1594-2640 TWh/yr
  - Tidal streams 445 TWh/yr
  - Ocean currents (more later) 200 TWh/yr
  - River currents 1381 TWh/yr
  - Ocean thermal gradients ?
- MHK technologies in an early stage of R&D
- <u>No</u> full-scale, commercial deployments in the United States
- DoE-$133M for 97 projects (2008-2014)
- Challenges
  - Commercial investments are high risk
  - Technology maturity too diverse
  - Cost competitiveness
  - Access to test facilities and demonstration sites difficult
  - Permit process not well understood

MHK Example

- CETO 6 Project located near Perth, Australia
  - Fully submerged buoys secured to pumps attached on the seafloor
  - The buoys react to the sea's wave energy and direct it to the pumps via and attached tether
  - The pumps are used to pressurize fluid and transmit it fluid to a hydroelectric power plant onshore
  - The pressurized fluid is used to drive a turbine to generate electricity
  - Cost $46M for 3 CETO 6 units-3MW

Tidal Lagoon Power

- Tidal Lagoon Power=Tides channeled into an artificially constructed lagoon
- First plant to be constructed at Swansea Bay in the UK
- An artificial harbor mouth is created within a semi-circular barrier built up from the seabed
- The inflow and outflow of water powers a set of turbines
- Operation: When tide is high water flows into the artificial lagoon generating power-when tide is low water flows out of the artificial lagoon running the turbines in the other direction
- Cost ~$1.5B
- Begin construction in 2017 if final approval is reached
- Plan is to build 6 more units providing 10% of the UK power need

Osmotic Power

- Osmotic power is the energy available from the difference in the salt concentration between sea water and river water. The two solvents are separated by a semipermeable membrane. Osmosis physics works to make the two solvents as equal in salt concentration as possible. If the solvents start with equal volumes osmosis will work to make the salty solvent grow in volume and pressure. As the pressure in the chamber is compensated it can enable a turbine to spin generating electricity.

- Estimates indicate as much as 1600 TWh/yr could be generated in this way.
- Netherlands Blue Energy pilot plant-50kW/hour
- Norway Statkraft worked on osmotic power for 10 years but stopped work in 2013
  - Norway Fjiords were viewed as excellent sources

- Problems
  - High membrane cost (Membranes are similar to those used in water desalination.)
  - Commercial Technology questionable
  - Brackish water residue is an environmental issue

What might be possible?

- Hydropower
  - DoE estimates >80 GW new Hydropower resource potential in the US (includes a lot of small hydro)
  - Issues: Growth has slowed: Lack of investments in infrastructure and constraints on water use
  - Int. Ctr. on Small Hydro estimates 173 GW potential worldwide
    - Over 50% in Asia and significant installations in Eurasia and Africa
    - China had 45,000 installations in 2010
- Geothermal
  - USGS estimates 9GW of identified and 30GW undiscovered in the US-Cost and Risk concerns.
- Marine and Hydrokinetic Power
  - DoE estimates U.S. continental technical resource potential of 500-750 TWh/yr (~90K homes can be powered by 1TWh each year) Only small projects currently exist in the US.
- Osmotic Power
  - Interest by Netherlands, Canada, Japan, U.S. NASA-Long way from commercial viability.
- Tidal Lagoon Power
  - 6 Installations could provide 10% of UK electricity demand (320MW)-Currently under Government review-start construction in Spring 2017 if approved. Cost ($1.5B) and maintenance concern.

Questions?

Russell Lefevre
r.Lefevre@ieee.org

Organization for Petra Nova CO_2 Capture Plant

Owner: Petra Nova

NRG Energy and JX Nippon Oil & Gas Exploration Corporation

Consortium

TIC
Construction
BOP E&P

MHI
CCS technology provider
CCS E&P

- Location: NRG WA Parish Power Plant in Thompsons, TX.
- CO_2 capture capacity: 4,776 TPD (240MW equivalent)
- CO_2 Use : CO_2 EOR
- Operation Start: 4th Quarter, 2016

[Reference] U.S. Department of Energy "W.A. Parish Post-Combustion CO2 Capture and Sequestration Project Final Environmental Impact Statement Volume I" (Feb, 2013), DOE/EIS-0473

Petra Nova Panel Discussion

nrg

Safe harbor statement

This presentation contains forward-looking statements within the meaning of Section 27A of the Securities Act of 1933 and Section 21E of the Securities Exchange Act of 1934. Forward-looking statements are subject to certain risks, uncertainties and assumptions and typically can be identified by the use of words such as "expect," "estimate," "should," "anticipate," "forecast," "plan," "guidance," "believe" and similar terms. Such forward-looking statements include our future growth and financial performance, Company operations, developments in renewables, and project development. Although NRG believes that its expectations are reasonable, it can give no assurance that these expectations will prove to have been correct, and actual results may vary materially. Factors that could cause actual results to differ materially from those contemplated above include, among others, general economic conditions, hazards customary in the power industry, weather conditions, competition in wholesale and retail power markets, the volatility of energy and fuel prices, failure of customers to perform under contracts, changes in the wholesale and retail power markets, changes in government regulation of markets and of environmental emissions, the condition of capital markets generally, our ability to access capital markets, unanticipated outages at our generation facilities, adverse results in current and future litigation, failure to identify or successfully implement acquisitions and repowerings, the inability to implement value enhancing improvements to plant operations and companywide processes, our ability to realize value through our commercial operations strategy, and our ability maintain successful partnering relationships.

NRG undertakes no obligation to update or revise any forward-looking statements, whether as a result of new information, future events or otherwise, except as required by law. The foregoing review of factors that could cause NRG's actual results to differ materially from those contemplated in the forward-looking statements included in this Investor Presentation should be considered in connection with information regarding risks and uncertainties that may affect NRG's future results included in NRG's filings with the Securities and Exchange Commission at www.sec.gov. Statements made in connection with the exchange offer are not subject to the safe harbor protections provided to forward-looking statements under Private Securities Litigation Reform Act.

nrg

Partners
and the Transaction Structure

Project Ownership Structure

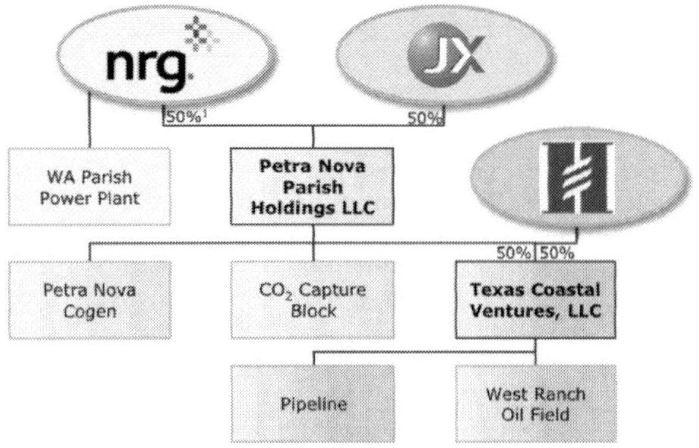

Partner Summary

Hilcorp Energy
- One of the largest privately-held oil and natural gas E&P companies in the US
- Strong track record of implementing new production techniques into mature reservoirs
- Specialized team that has extensive experience implementing CO_2 floods

JX Nippon Oil & Gas Exploration
- Currently conducting oil and natural gas business in 14 countries
- Parent company, JX Holdings, is a leading integrated energy, resources, and materials company

Japan Bank for Int'l Cooperation / NEXI
- Policy-based financial institution
- Wholly-owned by the Japanese government
- Over 25,000 loan and equity commitments valued at over $479 billion

US DoE
- Awarded $167 MM grant
- Funded through Clean Coal Power Initiative

Well-Structured Project with Strong and Experienced Partners

[1] Petra Nova will be deconsolidated from NRG's financial statements
[2] Includes investments already incurred during development of the project
[3] Includes costs associated with CCS system and contribution of Parish peaker

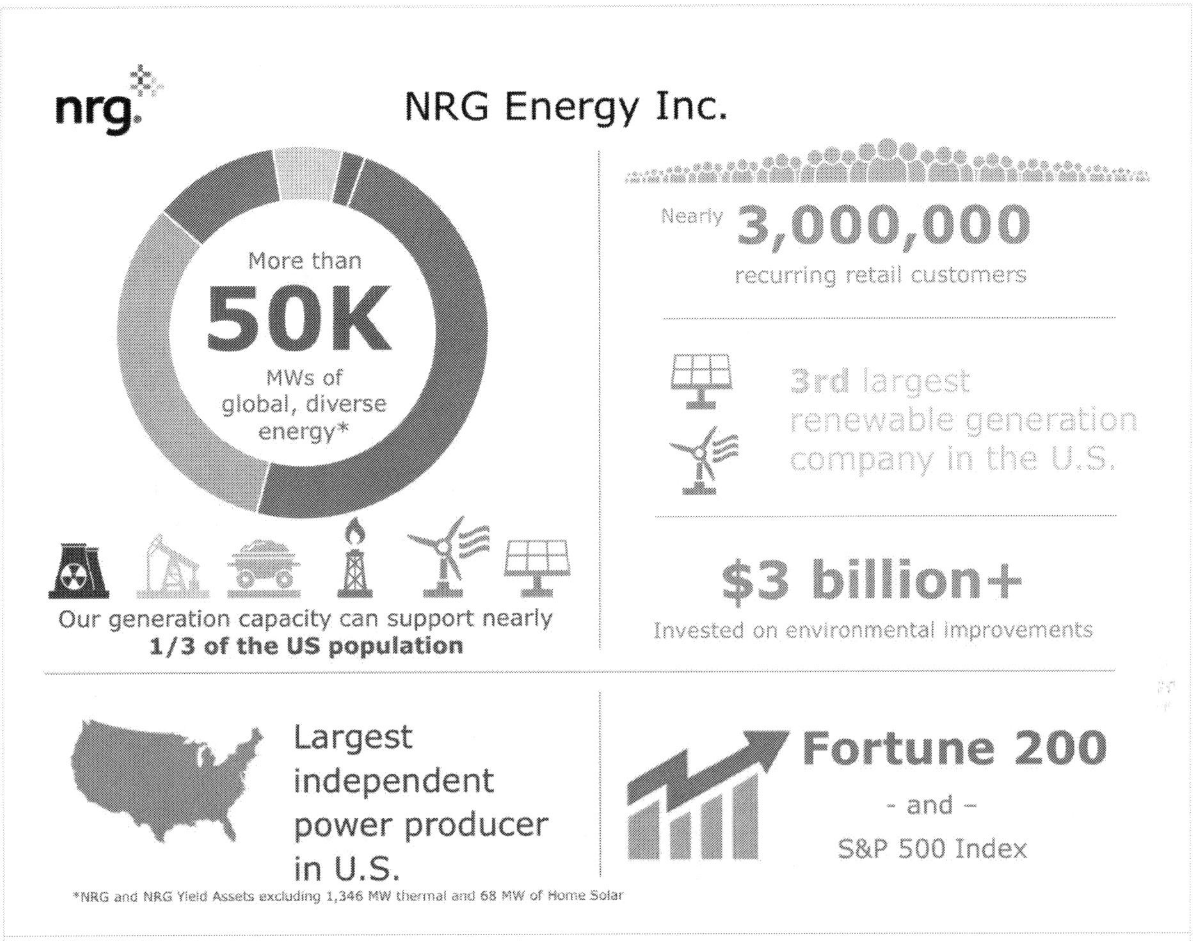
nrg.
NRG Energy Inc.
More than
50K
MWs of
global, diverse
energy*
Our generation capacity can support nearly
1/3 of the US population
Nearly 3,000,000
recurring retail customers
3rd largest
renewable generation
company in the U.S.
$3 billion+
Invested on environmental improvements
Largest
independent
power producer
in U.S.
Fortune 200
- and -
S&P 500 Index
*NRG and NRG Yield Assets excluding 1,346 MW thermal and 68 MW of Home Solar

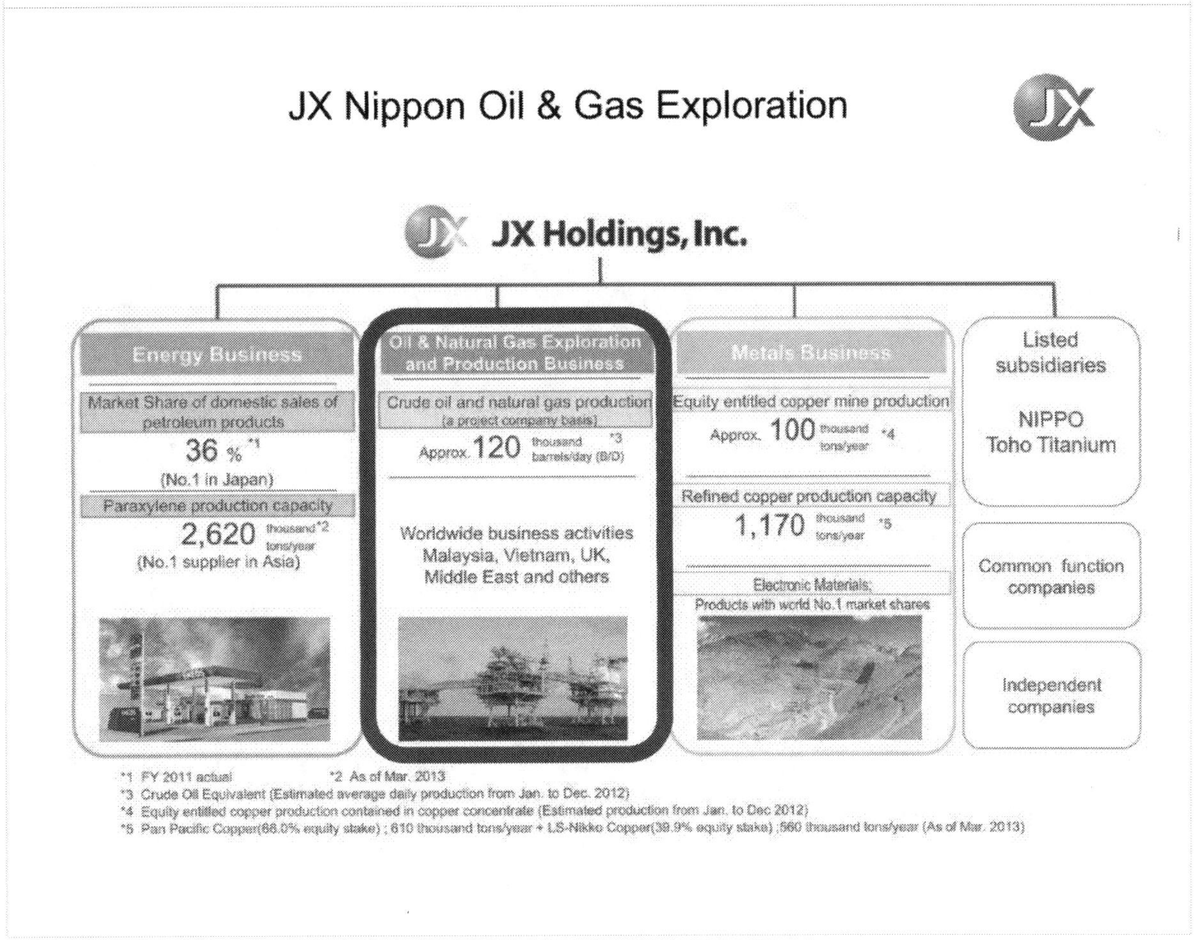
JX Nippon Oil & Gas Exploration
JX Holdings, Inc.
Energy Business
Market Share of domestic sales of petroleum products
36 % *1
(No.1 in Japan)
Paraxylene production capacity
2,620 thousand tons/year *2
(No.1 supplier in Asia)
Oil & Natural Gas Exploration and Production Business
Crude oil and natural gas production (a project company basis)
Approx. 120 thousand barrels/day (B/D) *3
Worldwide business activities
Malaysia, Vietnam, UK,
Middle East and others
Metals Business
Equity entitled copper mine production
Approx. 100 thousand tons/year *4
Refined copper production capacity
1,170 thousand tons/year *5
Electronic Materials;
Products with world No.1 market shares
Listed subsidiaries
NIPPO
Toho Titanium
Common function companies
Independent companies
*1 FY 2011 actual
*2 As of Mar. 2013
*3 Crude Oil Equivalent (Estimated average daily production from Jan. to Dec. 2012)
*4 Equity entitled copper production contained in copper concentrate (Estimated production from Jan. to Dec 2012)
*5 Pan Pacific Copper(66.0% equity stake) ; 610 thousand tons/year + LS-Nikko Copper(39.9% equity stake) ;560 thousand tons/year (As of Mar. 2013)

CCS Progress

CCS/BOP Progress

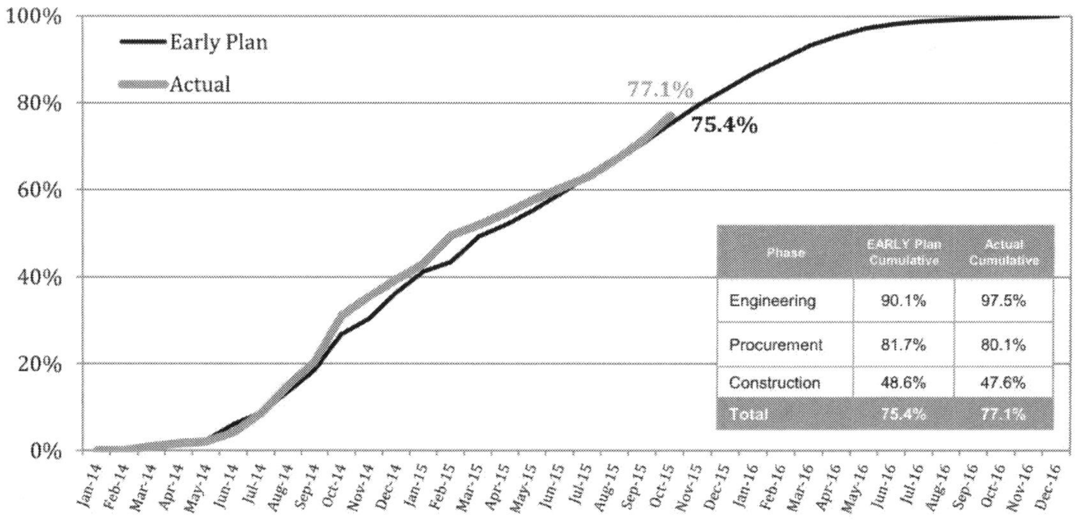

Phase	EARLY Plan Cumulative	Actual Cumulative
Engineering	90.1%	97.5%
Procurement	81.7%	80.1%
Construction	48.6%	47.6%
Total	75.4%	77.1%

On the Early Curve

5

CCS Progress

6

BOP Progress

7

TCV Pipeline

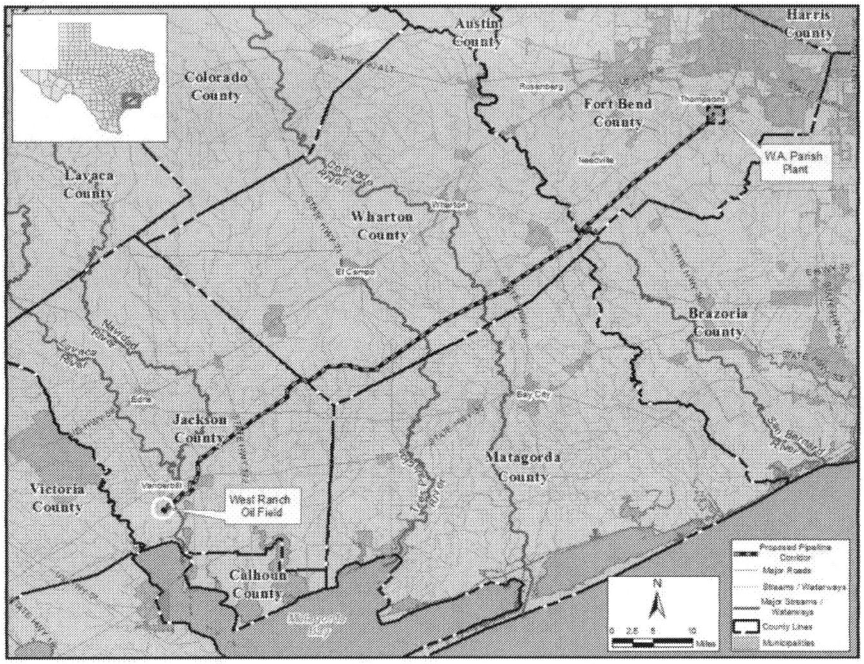

West Ranch

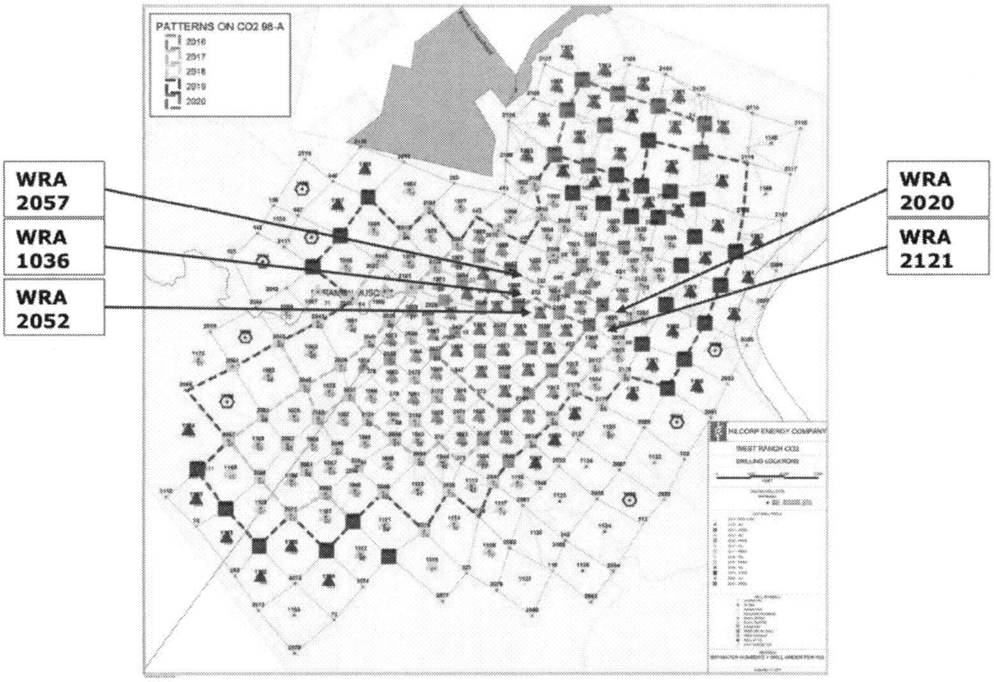

MHI's Scale-up Strategy

MITSUBISHI HEAVY INDUSTRIES, LTD.

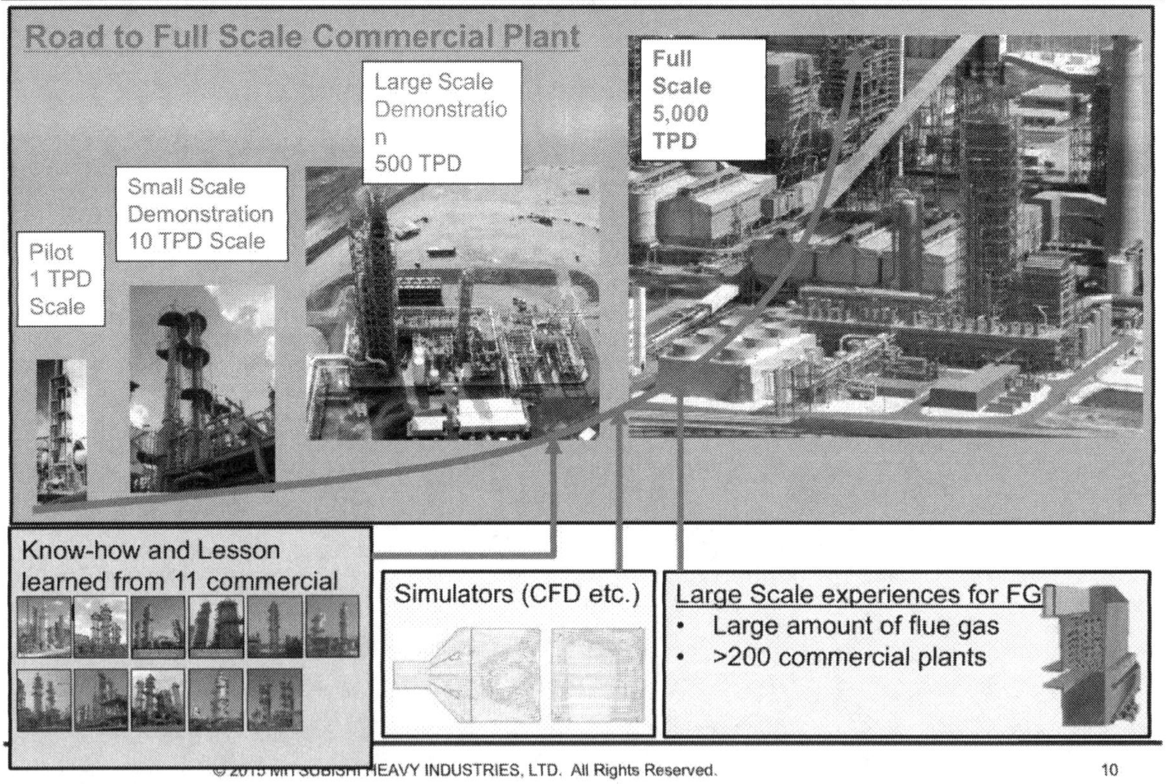

 10

Thank You

European Commission

European CCS Demonstration Projects

Michael Schuetz
European Commission, DG Energy
Michael.Schuetz@ec.europa.eu

European Commission

European CCS Demo Project Network

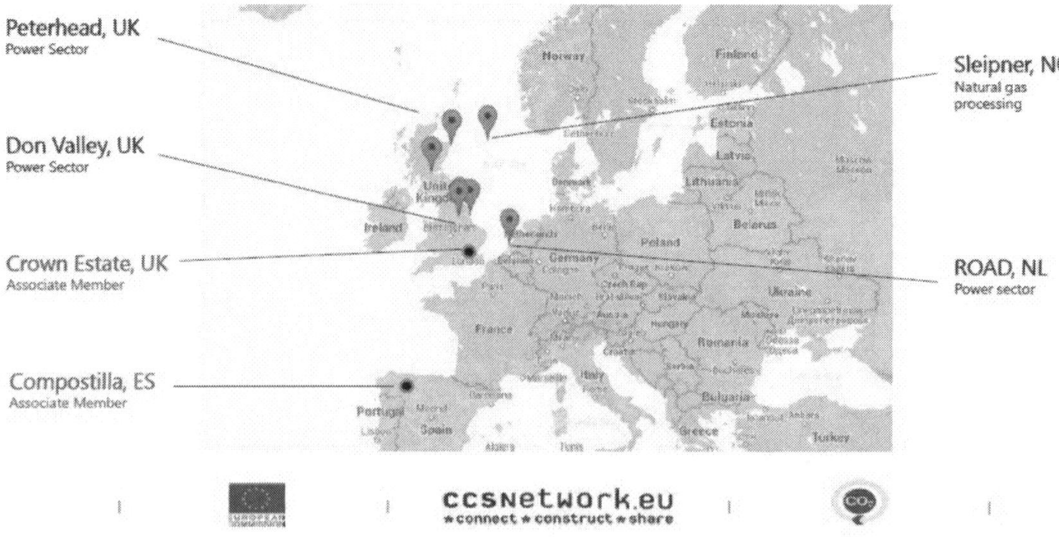

ccsnetwork.eu
*connect * construct * share

European Commission

Compostilla CO2 Capture Pilot

Fundación Ciudad de la Energía

Technologies Development Centre

- Total surface: **100,000 m^2**
- Circulating fludized bed (**CFB**) & pulverized coal boilers (**PC**) and **biomass gasifier** (BFB)
- **Air and oxycombustion** modes
- **Different fuels** (coals, pet coke, biomass and/or blends)
- **Uses** of CO_2

Further information:
Carlos de Dios, jc.dedios@ciuden.es

Disclaimer: the European Commission cannot guarantee accuracy of the above information which was provided by Ciuden

European Commission

Compostilla CO2 Transport Pilot

Fundación Ciudad de la Energía

Transport experimental facility

- **3,000 meters** piping length
- 2" pipe diameter
- Operating pressure 80 to 110 barg
- Operating temperature 10ºC to 31ºC

- **6 experimental areas**:
Despressurization, leakages, fractures, corrosion, instrumentation and doping

Further information:
Carlos de Dios, jc.dedios@ciuden.es

Disclaimer: the European Commission cannot guarantee accuracy of the above information which was provided by Ciuden

European Commission

Compostilla CO2 Storage Pilot

Fundación Ciudad de la Energía

Technical achievement

+1,600 m
with light drilling equipment

- Total surface: **140,000 m²**
- **On-shore deep saline aquifer storage**
- 2 wells
- 1,600 m depth
- CO_2 injection strategies: liquid, supercritical and gas

Further information:
Carlos de Dios, jc.dedios@ciuden.es

Disclaimer: the European Commission cannot guarantee accuracy of the above information which was provided by Ciuden

CCS

Carbon Management Technology Conference
17th November 2015

The ROAD Project
Creating a Rotterdam CO_2 hub for Europe

Andy Read, Capture Director
ROAD Project

ccsnetwork.eu

Global CCS Institute | Government of the Netherlands | Co-financed by the European Union, European Energy Programme for Recovery

road
CCS

Rotterdam Port and Industrial Area

- Largest port of Europe (top 10 world ports)
- Direct added value € 15.5 billion
- Indirect added value € 6.5 billion
- Direct and indirect added value 3.3% of GNP
- Direct employment 90 000 people
- Indirect employment 55 000 people
- Annual average business invesment € 1.5 billion

- ...and 17% (2013: 28 Mt) of total CO_2 emissions in the Netherlands (2013: 166 Mt)

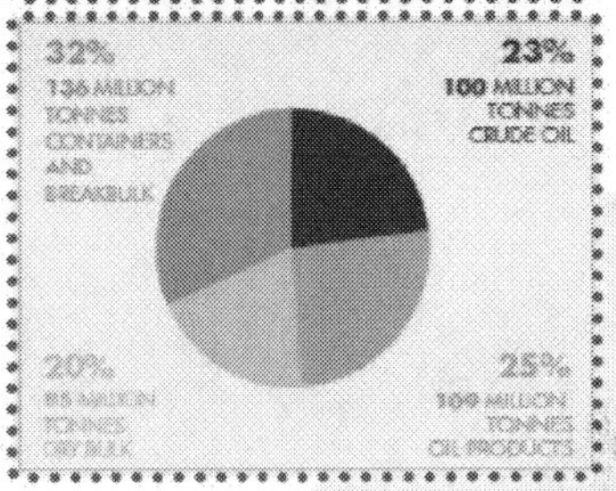

Rotterdam port and industrial complex is of great strategic value for (North-west) Europe
"Gateway to 350 million consumers"

road
CCS

Rotterdam CO_2 Hub: First Steps

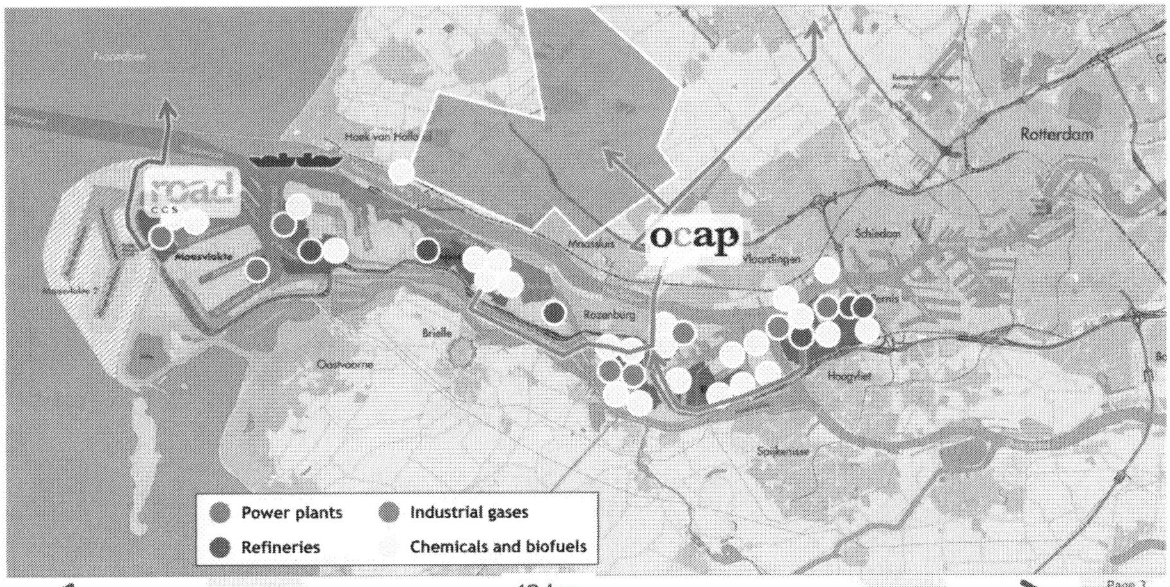

← 40 km →

road
CCS

Rotterdam CO_2 Hub: Vision 2030

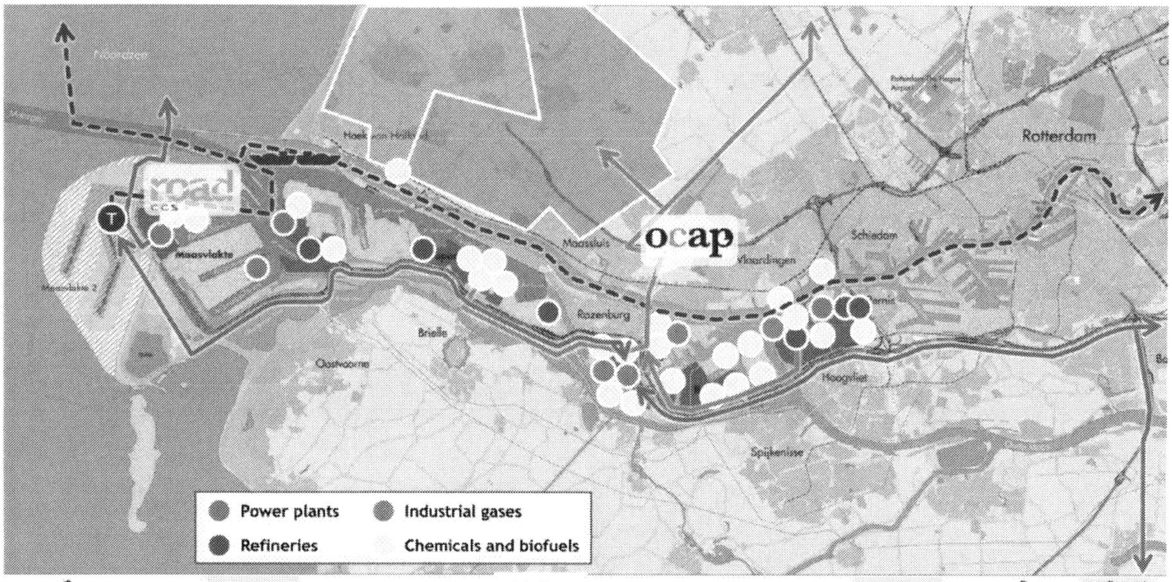

40 km

Page 4

road
CCS

The Rotterdam CO_2 Hub - Existing OCAP* System

OCAP System:

- Supplies 400 kt CO_2 to ≈600 greenhouses
- Shell Pernis CO_2 is from natural gas-based hydrogen production facility
- Abengoa is a first generation bio-ethanol plant
- OCAP claims CO_2 emission reduction of ≈200 kt/year through avoidance of natural gas use in the greenhouses
- There is additional CO_2 demand OCAP can't meet

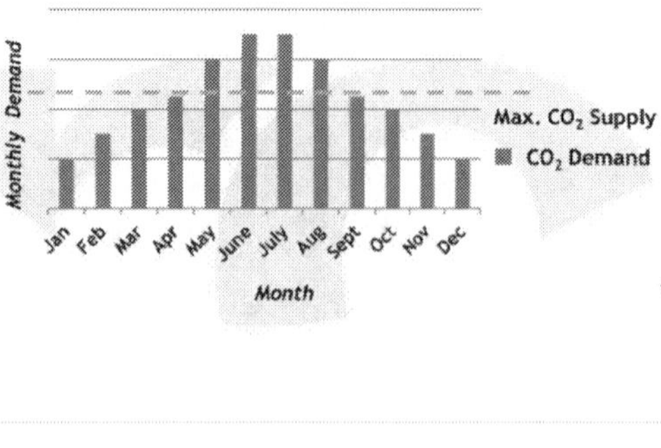

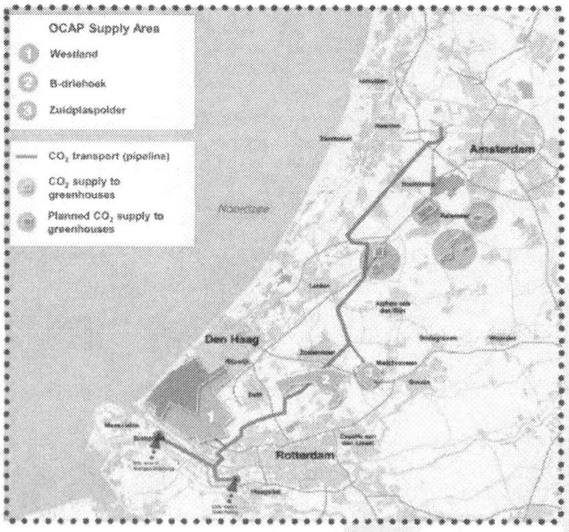

* *OCAP: Organic CO_2 for Assimilation by Plants*

Page 5

The Proposed Link between ROAD and OCAP

- OCAP cannot expand due to lack of CO_2 in summer peak (usually a few days only)
- ROAD could supply OCAP with summer peak CO_2
- Expansion is desirable to reduce CO_2 emissions by greenhouses
  - Estimate of greater than 200kt/yr further reduction achievable (with heat pipe - to be confirmed)
- In winter CO_2 from Shell and Abengoa (bio-CCS) could be sent for geological storage by ROAD

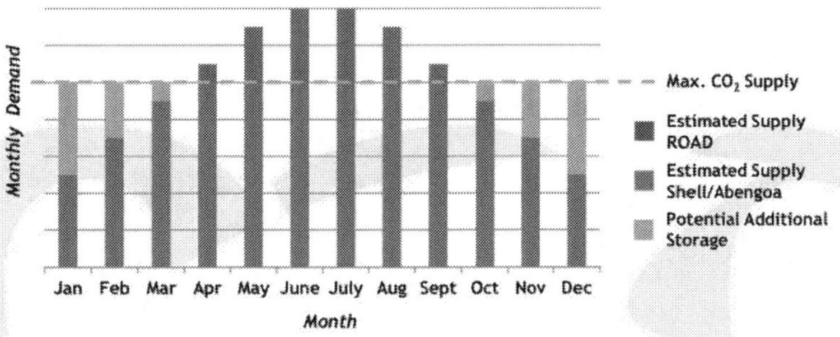

ROAD | Maasvlakte CCS Project C.V. is a joint venture of:

- E.ON Benelux
- ENGIE

In **co-operation** with intended storage partners:

- TAQA Energy

With **financial support** of:

- European Commission
- Government of the Netherlands
- Global CCS Institute

Co-financed by the European Union
European Energy Programme for Recovery

Government of the Netherlands

GLOBAL CCS INSTITUTE

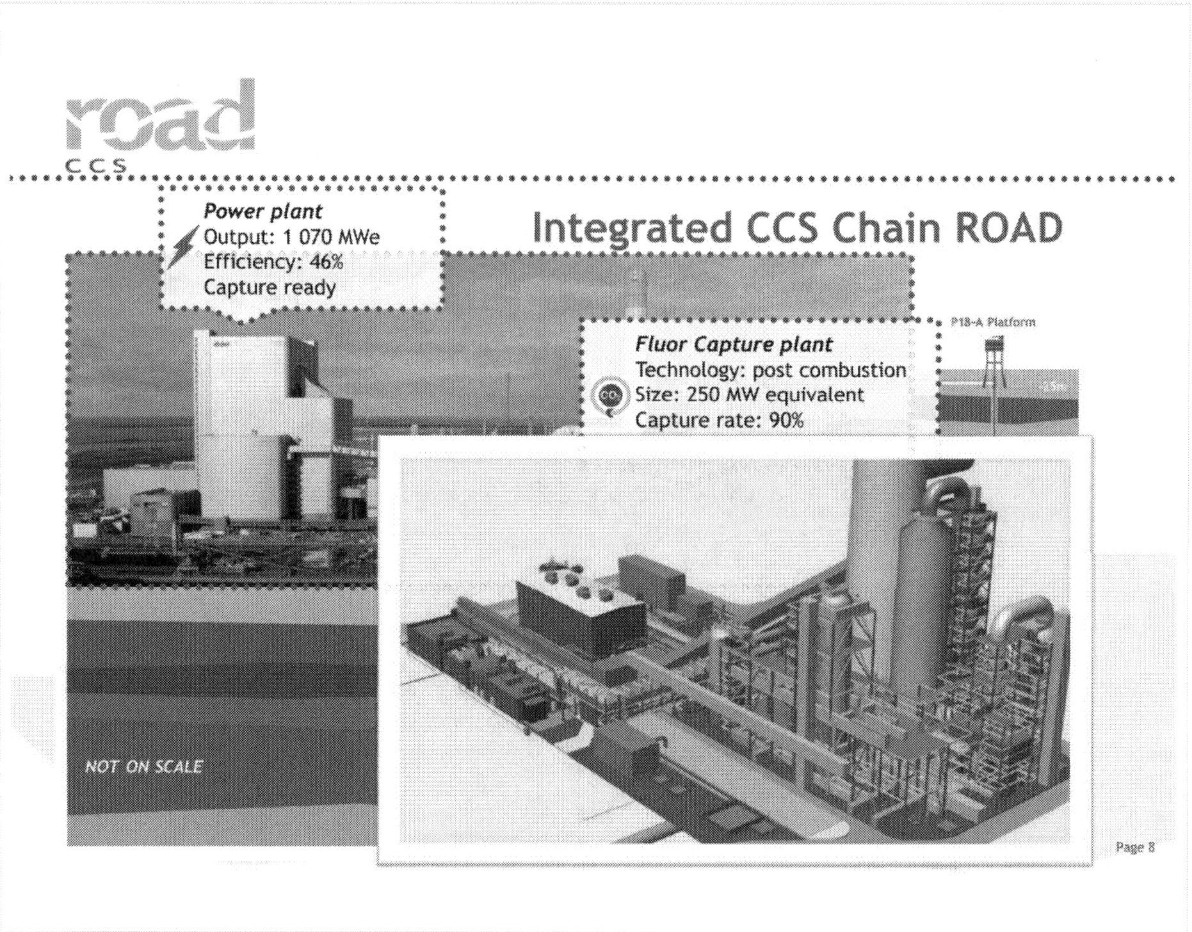
road
CCS
Integrated CCS Chain ROAD
Power plant
Output: 1 070 MWe
Efficiency: 46%
Capture ready
Fluor Capture plant
Technology: post combustion
Size: 250 MW equivalent
Capture rate: 90%
P18-A Platform
NOT ON SCALE
Page 8

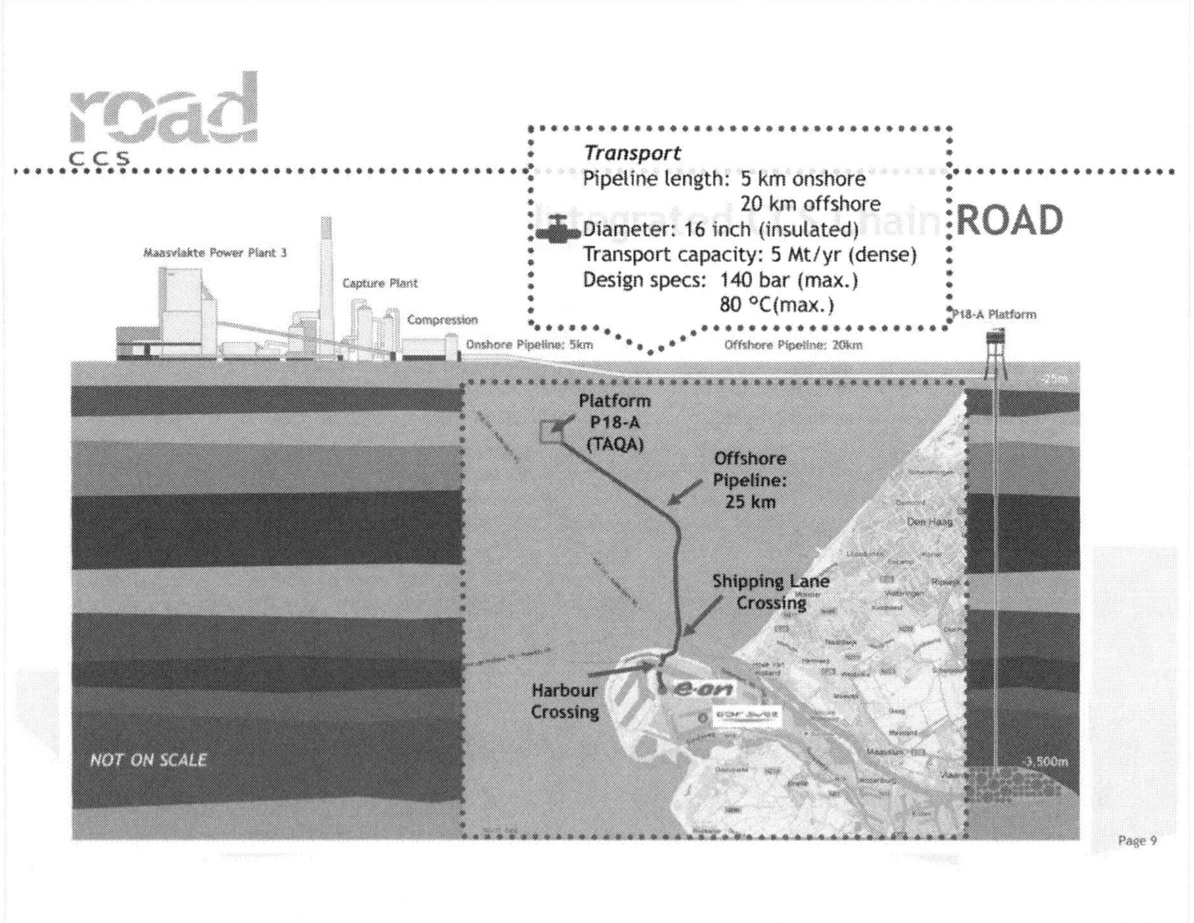
road
CCS
Integrated CCS Chain ROAD
Transport
Pipeline length: 5 km onshore
20 km offshore
Diameter: 16 inch (insulated)
Transport capacity: 5 Mt/yr (dense)
Design specs: 140 bar (max.)
80 °C(max.)
Maasvlakte Power Plant 3
Capture Plant
Compression
Onshore Pipeline: 5km
Offshore Pipeline: 20km
P18-A Platform
-25m
Platform
P18-A
(TAQA)
Offshore
Pipeline:
25 km
Shipping Lane
Crossing
Harbour
Crossing
Den Haag
e-on
NOT ON SCALE
-3,500m
Page 9

road CCS

Integrated CCS Chain ROAD

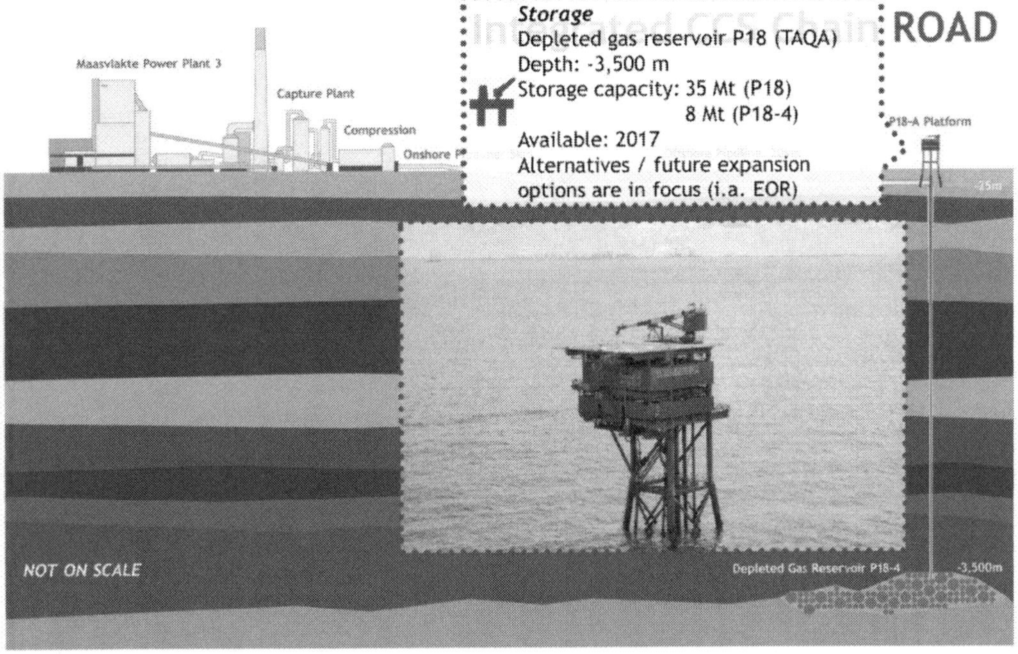

Page 10

road CCS

Status ROAD

- Engineering ✓
- Permits ✓
- Contracts ✓
- Finance —

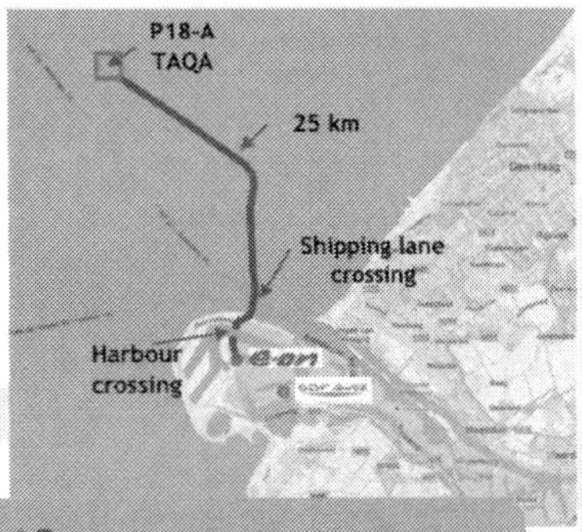

➡ Since summer 2012, ROAD focused on closing the funding gap

Page 11

Building a Business Case for ROAD

- Exploring alternative funding scenario's:
  - Reallocating EEPR funds
  - Private investors
  - NER300 (2nd round)
- Improving the project:
  - Capture: capture technologies
  - Transport: pipeline route
  - Storage: locations
  - Utilization: connection with OCAP - still being developed
- Phased approach:
  - Focus existing funding on Capex
  - Splitting CAPEX and OPEX funding
  - Horizon 2020/ERA-NET for operational funding

Page 12

Time Table ROAD

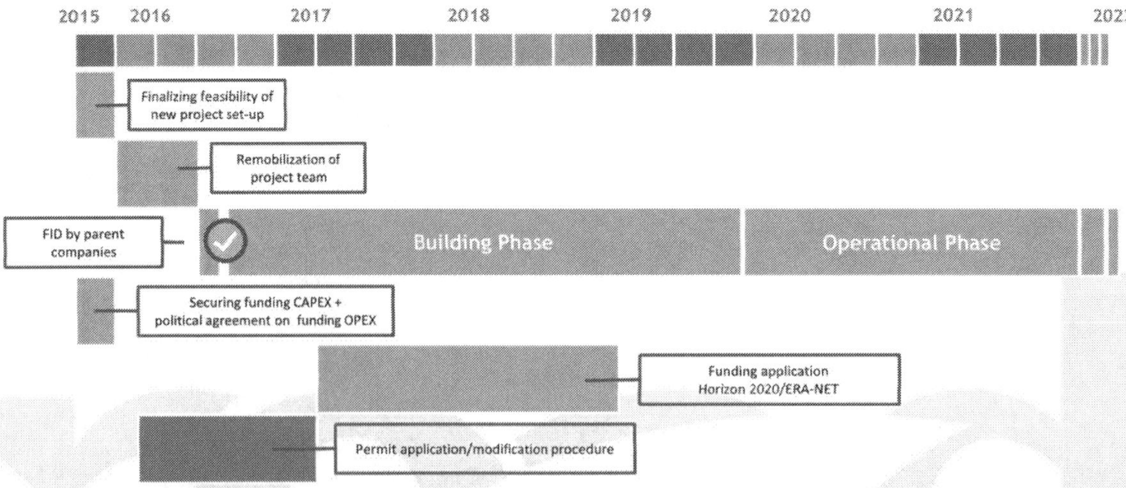

Page 13

road
CCS

ROAD | Maasvlakte CCS Project C.V.

Visit
Parallelweg 1
3112 NA Schiedam
The Netherlands

Post
P.O. Box 133
3100 AC Schiedam
The Netherlands

Contact
T: +31 (0)10 75 34 000
F: +31 (0)10 75 34 040
E: info@road2020.nl
W: www.road2020.nl

ccsnetwork.eu GLOBAL CCS INSTITUTE Government of the Netherlands Co-financed by the European Union European Energy Programme for Recovery

Page 14

road
CCS

Concentrations of CO_2 Emissions Europe

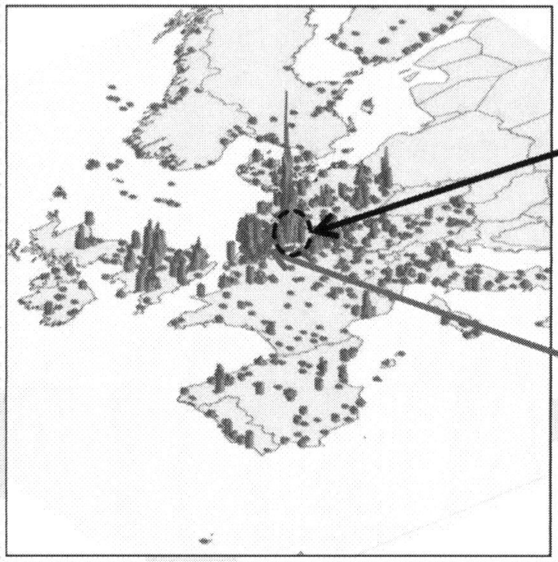

Ruhr area (GE)

Rotterdam area (NL)

Page 16

road
CCS

Rotterdam Vision: CO_2 Hub of Northwest Europe

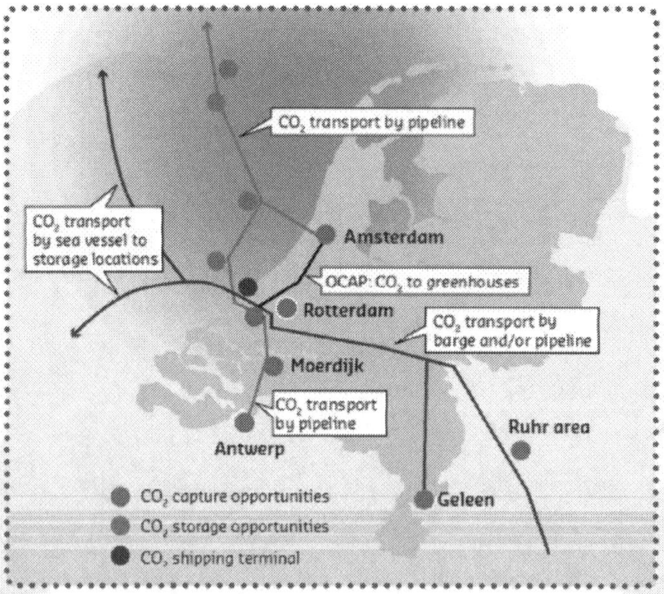

Page 17

road
CCS

Development of Total Cumulative Costs in Time

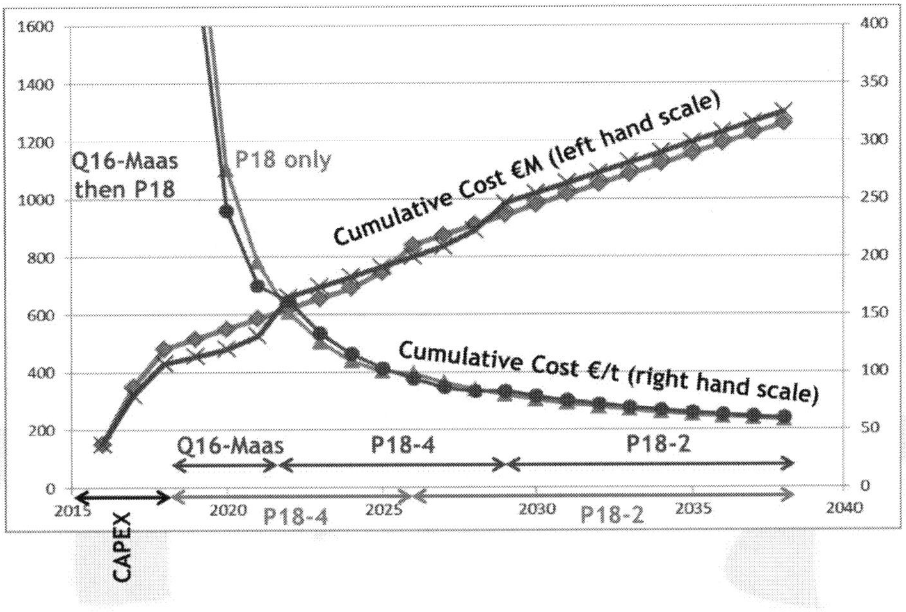

Page 18

road
CCS

The Rotterdam CO_2 Hub - Existing OCAP* System

OCAP System:

- Supplies 400 kt CO_2 to ≈600 greenhouses
- Shell Pernis CO_2 is from natural gas-based hydrogen production facility
- Abengoa is a first generation bio-ethanol plant
- CO_2 emission reduction of ≈200 kt/year through avoidance of natural gas use in the greenhouses
- There is additional CO_2 demand OCAP can't meet

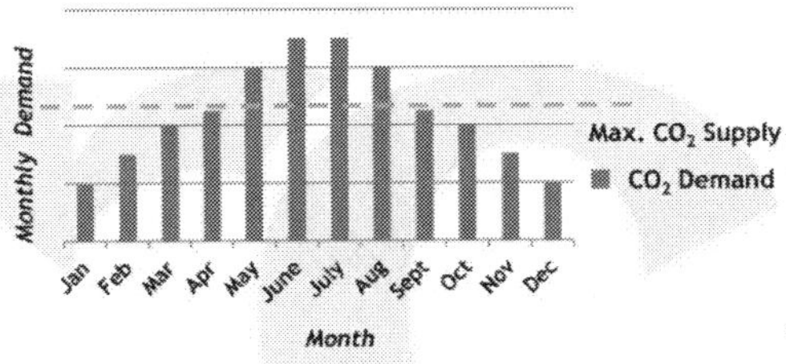

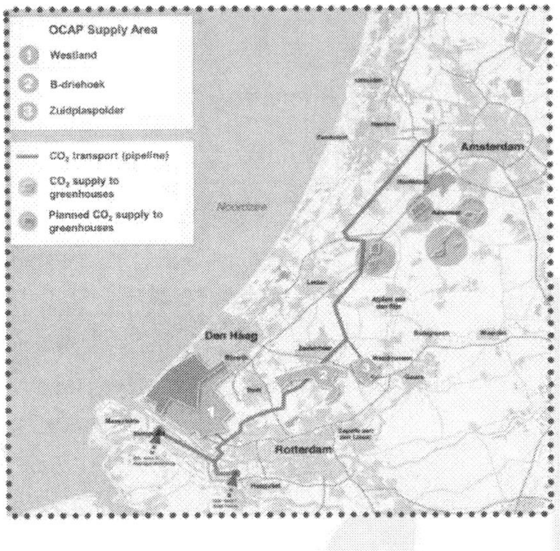

** OCAP: Organic CO_2 for Assimilation by Plants*

Page 19

road
CCS

The Proposed Link between ROAD and OCAP

- OCAP cannot expand due to lack of CO_2 in summer peak (usually a few days only)
- ROAD could supply OCAP with summer peak CO_2
- Expansion is desirable to reduce CO_2 emissions by greenhouses
  - Up to 200kt/yr further CO_2 reduction achievable by expanding the existing system
  - Much greater CO_2 reductions in long term if CO_2 supply is combined with heat supply
- In winter CO_2 from Shell and Abengoa (bio-CCS) could be sent for geological storage by ROAD

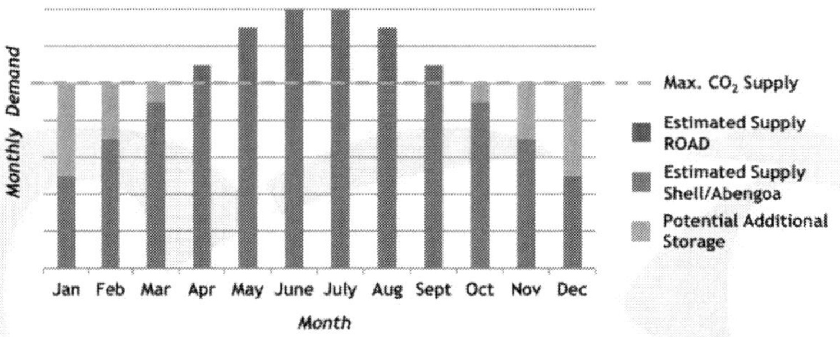

Page 20

SARGAS
Power

Delivering the Future of Clean Power Generation

Sargas Power's plans for Don Valley: the key activities and milestones

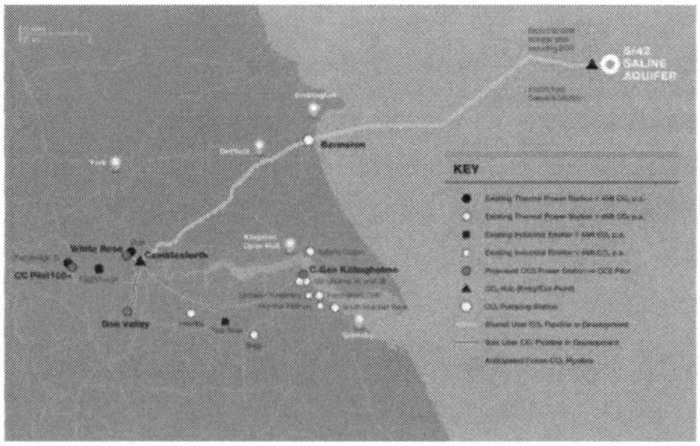

SARGAS Power

Don Valley Power Project, DVPP

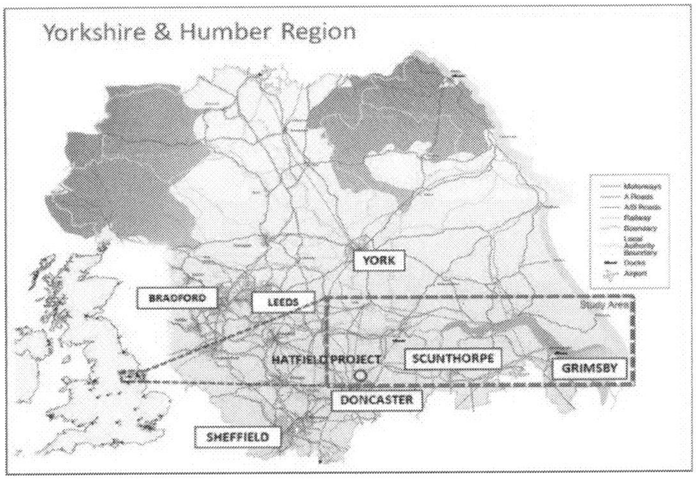

- Sargas Power Yorkshire Limited is an independent power generation technology developer and vendor, developing DVPP in the UK
- Development of the scheme commenced in 2006 and having secured the primary licences, permits and agreements, the project vehicle has now been acquired by Sargas Power Don Valley and will now be known as Sargas Power Yorkshire Limited ('SPYL').
- Sargas Power Yorkshire will complete pre-construction development ahead of securing the finance to commence construction of DVPP.
- The Scheme has and will be designed and developed to capture at least 90% of the CO2 from the fuel, and SPYL has signed a Co-operation Agreement with National Grid Carbon (NGC) for development of a shared user CO2 transportation solution and development of safe storage in the Southern North Sea.

SARGAS Power

The project location

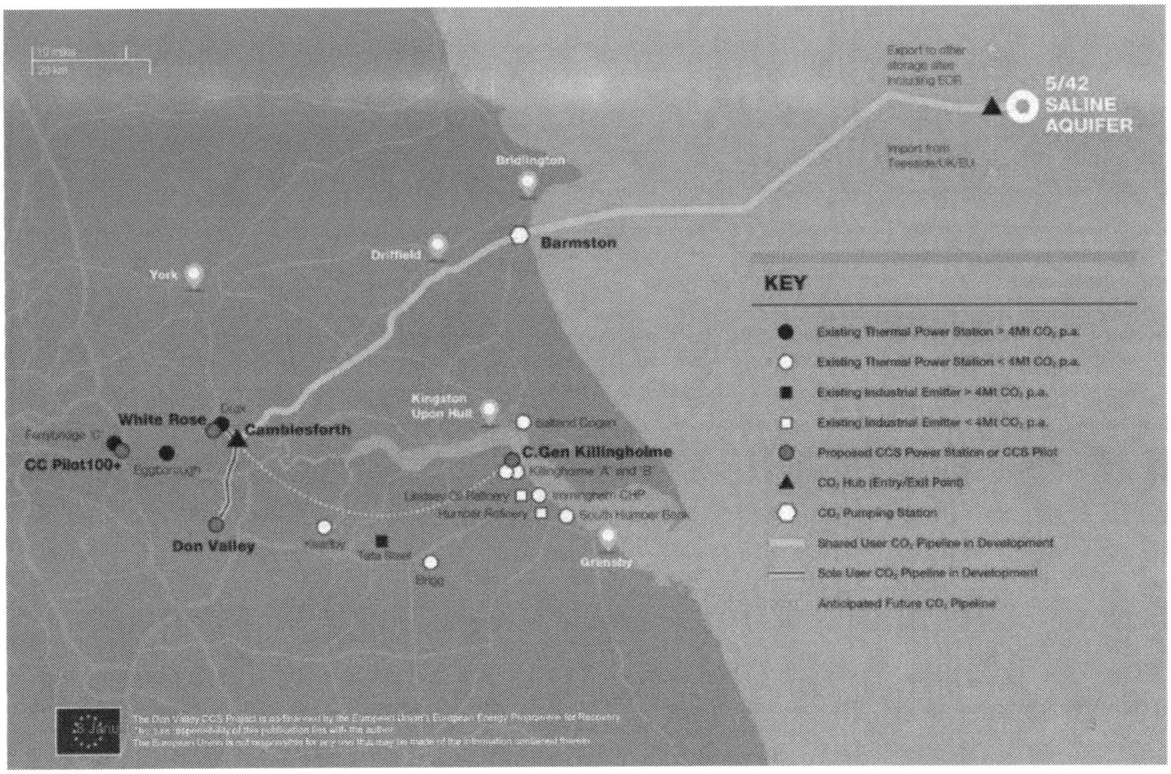

SARGAS Power

The full value chain as represented to the EU & DECC

Capture	Transport	Storage
SARGAS Power (Don Valley)	nationalgrid	nationalgrid

- The EEPR Grant was awarded in 2009 with all the grant to be committed by December 2015 now seeking an extension (via a 8th Amendment) to December 2017
- Cooperation Agreement between Don Valley and National Grid Carbon (co-beneficiaries) allocates €110M to Don Valley for CO_2 capture and €70m to National Grid Carbon Limited for safe and permanent CO_2 transportation and storage

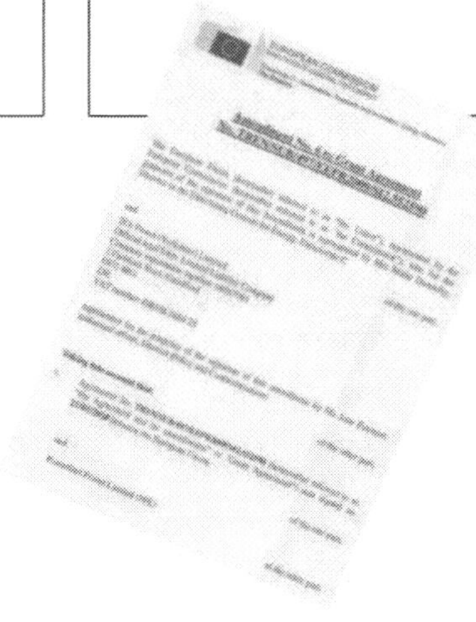

SARGAS Power

Sargas Technology Supports UK CCS Cost Trajectory

DECC & National Grid: White Rose infrastructure

Sargas Power: GE; DSME; Nooter/Eriksen; Catacarb

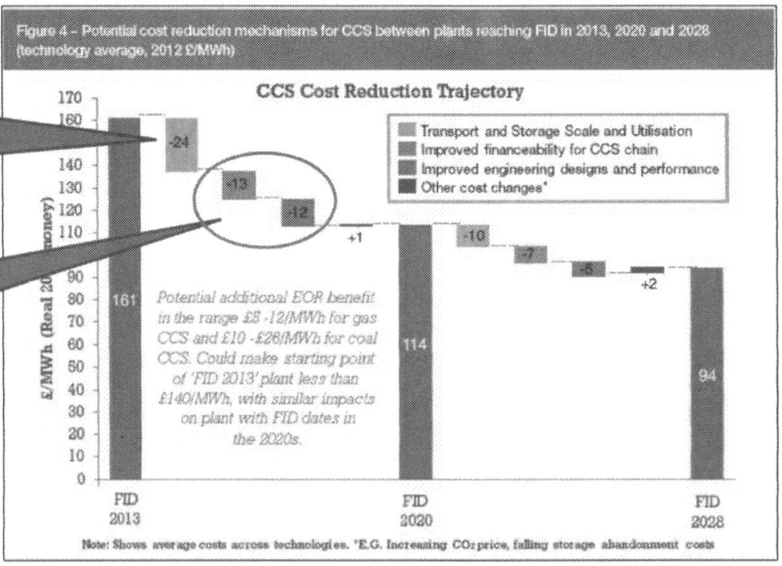

Source: UK CCS Cost Reduction Task Force

Sargas Power will work in partnership with DECC to deliver lower cost CCS over next 10 years

SARGAS Power

Consents & Licences already awarded or in-progress

Consent/Licence	Authority	Status / Award date	Complete
Section 36	**DECC**	**Awarded (February 2009)**	☑
Section 14	**DECC**	**Awarded (February 2009)**	☑
Water Abstraction	**Environment Agency**	**Awarded (February 2009)**	☑
Connection Agreement	**NGET**	**Awarded (April 2007)**	☑
Generation License	**OFGEM**	**Awarded (August 2009)**	☑
Section 37	**DECC**	**Awarded (May 2014)**	☑
Discharge consents	Environment Agency	On hold	☐
Local planning (gas)	Local Planning Authorities	On hold	☐
Local planning (water)	Local Planning Authorities	On hold	☐
COMAH (pre-construction)	Health & Safety Executive	On hold	☐
Environmental Permit	Environment Agency	On hold	☐

8 January 2016 Strictly Confidential 6

SARGAS Power

Multiple Sources of Value supported by major partners

Stargate 250 'tolls' natural gas into electricity and pressurized CO_2

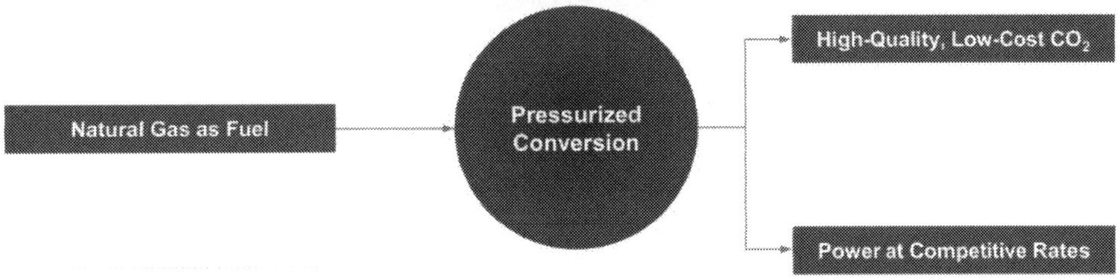

GE DSME SNC·LAVALIN

The Stargate 250 provides proven CO_2 capture & power technology for power and EOR companies

8 January 2016 Strictly Confidential

SARGAS Power

Integrated schedule: key activities & milestones to FID

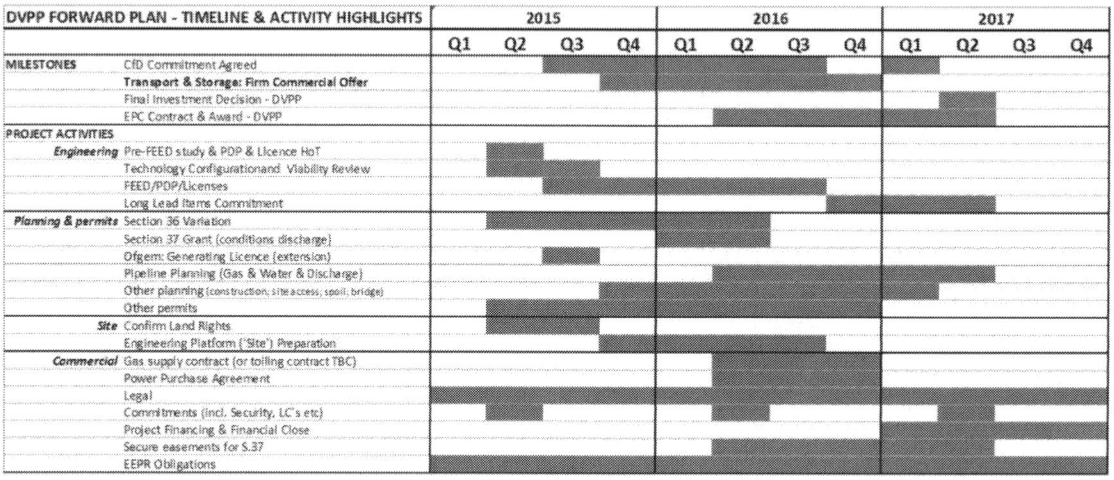

- By end 2015, Sargas Power will undertake a pre-FEED study to provide clarification and preliminary detail on the project design and economic business case
- This should enable DECC (amongst others) to understand the provisional CfD requirements and to endorse the Don Valley project and commit to negotiate a CfD with Sargas Power
- In parallel, DECC's commitment to a negotiation process for a CfD will underpin the EEPR investment
  - Current estimate is for Financial Close in late 2017
- Pre-FID development costs are approximately £70-75M - primarily financed by EU grant funding

SARGAS Power

Thank You!

NATIONAL COALCOUNCIL

Carbon Management Technology Conference
AIChE ASME AIME IEEE ASCE
November 18, 2015 – Houston, Texas

National Coal Council

Celebrating 30 years ~ 1984|2014

The National Coal Council provides advice and recommendations to the Secretary of Energy on general policy matters relating to coal and the coal industry.

NCC is a Federal Advisory Committee organized under FACA legislation

NCC »»FOSSIL FORWARD

>> STUDY REQUEST

Secretary Moniz's Charge to NCC – May 2014

... request the NCC conduct a study that assesses the value of DOE's Carbon Sequestration Program ... The assessment should address the question: "What is industry's assessment of the progress made by the DOE and others regarding cost, safety, and technical operation of CCS/CCUS?

... In other words, how does industry see and accept major technical findings from the CCS/CCUS community, and how do those relate to DOE programs and investments?

... an assessment based on technical soundness and results to date would provide a welcome perspective from leading companies with experience in CCS/CCUS technology."

NCC >>>> FOSSIL FORWARD

>> Study Leadership & Lead Authors

- NCC Chair – **Jeff Wallace**, VP Fuel Services, Southern Company Svcs
- Coal Policy Committee Chair – **Fred Palmer**, Senior VP, Peabody Energy
- NCC CPC Vice Chair– **Bill Brownell**, Chairman, Hunton & Williams

- Study Chair – **Amy Ericson**, President, Alstom Inc.
- Technical Chair – **Carl Bozzuto**, Alstom

- Lead Authors –
  - **Holly Krutka**, Shenhua Group
  - **Pam Tomski**, Global CCS Institute
  - **Shannon Angielski**, CURC
  - **Carl Bozzuto,** Alstom
  - **Jeff Phillips,** EPRI

NCC >>>> FOSSIL FORWARD

>> Fossil Forward Principal Theme

"While DOE is indisputably a world leader in the development of CCS technology, the DOE CCS/CCUS program has not yet achieved critical mass."

- "Without adequate demonstration there can be no commercialization."
- "There is no point in capturing CO2 if there is no place to use it or store it."

NCC >>>> FOSSIL FORWARD

>> Magnitude of the Problem

- Current # of demonstration projects in operation or under construction globally = 22

  Projected need by 2050 = 3,400
- The current global CO2 storage rate = 40 million tons/year

  Projected need = 10 billion tons/year
- Cumulative total CO2 emissions 2050 ~ 2,000 billion tons

  Projected "safe" level of emissions = 884 billion tons

NCC >>>> FOSSIL FORWARD

2,200+ Coal Units in Construction and Planned Globally

NATIONAL COAL COUNCIL

Major Growth of Fossil Fuel Plants in Global Markets

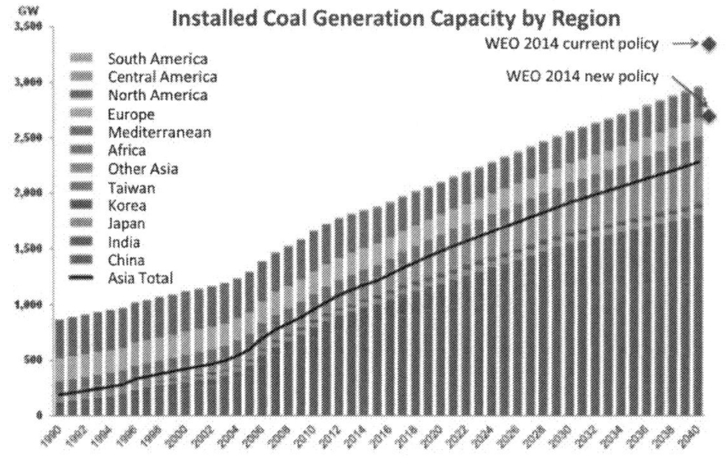

Source: World Coal Association.

7

>> The CCS/CCUS Imperative

CCS is the only large-scale technology that can mitigate CO_2 emissions from fossil fuel use for electricity generation and key industrial sectors.

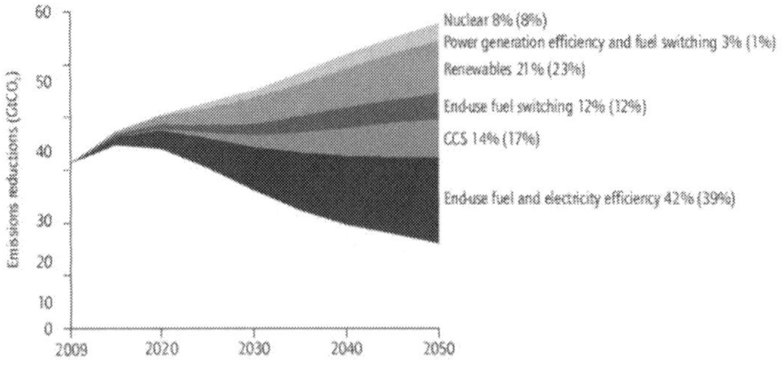

Note: numbers in brackets are shares in 2050. For example, 14% is the share of CCS in cumulative emission reductions through 2050, and 17% is the share of CCS in emission reductions in 2050, compared with the 6DS.

IEA Technology Road Map

NCC >>>> FOSSIL FORWARD

The World Must Have CCS to Achieve CO_2 Emissions Goals

NATIONAL COAL COUNCIL

CCS Needed to Achieve CO_2 Goals
Costs Increase 70% to 138%* Without CCS

- CCS provides most impactful opportunity to capture, use and store significant volumes of CO_2 from fossil fuels
- CCS reduces emissions from power and key industrial sectors: cement production, iron, steel, oil refining, chemicals manufacturing
- CCS maintains baseload power that enables reliable grid
- Commercializing CCS requires a level playing field

*Source: NCC Fossil Forward study, 2015.

9

>> Key Recommendation

In order to achieve CCS at commercial scale, policy parity with other low/no carbon technologies is required…

<u>The National Coal Council recommends that:</u>

- **DOE take a stronger position on the need for policy parity with respect to funding allocations**
- **DOE take a stronger position on the need for policy parity with respect to incentive mechanisms and subsidies applied to near zero emission energy technology**

NCC

>>>> FOSSIL FORWARD

NATIONAL COUNCIL

LEVELING THE PLAYING FIELD

Policy Parity for Carbon Capture and Storage Technologies

National Coal Council Report

Nov. 12, 2015

Environmental Policy Has Tilted The Energy Playing Field

NATIONAL COAL COUNCIL

- U.S. Dept. of Energy has stewarded successful R&D for early development of CCS
- Without sufficient government support, commercial demonstration has lagged
- Incentives too small to bridge divide between costs/risks of projects with CCS
- Commercial scale demonstration needed for broad CCS deployment
- Report follows policy parity path in NCC 2015 Fossil Forward study

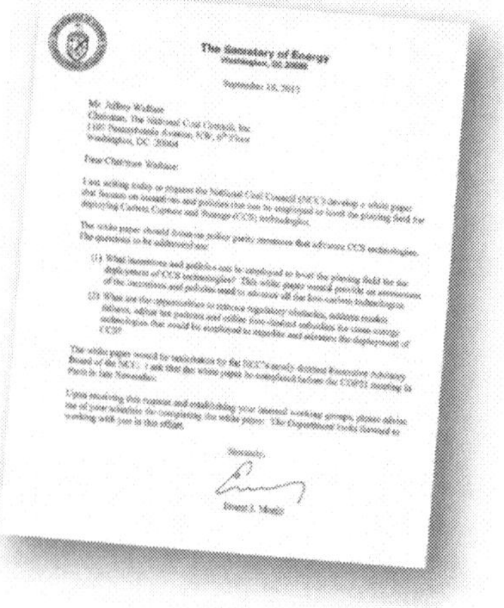

12

Eliminate Dis-Parity Between CCS and Renewables

NATIONAL COAL COUNCIL

- Renewables have received 100X more investment than CCS globally
- Current policies fail to adequately compensate for immaturity of CCS, which drives investment decisions due to increased cost and risk

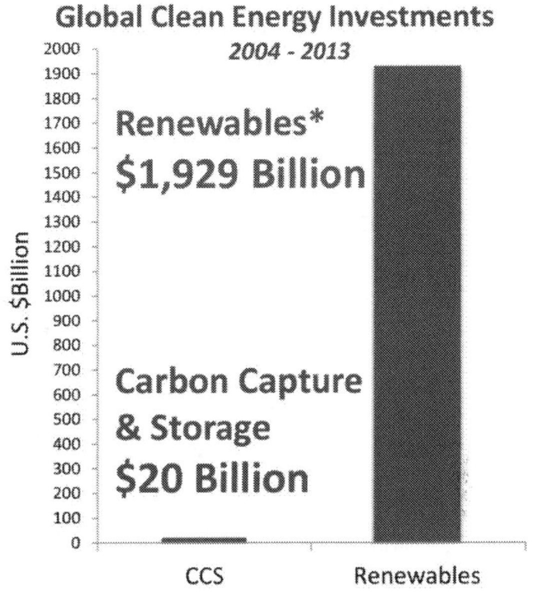

Source: International Energy Agency.

13

U.S. Renewables Received 12X Federal Incentives Versus Coal in 2013

NATIONAL COAL COUNCIL

Renewables Receive 72% of Total Incentives; Coal Just 6%

- Fossil fuels produced 78.5% of U.S. energy; renewables 11.4%, nuclear 10.1% in 2013
- Renewable subsidies increased from ~15% in 2007 to 72% in 2013
- Funding for coal declined from 12.7% in 2007 to 6% in 2013

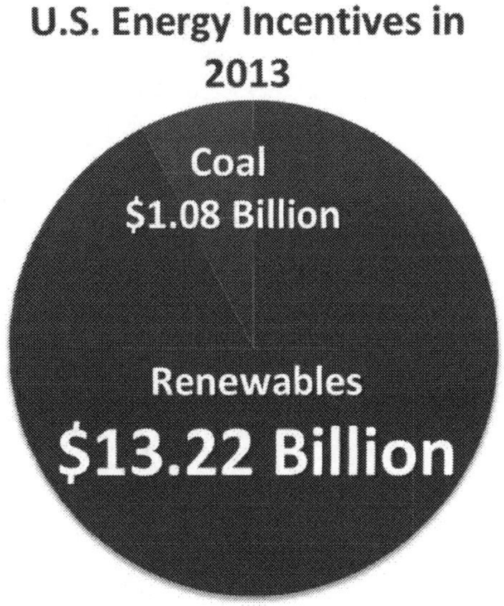

Source: Energy Information Administration, March 2015.

14

Recovery Act Funding: Coal and Renewables

NATIONAL COAL COUNCIL

ARRA Funding for Renewables $20 Billion, Coal $3.4 Billion in 2009

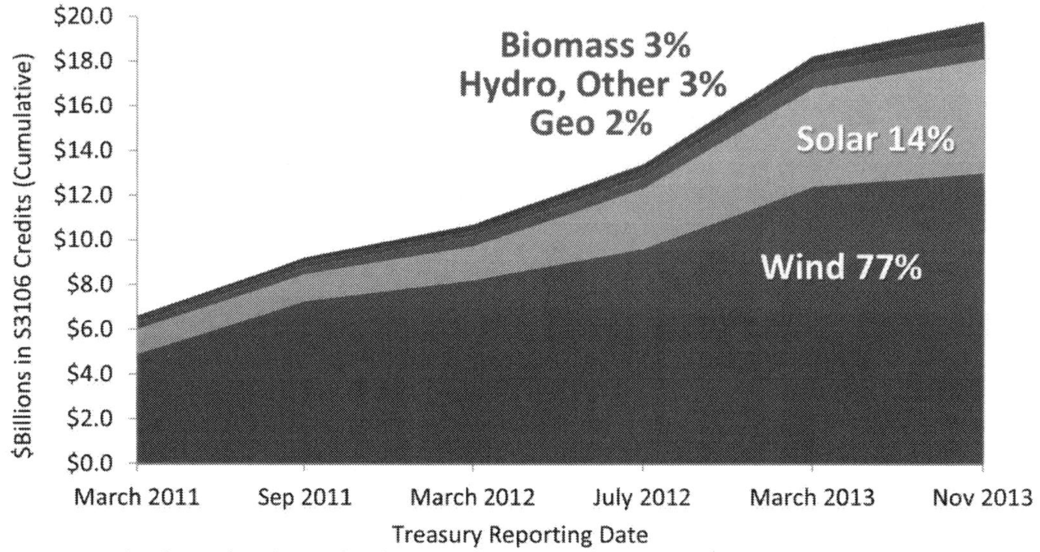

Source: National Coal Council Fossil Forward study, Treasury Grants American Recovery and Reinvestment Act Section 1603, 2009 - 20135AARAi.

15

>> Policy Dis-Parity

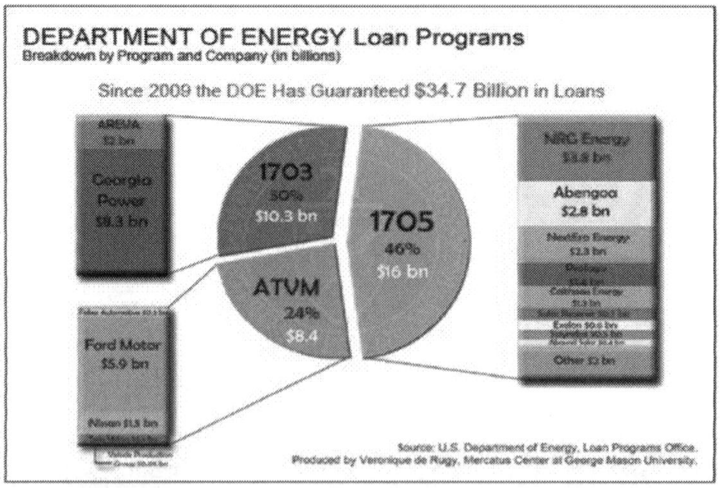

NCC

>>>> FOSSIL FORWARD

>> Potential Benefit from Policy Parity

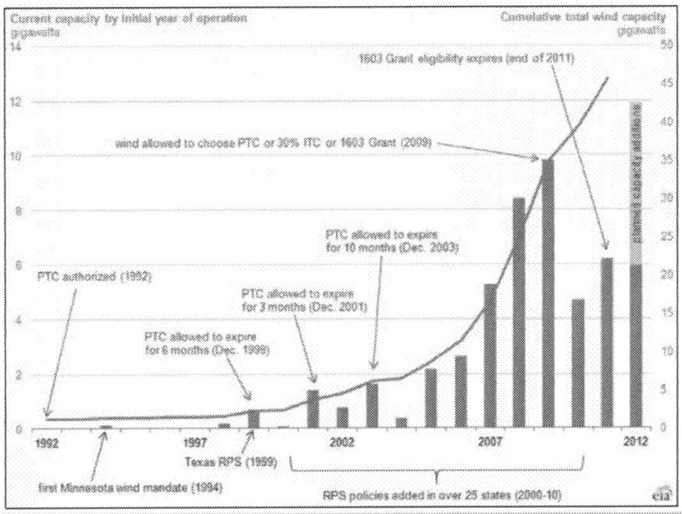

NCC

>>>> FOSSIL FORWARD

Recommendations for Leveling the Playing Field

NATIONAL COAL COUNCIL

Financial Incentives

- Broaden to include incentives available to other clean energy sources

Regulatory Improvements

- Regulatory blueprint needed to remove barriers to construction and development of CCS projects

Research Development and Demonstration

- U.S. Dept. of Energy must be a catalyst for commercial scale demonstrations to start immediately

Communications and Collaboration

- U.S. Dept. of Energy must assure policymakers that fossil fuels will continue to be used & CCS is essential

18

>> Key Recommendation

Funding for CCS RD&D is limited and must be enhanced and focused…

- **DOE continue fostering a portfolio of technologies for implementing CCS and "prime the pump" with early stage funding for promising concepts. NCC recommends that after technologies reach TRL 4, DOE cull its support to only those technologies which show a clear promise of meeting or exceeding DOE's CCS performance goals**
- **DOE continue to develop a plan for demonstrating second generation and transformational CCS technologies showing cost and performance advantage at a scale of 25-50 MW by 2020 and make subsequent budget request to carry out the plan**

NCC >>>> FOSSIL FORWARD

>> Key Recommendation

Technology and Funding Incentives must be significantly better coordinated to be effective…

The National Coal Council recommends that:

- **DOE develop a plan to have a total of 5-10 GW of CCS/CCUS demonstration projects in operation in the U.S. by 2025**
- **DOE expand the RCSP program to identify and certify at least one reservoir in each region that is capable of storing a minimum of 100 million tons of CO2 at a cost of less than $10/ton by 2025**
- **All federal incentives for CCS demonstration projects undergo a coordinated review for their combined adequacy and effectiveness in supporting CCS deployment in time to achieve the installation of storing a minimum of 100 million tons of CO2 at a cost of less than $10/ton by 2025**
- **Concerted effort be undertaken by DOE to identify and pursue creative mechanisms to finance CCS/CCUS projects**

NCC >>>> FOSSIL FORWARD

>> Current DOE CCS Program

**Funding for DOE programs is inconsistent with DOE goals.
DOE programs have been inadequately funded at levels that are insufficient to achieve the aggressive goals of the program.**

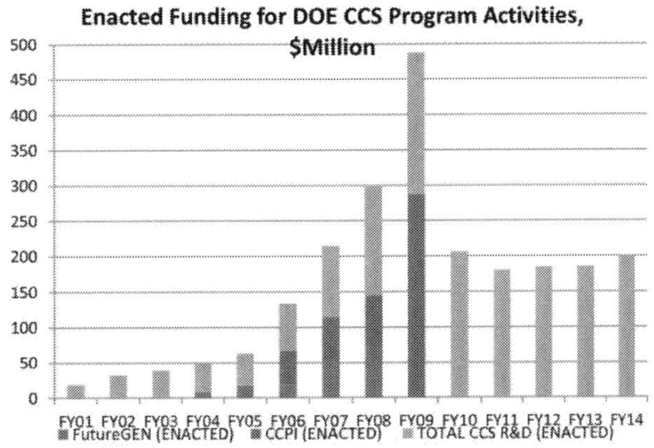

NCC >>>> FOSSIL FORWARD

>> DOE CCPI Program

**CCPI does not appear to have a high success rate.
Only a small number of projects have been selected for funding.
Ratio of Federal Grant to Total Project Cost = 5-18%**

Status	Number of Projects
Complete	4
Active	4
Withdrawn	7
Discontinued	2
Negotiations Ceased	1
Total	18

	Applications Submitted	Applications Selected
Round 1	36	8
Round 2	13	4
Round 3	36	6

Project	Total Federal Grant	Total Project Cost	Federal Cost Share
Hydrogen Energy California	$408 M	$4 B	10%
Summit Texas Clean Energy	$450 M	$2.5 B	18%
NRG Energy	$167 M	$1 B	17%
Southern Kemper Energy	$293 M	$6.1 B	5%
Totals	$1.752 B	$13.6 B	13%

NCC >>>> FOSSIL FORWARD

>> Current DOE CCS Program

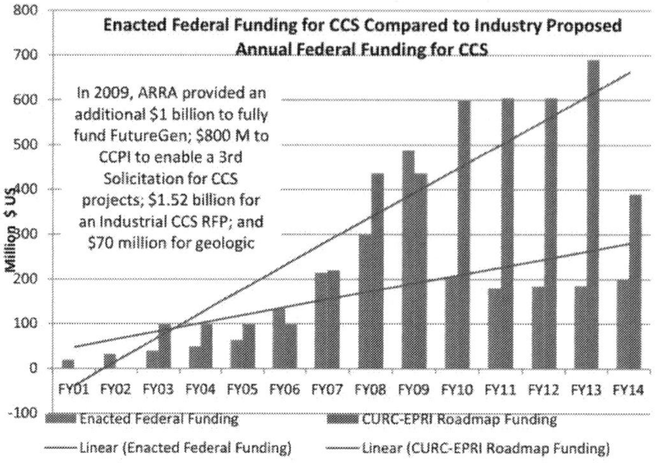

Federal Funding vs. Industry Recommended Funding

NCC >>>> FOSSIL FORWARD

>> Key Recommendation

DOE Program Goals need far greater clarity and alignment with Commercial Technology and Funding Approaches Used by Industry...

- **DOE and Industry prioritize projects critical to achieving goals consistent with the need to bring CCS technologies up to Technology Readiness Level 9**
- **DOE establish interim goals that are more amenable to testing for scale up of CCS technologies that show promise towards meeting the cost and performance goals**
- **A targeted number of projects or GW's be established with dates of operation that are consistent with overall emission reduction targets**
- **Future QER reports examine CCS infrastructure needs for a comprehensive nationwide CCS/CCUS system**
- **DOE undertake a general equilibrium model study to determine if the goal of CCS cost parity by 2035 is adequate and consistent with the overall CO2 reduction goals**

NCC >>>> FOSSIL FORWARD

>> Current DOE CCS Program

Stages of CO2 Capture Technology R&D

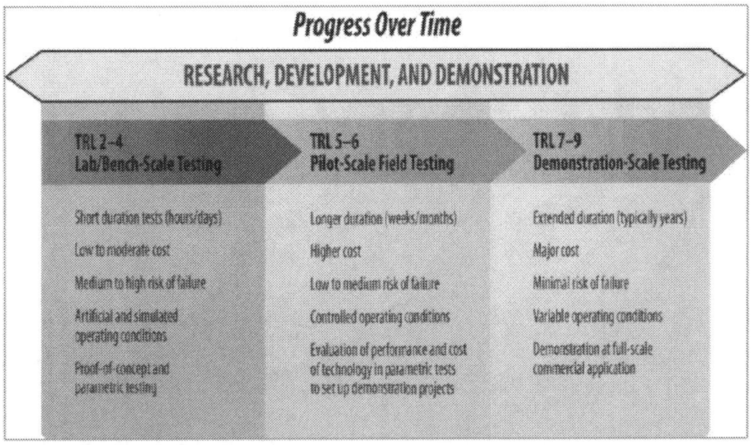

NCC >>>> FOSSIL FORWARD

>> Current DOE CCS Program

While DOE has enabled advancement of CCS technology, existing portfolio of 70 projects are predominantly small and in early stages of development.

Program Area	Key Technology	Number of R&D Projects					Total
		TRL 1	TRL 2	TRL 3-4	TRL 5-6	TRL 7	
Post-Combustion Capture	*Solvents*		3	9	5	-	17
	Sorbents		3	9	2	-	14
	Membranes		4	5	1	-	10
	Hybrid/Novel		5	3	1	1	10
Pre-Combustion Capture	*Solvents*		2	1	→	-	3
	Sorbents		2	1	1	-	4
	Membranes		2	5	-	-	7
	Hybrid/Novel		3	-	-	-	3
Compression	*Compression*		-	-	2	-	2
TRL Totals			24	33	12	1	70

Need for fresh, transformational ideas
"Wave" of bench scale projects approaching graduation (1/2 of portfolio)
Up to 12 candidate ≤1MW pilots progressing toward large pilot scale

NETL

NCC >>>> FOSSIL FORWARD

>> Current DOE CCS Program

Timeline of DOE Cost Goals

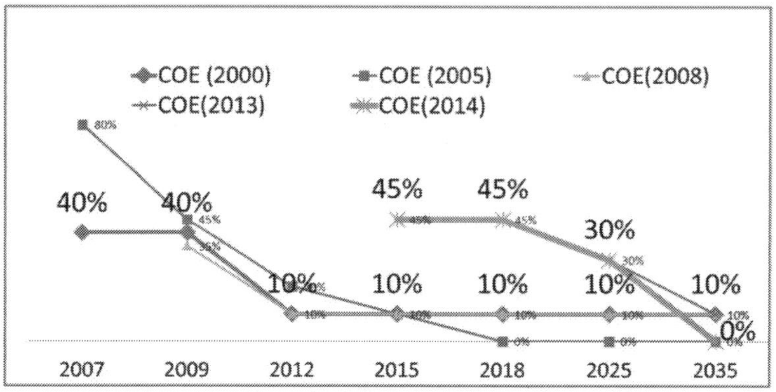

NCC >>> FOSSIL FORWARD

CCS Systems Must Be Demonstrated at Commercial Scale

NATIONAL COAL COUNCIL

Carbon Technology Development Highly Complex & In Early Stages

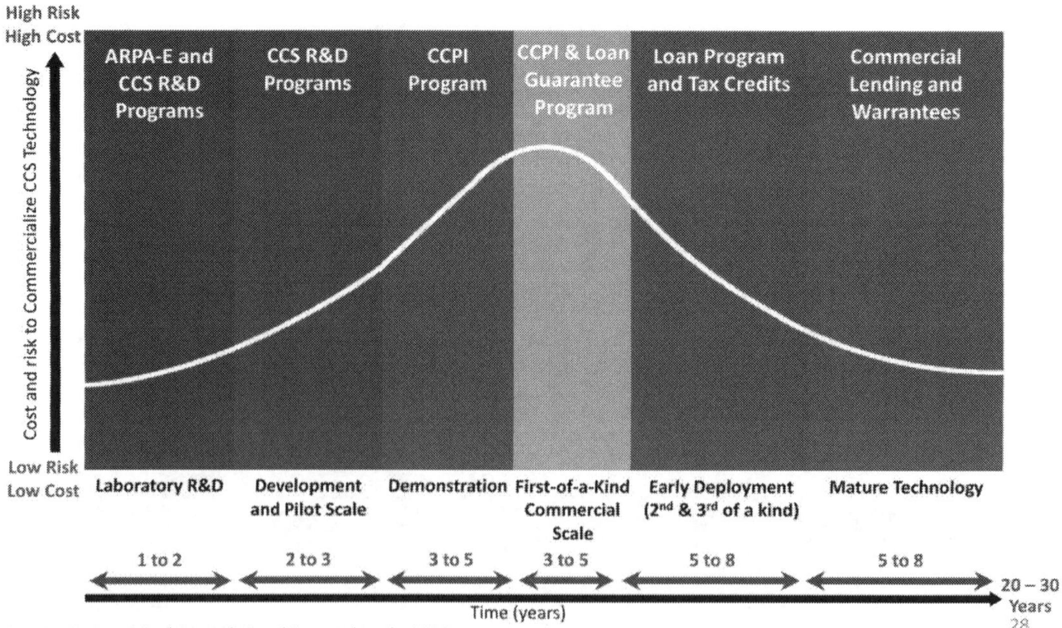

Source: National Coal Council, Fossil Forward study; 2015.

28

NATIONAL COALCOUNCIL

Janet Gellici, EVP & COO
jgellici@NCC1.org
www.nationalcoalcouncil.org

Recommendations for Leveling the Playing Field

NATIONAL COALCOUNCIL

Foundational Financial Structure: Contracts for Differences

- Establish a "contracts for differences" structure enabling greater competitiveness
- Allow developers to propose a menu of government incentives – e.g., CCPI grants to reduce capital, loan guarantees to reduce borrowing costs, tax credits to reduce cost of power over time
- Provide CFD structure for first 5-10 GW of projects with CCS on a competitive basis
- Use to create commercial demonstrations and large-scale pilot projects

30

Recommendations for Leveling the Playing Field

NATIONAL COAL COUNCIL

Financial Incentives

- **Limited Guaranteed Purchase Agreements**: Offer a limited number of pioneering CCS facilities to receive a guarantee that their output will be purchased in order to obtain financing
- **Market Set Aside**: Establish a market set-aside to provide parity with state RES requirements
- **Clean Energy Credits**: Allow CCS projects to receive credit under applicable programs for 100% of CO_2 emissions avoided

31

Recommendations for Leveling the Playing Field

NATIONAL COAL COUNCIL

Financial Incentives

- **Tax Credits and Price Interventions**: In addition to guaranteed purchase agreements and the ability to attract financing, offer other support through:
  - Production Tax Credit
  - Revised CO_2 Injection Credit
  - Electricity Price Stabilization
  - CO_2 Price Stabilization
- **Tax-Preferred Bonds**: Provide financing with tax-preferred and tax exempt bonds under Section 54 or Section 142 of Internal Revenue Code

32

Recommendations for Leveling the Playing Field

NATIONAL COAL COUNCIL

Financial Incentives

- **Master Limited Partnerships:** Provide MLP structure for projects with CCS so qualifying income is taxed at the individual level
- **Loan Guarantees:** Put in place a mechanism to pay the credit subsidy cost of loan guarantees similar to the Section 1705 program that helped renewables

33

Recommendations for Leveling the Playing Field

NATIONAL COAL COUNCIL

Regulatory Improvements

- **Regulatory Blueprint:** U.S. Dept. of Energy must take the lead in developing a regulatory blueprint to identify and remove barriers to construction and development of CCS projects
- **Remove Injection Barriers:** Eliminate barriers in EPA's Clean Power Plan and 111(b) rule to include beneficial reuse of CO_2 from regulated carbon capture facilities
- **New Source Review:** Eliminate uncertainty under NSR to encourage installation of CO_2 emission controls

34

Recommendations for Leveling the Playing Field

NATIONAL COAL COUNCIL

Regulatory Improvements

- **Infrastructure Siting:** Federal policymakers should consider providing backstop federal eminent domain authority for siting and construction of CO_2 pipelines, similar to authority provided under the Natural Gas act for gas pipelines
- **Storage Siting:** U.S. Dept. of Energy should identify at least one reservoir capable of storing at least 100 million tons of CO_2 at a cost of less than $10 per ton in each of the seven regions covered by the Regional Carbon Sequestration Partnership program

35

Recommendations for Leveling the Playing Field

NATIONAL COAL COUNCIL

Research, Development and Demonstration

- **Align RD&D Funding with Other Fuels:**
  - Substantially increase RD&D budget
  - Fully fund CCS RD&D recommended in CURC-EPRI Roadmap
  - Fund an 80% federal cost share for early stage RD&D
  - Fund 100% federal cost share for large-scale pilots
  - Fund 50% cost share for commercial demonstrations

36

Recommendations for Leveling the Playing Field

Communications and Collaboration

- **Vigorously Explain Reality:** U.S. Dept. of Energy must assure U.S. and global policymakers and other stakeholders that fossil fuels will be used to an even greater extent today, and there is a resulting need for CCS
- **Initiate Projects Immediately:** In addition to initiating 5 to 10 gigawatts of U.S. projects, the U.S. Dept. of Energy should advance collaboration of
  5 to 10 gigawatts of global demonstration projects in the next year, which will promote energy, environmental, and foreign policy interests

37

AUTHOR INDEX

AUTHOR INDEX

AUTHOR INDEX

Carbon Management Technology Conference 2015

Sustainable and Economical CCUS Options

Sugar Land, Texas, USA
17-19 November 2015

Volume 2 of 2

ISBN: 978-1-5108-1886-6

Printed from e-media with permission by:

Curran Associates, Inc.
57 Morehouse Lane
Red Hook, NY 12571

Some format issues inherent in the e-media version may also appear in this print version.

Printed by Curran Associates, Inc. (2016)

For permission requests, please contact AIChE
at the address below.

AIChE
120 Wall Street, FL 23
New York, NY 10005-4020

Phone: (800) 242-4363
Fax: (203) 775-5177

www.aiche.org

Additional copies of this publication are available from:

Curran Associates, Inc.
57 Morehouse Lane
Red Hook, NY 12571 USA
Phone: 845-758-0400
Fax: 845-758-2633
Email: curran@proceedings.com
Web: www.proceedings.com

TABLE OF CONTENTS

Carbon Capture at Coal-Fired Power Plants: SaskPower's Boundary Dam Project and Beyond

Reitenbach, G.,

Originally presented on: 11/18/2015 12:00:00 - 12:25:00

In October 2014, the world's first full-scale carbon capture plant at an operating coal-fired power plant went online at SaskPower's Boundary Dam Unit 3 in southern Saskatchewan. That landmark achievement earned the project *POWER*'s top award in 2015: Plant of the Year. This presentation will share details of the project drawn from a site visit made by *POWER*'s editor and from follow-up exchanges with project managers. It will also provide an overview of potential future technologies for use at power plants and why they may or may not prove successful.

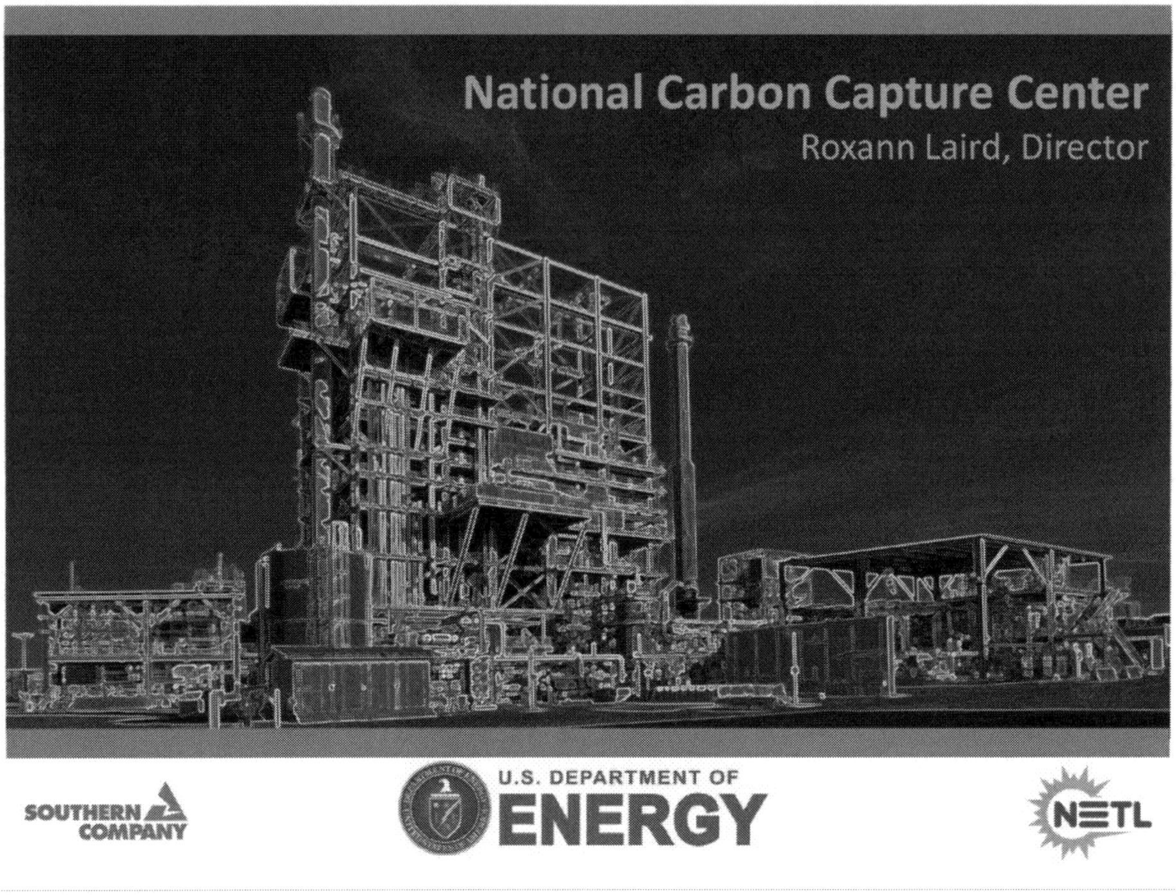
National Carbon Capture Center
Roxann Laird, Director
SOUTHERN COMPANY
U.S. DEPARTMENT OF ENERGY
NETL

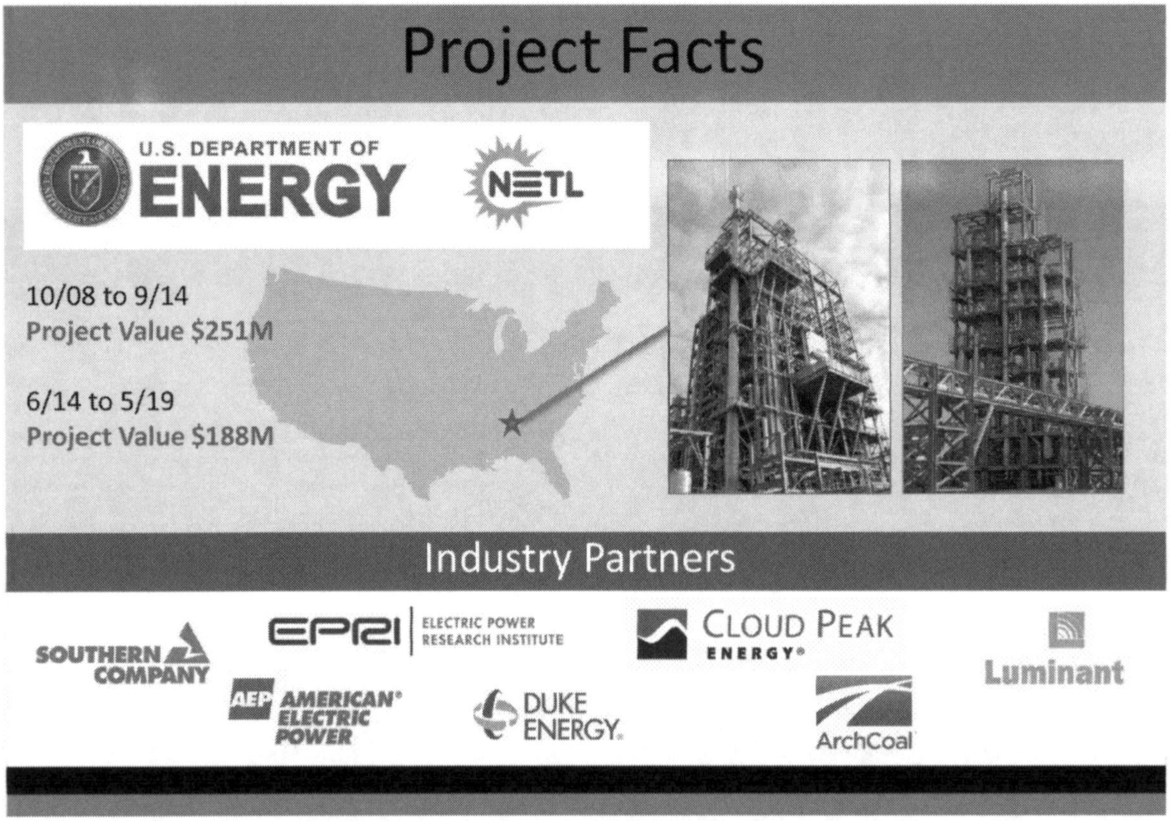
Project Facts
U.S. DEPARTMENT OF ENERGY
NETL
10/08 to 9/14
Project Value $251M
6/14 to 5/19
Project Value $188M
Industry Partners
SOUTHERN COMPANY
EPRI ELECTRIC POWER RESEARCH INSTITUTE
CLOUD PEAK ENERGY
Luminant
AEP AMERICAN ELECTRIC POWER
DUKE ENERGY
ArchCoal

Offering a **world-class neutral** test facility and a highly specialized staff, to **accelerate the commercialization** of advanced technologies and enable coal based power plants to achieve **near-zero emissions** **(low cost CO_2)**.

NCCC

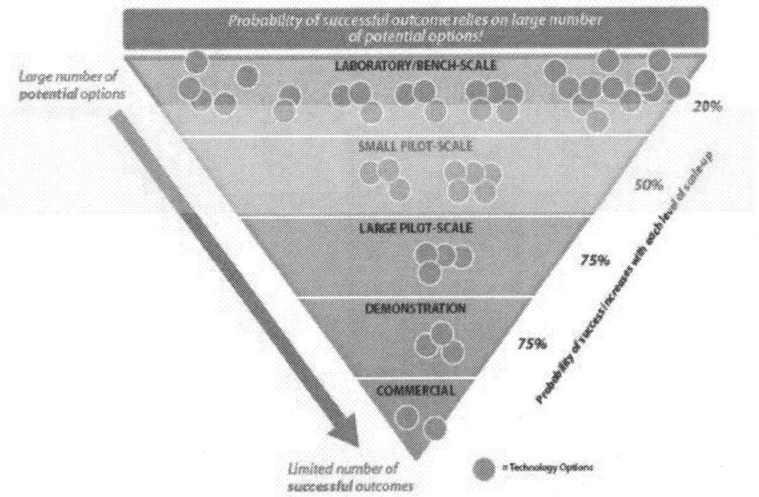

U.S. DEPARTMENT OF ENERGY

NETL

Technology Development Process

Evaluate and Screen Technologies

Define Scope of Work with Technology Developer

Design and Construct

Operate according to Test Matrix

Analyze Data and Report

Post Combustion Small Pilot Scale

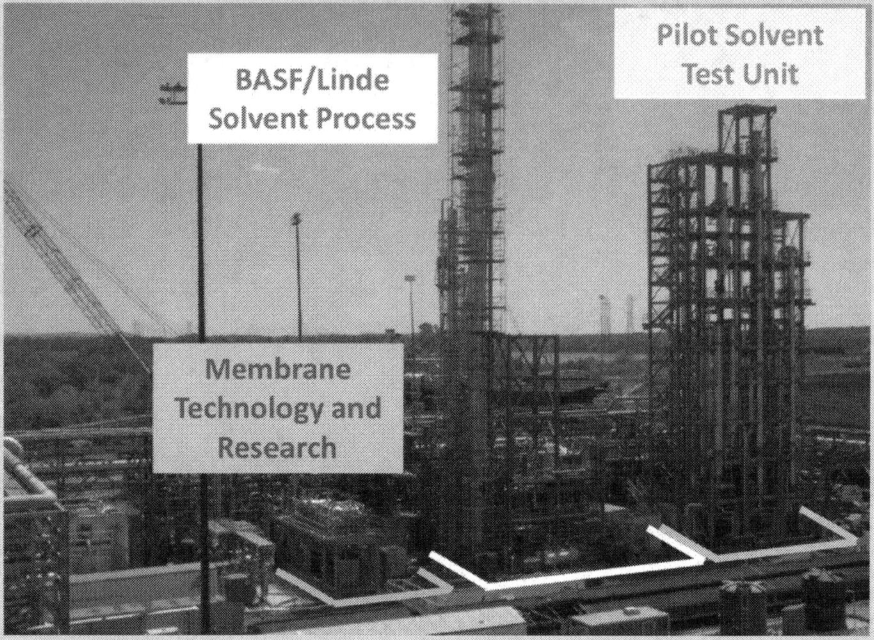

Post Combustion Bench Scale

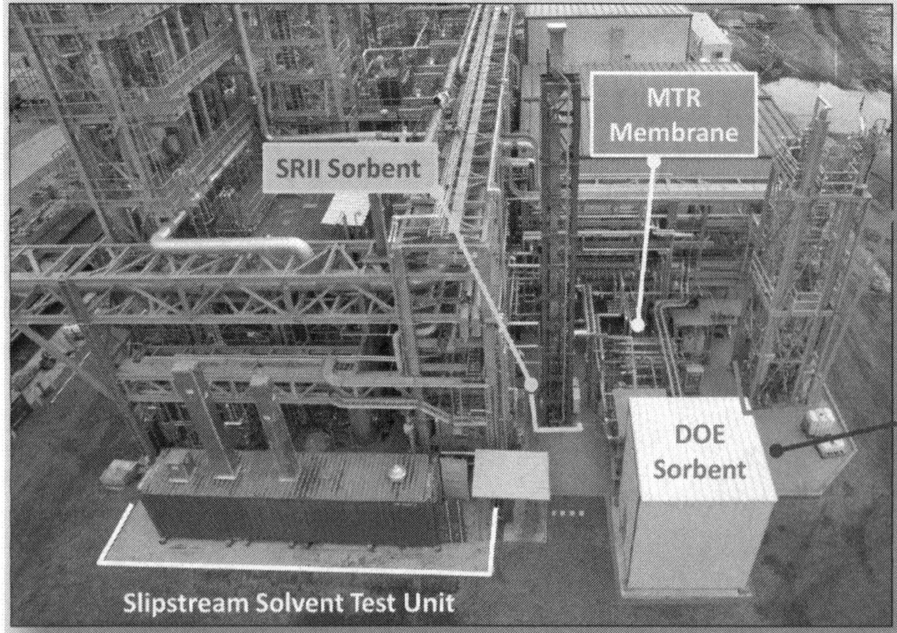

Akermin Enzyme Enhanced Solvent

CCSp Solvent (Carbon Capture Scientific)

Pre Combustion Carbon Capture

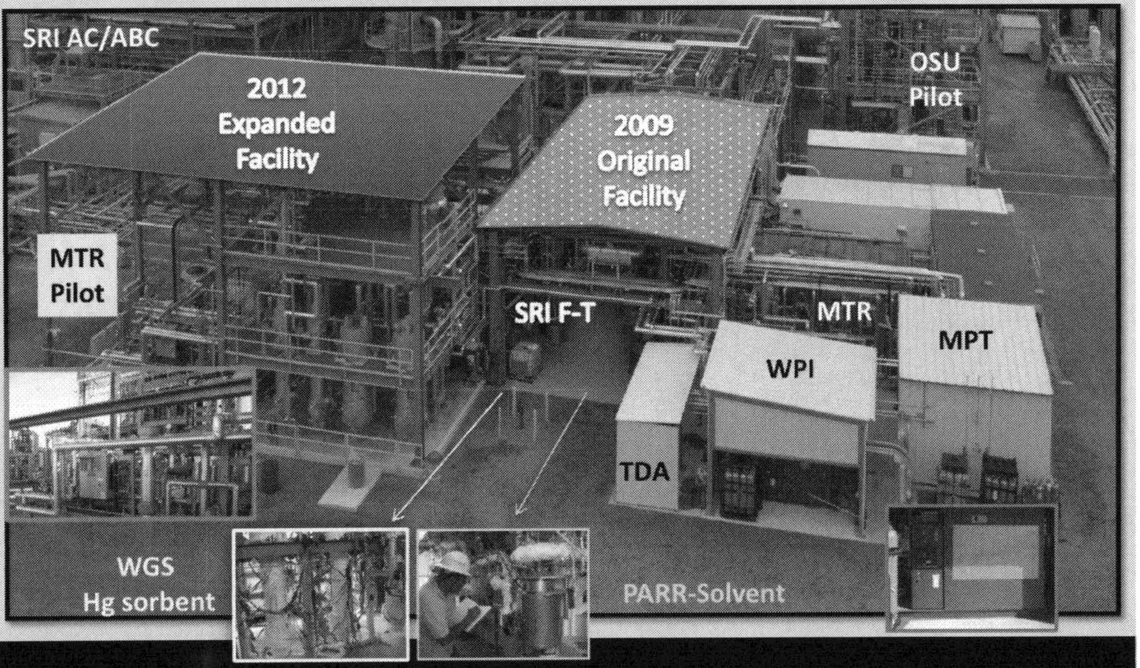

Successful Testing and Partnerships

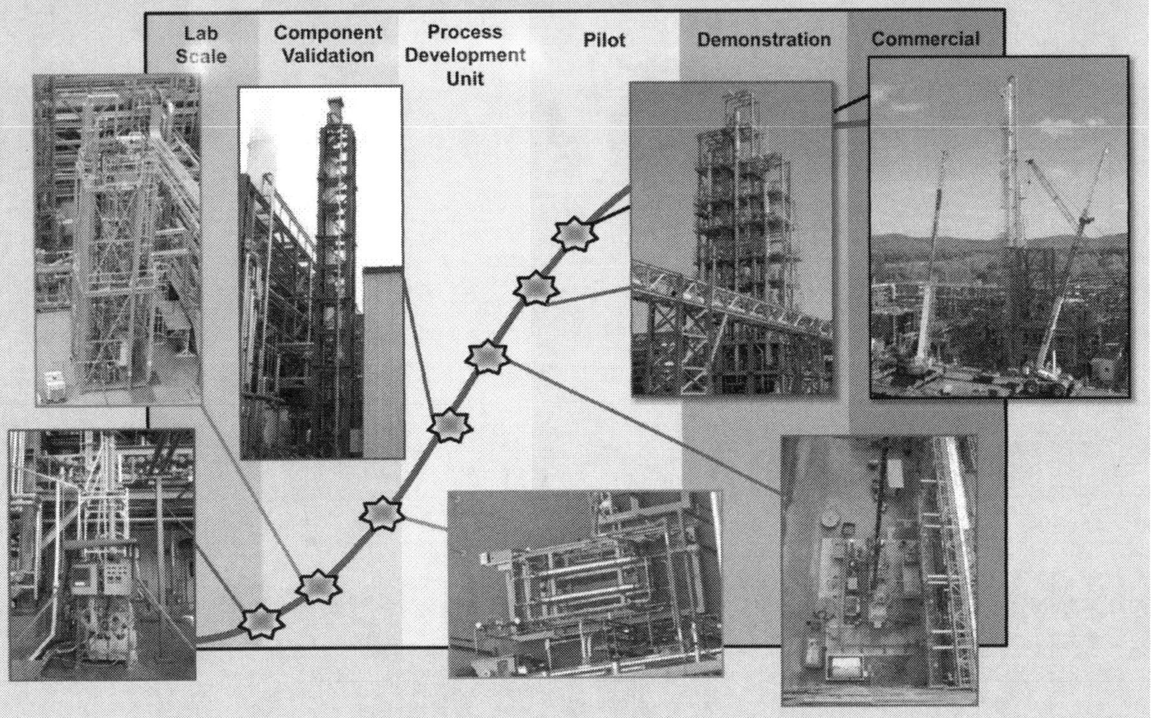

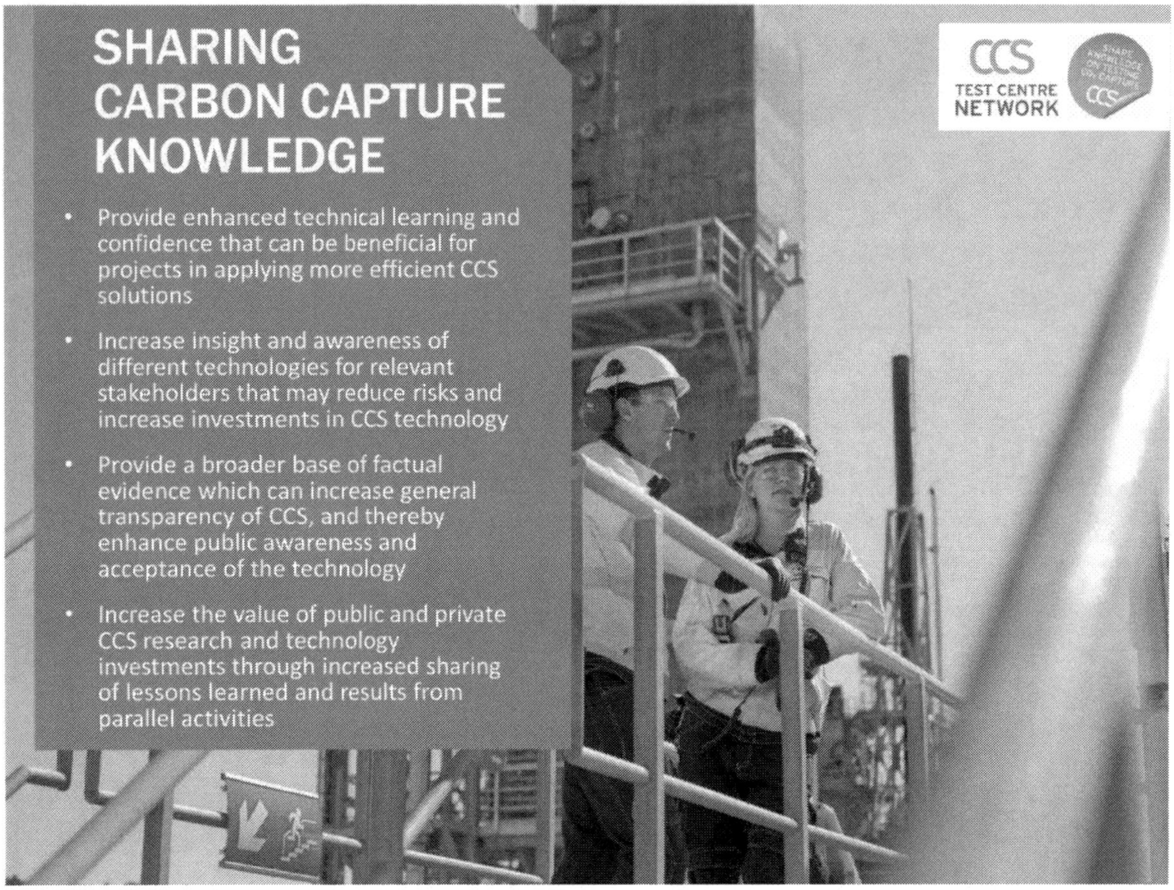

Acknowledgements

U.S. DEPARTMENT OF ENERGY

NETL

SOUTHERN COMPANY

EPRI ELECTRIC POWER RESEARCH INSTITUTE

ArchCoal

Luminant

AEP AMERICAN ELECTRIC POWER

DUKE ENERGY

CLOUD PEAK ENERGY

More information: *www.nationalcarboncapturecenter.com*

Project Management Guide

Goals:

- Answer FAQ's more efficiently
- Develop Consistency between Developers
- Identify Typical Pitfalls and offer Guidance

SOUTHERN COMPANY

Energy to Serve Your World

Project Management Guide
for
Technology Developers

At The

National Carbon Capture Center

Revision	Description	By	Date
1	Initial Issue	PWC	1/2/14
2	Updated	PWC	2/10/14
3	Updated Section 9.3	PWC	2/17/14

Accomplishments

2009

- DOE announces NCCC founding
- Implementation of technology screening process

Post-combustion

- PC4 design & construction began in earnest

Gasification & Pre-combustion

- 10-fold scale-up of CMS hydrogen membrane (first membrane tested)
- 1st biomass gasification

2010

Gasification & Pre-combustion

- Slipstream capacity increased 30-fold
- WGS study supported significant COE reduction for new IGCC plants
- Established solvent chemistry and sampling techniques

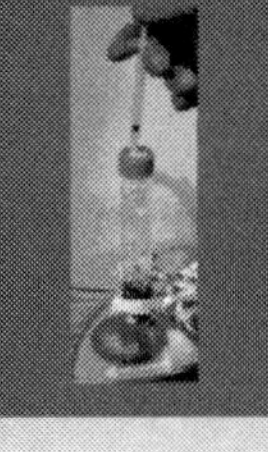

2011

Post-combustion

- Completion of PC4 construction
- Commissioning of PSTU with MEA solvent
- 1st pilot-scale testing (Aker)
- 1st technology developer testing in PSTU (B&W)
- 1st PC4 membrane testing (MTR 1-TPD/day unit)

Gasification & Pre-combustion

- 1st sorbent testing (TDA Research)

2012

Post-combustion

- 1st enzyme testing (Codexis)

Gasification & Pre-combustion

- 1st real-time syngas composition measurements (Stanford University TDL)
- Oxygen-blown biomass operation (DOD objective)
- 1st demonstration of membrane-assisted CO_2 liquefaction

2013

Post-combustion

- PC4 operation exceeds 25,000 hours
- 1st natural gas flue gas simulation
- 1st PC4 sorbent testing

Gasification & Pre-combustion

- Gasification operation exceeds 20,000 hours, with 30,000 hours of technology developer testing
- Expansion of TDL capabilities for additional syngas constituent measurements

2014

- NCCC awarded second DOE contract through competitive process

Post-combustion

- PC4 capacity doubled, allowing simultaneous pilot-scale testing
- 20-fold scale-up of MTR membrane
- Solvent testing for DOE Carbon Capture Simulation Initiative

Gasification & Pre-combustion

- 1st Fischer-Tropsch catalyst test

2015

Post-combustion

- 1st demonstration of gas pressurized stripping
- Slipstream solvent test unit commissioning

Gasification & Pre-combustion

- Largest scale slipstream testing to date (Ohio State syngas chemical looping)

Pre Combustion/Gasification Results

14 gasification runs – 10,920 hours

More than 30,000 hours combined testing of technologies

Technologies Tested

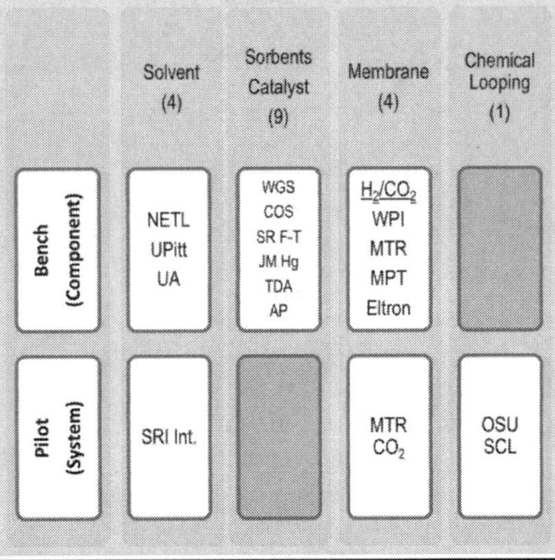

Accumulated Hours

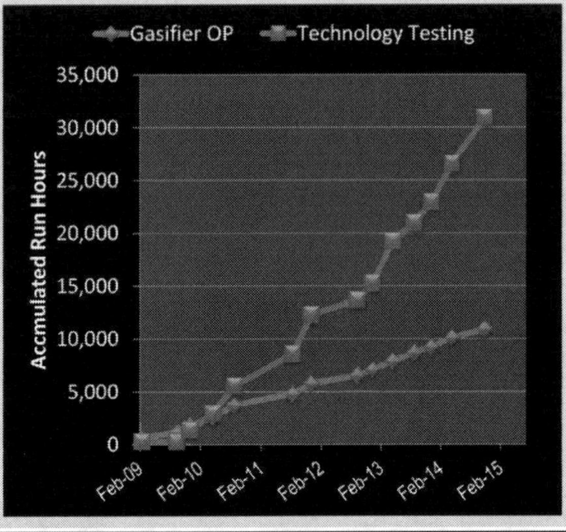

Post Combustion Results

- Developers
  - 8 pilot-scale (0.2 – 1.0 MW)
  - 6 bench-scale (0.001-0.05 MW)
- Oprating hours since 3/2011
  - Pilot 15,213 hrs
  - Bench 15,032 hrs
  - Cansolv & Carbon Clean Solutions of India (CCSi) tested w/ natural gas simulated flue gas
- Technologies
  - 10 amine-based solvent
  - 1 carbonate-based solvent
  - 2 catalytically assisted (enzymes)
  - 2 solid sorbents
  - 1 CO_2 membrane
- Others:
  - Amine carryover and aerosols
  - Trace metals accumulation in solvent
  - Analytical methods development & verification

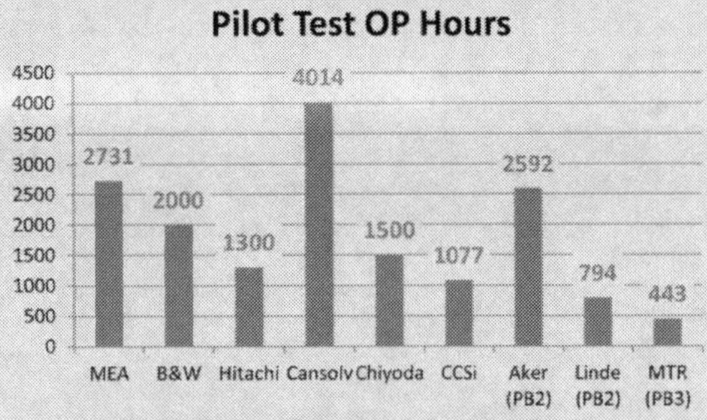

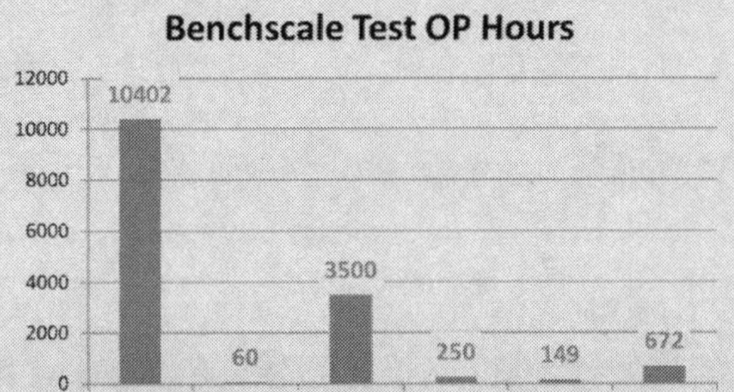

THE INTERNATIONAL CCS TEST CENTRE NETWORK (ITCN)
November 18, 2015
Roy Vardhejm, TCM, Norway
1
CCS
TEST CENTRE
NETWORK

• The United States and Norway announced their commitment to support the Carbon Capture and Storage (CCS) Test Centre Network at the CSLF Ministerial Meeting in Washington DC November 7, 2013.
08/01/2016
2
CCS
TEST CENTRE
NETWORK

Members of the International CCS Test Center Network as of June 2015

CO2 TECHNOLOGY CENTRE MONGSTAD

NCCC National Carbon Capture Center

e·on

SaskPower Powering the future

UKCCSRC PACT FACILITIES

CCS TEST CENTRE NETWORK

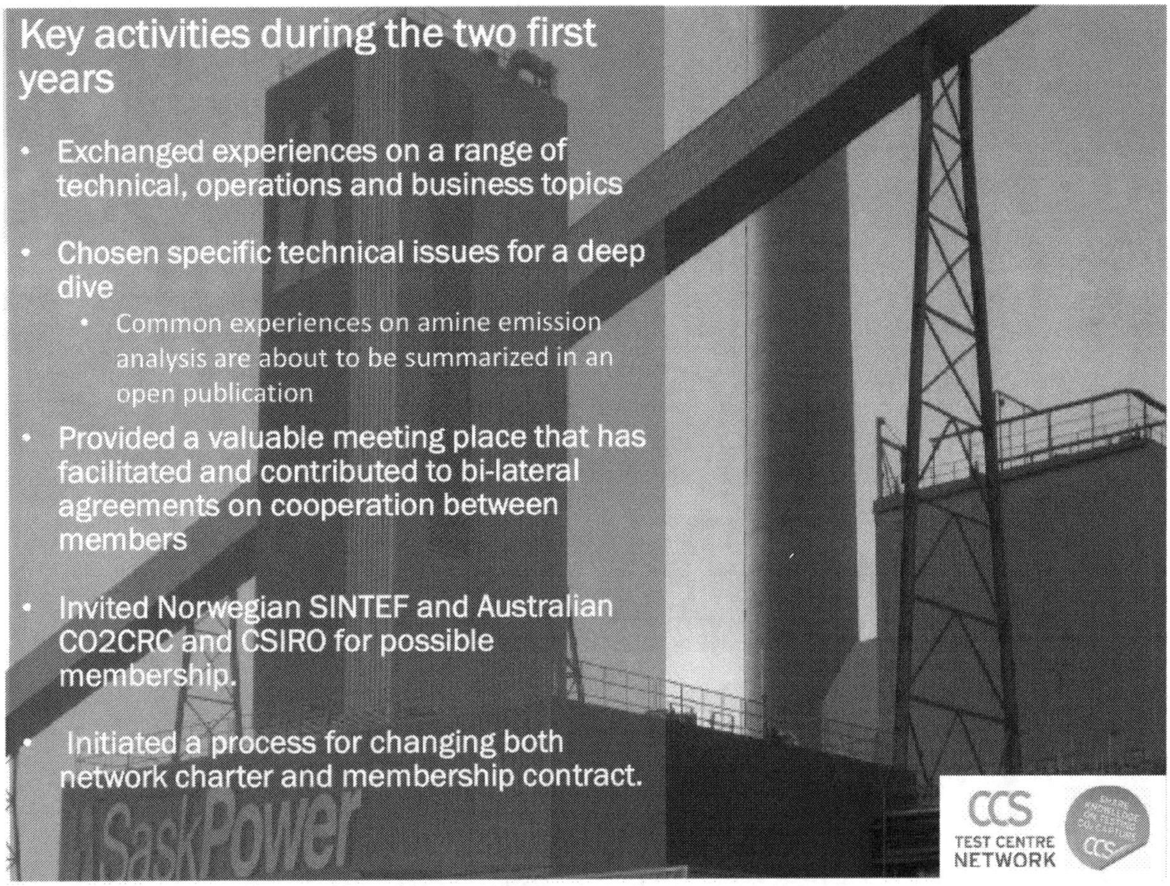
Key activities during the two first years
Exchanged experiences on a range of technical, operations and business topics
Chosen specific technical issues for a deep dive
Common experiences on amine emission analysis are about to be summarized in an open publication
Provided a valuable meeting place that has facilitated and contributed to bi-lateral agreements on cooperation between members
Invited Norwegian SINTEF and Australian CO2CRC and CSIRO for possible membership.
Initiated a process for changing both network charter and membership contract.
SaskPower
CCS
TEST CENTRE
NETWORK

Some benefits of ITCN membership:
Strengthen own competence through access to other's learnings and experiences
Enhanced opportunities for market collaboration
Bi- or multi-lateral agreements between facilities
Common promotion of technology testing, e.g. ability to offer testing at several scales in one package
Enhanced opportunities for cross-national funding
Bi- and multi-lateral national agreements on co-funding of CCS exist and the number is increasing
Enhanced opportunities to influence and improve technology qualification through common contributions to benchmarking and harmonisation of procedures
08/01/2016
6
CCS
TEST CENTRE
NETWORK

Technology Centre Mongstad (TCM)
Norway
GASSNOVA
Statoil
75.12%
20%
sasol
reaching new frontiers
2.44%
2.44%
TECHNOLOGY CENTRE MONGSTAD

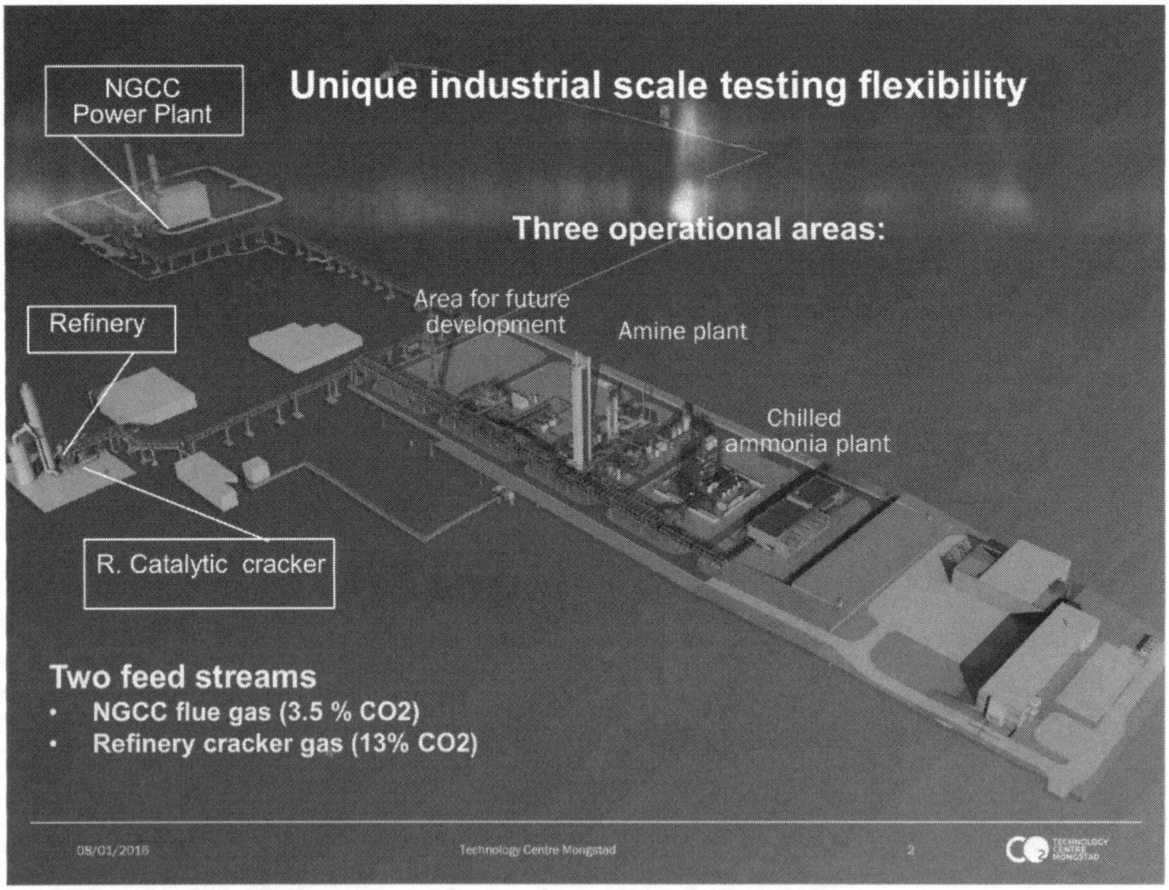
NGCC Power Plant
Unique industrial scale testing flexibility
Three operational areas:
Area for future development
Amine plant
Refinery
Chilled ammonia plant
R. Catalytic cracker
Two feed streams
• NGCC flue gas (3.5 % CO2)
• Refinery cracker gas (13% CO2)
08/01/2016
Technology Centre Mongstad
2
TECHNOLOGY CENTRE MONGSTAD

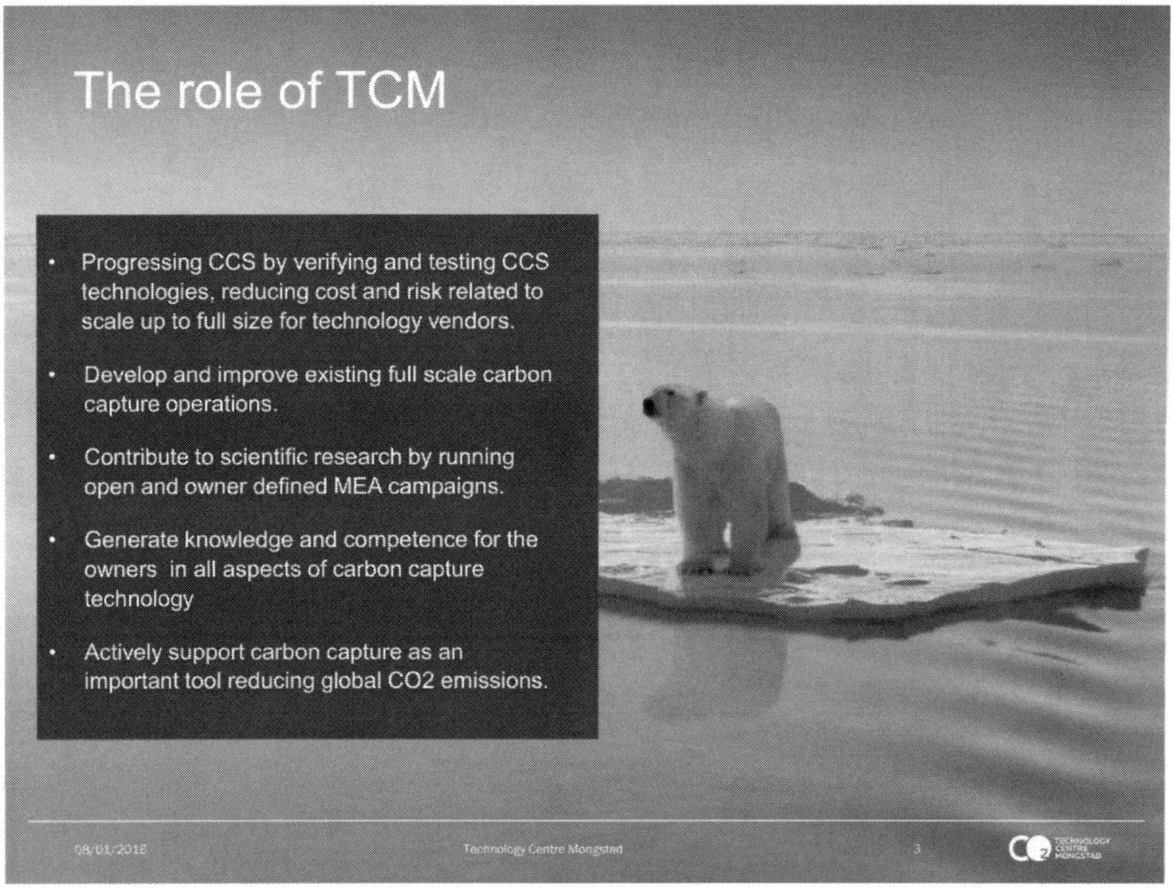
The role of TCM
Progressing CCS by verifying and testing CCS technologies, reducing cost and risk related to scale up to full size for technology vendors.
Develop and improve existing full scale carbon capture operations.
Contribute to scientific research by running open and owner defined MEA campaigns.
Generate knowledge and competence for the owners in all aspects of carbon capture technology
Actively support carbon capture as an important tool reducing global CO2 emissions.
08/01/2016
Technology Centre Mongstad
3
TECHNOLOGY CENTRE MONGSTAD

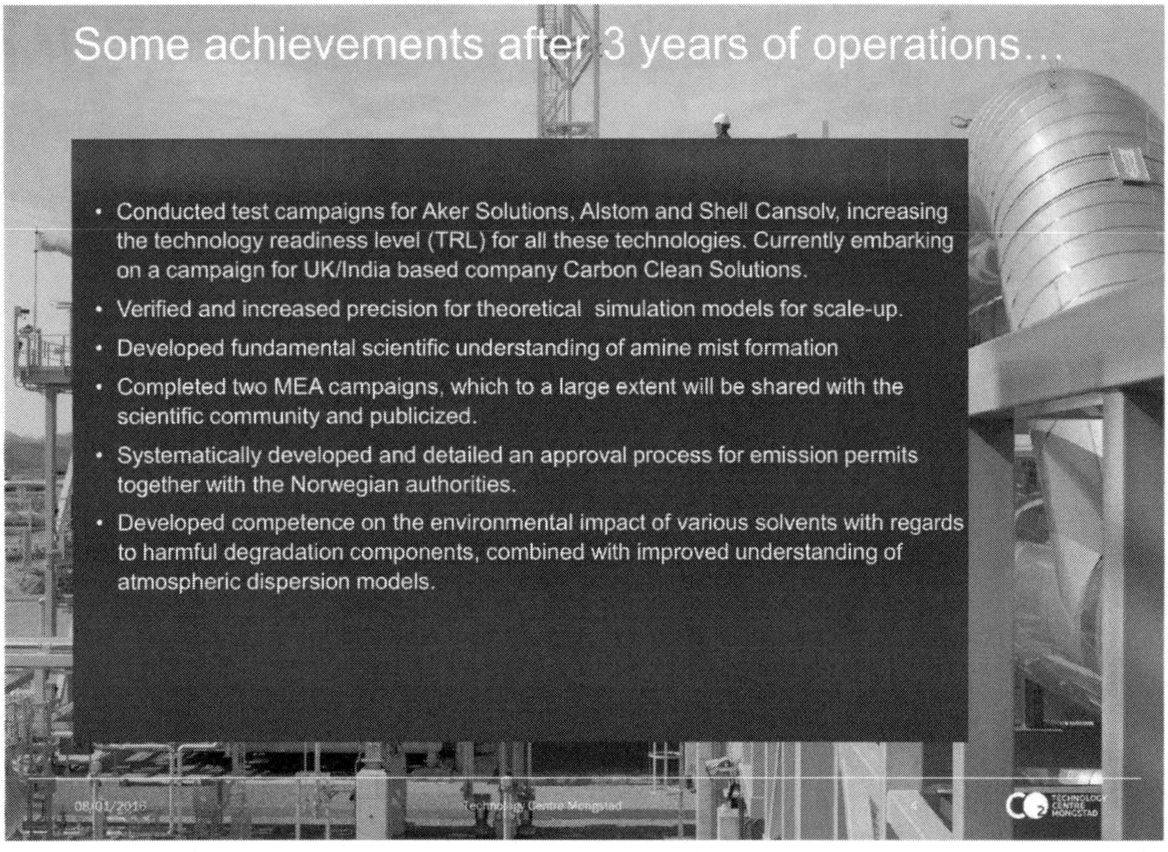
Some achievements after 3 years of operations...
Conducted test campaigns for Aker Solutions, Alstom and Shell Cansolv, increasing the technology readiness level (TRL) for all these technologies. Currently embarking on a campaign for UK/India based company Carbon Clean Solutions.
Verified and increased precision for theoretical simulation models for scale-up.
Developed fundamental scientific understanding of amine mist formation
Completed two MEA campaigns, which to a large extent will be shared with the scientific community and publicized.
Systematically developed and detailed an approval process for emission permits together with the Norwegian authorities.
Developed competence on the environmental impact of various solvents with regards to harmful degradation components, combined with improved understanding of atmospheric dispersion models.
08/01/2016
Technology Centre Mongstad
4
TECHNOLOGY CENTRE MONGSTAD

Access to TCM

Our business development process in brief.....

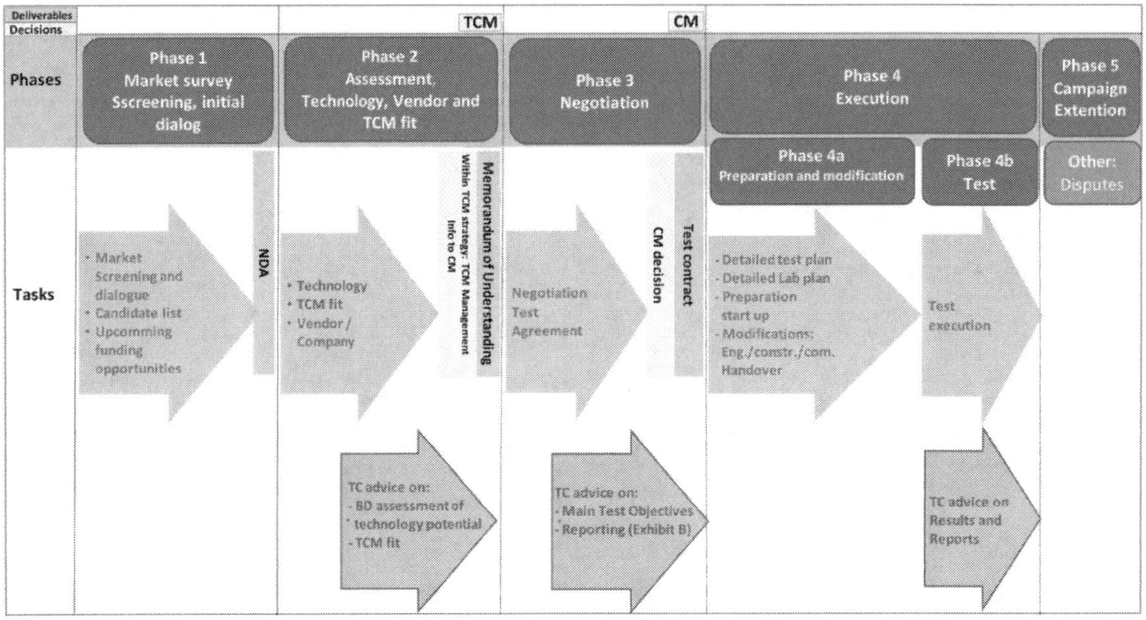

08/01/2016 Technology Centre Mongstad 5

UKCCSRC PACT FACILITIES

UKCCSRC PACT OVERVIEW

Dr. Bruce Adderley

ITCN Panel Session
CMTC 18th Nov 2015

The University Of Sheffield.

About UKCCSRC PACT

UKCCSRC PACT FACILITIES

- **P**ilot-scale **A**dvanced **C**apture **T**echnology facilities
  - Funded by: DECC and EPSRC
  - Partners: Cranfield, Edinburgh, Imperial, Leeds, Nottingham, Sheffield
  - Part of: UK Carbon Capture & Storage Research Centre (UKCCSRC)
- **Scope:** Specialist national facilities for research in advanced fossil-fuel energy, bioenergy and carbon capture technologies
  - Comprehensive range of pilot-scale facilities
  - Supporting specialist research and analytical facilities
  - Leading academic expertise
- **Aim:** Support and catalyse industrial and academic R&D to accelerate the development and commercialisation of novel low carbon technologies
- **Objectives**
  - Bridge gap between bench-scale R&D and industrial pilot trials
  - Provide shared access to industry and academia

Department of Energy & Climate Change

EPSRC Engineering and Physical Sciences Research Council

UKCCS RESEARCH CENTRE

Cranfield UNIVERSITY | THE UNIVERSITY of EDINBURGH | Imperial College London | UNIVERSITY OF LEEDS | The University of Nottingham | The University Of Sheffield.

UKCCSRC PACT Locations

UKCCSRC PACT FACILITIES

- 4 Facility Sites
- PACT Office

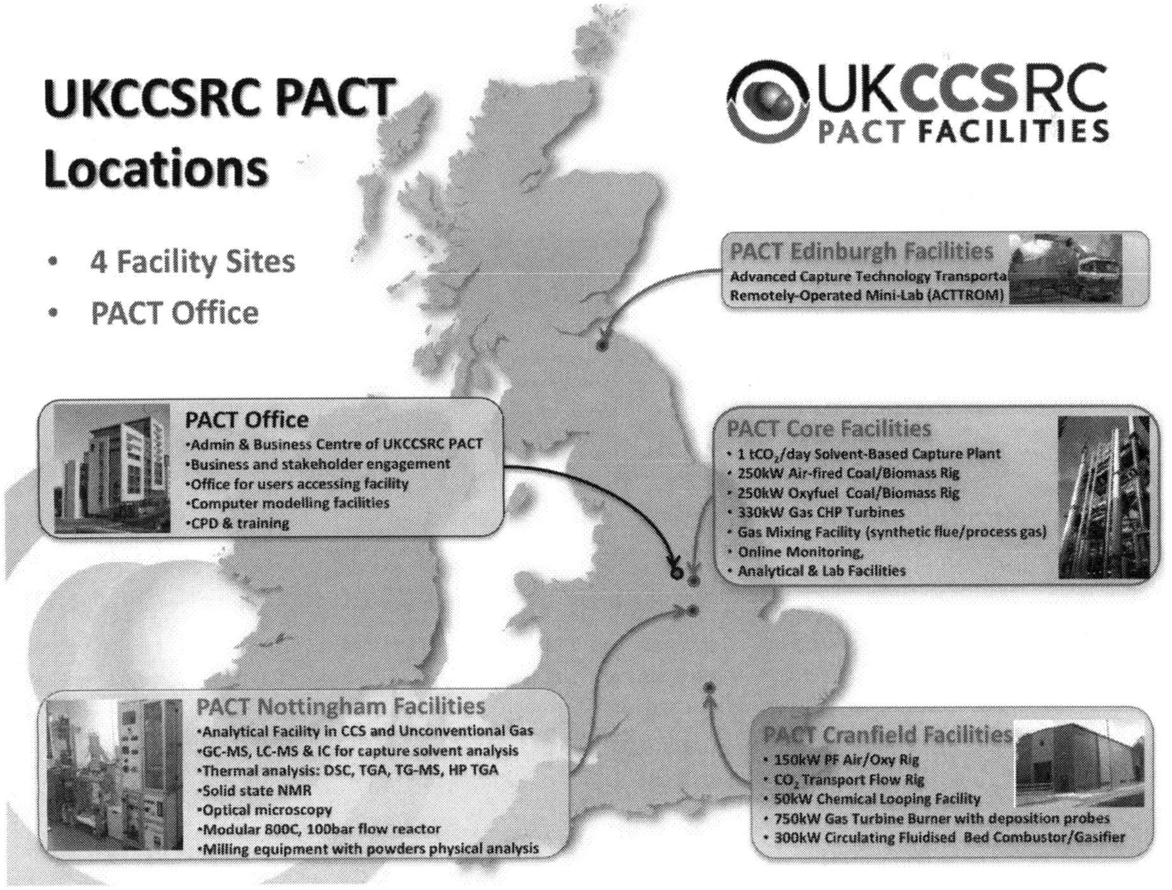

Core Facility Overview

UKCCSRC PACT FACILITIES

- Design & Installation from March 2012; operating since August 2013
- Officially opened in June 2014 by the Rt Hon Michael Fallon MP, DECC Minister of State for Energy

Cranfield UNIVERSITY | THE UNIVERSITY of EDINBURGH | Imperial College London | UNIVERSITY OF LEEDS | The University of Nottingham | The University Of Sheffield.

Core facility overview

UKCCSRC PACT FACILITIES

- **Setup costs**
  - Total setup cost £6.3m
  - 3.0m total funding from DECC
  - 3.3m total contribution from research groups
- **Operational income**
  - To date £3.2m of operational income till 2017-2018
  - Involved in £34m of research and training activity
    - 16 R&D projects (£26m - income to PACT of £2.9m)
    - 2 Doctoral Training Centres (£8m - £300k income to PACT)
- **Utilisation:**
  - To date 707 days allocated to current/committed projects till 2018
  - Currently 40% utilised with utilisation rate at 170% of planned
- **Visits/ presentations**
  - 1000+ visitors to the facility from 256 organisations including 65 academic and over 200 industrial and other organisations
  - Around 30 national/international presentations on PACT to a total estimated audience of 1100

Cranfield UNIVERSITY | THE UNIVERSITY of EDINBURGH | Imperial College London | UNIVERSITY OF LEEDS | The University of Nottingham | The University Of Sheffield.

PACT Core Facility

UKCCSRC PACT FACILITIES

1. $1tCO_2$/day Carbon Capture Plant

1. Gas Mixing Facility (GMF) synthetic flue/process

1. 300kW (CHP)Gas Turbines

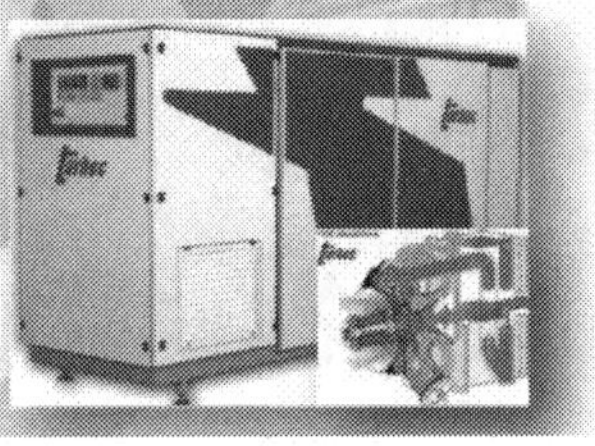

4. 250kW Air Combustion Plant

5. 250kW Oxyfuel Combustion Plant real/simulated EGR

6. Analytical facilities: online monitoring and analytical labs

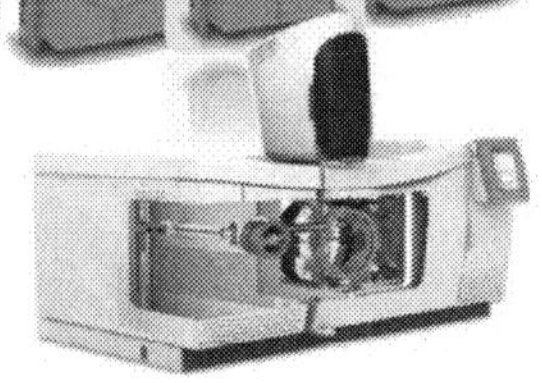

Integrated facilities

UKCCSRC PACT FACILITIES

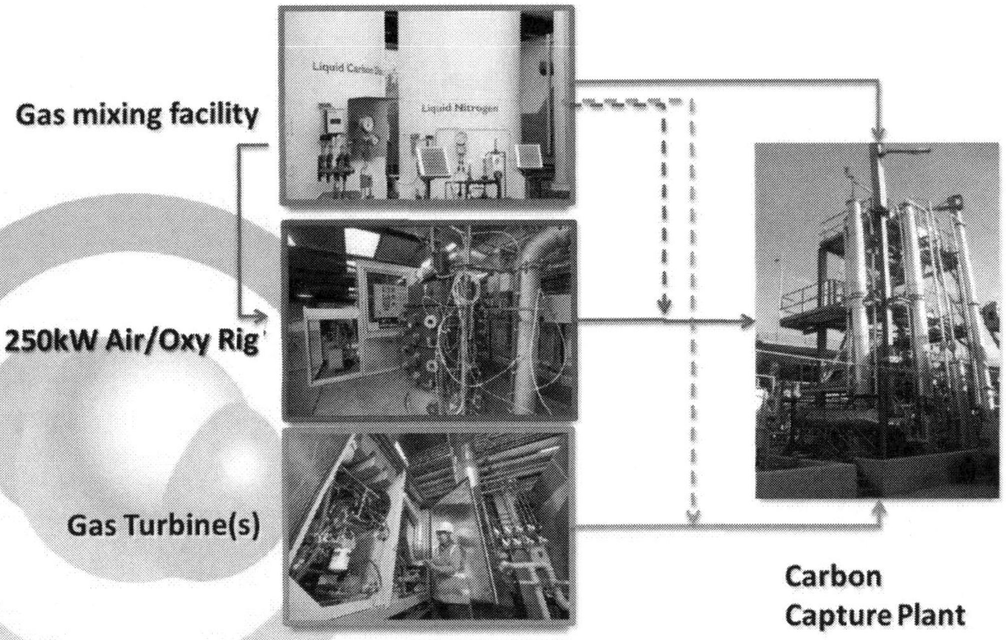

Gas Mixing Facility

UKCCSRC PACT FACILITIES

- Modulate real flue composition or generate synthetic flue /industrial process gas
- Connected directly to the Solvent-based Carbon Capture Plant
- Carbon capture from a variety of combustion, gasification processes and Industrial CCS

250kW Air/Oxy Plant

UKCCSRC PACT FACILITIES

- 250kWth, rig for coal/biomass (and gas)
- Advanced fossil-fuel energy research
- Bioenergy research
- Post combustion capture
- Oxyfuel capture (real/simulated)
- Fuel-testing: coal/biomass/co-firing
- Highly instrumented
  - Temperature and pressure monitoring
  - Combustion and deposition probes
  - 3D laser diagnostics and thermal imaging
  - Gas Analysis
  - Metals analysis (ICP)
- Modelling and model validation

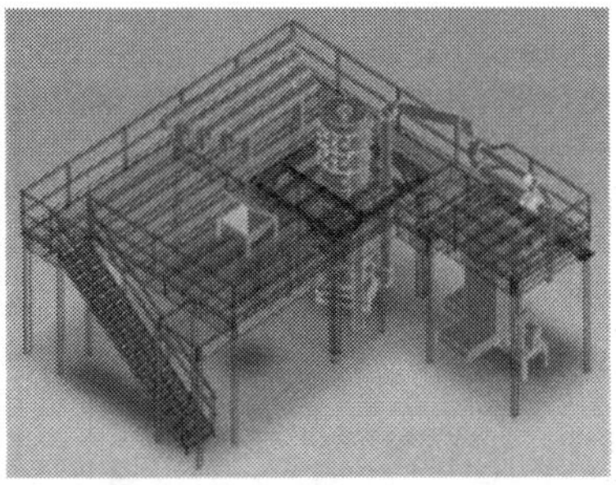

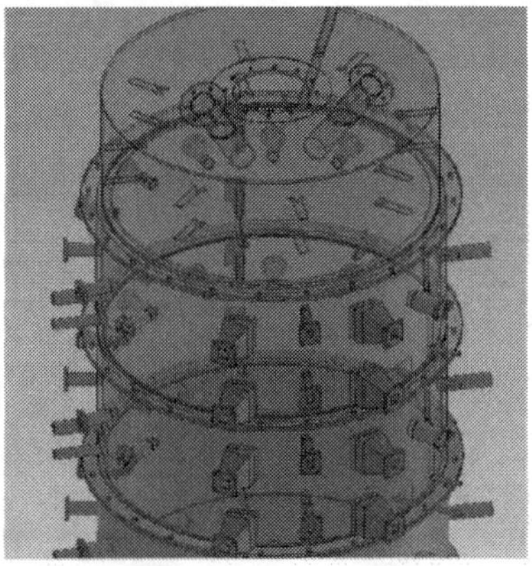

Gas Turbine System

UKCCSRC PACT FACILITIES

- Two Turbec T100 CHP Microturbines:
  - 100kWe, 170kWth
- Fuel: Natural gas, biogas, syngas, MeOH
- Analytical Facilities
  - GASMET FTIR flue gas analyser
  - HORIBA/SIGNAL Combustion Gas Analysers
  - CAMBUSTION DMS500 fast particle analyser
- Research applications
  - Humid Air Turbine (HAT) Cycle
  - Exhaust Gas Recirculation /selective EGR

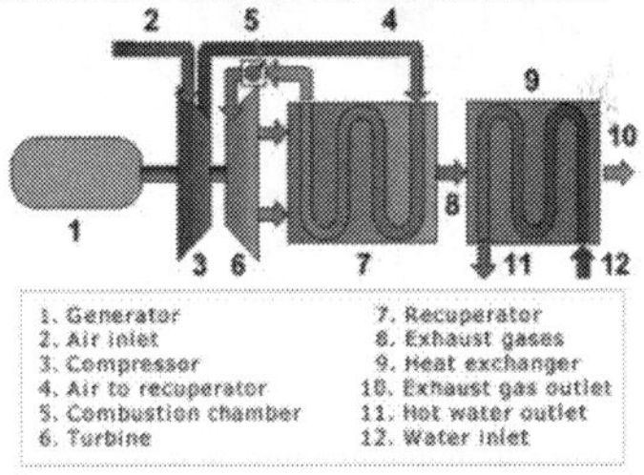

Carbon Capture Plant

UKCCSRC PACT FACILITIES

- 1 tonne of CO2 per day (MEA)
- Post combustion capture
  - Gas/coal/biomass
  - Industrial CCS
- Alternative solvent development
- Solvent degradation and emissions research
- Benchmarking & energy requirements
- Validation of baseline economics
- Integrated systems modelling and model validation

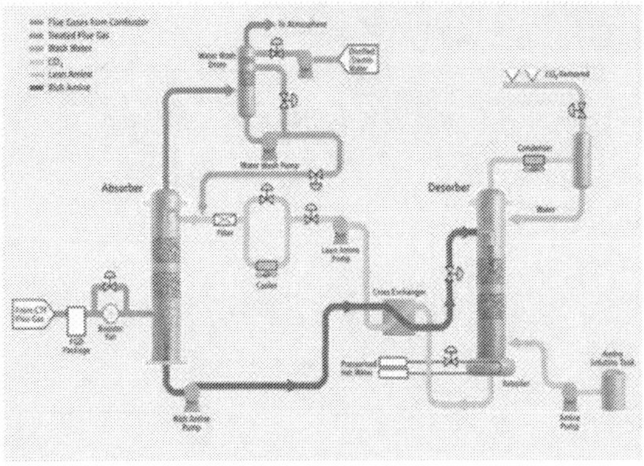

Analytical Facilities

UKCCSRC PACT FACILITIES

- Combustion MK2 fast particle size analyser
- Mobile ICP for 50+ metal analysis in flue gases
- 2 x Gasmet FTIR analysers for up to 25 compounds
- Signal/Horiba combustion gas analyser suite
- PerkinElmer® 2400 SII CHNS/O Elemental Analyser
- PerkinElmer® GC MS
- PerkinElmer® Pyris 1 TG and TG-MS
- PerkinElmer® FT-IR and TG-IR
- Portable SERVOFLEX CO_2 and O_2 analysers

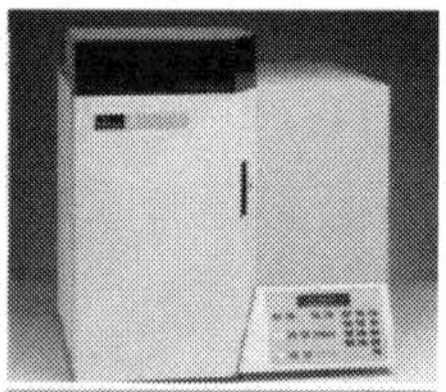

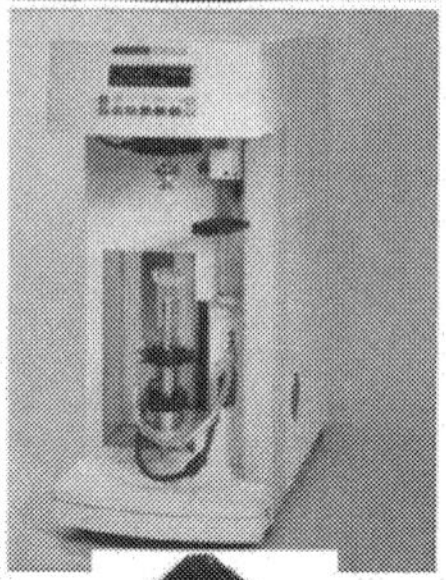

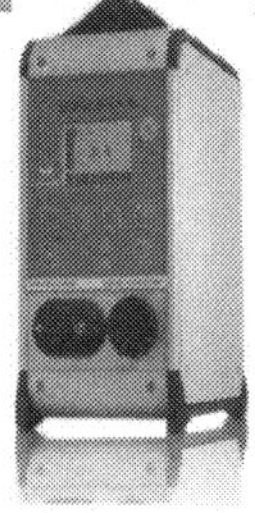

Online monitoring

UKCCSRC PACT FACILITIES

- New 5 point FTIR system
  1. Before FGD
  2. Before absorber
  3. After absorber
  4. After water wash
  5. Product gas stream (after pressure control valve)
- CO_2 column profiling
- Liquid sample analysis

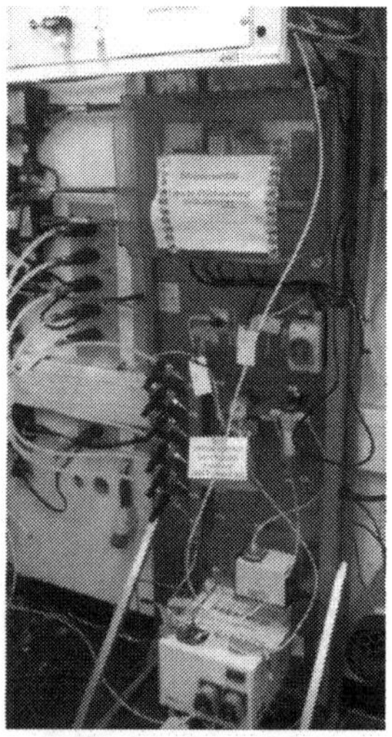

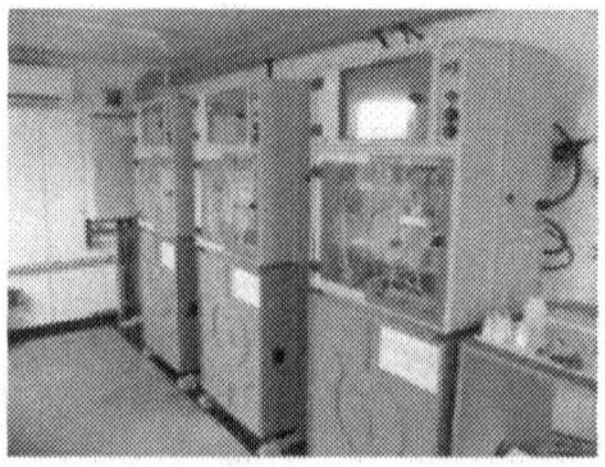

Edinburgh PACT Facilities

UKCCSRC PACT FACILITIES

- **A**dvanced **C**apture **T**esting **in a** **T**ransportable **R**emotely-**O**perated **M**ini-lab **(ACCTROM)**
- Integrated carbon capture media testing laboratory
- Facility is transported to the test site and connected to directly to onsite flue gasses for testing
- Long-term **on-site** testing of CO_2 capture media on real flue gases specific to a given process under a range of temperature and flue gas environments with a possibility of multiple parallel tests
- Remotely monitored and operated

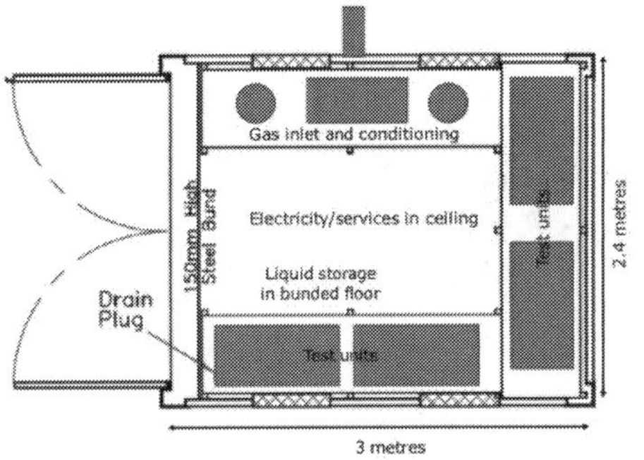

The ACTTROM System

UKCCSRC PACT FACILITIES

- ❑ Self-contained, standard chemical storage unit
- ❑ Capacity: 25-50 m³ of flue gas per day
- ❑ Meets DSEAR and COSHH regulations.
- ❑ Deployed and hosted on industrial site; host provide: flat site, supply of flue gas, electricity and internet connection.
- ❑ What's in the "Box"?
  - A number of small-scale units for carbon capture media testing; operated in parallel
  - Accelerated degradation testing units to examine system characteristics on exposure to **real** flue gases
  - Modular units for flue gas conditioning
  - Hot and cold water supplies for test rig temperature control.
  - Remote access control and data logging unit with video stream through a wireless internet link, allowing scientists anywhere in the country to monitor and control their experiments.

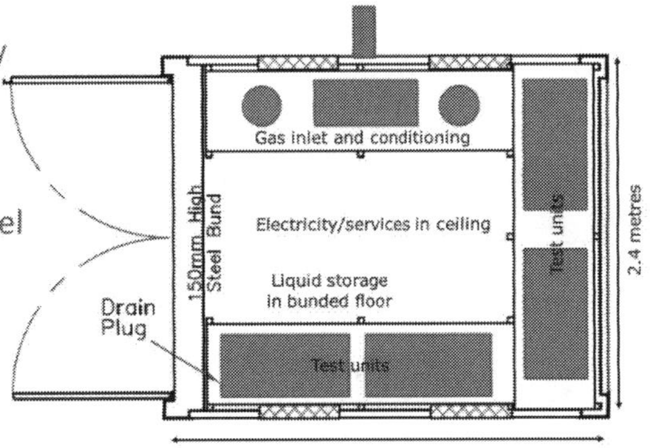

Cranfield PACT Facilities

UKCCSRC PACT FACILITIES

- 750kWth Burner Rig for Gas Turbine Hot Gas Path Research
- 350kWth Circulating Fluidised Bed Facility
- 50kWth Chemical/Ca Looping Facility
- 150kWth Pulverised Fuel
- Combustion Rig Dense-phase CO_2 Flow Loop

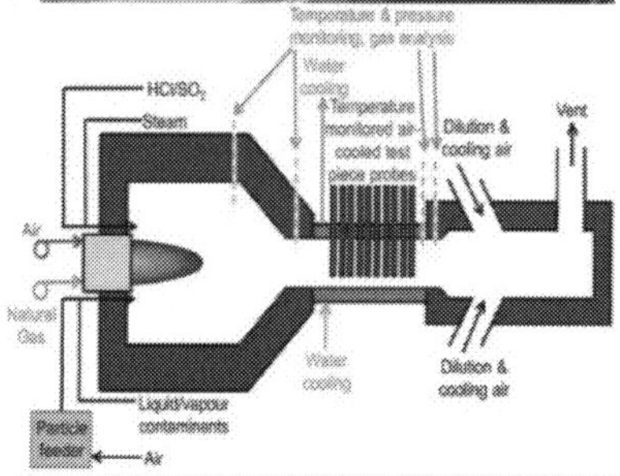

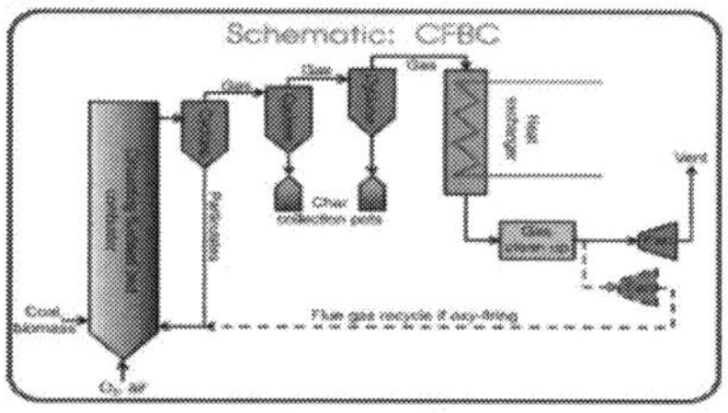

Cranfield PACT Facilities

UKCCSRC PACT FACILITIES

350kWth Circulating Fluidised Bed Facility

- ❑ UK's largest Circulating Fluidized Bed test facilities, operated flexibly in
  - combustion
  - gasification
  - oxy-fuel-fired mode, with full exhaust gas recycle (EGR), with both dry and wet flue gas.
- ❑ Sampling/feed ports permitting full solid, liquid and gas sampling with conditioning units
- ❑ Probing for in situ temperature, pressure and gas concentration measurements
- ❑ Automated, gas supply and mixing system.
- ❑ Hot and cold gas cleaning
- ❑ Gas leak safety detection system
- ❑ Fuel feed metering system,
- ❑ advanced flow meters
- ❑ continuous ash removal system.

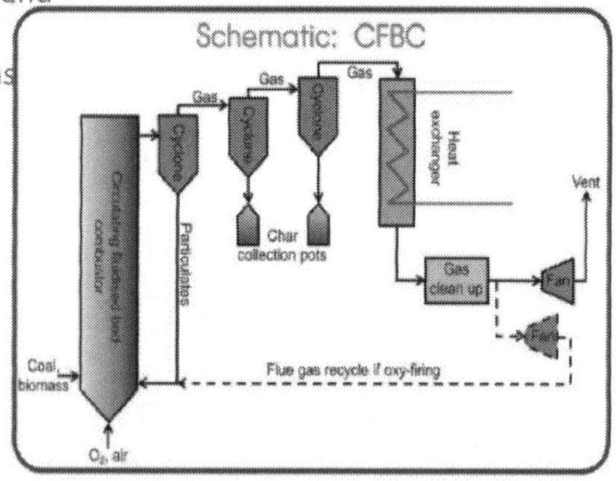

Nottingham PACT Facilities

UKCCSRC PACT FACILITIES

Analytics for CCS and Unconventional Gas

- GC-MS, LC-MS and ion chromatography to or the characterisation of post combustion capture capture solvents
- A range of thermal analysis equipment, including DSC, TGA coupled to GC-MS and high pressure TGA for investigating fundamentals of DE volatilisation and combustion
- Solid state NMR
- Optical microscopy
- A modular high pressure and temperature flow reactor system, enabling enable CO_2 utilisation and geological
- Milling equipment with continuous powder physical analysis.

Network Benefits

UKCCSRC PACT FACILITIES

www.pact.ac.uk

- Knowledge sharing and working together to establish best practice
- Forum for sharing research findings across the CCS system and advancing knowledge
- Operational and design experience and advice
- Developing process monitoring methods
- Sharing research to understand chemical and physical processes in the whole system

Funded by

Department of Energy & Climate Change

EPSRC

Engineering and Physical Sciences Research Council

UKCCS RESEARCH CENTRE

Cranfield UNIVERSITY · THE UNIVERSITY of EDINBURGH · Imperial College London · UNIVERSITY OF LEEDS · The University of Nottingham · The University Of Sheffield.

UKCCSRC
PACT FACILITIES

UKCCSRC PACT OVERVIEW

END

The University Of Sheffield.

Session 6
Panel Discussion: EOR & Storage

"State of U.S. EOR,
Introduction to ROZs and ROZ Science"

CMTC 2015
Sugar Land, Texas
November 17-19, 2015

Melzer CO2nsulting

2

State of U.S. EOR and an Introduction to ROZs and ROZ Science

- CO_2 EOR Projects and Production vs. Time
- Map of Where the Projects are Located
- Phases of CO_2 EOR
- What is a Residual Oil Zone (ROZ)?
- Why the Interest in Them?
- What is Happening Related to ROZ Exploitation?

Melzer CO2nsulting

HISTORY OF WORLDWIDE AND U.S. CO_2 EOR PROJECT GROWTH*

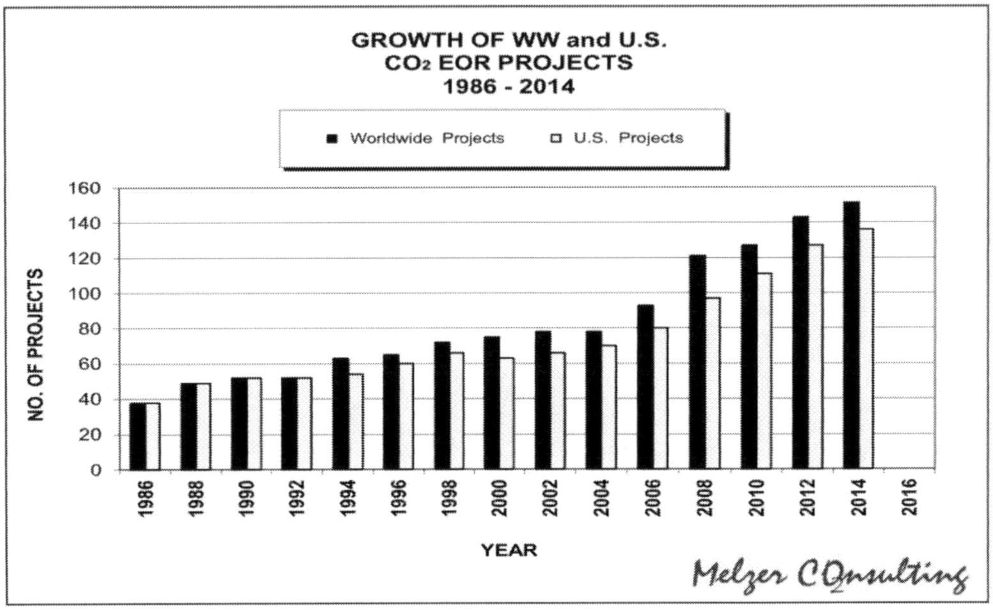

* Ref: O&GJ Biennial EOR Editions & UTPB Petr Industry Alliance

HISTORY OF WORLDWIDE AND U.S. CO_2 EOR PRODUCTION GROWTH*

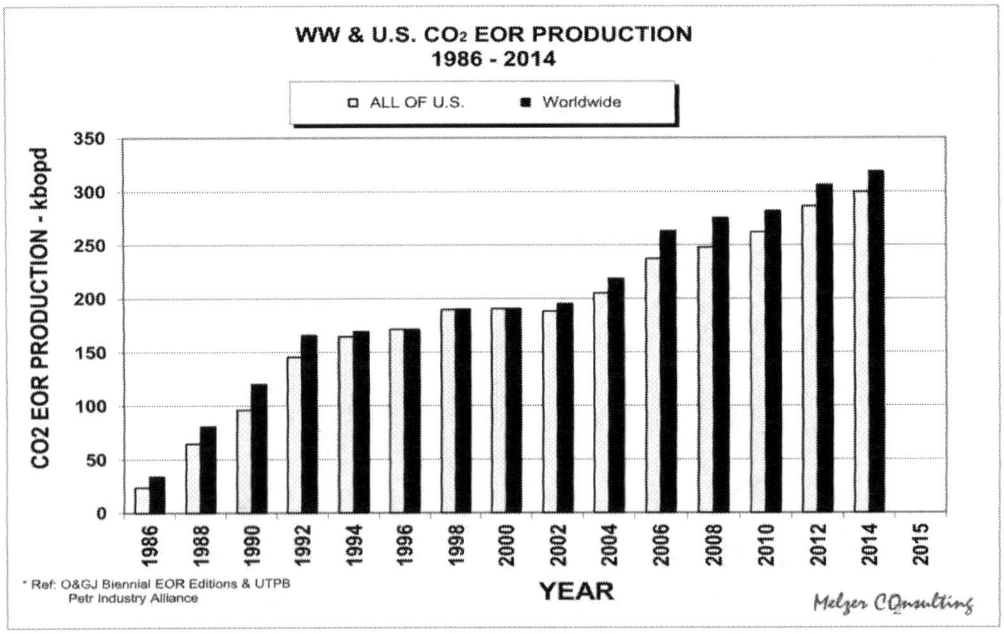

* Ref: O&GJ Biennial EOR Editions & UTPB Petr Industry Alliance

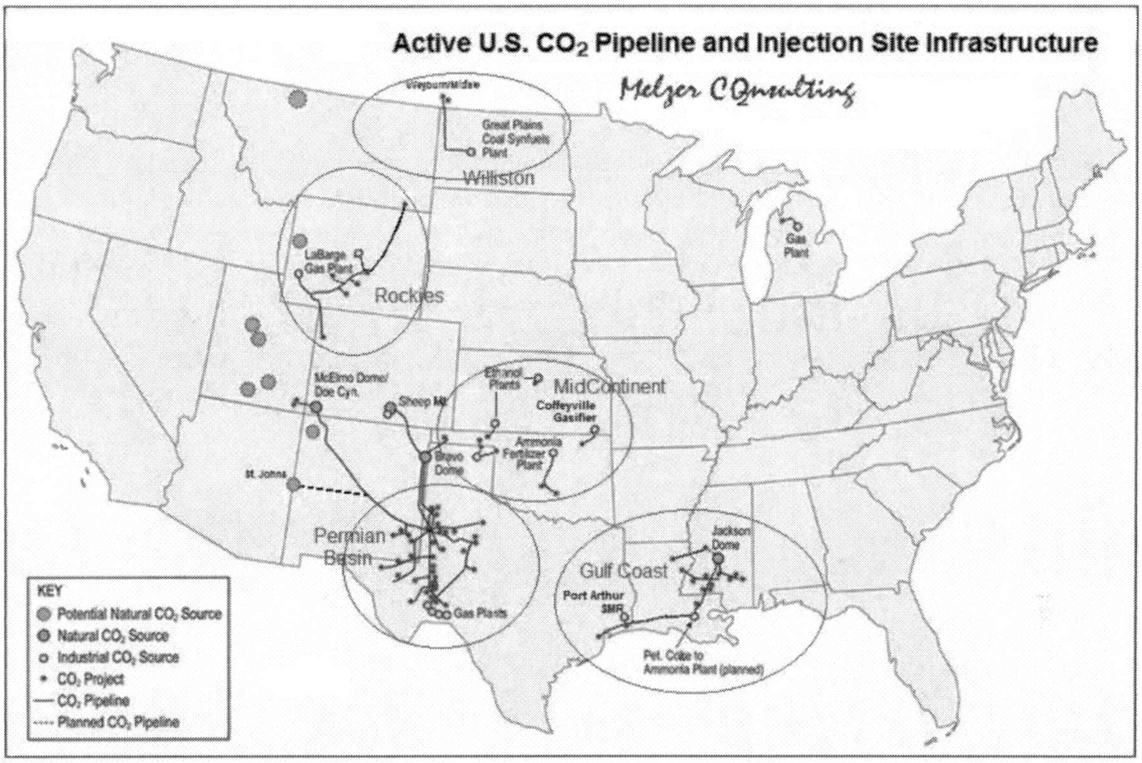

Carbon Management

Melzer CO2nsulting

The Phases of CO_2 Enhanced Oil Recovery

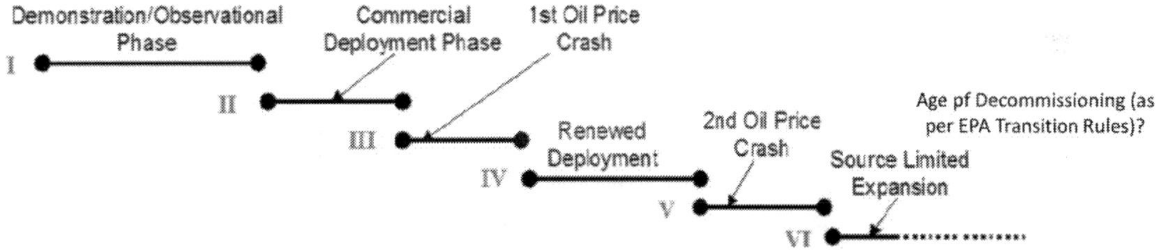

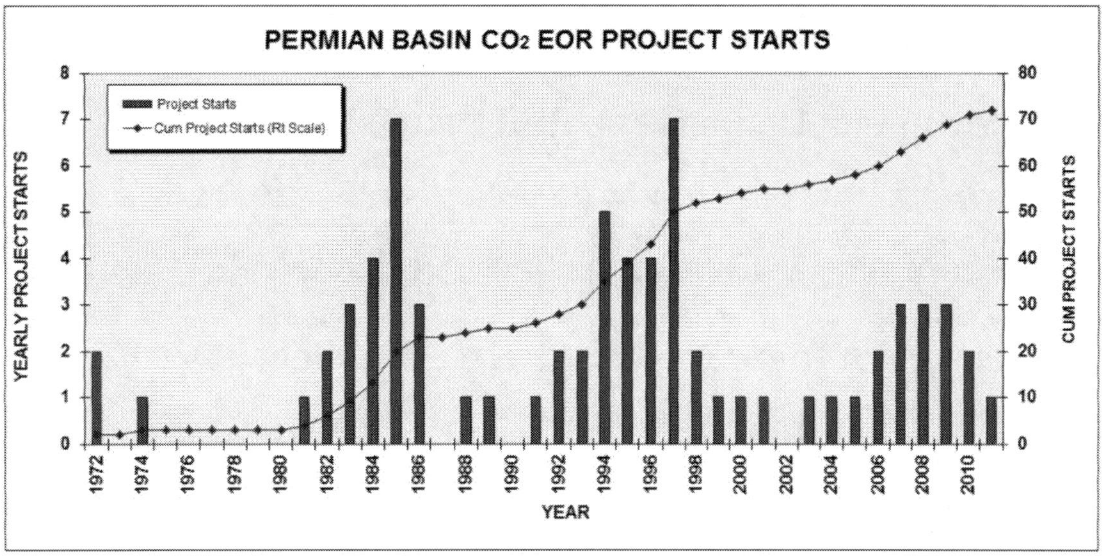

Residual Oil Zones (ROZs)

Key Concepts/Definitions for Residual Oil Zones (1)

- ROZ Definition: That Portion of a Reservoir Below the Oil/Water Contact Where Residual Oil Resides but the Mobile Phase is Water
- Brownfield: Defined as a ROZ that Exists Under a Developed Oil Field (Main Payzone {MPZ})
- Greenfield: A ROZ that has no Overlying MPZ
- Fairway: A Large Region Where Greenfield ROZs Exist and Align Geographically

A ROZ Brownfield: The Seminole Field (W. Tx) Original & Post Waterflood Saturation Profile

9

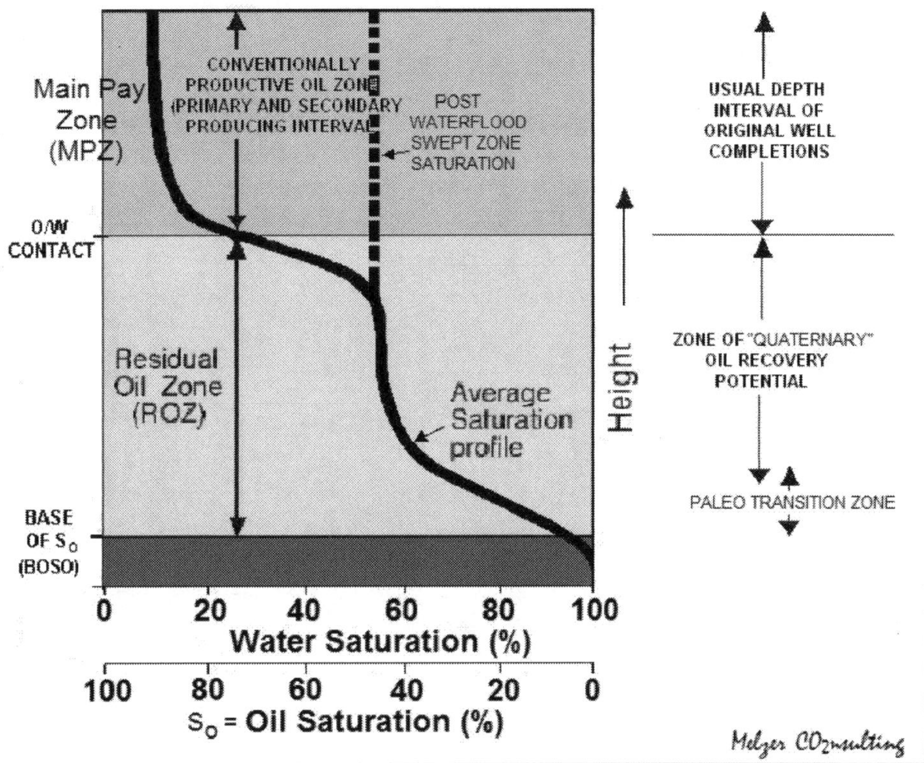

10

Key Concepts/Definitions for Residual Oil Zones (2)

- ROZ Types:
  - Type 1 ROZ – Post Oil Accumulation Tilt of a Basin or Area within a Basin, the OWC has readjusted and Water Has Encroached on a Portion of a Paleo Oil Trap
  - Type 2 ROZ – Water Encroachment into a Paleo Trap Due to Escape of Gas or Oil out of the Top (aka Breached Seal ROZ)
  - Type 3 ROZ – Laterally Flushed Paleo Trap Usually Due to Uplift on one Side of the Basin Exposing One or a Series of Reservoirs to Meteorically-derived Water Influx

CM Carbon Management

Melzer CQnsulting

11

Original Oil Accumulation Under Static Hydrological Conditions (a Hypothetic Example)

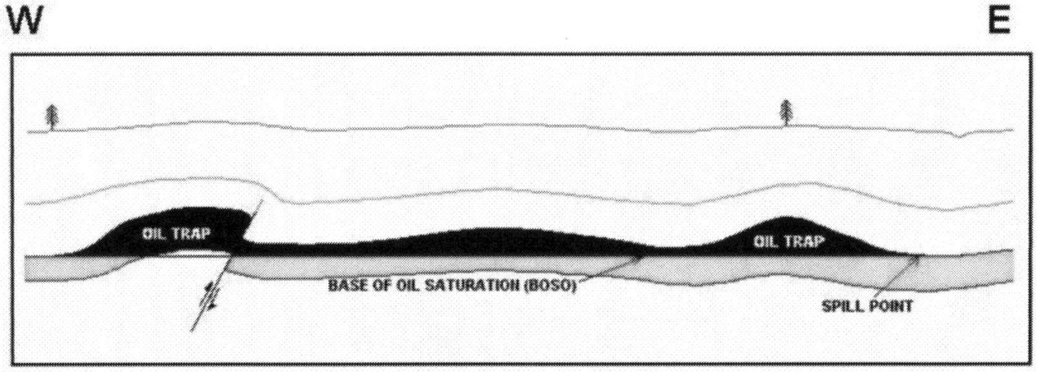

Melzer COnsulting

Carbon Management

Melzer COnsulting

12

RESIDUAL OIL ZONE: TYPE 1

Original Accumulation Subject to a Westward Regional Tilt & Forming a ROZ

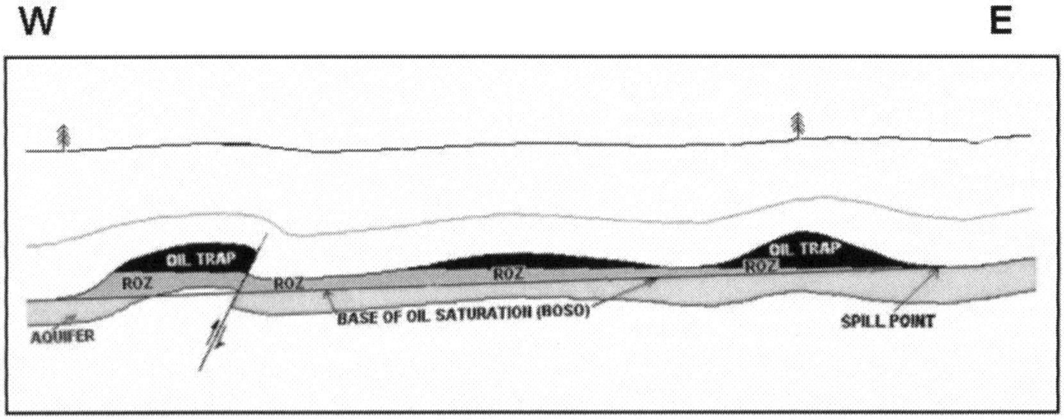

Melzer COnsulting

Carbon Management

13

RESIDUAL OIL ZONES: TYPE 2

Original Accumulation with a Breached then Repaired Seal & Forming a ROZ

ORIGINAL

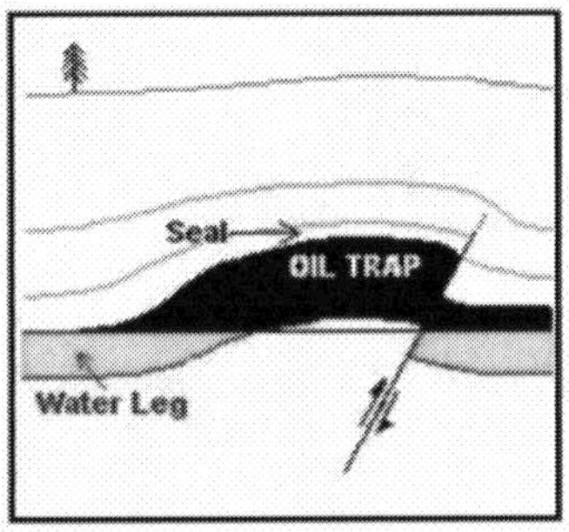

POST BREACH

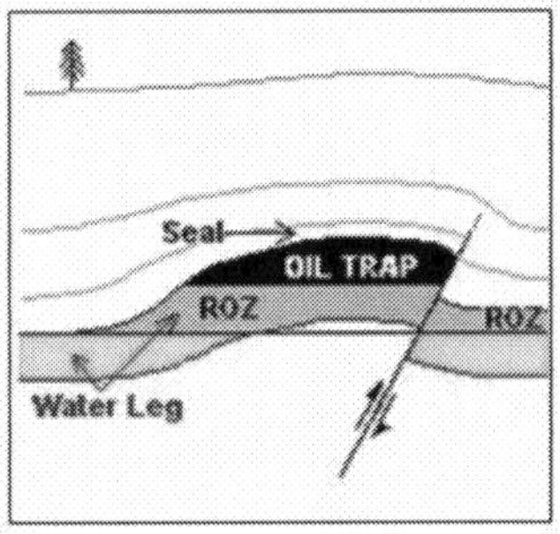

Melzer COnsulting

Carbon Management

14

RESIDUAL OIL ZONES: TYPE 3

Change in Hydrodynamic Conditions, Sweep of all or a Portion of the Oil Column, Oil/water Contact Tilt, and Development Of The Residual Oil Zone

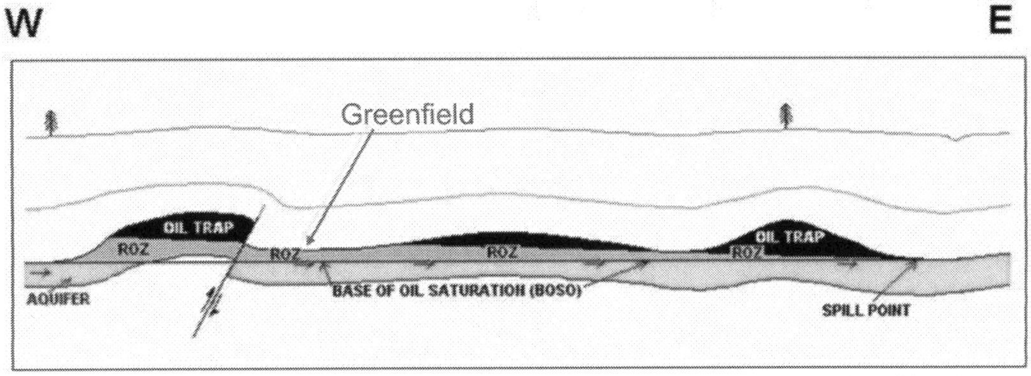

Melzer COnsulting

Carbon Management

Permian Basin ROZ Fairways*

* *Ref: RPSEA/DOE Report in Preparation*

15

Mapping of Active Major CO_2 EOR Projects in the Permian Basin and Residual Oil Zone Fairways

and Showing the First Pure ROZ Greenfield ROZ CO_2 Project

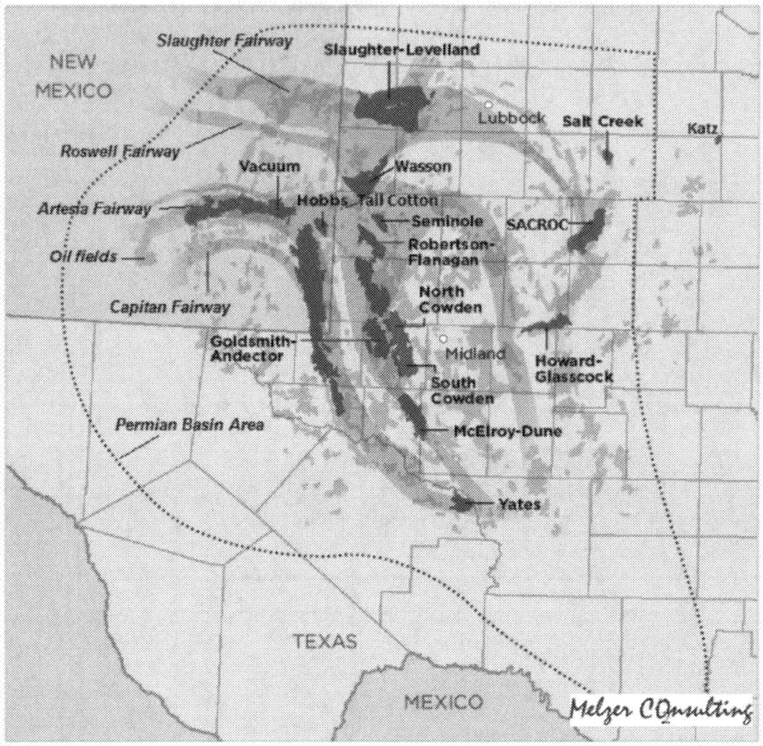

16

Remember this CO_2 EOR Phase Chart

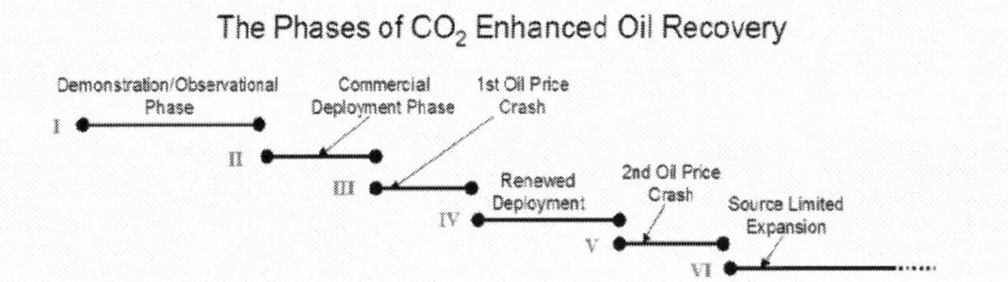

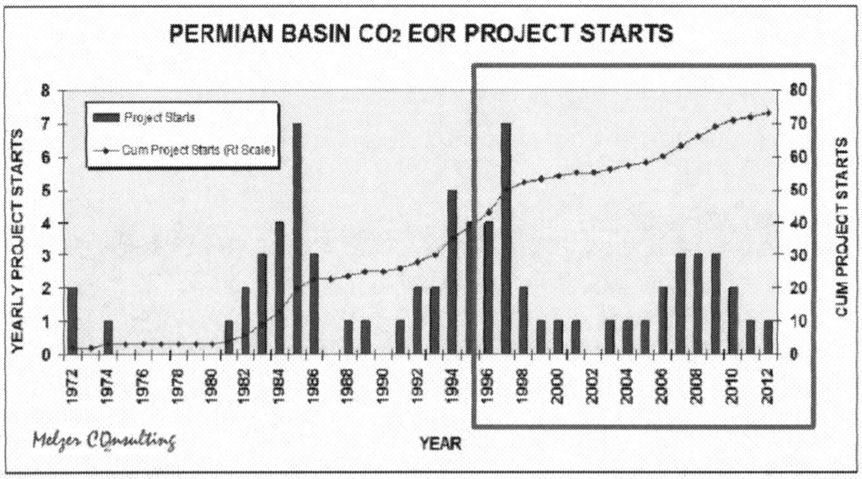

17

Residual Oil Zone Flooding: Project History and Phases of Development

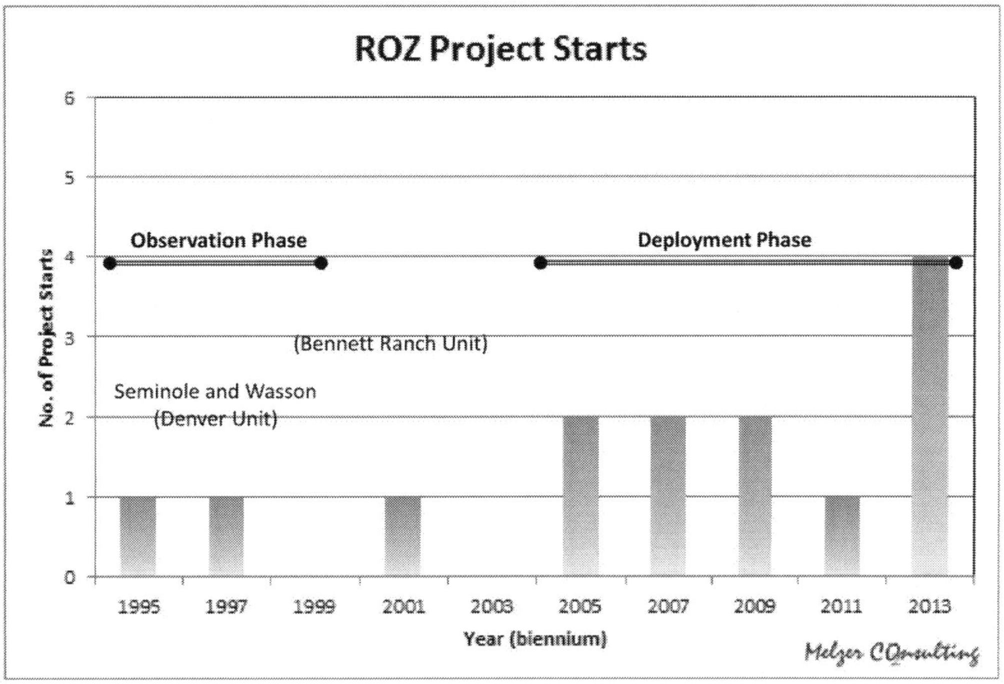

Carbon Management

18

The ROZ Projects Underway

Table 1

Melzer COnsulting
as of Jan 2015

ON-GOING ROZ CO_2 EOR PROJECTS
IN THE PERMIAN BASIN REGION OF THE U.S.

	Type and operator	Field	State	County	Top MPZ Depth, (ft)	Pay zone	Lith-ology	MPZ Start Date	ROZ Start Date
	Active CO_2 miscible								
1	Chevron	Vacuum San Andres Grayburg Unit	NM	Lea Co.	4,550	San Andres	Dolo.	2007	2007
2	Fasken	Hanford	Tex.	Gaines	5,500	San Andres	Dolo.	7/86	8/09
3	Hess	Seminole Unit-ROZ Phase 1	Tex.	Gaines	5,500	San Andres	Dolo.	7/83	7/96
4	Hess	Seminole Unit-ROZ Phase 2	Tex.	Gaines	5,500	San Andres	Dolo.	7/83	4/04
5	Hess	Seminole Unit-ROZ Stage 1 Full Field Dev	Tex.	Gaines	5,500	San Andres	Dolo.	7/83	10/07
6	Hess	Seminole Unit-ROZ Stage 2 Full Field Dev	Tex.	Gaines	5,500	San Andres	Dolo.	7/83	5/11
7	Hess	Seminole Unit-ROZ Stage 3 Full Field Dev	Tex.	Gaines	5,500	San Andres	Dolo.	7/83	7/13
8	Kinder Morgan CO_2	Goldsmith-Landreth Unit	Tex.	Ector	4,200	San Andres	Dolo.	8/09	8/09
9	Kinder Morgan CO_2	Tall Cotton Pilot Project *	Tex.	Gaines	5,250	San Andres	Dolo.	n/a	9/14
10	Occidental	Wasson Bennett Ranch Unit	Tex.	Yoakum	5,250	San Andres	Dolo.	6/95	2000
11	Occidental	Wasson Denver Unit	Tex.	Yoakum	5,200	San Andres	Dolo.	4/83	1995**
12	Occidental	Wasson ODC	Tex.	& Gaines	5,200	San Andres	Dolo.	Nov-84	2005?
13	Trinity CO_2	George Allen (BF&GF) *	Tex	Yoakum	4,900	San Andres	Dolo.	12/12	2012
14	Tabula Rasa	East Seminole	Tex	Gaines	5,400	San Andres	Dolo.	1/13	1/13
15	XTO/ExxonMobil	Salt Creek	Tex.	Kent	6,300	Canyon Lime	LS	Jun-05	1996
16	XTO/ExxonMobil	Means	Tex	Andrews	4,500	GrBrg/San Andres	Dolo.	Nov-83	1/12

Lithologies: Dolomite, Dolomite Anhydrite, except Salt Creek field (SC) which is a limestone
Pay Zones: San Andres (SC = Canyon Lime)
Completion Strategies: Commingle except in Projects # 4-7 (Wells were Deepened & Dedicated Injection into the ROZ) & Project # 13 (Wells were Deepened & Drilled into ROZ)
* **Brownfield/Greenfield:** All Projects except a portion of George Allen are Brownfields (possess an overlying MPZ); ROZ w/o an overlying MPZ = Greenfield
** Initial Dev called Transition Zone Sweetspot, now followed by Phases 1-4

Sources: SPE 167209, SPE 88770, SPE 81458, Melzer Consulting Database, CO_2 Flooding Conference Presentations (www.CO2Conference.net)

The Annual CO_2 Conference in Midland

www.CO2Conference.net

2015 CO_2 Conference Week: Midland, Texas

(All Events to be held in Downtown Midland Center (MC))

	Tues 12/8	Wed 12/9	Thur 12/10	Fri 12/11
7am	Check-in (MC)		Check-in (MC)	
8am	Session I Carbon Mgmt Workshop	Check-in (MC)	CO_2 Theme Session I: CO_2 EOR Policy and Flood Ops	Session III CO_2 Theme Session: Reservoir Case Histories
9am, 10am, 11am		Seminar entitled Case Histories and Science of Producing Oil from the ROZ		
12pm	Luncheon Keynote: Dan Cole, VP – Commercial Dvlpmnt & Gov't Relations	Check-in (MC)	Keynote Lunch Speaker: Kenneth A Golden, Commercial Adv for Exxon Mobil; "The Outlook for Energy: A View to 2040"	Light Lunch Midland Center
				Conf Conclusion
1pm, 2pm, 3pm, 4pm	Session II Carbon Mgmt Workshop	Field Trip to Seminole San Andres Unit CO_2 EOR Flood and Surface Facilities	CO_2 Theme Session II: CO_2 Flood Ops & Reservoir	
5pm				
6pm, 7pm, 8pm	Reception - Midland Center	Reception - Midland Center	Reception - Perm Basin Petroleum Museum	

LEGEND
EOR Carbon Management
CO2 Flooding Conference
STIPPLED EVENTS HAVE LIMITED SEATING AND REQUIRE SEPARATE RESERVATIONS

CARBON MANAGEMENT TECHNOLOGY CONFERENCE
ENHANCED OIL RECOVERY (EOR)/STORAGE

POSSIBLE IMPEDIMENTS TO EOR/CO_2 SEQUESTRATION

- Interstate CO_2 pipelines are not regulated by Federal Energy Regulatory Commission (FERC).
- Intrastate CO_2 pipelines are regulated by each state's constitution.
- The absence of State and Federal eminent domain authority.
- Greater Sage Grouse listing as endangered.
- Particulate matter air Quality non-attainment.
- State and Federal laws governing EOR and Carbon Sequestration.

CARBON MANAGEMENT TECHNOLOGY CONFERENCE
ENHANCED OIL RECOVERY (EOR)/STORAGE

WYOMING STATE LAWS GOVERNING CARBON SEQUESTRATION

- State of Wyoming paving the way on laws governing carbon sequestration
  - Specifies that the owner of the surface estate owns the subsurface pore space.
  - Specifies that the owner of subsurface minerals is dominant over the pore space.
  - Specifies that the injector of CO_2 into the subsurface pore space owns the CO_2 and is responsible for the possible consequences of the injection.
  - Adopts a procedure for unitizing subsurface pore space so that multiple owners can utilize the sequestration formation.
  - Directs the Wyoming Department of Environmental Quality to develop regulations, permits, and bonding requirements for the injection and sequestration of CO_2.
  - Establishes a special revenue account funded by sequestration permits to pay for the measurement, monitoring, and verification of CO_2 sequestration.
  - Similar and consistent laws are needed for carbon sequestration on public lands.

Opportunities for Enhanced Oil Recovery in Wyoming

David Mohrbacher, P.E.

Director – Wyoming Enhanced Oil Recovery Institute

ENHANCED OIL RECOVERY INSTITUTE

UNIVERSITY OF WYOMING

Centers of Excellence

- Enhanced Oil Recovery Institute
- Carbon Management Institute
- Wyoming Reclamation and Restoration Center
- Coal Bed Natural Gas Center
- Wind Energy Research Center
- Center for Energy Economics and Public Policy
- Center for Photoconversion and Catalysis
- Center for Fundamentals of Subsurface Flow

ENHANCED OIL RECOVERY INSTITUTE

UNIVERSITY OF WYOMING

Enhanced Oil Recovery Institute's Mission

- EORI was created by the Wyoming Legislature in 2004.
- Employs 22 full-time staff.
- Located at the University of Wyoming.
- Assists Wyoming operators with application of new technology and techniques.
- Completes field demonstrations.

ENHANCED OIL RECOVERY INSTITUTE

UNIVERSITY OF WYOMING

Producing Wyoming's Stranded Oil

- Oil Recovery
  - "Primary Recovery": wells flow or are pumped
- Improved Oil Recovery (IOR)
  - "Secondary Recovery": wells are flooded with water
- Enhanced Oil Recovery (EOR)
  - "Tertiary Recovery": wells are flooded with chemicals, CO_2, etc.

ENHANCED OIL RECOVERY INSTITUTE

UNIVERSITY OF WYOMING

CO_2 Flooding

ENHANCED OIL RECOVERY INSTITUTE

UNIVERSITY OF WYOMING

Historic Oil Production

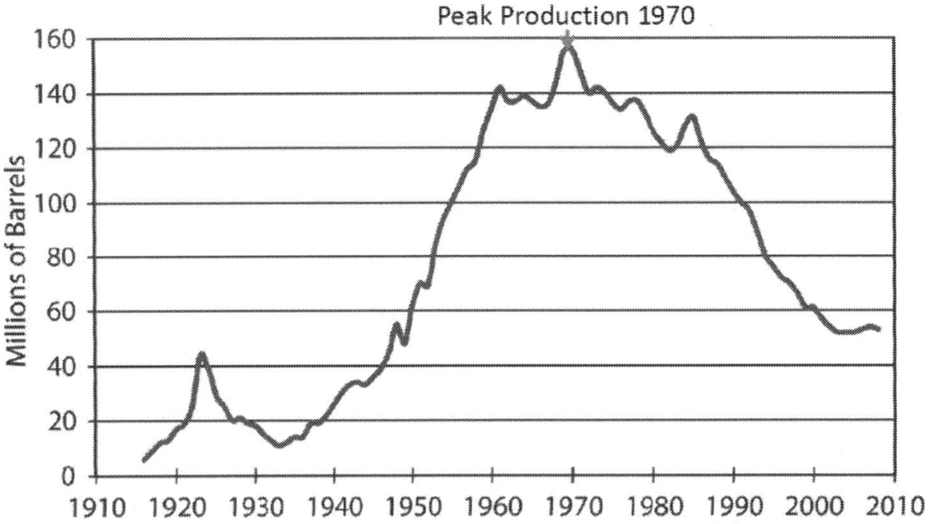

ENHANCED OIL RECOVERY INSTITUTE

UNIVERSITY OF WYOMING

CO_2-EOR in Wyoming

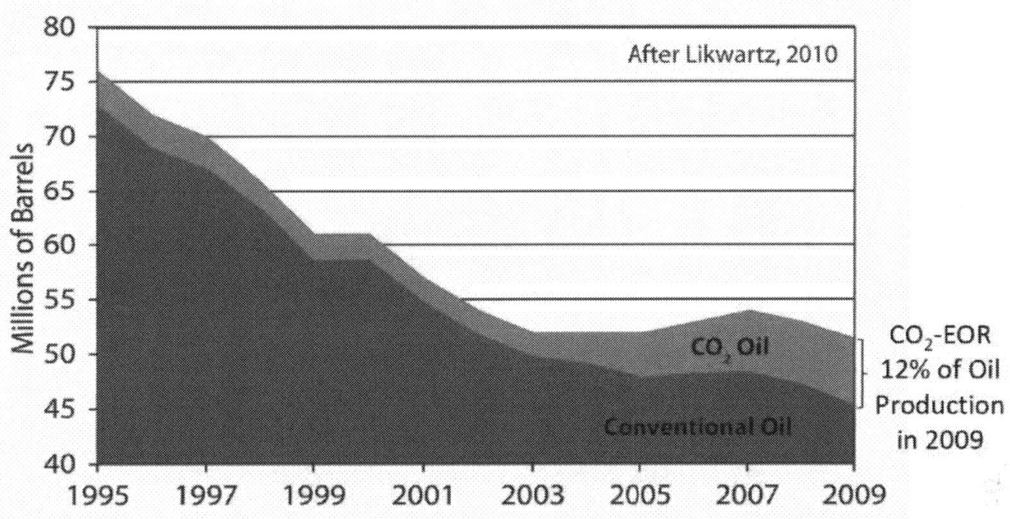

ENHANCED OIL RECOVERY INSTITUTE

UNIVERSITY OF WYOMING

Production of Wyoming's Stranded Oil

- Oil fields in Wyoming have produced about 7 billion barrels of oil since the industry first began.
- An additional 2 to 3 billion barrels of tertiary oil may be produced through EOR methods in Wyoming over the next 20 to 40 years.

ENHANCED OIL RECOVERY INSTITUTE

UNIVERSITY OF WYOMING

Economic Benefits for Wyoming

- A billion barrels of oil will result in an estimated **$9.2 billion** of severance and ad velorem taxes to the state of Wyoming and Wyoming counties.
- Additional **high paying energy jobs** will be sustained or created in oil producing portions of the state.
- **Additional income to the state** will include property taxes, mineral royalties paid to individual land owners and the state, and sales and use taxes.

ENHANCED OIL RECOVERY INSTITUTE

UNIVERSITY OF WYOMING

Current CO_2 Infrastructure

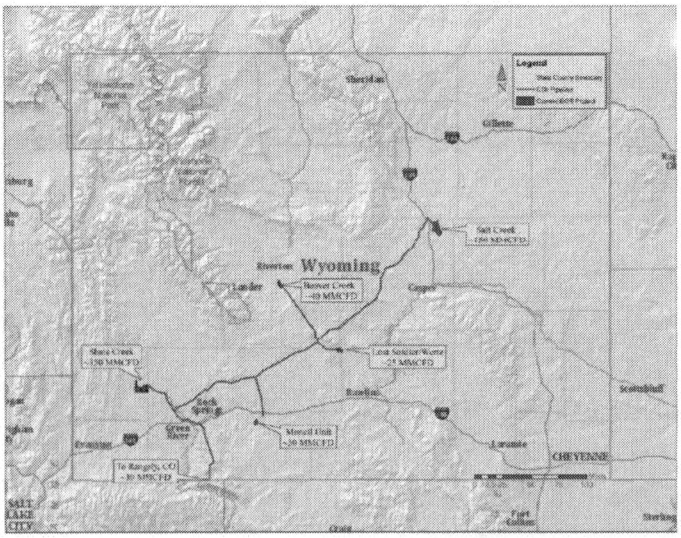

ENHANCED OIL RECOVERY INSTITUTE

UNIVERSITY OF WYOMING

Proposed CO_2 Infrastructure

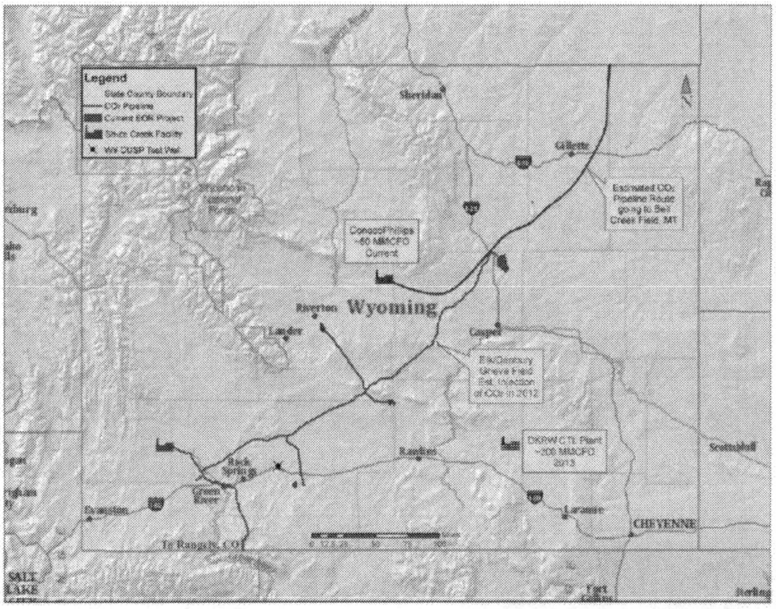

ENHANCED OIL RECOVERY INSTITUTE

UNIVERSITY OF WYOMING

Session 8 – Panel Discussion: Small Test Centers

CMTC 2015: Sustainable and Economical CCUS Options
Sugar Land, TX
November 18, 2015

Energy & Environmental Research Center, University of North Dakota

Center for Applied Energy Research, University of Kentucky

Research Institutes, Carbon Management Canada

The Institute for Clean and Secure Energy, University of Utah

Energy & Environmental Research Center (EERC) – Grand Forks, North Dakota

More than 254,000 square feet of state-of-the-art laboratory, demonstration, and office space.

EERC

EERC Carbon Management

- Plains CO_2 Reduction (PCOR) Partnership
- CO_2 Storage and Use in Oil Production
- CO_2 Capture Testing – Multiple Pilot Systems for Pre- and Post-Combustion
- Support of Industry Planning and Demonstration Projects.

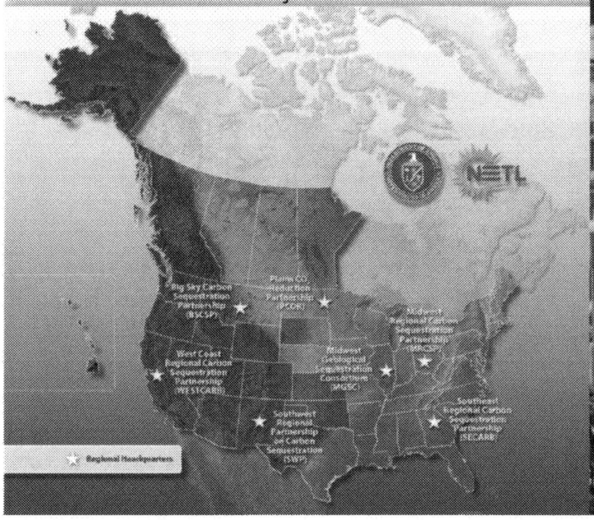

University of Kentucky CAER UK

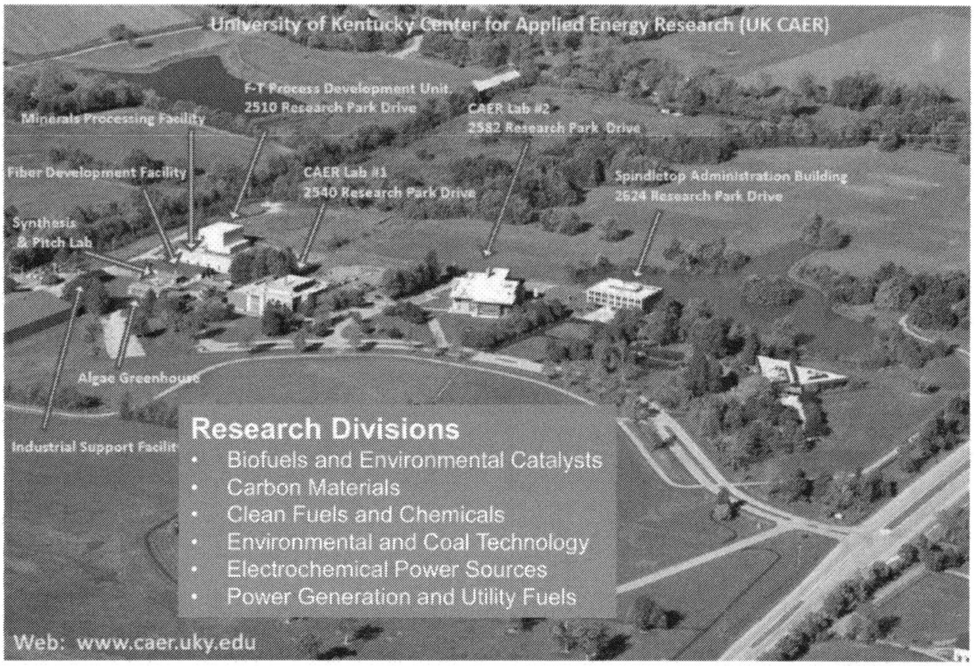

CO_2 Capture Technology Development Available at CAER

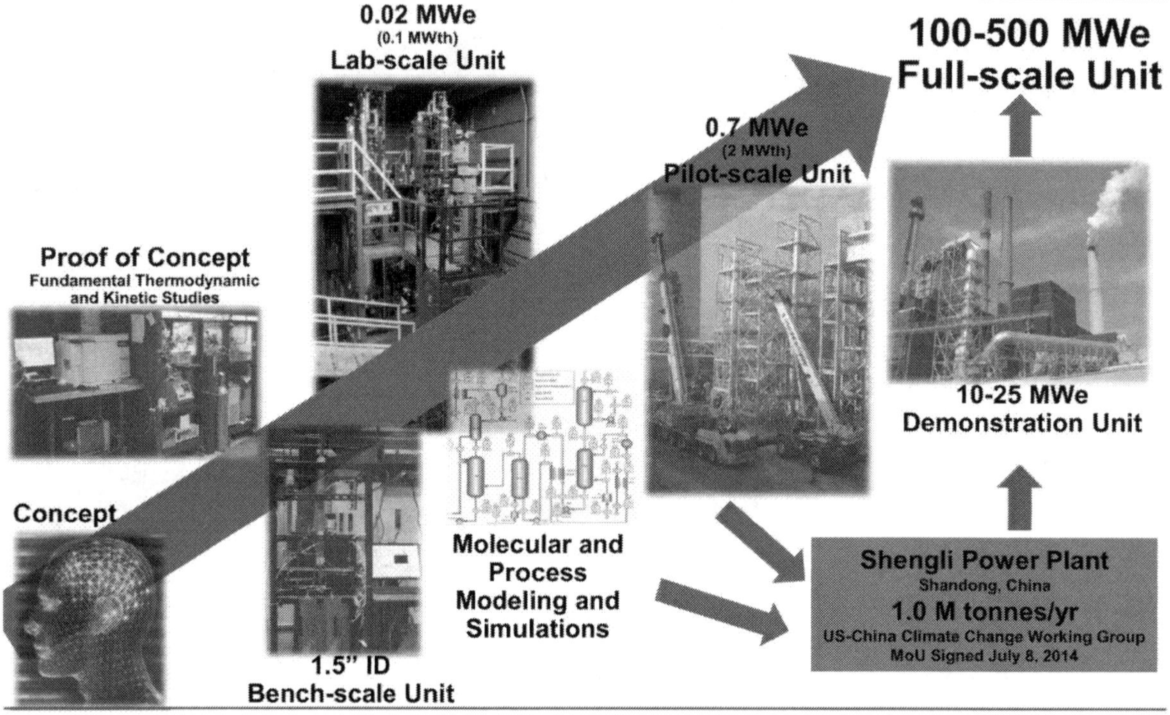

CARBON CAPTURE & CONVERSION INSTITUTE (CMC.CCCI)

Industrial process innovation:

- Integration
- Adaptation
- Scale-up
- Testing
- Performance verification

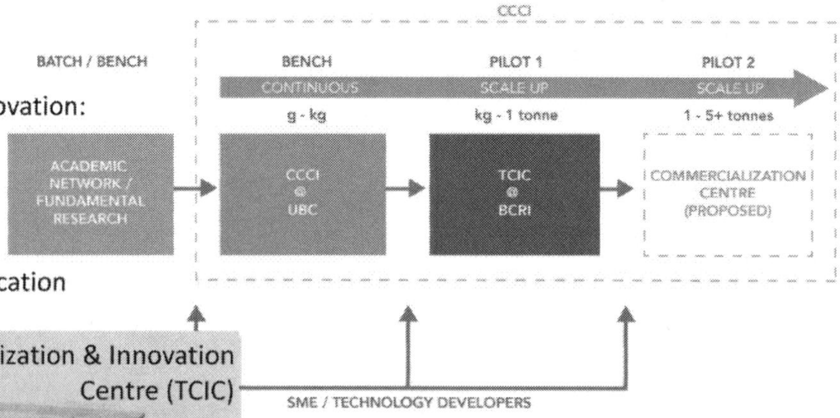

BC RESEARCH

a place of mind
THE UNIVERSITY OF BRITISH COLUMBIA

CMC | Carbon Capture & Conversion Institute

6

CONTAINMENT & MONITORING INSTITUTE (CMC.CaMI)

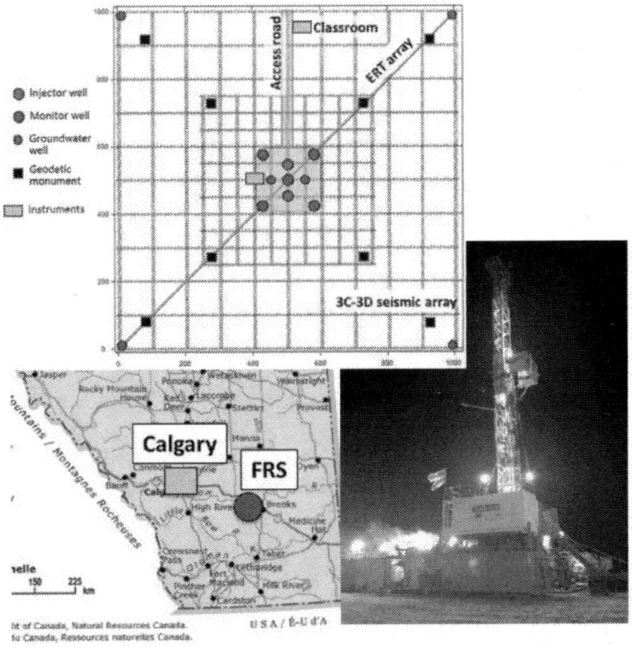

Field Research Station (FRS#1)

- ¾ section of land
  - Surface
  - Subsurface to 750m+
- Injection zones
  - 300m, 550m
- Monitoring
  - Shallow aquifer
  - Surface: fugitives, venting, soil, atmosphere
  - Injection zones
- Focus areas
  - CO_2 storage & containment
  - Methane management
  - Measurement, monitoring & verification

CMC | Containment & Monitoring Institute

7

CO_2LLABORATION

New pathways to reduce industrial greenhouse gas emissions.

CMC RESEARCH INSTITUTES

Contact:
Richard Adamson, PEng MSc
President
Richard.Adamson@cmcghg.com

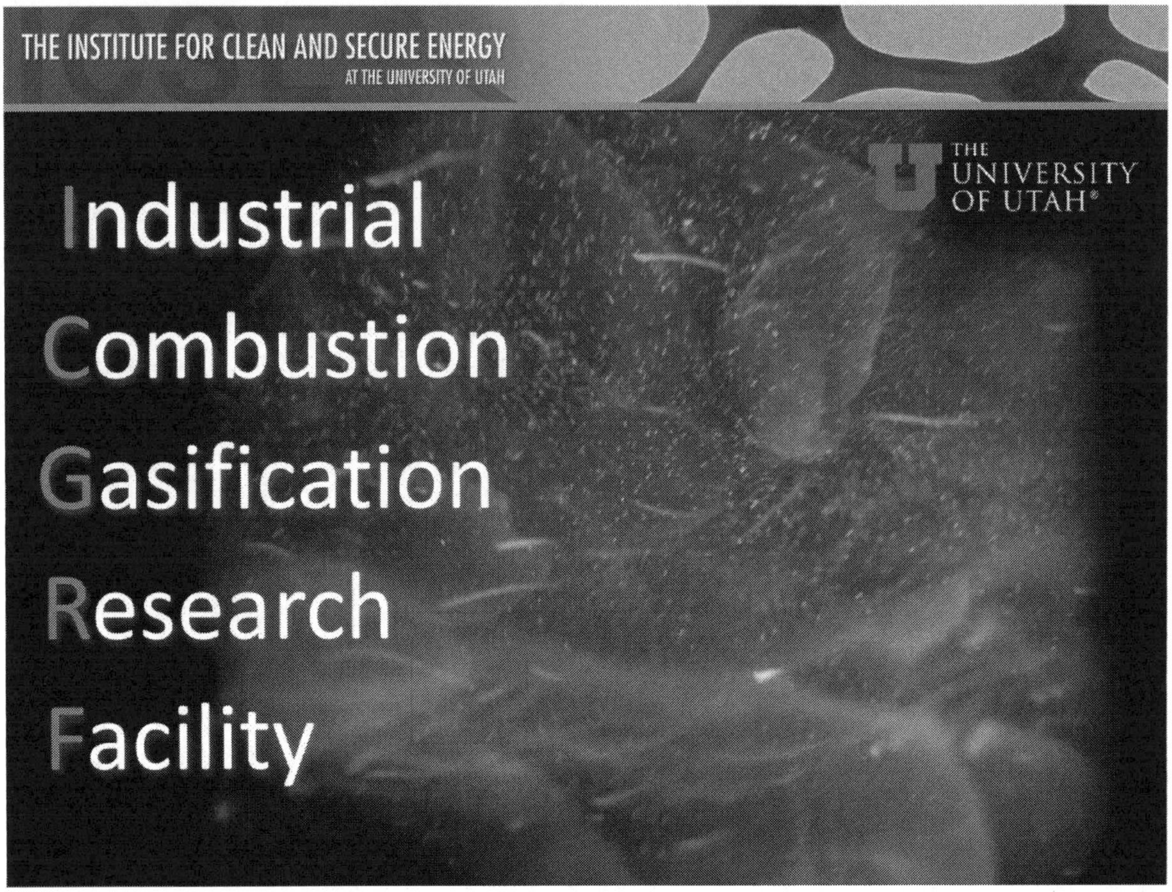
THE INSTITUTE FOR CLEAN AND SECURE ENERGY
AT THE UNIVERSITY OF UTAH
THE UNIVERSITY OF UTAH®
Industrial
Combustion
Gasification
Research
Facility

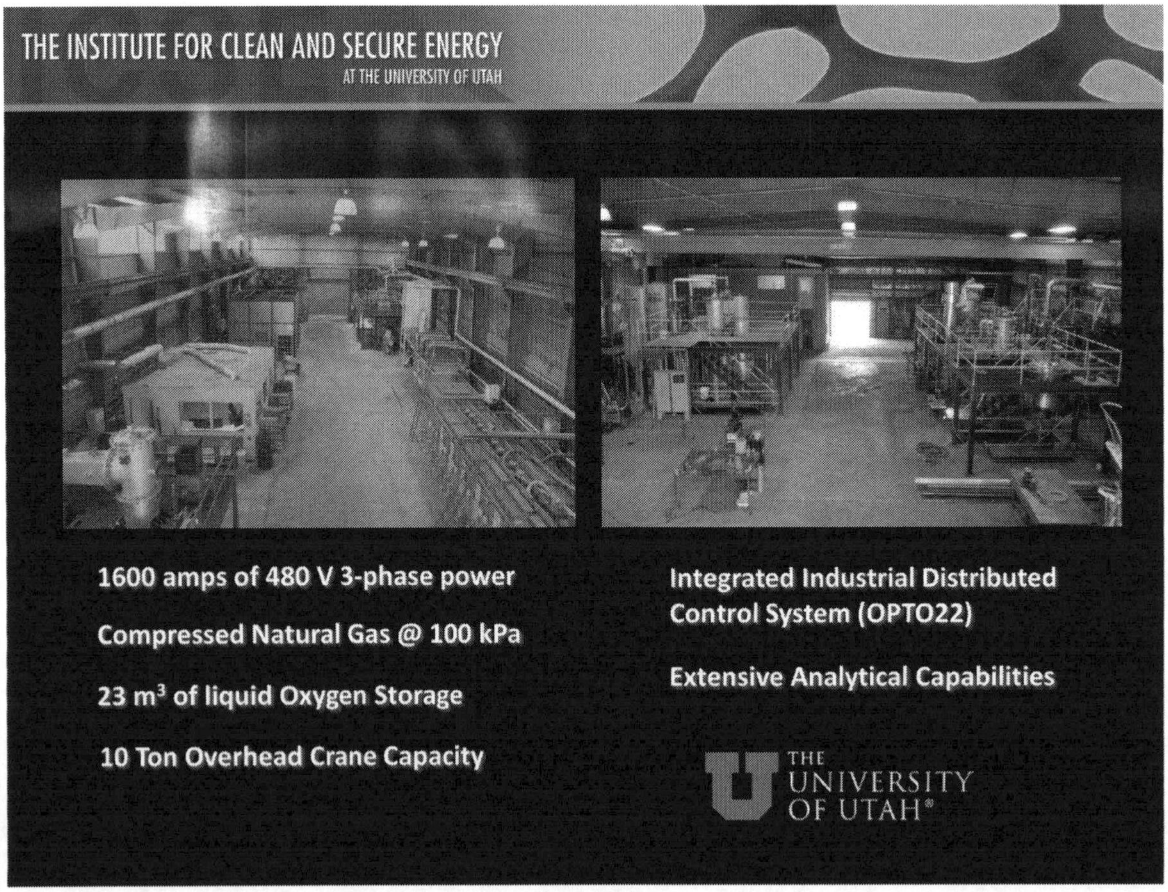
THE INSTITUTE FOR CLEAN AND SECURE ENERGY
AT THE UNIVERSITY OF UTAH
1600 amps of 480 V 3-phase power
Compressed Natural Gas @ 100 kPa
23 m³ of liquid Oxygen Storage
10 Ton Overhead Crane Capacity
Integrated Industrial Distributed Control System (OPTO22)
Extensive Analytical Capabilities
THE UNIVERSITY OF UTAH®

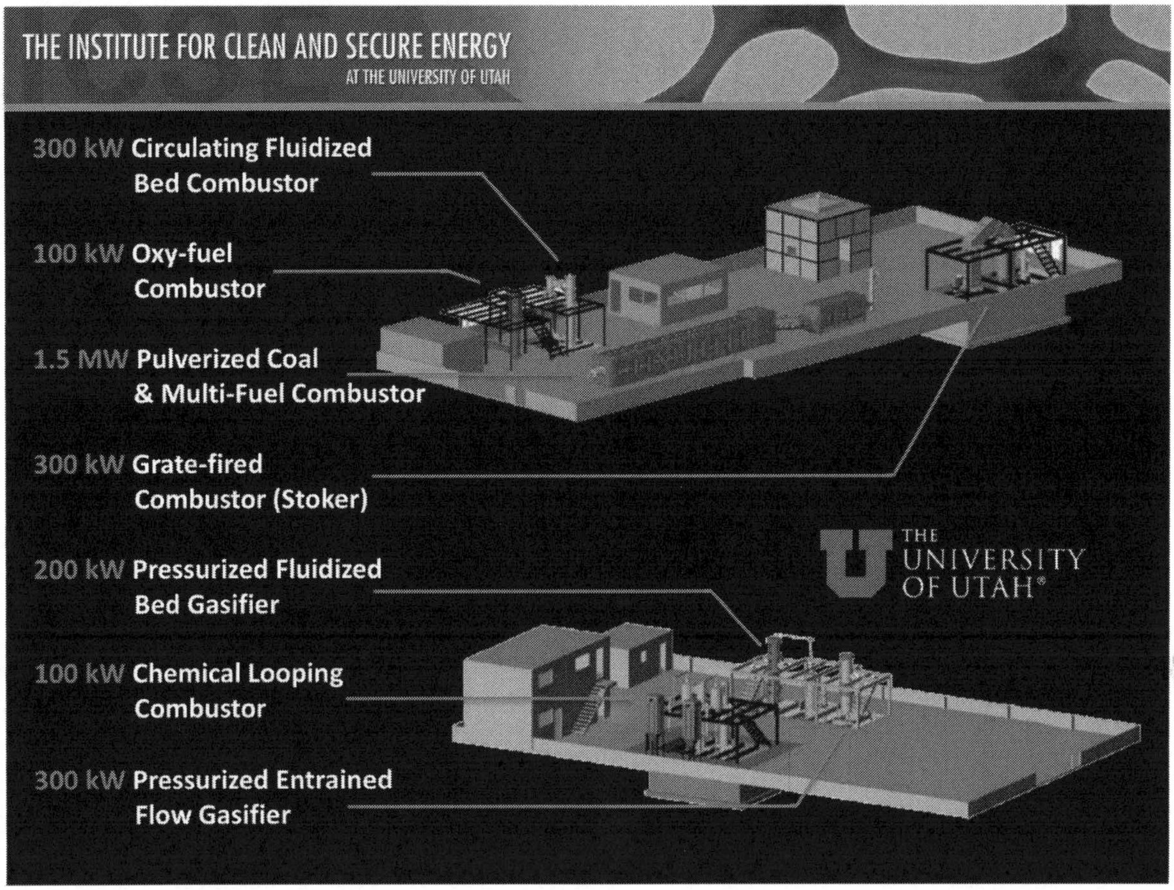

Session 8 – Panel Discussion: Small Test Centers

CMTC 2015: Sustainable and Economical CCUS Options
Sugar Land, TX
November 18, 2015

Michael J. Holmes
Director for Energy systems Development
Energy & Environmental Research Center

EERC
Energy & Environmental Research Center (EERC)

Energy & Environmental Research Center (EERC) – Grand Forks, North Dakota

More than 254,000 square feet of state-of-the-art laboratory, demonstration, and office space.

EERC

EERC Carbon Management

- Plains CO_2 Reduction (PCOR) Partnership
- CO_2 Storage and Use in Oil Production
- CO_2 Capture Testing – Multiple Pilot Systems for Pre- and Post-Combustion
- Support of Industry Planning and Demonstration Projects.

Pilot Facilities and Capabilities

THE UNIVERSITY OF NORTH DAKOTA

Combined Combustion Facilities

- Two "sister" pilot-scale units
- 550,000 (single) to 1.1 million Btu/hr (combined)
- 60 to 130 scfm (single), or 260 scfm (combined) flue gas
- Air- or O_2-fired
- Multifuel capability (coal, biomass, gas, liquid fuel, sludge, and municipal solid waste)

- Features
  - Air preheater and heat exchangers
  - Adjustable-swirl burner
  - Deposition section
  - Numerous ports
  - Selective catalytic reduction reactor
  - Baghouse
  - Electrostatic precipitator
  - Sulfur scrubber
- Testing of fuels and additives
  - Fouling and slagging
  - Corrosion
  - Hg
  - NO_x
  - SO_x
  - CO_2 capture
  - Particulates
  - Heat flux
  - Infrared flame characterization

EERC The International Center for Applied Energy Technology® 5

CO_2 Capture System

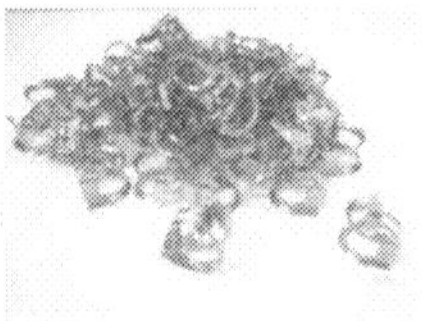

- Three 10-inch-diameter packed columns
  - Two absorbers and one stripper that are designed for maximum flexibility.
  - Able to evaluate different solvents by adjusting the column height.
  - Absorber columns are operated in series.
  - Packing type (both random and structured) can easily be changed.
- Water wash column used to monitor solvent slip.
- Very highly instrumented to allow for tight control and accurate, precise measurement of parameters.
- Can process up to 160 scfm of flue gas.
- Also performed solvent-based SO_2 scrubbing.

EERC The International Center for Applied Energy Technology®

6

Small Pilot-Scale Gasification Systems

- Entrained-flow gasifier (EFG)
  - ♦ 2750°F (1510°C)
  - ♦ 300 psi (20 atm)
- High-pressure fluid-bed gasifier (FBG)
  - ♦ 1800°F (980°C)
  - ♦ 1000 psi (68 atm)
- Process optimization, fuel behavior, ash and slag behavior, warm-gas cleanup, gas separation, and chemicals and liquid fuel production.
- Biomass to petcoke.
- Gasification, catalytic gasification, hydrogasification, methanation, and pyrolysis.
- Highly reconfigurable systems.

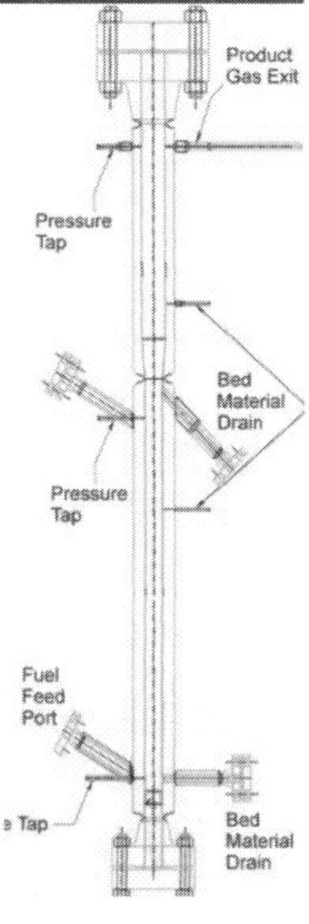

EERC The International Center for Applied Energy Technology®

Long-Term Demonstration of Hydrogen Production from Coal at Elevated Temperature – Hydrogen Membrane Consortium

- Project goal: Evaluate the performance of hydrogen separation membranes on coal-derived syngas produced from Powder River Basin (PRB) fuels.
- Over 800 membrane-hours of testing was completed with EERC gasifiers.
- Supports the U.S. Department of Energy (DOE) goals of low-cost CO_2 capture technologies.

EERC The International Center for Applied Energy Technology®

Gas Cleanup and Hydrogen Separation

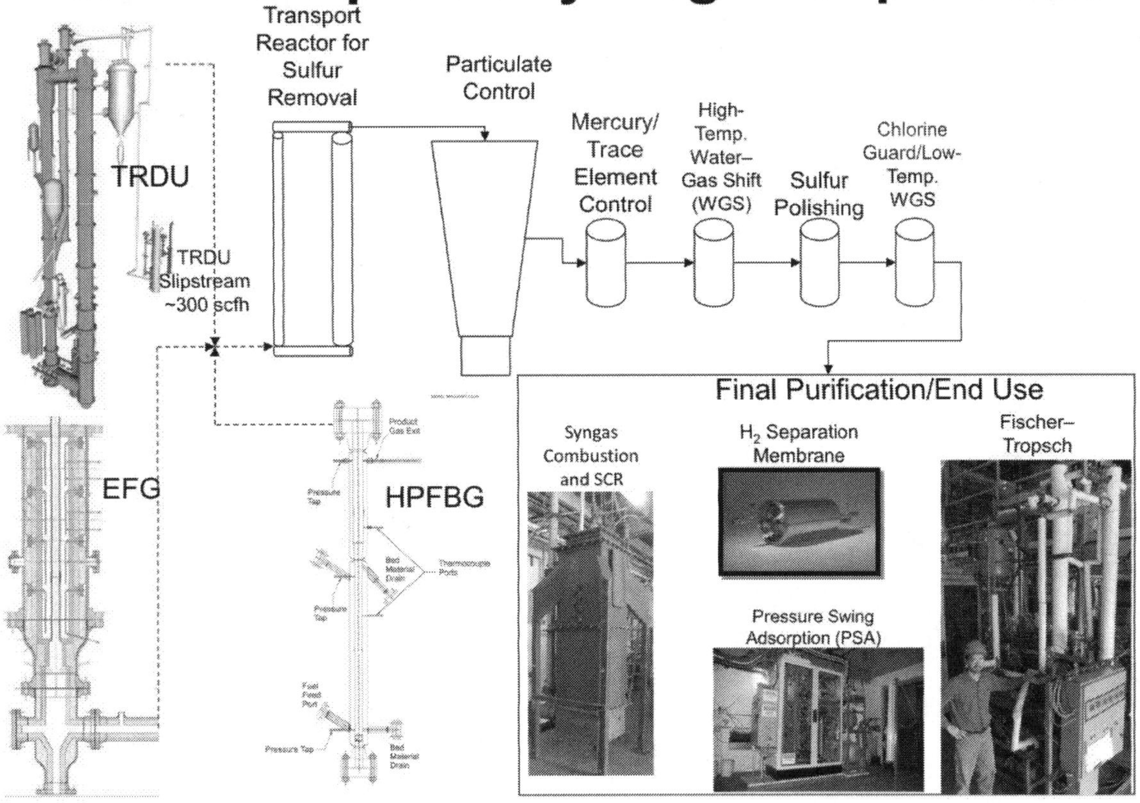

Systems Testing in Support of Liquid Fuels Development for Military Applications

- The Connecticut Center for Advanced Technology (CCAT) has teamed with the EERC to evaluate the cogasification of coal and biomass blends in pilot-scale gasification systems.
- Indirect liquefaction for coproduction of liquid fuels with electricity, hydrogen, and chemicals.
- Two systems were chosen for the test runs.
  - Transport reactor development unit (TRDU)—pilot-scale version of the full-scale system currently being installed in Kemper County, Mississippi.
  - Entrained-flow gasifier (EFG)—represents many commercially available entrained-flow systems.

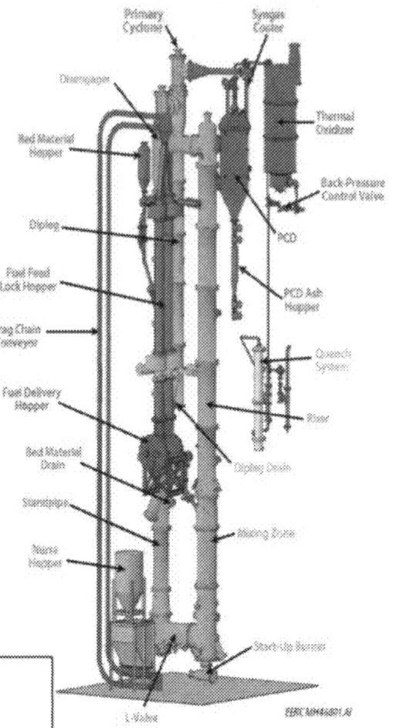

A U.S. Department of Defense (DoD) objective is to be prepared to procure 50% of domestic fuel from alternative sources by 2016.

EERC The International Center for Applied Energy Technology®

Aspen Modeling Capabilities

- Modeling gasification and combustion processes :
  - Combustion and steam generation
  - Gasification and gas separation
  - Steam cycle
  - CO_2 capture

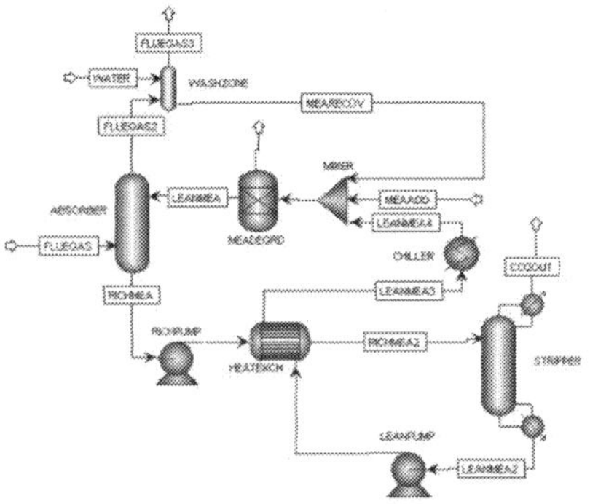

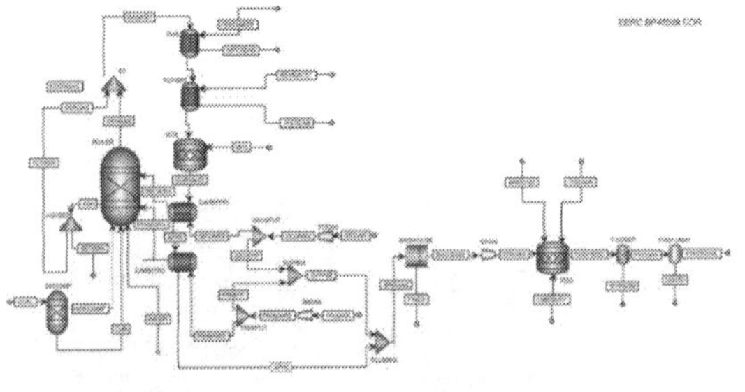

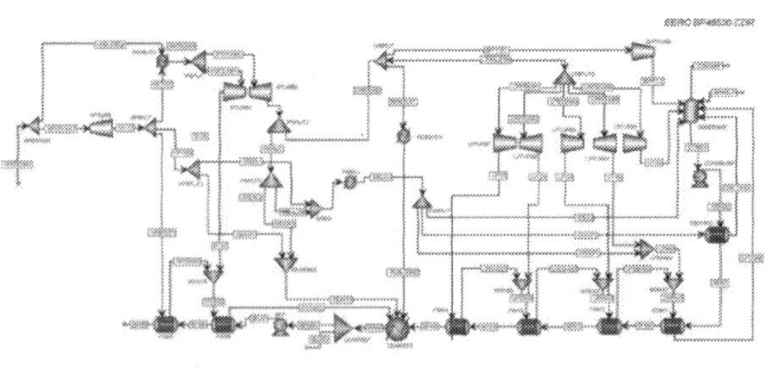

EERC The International Center for Applied Energy Technology®

11

Economic Evaluations

- All gasification, combustion, generation, and CO_2 capture modeling follows DOE NETL guidelines.
- Pilot-scale-derived data used as input for Aspen modeling.
- CO_2 compression to 2000 psi included but no transport (to the fence only).
- Pilot-scale-derived data used as input for Aspen modeling.

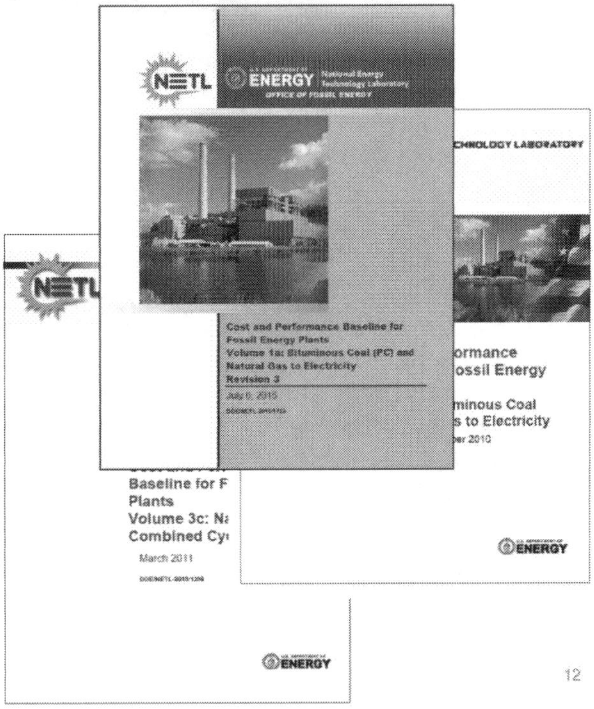

EERC The International Center for Applied Energy Technology®

12

CO_2 Capture Solvent Testing Results Coal-Fired System

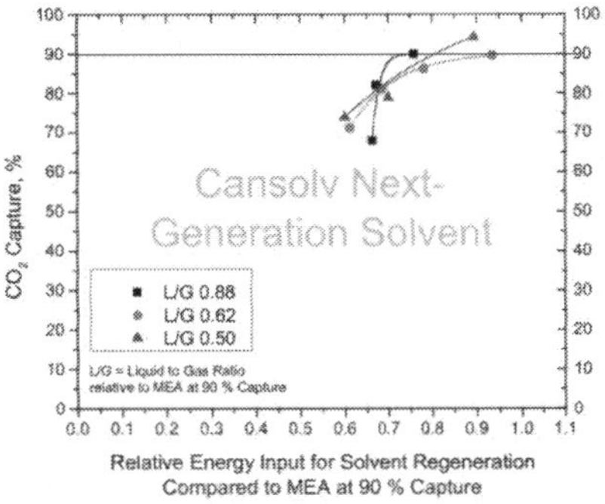

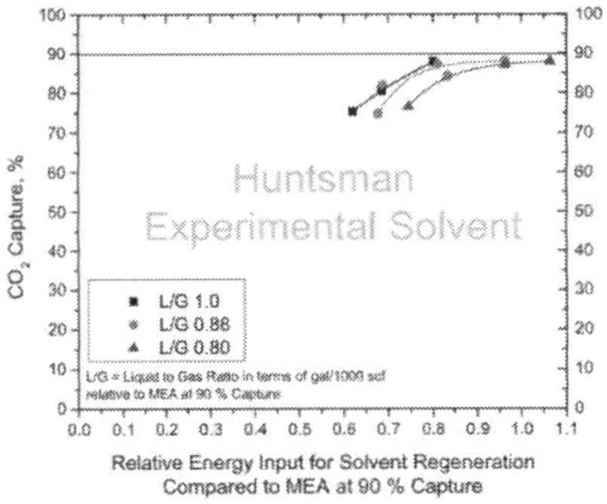

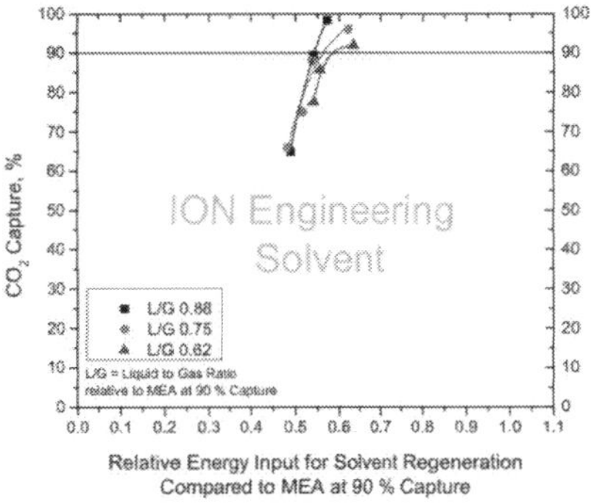

13

Economic Analysis – Tracking the Potential to Reduce Carbon Capture Costs

	DOE Report, Revision 2a		EERC Case 10 Models - Using Pilot-Scale Data for Advanced Solvents			
	Case 9 No Capture	Case 10 With Capture	MEA	ION	Huntsman	Cansolv
Gross Output, MW	582.6	672.7	660.3	623.8	646.0	641.7
Net Output, MW	550	550	550	550	550	550
Potential COE, $2011/MWh and Potential improvements	82	142	Baseline	0.87	0.95	0.92
Potential CO_2 Capture Cost, $2011/ton and Potential Improvements	–	59	Baseline	0.74	0.91	0.85

Potential costs projected here are based on pilot scale testing. Technologies must be verified through further pilot-scale evaluation, scale-up demonstrations, and long-term operation with a variety of coals and plant configurations.

EERC The International Center for Applied Energy Technology®

14

Contact Information

Energy & Environmental Research Center
University of North Dakota
15 North 23rd Street, Stop 9018
Grand Forks, North Dakota 58202-9018
www.undeerc.org

Mike Holmes, Director of Energy Systems Development
mholmes@undeerc.org
Telephone No. (701) 777-5276

EERC The International Center for Applied Energy Technology®

15

Acknowledgment

This material is based upon work supported by the U.S. Department of Energy National Energy Technology Laboratory under Award No. DE-FC26-08NT43291.

Disclaimer

This presentation was prepared as an account of work sponsored by an agency of the United States Government. Neither the United States Government, nor any agency thereof, nor any of their employees, makes any warranty, express or implied, or assumes any legal liability or responsibility for the accuracy, completeness, or usefulness of any information, apparatus, product, or process disclosed or represents that its use would not infringe privately owned rights. Reference herein to any specific commercial product, process, or service by trade name, trademark, manufacturer, or otherwise does not necessarily constitute or imply its endorsement, recommendation, or favoring by the United States Government or any agency thereof. The views and opinions of authors expressed herein do not necessarily state or reflect those of the United States Government or any agency thereof.

CAER UK

University of Kentucky Center for Applied Energy Research

Small test centers

Kunlei Liu

Power Generation and Utility Fuels Group

University of Kentucky, Center for Applied Energy Research

http://www.caer.uky.edu/powergen/home.shtml

Background on UKy-CAER

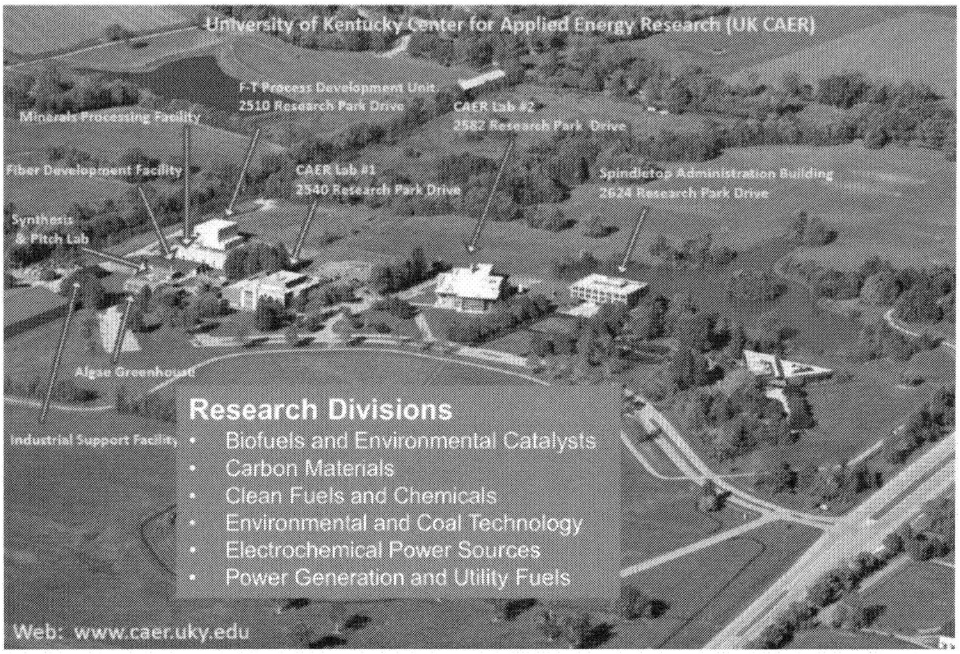

CMTC 2015 November 17 – 19, 2015

Facilities and Apparatus

Pilot Unit

Coal-fired FGG

Pre-concentrating Membrane

13.7 cfm Unit

Gas Sampling Unit

Emissions testing addresses EH&S concerns

- Integrated CO_2 capture process
- Solvent performance evaluation under real conditions with coal-derived flue gas
- Degradation studies for solvent degradation and make-up rates
- Flexibility for integration of other technologies (e.g., membranes) and process/method development (e.g., gas sampling)

1400 scfm CO_2 Capture Process

- Unique Uky-CAER CO_2 Capture Process
  - Heat integrated cooling tower and liquid desiccant air drying system for decreased energy consumption
  - Two-stage stripping for increased solvent working capacity
- Continuous gas stream composition monitoring
- Continuous steam usage measurement Continuous energy consumption measurement
- Comprehensive liquid sampling ports for solvent quality analyses
- Comprehensive gas sampling ports for emissions and degradation analyses
- Liquid/gas column profile sample ports for profile validation
- Corrosion coupon testing locations for material evaluation
- On-site analytical laboratory for routine liquid sample analysis

Gasification and Acid Gas Cleanup

Gasification

- Coal and Coal/Biomass Integrated Feed Preparation
- Entrained flow, oxygen blown, molten slag, OMB gasifier
- 1 ton dry coal/day
- 160 lb/hour syngas production

Acid Gas

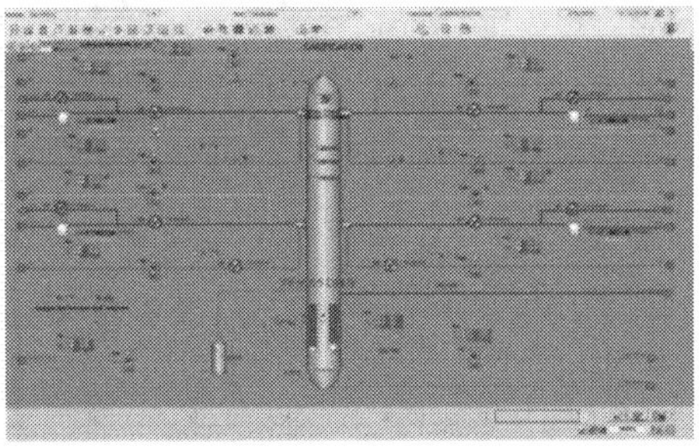

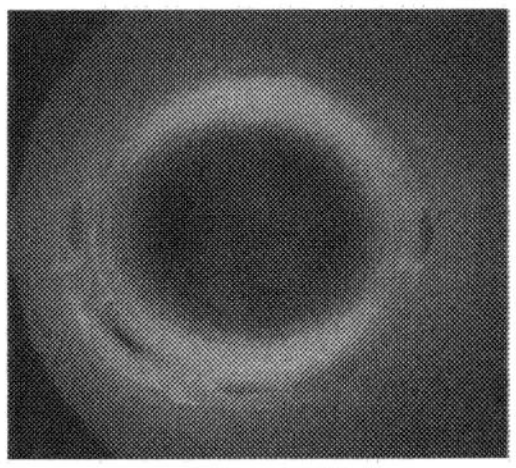

- Aqueous Amine system with absorber and stripper
- Pressurized system (400 psi)
- Sulfur treatment on both rejected and clean gas (<1ppmv sulfur)

Equipment and Instrumentation for Screening, Evaluation and Physical/Chemical Property Measurements

Equipment/Instrumentation	Type	Capacity	Function
Vapor-Liquid Equilibrium (VLE) with GC (high pressure)	Parr reactor	400 mL/charge	Determine chemical/physical properties such as equilibrium CO_2 pressure, capacity and heat of absorption
High pressure TGA	fixed bed	up to 10 gram/load	Evaluate sorbent stability, CO_2 absorption rate, carbon capacity, Vapor-Solid Equilibrium
Netzsch TG/QMS/DSC System	fixed bed	up to 30 gram/load	Evaluate sorbent stability, CO_2 absorption rate, carbon capacity, Vapor-Solid Equilibrium, heat of absorption.
Electrochemical Corrosion Cell	stirred reactor	200 mL/charge	Corrosion
Traditional Corrosion Cell	equilibrium cell	2 liter/run	Corrosion
Bench-Scale Mini-Scrubber	solvent-based, simulated flue gas	1.1 kg/hr gas feed rate	Evaluate foaming tendency, batch-mode CO_2 absorption for mass transfer measurement
Bench-Scale Mini-Stripper	random packing	20 mL/min liquid feed rate	Determine batch-mode CO_2 desorption rate
Wetted Wall Column (WWC)	-	2 L/run	Determine mass transfer/kinetic data
Vapor-liquid equilibrium (VLE) apparatus (Low pressure)	-	10 mL/charge	Determine Chemical/physical properties such as equilibrium CO_2 pressure, capacity and heat of absorption
Breakthrough Solvent Evaluation Apparatus	-	10 mL/charge	Determine kinetic data/equilibrium CO_2 pressure, capacity
Stop Flow Reactor	-	20 mL/charge	Reaction rate
Pilot-scale Extruder system with material preparation	thermo-compression	20 kg/day product	Catalyst/solid sorbent production
Freeze Granulator and Freeze Dryer	slurry	2 kg/day	Catalyst/solid sorbent production
EPA Method 5 Standard Gas Sampling Equipment	Apex Instruments (2009)		Monitoring emissions of particulate matter (PM) and solvent degradation compounds for secondary environmental monitoring including EPA stack emission methods

Structure to Prevent Confidentiality Issues or Potential Contamination

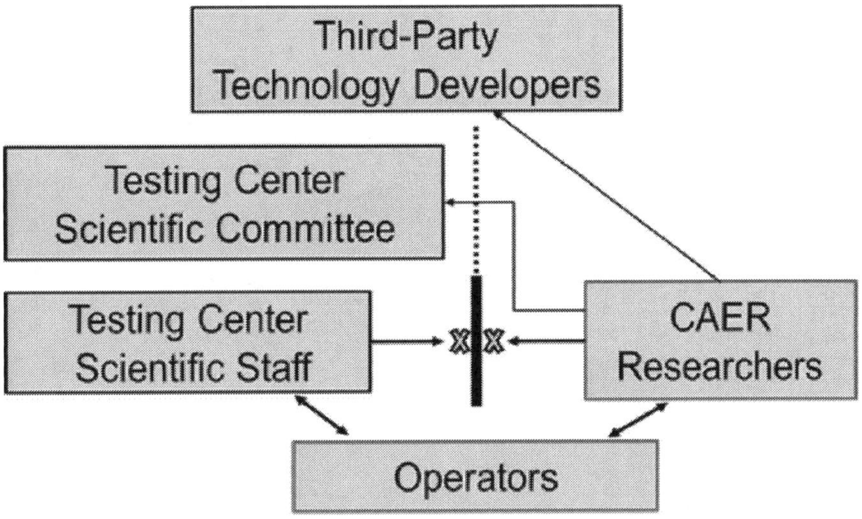

CMTC 2015 November 17 – 19, 2015

CONTAINMENT & MONITORING INSTITUTE (CMC.CaMI)

- Develop and validate performance of measurement, monitoring and detection technologies for location & movement of subsurface fluids
- Develop geological CO_2 & energy storage technologies
- Methane Management program – detection, fate & mitigation of loss-of-containment conditions

CMC Containment & Monitoring Institute

2

Field Research Station (FRS) : Location

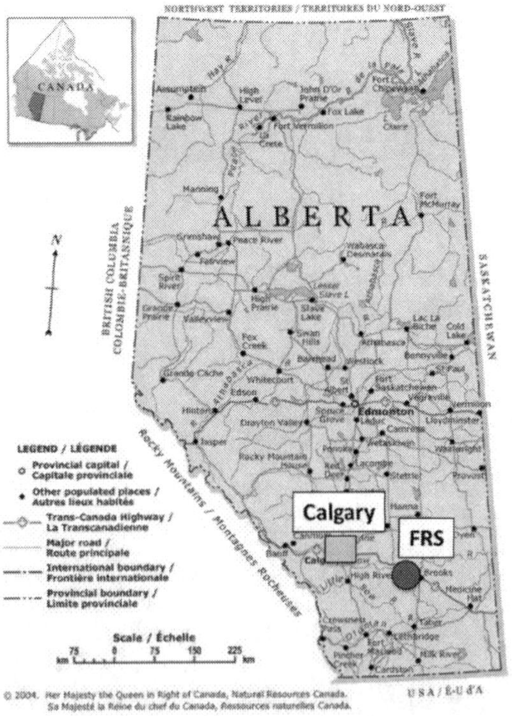

Land leased from Cenovus Energy

CMC Containment & Monitoring Institute

FRS monitoring layout

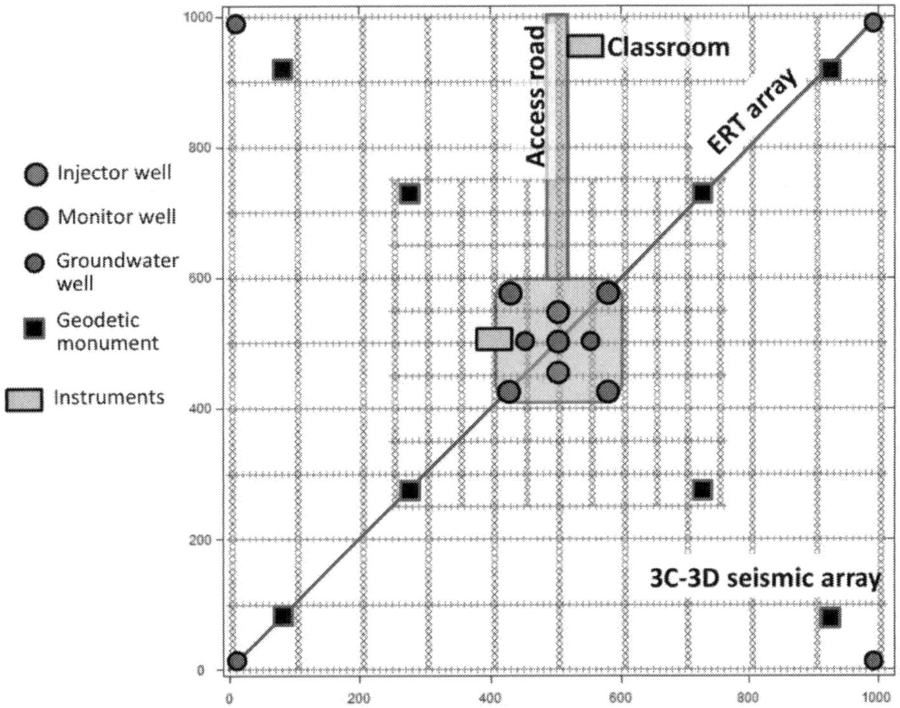

FRS walk-away and walk-around VSP

May, 2015

CMC | Containment & Monitoring Institute

Resourcing & Collaborations

- $4.40 M CMCRi, Inc. (capital & operating)
- $4.92 M from Western Economic Diversification (capital)
- Implementation through Schlumberger Carbon Services
- NRCan
- US Department of Energy
- UK Carbon Capture and Storage Research Centre
- Scottish Carbon Capture and Storage
- South Korea
- Industry subscriptions (JIP or programs)

CMC | Containment & Monitoring Institute

CARBON CAPTURE/CONVERSION INSTITUTE (CMC.CCCI)

- Develop, integrate, adapt, pilot and scale-up industrial process technologies for GHG emissions mitigation
- Reduce cost of CO_2 capture and convert to beneficial products
- Work with domestic & global researchers to convert bench/batch to continuous processes

CMC | Carbon Capture & Conversion Institute

7

CMC.CCCI ROLES

RESEARCH

- Sr. Research Director CCCI @ UBC
  - Dr. Naoko Ellis (Acting)
  - UBC is Academic Host institution
  - Develop & integrate early stage technologies & industrial processes
  - Prove at bench-scale continuous

DEVELOPMENT

- Technology Commercialization & Innovation Centre (TCIC)
  - Inst. Director TBD
  - Collaborative test & development centre
  - Linked to other test centres (Canada/global)
- Engineering/Design partners
  - NORAM Engineering
  - BC Research

CMC | Carbon Capture & Conversion Institute

8

CMC.CCCI KEY RELATIONSHIPS

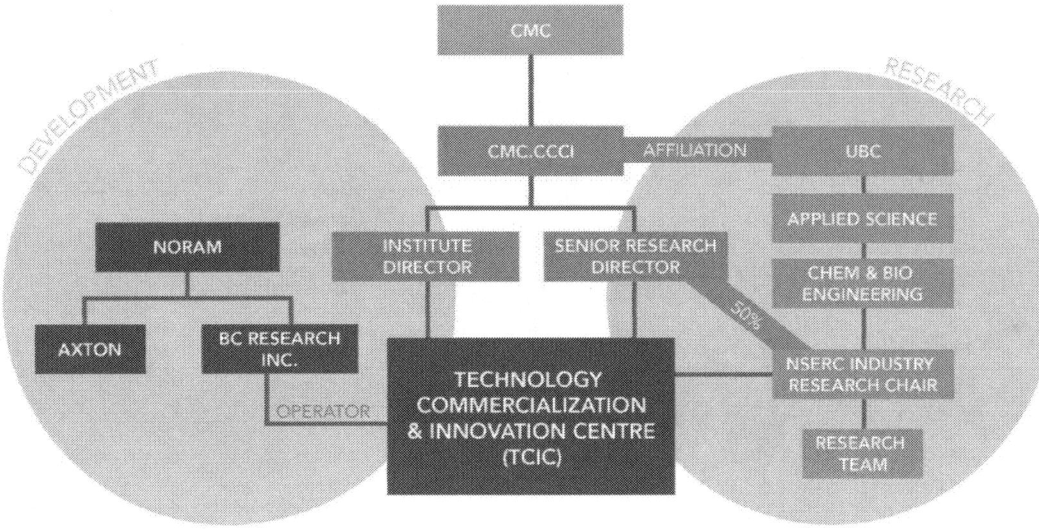

CMC | Carbon Capture & Conversion Institute

9

ACCELERATING INNOVATION

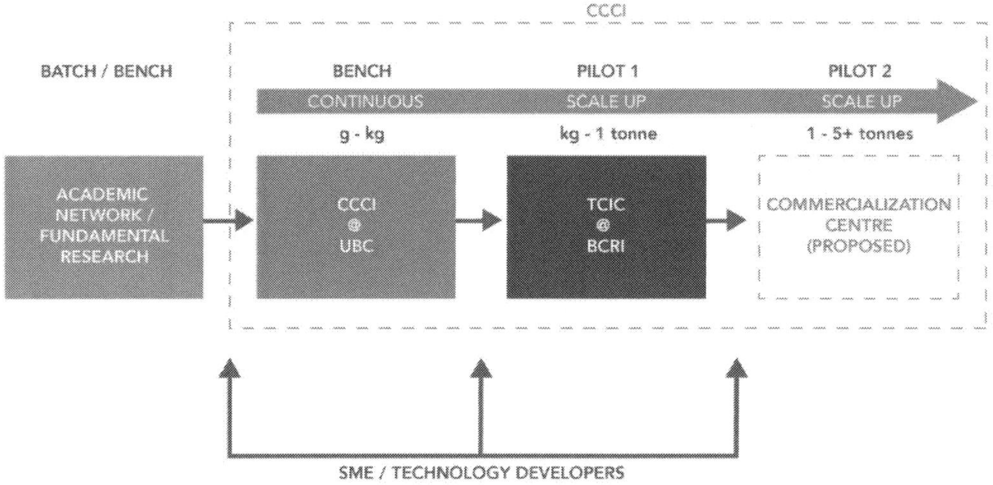

CMC | Carbon Capture & Conversion Institute

10

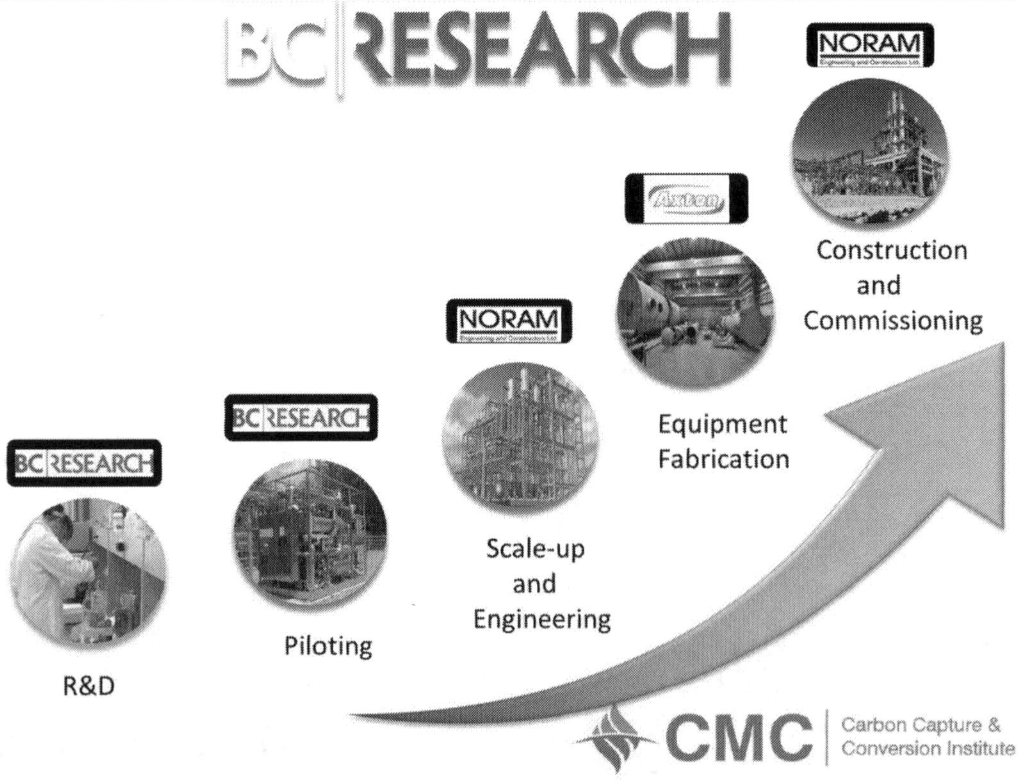

CMC | Carbon Capture & Conversion Institute

Technology Commercialization & Innovation Centre Mitchell Island, Greater Vancouver (BCRI/CCCI)

CMC | Carbon Capture & Conversion Institute

12

Technology Commercialization & Innovation Centre

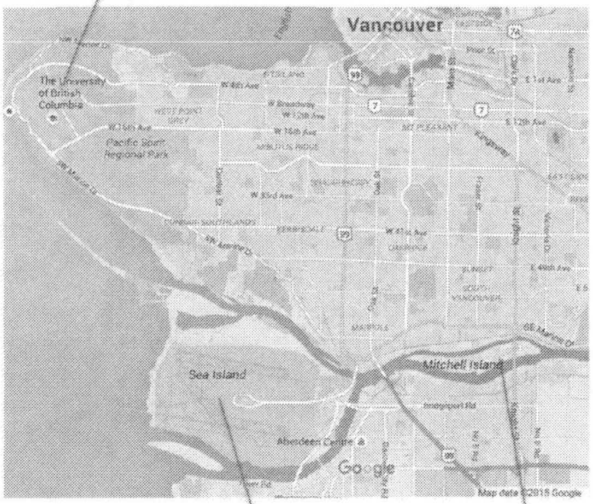

- Office Space 8400sqft
- Laboratory 3,500sqft
- Pilot Plant 25,000sqft (33ft height)
  - Class 1 Div 1 Area 2,000sqft

- 10 inch water supply, 3 inch distribution
- 2 inch natural gas line with 2 inch distribution
- 1 inch airline
- 1 inch hot water supply and return
- 30 KVA hot water boiler
- 1200 Amp 600 V electrical service
- 3 - 200 amp/600 V breakers
- 200 KVA low voltage transformer
- Grated trench to collection sump.

CMC | Carbon Capture & Conversion Institute

13

CO_2LLABORATION
CMC
RESEARCH
INSTITUTES
Contact:
Richard Adamson, PEng MSc
President
Richard.Adamson@cmcghg.com

THE INSTITUTE FOR CLEAN AND SECURE ENERGY
AT THE UNIVERSITY OF UTAH
THE
UNIVERSITY
OF UTAH
Industrial
Combustion
Gasification
Research
Facility

THE INSTITUTE FOR CLEAN AND SECURE ENERGY
AT THE UNIVERSITY OF UTAH

Industrial Combustion and Gasification Research Facility (ICGRF)

1600 amps of 480V 3-phase power

Compressed Natural Gas @ 100 kPa

23 m³ of liquid Oxygen Storage

10 Ton Overhead Crane Capacity

Integrated Industrial Distributed Control System (OPTO22)

Extensive Analytical Capabilities

THE UNIVERSITY OF UTAH®

THE INSTITUTE FOR CLEAN AND SECURE ENERGY
AT THE UNIVERSITY OF UTAH

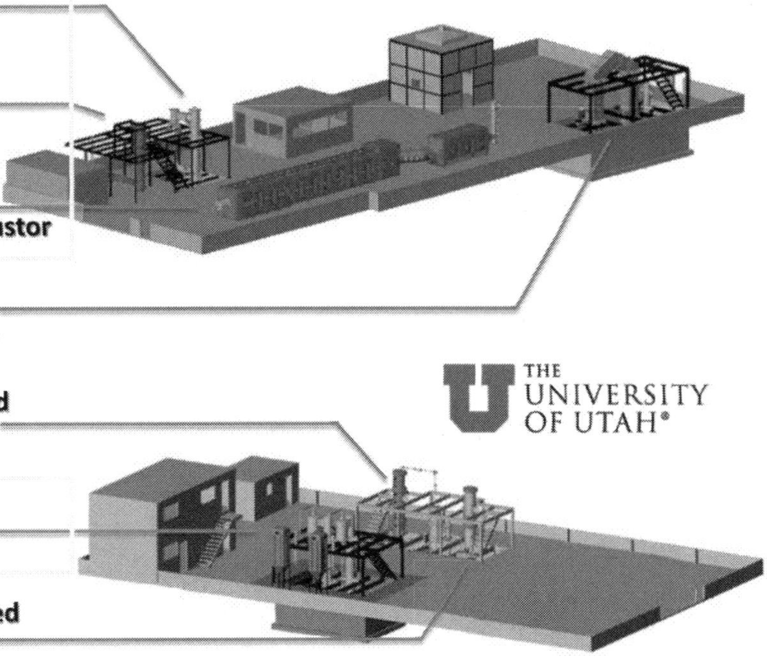

300 kW **Circulating Fluidized Bed Combustor**

100 kW **Oxy-fuel Combustor**

1.5 MW **Pulverized Coal & Multi-Fuel Combustor**

300 kW **Grate-fired Combustor (Stoker)**

200 kW **Pressurized Fluidized Bed Gasifier**

100 kW **Chemical Looping Combustor**

300 kW **Pressurized Entrained Flow Gasifier**

THE UNIVERSITY OF UTAH®

THE INSTITUTE FOR CLEAN AND SECURE ENERGY
AT THE UNIVERSITY OF UTAH

Circulating Fluidized Bed (CFB)

Specifications

- Dimensions - 6.5 m tall 0.27 m ID
- Circulating and Bubbling Bed Modes
- Air- and Oxy-fired with FGR

Typical Oxy-Fired Operation

Firing Rate	200	kW	
Coal Feed Rate	60	lb/hr	Ill. Bituminous
Oxygen Feed Rate	120	lb/hr	
Flue Gas Recycle	390	lb/hr	@ 27% O_2
Air Leakage	8	lb/hr	
Flue Gas Flow	34	SCFM	Wet

Typical Oxy-Fired Gas Composition, Vol. %

O_2	2	H_2O	31	N_2	5
CO_2	61	SO_2	1		

~ 87% on a dry basis

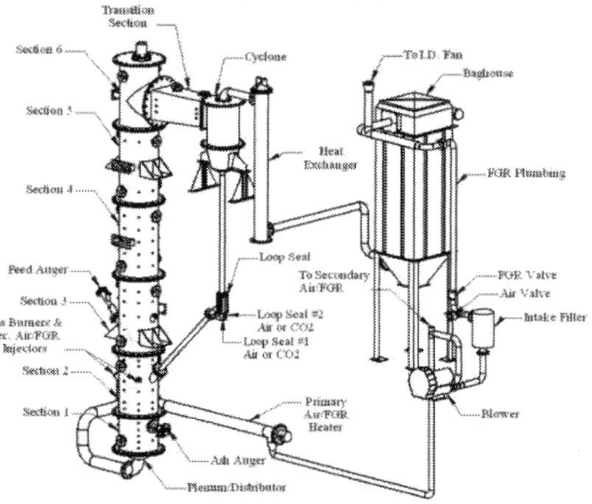

THE INSTITUTE FOR CLEAN AND SECURE ENERGY
AT THE UNIVERSITY OF UTAH

Oxy-Fuel Combustor (OFC)

Specifications

- Burner Zone Dimensions – 4.1 m tall x 0.5 & 0.27 m ID
- Convective Zone Dimensions – 3.7 m x 0.15m Square
- Air- and Oxy-fired with FGR or with CO_2

Typical Oxy-Fired Operation

Firing Rate	37	kW	
Coal Feed Rate	11	lb/hr	Ill. Bituminous
Oxygen Feed Rate	22	lb/hr	
Flue Gas Recycle	72	lb/hr	@ 27% O_2
Air Leakage	1	lb/hr	
Flue Gas Flow	6	SCFM	Wet

Typical Oxy-Fired Gas Composition, Vol. %

O_2	2	H_2O	32	N_2	4
CO_2	61	SO_2	1		

~ 90% on a dry basis

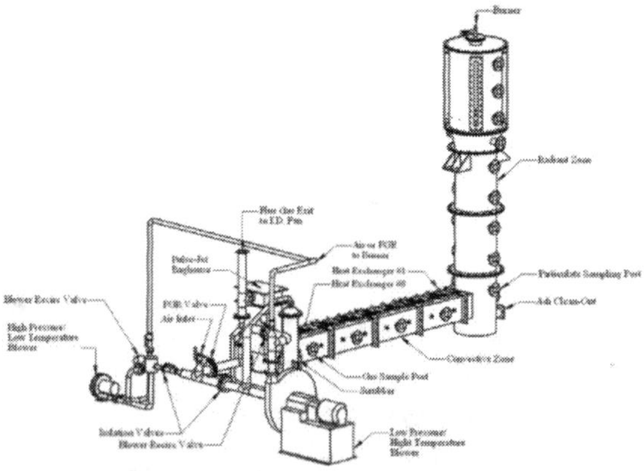

THE INSTITUTE FOR CLEAN AND SECURE ENERGY
AT THE UNIVERSITY OF UTAH

Multi-Fuel Combustor (L1500)

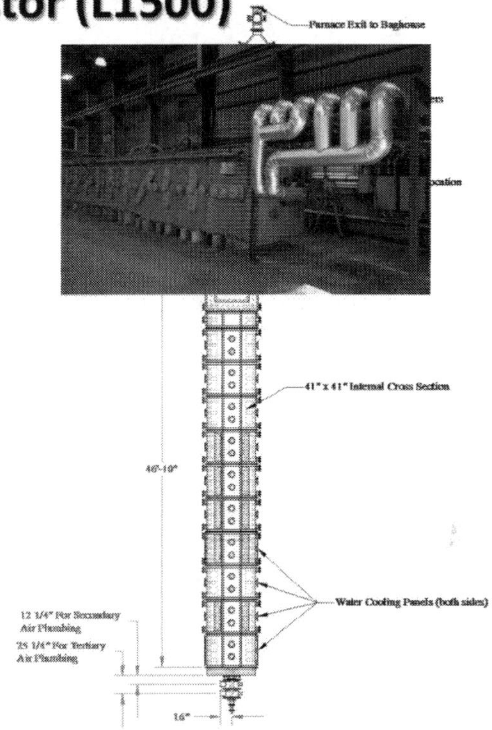

Specifications

- Radiative Zone Dimensions 14.1m long x 1.0 m square
- Pulverized Coal, Biomass, Natural Gas, Oil
- Air- and Oxy-fired with FGR

Typical Oxy-Fired Operation

Firing Rate	1.0	MW	
Coal Feed Rate	302	lb/hr	Ill. Bituminous
Oxygen Feed Rate	620	lb/hr	
Flue Gas Recycle	1990	lb/hr	@ 27% O_2
Air Leakage	58	lb/hr	
Flue Gas Flow	176	SCFM	Wet

Typical Oxy-Fired Gas Composition, Vol. %

O_2	2	H_2O	31	N_2	6
CO_2	60	SO_2	1		

~ 85% on a dry basis

THE INSTITUTE FOR CLEAN AND SECURE ENERGY
AT THE UNIVERSITY OF UTAH

Chemical Looping Combustor (CLOU)

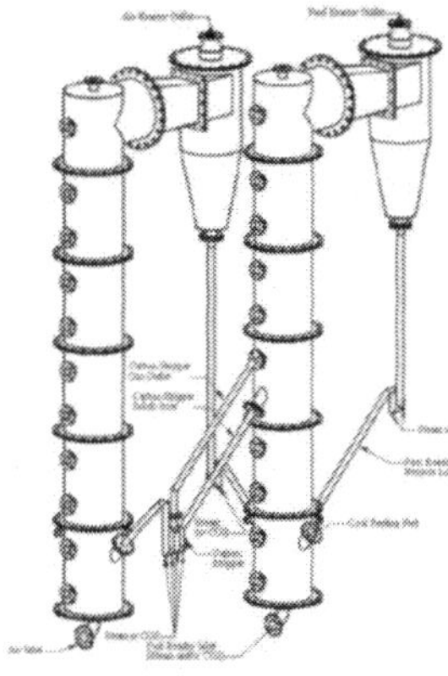

Specifications

- Air Reactor Dimensions 5.8 m tall x 0.27 m ID
- Fuel Reactor Dimensions 6.4 m tall x 0.27 m ID
- Capable of Producing Concentrated CO_2 without O_2

Expected Operation

Firing Rate	100	kW	
Coal Feed Rate	29	lb/hr	Ill. Bituminous
Steam Flow	250	lb/hr	
Air Leakage	0	lb/hr	
Flue Gas Flow	104	SCFM	Wet

Expected Gas Composition, Vol. %

O_2	1>	H_2O	90	N_2	1>
CO_2	10	SO_2	1>		

~ 96% on a dry basis

THE INSTITUTE FOR CLEAN AND SECURE ENERGY
AT THE UNIVERSITY OF UTAH

Synergistic Activities

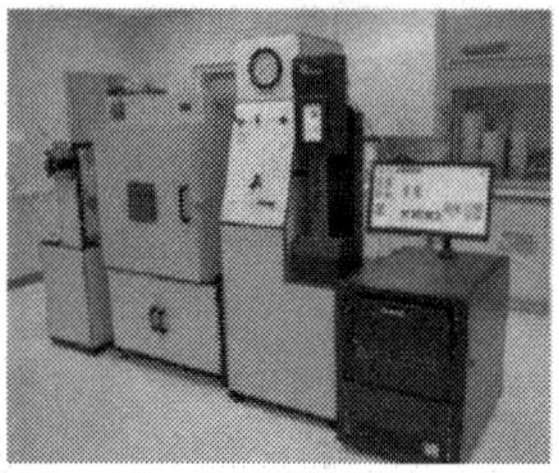

Carbon dioxide Enhanced Oil Recovery and Sequestration

- Oil-CO2 phase behavior and multiple contact miscibility
- High-pressure CO2 floods
- Geological characterization and screening of EOR and sequestration sites
- CO2-brine-rock reactions and kinetics
- Reservoir simulations of EOR and sequestration field studies

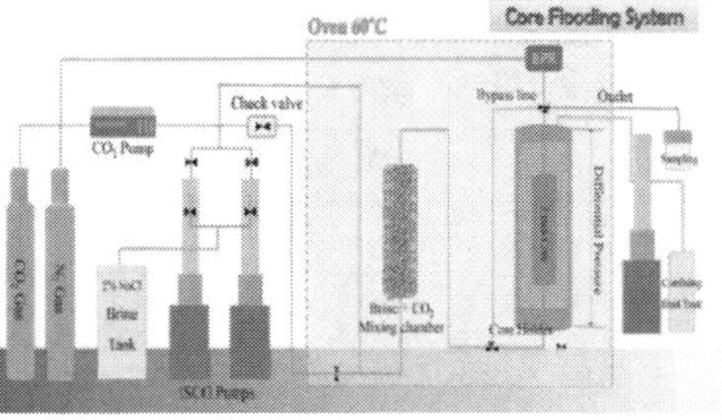

Department of
CHEMICAL ENGINEERING
THE UNIVERSITY OF UTAH

Energy & Geoscience Institute

THE INSTITUTE FOR CLEAN AND SECURE ENERGY
AT THE UNIVERSITY OF UTAH

Contractual and Legal Issues

1 Non-Disclosure Agreement → 2 Scope of Work and Budget → 3 Contract Negotiation → 4 Program Execution

- Work is performed according to the established scope and budget
- Periodic billing is performed by U of U Grants and Contracts Accounting. Some programs require an up-front investment to cover known expenditures like utilities

THE INSTITUTE FOR CLEAN AND SECURE ENERGY
AT THE UNIVERSITY OF UTAH

Andrew Fry
(801) 597-2647
andrew.fry@utah.edu
www.icse.utah.edu/combustion-gasification-facility

Carbon Capture and Storage (CCS) in Canada

Carbon Management Technology Conference

Sugar Land, Texas

November 17 - 19, 2015

Richard Adamson
CMC Research Institutes

WE'RE ALL IN THIS TOGETHER

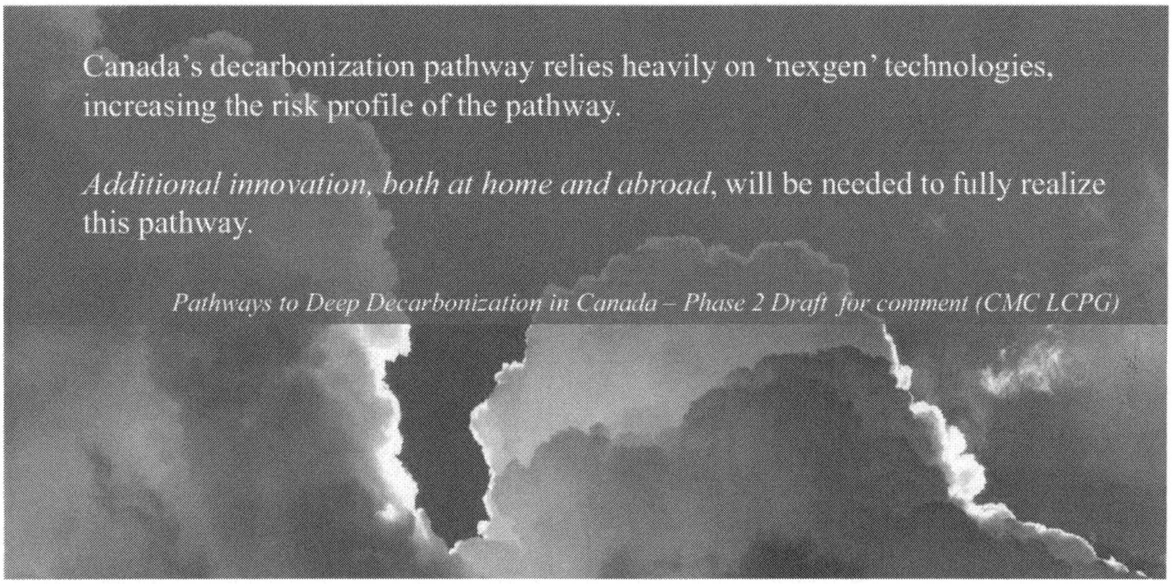

WHO IS CMC?

- **Neutral, independent** federally incorporated Canadian not-for-profit
  - Not government - no policy or jurisdictional priority conflicts
  - Not industry - no commercial/IP interest, longer-term perspective
  - Not Academic – driven by mission, not novelty
- **Single mission**: elimination of industrial greenhouse gas emissions
  - Accelerate innovation through integration, adaptation, application development, pilot testing, performance verification …
  - Leverage point:
    - Rapid & cost-effective reduction of uncertainty to enable earlier decision making
- Not an advocate – provide **facts/science-based** options
  - Elevate the conversation
  - Support policy makers/regulators:
    - Rapid, cost effective reduction of uncertainty to enable earlier decision making

CMC RESEARCH INSTITUTES

4

IDEAS TO IMPACT

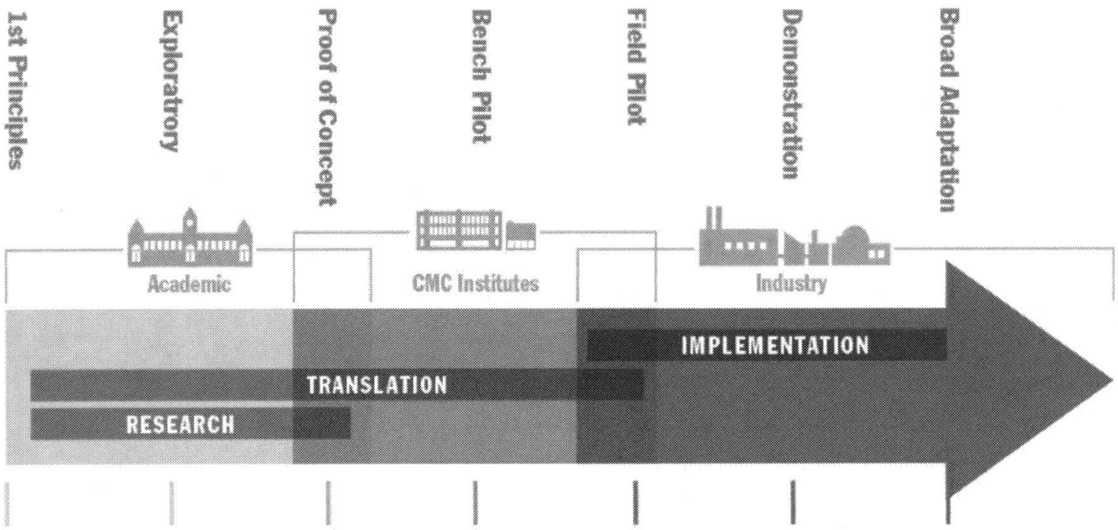

5

INNOVATION FUNNEL

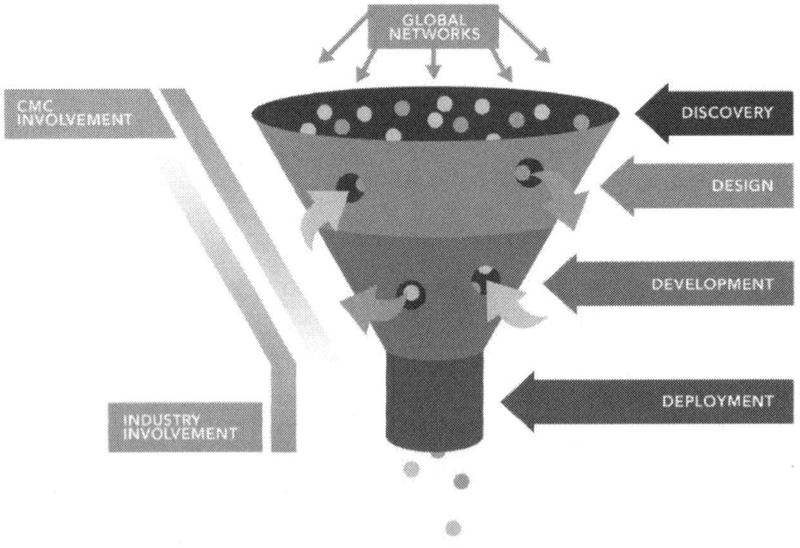

6

THE NEW PATH FORWARD

CONTAINMENT & MONITORING INSTITUTE (CMC.CaMI)

- Develop and validate performance of measurement, monitoring and detection technologies for location & movement of subsurface fluids
- Develop geological CO_2 & energy storage technologies
- Methane Management program – detection, fate & mitigation of loss-of-containment conditions

CMC | Containment & Monitoring Institute

8

CMC.CaMI KEY RELATIONSHIPS

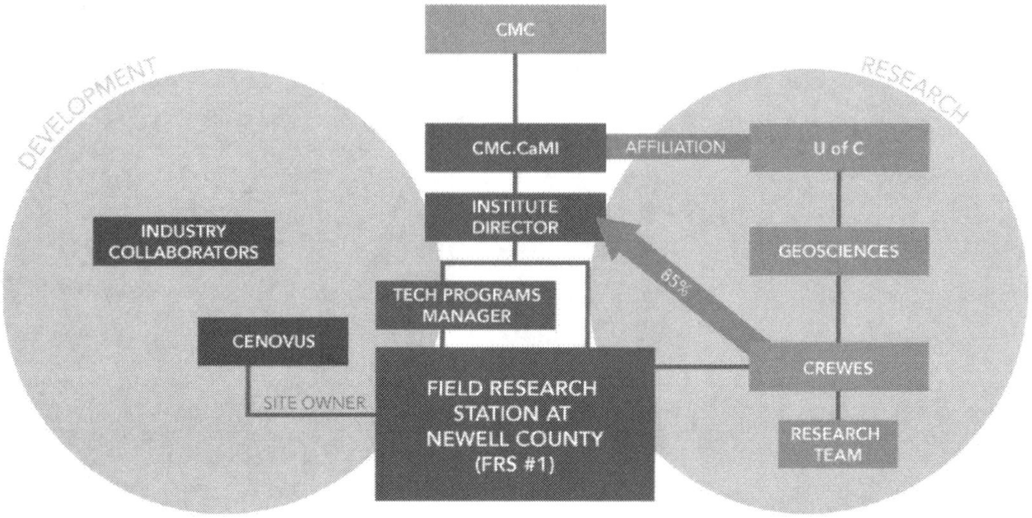

CMC Containment & Monitoring Institute

9

CaMI/UofC – Field Research Station (FRS)

- **Technical**
  - Undertake controlled CO_2 release at 300 m & 500 m depth; up to 1000 t/yr
  - Determine CO_2 detection thresholds
  - Improve monitoring technologies, cap rock assessment
  - Monitor gas migration at shallow to intermediate depths and impacts on groundwater (CO_2 and CH_4)
  - Determine fate of CO_2 (trapping/dissolution)
- **Non-technical**
  - University & industry field training & research
  - Integrating engineering and geoscience
  - Public outreach & education

CMC Containment & Monitoring Institute

Field Research Station (FRS) : Location

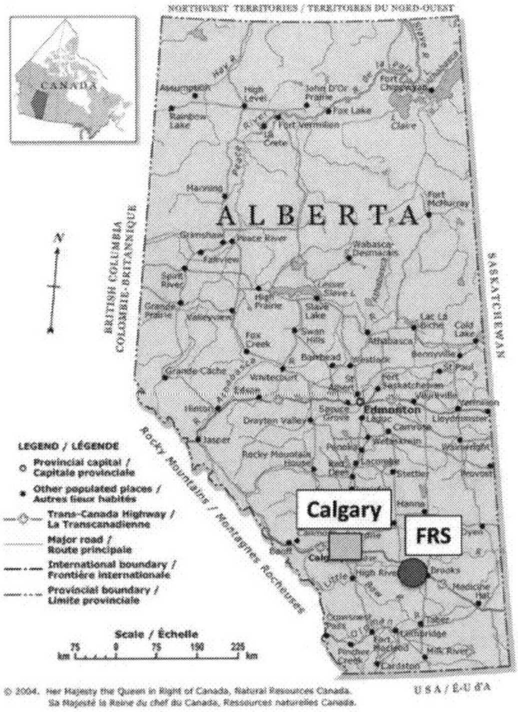

Land leased from Cenovus Energy

CMC Containment & Monitoring Institute

Mobile geochem laboratory

- Sondes for field measurements (pH, EC, T, DO, Eh)
- Soil gas flux chambers and soil gas collection probes
- Gas chromatographs for hydrocarbon and soil gas analyses
- Ion chromatograph (Dionex) for anion and cation concentration analyses on water samples
- Titrators for alkalinity and H_2S in water samples
- Portable H_2S gas analyzer
- Carbon isotope laser analyzer for methane
- Carbon and oxygen isotope laser

CMC Containment & Monitoring Institute

Dr. Bernhard Mayer

FRS walk-away and walk-around VSP

May, 2015

CMC | Containment & Monitoring Institute

FRS geostatic model - porosity

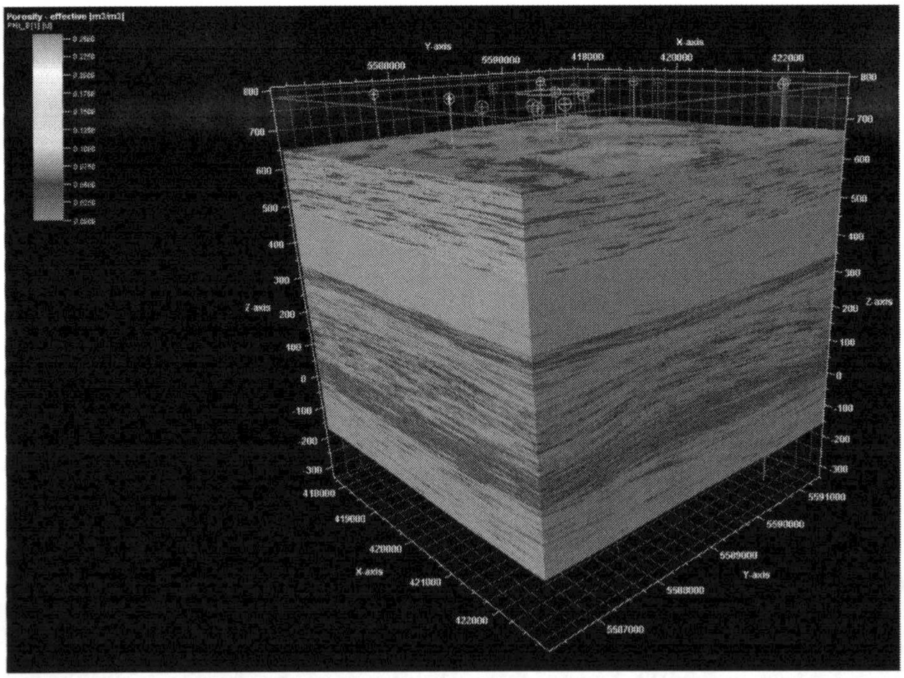

Schlumberger

Jessica Dongas: Univ. of Calgary

Resourcing & Collaborations

- $4.40 M CMCRi, Inc. (capital & operating)
- $4.92 M from Western Economic Diversification (capital)
- Implementation through Schlumberger Carbon Services
- NRCan
- US Department of Energy
- UK Carbon Capture and Storage Research Centre
- Scottish Carbon Capture and Storage
- South Korea
- Industry subscriptions (JIP or programs)

CMC | Containment & Monitoring Institute

CARBON CAPTURE/CONVERSION INSTITUTE (CMC.CCCI)

- Develop, integrate, adapt, pilot and scale-up industrial process technologies for GHG emissions mitigation
- Reduce cost of CO_2 capture and convert to beneficial products
- Work with domestic & global researchers to convert bench/batch to continuous processes

CMC | Carbon Capture & Conversion Institute

18

CMC.CCCI ROLES

RESEARCH

- Sr. Research Director CCCI @ UBC
  - Dr. Naoko Ellis (Acting)
  - UBC is Academic Host institution
  - Develop & integrate early stage technologies & industrial processes
  - Prove at bench-scale continuous

DEVELOPMENT

- Technology Commercialization & Innovation Centre (TCIC)
  - Inst. Director TBD
  - Collaborative test & development centre
  - Linked to other test centres (Canada/global)
- Engineering/Design partners
  - NORAM Engineering
  - BC Research

CMC | Carbon Capture & Conversion Institute

19

CMC.CCCI KEY RELATIONSHIPS

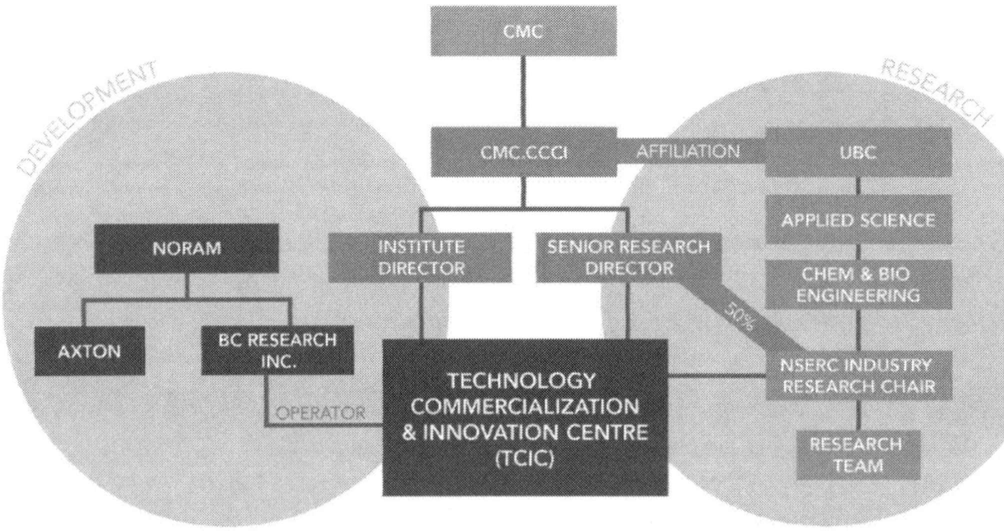

CMC | Carbon Capture & Conversion Institute

20

ACCELERATING INNOVATION

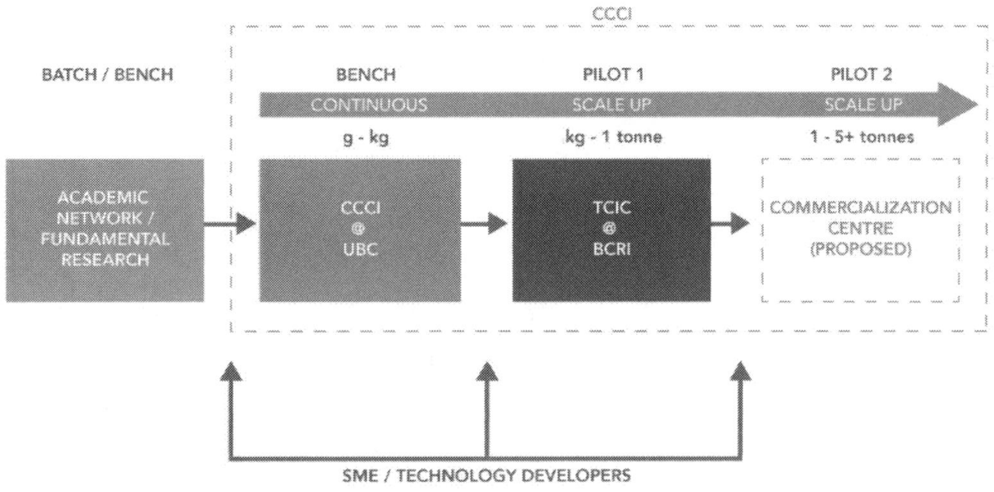

CMC | Carbon Capture & Conversion Institute

21

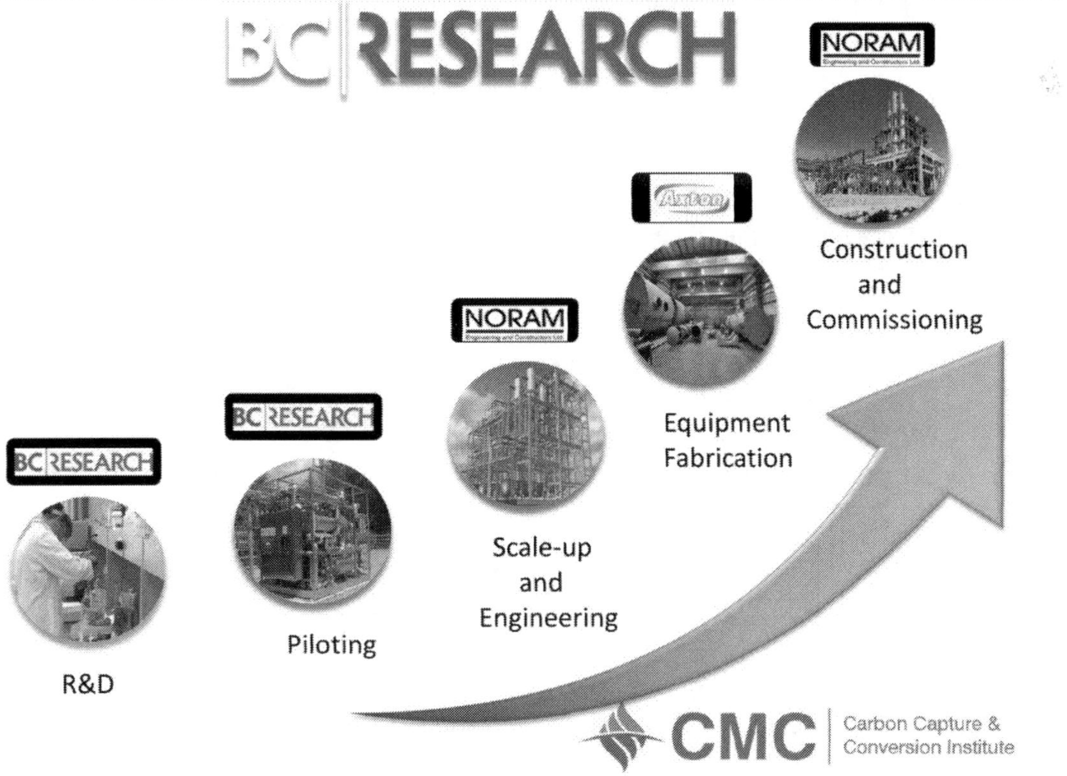

CMC | Carbon Capture & Conversion Institute

Technology Commercialization & Innovation Centre Mitchell Island, Greater Vancouver (BCRI/CCCI)

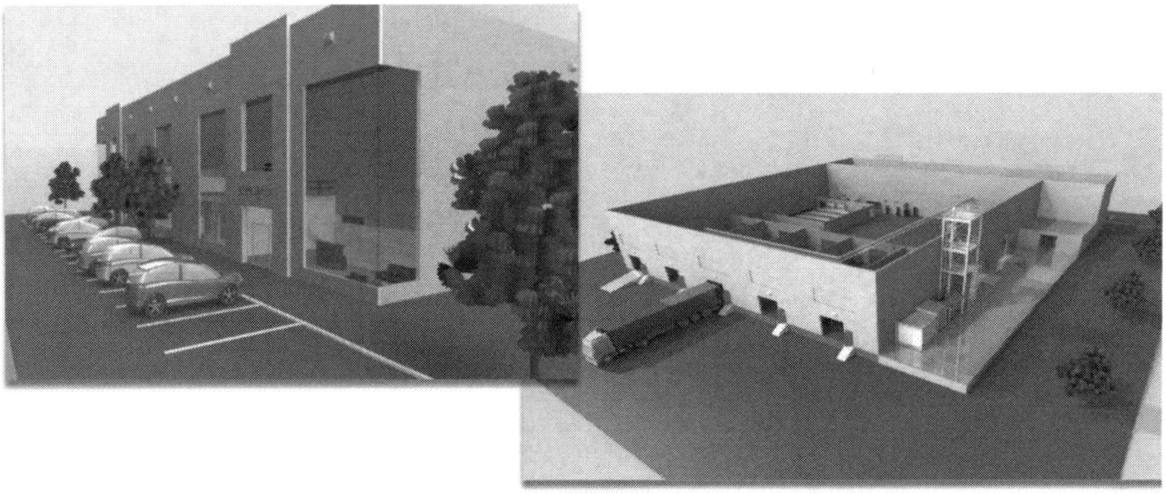

CMC | Carbon Capture & Conversion Institute

23

Technology Commercialization & Innovation Centre

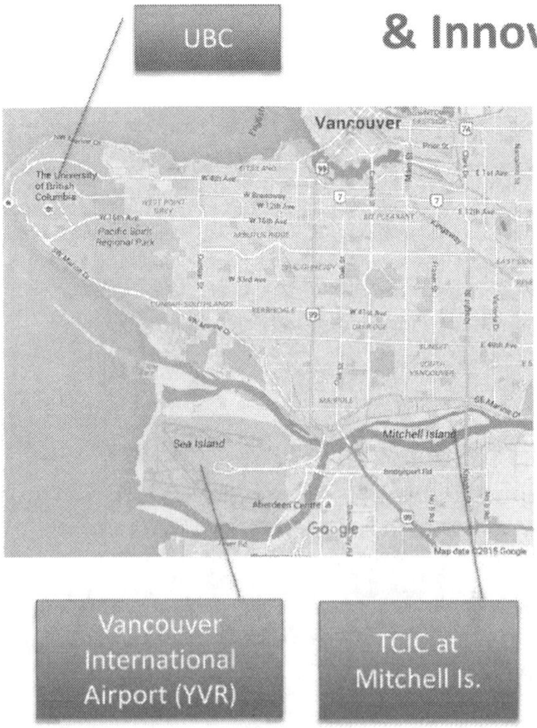

- Office Space 8400sqft
- Laboratory 3,500sqft
- Pilot Plant 25,000sqft (33ft height)
  - Class 1 Div 1 Area 2,000sqft

- 10 inch water supply, 3 inch distribution
- 2 inch natural gas line with 2 inch distribution
- 1 inch airline
- 1 inch hot water supply and return
- 30 KVA hot water boiler
- 1200 Amp 600 V electrical service
- 3 - 200 amp/600 V breakers
- 200 KVA low voltage transformer
- Grated trench to collection sump.

CMC | Carbon Capture & Conversion Institute

24

With thanks to:
Natural Resources Canada, SaskPower, Petroleum Technology Research Centre (PTRC), Northwest Refinery, Alberta Carbon Trunk Line (ACTL)/Enhance Energy, Shell Canada & Spectra Energy

CCS IN CANADA

27

Introduction to CCS in Canada

- Canada has three of the world's 15 operational large-scale projects.
  - Canada has significant underground CO_2 storage options.
  - CCS costs can be offset in Canada by selling CO_2 for use in enhanced oil recovery (EOR).
- NRCan is pursuing a variety of cutting-edge R&D in CCS, including innovation in CO_2 capture.
- The Global CCS Institute ranks Canada 2nd globally in its CCS regulatory environment, and 3rd globally in its supportive policy setting for CCS.

CMC RESEARCH INSTITUTES

Canada is implementing large-scale CCS demonstrations

1. **Weyburn-Midale Project** (in operation since 2000)
2. **SaskPower Boundary Dam** (in operation since 2014)
3. **Shell Quest project** (in operation since Nov. 6, 2015)
4. **Alberta Carbon Trunk Line** (will begin operations by end of 2017)

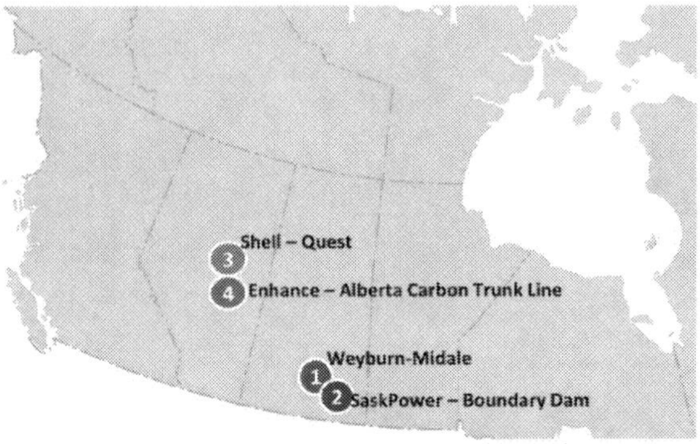

Federal-Provincial investments in CCS of over \$1.8 Billion with potentially up to \$4.5 Billion in public-private investment in RD&D CCS initiatives

CMC RESEARCH INSTITUTES

Support for CCS in Canada

Federal Government/NRCan (over \$580M)

Demo:
- Budget 2008 - \$240M for SaskPower's Boundary Dam CCS project
- Clean Energy Fund (CEF) – \$150M for 2 large-scale CCS projects in Alberta

R&D:
- Program for Energy Research and Development (PERD) - \$28M provided to federal laboratories for R&D related to CCS
- ecoENERGY Innovation Initiative (ecoEII) - \$27M for 2nd and 3rd generation capture technologies and CO_2 storage

Demo/R&D:
- ecoENERGY Technology Initiative (ecoETI) - \$112M for CCS initiatives led by industry, universities and federal laboratories

Provincial Support

Demo:
- Alberta CCS Fund – \$1.24B for 2 large-scale CCS projects in Alberta
- SaskPower, a provincial utility, invested over \$1.2B for Boundary Dam

R&D:
- Funding also provided through Alberta's Climate Change and Emissions Management Corporation (CCEMC), Saskatchewan's Go Green Fund, CCS Nova Scotia, etc.

CMC RESEARCH INSTITUTES

Recent Successes

2 ground-breaking CCS projects became operational since 2014

US Department of Energy Secretary, Ernest Moniz, on Quest: *"It will provide a new large-scale model to advance deep saline storage technologies that the world's CCS community can learn and benefit from. [As] the first CCS project in the oil sands industry, it is a significant milestone."*

Quest

Bellona President, Frederic Hauge: *"Finally, people cannot say that this is unproven technology. It will be much harder to reach climate targets without CCS."*

Boundary Dam

IEA Executive Director, Fatih Birol: *"Getting the Quest project up and running is another great example of how Canada is a leader in CCS. The launch of the Quest CCS project is remarkable, as it provides another excellent example of the fact that CCS is about so much more than just coal-fired power."*

Mike Monea, SaskPower, and Frederic Hauge, Bellona President

Next.. The Alberta Carbon Trunk Line

North West Redwater Partnership Sturgeon Refinery – Under construction

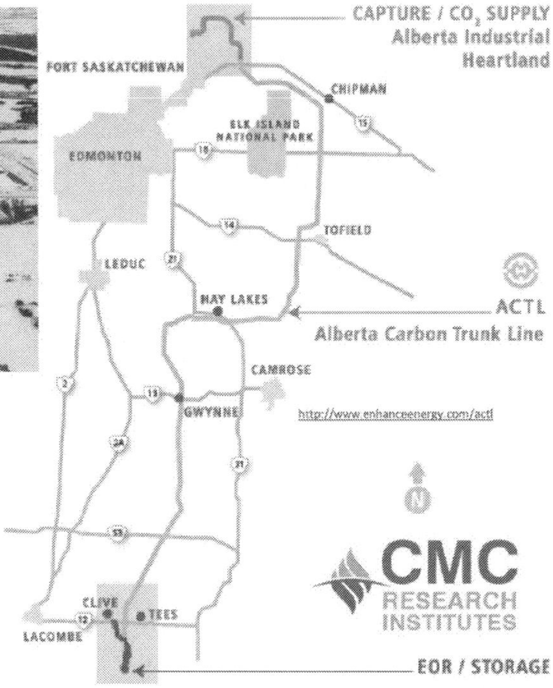

Petroleum Technology Research Centre projects

PTRC has two CO_2 focus areas.

1. CO_2-enhanced oil recovery:
   - The completed **Weyburn Midale** research project is the largest monitored CO_2 location in the world in association with a depleted oilfield.
   - The **SaskCO_2USER** project is exploring the use, shut-down and transition of CO_2-EOR fields to long-term storage.
2. CO_2 storage in deep saline formations:
   - **Aquistore**, Canada's first commercial dedicated CO_2 storage project, fits into the research and environmental side of CO_2 sequestration.

Spectra Energy North American Asset Portfolio
Fort Nelson CCS Project Overview

Spectra Energy.

- Fort Nelson Gas Plant processing capacity of 1.0 BCF/day (raw natural gas)
- At full capacity, Fort Nelson CCS has the potential to sequester up to 2.2 MT of CO_2 annually
- The Fort Nelson Gas Plant is designed to capture CO_2
- CCS design utilizes existing technology to transport and inject supercritical CO_2
- Close proximity from source to sink ~12 kilometers
- No net additional fuel or power requirements
- Storage:
  - Regional saline-filled reservoir
  - 2.1 kilometers below surface
  - Injection zone protected by >500 meters of impervious cap rock
  - Vast storage capacity (long term injection capability)

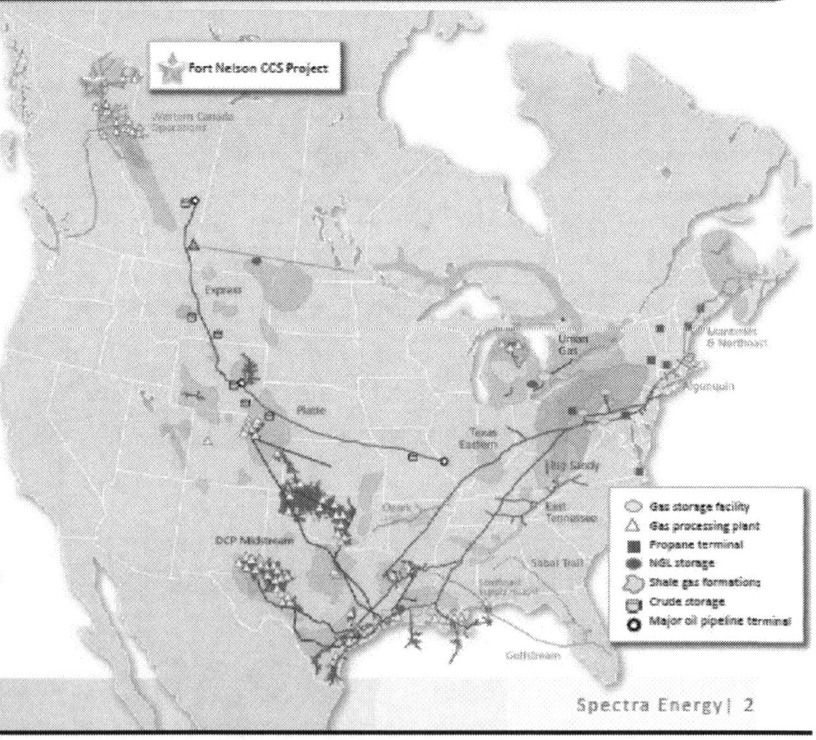

REAL WORLD
REAL RESULTS

Spectra Energy| 2

Fort Nelson CCS
2015 Project Status

Spectra Energy

Current Status:

Fort Nelson CCS is technically feasible:

- Exploratory phase completed
- Large storage capacity available
- Required injectivity available
- Competent containment
- Utilizes existing technology
- Technical risk assessment completed
- Preliminary Measurement, Monitoring and Verification completed

Future Advancement:

An investment decision on whether or not to advance large-scale CCS will require:

- Greater clarity as to the path forward for treatment of GHG reductions
- A clear economic incentive for GHG emission reductions within a competitive North American framework (an ability to quantify cost/benefit/risk)
- A price on carbon and ability to offset
- Comprehensive CCS regulations including treatment of long-term liability

Spectra Energy is looking forward to advancing this discussion at both the provincial and federal levels as we reduce our carbon footprint

REAL WORLD
REAL RESULTS

Spectra Energy| 3

Canada has world-class private sector expertise toward a new generation of lower-cost CCS

Company	Technology	Piloting Activities include
Saskatchewan-based HTC CO2 Systems Corp.	The HTC LCDesign™ advanced post combustion amine CO_2 capture technology	▪ Applying its capture technology and Delta Purification System™ technology at Husky Energy's Pikes Peak South heavy oil facility in Lashburn, Saskatchewan
Quebec-based CO2 Solutions Inc.	Enzyme-enabled CO_2 capture technology	▪ Successfully completed 8 tonne-CO_2/day demo in collaboration with Husky Energy, supported by NRCan's ecoEII ▪ Results of 1 tonne-CO_2/day pilot testing at the University of North Dakota show cost of $39/tonne including compression to 2250psi
BC-based Inventys Thermal Technologies Inc.	VeloxoTherm post-combustion CO_2 capture technology	▪ Funding of $ 1 million from US DOE for pilot project at one NRG's Gulf Coast coal plants

Canada has world-class private sector expertise toward a new generation of lower-cost CCS

Company	Technology	Piloting Activities include
Nova Scotia-based CarbonCure Technologies	Develops and licenses technology that sequesters CO_2 in concrete, resulting in improved performance	▪ CO_2 Utilization technology deployed widely in both US and Canada concrete sectors (25 licensed concrete manufacturing plants)
SaskPower Carbon Capture Test Facility (CCTF)	Facility is designed to provide evaluation of amine post-combustion technologies	▪ CCTF launched on June 18, 2015 ▪ After initial demonstration by Hitachi, CCTF will provide a unique platform to test other technologies ▪ Member of Int'l CCS Test Center Network

CMC RESEARCH INSTITUTES

Canada has also advanced CCS through regulatory development and public engagement

- Federal regulations for coal-fired power generation now in force
- Provincial regulations advancing - Alberta at the forefront globally
- Alberta co-leading a task force on developing ISO standards for CCS
- British Columbia introducing first legislation relating to use of CCS for GHG mitigation
- Public information and education efforts
  - ➢ Canadian project proponents, such as Shell, have been recognized as world-leading in this area.

As well as through knowledge-sharing and NRCan's ongoing international engagement

Multilateral

- Carbon Sequestration Leadership Forum (CSLF)
  - SaskPower participates in the International CCS Test Centre Network
  - Shell Quest participates in the new Large-Scale Saline Storage Project Network
  - NRCan and Norway co-lead Task Force on 2nd / 3rd Generation CCS Technologies
  - NRCan is member of CCS in the Academic Community Task Force
- IEA / IEA GHG R&D and Clean Energy Ministerial

Bilateral / Trilateral

- Canada-U.S. (Clean Energy Dialogue / Energy Cooperation MoU)
- Canada-UK Joint Statement on CCS
- Bilateral collaboration with governments in Mexico, Norway, Japan, China, Korea
- North American trilateral CCUS workshop on December 2-3, organized by US DOE and hosted by Bureau of Economic Geology at University of Texas in Austin.

CMC RESEARCH INSTITUTES

In addition to NRCan funding CCS technologies, NRCan's CanmetENERGY also performs CCS R&D

Our Focus in the Innovation Process

Conception → Implementation

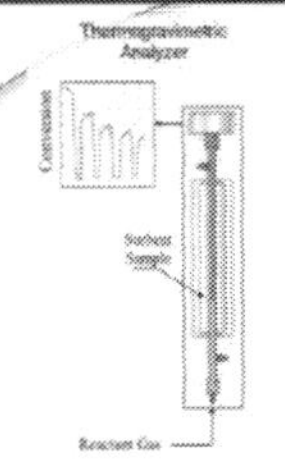

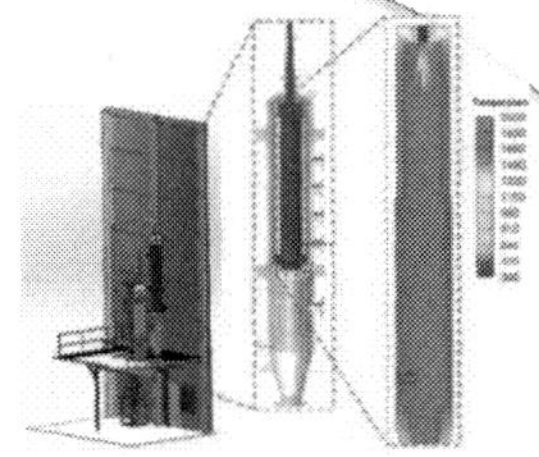

Pre-Commercial Demo

Formulation of ideas
Bench scale testing

Pilot-scale demo
Model development

Prototype demo
Scale-up modeling

Full-scale application

$10s to $100s K — ~$Millions — ~$10s

CanmetENERGY CFF

Other R&D Performers

Technology Providers

End Users

R&D on Oxy-Pressurized Fluid Bed Combustion (Oxy-PFBC)

Need in Power Sector and Oil Sands Producers

- Reliable power and steam production from solid fuels (coals or petroleum coke) that is cost competitive with natural gas technologies with near zero emission

Potential Benefits

- Provide steam and power at a lower price than natural gas by 2022 even under a low price gas scenario
- Lower GHG intensity
- Higher efficiency, fuel flexibility

Progress: 1 MW_{th} pilot plant under construction

Total Program scope: $20.5 M

AEROJET ROCKETDYNE
A GenCorp Company
CANADIAN CLEAN POWER COALITION
Alberta Innovates Energy and Environment Solutions
EPRI
Linde

R&D on G2 System using Supercritical CO_2 Turbine

Need in Power Sector, Oil Sands & Tight Resource Producers

- Highly efficient, mobile, water conserving power plants with near zero emission
- Enhanced recovery with reduced water demand

Potential Benefits

- Very compact, 350x smaller than conventional plants, suitable for remote applications
- 30% less CO_2, 40% drop in cost of electricity and 50% reduction in capital expenditure (CAPEX) with value-added by-products (pipeline-ready CO_2 for EOR and energized fracking, clean water at 9000 liter/MW_e)

Progress

- Engineering design of the 50 kW pilot-scale facility about 90% completed
- New IP and patent (WO/2012/159189 – PCT/CA2011/000593)

R&D on Pressurized Chemical Looping Combustion (P-CLC)

Need in Power Sector and Oil Sands Producers

- A competitive and promising technology in efficiency and cost to produce heat, hydrogen and power from solid fuels (vs. natural gas) with near zero emission

Approach
- Using Canadian oxygen carriers, such as ilmenite and iron ores
- Chemical looping steam reforming for hydrogen production, an advanced process integrated between conventional steam reforming and CLC

Benefits
- Very high efficiency 40 to 47% HHV for 96% CO_2 capture (Worley Parsons)
- Cost of CCS power reduced by 25% (Worley Parsons)
- Reduce oil sands emissions and increase productivity
- Suitable for current and potential future NO_X, SO_X, and CO_2 regulations

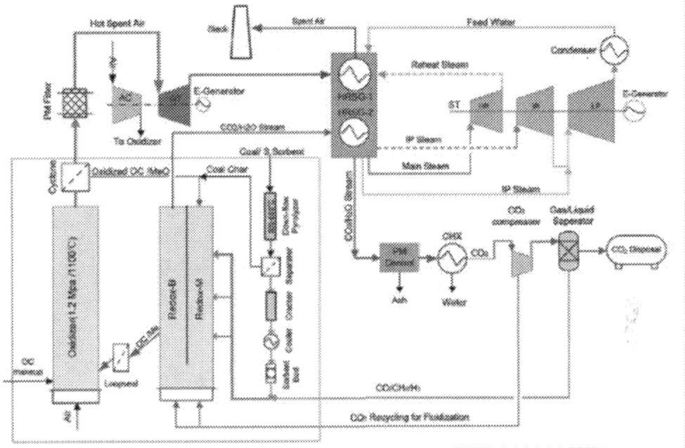

R&D on CO_2 Storage and Utilization

Needs for Tight Resource Producers and Power Sector

- Address gaps in knowledge related to injecting impurities into geologic reservoirs
- Improve recovery rates of tight resources

Focus

Effects of Impurities on CO_2 Storage Operations
- Determine the impact of impurities (H_2S, SO_2, O_2) in the injected flue stream on the reservoir and on wellbore integrity
- ISO Standards development for CCS

Enhanced Production in Shale and Tight oil and gas reservoirs
- Use of CO_2 for improved reservoir production and recovery efficiency

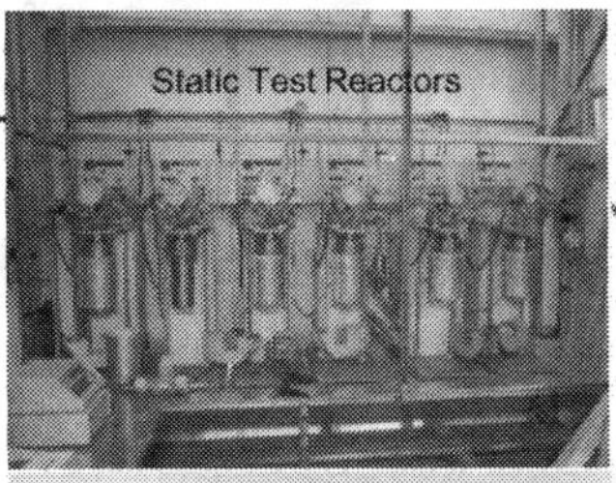

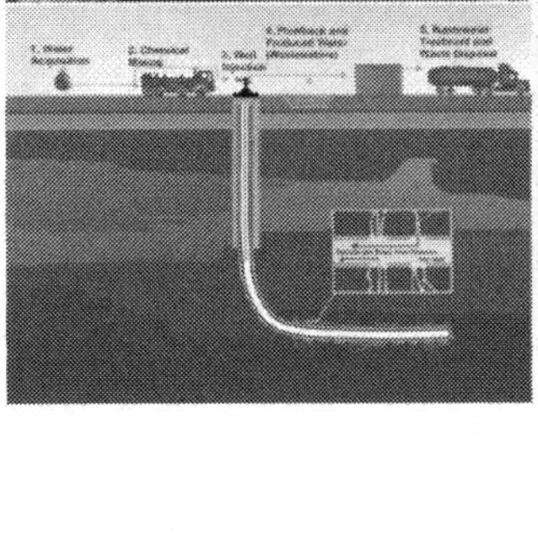

Core Flood Test Equipment

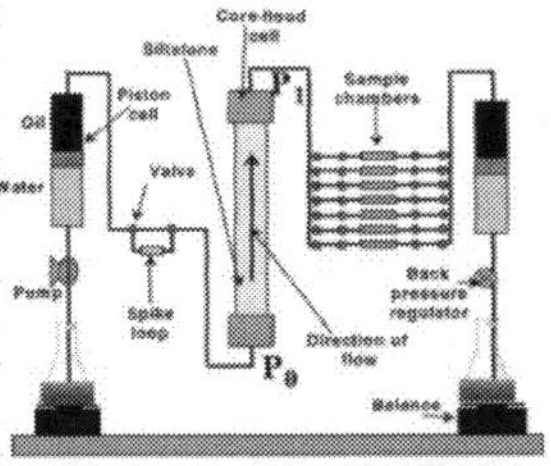

Basal Cambrian Sandstone

CanmetENERGY Pilot-Scale Facility

CO_2 R&D Consortium
1st of a kind Oxy-Fuel Pilot Facility (1994)

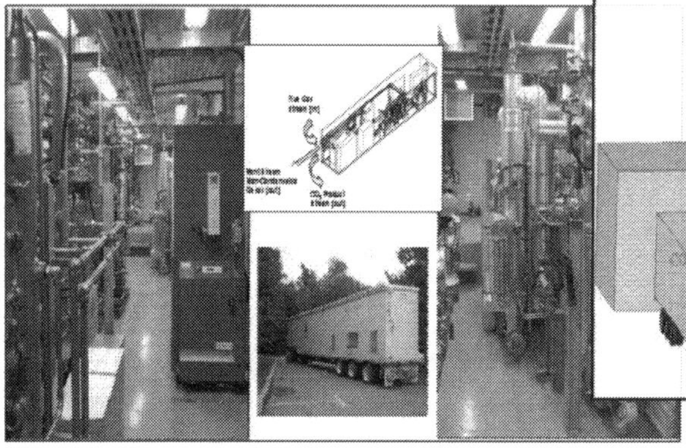

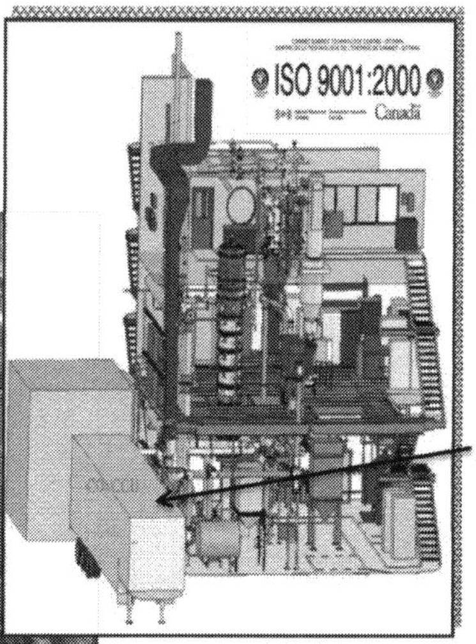

CO_2 Capture & Compression Unit ($CanCO_2$) added in 2008

CMC RESEARCH INSTITUTES

Challenge Awards/Prizes for Carbon Capture & Conversion to Beneficial Products

- CCEMC Grand Challenge
- COSIA/NRG Carbon X-Prize

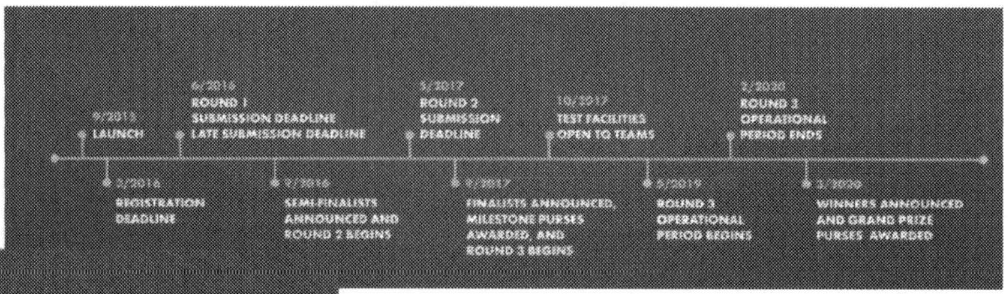

http://carbon.xprize.org/

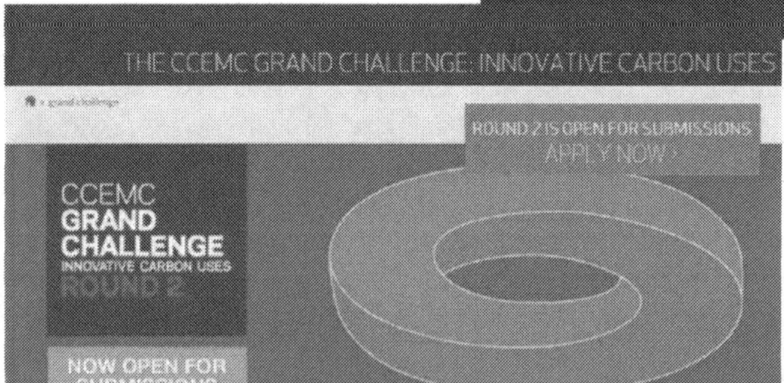

http://ccemc.ca/grand-challenge/

CMC RESEARCH INSTITUTES

46

In Summary:

- Canada has 3 large-scale demonstration projects in operation, with a 4th under construction (representing 1/5th of global CCS projects);
- Canada is proving CCS at scale while learning-from-doing;
- Canada is contributing to the global effort to advance CC(C)S by sharing its knowledge and expertise; and
- Canada is improving the economics by reducing technology costs through continued research and development.

CMC RESEARCH INSTITUTES

THE LOW CARBON FUTURE STARTS NOW.

We envision a leading global collaborative network developing solutions to industrial GHG emissions.

We anticipate industries known for innovative low-carbon technologies.

We foresee low net carbon footprint products welcomed in a carbon constrained world.

CO_2LLABORATION

CMC RESEARCH INSTITUTES

Contact:
Richard Adamson, PEng MSc
President
Richard.Adamson@cmcghg.com

SINOPEC CCS Technology and Project

Yuanyuan Zhang zhangyuanyuan.osec@sinopec.com
Sinopec Energy and Environmental Engineering Corporation
Sugar Land 2015

Content

- Introducing Sinopec
- Sinopec CCS Technology
- Sinopec CCS Projects
- Demands for CCS Development

Content

✓ Introducing Sinopec

- ☐ Sinopec CCS Technology
- ☐ Sinopec CCS Projects
- ☐ Demands for CCS Development

Sinopec

Sinopec is a state-owned company, as the largest integrated energy and chemical company.

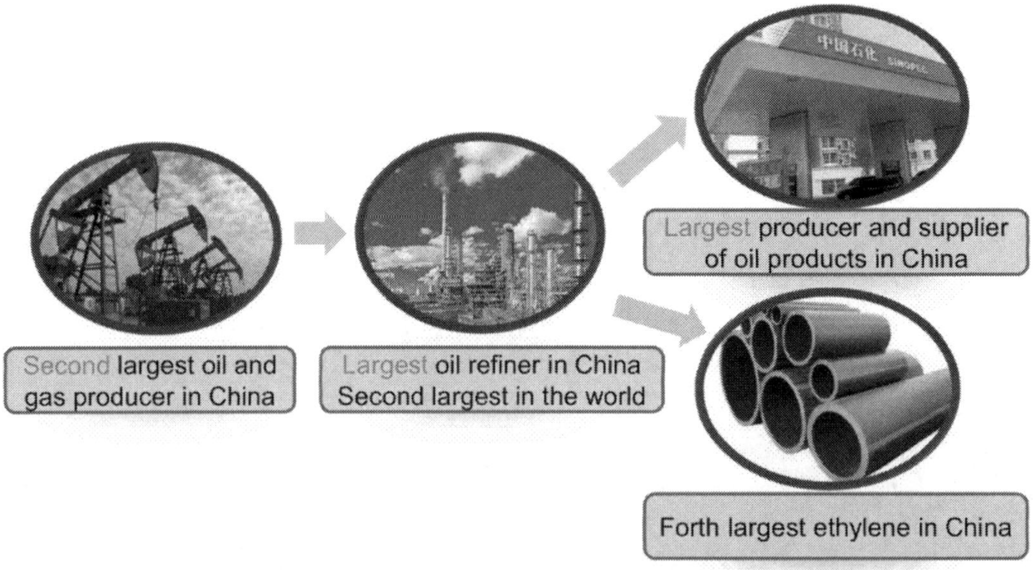

Background of Sinopec CCS

Global warming

The great threaten for environment
- **CCS is an effective way to reduce greenhouse gas emission**

China

- **China is going to reduce the intensity of CO_2 emissions per unit of GDP in 2020 by 40 to 45 percent compared with the level of 2005.**
- **China carries out a series of policies to promote CCS**

Sinopec

- **The social and environmental responsibilities**
- **Having CO_2 capture resources(refineries, natural gas with high CO_2 content, power plant, etc.)**
- **Having a vast amount of reservoirs suitable for CO_2–EOR and storage, CCS+EOR**

Content

Sinopec CCS Technology

Sinopec CCS Technology

Main Research Fields

CO_2 Capture and Purification
- Power Plant PCC
- Ammonia tail gas
- Natural CO_2 source
- Natural gas with high CO_2 content

CO_2 Pipeline
- Calculation of process
- Security control

CO_2 -EOR Reservoir
- Simulation
- Storage security assessment
- Selection Criteria for EOR & storage
- Development Scheme

CO_2 Corrosion
- Material
- Corrosive inhibiter
- Anticorrosive coating
- Cathodic protection

CO_2 capture simulation equipment (10Nm3/h)

Non homogeneous long core physical simulation

Simulation of CO_2 pipeline leakage

Drilling Tech.
- Special gas injection tube
- Special production pipe string
- Channeling prevention

Environmental Monitoring
- Monitoring tech.
- Data analysis
- Environmental impact
- Early warning scheme

Multi-phase flow loop for field corrosion modeling equipment

Post-combustion CO_2 Capture Technology

Problems

- High energy consumption
- Solvent degradation and corrosion
- Secondary pollution

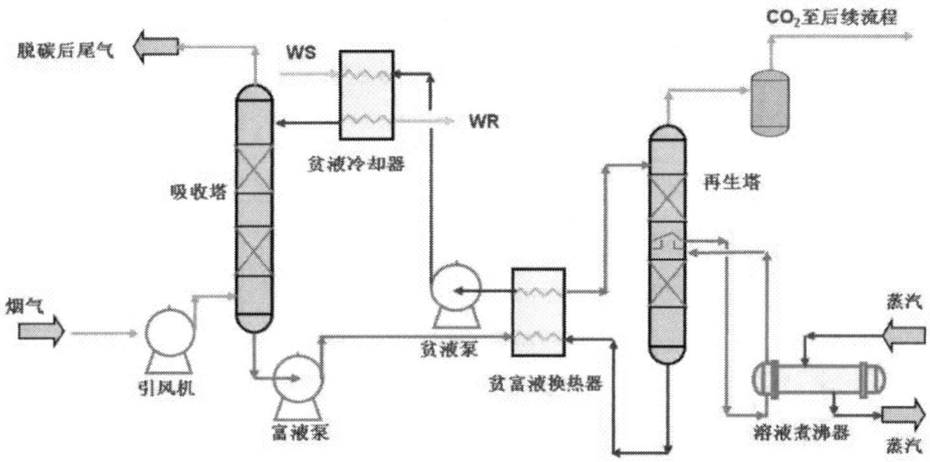

Post-combustion CO_2 Capture Technology

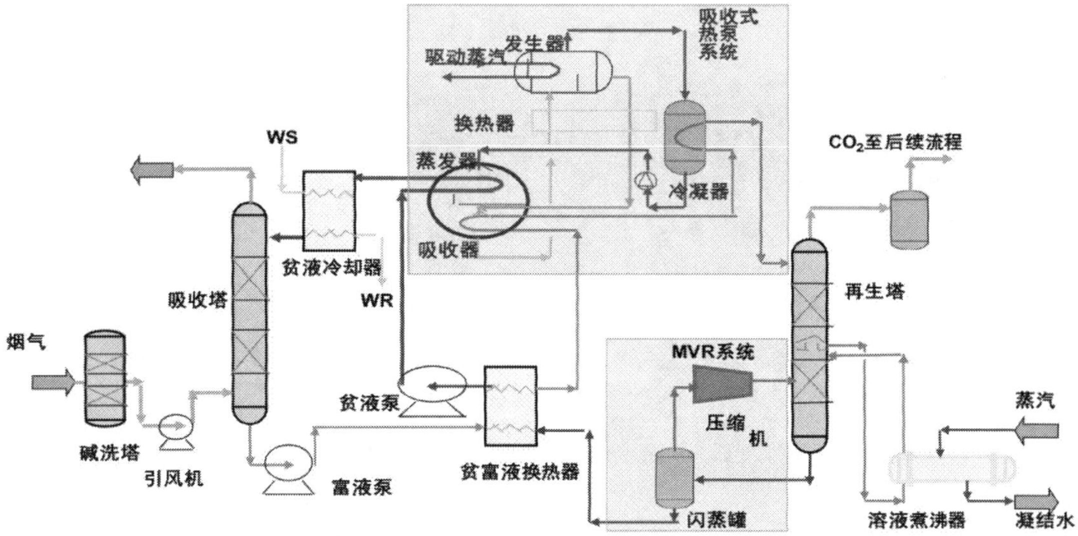

Approaches

- Improved process by thermal energy saving
- Developed new solvents-MSA
- Flue gas pretreatment
- Harmless treatment for secondary pollution

Post-combustion CO_2 Capture Technology

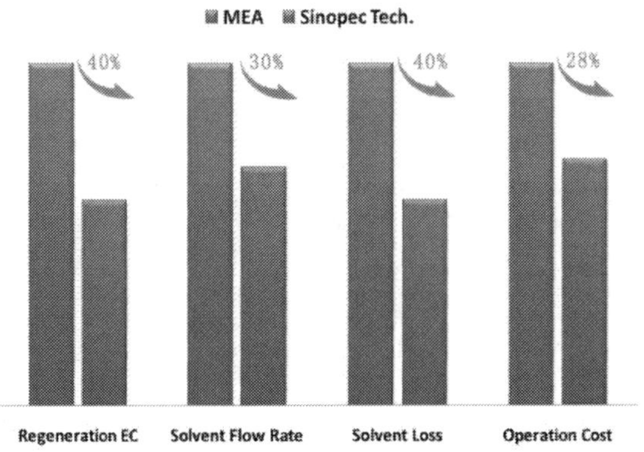

Approaches

- Improved process by thermal energy saving
- **New solvents-MSA**
- Flue gas pretreatment by $NaHCO_3$ solution
- Harmless treatment for secondary pollution

Post-combustion CO_2 Capture Technology

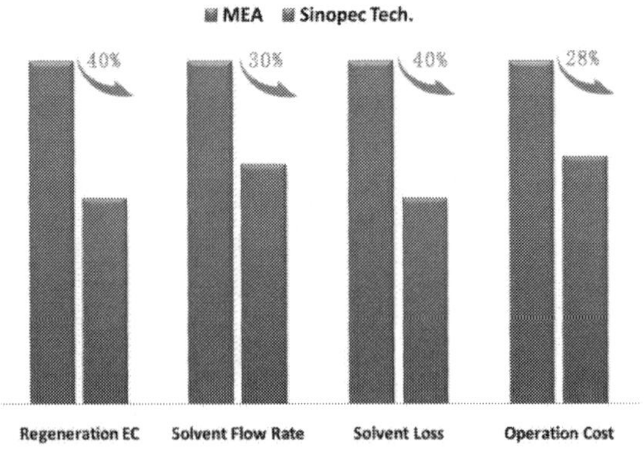

Approaches

- Improved process by thermal energy saving
- New solvents-MSA
- **Flue gas pretreatment by alkaline solution**
- **waste treatment with University of Kentucky**

CO_2 Pipeline Technology

Sinopec has done：

- **Effect of impurities on the phase properties of CO_2 and determine the limited content of impurities**
- **calculate hydraulic and thermal characteristics of CO_2 pipeline containing impurities**

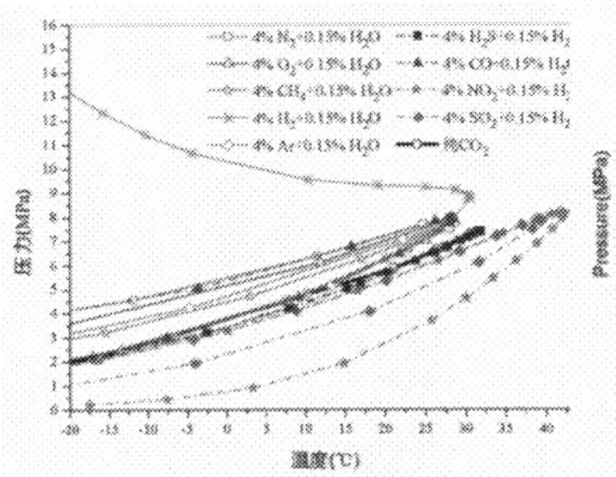

杂质对相平衡影响

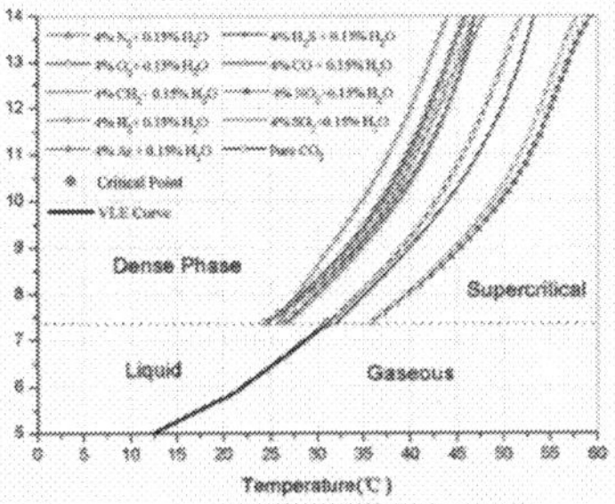

杂质对准临界区影响

Impurity	Concentration Limitation
CO_2	≥95%
N_2、CH_4、Ar	≤4%
O_2	≤10 ppm
H_2S	≤200 ppm
SOx	≤100 ppm
CO	≤200 ppm
H_2O	水露点比最低环境温度低5°C

CO_2 Pipeline Technology

Sinopec is doing

- **Investigate the dynamic transformation rules and the transient characteristics of CO_2 when emergency shutdown or relief happen**
- **Calculate the CO2 dispersion area and concentration during pipeline leakage**

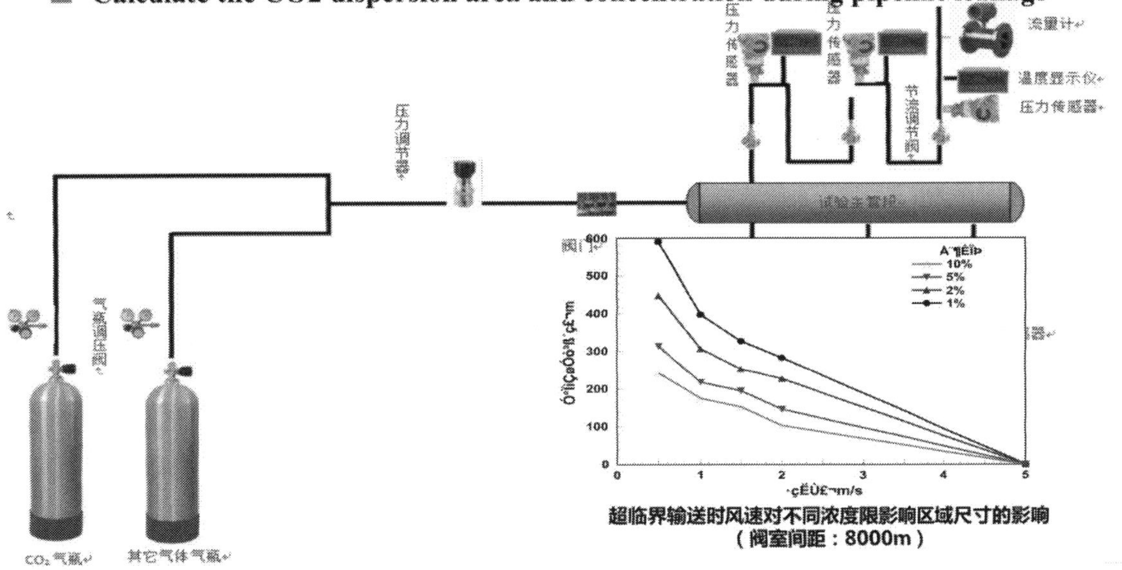

超临界输送时风速对不同浓度限影响区域尺寸的影响
（阀室间距：8000m）

CO_2 Pipeline Technology

Sinopec is doing

We cooperate with certain university to establish an industrial-scale platform for pipeline. Major research fields are the discharge flow state, pressures and temperatures on the pipeline, the dynamic pressures, temperatures and CO2 concentration on the near field of the discharge area, and so on.

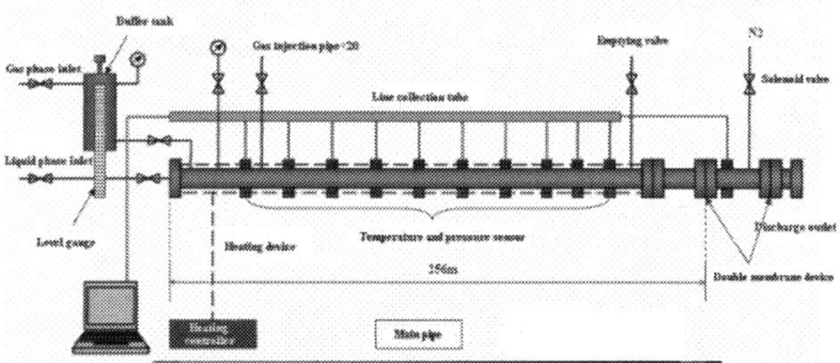

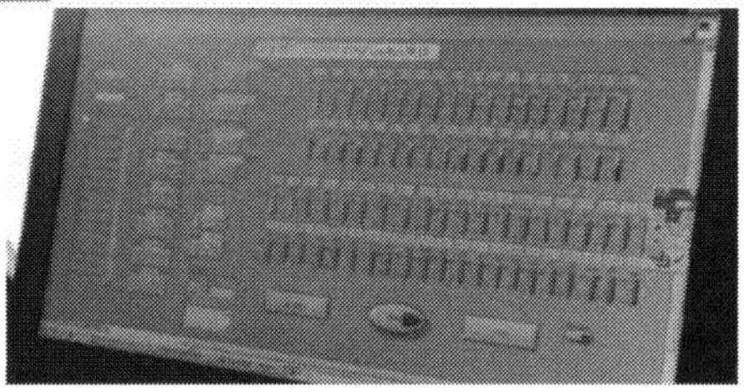

CO_2 EOR & Storage Technology

- ☐ **Block screening method of reservoir suitable CO_2-EOR and storage**
- ☐ **The effect of CO_2 purity on oil production**
- ☐ **Co-optimization of EOR and storage**

非均质长岩心物理模拟装置

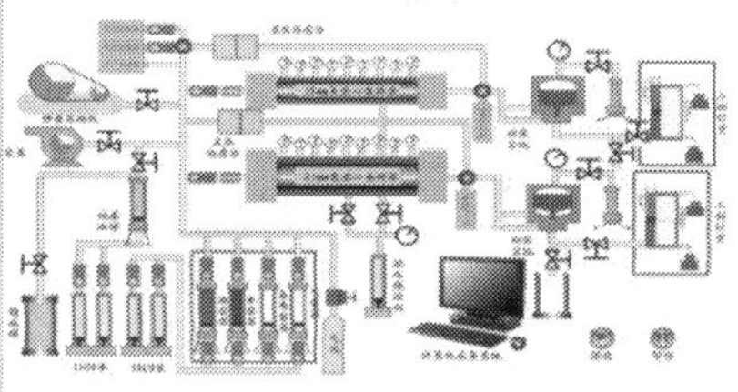

价格为同类进口产品的1/2

Block Screening method

评价参数		评价标准
关键参数	混相能力	≧1（混相驱）
		0.8～1（近混相驱）
	渗透率，mD	>0.5
参考参数	油藏条件下原油粘度，mPa.s	<12
	油藏条件下原油密度，$10^3 kg/m^3$	<0.8762
	剩余油饱和度，%	>25
	单储系数，$m^3/(km^2.m)$	>39000
	油层深度，m	>2000
	地层温度，℃	<145

The affection of CO2 purity on oil production

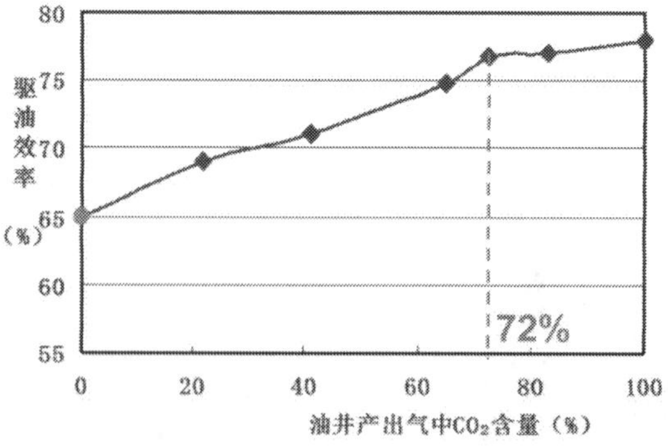

Multiphase Flow Corrosion Control

A oil-gas-water multiphase flow corrosion simulation device was built up. The device is mainly used to simulate the in-situ corrosion mechanisms. The pipe diameter is DN80 and the total length is about 130 meters.

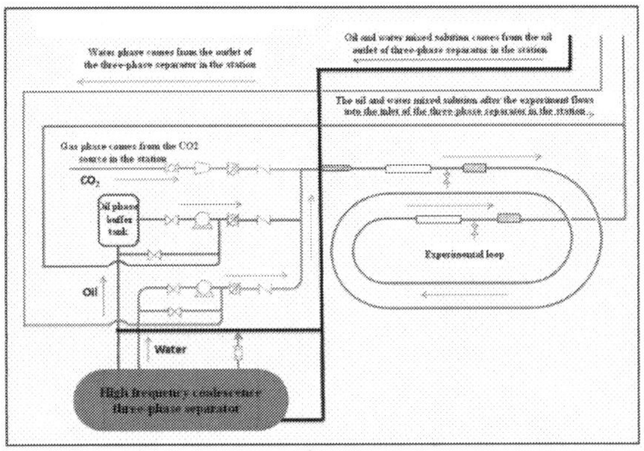

Site photo of corrosion annular channel.

Multiphase Flow Corrosion Control

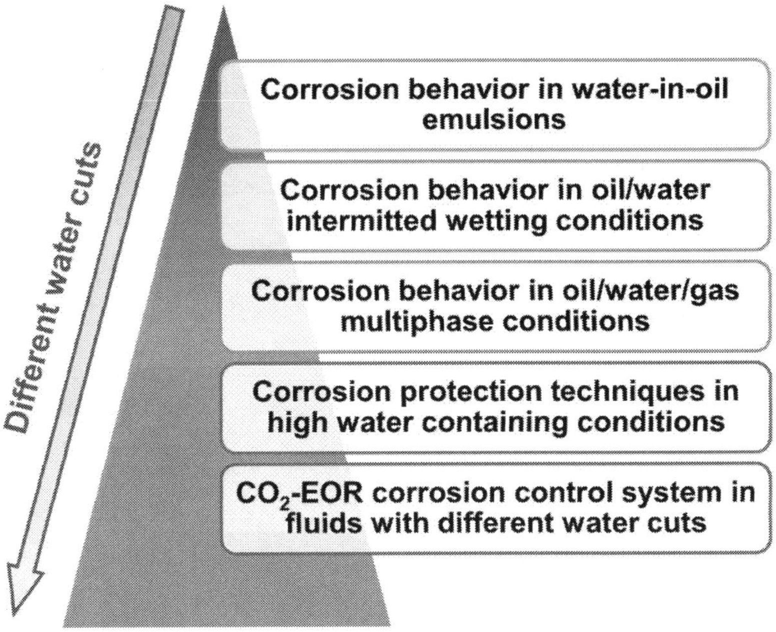

we did a series of works on the understanding of corrosion behavior of steels in different water cut conditions. Finally, we advised a controlling system for corrosion design in CO_2-EOR conditions.

Environmental monitoring Technology

- ☐ **Environmental monitoring plan for baseline has been finished.**
- ☐ **The baseline monitoring is on-going for 3 years.**
- ☐ **Mainly on surface environmental monitoring.**

4D seismic

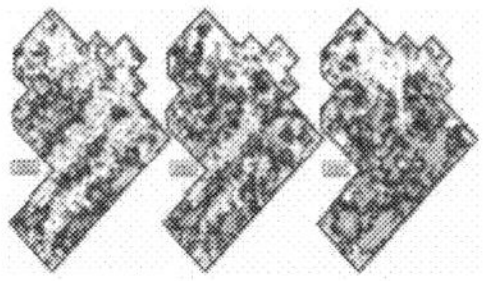

soil gas

Vegetation

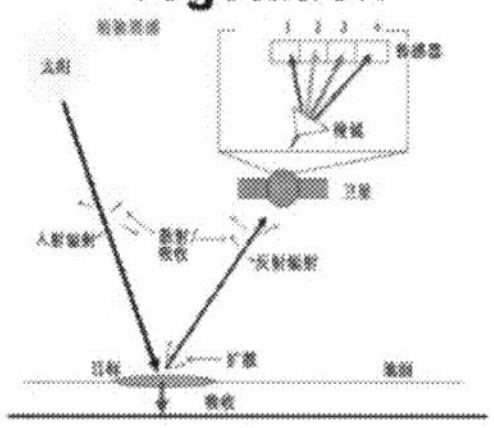

ground deformations

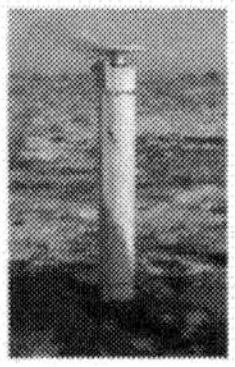

water

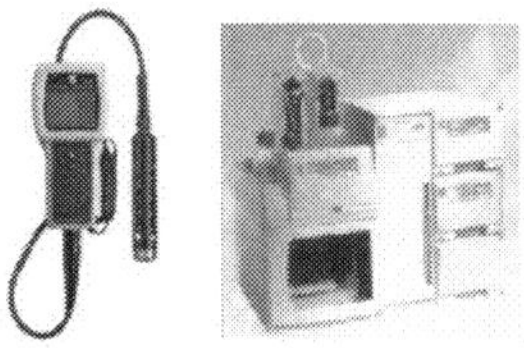

Air

Content

- ☐ Introducing Sinopec
- ☐ Sinopec CCS Technology Progress
- ✓ Sinopec CCS Projects
- ☐ Demands for CCS Development

Sinopec CCS Projects

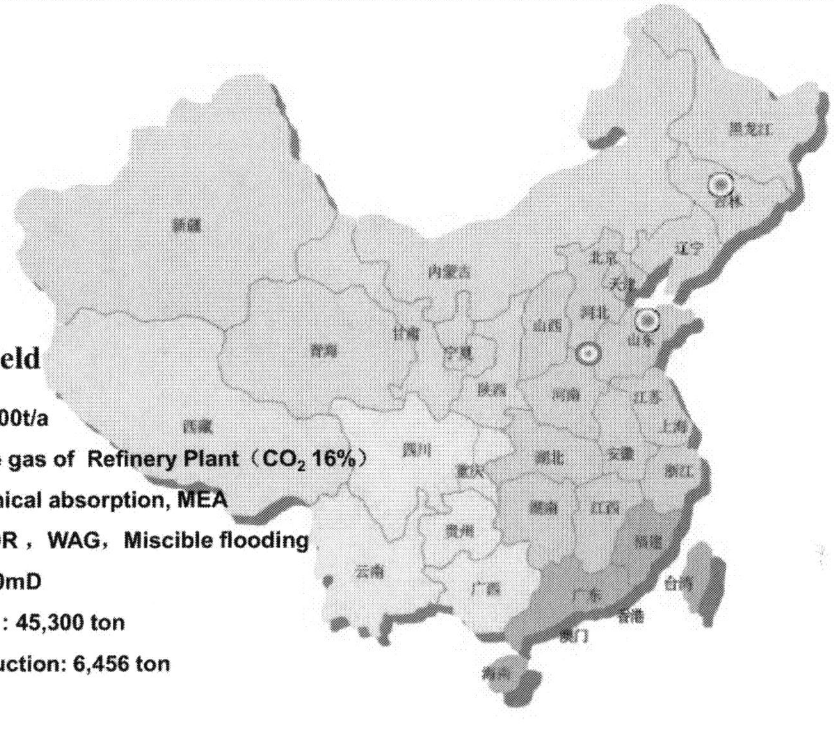

- **Zhongyuan Oilfield**
  - ✓Project Scale: 20,000t/a
  - ✓CO_2 resource: Flue gas of Refinery Plant（CO_2 16%）
  - ✓CO_2 capture: Chemical absorption, MEA
  - ✓Utilization: CO_2-EOR， WAG， Miscible flooding
  - ✓ Permeability： 590mD
  - ✓Cum CO_2 injection : 45,300 ton
  - ✓Increased oil production: 6,456 ton

Sinopec CCS Projects

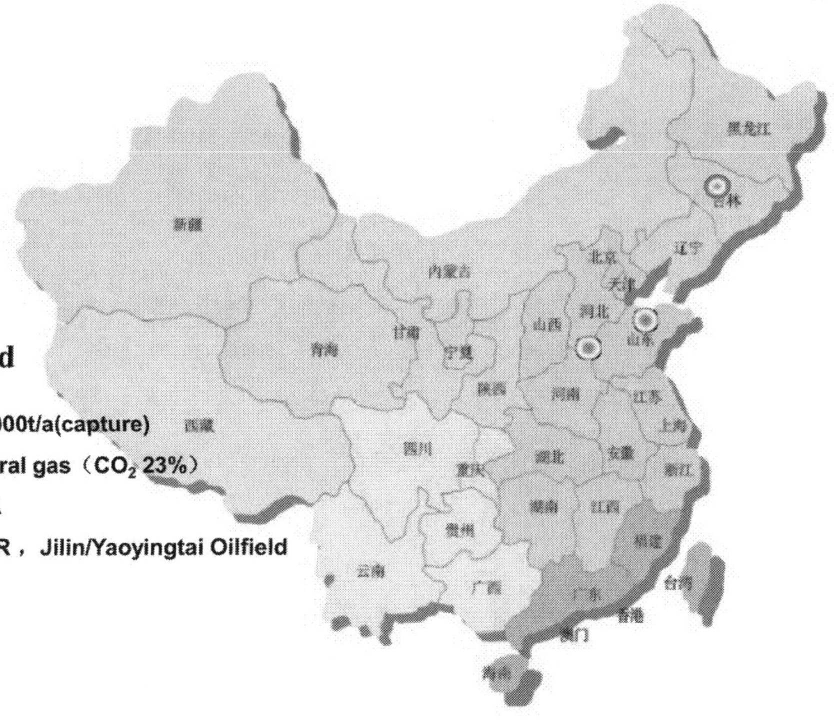

- **Songnan Gasfield**
  - ✓Project Scale: 500,000t/a(capture)
  - ✓CO_2 resource: Natural gas（CO_2 23%）
  - ✓CO_2 capture: MDEA
  - ✓Utilization: CO_2-EOR， Jilin/Yaoyingtai Oilfield

Sinopec CCS Projects

- **Shengli Oilfield**

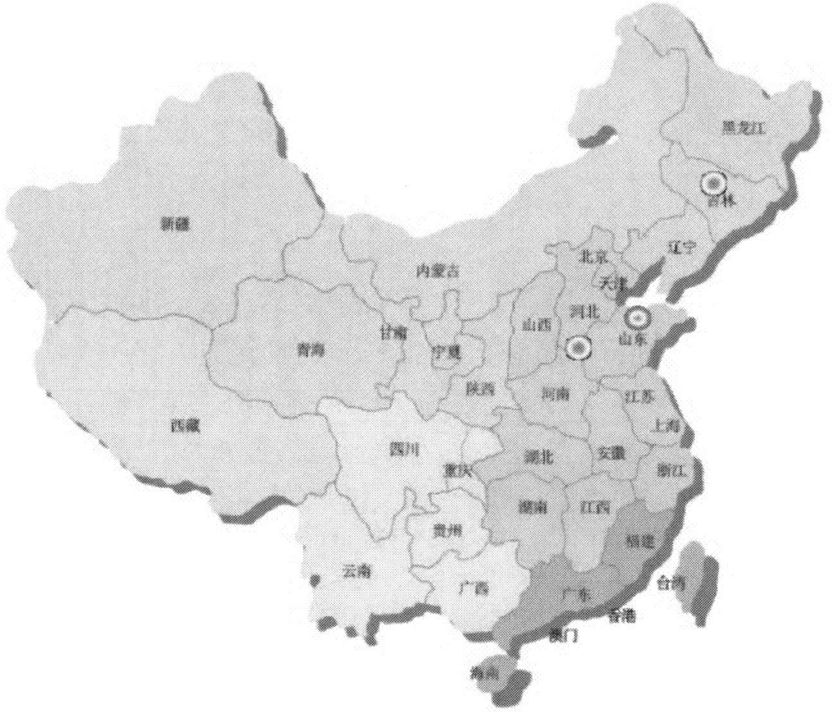

Shengli CCS Projects

Pilot platform

In 2008, a demonstration project for CO_2 capture, purification, flooding and storage of coal-fired power plant in Shengli Oilfield was built up. The scale 100 t/d, 99.5% purity CO_2.

Shengli CCS Projects

Pilot platform

- For CO_2-EOR pilot block, there are 14 oil wells, and 10 CO_2 injection wells. The permeability is 4.7 mD.
- Until now, 228 thousand tons of CO_2 is injected and 52 thousand tons of oil are produced.
- The dynamic CO_2 storage rate is 86%. Thus the oil recovery effectively enhanced as well as CO_2 emission reduction.

Geometric arrangement of wells for CO_2 pilot test in Gao 89-1 Block

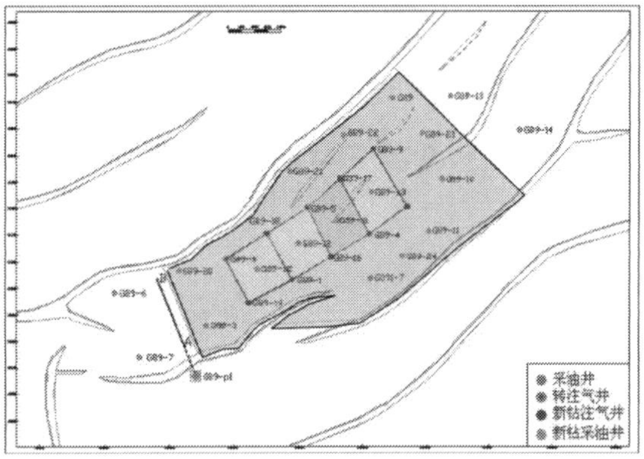

Shengli CCS Projects

Shengli has been launching two large-scale CO_2 capture, transport, oil enhanced recovery and sequestration projects under the strong supports of the Ministry of Science and Technology:

Part 1: 1 Mton/year CCS-EOR project from Shengli Coal-fired Power Plant flue gas

Part 2: 0.5 Mton/year CCS-EOR project from Sinopec Qilu Petrochemical coal gas

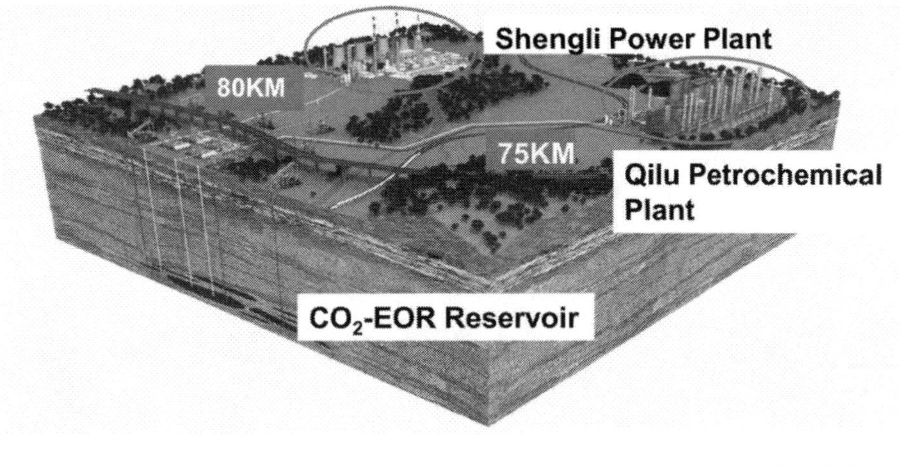

Shengli CCS Projects

Goals

- Testing and optimization of the developed technology
- Economy assessment of CCS-EOR project
- Best practice for the construction, operation and management
- Promotion of the commercial application of CCS-EOR

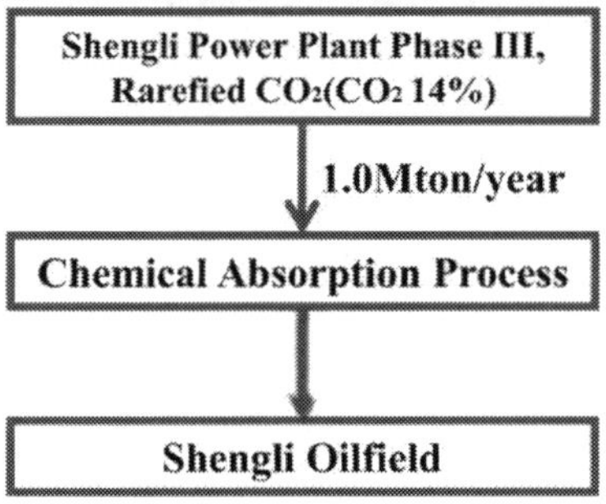

Shengli CCS Projects

- 1 Mton/year CCS-EOR project from Shengli Coal-fired Power Plant flue gas-----finish pre-feasibility
  - The pre-feasibility studies on CO_2 capture (the first version) and pipeline have been finished
  - The CO_2 flooding reservoir have been selected(0.35Mt/a)
  - Finish the design schemes of CO_2 injection, drilling and production engineering, produced fluid gathering and transportation system
  - Project investment estimation
  - Baseline for CO_2-EOR environmental monitoring

Shengli CCS Projects

In recent years, the Project Group has published more than 70 papers, declared more than 40 patents, obtained 5 software copyrights, published 2 translated books, and drafted 1 national carbon capture standard, 1 industrial standard for CO_2 pipeline transportation and 1 enterprise standard of Shengli Oilfield enterprises for CO_2 flooding and corrosion prevention in the field of CCUS.

Name	Type	Patent No./ Registration No.
Device for reducing the energy consumption of flue gas CO_2 capture system and improving the recovery rate of CO_2	Invention	ZL201210304297
Flue gas pretreatment and CO_2 capture and purification and recovery device	Invention	ZL201310599678
Electrolytic cell used in the in situ corrosion analysis of stereo microscope	Invention	ZL20121006793.0
Multi-functional annular oil-gas -water multiphase flow corrosion simulation test device	Invention	ZL20121006772.9
Decarbonization solution for produced gas recovery and CO_2 removal	Invention	ZL201310070382.0
CCS-EOR economic costs and carbon emissions accounting software	Software copyright	2015SR033370
Code for design of flue gas CO_2 capture and purification engineering	Standard	Under preparation
CO_2 transportation pipeline project-engineering design specification	Standard	Under preparation
....	......	......

Shengli CCS Projects

International cooperation & exchange

Sinopec carries out strategic cooperation and joint research with University of Kentucky, Ohio University, Rice University, ASPEN, PowerSpan, Saskpower, DNV, Shell Cansolv, SIEMENS, Virginia Commonwealth University,.

Exchange with Rice University

Schlumberger

DNV

RICE

Exchange with University of Kentucky

OHIO UNIVERSITY

SaskPower

aspentech

Cansolv

SIEMENS

JUPITER OXYGEN CORPORATION

VirginiaTech

Content

- Introducing Sinopec
- Sinopec CCS Technology Progress
- Sinopec CCS Projects
- ✓ Demands for CCS Development

Demands for CCS Development

Nature of a successful CCS-EOR project

Benefits

- Enhanced oil recovery
- Government investment support
- Tax allowance and exemption
- CO_2 credit

Costs

- CO_2 source/capture cost
- CO_2 transportation cost
- Oil processing facility modification cost
- New injection wells cost
- Additional operation cost

Demands for CCS Development

I. Establish the financial incentives and support.

-tax reduction

-low-interest loads

II. CCUS-related law, regulation and standard are urgently needed.

- Develop CCUS-related technical specifications as early as possible

safe for carbon storage, no leakage

III. To strengthen the sharing of existing knowledge and experience.

Thanks !

CCS in Australia

Prof. Paul A. Webley
Capture Program Manager
CO2CRC

Carbon Management
Technology Conference 2015
November 17-19, USA

CO2CRC

Outline

- CCS in Australia
  - Political and Economic Context
  - Legal Framework
  - Projects
  - R&D Programs
- CO2CRC
  - Capture, Otway Capture Project
  - Otway Storage Capabilities
  - Outreach
- Future Directions

CO2CRC

Political and economic context supporting a low emissions future in Australia

- Momentum for engagement on global targets, ETS, hard cuts to emissions have gone backwards
- Australian Government is primarily focused on economic growth. Australia is highly reliant on fossil fuel exports
- Some movement towards further environmental engagement now becoming evident.
- Mandatory CCS or low emissions technology inclusion at an industrial scale is regulated on a project by project basis.
- Differences exist between State and Australian Government policies with some States progressive, but the Australian Government is the key influencer
- Significant reliance on electricity from aging coal fired power plants
- Dynamic policy context with lack of incentives for investment in renewables or low emissions technology

CO2CRC

Sectoral trends in Australian GHG emissions

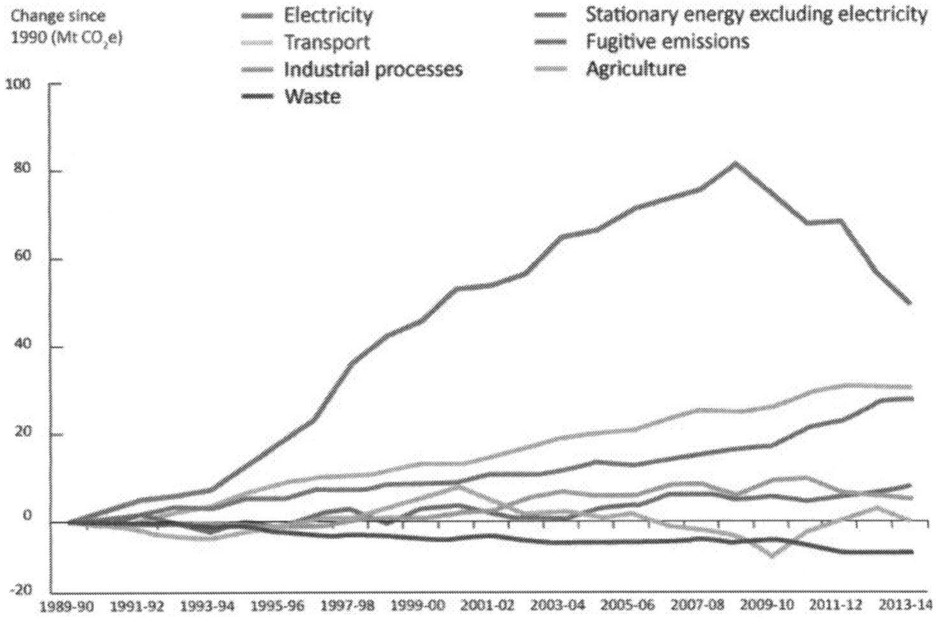

CO2CRC

Modified after Department of Environment (2014)

Drivers for CCS in Australia

- Maintain Australian Exports (60 billion)
- Stay within 2 degree scenario
- Greenhouse gas reduction 5% below 2000 level by 2020
- Heavy reliance on coal

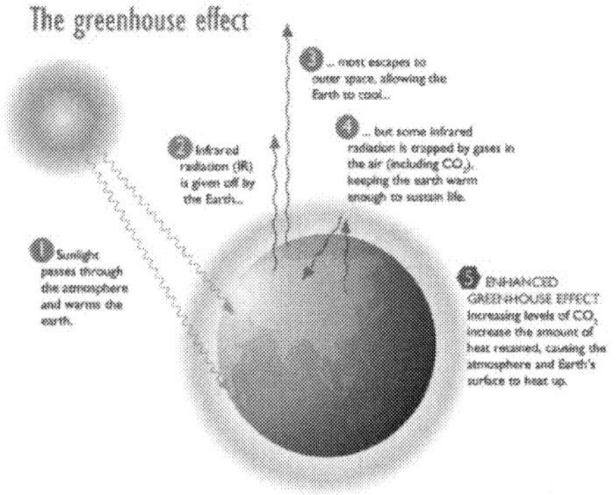

CO2CRC

CCS in Australia – Legal Framework

Commonwealth and State-level legal frameworks have been developed to regulate GHG storage activities:

- Commonwealth offshore Act;
- State Acts and amendments in Victoria, Queensland and South Australia;
- Western Australia's Barrow Island Act.

- Legislation has adopted three largely consistent regulatory models, which address both onshore and offshore storage:

I. Amended existing petroleum regimes to include provisions which regulate GHG storage;

II. Developed stand-alone CCS-specific regulatory models (albeit predicated upon aspects of petroleum legislation);

III. Project-specific legislation to regulate GHG storage as part of the Gorgon project.

CO2CRC

Major CCS work to date in Australia

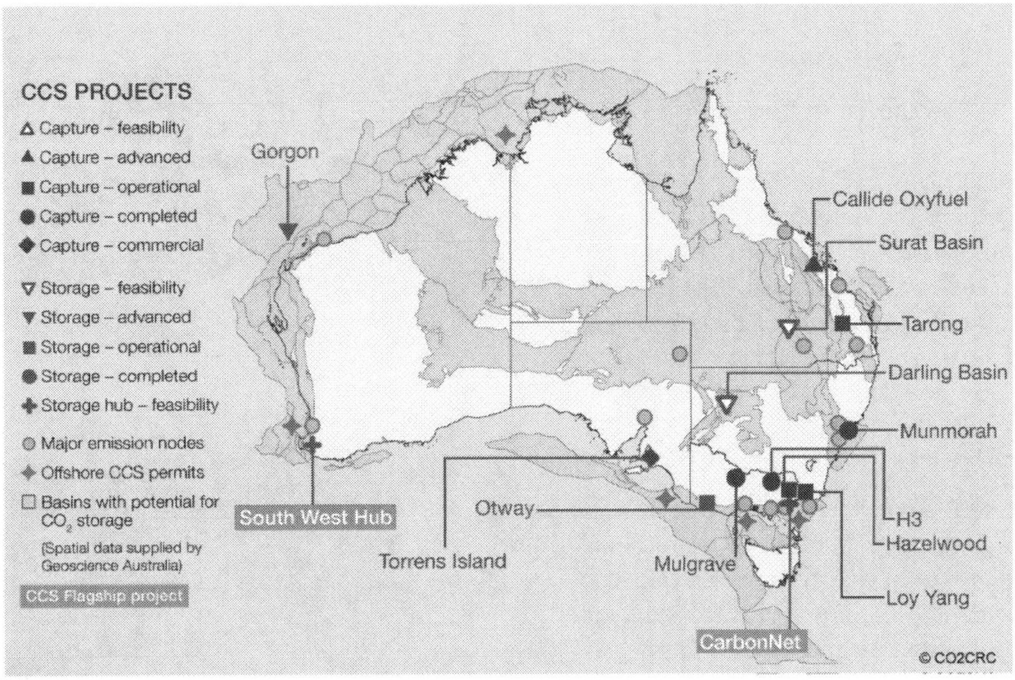

CO2CRC

CCS large scale projects in Australia

Gorgon Project

The Chevron-operated Gorgon Project is one of the world's largest natural gas projects (3.4-4Mtpa CO2) and the largest single-resource development in Australia. Construction operations are over 90% complete.

Source: Chevron Corporation

Callide Oxyfuel

The Callide Oxyfuel Project is world leading project that aims to demonstrate how carbon capture technology can be applied to a coal-fired power station to generate electricity with low emissions.

Source: CS Energy

CO2CRC

CCS large scale projects - Flagships

CarbonNet Project

The CarbonNet Project will implement a multi-user CO_2 capture and storage network in the Latrobe Valley (1-5Mtpa). The network could integrate multiple CO_2 capture projects in the Latrobe Valley, transporting CO_2 via a common-use pipeline and injecting it deep into offshore storage sites in the Gippsland Basin.

Source: DEDJTR, State Government Victoria

South West Hub Project

Potential industrial capture source, initial CO2 capture capacity of 2.5 Mtpa but longer term potential to include power facilities in a Hub concept, the project is at the stage of conducting seismic surveys/early planning

CCS large scale projects– CO2 CRC Otway Project

CO2CRC Otway Project

The CO2CRC Otway Project is the country's first demonstration of the deep geological storage or geosequestration of carbon dioxide (CO_2), the most common greenhouse gas.

CCS Pilot Scale Projects

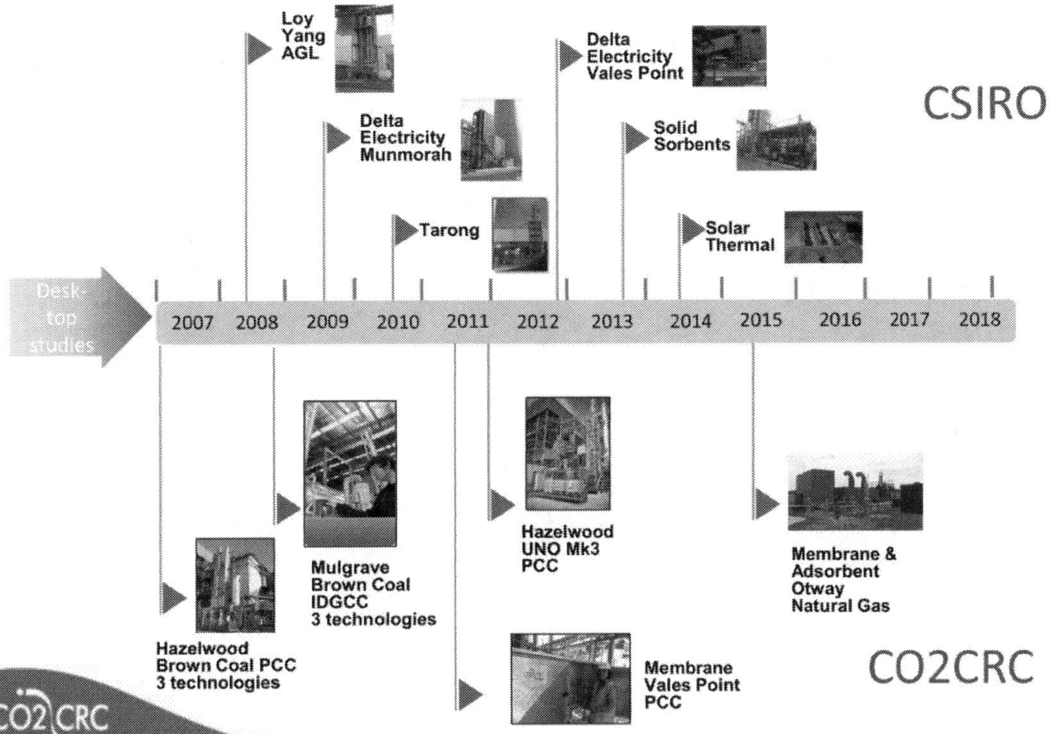

CO2CRC

What is CO2CRC?

- CO2CRC is a world leading carbon capture, utilisation & storage (CCUS) research organisation that works closely with major Australian and international industry to find solutions to reduce large scale greenhouse gas emissions through:
  - Investigating carbon capture, transport, utilisation and storage technologies
  - Examining the costs and drivers across the CCUS chain
  - Conducting research, development and demonstration across the CCUS value chain
  - Providing efficient collaboration amongst 150 researchers in capture and storage
  - Building experience in CCUS technology by addressing knowledge gaps

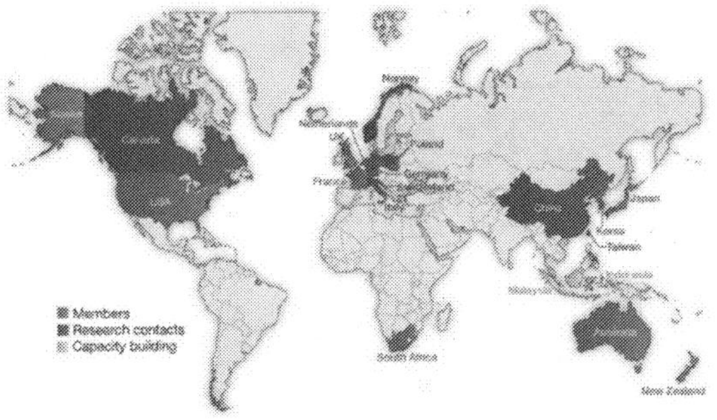

International Connections

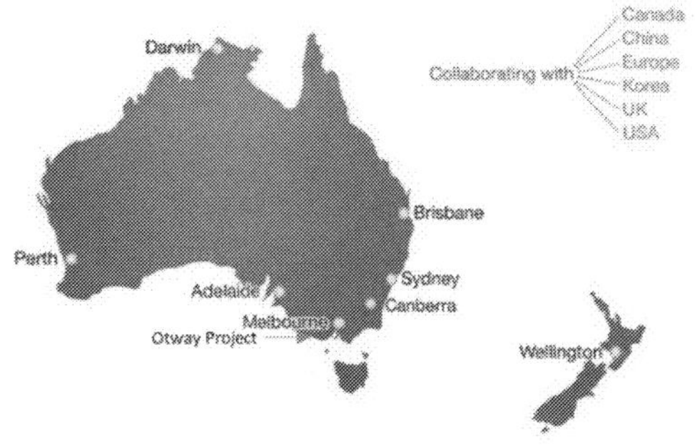

Research Providers

CO2CRC

CO2CRC's high level mandate

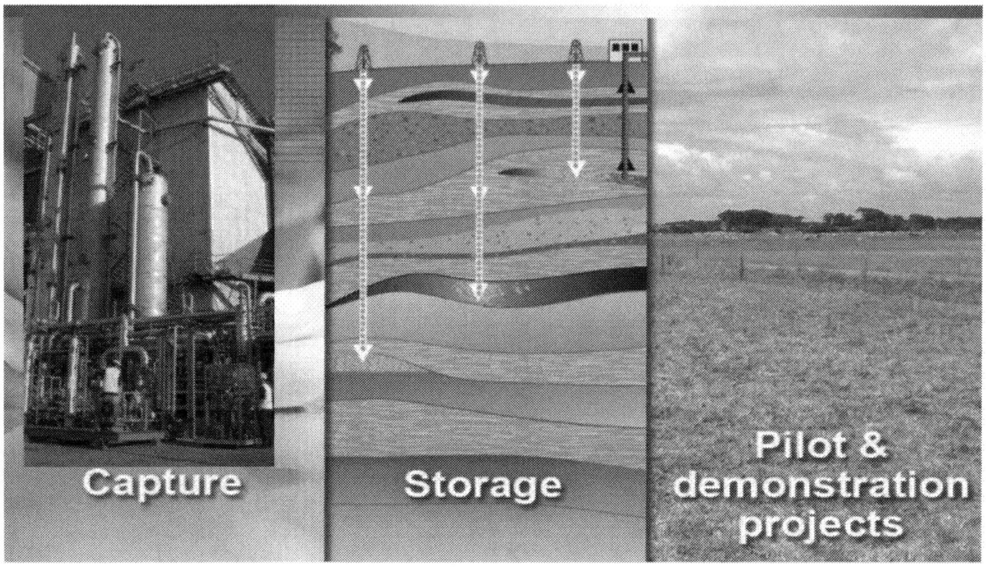

- 31 partner organisations with more than 100 researchers

Government, Industry and Research Partners

Australian Government Department of Industry and Science
Australian Government Department of Education
Australian Government Geoscience Australia
NSW Department of Industry
Government of South Australia
Victoria
Government of Western Australia Department of Mines and Petroleum
anlecr&d
Global CCS Institute
Chevron
INPEX
CSIRO
The University of Melbourne
The University of Adelaide
Australian National University
Curtin University
The University of Western Australia
UNSW Australia
The University of Queensland
Charles Darwin University
Federation University Australia
COAL21
KIGAM Korea Institute of Geoscience and Mineral Resources

Field Capture Projects

- Attained capability of scaling up of capture plants from concept till large scale (TRL level 1 till 9) with economic feasibility
- First application of 3 different capture technologies in parallel and their comparison for large scale implementation

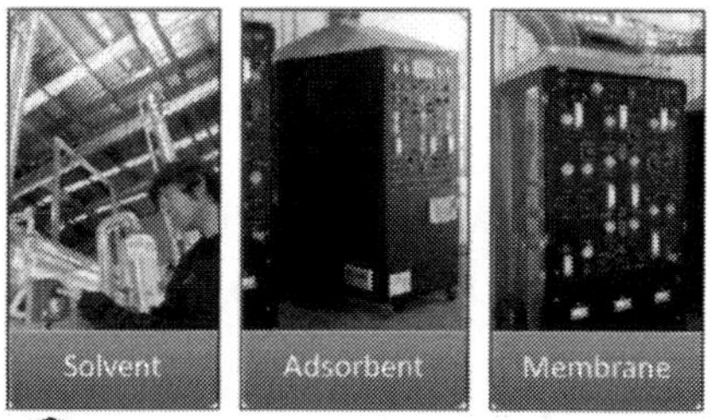

3 pre-combustion technologies at HRL, VIC : 2007-2011, ETIS/BCIA funded

3 post-combustion technologies at Hazelwood, VIC : 2007-2011 ETIS/BCIA funded

BCIA funded UNO MK 3 post-combustion at Hazelwood, VIC : 2011-2014

Membrane plant at Delta's Vales Point, NSW : ANLEC R&D funded

CO2CRC

Capture research in CO2CRC

Solvent Absorption	Membranes	Adsorption	Hydrates/ Cryogenics	Capture Demo Facilities	Engineering Integration & Large-scale Design
Melbourne Uni	*Melbourne Uni UNSW UQ*	*Melbourne Uni Monash*	*Curtin Uni*	*ALL*	*ALL*

Since 2003 — Since 2007

Economic evaluation of capture and overall CCS by *UNSW*

Applications: Post-combustion capture, pre-combustion capture, natural gas separation

CO2CRC

World class skills and knowledge

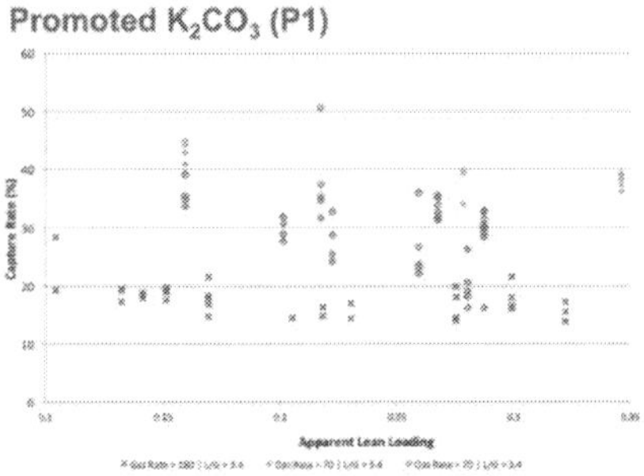

Solvents

- Physical & chemical properties affecting K_2CO_3 capture
- Field operation of ppt carbonate system
- Research into alternate solvents
- Spin-off of UNO Technology

Membranes

- Manufacture highly permeable & selective hollow fibres and flat sheet membranes for post-combustion capture, NG
- Prepare inorganic membrane for CO_2/H_2 separation
- Membrane modelling for pre/post-combustion

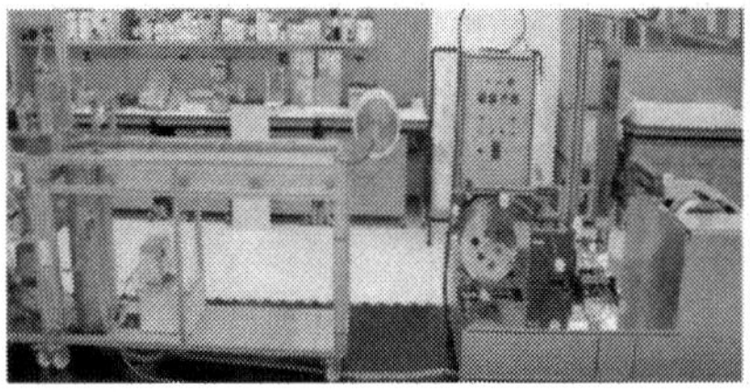

CO2CRC

World class skills and knowledge

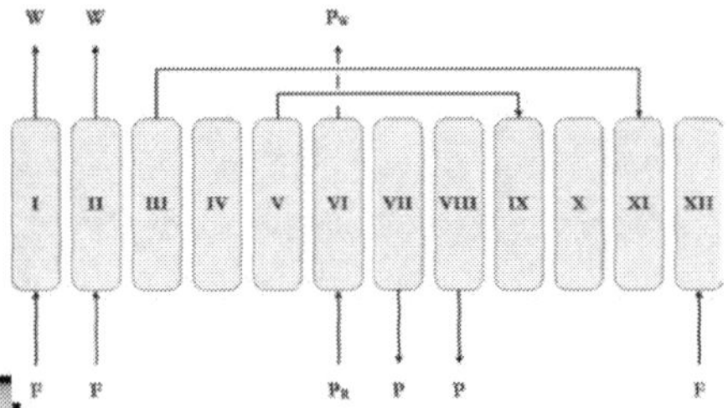

Adsorbents

- Development of adsorbent process cycles
- Rapid screening of adsorbents
- Adsorbent production and development
- Pilot CO2VSA plants

Cryogenics

- Process designs for IGCC and oxyfiring
- VSA-cryogenic hybrid for post-combustion capture

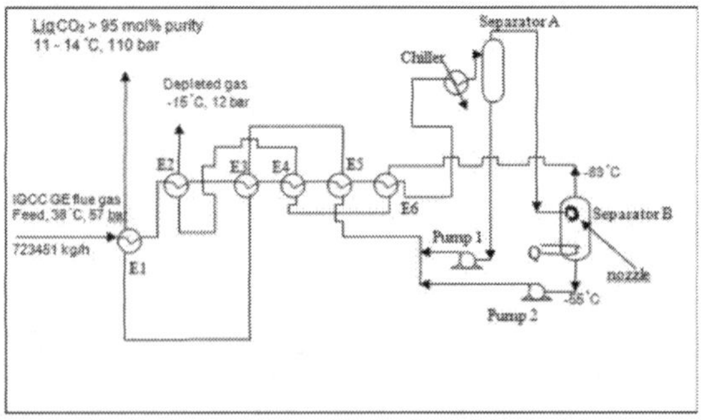

CO2CRC

Otway Capture Project

Main Objectives

- To develop cost effective, compact technologies to capture CO_2 mainly from high CO_2 content wells.
- To test new capture materials (membranes and adsorbents) and develop new capture processes over a range of adjusted CO_2 concentrations which mimic various gas field conditions of composition, temperature, and pressure.

Location: left side of the above existing facility

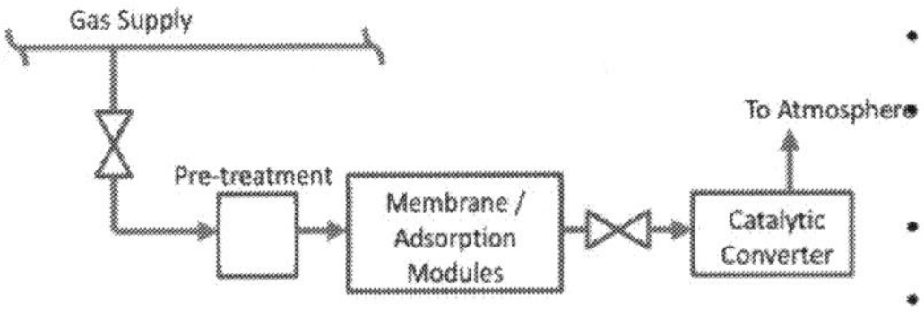

Main Features

- Feed Pressure: 90 to 20 bar
- Feed CO_2 Concentration: 5% to 80% achieved by CH_4 addition (existing is ~80% CH_4 and 19% CO_2)
- H_2S addition for impurities effect tests
- Rig on a common skid delivered on site for installation and commissioning (flow diagram shown left) 2016
- Parallel and Long term testing of membranes and adsorbents with laboratory supports (2016-19)
- Opportunities for vendors to conduct long term testing of capture technologies for CO2/CH4 separation at a fully instrumented site

CO2CRC

Location of CO2CRC Otway Project

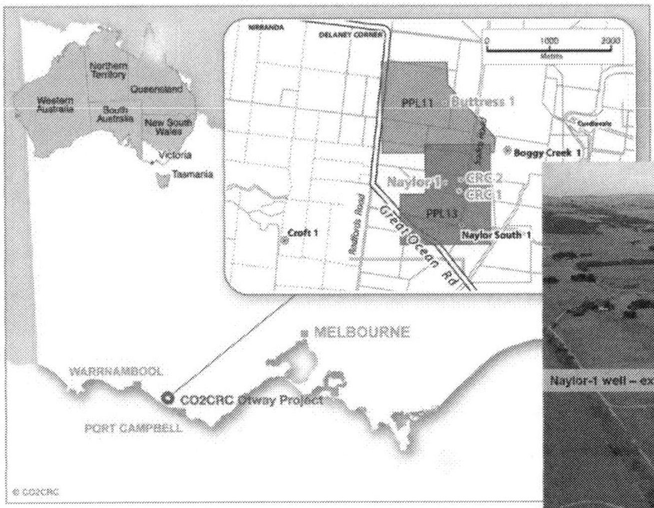

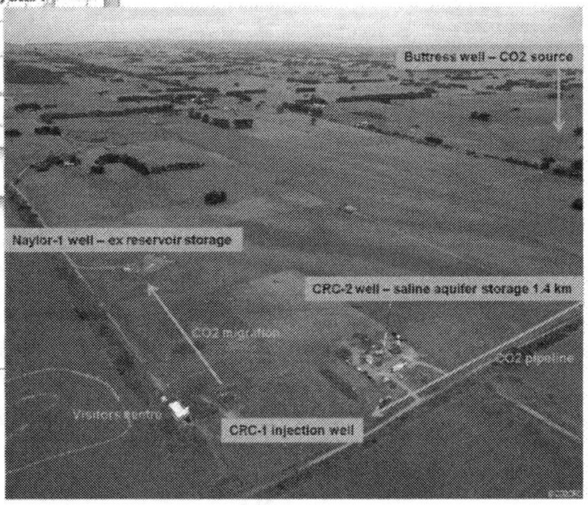

CO2CRC

The CO2CRC Otway Project

Stage 1: 2004 – 2009

Depleted Field Storage

✓ Demonstrate safe transport, injection and storage of CO_2 in a structural trap

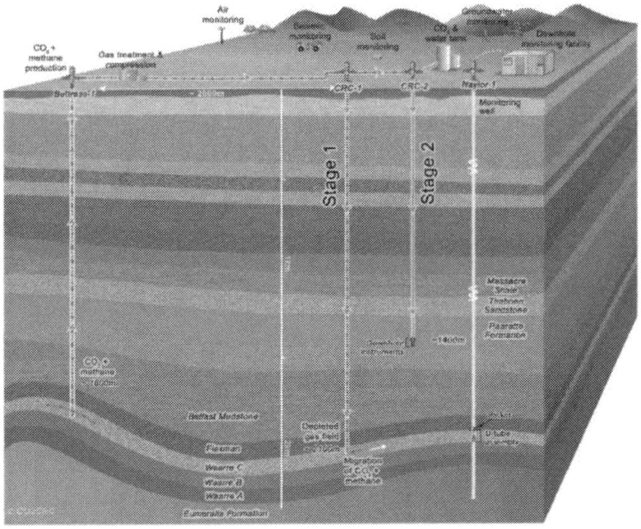

Otway Basin Schematic

CO2CRC

Stage 1 Modelling

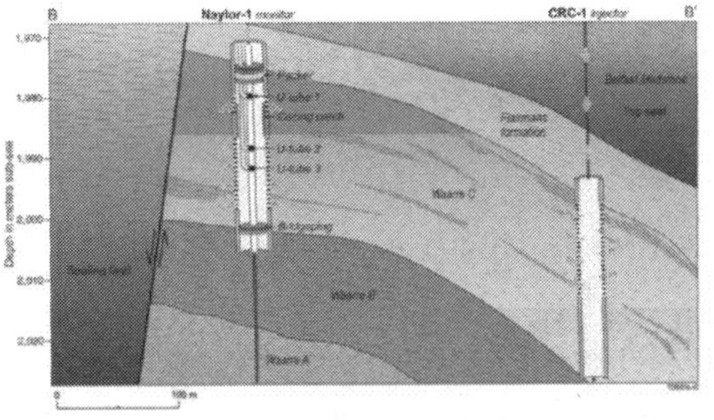

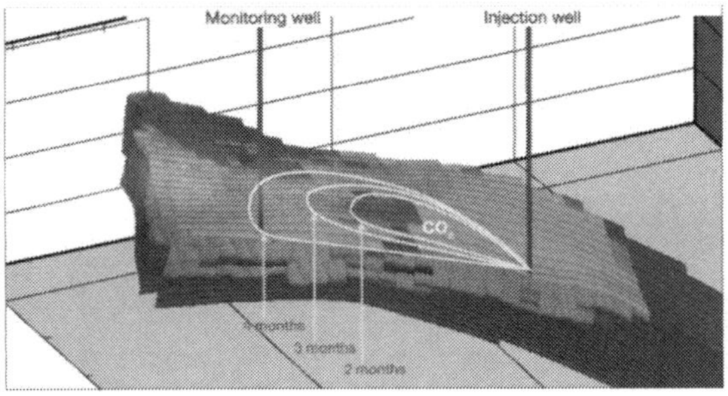

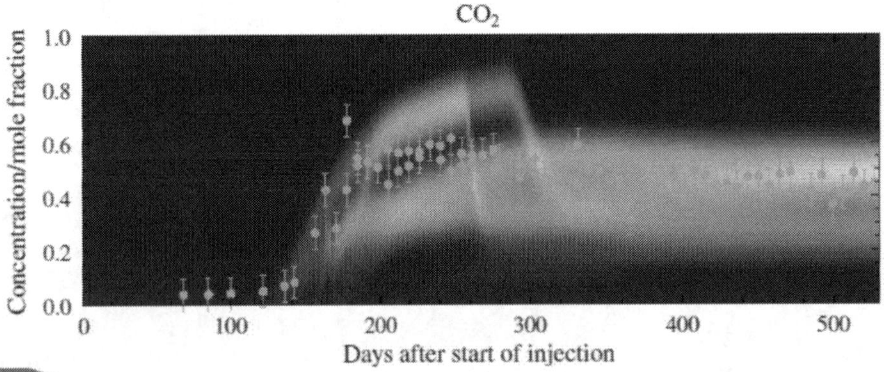

CO2CRC

The CO2CRC Otway Project: Stage 2

Stage 1: 2004 – 2009

Depleted Field Storage

- ✓ Demonstrate safe transport, injection and storage of CO_2 in a structural trap

Stage 2: 2009 – 2018

Saline Formation Storage

- ✓ 2A :Drill CRC-2
- ✓ 2B: Measure parameters affecting residual and dissolution trapping in a saline formation

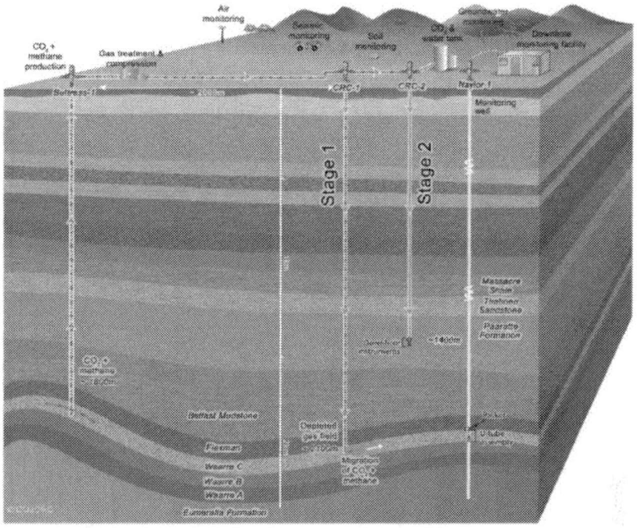

Otway Basin Schematic

CO2CRC

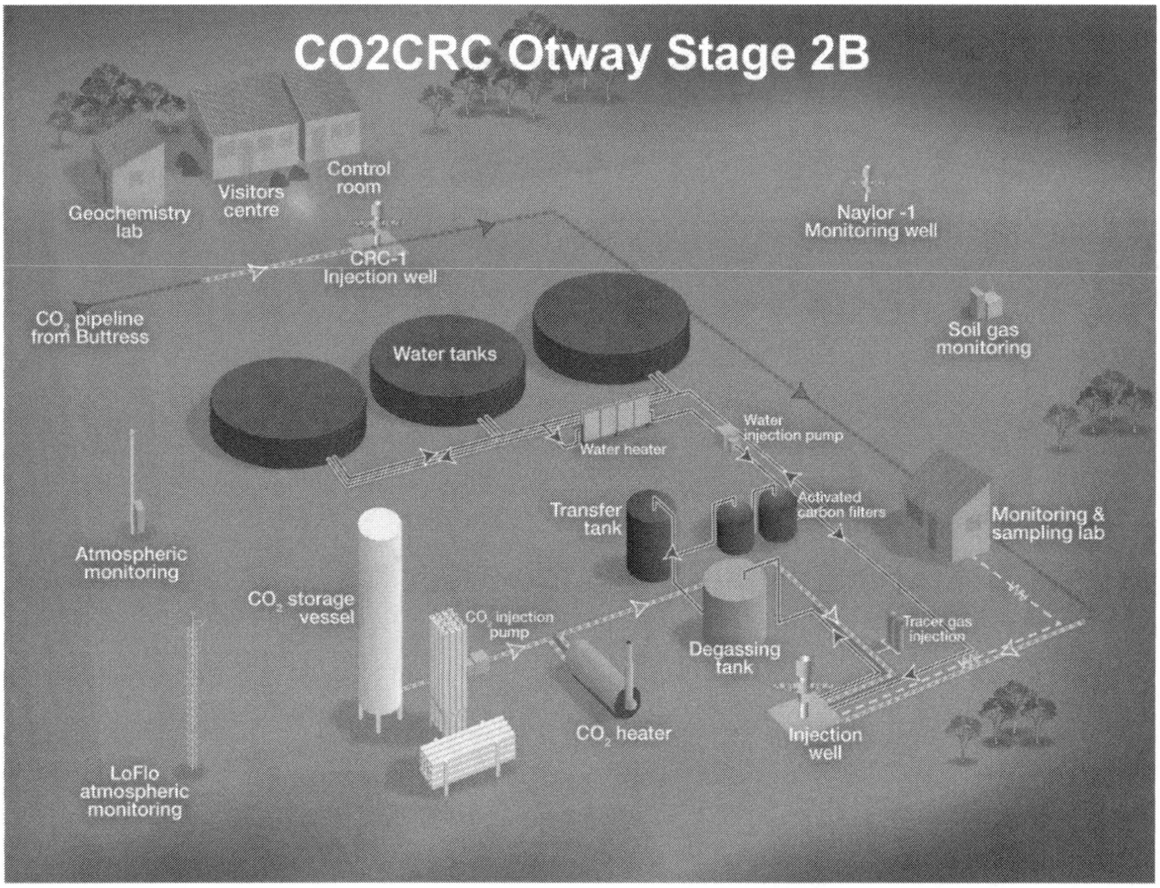

Measurements applied in Stage 2B

- Five (5) independent measurement approaches to determining residual trapping:
- The methods deployed have different depths of investigations, supply different information, and have difference benefits and limitations

Pressure response (<20 m)

Pulsed neutron logging tool (RST) (<1m)

Thermal testing (Fibre Optics) (1 - 2 m)

Noble gas tracer testing (KR-Xe) (4 - 10 m)

Liquid tracer (reactive ester tracer partitioning) (4 - 10 m)

Dissolution testing (<1 m)

Laboratory core testing

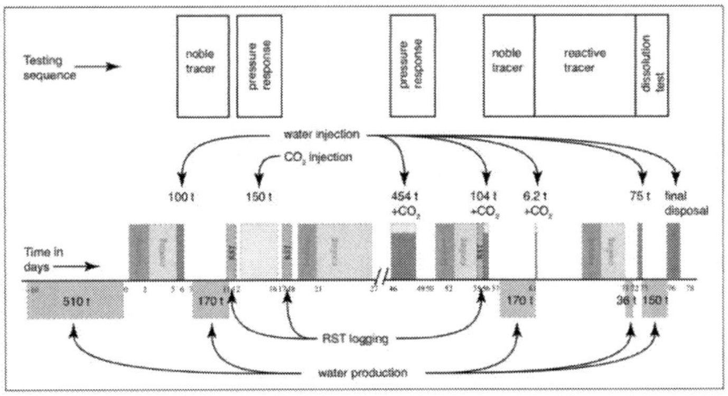

Monitoring to ensure security of storage

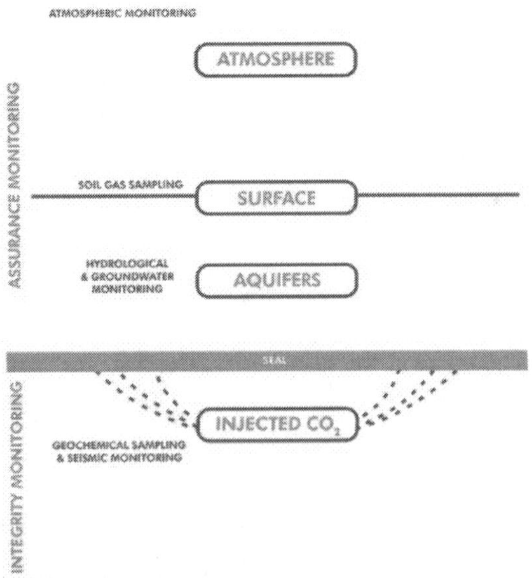

- Integrity monitoring:
  - Confirm containment of injected CO2
- Assurance monitoring:
  - Provide assurance that groundwater, soil and air are unaffected
- Data used in reporting to regulators
- Pertinent data provided to landowners

CO2CRC

Monitoring to ensure security of storage

Downhole fluid sampling

Groundwater sampling

Soil gas sampling

Seismic surveys

Stage 2C – Buried Seismic Array

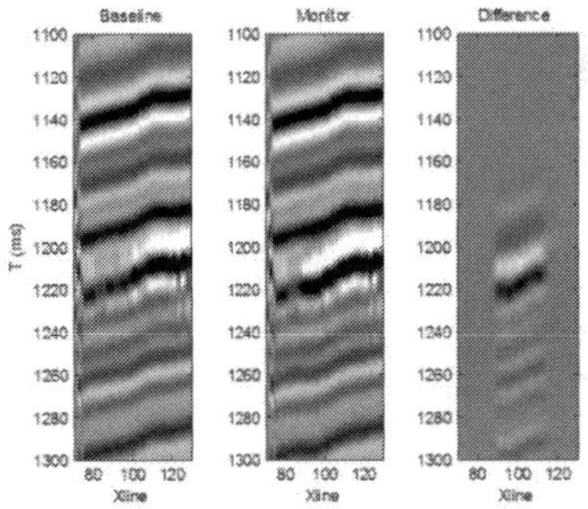

A: 2009/2010 TL noise level (Surface acquisition)

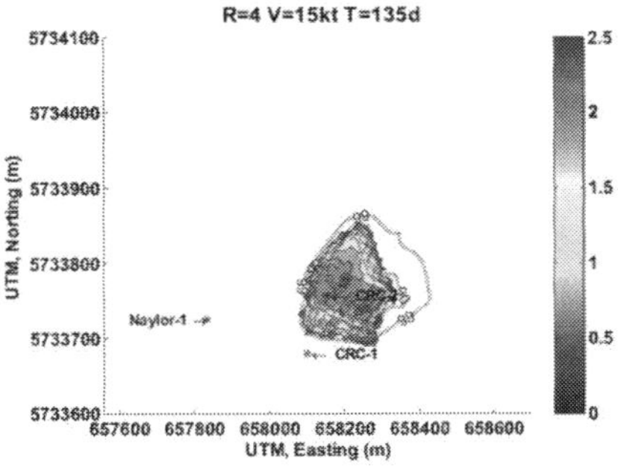

B: 2009/2010 TL noise level discounted by 6 dB (permanent installation conditions)

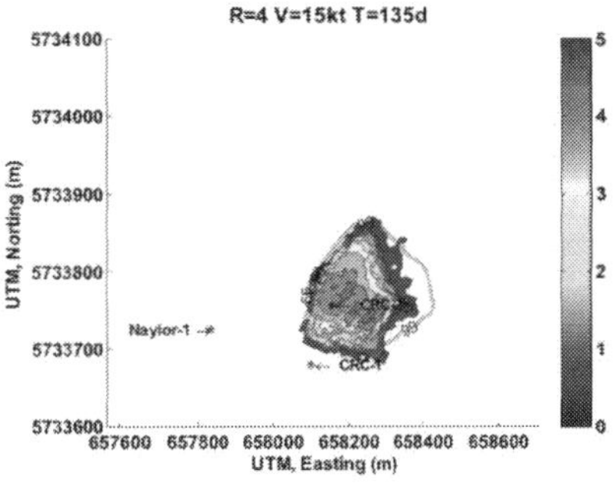

The CO2CRC Otway Project: Stage 3

Stage 1: 2004 – 2009

Depleted Field Storage

- ✓ Demonstrate safe transport, injection and storage of CO_2 in a structural trap

Stage 2: 2009 – 2018

Saline Formation Storage

- ✓ 2A :Drill CRC-2
- ✓ 2B: Measure parameters affecting residual and dissolution trapping in a saline formation
- 2C: Spatially track injected CO_2 in a saline formation

Stage 3: 2014 – 2021+

Otway Subsurface Laboratory

- Validate subsurface monitoring
- Trial storage management processes and technologies

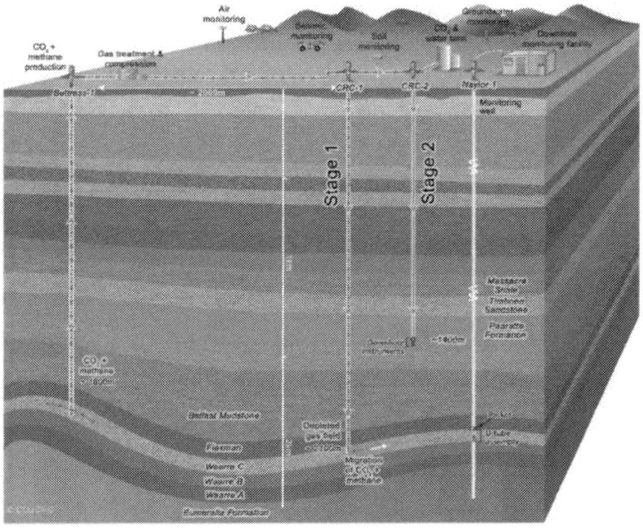

Otway Basin Schematic

Utilising the capital at the Otway – A National Deep Earth Research Facility

CO2CRC's aims to develop innovative clean energy technologies, over the next 10 years, by further understanding and exploiting deep earth processes.

Through international collaborations we wish to establish a **National Deep Earth Research Facility** that will continue to undertake leading CCS research while pursuing opportunities in:

- Geothermal
- Energy storage
- CCS Water Co-Production Management

National Deep Earth Research Facility

The Otway site and its operating environment is ideal because it has a:

1. Sedimentary basin with several kilometres of sediments
2. Variety of sediment types including several reservoir and seals
3. Several fresh aquifers, saline formations and depleted gas structures
4. Structurally simple with faulted structure at depth
5. Presence of earth resources
6. Comprehensive geological data set
7. Location with multiple CO2 sources and pipeline to site
8. Supportive local community
9. Politically stable and supportive government

CO2CRC

Education and Training

- World class educational resources and presenters
- Researchers are highly sought-after by industry
- From 2010 to 2014, CO2CRC supported:
    - 53 Honours theses
    - 61 PhD/Masters students graduated in this period
- From 2002 to 2009, CO2CRC supported:
    - 29 undergraduates Honours theses
    - 45 PhD/Masters students with 17 Masters and PhDs awarded by December 2009
    - 34 early-career post-doctoral researchers appointed to the CRC
- CO2CRC has also engaged and expanded the research skills and knowledge of more than 30 postdoctoral researchers
- Over 300 participants in CCS courses (run in collaboration with Global CCS Institute) in Australia and internationally

Students at Otway site

CO2CRC

CO2CRC and the Community

- Dedicated locally-based Community Liaison Officer
- Maintaining good relationships with landowners
- Regular community reference group meetings
- Annual public Open Days
- Site tours for special interest groups
- Community newsletter delivered to 1300 nearby residents, schools and Councils
- Market research into community attitudes
- Sponsorship of the Nirranda Football Netball Club
- Website, social and traditional media

CO2CRC

Future directions: Capture

Breakthrough solvents and UNO

Improving kinetics of absorbents for CH_4/CO_2 separation

Improving PEI adsorbents for post-combustion capture

Making improved hollow fibres for CH_4/CO_2 separation and post-combustion capture

Making better CAP membranes at scale for post-combustion capture, NG Separation

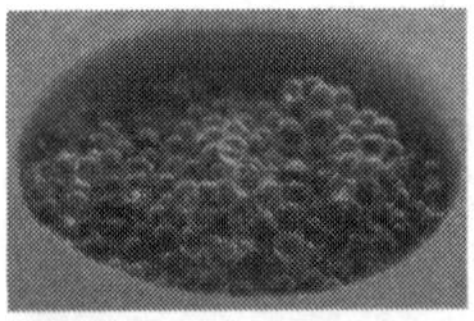

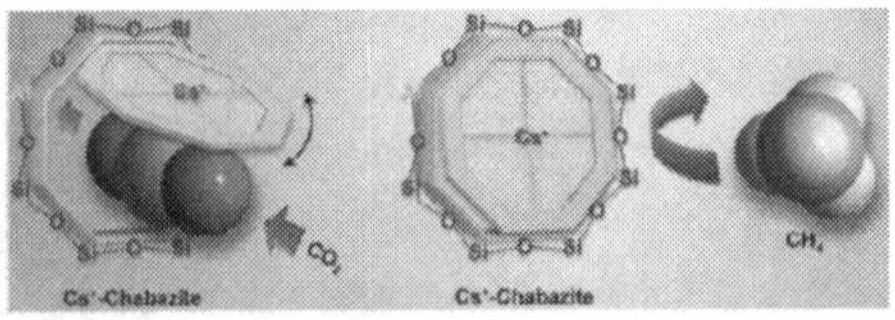

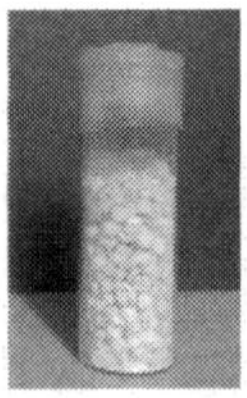

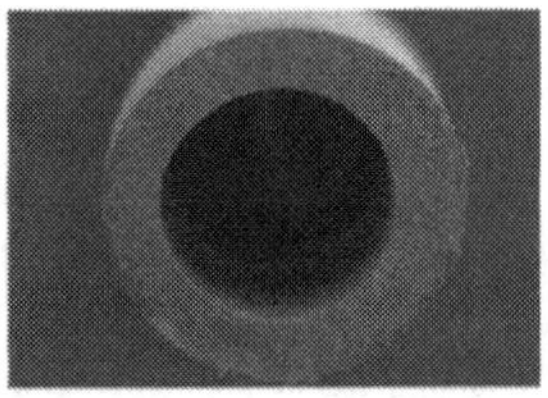

CO2CRC

CO2CRC Portfolio Assessment at 2012

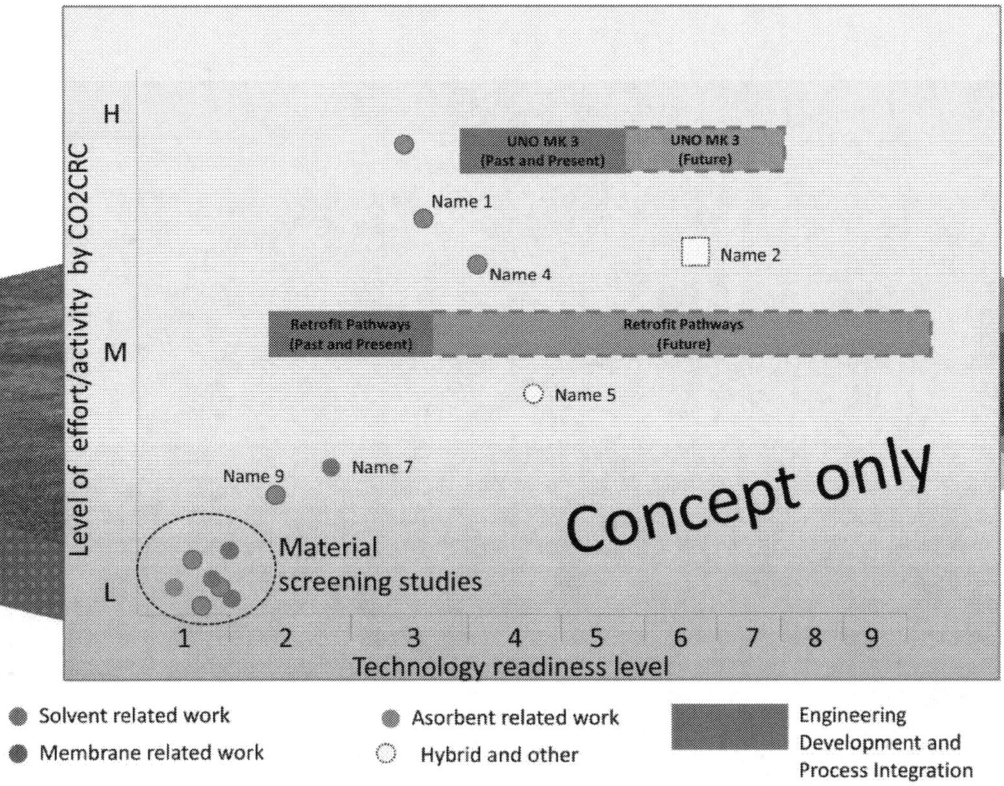

Future directions: Storage

Otway Stage 2C and 3

Alternate Storage assessment – Flagship Projects

National Deep Earth Research Facility

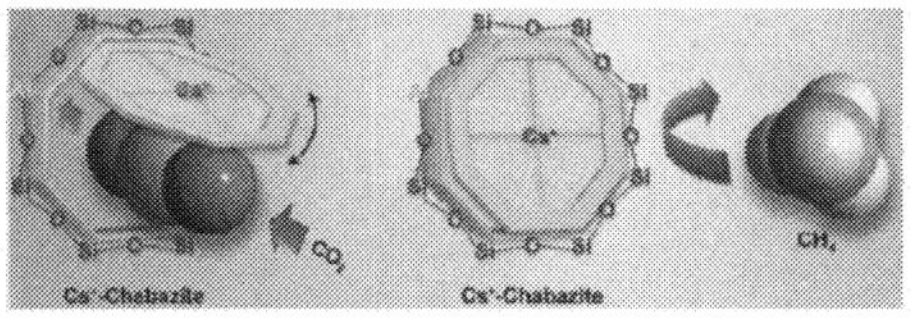

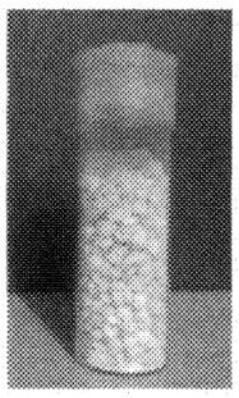

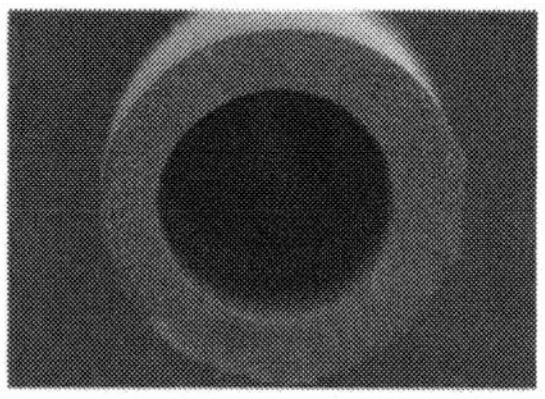

CO2CRC

Thank you
© CO2CRC 2015

Carbon Management Technology Conference
Houston
19th November 2015

UKCCS RESEARCH CENTRE

A View of the UK CCS Landscape

Dr Bruce Adderley
UK CCS Research Centre
University of Sheffield
www.ukccsrc.ac.uk
bruce@ukccsrc.ac.uk

The UKCCSRC is supported by the Engineering and Physical Sciences Research Council as part of the Research Councils UK Energy Programme

EPSRC
Pioneering research and skills

This Presentation

UKCCS RESEARCH CENTRE

- 'Leveling the Playing Field ?' – how has CCS landscape developed in the UK over the last 10 years
- Some Lessons Learnt
- Where Might the UK CCS Landscape be Headed?
- Where the UKCCSRC Fits In

UKCCS RESEARCH CENTRE

Leveling The Playing Field?

- With hindsight in 2005 CCS related legislation and regulation in the UK was still in its infancy
- Since then there have been many changes and additions

A Brief History of ……… UK CCS initiatives (2005-2015)

UKCCS RESEARCH CENTRE

Demonstrate CCS asap for G8 initiative, Peterhead 1	2005
Response to a dash for coal, Kingsnorth/Longannet	2007/2008
DECC CCS Competition launched	November 2007
New Kingsnorth power plant plans shelved	7 October 2009
Energy Act 2010, levy for CCS on all fuels	8 April 2010
Peterhead post-combustion project	14 July 2010
E.ON withdraws Kingsnorth from Competition	20 October 2010
Longannet CCS project cancelled	19 October 2011
National Audit Office report on Competition	16 March 2012
DECC CCS Commercialisation Programme (+£1bn)	3 April 2012
Four projects shortlisted	31 October 2012
Two preferred bidders announced	20 March 2013
Energy Bill (for EMR & FiT CfD) Royal Assent	18 December 2013
White Rose FEED funding awarded	9 December 2013,
Peterhead FEED funding awarded	24 January 2014
FID by the Government	2016 - up to two projects

The Committee on Climate Change

"In our December 2008 report, we set out a range of scenarios to meet our 80% emissions reduction target in 2050. The common theme running through these scenarios was the need for early decarbonisation of the power sector, with the application of low-carbon electricity to transport and heat. We showed therefore that the carbon intensity of power generation should decline over time, whilst at the same time electricity demand could increase."

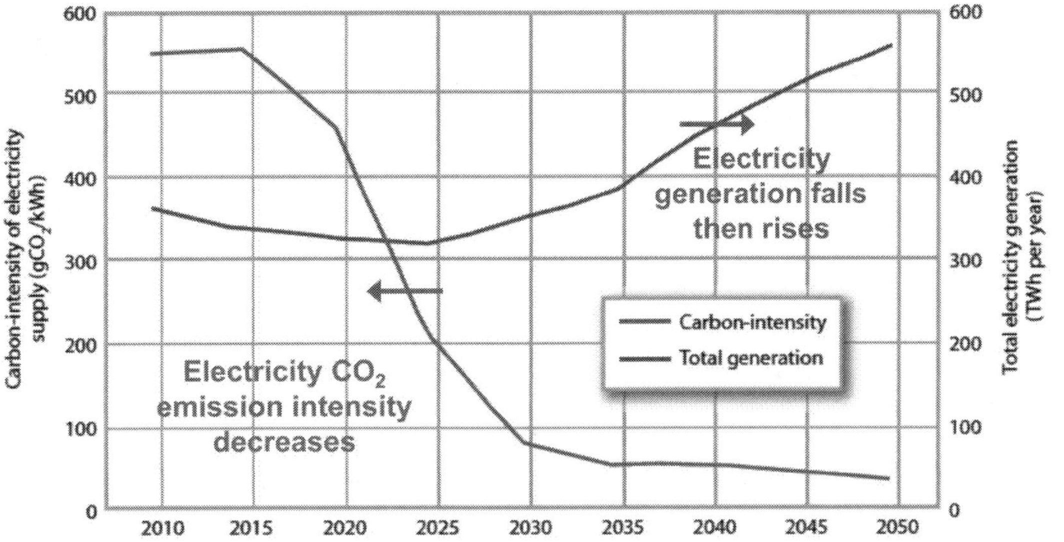

All New Power Plants above 300MW Must be Capture Ready

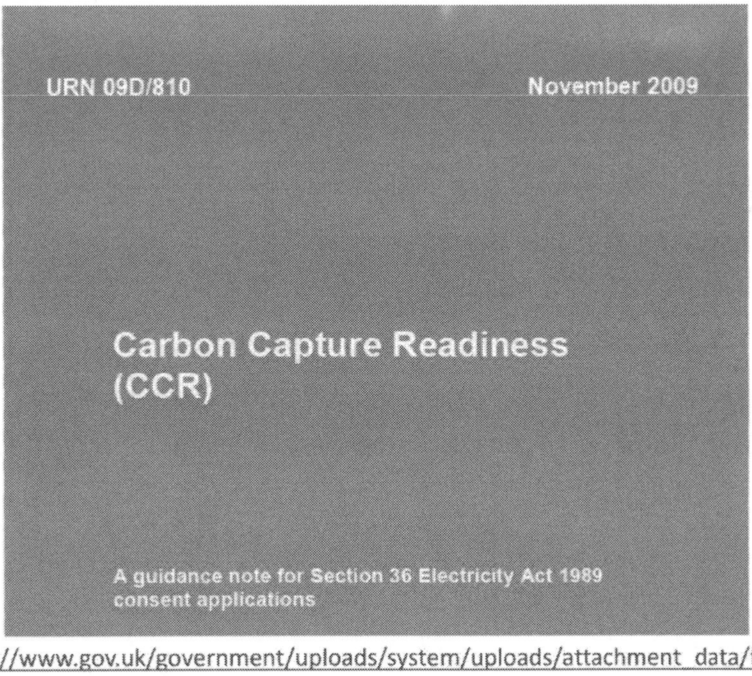
URN 09D/810

November 2009

Carbon Capture Readiness (CCR)

A guidance note for Section 36 Electricity Act 1989 consent applications

https://www.gov.uk/government/uploads/system/uploads/attachment_data/file/43609/Carbon_capture_readiness_-_guidance.pdf

DECC – The Energy Act 2010

UKCCS RESEARCH CENTRE

- Queen's Speech Nov 18 2009 - Energy Bill with funding for CCS
- Initially to be CCS on coal only but fuel neutrality adopted by the Labour Government.
- The Bill was passed just before the 2010 General Election.

- The Act introduced a new levy of up to £11bn on electricity
- To fund all types of power station CCS demonstration projects
- To support the retrofit of additional CCS capacity should it be required at a later date.

- But the CCS measures were never put into force when a Conservative/Liberal Democrat coalition government came to power

- Plans for all new coal power stations by utility companies were abandoned as a result of the recession.

Electricity Market Reform 2011

UKCCS RESEARCH CENTRE

- 'Planning Our Electric Future: A White Paper for Secure, Affordable and Low-carbon Electricity' - 12 July 2011

- The White Paper sets out key measures to attract investment, reduce the impact on consumer bills, and create a secure mix of electricity sources including gas, new nuclear, renewables, and carbon capture and storage.

"Creates a level playing field for low-carbon electricity" paid for by Feed-in Tariffs with a Contract for Difference.

http://www.decc.gov.uk/en/content/cms/legislation/white_papers/emr_wp_2011/emr_wp_2011.aspx

The Emissions Performance Standard 2013

- The Energy Act 2013 established an EPS to limit carbon dioxide emissions from new fossil fuel power stations.
- The Government has now consulted on draft regulations that will fully implement the EPS.

- The consultation, titled 'Implementing the Emissions Performance Standard: Further Interpretation and Monitoring and Enforcement Arrangements in England and Wales'
  - Launched on 25 September 2014
  - Ran until the 6 November 2014
  - On 15 January 2015 the government published their response to the consultation, together with an impact assessment.

https://www.gov.uk/government/publications/2010-to-2015-government-policy-uk-energy-security/2010-to-2015-government-policy-uk-energy-security

The UK CCS Commercialisation Competition

- **Makes available £1 billion capital funding**
- **Together with additional support through EMR addresses design, construction and operation of commercial-scale CCS.**
  - Opened in April 2012, closed in July 2012
  - Four projects shortlisted in October 2012
  - Revised proposal 14 January 2013
  - Two preferred bidders 20 March 2013
    - Peterhead Project in Aberdeenshire, Scotland - Shell and SSE.
    - White Rose Project in Yorkshire, England - Alstom, Drax Power, BOC and National Grid.
  - FEED funding awarded 9 December 2013, Peterhead 24 January 2014
- **FID by the Government in Q1 2016 - up to two projects.**
- **Two other projects, Caledonia Clean Energy and Teesside Low Carbon in reserve**

https://www.gov.uk/uk-carbon-capture-and-storage-government-funding-and-support

Industrial Decarbonisation and Energy Efficiency Roadmaps to 2050

UKCCS RESEARCH CENTRE

- 8 reports that assess the potential for a low-carbon future across the most heat-intensive industrial sectors in the UK.
- Emissions-abatement potential of individual industrial sectors, the relative costs of alternative abatement options and the related business environment including investment decisions, barriers and issues of competitiveness;
- Evidence base to inform future policy, and identify strategic conclusions and potential next steps to help deliver cost-effective decarbonisation in the medium to long term
- **Significant direct CCS in four sectors: Chemicals, Iron & Steel, Cement, Oil Refining.**
- **Indirect CCS through decarbonised electricity in all sectors**

https://www.gov.uk/government/publications/industrial-decarbonisation-and-energy-efficiency-roadmaps-to-2050

UKCCS RESEARCH CENTRE

Some Lessons Learnt

- A selection for the purposes of today's discussion
- There are many many more

Some Lessons Learnt

- Multiple UK East Coast sites available for CCS projects with direct access to offshore storage, notably Teesside with 0.5MtCO_2/yr high purity CO_2 already available from fertiliser plant to boost cluster output.
- Ability to access wider range of CO_2 storage sites, especially if already proven, would help to limit risks for developing capture/transport/storage systems from scratch.
- Stated DECC preference for long-term fixed price strike price values for the FiT with CfD under EMR, would require ability to handle fuel price fluctuations..

Some Lessons Learnt

- UK electricity market also likely to evolve to value firm and controllable low carbon generation capacity more than fluctuating renewables.
- Utilities happy to operate gas turbine combined cycle (GTCC) power plants and electricity sales, possibly OK with post-combustion capture units, not prepared or able (generally) to deal with CO_2 transport and storage.
- UK looking to build up supply chain and reduce capture capital costs for liquid solvent capture from GTCC– scope for collaboration.
- UK Govt and others (finance, utilities, equipment suppliers etc.) are open to further development of EMR legislation that supports credible players entering the UK CCS market.

Where Might the UK CCS Landscape Be Headed?

Future Carbon Budgets

Figure 2. UK GHG emissions against legislated budgets and 2050 target ($MtCO_2e$)

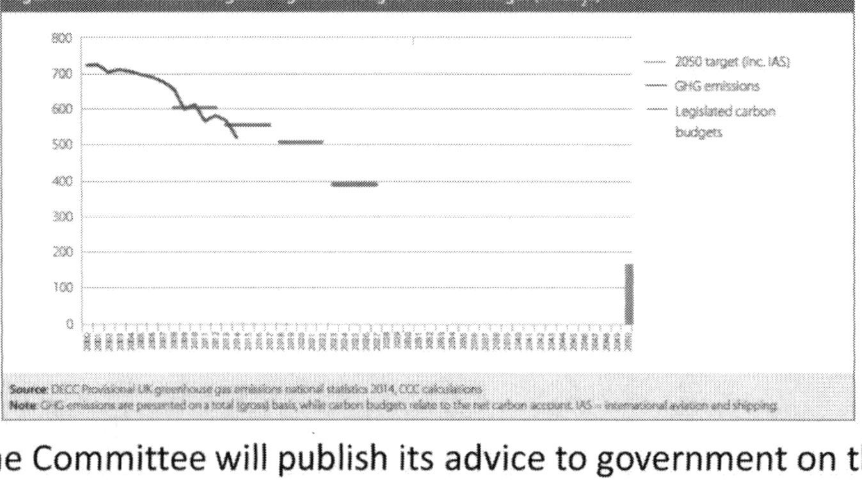

Source: DECC Provisional UK greenhouse gas emissions national statistics 2014, CCC calculations
Note: GHG emissions are presented on a total (gross) basis, while carbon budgets relate to the net carbon account. IAS = international aviation and shipping.

- The Committee will publish its advice to government on the fifth carbon budget in December 2015, covering the period 2028-2032
- The government will propose draft legislation for the fifth budget in 2016.

https://www.theccc.org.uk/publication/reducing-emissions-and-preparing-for-climate-change-2015-progress-report-to-parliament/

Planning for 2 More Phases

UKCCS RESEARCH CENTRE

DECC – August 2014

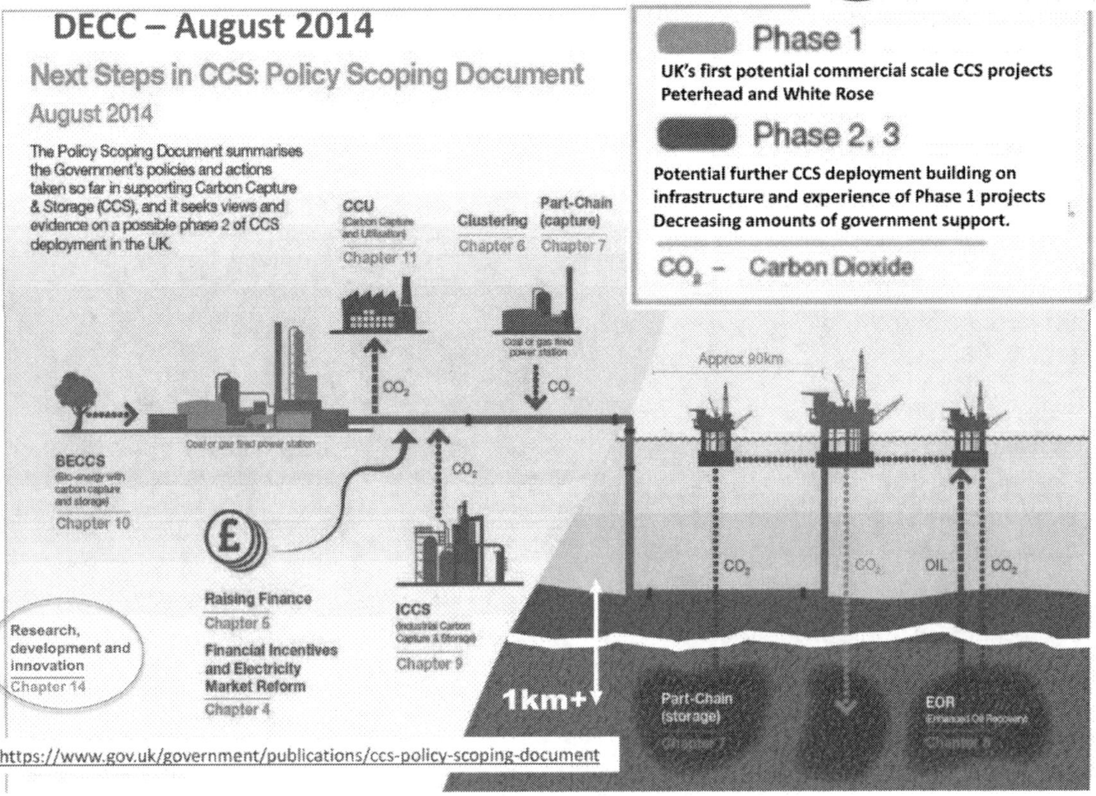

https://www.gov.uk/government/publications/ccs-policy-scoping-document

ETI Models suggest high £ without CCS

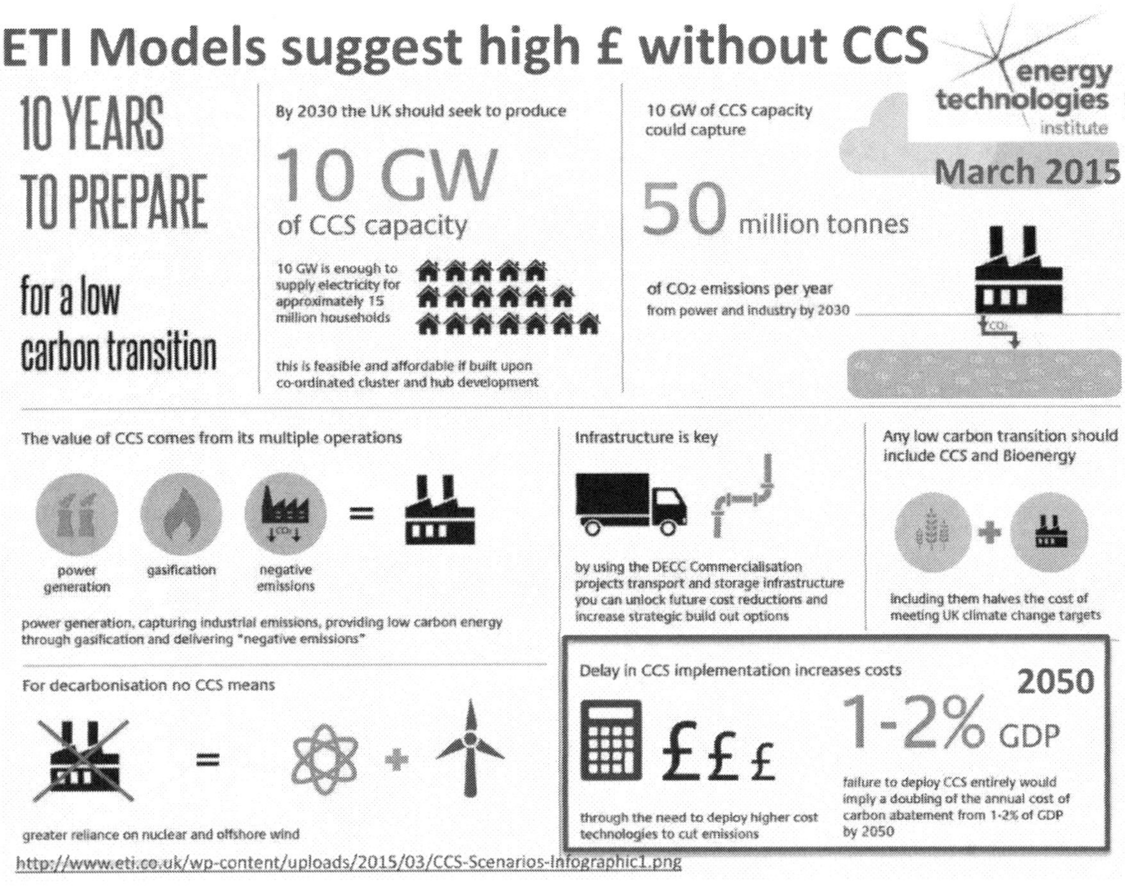

http://www.eti.co.uk/wp-content/uploads/2015/03/CCS-Scenarios-Infographic1.png

2030: ~5GW nat gas CCS (+ coal + ind)

UKCCS RESEARCH CENTRE

energy technologies institute   elementenergy   PÖYRY

CCS Sector Development Scenarios in the UK, May 2015

http://www.eti.co.uk/wp-content/uploads/2015/05/2015-04-30-ETI-CCS-sector-development-scenarios-Final-Report.pdf

- **Deployment of CCS capacity at scale (i.e. ~10 GW electricity) and infrastructure capable of capturing 40-50 $MtCO_2$/year from power and industry by 2030 .**
- **Three scenarios varying cluster and storage locations**
- **Storage ~ 100 $MtCO_2$/year by 2050.**

Cost reductions

- **Economies of scale for T&S**
- **EOR if it happens**

Learning by doing for capture technology

- **Always some learning for gas**
- **Coal most in EOR scenario**
- **Limited across Balanced scenario…**
- **but most opportunity for technology transfer from overseas**

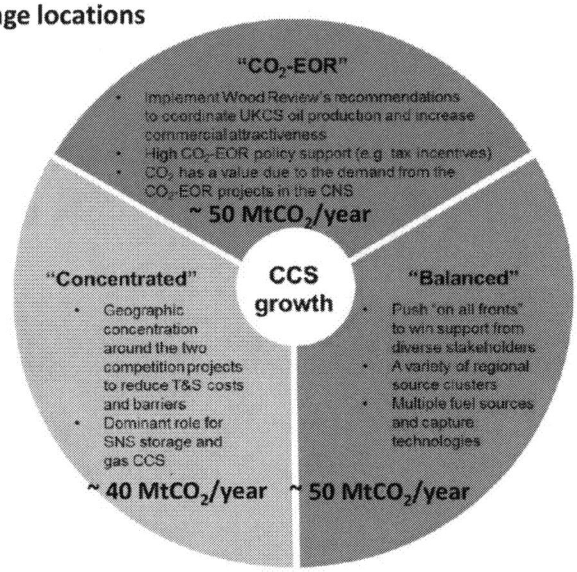

> But coordination probably needed to maximise learning by doing with a very limited number of projects, globally as well as in the UK.

Projects in These Scenarios

elementenergy   PÖYRY   energy technologies institute   UKCCS RESEARCH CENTRE

Deployment of CCS capacity at scale (i.e. ~10 GW electricity) and infrastructure capable of capturing 40-50 $MtCO_2$/year from power (as part of <100 $kgCO_2$/MWh) and industry by 2030.

Eventual storage target for 2050 scenarios (80% cut in UK emissions) ~ 100 $MtCO_2$/year.

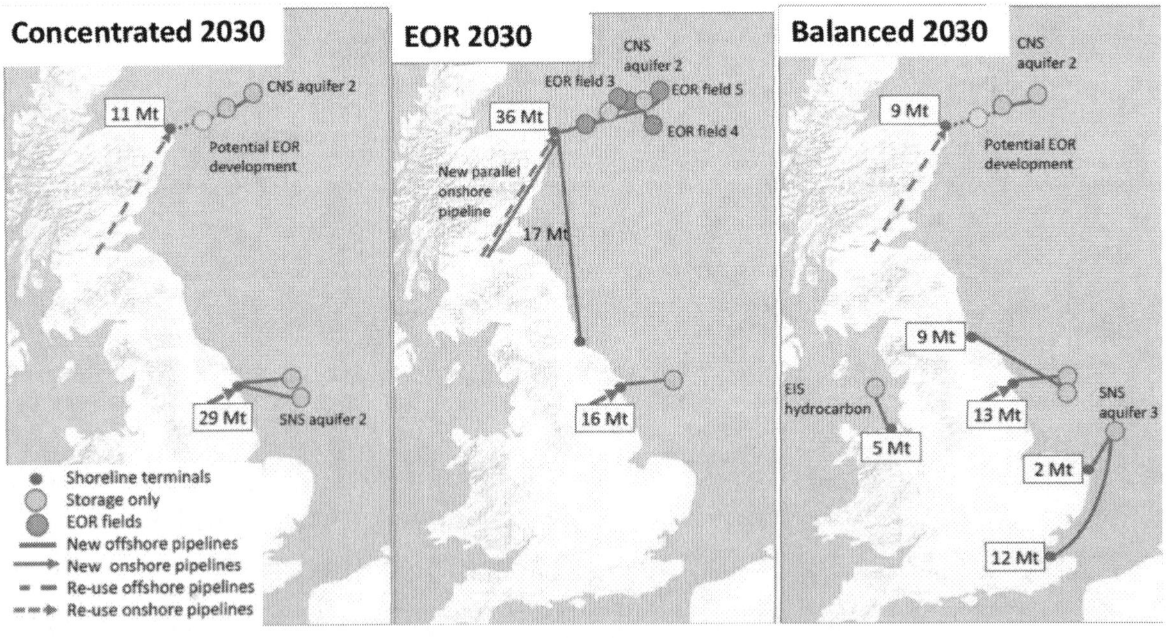

Gas CCS Cheaper Than Coal CCS?

UK CCS RESEARCH CENTRE

From: Decarbonising British industry why industrial CCS clusters are the answer by Dustin Benton, Green Alliance

http://www.green-alliance.org.uk/decarbonising_british_industry.php

Cost of CO2 abatement by different low carbon technologies[13]

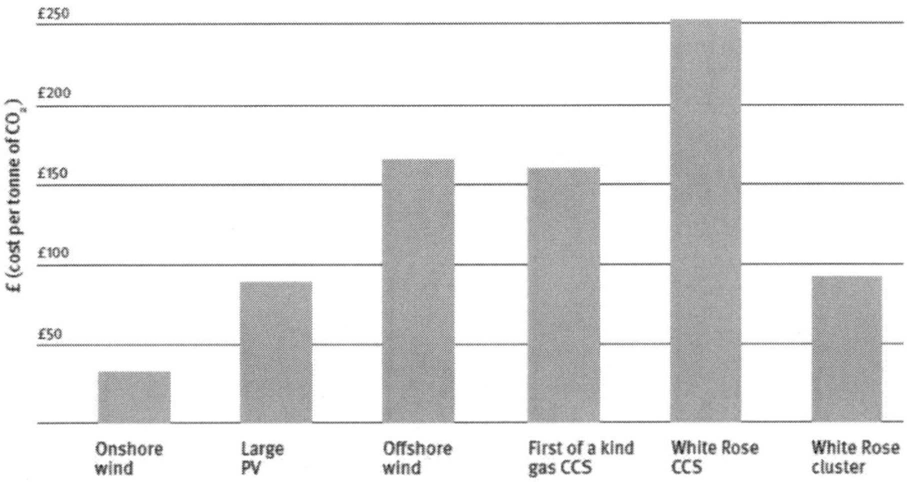

[13] These costs are based on CfD strike prices for onshore wind, large PV and offshore wind. Power sector CCS costs are based on CCS Cost Reduction Task Force analysis for 'first of a kind' CCS power projects. Industrial CCS costs are based on metanalysis of CCS costs by Element Energy and Imperial College London. The costs are presented in £/tCO2 , abated solely to enable comparison between industrial and power sector technologies.

System-wide Costs cf £/MWh ?

UK CCS RESEARCH CENTRE

From: Jon Rees, Financial Mail on Sunday, 22 August 2015

- Energy Secretary Amber Rudd has launched a groundbreaking project to examine the actual cost of electricity generation in a move which could spell the end of billions of pounds of subsidies for green energy.
- The Department of Energy & Climate Change said its aim is to look at every aspect of the cost of generating energy.
- It will include not just the cost of constructing offshore wind farms, for instance, but also of connecting them to the national grid – something which critics of the green energy industry say is often overlooked by its supporters. It will also examine nuclear power and conventional energy.
- The study is being conducted by Frontier Economics, the consultancy chaired by former Cabinet Secretary Lord O'Donnell.
- The review comes amid opposition to the Government's plan to build two new nuclear reactors – Hinkley Point C – in Somerset.
- The complainants say the UK Government's subsidies may reach £76billion and are in breach of European rules relating to state aid.

System-wide Costs cf £/MWh ?

August 2015, Energy Research Partnership
Managing Flexibility Whilst Decarbonising the GB Electricity System

The chart starts with an assumed 2030 mix of 10 GW nuclear, 28 GW wind and 5 GW gas-CCS resulting in a system close to 100 g/kWh. The technology tracks have the same shape but sometimes curve more sharply. Carbon price is set at £70/t. The earlier upward curve from CCS is due to the residual emissions assumed in the modelling.

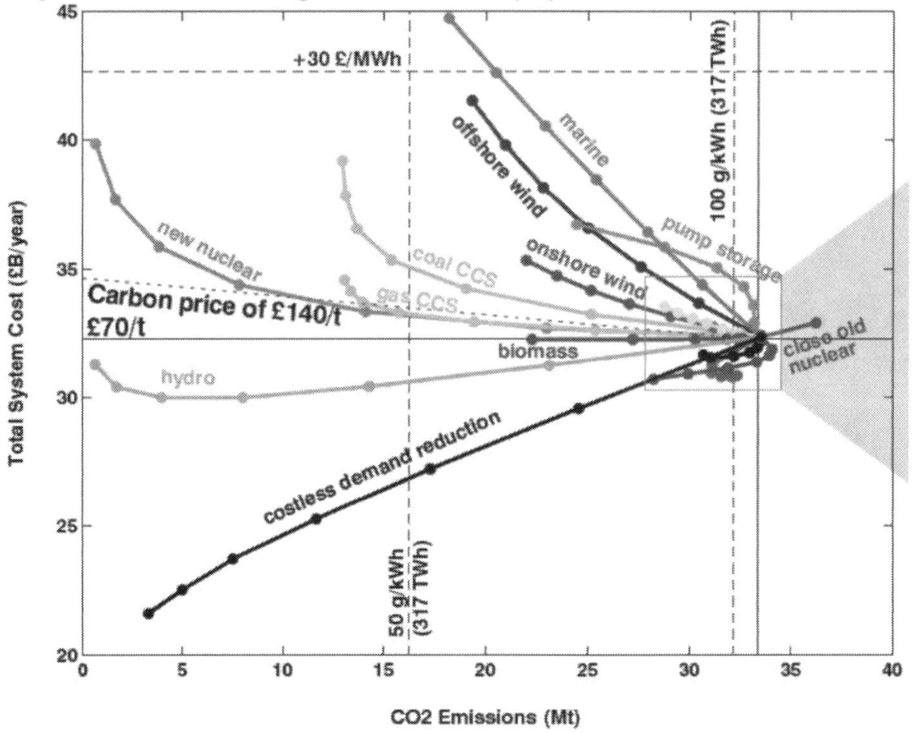

Where the UKCCSRC Fits In

- It is crucial that the research community is flexible enough to run alongside the inevitable ongoing changes in the CCS landscape
- I am sure that many of you can relate to this

UKCCS RESEARCH CENTRE

The UK Carbon Capture and Storage Research Centre

- Leads and coordinates a programme of underpinning research on all aspects of CCS
- Brings together over 260 of the UK's world-class CCS academics to provide a national focal point for CCS research and development
- Supports the UK economy by driving an integrated research programme and building research capacity that is focused on maximising the contribution of CCS to a low-carbon energy system

UKCCSRC funders

UKCCS RESEARCH CENTRE

- The UKCCSRC grant started on 1 April 2012
- Initial core funding for the UKCCSRC is provided by **£10M grant** from the Engineering and Physical Sciences Research Council (EPSRC) as part of the RCUK Energy Programme
- This is complemented by **£3M** in additional funding from the Department of Energy and Climate Change (DECC) to establish new capital facilities that will support innovative research
- Research institutions receiving funding have contributed **£2.5M**
- Research co-funders have contributed an additional **£2.5M** to Calls 1 and 2 projects

EPSRC Engineering and Physical Sciences Research Council

Energy For a Low Carbon Future

Department of Energy & Climate Change

Building A Virtual Research Centre

UKCCS RESEARCH CENTRE

UKCCSRC member type	Number of members, April 2015
Academic	264
Associate	154 (134 UK, 20 International)
Early Career Researcher	263
Network	538 (389 UK, 149 International)
Total members	**1219**

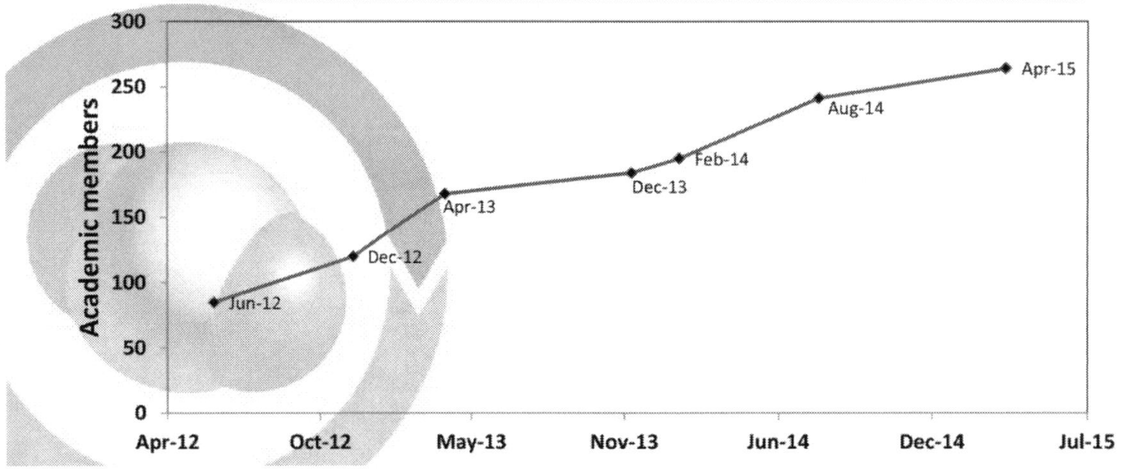

UKCCSRC Research Groups

UKCCS RESEARCH CENTRE

- Members belong to at least one of three research groups
  - Capture
  - Storage
  - Cross-cutting issues
- Groups provide a focal point for the development of new project ideas and collaborations.
- They organise specialist meetings and are key communication channels for the Centre and for external stakeholders

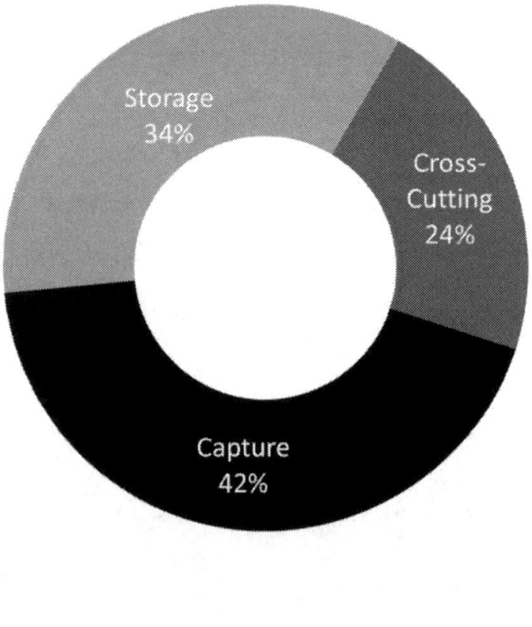

Research and Pathways to Impact Delivery (RAPID)

UKCCS RESEARCH CENTRE

- Academics and CCS practitioners work closely together to help identify, plan, execute and implement relevant, high-impact research.
- Now being developed and implemented with support from Independent Advisory Panel members and industry through the CCSA Technical Working Group.

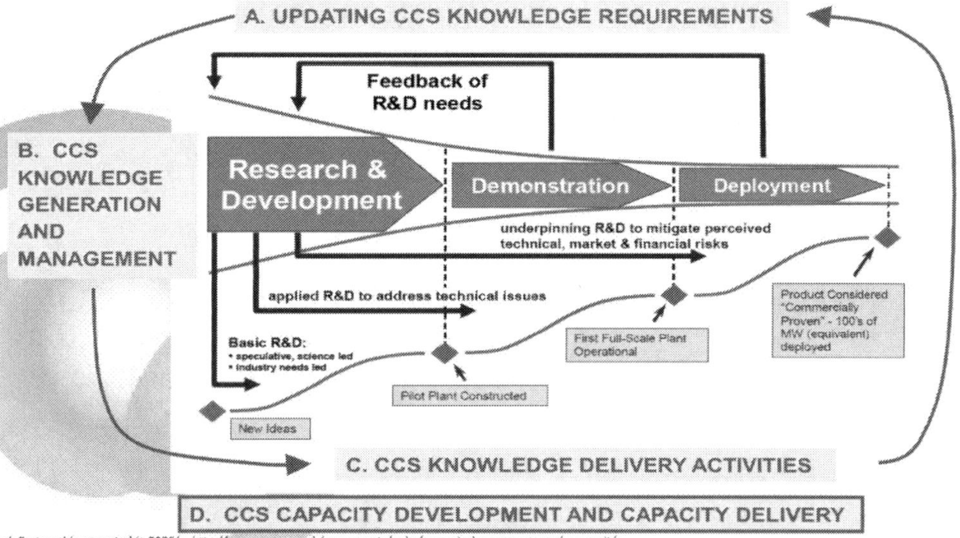

Energy Research Partnership, reported in 2006 in http://www.epsrc.ac.uk/newsevents/pubs/second-uk-energy-research-summit/
(and subsequently reproduced in various forms including , www.ukerc.ac.uk/support/tiki-download_file.php?fileId=3619)

Working with the world to progress CCS

UKCCS RESEARCH CENTRE

- Essential to draw on **world-wide experience, particularly full-scale projects for learning by doing** to maximise relevance and impact of research given limited time and resources
- MOUs with **Carbon Management Canada, SaskPower, Guangdong (+ Shanghai), CATO, CO2CRC, RCN-CCUS**, links with the **Research Council of Norway**
- Supported FCO links to **USA, Canada, South Korea, Mexico**
- Strong links with **IEAGHG, GCCSI** and other international bodies
- Funding to support UKCCSRC members' efforts to develop significant exchanges and **'best with best' collaborations** with overseas academics and research institutions through links between existing projects
- **ERA-NET (ACT) Co-fund for CCS** : Norway, Germany, UK, The Netherlands, Switzerland, Romania, Turkey , Spain, and Greece

What R&D could reduce CCS costs in the 2020s?

UKCCS RESEARCH CENTRE

- This must be R&D that actually gets applied
- So will have to be R&D that will evolve 'current' technologies – not inventing completely novel approaches

Why current technologies? - Industry 'clockspeed' is the time to complete an iteration of design-build-test-market-learn, e.g:

- Clothing fashions – weeks
- Consumer goods - months
- Automotive - years,
- Pharmaceuticals – decade
- Big energy - CCGT, nuclear, coal or CCS – one or more decades

- Also, the organisations who will build the series of Phase 2 and Phase 3 projects need to shape and influence the R&D agenda and must apply the results

Acknowledgement: J. Carey

UKCCSRC industry working group conclusions for UK Phase 2 and Phase 3 post-combustion

a) Natural gas only – NGCC+CCS

b) Conventional solvents and compatibles

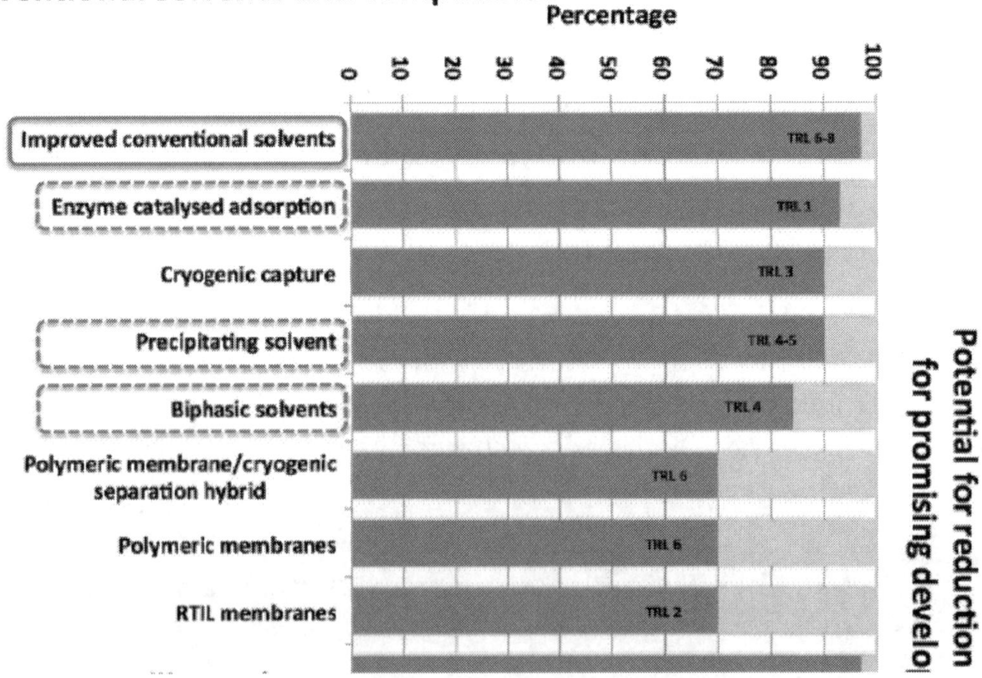

How can research contribute to R&D that will evolve 'current ' CCS technologies

UKCCS RESEARCH CENTRE

- Only 'current generation' CCS with reference plants at TRL 9 now, or in the next few years, will be bankable in the 2020s; but isn't academic research more appropriate at lower TRLs?
- Need to consider system level TRL vs. sub-system or component TRL
- TRLs for evolving current technologies should be applied to innovation in sub-systems not across a whole system; improvements to sub-systems can start at TRL 1 while the overall technology is at TRL 9

E.g. NASA Chevron
http://www.nasa.gov/topics/aeronautics/features/trl_demystified.html

Future Roles for UKCCSRC

UKCCS RESEARCH CENTRE

A. Phase 1 project(s) approved in 2016 and Phase 2/3 next objective
B. No Phase 1 projects approved but reserve projects being developed
C. No Phase 1 projects approved, CCS deployment postponed until 2020's

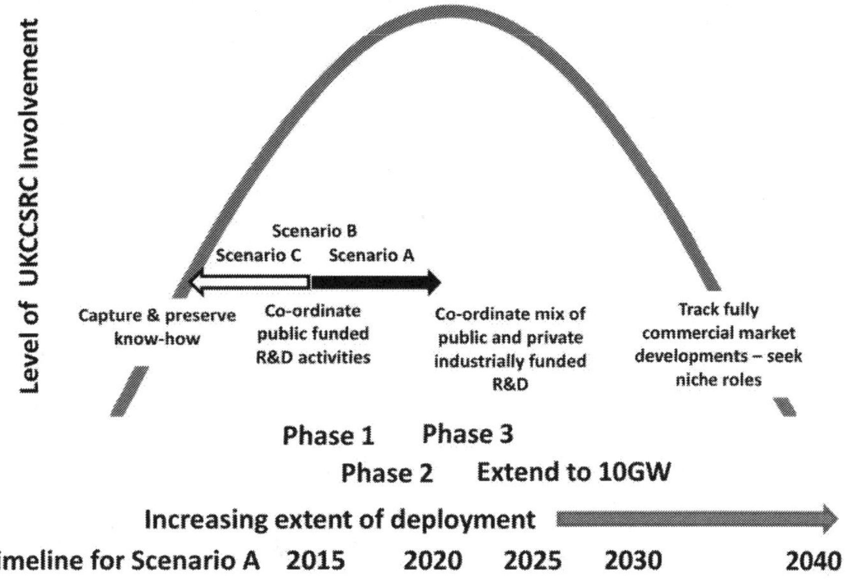

Acknowledgement: R. Irons

END

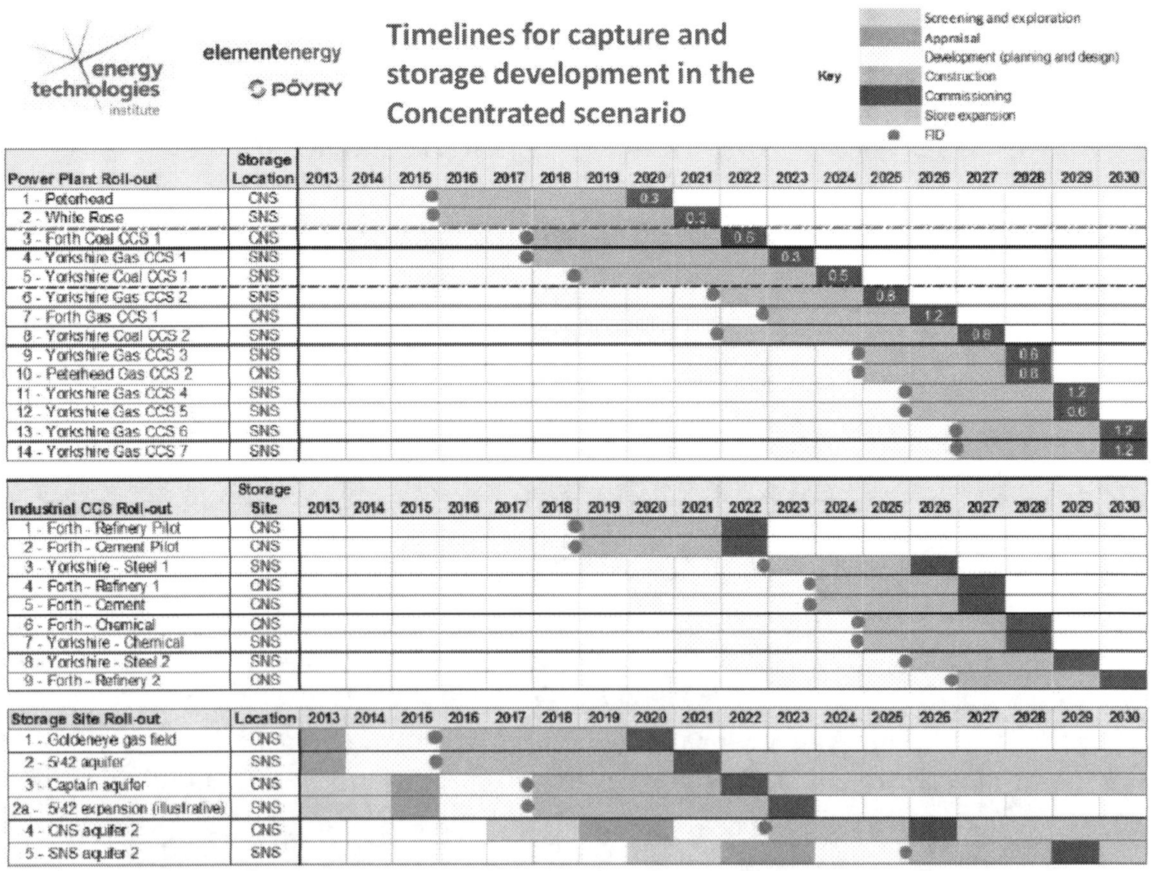
energy technologies institute
elementenergy
PÖYRY
Timelines for capture and storage development in the Concentrated scenario
Key
Screening and exploration
Appraisal
Development (planning and design)
Construction
Commissioning
Store expansion
FID
Power Plant Roll-out
Storage Location
2013 2014 2015 2016 2017 2018 2019 2020 2021 2022 2023 2024 2025 2026 2027 2028 2029 2030
1 - Peterhead CNS 0.3
2 - White Rose SNS 0.3
3 - Forth Coal CCS 1 CNS 0.5
4 - Yorkshire Gas CCS 1 SNS 0.3
5 - Yorkshire Coal CCS 1 SNS 0.5
6 - Yorkshire Gas CCS 2 SNS 0.8
7 - Forth Gas CCS 1 CNS 1.2
8 - Yorkshire Coal CCS 2 SNS 0.8
9 - Yorkshire Gas CCS 3 SNS 0.8
10 - Peterhead Gas CCS 2 CNS 0.8
11 - Yorkshire Gas CCS 4 SNS 1.2
12 - Yorkshire Gas CCS 5 SNS 0.8
13 - Yorkshire Gas CCS 6 SNS 1.2
14 - Yorkshire Gas CCS 7 SNS 1.2
Industrial CCS Roll-out
Storage Site
1 - Forth - Refinery Pilot CNS
2 - Forth - Cement Pilot CNS
3 - Yorkshire - Steel 1 SNS
4 - Forth - Refinery 1 CNS
5 - Forth - Cement CNS
6 - Forth - Chemical CNS
7 - Yorkshire - Chemical SNS
8 - Yorkshire - Steel 2 SNS
9 - Forth - Refinery 2 CNS
Storage Site Roll-out
Location
1 - Goldeneye gas field CNS
2 - 5/42 aquifer SNS
3 - Captain aquifer CNS
2a - 5/42 expansion (illustrative) SNS
4 - CNS aquifer 2 CNS
5 - SNS aquifer 2 SNS

energy technologies institute

elementenergy

PÖYRY

CCS Sector Development Scenarios in the UK

http://www.eti.co.uk/wp-content/uploads/2015/05/2015-04-30-ETI-CCS-sector-development-scenarios-Final-Report.pdf

Concentrated scenario

4 coal, 10 gas, 2 steel, 2 chemical, 2 refinery, 1 cement

Power Plant Roll-out	Storage Location
1 - Peterhead	CNS
2 - White Rose	SNS
3 - Forth Coal CCS 1	CNS
4 - Yorkshire Gas CCS 1	SNS
5 - Yorkshire Coal CCS 1	SNS
6 - Yorkshire Gas CCS 2	SNS
7 - Forth Gas CCS 1	CNS
8 - Yorkshire Coal CCS 2	SNS
9 - Yorkshire Gas CCS 3	SNS
10 - Peterhead Gas CCS 2	CNS
11 - Yorkshire Gas CCS 4	SNS
12 - Yorkshire Gas CCS 5	SNS
13 - Yorkshire Gas CCS 6	SNS
14 - Yorkshire Gas CCS 7	SNS

Industrial CCS Roll-out	Storage Site
1 - Forth - Refinery Pilot	CNS
2 - Forth - Cement Pilot	CNS
3 - Yorkshire - Steel 1	SNS
4 - Forth - Refinery 1	CNS
5 - Forth - Cement	CNS
6 - Forth - Chemical	CNS
7 - Yorkshire - Chemical	SNS
8 - Yorkshire - Steel 2	SNS
9 - Forth - Refinery 2	CNS

EOR scenario

9 coal, 6 gas, 2 steel, 2 chemical, 2 refinery, 1 cement

Power Plant Roll-out	Storage Location
1 - Peterhead	CNS
2 - White Rose	SNS
3 - Forth Coal CCS 1	CNS
4 - Yorkshire Gas CCS 1	SNS
5 - Yorkshire Coal CCS 1	SNS
6 - TeessideCoal CCS 1	CNS
7 - Forth Coal CCS 2	CNS
8 - Yorkshire Coal CCS 2	SNS
9 - Peterhead Gas CCS 2	CNS
10 - Yorkshire Gas CCS 2	SNS
11 - Yorkshire Gas CCS 3	SNS
12 - TeessideCoal CCS 2	CNS
13 - Forth Coal CCS 3	CNS
14 - TeessideGas CCS 1	CNS
15 - Scotland Coal CCS	CNS

Industrial CCS Roll-out	Storage Site
1 - Forth - Refinery Pilot	CNS
2 - Forth - Cement Pilot	CNS
3 - Teesside - Chemical	CNS
4 - Teesside - Steel	CNS
5 - Yorkshire - Steel	SNS
6 - Forth - Refinery 1	CNS
7 - Forth - Cement	CNS
8 - Forth - Chemical	CNS
9 - Forth - Refinery 2	CNS

Balanced scenario

7 coal, 7 gas, 1 biomass, 2 steel, 2 chemical, 2 refinery, 1 cement

Power Plant Roll-out	Storage Location
1 - Peterhead	CNS
2 - White Rose	SNS
3 - Forth Coal CCS 1	CNS
4 - Yorkshire Gas CCS 1	SNS
5 - Yorkshire Coal CCS 1	SNS
6 - TeessideCoal CCS 1	SNS
7 - Liverpool Gas CCS 1	EIS
8 - Bacton Gas CCS	SNS
9 - Thames Coal CCS 1	SNS
10 - Yorkshire Gas CCS 2	SNS
11 - Yorkshire Biomass	SNS
12 - TeessideCoal CCS 2	SNS
13 - Liverpool Gas CCS 2	EIS
14 - Thames Coal CCS 2	SNS
15 - Scotland Gas CCS	CNS

Industrial CCS Roll-out	Storage Site
1 - Forth - Refinery Pilot	CNS
2 - Forth - Cement Pilot	CNS
3 - Teesside - Chemical	SNS
4 - Teesside - Steel	SNS
5 - Yorkshire - Steel	SNS
6 - Forth - Refinery 1	CNS
7 - Forth - Cement	CNS
8 - Forth - Chemical	CNS
9 - Forth - Refinery 2	CNS

Essential Requirements of a CCSR facility

https://www.iea.org/publications/freepublications/publication/ccs_g8.pdf

The essential requirements represent the minimum criteria that should be met before a facility can be considered CCSR. The project developer should:

- Carry out a site-specific study in sufficient engineering detail to ensure the facility is technically capable of being fully retrofitted for CO_2 capture.
- Demonstrate that retrofitted capture equipment can be connected to the existing equipment effectively and without an excessive outage period and that there will be sufficient space available to construct and safely operate additional capture and compression facilities.
- Identify realistic pipeline or other route(s) to storage of CO_2.
- Identify one or more potential storage areas which have been appropriately assessed and found likely to be suitable for safe geological storage of projected full lifetime volumes and rates of captured CO_2.
- Identify other known factors, including any additional water requirements that could prevent installation and operation of CO_2 capture, transport and storage, and identify credible ways in which they could be overcome.
- Estimate the likely costs of retrofitting capture, transport and storage.
- Engage in appropriate public engagement and consideration of health, safety and environmental issues.
- Review CCSR status and report on it periodically.

Incentivizing Innovation To Convert Carbon Dioxide Emissions

Carbon Management Technology Conference
19 November 2015

Marcius Extavour, PhD
Director of Technical Operations, Energy & Environment, XPRIZE

Today CO2
emissions are viewed
primarily as a
waste...

... But CO_2
is also a
building
block for
many types
of materials
and
products

INTRODUCING THE

nrg | cosia CARBON XPRIZE®

The Carbon XPRIZE is a US$20 million competition to convert the most CO_2 into one or more products with the highest net value

- CONVERSION, not capture
- Using standard capture not disallowed...
- ... direct conversion from flue also encouraged

SPONSORS

A sustainable, renewable energy-driven future is no longer the future. It's here and now. We are embracing the possibilities and so are hospitals, pro football stadiums, downtown business districts, and many others.

Canada's Oil Sands Innovation Alliance (COSIA) is an alliance of oil sands producers focused on accelerating the pace of improvement in environmental performance in Canada's oil sands through collaborative action and innovation.

Snapshot of the NRG COSIA Carbon XPRIZE

THE WINNING TEAM WILL...	Convert the most CO_2 into one or more products with the highest net value
PRIZE PURSE	• US$2.5M milestone prize purse for each track • US$7.5M grand prize purse for each track
TIMELINE	4.5 years (54 months)
STRUCTURE	Teams register for one or both of two tracks: • Track A - tested on coal flue gas (Wyoming) • Track B - tested on natural gas flue gas (to be announced) Teams compete in three rounds: • Round 1: Technical and Business Viability Assessment • Round 2: Pilot Scale Competition • Round 3: Demonstration Scale Competition
JUDGING CRITERIA	**Scoring Criteria:** • Amount of CO_2 converted into products • Net value of products **Performance Thresholds:** • > 30% of CO2 converted • < 4 m^3 water per ton CO2 converted • Maximum land footprint in Round 3 demonstration

Snapshot of Competition Rounds

Round 1: Paper Down-select

Round 2: Pilot-Scale
30 teams 200 kg CO_2 / day Demonstrate Anywhere

Round 3: Demonstration-Scale
10 teams 2T CO_2 / day Test Facilitis

nrg | cosia CARBON XPRIZE
Reimagine CO_2

GET INVOLVED

carbon.xprize.org

#carbonXPRIZE

Marcius Extavour, PhD
Director of Technical Operations, Energy & Environment
xprize.org | @xprize

Carbon Management Technology Conference

November 2015

About Forward-Looking Statements

The data contained in this presentation that are not historical facts are forward-looking statements that involve a number of risks and uncertainties. Such statements may relate to, among other things: long-term strategy; anticipated levels of future dividends and their rate of growth and sustainability; the length or severity of the oil price downturn in late 2014 and early 2015; forecasts of capital expenditures, drilling activity and developmental activities; timing of carbon dioxide (CO_2) injections and production response to such tertiary flooding projects; estimated timing of pipeline construction or completion or the cost thereof; anticipated dates of completion of industrial plants to be constructed or under construction and the initial date of capture and amount of anthropogenic CO_2; estimates of liquidity, costs, forecasted production rate or peak production rates and the growth thereof; estimates of hydrocarbon reserve quantities and values, including potential and recoverable reserves, CO_2 reserves, and helium reserves; projected future hydrocarbon prices or costs; estimated future cash flows, including from our hedging positions, or uses of cash; availability of capital or borrowing capacity; estimated rates of return and overall economics; and anticipated availability and cost of equipment and services. These forward-looking statements are generally accompanied by words such as "believe", "estimated", "preliminary", "projected", "potential", "anticipated", "forecasted", "expected", "assume" or other words that convey the uncertainty of future events or outcomes. These statements are based on management's current plans and assumptions and are subject to a number of risk and uncertainties as further outlined in our most recent Form 10-K filed with the SEC. Therefore, actual results may differ materially from the expectations, estimates, forecasts, projections, or assumptions expressed in or implied by any forward-looking statement herein made by or on behalf of the Company.

Cautionary Note to U.S. Investors – Current SEC rules regarding oil and gas reserves information allow oil and gas companies to disclose in filings with the SEC not only proved reserves, but also probable and possible reserves that meet the SEC's definitions of such terms. We disclose only proved reserves in our filings with the SEC. Denbury's proved reserves as of December 31, 2014 were estimated by DeGolyer and MacNaughton, an independent petroleum engineering firm. In this presentation, we make reference to probable and possible reserves, some of which have been estimated by our independent engineers and some of which have been estimated by Denbury's internal staff of engineers. In this presentation, we also refer to estimates of original oil in place, resource or reserves "potential", barrels recoverable, or other descriptions of volumes potentially recoverable, which in addition to reserves generally classifiable as probable and possible (2P and 3P reserves), include estimates of reserves that do not rise to the standards for possible reserves, and which SEC guidelines strictly prohibit us from including in filings with the SEC. These estimates, as well as the estimates of probable and possible reserves, are by their nature more speculative than estimates of proved reserves and are subject to greater uncertainties, and accordingly the likelihood of recovering those reserves is subject to substantially greater risk.

A Different Kind of Oil Company

Long-Term Visibility

- **Proven process**
- **Lower-risk & long-life assets**
- **Tremendous resource potential**

Capital Flexibility

- **Fund capex & dividends with cash flow**
- **Relatively low capital intensity**
- **Adjust to oil price environment**

Competitive Advantages

- **Strategic CO_2 supply**
- **>1,100 miles of CO_2 pipelines**
- **Large inventory of oil fields**

Denbury at a Glance

Total 3P Reserves (12/31/14)	~1.2 BBOE
Total Proved Reserves (12/31/14)	437.7 MBOE
% Oil Production (2Q15)	95%
Total Daily Production – BOE/d (2Q15)	73,716
Proved PV-10 (12/31/14) $94.99 NYMEX Oil Price	$8.7 billion
Market Cap (8/31/15)	$1.5 billion
Total Debt (6/30/15)	$3.5 billion
CO_2 Supply 3P Reserves (12/31/14)	~17 Tcf
CO_2 Pipelines Operated or Controlled	>1,100 miles
Credit Facility Availability (6/30/15)	$1.2 billion
Anticipated Annual Dividend per Share	2015E - $0.25

Core Focus: CO_2 EOR

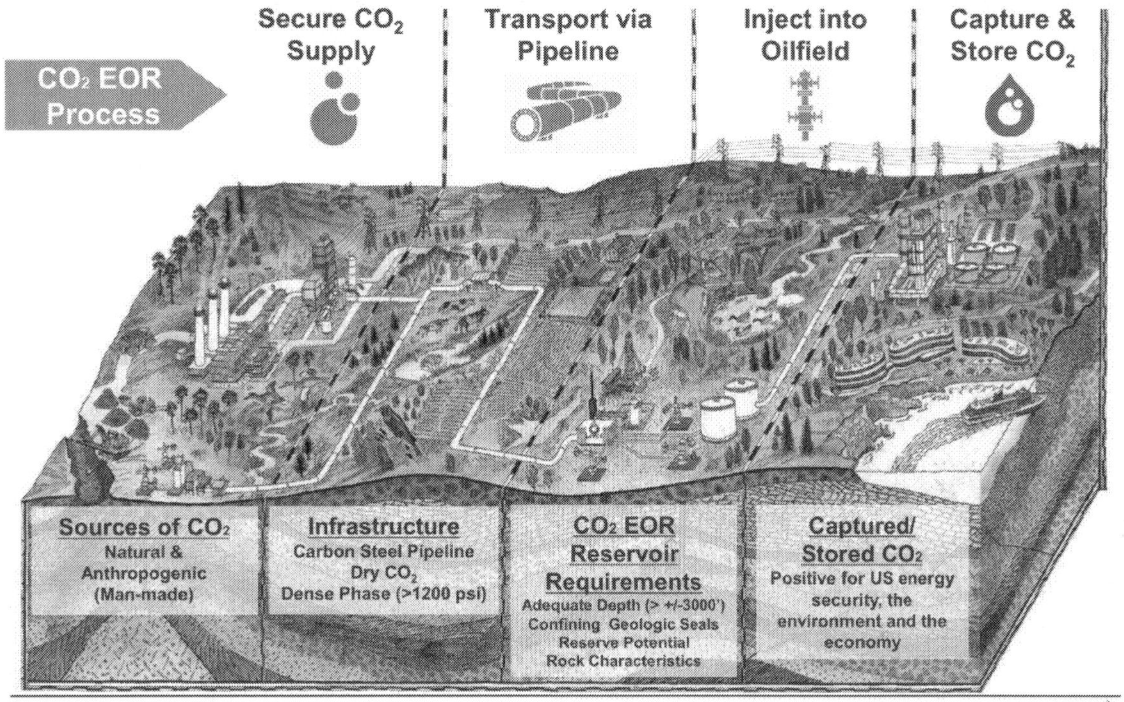

CO_2 EOR – A Brief History

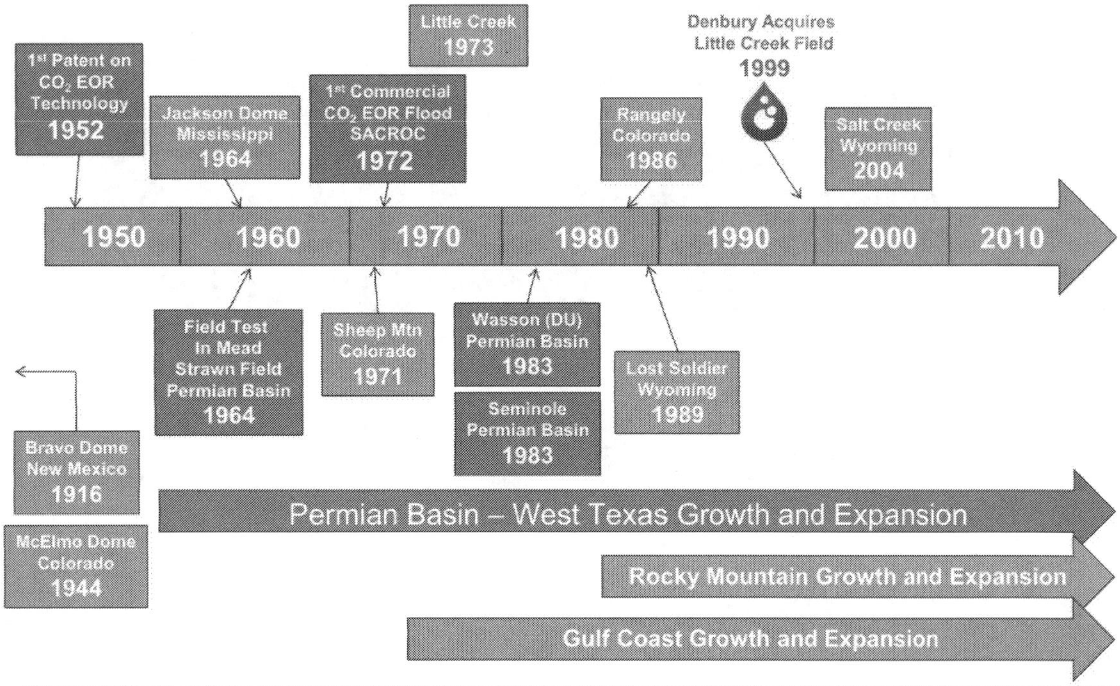

CO_2 EOR Potential – 2008 & 2011 Reports

2008 - DOE/NETL Report

- "CO_2 enhanced oil recovery (CO_2 EOR) offers the potential for storing significant volumes of carbon dioxide emissions while increasing domestic oil production"
- Next generation technology offers potential for recovering more stranded oil and storing significantly more CO_2

2011 - DOE/NETL Report

- "Next Generation" CO_2 EOR can provide 137 billion barrels of additional technically recoverable domestic oil, with about half (67 billion barrels) economically recoverable at an oil price of $85 per barrel. Technical CO_2 storage capacity offered by CO_2 EOR would equal 45 billion metric tons
- The market for captured CO_2 emissions from power plants created by economically feasible CO_2 EOR projects would be sufficient to permanently store the CO_2 emissions from 93 large one GW size coal-fired power plants operated for 30 years

Current U.S. CO_2 Sources & Pipelines

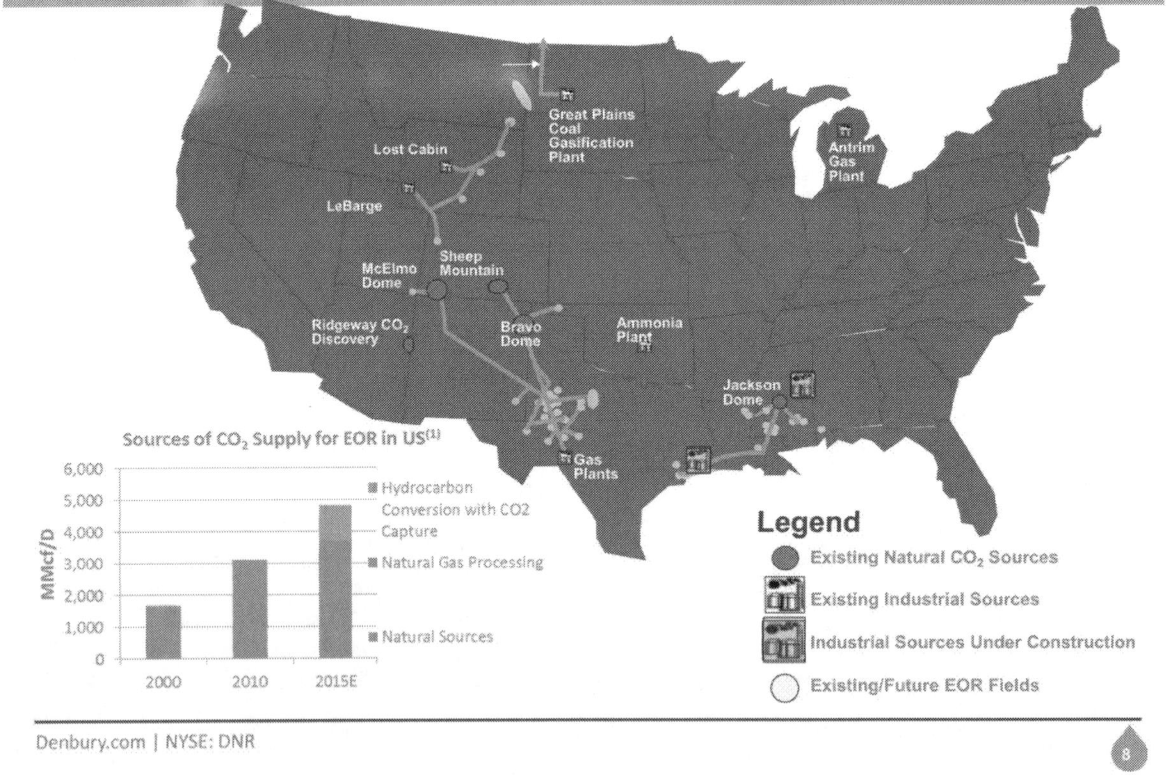

What is CO_2 EOR & How Much Oil Does it Recover?

CO_2 EOR Delivers Almost as Much Production as Primary or Secondary Recovery[(1)]

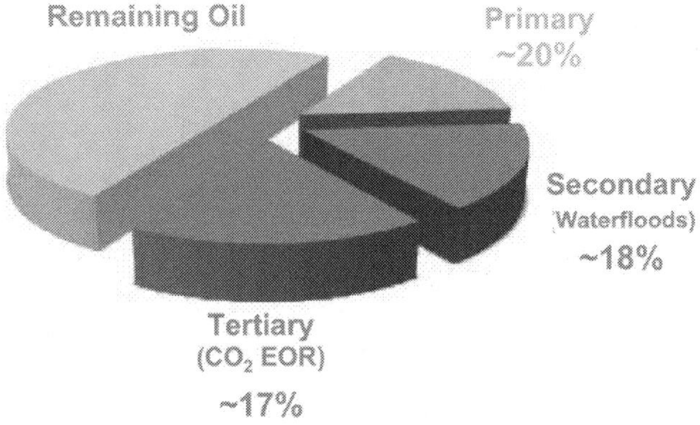

(1) Recovery of original oil in place based on history at Little Creek Field.

How much oil remains in an old oil field?

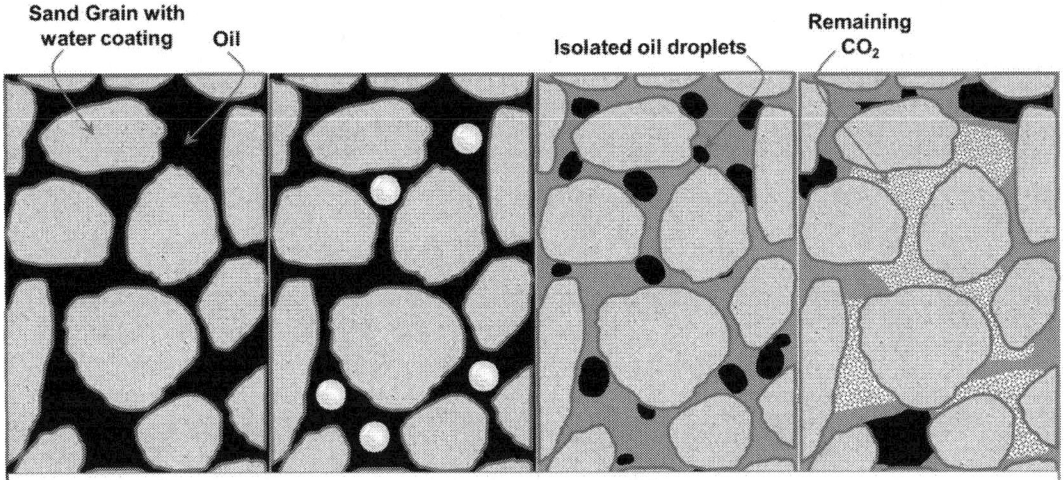

At Microscopic Level

Initial Discovery Conditions	After Primary Recovery	After Secondary Recovery (Waterflooding)	After Tertiary Recovery (CO_2 EOR)
Oil Saturation ~70%	Oil Saturation ~50%	Oil Saturation ~30%	Oil Saturation ~15%

Up to 16 Billion Gross Barrels Recoverable[1] in Our Two CO_2 EOR Target Areas

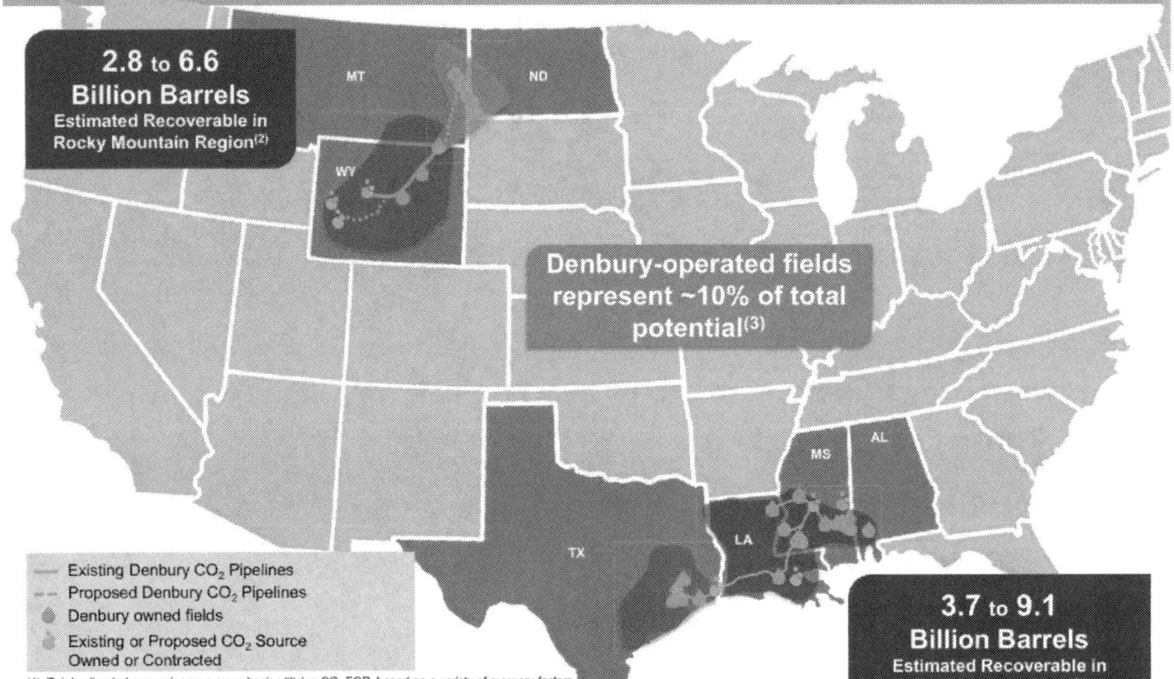

(1) Total estimated recoveries on a gross basis utilizing CO_2 EOR, based on a variety of recovery factors.
(2) Source: 2013 DOE NETL Next Gen EOR
(3) Using approximate mid-points of ranges, based on a variety of recovery factors.

CO_2 EOR in Gulf Coast Region:

Control of CO_2 Sources & Pipeline Infrastructure Provides a Strategic Advantage

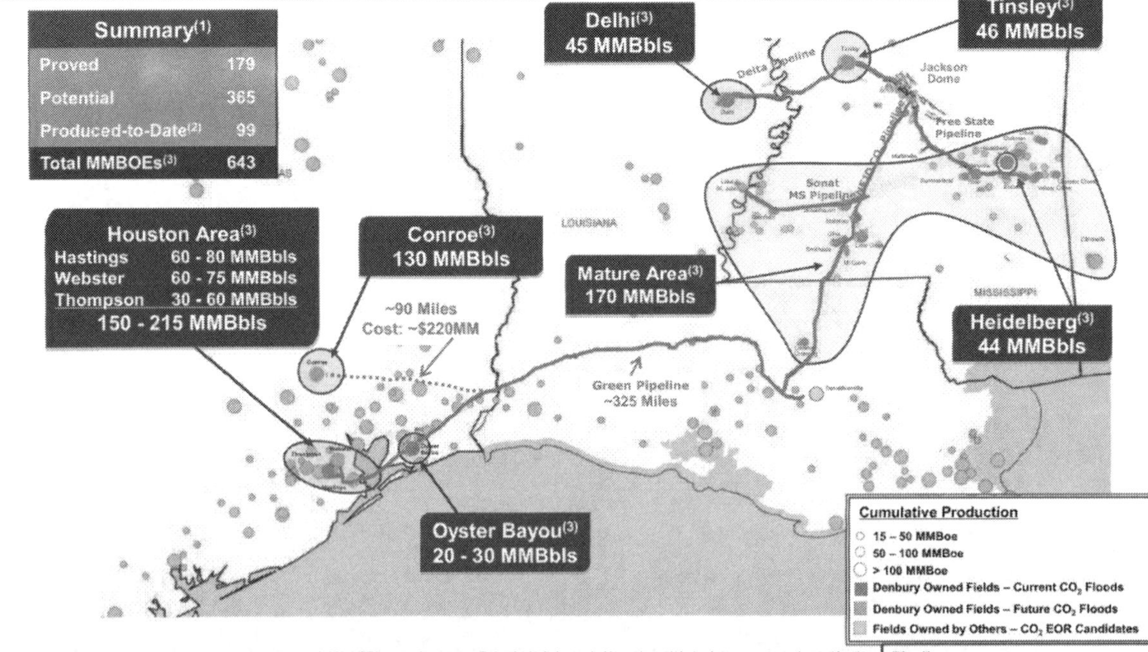

(1) Proved tertiary oil reserves based on year-end 12/31/14 SEC proved reserves. Potential includes probable and possible tertiary reserves estimated by the Company as of 12/31/14, using mid-point of ranges, based on a variety of recovery factors.
(2) Produced-to-date is cumulative tertiary production through 12/31/14.
(3) Field reserves shown are estimated total potential tertiary reserves, including cumulative tertiary production through 12/31/14.

CO_2 EOR in Rocky Mountain Region:

Control of CO_2 Sources & Pipeline Infrastructure Provides a Strategic Advantage

Summary[1]	
Proved	37
Potential	313
Produced-to-Date[2]	<1
Total MMBbls	351

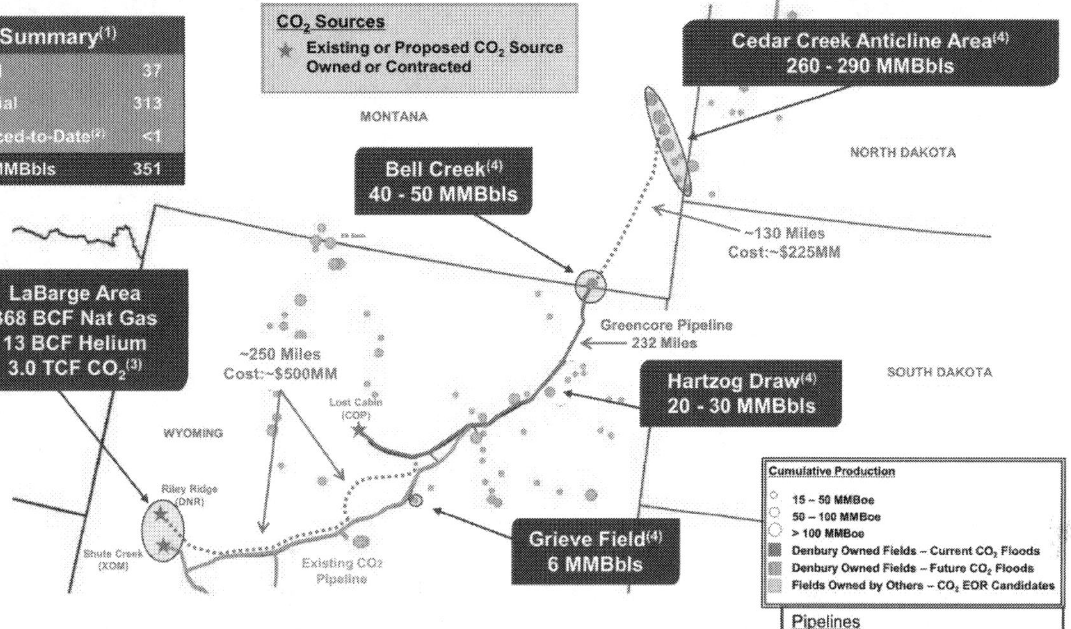

(1) Proved tertiary oil reserves based on year-end 12/31/14 SEC proved reserves. Potential includes probable and possible tertiary reserves estimated by the Company as of 12/31/14, using approximate mid-points of ranges, based on a variety of recovery factors.
(2) Produced-to-date is cumulative tertiary production through 12/31/14.
(3) Reported on a gross working interest or 8/8ths basis, except for overriding royalty interest in LaBarge Field.
(4) Field reserves shown are estimated total potential tertiary reserves, including cumulative tertiary production through 12/31/14.

Gulf Coast Industrial Partners

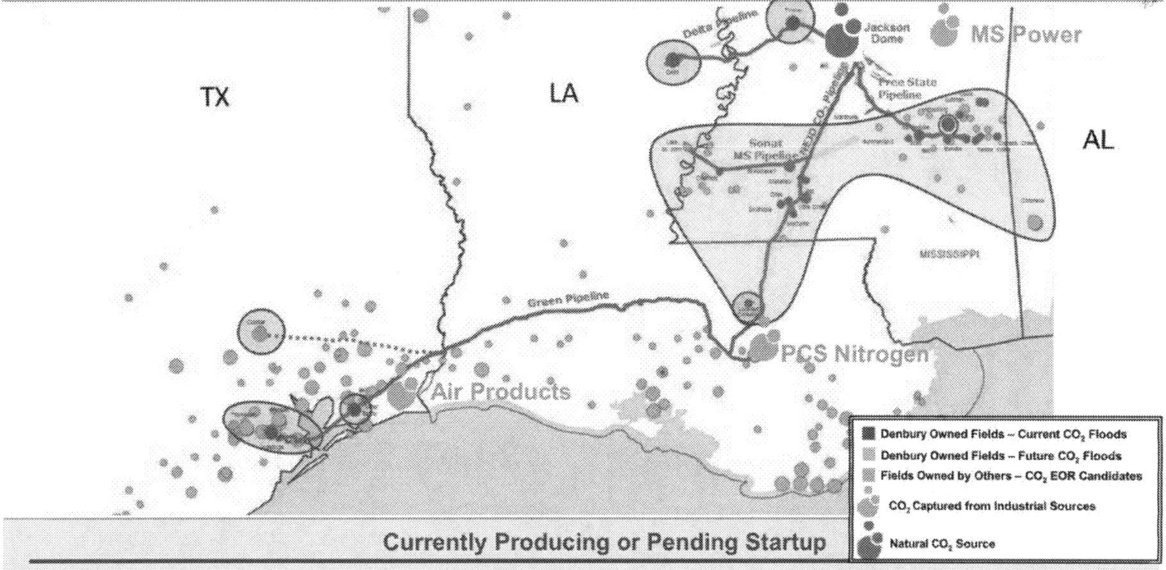

Currently Producing or Pending Startup

Air Products
- Port Arthur, Texas
- Hydrogen Plant
- Producing Since: 1Q 2013
- Quantity: ~50 MMcf/d

PCS Nitrogen
- Geismar, Louisiana
- Ammonia Products
- Producing Since: 2Q 2013
- Quantity: ~20 MMcf/d

Mississippi Power (Pending Startup)
- Kemper County, MS
- Gasifier
- Estimated Capture Date: ~2016
- Quantity: ~115 MMcf/d

Simplified CO_2 EOR Closed Loop Recycle Facility

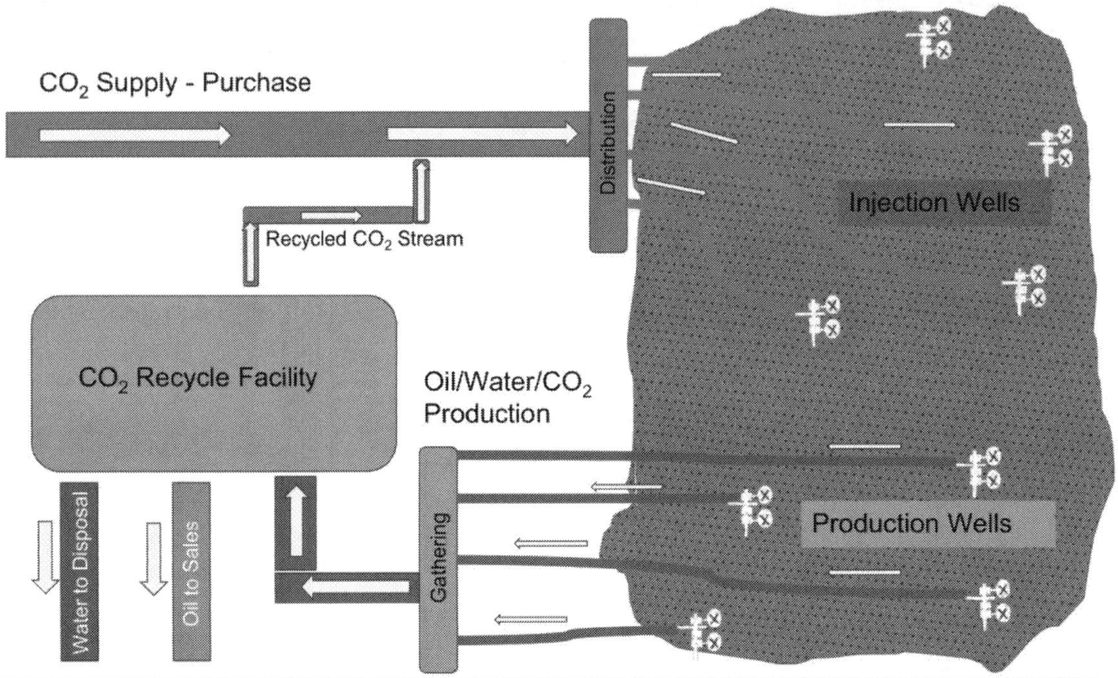

CO_2 EOR Generalized Type Curve

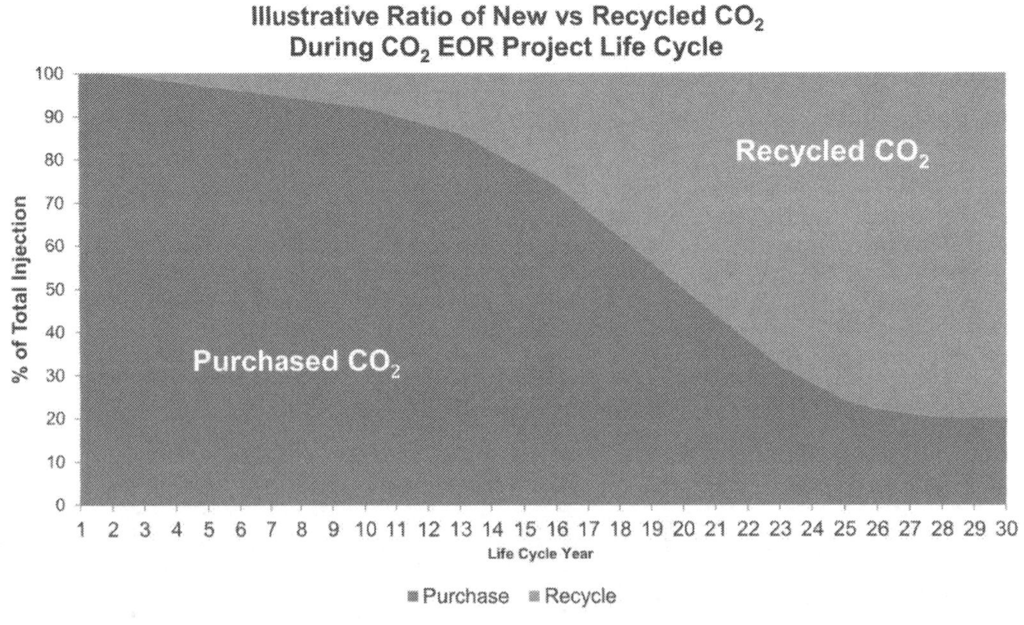

Hastings Aerial View

CO_2 EOR Storage

How CO_2 EOR to Storage Works

When CO_2 comes into contact with oil, a significant portion of the CO_2 dissolves into the oil, reducing oil viscosity and increasing the oils mobility. This, combined with the Increased pressure, can result in increased oil production rates as well as an extension of the operational lifetime of the oil reservoir.

In an oil field, this EOR method is called CO_2 Flooding. CO_2 floods are designed to be active for decades. Over the years there are many cycles of CO_2 injection. With each cycle, another portion of injected CO_2 becomes permanently trapped, or stored, in the oil reservoir. As a result of ongoing CO_2 EOR projects since the 1970s, hundreds of millions of tons of CO_2 is now permanently stored in oil fields.

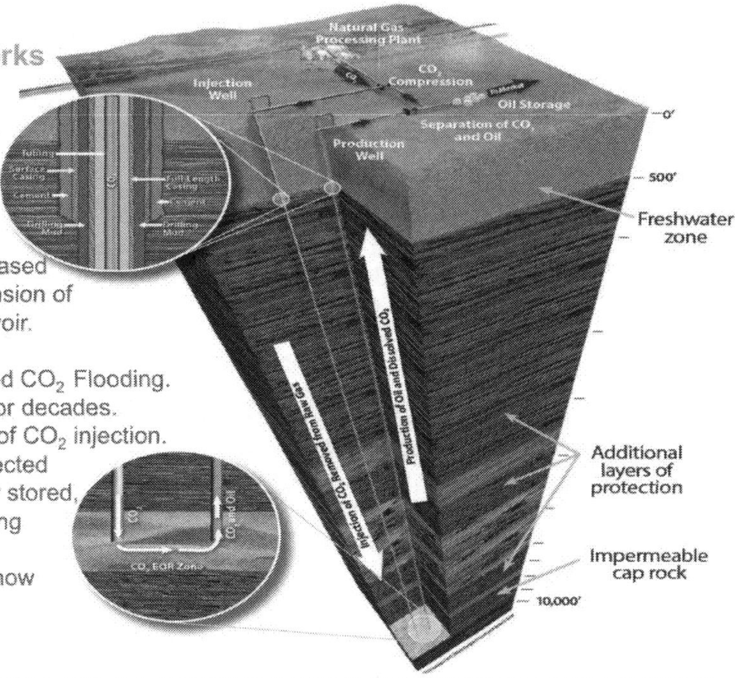

CO_2 Injections, Oil Recovery & Associated CO_2 Storage

- **Commercial CO_2 Operations Have Been Ongoing Continuously in the United States since 1972**
  - Current Annual injections: Approximately 63-68[(1)] Million Metric Tonnes Per Year
  - Cumulative CO_2 Injected (Net of Recycle) from 1984-2014: approximately 1,000 Million Metric Tonnes (i.e. One Gigaton)
  - Oil Production via CO_2 EOR 1984-2014: approximately 2 Billion Barrels
    - Annual Recovery of Oil via CO_2 EOR Currently Exceeds 100,000,000 bbls/yr

- **"All of the injected CO2 is retained within the subsurface formation. . . . Or recycled to subsequent projects"; "When a CO_2 EOR flood is finished, the CO_2 that remains underground stays there."[(2)]**

(1) US DOE estimated 63 million metric tonnes, US EPA estimated 64 million metric tonnes, Advanced Resources International estimated 68 million metric tonnes

(2) U.S. National Energy Technology Laboratory(NETL)

Federal Government Determines CO_2 is a Pollutant

Under Clean Air Act and Massachusetts vs. EPA (2007)

- The atmospheric release of Greenhouse Gases (CO_2)
  "fit well within the [Clean Air] Act's … definition of air pollutant"

- 2009 EPA issues the "*Endangerment*" finding – prerequisite for implementing GHG emission standards

- EPA issued the "*Tailoring Rule*" in 2010; a phased in approach for GHG emissions for stationary sources and Title V operating permitting

- As a regulated New Source Review pollutant (NSR) CO2 become subject to requirements that major emitters apply "Best Available Control Technology" (BACT); in 2011 EPA issued guidance discussing emission control technologies that should be evaluated by permitting authorities on applying the BACT requirement
  - Under Federal Law CO_2 is now a regulated air pollutant for all major emitters
  - EPA determines CCS to be a pollution control technology for Greenhouse CO_2
  - EPA recognized a CO_2 pipeline as a "main component" of CCS Control System

Federal Government Determines CO_2 is a Pollutant

- 2012 U.S. Court of Appeals D.C. Circuit rules EPA was "*unambiguously correct*" in its effort to address global warming through regulatory programs
- 2013 Supreme Court agrees to hear if prior legal determination in MA vs. EPA as applied to mobile sources can be extended to stationary sources governed under separate programs
- 2014 US Supreme Court substantially upholds EPA GHG regulatory authority under the CAA. EPA may not treat GHG's as an air pollutant for purposes of determining whether it is a major source required to obtain a PSD or a Title V permit; however, PSD permits that are otherwise required may continue to require limitations on GHG's based on BACT

U.S. Federal Regulation Distinguishes Role of CO_2 EOR[(1)]

- Geologic storage of CO_2 can continue to be permitted under the UIC Class II program

  "CO_2 storage associated with Class II wells is a common occurrence, and CO_2 can be safely stored where injected through Class II-permitted wells for the purpose of oil or gas-related recovery."
- Use of anthropogenic CO_2 in ER operations does not necessitate a Class VI permit

  "ER operations can continue to be permitted as Class II wells, regardless of the source of CO_2. An owner or operator of an ER operation can switch from using a natural source to an anthropogenic source of CO_2 without triggering the need for a Class VI permit."
- Class VI site closure requirements are not required for Class II CO_2 injection operations

  "The most direct indicator of increased risk to USDW's is increased pressure in the injection zone related to the significant storage of CO_2. Increases in pressure with the potential to impact USDWs should first be addressed using tools within the Class II program. Transition to Class VI should only be considered if the Class II tools are insufficient to manage the increased risk."

(1) EPA Office of Ground Water and Drinking Water Memorandum, April 2015

Carbon Pollution Standards/Clean Power Plan

EPA Final Rule and Plan Creates Obstacles for EOR

- Conflicting objectives of resource conservation and waste disposal
  - Subpart RR will transform EOR operations from resource recovery operations to waste disposal operations
- Subpart RR compliance will conflict with state mandates to conserve natural resources, prevent waste and protect correlative rights
  - Classifying CO_2 as a waste will preclude future timely access to any future technologies and access to the remaining oil at the end of EOR operations.
- Subpart RR reporting is a vehicle for litigation and substantive regulation under the yet undefined Monitoring, Reporting and Verification (MRV) plans.
  - CO_2 injected as a waste will require the operator to obtain approvals by the EPA for a MRV plan. The MRV plans are open for public comment, debate and litigation.
  - The EPA will control MRV plan not the oil operator or the developer of the generating project

Texas Adopts CO_2 Management Rules

ADOPTED RULES

Adopted rules include new rules, amendments to existing rules, and repeals of existing rules. A rule adopted by a state agency takes effect 20 days after the date on which it is filed with the Secretary of State unless a later date is required by statute or specified in the rule (Government Code, §2001.036). If a rule is adopted without change to the text of the proposed rule, then the *Texas Register* does not republish the rule text here. If a rule is adopted with change to the text of the proposed rule, then the final rule text is included here. The final rule text will appear in the Texas Administrative Code on the effective date.

TITLE 16. ECONOMIC REGULATION

PART 1. RAILROAD COMMISSION OF TEXAS

CHAPTER 5. CARBON DIOXIDE (CO_2)
SUBCHAPTER C. CERTIFICATION OF GEOLOGIC STORAGE OF ANTHROPOGENIC CARBON DIOXIDE (CO_2) INCIDENTAL TO ENHANCED RECOVERY OF OIL, GAS, OR GEOTHERMAL RESOURCES

16 TAC §§5.301 - 5.308

Bottom Line

CO_2-EOR is a viable, economical and technologically feasible way to encourage CCS of anthropogenic CO_2 in a safe and secure manner under a known and proven regulatory system

Denbury.com | NYSE: DNR

25

Bell Creek Central Facility

Denbury.com | NYSE: DNR

26

Bell Creek High Pressure Compressors

Denbury.com | NYSE: DNR

27

Bell Creek Low Pressure Compressors

Denbury.com | NYSE: DNR

28

Bell Creek Process Building

Bell Creek Pump Building

Bell Creek Manifold Building

Corporate Information

Corporate Headquarters:
Denbury Resources Inc.
5320 Legacy Drive
Plano, Texas 75024
(972) 673-2000
denbury.com

Contact Information:
Greg Schnacke
Executive Director Governmental Relations
(972) 673-2324
greg.schnacke@denbury.com

Denbury NYSE: DNR

Using Nature's CO2 Sink to Convert Carbon from Challenge to Economic Opportunity:
Are Biological Capture and Utilization Technologies Ready to Deliver?

Carbon Technology Management Conference
November 19, 2015

Matt Carr
Executive Director
mcarr@algaebiomass.org

ALGAE BIOMASS ORGANIZATION

ALGAE BIOMASS ORGANIZATION

The Challenge

- Reduce GHG emissions from fossil power generation and other industrial sources…

Economically

What are Algae?

Algae

- ➢ Some of earliest life on Earth
- ➢ Highly diverse set of organisms
  - Estimated 50,000+ species
- ➢ Size: microscopic (microalgae) to large seaweeds (macroalgae) e.g. giant kelp
- ➢ Microalgae include:
  - Cyanobacteria (blue-green)
  - Green, brown and red algae

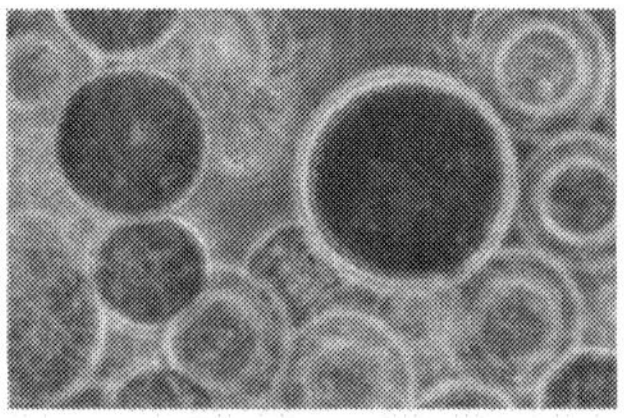

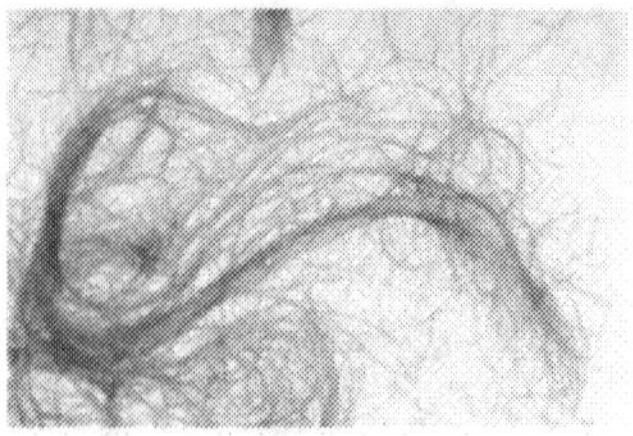

Algae Biomass Organization | www.algaebiomass.org

What are Algae?

Algae

- Mostly photosynthetic
  - Sunlight + CO_2 (autotrophic)
- Some feed on sugars…
  - Heterotrophic
- …or do both
  - Mixotrophic

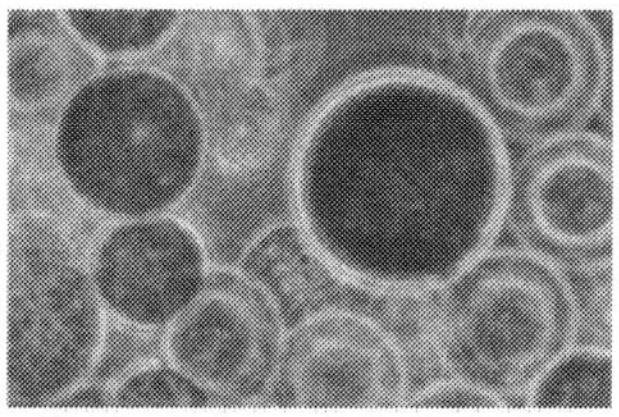

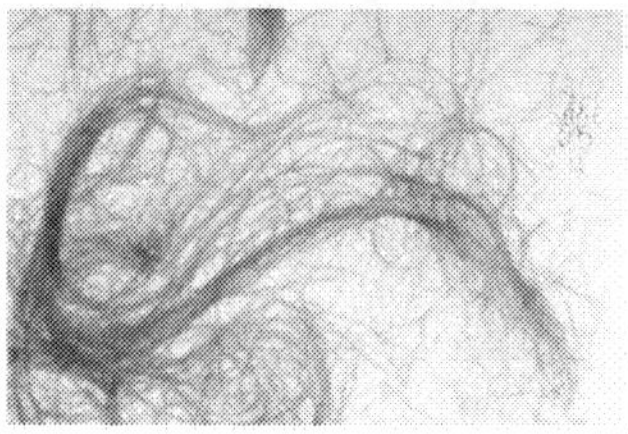

Algae Biomass Organization | www.algaebiomass.org

What are Algae?

Algae

- Exceptionally fast-growing
  - Mature in days vs. months/years

- Exceptionally productive

Algae Biomass Organization | www.algaebiomass.org

Why Algae: Yield

Comparison of Feedstock Crops					
Plant	**Average Biomass Yields** (MT / ha / yr)	**Oil Content** (% dry mass)	**Sugar / Starch Content** (% dry mass)	**Energy Content of Oil / Sugar / Starch** (boe / 1000ha / day)	**Protein Content** (% dry mass)
Soy/Soy oil	1 - 2.5	20%	18%	3 - 8	37%
Rapeseed	3	40%	NA	22	23%
Palm/Palm oil	19	20%	NA	63	15%
Jatropha	7.5 - 10	30 - 50%	NA	40 - 100	24 - 28%
Corn	10 - 12	4%	75%	240 - 300	4 - 14%
Sugarcane	60 - 70	NA	12 - 16%	230 - 370	3 - 4%
Microalgae	40 - 100+	**25 - 50%+**	**15 - 25%**	**330 - 785+**	**25 - 60%+**

Why Algae: Sustainability

Lowest Carbon, Water, & Arable Land Footprints of Any Crop

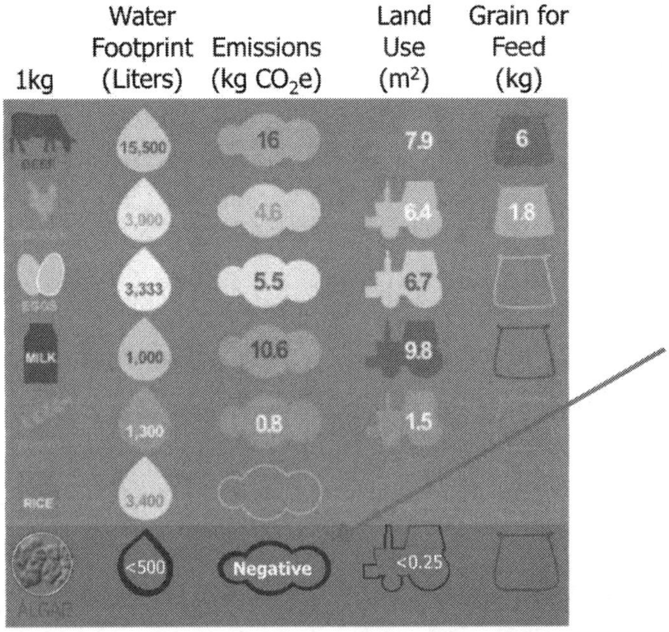

Sources: Water www.waterfootprint.org/?page=files/productgallery; *emissions and land use UK DEFRA* (2006), http://goo.gl/T12.hq; *grain* National Geographic, http://goo.gl/4CgFB; *Algae,* Cellana estimates

The Biological Capture and Utilization Opportunity

Algae

- ➢ CO2 eating machines…

Algae Biomass Organization | www.algaebiomass.org

The Algae CO2 Utilization Opportunity

Algae

- Factory for production of:
  - oils
  - proteins
  - other high-value products
- Can produce …
  100 gallons biofuel per ton CO_2

Algae Biomass Organization | www.algaebiomass.org

The Algae CO2 Utilization Opportunity

- Product value from algae conversion:
  > \$200 / ton CO_2
- Today: Algae developers buy CO_2
  (at \$40+ / ton) as a feedstock
- Industrial CO_2 Challenge + Algae =
  Opportunity

Algae Biomass Organization | www.algaebiomass.org

ALGAE BIOMASS ORGANIZATION

Diamond Members

ALGENOL
bioprocessalgae
solazyme

Platinum Members

ACCELERGY CORPORATION
BOEING
heliae
JOULE
PHITEC SUSTAINING EFFICIENCY
Sapphire Energy
Wilson Sonsini Goodrich & Rosati PROFESSIONAL CORPORATION

Gold Members

algae systems
ALGIX
BYRNE & COMPANY LIMITED
cellana
LanzaTech
MATRIX GENETICS
The Gas Company A Sempra Energy utility
subitec

ALGAE BIOMASS ORGANIZATION

Corporate Members

APPLIED CHEMICAL TECHNOLOGY The World Leader in Process Equipment Design
ATP³ Algae Testbed Public-Private Partnership
Aurora algae
AzCATI Arizona Center for Algae Technology and Innovation
cleanTECH san diego
CHURCH & DWIGHT CO., INC.
COLORADO LINING INTERNATIONAL
DIVERSIFIED TECHNOLOGIES, INC.
DSM BRIGHT SCIENCE. BRIGHTER LIVING.
DUKE ENERGY
EARTHRISE
evodos
FedEx Express
gti
GICON Großmann Ingenieur Consult GmbH
GENERAL ATOMICS
+GF+ GEORG FISCHER PIPING SYSTEMS
GLOBAL ALGAE INNOVATIONS
AIIC JAPAN
Parker
Kimberly-Clark Corporation
Linde
NCMA
PETRIXO
MBIO MicroBio Engineering
MURADEL
NESTE OIL
OpenAlgae
OriginOil
RAE RENEWABLE ALGAL ENERGY
POS BIO-SCIENCES
SYNTHETIC GENOMICS
SCHOTT glass made of ideas
STOEL RIVES LLP ATTORNEYS AT LAW
TASA
TEXAS A&M AGRILIFE RESEARCH
UK Center for Applied Energy Research
PRANA
سابك sabic
The Scoular Company
Valicor Renewables

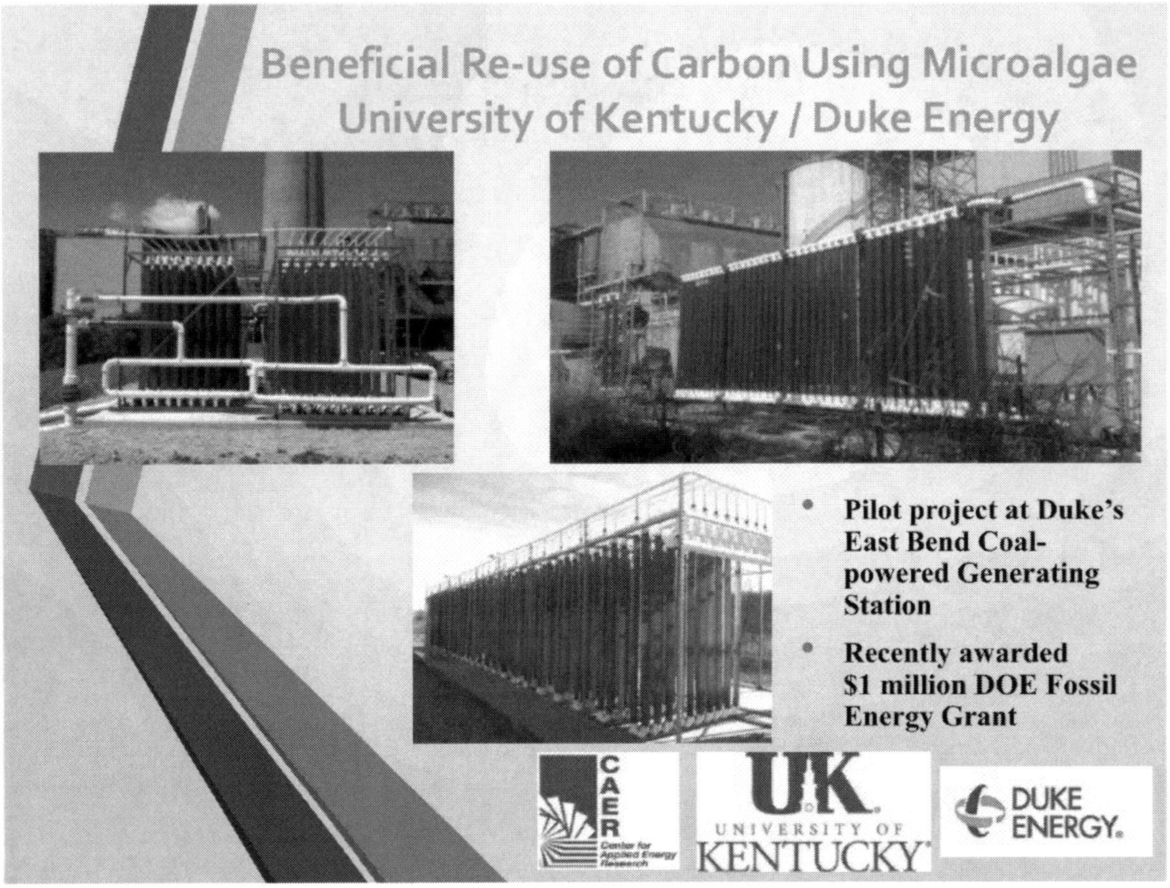
Beneficial Re-use of Carbon Using Microalgae
University of Kentucky / Duke Energy
Pilot project at Duke's East Bend Coal-powered Generating Station
Recently awarded $1 million DOE Fossil Energy Grant
CAER
Center for Applied Energy Research
UK
UNIVERSITY OF KENTUCKY
DUKE ENERGY

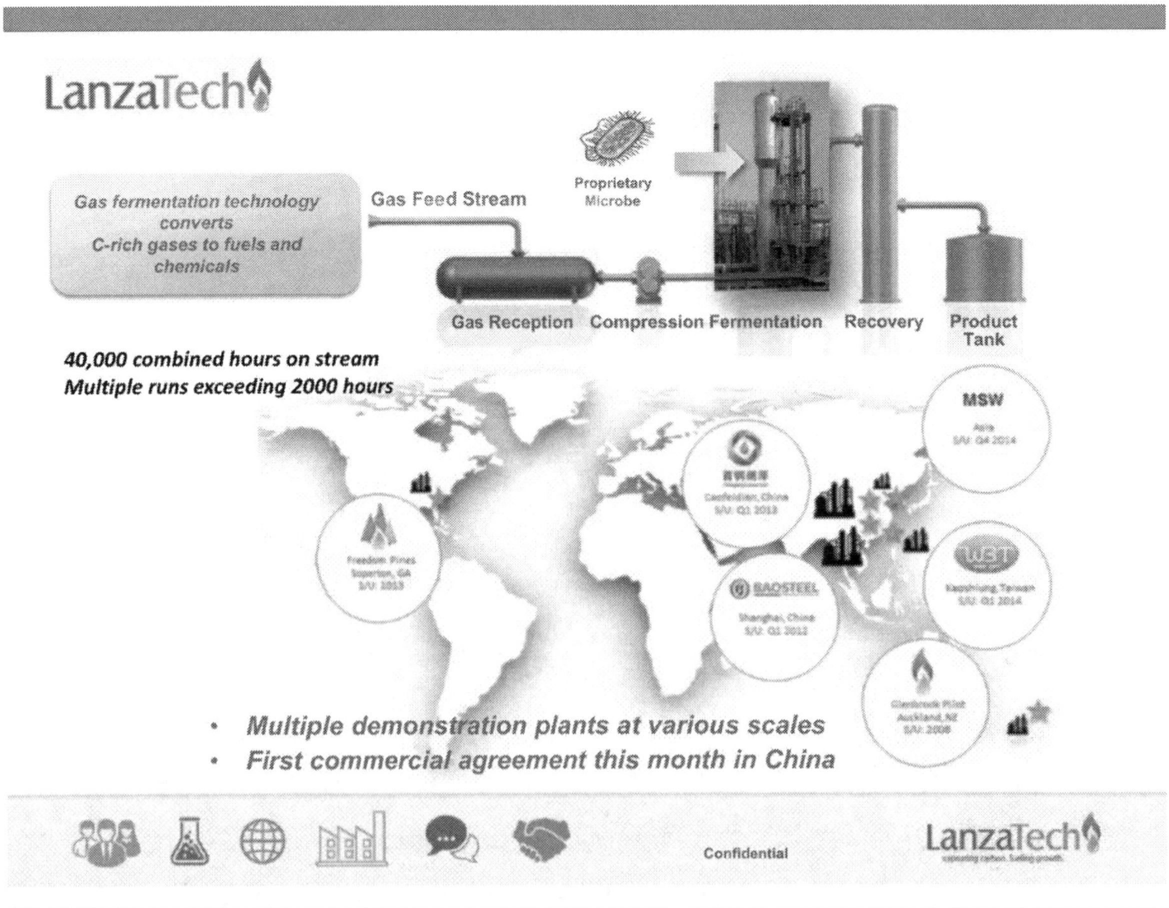
LanzaTech
Gas fermentation technology converts C-rich gases to fuels and chemicals
Gas Feed Stream
Proprietary Microbe
Gas Reception
Compression
Fermentation
Recovery
Product Tank
40,000 combined hours on stream
Multiple runs exceeding 2000 hours
MSW
BAOSTEEL
Multiple demonstration plants at various scales
First commercial agreement this month in China
Confidential
LanzaTech

33-Acre Kauai Algae Facility

GLOBAL ALGAE INNOVATIONS

Distribution Statement "A" (Approved for Public Release, Distribution Unlimited)

AB ALGAE BIOMASS ORGANIZATION

State of Algae Carbon Utilization Technology

➢ Algenol Biofuels

- 2-acre commercial demonstration in Lee County, FL
- Pilot demo at Reliance **petroleum refinery** in India
- Received EPA approval as advanced biofuel with 69% GHG reduction vs. gasoline

ALGENOL BIOFUELS

"Harnessing the sun to fuel the world"

Algae Biomass Organization | www.algaebiomass.org

State of Algae Carbon Utilization Technology

➢ BioProcess Algae

- 5-acre commercial demonstration in Shenandoah, IA, co-located with Green Plains **corn ethanol plant**
- Continuous operation since 2009 utilizing waste CO2, waste heat, waste nutrients
- Markets: Fuel, fish/animal feed

bioprocessalgae
Valuing Carbon

Algae Biomass Organization | www.algaebiomass.org

State of Algae Carbon Utilization Technology

➢ Technology Maturation / Demonstration

- **DOE-funded integrated algae biorefinery commercial demonstration projects in**
    - Florida (Algenol Biofuels)
    - New Mexico (Sapphire Energy)
- **Pilot demonstrations with flue gas from variety of industrial sources**
    - Coal, other power plants
    - Cement, steel
    - Ethanol

Algae Biomass Organization | www.algaebiomass.org

ALGAE BIOMASS ORGANIZATION

State of Algae Carbon Utilization Technology

➢ Technology Maturation / Demonstration

- Contracts with major refiners

cellana

Sapphire Energy™

TESORO

ALGENOL BIOFUELS
Cleaner - Greener - Cheaper

Reliance Industries Limited

ALGAE BIOMASS ORGANIZATION

State of Algae Carbon Utilization Technology

➢ Product Diversification

➢ Bioplastics
- Algix manufacturing plant opens in Mississippi

➢ Biofertilizers
- Accelergy validating Terrasync terrestrial sequestration platform

➢ Human / Animal Nutrition
- Synthetic Genomics , ADM to produce omega-3s

ALGIX™ Formulating the Potential of Algae

ACCELERGY CORPORATION

SYNTHETIC GENOMICS

ADM

Algae Biomass Organization | www.algaebiomass.org

ALGAE BIOMASS ORGANIZATION

State of Algae Carbon Utilization Technology

- Strategic Partnerships / Globalization
- Sapphire / Sinopec
  - Validate and deploy in China
- Heliae / Sincere Corp
  - Build first commercial facility in Japan
- Solazyme / Bunge
  - Manufacturing plant online in Brazil

Sapphire Energy | SINOPEC | heliae | sincere | solazyme | BUNGE

Algae Biomass Organization | www.algaebiomass.org

ALGAE BIOMASS ORGANIZATION

Algae Facilities in U.S. and Canada

Algae Biomass Organization | www.algaebiomass.org

Clean Power Plan as Economic Opportunity

- **CPP Final Rules**
  - Include Carbon Capture and Use (CCU) as a compliance option
  - Specifically recognize algae CCU as a potential platform
  - Establish procedure to qualify CCU projects

Algae Biomass Organization | www.algaebiomass.org

What Needs to Happen

- **State(s) include CCU in their compliance plan**
  - Perhaps as part of an "innovative technologies" bucket?
- **Establish approved accounting methodology**
  - Have a key state take the lead?
- **Increased federal funding for R&D and scale-up**
  - FY16 U.S. Senate Energy Appropriations includes $10 million for CCU

Algae Biomass Organization | www.algaebiomass.org

ALGAE BIOMASS ORGANIZATION

Conclusions

- Carbon Capture and Utilization (CCU) via algae and other micro-organisms offers an **economically attractive CO2 mitigation option** for new and existing sources
- Clean Power Plan lays groundwork for commercial deployment
- Opportunity for states to **substantially reduce, or even offset, portion of compliance cost**

Algae Biomass Organization | www.algaebiomass.org

ALGAE BIOMASS ORGANIZATION

Learn More – RecycleCarbon.org

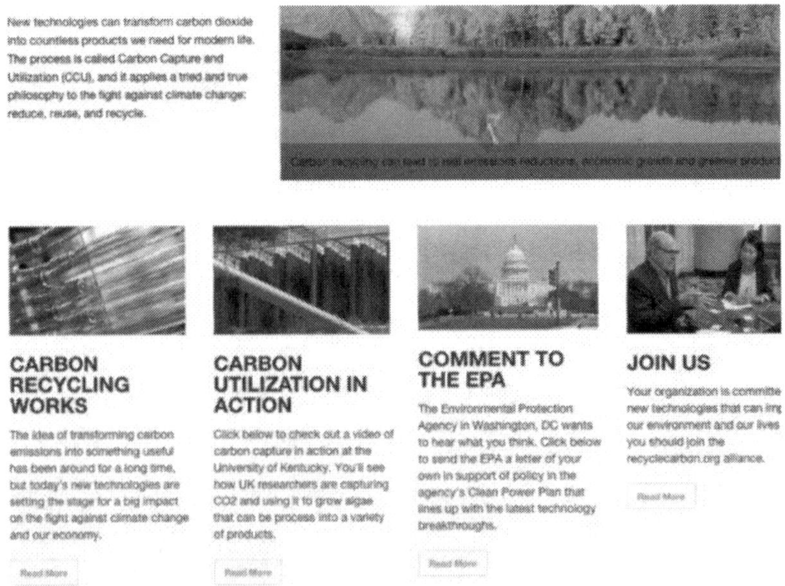

Algae Biomass Organization | www.algaebiomass.org

2016 Algae Biomass Summit

Phoenix, Arizona
October 23 - 27, 2016

Industry Panel

- **Tom Jensen**, Executive Vice President, Joule
- **Laurel Harmon**, Vice President Government Relations, LanzaTech
- **Mark Allen**, Vice President - Integrated Carbon Solutions, Accelergy
- **Michael Wilson**, Research Senior Engineer, University of Kentucky Center for Applied Energy Research
- **Noah Deich**, Executive Director, Center for Carbon Removal

Center for Carbon Removal

THE CENTER
FOR CARBON
REMOVAL

CMTC 2015
November 19, 2015
Sugar Land, Texas
www.centerforcarbonremoval.org

About Us

Noah Deich
Executive Director
Center for Carbon Removal

BECI
BERKELEY ENERGY &
CLIMATE INSTITUTE
LBNL & UC BERKELEY

Mission:

To ***accelerate the development of carbon removal solutions*** to assist in the fight against climate change.

CCS 1.0: Plants

Challenge:

Plants are "catch and release"…
…we need "capture and reliable sequestration"

CCS 2.0: Fossil Energy

Challenge:

Seen as more expensive / dirtier substitute for renewables → struggles to gain financial / regulatory support

CCS 3.0: Bio-Capture & Storage?

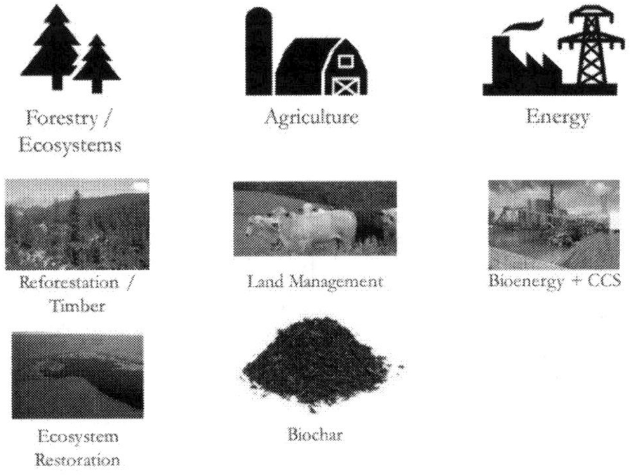

Challenges:

- Costs
- Sustainability
- Scalability
- Accounting / Reliability

The Need for Negative Emissions

The Climate Modeling Math

IPCC integrated modeling scenarios consistent with 2°C of warming:

- 87% have net-negative emissions globally by 2100.
- Up to 25 GW of negative emissions energy systems annually by 2040
- Average GHG budget overshoot of >160 Gt (>15% of total)

The Climate Negotiations Math

INDCs for COP21 *not* expected to peak emissions by 2030—~75% of GHG emissions budgets exhausted by 2030.

Starting Points:

Commercialize "negative-emissions" energy today—not fossil+CCS alone

1. Test and demonstrate geologic sequestration with inexpensive (but not scalable) CO_2 from biofuels.

2. Co-fire with sustainable biomass and/or retrofit bioenergy facilities

Enables testing of:
- Sustainability of feedstocks
- Regulatory and consumer reaction to geologic storage

The price for better politics: increased short-term capital costs.

Moving Beyond CO_2 Neutrality

Bioenergy needs fossil + CCS to:

- Prove technology
- Reduce costs
- Establish regulatory structures and protocols

Fossil needs bioenergy + CCS to:

- Provide a credible commitment to negative emissions pathway in the future to gain the political support it needs

Our emissions budget for mitigating climate change is rapidly approaching; we are on track to breach the 2° C budget by 2020-2030

JOULE

 www.jouleunlimited.com

The Smartest Way To Sequester Carbon: Industrializing Photosynthesis

JOULE

 2 www.jouleunlimited.com

Joule: The Ultimate Low Carbon Fuels Company

JOULE

Hydrocarbon Based Fuels

One way carbon movement:
From underground to the atmosphere

VS

Recycled Carbon Fuels

Closed carbon loop

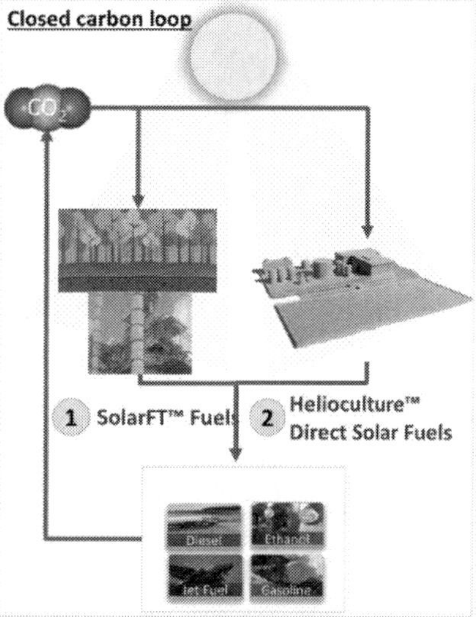

Pillars of Commercial Success: Productivity, Modularity, Scalability

JOULE

Pillars Of Commercial Success: Global Scalability

JOULE

Global Above- and Below-ground Living Biomass Carbon Density

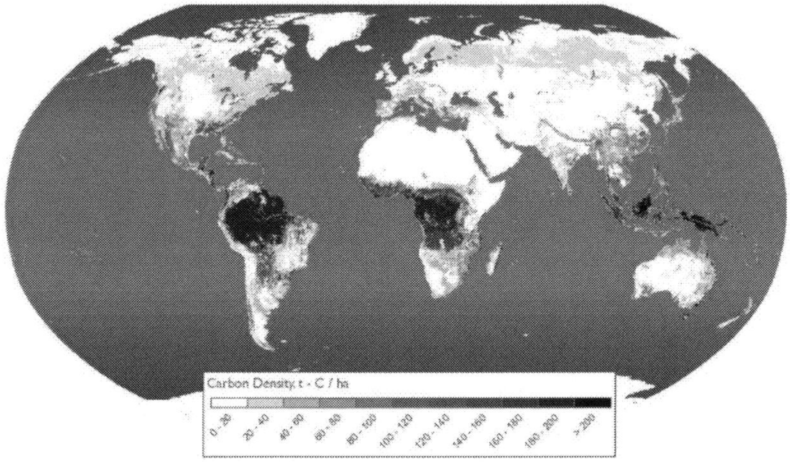

Store Sunlight In Liquid Fuels Through Recycling Carbon Cost Effectively

JOULE

LanzaTech
Creating a Carbon Smart Future!

Houston, TX
Laurel Harmon
VP Government Relations

LanzaTech

STAYING BELOW 2°C: THE CHOICES WE FACE

With current pledges on the table to cut emissions, we are heading to a 3.3° C warming future. No further action before 2020 will limit society's choices. As temperatures rise, so do the impacts.

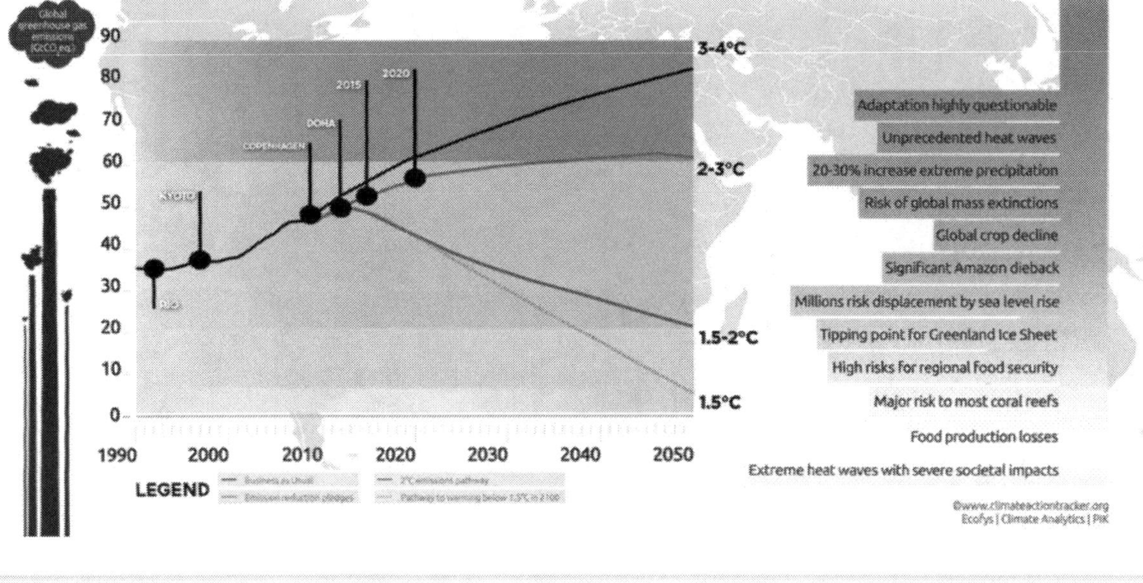

©www.climateactiontracker.org
Ecofys | Climate Analytics | PIK

LanzaTech

2

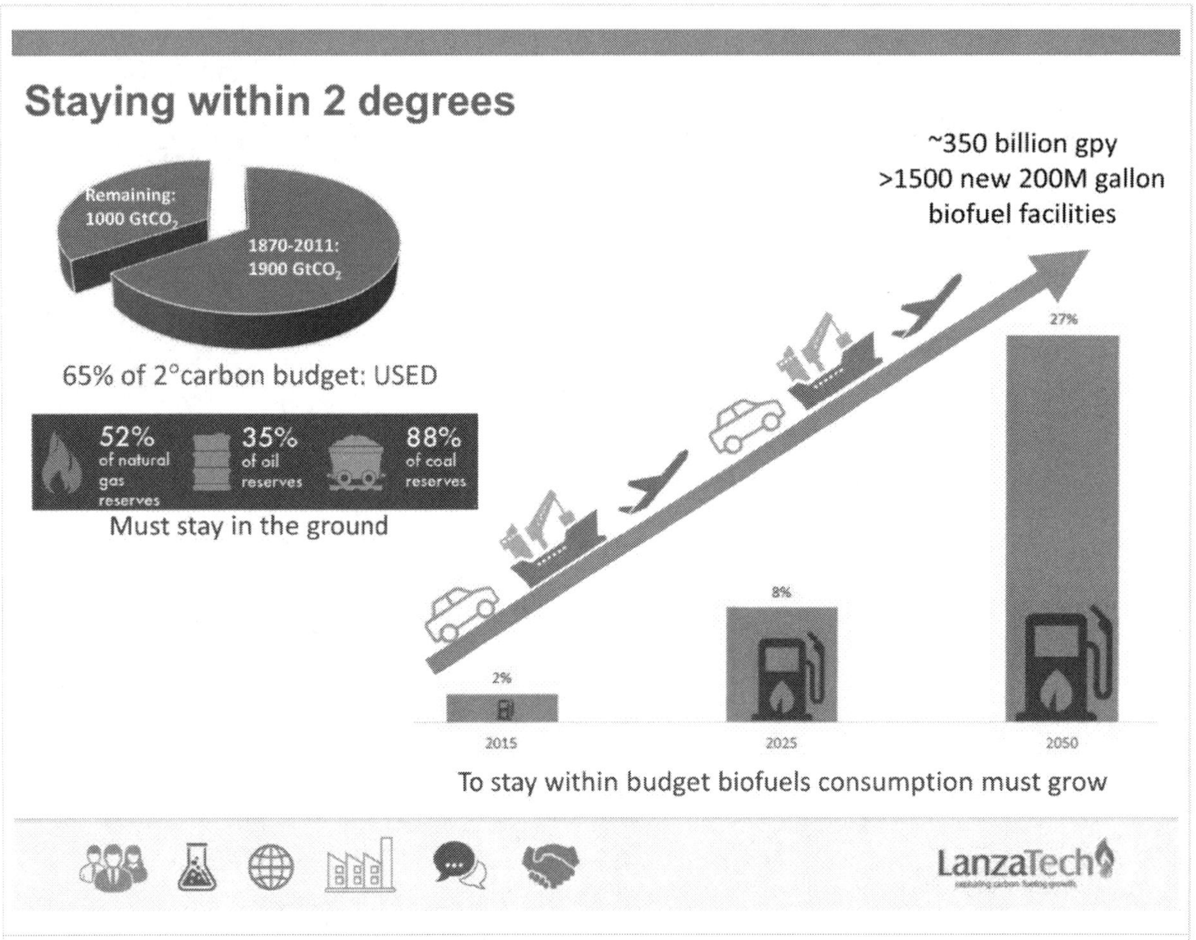
Staying within 2 degrees
Remaining:
1000 GtCO2
1870-2011:
1900 GtCO2
65% of 2°carbon budget: USED
52%
of natural gas reserves
35%
of oil reserves
88%
of coal reserves
Must stay in the ground
~350 billion gpy
>1500 new 200M gallon
biofuel facilities
2%
8%
27%
2015
2025
2050
To stay within budget biofuels consumption must grow
LanzaTech

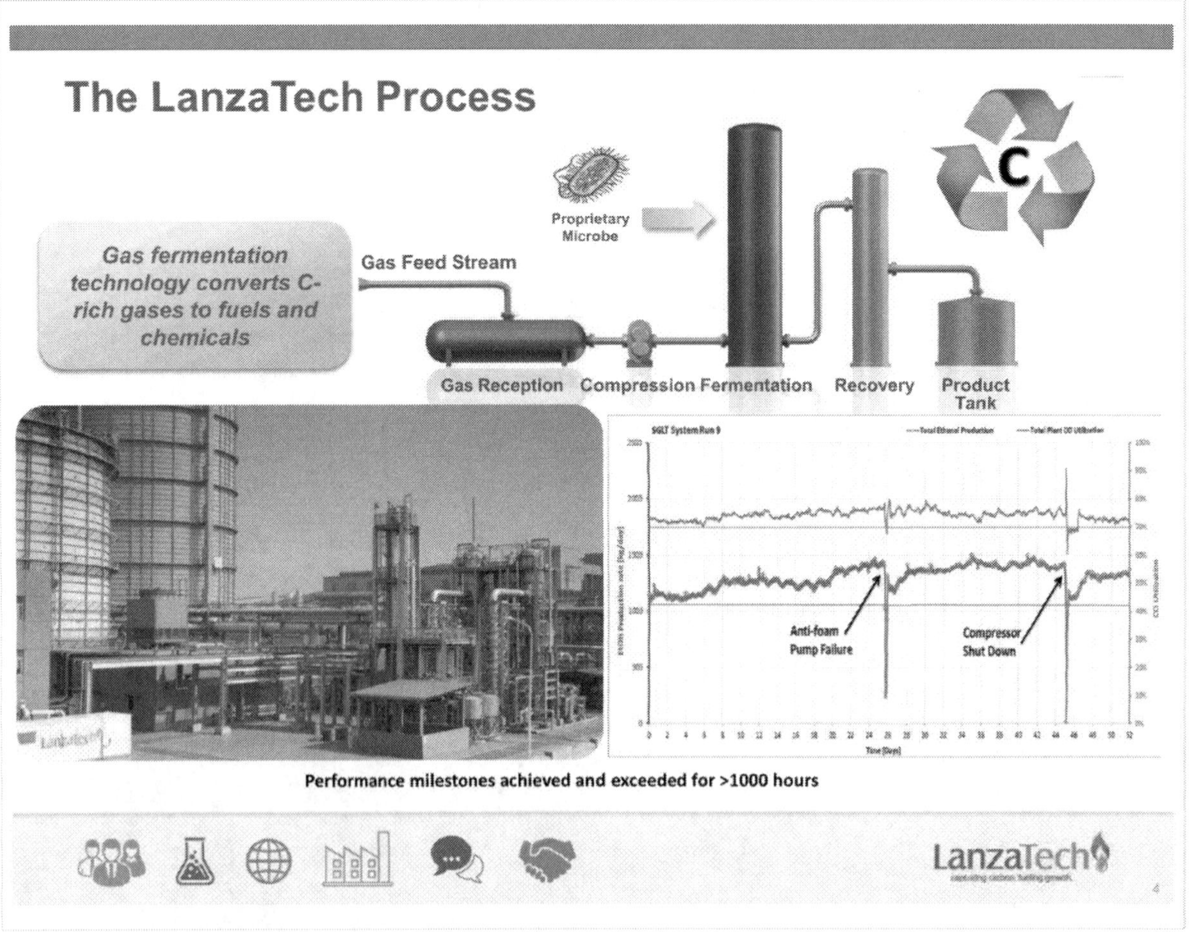
The LanzaTech Process
Gas fermentation technology converts C-rich gases to fuels and chemicals
Gas Feed Stream
Proprietary
Microbe
C
Gas Reception
Compression
Fermentation
Recovery
Product
Tank
Anti-foam
Pump Failure
Compressor
Shut Down
Performance milestones achieved and exceeded for >1000 hours
LanzaTech
4

Waste carbon streams as a Resource

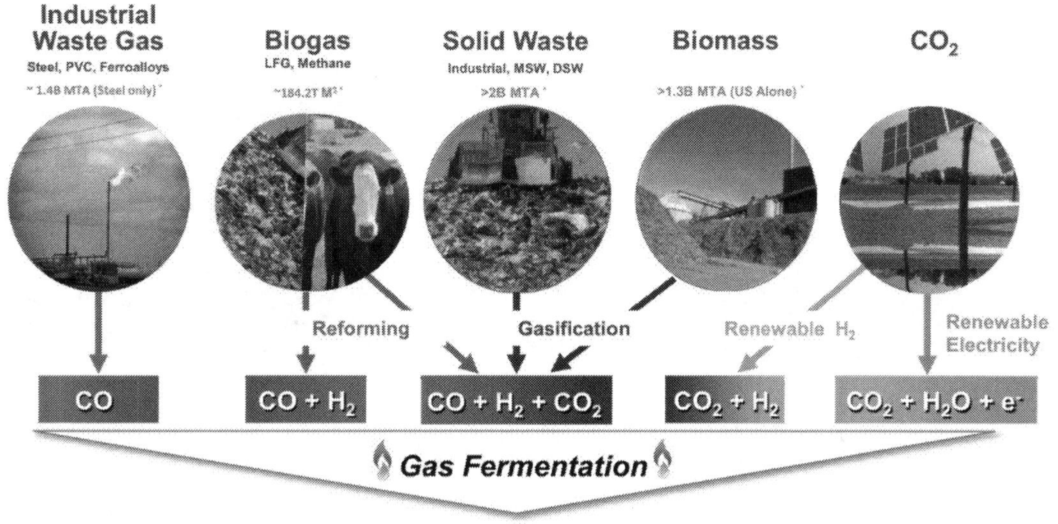

✓ *Available* ✓ *High Volume/Low Intrinsic Value* ✓ *Most Point Sourced*

✓ *Non-Food*

*2010 global production, 2012 proven gas reserves data (IEA, UNEP, IndexMundi, US DOE Billion Ton Update)

LanzaTech

Global Technology "Lab"
Data, Data, Data

40,000 combined hours on stream
Multiple runs exceeding 2000 hours

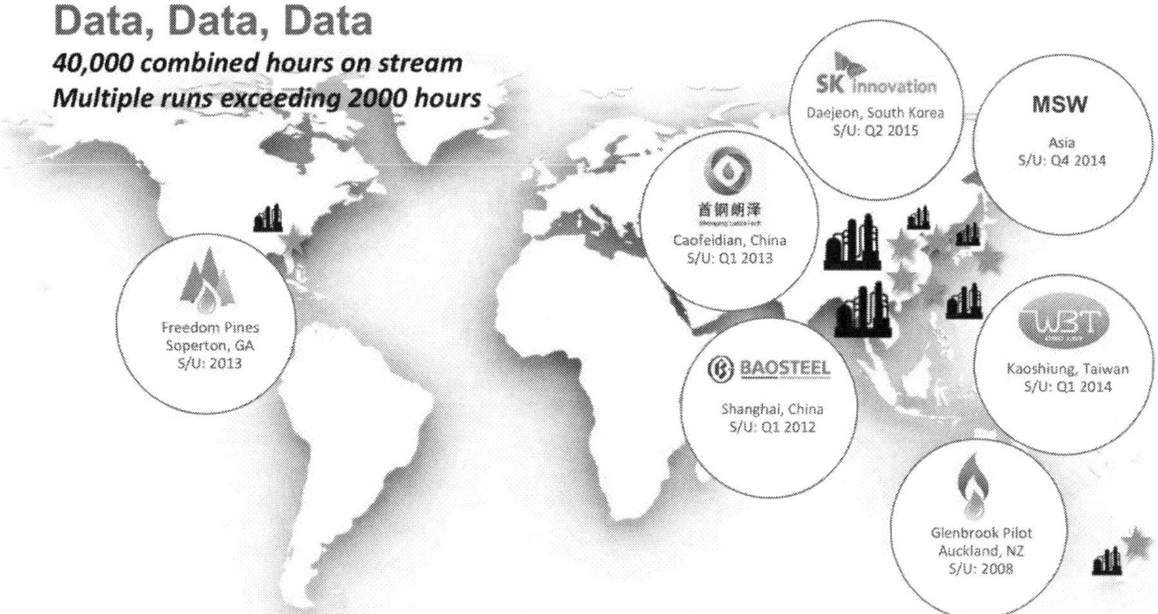

Multiple Demo plants at various scales all demonstrating <u>different</u> key aspects of process

LanzaTech

Pre commercial steel mill demonstrations

- ✓ Performance milestones exceeded
- ✓ First commercial in design; fully financed in China

Exceeded design capacity
Local chemicals, water

Mitigating Scale up Risk through Successful Technology Demonstration

LanzaTech

CSC Board Approval

CHINASTEEL

最新消息

27 Mar

Published Board Meeting Agenda

5) Capital Injection to participate WBT concerting BOF to EtOH Project

CSC diversified investment to WBT by applying LanzaTech (USA) fermentation technology to erect an EtOH demonstration plant. In accordance with the results of the demonstration plant, CO utilization, broth concentration and robust of bacteria are all way exceed the targets. It approves the technology has commercial value, it will proceed commercial operation plant.

五、參與新能公司增資建置轉爐氣產製乙醇商轉工廠案
中鋼公司轉投資事業新能公司應用美國Lanzatech公司厭氧菌發酵技術，建置乙醇示範工場，依該示範工場實際運轉結果，其CO利用率、發酵液乙醇濃度及菌體耐受度等指標，皆大幅超越預定目標，證明此技術具進行商業化發展價值，乃進行建置商轉工廠。

http://www.csc.com.tw/indexe.html

Public Announcements

China Steel Invest 7000 MM TWD Launch Two Projects

2015 -03-28 Economic Daily News

CSC board meeting held yesterday (27) has approved two investment project. It will spend 5600 MM TWD to overhaul No. 2 blast furnace of the third train; meanwhile investment 1400 MM TWD to erect a 100,000 ton per annum ethanol plant.

CSC states, trans-invested to WBT using LanzaTech (USA) fermentation technology, built an ethanol demonstration plant, in accordance with actual operation result, all indices were way exceed the targets, approve this technology worth for commercialization, will begin commercial plant at the year end.

Per the decision of CSC board, it will use the land near Xiao Gang Zhong Lin Road purchased from LCY to build the plant. The investment amount 1400 MM TWD. Starts production after estimated 2~3 years later. Per understanding, the plant will be in two phases, it will erect a 100,000 MT per annum production line, main application is gasoline additives and industrial ethanol.

3大投資案 中鋼砸76.28億元

"3 Large Investments Hit 7.628 Billion Yuan Steel Case"

http://www.chinatimes.com/newspapers

LanzaTech

8

ArcelorMittal Gent

ArcelorMittal

Electricity generation

- €300M investment in electricity from CO in Gent
- Fully integrated Flagship mill

Carbon recycling with LanzaTech

- First of a kind project in EU Q2 2017
- Aggressive deployment across EU post Gent
- Power from renewables not carbon

ArcelorMittal supports a carbon smart future

- 120,000 tons/annum CO_2 reduction
- 300,000 tons/annum CO2 captured for future reuse
- 60k MTA ethanol capacity
- Adding value to CO

LanzaTech

Commercialization of Aviation Fuel

HSBC · virgin atlantic · BOEING · U.S. DEPARTMENT OF ENERGY | Energy Efficiency & Renewable Energy · GE · Pacific Northwest NATIONAL LABORATORY · Imperium renewables

- **World First Proving Flight using sustainable ATJ from steel mill off gases**
- **Flight will provide fuel performance data to help accelerate ASTM certification of ATJ production pathway**

LanzaTech

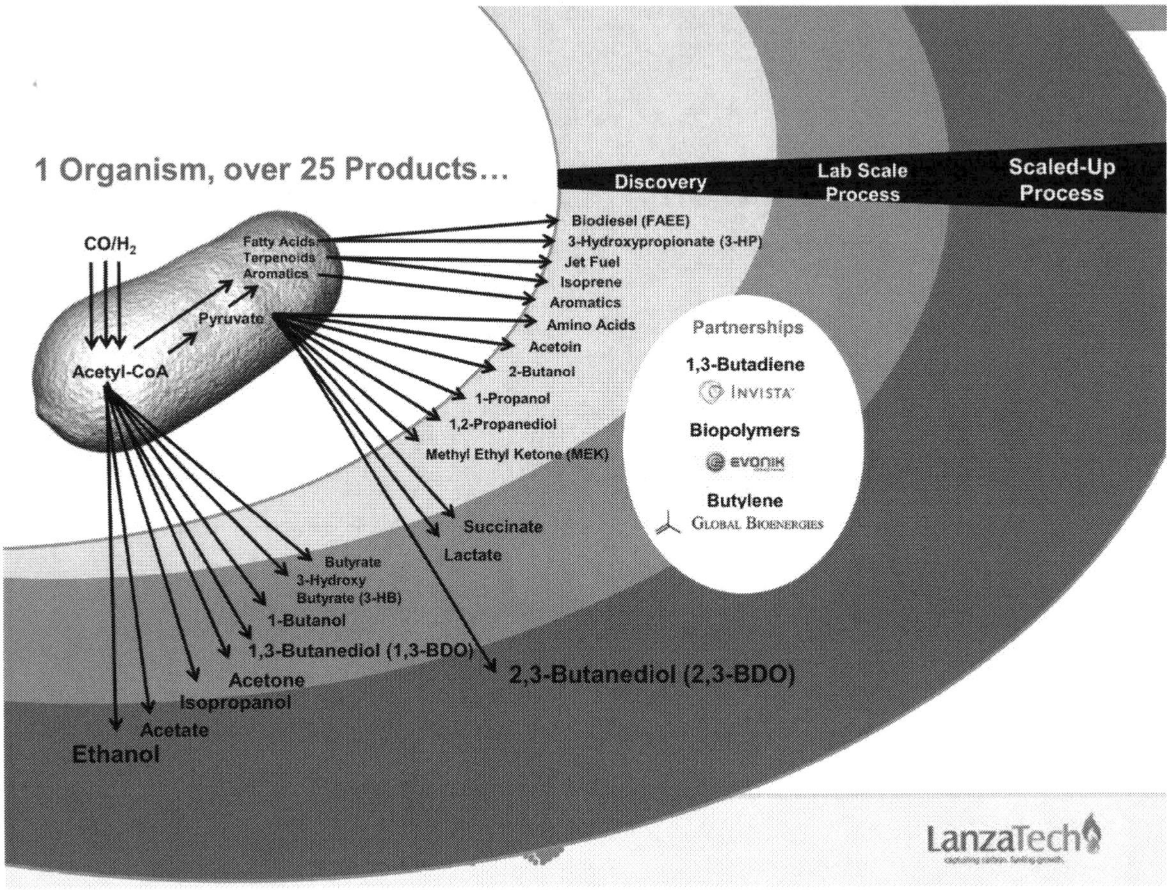

Fixing Carbon into Products. A Carbon Smart Future.

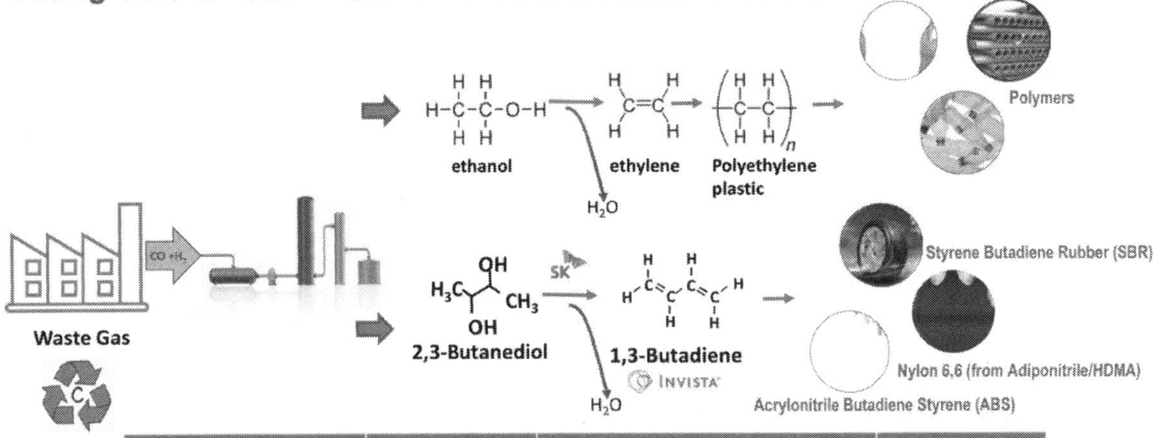

		Fossil Energy (toe/t)			GHG
		Fuel	Feedstock	Total	(kg CO2e/kg)
Plastic bags (LDPE)	Conventional	0.8	1.1	1.9	1.9
	LanzaTech	0.8	0.0	0.8	-1.8
Plastic bottles (PET)	Conventional	0.7	0.8	1.6	2.2
	LanzaTech	0.7	0.7	1.4	1.6
Tires (SBR)	Conventional	0.8	1.2	2.0	2.3
	LanzaTech	0.8	0.3	1.0	0.1

LanzaTech

Reduce Pressures on Reserves: Recycle Carbon
Steel
Aluminum
Glass
Plastic
Carbon
C
Ore
Alum
Sand
Oil
Oil
Carbon Reduction through Re-use and Recycling
LanzaTech
13

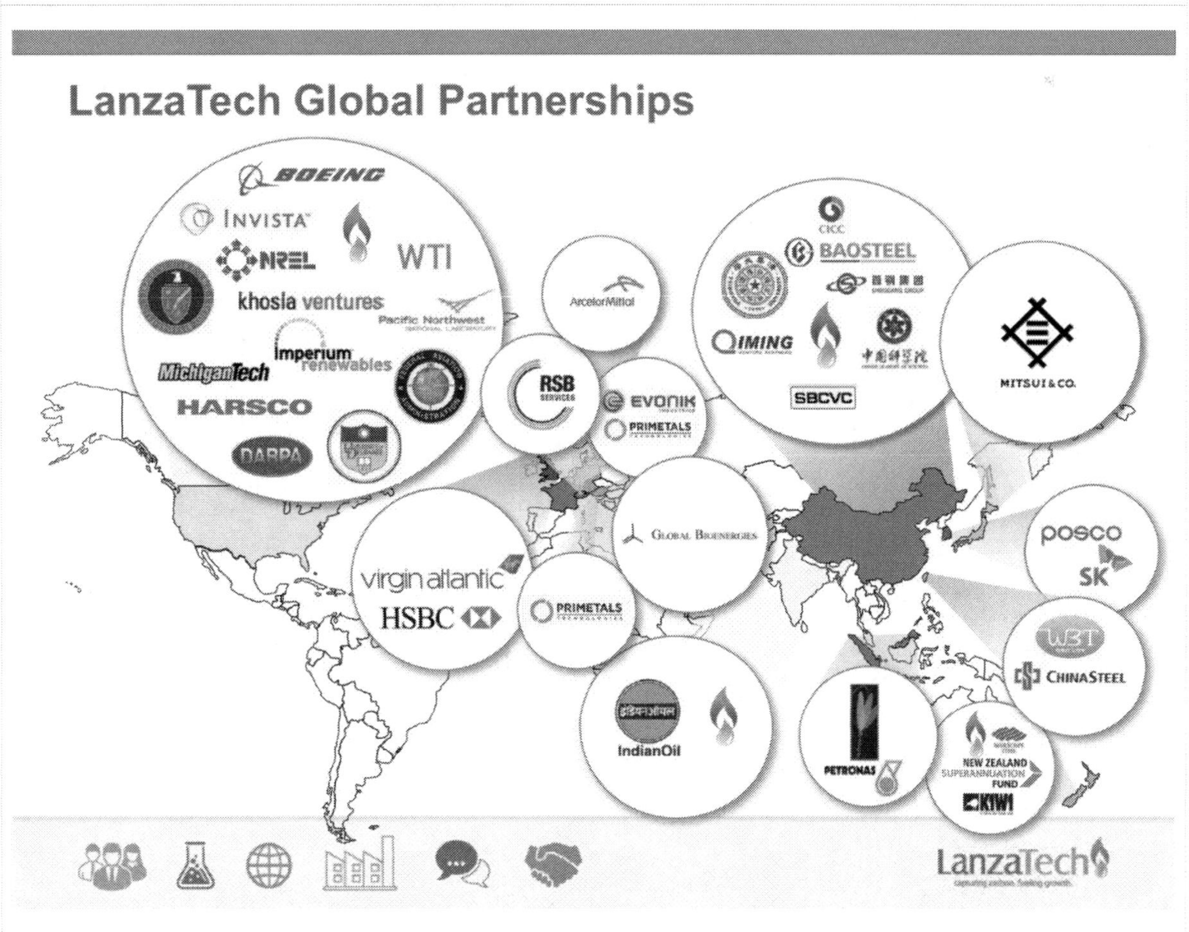
LanzaTech Global Partnerships
BOEING
INVISTA
NREL
WTI
khosla ventures
Pacific Northwest
Imperium renewables
MichiganTech
HARSCO
DARPA
ArcelorMittal
RSB
EVONIK
PRIMETALS
CICC
BAOSTEEL
QIMING
SBCVC
MITSUI&CO.
GLOBAL BIOENERGIES
virgin atlantic
HSBC
PRIMETALS
posco
SK
CHINASTEEL
IndianOil
PETRONAS
NEW ZEALAND SUPERANNUATION FUND
KIWI
LanzaTech

ACCELERGY CORPORATION

TerraSync Biofertilizer

Atmospheric CO_2 & N ↓

Process CO_2 → Algae → Biofertilizer → Terrestrial Carbon Sequestration

1

Nitrogen Fixing Cyanobacteria – Old as Dirt

ACCELERGY CORPORATION

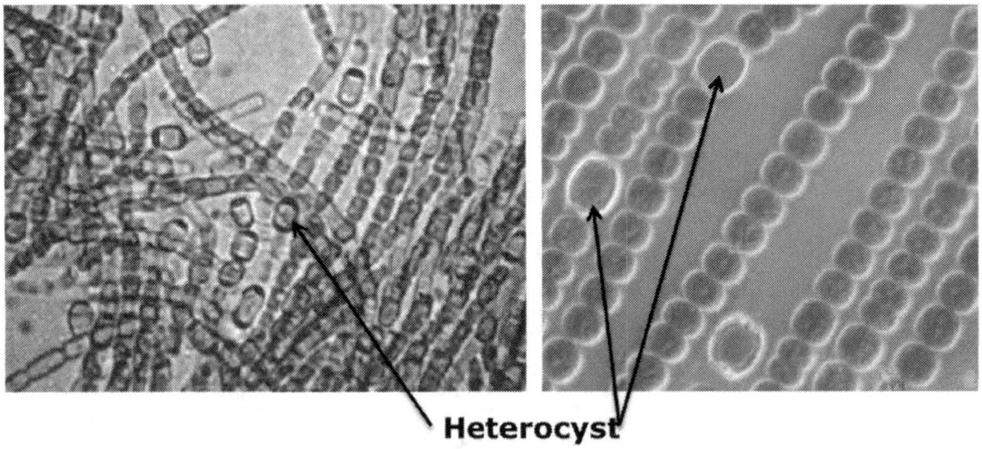

2

TerraSync Soil Inoculation

ACCELERGY CORPORATION

0 day | 2 weeks | 4 weeks | 6 weeks | 8weeks | 12 weeks | 16 weeks

	Control	TerraSync
N Content, mg/m^2	---	500
NO_3^-	19.2	452
NH_4^+	9.0	121
NO_2^-	19.0	9.7
Total N	47.2	583
Chl. A	Below the lower limit of detection	12.4
Microalgae Biomass	Below the lower limit of detection	4788

Accelergy TerraSync Biofertilizer Development Program

US Department of Energy – Montana State University Plant Trials

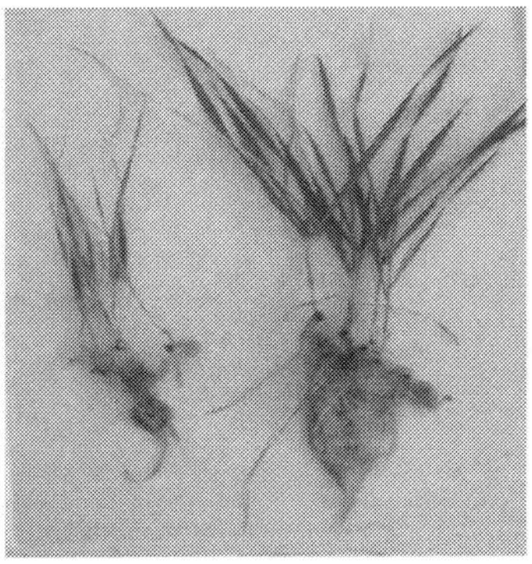

- Yellowstone Wheat is shown growing in Crow Reservation soil without any fertilizer (control) and two normal applications of standard fertilizer, and Strain B1611 nitrogen fixing cyanobacteria (blue-green algae) at eqv. standard nitrogen application rate
- Strain B1611 was selected from algae cultivation trials funded by DOE under Accelergy supervision

- Root development, (control on left – Strain B1611 biofertilizer on right) is indicative of overall healthy plant development

Camilena Seed Oil Energy Crop

Accelergy & Elion Corporation

Desert Drought Resistant Science
& Technology Park, DuGui, Inner Mongolia, China

Carbon Depleted Kubuqi Desert Soil

200 Acre Field Trial – Abandoned Farm Land

ACCELERGY

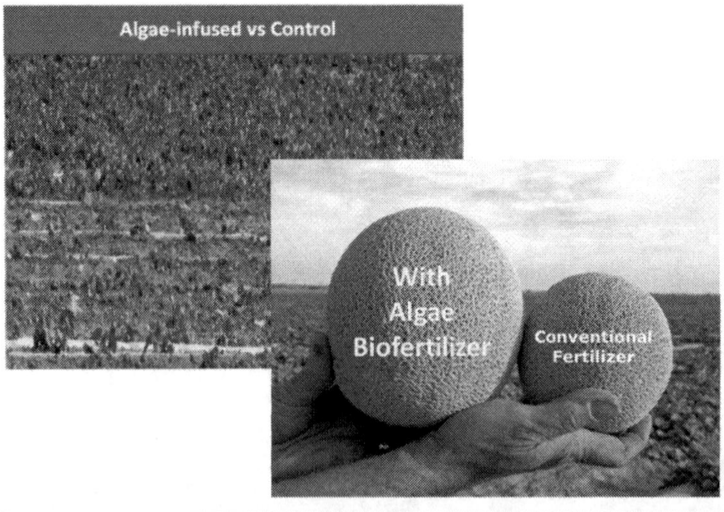

Del Monte Quality

Yosemite Melons	
Germination	2X
Early Growth	2.5X
Plant Size & Density	3X
Early Melon Maturity	2X
Mature Melon Size	1.4X
Mature Melon Weight	1.5X
Taste (blind panel test)	14:1

Yielding Superior Crops and Price for Farmers

ACCELERGY CORPORATION

Cantaloupe WITHOUT Algae

Nutrition Facts

Serving Size (100g)
Servings Per Container varies

Amount Per Serving	
Calories 40	Calories from Fat 5
	% Daily Value*
Total Fat 0g	0%
Saturated Fat 0g	0%
Trans Fat 0g	
Cholesterol 0mg	0%
Sodium 25mg	1%
Total Carbohydrate 8g	3%
Dietary Fiber 2g	8%
Sugars 2g	
Protein 1g	
Vitamin A 70%	Vitamin C 25%
Calcium 2%	• Iron 2%

*Percent Daily Values are based on a 2,000 calorie diet. Your daily values may be higher or lower depending on your calorie needs:

	Calories:	2,000	2,500
Total Fat	Less than	65g	80g
Saturated Fat	Less than	20g	25g
Cholesterol	Less than	300mg	300mg
Sodium	Less than	2,400mg	2,400mg
Total Carbohydrate		300g	375g
Dietary Fiber		25g	30g

Calories per gram:
Fat 9 • Carbohydrate 4 • Protein 4

+ 300%
+ 60%

Cantaloupe WITH Algae

Nutrition Facts

Serving Size (100g)
Servings Per Container varies

Amount Per Serving	
Calories 40	Calories from Fat 0
	% Daily Value*
Total Fat 0g	0%
Saturated Fat 0g	0%
Trans Fat 0g	
Cholesterol 0mg	0%
Sodium 20mg	1%
Total Carbohydrate 9g	3%
Dietary Fiber 2g	8%
Sugars 6g	
Protein 1g	
Vitamin A 70%	Vitamin C 40%
Calcium 0%	• Iron 0%

*Percent Daily Values are based on a 2,000 calorie diet. Your daily values may be higher or lower depending on your calorie needs:

	Calories:	2,000	2,500
Total Fat	Less than	65g	80g
Saturated Fat	Less than	20g	25g
Cholesterol	Less than	300mg	300mg
Sodium	Less than	2,400mg	2,400mg
Total Carbohydrate		300g	375g
Dietary Fiber		25g	30g

Calories per gram:
Fat 9 • Carbohydrate 4 • Protein 4

TerraSync - Organic Rice Production in China

ACCELERGY CORPORATION

- **Local strain of blue-green algae**
- **Increased Yields of 15+% Per Acre**
- **High Quality Organic Rice for Premium Market**
- **Reduced CO2 emissions from conventional N fertilizer production and reduced NO3 emission from field**

Algae Based Platform for the Beneficial Re-use of Industrial Carbon Emissions

Michael Hayes Wilson
Senior Research Engineer
University of Kentucky Center for Applied Energy Research

1

Introduction:
University of Kentucky Center for Applied Energy Research

Introduction:
University of Kentucky Center for Applied Energy Research

RESEARCH AREAS

- Biofuels and Environmental Catalysis
  - Catalytic Converters, Catalytic Upgrading for Fuels, Algae based CO2 utilization.
- Carbon Materials
  - Carbon Fibers, Nanotubes
- Clean Fuels and Chemicals
  - Fischer-Tropes, Indirect Coal Liquefaction
- Electrochemical Power Sources
  - Low cost capacitors, batteries, and hybrids
- Environmental and Coal Technologies
  - Post Combustion Byproduct Beneficiation, Advanced Concretes
- Environmental Remediation and Restoration Group
  - Paducah Gaseous Diffusion Plant
- Power Generation and Utility Fuels
  - Chemical Carbon Capture, Chemical Looping, Demonstration Projects

3

Project Timeline

- 2008 – UK Approached by Kentucky State Department of Energy Development and Independence to investigate the techno-economic feasibility of algae based CO2 mitigation
- 2011-2012:Initial Demonstration Work started at EKPC's Dale Station
- **2012-Present: Demonstration Project at Duke Energy's East Bend Station**
- 2011-Present: Part of US-China Clean Energy Research Center (CERC)
- 8/2015: NETL Biological CO_2 Utilization Award - $1,257,415

Research Focus

Power plant integration, PBR design, low cost/low energy dewatering, utilization studies, techno economic modeling

Utilization Studies

Anaerobic digestion, lipid extraction, catalytic upgrading, bio polymers, HTL, Pyrolysis, Aquaculture, Aquaponics, etc.

4

Current Research Focus

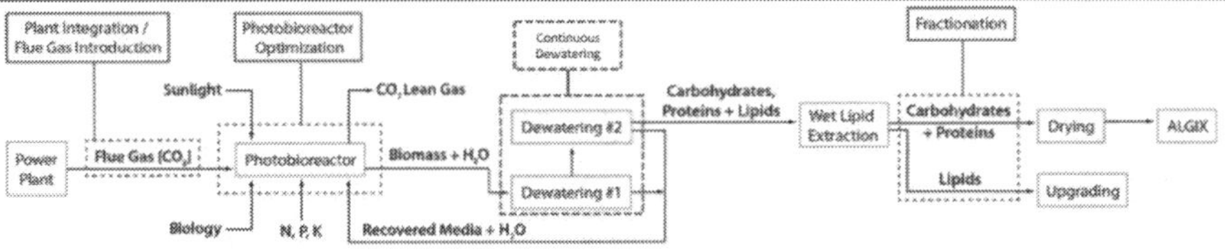

Flow Chart of Process With Research Focus Areas Highlighted in Blue

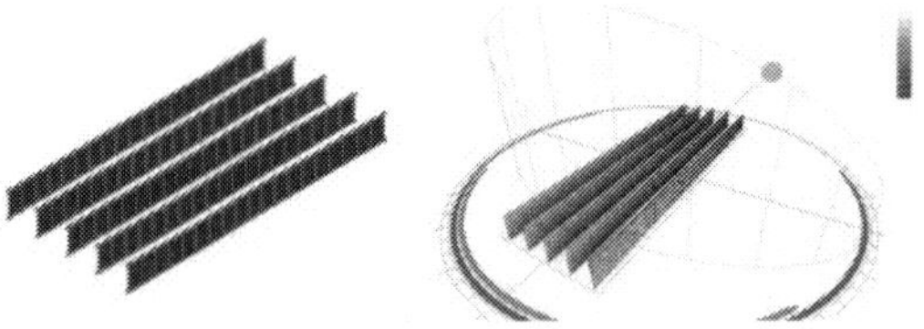

Solar modeling for optimum photobioreactor spacing

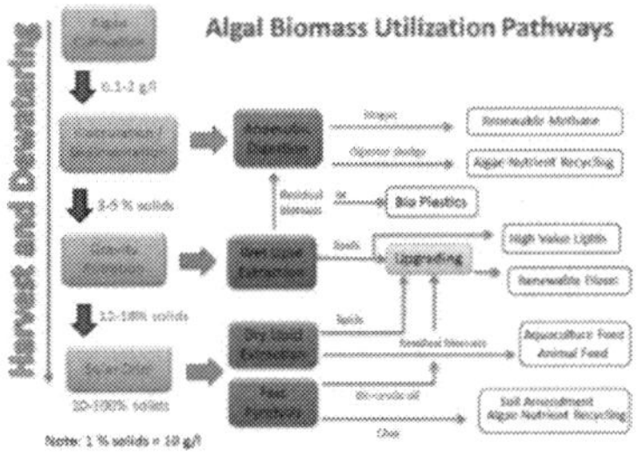

Low-Cost/Low-Energy Dewatering Process Leverages Existing Technologies to Deliver Biomass to a Variety of Utilization Pathways

5

Field Demonstration of Algae Technology

- Pilot algae facility operated at Duke Energy's East Bend Station in 2012/13, incorporating 1st generation U. KY photobioreactor (see top right)
- East Bend is a scrubbed unit with ESP, FGD, and SCR emission control
- Areal productivity of routinely ≥ 30 $g/m^2/day$ (summer) and ≥ 10 $g/m^2/day$ (winter) demonstrated at significant scale (18,000 L)
- New "cyclic flow" photobioreactor deployed in 2014 (top left, bottom) – promises lower cost, higher productivity, and more robust operation

6

System Productivity

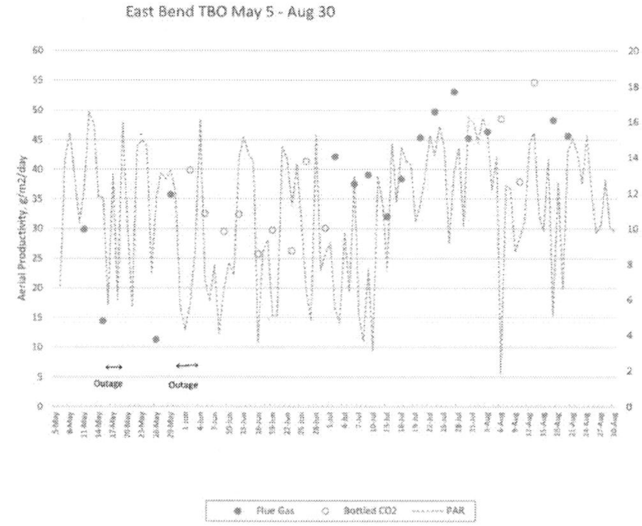

80% (240 Gallons) harvested 2 x week

Clarified water returned to system via UV sterilizer.

Productivity calculated based upon measured harvest weights.

T, dO_2, PAR, pH data measured and stored automatically.

7

Typical Data Set

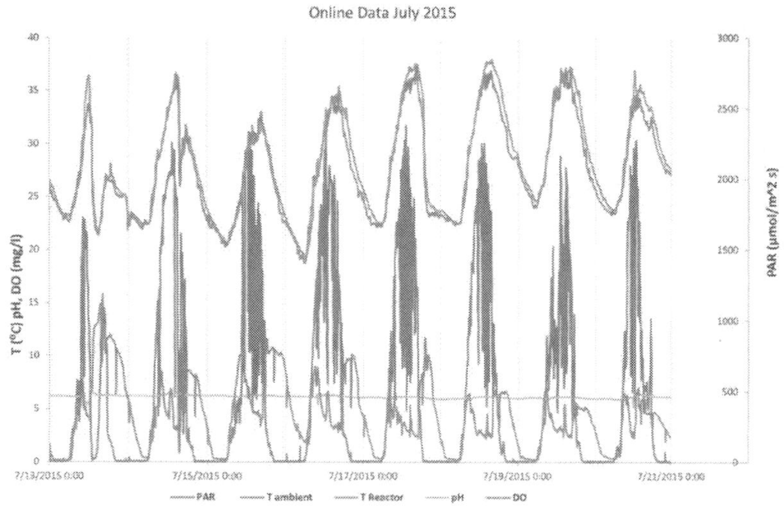

8

Mass Balance Set Up

Temperature, dissolved O_2, and pH measured in reactor

MRU Flue Gas analyzers in inlet and outlet streams measure T, CO_2, O_2, NOx, SOx, CO, and CH_4.

Data measured and stored automatically.

Mass Balance Data

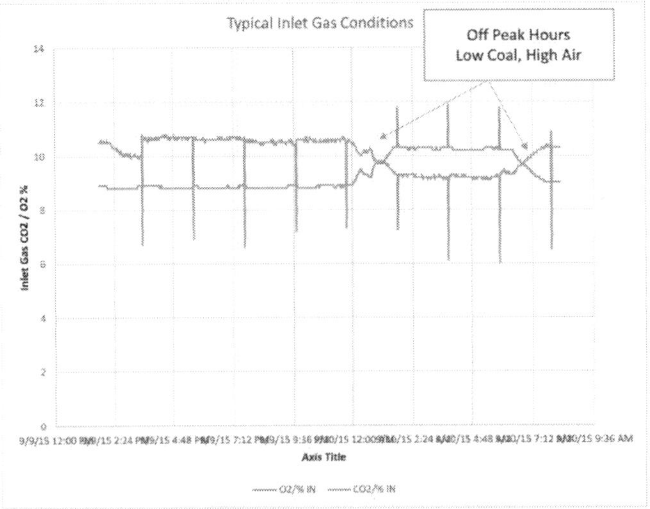

Fairly consistent and steady concentration

As plant ramps down overnight to match lower demand, CO_2 concentration decreases while O_2 increases.

Mass Balance Data 2

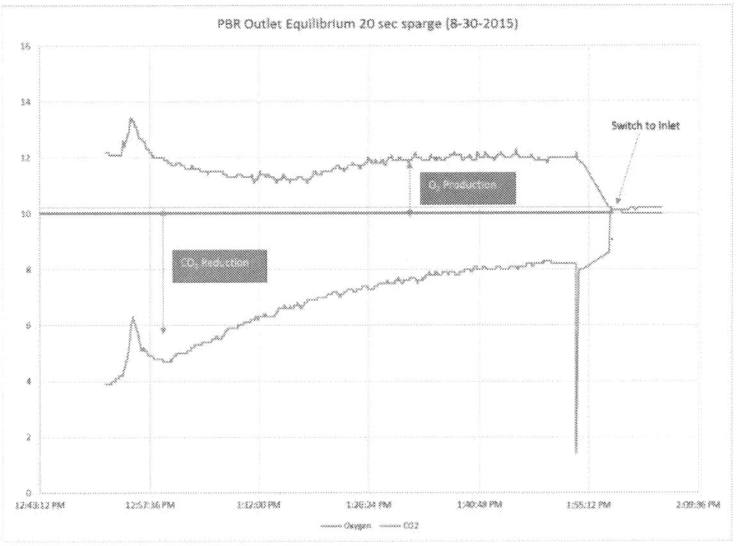

CO_2 reaches equilibrium below inlet concentration.

Indicates CO_2 conversion to O_2 via photosynthesis.

Highlights opportunity to optimize CO_2 conversion

Mass Balance Data

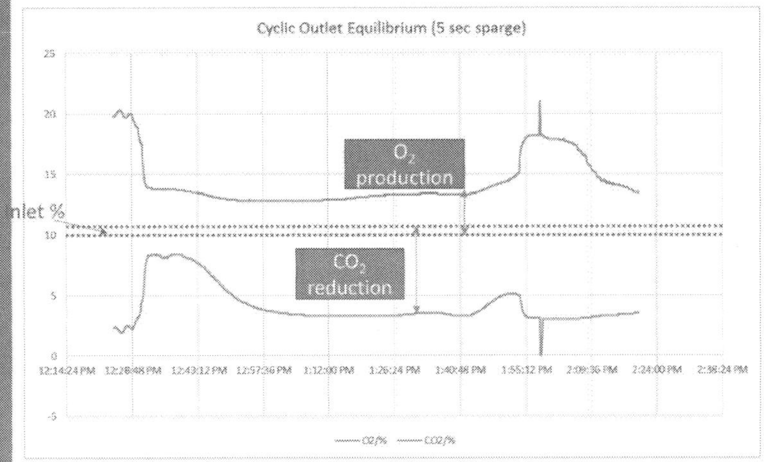

CO_2 reaches equilibrium at lower concentration.

Indicates CO_2 conversion to O_2 via photosynthesis.

Highlights opportunity to optimize CO_2 conversion. Targeting 75%

CO2

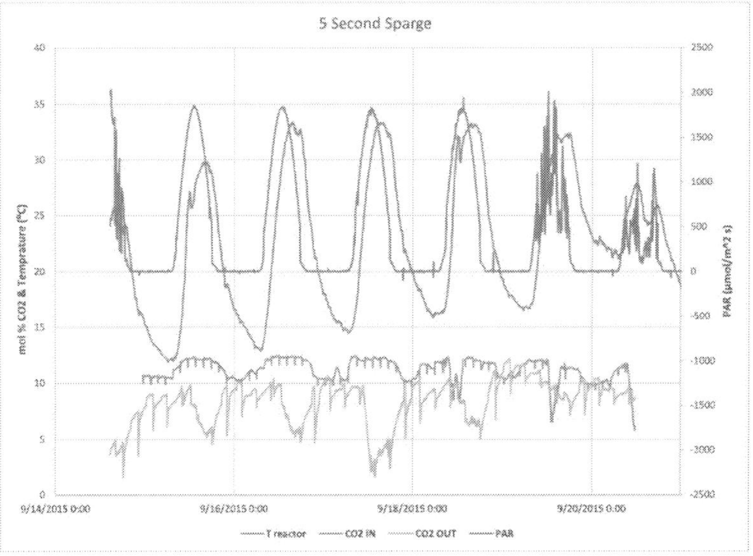

CO_2 uptake a function of sunlight

Indicates CO_2 conversion to O_2 via photosynthesis.

Limited night loss via respiration observed.

13

Oxygen

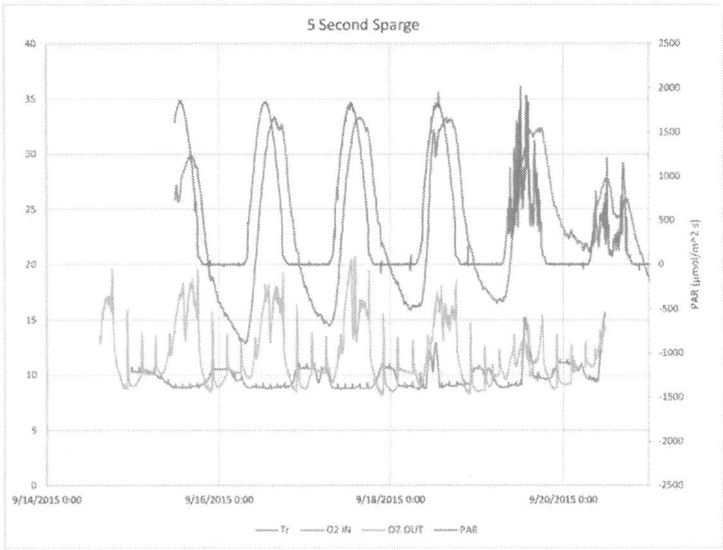

O_2 production a function of available sunlight

Indicates CO_2 conversion to O_2 via photosynthesis.

Lower partial pressure of O2 in flue gas beneficial to algae growth.

14

SOx in vs out

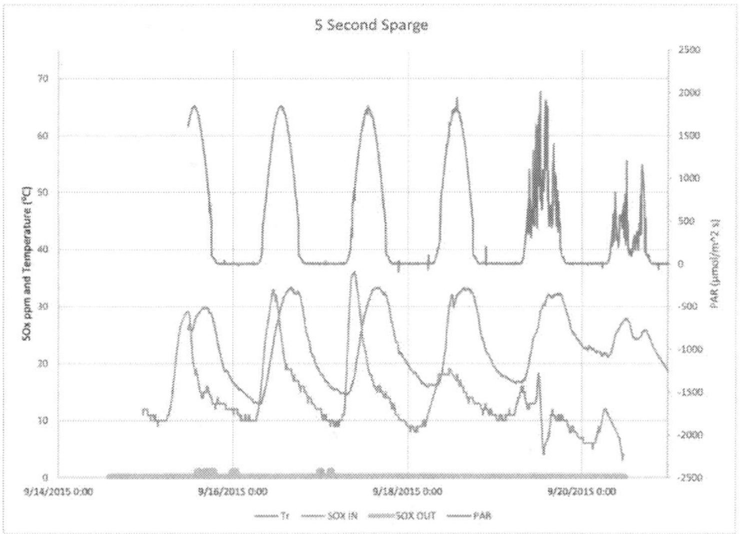

SO_x in inlet stream a function of temperature.

High uptake efficiency with limited effect on system pH

Ion Chromotagraphy used to monitor accumulation of sulfates in media.

15

NOx in vs out

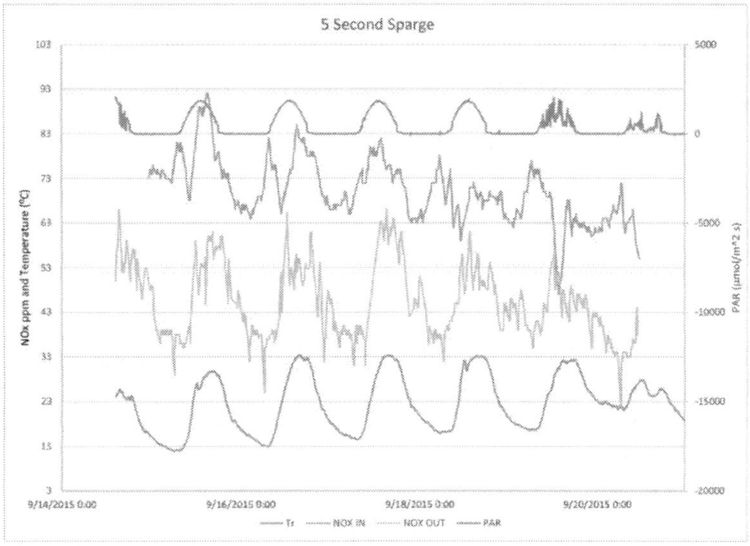

NO_x in inlet stream a function of temperature.

Opportunity to use as Nitrogen source for algae production.

Ion Chromotagraphy used to monitor accumulation of nitrates in media.

16

UK UNIVERSITY OF KENTUCKY

Harvesting/Dewatering Flowsheet

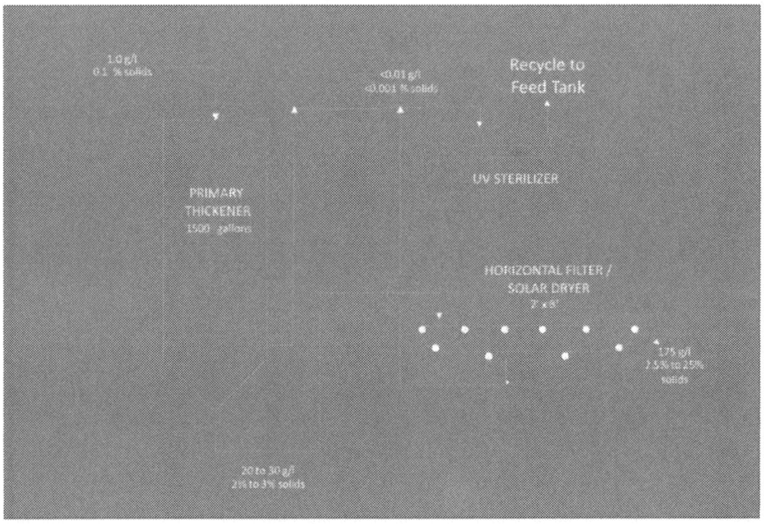

17

UK UNIVERSITY OF KENTUCKY

Prototype Gravity Filter/Solar Dryer

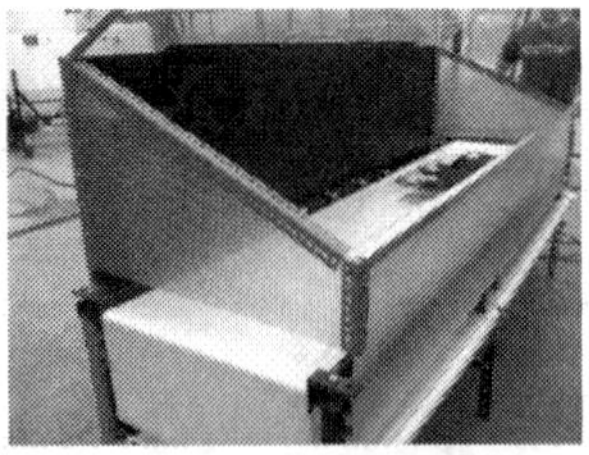

- Multifilament nylon fiber weave allows for cake formation
- Allows separation and recycling of all free water containing unused nutrients
- Short vacuum pulse can improve throughput
- Can produce 10-25% solids for utilization
- Solar oven can reach 60 degrees C in summer
- Energy efficient based off of existing tech

18

Algal Biomass Utilization

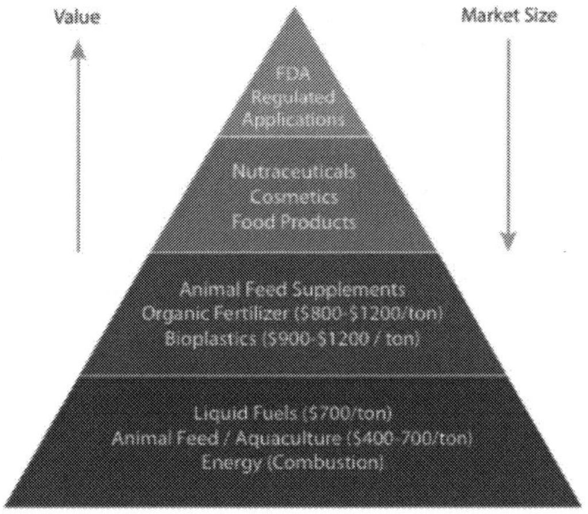

19

Algal Biomass Utilization Pathways

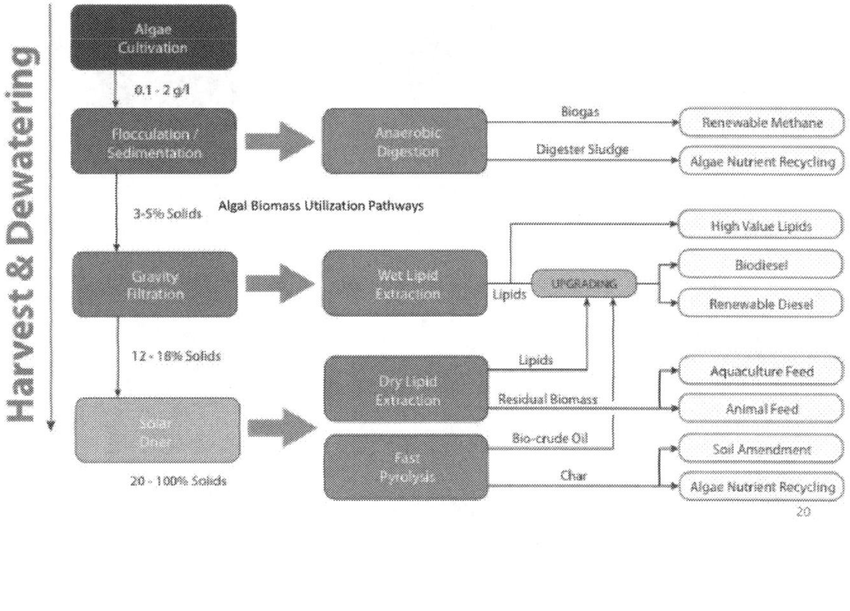

20

Current UK Concept for CO_2 Capture / Bioproduct Production

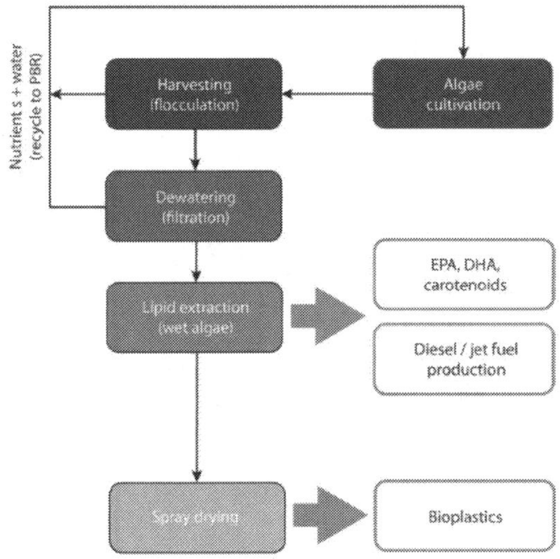

21

From Flue Gas to Fungible Fuels

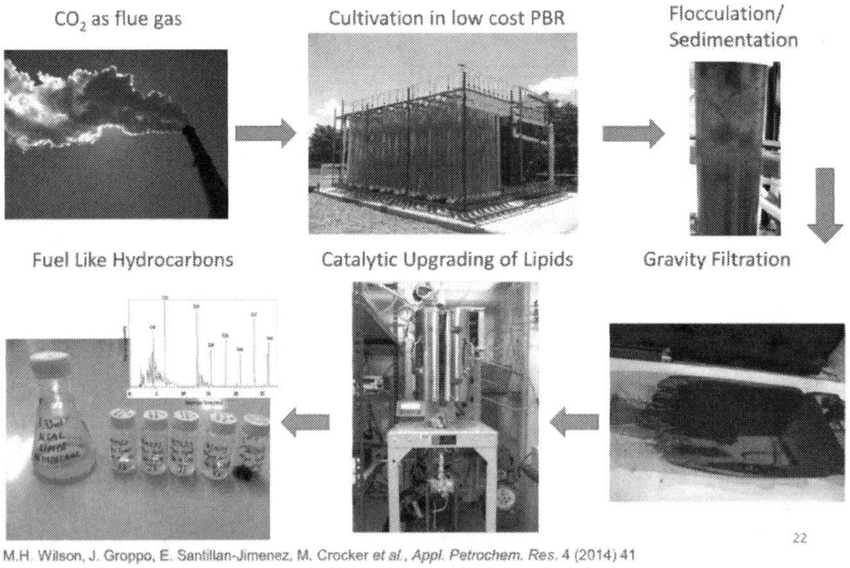

M.H. Wilson, J. Groppo, E. Santillan-Jimenez, M. Crocker *et al.*, *Appl. Petrochem. Res.* 4 (2014) 41

22

Techno-economic Analysis

Preliminary TEA highlights effect of amortization period (10 vs. 30 years), capital cost reduction and algae growth rate
***Does NOT account for utilization**

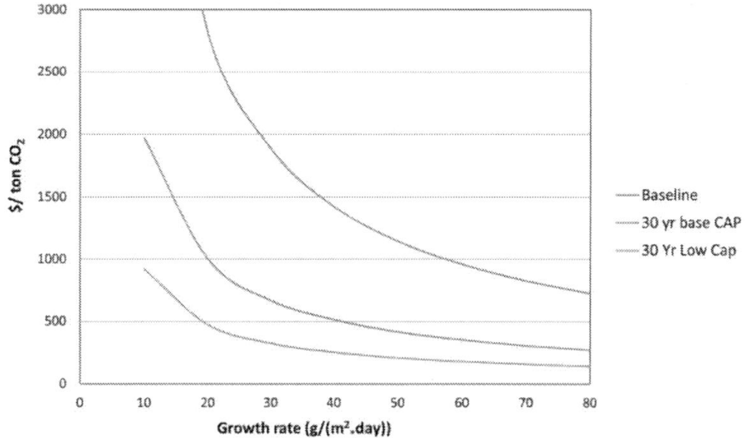

23

SYSTEM SIZING

- 1 MW–hr of coal based power generation produces 1 ton of CO_2/hr
- Elemental composition of algae shows that 1 ton of algae will consume ca. 2 tons of CO_2

Areal Productivity	Land Required	Production
(g algae/m²/day)	*(Acres)*	*(Tons algae/acre/day)*
10	269	0.04
20	134	0.09
30	89.5	0.13
40	67.3	0.18
50	53.8	0.22
60	44.8	0.27
100	27	0.45

- The dramatic effect of areal productivity on required land drove the team toward the development of a low-cost photobioreactor

24

UK UNIVERSITY OF KENTUCKY CAER Center for Applied Energy Research

Proposed Layout of a 3 Acre Cyclic Flow Photobioreactor System

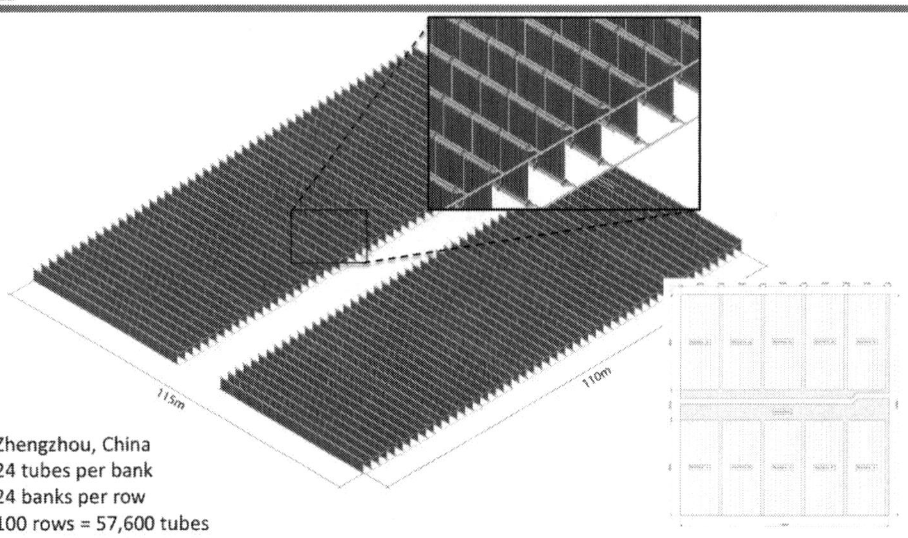

- Zhengzhou, China
- 24 tubes per bank
- 24 banks per row
- 100 rows = 57,600 tubes
≈ **1,000,000L**

25

UK UNIVERSITY OF KENTUCKY CAER Center for Applied Energy Research

Student Engagement

Student involvement is an important focus of the project, leveraging creative problem solving and enthusiasm to solve real world research problems while developing the scientists and engineers of tomorrow.

- Student Employment / Experiential Learning
  - Undergraduate Engineers, Scientists, and Architects contribute to day to day research activities
- Senior Design Projects
  - CAER researchers act as customer/advisor to provide real world projects for student teams in Mechanical, Electrical, and Chemical Engineering
  - Students get exposed to research and researchers at CAER get prototype equipment and/or models to aid research
- College of Design Studios
  - Architecture and/or Interior Design students work on developing forward thinking designs, large scale instillations, next generation research facilities, and creative applications of current research
- Graduate Students / Postdocs

26

Ongoing Collaborations

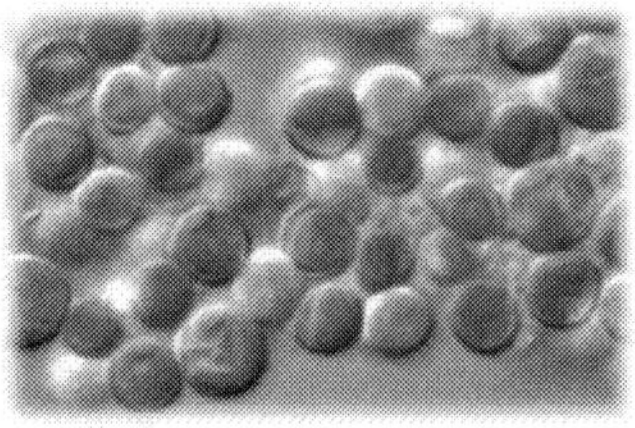

Cyanidium merolae

PBR installed at University of Delaware

Future Work

- Reduce Cost / Increase Productivity
  - Optimized photobioreactor design and operation
  - Batch→ continuous dewatering process
- Conceptual Design of System Integrated with Power Plant
  - Mass and Energy Balances
  - Power plant integration (heat, flue gas, etc.)
  - Life Cycle Assessment
  - Techno Economic Analysis
- Biomass Utilization / Valorization
  - Investigate alternative / multi-product utilization pathways
  - Fate of NOx, SOx, heavy metals
  - Focus on fuels/chemicals and biopoloymers
- Systems Biology
  - Power plant outage mitigation system
  - Flue gas constituents on biomass composition
  - Abiotic Parameter Optimization

Acknowledgments

KY Department of Energy Development and Independence

Duke Energy

Department of Energy: U.S.-China Clean Energy Research Center

National Energy & Technology Lab

The UK algae team:

- Dr. Mark Crocker
- Dr. Jack Groppo
- Stephanie Kesner
- Thomas Grubbs
- Robert Pace
- Dr. Czarena Crofcheck
- Aubrey Shea
- Daniel Mohler
- Dr. Eduardo Santillan-Jimenez
- Tonya Morgan

And..... ca. 30 students

Questions?

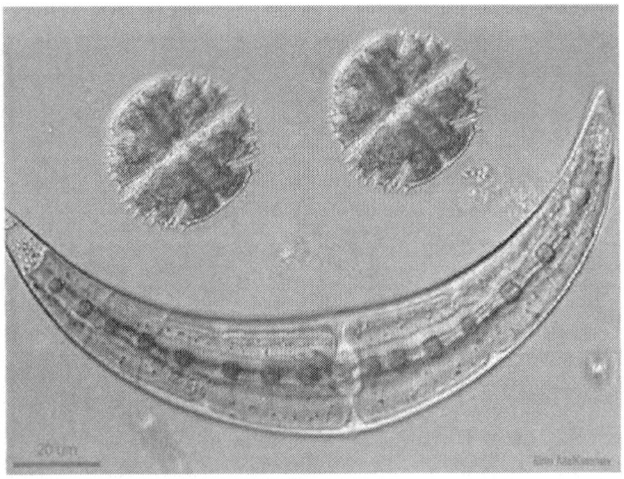

30

UK UNIVERSITY OF KENTUCKY CAER

nversion of Algal Lipids to Fuel-Like Hydrocarbons

Catalyst	Conversion (%)	Selectivity to C10-C17 (%)	Selectivity to C17 (%)
Ni-Al LDH	95	73	7

300 °C, 580 psi H_2, fixed bed reactor

Gas chromatogram of product

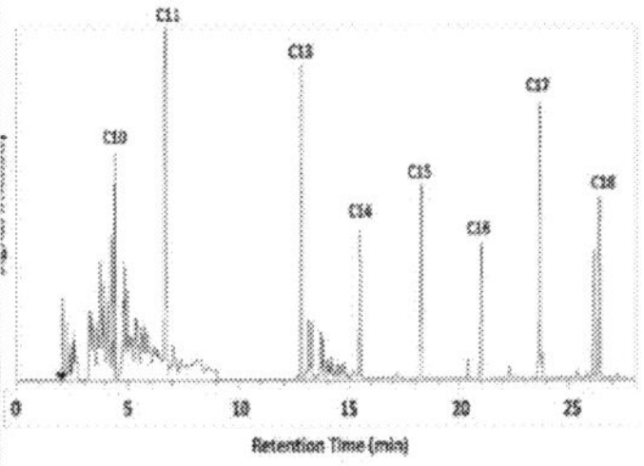

Boiling point distribution plot

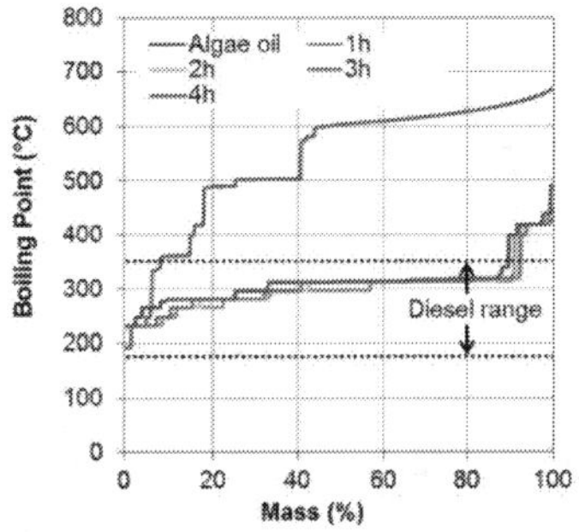

T. Morgan, D. Grubb, E. Santillan-Jimenez, M. Crocker, *Top. Catal.* 53 (2010) 820;
B. Peng, X. Yuan, C. Zhao, J.A. Lercher, *J. Am. Chem. Soc.* 134 (2012) 9400;
T. Morgan, E. Santillan-Jimenez, A.E. Harman-Ware, M. Crocker, *Chem. Eng. J.* 189-190 (2012) 346;
E. Santillan-Jimenez, T. Morgan, J. Lacny, S. Mohapatra, M. Crocker, *Fuel* 103 (2013) 1010.

31

UK UNIVERSITY OF KENTUCKY CAER

ental Analysis of Algae Harvested at East Bend

	6/21/2013	6/28/2013	7/2/2013	7/8/2013	7/22/2013	7/29/2013	8/6/2013
% Ash	29.88	22.63	19.52	10.97	14.01	14.52	23.98
% Moisture	5.66	4.49	5.11	6.14	5.14	4.05	4.51
% Carbon	35.16	40.36	42.15	47	45.07	46.61	40.93
% Hydrogen	5.74	6.15	6.57	7.28	6.77	6.96	6.03
% Nitrogen	5.64	6.52	6.81	7.61	7.05	7.51	6.24
% Total Sulfur	0.42	0.48	0.53	0.59	0.6	0.6	0.44
% Oxygen	23.16	23.86	24.42	26.55	26.5	23.8	22.38
% Volatile Matter	60.16	64.05	65.86	71.63	70.57	70.22	63.27
% Fixed Carbon	4.3	8.83	9.51	11.25	10.28	11.21	8.24
ppm As	<0.1	<0.1	<0.1	<0.1	<0.1	<0.1	<0.1
ppm Se	<0.1	<0.1	<0.1	<0.1	<0.1	<0.1	<0.1
ppm Cd	<0.1	<0.1	<0.1	<0.1	<0.1	<0.1	<0.1
ppm Hg	<0.1	<0.1	<0.1	<0.1	<0.1	<0.1	<0.1

32

Illinois Industrial Carbon Capture & Storage Project

Eliminating CO_2 Emissions from the Production of Bio Fuels - A 'Green' Carbon Process

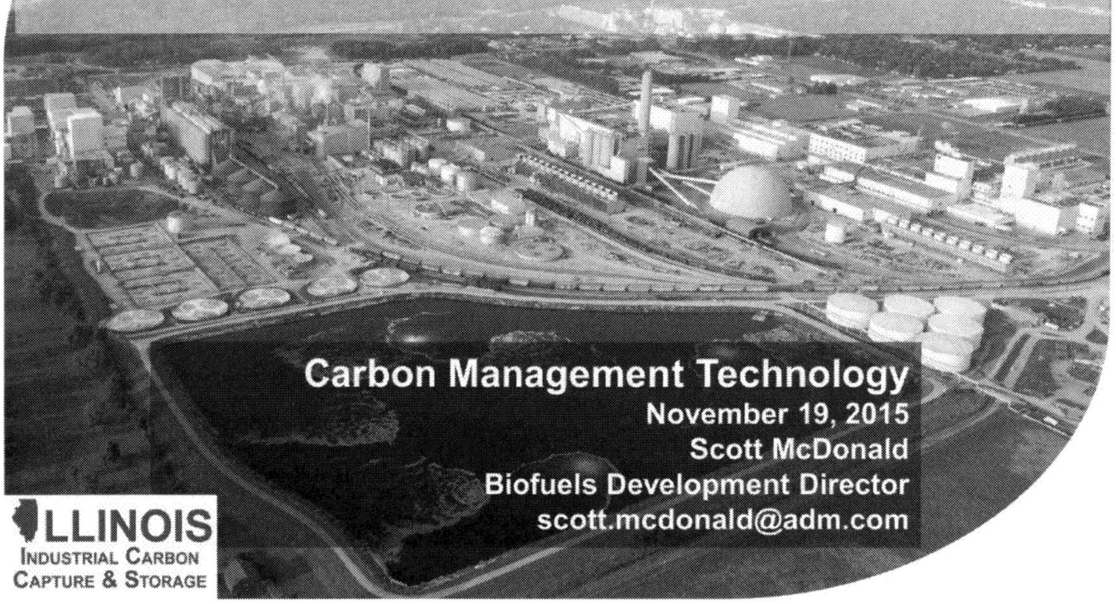

Carbon Management Technology
November 19, 2015

Scott McDonald
Biofuels Development Director
scott.mcdonald@adm.com

ILLINOIS
INDUSTRIAL CARBON
CAPTURE & STORAGE

ADM

Illinois Basin Decatur Project (IBDP)

ILLINOIS
INDUSTRIAL CARBON
CAPTURE & STORAGE

Program Objective

Large scale geologic test to inject 1.0 million tons of CO_2 over a three year period (1,000 MT/day).

Project Team Members

ADM | Schlumberger Carbon Services | NETL

Knowledge Base

- Site Geological Characterization
- Risk Assessment & Reservoir Modeling
- Engineering Design & MVA

Nov 2014-Completed goal of injecting and storing 1,000,000 tons of CO_2.

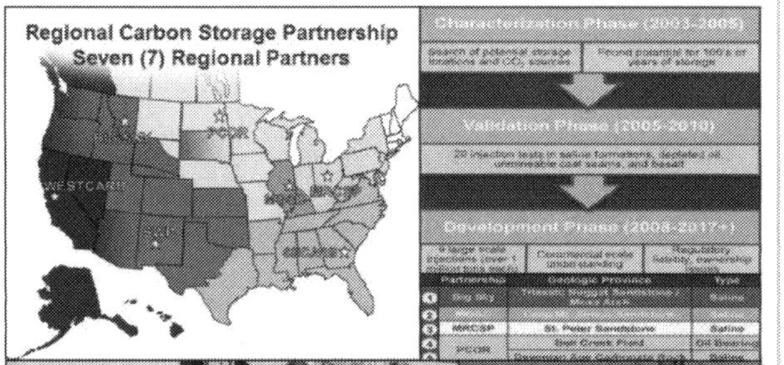

ARCHER DANIELS MIDLAND COMPANY

Illinois Industrial CCS Project (IL-ICCS)

ILLINOIS
INDUSTRIAL CARBON
CAPTURE & STORAGE

Program Objectives

- Target & Demonstrate Advanced CCS Technologies at Industrial Scale Facilities
- Inject and Store One Million Tons of CO_2 Annually (3,000 tons/day)

Project Team Members

ADM · Schlumberger Carbon Services · Richland · NETL

Knowledge Base

- Site Geological Characterization
- Risk Assessment & Reservoir Modeling
- Engineering Design & MVA
- Education and Public Outreach

Study the interaction between the CO_2 plumes from two injection wells within the same formation.

Intelligent Monitoring System (IMS)

ILLINOIS
INDUSTRIAL CARBON
CAPTURE & STORAGE

Program Objectives

- Develop and validate software tools that advance CCS-specific IMS by enabling access, integration and analysis of real-time surface and subsurface data for decision-making and automation of process
- Demonstrate integration of system components to validate feasibility of real-world application to CCS.

Project Team Members

ADM · BERKELEY LAB · SILIXA · NETL · USGS science for a changing world · Richland · Schlumberger Carbon Services

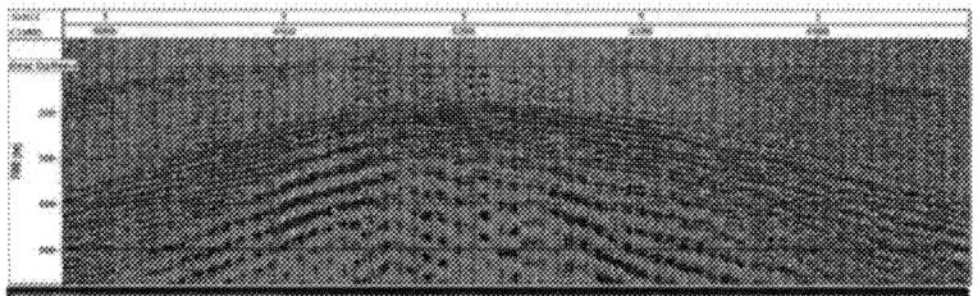

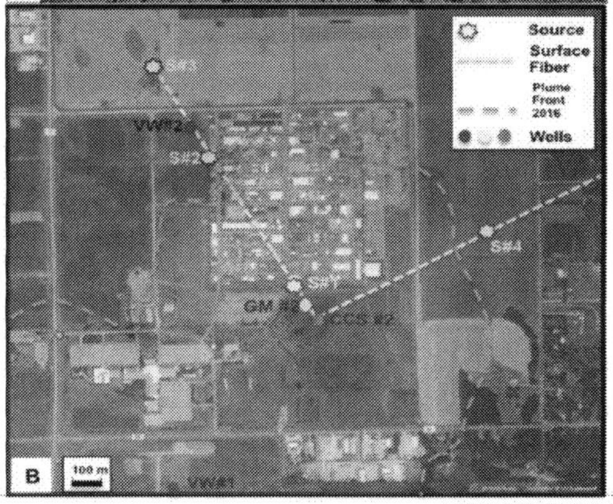

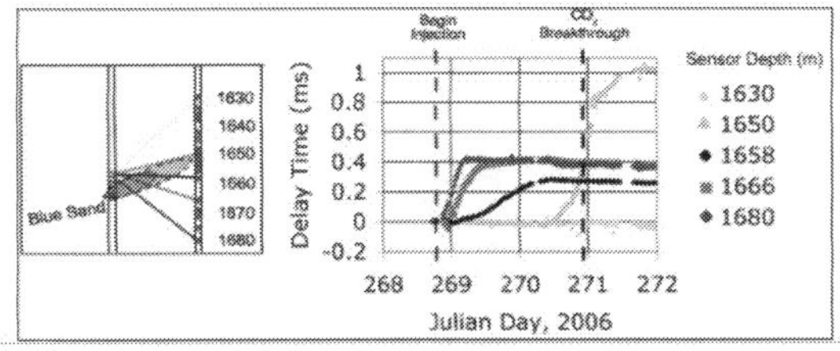

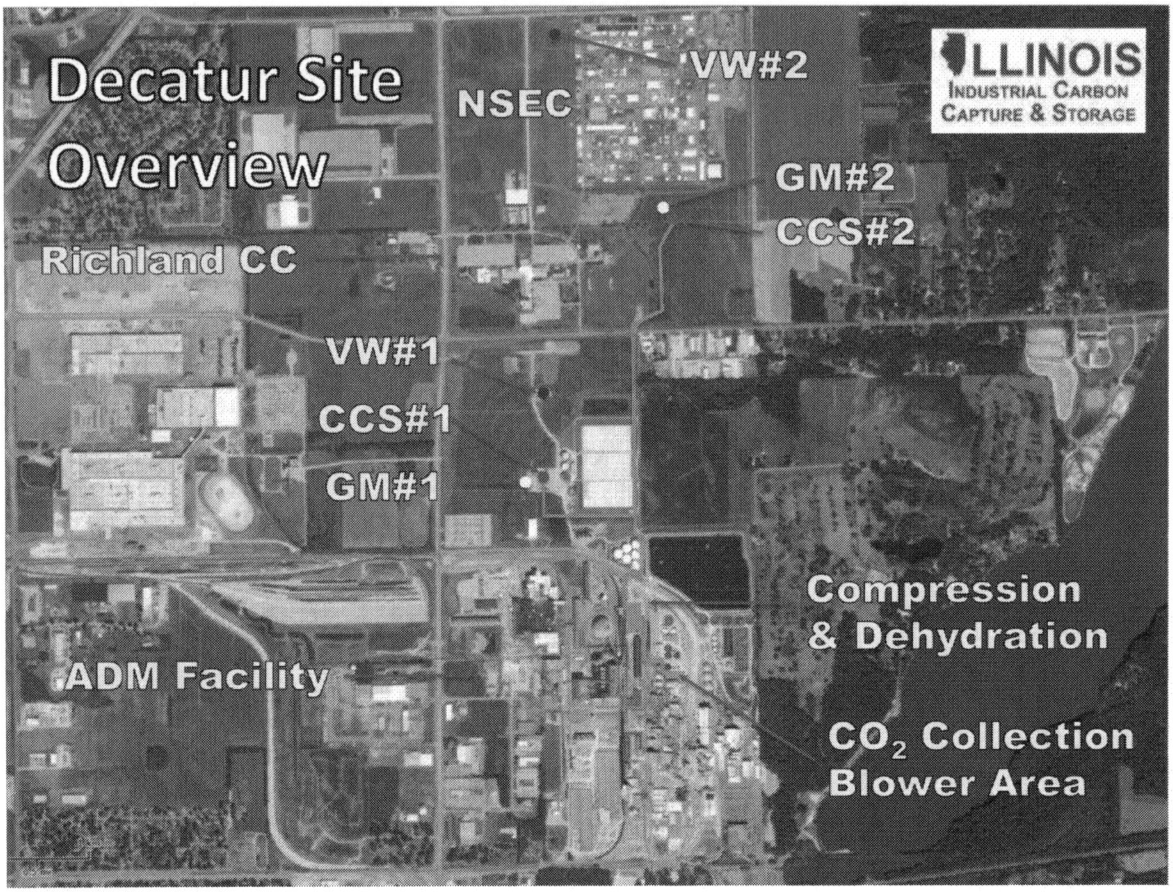
Decatur Site Overview
ILLINOIS
Industrial Carbon Capture & Storage
NSEC
VW#2
GM#2
CCS#2
Richland CC
VW#1
CCS#1
GM#1
Compression & Dehydration
ADM Facility
CO_2 Collection Blower Area

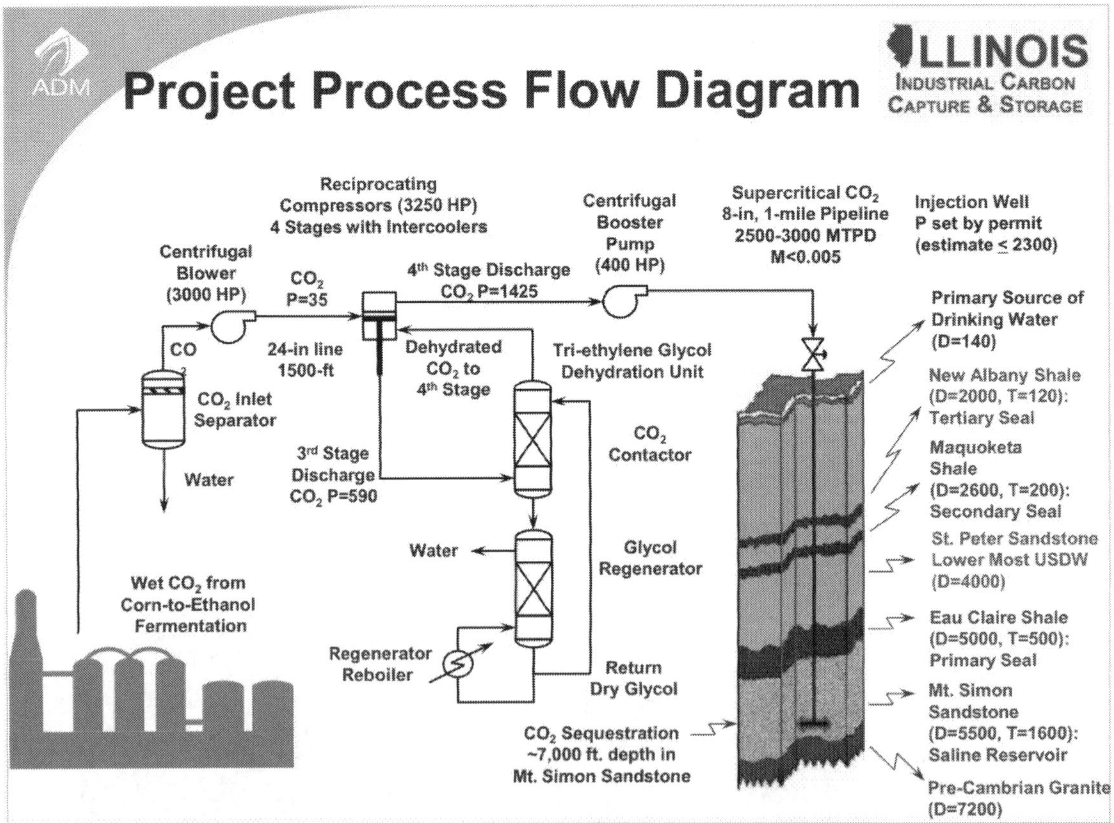
ADM
Project Process Flow Diagram
ILLINOIS
Industrial Carbon Capture & Storage
Reciprocating Compressors (3250 HP) 4 Stages with Intercoolers
Centrifugal Booster Pump (400 HP)
Supercritical CO_2 8-in, 1-mile Pipeline 2500-3000 MTPD M<0.005
Injection Well P set by permit (estimate ≤ 2300)
Centrifugal Blower (3000 HP)
CO_2 P=35
4th Stage Discharge CO_2 P=1425
Primary Source of Drinking Water (D=140)
CO
24-in line 1500-ft
Dehydrated CO_2 to 4th Stage
Tri-ethylene Glycol Dehydration Unit
CO_2 Inlet Separator
New Albany Shale (D=2000, T=120): Tertiary Seal
CO_2 Contactor
Maquoketa Shale (D=2600, T=200): Secondary Seal
Water
3rd Stage Discharge CO_2 P=590
St. Peter Sandstone Lower Most USDW (D=4000)
Water
Glycol Regenerator
Wet CO_2 from Corn-to-Ethanol Fermentation
Eau Claire Shale (D=5000, T=500): Primary Seal
Regenerator Reboiler
Return Dry Glycol
Mt. Simon Sandstone (D=5500, T=1600): Saline Reservoir
CO_2 Sequestration ~7,000 ft. depth in Mt. Simon Sandstone
Pre-Cambrian Granite (D=7200)
ARCHER DANIELS MIDLAND COMPANY

ADM

Site Selection

Regional Geologic Characterization

ILLINOIS
INDUSTRIAL CARBON
CAPTURE & STORAGE

- Cratonic basin
- 60,000 square mile area
- Structurally complex to the south with faulting and seismicity
- ADM Decatur facility is located near the center of this geologic formation
- Estimated CO_2 storage capacity between 27 to 109 billion metric tons

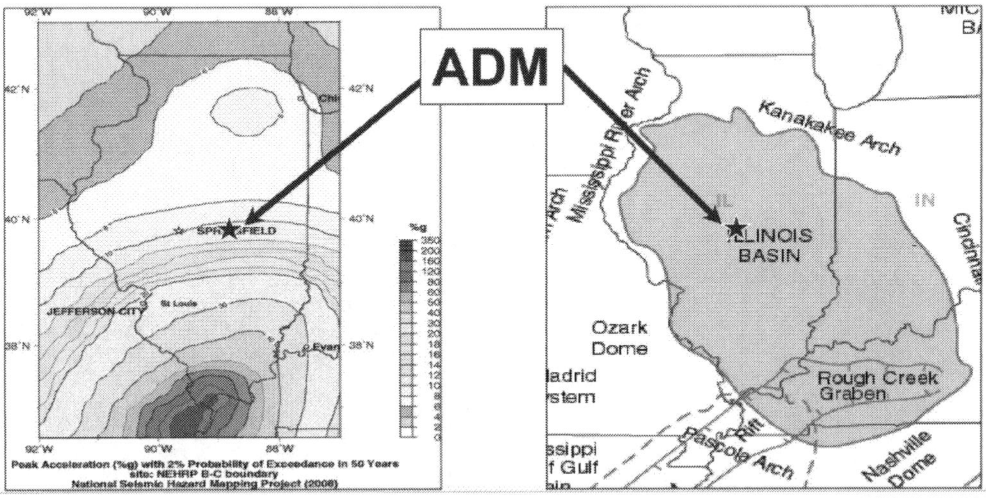

ARCHER DANIELS MIDLAND COMPANY

ADM

Site Characterization

Seismic Acquisition

ILLINOIS
INDUSTRIAL CARBON
CAPTURE & STORAGE

ARCHER DANIELS MIDLAND COMPANY

ADM

Test Well Construction

ILLINOIS
INDUSTRIAL CARBON
CAPTURE & STORAGE

ARCHER DANIELS MIDLAND COMPANY

ADM

Coring and Well Logging

ILLINOIS
INDUSTRIAL CARBON
CAPTURE & STORAGE

ARCHER DANIELS MIDLAND COMPANY

Geophysical Model Development

ILLINOIS INDUSTRIAL CARBON CAPTURE & STORAGE

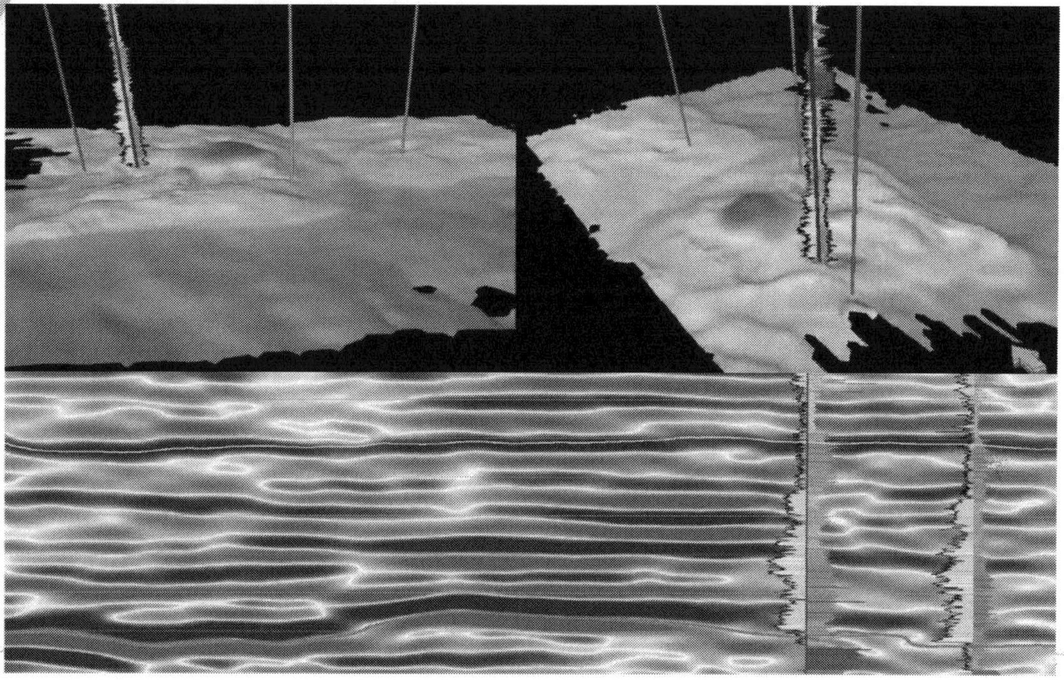

Site Permitting

USEPA: UIC Class VI Permit

1st UIC Class VI Permit Application Reviewed by the USEPA

- UIC Class VI permit application submitted on July 25, 2011.
- US EPA Region V issued a draft permit on April 15, 2014.
- Public hearing conducted on May 21, 2014.
- Public Comment Period concluded May 31, 2014.
- Final permit issued December 28, 2014.

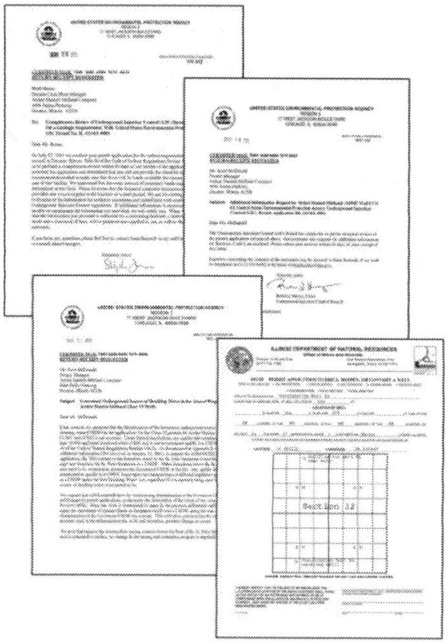

ARCHER DANIELS MIDLAND COMPANY

ADM

Geophysical Model Framework

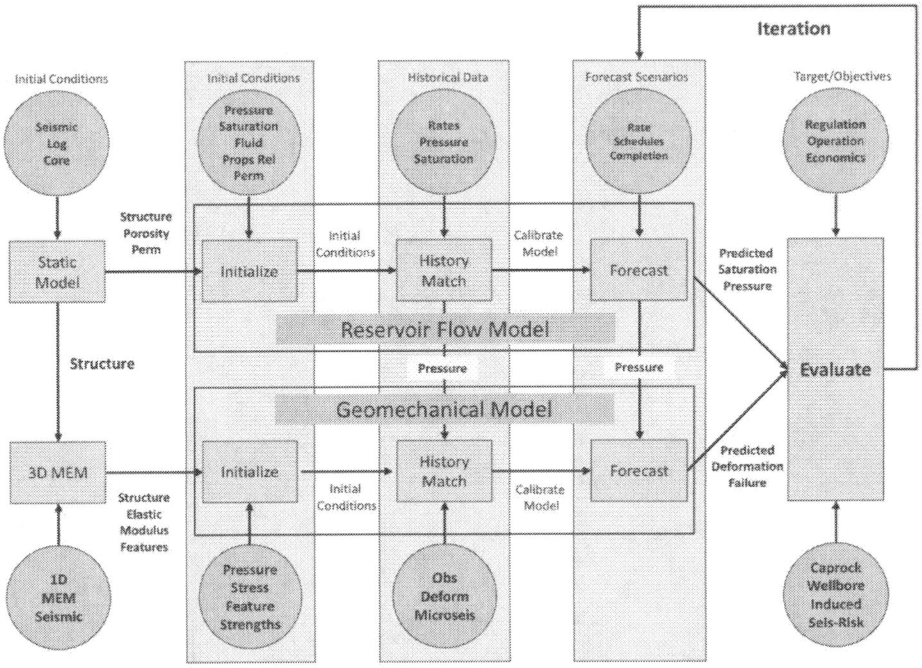

ADM

ILLINOIS
INDUSTRIAL CARBON
CAPTURE & STORAGE

Geophysical Modeling

CO_2 Plume Position - 2012

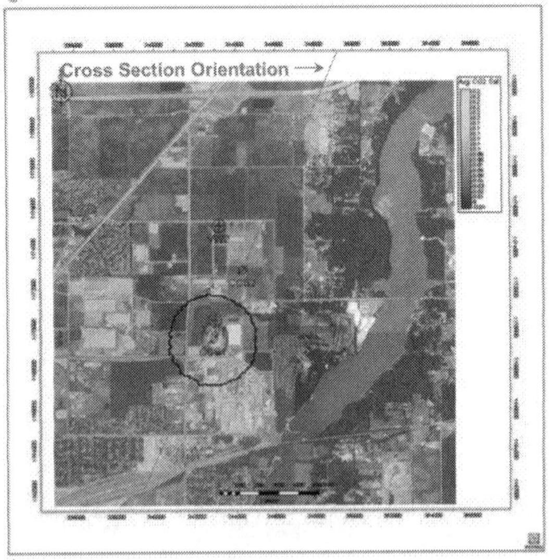

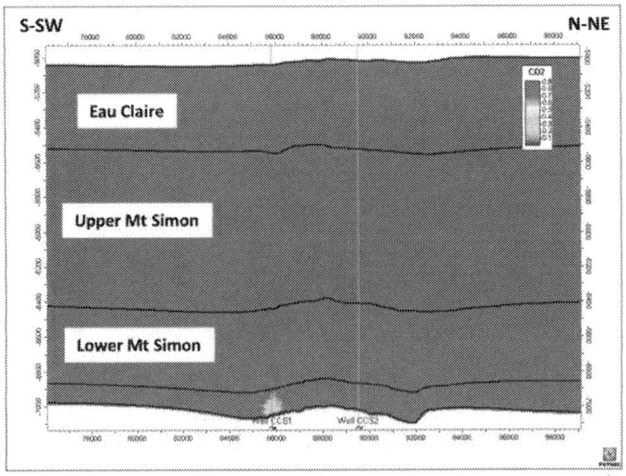

ADM

ILLINOIS
INDUSTRIAL CARBON
CAPTURE & STORAGE

Geophysical Modeling

CO_2 Plume Position - 2013

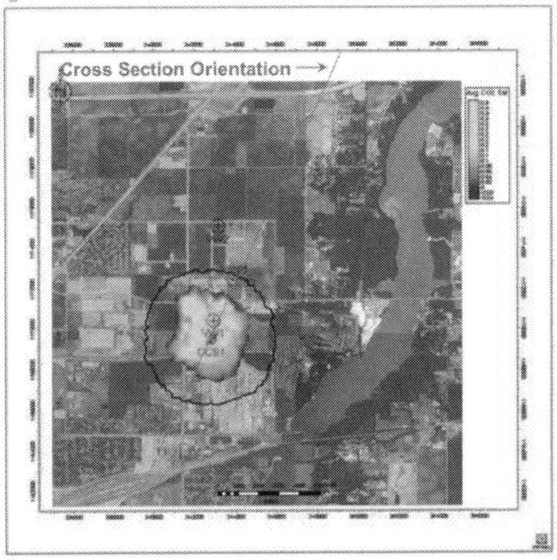

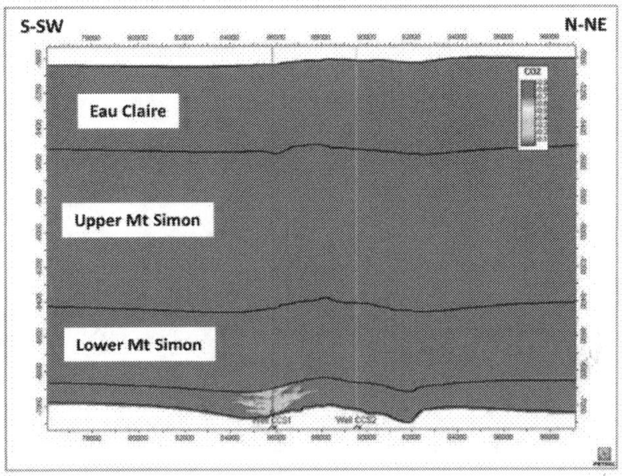

ADM

ILLINOIS
INDUSTRIAL CARBON
CAPTURE & STORAGE

Geophysical Modeling

CO_2 Plume Position - 2014

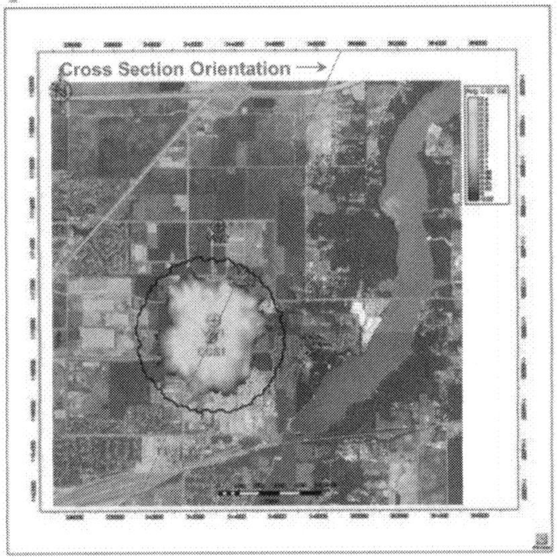

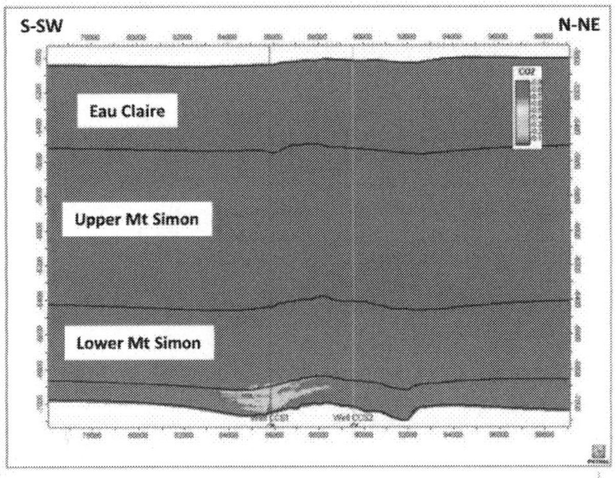

ADM

Geophysical Modeling

CO_2 Plume Position - 2015

ILLINOIS
INDUSTRIAL CARBON
CAPTURE & STORAGE

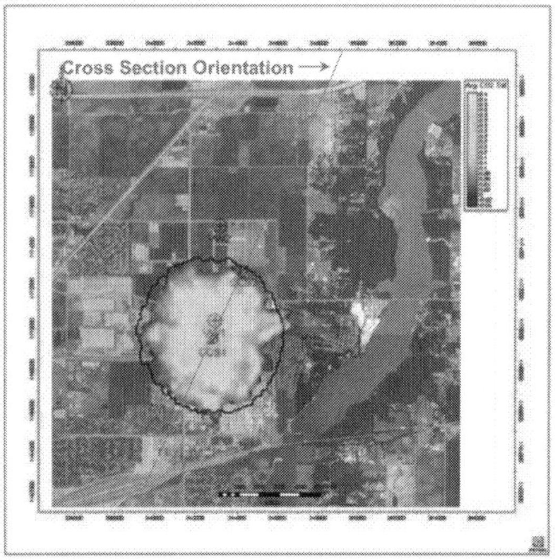

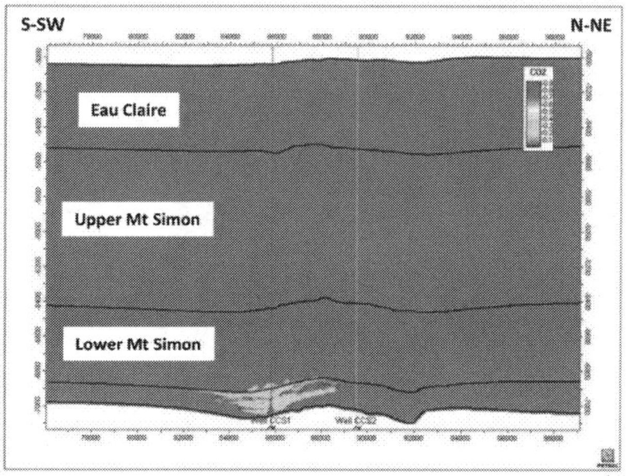

ADM

Geophysical Modeling

CO_2 Plume Position - 2016

ILLINOIS
INDUSTRIAL CARBON
CAPTURE & STORAGE

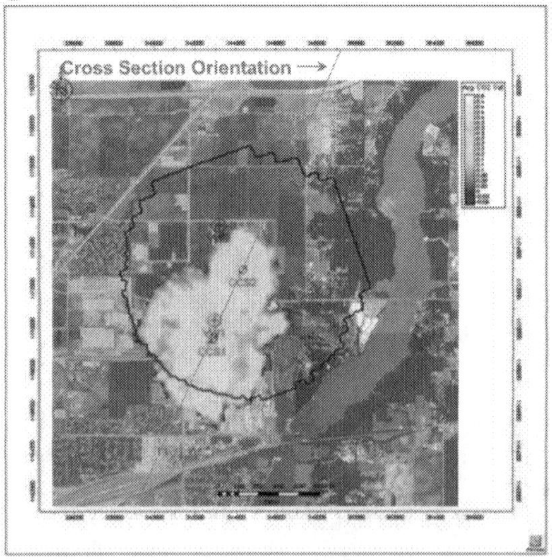

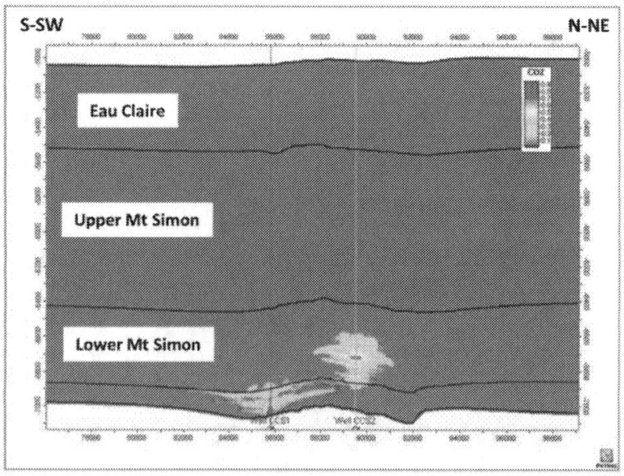

ADM

ILLINOIS
INDUSTRIAL CARBON
CAPTURE & STORAGE

Geophysical Modeling

CO_2 Plume Position - 2017

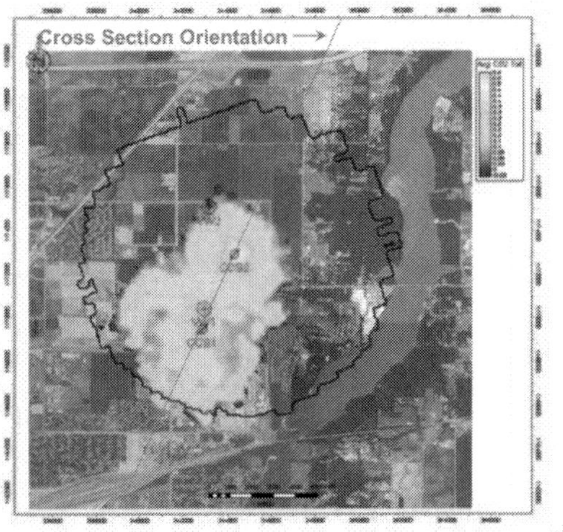

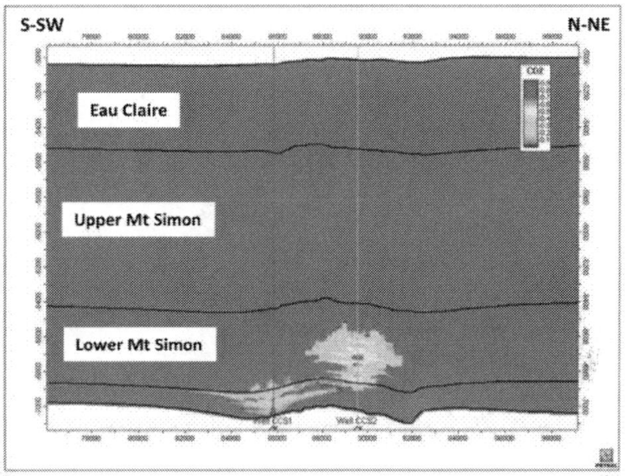

ARCHER DANIELS MIDLAND COMPANY

ADM

ILLINOIS
INDUSTRIAL CARBON
CAPTURE & STORAGE

Geophysical Modeling

CO_2 Plume Position - 2018

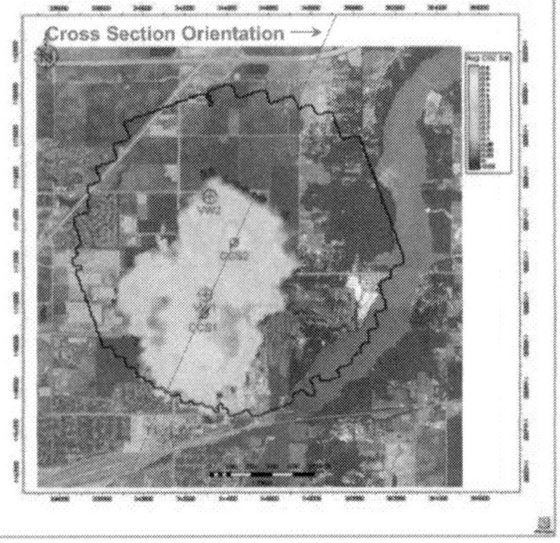

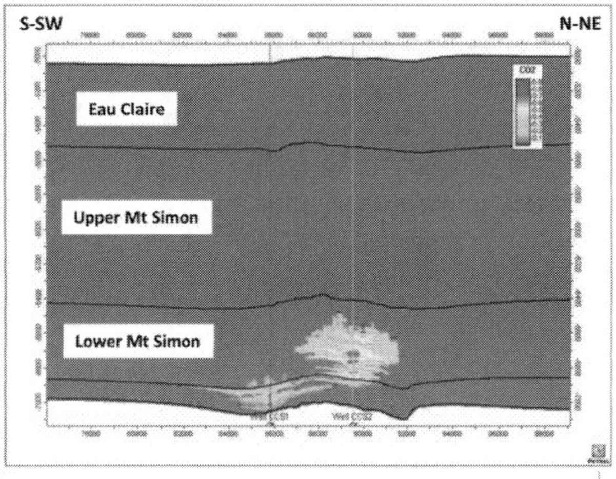

ARCHER DANIELS MIDLAND COMPANY

ADM

Geophysical Modeling
CO_2 Plume Position - 2019

ILLINOIS INDUSTRIAL CARBON CAPTURE & STORAGE

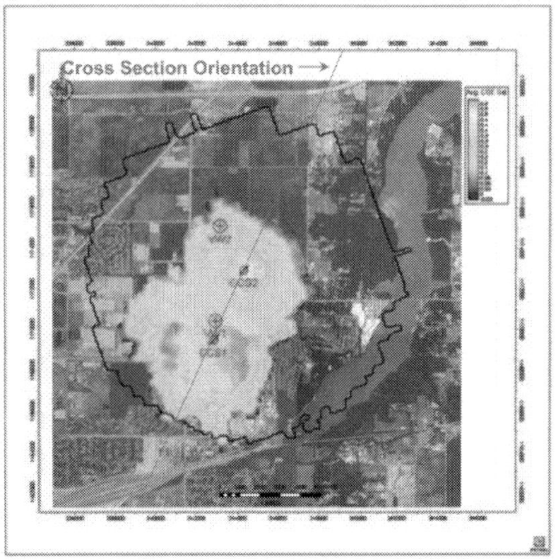

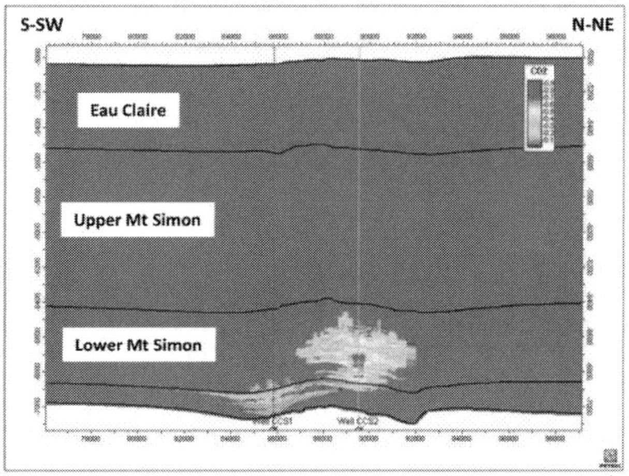

ARCHER DANIELS MIDLAND COMPANY

ADM

Geophysical Modeling
CO_2 Plume Position - 2020

ILLINOIS INDUSTRIAL CARBON CAPTURE & STORAGE

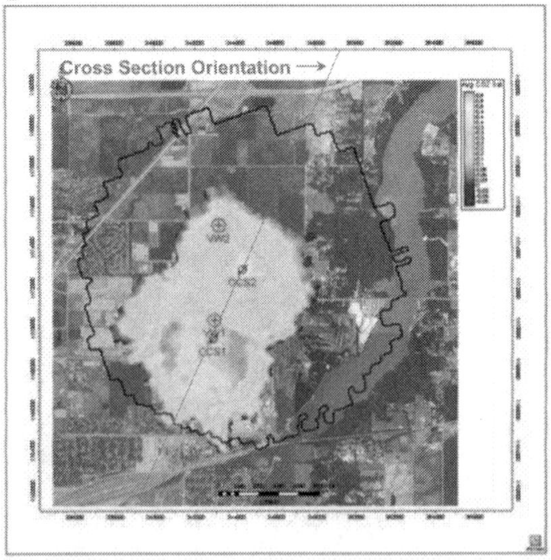

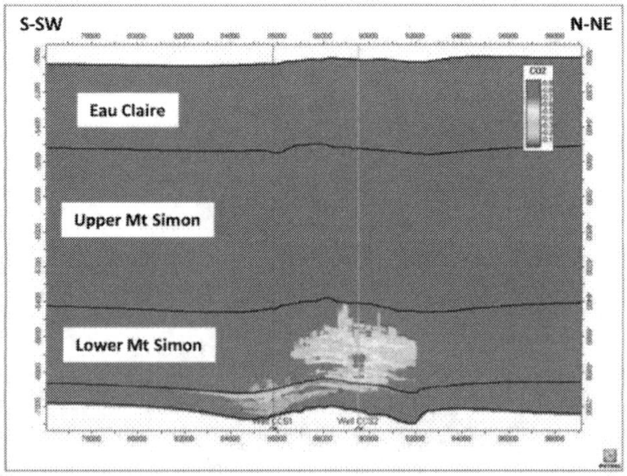

ARCHER DANIELS MIDLAND COMPANY

ADM

ILLINOIS
INDUSTRIAL CARBON CAPTURE & STORAGE

Environmental Monitoring (MVA)
Conceptual Framework

Near Surface
- **Soil and Vadose Zone**
  - Aerial Imagery
  - Soil CO_2 Flux
- **Ground Water**
  - Geochemical Sampling
  - P/T Monitoring

Deep Subsurface
- **Above Seal**
  - Geophysical Surveys
  - Seismic Monitoring
  - P Monitoring
- **Injection Zone**
  - Geophysical Surveys
  - Geochemical sampling
  - P/T Monitoring

ARCHER DANIELS MIDLAND COMPANY

ADM

ILLINOIS
INDUSTRIAL CARBON CAPTURE & STORAGE

Environmental Monitoring
Near Surface Monitoring

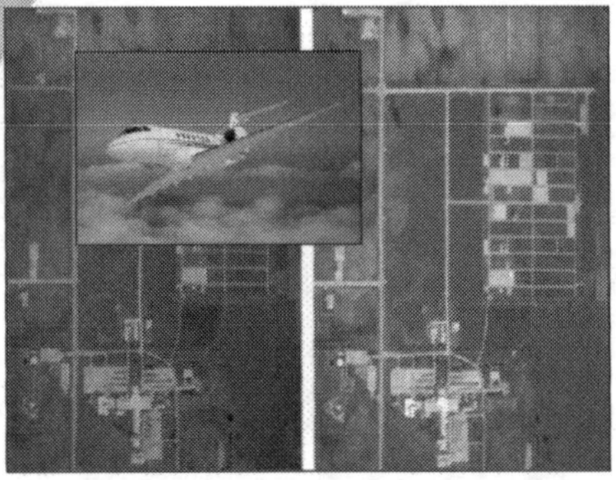

- Near infrared aerial imagery will be used to evaluate plant stress
- Soil resistivity characterized shallow depths for identification of optimum GWM locations
- GWM for baseline conditions and operational surveillance
- Surface soil CO_2 flux monitoring

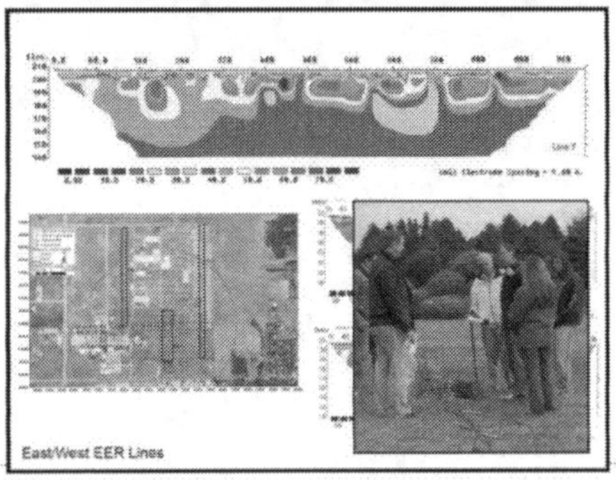

ARCHER DANIELS MIDLAND COMPANY

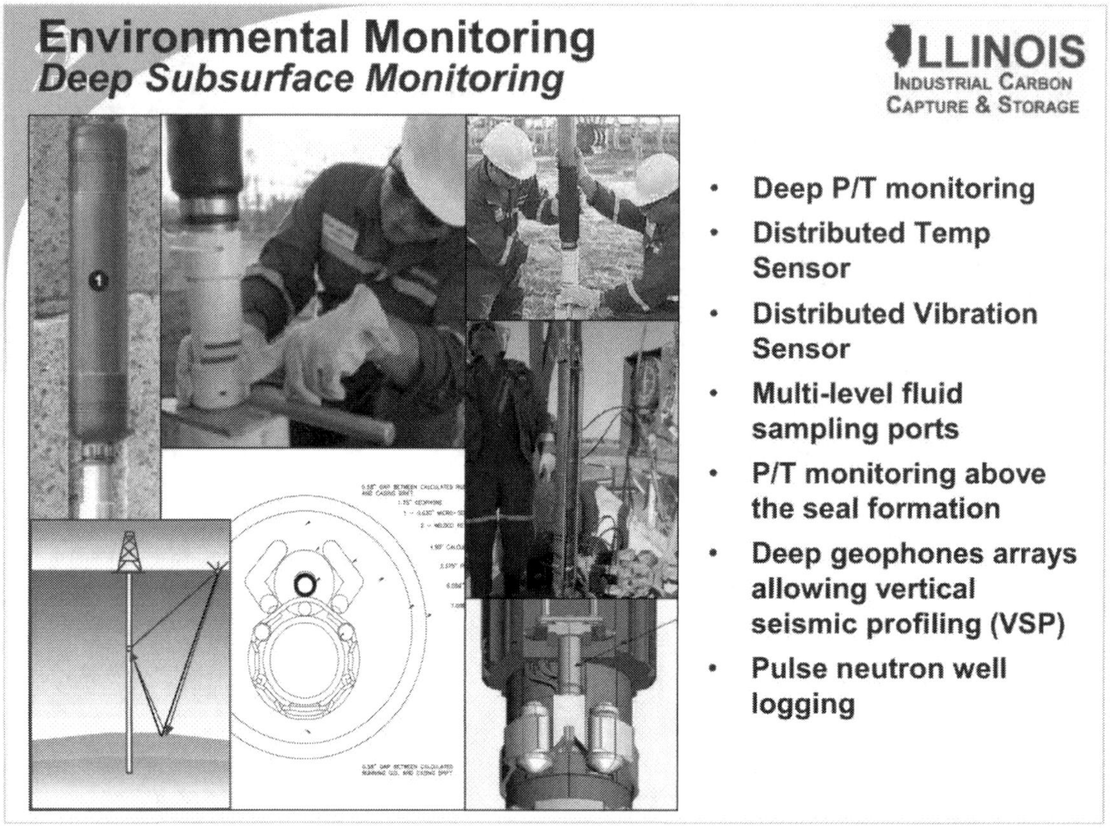
Environmental Monitoring
Deep Subsurface Monitoring
ILLINOIS
INDUSTRIAL CARBON
CAPTURE & STORAGE
• Deep P/T monitoring
• Distributed Temp Sensor
• Distributed Vibration Sensor
• Multi-level fluid sampling ports
• P/T monitoring above the seal formation
• Deep geophones arrays allowing vertical seismic profiling (VSP)
• Pulse neutron well logging
ARCHER DANIELS MIDLAND COMPANY

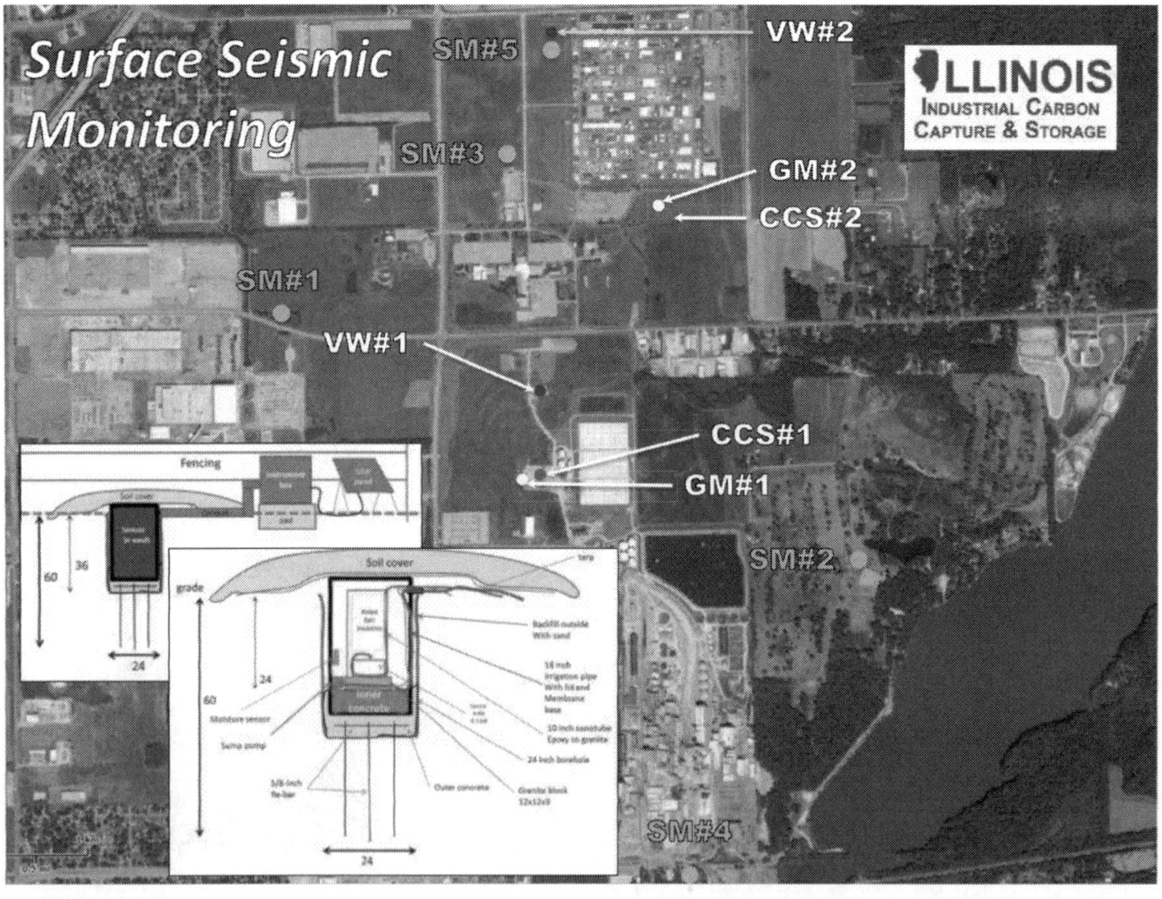
Surface Seismic Monitoring
ILLINOIS
INDUSTRIAL CARBON
CAPTURE & STORAGE
SM#5
VW#2
SM#3
GM#2
CCS#2
SM#1
VW#1
CCS#1
GM#1
SM#2
SM#4
Fencing
Soil cover
60
36
24
grade
Backfill outside With sand
Moisture sensor
Sump pump
24 inch borehole
Outer concrete
Granite block

Seismic Monitoring

USGS Site Monitoring

USGS
science for a changing world

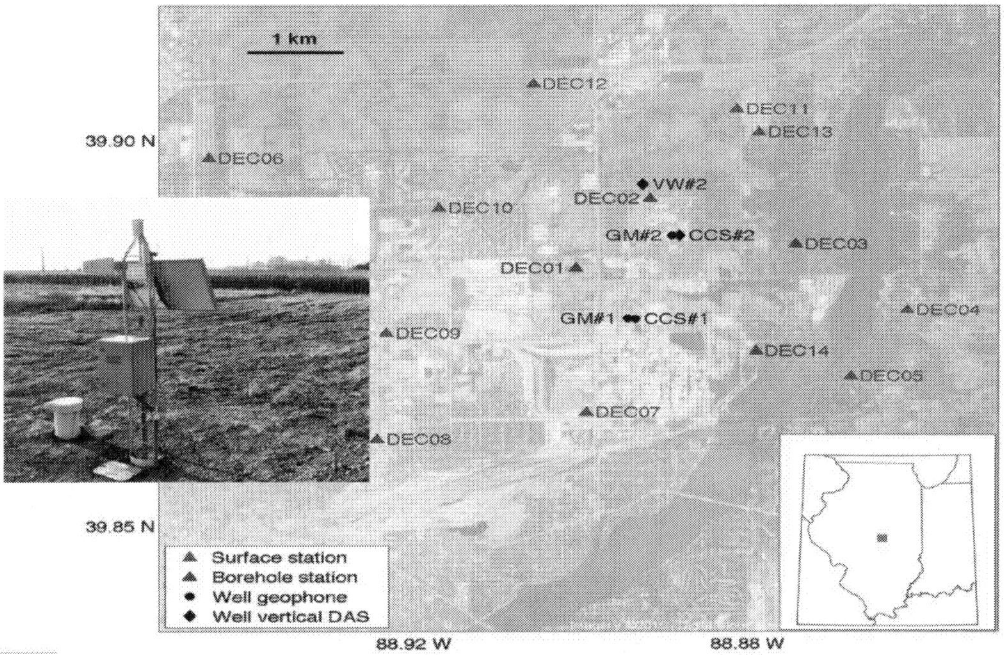

Deep Seismic Monitoring

ILLINOIS INDUSTRIAL CARBON CAPTURE & STORAGE

- CCS#1: 3 Level Geophone
- GM#1: 24 level Geophone
- CCS#2: Fiber Optic DAS
- VW#2: Fiber Optic DAS
- GM#2: 5 Level Geophone
- 6 Borehole Stations
- 14 Surface Stations

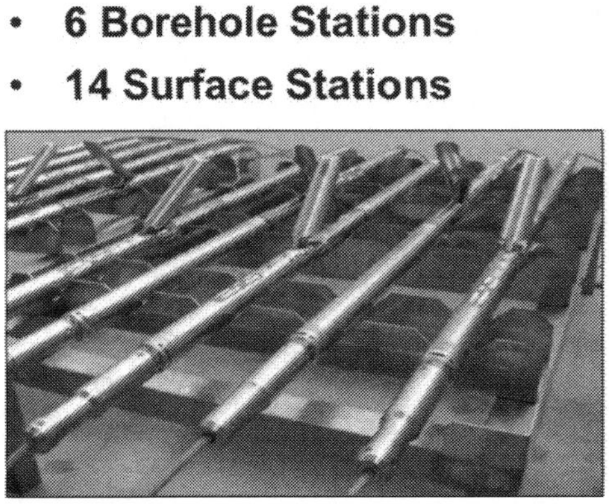

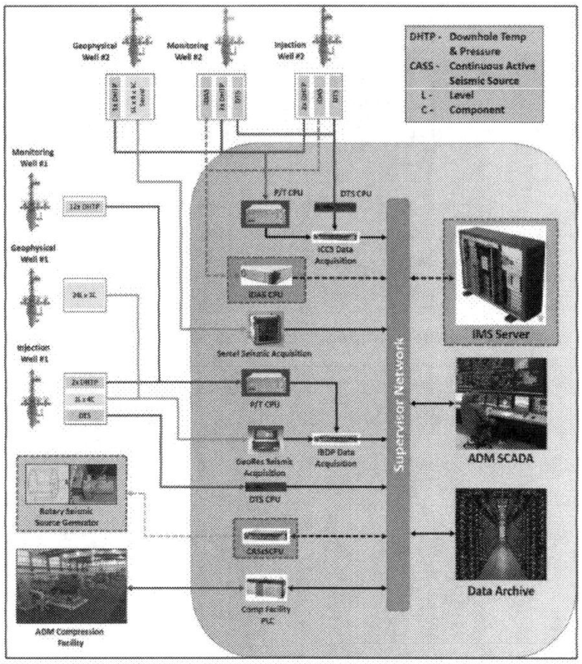

ADM

Seismic Monitoring Data

ILLINOIS
INDUSTRIAL CARBON
CAPTURE & STORAGE

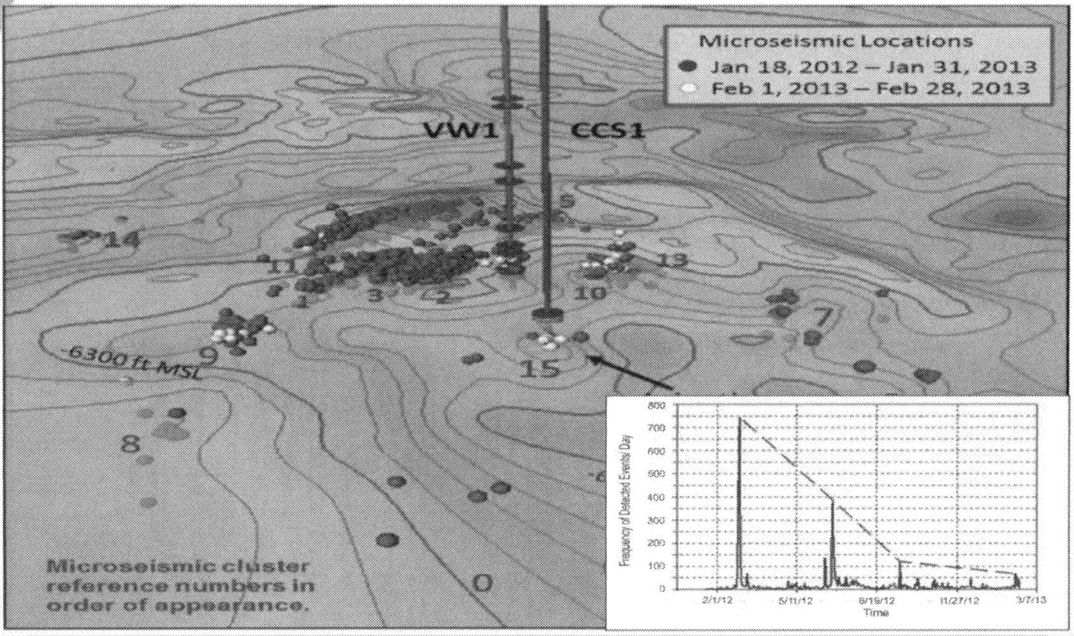

ARCHER DANIELS MIDLAND COMPANY

ADM

Time Lapse VSP Surveys

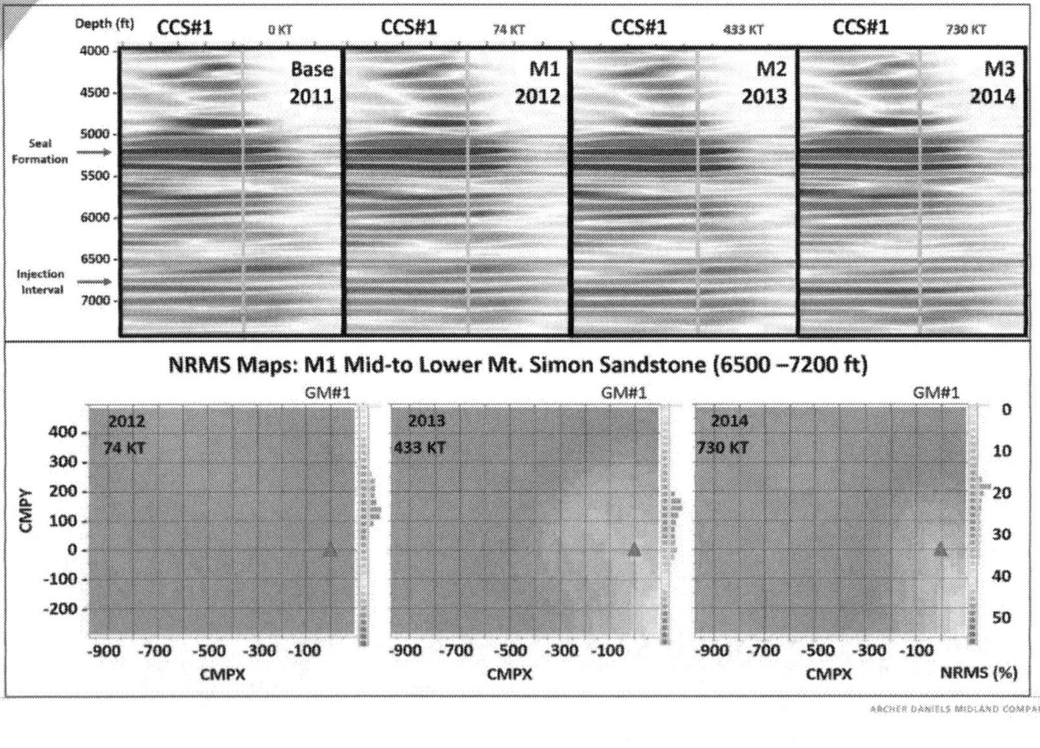

ARCHER DANIELS MIDLAND COMPANY

CO_2 Based Chemicals

- •Carbonates
  - Glycerol Carbonate
  - Propylene Carbonate
  - Dimethyl Carbonate
- •Fertilizers
- •Alcohols
- •Fuels
- •Acids
- •Others

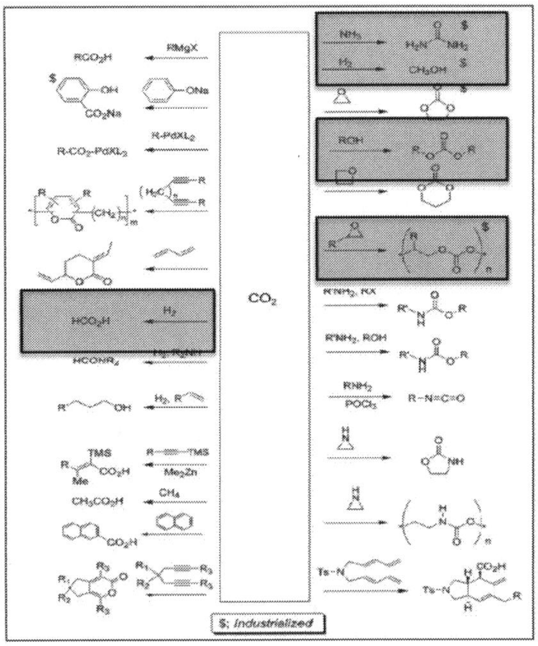

Transformation of Carbon Dioxide, Sakakura, Choi, & Yasuda, 2007

Enhanced Oil Production

http://www.youtube.com/watch?v=azLVjYij5U4

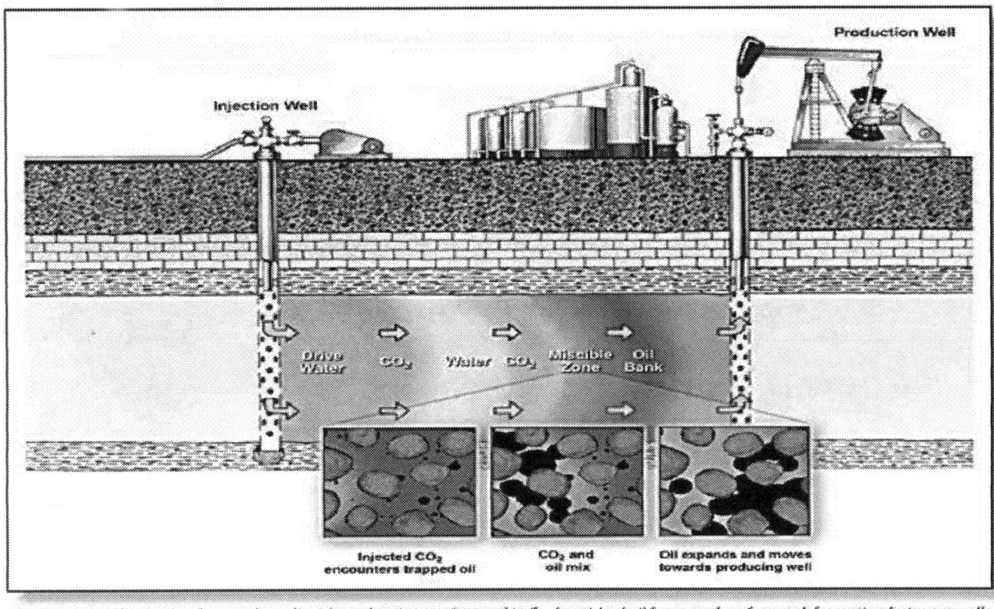

Cross-section illustrating how carbon dioxide and water can be used to flush residual oil from a subsurface rock formation between wells

Illinois Basin Potential

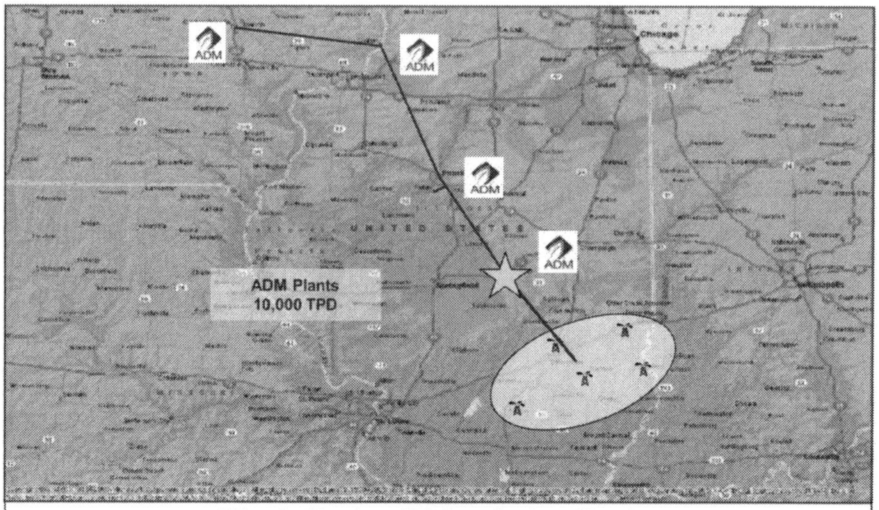

•Illinois Basin - Oil Producing Region
•Est. Recoverable Oil = 700 million bbls[(1)]
•Est. CO_2 Requirements = 150 million tons

(1) BASIN ORIENTED STRATEGIES FOR CO2 ENHANCED OIL RECOVERY: ILLINOIS AND MICHIGAN BASIN OF ILLINOIS, INDIANA, KENTUCKY AND MICHIGAN; Advanced Resources International, February 2006

ARCHER DANIELS MIDLAND COMPANY

Thank You!

ILLINOIS INDUSTRIAL CARBON CAPTURE & STORAGE

Industrial Carbon Capture and Storage Project:

- U.S. Department of Energy Award No. DE-FE-0001547
- Administered by the DOE's Office of Fossil Energy
- Managed by the National Energy Technology Laboratory
- DOE cost share from American Recovery and Reinvestment Act of 2009

Cost Share Agreements:

- Archer Daniels Midland Company
- University of Illinois through the Illinois State Geological Survey
- Schlumberger Carbon Services
- Richland Community College

Project Team Members Contacts:

- Dr. Sai Gollakota (NETL-DOE) Sai.Gollakota@NETL.DOE.GOV
- Dr. Robert Finley, (ISGS) finley@isgs.illinois.edu
- John Medler, (Schlumberger Carbon Services) jmedler@slb.com
- Dr. Douglas Brauer (RCC) dbrauer@richland.edu

ARCHER DANIELS MIDLAND COMPANY

Demonstration Project for the Capture, Transportation, Flooding and Storage of CO_2 from Shengli Coal-fired Power Plant

ZHANG Yuanyuan
Sinopec Energy and Environment Engineering Corporation
Nov. 19th, 2015

Sinopec 1.5 Mton/year CCS-EOR Project

Sinopec is undertaking two large-scale CO_2 capture, transport, enhanced oil recovery and sequestration projects under the strong support of the Ministry of Science and Technology:

Part 1: 1 Mton/year CCS-EOR project from Shengli Coal-fired Power Plant flue gas

Part 2: 0.5 Mton/year CCS-EOR project from Sinopec Qilu Petrochemical coal gas

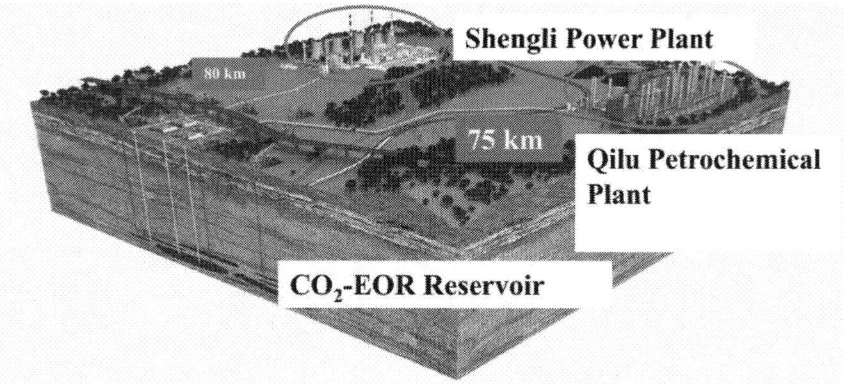

Sinopec 1.5 Mton/year CCS-EOR Project

Shengli Power Plant Phase III, (CO_2 14%)

↓ 1.0Mton/year

Chemical Absorption Process

↓

Shengli Oilfield

Goals

- Testing and optimization of the developed technology
- Economy assessment of CCS-EOR project
- Best practice for the construction, operation and management
- Promotion of the commercial application of CCS-EOR

QPEC Coal Gas, (CO_2 90%)

↓ 0.5Mton/year

Cryogenic Distillation Process

↓

Shengli Oilfield

Overview of the Project

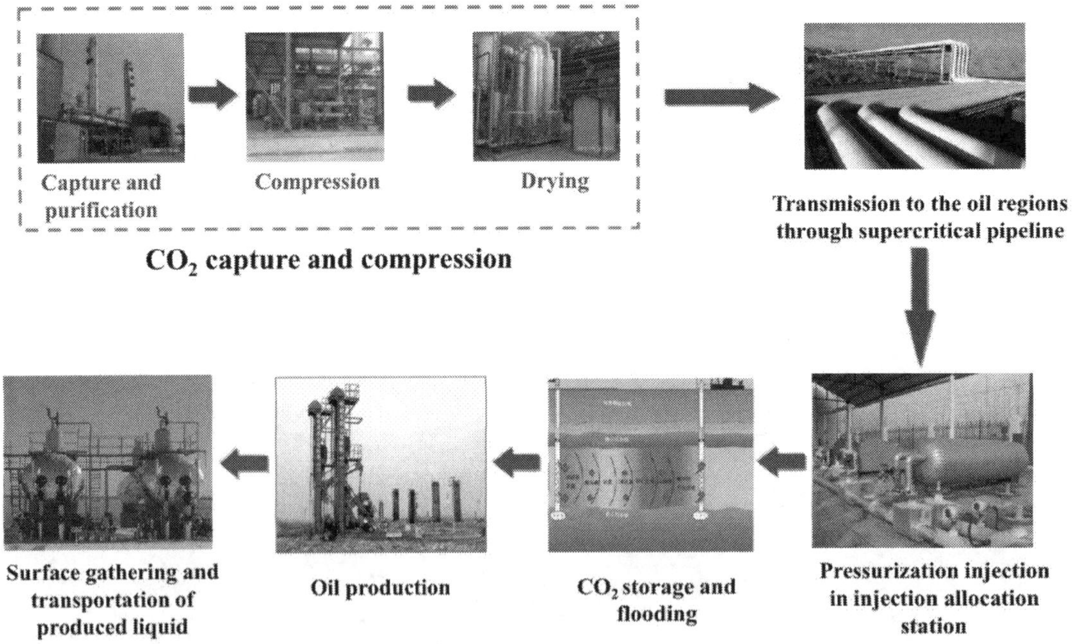

Progress Up to Day

- 1 Mton/year CCS-EOR project from Shengli Coal-fired Power Plant flue gas——finish the pre-feasibility
  - The feasibility studies on CO_2 capture (the first version) and pipeline have been finished
  - The CO_2 flooding reservoir have been selected(0.35Mt/a)
  - The feasibility studies of CO_2 injection, drilling and production engineering, produced fluid gathering and transportation system are done
  - Project investment estimation
  - Baseline for CO_2-EOR environmental monitoring

Progress Up to Day

Capture

The flue gas comes from part of the outlet gas after desulfuration and denitrification from the Phase III 600MW units in Shengli Power Plant (gas flow for capture is 600,000Nm^3/h, accounting for 35% of the total flue gas from Phase III). CO_2 concentration is 12-14%. The process will include three units, capture, compression and drying, to obtain the product CO_2 with the purity of 99.5% (V%), which will be transported via pipeline to Shengli oilfield.

The process of carbon capture

Capture and purification

Compression

Drying

Progress Up to Day

Capture scheme

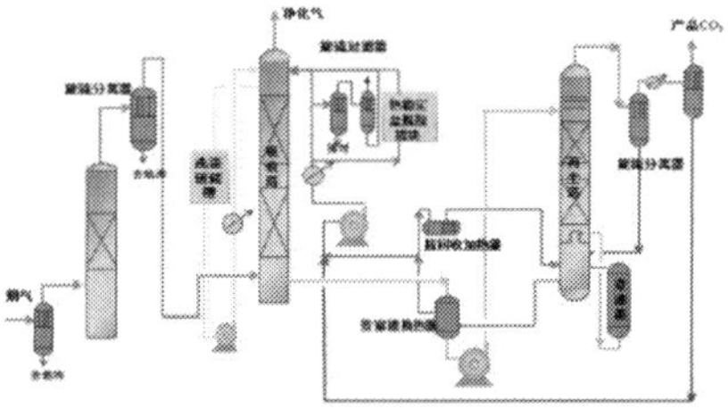

Sinopec MAS carbon capture technique for low partial pressure flue gas

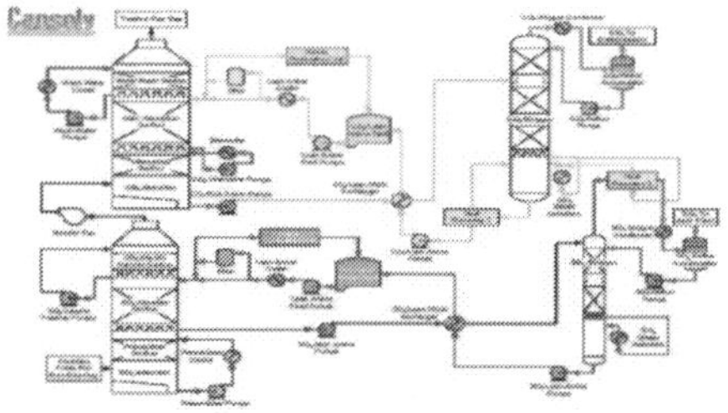

Shell Cansolv carbon capture technology

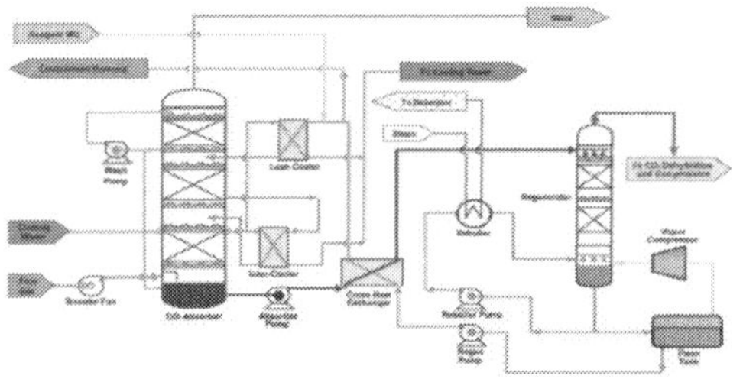

SIEMENS Carbon capture technology based on the amino acid salts

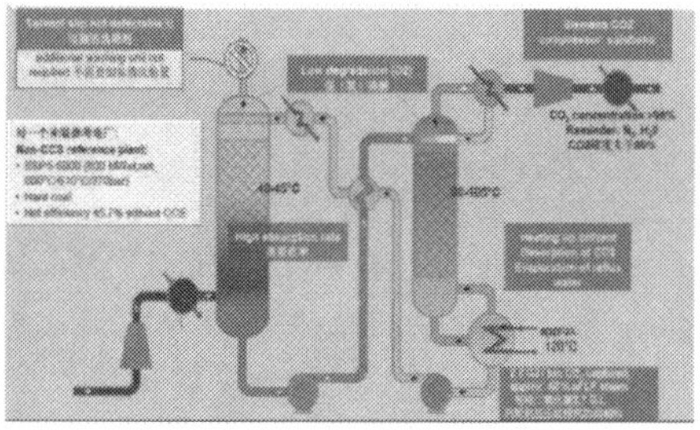

PowerSpan ECO_2 technology

Progress Up to Day

- CO_2 capture

The estimated land area : 21,438m^2
The reserved land area : 34,400m^2

So the reserved lands can meet the demand.

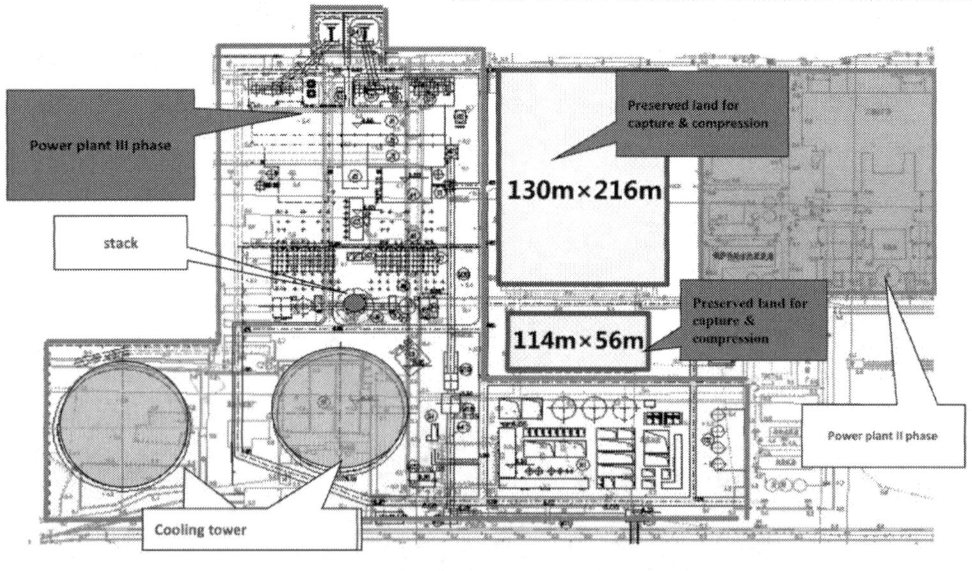

Reserved land for CO_2 capture and compression

Progress Up to Day

- CO_2 pipeline
  - 80km
  - Supercritical
  - DN250,12MPa
  - Seamless steel pipe with anti-corrosion measures

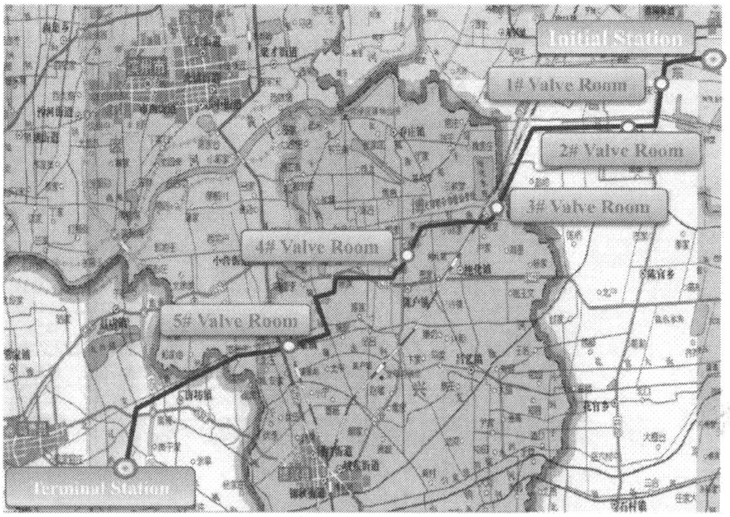

Progress Up to Day

- CO_2 flooding and storage zones

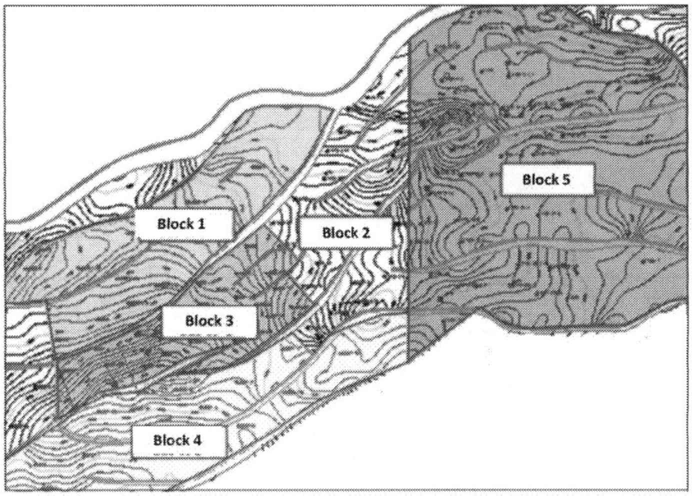

- **CO_2 Injection Scale: 350,000t/a**
- **Both new and old oil block**
- **Elastic development is used in old block**
- **Depth: 3000m**
- **Permeability : 2-3mD**
- **Oil-bearing area:40km^3**
- **Comprehensive watercut:6%**
- **Area: 10.2*7.8km^2**

Progress Up to Day

- Reservoir development program for CO_2 flooding and storage

There are 180 wells totally, including 110 oil wells and 70 gas wells; 60 wells are newly drilled wells, including 26 oil wells and 34 gas-injection wells; there are 25 conversion wells and 11 existing gas-injection wells.

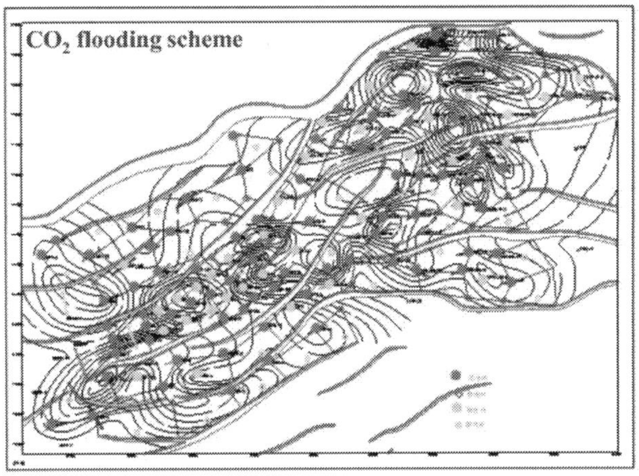

1) A set of reservoir development by CO_2 flooding

2) Well pattern with irregular areas

3) Well spacing of 350-600m

Progress Up to Day

- Project investment estimation

Project investment :300 million USD

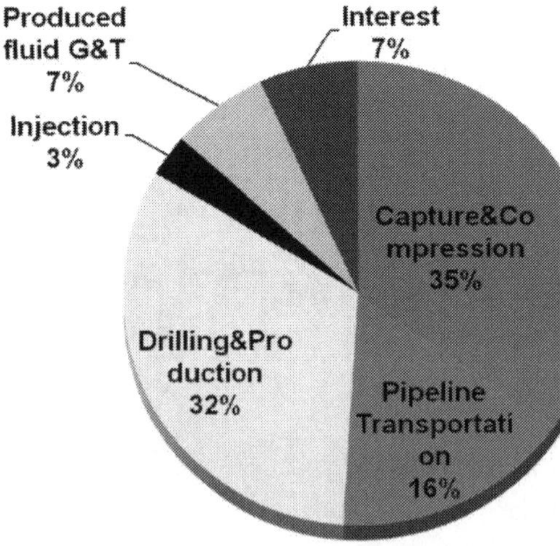

Progress Up to Day

Proposed environmental monitoring scheme

Background monitoring framework for Co_2-EOR project in Shengli Oilfield

Monitoring category		Monitoring object/technology	Monitoring instrument/method	Monitoring frequency
Atmosphere		Weather: temperature, humidity, wind speed, atmosphere stability	Eddy correlation method	
		Co_2 flux	Eddy correlation method	Real time
		Co_2 concentration	Infrared diode laser	
		Stable isotope 13C in air	Isotope analyzer	Per month
		Co_2 emission source survey	The ecological system CO_2 and existing industrial, agricultural CO_2 source model are established	Once
Soil gas		Soil temperature, matrix potential, water cut	Underground sensor	
		Soil surface CO_2 flux	Soil respiration measurement system	
		Soil gas CO_2 concentration at a certain depth	Soil respiration measurement system	Per month
		Other soil gas compositions: CH_4, O_2, N_2	Portable meteorological chromatograph	
		Isotope ratio at soil air temperature 13C	Isotope analyzer	
Vegetation ecology		Flora and fauna survey	Sampling surveys	Once
		Vegetation index	Airborne spectral imaging	Quarterly
Surface deformation		Vertical direction	Electronic digital level	Quarterly
		Horizontal direction	High precision total station	Quarterly
Water quality	Surface water Shallow groundwater	(1) Temperature, pH, electrical conductivity, total salinity, total organic carbon, alkalinity (2) Main anions and cations	(1) Glass electrode method, titration method, combustion oxidation, non dispersive infrared absorption method (2) Ion chromatograph (3) Meteorological chromatograph	Per month

Progress Up to Day

- Baseline for CO_2-EOR environmental monitoring

4D seismic

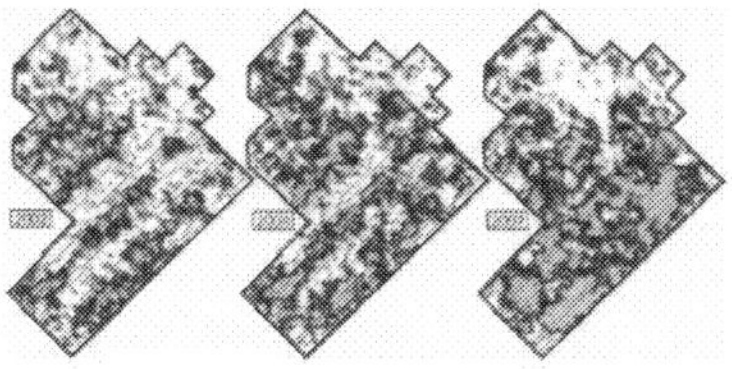

soil gas

Vegetation

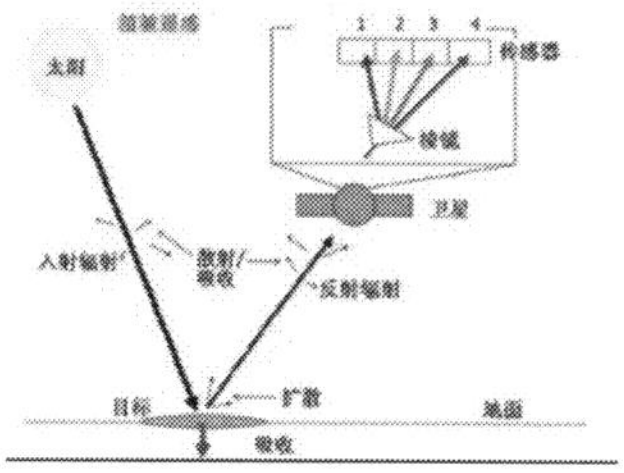

ground deformations

GPS

Groundwater

便携式水质分析仪

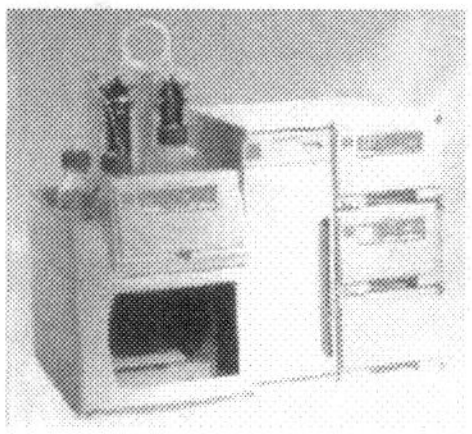

离子色谱

Air

Progress Up to Day

Public acceptance survey

Questionnaire survey

Respondents:

1. **Random**
2. **Different ages**
3. **Different occupations**
4. **Different cultural levels**

Progress Up to Day

Public acceptance survey

Enter into communities, villages, wholesale markets for public survey.

Public survey results show that CCUS technology has been strongly supported by the public, with the expectant attitude for the CO_2 emission reduction and environmental protection effect brought out by the large-scale promotion of CCUS technology.

Thanks!

Global Opportunities for Offshore CCS on Continental Shelves

AIChE – Carbon Management Technology Conference

Houston, Texas

November 18, 2015

Dr. Tip Meckel, Research Scientist

Gulf Coast Carbon Center

TEXAS Geosciences

Bureau of Economic Geology

Jackson School of Geosciences

The University of Texas at Austin

Bureau of Economic Geology · Gulf Coast Carbon Center · ieaghg

Three main messages:

1. **Scale of issue eventually eclipses EOR. CO2-EOR is great driver and will work for many (but not all) places globally for some time. Then what? What about places where CO2-EOR is not an option?**
2. **For CCS to be a globally-deployed technology with significant atmospheric benefit, additional focus on <u>offshore</u> use and storage is needed. Why offshore?**
   - **6 Gt by 2050 (6,000 'Sleipners')**
   - **2/3 of CCS potential will need to come from non-OECD countries (IEA)**
3. **Focus needs to be on storage capacity validation and knowledge transfer from experienced countries and companies to those aspiring to undertake projects.**

Outline of Topics

1. **Offshore potential**

2. **Provide summary of Carbon Sequestration Leadership Forum - Offshore CCS Task Force report.**

3. **Provide information on planned international workshop.**

Global Offshore CO_2 Storage Potential

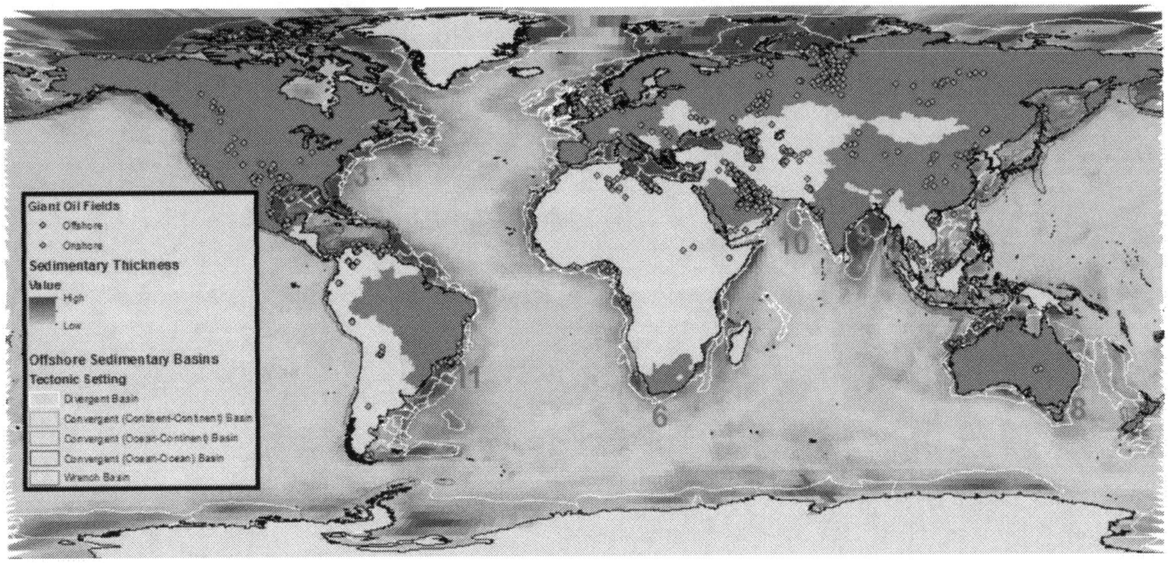

One prior and five new US Offshore Storage Assessment Projects: 3 GoM; 2 Atlantic

- **Offshore CO_2 Storage Resource Assessment of the Northern Gulf of Mexico (Upper Texas-Western Louisiana) ($3.3M) – *GCCC, UT-Austin*.**
- **Assessment of CO2 Storage Resources in Depleted Oil and Gas Fields in the Ship Shoal Area, Gulf of Mexico ($1M) – *Geomechanics Technologies***
- **Offshore GoM Storage Resource Assessment – *NITEC LLC***

- **Southeast Offshore Storage Resource Assessment ($3.9M) - *SSEB***
- **Mid-Atlantic U.S. Offshore Carbon Storage Resource Assessment Project ($3.7M) - *Batelle***

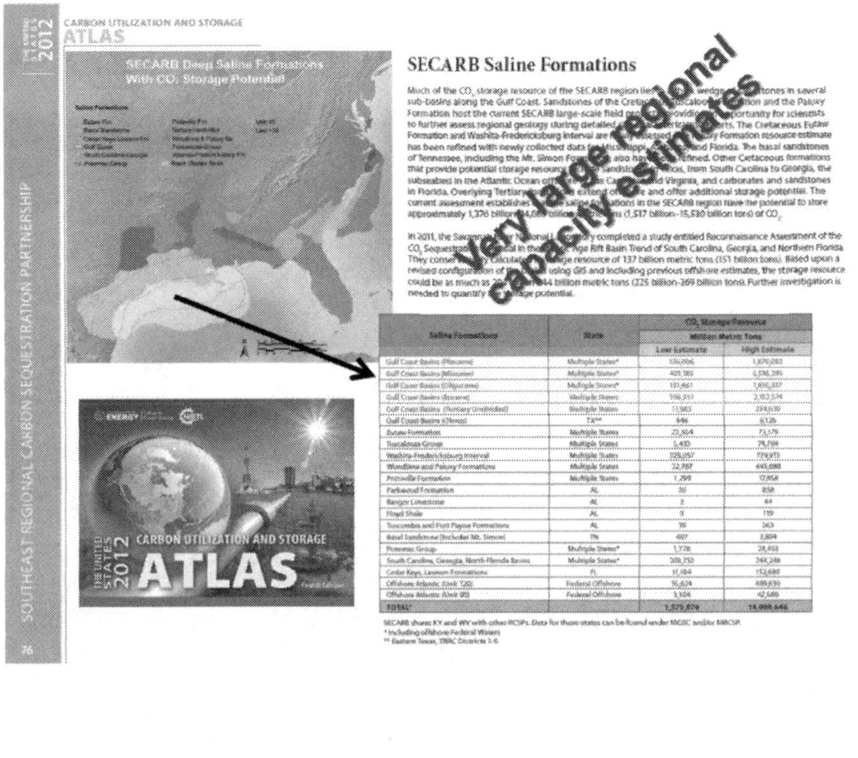

Static Regional Capacity

- NETL Methodology
- 40,000 sq. km.
- 3,300 logs
  - Tops, net sand, porosity
- 172 Gt total (TX State Waters)

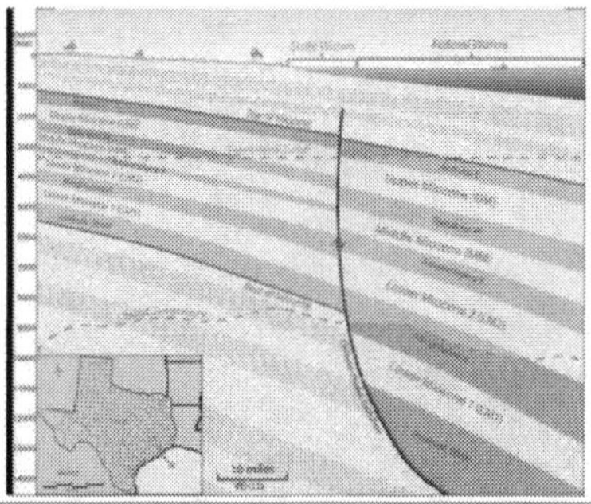

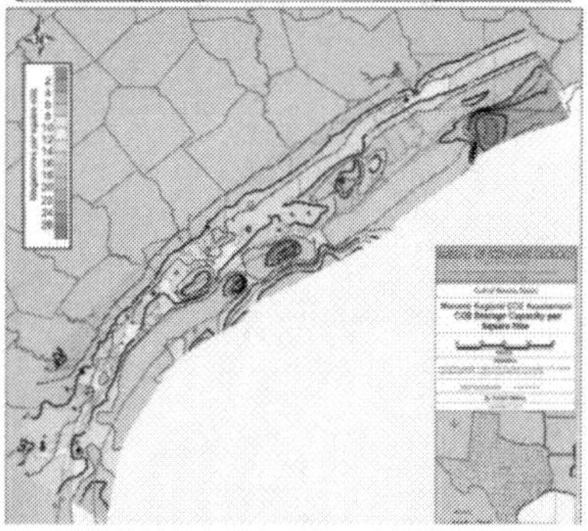

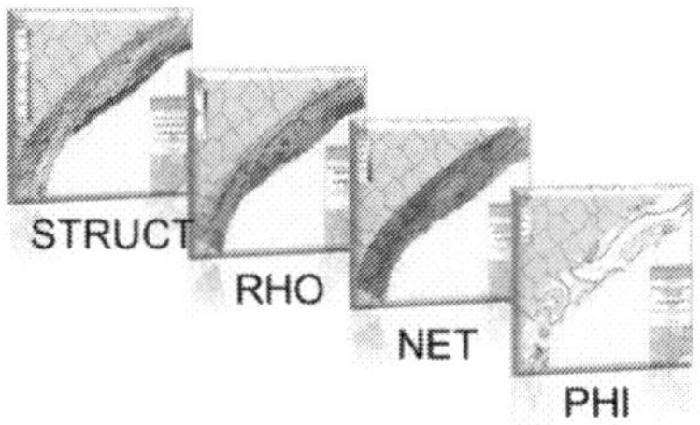

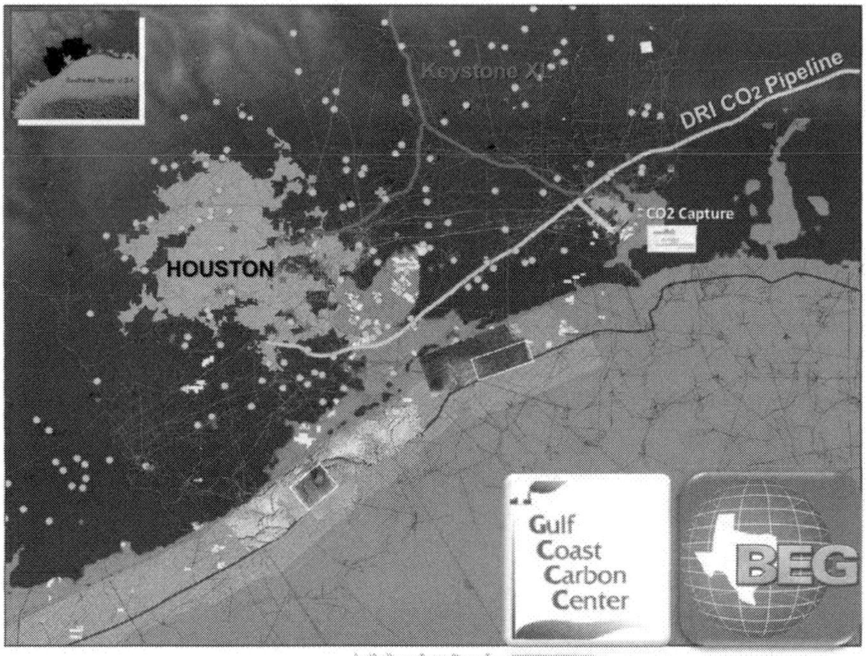

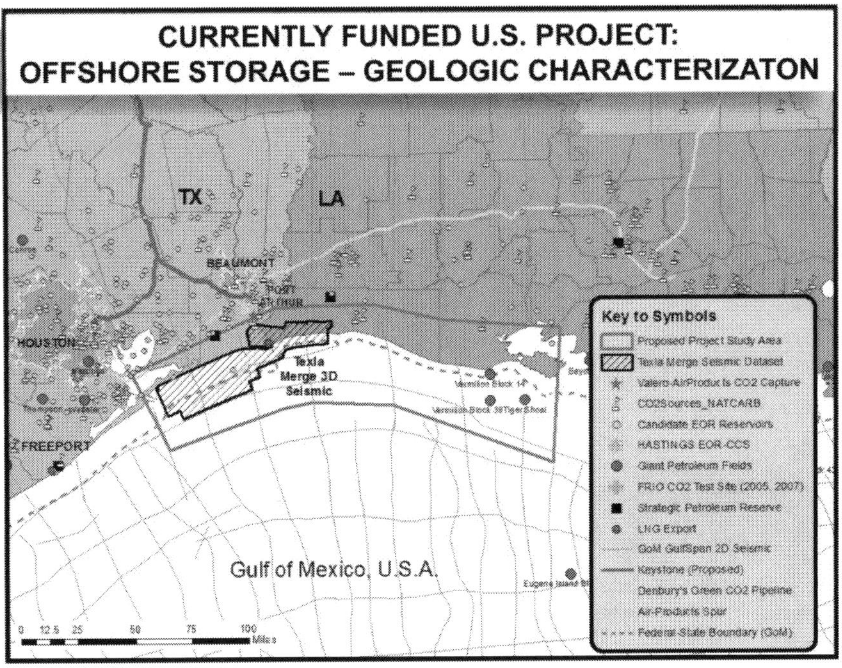
CURRENTLY FUNDED U.S. PROJECT:
OFFSHORE STORAGE – GEOLOGIC CHARACTERIZATON
TX
LA
BEAUMONT
PORT ARTHUR
HOUSTON
FREEPORT
Texla Merge 3D Seismic
Gulf of Mexico, U.S.A.
Key to Symbols
Proposed Project Study Area
Texla Merge Seismic Dataset
Valero-AirProducts CO2 Capture
CO2Sources_NATCARB
Candidate EOR Reservoirs
HASTINGS EOR-CCS
Giant Petroleum Fields
FRIO CO2 Test Site (2005, 2007)
Strategic Petroleum Reserve
LNG Export
GoM GulfSpan 2D Seismic
Keystone (Proposed)
Denbury's Green CO2 Pipeline
Air-Products Spur
Federal-State Boundary (GoM)
0 12.5 25 50 75 100 Miles

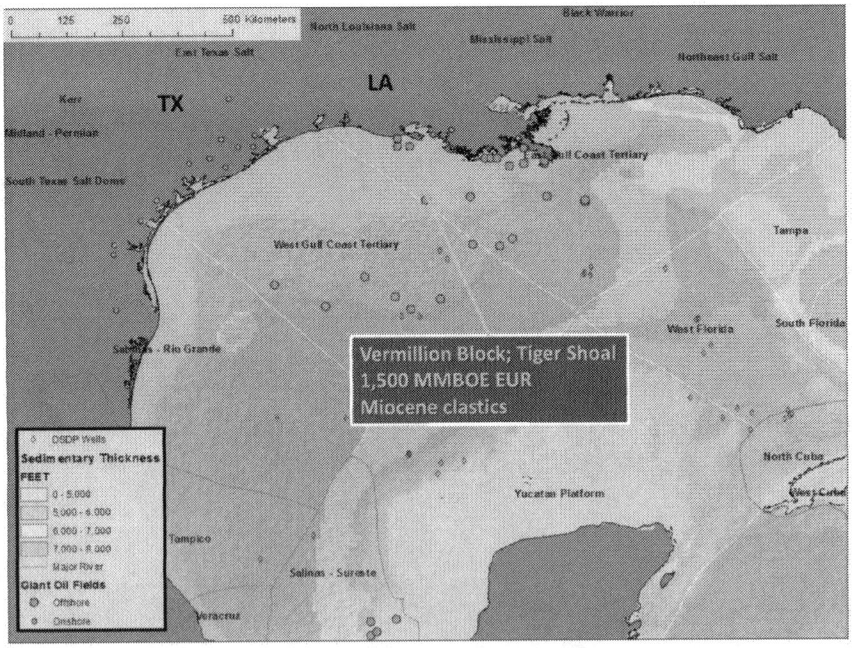
0 125 250 500 Kilometers
North Louisiana Salt
Mississippi Salt
Black Warrior
East Texas Salt
Northeast Gulf Salt
LA
TX
Kerr
Midland - Permian
South Texas Salt Dome
East Gulf Coast Tertiary
West Gulf Coast Tertiary
Tampa
West Florida
South Florida
Sabinas - Rio Grande
Vermillion Block; Tiger Shoal
1,500 MMBOE EUR
Miocene clastics
North Cuba
Yucatan Platform
Tampico
Salinas - Sureste
Veracruz
DSDP Wells
Sedimentary Thickness
FEET
0 - 5,000
5,000 - 6,000
6,000 - 7,000
7,000 - 8,000
Major River
Giant Oil Fields
Offshore
Onshore

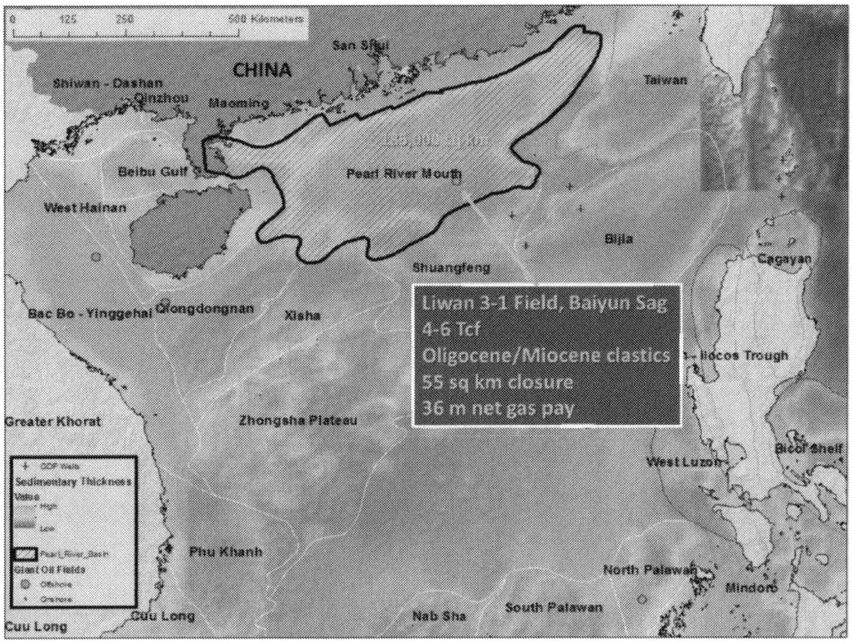

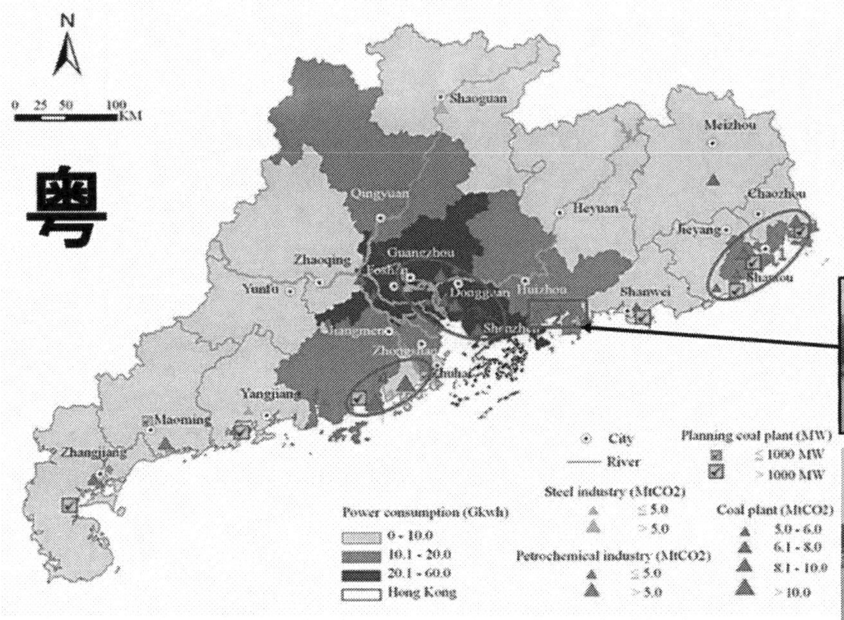

The most industrialized Province in China
Has no onshore storage potential.

Figure 5-1 CO_2 capture opportunities of Guangdong Province

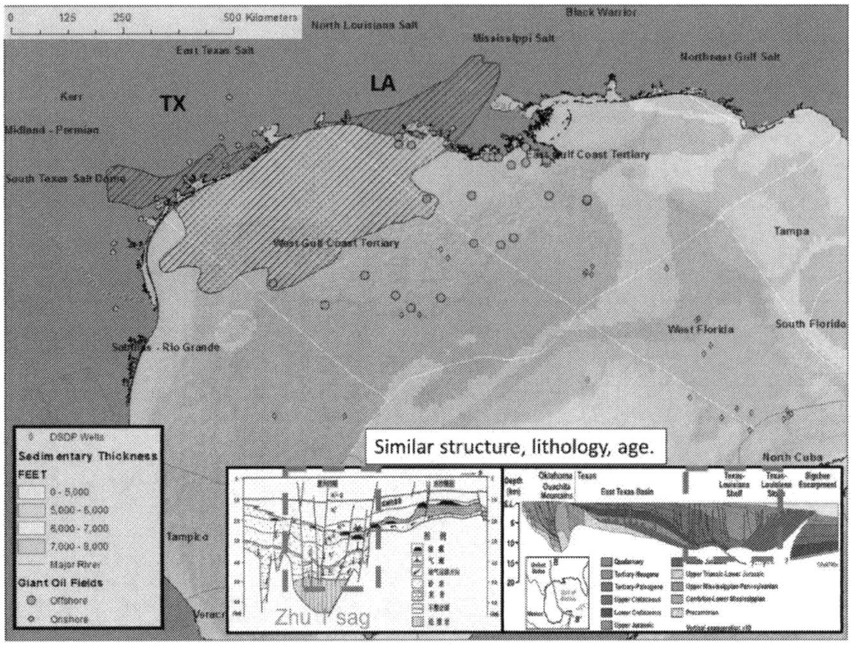
TX
LA
Similar structure, lithology, age.
Zhu I sag
Sedimentary Thickness
FEET
Giant Oil Fields
Offshore
Onshore

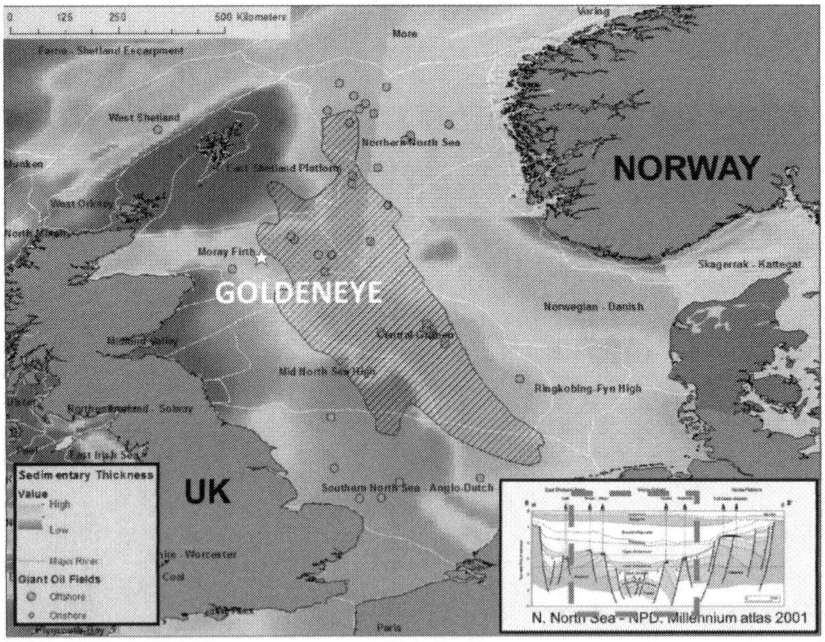
NORWAY
UK
GOLDENEYE
Moray Firth
Sedimentary Thickness
Giant Oil Fields
N. North Sea - NPD, Millennium atlas 2001

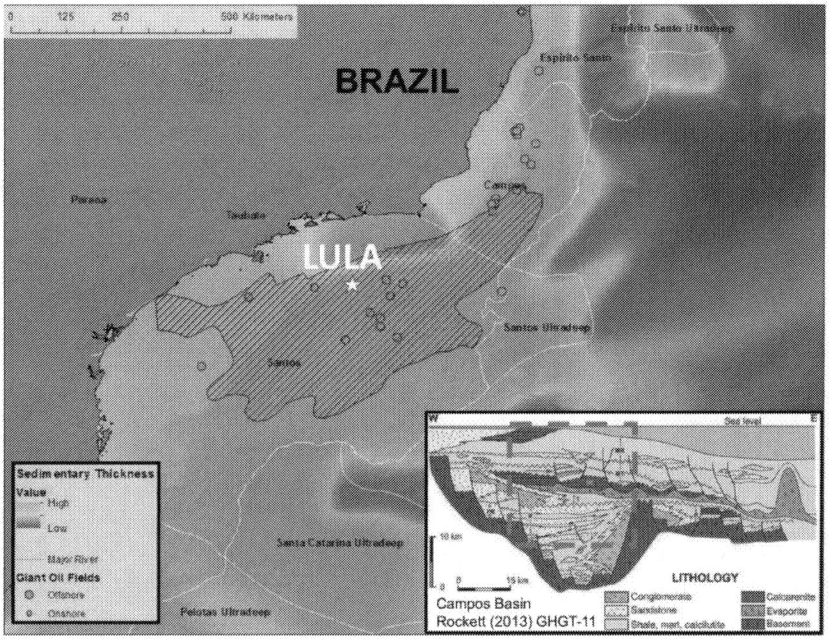
0 125 250 500 Kilometers
BRAZIL
Espirito Santo Ultradeep
Espirito Santo
Parana
Taubate
Campos
LULA
Santos Ultradeep
Santos
Santa Catarina Ultradeep
Pelotas Ultradeep
Sedimentary Thickness
Value
High
Low
Major River
Giant Oil Fields
Offshore
Onshore
W
Sea level
E
10 km
0
15 km
LITHOLOGY
Campos Basin
Rockett (2013) GHGT-11
Conglomerate
Sandstone
Shale, marl, calcilutite
Calcarenite
Evaporite
Basement

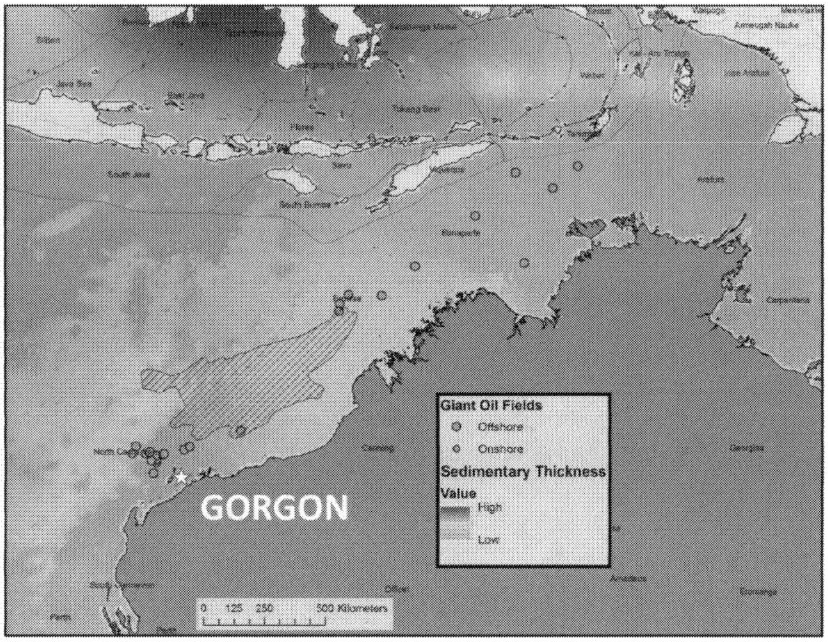
South Java
South Bonaparte
Bonaparte
Arafura
Carpentaria
Canning
Georgina
GORGON
Giant Oil Fields
Offshore
Onshore
Sedimentary Thickness
Value
High
Low
0 125 250 500 Kilometers

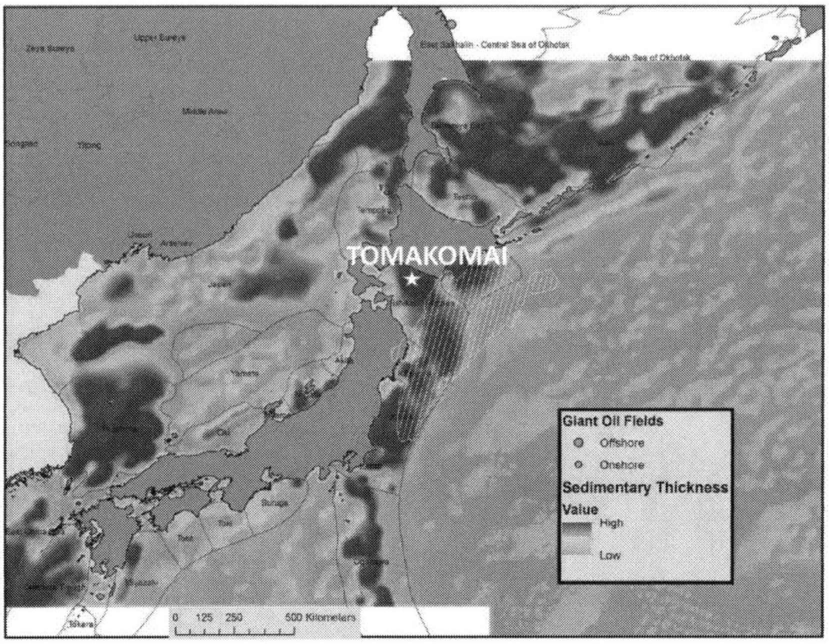

Carbon Sequestration Leadership Forum

CSLF-T-2015-06

20 October 2015

www.cslforum.org

- Total task force members: 28
- 6 countries, 4 continents
- Government agencies, universities, research laboratories, industry, non-governmental organizations

TECHNICAL GROUP

Final Report from Task Force on Technical Barriers and R&D Opportunities for Offshore, Sub-Seabed Geologic Storage of CO_2

Identify technical barriers and R&D needs/opportunities for offshore, sub-seabed storage of carbon dioxide.

Report Outline and Structure

High-Level Report Outline

1. Introduction
2. Status and barriers of existing and proposed offshore CO_2 storage and EOR Projects
3. Offshore CO_2 Storage and EOR Resource Assessments
4. CO_2 Transport for Offshore Storage
5. Risk Analysis for Offshore CO_2 Storage
6. Wellbore Management
7. Monitoring, Verification, and Assessment Tools for Offshore Storage
8. Summary of Regulatory Requirements for Offshore Storage
9. Summary and Recommendations

General Structure of Chapters

- Status/Overview
- Barriers/Technical Challenges
- R&D Opportunities
- Recommendations

http://www.cslforum.net/publications/documents/Riyadh2015/tg_OffshoreSubSeabedStorageTaskForceFinalReport-Riyadh1115.pdf

Recommendations

Topic	Status/Description	Recommendation
Storage Capacity Assessments	Currently inadequate for many regions.	Basin assessment; Pre-qualify storage locations; Knowledge sharing and int'l collaboration.
Transport Infrastructure	Limited and potentially expensive, but less exposure to issues around routing.	Optimization of current practices and infrastructure; take advantage of pilots and demos.
Offshore CO2-EOR	Only one project - Lula in Brazil. Possible to catalyze storage opportunities and infrastructure.	Recent advances in subsea separation and processing could extend the current level of utilization of sea bottom equipment to also include the handling of CO2 streams. Explore opportunities to leverage existing infrastructure and field tests.
Understanding of CO2 Impacts on the Subsea Environment	Significant body of research exists, but complexity of impacts and the challenges to efficient monitoring, particularly natural variability to correctly identify and quantify non-natural change.	Leverage existing work. Understand buffering potential of sediments, and the impact of longer term exposures. Modeling: CO2 dispersion and influencing factors, marine systems.
Monitoring Technology Development	Technology exists but room for improvements. Cover large areas and lengthy periods.	Data processing and interpretation for CO2 storage. The quantification of CO2 within a reservoir still remains a challenge. Real-time data retrieval and navigation. Further development in integrated in situ sensors.

WORKSHOP Planning: Offshore CCS

- **CSLF Task Force final report general recommendation:** knowledge sharing and capacity building.
- **Seeks to build international awareness of global potential.**
  - **Focus on knowledge transfer from experienced countries to those that may want to pursue offshore storage.**
  - **CSLF and UNFCC CTCN support.**
- **Provide forum for outlining collective opportunities and barriers.**
  - **Find commonalities and areas of focus.**
  - **Industry and research participation.**
- **Initiate dialog toward strategies for deploying (funding) a collaborative and co-funded offshore demonstration project(s).**

WORKSHOP Planning: Current Timeline

✓Introduce workshop at CSLF Ministerial, Riyadh.

- Introduce workshop at COP21 side event and engage CTCN.
- January 2016: Form steering committee and begin planning.
- February – Participant contact and survey.
- Workshop: *proposed* host at Gulf Coast Carbon Center, The University of Texas at Austin, dates TBD.

- Parallel activity: New CSLF Task force on Offshore EOR
  - Led by Norway and USA.

Role of Subsea Geologic Carbon Storage in the U.S. and Worldwide

The offshore holds great promise for global CCS deployment.

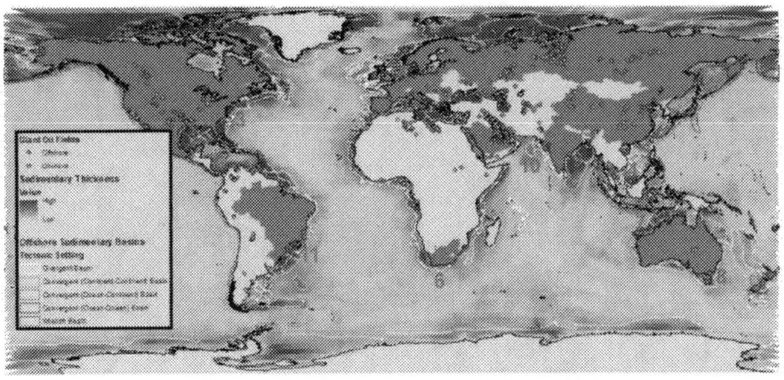

- **Keeps discussion going**:
  - *reduces geopolitical aspects (non-OECD)*
  - *nationally-owned (mandate but provide sites)*
- **Reduced barriers to rapid large-scale deployment:** *Single owner; Public acceptability. Monitoring fairly mature, and arguably more cost-effective. Reuse of infrastructure.*
- **Protects economic positions:**
  - *Future energy: e.g. Offshore Gas (LNG)*

CMTC 2015 November 2015 Sugar Land, TX

EFFC-DFI GEOTECHNICAL CARBON CALCULATOR PROJECT

Gerald Verbeek
DFI Testing & Evaluation Committee Chair

Luca Bruni
Member EFFC Sustainability Committee

Presentation Objectives

- Sustainability Awareness
- Geotechnical Carbon Calculator
- DFI Carbon Database
- Input Forms

Sustainability - Awareness

- What is Sustainability?
  - **UN, Brundtland Commission** – "*Development that meets the needs of the present without compromising the ability of future generations to meet their own needs.*"
  - **ASCE** – "Sustainability is a set of environmental, economic and social conditions in which all of society has the capacity and opportunity to maintain and improve its quality of life indefinitely without degrading the quantity, quality or availability of natural resources and ecosystems."
  - **EFFC** - "*Maintenance of a healthy economic environment; prudent use of natural resources; social progress which recognizes the needs of all people; and effective protection of the environment.*"

Sustainability - Awareness

- What is Sustainability?
  - **UN, Brundtland Comission** – "*Development that meets the needs of the present without compromising the ability of future generations to meet their own needs.*"

Sustainability - Awareness

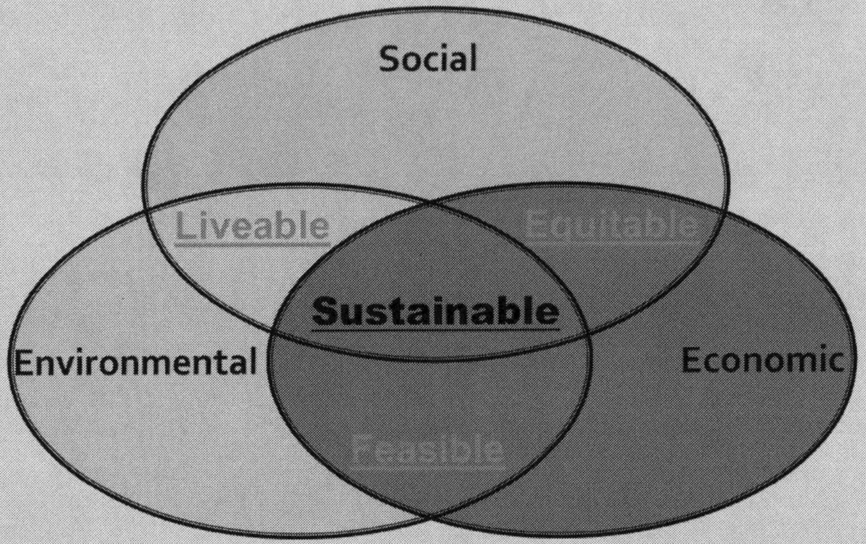

Sustainability - Awareness

Balance of the Economic, Social, and Enviromental Factors

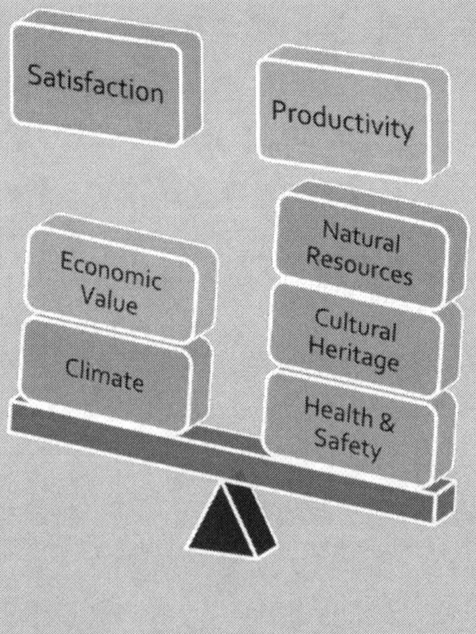

Sustainability Awareness

Natural Resources

The World's natural oil supply is fixed because petroleum is naturally formed far too slowly to be replaced at the rate at which it is being extracted

Sustainability Awareness

Climate

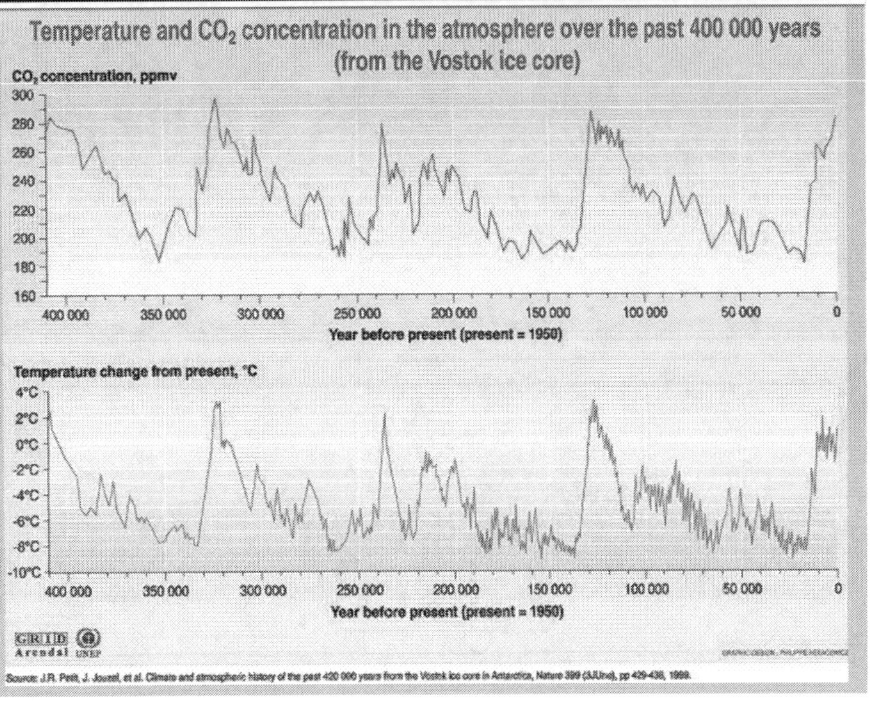

Sustainability Awareness

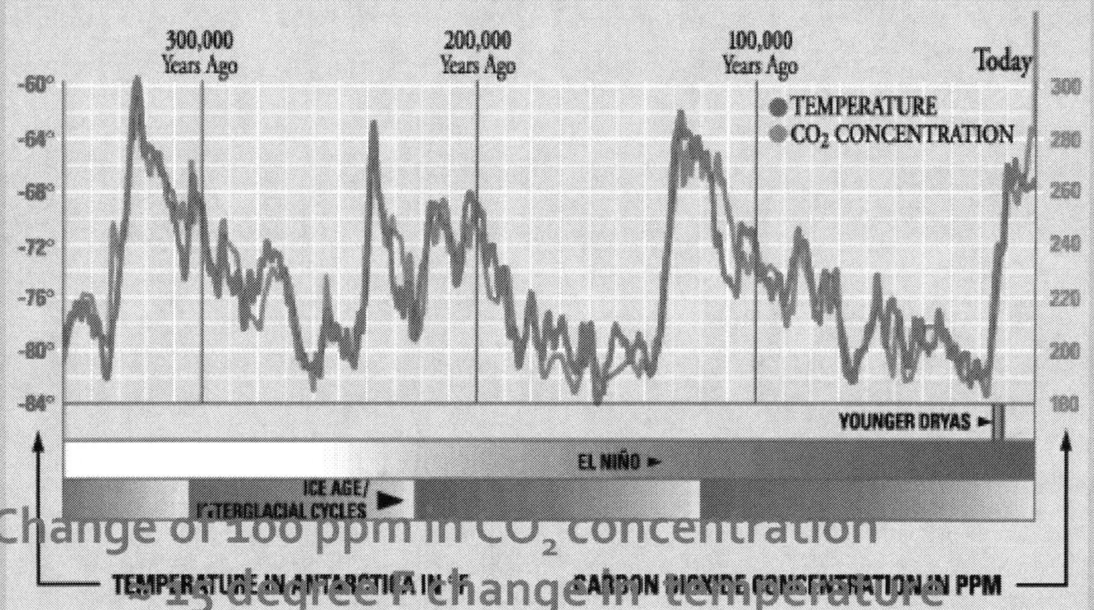

Change of 100 ppm in CO_2 concentration

≈ 15 degree F change in temperature

Sustainability Awareness

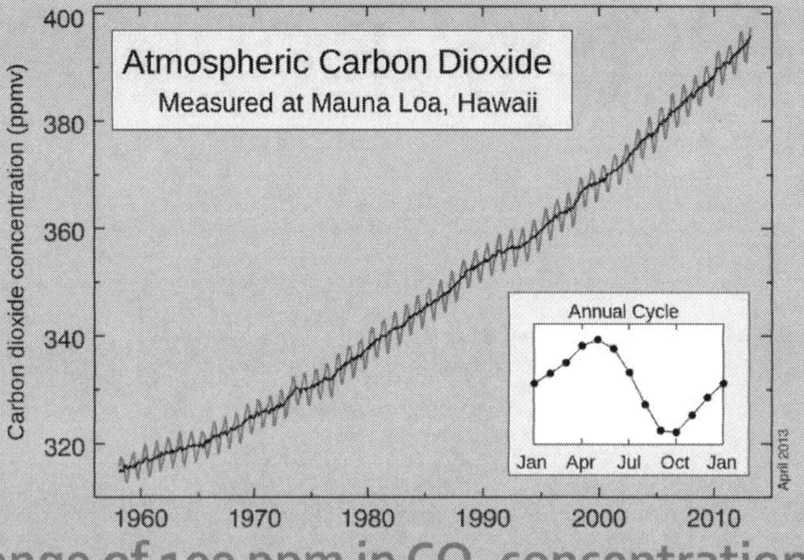

Change of 100 ppm in CO_2 concentration

≈ 15°F or 8°C change in temperature

Sustainability Awareness

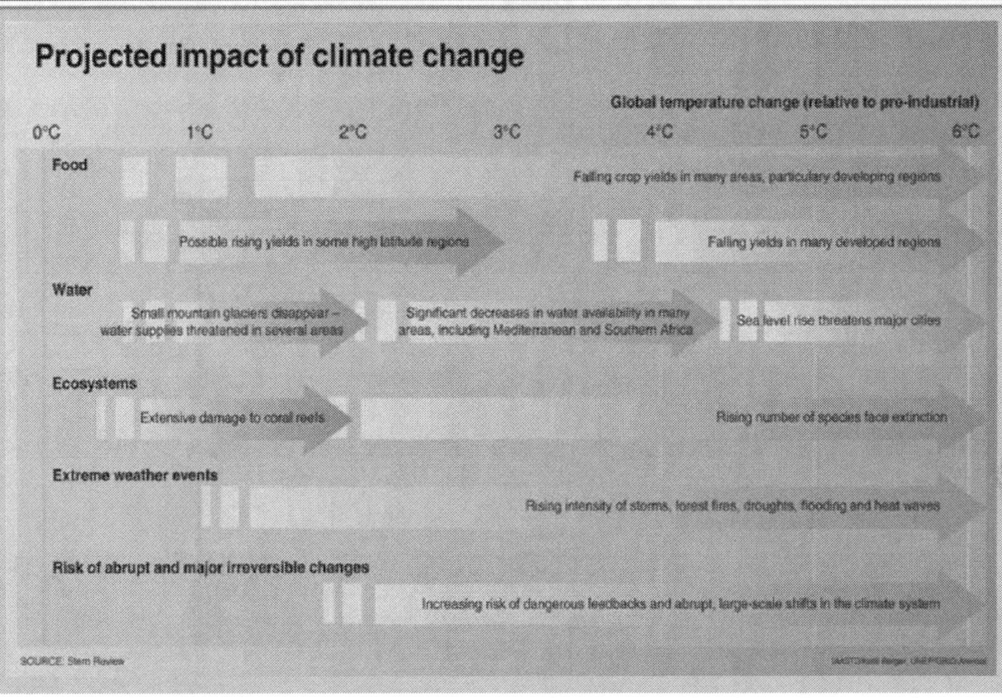

Sustainability - Awareness

- What is Sustainability?

"Development that meets the <u>needs of the present</u> without compromising the ability of <u>future</u> generations to meet their own <u>needs</u>."

Sustainability Awareness

- **Carbon footprint is defined as:** The total amount of greenhouse gases produced to directly and indirectly support human activities, usually expressed in equivalent tons of carbon dioxide (CO2).

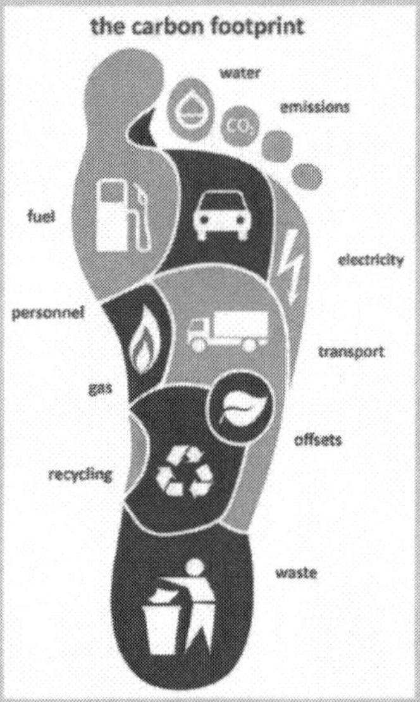

Sustainability Awareness

Why should we foster the movement towards low carbon processes?

- Help preventing uncontrollable climate change
- Reduce dependency on Energy
- Promote innovation
- Develop marketing arguments
- Motivate teams working on values
- Reduce operational costs
- Improve image
- Answer (or initiate new) customers needs

Sustainability Awareness

- **Carbon footprint – Fly Ash vs. Cement**
  Fly ash is a waste product from coal-fired power plants.
  The carbon footprint is attributed to the power industry and not the concrete industry that uses this waste product.
  By replacing 20% of the cement in a typical mix with fly ash, the carbon footprint of the concrete is reduced by 15 %.
  Fly ash is cheaper than cement.

Sustainability Awareness

- **Carbon footprint – more testing under LRFD**
  More testing allows the use of higher resistance factors.
  Higher resistance factors allow smaller foundations, which reduce the required amount of foundation materials.
  Doing this reduces the carbon footprint of the foundation
  Doing this generally leads to cost savings.

Sustainability Awareness

- **What is the carbon footprint of our industry?**

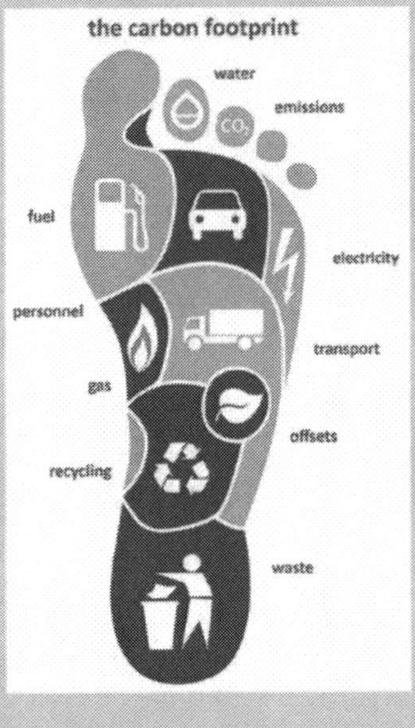

Geotechnical Carbon Calculator

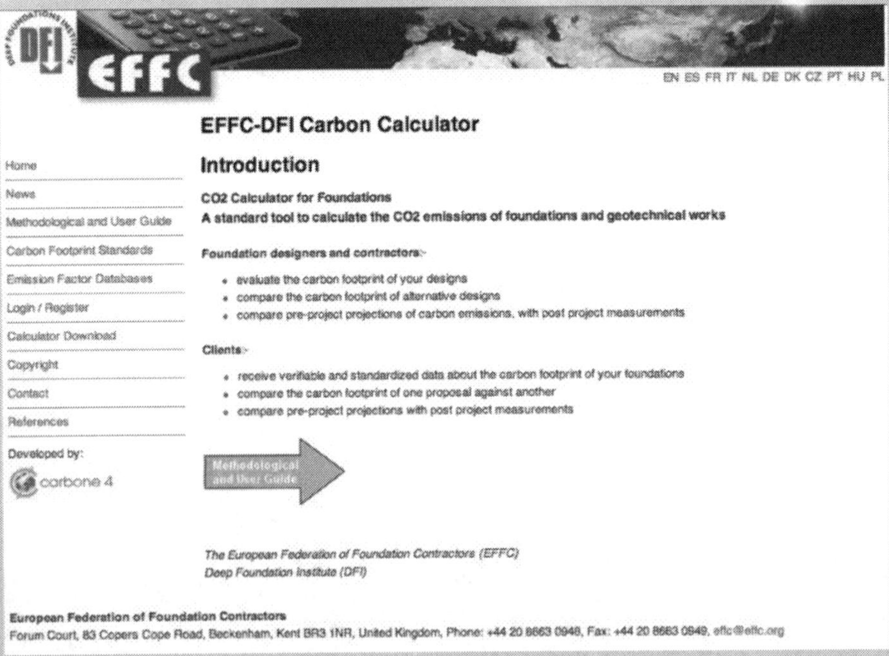

EN ES FR IT NL DE DK CZ PT HU PL

EFFC-DFI Carbon Calculator

Home
News
Methodological and User Guide
Carbon Footprint Standards
Emission Factor Databases
Login / Register
Calculator Download
Copyright
Contact
References

Developed by:
carbone 4

Introduction

CO2 Calculator for Foundations
A standard tool to calculate the CO2 emissions of foundations and geotechnical works

Foundation designers and contractors:-

- evaluate the carbon footprint of your designs
- compare the carbon footprint of alternative designs
- compare pre-project projections of carbon emissions, with post project measurements

Clients:-

- receive verifiable and standardized data about the carbon footprint of your foundations
- compare the carbon footprint of one proposal against another
- compare pre-project projections with post project measurements

Methodological and User Guide

The European Federation of Foundation Contractors (EFFC)
Deep Foundation Institute (DFI)

European Federation of Foundation Contractors
Forum Court, 83 Copers Cope Road, Beckenham, Kent BR3 1NR, United Kingdom, Phone: +44 20 8663 0948, Fax: +44 20 8663 0949, effc@effc.org

Geotechnical Carbon Calculator

- **Context**
  Until recently there was no carbon calculator for Deep Foundation / Ground Improvement Works
- **EFFC and DFI Objectives**
  Provide a Carbon calculator for Deep Foundation and Ground Improvement Works, allowing for absolute calculation and for comparisons
- **The tool should be simple, open, though comprehensive, methodologically sound and usable by all**

Geotechnical Carbon Calculator

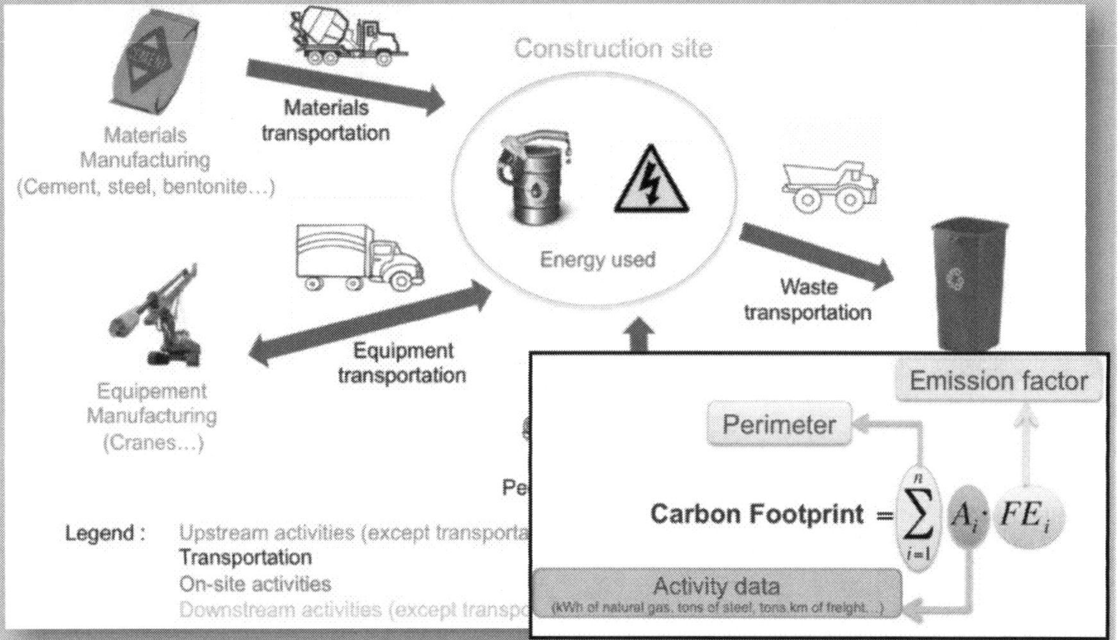

Geotechnical Carbon Calculator

- Emission Factor Database
  - A specific EFFC-DFI recommended database has been built from different databases
  - The tool contains also French, UK and US specific databases

Activity	EFFC-DFI recommended database
Materials	• Ecoinvent v2.2 • Bilan Carbone V7 • Sustainableconcrete • ICE v2 • DEFRA 2012
Energy	• DEFRA 2012 • IEA 2012
Transportation	• DEFRA 2012 • EcoTransit
Assets	• Bilan Carbone V7

Geotechnical Carbon Calculator

- CO_2 Emission Distribution

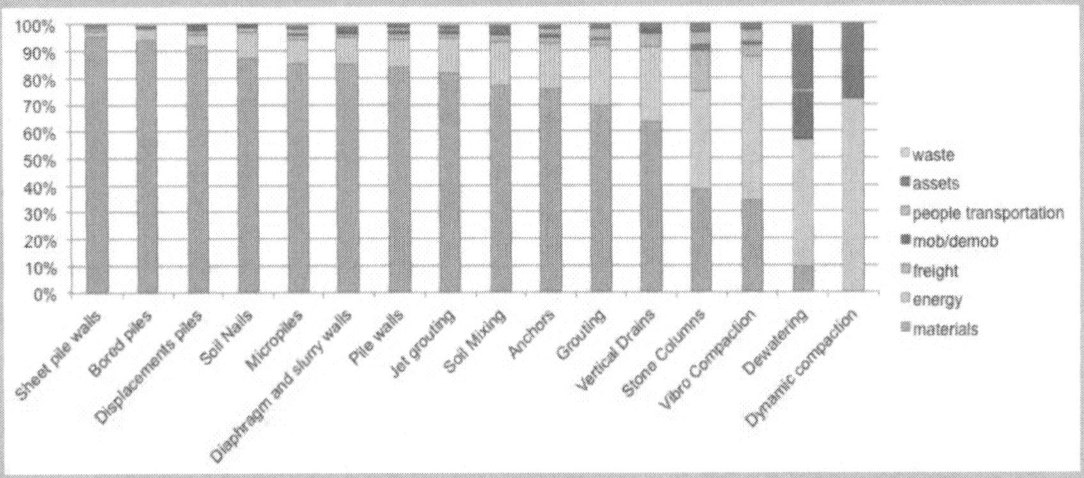

Geotechnical Carbon Calculator

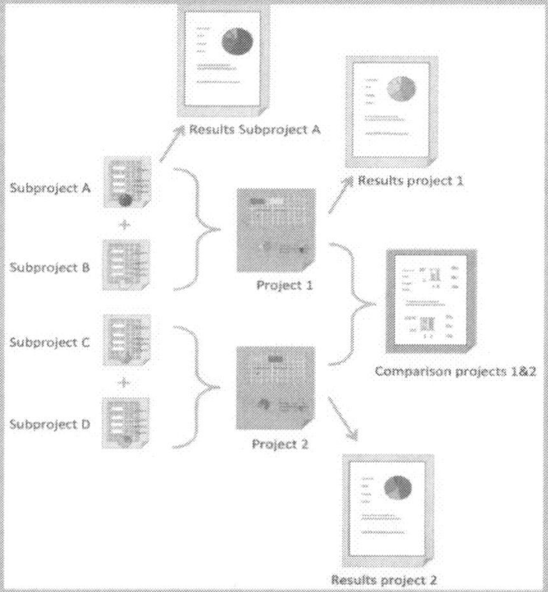

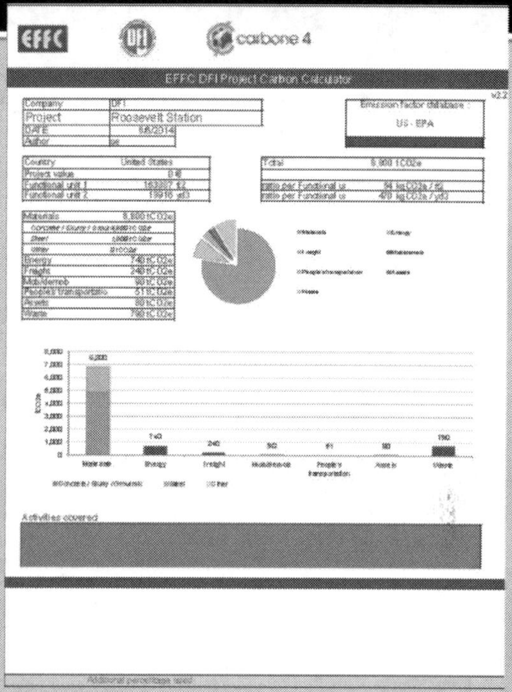

Geotechnical Carbon Calculator

- Bored Piles
  - 9 projects
    - Parker Steel Warehouse
    - Kühnsdorf
    - Graz
    - Wembley
    - Radlett Place
    - Barrow
    - A465 Dualling Section 3
    - Wigmore
    - IKEA

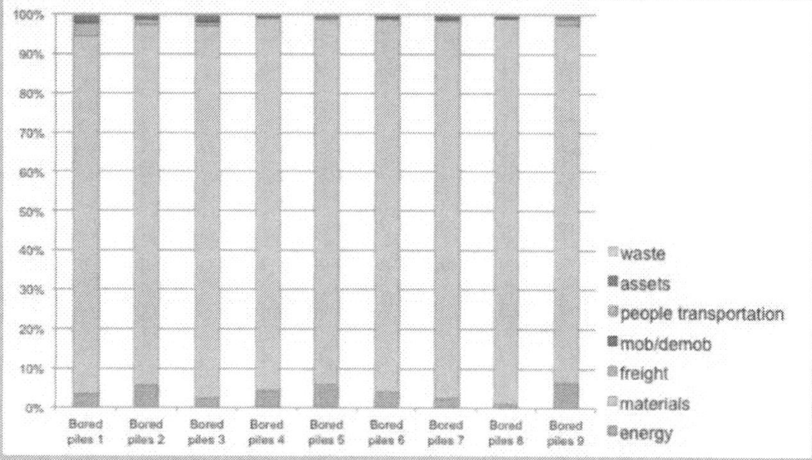

Geotechnical Carbon Calculator

- Displacement Piles
  - 8 projects
    - Southampton Dock
    - Houses Rotterdam
    - Burnley
    - DP World Welfare and Workshop Block
    - Rochdale
    - BridgeWater
    - Hoe Valley Leisure Centre
    - Riverside Museum

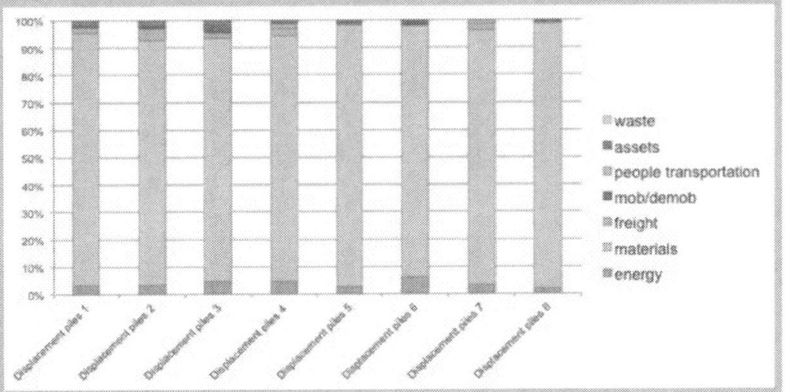

Geotechnical Carbon Calculator

- Sheet Piles
  - 3 projects
    - Berges de Varennes sur Seine
    - Highway A4
    - FBH Klagenfurt

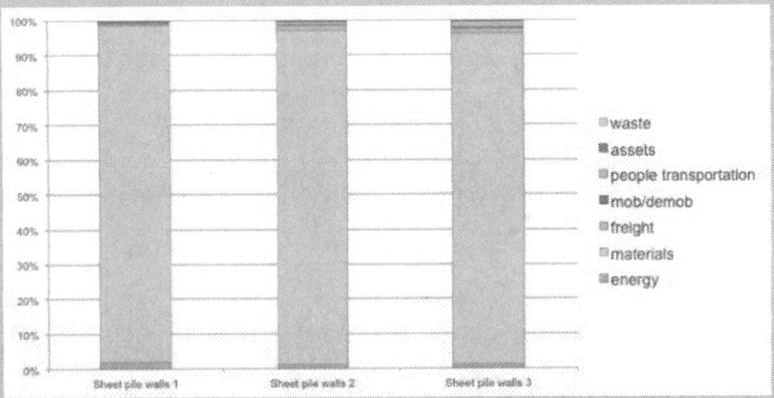

Geotechnical Carbon Calculator

- Diaphragm Walls
  - 11 projects
    - PML
    - Barangaroo
    - Crossrail
    - Toulon
    - Paddington
    - TourD2
    - Reservoir Enhancement
    - Singapour
    - ROP
    - Shoreham Adur Tidal Walls East Bank
    - Valenton

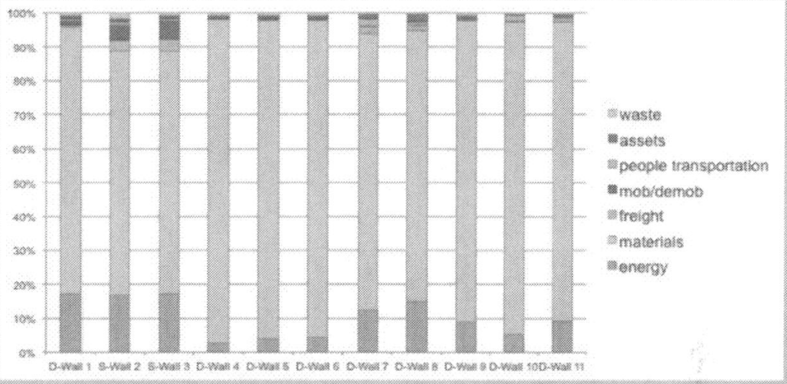

Geotechnical Carbon Calculator

- Soldier Pile and Secant Shaft Walls
  - 7 Projects
    - Prochazkova
    - Mesdag
    - Essex Waste
    - Maasvlakte2
    - Brussel
    - Holland Park
    - Metro Ceveny vrch

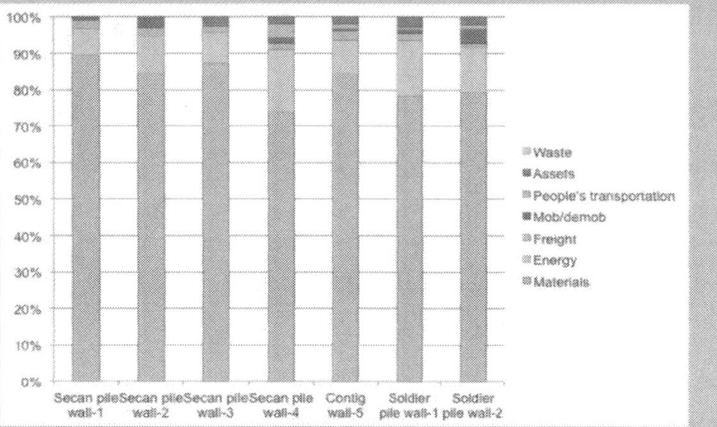

Geotechnical Carbon Calculator

- Anchors
  - 12 Projects
    - Voltaire
    - LKH Leoben
    - "Aurelia" Bergeggi
    - Bypass Bad St. Leonhard
    - Zwijndrecht
    - ROSAPARKS
    - Terneuzen
    - Schiedam
    - BARANGAROO
    - ALWA Donnersbach
    - Anchors – 32mm – 4 rigs
    - PRESSENSE

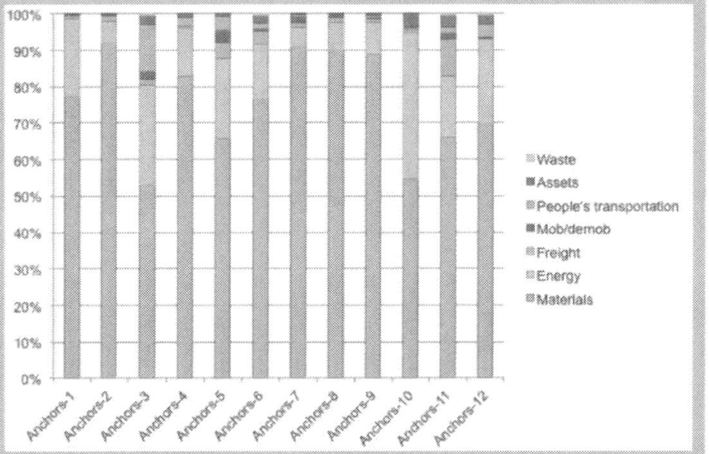

DFI Carbon Database Project

- Carbon Database Project
  - Collect actual US project data
    - Confirm Primary and Secondary emission sources for US based practice
    - Obtain standard industry emissions per Functional Units (i.e. per ft^2, yd^3, linear ft, etc)
      - Compare between different techniques
      - Understand and identify areas for improvement
      - Be ready to provide data to policy makers and avoid imposition of unrealistic benchmarks

DFI Carbon Database Project

- Database Input Forms
  - Diaphragm Walls
  - Secant, Tangent and Soldier Pile Walls
  - Anchors, others to come
  - Information Needed:
    - Project Description (Name, Year, State, Type)
    - Project Quantities (Area, Length, Volume)
    - Project Duration
    - Project Total Manpower Hours
    - Project Materials (Type, Quantities)
    - Project Energy Consumption (Gallons of Diesel, kWh)
    - The more and detailed information the better, but any information is welcome
  - Contact Carlos Englert (cenglert@schnabel-eng.com) for help or comments

DFI Carbon Database Project

DFI CARBON DATABASE
ANCHOR INPUT FORM

DEEP FOUNDATIONS INSTITUTE DFI

GENERAL PROJECT INFORMATION

Location (Country):

Location (City):

Total Linear Feet of Tiebacks (ft):

Tieback Type (Select One):
- ○ Temporary
- ○ Permanent
- ○ Removable
- ○ SBMA
- ○ Compression

Tieback Reinforcement (Select One):
○ Strands ○ Bars

Tieback Grouting Method (Select One):
- ○ Tremie
- ○ Pressure Grouted
- ○ Post-Grouted

Other Information of Interest:

Project Name:

Location (State):

Date:

Total Number of Tiebacks (No.):

Tieback Drilling Method (Select One):
- ○ Cased Hole
- ○ Open Hole
- ○ Auger
- ○ Other (Specify):

Corrosion Protection System (Select One):
- ○ Single
- ○ Double
- ○ Other (Specify):

Project Duration (Days):

Workforce (Full-Time Equivalent):

Average Commute Distance (miles):

DFI CARBON DATABASE
ANCHOR INPUT FORM

DEEP FOUNDATIONS INSTITUTE DFI

PRIMARY EMISSION SOURCES
Materials and Energy

Energy
Diesel (Gallons):

Network Electricity (kWh):

Tendon
Fabrication Location (Select One):
○ Onsite ○ Outside, indicate distance to Site (miles):

Total Weight of Steel (ton):

Steel Source (Select One):
- ○ New
- ○ Reused
- ○ Recycled, indicate Recycled Content (%):

Distance to Site if fabricated Onsite (miles):

Type of Encapsulation Material (Select One):
- ○ PVC
- ○ HDPE
- ○ Other, Specify

Total Weight of Encapsulation (tons):

Distance to Site if fabricated Onsite (miles):

Grout
Mixing Location (Select One):
○ Onsite ○ Outside, indicate distance to Site (miles):

Total Volume of Grout (cy):

Cement Type (Select One):
○ Type I ○ Type II
○ Type III ○ Type IV
○ Type V ○ Type IS
○ Type IP ○ Type P
○ Type I(PM) ○ Type S
○ Type I(SM) ○ Other (Specify):

Distance to Site if mixed Onsite(miles):

Weight of Cement (lbs/cy):

Weight of Aggregate (lbs/cy):

Distance to Site if mixed Onsite (miles):

Weight of Water (lbs/cy):

Distance to Site if mixed Onsite (miles):

Weight of Additive (lbs/cy):

Distance to Site if mixed Onsite (miles):

Closing Remarks

- Sustainability - Balance of the Economic, Social, and Enviromental factors to meet present and future needs
- The Carbon Footprint is a good parameter to quantify the sustainability
- The Carbon Calculator is a useful tool to determine the Carbon Footprint
- Please collaborate with the Carbon Database Project

Accelerating CCUS Deployment for Non-coal Energy Chains & Offshore Storage

Tip Meckel & GCCC

AIChE – Carbon Management Technology Conference
Houston, Texas
November 18, 2015

TEXAS Geosciences
Bureau of Economic Geology
Jackson School of Geosciences
The University of Texas at Austin

Gulf Coast Carbon Center

What Currently Limits Success of Global-Scale CO_2 Storage?

UTILIZATION

- **Costs are high**: These relate primarily to **capture, but also infrastructure**. Focus should be put on capture from lowest cost / lowest risk sources. Start easy, transfer gains to difficult.
- **Market immaturity**: Lack of a carbon price prevents market efficiencies and risk-management of investments.

OFFSHORE

- **Public Scrutiny**: HSE: EU experience, US shale gas experience.
- **Geographic distribution of onshore storage is not uniform**. Suitable onshore storage sites are geologically diverse but unequally geographically distributed, presenting potential geopolitical challenges for consensus on CCUS.

Mostly, CCS has been pursued as an emissions reduction strategy (coal, electricity) with a dominant focus local (onshore) storage. Early international projects are associated with gas, or EOR.

What other lenses can this technology be seen through?

CCS has a role in diverse energy chain development

CCS can/should be more than just post-combustion capture from coal-fired electric utilities.

- **Natural Gas** Most existing CCS projects are gas related.
  - *Global Production: CO_2 liability on reserves.*
- **Refining**
  - *Hydrogen generation (SMR)*
- **Enhanced oil recovery**
  - *Utilization drives infrastructure*
- **Methanol**
- **LNG**
  - *Compression Permitting: CCS as BACT*

World Gas Reserves

Gas quality problems are holding back investment.

Approximately 40% (2600 Tcf) of reserves are estimated to be sour

Malaysia, Qatar, Saudi Aramco, Canada, China, Venezuela, Brazil and many other countries have a high demand to treat sour gas fields.

Table 12.2 • World proven sour gas reserves, end- 2006

	High H_2S only (tcm)	High CO_2 only (tcm)	High H_2S and CO_2 (tcm)	Total (tcm)	Total % of total reserves
Mexico & Latin America	0.3	1.1	0.3	1.7	21
Europe	0.1	0.7	0.3	1.1	19
Former Soviet Union	0.8	10.1	7.3	18.2	34
Africa	0.0	0.5	0.5	1.0	8
Middle East	2.6	0.4	40.9	44.0	60
Asia-Pacific	0.3	4.4	2.3	7.1	46
World	4.2	17.2	51.6	73.1	43

Note: Excludes North America. High H_2S is more than 100 parts per million; high CO_2 is more than 2%.
Source: Bourdarot (2007).

SOURCE: US EIA

Challenges that high CO_2 fields present

INFRASTRUCTURE: Material costs mostly.

MARKETABILITY: must meet specs

Natuna (70% CO_2; >30 Tcf CO_2)
Kuala Langsa (82% CO_2; >20 Tcf CO_2)

DISPOSAL (CCS): cost of capture of CO_2 from natural gas production streams
is significantly lower than from the flue gases of power stations (IEA, 2008).

EOR: Various & vigorous global offshore CO_2-EOR projects: Malaysian Developments

Natuna D-Alpha

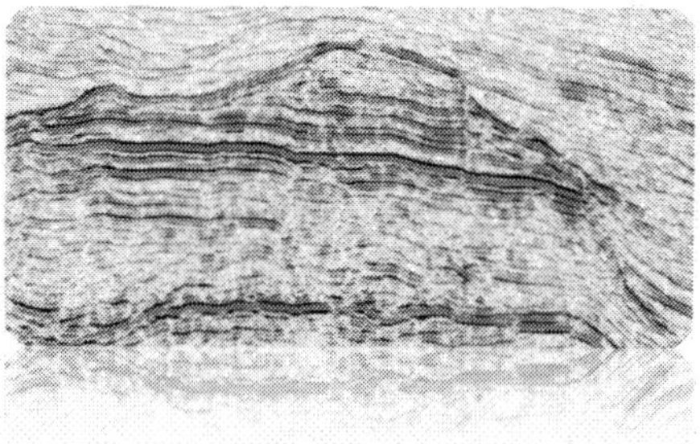

Southeast Asia's largest gas reservoir is the Indonesian Natuna field, with 1.3 tcm (46 TCF) of gas reserves, but it has a CO_2 content as high as 71%.

Technical difficulties have delayed Natuna's development for more than two decades.

Sarawak K5, Malaysia

Petronas/Shell Baram Delta Operations

World's biggest offshore EOR project
\$12B over 30 years

- Gas from the shallow-water discovery contains about ***<u>70% CO_2</u>***
- CO_2 extracted from K5 is understood to be intended for ***<u>re-injection</u>*** into three fields — Jintan, Seria and M4.
- M4 is a depleted gas field that has been flagged up as a potential candidate for CO_2 sequestration.
- Studies will look at reinjecting CO_2 in an unidentified field about 50 kilometres away from Bintulu to boost hydrocarbon recovery.

- **<u>Global Technology Race</u>: pioneering high CO_2 gas field development + CCS**
  - **Lula, Brazil = EOR.**

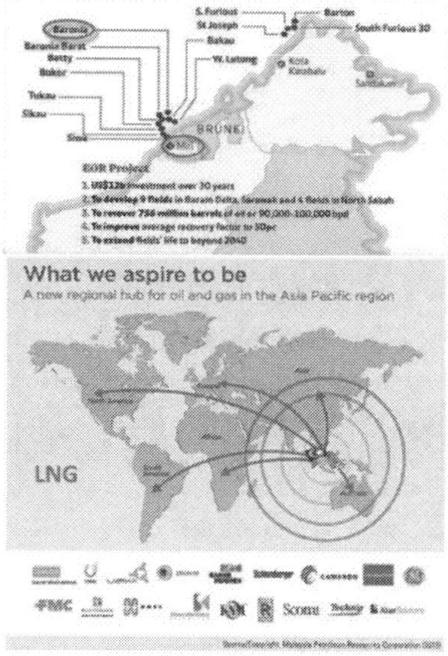

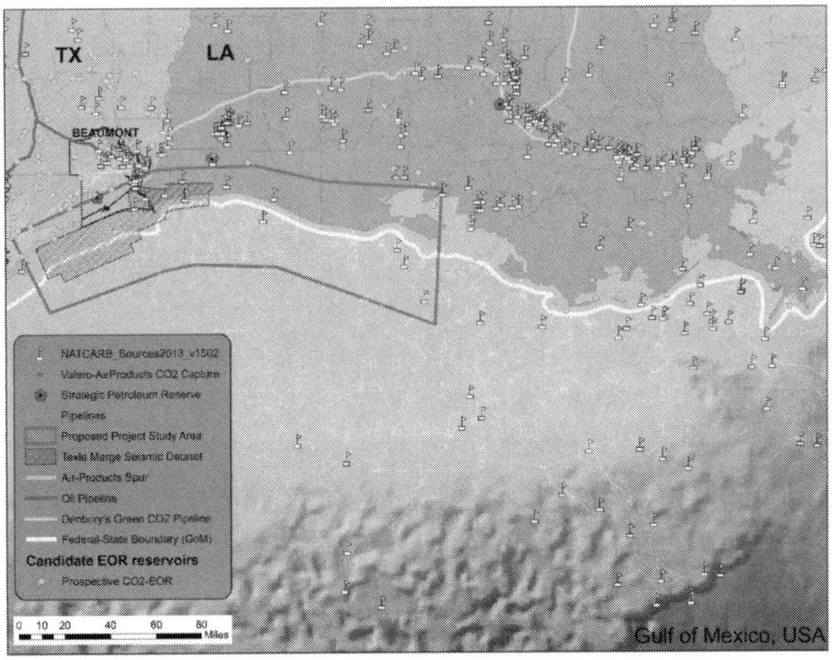

REFINING:

Generate tremendous CO_2 (H2 production from CH4)

Solving CO_2 emissions allows development of those resources in more publically acceptable and environmentally mitigated way.

Alberta, CANADA

Quest Carbon Capture and Storage Project

- Goal to reduce CO_2 emissions (35% capture, Amine) from Shell's oil sands operations by more than ***one million tonnes a year*** by capturing CO_2 from its Scotford bitumen upgrader and permanently storing it deep underground.

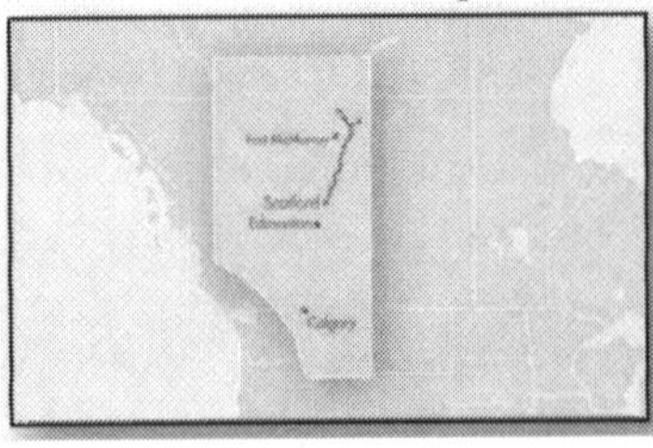

Texas, USA

Keystone Pipeline: proposed termination in east TX

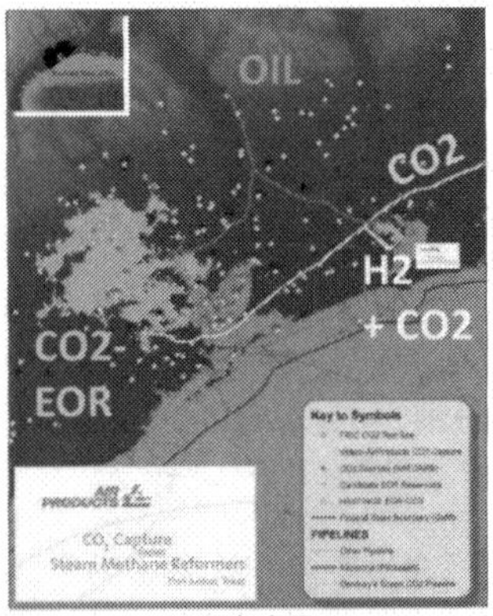

US CO_2 Networks for EOR

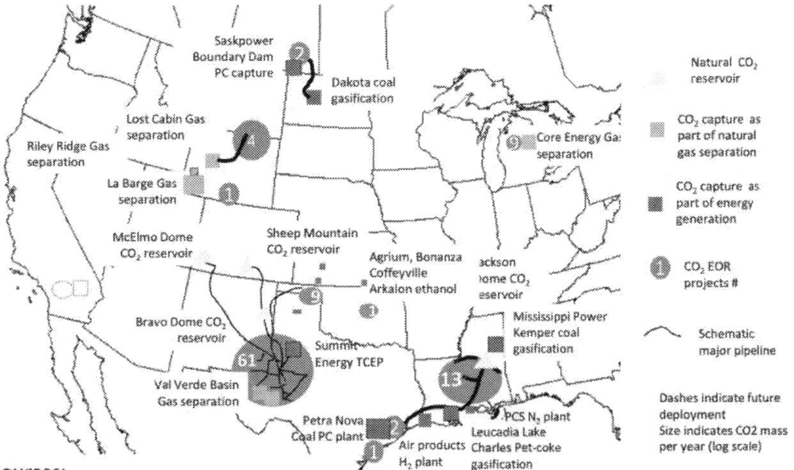

Sources:
O&G J 2014; ARI 2014; Melzer 2012,
Corporate web pages

Methanol / Ammonia

- **Conventional steam reforming is the simplest and most widely practiced route to synthesis gas production:**

  $2\ CH_4 + 3\ H_2O \rightarrow CO + CO_2 + 7\ H_2$ (Synthesis Gas)

  $CO + CO_2 + 7\ H_2 \rightarrow 2\ CH_3OH + 2\ H_2 + H_2O$

  This process results in a considerable hydrogen surplus, as can be seen.

  If an external source of CO2 is available, the excess hydrogen can be consumed and converted to additional methanol. The most favorable gasification processes are those in which the surplus hydrogen is "burnt" to water, during which steam reforming is accomplished through the following partial oxidation reaction:

  $CH_4 + ½O_2 \rightarrow CO + 2\ H2 \rightarrow CH3OH$

  $CH4 + O2 \rightarrow CO2 + 2\ H2$

  The carbon dioxide and hydrogen produced in the last equation would then react with an additional hydrogen from the top set of reactions to produce additional methanol. This gives the highest efficiency, but may be at additional capital cost.

  Unlike the reforming process, the synthesis of methanol is highly exothermic, taking place over a catalyst bed at moderate temperatures. Most plant designs make use of this extra energy to generate electricity needed in the process.

LNG

"Carbon Capture and Storage (CCS) is "available" and should be considered in Step 1 of a BACT analysis for high CO2-emitting sources"

Typical global feed gas composition up to 12% CO2

EPA

GHG Guidance: BACT Step 2 (cont.)

- For Carbon Capture and Storage:
  - CCS technology is composed of 3 main components: (1) CO_2 capture and/or compression, (2) transport, and (3) storage.
  - CCS may be eliminated if any of the 3 components working together are deemed technically infeasible for the proposed source.
    - *e.g.*, no space available for CO_2 capture equipment at an existing facility; right-of-ways prevent building a pipeline or access to an existing CO_2 pipeline; no access to suitable geologic reservoirs for sequestration or other storage options.
  - CCS may be eliminated if it can be shown that there are significant differences pertinent to the successful operation for each of these three main components from what has already been applied to a different source type.

21

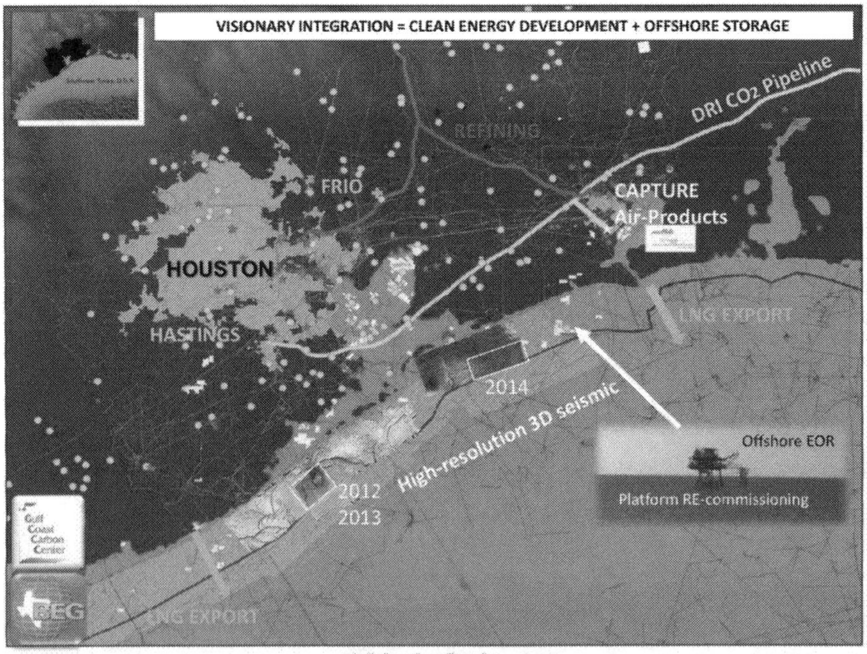

Accelerating sustainability of diverse future clean energy developments through CCUS

CCS can/should be more than just post-combustion capture from coal-fired electric utilities.

- Integration of CCUS into diverse clean energy chains will accelerate deployment.
- Kick off technology competition: (Jobs, patents)
- Pioneering high CO_2 gas field development+CCS;
- EOR; LNG; REFINING

The offshore holds great promise for global CCS deployment.

- Offshore storage suits most nations.
- With collective access to adequate storage resources many countries may feel more willing to enter into future binding emissions negotiations.

International Organization for Standardization (ISO) Technical Committee 265 CCUS, including CO2-EOR: International Panel Discuss – the what, the why, and the how.

Prepared By:

Steven M. Carpenter, Vice, President
ADVANCED RESOURCES INTERNATIONAL, INC.
Arlington, VA

Unconventional Resources • Enhanced Recovery • Carbon Sequestration

Advanced Resources International, Inc.

Agenda – an open conversation

- What is being done – overview of each of the working groups within TC-265
- Why is this work important? How is it viewed as a best practice and how is the work completed by ISO planned to be used by the developing world (non-OECD economies)?

Advanced Resources International, Inc.

2

Agenda – an open conversation

- What are the real (granular, technical) issues that are being wrestled with and why are they an issue?

didaulamgi24h.com

mostholyplace.com

Advanced Resources International, Inc.

3

Introduction of Panel

- Jim Ekkman, LTI (USA): WG1 Capture subject matter expert, member of the US TAG, engineer and economist for Leonardo Technologies, as a prime contractor to the US Department of Energy
- Greg Schnacke, Denbury (USA): WG2 & 6 Transportation & CO2-EOR subject matter expert, member of the US TAG, Executive Director, Governmental Relations for Denbury Resources, former Chief Executive of the Colorado Oil & Gas Association, a graduate of the University of Kansas and the University of Tulsa, College of Law

Advanced Resources International, Inc.

4

Introduction of Panel

- Stefan Bachu, Alberta Innovates (CAN), WG3 Storage, subject matter expert and member of the Canadian Mirror Committee, holds an engineering degree in Water Resources and a Ph.D. in Flow and Transport Processes, currently he is the Distinguished Scientist at Alberta Innovates, and is the Associate Editor for CO2 storage for the International Journal of Greenhouse Gas Control

- Peter Andersen, SGS (UK), WG4 Quantification & Verification subject matter expert to the UK Mirror Committee, Global CCS Director for SGS, extensive experience as a Greenhouse Gas Verifier in the EU Emissions Trading Scheme and many other GHG reporting programs.

Advanced Resources International, Inc.

5

Introduction of Panel

- George Koperna, ARI (SA), WG6 CO2-EOR subject matter expert to US TAG, VP & Director of Reservoir Engineering for ARI, and the Co-chair of the Conference, George holds Bachelor's and Master's degrees in Petroleum Engineering from WVU

- Steve Carpenter, ARI (USA), WG2, 5 & 6 subject matter expert, Chair of United States TAG, VP at ARI, moderator for today's panel.

Advanced Resources International, Inc.

6

ISO = A Global System

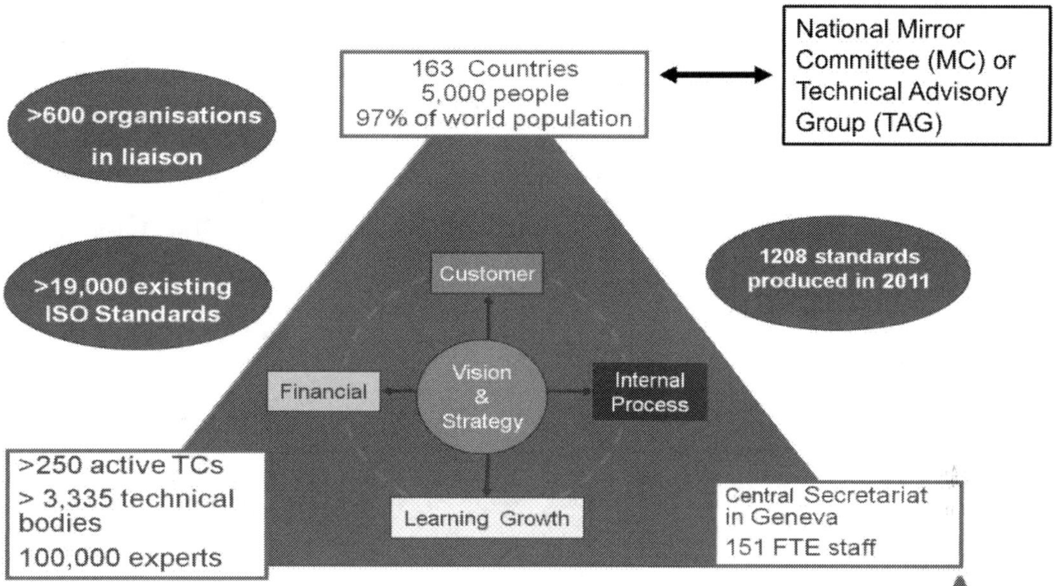

Advanced Resources International, Inc.

7

ISO Standards Development

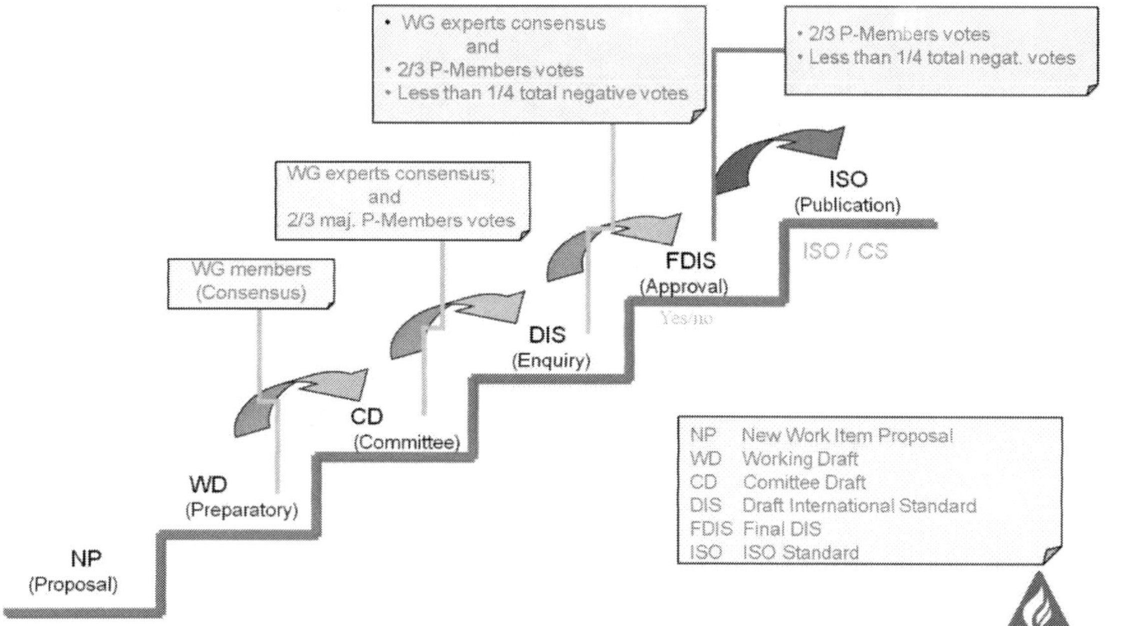

Advanced Resources International, Inc.

8

ISO Standards Development

- ISO does not write standards
- Technical Committees write standards
- P-Member countries approve standards
- Nations adopt ISO standards (MC)
- ISO does not influence the technical content

Advanced Resources International, Inc.

9

TC-265 Working Groups

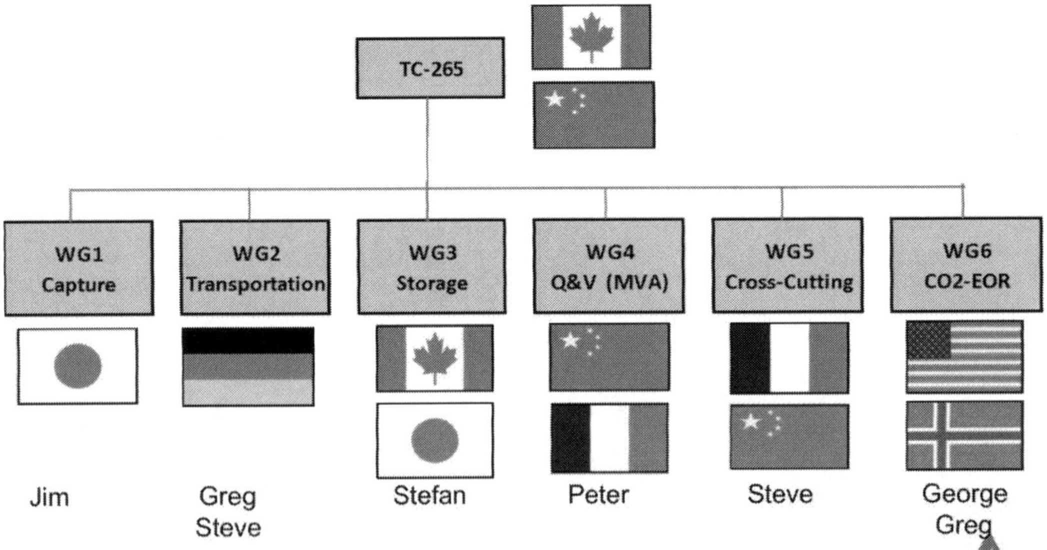

Advanced Resources International, Inc.

10

Issues

Example of harmonizing cross-cutting terms among WGs: CO_2 stream

- **WG5**: *a stream consisting overwhelmingly of carbon dioxide*
- **WG2:** stream consisting overwhelmingly of carbon dioxide *with a limited fraction of other chemical substances*
- **WG3**: a stream of carbon dioxide *that has been captured from an emission source (e.g., a fossil fuel power plant) and meets applicable regulatory requirements for CO_2 storage*

  Note: It may include any incidental associated substances derived from the source materials or the capture process, added as a result of commingling for transportation, added to the stream to enable or improve the injection process and/or trace substances added to assist in CO_2 migration detection.

11

Advanced Resources International, Inc.

Forward Looking Statements

All statements in this presentation that are other than statements of historical facts are forward-looking statements which contain our current expectations about our future results. Forward-looking statements involve numerous risks and uncertainties. We have attempted to identify any forward-looking statements by using words such as "anticipates", "believes", "could", "expects", "intends", "may", "should" and other similar expressions.

Although we believe that the expectations reflected in all of our forward-looking statements are reasonable, we can give no assurance that such expectations will prove to be correct. A number of factors may affect our future results and may cause those results to differ materially from those indicated in any forward-looking statements made by us or on our behalf. Such factors include our early stage of technology development; our need for capital to finance necessary research and product development; our ability to attract and retain key employees and strategic partners; our ability to achieve and maintain profitability; fluctuations in the trading price and volume of our stock; competition from other providers of similar products and services; and other unanticipated future events and conditions. For further information concerning risks and uncertainties that may affect our future results, please review the disclosures as may be contained from time to time in our filings with SEDAR. Other than as required by applicable securities laws, we undertake no obligation to publicly update or revise any of our forward-looking statements, whether as a result of changed circumstances, new information, future events, or for any other reason occurring after the date of this presentation. *This presentation does not constitute an offer to sell or solicitation of an offer to buy securities in any jurisdiction.*

2

About CO_2 Solutions Inc.

- Leader in enzyme-based CO_2 capture
- Based in Quebec, Canada
- $50 million invested to date
- 22 employees
  - 17 R&D, including 7 PhDs
- 48 issued and 40 pending patents
- Entering commercial phase
- Publicly traded on TSX Venture Exchange (Symbol: CST)

3

Amine-Based CO_2 Capture

- High costs
  - Steam for solvent regeneration = high operating costs
  - $60-90/tonne of CO_2 for flue gas application[1,2]
  - Cost is critical issue to be resolved for EOR and other CO_2 reuse and sequestration to be expanded[3,4]
- Environmental issues[5]
  - Solvent degradation and stability issues; extensive flue gas pre-treatment required
  - Toxic aerosol emissions
  - Solvent losses
  - Waste products to handle

Sources:
1) http://ccemc.ca/wp-content/uploads/2012/12/C101033-HTC-DEV-JA-03-001-1-R1-+-FEED-REPORT-CCEMC.pdf, Page 19 (~$70/tonne cost)
2) https://www.netl.doe.gov/File%20Library/Events/2014/2014%20NETL%20CO2%20Capture/A-Bhown-EPRI-CO2-Capture-RD-EPRI.pdf, Page 6
3) http://www.ai-ees.ca/media/10958/2010_barriers_to_co2_eor_report_final_june4-13.pdf, Page 44
4) http://www.energy.gov.ab.ca/Org/pdfs/CCS_Implementation.pdf, Page 10
5) http://www.sciencedirect.com/science/article/pii/S1750583614001777

4

Our Process – An 'Industrial Lung'

- Benign aqueous carbonate salt solvent
- Low temperature stripping using low-grade (75-85°C) heat
- Absent
  - Solvent reclamation
  - Water wash
  - Intercooling

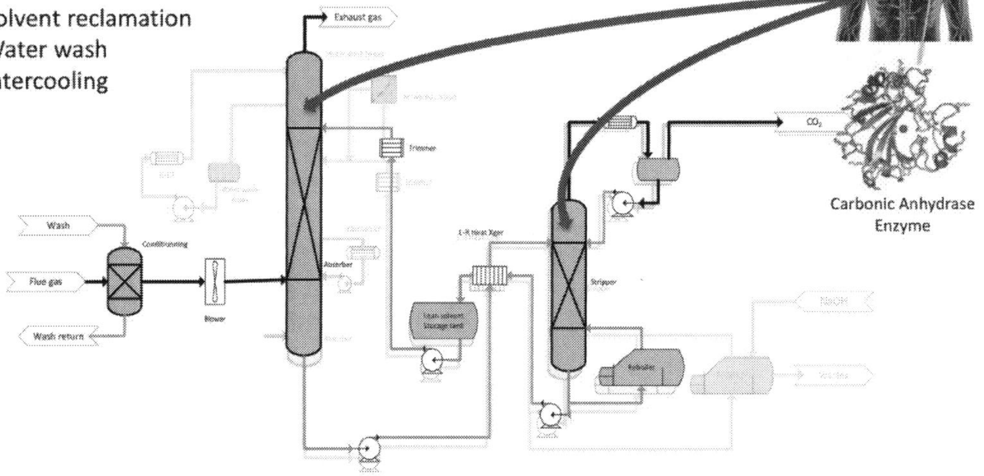

5

Enzymes Today

- Enzymes are nano-sized protein structures with an active site(s) which catalyze / accelerate a specific biochemical reaction; enzymes are not living organisms
- Used to optimize dozens of existing processes
  - Biofuels production
  - Detergents
  - Food production
  - Pharmaceuticals production
  - Beer making
  - And many others!
- Enzymes save energy, chemicals, raw materials and reduce waste
- Safe and environmentally benign
- A \$3.5 billion market in 2011[1]

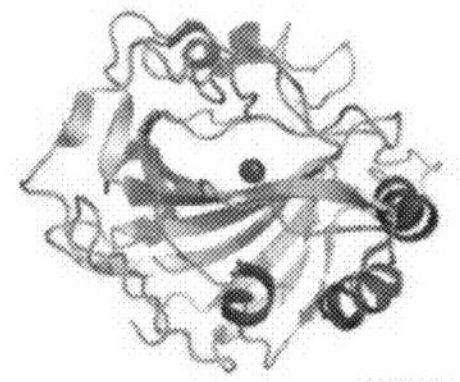

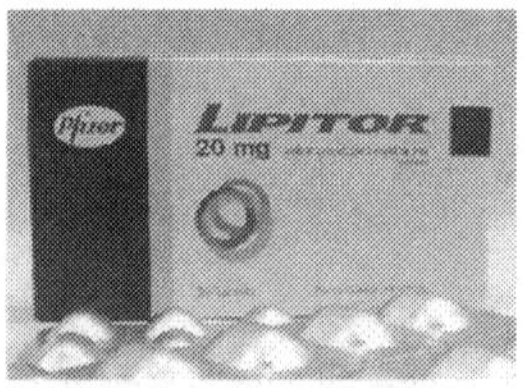

6

Sources: (1) ReportsnReports.com

Intellectual Property

CO2 SOLUTIONS

BROAD PATENT PORTFOLIO FOR USE OF **CARBONIC ANHYDRASE** IN CANADA, U.S., EU, CHINA, AUSTRALIA AND OTHER MARKETS

48 ISSUED **40** PENDING*

vs. 13 and 19 respectively in 2008

SOLVENTS
Carbonates
Amino Acids
Amines
Combinations

INDUSTRIAL SECTORS
Power
Steam
'Drop-in' Applications

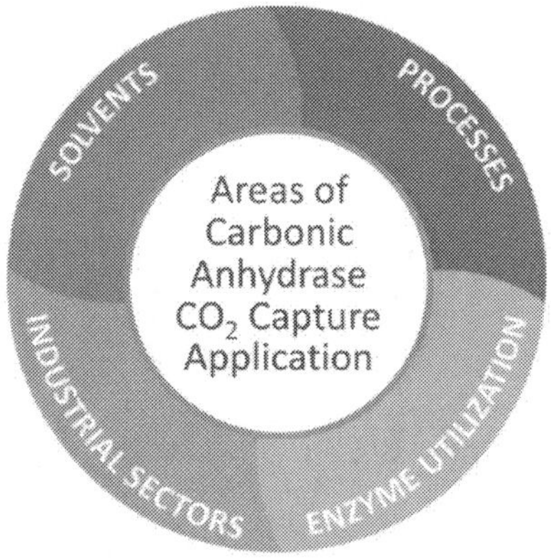

PROCESSES
Packed Tower
High-Intensity Contactor
Spray Scrubber
Universal

ENZYMES AND THEIR UTILIZATION
Enzyme Strains
Soluble
Particle-Based
Analogs

*at March 31, 2015

7

Completed: 1 tonne/day Pilot Testing

CO2 SOLUTIONS

- Conducted at Energy & Environmental Research Center (EERC), Jan. 2015
- Coal and natural gas flue gases
- Stripping heat was provided by hot water; steam never used
- Cost of capture incl. compression to 2250psi estimated at $39/tonne by EERC
  - Use of low-grade, nil value heat from outside power plant steam cycle is the key
- Constant capture performance
  - No enzyme activity loss during the test
- No solvent make-up
- Zero waste produced
- Detailed results presented April 29, 2015 at Pittsburgh CCUS conference (http://tinyurl.com/nkuabns)

EERC Energy & Environmental Research Center® Putting Research into Practice

DEPARTMENT OF ENERGY UNITED STATES OF AMERICA

8

Completed: 8 tonne/day Demo Testing

CO2 SOLUTIONS

- Located in Salaberry-de-Valleyfield, near Montreal
- 2,500 operational hours completed November, 2015
- CO_2 capture from flue gases of natural gas fired boiler (~8.5% CO_2 content)
- Uses aqueous carbonate salt solution accelerated by proprietary CA enzyme
- Enzyme neat in solvent flowing freely throughout process including in reboiler
- Stripping uses hot water (82.5°C)

Canada ecoENERGY

CCEMC

9

Demo Testing: Some Facts

CO2 SOLUTIONS

- From sketches to operation in less than 14 months
- On schedule and on budget
  - Built for $2.1M
- 1 week of training of third-party operators
- 3 weeks from power up to normal operation at design capacity
- Objective was to test drive the process
  - Execute start-ups and shut-downs, simulate power failure and restart
  - Extended time with enzyme at high temperature; challenge enzyme stability
  - Determine impact of operating conditions on energy consumption and capture performance
- Largest and longest ever enzyme-based demo project built and operated under realistic conditions

Canada ecoENERGY

CCEMC

10

Performance Observed – Enzyme Stability

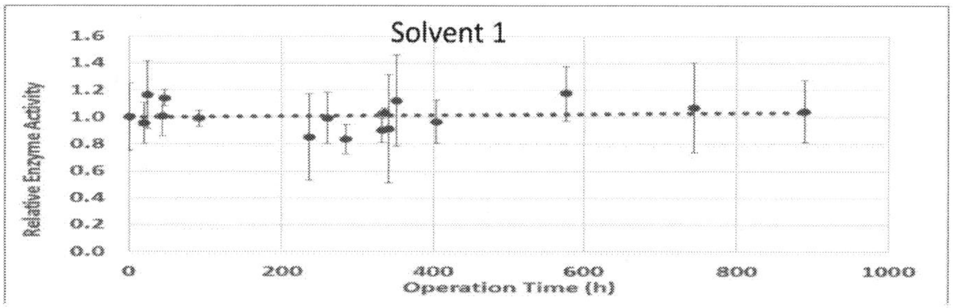

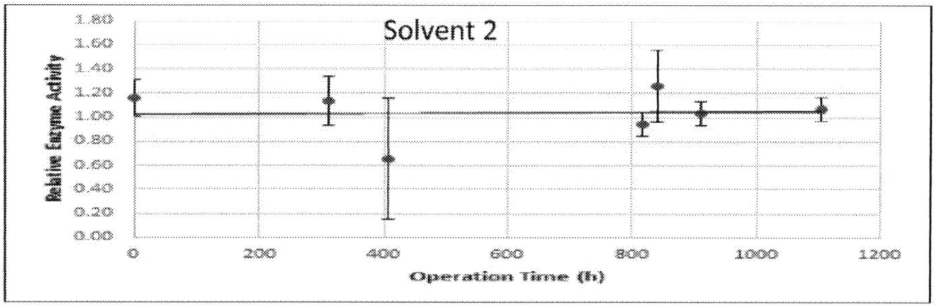

Enzyme activity measured from solvent samples in standardized kinetic measurement cell

11

Performance Observed – Enzyme Dosing

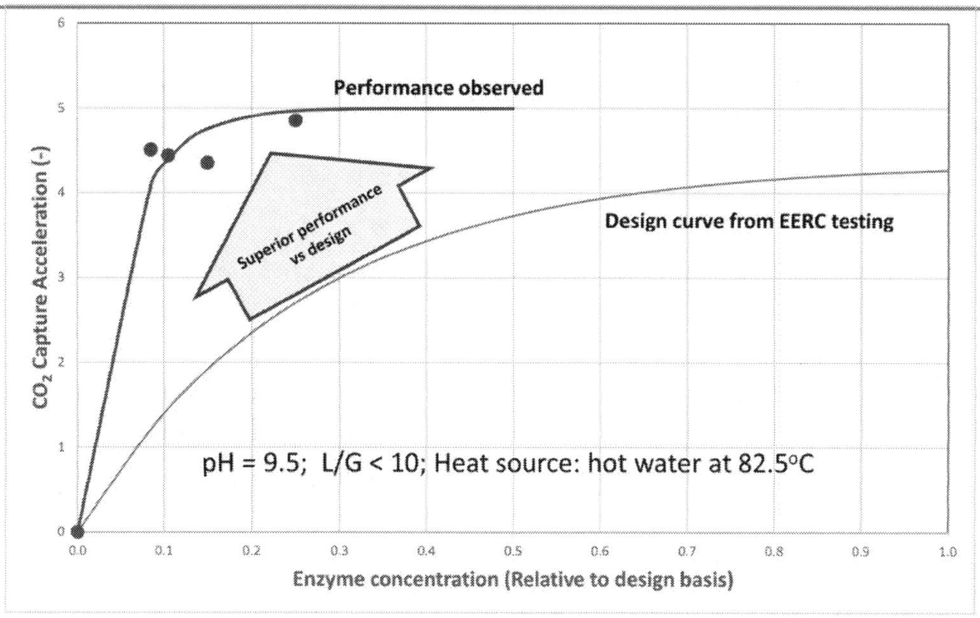

- Reached unit capture capacity with enzyme dosing ~¼ of design value
- Direct benefit to operating costs

12

Performance Observed - Capture

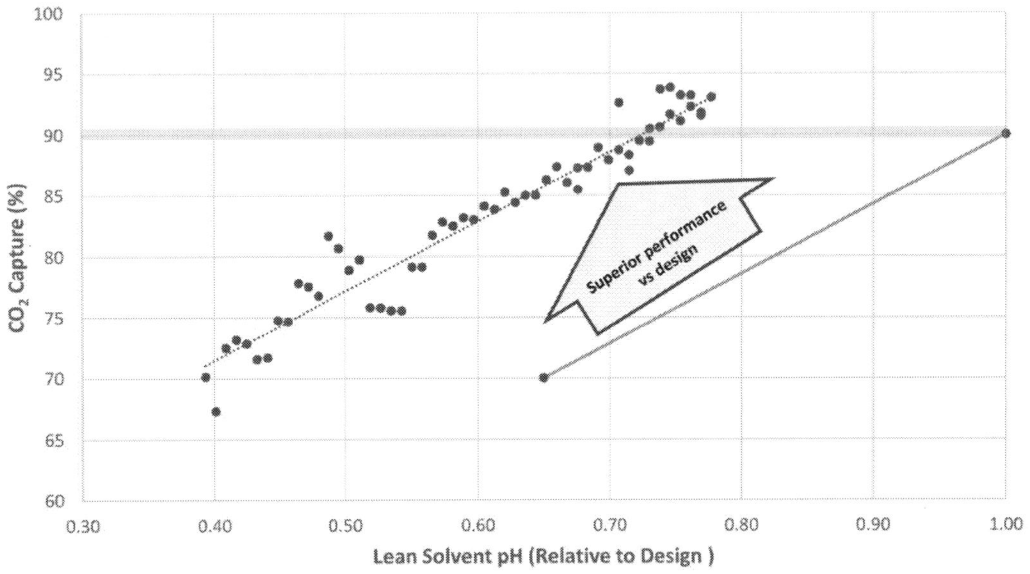

- Lower pH required than expected = longer enzyme stability

13

Environmental and Other Performance

- No solvent degradation
- No solvent makeup
- No aerosol emissions
- No waste products generated
- Simple, autonomous operation
  - Pilot operation was with day shift only
  - Commercial operation does not require full-time operators; est. 0.25 FTE

14

CO_2 Purity

- 99.5% CO_2 purity achieved (dry basis) by independent lab verification with only O_2 and N_2 remaining to be removed
- Can meet requirements for food and beverage use (ISBT)[1] without extensive downstream purification
- Exceeds Enhanced Oil Recovery (EOR) and other reuse application purity requirements
- Eliminates typical additional cost of post-capture purification

[1]*http://www.bevtech.org/*

Initial Conclusions and Next Steps

- **>2,500 hours** of operation at design capacity with **stable enzyme performance**
- **No steam used**; confirmed ability to use of low-grade (75-85°C) heat for stripping
  - Total energy consumption similar to amines, but energy is far lower quality/value
- **No aerosols or wastes** generated
- **No degradation** of solvent
- **High purity CO_2** produced
- **Simple**, autonomous operation
- Mass and energy balance data now being independently validated by leading engineering firm
  - Will be used for costing and engineering of commercial units at various scale
  - **Expected to confirm potential <$40/tonne at large scale**

Further Innovation: Rotating Packed Bed

CO2 SOLUTIONS

- Exclusive collaboration with GasTran Systems (GTS) of Cleveland (OH) - leader in RPB technology
- Process intensification in action
- Generates extremely fast G-L interface renewal; Mass transfer is orders of magnitude greater than packed column
- Enables maximum performance from enzyme
- RPB never used for CO_2 capture previously because of slower kinetics of conventional amine solvents
- RPB is proven commercially in beverage deaeration and chemical production

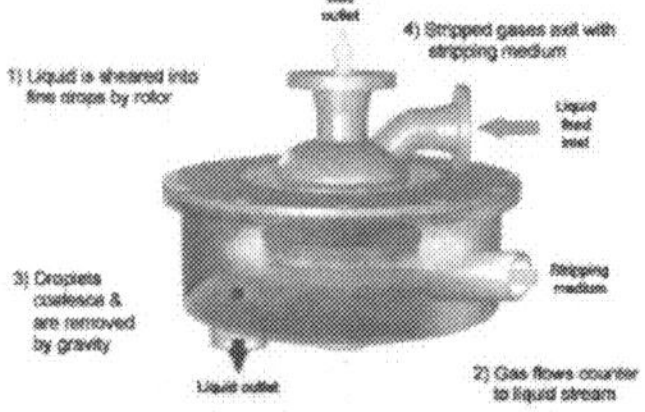

RPBs in Use at Dow Chemical

17

RPB vs. Packed Column

CO2 SOLUTIONS

90% CO_2 Capture from Natural Gas-Fired Boiler Flue Gases; 300 tonnes/day CO_2 captured

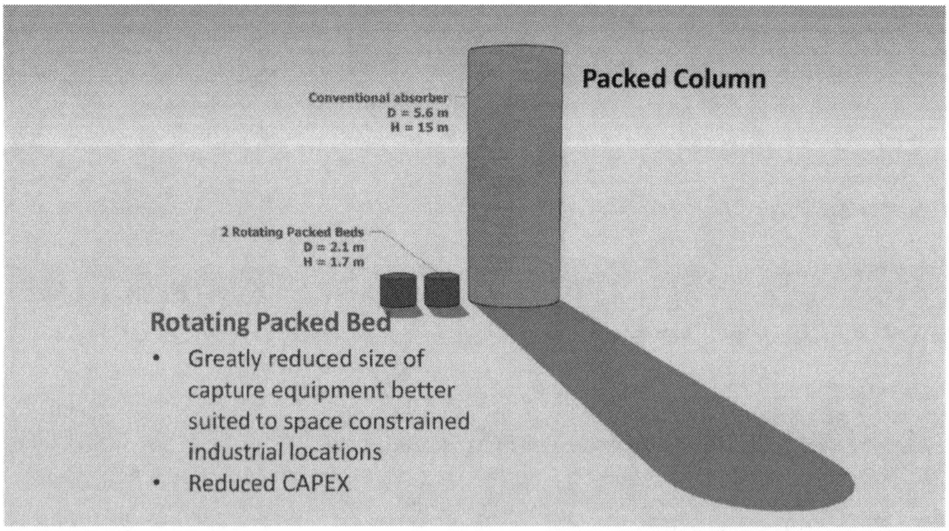

18

Commercial Application for CCUS

Enhanced Oil Recovery (EOR) with Geologic Sequestration

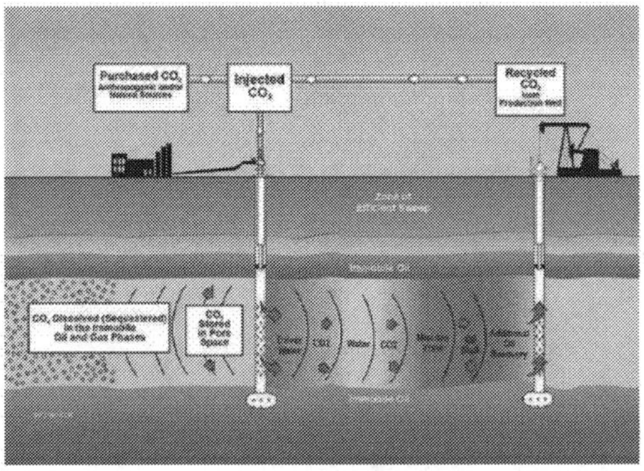

Carbonation

Pulp and Paper

Greenhouses

Water Treatment

Algae Biofuels

Mineralization

- >$1.5 billion existing market for CO_2 as industrial input/commodity(1)
- CCUS provides 'bridge' to CCS

19 (1) Esprit Associates and CO_2 Solutions, 2014

Commercial Activities

- Full techno-economics (Q4 2015)
- Relocation of **8 tonne/day** demo unit for commercial operation (H1 2016)
- Pilot testing of RPB (Q1 2016)
- **30 tonne/day** commercial CO_2 reuse project at greenhouse incorporating RPB (2016-2017) and **300 tonne/day** commercial Enhanced Oil Recovery (EOR) project incorporating RPB (2016-2018)
- Larger scale broader commercial rollout (2017+)

20

CO2 SOLUTIONS
THANK YOU
CO2 Solutions Inc.
Jonathan A. Carley
Vice-President, Business Development
jonathan.carley@co2solutions.com
M: 905-320-6260
www.co2solutions.com
© 2014 CO2 Solutions Inc. and CO2 Solution Technologies Inc. All Rights Reserved.

CAER
CENTER FOR APPLIED ENERGY RESEARCH

UK

SECONDARY AIR STRIPPER PERFORMANCE IN A PCCC PILOT PLANT INTEGRATED WITH A COAL FIRED POWER PLANT

Jonathan Pelgen[a], Kunlei Liu[a], Heather Nikolic[a], Jesse Thompson[a], Reynolds Frimpong[a], Michael Manahan[b], David Link[b]

[a] *Center for Applied Energy Research, University of Kentucky, Lexington, KY, USA*
[b] *LG&E and KU Energy LLC, Louisville, KY USA*

CAER
CENTER FOR APPLIED ENERGY RESEARCH

Agenda

UK

1. **Background**
2. **Purpose of the Air Stripper**
4. **Sensitivity of the air stripping rate to:**
   a. **Air flow**
   b. **Feed C/N**
   c. **Temperature**
5. **Absorber Height Advantage**
6. **TEA Comparison**
7. **Oxidative Degradation and Solvent Emissions**
8. **Conclusions**
9. **Acknowledgements**

University of Kentucky
Center for Applied Energy Research

www.caer.uky.edu

CMTC 2015, Sugar Land, TX

November 17 - 19, 2015

The Unique UKy-CAER Carbon Capture Facility

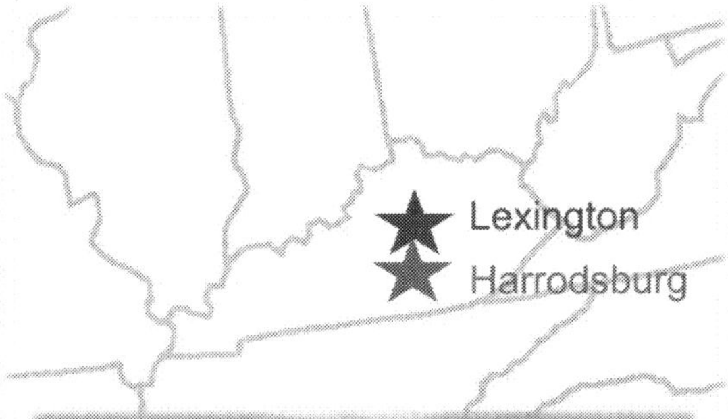

CMTC 2015, Sugar Land, TX

November 17 - 19, 2015

Background: UKy-CAER Process

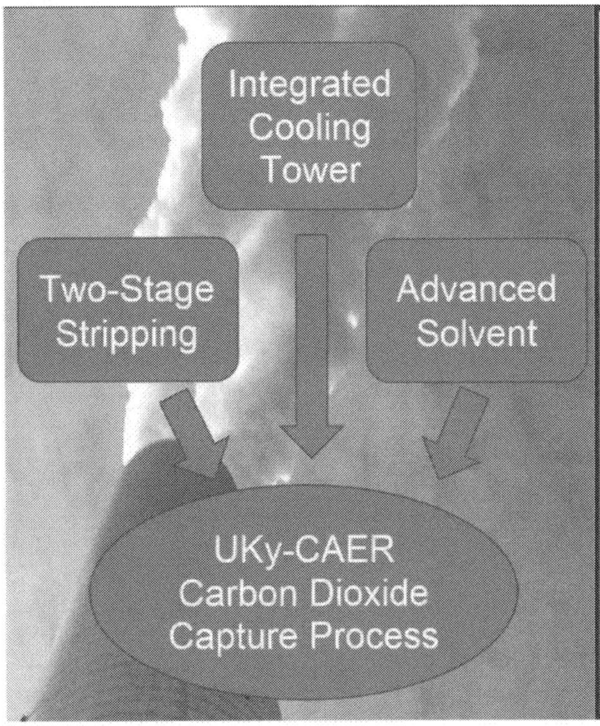

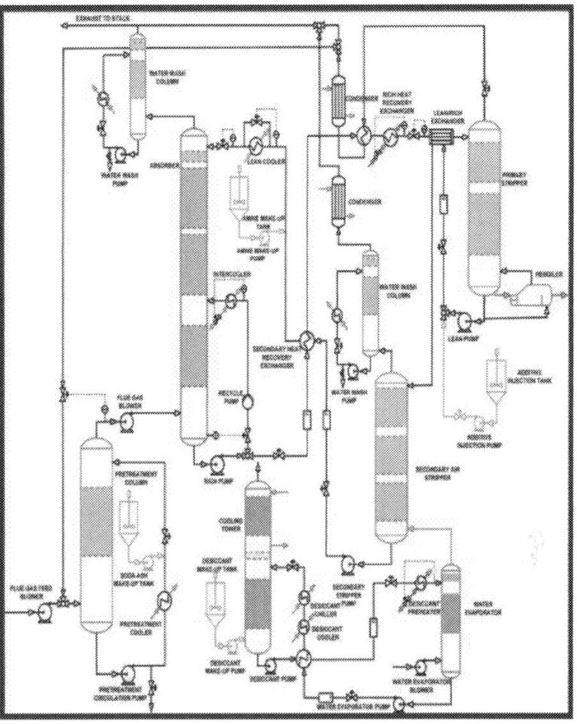

Background:TEA

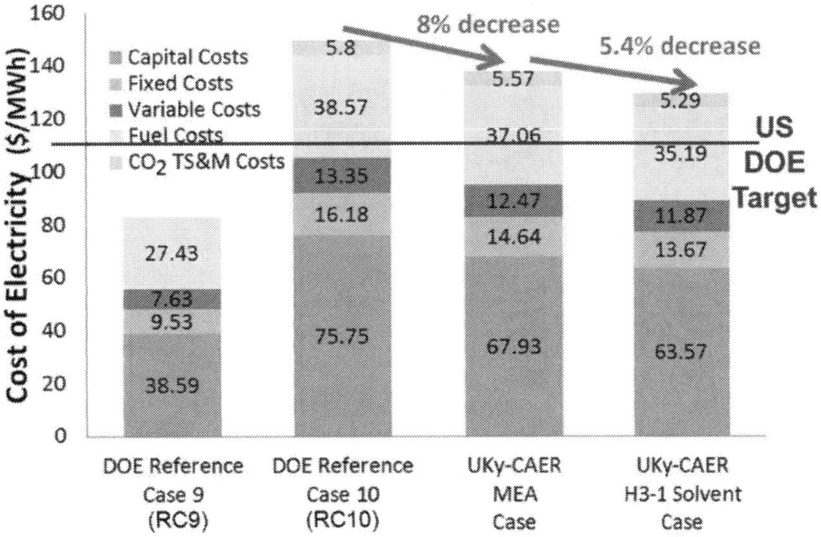

All at 90% capture rate and CO_2 compression to 2200 psia.

55.8% increase in Cost of Electricity from RC9

$46.93/tonne CO_2 Captured, lower than RC10 by $14.38/tonne CO_2

Background: Project Progression

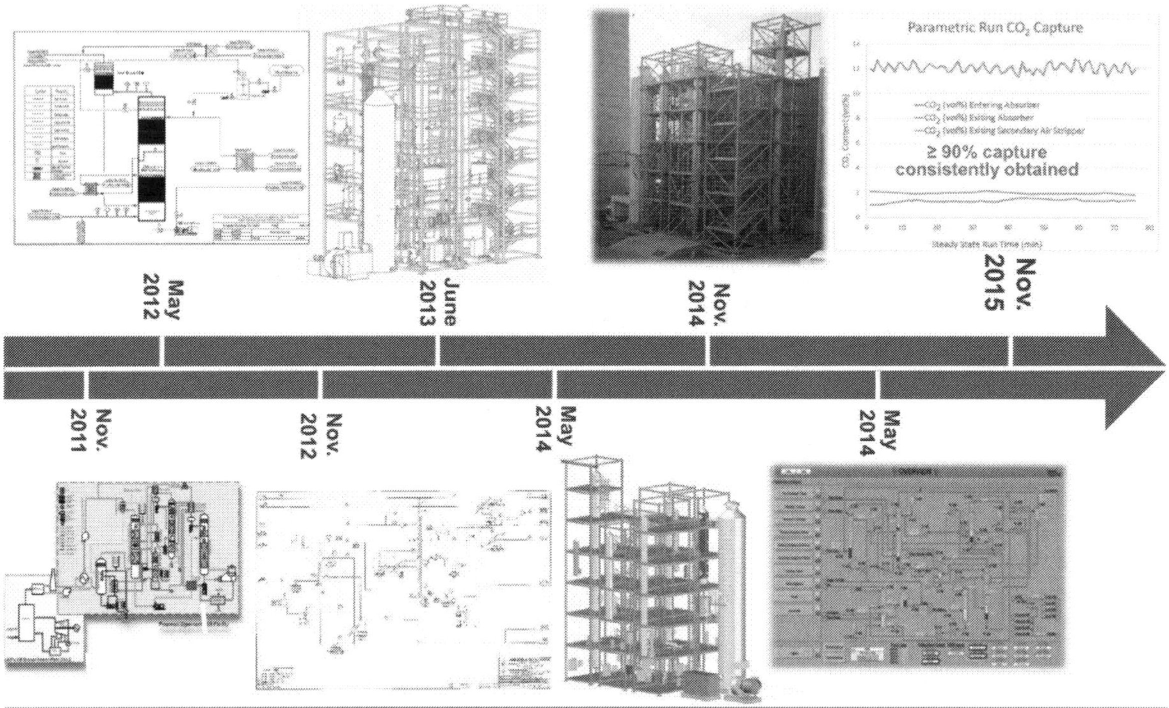

CMTC 2015, Sugar Land, TX

November 17 - 19, 2015

Air Stripper Focus

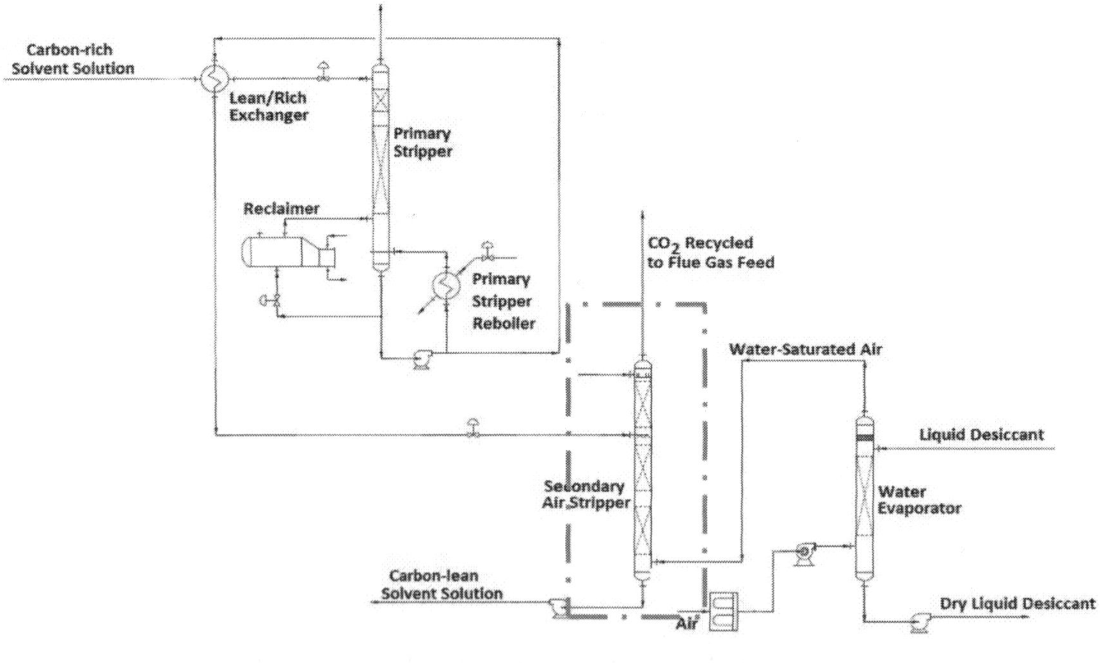

CMTC 2015, Sugar Land, TX

November 17 - 19, 2015

Air Stripper Focus

- Considerations:
  1. Inlet air flow rate
  2. Inlet solvent carbon loading
  3. Inlet solvent and air temperatures
  4. Outlet carbon loading
  5. Outlet CO_2 concentration

The point is to increase the solvent working capacity in the absorber.

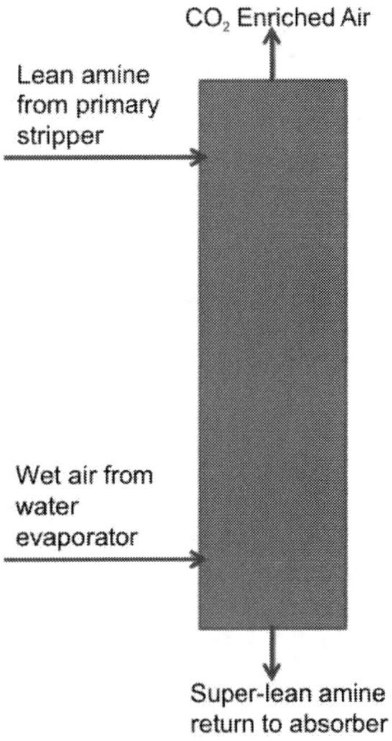

CMTC 2015, Sugar Land, TX

November 17 - 19, 2015

Air Flow Sensitivity

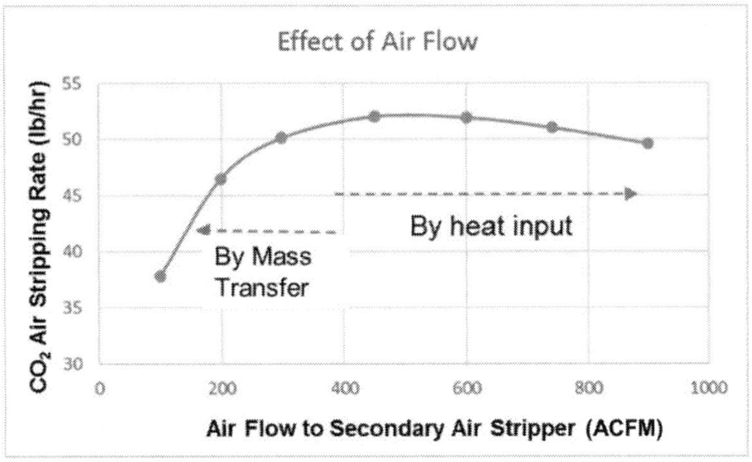

- Three ways to improve $P^*(CO_2)$ in the air stripper:
  1. Increase lean C/N from primary stripper (lower reboiler duty)
  2. Increase lean solvent temperature from primary stripper
  3. Increase water evaporator air exit temperature

CMTC 2015, Sugar Land, TX

November 17 - 19, 2015

Air Flow Sensitivity

- Given:
  - Flue Gas: 1400 acfm
  - CO_2: 16.0%
  - Stripper pressure: 36 psia
  - Solvent feed to air stripper temp: 195 F
  - Alkalinity = 3.26 mol/kg

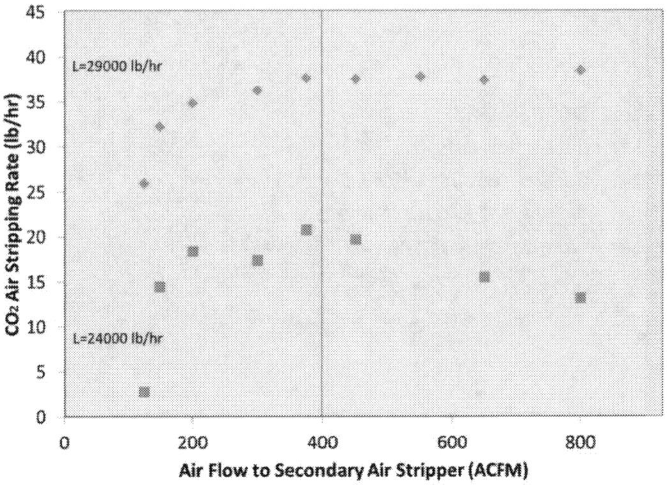

- Stripping does not increase in a linear fashion with increased air flow but is limited by the equilibrium vapor pressure of CO_2 in the air stripper.
- Beyond 400 acfm, stripping can be increased with additional heat input
- Model agrees with the data

CMTC 2015, Sugar Land, TX November 17 - 19, 2015

Feed C/N Sensitivity

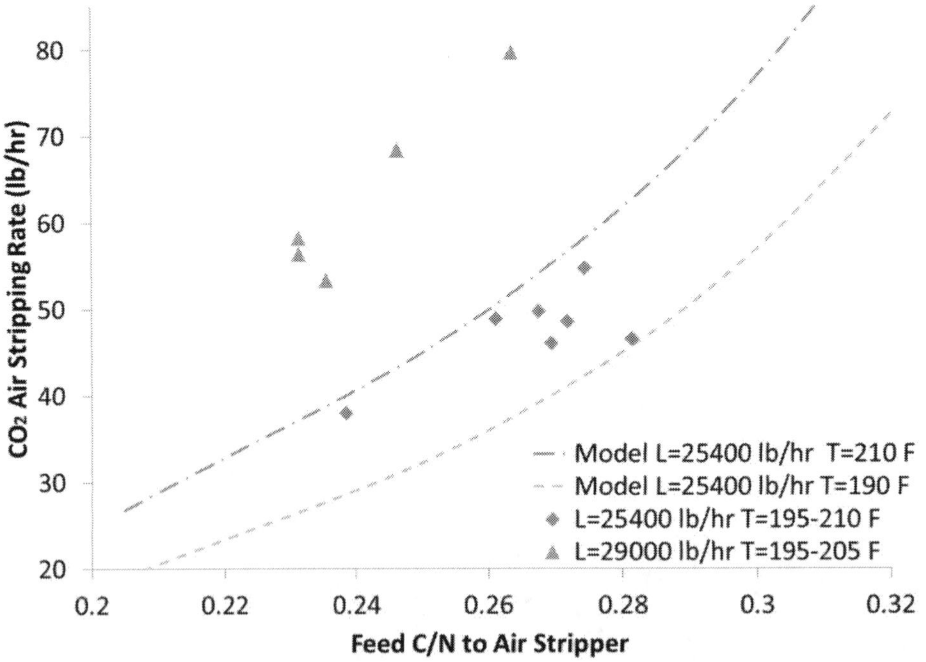

CMTC 2015, Sugar Land, TX November 17 - 19, 2015

Temperature Sensitivity

- Lean Feed T
  - Provide more heat
  - Raise air exit T

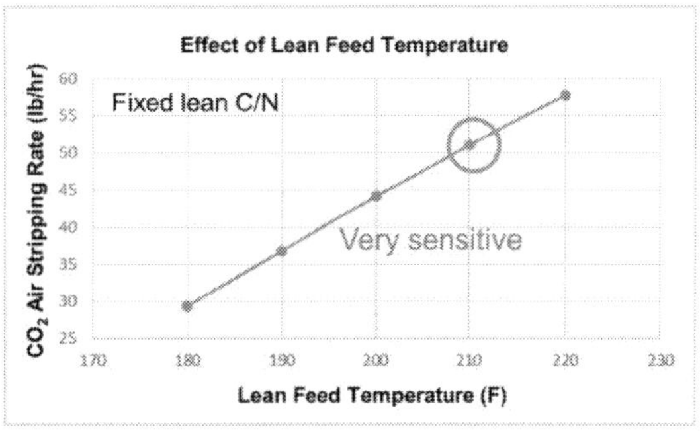

- Wet air T
  - High flow (more H_2O) from high air T
  - Insensitive (< 170 F)

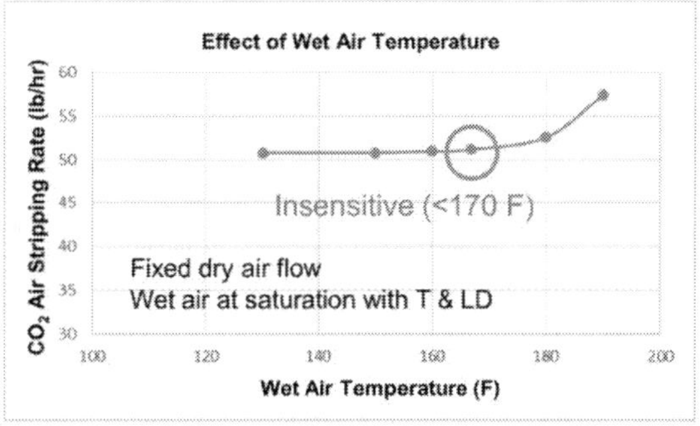

Temperature Sensitivity

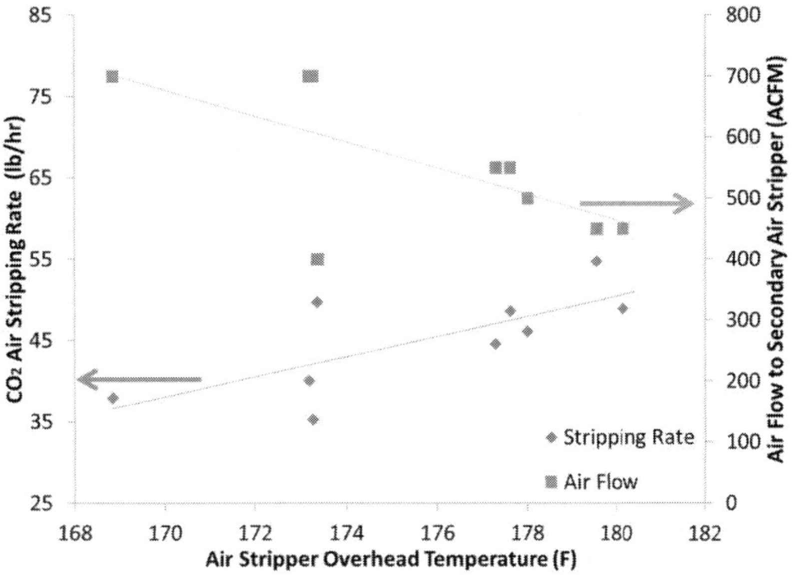

- Low blower flow rate = less air cooling effect = higher $P^*(CO_2)$= more CO_2 stripped
- Conditions = 16% CO_2 to the absorber; L=29000 lb/hr

CMTC 2015, Sugar Land, TX November 17 - 19, 2015

Oxidative Degradation?

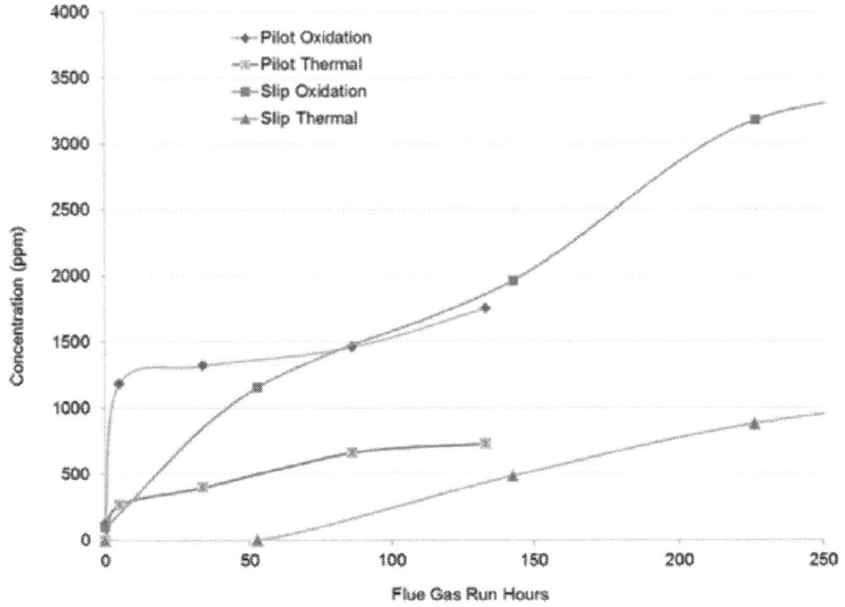

No excessive oxidation products were observed as a result of amine in contact with air

Conclusions

- Air stripper max overhead CO_2 concentration is limited by $P^*(CO_2)$
  - Equilibrium Controlled
- The air stripper increases the solvent working capacity
- The air stripper does not contribute to any detrimental solvent degradation

Technology Development Pathway

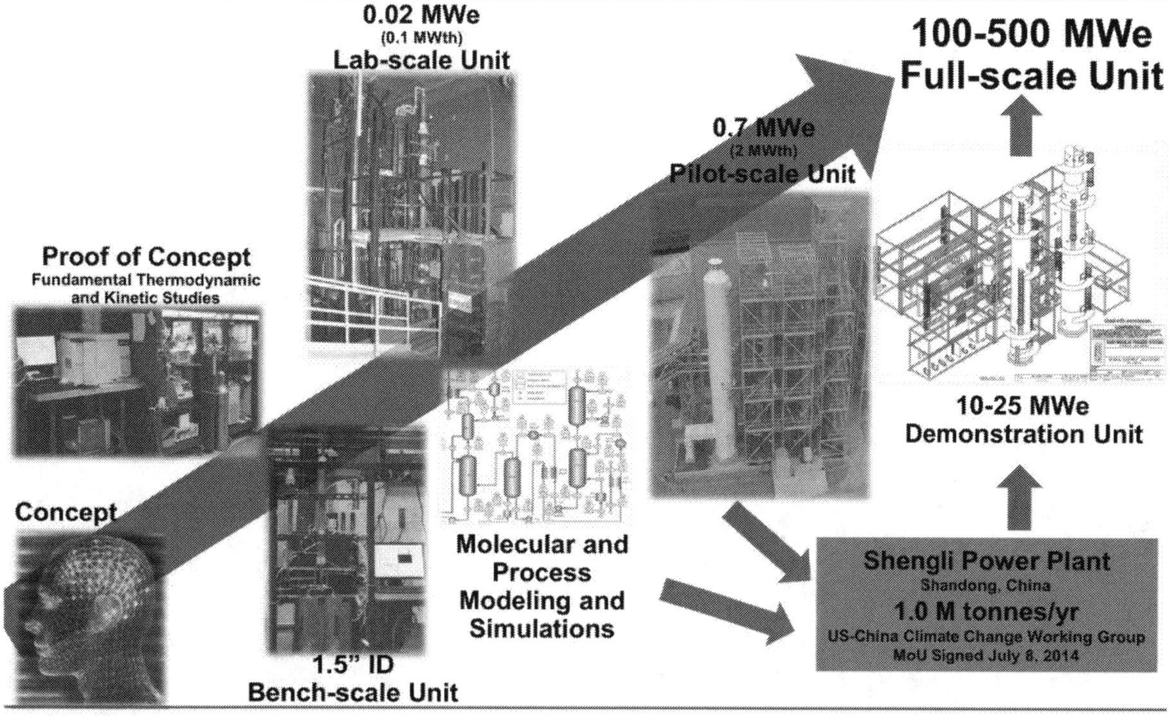

CMTC 2015, Sugar Land, TX

November 17 - 19, 2015

Team Members

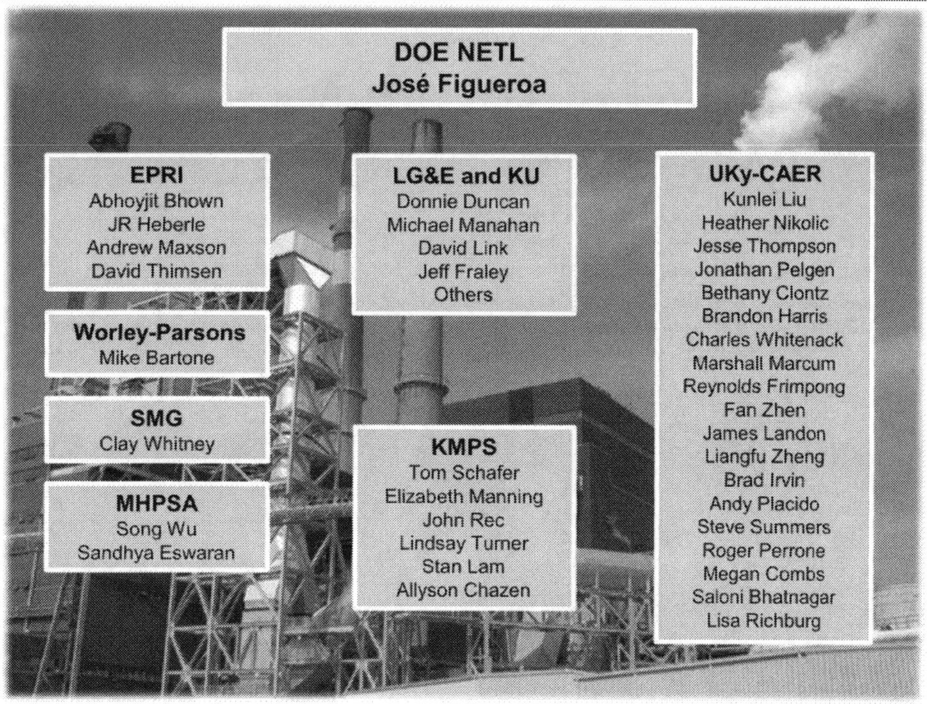

CMTC 2015, Sugar Land, TX

November 17 - 19, 2015

DRESSER-RAND
A Siemens Business

DRESSER-RAND
A Siemens Business

Table of Contents

- Brief overview of the CCS power penalty
- Multi-stage centrifugal solutions
  - In-line
  - Integrally geared
- Intercooled centrifugal (liquid cooled diaphragm)
- Supersonic – high ratio compression
- Conclusions

Dresser-Rand Thanks the DOE for Supporting This Research

DE-FC26-05NT42650 Liquid Cooled Diaphragms

DOE Project DE-FE-0000493 - "Ramgen Supersonic Shock Wave Compression and Engine Technology"

DE-FOA-0001190 "Pilot proposal for reducing the cost of CO_2 Capture and Compression - Advanced CO_2 Compression with Supersonic Technology"

CO_2 Compression

DRESSER-RAND
A Siemens Business

- Compression is an integral part of any CO_2 capture system.
- Separation systems deliver CO_2 near 1 bara and compression is required to increase pressure and reduce volume, making transport more practical.
  - Pipelines typically require ~ 2215 PSIA (153 bara).
  - EOR Injection pressures vary (50 bar – 600 bar).
  - On the capture side, given the high flow volumes, centrifugal compressors are generally employed instead of positive displacement screw or reciprocating compressors.
    - Small pilots (1-25 MWe) could utilize positive displacement compressors.
- Coal-based electric generating plants that are fitted with CO_2 capture systems have a compression penalty.
  - This penalty can range from 8–12% depending on the starting and ending CO_2 state points (temperature, pressure).
- With 90% capture, typical compression flow rates are:
  - 1,000,000 lbm/hr (454,000 kg/hr) of CO_2 for a 550 MW PC unit
  - 700,000 lbm/hr of (318,000 kg/hr) of CO_2 for a 400 MW PC unit

CO_2 Compression Experience

DRESSER-RAND
A Siemens Business

- Dresser-Rand has more than 450 units in carbon dioxide service, totaling more than 1.1 Million HP (800 MW)
  - ~ 100 in-line centrifugal compressors
  - ~ 300 reciprocating compressors
  - ~ 30 integrally geared compressors
- In-line centrifugal
  - First shipped in 1948
  - Max discharge pressure 8,000 psia (552 bara)
    - Delivery in 2011
  - Maximum inlet flow greater than 48,000 acfm (82,000 m^3/hr)
  - Total installed power > 400,000 HP (>300MW)
- Integrally Geared
  - First shipped in 1996
  - Max discharge pressure 2,277 psia (157 bara)
    - Delivery in 2010
  - Maximum inlet flow greater than 80,000 acfm (136,000 m^3/hr)
  - Total installed power > 173,000 HP (>129 MW)

Page 5 2015-11-19 Mark J. Kuzdzal/ PG DR GS SUP

Typical In-Line Centrifugal Compressor

DRESSER-RAND
A Siemens Business

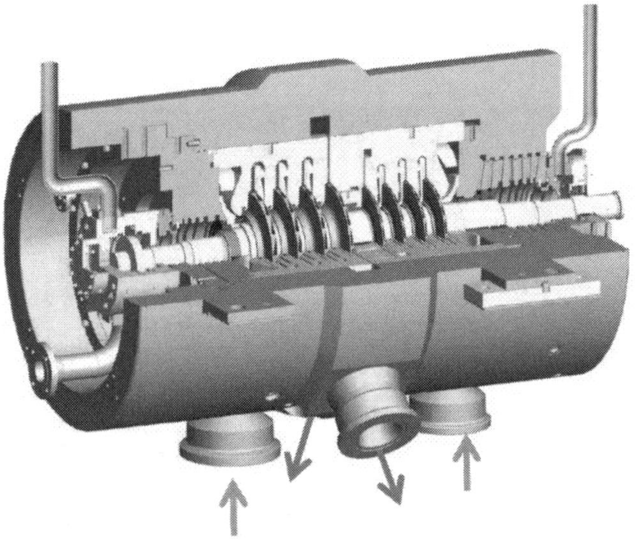

•Research continues to focus on developing new aerodynamic stages capable of higher efficiency over a broad operating flow map and higher head, enabling smaller, higher speed equipment.

Page 6 2015-11-19 Mark J. Kuzdzal/ PG DR GS SUP

Typical Integrally Geared Centrifugal Compressor

DRESSER-RAND
A Siemens Business

CO2 has a relatively low speed of sound and traditionally this class of equipment has been designed to operate sub-sonically, limiting impeller inlet relative Mach number to less than 1

Page 7 2015-11-19 Mark J. Kuzdzal/ PG DR GS SUP

In-Line Compressor Example: North Sea CO_2 Injection Compressor

DRESSER-RAND
A Siemens Business

- First CO2 re-injection project for the purpose of mitigating greenhouse emissions
- As of January 2015, 15.2 million tonnes CO2 injected in 20 years
- Objective: reduce the CO_2 content in methane from 9% to 2.5% (sale specification)
- Capture the CO_2 from an amine plant
- CO_2 storage in an aquifer
- Start-up: Aug 1996
- Injection: ~ 0.75 million tonne CO_2/year
- Availability: 99%

Page 8 2015-11-19 Mark J. Kuzdzal/ PG DR GS SUP

Liquid Cooled Diaphragms – Internal to Centrifugal

DRESSER-RAND
A Siemens Business

- DOE, SwRI and Dresser–Rand: Successfully completed DOE award DE-FC26-05NT42650
- Internal heat exchangers were developed and placed inside a in-line compressor casing
  - Integrated into the aerodynamic flow path
- The red flow path is the CO_2 gas path and blue region represents a cooling medium
- This approach eliminates the need for the CO_2 to exit the casing, travel through piping to a external cooler, then re-enter through another nozzle leading to the next stage
- The internal heat exchanger results in no pressure drop and fits within the same envelope of the standard internal diaphragm (Moore, et al., 2007)

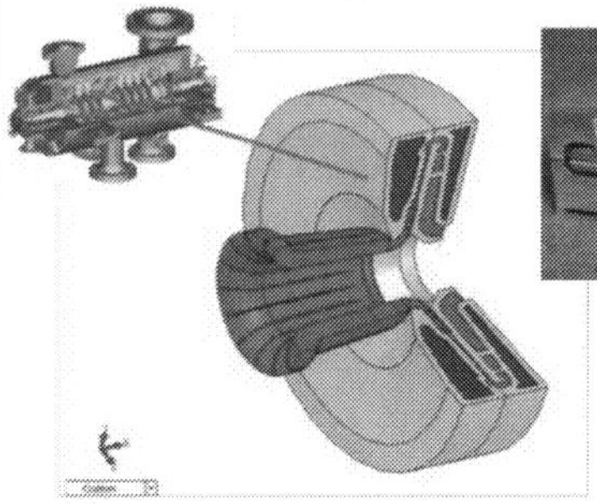

Photograph of Generation-1 Cooled Diaphragm Test Rig

Reference; J. Moore, M. Nored and K. Brun, "Researchers Seek to Economically Compress Large Volumes of Carbon Dioxide," Pipeline and Gas Journal, 2007.

Liquid Cooled Diaphragms – Internal to Centrifugal

DRESSER-RAND
A Siemens Business

- Test results show cooled diaphragm can remove up to 55% of the heat of compression in each stage
- The technology led to the development of a pilot scale demonstration of a 6-stage back-to-back, 25:1 pressure ratio compressor in a CO_2 test loop constructed and operated at SwRI.

Dresser-Rand 6-stage CO2 compressor with generation-2 cooled diaphragms

Liquid Cooled Diaphragms – Internal to Centrifugal

DRESSER-RAND
A Siemens Business

- The D-R compressor installed at SwRI is a D12R6B, 37 inch (0.93 m) case bore, 20 inch (0.5 m) diameter impellers.
- Diaphragms were successfully developed with the same pressure recovery as a standard compressor flow path.
- The compressor was operated first without cooling and compared to identical operating conditions with cooling allowing for an apples-to-apples comparison.

Internally-cooled inline centrifugal test results show up to a 8% power savings over the same non-cooled inline centrifugal, but with slight increase of compressor costs due to manufacture of liquid-cooled diaphragms.

Page 11 2015-11-19 Mark J. Kuzdzal/ PG DR GS SUP

Ramgen → DATUM® S → History

DRESSER-RAND
A Siemens Business

- Shock waves generate compression and heat
  - Shock compression has been applied in supersonic engine inlet designs for decades
- In 2008, D-R invested in Ramgen's supersonic compression technology
  - DOE Project DE-FE-0000493 - "Ramgen Supersonic Shock Wave Compression and Engine Technology"
- Late 2012 in Olean, NY, the team successfully tested a supersonic CO_2 compressor achieving a 9:1 pressure ratio in a single rotating disk
  - Proof of concept – successful !
- On Aug 8th, 2014, D-R announces the opening of the Seattle Technology Center along with a 11,000 sq ft testing lab
  - Initial staffing is comprised of 19 engineers and support staff, all of whom are former Ramgen Power Systems employees

Page 12 2015-11-19 Mark J. Kuzdzal/ PG DR GS SUP

CO_2 Supersonic Compressor

DRESSER-RAND
A Siemens Business

- 10:1 pressure ratio with single rotating disk on CO_2
  - Enables a 100:1 pressure ratio in a low-pressure (LP) and high-pressure (HP) case configuration
    - 22 PSIA (1.52 bara) → 2215 psia (153 bara)
- Compression equipment
  - ~ 70% weight and size of traditional in-line centrifugal offering
  - ~ 50% weight and size of an integrally geared centrifugal offering

Heart of the value proposition is integration of waste heat for amine regeneration or boiler feed water preheat

DATUM® S – Shock Wave Compression

DRESSER-RAND
A Siemens Business

- HP Demonstration Unit size 13,400 hp (10 MW) shaft power level
- Mass flow 100 lbm/s (46 kg/s) (76 MMSCFD)
  - ~ 90% capture of a 230 MW coal-fired plant
- 220 psig suction, 2200 psig discharge pressure – HP stage
- Program plan includes:
  - Design, fabrication, assembly and test
  - Test phase conclusion late November 2015
- Follow-on compressor pilot demonstration program at a client site targeted for 2016

Nov 5th, 2014 in Olean, NY, the team successfully tested a supersonic CO2 compressor achieving a 10.1:1 pressure ratio in a single rotating disk

HP DATUM S - CO_2 Test Facility

DRESSER-RAND
A Siemens Business

- CO_2 and Inert Test Gas
  - 220 PSIG Inlet
  - 2200 PSIG Discharge
  - 3363 PSIG Discharge Design Rating
  - 1500 # ANSI, 5000 PSIG Helium and 1000 PSIG Nitrogen utility gas
- 10 MW VFD Driven Motor
- CO_2 Gas collection system
- Cooling Tower

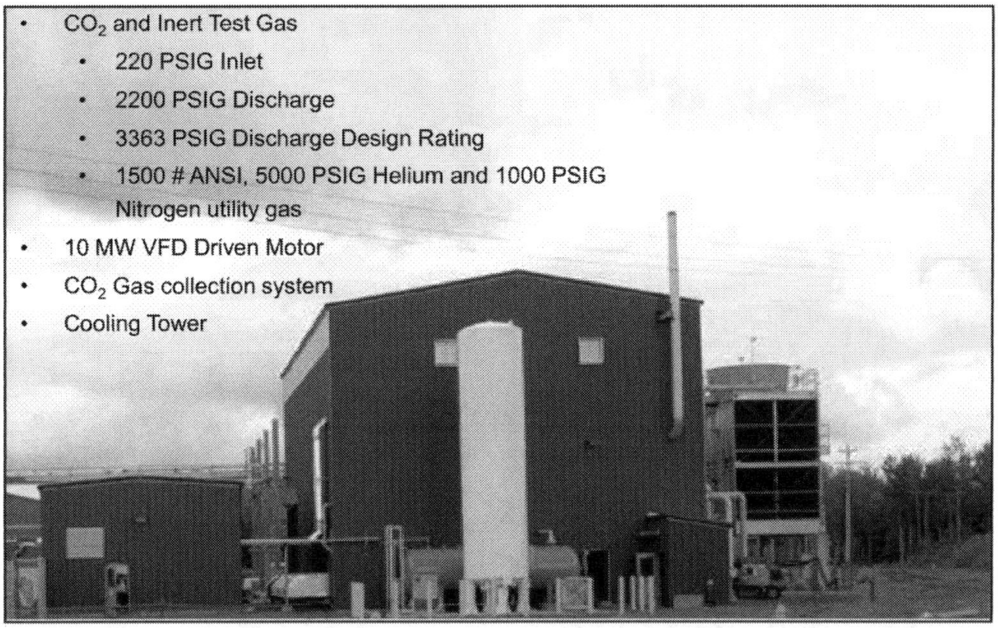

DATUM® S – Shock Wave Compression

DRESSER-RAND
A Siemens Business

- LP Demonstration Unit
  - Rig size 24,100 hp (18 MW) shaft power level
  - Mass flow 200 lbm/s (91 kg/s) (152 MMSCFD)
  - ~ 90% capture of a 440 MW coal-fired plant
  - 22 psig suction, 220 psig discharge pressure – LP stage
  - Program plan includes:
    - Design, fabrication, assembly and test
    - Test phase expected conclusion 1Q2018

- DE-FOA-0001190 "Pilot proposal for reducing the cost of CO_2 Capture and Compression - Advanced CO_2 Compression with Supersonic Technology"

Summary

DRESSER-RAND
A Siemens Business

- The integrally geared (IG) compressor is the current benchmark for overall shaft power, but is large and incorporates many coolers, bearings and seals.
- In-line centrifugal have less bearings and seals, but will consume more power:
  - Will have a reduced footprint relative to IG
  - Much of power penalty mitigated with water cooled diaphragms.
  - Internally-cooled inline centrifugal test results show up to a 8% power savings over the same non-cooled inline centrifugal
    - Slight increase of compressor costs due to manufacture of liquid-cooled diaphragms.
- DATUM® S:
  - High ratio results in increase of shaft horsepower
  - Single rotating disk results in smaller footprint, reduced CAPEX and commissioning costs
  - Size reduction of 33% percent compared to inline centrifugal
  - Delivers usable heat which, when integrated into the system, will produce the lowest overall power for the compression service - lowest OPEX solution if waste heat can be integrated.

Dresser-Rand is committed to realizing the full potential of supersonic compression technology and deploying a viable commercial product to serve CCS applications and other CO2 compression markets.

Page 17 2015-11-19 Mark J. Kuzdzal/ PG DR GS SUP

Contact

DRESSER-RAND
A Siemens Business

Mark J. Kuzdzal

Director Business Development

500 Paul Clark Drive

Olean, New York 14760

Phone:+1 (716) 375-3573
Mobile: +1 (716) 378-5483

E-mail:

mkuzdzal@dresser-rand.com

Dresser-Rand.com

Page 18 2015-11-19 Mark J. Kuzdzal/ PG DR GS SUP

the Energy to Lead

Pilot Test of a Nanoporous, Super-hydrophobic Membrane Contactor Process for Post-combustion CO_2 Capture

S. James Zhou, Shiguang Li, Travis Pyrzynski, and Howard Meyer, *GTI*
Yong Ding and Ben Bikson, *PoroGen*
Katherine Searcy, *Trimeric*

Carbon Management Technology Conference 2015 (CMTC 2015)

November 17-19, 2015, Sugar Land, TX

gti PoroGen

Introduction to GTI and PoroGen

gti

- Not-for-profit research company, providing energy and natural gas solutions to the industry since 1940s
- Facilities
  - 18 acre campus near Chicago
  - 200,000 ft², 28 specialized labs

AIR LIQUIDE ADVANCED BUSINESS & TECHNOLOGIES PoroGen INNOVATIVE MEMBRANE PRODUCTS

- Materials technology company commercially manufacturing products from high performance plastic PEEK (poly (ether ether ketone))
- Products ranging from membrane separation filters to heat transfer devices

2

Project overview

- **Performance period**: Oct. 1, 2013 – June 30, 2018
- **Total funding**: \$13.3 MM (DOE: \$10.2MM, Cost share: \$3.1MM)
- **Objectives**:
  - Build a 0.5 MW_e pilot-scale CO_2 capture system and conduct tests on flue gas at the National Carbon Capture Center (NCCC)
  - Demonstrate a continuous, steady-state operation for ≥ 2 months
- **Goal:** Achieve DOE's goal of 90% CO_2 capture rate with 95% CO_2 purity at a cost of \$40/tonne of CO_2 captured by 2025
- **Team**:

Member	Roles
gti	• Project management and planning • Process design and testing
AIR LIQUIDE, PoroGen	• Membrane and module development
TRIMERIC CORPORATION	• Techno-Economic Analyses (TEA)
NCCC	• Site host

3

What is a membrane contactor?

- High surface area membrane device that facilitates mass transfer

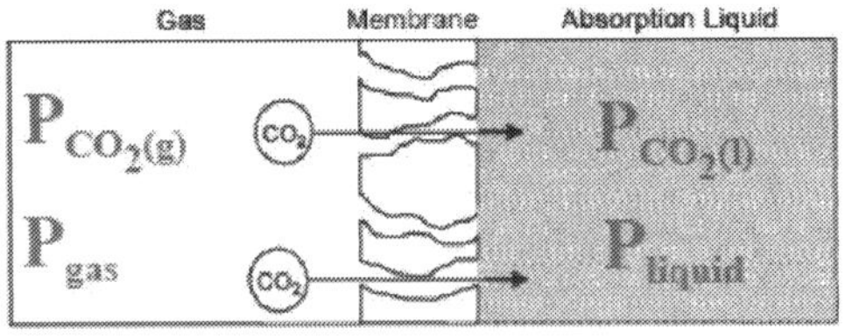

- **Separation mechanism**: CO_2 permeates through membrane and reacts with the solvent; N_2 does not react and has low solubility in solvent
- Why **PEEK HFMC**?

Polymer	Max service temperature (°C)
Teflon™	250
PVDF	150
Polysulfone	160
PEEK	**271**

PEEK= Polyether ether ketone

HFMC= Hollow Fiber Membrane Contactor

4

Process description

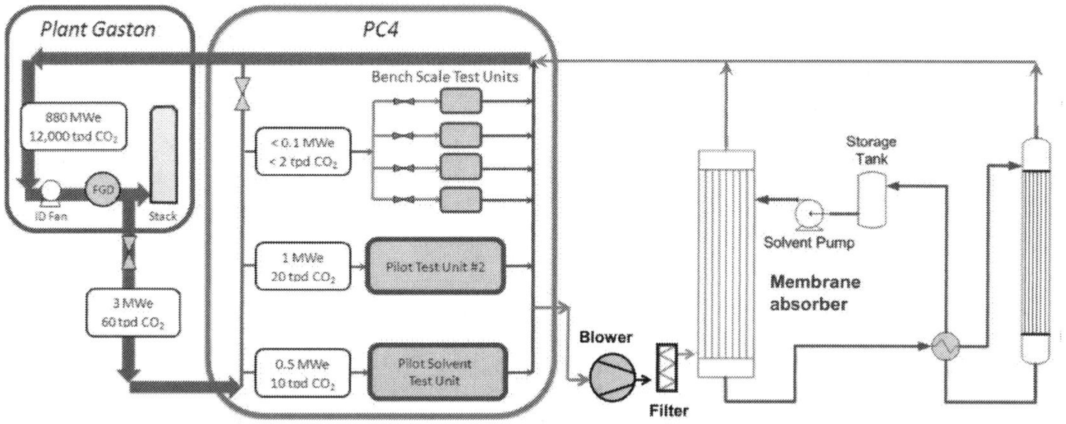

NCCC's PC4 **HFMC plant**

5

Technical and economic <u>challenges</u> of applying membrane contactor to existing PC plants

- **<u>Performance</u>**
  - Maximize overall mass transfer coefficient to reduce absorption system size
- **<u>Durability</u>** – Long-term membrane life in contact with solvent
  - Improve membrane hydrophobicity
- Contactor **<u>scale-up</u>** and **<u>cost reduction</u>**
  - Make larger diameter module, module packaging to reduce cost

6

Timeline and scope

Oct., 2013

BP1: Design of the 0.5 MW equivalent scale system

Nov., 2015

BP2: Procurement of parts, and construction

Oct., 2016

BP3: System installation and shakedown

Oct., 2017

BP4: Parametric and continuous tests

June 2018

BP2 Scope:

- 8-inch diameter commercial-sized module fabrication
- Parts and equipment procurement
- 0.5 MW_e CO_2 capture pilot plant construction

- **BP1 scope:**
    - Preliminary TEA and EH&S study
    - Bench-scale testing in support of the pilot-scale design
    - Design and costing of the pilot plant

7

Preliminary TEA shows HFMC technology with aMDEA solvent can cost 25% less than DOE Case 12

Item	Unit	[2]Case 12	[1]HFMC with aMDEA solvent	
Mass transfer coefficient	*(sec)*$^{-1}$		bench-scale tested, 1.2	Target of the current program, 1.7
LCOE - no TS&M	*mills/kWh*	137.3	127.2	122.3
LCOE - total	*mills/kWh*	147.3	137.2	132.3
Incremental cost of CO_2 capture - No TS&M	*mills/kWh*	56.3	46.3	41.4
Increase in LCOE - No TS&M	%	69.6%	57.2%	51.19%
Increase in LCOE - total	%	81.9%	69.5%	63.5%
[3]Cost of CO_2 capture - no TS&M	*$/tonne*	56.47	47.53	42.48

[1] The field test module had a CO_2 permeance of 2,000 GPU
[2] "Updated Costs (June 2011 Basis) for Selected Bituminous Baseline Cases." August 2012. DOE/NETL-341/082312.
[3] In 2011$

8

PEEK membranes continue to improve, larger inner diameter fibers with high CO_2 permeance developed

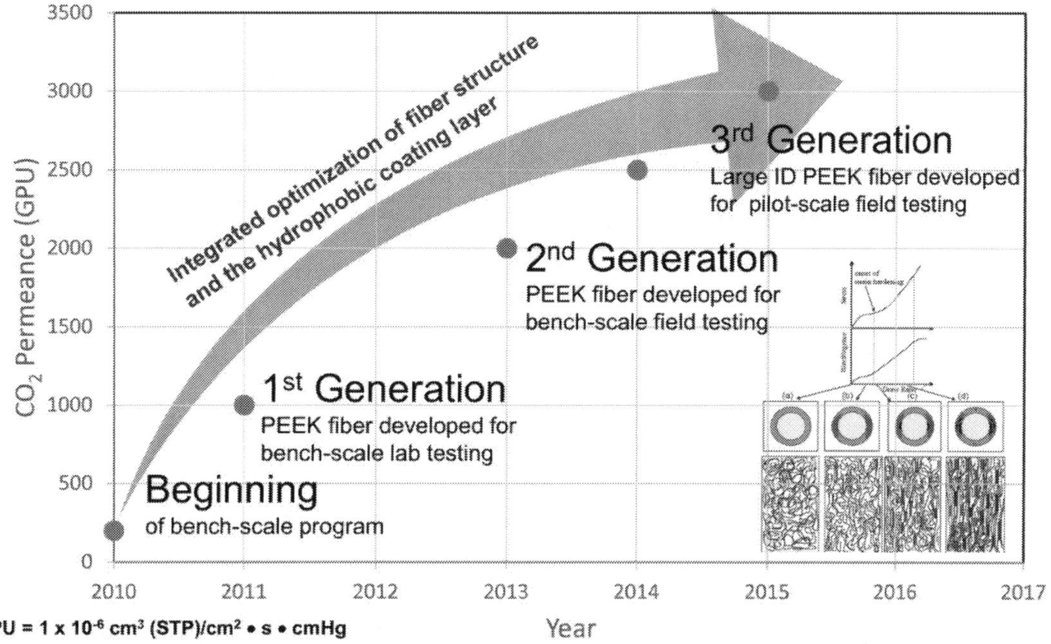

1 GPU = 1 x 10^{-6} cm^3 (STP)/cm^2 • s • cmHg

New larger ID fibers have intrinsic CO_2 permeance as high as 3,500 GPU, surpassing our <u>target</u> of 2,000 GPU

- Achieved in October 2015 with testing conditions:
  - 2-inch diameter modules
  - Pure CO_2 gas
  - ~3 psig feed pressure
  - Room temperature (22 °C)

Cartridge No.	Number of fibers	Active fiber area (cm^2)	Pure CO_2 permeance (GPU)
2PG-664	448	3,161	2,580
2PG-665	448	3,161	2,500
2PG-3035	1,088	18,417	3,500

Why larger inner diameter (ID) fibers?

- During bench-scale field test, pressure drop (ΔP) between flue gas Inlet and outlet was 5 psi
- Current pilot program target: decrease ΔP to $\leq$ 2 psi because the lower the ΔP the more saving on operating costs
- The Hagen–Poiseuille equation: ΔP very sensitive to ID

$$\Delta P = \frac{8Q\eta L}{\pi \cdot r^4}$$

  - *Q:* volumetric flow rate, η*:* absolute viscosity of the fluid, *L:* length of the hollow fiber, and *r:* radius of the hollow fiber
- Fiber ID is being increased from 13 mil to 20 mil

High performance for the larger ID fibers confirmed

- Intrinsic CO_2 permeance as high as 3,500 GPU (our target: 2,000 GPU)

Cartridge No.	Number of Fibers	Active Fiber Area (inside, cm^2)	Pure CO_2 Permeance (GPU)
2PG-664	448	3,161	2,600
2PG-665	448	3,161	2,500
2PG-3035	1,088	18,417	3,500
8PG-3034	18,756	317,504	2,600

- Contactor testing with aMDEA solvent: mass transfer coefficient of 2.0 sec^{-1} (surpassing our target of 1.7 sec^{-1}) at 90% CO_2 removal
- Low gas-side ΔP of 0.28 psi was observed for the 2-inch module
  - ΔP as low as 0.74 psi (our target: less than 2 psi) is predicted for the 8-inch diameter module

Preliminary design completed, HAZOP reviewed at NCCC in February 2015

- **Major HAZOP review outcome**
  - No "show-stopper" or "major" modifications were identified
  - Valuable insights from NCCC engineers ⟶ revised design

HFMC pilot system – layout

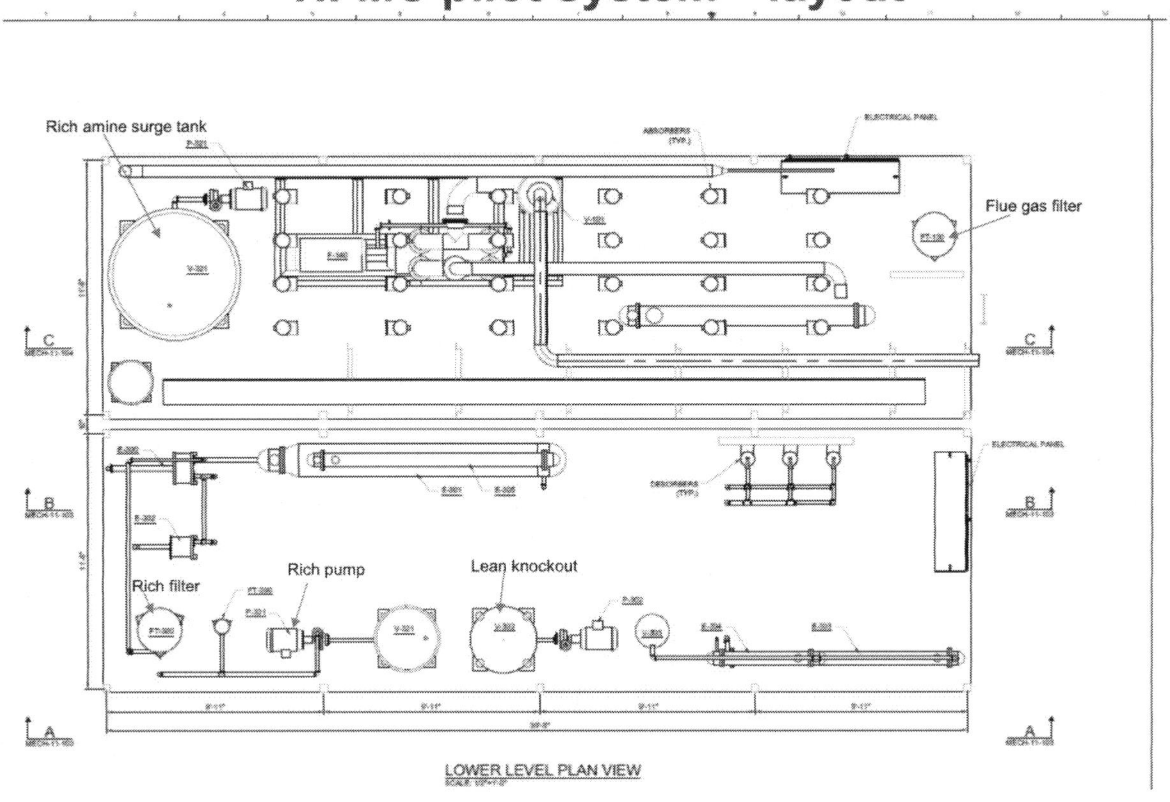

HFMC pilot system – layout

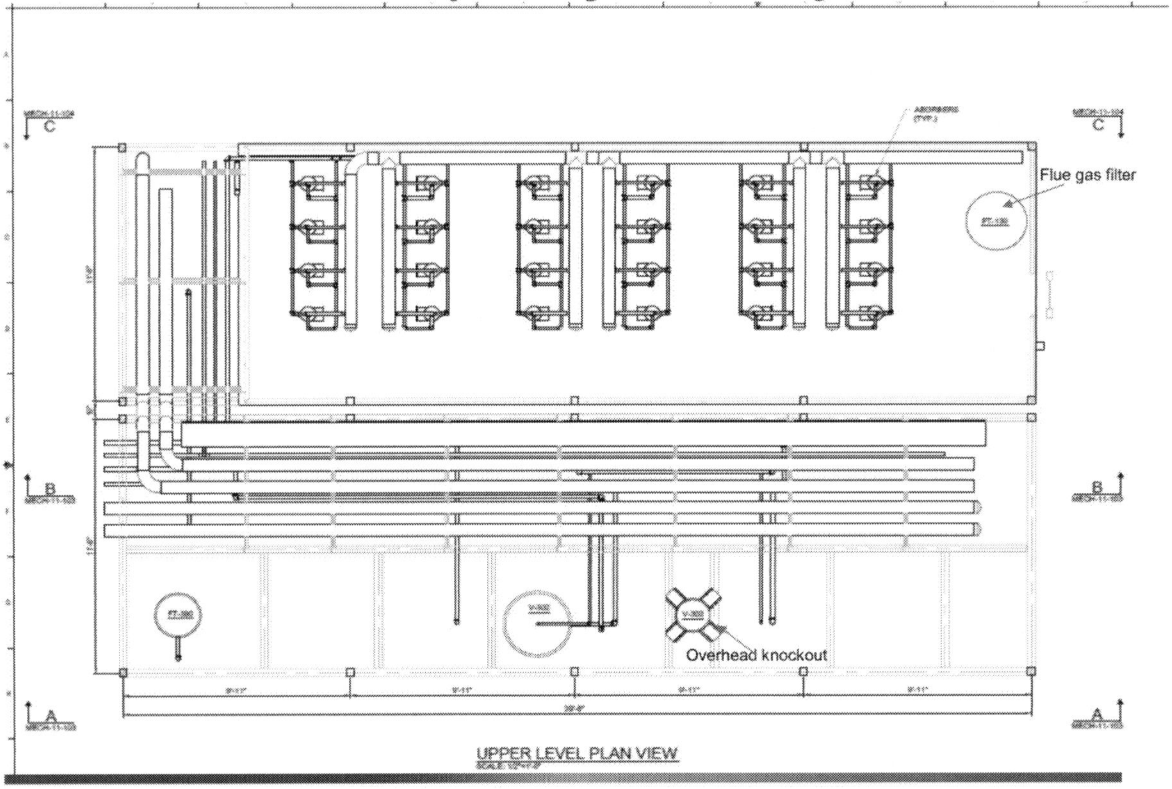

HFMC pilot system – skid sections

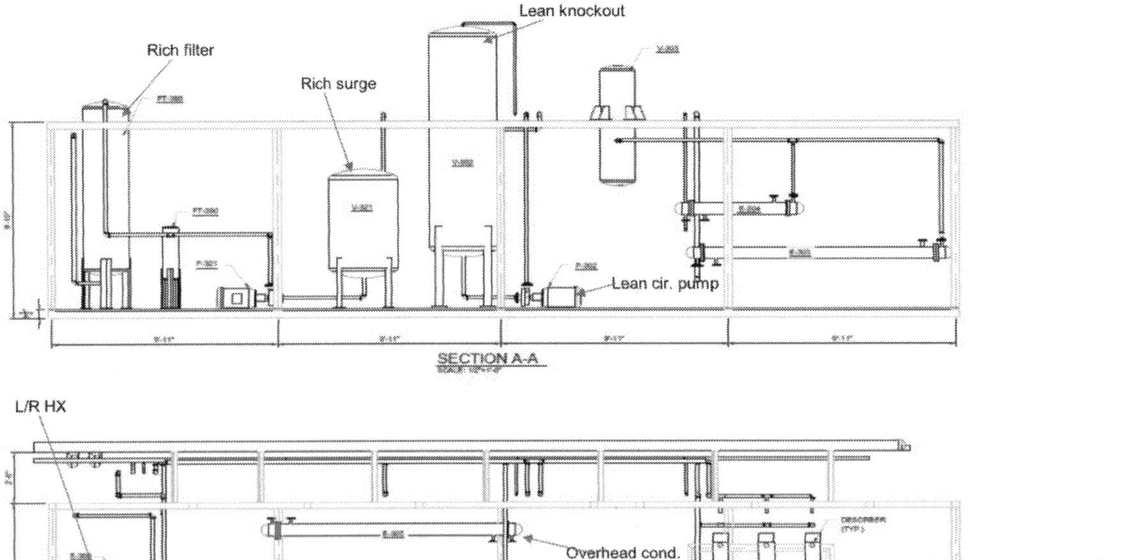

HFMC pilot system – skid sections

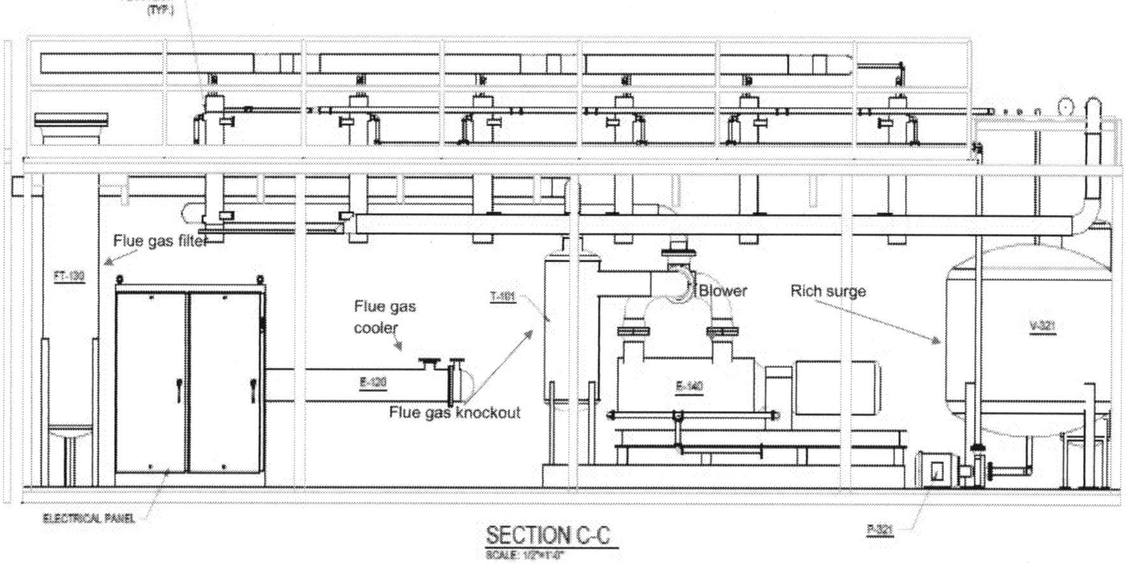

17

HFMC pilot system – isometric

2 Skids

40'W x 15'H x 12'L

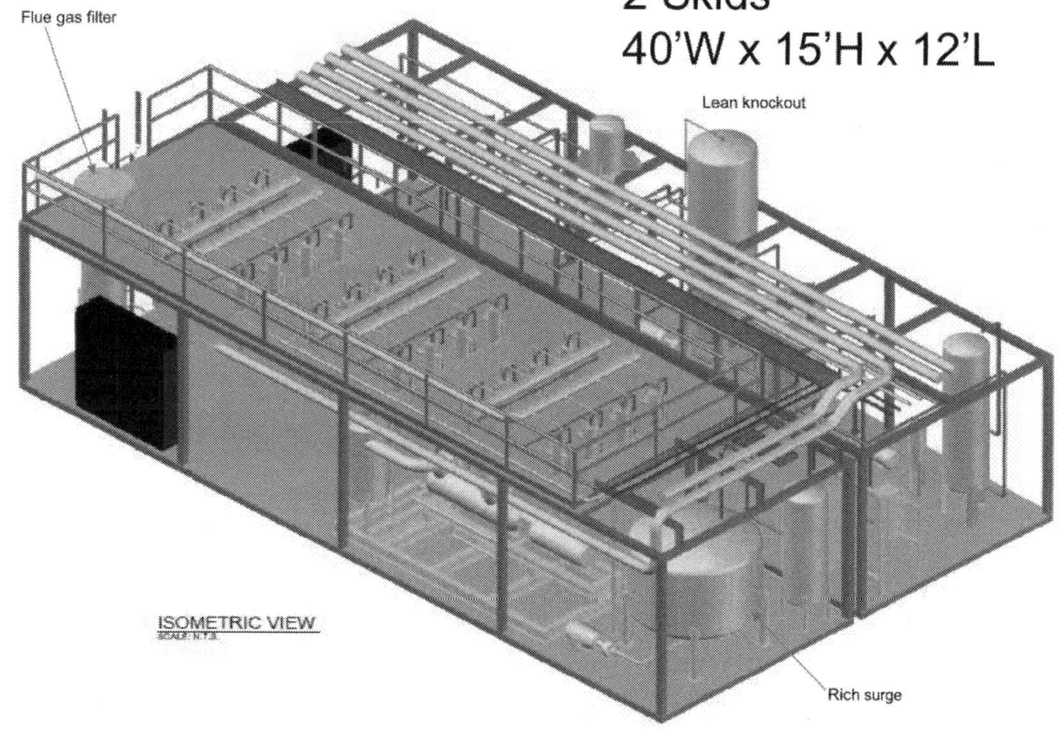

HFMC pilot system – isometric

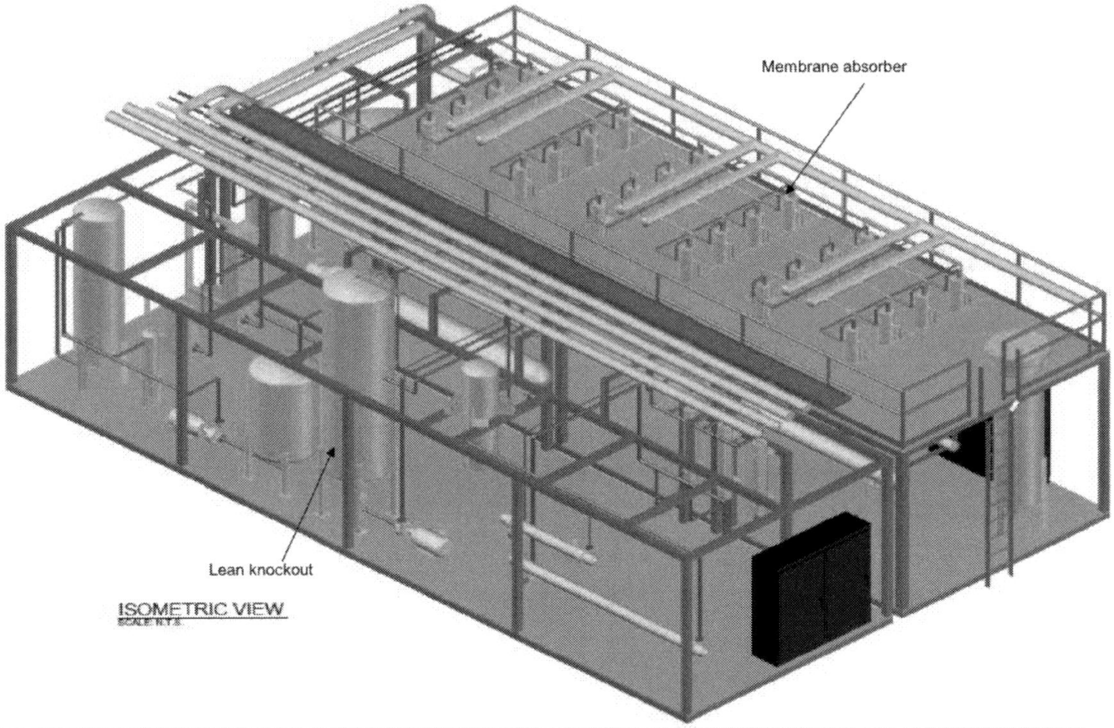

Plans for future testing/development/commercialization

- **In this project**
  - BP3
    - Site preparation and system installation at the NCCC
    - Procure aMDEA solvent for the pilot testing
    - Test system shake down at NCCC
    - Parametric testing at NCCC performed prior to continuous testing
  - BP4
    - Identify operational conditions for the continuous steady-state run at NCCC
    - Run continuous steady-state tests for a minimum of two months
    - Gather data necessary for further process scale-up
    - Final Techno-Economic Analysis and EH&S study
- **After this project**

Time	Development	CO_2 capture, Ton/day
By 2022	10 - 25 MWe demonstration	200 - 500

20

Summary

- HFMC technology field test system at 25kW scale was installed at Midwest Generation's coal-fired power plant for one month and tested with aMDEA and other solvents for CO_2 capture and start-up/shutdown stability
- Field test results were used in the Preliminary TEA and EH&S study, the CO_2 capture cost of HFMC with aMDEA solvent can be 22% lower than the DOE Case 12
- New PEEK fibers targeting low gas side pressure has been developed
  - Intrinsic CO_2 permeance as high as 3,500 GPU
  - Gas side ΔP as low as 0.74 psi predicted for 8-inch module
- Pilot unit has been designed and fabrication is underway
- The pilot system will be tested at NCCC from Oct./Nov., 2016 to July/Aug., 2017
- Test results at NCCC with commercial PEEK membrane contactors from PoroGen will provide
  - Validation and scale-up information
  - Information to support decision for further scale-up
  - Information for continuous operation and start-up/shutdown

21

Acknowledgements

- **Financial support**

U.S. DEPARTMENT OF ENERGY

NETL

ICCI Illinois Clean Coal Institute

gti JIP Partners

- **DOE NETL José Figueroa**
- **ICCI Dr. Debalina Dasgupta**

22

CARBON MANAGEMENT TECHNOLOGY CONFERENCE

CMTC-437529-MS

Carbon Capture and Storage - Using CO2 EOR Method for Economic Viability.

Md Imtiaz, Sanchita Dei, Indian School of Mines.
Manas Pathak, University of Utah.
Manika Tiwari, Drilling Mentor Pvt Ltd.

This paper was prepared for presentation at the Carbon Management Technology Conference held in Sugarland, Texas, USA, 17–19 November 2015.

Abstract

Every day, we use energy and every day we ask for more. And with the global population set to rise from 7 to 9 billion by 2050, world energy demand is expected to increase by 50% over the next 20 years alone. Fossil fuels still supply 80% of our energy and we emit enormous quantities of CO2 when we burn them. Carbon dioxide, though not the most potent of greenhouse gases, is the most significant one. Rising CO2 concentrations in the atmosphere has led to evolution of many problems; two important of that are - ocean acidification and rising of global temperatures. In order to prevent this large volume of CO2 reaching the atmosphere, there is a need to capture and dispose CO2 in an eco-friendly and economically feasible method.

Carbon Capture and Storage (CCS) method which is essentially a three stage technology where CO2 is captured from large man-made CO2emission sources, transported via a network of pipelines and stored in deep subsurface geological formations. This paper dissertates on the process of CCS which includes three stages which are:

Capture CO2 emissions from large industrial facilities before the CO2 is emitted to the atmosphere. Transport the CO2 through pipelines, and permanently store the CO2 in deep, secure underground rock formations. And also it cogitates on the mutation taking place in this field so as to make this process more expeditious.

Here in this paper we have reckoned on the use of carbon dioxide in this process of Enhanced Oil recovery from the oil fields. The conditions under which this EOR method is executable and the effect CO2 has on the underground oil which led to an increase in the recoverable oil has been highlighted. EOR can create a market for CO2 that can generate revenues that could play a vital part in contributing to the large capital investment that is required for Carbon Capture and Storage deployment.

The effects of CCS method on environment and loss in the amount of CO2 has also been discussed with referring to a case study which mainly focus on the cost estimate, reduction of CO2 in the environment and the risks assessment of the Carbon Capture and Storage project.

Introduction

Fossil fuels still supply 80% of our energy and we emit enormous quantities of CO2 when we burn them. Rising CO2 concentrations in the atmosphere has led to evolution of many problems and in order to prevent this large volume of CO2 reaching the atmosphere, there is a need to capture and dispose CO2 in an eco-friendly and economically feasible method. In this paper we discuss a method for conversion of Carbon dioxide to carbonates or bicarbonates which is economically viable and who's kinetic we can control.

Storing CO_2 in geologic strata over a long period of time includes three mechanisms; capillary trapping, dissolution trapping, and mineralization. Due to its low density and high viscosity, mobile CO_2 will rise in the geological strata. The current estimations suggest that it will take more than 1000 years for CO_2 to be permanently trapped in the formation due to solubilization or mineral trapping of CO_2. Sequestered may rise through the geologic strata and be emitted back into the atmosphere.

The CCS (Carbon Storage and Capture process as shown in figure 1) is extensively being put to use across the globe, with the Northern American countries and the European countries putting it to extensive use. The middle eastern countries als have started to apply the technology with a major chunk of the projects concentrated in Saudi Arabia.Other countries such as China and Australia also have a significant number of operational projects going on . All in all the technology is being use in a widespread scale on the industrial front.

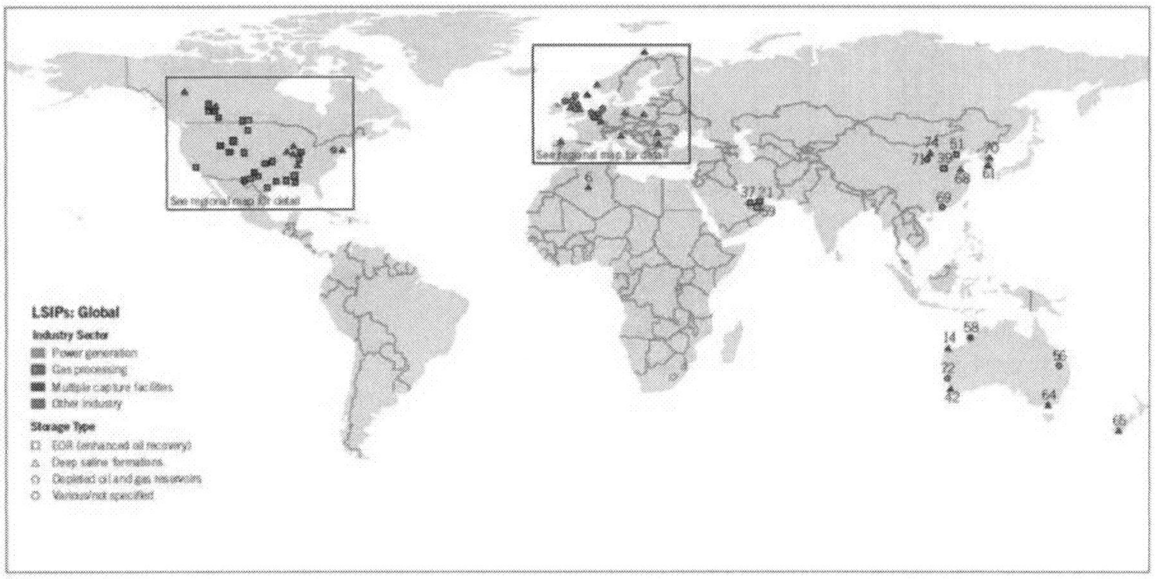

Figure 1: The Global Status of CCS

Prevailing Sequestration Methods

After the CO2 has been captured compressed, and transported, it must be sequestered.As illustrated in the figure 2. The company can either directly sequester the CO2 or treat it as a commodity for industrial use. The value of the CO2 is dependent upon its level of contamination and the purpose for which it is intended. In this section, the key sequestration method are described.

- Sequestration in Depleted Oil and Gas Reservoirs
- Sequestration in Deep Saline Aquifer
- Sequestration in Oceanbed
- Food Grade & Industrial Grade CO2
- Enhanced Oil Recovery (EOR)
- Gas Bearing Sandstone
- Gas Bearing Shale
- Coal Bed Methane
- Gas Storage

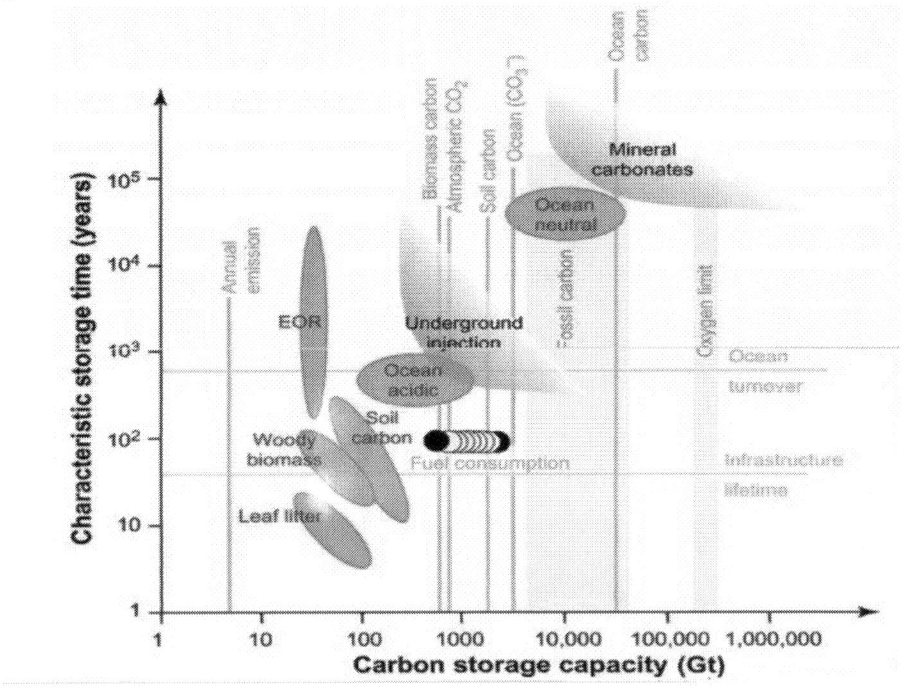

Figure 2: Current Sequestration Methods

Benefits of CO2 EOR

Worldwide we have about 2100 reservoirs which have the basic amenities required for supporting CO2 EOR. A large chunk of these reservoirs have the capability to economically run these projects at current oil prices or higher. But the real problem lies in the availability of CO2 gas, which greatly limits the viability of these projects. One of the most plausible ways of ensuring sufficient availability of CO2 gas is through widespread capture of industrial CO2 for EOR which would vastly increase the volume of CO2 available from a meagre 3Bcf per day (that is obtained mainly from natural sources) to a whopping 70 Bcf per day , thereby eliminating the main problem of limited CO2 supply. Moreover the capture of industrial CO2 ensures that there is a steady supply available at all times. With a stable supply of CO2 ensured more than 200 CO2 EOR projects with increasing production reaching nearly 1 mn barrels per day can be realised, assuming that the prices in USA remain between \$1 and \$3 Mcf day. In addition to this, these projects would provide a platform to sequester nearly 5 Bcf of CO_2 each day. With estimates and assumptions, it can be predicted that production as shown in figure 3 may boom to a volume of more than 5.5 billion barrels of oil and the sequestration of nearly Tcf CO_2 in a period of 25 years.

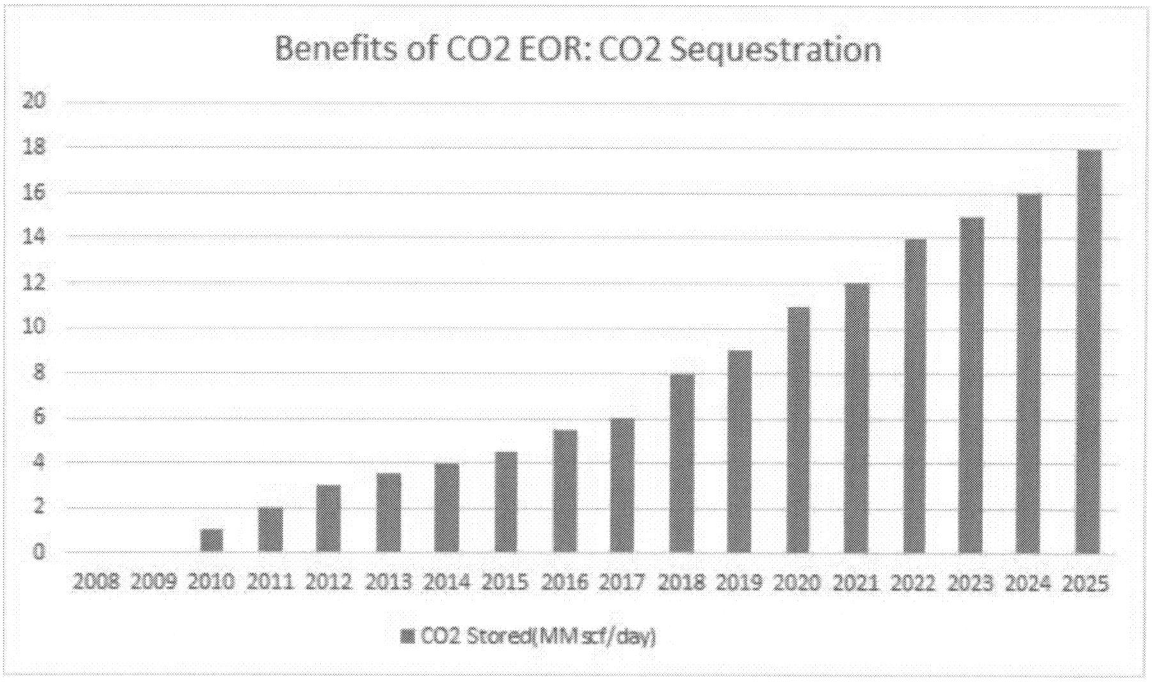

Figure 3: Benefits of CO2 EOR: CO2 Sequestration

The graph below (figure 4) shows the increasing oil production from the new CO_2 projects over two decades ranging from 2010 to 2030. It is estimated that by 2030 an excess of 800 thousands barrels of daily incremental production could be achieved. The same estimates show that by that time nearly 5 Bcf of CO_2 would be sequestered on a day to day basis. By definition, CO_2 sequestered is the difference in volumes between CO_2 injected and CO_2 produced.

As per the estimated data in the graph, the two humps created (one in 2021-22 and the other in 2027-28) are a result of an additional CO_2 produced from refineries and power plants that are going to be operational after 2021 for additional projects.

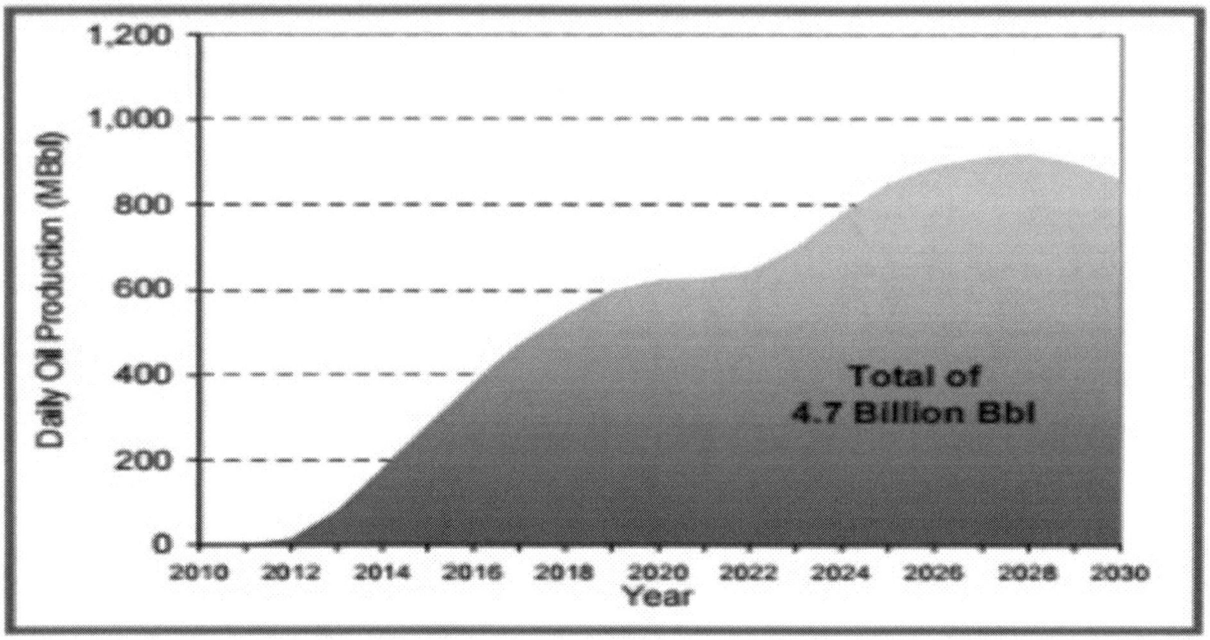

Figure 4: Benefits of CO2 EOR: Incremental Production

COMPARISON OF THE COSTS OF CCS FOR DIFFERENT POWER

Power Plant	Natural Gas Combined Cycle(U.S.$/kW-h)[a]	Pulverised Coal (U.S.$/kW-h)[b]	Integrated Gasification Combined Cycle (U.S.$/kW-h
Without Capture	0.03-0.05	0.04-0.05	0.04-0.06
With Capture and Geological Storage[c]	0.04-0.08	0.06-0.1	0.05-0.09
With Capture and EOR[d]	0.04-0.07	0.05-0.08	0.04-0.07

Table 1: Comparison of Cost of CCS

With these assumptions and the present data available it is clearly seen that the cost prices of CCS in the Capture and EOR sector is slightly higher in all fields with respect to the cost price of the cycles during operation of the power plant without capture. The difference in cost comes out about 0.01-0.003 for each cycle (natural gas combined cycle, pulverized coal cycle, and integrated gasification combined cycle), the margin of difference is small compared to the advantage that is obtained by operating each of these cycles in a power plant with capture and EOR of CO2 than without its capture.

*A Gas prices are assumed to be 2.8–4.4 U.S. $/GJ.
*B Coal prices are assumed to be 1–1.5 U.S. $/GJ.

*C The cost of a full CCS system for electricity generation from a newly built, large-scale fossil fuel-based power plant depends on a number of factors, including the characteristics of both the power plant and the capture system, the specifics of the storage site, the amount of CO2, and the required transport distance. The numbers assume experience with a large-scale plant.
*D Enhanced oil recovery.

Costs and Benefits of CO2 Capture and Sequestration

The total cost of carbon capture and storage includes the cost of capture, the cost of transportation, and the cost of either sequestering the CO2 (as shown in figure 5) or the value which can be realized through its sale. The sequestration costs can range from $1/ton to $41/ton depending upon the geologic media. The costs for sequestering CO2 in depleted oil and gas reservoirs and brine aquifers range from $1 to $10 per ton depending upon the depth of the reservoir or aquifer. The additional technical difficulties in ocean bed sequestration raise the costs to between $7 and $41 per ton of CO2 sequestered. These costs include transportation of the CO2.

If the CO2 is sold for commercial or industrial use, the value will depend upon the purpose of the CO2. The industrial and food grade CO2 have the highest values ranging between $50 and $120 per ton. The values of CO2 for enhanced oil or gas recovery can range between $36 and $106 dollars per metric ton.

In addition, the food grade and industrial grade CO2 require liquefaction and transportation. The CO2 used in the enhanced oil and gas recovery processes as well as the gas storage does not include transportation costs. It is assumed that these costs will be paid by the operator of the reservoir. The final cost of sequestration including capture costs between $31 and $55 per ton range from nearly $36 to $106 per ton of CO2. The range of cost varies based on the sequestration technology applied. The food grade CO2 continues to be profitable for the seller of CO2. The industrial grade and other value options are profitable and range from value of $4 to 41 per ton of CO2 captured and sold. The costs provided do not reflect monitoring, liability, or other costs required by future regulations.

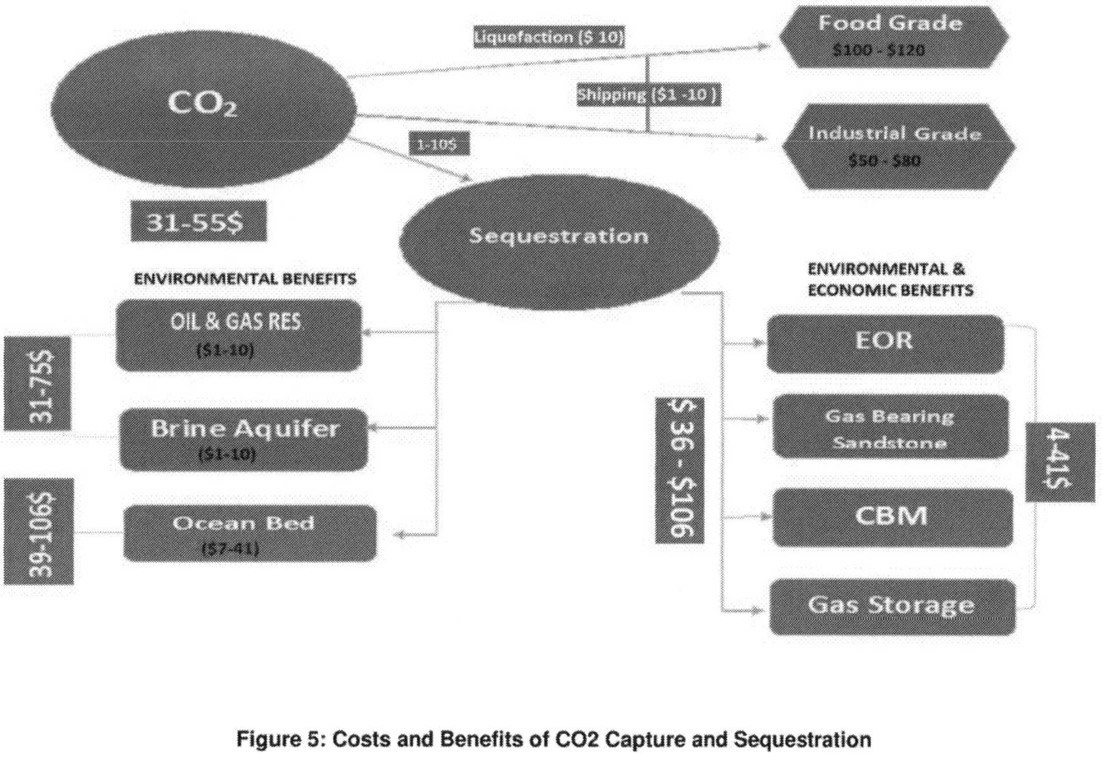

Figure 5: Costs and Benefits of CO2 Capture and Sequestration

Conclusion

There are several key challenges that need to be resolved to fully demonstrate carbon storage, including:

- Identifying optimal storage sites.
- Understanding corrosion and long-term sealing of wells.
- Understanding the behaviour, interactions, and physical properties of CO_2 under storage conditions.
- Maximizing storage potential.
- Monitoring storage sites in the long term.
- Demonstrating successful projects to build confidence in the technology.

References

1. EIA Historical Data. EIA website http://eia.doe.gov
2. The National Petroleum Council. Reference Report #36: Capturing the Gains from Carbon Capture. July 18, 2007. NPC website www.npc.org
3. UOP "Amine Guard ™ FS Process", 2000.
4. Edited by Metz, Bert, Davidson, Ogunlade, et. al., Special Report on Carbon dioxide Capture and Storage, Intergovernmental Panel on Climate Change (IPCC), 2005.
5. NATCARB Website, http://www.natcarb.org and INTEK, Inc. 2009.
6. INTEK Inc.
7. Renewable Fuels Association, http://www.ethanolrfa.org
8. INTEK Inc.
9. Dooley, J.J, et. al., Carbon Dioxide Capture and Geologic Storage, Global Energy Technology Strategy Program, April 2006.
10. Edited by Metz, Bert, Davidson, Ogunlade, et. al., Special Report on Carbon dioxide Capture and Storage, Intergovernmental Panel on Climate Change (IPCC), 2005

CARBON MANAGEMENT TECHNOLOGY CONFERENCE

CMTC-439516-MS

Geothermal Exploitation with Considering CO2 Mineral Sequestration in High Temperature Depleted Gas Reservoir By CO2 Injection

Guodong Cui, Liang Zhang, China University of Petroleum (Huadong),Bo Ren, The University of Texas at Austin, Yuan Zhuang, Xin Li, Shaoran Ren, China University of Petroleum (Huadong), Xiaohui Wang, Shengli Engineering & Consulting Co., Ltd, SINOPEC.

This paper was prepared for presentation at the Carbon Management Technology Conference held in Sugarland, Texas, USA, 17–19 November 2015.

Abstract

High temperature depleted gas reservoir has the huge potential of heat mining. In this article, the method of geothermal energy exploitation in high temperature depleted gas reservoir by injection of supercritical CO_2 was proposed based on the analysis of thermal properties of supercritical CO_2 and the conventional heat transmission fluid-H_2O. Meanwhile, reservoir numerical simulations were conducted to analyze the heat mining capacity of supercritical CO_2 and related influence factors in high temperature depleted gas reservoir. The results show that this method not only can exploit geothermal energy, but also enhance the gas ultimate recovery and achieve the CO_2 geological storage. Compared with the geothermal energy exploitation in water reservoir using water (100 ℃, 35MPa), the method of geothermal energy exploitation in high temperature depleted gas reservoir (100℃, 35MPa) using supercritical CO_2 can achieve high-speed geothermal energy exploitation (the heat mining capacity of supercritical CO_2 is one and half times heat mining capacity of water), and this method can drastically reduce initial investment using the existing well pattern so that the high-efficient development of high temperature gas reservoir can be achieved. Reaction flow simulations in typical gas reservoir were also conducted to analyze the CO_2-brine-rock geochemical reactions. The results indicate that CO_2 mineral sequestration cannot be ignored in the high temperature gas reservoir, which has a certain effect on CO2 storage and heat transmission. The technology of geothermal energy exploitation in high temperature depleted gas reservoir by injection of supercritical CO_2 can further enhance the development value of the gas reservoir and prolongs the gas reservoir working life, which is favorable for the technology progress about geothermal development and CO_2 geological storage.

Introduction

Geothermal energy is one of the most promising renewable energy, which has the following advantages: wide distribution, little affected by the outside world(such as day and night, wind speed, temperature difference), low carbon emission and low maintenance costs, compared with the other new energy such as solar energy, wind energy and biomass energy. China is located between the circum Pacific geothermal belt and the Mediterranean-Himalayan Geothermal Belt, and is rich in geothermal energy.

Water is used as the heat transmission fluid to extract the heat in the traditional geothermal energy exploitation. Traditional geothermal reservoir is usually located in the geologically volatile zone or the region where surface temperature is higher or near the volcano, and the abundant geothermal energy can't reasonable be used because of the limit of the reservoir location[1]. Recent studies show that China has the hot dry rock geothermal recoverable reserves of 1.49×10^{21} J, saline aquifer geothermal recoverable reserves of 3.28×10^{19} J, geopressured reservoir geothermal recoverable reserves of 2.74×10^{19} J[2]. The recoverable geothermal reserves are up to 2×10^{23} J in America[3]. The abundant geothermal energy is far more than 2011 global energy consumption (5.4×10^{20} J)[4].

The concept of CO_2-Enhanced Geothermal System is first suggested by Brown (2000)[5]. In this concept $SCCO_2$ is used as the fracturing fluid for reservoir creation as well as the heat transfer fluid[6]. But the super deep well drilling technique and fracturing technique used in the EGS will cost a great deal of money, which decline economic benefits[7]. To overcome these problems, Randolph (2011) et al. proposed to inject $SCCO_2$ into deep brine reservoir to extract the heat. Thy differentiated this approach from EGS and name it as CO_2-plume geothermal-CPG[8].

Geothermal extracting capacity and advantage in high temperature depleted gas reservoir

Geothernal extracting capacity in high temperature depleted gas reservoir

Geothermal resource is very rich in China oil and gas fields. More and more high temperature gas fields are found with the development of petroleum exploration such as Puguang gas field (120～133 ℃), Dagang Qianmiqiao condense gas field(168 ℃), Daqing Xujiawenzi gas field (145 ℃～170 ℃) , and so on[9-13]. The rational development of gas reservoir geothermal resource at the late period of development is conducive to the oil gas industry reconversion. Table 1 lists geological characteristics and geothermal exploitation potential of high temperature gas reservoir in China.

Tab 1 Geological characteristics and geothermal extracting capacity of high temperature gas field in China

Name	Type	Volume ,10^8m^3	Temperature ,℃	Pressure ,MPa	Porosity	Permeability ,md	Geothermal potential, 10^{16} J	Coal equivalent, 10^6t
Dagang qianmiqiao	Carbonate	2.21	168	43	0.1	0.5	4.01	1.37
Daqing xushen	Volcanic	62.8	150	35	0.07	5	56.4	19.27
Eerduosi Sulige	sandstone	100	115	29	0.086	0.73	66.7	22.79
Puguang Feixianzu	Dolomite	53.6	130	55	0.0817	94.4	40	13.37

From Table 1 we can see that the temperature of high temperature gas reservoir usually higher than 110℃, and the initial reservoir pressure usually higher 30 MPa, the energy equivalent of four high temperature gas field all more than one million tons of standard coal.

Geothernal extracting advantage in high temperature depleted gas reservoir

Compared with the conventional geothermal resource, the geothermal exploitation of high temperature depleted gas reservoir has the following advantages: 1) High temperature depleted gas reservoir has been established a relatively perfect pattern, which can decrease drilling cost by using the existing wells during geothermal exploitation; 2) The detailed geological date of gas reservoir can accurately predict the geothermal reserves and reduce the reduce the exploration cost; 3) The natural pore structure of gas reservoir is beneficial to the heat exchange between heat transmission fluid and geothermal reservoir; 4) The perfect trap structure of gas reservoir can reduce the risk of leakage.

Gothermal extracting mechanism of $SCCO_2$ in high temperature depleted gas reservoir

The method to extract heat from high temperature reservoir using $SCCO_2$ as transmission fluid is a new type of geothermal exploitation. This method has been applied to hot dry rock, deep brine reservoir and geopressure geothermal reservoirs. But few researches for the geothermal exploitation of high temperature gas reservoirs at present, especially the study of using $SCCO_2$. The method of geothermal energy exploitation in high temperature depleted gas reservoir by injection of supercritical CO_2 was proposed based on the analysis of thermal properties of supercritical CO_2, and will be applied to the high temperature gas reservoir (150℃).

Advantage of $SCCO_2$ in heat transmission

Under the condition of typical high temperature gas reservoir (150 ℃, pressure>7.4 MPa), CO_2 can reach supercritical state. Pruess et al. indicated $SCCO_2$ will shows different from the conventional transmission fluid because of the sensitivity of thermophysical property vary with temperature and pressure.

The heat capacity of CO_2 and H_2O are calculated by combining PR equation of state and Passut-Danner thermal enthalpy coefficient[14]. And the mass flow evaluation parameters I_Q is calculated by Pedersen empirical viscosity formula[15,16]. As you can see from the dotted path in Fig 1a and 1c, under the condition of 150℃, the heat mining capacity of $SCCO_2$ increases with the increases of pressure when the pressure is lower than 35MPa, and maintains constant when the pressure is higher than 35MPa. Therefore, the 35 MPa is the best choice to extract heat of high temperature gas reservoir in 150℃. The changes of thermo-physical property about transmission fluids are shown as the red line in Fig 1 under the development of gas reservoir geothermal resource. Although the mass heat capacity of CO_2 is about 1.75 kJ/kg/℃ which is the 45% of the heat capacity of H2O, the mass flow of CO_2 is 2.5 times of mass flow of H_2O, so the heat mining capacity of CO_2 is 1.5 times of that of H_2O. The mass heat capacity of CO_2 and the ratio mass flow of CO_2 to mass flow of H_2O increase with the further development of the gas reservoir geothermal resource, which shows the CO_2 is better as the transmission fluid.

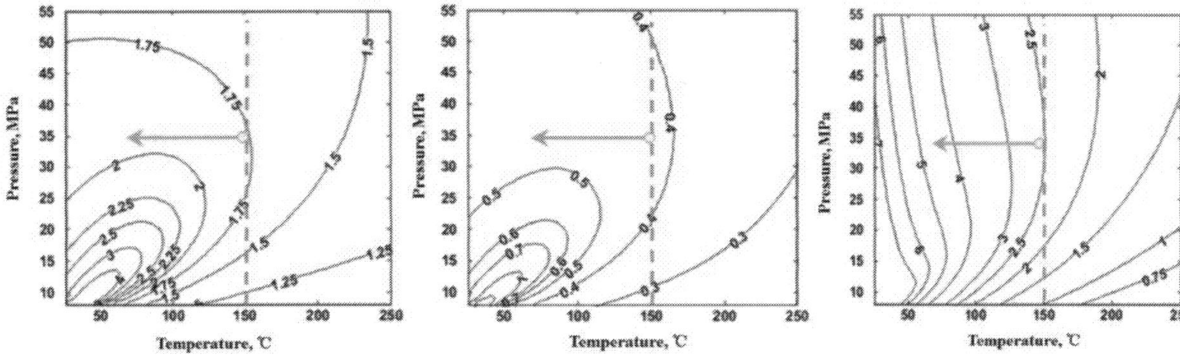

a mass heat capacity of CO_2 b ratio of heat capacity between CO_2 and H_2O c ratio of flow rate between CO_2 and H_2O

Fig.1 Heat capacities and mass flow indexes of CO_2 and H_2O

The method of geothermal energy exploitation using $SCCO_2$

In addition to the advantages mentioned above, geothermal energy exploitation using CO_2 can also lower injection-production pump power and avoid the scaling[5,17,18,19]. the method of geothermal energy exploitation in high temperature depleted gas reservoir by injection of supercritical CO_2 was proposed considering the CO_2 geological storage technology. In this paper, the gas high temperature depleted under 150℃ and 5MPa (abandonment pressure is estimated using Meike empirical forumla) is chosen as the typical geothermal exploration object.

In this paper, the gas high temperature depleted under 150℃ and 5MPa (abandonment pressure is estimated using Meike empirical forumla[20]) is chosen as the typical geothermal exploration object. Firstly, the ultimate recovery of natural gas can be improved through the gravitational differentiation.by injection of CO_2; secondly, the CO_2 is used as the transmission fluid to extract heat of gas reservoir after adjusting the working system (reservoir pressure has changed, as shown the yellow circle in Fig 1); lastly, the CO_2 can be stored in the gas reservoir when the gas reservoir no longer has the geothermal exploitation value.

Analysis of Geothernal extracting capacity about $SCCO_2$

Simulation models

A reservoir simulation study has been conducted to analyze the heat mining capacity of different transmission fluids.by using the simulator CMG which has excellent performance on calculating heat exchange. Basic parameters of the simulation model are adopted according to the published literatures. Temperature of gas reservoir is set to 150℃, reservoir pressure is set to 35 MPa. And the stability of reservoir pressure is obtained by setting constant injection-production pressure difference. The geological structure of the model is an anticline which has angle of 5 ℃. Meanwhile, five simulation cases are set to analyze the effect of transmission fluids and reservoir fluids, the simulation cases and results as shown in table 3 and figure 4-6.

Table 2 Parameters of geothermal extraction simulation model in gas reservoir

Parameters	Values	Parameters	Values
Deep, m	4000	Angle, (°)	5
Volume, m^3	1000×500×50	Heat capacity of CO_2, J/(mol·K) [b]	77 [b]
Temperature, ℃	150	Heat capacity of H_2O, J /(mol·K) [c]	80 [c]
Initial pressure, MPa	35[a]	Heat capacity of reservoir rock, J/(m^3·℃)	2.65×10^6
Reservoir pressure, MPa	35	Thermal conductivity reservoir rock, J/(m·day·℃)	1.814×10^5
porosity	0.1	Heat capacity of cap rock, J/(m^3·℃)	2.347×10^6
Permeability, md	20	Thermal conductivity cap rock, J/(m·day·℃)	1.496×10^5
Injection temperature, ℃	20	Distribution of injection-production wells	Top, bottom
Injection-production pressure, MPa	4	Production, a	40

Note: a, the phase of build-up pressure is ignored in this simulation. b, c, heat capacity of transmission fluids in different temperature are calculated in following formula : C_{CO2}=-329.71+3.375K-0.0076K^2+6x$10^{-6}$$K^3$, H_2O=20.403+0.6662K-0.0025K^2+3x10^{-6}K, where K is kelvin degree.

Table 3 Simulation schemes and corresponding results

	Simulation cases description			Simulation results					
	Description	Reservoir fluid	Injection fluid	Mining rate g/s	Heat mining rate W	Temperature decrease, ℃	Accumulated heat J	Coal Equivalent10^6t
1	Ideal gas reservoir	CO_2	CO_2	6.3	1.82	21	2.34×10^{15}	0.079
2	Ideal water reservoir	H_2O	H_2O	2.4	1.15	13.3	1.5×10^{15}	.0.051
3	Reference	H_2O	CO_2	a	1.65	15.3	1.68×10^{15}	0.057
4	Actual as reservoir-1	CH_4	CO_2	6.3	1.81	22	2.44×10^{15}	0.083
5	Actual gas reservoir -2	50%CH_4+50%H_2O	CO_2	a	1.67	17	1.91×10^{15}	0.065

Note: the values of mining rate and heat mining rate et al. are the results of simulation in the fortieth year. a, the values can be seen in figure 5 and 6.

Advantage of $SCCO_2$ in high temperature depleted gas reservoir

Cases 1 and 2 are set to analyze the seepage characteristics of gas reservoir geothermal exploitation using CO_2 and water reservoir geothermal exploitation using water. As shown in figure 2, the mining rate of CO_2 is 6.3 kg/s and mining rate of H_2O is 2.5 kg/s (the former is 2.5 times of latter, which is consistent with the result of figure 1c) under the same injection-production pressure; the heat mining rate of CO_2 is about 1.82 MW, however the heat mining of H_2O is about 1.15 MW (the former is 1.58 times of latter). These values indicate that CO_2 is better than H_2O as the transmission because CO_2 has better seepage ability. As you can see in the figure 3, the cumulative heat production of CO_2 is double the cumulative heat production of H_2O after 40 years exploration, which also indicates CO_2 is better than H_2O.

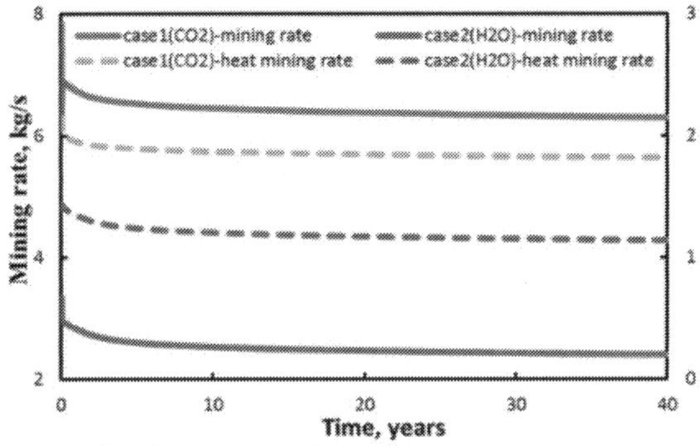

Fig.2 Mining rate and heat mining rate in cases 1,2

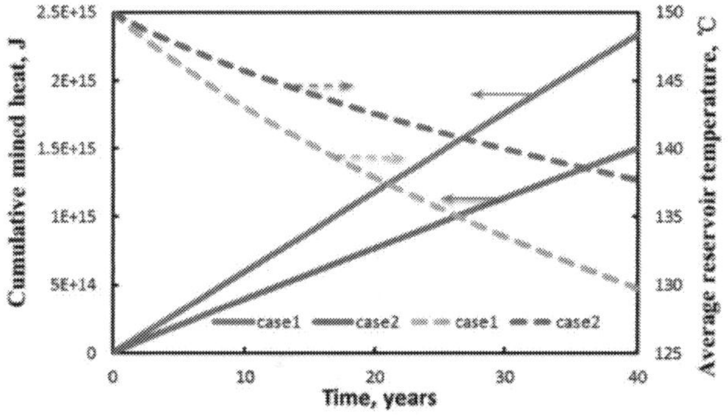

Fig.3 Accumulative mined heat and average reservoir temperature in cases 1,2

Ressure distribution and temperature distribution of gas reservoir and water reservoir can be seen in the figure 4. As shown, pressure drop of water reservoir near the injection well is bigger than that of gas reservoir, however, pressure drop of gas reservoir near the production well is bigger than that of gas reservoir. The difference of pressure distribution is caused by the thermo-physical property of transmission fluids. The viscosity of H_2O near the injection well where the temperature is lower than the average temperature is higher than the average viscosity, which consumes more displacement pressure. The higher pressure drop near the production well in gas reservoir can be attributed to the compressibility of CO_2 which offset some pressure drop.

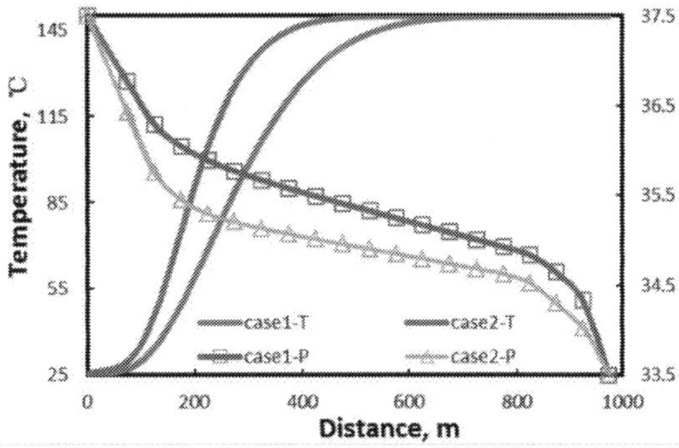

Fig.4 Pressure distribution and temperature distribution in cases 1,2 (40 years)

The influence of reservoir fluid to geothermal extracting capacity

As shown in figure 5-8, the reservoir fluids can affect the mining rate and heat mining rate. When the reservoir is full of CH_4, the simulation results show higher heat mining rate can be obtained because of the heat capacity of CH_4. When the reservoir is full of 50%CH_4+50%H_2O, the heat mining rate firstly gradually increases, then gradually decreased, lastly gradually increases, in the early production period. Combined with the result of case 3, the existence of H_2O in the reservoir can impact the seepage velocity and reduce the heat mining rate ultimately. So it is best to shorten the water production time to improve the heat mining rate.

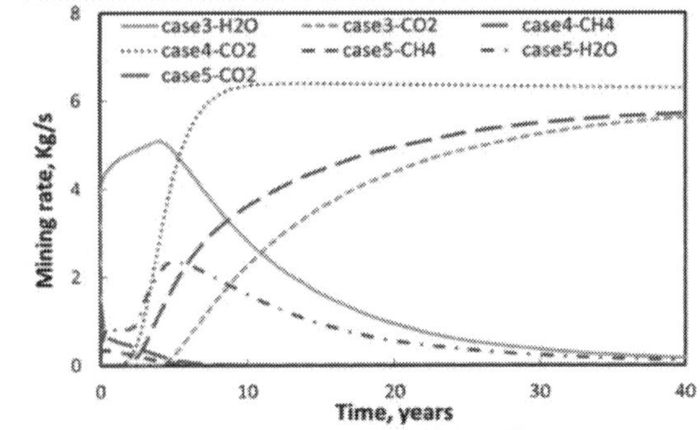

Fig.5 Mining rate in cases 3,4 and5

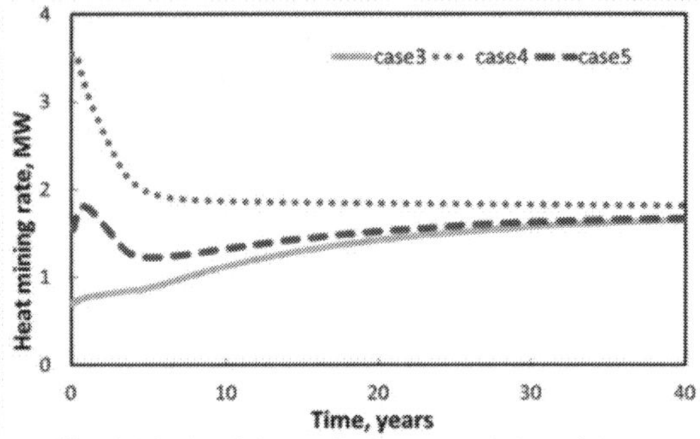

Fig.6 Heat mining rate in cases 3,4 and5

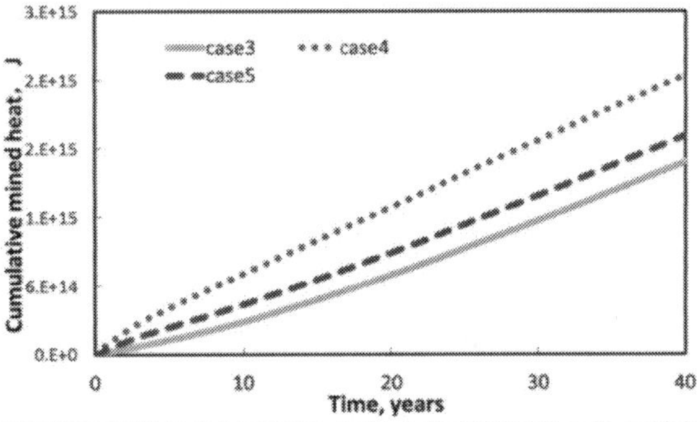

Fig.7 Accumulative total of heat mining in cases 3, 4 and 5

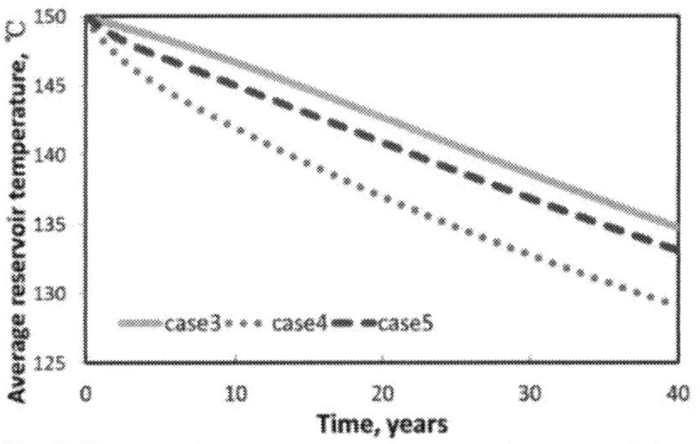

Fig.8 Reservoir average temperature in cases 3,4 and5

The analysis of geochemical reaction in high temperature depleted gas reservoir

On the basis of case 5 above, simulation models of sandstone gas reservoir and carbonate gas reservoir considering CO_2-brine-rock chemical reaction have been established for preliminary assessment of the CO_2 mineral sequestration contribution. The sandstone mineral composition include 3.6% anorthite, 15.3% calcite , 13.5% kaolinite and quartz, coming from Utsirf sandstone reservoir. The carbonate is full of calcite in the simulation model. The mineral reactions and related kinetic parameters in this study are given in table 4. The results of simulation are given in figure 9-12.

Table 4 Mineral reactions and kinetic parameters

Mineral	Reaction equation	$\log K_{25℃}$ mol/(m^{2}s)	Activation energy J/mol	Surface area m^2/m^3
Calcite	Calcite + H^+ = Ca^{2+} + HCO_3^-	-8.8	41870	88
Anorthite	Anorthite + $8H^+$ = $4H_2O$ + Ca^{2+}+$2Al^{3+}$+$2SiO_2$(aq)	-13	67830	88
Kaolinite	Kaolinite + $6H^+$ =$5H_2O$ +$2Al^{3+}$ +$3SiO_2$(aq)	-13	62760	1760

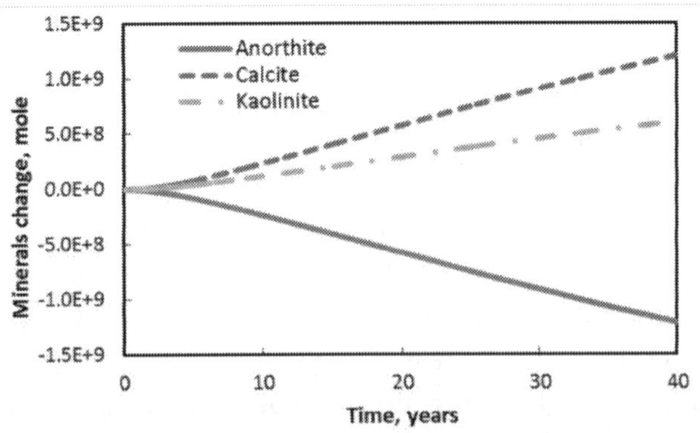

Fig.9 Minerals change in sandstone reservoir

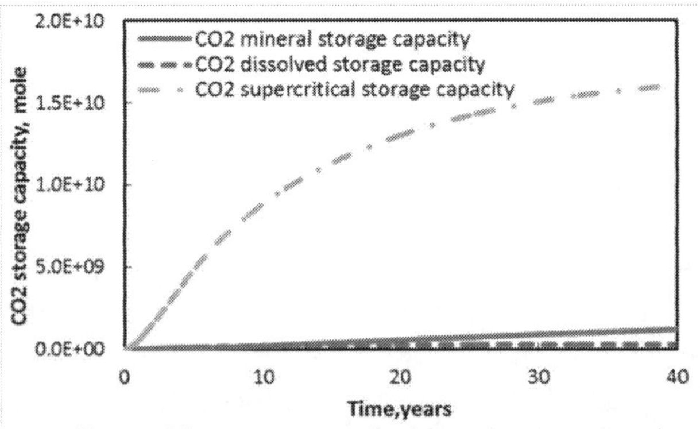

Fig.10 CO_2 storage capacity in sandstone reservoir

The dissolved CO_2 continually react with the reservoir minerals along with the sustained injection of CO_2 (figure 9, 10). And these chemical reactions rate is faster than that of conventional CO_2 geological storage because the temperature of sandstone gas reservoir is very high. Therefore, the CO_2 mineral sequestration can't be ignored. As shown in figure 9, the dissolution of anorthite can increase the concentration of Ga^{2+}, $Al^{3+}$and SiO_2, which will result in precipitate of calcite and kaolinite. The contribution of mineral sequestration to CO_2 storage can't be ignored because the fraction of mineral sequestration in CO_2 storage is large (figure 10).

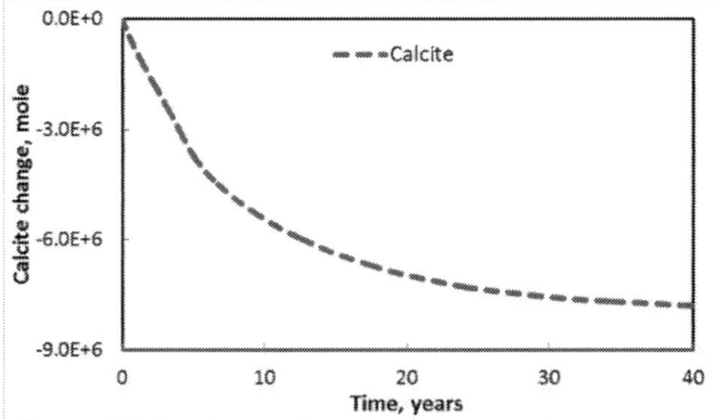

Fig.11 Calcite changes in carbonate reservoir

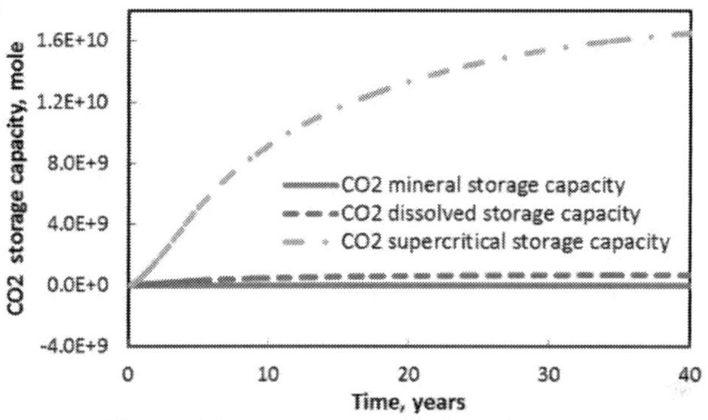

Fig.12 CO_2 storage capacity in carbonate reservoir

The results of carbonate gas reservoir simulation can be seen from figure 11, 12. The carbonate dissolves gradually with the injection of CO_2, but the dissolved content is low because of the ionic composition formation water. The contribution of mineral sequestration to CO_2 storage can be ignored.

Conclusions

There is enormous geothermal exploration potential in China. Compared with the conventional geothermal exploration, it has the following advantages: decreasing drilling cost, detailed geological date of gas reservoir and low risk of leakage.to extract heat from high temperature depleted gas reservoirs by injection of supercritical CO_2.and it can promote the rational utilization of gas reservoir geothermal, achieve energy saving and emission reduction, and prolongs the gas reservoir working life through the use of gas reservoir geothermal.

The heat mining rate of high temperature depleted gas reservoir using CO_2 as transmission fluid is higher than that of high temperature water reservoir using H_2O as transmission fluid. The advantage of CO_2 is not only due to its excellent flowing ability but it can also be attributed to its good pressure distribution. The research results shows that the heat mining rate of CO_2 is 1.5 times of that of H_2O in the typical high temperature condition (150℃).

The existence of initial reservoir fluids can affect the heat mining rate when using CO_2 to extract the heat of high temperature gas reservoir. When the reservoir is full of CH_4, the simulation results show higher heat mining rate can be obtained because of the heat capacity of CH_4. When the reservoir is full of 50%CH_4+50%H_2O, the heat mining rate is lower because of the existence of water. So it is best to shorten the water production time to improve the heat mining rate.

The CO_2-brine-rock chemical reaction and induced CO_2 mineral sequestration can't be ignored because of the high reservoir temperature. In this paper, for the sandstone gas reservoir, the contribution of mineral sequestration to CO_2 storage is positive. However, for the carbonate gas reservoir, the contribution of mineral sequestration to CO_2 storage is negative.

Acknowledgement

This research is supported by the Qingdao basic research program for young scholars (No. 13-1-4-254-jch), the Shandong Natural Science Foundation (No. ZR2013EEQ032) and the Fundamental Research Funds for the Central Universities (No. 15CX05036A).

Reference

1 Chapman D S. Geothermal Map of North America[M]. Geological Society of America, 1992.
2 Zhang L, Ezekiel J, Li D, et al. Potential assessment of CO2 injection for heat mining and geological storage in geothermal reservoirs of China[J]. Applied Energy, 2014, 122(2):237–246.
3 Tester J W, Anderson B, Batchelor A, et al. The future of geothermal energy: Impact of enhanced geothermal systems (EGS) on the United States in the 21st century[J]. Massachusetts Institute of Technology, 2006, 209.
4 Morlan T H. Energy Information Administration[J]. Energy Policy Planning, 2012, 9: 187.
5 Brown D W. A hot dry rock geothermal energy concept utilizing supercritical CO2 instead of water[C]//Proceedings of the twenty-fifth workshop on geothermal reservoir engineering, Stanford University. 2000: 233-238.
6 Pruess K. Enhanced geothermal systems (EGS) using CO 2 as working fluid—a novel approach for generating renewable energy with simultaneous sequestration of carbon[J]. Geothermics, 2006, 35(4): 351-367.
7 Salimi H, Wolf K. Integration of heat-energy recovery and carbon sequestration[J]. International Joural of Greenous Gas Control. 2012,6:56-68.
8 Randolph J B, Saar M O. Coupling carbon dioxide sequestration with geothermal energy capture in naturally permeable, porous geologic formations: Implications for CO 2 sequestration[J]. Energy Procedia, 2011, 4: 2206-2213.
9 Tang Y, Sichuan C, China, et al. Formation water vaporization and salt out at near well bore zone in high temperature gas reservoir [J]. Journal of Southwest Petroleum University, 2007, 29(2):96-99.
10 Yang S, Wang S, et al. Evaluation and development of fractured buried-hill condensate gas reservoir——taking Qianmiqiao buried-hill condensate gas reservoir as an example [J]. Natural Gas Geoscience, 2006, 17:857-861.
11 Shao Rui. Evaluation and optimization of development program for volcanic gas reservoir in Xushen gas feld[D].Northeast Petroleum University,2011.
12 He G, Li J, Wang J, et al. New progress and outlook of development technologies in the Sulige Gas Field[J]. Natural Gas Industry, 2011, 31(2):12-16.
13 hang Bowen. Resarch of main controlling factor and reservoir characteristics in Guanxian group of puguang gas field [D]. Southwest University,2012.
14 Passut C A, Danner R P. Correlation of Ideal Gas Enthalpy, Heat Capacity and Entropy[J]. Ind.eng.chem.process Des.dev, 1972, 11(4):543-546.
15 Pedersen K S, Fredenslund A, Christensen P L, et al. Viscosity of crude oils[J]. Chemical Engineering Science, 1984, 39(6):1011–1016.
16 Pedersen K S, Fredenslund A. An improved corresponding states model for the prediction of oil and gas viscosities and thermal conductivities[J]. Chemical Engineering Science, 1987, 42:182–186.
17 Wilke C R, Chang P. Correlation of diffusion coefficients in dilute solutions[J]. AIChE Journal, 1955, 1(2): 264-270.
18 Remoroza A, Doroodchi E, Moghtaderi B. CO2-EGS in hot dry rock: preliminary results from CO2-rock interaction experiments[C]//Proceedings of 37th Workshop on Geothermal Reservoir Engineering. 2012, 30.
19 Zhang L, Pei J, et at. Advantages of supercritical CO_2 as heat transmission fluid and its application potential in development of geothermal resources[J] Renewable energy sources,2014,32(3):330-334.

20 Chen Wei. The discussion of several methods to calculate the abandonment pressure of gas reservoirs[J]. Liaoning Chemical Industry,2013,08.

CARBON MANAGEMENT TECHNOLOGY CONFERENCE

CMTC-439520-MS

Analysis of CO_2 Swelling and Modeling of PVT Properties of Reservoir Fluid of an Onland Field of Gujarat, India

N. K. Jha and U. Ojha, RGIPT; A. Kumar and A. Mathur, ONGC Ltd.

This paper was prepared for presentation at the Carbon Management Technology Conference held in Sugarland, Texas, USA, 17–19 November 2015.

Abstract

With the advancement in technology, interest in different enhanced oil recovery (EOR) techniques to recover this untapped oil has increased in recent years. The paper presents a fluid with extensive PVT and carbon dioxide (CO_2) Swelling Test data from a major on-land field of Gujarat, India. The reservoir is at pressure of 2,860.7 psia and temperature of 192.2 °F. The reservoir is on depletion drive mechanism with very low gas/oil ratio of 16.2 (v/v). 12-component equation of state (EoS) models were developed for the fluid using the volume corrected Peng-Robinson EoS, which shows a good match between experimental and simulated data. Simulation of multi-contact process was carried out to predict the minimum miscibility pressure (MMP). Purely predictive MMP is considerably lower than the reservoir pressure which is the most important criteria for a miscible gas injection project and the simulation result shows a forward contact vaporizing gas drive mechanism of miscibility. The representative reservoir fluid sample was experimentally analyzed and the resulting data were used to obtain parameters for EoS model. The tuning procedure for EoS was carried out systematically by matching the volumetric and phase behavior results with that of the experimental results. Splitting method by Whitson et al was used to split pentane plus, C_{5+} fraction into 5 pseudo-components to increase the predictive capability of EoS. The EoS model developed for this field may be used in compositional reservoir simulation studies and well optimization for a miscible CO_2 EOR.

Introduction

Worldwide the numbers of mature oil field are continuously growing. Two-third of the resources is left in reservoirs after traditional recovery methods, such as primary production and water flooding. The world oil production is 90 million barrels per day (bpd) in which production through enhanced oil recovery (EOR) methods is less than 3 million bpd.[1] Oil companies have stepped up efforts in accessing CO_2 as an effective EOR agent. The injection of this greenhouse gas has established itself as a very efficient mechanism for increasing oil recovery. Moreover, worldwide environmental concerns are also driving the use of CO_2 for EOR. Presently, 135 CO_2 based EOR projects are active worldwide. 112

miscible CO_2 EOR projects in USA are accounting for 41% of oil production from EOR methods, which is higher than any other EOR projects.

CO_2 injection has various advantages in a flood process for improving oil recovery such as sweep improvement, viscosity reduction, oil swelling, miscibility development at lower pressures and high incremental recovery. Carbon dioxide mixes in oil mainly as a result of three mass transfer mechanisms; solubility, diffusion and dispersion.[2] Out of these three mass transfer mechanisms, solubility accounts for the greater part of the mixing. Solubility is affected greatly by pressure and to a lesser degree by temperature and composition. Compositional reservoir simulation studies are necessary to evaluate the possible incremental recovery by applying miscible gas injection. The accuracy of results from the reservoir simulation depends on the EoS models for oils as well as their ability to reproduce the phase behavior during miscible gas injection. Tuning the selected EoS is an important pre-requisite to provide meaningful and reliable predictions. It refers to adjusting the parameters of the EoS to achieve a satisfactory match between the experimental PVT data and EoS results. Manual adjustments through trial and error or by an automatic nonlinear regression approach are used to adjust the EoS parameters.[3] The regression variables are a selected number of EoS parameters that may be adjusted to achieve a match between the two results. Christensen has shown that adjusting the critical temperatures and critical pressures of the heavy end pseudo-components could result into the required match.[4] Coats and Smarts have recommended the adjustment of critical temperature T_c, critical pressure p_c and accentric factor ω of the heavy end pseudo-components.[5] They also suggested the adjustment of binary interaction co-efficient (BIC) between methane and pseudo-components and between methane and CO_2 respectively.

The design of gas or solvent injection processes involves a multiple-contact calculation to study the vaporization or extraction process. Miscibility between fluids can be either first contact miscible or multi-contact miscible (also known as dynamic miscibility). If carbon dioxide is first contact miscible with a reservoir fluid then it will mix in all proportions as soon as the two fluids contact one another, no matter what amounts of each component is used for the mixture [6]. Multi-contact miscibility can be achieved through two techniques; the vaporizing gas drive and condensing gas-drive.[7-8] In the vaporizing-gas drive, a lean gas or dry gas is injected. The leading edge of the injection gas will first vaporize the lighter components present in the reservoir fluid, such as, methane through to hexane. Then this mixture will progressively mix with heavier components until the leading edge of the injection gas is enriched enough such that it is miscible with the virgin reservoir fluid.[7-8] In a condensing-gas drive, an enriched gas with C_{2+} components is injected and the heavier components are given up into the oil. When the oil becomes sufficiently enriched with these light hydrocarbon components it becomes miscible with the injection gas.[7-8] In the condensing gas drive miscibility is achieved at the trailing edge. The enriched reservoir fluid becomes miscible with fresh injection gas. Since, the miscibility in an oil reservoir undergoing miscible gas injection develops through a critical point, it is required for an EoS model to be used for simulating miscible gas injection processes that the critical point on the Swelling Test saturation point curve is accurately matched [9]. In the present study an Indian onland field of Gujarat has been identified as suitable for CO_2 injection for EOR. The field is having sandstone formation in the depth interval of 5,272.3-5,339.5 ft and °API gravity of reservoir fluid is 37.7. The reservoir under consideration is under saturated and on depletion drives mechanism. Reservoir pressure and temperature are 2,860.7 psia and 192.2 °F respectively. The reservoir is having permeability in the range of <5 to 150 md with oil saturation greater than 70%. Neutron porosity value averages to the value of 26%. Various properties such as relative volume, viscosity and density in presence and absence of CO_2 were measured as a function of pressure and the data were compared with that of the simulated results. Many pseudo-ternary diagrams are generated from the calculations to help with the interpretation of the results.

Experimental

Compositional Analyses

The reservoir fluid composition was analyzed using gas chromatography and C_{5+} fraction molecular weight and liquid densities have been determined for the liquid from a flash of the sub-surface sample to standard conditions. The composition of the fluid is shown in Table 1. The reservoir fluid in the shipping bottle was rocked several times at 5,000 psia for 5-7 days to make it in single phase before charging into an ultra-high pressure PVT cell. The process of high pressure rocking for several days is necessary to ensure the integrity and homogeneity of the reservoir fluid. The single phase reservoir fluid is flashed to standard condition from reservoir conditions and determines the oil formation volume factor B_o, gas/oil ratio GOR. The compositions of evolved gas and dead oil and their respective molecular weights were also measured. The calibration of gas chromatograph was done by injecting standard gases in advance. Gas samples were analyzed at least two times, and the values of the composition were averaged. The dead oil was analyzed separately. Overall composition of the fluid was determined by combining the compositions of gas and dead oil in GOR in terms of their weight. C_{5+} fraction molecular weight is determined by the correlation developed by US DOE's NETL [10] given by,

$$M_{C_{5+}} = 4247.98641\, X^{-0.87022}$$

Where, X is dead oil gravity in °API at STP.

Constant Mass Expansion (CME) Test

A versatile PVT instrument capable to operate up to a pressure of 15,000 psia and temperature up to 175°C was used for the study of thermodynamic and phase behavior properties of the fluid. The volume capacity of the cell was 500 cc. The digital camera which displayed the view of inside PVT cell through a sapphire window on a personal computer screen provide the direct viewing of phase separation during studies. The Constant Mass Expansion test (CME) was performed by charging a 71.435 cc of representative reservoir fluid sample in to the PVT cell at reservoir pressure of 2,860.7 psia and temperature of 192.2 °F. The single phase fluid was deep black in color. The composition of the fluid was kept constant throughout the test. This test was performed to determine saturation or bubble point pressure at reservoir temperature. At 414.7 psia, it was observed that some bubble of foggy nature started accumulating at the top of the cell which was clearly visible on computer screen. From the observation it was recorded as the saturation pressure of the reservoir fluid at constant reservoir temperature. After determining the saturation pressure, differential liberation studies were carried out to determine the relative volume-pressure relationship, density and compressibility at different stages of pressure depletion. An embedded pump was programmed to decrease the pressure in 10 steps up to 214.7 psia. Below saturation pressure two phases occurred and a large amount of gas evolved in the process. The relative volume is plotted against the cell pressure and is illustrated in Figure 1. The relative volume was determined from the ratio of the cell volume at a pressure over the cell volume at saturation pressure.[11] Thus at saturation pressure; the value of relative volume is 1. The oil formation volume factor B_o is defined as the ratio of the total volume of the hydrocarbon mixture, i.e., oil and gas in single phase at the prevailing pressure and temperature per unit volume of the stock-tank oil. Constant Mass Expansion Test data for the fluid is shown in Table 2.

Viscosity measurement

The capillary viscometer of 500 mm length was used to determine the viscosity of reservoir fluid in the range of 1 to 10 cp. One end of capillary viscometer was connected to the PVT cell and the other end to a piston cell and two pressure transducers measured the pressure at both the ends. A constant flow rate was maintained through the capillary and when the flow was continuous; the pressure drop Δp across the capillary tubing of length L and internal radius r was measured.

The viscosity data is based on Poiseuille's law for dynamic viscosity.

$$\eta = K\frac{\pi r^4 \Delta p}{8LQ}$$

Where, K is a constant factor relative to the capillary tubing. The kinematic viscosity can be determined by the following expression,

$$v = \eta / \rho$$

Where, ρ is density of sample. The fluid is supposed to be Newtonian (non-dependent to shear rate). The flow is supposed to be laminar according to Reynold's number criteria. Referring to Poiseuille's law, the internal radius 'r' should be known with a sharp accuracy, determinant on the viscosity accuracy. Since it is not possible to measure the internal radius of capillary tubing along its whole length, and neither the losses in fittings, the viscosity measures will be determined as a ratio to a standard sample of known viscosity.

$$\eta_{sample\,(T)} = \eta_{standard\,(T)}\, X\, \frac{\Delta p_{sample\,(T)}}{\Delta p_{standard\,(T)}}$$

at a given temperature T.

with $\left[\frac{\eta_{standard\,(T)}}{\Delta p_{standard\,(T)}}\right]$ = constant calibrated with the PVT instrument for each capillary tubing. This requires to measure Δp sample and $\Delta p_{standard}$ and to know η_{sample} exactly at the same temperature T. The measured viscosity data at different pressures is shown in Table 2

Swelling Test

The purpose of this laboratory test at different pressure between reservoir pressure and bubble point pressure was to determine the degree to which CO_2 will dissolve in the reservoir fluid. The data obtained from the experiment gave the relationship of saturation pressure and volume of gas injected and changes in various fluid properties at different saturations of CO_2. The representative sample of the hydrocarbon mixture with a known overall composition was charged in to the PVT cell at working pressure (typically 1,000 psia above the reservoir pressure) and reservoir temperature (192.2 °F). Initially a predetermined volume of CO_2, calculated to be 0.186 mole fraction of the reservoir fluid/CO_2 mixture was added to the reservoir fluid. The cell was pressurized up till a black colored homogenous mixture was obtained. The new saturation pressure and volume were recorded and designated as p_s and v_s, respectively. The original saturation volume determined from constant mass expansion test was used as a reference value and the results were presented as relative total volumes. The amount of CO_2 was increased subsequently and the process was repeated up to 0.49 mole fraction of CO_2. The liquid which was originally black oil changed its properties to first, become a volatile oil and finally a critical fluid. At this point experiment was stopped. The identification of the appearance of the critical fluid mixture were carefully carried out as the slightest variation in pressure would have either lead to major deviations in volume measurement as the liquid would have either developed from the bottom of the cell (a case for a dew point) or shrunk from the top (a case of a bubble point). The density and viscosities of the swelled oil mixtures were also determined. Capillary viscometer used for the reservoir fluid/CO_2 mixture was of 2,000 mm length with a viscosity range from 0.1 to 1 cP. The process followed was same as described in "Viscosity Measurement" section. Table 3 shows Swelling Test data using pure CO_2 injection gas. The critical composition was found for 0.45 mole fraction of CO_2 in the reservoir fluid with a swelling factor of 1.39 v/v ratio.

EoS Model Development

Pre-requisite of EoS modeling is matching saturation point which would simulate the correct number of phases at pressures near the saturation point. Liquid densities, relative volumes and viscosities are the most important property to match for a CME Test. For matching densities and viscosities rigorous adjustment in tuning weight is required in regression. For Swelling Test experiment, accurate match of critical point after a single contact between reservoir fluids and injection gas is necessary [9]. The miscibility develops through this critical point for which minimum pressure is required to develop a miscible zone after multiple contacts between reservoir oil and injection gas. This minimum pressure is called minimum miscibility pressure (MMP) and it should be much below reservoir pressure in miscible EOR. The Peng-Robinson EoS [12] with volume correction was used for the development of the model. In 1976, Peng and Robinson introduced a new two-parameter equation of state. The binary interaction coefficients were used to model the intermolecular interaction through empirical adjustment of the term. The interaction between hydrocarbon components increases as the relative difference between their molecular weights increases. Earlier Jhaveri and Youngren [13] found the error associated with the equation in the prediction of gas-phase *z*-factors ranged from 3 to 5% and the error in the liquid density predictions ranged from 6 to 12%. They followed the procedure proposed by Peneloux and co-workers [14] and introduced the volume correction parameter c_i to the PR-EoS. This third parameter has the same unit as second parameter b_i of unmodified PR-EoS and given as:

$$c_i = S_i b_i$$

Where, S_i is a dimensionless parameter, called the shift parameter, and b_i is the Peng-Robinson covolume. Whitson and Brule [15] extended the work of Jhaveri and Youngren [13] and tabulated the shift parameters S_i for a selected number of pure components thereby eliminating the volumetric deficiency associated with the EoS application. Splitting method by Whitson et al [16-18] is used here to split C_{5+} fraction into 5 pseudo-components. This splitting method uses gamma distribution function for calculating molecular weights of pseudo-components and their respective molar composition. The three-parameter gamma function that was first proposed by Whitson is the probability density function used for the description of the method.

The equations represent the basic application of the three-parameter gamma function for splitting of plus fraction into many single carbon number (SCN) groups. Whitson *et al.* adapted the gamma distribution model with Gaussian quadrature using the principles of semi-continuous thermodynamics to characterize multiple fluid samples from the same reservoir simultaneously. Each fluid sample can have its own molecular weight and specific gravity for the C_{n+} fraction before the split. After the split, each split fraction (pseudo-component) will have the same molecular weight and specific gravity while the mole fraction distribution will be different for each fluid sample. Critical properties and accentric factors of the pseudo-components were calculated using Kesler and Lee [19] correlations. The viscosity model Jossi-Stiel-Thodos [20] was selected to model the viscosity experiments for the fluid. The JST correlation determined the mixture viscosity from the low-pressure mixture viscosity. Regression procedure of Agarwal *et al.* [21] was used to tune the EoS to match experimental measurements. The parameters used in the regression were component properties and binary interaction co-efficient (BIC). Regression procedure was based on a non-linear mathematical model that places global upper and lower limits on each regression variable. The non-linear mathematical model determined the optimum values of these regression variables that minimize the objective function.

Results and Discussion

Results of EoS modeling are presented in Table 4. Figure 1-3 shows graphical representation of experimental PVT parameters measured at simulated reservoir conditions. Figure 1 shows the plot of relative volume (red circles) with pressure and Figure 2 shows the density (red circles) with pressure. An excellent match of PVT data with that of the simulated data was obtained by tuning parameters of PR-EoS. Saturation pressure of the relative volume plot in Figure 1 is matched accurately. Whitson splitting method was used to split C_{5+} fraction into five pseudo-components and Lee-Kesler correlation was used to characterize the resulting pseudo-components. The tuning parameters such as critical temperatures T_c, critical pressures p_c and accentric factors ω have the largest impact on simulation of PVT data. Tuning weight was adjusted to override value of 10 from a default value of 1 for better density simulation. This can be seen from the good match of simulated and experimental plot of density in Figure 2. Viscosity is another sensitive parameter as it affects the mobility of reservoir fluids through porous media. Experimental results indicate there is no dissolved gas in the oil at atmospheric pressure and reservoir temperature and therefore the oil has its highest viscosity value (Table 2). The plot of fluid viscosity (red circle) with pressure is shown in Figure 3. Viscosity data obtained by JST correlation was matched with experimental data by varying correlation coefficients (α, a_i) and critical volume v_c. Density match by tuning EoS parameters is very critical to the use of JST correlation for viscosity simulation [22]. Thus a good match of density was ensured prior to the viscosity simulation by this method.

Swelling test experiment determined the degree to which CO_2 dissolved into the reservoir fluid. At each step of CO_2 injection, the new saturation volume and saturation pressure were recorded. The saturation pressure increased with increase in CO_2 mole fraction as shown in Figure 4. Critical saturation was recorded at 0.45 mole fraction of CO_2, beyond which phase separation occurred and to distinctly different phase visibly separated out.

The experimental data (red line) matched well with the data obtained from that of the simulation (blue line) as can be seen in Figure 4. Simulation is mostly dependent on the critical properties of components (T_c, p_c, v_c), accentric factor (ω) as well as BIC. It was found that tuning of BIC between methane and pseudo-components and between methane and CO_2 was in agreement with the suggestions by Coats and Smarts [5]. The critical point on the Swelling Test saturation point curves matched accurately for EoS model.

This is important for the developed EoS model to describe gas injection effect which shows forward contact vaporizing gas drive in simulation of multi-contact process. The MMP have been calculated by the simulation of multi-contact process [23-24] which is purely predictive. The MMP is predicted to be 1980 psia which is much lower than the reservoir pressure at temperature of 192.2 °F.

The condensing / vaporizing process Zick [25] refers to a process in which there is a region where condensation of intermediates from the solvent into the oil and another region where vaporization of components up to C_{30} from the oil into the vapor phase takes place. The phase envelope for the immiscible condensing / vaporizing process does not close which indicates that multiple-contact miscibility is not achieved.

Novosad and Costain [26-27] have shown that for some condensing/vaporizing processes upper phase extraction (vapor or liquid) plays an important role in recovering oil. It could result into high recovery by efficient extraction of components when the upper phase becomes liquid-like in density. Under these conditions the IFT between the phases becomes low. At high mole fraction of CO_2, the process goes from a combined condensing / vaporizing process to a process mostly dominated by the vaporizing or upper phase extraction mechanism. At these conditions the solvent is multiple contacts miscible with the oil. Miscibility is generated by forward contacts of the equilibrium vapor phase with fresh oil.

A p/T phase envelope diagram for the reservoir fluid is generated and shown in Figure 5. 2-phase boundary shows the saturation curve in which upper part up to critical point (green dot) is bubble point curve and the lower part is dew point curve. Quality lines were generated representing the percentage of liquid volume. Quality lines that are approximately equally spaced characterize this black oil phase diagram and converge to the critical point. However, this diagram represents the possible phase behavior of reservoir fluid under primary recovery or under depletion drive mechanism

Conclusion

Extensive PVT and Swelling Test data of reservoir fluid from the major field of Gujarat, India have been presented. Twelve components EoS models of the fluid is developed using volume corrected Peng-Robinson EoS. In EoS model development, saturation point of relative volume curve and critical point of the Swelling Test saturation point curves are mostly focused. Density and viscosity curves also matched well. Tuning of parameters through regression contributed to larger extent for the achievement of EoS models. It is found that critical temperatures, critical pressures, accentric factors, binary interaction co-efficient and selection of tuning weights are the most important and sensitive parameters for EoS model development both for PVT data and Swelling Test data. Both PVT and Swelling Test data matched accurately with the consistent Peng-Robinson EoS models. Predicted MMP is found to be much less than the reservoir pressure through the simulation of multi-contact process. Prepared data can be used in EoS compositional simulator for simulating various important mechanisms of the miscible CO_2 injection process for reservoir performance prediction. These data can also be used for the design of facilities in miscible CO_2 injection process.

Acknowledgements

Nilesh K. Jha acknowledges KDMIPE, ONGC Academy, ONGC Limited, India for providing the infrastructure and facility to conduct the above research. Department of Science & Technology (DST), India (Grant: SERB/F/5311/2012-13) is acknowledged for providing financial assistance.

References

1. Schenwerk, P. EOR can extend the promise of unconventional oil and gas. *Oil & Gas J.* **2012,** *110,* 48–69.

2. Bybee, K. Challenges for Offshore Heavy-Oil Field Development. *Journal of Petroleum Technology.* **2003,** *55*, 66-67.

3. Ahmed, T. *Equations of State and PVT Analysis*; Gulf Publishing Company: Houston, Texas, 2010; pp 331-456

4. Christensen, P. L. Regression to Experimental PVT Data. *Journal of Canadian Petroleum Technology*. **1999,** *38*, 1-9.

5. Coats, K.; Smart G. Application of Regression-Based EOS PVT Program to Laboratory Data. *SPE Reservoir Engineering*. **1986**, 1, 277-299

6. Stalkup Jr., F. I. *Miscible Displacement*; Society of Petroleum Engineers, AIME, Monograph: Texas, 1984; 8, pp 1-3.

7. Holm, L.W.; Josendal, V.A. Mechanisms of Oil Displacement by Carbon Dioxide. *Journal of Petroleum Technology*. **1974**, *26*, 1427-1438.

8. Holm, L.W. Miscibility and Miscible Displacement. *Journal of Petroleum Technology*. **1986**, *38*, 817-818.

9. Negahban, S.; Pedersen, K.S.; Baisoni, M.A.; Sah, P.; Azeem, J. An EoS Model for a Middle East Reservoir Fluid with an Extensive EOR PVT Data Material. *Proceedings of the Abu Dhabi International Petroleum Exhibition & Conference*, Abu Dhabi, UAE, November 1-4, 2010.

10. *Storing CO_2 and Producing Domestic Crude Oil with Next Generation CO_2-EOR Technology: An Update*; DOE/NETL-1417, Office of Systems, Analyses and Planning, U.S., 2010.

11. Pedersen, K.S.; Christensen, P.L. *Phase Behavior of Petroleum Reservoir Fluid;* Taylor & Francis: Boca Raton, 2006; pp 41-62.

12. Peng, D. Y.; Robinson, D. B. A New Two Constant Equation of State. *Ind. Eng. Chem. Fundam.* **1976**, *15*, 59–64.

13. Jhaveri, B. S.; and G. K. Youngren. Three-Parameter Modification of the Peng-Robinson Equation of Sate to Improve Volumetric Predictions. *Proceedings of the SPE Annual Technical Conference*, Houston, September 16–19, 1984.

14. Peneloux, A.; Rauzy, E.; Freze, R. A Consistent Correlation for Redlich-Kwong-Soave Volumes. *Fluid Phase Equilibria.* **1982**, *8*, 7–23.
15. Whitson, C. H.; Brule M. R. *Phase Behavior;* Society of Petroleum Engineers: Richardson, TX, 2000.
16. Whitson, C. H.; Anderson, T. F.; Soreide, I. C_{7+} Characterization of Related Equilibrium Fluids Using the Gamma Distribution. In *C_{7+} Characterization*; Chorn, L. G., Mansoori, G. A., Eds; Taylor & Francis, New York, 1989; pp 35-56
17. Whitson, C. H. Characterizing Hydrocarbon Plus Fractions. *SPE J.* **1983**, 4, 683–694.
18. Whitson, C. H.; Rogaland, R. C. Effect of C_{7+} Properties on Equation-of-State Predictions. *SPE J.* **1984**, 6, 685–696.
19. Kesler, M.G.; Lee, B. I. Improve Predictions of Enthalpy of Fractions. *Hydro. Proc.* **1976**, pp. 153-158.
20. Reid, R.C.; Prausnitz, J.M.; Sherwood, T.K. *The Properties of Gases and Liquids*; McGraw-Hill: New York, 1977.
21. Agarwal, R.K.; Li, Yau-Kun.; Nghiem, L. A Regression Technique With Dynamic Parameter Selection for Phase-Behavior Matching. *SPE Reservoir Engineering,* **1990**, *5*, 115-120.
22. *User Guide, WinProp, Phase Property Program*; Version 2009, Computer Modeling Group: Calgary, Canada, 2009, pp. 35-51.
23. Nutakki, R.; Nghiem, L.; Li, Yau-Kun.; George, A. Optimal Representation of Heavy Fractions in the Simulation of Multiple-Contact Processes. *Proceedings of the CIM/AOSTRA Technical Conference*, Banff, April 21-24, 1991.
24. Nutakki, R.; Hamoodi, A. N.; Li, Yau-Kun; Nghiem, L. X. Experimental Analysis, Modelling, and Interpretation of Recovery Mechanisms in Enriched-Gas Processes. *Proceedings of the 66th Annual Technical Conference and Exhibition of SPE*, Dallas, Texas, October 6-9, 1993.
25. Zick, A. A. A Combined Condensing/Vaporizing Mechanism in the Displacement of Oil by Enriched Gases. *Proceedings of 61st Annual Conference*, New Orleans, Louisiana, October 5-8, 1996.
26. Novosad, Z.; Costain, T. New Interpretation of Recovery Mechanisms in Enriched Gas Drives. *Journal of Canadian Petroleum Technology*, **1988**, *27*, 54-60.
27. Novosad, Z.; Costain, T. Mechanisms of Miscibility Development in Hydrocarbon Gasdrives: New Interpretation. *SPE Reservoir Engineering,* **1989**, *4*, 341-347.

Tables

Table 1. Molar composition of the fluid

Component	Mole%
H_2S	0.00
N_2	0.20
CO_2	0.16
C_1	10.19
C_2	1.11
C_3	1.47
i-C_4	0.74
n-C_4	0.70
C_{5+}*	85.43

*Molecular weight and density of C_{5+} is 180.2g/mol and 0.826g/cc respectively.

Table 3. Swelling Test data

p_{sat} (psia)	v_{rel} (v/v)	CO_2 Mole Fraction	ρ(g/cc)	μ (cP)
2860.7	1.54	0.493	0.593	0.28
2514.7	1.49	0.489	0.612	0.49
2014.7	1.39	0.445	0.656	0.72
1514.7	1.31	0.378	0.698	1.03
1014.7	1.23	0.301	0.741	1.21
514.7	1.17	0.186	0.781	1.29

Table 2. Constant Mass Expansion data

p (psia)	v_{rel} (v/v)	Bo (v/v)	Bg (v/v)	Sol.GOR (v/v)	ρ (g/cc)	μ (cP)	Z-factor	Gas at STP (cc)
2860.7	0.982	1.074		16.187	0.834	0.28		
2514.7	0.984	1.077		16.187	0.832	0.49		
2014.7	0.987	1.080		16.187	0.829	0.72		
1514.7	0.991	1.084		16.187	0.826	1.03		
1014.7	0.995	1.088		16.187	0.823	1.21		
814.7	0.997	1.090		16.187	0.822			
714.7	0.997	1.091		16.187	0.821			
614.7	0.998	1.092		16.187	0.820			
514.7	0.999	1.094		16.187	0.819	1.29		
414.7	1	1.094		16.187	0.819			
214.7	1.595	1.077	0.087	8.506	0.822		0.069	510.673
14.7				0	0.873	2.37		565.507

Table 4. EoS model developed for the fluid using volume corrected Peng-Robinson EoS.

Component	Mol%	Mol Wt (g/mol)	p_c (psia)	T_c (°F)	ω	Volume Shift
CO_2	0.0016	44.010	1070.16	87.89	0.225	-0.094
N_2	0.0020	28.013	492.45	-232.51	0.040	-0.128
C_1	0.1020	16.043	667.38	-116.59	0.008	-0.154
C_2	0.0111	30.070	708.54	90.05	0.098	-0.102
C_3	0.0147	44.097	615.93	205.97	0.152	-0.073
i-C_4	0.0074	58.124	529.20	274.91	0.176	-0.058
n-C_4	0.0070	58.124	551.25	305.69	0.193	-0.057
C5_1	0.2697	76.697	429.33	359.13	0.222	-0.145
C5_2	0.2680	127.727	423.61	533.02	0.468	0.256
C5_3	0.1880	224.610	302.13	771.22	0.702	0.274
C5_4	0.0961	379.469	140.71	999.48	0.716	0.185
C5_5	0.0326	626.000	93.01	1207.35	1.422	0.169

Figures

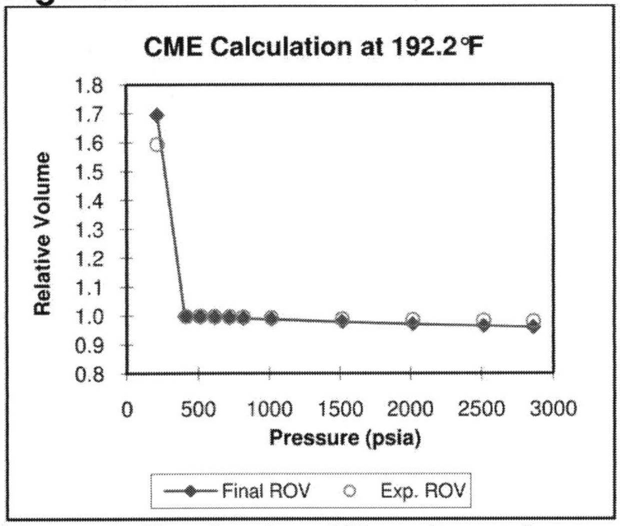

Figure 1. Simulated and Experimental Relative Volume of the Fluid in CME Test

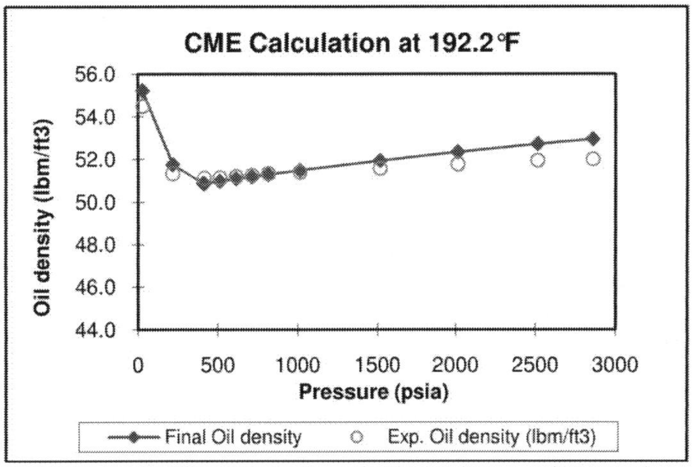

Figure 2. Simulated and Experimental Density of the Fluid in CME Test

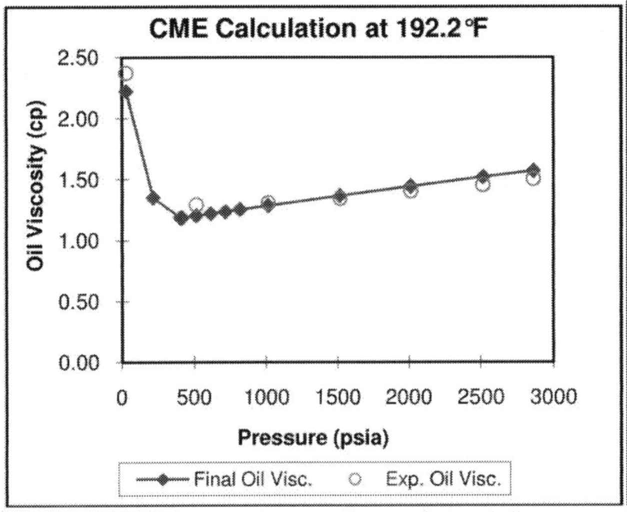

Figure 3. Simulated and Experimental Viscosity of the Fluid in CME Test

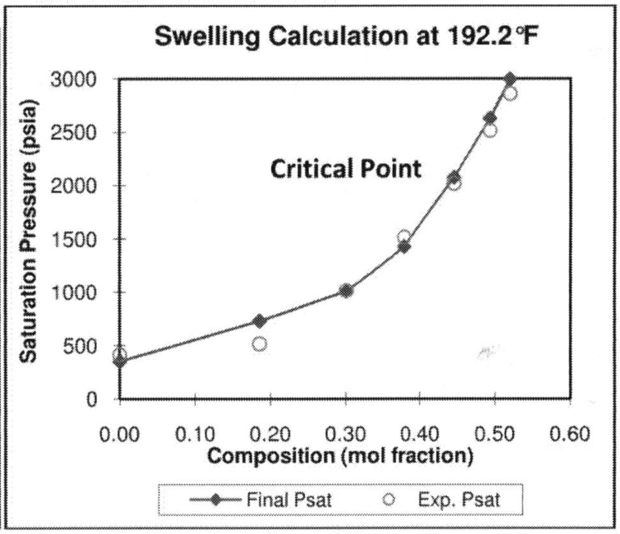

Figure 4. Simulated and Experimental Saturation Pressure of the Fluid in Swelling Test

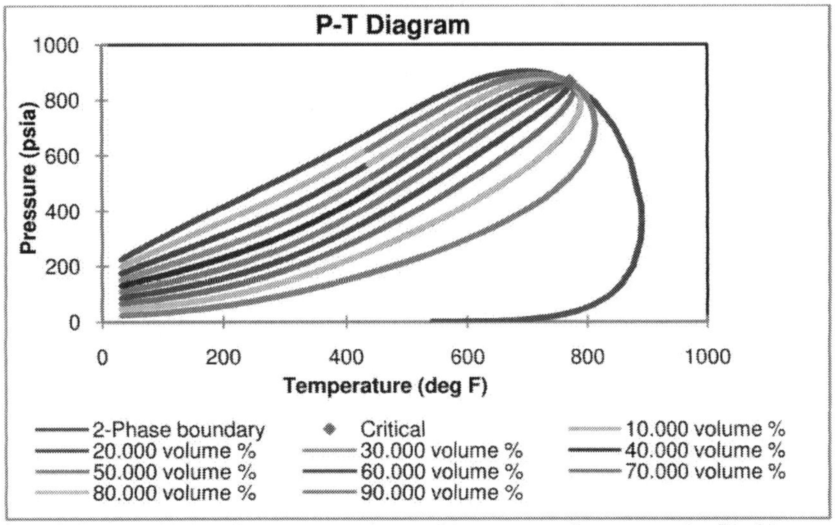

Figure 5. 2-phase envelope diagram of reservoir fluid

CMTC-439523-MS

The Pore Scale Description of Carbon Dioxide Storage into High Asphaltene Content Reservoirs

Raphael Coelho and Berna Hascakir, Texas A&M University

This paper was prepared for presentation at the Carbon Management Technology Conference held in Sugarland, Texas, USA, 17–19 November 2015.

Abstract

The knowledge of how CO_2 interacts with reservoir fluids and rock is essential for effective and reliable CO_2 storage. This work investigates CO_2 storage in unconventional oil reservoirs with high asphaltene content and explains the CO_2 storage mechanism in pore-scale. Considering that asphaltenes are insoluble in CO_2, asphaltene precipitation is induced during CO_2 injection, controlling CO_2 solubility and capillary trapping mechanisms. In this study, CO_2-oil interaction is examined through core flooding experiments in a high asphaltene content (34.3 wt%) Canadian bitumen sample with 8.8° API. To study how irreversible clay-asphaltene interaction affects CO_2 capillary trapping, the reservoir rock is prepared with and without clay addition. The role of CO_2 injection rate on solubility trapping with experiments at varying injection flow rates. Displaced fluids and postmortem samples are subjected to several analyses to observe the CO_2 storage mechanisms in pore scale due to CO_2-asphaltene and CO_2-clay-asphaltene interactions. It was found that CO_2-clay-asphaltene interaction may favor CO_2 capillary trapping into high asphaltene content reservoirs. Reservoir clays play important roles in porosity and permeability reduction due to clay interactions with asphaltenes. The low CO_2 flow rate was found to favor solubility trapping. Therefore, our results suggest that the presence of clays and the CO_2 injection rate are critical parameters controlling the effectiveness of CO_2 storage in high asphaltene content reservoirs.

Introduction

Underground carbon dioxide (CO_2) storage projects usually target the depleted light oil and natural gas fields due to operational expertise, favorable geological features, and existence of infrastructure [1; 2]. However, as CO_2 emissions are projected to increase in the next 35 years [3], other storage alternatives are required to mitigate environmental impacts. In this context, CO_2 storage in unconventional oil reservoirs with low API gravity may be an attractive candidate on the short to medium term [4; 5].

These low API gravity oils usually have high asphaltene content, which can represent up to 45.3 wt% of the crude oil [6]. CO_2 injection can cause asphaltenes precipitate in the pore space as CO_2 is insoluble in asphaltenes [7; 8]. Therefore, the effectiveness of CO_2 microscopic storage mechanisms will be affected, particularly for capillary (or residual) and solubility trapping.

In capillary trapping, due to capillary forces, residual CO_2 droplets become immobile in the pore spaces [9; 10]. In high asphaltene content oil reservoirs under CO_2 injection, this mechanism is expected to be highly influenced by mutual interactions of asphaltenes, CO_2, and reservoir fines (clays) in the pore-scale [11; 12; 13].

Solubility trapping occurs due to solubility and swelling of CO_2 in reservoir fluids [2; 9; 14; 15]. The solubility of CO_2, which is characteristically limited in low API gravity oils and brines [15; 16; 17], is also predicted to alter during CO_2 flooding as asphaltenes leave the oil phase by precipitation and clays interact with oil [7; 18].

Still, during immiscible CO_2 flooding, high CO_2 solubility in low API gravity oils may be achieved at reservoir pressures [15; 19; 20]. As solubility is developed, the net oil volume expands substantially and the reservoir oil swells (10-20%) [21; 22; 23]. This swelling effect is believed to be one of the main oil displacing mechanisms for immiscible CO_2 injection, along with viscosity reduction [22; 24].

This work investigates the feasibility of CO_2 in high asphaltene content reservoirs. The experimental work is interested in developing quantitative and qualitative knowledge of the CO_2 capillary and the solubility trapping mechanisms in pore scale.

Materials and Methods

Experimental studies to determine the pore scale CO_2 storage mechanisms in high asphaltene content reservoirs were conducted using a Canadian bitumen sample from the Peace River region, northwestern Alberta. The original bitumen sample has 8.8° API, 54,000 cP (at room temperature), and 34.3 wt% asphaltene content [25].

The reservoir rock was prepared by mixing 85 wt% of 20-40 mesh-size Ottawa sand with 15 wt% clay. The clay was a 90 wt% kaolinite and 10 wt% illite mixture with 2.3 μm average grain size. This mixture with 32 % porosity mimics the Peace River reservoir rock [26]. To investigate the impact of clay-asphaltene interactions on the CO_2 trapping mechanisms, a reservoir rock was prepared without clay by using only Ottawa sand (39.1% porosity). For all experiments, 84 wt% bitumen and 16 wt% water saturations are used [27]. Then, the oil sand mixture was packed in a stainless steel cylinder core holder cell. The outlet was equipped with a screen to avoid sand production. A schematic diagram of the experimental setup and the picture of core holder are presented in Figure 1.

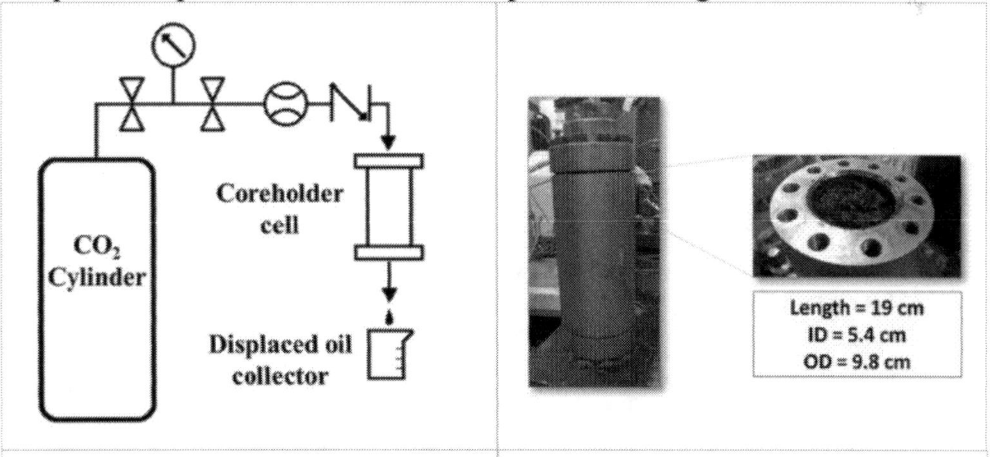

A-Schematic diagram of Experimental Set-up	B-Photograph of the core holder

Figure 1- Experimental setup (The core holder has ID = 5.4 cm and OD = 9.8 cm).

Piston-like displacement was achieved by conducting core flooding experiments vertically. Experiments were performed by injecting high purity CO_2 (≥99.9% purity CO_2). The outlet was kept at atmospheric pressure and connected to a sampling bottle, which collected the displaced oil and water. The experiments were conducted at room temperature (approximately 20°C).

The displaced oil was analyzed by measuring its viscosity at varying temperature values using a Brookfield HBDV-III rheometer. Asphaltene content was determined by separating de-asphalted oil using *n*-pentane dissolution [28]. Asphaltene-clay interactions were investigated by measuring the clay content of the asphaltenes by washing samples with toluene (a strong asphaltene soluble solvent) [29; 30; 31]. Fourier Transform Infrared (FTIR) spectroscopy was used to verify if the entire clay content of asphaltenes was removed from the asphaltenes.

Similar separation technique was also implemented to the residual oil samples. The postmortem samples both from injection and production sides were subjected to toluene extraction [29; 30].

Results and Discussions

Three CO_2 core flooding experiments were conducted to investigate the pore-scale CO_2 storage mechanisms on a high-asphaltene content bitumen sample (Table 1). Experiments E1 and E2 were conducted to investigate the effect of CO_2 injection flow rate; thus, low and high injection rates were tested. E3 was conducted to study the effect of clay on CO_2 storage mechanisms. Therefore, while the reservoir rock for E2 was prepared with clay, for E3 the reservoir rock was prepared without clay, with other experimental and initial conditions kept constant.

CO_2 is a well-known asphaltene non-solvent, hence, during CO_2 injection asphaltenes precipitate [7]. Also, it is known that clay and asphaltene interactions may induce the precipitation of asphaltenes by forming clay-asphaltene clusters in the reservoir pores [13; 18], which may affect the CO_2 storage.

Therefore, both the effects of CO_2-asphaltene and CO_2-clay-asphaltene interactions on CO_2 storage were investigated in this study.

Table 1 – Experimental conditions

Experiment	Rock Sample	CO_2 flow rate, cc/min	Tested effect	Experimental time, h	Gas breakthrough, h
E1	Sand + clay	2,540	CO_2 flow rate	5.3	3.2
E2	Sand + clay	18 - 101.7	CO_2 flow rate and clay presence	8.5	7.7
E3	Sand	18 - 101.7	Clay presence	6.2	~0.1

Displaced Oil Analysis

The reservoir oil is displaced by CO_2 through the combination of swelling and viscosity reduction mechanisms [22; 31]. It was expected that the CO_2-asphaltene interactions would cause asphaltene to leave the oil phase and deposit on reservoir rock, while clay-asphaltene irreversible interactions would cause clay migration into the oil. To quantify such interactions during each of the three experiments, asphaltene and clay contents in the displaced oil were measured (Figure 2).

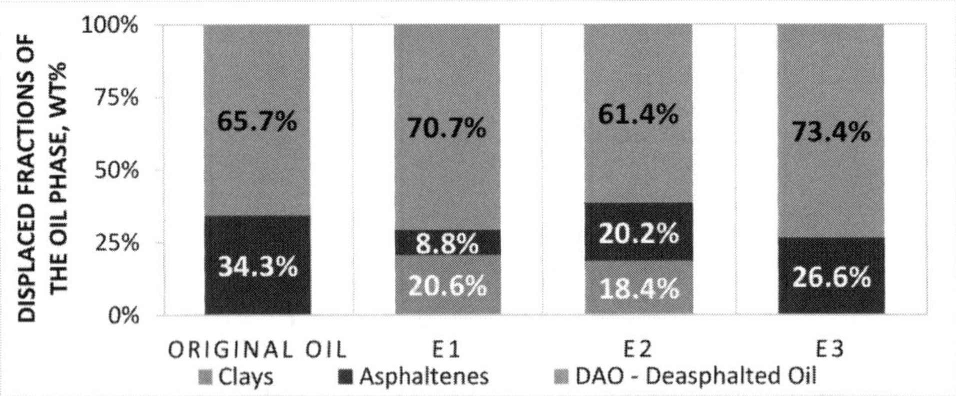

Figure 2 – Asphaltene and clay content in displaced oil.

CO_2 injection flow rate is found to greatly affect asphaltene precipitation, with asphaltene content in the displaced oil being 56.4% lower in E1 than in E2. However, the presence of clay caused less impact by having 31.7% more asphaltenes in E3 than E2.

The viscosity of the displaced oil samples as a function of temperature was also measured (Figure 3A). For E1 and E2 (with clays), the displaced oils are more viscous than the original oil. This high viscosity is justified by the existence of dispersed clay-asphaltene solids. This is confirmed by the lower viscosity of E3 samples, which do not have clays and allow the solvent action of CO_2 to be perceived.

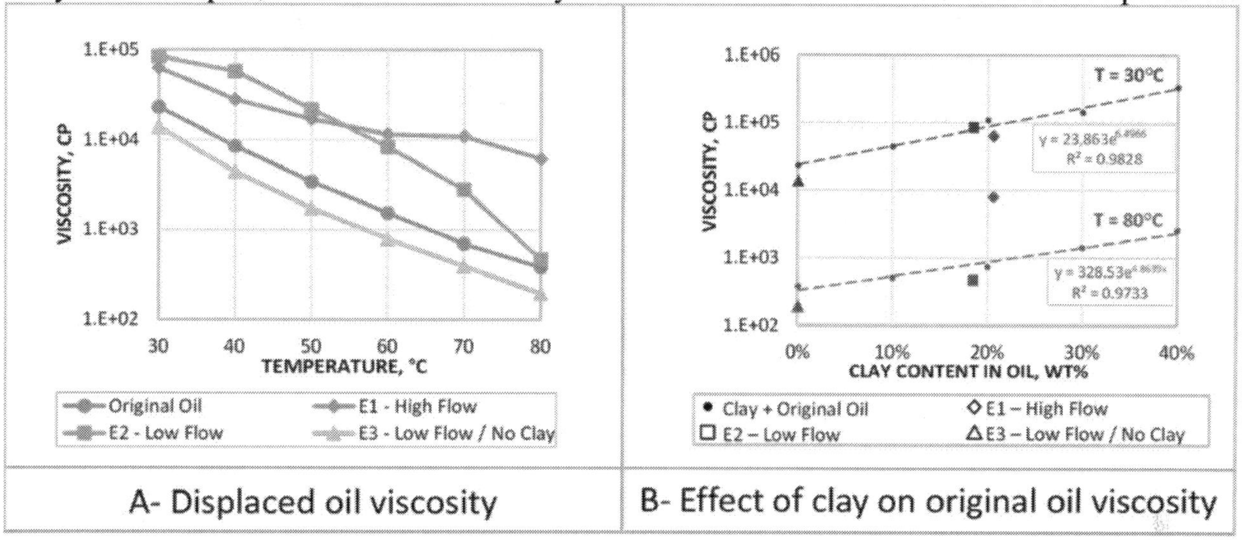

Figure 3 – Viscosity readings of original oil, displaced oil, and clay-bitumen pseudo-mixtures.

To better understand the contribution of clay in viscosity increase, original bitumen samples were mixed with clay at different amounts [0% (original bitumen); 10 wt%; 20 wt%; 30wt; and 40 wt%] and viscosity of each sample was measured at 30 °C (purple circle points) and at 80 °C (red circle points) (Figure 3B). It has been observed that there is a strong correlation between clay content and viscosity of the bitumen (purple and red dashed lines in Figure 3B for 30 °C and 80 °C, respectively).

The viscosities of the displaced oil samples are compared to the viscosity of the clay-bitumen mixtures both at 30 °C and at 80 °C. Those measurements are given in Figure 3B purple markers for the measurements taken at 30 °C and red markers for the measurements taken at 80 °C. In the same figure, the diamond markers represent the produced oil originated from E1, the square markers represent the produced oil originated from E2, and the triangular markers represent the produced oil originated from E3. None of the produced oil samples show the similar trend observed for original bitumen and clay mixtures. This result shows that clay-asphaltene interaction is the major factor controlling the displaced oil viscosity. Hence, E3 (no clay addition case) had the lowest viscosity value.

The rheology of the displaced oil from E2 and E3, in which the lower rate of CO_2 was used, showed shear thickening behavior due to foaming and/or swelling feature of CO_2-bitumen interaction. On the other hand, the displaced oil from E1 exhibited shear thinning behavior like for original bitumen (Figure 3A).

The foaming and swelling effects resulted increase in viscosity (Figure 3A). However, this behavior changed after 55 °C because the solubility of CO_2 in Peace River bitumen decreases at high temperatures [20]. The authors believe that at high temperatures more CO_2 is released from the displaced oil from E2 sample than from E1, causing the greatest viscosity drop on E2 at high temperatures. Similar interpretation is obtained from Figure 3B: at low temperature, both E1 and E2 present viscosities close to the oil/clay pseudo-mixture trend; at high temperature, however, E1 stays above the trend, while E2 falls below it.

Such observations are strong indicators that the injection flow rate of CO_2 plays a major role on CO_2-oil interactions due to the slow diffusion of CO_2 in low API gravity oil [32; 33]. Therefore, lower injection flow rates favors the dissolution of CO_2 in the oil and storage through solubility trapping and swelling. These results agree with the work in which hydrocarbon (benzene, toluene, and naphtha) and non-hydrocarbon (carbon tetrachloride) solvents were injected to displace Athabasca bitumen (with 40.1 wt% asphaltene content) [33].

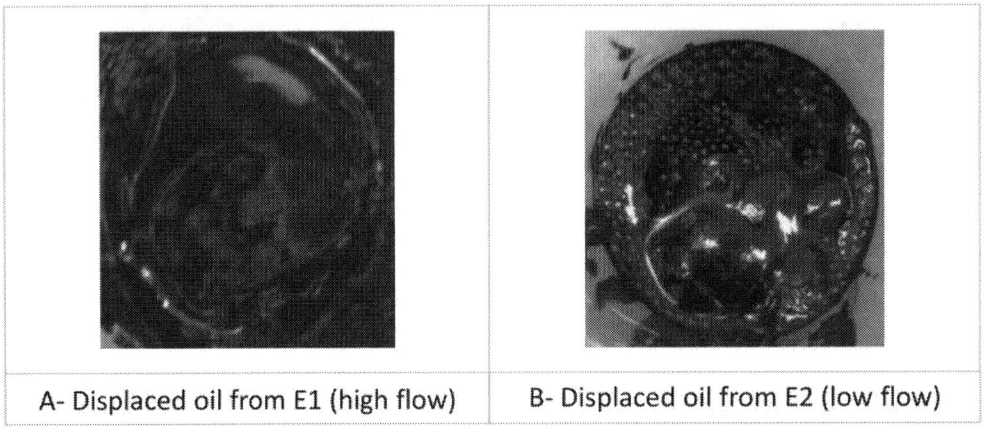

Figure 4 – Displaced oil samples (Only E2 displaced oil displayed gas bubbles).

Additionally, clay migration to the oil phase may change the CO_2 solubility in oil, which directly affects the CO_2 solubility trapping. Our experiments showed gas bubbles in both low flow rate cases (E2 and E3-Figure 4B), indicating CO_2 solubility in oil present both with and without clay cases, and more significantly at lower CO_2 rates. Therefore, further studies are required to quantify how clay content in oil affects CO_2 solubility.

Postmortem Analysis

Residual oil saturations were extracted from the postmortem samples. Postmortem pictures are given in Figure 5. The top three pictures show the entire postmortem samples and the bottom pictures show the cross-sectional view of the outlet of the postmortems. Note that all experiments were conducted vertically and the CO_2 injection direction of the top pictures are from left to right, hence, the left hand side of the pictures are the injection side and the right hand sides show the production end of the cores.

Figure 5 – Postmortem pictures at the top and the cell outlet appearance after the experiments at the bottom. Red circles (C-bottom pictures) indicate CO_2 bubbles, observed only for E3 (low flow, no clay case).

In the experiments conducted with clays (Figure 5A and Figure 5B-top pictures), the postmortem displays more consolidation than E3 sample. The higher consolidation in E1 and E2 indicates that porosity and permeability were lower for such samples due to clay-water-crude oil interactions. Therefore, those interactions are expected to favor CO_2 trapping due to higher capillary forces. Gas bubbles were observed only on E3 postmortem (prepared without clays) (Figure 5C-bottom image). The presence of such gas bubbles indicates that CO_2 capillary trapping did not occur for experiment E3.

After visual inspection of postmortem samples, residual oil samples both from the inlet and outlet of the cores were extracted by washing spent rock samples with toluene [30; 31]. Then, the asphaltenes of residual oil samples were separated by following an ASTM method [28]. The asphaltenes, clays, and deasphalted oil fractions of the residual oil samples along with the original bitumen asphaltene and deasphalted oil contents are given in Figure 7.

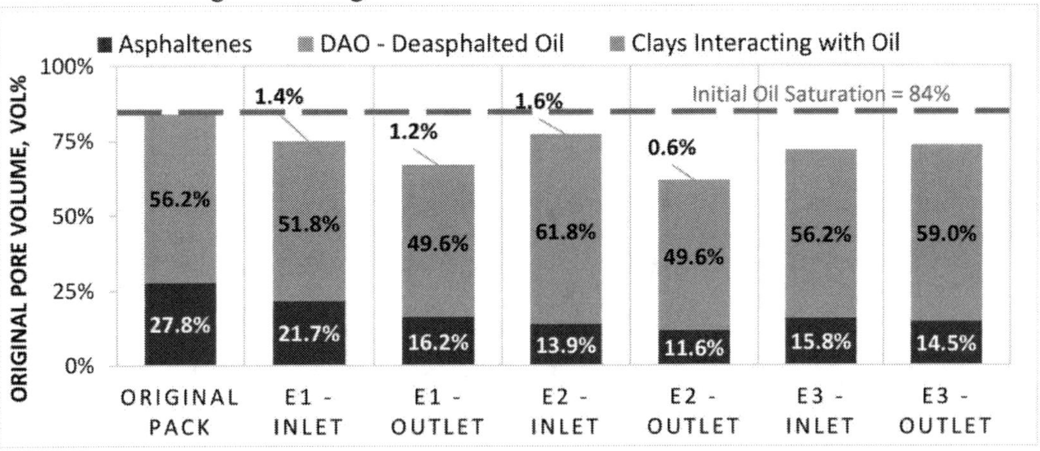

Figure 7 – The distribution of residual oil fractions in reservoir pore

Note that while the asphaltenes given in Figure 2 are the movable asphaltenes, the asphaltenes in Figure 7 are the precipitated asphaltenes. The greatest asphaltene precipitation is observed for E1. This is attributed to the exposure of oil to more CO_2 during E1 due to higher CO_2 rate used in this experiment.

During CO_2 injection, asphaltene-clay interaction was observed both in displaced and residual oil (Figure 2 and Figure 7). The effect of CO_2 on asphaltene precipitation was higher for E3 than E2, demonstrating that CO_2 diffusion becomes difficult with clay presence.

E3 was the only experiment to present water displacement as a separate phase, indicating that the clay-water interactions occurring in E1 and E2 also affect solubility of CO_2 in the water phase. This can be associated to immobile connate water due to clay-water interactions, which may cause cementing in the pores, particularly when illite is present [18]. The reduction in permeability due to the cementing behavior of clay causes gas breakthrough later in E1 and E2 than E3 (Table 1).

Such phenomenon is therefore expected to affect CO_2 storage effectiveness as CO_2 trapping in the water phase may be reduced due to clay-water cementing. Further studies are required to investigate and quantify the effect of CO_2-clay-water interactions on CO_2-water solubility.

Proposed CO_2 Storage Mechanism due to Clay-Asphaltene-CO_2 Interactions

Based on the previous discussions, we propose a novel CO_2 storage capillary trapping mechanism, which we believe to occur on high asphaltene content reservoirs due to clay-asphaltene-CO_2 interactions.

The presence of clays is by itself expected to cause to porosity and permeability reduction due to clay pore lining, pore bridging, and pore filling features [11]. Also, CO_2 triggers asphaltene precipitation, causing clay-asphaltene association due to the presence of mainly illite. Such interactions are expected to reduce porosity and to cause pore-throat clogging, therefore, enhancing CO_2 capillary trapping.

This novel capillary trapping mechanism is depicted in Figure 8, where the sandstone's (yellow grains) pore-throat is blocked by kaolinite (light blue hexagons), illite (dark blue strips), and asphaltenes (dark-gray discs), which trap the CO_2 bubble.

Another phenomenon that may affect CO_2 storage through capillary trapping is the high CO_2 adsorption capacities of illite and kaolinite clays [12]. Though our experiments cannot quantify the contribution of adsorption, such effect might help explain why only experiment E3 presented CO_2 gas bubbles in the postmortem sample.

Furthermore, CO_2 injection flow rate was recognized to control asphaltene precipitation in the reservoir, which will affect clay-asphaltene interactions, ultimately controlling the development and effectiveness of the proposed CO_2 capillary trapping.

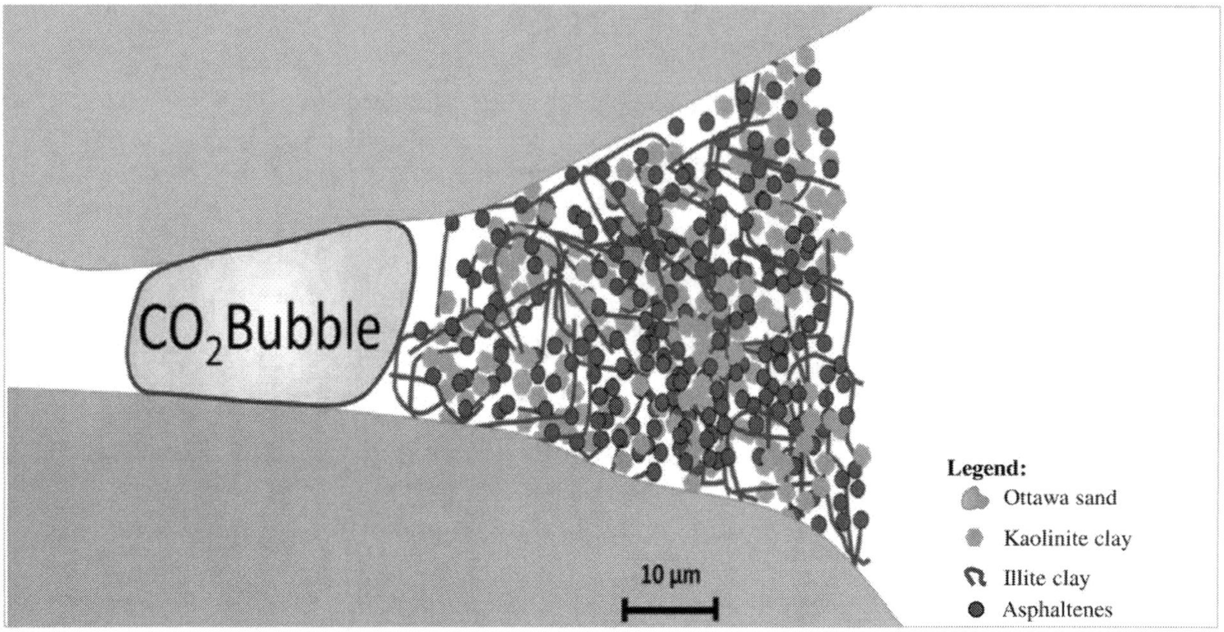

Figure 8 – Proposed CO_2 capillary trapping due to clay-asphaltene-CO_2 interactions.

Conclusions

Low CO_2 injection flow rates resulted in lower asphaltene precipitation and higher CO_2 solubility in the oil phase than higher CO_2 rates. Thus, the CO_2 flow rate is recognized as a critical parameter controlling CO_2 storage through solubility trapping into high asphaltene content reservoirs. CO_2-clay-asphaltene interactions greatly impact the viscosity of the displaced oil. Presence of clays in the high asphaltene content crude oils is increased the viscosity exponentially. Clay-water interactions also affect CO_2 storage due to cementing features of clays by reducing significantly the porosity and permeability of the system.

To depict all major factors acting upon effective CO_2 storage, a novel CO_2 capillary trapping mechanism due to clay-asphaltene-CO_2 interactions is proposed. The combined effects of asphaltene precipitation, formation of clay-asphaltene fines, and CO_2 adsorption are believed to enhance CO_2 storage through capillary trapping. Clay migration into the oil phase may change CO_2 solubility in oil, affecting CO_2 solubility trapping. Further investigation is needed to quantify this effect.

Acknowledgments

The authors acknowledge the financial support and the opportunity provided by CAPES, Coordenação de Aperfeiçoamento de Pessoal de Nível Superior – Brasil. We also acknowledge the members of the Heavy Oil, Oil shales, Oil sands, & Carbonate Analysis and Recovery Methods Research Team at the Texas A&M University Petroleum Engineering Department for their help.

References

1 Gielen, D., and Podkanski, J. 2004. Prospects for CO_2 Capture and Storage. Paris, OECD/IEA

2 IPCC Special Report on Carbon Dioxide Capture and Storage. In: Metz, B., Davidson, O., de Coninck, H.C., Loos, M., Meyer, L.A. (Eds.). Prepared by Working Group III of the Intergovernmental Panel on Climate Change. Cambridge University Press, Cambridge, United Kingdom and New York, NY, USA.

3 OECD, 2011- Environmental Outlook to 2050: Climate Change - Chapter 3. Pre-Release Version. November 2011. http://www.oecd.org/env/cc/49082173.pdf (Accessed in 28/10/2015)

4 Alboudwarej, H., Felix, J., and Taylor, S. 2006. Highlighting Heavy Oil. Oilfield Review. Summer 2006.

5 Cavallaro, A.N., Galliano, G. R., Moore, R.G., Mehta, S.A, Ursenbach, M.G., Zalewski, E., Pereira, P. 2008. In Situ Upgrading of Llancanelo Heavy Oil Using In Situ Combustion and a Downhole Catalyst Bed. *J. Can. Pet. Technol.* **47** (9): 23-13. doi:10.2118/08-09-23.

6 Prakoso, A., Punase, A., and Hascakir, B. 2015. A Mechanistic Understanding of Asphaltene Precipitation from Varying Saturate Concentration Perspective, SPE Latin American and Caribbean Petroleum Engineering Conference, 18-20 November 2015, Quito, Ecuador, SPE-177280-MS.

7 Mungan, N. 1981. Carbon Dioxide Flooding-fundamentals. Petroleum Society of Canada. 81-01-03.

8 Deo, M., and Parra, M. 2012. Characterization of carbon-dioxide-induced asphaltene precipitation. *Energy Fuels* **26**: 2672−2679.

9 Juanes, R., Spiteri, E.J., Orr Jr., F.M., and Blunt, M.J. 2006. Impact of Relative Permeability Hysteresis on Geological CO_2 Storage. *Water Resour. Res.* **42:** W12418.

10 Pentland, C.H., El-Maghraby, R., Georgiadis, A., Iglauer, S., and Blunt, M.J. 2011. Immiscible Displacements and Capillary Trapping in CO_2 Storage. *Energy Procedia* **4:** 4969-4976.

11 Neasham, J. W. 1977. The Morphology of Dispersed Clay in Sandstone Reservoirs and Its Effect on Sandstone Shaliness, Pore Space and Fluid Flow Properties. Society of Petroleum Engineers. doi:10.2118/6858-MS.

12 Heller, R., and Zoback, M. 2014. Adsorption of Methane and Carbon Dioxide on Gas Shale and Pure Mineral Samples. *Journal of Unconventional Oil and Gas Resources* **8:** 14-24.

13 Kar, T., Williamson, M., and Hascakir, B. 2014.The Role of Asphaltenes in Emulsion Formation for Steam Assisted Gravity Drainage (SAGD) and Expanding Solvent - SAGD (ES-SAGD). Society of Petroleum Engineers. doi:10.2118/171076-MS.

14 Bachu, S., and Shaw, J. 2003. Evaluation of the CO Sequestration Capacity in Alberta's Oil and Gas Reservoirs at Depletion and the Effect of Underlying Aquifers. Petroleum Society of Canada. doi:10.2118/03-09-02.

15 Li, H., Zheng, S., Yang, D., 2013. Enhanced Sweeling Effect and Viscosity Reduction of Solvent(s)/CO_2/Heavy-Oil Systems, SPE Journal **18-04**:695-707.

16 Donaldson, E., Chilingar, G., and Yen, T.F, 1989. Enhanced Oil Recovery, II: Processes and Operations. Amsterdam: Elsevier. Print.

17 Pruess, K., and García, J. 2002. Multiphase Flow Dynamics during CO_2 Disposal into Saline Aquifers. *Environ. Geol.,* **42** (2-3): 282-295.

18 Unal, Y., Kar, T., Mukhametshina, A., and Hascakir, B. 2015. The Impact of Clay Type on the Asphaltene Deposition during Bitumen Extraction with Steam Assisted Gravity Drainage. Society of Petroleum Engineers. 173795-MS.

19 Holm, L. W. 1982. CO_2 Flooding: Its Time Has Come. Society of Petroleum Engineers. doi:10.2118/11592-PA.

20 Svrcek, W. Y. and A. K. Mehrotra. 1989. Properties of Peace River Bitumen Saturated with Field Gas Mixtures. *J Can Pet Technol* **28** (2): 50-56.

21 Simon, R., and Graue, D. J. 1965. Generalized Correlations for Predicting Solubility, Swelling and Viscosity Behavior of CO_2 -Crude Oil Systems. Society of Petroleum Engineers. doi:10.2118/917-PA

22 Rojas, G. A., and Farouq Ali, S. M. 1988. Dynamics of Subcritical CO_2/Brine Floods for Heavy-Oil Recovery. Society of Petroleum Engineers. doi:10.2118/13598-PA

23 Martin, D. F., and Taber, J. J. 1992. Carbon Dioxide Flooding. Society of Petroleum Engineers. doi:10.2118/23564-PA

24 Naderi, K., and Babadagli, T. 2014. Use of Carbon Dioxide and Hydrocarbon Solvents During the Method of Steam-Over-Solvent Injection in Fractured Reservoirs for Heavy-Oil Recovery From Sandstones and Carbonates. Society of Petroleum Engineers. doi:10.2118/169815-PA

25 Mukhametshina, A., Kar, T., and Hascakir, B. 2015. Asphaltene Precipitation during Bitumen Extraction with Expanding-Solvent Steam-Assisted Gravity Drainage: Effects on Pore-Scale Displacement. Society of Petroleum Engineers. doi:10.2118/170013-PA.

26 Bayliss, P., and Levinson, A.A. 1976. Mineralogical Review of the Alberta Oil Sand Deposits (Lower Cretaceous, Mannville Group). *Bulletin of Canadian Petroleum Geology* **24** (2): 211-224.

27 Hamm, R.A., and Ong, T.S. 1995. Enhanced Steam-Assisted Gravity Drainage: A New Horizontal Well Recovery Process for Peace River, Canada. *J Can Pet Technol* **34** (4). SPE- 95-04-03-PA.

28 ASTM. 2011. ASTM D2007-11: Standard Test Method for Characteristic Groups in Rubber Extender and Processing Oils and Other Petroleum-Derived Oils by the Clay-Gel Absorption Chromatographic Method.

29 Kar, T., Hascakir, B., The Role of Resins, Asphaltenes, and Water in Water-Oil Emulsion Breaking with Microwave Heating, Energy and Fuels, 29, 3684-3690, 2015.

30 Amyx, J. W., Bass, D. M., Whiting, R. L., 1960. "Petroleum Reservoir Engineering" McGraw Hill, ISBN 07-001600-3: 103-104.

31 Speight, J. G. 2014. The Chemistry and Technology of Petroleum. CRC Press, Taylor and Francis.

32 Yang, C. and Gu, Y. 2006. Diffusion Coefficients and Oil Swelling Factor of Carbon Dioxide, Methane, Ethane, Propane, and Their Mixtures in Heavy Oil. *Fluid Phase Equilibr.* **243** (1-2): 64-73.

33 Hernandez, O. E., and Farouq Ali, S. M. 1972. Oil Recovery from Athabasca Tar Sand by Miscible-Thermal Methods. Petroleum Society of Canada. doi:10.2118/7249.

CMTC-439524-MS

Effect of Impurities in Carbon Dioxide Stream on Phase Behavior for Geological Storage of Carbon Dioxide in Low API Gravity Oil Reservoirs

Raphael Coelho, Maria Barrufet, and Berna Hascakir, Texas A&M University

This paper was prepared for presentation at the Carbon Management Technology Conference held in Sugarland, Texas, USA, 17–19 November 2015.

Abstract

While previous studies investigate the effect of impure CO_2 streams in high and medium API gravity oils, this work focus on reservoirs with low API. This analysis provides additional alternatives for CO_2 storage due to the uneconomical exploration and the vast extension of heavy oil resources. CO_2 streams (re)injected for storage may contain impurities such as nitrogen, water, and light hydrocarbons, particularly for flue gases and CO_2-rich natural. These impurities greatly affect CO_2 solubility trapping mechanisms and storage effectiveness. This work investigates the phase behavior of reservoir fluids during injection of CO_2 with impurities for storage into two low API gravity oils; 12° (San Ardo heavy oil) to 8.2° (Athabasca extra-heavy bitumen). Numerical studies are conducted using the Peng-Robinson Equation-of-State with Volume-Translation and evaluating Vapor-Liquid and Vapor-Liquid-Liquid equilibria, with rigorously implementation of mutual solubility water-hydrocarbons-CO_2. Simulations are conducted at typical reservoir conditions and at high pressure/high temperature synthetic scenarios. The effect of impurity concentration is represented constructing pseudo-ternary diagrams. Results show that C_3-CO_2 mixtures may achieve miscibility at low pressures. In the presence of N_2 and C_1 as impurities, miscibility with crude oil was found to be affected negatively. The water presence also affects the solubility.

Introduction

Carbon dioxide (CO_2) storage in oil reservoirs is a well-known carbon capture and sequestration (CCS) alternative to reduce the environmental footprint of CO_2 emissions [1, 2]. Though high purity CO_2 injection is ideal to maximize storage. The industrial activities (e.g. flue gas and CO_2-rich natural gas) or carbon capture plants are usually the sources of impure CO_2, containing N_2, H_2, O_2, H_2S, water, and light hydrocarbons [2, 3, 4, 5, 6]. Such impurities present an economic challenge to CCS as current CO_2 capturing technologies to achieve high-purity streams are expensive, costing 12-60 US$/t$CO_2$ and accounting for up to two-thirds of the total CCS project cost [2, 7, 8]. Hence, the impure CO_2-rich streams may be disregarded for storage and contribute even further to CO_2 emissions if vented or flared.

An alternative is storing these impure gases in their raw state in uneconomic unconventional oil reservoirs. The low API gravity crude oils account for approximately 70% of the world's oil reserves [9]. Mature heavy oil fields not subjected to thermal recovery methods could be a good potential for CCS [10].

Only for the Alberta's (Canada) heavy oil reservoirs, the combined theoretical storage capacity can be up to 120 Mton CO_2 [11].

To maximize CO_2 storage in these reservoirs, the development of a single-phase mixture (miscible conditions) in the reservoir is targeted. This can be achieved through solubility trapping mechanisms [2, 11]. Though CO_2 is not completely miscible with heavy oils rich in C_{30+} fractions, high solubility may occur [12], which favors CO_2 trapping. The impurities in the CO_2 stream may contribute the formation of the desired single-phase mixture [6, 13]. Numerous studies have investigated the phase behavior of impure CO_2 streams injected into light oils for CO_2 Enhanced Oil Recovery [13, 14, 15, 16, 17, 18]. This work extends this analysis to CO_2-rich impure streams interacting with heavy oils and bitumen through phase behavior simulations considering C_1, C_3, N_2, and water as impurities. Simulations are conducted considering the addition of a single impurity to the CO_2 stream injected into the oil reservoir.

Methodology

1st Step- Oil Samples Composition: To mimic real low API reservoir fluids under commercial exploitation, the necessary data to characterize a heavy oil and a bitumen samples were compiled from the literature [19, 20, 21]. A 12° API gravity San Ardo crude sample and an 8.2° API gravity Athabasca bitumen were chosen for simulations. The San Ardo reservoirs are sandstones bearing heavy oils with 10 to 12° API [22]. Typical reservoir temperature is 100°F, with reservoir pressure from 115 to 315 psi [23]. Reservoir conditions for the Athabasca reservoirs are reported 50°F with up to 174 psi [24, 25].

2nd Step- Simulate Binary P-T Diagram: A pseudo-binary Pressure-Temperature (P-T) diagram was initially simulated. Phase diagrams for the San Ardo and Athabasca oil samples, as well as single component saturation vapor curves for CO_2 and impurities (N_2, C_1, C_3, and H_2O) were simulated using thermodynamically consistent Vapor-Liquid Equilibrium (VLE). Simulations were performed using PVTSim Nova commercial software [26]. VLE calculations were carried out for each component using the Peng-Robinson Equation-of-State (PR-EOS) [27, 28]. At specified pressure and temperature conditions, VLE was achieved for single components at equal fugacities of vapor and liquid phases [29, 30]. For mixtures like the oil samples used in this study, the equality of fugacities principle is also valid. However, the VLE is only attained when the fugactities of each (pseudo-) component are equal in both vapor and liquid phases [30].

3rd Step- VLE and VLLE Flash Calculations: To simulate the phase behavior of the crude oil/CO_2/impurity mixtures, flash calculations are performed at fixed temperatures and pressures, but with varying molar compositions. To mimic reservoir conditions, initial simulations were run at 315 psi and 100°F (for San Ardo heavy oil) and at 400 psi (fracture pressure for Athabasca reservoir [25]) and 50°F (for Athabasca bitumen). Afterwards, simulations at 1200 psi and 122°F were conducted to achieve CO_2 supercritical fluid conditions [31], which is preferred for CO_2 geological storage [2].

While liquid, vapor, and/or solid phases may exist in the system, this work does not consider the solid phases. Though heavy oils and bitumen may have a considerable amount of asphaltenes, which may precipitate when contacted by CO_2, C_1, and C_3 and form a solid phase [32, 33, 34]. Multiphase flash calculations for the crude oil/CO_2/impurity mixtures are solved by minimization of Gibbs free energy of the system in order to achieve the equilibrium state.

When water is present in the mixtures, an aqueous liquid phase may be formed additionally to the hydrocarbon-rich liquid phase. Hence, the phase equilibria calculations must consider both Vapor-Liquid and Vapor-Liquid-Liquid (VLLE) equilibria [30].

In this work, all VLE and VLLE flash calculations were performed using PR-EOS with Peneloux volume translation to enhance results on near-critical simulations [33, 35]. The n (pseudo-) components from each of the oil mixtures were grouped in a pseudo-component before the VLE and VLLE calculations by using the following three material balance rules [28, 36]:

$\sum_{i=6}^{n} z_i = z_{C+}$ (Eq^{n}1) $\sum_{i=6}^{n} z_i M_i = z_{C+} \cdot M_{C+}$ (Eq^{n}2) $\sum_{i=6}^{n} \frac{z_i M_i}{\gamma_i} = \frac{z_{C+} M_{C+}}{\gamma_{C+}}$ (Eq^{n}3)

A single grouped pseudo-component represents the low API gravity nature of the oil samples. However, the procedure would not be successful if applied to heavy oils with high gas content [37].

4th Step- Develop Pseudo-Ternary Diagrams: Finally, using the results from the multiphase, multicomponent flash calculations, pseudo-ternary diagrams are built for each crude oil/CO_2/impurity mixture both at reservoir and supercritical conditions. This is achieved by using the molar composition of each phase at the calculated equilibrium states.

Results and Discussions

The miscibility behavior of each oil/CO_2/impurity mixture at given pressure and temperature conditions is summarized in Table 1.

Table 1 – The miscibility behavior of each component with reservoir fluids at reservoir conditions and at supercritical CO_2 conditions with CO_2.

Impurity added to CO_2 stream	Reservoir Conditions (San Ardo: 315 psi & 100ºF / Athabasca: 400 psi & 50ºF)	Supercritical CO_2 Conditions (1200 psi & 122ºF)
N_2	Predominantly immiscible	Predominantly immiscible
C_1	Predominantly immiscible	Predominantly immiscible
C_3	Miscibility at intermediate to high C_3 concentrations	Miscible
H_2O	Immiscible	Immiscible

Pressure-Temperature (P-T) diagram: A P-T diagram of CO_2 and impurities, as well as the simulated phase diagrams for the two low API oil samples under study, is presented in Figure 1.

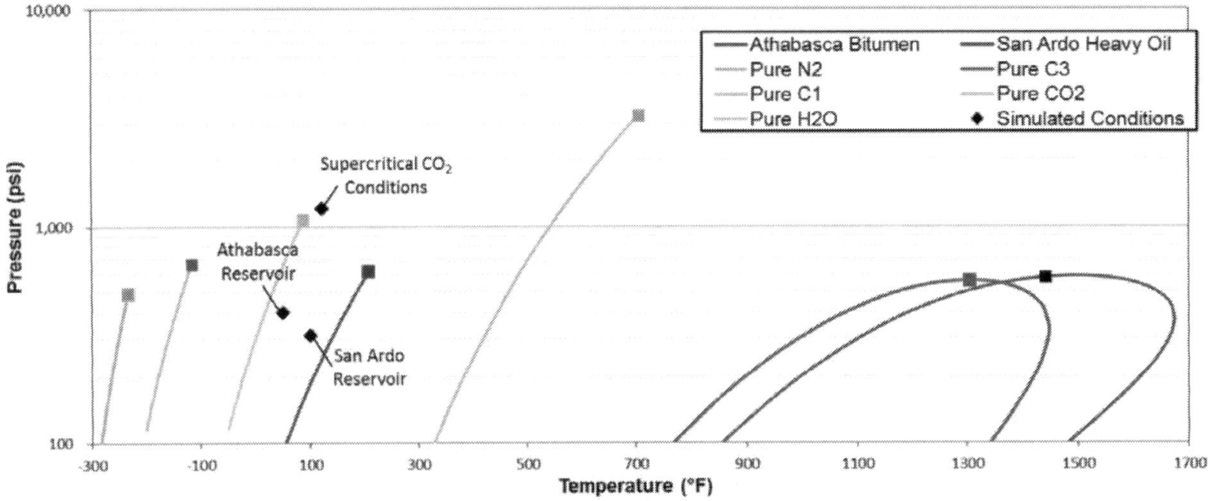

Figure 1 –Single component vapor pressure lines [31], and the simulated P-T diagram of San Ardo Heavy Oil [19] and Athabasca Bitumen [20].

At the reservoir conditions, both the San Ardo heavy oil and Athabasca bitumen samples are in single phase. At supercritical CO_2 conditions (112°F - 1,200 psi), both N_2 and C_1 are also supercritical fluids, whereas C_3 and H_2O are in the liquid state [31]. However, when these components are mixed with crude oils, complex phase behavior may occur due to the VLE and the VLLE, which will be discussed in the following sections for CO_2 streams with varying impurities.

1. N_2 as impurity (Figure 2-A): Results of both oil samples indicate that N_2/CO_2 mixtures fall predominantly in the undesired two-phase region at simulated conditions. For all simulations, the single-phase is limited to extremely low CO_2 and N_2 compositions or nearly pure oil. Additionally, as the CO_2 purity is increased in the stream, more miscibility is achieved. Therefore, N_2 is found to negatively affect CO_2 storage in low API gravity oils.

For the injection of a flue gas with 85 mol% N_2 concentration [3], the single-phase conditions can be maintained by the maximum flue gas concentration and 10 mol% of the heavy oil and 15 mol% for the bitumen samples, both at the supercritical CO_2 scenario (112°F - 1,200 psi). At reservoir conditions, the maximum flue gas concentration decreases to 3-4 mol% for both samples.

Operationally, such results indicate that effective CO_2 storage through solubility trapping in the low API gravity oil would only be feasible for low injected volumes of the N_2/CO_2 mixture.

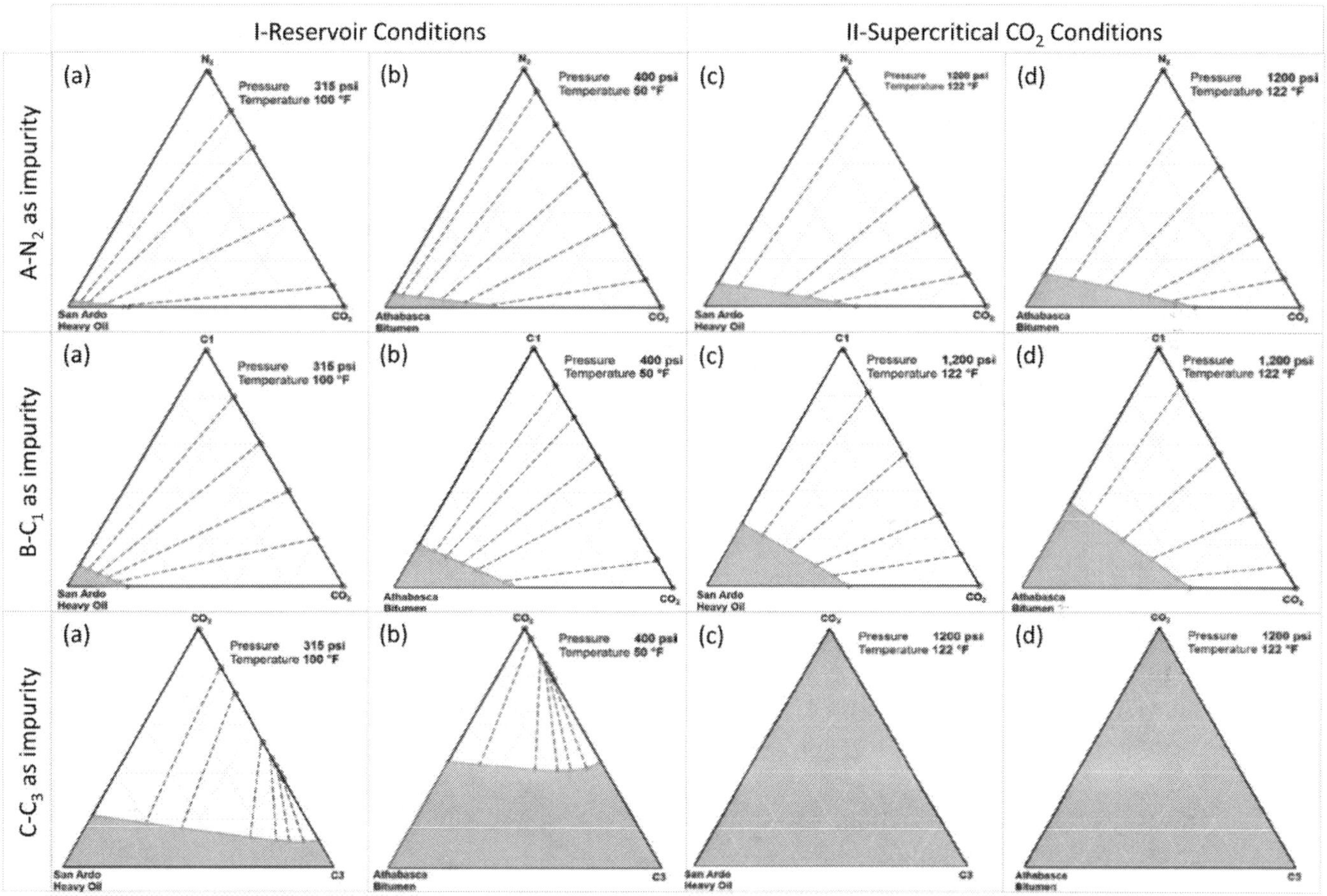

Figure 2- Simulated pseudo-ternary diagrams using A-Nitrogen (at the top), B-Methane (in the middle), and C-Propane (at the bottom) as impurity at I-Reservoir Conditions and at II-Supercritical CO_2 Conditions. The green regions indicate the targeted single-phase liquid.

C_1 as impurity (Figure 2-B): Results under reservoir and CO_2 supercritical conditions demonstrate that C_1/CO_2 mixtures will also have a limited single-phase liquid region, similar to the of N_2/CO_2 mixtures, though in the presence of C_1 the liquid region is larger. At reservoir conditions, the maximum concentration that allows single-phase is 20 mole % for C_1-rich mixtures and 40%molar for CO_2-rich mixtures injected to the bitumen sample, whereas for the heavy oil sample these concentrations decrease approximately 50%.

Even at CO_2 supercritical conditions, similar behavior is observed, though the single-phase liquid region is expanded, allowing higher molar concentrations of the C_1/CO_2 mixtures to be injected. These results ultimately mean that the presence of C_1 as impurity also negatively impacts CO_2 storage at the simulated reservoir conditions, though solubility trapping in the lower API gravity may be enhanced compared to N_2/CO_2 streams due to higher solubility of C_1 in the reservoir fluids.

C_3 as impurity (Figure 2-C): Already at the typical reservoir conditions (i.e. 315 psi - 100 °F for San Ardo, and 400 psi and 50°F for Athabasca), first contact miscibility (FCM) with the reservoir fluid could be achieved for intermediate to high C_3 concentration in the CO_2 stream both for the heavy oil and bitumen samples [Figure 2-C (a) and Figure 2-C (b)]. For supercritical CO_2 conditions, full miscibility is achieved with both oils [Figure 2-C (c) and Figure 2-C (d)].

However, investigation of Figure 2-C reveals that the presence of C_3 as impurity has negative impact on the development of miscibility with the San Ardo sample. For this oil, pure CO_2 injection would allow single-phase mixtures with CO_2 concentrations as high as 22 mole %, whereas CO_2-C_3 mixtures would only be miscible with concentrations up to 12 mole % (90 mole % of C_3).

For the Athabasca bitumen, the occurrence of C_3 in the CO_2 stream would not affect how miscibility is developed. For both pure CO_2 and CO_2-C_3 streams, FCM would be guaranteed only to a maximum of 55 mole % concentration of the injected fluid.

Water as impurity: The presence of water induces the multiphase conditions with either two (aqueous and liquid) or three phases (aqueous, liquid, and vapor). For all simulated cases, when three phases are formed, miscibility among three phases is extremely limited, with predominant binary miscibility among CO_2-water (vapor phase), CO_2-oil (liquid phase), and water-CO_2 (aqueous phase). Therefore, pseudo-ternary diagrams are ineffective to represent the phase behavior of such mixtures.

Alternatively, to investigate the phase behavior effects on CO_2 storage, four multiphase flash results for each low API gravity oil sample are presented in Tables A1 through A4 (Appendix A). All simulated cases presented similar phase behavior, thus only the results for Athabasca bitumen at reservoir conditions (Table A1) will be discussed to illustrate the water-CO_2-crude oil interactions.

Results in Table A1 indicate that for high molar concentrations of CO_2, the formation of a third vapor phase composed by nearly pure CO_2 with 0.06% water is triggered (mixtures 1 and 3). This vapor phase diminishes CO_2 storage efficiency due to viscous fingering and to buoyance effects, as CO_2 gas is much less viscous and dense than the reservoir fluids. In such cases, CO_2 storage would be achieved by solubility trapping both in the aqueous and in the hydrocarbon-rich liquid phases. The aqueous phase was found to solubilize 2.17 mole % CO_2, while the liquid (hydrocarbon-rich) phase solubilizes 44.37 mole % CO_2. Comparison of results for mixtures 1 and 3 shows that the same equilibrium compositions for each phase will sustain for both equimolar and CO_2-rich mixture, but the molar distribution varies among each phase.

For the oil-rich and water-rich cases (mixtures 2 and 4), only liquid and aqueous phases are formed. For the oil-rich mixture, the liquid phase accounts for 95 mole % of the whole mixture. In this liquid phase, limited three-phase miscibility is achieved, with 0.02 mole % of water, 5.25 mole % of CO_2, and 94.73 mole % of Athabasca oil. The aqueous phase will have only 0.25 mole % of CO_2 solubilized in water. The water-rich mixture 4 results indicates that the CO_2 solubility dramatically increases both in the liquid (from 5.25 to 61.05 mole %) and aqueous (from 0.25 to 1.97 mole %) phases. When compared to the oil-rich mixture 2, however, the water-rich mixture was composed of 91.81 mole % of the aqueous phase. Accounting for the total CO_2 moles solubilized in each case, the water-rich mixture yielded 6.8 mole % total CO_2 solubility, whereas the oil-rich mixture resulted in 6.24 mole % CO_2 solubility.

All previous results were achieved by considering the phase behavior of fresh water as the CO_2 stream impurity. In perspective, this scenario would allow 3.5 - 4 w/w CO_2 at 1,160 psi and 122°F [38]. However, supercritical CO_2 is insoluble in saline waters, which may be the case for connate water from aquifers [2]. Therefore, operationally, the solubility of CO_2 in brine is considered negligible in the short term due to slow dissolution mechanisms and to the higher solubilization in oil [6, 39]. From the previous discussions, it is evident that when water is present in the reservoir or in the injected stream, prior knowledge of water and oil saturations, as well as water salinity are desirable to predict the feasibility of CO_2 storage in the liquid and aqueous phases, as well as to address the storage effectiveness in the long term.

Conclusions

1. Presence of N_2 or C_1 as impurities in the CO_2 stream negatively affects storage, as CO_2-rich mixtures may be 2 to 5 times more miscible than N_2- or C_1-rich mixtures. For both impurities, immiscible conditions are predominant, with maximum 10-15 mole % concentration of flue gas and 40-45 mole % of CO_2–C_1 streams.
2. Presence of C_3 negatively affects CO_2 storage into San Ardo heavy oil. Pure CO_2 streams allow 83% more miscibility with this oil when compared to C_3-rich streams. For Athabasca bitumen, CO_2-C_3 and CO_2-rich streams have similar miscibilities in oil: approximately 55 mole %.
3. Presence of fresh water in the CO_2 stream triggers the formation of an aqueous phase, which may induce CO_2 storage by solubility trapping in both oil and water. It is desirable to know water saturations and salinity beforehand to quantify CO_2 solubility in the liquid and aqueous phases.

Acknowledgments

The authors acknowledge the financial support and the opportunity provided by CAPES, Coordenação de Aperfeiçoamento de Pessoal de Nível Superior – Brasil. We also acknowledge the members of the Heavy Oil, Oil shales, Oil sands, & Carbonate Analysis and Recovery Methods Research Team at the Texas A&M University Petroleum Engineering Department for their help.

Nomenclature

C_1	Methane	H_2O	Water
C_3	Propane	N_2	Nitrogen
CCS	Carbon Capture and Sequestration	PR-EOS	Peng-Robinson Equation of State
CO_2	Carbon Dioxide	VLE	Vapor-Liquid Equilibria
FCM	First Contact Miscibility	VLLE	Vapor-Liquid-Liquid Equilibria

References

[1] Gielen, D., and Podkanski, J. 2004. Prospects for CO_2 Capture and Storage. Paris, OECD/IEA.

[2] IPCC. 2005. IPCC Special Report on Carbon Dioxide Capture and Storage. In: Metz, B., Davidson, O., de Coninck, H.C., Loos, M., Meyer, L.A. (Eds.). Cambridge University Press, Cambridge.

[3] Nasr, T. N., Prowse, D. R., and Frauenfeld, T. 1987. The Use of Flue Gas with Steam in Bitumen Recovery from Oil Sands. *Petroleum Society of Canada*. doi:10.2118/87-03-06.

[4] Li, H., and Yan, J. 2008. Evaluating cubic equations of state for calculation of vapour-liquid equilibrium of CO2 and CO2-mixtures for CO2 capture and storage processes. Applied Energy, **86** (2008), pp. 826–836.

[5] Chapoy, A., Burgass, R., Tohidi, B., Austell, J. M., and Eickhoff, C. 2011. Effect of Common Impurities on the Phase Behavior of Carbon-Dioxide-Rich Systems: Minimizing the Risk of Hydrate Formation and Two-Phase Flow. Society of Petroleum Engineers. doi:10.2118/123778-PA.

[6] Valbuena, E., Barrufet, M., & Falcone, G. 2012. Analytical Estimation of CO2 Storage Capacity in Depleted Oil and Gas Reservoirs Based on Thermodynamic State Functions. Society of Petroleum Engineers. doi:10.2118/153307-MS.

[7] Dooley, J., Davidson, C., Dahowski, R., Wise, M., Gupta, N., Kim, S., and Malone, E. 2006. Carbon Dioxide Capture and Geologic Storage: A Key Component of a Global Energy Technology Strategy to Address Climate Change. Joint Global Change Research Institute, College Park, MD.

[8] Bybee, K. 2008. The CO_2 Pilot at Lacq: An Integrated Oxycombustion CO_2-Capture and -Storage Project. Society of Petroleum Engineers. doi:10.2118/0708-0090-JPT.

[9] Alboudwarej, H., Felix, J., and Taylor, S. 2006. Highlighting Heavy Oil. Oilfield Review. Summer 2006.

[10] Huc, A. Y. 2011. Heavy Crude Oils - From Geology to Upgrading: an overview. Paris: Editions Technip.
[11] Bachu, S., and Shaw, J. 2003. Evaluation of the CO_2 Sequestration Capacity in Alberta's Oil and Gas Reservoirs at Depletion and the Effect of Underlying Aquifers. Petroleum Society of Canada. 03-09-02.
[12] Donaldson, E., Chilingar, G., and Yen, T.F, 1989. Enhanced Oil Recovery, II: Processes and Operations. Amsterdam: Elsevier. Print.
[13] Metcalfe, R. 1982. Effects of Impurities on Minimum Miscibility Pressures and Minimum Enrichment Levels for CO2 and Rich-Gas Displacements. Society of Petroleum Engineers. doi:10.2118/9230-PA.
[14] Jacobson, H. A. 1972. Acid Gases and Their Contribution to Miscibility. Petroleum Society of Canada. doi:10.2118/72-02-03.
[15] Sebastian, H. M., Wenger, R. S., and Renner, T. A. 1985. Correlation of Minimum Miscibility Pressure for Impure CO2 Streams. Society of Petroleum Engineers. doi:10.2118/12648-PA.
[16] Alston, R. B., Kokolis, G. P., and James, C. F. 1985. CO_2 Minimum Miscibility Pressure: A Correlation for Impure CO_2 Streams and Live Oil Systems. Society of Petroleum Engineers. 11959.
[17] Glass, O. 1985. Generalized Minimum Miscibility Pressure Correlation (includes associated papers 15845 and 16287). Society of Petroleum Engineers. doi:10.2118/12893-PA.
[18] Nguyen, T. A., and Farouq Ali, S. M. 1998. Effect of Nitrogen on the Solubility and Diffusivity of Carbon Dioxide into Oil and Oil Recovery by the Immiscible WAG Process. Petroleum Society of Canada. doi:10.2118/98-02-02.
[19] Jaiswal, N., and Mamora, D. D. 2007. Distillation Effects in Heavy-Oil Recovery under Steam Injection with Hydrocarbon Additives. Society of Petroleum Engineers. doi:10.2118/110712-MS.
[20] Ashrafi, M., Souraki, Y., Karimaie, H., Torsaeter, O., and Bjorkvik, B. J. A. 2011. Experimental PVT Property Analyses for Athabasca Bitumen. Society of Petroleum Engineers. 147064-MS.
[21] Katz, D. L., and Firoozabadi, A. 1978. Predicting Phase Behavior of Condensate/Crude-Oil Systems Using Methane Interaction Coefficients. Society of Petroleum Engineers. doi:10.2118/6721-PA.
[22] Tennyson, M.E. 2005. Growth History of Oil Reserves in Major California Oil Fields During the Twentieth Century. U.S. Geological Survey Bulletin 2172-H.
[23] Traverse, E. F., Deibert, A. D., and Sustek, A. J. 1983. San Ardo - A Case History of a Successful Steamflood. Society of Petroleum Engineers. doi:10.2118/11737-MS.
[24] Bell, J. S., Price, P. R., and McLellan, P. J., 1994. In-situ stress in the western Canada Sedimentary Basin. In Geological atlas of the western Canada sedimentary Basin. Mossop, G.D., Shetsen, I. *Canadian Society of Petroleum Geologists and Alberta Research Council*, 439 – 446.
[25] Badamchi-Zadeh, A., Yarranton, H. W., Maini, B. B., and Satyro, M. A. 2009. Phase Behavior and Physical Property Measurements for VAPEX Solvents: Part II. Propane, Carbon Dioxide and Athabasca Bitumen. Petroleum Society of Canada. 09-03-57.
[26] PVTSim Nova software. 2015. Calsep A/S, Lyngby, Denmark.
[27] Peng, D. Y., and Robinson, D. B. 1976. "A New Two-Constant Equation of State". *Industrial and Engineering Chemistry: Fundamentals* **15**: 59–64. doi:10.1021/i160057a011.
[28] Al-Meshari, A. A., and McCain, W. 2007. Validation of Splitting the Hydrocarbon plus Fraction First step in tuning Equation-of-State. Society of Petroleum Engineers. 104631-MS.
[29] Firoozabadi, A. 1999. *Thermodynamics of Hydrocarbon Reservoirs*. New York: McGraw-Hill. Print.
[30] Elliott, J. R., and Lira, C. T. 2012. Introductory Chemical Engineering Thermodynamics. 2nd Edition. Prentice Hall. 346 – 347.
[31] McCain, W.D. 1990/ *Properties of Petroleum Fluids*. Tulsa: Pennwell Books. Print.
[32] Nghiem, L.X., and Coombe, D.A. 1997. Modeling Asphaltene Precipitation during Primary Depletion. *SPE Journal*, **2**, 170-176.
[33] Pedersen, K.S., Christensen, P.L., and Shaikh, J.A. 2006. Phase Behavior of Petroleum Reservoir Fluids. 2nd Edition. Boca Raton, Florida. CRC Press.

[34] Gonzalez, K. G., Barrufet, M. A., and Nasrabadi, H. 2014. Development of a Compositional Reservoir Simulator Including Asphaltene Precipitation from a Thermodynamic Consistent Model. Society of Petroleum Engineers. doi:10.2118/169401-MS.

[35] Peneloux, A., Rauzy, E., and Fréze, R. 1982. A Consistent Correction for Redlich-Kwong-Soave volumes. *Fluid Phase Equilib* **8**.

[36] Danesh, A. 1998. PVT and Phase Behavior of Petroleum Reservoir Fluid. Elsevier, Amsterdam.

[37] Nasrabadi, H., Firoozabadi, A., Esposito, R. O., and Vieira, A. J. M. 2008. Interpretation of an Unusual Bubblepoint Pressure Variation in an Offshore Field. Society of Petroleum Engineers. 113574-MS.

[38] Rochelle, C.A., and Moore, Y.A. 2002. The solubility of supercritical CO_2 into pure water and synthetic Utsira porewater. British Geological Survey. Commissioned report CR/02/052.

[39] Sifuentes, W. F., Giddins, M. A., and Blunt, M. J. 2009. Modeling CO_2 Storage in Aquifers: Assessing the key contributors to uncertainty. Society of Petroleum Engineers. doi:10.2118/123582-MS.

Appendix: VLLE flash results for the H2O-CO_2-Crude oil mixtures

Table A1 –Flash results for the H_2O-CO_2-Athabasca mixture at reservoir conditions (400 psi - 50°F).

	Mixture component	Mixture composition, mol%	Mixture composition, wt%	Vapor phase, mol%	Liquid Phase, mol%	Aqueous phase, mol%
Mixture 1: Equimolar	Athabasca Bitumen	33.33	89.59	0	55.61	0
	CO_2	33.33	7.38	99.94	44.37	2.17
	H_2O	33.33	3.02	0.06	0.02	97.83
	Phase total	100		6	59.95	34.06
Mixture 2: Oil-rich	Athabasca Bitumen	90	99.36	0	94.73	0
	CO_2	5	0.45	0	5.25	0.25
	H_2O	5	0.19	0	0.02	99.75
	Phase total	100		0	95	5
Mixture 3: CO_2-rich	Athabasca Bitumen	5	39.73	0	55.61	0
	CO_2	90	58.93	99.94	44.37	2.17
	H_2O	5	1.34	0.06	0.02	97.83
	Phase total	100		85.95	8.99	5.05
Mixture 4: H_2O-rich	Athabasca Bitumen	5	59.18	0	38.93	0
	CO_2	5	4.88	0	61.05	1.97
	H_2O	90	35.94	0	0.02	98.03
	Phase total	100		0	8.19	91.81

Table A2. Flash results for the H_2O-CO_2-Athabasca mixture at supercritical conditions (1,200 psi - 122°F).

	Mixture component	Mixture composition, mol%	Mixture composition, wt%	Vapor phase, mol%	Liquid Phase, mol%	Aqueous phase, mol%
Mixture 1: Equimolar	Athabasca Bitumen	33.33	89.59	0	50.37	0
	CO_2	33.33	7.38	0	49.49	1.73
	H_2O	33.33	3.02	0	0.14	98.27
	Phase total	100		0	66.18	33.82
Mixture 2: Oil-rich	Athabasca Bitumen	90	99.36	0	94.64	0
	CO_2	5	0.45	0	5.25	0.18
	H_2O	5	0.19	0	0.11	99.82
	Phase total	100		0	95.09	4.91
Mixture 3: CO_2-rich	Athabasca Bitumen	5	39.73	0.53	38.39	0
	CO_2	90	58.93	99.46	61.45	2.06
	H_2O	5	1.34	0.01	0.16	97.94
	Phase total	100		82.36	12.99	4.65
Mixture 4: H_2O-rich	Athabasca Bitumen	5	59.18	0	57.71	0
	CO_2	5	4.88	0	42.16	1.48
	H_2O	90	35.94	0	0.13	98.52
	Phase total	100		0	8.66	91.34

Table A3. Flash results for the H_2O-CO_2-San Ardo mixture at reservoir conditions (400 psi-100°F).

	Mixture component	Mixture composition, mol%	Mixture composition, wt%	Vapor phase, mol%	Liquid Phase, mol%	Aqueous phase, mol%
Mixture 1: Equimolar	Athabasca Bitumen	33.33	87.05	0.01	77.56	0
	CO_2	33.33	9.19	99.64	0.07	0.88
	H_2O	33.33	3.76	0.35	22.38	99.12
	Phase total	100		23.51	42.97	33.52
Mixture 2: Oil-rich	Athabasca Bitumen	90	99.15	0	94.69	0
	CO_2	5	0.58	0	5.25	0.20
	H_2O	5	0.24	0	0.06	99.80
	Phase total	100		0	99.05	4.95
Mixture 3: CO_2-rich	Athabasca Bitumen	5	33.97	0.01	77.56	0
	CO_2	90	64.56	99.64	22.37	0.88
	H_2O	5	1.47	0.35	0.07	99.12
	Phase total	100		88.84	6.44	4.72
Mixture 4: H_2O-rich	Athabasca Bitumen	5	53.09	0.01	77.56	0
	CO_2	5	5.61	99.64	22.37	0.88
	H_2O	90	41.30	0.35	0.07	99.12
	Phase total	100		2.77	6.45	90.78

Table A4. Flash results for the H_2O-CO_2-San Ardo mixture at supercritical conditions (1,200 psi - 122°F).

	Mixture component	Mixture composition, mol%	Mixture composition, wt%	Vapor phase, mol%	Liquid Phase, mol%	Aqueous phase, mol%
Mixture 1: Equimolar	Athabasca Bitumen	33.33	87.05	0	50.41	0
	CO_2	33.33	9.19	0	49.45	1.89
	H_2O	33.33	3.76	0	0.14	98.11
	Phase total	100		0	66.12	33.88
Mixture 2: Oil-rich	Athabasca Bitumen	90	99.15	0	94.65	0
	CO_2	5	0.58	0	5.25	0.19
	H_2O	5	0.24	0	0.10	99.81
	Phase total	100		0	95.09	4.91
Mixture 3: CO_2-rich	Athabasca Bitumen	5	33.97	0.04	45.68	0
	CO_2	90	64.56	99.44	54.17	2.06
	H_2O	5	1.47	0.52	0.15	97.94
	Phase total	100		84.49	10.88	4.64
Mixture 4: H_2O-rich	Athabasca Bitumen	5	53.09	0	58.43	0
	CO_2	5	5.61	0	41.45	1.59
	H_2O	90	41.30	0	0.12	98.41
	Phase total	100		0	8.56	91.44

CARBON MANAGEMENT TECHNOLOGY CONFERENCE

CMTC-439561-MS

Modeling Study on CO_2 Capture and Storage in Organic-Rich Shale

Yinghui Li*, Hui Pu, InPetro Technologies Inc.

This paper was prepared for presentation at the Carbon Management Technology Conference held in Sugarland, Texas, USA, 17–19 November 2015.

Abstract

Shale reservoirs, with huge oil and gas resources in place, play an important role as energy resources in the world. In the meantime, widely-distributed organic-rich shales are being considered as one of the important carbon sequestration targets. Some studies have been done to characterize rock-fluid properties, and flow mechanisms in the shale reservoirs. Most of these studies, through modifying methods used for conventional reservoirs, fail to capture dynamic features of shale rock and fluids. In confined nano-pore space in shale reservoirs, interaction between the wall of shale and the contained fluid significantly affects fluid properties. The incapability to model capillarity with the consideration of imperative pore size distribution characteristics using commercial software may lead to inaccurate results for Carbon Capture, Utilization, and Storage (CCUS). This paper presents a novel formulation that consistently evaluates capillary force in the light of using pore size distribution (PSD) directly from core measurements, and develops an adsorption model to achieve a realistic representation of CO_2 storage mechanisms. The assumptions in Langmuir isotherm model affect the validity when it is applied to unconventional reservoirs, especially CCUS. An adsorption model was developed to incorporate adsorption behavior using a local density optimization algorithm. Our advances were also designed for multi-component interactions to adsorption sites for a full spectrum of reservoir pressures of interests. This feature allowed us to fully understand the storage and sieving capabilities for CO_2 system with remaining reservoir fluids.

Introduction

It is widely regarded that the increasing emission of CO_2 and other Greenhouse gases can cause Greenhouse effect and CO_2 has been a major challenge to the global climate change. Carbon capture and storage is one of the options under investigation to reduce CO_2 emissions in the atmosphere (House *et al.*, 2006; Stefan, 2008). The options for CO_2 geological sequestration have been widely investigated and the technical papers and reports are well-documented for almost every aspect of CO_2 geological sequestration. Geologic formations have attracted much attention for the purpose of reducing of emission of carbon dioxide to the atmosphere through storing and sequestration process in recent years. The U.S. Department of Energy has identified several categories of geologic reservoirs for potential

CO_2-sequestration (U.S. Department of Energy, 1999, 2004, 2005). Carbon sequestration options include oil and gas reservoirs, un-mineable coal seams, deep saline reservoirs, basalts, and shales. It is beyond the scope of this paper to focus on every type of geologic formations for CO_2 storage. Instead, oil and gas reservoirs, specifically organic-rich shales, are the focus of this paper.

One of the advantages of oil and gas reservoirs are that there are so much important data and information about the reservoirs already available that they represent known geologic (structural or stratigraphic) traps containing hydrocarbons within one or more reservoirs that are confined by a known cap or seal. We know that depleted oil and gas reservoirs are natural traps for liquids and gas. After all, that is where the oil or natural gas was safely "stored" for millions of years before being produced and brought to the surface by humans. The development of unconventional reservoirs due to the advances obtained in drilling and hydraulic fracturing not only contributes to the rapid rise of the critical energy supply, but also it presents the opportunity for Carbon Capture and Sequestration (CCS). Besides many differences, one simple difference between conventional and unconventional reservoirs is the fact that unconventional reservoirs are widely present around the world but conventional reservoirs exist just in some parts of the world. This strongly means that since unconventional reservoirs are everywhere on the map then the CO_2 sequestration in such reservoirs is very reasonable and economically feasible (Li *et al.*, 2006; DOE NETL report, 2010; Eshkalak and Aybar, 2015). Shale is able to trap significant amount of gas permanently because of adsorption in its finely dispersed organic matter (Kang *et al.*, 2011; Li and Elsworth, 2014). The total organic content (TOC) in shale could be up to 10% of the total shale weight for many North American shale plays. It is reported that the ratio of organic-pore-volume to total-pore-volume could be up to 70%, i.e., for Barnett shale samples (Kang *et al.*, 2011). This allows CO_2 being stored in shale reservoirs for much longer time, often in millions of years, than conventional mechanisms without the concerns of traps seals, and leakage. In addition to the large storage capacity in organic matter in shale and its eternity, storage in kerogen-rich shale has the advantage that these nano-pores may act as molecular sieve thanks to much higher affinity of CO_2 compared with other gas components with the organic matter in kerogen.

In spite of great success in the development of unconventional reservoirs, the fluid flow mechanisms in shale, fluid properties in confined pore space and interactions between the wall of shale and the contained fluid are still not well understood. One of unique features of shale rocks are their extremely small pore-sizes. According to Kuila and Prasad (2011), the matrix in shale reservoirs consists of micropores smaller than 2nm in diameter, mesopores with diameters in the range of 2 to 50nm, and macropores larger than 50nm in diameter. So pore sizes can be on the order of nanometers for formations. Such small pores lead to the increase of capillary pressures. At such small pores the confined hydrocarbon phase behavior deviates from the bulk measurements where vapor-liquid interface is assumed as a flat surface. This assumption is generally valid when the vapor-liquid equilibrium takes place in PVT (Pressure-Volume-Temperature) cells. But when the fluid is confined in pore spaces of nano-size, the significant interfacial curvature may cause a large capillary pressure difference between liquid and vapor phases (Wang *et al.*, 2013; Jin *et al.*, 2015). The effect of capillary pressure on vapor-liquid equilibrium in conventional reservoirs had been the subject of study for long time, and it is generally negligible for conventional reservoirs (Leverett 1941; Sigmund *et al.* 1973; Shapiro and Stenby 1997; Shaprio *et al.* 2000; Wang *et al.*, 2013). Perhaps for this reason, the current commercial reservoir simulators assume no pressure difference between vapor and liquid phases during flash calculations. Not accounting for increased capillary pressure in small pores can lead to inaccurate estimates of ultimate recovery, and of saturation pressures (Nojabaei *et al.*, 2013; Jin *et al.*, 2015; Kazi *et al.*, 2015). Aiming at improving the characterization of capillary pressure for simulation of unconventional reservoir, this paper presents a novel formulation that consistently evaluates capillary force in the light of using pore size distribution (PSD) directly from core measurements and the capillary pressure model is incorporated into a compositional reservoir simulator. Therefore, how PSD would alter phase and flow behavior under nano-pore confinements is examined in the paper.

Adsorption is the attraction of gas or liquid molecules to the surface of a solid, creating a high-density phase that is distinct from the surrounding free gas or liquid molecules. Organic-rich shales are composed of several percent of microporous organic kerogen, which commonly exhibits high adsorption potential due to its significant internal surface area. Adsorption is an important storage mechanism in shales. Quantification of adsorption for a material involves the measurement of amount of adsorption at many different pressures at a constant temperature to obtain an adsorption isotherm. In general, the adsorption isotherm will be controlled by the distribution of pore sizes in the material, the effective molecular size of the adsorbing gas, and the combined attraction energy of the surface material and the adsorbing gas (Langmuir, 1916; Brunauer *et al.*, 1938). In an effort to better understand the adsorption, numerous authors have made valuable contributions to the literature through both laboratory studies and models to characterize adsorption. Many isotherm models have been developed for gas adsorption on solids surface (Kohlmann and Zevenhoven, 2001). The Langmuir isotherm, which is widely used, represents the adsorption of a monolayer of molecules on the inner surfaces of the material (Langmuir, 1916). In existing commercial software package, adsorption was modeled using Langmuir isotherm adsorption. The assumptions of the Langmuir model (Do, 1998): 1). The surface is homogeneous, that is the adsorption energy is constant over all sites; 2). Adsorption on surface is localized, that is adsorbed atoms or molecules are adsorbed at definite, localized sites; 3). Each site can accommodate only one molecule or atom. The assumptions in Langmuir isotherm, however, are not held for wet or liquid-rich shale formations of interests, and CO_2 storage purpose in particular. This paper presents a new adsorption model that incorporates adsorption behavior using a local density optimization algorithm.

The characteristics of unconventional reservoirs make them a preferable alternative over CCS in depleted conventional reservoirs and saline aquifers, given its storage capacity, geological presence, and unique in-situ compositional sieving. It is a challenge to a better representation of these features in organic-rich shales using existing commercial software for studying CO_2 storage mechanisms, and guiding its application optimization. The objective of this work is to promote the advances of simulation of CO_2 storage in shale formation aiming to achieve a realistic representation that differentiates CO_2 storage in unconventional reservoirs vs. conventional ones.

Pore Size Distribution in Shales and Capillary Pressure

Shale, as the most abundant sedimentary rock, is a fine-grained, fissile, detrital sedimentary rock formed by consolidation of clay- and silt-sized particles into thin, relatively impermeable layers. In contrast to conventional reservoir rocks, shales have the following characteristics: fine-grained rock texture (dominant grain size ≤62.5 μm), low porosity (≤10%) and very low permeabilities (in nanodarcy range) (Kuila and Prasad, 2011). Fine-grained nature of shale rocks results in their extremely small pore-sizes. Shale rocks have predominantly micro (pore diameter<2nm) to meso-pores (2nm<pore diameter<50nm). Even tough there were significant advancement in characterization techniques, quantification of pore structure attributes in shales still remains a challenge. **Figure 1** shows the pore size distribution of different shale rock samples obtained from gas (N_2/Ar) adsorption isotherms (Wilcox, 2012). It clearly shows that the majority of pore diameter is in the range of 2 nm to 10 nm. Due to the extremely small pores, shale reservoirs exhibit flow and interaction characteristics that are uniquely related with nano-scale pore throats and pore size distribution. The tiny pore can act as a confinement to change the fluid behavior and properties because of smalle pores' geometric constraint upon fluid molecules and their influence on the interaction of these molecules with the pore walls (the molecule–wall interaction). Experimental results indicate that a confined fluid has a depressed critical temperature and an increased critical density compared to its bulk properties (Wong *et al.*, 1993, Thommes *et al.*, 1997, Hiejima and Yao, 2004, Arakcheev *et al.*, 2008). Sapmanee (2011) reported pore proximity effects in which confined pores could change fluid properties and flow conditions. For conventional reservoirs, the phase interface curvature is negligible, so this simplification is appropriate. In current available compositional simulators, the difference between vapor and liquid phase pressures is ignored. For unconventional reservoirs, small pores increase capillary pressure. More and more findings show the non-negligible influence of pore proximity effect and increased capillary pressure in unconventional reservoirs, the continuous ignorance of the effect of capillary pressure on vapor-liquid equilibrium is inappropriate. Historically, mercury injection techniques are used to perform pore-size analysis for conventional reservoir rock samples. However, when it comes to unconventional shale reservoirs, very high pressures (> 60,000 psi) would be required for mercury to access the full pore-structure (Kuila and Prasad, 2011). Elgmati *et al.* (2011) conducted mercury injection capillary pressure (MICP) experiment on Haynesville shale rock sample (**Figure 2**). We can observe from the figure that the major intrusion occurred between 10,600 and 60,000 psi, which indicates that the nano-scale pores lead to very high mercury injection pressure.

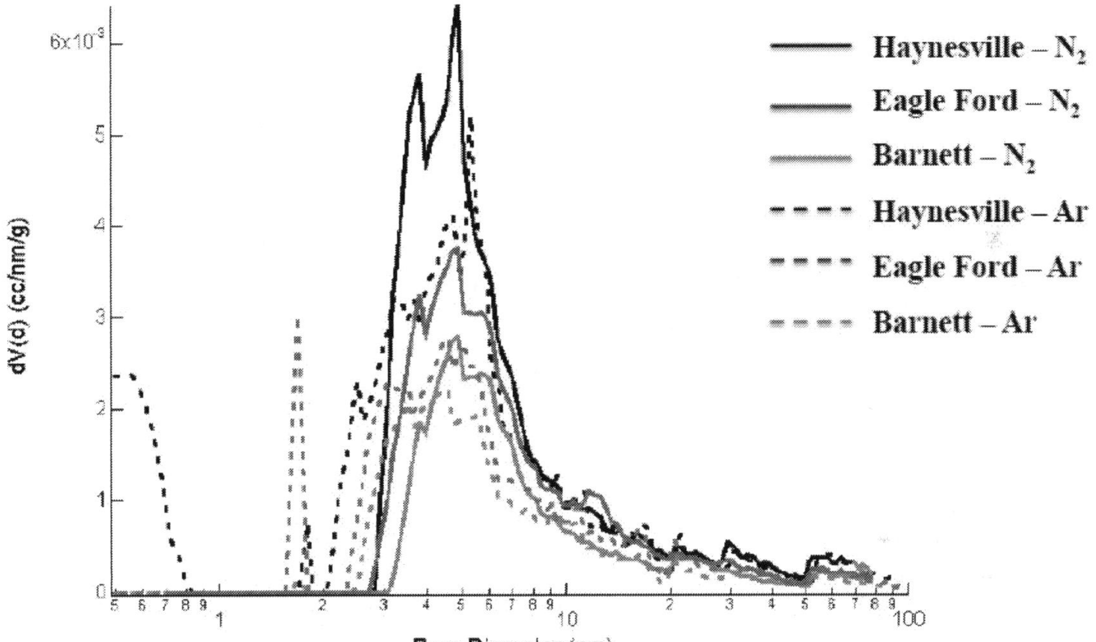

Figure 1. Pore size distribution of different shale rock samples (Wilcox, 2012)

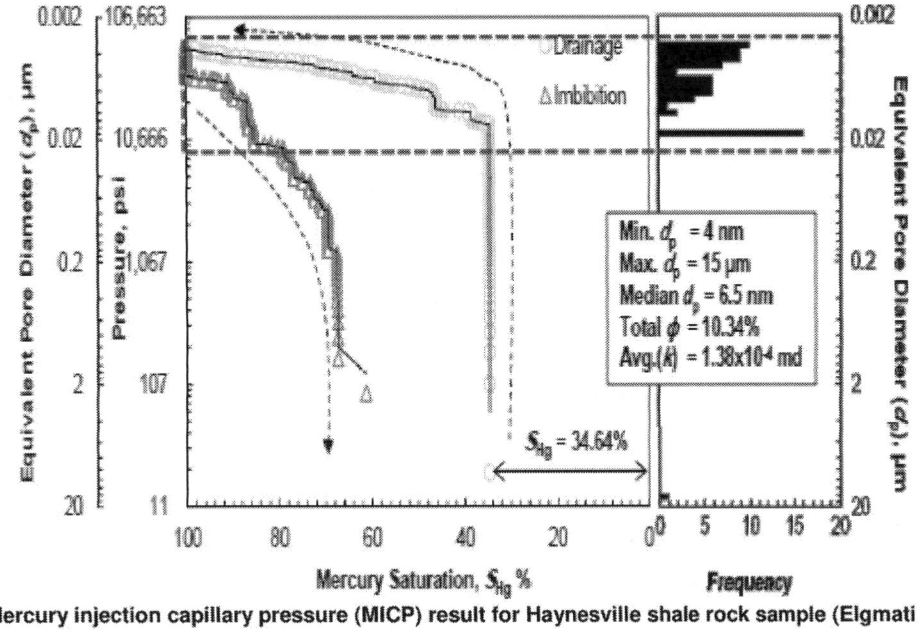

Figure 2. Mercury injection capillary pressure (MICP) result for Haynesville shale rock sample (Elgmati *et al.*, 2011)

Capillary pressure is usually calculated either by 1) Young-Laplace Equation (Eq. 1), where σ is the interfacial tension between phases j and i in dynes per cm, and r is the pore radius and contact angle is assumed to be 0 here. σ is calculated using Macleod-Sugden correlation, where ψ_i is the parachor of component i , ρ_i is the molar density of phase j in lb-mole per cubic feet.

$$P_c = \frac{2\sigma}{r},\ \sigma_{jl}^{0.25} = 0.016018 \sum_{i=1}^{n_c} \Psi_i (\rho_j x_{ij} - \rho_l x_{il}) \quad (1)$$

or 2) J-Function (Eq. 2), where the parameters C_{pc} and E_{pc} need to be determined by matching gas/oil experimental capillary pressure curves. ϕ is the porosity and K is the permeability. S_{on} and S_{gn} is the normalized oil and gas saturation.

$$P_c = C_{pc}\sigma\sqrt{\frac{\emptyset}{K}}\left(\frac{S_{gn}}{S_{on}+S_{gn}}\right)^{E_{pc}} \quad (2)$$

Vapor-liquid equilibrium (VLE) calculation is an important part to characterize the phase behavior using equation-of-state (EOS) model. As pointed out in the previous section that the capillary pressure in unconventional shale reservoirs affects the vapor-liquid equilibrium so significantly that it has to be considered in order to have a correct representation of flow mechanisms in shales. Vapor-liquid equilibrium is calculated considering the effect of capillary pressure, and adsorption for the vapor phase is superimposed to account for molecular interactions with the surface and changes in storage properties, which was described in a later session. The condition of equilibrium of the two phases can be written in terms of the chemical potential, μ, at a given pressure, P:

$$\mu_i^L(T, P^L, x) = \mu_i^V(T, P^V, y) \qquad i = 1,2, \ldots, m \quad (3)$$

Under the action of capillary forces, the pressure exerted at the curved liquid-vapor interface, P_c, is defined as

$$P^V - P^L = P_c \quad (4)$$

Where P^L and P^V are the phase pressures of liquid and vapor phases, respectively. u_i is the chemical potential of component i at the respective temperature, phase pressure, and mole fraction. m is the number of components in the system.

Optimized Simplified Local Density (OSLD) Model and Adsorption

The trapping of gas in an adsorbed state within organic-rich shales' finely dispersed matter (i.e.: kerogen) leads to their ability to store significant amount of gas permanently. Organic-rich shales are ideal targets for sequestration of significant volumes of anthropogenic CO_2. The extensive study of gas adsorption, e.g. methane adsorption, contributed to the early recognization of importance of adsorption of methane has for gas production in unconventional reservoir. The study of CO_2 adsorption focused on enhanced gas/oil recovery and sequestration. CO_2 adsorption in coal has been studied extensively, with researchers finding significant preferential adsorption of CO_2 over CH_4 and thus the potential for enhanced recovery and sequestration (Kovscek *et al.*, 2005). A study by Nuttall *et al.* (2005) showed that Langmuir-like adsorption of CO_2 in gas shales was more than five times greater than the amount of CH_4 adsorption. The results of Kang *et al.* (2010) showed that adsorption of CO_2 is five to ten times higher than that of CH_4. It is well-known that Langmuir isotherm (1916) is the classic theory usually used to describe adsorption isotherm. But several assumptions in Langmuir isotherm are not valid for organic-rich shale formations, and CO_2 storage purpose in particular. In this work, a Simplified Local Density (SLD) theory model is further improved to include various pore sizes of shale samples, then is utilized to modify Peng-Robinson Equation of State (PR EOS), and ultimately the new Optimized Simplified Local Density (OSLD)-PR EOS is integrated to a compositional simulator for study of mechanisms of CO_2 sequestration in organic-rich shales.

Rangarajan *et al.* (1995) originally developed the SLD-PR model with a goal to correlate adsorption isotherms of a pure component on a flat surface over a large pressure range. Through a series of experiments, Chen *et al.* (1997) investigated adsorption and desorption isotherm of carbon dioxide adsorbed on activated carbon at subcritical conditions, and found that the simplified local density model was able to characterize adsorption very well. The SLD-PR model was further modified by Fitzgerald *et al.* (2003) to investigate the adsorption phenomena of pure gases (such as methane, nitrogen, and CO_2) on coals at near-critical and supercritical conditions, and their research results showed that the modified SLD-PR model can correlate accurately with data for supercritical adsorption systems. The SLD-PR model was used by to Ma and Jamili (2014) to study the effect of confinement in shale formations on phase behavior.

Some basic assumptions were made in SLD-PR model: 1) the chemical potential at any point near the absorbent surface is equal to the bulk phase chemical potential, and 2) the chemical potential at any point above the surface is sum of the fluid-fluid and fluid-solid interactions. The equilibrium chemical potential is calculated by considering the contributions from these fluid-fluid and fluid-solid interactions. Our OSLD-PR EOS is based on the work of Chen *et al.* (1997). The pore geometry used in local density model is slit with a width of *L*. Let distance *z* be the normal distance from the point of interest to the surface of the solid. The chemical potential of fluid inside a porous medium at a distance z is the result of fluid-fluid and fluid-solid interactions:

$$\mu_{bulk} = \mu(z) = \mu_{ff}(z) + \mu_{fs1}(z) + \mu_{fs2}(L - z) \quad (5)$$

where the subscript "*bulk*" refers to the bulk fluid, "*ff*" refers to fluid-fluid interactions, and "*fs*" refers to the fluid-solid interactions.

The chemical potential of bulk can be written in terms of fugacity:

$$\mu_{bulk} = \mu_0(T) + RTln(f_{bulk}/f_o) \quad (6)$$

where f_{bulk} is the bulk fugacity and f_o is the fugacity at standard state.

At d distance *z* normal to the surface of pore wall, the chemical potential of fluid-fluid interactions can be defined in terms of fugacity due to fluid-fluid interaction $f_{ff}(z)$ as:

$$\mu_{ff}(z) = \mu_0(T) + RTln\left(f_{ff}(z)/f_o\right) \quad (7)$$

The chemical potential due to fluid-solid interaction can be written as

$$\mu_{fs}(z) = N_A[\Psi(z) + \Psi(L - z)] \quad (8)$$

where $\Psi(z)$ is the fluid-solid potential energy function. Lee's partially integrated 10-4 Lennard-Jones potential was used (Lee, 1988):

$$\Psi(z) = 4\pi\rho_{atoms}\varepsilon_{fs}\sigma_{fs}^2\left(\frac{\sigma_{fs}^{10}}{5(z')^{10}} - \frac{1}{2}\sum_{i=1}^{4}\frac{\sigma_{fs}^4}{(z'+(i-1)\cdot\sigma_{ss})^4}\right) \quad (9)$$

where ρ_{atoms} is the number of carbon plane atoms per unit area (38.2 atoms/nm^2), ε_{fs} is the fluid-solid interaction energy parameter, σ_{fs} is the molecular diameter of the adsorbate, $\sigma_{fs} = \left(\sigma_{ff} + \sigma_{ss}\right)/2$, σ_{ss} is the carbon interplanar distances which is taken to be that of graphite (0.335nm) (Subramanian *et al.*, 1995), The distance z' is the perpendicular distance between the center of a fluid molecule to the center of the first plane of solid atoms, i.e., $z' = z + \sigma_{ss}/2$.

Through combining equations (5)-(8), the relations of the bulk fugacity and fluid-fluid fugacity can be expressed as:

$$f_{ff}(z) = f_{bulk}exp\left(-\frac{\mu_{fs1}(z)+\mu_{fs2}(L-z)}{RT}\right) \quad (10)$$

The Peng-Robinson equation of state was used to calculate the fugacity of a bulk fluid:

$$lnf_b = \frac{b\rho_b}{1-b\rho_b} - \frac{a_b\rho_b}{RT(1+2b\rho_b-b^2\rho_b{}^2)} - ln\left[\frac{(1-b\rho_b)}{RT\rho_b}\right] - \frac{a_b}{2\sqrt{2}RT}ln\left[\frac{1+(1+\sqrt{2})\rho_b}{1+(1-\sqrt{2})\rho_b}\right] \quad (11)$$

Similarly, the expression for fugacity of fluid-fluid interaction can be written:

$$lnf_{ff}(z) = \frac{b\rho(z)}{1-b\rho(z)} - \frac{a(z)\rho(z)}{RT\{1+2b\rho(z)-b^2[\rho(z)]^2\}} - \ln\{\frac{[1-b\rho(z)]}{RT\rho(z)}\} - \frac{a(z)}{2\sqrt{2}RT}ln\left[\frac{1+(1+\sqrt{2})\rho(z)}{1+(1-\sqrt{2})\rho(z)_b}\right] \quad (12)$$

The equations for $a(z)$ depend on the ratio of the slit width L to the molecular diameter σ_{ff}. The details for calculations of are $a(z)$ are described in paper of Chen *et al.* (1997), and are shown in Appendix A. From equations (6)-(12), the local density $\rho(z)$ at any point can be calculated.

In our model, the adsorption per unit weight area of adsorbent is calculated as follows:

$$\Gamma = A\int_{\sigma_{ff}/2}^{L-\sigma_{ff}/2} [\rho(z) - \rho_b(z)]dZ$$

where A is the surface area per unit weight of the adsorbent.

Through characterizing the adsorption behavior using an optimized simplified local density optimization function, we successfully developed the OSLD-PR model, which allows the characterization and understanding of multi-component interactions at adsorption sites for a full spectrum of reservoir pressures.

Simulation Model Description

The capillarity model with consideration of PSD characteristics and adsorption model using a local density function were integrated to our in-house compositional simulator aiming at characterizing the flow mechanisms in unconventional reservoirs. Our simulator is then used to model the process of injection of CO_2 in Bakken Formation in this study. An 1-D model was built in this work. The model size is 6.25 ft × 0.25 ft × 0.5 ft, which corresponds to length, width, and thickness, respectively. There are 25 cells in *x*-direction, 1 cell in *y*-direction and 1 cell in *z*-direction. There are two wells at the either end of model. The basic reservoir properties data of the model is shown in **Table 1**. The reservoir is assumed to be homogeneous with a permeability of 1 μD. The posority is 20%.

Table 1. Parameters for Bakken Simulation Model

Parameter	Value
Initial reservoir pressure, psi	3200
Reservoir temperature, ºF	240
Reservoir permeability, μD	1
Initial water saturation	0.1

The pressure, volume, and temperature (PVT) data for Bakken crude oil samples are available and were used to define PVT relationships under reservoir condition. Constant composition expansion (CCE), differential liberation (DL) analysis, separator, swelling test and fluid compositional analysis data were also available for oil samples. Our SLD-PR EOS model was tuned based on the available experimental PVT data to perform phase equilibrium and property calculations. The Bakken reservoir oil is carefully grouped into nine pseudocomponents, i.e., CO_2, C_1, C_2, C_3, C_4, C_5-C_6, C_7-C_{12}, C_{13}-C_{21}, C_{22}-C_{80}, and their molar fractions are 0.0001, 0.36726, 0.14885, 0.09334, 0.05751, 0.06406, 0.15854, 0.0733, and 0.03704. The SLD-PR EOS model was tuned to match with experimental data.

Flow Simulation Results

After the reservoir model was built, different cases were designed and run to study and evaluate both primary production and CO_2 EOR, and CO_2 storage in the Bakken Formation.

Primary Recovery

For primary recovery cases, the reservoir models are simulated for 100 days, and the results are used as base case for comparison with CO_2 EOR cases. The reservoir is produced by one production well which is under constant bottom-hole pressure (BHP) control of 1,100 psi. Four cases, as shown below, were designed by choosing the different options in the simulator for capillary pressure and PSD:

1)."No Pc" case means the effect of capillary pressure is not considered in this scenario;

2)."Fixed Pore Radius" case in which a uniform pore size of 10 nm is used and Young-Laplace equation is used to calculate the P_c;

3)."J-function" case in which J-function is used to calculate the P_c while a uniform pore size of 10 nm is used;

4)."PSD" case in which pore size distribution data of Bakken sample is used and Young-Laplace equation is used to calculate the P_c.

Figure 3 shows the results of oil recoveries of four primary production cases. The recoveries of primary production vary from 11%OOIP to 12.4%OOIP. The oil recovery increased very fast during the first 25 days of production, and after that the recovery plateaued. The average reservoir pressure profile of primary production cases is shown in **Figure 4**, which represents pressure decline of a typical natural depletion process.

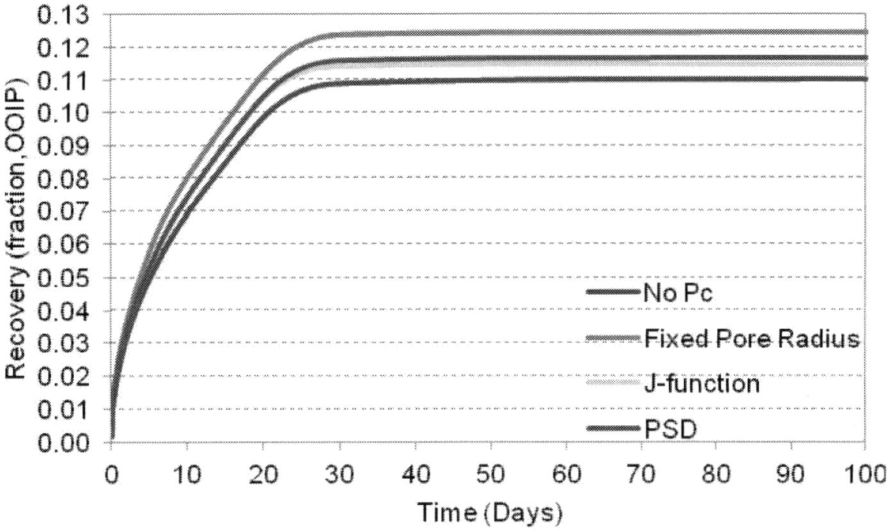

Figure 3. Recovery Factors of Primary Production Cases

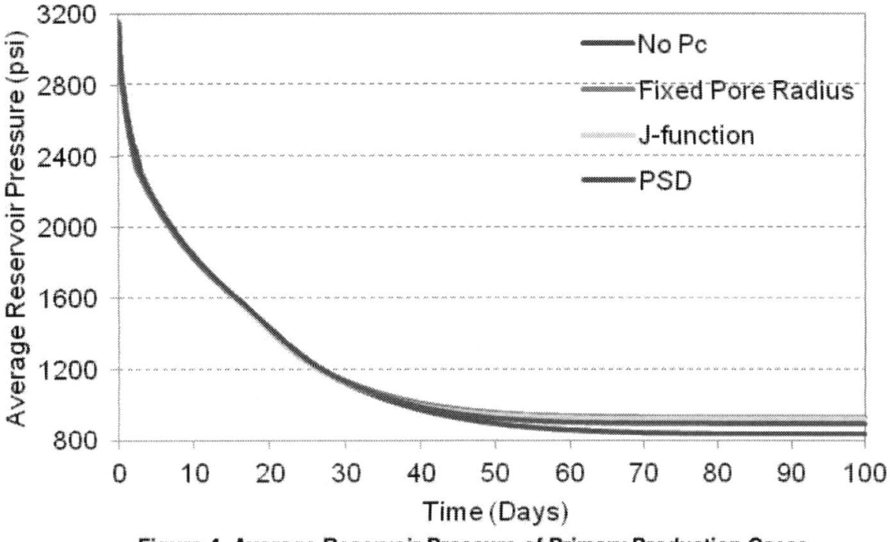

Figure 4. Average Reservoir Pressure of Primary Production Cases

Continuous CO_2 Flooding

During the CO_2 flood phase, there are two wells in the model, one injection well and one production. Initially the production well produces for 30 days, and then it is followed by a 70 days of continuous CO_2 injection for four cases of primary recovery. The BHP control for CO_2 injection well is 3,200 psi and 1,100 psi for production well. The oil recovery of four cases is shown in **Figure 5**. As shown in the figure, the incremental oil recovery by CO_2 injection is significant compared to the base case of primary recovery. The oil recovery and incremental oil recovery of CO_2 injection cases and primary recovery are summarized in **Table 2**. The incremental oil recovery of CO_2 flooding ranges from 21.54%OOIP to 25.29%OOIP. The relationship of oil recovery with the volume of CO_2 injected is illustrated in **Figure 6**. The oil recovery profile shows that the methods for capillary pressure calculation and the PSD do affect the oil recovery by almost 4%OOIP. The daily oil rate and cumulative oil rate of continuous CO_2 flooding cases are shown in **Figures 7** and **8**, respectively. The injection of CO_2 after primary recovery resulted in a huge increase of oil production for CO_2 injection cases.

Table 2. Summary of Oil Recovery and Incremental Recovery of Continuous CO_2 Injection Cases

Cases	Oil recovery@100 Days,%OOIP	CO_2 Incremental Oil Recovery, %OOIP
"No Pc"	32.56	21.54
"Fix Pore Radius"	36.31	25.29
"J-function"	35.93	24.91
"PSD"	33.89	22.87
Base Case (Primary)	11.02	N/A

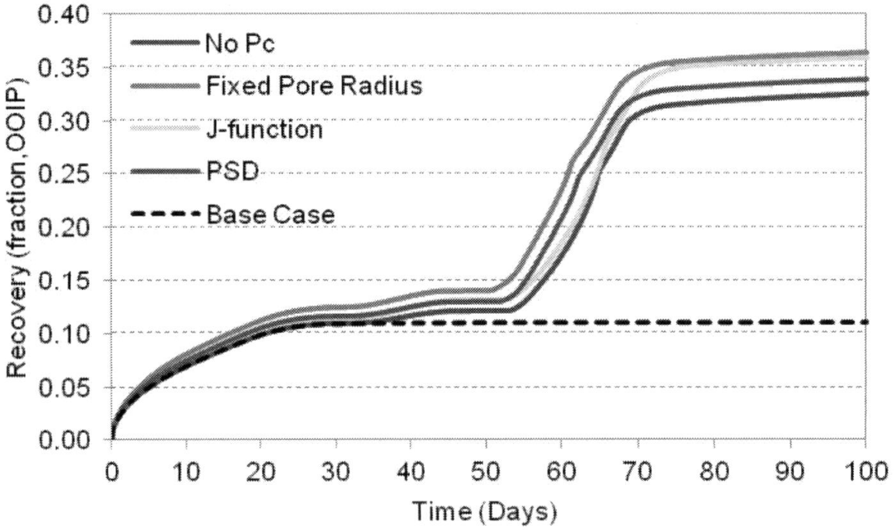

Figure 5. Oil Recoveries of Four Continuous CO2 Injection Cases

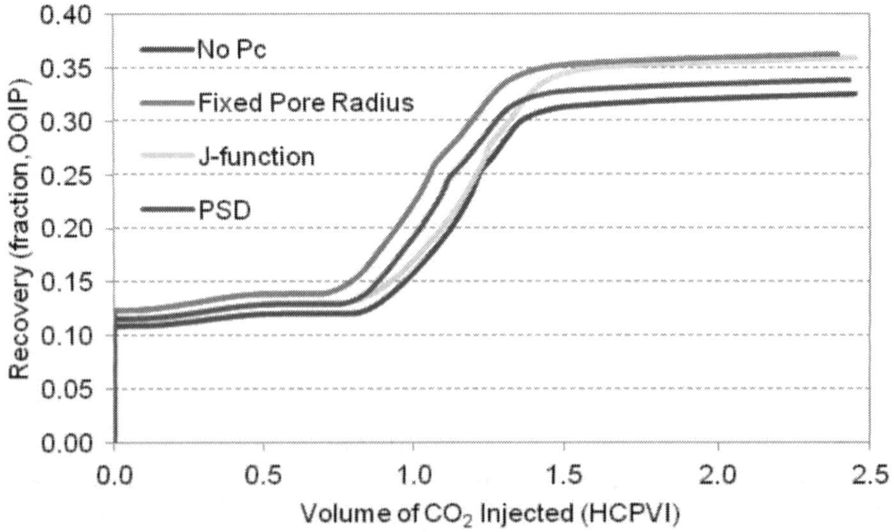

Figure 6. Oil Recoveries vs Volume of CO2 Injected for Continuous CO2 Injection Cases

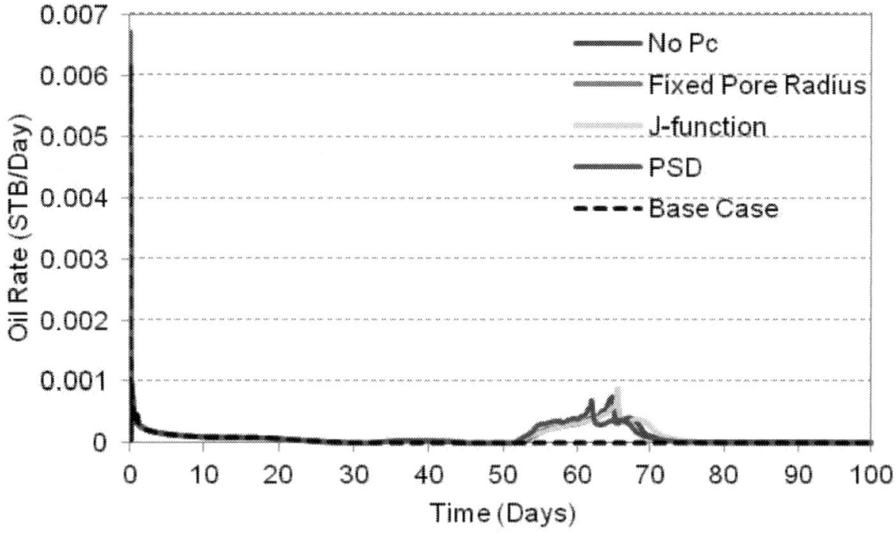

Figure 7. Daily Oil Rates of Four Continuous CO2 Injection Cases

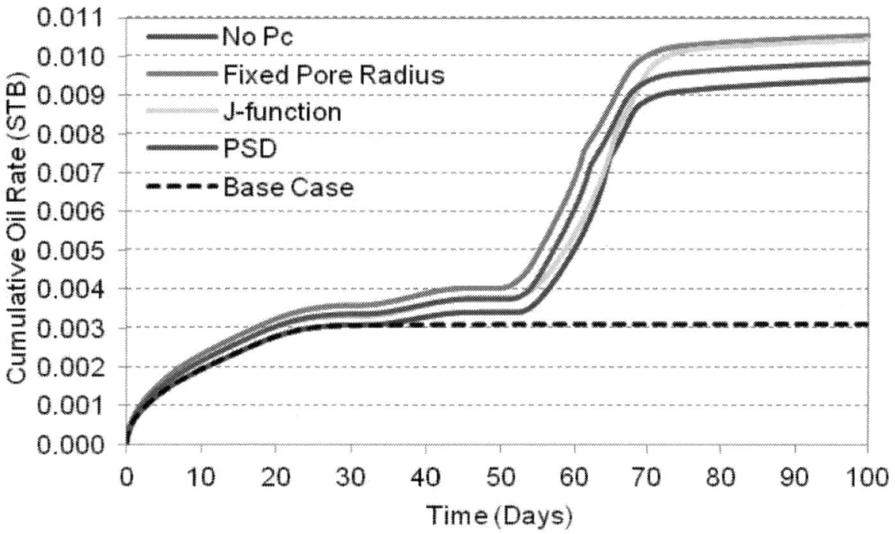

Figure 8. Cumulative Oil Rates of Four Continuous CO2 Injection Cases

CO_2 Storage

Based on the cases of continuous CO_2 injection, two CO_2 storage cases were run through continuing the CO_2 injection case processes ("No Pc" case, and "Fixed Pore Radius" case), but shut in the production well. Both "No Pc" case and "Fixed Pore Radius" case were continued to run from the end of continuous CO_2 injection for another 100 days as the phase of CO_2 storage. The BHP control for injection well during CO_2 storage was 6,500 psi. **Figure 9** shows the average reservoir pressure of two cases during primary recovery, CO_2 flood and CO_2 storage. The reservoir pressure increased steadily during the storage phase to a maximum pressure of 6,500 psi. The volume of the injected and stored CO_2 for "No Pc" case and "Fixed Pore Radius" case are illustrated in **Figures 10** and **11**. As shown in these two figures, the volume of stored CO_2 increased before the breakthrough of CO_2 during CO_2 flood, after the front of CO2 moved to the producing well, the volume of CO_2 contained in the reservoir was produced to the surface, so the stored CO_2 decreased. During CO_2 storage phase, the amount of CO_2 stored increased linearly with the increase of injected CO_2. Also the stored and injected CO_2 profiles of two cases are very similar.

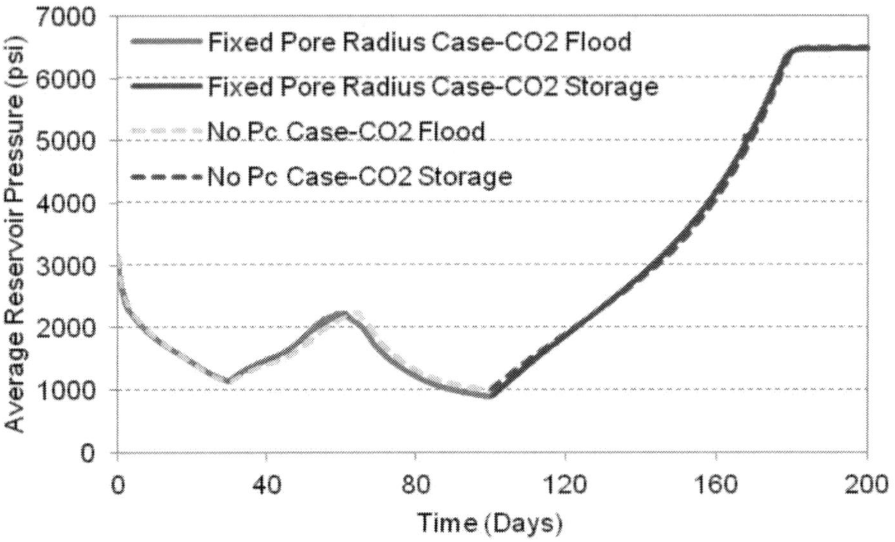

Figure 9. Average Reservoir Pressure for "No Pc" Case

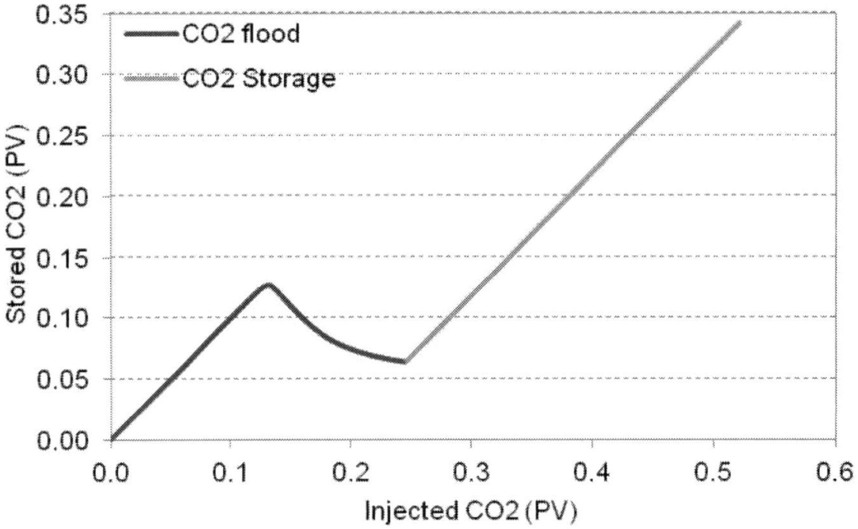

Figure 10. Relation of Stored CO_2 with Injected CO_2 for "No Pc" Case

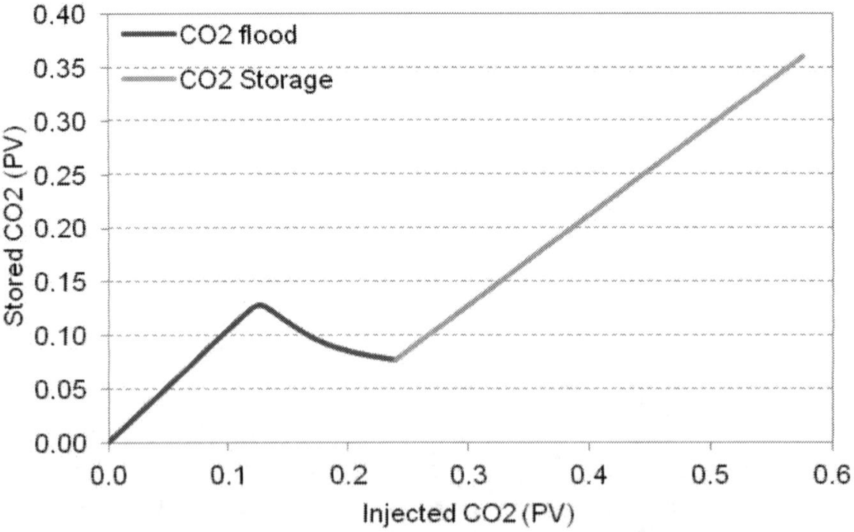

Figure 11. Relation of Stored CO_2 with Injected CO_2 for "Fixed Pore Radius" Case

Conclusions

Sequestration of CO_2 in organic-rich shale formations has significant potential. Shales are widely distributed. The proximity of shales and CO_2 emissions from the coal-firing power plants and other industrial complexes, e.g. in Bakken formation, provides an excellent opportunity for carbon capture and storage to mitigate the impact of CO_2 emissions. In this paper, we have developed methods that allow us to characterize and model two key features of adsorption and capillarity, our simulator allows a realistic representation that differentiates CO_2 storage in unconventional reservoirs vs. conventional ones. The following observations and conclusions can be made.

1).Representation of capillary pressure through including PSD features of shale rocks enables us to characterize the phase and flow mechanisms in the nano-sized pores of shale formations;

2).Our Optimized Simplified Local Density (OSLD) PR EOS Model presents a method to describe adsorption of more complex multilayer, multicomponent system, and is a significant advancement with respect to Langmuir isotherm model;

3). It is found that should capillarity be consistently formulated with PSD, difference in retaining of CO_2 is captured;

Due to the effect of molecular sieving and higher affinity to CO_2 by shales, the application of flue gas (mostly N_2 and CO_2) for CCUS may show a great economic advantage for large-scale CCUS projects. It is of particular interests for studying that application, and the cost for CCUS could be significantly reduced because of much less infrastructure requirement for CO_2 capture thanks to selectivity of shale formations for CO_2/N2 mixtures.

References

Arakcheev, V.G., Valeev, A.A. Morozov, V.B. Olenin, A.N., CARS Diagnostics of Molecular Media under Nanoporous Confinement, Laser Physics 18 (2008) 1451–1458.

Brunauer, S., Emmett, P.H., and Teller, E., 1938, Adsorption of Gases in Multimolecular Layers: Journal of the American Chemical Society, v. 60, p. 309-319.

Chen, J. H., Wong, D. S. H., Tan, C. S., Subramanian, R., Lira, C. T., & Orth, M. 1997, Adsorption and Desorption of Carbon Dioxide onto and from Activated Carbon at High Pressures. Industrial & Engineering Chemistry Research, 36(7), 2808-2815.

Do, D., Duong, 1998, "Adsorption Analysis: Equilibria and Kinetics".

Elgmati, M., Zhang, H., Bai. B., Flori, R., Qu, Q., 2011, Submicron-Pore Characterization of Shale Gas Plays, Paper SPE 144050 presented at North American Unconventional Gas Conference and Exhibition, 14-16 June.

Eshkalak, M.O., Aybar,U., Carbon Dioxide Storage and Sequestration in Unconventional Shale Reservoirs, Journal of Geoscience and Environment Protection, 2015, 3, 7-15.

Fitzgerald, J.E., Sudibandriyo, M., Pan, Z., Robinson R.L. Jr., Gasem, K.A.M. Modeling the Adsorption of Pure Gases on Coals with The SLD Model, Carbon 41 2203-2216 (2003).

Hiejima,Y., Yao, M., Phase Behaviour of Water Confined in Vycor Glass at High Temperatures and Pressures, Journal of Physics: Condensed Matter 16 (2004) 7903–7908.

House, K. Z.; Schrag, D. P.; Harvey, C. F.; Lackner, K. S., 2006, Permanent Carbon Dioxide Storage in Deep-Sea Sediments. Proceedings of the National Academy of Sciences of the United States of America, 103 (33): 12291–12295.

Jin, L., Pu, H., Wang, Y., Li, Y., 2015, The Consideration of Pore Size Distribution in Organic-rich Unconventional Formations May Increase Oil Production and Reserve by 25%, Eagle Ford Case Study.

Paper SPE 178507/URTeC-2148314, Proceedings of the Unconventional Resources Technology Conference, San Antonio, TX, Jul. 20-22.

Kang, S.M., Fathi, E., Ambrose, R.J., Akkutlu, I.Y., and Sigal, R.F., 2010, CO_2 Storage Capacity of Organic-Rich Shales, paper SPE 134583 presented at SPE Annual Technical Conference and Exhibition.

Kang, S.M., *et al.* 2011, Carbon Dioxide Storage Capacity of Organic-Rich Shales. SPE Journal

Kazi, T., Li, Y., Jin, L., Wu, X., 2015, How Typical Langmuir Isotherm Curves Underestimate OGIP and Production by 20%, a Barnett Case Study. AAPG Datapages/Search and Discovery Article #90214, Proceedings of Southwest Section AAPG Annual Convention, Wichita Falls, TX, April 11-14.

Kohlmann, J.; Zevenhoven, R., 2000, The Removal of CO_2 from Flue Gases using Magnesium Silicates in Finland. In Proceedings of the 11th International Conference on Coal Science, San Francisco, CA, USA, 29 September–5 October.

Kovscek, A.R., Tang, G.-Q., and Jessen, K., 2005, Laboratory and Simulation Investigation of Enhanced Coalbed Methane Recovery by Gas Injection: Paper SPE 95947 presented at SPE Annual Technical Conference and Exhibition.

Kuila, U. and Prasad, M., 2011, Surface Area and Pore-Size Distribution in Clays and Shales. Paper SPE 146869 presented at the SPE Annual Technical Conference and Exhibition, Denver, Colorado, 30 October–2 November

Langmuir, I., 1916, The Constitution and Fundamental Properties of Solids and Liquids. Part I. Solids: Journal of the American Chemical Society, v. 38, p. 2221-2295.

Lee, L. L. Molecular Thermodynamics of non-Ideal Fluids; Butterworth: Stoneham, MA, 1988.

Leverett, M.C., 1941, Capillary Behavior in Porous Solids. Transactions of the AIME

Li, X., Elsworth, D., 2014, Geomechanics of CO_2 Enhanced Shale Gas Recovery. Journal of Natural Gas Science and Engineering, 2014 1-13.

Li, Z.W., Dong, M.Z., Li, S.L. and Huang, S., 2006, CO_2 Sequestration in Depleted Oil and Gas Reservoirs—Caprock Characterization and Storage Capacity. Energy Conversion and Management, 47, 1372-1382

Ma, Y. and Jamili, A., 2014, Using Simplified Local Density/ Peng-Robinson Equation of State to Study the Effects of Confinement in Shale Formations on Phase Behavior, Paper SPE 168986 presented at SPE Unconventional Resources Conference,The Woodlands, Texas, 1-3 April.

National Energy Technology Laboratory, 2010, Carbon Sequestration Atlas of the United States and Canada. 3rd Edition, the U.S. Department of Energy, Washington DC, p162

Nojabaei, B, Johns, R.T. and Chu, L., 2013, Effect of Capillary Pressure on Phase Behavior in Tight Rocks and Shales, SPE Reservoir Evaluation & Engineering Journal, August

Nuttall, B.C., Eble, C.F., Drahovzal, J.A., and Bustin, R.M., 2005, Analysis of Devonian Black Shales in Kentucky for Potential Carbon Dioxide Sequestration and Enhanced Natural Gas Production.

Rangarajan, B., Lira, C. T. and Subramanian R., 1995, Simplified Local Density Model for Adsorption over Large Pressure Ranges. AIChE Journal. 41.4: 838-845.

Sapmanee, K., 2011, Effects of Pore Proximity on Behavior and Production Prediction of Gas/ Condensate. Master Thesis, University of Oklahoma, 2011.

Shapiro, A.A., and Stenby, E.H. 1997, Kelvin Equation for a Non-ideal Multicomponent Mixture. Fluid Phase Equilibria 134(1-2): 87-101.

Shapiro, A.A., Potsch, K., Kristensen, J.G., Stenby, E.H., 2000, Effect of Low Permeable Porous Media on Behavior of Gas Condensates. Paper SPE 65182 presented at SPE European Petroleum Conference, Paris, France, 24-25 October.

Sigmund, P. M., Dranchuk, P. M., Morrow, N. R., *et al.* 1973. Retrograde Condensation in Porous Media. SPE Journal 13 (2): 93–104.

Stefan, B., 2008, CO_2 Storage in Geological Media: Role, Means, Status and Barriers to Deployment.

Subramanian R., Pyada H., Lira C.T., An Engineering Model for Adsorption of Gases onto Flat Surfaces and Clustering in Supercritical Fluids, Ind. Eng. Chem. Res., 34 (11) 3830-37 (1995).

Thommes, M., Findenegg, G.H., Schoen, M., Critical Depletion of a Pure Fluid in Controlled-Pore Glass: Experimental Results and Grand Canonical Ensemble Monte Carlo Simulation, Langmuir 11 (1995) 2137–2142.

U.S. Department of Energy, 1999, Carbon Sequestration Research and Development: Office of Fossil Energy, Office of Science, Washington, D.C., December, p195

U.S. Department of Energy, 2004, Carbon Sequestration Technology Roadmap and Program Plan: National Energy Technology Laboratory, P24.

U.S. Department of Energy, 2005, Carbon Sequestration, Technology Roadmap and Program Plan 2005: National Energy Technology Laboratory, P26.

Vermylen, J.P., 2011, Gemechanical Studies of the Barnett, Texas, USA. A dissertation submitted to the Department of Geophysics and the Committee on Graduate Studies of Stanford University

Westrich, H., *et al.*, 2002, Sequestration of CO_2 in a Depleted Oil Reservoir: an Overview: Journal of Energy and Environmental Research, v. 2, no. 1. p64-74

Wang, Y., Yan, B., Killough, J., 2013, Compositional Modeling of Tight Oil Using Dynamic Nanopore Properties, Paper SPE 166267 presented at Annual Technical Conference and Exhibition, New Orleans, Louisiana, 30 September–2 October.

Wilcox, J., 2012, Pore Characterization and Modeling of Gas Adsorption and Transport in Shales, 29 October

Wong, A.P.Y, Kim, S.B., Goldburg, W.I. Chan, M.H.W. Phase-Separation, Density Fluctuation, and Critical-Dynamics of N_2 in Aerogel, Physical Review Letters 70 (1993) 954–957.

Appendix A

The calculation of $a(z)$is described in the paper of Chen *et al.* (1997). For $L/\sigma_{ff} \geq 3$, $a(z)$ is given by

$$\frac{a(z)}{a_b} = \frac{3}{8}\left[\frac{z}{\sigma_{ff}} + \frac{5}{6} - \frac{1}{3\left(\frac{L-z}{\sigma_{ff}}-0.5\right)^3}\right] \quad \text{for} \quad 0.5 \leq \frac{z}{\sigma_{ff}} \leq 1.5$$

$$\frac{a(z)}{a_b} = \frac{3}{8}\left[\frac{8}{3} - \frac{1}{3\left(\frac{z}{\sigma_{ff}}-0.5\right)^3} - \frac{1}{3\left(\frac{L-z}{\sigma_{ff}}-0.5\right)^3}\right] \quad \text{for} \quad 1.5 \leq \frac{z}{\sigma_{ff}} \leq \frac{L}{\sigma_{ff}} - 1.5$$

$$\frac{a(z)}{a_b} = \frac{3}{8}\left[\frac{L-z}{\sigma_{ff}} + \frac{5}{6} - \frac{1}{3\left(\frac{z}{\sigma_{ff}}-0.5\right)^3}\right] \quad \text{for} \quad \frac{L}{\sigma_{ff}} - 1.5 \leq \frac{z}{\sigma_{ff}} \leq \frac{L}{\sigma_{ff}} - 0.5$$

When $2 < L/\sigma_{ff} < 3$, $a(z)$ is given by

$$\frac{a(z)}{a_b} = \frac{3}{8}\left[\frac{z}{\sigma_{ff}} + \frac{5}{6} - \frac{1}{3\left(\frac{L-z}{\sigma_{ff}}-0.5\right)^3}\right] \quad \text{for} \quad 0.5 \leq \frac{z}{\sigma_{ff}} \leq \frac{L}{\sigma_{ff}} - 1.5$$

$$\frac{a(z)}{a_b} = \frac{3}{8}\left(\frac{L}{\sigma_{ff}} - 1\right) \quad \text{for} \quad \frac{L}{\sigma_{ff}} - 1.5 \leq \frac{z}{\sigma_{ff}} \leq 1.5$$

$$\frac{a(z)}{a_b} = \frac{3}{8}\left[\frac{L-z}{\sigma_{ff}} + \frac{5}{6} - \frac{1}{3\left(\frac{z}{\sigma_{ff}}-0.5\right)^3}\right] \quad \text{for} \quad 1.5 \leq \frac{z}{\sigma_{ff}} \leq \frac{L}{\sigma_{ff}} - 0.5$$

When $1.5 \leq L/\sigma_{ff} \leq 2$, $a(z)$ is given by

$$\frac{a(z)}{a_b} = \frac{3}{8}\left(\frac{L}{\sigma_{ff}} - 1\right)$$

CMTC-439562-MS

Mechanical Integrity of Well Cement under Geologic Carbon Sequestration Conditions

Omotayo Omosebi, University of Oklahoma
Ramadan Ahmed, University of Oklahoma
Subhash Shah, University of Oklahoma
Samuel Osisanya, The Petroleum Institute, Abu Dhabi, UAE

This paper was prepared for presentation at the Carbon Management Technology Conference held in Sugarland, Texas, USA, 17–19 November 2015.

Abstract

Carbon dioxide (CO_2) is injected into subsurface rocks either to improve hydrocarbon recovery or for permanent storage in geologic formations. At storage sites, CO_2 injection wells are drilled and completed with multiple-string casings, which are cemented to the host rock. Cement is the primary means of protecting these casings from corrosive fluids and isolating storage zones from overlying fresh water aquifers by effectively sealing the casing-rock annulus. Over the life of the well, dissolved CO_2 interacts with the casing-cement-rock system thereby degrading the hardened cement and resulting in loss of zonal isolation and structural integrity required to support the casing and to ensure long-term integrity of the well. Cement performance is often assessed through laboratory measurement of compressive strength, porosity, and permeability. These parameters are indicators of mechanical integrity and seal capacity of cement. Although compressive strength in the range of 0.7 to 5 MPa is generally sufficient to continue drilling after the casing is cemented, further hydration and chemical degradation occur throughout the operational life of the well. In CO_2 injection wells, environmental conditions (i.e. CO_2 concentration, pressure, and temperature) around the cement result in further alteration in physical and mechanical properties. This hostile environment causes severe mechanical damage and ultimate failure of cement sheath, potentially leading to micro-channelling and formation of micro-annuli.

In this study, experiments were conducted to investigate the effect of CO_2 concentration on mechanical integrity of well cements (Class G and H) after exposure to carbonic acid environment. Cement cores were prepared and aged for 14 days in autoclave filled with 2% NaCl solution saturated with methane gas containing carbon dioxide. Aging tests were performed at 38°C and 42 MPa, varying CO_2 concentration. To assess the level of degradation and describe the process of acid attack, compressive strengths of unexposed and exposed samples were measured and compared. In addition, other properties of cement cores including porosity, permeability, stiffness (i.e. elasticity) and Fourier Transform Infrared Spectroscopy (FTIR) measurements were obtained and used in the assessment.

After 14 days of exposure, increase in CO_2 concentration did not result in major change in compressive strength due to combined effect of carbonation of cement hydrates (calcium silicate hydrates and calcium hydroxide) and subsequent leaching, with each process offsetting the other. Porosity, permeability and FTIR measurements were consistent with this observation. Under the experimental conditions adopted in this study, durability of Classes G and H cement were comparable.

However, Class H cement became more elastic (i.e. less stiff) than Class G cement as CO_2 concentration increased. This could be attributed to the change in material composition and mineralogy. FTIR mineralogy provides evidence that the mechanical behavior of well cement after exposure to CO_2-saturated brine is a direct consequence of interrelated chemical processes. Under carbon sequestration conditions, these processes counterbalance one another to ensure long-term integrity of the well for CO_2 storage and containment.

Introduction

Carbon dioxide (CO_2) is injected into subsurface rocks to improve hydrocarbon recovery or store the gas. Permanent storage of CO_2 is achieved through Carbon Capture and Storage (CCS) or Carbon sequestration, which is an emerging technology that helps to reduce the effect of industrial carbon emission on global climate change. It involves the capture and perpetual storage of CO_2 in subsurface depleted petroleum reservoirs or deep saline aquifers. To achieve this purpose, CO_2 injection wells are drilled up to thousands of feet of depth where downhole conditions exist. As of November 2014, there are 274 CCS projects in thirty countries across the six continents of the world. This is according to the National Energy Technology Laboratory's (NETL's) Carbon Capture and Storage Database. These projects are in either the planning and development stage or actively capturing and injecting CO_2. At storage sites, CO_2 injection wells are drilled and completed with multiple strings casings, which are cemented to the host rock. Cement is the primary means of protecting these casings from corrosive fluids and isolating storage zones from overlying fresh water aquifers. Over the life of the well, the injected CO_2 and formation fluids (i.e., formation brine, hydrocarbons etc.) interact with the casing-cement-rock system thereby degrading the hardened cement, leading to loss of zonal isolation and structural integrity required to support the casing. After cement degrades, fluid migration occurs through potential leakage paths (**Fig. 1**) thereby contaminating drinkable water and ultimately compromising safety of lives and the environment.

Fig. 1: CO2 leakage pathways along a well: (a) casing-cement interface; (b) deformed cement; (c) cement-rock interface; (d) corroded/deformed casing; (e) cement pores or mico-fractures; (f) cement plug-casing interface; (g) cement-perforation interface

In the industry today, baseline API Class G and H cements are most commonly used. Standard practice requires lesser water-to-cement ratio for Class H (i.e., 38% leading to 1965 kg/m^3 density) than Class G (i.e., 44% leading to 1893 kg/m^3 density) due mainly to their differences in particle size distribution. Studies (Bruckdorfer, 1986; Akkaya et al., 2003; Voigt et al., 2003; Rahmani et al., 2012; Alawode and Idowu, 2011) have shown that compressive strength decreases with increase in water content. Therefore, Class H is typically stronger than Class G when the standard formulations are compared. The primary components in these cements are tricalcium silicate, (Ca_3SiO_5 or C_3S), dicalcium silicate (Ca_2SiO_4 or C_2S), tricalcium aluminate ($Ca_3Al_2O_6$ or C_3A) and tetracalcium aluminoferrite ($Ca_4Al_2Fe_2O_{10}$ or C_4AF). Hydration of C_3S and C_3A is very fast while C_2S and C_4AF hydrate slowly. C_3S and C_2S control short and long-term strength development, respectively while C_3A and C_4AF control rheology of the slurry. In field operations, most of the C_3S and C_2S components are hydrated during "waiting on cement (WOC)" period, through a curing process, thereby producing calcium silicate hydrate ($3CaO.2SiO_2.3H_2O$ or CSH) and calcium hydroxide or portlandite ($Ca(OH)_2$, both of which aid early strength development. At elevated temperature, CSH transforms to crystalline forms which leads to strength retrogression. To prevent this from happening, standard field practice in slurry design involves adding 35 to 40% (BWOC) silica to the base slurry (Bensted and Barnes, 2002; Bourgoyne Jr., 1986; Bruckdorfer, 1986).

In wells where CO_2 is either injected (e.g. CO_2-EOR and CCS, etc.) or produced as associated gas (e.g. deep sour gas condensate and volatile oil fields), environmental conditions around the cement result in further alteration in mechanical properties, especially if it is aggressive. This hostile environment results in severe mechanical damage and ultimate failure of cement sheath, potentially leading to micro-channelling and formation of micro-annuli. Review of previous studies (Bruckdorfer, 1986; Brandl et al., 2011; Eilers et al., 1980; Lesti et al., 2013; Neuville et al., 2012; Santra et al., 2009; Barlèt-Gouédard et al., 2006, 2009; Laudet et al., 2011; Condor and Asghari, 2009; Carey et al., 2007; Scherer et al., 2011; Takase et al., 2010) indicates that carbonation and leaching are principal mechanisms responsible for physical and mechanical degradation of well cement after exposure to carbonic acid under geologic sequestration conditions. Minerals that are typically formed during the degradation process include: calcium carbonate (calcite, aragonite, vaterite, etc.) and calcium silicate carbonate group which comprises scawtite ($Ca_7(Si_6O_{18})(CO_3).2H_2O$), fukalite ($Ca_4Si_2O_6(OH)_2CO_3$), spurrite ($Ca_5(SiO_4)_2CO_3$), tilleyite ($Ca_5(Si_2O_7)(CO_3)_2$) and galuskinite ($Ca_7(SiO_4)_3CO_3$) (Zhang et al., 2013). At ambient temperature and pressure, calcite is the most thermodynamically stable form of calcium carbonate; hence, it is the most expected under the operational conditions frequently encountered in CCS. Generally, portlandite is attacked at faster rate and is the first to be completely leached out before CSH because it is more reactive. Field results of cement and rock samples retrieved from CO_2 injection wells confirm this phenomenon (Carey et al., 2007; Mito et al., 2008; Scherer et al., 2011).

According to the MIT CCS project database (MIT 2015), CO_2 concentration from producing fields can be between 5-65% (e.g. Greater Gorgon Fields, Snohvit field, Salah Oil field, Sleipner West field, LaBarge field). In active CCS projects, it could be as high as 98% (e.g. Port Arthur Project, Wasatch Plateau). Qualitative and mineralogy studies on cement degradation have been conducted previously, considering the effect of CO_2 (Duguid et al., 2005, 2008, 2009; Kutchko et al., 2007, 2008; Matteo and Scherer, 2012). However, data on mechanical integrity of well cement, after exposure to CO_2-containing fluid is limited. Besides, the impact of variation in CO_2 concentration is not well understood. In this study, Class G and H Portland-based cements are evaluated to assess the level of mechanical degradation after exposure to CO_2-saturated brine at 38°C (100°F) and 42 MPa (6000 psi).

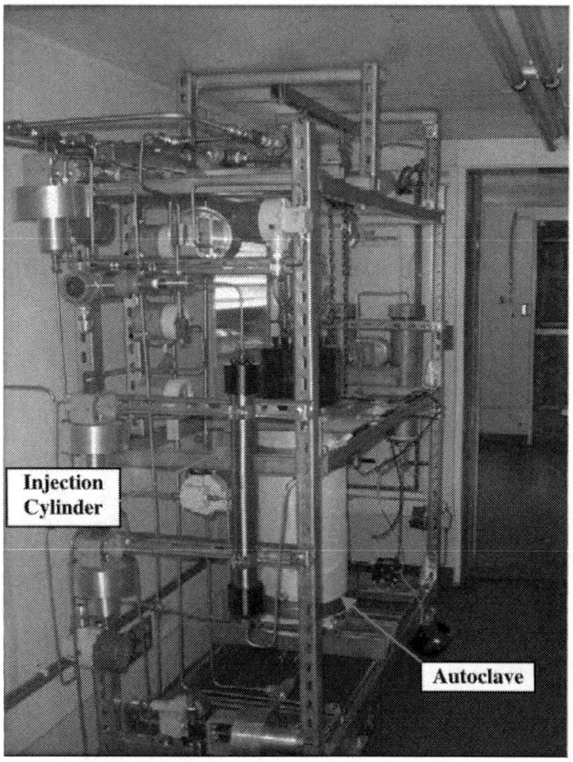

Fig. 2: Photo of sample aging setup

Materials and Methods

Materials

High sulfate resistant Class G and H cements were used in the experiment. Cement slurries were prepared according to API RP 10B (1997) procedure. To prevent strength deterioration at elevated temperature, 35% silica flour was added. After mixing, slurry properties (thickening time, density, fluid loss, and rheology) were measured. Slurry composition and properties are presented elsewhere (Omosebi et al. 2015).

Methods

To prepare samples, cement slurry was poured into a rubber mould and cured in pressure vessel placed in an oven for 5 days in 2% (by weight) sodium chloride solution at 93°C. After curing, cylindrical cores (38.10 mm in length and 25.4 mm in diameter) were prepared and aged in a HPHT autoclave (**Fig. 2**) for 14 days. Unaged samples (i.e. control samples) were kept in 2% NaCl solution for 14 days under atmospheric condition. Methane and carbon dioxide were injected into the autoclave using injection cylinder. Detailed descriptions of the experimental setup and test procedure are presented elsewhere

(Omosebi et al. 2015). CO_2 content of the gas was varied. **Table 1** shows the test matrix for the aging experiments.

After aging, destructive and non-destructive tests were conducted on samples. Using laboratory testing apparatus (Omosebi et al. 2015), compressive strength and axial strain were measured according to ASTM C39/C39M (2012) standard procedure. The tests were repeated to ensure repeatability. Furthermore, porosity, permeability and FTIR measurements were obtained to analyze and interpret results obtained from mechanical tests. Porosity and gas permeability of unaged and aged cores were measured using Coretest AP-608 Permeameter (Maheshwari, 2014). Before measurement, the cores were dried in an oven at 43°C for 24 hours until weight loss became negligible with time. Dried cement cores were cut transversely and FTIR samples were collected from the inner, middle and outmost zones. Using FTIR equipment (Nicolet 6700), mineralogical analysis was conducted on the samples to measure the degree of carbonation and leaching in each of these degraded regions.

Table 1: Test matrix during CO_2 aging experiment

Temperature (°C)	Pressure (MPa)	CO_2 (%)	CH_4 (%)
38	42	10	90
38	42	40	60
38	42	100	0

Results and Discussion

Results of the effect of CO_2 concentration and pressure on compressive strength are presented in this section. Compressive strength and stiffness data are analyzed and supported with porosity, permeability and FTIR measurements.

Effect of CO_2 Concentration on Mechanical Integrity

Figure 3 shows the compressive strength, stiffness, porosity and permeability of the degraded samples after immersion in 2% NaCl solution and exposure to varied CO_2 gas composition at 38°C and 42 MPa for 14 days. In both classes of cement, strength was preserved as CO_2 concentration increased (Fig. 3a). Similarly, porosity slightly decreased (Fig. 3b) while permeability slightly increased (Fig. 3c). In addition to the rapid dissolution of portlandite, this can be attributed to a sequence of reactions that leads to chemical degradation through carbonation and leaching.

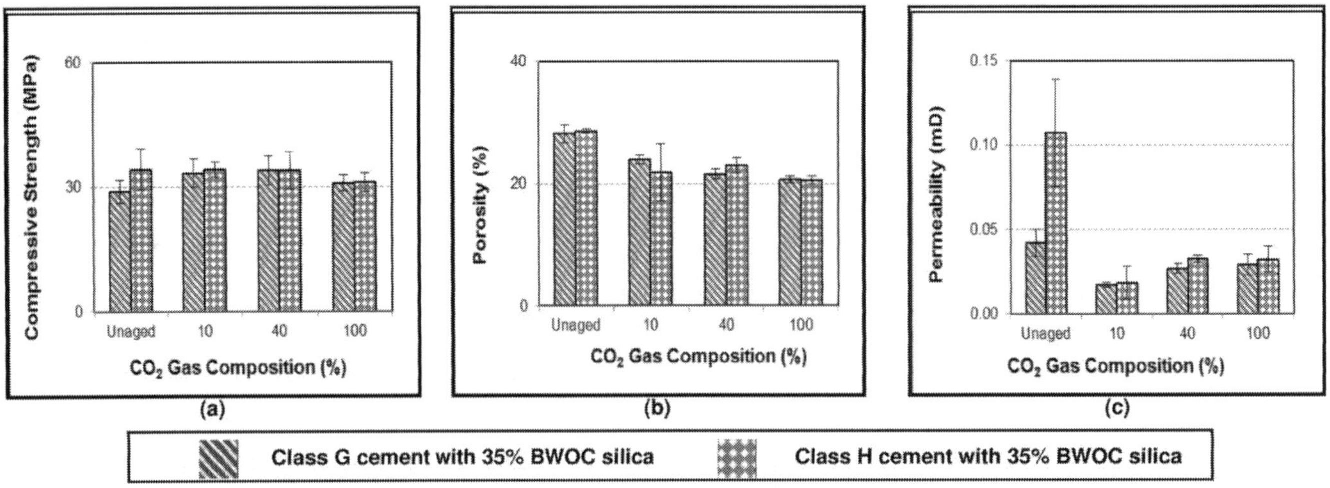

Fig. 3: Effect of CO_2 concentration on: (a) Compressive strength; (b) Porosity; (c) Permeability

Solubility of portlandite is high at low temperature (Nakarai et al., 2006). High rate of dissolution contributes to mechanical degradation. However, this mode of degradation is compensated by carbonation of cement hydrates (CH and CSH) which is known to reduce porosity and reinforce cement matrix due to calcite precipitation in the pores. Immediately carbonation occurs, leaching follows negating the effect of carbonation. It reduces strength but increases porosity and permeability, thereby allowing further CO_2 ingress. These mechanisms of cement degradation occur simultaneously and the dominant process dictates overall effect on cement integrity. The trends in both Class of cements suggest an even effect since aggregate strength is approximately constant. However, the slight reduction in porosity (Fig. 3b) underscores the impact of carbonation. Ideally, calcium carbonate ($CaCO_3$) is produced during carbonation of cement hydrates. It fills the pores of cement, thereby reducing porosity and compensating for strength loss due to increased solubility of portlandite in the penetrated portion of cement. The rate of carbonation is controlled by several parameters such as temperature, pressure, brine, and CO_2 concentration. Increase in CO_2 concentration leads to further formation of $CaCO_3$, which reduces porosity further (Fig. 3b). The decrease in porosity suggests that carbonation maintains a slight edge over leaching as CO_2 concentration increases. The slight increase in permeability could be due to the development of micro-cracks, which were observed during visual inspection of the degraded sample. Furthermore, the progress of reaction fronts is expected to be slow at low temperature, thereby minimizing the loss of mechanical strength.

At 38°C, strength retrogression, due to structural transformation of calcium silicate hydrate (CSH), is not expected since sufficient amount (i.e. 35% BWOC) of silica flour was added to the baseline cement. This mechanical deterioration normally occurs at elevated temperature (above 90°C, approximately). At best, the added silica acts as a pozzolanic material, which aids further strength development without risk of compromising integrity. In addition, the presence of calcium silicate carbonates (i.e. scawtite, fukalite, spurrite, tilleyite and galuskinite) is not anticipated, since they are formed predominantly under high temperature when structural polymorphs of CSH (e.g. xonotlite, pectolite, truscottite, etc.) are incompletely carbonated (Eilers et al., 1980).

Effect of CO_2 Concentration on Cement Elasticity

Elasticity of cement provides an understanding of its behavior when subjected to loading. Mechanical loading arise from pressure and/or thermal related loadings during well operation. Thermal related loading conditions develop during drilling, completion, and intervention operations. Mechanical loads are often caused by high external pressure, cyclic loading resulting from intermittent production (ballooning/reverse ballooning), pressure decline due to gas production, and tensile/compression strain induced by reservoir compaction during gas production (Teodoriu et al. 2010a, 2010b; Yuan et al. 2013; Shadravan et al. 2015). Extreme loading during production or after abandonment can cause potential CO_2 leakage pathways within the storage zone. When chemical attack is involved, these loading conditions change cement properties significantly with time.

Figure 4 presents a plot of differential stress (σ_1-σ_3) versus axial strain for aged Class H sample. The linear portion of the curve depicts the elastic regime while non-linear portion is the post-peak behavior. Like rocks, when the sample is initially loaded it compresses but commencement of dilation (i.e. increase in volume due to initiation of internal cracks) is expected before the peak stress is reached and this may continue in the post-peak region. Post-peak non-linear behavior suggests a residual strength of approximately 12 MPa after failure, which is still acceptable for casing support and well integrity. However, the conductive path created after failure compromises sealing integrity. The linear potion of the stress-strain curve has a slope of 1395 MPa, which corresponds to the elastic modulus (sometimes called stiffness or Young's modulus). This slope is estimated from the ratio of change in axial stress at failure (i.e. compressive strength) and corresponding change in axial strain. Stiff cement is less elastic and vice versa; hence, a lower value of this parameter would be preferred. Modulus of elasticity of oil well cement depends on cement formulation, curing and aging conditions. Previous studies (Baumgarte et al., 1999; Stiles, 2006; Takase et al., 2010; Prisca and Mahmood, 2012) show variation from 1,000 to 20,000 MPa.

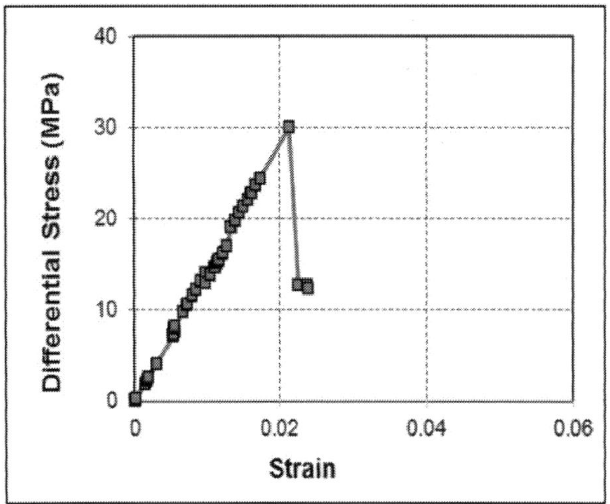

Fig. 4: Stress-strain response of Class H cement aged in brine saturated with 100% CO_2

Averaged stiffness of samples aged at different CO_2 concentration is presented in **Fig. 5.** Class H cement stiffness decreased with concentration. However, for Class G cement, stiffness increased up to 40% CO_2 concentration, then decreased. This is also attributed to disparity in material properties of the multi-component minerals (CH, CSH, calcium carbonate, etc.) that were formed because of hydration and acid attack. As CO_2 concentration increases, Class H cement demonstrates better elasticity than Class G cement.

FTIR Spectra Analysis

FTIR-based mineralogy analysis was conducted on Class H samples to assess the degree of degradation and alteration due to carbonation and/or leaching after the samples were aged. **Figure 6** shows a photo of actual degraded cement core and schematic indicating three locations on the core selected for FTIR analysis: virgin, middle, and outer zones. Each peak of absorbance at a certain wavenumber in the FTIR spectra denotes a specific mineral. Calcium carbonate, calcium hydroxide, and calcium silicate hydrate exhibit peaks at 1480, 3645 and 950 cm^{-1}, respectively (Matsuda and Tsukada, 1995; Zaki et al., 2006; Baciu and Simitzis, 2007).

Figure 7 compares FTIR spectra of unaged and aged samples. At wavenumber 3645 cm^{-1}, only the unaged sample shows a peak that indicates the presence of calcium hydroxide. All three regions in the aged samples show the absence of calcium hydroxide, which suggests that calcium hydroxide has been consumed during carbonation and/or leaching. At 950 cm^{-1}, the unaged sample and virgin zone of the aged sample show this peak indicating the preservation of calcium silicate hydrate. All the calcium hydroxide and calcium silicate hydrate were carbonated in the middle and outer regions; hence, the absorbance peak of calcium carbonate is maximum in these zones but minimum in the unaged sample and virgin zone of the aged sample. The qualitative descriptions of minerals in these zones are similar in characteristics to the zones identified previously (Kutchko et al. 2007, 2008; Santra et al., 2009). Since CO_2-saturated brine penetrates the aged sample radially inward from the surface, this mineralogical evidence verifies the occurrence of maximum carbonation in the outer and middle zones of the aged sample.

To investigate the effect of CO_2 concentration on mineralogy at 38^0C and 42 MPa, FTIR absorbance peaks of calcium hydroxide (CH), calcium silicate hydrate (CSH) and carbonates were retrieved, normalized and presented in **Fig. 8**. For the three degraded zones illustrated in Fig. 6, effect of CO_2 concentration was investigated. FTIR data shows that CH and CSH were completely consumed in the outer zone through reaction with carbonated brine. The products of this reaction are calcium carbonate and carbonates. Therefore, the outer zone is the most carbonated as CO_2 concentration increases. In the virgin and middle zones, CH and CSH were preserved due to slow propagation of reaction fronts at low temperature (38°C). In addition, the carbonates

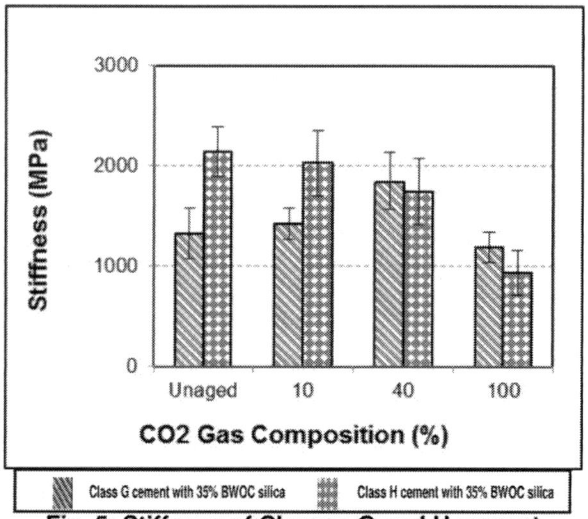

Fig. 5: Stiffness of Classes G and H cement

Fig. 6: Three zones of degraded cement used for FTIR analysis: virgin (red), middle (yellow) and outer (green) zones

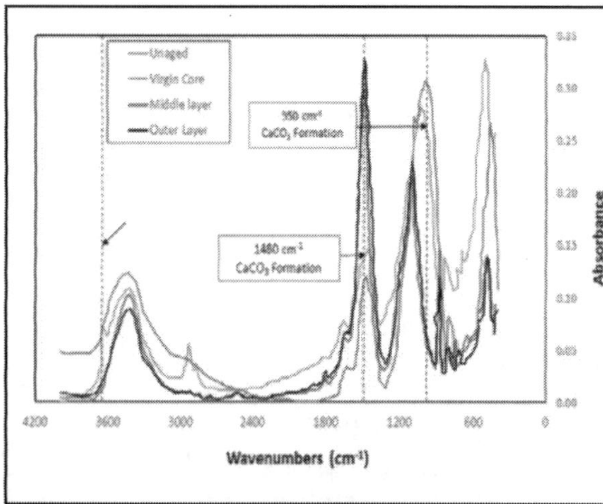

Fig. 7: FTIR spectra of unaged and aged cores

produced in the outer zone form a protective layer which limits further ingress of dissolved CO_2, thereby preserving CH and CSH.

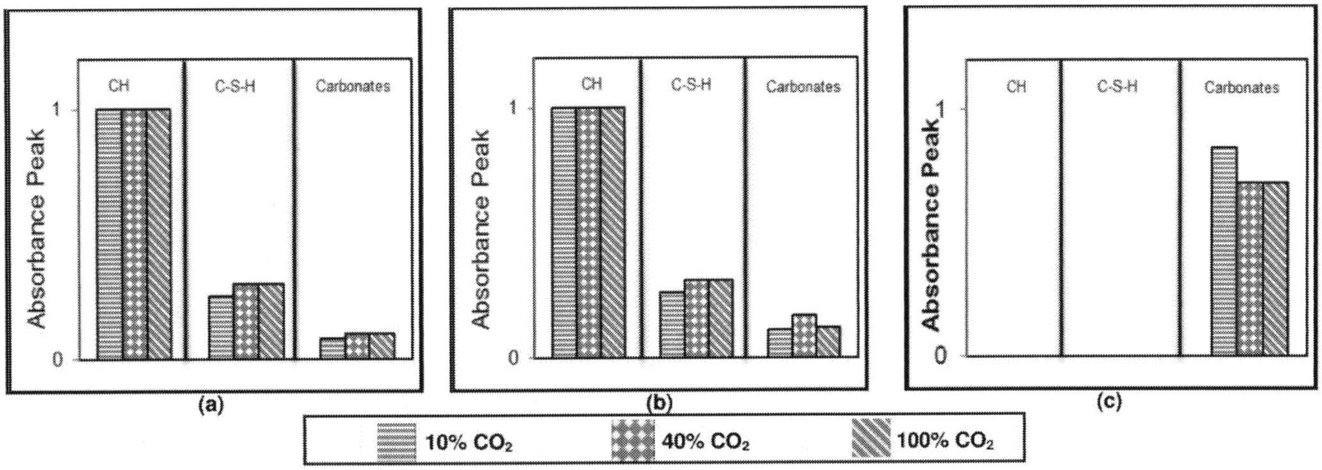

Fig. 8: Normalized absorbance peaks extracted from FTIR spectra of the: (a) Virgin zone; (b) Middle zone; and (c) Outer zone of Class H cement aged at 38°C and 42 MPa

Photos of the degraded samples are shown in **Fig. 9**. In both Class of cements, CO_2 concentration is varied from left to right. Class G cement shows more discoloration than Class H cement through progressive degradation as CO_2 concentration increases. As discussed previously, this is due to carbonation of cement hydrates (CH and CSH). Visual observation shows increasing depth of penetration as the concentration of CO_2 increases, especially in Class G cement. Discoloration in Class H cement is not visually noticeable, which explains the trends in mechanical strength and elasticity. However, the absence of observable discoloration does not indicate absence of chemical degradation. Obviously, the FTIR data (Fig. 7) confirms degradation of Class H cement.

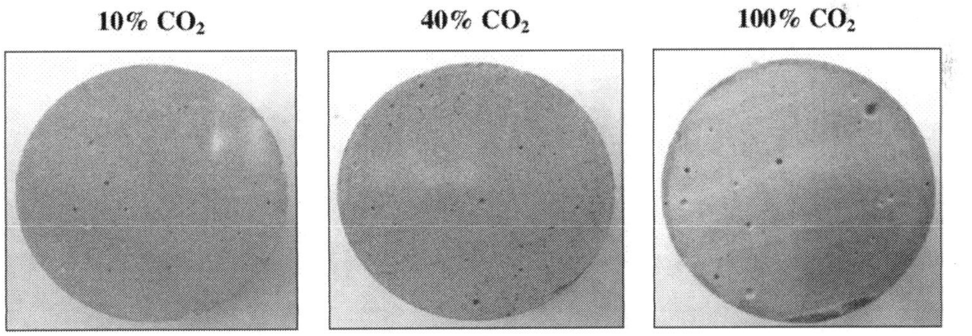

(a)

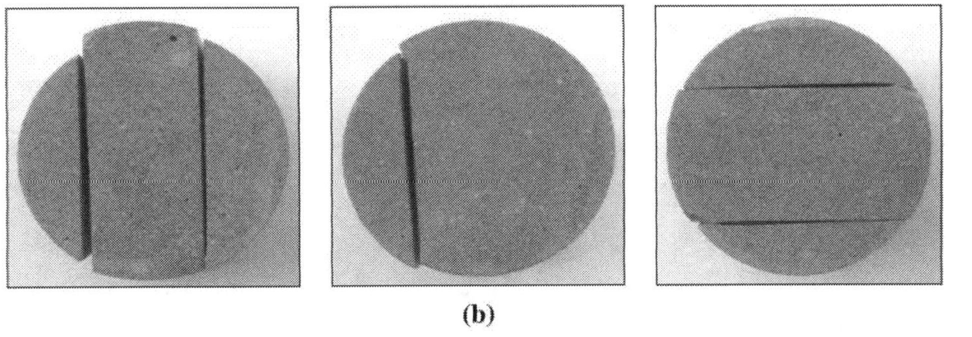

(b)

Fig. 9: Degraded zones due to variation in CO_2 concentration at 38°C and 42 MPa: (a) Class G; (b) Class H

Conclusions

The compressive strength, porosity, permeability, stiffness, and FTIR measurements indicate that mechanical behavior of well cement after exposure to CO_2-saturated brine is a direct consequence of interrelated physical and chemical processes. Some of these processes (such as carbonation) improve cement performance while others (such as leaching) compromise cement integrity. The dominant process dictates overall mechanical behavior. Under the geologic carbon sequestration conditions adopted in this study, Class G and H cements demonstrated acceptable mechanical integrity required for safe well operations after two weeks of exposure. This study is focused on the effect of CO_2 concentration on mechanical integrity of conventional Portland-based well cement under geologic carbon storage conditions. Major outcomes of the study are:

- Increasing CO_2 concentration at 38°C and 42 MPa results in minor change in compressive strength of Class G and H cements, although the samples were chemically degraded. This is due to simultaneous occurrence of carbonation and leaching, each process offsetting the effect of each other. Porosity, permeability, and FTIR measurements are consistent with this observation.
- FTIR mineralogical analyses confirm that maximum degradation occurs in the outer zone of the cylindrical core because it is directly exposed to acid attack.
- Generally, Class H cement demonstrates slight edge over Class G cement in mechanical performance because it becomes more elastic as CO_2 concentration increases.

Acknowledgement

The authors acknowledge Bureau of Safety and Environmental Enforcement (BSEE) for sponsoring this project (E12PC00035). Technical contributions from industry advisory board members are also appreciated. Special thanks to Halliburton Energy Services, Mountain Cement Company, Lone Star Portland Cement, Fritz Industries and Schlumberger Limited for donating materials for this research.

Nomenclature

API	–	*American Petroleum Industry*
ASTM	–	*American Society for Testing and Materials*
BWOC	–	*By Weight of Cement*
EOR	–	*Enhanced Oil Recovery*
MIT		*Massachusetts Institute of Technology*
σ_1-σ_3		*Differential stress*

References

Akkaya, Y., Voigt, T., Subramaniam, K. V. and Shah, S. P., 2003. Non-destructive measurement of concrete strength gain by an ultrasonic wave reflection method. 1-10.

Alawode, O., Idowu, O. I., 2011. Effects of water-cement ratios on the compressive strength and workability of concrete and lateritic concrete mixes. The Pacific Journal of Science and Technology 12 (2), 99-105.

API RP 10B, 1997. Recommended practice for testing well cements. American Petroleum Institute.

ASTM C39/C39M-12, 2012. Standard test method for compressive strength of cylindrical concrete specimens. ASTM International, West Conshohocken, PA.

Baciu, D., Simitzis, J., 2007. Synthesis and characterization of a calcium silicate bioactive glass. Journal of Optoelectronics and Advanced Materials 9, 3320-3324.

Barlèt-Gouédard, V., Rimmelé, G., Goffé, B., Porcherie, O., 2006. Mitigation strategies for the risk of CO_2 migration through wellbores. IADC/SPE 98924 presented at the IADC/SPE Drilling Conference held in Miami, Florida, 21–23 February.

Barlèt-Gouédard, V., Rimmelé, G., Porcherie, O., Quisel, N., Desroches, J., 2009. A solution against well cement degradation under CO_2 geological storage environment. International Journal of Greenhouse Gas Control 3, 206-216.

Baumgarte, C., Thiercelin, M., Klaus, D., 1999. Case studies of expanding cement to prevent microannular formation. SPE 56535, Houston, Texas, 3–6 October.

Bensted, J., Barnes, P., 2002. Structure and performance of cements, Second Edition. Spon Press, London and New York.

Bourgoyne Jr., A. T., Millheim, K. K., Chenevert, M. E., Young Jr., F. S., 1986. Applied drilling engineering, Volume 2. SPE Richardson, Texas.

Brandl, A., Cutler, J., Seholm, A., Sansil, M., Braun, G., 2011. Cementing solutions for corrosive well environments. SPE Drilling & Completions, 208-219.

Bruckdorfer, R. A., 1986. Carbon dioxide corrosion in oil well cements. SPE 15176 presented at the Rocky Mountain Regional Meeting of the Society of Petroleum Engineers, Billings, MT, 19-21 May.

Carey, J. W., Wigand, M., Chipera, S. J., WoldeGabriel, G., Pawar, R., Lichtner, P. C., Wehner, S. C., Raines, M. A., Guthrie Jr., G. D., 2007. Analysis and performance of oil well cement with 30 years of CO_2 exposure from the SACROC Unit, West Texas, USA. International Journal of GreenHouse Gas Control 1 (1), 75-85.

Condor, J., Asghari, K., 2009. Experimental study of stability and integrity of cement in wellbores used for CO_2 storage, Energy Procedia 1, 3633-3640.

Duguid, A., Radonjic, M., Scherer G., 2005. Degradation of well cements exposed to carbonated brine. Fourth Annual Conference on Carbon Capture and Sequestration DOE/NETL, May 2-5.

Duguid, A., 2008. The effect of carbonic acid on well cements as identified through lab and field studies. SPE 119504 presented at the 2008 SPE Eastern Regional/AAPG Eastern Section Joint Meeting, Pittsburgh, Pennsylvania, USA, 11-15 October.

Duguid, A., 2009. An estimate of the time to degrade the cement sheath in a well exposed to carbonated brine. Energy Procedia 1, 3181-3188.

Eilers, L. H., Nelson, E. B., Moran, L. K., 1980. High temperature cement compositions-pectolite, scawtite, truscottite or xonotlite: which do you want? SPE 9286 presented at SPE Annual Technical Conference and Exhibition, Dallas, Texas, 21-24 September.

Kutchko, B. G., Strazisar, B. R., Dzombak, D. A., Lowry, G. V., Thaulow, N., 2007. Degradation of well cement by CO_2 under geologic sequestration conditions. Environmental Science and Technology 41, 4787-4792.

Kutchko, B. G., Strazisar, B. R., Lowry, G. V., Dzombak, D. A., Thaulow, N., 2008. Rate of CO_2 attack on hydrated class h well cement under geologic sequestration conditions. Environmental Science and Technology, 42, 6237-6242.

Laudet, J. B., Garnier, A., Neuville, N., Le Guen, Y., Fourmaintraux, D., Rafai, N., Burlion, N., Shao, J. F., 2011. The behavior of oil well cement at downhole CO_2 storage conditions: static and dynamic laboratory experiments. Energy Procedia 4, 5251-5258.

Lesti, M., Tiemeyer, C., Plank, J., 2013. CO_2 stability of portland cement based well cementing systems for use on carbon capture & storage (CCS) wells. Cement and Concrete Research 45, 45-54.

Maheshwari, H., 2014. Effects of CO_2 on Degradation of Oilwell Cement under HPHT Environment. MS Thesis, University of Oklahoma.

Matsuda, Y., Tsukada, M., 1995. Identification of calcium carbonate contained as body in modern paints by FTIR spectroscopy. IRUG 2 Postprints, 25-34.

Matteo, E. N., Scherer, G. W., 2012. Experimental study of the diffusion-controlled acid degradation of Class H Portland cement. International Journal of Greenhouse Gas Control 7, 181-191.

Mito, S., Xue, Z., Ohsumi, T., 2008. Case study of geochemical reactions at the Nagaoka CO_2 injection site, Japan. International journal of greenhouse gas control 2, 309-318.

MIT 2015. Carbon Capture and Sequestration Project Database, Massachusetts Institute of Technology, https://sequestration.mit.edu, last accessed on 10/31/2015.

Nakarai, K., Ishida, T., Maekawa, K., 2006. Modeling of calcium leaching from cement hydrates coupled with micro-pore formation. Journal of Advanced Concrete 4, 395-407.

Neuville, N., Aouad, G., Lecolier, E., Damidot, D., 2012. Innovative leaching tests of an oilwell cement paste for CO_2 storage: Effect of the pressure at 80°C. Energy Procedia 23, 472-479.

Omosebi, O., Maheshwari, H., Ahmed, R., Shah, S., Osisanya, S., Santra, A., Saasen, A., 2015. Investigating temperature effect on degradation of well cement in HPHT carbonic acid environment. Journal of Natural Gas Science and Engineering 26, 1344-1362.

Prisca, S., Mahmood, A., 2012. Special considerations in cementing high pressure high temperature wells. International Journal of Engineering and Applied Sciences 1, 2305-8269.

Rahmani, K., Shamsai, A., Saghafian, B., Peroti, S., 2012. Effect of water and cement ratio on compressive strength and abrasion of microsilica concrete. Middle-East Journal of Scientific Research 12 (8), 1056-1061.

Santra, A., Reddy, B. R., Liang, F., Fitzgerald, R., 2009. Reaction of CO_2 with portland cement at downhole conditions and the role of pozzolanic supplements. SPE 121103 presented at the SPE International Symposium on Oilfield Chemistry, The Woodlands, Texas, 20-22 April.

Scherer, G. W., Kutchko, B., Thaulow, N., 2011. Characterization of cement from a well at teapot dome oil field: implications for geological sequestration. International Journal of Greenhouse Gas Control 5, 115-124.

Shadravan, A., Schubert, J., Amani, M., Teodoriu, C., 2015. Utilizing fatigue failure envelope for cement sheath integrity evaluation. SPE Journal Drilling and Completions.

Stiles, D., 2006. Effects of long-term exposure to ultrahigh temperature on the mechanical parameters of cement. IADC/SPE 98896 presented at the IADC/SPE Drilling Conference, Miami, Florida, 21 – 23 February.

Takase, K., Barhate, Y., Hashimoto, H., Lunkad, S. F., 2010. Cement-sheath wellbore integrity for CO_2 injection and storage wells. SPE 127422 presented at the SPE Oil and Gas India Conference and Exhibition, Mumbai, India, 20-22 January.

Teodoriu, C., Ugwu, I., Schubert, J., 2010a. Estimation of casing-cement-formation interaction using a new analytical model. SPE131335, SPE EUROPEC/EAGE Annual Conference and Exhibition held in Barcelona, Spain, 14–17 June 2010.

Teodoriu, C., Reinicke, K.M., Fichter, C., Wehling, P., 2010b. Investigations on casing-cement interaction with application to gas and CO_2 storage wells. SPE EUROPEC/EAGE Annual Conference and Exhibition held in Barcelona, Spain, 14–17 June 2010.

Voigt, T., Shah, S. P. 2003. Nondestructive monitoring of setting and hardening of portland cement mortar with sonic methods. Northwestern University, Evanston, USA, 1-8.

Yuan, Z.; Teodoriu, C., Schubert, J., 2013. Low cycle cement fatigue experimental study and the effect on HPHT well integrity. Journal of petroleum science and engineering. Accepted on March 22, 2013.

Zaki, M. I., Knözinger, H., Tesche, B., Mekhemer, G. A. H., 2006. Influence of phosphonation and phosphation on surface acid–base and morphological properties of CaO as investigated by in situ FTIR spectroscopy and electron microscopy. Journal of Colloid and Interface Science 303, 9-17.

Zhang, Y. Q., Radha, A. V., Navrotsky, A., 2013. Thermochemistry of two calcium silicate carbonate minerals: scawtite, $Ca_7(Si_6O_{18})(CO_3).2H_20$, and spurrite, $Ca_5(SiO_4)_2(CO_3)$. Geochimica et Cosmochimica Aeta 115, 92-99.

CMTC-439577-MS

A combined cooling and power system based on compressed CO_2

ZHENG Dewen[1,2], TANG Ligen[1,2,*], WANG Jieming[1,2], GUO Kai[3], BAI Fengjuan[4]

1. PetroChina Research Institute of Petroleum Exploration & Development, Langfang 065007, China
2. CNPC Key Laboratory of Oil and Gas Underground Storage, Langfang 065007, China
3. Petro China Company Limited Natural Gas & Pipeline Company, Beijing 10001, China
4. China Kunlun Constracting & Engeering Corporation, Beijing 100037, China

This paper was prepared for presentation at the Carbon Management Technology Conference held in Sugarland, Texas, USA, 17–19 November 2015.

Abstract

In order to use and store greenhouse gas CO_2, compressed CO_2 energy storage system(CCSE) was designed by using model of Compressed Air Energy Storage(CASE) and with the advantage of high compressibility and big specific heat capacity. The system CCSE in question, located in the seaside and consisted of wind power generator/Compressors/Underground salt cave/Expander and power generator/ Refrigerator for fish, was designed to realize the function of storage wind power and reuse cold energy. When power grid was in low load, such as in the nighttime, power generated from wind was used to compress CO_2 from low pressure into high pressure, and CO_2 with high pressure was injected into underground salt cave for power grid becoming peak load, then CO_2 was withdrew to generate power to peak shaving and generate cold energy to maintain fish refrigerator in low temperature. CO_2 carried cold energy have another name as dry ice, solid state of CO_2 is easy to preservation and transportation, when transfer the cold energy to fish refrigerator, dry ice become as CO_2 in gas phase and flow into another underground salt caves to prepare for next CO_2 cycle. Energy efficiency and economic feasibility of the CCSE above was calculated using thermal physics knowledge, and results show that, total energy efficiency exceeds 50%, moreover, the output power price is about 1/3 of peak price when input price is 0.2CHY/kw.h.

Introduction

Now days more and more attention has been payed to environment, wind power would have a great potential development as a clean energy. However, wind power, with the characteristics of intermittent, cannot maintain continuous supply. So utilization potential of wind energy is reduced dramatically. Otherwise, power load is uneven in that high consumption in daytime cases a peak load power grid and low consumption at night cases a load valley. Especially in the seaside, the contrary phenomenon between more abundant wind resources and much smaller consumption at night than that in daytime, results serious waste of wind power. Therefore, a system of energy storage was designed to store wind at night and release in daytime.

1 System description

In order to use and store greenhouse gas CO_2, compressed CO_2 energy storage system(CCSE) was designed by using model of Compressed Air Energy Storage[1,2](CASE) and with the advantage of high compressibility and big specific heat capacity. The system (CCSE) in question located in the seaside and consisted of Wind power generator, Compressors, Underground salt cave, Expander and power generator, Refrigerator for fish, was designed to realize the function of storage wind power and reuse cold energy. The schematic of CCSE worked as follows: When power grid was in low load, such as in the nighttime, power generated from wind was used to compress CO_2 from low pressure into high pressure, and CO_2 with high pressure was injected into underground salt cave for power grid becoming peak load, then CO_2 was withdrew to generate power to peak shaving and generate cold energy to maintain fish refrigerator in low temperature. CO_2 carried cold energy have another name as dry ice, solid state of CO_2 is easy to preservation and transportation, when transfer the cold energy to fish refrigerator, dry ice become as CO_2 in gas phase and flow into another underground salt caves[3,4] to prepare for next CO_2 cycle.

2 Efficiency of energy storage

2.1 Assumptions

CCSE used greenhouse gas CO_2 as working medium, seawater as heat storage medium. CO_2 was assumed to meet ideal gas state equation and the heat capacity of the gas and water was a constant during the process of energy conversion and the system cycle. Loss of pressure in pipeline, salty cavity and heat exchanger was ignored, as well as loss of energy between compressor-motor and turbine-generator. A complete cycle consisted of one energy-storage and one following energy-release, and there was no energy interference with any other cycles.

2.2 Transmission of energy

A. Compressor

Compressor worked and operated isentropically during the process of cycle, when CO_2 flowed through compressor with quality of ma, the quantity of electricity power consumed by compressor Wc was obtained as shown:

$$W_c = \frac{m_a c_p T_1 (\pi_c^{\frac{\gamma-1}{\gamma}} - 1)}{\eta_c}$$

Then the temperature of compressor outlet T2 was obtained:

$$T_2 = T_1 \left[\frac{\pi_c^{\frac{\gamma-1}{\gamma}} - 1}{\eta_c} + 1 \right]$$

Among them, the temperature of compressor inlet T1 was assumed as environment temperature T0.

B. First heat-exchanger

Compressed CO_2 exchanges heat with cold seawater in first heat-exchanger, according to the definition of energy efficiency of heat-exchanger, outlet temperature of seawater and CO_2 would be received:

$$\varepsilon_1 = \frac{m_a c_p (T_2 - T_3)}{(mc)_{min} (T_2 - T_0)} = \frac{m_w c_w (T_w - T_0)}{(mc)_{min} (T_2 - T_0)}$$

Based on the assumption that there was no energy loss in the heat exchanger, outlet temperature of the seawater and CO_2 can be obtained as Tw and T3 respectively.

C. Heat storage

Cold seawater became hot seawater and stored in heat storage when flowed away from first heat exchanger, some heat might loose when withdrew the next day, and the loss proportion was ε_0 , so the effective heat stored in the heat storage was got:

$$Q_{TES} = m_w c_w (T_w - T_0) \varepsilon_0$$

D. Underground salt cave

Compressed CO_2 was stored in isothermal condition because storage time was short, generally overnight, as a result, the temperature of the gas from salt cavity equaled to the values for environment temperature:

$$T_4 = T_0$$

E. Second heat-exchanger

In order to get more energy, compressed CO_2 was heated in the Second heat-exchanger by hot seawater from heat storage before flowed into turbine motor. The CO_2 temperature T5 of Second heat-exchanger outlet was obtained based on the definition of heat exchanger 2 energy efficiency.

$$T_5 = \varepsilon_2(T_w - T_4)\varepsilon_0 + T_4$$

F. Turbine motor

Turbine motor worked and operated isentropically during the process of cycle, when CO_2 flowed through compressor with quality of ma, the quantity of electricity power generated by compressor Wt was obtained:

$$W_t = m_a c_p T_5 (1 - \pi_t^{\frac{1-\gamma}{\gamma}})\eta_t$$

Then the temperature of turbine motor outlet T6 was obtained:

$$T_6 = T_5\left[1 - \eta_t(1 - \pi_t^{\frac{1-\gamma}{\gamma}})\right]$$

G. Refrigerator for fish

In order to keep the fish fresh, temperature of refrigerator was maintained at T7=-10 ℃, therefore, total cold energy released by CO_2 from turbine motor was calculated as Qc:

$$Q_C = m_a c_p (T_7 - T_6)$$

2.3 Energy efficiency

As a result of energy transmission among several parts above, efficiency of electricity storage and efficiency of cooling storage were got.

$$\eta_{work} = \frac{W_t}{W_c}$$

$$\eta_{cool} = \frac{Q_C}{W_c}$$

It was obvious that many factors could impose efficiency of electricity storage and efficiency of cooling storage dramatically. Sensitive analysis was studied on different ratio of compression and heat loss proportion.

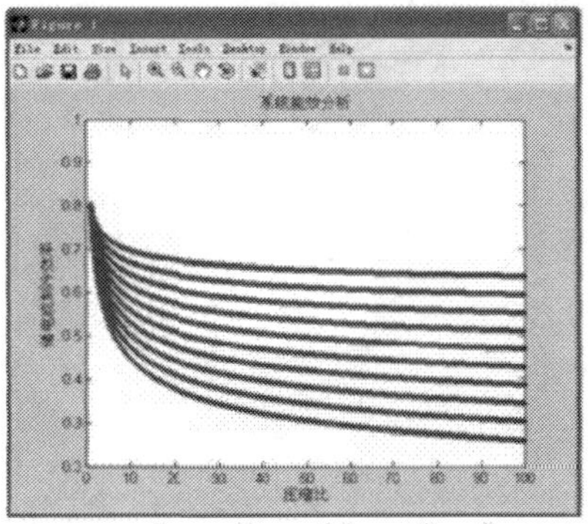

Figure1. Efficiency of electricity storage changing with ratio of compression and heat loss proportion

Results show that efficiency of electricity storage decrease gradually when ratio of compression increase and heat loss proportion decrease.

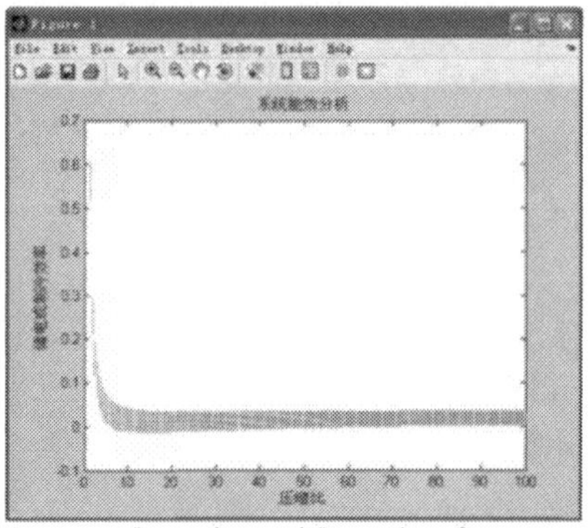

Figure2. Efficiency of cooling storage changing with ratio of compression and heat loss proportion

Results show that efficiency of cooling storage decrease gradually when ratio of compression increase and heat loss proportion decrease.

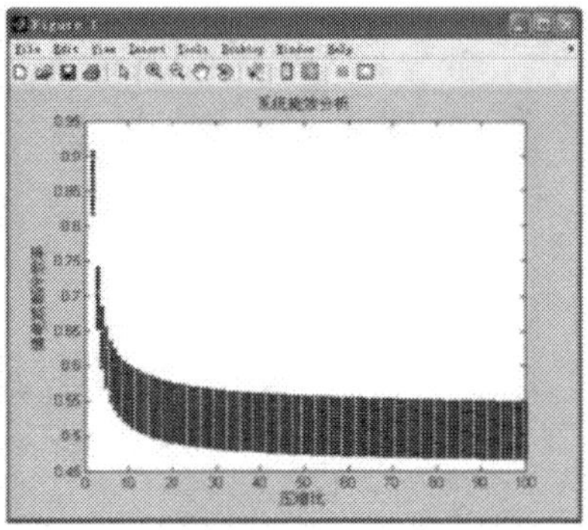

Figure3. Total efficiency of cooling storage changing with ratio of compression and heat loss proportion

In order to enhance efficiency, power generated from turbine motor was totally used for cooling, so total efficiency of cooling storage consisted of both cooling and electricity from CCSE was got. Results show that total efficiency of cooling storage decrease gradually when ratio of compression increase and heat loss proportion decrease.

3 Economic analysis

Case was set with parameters as: compressor efficiency 0.9, heat-exchanger efficiency 0.8, turbine motor efficiency 0.9, heat storage efficiency 0.9, ratio of compression 5-100, electricity price in daytime was 0.9CHY/kv.h, electricity price at night was 0.1 CHY/kv.h, 0.2 CHY/kv.h,0.3 CHY/kv.h,0.4 CHY/kv.h.

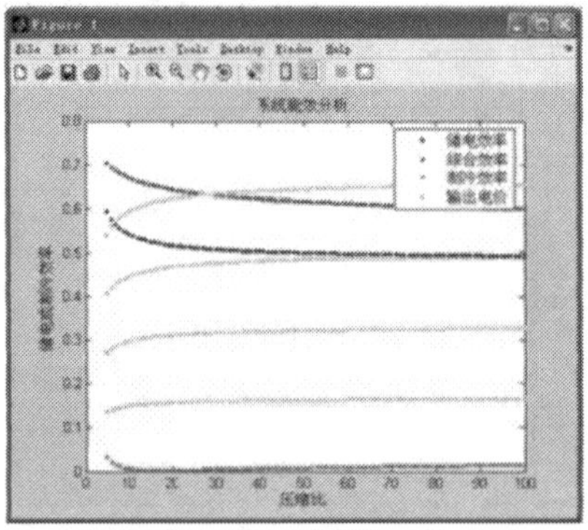

Figure4. Efficiency and electricity price from CCSE with different ratio of compression and electricity price at night

Efficiency and electricity price from CCSE were forecasted with help of parameters above, efficiency remained the same level and electricity price from CCSE increased when electricity price at night increased. The range of electricity price at night to electricity price from CCSE might be 1.35 and 1.65, and the range of electricity price from CCSE might be 0.27 CHY/kv.h and 0.33 CHY/kv.h when electricity price at night was 0.2 CHY/kv.h,, this price was about 30% of electricity price in daytime. Case showed that CCSE in question had strong market competitive ability.

4 Conclusion

The system of energy storage was designed to store wind at night and release in daytime with CO2 as working medium. Efficiency of storage in CCSE decreases slightly with much higher energy density. CCSE had strong market competitive ability when electricity price at night was 0.2 CHY/kv.h.

References

[1] Chen Haisheng, Gong T N, Yang Wei, et al. Progress in Electrical Energy Storage System: a Critical Review [J].Progress in Natural Science, 2009, 19(3):291-312.

[2] Lund H, Salgi G.The Role of Compressed Air Energy Storage (CAES) in Future Sustainable Energy Systems [J]. Energy Conversion and Management, 2009, 50(5):1172-1179.

[3]Tang Ligen, Wang Jieming. Inventory forecast in underground gas storage based on modified material balance equation [J]. Petroleum Exploration and Development, 2014,41(4): 480~484

[4]Tang Ligen, Wang Jieming. Forecast Inflow-performance in Gas Storage Based on Datas from Field Exploration & Development [J]. Petroleum Exploration and Development, 2016,43(1): 12~17

CMTC-438009-MS

Optimization of CO2 Sequestration in Reservoirs under Geochemistry and Thermal Effects

Cenk Temizel, Aera Energy; Aditya Tiwari, Consultant; Tayfun Tuna, University of Houston; Sinem Setenay Erdogan, Turkish Petroleum, Mahmut Unan, University of Houston

This paper was prepared for presentation at the Carbon Management Technology Conference held in Sugarland, Texas, USA, 17–19 November 2015.

Abstract

Disposal and long-term sequestration of anthropogenic "greenhouse gases" such as CO_2 is a proposed approach to reducing global warming. Deep, regional-scale aquifers in sedimentary basins are possible sites for sequestration, given their ubiquitous nature. In order for CO_2 to be stored in aquifers responsibly, it is essential to identify key concepts that need to be considered for potential implementation. Ideally, injected CO_2 will migrate through an aquifer from injection wells to remote storage sites, and remain isolated from the atmosphere for a considerable period of time. Fundamental topics of interest in sequestration research have concerned not just scientific and technical aspects, but practical concerns such as the economic feasibility of storage, safety, and the maximum possible amount of CO_2 storage globally and for specified regions. Thus, it is crucial to have a robust understanding of this important process not only in theory but in practice too through illustration with solid examples as in this study.

A robust commercial optimization and uncertainty software is coupled with a full-physics commercial simulator that models the phenomenon in order to investigate the significance of the major parameters on performance of wells in CO_2 sequestration under geochemistry and thermal effects. CO_2 injection is done for first 25 years. The injector is shut-in thereafter and the fate of CO_2 is modeled for the next 225 years. Chemical equilibrium constants are functions of temperature. Trace gas is used. Molecular diffusion of CO_2 in water is modeled. Thermal effects due to injection of CO_2 at are modeled.

Sensitivity and optimization have been done on major reservoir parameters such as fluid and rock properties and well operational parameters and then significance of each has been illustrated in tornado diagrams. It is observed that a robust approach on handling of uncertainties in reservoir are as important as management of well operational parameters in the scope of reservoir management. Presence of impact of geochemistry and temperature effects have proven to play an important role in the process.

Introduction

In past few years CO_2 capture from various combustion sources like power plants, refineries, petrochemical industries etc. has gained widespread interest to control emission and control change in climatic conditions due to global warming. Worldwide emissions of CO_2 are listed in Table 1 as per the IPCC report on Carbon dioxide capture and storage.

Table 1: Carbon dioxide emission in different regions around world

Region	CO_2 emission in 10^3 metric tons	
	1990	2011
East Asia & Pacific	3,090,194	10,486,424
Europe & Central Asia	1,663,171	1,400,690
Latin America & Caribbean	700,239	1,210,407
Middle East & North Africa	559,388	1,309,434
South Asia	782,138	2,328,351
Sub-Saharan Africa	447,994	738,901
Euro area	2,567,009	2,403,352
World	**22,200,872**	**34,649,483**

Carbon dioxide capture can play a vital role in controlling volume of greenhouse gas in atmosphere and has been adopted by around 120 facilities in United States as per Greenhouse gas reporting program (GHGRP, 2012). Carbon Sequestration is defined as process of depositing CO_2 emitted from various anthropogenic sources in geologic formation after capturing it from its source point and transporting it to disposal site. Disposal of emitted CO_2 is considered as effective means of mitigating the contribution of fossil fuel to greenhouse effect and ocean acidification.

CO_2 has been used in past for enhanced oil recovery (EOR) projects but sequestration projects are different because in case of EOR process keeping CO_2 underground permanently is not ensured. Sequestration was initially believed as a simple process similar to CO_2 injection for EOR but with analysis of technical issues and reservoir simulation studies it has been identified that fluid and pressure conditions are critical in a reservoir subjected to sequestration and due to long term injection process well integrity is a major challenge. Life of a project can be classified mainly in four stages: construction, operation, suspension and abandonment. A design life of a sequestration well is around 40+ years with a need for continued integrity for an abandonment period of thousands of year. In order to accept a sequestration project as safe and reliable, comprehensive monitoring and testing of site with efficient warning system to identify leakage are required.

Economic viability of projects is hindered by extensive investments required for separating and capturing greenhouse gas as well as assessment and development of long term integrity of sequestration sites. Challenges in sequestration projects can be classified as Near Wellbore challenges and reservoir associated challenges. This paper briefly explains near wellbore challenges and presents extensive study of reservoir-associated challenges with focus on geochemistry and thermal effects.

Reservoir Associated Challenges

Injected CO_2 migrates in the reservoir and monitoring the fate of gas is based on oil-gas related geophysical measurements. Since storage capability of a reservoir is directly associated with caprock, depletion and pressurization of formation can directly impact integrity of caprock. Modelling of a sequestration reservoir needs evaluation of adsorption mechanism, solubility trapping, capillary trapping and mineralization with calibration based on laboratory measurements. Mechanism dominating storage capacity of a reservoir is based on pore structure, minerology, rock and fluid properties (Rao, 2011).

CO_2 is miscible in water and sea water and forms various species of carbonate components depending on temperature and pressure conditions in the reservoir. In ocean water equilibrium condition can be mainly given by following equation:

$$CO_2(aq) + H_2O \rightleftharpoons H_2CO_3 \rightleftharpoons HCO_3^- + H^+ \rightleftharpoons CO_3^-$$

Density of CO_2 is greatly influenced by pressure and temperature conditions and impact of water depth on density of CO_2 is shown in Figure 1 below. At depth of around 9400 feet it can be seen that density of CO_2 exceeds density of sea water and becomes gravitationally stable. With increase in depth difference in density increases but starts decreasing after 11200 feet as shown in Figure 1. This plot shows that low depth reservoirs depending on geologic trap and type of reservoir rocks in the region will be effective for CO_2 sequestration in the region.

Injection of CO_2 displaces formation fluids and thus developing a pressure gradient due to injection, immiscible CO_2 will rise towards the caprock. Based on the relative permeability and difference in density of formation fluid and CO_2 injection overpressure required for given influx rate and thus estimates the pressure around caprock to ensure caprock integrity. Plume formation around caprock is associated with relative permeability and volumetric expansion of CO_2 during ascent towards caprock and has been covered in various studies (Johnson et al., 2004; Kumar et al., 2007).

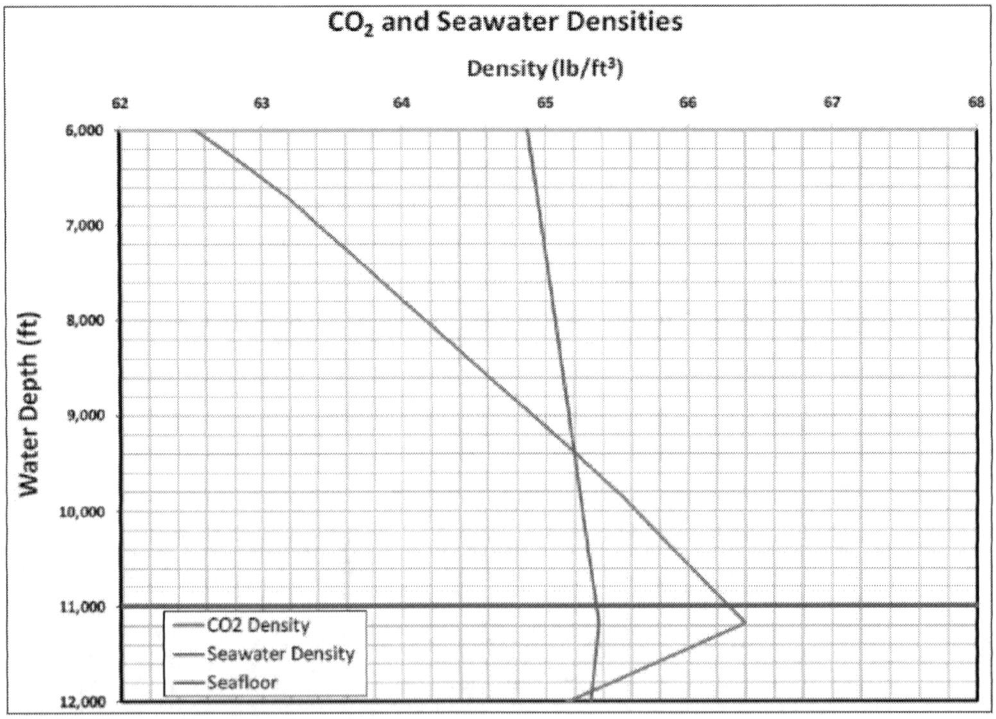

Fig 1: CO_2 and sea water densities above and below sea floor (Pilisi et.al, 2010)

In order to inject large volume of CO_2 in a reservoir, more than one injection wells are required hence; interference of flow between wells is expected. Shembre et.al (2007) studies the importance of pressure interference of wells and effect on injectivity. Simulation studies suggest that reservoir size and presence of interfering wells can have great impact on well count needed and injectivity of each well. It has also been observed that even in case of infinite reservoirs interference from neighboring wells results in decrease of injectivity.

Near Wellbore Challenges

Along with caprock integrity, well integrity is also critical and near wellbore stress study is important to avoid migration of fluid into water table through debonding of cement from casing or debonding of cement and formation or failure of casing/cement due to excessive tensile/shear stress. Well integrity data management and efficient well testing methodologies need to be adopted in order to identify possible gas leakage issues. Ensuring wellbore integrity is dependent on understanding of well construction, cement chemistry and cement evaluation.

During injection phase of a sequestration corrosion is rarely an issue as injected CO_2 is dry and non-corrosive making standard carbon steel alloy as feasible material for well construction. This ideal scenario doesn't exist in long term as brine in the reservoir dissolves CO_2 making it more corrosive as its pH drops. Once the injection is stopped corrosive brine might migrate towards wellbore in no flow condition and exposed section of tubing below packer will corrode. In order to avoid this after completion of injection phase it is recommended to remove the tubing and cap the well. It is necessary to continuously monitor pressure around annulus in case tubing needs to be in place after stopping injection activities. In such case use of corrosion resistance alloy is recommended for tubing material. Material selection for downhole equipment and cement are based on bottomhole pressure, temperature, injection fluid composition and brine composition.

Concerns associated with CO_2 Sequestration

Injection of CO_2 is an efficient way to trap greenhouse gas but like hydraulic fracturing and disposal wells, it is also associated with groundwater contamination and induced seismicity if not done efficiently.

Groundwater contamination can be due to caprock integrity, well leakage or fault leakage. Several studies have discussed risk associated with trapping of CO_2 (*Price et.al, 2007; Wilson et.al 2007 etc.*).Since CO_2 is not flammable these risks can be considered relatively less hazardous as compared to oil and gas related leakages unless in case of high concentration leaks (>15000 ppm). Experience from leaks in oil and gas fields can be used to prevent and mitigate such leaks. Most of such leaks occur via fractured zones or unsealed faults in the region or through poor well completion in nearby region. Primary concern in such cases is due to potential mobilization of hazardous inorganic constituents into ground water and possibility of increased acidity as studied by *Birkholzer et.al* (*2008*) and others.

With increase in seismicity in Central USA since 2008 (*Ellsworth, 2013*) and its association with injection activities concerns it is important to understand impact of sequestration on seismic events. Deformation of rock can result in release of seismic energy and can activate faults which might result in high magnitude seismic events. Large faults can be avoided using geologic models during site screening but faults which are not evident in geophysical survey can't be avoided easily thus resulting in potential hazards with sequestration.

Geochemistry and Thermal Effects in C02 Sequestration

CO_2 sequestration in saline aquifers involves different trapping mechanisms: geological trapping, hydrodynamic trapping and geo-chemical trapping. Geo-chemical trapping is consists of solubility trapping and mineral trapping. According to Silva et al., (2015) upon these trapping mechanisms geological trapping an d solubility trapping are more effective in short term but mineral trapping is safer and more economical in long term.

CO_2 preferably injected in super critical state ($scCO_2$) since in super critical state CO_2 is denser than gaseous CO_2. The critical point is 31.1 ˚C and 7.38 Mpa at which CO_2 becomes a supercritical fluid (Pruess & Garcia, 2002). Silva et al., (2015) mentioned that super critical CO_2 has liquid like density and gas like compressibility. Injecting super critical CO_2 reduces the buoyant forces, which mean it can be stored in great amounts and easily occupy available space entirely.

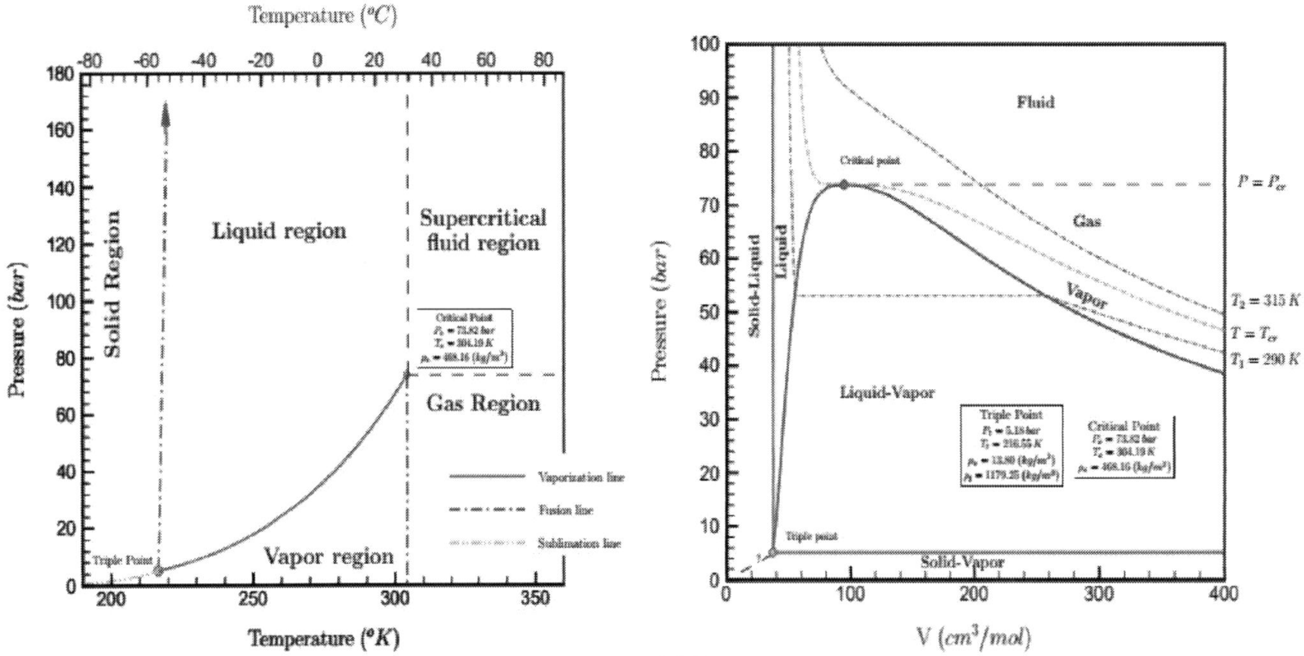

Figure 2: PVT schemes of CO2

Through geo-chemical trapping processes, injected CO_2 is either dissolved into the brine (solubility trapping) or reacted with the rock minerals (mineral trapping).

CO_2 injection can trigger some geochemical reactions between CO_2, aquifer water and formation rock that can alter the mineral composition of the host rock and can shift the thermodynamic equilibrium existed before the injection. Reaction dynamics depend on the mineralogical composition of the rock mass and the chemical composition of the aquifer brine. These reactions have an undeniable effect on the mineralogical structure and thermodynamic equilibrium of the aquifer (Benson & Cole, 2008).

When CO_2 is injected into the formation, the first process is the dissolution of CO_2 into the aquifer brine. As a result of this process, the aqueous CO_2 is formed (Eq.1).

$$CO_2(g) \rightarrow CO_2(aq) \qquad \text{Eq. 1}$$

Then an equilibrium is reached between aqueous CO2 and carbonic acid (Eq.2).

$$CO_2\,(g) + H_2O(\iota) \leftrightarrow H_2CO_3(aq) \qquad \text{Eq. 2}$$

The formed weak carbonic acid,H_2CO_3, which eventually dissociates into hydrogen and bi-carbonate ions (Eq.3).

$$H_2CO_3(aq) \leftrightarrow \; H^{+}(aq) + \; CO_3^{-}(aq) \qquad \text{Eq. 3}$$

Consequently, bi-carbonate ions partially dissociate into H^+ and carbonate $[CO_3^{2-}]$

$$HCO_3^{-}(aq) \leftrightarrow \; H^{+}(aq) + \; CO_3^{2-}(aq) \qquad \text{Eq. 4}$$

(Silva et al., 2015)

It also should be noted that at the beginning of the process 99% of CO2 exists as dissolved gas state rather than carbonic acid.

The solubility of CO_2 is mainly depends on the temperature, pressure, salinity of the formation brine and the type and the composition of the hosting rock.

According to the various authors' studies, the CO_2 solubility in saline water mainly increases with increasing pressure and reducing temperature and reduces with increasing salinity of the formation brine. The studies and their findings on CO_2 solubility in brine are summarized in Table 2 and Table 3.

Table 2: Solubility of CO_2 in saline water funder geo-sequestration conditions- experimental data (retrieved from Silva et al., 2015). It also should be noted that at the beginning of the process 99% of CO_2 exists as dissolved gas state rather than carbonic acid.

The solubility of CO_2 is mainly depends on the temperature, pressure, salinity of the formation brine and the type and the composition of the hosting rock.

Table 2: Solubility of CO_2 in saline water funder geo-sequestration conditions- experimental data (retrieved from Silva et al., 2015)

Referenc e	Fluid Type	Salinity	Temp.	Press . (Mpa)	Max. CO2 solubility	Comments
Nighswander et al. ; 1989	Synthetic brine solution	1wt. % NaCl (0.17 mol/kg)	80-200 ˚C (at 40 ˚C intervals)	4-10	0.86 mol/kg at 80 ˚C and at 10 Mpa	CO2 solubility in brine is lower than in pure water due to the salting-out effect. Salinity levels considered are very low compared to salinity levels in deep saline aquifers
Rumpf et al. ; 1994	NaCl solution	4 and 6 mol/kg NaCl	40-160 ˚C	Up to 10	0.5497 mol/kg at 40 ˚C, 10 Mpa am 4 mol/kg salinity	Most of the temperatures are well above those of the actual deep aquifer and therefore solubilities may be considerably lower
Bando et al. ; 2003	NaCl solution	1-3 wt. % NaCl (0.17-0.5 mol/kg)	30-60 ˚C	10-20	2.4 mol/kg at 30 ˚C, 20 Mpa and 1 wt. % NaCl	Test conditions are in good agreement with actual field conditions
Zhaowen et al.; 2004	Weyburn formation brine	92.59 gr/lt (~1.6 mol/kg)	59 ˚C	Up to 20	0.958x10^-3 mol cm3 at 20 Mpa	Real formation water sample was used instead of synthetic solutions. 25-30 % of solubility reduction occured compared to the solubility in water.
Portier and Rochelle; 2005	Synthetic Utisara formation water	Up to 3 mol/kg	35-80 ˚C	8-10	1.087mol/kg at 37 ˚C, 10 Mpa and 1 mol/kg NaCl	A synthetic brine solution that represented various ionic concentrations was used

According to the various authors' studies, the CO_2 solubility in saline water mainly increases with increasing pressure and reducing temperature and reduces with increasing salinity of the formation brine. The studies and their findings on CO_2 solubility in brine are summarized in Table 2 and Table 3.

Table 3: Solubility of CO_2 in saline water under geo-sequestration conditions modelling studies (retrieved from Silva et al., 2015)

Reference	Fluid type	Salinity	Temp.	Press. (Mpa)
Duan and Sun ; 2003	$NaCl_{(aq)}$	0-4.3 mol/kg	273-533 K	0-200
Portier and Rochelle; 2005	$NaCl_{(aq)}$	0-3 mol/kg	0-150 °C	0-30
Darwish and Hilal; 2010	$NaCl_{(aq)}$	1-4 mol/kg	300-500 K	5-200
Akinfiev and Diamond; 2010	$NaCl_{(aq)}$	1-6 mol/kg	-22 to 100 °C	0.1-100
Spycher and Pruess; 2010	$NaCl_{(aq)}$	1-6 mol/kg	12-300 °C	0.1-60

For further interest on the solubility studies of CO_2 the study of Silva et al. (2015) can be referenced. After dissolution of CO_2 into brine, the mineralogical reactions between dissolved CO_2 and formation rock. These reactions led a considerable amount of CO_2 to be stored in the saline aquifers. This process is commonly called as mineral trapping. This is a slow process that occurs in geological time scales and needed reactive minerals such as feldspar, magnesium and iron-rich clays, micas and Fe-oxides (Silva et al., 2015).

According to Silva et al. (2015); after injecting CO_2 mainly two types of reactions occur. The first kind of reactions are pH-buffered reactions. These reactions change the acidity level of the formation.

According to the researchers, in these reactions, because of injecting CO2 into brine, the aquifer pH decreased from 7 to 3 and consequently the reactions start between the co2 and carbonate minerals. This reactions led to increase aquifer pH to around 5. Therefore these pH-buffered reactions increase the solubility trapping through the formation of HCO_3^- ions and complexes. Moreover, reactions with the formation rock minerals led to increase in porosity and permeability near the injection wells. This also enhances the trapping process. On the other hand, with the precipitation of carbonate minerals, the permeability can also decrease significantly at the early stages of geochemical reactions and also negatively impact the trapping process.

According to various studies(Rosenbauer, Koksalan, & Palandri, (2005);Lagneau, Pipart, & Catalette, (2005);Goldberg, Takahashi, & Slagle, (2008);Darwish & Hilal, (2010);Gislason et al., (2010);Busch et al., (2008);Izgec, Demiral, Bertin, & Akin, (2008)), different types of formations studied for the suitability of the process. Sandstones, carbonates, shale and basalt formations are considered as potential saline aquifers for CO2 sequestration. Among these formations, sandstone formations presents more potential of pH buffering reactions than the other formations. Mineral-CO_2-brine interactions that can occur in sandstone reservoirs are summarized by Silva et al., (2015) at Table 4.

Table 4: Mineral-CO2-Brine interactions that can occur (Silva et al., 2015)

Primary mineral	Reaction	Secondary minerals
Dissolution reactions		
	$CO_2(g) \rightarrow CO_2(aq)$	
	$CO_2(g) + H_2O(l) \leftrightarrow H_2CO_3(aq)$	
	$H_2CO_3(aq) \leftrightarrow H^+(aq) + HCO_3^-(aq)$	
	$HCO_3^-(aq) \leftrightarrow H^+(aq) + CO_3^{2-}(aq)$	
Kaolinite	$Al_2Si_2O_5(OH)_4(s) + 6H^+(aq) \rightarrow 5H_2O + 2SiO_2(aq) + 2Al^{3+}(aq)$	Complete dissolution
Illite	$Illite + 8H^+(aq) \rightarrow 5H_2O(l) + 0.6K^+(aq) + 0.25Mg^{2+}(aq) + 3.5SiO_2(aq) + 2.3Al^{3+}(aq)$	Complete dissolution
Arnorthite	$CaAl_2Si_2O_8(s) + CO_2(g) + 2H_2O(l) \rightarrow CaCO_3(s) + Al_2Si_2O_5(OH)_4(s)$	Calcite, kaolinite
Labradorite	$Ca_{0.6}Na_{0.4}Al_{1.6}Si_{2.4}O_8 + 5.4H^+(aq) + CO_2(aq) \rightarrow 0.6Ca^{2+}(aq) + HCO_3^-(aq) + 2.2H_2O(l) + 0.4Na^+(aq) + 1.6Al^{3+} + 2.4SiO_2(aq)$	Complete dissolution
Albite	$NaAlSi_3O_8(s) + CO_2(g) + H_2O(l) \rightarrow NaAl(CO_3)(OH)_2(s) + 3SiO_2(s)$	Dawsonite, quartz
K-feldspar	$KAlSi_3O_8(s) + Na^+(aq) + CO_2(g) + 2H_2O(l) \rightarrow NaAl(CO_3)(OH)_2(s) + 3SiO_2(s) + K^+(aq)$	Dawsonite, quartz
Calcite	$CaCO_3(s) + H^+(aq) \rightarrow Ca^{2+}(aq) + HCO_3^-(aq)$, *at high PH values only*	Complete dissolution
Glauconite	$Glauconite + 14H^+(aq) \rightarrow 1.5K^+(aq) + 2.5Fe^{3+}(aq) + 0.5Fe^{2+}(aq) + Mg^{2+}(aq) + 1.0Al^{3+}(aq) + 7.5SiO_2(aq) + 9H_2O(l)$	Quartz
Annite	$annite + 3CO_2 \leftrightarrow 3siderite + K\text{-}feldspar$	Siderite, K-feldspar
Chlorite	$Chlorite + 20H^+(aq) \rightarrow 5Fe^{2+}(aq) + 5Mg^{2+}(aq) + 4Al(OH)_3(aq) + 6H_4SiO_4(aq)$	Aluminium hydroxide
Dolomite	$CaMg(CO_3)_2(s) + 2H^+(aq) + Mg^{2+}(aq) + HCO_3^-(aq)$	Complete dissolution
Precipitation reactions		
	$Ca^{2+}(aq) + CO_3^{2-}(aq) \rightarrow CaCO_3(s)$	Calcite
	$Fe^{2+}(aq) + CO_3^{2-}(aq) \rightarrow FeCO_3(s)$	Siderite
	$Mg^{2+}(aq) + CO_3^{2-}(aq) \rightarrow MgCO_3(s)$	Magnesite
	$Ca^{2+}(aq) + SO_4^{2-}(aq) \rightarrow CaSO_4(s)$	Anhydrite
	$K^+(aq) + 3Al^{3+}(aq) + 2SO_4^{2-}(aq) + 6H_2O(l) \rightarrow KAl_3(SO_4)_2(OH)_6(s) + 6H^+(aq)$	Alunite
	$Ca^{2+}(aq) + Mg^{2+}(aq) + 2HCO_3^-(aq) \rightarrow CaMg(CO_3)_2(s) + 2H^+(aq)$	Dolomite

CO_2 Sequestration Projects

Department of Energy (DOE) has listed a network of seven Regional Carbon Sequestration Partnerships (RCSPs) to help develop the technology, infrastructure, and regulations to implement large-scale carbon sequestration in different regions and geologic formations within the Nation. All in all, the seven RCSPs represent regions surrounding 97 percent of coal-fired CO2 emissions, 97 percent of industrial CO2 emissions, 96 percent of the total land mass, and essentially all the geologic sequestration sites in the U.S. potentially available for carbon storage. This initiative is launched in 2003. It's being completed in phases (I, II, and III) and forms the centerpiece of national efforts to develop the infrastructure and knowledge base needed to place carbon storage and utilization technologies on the path to commercialization.

Phase I is the characterization the potential for CO2 storage in deep oil-, gas-, coal-, and saline-bearing formations. Phase II is validation of regional sequestration opportunities through small-scale geological storage tests. In this phase, twenty small-scale geologic storage tests were conducted that collected additional data to confirm storage and injectivity estimates; to validate reservoir simulation models; to demonstrate effective monitoring, verification and accounting methods; and to develop guidelines for well completion, operations, and closure. Phase III is development and the RCSPs are working to implement eight large-scale (one million metric tons or more total) geologic storage and utilization projects that will demonstrate the long-term, effective, and safe storage and utilization of CO_2 in the major geologic formations throughout the United States and portions of Canada.

The seven RCSPs are consist of more than 400 diverse organizations which covers 43 states and four Canadian provinces.

- Big Sky Regional Carbon Sequestration Partnership (Big Sky)
- Plains CO2 Reduction Partnership (PCOR)
- Midwest Geological Sequestration Consortium (MGSC)
- Midwest Regional Carbon Sequestration Partnership (MRCSP)
- Southeast Regional Carbon Sequestration Partnership (SECARB)
- Southwest Regional Partnership on Carbon Sequestration (SWP)
- West Coast Regional Carbon Sequestration Partnership (WESTCARB)

The Success of Mitsubishi Heavy Industries:

In Jan 2014, Mitsubishi Heavy Industries and Southern Company Services announced that they have successfully tested a technology that captures and sequesters carbon dioxide from coal plants. U.S. Energy Department pledged to come up with about $US 1 billion for the $1.65-billion FutureGen 2.0 carbon capture and storage project in Illinois.

Photo 1: Mitsubishi Heavy Industries/PRNewsFoto

Based on the results released, Mitsubishi Heavy seems will accelerate its program to commercialize the technology. Southern Company owns the 2657 megawatt power plant in Alabama. The DOE and the power industry's R&D arm, the Electric Power Research Institute is involved in this project.

The project demonstration was done in June 2011. The plant consisted mostly of a flue-gas scrubber, flue-gas carbon capture system, as well as carbon compression machinery and electrical components. It recovered 90 percent of the carbon dioxide generated and has a capacity to catch 150 000 metric tons of carbon dioxide a year. Mitsubishi Heavy did not reveal how this system affected the plant's efficiency or its water use.

Besides Southern Company's work with Mitsubishi Heavy in Alabama, it also has a carbon capture project going in Kemper County, Mississippi. That one would capture the carbon dioxide and use it to enhance oil recovery. DOE contributed $290 million to help Southern finish the job.

Characteristics of Successful CO2 Sequestration Projects

During the past decade, projects in support of the development and deployment of the technology have been initiated across the globe. However, a considerable number of these projects have later been put on hold or cancelled as shown in Table 3. Currently, there is little understanding of what characteristics may have led to these undesirable outcomes (Thronicker et al, 2014). Thronicker aims to elicit technological, economic and policy characteristics that render CCS projects more or less likely to become operational.

Their results consistently find that storage site confirmed statistically improves the likelihood of project success while public funding, being located in an area with a carbon price, and previous CCS experience statistically reduces the likelihood of project success. The coefficients and statistical significance of the two size variables imply a non-linear relationship between project success and size. The coefficients imply that as size increases, the chance of project success decreases at a decreasing rate. The size that would minimize the chance of success, given the coefficients, is larger than any size in our sample, implying a convex relationship between size and chance of success. The storage of CO2 in saline formations, compared to enhanced oil recovery (EOR), is detrimental to project success while the capture of CO2 by pre-combustion, industrial separation or natural gas processing, compared to post-combustion capture, is advantageous to project success. Another finding is CCS at oil-fueled plants to be more likely to be successful when compared to coal feedstock. When excluding pilot projects from the observations, gas feedstock becomes less likely to be success compared to coal.

Table 3 - Canceled and Inactive Projects(retrieved from MIT website [20])

Project Name	Leader	Location	Region	Feedstock	Size	Capture Process	CO2 Fate
FutureGen	FutureGen Alliance	Illinois	USA	Coal	229	Oxy	Saline
Lake Charles	Leucadia Energy	Louisiana	USA	Methanol Production	4.5	EOR	Planning
Kimberlina	Clean Energy Systems	California	USA	Coal	50 MW	Oxy	Saline
Taylorville	Tenaska	Illinois	USA	Coal	602 MW	Pre	Saline
Indiana Gasification	Leucadia	Indiana	USA	Coal	5 Mt/yr	Pre	EOR
Antelope Valley	Basin Electric	North Dakota	USA	Coal	120 MW	Post	EOR
Trailblazer	Tenaska	Texas	USA	Coal	765 MW	Post	EOR
ZENG Worsham-Steed	Worsham-Steed	Texas	USA	Coal	70 MW	Oxy	EOR
Wallula	Wallula Resource	Washington	USA	Coal	600 MW	Pre	Saline
AEP Mountaineer	AEP	West Virginia	USA	Coal	235 MW	Post	Saline
Appalacian Power	AEP	West Virginia	USA	Coal	629 MW	Pre	Saline
Project Pioneer	TransAlta	Alberta	Canada	Coal	450 MW	Post	Saline/ EOR
Compostilla	ENDESA	Spain	European Union	Coal	323	Oxy	Saline
Goldenbergwerk	RWE	Germany	European Union	Coal	450 MW	Pre	Saline
Janschwalde	Vattenfall	Germany	European Union	Coal	250 MW	Oxy	Saline
Barendrecht	Shell	Netherlands	European Union	Oil Refinery	0.4 Mt/Yr	Post	Depleted Gas
Belchatow	PGE	Poland	European Union	Coal	260	Post	Saline
Longannet	Scottish Power	UK	European Union	Coal	300 MW	Post	Saline
FINNCAP	Fortum	Finland	European Union	Coal	565 MW	Post	Depleted Gas
Hunterston	Ayrshire Power	UK	European Union	Coal	1,600 MW	Post	Depleted Gas
Zeng Risavika	CO2 Norway	Norway	Norway	Gas	50 MW	Oxy	Saline
Halten	Statoil	Norway	Norway	Gas	860 MW	Post	EOR
ZeroGen	ZeroGen	Australia	Other	Gas	2 MT/Yr	Pre	Saline
Kwinana	BP	Australia	Other	Coal	500 MW	Pre	Saline

Abbreviations used:

Oxy = Oxyfuel Combustion Capture
Pre = Pre Combustion Capture
Post = Post Combustion Capture
EOR = Enhanced Oil Recovery
TBD = To Be Decided
EGR = Enhanced Gas Recovery
Saline = Saline Formation
Depleted Gas = Depleted Gas Reservoir
Depleted Oil = Depleted Oil Reservoir

Simulation, Optimization & Model Details

The simulation model consists of a Cartesian Grid of 9x9x4 of 324 blocks. It is a single-porosity system. This data set is built to model gas sequestration into an aquifer. CO2 injection is done for first 25 years. The injector is shutin thereafter and the fate of CO2 is modelled for next 225 years. Chemical equilibrium constants are functions of temperature. Trace gas is used. Molecular diffusion of CO2 in water is modelled. Thermal effects due to injection of CO2 at 50 C are modelled.

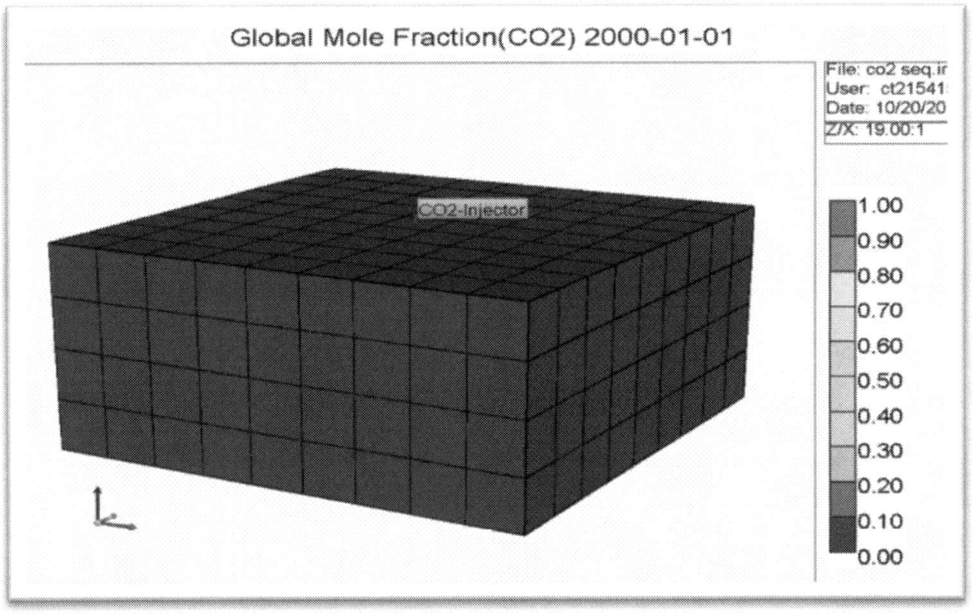

Figure 3: **Initial Global Mole Fraction of CO2**

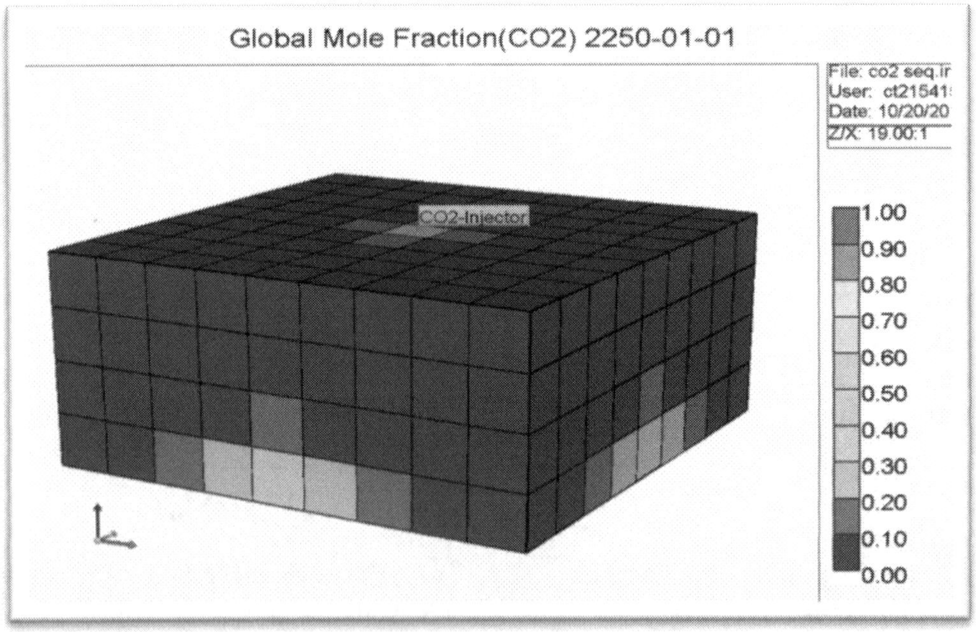

Figure 4 Final **Global Mole Fraction of CO2**

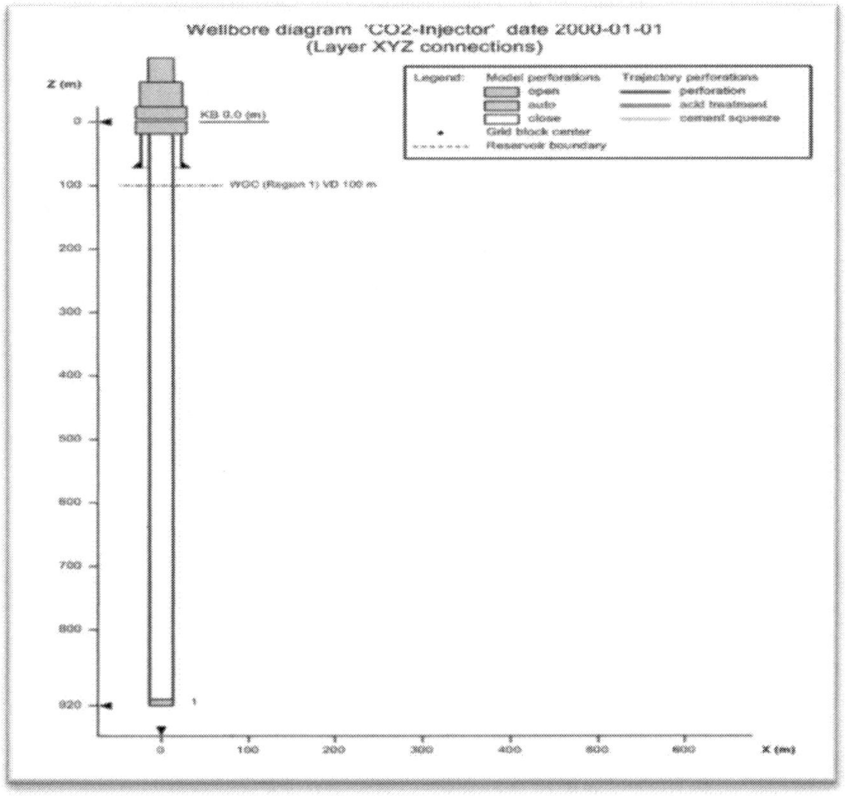

***Figure 5:** **Cr**oss Section of CO2 Injection Well*

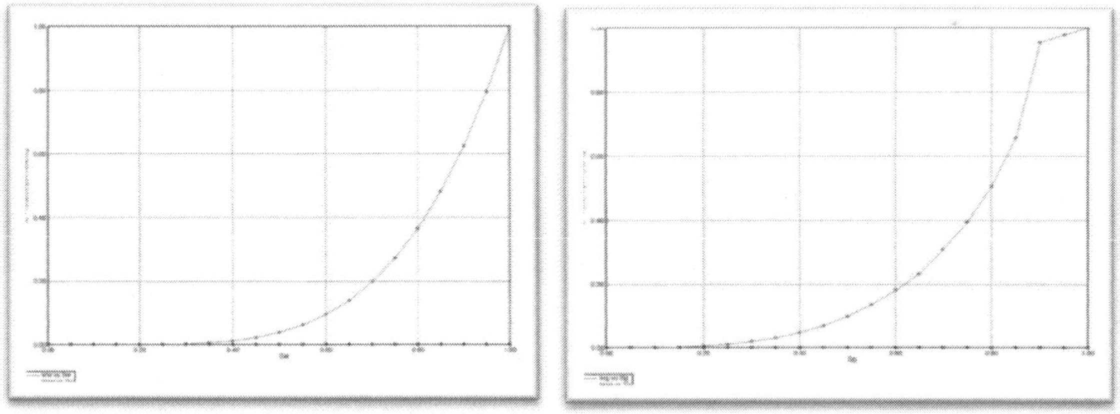

Figure 6: **Relative Pemeability Curves**

Table 4 – Interaction parameters

		CO2	C1
		non-HC	non-HC
1	CO2		
2	C1	1.0300000E-01	

Table 5 – Aqueous Phase Solubility

	Component	Henry's constant	Ref. pressure	Vinfinity
		kPa	kPa	m3/kgmole
1	CO2	1.9510547E+05	9.4000000E+03	3.5089333...
2	C1	0.0	9.4000000E+03	3.5242646...

Table 6 – Component Definition

#	Component	Hydrocarbon?	Crit. pressure	Crit. temperat...	Acentric factor	Mol. weight	Volume shift	Compress. fa...
			atm	K		g/gmole		
1	CO2	No	7.2800000E+01	3.0420000E+02	2.2500000E-01	4.4010000E+01	0.0000000E+00	
2	C1	No	4.5400000E+01	1.9060000E+02	8.0000000E-03	1.6043000E+01	0.0000000E+00	

#	Component	Crit. volume (...	Omega A	Omega B	Specific gravi...	Average nor...	Parachor
		m3/kgmole				C	
1	CO2	9.4000000E-02	4.5723553E-01	7.7796074E-02	8.1800000E-01	-7.8450000E+01	7.8000000E+01
2	C1	9.9000000E-02	4.5723553E-01	7.7796074E-02	3.0000000E-01	-1.6145000E+02	7.7000000E+01

Table 7 – Sensitivity and Optimization Parameters

	Name	Comment	Active	Default Value
1	CP		✓	1000
2	CW		✓	4.35E-07
3	GEOMETRY		✓	0.2
4	HLPROP		✓	2.0
5	OPERATE		✓	1.4E+06
6	SALINITY		✓	0.1
7	THCONR0		✓	2.0

Where:

HLPROP
It consists of the following parameters:

- **rock_den**

mass density of surrounding rock [kg/m3 or lb/ft3].

- **rock_conductivity**

thermal conductivity of surrounding rock [J/m·s·K | Btu/ ft·hr·oF].

OPERATE
This sub-keyword identifies a surface gas rate (m3/day | SCF/day) constraint.

CP-ROCK
cR, heat capacity of rock [J/kg·K | Btu/lb·oF].

THCONR0
τc, thermal conductivity of rock and fluid [J/m·s·K | Btu/ ft·hr·oF].

CW
Water component isothermal compressibility (1/kPa | 1/psia).

GEOMETRY
It specifies the well geometric characteristics to be used by the simulator to calculate the well index internally. A real number between 0. and 1. specifying the fraction of a circle that the well models. (Usually one; set to one half or one quarter if the well is at the edge or corner of the grid block on the grid boundary.) Dimensionless.

SALINITY
Specifies the salinity of the aqueous phase in terms of equivalent NaCl concentration for use with the AQUEOUS-VISCOSITY model.

Table 8 – Coefficients

Coefficients in Terms of Actual Parameters

Term	Coefficient
Intercept	7.93115E+09
CP	1.9602E+06
CW	1.17566E+15
GEOMETRY	5.4275E+09
HLPROP	6.94969E+08
SALINITY	-2.15294E+09
THCONR0	2.62616E+08
CW*CW	-1.25026E+21
GEOMETRY*GEOMETRY	-1.37877E+10
HLPROP*HLPROP	-1.21133E+08
THCONR0*THCONR0	-6.57883E+07

Table 9 – Summary of Fit

Summary of Fit

R-Square	0.999114
R-Square Adjusted	0.998432
R-Square Prediction	0.997335
Mean of Response	1.15865E+10
Standard Error	1.22398E+07

Analysis of Variance

Source	Degrees of Freedom	Sum of Squares	Mean Square	F Ratio	Prob > F
Model	10	2.19534E+18	2.19534E+17	1465.38	<0.00001
Error	13	1.94758E+15	1.49814E+14		
Total	23	2.19729E+18			

Effect Screening Using Normalized Parameters (-1, +1)

Term	Coefficient	Standard Error	t Ratio	Prob > |t|	VIF
Intercept	1.16524E+10	1.05516E+07	1104.32	<0.00001	0.00
CP(750, 1250)	4.90051E+08	4.33686E+06	112.997	<0.00001	1.20
CW(3.2625E-07, 5.4375E-07)	9.5625E+06	4.3778E+06	2.18431	0.04786	1.22
GEOMETRY(0.15, 0.25)	-4.37943E+06	4.5443E+06	-0.963719	0.35277	1.22
HLPROP(1.5, 2.5)	1.05219E+08	4.29117E+06	24.5198	<0.00001	1.16
SALINITY(0.075, 0.125)	-5.38234E+07	4.83833E+06	-11.1244	<0.00001	1.63
THCONR0(1.5, 2.5)	-268571	4.18995E+06	-0.0640988	0.94987	1.27
CW*CW	-1.47863E+07	8.53622E+06	-1.73218	0.10688	1.39
GEOMETRY*GEOMETRY	-3.44693E+07	8.80871E+06	-3.91309	0.00178	1.56
HLPROP*HLPROP	-3.02832E+07	1.06684E+07	-2.83859	0.01396	2.25
THCONR0*THCONR0	-1.64471E+07	8.72828E+06	-1.88434	0.08207	1.73

Results

As seen in Figure 7 below, the proxy model estimation and the actual simulation results match very closely.

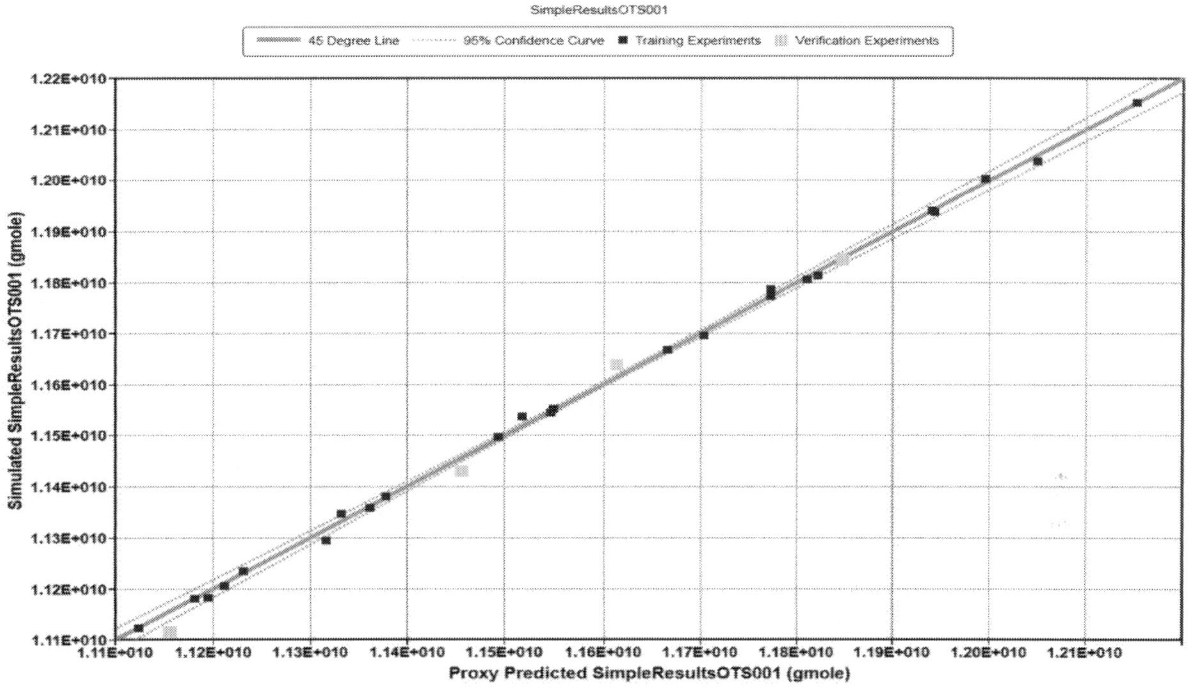

Figure 7: **Simulation vs. Proxy Results**

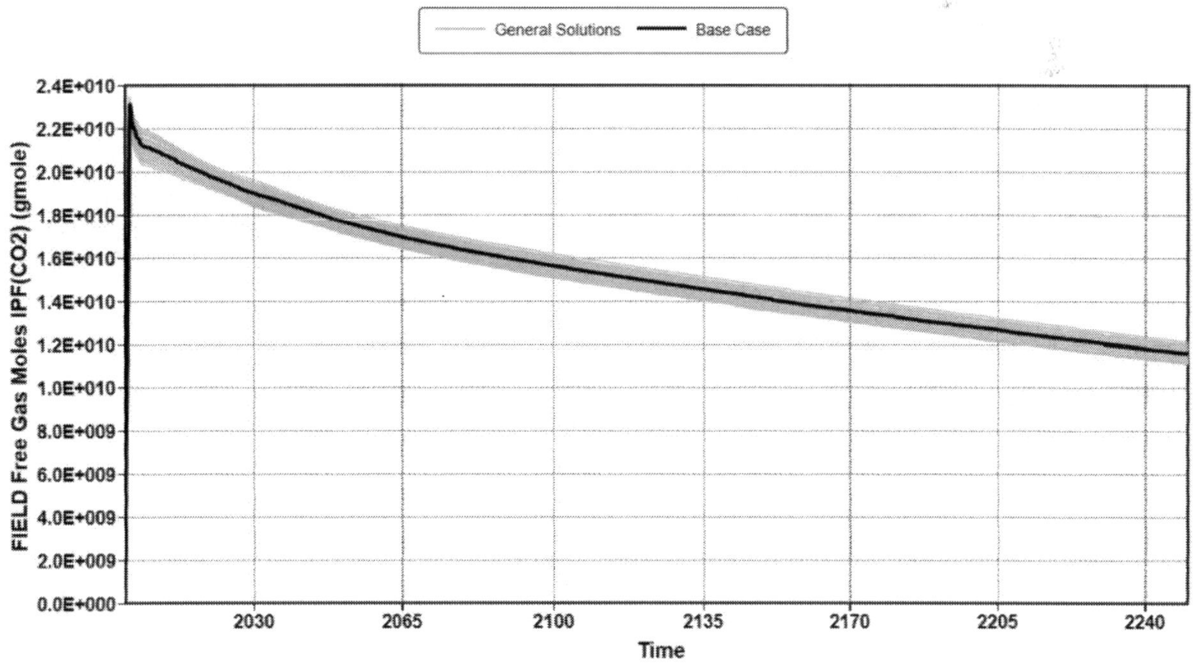

Figure 8: **Free CO2 Gas Molecules vs Time**

Figure 8 shows the comparison of CO2 free gas moles over time for different cases run in this study. Black curve indicates the base case and green curves present the cases where the optimization and sensitivity parameters have been changed in the input.

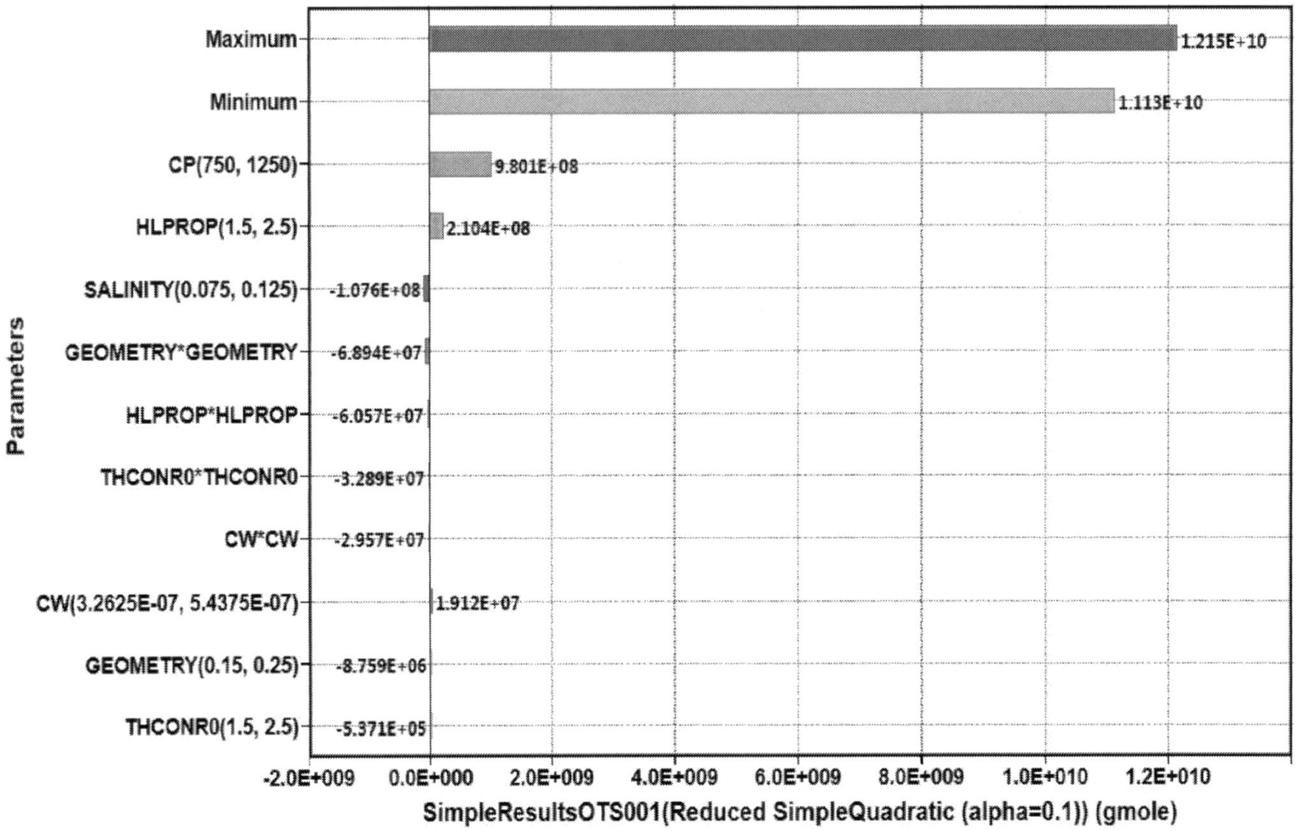

Figure 9: **Sensitivity Plot**

The sensitivity to heat capacity of the surrounding rock is the highest, and then rock density and conductivity follow under the parameter of heat loss (HLPROP) property. Sensitivity to salinity is much lower but relatively higher than that of other following factors. Thus, it is crucial to have a better understanding of the most significant parameters analyzed above to have a more robust plan for long - term CO2 sequestration.

Acknowledgement

Authors would like to thank Aera Energy (a Shell-ExxonMobil Affiliate), Turkish Petroleum and University of Houston.

References

1. Green House Gas Reporting Programme, 2012. EPA Greenhouse Gas Reporting Data–*Subpart PP–Suppliers of Carbon Dioxide.* Based on 2011 data.
2. Smith, L., Billingham, M., Lee, C. H., Milanovic, D. Z., & Lunt, G. (2010, January 1). CO2 Sequestration Wells - the Lifetime Integrity Challenge. Society of Petroleum Engineers. doi:10.2118/136160-MS

3. Pilisi, N., Ceyhan, I., & Vasantharajan, S. (2010, January 1). CO2 Sequestration in Deepwater Subseabed Formations. Society of Petroleum Engineers. doi:10.2118/139498-MS
4. Al-Ameri, W. A., Abdulraheem, A., Mahmoud, M. A., Abdullatif, O. M., & Adebayo, A. A. (2014). Effect of CO2 Sequestration Period on the Mechanical Properties of Carbonate Aquifers. Society of Petroleum Engineers. doi:10.2118/171702-MS
5. Rao, Dandina N., and Richard G. Hughes. "Current Research and Challenges Pertaining to CO2 Flooding and Sequestration." The Way Ahead Vol. 7, 2011
6. Silva, G. P. D. De, Ranjith, P. G., & Perera, M. S. A. (2015). Geochemical aspects of CO 2 sequestration in deep saline aquifers : A review. *Fuel*, *155*, 128–143. http://doi.org/10.1016/j.fuel.2015.03.045
7. Pruess, K., & Garcia, J. (2002). Multiphase flow dynamics during CO 2 disposal into saline aquifers. *Environmental Geology*, *42*(2-3), 282–295. http://doi.org/10.1007/s00254-001-0498-3
8. Benson, S. M., & Cole, D. R. (2008). CO2 Sequestration in Deep Sedimentary Formations. *Elements*, *4*(5), 325–331. http://doi.org/10.2113/gselements.4.5.325
9. Rosenbauer, R. J., Koksalan, T., & Palandri, J. L. (2005). Experimental investigation of CO2-brine-rock interactions at elevated temperature and pressure: Implications for CO2 sequestration in deep-saline aquifers. *Fuel Processing Technology*, *86*(14-15), 1581–1597. http://doi.org/10.1016/j.fuproc.2005.01.011
10. Lagneau, V., Pipart, a., & Catalette, H. (2005). Reactive transport modelling of CO2 sequestration in deep saline aquifers. *Oil Gas Sci. Tech.*, *60*(2), 231–247. http://doi.org/10.2516/ogst:2005014
11. Goldberg, D. S., Takahashi, T., & Slagle, A. L. (2008). Carbon dioxide sequestration in deep-sea basalt. *Proceedings of the National Academy of Sciences of the United States of America*, *105*(29), 9920–9925. http://doi.org/10.1073/pnas.0804397105
12. Darwish, N. a., & Hilal, N. (2010). A simple model for the prediction of CO2 solubility in H2O-NaCl system at geological sequestration conditions. *Desalination*, *260*(1-3), 114–118. http://doi.org/10.1016/j.desal.2010.04.056
13. Gislason, S. R., Wolff-Boenisch, D., Stefansson, A., Oelkers, E. H., Gunnlaugsson, E., Sigurdardottir, H., … Stute, M. (2010). Mineral sequestration of carbon dioxide in basalt: A pre-injection overview of the CarbFix project. *International Journal of Greenhouse Gas Control*, *4*(3), 537–545. http://doi.org/10.1016/j.ijggc.2009.11.013
14. Busch, A., Alles, S., Gensterblum, Y., Prinz, D., Dewhurst, D. N., Raven, M. D., … Krooss, B. M. (2008). Carbon dioxide storage potential of shales. *International Journal of Greenhouse Gas Control*, *2*(3), 297–308. http://doi.org/10.1016/j.ijggc.2008.03.003
15. Izgec, O., Demiral, B., Bertin, H., & Akin, S. (2008). CO2 injection into saline carbonate aquifer formations I: Comparison of numerical simulations to experiments. *Transport in Porous Media*, *72*(1), 1–24. http://doi.org/10.1007/s11242-007-9160-1
16. Thronicker, Dominique, and Ian Lange. "Determining the success of carbon capture and storage projects." (2015).
17. http://energy.gov/fe/science-innovation/carbon-capture-and-storage-research/regional-partnerships
18. http://www.undeerc.org/pcor/CO2SequestrationProjects/default.aspx
19. http://spectrum.ieee.org/energywise/energy/fossil-fuels/carbon-capture-success-in-alabama-1-billion-pledge-in-illinois
20. https://sequestration.mit.edu/tools/projects/index_cancelled.html

CARBON MANAGEMENT TECHNOLOGY CONFERENCE

CMTC-438055-MS

Robust and Cost Effective Distributed Coaxial Cable Temperature Sensor Verified as Real-Time Permanent Downhole Monitoring for Groundwater Safety in Geological CO_2 Storage

Yurong Li, Runar Nygaard, Missouri University of Science and Technology, and Wenge Zhu, Hai Xiao, Clemson University

This paper was prepared for presentation at the Carbon Management Technology Conference held in Sugarland, Texas, USA, 17–19 November 2015.

Abstract

Downhole monitoring plays a crucial part in a geological carbon dioxide (CO_2) sequestration project, especially in providing early warnings of failure. However, the current downhole monitoring technologies are either low in spatial resolution and time-consuming, or expensive and have system longevity issues. To address this issue a robust and cost effective real-time downhole monitoring system, distributed coaxial cable temperature sensor is proposed using the Fabry-Perot interferometer technology.

The coaxial cable sensor (CCS) is made in house and tested with a high pressure high temperature (HPHT) testing apparatus with water to study the sensor accuracy, sensitivity, long-term stability and crosstalk effect in simulated downhole conditions. The laboratory test results indicate that the sensor can work under simulated downhole conditions of pressure up to 1000psia and temperature up to 110 °C. At 1 ATM, the sensor has an accuracy more than 1%. At 1000 psia, the sensor sensitivity tends to be stable after repeated heating and cooling treatment. The hysteresis phenomenon is observed, but it is reduced after several treatment cycles. The pressure crosstalk effect is observed during the test.

The temperature and pressure range of the distributed CCS allows a long-term in-situ monitoring for a surface casing well section up to 2500 feet, which would prove great value in detecting wellbore leakage that will contaminate the ground water.

Introduction

A program for monitoring of CO_2 distribution is required once injection begins in order to manage the injection process, delineate and identify leakage risk or actual leakage that may endanger underground source of drinking water, verify and provide input into reservoir models, and provide early warnings of failure. Monitoring of the wells, deep subsurface, shallow subsurface and ground surface is expected to continue for long periods after the injection is terminated for safety and to confirm predictions of storage behavior (United States Environmental Protection Agency 2010).

To ensure the public safety as well as obtaining carbon credits in a future cap and trade system, monitoring and modeling of sequestration projects have to reach a high degree of accuracy. The objective is to reach 99% accuracy in a monitoring and verification program (The National Energy Technology Laboratory 2009). However, the predictions based on current methodology are far too uncertain to achieve the goal to account for 99% of the injected CO_2 (The National Energy Technology Laboratory 2012).
Various sensing technologies have been explored for characterization and monitoring of subsurface geologic environments. However, the current monitoring technologies have the following limitations: (1) Time-consuming. Some of the monitoring technologies require time-consuming field sample collection work, which loses the time effectiveness of the collected data, such as geochemical sampling and coring. (2) High cost. Although in part due to the success at Sleipner, 4-D seismic has emerged as the standard for comparison, this technology requires high cost for implementation. A monitoring cost estimate of the Wabamun Area Sequestration Project (WASP) showed that 4D seismic to be the most expensive monitoring methods when well cost was not included (Nygaard & Lavoie 2009). (3) Low sensitivity. The smallest detectable amount of CO_2 using seismic surveys depends on the fluid distribution, if homogeneously distributed, 1% CO_2 is detectable, however if the CO_2 is patchily distributed anything below 18% CO_2 is indistinguishable from brine (Cairns et al. 2010). (4) Low spatial resolution. Depending on the relative position of the observation well, the resolution of the microseismic results might be a few meters, which is not sufficient to answer detailed questions regarding points of fracture initiation from the wellbore (Holley et al. 2010).
Optical fiber sensors (OFS) are able to perform efficient monitoring with their multiplexing ability and reduced size compared with other wire-connected downhole sensors. Fiber optic based downhole temperature, pressure, strain and acoustic sensors for petroleum industry application are currently available (Tardy et al. 2011; Koelman et al. 2011; Costello et al. 2012; Medina et al. 2012; Molenaar et al. 2012). Despite its significant advantages and progressively increasing application for downhole monitoring in oil industry, optical fiber sensors do have some inherent drawbacks that have to be evaluated when they are considered as an alternative to conventional electronic transducers. Attenuation and longevity of the system are the major concerns. Hydrogen (H_2) attenuation, liquid ingress and micro-bending effects are the three factors that will give rise to either intrinsic or extrinsic energy loss (Williams et al. 2000). To address these concerns this paper studies the potential of using CCS for permanent installation in the wellbore.

Coaxial cable Fabry-Perot interferometer (CCFPI) temperature sensor

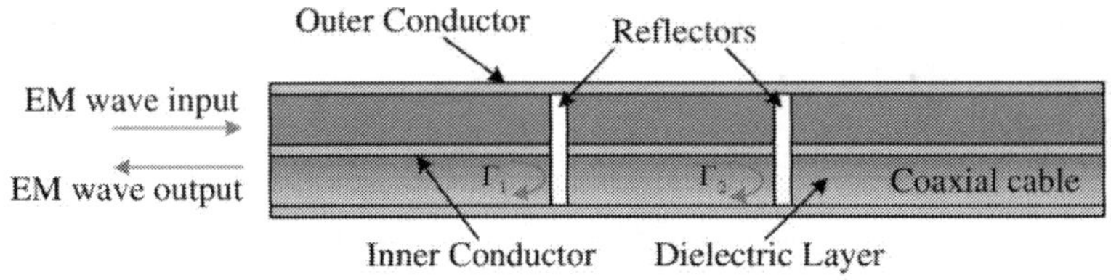

Figure 1 Schematic (left) and picture (right) of the CCFPI temperature sensor.

As Figure 1 (left) shows, two reflectors are generated on a coaxial cable, constructing a Fabry-Perot interferometer. An electro-magnetic (EM) wave propagating inside the coaxial cable is partially reflected at the first reflector due to characterisitic impedance discontinuity. The remaining wave transmitting through the first reflector is once agin partially reflected. The two reflected waves interfere coherently to generate a constructive or destructive interference signal. When the frequency of EM wave is swept, an interferogram in frequency domain is observed (Huang et al. 2013).

Since the Fabry-Perot interferometer is generated by two reflection points, the CCFPI temperature sensor is fabricated by constructing two structural reflectors, as shown in Figure 1 (right). Two copper crimp rings are firmly compressed onto the cable separated at a 4 inch distance to deform the outer conductor and generate the characteristic impedance discontinuities. The copper crimp rings also provide higher survivability for application under harsh environment. When a radio frequency (RF) waveform is transmitted into the cable, the deformed locations would generate two reflections and form the interference pattern in the frequency domain.

Experimental setup

A high pressure high temperature (HPHT) sensor testing system has been completed for sensor testing under various temperature, pressure and strain conditions. A picture of the whole system set up is shown in Figure 2. The HPHT testing cell is fabricated with stainless steel as container for fluid (water was used in all the tests), thermocouple and coaxial cable temperature sensor. The cell has a length of 30 cm and an inner diameter of 5 cm with a maximum pressure capability of 5,000 psia. O-rings are used to seal the cable on the end caps of the testing cell. A flexible silicone-rubber heating pad with a nominal maximum temperature of 232 °C is attached to the outer wall of the testing cell. It is then connected to a temperature controller (Omega CN7523, Stamford, CT, USA) which uses the thermocouple reading as input to control the fluid temperature. Flexible fiberglass insulation is wrapped around the testing cell to prevent heat loss. The pressure within the testing cell is controlled by a pump (Teledyne Isco 100DX, Lincoln, NE, USA) which has a maximum pressure rating of 10,000 psia and a specification of ±1 psia. A linear variable displacement transformer (LVDT, Omega LD630-10, Stamford, CT, USA) is mounted on the base to measure the displacement of the cable clamp which is assumed equal to the cable strain. The cable is connected to a vector network analyzer (VNA, HP 8753ES, Santa Clara, CA, USA) which generates the RF signal and monitors the reflection spectrum.

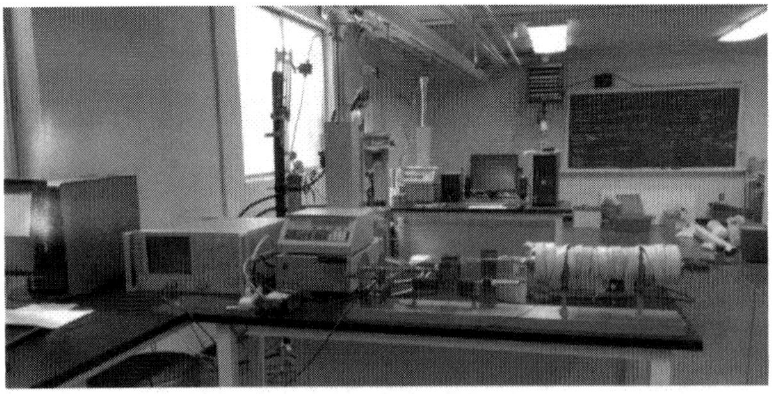

Figure 2 Picture of the HPHT Sensor Testing System.

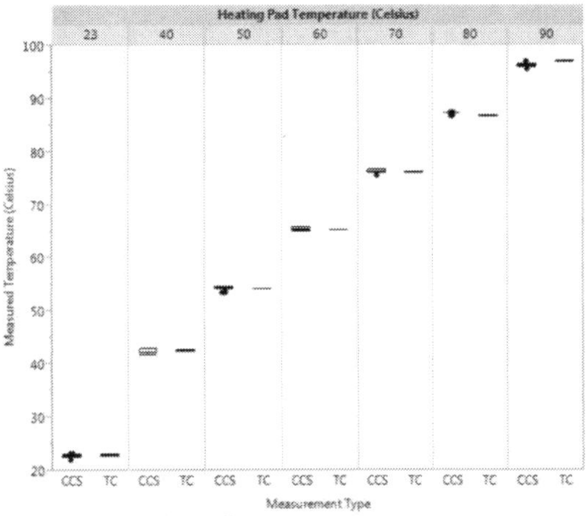

Figure 3 Boxplot comparison between CCS and TC measured temperature at 1 ATM.

Results

Sensor performance at atmospheric pressure.

The sensor is first tested under 1 ATM with temperature step increasing from room temperature (RMT) to 100 °C then air cooled to RMT. Figure 3 shows a boxplot comparison between coaxial cable sensor (CCS) measured temperature and thermocouple (TC) measured temperature when the testing cell is under a relatively constant temperature condition. The temperature measured by the two methods shows a very good consistency, and during the test the sensor shows an instantaneous response to the temperature change.

Sensor performance at 1000 psia.

To investigate the sensor response under a higher pressure condition, the testing cell is then pressurized to 1,000 psia with temperature step increasing from RMT to 110 °C and then air cooled to RMT. The heating and cooling cycle is repeated for 5 times and the boxplot of sensor measurement deviation (sensor measured temperature minus thermocouple measured temperature) for the 5 cycles is shown in Figure 4 (a)-(c). The results indicate that for the 1st cycle, there is a poor consistency between the two measurement methods, and the hysteresis is very large. However, as the cycle number increases, the consistency tends to be better. And further analysis of the sensor sensitivity and hysteresis change shows that the sensor tends to have a stable sensitivity and the hysteresis tends to become close to zero as the cycle number increases.

To verify the test results, the same tests are conducted on a second coaxial cable temperature sensor, and the same conclusion can be made from the results shown in Figure 4 (d)-(f). These two sets of tests indicate that the sensor needs to be pre-stressed (multiple heating-cooling cycles) to have a stable performance under a high pressure condition.

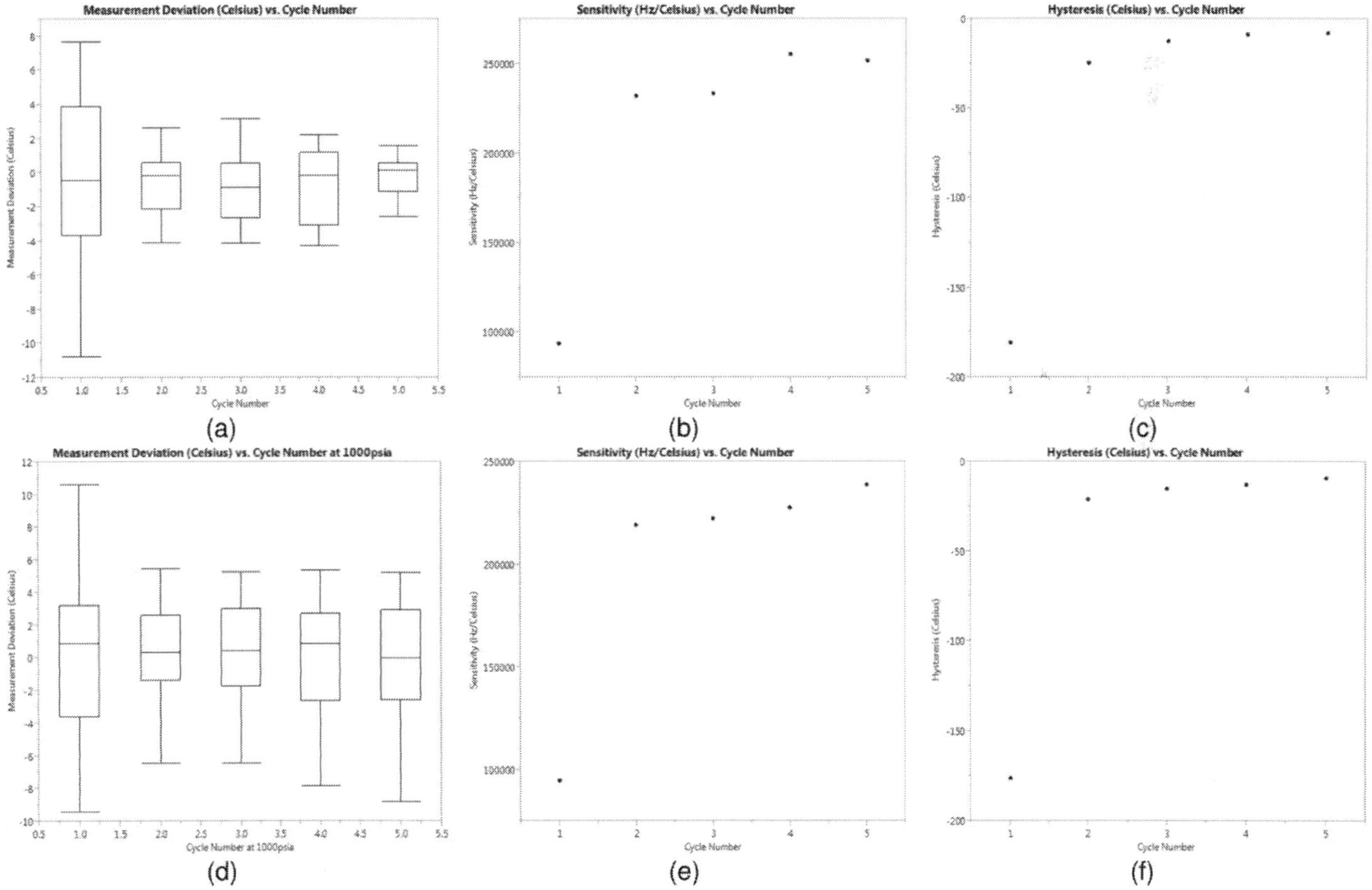

Figure 4 Test results of the sensor at 1000 psia for five cycles where (a) Sensor 1 measurement deviation boxplot (b) Sensor 1 sensitivity change (c) Sensor 1 hysteresis change (d) Sensor 2 measurement deviation boxplot (e) Sensor 2 sensitivity change (f) Sensor 2 hysteresis change.

Sensor long-term stability.

To test the sensor stability over a long period, the testing cell is pressurized to 1000 psia and the temperature is hold constant at both 40 °C and 90 °C each for more than 40 hours. Figure 5 shows the boxplot comparison of the pre-stressed CCS and TC measured temperature at 40 °C and 90 °C. At 40 C, the CCS measured temperature is 38 °C ± 0.6 °C, compared to the TC measured temperature of 38.1 °C ± 0.2 °C. While at 90 °C, the CCS measured temperature is 88.2 °C ± 1.3 °C, compared to the TC measured temperature of 86.9 °C ± 0.2 °C. The result shows that the sensor has a more stable performance at lower temperature, with higher accuracy and smaller measurement deviation.

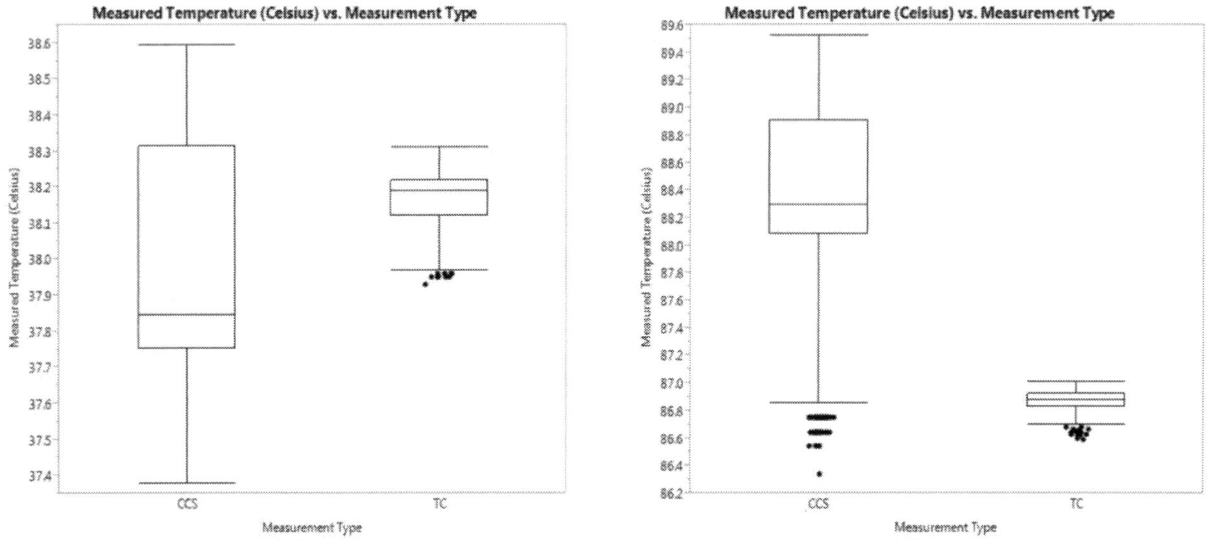

(a) (b)

Figure 5 Boxplot comparison of CCS and TC measured temperature at 1000 psia over a long period where (a) at 40 °C (b) at 90 °C.

Sensor pressure crosstalk.

The pressure crosstalk phenomenon has been observed during the test while decreasing the pressure from 1000 psia to 1 ATM at a constant temperature at 90 °C, as shown in Figure 6. To study the influence of pressure on the pre-stressed temperature sensor, the pressure is step increased from 1 ATM to 1000 psia and then step decreased back to 1 ATM at both 40 °C and 80 °C. Figure 7 shows the sensor frequency change with respect to changing pressure at 40 °C and 80 °C. It can be observed that the sensor has a constant response ratio to pressure at a constant temperature regardless of increasing or decreasing pressure. However, at different temperatures, the response ratios are different.

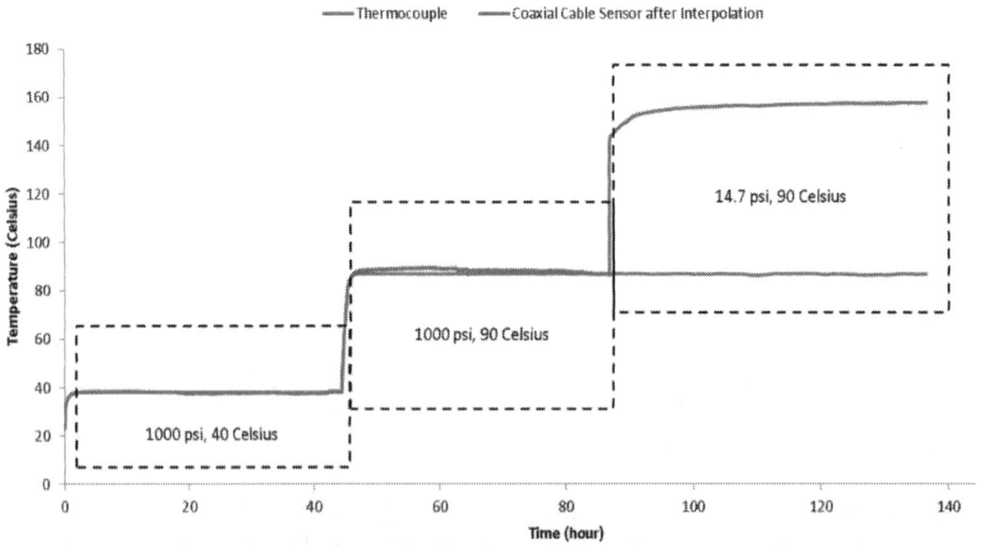

Figure 6 Observation of pressure crosstalk when decreasing pressure from 1000 psia to 1 ATM at 90 °C.

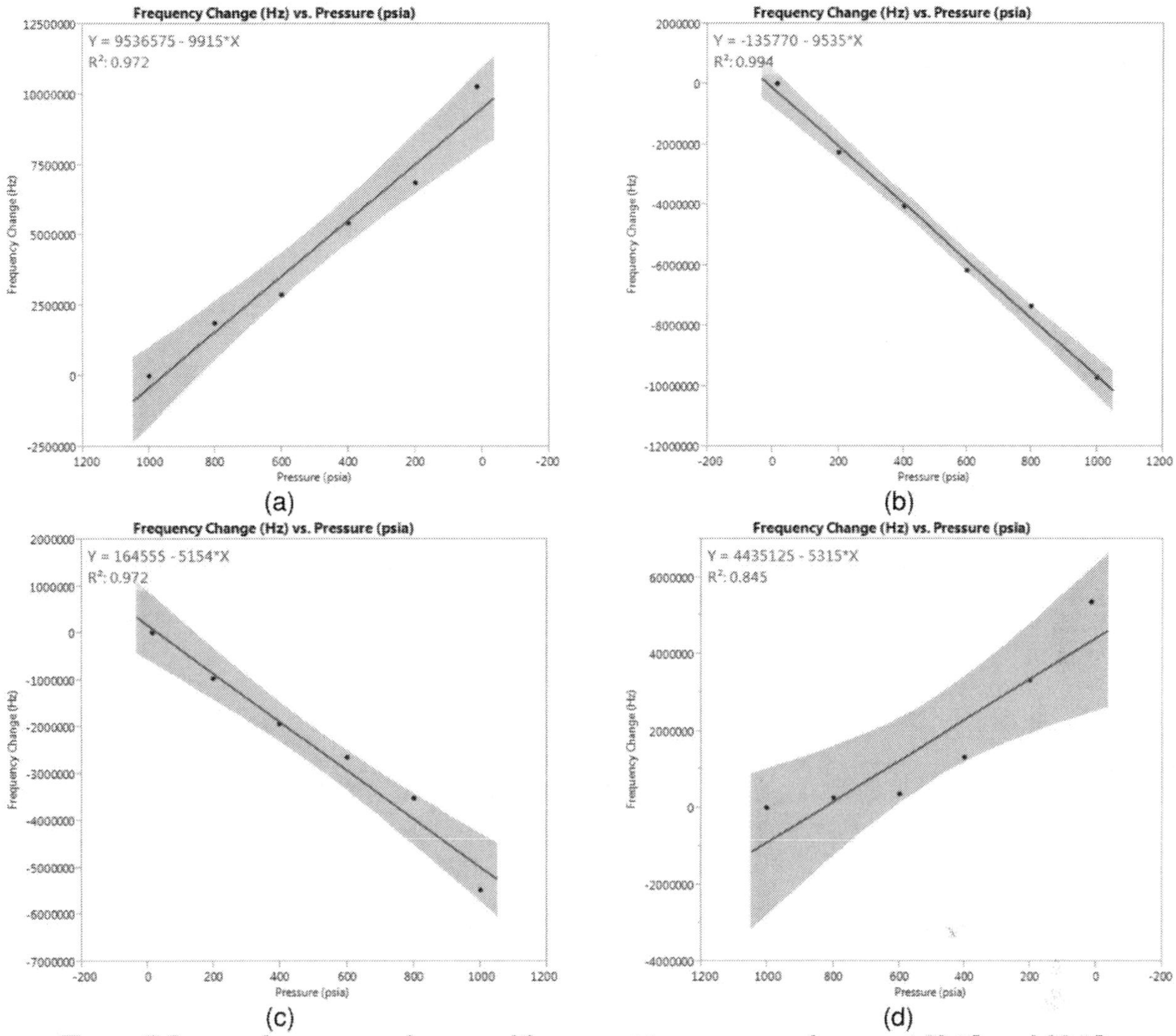

Figure 7 Sensor frequency change with respect to pressure change at 40 °C and 80 °C.

To quantify the pressure crosstalk, a central composite experiment is designed as shown in Table 1. A pre-stressed sensor is tested under the testing scheme with 3 repetitions. The analysis of variance (ANOVA) of the testing results show that temperature*pressure has a P-value smaller than 0.05, as shown in Table 2, which means that temperature and pressure interaction exists. The regression analysis gives a predictive model as;

$$\Delta F = 817485 + 10429817 \times \left(\frac{T - 67.5}{42.5}\right) - 6239952 \times \left(\frac{P - 507.35}{492.65}\right) - 2447524 \times \left(\frac{T - 67.5}{42.5}\right) \times \left(\frac{P - 507.35}{492.65}\right) + 1856260 \times \left(\frac{T - 67.5}{42.5}\right) \times \left(\frac{T - 67.5}{42.5}\right) + 1621496 \times \left(\frac{P - 507.35}{492.65}\right) \times \left(\frac{P - 507.35}{492.65}\right) \quad \text{Equation 1}$$

Where ΔF is sensor frequency change, Hz; T is temperature, °C; P is pressure, psia. With the predictive equation, for a known pressure and sensor frequency change, the sensor measured temperature can be calculated. Two verification tests are performed to compare the sensor measured temperature based on the predictive equation and the actual temperature. Table 3 shows the test results for the verification tests. As it indicates, the sensor measured temperature is around 10 °C lower than the actual temperature with the predictive equation.

Table 1 Central composite experimental design.

Reference Parameters	
Temperature (°C)	Pressure (psia)
25	14.7
Testing Parameters	
Temperature (°C)	Pressure (psia)
37.45	158.99
37.45	855.71
97.55	158.99
97.55	855.71
25	507.35
110	507.35
67.5	14.7
67.5	1000
67.5	507.35
67.5	507.35

Table 2 ANOVA table.

Analysis of Variance					
Source	DF	S.S.	M.S.	F Ratio	
Model	5	1.2035e+15	2.407e+14	580.8188	
Error	14	5.8019e+12	4.144e+11	Prob>F	
C. Total	19	1.2093e+15		<0.001*	
Effect Tests					
Source	Nparm	DF	S.S.	F Ratio	Prob>F
T	1	1	8.7025e+14	2099.925	<0.0001*
P	1	1	3.115e+14	751.6453	<0.0001*
T*P	1	1	1.1981e+13	28.9098	<0.0001*
T*T	1	1	7.8759e+12	19.0047	<0.0007*
P*P	1	1	6.0097e+12	14.5015	<0.0019*

Table 3 Comparison of sensor measured temperature and actual temperature based on the predictive equation.

Actual temperature (°C)	Actual pressure (psia)	Measured temperature (°C)
40	200	30.91
80	800	69.31

Discussion

Several factors can cause the reduction of sensor accuracy. The VNA used in the tests has a resolution of 1601 points for a frequency up to 6 GHz, while VNA with much higher resolution is available in market. Although interpolation is applied on the VNA frequency reading, higher VNA resolution is still required for accurate frequency change detection. Besides, the pressure and temperature maintaining of the testing cell is controlled by the pump and temperature controller. Pump pulsation and thermal conduction between the testing cell and the environment may reduce the system stability during the test.

The hysteresis phenomenon might be due to the residual stress on the cable. Although the sensor hysteresis tends to be close to zero after a few heating-cooling cycles, it still remains a problem. A small hysteresis can cause the sensor measurement error build up with increasing testing cycles. In addition, to solve the sensor pressure crosstalk problem, a pressure insensitive package is required for higher sensor measurement accuracy.

Conclusion

1. The distributed coaxial cable temperature sensor can work under conditions of temperature up to 110 °C and pressure up to 1000 psia.
2. At 1 ATM the sensor has an accuracy higher than 1%, and the sensor has instantaneous response to temperature change.
3. At 1000 psia the sensor measurement deviation tends to reduce with increase of heating-cooling cycles. The sensor sensitivity becomes stable after 5 cycles and the hysteresis becomes almost 0.
4. The sensor is more stable at lower temperatures than at higher temperatures.
5. Pressure has a very large effect on the sensor. Pressure insensitive package is required to remove the pressure crosstalk effect.

Acknowlegement

The experimental setup was completed with the help of the research technicians Jeffrey Heniff and John Tyler from the Rock Mechanics and Explosives Research Center of Missouri University of Science and Technology. The research is supported by the Department of Energy under Award Number DE-FE0009843. Neither the United States Government nor the Department of Energy, nor any of their employees, makes any warranty, express or implied, or assumes any legal liability or responsibility for the accuracy, completeness, or usefulness of any information, apparatus, product, or process disclosed, or represents that its use would not infringe privately owned rights. Reference herein to any specific commercial product, process, or service by trade name, trademark, manufacturer, or otherwise does not necessarily constitute or imply its endorsement, recommendation, or favoring by the United States Government or any agency thereof. The views and opinions of authors expressed herein do not necessarily state or reflect those of the United States Government or any agency thereof.

References

Cairns, G., Jakubowicz, H., Lonergan, L. et al (2010, January). Issues regarding the use of time-lapse seismic surveys to monitor CO_2 sequestration. *Society of Exploration Geophysicists.*

Costello, C., Sordyl, P., Hughes, C. T. et al (2012, January). Permanent distributed temperature sensing (DTS) technology applied in mature fields - a Forties field case study. *Society of Petroleum Engineers*. DOI:10.2118/150197-MS.

Holley, E. H., Zimmer, U., Mayerhofer, M. J. et al (2010, January). Integrated analysis combining microseismic mapping and fiber-optic distributed temperature sensing (DTS). In *Canadian Unconventional Resources and International Petroleum Conference*. Society of Petroleum Engineers.

Huang, J., Wang, T., Hua, L. et al (2013). A Coaxial Cable Fabry-Perot Interferometer for Sensing Applications. *Sensors, 13*(11), 15252-15260.

Koelman, J. V. V., Lopez, J. L., and Potters, H. (2011, January). Fiber optic technology for reservoir surveillance. *International Petroleum Technology Conference*. DOI:10.2523/14629-MS.

Medina, M., Torres, C. E., Sanchez, J. et al (2012, January). Real-Time downhole monitoring of electrical submersible pumps rated to 250 degree C using fiber optic sensors: case study and data value in the Leismer SAGD project. *Society of Petroleum Engineers*. DOI:10.2118/153984-MS.

Molenaar, M. M., Fidan, E., and Hill, D. J. (2012, March). Real-time downhole monitoring of hydraulic fracturing treatments using fibre optic distributed temperature and acoustic sensing. In *SPE/EAGE European Unconventional Resources Conference & Exhibition-From Potential to Production.*

Nygaard, R. and Lavoie, R. (2009). Project cost estimate-Wabamun area CO_2 sequestration project (WASP). University of Calgary.

Tardy, P., Chang, F., and Qiu, X. (2011, November). Determining Matrix Treatment Performance From Downhole Pressure And Temperature Distribution: A Model. In *International Petroleum Technology Conference.*

The National Energy Technology Laboratory. (2009). Project solicitation innovative and advanced technologies and protocol for monitoring/verification/accounting (MVA), simulation, and risk assessment of carbon dioxide (CO_2) sequestration in geologic formations. United States Department of Energy.

The National Energy Technology Laboratory. (2012). Project S national energy technology laboratory, technologies to ensure permanent geologic carbon storage. United States Department of Energy.

United States Environmental Protection Agency. (2010, November). Geologic CO_2 sequestration technology and cost analysis (Report No. 816-B-08-009). Retrieved from http://water.epa.gov/type/groundwater/uic/class6/upload/

Williams, G. R., Brown, G., Hawthorne, W. et al (2000, December). Distributed temperature sensing (DTS) to characterize the performance of producing oil wells. In *Environmental and Industrial Sensing* (pp. 39-54). International Society for Optics and Photonics.

CMTC-438081-MS

Low-carbon Development Strategy and CCS/CCUS Practices of Sinopec

XianLing Meng, EDRI of Sinopec

This paper was prepared for presentation at the Carbon Management Technology Conference held in Sugarland, Texas, USA, 17–19 November 2015.

Abstract

In responding to the challenges of climate change and in order to meet the carbon emissions peak goal around 2030, the petroleum and petrochemical industries in China have been taking great efforts in carbon reduction and achieving sustainable development. As one of the major operators in China's petroleum and petrochemical industries, Sinopec has been carried out low-carbon development strategy and tactics, such as optimizing industries structure, developing low-carbon energy, improving energy efficiency, carbon asset management and CCU/CCUS practicing etc.. Besides the low-carbon development strategy, the paper also introduces the progress and effects of CCS/CCUS practices in Sinopec, such as CO_2-based EOR, conversion of CO_2 into fuels and chemicals, mineralization of CO_2 , bio-utilization of CO_2 etc..

1. Overview of Sinopec

Sinopec Group is one of the largest integrated energy and chemical companies in China. Its principal operations include the exploration and production, pipeline transportation and sale of petroleum and natural gas; the sale, storage and transportation of petroleum products, petrochemical products, coal chemical products and other chemical products; the import and export of petroleum, natural gas, chemical products, and other commodities and technologies; and research, development and application of technologies and information.

Sinopec is the largest oil refiner, the largest producer and supplier of oil products, and the largest manufacturer and distributor of petrochemicals in China , it is also the second largest oil and gas producer in China and the fourth largest ethylene producer in the world. In 2014, Sinopec's revenue reached 2.89 trillion yuan, ranked 3rd on *Fortune*'s Global 500 list.

In 2014, Sinopec produced 78.1 million tonnes of crude oil and 28.2 billion cubic meter of natural gas, of which 34.3 million tonnes oil and 8 billion cubic meter natural gas were from overseas. Moreover,

Sinopec has constantly fueled the economic growth in China. In 2014, Sinopec's processing capacity of crude oil reached 294.62 million tonnes per year and accounted for around 40% of the total refining capacity in China(FIG.1), its refining throughput in 2014 was 237 million tonnes of crude oil. Its Ethylene capacity was 10.52 million tonnes and accounted for more than 51% of the total ethylene capacity in China(FIG.2). The company distributes most of its products in domestic market.

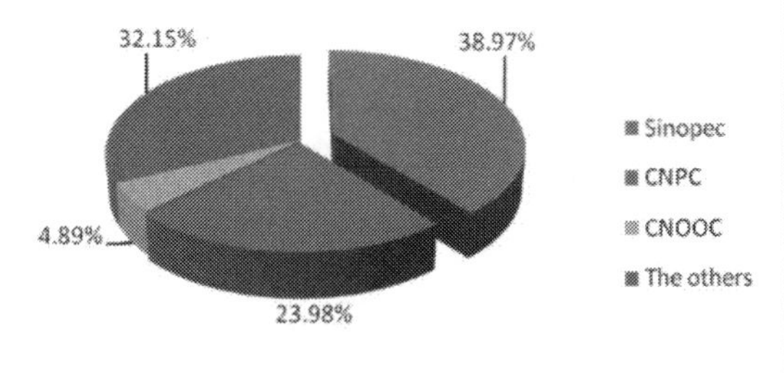

FIG.1. Refining capacity distribution in China

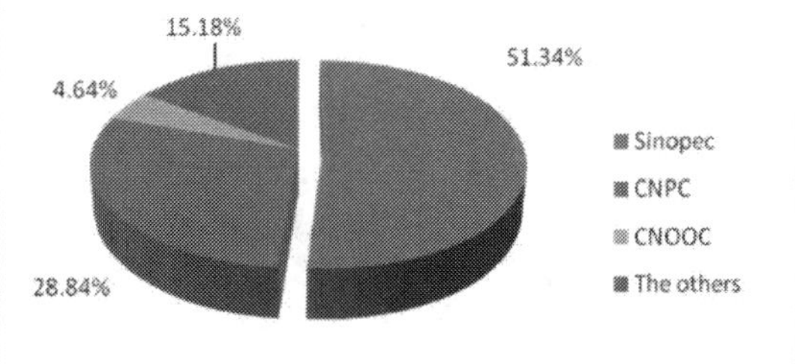

FIG. 2. Ethylene capacity distribution in China

As a responsible energy and petrochemical company, Sinopec sets 'fueling beautiful life' as its corporate mission, puts 'people, responsibility, integrity, precision and win-win' as its corporate core values. It always pays great attention to environmental protection and keeps striving to achieve green growth and contribute to clear waters, green land and a blue sky.

2. Low-carbon Development Strategy of Sinopec

2.1 Adding low-carbon strategy to the company development strategy

In August 2011, Sinopec set the development vision of building a world leading energy and chemical company. The company has also fine-tuned the sustainable development strategies, and low carbon development was added into the company development strategies(FIG.3)

FIG.3. Development strategies of Sinopec

In 2013, the Energy Management and Environment Protection Department was established, which is responsible for implementing the low-carbon development strategy of Sinopec. its main responsibilities include planning of low-carbon development of Sinopec, management of energy saving and emissions reduction, management of carbon assets of Sinopec etc.. The construction of low-carbon management system is enhanced further.

2.2 Main tactics on low-carbon development

Climate change is a critical global issue that challenges us all. As a responsible energy and chemical company, Sinopec takes it a major responsibility to cope with climate change, and is working hard to reduce GHG emissions by speeding up structural adjustment, promoting energy saving, reducing energy consumption, advancing carbon asset management and developing circular economy. The main tactics taken by Sinopec are shown in FIG. 4.

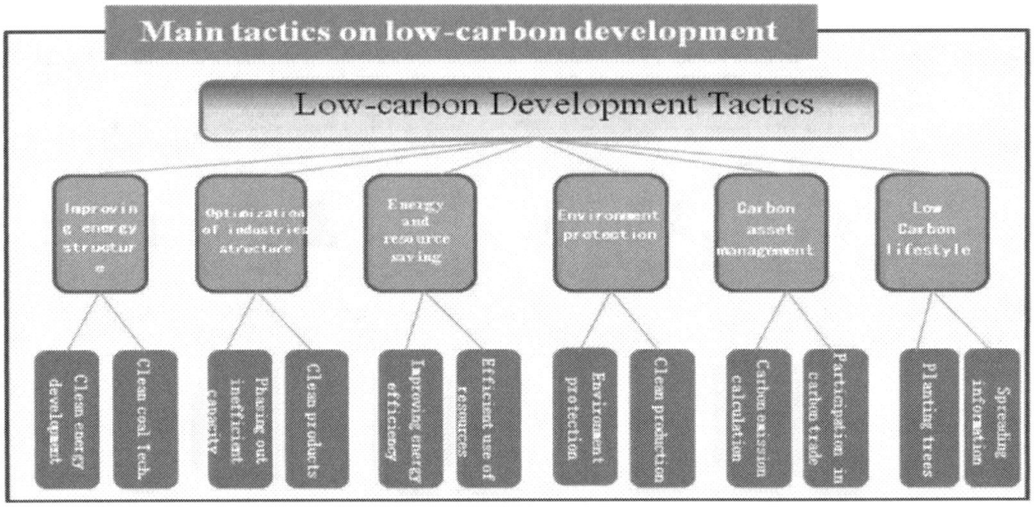

FIG.4. Illustration of main tactics of low-carbon development of Sinopec

2.2.1 Speeding up structural adjustment

Adjustment of industrial structure and energy mix is one of the important measures to reduce CO_2 emission. We can significantly cut CO_2 emission through development of low-carbon industries and utilization of low-carbon energy.

From 2007 to 2014, Sinopec has shut down 21.6 million tonnes of outdated refining capacity and increased refinery's average scale by 43%. Also, Sinopec has exited from polyester filament yarn and urea businesses, shut down two sets of 140,000-tonnes-per-year ethylene facilities. The technical and economic indicators of refining and chemical business have improved continuously.

Sinopec endeavors to develop and utilize low-carbon energy, raise the weight of low-carbon energy in production and consumption, so as to cut down CO_2 emission from the source. Therefore, Sinopec continues to increase natural gas supply, pace up shale gas development, develop and utilize bio-fuels and geo-thermal resources positively. For example, by the end of 2014, Sinopec has built up geo-thermal heating capacity of 23.23 million square meters, accounting for 30% of the national total geo-thermal heating areas. The geo-thermal business can provide clean heating for nearly 200,000 families, replace 320,000 tonnes of standard coal, and reduce CO_2 emission by 850,000 tonnes.

2.2.2 Enhancing energy saving

Energy saving is the most direct and effective measure to reduce CO_2 emission at the current phase. Sinopec always attaches great importance to energy saving, persists in the priority of energy saving along with the growing of the company, and endeavor to build Sinopec into an energy-saving company. On 26 June 2014, Sinopec initiated "Doubling Energy Efficiency" program on 2014 China Summit on Caring for Climate held by United Nations Global Compact China Network. It is another important move of Sinopec on green and low-carbon development and ecological civilization in the wake of the "Clear Water and Blue Sky" program launched in 2013. Moreover, Implementing energy performance contracting regulations were modified and implemented. By the end of 2014, "Doubling Energy Efficiency" initiative accumulatively saved 2.47 million tonnes of coal equivalent; energy performance contracting projects saved 57,000 tonnes of coal equivalent.

2.2.3 Strengthening carbon assets management

Establishing sound carbon asset management system, strengthening carbon assets management and forming economic methods to restrain GHG emissions will not only help to reduce GHG emissions, but also increase the value of carbon assets. Sinopec has built up regulations on carbon assets management, implemented carbon assets inspection within the company and participated in pilot carbon trade in China. In 2014, Sinopec traded 2.1 million tonnes of CO_2 with a total transaction value of 86.34 million yuan. Most of Sinopec subsidiaries made the first deal in several pilot carbon exchanges, helping nurture a domestic carbon trade market. In addtion, Sinopec continues to push forward CCUS tests so as to accumulate technologies and experiences.

2.2.4 Circular economy

Sinopec is actively developing circular economy, pushing forward high efficiency and cyclic utilization of resources and enhancing comprehensive utilization of by-products and three wastes. For example, in 2014, Northwest Oilfield of Sinopec increased light hydrocarbon production by 13,300 tonnes, LNG 25,000 tonnes, and power generation 50 MW, which equals saving 179,000 tonnes of standard coal. moreover, Yangzi Petrochemical Company's syngas unit realized utilization of waste by using 62,000 tonnes of high-purity CO_2 from ethylene glycol unit to produce CO.

2.3 Strategic targets of low-carbon development of Sinopec

Firstly, with the goal of low-carbon development, Sinopec will take the world-class energy chemical companies as the benchmark and become a leader of green and low-carbon development in China. Moreover, turning from the petrochemical industry to energy chemical industry, Sinopec pursues to realize the diversification of energy structure, solve the problem of traditional fossil energy carbon emissions, realize the green and low-carbon development.
In 2011, low-carbon development became the company's development strategy. In 2013, the company began to disclose information on carbon emissions, and took activities to reduce carbon emissions. In 2020, the company will contribute to reducing the intensity of CO_2 emission per unit of GDP in 2020 by 40-45% compared with the level of 2005.

3. CCS/CCUS Practices in Sinopec

3.1 CO_2-based EOR practices

The combined technology of CCS and EOR is to utilize captured CO_2 to enhance the recovery of crude production. It will be an important measure to reduce CO_2 emission. In recent years, Sinopec is vigorously carrying out CCUS research and tests in Shengli, Zhongyuan, East China, Northeast oilfields. So far, about 1 million tonnes of CO_2 was accumulatively injected and 188,000 tonnes of crude production was increased. In 2014, 345,000 tonnes of CO_2 was injected and 43,000 tonnes of crude production was increased. Table 1 lists the general profile of the Industrial Demo projects.

TABLE 1 General parameters of the Demo projects

Items	Demo projects NO.1	Demo projects NO.1	Demo projects NO.1
Location	Zhongyuan	Northeast oilfield	Shengli
CO_2 capture scale,tpy	20000	500000	30000
CO_2 resource	Refinery flu gas	CO_2 from nature gas	Coal-fired power plant
CO_2 purity, v%	17	23	14

CO_2 Capture and Purification Project in Shengli Power Plant is a fully integrated project which was implemented in 2010. The main procedures include CO_2 capture, compressing, dehydrating, liquefaction and storage, as shown in FIG.5.

FIG.5. Flow diagram of CO_2-EOR procedures in Shengli oil field

In addtion, there are two planning large scale CO_2 capture projects being planned in Sinopec. The first one is the 0.5 million tpy CO_2 capture project in Qilu Petrochemical Plant, Shandong province, the original CO_2 purity is about 91v%. Assessment of feasibility study has been finished. The other is the 1.0 million tpy CO_2 capture project in Shengli Power Plant, Shandong province, the original CO_2 purity is 12-14v%. Feasibility study report is being assessed. Pipeline route map of the planning projects is shown in FIG.6.

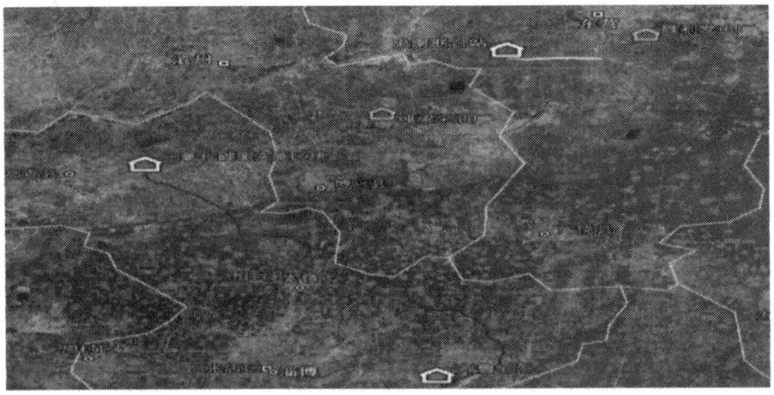

FIG.6. Illustration of the planning pipelines' route map

3.2 Conversion of CO_2 into Chemicals

As for the CO_2 resources which are located far away from the oil fields, what can be done with the enormous amount of CO_2 once captured? Fortunately, as medium-term consideration, a lot of efforts have been put to develop various CO_2 conversion technologies in recent years.

The major research institutes of Sinopec have focused on the following conversion technologies. For example, the research of the transformation of biomass and CO_2 into chemicals such as butanedioic acid or butanediol, the research of CO_2 reforming of CH_4 to produce syngas have achieved notable progress.

In addition, Sinopec and Sichuan University jointly initiated the research on direct mineralization process of CO_2 and phosphogypsum and co-production of sulfur-based compound fertilizer. The research has entered pilot phase.

Although there are still many notable economic hurdles for the listed CO_2 conversion technologies according to the overall assessment and products market investigation, opportunities exist for technology developers to yield efficiency gains.

3.3 Bio-utilization of CO_2

In the long run, CO_2 -based bio-fuels will be one of the important utilization ways. Currently there are many researchers devoted to this field. Sinopec has also started the research of fuels production from algae since 2009. In 2010, the algae cultivation lab was built in one of the refineries of Sinopec and a series of researches on cultivation of algae and fuel production have been operated.

4. Summary

Sinopec is committed to green and low-carbon development, it has contributed to the planning and deployment of CCS/CCUS in China.

There are still many challenges faced by the development and demonstration of CCS/CCUS projects in Sinopec, such as technology maturity, high cost, financial requirements, operation and monitoring, assessment of environmental impact and safety, public confidence etc..

It is essential for Sinopec to enhance further international cooperation in implementing both the low-carbon strategy and the complex CCS/CCUS projects.

CMTC-438891-MS

Chemosequestration in Combustion Engines

Nnamdi A. Nwaka and Olukunle O. Owolabi, University of Ibadan.

This paper was prepared for presentation at the Carbon Management Technology Conference held in Sugarland, Texas, USA, 17–19 November 2015.

Abstract

Climate change, resulting from combustion of fossil fuels have become of major concern to the Global Community, especially in the last two decades. Fossil fuels, no doubt are indispensable to the economic development of most countries and facilitate better living conditions for teeming populations. Improved efficiencies of combustion engines (internal and external) also present realistic and cost effective opportunities to reduce emissions and complement global efforts that is currently largely focused on the adoption of renewable energy.

This paper examines CO_2 reductions in combustion engines by the process of Chemical Carbonation through the application of the Law of Constant Composition. We evaluate the mechanical and chemical processes involved in chemosequestration using aqueous solution of Magnesium Silicate (naturally occurring Serpentinite) while defining clearly the conditions for reaction. An initial design and the method of operation for the carsequestor (the sequestration device) are presented after a proper volumetric analysis of a variable case performed under specified conditions. Cost analysis, global emissions reduction estimation and limitations are also discussed. This paper shows that the entire process is feasible and will reduce CO_2 emissions significantly while maintaining fuel diversity for sustainable growth.

Keywords: Carbonation, Serpentinite, Emissions reduction, Carsequestor, Sequestration.

Introduction

Heat engines have contributed immensely to global economic and social development since its invention. Could you think about the state of the world this moment without heat engines? In as much as the losses would be unbearable, it is fast becoming an option since the aftermaths of their use pose a threat to life and our environment through global warming and climate change. Global warming is the continuous increase in the average temperature of the Earth's atmosphere and its oceans, a change that is seen to be permanently changing the Earth's climate. The rise in the volumes of carbon dioxide and other greenhouse gases (GHGs) released by the use of fossil fuels, land clearing, agriculture, and other human activities, are believed to be the principal sources of the global warming that has occurred over the past 50 years [Intergovernmental Panel on Climate Change, 2007]. Evident changes resulting from global warming include: sea levels rise due to melting of the polar ice caps, change in the amount and pattern of precipitation as well a probable desert expansion, frequent and severe storms, ocean acidification and species extinction due to shifting temperature region, decreasing crop yield - a threat to food security, inundation leading to habitat loss, drought and other severe events.

Over the past three decades, efforts to avert the situation have dwelt greatly on carbon sequestration and adoption of renewable (cleaner) sources of energy. Through the Kyoto Protocol (an international agreement), tremendous results have been achieved in terms of global awareness, energy diversification and global reductions of emissions through its mechanisms. Presently, the major contributors of CO_2 to the atmosphere are fossil fuel use which account for about 62% of CO_2 emissions according to International Energy Agency [2014]. Conversely, the numerous benefits from fossil fuel use is something to put to mind. Employment opportunities, raw materials generation, cheapness and availability of this energy source have kept it in the game.

Chemosequestration denotes capturing and storage by the use of a chemical. Usually, the CO_2 captured is converted to a stable compound that could be a raw material. It is an attempt to sequester CO_2 from the release point before it gets out to the atmosphere. After all, who says we cannot capture what we create? Bulk sequestrations are most difficult in transportation systems. But with significant emissions from transportation sources, unit chemosequestrations become the best option. In order to chemosequester in a vehicle for instance, there is need for the design of a sequestration device particular to that vehicle. But first, we have to be able to determine the average power requirement of a sample engine, then we determine the rate of production of CO_2. The rate of CO_2 production, the rate of sequestration as well as an optimum refill time will enable us design the carsequestor - the chemosequestration device.

Case Description

The hydrocarbon fuel used in most combustion engines are in most cases, not pure compounds but complex mixtures of isooctanes and other additives. Consider an engine X which is a four stroke gasoline engine that powers a car. Engine X is without a secondary condenser and is cooled by a water jacket. Assuming, for analysis, the gasoline to be of pure isooctane, the combustion equation is simply:

$$C_8H_{18} + 12.5O_2 \rightarrow 8CO_2 + 9H_2O$$

Considering the Nitrogen (78% of air), the stoichiometric equation becomes:

$$C_8H_{18} + 12.5O_2 + 12.5(3.76)_2 \rightarrow 8CO_2 + 9H_2O + 12.5(3.76)_2$$

The fuel-air mixture could be for example 150% of stoichiometric with an excess air. This is called a lean mixture as a result of the fuel deficiency and there would be complete combustion. If on the other hand, the fuel-air mixture is 80% of stoichiometric, there is excess fuel and carbon monoxide will be produced and this mixture is called rich mixture. For engine X, in all cases, there is excess air and complete combustion.

Determination of Rate of CO_2 Production from Engine X

By the law of conservation of mass and assuming negligible CO_2 content of air, the amount of carbon (in the gasoline) injected into engine X should be equal to the amount ejected at exhaust. The rate of fuel intake is however determined by the total power consumption (fuel power) of the engine and is given by:

$$FR = P_f/LHV$$

Where:

FR = Fuel Rate in grams per second
P_f = Fuel Power in Watts
LHV = Lower Heating Value of Gasoline (approximately 43.7 kJ/g for gasoline)

Note: The lower heating value is used since engine X has no secondary condenser and the water vapor generated from combustion leaves with the exhaust stream. The fuel power depends on the power demand and the overall (thermal) efficiency of the engine. The power demand, according to Nam and Giannelli [2005], is based on overcoming inertia, road grade, tire friction, and aerodynamic loss. This road-load methodology was mainly introduced by Sovran and Bohn [1981].

The power demand (in Watts) is the brake (or tractive) power:

$$P_b = mv\,[a\,(1+f) + g\,{*}grade + gC_R] + 0.5\rho C_D A_F v^3$$

Where:

v = vehicle speed (assuming no headwind) in m/s (or mps)
a = vehicle acceleration in m/s^2
f = mass factor accounting for the rotational masses (~0.1)
g = acceleration due to gravity (9.8 m/s^2)
$Grade$ = road grade
C_R = rolling resistance (~0.009)
ρ = air density (~1.2 kg/m^3)
C_D = aerodynamic drag coefficient (~0.3)
A_F = frontal area in meters^2 (~2 m^2)
m = vehicle mass in metric tonnes.

The rotating mass term f, is assumed to be 0.1 [Jimenez, 1999]. However it increases at lower gears [Gillespie, 1992]. The rolling resistance for radial tires, C_R, can range from 0.008 – 0.013 for a majority of the on- road passenger car tires. This value can be larger depending on state of inflation, temperature, ground surface type, and speed (though this effect is minor) [Bosch, 2000; Gillespie, 1992]. Heavy-duty trucks tend to have lower resistance. The aerodynamic drag, C_D, varies according to the body shape, and is sometimes provided by the manufacturers [Nam and Giannelli, 2005]. The frontal area, A_F, of the vehicle is also sometimes supplied and it is unknown, 93% of the box frontal area is used based on a number of vehicles examined by the Nam and Giannelli [2005].

This is calculated using the dimensions available for the vehicles:

$$A_F = (H - GC) * W * 0.93$$

Where:

H = vehicle height in meters
GC = ground clearance in meters
W = width in meters

The thermal efficiency of the Otto cycle (the ideal cycle for gasoline engines) under the cold air standard assumptions is:

$$\eta_{th,otto} = 1 - \frac{1}{r^{k-1}}$$

Where:

η_{th} = thermal efficiency of the Otto cycle
r = compression ratio
k = specific heat ratio

However, for a given compression ratio, the thermal efficiency of engine X is less than that of an ideal Otto cycle because of the irreversibilities, such as friction and the thermal efficiencies of actual sparkignition engines range from about 25 to 30 percent [Çengel and Boles, 2006]. The fuel power is then determined thus:

$$P_f = \frac{P_b}{\eta_{th,otto}}$$

From, the stoichiometric equation, 114g of C_8H_{18} produces 352g of CO_2. Assuming complete combustion, the rate of CO_2 (in grams/s) that will be produced from the already determined fuel rate is:

CO_2 rate = 3.09FR

Note: This process can be used to determine the CO_2 rate for any combustion engine provided the required expressions are used and velocity-time graphs can be used to account for the different stages in motion for the vehicle.

Reaction between Aqueous Serpentine and Carbon dioxide

Direct aqueous mineral carbonation involves the reaction of CO_2 with alkaline minerals in an aqueous suspension in a single stage [Sipilä et al., 2008]. CO_2 reacts with water to form bicarbonate and a proton, the proton liberates the metal ion from the mineral, and the metal ion reacts with bicarbonate to produce a carbonate precipitate [O'Connor et al., 2000a]. The direct aqueous carbonation of serpentine is shown in the following reactions:

$$CO_2 + H_2O \rightarrow H_2CO_3 \rightarrow H^+ + HCO_3^-$$

$$Mg_3Si_2O_5(OH)_4 + 6H^+ \rightarrow 3Mg^{2+} + 2SiO_2 + 5H_2O$$

$$Mg^{2+} + HCO_3^- \rightarrow MgCO_3 + H^+$$

Elevated pressure and temperature enhance direct aqueous mineral carbonation. Increased pressure facilitates the dissolution of CO_2, thereby driving the reaction forward. Increases in temperature decrease the solubility constant of $MgCO_3$ and increase the dissociation of aqueous CO_2 to bicarbonate, and then to carbonate. Increasing temperature, however, decreases the solubility of CO_2, placing an upper limit on the degree to which increased temperature can enhance the reaction [Erin Rae Bobicki, 2014].

The reaction rate can be further increased by using a bicarbonate/salt ($NaHCO_3$/NaCl) mixture as the slurry supernatant [Metz et al., 2005; Huijgen and Comans, 2003; Maroto-Valer et al., 2005; Sipilä et al., 2008; O'Connor et al., 2000a], adding acids or other chemical activators [Maroto-Valer et al., 2004], and employing various pre-treatment techniques, such as crushing to small particle sizes (typically < 75 μm) [O'Connor et al., 2000a; Gerdemann et al., 2003; Alexander et al., 2007], magnetic separation to remove any magnetite liberated during mineral comminution (iron compounds form a passivating layer on silicate particles and inhibit carbonation) [Alexander et al., 2007], and heat treatment to remove hydration water in the case of serpentine [Gerdemann et al., 2007; Gerdemann et al., 2003; O'Connor et al., 2000b].

Serpentine Pre-treatment Route

Size reduction

The first step In order to achieve a reasonable reaction rate is to grind the minerals. The reaction rate increases as the surface area increases. O'Connor et al. [2000b] found that a reduction from 106-150μm to <37μm increased the conversion in their experiments from 10% to 90%.

High-energy attrition grinding induces imperfections into the crystal lattice [Gerdemann et al., 2002]. This results in a higher conversion than size reduction to the same diameter using 'normal' grinding [Huijgen & Comans, 2003].

Magnetic separation

The oxidation of iron (magnetite) slows down the carbonation of serpentine due to the formation of a layer of hematite on the mineral surface [Fauth et al., 2000]. Magnetic separation of the iron compounds prior to the carbonation process resolves this complication [NETL, 2001]. Furthermore, a potentially marketable iron ore by product is formed [O'Connor et al., 2001a]. When a combination of magnetic separation and thermal treatment is used, it is more effective first to conduct the magnetic treatment step [O'Connor et al., 2000a].

Thermal treatment

Serpentine contains up to 13wt% chemically-bound water. By heating the serpentine to 600- 650°C the water is removed and an open structure is created [O'Connor et al., 2000b]. This significantly improves the reaction kinetics owing to the increase reactive surface. The mineral porosity can also be increased by treatment with steam [NETL, 2001] or supercritical water (T=385°C, p=272atm) [O'Connor et al., 2000b].

It is worthy to note that research has been focused on understanding the reaction mechanisms and kinetics [Prigiobbe et al., 2009, Dufaud et al., 2009], improving the kinetics by varying the process conditions and contents of the aqueous phase [Krevor and Lackner, 2009; Zhao et al., 2010], developing improved pre-treatment techniques (Haug et al., 2010; Jang et al., 2010) and studying the applicability of different kinds of waste products as source materials [Uibu et al., 2010; Baciocchi et al., 2009; Wang et al., 2010].

As seen previously, the products of the combustion reaction are carbon dioxide, water and Nitrogen. Since nitrogen is an unreactive gas, it is not needed in the carbonation process. The exhaust gases are then passed through a selective membrane in which the CO_2 is allowed to pass through more easily under high pressure. The active reactant therefore becomes the CO_2 alongside the pre-treated serpentine in aqueous solution. The suggested pre-treatment route for the serpentine is same as written above:

SIZE REDUCTION → MAGNETIC SEPARATION → THERMAL TREATMENT

The thermally heated and grinded serpentine is then passed into a chamber in which it mixes with the product of the combustion reaction (exhaust). The reaction of carbon dioxide in the chamber is given below:

$$Mg_3Si_2O_5\,(OH)_4 + 3CO_2 + H_2O \rightarrow 3MgCO_3 + 2SiO_2 + 3H_2O$$

Amount of Carbon Captured

Stoichiometric requirement

From, the stoichiometry of the above equation, ***276g of*** $\boldsymbol{Mg_3Si_2O_5(OH)_4}$ on reaction with ***132g of*** $\boldsymbol{CO_2}$ will ideally produce ***252g of*** $\boldsymbol{MgCO_3}$ (this is correct if all the CO_2 is captured). This means that under appropriate conditions for every 1g of CO_2 captured, 1.9g of $MgCO_3$ will be produced.

In rate condition, every 1g/min of gasoline expended, 3.09g/min of CO_2 will be produced and 3.09*1.9= 5.871g/min of CO_2 will be obtained in form of $MgCO_3$. On the operational side this implies that the volume required for CO_2 storage will exceed that displaced by gasoline.

Amount of Serpentine Required

Stoichiometric requirement

Starting with serpentine ore, $\boldsymbol{Mg_3Si_2O_5(OH)_4}$, we can calculate the stoichiometric amount required per g of CO_2 based on the following reaction:

$$1/3Mg_3Si_2O_5(OH)_4 + 1/3H_2O + CO_2 \rightarrow MgCO_3 + 2/3SiO_2 + 2/3H_2O \qquad (\alpha)$$

(? mol serpentine/ mol CO_2)*(mol CO_2/44 g CO_2)*(276 g serpentine/ mol serpentine) = **2.09g serpentine per g of CO_2**

Assuming a carbonation efficiency of 70% which is reasonably close to the 73.5% determined by Sipilä et al [2007], the actual serpentine requirement becomes 2.09g/0.7 **=2.99g serpentine per g of CO_2 sequestered**

Amount of Water Required

Stoichiometric requirement

From the carbonation reaction, we can calculate the stoichiometric amount of water required per g of CO2 based the reaction (α) to be:

(? mols water/ mol CO_2)*(mol CO_2/44 g CO_2)*(18g water/ mol water) **=0.14g water per g of CO_2**

The actual amount of water required is therefore 0.14/0.7= **0.2g water per g of CO_2 sequestered**

Amount of Solid Products

Amount of $MgCO_3$ and SiO_2 produced per g of CO_2 reacted = (84g/mol + 2/3 x 32g/mol)/44g/mol = **2.4g solids per CO_2 reacted**

Actual solids produced = 0.7 x 2.4 = **1.68g solids produced per CO_2 reacted**

The size of the precipitate obtained after direct aqueous carbonation using serpentine mineral has not been widely discussed but for this study we shall use the experimental result obtained by Park et al. [2004] where the size of $MgCO_3$ precipitate is 10μm.

Carsequestor Design

The inlet pipe allows for a higher flow velocity and directs the fumes to the inlet diverter. The inlet diverter has a down-comer that directs the fumes downwards below the liquid interface. The diverter makes use of the higher velocity of the fumes to achieve this. The maximum liquid level is somewhat below the level of the inlet pipe. As the exhaust gases are washed through the liquid section, the CO_2 is trapped back by the CO_2-philic solution in the liquid chamber which is in constant agitation to promote the reaction.

Figure 1: Carsequestor Cross-Section

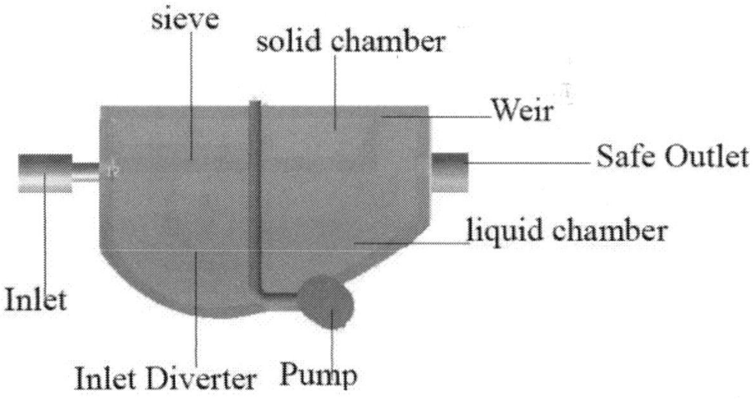

The electric pumps pump the liquid/solid mixture upwards unto the sieve in the solid section. The solid section is bounded below by a <10microns mesh that sieves the solid out. The weir allows for over flow while providing sufficient retention time for the solid particles to settle should in case the sieving rate slows down. The weir height is determined by the sum of the solid chamber height and the solid travel height (depending on the terminal velocity and desired retention time). The safe outlet, which contains a mist extractor, allows for passage of unreacted gases.

Figure 2: Carsequestor in 3D

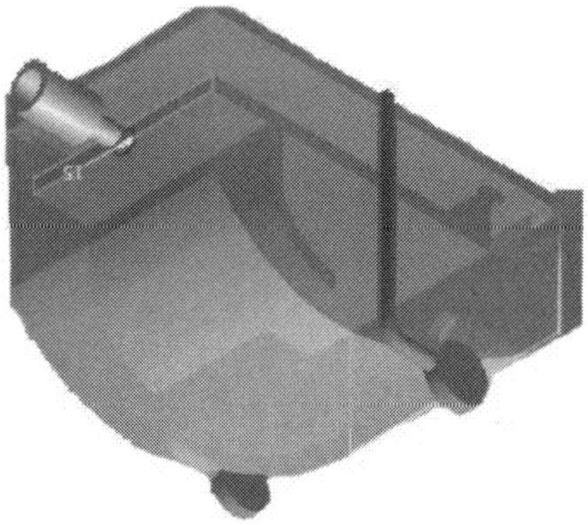

Carsequestor Functional Requirements

Several requirements were considered before designing a carsequestor for a specific application. These functional requirements are detailed below:

Backpressure

Backpressure indicates the extra static pressure that will be exerted by the carsequestor on the engine through the restriction in flow of exhaust gases. The gases will be restricted by the diverter as well as the liquid surface in an attempt to break the surface tension. The agitation of the liquid is therefore essential to reduce back pressure and promote reaction.

Size

The available space will have a great influence on the size of a carsequestor. But this has to be included in vehicular design. The size of the carsequestor will have to be optimized while considering space availability and optimum refill time for the carsequestor. The available space has a great influence on the size and therefore the carsequestor that may be used. A carsequestor may have its geometry designed for optimum performance however if it does not meet the space constraints, it is useless.

Cost

The effectiveness of this device is not the only concern but also the cost. Ideally, carsequestors are required to be made from corrosion resistive materials such as stainless steel or aluminium. The overall cost will be determined by the cost of the materials, cost of the refill solution and labor costs.

Refill and Solid Disposal

The viability of the carsequestor depends greatly of the reasonability of the refill life. Proper disposal channels are also required for efficient operation.

Global Emissions Reduction and Significance

In 2012, global CO_2 emissions were estimated by IEA [2014] to be at 31.7GtCO_2. Transportation alone expended about 23% while electricity and heat accounted for 42%. (Figure 3) The use of the carsequestor in the entire transport system alone, with an estimated efficiency of 70%, will reduce emissions by 5.1 GtCO_2 annually. With further use in industrial systems as well as electricity and heat generation, more significant reductions could be seen. The use of the carsequestor will inevitably make significant impacts in societal environment through fuel diversification, employment generation and more resource utilization.

Figure 3: World CO_2 emissions by sector in 2012

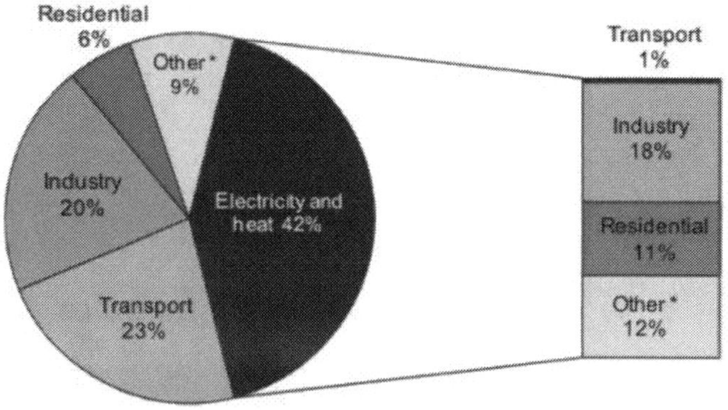

Source: IEA (2014). "CO2 emissions from fuel combustion highlights"

Conclusion

The term carsequestor is evidently coined from carbon and sequestration. This clearly reflects the principal aim of works done in this respect which is to reduce emissions and protect our environment. Amidst the wide campaigns for adoption of renewables, the numerous benefits from fossil fuels (petroleum especially) cannot be dismissed. Bearing in mind that this publication aims to illuminate conceptions, there is great need for more research work to improve the performance of the proposed carsequestor. However, the mobility in transportation systems makes the carsequestor the only available option for it as long as it is driven by heat engines. For engine X and for any other heat engine, conscious individual as well as collective responsibility must be put in to see to the success of the carsequestor.

References

Alexander G., Maroto-Valer M. M. & Gafarova-Aksoy P. (2007). *Evaluation of reaction variables in the dissolution of serpentine for mineral carbonation*. Fuel. 86, 273-281.

Baciocchi R., Costa G., Polettini A., Pomi R. & Prigiobbe V. (2009). *Comparison of different reaction routes for carbonation of APC residues*. Energy Procedia. 1, 4851-4858.

Bobicki E. R. (2014). *Pre-treatment of Ultramafic Nickel Ores for Improved Mineral Carbon Sequestration*. Department of Chemical and Materials Engineering, University of Alberta, Canada.

Bosch. (2000). *Automotive Handbook, (5th Ed)*. Stuttgart, Germany: Robert Bosch GmbH.

Çengel Y. A. & Boles M. A. (2006). *Thermodynamics: An Engineering Approach*, (5th ed). New York: McGraw-Hill.

Dufaud F., Martinez I. & Shilobreeva S. (2009). *Experimental study of Mg-rich silicates carbonation at 400 and 500 ℃ and 1 kbar*. Chemical Geology. 265, 79-87.

Fauth D. J., Baltrus J. P., Knoer J. P., Soong Y., Howard B. H. Graham W. J. Maroto-Valer M. M.& Andresen, J. M. (2001). *Conversion of silicate minerals with carbon dioxide producing environmentally benign and stable carbonates*, Preprints of papers, Division of Fuel Chemistry, American Chemical Society 46/1, pp. 278-279.

Gerdemann S. J., Dahlin D. C. & O'Connor W. K. (2003). *Carbon Dioxide Sequestration by Aqueous Mineral Carbonation of Magnesium silicate Minerals*. Conference Proceedings GHGT-6 Kyoto Japan Oct 2003.

Gillespie, T. (1992). *Fundamentals of Vehicle Dynamics*. Warrendale, Pennsylvania: Society of Automotive Engineers.

Haug, T. A. & Kleiv, R.A., Munz, I.A. (2010). *Investigating dissolution of mechanically activated olivine for carbonation purposes*. Applied Geochemistry. 25(10), 1547-1563

Huijgen W. J. J. & Comans R. N. J. (2003) *Carbon dioxide sequstration by mineral carbonation: Literature review update 2003-2004*. ECN-C--05-022, Energy Research Centre of The Netherlands, Petten, The Netherlands.

Intergovernmental Panel on Climate Change. (2007). *Climate Change 2007: IPCC Fourth Assessment Report*. Retrieved from https://www.ipcc.ch/ipccreports/far/wg.../ipcc_far_wg_II_full_report.pdf

International Energy Agency. (2014). *CO_2 Emissions from Fuel Combustion, Highlights*. Paris, France: Author.

Jimenez-Palacios, J. (1999). *Understanding and Quantifying Motor Vehicle Emissions with Vehicle Specific Power and TILDAS Remote Sensing*, PhD Thesis, Massachusetts Institute of Technology, Cambridge, MA.

Krevor, S. C. & Lackner, K. S. (2009). *Enhancing process kinetics for mineral carbon sequestration*. Energy Procedia. 1, 4867-4871.

Maroto-Valer M. M., Kuchta M. E., Zhang Y., Andrésen J. M. & Fauth D. J. (2005). *Activation of magnesium rich minerals as carbonation feedstock materials for CO2 sequestration*. Fuel Processing Technology. 86, 1627-1645.

Metz B., Davidson O., De Coninck H. C., Loos M. & Meyer L. A. (Eds.) (2005). *IPCC Special Report on Carbon Dioxide Capture and Storage*. Prepared by Working Group III of the Intergovernmental Panel on Climate Change. Cambridge, UK and New York, NY: Cambridge University Press. p.443.

Nam E. K. & Giannelli R. (2005). *Fuel Consumption Modeling of Conventional and Advanced Technology Vehicles in the Physical Emission Rate Estimator (PERE), Draft*. Retrieved from www3.epa.gov/otaq/models/ngm/420p05001.pdf

National Energy Technology Laboratory (2001). *Proceedings of Workshop NETL Mineral CO_2 Sequestration*. Pittsburgh, Pennsylvania: Author.

O'Connor, W. K., Dahlin, D. C., Rush, G. E., Dahlin C. L., & Collins W. K. (2000) "*Carbon Dioxide Sequestration by Direct Mineral Carbonation: Process mineralogy of Feed and Products*", Minerals & Metallurgical Processing, Vol. 19 No. 2, pp. 95-101.

O'Connor W. K., Dahlin D. C., Nilsen D. N., Rush G. E., Turner P. C. & Walters R. P. (2000) "*CO_2 Storage in Solid Form: A Study of Direct Mineral Carbonation*," Proc. of the 5th Inter. Conference on Greenhouse Gas Technologies, Cairns, Australia, August 14-18, 2000, p 7.

Park A. & Fan L. (2004). *CO2 mineral sequestration: physically activated dissolution of serpentine and pH swing process*. Department of Chemical Engineering, The Ohio State University, 140 W. 19th Ave. Columbus, OH 43210, USA.

Prigiobbe V., Hänchen M., Werner M., Baciocchi R. & Mazzotti, M. (2009). *Mineral carbonation process for CO_2 sequestration*. Energy Procedia. 1, 4885-4890.

Sipilä J., Teir S. & Zevenhoven R. (2007). *Carbon dioxide sequestration by mineral carbonation: Literature review update 2005-2007*. Åbo Akademi University, Heat Engineering Laboratory, Turku, Finland.

Sovran G. & Bohn M. S. (1981). "Formulation for the Tractive Energy Requirement of Vehicles during the EPA Schedules. SAE810184.

Uibu M., Velts O. & Kuusik R. (2010). *Developments in CO2 mineral carbonation of oil shale ash*. Journal of Hazardous Materials. 174, 209-214.

Zhao L., Sang L., Chen J., Ji J. & Teng H. H. (2010). *Aqueous carbonation of natural brucite: Relevance to CO_2 sequestration*. Environmental Science and Technology. 44, 406-411.

Chemosequestration in Combustion Engines

Nnamdi A. Nwaka and Olukunle O. Owolabi

Department of Petroleum Engineering, University of Ibadan, Nigeria.

ABSTRACT

Our environment has been significantly marred by climate change and global warming. These arise from increasing volumes of GHGs among which CO_2 is the chief culprit. The principal sources of CO_2 are however our principal artificial sources of energy which are the combustion engines.

This work examines chemosequestrations in combustion engine systems by the use of a proposed device called the Carsequestor. The carsequestor is meant to facilitate reaction between the generated CO_2 and aqueous treated serpentine.

INTRODUCTION

- Heat Engines have been very beneficial to mankind since time past.
- In the midst of these benefits, its use and most importantly our environment are threatened by the emission of CO_2.
- Carbon sequestrations and adoption of renewables have, for some time, become viable means to reduce global emissions.
- Bulk sequestrations in transportation systems are however difficult due to the mobility.

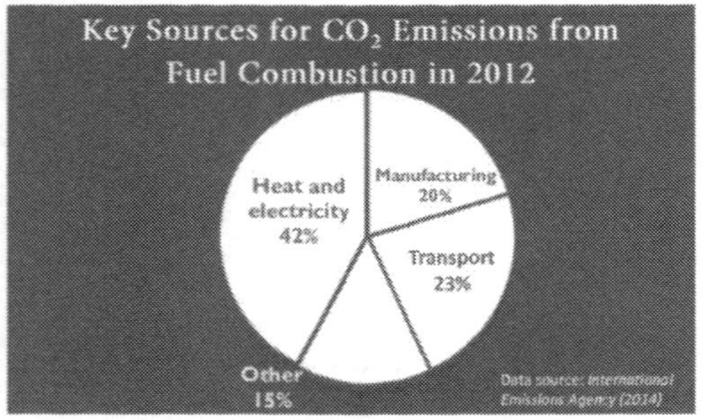

- With significant CO_2 emissions from transportation systems, how can these emissions be reduced?
- The carsequestor (the chemosequestration device) is therefore proposed to facilitate unit chemosequestrations in such engines.
- The carsequestor facilitates the reaction:

$$Mg_3Si_2O_5(OH)_4 + 3CO_2 + H_2O \rightarrow 3MgCO_3 + 2SiO_2 + 3H_2O$$

- The soilid products are separated from the solution.

METHODS

Determination of the Power Demand of the system using the Road-Load methodology.

⇓

Conversion of the Power Demand to Fuel Power (power input) using the engine thermal efficiency.

⇓

Conversion of the Fuel Power to Fuel Rate using the heating value of the fuel used.

⇓

Determination of the CO_2 generation rate from the Fuel Rate and a conversion factor obtained from the stoichiometric combustion equation

⇓

Calculation of the stoichiometric amount of Serpentine and water required per gram of CO_2 captured.

⇓

Calculation of the stoichiometric amount of solids produced per gram of CO_2 captured.

⇓

The reactant depletion rate, the solids production rate and optimum reactant refill time will allow for the carsequestor sizing.

CARSEQUESTOR DESIGN

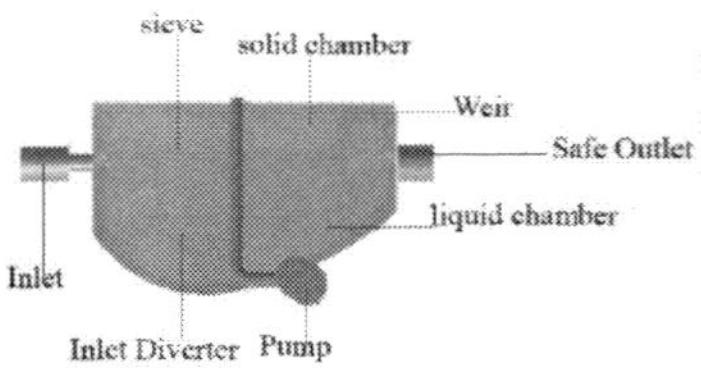

Carsequestor Cross-Section

Carsequestor in 3D

- The inlet pipe directs the fumes onto the inlet diverter.
- The inlet diverter with its down-comer directs the fumes downwards below the liquid interface.
- The liquid chamber contains the CO2-philic serpentine solution which is supposedly in constant agitation to break surface tension and facilitate reaction.
- The electric pumps pump the liquid-solid mixture continuously onto the sieve in the solid chamber.
- The solid chamber houses the sieve and weir. The volume is determined to be about half the size of the liquid chamber volume.

CARSEQUESTOR DESIGN

- The weir is supposedly a thin plate that allows for overflow in the case of a slow sieving rate.
- The sieve is < 10µm mesh screen that discriminates between the liquid and solid.
- The safe outlet allows for passage of unreacted gases. The mist extractor in the safe outlet traps back the liquid droplets.

SIGNIFICANCE

- With 70% carsequestor efficiency, there is an estimated emissions reduction of 5.1GtCO$_2$
- More significant reductions could be seen with further use in industrial systems as well as electricity and heat generation.
- The use of the carsequestor will inevitably make significant impacts in societal environment.

FUTURE WORK

- Research for a non-refill carsequestor.

CONCLUSIONS

- There is undoubtedly reduced emissions from bulk carbon sequestrations and use of renewables.
- The carsequestor is the only available option for mobile engines.
- Conscious individual and collective efforts are required to see to its success.

REFERENCES

- International Energy Agency. (2014). *CO2 Emissions from Fuel Combustion, Highlights*. Paris, France: Author.
- Nam E. K. & Giannelli R. (2005). *Fuel Consumption Modeling of Conventional and Advanced Technology Vehicles in the Physical Emission Rate Estimator (PERE), Draft*. Retrieved from www3.epa.gov/otaq/models/ngm/420p05001.pdf

CARBON MANAGEMENT TECHNOLOGY CONFERENCE

CMTC-439293-MS

Effect of Depth and Leakage-Pathway Flow Properties on Thermal Response to Leakage from CO_2 Storage Zone

Mehdi Zeidouni, Louisiana State University, JP Nicot, University of Texas at Austin, Susan D. Hovorka, University of Texas at Austin, Vanessa Nunez-Lopez, University of Texas at Austin

This paper was prepared for presentation at the Carbon Management Technology Conference held in Sugarland, Texas, USA, 17–19 November 2015.

Abstract

Temperature can be used as a tracer to detect leakage of fluids from a CO_2 storage zone. Brine leakage from the injection zone to an above-zone interval will induce a temperature increase as a result of geothermal gradient. Leakage of CO_2 can induce a temperature decrease owing to the Joule-Thompson effect associated with pressure drop toward shallow zones. A larger pressure drop at shallower depths is associated with more CO_2 expansion upon leakage and could induce more cooling and, hence, a stronger temperature signal. We investigate the strength of the temperature signal as a function of depth for two scenarios in which either a well or a fault acts as leakage pathway. The hydraulic properties of the leakage pathway, modeled as a fractured medium or as a porous medium, also impact the results. Using dual-porosity and dual-permeability models for which CO_2 relative permeability and average absolute permeability are modeled higher than in the simple porous medium case, we study the effect of fractures on the temperature signal. The leakage rate increases significantly for dual medium models. However, the temperature change in above-zone interval does not change much as it depends on the pressure gradient which is reduced compared to the single-porosity medium case.

1. Introduction

The value of CO_2 geological storage sites may be limited when defective wells and faults create pathways allowing migration out of the CO_2 injection zone (IZ). Various monitoring techniques are available to assure storage quality and to detect and characterize leakage pathways. Leakage of CO_2, brine, or their mixture can induce pressure and temperature changes in an above-zone monitoring interval (AZMI) that is separated from the IZ by a sealing confining layer. Monitoring of the pressure signal in the AZMI has provided information about leakage pathways (Zeidouni and Pooladi-Darvish 2012a, b, Sun et al. 2013, Jung, Zhou, and Birkholzer 2013, Haghighat et al. 2013). More recent work by Zeidouni et al. (Zeidouni, Nicot, and Hovorka 2014) investigated the potential for leakage detection on the basis of the AZMI's temperature signal. Results showed that the thermal signal is controlled by several processes, including

Joule-Thompson (JT) effect, heat of dissolution/vaporization of CO_2/water, temperature discrepancy between the injected and native fluids, geothermal gradient, and heat exchange with the surrounding rock-fluid system. Results further showed that the temperature pulse caused by CO_2 leakage is local and restricted to the immediate vicinity of the leakage pathway. However, leakage can result in large temperature changes along the leakage pathway. Running temperature logs along suspicious wells and/or wells intersecting suspicious faults may provide useful information on the leakage potential and amount.

With careful site characterization, it is likely that the injection well will be located far from potential leakage pathways. When CO_2, traveling in the IZ reaches the leakage pathway distant from the injection well, it is already in local thermal equilibrium, owing mainly to the high heat capacity of the rock-fluid system, which maximizes heat adsorption from the injected CO_2. CO_2 dissolution in the brine may cause minor temperature increase at the CO_2 front, which is likely to be negligible compared to the leakage-induced thermal signal, controlled mainly by the JT effect. The CO_2 front is therefore ahead of the thermal signal front. Brine leakage, which precedes the CO_2 leakage, will generate a temperature increase (owing to geothermal gradient) that could be observed in the AZMI. After CO_2 arrival at the leak location, both brine and CO_2 leak. As a result of JT effects, CO_2 cools down upon pressure decline and expansion caused by leakage. The cooling signal corresponding to CO_2 leakage is stronger than and can mask the effect of brine leakage. The cooling signal caused by JT effects is a direct function of the pressure gradient. However, the cooling effect also depends on the magnitude of the pressure because, for the same pressure gradient, the expansion is more significant at lower pressure. Hence, the temperature signal could be weakened with increasing reservoir depth.

In this study, we perform numerical simulations for a range of depths and, thus, initial pressure and temperature conditions, for the IZ and AZMI along with various geothermal gradients to further investigate the thermal signal in the AZMI. The thermal response of leakage is investigated for both leaky wells and leaky faults. In terms of fluid-flow properties, leakage pathways are generally a different medium than the reservoir and may have cracks and fractures. We investigate the effect of assigning different relative permeability and flow behavior to the fractures through the use of a dual-porosity and dual-permeability modelling approach.

The effects are investigated by extending a base-case problem defined in an earlier work (Zeidouni, Nicot, and Hovorka 2014). CO_2 is injected at a rate of 50 kton/year (69,185 m^3/day) into a 55-m-thick brine aquifer (IZ) for two years. A 5-m-thick impermeable caprock separates the IZ from a 10-m-thick AZMI. Both IZ and AMZI have identical flow properties (Table 1). The depth to top of the IZ is 1000 m. The IZ and AZMI are connected by either (a) a leaky well or (b) a leaky fault. The leaky well and leaky fault properties are given in Table 1. The same relative permeability curves (Table 2) are considered for all the fluid-flow media of the system. Injected CO_2 temperature is identical to the initial reservoir temperature (41.7°C), and the geothermal gradient is set at 0.03°C/m. Rock density and heat capacity are 2,650 kg/m^3 and 1,000 J/(kg.K), respectively. Total thermal conductivity is 2.51 W/(m.K). Simulations are run in CMG-GEM (2014) using Cartesian gridding. Because of the system symmetry and in order to reduce computational cost, only a quadrant of the system is simulated, with the injection well at the corner of the model.

2. Effect of depth on the leakage thermal signal

In this section the effect of depth on the strength and trend of the thermal signal is investigated. We take the IZ-Seal-AZMI storage complex to various depths and the investigate temperature signal in the AZMI. We varied depth to top of IZ (seal base), modeling it at 1000, 1500, and 2000 m. We varied initial temperature, considering the temperature gradient to be 0.03°C/m. As a result, temperature at the base of the1500-m IZ is 56.7°C and that for 2000-m IZ is 71.7°C. Also, the injection temperature is considered the same as the base of the IZ. All other properties of the system and the injection temperature are kept constant.

At greater depths, both temperature and pressure are higher. CO_2 density decreases as a function of increasing temperature and increases as a function of increasing pressure. The net effect depends highly on the temperature gradient (Ennis-King and Paterson 2001, Bachu 2003). For the conditions used in this study, the density increases with depth. The increased density means that the injected CO_2 takes less volume upon injection, and thus the breakthrough time for CO_2 at the leak will increase with increasing depth. CO_2 viscosity also increases with density, as shown by the correlations of Fenghour et al. (Fenghour, Wakeham, and Vesovic 1998). Therefore, it takes longer to sense the cooling (caused by JT effect upon CO_2 leakage) when depth is increased. Fig. 1 (a, b) illustrates that the breakthrough time varies from 0.48 to 0.55 to 0.61 years for 1000-, 1500-, and 2000-m-IZ for the leaky-well case. The breakthrough times for the leaky-fault case are slightly shorter as a result of higher leakage rate through the fault (compared to the well case), which reduces resistance to flow.

Cooling will be preceded by brine leakage (and accompanying increasing temperature signal) over some period of time. We can infer that higher temperature increase should be observed as a result of the larger amount of leaked brine preceding the CO_2 leakage when depth is increased. This postulated pattern proves correct for the leaky well (Fig. 2a). However, for the leaky-fault case, the brine leakage rate will be less, owing to lower pressure change at the IZ leading to less cumulative brine leakage and thus a smaller temperature increase (Fig. 2b).

After CO_2 breakthrough in the leaky-well case, the temperature signal will be driven by the JT effect. The larger pressure gradient at shallower depths implies a larger temperature signal. The strength of the cooling signal in the AZMI after 2 years reduces from 5°C in the 1000-m case to 1.8 and 0.8°C in the 1500- and 2000-m-IZ cases, respectively.

In the leaky-fault case, after CO_2 leakage starts, the CO_2 leakage rate rises quickly until it reaches a maximum. The maximum is reached as a result of the fast pressure increase in the AZMI causing a reduced pressure differential (the driving force of leakage). The maximum leakage rate occurs earlier for larger rates and larger depths. After the maximum leakage rate is reached, the cooling signal owing to CO_2 leakage will decrease. However, a larger pressure gradient at lower depth will lead to a larger cooling signal.

3. Effect of leakage-pathway flow properties

The leakage pathway is likely to have a pore structure system different from that of the IZ and AZMI media. In a leakage pathway, porosity is more likely to be secondary porosity resulting from processes such as fracturing and/or dissolution, whereas the porosity of the IZ and AZMI is more likely to be intergranular primary porosity (especially in clastic reservoirs). In this section we investigate the effect of such difference on the thermal signal. We consider the leakage pathway as a fractured medium and model the flow within it using a dual-porosity and dual-permeability approach (Hill and Thomas 1985, Blaskovich et al. 1983) to honor different flow characteristics of the fracture and matrix systems. The dual-permeability approach is an extension of the dual-porosity model presented by Warren and Root (Warren and Root 1963) and Kazemi et al. (Kazemi et al. 1979). Fluid exchange between fracture blocks and between matrix and fracture blocks are accommodated using the dual-porosity approach. The dual-permeability model also accounts for the matrix-to-matrix fluid exchange. The matrix and fracture-flow domains are coupled using a shape factor transfer term. Based on this model, a fracture cell is assigned for every matrix grid block. Therefore, to make this case comparable to the single-porosity problem, the volume of the leakage pathway should be reduced by half. In other words, the cross-sectional area of the leak is reduced to one half the original area. To ensure that the pore volume is the same as in the single-porosity case, the arithmetic mean of the matrix and the fracture porosities must be equal to the porosity used in the single-porosity model (0.1). We consider the porosity of the matrix and fracture to be 0.18 and 0.02, respectively, for an average of 0.1. A simple straight-line relative permeability is used for the fracture medium and the absolute permeability of the fracture medium is considered 100 mD. The matrix permeability is identical to that of the single-porosity case, 10 mD.

With the change in the nature of the fluid-flow within the leakage pathway and the increased average permeability, more leakage of brine is expected prior to CO_2 breakthrough compared to the single-porosity problem, in spite of the reduced leakage-pathway area. The temperature change at the base of the AZMI is shown in Fig 3. More temperature increase prior to CO_2 breakthrough is observed for both the leaky-well and leaky-fault cases, as a result of the larger amount of leaked brine as compared to the full single-porosity model. The breakthrough of CO_2 also occurs earlier owing to the reduced resistance to flow permitted by the fractured leak. After CO_2 breakthrough, the temperature quickly drops in the leaky-well case. Because of the high leakage rate of CO_2, the pressure builds up quickly in the AZMI. Consequently, the temperature signal weakens as a result of reduced pressure differential. Therefore, the temperature starts rising after reaching a minimum (Fig. 3a). This contrasts with the single-porosity case, in which the temperature declines monotonically. For completeness, the dual-porosity model is extended to account for the interaction between the matrix blocks using a dual-permeability model and the results are very similar to those of the dual-porosity model (Fig. 3a). However, the enhanced heat exchange as a result of increased flow dynamics within the leak further reduced the thermal signal for the dual-permeability model.

Similar results are obtained for the leaky fault (Fig. 3b). The temperature increase before CO_2 breakthrough is slightly larger for the dual-porosity and dual-permeability models compared to the single-porosity case. Temperature drops after the start of CO_2 leakage but starts increasing 1 month later as a result of reduced pressure change caused by increase leakage rate.

Conclusions

Sensitivity of the leakage thermal signal to depth and considering the leak as a fractured medium were investigated in this study. Assuming constant geothermal gradient at different depths, the temperature increase due to brine leakage preceding the CO_2 leakage becomes stronger with depth, owing to larger CO_2 breakthrough time at the leak. The strength of the cooling signal after the start of CO_2 leakage decreases with depth. The cooling signal may also decrease with time as pressure builds up in the AZMI, creating resistance to CO_2 leakage.

The effect of treating the leakage path as a fractured medium was investigated using dual-porosity and dual-permeability models. We showed that, as a result of increased absolute and relative permeability of the CO_2-rich phase, CO_2 leakage increases significantly. This leakage results in a rapid pressure buildup in the AZMI and reduces the pressure gradient, leading to a weaker cooling signal. Using a dual-permeability model instead of a dual-porosity model to allow for fluid-flow interactions between matrix blocks resulted in more efficient heat exchange and slightly less cooling in AZMI.

References

Bachu, Stefan. 2003. Screening and ranking of sedimentary basins for sequestration of CO2 in geological media in response to climate change. *Environmental Geology* **44** (3): 277-289.

Blaskovich, F. T., Cain, G. M., Sonier, Fernand et al. 1983. A Multicomponent Isothermal System for Efficient Reservoir Simulation. Proc., 10.2118/11480-MS.

Ennis-King, J, Paterson, L. Reservoir engineering issues in the geological disposal of carbon dioxide. Vol. 1, 290-295.

Fenghour, A., Wakeham, W. A., Vesovic, V. 1998. The viscosity of carbon dioxide. *Journal of Physical and Chemical Reference Data* **27** (1): 31-44.

Haghighat, S. Alireza, Mohaghegh, Shahab D., Gholami, Vida et al. 2013. Using Big Data and Smart Field Technology for Detecting Leakage in a CO2 Storage Project. Proc., SPE-166137-MS, SPE Annual Technical Conference and Exhibition, New Orleans, Louisiana, 10.2118/166137-MS.

Hill, A. C., Thomas, G. W. 1985. A New Approach for Simulating Complex Fractured Reservoirs. Proc., 10.2118/13537-MS.

Jung, Y. J., Zhou, Q. L., Birkholzer, J. T. 2013. Early detection of brine and CO2 leakage through abandoned wells using pressure and surface-deformation monitoring data: Concept and demonstration. *Advances in Water Resources* **62**: 555-569. DOI 10.1016/j.advwatres.2013.06.008.

Kazemi, H., Merrill, L. S., Jr., Porterfield, K. L. et al. 1979. Numerical Simulation of Water-Oil Flow in Naturally Fractured Reservoirs. 10.2118/5719-PA.

Sun, Alexander Y., Zeidouni, Mehdi, Nicot, Jean-Philippe et al. 2013. Assessing leakage detectability at geologic CO2 sequestration sites using the probabilistic collocation method. *Advances in Water Resources* **56** (0): 49-60. http://dx.doi.org/10.1016/j.advwatres.2012.11.017.

Warren, JE, Root, P Jj. 1963. *The behavior of naturally fractured reservoirs*, SPE Journal.

Zeidouni, Mehdi, Nicot, Jean-Philippe, Hovorka, Susan D. 2014. Monitoring above-zone temperature variations associated with CO2 and brine leakage from a storage aquifer. *Environmental Earth Sciences*: 1-15. 10.1007/s12665-014-3077-0.

Zeidouni, Mehdi, Pooladi-Darvish, Mehran. 2012a. Leakage characterization through above-zone pressure monitoring: 1—Inversion approach. *Journal of Petroleum Science and Engineering* **98–99** (0): 95-106. http://dx.doi.org/10.1016/j.petrol.2012.09.006.

Zeidouni, Mehdi, Pooladi-Darvish, Mehran. 2012b. Leakage characterization through above-zone pressure monitoring: 2—Design considerations with application to CO2 storage in saline aquifers. *Journal of Petroleum Science and Engineering* **98–99** (0): 69-82. http://dx.doi.org/10.1016/j.petrol.2012.09.005.

Table 1. Description of the base-case problem.

Porosity (fraction)	0.1	Leaky-pathway lateral permeability (m^2)	1e-14
Lateral permeability (m^2)	1e-14	Leaky-pathway vertical permeability (m^2)	1e-14
Vertical permeability (m^2)	1e-15	Leaky-pathway porosity (fraction)	0.1
Reservoir radial extent (km)	24	Leak-injector distance (m)	100
Brine salinity (wt%)	15	Leaky-well radius (m)	0.3
Initial pressure at 1000-m depth (MPa)	8.03	Leaky-fault width (m)	2

Table 2. Relative permeability of CO_2-rich phase. Relative permeability of the aqueous phase is identical to its saturation (straight line).

CO_2-rich phase: Corey curve (1954)

$$k_{rg} = \left(1-\hat{S}\right)^n \left(1-\hat{S}^2\right)$$

where:

$$\hat{S} = \frac{S_a - S_{wirr}}{1 - S_{wirr} - S_{gc}}, \quad n=2,\ S_{gc}=0.05$$

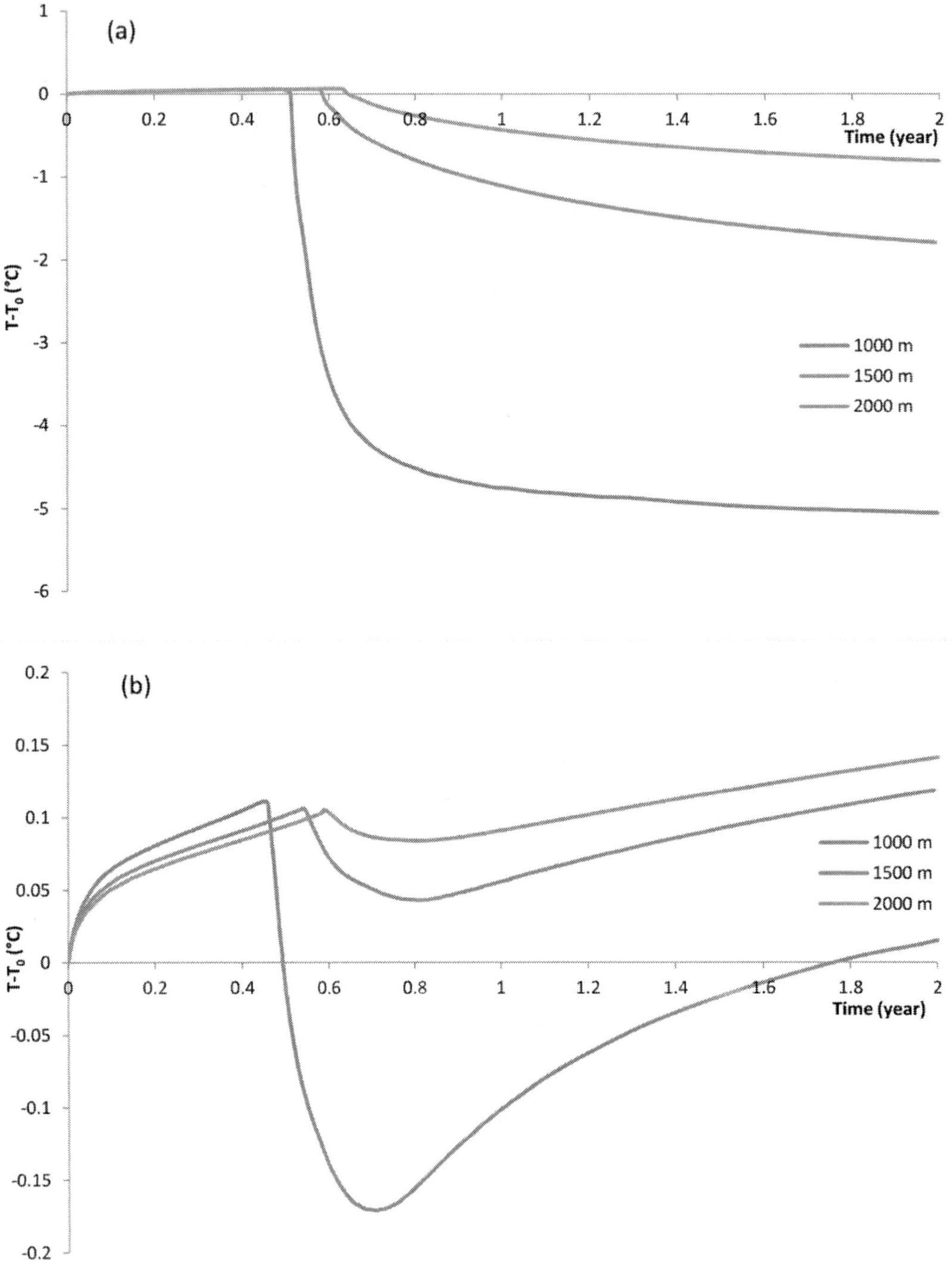

Fig 1. Temperature change in the lowermost layer of AZMI at the location of the leak for (a) leaky-well and (b) leaky-fault cases.

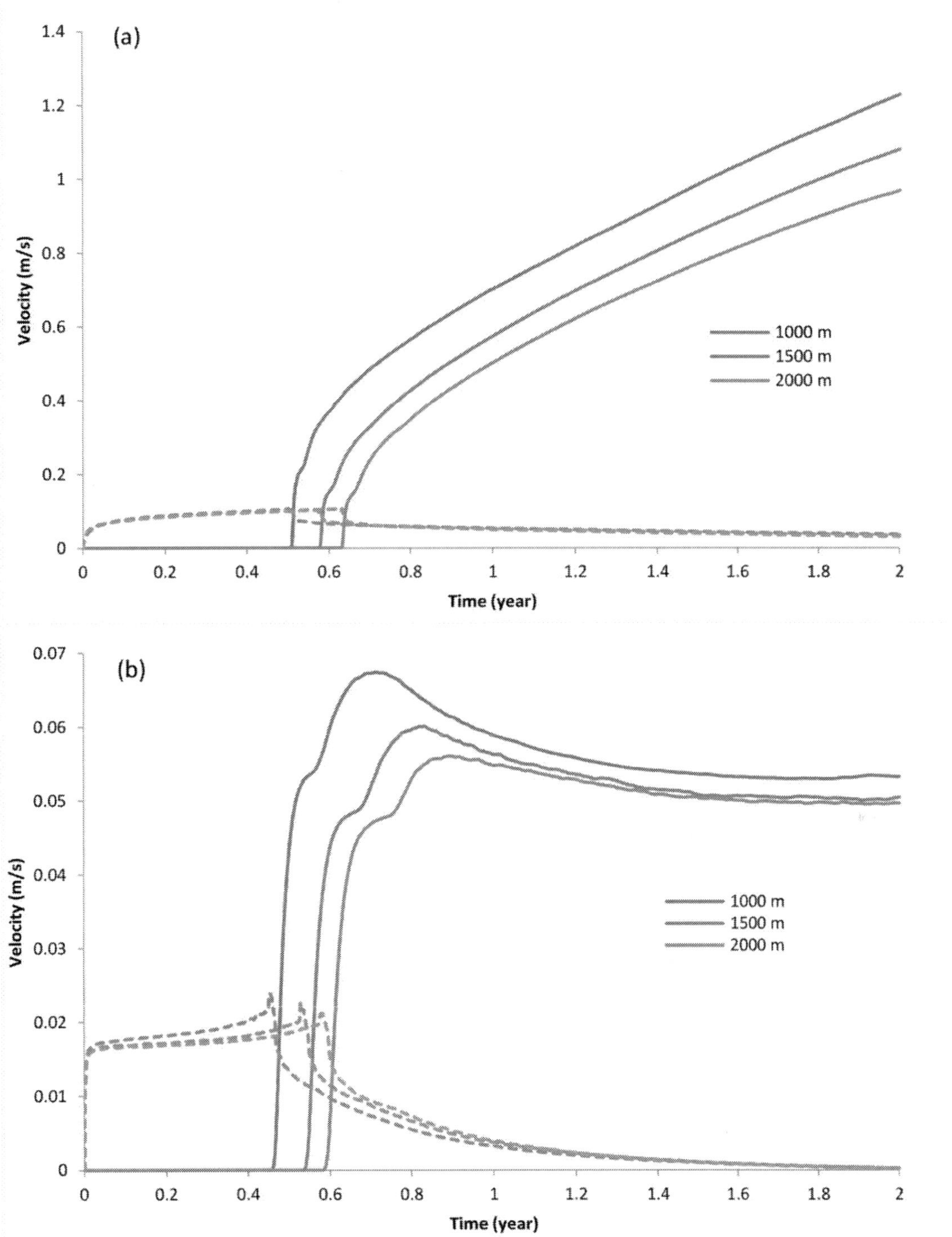

Fig 2. Darcy velocity of CO_2-rich phase (solid curves) and the aqueous phase (dashed curves) at the location of the leak for various depths of (a) leaky-well case and (b) leaky-fault case.

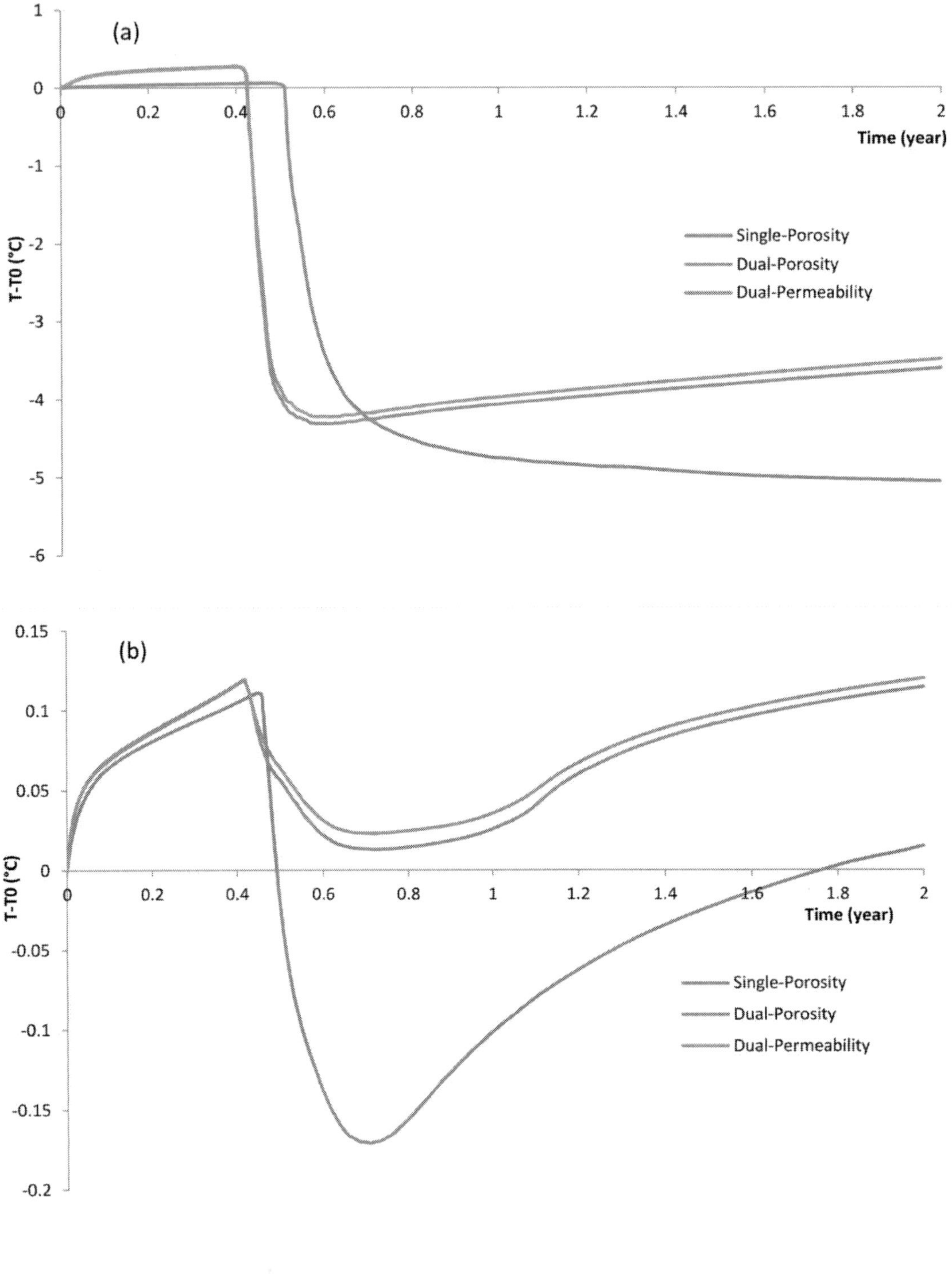

Fig 3. Temperature change in the lowermost layer of AZMI at the location of the leak for (a) leaky-well and (b) leaky-fault cases.

CARBON MANAGEMENT TECHNOLOGY CONFERENCE

CMTC-439377-MS

Reduce CO_2-Oil Miscible Pressure by Injection Man-Made Miscible-Slug to Enhance Oil Recovery

SUN Shasha[1], TANG Ligen[1,*], GUO Kai[2], BAI Fengjuan[3]

1. PetroChina Research Institute of Petroleum Exploration & Development, Langfang 065007, China
2. Petro China Company Limited Natural Gas & Pipeline Company, Beijing 10001, China
3 .China Kunlun Constracting & Engeering Corporation, Beijing 100037, China

Abstract

In order to reduce CO_2-oil minimum miscible pressure (MMP), a brave idea of injection miscible-slug instead of pure CO_2 into reservoir was proposed, and then several slim-tube tests with different pressures were deployed to acquire the composition of miscible-slug, and a consecutive series of output samples including oil and gas from slim-tube at different injection pore-volume CO_2 were collected, composition of every output samples was tested and analyzed by using equipment of liquid phase chromatograph and gas phase chromatograph.

Composition of consecutive output samples at different injection pore-volume CO_2 were obtained through the help of gas & oil phase chromatograph, and the result above shows the quit difference between miscible-slug and immiscible-slug. The components with molecular weight lighter than C18 are mainly extracted by CO_2 immiscible-slug, while some additional components with molecular weight between C18 and C25 are also extracted by CO_2 miscible-slug. Moreover, low-carbon molecules are initially extracted, with the ongoing flooding, higher –carbon molecules are gradually extracted. Based on the composition of consecutive output samples, man-made miscible-slug for the this oil reservoir was made up and the slug PV was amplified used for injection instead of CO_2, another slim-tube test using above man-made miscible-slug was deployed again with pressure lower than MMP obtained from above series of slim-tube tests, the result shows a CO_2 –oil miscible flooding and the idea of CO_2 –oil MMP reduction comes true.

A method of reduction CO_2 –oil MMP by injection man-made miscible-slug made up from slim-tube tests was proved, which would improve a lot of oil reservoir immiscible with CO_2 into miscible with CO_2 under reservoir temperature.

Introduction

Now days more and more attention has been payed to enhancement oil recovery and reduction emission of greenhouse gas CO_2, injection CO_2 into oil reservoir to enhance oil recovery has became one important action to solve the problem of both climate and recovery. As we all known, the value of oil recovery might achieve more than 90% when injection CO_2 becomes miscible with oil in place. So we always prefer to deploy the project of enhancement oil recovery by CO_2 in which injection CO_2 could be miscible with oil. However, it is not very easy find CO_2 miscible with any kind of oil in place, especially when the gravity of oil is too big. This problem makes more and more reservoirs have to quit enhancement oil recovery by CO_2 injection. To counter this problem, it would become urgent and necessary to test some method to reduce CO_2-oil miscible pressure.

1 Experiment

In order to reduce CO_2-oil minimum miscible pressure (MMP), a brave idea of injection miscible-slug instead of pure CO_2 into reservoir was proposed, and then several slim-tube tests with different pressures were deployed to acquire the composition of miscible-slug, and a consecutive series of output samples including oil and gas from slim-tube at different injection pore-volume CO_2 were collected, composition of every output samples was tested and analyzed by using equipment of liquid phase chromatograph and gas phase chromatograph.

2 Results

2.1 Immiscible-slug

As to immiscible-slug test, slim-tube saturated with oil in place was displaced by CO_2 with purity of 99.99% at 18.37MPa. During the test, six samples of output were collected when the amount of CO_2 injected were 0.5444 PV, 0.6184 PV, 0.7966 PV, 1.0433 PV, 1.2266 PV and 1.6102 PV respectively.

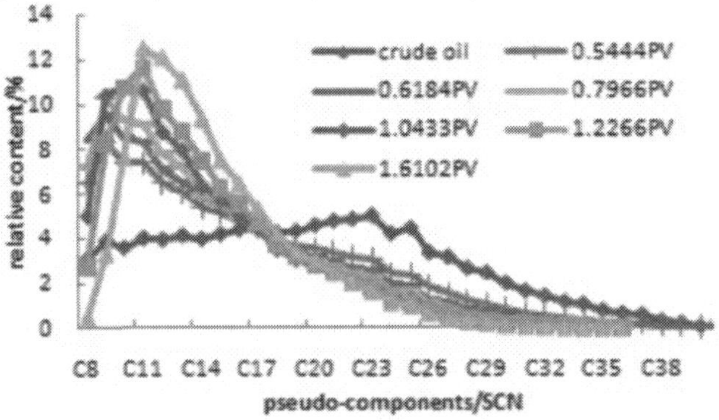

Figure 1 Law of relative content changing with different CO_2 Pore Volume in immiscible test

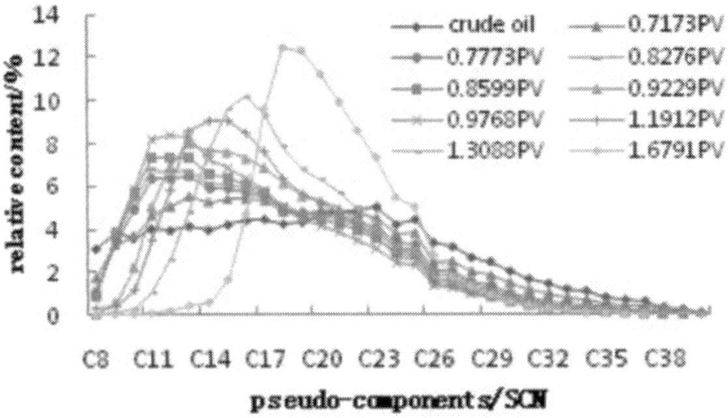

Figure 2 Law of relative content changing with different CO_2 Pore Volume in miscible test

The results of immiscible-slug test that, component curves at different injection CO_2 PVs intersect with crude oil curve at one point near C18, and this point separates each curve into left part and right part as figure 1 shows. Then the laws of component curves change opposite to each other between the left and the right part, such as relative content of the left one increase and even much higher than that of crude oil when inject more CO_2 into slim-tube, while the relative content of right part decrease and even slightly lower than that of the crude oil when inject more CO_2 into slim-tube. This different changing and intersection near C18 signify that, much of several components lighter than C18 especially for C10-C14 components have been extracted from original oil, while a small part of components heavier than C18 have been extracted from original oil in immiscible test.

2.2 Miscible-slug

In the case of miscible-slug test, slim-tube saturated with oil in place was displaced by CO_2 with purity of 99.99% at 25.18MPa. During the test, nine samples of output were collected when the amount of CO_2 injected were 0.7173 PV, 0.7773 PV, 0.8276 PV, 0.8599 PV, 0.9229 PV, 0.9768 PV, 1.1912 PV, 1.3088 PV and 1.6791 PV respectively.

By comparison, the results of miscible-slug test show not the same as those of immiscible-slug test. The points, at which component curves intersect with crude oil, move from light component C18 to heavier component C25 slowly with more injection CO_2 PVs. And also these points separate each curve into left part and right part as figure 2 shows. Then the laws of component curves change opposite to each other between the left and the right part, such as relative content of the left one increase and even much higher than that of crude oil when inject more CO_2 into slim-tube, while the relative content of right part decrease and even slightly lower than that of the crude oil when inject more CO_2 into slim-tube. These intersections from C18 to C25 signify that, much of several components not only lighter than C18 have been extracted from original oil but also the components of C18, C19, C20, C21, C22, C23, C24 and C25 have been extracted from original oil, while a small part of components heavier than components of intersection have been extracted from original oil in immiscible test in miscible test.

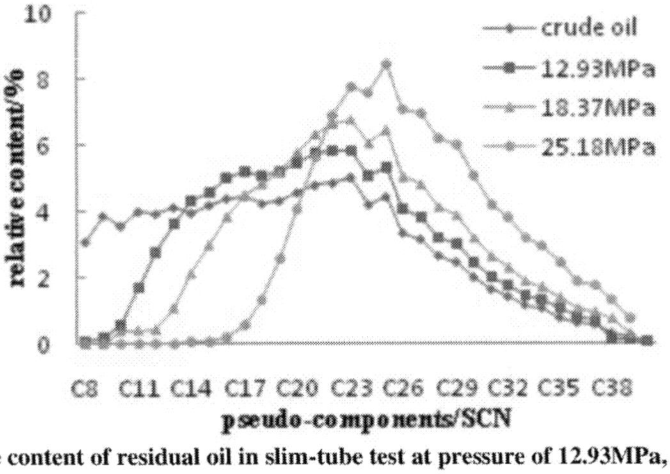

Figure 3 Law of relative content of residual oil in slim-tube test at pressure of 12.93MPa, 18.37MPa and 25.18MPa

2.3 Residual oil

What is more, residual oil in the slim-tube was also collected before tests finished. The results of residua oil component at three different pressures show that, the points, at which component curves intersect with crude oil, move from light component C12 to heavier component C20 slowly with more injection CO_2 PVs. And the law of components changing in residual oil opposite to that in sample collected during CO_2 flooding, this difference expresses as that left parts decrease and right parts increase when pressures increase. The difference signifies that much of components lighter than intersection components have been extracted from crude oil. Then it is obvious that the range of components extracted from original oil is much larger in miscible flooding than that in immiscible flooding. The different ranges between miscible test and immiscible test may become a good explain to interpret what cause oil recovery much larger in miscible flooding than that in immiscible flooding.

2.4 Man-made miscible-slug

Based on the composition of consecutive output samples, man-made miscible-slug for the this oil reservoir was made up and the slug PV was amplified used for injection instead of CO_2, another slim-tube test using above man-made miscible-slug was deployed again with pressure lower than MMP obtained from above series of slim-tube tests, the result shows a CO_2 –oil miscible flooding and the idea of CO_2 –oil MMP reduction comes true.

4 Conclusion

A method of reduction CO_2 –oil MMP by injection man-made miscible-slug made up from slim-tube tests was proved, which would improve a lot of oil reservoir immiscible with CO_2 into miscible with CO_2 under reservoir temperature.

References

[1] Yang Zhengming, Guo Hekun, Liu Xuewei, et al. Characteristic experimental technology of an extra- ultra low permeability reservoir [M]. Beijing: Petroleum Industry Press, 2012: 2-14.

[2] Bachu S, Adams J. Sequestration of CO_2 in geological media in response to climate change: Capacity of deep saline aquifers to sequester CO_2 in solution [J]. Energy Conversion and Management, 2003, 44(5):3151-3175.

[3] Metz B, Davison O, de Coninck H C, et al. IPCC Special report on carbon dioxide capture and storage[R/OL].[2012-08-31].http://library.wmo.int/pmb_ged/IPCC_CO2_e.pdf

[4] Tang Ligen, Wang Jieming. Inventory forecast in underground gas storage based on modified material balance equation [J]. Petroleum Exploration and Development, 2014, 41(4): 480~484

CARBON MANAGEMENT TECHNOLOGY CONFERENCE

CMTC-439470-MS

Technical and Economic Feasibility of Storing CO_2 in Mature Fractured Monterey Reservoirs of Offshore California

Xiaoxi Zhao and Iraj Ershaghi, University of Southern California

This paper was prepared for presentation at the Carbon Management Technology Conference held in Sugarland, Texas, USA, 17–19 November 2015.

Abstract

In this paper we present a study on the feasibility of utilizing the mature Monterey oilfields of offshore California for sequestering carbon dioxide while potentially enhancing crude oil recovery from the tight matrix.

An important characteristic of these fields is that, because of natural fractures, well productivities have been high and as a result, a significant amount of reservoir voidage has been generated in each field by production of more than 780 million barrels of oil and more than a TCF of natural gas during the last three decades. These productions translate into a computed voidage of about a billion barrels allowing storage of more than 90 Million tons of CO_2. This is roughly equivalent to storing 50 years of CO_2 from a 400 MW Power plant. Another important observation discussed in the paper is the cap rock sealing nature of these fields allowing preservation of the oil and initial pressure as evidenced from the analyses of the initial GOR data. Furthermore many of these fields will be at the verge of abandonment during the next two decades unless steps are taken to enhance recovery by extracting oil from the tight matrix. CO_2 extraction from tight rocks is a technically feasible process and our analysis indicate that CO_2 storage can indeed prolong the life of many of these fields and enhance matrix oil recovery. An important infrastructure requirement for CO_2 transportation into these offshore fields is the cost of building offshore pipelines. Availability of existing multiple pipelines connecting the fields to their corresponding onshore treatment facilities solves this transportation of CO_2 to the offshore fields. Finally, these fields are under active water influx, where aquifers may provide additional CO_2 storage possibilities.

Introduction

Power plants are a major contributor to CO_2 emissions. Fig. 1 is a map showing electric power plants in California[1]. Conceptual solutions to Carbon capture and storage have generated considerable interest as solution to reducing these emissions[2]. Various geological sites have been suggested for CO_2 storage, including depleted oil and gas reservoirs, deep saline aquifers and deep coal seams[3]. Currently, there are several successful projects for such sequestration concepts. Examples include Sleipner (Norway), In Salah (Algeria), Ketzin (Germany), and K12B (Netherlands).

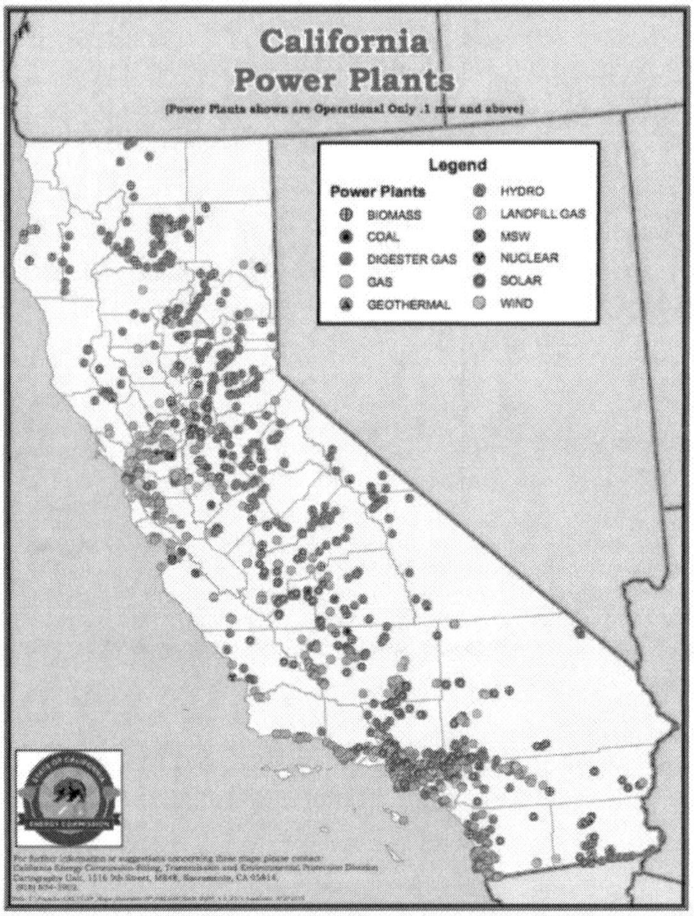

Figure 1 California Power Plants[1]

A concern for sequestration of CO_2 in onshore natural storage sites has been potential migration into groundwater leading to contamination[3,4] and risk of over pressuring, causing fracturing, possible leakage and potential hazards to human and the environment. A plausible alternative has been the consideration of CO_2 storage in offshore natural storage sites. This may minimize risks associated with such operations.

This paper focuses on some California federal offshore fields that have produced from the Monterey formation. These fields have a long history of production and considerable voidage has been developed because of their depletion history. Offshore Monterey intervals are siliceous and fractured and have demonstrated significantly high well productivities. Our study demonstrates the suitability of these producing intervals and reservoir compartmentalization as factors making such sequestration operation technically quite successful. Furthermore the availability of multiple exiting pipelines connecting the shore treatment facilities to the producing platforms make the CO_2 transport quite feasible.

Geology

Monterey Formation is a middle and upper Miocene biogenic siliceous rock deposit widespread in California. Reservoirs producing from the offshore fields generally consist of brittle rocks exhibiting the most mature form of the Monterey Formation. In some places the total thickness is greater than 2000ft. The formation while primarily consisting of fractured cherts, may include other digenetic phases of silica and also dolomites. As shown in Figure 2, some of the fields contained in this formation include Hondo, Point Pedernales, Point Arguello, Sacate and Rocky Point.

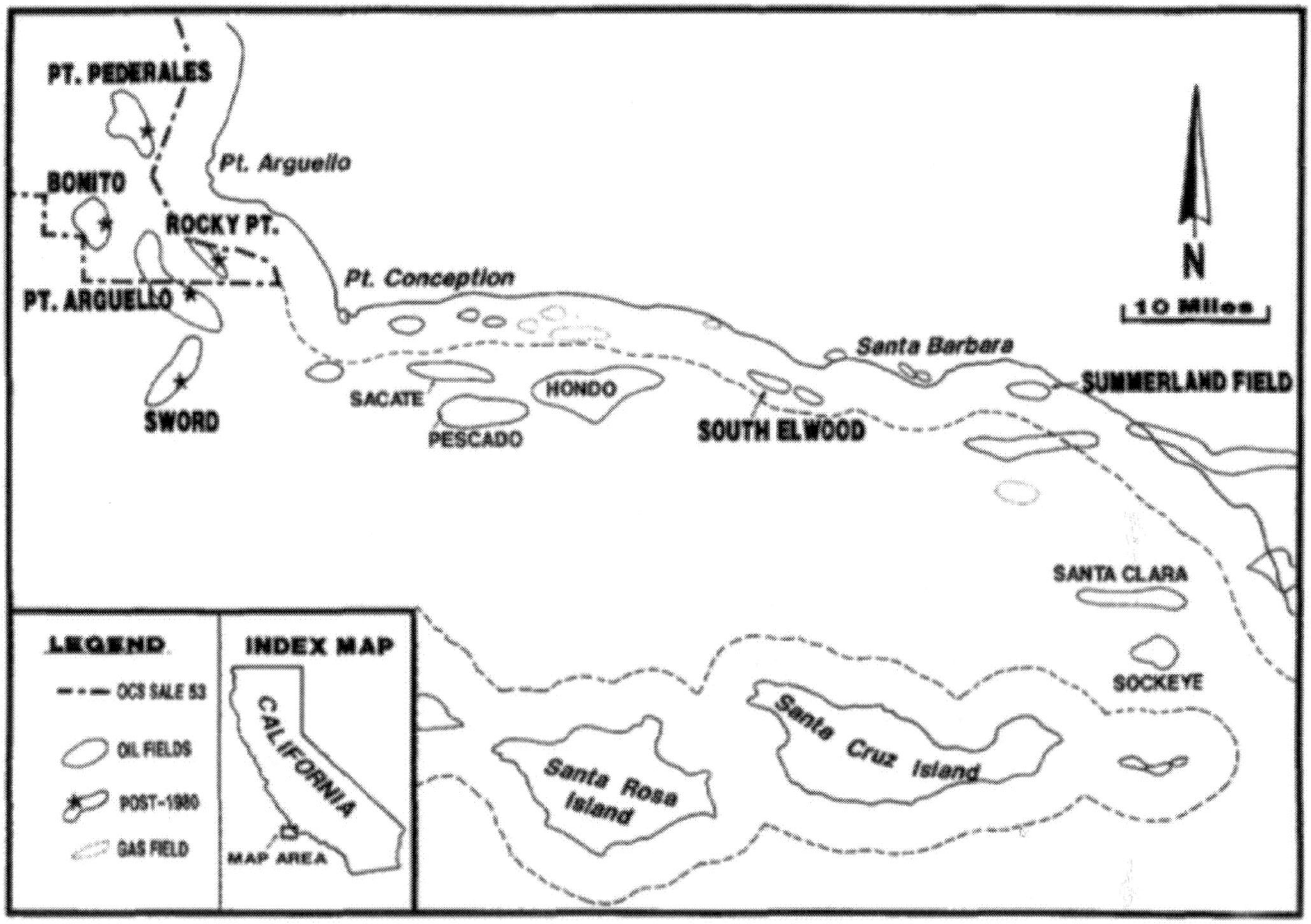

Figure 2 Map of Monterey Formation[5]

Fig. 3 shows a typical section of Monterey column. Intervals listed include upper siliceous, upper calcareous, transition, lower siliceous, lower calcareous, and other horizons with various proportions of detrital material.

Lithology varies considerably both vertically and laterally. Fig. 4 shows the nature of fracturing observed on some outcrops. Core recovery is poor in some parts of the reservoir. But the matrix is extremely tight. A sample of a recovered core is shown in Fig. 5, which indicates the rubblelized segments in the highly productive chert member of the Monterey Formation. Mud logs indicate chert sections are often sandwiched between other siliceous members). 2D seismic also indicates multiple faults in the reservoir. Evaluation of available geological data shows that stratigraphic chert breccia and crosscutting fault breccia are the most important conduits for fluid movement within the formation.

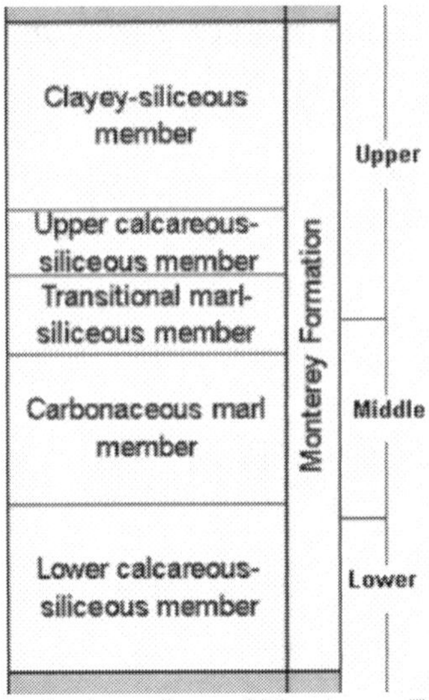

Figure 3 Typical section of Monterey Formation

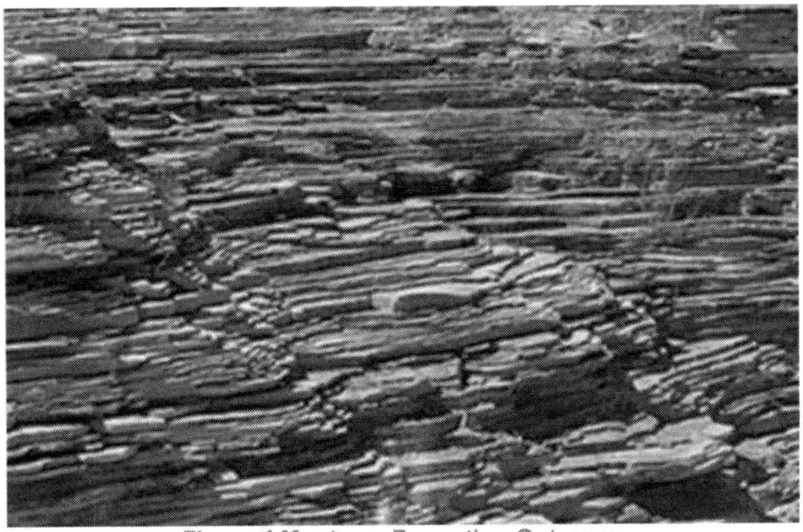

Figure 4 Monterey Formation Outcrops

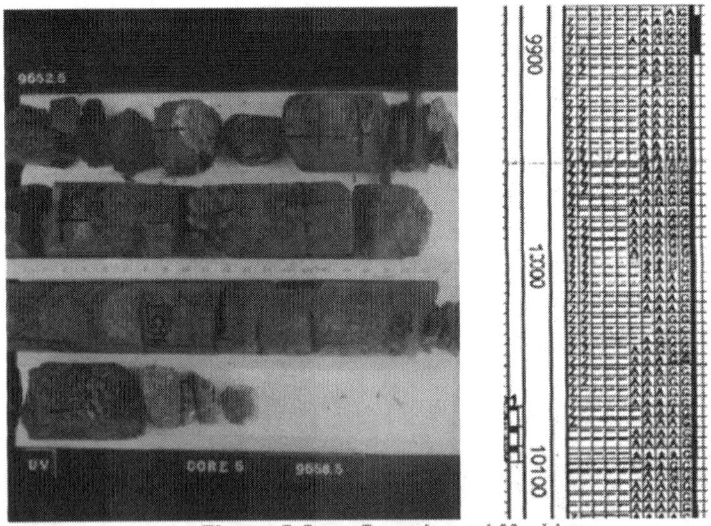

Figure 5 Core Sample and Mud Log

Production History

The first onshore Monterey oil field was discovered in 1901 in the Santa Maria basin. Productivity from onshore wells was not spectacular. Accidental discoveries of the offshore Monterey potential in the South Ellwood and the Hondo oil fields in Santa Barbara generated substantial interest for exploitation and development activities. Twenty years later, other fields such as the Point Arguello and the Point Pedernales oil fields were discovered.

Because of natural fractures, well productivities have been high and as a result a significant amount of reservoir voidage has been generated in each field. From 1987 to September 2014, more than 780 million barrels of oil and more than a TCF of natural gas have been produced from about 200 wells in these fields. Half of these wells have individually produced more than 3 million STB. One of these wells has produced over 31 million STB over the past three decades. Cumulative production histograms for the fields indicated above are shown in Fig. 6.

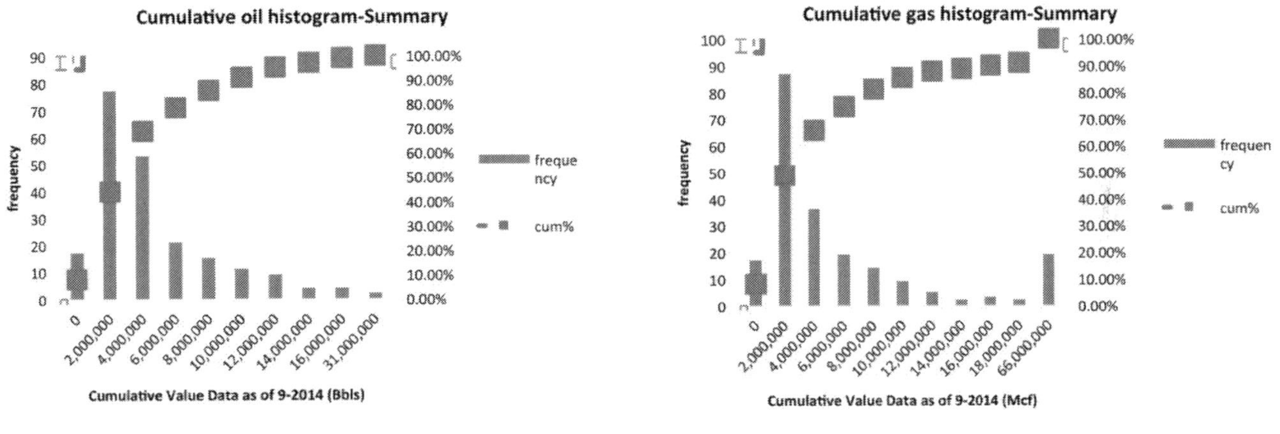

Figure 6a Cumulative Production Histogram Oil and Gas

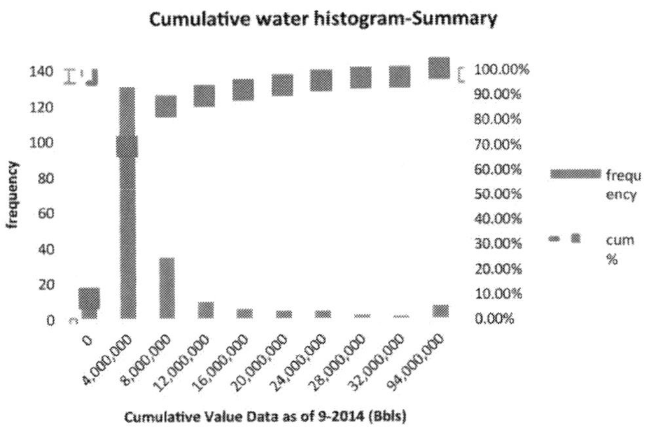

Figure 7b Cumulative Production Histogram for Water

Before we discuss the storage capacity and the voidage that can be used for CO_2 storage in these reservoirs, it is important to emphasize the unique structural conditions favoring these sites.

With the exception of a segment of the South Ellwood Field, an important observation in these fields is the sealing nature of the cap rocks and tar seal plugging of fractures resulting in preservation of the oil and initial pressure as evidenced from the analyses of the initial GOR data. One example is shown below. In Fig. 7, seismic information is combined with well directional path on a cross section. From the seismic, the pink line is top of Monterey formation, blue line is water oil contact, also black line indicates faults. Resistivity and gamma ray log are shown for each well.

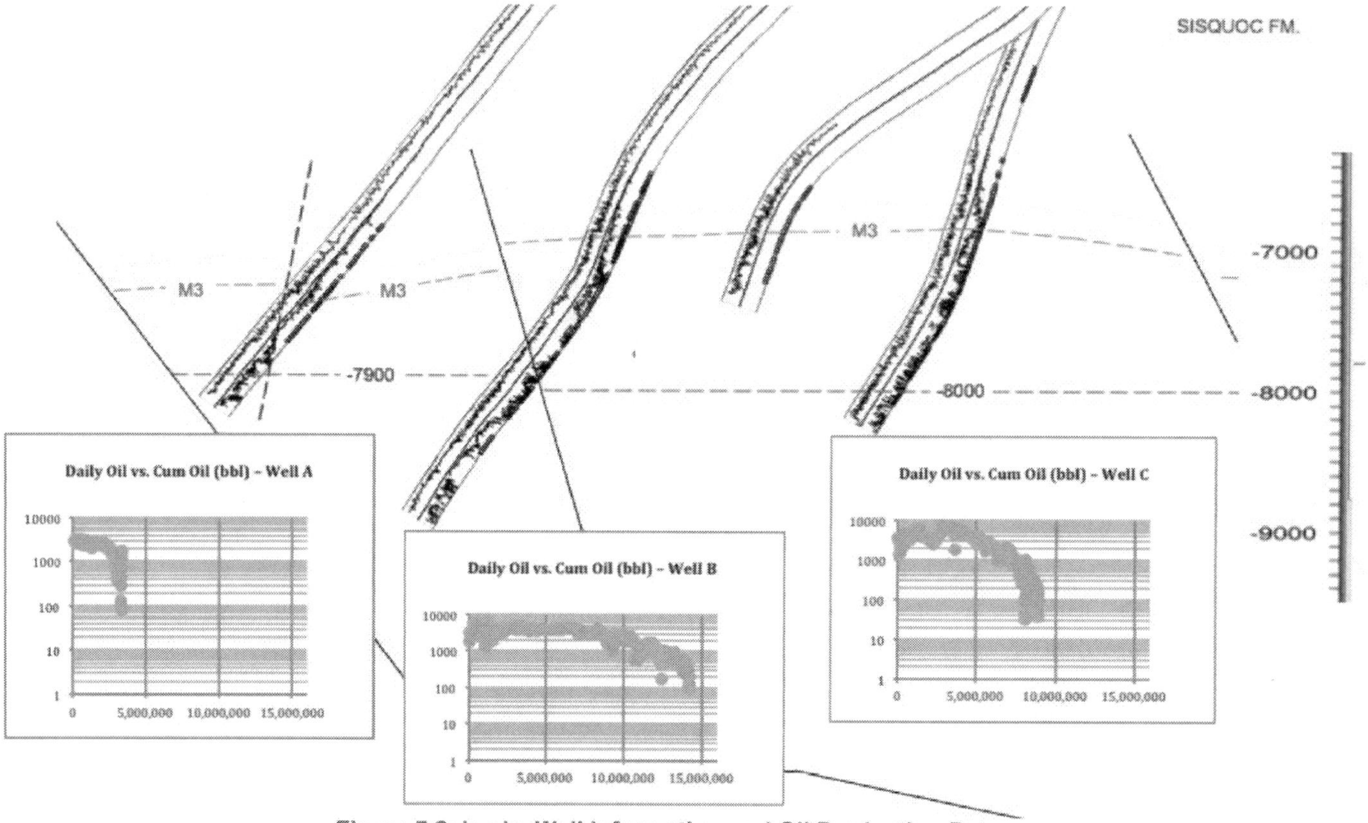

Figure 7 Seismic, Well information and Oil Production Data

In this cross section, three wells are compared with each other and there are many examples like this. Although these wells are close to each other, they have behaved differently. While cumulative production from wells, A and C, have ranged from 3 million STB to 8 million STB, Well B, which has a longer production interval and intersect with nearby fault has performed excellent with a cumulative production of 13 million STB.

Because there are not much static pressure measurements, gas oil ratio data was used as a proxy for pressure. The initial gas oil ratio in this field was about 400SCF/STB. As shown in Fig. 8, gas oil ratio history shows clearly that these blocks are separate from each other. Among these nearby wells, in one case, Well A produced for a little time before GOR jumped up. However, for the good well, which is Well C, it produced 6 million before the GOR showed the evidence of a drop in reservoir pressure. Also for Well B, after some production, pressure dropped and GOR increased a little. However, based on WOR data, water influx from the bottom of the formation has maintained the pressure. After cumulative production reached to 8 million STB, GOR started to increase. This information indicates examples of compartmentalization and different timing for pressure depletion.

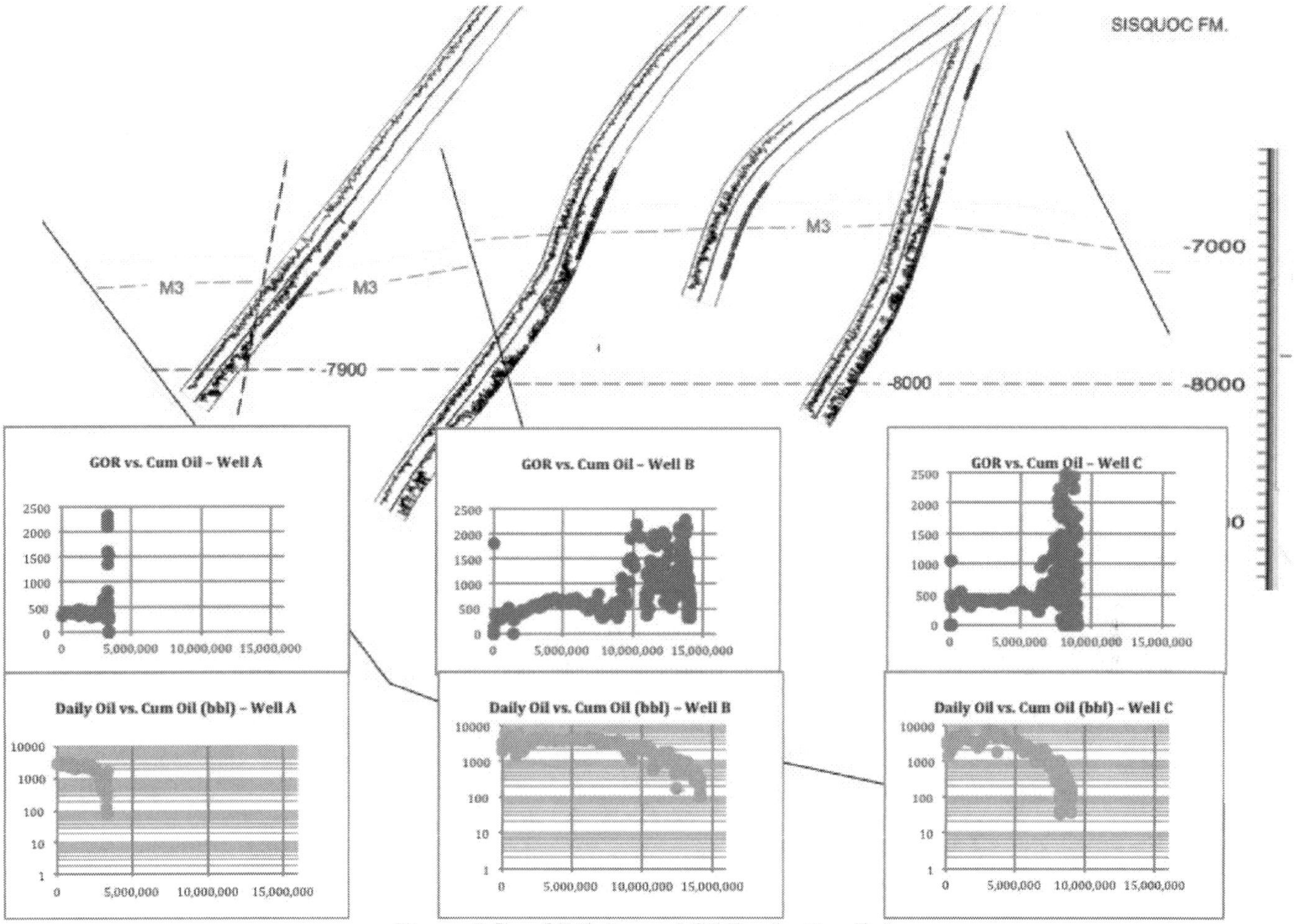

Figure 8 Gas Oil Ratio and Oil Production Data

Similar examples exist for other fields. There seems to be little communication among production blocks. If CO_2 is injected into one of the production blocks, no effect will occur in other production blocks. Also the sealing cap rock can contribute to preservation of injected carbon dioxide. As shown earlier, production histograms of other fields indicate similar compartmentalized systems from their reservoir geology.

Potential Opportunity

Although tremendous amounts of oil and gas have been produced, depending on the economics, many of these offshore fields will be at the verge of abandonment during the next two decades unless steps are taken to enhance recovery by extracting oil from the tight matrix. The tight matrix in some geologic units may hold substantial amounts of oil. Injection of the CO_2 gas in the fracture system and allowing the diffusion and extraction of oil from the tight matrix may help in extracting the light fractions of the oil from the matrix resulting in an overall increase in recovery factor[7].

Storage Capacity

Large voidage created by the past production in different fields means a huge CO_2 storage capacity. For illustrative purposes, we assumed a maximum storage reservoir pressure of 3000 psi, a temperature of 200°F; formation oil volume factor of 1.15RB/STB and formation gas volume factor of 0.0049 CF/SCF. Estimated density of carbon dioxide is 0.3 tons/m^3 under this condition. As shown in Table 1, oil and gas productions from 1987 to 2014 translate into a total computed voidage of about a 3 million cubic meters allowing storage of more than 90 Million tons of CO_2.

According to information published by the U.S. Energy Information Administration[8], total CO_2 emission by the electric power industry in year 2013 was 45.7 million tons in California. Voidage in the Monterey formation in the fields indicated is sufficient to store almost 2 years of CO_2 emitted in the entire California. According to California Energy Commission, in 2013, total usage of residential and commercial has been 3,334,000 of MWh in Santa Barbra County, which requires roughly a power plant equivalence of 400 MW at 60 Hz. Based on EPA's final "Carbon Pollution Standard for New Power Plants"[6], published in 2015, new natural gas power plants can emit no more than 1,000 pounds (lbs.) of carbon dioxide per megawatt-hour (MWh) of electricity produced. Thus, the total CO_2 released annually would be 1.728 million tons. Our computation based on voidage indicates a potential of 53 years storage capacity. For Santa Barbara County assuming no increase in power production.

Table 1 CO_2 Storage Capacity Calculation

Field	Pressure	Cumulative Oil (STB)	Cum Gas (MSCF)	Cum Water (STB)	Oil Void (STB)	Gas Void (MSCF	Cum Void (m^3)	CO2 injected (ton)
A	3000	310,032,646	681,458,529	191,976,202	356,537,543	3,339,147	151,239,049	4.54E+07
B	3000	145,057,024	223,062,135	117,683,159	166,815,578	1,093,004	57,471,998	1.72E+07
C	3000	187,162,982	173,614,339	434,146,529	215,237,429	850,710	58,309,450	1.75E+07
D	3000	94,561,407	34,001,890	423,281,669	108,745,618	166,609	22,007,021	6.60E+06
E	3000	46,515,332	46,040,251	38,908,874	53,492,632	225,597	14,892,851	4.47E+06
F	3000	2,789,485	2,259,712	6,132,094	3,207,908	11,073	823,557	2.47E+05
Cumulative		786,118,876	1,160,436,856		904,036,707	5,686,141	304,743,926	9.14E+07

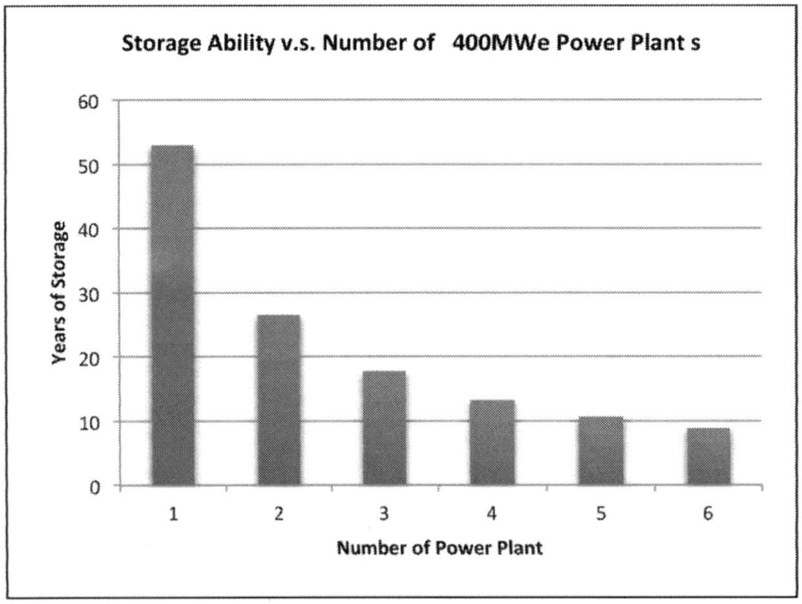

Figure 9 Storage Ability vs. Number of 400MWe Power Plants

On the other hands, an important infrastructure requirement for CO_2 transportation into these offshore fields is the cost of building offshore pipelines. The Monterey reservoirs cited are connected to the onshore via existing oil and gas pipeline system.

Conclusion

A study was conducted on the feasibility of letting the mature semi depleted Monterey oilfields of offshore California serve as storage for sequestering carbon dioxide while potentially enhancing their crude oil recoveries.

The voidage calculated for certain offshore fields in California until September 2014, based on a total production of 780 million barrels of oil as well as a TCF of natural gas, can provide a total storage for about 90 million tons of CO_2.

The compartmentalized nature of these fields makes them ideal for storage purposes. An exam cross-section shows production blocks are separated from each other and little communication exists among those blocks. Such compartmentalization will limit the impact of pressure changes beyond target storage volume, and also reduces the risk of CO_2 migration.

Future enhanced oil recovery potential of this storage CO_2 has been discussed. Preexistence of multiple pipelines can address economics associated with transportation of CO_2 from onshore to these offshore fields. Finally, these fields have all been under active water influx. The aquifers also can potentially provide additional storage possibilities.

References

1-California Energy Commission, CEC-1304 QFER Database, http://www.ca.gov

2-Holloway, S., Vincent, C. J., & Kirk, K. L., 2006, "Industrial carbon dioxide emissions and carbon dioxide storage potential in the UK", British Geological Survey

3-Gale, J, 2004, "Geological Storage of CO_2: What do we know, where are the gaps and what more needs to be done?" Energy 29 1329-1338

4-Takeshi Komai, Yasuhide Sakamoto, Atsuko Tanaka, 2011, "Risk Assessment for CO_2 Capture and Geological Storage in Offshore Environment", ISBN 978-1-880653-95-1; ISSN 1946-0066

5-W. E. Mero, "Point Arguello Field – U.S.A. Santa Maria Basin. Offshore California"

6-U.S. Energy Information Administration, http://www.eia.gov

7-A.R. Kovscek, G.Q. Tang, B. Vega, 2008, "Experimental Investigation of Oil Recovery From Siliceous Shale by CO_2 Injection" SPE 115679

8-U.S. Environmental Protection Agency (EPA), 2015, "Carbon Pollution Standard for New Power Plants"

CMTC-439486-MS

Fast Modeling of Local Capillary Trapping during CO_2 Injection into a Saline Aquifer

Bo Ren, The University of Texas at Austin; Steven L. Bryant, University of Calgary; Larry W. Lake, The University of Texas at Austin

This paper was prepared for presentation at the Carbon Management Technology Conference held in Sugarland, Texas, USA, 17–19 November 2015.

Abstract

When CO_2 migrates upwards under buoyancy in the subsurface saline aquifer and encounters local capillary barriers (regions of rock with large capillary entry pressure), CO_2 would accumulate beneath these small barriers, and these accumulations are called *local capillary trapping* (LCT). LCT benefits storage because locally trapped CO_2 has a much larger saturation than residual gas, and such trapped gases cannot escape from the formation even if leakage conduits (fractures or fault) in the seal develop during the long-term storage of CO_2. Thus predicting and maximizing LCT is valuable in design and risk assessment of geologic storage projects.

Modeling LCT is computationally expensive and may even be intractable by using a conventional reservoir simulator. In this work, we decouple the problem into two parts: permeability-based flow simulation and capillary entry pressure-based local capillary trapping phenomenon. The connectivity analysis originally developed for characterizing well-to-reservoir connectivity is adapted to the flow simulation by means of a newly defined edge weight property between neighboring grid blocks, which accounts for the multiphase flow properties, injection rate, and buoyancy effect. Then the connectivity was estimated from shortest path algorithm to predict the CO_2 migration behavior and plume shape during injection. A geologic criteria algorithm is developed to estimate the potential LCT only from the entry capillary pressure field. The latter is correlated to a geostatistical realization of permeability field.

The extended connectivity analysis shows a good match of CO_2 plume computed by the full-physics simulation. We then incorporate it into the geologic algorithm to quantify the amount of LCT structures identified within the entry capillary pressure field that can be filled during CO_2 injection. Several simulations were conducted in the reservoirs with different level of heterogeneity (measured by the Dykstra-Parsons coefficient) under various injection scenarios. We demonstrate the reservoir heterogeneity affects the optimal injection rate in maximizing the LCT during injection.

Both the geologic algorithm and connectivity analysis are very fast; therefore, the integrated methodology can be used as a quick tool to estimate LCT. It can also be used as a potential complement to the full-physics simulation to evaluate the total safe storage capacity.

Introduction

Geological carbon sequestration (GCS) has been widely accepted as the most promising technique in this century to reduce the carbon emission and global warming. The subsurface saline aquifer has been considered as the major storage media in terms of the large storage capacity and wide accessibility. One of the main tasks of GCS is to ensure the CO_2 will be permanently stored in a safe way.

After CO_2 injection, CO_2 would migrate upward because of the buoyant force (CO_2 is less dense than brine). When the buoyant CO_2 bulk phase encounters a region with high entry capillary pressure within a saline aquifer exhibiting spatially varying properties (permeability and capillary entry pressure), the CO_2 would accumulate beneath the barriers with locally large capillary entry pressure. This kind of accumulation is termed as local capillary trapping which is intrinsic to the capillary pressure heterogeneity (Saadatpoor, 2009). As shown in Fig. 1 which is a quasi-2D bench-scale experimental observation of buoyant plumes in countercurrent displacement, the nonwetting phase (CO_2 in GCS, oil in this experiment) is initially emplaced in a high permeability region at the bottom of the domain (filled with 5 mm grains) and then allowed to rise under buoyancy. The circled sections illustrate local capillary trapping of oil (red-dyed) beneath boundaries of regions of small beds (the yellow number) above larger beads. If the nonwetting phase is emplaced under the control of viscous forces, e.g. in reservoir volumes near the injection well, local capillary traps as well as lower-permeability regions in the near-well region can be fully filled during injection; the traps remain filled after post-injection buoyancy-driven flow ends (Ren *et al.*, 2014). This would greatly enhance the safe storage of CO_2 and reduce the risk of escape, as the CO_2 held in local capillary traps will not escape even if the overlying sealing structures lose their integrity.

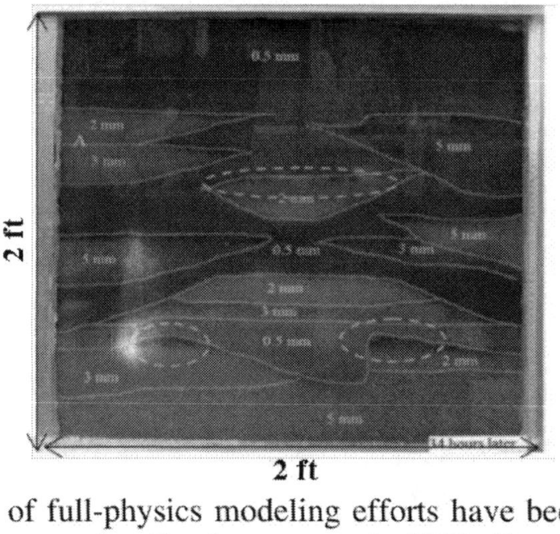

Fig. 1. End state after buoyant flow of non-wetting phase (red) which is initially emplaced at the bottom of a heterogeneous domain; yellow text indicates grain size in each region (Sun, 2014). The non-wetting phase accumulates beneath regions of higher capillary entry pressure (smaller grain size).

A number of full-physics modeling efforts have been directed towards exploration of the impact of capillary heterogeneity (Saadatpoor *et al.*, 2009; Krause *et al.*, 2011; Li and Benson 2015) on the CO_2 movement, saturation distribution and trapping in brine-saturated porous media. However, the full-physics simulation suffers from the drawbacks of computational intensity and time-step limitation. Resolving the discontinuity in capillary pressure and saturation that arises between grid cells with different capillary pressure-saturation curves introduces convergence problem in the numerical simulation, which requires the use of very small time-step size. Therefore, modeling this small-scale trapping phenomena using traditional reservoir simulation would be much more computationally intensive or even intractable at the field scale.

The inexpensive physics-proxy techniques that could rapidly evaluate flow response might provide a good alternative solution. The reported applications of the approximation methodologies to predict CO_2 migration in the GCS include streamline (Qi *et al.*, 2009; Tanaka *et al.*, 2014) and random walker particle tracking (Bhowmik 2014). The other category of approaches is based on modern methods of statistical physics of disordered media (Sahimi, 2012). One of them is the invasion percolation (Wilkinson and Willemsen, 1983) for modeling the two-phase immiscible flow. It allows for the high-resolution representation of subtle geologic features in a fast way and, therefore, has been employed to predict the CO_2 buoyant migration behavior (Cavanagh and Haszeldine 2014; Meckel *et al.*, 2015) and evaluate wellbore leakage risk (Cavanagh and Rostron, 2013) in the GCS. However, the invasion percolation can only be physically justified in the capillary-dominated flow regime. It is significantly different in behavior from the above traditional Darcy flow solvers in that the simulations are sensitive to threshold pressures and interfacial tension rather than permeability and viscosity. However, both the IP models and conventional reservoir simulations need to close the gap in terms of addressing both the final detail and the large scale of reservoir simulation models.

Another important endeavor for the safe CO_2 storage is to optimize the trapping process and enhance the safe trapping mechanisms, namely dissolution, residual and mineral trapping. In designing the CO_2 process, different injection strategies have been proposed to accelerate the solubility trapping (Leonenko and Keith 2008; Hassanzadeh *et al.*, 2009) or residual gas trapping (Qi *et al.*, 2009; Javaheri and Jessen, 2011) or both of them (Kumar *et al.*, 2005; Nghiem *et al.*, 2010; Anchliya *et al.*, 2012). Considering LCT can be regarded as one of the safe trapping mechanisms, it is very necessary to maximize the LCT storage capacity.

The main objective of this work is to develop a fast simulation technique that can be used to model the LCT and optimize its storage capacity in a given storage formation. We firstly employ the connectivity analysis to predict the CO_2 flow path and plume behavior. After that, we use a geologic criteria algorithm to identify the local capillary traps filled by CO_2 as determined from the connectivity analysis.

Connectivity Analysis Algorithm

According to Hirsch and Schuette (Hirsch and Schuette, 1999), the geologic model can be considered as a graph, and the grid block in the geologic model is equivalent to the node in the graph. These nodes are connected with edges which are weighted by reservoir parameters that impact connectivity such as porosity and permeability. The original definition of the edge is for the single phase flow. Here, we adapt it to the two-phase immiscible flow problems based on the Darcy's law.

(1) Edge weight definition

Eq. 4 shows the original definition of edge weight (ED1) for the single phase flow. The physical meaning of edge is the time needed to fill a given pore volume with a fluid of unit viscosity under a unit pressure gradient. In this work, we extend this definition by incorporating the buoyancy effect, relative permeability, and viscous pressure. The new definition of edge weight (ED2) is shown in Eq. 6, and its derivation based on the Darcy's law is shown below. In Eq. 6, the viscous pressure difference is approximated by the analytical solution of steady state radial flow in the homogeneous media (Eq. 7). After identifying the connectivity from the wellbore to the reservoir grids, these grid blocks will be filled with CO_2 to a specific saturation, which is same as average saturation before the breakthrough in the 1D immiscible displacement determined from the fractional flow curve (Lake, 1989).

The new edge weight is derived based on the Darcy's law as shown below. The subscript 'g' represents CO_2.

$$q_g = \frac{kk_g A(\Delta P + \rho_g gh)}{\mu_g L} \quad \text{.......} (1)$$

According to the definition of transmissivity (Eq. 2), we can write the above Eq. 1 into 3

$$T = k\frac{A}{L} \quad \text{.......} (2)$$

$$q_g = \frac{Tk_g(\Delta P + \rho_g gh)}{\mu_g} \quad \text{.......} (3)$$

Original definition of edge weight for single phase flow is shown in Eq. 4. Here we consider two-phase immiscible flow and assume the grid is filled by a given fluid to a specific saturation, so the equation of edge weight can be written as in Eq. 5

$$ED1 = \frac{\sqrt{Vp_i * Vp_j}}{T_{ij}} \quad (4)$$

$$ED2 = \frac{\sqrt{Vp_i * Vp_j} * \overline{S_g}}{q_g} \quad (5)$$

Since we assume the fluid is filled to a specific saturation, the corresponding relative permeability of gas which affects the filling speed will be calculated using this specific saturation. Substituting Eq. 3 into Eq. 5, we get the final form of edge weight (Eq. 6).

$$ED2 = \frac{\sqrt{Vp_i * Vp_j} * \overline{S_g} * \mu_g}{T\overline{k_g}(\Delta P + \rho_g gh)} \quad (6)$$

$$\Delta P = \frac{Q\mu_g}{2\pi \overline{k} H_{re}}[Ei(-\frac{\phi\mu_g C_t r_2^2}{4\overline{k}t}) - Ei(-\frac{\phi\mu_g C_t r_1^2}{4\overline{k}t})] \quad (7)$$

$$Ei(-x) = -\int_x^\infty \frac{e^{-y}}{y} dy \quad (8)$$

(2) Connectivity Analysis Flow Chart

As shown in Fig. 2, firstly the edge weight between neighboring grid blocks is calculated using the new definition. After that the edge weight is scaled by the local minimums to enhance the buoyancy effect (Jeong, 2014). Then we calculate the connectivity from the perforation grids to any reservoir grids by summing the edge weight, and then rank these connectivities using the Dijkstra's algorithm (Dijkstra, 1959). Finally, we fill the grid blocks with CO_2 to a specific saturation in the order of connectivity magnitude until the cumulative volume in the filled blocks reaches the desired injected CO_2 volume.

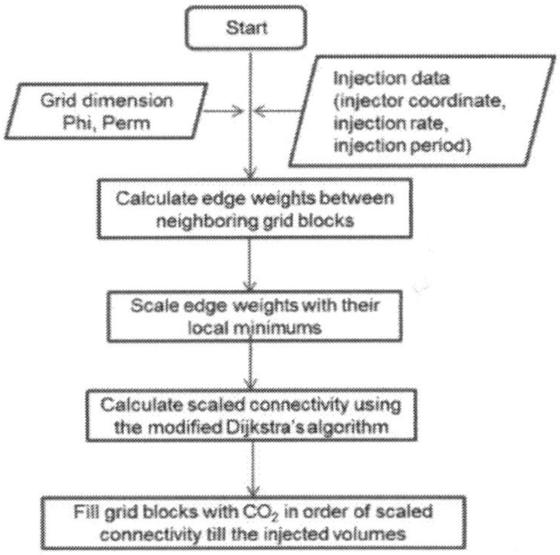

Fig. 2. Flow chart of connectivity analysis

Verification of Connectivity Analysis (CA) using CMG-GEM

We employ CMG-GEM simulator to verify the CA algorithm. The simulation model and injection parameters settings are shown in Table 1. The permeability field is shown in Fig. 3; it is a log-normal distribution with the average of 194 mD. The porosity is correlated with permeability by the Holtz equation (Holtz, 2002), and its distribution is shown in Fig. 4. The relative permeability and capillary pressure curve are shown in Fig. 5 and Fig. 6, respectively. The relevant fluid properties are the same as in Kumar (Kumar, 2005).

In the connectivity analysis, we calculate the average saturation to be 0.56 from the fractional flow curve constructed by the relative permeability curve. The average gas relative permeability shown in Eq. 6 is set to be the endpoint relative permeability of CO_2 rather than that corresponding to the average saturation to fast fill the pore volumes of grid blocks. The other fluids properties parameters for the CA algorithm are shown in the lower part of Table1.

Table 1. Parameters settings in the CMG-GEM and CA

Grid dimensions	100*100*4
Grid size, ft	1*1*1
Auto correlation length (L_x, L_y, L_z), ft	(4, 4, 0)
Well type	Vertical injector
Injection rate (RC ft^3/d)	60
Injection time, day	20
Perforated grid	(50,50,4)
Simplified multiphase properties in the CA	
CO_2 volume ratio RC/SC	3.02E-3
CO_2 viscosity (RC, cp)	8.61E-2
CO_2 density (RC, kg/m^3)	618.70
Water density (RC, kg/m^3)	1024.60
Average CO_2 saturation	0.56
Endpoint CO_2 relative permeability	0.86

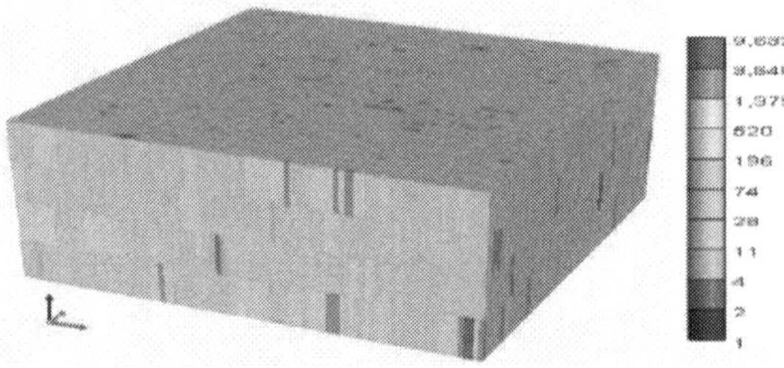

Fig. 3. Permeability field (mD)

Fig. 4. Porosity field

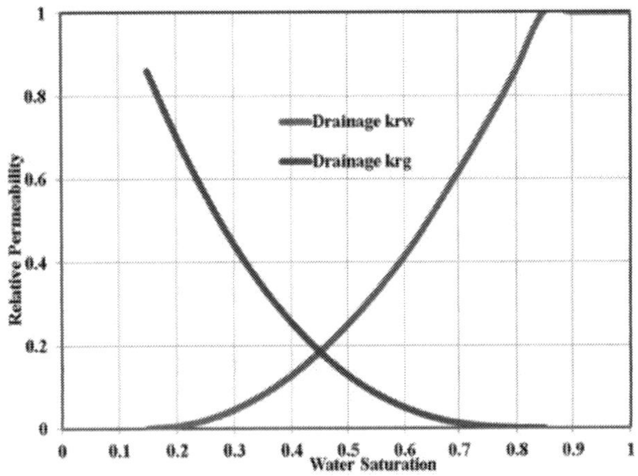

Fig. 5. Relative permeability curves in CMG and CA

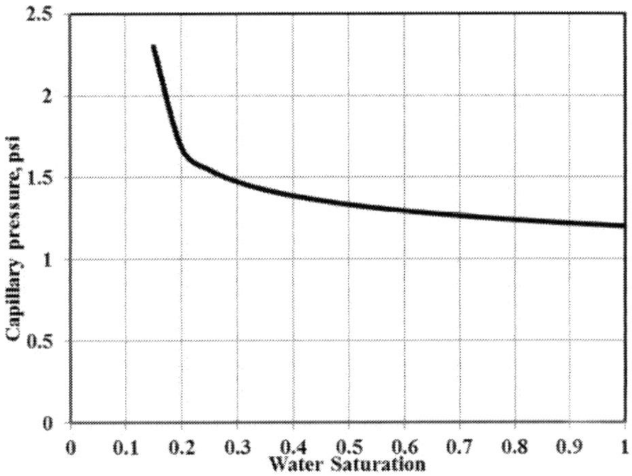

Fig. 6. Capillary pressure curve in CMG

The comparisons of CO_2 plume behavior between CMG-GEM and CA in the bottom layer, top layer and cross-section along perforation of the storage domain are shown in Fig. 7a, 7b, and 7c respectively. Different rows represent different time periods of injection which are shown in the left column. It can be seen that the CA could give a good approximation on the CO_2 plume behavior and flow path. Also, it accounts for the buoyancy effect very well as demonstrated from the cross-section profile comparison. Therefore, the CA could capture the multiphase flow dynamics during CO_2 injection into the storage domain.

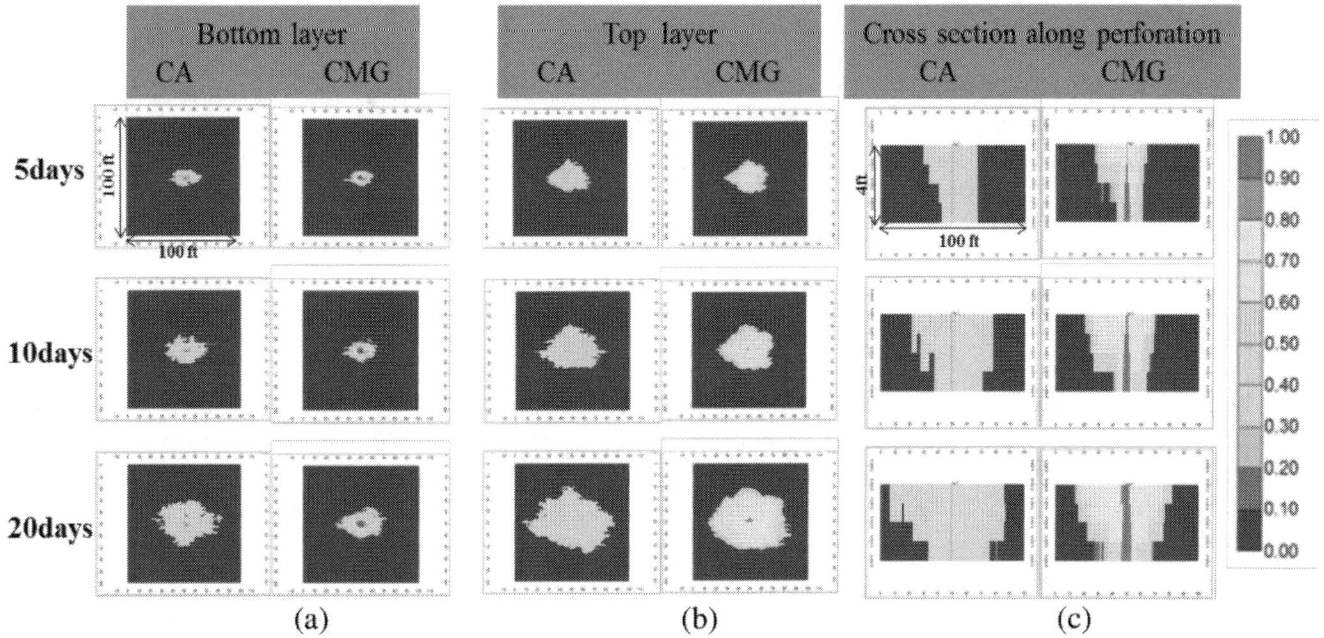

Fig. 7. Comparison of CO_2 plume behavior between CMG-GEM and CA at the different time periods of injection: (a) bottom layer; (b) top layer; (c) cross section along the perforation. Note: the scale aspect ratios for X/Y (a and b) and Z/X (c) is not the same as those in the geologic model.

Geologic Criteria Method to Predict the LCT Distribution

In a companion paper (Ren *et al.*, 2015), we present a geologic criteria (GC) algorithm to predict the local capillary traps only based on the capillary entry pressure field. The capillary entry pressure field is generated from the permeability realization using the Leverett J-function (Leverett, 1941) which is shown in Eq. 9. The detailed procedures of generating capillary entry pressure fields are presented in the companion paper. The capillary entry pressure field and its histogram for the geologic model employed in the verification part are shown in left and right plots of Fig. 8, respectively.

$$J(S_w) = \frac{p_c(S_w)}{\sigma \cos\theta}\sqrt{\frac{k}{\phi}} \qquad (9)$$

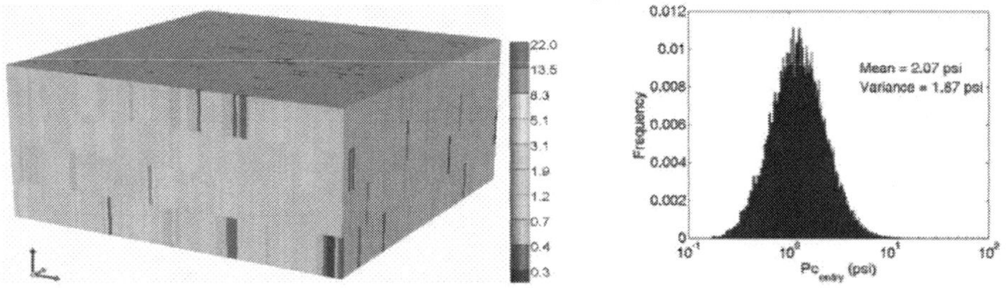

Fig. 8. Left: Capillary entry pressure field (psi); Right: Histogram of capillary entry pressure field.

Here, we give a simple description on how the GC works to identify the LCT on the capillary entry pressure field. Firstly, we select a critical capillary entry pressure (CCEP) which is used to define a given grid block to be the barrier or flow path. If the capillary entry pressure of the grid block is smaller than CCEP, it will be considered as the flow path, and otherwise, it will be barriers. Then we find the connected clusters of barrier grid blocks. After that, we search for the holes surrounded by those connected clusters of barriers. These holes are the local capillary traps.

The success of determining local capillary traps depends on the appropriate selection of CCEP. In the companion paper (Ren *et al.*, 2015), we show that the value of CCEP at which a percolating cluster is established gives a good representation of the LCT distribution. Thus we will use this choice of CCEP in the following part of the model application.

Model Application

In this part, we will combine the CA and GC to predict the local capillary traps filled during the CO_2 injection in the heterogeneous storage formation. The main objective is to explore how to maximize the LCT in the different heterogeneous reservoirs through changing injection rates. We generate a series of 3D storage domains with different level of heterogeneity. The heterogeneity is measured by the Dykstra-Parsons variation coefficient (Dykstra and Parsons, 1950), V_{dp}, which is defined in the following equation.

$$V_{dp} = \frac{k_{50} - k_{84.1}}{k_{50}} \quad \text{...(10)}$$

Where $k_{50}$and $k_{84.1}$is defined on the log-normal probability graph. $\log(k_{50})$ is less than 50% of the samples in the log permeability spectrum, whereas $\log(k_{84.1})$ is less than 84.1% of the samples, which is one standard deviation from the mean. The Dykstra-Parsons coefficient of variation is a dimensionless number that ranges from 0 to 1. A homogeneous reservoir has a coefficient of permeability variation that approaches 0 whereas an extremely heterogeneous reservoir has a coefficient of permeability variation that approaches 1.

The Dykstra-Parson variation coefficients of the nine generated isotropic permeability field are shown in Table 2, it covers a wide range from 0.13 through 0.91. The dimensions of these 3D models are the same of 64*64*32, and the grid size is 1*1*1 ft. The auto correlation in the x, y and z-direction is 4, 4, and 1 ft, respectively. The mean of the permeability for all the fields keeps the same as 500 mD with different standard deviations. Following the same procedures described above, we correlate permeability with porosity, and then correlate the above two parameters with the capillary entry pressure by Leverett J-function. The detailed properties of these permeability, porosity and capillary entry pressure fields are shown in Table 2.

Table 2. Heterogeneous permeability, porosity and capillary entry pressure fields

Model #	k Mean, mD	k Standard deviation, mD	V_{dp}	ϕ Mean	P_c^{entry} Mean, psi	P_c^{entry} Standard deviation, psi	Percolated CCEP, psi
Hete01	500	70	0.13	0.29	1.21	0.09	1.1
Hete02	500	140	0.24	0.29	1.23	0.17	1.05
Hete03	500	220	0.34	0.29	1.28	0.27	1.00
Hete04	500	299	0.43	0.29	1.35	0.38	0.95
Hete05	500	447	0.54	0.28	1.50	0.60	0.95
Hete06	500	798	0.68	0.27	1.94	1.19	0.95
Hete07	500	1336	0.76	0.26	2.60	2.12	0.95
Hete08	500	3783	0.86	0.24	5.04	6.22	1.15
Hete09	500	9574	0.91	0.22	10.09	16.97	1.45

For the connectivity analysis, the fluid property values keep the same as shown in Table 1. We select six injection rates to be 8E+2, 8E+3, 8E+4, 8E+5, 8E+6, and 8E+7 MMScf/d, which cover a wide range of the reported industrial GCS CO_2 injection rates. The corresponding buoyancy number is calculated based on the following equation 6 (Zhou *et al.*, 1997). In the equation, N_{gr} is the buoyancy number, $\Delta\rho$ is density difference between brine and CO_2 (405.9 kg/m^3), g is buoyancy constant (9.8 m/s), k_v is the average vertical permeability (refer to the 2nd column of Table 2), u is the Darcy velocity of CO_2 along the wellbore, μ is the viscosity of CO_2 (0.049 mPa·s). Table 3 shows the injection parameters settings and buoyancy numbers in detail.

$$N_{gr} = \frac{\Delta \rho g k_v L_{pr}}{u \mu_g H_{re}} \quad (11)$$

Table 3. The injection parameter settings and buoyancy numbers in the CA

Well type	Vertical injector					
Perforation grid	(32, 32, 32)					
Injection rate, Scf/d	8E+2	8E+3	8E+4	8E+5	8E+6	8E+7
Injection duration, day	2000	200	20	2	0.2	0.02
Buoyancy number	6E+1	6E+0	6E-1	6E-2	6E-3	6E-4

Here, we define two different terminologies: static LCT volume ratio and dynamic LCT volume fraction. The static LCT volume ratio is defined as the number ratio of grid blocks in LCT to those in the whole storage domain. The dynamic LCT volume fraction means the fraction of LCT grid blocks in the CO_2 swept grid blocks under the scenario of injection. We show the difference between these two definitions in the following sketch of Fig. 9. Based on the definitions, we can see that the physical implication of the static LCT volume ratio is the fraction of the pore space occupied by LCT in the storage domain, whereas the dynamic LCT volume fraction is the fraction of CO_2 stored in the safe LCT mode relative to the cumulatively injected CO_2. Therefore, the former is definite while the latter dynamically changes as the injection proceeds. As the injected CO_2 fully fill the whole domain, the dynamic LCT volume fraction will be the same as static LCT volume ratio.

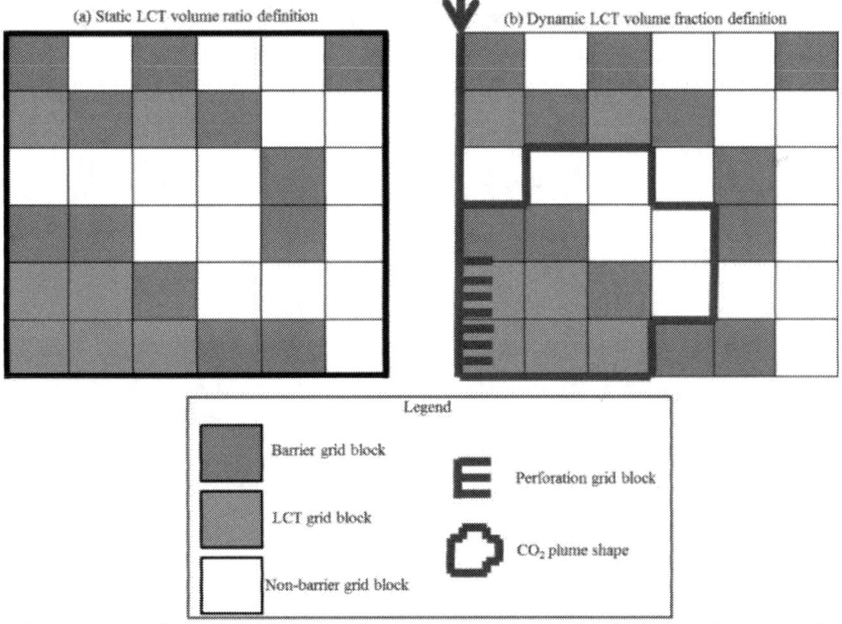

Fig. 9. Schematic explanation of the difference between the static LCT volume ratio and the dynamic LCT volume fraction. (a) the static LCT volume ratio is the ratio of the number of green LCT grid blocks over the total number of grid blocks in the whole domain (the black square); (b) the dynamic LCT volume fraction is the ratio of the number of green LCT grid blocks in the CO_2 swept zone (the connected blue line) over the total number of grids blocks in the CO_2 swept zone.

In the geologic criteria, we choose the CCEP to be the percolated capillary entry pressure values. The percolated CCEP corresponds to the peaked static LCT volume ratio as shown in the Fig. 10. In the left and right plots of Fig. 10, we show the impact of CCEP on the static LCT volume ratio in the two different ranges of Dykstra-Parsons variation coefficients, i.e., 0.13~0.68 and 0.68~0.91, respectively. The former represents the homogenous and intermediately heterogeneous reservoirs while the latter the highly heterogeneous domains. In the reservoirs with the V_{dp} ranging from 0.13 to 0.68, the peaked static LCT volume ratio moves to the left side which means that the percolated CCEP decrease with the increase of V_{dp}. The highly heterogeneous reservoir shows the opposite trends. The percolated CCEP for the different heterogeneous reservoirs is shown in Table 2.

Also, we plot the peaked static LCT volume ratio versus Dykstra-Parsons variation coefficient in the left plot of Fig. 11. As the increase of the Dykstra-Parsons variation coefficient, the peaked static LCT volume ratio firstly increases fast followed by marginal augment. The inflection point corresponds to the V_{pd} around 0.3. As observed from Table 2, when the V_{dp} is not higher than 0.3, the percolated CCEP is very close to the mean of the entry capillary pressure field. In this region, the small increase of standard deviation of the entry capillary pressure fields would increase the possibility of entrapment of gas in the form of local capillary trapping; this explains the fast increase of static LCT volume ratio. However, as the V_{dp} is higher than 0.3, the percolated CCEP become much smaller than the mean. This will bring about more barriers in the storage domain. On one hand, more barriers would more likely to be connected and surrounds more LCT, so we see some marginal increase of static LCT volume ratio. On the other hand, more barriers mean less flow path grids blocks, which decease of the possibility of forming LCT. Therefore, the peaked static LCT volume ratio decreases a little bit when the V_{dp} is higher than 0.76.

The V_{dp} for most of the reservoirs ranges from 0.50 to 0.90 (Lake, 1989). We can observe that the peaked static LCT volume ratio under this typical range of V_{dp} is around 11%, which can be considered as the upper most storage capacity of LCT. It should be indicated that the heterogeneity here is indicated only through different V_{dp} under a constant correlation length.

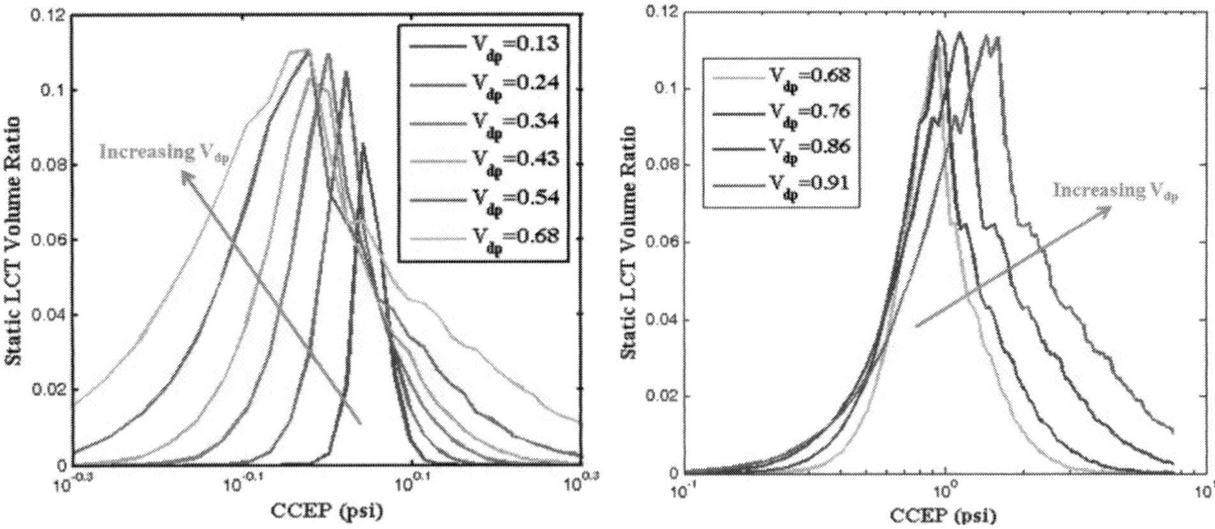

Fig. 10. Evolution of the LCT with the CCEP for the different heterogeneous models. Left: V_{dp}=0.13~0.68; Right: V_{dp}=0.68~0.91.

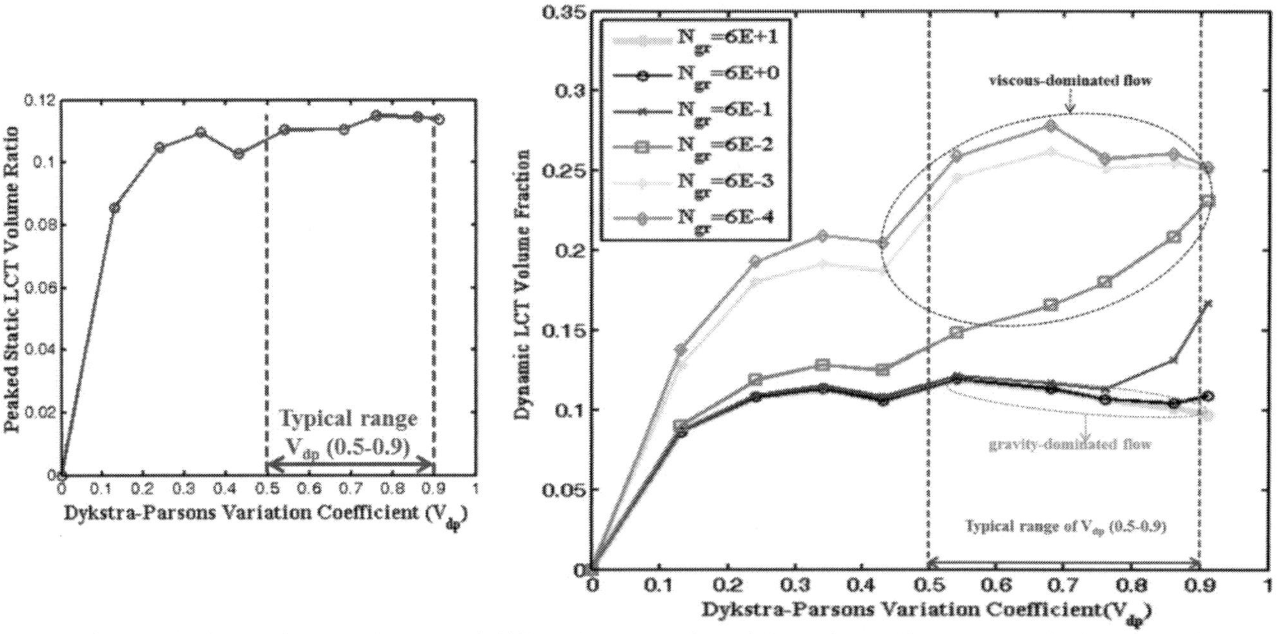

Fig. 11. Left: evolution of peaked static LCT volume ratio with Dykstra-Parsons variation coefficient. Right: change of dynamic LCT volume fraction with the Dykstra-Parsons variation coefficient (V_{dp}) under different buoyancy numbers (injection rates).

We firstly run the CA to predict the CO_2 plume shape under different injection rates and then combine the GC to predict the dynamic LCT volume fraction. The right plot of Fig. 11 shows the impact of injection rate on the dynamic LCT volume fraction in the different level of heterogeneous models at the end of injection. In this figure, we categorize the CO_2/brine displacement into three flow regimes using the buoyancy number: viscous-dominated flow (N_{gr}<0.6), buoyancy-dominated flow (N_{gr}>6), and transient regime (0.6≤N_{gr}≤6). Several observations can be obtained from this figure:

(1) Impact of heterogeneity on the LCT

We describe the impact of heterogeneity on the dynamic LCT volume ratio in the above three separate flow regimes. In the viscous-dominated flow, the dynamic LCT volume fraction generally increases with the V_{dp}; this trend is very similar to that for static LCT ratio versus V_{dp} as shown in left plot of Fig. 11. However, in the buoyancy-dominated and transient regimes, the dynamic LCT volume fraction is almost independent of the heterogeneity for the V_{dp} ranging from 0.25 to 0.68. When the V_{dp} is higher than 0.68, the dynamic LCT fraction increases with the V_{dp} in the transient flow regime while it decreases in the buoyancy-dominated flow regime.

(2) Impact of injection rate (buoyancy number) on the LCT

When the reservoir is relatively homogenous (V_{dp} <0.45), the dynamic LCT volume fraction is independent of injection rate in the buoyancy-dominated and transient flow regimes. Whereas, increasing the injection rate in the viscous-dominant regime would give rise to higher dynamic LCT volume fraction, and this effect is pronounced in the more heterogeneous reservoirs. However, when the buoyancy number is less than 6E-4, the dynamic LCT volume fraction is not sensitive to the injection rate. This is consistent with the coreflooding observation (Shi *et al.*, 2011) that the mean CO_2 saturation profiles along the core become gradually diminished as the injection rate was increased in step to 3 cc/min (N_{gr}=6.4E-3).

For the further analysis and explanation, we show the visualized flooded LCT (LCT in the CO_2 plume), flooded non-LCT, and non-flooded LCT of the storage domain in Fig. 12. Here, the flooded LCT and flooded non-LCT can be considered to compose the CO_2 plume. So the dynamic LCT volume fraction can be interpreted as the ratio of flooded LCT over the sum of flooded LCT and non-LCT. Fig. 12 compares these three types of LCT in the selected layers # 8, 16, 24, and 32 and cross section along the wellbore for two extreme reservoirs, one is almost homogeneous with the V_{dp} of 0.13, the other with the largest V_{dp} (the most heterogeneous). The layer numbering is from the top to the bottom of the reservoir.

As we can see, the homogenous reservoir shows good uniform sweeping profile under small injection rate (N_{gr}=60, the upper row of Fig. 12), similar behavior is observed for the large injection rate (N_{gr}=6E-4). This uniform profile gives less chance of CO_2 filling the sparsely-distributed LCT. Therefore, we see the smaller dynamic LCT volume fraction compared with that for the heterogeneous reservoirs. Specifically, under the large injection rate in the homogeneous reservoirs, the CO_2 could sweep more grid block in the bottom of the domain, and the CO_2 would more likely fill the LCT under the buoyancy far away from the wellbore. Therefore, the increase of injection rate could slowly enhance the dynamic LCT volume fraction when the reservoir is relatively homogeneous (V_{dp} <0.45) as shown in the right plot of Fig. 11.

For the heterogeneous reservoirs (V_{dp}=0.91), injected CO_2 sweeps most of the grids in the bottom of the storage domain under high injection rate (N_{gr}=6E-4, the lower row of Fig. 12). As CO_2 moves far away from the wellbore, it would move along the specific paths with good connectivities under the buoyancy. This kind of channeling increases the possibility of avoiding barriers and filling of local capillary traps. Therefore, the more heterogeneity the reservoir becomes the higher dynamic LCT volume fraction achieved under higher injection rate (N_{gr}<0.6). However, when the small injection rate (N_{gr}=60, the middle row of Fig. 12) is employed, the CO_2 moves quickly to the top layer and cause less filling of LCT.

(3) For the typical range of Dykstra-Parsons coefficients, the dynamic LCT volume ratio varies from 8% to 28% depending on the V_{dp} and injection rates (refer to Fig. 11). This demonstrates that LCT accounts for an important portion of the CO_2 plume. Also, considering the LCT trapped saturation is typically 50-70% higher than residual saturation, it can be expected that the trapped LCT CO_2 mass fraction would be 3-6% higher than the dynamic LCT volume fraction. It should be noted that a constant correlation length is assumed here, the impact of this heterogeneity indicator on the LCT will be reported in the future publications.

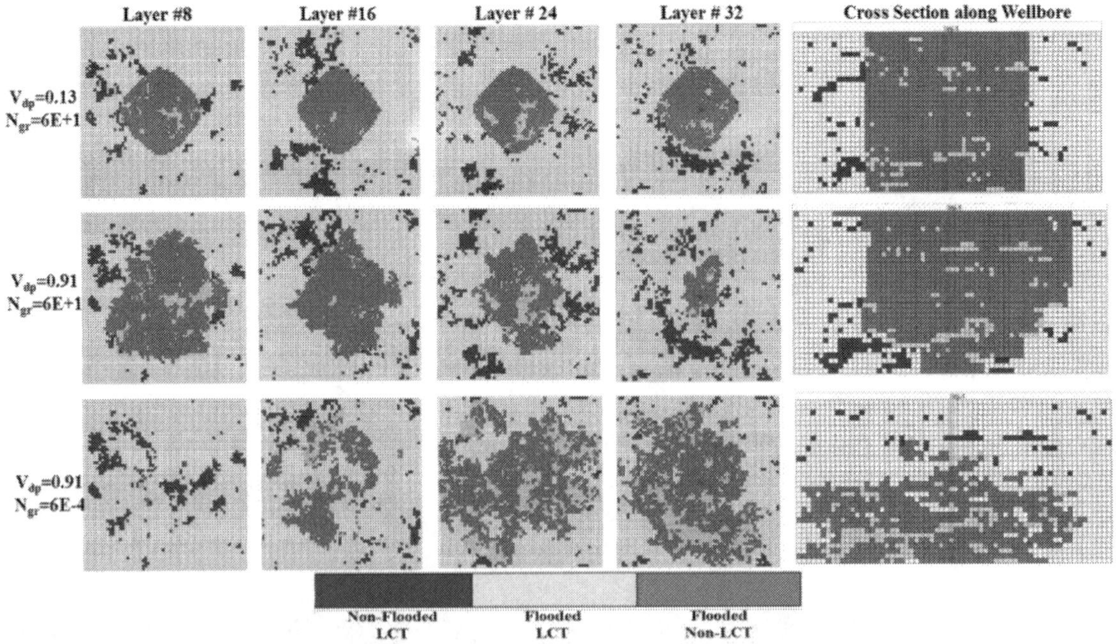

Fig. 12. The distribution of flooded LCT, flooded non-LCT, and Non-flooded LCT in the different layers (#8, 16, 24, 32) and cross section along the wellbore (XZ) for different heterogeneous reservoirs under different injection rates (buoyancy numbers). Upper: V_{dp} =0.13, N_{gr}=6E+1; Middle: V_{dp} =0.91, N_{gr}=6E+1; Lower: V_{dp} =0.91, N_{gr}=6E-4.

Summary and Conclusions

The full-physics simulation of displacement incorporating local capillary heterogeneity is computationally intensive. We decouple the flow simulation under the effect of capillary pressure heterogeneity into two parts: one is the permeability-based CO_2 plume behavior prediction by the connectivity analysis; the other is the capillary entry pressure heterogeneity-based local capillary trapping (LCT) distribution identification by the geologic criteria. The main conclusions from this work are summarized as below;

(1) Final CO_2 distribution (post-injection and post-buoyancy-driven migration) is controlled by the local capillary traps intrinsic to the capillary pressure heterogeneity.

(2) The CO_2 plume behavior calculated from the extended connectivity analysis shows a good match with that computed by the full-physics simulation of CMG-GEM.

(3) Statically, for a fixed correlation length of geologic features, the increase of reservoir heterogeneity (measured by Dykstra-Parsons coefficient) would increase the LCT storage capacity when the reservoir is relatively homogenous with the Dykstra-Parson coefficient smaller than 0.3. The static LCT volume ratio for the typical range of V_{dp} (0.5~0.9) is around 11%.

(4) After incorporating the flow dynamics using connectivity analysis, increasing injection rates in a specific range of buoyancy number (6E-4<N_{gr}<6E-1) would bring higher volume fraction of CO_2 stored in the safe LCT mode and this effect is enhanced for the heterogeneous reservoir. The buoyancy dominated flow ($N_{gr}\geq 6$) shows almost constant fraction (11%) of CO_2 trapped in the LCT mode which is independent of the reservoir heterogeneity when the V_{dp} is larger than 0.25.

Nomenclature

Roman Symbols

A	Contact area of the adjacent grid blocks, cm^2
C_t	Reservoir total compressibility, 1/atm
$ED1$	Original edge weight, []
$ED2$	New edge weight, sec
g	Gravity acceleration, 9.8m/s^2
H_{re}	Reservoir thickness, cm
h	Height of CO_2 column, cm
$J(S_w)$	Leverett J-function, []
k	Permeability, Darcy
k_v	Average vertical permeability, Darcy
$\bar{k}$	Reservoir geometric average, Darcy
k_i	Permeability in grid i, Darcy
k_j	Permeability in grid j, Darcy
k_g	Relative permeability of CO_2, []
$\overline{k_g}$	Average relative permeability of CO_2, []
k_{50}	50 percentile permeability value, Darcy
$k_{84.1}$	84.1 percentile permeability value, Darcy
L	Length of grid block, cm
L_{pr}	Reservoir perforation length, cm
L_x	Correlation length in x-direction, ft
L_y	Correlation length in y-direction, ft

L_z Correlation length in z-direction, ft
$p_c(S_w)$ Capillary pressure, psi
$p_{ci}(S_w)$ Capillary pressure for grid i, psi
$p_{cj}(S_w)$ Capillary pressure for grid j, psi
p_c^{entry} Capillary entry pressure, psi
Q CO_2 injection rate, cm^3/s
q_g CO_2 flow flux, cm^3/s
r_1 The radial distance of grid i from wellbore, cm
r_2 The radial distance of grid j from wellbore, cm
$\overline{S_g}$ Specific saturation at which the given grid will be filled, []
T Transmissivity between the adjacent grid blocks, cm^3
t CO_2 injection time period, sec
u Total or Darcy velocity of CO_2, ft/day
V_{dp} Dykstra-Parsons variation coefficient, []
Vp_i Pore volume in grid i, cm^3
Vp_j Pore volume in grid j, cm^3

Greek Symbols
μ_g Viscosity of gas (CO_2), cp
μ_w Viscosity of brine, cp
ρ_g Density of gas (CO_2), kg/m^3
σ Interfacial tension between CO_2 and brine, N/m
θ Contact angle, degree
ϕ Porosity, []
ϕ_i Porosity in grid i, []
ϕ_j Porosity in grid j, []
ΔP Viscous pressure difference, atm
$\Delta\rho$ Density difference between brine and CO_2, kg/m^3

Acronyms
CA Connectivity analysis
CCEP Critical capillary entry pressure
GC Geologic criteria
CMG Computer modeling group
GCS Geological carbon sequestration
LCT Local capillary trapping
RC Reservoir condition
SC Surface condition

Acknowledgement

We would like to thank the sponsors of the Geological CO_2 Storage Industrial Affiliates Project at The University of Texas at Austin. Part of this work was supported by the Office of Fossil Energy, National Energy Technology Laboratory of the United States Department of Energy under DOE Award

Number DE-FE0004956. Larry W. Lake holds the W.A. (Monty) Moncrief Centennial Chair in the Department of Petroleum and Geosystems Engineering at The University of Texas at Austin. Steven Bryant holds the Canada Excellence Research Chair in Materials Engineering for Unconventional Oil Resource at the University of Calgary.

References

Anchliya, A., Christine, A., Ehlig-Economides, C., Jafarpour, B. 2012. Aquifer Management To Accelerate CO_2 Dissolution and Trapping. *SPE Journal* **17** (3): 805-816.

Bhowmik, S. 2014. *Particle Tracking Proxies for Predication of CO_2 Plume Migration within a Model Selection Framework*. Ph.D. Dissertation, The University of Texas at Austin, Austin, TX.

Bryant, S.L., Lakshminarasimhan, S., Gary, A.P. 2008. Buoyancy-Dominated Multiphase Flow and Its Effect on Geological Sequestration of CO_2. *SPE Journal* **13** (4): 447-454.

Cavanagh, A.J., Haszeldine, R.S. 2014. The Sleipner Storage Site: Capillary Flow Modeling of a Layered CO_2 Plume Requires Fractured Shale Barriers within the Utsira Formation. *International Journal of Greenhouse Gas Control* **21**: 101-112.

Cavanagh, A.J. and Rostron, B. 2013. High-resolution Simulations of Migration Pathways and the Related Potential Well Risk at the IEAGHG Weyburn–Midale CO_2 Storage Project. *International Journal of Greenhouse Gas Control* **16** (S1): S15-S24.

Dijkstra, E.W. 1959. A Note on Two Problems in Connexion with Graphs. *Numerische Mathematik* **1** (1): 269-271.

Dykstra, H., Parsons, R.L. 1950. The Prediction of Oil Recovery by Waterflooding. In Secondary Recovery of Oil in the United States, second edition, 160-174. *Washington, DC*: API.

Hassanzadeh, H., Pooladi-Darvish, M., Keith, D.W. 2009. Accelerating CO_2 Dissolution in Saline Aquifers for Geological Storage - Mechanistic and Sensitivity Studies. *Energy & Fuels* **23** (6): 3328-3336.

Hirsch, L.M., Schuette, J.F.1999. Graph Theory Applications to Continuity and Ranking in Geologic Models. *Computers & Geosciences* **25** (2): 127-139.

Holtz, M.H. 2002. Residual Gas Saturation to Aquifer Influx: A Calculation Method for 3-D Computer Reservoir Model Construction. Presented at the SPE Gas Technology Symposium, Calgary, Alberta, Canada, 30 April-2 May. SPE-75502-MS.

Javaheri, M, Jessen, K. 2011. Residual Trapping in Simultaneous Injection of CO_2 and Brine in Saline Aquifers. Presented at the SPE Western North American Region Meeting, Anchorage, Alaska, USA, 7-11 May. SPE-144613-MS.

Jeong, H. 2014. *Uncertainty Quantification of Fluid Displacements in Heterogeneous Media Using a Connectivity Based Proxy*. Ph.D. Proposal, unpublished, The University of Texas at Austin, Austin, Texas.

Krause, M.H., Perrin, J.C., Benson, S.M. 2011. Modeling Permeability Distributions in a Sandstone Core for History Matching Coreflood Experiments. *SPE Journal* **16** (4): 768-777.

Kumar, A., Noh, M.H., Ozah, R.C., Pope, G.A., Bryant, S.L., Sepehrnoori, K. and Lake, L.W. 2005. Reservoir Simulation of CO_2 Storage in Deep Saline Aquifers. *SPE Journal* **10** (3): 336-348.

Lake, L.W. 1989. *Enhanced oil recovery*, second edition. New Jersey: Prentice Hall.

Leonenko, Y., Keith, D.W. 2008. Reservoir Engineering To Accelerate the Dissolution of CO_2 Stored in Aquifers. *Environmental Science & Technology* **42** (8): 2742-2747.

Leverett, M. C. 1941. Capillary Behavior in Porous Solids. *AIME Petroleum Transactions* **142**: 152-169.

Li, B.X., Benson, S.M. 2015. Influence of Small-scale Heterogeneity on Upward CO_2 Plume Migration in Storage Aquifers. *Advances in Water Resources* **83**: 389-404.

Meckel, T.A., Bryant, S.L., Ganesh, P.R. 2015. Characterization and Prediction of CO_2 Saturation Resulting from Modeling Buoyant Fluid Migration in 2D Heterogeneous Geologic Fabrics. *International Journal of Greenhouse Gas Control* **34**: 85-96.

Nghiem, L., Shrivastava, V., Kohse, B., Hassam, M., Yang, C. 2010. Simulation and Optimization of Trapping Processes for CO_2 Storage in Saline Aquifers. *Journal of Canadian Petroleum Technology* **49** (8): 15-22.

Qi, R., Laforce, T., Blunt, M. 2009. Design of carbon dioxide storage in aquifers. *International Journal of Greenhouse Gas Control* **3** (2): 195-205.

Ren, B., Lake, L.W., Bryant, S.L. 2015. Quantifying Local Capillary Trapping Storage Capacity Using Geologic Criteria. Presented at the AIChE Carbon Management Technology Conference, Houston, Texas, USA, 17-19 November. CMTC-439489-MS.

Ren, B., Sun, Y.H., Bryant, S.L. 2014. Maximizing Local Capillary Trapping During CO_2 Injection. *Energy Procedia* **63**: 5562-5576.

Saadatpoor, E. 2009. *Effect of Capillary Heterogeneity on Buoyant Plumes: New Trapping Mechanism in Carbon Sequestration*. MS thesis, The University of Texas at Austin, Austin, Texas.

Saadatpoor, E. 2012. *Local Capillary Trapping in Geological Carbon Storage*. Ph.D. Dissertation, The University of Texas at Austin, Austin, Texas.

Saadatpoor, E., Bryant, S.L., Sepehrnoori, K. 2009. New Trapping Mechanism in Carbon Sequestration. *Transport in Porous Media* **82** (1): 3-17.

Sahimi, M. 2012. *Flow and Transport in Porous Media and Fractured Rock: From Classical Methods to Modern Approaches*, second edition. New Jersey: Wiley.

Shi, J.Q., Xue, Z.Q., Durucan, S. 2011. Supercritical CO_2 Core Flooding and Imbibition in Tako Sandstone-Influence of Sub-core Scale Heterogeneity. *International Journal of Greenhouse Gas Control* **5**(1): 75-87.

Zhou, D.E, Fayers, F.J., Orr, F.M. 1997. Scaling Multiphase Flow in Simple Heterogeneous Porous Media. *SPE Reservoir Engineering* **12**(03): 173-178.

Sun, Y.H. 2014. *Investigation of Buoyant Plumes in a Quasi-2D domain Characterizing the Influence of Local Capillary Trapping and Heterogeneity on Sequestered CO_2: A Bench Scale Experiment*. MS thesis, The University of Texas at Austin, Austin, Texas.

Tanaka, S., Datta-Gupta, A., King, M.J. 2014. Compositional Streamline Simulation of CO_2 Injection Accounting for Gravity and Capillary Effects Using Orthogonal Projection. Presented at the SPE Improved Oil Recovery Symposium, Tulsa, USA, 12-16 April. SPE-169066-MS.

Wilkinson, D. and Willemsen, J.F. (1983). Invasion Percolation: a New Form of Percolation Theory. *Journal of Physics A: Mathematical and General* **16**(14): 3365-3376.

CMTC-439497-MS

Preliminary Assessment of CO_2 Storage Potential in the H-59 Block in Jilin Oilfield CCS Project

L. Zhang, X. Li, China University of Petroleum (huadong); B. Ren, University of Texas at Austin; G. D. Cui, S. R. Ren, China University of Petroleum (huadong); and G. L. Chen, PetroChina

This paper was prepared for presentation at the Carbon Management Technology Conference held in Sugarland, Texas, USA, 17–19 November 2015.

Abstract
The H-59 pilot test has been conducted for several years since 2008. There are enough available data for the operator to track the CO_2 storage and distribution in the H-59 block, which is important for the storage potential verification, storage safety assessment, and monitoring scheme adjustment in the next phase of the project. In this paper, the storage capacity and distribution of CO_2 trapped at different states in the H-59 block are investigated. An assessment method for CO_2 storage potential is established to calculate the theoretical and effective capacities of CO_2 storage in the target oil reservoir. Then the actual amount of CO_2 buried in the H-59 block up to June 2014 is calculated, and its distribution and trapping states are predicted using reservoir simulation method based on the field data. At last, the reliability of the assessment method and the prediction accuracy of reservoir simulation as well as the remaining potential for CO_2 storage in the H-59 block are discussed. The assessment results show that the theoretical and effective CO_2 storage capacities in the H-59 block are 59.04×10^4 tons and 26.37×10^4 tons respectively. Until June 2014, 20.12×10^4 tons CO_2 has been stored in the block, and a potential of about 6.25×10^4 tons CO_2 is remained. Most of injected CO_2 is distributed within the well pattern and its shape and cover area are determined by the reservoir properties and the amount of CO_2 injected in each well. 60.15% of stored CO_2 is trapped at supercritical state, and the CO_2 dissolved in residual oil and water take 24.85% and 15.00% respectively. In addition, the accuracy of geological model has a significant effect on the prediction of CO2 distribution in the oil reservoir. More efforts should be taken to improve the understanding of the geological conditions of the target reservoir.

Introduction
Carbon Capture and Storage (CCS) technology has been proven an effective option to sequester CO_2 underground permanently at a large scale. CO_2 in the industrial waste gas can be separated, transported and injected into suitable geological structures for storage, to achieve a long-term trapping of CO_2 away from the atmosphere [1]. Since 1990s, many CO_2 storage demonstration projects have been conducted, such as Sleipner, Weyburn, In salah, K-12B, Snowhit projects [2-7]. Oil and gas reservoirs, deep saline aquifers are the preferable sites for CO_2 storage. However, at present and for the near future, oil reservoirs are the most favorable choices because of their proven sealing conditions and economic potentials combined with EOR. Especially for the reservoirs associated with large underlying or lateral aquifers, they can provide a greater capacity for CO_2 storage [8].

In recent years, China petroleum industry has been paying more attentions on CO_2 EOR and storage. The field operation of CO_2 injection is gradually scaled up. The CO_2 EOR and storage project conducted in Jilin oilfield is the first large-scale CCS project in China, which is located in the Jilin province section of the Songliao Basin in the northeast China (see Figure 1a). The Daqingzijing region was selected as the target reservoir for CO_2 injection, aimed at demonstrating the technical and economic feasibility of CO_2 EOR and storage in the widespread tight and light oil reservoirs in China at an industrial scale. The project contains three phases (see Figure 1b). The first phase is a small pilot test, located at the H-59 block including 5 injectors and 19 producers forming an inverted seven-point well pattern with a spacing of 250-300 m (see Figure 1c). The injected CO_2 is from the nearby Changshen 4 well which produces natural CO_2 gas with a high purity up to 97 mol%. The second phase is an enlarged test which will be conducted in the H-79 block. 18 injectors and 60 producers will be involved. Then CO_2 injection will be extended to other blocks of Daqingzijing region with more than 200 well groups covered [9]. The captured CO_2 from Changling gas field close to the H-79 block which has a CO_2 concentration of 21-23 mol% will be used for the second and full-scale phases of CO_2 EOR and storage in the Jilin oilfield. The Jilin CCS project is expected to provide valuable experiences for the future CO_2 EOR projects in China.

Geological Characteristics of the H-59 Block

The Daqingzijing region is located in the center of the Changling sag in the south of the Songliao Basin. Changling sag is composed of two sub-sags, including the southern Heidimiao sub-sag and the northern Qian'an sub-sag, while the Daqingzijing region is located in the uplift between the two sub-sags. Daqingzijing region consists of more than 20 individual blocks mainly divided by faults. Most of these blocks are low permeability oil reservoirs, buried at depth from 1800 m to 2500 m, with net thickness between 2 m and 20 m. The main oil-bearing formations are located in the Qingshankou Group which was formed in the age of Cretaceous and deposited in a delta front environment. Mature sand bodies are well developed in these oil reservoirs. The above cap rock is 500-550 m thick and distributed stably, composed of mudstone layers and shale intercalations, which divide the upper and lower groups into different hydrodynamic systems [10].

The H-59 block is located at the west side of the central fault belt of Daqingzijing region. It covers an area of approximately 3.1 km^2 and contains about 160×10^4 tons of light oil. The five well groups in the north of the block were selected for the pilot test, which covers an area of about 1.8 km^2 and contains about 86×10^4 tons of crude oil. The oil reservoir is buried at depth of 2000-2400 m, with a small downward inclination angle of 2-4 degrees from west to east, and surrounded by four normal faults extending from north to south with a length of 1-3 km. The total thickness of oil-bearing layers ranges from 11.2 m to 18.2 m, including main layers 7, 12, 14, 15 in the Qing-I formation of the Qingshankou Group, which take 78% of OOIP. The average porosity and permeability of the reservoir are 12.7% and 3.5 md, respectively. The reservoir temperature is 98.9 °C, and the original reservoir pressure is 24.15 MPa. The CO_2 minimum miscible pressure (MMP) of the crude oil is 22.3 MPa. The oil viscosity and density at reservoir are 1.85-2.18 mPa·s and 0.7615-0.7686 g/cm^3, respectively. The original gas-oil ratio is 35 m^3/ton corresponding to a bubble point of 7.31 MPa. The salinity of the formation water is between 10229-19320 mg/L (Zhang et al., 2014). In addition, the core analysis shows that the reservoir lithology of the H-59 block is mainly siltstone associated with a small amount of fine sandstone and muddy siltstone. The main mineral components of sandstone grains are quartz (18%-29%), feldspar (24.5%-37.5%) and debris (26.8%-30.7%). The main mineral components of cements between the sandstone grains are mudstone (6%-21%) and limestone (3%-34%) as well as a small amount of siliceous matters (1%-3%) [11].

CO_2 Storage Potential in the H-59 Block

Generally CO_2 are stored in geological structures through structure trapping, residual gas trapping, dissolution trapping and mineral trapping (Chadwick et al., 2008). The amounts of CO_2 stored by structure and residual gas trapping can be regarded as displacement trapping capacity for convenience and simplicity, in which the CO_2 trapped at supercritical state. Hence, the storage capacity of CO_2 in the H-59 block is composed of the following three parts (see formula 1).

$$M_{CO_2}^{theory} = M_{CO_2}^{displacement} + M_{CO_2}^{dissolution} + M_{CO_2}^{mineral} \quad (1)$$

where M_{CO2}^{theory} is the theoretical capacity of CO_2 storage in the reservoir; $M_{CO2}^{displacement}$ is the displacement trapping capacity ; $M_{CO2}^{solution}$ is the dissolution trapping capacity; $M_{CO2}^{mineral}$ is the mineral trapping capacity.

Displacement Trapping Capacity

The H-59 block has only carried out the primary recovery and a short-term water flooding before CO_2 injection. The recovery factor was only 5% from 2004-2008. Due to the bad water flooding performance, CO_2 injection was adopted and WAG (Water-Alternating-Gas) injection was considered to improve the sweeping efficiency of CO_2 after CO_2 breakthrough in the production wells. Therefore, the H-59 block can be treated as an original reservoir only with primary recovery starting CO_2 injection, and the method of Shaw and Bachu (2002) was used to estimate the displacement trapping capacity of CO_2 in the target reservoir [12]. This method is based on the material balance principle which assumes that CO_2 will occupy all the reservoir volume of produced crude oil before CO_2 breakthrough in the production wells, and only a part of produced oil volume is used to store CO_2 after CO_2 breakthrough in the production wells when the WAG flooding is adopted. The calculation formula is as follows:

$$M_{CO_2}^{displacement} = \rho_{CO2res}\left[RF_{BT}+0.6\left(RF_{\%HCPV}-RF_{BT}\right)\right]OOIP\times B_o/\rho_{osurf} \quad (2)$$

where $M_{CO2}{}^{displacement}$ is the displacement trapping capacity, 10^4 tons; ρ_{CO2res} is the CO_2 density at reservoir condition, ton/m^3; RF_{BT} is the oil recovery factor when the CO_2 breakthrough in the production wells, fraction; $RF_{\%HCPV}$ is the oil recovery factor after CO_2 breakthrough in the production wells, fraction; OOIP is the original oil in place, 10^4 tons; B_o is the oil volume efficiency; ρ_{osurf} is the oil density at surface, ton/m^3.

The main reservoir and fluid properties of the H-59 block (referring to the five well groups in the north for CO_2 injection) are summarized in Table 1. Based on the preliminary reservoir simulation results before CO_2 injection, the oil recovery factor before CO_2 breakthrough was expected to be 0.17 (2008-2013) while the finial recovery factor was 0.28 at the end of WAG flooding (2013-2018). According to the formula (2), the displacement trapping capacity of CO_2 in the H-59 block was calculated to be about 15.53×10^4 tons.

In addition, according to the material balance equation and assuming that the reservoir is incompressible, the volumes of various reservoir rocks and fluids in the H-59 block at reservoir conditions can be further calculated, as shown in Table 2. The relevant formulas adopted are also remarked in the table. The values of different parameters used in these formulas refer to Table 1.

Dissolution Trapping Capacity

The theoretical CO_2 dissolution trapping capacity is composed of two parts: the amount of CO_2 dissolved in residual water and the amount of CO_2 dissolved in the residual oil.

CO_2 solubility in formation water. The average salinity of the formation water in H-59 block is 14607.8 mg/L. Duan & Sun model [13, 14] was used to calculate the CO_2 solubility in the formation water of H-59 block. This model can take into account the effect of different mineral ions on the solubility of CO_2 in water, and has a wide application condition of temperature within 0-260 °C, pressure within 0-200 MPa, and salinity within 0-4 mol NaCl/kg H_2O (Duan and Sun, 2003; Duan et al., 2006). At typical reservoir conditions, CO_2 solubility in water is between 0.5-2.0 mol/kg while at the original reservoir conditions of H-59 block, the solubility of CO_2 in water ($S_{CO2}{}^{water}$) is 1.1398 mol/kg or about 0.0502 ton/m^3.

CO_2 solubility in crude oil. The prediction model of N_2, CO_2 and CH_4 solubility in the oil phase proposed by Xue et al (2005) [15] was selected to calculate the CO_2 solubility in the oil ($S_{CO2}{}^{oil}$), which is simple and only needs several easily available parameters. CO_2 solubility in the crude oil increases as pressure increases, but decreases as temperature rises. At typical reservoir conditions, CO_2 solubility in oil is generally between 0.05-0.35 ton/m^3 while at the original reservoir conditions of H-59 block, the solubility of CO_2 in oil ($S_{CO2}{}^{water}$) is 0.2156 ton/m^3.

CO_2 dissolution trapping capacity. Therefore, the theoretical dissolution trapping capacity of CO_2 in the H-59 block can be calculated based on the solubility data of CO_2 in formation water and crude oil, and the residual water and oil volumes in the block (see the table 2) using the following formula (3). About 25.93×10^4 tons CO_2 can be dissolved in the residual fluids in the H-59 block, of which the contributions of residual water and oil are 7.53×10^4 tons and 18.40×10^4 tons respectively.

$$M_{CO_2}^{dissolution} = M_{CO_2}^{dissolution\ water} + M_{CO_2}^{dissolution\ oil} = V_{wr}\times S_{CO_2}^{water} + V_{or}\times S_{CO_2}^{oil} \quad (3)$$

Mineral Trapping Capacity

$CaCO_3$, $MgCO_3$ and $FeCO_3$ are regarded as the main mineral trapping types of CO_2, and those non-carbonate minerals which can release Ca^{2+}, Mg^{2+} and Fe^{2+} through dissolution reactions are called the minerals with carbon trapping potential. The carbon trapping potential of minerals (R_{ctp}) can be assessed by calculating the amount of CO_2 consumed when all the Ca^{2+}, Mg^{2+} and Fe^{2+} in non-carbonate minerals in unit volume of rock are transferred to carbonate precipitations [16, 17]. Table 3 lists the mineral compositions and carbon trapping potentials of the rock in the H-59 block. The reservoir rock of H-59 block contains a large number of quartz, feldspar (including K-feldspar and plagioclase), a certain amount of calcite, ankerite, and a little pyrite and clay minerals (including kaolinite, illite, chlorite and montmorillonite). Although clay minerals just take a little part of rock minerals, they usually have a great impact on reservoir and formation water characteristics. As shown in Table 3, only the chlorite and illite have carbon trapping potentials in the H-59 block. The carbon trapping potential per unit volume of rock is about 8.4744×10^{-3} ton /m^3. Then the mineral trapping capacity of CO_2 in the H-59 block was calculated by the formula (4) and a capacity of 17.58×10^4 tons CO_2 was gotten.

$$M_{CO_2}^{mineral} = V_r\times R_{ctp} \quad (4)$$

Theoretical and Effective Storage Capacities

Theoretical storage capacity. Based on the above analysis, the theoretical CO_2 storage capacity of H-59 block (M_{CO2}^{theory}) is 59.04×10^4 tons by summing the capacities of displacement, dissolution and mineral trapping. The displacement trapping occupies a part of 26.30%, the dissolution trapping contributes 43.92% (12.75% and 31.17% for dissolution in residual water and oil respectively), and the mineral trapping takes the rest of 29.78%, as shown in Table 4. The Table 4 also lists the effective CO_2 storage capacities of different trapping mechanisms which will be analyzed below.

Effective storage capacity. However, considering the field operation during CO_2 EOR and storage process, the practical storage capacity of CO_2 in the target reservoir will be limited by time and space factors such as the project duration and well patterns. Only the CO_2 swept area can be used to store the supercritical CO_2 in the pore spaces, as well as only the residual fluids and rock minerals in the area used for CO_2 dissolution and geochemial reactions respectively in a project life. Therefore, on the basis of the theoretical CO_2 storage capacity, the effective storage capacity of CO_2 takes area efficiency of well pattern k_1 (cover area of well pattern / cover area of OOIP) and CO_2 sweep efficiency k_2 (within the cover area of well pattern) into consideration. The effective CO_2 storage capacity can be calculated as follows.

$$M_{CO_2}^{effective} = M_{CO_2}^{eff\ displacement} + M_{CO_2}^{eff\ dissolution} + M_{CO_2}^{eff\ mineral} \quad (5)$$

Where $M_{CO2}^{effective}$ is the effective CO_2 storage capacity, 10^4 tons; $M_{CO2}^{eff\ displacement}$ is the effective CO_2 displacement trapping capacity, 10^4 tons; $M_{CO2}^{eff\ dissolution}$ is the effective CO_2 dissolution trapping capacity, 10^4 tons; $M_{CO2}^{eff\ mineral}$ is the effective CO_2 mineral trapping capacity, 10^4 tons.

- *Effective displacement trapping capacity.* The oil recovery factor has been put into use for the calculation of theoretical displacement trapping capacity of CO_2. That is to say that it has already taken into account the effect of area efficiency of well pattern and sweep efficiency of CO_2 implicitly. Thus, the effective CO_2 displacement trapping capacity is the same as the theoretical calculation in this study ($M_{CO2}^{displacement}=M_{CO2}^{eff\ displacement}=15.53\times10^4$ tons).

- *Effective dissolution trapping capacity.* The theoretical dissolution trapping capacity assumes that all the residual water and oil can be used to dissolve CO_2. However, in practice, the injected CO_2 only can dissolve in the residual water and oil in the sweep area of CO_2 flooding. The specific calculation formulas (6-8) are as follows.

$$M_{CO_2}^{eff\ dissolution} = M_{CO_2}^{eff\ dis\ water} + M_{CO_2}^{eff\ dis\ oil} = V_{wr}^{eff} \times S_{CO_2}^{water} + V_{or}^{eff} \times S_{CO_2}^{oil} \quad (6)$$

$$V_{or}^{eff} = V_{oi} \times (k_1 k_2 - RF_{\%HCPV}) \quad (7)$$

$$V_{wr}^{eff} = V_{\varphi} \times k_1 k_2 - V_{CO_2}^{dis} - V_{or}^{eff} \quad (8)$$

Where $M_{CO2}^{eff\ dis\ water}$ and $M_{CO2}^{eff\ dis\ oil}$ are the effective CO_2 dissolution trapping capacities in residual water and oil respectively, 10^4 tons. V_{wr}^{eff} and V_{or}^{eff} are the volumes of residual water and oil in the sweep area of CO_2 respectively. Parameters k_1 and k_2 (have been mentioned above) are important factors for calculating residual water and oil volumes in sweep area of CO_2 flooding. In this study of H-59 block, 0.9 is estimated for k_1 based on the ratio between the internal cover area of the five well groups and the cover area of OOIP. For the k_2, a larger and ideal CO_2 sweep efficiency of 60% for k_2 is considered assuming a perfect miscible flooding and possible necessary measures used to reach such a sweep efficiency of CO_2 flooding. Based on the above formulas, the effective dissolution trapping capacity of CO_2 in the H-59 block is 10.84×10^4 tons, of which residual water contributes 4.19×10^4 tons while residual oil contributes 6.65×10^4 tons.

- *Effective mineral trapping capacity.* Geochemical reactions in reservoirs induced by CO_2 injection is a complicated process. A more accurate assessment on mineral trapping capacity of CO_2 in the rock of H-59 block was conducted using reservoir simulation method (CMG, 2012) which only considered the interactions between CO_2, water, and minerals (oil phase is neglected). Based on the reservoir properties of H-59 block, a conceptual homogeneous model was built with one injection well and one production well, and assuming that the initial formation fluid is of 100% water. The bottom-hole flowing pressures in injector and producer are defined to be 28 MPa and 20 MPa respectively. CO_2 injection lasts for 10 years, and two wells shut in for 1000 years. The geochemical reactions involved are listed in Table 5.

As shown in Figure 2a, when injected CO_2 dissolves into formation water, the decrease of pH value of the water induces the dissolution of calcite, illite and chlorite, releasing Ca^{2+}, Mg^{2+}, Fe^{2+} and other ions, meanwhile mainly five minerals are generated, including albite, k-feldspar, ankerite, siderite and kaolinite. Part of released Ca^{2+}, Mg^{2+}, and Fe^{2+} react with CO_3^{2-} and change to siderite and ankerite. After a long-time reaction, the interaction between various ions and the amount of each mineral gradually tend to be stable. The kaolinite has the biggest accumulative generation, while the illite and calcite have the biggest dissolution. As shown in Figure 2b, although the mineral compositions change a lot due to geochemical reactions, only a very little CO_2 is stored by the mineral trapping after CO_2 is injected into the reservoir for a very long time. Most of CO_2 mainly exists at the supercritical state, and the secondary is trapped at dissolution state. The proportion of CO_2 trapped at mineral state is very small. That is to say, although the carbon trapping potential per unit volume of rock is great in the H-59 block, however, at the real reservoir conditions, both the complicated geochemical reactions and the short time of project life make the mineral trapping mechanism playing a very small effect in the H-59 block which can be neglected.

CO_2 Storage Amount and Trapping States in the H-59 Block

Amount of CO_2 Storage in the H-59 Block

Injection and production history. The H-59 block started oil production in 2004, mainly depending on the natural energy of the reservoir. The reservoir kept void status during the first 4 years although water injection was adopted later but its performance was bad. The primary oil recovery factor only reached around 5%. In April 2008, the selected five well groups in the north of H-59 block started to inject CO_2. In order to recover the reservoir pressure quickly, most of the production wells were shut down from October 2008 while all the injection wells continued to inject CO_2 at a rate of 30-50 t/d for single well. 5 months later, the reservoir pressure was restored to 20.3 MPa (in March 2009). Then all the production wells resumed production. According to the initial designed injection scheme, WAG operation or water injection would only start after 5 years of pure CO_2 injection. However, due to an early CO_2 breakthrough in some production wells, from May 2009, aperiodic and non-simultaneous WAG operations were initiated in some injectors to improve the sweep efficiency associated with a water-gas ratio between 1:2 and 1:5 (subsurface volume ratio). Water injection usually lasted for 1-4 months at a rate of about 30-40 m^3/d (1000-4800 m^3 pore volume per water slug), then changed to CO_2 injection for 2-12 months (4000-24000 m^3 pore volume per CO_2 slug)

Up to June 2014, more than 21.08×10^4 tons of CO_2 has been injected into the target oil reservoir which is equivalent to about 32.03% of hydrocarbon pore volume (HCPV) (118.6×10^4 m^3). Nearly 95.45% of the injected CO_2 (30.57% of HCPV, 20.12×10^4 tons) has been stored in the oil reservoir with the rest of the injected CO_2 breakthrough in the production wells. About 12.64×10^4 tons of crude oil has been produced by CO_2 injection, increasing the oil recovery factor from 5% to 14.70%.

CO_2 Distribution and Trapping States in the H-59 Block

Reservoir Simulation Model. The geological model of the target reservoir in the H-59 block was built based on the geological data using corner point grid with a grid number of 77×80×25=154000, and an average grid size of 25 m×25 m. The layers 5, 10, 12, 13 and 21 in the geological model are corresponding to the real reservoir layers 7, 12, 14, 15 and 23. The porosity, permeability and initial oil saturation distributions were generated through stochastic modeling based on the well data and previous understanding. Four components were defined in the fluid model including CO_2, N_2-C1, C2-C6 and C7+. Due to the above geochemical study indicating that the contribution of mineral trapping to CO_2 storage capacity was very small in the H-59 block, mineral reactions were neglected in this model, which also can make the simulation run faster and achieve a good iterative convergence.

Production History Matching. Production history matching is an important basis for CO_2 distribution prediction in oil reservoir. The oil reserve and production data of the H-59 block have been fitted carefully by mainly adjusting the relative permeability curves and local reservoir physical properties based on the current understanding to the target reservoir. Finally, the reserve fitting error is only 1.17%. The injection data of the target reservoir are easily fitted, and all the CO_2 and water can be injected according to field data of each injection well during the simulation. The cumulative and daily production data of the reservoir are also matched well, and nearly 90% production wells have a fitting error within 5%. Although the fitted final cumulative production of CO_2 is very close to the field data, the fitted breakthrough time of CO_2 in production wells falls a little behind the real time. It's hard to match the CO_2 production especially for thin-layer oil reservoir with serious heterogeneity, which will be further discussed later.

Distribution of CO_2 trapped at Different States. The injected CO_2 in the target reservoir is mainly kept at supercritical and dissolution states. The distribution of supercritical CO_2 can be described in term of the size and shape of the gas bearing area with different gas saturations. As shown in Figure 3, the injected CO_2 mainly accumulates around the injection wells with a maximum CO_2 gas saturation of 0.55. The relative location of injection and production wells has a great effect on the shape of CO_2 gas bearing area. The CO_2 along the main stream line between injection and production wells usually migrates fasters than any other areas. This process is also affected by the reservoir heterogeneity. The reservoir layer with a good permeability usually has a strong gas suction capacity, often leading to a larger CO_2 bearing area. In addition, the gas bearing area is also affected by the CO_2 injection volume. The greater the amount of CO_2 gas is injected, the larger the gas bearing area is. It is obvious that the distribution of CO_2 in the reservoir is mainly the interaction results of the well relative position, the reservoir heterogeneity and the injection amount of CO_2.

Apart from trapped at supercritical state, CO_2 can also dissolve into the oil and water. Mole fraction of CO_2 in crude oil and formation water can be used to assess the performance of CO_2 dissolution underground. The distribution of CO_2 dissolved in the oil and water are similar as that of CO_2 at supercritical state, but the size of the former is a little larger than the latter due to CO_2 dissolution and diffusion in the water and oil, as shown in Figure 4, taking the CO_2 distribution in the layer 12 for example. In addition, the mole fraction of CO_2 in water of layer 12 is no more than 0.0227, while the mole fraction of CO_2 in crude oil is usually between 0.4-0.8, which indicates that CO_2 dissolution in oil is more significant than that in water.

Amounts and Proportions of CO_2 Trapped at Different States. In order to further confirm the storage state and distribution of CO_2 underground, the amounts of CO_2 trapped at different states and their proportions were calculated, as shown in Figure 5. As CO_2 flooding conducted from 2008 to 2014, the amounts of CO_2 trapped at different states increased nearly linearly. But in 2012, the amount of CO_2 trapped at supercritical state started to increase slowly while the trapped CO_2 mostly dissolved in the residual water and oil. It was mainly due to the CO_2 breakthrough in production wells at a relatively large scale. However, the proportions of CO_2 trapped at different states stayed at a stable level. Up to June 2014, there had been 21.08×10^4 tons CO_2 injected underground and 20.12×10^4 tons CO_2 had been buried in the oil reservoir. Most of CO_2 was constrained by displacement trapping at supercritical state, which takes about 60.15%. The amounts of CO_2 trapped in residual oil and water account for 24.85% and 15.00% respectively.

Discussions

Reliability Analysis of CO_2 Storage Capacity Assessment Method

The established CO_2 storage capacity assessment method has considered the way of gas injection, the effective sweep volume of gas flooding, and the different CO_2 trapping mechanisms. The proportions of CO_2 trapped at various states calculated using the above method (on effective CO_2 storage capacity) are similar to the results of reservoir simulation with an error within 2%, as shown in Table 6. It indicates that this method can reflect the contributions of different trapping mechanisms to CO_2 storage capacity, and has good reliability and accuracy. It is enough for the preliminary assessment of CO_2 storage potential in oil reservoir. But it should be noted that the effective CO_2 storage capacity is affected greatly by the parameters k_1 and k_2 which deserve for further discussion based on specific cases.

Prediction Accuracy of CO_2 Distribution and Trapping State by Reservoir Simulation

The description accuracy of reservoir heterogeneity has a great effect on CO_2 production fitting especially for the thin-layer oil reservoir with serious heterogeneity, although it's usually very hard to build an exact geological model for reservoir simulation. H-59 block is such a reservoir, which consists of many thin oil-bearing layers in vertical. It has a ultra-low heterogeneous permeability, and many wells have been fractured at the beginning of development. The current geological model of the H-59 block used for simulation is not accurate enough especially in the aspect of reservoir heterogeneity. In the next step, the geological model of the H-59 block should be updated periodically as the understanding of reservoir properties is deepened, in order to improve continually the prediction accuracy of CO_2 distribution and trapping state in the target oil reservoir using simulation method. Some monitoring data can be used to improve the description of the permeability heterogeneity in the reservoir, such as the gas tracer test which can indicate the connectivity between injection and production wells and identify the distribution of high permeability channels or fractures in the reservoir.

CO_2 Storage Capacity, Actual amount and the Remaining Potential

The current amount of CO_2 stored in the H-59 block (up to June 2014) is still 6.25×10^4 tons less than the effective CO_2 storage capacity (see Table 6). This remaining potential can be used for the future CO_2 storage. But it should be pointed out that the calculation of effective storage capacity is based on a relatively ideal gas flooding in the first several years and then a WAG operation adopted after gas channeling in production wells. The CO_2 storage efficiency before gas channeling is higher than that after gas channeling. However, in the field, in the second year of CO_2 injection, CO_2 channeling had occurred in the production wells. This decreases the storage efficiency of CO_2 in the H-59 block, although WAG operation has a good performance on controlling the gas channeling and keeping the GOR in the production wells at a level no more than 200 m^3/t. Even so, it is thought that the H-59 block still has a potential of about 6.25×10^4 tons CO_2 for storage which may be achieved through taking some necessary measures to improve the CO_2 storage amount in the target reservoir.

Conclusions

(1) The theoretical storage capacity of CO_2 in the H-59 block is estimated to be 59.04×10^4 tons. During CO_2 EOR and storage process, the mineral trapping capacity of CO_2 can be neglected. The coverage factor of well pattern (k_1) and the sweep coefficient of CO_2 within the well pattern (k_2), were introduced for the calculation of effective CO_2 storage capacity. The effective storage capacity of CO_2 in the H-59 block is estimated to be 26.37×10^4 tons, which has a more significant meaning for the field operation.

(2) Based on actual production data and reservoir simulation results, it is obtained that the amount of CO_2 stored in the block H-59 was 20.12×10^4 t up to June 2014, of which structure trapping, dissolution trapping in oil and water account for 60.15%, 24.85% and 15.00% respectively. Most of trapped CO_2 maintains at supercritical state. The distribution of CO_2 in the reservoir is determined by the combined effects of CO_2 injection amount, reservoir heterogeneity and the relative position between injection and production wells.

(3) Compared with the effective storage capacity of CO_2, it is thought that the H-59 block still has a potential of about 6.25×10^4 tons CO_2 for further storage which may be achieved through taking necessary measures to improve the CO_2 storage amount in the target reservoir. But it should be noted that the calculation of effective CO_2 storage capacity is affected significantly by parameters k_1 and k_2 which deserve for further discussion. In addition, the accuracy of CO_2 distribution predicted by reservoir simulation greatly depends on the accuracy of geological model. More efforts should be taken to improve the understanding of the target reservoir geological conditions.

Acknowledgement

This research is supported by the National Major S&T Project (No. 2011ZX05016-005), the Qingdao basic research program for young scholars (No. 13-1-4-254-jch), the Shandong Natural Science Foundation (No. ZR2013EEQ032) and the Fundamental Research Funds for the Central Universities (No. 15CX05036A).

References

1. IPCC (Intergovernmental Panel on Climate Change), 2005. IPCC Special Report on Carbon Dioxide Capture Storage. Cambridge University Press. Cambridge, UK, pp. 51-72.
2. Ren, S.R., Zhang, L., Zhang, L., 2010. Geological Storage of CO_2: Overseas Demonstration Projects and Its Implications to China. Journal of China University of Petroleum 34(1), 93-98. (in Chinese with English abstract)
3. Arts, R., Eiken, O., Chadwick, A., Zweigel, P., van der Meer, L., Zinszner, B., 2004. Monitoring of CO_2 injected at Sleipner using time-lapse seismic data. Energy 29(9-10), 1383-1392.
4. Maldal, T., Tappel, I. M., 2004. CO_2 underground storage for Snøhvit gas field development. Energy 29(9-10), 1403- 1411.
5. Preston, C., Monea, M., Jazrawi, W., Brown, K., Whittaker, S., White, D., Law, D., Chalaturnyk, R., Rostron, B., 2005. IEA GHG Weyburn CO_2 Monitoring and Storage Project. Fuel Process Technol 86(14-15), 1547-1568.
6. Van der Meer, L.G.H., Kreft, E., Geel, C., 2005. K12-B: A test site for CO_2 storage and enhanced gas recovery. SPE paper 94128, SPE Europec/EAGE Annual Conference, Madrid, Spain, 13-16 June.
7. Rutqvist, J., Vasco, D.W., Myer, L., 2010. Coupled reservoir-geomechanical analysis of CO_2 injection and ground deformations at In Salah, Algeria. International Journal of Greenhouse Gas Control 4(2), 225-230.
8. Zhang, L., Ren, S.R., Ren, B., Zhang, W.D., Guo, Q., Zhang, L, 2011. Assessment of CO_2 Storage Capacity in Oil Reservoirs Associated with Large Lateral/Underlying Aquifers: Case Studies from China. International Journal of Greenhouse Gas Control 5(4), 1016-1021.
9. Zhang, L., Huang, H.D., Wang, Y.Q., Ren, B., Ren, S.R., Chen G.L., Zhang, H., 2014. CO_2 Storage Safety and Leakage Monitoring in the CCS Demonstration Project of Jilin Oilfield, China. Greenhouse Gases: Science and Technology 4(4): 425-439.
10. Xu, Z.G., Chen, D.Z., Zeng, R.S., Guo, K., Li, Y.P., Xiao, B.,2009. Geological Storage Framework of CO_2 Subsurface Burial Trial Area of Daqingzi Block in the Jilin Oilfield. Acta Geol Sin 83(6), 875–884. (In Chinese with English abstract)
11. Cheng, M.H., 2012. Characteristic Analysis of Typical Block Reservoirs in the Daqingzhijing Oilfield.: Northeast Petroleum University, Daqing, China, pp.26-51. (in Chinese with English abstract)
12. Shaw, J.C., Bachu, S., 2002. Screening, Evaluation and Ranking of Oil Resevoirs Suitable for CO_2-flood EOR and Carbon Dioxide Sequestration. Journal of Canadian Petroleum Technology 41(9), 51-61.
13. Duan, Z.H., Sun, R., 2003. An Improved Model Calculating CO_2 Solubility in Pure Water and Aqueous NaCl Solutions from 273 to 533 K and from 0 to 2000 bar. Chem Geol 193(3-4), 257-271.
14. Duan, Z.H., Sun, R., Zhu, C., Chou, I.M., 2006. An Improved Model for the Calculation of CO_2 Solubility in

Aqueous Solutions Containing Na^+, K^+, Ca^{2+}, Mg^{2+},Cl-, and SO_4^{2-}. Mar Chem 98(2-4), 131-139.
15. Xue, H.T., Lu, S.F., Fu, X.T., 2005. Forecasting model of solubility of CH_4. CO_2 and N_2 in crude oil. Oil & Gas Geology 26 (4), 444–449. (in Chinese with English abstract)
16. Xu, T.F., Apps J.A., Pruess, K., 2001. Analysis of Mineral Trapping for CO_2 Disposal in Deep Aquifers. Lawrence Berkeley National Laboratory, University of California, Berkeley, CA, http://escholarship.org/uc/item/59c8k6gb.
17. Zhang, L., 2011. Saline Aquifer Storage of CO_2 from Natural Gas Reservoirs in the South China Sea: Trapping Me chanisms and Project Design. PhD dissertation, China University of Petroleum (huadong), Qingdao, China, pp.77-86.

Table 1 Reservoir and fluid properties in the H-59 block (the five well groups)

Parameter	Value	Parameter	Value
Cover area A, km^2	1.8	Initial average water saturation S_{wi}, fraction	0.55
Buried depth h, m	2000-2400	Oil viscosity at reservoir μ_{ores}, mPa·s	1.85-2.18
Reservoir thickness H, m	11.2-18.2	Oil density at reservoir ρ_{ores}, ton/m^3	0.7615-0.7686
Average thickness H_a, m	14.3	Oil density at surface ρ_{osurf}, ton/m^3	0.8503
Dip angle α, °	2-4	Oil volume factor B_o	1.1723
Reservoir temperature T, °C	98.9	Gas-oil ratio R_{si}, m^3/ton	35
Reservoir pressure P, MPa	24.15	Salinity, mg/L	10229-19320
Porosity φ, %	8-15	Average salinity, mg/L	14607.8
Average porosity φ_a, %	12.7	pH value of formation water	6.8
Permeability k, md	0.24-9.85	CO_2 density at reservoir ρ_{CO2res}, ton/m^3	0.555
Average permeability k_a, md	3.5	CO_2-oil MMP, MPa	22.3
OOIP (the five well groups), tons	86×10^4	RF_{BT} (2008-2013), fraction	0.17
Initial average oil saturation S_{oi}, fraction	0.45	$RF_{\%HCPV}$ (2013-2018), fraction	0.28

Table 2 Volumes of various reservoir rocks and fluids in the H-59 block at reservoir conditions

Volumes of rock and fluid	value	Calculation formula
Rock volume V_r, m^3	2.0747×10^7	$V_r=V_\varphi/\varphi_a$
Pore volume V_φ, m^3	2.635×10^6	$V_\varphi=V_{oi}/S_{oi}$
Initial oil V_{oi}, m^3	1.186×10^6	$V_{oi}=OOIP\times B_o/\rho_{osurf}$
Initial water V_{wi}, m^3	1.449×10^6	$V_{wi}=V_\varphi\times S_{wi}$
Oil recovered by CO_2 injection V_{op}, m^3	3.321×10^5	$V_{op}=V_{oi}\times RF_{\%HCPV}$
Residual oil V_{or}, m^3	8.537×10^5	$V_{or}=V_{oi}\times(1-RF_{\%HCPV})$
CO_2 trapped at supercritical state V_{CO2}^{dis}, m^3	2.798×10^5	$V_{CO_2}^{dis}=M_{CO_2}^{displacement}/\rho_{CO2res}$
Residual water V_{wr}, m^3	1.501×10^6	$V_{wr}=V_\varphi-V_{or}-V_{CO_2}^{str}$

Table 3 Mineral compositions and carbon trapping potentials of the rock in the H-59 block

Mineral	Composition, vol% (based on rock matrix)		Composition, vol% (based on rock)	Carbon trapping potential, 10^{-3} ton CO_2/m^3 mineral	Carbon trapping potential, 10^{-3} ton CO_2/m^3 rock	Chemical formula
	range	average				
Quartz	30.17-54.67	43.33	37.83	0	0	SiO_2
Plagioclase	27.64-53.15	39.91	34.84	0	0	$NaAlSi_3O_8$
K-feldspar	1.57-10.10	5.22	4.56	0	0	$KAlSi_3O_8$
Ankerite	0-25.67	5.32	4.64	0	0	$Ca[Fe/Mg](CO_3)_2$
Calcite	0-19.24	3.25	2.84	0	0	$CaCO_3$
Siderite	0-2.08	0.10	0.09	0	0	$FeCO_3$
Chlorite	0.19-2.68	0.78	0.68	1049.20	7.1346	$[Fe/Mg]_5Al_2Si_3O_{10}(OH)_8$
Kaolinite	0.03-0.48	0.14	0.12	0	0	$Al_2(Si_2O_5)(OH)_4$
Illite	0.49-6.71	1.95	1.70	78.81	1.3398	$K_{0.6}Mg_{0.25}Al_{1.8}(Al_{0.5}Si_{3.5}O_{10})(OH)$
Carbon trapping potential R_{ctp}, 10^{-3} ton CO_2/m^3 rock				8.4744		

Table 4 Composition of CO_2 storage capacity in the H-59 block

Trapping mechanism	Displacement	Residual water	Residual oil	Mineral	Total
Theoretical capacity, 10^4 tons	15.53	7.53	18.40	17.58	59.04
Effective capacity, 10^4 tons	15.53	4.19	6.65	0	26.37

Table 5 Aqueous and mineral reactions involved in the simulation

No	Chemical reaction equation	$lgK_{eq}^{a/m}$@50 °C	lgk_β, mol/(m^2·s) @25°C	A_β, m^2/m^3	$E_{a\beta}$, J/mol
a1	$H_2O = H^+ + OH^-$	-13.263			
a2	$CO_2(aq) + H_2O = H^+ + HCO_3^-$	-6.322			
a3	$CO_2(aq) + H_2O = 2H^+ + CO_3^{2-}$	-16.556			
m1	Albite+$4H^+$ = $2H_2O$+ Na^++Al^{3+}+$3SiO_2(aq)$	-0.344	-12.0	250	67830
m2	K-feldspar+$4H^+$ = $2H_2O$+ K^++Al^{3+}+$3SiO_2(aq)$	-0.344	-12.0	176	67830
m3	Ankerite+$2H^+$=$2HCO_3^-$+ Ca^{2+}+$0.3Mg^{2+}$+$0.7Fe^{2+}$	1.673	-9.2	88	41870
m4	Calcite+ H^+ = Ca^{2+} + HCO_3^-	1.360	-8.8	88	41870
m5	Siderite+ H^+= Fe^{2+} + HCO_3^-	-0.37@33°C	-9.2	88	41870
m6	Chlorite+$16H^+$=5[Fe/Mg]$^{2+}$+$2Al^{3+}$+$3SiO_2$+ $12H_2O$	-1.466@100°C	-14.0	88000	8870
m7	Kaolinite+6 H^+ = 5 H_2O +$2Al^{3+}$+$2SiO_2$	5.470	-13.0	17600	62760
m8	Illite+$8H^+$ = $5H_2O$+$0.6K^+$+$0.25Mg^{2+}$+$2.3Al^{3+}$+$3.5SiO_2(aq)$	7.486	-14.0	26400	58620

Table 6 Comparison of current CO_2 storage amount and effective storage capacity

CO2 capacity and amount	Value, 10^4 tons	Proportion, %		
		Structure	Residual oil	Residual water
Effective storage capacity	26.37	58.89	25.22	15.89
Current storage amount (June 2014)	20.12	60.15	24.85	15.00

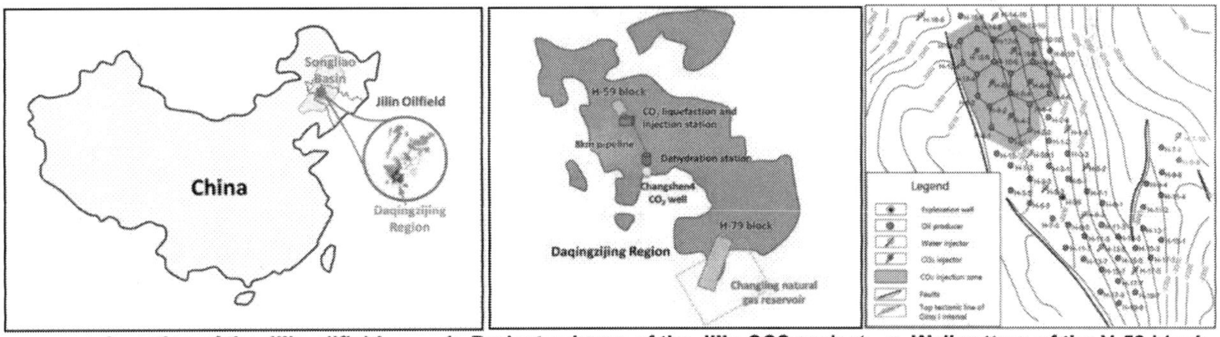

a. Location of the Jilin oilfield b. Project scheme of the Jilin CCS project c. Well pattern of the H-59 block

Fig. 1 CO_2 EOR and storage project in the Jilin oilfield in the northeast China

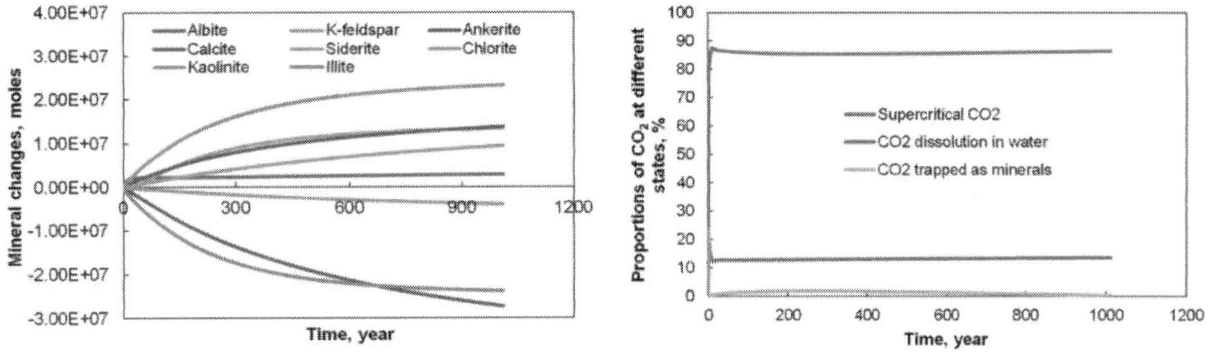

a. Amount changes of various minerals b. Proportion of CO_2 trapped at mineral state

Fig. 2 Simulation results of CO_2 geochemical reactions at reservoir conditions of H-59 block during 1000-year storage

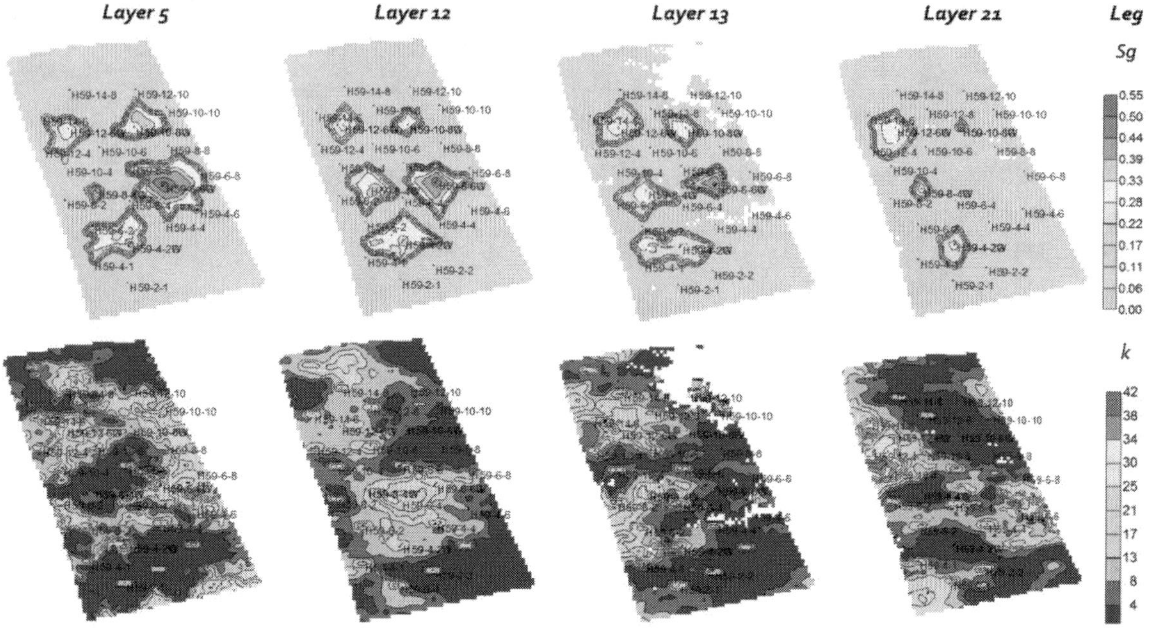

Fig. 3 CO_2 saturation (first line) and permeability (second line) distribution in different layers (up to June 2014)

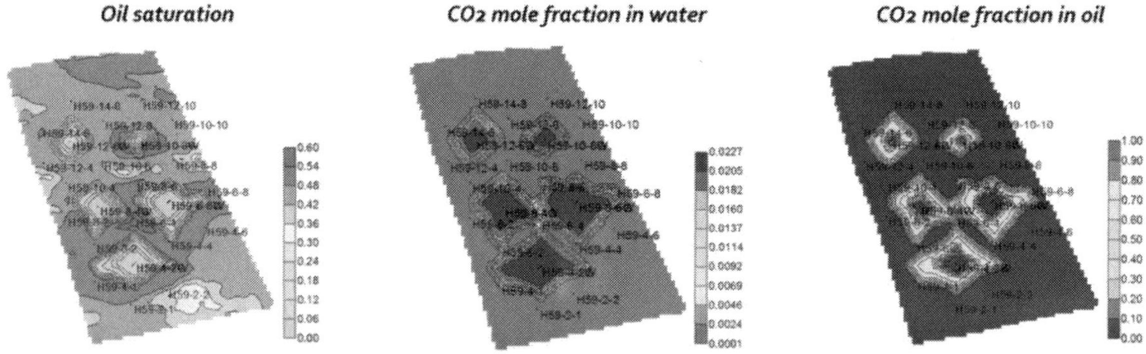

Fig. 4 CO_2 mole faction distribution in water and oil (layer 12, up to June 2014)

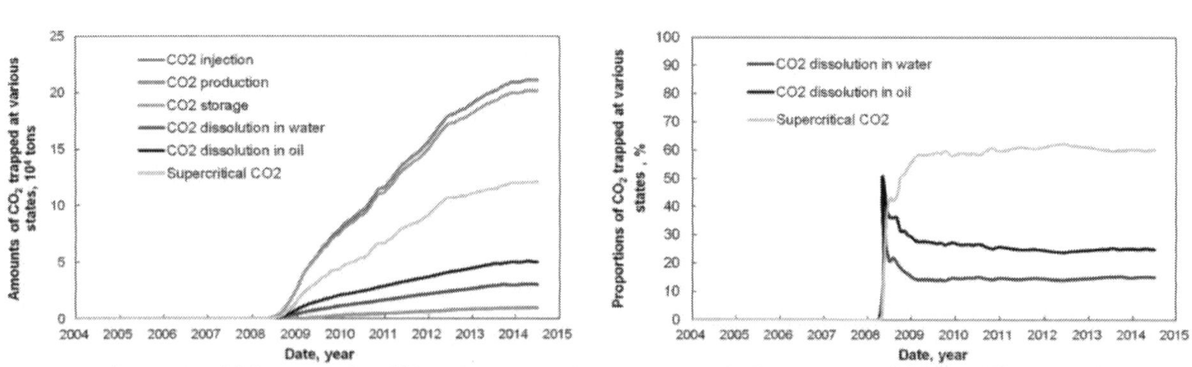

a. Amounts of CO_2 trapped at different states **b. Proportions of CO_2 at different states**

Fig. 5 Amounts and proportions of CO_2 trapped at different states

CMTC-440237-MS

Optimization of Geological Carbon Sequestration in Heterogeneous Saline Aquifers through Managed Injection for Uniform CO_2 Distribution

Mohammadali Tarrahi, SPE, Shell Global Solutions (US), Inc., Sardar Afra, SPE, Texas A&M University

This paper was prepared for presentation at the Carbon Management Technology Conference held in Sugarland, Texas, USA, 17–19 November 2015.

Abstract

To mitigate continued release of CO_2 from anthropogenic sources, underground storage of carbon dioxide (CO_2) is accepted worldwide as a promising and well-established technology. Due to increased concerns on CO_2 emissions, there has been foremost importance in recent years on the development and establishment of safe, technologically feasible and economic geological carbon sequestration (GCS) technology. GCS has been practiced widely as a potentially practical climate change mitigation decision. For this purpose to dispose large amount of CO_2 in economic and safe fashion for long term periods, deep ocean and geologic sequestration are considered viable options. Geologic sequestration is a potential technology to decrease released CO_2 into the atmosphere through capturing CO_2 from hydrocarbon emissions, transporting compressed CO_2 from the source to the field, and injecting and storage of CO_2 into the underground formation. While injecting CO_2 into the formation, residual gas trapping and solubility trapping are the most important trapping mechanisms for CO_2 storage in saline aquifers. The lack of information about the geological formation results in uncertainties in specifying the storage capacity of the formation and the safety of sequestered CO_2 caused by leakage. These uncertainties influence the sequestration capacity and CO_2 plume migration. Moreover, the sequestration efficiency is highly dependent on the injection strategy which includes injection rate, injection pressure, type of injection well employed and its trajectory. The goal of GCS is to maximize the sequestration capacity and minimize the CO_2 plume migration by optimizing the GCS operation before progressing with its large scale deployment. Hence the objective of CO_2 sequestration is to achieve a more uniform sweep efficiency and more extensive CO_2 sweep of the geological formation for a fixed amount of injected CO_2. To accomplish this objective, we propose a novel formulation for CO_2 sequestration optimization problem to promote uniform distribution of CO_2 in geological formation through minimizing the gas saturation differences in cells with the same distance to CO_2 injection wells.

We propose a new objective function based on uniform distribution of injected CO_2 in the heterogeneous formation which optimizes CO_2 flooding sweep efficiency through controlling operating conditions of CO_2 injection wells. The objective of the proposed method is to equalize the gas breakthrough time to the equidistant pair of cells from the injection wells. We first determine all equidistant pairs of cells measured from all the injection wells and then we set up the objective function in a way to minimize the difference in CO_2 breakthrough times of each pair of cells. Minimizing this objective function and finding the optimal injection scenario guarantees CO_2 uniform sweep efficiency in the geological formation.

We show the suitability of the proposed method for optimized CO_2 sequestration by performing a numerical case study on PUNQ-S3 benchmark model. The results of applying the proposed method to geological carbon storage numerical example illustrate the efficiency of this novel objective function to uniformly distribute and spread the injected CO_2 in the heterogeneous aquifer and consequently enhance solubility and residual trapping. It is demonstrated that by employing the new objective function for promoting uniform distribution of CO_2 comparing to unmanaged CO_2 injection and the traditional sequestration optimization techniques, the amount of total trapped gas substantially increases.

Introduction

Deep abundant saline aquifers are accepted widely for Geologic CO_2 storage (GCS) due to the associated safe removal mechanism and sufficient potential storage capacity of such aquifers for holding a large volume of CO_2 emissions for decades (McPherson and Cole 2000; Lindeberg and Wessel-Berg 1997). Low density and relatively high viscosity of injected supercritical CO_2 decrease sequestration's cost and control buoyancy forces. Should the mobility of CO_2 decreases, the storage security will increase accordingly. Thus, employing trapping techniques with capability of reducing CO_2 mobilization in the aquifer such as brine injection, mineral trapping and controlled injection rate to name a few, may result in a lower leakage risk especially from leaky cap rocks (Shamshiri and Jafarpour 2012). The lack of information about the geological formation results in uncertainties in specifying the storage capacity of the formation and the safety of sequestered CO_2 caused by leakage. These uncertainties influence the sequestration capacity and CO_2 plume migration. Moreover, the sequestration efficiency is highly dependent on the injection strategy which includes injection rate, injection pressure, type of injection well employed and its trajectory. The goal of GCS is to maximize the sequestration capacity and minimize the CO_2 plume migration by optimizing the GCS operation before progressing with its large scale deployment (Weir et al. 1996; Intergovernmental Panel on Climate Change (IPCC) 2005).

Furthermore, one may be interested in conducting a number of investigations to understand the effect of various injection parameters like injection rate, injection duration, and injection well orientation and displacement on CO_2 storage efficiency and plume migration in a given reservoir (Zhang and Agarwal 2013; IPCC 2005). We propose and test a new objective function considering a uniform distribution for injecting CO_2 in the heterogeneous formation. Controlling operating conditions of CO_2 injection wells is the key to optimize CO_2 flooding sweep efficiency. The objective of the proposed method is to equalize the gas breakthrough time to the equidistant pair of cells from the injection wells and improve overall CO_2 storage (McMillan and Bryant 2007; Keith et al. 2005).

It is essential to consider reservoir properties uncertainty in history matching procedure for the purpose of field development (Zafari and Reynolds 2005; Afra and Gildin 2013; Tarrahi et al. 2015; Afra et al. 2014). Since the aquifer heterogeneous properties such as permeability and porosity are the major factors in determining the spread and extent of CO_2 plume in the reservoir and they always have associated uncertainties, it is also recommended to consider the aquifer properties uncertainty in optimization process (Shamshiri and Jafarpour 2012). Parallel to geological sequestration projects, CO_2 flooding is also a widespread method for enhanced oil recovery a tertiary mechanism to improve oil recovery factor (Shaw and Bachu 2002; Tarrahi et al. 2015).

The paper is organized as follows. In the next section, the gas sequestration optimization problem is defined and the new objective function for enhancing sweep efficiency is introduced. In case study section, the proposed optimization workflow is applied to PUNQ-S3 benchmark model with 6 CO_2 injectors to optimize gas sweep in the aquifer and consequently improve dissolved and trapped CO_2. Finally, last section presents the conclusions and the important remarks regarding to the proposed objective function and optimization workflow and its advantages and applicabilities.

Gas Storage Optimization Methodology

The injection strategy which includes injection rate, injection pressure, type of injection wells employed and their trajectory control CO_2 sequestration performance. In GCS we aim to maximize the sequestration capacity and minimize the CO_2 plume migration by numerical reservoir simulation as well as numerically optimizing the CO_2 injection procedure before advancing with its large scale deployment in the real field (Bryant 2006; Shamshiri and Jafarpour 2010; Gunter et al. 1997). Achieving a more uniform sweep efficiency and more extensive CO_2

sweep of the geological formation for a fixed amount of injected CO_2 are the main objectives of CO_2 sequestration optimization (Shamshiri and Jafarpour 2012). To accomplish these objectives, we assemble a novel objective function for CO_2 sequestration optimization problem to promote uniform distribution of CO_2 in geological formation through minimizing the gas saturation differences in cells with the same distance to all CO_2 injectors. We seek to optimize CO_2 flooding sweep efficiency through controlling operating conditions of CO_2 injection wells. The objective of the proposed sequestration optimization algorithm is to equalize the gas breakthrough time to the equidistant pair of cells to all the injection wells. Firstly we determine all equidistant pairs of cells measured from all the injectors and then we construct the objective function in a way to minimize the differences in CO_2 breakthrough times for each pairs of cells. Minimizing this objective function and finding the optimal injection scenario guarantees CO_2 uniform sweep efficiency in the geological formation. The proposed objective function in this study for equalizing the gas breakthrough times is inspired by the work done by *Shamshiri and Jafarpour 2012* where a few pseudo production wells or representative cells are considered for gas production rate equalization. While in this paper to promote uniform CO_2 spread and equalize gas breakthrough times, we first find all the equidistant pair of cells from all the wells then we consider all these pairs in calculating the objective function to widen the objective of uniform gas front to all the aquifer.

The aquifer flow properties heterogeneity determines the extent and shape of CO_2 plume in the aquifer and the uniformity of injected gas spread. Permeability and porosity distributions are the key players in guiding the injected gas into different regions of the aquifer. The other important factor is the injection scenario or the injection well controls. Having a heterogeneous aquifer we should be able to manage and determine the injection rates or control variables to achieve a uniform spread of CO_2 in the field and consequently maximum residual trapping and dissolved gas as the prime objectives of geological gas storage. Aquifer and reservoir properties are always associated with significant levels of uncertainty that should be handled in reservoir characterization problems (Oliver 2008; Afra 2014; Jafarpour and Tarrahi 2011; Tarrahi et al. 2013) likewise it is recommended to develop a robust optimization procedure to consider aquifer properties uncertainty.

The full objective function for CO_2 sequestration optimization can be the combination of Net Present Value (NPV) and sweep efficiency term. For instance when NPV and sweep efficiency (equalizing breakthrough times) objective functions are named $F_1(u)$ and $F_2(u)$, respectively, we can setup the full objective function as:

$$F(u) = \frac{F_2(u)}{F_1(u)} \qquad (1)$$

where u is the vector of controllable variables such as injection well controls. By minimizing $F(u)$, we can guarantee improved NPV (reduce the overall cost) and uniform CO_2 sweep in the reservoir. In CO_2 sequestration project, NPV term may contain the cost of gathering, compressing and injecting gas into the aquifer as well as the capital cost of drilling injection wells. In this study we only consider the second term i.e. sweep efficiency optimization as the objective function. We first need to find the set of pairs with equidistant cells that collectively cover the whole aquifer. For each cell, we calculate the distance to all the injectors (a set of distances) and then we find pair of cells with the same set of distance (while there is no order in this distance set). The set of all pairs of equidistant cells are defined as:

$$pair_i = \{cell^{i,1}, cell^{i,2}\};\ i = 1, \ldots, N_{pair} \qquad (2)$$

where $cell^{i,1}$ and $cell^{i,2}$ belongs to i^{th} pair and have the same set of distances to all the injectors (order of distances are not important). Having defined the representative pairs of cells, the sweep efficiency objective function is defined as follows:

$$F(u) = \sum_{i=1}^{N_{pair}} \left(t^{i,1}_{gas\ B.T.} - t^{i,2}_{gas\ B.T.}\right)^2 \qquad (3)$$

$t^{i,1}_{gas\ B.T.}$, $t^{i,2}_{gas\ B.T.}$ = gas breakthrough times of cells # 1 and # 2 of ith pair
N_{pair} = total number of equidistant pairs of cells
u = control variables (CO_2 injection rates)
$F(u)$ = CO_2 sequestration objective function to be minimized (to promote uniform CO_2 spread in the field)

This objective function contains all the mismatches of gas breakthrough times for equidistant cells to all injectors. Minimizing this objective function results in improved sweep efficiency. In this study to spread injected gas uniformly, we only consider and calculate top most layer's cells gas breakthrough times. The constraints of this optimization problem are minimum and maximum values for injection rates, total injected CO_2 volume (in standard condition) and maximum injection bottom hole pressure.

As an example to demonstrate the procedure of finding the target pairs of cells, a color-coded distance map for PUNQ-S3 benchmark model (top most layer) is plotted in Figure 1. Each pair of cells with the same color have the same set of distances to injectors (unlabeled injectors) while the color-scale does not represent the distance value and it only shows the similarity of cells in terms of cell to well distance.

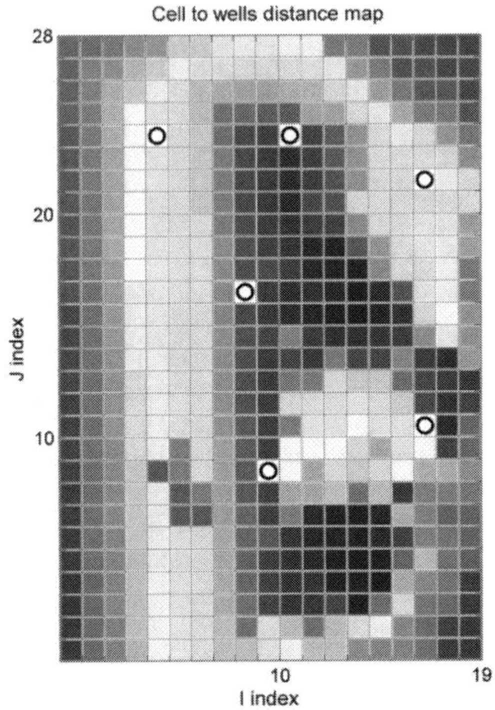

Figure 1. The map of distance of cells to all wells in PUNQ-S3 model: CO2 injection wells are indicated by black circles. Cells with the same color have the same distance to all the wells (calculated for the top most layer).

In Figure 1 constructed for PUNQ-S3 model, any pair of cells with the same color have the same set of distances from the injection wells and should have similar (close enough) gas breakthrough time. In the proposed algorithm, we pick pair of cells with the same color (distance to wells) and extract their gas breakthrough times and through the optimization procedure, we minimize the difference in their gas breakthrough times. In this study to spread injected gas uniformly, we only consider and calculate gas breakthrough times of the cells in the top most layer. For instance in PUNQ-S3 model (in top most layer), there are in total 263 equidistant pairs of cells considering all the grids and 166 equidistant pairs considering the active grid blocks.

The control variables for optimization are decided to be gas injection rates in each control time step. In PUNQS-3 model, since there are 6 injection wells and each control time step is 5 years and the total injection time is 25 years, there are 30 control variables (gas injection rates).

A gradient-based optimization procedure with the proposed objective function is implemented using MATLAB optimization toolbox (MATLAB 2010) while interior-point algorithm is used for optimization and BFGS (Liu and Nocedal 1989) method is utilized for hessian approximation. As alternative optimization methods, random and evolutionary search methods as well as machine learning based optimization methods can also employed for this problem (Angeline, P. J. 1998; Afra and Braga-Neto 2011; Shadravan et al. 2015).

Case study and Results

To show the utility of the proposed objective function for CO_2 sequestration optimization, we perform numerical examples with PUNQ-S3 model which is a 3-D synthetic benchmark model (available at: https://www.imperial.ac.uk/engineering/departments/earth-science/research/research-groups/perm/standard-models). We adapt the original black oil three phase reservoir model to a compositional gas-brine aquifer model. We model CO_2 sequestration into a heterogeneous aquifer using CMG-GEM (CMG-GEM 2014). CO_2 is injected for first 25 years and then the injector is shut-in and the fate of CO_2 is modelled for next 225 years. The storage mechanisms include dissolving gas and residual trapping because of gas relative permeability hysteresis. The important reservoir and fluid properties of the employed model is reported in Table 1.

Table 1. Reservoir and fluid properties for CO_2 sequestration case study

Property	**Value**
Reservoir reference pressure	3550.0 psi
Reference depth	7700 ft
Reservoir temperature	200 F
Maximum residual gas saturation	0.3
Residual water saturation	0.16
Reservoir dimensions	19 x 28 x 5
Total injection time	25 years
Total simulation time	250 years
Number of injectors	6
Number of control time steps	5
Number of control variables	30
Injection wells control	CO_2 injection rate
Rock compressibility	4.0e-6 psi^{-1}

The major computational effort of optimization procedure is spent in performing reservoir simulation using CMG-GEM, therefore to accelerate the optimization we used CMG-GEM parallel run capability (CMG-GEM 2014). The original permeability and porosity distributions of PUNQ-S3 model are shown in Figure 3 and these are the major factors that affect CO_2 plume spread and sweep performance in the aquifer.

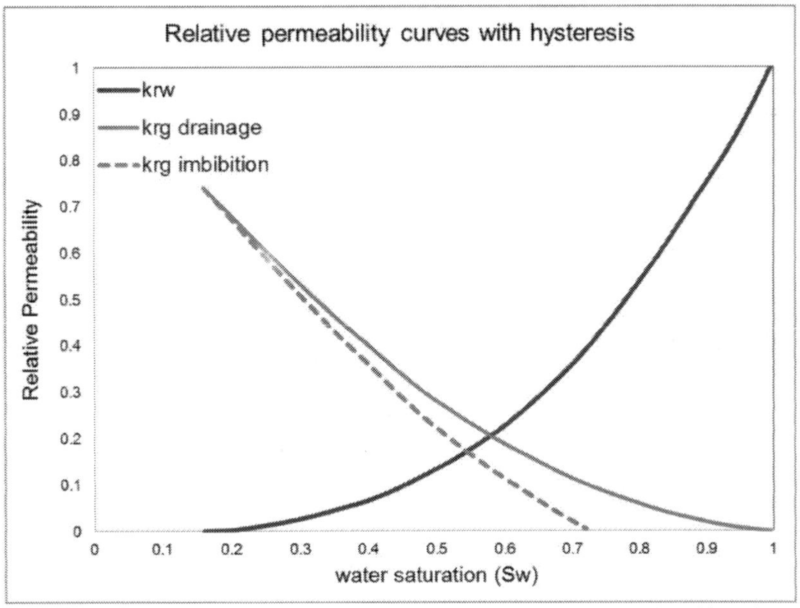

Figure 2. Relative permeability curves of CO_2 and brine with hysteresis effect for gas phase

Hysteresis effect in gas relative permeability is modeled and the maximum residual gas saturation is 0.3 (CMG-GEM 2014). The residual or trapped gas is resulted by the hysteresis effect in gas phase.

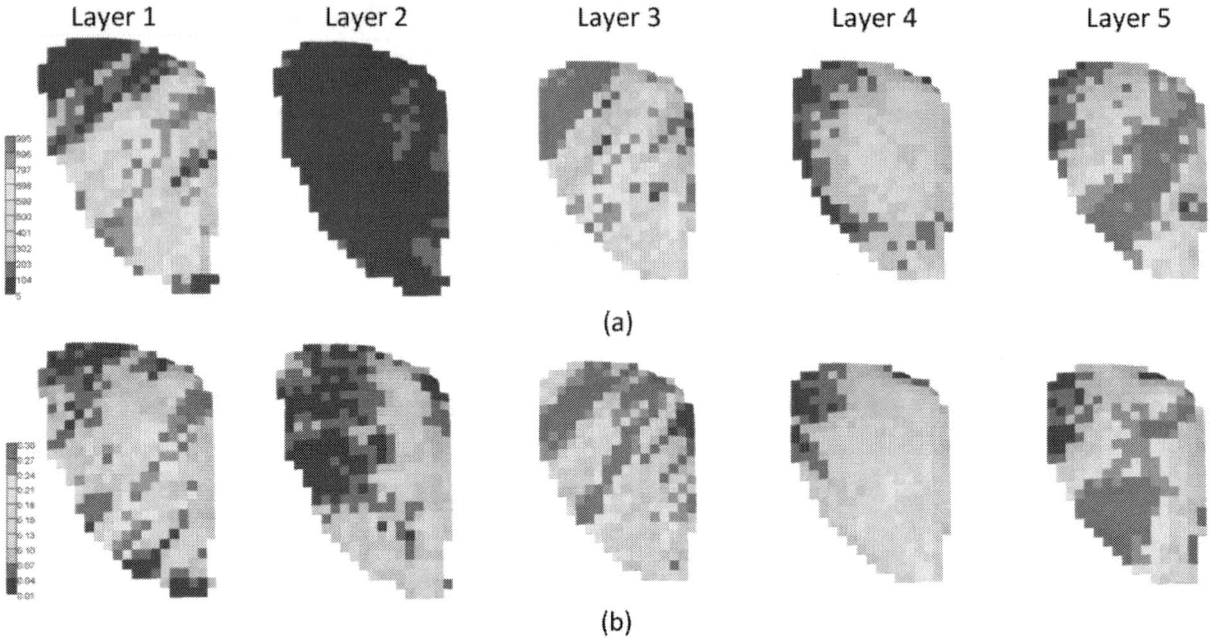

Figure 3. Heterogeneous (a) permeability and (b) porosity distributions of PUNQ-S3 model in different layers

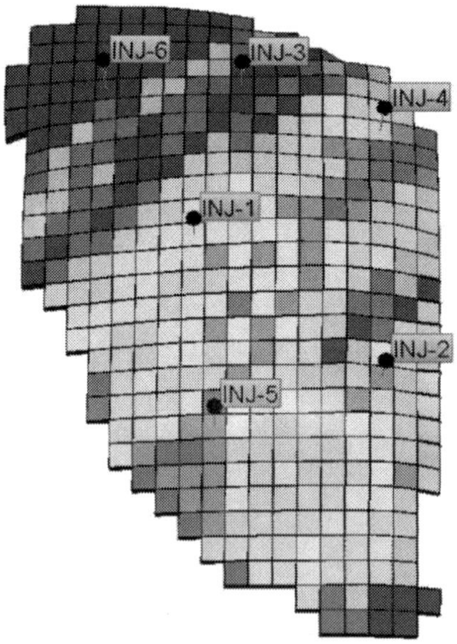

Figure 4. Well configuration, 6 CO_2 injection wells all perforated in bottom most layer (top layer of permeability is shown)

We placed 6 injectors in the aquifer. We chose their locations to be in a wide range of permeability and porosity in order to clearly see how important reservoir properties are in injected gas distribution. To maximize the contact of injected gas in the aquifer, all injectors are perforated in the bottom most layer and while CO_2 moves upward to the top most layer the breakthrough times of the cells in the top most layer are recorded and analyzed. In the base case simulation all the control variables (CO_2 injection rates) are equal to 2 MMscf and the injected pore volume (in reservoir condition) is 0.17 PV.

After performing the optimization by the proposed objective function, the resulted optimal injection scenario is presented in Figure 5. It should be noted that the initial guess or the base case had uniform injection rate in all the wells. This optimal injection rate scenario obtained by equalizing gas breakthrough times enhanced trapped and dissolved gas as shown in the next plots. The optimal injection scenario as shown in Figure 5 has clear correlation with aquifer hydraulic properties and connectivities. For instance as the proposed optimal injection rates, it is seen that well INJ-6 injection rates are lower initially and increased to high values by time which can be linked to the location of the well in the low permeability area of the aquifer. And the injection rates of well INJ-2 are generally low values that can be related to the high permeability area which the well is placed in.

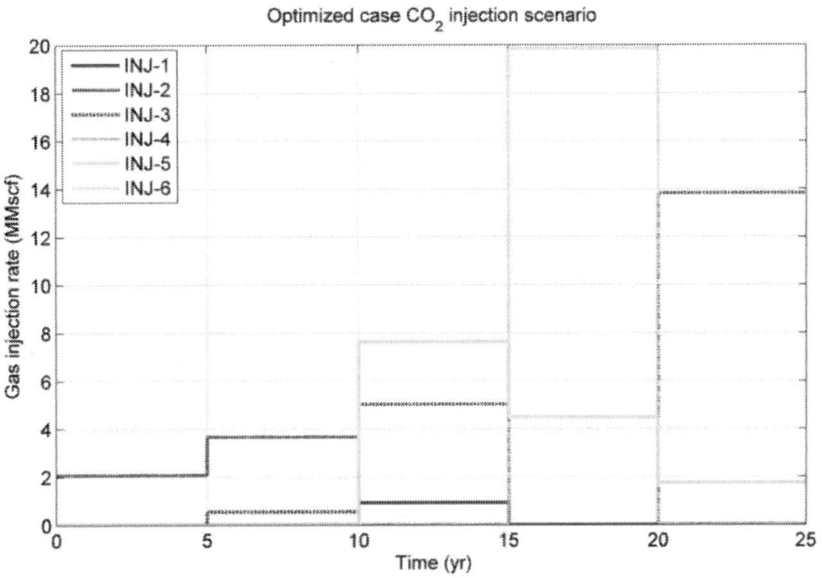

Figure 5. Gas injection rates for optimized case

Figure 6 compares the gas saturations of the representative pair of cells for the base case and optimized case. As shown, after optimization cell gas breakthrough times in each representative pair are more similar which consequently results in more uniform spread of injected CO_2 in the aquifer. As shown in Figure 6, the proposed optimization workflow was able to equalize the gas breakthrough times for equidistant cells. Two representative pairs of cells are demonstrated in Figure 6. It should be noted again that gas injection is performed for the first 25 years and then injectors are shut-in and the simulation is continued for 225 years to model CO_2 plume migration in the aquifer.

To show the performance of the sweep efficiency optimization procedure, base case and optimized case are compared in terms of the total (after 250 years) trapped and dissolved gas in Table 2. Also Figure 7 shows the change in dissolved and trapped gas by time for base case and optimized case. As offered by these results, sweep efficiency enhancement via gas breakthrough times equalization is able to improve gas storage (through increasing trapped and dissolved gas) in aquifer.

The proposed objective function as well as the optimization procedure can also be employed for optimal well placement to achieve better sweep efficiency. Well placement along with optimal injection scenario determination can be done in an iterative optimization loop using the proposed objective function to enhance gas storage in the aquifers.

Table 2. CO_2 storage amounts in reservoir

Simulation case	Trapped or residual gas (lb)	Dissolved gas (lb)	Total stored gas (lb)
Base case	4.6618e9	2.2161e9	6.8779e9
Optimized case	4.9727e9	2.6860e9	7.6588e9

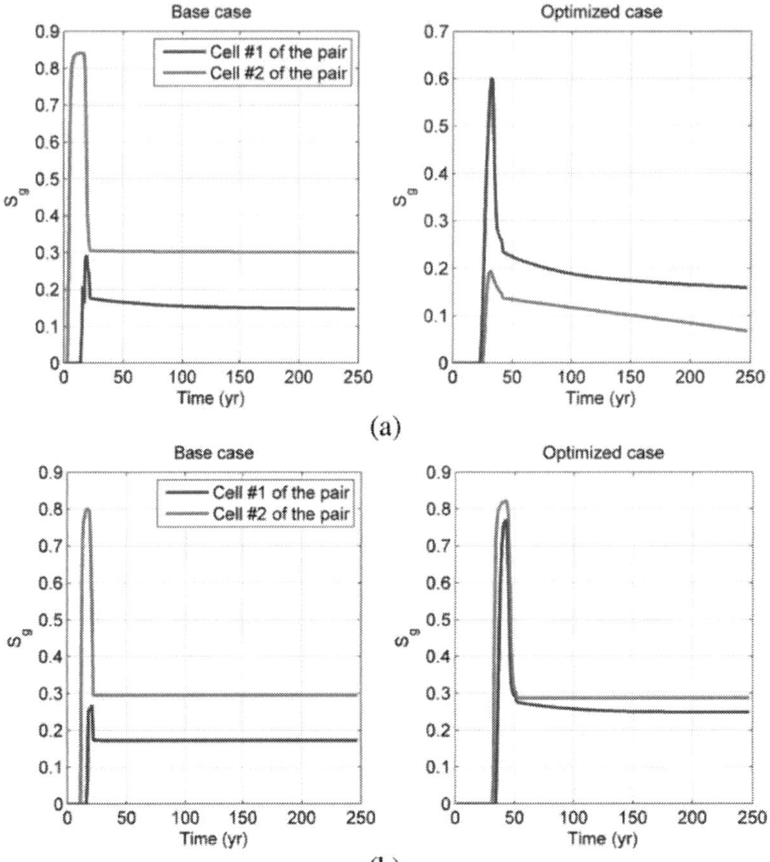

Figure 6. Comparison of gas breakthrough times and gas saturation in the representative pairs of cells between the base case and the optimized case, two pairs of cells are shown in (a) and (b)

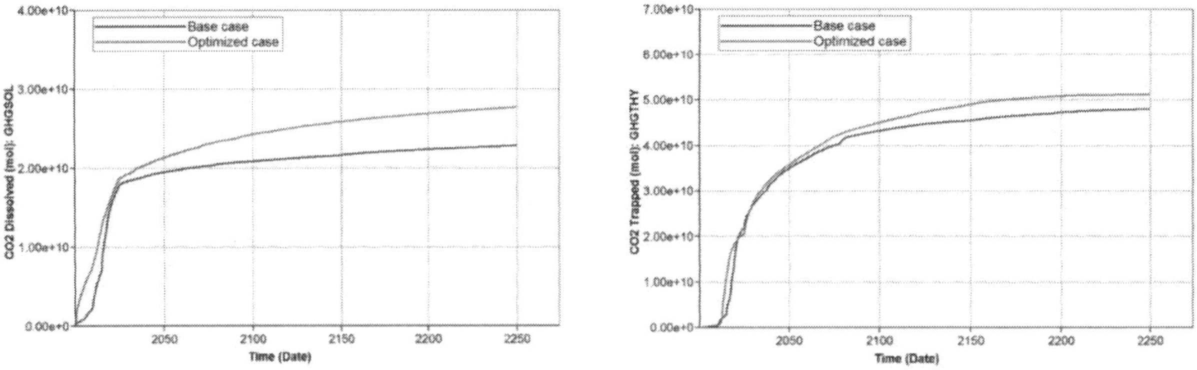

Figure 7. (a) The dissolved and (b) trapped CO_2 versus time for the base case and optimized case

Conclusion

Subsurface storage of CO_2 is a widely accepted and applicable method to lessen continued release of CO_2 from anthropogenic sources. To dispose large amount of CO_2 in economic and safe manner for long term periods, deep ocean and geologic sequestration are considered viable options. Geologic gas sequestration method decreases released CO_2 into the atmosphere through capturing CO_2 from hydrocarbon emissions, transporting compressed CO_2 from the source to the field, and injecting and storage of CO_2 into the underground formation. The most important trapping mechanisms for CO_2 storage in saline aquifers are residual gas trapping and solubility trapping. The efficiency of the sequestration process highly depends on CO_2 injection scenario such as injection rates. Therefore the prime goal of CO_2 sequestration is to achieve more uniform sweep efficiency and improved CO_2 sweep of the geological aquifer for a fixed amount of injected CO_2. To accomplish this objective in this paper, we proposed and implemented a novel formulation for CO_2 sequestration optimization problem to promote uniform distribution of CO_2 via minimizing the differences in gas breakthrough times in cells with the same distance to CO_2 injection wells. The proposed method equalizes the gas breakthrough time to the equidistant (cell to well distance) pair of cells. Minimizing this objective function and finding the optimal injection scenario guarantees CO_2 uniform sweep efficiency in the geological formation. We demonstrated the suitability of the proposed method for optimized CO_2 sequestration by performing a numerical case study on PUNQ-S3 benchmark model. The results showed improved gas storage via increasing trapped and dissolved gas in the aquifer. The results of numerical example illustrate the efficiency of this novel objective function to uniformly distribute and spread the injected CO_2 in the heterogeneous aquifer and consequently enhance solubility and residual trapping.

References

Afra, S., and Braga-Neto, U. 2011. Studying the Possibility of Peaking Phenomenon in Linear Support Vector Machines with non-Separable Data. Presented at IEEE International Workshop on Genomic Signal Processing and Statistics (GENSIPS), San Antonio, December 4-6. doi: 10.1109/GENSiPS.2011.6169.484.

Afra, S., and Gildin, E. 2013. Permeability Parametrization using Higher Order Singular Value Decomposition (HOSVD). Presented at the 12th International Conference on Machine Learning and Applications (ICMLA), Miami, December 4-7. doi: 10.1109/ICMLA.2013.121.

Afra, S. and Gildin, E. 2014. Efficient inference of reservoir parameter distribution utilizing higher order SVD Reparameterization. Presented at the 14th European conference on the mathematics of oil recovery (ECMOR XIV), Sisil, Italy, August 7-9. doi: 10.3997/2214-4609.20141826.

Afra, S., Gildin, E., and Tarrahi, M. 2014. Heterogeneous Reservoir Characterization using Efficient Parameterization through Higher Order SVD (HOSVD). Presented at the American Control Conference (ACC), Portland, 4-6 June. doi: 10.1109/ACC.2014.6859246.

Alfi, M., Hosseini, S. A., Alfi, M., Shakiba, M. 2015. Effectiveness of 4D Seismic Data to Monitor CO2 Plume in Cranfield CO2-EOR Project, Carbon Management Technology Conference, Sugarland, Texas, USA, CMTC-439559-MS

Angeline, P. J. 1998. Evolutionary optimization versus particle swarm optimization: Philosophy and performance differences. In Evolutionary Programming VII (pp. 601-610). Springer Berlin Heidelberg.

Bryant, S. L., S. Lakshiminarasimhan, and G. A. Pope 2006, Buoyancy dominated multiphase flow and its impact on geological sequestration of CO2, SPE 99938, SPE/DOE Symposium on Improved Oil Recovery, Tulsa, Okla.

CMG-GEM 2014. Advanced Compositional and Unconventional Reservoir Simulator. Version 2014 User's Guide. Computer Modeling Group Ltd., Calgary, Alberta.

Gunter, W. D., B. Wiwchar, and E. H. Perkins 1997, Aquifer disposal of CO2-rich greenhouse gases: Extension of the time scale of experiment for CO-sequestering reactions by geochemical modeling. Mineral. Petrol., 59, 121–140.

Intergovernmental Panel on Climate Change (IPCC) 2005. IPCC special report on carbon dioxide capture and storage, summary for policy makers, Eighth Session of IPCC Working Group III, Montreal, Canada, 22–24 September 2005.

Jafarpour, B., and Tarrahi, M. 2011. Assessing the performance of the ensemble Kalman filter for subsurface flow data integration under variogram uncertainty. *Water Resources Research*, **47**(5).

Keith, D., H. Hassanzadeh, and M. Pooladi-Darvish 2005. Reservoir engineering to accelerate dissolution of stored CO2 in brines, paper presented at Proceedings of the 7th International Conference on Greenhouse Gas Control Technologies (GHGT-7), Vancouver, Canada, September 5–9, 2004, V.II, pp. 2163–2168.

MATLAB 2010. The MathWorks, Natick, Mass.

Lindeberg, E., and D. Wessel-Berg 1997. Vertical convection in an aquifer column under a gas cap of CO2, Energy Convers. Manage., 38(Suppl.), S229–S234.

Liu, D. C., and Nocedal, J. 1989. On the limited memory BFGS method for large scale optimization. Mathematical programming, 45(1-3), 503-528.

McMillan, B., and S. Bryant 2007. Surface mixing as method for minimizing risk in CO2 sequestration, paper presented at Proceedings of the Sixth Annual Conference on Carbon Capture and Sequestration, Dept. of Energy/NETL, Austin, TX.

McPherson, B. J. O. L., and B. S. Cole 2000. Multiphase CO2 flow, transport and sequestration in the Powder River Basin, Wyoming, U.S.A., J. Geochem. Explor., 69, 65–69.

Oliver, D. S., Reynolds, A. C., and Liu, N. 2008. Inverse theory for petroleum reservoir characterization and history matching. Cambridge University Press.

PUNQ-S3 benchmark model:
https://www.imperial.ac.uk/engineering/departments/earth-science/research/research-groups/perm/standard-models/

Shadravan, A., Tarrahi, M., and Amani, M. (2015). Intelligent Tool to Design Fracturing, Drilling, Spacer and Cement Slurry Fluids Using Machine Learning Algorithms. In SPE Kuwait Oil and Gas Show and Conference. Society of Petroleum Engineers.

Shamshiri, H., and Jafarpour, B. 2012. Controlled CO_2 injection into heterogeneous geologic formations for improved solubility and residual trapping. Water Resources Research, 48(2).

Shamshiri, H., and Jafarpour, B. 2010. Optimization of geologic CO_2 storage in heterogeneous aquifers through improved sweep efficiency. In SPE International Conference on CO_2 Capture, Storage, and Utilization. Society of Petroleum Engineers.

Shaw, J., and Bachu, S. 2002. Screening, evaluation, and ranking of oil reservoirs suitable for CO2-flood EOR and carbon dioxide sequestration.Journal of Canadian Petroleum Technology, 41(09).

Tarrahi, M., B. Jafarpour, and A. Ghassemi 2015. Integration of microseismic monitoring data into coupled flow and geomechanical models with ensemble Kalman filter, Water Resour. Res., 51, doi:10.1002/2014WR016264.

Tarrahi, M., Jafarpour, B., and Ghassemi, A. 2013. Assimilation of Microseismic Data into Coupled Flow and Geomechanical Reservoir Models with Ensemble Kalman Filter. Society of Petroleum Engineers. September 30. doi:10.2118/166510-MS.

Tarrahi, M., Afra, S., & Surovets, I. 2015. A Novel Automated and Probabilistic EOR Screening Method to Integrate Theoretical Screening Criteria and Real Field EOR Practices Using Machine Learning Algorithms. Society of Petroleum Engineers. doi:10.2118/176725-MS

Weir, G. J., S. P. White, and W. M. Kissling 1996. Reservoir storage and containment of greenhouse gases, Transp. Porous Media 23, 37–60.

Zafari, M., and Reynolds, A. C. 2005. Assessing the uncertainty in reservoir description and performance predictions with the ensemble Kalman filter. In SPE Annual Technical Conference and Exhibition. Society of Petroleum Engineers.

Zhang, Z., and Agarwal, R. K. 2013. Numerical simulation and optimization of CO_2 sequestration in saline aquifers for enhanced storage capacity and secured sequestration. International Journal of Energy and Environment, 4(3), 387-398.

CMTC-440236-MS

Intelligent Cement Design: Utilizing Machine Learning Algorithms to Assure Effective Long-term Well Integrity

Arash Shadravan, ReservoirFocus LLC, Mohammadali Tarrahi, Shell, Mahmood Amani, Texas A&M University at Qatar

This paper was prepared for presentation at the Carbon Management Technology Conference held in Sugarland, Texas, USA, 17–19 November 2015.

Abstract

Designing drilling fluids, spacers and cement slurries are all often done by trial and error in the laboratory. There are several test data that would not be used in this method and it is hard to digest a plethora of information for users and take intelligent and cost-effective decision to design a fluid with the desired properties. Therefore, trial and error method is considered to be time consuming, very costly and misleading. Today, there is a need for an intelligent system which uses all the available fluid design data stored in a database by which we can benefit from its insights for smart fluid designs. This predictive tool suggest a composition for drilling fluids, spacer fluid or cement slurry by implementing a machine learning algorithm on imported experimental data.

This paper investigates and implements a data-driven predictive tool which uses Gaussian Process Regression (GPR). GPR as a novel machine learning method considerably reduces the costs of testing, optimizes the material use, integrates available experimental data and eliminates the user bias. This practical nonlinear regression method fosters an efficient and fast prediction analysis which do not require including complex physics of the underlying intricate chemical fluid behavior while integrating all available data from different databases. GPR has exceptional advantages over traditional regression methods since it does not require a known form for regression function. Also its capability of determining estimation error and confidence interval is unique. This machine learning based tool offers comprehensive insights for intelligent fluid design and considerably reduces the experiment cost. This study showcases an example through which of GPR predicts rheological properties and helped engineers to maintain the fluid rheological hierarchy for better cement jobs and well integrity. Zonal isolation has the utmost importance for carbon sequestration projects.

Introduction

Machine learning methodologies and expert systems have proved to be conducive to digest big data sets and make intelligent decisions for agile productivity and more accurate measures. These algorithms have been successfully applied to petroleum engineering fields such as production forecasting, history matching, reservoir characterization and EOR screening (Alvarado *et al.* 2002; Parada *et al.* 2012; Tarrahi *et al.* 2015). The proposed workflow in this paper includes the following tools:

1- ***Data preparation tool:*** preparing the training data through guided experiments

- Determining the most sensitive and effective design parameters (e.g. temperature, density, etc.)
- Design of experiment (DOE), Latin Hypercube Sampling (LHS), or Monte Carlo sampling to produce the most representative and wide enough sample cases

2- ***Prediction tool:*** building an intelligent data-driven prediction tool

- Build a reliable prediction tool (regressor) through Machine Learning algorithms
- Predict (and assess the uncertainty of) any new samples properties (e.g. rheological properties)

3- ***Design tool:*** intelligent fluid design tool

- Designing the desired fluid system based on the requested sample properties
  a. Performing the opposite of prediction tool and building a regressor to predict corresponding parameters of a suitable fluid system (through machine learning algorithms)
  b. Applying inverse problem methods to prediction tool and obtaining the desired fluid system

In the machine learning literature some of the most well-known regression methods are: Linear Regression (Ordinary Least Square), Nonlinear Regression or Curve Fitting (e.g. Polynomial Regression), Logistic Regression, Ridge regression, Artificial Neural Network, Bayesian Network, Radial Basis Function (RBF) Network, Support Vector Machine (SVM), Gaussian Process Regression and Kernel Regression (Duda *et al.* 2012; Theodoridis *et al.* 2010). In this paper we choose Gaussian Process Regression (GPR) algorithm to investigate its application in fluid characteristics estimation and fluid design and build an intelligent multi-purpose prediction and design tool (Shadravan *et al.* 2015). Predicting the rheological profiles mud, spacer and cement slurries and maintaining proper rheological hierarchy will benefit long term zonal isolation, Shadravan *et al.* (2015). **Figure 1** lists various types of fluid designs experiments which usually are pursued by a trial and error methods.

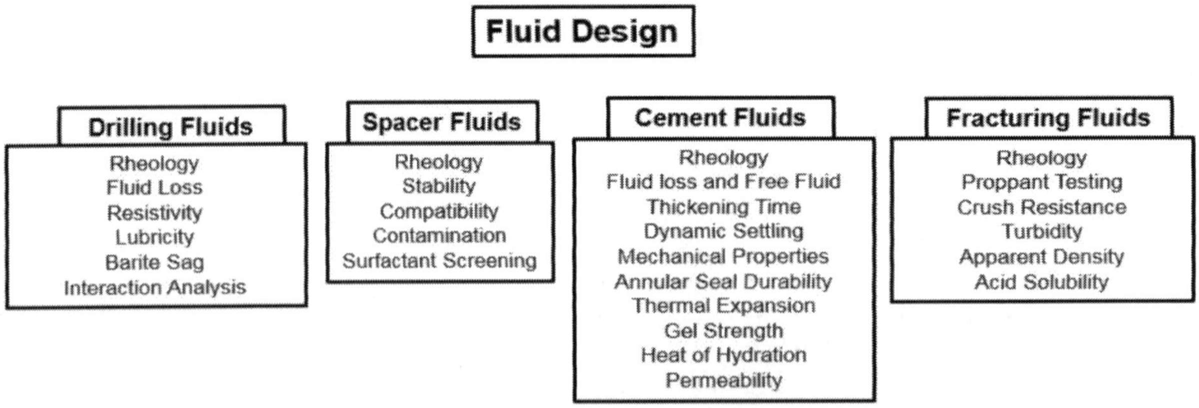

Figure 1: Applications of the intelligent design tool for various experiments

Lee *et al.* (2013) studied the rheological properties of an invert emulsion drilling fluid under extreme high pressure high temperature conditions, 300°F to 600°F. Shadravan *et al.* (2014) presented a model which could predict the rheological profiles of a micro-emulsion spacer fluid up to 300 °F. This model was built based on Design of Experiments (DOE). DOE relies heavily on individual and joint statistics of fluid system parameters, and its implementation in fluid systems with many components requires huge investment in time and cost. DOE along with response surface methodology (RSM) are also not able to control certain variables, problems identifying all the variables that affect the process, and a lack of linearity in variables that makes it hard to identify optimum settings. These drawbacks tend to limit the frequency with which DOE is used to solve problems. Therefore in this paper, GPR method shows how to predict such properties for desired fluid design.

The complexity and variations of downhole conditions, geology, additives (water, polymers, cement types, weighting agents, crosslinker, breaker, etc.), testing procedures, and experiment schedule could be all properly taken into account to construct a useful laboratory database for an intelligent fluid design tool. In this study, separate data silos for drilling fluids, spacers and cement slurries were created. Proper equipment calibration and competent laboratory personnel are also evidently valuable elements of successful fluid design. **Figure 2** shows traditional trial and error fluid design and **Figure 3** demonstrates the comprehensive intelligent machine learning fluid design workflow.

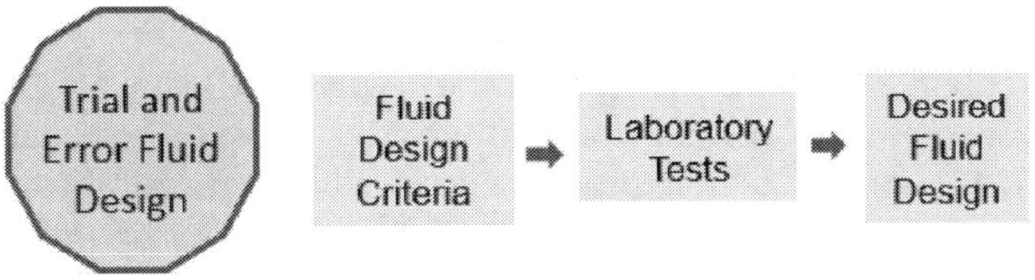

Figure 2: Traditional arbitrary and trial-and-error-based fluid design procedure

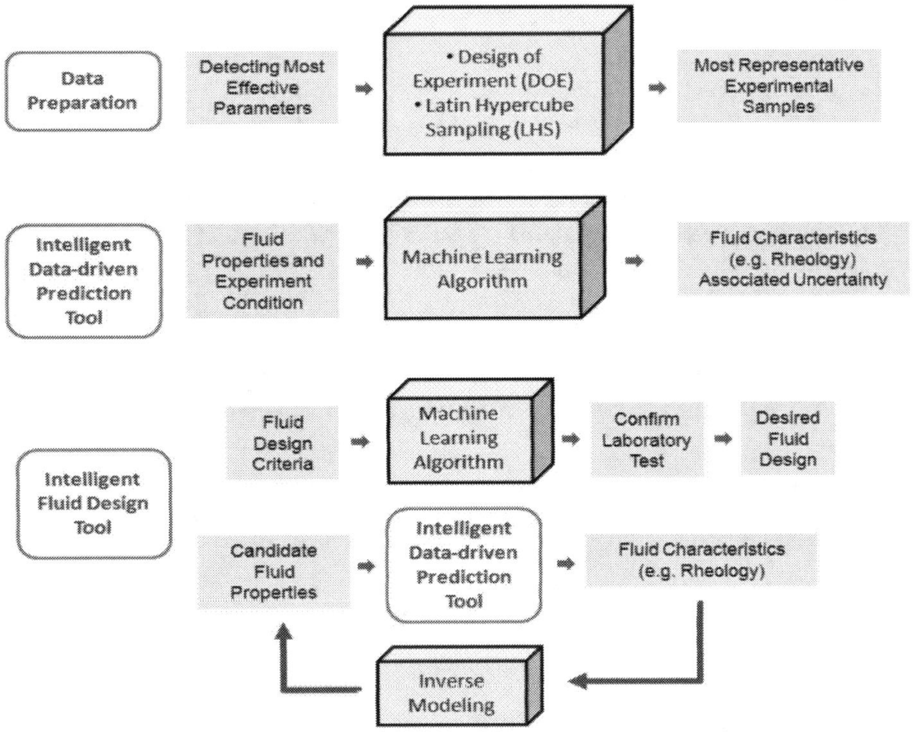

Figure 3: Schematic of the proposed intelligent integrated tools for guided experiment design, characteristic prediction and fluid system design

Data-Driven Analytics for Fluid Characteristics Prediction and Optimal Design

Rheology is a common experiment which is conducted on drilling, spacer and cement slurry fluids. This paper delineates the benefits of GPR as a powerful and robust machine learning algorithm to predict the rheological properties of a fluid as a simple example of the breath of Machine Learning methods application. Predicting such properties can be viewed as a regression analysis (a data-driven approach) where the goal is to design the best linear or nonlinear regression function (or rule).

We solved this problem by GPR machine learning method, which is well-established in computer science literature. The proposed algorithm eliminate arbitrary approach in making decisions, and provide accuracy and fast computation. This approach substitutes the need for days and weeks of testing in the laboratory with an intelligent and agile fluid design which could require only a few number of tests to confirm the proposed compositions (additives) for the desired properties.

In the parametric regression methods, a known regression function (e.g. polynomial, exponential, etc.) is defined in terms of finite number of unknown parameters which are estimated from the training data. In general since the nature of the underlying physical process (e.g. relation of rheological properties and fluid characteristics) is very complex, assuming a known and closed form for the regression function is a rigorous task. Therefore nonparametric regression methods are proposed that have no (or very little) a-priori knowledge of the form of the function that is being estimated. These methods allow the class of functions which the model can represent to be very broad (Bishop, 2006, Duda, *et al.* 2012, Theodoridis *et al.* 2010). Here we implement and apply Gaussian Process Regression that counts as one the most practical nonparametric regression methods. One of the advantages of Gaussian Processes application for machine learning is that the prediction or estimated value is probabilistic (Gaussian) so empirical confidence intervals and exceedance probabilities can be computed that might be used to refit (online fitting, adaptive fitting) the prediction in some regions of interest. Also the estimation and regression procedure is versatile i.e. different linear regression models and correlation models can be specified.

Gaussian Process Regression for Optimal Interpolation

Gaussian process regression is a nonlinear regression or interpolation technique that models the new estimated (interpolated) values based on Gaussian process determined by a covariance function (Williams & Rasmussen 2006). GPR is a statistical machine learning method which is also known as Kriging and is well established in Geostatistics and computer science literature (Deutsch & Journel 1992, Shi & Choi 2011). To estimate the corresponding value for the new input, GPR calculates the weighted average (linear combination) of the known values (training data) based on the correlation (or proximity) of the new input and the training data governed by the covariance function (Rasmussen, 2006). Providing the training data and the prior assumptions (e.g. covariance function) are suitable (Deutsch & Journel 1992), GPR or kriging is the best linear unbiased estimator (BLUE). In Geostatistics literature the spatial dependencies are represented by variaogram function instead of covariance function. In signal processing literature this estimation procedure is known as Kalman filtering which have been also applied to history matching problems (Tarrahi *et al.* 2013, 2015).

In our current experiment, there are 3 input parameters (fluid density, ingredient A content and temperature) and 7 output values (300, 200, 100, 60, 30, 6, 3 RPM viscosities) and we have a set of experimental data points (training samples).

$$training\ data: \{X_i, Y_i\}, \quad i = 1,2,\dots,N$$

$$X_i = i^{th}\ input\ vector\ (n \times 1)$$

$$Y_i = i^{th}\ output\ vector\ (m \times 1)$$

Where N represents the number of experiments or the number of data points. In this study, n and m are 3 and 7, respectively.

To perform Gaussian process regression, we consider one output at a time i.e. we construct seven GPR models for seven dial reading outputs. So the training data for each GPR model design is:

$$training\ data\ D\colon \{X_i, y_i\}, \quad i = 1,2, \dots, N$$

Where y_i is a scalar and represents one of the dial reading outputs.

Each nonlinear regression function is a random sample (parametrized by the input variable) drawn from a joint Gaussian probability function given the training data set D. Our goal in GPR is to train a function f from data D. A Gaussian process is a prior (information) over functions $p(f)$ that can be utilized for Bayesian regression (inference):

$$p(f|D) = \frac{p(f)p(D|f)}{p(D)}$$

Where $p(f|D)$ shows the posterior probability of all the regression functions given the training data. $p(f)$ represents the prior probability on all the possible regression functions and $p(D|f)$ is the likelihood function. In Bayesian inference context, $p(D)$ is only a scaling parameter which does not affect the intended posterior probability density function $p(f|D)$.

The nonlinear regression function in GPR is a set of random variables indexed by a continuous parameter (e.g. time, space, temperature, etc.) which is also called random function $f(X)$. The assumption in GPR is that any set of regression functions has jointly Gaussian distribution with zero mean:

$$p(f|X) = N\big(0, C(X)\big) = (2\pi)^{-\frac{k}{2}}|C(X)|^{-\frac{1}{2}}e^{-\frac{1}{2}fC(X)^{-1}f}$$

Where $C(X)$ is the covariance matrix and k represents the dimension of function f with respect to continuous parameter. Estimating the unknown target value y_0 given the set of training y presents a conditional probability function:

$$p(y_0|y) = N(\overline{y_0}, \sigma_{y_0})$$

The GPR estimation value is the mean of the above conditional probability density function. Basically GPR proposes a Gaussian probability density function (PDF) for the new estimated value which can be fully characterized by a mean and a standard deviation value.

As a necessary preprocessing step to avoid scaling issues due to different input and output ranges, we apply data normalization and scaling to input and output values. Target or output values are assumed to be normal standard variables (resulted from a Gaussian distribution with zero mean and unit variance). Moreover the input values (input parameters) are shifted and scaled to [0,1] interval.

$$X_s = \frac{X - X_{min}}{X_{max} - X_{min}}$$

$$Y_{sn} = \frac{Y - \bar{Y}}{\sigma_Y}$$

To define the spatial relation of the samples in the input space (in this study a 3D space), a covariance function is established:

$$C\big(X_i, X_j\big) = vexp(-h^p)$$

$$h = \sqrt{g^T L^{-1} g}$$

$$g = R\big(X_i - X_j\big)$$

$$L = \begin{bmatrix} l_1 & 0 & 0 \\ 0 & l_2 & 0 \\ 0 & 0 & l_3 \end{bmatrix} = scaling\ matrix$$

$$X_i, X_j = experiment\ input\ values\ (3 \times 1)$$

l_1, l_2 and l_3 are correlation lengths in three input directions. Matrix R represent the rotation matrix in 3D which can be calculated based on the rotation angles around different axes (D'Orangeville et al. 2003). The combination of L and R matrices determines the spatial correlation of fluid properties in matrix X. v represents the variance parameter. In specific cases, where the parameter p is equal 2, the resulted covariance is the squared exponential covariance function (or Gaussian) and if $p = 1$, the exponential covariance function is reproduced. The parameters involved in defining the covariance function are called hyper parameters. By changing the hyper parameters the GPR estimator quality will change. In this study we tune 7 hyper parameters (three correlation lengths, three rotation angles and the power) through leave-one-out cross validation (LOOCV) to obtain the optimal GPR estimator. In LOOCV, we assume one of the training data points is unknown and use the other 8 training data to design GPR estimator and estimate the unknown data point and then compare the estimated value with the true value. This procedure is repeated for all the outputs (7 viscosity readings) and for all the training data. In cross validation procedure, the performance measure of the GPR estimator is the average relative error.

The following is the relation between semivariogram (popular spatial correlation representation in Geostatistics) and covariance function:

$$\gamma(X_i, X_j) = v - C(X_i, X_j)$$

To estimate the corresponding output y_0 of the new input parameter X_0, we use the Gaussian process interpolation (also called simple Kriging):

$$y_0 = \begin{bmatrix} y_1 \\ \vdots \\ y_N \end{bmatrix}^T C_{nn}^{-1} C_{nu} = \sum_{i=1}^{N} \lambda_i y_i$$

$$C_{nn} = \begin{bmatrix} C(X_1, X_1) & \cdots & C(X_1, X_N) \\ \vdots & \ddots & \vdots \\ C(X_N, X_1) & \cdots & C(X_N, X_N) \end{bmatrix} = covariance\ matrix\ of\ training\ data$$

$$C_{nu} = \begin{bmatrix} C(X_1, X_0) \\ \vdots \\ C(X_N, X_0) \end{bmatrix} = cross\ covariance\ of\ training\ data\ and\ new\ input$$

The estimated output is basically a linear combination of the training outputs where the coefficients are determined by the spatial configuration of the input (with respect to the training data) and the covariance function shape. One of the unique advantages of GPR for instance comparing to Artificial Neural Networks (ANN) is the ability to obtain the estimation error. GPR not only provides us with the estimated output at the new input value but also it is able to calculate its associated error or variance. With this capability we can specify how good each estimation is or determine the confidence of each estimation and decide if we need to perform new laboratory experiments.

$$estimation\ variance\ of\ y_0 = v - C_{nu}^T C_{nn} C_{nu}$$

This provides us with the estimation standard deviation and consequently the confidence interval associated with each estimation. It should be noted that GPR is an absolute estimator i.e. estimated values for the training data are equal to true outputs. After tuning using LOOCV the average minimum cross validation error is 15%. We obtained different hyper parameters (different covariance functions) for different viscosity outputs. For all the output viscosity readings, the obtained power parameter p through cross validation is 2 so the preferred covariance function is the squared exponential covariance function.

Since there are 3 input parameters and plotting each of the viscosity readings with respect to 3 independent input parameters is not very informative, for illustration purposes, we plotted the viscosity with respect to only two input parameters and kept the third one constant. In **Figure 4**, all of the predicted output values are plotted versus two pairs of the input parameters for mud, spacer and cement. For instance, first plot in **Figure 4** shows the 3 RPM predicted viscosity of cement, mud and spacer with respect to temperature and ingredient of different material contents while their density is constant (lb/gal). The plotted surfaces are the nonlinear regression models resulted from GPR. Having this nonlinear regression function, we are able to predict the rheological properties of the new mud, spacer and cement samples prior to performing the confirmation laboratory experiment.

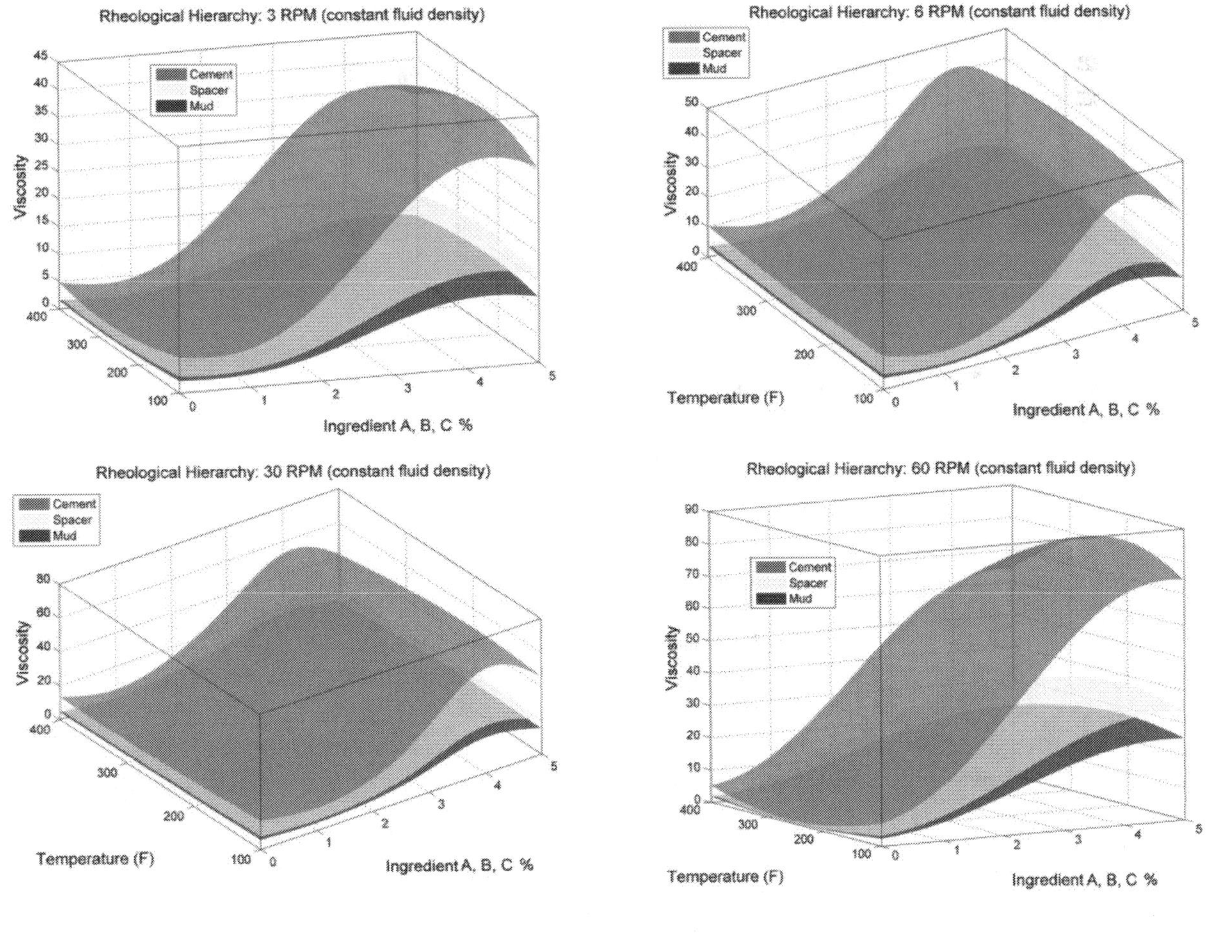

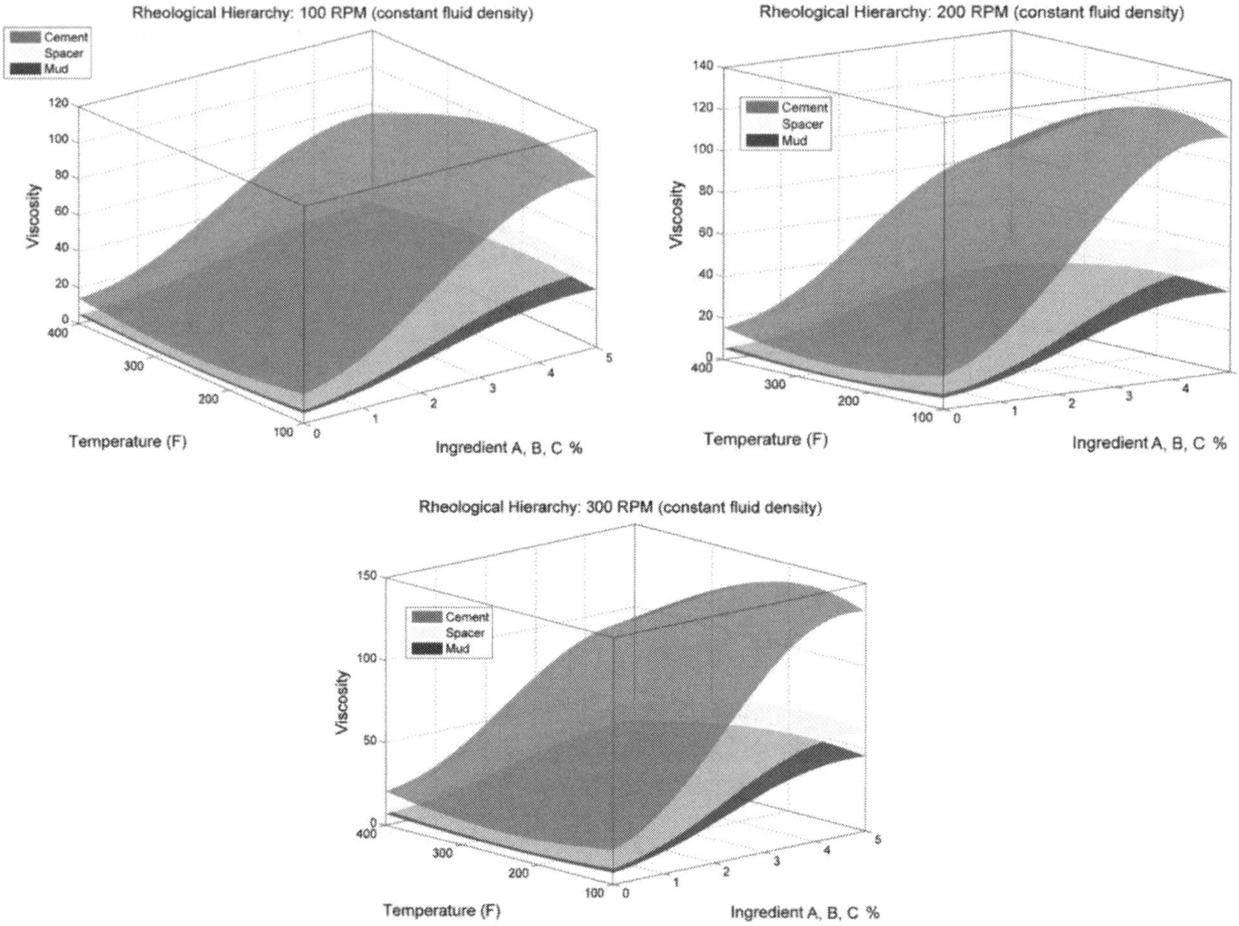

Figure 4: Nonlinear estimator resulted from GPR for different RPMs

Work flow excecution for predicting mud, spacer and cement properties

The following figures show the comprehensive procedure of experimental data importing, prediction with machine learning technique, visualization and finally fluid characteristics design. As shown in **Figure 5**, experimental fluid data can be manually or automatically imported to the design tool in the first step. Then in the prediction step (**Figure 6**), GPR based data-driven tool will be trained by the imported data and customized fluid characteristics plots can be visualized and their data can be exported. In the final step in **Figure 7**, the rheological characteristics of mud, spacer and cement are predicted and plotted based on their fluid system properties.

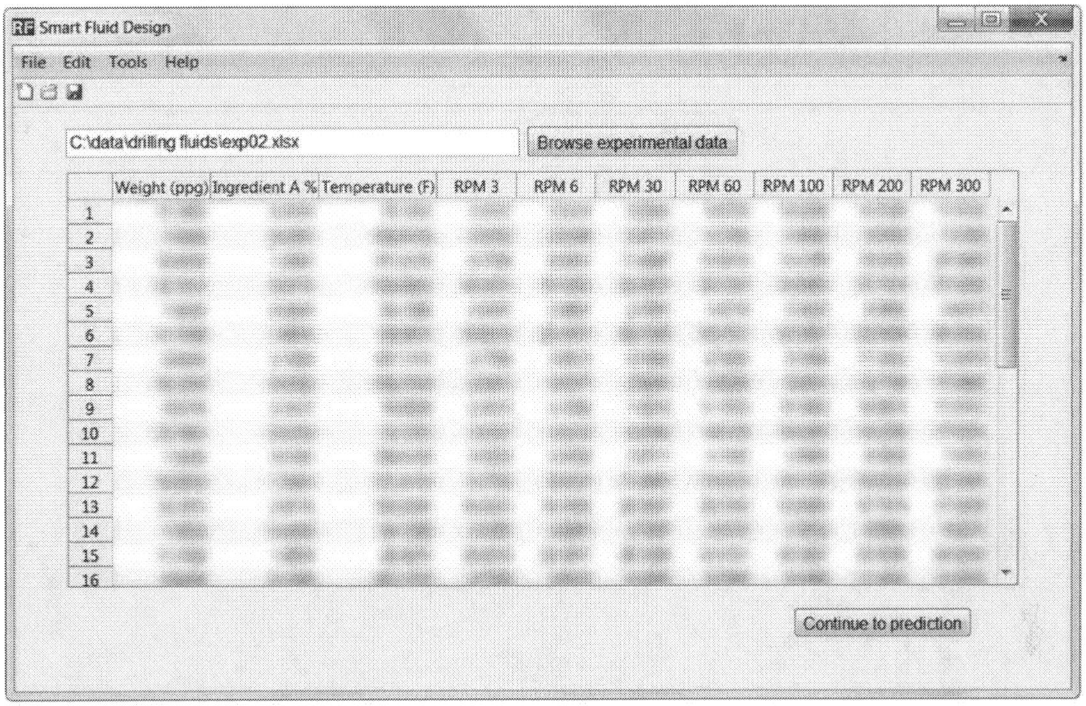

Figure 5: Importing experimental data (training data) to smart data- driven fluid design tool

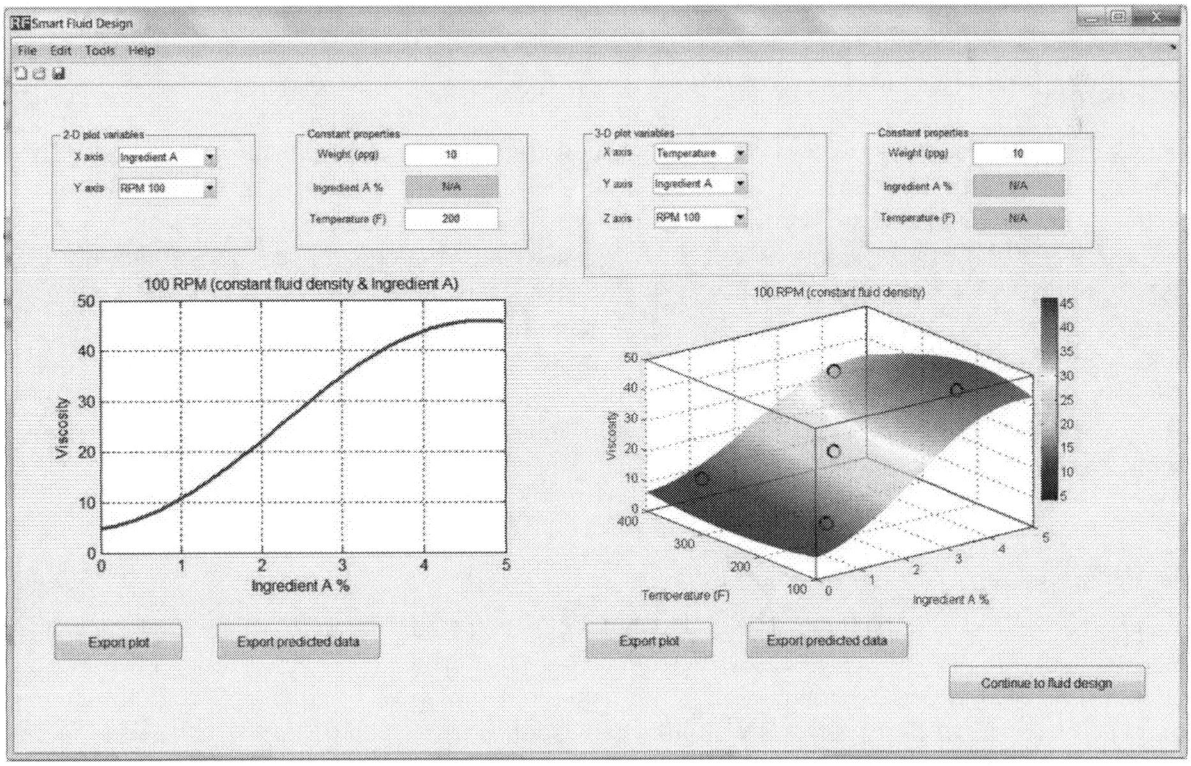

Figure 6: Data-driven rheological properties estimation and visualization

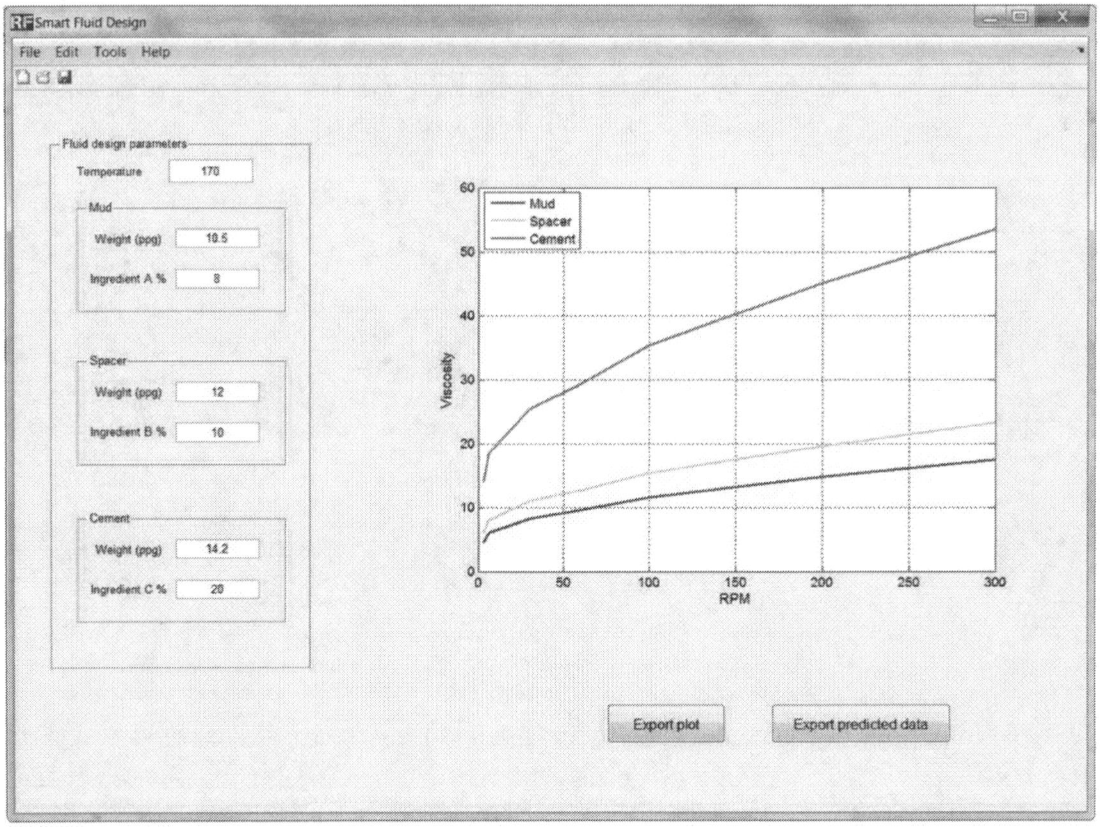

Figure 7: Predicting rheological properties of mud, spacer and cement: smart fluid system design

Discussion

Predicting the fluid properties is very beneficial to the operations cost effectiveness and success. Trial and error methods are incapable of offering an agile comprehensive computation capability which is provided by machine learning algorithms. DOE along with RSM is a good method for data point selection to learn more about R&D and new products however it is limited and time consuming method and not flexible enough for agile decision making. GPR is a strong regressor which can suitably estimate the mud, spacer and cement properties. This paper showcased the strength of GPR in predicting the rheological properties of mud, spacer and cement. This is an important example through which GPR assisted maintaining the rheological hierarchy at different RPMs. A good rheological hierarchy would be a contributor to better friction pressure hierarchy and therefore less fluid intermixing and viscous fingering, better cement jobs and well integrity excellence.

Conclusion

Trial and error fluid design is not cost effective and lacks long term vision to accumulate knowledge on mud, spacer and cement fluid systems. There is a need for a tool which is empowered by machine learning methodologies. This paper investigated an intelligent tool which is equipped with flexible machine learning algorithms. GPR unlike ANN honors the training data and offers a prior-free regression implementation. Furthermore GPR is unique in the sense of providing the estimation error and uncertainty assessment while predicting the fluid characteristics and designing a fluid system. The developed intelligent prediction and design tool is able to extract and relate any set of input parameters to output variables based on the provided training samples that can be both experiment and simulation results.

Acknowledgment

Authors would like to thank the management of ReservoirFocus LLC for the permission to publish this work.

References

Alfi, M., Hosseini, S. A., Alfi, M., & Shakiba, M. (2015, November 17). Effectiveness of 4D Seismic Data to Monitor CO_2 Plume in Cranfield CO_2-EOR Project. Carbon Management Technology Conference. doi:10.7122/439559-MS

Alfi, M., and Park et al., "Simulation of Formwork Filling by Cement Fluid: the Effect of the Formwork Structure on Yield-stress Fluid", COMSOL Conference Boston, 2013

Alvarado, V., Ranson, A., Hernandez, K., Manrique, E., Matheus, J., Liscano, T., & Prosperi, N. (2002, January 1). Selection of EOR/IOR Opportunities Based on Machine Learning. Society of Petroleum Engineers. doi:10.2118/78332-MS

Amani, M., Al-Jubouri, M. and Shadravan, A., (2012), Comparative study of using oil-based mud versus water-based mud in HPHT fields, Advances in Petroleum Exploration and Development, Vol. 4, No. 2, 2012, pp. 18-27 DOI:10.3968/j.aped.1925543820120402.987

Bishop, C. M. (2006). Pattern recognition and machine learning. springer.

Deutsch, C. V., & Journel, A. G. (1992). Geostatistical software library and user's guide. New York.

D'Orangeville, C., & Lasenby, A. N. (2003). Geometric algebra for physicists. Cambridge University Press.

Duda, R. O., Hart, P. E., & Stork, D. G. (2012). Pattern classification. John Wiley & Sons.

Lee, J., Shadravan, A., & Young, S. (2012, January 1). Rheological Properties of Invert Emulsion Drilling Fluid under Extreme HPHT Conditions. Society of Petroleum Engineers. doi:10.2118/151413-MS

Parada, C. H., & Ertekin, T. (2012, January 1). A New Screening Tool for Improved Oil Recovery Methods Using Artificial Neural Networks. Society of Petroleum Engineers. doi:10.2118/153321-MS

Rasmussen, C. E. (2006). Gaussian processes for machine learning.

Rad, A., Shadravan, A., Soleymani, A., Motaghedi, N., "Lewis Acid-base Surface Interaction of Some Boron Compounds with N-doped Graphene; First Principles Study", Current Applied Physics, Volume 15, Issue 10, October 2015, Pages 1271–1277

Shadravan, A., Alegria, A., & Castanedo, R. (2015, September 28). Rheological Hierarchy Optimization Improves Fluid Displacement and Well Integrity, 3 Worldwide Case Studies. Society of Petroleum Engineers. doi:10.2118/174773-MS

Shadravan, A., Narvaez, G., Alegria, A., Carman, P., Perez, C., & Erger, R. (2015, March 16). Engineering the Mud-Spacer-Cement Rheological Hierarchy Improves Wellbore Integrity. Society of Petroleum Engineers. doi:10.2118/173534-MS

Shadravan, A., Tarrahi, M., & Amani, M. (2015). Intelligent Tool to Design Fracturing, Drilling, Spacer and Cement Slurry Fluids Using Machine Learning Algorithms. Society of Petroleum Engineers. SPE-175238-MS

Shadravan, A., Ghasemi, M. and Alfi, M. "Zonal Isolation in Geothermal Wells", Fortieth Workshop on Geothermal Reservoir Engineering, Stanford University, Stanford, California, January 26-28, 2015

Shi, J. Q., & Choi, T. (2011). Gaussian process regression analysis for functional data. CRC Press.

Tarrahi, M., Jafarpour, B., & Ghassemi, A. (2015). Integration of microseismic monitoring data into coupled flow and geomechanical models with ensemble Kalman filter, Water Resour. Res., 51, doi:10.1002/2014WR016264.

Tarrahi, M., Jafarpour, B., & Ghassemi, A. (2013, September 30). Assimilation of Microseismic Data into Coupled Flow and Geomechanical Reservoir Models with Ensemble Kalman Filter. Society of Petroleum Engineers. doi:10.2118/166510-MS

Tarrahi, M., Afra, S., & Surovets, I. (2015). A Novel Automated and Probabilistic EOR Screening Method to Integrate Theoretical Screening Criteria and Real Field EOR Practices Using Machine Learning Algorithms. Society of Petroleum Engineers. SPE-176725-MS

Theodoridis, S., Pikrakis, A., Koutroumbas, K., & Cavouras, D. (2010). Introduction to Pattern Recognition: A Matlab Approach: A Matlab Approach. Academic Press.

Williams, C. K., & Rasmussen, C. E. (2006). Gaussian processes for machine learning. the MIT Press, 2(3), 4.

CARBON
MANAGEMENT
TECHNOLOGY CONFERENCE

CMTC-440237-MS

Assisted EOR Screening Approach for CO_2 Flooding with Bayesian Classification and Integrated Feature Selection Techniques

Sardar Afra, SPE, Texas A&M University, Mohammadali Tarrahi, SPE, Shell Global Solutions (US), Inc.

This paper was prepared for presentation at the Carbon Management Technology Conference held in Sugarland, Texas, USA, 17–19 November 2015.

Abstract

Predicting the performance of different enhanced oil recovery methods is of high importance due to the need for making feasible and practical operational decisions in field development. A proper recovery strategy has to be selected in early stage of field development planning. EOR screening criteria are widely employed in order to provide engineers with an appropriate recovery technique corresponding to specific reservoir and fluid properties. A dependable first order screening evaluation algorithm enables critical decision making on potential enhanced oil recovery strategies with the limited reservoir information.

We present and test a novel screening techniques through the means of machine learning and pattern recognition techniques, which are well-established in computer science literature, in order to discriminate between various EOR projects with a special focus on CO_2 flooding. A comprehensive study is conducted utilizing various classification rules to solve EOR screening problem. Also, the effects of various features which are globally shared within different reservoirs are investigated as well. Well known feature selection methods like data whitening, feature ranking and dimensionality reduction have been integrated to the process to improve the performance of the classifiers. In order to perform feature selection, some important fluid and reservoir characteristics such as permeability, depth, API, temperature, oil saturation and viscosity are taken into account.

The proposed data-driven screening algorithm is a high-performance tool to select an appropriate EOR method such as steam injection, combustion, miscible injection of CO_2 and N_2, based on aforementioned reservoir and fluid properties. In this innovative approach, we integrate both theoretical screening principles such as Taber criteria and successful field EOR practices worldwide. Not only this algorithm proposes an appropriate EOR method for a specific reservoir condition but it also gives the probability of success or success rate corresponding to each EOR method. In addition, the proposed algorithm is able to address environmental, economical, geographical and technological limitations.

The proposed algorithm permits integration of different types of data, eliminates arbitrary approach in making decisions, and provides accuracy and fast computation. The suitability of the proposed method is demonstrated by different synthetic and real field EOR cases. This novel EOR screening method is capable of evaluating the effectiveness of different EOR scenarios given a specific reservoir condition. We showed that the proposed EOR screening algorithm is able to predict the appropriate EOR method correctly in more than 90% of cases. We also ranked the proposed screening algorithms based on their screening performance.

Introduction

Enhanced oil recovery (EOR) techniques have been widely employed to exploiting remaining oil or natural gas within the reservoir, which cannot be retrieved using primary and secondary methods, usually by injecting non-reservoir materials into the rock. Continuous fall in the conventional reservoirs' production and low probability of new reservoirs discovery make the EOR practices important. Engineers also must consider several important factors while performing EOR projects such as capital sensitivity, the high risk of unfavorable instances, and the fact that not all reservoirs can be recovered through EOR methods. Hence, selecting the best recovery technique becomes the choice of interest for oil and gas industry and reservoir engineers. Furthermore, expertise along with many assumptions, structures and relations among reservoir and fluid characteristics and features are needed to obtain an efficient EOR technique in each EOR screening scenario.

EOR screening has been first introduced by Taber et al. (Taber et al. 1997) in which all EOR screening criteria have been briefly summarized in a table and each EOR method spectrum has been graphically determined in order to obtain and execute an efficient EOR technique for all types of reservoir with various fluid and rock properties. Although, automated algorithms have attracted researchers from various engineering and science fields such as chemical engineering (Emrani et al. 2011 and 2012), little work has been done in developing automated EOR screening. While introducing new EOR techniques is very critical to improve the existing EOR schemes, a need for providing a more reliable and structured screening techniques becomes inevitable (Emrani and Nasr-El-Din 2015; Afra and Nasr-El-Din 2015).

Reservoir engineering has recently advanced to a more structured paradigm, in which reservoir production strategies can be optimized simultaneously with assimilation of new production information in a real-time fashion (Jafarpour and Tarrahi 2011; Tarrahi et al. 2013 and 2015). Employing machine learning and big data analytics helped petroleum engineers to obtain more reliable, accurate and cost efficient methods in reservoir management including EOR screening. Many works have been done in literature to apply such artificial intelligence techniques in many aspects of petroleum engineering like reservoir simulation, production optimization, field development and history matching. For instance, a very comprehensive set of studies have been conducted by authors in the field of automated history matching and reservoir parameterization by introducing a novel parameterization method through tensor algebra (Afra et al. 2013 and 2014; Gildin et al. 2014). Another work by authors (Afra and Braga-Neto 2011) also showed the suitability of supervised classification algorithms for high dimensional small sample data sets.

Automated EOR screening through artificial intelligence has recently been addressed in the literature. Various machine learning techniques have been proposed to facilitate the automated EOR screening procedures including artificial neural networks (ANN), fuzzy classification and Bayesian networks (BN). Alvarado et al. (Alvarado et al. 2002) employed machine learning algorithms, e.g., clustering, and dimensionality reduction methods to extract information from a worldwide EOR/IOR data set to acquire desired screening. Parada and Ertekin (Parada and Ertekin 2012) developed a neural network based simulation tool to deliver general EOR screening criteria for variety of recovery schemes which provided a new design for steam injection recovery projects. Zerafat et al. (Zerafat et al. 2011) used Bayesian network analysis to come up with a unified expert system that assists further economical and environmental assessments through classifying suitable EOR methods. Of special interest in this paper is to study the performance of different pattern recognition algorithms in solving EOR screening problem. A comprehensive comparison is also provided on the performance of the classifiers with respect to the corresponding average correct classification rate (ACCR).

The paper is organized as follows. In the next section, the classification framework and designing proper classifiers are briefly introduced. Data collection section describes how EOR screening data collected and integrated into the proposed automated EOR screening schemes. Also, the probabilistic process of synthetic data generation is explained in details. In results and discussion section experiment results are presented and explained in details for 9 different EOR methods in a 9-dimentional feature space of reservoir properties and then all results are compared with respect to the ACCR. Finally, last section concludes several remarks of the work and proposes best classification method along with some guidelines for the future work by authors regarding the effect of feature selection and feature conditioning on the performance of automated EOR screening.

Methods

In order to demonstrate the main concepts developed in the present work, we will briefly describe the classification problem. Eager readers may refer to the machine learning and pattern recognition resources to get more thorough understanding of the classification paradigm and its application in energy sector. We start by describing classification rule and then continue by introducing the most applicable classifiers in EOR screening.

Classification

In machine learning and statistics, the problem of recognizing the label or class of a new incoming unknown input given a training set of observations with known labels is referred to as classification. In this case, the classification problem would be based on a supervised learning in which a set of properly labeled and known training data is available. On the other hand, in the lack of such labeled observation, the unknown data set is categorized based on a specific measure like distance or data similarity which is known as clustering or unsupervised learning (Alpaydin 2010; Devroye et al. 1996; Webb and Copsey 2011).

Classifier

In two-group statistical pattern recognition, there is a feature vector $X \in \Re_p$ and a label $Y \in \{-1, 1\}$. The pair (X, Y) has a joint probability distribution F, which is unknown in practice. Hence, one has to resort to designing classifiers from training data, which consists of a set of n independent points $S_n = \{(X_1, Y_1), \ldots, (X_n, Y_n)\}$ drawn from F. A classification rule is a mapping $g: \{\Re_p \times \{0,1\}\}_n \times \Re_p \to \{0,1\}$. A classification rule maps the training data S_n into the designed classifier $g(S_n, .): \Re_p \to \{0,1\}$. A linear discriminant classifier is given by **Eq. 1**:

$$g(S_n, x) = \begin{cases} 1, & a^T x + a_0 > 0 \\ 0, & otherwise \end{cases} \quad (1)$$

Therefore, all training points are correctly classified if $Y_i(a^T X_i + a_0) > 0, i = 1, \ldots, n$ where$y_i \in \{-1,1\}$ (Afra and Braga-Neto 2011).

Of special interest in the present work is to employ pattern recognition tools specifically Bayesian classifiers to rank the probability of success for all possible EOR schemes for a given reservoir with specified fluid and rock properties. One may choose the best EOR method for a given reservoir based on the outcome of this automated screening, known as expert system, and considering the non-learning parameters outside the training set, e.g., reservoir heterogeneity, production history to name a few, which either are not a quantity or not included in Taber table. In this paper, in order to perform a computer assisted EOR screening, we utilized 9 different classification rules including Bayes optimal classification, neural network, support vector machines, k-nearest neighbors, Gaussian mixture models, Naïve Bayes classifier, decision tree and RBF. Here we only describe two methods. Eager readers may refer to (Devroye et al. 1996; Grubbs 1969; Webb and Copsey 2011) for more details on the different classification methods as well as error estimation analysis.

Neural Network Classifier

Neural network technique processes data by mathematically reproducing the biological neural networks function. Kolmogorov-Lorentz theorem states that every continuous function $f: [0,1]^d \to \mathbb{R}$ can be rewritten as follows (**Eq. 2**):

$$f(x) = \sum_{i=1}^{2d+1} P_i\left(\sum_{j=1}^{d} Q_{ij}(x_j)\right) \quad (2)$$

Wherein $P_i: \mathbb{R} \to \mathbb{R}$ and $Q_{ij}: \mathbb{R} \to \mathbb{R}$ are continuous functions as well. The theorem proposed that a multivariate function can be computed as a finite sum of univariate functions of sums of univariate functions of the coordinates. It is not obvious at all how to find functions P_i and Q_{ij} required to compute exactly a general function f. Instead, it is possible to use simpler functions such as linear functions and sigmoid function. A two-layer network with k hidden neurons is determined by (**Eq. 3**):

$$\zeta(x) = c_0 + \sum_{i=1}^{k} c_i \sigma(\phi_i(x)) = c_0 + c^T \zeta(x) \quad (3)$$

Where $\xi(x) = \left(\sigma(\phi_1(x)), \cdots, \sigma(\phi_d(x))\right)^T$ is the output of hiddin node i and

$$\phi_i(x) = b_i + \sum_{j=1}^{d} a_{ij}x_j \quad (4)$$

And consequently a neural network classifier can be obtained by thresholding the output of the network as follows (**Eq. 5**):

$$\psi(x) = \begin{cases} 1, & \zeta(x) > 0 \\ 0, & otherwis \end{cases} \quad (5)$$

The neural network approach is similar to the non-linear support vector machine (SVM) approach. The hidden layers nonlinearly map the original feature space into a different space, and the output layer acts on the transformed features by means of a linear decision (hyperplane). The decision in the original feature space is nonlinear. If the first hidden layer is composed of threshold sigmoids, then no matter what follows, the decision regions will be piecewise linear, given by the intersection of k hyperplanes which is called an arrangement classifier. The decision regions for an arrangement classifier are convex polytopes, which are determined by intersections of half-spaces. The classifier is therefore piecewise linear. The decision regions in the arrangement classifier are determined by the first hidden layer of k threshold sigmoids. Each of the regions corresponds to a binary vector $b = [b_1, \cdots, b_k]$. All other subsequent hidden layers can only alter the class attributed to each of these regions. Finally, by minimizing an error criterion like empirical error, absolute error or mean square error one may select parameters. By using *augmented vectors* to represent the bias coefficients, we can write the output of a two-layer network with k hidden nodes as (**Eq. 6**):

$$\zeta(x) = \sum_{i=0}^{k} c_i\sigma\left(\sum_{j=1}^{d} a_{ij}x_j\right) \quad (6)$$

The vector of parameters for this equation, also known as the vector of weights w, is given by (**Eq.7**):

$$w = (c_0, \cdots, c_d, a_{10}, \cdots, a_{1d}, \cdots, a_{k1}, \cdots, a_{kd})^T \quad (7)$$

given a training pattern (x, y), and considering mean square error, one may determin the total error on the entire training set as follows (**Eq. 8**):

$$\epsilon_T = \sum_{i=1}^{n}\left(t_i - \zeta(x_i)\right)^2 \quad (8)$$

And the goal is to find the vector of weights w that minimizes the error criterion whis is a non-linear optimization problem. Following the gradient descent, i.e., backpropagation algorithm will be described to solve the optimization problem in hand. The basic iteration step in gradient descent is given by (**Eq. 9**):

$$\Delta w = -\ell\nabla\epsilon_T \quad (9)$$

That is, the vector of weights w is updated in the direction that decreases the error criterion, according to the step length ℓ. Now, one can apply chain rule to the componentwise form of the above difference (**Eq. 10**).

$$\frac{\partial\epsilon}{\partial w_{jk}^i} = \frac{\partial\epsilon}{\partial\alpha_k^i}\frac{\partial\alpha_k^i}{\partial w_{jk}^i} = -\delta_k^i\beta_j^{i-1}(x) \quad (10)$$

where

$$\delta_k^i = -\frac{\partial\epsilon}{\partial\alpha_k^i} \quad (11)$$

So that δ_k^i can be determined as follows:

$$\delta_k^i = -\frac{\partial\epsilon}{\partial\alpha_k^i} = -\sum_{l=1}^{n_{l+1}}\frac{\partial\epsilon}{\partial\alpha_l^{i+1}}\frac{\partial\alpha_l^{i+1}}{\partial\alpha_k^i} = \sum_{l=1}^{n_{l+1}}\delta_l^{i+1}\frac{\partial\alpha_l^{i+1}}{\partial\alpha_k^i} \quad (12)$$

and

$$\frac{\partial\alpha_l^{i+1}}{\partial\alpha_k^i} = w_{kl}^{i+1}\sigma'(\alpha_k^i(x)) \quad (13)$$

Whis yeailds the backpropagation equation as follows (**Eq.14**):

$$\delta_k^i = \sigma'(\alpha_k^i(x))\sum_{l=1}^{n_{l+1}} w_{kl}^{i+1}\delta_l^{i+1} \quad (14)$$

Bayesian Classification

As mentioned previously, in many classification approaches the goal is to construct a map from the feature space to the space of labeled class. In fact, in such cases a classification rule explicitly maps training data into a designed classifier. Decision trees and neural networks are good examples of such classifiers. However, Bayesian classification approach slightly differs from others. As a matter of fact, in the Bayesian approach, the joint probability distribution of the class and feature have to be estimated.

Given an n-dimensional random variable describing the feature space, **X**, and a random variable, **C**, representing class space, the problem is to estimate the following conditional probability function utilizing Bayes rule in (**Eq. 15**):

$$P(\boldsymbol{C}|\boldsymbol{X}) = \int f(\boldsymbol{X};\boldsymbol{\theta})\, P(\boldsymbol{\theta}|D)d\boldsymbol{\theta} \quad (15)$$

Wherein D is the training set. Therefore, learning process in Bayesian classification turns to a joint probability distribution function approximation. Then, a new incoming data can be classified only by calculating the conditional probability of class space variables vector given the corresponding features vector (reservoir and fluid properties) which obtains the most probable class (proper EOR method) through the constructed estimation in the learning process.

For the sake of simplicity and without losing generality, we assume a two-class classification problem. In this case we are interested in classifying any incoming data through identifying its label and consequently the correct class of the data based on the joint probability distribution and feature space probability. The Bayes rule suggests to calculate a-posteriori probabilities, $P(c_i|X)$, and then assigns the data to the class with highest probability (**Eq. 16**).

$$P(c_i|X) = \frac{P(X|c_i)P(c_i)}{P(X)} \quad (16)$$

Wherein, $P(c_i)$ is the prior probability of the class lables that can be easily estimated from the training data. Now, Bayes classification rules can be stated as follows. An incoming instance is assigned to class c_i if the posterior probability of that class given the features is greater than of the second class. Thus, considering Bayes rule the problem of classification shrinks to estimating class-conditional probability density function, $P(X|c_i)$. The most important point to be noted here is to minimize the misclassification probability or the classification error probability. Misclassification probability consists of two terms in case of two-class problem. First term is the probability of classifying unknown data to c_1 although it belongs to c_2 and the second term would be stated vice versa. Thus (**Eq. 17**):

$$P_{error} = P(x \in c_1;\, c_2) + P(x \in c_2;\, c_1). \quad (17)$$

Now, employing *chain rule*, one can write the following (**Eq. 18**):

$$P_{error} = P(x \in c_1|c_2)p(c_2) + P(x \in c_2|c_1)P(c_1). \quad (18)$$

And utilizing Bayes rule results in (**Eq. 19**):

$$P_{error} = \int_{C_1} P(c_1|x)p(x)dx + \int_{C_2} P(c_2|x)p(x)dx. \quad (19)$$

Considering equation 8 one may select the decision regions in feature space as follows (**Eq. 20**).

$$\begin{cases} C_1: & P(c_1|x) > P(c_2|x) \\ C_2: & P(c_2|x) > P(c_1|x) \end{cases} \quad (20)$$

Hence, the Bayesian classifier is optimal in minimizing classification error probability. For more details on Bayesian classification and probability density function estimation see (Alpaydin 2010; Devroye et al. 1996; Webb et al. 2011).

Confusion Matrix

In the field of machine learning, the performance of a supervised classifier is monitored by means of confusion matrix. A confusion matrix actually is a counter array of size $C \times C$ where C is the size of the class space, at which every classification step is recorded. Each row represents the true class while each column represents a predicted class. Once a classification proceeds to assign an unknown input to a class, the corresponding entry in the confusion matrix increases by one. The ideal classifier is the one that its corresponding confusion matrix is a diagonal matrix in which all entries outside the main diagonal are zero. Considering confusion matrix, correct classification rate can be determined as follows (**Eq. 21**).

$$CCR = \frac{summation\ on\ the\ diagonal\ elements}{Number\ of\ all\ test\ samples} = 1 - P_{error} \quad (21)$$

One has to note that a legitimate error estimator always should be calculated based on the test data sets not the training data. The next session describes data collection process and explains how synthetic data are generated based on real reservoir data sets and Taber criteria table.

Feature Selection

In many instances, it is often necessary to reduce the number of measurements (features) in a data set to make the problem tractable. As a matter of fact, dimensionality reduction is of special interest for number of reasons. First, it improves classification performance by avoiding peaking phenomenon (Afra and Braga-Neto 2011). Second, it may reduce computational costs with respect to execution time and data storage. Third, it assists visualization of high dimensional data, e.g., big data. It also enables classifiers to handle categorical and nominal features in an efficient way. Dimensionality reduction will generally involve loss of information. One typically wants to reduce the number of features in such a way that this loss is minimized. One may be interested in dimestionality reduction, e.g., feature selection to eliminate correlation between features, and those that are nearly constant, i.e., small variance. Overall, dimensionality reduction help to capture noise and redundancies and to retain most informative features.

Results and Discussions

This section explains how the required training and test data sets are generated utilizing real EOR data sets. Here, in automated EOR screening, a combination of Taber table, successful EOR projects and prior knowledge about the reservoir are employed to generate training data in a 9-dimentional feature space. All the settings and setups are similar to (Tarrahi and et. al. 2015). We choose oil gravity (API), oil viscosity (cp), oil composition, oil saturation, formation type, net thickness (ft), permeability (md), reservoir depth (ft), temperature (°F) as the selection properties or the members of feature space. Also, the classes of the classification problem are chosen to be the EOR techniques including nitrogen and flue gas, hydrocarbon, CO_2, immiscible gases, micellar/polymer, ASP and Alkaline flooding, polymer flooding, combustion, steam, and surface mining. A multivariable joint probability density function (PDF) is employed in order to generate the synthetic realistic training data set based on a real data set of 7 features (continuous features) and 8 classes.

In the present work, 9 data sets each including 1000 samples are generated for all 9 classes where EOR methods represent the classes here. In order to create training and test data sets, one third of the samples, 300 samples, used as test samples and the rest utilized to train classifiers. We utilized uniform and triangular distribution to generate data sets though a Monte Carlo simulation. The Latin hypercube sampling (LHS) is employed to perform sampling step in an efficient Monte Carlo simulation and to achieve reasonably accurate random distribution. After generating data points, the calculated correlation coefficients (calculated from field EOR projects data) have to be applied to make the final samples that are correlated random variables.

In the following section EOR screening results of all classifiers are presented and the performances of these classifiers are compared with respect to average correct classification rate (ACCR) defined earlier. Based on the results, Bayes optimal classifier with Gaussian parametric PDF estimator tends to have the best performance to classify unknown data samples from test data set. Bayes optimal classifier aims at minimizing the probability of classification error. We also run an exhaustive feature selection approach to find the best feature sets. We find the best feature set which its error is minimum over all possible feature sets. As expected, corresponding errors decrease when the feature size increases. However, we should consider the peaking phenomenon, that is, increasing the feature size does not always lead to monotonically decreasing the errors and in a specific feature size, may decreasing stops or even worse start to increase. Table one summarizes all results.

Conclusion

We performed a comprehensive study on the application of different classifiers including Bayesian classification and artificial neural networks along with exhaustive search as feature selection technique, to deliver a better understanding of automated EOR screening based on real EOR data sets. We also presented a complete performance comparison between the classifiers with respect to average correct classification rate criteria. The analysis showed that Bayesian optimal classifier is the most reliable one among all others that not only proposes an appropriate EOR method for a specific reservoir condition but also gives the probability of success or success rate corresponding to each EOR method. The proposed algorithm is able to address environmental, economical, geographical and technological limitations. Furthermore, the proposed technique permits integration of different types of data, eliminates arbitrary approach in making decisions, and provides accuracy and fast computation. Corresponding errors decrease when the feature set size increases.

Number of Features	Average Correct Classification Rate (ACCR)	Number of Misclassifications	Probability of Classification Error (PCE)
4	0.8089	516	0.1911
5	0.8459	416	0.1541
6	0.8715	347	0.1285
7	0.8859	308	0.1141

Table 1. Classification error with feature selection.

Feature rank	1	2	3	4	5	6	7	8	9
Feature	μ	Depth	API	Comp.	Form.	K	S_o	Thickness	T
ACCR	0.7385	0.3993	0.3715	0.3541	0.2567	0.2496	0.2204	0.2163	0.2067
Number of Misclassifications	706	1622	1697	1744	2007	2026	2105	2116	2142
Probability of Classification Error (PCE)	0.2615	0.6007	0.6285	0.6459	0.7433	0.7504	0.7796	0.7837	0.7933

Table 2. Feature ranking by classification performance.

References

Afra, S., and Gildin, E. 2013. Permeability Parametrization using Higher Order Singular Value Decomposition (HOSVD). Presented at the 12th International Conference on Machine Learning and Applications (ICMLA), Miami, December 4-7. doi: 10.1109/ICMLA.2013.121.

Afra, S., and Braga-Neto, U. 2011. Studying the Possibility of Peaking Phenomenon in Linear Support Vector Machines with non-Separable Data. Presented at IEEE International Workshop on Genomic Signal Processing and Statistics (GENSIPS), San Antonio, December 4-6. doi: 10.1109/GENSiPS.2011.6169.484.

Afra, S., Nasr El-Din, H.A., Socci, D. et al. 2015. A novel viscosity reduction plant-based diluent for heavy and extra heavy oil. Presented at the 65th Canadian Chemical Engineering Conference (CSChE 2015), Calgary, October 4-7.

Alpaydin, E. 2010. Introduction to Machine Learning. MIT Press. p. 9. ISBN 978-0-262-01243-0.

Alvarado, V., Ranson, A., Hernandez, K. et al. 2002. Selection of EOR/IOR Opportunities Based on Machine Learning. Society of Petroleum Engineers. January 1. doi:10.2118/78332-MS.

Emrani, A.S., Saber, M., and Farhadi, F. 2012. Modeling and Optimization of Fixed-Bed Fischer-Tropsch Synthesis Using Genetic Algorithm. *Journal of Chemical and Petroleum Engineering* **46**(1): 1-11.

Emrani, A.S., Saber, M., and Farhadi, F. 2011. A Decision Tree for Technology Selection of Nitrogen Production Plants. *Journal of Chemical and Petroleum Engineering* **45**(1): 1-11.

Emrani, A.S., and Nasr-El-Din, H.A. 2015. Stabilizing CO_2-Foam using Nanoparticles. Presented at the SPE European Formation Damage Conference and Exhibition, Budapest, Hungary, 3-5 June. SPE-174254-MS. doi:10.2118/174254-MS.

Devroye, L., Gyorfi, L., and Lugosi, G. 1996. A probabilistic theory of pattern recognition. Applications of mathematics. Springer.

Gildin, E., and Afra, S. 2014. Efficient inference of reservoir parameter distribution utilizing higher order SVD Reparameterization. Presented at the 14th European conference on the mathematics of oil recovery (ECMOR XIV), Sisil, Italy, August 7-9. doi: 10.3997/2214-4609.20141826.

Gildin, E., Afra, S., and Tarrahi, M. 2014. Heterogeneous Reservoir Characterization using Efficient Parameterization through Higher Order SVD (HOSVD). Presented at the American Control Conference (ACC), Portland, 4-6 June. doi: 10.1109/ACC.2014.6859246.

Grubbs, F.E. 1969. Procedures for Detecting Outlying Observations in Samples. *Technometrics* **11** (1): 1–21. doi:10.1080/00401706.1969.10490657.

Jafarpour, B., and Tarrahi, M. 2011. Assessing the performance of the ensemble Kalman filter for subsurface flow data integration under variogram uncertainty. *Water Resources Research*, **47**(5).

Parada, C.H., and Ertekin, T. 2012. A New Screening Tool for Improved Oil Recovery Methods Using Artificial Neural Networks. Society of Petroleum Engineers. January 1. doi:10.2118/153321-MS.

Taber, J.J., Martin, F.D., and Seright, R.S. 1997. EOR Screening Criteria Revisited - Part 1: Introduction to Screening Criteria and Enhanced Recovery Field Projects. Society of Petroleum Engineers. August 1. doi:10.2118/35385-PA.

Tarrahi, M., B. Jafarpour, and A. Ghassemi 2015. Integration of microseismic monitoring data into coupled flow and geomechanical models with ensemble Kalman filter, Water Resour. Res., 51, doi:10.1002/2014WR016264.

Tarrahi, M., Jafarpour, B., and Ghassemi, A. 2013. Assimilation of Microseismic Data into Coupled Flow and Geomechanical Reservoir Models with Ensemble Kalman Filter. Society of Petroleum Engineers. September 30. doi:10.2118/166510-MS.

Tarrahi, M., Afra, S., & Surovets, I. 2015. A Novel Automated and Probabilistic EOR Screening Method to Integrate Theoretical Screening Criteria and Real Field EOR Practices Using Machine Learning Algorithms. Society of Petroleum Engineers. doi:10.2118/176725-MS

Webb, A.R., Copsey, K.D. 2011. *Statistical Pattern Recognition*. John Wiley & Sons.

Zerafat, M.M., Ayatollahi, S., Mehranbod, N. et al. 2011. Bayesian Network Analysis as a Tool for Efficient EOR Screening. Society of Petroleum Engineers. January 1. doi:10.2118/143282-MS.

Speciation of CO_2 in Amine and Alkaline Containing Solutions Using Raman Spectroscopy

Venkatesh R. Naik and Dr. John Kitchin

Department of Chemical Engineering, Carnegie Mellon University, Pittsburgh, PA

Motivation

- Post-combustion carbon capture remains an important option to mitigate green house gas emissions.
- The most mature technology for CO_2 capture is reversible chemical absorption into an aqueous amine solution. [2]
- Piperazine has been identified as a promising replacement to monoethanolamine(MEA).
- Piperazine has a much faster reaction rate compared to MEA (the most commonly used solvent). [3]

Objective of the Experiment

Investigated if any functionalized piperazine is better at CO_2 capture.

Samples used in the experiment

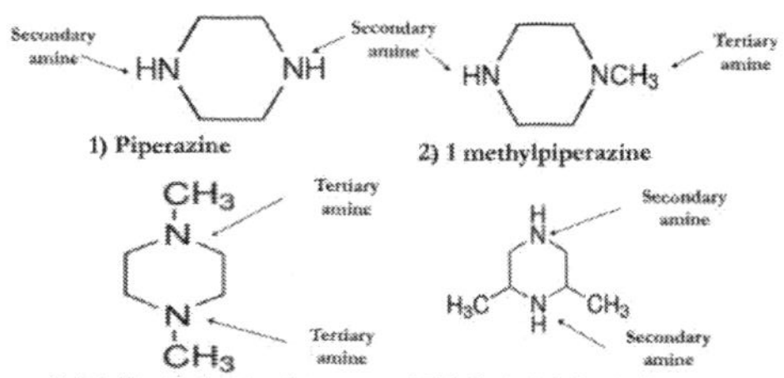

1) Piperazine
2) 1 methylpiperazine
3) 1,4 dimethylpiperazine
4) 2,6 dimethylpiperazine

Primary and Secondary amines reaction mechanism

$$CO_2 + R_1R_2NH \rightleftharpoons R_1R_2NCOOH$$

$$R_1R_2NCOO^- + H^+ \rightleftharpoons R_1R_2NCOOH$$

$$R_1R_2NH + H^+ \rightleftharpoons R_1R_2NH_2^+$$ [2]

Tertiary amine reaction mechanism

$$CO_2 + H_2O \rightleftharpoons HCO_3^- + H^+$$

$$R_1R_2R_3N + H^+ \rightleftharpoons R_1R_2R_3NH^+$$ [2]

Experimental Results

Maximum carbon dioxide (CO_2) loading capacity for unfunctionalized piperazine was found as compared to other functionalized piperazines.

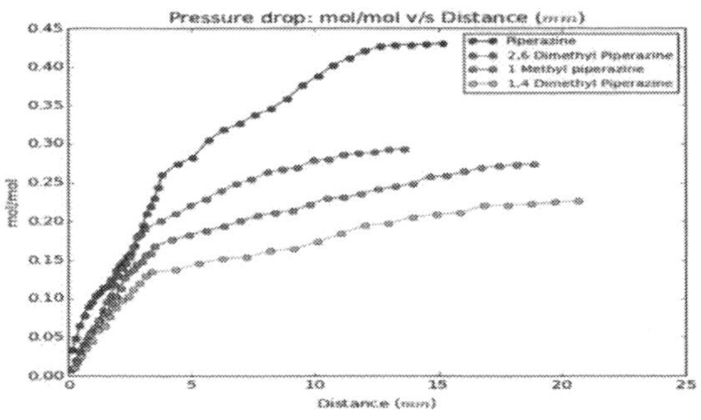

The order of CO_2 loading capacity in this experiment as a function of distance(mm):

Sr. No	Sample	CO_2 loading (mol_{CO2}/mol_{amine})
1	Piperazine	0.43
2	2,6 dimethyl piperazine	0.30
3	1 methyl piperazine	0.27
4	1,4 di methyl piperazine	0.22

Image Analysis Approach

The length of Gas Plug decreases as we move down the microfluidic reactor.

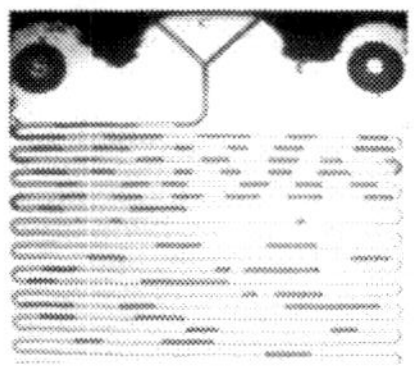

Microfluidic reactor

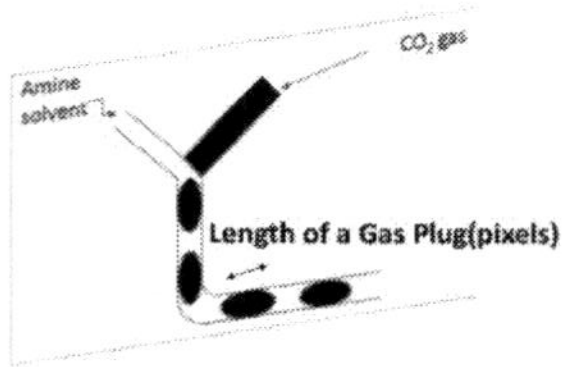

- **Calculate the length of gas plug in pixels.**
- **Pixel ⟶ millimeter (using the reference).**
- **Plug length ⟶ Volume ⟶ Concentration ⟶ mol $_{CO2}$/mol$_{amine}$**

Raman Spectroscopy

The speciation of CO_2 in the reaction product was validated using Raman Spectroscopy.

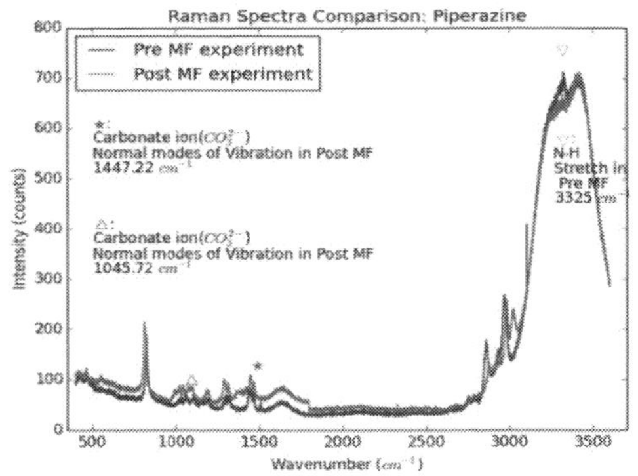

- In the pre microfluidic experiment spectrum we see the N-H stretch at 3325 cm^{-1} for secondary amine.
- In the post microfluidic experiment spectrum we observe two things:
  N-H stretch at 3325 cm^{-1} is not present.
  Two new peaks at 1045.72 cm^{-1} and 1447.22 cm^{-1} are observed which represent carbonate ion (CO_3^{2-}).

These observations validate the speciation of CO_2 in amine.

Discussions

- **Secondary amines are stronger bases than their corresponding tertiary amines [1].**
- **Unfunctionalized piperazine is the most basic amine among the four samples used in this experiment.**
- **1,4 dimethyl piperazine is the weakest base.**
- **Mass transfer coefficient of piperazine is usually two to three times greater than MEA.[3]**

References

1. Khalili, Farhad, Amr Henni, and Allan L. L. East. "P K a Values of Some Piperazines at (298, 303, 313, and 323) K." *Journal of Chemical & Engineering Data J. Chem. Eng. Data* 54.10 (2009): 2914-917. Print.
2. Carbon Dioxide Postcombustion Capture: A Novel Screening Study of the Carbon Dioxide Absorption Performance of 76 Amines Graeme Puxty, Robert Rowland, Andrew Allport, Qi Yang, Mark Bown, Robert Burns, Marcel Maeder, and Moetaz Attalla *Environmental Science & Technology* 2009 *43* (16), 6427-6433 DOI: 10.1021/es901376a.
3. Cullinane, J. T.; Rochelle, G. T. Kinetics of Carbon Dioxide Absorption into Aqueous Potassium Carbonate and Piperazine. *Ind. Eng. Chem. Res.* 2006, *45*, 2531-2545.

SAVE US ENERGY AND REDUCE WORLD POLLUTION

PRESENT SITUATION: BY-PRODUCT COKE >25% OF CARBON IN Sour CRUDE OIL

1.GREEN PETROLEUM COKE, REFINERY BY-PRODUCT, IS SHIPPED FROM US TO FAR EAST.

2.FOR EACH 100,000 TONS SHIPPED WE LOSE 28MM THERMS OF ENERGY

3.WORLD ATMOSPHERE RECEIVES

11,000 TONS OF SOX

4,000 TONS OF NOX,

385,000 TONS OF CO2

SOLUTION: REDUCE SULFUR FROM 5.5% TO 0.5% FOR US USE.

COST TO DESULFURIZE 100,000 TONS/DAY OF PETROLEUM COKE?

AMINE SCRUBBER	RAW MATERIAL:		PER TON
CLAUS UNIT	Amine + Catalyst		$ 1.00
	OXYGEN (at 8 cents/lb)		$ 7.33
(ALLOWED CAPITOL TEST)	$MM	TOTAL	$ 8.33
DEPRECIATION allowed 5yr	$30		
OVERHEAD	$2		
RAW MATERIAL	$292		
_TOTAL COST	$ 324		
CLEAN FUEL COKE	29.2MM TON/YR		

IF COST PER TON $11
CAPITAL AVAILABLE $150MM
SEE US PATENT 8,372,171, B2

IF REFINERY DEMANDS $40 PER TON FOR WET GREEN COKE THE COST GOES TO $77/TON

IF CLEAN COKE IS USE AT WELL HEAD IT HAS A VALUE OF $121 AT $2 PER MCF FOR CO2. USE MEA TO EXTRACT

DESULFURIZER OPERATES UNDER PRESSURE PERHAPS SUITABLE FOR DOWN HOLE COMPRESSER

AUTO BRAKE DRUMS NEED LOW SULFUR CARBON

ANNODES NEED LOW SULFUR COKE

CO GENERATORS CAN MAKE ULTRA PURE CO WITH LOW SULFUR COKE FOR CHEMICAL INDUSTRY

HEAT OF COMBUSTION FOR LOW SULFUR COKE IS EQUILIVANT TO $49.5 COAL

$150MM IS QUITE SUFFICIENT TO BUILD THIS PLANT

WET GREEN COKE CAN BE DESULFURIZED FOR PRODUCT IN EXCHANGE FOR HYDROGEN WITH A REFINERY

WE CANNOT SEQUESTER CO2 FROM EXPORTED COKE

MY SAE GROUP PLANS TO BUILD & TEST THE DESULFURIZER IN CONJUNCTION WITH MICROORGANISM & ETHANOL MANUFACTURE

DOE MIGHT WISH TO OBSERVE OUR TEST WHEN CAPITAL IS SECURED

SAE HAS A P.E. ChE, PhD BIOLOGIST, WOOD SPECIALIST, COKE MANAGER, CONSTUCTION OWNER, ALGAE SPECIALIST, VAST BUSINESS KNOWLEDGE

CMTC-439585-MS

The Road to Energy Autarky in Municipal Water Resource Recovery Facilities

Paul Wood P.E., Lockwood,Andrews & Newnam, Inc.

This paper was prepared for presentation at the Carbon Management Technology Conference held in Sugarland, Texas, USA, 17–19 November 2015.

Abstract

Wastewater treatment has traditionally been one of the largest power users in a municipality's energy budget. New technologies and operating methods have allowed some water resource recovery facilities (WWRF) to actually achieve energy autarky (the ability to recover and generate as much energy as they consume).

General wastewater treatment strategies used around the world involve the growth of bacteria to consume carbon and generate biomass or methane. Treatment is typically divided into physical processes (clarification, etc.), and biological processes (aerobic, anoxic and anaerobic). Typically, 35% of the carbon in raw sewage can be separated with primary clarification and digested anaerobically. The remaining carbon is treated by anaerobic, aerobic, and anoxic biological processes which converts carbon to biosolids and CO_2, and ammonia to nitrate and nitrogen. The resulting biosolids are separated from the treated wastewater through the secondary clarification process and are added to the primary solids prior to digestion.

The anaerobically digested primary and secondary solids produce a biogas, which contains approximately 60% to 70% methane. The generated biogas can heat anaerobic digesters, dry the dewatered biosolids (reducing transportation costs), and power engines to generate electricity.

Large quantities of air must be supplied to allow the secondary process to work correctly. Typically, it takes 1.2 pounds of oxygen for every pound of biochemical oxygen demand (BOD) and 4.3 pounds of oxygen for every pound of ammonia, including organic ammonia, that is present in the wastewater entering the secondary treatment process. Inefficiencies of mass transfer require more oxygen to be supplied. Because air is only 21% oxygen, large quantities of air must be compressed and bubbled through the wastewater to provide the needed oxygen. Because of the inefficiencies of the oxygen transfer, energy consumption estimates for secondary wastewater treatment can range from 0.4 to 2.6 kWh per lb BOD removed[1].

Improvements in gas transfer methods continue to be made. A shift has slowly been taking place from the use of course bubble aeration systems to fine bubble and micro-bubble aeration systems. The increase in surface area due to smaller bubbles increases the efficiency of gas transfer. The use of deeper tanks for aeration also improves transfer efficiency since the pressure at the deeper depths allows more oxygen to be dissolved into the liquid phase. There is, however, a tradeoff to deeper tank depths. Increased depths equate to higher discharge pressures for air compression equipment, which means more power must be used to provide this increased compression. Typical depths of aeration have increased from about 12 feet to between 20 and 30 feet. Even with these improvements, a typical wastewater plant remains a net consumer of electrical power.

Several new developments in wastewater treatment have emerged in recent decades that have converged to allow the power required for wastewater treatment to be substantially reduced, and the capacity for biogas generation to increase greatly. One of these developments was the discovery of anaerobic bacteria that consumes ammonia as its food source, and uses nitrite as an electron acceptor[2]. These bacteria, called anammox, were first used at WWRFs in side-stream processes, to reduce the total ammonia returning to the plant from sludge handling processes.

Research into the mechanism of the nitrification/denitrification process led to a discovery that the traditional process path of nitrification and denitrification had a shortcut path, referred to as nitrite shunt. The nitrite shunt pathway reduces the total amount of oxygen needed to convert ammonia to nitrogen. Concerns with energy use for wastewater treatment combined with goals for nutrient limits, led to the realization that these anammox bacteria and the nitrite shunt pathway could be utilized in the mainstream treatment process, and an ideal configuration could be established whereby:

- Additional carbon can be removed in the primary treatment process though either chemical co-precipitation or through biological adsorption in a High Rate Activated Sludge (HRAS) process. This diversion of carbon away from the mainstream treatment process provides additional feed to the anaerobic digesters
- The additional feed to the anaerobic digesters allows greater biogas production
- Improvements in the digestion process, along with co-digestion of highly digestible organic feed stock (i.e. Fats, Oils and Greases or FOG) allows yet further increases in biogas production
- The decrease in carbon to the secondary treatment process (through improved primary treatment or A-Stage/HRAS treatment) reduces the amount of oxygen that must be supplied for secondary treatment
- Ammonia treatment by the nitrite shunt pathway and anammox bacteria further reduces the amount of air required for secondary treatment[3]

The convergence of these process changes and discoveries has allowed several plants to achieve energy autarky. Even if not fully implemented to achieve energy independence, the partial implementation of some of these changes can allow significant energy savings.

Introduction

Carbon management is not just about CO_2 sequestration. In wastewater treatment, raw sewage includes carbon in the form of particulates, colloids, and solubles that must be removed or mineralized as part of treatment before discharge to a receiving stream. Historically, two types of treatment have been used to accomplish this removal and mineralization: aerobic treatment and anaerobic treatment.

In aerobic treatment, oxygen is provided to the bacteria, which utilize the available carbon as a food source. As they utilize the carbon and oxygen, they produce CO_2, and reproduce. In anaerobic treatment, four reactions are typically desired: hydrolysis, acidogenesis, acetogenisis, and methanogenesis. Aerobic treatment produces a large quantity of biosolids. Anaerobic treatment for mainstream treatment is not common in non-tropical latitudes. However, it is extensively used for sludge treatment.

Typically, settleable solids are removed from the main raw sewage stream and diverted to anaerobic digesters. This settling process is known as primary treatment. Approximately 35% of the incoming carbon can be removed from the main stream by primary treatment. This diversion of carbon by primary treatment therefore reduces the amount of energy that needs to be provided by aerobic, or secondary treatment, by the same 35%. In addition to the primary solids that are diverted to the anaerobic digesters, waste activated sludge (WAS) is also sent to the anaerobic digesters. WAS is essentially the biomass growth that has occurred during the consumption of the remaining carbon in the secondary treatment process.

Aerobic treatment requires significant energy input to provide oxygen from air. Anaerobic treatment only requires energy for mixing and heat maintenance, which is substantially less than that required for aerobic treatment. A benefit of anaerobic digestion is the production of biogas. Biogas contains between 60 to 70% methane by volume. This methane can be used to power engines and generated electricity. Carbon that has not been aerobically treated is more easily digested. Therefore, if more carbon can be diverted to sludge treatment, prior to mainstream secondary treatment, more methane can be produced, and less energy for aeration is required.

Carbon Diversion

Several methods are available which will allow this carbon diversion; these include chemically enhanced primary treatment, and bioflocculation optimized for carbon concentration and separation with minimal energy use. Both processes have the ability to remove approximately 50-65% of the incoming carbon, allowing it to be diverted to the anaerobic digesters in the sludge treatment train. The A/B process is a two-stage process that uses bioflocculation to divert carbon. By diverting carbon from the mainstream process, two important goals are achieved: energy required for secondary treatment is greatly reduced, and generated solids, which must be transported (disposed of or processed for beneficial reuse) are minimized.

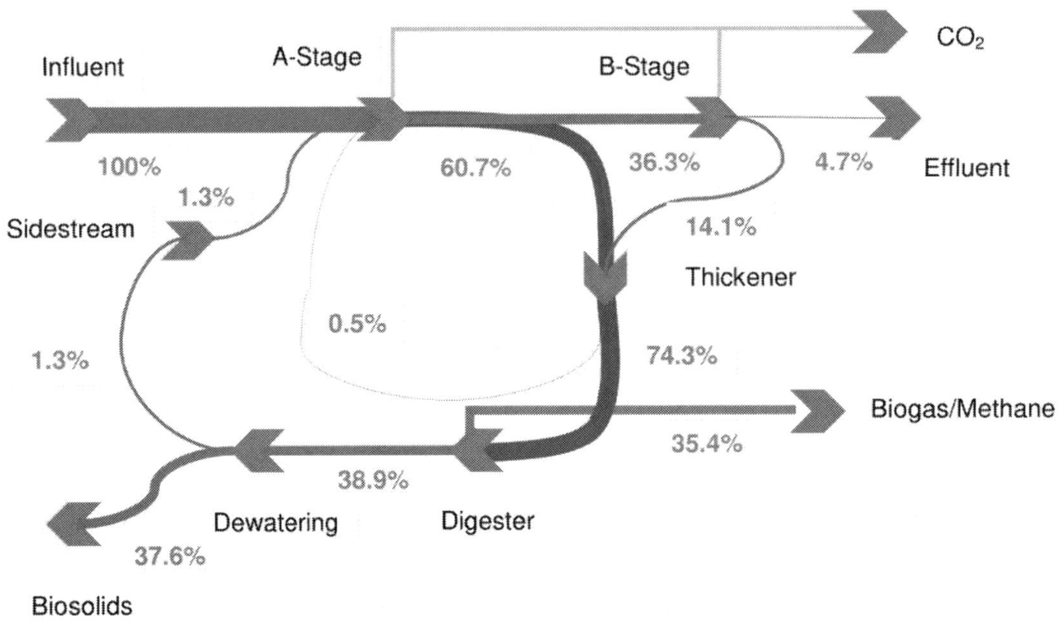

Figure 1. Simulated COD-balance of the Strass WWTP [4]

By concentrating and diverting a higher percentage of carbon to anaerobic digesters from the raw influent stream, methane production is increased. Because the anaerobic process converts carbon to methane, and carbon which has not be biologically converted to biomass is easier to digest, the total quantity of solids generated by the waste treatment process is reduced. When the solids from the anaerobic digestion process are dewatered, a recycle stream high in ammonia is produced. The ammonia rich recycle stream presents an additional load: either it must be pretreated prior to returning to the main train (sidestream treatment), or it must be treated in the main train by the provision of additional air.

Other Associated Benefits of Carbon Diversion

When the carbon content of the wastewater being treated in a secondary aerobic treatment process is reduced, the secondary process can be better optimized for nitrogen removal. Normally, when carbon diversion is not practiced, nitrogen removal occurs by the nitrification/denitrification process. In the nitrification process, ammonia is oxidized to nitrite, and nitrite is oxidized to nitrate in sequential steps. In denitrification, nitrate is reduced to nitrite, and nitrite is reduced to nitrogen gas.

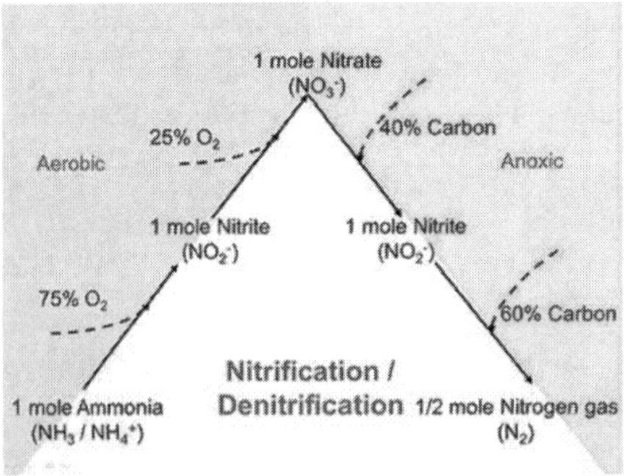

Figure 2. The Nitrification/Denitrification Pathway

Nitrification is an aerobic process and denitrification is an anoxic process. In the anoxic process, heterotrophic organisms utilize nitrate and nitrite as an oxygen source for respiration.

In the nitrification process, there are two sets of bacteria at work: ammonia oxidizing bacteria (AOBs) and nitrite oxidizing bacteria (NOBs). Normally NOBs grow faster than AOBs and therefore nitrite concentration is minimized.

By controlling a few of the operational parameters in the secondary treatment unit (principally dissolved oxygen (DO), sludge retention time (SRT), and ammonia and NOx concentrations), an enhanced environment for AOBs can be created, and a shortcut path for nitrification can be obtained.

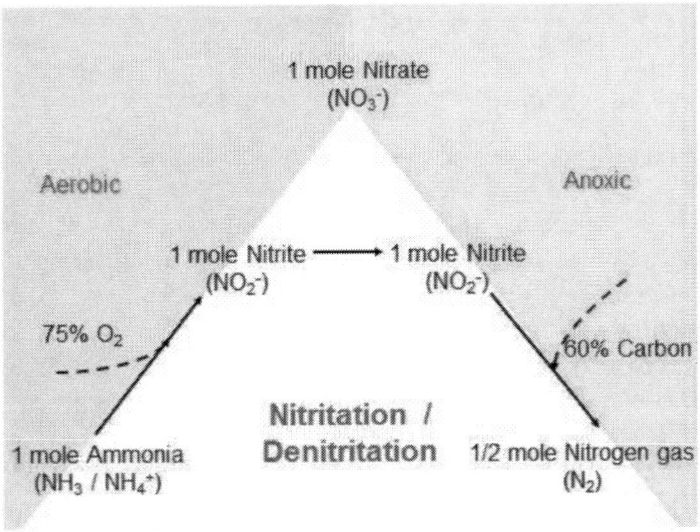

Figure 3. The Shortcut Nitritation/Denitritation Pathway

By wasting at a high rate and keeping the SRT low, NOBs tend to preferentially wash out due to their slow growth rate. In addition, if the concentration of NO_x is maintained approximately equal to the ammonia concentration by limiting the dissolved oxygen concentration, an enhanced environment for ammonia oxidizing bacteria is formed, the complete oxidation of ammonia is minimized, and an increased quantity of nitrite is available for reduction. This shortcut process is called Nitritation/Denitritation. The Nitritation/Denitritation process uses less oxygen and requires less carbon than traditional Nitrification/Denitrification. The lower dissolved oxygen requirements of this optimized secondary treatment unit also increase the oxygen transfer efficiency by increasing the driving force of diffusion for the oxygen molecules traveling to a lower concentration environment than in a normal secondary treatment unit.

Side Stream Treatment

As discussed above, when dewatering anaerobically treated solids an ammonia rich stream is generated. This side stream can be pretreated by either of the process noted above (nitrification/denitrification or shortcut nitritation denitritation). However a third option exists which requires even less energy input.

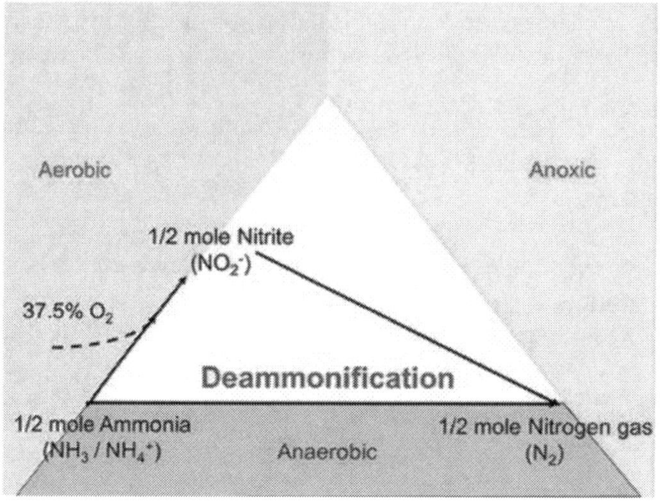

Figure 4. The Deammonification Pathway Using Anammox Bacteria

Anaerobic bacteria, which can reduce ammonia without carbon exist, and are classified as anammox. Processes for side stream treatment of high ammonia wastewaters utilizing these bacteria have been developed. Anammox processes require half of the incoming ammonia be oxidized by AOBs to nitrite, and then the anammox bacteria use the nitrite directly to convert the remaining ammonia to nitrogen gas. Most processes being developed rely on sequenced batch reactor (SBR) technology, moving bed bioreactor (MBBR) technology, or granular sludge technology. All these processes rely on providing an environment in which AOBs can thrive, NOBs are suppressed, and sludge residence time is decoupled from hydraulic residence time. Anammox grows so slowly that methods must be developed whereby it is preferentially retained and other bacteria can be wasted. Fortunately, anammox bacteria are slightly denser than other bacteria. Separation has been accomplished by hydrocyclones and fine sieves. Anammox can also be retained as a granule under the correct conditions. Anammox also colonizes the structured media used in MBBR configurations. Unlike most other anaerobic bacteria, anammox bacteria can survive in low DO environments. Because of their slow growth and independence from carbon, sludge production from deammonification process is extremely low.

Main Stream Treatment

Mainstream treatment involves the reduction of remaining carbon, and ammonia from the wastewater. Phosphorus must sometimes also be reduced. Biological pathways to accomplish all of these exist. If a large portion of the raw sewage incoming carbon is diverted to anaerobic treatment, considerations for ammonia removal become more important if energy optimization is a priority. Nitrification will generally occur in warm climates if enough oxygen is provided. Total nitrogen limits require controlled aeration so that denitrification can occur. As a best practice, denitrification should be utilized even if total nitrogen limits are not imposed because it reduces the total energy requirements. Nitrate and nitrite can be used as an oxygen source by heterotrophic bacteria, reducing the total amount of energy required for aeration.

Improved energy performance can be accomplished by designing for shortcut nitrification/denitrification. This process requires greater operational attention and instrumentation. Solids retention time must be kept low so that NOBs are preferentially washed out, but the AOBs are retained. Maintaining some ammonia residual levels is inhibitory to NOBs, but encourages the growth of AOBs. Transient anoxia also encourages NOB out selection, but the transition from aerobic to anoxic conditions must be rapid, which can be enhanced by organic addition when aeration is turned off.

Recent projects have incorporated the deammonification process in mainstream treatment. This is slightly more complex than the shortcut processes. Typically, the waste activated sludge (WAS) stream is run though a hydrocyclones or vibrating sieves to separate and reclaim the anammox bacteria. Additionally, the mainstream treatment unit can be continuously seeded with anammox bacteria if a plant also has a side stream process that utilizes a deammonification process.

Conclusions

Carbon management in wastewater treatment can be viewed from several perspectives. Carbon diversion allows a wastewater treatment plant to operate more economically by shifting organic load from energy intensive aerobic treatment processes to less energy intensive anaerobic processes. Increased biogas production from anaerobic digestion, due to enhanced carbon diversion, allows a plant to increase production of a sustainable fuel source, and reduce the plants total carbon footprint. Utilization of advanced ammonia treatment processes can further reduce both energy requirements and total carbon footprint. Wastewater plants that have aggressively undertaken all these steps as an integrated program have already achieved energy autarky.[3,5]

References

1. Shahid Chaudhry, California Energy Commission, Dec. 16, 2008: Energy Management Opportunities at Wastewater Facilities, Innovative Energy Management Workshop –Sacramento
2. Mulder et al., FEMS Microbiol. Ecol, 1995: Anaerobic Ammonium Oxidation Discovered in a Denitrifying Fluidized Bed Reactor
3. Shortcut Nitrogen Removal Task Force, WEF/WERF joint publication, 2015: Shortcut Nitrogen Removal – Nitrite Shunt and Deammonification
4. Wett et al, IWA, Jan. 2007: Energy Self-Sufficiency as a Feasible Concept for Wastewater Treatment Systems, Leading Edge Technology Conference
5. WERF Progress, Spring 2010: Independent Thinking – How One Treatment Facility Achieved Energy Independence through Energy Optimization

The Road to Energy Autarky in Municipal Water Resource Recovery Facilities

Benefits of Carbon Diversion

- ✓ Energy autonomy
- ✓ Conversion to biogas/methane
- ✓ Biosolids reduction
- ✓ Reduced energy costs
- ✓ Ammonia treatment

Simulated COD-Balance

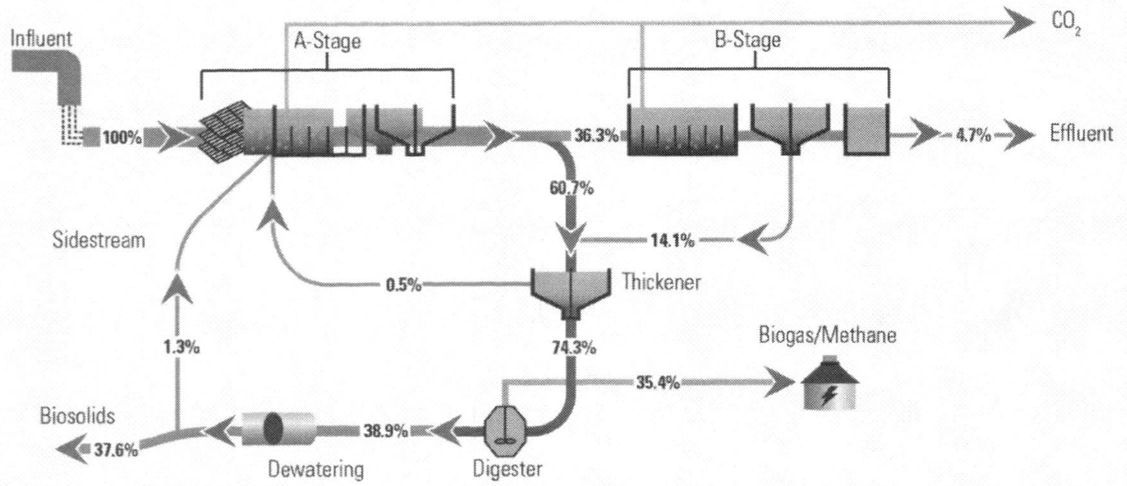

Benefits of Alternate Ammonia Reduction Pathways

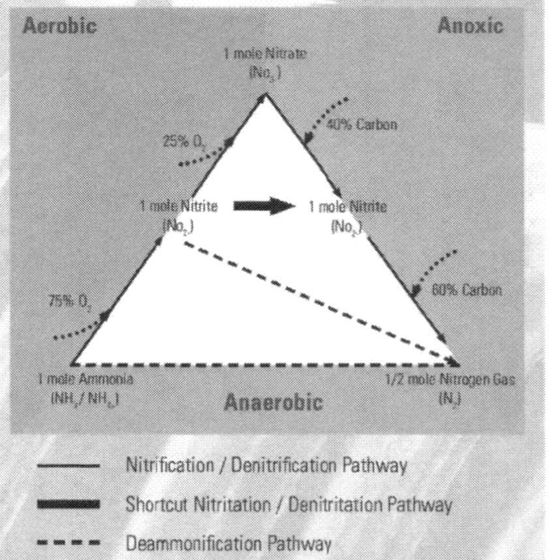

LAN is currently implementing this process on the 6-mgd, South Conroe Wastewater Treatment Plant in Texas.

Lockwood, Andrews & Newnam, Inc.
A LEO A DALY COMPANY

CMTC-440075-MS

SPE-139516 "Life beyond 80 – A Look at Conventional WAG Recovery beyond 80% HCPV Injection in CO2 Tertiary Floods"

David Merchant, Merchant Consulting

This paper was prepared for presentation at the Carbon Management Technology Conference held in Sugarland, Texas, USA, 17–19 November 2015.

Abstract

During the past 43 years, CO_2 flood technology for Enhanced Oil Recovery projects evolved from a partially understood process filled with uncertainties to a process based on proven technology and experience. Many questions involved with CO_2 flooding have been thoroughly analyzed and answered. This knowledge is currently being used by a limited number of companies that actually know how to design, implement, and manage a CO_2 flood for long term profit. The purpose of this report is to help disseminate this knowledge to operating companies interested in EOR flooding or to CO_2 Sequestration Communities interested in storing CO_2 in EOR projects.

In 2010, Merchant Consulting published SPE 139516 "Life beyond 80 – A look at Conventional WAG Recovery beyond 80% HCPV Injection in CO_2 Tertiary Floods". The primary objective of the report was to target "Conventional WAG Techniques" which have been used in over 90% of all the Enhanced Oil Recovery projects implemented in the Permian Basin in Texas, Colorado, Oklahoma, and Wyoming. The paper presents answers to the question "What is life after 80% HCPV Injected?" And "What effect does life after 80% HCPV have on Tertiary Oil Recovery, CO_2 Utilization and CO_2 Retention in different producing formations?" Results of this study show Tertiary Oil Recovery can be as high as 26% OOIP when slug sizes exceed 190% HCPV injected.

Carbon Sequestration Options for EOR:

The purpose of this report is to provide the CO_2 Sequestration Community an update to the original SPE 139516 paper presented in 2010 at the SPE International Conference on CO_2 Capture, Storage, and Utilization held in New Orleans, Louisiana. The original paper covered only 5 of the 10 methods currently used today to recover tertiary oil with the injection of CO_2. This report will provide insight to all 10 methods used today in EOR operations.

Introduction – CO_2 Flooding

CO_2 Flood History

The Permian Basin has had over 85+ years of oil production history and produced over 29 billion barrels of oil and represents the 3rd largest petroleum producing area in the United States after the Gulf of Mexico and Alaska. Today, it produces 20% of the U.S. oil production excluding Alaska. In 1972, CO_2 injection was introduced, first in the Sacroc Field. Today, over 1.1 billion barrels have been produced with CO_2 and accounts for over 237,000 BOPD or 15% of all Permian Basin oil production from over 15,000 production wells and 12,000 CO_2 injection wells.

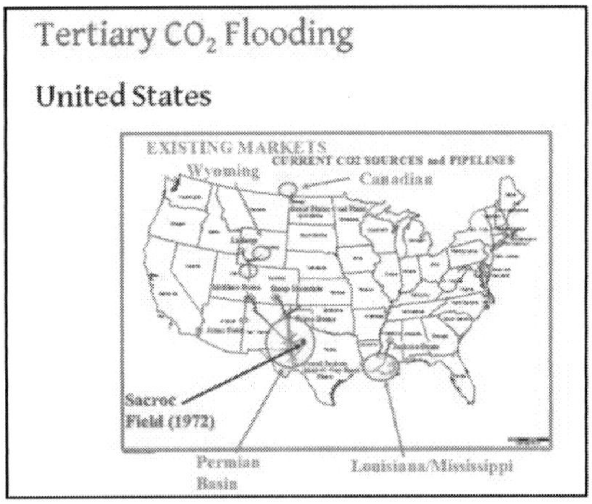

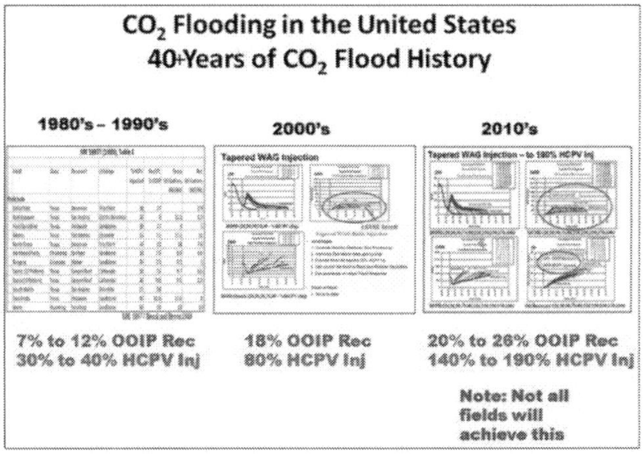

What makes CO_2 Tertiary Oil Recovery Work?

The reason why CO_2 works is simple. The CO_2 acts as a solvent when injected into the reservoir and swells the Residual Oil (oil left after water flood) and reduces its viscosity. This causes the Residual Oil to swell and become mobile and be produced.

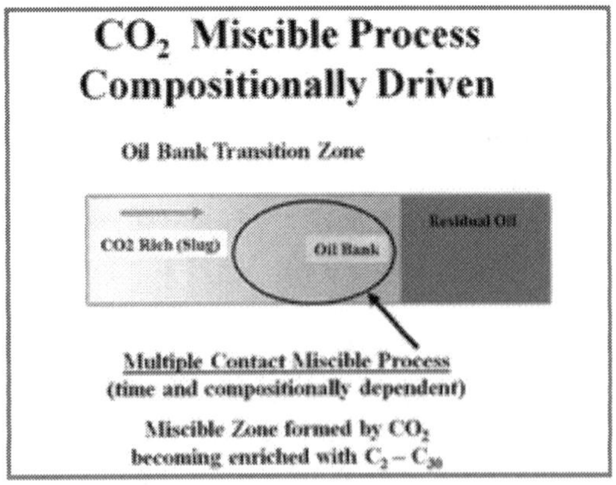

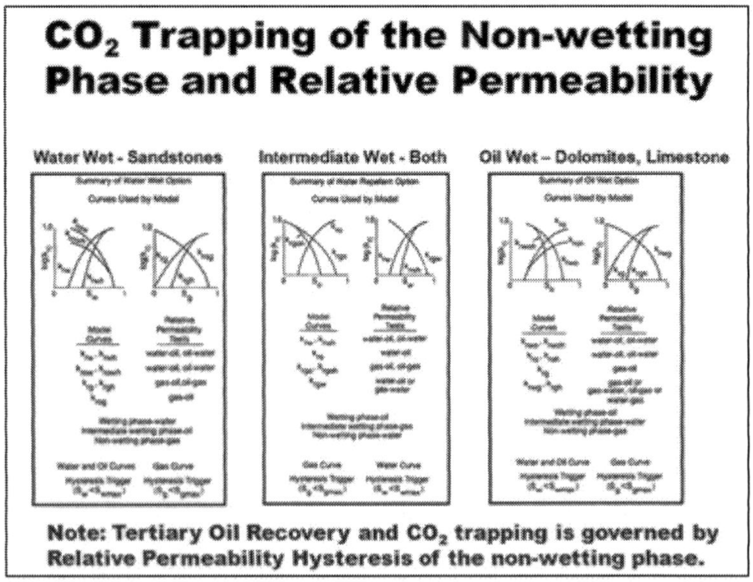

The amount of Tertiary Oil Recovered and the amount of CO_2 Trapped or Sequestered is dependent on the rock type. In addition to mobilizing residual oil saturation, a portion of the CO_2 becomes trapped. In the Petrophysical world, this is known as Relative Permeability Hysteresis or Trapping of the Non-wetting Phase. In this case CO_2 is the non-wetting phase. As demonstrated above, this phenomenon occurs in Water Wet Rocks (Sandstones), Intermediate Wet Rocks (Sandstones or Carbonates), and in Oil Wet Rocks (Dolomites and Limestone formations).

CO_2 Tertiary Recovery Methods

Introduction

CO_2 Tertiary Recovery Processes to date encompass "Ten" Recovery Methods. Four of these methods are used in the Permian Basin. Conventional WAG Techniques have been used in over 90+% of the CO_2 floods around the World. The Seminole field example presented in this report is an example of a field operated under Conventional WAG that targets both the main pay and ROZ zones. The Yates field example has operated under both Gravity Drainage and Double Displacement recovery methods.

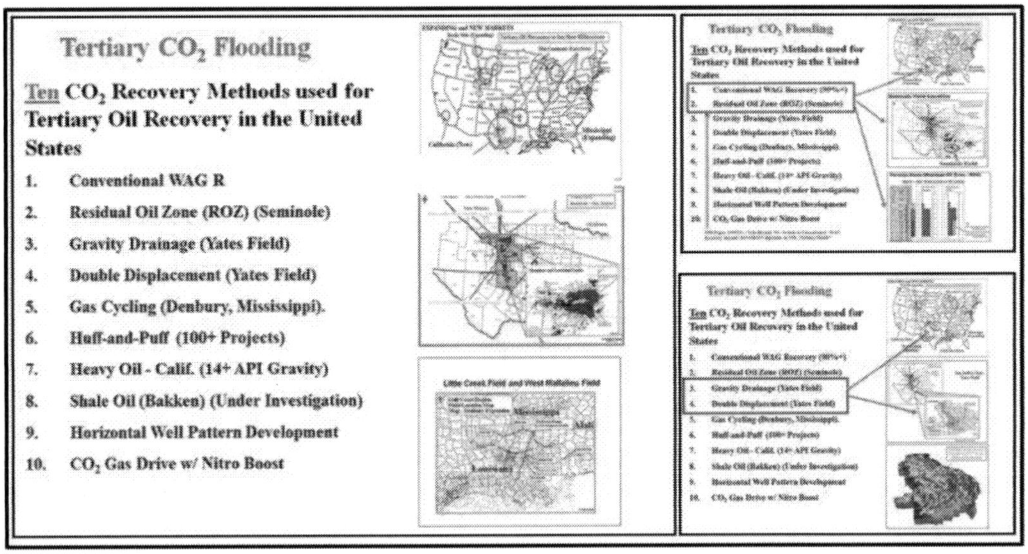

Additional information can be found: SPE Paper 139516 – "Life Beyond 80 – A look at Conventional WAG Recovery beyond 80% HCPV Injection in CO_2 Tertiary Floods"

Conventional WAG Techniques

Conventional WAG Techniques have been used in over 90% of the CO_2 Floods implemented today. They are designed to work well in reservoirs that are developed on pattern spacing. These include: Five-spot, Nine-spot, and Chickenwire pattern development. The objective is to "Level Load" produced gas production to a CO_2Recovery Plant inlet gas rate, which extends field life and recovers more tertiary oil.

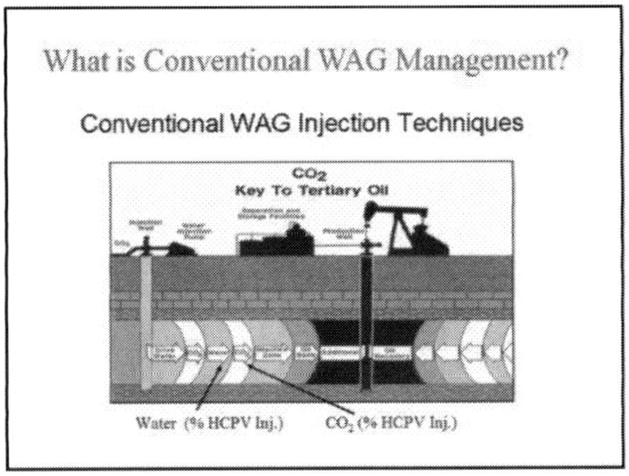

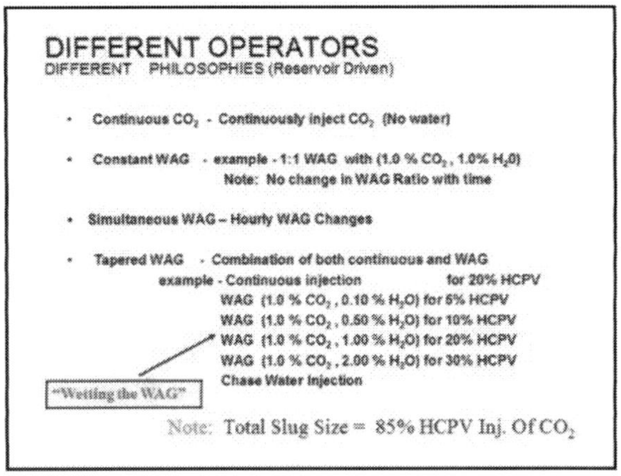

Today, most operators have adopted the Tapered WAG approach to optimize WAG management.

Conventional WAG Management – Seminole San Andres Unit

The best example of CO_2 Tertiary Recovery is the Seminole San Andres Unit in the Permian Basin. The field has undergone Primary, Secondary, CO_2 Tertiary operations in the Main Pay, and CO_2 Tertiary operations in the Residual Oil Zone (ROZ).

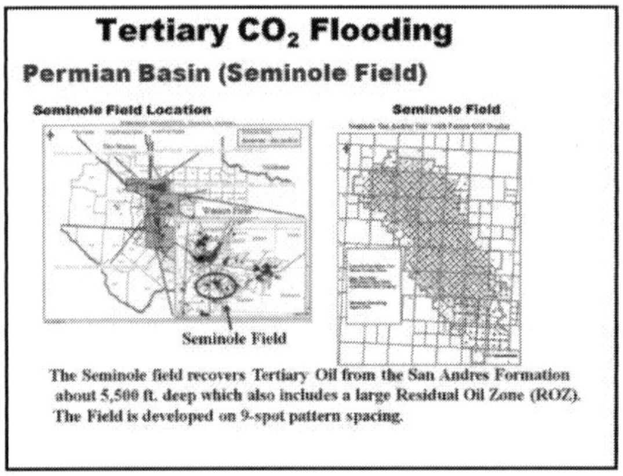

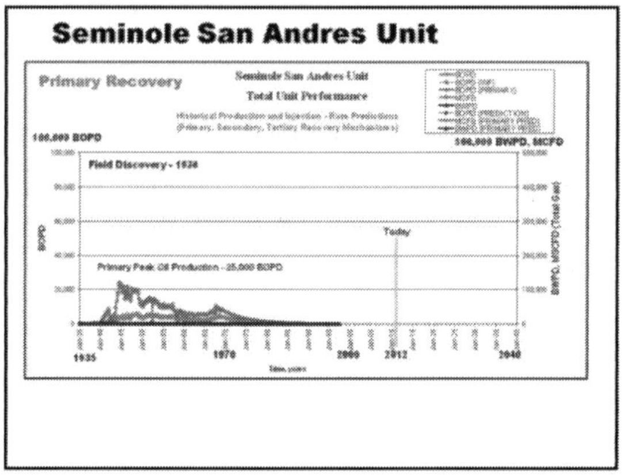

The Seminole field was discovered in 1936 with water injection operations initiated in 1971. Under Primary Recovery the field would have only recovered 12.8% of its Original Oil-in-Place (OOIP). With Secondary Water flood operations, the field would have recovered 42.7% of its Original Oil-in-Place.

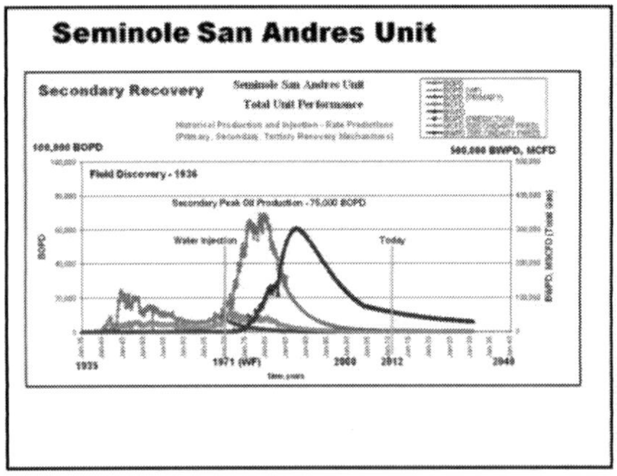

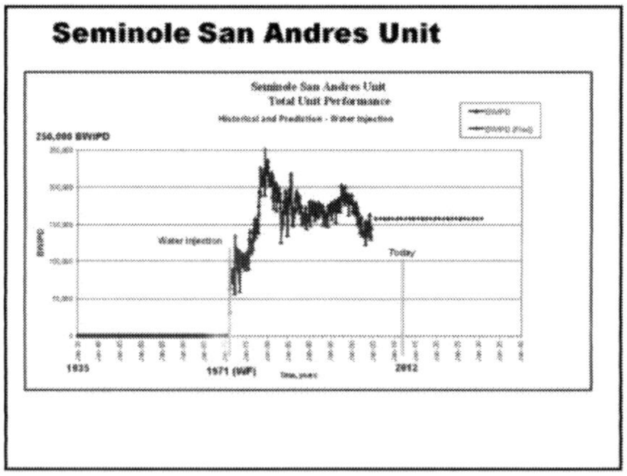

Tertiary CO_2 Injection operations into the <u>Main Pay</u> Zone was initiated in 1983. Under CO_2 Injection the field is expected to recover an additional 24% of its Original Oil-in-Place (OOIP).

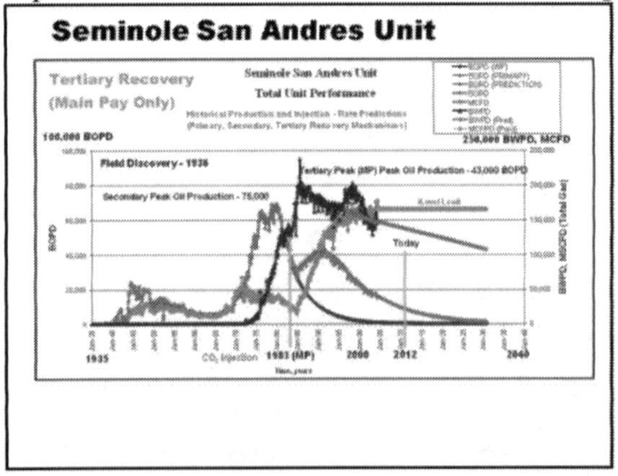

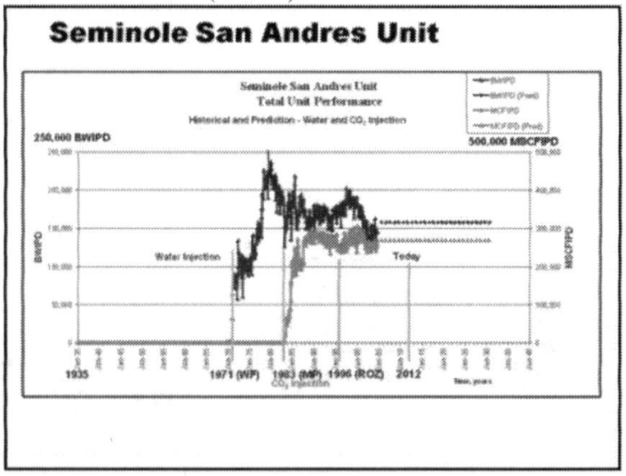

Seminole San Andres Unit – (Main Pay Oil Zone Example)

Primary, Secondary, and CO_2 Main Pay Oil Recovery as a percent of Original Oil-in-Place (OOIP) are shown below. Under Primary Operations, the field would have recovered 12.8% of its OOIP before abandoning operations. Water flood operations would have increased oil recovery to 42.7% OOIP (Primary plus Secondary). Tertiary operations with CO_2 in the Main Pay zone should increase oil production to 24% OOIP. Total Primary+Secondary+Tertiary (MP) = 66.6% OOIP.

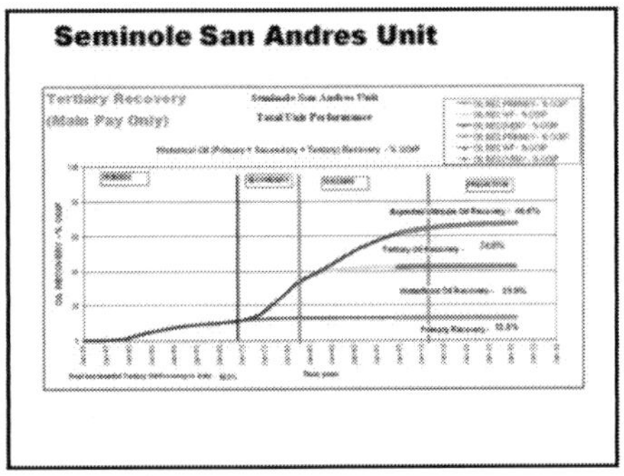

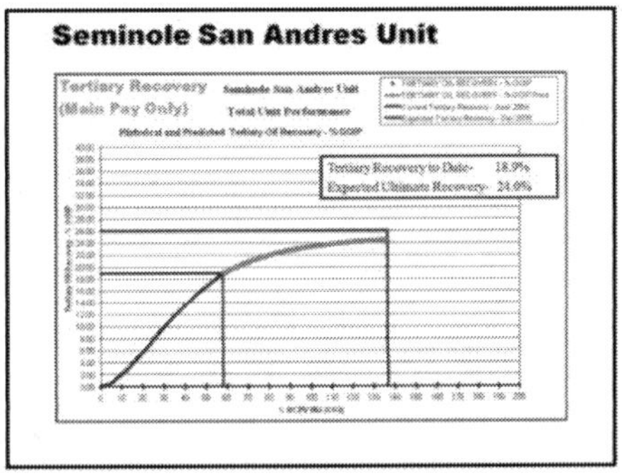

Seminole San Andres Unit – (Residual Oil Zone (ROZ) Example)

A Residual Oil Zone (ROZ) is created when oil within the original oil column migrates away from an oil field over geologic time creating a ROZ interval. In addition to the Main Pay Zone, the Seminole field contains a very large Residual Oil Zone (ROZ). The size of the ROZ is about the same size at the residual oil remaining in the Main Pay. CO_2 Injection commenced in 1996 into the ROZ. Injection into the ROZ will extend field life beyond the 2050's.

The Residual Oil Zone is responding well to CO_2 Injection. Since the ROZ reserves have never been tied to any Basin Study Estimates, these reserves are "NEW" bookable reserves. Tertiary Performance is expected to be similar to that of the Main Pay, but depends on the quality of the ROZ pay section.

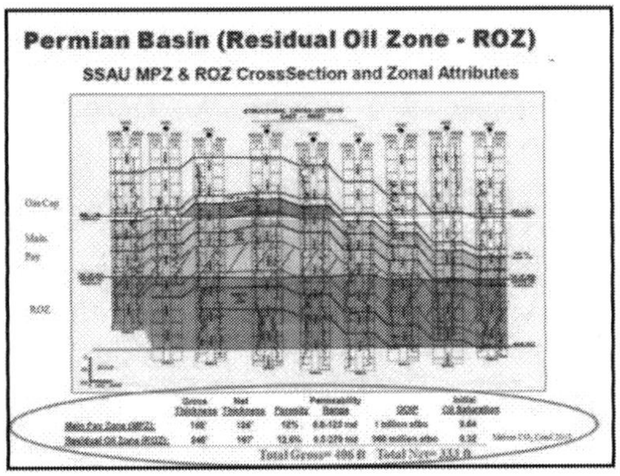

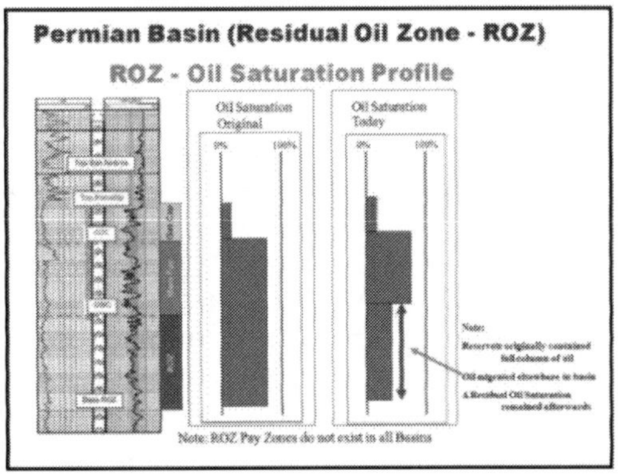

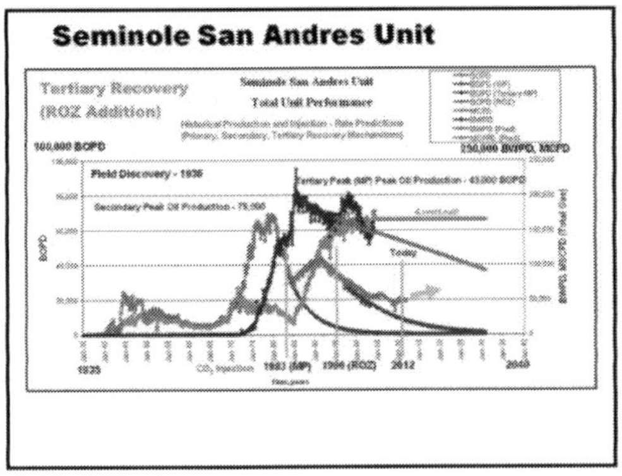

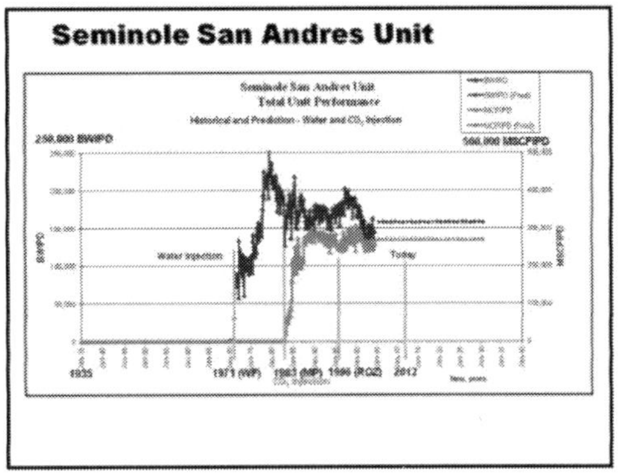

Gravity Drainage and Double Displacement Techniques

Gravity Drainage and Double Displacement

Gravity Drainage and Double Displacement CO_2 Flood Techniques work in reservoirs that have structure. The Hc-gas, Flue Gas, Nitrogen, or CO_2 is usually injected at the top of the reservoir and oil is produced at the bottom. The Yates field in the Permian Basin is an example where CO_2 has used both Gravity Drainage and Double Displacement Recovery Mechanisms to recover tertiary oil in this heavily fractured and karsted reservoir.

Yates Field (Gravity Drainage and Double Displacement Example)

The Yates field has undergone a long history of Gravity Drainage, Double Displacement, and Contact Stabilization operations to recover oil. This included the injection of Flue Gas between 1976 and 1986, 100% CO_2 from 1986 to 1994, Nitrogen Injection from 1994 to 2003, and from 2003 a continuation back to CO_2 injection.

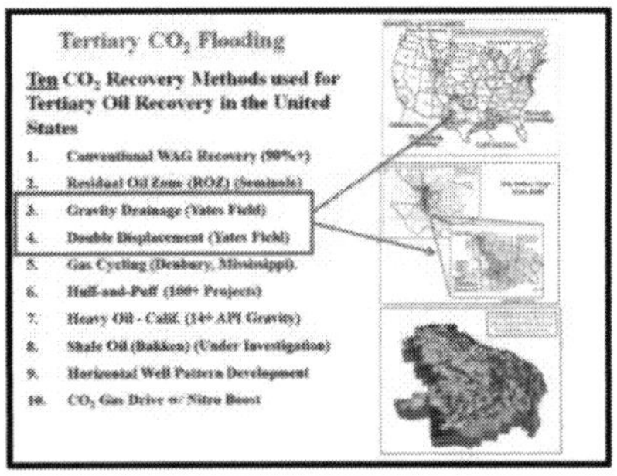

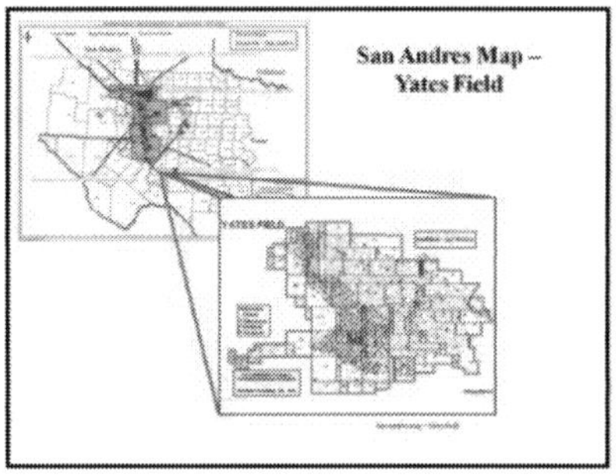

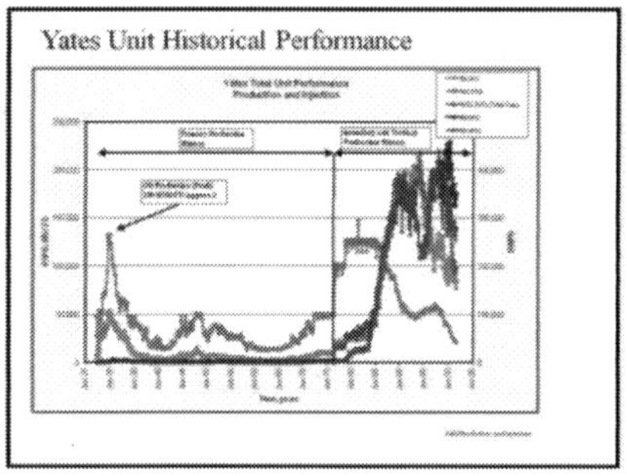

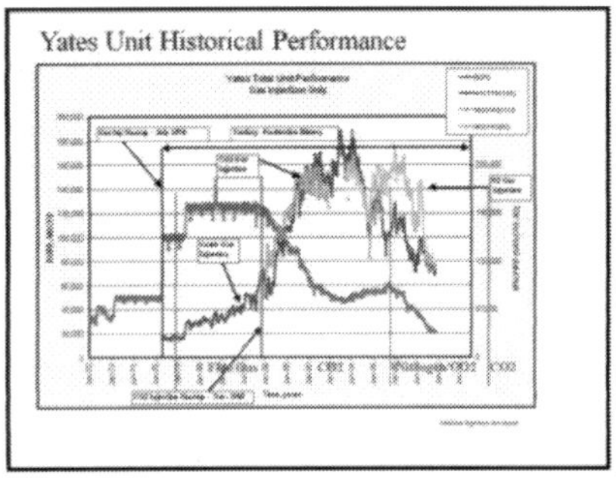

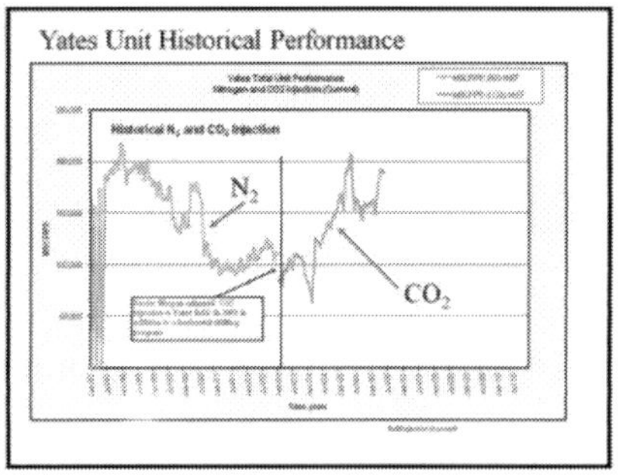

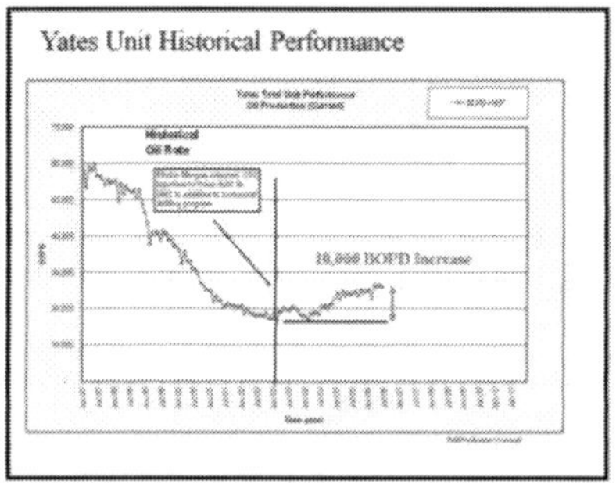

The Yates field is a highly fractured San Andres Dolomite formation. Gravity Drainage in a fractured system involves movement of the CO_2 from the fracture to the matrix displacing oil from the matrix back into the fracture system. This "Counter-flow" movement provides the driving force behind recovery in fractured formations. The degree of oil recovery depends on the permeability and relative permeability of the matrix and the miscible interaction between the injected gas and the residual oil saturation. Since CO_2 is more miscible to residual oil than N_2, the CO_2 has the capability of recovering more oil than with nitrogen injection.

This was proven in 2003 when Kinder Morgan acquired the Yates field from Marathon Oil Company and replaced the nitrogen (1994+) that was being injected back to 100% CO_2 injection (1986-1993). The results are shown above. The field experienced a 10,000 BOPD increase in oil rate and continues to be strong today.

The primary force behind "Gravity Drainage" is Gravity. Today, the field still produces over 20,000 BOPD. The CO_2 provides "Mobilization" of the remaining tertiary oil in the gas cap. Horizontal producing wells have been drilled in the 30 foot oil column at the bottom to "Capture" the remaining oil as it drains from the gas cap.

Gas Cycling CO_2 Project Techniques

Gas Cycling CO_2 projects are a common flooding technique used in Mississippi and Louisiana CO_2 flood projects by Denbury Resources. The objective is to inject continuous CO_2 through the reservoir and recycle CO_2 with many pore volumes of injection. The 17% OOIP tertiary recovery predicted by Denbury will be achieved with 5 pore volumes of re-cycled CO_2.

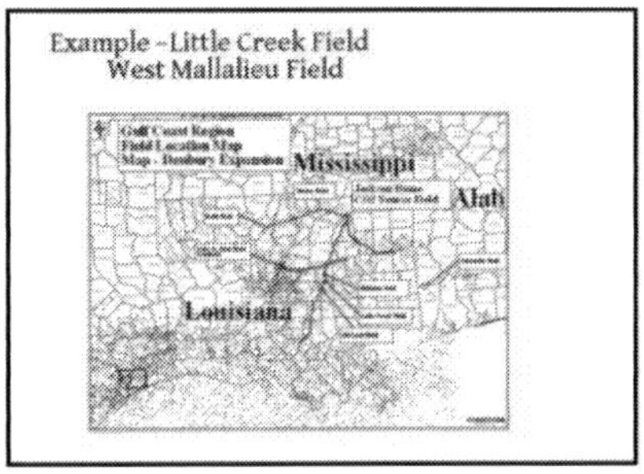

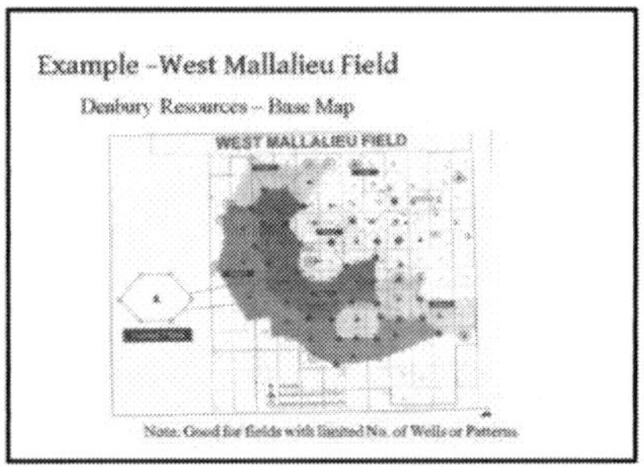

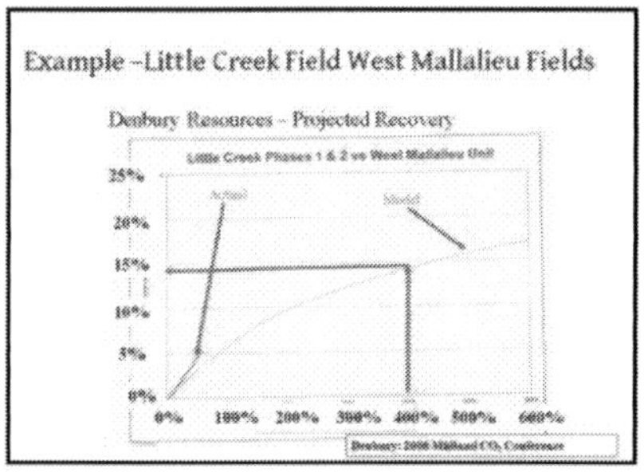

This technology works well in sandstone reservoirs that have a high injection flow capacity. Under this type of operating philosophy, pumping units and submersibles are removed from the production wells which are allowed to flow on their own due to the lifting capacity of the CO_2. In this case, these savings offset the additional costs to compress and re-cycle CO_2 through the reservoir.

Huff-n-Puff CO_2 Single Well Techniques

There have been over 100+ Single Well CO_2 Injection projects implemented since 1972 in most regions of the United States. Most of these projects were implemented as pilot projects to understanding reservoir parameters such as injection well injectivity and tertiary oil recovery response. There have been reported commercial projects.

The technique involves injecting CO_2 into individual wells, allow a certain soak time for the CO_2 to work, and then back-producing the well to recover additional tertiary oil. The objective is strictly for EOR purposes and little CO_2 is stored.

Heavy Oil (14+ API Gravity) CO_2 Projects

Heavy Oil (14+ API) CO_2 Projects

In California during the 1980's, three CO_2 Pilots were conducted to test the feasibility of recovering tertiary oil from the Wilmington field in Los Angeles, California. The pilots proved tertiary oil could be recovered with some wells increasing rate from 30 BOPD to over 300 BOPD without severe CO_2 breakthrough. With a severe oil price drop in 1986, all of these projects were abandoned. However, the pilots did prove the CO_2 recovery process works well in moderately heavy oils.

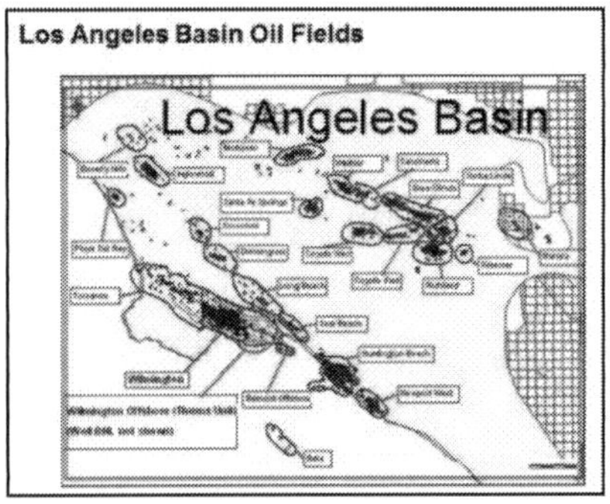

Wilmington CO_2 Pilot Results

1. Pilots were Demonstration Projects (Not Oil in the Tank Pilots)
2. Wilmington Oil – 10 fold decrease in Viscosity (300 cp to 30 cp)
3. Wilmington Oil – 1.15 fold increase in Oil Swelling
4. Wilmington Oil – Immiscible with 85% CO_2 and 15% N_2
5. Wilmington Oil Response – Single Well Response 30 BOPD to 300 BOPD

Shale Oil (Bakken) CO_2 Injection Techniques

Shale Oil (Bakken) CO_2 Projects

Shale Oil reservoirs, such as the Bakken in the Williston basin are extremely tight. Horizontal wells with extensive fracking are the only way to recover oil. Single well recoveries under primary recovery are typically in the range of 4% to 6% OOIP. Oil response rate is high initially but severely falls off. Development plans include in certain cases hundreds of wells in order to maintain oil rate. Several CO_2 injectivity tests have been conducted in the Bakken formation in Canada, with little documented results.

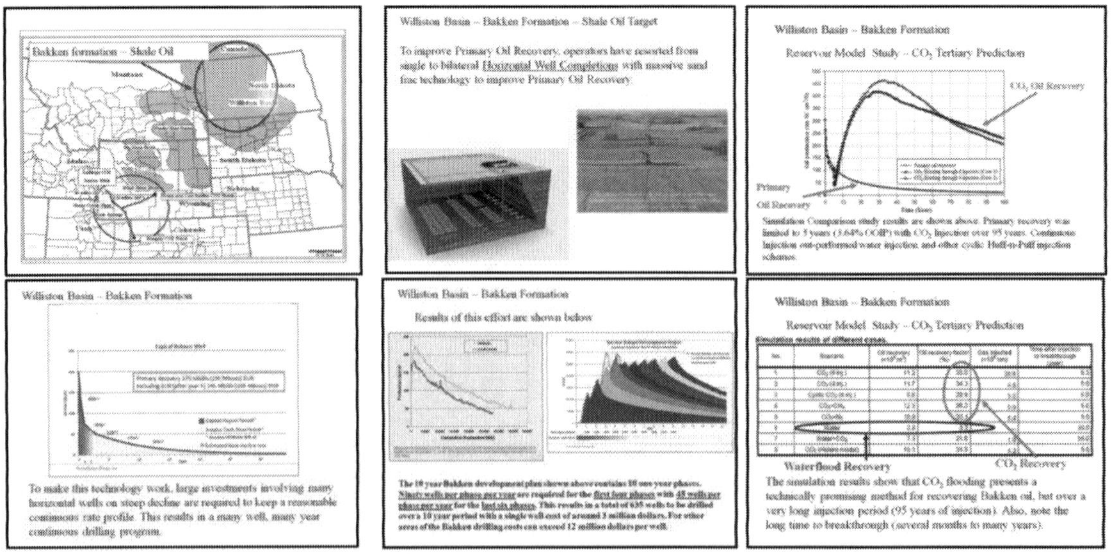

Reservoir model studies show high CO_2 recoveries can be achieved in excess of 30% OOIP compared to today's expected 4% to 6% per well oil recovery demonstrated under Primary Recovery. However, this high oil recovery can only be achieved with over 100 years of CO_2 injection. First tertiary response may not occur until after 10 years of CO_2 injection. At this time, the technology may <u>not</u> be here today, but the <u>Tertiary Target</u> is certainly large enough that it will supply both the 21st and 22nd Centuries with oil and other hydro-carbon products, once these technology are developed.

CO_2 Injection with Nitro Boost Techniques

Gravity Drainage and Nitro Boost

Gravity Drainage and Double Displacement CO_2 Flood Techniques work in reservoirs that have structure. The Hc-gas, Flue Gas, N_2, or CO_2 is usually injected at the top of the reservoir and oil is produced at the bottom. In Oxy-fuel Power Plant operations, which also require a separate Air Separation Plant to extract O_2 from the air, recovers both a clean stream of CO_2, along with a clean stream of Nitrogen. The Question has always been: "Does the six molecules of Nitrogen generated with the one molecule of CO_2 have any use except to be vented to the atmosphere?"

The Rio Bravo field in California is a field that has under-gone multiple types of recovery mechanisms over the years. In the 1940's, Conventional Gravity Drainage Double Displacement techniques were implemented in the field where Hc-gas was injected at the top in a limited number of wells. Production occurred as the front advanced down-dip towards the original oil water contact. The reservoir was then back flooded with water and then pressure depleted. Current performance has shown this type of recovery method left a large tertiary oil target with many millions of barrels of tertiary oil remaining.

With the Nitro-boost concept, an additional recovery stage is introduced into the reservoir. This last stage of recovery uses CO_2 to mobilize the remaining oil and Nitrogen to accelerate the process. The CO_2 is first introduced at the crest to build a bank. Once built, N_2 is injected at the crest to create a secondary gas cap and displace CO_2 down-dip through a pattern development. The re-cycled CO_2 is recovered to be injected down-dip again at the front of the developing oil bank. The objective is to recover more tertiary oil through better Reservoir Management Practices through improved accelerate response and better CO_2 Utilization Managment.

Conventional Gravity Drainage Recovery

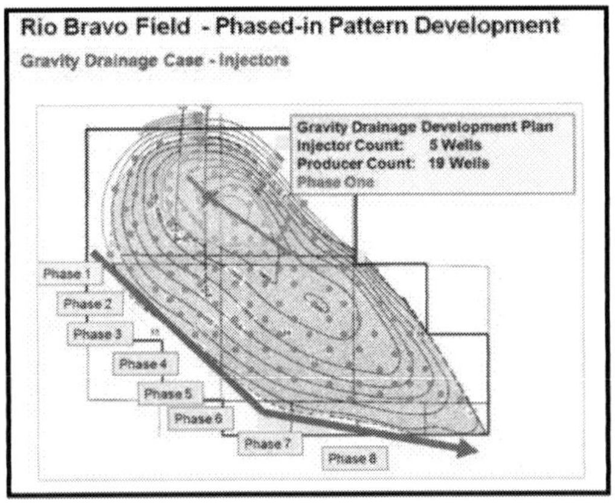

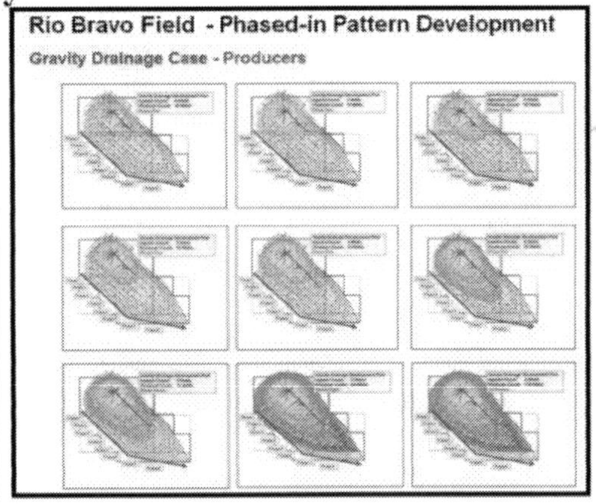

Enhanced Gravity Drainage w/ Nitro Boost

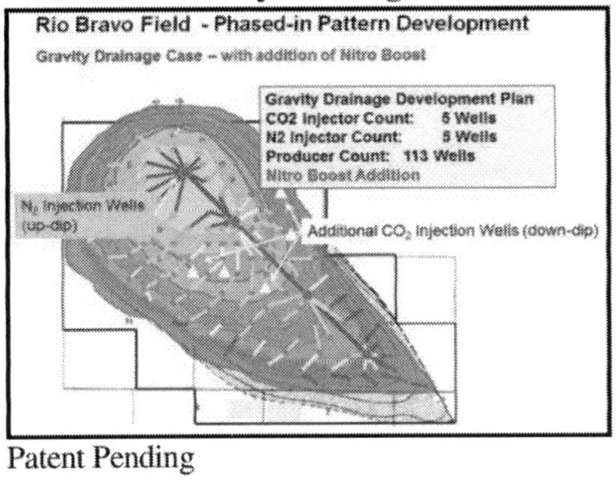

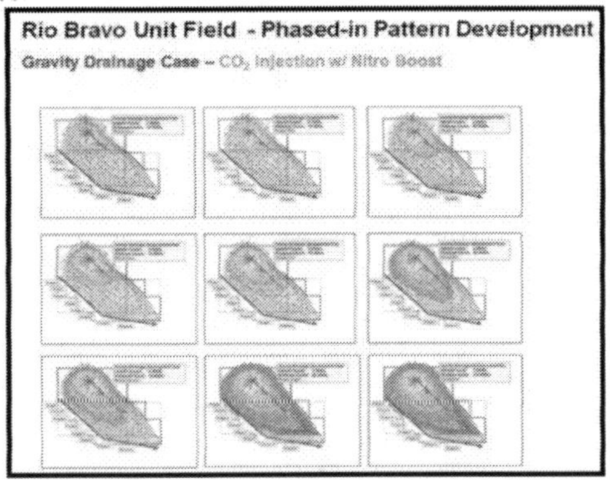

Patent Pending

Horizontal Drilling w/CO_2 WAG Techniques

Gravity Drainage and Nitro Boost

Horizontal Drilling in CO_2 Flood projects are not new to the oil Industry. They have been successfully operated for many years in Canada at the Weyburn project and at the Aneth CO_2 flood in Utah. Future CO_2 flood projects using Horizontal Well Technology exist at Cedar Creek Anticline in Montana and the Dan field, located in the North Sea, Offshore Denmark.

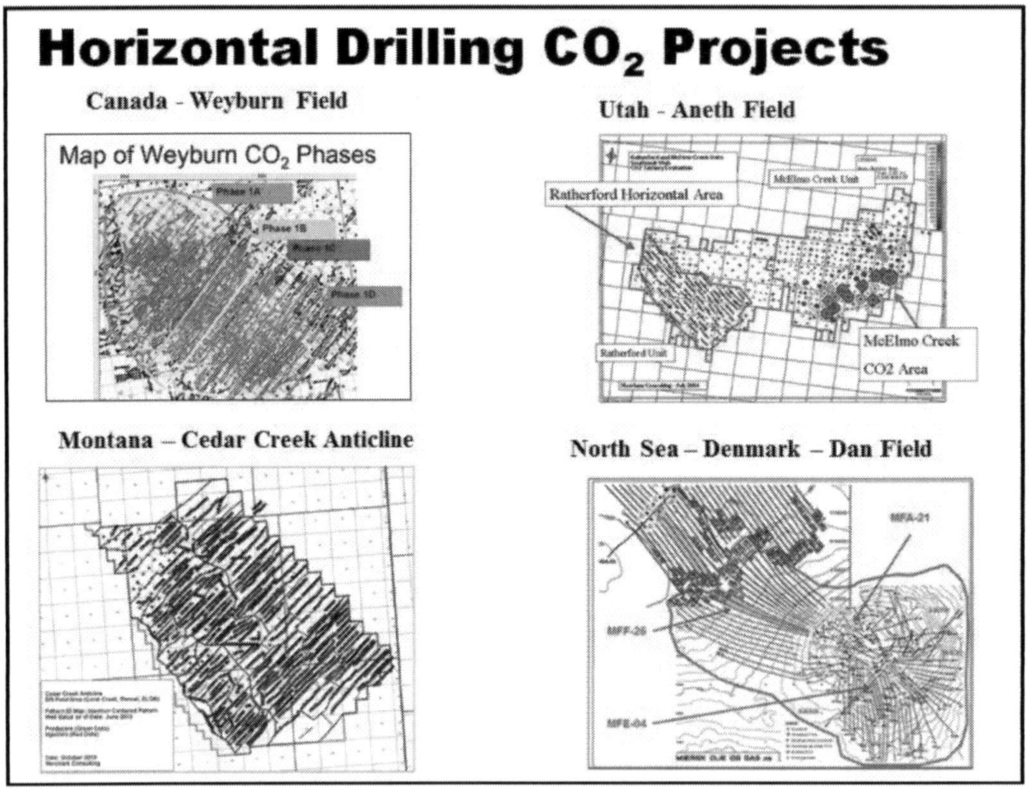

Acknowledgements

I would like to express my gratitude to the members of the Carbon Management Technology Conference (CMTC) for allowing me the opportunity to be part of this CO_2 Conference. I would especially like to thank George Koperna of Advanced Resources International for his support in making this effort possible.

David H. Merchant
Merchant Consulting

CARBON MANAGEMENT TECHNOLOGY CONFERENCE

CMTC-440176-MS

Reservoir Fluid Characterization for Miscible Enhanced Oil Recovery

Dhiraj Gunda, William Ampomah, Reid Grigg and Robert Balch, Petroleum Recovery Research Center, New Mexico Institute of Mining and Technology, Socorro, NM

This paper was prepared for presentation at the Carbon Management Technology Conference held in Sugarland, Texas, USA, 17–19 November 2015.

Abstract

CO_2 injection is an effective method for enhanced oil recovery (EOR). The Pennsylvanian Morrowan sequence of the panhandle of Texas through Southeast Colorado in the United States currently presents an opportunity for CO_2 EOR and carbon sequestration. The Farnsworth Unit (FWU) of Ochiltree County, Texas operated by Chaparral Energy L.L.C. is the site of a CO_2-EOR project using anthropogenic CO_2. The Southwest Regional Partnership on Carbon Sequestration is working closely with Chaparral Energy to study the effectiveness of large-scale CO2 injection into the FWU.

This paper presents the reservoir fluid model for the FWU. It includes a comprehensive PVT analysis of fluid that has been tuned to an equation of state (EOS) with 9 components. After tuning, the experimental and simulated data had an excellent match. A series of slim tube simulated experiments were carried out to determine the minimum miscibility pressure (MMP). The C7+ fractions were split into two pseudo-components with appropriate critical properties without affecting the actual fluid properties. This aided in optimizing the predictions even while using fewer pseudo components to characterize the fluid.

Simulation of the MMP using the EOS tuned fluid was conducted and the predicted MMP was close to the experimental value with an error less than five percent. A sensitivity analysis was also conducted with varying impurities in the CO_2 injection fluid to analyze the effect of recycle gas on the MMP.

The approach used in this study demonstrated an improved strategy in lumping and splitting of compositions to ensure accuracy in reservoir fluid characterization.

Introduction

Fluid characterization plays an important role in the compositional simulation of enhanced oil recovery processes (EOR) like miscible gas flooding. Black oil models do not efficiently deal with the complexities of phase behavior in the case of compositional modelling. This makes the use of compositional fluid characterization essential.

Equations of state (EOS) are used extensively in the fluid characterization process to describe the volumetric and phase behavior and other thermodynamic properties of pure substances and mixtures. The use of an EOS helps in establishing consistency over all phases in the reservoir processes. Van der Waals EOS (van der Waals, 1873) is considered to be the pioneer in the terms that he proposed that gases do not behave in the ideal way and developed the first cubic EOS. Numerous EOS's have been proposed since then including Redlich and Kwong (Redlich and Kwong, 1949), Peng and Robinson (Peng and Robinson, 1976, 1978), Abbott (Abbott, 1973, 1978), Yarborough (Yarborough, 1978). The petroleum industry uses the Peng and Robinson EOS and the Soave-Redlich-Kwong EOS (Soave, 1972) extensively for reservoir fluid modelling studies. Another concept of volume translation was proposed to improve the volumetric capabilities of the above mentioned EOS by Peneloux (Peneloux et al, 1982). The EOS was proposed for pure substances and was extended to mixtures by providing mixing rules. The critical properties and the acentric factor of the mixture components are the requirements for using the EOS. These are then used in the numerical modelling of the fluids to evaluate the consistency of measured PVT data or to generate such data for reservoir simulation.

The time required for simulations in compositional modeling is much higher as compared to black oil simulation. This increases the importance of having fewer components to make the simulation process efficient but without compromising on accuracy. Reservoir hydrocarbon fluids are mixtures of a large number of components. This presents us with the task of limiting the number of components to save substantial time in reservoir simulation while also making sure the fluid properties are not affected. This is done by lumping the various components of the original fluid. Several methods for lumping of components are available (Lee et al. 1982; Whitson 1983; Pedersen et al. 2002; Li et al. 1985). The lumped fluid model should be able to reproduce the results of the original fluid before it can be used for reservoir simulations. This is done by a technique known as tuning or regression. There are several regression methods proposed in the literature; the primary objective is to change the uncertain parameters in order to tune the EOS so that the new model replicates the original fluid and matches the experimental PVT data.

Minimum miscibility pressure (MMP) is the minimum pressure at which an injected gas develops miscibility with the hydrocarbon fluid. The accurate estimation of the MMP is important for a successful injection process. The MMP is determined by both experimental and analytical methods. The slim tube process is a commonly used experiment to determine the MMP and also to validate the tuned EOS model. The oil recovery is calculated for 1.2 pore volumes (PV) and plotted against different displacement pressures. The curve starts to flatten out when the miscible displacement starts. This break point on the curve is considered as the MMP.

There are two widely used analytical methods for estimating MMP: 1-D slim tube simulation and the key tie line method or the Method of characteristics (MOC) (Yuan and Johns, 2005; Jensen et al., 1998; Wang and Orr, 1998). The slim tube compositional simulation replicates the experimental process by using 1-D grid blocks. The MMP is determined similar to the experimental process using the recovered oil and pressure curve.

Our study focuses on an improved strategy for fluid characterization of the complex hydrocarbon fluids. This is required for reducing the number of pseudocomponents for compositional reservoir simulation. This approach was developed by attaining a match between both the experimental and simulated PVT data by tuning the EOS and further validating it with the MMP data. The fluid model was then used to study the effect of impurities in the CO_2 injection gas to understand the impact of recycle gas. Results from this study have been used for numerical simulations for Farnsworth Unit (Ampomah et al., 2015a; Ampomah et al., 2015b).

Experimental database

The original reservoir fluid composition for the Farnsworth Unit (FWU) is given below in Table 1:

Table 1: Reservoir fluid composition for Farnsworth field

Component	Weight %	Mol%	Density	API	Molecular weight
C1	3.98	38.44			
C2	0.75	3.85			
C3	0.7	2.46			
IC4	0.11	0.29			
NC4	0.62	1.66			
IC5	0.25	0.53			
NC5	0.59	1.26			
C6	1.57	2.83			
C7+	91.43	48.68	0.8477	35.3	291

The reservoir pressure for the field is estimated to be 2203 psig at the reservoir temperature of 168 F. The saturation pressure is 2059 psig. This study also uses data from PVT experiments including Constant composition expansion (CCE), Differential liberation (DL), and multi stage separator tests acquired from Farnsworth field fluid analysis.

Methods

The process of EOS fluid characterization in this study primarily involved splitting and lumping of heavier components, tuning simulation results to EOS and validation.

Fluid system setup

The C7+ fractions were split and consequently lumped into two pseudocomponents in accordance with the data available for the average molecular weight, average specific gravity and the total mole percent using the mixing rule by Pedersen (Pedersen et al, 1989). Using the same methodology the isomers of C4 (butane) and C5 (pentane) were lumped in order to further reduce the number of components. This required a comprehensive iterative process performed manually in order to achieve convergence with the available data.

The EOS parameters for the pseudocomponents (HC-1 & HC-2) such as the critical temperature (T_c), critical pressure (P_c), acentric factor (ω) have been calculated based on the approach suggested by Pedersen (Pedersen et al, 2002) using a compositional PVT software. All the calculations were performed using the 3 parameter Peng Robinson EOS (Peng and Robinson, 1976) with Peneloux volume correction (Peneloux et al. 1982). The Lohrenz-Bray-Clark correlation is used for the calculations of viscosities (Lohrenz et al. 1964).

Experimental PVT data

The data from experimental PVT tests has been used for accurately tuning the EOS. The PVT experiments in this study include the constant composition expansion (CCE), saturation pressure, differential liberation (DL), multi stage separator tests, and viscosity experiments.

Tuning the EOS

The objective of tuning the EOS is to match the simulated PVT data to the experimental data. The tuning process was performed using the regression option in the PVT software. In general it is very unlikely that properties for the components up to C6 or the lower components will differ from the library values. These values were not used during the regression. The critical properties of the heavier components are not as accurately determined due to the mixture of components. The fluid characterization studies have also indicated that the C7+ fraction is a key factor for obtaining a match between the experimental and simulation data. As the plus fraction is a mixture of a large number of hydrocarbons usually the properties determined by the correlations are just as good as the characterization method used to generate them. The P_c, T_c, ω of the plus fraction are determined from the plus fluid characterization making them the ideal candidates for tuning (Whitson et al. 2000). These properties along with the molecular weights were subsequently used in the regression process.

The binary interaction parameters (BIP's) are introduced into the cubic equation of state to account for the unlike pair interaction between the components. The BIP's are generally considered to be zero for hydrocarbon pairs except probably for C1/C7+ pairs. There are numerous binary coefficients and most of the time there is insufficient PVT data to justify the tuning of all of them. Tuning the data more closely than necessary can lead to convergence issues in compositional simulations. In this case binary interaction coefficients were not used to avoid over tuning.

Additionally, weights were given to certain variables such as saturation pressure, saturation densities and other data to enhance the tuning process. In the case of some variables the approach suggested by Coats and Smart (Coats et al., 1986) was used and for other variables weights were allocated on the basis of their impact on the whole tuning process. The acceptable mole fractions obtained after tuning are shown in Table 2:

Table 2: Fluid composition with pseudocomponents

Components	ZI	Weight fraction	Mol Weight	Spec Gravity	Critical Temperature	Critical Pressure
	(percent)	(percent)			°C	bar
C1	38.491	3.873	16.043		-82.55	46.00
C2	3.855	0.727	30.07		32.55	48.84
C3	2.463	0.681	44.097		96.65	42.46
C4's	1.952	0.712	58.124		149.507	37.77
C5's	1.793	0.811	72.151		193.726	33.78
C6's	2.834	1.532	86.178	0.664	234.25	29.69
HC1	33.484	39.891	189.952	0.804	428.305	22.49
HC2	15.128	51.773	545.65	0.892	580.97	16.25

Validation of the EOS model

The PVT data from the lumped fluid was compared to the experimental data for the original fluid. This is the final step to ensure the validity of the EOS model. In addition, 1-D simulation of the MMP using PVT software was also performed to further evaluate the performance of the fluid model.

Sensitivity analysis

It is important to analyze the long-term effects of using recycled CO_2 in the reservoir. Recycling reduces dependence on external sources for CO_2 and thereby reduces cost. However, due to mixing in the reservoir and the use of recycled gas, injected gas composition can very with time. Therefore a sensitivity analysis was done using the PVT software to understand the impact of impurities and varying composition of injection gas on reservoir behavior.

Results and Discussions

The results from the tuned fluid model showed much more accurate predictions as compared to the original fluid. Figures 1-13 show the simulated results compared to the observed results. Figures 1-6 show results from the default Peng-Robinson three parameter EOS without regression. Figure 7 shows the phase envelope for the FWU fluid. It can be seen that the simulated saturation pressure has been closely matched to the observed reservoir saturation pressure after regression. Figures 8-13 show comparative results after tuning the EOS to the experimental data. The adjustment of critical properties and the molecular weights of the pseudocomponents were crucial for obtaining a precise match between the simulated and the experimental results. The regression variables were able to keep the results in the permissible error range of 5-10%. The results are more significant due to the fact that a very few variables were adjusted to tune the EOS as it reduces the convergence problems.

Table 3 summarizes a few results comparing the predicted and the experimental values for some of the properties for the lumped fluid before regression. Table 4 compares the values for the same properties for the lumped fluid after the tuning has been done.

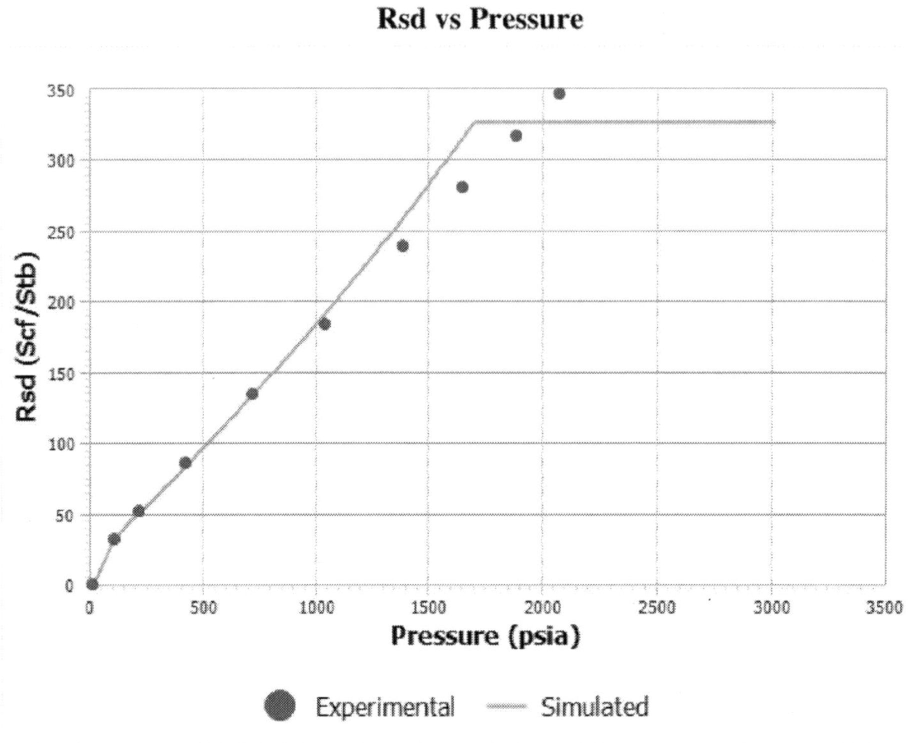

Figure 1: Experimental & Simulated GOR from differential liberation experiments

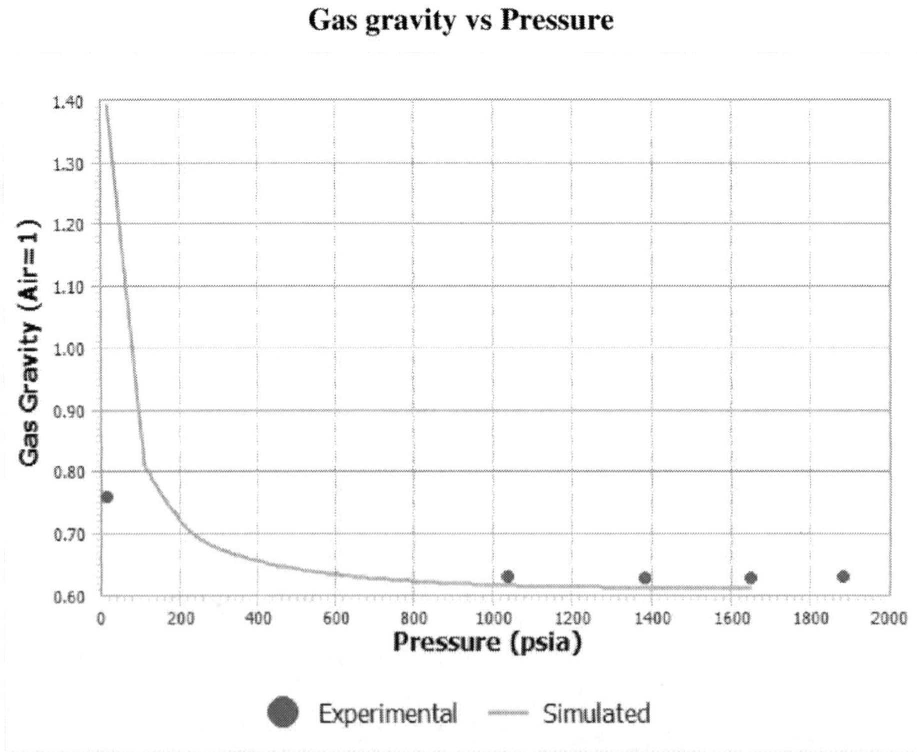

Figure 2: Experimental & Simulated gas gravity from differential liberation experiments

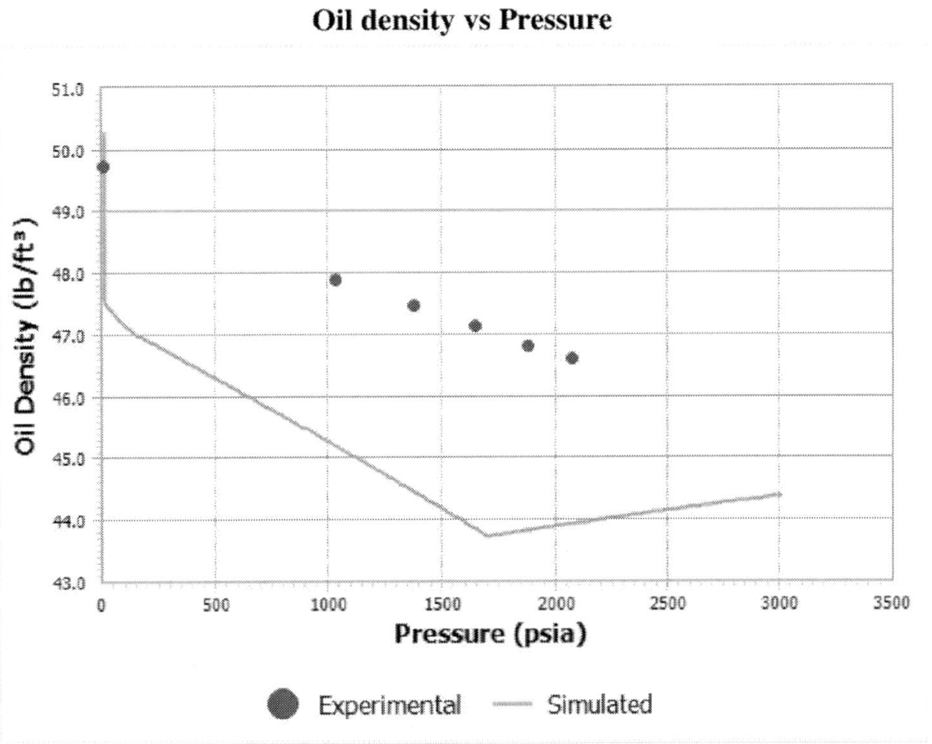

Figure 3: Experimental & Simulated oil density from differential liberation experiments

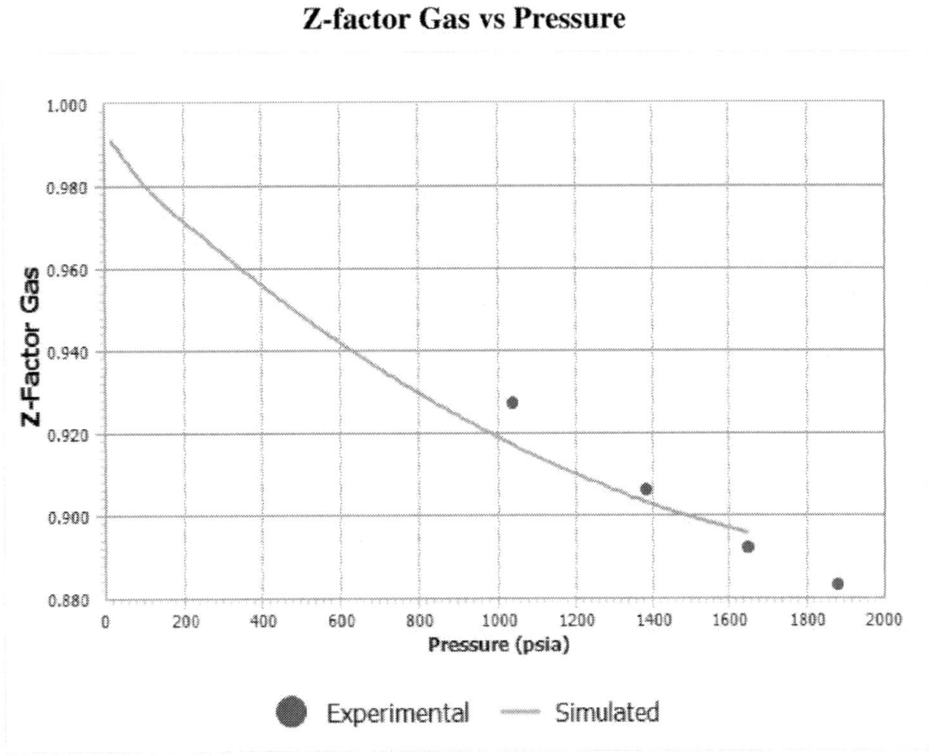

Figure 4: Experimental & Simulated Z factor from differential liberation experiments

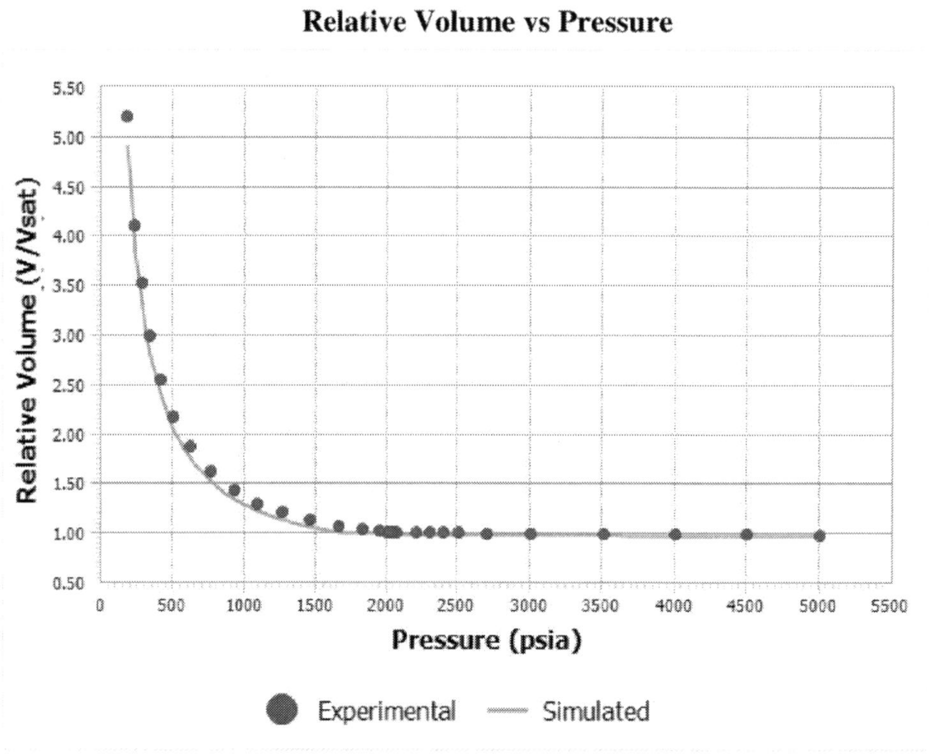

Figure 5: Experimental & Simulated relative volume from constant mass expansion experiments

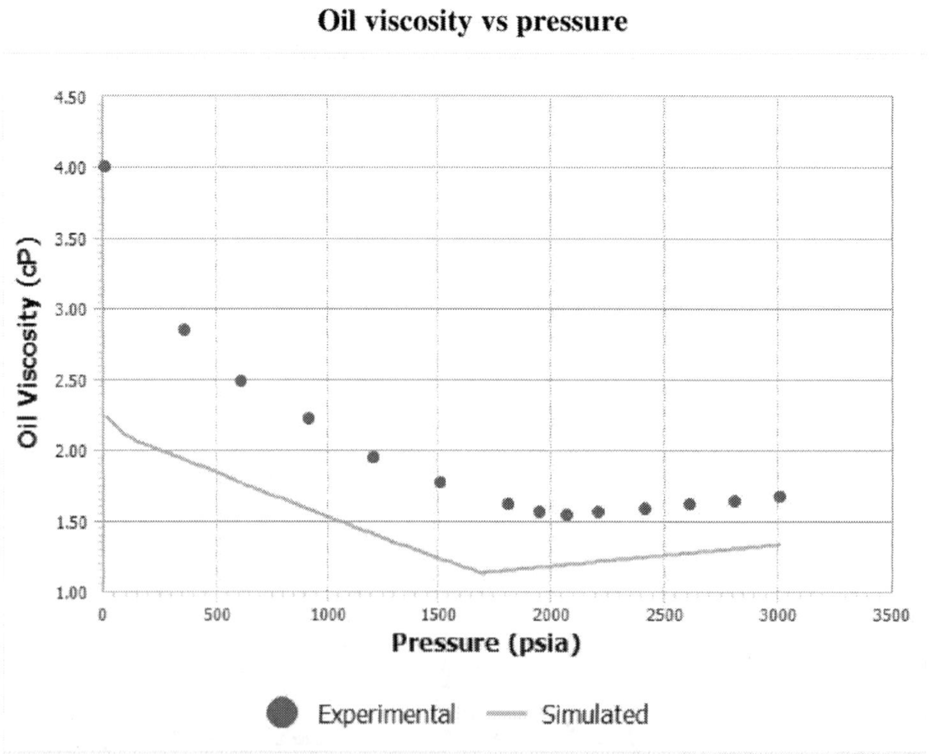

Figure 6: Experimental & Simulated oil viscosity from differential liberation experiments

Table 3: Comparison of predicted and observed values for some properties before regression

Properties	Units	Observed	Calculated	Reference Pressure (psig)	%Error
Saturation Pressure (Pb)	psia	2073.7	1701.3		17.95
Oil density	g/cc	0.7466	0.7032	Observed Pb	5.813
Vapor Z-factor		0.906	0.903	1372	0.33
GOR	MSCF/STB	0.347	0.326	Observed Pb	6.05
Gas Gravity		0.627	0.613	1372	2.23
Liquid Viscosity	CP	1.536	1.135	Observed Pb	26.11

There are large errors observed in some properties especially the saturation pressure, gas oil ratio and the liquid viscosity. The error in saturation pressure, one of the most important parameters was close to 18 percent, which is unacceptable. The plots for the oil density and the viscosity follow a continuous error pattern for all data points. These results make it imperative to tune the EOS in order to reduce the errors.

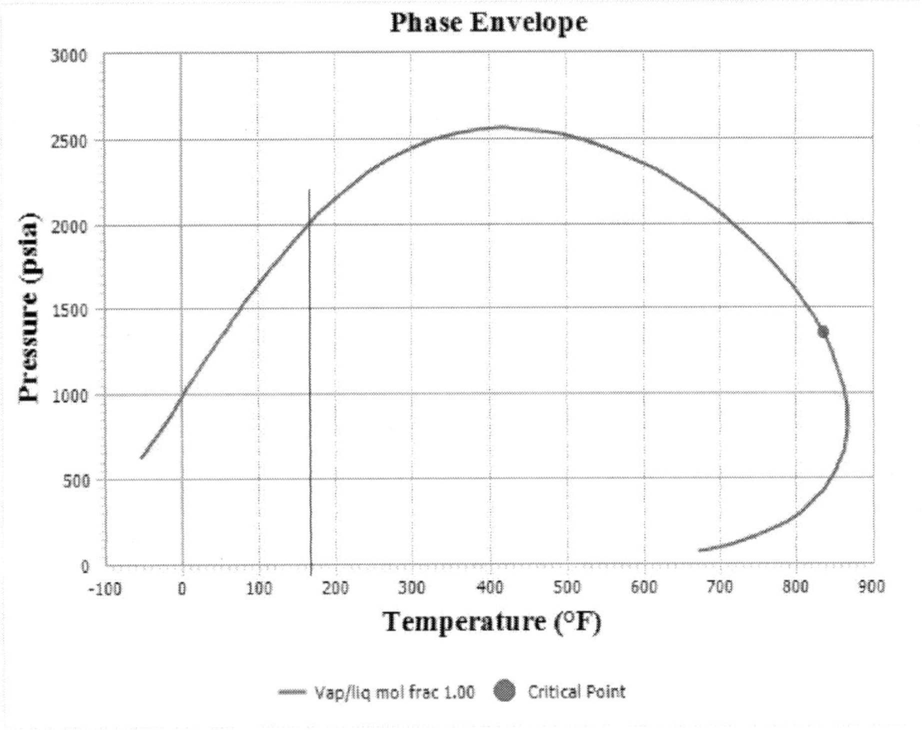

Figure 7: Phase diagram after regression

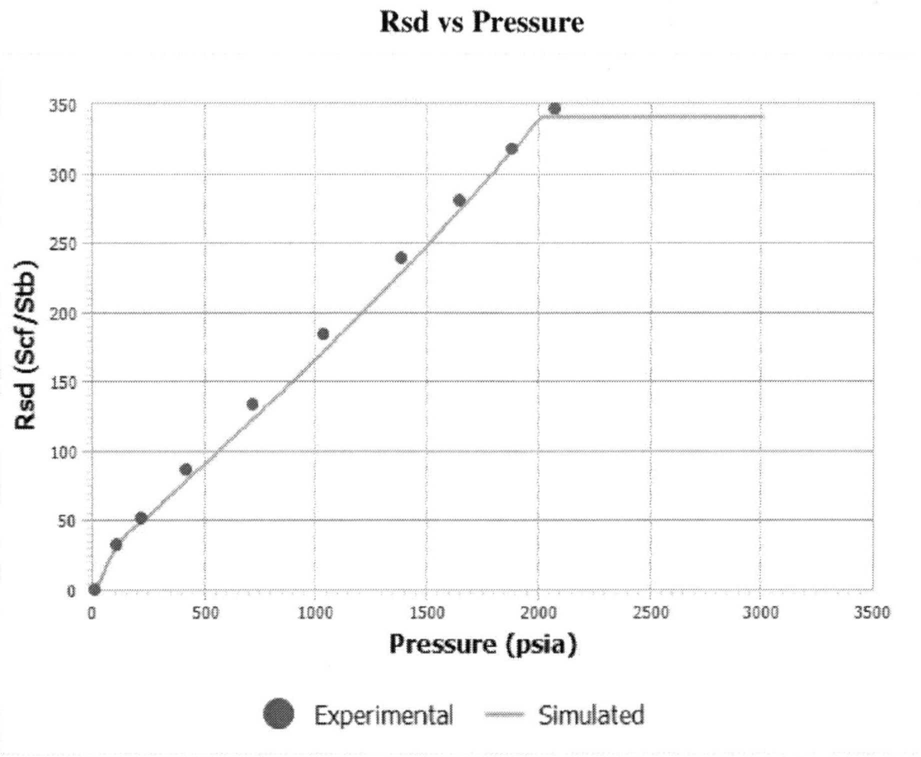

Figure 8: Experimental & Simulated GOR from differential liberation experiments

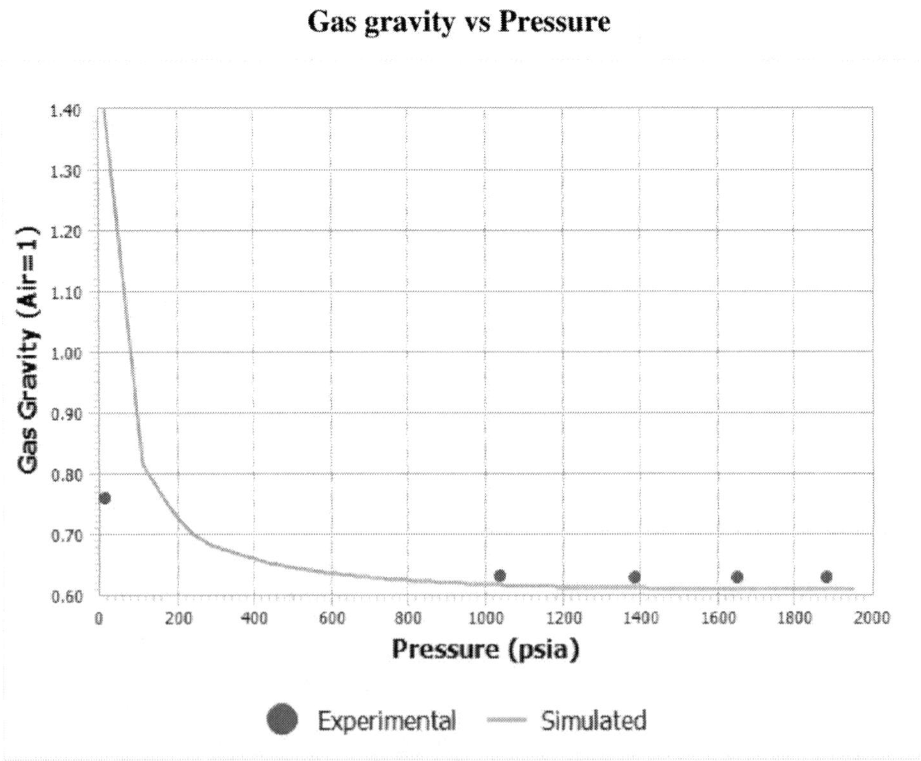

Figure 9: Experimental & Simulated gas gravity from differential liberation experiments

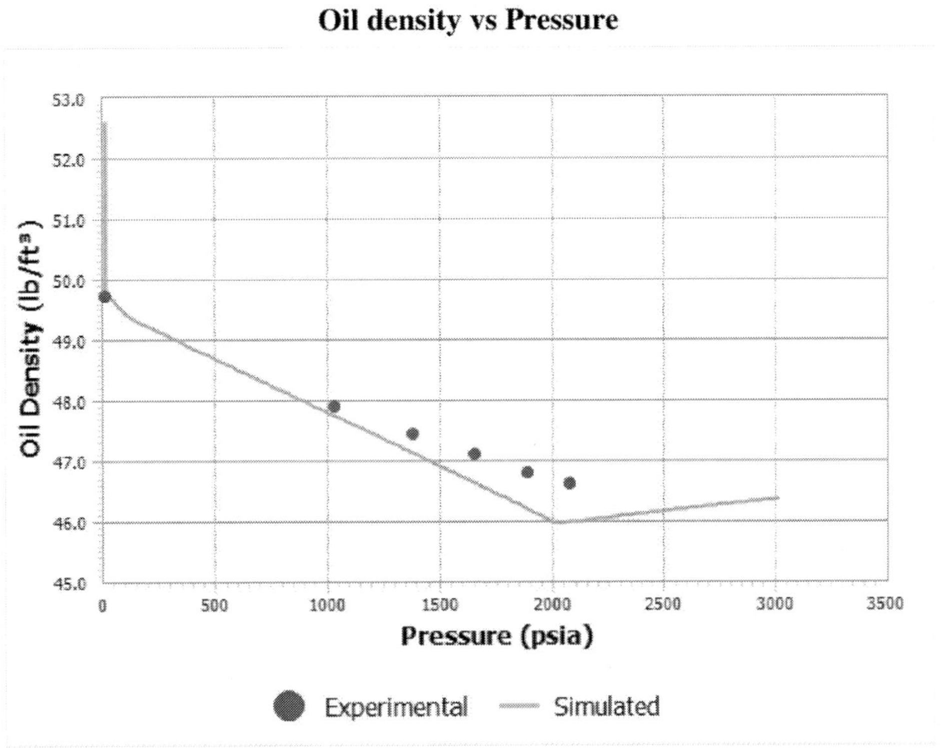

Figure 10: Experimental & Simulated oil density from differential liberation experiments

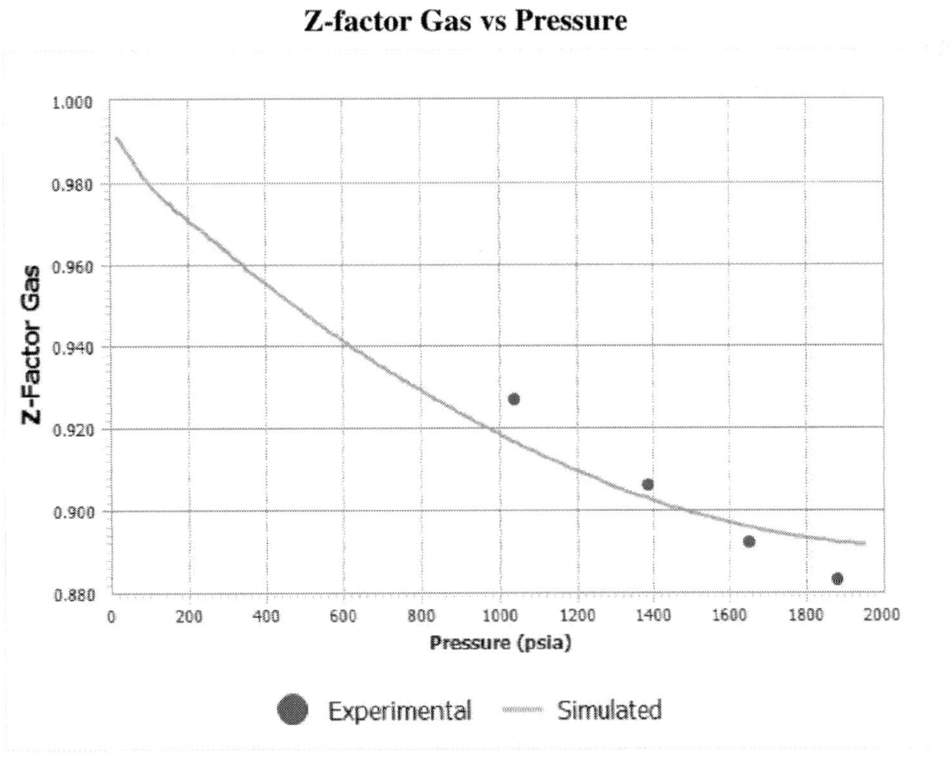

Figure 11: Experimental & Simulated Z factor from differential liberation experiments

Relative Volume vs Pressure

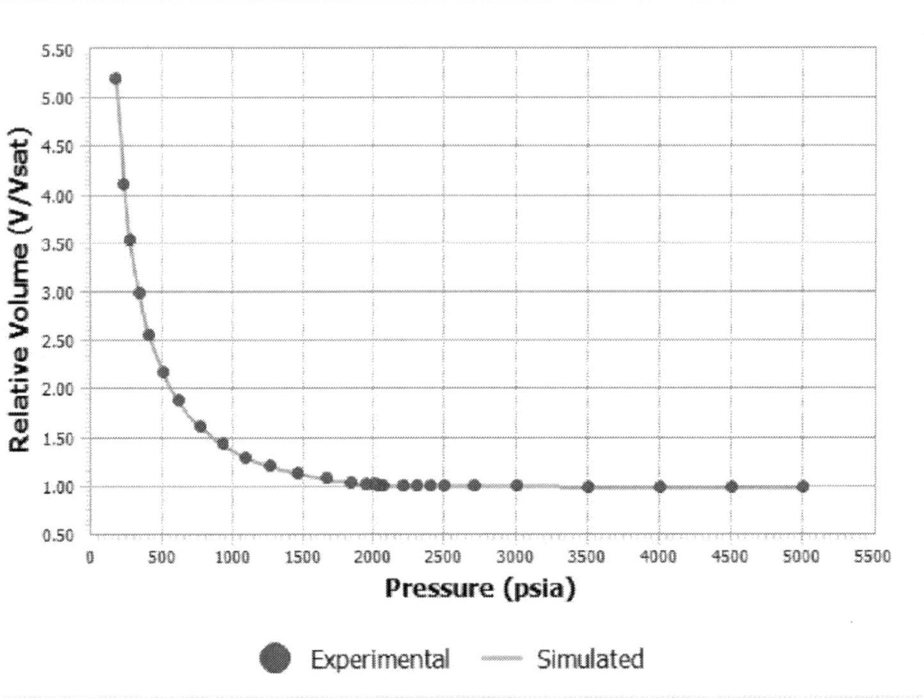

Figure 12: Experimental & Simulated relative volume from constant mass expansion experiments

Oil viscosity vs Pressure

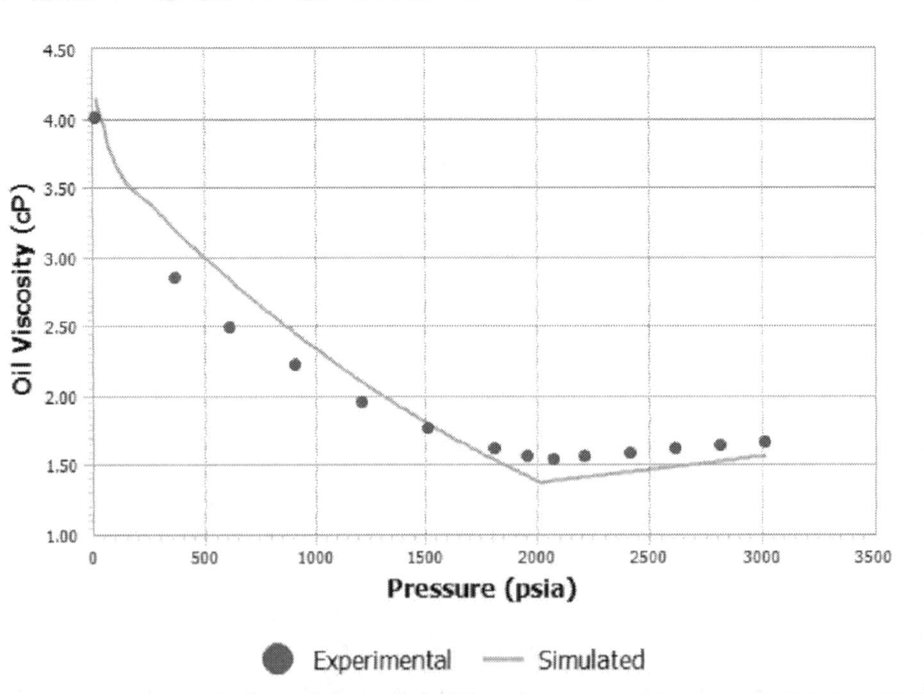

Figure 13: Experimental & Simulated oil viscosity from differential liberation experiments

Table 4: Comparison of predicted and observed values for some properties after regression

Properties	Units	Observed	Calculated	Reference Pressure (psig)	%Error
Saturation Pressure (Pb)	psia	2073.7	2014.8		2.84
Oil density	g/cc	0.7466	0.7367	Observed Pb	1.3
Vapor Z-factor		0.906	0.904	1372	0.22
GOR	MSCF/STB	0.347	0.3415	Observed Pb	1.58
Gas Gravity		0.627	0.612	1372	2.39
Liquid Viscosity	CP	1.536	1.387	Observed Pb	9.7

The results in Table 4 demonstrate the impact of tuning the lumped fluid model. The error in the saturation pressure was reduced from 17.95 percent to 2.84 percent, which is significant. The liquid viscosity error was also reduced to a 10 percent error versus over 26 percent in the untuned model.

This improved fluid model was then used to predict the minimum miscibility pressure (MMP), and the results were within 5 percent of experimental data. This further validates the lumped fluid model and the lumping technique used for the pseudocomponents. Figure 14 presents the oil recovery plot from the 1-D slim tube test.

Table 5: Comparison of MMP values before and after regression

Properties	Units	Observed	Before Regression	After Regression	%Error
MMP	psia	4200	3038.4	4008.8	4.5

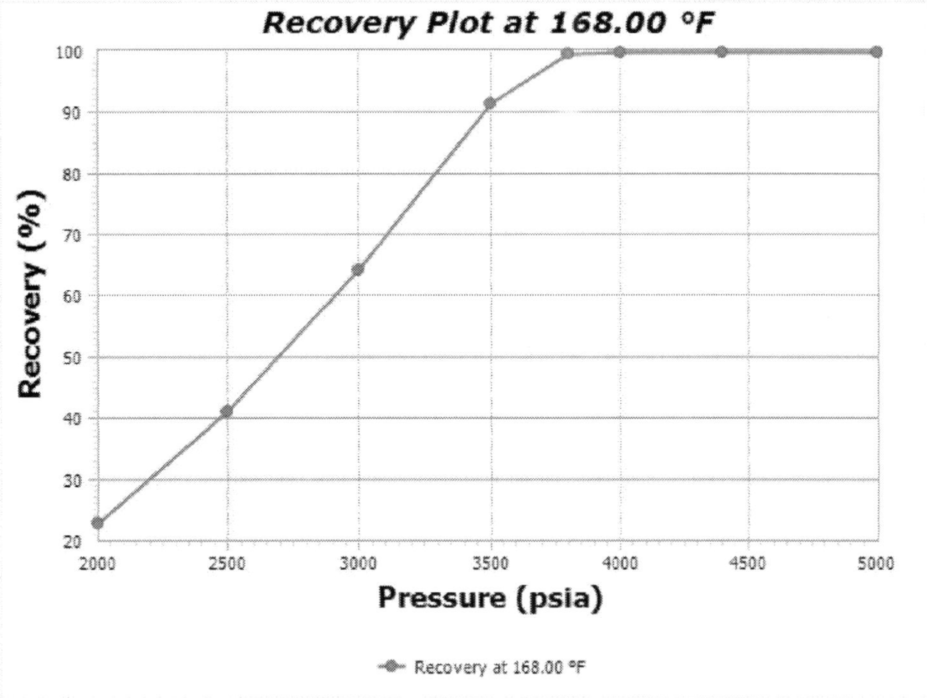

Figure 14: Oil recovery plot for 1-D slim tube test

The results from the sensitivity analysis of MMP showed an increase in the MMP with increasing amount of methane impurity in the injection gas. A decrease in the MMP was observed with increase in the percentage of ethane in the injection gas. Nitrogen showed a higher increase in the MMP as compared to the methane. Table 6 shows the results observed for various amounts of impurities in the CO_2 injection stream. These results reiterate the findings available in the literature for injection of impure CO_2 (Alston et al 1985).

Table 6: MMP for CO_2 injection gas with impurities

CH_4	CO_2	C_2H_6	N_2	MMP(psia)
0	100	0	0	4008.8
5	95	0	0	4025.6
10	90	0	0	4094.6
15	85	0	0	4196.6
20	80	0	0	4321.4
0	95	5	0	3635
0	90	10	0	3389.9
0	85	15	0	3211.3
0	80	20	0	3070.5
0	95	0	5	4641.5
0	90	0	10	5336.1
0	85	0	15	6102.3
0	80	0	20	6952

Conclusion

The approach used in this study provides an improved strategy in lumping and splitting of compositions to ensure accuracy in reservoir fluid characterization.

The EOS-based phase behavior models present the possibility of estimating erroneous results especially for near critical fluids. The heavier fractions of the reservoir fluids are in general described by a limited number of components. The generalized correlations present in the literature are used to estimate the properties for these components. But the presence of so many uncertainties increases the deficiencies in the predictions. This gives rise for a need to tune the EOS model by using the experimental data available, which is done by varying the uncertain factors in the fluid model to decrease the errors between the predicted and the experimental values.

In this study the C7+ fractions have been successfully lumped to 2 pseudocomponents and the fluid model was then tuned to the EOS. The data with greater influence on the phase behavior was tuned with a higher preference. The predicted results for the PVT experiments support the efficiency of this model. The fluid model was further validated by using the 1-D slim tube experiment in the PVT software package to predict the MMP, which was found to be close to the available MMP with less than a 5 percent error.

This study has further helped in improving the compositional reservoir model constructed for the FWU (Ampomah et al., 2015a; Ampomah et al., 2015b).

Acknowledgements

Funding for this project is provided by the U.S. Department of Energy's (DOE) National Energy Technology Laboratory (NETL) through the Southwest Regional Partnership on Carbon Sequestration (SWP) under Award No. DE-FC26-05NT42591. Additional support has been provided by site operator Chaparral Energy, L.L.C. and Schlumberger Carbon Services.

References

Abbott, M.M.: "Cubic Equation of State," AIChE J. (May 1973) 596.

Abbott, M.M.: "Cubic Equations of State: An Interpretive Review," Equations of State in Engineering and Research, K.C. Chao and R.L.Robinson Jr. (eds.), Advances in Chemistry Series, Amer. ChemicalSoc., Washington, DC (1978).

Ampomah, W., Balch, R. S., and Grigg, R. B., Dai Z., and Pan F. (2015) Compositional Simulation of CO2 Storage Capacity in Depleted Oil Reservoirs. Carbon Management Technology Conference November 16-19 2015, Sugarland, Houston-Texas USA. CMTC-439476-MS

Ampomah, W., Balch, R. S., and Grigg, R. B. (2015) Analysis of Upscaling Algorithms in Heterogeneous Reservoirs with Different Recovery Processes. SPE Production and Operations Symposium March 1-5 2015, Oklahoma USA. doi:10.2118/173588-MS

Alston, R. B., Kokolis, G. P., & James, C. F. (1985, April 1). CO2 Minimum Miscibility Pressure: A Correlation for Impure CO2 Streams and Live Oil Systems. Society of Petroleum Engineers. doi:10.2118/11959-PA.

Coats, K. H., & Smart, G. T. (1986, May 1). Application of a Regression-Based EOS PVT Program to Laboratory Data. Society of Petroleum Engineers.

Jessen, K., Michelsen, M.L. and Stenby, E.H.: "Effective Algorithm for Calculation of Minimum Miscibility Pressure", SPE Paper 50632, 1998.

Johns, R.T., Yuan, H. and Dindoruk, B., "Quantification of Displacement Mechanisms in Multicomponent Gasfloods", SPE 77696 presented at the SPE ATCE in San Antonio, Tx September 29-October 2, 2002.

Katz, D.L. and Firoozabadi, A., "Predicting Phase Behavior of Condensate/Crude-Oil Systems Using Methane Interaction Coefficients", J. Pet. Technol. 20, 1978, pp. 1649-1655.

Lee, S.T.et al.: "Experiments and Theoretical Simulation on the Fluid Properties Required for Simulation of Thermal Processes," SPEJ (October 1982) 535.

Li, Y.-K., Nghiem, L. X., & Siu, A. (1985, November 1). Phase Behavior Computations For Reservoir Fluids: Effect Of Pseudo-Components On Phase Diagrams And Simulation Results. Petroleum Society of Canada. doi:10.2118/85-06-02.

Lohrenz, J., Bray, B. G., & Clark, C. R. (1964, October 1). Calculating Viscosities of Reservoir Fluids From Their Compositions. Society of Petroleum Engineers. doi:10.2118/915-PA.

Pedersen, K.S., Thomassen, P., and Fredenslund, A.: "Characterization of Gas Condensate Mixtures,"C7+ Fraction Characterization, L.G.Chorn and G.A. Mansoori (eds.), Advances in Thermodynamics, Taylor & Francis, New York City (1989).

Pedersen, K. S., Calsep, A. S., Milter, J., Sørensen, H., & Calsep, A. S. (n.d.). SPE 77385 Cubic Equations of State Applied to HT / HP and Highly Aromatic Fluids (2002).

Pedersen, K.S., Fredenslund, Aa. and Thomassen, P.,"Properties of Oils and Natural Gases", Gulf Publishing Inc., Houston, 1989a.

Peng, D.Y. and Robinson, D.B.: "A New-Constant Equation of State," Ind. & Eng. Chem. (1976).

Peneloux, A., Rauzy, E., and Freze, R.: "A Consistent Correction for Redlich-Kwong-Soave Volumes," Fluid Phase Equilibria (1982).

Redlich, O. and Kwong, J.N.S.: "On the Thermodynamics of Solutions'. An Equation of State. Fugacities of Gaseous Solutions,"Chem. Rev. (1949)

Robinson, D.B. and Peng, D.Y.: "The Characterization of the Heptanes and Heavier Fractions," Research Report 28, Gas Producers Assn., Tulsa, Oklahoma (1978).

Soave, G., "Equilibrium Constants from a Modified Redlich-Kwong Equation of State", Chem. Eng. Sci. 27, 1972, pp. 1197-1203.

Van der Waals, J.D.: Continuity of the Gaseous and Liquid State of Matter (1873).

Wang, Y., and Orr, F.M.,"Calculation of Minimum Miscibility Pressure", SPE paper 39683, 1998.

Whitson, C.H and Brule, M.R, 2000. Phase Behavior. Monograph Series. SPE, Richardson, Texas.

Whitson, C.H.: "Effect of C7+ Properties on Equation-of-State Predictions," SPEJ (December 1984) 685; Trans., AIME, 277.

Yarborough, L.: "Application of a Generalized Equation of State to Petroleum Reservoir Fluids," Equations of State in Engineering and Research, K.C. Chao and R.L. Robinson Jr. (eds.), Advances in Chemistry Series, Amer. Chemical Soc., Washington, DC (1978).

Yuan, H., & Johns, R. T. (2005, December 1). Simplified Method for Calculation of Minimum Miscibility Pressure or Enrichment. Society of Petroleum Engineers. doi:10.2118/77381-PA

CMTC-440179-MS

Synthetic Coal Cycle Technology(TM): A Novel Carbon Utilization Technology

D. A. Zietlow, Zietlow Innovative Engineering and Technology, LLC

This paper was prepared for presentation at the Carbon Management Technology Conference held in Sugarland, Texas, USA, 17–19 November 2015.

Abstract

This paper discusses Synthetic Coal Cycle Technology™ (synCCT™), a novel, patent pending, energy conversion and storage technology that uses carbon dioxide (CO_2) to make grid-scale renewable energy dispatchable. synCCT™ uses renewable energy to dissociate CO_2 into carbon, a renewable fuel named Synthetic Coal™ or synC™, and oxygen. synC™ is oxidized to generate electricity and heat on demand, producing CO_2, which is captured and reused in the process.

Both synC™ and CO_2 are feasibly and economically stored for long periods and/or transported long distances by conventional technologies. The synCCT™ process is carbon neutral and sustainable. In addition, synCCT™ has a number of advantages over competitive technologies, including:

- No greenhouse gas (GHG) emissions
- Simplified integration of grid-scale renewable electricity
- Virtual transmission of renewable electricity without wires
- Simplicity
- No pollution or waste products
- Minimal water use
- Relative safety
- Low toxicity
- Minimal infrastructure development required
- Potential use and repurposing of existing infrastructure

This paper discusses the rationale for synCCT™, how it works and how it compares with competitive technologies for producing renewable fuels from CO_2.

Introduction

The embodiments of Synthetic Coal Cycle Technology™ (synCCT™) discussed in this paper simultaneously address two major issues for the production of electricity: carbon dioxide (CO_2) emissions and the limited availability of solar and wind energy to reduce these emissions (Zietlow 2014). Although renewable energy sources, such as wind and solar power, reduce CO_2 emissions by replacing fossil fuels, this reduction is limited by their availability. Because most renewable energy sources are either intermittent, geographically fixed or both, they are often not available when and where needed. Although intermittency can be mitigated to some extent by using peaking generation or conventional energy storage, both have significant limitations. Peaking generation emits CO_2 and most commercially available energy storage only works over short time frames. Both require capital investment beyond that for renewable energy generation, often including new, expensive electrical transmission lines to transport the electricity they provide.

synCCT™ converts renewable energy to a renewable fuel, synC™, that stores renewable energy indefinitely and can be transported significant distances by conventional transportation, thereby, solving the problems of intermittency and fixed location without the need for new electrical transmission lines. Although synCCT™ requires additional capital investment beyond that for renewable energy generation, the added value of greater location and time flexibility and the avoided cost of new electrical transmission offsets its cost relative to peaking generation and conventional energy storage.

Synthetic Coal Cycle Technology™ (synCCT™)

The synCCT™ process is patterned after the biological carbon cycle of photosynthesis and respiration, hereinafter referred to simply as the carbon cycle, and is analogous to the renewable hydrogen process for storing and transporting renewable energy. All three processes produce renewable fuels and are carbon neutral. Both synCCT™ and the carbon cycle produce renewable fuel from CO_2.

Figure 1 presents a block flow diagram for one embodiment of the synCCT™ process, the main steps of which are synC™ production and electricity generation. The synCCT™ process also includes the additional steps discussed below.

- CO_2 capture
- CO_2 storage
- CO_2 transportation (optional)
- synC™ storage
- synC™ transportation (optional)

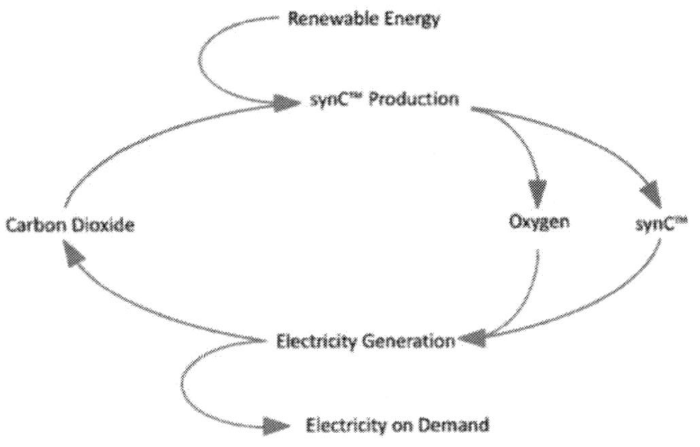

Figure 1 – synCCT™ Flow Diagram

synC™ Production

synC™ production produces carbon (named Synthetic Coal™ or synC™), which is a renewable fuel, and oxygen (O_2) from CO_2 using renewable energy. For the remainder of this paper, the terms carbon and synC™ are used interchangeably when referring to any aspect of synCCT™. Carbon is storable for long times, easily transported, non-toxic, relatively safe, environmentally benign and convertible to usable energy on demand with commercial or near-commercial technology. In addition, most of the technology needed to store and transport carbon safely and economically is commercially developed for handling carbon black and coal. All of these factors make synC™ a strong contender for a renewable, carbon-neutral fuel.

The overall reaction for synC™ production is:

$$CO_2 + renewable\ energy \rightarrow \mathrm{C\ (synC^{TM})} + O_2 \tag{1}$$

synC™ production is the one step of the synCCT™ process that needs significant development to be commercially viable. However, the dissociation of carbon oxide has been demonstrated at lab and pilot scale by a number of processes (Zietlow 2014, Walker 1980, Schora 1975, Fukuda, et al. 2007, Hosback, et al., 2011, Cundliffe 2011).

Electricity Generation

The overall reaction for electricity generation is:

$$\mathrm{C\ (synC^{TM})} + O_2 \rightarrow CO_2 + electricity + heat \tag{2}$$

The energy of oxidation produces both electricity and heat, because conversion of the heat of reaction to electricity is less than 100%. The means of oxidation can be combustion in a conventional coal-fired power plant or electrochemical reaction in a fuel cell. In either case, the use of oxygen, rather than air, produces a concentrated stream of CO_2 facilitating capture of the CO_2 produced. In the case of combustion, chemical looping combustion could also provide the same benefit.

CO_2 Capture, Storage and Transportation

As noted above, the difficulty of CO_2 capture depends on the means of electrical generation, specifically the means of carbon oxidation. Although the synCCT™ process requires CO_2 capture, the location and means of doing so depends on the specific embodiment of the process. If carbon oxidation produces a concentrated CO_2 stream, conventional means of CO_2 capture exist. Economical, means of capturing CO_2 from a dilute CO_2 stream or the atmosphere are being developed, but are not yet commercialized.

Storage and transportation of bulk CO_2 as an industrial chemical are fully commercialized and require little or no development for the synCCT™ process. Transportation is not required is synCCT™ is used for onsite energy storage.

Comparison with Other Renewable Fuels

In addition to its similarities to the carbon cycle and the renewable hydrogen process, both of which produce renewable fuels, synCCT™ is similar to a class of chemical processes that convert CO_2 to carbon-based fuels. Some of these fuels are conventionally produced from petroleum, in which case they are not renewable. However, when produced from biomass, CO_2 that is naturally occurring or CO_2 that results from oxidizing renewable carbon-based fuels, and the reaction energy for fuel production is provided by renewable energy, these fuels are renewable.

Generally, the overall reaction for producing renewable carbon-based fuels from CO_2 is:

$$a\,CO_2 + \frac{b}{2}\,H_2O + energy \rightarrow C_aH_bO_c + \left[a + \frac{b}{4} - \frac{c}{2}\right] O_2 \tag{3}$$

The overall reaction for the use of these fuels is simply the reverse reaction.

Renewable carbon-based fuels are storable for long periods and most have high enough specific energy and energy density to be feasibly transported significant distances by conventional means. Although these fuels produce CO_2 when oxidized to generate electricity or other usable energy on demand, they are effectively carbon neutral because the original CO_2 was naturally occurring or the produced CO_2 is captured and recycled to produce more fuel.

Table 1 compares select renewable fuels on a number of criteria that affect the feasibility, cost and sustainability of renewable fuels. The last column of the table presents a composite score for ranking the fuels, which is an attempt to summarize all of the factors in the table in one number for overall comparison and screening. The factors are designed so that a smaller value is "better" than a larger value, which is why reciprocals of specific energy and energy density are used. The composite score is a weighted sum of the individual factors. Weighting is only used to make the effective numerical range of each factor one (1). Other than this, no attempt is made to weight individual factors.

Although some of the ranking factors are based on objective information, all are somewhat subjective. Consequently, exact values of individual factors and weighting factors for calculating a composite value may be debatable. However, the factors and the composite score are still helpful for comparing the feasibility, cost and sustainability of the selected fuels.

			Specific Energy		Energy Density		Water Consumption			Production Complexity			
Fuel (As Pure Chemical)	Chemical Formula	Physical State @ STP	MJ/ kg of fuel	Reciprocal Ratio Relative to H_2	MJ/ m^3 of fuel	(Reciprocal Ratio Relative to synC™)/ 1000	kg H_2O/ kg fuel produced	kg H_2O/ GJ LHV	Water Consumed per Unit of LHV Relative to H_2	# of Reactants for Overall Reaction	Minimum Theoretical Reaction Steps	Energy Intensity Relative to Biomass (as Cellulose)	Sum of Production Complexity Factors
synC™ (Carbon)	C	solid	32.8	3.7	63,885	0.001	0.00	0.0	0.00	1	1	2.02	4.0
Gasoline (Octane)	C_8H_{18}	liquid	44.3	2.7	33,471	0.002	1.42	32.0	0.44	2	6	2.74	10.7
Biomass (Cellulose)	$n(C_6H_{10}O_5)$	solid	16.2	7.5	24,282	0.003	0.56	34.3	0.47	1	2	1.00	4.0
Dimethyl Ether	C_2H_6O	gas	29.1	4.2	61	1.040	1.17	40.4	0.55	2	4	1.79	7.8
Methane Gas	CH_4	gas	50.5	2.4	36	1.769	2.25	44.5	0.61	2	3	3.12	8.1
Methanol	CH_3OH	liquid	19.8	6.2	15,664	0.004	1.12	56.9	0.77	2	3	1.22	6.2
Hydrogen Gas	H_2	gas	121.7	1.0	11	5.840	8.94	73.4	1.00	1	1	7.52	9.5

	Hazards 1 = Low, 3 = High				Development Status 1 = Exists/Commercial, 2 = Needs Some Development, 3 = Needs Significant Development				
Fuel (As Pure Chemical)	Safety	Health	Environ-ment	Sum of Hazard Factors	Production as Renewable Fuel	Storage & Transport	Use as Fuel for Grid-Scale Electricity Generation	Sum of Development Status Factors	Composite Score for Ranking
synC™ (Carbon)	2	2	1	5	3	1	2	6	2.1
Gasoline (Octane)	3	3	3	9	1	1	3	5	3.4
Biomass (Cellulose)	1	1	1	3	1	1	1	3	2.5
Dimethyl Ether	3	3	3	9	1	1	3	5	3.6
Methane Gas	3	2	3	8	2	1	1	4	3.3
Methanol	3	3	3	9	1	1	3	5	3.7
Hydrogen Gas	3	2	1	6	2	2	3	7	4.5

Table 1 - Comparison of Select Renewable Fuels

Table 1 presents the physical state of each fuel at standard temperature and pressure (STP), 0 °C and 1 atm, for reference. Most renewable fuels are liquid or gases, but synC™, like coal, is solid. Although bulk liquids and gases may have some storage and handling advantages over bulk solids, storage and handling of bulk solids is well developed commercial technology. Though a solid, coal has been the largest single source of energy for electrical generation in the world for many years.

High specific energy and high energy density are necessary for a transportable fuel. Both properties affect the cost of transportation: specific energy, the energy cost of transporting the fuel itself; energy density, the size and, therefore, capital cost of transportation vehicles and the energy cost of operating them.

For purposes of this paper, specific energy is defined as the lower heating value (LHV) per unit mass of fuel. The reciprocal ratio relative to H_2, which has the highest specific energy, is used in the composite score to rank renewable fuels, the higher the number the lower specific energy.

Similarly, energy density is defined as the LHV per unit volume of fuel. For synC™ and biomass as cellulose, which both have variable volumetric mass density, the highest value of volumetric mass density is used for calculation of energy density. The reciprocal ratio relative to synC™ divided by 1000 is used to compare renewable fuels, the higher the number the lower energy density. synC™ is used for reference because it has the highest energy density of the renewable fuels considered.

Table 1 presents and Figure 2 plots the specific energy and energy density of select renewable fuels. Note that Figure 2 plots energy density on a logarithmic scale because the energy density of gases is several orders of magnitude less than that of liquids and solids.

Hydrogen gas has the highest specific energy by a factor of more than two. It also has the lowest energy density by a factor of about three. In contrast, synC™ has a reasonably high specific energy and the highest energy density, depending on which allotrope of carbon is used.

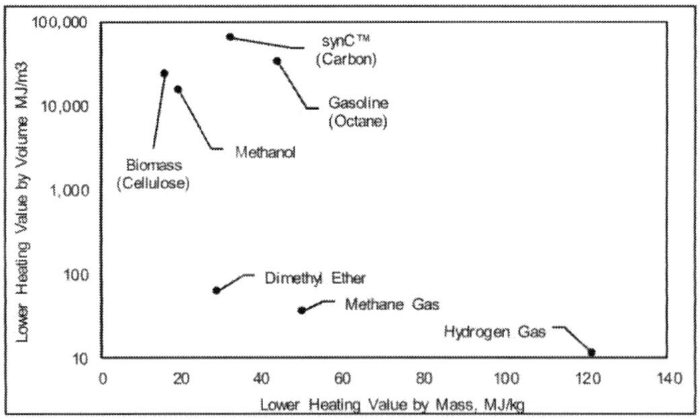

Figure 2 – Specific Energy and Energy Density of Select Renewable Fuels

All renewable fuels, except synC™, require water as a feedstock for production. Minimizing water consumption is an important factor for the feasibility and sustainability of a renewable fuel. Water is an increasingly scarce resource, and grid-scale electricity generation with carbon-based fuels is already water intensive. In addition, water may be especially scarce at the location of renewable fuel production due to arid conditions for solar and competition with agriculture for wind. Given that abundant water is not necessarily available where renewable energy is available for renewable fuel production, lower water consumption is important.

Table 1 and Figure 3 compare water consumption for the production of renewable fuels. Both present stoichiometric water consumption, which is the theoretical minimum. Table 1 presents water consumption by both mass and LHV; Figure 3, only by LHV, which is the more important value because the objective of renewable fuels is energy storage. Because water consumption for synC™ production is zero, H_2 is used as the reference for the water consumption factor.

synC™ is the only carbon-based renewable fuel that consumes no water for production. Hydrogen consumes the most water of any renewable fuel, carbon-based or otherwise, both per unit mass and unit heating value of fuel produced. Next to synC™, gasoline has the lowest water consumption of the fuels compared, consuming 44% of the water per unit energy of hydrogen. Next to hydrogen, methanol has the highest water consumption of the fuels compared, consuming 77% of the water per unit energy of hydrogen.

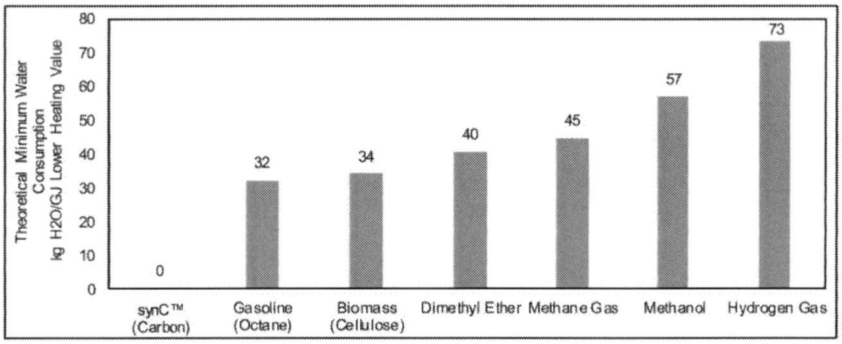

Figure 3 – Water Consumption of Select Renewable Fuels

The production complexity factors presented in Table 1 compare the likely complexity, cost and efficiency of renewable fuel production. The number of reactants for the overall reaction, number of reactions and energy intensity all impact capital and operating costs. The number or reactants for the overall reaction corresponds to the number of feedstock chemicals required, each requiring storage and handling systems. The theoretical minimum number of chemical reactions is a surrogate for the number and complexity of reactors and other unit operations. Relative energy intensity is expressed as the ratio of specific energy relative to biomass, which has the lowest specific energy of the fuels compared. Greater energy intensity requires more/larger heat exchangers, and/or higher temperature and/or pressure, either of which requires more expensive equipment and materials of construction, and tends to result in lower energy efficiency due to greater thermal losses to the environment. The metric for production complexity is simply the sum of individual factors.

Note that, although biomass actually requires two reactants, CO2 and H2O, they often are provided naturally by the atmosphere with little or no material handling required. However, ground and surface water is often applied to agricultural crops to increase productivity. Therefore, water is considered a feedstock for biomass. Likewise, plants naturally perform photosynthesis with little or no human intervention. However, plants do require cultivation and harvesting each of which is considered a reaction step.

As might be expected, biomass has the lowest value for production complexity. However, synC™ is virtually tied with biomass. Gasoline and hydrogen gas have the highest complexity factors.

Hazard factors compare environment, health and safety issues. The hazards factor metric is the sum of individual hazard factors.

Development status factors compare the current status of technological and commercial development for the production, handling and use of renewable fuels for grid-scale electricity generation. The development status metric is simply the sum of these factors.

Figure 4 presents the composite score for ranking renewable fuels on overall feasibility, cost and sustainability. As discussed above, lower scores are better. synC™ has the lowest score, 2.1, followed by biomass, 2.5, and methane gas, 3.3. Hydrogen gas has the highest score, 4.5.

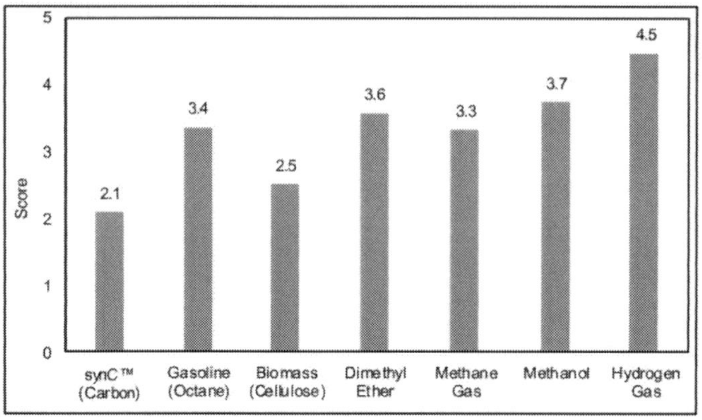

Figure 3 – Composite Ranking Scores of Select Renewable Fuels

Conclusion

Synthetic Coal Cycle Technology™, which produces the renewable fuel synC™, is a promising means for carbon dioxide use for electricity generation. synC™ compares favorably with other potential renewable fuels from carbon dioxide. More research and development is required for synCCT™ to become a commercial reality.

Acknowledgments

The author thanks all who have supported this work, especially his wife, family and those who contributed to the Indiegogo crowdfunding campaign "Increase the Four Percent" and God who makes all things possible.

References

Cundliffe, 2011. Carbon, Nitrogen and Oxygen Separator and Method of Use Thereof, World Intellectual Property Office, PCT application WO 2011/050437

Fukuda, et al., 2007. Dissociation of carbon dioxide and creation of carbon particles and films at room temperature, New Journal of Physics, 9 (321)

Hosback, et al., 2011. Physical and technical method for splitting carbon dioxide into carbon and oxygen, in which the binding energy of the carbon dioxide molecule is overcome using pulsed energy input, German patent publication DE 10 2009 048 541 A1

Schora, 1975, Carbon Black Fuel Production, US patent 3,861,885

Walker, 1980. Manufacture of Finely Divided Carbon, US patent 4,185,083

Zietlow, D. A., 2014. Processes and Systems for Storing, Distribution and Dispatching Energy on Demand Using and Recycling Carbon, US patent application 2014/0272639 A1

CMTC- 440210-MS

Restoring Gases Emission in Aspect of CO2 Capture :(Case Study: Borzouyeh Petrochemical Complex (BPC) in Pseez)

Peyman Razmjoo Sr., Senior Process Engineering, Borzouyeh Petrochemical Complex (BPC), Tehran, Iran and
Ali Rahneshin Sr., Business Dept, Borzouyeh Petrochemical Complex (BPC), Bandar abbas, Iran

Abstract

The industry provides significant opportunities for a reduction in energy demand and CO2 emissions. It could be achieve through a variety of measures and technology advances in equipment and materials. The Refineries & Petrochemical industries are the third largest source of CO2 emissions takes an account for 30 % of the final energy use.

This huge producers are burning of thousands Tones of gases in the flaring system, the most amount of Carbon compounds (i.g.hydrocarbons) and valuable gases such as hydrogen burn in flaring system. Major emissions are Co2, Nox. This Losses in addition to the waste of a large amount of the nation's capital, suffered the ecosystem imported and the cost of many diseases such as respiratory as well as skin will be followed.**Second**, BPC (same as a Petro-Refinery) is followed by macro policy in the environmental debate establish the three projects for management of Restoring the Gases from Flaring system.

With due these projects, it could be to prevent the release of more than 380 tons of Co2 gas daily and in other side, it will be in access as a revenue from the sale of fresh gases with margin of income around 500,000 $ per year. These projects were defined in our company.

- **Construction & Commissioning of the Hydrotreating(Sweetening) process of light End(LTE) as by-product in Complex**
- **Flare Gas Recovery plan**
- **Pilot unit of Sour gas conversion to DMDS (Di-methyl Di-Sulfide) as new product.**

BPC could be to prevent the loss of hundred tons of hydrocarbons, leading in carbon management requirements in terms of Capture and Storage of Carbon. It should be noted that another important aspect is commitment to environmental issues and the implementation of relevant considerations. Due to the increasing sensitivity to global environmental issues, is in particular importance.

❖ Process of light End(LTE) as by-product in Complex

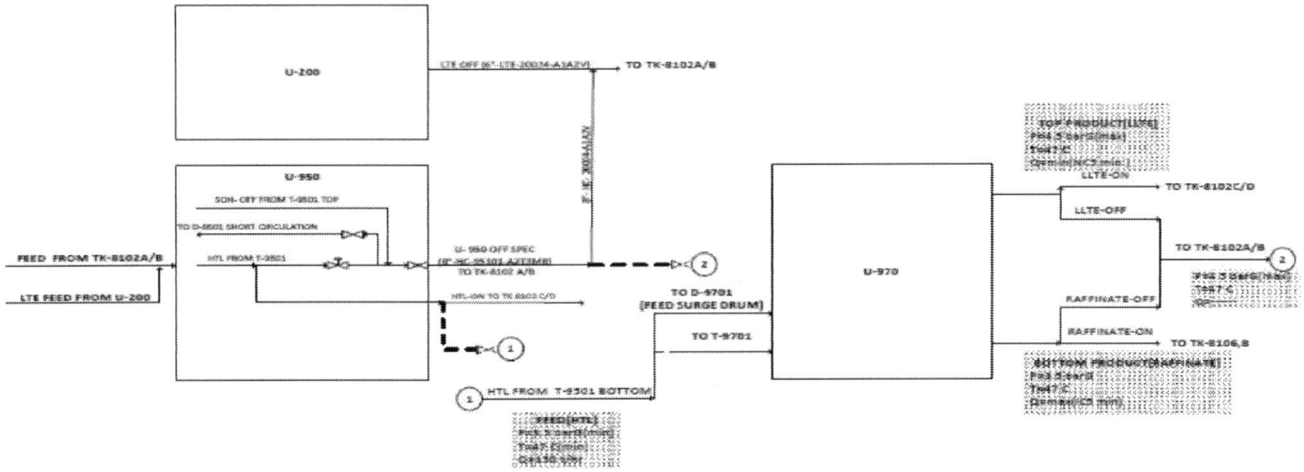

Fig No.1 Block Flow Diagarm

Aspects of Plan:

- Reduce the losses of Flaring because of new deffination of source for using the end product based on the Market scan of World Petrochemical & Refineries demands.
- Enviromental care because of reducing flaring by establishing this unit.More than thousands Tones of gases in the flaring system, the most amount of Carbon compounds (i.g.hydrocarbons) and valuable gases such as hydrogen burn in flaring system
- In short time turn over of fixed cost of project and made a profit for the plant in both side as an one of pioneers of Enviromental friendly plants and produce new product based on the existing demand.
- In optimistic manner, it is a good opportunity for the PSEEZ to promote this figure to reduce the CO_2 and other un-favor gases Emmision to the Atmosphere.
- Many projects are under investigation and study for doing well this world plan nearer to the pioneers.

❖ **Flare Gas Recovery plan**

The establishing in pilot phase for recovery of valuable gases are another project of BPC Plant for scale up to the industerieal phase.The field monitoring are comprehensive for investigation gainst the Industerial scale and the Plant is proceede to establish the industrial scale of Flare Gas Recovery by aspect of own mission of statement repeect to Enviroment and Global Climate change plan.

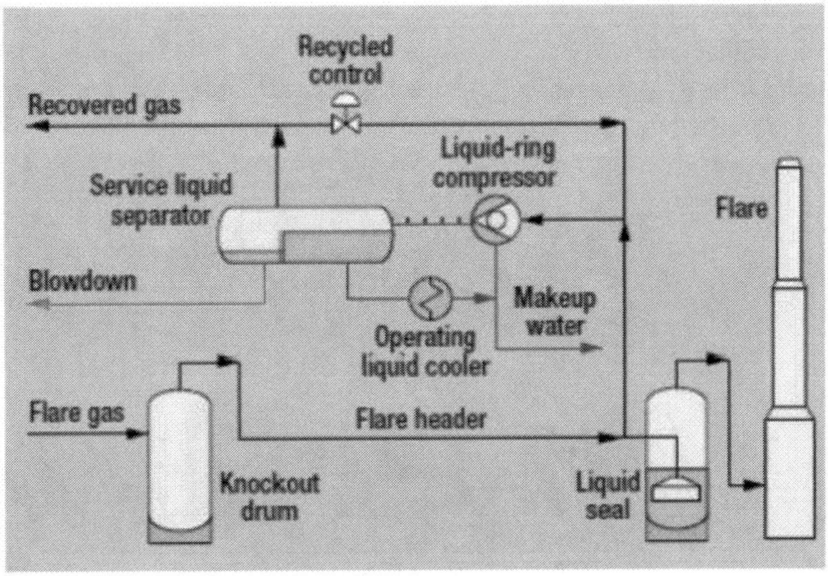

Fig No.2 Flare Gas Recover Schematic

Item No.	Major Companies	Number of oil fields	Faring gas amount	Price (Cent/cm)
1	National Iranian South Oil Company	21 fields	6.186 mcm/d	3.5 to 4.2
2	Iranian Offshore Oil Company	11 fields	10.173 mcm/d	1.75 to 3.5
3	Arvandan Oil & Gas Company	4 fields	2.778 mcm/d	3.5
4	Iranian Central Oil Fields Company	4 fields	0.3 mcm/d	3.5
5	South Pars gas field (phases 1 to 10)	10 phases	0.39 mcm/d	3.5

Table No.1 Four major companies as an affiliated of Iran Oil Ministry

❖ **Pilot unit of Sour gas conversion to DMDS (Di-methyl Di-Sulfide)**

Use the new methods same as GTL Process to recover and convert the Sulfure compounds for producing DMDS is a FS(Fiesibility study) that it was proceed in R&D Department.This is a new plan for in pilot phase by biochemical catalysts to do this job.The apparatuse is skid mounted with Up/Down stream connection that mounted in process connections of unit.

Now a day's, A suggested method forcontrolling the level of greenhouse gases in the atmosphere is prevention of flaring gas. In this work,three methods are proposed to recover gas instead of conventional burning in a flare at the Asalooye GasRefinery. These methods aim to minimize environmental and economical disadvantages of burning flaregas. The proposed methods are: 1) Gas-to-Liquid (GTL) production, 2) electricity generation with a gasturbine and, 3) compression and injection into the refinery pipelines. In order to find the most suitablemethod the required equipments for the three aforementioned methods are simulated.

Reffrences:

1. Resaerch & Development Investments Reports of Borzouyeh Petrochemical Complex (BPC)

2. National Iranian Oil Company Research & Development, Technical issues 20110-2015

3. Overall Conferences reports of Climate Change and the role of Iran to made monitor and control own situation

4. A comparative study of three different methods for flare gas recovery of Asalooye Gas Refinery, M.R. Rahimpour , Z. Jamshidnejad , S.M. Jokar , G. Karimi , A. Ghorbani , A.H. Mohammadi,Elsewire

5. Study on Falre Gas Recovery (FGR) to minimize wastes and economics benefits, Soroush Younessi, IAU Facualty of Engineering, Tehran, Iran.

6. FGRS the Key Step to Protect Environment in Petrochemical Plants, Omid Zadakbar, Ali Vatani, Armin Zadakbar,Faculty of Chemical Engineering, the University College of Engineering, University of Tehran

CMTC-440223-MS

The Energy Gap – How CO_2 Tertiary Oil Recovery will mark its place in the 21st Century

David Merchant, Merchant Consulting

This paper was prepared for presentation at the Carbon Management Technology Conference held in Sugarland, Texas, USA, 17–19 November 2015.

Abstract

The Economic Sustainability of the 21st Century is based largely on how well Mankind can best fill its World Oil Energy needs. Back in 2005, the World was consuming around 80 million barrels of oil a day. Today, that number has risen to 86.8 million barrels a day. Based on straight line projections, the World would require between 160 million barrels a day and 200 million barrels a day in the year 2100 to support an ever increasing population growth.

In 2005, the World thought we were running out of oil. What we learned is that Peak Oil is a function of how fast we can keep up with technology to help "Fill" the "Energy Gap" that exists between Consumption and Energy Supply.

For the World to maintain Economic Sustainability in the 21st Century and lower our CO_2 impact on the Planet, we must at some point transition ourselves between a Hydro-Carbon Based Economy to a Hydrogen Based Economy without sacrificing all of what we have accomplished as a society to date. The Hydrogen Economy currently contains Wind, Solar, and Nuclear (Fission) that by themselves will add to the curve, but will still be small in comparison to our total energy needs. In addition, Hydrogen Power along with Nuclear (Fusion) are only in their infant stages of development and will not see any impact until the technology can compete economically with conventional sources of power.

To achieve Economic Sustainability requires filling the total area under the curve all the way through the 21st Century. "Gap Fillers" such as Conventional Oil Recovery (Primary, Secondary, and Tertiary) along with Unconventional Methods (Shale Oil and Shale Gas) will help fill our energy needs in the 21st Century.

The question becomes "How will CO_2 for Enhanced Oil Recovery mark its place in the 21st Century?"

Peak Oil in 2005 – Fact or Fiction

M. King Hubbert was a distinguished scientist who published on many aspects of petroleum geology. In 1974, he predicted World Oil Supply would Peak around 1995 after he had successfully demonstrated he could predict the Continental USA Oil Production would Peak in 1970, in which it did, in a prediction he made in the 1950's. His methods still hold true today.

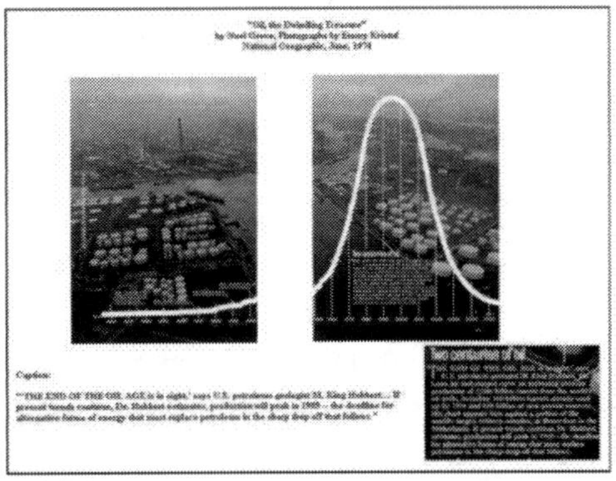

1974

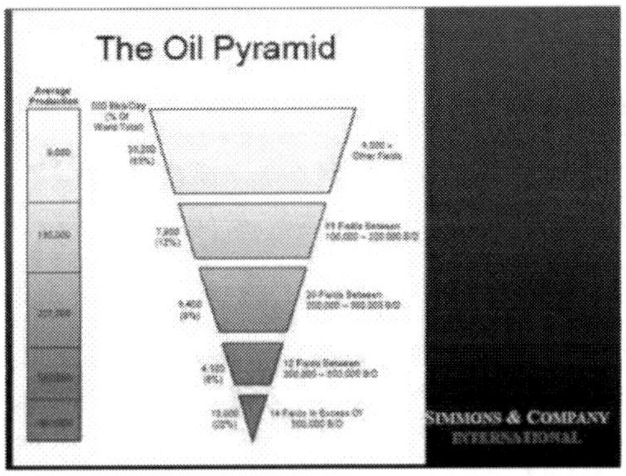

2005

In 2005, Mathew Simmons, a respected Consultant in Houston, wrote the Book "Twilight in the Desert" which challenges whether or not Saudi Arabia can meet today's World Oil Demand. The World's Top 14 fields supply 20% of the World's Current Oil Producing Capacity. All these fields are 60 years or older and based on historical basin performance, are expected to decline sharply.

In 2015, Helge Hove Haldorsen, President of SPE, along with Scott Tinker, announced the G7 members' ambition is to "Switch" from a "Hydro-carbon" based economy to a total "Hydrogen" based economy within the next 85 years (2100).

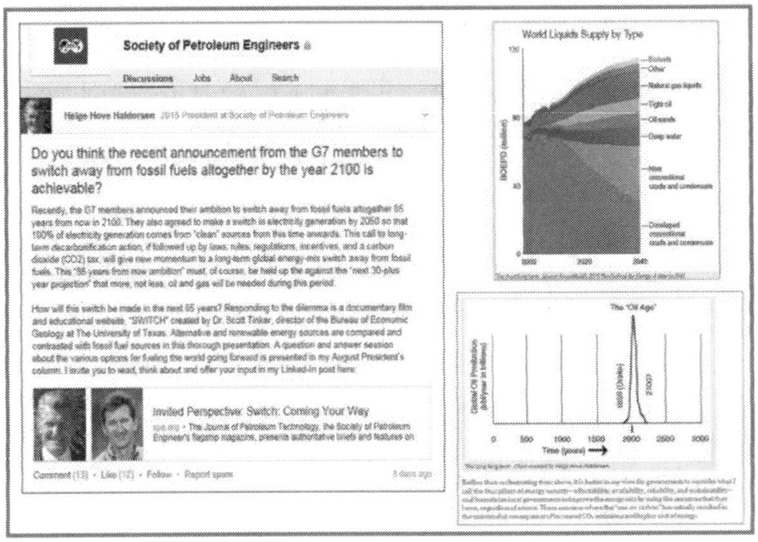

http://www.switchenergyproject.com/

What that means is an end to the "Hydro-carbon" Revolution that has spurred both Domestic and International Economic Sustainability for over a hundred and fifty years.

The Energy Gap – Future Economic Sustainability through CO_2 Injection in Enhanced Oil Recovery Projects

The objective of this paper is to present "The Energy Gap – Future Economic Sustainability through CO_2 Injection in Enhanced Oil Recovery (EOR) Projects". The paper addresses the current World Status Quo in regards to Future Oil Recovery and how Anthropogenic CO_2 from IGCC Plants, Refineries, and other plants will be used to fill the "Energy Gap" that exists between oil depletion of our current Hydrocarbon Economy and full implementation of the future Hydrogen Economy. New Technologies are currently being developed to fill this gap such as gas-to-liquids and Clean Coal Technology such as IGCC (Integrated Gasification Combined Cycle). IGCC technology converts biomass (Coal, Petroleum Coke, etc.) to a gaseous state whereby the gas can be reformed into various components, creating a variety of different products. Electricity has been the central driver for economic purposes. However, Ammonia and Urea can be output for the creation of fertilizer. The IGCC plant also outputs a high quality CO_2 stream at high pressure, which makes the CO_2 economically attractive when compared to Post-combustion CO_2 extraction stack gas processes. The availability of CO_2 at high pressure can "Revitalize the Oil Industry" in many Mature Basins around the World. Hydrogen is also output. In the future, the Hydrogen from these plants will provide the necessary fuel to drive the World from coal producing basins for the next thousand years.

Most countries in the World recognize CO_2 is a major contributing factor to Global Warming. In addition, most countries in the World also recognize we are running out of light oil and expect Saudi Arabia and other major Middle East countries to supply the World with enough oil to meet our needs until the Hydrogen Economy can take over. In 2005, the World contained approximately 1,106 BBO (Billion Barrels of Crude Oil) as reported by the BP Statistical Annual Review, World Oil, and Oil & Gas Journal. Of this total, two-thirds of the World's Proved Oil Reserves lie in the Middle East. In 2005, the World consumed around 80 MMBOPD (Million Barrels of Crude Oil a day) of which 10 MMBOPD was supplied by Saudi Arabia. Today, the World consumes around 86 MMBOPD (Million Barrels of Crude Oil a day). To maintain market share, Saudi Arabia increased oil production in 2014 to 11.5 MMBOPD, bringing oil price down to historically low levels less than $50.00 per barrel. The Saudis have also provided predictions for a sustained 10 MMBOPD or 12 MMBOPD case for years to come. However, as current World Oil Demand increases to 100 MMBOPD by 2030, the question becomes "How long can the current low oil prices be sustained?" And, "At what oil price will CO_2 for Enhanced Oil Projects and CO_2 for Sequestration Projects be profitable?" And, from competing technologies for funding, "At what Oil Price will "Horizontal Fracture Drilling and Heavy Oil Extraction Recovery Techniques come back to market?" In 2005, the World was concerned with the question of "Peak Oil" and its impact on global oil supply. Today, in the "New Millennium" the question becomes "What new technology will fill the energy gap? And, "What place will CO_2 for Enhanced Oil Recovery and CO_2 for Sequestration have to help fill the energy gap?"

During the past 43 years, CO_2 flood technology for enhanced oil projects evolved from a partially understood process filled with uncertainties to a process based on proven technology and experience. Many questions involved with CO_2 flooding have been thoroughly analyzed and answered. This knowledge is currently being used by a limited number of companies that actually know how to design, implement, and manage a CO_2 flood for long term profit. The purpose of this report is to help disseminate this knowledge to operating companies interested in EOR flooding or to CO_2 Sequestration Communities interested in storing CO_2 in EOR projects.

The Old Millennium - Enhanced Oil Recovery

With more than forty-three years of successful Enhanced Oil Recovery (EOR) Projects in the Permian Basin, Rocky Mountains, and Louisiana, Carbon Dioxide (CO_2) Flooding is a proven method for recovering oil not recoverable by either Primary or Secondary Recovery Methods. In the **Previous Millennium**, CO_2 was produced from Natural CO_2 Source Fields located in Colorado, Wyoming, New Mexico, and Louisiana and delivered to oil fields with over 4,500 miles of pipeline infra-structure. Today, the Permian Basin produces over 900,000 BOPD, of which 237,000 BOPD is a direct result of CO_2 Injection. Since 1972, over 1.1 Billion Barrels of Tertiary Oil have been recovered with CO_2 Injection.

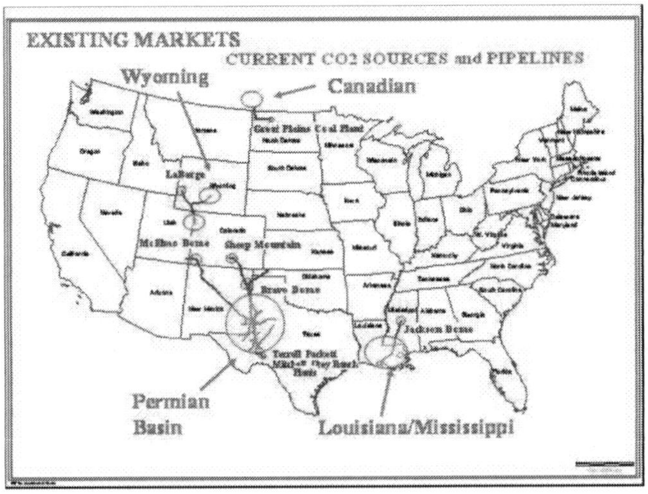

The New Millennium - Enhanced Oil Recovery and CO_2 Sequestration

In the "**New Millennium**" CO_2 produced from IGCC Plants, Post Combustion, Oxy-fuel Plants, and other CO_2 Industrial Sources will provide the oil we so desperately need in tomorrow's hungry Energy Environment. The availability of CO_2 from these plants will provide the ability to "Revitalize" the Oil Industry in many Mature Oil Basins in the United States and throughout the World. Tertiary Oil Recovery using CO_2 will be used to recover an additional 10% to 26% of the Original Oil-in-Place, not recoverable with any other type of Oil Recovery Mechanism.

World Oil Reserve Determination

M. King Hubbert was a distinguished scientist who published on many aspects of petroleum geology. In 1974, he predicted World Oil Supply would Peak around 1995 after he had successfully demonstrated he could predict the Continental USA Oil Production would Peak in 1970, in which it did, in a prediction he made in the 1950's. His methods still hold true today.

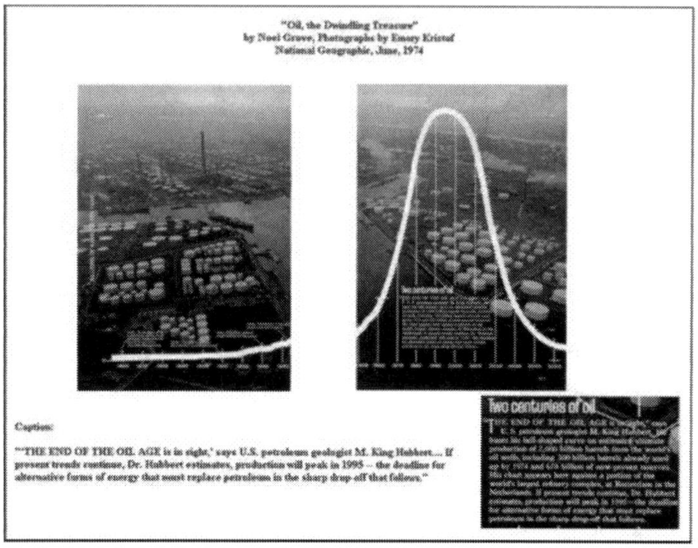

In 2005, Mathew Simmons, a respected Consultant in Houston, wrote the Book "Twilight in the Desert" which challenges whether or not Saudi Arabia can meet today's World Oil Demand. The World's Top 14 fields supply 20% of the World's Current Oil Producing Capacity. Based on historical basin performance, all these fields are expected to decline sharply.

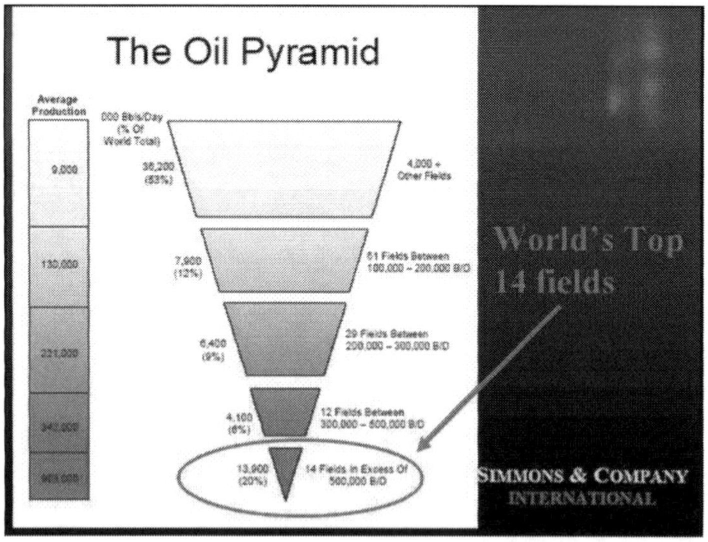

The Top 14 fields (500,000+ BOPD) in the World produce 20% of the Worlds Current Oil Producing Capacity. These fields are 60 years or older. No new Giant Oil fields have been found in over 30 years. New Discoveries today are much smaller in size and rate when compared to these Giants.

A Hubbert Curve (Merchant) was developed based on historical World oil production and current proved reserves. Current World Oil Reserves at the end of December 2004 were approximately 1,106 Billion Barrels according to the BP Statistical Review, World Oil, and the Oil & Gas Journal.

World Oil Production
Current Proved Reserves

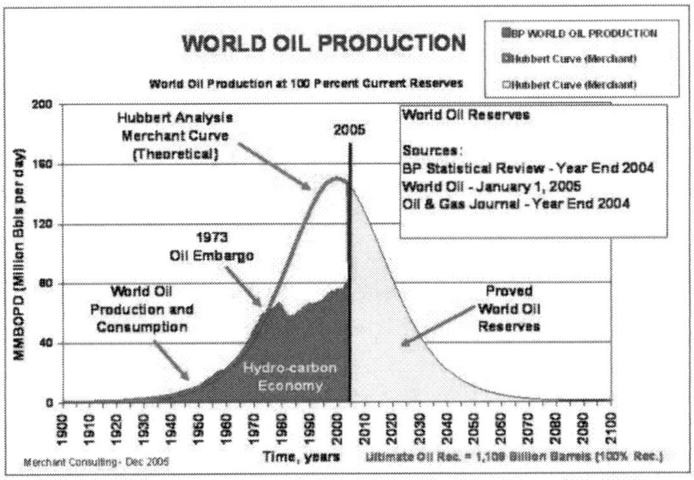

Current World Oil Reserves are approximately 1,106 Billion Barrels according to the BP Statistical Review, World Oil, and the Oil & Gas Journal.

In 2004, Middle East Reserves represented 734 Billion Barrels of the total 1,106 Billion Barrels or 66 % of Total World Oil Reserves.

Hydro-carbon Economy versus Hydrogen Economy
World Oil Consumption

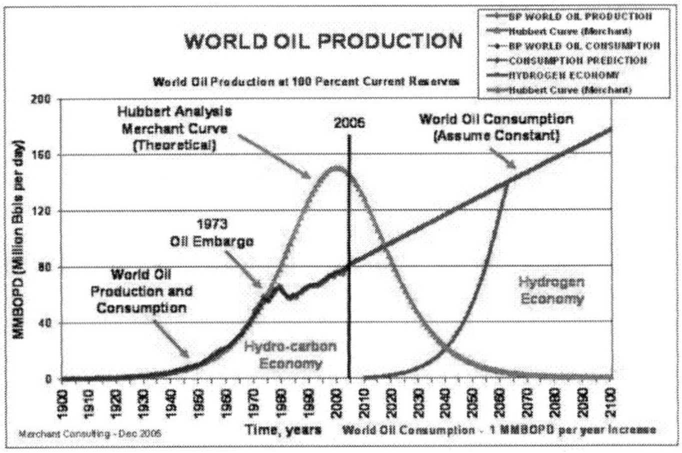

World Oil Consumption is assumed to increase at a rate of 1 Million Barrels per Day on an Annual Basis even though World Population is increasing Exponentially.

World Oil Consumption (Blue Curve) will continue to increase. Here we assume the Oil Rate Consumption will grow at a constant rate of 1 Million BOPD per year even though population is growing exponentially. After 10 years of history, the Oil Consumption Rate was 86 MMOBPD in 2014, which is still on trend with the original 2005 prediction. The Hydrogen Economy will have to grow quickly in order to fill the Energy Gap as oil rate falls in future years unless suitable alternative sources are found.

World Oil Rate Prediction

The ability to predict future World Oil Rate Performance depends on many factors, the first factor being the amount of oil remaining to be produced. Proved Oil Reserves by definition are estimated quantities that analysis of geologic and engineering data demonstrates with reasonable certainty are recoverable under existing economic and operating conditions. With the inclusion of Saudi Arabian Reserves, the World certainly contains a sufficient supply of oil reserves. Therefore, the World doesn't have a RESERVE SUPPLY problem. The second controlling factor is World Oil Producing Rate. The World's current Producing Oil Rate Capacity is controlled by the top producing fields in the World, along with new oil recovery technologies. To date, Oil Supply has been able to meet all our Consumption needs, new discoveries and technology was found to help fill the gap. A graph showing how the prediction was adjusted is shown below. Each of the three predictions is tied to current World Oil Reserves of 1,106 MMBO.

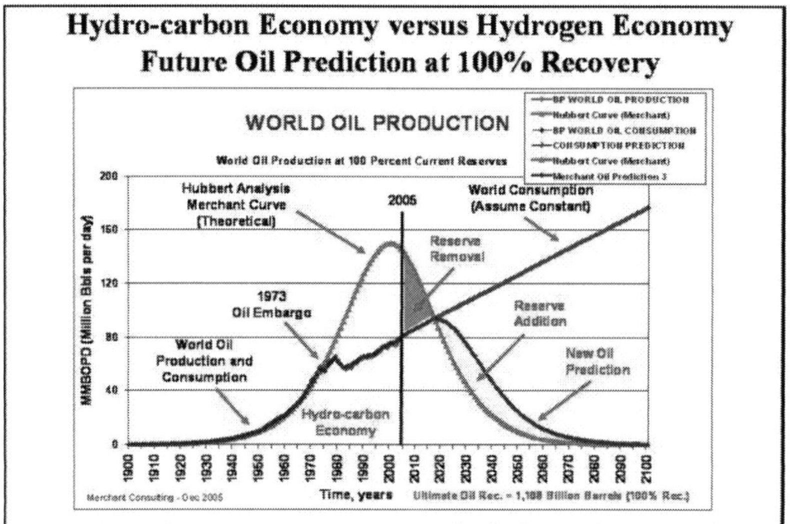

Even with the 160 Billion Reserve Inclusion added to the Base Prediction, an Energy Gap will exist in the Future between today's Current Oil Rate Producing Capacity and the start of the Hydrogen Economy. These reserves include Heavy Canadian Oil Reserves.

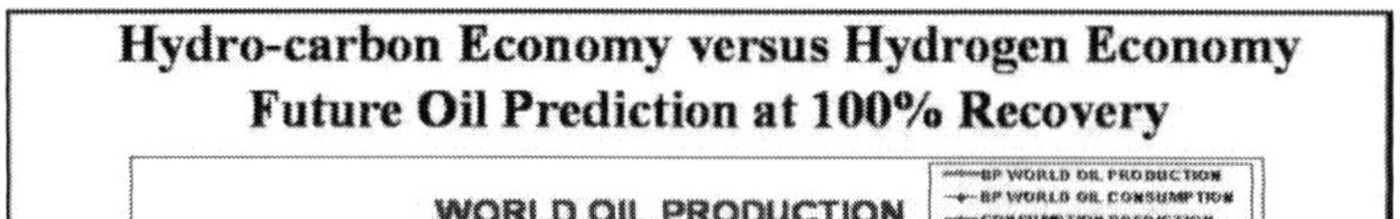

The Three Oil Predictions presented above all recover 1,108 Billion Barrels of Proved Reserves. The Timing as to when World Oil will Peak will depend on how fast the top 14 fields in the World decrease in production and how fast we can add New Production. Even with projected New Reserves being added, there will be an "Energy Gap" that will have to be filled with New Oil.

Where in the 21st Century will CO_2 for Tertiary Oil Recovery fit-in?
The Energy Gap:

For the World to maintain Economic Sustainability in the 21st Century and lower our CO_2 impact on the Planet, we must at some point transition ourselves between a Hydro-Carbon Based Economy to a Hydrogen Based Economy without sacrificing all of what we have accomplished as a society to date. The Hydrogen Economy currently contains Wind, Solar, and Nuclear (Fission) that by themselves will add to the curve, but will still be small in comparison to our total energy needs. In addition, Hydrogen Power along with Nuclear (Fusion) are only in their infant stages of development and will not see any impact until the technology can compete economically with conventional sources of power.

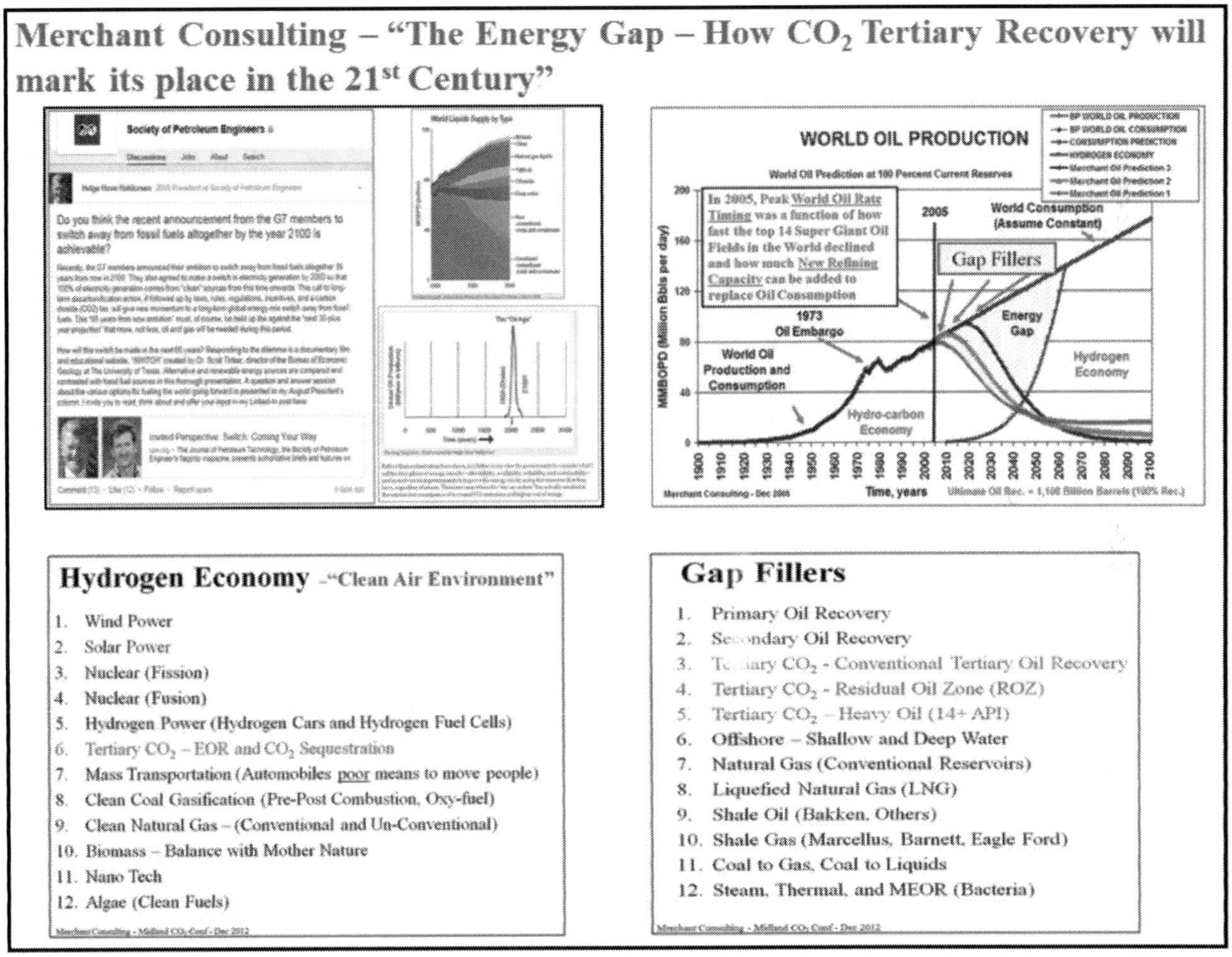

To achieve Economic Sustainability requires filling the total area under the curve all the way through the 21st century. "Gap Fillers" such as Conventional Oil Recovery (Primary, Secondary, and Tertiary) along with Unconventional Methods (Shale Oil and Shale Gas) will help fill our energy needs in the 21st Century.

BP Statistical Review - 2015

The BP Statistical Review for Total Proved Reserves in 2015 demonstrates how the "Energy Gap" theory works when New Technologies are added to help "Fill the Gap". The report provides a basic understanding of how New Technologies including CO_2 for Enhanced Oil Recovery will mark its place in the 21st Century. Here, current high oil prices over the past ten years advanced Heavy Oil Sands development and Horizontal Frack Drilling. The result was an improvement in total proved reserves.

BP 2015 Statistical Review

Oil: Total proved reserves

	at end 1994 Thousand million barrels	at end 2004 Thousand million barrels	Inc/Dec over 1994 at end 2004 Thousand million barrels	at end 2014 Thousand million barrels	Inc/Dec over 2004 at end 2014 Thousand million barrels	Share of total	R/P ratio	
Total World	1118.0	1366.2		1700.1		100.0%	52.5	
Venezuela	64.9	79.7	22.9%	298.3	274.2%	17.5%	100.0	Orinoco Belt/Heavy Oil
Saudi Arabia	261.4	264.3	1.1%	267.0	1.0%	15.7%	63.6	
Canada	48.1	179.6	273.0%	172.9	-3.7%	10.2%	100.0	Canadian Oil Sands
Iran	94.3	132.7	40.8%	157.8	18.9%	9.3%	100.0	
Iraq	100.0	115.0	15.0%	150.0	30.4%	8.8%	100.0	
Russian Federation	115.1	105.5	-8.4%	103.2	-2.2%	6.1%	26.1	
Kuwait	96.5	101.5	5.2%	101.5	0.0%	6.0%	89.0	
United Arab Emirates	98.1	97.8	-0.3%	97.8	0.0%	5.8%	72.2	
United States	29.6	29.3	-1.1%	48.5	65.4%	2.9%	11.4	Horizontal Fracking
Libya	22.8	39.1	71.6%	48.4	23.6%	2.8%	100.0	
Nigeria	21.0	35.9	70.9%	37.1	3.3%	2.2%	43.0	
Kazakhstan	5.3	9.0	69.2%	30.0	233.3%	1.8%	48.3	
Qatar	3.5	26.9	667.6%	25.7	-4.3%	1.5%	35.5	
China	16.3	15.5	-4.5%	18.5	19.0%	1.1%	11.9	
Brazil	5.4	11.2	109.2%	16.2	43.7%	1.0%	16.9	
Angola	3.0	9.0	204.8%	12.7	40.2%	0.7%	20.3	
Algeria	10.0	11.8	18.2%	12.2	3.4%	0.7%	21.9	
Mexico	49.8	14.8	-70.3%	11.1	-25.2%	0.7%	10.9	
Ecuador	3.5	5.1	44.9%	8.0	58.1%	0.5%	39.4	
Azerbaijan	1.2	7.0	505.0%	7.0	0.0%	0.4%	22.6	
Norway	9.7	9.7	0.2%	6.5	-32.7%	0.4%	9.5	
India	5.8	5.6	-4.2%	5.7	3.2%	0.3%	17.6	
Oman	5.1	5.6	9.3%	5.2	-7.6%	0.3%	15.0	
Australia	3.8	3.9	1.9%	4.0	2.4%	0.2%	24.3	
Malaysia	5.2	5.2	-0.8%	3.8	-27.3%	0.2%	15.4	
Indonesia	5.0	4.3	-13.7%	3.7	-14.1%	0.2%	11.9	
Colombia	3.1	1.5	-52.9%	2.4	65.5%	0.1%	6.8	
Argentina	2.3	2.5	10.0%	2.3	-6.1%	0.1%	10.1	
Other Europe & Eurasia	2.3	2.2	-1.5%	2.0	-11.7%	0.1%	14.0	
Other S. & Cent. America	1.0	1.5	53.3%	0.5	-64.3%	•	9.6	

Gap Fillers

Note:

1. **Canadian Heavy Oil Sands Development Increase**
2. **Venezuelian Orinoco Heavy Oil Increase**
3. **United States Horizontal Fracking Increase**

Acknowledgements

I would like to express my gratitude to the members of the Carbon Management Technology Conference (CMTC) for allowing me the opportunity to be part of this CO_2 Conference. I would especially like to thank George Koperna of Advanced Resources International for his support in making this effort possible.

David H. Merchant
Merchant Consulting

NETL

Driving Innovation ♦ Delivering Results

The U.S. DOE Carbon Storage Program: Status and Future Directions

Bruce Brown
Infrastructure Coordinator
Carbon Management Technology Conference
November 19, 2015

U.S. DEPARTMENT OF ENERGY | National Energy Technology Laboratory

Carbon Storage Program
What is the Carbon Storage Program

NETL

- **Integrated Research Program**
  - Applied Laboratory and Field Testing
- **Focused on Technical Challenges**
  - Research pathways focus on needs identified by "technical experts" to move toward CCS deployment

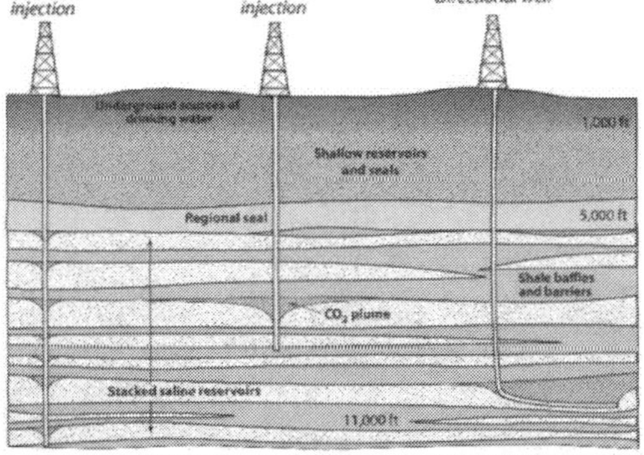

- **Technology Development**
  - Tools and techniques are ready for commercial application by both industry and regulators
    - *Do not develop/enforce Regulations*
    - *Do not develop Policy*
- **Research Mechanisms**
  - Funding Opportunity Announcements, National Laboratory Network, NRAP and SubTER

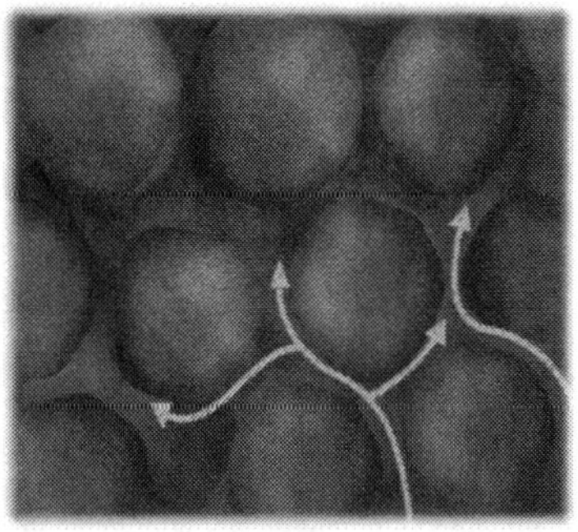

U.S. DEPARTMENT OF ENERGY | National Energy Technology Laboratory

2

Carbon Storage Program
Technical Priorities to "Master the Subsurface"

- Predicting and monitoring CO_2 plume and brine pressure front movement, stabilization, and impacts
- Optimization of reservoirs for CO_2 storage capacity
- Developing and validating risk assessment strategies
- Mitigating risks such as the risk of leakage from old wells and induced seismicity
- Carrying out (Large-volume and Fit-for-Purpose) field tests for different storage types and depositional environments

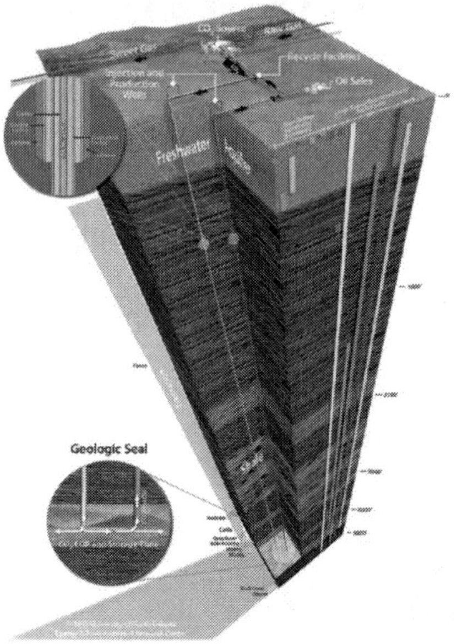

Carbon Storage Programmatic Structure

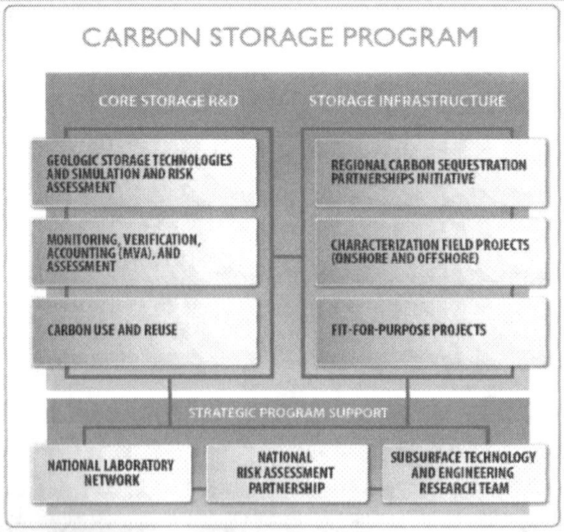

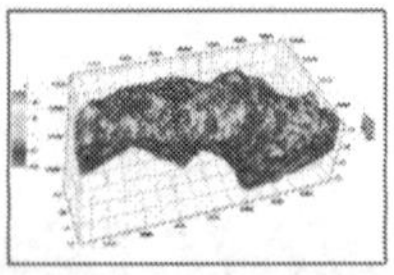

Storage Infrastructure
Regional Carbon Sequestration Partnerships (RCSP)

NETL

Seven Regional Partnerships

400+ distinct organizations, 43 states, 4 Canadian Provinces

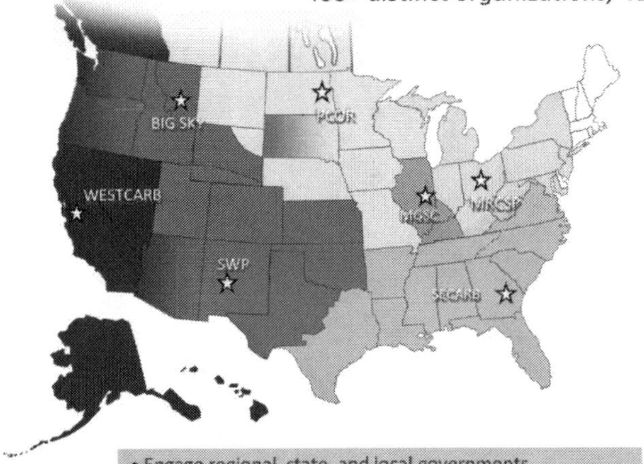

Characterization Phase (2003-2005)
- Search of potential storage locations and CO_2 sources
- Found potential for 100s of years of storage

Validation Phase (2005-2011)
- 20 injection tests in saline formations, depleted oil, unmineable coal seams, and basalt

Development Phase (2008-2018+)
- 8 large scale injections (over 1 million tons each)
- Commercial scale understanding
- Regulatory, liability, ownership issues

- Engage regional, state, and local governments
- Determine regional sequestration benefits
- Baseline region for sources and sinks
- Establish monitoring and verification protocols
- Address regulatory, environmental, and outreach issues
- Validate sequestration technology and infrastructure

Storage Infrastructure
RCSP Development Phase CO_2 Injection Volumes

NETL

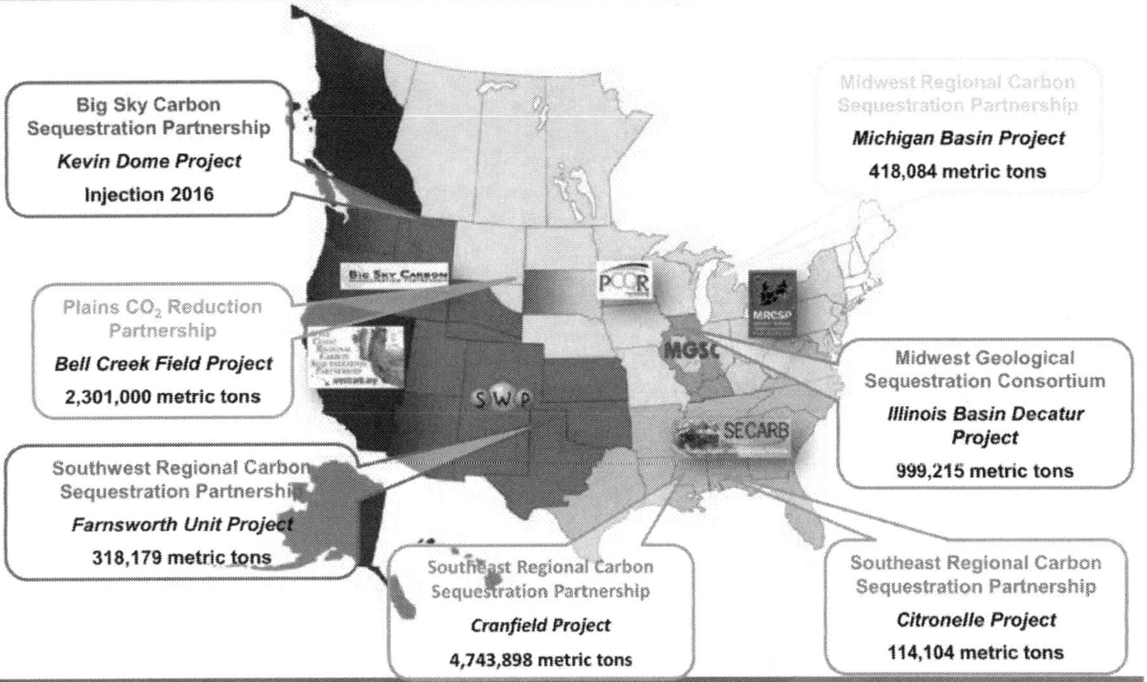

Citronelle Project
Southeast Regional Carbon Sequestration Partnership (SECARB)

Project Summary
Project Summary

Objective

Large scale field validation of saline formation storage of CO_2 in the Paluxy formation, Citronelle Dome, Alabama.

Status

Completed injection of 114,000 metric tons of CO_2 from Alabama Power Plant Barry. Conducting post injection monitoring. Will extract lessons learned and conduct tech transfer.

Key Findings/Accomplishments

- First integration of a CO_2 capture plant on a coal plant with pipeline transportation and injection for geologic storage
- Crosswell seismic shows CO_2 plume contained within reservoir
- Implemented unique wellbore–based monitoring instrumentation

Challenges

- Monitoring of individual reservoirs in a stacked storage system

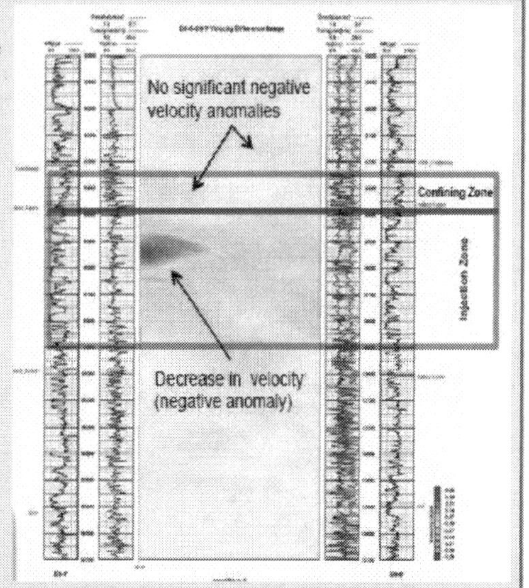

Time-lapse image of CO_2 plume

Cranfield Project
Southeast Regional Carbon Sequestration Partnership (SECARB)

Project Summary
Project Summary

Objective

Large scale field validation of associated CO_2 storage in conjunction with EOR. Injection into the Lower Tuscaloosa formation, near Cranfield, Mississippi.

Status

Field work complete; more than 4,743,000 metric tons CO_2 injected. Data analysis continues.

Key Findings/Accomplishments

- Validation and comprehensive analysis of above zone pressure monitoring
- Multiple subsurface monitoring methods tested in Detailed Area of Study
- Heterogeneity in properties/structure significantly influences plume movement
- Comprehensive measurement and analysis of groundwater and near surface; no leakage signal detected

Challenge

- Defining plume extent in a domal structure where the plume varies in thickness

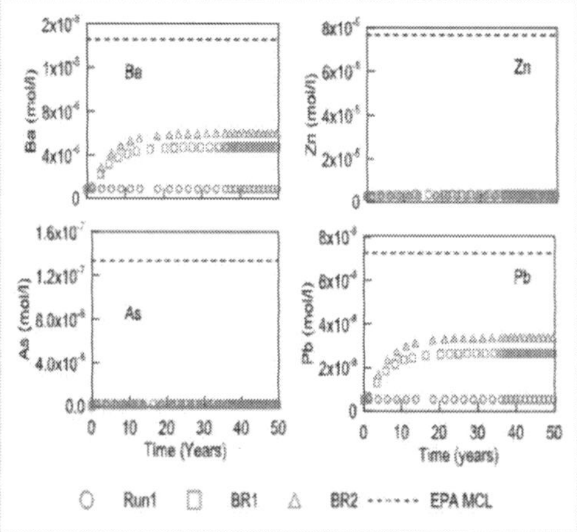

Groundwater simulation results

Core R&D Accomplishments

Wellbore Integrity and Mitigation

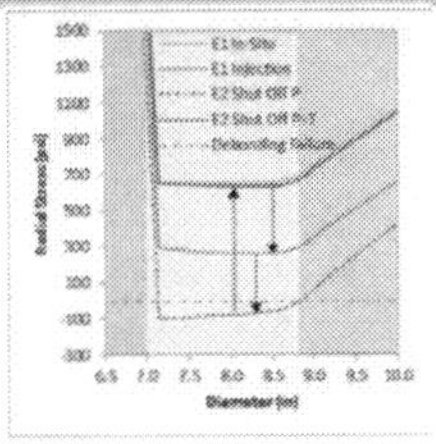

Statistical Analysis of CO_2 Exposed Wells to Predict Long Term Leakage – **University of Louisiana at Lafayette**

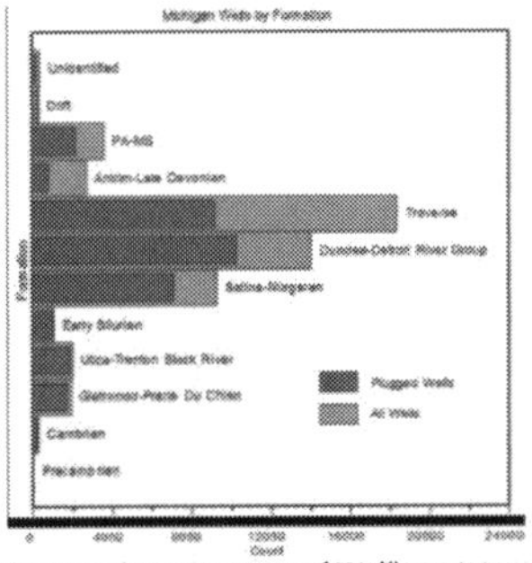

Systematic Assessment of Wellbore Integrity for Geologic Carbon Storage Projects Using Regulatory and Industry Information - **Battelle**

- **Wellbore integrity modeling**
  - Finite element model developed to predict integrity of CO_2 injection wells
  - Detailed modeling completed of wellbore deformation upon heating and cooling
- **Impact of integrity of existing wells on site suitability for CO_2 Storage**
  - Regional database on wellbore integrity indicators for MI and OH
  - Neural-Genetic Model developed for leak probability of CO_2-exposed wells
- **Mitigation/Remediation**
  - Successful field test of biomineralization sealing of a fracture in the near-wellbore region
  - Evaluated rheology and bench tested placement of pH-triggered polymer gelant for sealing potential release pathways
  - Synthesized and characterized nanocomposite materials and developed methods for testing their use in wellbore sealing systems

Other awardees not featured:, Lawrence Livermore National Laboratory, University of Texas at Austin, Los Alamos National Laboratory

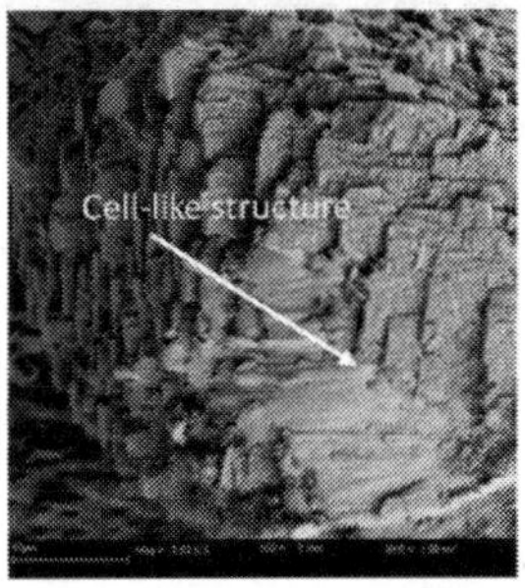

Field Test and Evaluation of Engineered Biomineralization Technology for Sealing Existing Wells – **Montana State University**

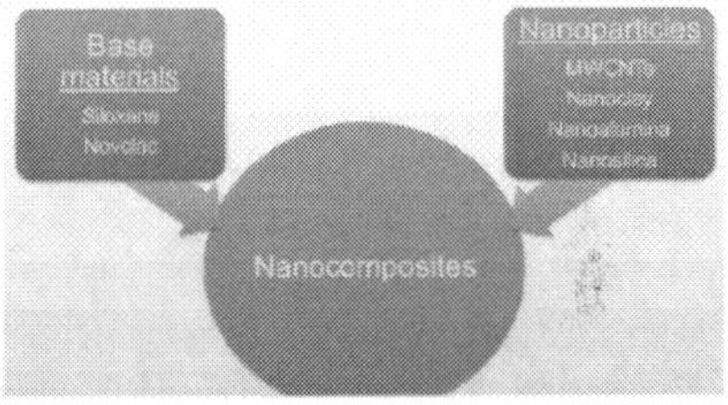

Wellbore Seal Repair Using Nanocomposite Materials – **University of New Mexico**

Core Storage R&D

Advanced Wellbore Integrity and Mitigation

- **Develop diagnostic tools and methods capable of characterizing borehole leakage pathways or fluid flow in existing wells**
- **Develop next-generation materials and methods for mitigating wellbore leakage**

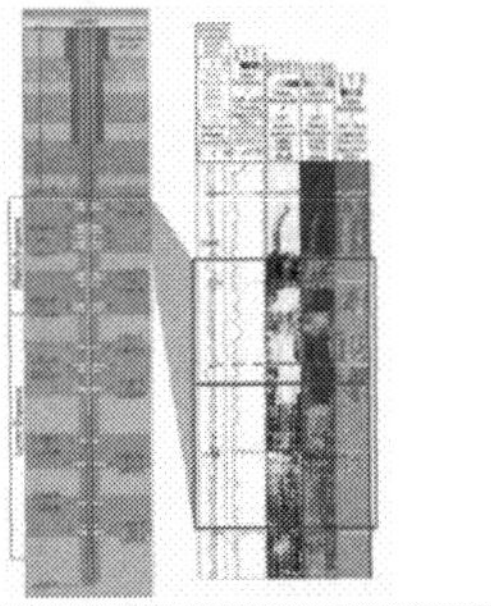

Montana State University– Wellbore Leakage Mitigation using Advanced Mineral Precipitation Strategies

Battelle Memorial Institute– Integrated Wellbore Integrity Analysis Program for CO_2 Storage Applications

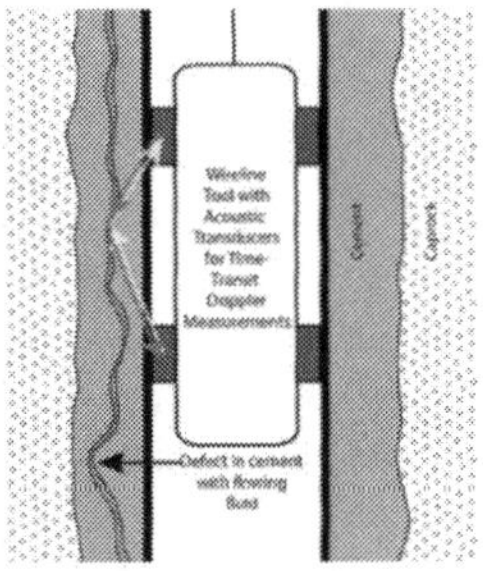

Los Alamos National Laboratory – Time reversal methods for the detection and monitoring of CO_2/brine leakage pathways in wellbore systems

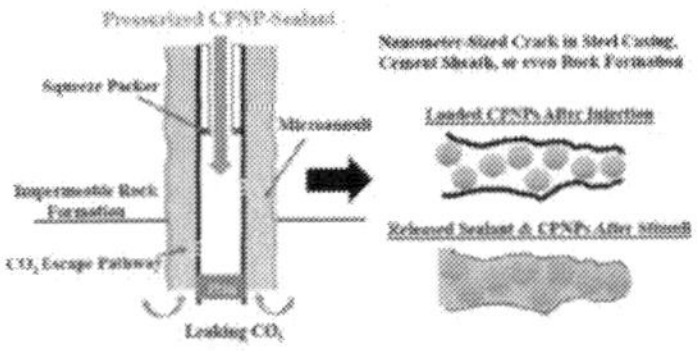

C-Crete Technologies, LLC - Programmable sealant-loaded mesoporous nanoparticles for gas/liquid leakage mitigation

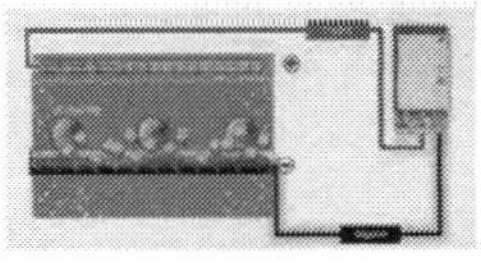

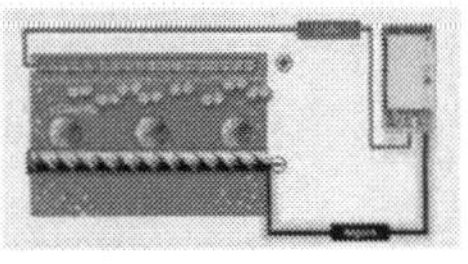

University of Colorado – Nanoparticle Injection Technology for Remediating Leaks of CO_2 Storage Formations

Core R&D Accomplishments

MVA Geophysics

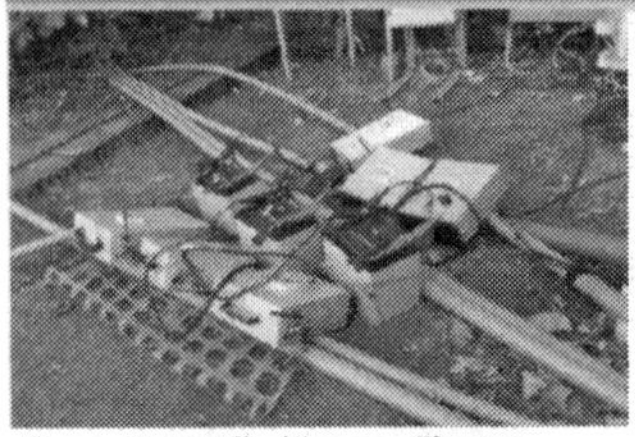

Deep, Controlled Source Electro--Magnetic Sensing: A Cost Effective, Long-term Tool for Sequestration Monitoring – Multi-Phase Technologies

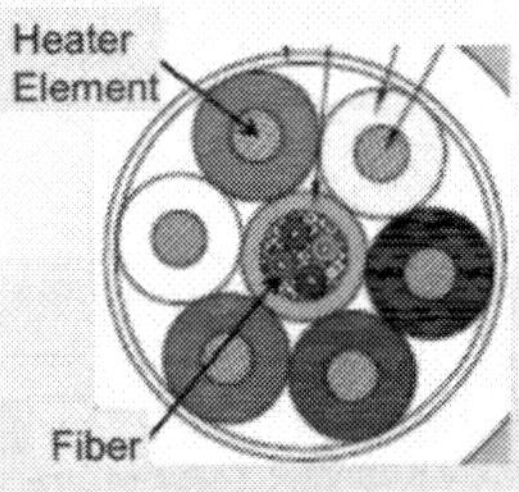

Distributed Fiber Optic Arrays: Integrated Temperature and Seismic Sensing for Detection of CO_2 Flow, Leakage and Subsurface Distribution - Electric Power Research Institute

- **Successful performance field testing of Distributed Acoustic Sensor (DAS) arrays in vertical seismic profiling (VSP) and crosswell seismic survey configurations**
- **Equipment procured and modeling carried out to support field testing of a new seismic method (scalable, automated, semi-permanent seismic array) for tracking a CO_2 plume**
- **Initial field validation testing of a hydrologic pulse testing technology for leakage detection**
- **Initial field testing of a controlled source electromagnetic (CSEM) system incorporating a borehole source for monitoring changes in electrical properties of CO_2 reservoirs**
- **Software using geostatistical and Kalman Filter methods has been tested for fast and reliable joint inversion of large datasets for monitoring**

Other awardees not featured: National Energy Technology Laboratory, Lawrence Livermore National Laboratory, Lawrence Berkley National Laboratory

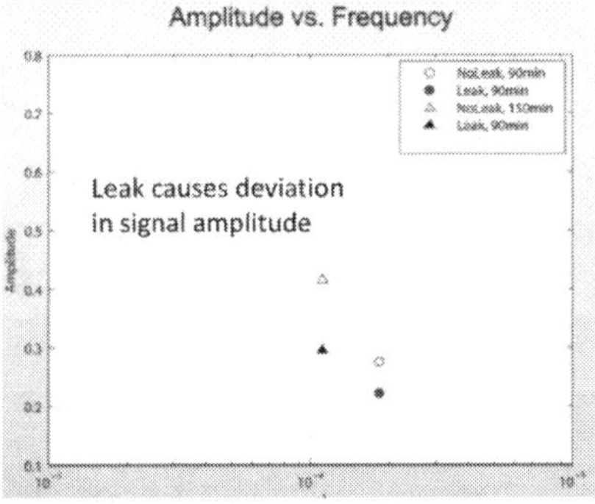

Pressure-Based Inversion and Data Assimilation System for CO_2 Leakage Detection – University of Texas at Austin

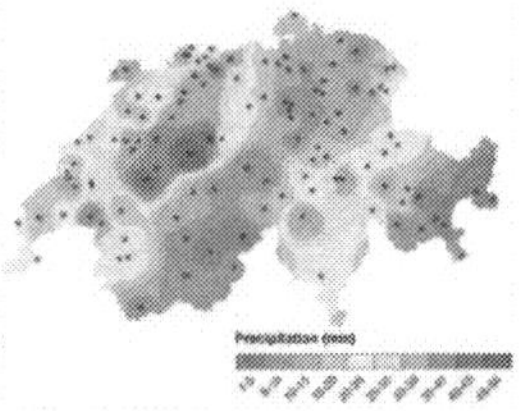

SASSA for Detecting CO_2 Plume Extent During Geological CO_2 Injection – Univ. of North Dakota

Core Storage R&D

Intelligent Monitoring Systems

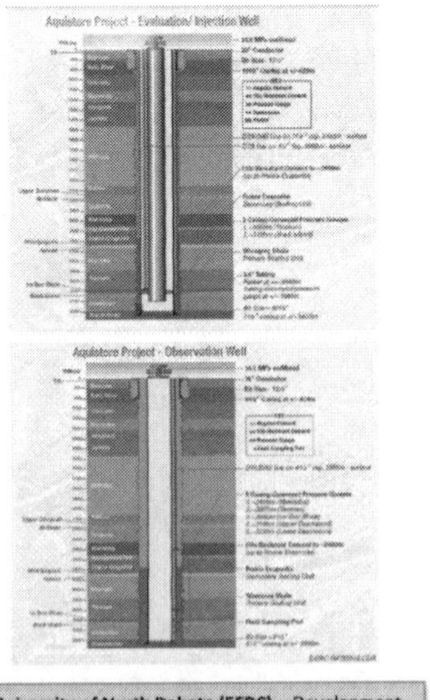

University of North Dakota (EERC) – Development of an Intelligent Monitoring System (IMS) Modules

- **Develop technological advancements for CCS-specific intelligent systems for monitoring, controlling, and optimizing CO_2 injection operations**

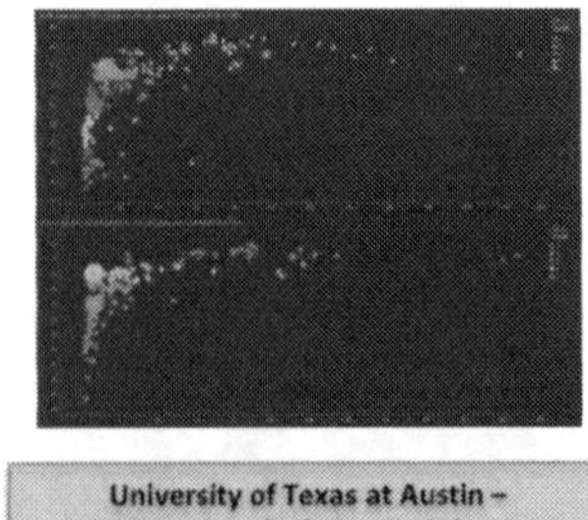

University of Texas at Austin – Development of a framework for data integration, assimilation, and learning for geological carbon sequestration

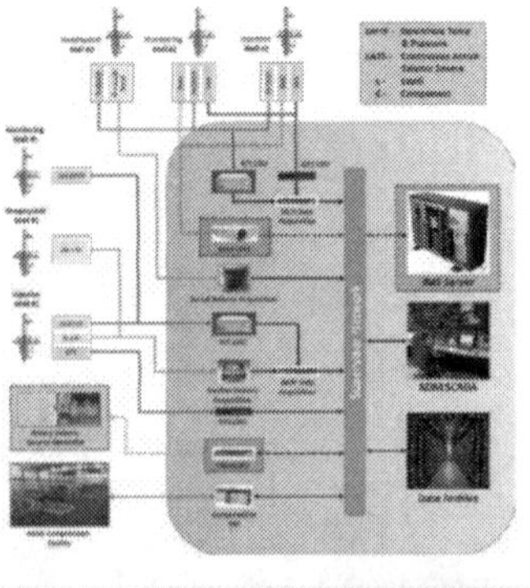

Archer Daniels Midland – Intelligent monitoring system for real-time geologic storage, optimization, and reservoir management

Fit-for-Purpose Field Project
Brine Extraction Storage Tests

Benefits

- Potential reduction in the risk of induced seismicity
- Improve reservoir storage efficiency while ensuring containment effectiveness

R&D Activities

- Lawrence Livermore National Laboratory simulations have shown that
  - Induced seismicity is related to saline formation pressure increases due to CO_2 injection
  - Localized pressure reduction can be used to steer the plume/pressure front
- BEST will be conducted to confirm simulation results

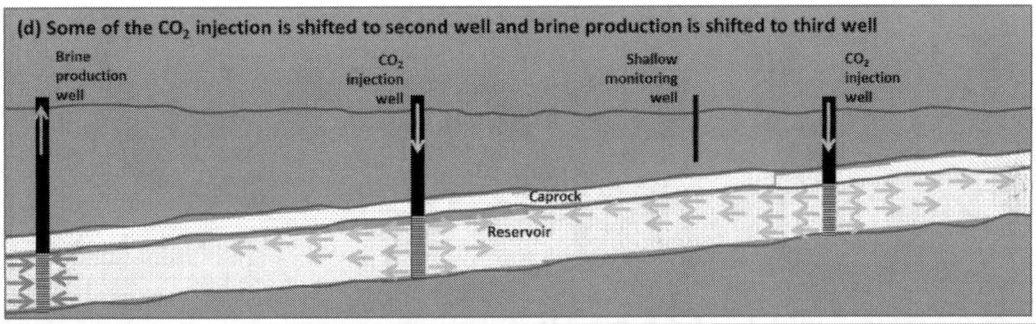

U.S. DEPARTMENT OF ENERGY | National Energy Technology Laboratory

13

Fit-for-Purpose Field Project
Brine Extraction Storage Tests

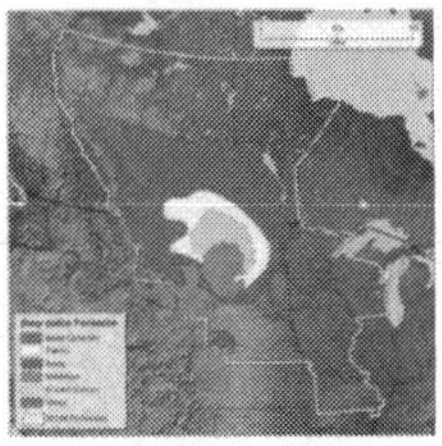

University of North Dakota – Developing and Validating Pressure Management and Plume Control Strategies in the Williston Basin through a Brine Extraction and Storage Test

- Managing formation pressure plumes as well as measuring/monitoring the movement of the differential pressure and CO_2 plumes in the subsurface for future saline CO_2 storage projects.
- Brines extracted shall be utilized as a part of a test-bed for brine treatment technologies

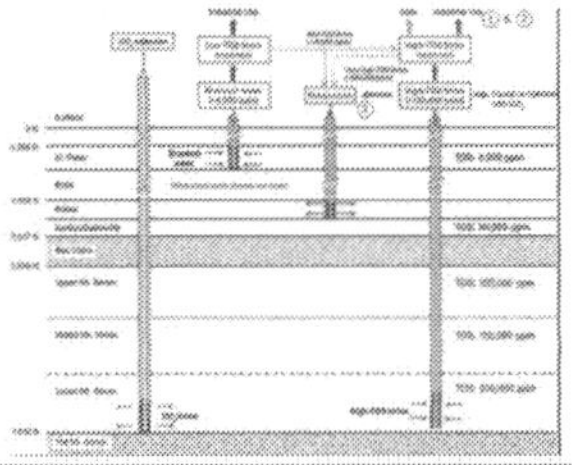

Illinois State Geological Survey – Brine extraction and treatment strategies to enhance pressure management and control of CO_2 plumes in deep geologic formations

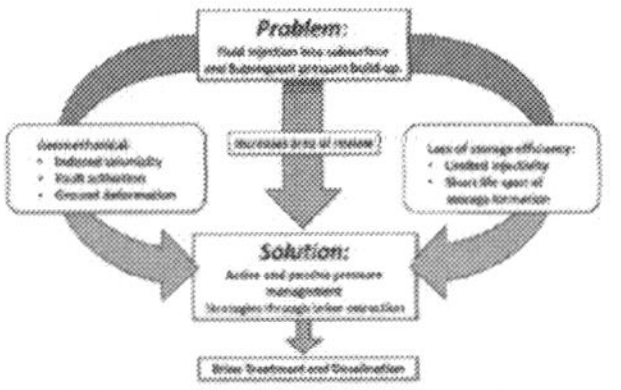

University of Texas at Austin– Pressure management and plume control strategies through a brine extraction storage test (BEST) at the Devine Test Site (DTS) in Texas

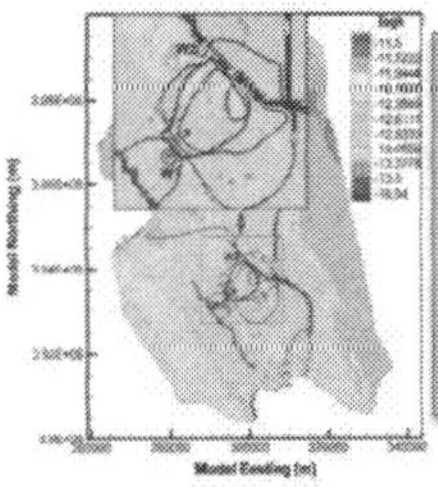

Electric Power Research Institute – Gulf Coast field demonstration at a flagship power plant site: Assessment of opportunities for optimal reservoir pressure control, plume management and produced water strategies

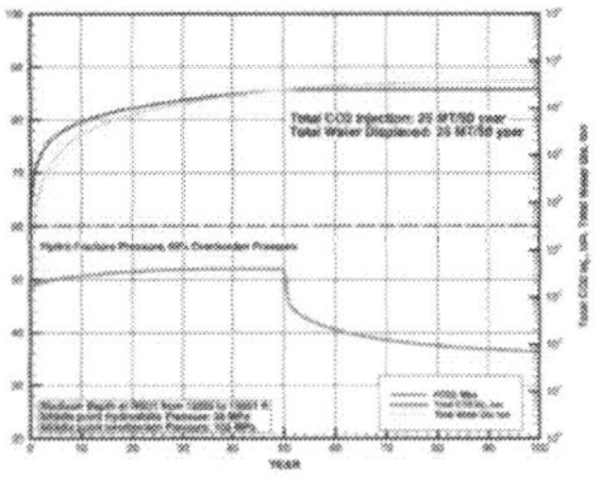

University of Wyoming – Field demonstration of an active reservoir pressure management through fluid injection and displaced fluid extraction at the Rock Springs Uplift

U.S. DEPARTMENT OF ENERGY | National Energy Technology Laboratory

14

Offshore Storage Resource Assessment

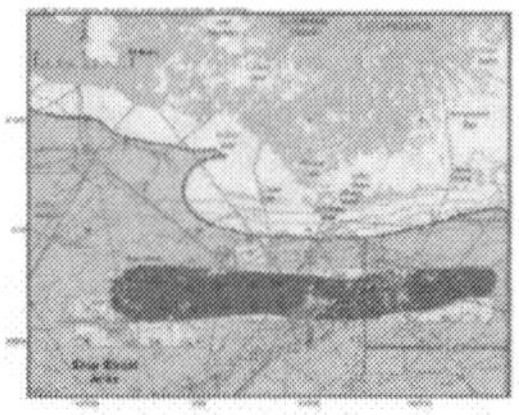

GeoMechanics Technologies – Assessment of CO_2 Storage Resources in Depleted Oil and Gas Fields in the Ship Shoal Area, Gulf of Mexico

- **Develop preliminary CO_2 storage resources in the offshore:**
  - Subsurface depleted oil and gas fields and saline formations in State and Federal waters on the East Coast and Gulf of Mexico.
  - Utilize existing geologic and geophysical data (well logs, records and sample descriptions) from existing or plugged/abandoned wells, seismic data, and existing core samples.

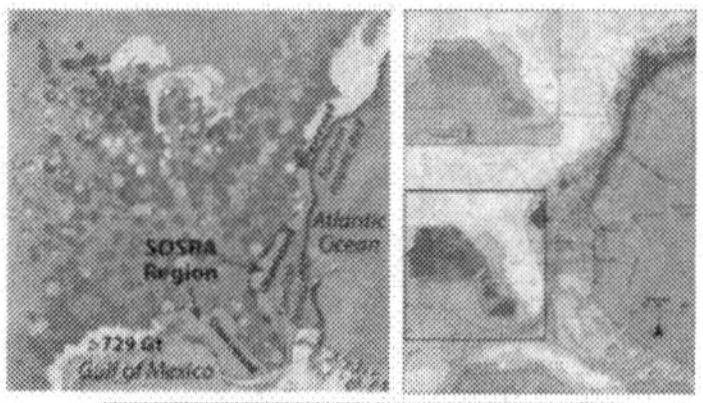

Southern States Energy Board – Southeast Offshore Storage Resource Assessment

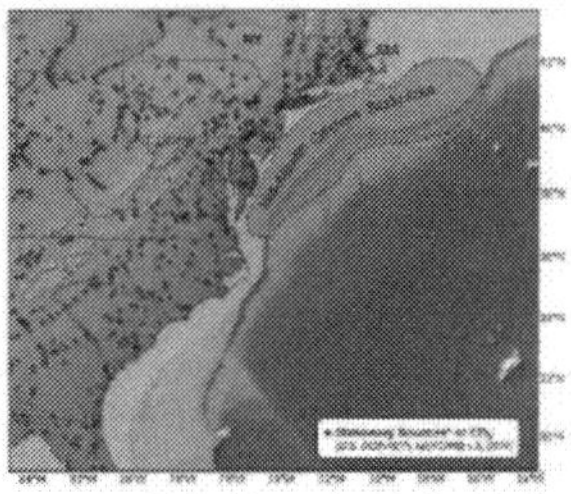

Battelle Memorial Institute – Mid-Atlantic U.S. Offshore Carbon Storage Resource Assessment Project

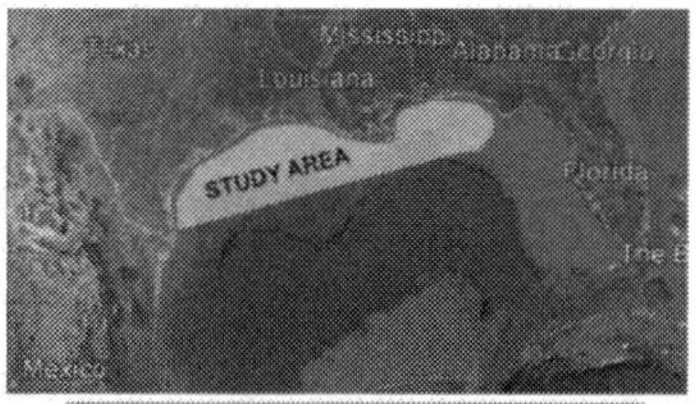

NITEC, LLC.– Offshore Storage Resource Assessment (Federal offshore – Gulf of Mexico)

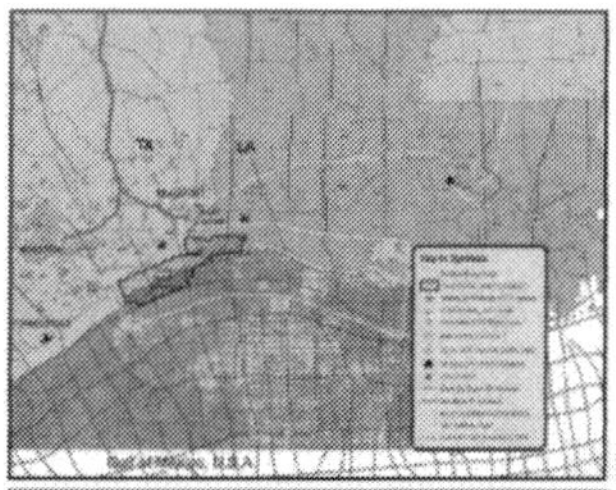

The University of Texas at Austin – Offshore CO_2 Storage Resource Assessment of the Northern Gulf of Mexico (Upper Texas – Western Louisiana Coastal Areas)

National Risk Assessment Partnership
www.edx.netl.doe.gov/nrap

NRAP leverages DOE's capabilities to help quantify uncertainties and risks necessary to remove barriers to full-scale CO_2 storage deployment.

Building toolsets and improving the science base to address...

- Potential impacts related to release of CO_2 or brine from the storage reservoir
- Potential ground-motion impacts due to injection of CO_2

Technical Team

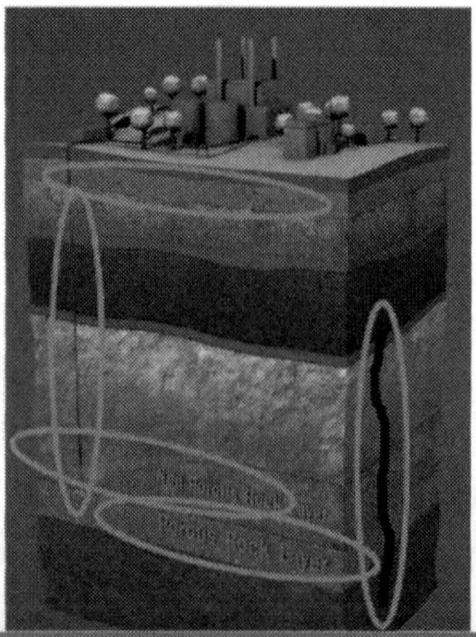

Stakeholder Group

Subsurface Technology and Engineering RD&D (SubTER)

- Discovering, characterizing and predicting subsurface conditions
- Accessing through wells
- Engineering and permeability control
- Sustained production and sustainable environment
- Monitoring technologies

Who's Involved?

Representing the geosciences, research, modeling, technology development, policy, and stakeholders, the participating DOE program and staff office include:

Fossil Energy	ARPA-E
Energy Efficiency & Renewable Energy	Electricity Delivery & Energy Reliability
Nuclear Energy	Energy Policy & Systems Analysis
Environmental Management	Congressional & Intergovernmental Affairs
Science	Energy Information Administration

Nuclear Waste Disposal

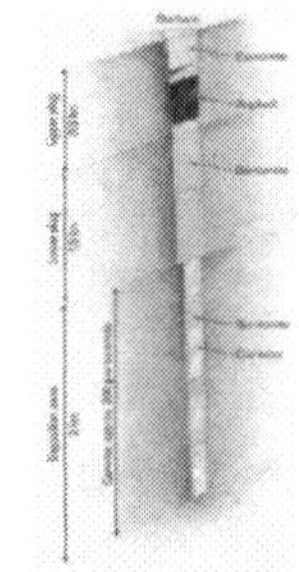

Geothermal

For More Information

National Energy Technology Laboratory
www.netl.doe.gov

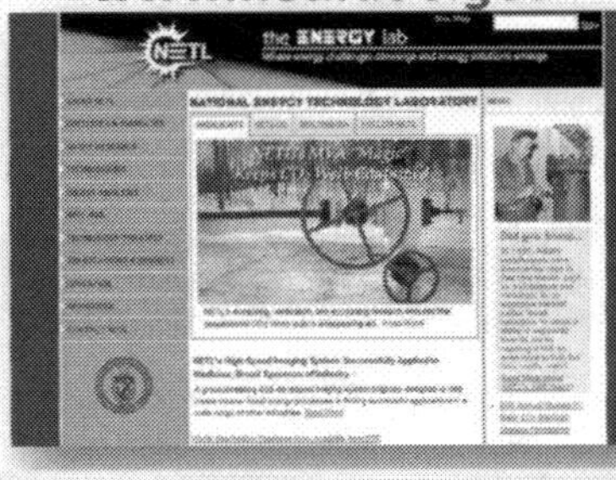

Traci D. Rodosta
304-285-1345
Traci.Rodosta@netl.doe.gov

Office of Fossil Energy
www.fe.doe.gov

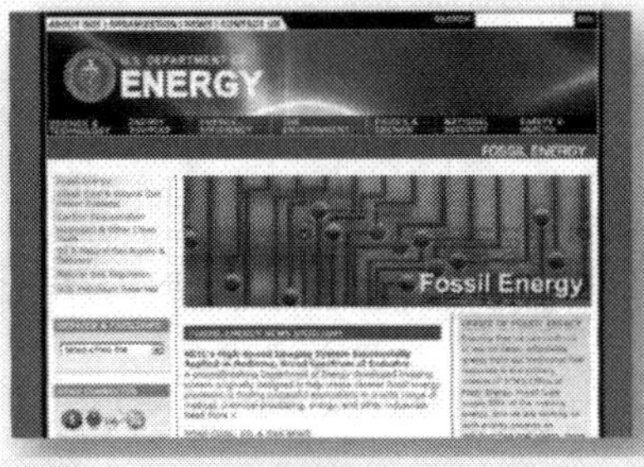

Darin Damiani
202-287-5282
Darin.Dmiani@hq.doe.gov

National Energy Technology Laboratory

@NETL_News

Fossil Energy

@Fossil Energy Gov

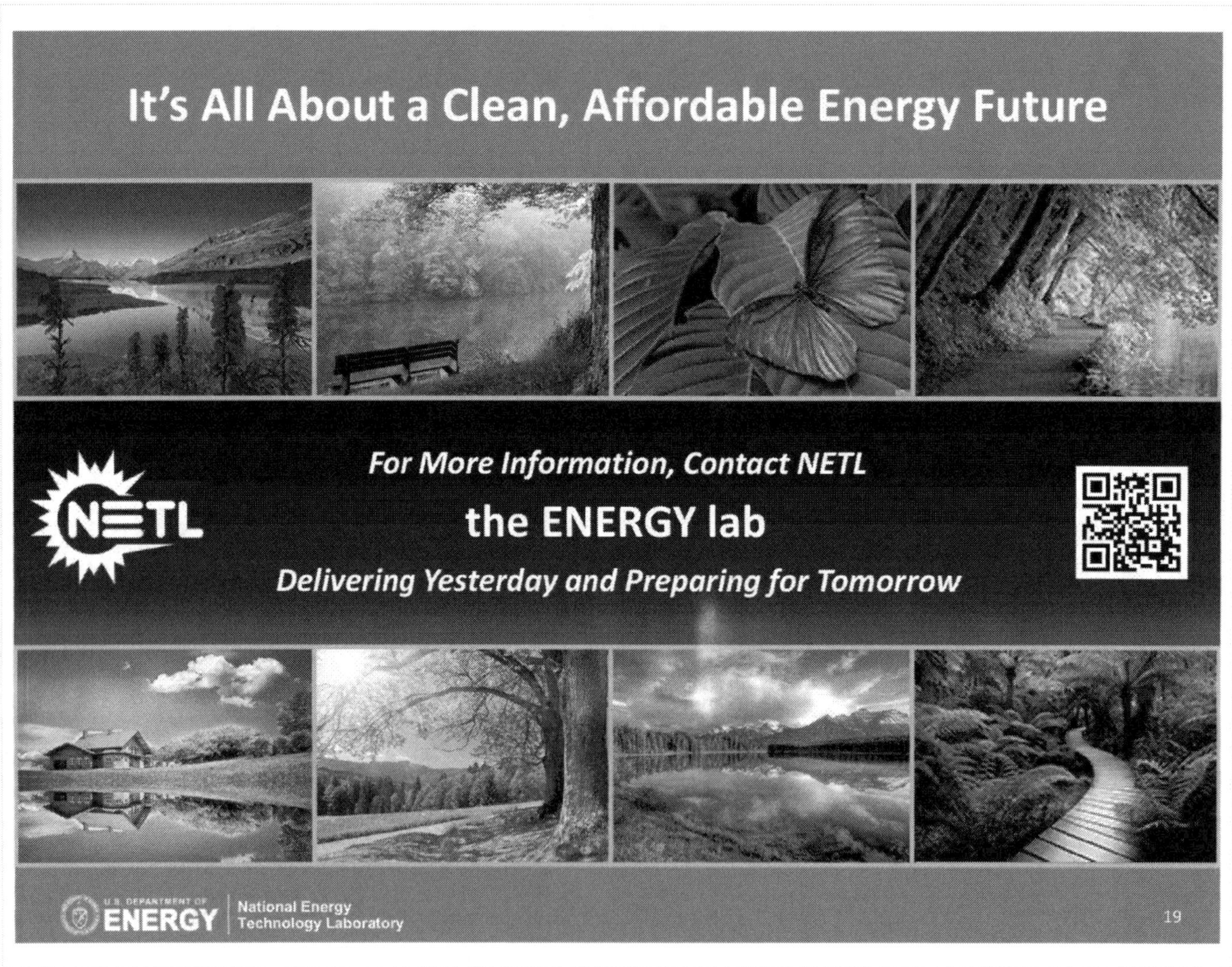
It's All About a Clean, Affordable Energy Future
For More Information, Contact NETL
NETL
the ENERGY lab
Delivering Yesterday and Preparing for Tomorrow
U.S. DEPARTMENT OF ENERGY
National Energy Technology Laboratory
19

ILLINOIS SUSTAINABLE TECHNOLOGY CENTER
PRAIRIE RESEARCH INSTITUTE

ILLINOIS STATE GEOLOGICAL SURVEY
PRAIRIE RESEARCH INSTITUTE

ILLINOIS
1867 UNIVERSITY OF ILLINOIS AT URBANA-CHAMPAIGN

Sustainable Power Plants

Kevin C. OBrien, Vinod Patel, Nandakishore Rajagopalan, Yongqi Lu, Hong Lu, Ashlynn S. Stillwell

CARBON MANAGEMENT TECHNOLOGY CONFERENCE

Carbon Management Technology Conference
Sustainable and Economical CCUS Options
November 2015

Prairie Research Institute: Global Reach

Able to address holistic solutions

I 1867

↓

PRAIRIE
RESEARCH INSTITUTE

↓

ILLINOIS NATURAL HISTORY SURVEY
PRAIRIE RESEARCH INSTITUTE

ILLINOIS SUSTAINABLE TECHNOLOGY CENTER
PRAIRIE RESEARCH INSTITUTE

ILLINOIS STATE GEOLOGICAL SURVEY
PRAIRIE RESEARCH INSTITUTE

ILLINOIS STATE ARCHAEOLOGICAL SURVEY
PRAIRIE RESEARCH INSTITUTE

ILLINOIS STATE WATER SURVEY
PRAIRIE RESEARCH INSTITUTE

ILLINOIS SUSTAINABLE TECHNOLOGY CENTER
PRAIRIE RESEARCH INSTITUTE

ILLINOIS STATE GEOLOGICAL SURVEY
PRAIRIE RESEARCH INSTITUTE

2

Sustainability and The Triple Bottom Line

Actions must account for impact on planet and people as well as profit

ISTC : Driving Economic Growth Through Sustainability

Catalyze job growth, improve quality of life, while preserving natural resources

- *Reduced operational & maintenance costs*
- *Pollution prevention*
- *Industrial "best practices" in sustainability*

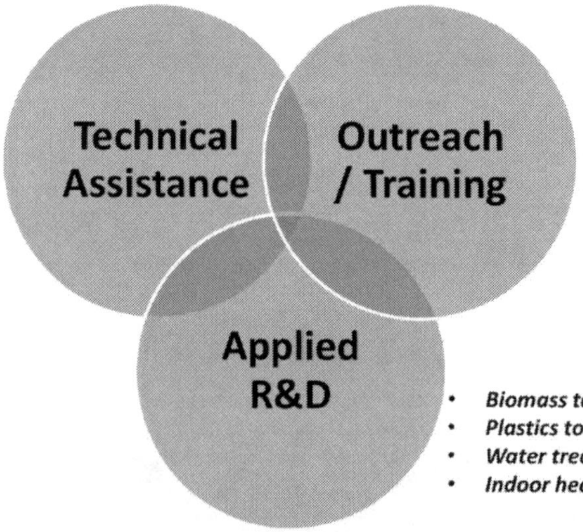

- *Energy efficiency training*
- *Sustainable electronics programs*
- *Great Lakes Pollution Prevention program*

- *Biomass to electricity*
- *Plastics to fuel*
- *Water treatment*
- *Indoor health and safety*

29th Annual Illinois Governor's Sustainability Award

Technical Symposium & Award Ceremony

October 27th, 2015

Abbott Laboratories	AbbVie, Inc.
Argonne National Laboratory	Caterpillar – Morton PDC
Clarke	ComEd
Cook County	Golden State Foods
Griffith Laboratories	Hoffer Plastics Corp.
Illinois Tollway	JL Clark
J. G. Shedd Aquarium	McHenry County Government
Public Building Commission	Saratoga Food Specialties
Silgan Containers Manuf. Corp.	University Housing at UIUC
Western Illinois University	

Did you know....

- 547 companies have won Awards over 28 years?
- The Awards is the longest running Award program in the nation?
- 50-55% of companies that apply win Awards?

ILLINOIS SUSTAINABLE TECHNOLOGY CENTER
PRAIRIE RESEARCH INSTITUTE

NOAA Report Highlights Global Need for Sustainability

Links extreme events of 2014 from a climate perspective

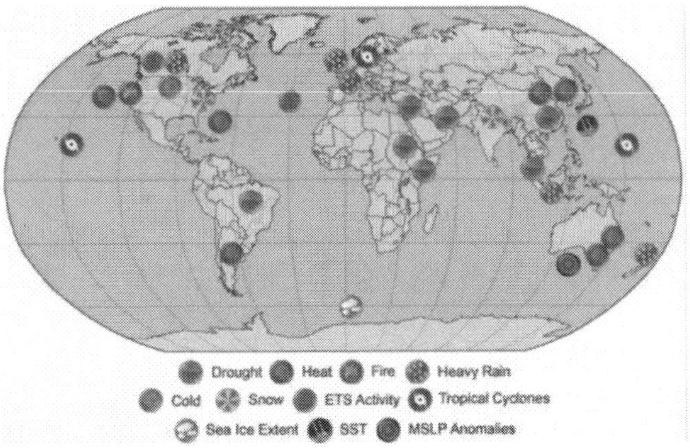

- For heat events the vast majority find a climate change signal: ~95% found a human influence
- For precipitation events it is more mixed: ~40% found a human influence

ILLINOIS SUSTAINABLE TECHNOLOGY CENTER
PRAIRIE RESEARCH INSTITUTE

ILLINOIS STATE GEOLOGICAL SURVEY
PRAIRIE RESEARCH INSTITUTE

6

Some common themes emerge

EFFORTS TO ADDRESS SUSTAINABILITY FOR POWER PLANTS

ILLINOIS SUSTAINABLE TECHNOLOGY CENTER
PRAIRIE RESEARCH INSTITUTE

ILLINOIS STATE GEOLOGICAL SURVEY
PRAIRIE RESEARCH INSTITUTE

Life Cycle Assessment Performed on Power Plants

Based on EPA's TRACI framework

Indicator	Abbr.	Summary	Relevant Emissions	Characterization Factor
Global Warming Potential	GWP	Impact of anthropogenic emissions which enhance the radiative forcing of the atmosphere.	CO_2, CH_4, N_2O, halocarbons	kg CO_2-eq/MWh
Ozone Depletion Potential	ODP	Impact on stratospheric ozone layer due to anthropogenic emissions, which causes a greater level of UV-B radiation to reach the earth's surface	CFCs, HCFCs, halons, methyl bromide	kg CFC11-eq/MWh
Acidification Potential	AP	Impact of acidifying pollutants on soil, groundwater, surface waters, and ecosystems.	SO_x, NO_x, HCl, HF, NH_3	kg SO_2-eq/MWh
Eutrophication Potential	EP	Impact of excessive macronutrients in terrestrial and marine ecosystems	PO_4, NO_x, nitrates, NH_3	kg PO_4^{-3} –eq/MWh
Photo-oxidant Formation Potential	POFP	Propensity to produce certain air pollutants which react with sunlight to form reactive chemical compounds, such as ozone, which negatively impact on the terrestrial ecosystem	Non-methane-hydrocarbon. Note: most LCAs do not account for impact of NO_x on ozone forming reactions	kg C_2H_4 –eq/MWh
Ecotoxicity Potential	ETP	Impacts of toxic substances on aquatic and terrestrial ecosystems	Mercury, chromium, cadmium, dioxins, arsenic, other toxic compounds	kg DCB-eq/MWh
Human Toxicity	HTP	Impacts of toxic substances on human health	PM_{10}, $PM_{2.5}$, soot, XO_2, NO_x, CH_4,	kg DCB-eq/MWh

SH Widder; Elliott, ML; Butner, RS; Freeman, CJ, "Sustainability Assessment of Coal-Fired Power Plants with Carbon Capture and Storage", PNNL 20933, October 2011

ILLINOIS SUSTAINABLE TECHNOLOGY CENTER
PRAIRIE RESEARCH INSTITUTE

ILLINOIS STATE GEOLOGICAL SURVEY
PRAIRIE RESEARCH INSTITUTE

Components of the LCA

Processes used in PNNL study to define boundaries for LCA

- *Coal Mining and Transportation.* Mining and transportation of coal to the boiler, including pulverizing the coal. This also includes the coal plant construction impacts.
- *Natural Gas Production and Transportation.* Production and transportation of natural gas. This does not include combustion of natural gas; it is included under Power Production.
- *MEA Production and Disposal.* Production and disposal of MEA. This does not include additional coal requirements.
- *Other Emissions Control Technologies.* Combined impacts of sulfur dioxide (SO_2), NO_x, and particulate removal technologies, including production of raw materials to disposal of waste streams. There is significantly reduced confidence in the estimated contributions from these processes due to the lack of specific data.
- *Power Production.* Combustion of coal in the boiler, resulting air emissions and waste generation
- *Sequestration.* CO_2 transportation and storage. Studies typically assume sequestration strategies do not leak or fail.

LCA Indicates Tradeoffs When Using Carbon Capture

Increased eutrophication and acidification from ammonia emissions while GHG emissions decrease

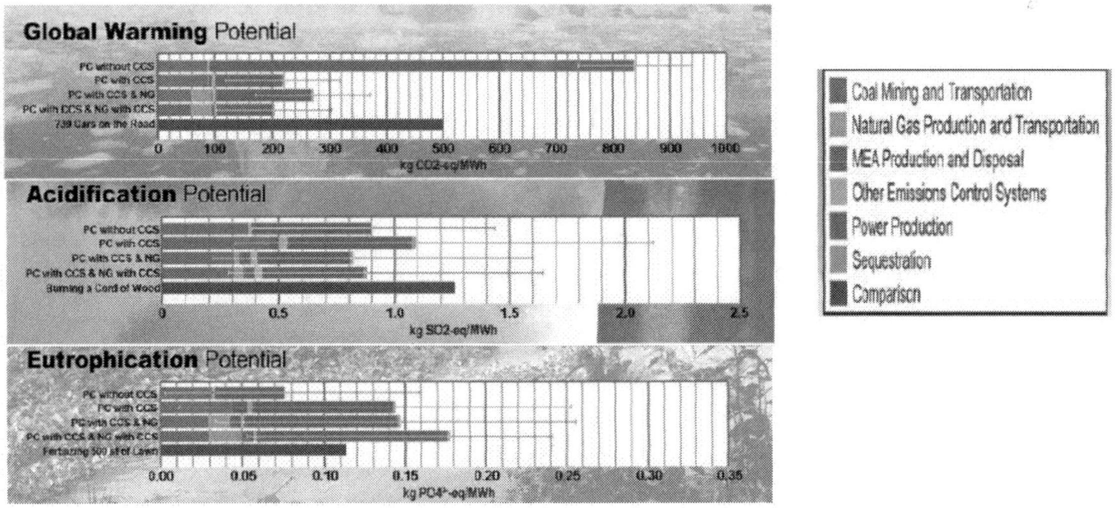

- Impact of MEA generation on increased eutrophication and acidification from ammonia emissions and increased toxicity from MEA production
- Impact of increased coal use - increased generation of NOx from combustion and transportation, impacts of increased mining of coal and limestone, and the disposal of toxic fly ash and boiler ash waste streams.

EPRI Study Identifies Sustainability Metrics

Based on sustainability interest group = 337 GW of generating capacity

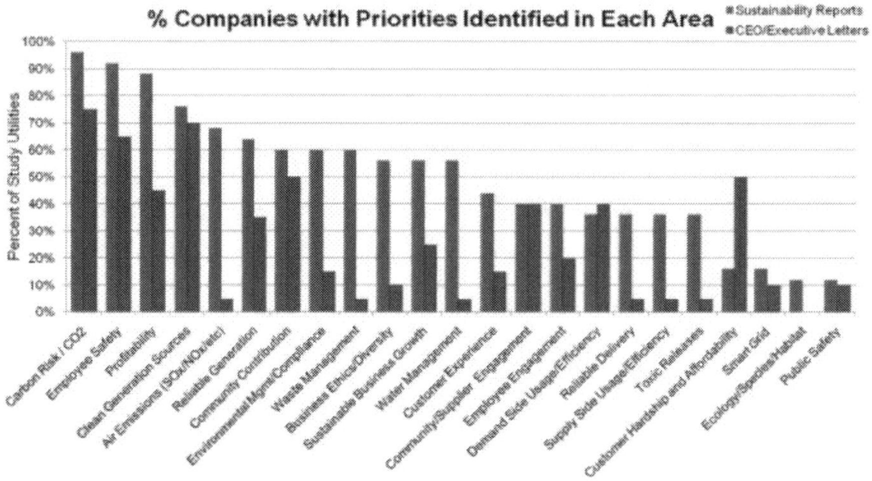

Jessica Fox, "Making Sustainability Decisions in the Electric Power Sector", EPRI, Portlands Energy Centre November 29, 2012

ILLINOIS SUSTAINABLE TECHNOLOGY CENTER
PRAIRIE RESEARCH INSTITUTE

ILLINOIS STATE GEOLOGICAL SURVEY
PRAIRIE RESEARCH INSTITUTE

11

EPRI Survey Rates Priorities of Sustainability Parameters

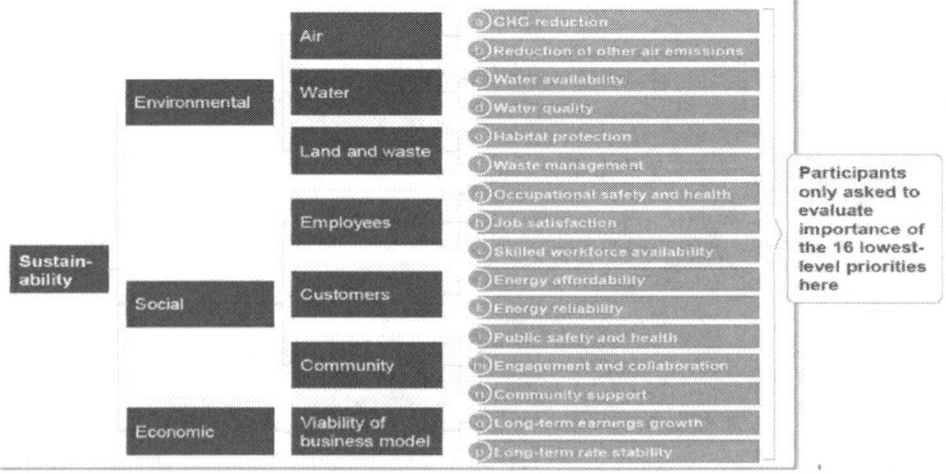

Jessica Fox, "Making Sustainability Decisions in the Electric Power Sector", EPRI, Portlands Energy Centre November 29, 2012

ILLINOIS SUSTAINABLE TECHNOLOGY CENTER
PRAIRIE RESEARCH INSTITUTE

ILLINOIS STATE GEOLOGICAL SURVEY
PRAIRIE RESEARCH INSTITUTE

12

Sustainability Reporting for Coal Operations

SASB Accounting for a Sustainable Future.

Sustainability Disclosure Topics	EVIDENCE OF INTEREST				EVIDENCE OF FINANCIAL IMPACT				FORWARD-LOOKING IMPACT		
	HM (1-100)	IWGs %	IWGs Priority	EI	Revenue & Costs	Assets & Liabilities	Cost of Capital	EFI	Probability & Magnitude	Externalities	FLI
Greenhouse Gas Emissions	80*	82	1t	High	•		•	High	•	•	Yes
Water Management	30	91	2	Medium	•	•	•	High	•		Yes
Waste Management	75*	91	3	High	•	•	•	High			No
Biodiversity Impacts	50*	100	1t	High	•	•	•	High	•		Yes
Community Relations & Rights of Indigenous Peoples	23	91	5t	Medium	•	•	•	Medium			No
Workforce Health, Safety, and Well-being	80*	91	1t	High	•	•		High			No
Labor Relations	45	91	5t	Medium	•	•	•	Medium			No
Reserves Valuation & Capital Expenditures	N/A	82	4	Medium	•	•	•	High	•		Yes

HM: Heat Map, a score out of 100 indicating the relative importance of the topic among SASB's initial list of 43 generic sustainability issues; asterisks indicate "top issues." The score is based on the frequency of relevant keywords in documents (i.e., 10-Ks, shareholder resolutions, legal news, news articles, and corporate sustainability reports) that are available on the Bloomberg terminal for the industry's publicly-listed companies; issues for which keyword frequency is in the top quartile are "top issues."

IWGs: SASB Industry Working Groups

%: The percentage of IWG participants that found the disclosure topic to likely constitute material information for companies in the industry. ( ) denotes that the issue was added after the IWG was convened.

Priority: Average ranking of the issue in terms of importance. One denotes the most important issue. (-) denotes that the issue was added after the IWG was convened.

EI: Evidence of Interest, a subjective assessment based on quantitative and qualitative findings.

EFI: Evidence of Financial Impact, a subjective assessment based on quantitative and qualitative findings.

FLI: Forward Looking Impact, a subjective assessment on the presence of a material forward-looking impact.

Financial Impacts of Sustainability Factors

Coal Operations

SASB Accounting for a Sustainable Future.

Evidence of Financial Impact	REVENUE & EXPENSES: Revenue – Market Size	Revenue – Pricing Power	Operating Expenses – COGS	Operating Expenses – R&D	Non-operating Expenses – Capex	Non-operating Expenses – Extra-ordinary Expenses	ASSETS & LIABILITIES: Assets – Tangible Assets	Assets – Intangible Assets	Liabilities – Contingent Liabilities & Provisions	Liabilities – Pension & Other Liabilities	COST OF CAPITAL: Risk Profile	Industry Divestment Risk
Greenhouse Gas Emissions			• (High)		• (High)						• (High)	• (High)
Water Management	• (High)		• (High)		• (Medium)	• (Medium)			• (Medium)		• (High)	
Waste Management	• (Medium)		• (High)		• (Medium)	• (Medium)		• (Medium)	• (Medium)		• (Medium)	
Biodiversity Impacts	• (High)		• (High)			• (High)		• (Medium)	• (High)		• (Medium)	
Community Relations & Rights of Indigenous Peoples	• (High)		• (High)			• (Medium)	• (High)	• (Medium)	• (Medium)		• (High)	
Workforce Health, Safety, and Well-being	• (Medium)		• (Medium)			• (High)		• (High)	• (High)			
Labor Relations	• (High)		• (Medium)					• (Medium)			• (High)	
Reserves Valuation & Capital Expenditures	• (High)						• (High)				• (Medium)	• (High)

MEDIUM IMPACT   HIGH IMPACT

ILLINOIS SUSTAINABLE TECHNOLOGY CENTER
PRAIRIE RESEARCH INSTITUTE

ILLINOIS STATE GEOLOGICAL SURVEY
PRAIRIE RESEARCH INSTITUTE

Requires a suite of activities – should be coordinated for maximum impact

ADDRESSING SUSTAINABILITY ISSUES

ILLINOIS SUSTAINABLE TECHNOLOGY CENTER
PRAIRIE RESEARCH INSTITUTE

ILLINOIS STATE GEOLOGICAL SURVEY
PRAIRIE RESEARCH INSTITUTE

Capturing Carbon from Abbott Power Plant

Team set and Phase I accepted by DOE for 25 MWe Capture Facility

- Strong Illinois team led by University: University of Illinois, Linde, Burns & McDonnell, Affiliated Engineering
- Phase I awarded to University in August 2015
- Phase II (build & test) is a $75 Million project with $60 Million from DOE
- Phase II cost share is $15 Million

ILLINOIS SUSTAINABLE TECHNOLOGY CENTER
PRAIRIE RESEARCH INSTITUTE

ILLINOIS STATE GEOLOGICAL SURVEY
PRAIRIE RESEARCH INSTITUTE

SoyFace: CO_2 User Close to Abbott Plant

Attractive use for captured CO_2

Only 5 miles from Power Plant

Utilizing Captured CO_2 for Enhanced Oil Recovery (EOR)

EOR at Mattoon reasonable proximity and have been characterized previously

- ***Abbott Power Plant to Mattoon ~ 50 miles***
- ***Previously characterized field***
- ***Ability to enhance oil production within Illinois***
- ***Potential rail accessibility***

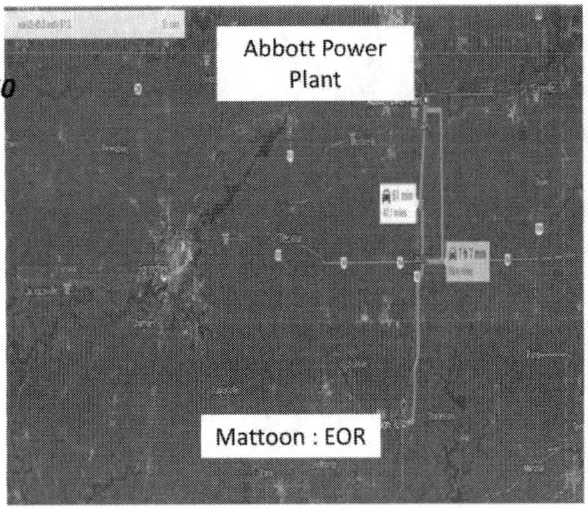

Illinois: A Regional & Global Test Bed for CCUS

Concentration of natural resources and intellectual capital

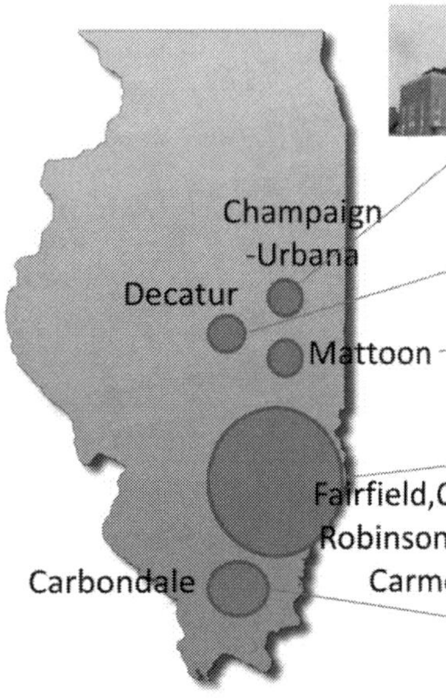

- Capture of CO_2 : Abbott Power Plant UIUC

ADM MGSC Richland

- Storage of CO_2 : ADM Project

MATTOON

- Utilization of CO_2 : Enhanced Oil Recovery (EOR)

Illinois Eastern Community Colleges

- Operator Training

Southern Illinois University Carbondale

Phase I acceptance by DOE: A Step Closer to the Goal

- Coal combustion

Life Cycle Assessment of Mercury in Illinois Coals

Consider physical cleaning, combustion, flue gas cleaning, and byproducts disposal

Results

- Same batch of coal followed from coal prep plant to power plant during both ozone and non-ozone seasons
- Coal cleaning process removed 40 to 50% (lbs/trillion BTU basis) of Hg from the as-mined raw coal.
- Average Hg reduction at the power plant was around 63% in non-ozone season (when selective catalytic reduction unit was not in operation) and around 40% in ozone season.
- Fly ash had comparable Hg content in ozone and non-ozone seasons.

Fired Clay Bricks with Class F Fly Ash from Illinois Basin Coals

Replacing Conventional Clay and Shale with Class F Fly Ash from Illinois Basin Coals

E1 E2 E3 E4

Fly ash inputs at 0% (E1), 20% (E2), 30% (E3), and 40% (E4) by volume

Engineering properties of building bricks from commercial-scale test runs

Brick Sample ID		E1	E2	E3	E4
Max. water absorption	Cold water 24 hr soak, % (< 8 wt%)	3.75	4.00	3.93	4.05
	Boiling water 5 hr, %	(5.30)	(6.06)	(5.89)	(4.30)
	Saturation coefficient	(0.71)	(0.66)	(0.67)	(0.94)
Fired compressive strength, psi (Severe weather grade > 3000 psi)		16905	16410	16476	20980
Suction, gm wt gain/minute		5.02	5.62	4.91	2.92
Scum		No	No	No	No

E1, E2, E3, and E4 refer to the brick samples from Run E1, Run E2, Run E3, and Run E4 respectively. The data indicated in the parentheses are not needed because all the cold water absorption data were less than 8%.

Building bricks far exceed ASTM building brick specifications and are marketable for the severe weather grade.

Dewatering Coal Fines by Forward Osmosis

Directed transport of water from the suspension to an osmotic agent

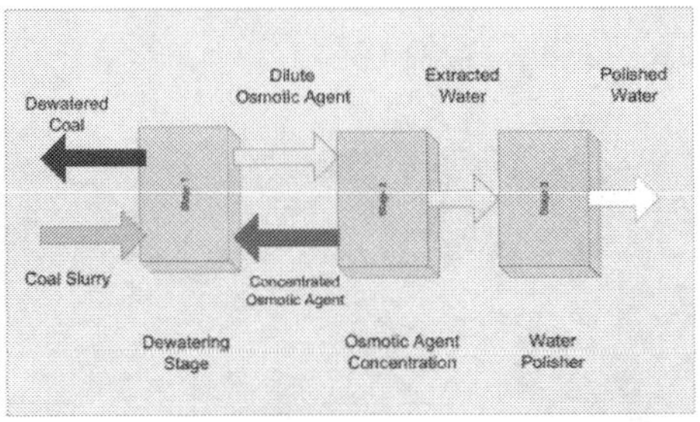

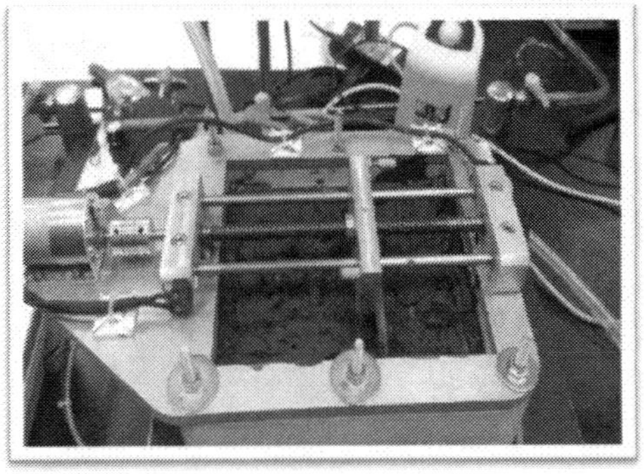

Technology merits

- **Passive dewatering technology**
- **Low pressure (atmospheric) process**
- **Low emissions – no phase change**
- **Dewatered solids content to 80%**

Major activities and achievements

- **Proof-of-concept completed**
- **Process scale-up to kg scale completed**
- **Critical process characteristics identified**
- **Estimated op. costs $3/ton solids**

Status

- **Design concepts for continuous processing initiated – prototyping in progress**

Power Plants Can Use Reclaimed Water for Cooling

Leveraging the proximity of waste water treatment facilities (WWTF)

- Many thermoelectric power plants use non-fresh water for cooling
- In 2010, 46 U.S. power plants used reclaimed water for cooling
- Reclaimed water has advantages
  - Drought-resistant
  - Can be abundant
  - Can be safe
- Reclaimed water can pose operational challenges

ILLINOIS SUSTAINABLE TECHNOLOGY CENTER
PRAIRIE RESEARCH INSTITUTE

ILLINOIS STATE GEOLOGICAL SURVEY
PRAIRIE RESEARCH INSTITUTE

Collaboration Between WWTPs and Power Plants

Analysis indicates promise

- Study area average flow from WWTPs: 1.6 billion gal/d
- Study area withdrawal for power plants: 5.1 billion gal/d
- Study area consumption for power plants: 0.05 billion gal/d

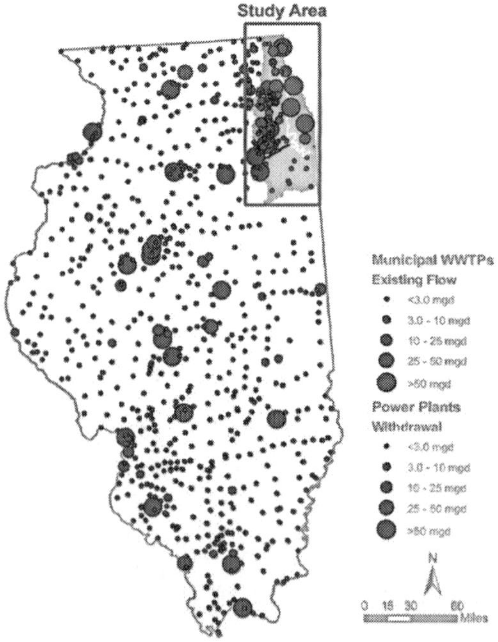

ILLINOIS SUSTAINABLE TECHNOLOGY CENTER
PRAIRIE RESEARCH INSTITUTE

[Barker & Stillwell, in preparation]

ILLINOIS STATE GEOLOGICAL SURVEY
PRAIRIE RESEARCH INSTITUTE

Engineered Water Reuse vs De Facto Reuse

Need to determine tradeoffs between both approaches

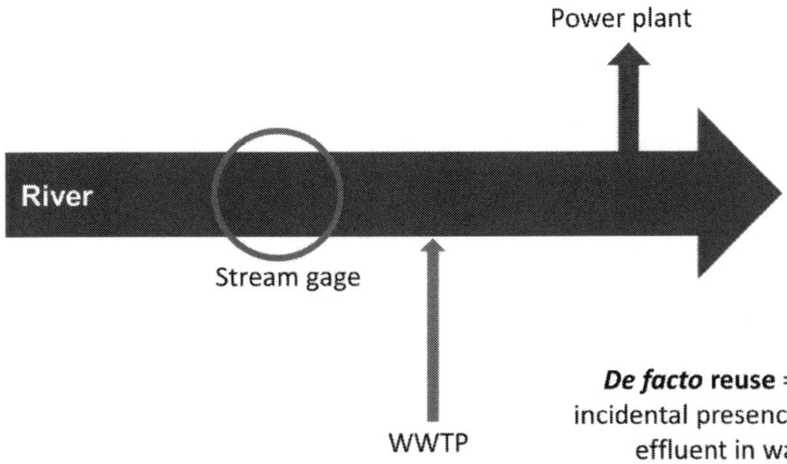

***De facto* reuse** = unintended, incidental presence of wastewater effluent in water supply

ILLINOIS SUSTAINABLE TECHNOLOGY CENTER
PRAIRIE RESEARCH INSTITUTE

[Barker & Stillwell, in preparation]

ILLINOIS STATE GEOLOGICAL SURVEY
PRAIRIE RESEARCH INSTITUTE

Engineered Reuse: Major Advantage

Decreases generation capacity loss from inefficient cooling

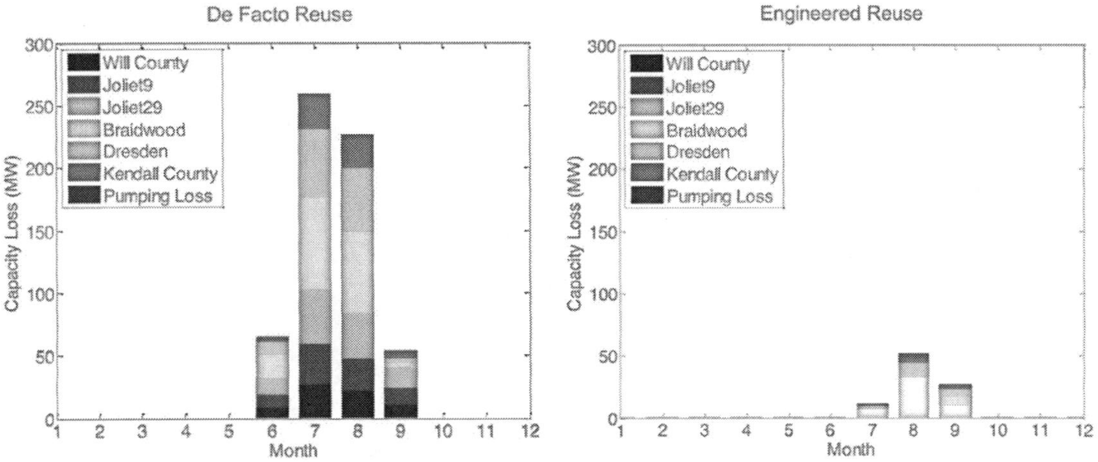

ILLINOIS SUSTAINABLE TECHNOLOGY CENTER
PRAIRIE RESEARCH INSTITUTE

ILLINOIS STATE GEOLOGICAL SURVEY
PRAIRIE RESEARCH INSTITUTE

Reclaimed Water for Power Plant Cooling

Significant advantages to this approach

- In study area of Chicago area, engineered water reuse with reclaimed water to cool power plants has notable advantages
  - ***Decreased withdrawals*** from surface water resources
  - Increased power generation ***reliability and performance***
- Tradeoffs include substantial infrastructure investment and operating costs

Secondary Revenue Stream for Power Plants

Utilizing solid and liquid waste for WWTFs

- **10 ton/day mobile pilot plant converting municipal, manure & food wastes to bio-oil**
- **Refining to bio-jet & bio-diesel**
- **Algae amplifies biofuels and purifies wastewater**
- **Potential to replace petroleum imports using waste inputs**

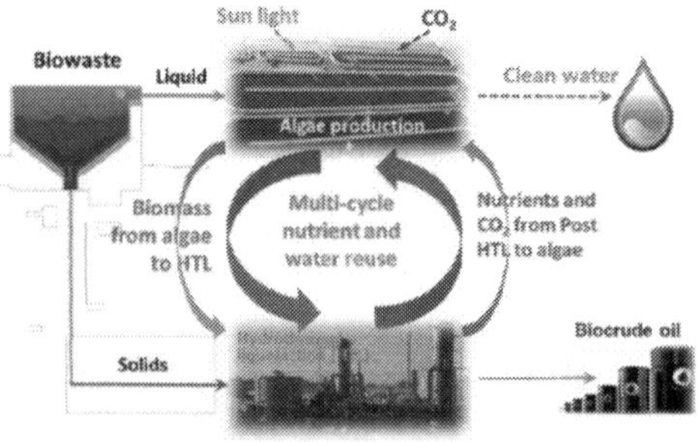

Future Power Plant Design

http://www.arch2o.com/coal-power-plant-mutation-bogdan-chipara/

ILLINOIS SUSTAINABLE TECHNOLOGY CENTER
PRAIRIE RESEARCH INSTITUTE

ILLINOIS STATE GEOLOGICAL SURVEY
PRAIRIE RESEARCH INSTITUTE

29

Discussion Questions

- What is the value of standardizing a means to evaluate sustainability aspects in power plants and their supply chains?
- Can a point system be developed as a metric? Perhaps a system that parallels LEED rating of buildings?
- Current systems only PENALIZE power plants. Goal is to develop a system that REWARDS sustainable behavior.

ILLINOIS SUSTAINABLE TECHNOLOGY CENTER
PRAIRIE RESEARCH INSTITUTE

ILLINOIS STATE GEOLOGICAL SURVEY
PRAIRIE RESEARCH INSTITUTE

30

Process Level Simulations of Calcium Looping Combustion

Wei Dai and Ramesh K. Agarwal
Department of Mechanical Engineering and Material Science
Washington University in St. Louis

Washington University in St. Louis
School of Engineering & Applied Science

CMTC 2015, Sugarland, TX, 17-19 November 2015

Two Recent Promising Carbon Capture Technology

- Chemical Looping Combustion (CLC)
- Calcium Looping Combustion (CaL)

2

What is Calcium Looping (CaL)?

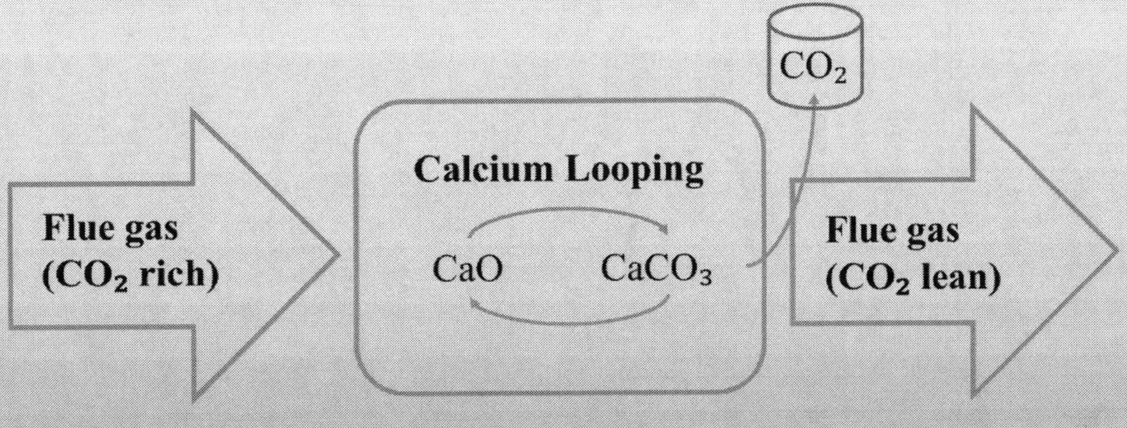

3

Post- & Pre- Combustion CaL

- There are two approaches to Calcium Looping:
- Post-combustion CaL to capture CO_2 after full combustion of fuel
- Pre-combustion CaL to remove carbon from fuel before combustion so that CO_2 will not form later

4

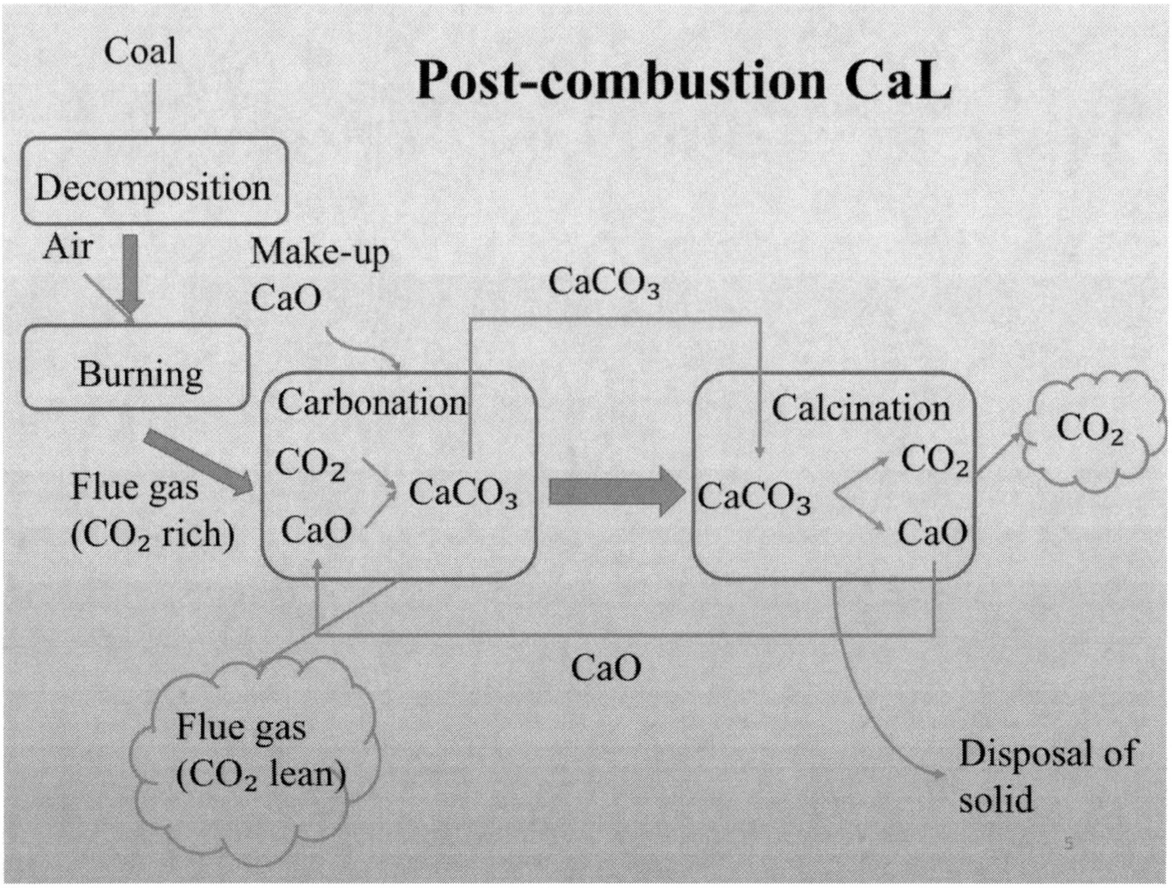
Post-combustion CaL
Coal
Decomposition
Air
Burning
Make-up
CaO
CaCO3
Carbonation
CO2
CaO
CaCO3
Flue gas
(CO2 rich)
Calcination
CaCO3
CO2
CaO
CO2
CaO
Flue gas
(CO2 lean)
Disposal of
solid
5

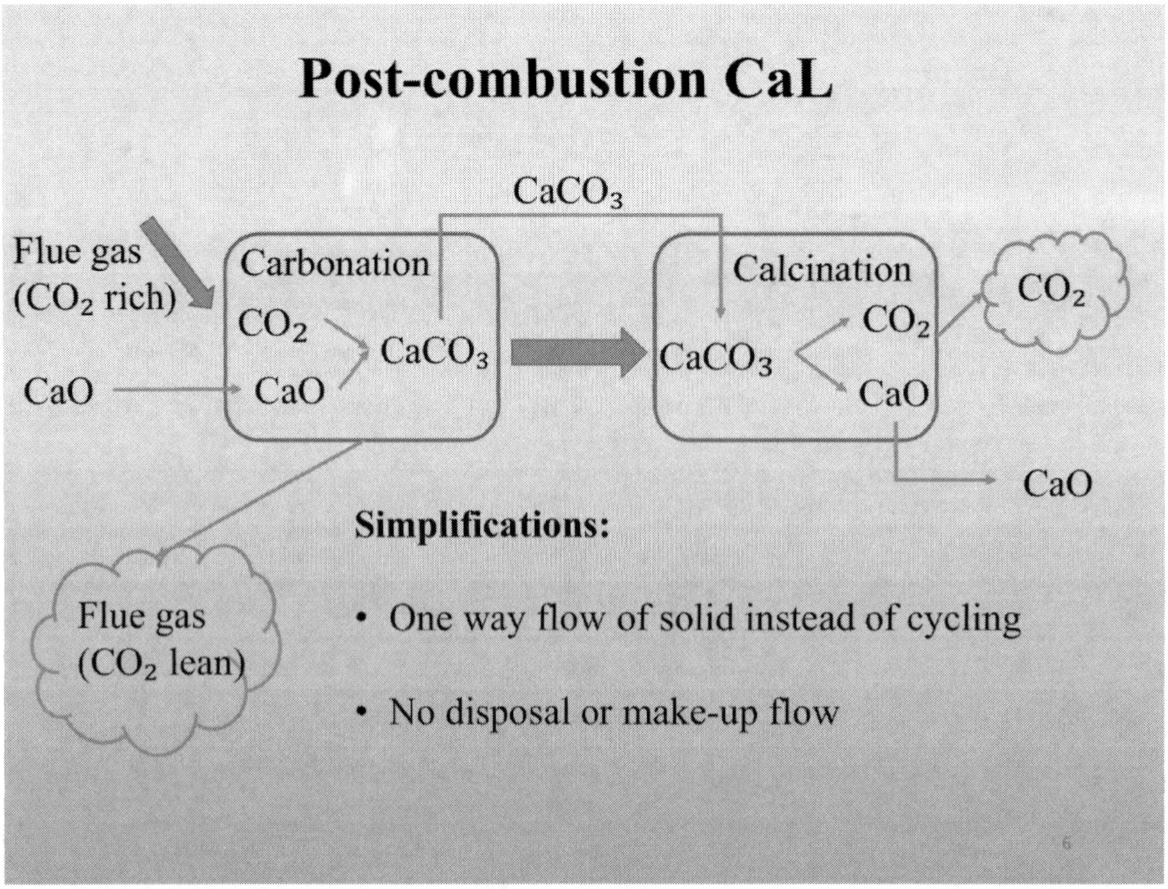
Post-combustion CaL
CaCO3
Flue gas
(CO2 rich)
Carbonation
CO2
CaCO3
CaO
CaO
Calcination
CO2
CaCO3
CaO
CO2
CaO
Simplifications:
• One way flow of solid instead of cycling
• No disposal or make-up flow
Flue gas
(CO2 lean)
6

Post-combustion CaL

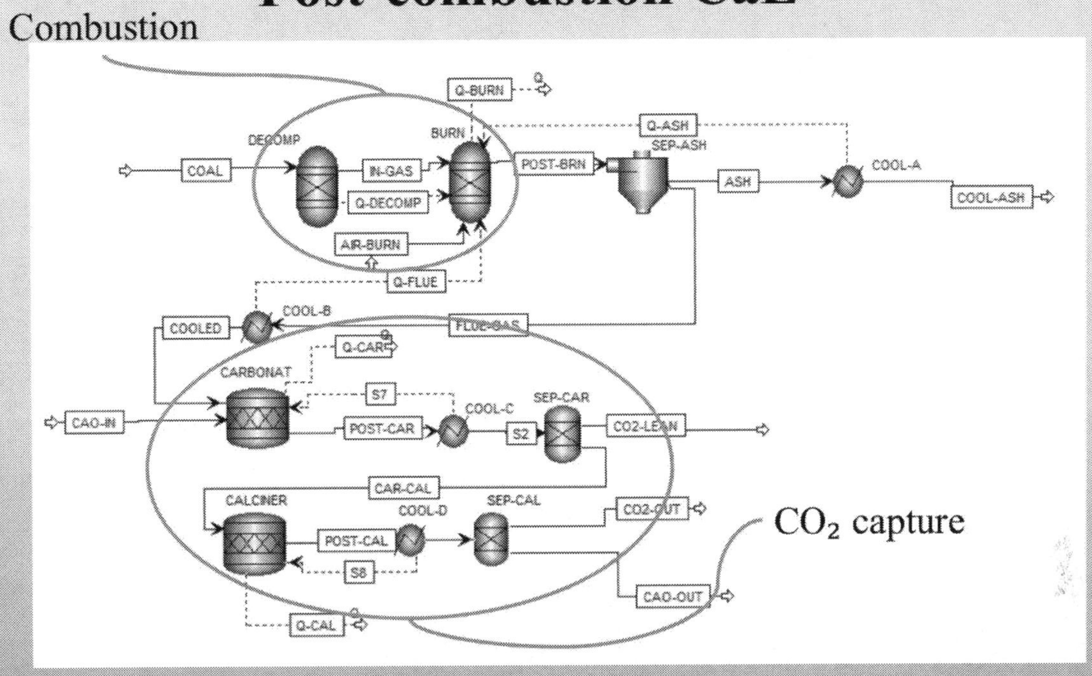

The flow sheet setup in ASPEN Plus for post-combustion capture

7

Process Models used ASPEN Plus in Post-Combustion CaL

Name	Model	Function	Reaction formula
DECOMP	RYIELD	Turn non-conventional into conventional	Coal → char + simple substances
BURN	RGIBBS	Coal burns with air	Char + simple substances+ $O_2 \rightarrow CO_2 + H_2O$
CARBONATOR	RSTOIC	Carbonation	CaO + CO2 → CaCO3
CALCINER	RSTOIC	Calcination	$CaCO_3 \rightarrow CaO + CO_2$

8

Process models used in ASPEN Plus in post-combustion CaL (cont.)

Name	Model	Function	Reaction formula
SEP-ASH	SSPLIT	Flue gas and ash separation	-
SEP-CAR	SEP	Flue gas (CO_2 lean) and Ca solids separation	-
SEP-CAL	SEP	CO_2 and Ca solids separation	-
COOL-A	HEATER	Ash cooler	-
COOL-B	HEATER	Flue gas cooler	-
COOL-C	HEATER	Cooler downstream of carbonator	-
COOL-D	HEATER	Cooler downstream of calciner	-

9

Post-combustion CaL

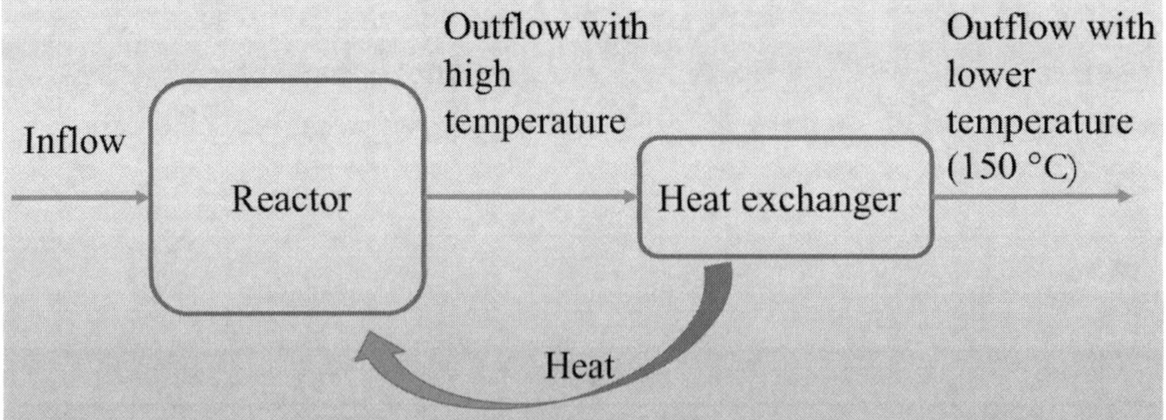

Add heat exchangers to release heat t lower temperature

10

Post-combustion CaL

Temperature and pressure specification for each block

Block name	Temperature (°C)	Pressure (bar)
DECOMP	925	1
BURN	1300	1
CARBONATOR	650	1
CALCINER	900	1
All SEP's	-	-
All COOL's	150	1

11

CO2 Capture Efficiency under Different Flow Rates of CaO and CO_2 (Sivalingam 2013)

Make-up CaO flow

Circulating CaO flow

CO_2 flow

Reaction: $CaO + CO_2 \rightarrow CaCO_3$

	E_{CO_2}				
F_O/F_{CO_2}	$F_R/F_{CO_2}=1$	$F_R/F_{CO_2}=2$	$F_R/F_{CO_2}=3$	$F_R/F_{CO_2}=4$	$F_R/F_{CO_2}=5$
0.05	0.28	0.46	0.63	0.81	0.99
0.1	0.36	0.57	0.76	0.95	0.99
0.2	0.46	0.72	0.95	0.99	0.99
0.3	0.52	0.84	0.98	0.99	0.99
0.4	0.58	0.95	0.99	0.99	0.99
0.5	0.61	0.97	0.99	0.99	0.99

Capture efficiency

12

- Reaction: $CaO + CO_2 \rightarrow CaCO_3$
- Use the flow ratio required for certain capture efficiency to define the conversion fraction of CaO in carbonator
- Try a series of CaO flow rates until the experimental efficiency is achieved

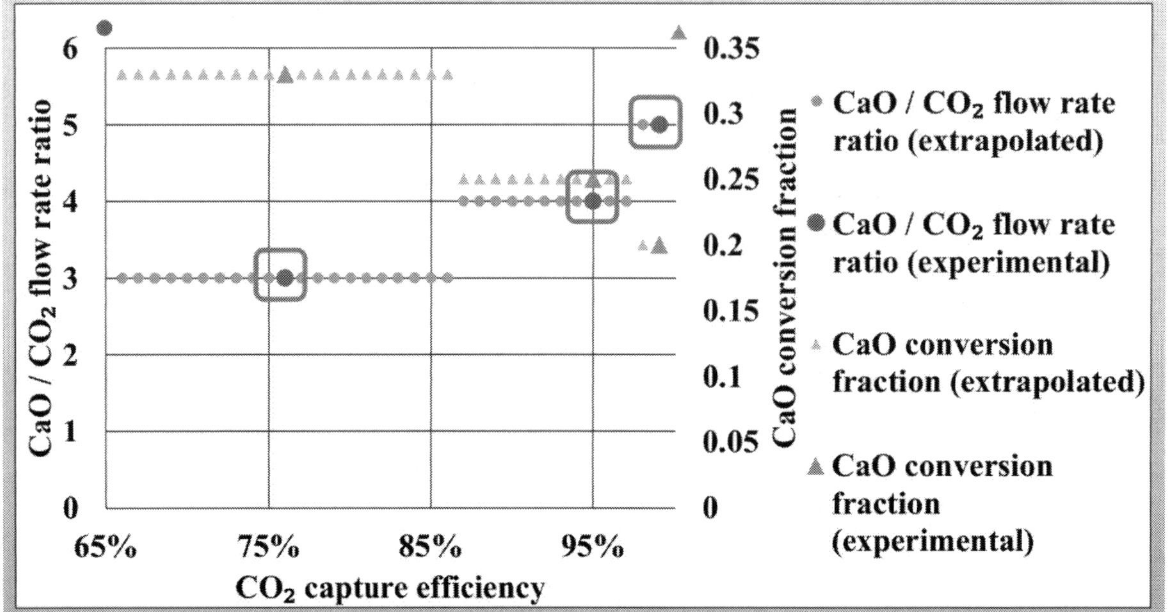

Range of CO_2 capture efficiency for each CaO conversion fraction 13

Results of Post-combustion CaL

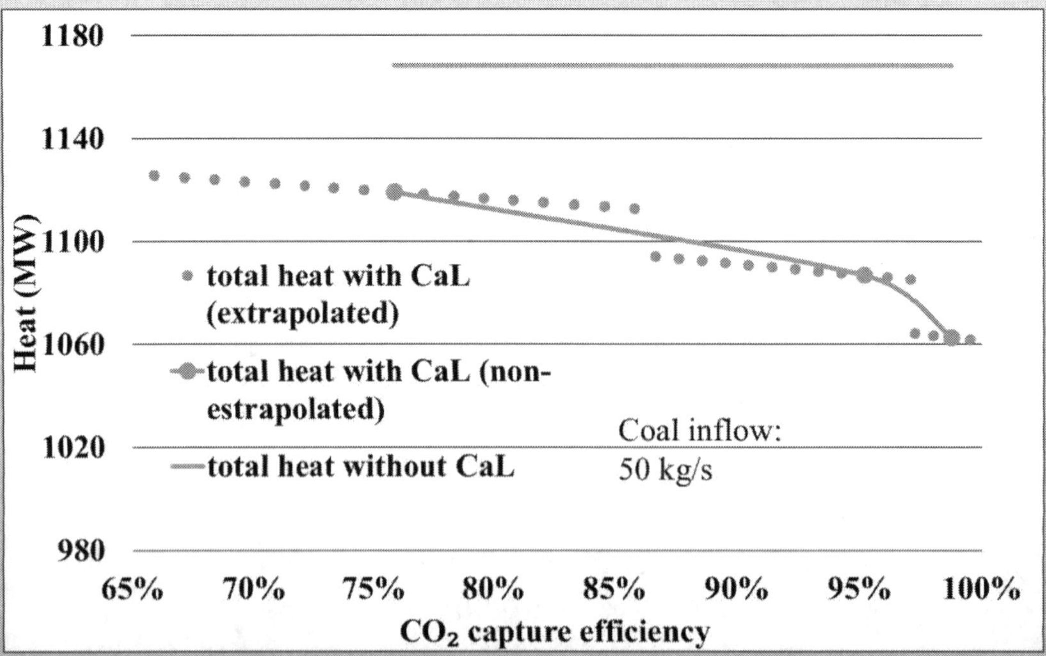

Total heat output for post-combustion setup with and without CaL 14

Results of Post-combustion CaL

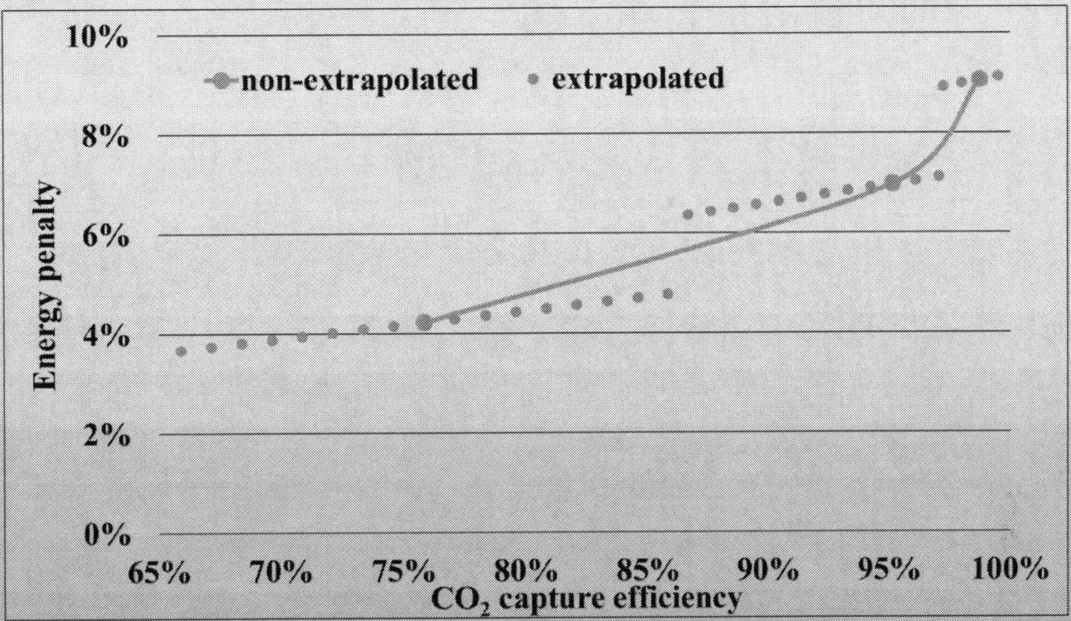

Energy penalty vs. CO_2 capture efficiency

15

Process Level Simulation

- Post-combustion CaL
- Pre-combustion CaL

16

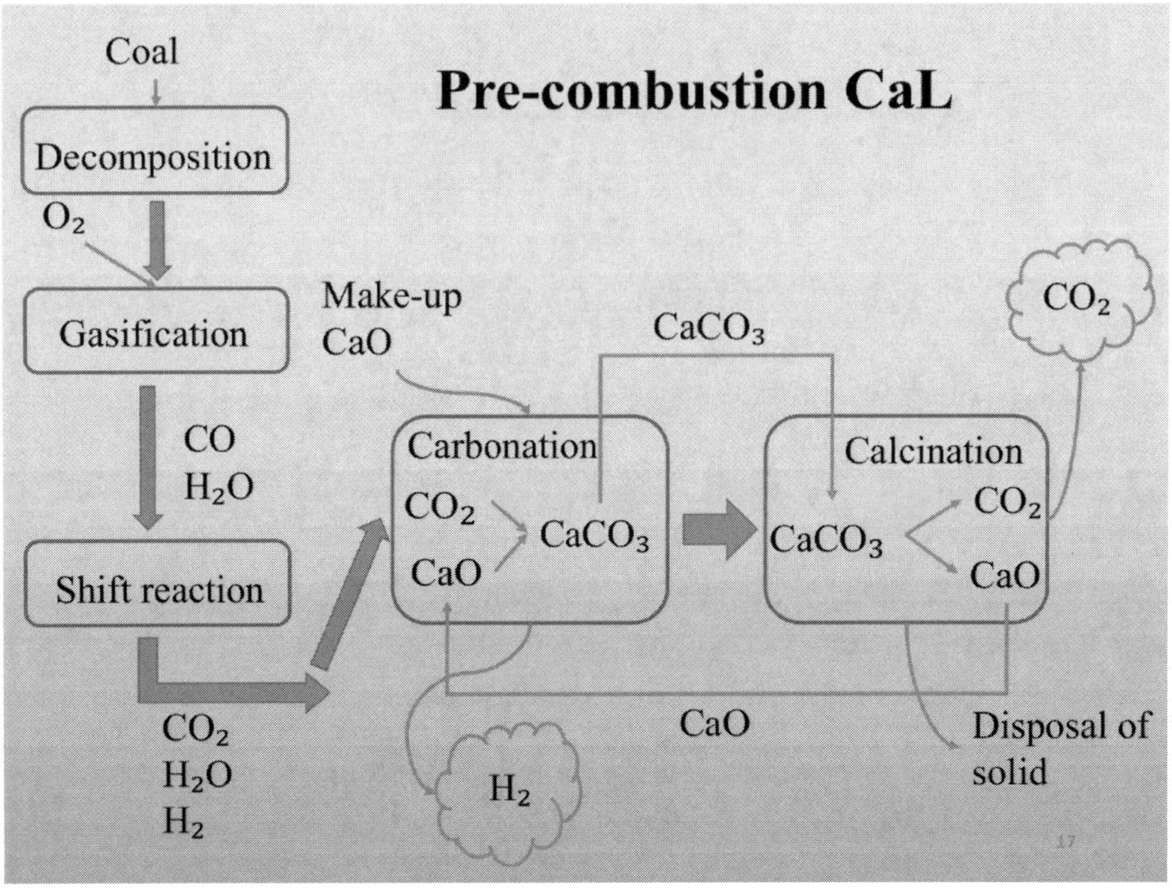
Pre-combustion CaL
Coal
Decomposition
O_2
Gasification
CO
H_2O
Shift reaction
CO_2
H_2O
H_2
Make-up
CaO
$CaCO_3$
Carbonation
CO_2
CaO
$CaCO_3$
Calcination
$CaCO_3$
CO_2
CaO
CO_2
CaO
H_2
Disposal of
solid
17

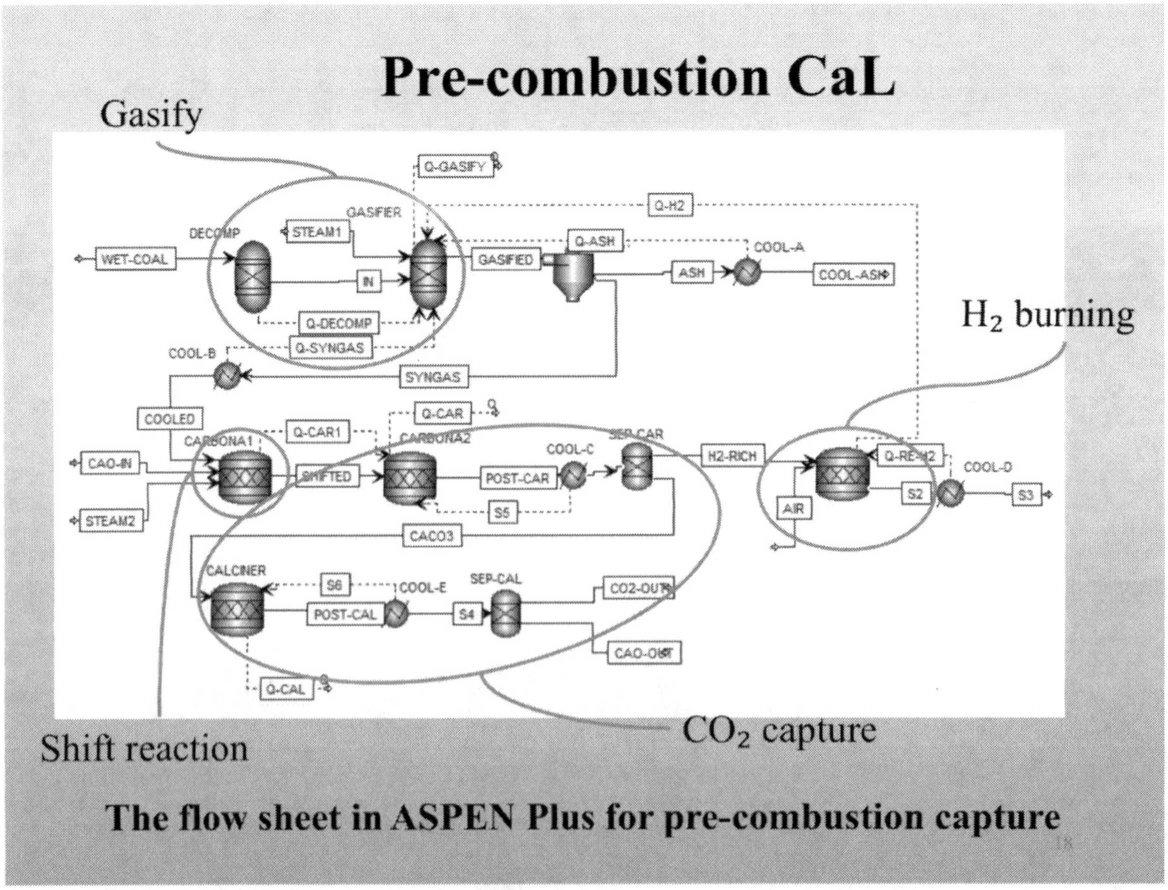
Pre-combustion CaL
Gasify
Q-GASIFY
Q-H2
GASIFIER
DECOMP
STEAM1
WET-COAL
GASIFIED
Q-ASH
COOL-A
IN
ASH
COOL-ASH
Q-DECOMP
Q-SYNGAS
COOL-B
SYNGAS
H₂ burning
COOLED
Q-CAR
CARBONA1
Q-CAR1
CARBONA2
COOL-C
SEP-CAR
H2-RICH
CAO-IN
SHIFTED
POST-CAR
COOL-D
S5
S2
S3
STEAM2
AIR
CACO3
CALCINER
S6
COOL-E
SEP-CAL
CO2-OUT
POST-CAL
S4
CAO-OUT
Q-CAL
Shift reaction
CO_2 capture
The flow sheet in ASPEN Plus for pre-combustion capture
18

Process models used in ASPEN Plus for pre-combustion CaL

Name	Model	Function	Reaction formula
DECOMP	RYIELD	Turn non-conventional into conventional	Coal → char + simple substances
GASIFIER	RGIBBS	Coal gasifies with steam	Char + simple substances + H_2O → $CO + H_2$ + volatile matter
CARBONA1	RSTOIC	Shift CO to be CO_2 and produce H_2	CO + volatile matter + H_2O → $CO_2 + H_2$
CARBONA2	RSTOIC	Carbonation	$CaO + CO_2 \rightarrow CaCO_3$
H2-BURN	RSTOIC	Combustion of H_2	$H_2 + O_2 \rightarrow H_2O$
CALCINER	RSTOIC	Calcination	$CaCO_3 \rightarrow CaO + CO_2$

19

Process models used in ASPEN Plus for pre-combustion CaL (cont.)

Name	Model	Function	Reaction formula
SEP-ASH	SSPLIT	Flue gas and ash separation	-
SEP-CAR	SEP	Flue gas (CO_2 lean) and Ca solids separation	-
SEP-CAL	SEP	CO_2 and Ca solids separation	-
COOL-A	HEATER	Ash cooler	-
COOL-B	HEATER	Flue gas cooler	-
COOL-C	HEATER	Carbonator downstream cooler	-
COOL-D	HEATER	Downstream cooler for H_2 burner	-
COOL-E	HEATER	Calciner downstream cooler	-

20

Pre-combustion CaL

Temperature and pressure specification for each block (pre-combustion)

Block name	Temperature (°C)	Pressure (bar)
DECOMP	925	1
GASIFIER	1300	1
CARBONA1	650	1
CARBONA2	650	1
H2-BURN	1300	1
CALCINER	900	1
All SEP's	-	-
All COOL's	150	1

21

Results Comparison between the Two Cases

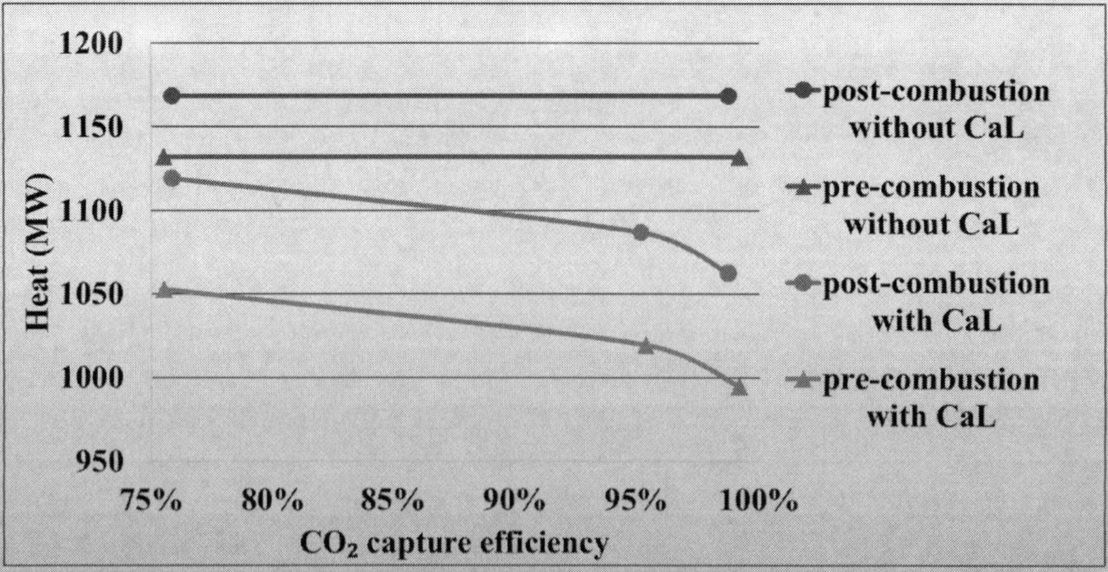

Total heat output with and without CaL for post- & pre-combustion

22

Results Comparison between the Two Cases

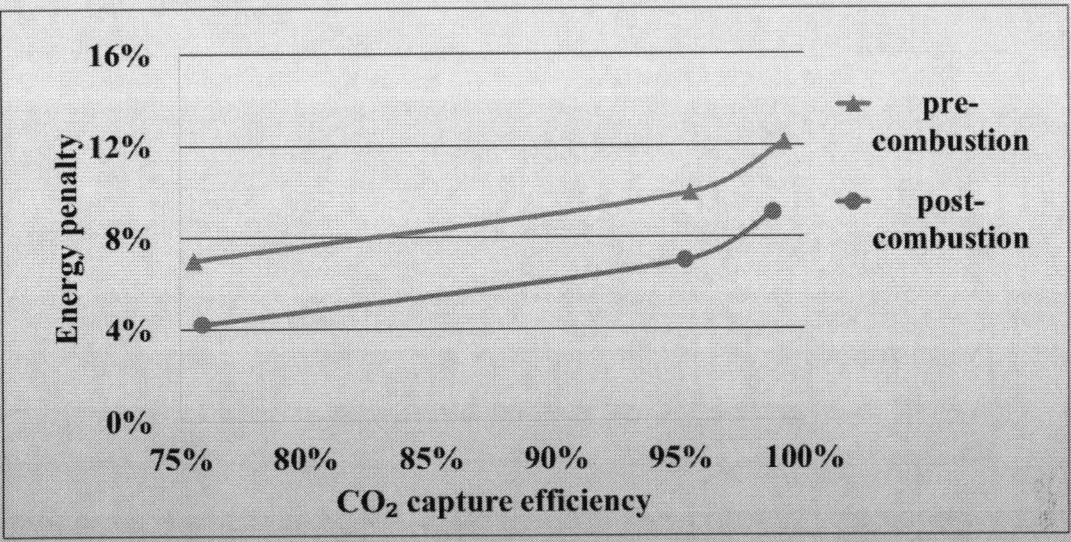

Energy penalty vs. CO_2 capture efficiency for post- & pre-combustion CaL

23

Scaled Cases for Post- & Pre-combustion CaL

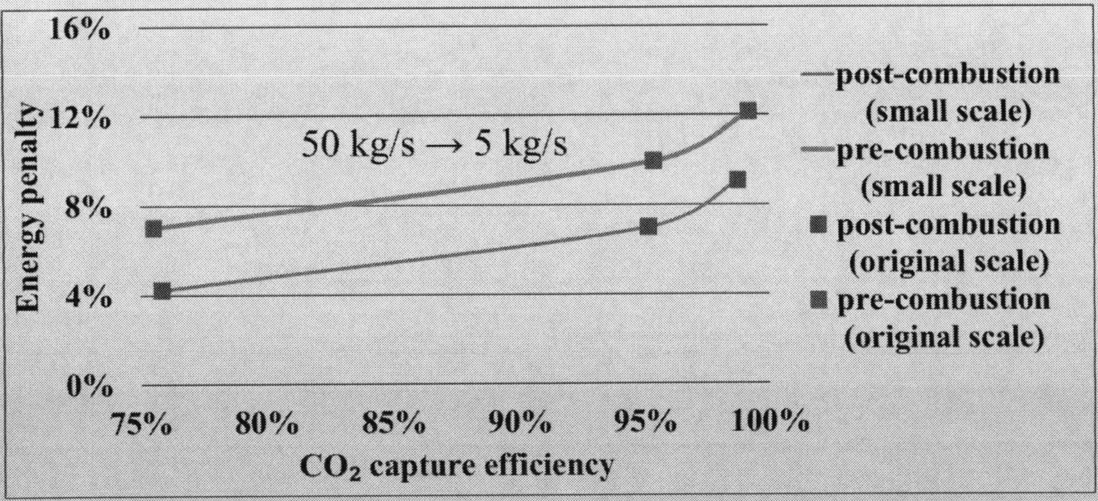

Energy penalty vs. CO_2 capture efficiency for post- & pre-combustion CaL at different scales

24

Conclusions

- Energy penalty is 7-9% for post-combustion CaL and 10-12% for pre- combustion CaL for high CO_2 capture (95-99%).
- 99% is a reasonable limit for CO_2 capture since marginal energy penalty increases rapidly after reaching 95% capture.

25

Techno-Economic Merit Analysis of Conversion of Captured CO_2 to Value-Added Products

C.B. Panchal, John C. Prindle,
Richard Doctor and Ajay Madgavkar

E3Tec Service, LLC

AIChE CMTC Conference, Sugar Land, TX
Session 23, Transformational Technology Advancements within the CCUS Value Chain

November 17-19, 2015

Presentation Outline

- Overview of CO_2 capture and sequestration
- Overview of CO_2 emission management
- Conversion processes and products
- Techno-economic merit criteria
- Conversion of CO_2 to alkyl carbonates
- Summary and path forward

Overview of CO_2 Sequestration – Fossil Electric with CO_2 capture

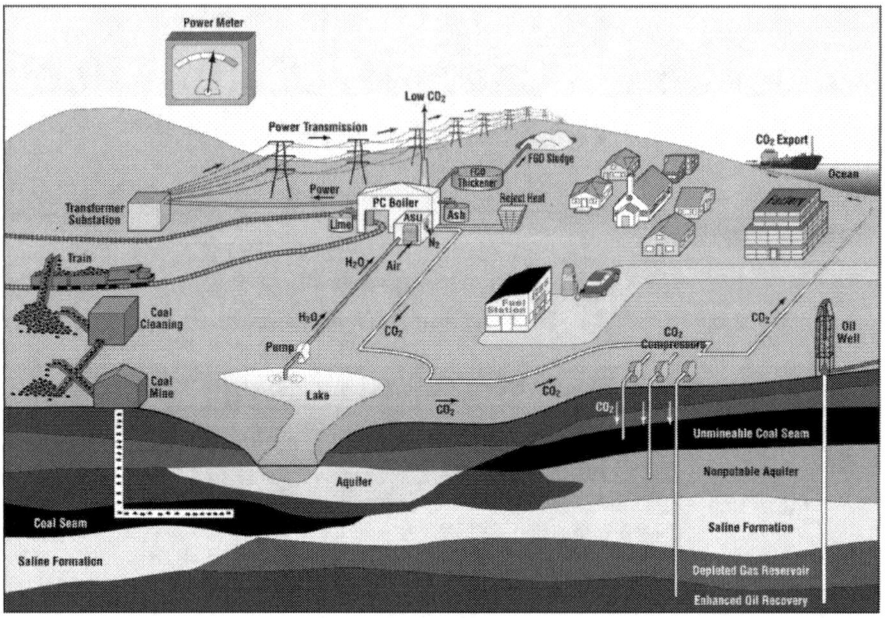

Overview of CO2 Sequestration

- Costs and programmatic barriers of CO_2 capture, transport and storage
- Major industry sources of concentrated CO_2 not located close to storage sites
- Local storage sites may or may not have capacities for long-term sequestration

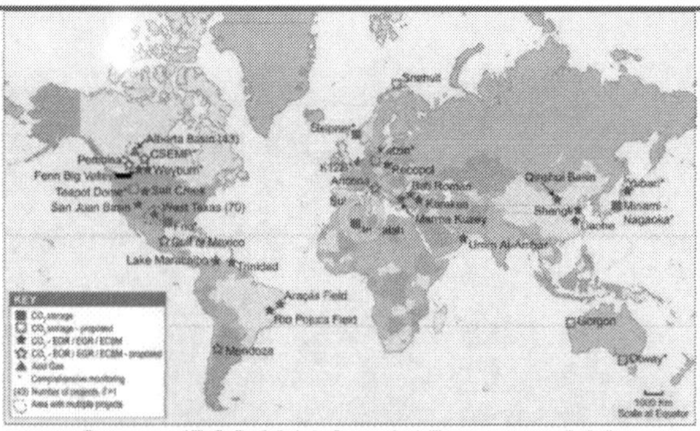

Source: IPCC 2014 Special Report on CCS

In Salah cost US$2.7 billion. Injection started in 2004 and injection suspended in 2011.

Alberta's 2008 Climate Change Strategy

Defines Long Term Goal (Source CCEMC Grand Challenge)

E3Tec

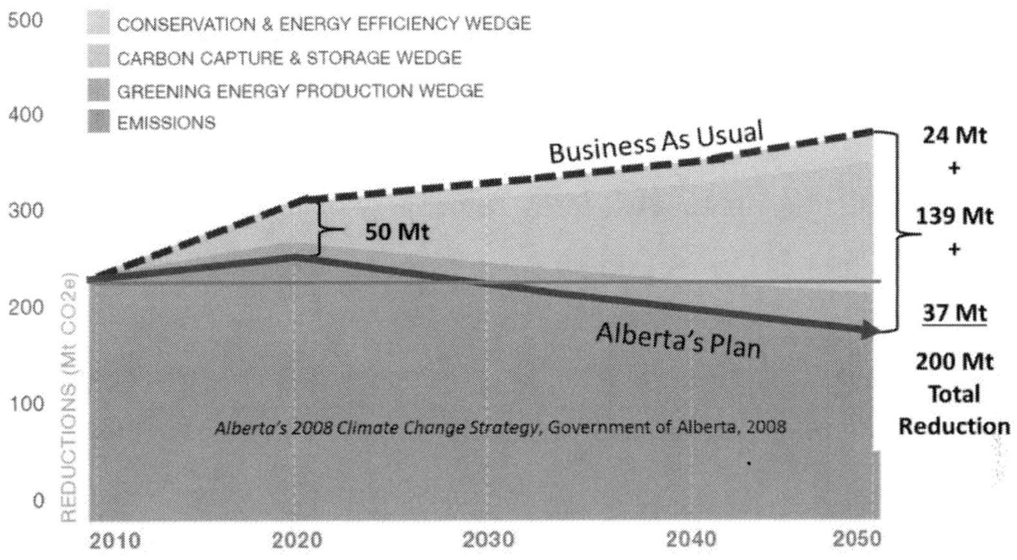

Logistics of Converting CO_2 to High-Value Products

E3Tec

- Logistics

CO_2 Generation → Transport →  **CO_2 Sinks** 

CO_2 Generation → Conversion → **Liquid Product** →  **Global Market**

- Product Economics

High Value Proposition

High CO_2 Merit Ratio $\left[\frac{CO_2 \text{ Consumed}}{CO_2 \text{ Generated}}\right]$

- Stakeholders

Partners	Government	Public
Industry with CO_2 Sources or DMC Manufacturer or Feedstock Company	Regulations Encouraging New Technology Commercialization	Local & Visible Economic Benefits

Alberta is an Ideal Candidate for Commercialization of Our Technology

Alberta's 2008 Climate Change Strategy

Three key messages to be derived from the strategic plan

E3Tec

1. CO_2 capture and storage from large sources would be the major part of emission control

2. Alternate fuel, renewable energy and energy efficiency would play key role in reducing CO_2 emission

3. Conversion of CO_2 to value-added products should be in the portfolio for distributed CO_2 sources of small utilities and industrial sites

Overview of Conversion of CO2 to Value-Added Products

Four Major Categories (Source CCEMC Grand Challenge)

E3Tec

Fuels

- Conversion to methanol with renewable or nuclear hydrogen
- SynGas by catalytic reforming of CO_2 with NG

Bio-Fixation

- Microalgae
- Bioconversion of CO_2 to butanol

Solid Products

- Inorganic carbonates and concrete
- Graphene

Chemicals

- Methanol
- Ethylene oxide
- Acetic acid
- Alkyl carbonates

Techno-Economic Merit Criteria

Prioritized Merit Criteria

1. Technical viability of the process and commercial use of products
2. Energy-efficient synthesis process for low C-Footprint preferably net CO_2 consumption
3. Competitive product margin for offsetting costs associated with capturing CO_2
4. Market potentials of the product should be clearly identified with potential high growth for sequestration of large-scale CO_2 from industry sources
5. Service life of the products and end products at the end of service life
6. Modular design for integration with CO_2 source sites and near-term commercialization

E3Tec Service, LLC 9

Techno-Economic Merit Criteria

Applied to CO_2 Conversion to Di-Methyl Carbonate (DMC)

PHOSGENE ROUTE:

Phosgenation of Methanol

$COCl_2$ + 2 CH_3 OH → CH_3 $OCOOCH_3$ + 2 HCL
Phosgene, Methanol, DMC

NON-PHOSGENE ROUTES:

Oxidative Carbonylation of Methanol

2 CH_3OH + CO + 0.5 O_2 → CH_3 $OCOOCH_3$ + H_2O
Methanol, DMC

E^3TEC PROCESSES

Methanolysis of Urea

2 NH_3 + CO_2 → H_2NCONH_2 + H_2O
H_2NCONH_2 + 2 CH_3 OH ↔ CH_3 $OCOOCH_3$ + 2 NH_3
Urea, Methanol, DMC

Epoxidation of Ethylene Oxide by CO_2 followed by Transesterification of Ethylene Carbonate

CH_2OCH_2 + CO_2 → $CH_2OCOOCH_2$
Ethylene Oxide, Ethylene Carbonate (EC)

$CH_2OCOOCH_2$ + 2 CH_3 OH ↔ $CH_3OCOOCH_3$ + CH_2OHCH_2OH
Ethylene Carbonate (EC), Methanol, DMC, Ethylene Glycol (EG)

E3Tec Service, LLC 10

Di-Methyl Carbonate (DMC) Synthesis Process

Heat Integrated Reactive Distillation (HIRD)

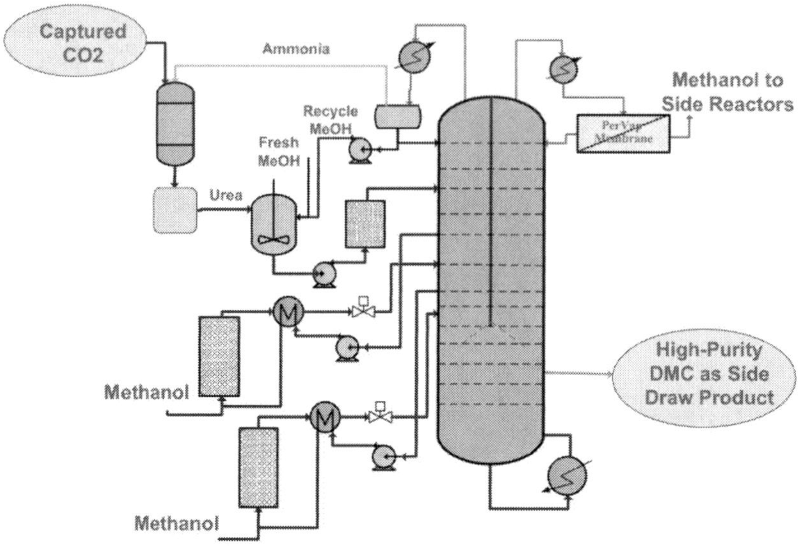

Patent Pending

Techno-Economic Merit Criteria

Applied to CO_2 Conversion to Dimethyl Carbonate (DMC)

1. **Technical viability of the process and commercial use of products**
   - ***Proven chemical reactions using prototype test units***
   - ***ASPEN Plus™ process simulation to establish scale-up criteria***
   - ***Three major use of DMC and large scale potential use***
2. **Energy-efficient synthesis process for low C-Footprint preferably net CO_2 consumption**
   - ***Process analysis shows significantly lower C-Footprint compared to commercial SynGas based DMC process***
   - ***Co-production of Mono-Ethylene Glycols (Mono-EG) further improves combined C-Footprint***

Techno-Economic Merit Criteria

C-Footprint Process Block Diagram

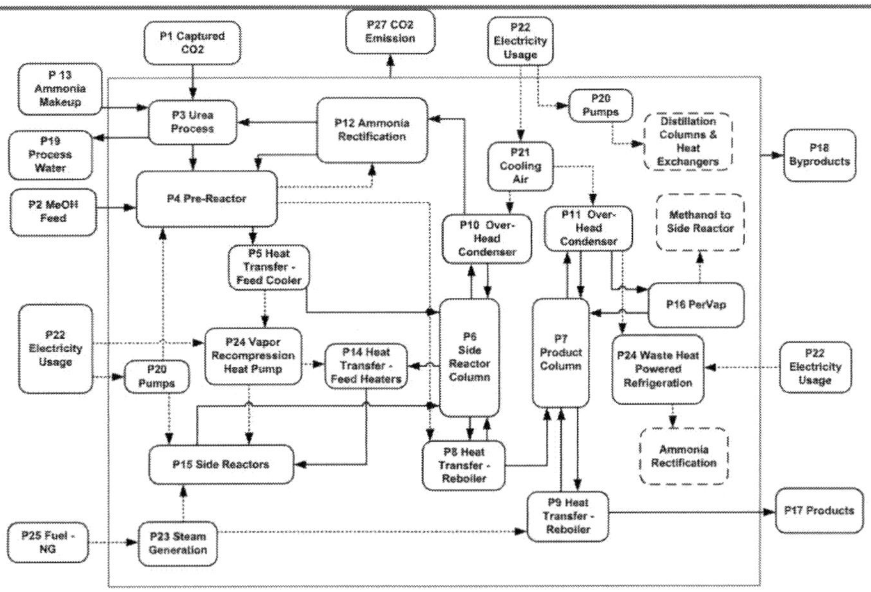

Dimethyl Carbonate (DMC) Process Block Diagram
Mass Flow ⟶ Energy Flow ┈┈►

Techno-Economic Merit Criteria

Applied to CO_2 Conversion to Dimethyl Carbonate (DMC)

3. **Competitive product margin for offsetting costs associated with capturing CO_2**
   - *Market price of high-purity DMC is about $1,900/MT, while the cost of methanol is about $395/MT*
   - *Co-production of Mono-EG (~ $1,300/MT) further enhances competitive margin*
   - *This provides a competitive margin for capital and operative costs*
4. **Market potentials of the product should be clearly identified with potential high growth for sequestration of large-scale CO_2 from industry sources**
   - *Major market potentials: a) polycarbonates; b) solvents; c) electrolyte solvent in lithium-ion batteries; d) intermediate chemicals; and e) potential use as diesel additive*

Di-Methyl Carbonate (DMC) as Value-Added Product

E3Tec

- DMC provides permanent CO_2 sequestration
- Present applications
  - Polycarbonate
  - Solvents
  - Chemical intermediates
- Emerging applications
  - Solvent with low VOC replacing MIBK
  - Electrolyte solvent in Lithium-Ion batteries
  - Potential oxygenated diesel fuel additive

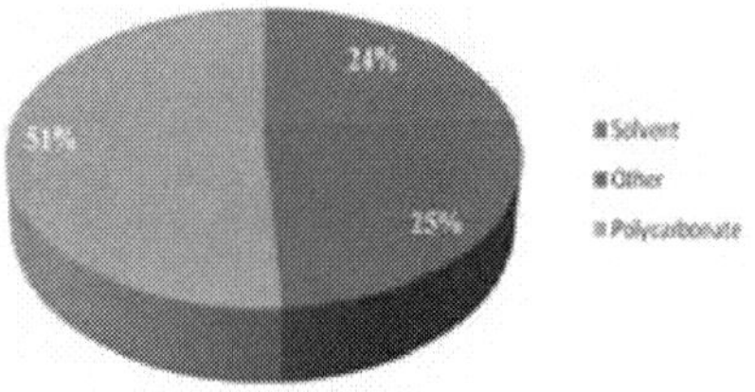

Di-Methyl Carbonate (DMC) as Value-Added Product

CO_2 Sequestration Potentials

E3Tec

Application	Global Demands - 2012 KTA	Growth	DMC in Product $T_{DMC}/T_{product}$	CO2 Sequestration T_{CO2}/T_{DMC}	CO2 Sequestration by Replacement $T_{CO2}/T_{product}$	CO2 Sequestration Potentials KTA Growth	25% Market
Polycarbonate	4,925	6.9%	1.0	0.29		99	357
Solvents	1,000	2.0%	1.0		0.80	16	200
Fuel Additives - Diesel	1,580,000	2.0%	0.076	0.29		700	8,752
Li-Ion Battery in 2017	$25 billion/yr		Electrolyte market $5.23 billion/yr	0.29		227	

Techno-Economic Merit Criteria

Applied to CO_2 Conversion to Dimethyl Carbonate (DMC)

E3Tec

5. Service life of the products and end products at the end of service life
   - *Carbonates and solvents provide long-term sequestration*
   - *Lithium-ion battery electrolytes/solvents would be recycled*
   - *Diesel fuel additive would produce CO_2, however it displaces petroleum*
6. Modular design for integration with CO_2 source sites
   - *Heat Integrated Reactive Distillation (HIRD) representing process intensification with low foot print should be easy to integrate with CO_2 source sites*

E3Tec

Summary and Path Forward

- Techno-economic viability of CO_2 conversion to alkyl carbonates is established
- C-Footprint analysis shows potentials of CO_2 abatement
- Competitive process using process intensification of HIRD
- Expanding global market of alkyl carbonates
- Applicable to distributed CO_2 sources

CO2 Conversion to Value-Added Products Should be in the Portfolio of the Long-Term CO_2 Emission Strategy

Thank You

Acknowledgements

The effort was sponsored by the U. S. Department Energy's (DOE's) through the SBIR-STTR program. The authors also acknowledge funding support from CCEMC Grant Challenge, Alberta, program.

AUTHOR INDEX

AUTHOR INDEX

AUTHOR INDEX